2016 IEEE Applied Power Electronics Conference and Exposition (APEC 2016)

Long Beach, California, USA
20-24 March 2016

Pages 1-714

IEEE Catalog Number: CFP16APE-POD
ISBN: 978-1-4673-9551-9

Copyright © 2016 by the Institute of Electrical and Electronic Engineers, Inc
All Rights Reserved

Copyright and Reprint Permissions: Abstracting is permitted with credit to the source. Libraries are permitted to photocopy beyond the limit of U.S. copyright law for private use of patrons those articles in this volume that carry a code at the bottom of the first page, provided the per-copy fee indicated in the code is paid through Copyright Clearance Center, 222 Rosewood Drive, Danvers, MA 01923.

For other copying, reprint or republication permission, write to IEEE Copyrights Manager, IEEE Service Center, 445 Hoes Lane, Piscataway, NJ 08854. All rights reserved.

***This publication is a representation of what appears in the IEEE Digital Libraries. Some format issues inherent in the e-media version may also appear in this print version.*

IEEE Catalog Number: CFP16APE-POD
ISBN (Print-On-Demand):   978-1-4673-9551-9
ISBN (Online):   978-1-4673-9550-2
ISSN: 1048-2334

Additional Copies of This Publication Are Available From:

Curran Associates, Inc
57 Morehouse Lane
Red Hook, NY 12571 USA
Phone: (845) 758-0400
Fax: (845) 758-2633
E-mail: curran@proceedings.com
Web: www.proceedings.com

2016 IEEE Applied Power Electronics Conference and Exposition (APEC 2016)

Long Beach, California, USA
20-24 March 2016

Pages 1-714

IEEE Catalog Number: CFP16APE-POD
ISBN: 978-1-4673-9551-9

TECHNICAL PAPERS

Session T01: Three-Phase AC-DC Converters
Location: 101A
March 22, 2016 8:30 - 12:00
Session Chairs: Gerry Moschopoulos, *Western University, Canada*
Patrick Wheeler, *University of Nottingham*

Hardware Implementation and Characterization of SiC-Based Hybrid Three-Phase Rectifier Employing Third Harmonic Injection 1
M. Makoschitz, *Technische Universität Wien, Austria*
M. Hartmann, *Schneider Electric SE, Austria*
H. Ertl, *Technische Universität Wien, Austria*

Voltage Oriented Control of the Three-Level Vienna Rectifier using Vector Control Method 9
Jeevan Adhikari, *National University of Singapore, Singapore*
Prasanna IV, *National University of Singapore, Singapore*
S.K. Panda, *National University of Singapore, Singapore*

Compensation of Neutral Point Deviation in 3-Level NPC Converter Under Unbalanced Grid Conditions 17
Kyungsub Jung, *Chungbuk National University, Korea, South*
Yongsug Suh, *Chungbuk National University, Korea, South*

High Power Factor Modular Polyphase AC/DC Converters with Galvanic Isolation based on Resistor Emulators 25
Javier Sebastián, *Universidad de Oviedo, Spain*
Ignacio Castro, *Universidad de Oviedo, Spain*
Diego G. Lamar, *Universidad de Oviedo, Spain*
Aitor Vázquez, *Universidad de Oviedo, Spain*
Kevin Martín, *Universidad de Oviedo, Spain*

Reduced Duty-Cycle Loss and Output Inductor Current Ripple in a ZVS Switched Three-Phase Isolated PWM Rectifier 33
Jahangir Afsharian, *Ryerson University, Canada*
Dewei David Xu, *Ryerson University, Canada*
Tao Zhao, *Ryerson University, Canada*
Bing Gong, *Murata Power Solution, Canada*
Zhihua Yang, *Murata Power Solution, Canada*

Analysis, Design, and Evaluation of Three-Phase Three-Wire Isolated AC-DC Converter Implemented with Three Single-Phase Converter Modules 38
Laszlo Huber, *Delta Products Corporation, United States*
Misha Kumar, *Delta Products Corporation, United States*
Milan M. Jovanović, *Delta Products Corporation, United States*
Dinggang Ping, *Delta Electronics Shanghai Co., Ltd., China*
Gang Liu, *Delta Electronics Shanghai Co., Ltd., China*

Startup Procedure for Three-Phase Three-Wire Isolated AC-DC Converter Implemented with Three Single-Phase Converter Modules 46

Misha Kumar, *Delta Products Corporation, United States*
Laszlo Huber, *Delta Products Corporation, United States*
Milan M. Jovanović, *Delta Products Corporation, United States*
Dinggang Ping, *Delta Electronics Shanghai Co., Ltd., China*
Gang Liu, *Delta Electronics Shanghai Co., Ltd., China*

Control of a Single-Stage Three-Phase Boost Power Factor Correction Rectifier 54

Ayan Mallik, *University of Maryland, United States*
Bryan Faulkner, *Virginia Polytechnic Institute and State University, United States*
Alireza Khaligh, *University of Maryland, United States*

A Bidirectional Single-Stage Three-Phase Rectifier with High-Frequency Isolation and Power Factor Correction 60

Bruno Ricardo de Almeida, *Universidade Federal do Ceará, Brazil*
Demercil de Souza Oliveira Jr., *Universidade Federal do Ceará, Brazil*
Paulo P. Praça, *Universidade Federal do Ceará, Brazil*

Session T02: High Frequency and Fast-Response DC-DC Converters
Location: 104A
March 22, 2016 8:30 - 12:00
Session Chairs: Olivier Trescases, *University of Toronto*
Jeff Nilles, *Texas Instruments*

A 5 MHz, 12 V, 10 A, Monolithically Integrated Two-Phase Series Capacitor Buck Converter 66

Pradeep S. Shenoy, *Texas Instruments Inc., United States*
Orlando Lazaro, *Texas Instruments Inc., United States*
Ramanathan Ramani, *Texas Instruments Inc., United States*
Mike Amaro, *Texas Instruments Inc., United States*
Wlodek Wiktor, *Texas Instruments Inc., United States*
Joseph Khayat, *Texas Instruments Inc., United States*
Brian Lynch, *Texas Instruments Inc., United States*

A 10-MHz Isolated Class-Φ_2 Synchronous Resonant DC-DC Converter 73

Yuan Zhou, *Nanjing University of Aeronautics and Astronautics, China*
Zhiliang Zhang, *Nanjing University of Aeronautics and Astronautics, China*
Xue-Wen Zou, *Nanjing University of Aeronautics and Astronautics, China*
Zhou Dong, *Nanjing University of Aeronautics and Astronautics, China*
Xiaoyong Ren, *Nanjing University of Aeronautics and Astronautics, China*

865 MHz Switching-Speed Step-Down DC-DC Power Converter for Envelope Tracking 79

Vivek Mehrotra, *Teledyne Scientific Company, United States*
Andrea Arias, *Teledyne Scientific Company, United States*
Joshua Bergman, *Teledyne Scientific Company, United States*
Charles Neft, *Teledyne Scientific Company, United States*
Miguel Urteaga, *Teledyne Scientific Company, United States*
Berinder Brar, *Teledyne Scientific Company, United States*

Current Parking Regulator for Zero Droop/Overshoot Load Transient Response 86

Sudhir S. Kudva, *Nvidia Corporation, United States*
William J. Dally, *Nvidia Corporation, United States*
Thomas H. Greer III, *Nvidia Corporation, United States*
C. Thomas Gray, *Nvidia Corporation, United States*

A 5MHz, 24V-to-1.2V, AO^{2}T Current Mode Buck Converter with One-Cycle Transient Response and Sensorless Current Detection for Medical Meters 94

Xugang Ke, *University of Texas at Dallas, United States*
Joseph Sankman, *Texas Instruments Inc., United States*
Dongsheng Ma, *University of Texas at Dallas, United States*

Capacitively-Aided Switching Technique for High-Frequency Isolated Bus Converters 98

Seungbum Lim, *Massachusetts Institute of Technology, United States*
Alex J. Hanson, *Massachusetts Institute of Technology, United States*
Juan A. Santiago-González, *Massachusetts Institute of Technology, United States*
David J. Perreault, *Massachusetts Institute of Technology, United States*

A 10 MHz, 48-to-5V Synchronous Converter with Dead Time Enabled 125 ps Resolution Zero-Voltage Switching 106

Alexander Barner, *Robert Bosch GmbH, Germany*
Jürgen Wittmann, *Hochschule Reutlingen, Germany*
Thoralf Rosahl, *Robert Bosch GmbH, Germany*
Bernhard Wicht, *Hochschule Reutlingen, Germany*

Plug-and-Play Electronic Capacitor for VRM Applications 111

Or Kirshenboim, *Ben-Gurion University of the Negev, Israel*
Alon Cervera, *Ben-Gurion University of the Negev, Israel*
Bar Halivni, *Ben-Gurion University of the Negev, Israel*
Eli Abramov, *Ben-Gurion University of the Negev, Israel*
Mor Mordechai Peretz, *Ben-Gurion University of the Negev, Israel*

Adaptive Voltage Positioning (AVP) Design of Multi-Phase Constant On-Time I^2 Control for Voltage Regulators with Ramp Compensations 118

Kuang-Yao Cheng, *Texas Instruments Inc., United States*
Yipeng Su, *Texas Instruments Inc., United States*

Session T03: Microgrids and Hybrid Systems
Location: 104B
March 22, 2016 8:30 - 12:00
Session Chairs: Yunwei Li, *University of Alberta*
Joesep Guerrero, *Aalborg University*

Reactive Power Support Capabilities of Nonsynchronous Interconnection Systems in Microgrid Applications 125

Yong-Duk Lee, *University of Connecticut, United States*
Sung-Yeul Park, *University of Connecticut, United States*

Zero Standby Power High Efficiency Hot Plugging Outlet for 380VDC Power Delivery System 132

Kai Tan, *North Carolina State University, United States*
Chang Peng, *North Carolina State University, United States*
Pengkun Liu, *North Carolina State University, United States*
Xiaoqing Song, *North Carolina State University, United States*
Alex Q. Huang, *North Carolina State University, United States*

Design of Control System for Smooth Mode Transfer in Smart Microgrid Application 138
Mingzhi Gao, *Zhejiang University, China*
Canhui Zhang, *Zhejiang University, China*
Maohang Qiu, *Zhejiang University, China*
Min Chen, *Zhejiang University, China*
Aron Levy, *Technology Dynamics Inc., United States*

Resonance Propagation Modeling and Analysis of AC Filters in a Large-Scale Microgrid ... 143
Yusi Liu, *University of Arkansas, United States*
Chris Farnell, *University of Arkansas, United States*
H. Alan Mantooth, *University of Arkansas, United States*
Juan Carlos Balda, *University of Arkansas, United States*
Roy A. McCann, *University of Arkansas, United States*
Cheng Deng, *University of Arkansas, United States*

A New Bidirectional DC-DC Converter for Fuel Cell, Solar Cell and Battery Systems 150
Ankur Patel, *Vicor Corporation, United States*

A Multiport Isolated DC-DC Converter 156
Yan-Kim Tran, *École Polytechnique Fédérale de Lausanne, Switzerland*
Drazen Dujic, *École Polytechnique Fédérale de Lausanne, Switzerland*

A Seamless Transfer Control Method with High Load Sharing Performance for Modular ESS ... 163
Jung-Hoon Ahn, *Sungkyunkwan University, Korea, South*
Won-Yong Sung, *Sungkyunkwan University, Korea, South*
Chang-Yeol Oh, *Sungkyunkwan University, Korea, South*
Byoung-Kuk Lee, *Sungkyunkwan University, Korea, South*
Yun-Sung Kim, *Dongahelecomm Corporation, Korea, South*

A Plug-and-Play Ripple Mitigation Approach for DC-Links in Hybrid Systems 169
Sinan Li, *University of Hong Kong, Hong Kong*
Albert T.L. Lee, *University of Hong Kong, Hong Kong*
Siew-Chong Tan, *University of Hong Kong, Hong Kong*
S.Y. Ron Hui, *University of Hong Kong, Hong Kong*

Active Control of Low Frequency Common Mode Voltage to Connect AC Utility and 380 V DC Grid 177
Fang Chen, *Virginia Polytechnic Institute and State University, United States*
Rolando Burgos, *Virginia Polytechnic Institute and State University, United States*
Dushan Boroyevich, *Virginia Polytechnic Institute and State University, United States*
Xuning Zhang, *Virginia Polytechnic Institute and State University, United States*

Session T04: Control Strategies for Inverters and Motor Drives
Location: 103C
March 22, 2016 8:30 - 12:00
Session Chairs: Bilal Akin, *Univeristy of Texas, Dallas*
                    Babak Nahid-Mobarakeh, *University of Lorraine*

A Three-Level Space Vector Modulation Scheme for Paralleled Two Converters to Reduce Zero-Sequence Circulating Current and Common Mode Voltage 185
Zhongyi Quan, *University of Alberta, Canada*
Yunwei Li, *University of Alberta, Canada*

Nonlinearity Analysis and Linear Modulation Method for Two Level Voltage Source Inverter with Low Switching to Operating Frequency Ratio 193

Yongjae Lee, *Seoul National University, Korea, South*
Jung-Ik Ha, *Seoul National University, Korea, South*

Synchronization Strategies in Cascaded H-Bridge Multi Level Inverters for Carrier based Sinusoidal PWM Techniques 199

Saroj Kumar Sahoo, *Indian Institute of Technology Kharagpur, India*
Tanmoy Bhattacharya, *Indian Institute of Technology Kharagpur, India*

Design and Implementation of a Sinusoidal Flux Controller for Core Loss Measurements 207

Burak Tekgun, *University of Akron, United States*
Ali R. Boynuegri, *University of Akron, United States*
Md Asif Mahmood Chowdhury, *University of Akron, United States*
Yilmaz Sozer, *University of Akron, United States*

Implementation of Deadbeat-Direct Torque and Flux Control for Synchronous Reluctance Machines to Minimize Loss Each Switching Period 215

Michael Saur, *Universität der Bundeswehr München, Germany*
Francisco Ramos, *Universität der Bundeswehr München, Germany*
Aday Perez, *Universität der Bundeswehr München, Germany*
Dieter Gerling, *Universität der Bundeswehr München, Germany*
Robert D. Lorenz, *University of Wisconsin at Madison, United States*

Addressing the Unbalance Loading Issue in Multi-Drive Systems with a DC-Link Modulation Scheme for Harmonic Reduction 221

Yongheng Yang, *Aalborg University, Denmark*
Pooya Davari, *Aalborg University, Denmark*
Firuz Zare, *Danfoss Power Electronics A/S, Denmark*
Frede Blaabjerg, *Aalborg University, Denmark*

Input Current Interharmonics in Adjustable Speed Drives Caused by Fixed-Frequency Modulation Techniques 229

Hamid Soltani, *Aalborg University, Denmark*
Pooya Davari, *Aalborg University, Denmark*
Poh Chiang Loh, *Aalborg University, Denmark*
Frede Blaabjerg, *Aalborg University, Denmark*
Firuz Zare, *Danfoss Power Electronics A/S, Denmark*

Low-Frequency Voltage Ripples in the Flying Capacitors of the Nested Neutral-Point-Clamped Converter 236

Amer M.Y.M. Ghias, *University of Sharjah, U.A.E.*
Josep Pou, *University of New South Wales, Australia*
Salvador Ceballos, *TECNALIA, Spain*
Vassilios G. Agelidis, *University of New South Wales, Australia*

DC Bus Capacitor Discharge of Permanent Magnet Synchronous Machine Drive Systems for Hybrid Electric Vehicles 241

Ziwei Ke, *Oregon State University, United States*
Julia Zhang, *Oregon State University, United States*
Michael W. Degner, *Ford Motor Company, United States*

Session T05: Si Devices and Power Module Packaging
Location: 101B
March 22, 2016 8:30 - 12:00
Session Chairs: Iulian Nistor, *Corporate Research, ABB Inc.*
Brian Rowden,

C_{OSS} Hysteresis in Advanced Superjunction MOSFETs 247
J.B. Fedison, *Enphase Energy, Inc., United States*
M.J. Harrison, *Enphase Energy, Inc., United States*

Compact Electrothermal Models for Unbalanced Parallel Conducting Si-IGBTs 253
Roozbeh Bonyadi, *University of Warwick, United Kingdom*
Olayiwola Alatise, *University of Warwick, United Kingdom*
Ji Hu, *University of Warwick, United Kingdom*
Zarina Davletzhanova, *University of Warwick, United Kingdom*
Yeganeh Bonyadi, *University of Warwick, United Kingdom*
Jose Ortiz-Gonzalez, *University of Warwick, United Kingdom*
Li Ran, *University of Warwick, United Kingdom*
Philip Mawby, *University of Warwick, United Kingdom*

**General 3D Lumped Thermal Model with Various Boundary Conditions for High Power
IGBT Modules** 261
Amir Sajjad Bahman, *Aalborg University, Denmark*
Ke Ma, *Aalborg University, Denmark*
Frede Blaabjerg, *Aalborg University, Denmark*

Improved 6.5kV FREEMD-Pair based on SiC JFET and Si IGBT 269
Xiaoqing Song, *North Carolina State University, United States*
Alex Q. Huang, *North Carolina State University, United States*
Chang Peng, *North Carolina State University, United States*
Liqi Zhang, *North Carolina State University, United States*

**On the Comparative Assessment of 1.7 kV, 300 a Full SiC-MOSFET and Si-IGBT
Power Modules** 276
Muhammad Nawaz, *ABB Corporate Research, Sweden*
Kalle Ilves, *ABB Corporate Research, Sweden*

**Suppression of Reverse Recovery Ringing 3.3kV/450A Si/SiC Hybrid in Low Internal
Inductance Package Next High Power Density Dual; nHPD[2]** 283
Katsuaki Saito, *Hitachi Europe Ltd., United Kingdom*
Daisuke Kawase, *Hitachi Power Semiconductor, Ltd., Japan*
Masamitsu Inaba, *Hitachi Power Semiconductor, Ltd., Japan*
Keiichi Yamamoto, *Hitachi Power Semiconductor, Ltd., Japan*
Katsunori Azuma, *Hitachi Power Semiconductor, Ltd., Japan*
Seiichi Hayakawa, *Hitachi Power Semiconductor, Ltd., Japan*

**New Layout Concepts in MW-Scale IGBT Modules for Higher Robustness during Normal
and Abnormal Operations** 288
Paula Diaz Reigosa, *Aalborg University, Denmark*
Francesco Iannuzzo, *Aalborg University, Denmark*
Stig Munk-Nielsen, *Aalborg University, Denmark*
Frede Blaabjerg, *Aalborg University, Denmark*

Design, Package, and Hardware Verification of a High Voltage Current Switch 295

Ankan De, *North Carolina State University, United States*
Adam Morgan, *North Carolina State University, United States*
Vishnu Mahadeva Iyer, *North Carolina State University, United States*
Haotao Ke, *North Carolina State University, United States*
Xin Zhao, *North Carolina State University, United States*
Kasunaidu Vechalapu, *North Carolina State University, United States*
Subhashish Bhattacharya, *North Carolina State University, United States*
Douglas C. Hopkins, *North Carolina State University, United States*

Investigation of Short Circuit in a IGBT Power Module with Three-Level Neutral Point Clamped Type 2 (NPC2, T-NPC, Mixed Voltage) Topology 303

Kevin Lenz, *Danfoss Silicon Power, Germany*
Vladan Jerinic, *Danfoss Silicon Power, Germany*
Reiner Hinken, *Danfoss Silicon Power, Germany*

Session T06: DC-DC Converter Control
Location: 102AB
March 22, 2016 8:30 - 12:00
Session Chairs: Sombuddha Chakraborty, *Texas Instruments*
Rafael Pena Alzola, *University of British Columbia*

Closed-Loop Design and Time-Optimal Control for a Series-Capacitor Buck Converter 308

Timur Vekslender, *Ben-Gurion University of the Negev, Israel*
Ofer Ezra, *Ben-Gurion University of the Negev, Israel*
Yevgeny Bezdenezhnykh, *Ben-Gurion University of the Negev, Israel*
Mor Mordechai Peretz, *Ben-Gurion University of the Negev, Israel*

Unified Constant On/Off-Time Hybrid Compensation for Fast Recovery in Digitally Current-Mode Controlled Point-of-Load Converters 315

K. Hariharan, *Indian Institute of Technology Kharagpur, India*
Santanu Kapat, *Indian Institute of Technology Kharagpur, India*
Siddhartha Mukhopadhyay, *Indian Institute of Technology Kharagpur, India*

Digital Implementation of Adaptive Synchronous Rectifier (SR) Driving Scheme for LLC Resonant Converters 322

Chao Fei, *Virginia Polytechnic Institute and State University, United States*
Fred C. Lee, *Virginia Polytechnic Institute and State University, United States*
Qiang Li, *Virginia Polytechnic Institute and State University, United States*

Digital Synchronous Rectification Controller for LLC Resonant Converters 329

Maryam S. Amouzandeh, *University of Toronto, Canada*
Behzad Mahdavikhah, *University of Toronto, Canada*
Aleksandar Prodić, *University of Toronto, Canada*
Brent McDonald, *Texas Instruments Inc., United States*

A Novel Adaptive Synchronous Rectification Method for Digitally Controlled LLC Converters ... 334

Fan Wang, *Texas Instruments Inc., United States*
Brent A. McDonald, *Texas Instruments Inc., United States*
Jeff Langham, *Texas Instruments Inc., United States*
Bo Fan, *Texas Instruments Inc., China*

Influence of the ADC Zero Bin on the Performance of an Integrated DC-DC Converter 339
S. Vesti, *Infineon Technologies Austria AG, Austria*
M. Agostinelli, *Infineon Technologies Austria AG, Austria*
H. Koltsov, *Infineon Technologies Austria AG, Austria*
S. Marsili, *Infineon Technologies Austria AG, Austria*

Improved Current-Mode Control with Single-Cycle Load Transient 343
Virginia Li, *Virginia Polytechnic Institute and State University, United States*
Pei-Hsin Liu, *Virginia Polytechnic Institute and State University, United States*
Qiang Li, *Virginia Polytechnic Institute and State University, United States*
Fred C. Lee, *Virginia Polytechnic Institute and State University, United States*

A Mixed-Signal Ripple-Based Controller for a 16 V, 10 MHz Integrated Buck Converter 350
Sergii Tkachov, *Infineon Technologies Austria AG, Austria*
Matteo Agostinelli, *Infineon Technologies Austria AG, Austria*

New Control Concept for Soft-Switching Flyback Converters with Very High Switching Frequency 355
A.M. Connaughton, *Technische Universität Graz, Austria*
K. Krischan, *Technische Universität Graz, Austria*
K.K. Leong, *Infineon Technologies AG, Austria*
A. Muetze, *Technische Universität Graz, Austria*

Session T07: Solar Energy Systems
Location: 104C
March 22, 2016 8:30 - 12:00
Session Chairs: Babak Fahimi, *UT- Dallas*
Morgan Kiani, *Texas Christian University*

Analysis, Modeling and Control of an Interleaved Isolated Boost Series Resonant Converter for Microinverter Applications 362
Luciano A. Garcia-Rodriguez, *University of Arkansas, United States*
Cheng Deng, *University of Arkansas, United States*
Juan Carlos Balda, *University of Arkansas, United States*
Andrés Escobar-Mejía, *Universidad Tecnologica de Pereira, Colombia*

Benchmarking of Constant Power Generation Strategies for Single-Phase Grid-Connected Photovoltaic Systems 370
Ariya Sangwongwanich, *Aalborg University, Denmark*
Yongheng Yang, *Aalborg University, Denmark*
Frede Blaabjerg, *Aalborg University, Denmark*
Huai Wang, *Aalborg University, Denmark*

Advanced Slip Mode Frequency Shift Islanding Detection Method for Single Phase Grid Connected PV Inverters 378
Bahador Mohammadpour, *Queen's University, Canada*
Majid Pahlevani, *Queen's University, Canada*
Sajjad Makhdoomi Kaviri, *Queen's University, Canada*
Praveen Jain, *Queen's University, Canada*

Direct MPPT Control of PWM Converters for Extreme Transient PV Applications 386

Ignacio Galiano Zurbriggen, *University of British Columbia, Canada*
Francisco Paz, *University of British Columbia, Canada*
Martin Ordonez, *University of British Columbia, Canada*

Feeding Partial Power into Line Capacitors for Low Cost and Efficient MPPT of Photovoltaic Strings 392

Ali Elrayyah, *Qatar Environment and Energy Research Institute, Qatar*
Mohammed Badawey, *University of Akron, United States*
Yilmaz Sozer, *University of Akron, United States*

Single Phase Cascaded H5 Inverter with Leakage Current Elimination for Transformerless Photovoltaic System 398

Xiaoqiang Guo, *Yanshan University, China*
Xiaoyu Jia, *Yanshan University, China*
Zhigang Lu, *Yanshan University, China*
Josep M. Guerrero, *Aalborg University, Denmark*

Optimal Low Switching Frequency Pulse Width Modulation of Current-Fed Three-Level Inverter for Solar Integration 402

Gnana Sambandam Kulothungan, *National University of Singapore, Singapore*
Akshay K. Rathore, *National University of Singapore, Singapore*
Amarendra Edpuganti, *National University of Singapore, Singapore*
Dipti Srinivasan, *National University of Singapore, Singapore*

Low Leakage Current Single-Phase PV Inverters with Universal Neutral-Point-Clamping Method 410

Liwei Zhou, *Shandong University, China*
Feng Gao, *Shandong University, China*

Modular Subpanel Photovoltaic Converter System: Analysis and Control 417

Yuan Li, *Sichuan University / Northeastern University, China*
Yue Zheng, *Northeastern University, United States*
Su Sheng, *Northeastern University, United States*
Brad Scandrett, *PowerFilm, Inc., United States*
Brad Lehman, *Northeastern University, United States*

Session T08: Advanced Converter for Power Systems used in Transportation
Location: 103AB
March 22, 2016 8:30 - 12:00
Session Chairs: Omer Onar, *Oak Ridge National Laboratory*
Khurram Afridi, *University of Colorado, Boulder*

Integrated DC-DC Converter Design for Electric Vehicle Powertrains 424

Saeed Anwar, *University of Tennessee, United States*
Weimin Zhang, *University of Tennessee, United States*
Fred Wang, *University of Tennessee, United States*
Daniel J. Costinett, *University of Tennessee, United States*

A 1 MHz Bi-Directional Soft-Switching DC-DC Converter with Planar Coupled Inductor for Dual Voltage Automotive Systems 432

Chenhao Nan, *Arizona State University, United States*
Raja Ayyanar, *Arizona State University, United States*

A Bridgeless Totem-Pole Interleaved PFC Converter for Plug-In Electric Vehicles 440
Yichao Tang, *University of Maryland, United States*
Weisheng Ding, *University of Maryland, United States*
Alireza Khaligh, *University of Maryland, United States*

Stability Analysis of Hybrid AC/DC Power Systems for More Electric Aircraft 446
Mehdi Karbalaye Zadeh, *Norwegian University of Science and Technology, Norway*
Roghayeh Gavagsaz-Ghoachani, *Université de Lorraine, France*
Babak Nahid-Mobarakeh, *Université de Lorraine, France*
Serge Pierfederici, *Université de Lorraine, France*
Marta Molinas, *Norwegian University of Science and Technology, Norway*

On the Concept of the Multi-Source Inverter 453
Lea Dorn-Gomba, *McMaster University, Canada*
Pierre Magne, *McMaster University, Canada*
Clement Barthelmebs, *McMaster University, Canada*
Ali Emadi, *McMaster University, Canada*

Time-Domain Analysis of a Wide-DC-Range Series Resonant Dual-Active-Bridge Bidirectional Converter with a New Passive Auxilliary Circuit 460
Alireza Safaee, *Queen's University, Canada*
Praveen Jain, *Queen's University, Canada*
Alireza Bakhshai, *Queen's University, Canada*

A New High Capacity Compact Power Modules for High Power EV/HEV Inverters 468
Seiichiro Inokuchi, *Mitsubishi Electric Corporation, Japan*
Shoji Saito, *Mitsubishi Electric Corporation, Japan*
Arata Izuka, *Mitsubishi Electric Corporation, Japan*
Yuki Hata, *Mitsubishi Electric Corporation, Japan*
Shinji Hatae, *Mitsubishi Electric Corporation, Japan*
Toshiya Nakano, *Powerex, Inc., United States*
Eric R. Motto, *Powerex, Inc., United States*

Modular Pet, Two-Phase Air-Cooled Converter Cell Design and Performance Evaluation with 1.7kV IGBTs for MV Applications 472
Frederick Kieferndorf, *ABB Switzerland Ltd, Switzerland*
Uwe Drofenik, *ABB Switzerland Ltd, Switzerland*
Francesco Agostini, *ABB Switzerland Ltd, Switzerland*
Francisco Canales, *ABB Switzerland Ltd, Switzerland*

A Phase Shift Full Bridge based Reconfigurable PEV Onboard Charger with Extended ZVS Range and Zero Duty Cycle Loss 480
Haoyu Wang, *ShanghaiTech University, China*

Session T09: Gate Drives, Failure Analysis, and Protection
Location: 102C
March 22, 2016 8:30 - 12:00
Session Chairs: Zhiliang Zhang, *Nanjing University of Aeronautics and Astronautics*
Indumini Ranmuthu, *Texas Instruments*

Series Arc Fault Detection Method based on Statistical Analysis for DC Microgrids 487
Gab-Su Seo, *Seoul National University, Korea, South*
Jung-Ik Ha, *Seoul National University, Korea, South*
Bo-Hyung Cho, *Seoul National University, Korea, South*
Kyu-Chan Lee, *Smart Power Supply Co., Ltd., Korea, South*

Arc Welding Inverter with Embedded Digital Active EMI Controller 493
Junpeng Ji, *Xi'an Jiaotong University, China*
Wenjie Chen, *Xi'an Jiaotong University, China*
Xu Yang, *Xi'an Jiaotong University, China*

A Thermo-Sensitive Electrical Parameter with Maximum dI_C/dt during Turn-Off for High Power Trench/Field-Stop IGBT Modules 499
Yuxiang Chen, *Zhejiang University, China*
Haoze Luo, *Zhejiang University, China*
Wuhua Li, *Zhejiang University, China*
Xiangning He, *Zhejiang University, China*
Jun Ma, *Shanghai Electric, China*
Guodong Chen, *Shanghai Electric, China*
Ye Tian, *Shanghai Electric, China*
Enxing Yang, *Shanghai Electric, China*

A Software Frequency Response Analysis Method to Monitor Degradation of Power MOSFETs in Basic Single-Switch Converters 505
Serkan Dusmez, *University of Texas at Dallas, United States*
Manish Bhardwaj, *Texas Instruments Inc., United States*
Lei Sun, *University of Texas at Dallas, United States*
Bilal Akin, *University of Texas at Dallas, United States*

A New Capacitance Estimation Method of Supercapacitor Bank using a Bank Impedance and Current Injection 511
Junwon Lee, *Chungnam National University, Korea, South*
Hyunsik Jo, *Chungnam National University, Korea, South*
Hanju Cha, *Chungnam National University, Korea, South*

Gate Driver Design for 1.7kV SiC MOSFET Module with Rogowski Current Sensor for Shortcircuit Protection 516
Jun Wang, *Virginia Polytechnic Institute and State University, United States*
Zhiyu Shen, *Virginia Polytechnic Institute and State University, United States*
Christina Dimarino, *Virginia Polytechnic Institute and State University, United States*
Rolando Burgos, *Virginia Polytechnic Institute and State University, United States*
Dushan Boroyevich, *Virginia Polytechnic Institute and State University, United States*

2 MHz High-Density Integrated Power Supply for Gate Driver in High-Temperature Applications 524

Remi Perrin, *Université Claude Bernard Lyon 1, France*
Bruno Allard, *Université Claude Bernard Lyon 1, France*
Cyril Buttay, *Université Claude Bernard Lyon 1, France*
Nicolas Quentin, *Université Claude Bernard Lyon 1, France*
Wenli Zhang, *Virginia Polytechnic Institute and State University, United States*
Rolando Burgos, *Virginia Polytechnic Institute and State University, United States*
Dushan Boroyevich, *Virginia Polytechnic Institute and State University, United States*
Philippe Preciat, *Labinal Power Systems, France*
Donatien Martineau, *Labinal Power Systems, France*

Design Consideration of Gate Driver Circuits and PCB Parasitic Parameters of Paralleled E-Mode GaN HEMTs in Zero-Voltage-Switching Applications 529

Juncheng Lu, *Kettering University, United States*
Hua Bai, *Kettering University, United States*
Alan Brown, *Hella Corporate Center USA Inc., United States*
Matt McAmmond, *Hella Corporate Center USA Inc., United States*
Di Chen, *GaN Systems Inc., Canada*
Julian Styles, *GaN Systems Inc., Canada*

A Gate Driver of SiC MOSFET for Suppressing the Negative Voltage Spikes in a Bridge Circuit 536

Qi Zhou, *Shandong University, China*
Feng Gao, *Shandong University, China*

Session T10: Control of AC-DC Converters
Location: 102AB
March 23, 2016 8:30 - 10:10
Session Chairs: Tsorng-Juu Liang, *National Cheng-Kung University (Taiwan)*
Laszlo Balogh, *Fairchild Semiconductor*

Interleaved Boost based AC/DC Bidirectional Converter with Four Quadrant Power Control based on One-Cycle Controller (OCC) 544

Snehal Bagawade, *Queen's University, Canada*
Praveen Jain, *Queen's University, Canada*

A New Control Scheme to Improve Load Transient Response of Single Phase PWM Rectifier with Auxiliary Current Injection Circuit 552

Naga Brahmendra Yadav Gorla, *National University of Singapore, Singapore*
Sandeep Kolluri, *National University of Singapore, Singapore*
Pritam Das, *National University of Singapore, Singapore*
Sanjib Kumar Panda, *National University of Singapore, Singapore*

Active Capacitor with Ripple-Based Duty Cycle Modulation for AC-DC Applications 558

Ching-Chieh Yang, *National Taiwan University, Taiwan*
Yang-Lin Chen, *National Taiwan University, Taiwan*
Yaow-Ming Chen, *National Taiwan University, Taiwan*

Novel Approach to Current-Mode Control in DCM/CCM Boundary Boost PFC 564

Giovanni Gritti, *STMicroelectronics, Italy*
Claudio Adragna, *STMicroelectronics, Italy*

Reducing the Switching Frequency Variation Range for CRM Buck PFC Converter by Variable On-Time Control 572

Xiaoping Wang, *Nanjing University of Science and Technology, China*
Kai Yao, *Nanjing University of Science and Technology, China*
Junfang Zhang, *Nanjing University of Science and Technology, China*

Session T11: GaN-Based DC-DC Converters
Location: 104A
March 23, 2016 8:30 - 10:10
Session Chairs: Alexis Kwasinski, *University of Pittsburgh*
Regan Zane, *Utah State*

High Efficiency 20-400 MHz PWM Converters using Air-Core Inductors and Monolithic Power Stages in a Normally-Off GaN Process 580

Alihossein Sepahvand, *University of Colorado at Boulder, United States*
Yuanzhe Zhang, *University of Colorado at Boulder, United States*
Dragan Maksimović, *University of Colorado at Boulder, United States*

Thermal Evaluation of Chip-Scale Packaged Gallium Nitride Transistors 587

David Reusch, *Efficient Power Conversion Corporation, United States*
Johan Strydom, *Efficient Power Conversion Corporation, United States*
Alex Lidow, *Efficient Power Conversion Corporation, United States*

Over 300kHz GaN Device based Resonant Bidirectional DCDC Converter with Integrated Magnetics 595

Gang Liu, *Fudan University, China*
Dan Li, *Fudan University, China*
Yungtaek Jang, *Delta Products Corporation, United States*
Jianqiu Zhang, *Fudan University, China*

Effective Control & Software Techniques for High Efficiency GaN FET based Flexible Electrical Power System for Cube-Satellites 601

Ashish Shrivastav, *North Carolina State University, United States*
Shikhar Singh, *IBM, United States*
Anirudh Mahajan, *North Carolina State University, United States*
Subhashish Bhattacharya, *North Carolina State University, United States*

A 98.8% Efficient Bidirectional Full-Bridge Isolated DC-DC GaN Converter 609

Rakesh Ramachandran, *University of Southern Denmark, Denmark*
Morten Nymand, *University of Southern Denmark, Denmark*

Session T12: Electric Machines
Location: 101A
March 23, 2016 8:30 - 10:10
Session Chairs: Bilal Akin, *Univeristy of Texas, Dallas*
Bulent Sarlioglu, *University of Wisconsin - Madison*

Comparison of Lateral- and Cylindrical-Stator Electrical Machines for High-Speed Direct-Drive Applications in Confined Spaces 615

Arda Tüysüz, *ETH Zürich, Switzerland*
Johann W. Kolar, *ETH Zürich, Switzerland*

Novel Contactless Axial-Flux Permanent-Magnet Electromechanical Energy Harvester 623
Michael Flankl, *ETH Zürich, Switzerland*
Arda Tüysüz, *ETH Zürich, Switzerland*
Ivan Subotic, *Liverpool John Moores University, United Kingdom*
Johann W. Kolar, *ETH Zürich, Switzerland*

Design of Rare-Earth Free Five-Phase Outer-Rotor IPM Motor Drive for Electric Bicycle 631
Md. Zakirul Islam, *University of Akron, United States*
Seungdeog Choi, *University of Akron, United States*

Transverse Flux Machines with Rotary Transformer Concept for Wide Speed Operations without using Permanent Magnet Material 638
Iftekhar Hasan, *University of Akron, United States*
Md Wasi Uddin, *University of Akron, United States*
Yilmaz Sozer, *University of Akron, United States*

Field Oriented Modeling and Control of Six Phase, Open-Delta Winding, Interior Permanent Magnet Synchronous Machines Considering Current Unbalance and Zero Sequence Currents 643
Murat Senol, *RWTH Aachen University, Germany*
Michael Schubert, *RWTH Aachen University, Germany*
Georges Engelmann, *RWTH Aachen University, Germany*
Rik W. De Doncker, *RWTH Aachen University, Germany*
Thorben Grosse, *RWTH Aachen University, Germany*
Kay Hameyer, *RWTH Aachen University, Germany*

Session T13: Advances in Magnetics
Location: 101B
March 23, 2016 8:30 - 10:10
Session Chairs: Matthew Wilkowski, *Enpirion*
Charles Sullivan, *Dartmouth*

Passive Integration using FMLF Technique for Integrated Boost Resonant Converters 651
Cheng Deng, *University of Arkansas, United States*
Luciano Andres Garcia Rodriguez, *University of Arkansas, United States*
Juan Zou, *Xiangtan University, China*
Juan Carlos Balda, *University of Arkansas, United States*

Magnetic Characterization Technique and Materials Comparison for Very High Frequency IVR 657
Dongbin Hou, *Virginia Polytechnic Institute and State University, United States*
Fred C. Lee, *Virginia Polytechnic Institute and State University, United States*
Qiang Li, *Virginia Polytechnic Institute and State University, United States*

Large-Signal Power Circuit Characterization of On-Silicon Coupled Inductors for High Frequency Integrated Voltage Regulation 663
S. Kulkarni, *Tyndall National Institute, Ireland*
Z. Pavlovic, *Tyndall National Institute, Ireland*
S. Kubendran, *Tyndall National Institute, Ireland*
C. Carretero, *Universidad de Zaragoza, Spain*
N. Wang, *Tyndall National Institute, Ireland*
C. O'Mathuna, *Tyndall National Institute / University College Cork, Ireland*

Point-of-Load Inductor with High Swinging and Low Loss at Light Load 668
Ting Ge, *Virginia Polytechnic Institute and State University, United States*
Khai Ngo, *Virginia Polytechnic Institute and State University, United States*
Jim Moss, *Texas Instruments Inc., United States*

Iron Loss Evaluation of Three-Phase Inductor for Three-Phase PWM Inverter 676
Hiroaki Matsumori, *Tokyo Metropolitan University, Japan*
Toshihisa Shimizu, *Tokyo Metropolitan University, Japan*
Koushi Takano, *Iwatsu Test Instrument Corporation, Japan*
Ishii Hitoshi, *Iwatsu Test Instrument Corporation, Japan*

Session T14: System Design and Layout for Improved Performance
Location: 102C
March 23, 2016 8:30 - 10:10
Session Chairs: Jeff Nilles, *Texas Instruments*
Ernie Parker, *Crane Aerospace & Electronics*

CMOS Gate Drive IC with Embedded Cross Talk Suppression Circuitry for SiC Devices 684
Jeffery Dix, *University of Tennessee, United States*
Zheyu Zhang, *University of Tennessee, United States*
Benjamin J. Blalock, *University of Tennessee, United States*

Optimal Design of a Voltage Regulator based Resonant Switched-Capacitor Converter IC 692
Eli Abramov, *Ben-Gurion University of the Negev, Israel*
Alon Cervera, *Ben-Gurion University of the Negev, Israel*
Mor Mordechai Peretz, *Ben-Gurion University of the Negev, Israel*

Novel Highly Integrated Current Measurement Method for Drive Inverters 700
N. Langmaack, *Technische Universität Braunschweig, Germany*
G. Tareilus, *Technische Universität Braunschweig, Germany*
M. Henke, *Technische Universität Braunschweig, Germany*

**A Novel DBC Layout for Current Imbalance Mitigation in SiC MOSFET Multichip
Power Modules** 704
Helong Li, *Aalborg University, Denmark*
Stig Munk-Nielsen, *Aalborg University, Denmark*
Szymon Bęczkowski, *Aalborg University, Denmark*
Xiongfei Wang, *Aalborg University, Denmark*

**A Double-End Sourced Multi-Chip Improved Wire-Bonded SiC MOSFET Power
Module Design** 709
Miao Wang, *Ohio State University, United States*
Fang Luo, *Ohio State University, United States*
Longya Xu, *Ohio State University, United States*

Session T15: Modeling of AC Energy Converters and Systems
Location: 104B
March 23, 2016 8:30 - 10:10
Session Chairs: Jaber Abu Qahouq, *The University of Alabama*
Xiongfei Wang, *Aalborg University*

Comparing Extended Kalman Filter and Particle Filter for Estimating Field and Damper Bar Currents in Brushless Wound Field Synchronous Generator for Stator Winding Fault Detection and Diagnosis 715
Sivakumar Nadarajan, *National University of Singapore, Singapore*
S.K. Panda, *National University of Singapore, Singapore*
Bicky Bhangu, *Rolls-Royce Singapore Pte. Ltd., Singapore*
Amit Kumar Gupta, *Rolls-Royce Singapore Pte. Ltd., Singapore*

Analytical Determination of Conduction Power Losses for Active Neutral-Point-Clamped Multilevel Converter 720
Vahid Dargahi, *Clemson University, United States*
Arash Khoshkbar Sadigh, *Extron Electronics, United States*
Keith Corzine, *Clemson University, United States*

Multifrequency Small-Signal Model of Voltage Source Converters Connected to a Weak Grid for Stability Analysis 728
Xing Li, *Huazhong University of Science and Technology, China*
Hua Lin, *Huazhong University of Science and Technology, China*

A New Approach to Control the Modified LinVerter for High Frequency Applications 733
Peyman Farhang, *University of Southern Denmark, Denmark*
Stefan Mátéfi-Tempfli, *University of Southern Denmark, Denmark*

Small-Signal Terminal Characteristics Modeling of Three-Phase Boost Rectifier with Variable Fundamental Frequency 739
Zeng Liu, *Xi'an Jiaotong University, China*
Jinjun Liu, *Xi'an Jiaotong University, China*
Dushan Boroyevich, *Virginia Polytechnic Institute and State University, United States*

Session T16: Manufacturing, Test, and Reliability
Location: 103C
March 23, 2016 8:30 - 10:10
Session Chairs: Jim Marinos, *Payton Group*
Brian Narveson, *Narveson Innovative Consulting*

Reliability Analysis of a High-Efficiency SiC Three-Phase Inverter for Motor Drive Applications 746
Juan Colmenares, *KTH Royal Institute of Technology, Sweden*
Diane-Perle Sadik, *KTH Royal Institute of Technology, Sweden*
Patrik Hilber, *KTH Royal Institute of Technology, Sweden*
Hans-Peter Nee, *KTH Royal Institute of Technology, Sweden*

RCP Evaluation of Electrolytic Capacitor Degradation for SMPS Failure Prediction 754
Hiroshi Nakao, *Fujitsu Laboratories Ltd., Japan*
Yu Yonezawa, *Fujitsu Laboratories Ltd., Japan*
Yoshiyasu Nakashima, *Fujitsu Laboratories Ltd., Japan*
Fujio Kurokawa, *Nagasaki University, Japan*

Modular Test System Architecture for Device, Circuit and System Level Reliability Testing 759
Roland Sleik, *Kompetenzzentrum Automobil- und Industrieelektronik GmbH, Austria*
Michael Glavanovics, *Kompetenzzentrum Automobil- und Industrieelektronik GmbH, Austria*
Sascha Einspieler, *Kompetenzzentrum Automobil- und Industrieelektronik GmbH, Austria*
Annette Muetze, *Technische Universität Graz, Austria*
Klaus Krischan, *Technische Universität Graz, Austria*

EMI Noise Cancelation by Optimizing Transformer Design without Need for the Traditional Y-Capacitor 766
Yongjiang Bai, *Xi'an Jiaotong University, China*
Wenjie Chen, *Xi'an Jiaotong University, China*
Ruirui He, *Xi'an Jiaotong University, China*
Dan Zhang, *Silergy Corp., China*
Xu Yang, *Xi'an Jiaotong University, China*

Manufacturing, Assembly and Production Qualifications of High Density, High Reliability POL DC-DC Converters 772
Fariborz Musavi, *CUI Inc., United States*

Session T17: Soft-Switching Converters in Renewable Energy Systems
Location: 104C
March 23, 2016 8:30 - 10:10
Session Chairs: Khurram Afridi, *University of Colorado at Boulder*
Katherine Kim, *Ulsan NIST*

Power Flow Control and ZVS Analysis of Three Limb High Frequency Transformer based Three-Port DAB 778
Ritwik Chattopadhyay, *North Carolina State University, United States*
Subhashish Bhattacharya, *North Carolina State University, United States*

A Novel Multi-Input Converter using Soft-Switched Single-Switch Input Modules with Integrated Power Factor Correction Capability for Hybrid Renewable Energy Systems 786
Sanjida Moury, *York University, Canada*
John Lam, *York University, Canada*
Vineet Srivastava, *Cistel Technology Inc., Canada*
Ron Church, *Cistel Technology Inc., Canada*

Analysis and Design of Impulse Commutated ZCS Three-Phase Current-Fed Push-Pull DC/DC Converter 794
Radha Sree Krishna Moorthy, *National University of Singapore, Singapore*
Akshay Kumar Rathore, *National University of Singapore, Singapore*

ZCS Resonant Converter based Parallel Balancing of Serially Connected Batteries String 802
Ilya Zeltser, *Rafael Advanced Defense Systems Ltd., Israel*
Or Kirshenboim, *Ben-Gurion University of the Negev, Israel*
Nadav Dahan, *Ben-Gurion University of the Negev, Israel*
Mor Mordechai Peretz, *Ben-Gurion University of the Negev, Israel*

A Novel Topology of High Voltage and High Power Bidirectional ZCS DC-DC Converter based on Serial Capacitors 810
Lejia Sun, *Xi'an Jiaotong University, China*
Fang Zhuo, *Xi'an Jiaotong University, China*
Feng Wang, *Xi'an Jiaotong University, China*
Tianhua Zhu, *Xi'an Jiaotong University, China*

Session T18: Solid State Lighting
Location: 103AB
March 23, 2016 8:30 - 10:10
Session Chairs: Jim Spangler, *Spangler Prototype Inc*
Nan Chen, *ABB*

Control Scheme for TRIAC Dimming High PF Single-Stage LED Driver with Adaptive Bleeder Circuit and Non-Linear Current Reference 816
Weizhong Ma, *Hangzhou Dianzi University, China*
Xiaogao Xie, *Hangzhou Dianzi University, China*
Yang Han, *Hangzhou Dianzi University, China*
Hao Deng, *Hangzhou Dianzi University, China*

Three Phase Converter with Galvanic Isolation based on Loss-Free Resistors for HB-LED Lighting Applications 822
Ignacio Castro, *Universidad de Oviedo, Spain*
Diego G. Lamar, *Universidad de Oviedo, Spain*
Manuel Arias, *Universidad de Oviedo, Spain*
Javier Sebastián, *Universidad de Oviedo, Spain*
Marta M. Hernando, *Universidad de Oviedo, Spain*

A ZV-ZCS Electrolytic Capacitor-Less AC/DC Isolated LED Driver with Continous Energy Regulation 830
John Lam, *York University, Canada*
Nader A. El-Taweel, *York University, Canada*

High Efficiency and Power Density GaN-Based LED Driver 838
Eric Faraci, *Texas Instruments Inc., United States*
Michael Seeman, *Texas Instruments Inc., United States*
Bin Gu, *Texas Instruments Inc., United States*
Yogesh Ramadass, *Texas Instruments Inc., United States*
Paul Brohlin, *Texas Instruments Inc., United States*

A Novel LED Drive System based on Matrix Rectifier 843
Baoping Shi, *Nanjing University of Aeronautics and Astronautics, China*
Bo Zhou, *Nanjing University of Aeronautics and Astronautics, China*
Jiadan Wei, *Nanjing University of Aeronautics and Astronautics, China*
Xianhui Qin, *Nanjing University of Aeronautics and Astronautics, China*
Yuanyu Yang, *Nanjing University of Aeronautics and Astronautics, China*
Bing Liu, *Nanjing University of Aeronautics and Astronautics, China*

Session T19: Resonant and Soft Switching DC-DC Converters
Location: 101A
March 23, 2016 14:00 - 17:30
Session Chairs: Mahshid Amirabadi, *Northeastern University*
Ray Orr, *Solantro*

LLC Synchronous Rectification using Coordinate Modulation 848
Mehdi Mohammadi, *University of British Columbia, Canada*
Navid Shafiei, *University of British Columbia, Canada*
Martin Ordonez, *University of British Columbia, Canada*

Low Parasitics Planar Transformer for LLC Resonant Battery Chargers 854
Mohammad Ali Saket, *University of British Columbia, Canada*
Navid Shafiei, *University of British Columbia, Canada*
Martin Ordonez, *University of British Columbia, Canada*
Marian Craciun, *Delta-Q Technologies Corporation, Canada*
Chris Botting, *Delta-Q Technologies Corporation, Canada*

New Symmetrical Bidirectional L3C Resonant DC-DC Converter with Wide Voltage Range 859
Minjae Kim, *Seoul National University of Science and Technology, Korea, South*
Shinyoung Noh, *Seoul National University of Science and Technology, Korea, South*
Sewan Choi, *Seoul National University of Science and Technology, Korea, South*

Influence of the Junction Capacitance of the Secondary Rectifier Diodes on Output Characteristics in Multi-Resonant Converters 864
Stefan Ditze, *Fraunhofer Institute for Integrated Systems and Device Technology, Germany*
Thomas Heckel, *Fraunhofer Institute for Integrated Systems and Device Technology, Germany*
Martin März, *Fraunhofer Institute for Integrated Systems and Device Technology, Germany*

A Triple Active Bridge DC-DC Converter Capable of Achieving Full-Range ZVS 872
Ling Jiang, *University of Tennessee, United States*
Daniel Costinett, *University of Tennessee, United States*

A Novel High Gain Step-Up Resonant DC-DC Converter for Automotive Application 880
Fei Shang, *Illinois Institute of Technology, United States*
Mahesh Krishnamurthy, *Illinois Institute of Technology, United States*
Alexander Isurin, *Vanner Inc., United States*

Series Injection Enabled Full ZVS Light Load Operation of a 15kV SiC IGBT based Dual Active Half Bridge Converter 886
Awneesh Tripathi, *North Carolina State University, United States*
Sachin Madhusoodhanan, *North Carolina State University, United States*
Krishna Mainali, *North Carolina State University, United States*
Kasunaidu Vechalapu, *North Carolina State University, United States*
Subhashish Bhattacharya, *North Carolina State University, United States*

Soft Switching for Half Bridge Current Doubler for High Voltage Point of Load Converter in Data Center Power Supplies 893
Yutian Cui, *University of Tennessee, United States*
Weimin Zhang, *University of Tennessee, United States*
Leon M. Tolbert, *University of Tennessee, United States*
Daniel J. Costinett, *University of Tennessee, United States*
Fred Wang, *University of Tennessee, United States*
Benjamin J. Blalock, *University of Tennessee, United States*

An Algorithm to Analyze Circulating Current for Multi-Phase Resonant Converter 899
Hongliang Wang, *Queen's University, Canada*
Yang Chen, *Queen's University, Canada*
Zhiyuan Hu, *Queen's University, Canada*
Laili Wang, *Queen's University, Canada*
Tianshu Liu, *Queen's University, Canada*
Wenbo Liu, *Queen's University, Canada*
Yan-Fei Liu, *Queen's University, Canada*
Jahangir Afsharian, *Murata Power Solutions, Canada*
Zhihua Yang, *Murata Power Solutions, Canada*

Session T20: Control Applications and Modulation Schemes
Location: 102C
March 23, 2016 14:00 - 17:30
Session Chairs: Masoud Karimi Ghartemani, *Mississippi state University*
Paul Bauer, *University of Lorraine*

A Simple Active Damping Method for Active Power Filters 907
Huawei Yuan, *Tsinghua University, China*
Xinjian Jiang, *Tsinghua University, China*

Simultaneous Voltage and Current Compensation of the 3-Phase Electric Spring with Decomposed Voltage Control 913
Shuo Yan, *University of Hong Kong, Hong Kong*
Tianbo Yang, *University of Hong Kong, Hong Kong*
C.K. Lee, *University of Hong Kong, Hong Kong*
Siew-Chong Tan, *University of Hong Kong, Hong Kong*
S.Y. Ron Hui, *University of Hong Kong / Imperial College London, Hong Kong*

Self-Synchronization Operation of Global Synchronous Pulsewidth Modulation with Communication Fault Tolerant and Simplified Calculation Capabilities 921
Tao Xu, *Shandong University, China*
Feng Gao, *Shandong University, China*

Design Considerations and Predictive Direct Current Control of Active Regenerative Rectifiers for Harmonic and Current Ripple Reduction 928
Alberto Berzoy, *Florida International University, United States*
A.A.S. Mohamed, *Florida International University, United States*
Osama Mohammed, *Florida International University, United States*

A Robust Controller for Medium Voltage AC Collection Grid for Large Scale Photovoltaic Plants based on Medium Frequency Transformers 936
Bahaa Hafez, *Texas A&M University, United States*
Prasad Enjeti, *Texas A&M University, United States*
Shehab Ahmed, *Texas A&M University at Qatar, Qatar*

Optimal Low Switching Frequency Pulse Width Modulation of Current-Fed Five-Level Inverter for Solar Integration 943
Gnana Sambandam Kulothungan, *National University of Singapore, Singapore*
Akshay K. Rathore, *National University of Singapore, Singapore*
Amarendra Edpuganti, *National University of Singapore, Singapore*
Dipti Srinivasan, *National University of Singapore, Singapore*

Design and Implementation of D-Σ Digital Controlled Multi-function Inverter to Achieve APF, Active Power Injection and Rectification.. 951
T.-F. Wu, *National Tsing Hua University, Taiwan*
H.-C. Hsieh, *National Chung Cheng University, Taiwan*
L.-C. Lin, *National Tsing Hua University, Taiwan*
C.-H. Chang, *National Tsing Hua University, Taiwan*

Operation and Analysis of an Improved Transformerless Unified Power Flow Controller 959
Yang Liu, *Michigan State University, United States*
Shuitao Yang, *Michigan State University / Ford Motor Company, United States*
Fang Zheng Peng, *Michigan State University, United States*

Design Consideration of Converter based Transmission Line Emulation 966
Bo Liu, *University of Tennessee, United States*
Shuoting Zhang, *University of Tennessee, United States*
Sheng Zheng, *University of Tennessee, United States*
Yiwei Ma, *University of Tennessee, United States*
Fred Wang, *University of Tennessee, United States*
Leon M. Tolbert, *University of Tennessee, United States*

Session T21: Advances in Wide BandGap Devices
Location: 104A
March 23, 2016 14:00 - 17:30
Session Chairs: Doug Hopkins, *North Carolina State University*
Alex Huang, *North Carolina State University*

Short-Circuit Characterization of 10 kV 10A 4H-SiC MOSFET 974
Emanuel-Petre Eni, *Aalborg University, Denmark*
Szymon Bęczkowski, *Aalborg University, Denmark*
Stig Munk-Nielsen, *Aalborg University, Denmark*
Tamas Kerekes, *Aalborg University, Denmark*
Remus Teodorescu, *Aalborg University, Denmark*

Record-Low 10mΩ SiC MOSFETs in TO-247, Rated at 900V 979
Vipindas Pala, *Wolfspeed, A Cree Company, United States*
Gangyao Wang, *Wolfspeed, A Cree Company, United States*
Brett Hull, *Wolfspeed, A Cree Company, United States*
Scott Allen, *Wolfspeed, A Cree Company, United States*
Jeffrey Casady, *Wolfspeed, A Cree Company, United States*
John Palmour, *Wolfspeed, A Cree Company, United States*

Performance Evaluation of Multiple Si and SiC Solid State Devices for Circuit Breaker Application in 380VDC Delivery System 983
Kai Tan, *North Carolina State University, United States*
Pengkun Liu, *North Carolina State University, United States*
Xijun Ni, *North Carolina State University, United States*
Chang Peng, *North Carolina State University, United States*
Xiaoqing Song, *North Carolina State University, United States*
Alex Q. Huang, *North Carolina State University, United States*

Evaluation of High Voltage Cascode GaN HEMTs in Parallel Operation 990

He Li, *Ohio State University, United States*
Xuan Zhang, *Ohio State University, United States*
Lucheng Wen, *Ohio State University, United States*
John Alex Brothers, *Ohio State University, United States*
Chengcheng Yao, *Ohio State University, United States*
Ke Zhu, *Ohio State University, United States*
Jin Wang, *Ohio State University, United States*
Liming Liu, *ABB Inc., United States*
Jing Xu, *ABB Inc., United States*
Joonas Puukko, *ABB Inc., United States*

A New Driving Concept for Normally-On GaN Switches in Cascode Configuration 996

Bernhard Zojer, *Infineon Technologies Austria AG, Austria*

Avoiding Divergent Oscillation of Cascode GaN Device Under High Current Turn-Off Condition 1002

Weijing Du, *Virginia Polytechnic Institute and State University, United States*
Xiucheng Huang, *Virginia Polytechnic Institute and State University, United States*
Fred C. Lee, *Virginia Polytechnic Institute and State University, United States*
Qiang Li, *Virginia Polytechnic Institute and State University, United States*
Wenli Zhang, *Virginia Polytechnic Institute and State University, United States*

Temperature-Dependent Turn-On Loss Analysis for GaN HFETs 1010

Edward A. Jones, *University of Tennessee, United States*
Fred Wang, *University of Tennessee, United States*
Daniel Costinett, *University of Tennessee, United States*
Zheyu Zhang, *University of Tennessee, United States*
Ben Guo, *United Technologies Research Center, United States*

Analysis of Parasitic Elements of SiC Power Modules with Special Emphasis on Reliability Issues 1018

Diane-Perle Sadik, *KTH Royal Institute of Technology, Sweden*
Juan Colmenares, *KTH Royal Institute of Technology, Sweden*
Hans-Peter Nee, *KTH Royal Institute of Technology, Sweden*
Konstantin Kostov, *Acreo Swedish ICT AB, Sweden*
Florian Giezendanner, *Alstom Power Sweden AB, Sweden*
Per Ranstad, *Alstom Power Sweden AB, Sweden*

Static and Dynamic Characterization of GaN HEMT with Low Inductance Vertical Phase Leg Design for High Frequency High Power Applications 1024

Nidhi Haryani, *Virginia Polytechnic Institute and State University, United States*
Xuning Zhang, *Virginia Polytechnic Institute and State University, United States*
Rolando Burgos, *Virginia Polytechnic Institute and State University, United States*
Dushan Boroyevich, *Virginia Polytechnic Institute and State University, United States*

Session T22: Motor Drive Design and Inverter Topologies

Location: 101B
March 23, 2016 14:00 - 17:30
Session Chairs: Yingying Kuai, *Caterpillar Inc.*
Jin Wang, *The Ohio State University*

A Family of Single-Phase Current Source Converters with Double Outputs 1032
Louelson A. Costa, *Universidade Federal de Campina Grande, Brazil*
Maurício B.R. Corrêa, *Universidade Federal de Campina Grande, Brazil*
Montiê A. Vitorino, *Universidade Federal de Campina Grande, Brazil*
Gutemberg G. Dos Santos, *Universidade Federal de Campina Grande, Brazil*
Darlan A. Fernandes, *Universidade Federal da Paraíba, Brazil*

Multiple-Output Boost Resonant Inverter for High Efficiency and Cost-Effective Induction Heating Applications 1040
Hector Sarnago, *Universidad de Zaragoza, Spain*
Oscar Lucia, *Universidad de Zaragoza, Spain*
José M. Burdío, *Universidad de Zaragoza, Spain*

Development of 2-kW Interleaved DC-Capacitor-Less Single-Phase Inverter System 1045
Runruo Chen, *Michigan State University, United States*
Hulong Zeng, *Michigan State University, United States*
Deepak Gunasekaran, *Michigan State University, United States*
Yunting Liu, *Michigan State University, United States*
Fang Z. Peng, *Michigan State University, United States*

Single Stage Transformer Isolated High Frequency AC Link based Open End Drive 1051
Srikant Gandikota, *University of Minnesota, United States*
Ned Mohan, *University of Minnesota, United States*

A Quasi-Z-Source Integrated Multi-Port Power Converter with Reduced Capacitance for Switched Reluctance Motor Drives 1057
Fan Yi, *University of Texas at Dallas, United States*
Wen Cai, *University of Texas at Dallas, United States*

A Fault-Tolerant Topology of T-Type NPC Inverter with Increased Thermal Overload Capability 1065
Jiangbiao He, *Marquette University, United States*
Nathan Weise, *Marquette University, United States*
Lixiang Wei, *Rockwell Automation, United States*
Nabeel A.O. Demerdash, *Marquette University, United States*

A Novel Analysis and Design Method of Phase Lead Filters in Repetitive Controllers for Pulse-Width Modulated Inverters 1071
Shunfeng Yang, *Nanyang Technological University, Singapore*
Peng Wang, *Nanyang Technological University, Singapore*
Yi Tang, *Nanyang Technological University, Singapore*
Michael Zagrodnik, *Rolls-Royce Singapore Pte. Ltd., Singapore*
Xiaolei Hu, *Nanyang Technological University, Singapore*
King Jet Tseng, *Nanyang Technological University, Singapore*

Research on the Filter of Load Side Converter in BDFG based Ship Shaft Power Generation System 1078
Meilin Wang, *Huazhong University of Science and Technology, China*
Hua Lin, *Huazhong University of Science and Technology, China*
Hongbin Yang, *Huazhong University of Science and Technology, China*
Xingwei Wang, *Huazhong University of Science and Technology, China*

Investigation of Common Mode Current Related DC-Bus Overvoltage in Multiple Converter Systems 1084
Jiangbiao He, *Rockwell Automation, United States*
Zoran Vrankovic, *Rockwell Automation, United States*
Patrick E. Ozimek, *Rockwell Automation, United States*
Craig Winterhalter, *Rockwell Automation, United States*

Session T23: Modeling of Magnetic Circuits and Systems
Location: 102AB
March 23, 2016 14:00 - 17:30
Session Chairs: Ed Herbert,
Jin Ye, *San Francisco State University*

High Frequency AC Inductor Analysis and Design for Dual Active Bridge (DAB) Converters .. 1090
Zhe Zhang, *Technical University of Denmark, Denmark*
Michael A.E. Andersen, *Technical University of Denmark, Denmark*

A Comprehensive Assessment of PM Motor Topology Impact on Magnet Defect Fault Signatures 1096
Mohsen Zafarani, *University of Texas at Dallas, United States*
Taner Goktas, *University of Texas at Dallas, United States*
Bilal Akin, *University of Texas at Dallas, United States*

High Frequency Modeling for Transformer Common Mode Noise Coupling Path based on Multiconductor Transmission Line Theory 1102
Peipei Meng, *Wuhan University of Technology, China*
Xiangming Zhang, *Naval University of Engineering, China*

Leakage Flux Modelling of Multi-Winding Transformer using Permeance Magnetic Circuit ... 1108
Min Luo, *École Polytechnique Fédérale de Lausanne, Switzerland*
Drazen Dujic, *École Polytechnique Fédérale de Lausanne, Switzerland*
Jost Allmeling, *Plexim GmbH, Switzerland*

Modeling Magnetic Devices using SPICE: Application to Variable Inductors 1115
J. Marcos Alonso, *Universidad de Oviedo, Spain*
Gilberto Martínez, *Continental Automotive R&D, Mexico*
Marina Perdigão, *Universidade de Coimbra, Portugal*
Marcelo Cosetin, *Universidade Federal de Santa Maria, Brazil*
Ricardo N. do Prado, *Universidade Federal de Santa Maria, Brazil*

Investigation of a Thermal Model for a Permanent Magnet Assisted Synchronous Reluctance Motor 1123
Joseph Herbert, *University of Akron, United States*
A.K.M. Arafat, *University of Akron, United States*
Guo-Xiang Wang, *University of Akron, United States*
Seungdeog Choi, *University of Akron, United States*

Design Procedure for Multi-Phase External Rotor Permanent Magnet Assisted Synchronous Reluctance Machines 1131
Sai Sudheer Reddy Bonthu, *University of Akron, United States*
Seungdeog Choi, *University of Akron, United States*

Applicability and Limitations of an M2Spice-Assisted "Planar-Magnetics-in-the-Circuit" Simulation Approach 1138
Samantha J. Gunter, *Massachusetts Institute of Technology, United States*
Minjie Chen, *Massachusetts Institute of Technology, United States*
Stephanie A. Pavlick, *Massachusetts Institute of Technology, United States*
Rose A. Abramson, *Massachusetts Institute of Technology, United States*
Khurram K. Afridi, *University of Colorado at Boulder, United States*
David J. Perreault, *Massachusetts Institute of Technology, United States*

Session T24: Inverter/Converter Control
Location: 103C
March 23, 2016 14:00 - 17:30
Session Chairs: Siavash Pakdelian, *UMass Lowell*
Behrooz Mirafzal, *Kansas State University*

Solution of Input Double-Line Frequency Ripple Rejection for High-Efficiency High-Power Density String Inverter in Photovoltaic Application 1148
Xiaonan Zhao, *Virginia Polytechnic Institute and State University, United States*
Lanhua Zhang, *Virginia Polytechnic Institute and State University, United States*
Rachael Born, *Virginia Polytechnic Institute and State University, United States*
Jih-Sheng Lai, *Virginia Polytechnic Institute and State University, United States*

Fractional-Order Phase Lead Compensation for Multi-Rate Repetitive Control on Three-Phase PWM DC/AC Inverter 1155
Zhichao Liu, *University of South Carolina, United States*
Bin Zhang, *University of South Carolina, United States*
Keliang Zhou, *University of Glasgow, United Kingdom*

A Robust Modified Model Predictive Control (MMPC) based on Lyapunov Function for Three-Phase Active-Front-End (AFE) Rectifier 1163
M. Parvez, *University of Malaya, Malaysia*
S. Mekhilef, *University of Malaya, Malaysia*
Nadia M.L. Tan, *Universiti Tenega Nasional, Malaysia*
Hirofumi Akagi, *Tokyo Institute of Technology, Japan*

Adaptive Reference Model Predictive Control for Power Electronics 1169
Yun Yang, *University of Hong Kong, Hong Kong*
Siew-Chong Tan, *University of Hong Kong, Hong Kong*
Shu-Yuen Ron Hui, *Imperial College London, United Kingdom*

Power Switch Lifetime Extension Strategies for Three-Phase Converters 1176
Serkan Dusmez, *University of Texas at Dallas, United States*
Enes Ugur, *University of Texas at Dallas, United States*
Bilal Akin, *University of Texas at Dallas, United States*

Current Controller Modeling for an Interleaved Boost with Voltage Multiplier Cells for PV Applications 1183

Alessandro Pevere, *Katholieke Universiteit Leuven, Belgium*
Urmimala Chatterjee, *Katholieke Universiteit Leuven, Belgium*
Johan Driesen, *Katholieke Universiteit Leuven, Belgium*

New Active Capacitor Voltage Balancing Method for Five-Level Stacked Multicell Converter 1191

Arash Khoshkbar Sadigh, *Extron Electronics, United States*
Vahid Dargahi, *Clemson University, United States*
Keith Corzine, *Clemson University, United States*

Gate Signal Jitter Elimination and Noise Shaping Modulation for High-SNR Class-D Power Amplifiers 1198

M. Mauerer, *ETH Zürich, Switzerland*
A. Tüysüz, *ETH Zürich, Switzerland*
J.W. Kolar, *ETH Zürich, Switzerland*

Analysis and Compensation of Inverter Nonlinearity for Three-Level T-Type Inverters 1206

Hyeon-Sik Kim, *Seoul National University, Korea, South*
Yong-Cheol Kwon, *Seoul National University, Korea, South*
Seung-Jun Chee, *Seoul National University, Korea, South*
Seung-Ki Sul, *Seoul National University, Korea, South*

Session T25: Topics in Renewable Energy Systems I
Location: 104B
March 23, 2016 14:00 - 17:30
Session Chairs: Fei Gao, *University of Technology of Belfort-Montbéliard*
　　　　　　　Kent Wanner, *John Deere*

Front-End Isolated Quasi-Z-Source DC-DC Converter Modules in Series for Photovoltaic High-Voltage DC Applications 1214

Yushan Liu, *Texas A&M University at Qatar, Qatar*
Haitham Abu-Rub, *Texas A&M University at Qatar, Qatar*
Baoming Ge, *Texas A&M University, United States*

Analysis of Non Detection Zone for Multiple Distributed PCS based on Equivalent Single PCS using Reactive Power Approach 1220

Byeong-Heon Kim, *Seoul National University, Korea, South*
Seung-Ki Sul, *Seoul National University, Korea, South*

Optimal Power Scheduling for a Grid-Connected Hybrid PV-Wind-Battery Microgrid System .. 1227

Adriana Luna, *Aalborg University, Denmark*
Nelson Diaz, *Aalborg University, Denmark*
Mehdi Savaghebi, *Aalborg University, Denmark*
Juan C. Vásquez, *Aalborg University, Denmark*
Josep M. Guerrero, *Aalborg University, Denmark*
Kai Sun, *Tsinghua University, China*
Guoliang Chen, *Shanghai Solar Energy & Technology Co., Ltd., China*
Libing Sun, *Shanghai Solar Energy & Technology Co., Ltd., China*

High Efficiency Power Converter for a Doubly-Fed SOEC/SOFC System 1235
Kevin Tomas-Manez, *Technical University of Denmark, Denmark*
Alexander Anthon, *Technical University of Denmark, Denmark*
Zhe Zhang, *Technical University of Denmark, Denmark*

A Hierarchical Active Balancing Architecture for Li-Ion Batteries 1243
Han-Dong Gui, *Nanjing University of Aeronautics and Astronautics, China*
Zhiliang Zhang, *Nanjing University of Aeronautics and Astronautics, China*
Dong-Jie Gu, *Nanjing University of Aeronautics and Astronautics, China*
Yang Yang, *Nanjing University of Aeronautics and Astronautics, China*
Zhouyu Lu, *Nanjing University of Aeronautics and Astronautics, China*
Yan-Fei Liu, *Queen's University, Canada*

A Series-DG based Autonomous Islanding Microgrid 1249
Beihua Liang, *Tianjin University, China*
Yun Wei Li, *University of Alberta, Canada*
Jinwei He, *Tianjin University, China*
Chengshan Wang, *Tianjin University, China*

**An Enhanced Droop Control Scheme for Resilient Active Power Sharing in Paralleled
Two-Stage PV Inverter Systems** 1253
Hongpeng Liu, *Harbin Institute of Technology, China*
Yongheng Yang, *Aalborg University, Denmark*
Xiongfei Wang, *Aalborg University, Denmark*
Poh Chiang Loh, *Aalborg University, Denmark*
Frede Blaabjerg, *Aalborg University, Denmark*
Wei Wang, *Harbin Institute of Technology, China*
Dianguo Xu, *Harbin Institute of Technology, China*

**Voltage Closed-Loop Virtual Synchronous Generator Control of Full Converter Wind
Turbine for Grid-Connected and Stand-Alone Operation** 1261
Yiwei Ma, *University of Tennessee, United States*
Liu Yang, *University of Tennessee, United States*
Fred Wang, *University of Tennessee, United States*
Leon M. Tolbert, *University of Tennessee, United States*

**DC Voltage Ripple Quantification for a Flywheel-Battery based Hybrid Energy
Storage System** 1267
Christopher R. Lashway, *Florida International University, United States*
Ahmed T. Elsayed, *Florida International University, United States*
Osama A. Mohammed, *Florida International University, United States*

Session T26: Electric Vehicle Charging Systems
Location: 104C
March 23, 2016 14:00 - 17:30
Session Chairs: Jim Spangler, *Spangler Prototype Inc*
Hadi Malek, *Ford*

**Adaptive Loss Reduction Charging Strategy Considering Variation of Internal
Impedance of Lithium-Ion Polymer Batteries in Electric Vehicle Charging Systems** 1273
Nari Kim, *Sungkyunkwan University, Korea, South*
Jung-Hoon Ahn, *Sungkyunkwan University, Korea, South*
Dong-Hee Kim, *Sungkyunkwan University, Korea, South*
Byoung-Kuk Lee, *Sungkyunkwan University, Korea, South*

A Pulse Width Modulated LLC Type Resonant Topology Adpated to Wide Output Voltage Range 1280
Haoyu Wang, *ShanghaiTech University, China*

A Series Resonant Circuit for Voltage Equalization of Series Connected Energy Storage Devices 1286
Yanqi Yu, *University of British Columbia, Canada*
Raed Saasaa, *University of British Columbia, Canada*
Wilson Eberle, *University of British Columbia, Canada*

Implementation of 3.3-kW GaN-Based DC-DC Converter for EV On-Board Charger with Series-Resonant Converter that Employs Combination of Variable-Frequency and Delay-Time Control 1292
Yungtaek Jang, *Delta Products Corporation, United States*
Milan M. Jovanović, *Delta Products Corporation, United States*
Juan M. Ruiz, *Delta Products Corporation, United States*
Misha Kumar, *Delta Products Corporation, United States*
Gang Liu, *Delta Electronics Shanghai Co., Ltd., China*

Dual Active Bridge-Based Full-Integrated Active Filter Auxiliary Power Module for Electrified Vehicle Applications with Single-Phase Onboard Chargers 1300
Ruoyu Hou, *McMaster University, Canada*
Ali Emadi, *McMaster University, Canada*

All-SiC Inductively Coupled Charger with Integrated Plug-In and Boost Functionalities for PEV Applications 1307
M. Chinthavali, *Oak Ridge National Laboratory, United States*
O.C. Onar, *Oak Ridge National Laboratory, United States*
S.L. Campbell, *Oak Ridge National Laboratory, United States*
L.M. Tolbert, *Oak Ridge National Laboratory, United States*

Switching Condition and Loss Modeling of GaN-Based Dual Active Bridge Converter for PHEV Charger 1315
Lingxiao Xue, *Virginia Polytechnic Institute and State University, United States*
Dushan Boroyevich, *Virginia Polytechnic Institute and State University, United States*
Paolo Mattavelli, *Università degli Studi di Padova, Italy*

Analysis of Cascaded Multi-Output-Port Converter for Wireless Plug-In Hybrid/On-Board EV Chargers 1323
Erdem Asa, *Hevo Power Inc. / New York University, United States*
Kerim Colak, *Istanbul Ulasim A.S., Turkey*
Dariusz Czarkowski, *New York University, United States*

Comparative Analysis of High Step-Down Ratio Isolated DC/DC Topologies in PEV Applications 1329
Zhiqing Li, *ShanghaiTech University, China*
Haoyu Wang, *ShanghaiTech University, China*

Session T27: Utility Interface and Inverter Applications
Location: 103AB
March 23, 2016 14:00 - 17:30
Session Chairs: Akshay Kumar Rathore, *Concordia University*
Yichao Tang, *Texas Instruments*

DC to Single-Phase AC Voltage Source Inverter with Power Decoupling Circuit based on Flying Capacitor Topology for PV System 1336
Hiroki Watanabe, *Nagaoka University of Technology, Japan*
Keisuke Kusaka, *Nagaoka University of Technology, Japan*
Keita Furukawa, *Nagaoka University of Technology, Japan*
Koji Orikawa, *Nagaoka University of Technology, Japan*
Jun-Ichi Itoh, *Nagaoka University of Technology, Japan*

GaN FET and Hybrid Modulation based Differential-Mode Inverter 1344
Sudip K. Mazumder, *NextWatt LLC, United States*
Ankit Gupta, *University of Illinois at Chicago, United States*
Shirish Raizada, *University of Illinois at Chicago, United States*
Harshit Soni, *University of Illinois at Chicago, United States*
Nikhil Kumar, *University of Illinois at Chicago, United States*
Paromita Mazumder, *NextWatt LLC, United States*
Parijat Bhattachaarjee, *NextWatt LLC, United States*

Thermal and Electrical Co-Design of a Modular High-Density Single-Phase Inverter using Wide-Bandgap Devices 1350
Steven Chung, *University of Toronto, Canada*
Miad Nasr, *University of Toronto, Canada*
David Guirguis, *University of Toronto, Canada*
Masafumi Otsuka, *University of Toronto, Canada*
Shahab Poshtkouhi, *University of Toronto, Canada*
David K.W. Li, *University of Toronto, Canada*
Vishal Palaniappan, *University of Toronto, Canada*
David Romero, *University of Toronto, Canada*
Cristina Amon, *University of Toronto, Canada*
Ray Orr, *Solantro Semiconductor, Canada*
Olivier Trescases, *University of Toronto, Canada*

Reactive Power Compensation with Improvement of Current Waveform Quality for Single-Phase Buck-Type Dynamic Capacitor 1358
Xinwen Chen, *Huazhong University of Science and Technology, China*
Ke Dai, *Huazhong University of Science and Technology, China*
Chen Xu, *Huazhong University of Science and Technology, China*
Ziwei Dai, *Huazhong University of Science and Technology, China*
Li Peng, *Huazhong University of Science and Technology, China*

Circulating Current Reduction for a D-Σ Digital Controlled Transformerless UPS 1364
T.-F. Wu, *National Tsing Hua University, Taiwan*
T.-H. Shiu, *National Tsing Hua University, Taiwan*
P.-H. Lin, *National Tsing Hua University, Taiwan*
L.-C. Lin, *National Tsing Hua University, Taiwan*
J.-W. Huang, *Industrial Technology Research Institute, Taiwan*

A Multi-Function Three-Level Dynamic Voltage Corrector with Wide Correction Range and Short Circuit Fault Isolation 1371
Jiankun Cao, *Nanjing University of Aeronautics and Astronautics, China*
Pengling Ding, *Nanjing University of Aeronautics and Astronautics, China*
Haichun Liu, *Nanjing University of Aeronautics and Astronautics, China*
Shaojun Xie, *Nanjing University of Aeronautics and Astronautics, China*

Effects and Analysis of Minimum Pulse Width Limitation on Adaptive DC Voltage Control of Grid Converters 1376
Bo Sun, *Aalborg University, Denmark*
Ionut Trintis, *Aalborg University, Denmark*
Stig Munk-Nielsen, *Aalborg University, Denmark*
Josep M. Guerrero, *Aalborg University, Denmark*

Improved Three-Phase Micro-Inverter using Dynamic Dead Time Optimization and Phase-Skipping Control Techniques 1381
S. Milad Tayebi, *University of Central Florida, United States*
Xianmin Mu, *University of Central Florida, United States*
Issa Batarseh, *University of Central Florida, United States*

Correcting Current Imbalances in Three-Phase Four-Wire Distribution Systems 1387
Vinson Jones, *University of Arkansas, United States*
Juan Carlos Balda, *University of Arkansas, United States*

Session T28: Isolated DC-DC Converters
Location: 104A
March 24, 2016 8:30 - 11:20
Session Chairs: Dragan Maksimovic, *UC Boulder*
                     Zhong Ye, *Texas Instruments*

New Design Methdology for Megahertz-Frequency Resonant DC-DC Converters using Impedance Control Network Architecture 1392
Yushi Liu, *University of Colorado at Boulder, United States*
Ashish Kumar, *University of Colorado at Boulder, United States*
Jie Lu, *University of Colorado at Boulder, United States*
Dragan Maksimovic, *University of Colorado at Boulder, United States*
Khurram K. Afridi, *University of Colorado at Boulder, United States*

Dual Voltage Regulations of Single Switch Flyback Converter using Variable Switching Frequency 1398
Jin-Woong Kim, *Seoul National University, Korea, South*
Jung-Ik Ha, *Seoul National University, Korea, South*

On-Chip PLL-Based Methods for Synchronizing Active Switches Across the Isolation Boundary in DC-DC Converters 1403
Shahab Poshtkouhi, *University of Toronto, Canada*
Miad Fard, *University of Toronto, Canada*
Olivier Trescases, *University of Toronto, Canada*

An Isolated Soft-Switching Buck-Boost Converter Utilizing Two Transformers and Embedded Bidirectional Switches on Secondary-Side for Wide Voltage Applications 1410

Tingting Liu, *Nanjing University of Aeronautics and Astronautics, China*
Hongfei Wu, *Nanjing University of Aeronautics and Astronautics, China*
Yan Xing, *Nanjing University of Aeronautics and Astronautics, China*
Kai Sun, *Tsinghua University, China*

Effect of Transformer Design on Operation of Fundamental Duty Modulation for Dual-Active-Bridge Converter 1416

Wooin Choi, *Seoul National University, Korea, South*
Moonhyun Lee, *Seoul National University, Korea, South*
Bo-Hyung Cho, *Seoul National University, Korea, South*

A High Step-Up Bidirectional Isolated Dual-Active-Bridge Converter with Three-Level Voltage-Doubler Rectifier for Energy Storage Applications 1424

Xiaohai Zhan, *Nanjing University of Aeronautics and Astronautics, China*
Hongfei Wu, *Nanjing University of Aeronautics and Astronautics, China*
Yan Xing, *Nanjing University of Aeronautics and Astronautics, China*
Hongjuan Ge, *Nanjing University of Aeronautics and Astronautics, China*
Xi Xiao, *Tsinghua University, China*

Digitized Self-Oscillating Loop for Piezoelectric Transformer-Based Power Converters 1430

Marzieh Ekhtiari, *Technical University of Denmark, Denmark*
Thomas Andersen, *Technical University of Denmark, Denmark*
Zhe Zhang, *Technical University of Denmark, Denmark*
Michael A.E. Andersen, *Technical University of Denmark, Denmark*

Session T29: Multilevel Converters
Location: 101A
March 24, 2016 8:30 - 11:20
Session Chairs: Maryam Saeedifard, *Georgia Tech*
Julia Zhang, *Oregon State University*

An Isolated Topology for Reactive Power Compensation with a Modularized Dynamic-Current Building-Block 1437

Hao Chen, *Georgia Institute of Technology, United States*
Anish Prasai, *Varentec, Inc., United States*
Deepak Divan, *Georgia Institute of Technology, United States*

Design and Control of a Compact MMC Submodule Structure with Reduced Capacitor Size using the Stacked Switched Capacitor Architecture 1443

Yuan Tang, *University of Warwick, United Kingdom*
Minjie Chen, *Massachusetts Institute of Technology, United States*
Li Ran, *University of Warwick, United Kingdom*

Fundamental Frequency Sorting Strategy for Capacitor Voltage Balance of Modular Multilevel Converters with Phase Disposition PWM 1450

Kun Wang, *Zhejiang University, China*
Yan Deng, *Zhejiang University, China*
Wenyu Li, *Zhejiang University, China*
Hao Peng, *Zhejiang University, China*
Guipeng Chen, *Zhejiang University, China*
Xiangning He, *Zhejiang University, China*

Active Voltage Balancing Control for 10kV Three-Level Converter using Series-Connected HV-IGBTs 1456

Shiqi Ji, *Tsinghua University, China*
Ting Lu, *Tsinghua University, China*
Zhengming Zhao, *Tsinghua University, China*
Hualong Yu, *Tsinghua University, China*
Fred Wang, *University of Tennessee, United States*

Average-Value Model of Modular Multilevel Converters Considering Capacitor Voltage 1462

Heya Yang, *Zhejiang University, China*
Yuxiang Chen, *Zhejiang University, China*
Wuhua Li, *Zhejiang University, China*
Xiangning He, *Zhejiang University, China*
Wei Sun, *China Electric Power Research Institute, China*
Yongning Chi, *China Electric Power Research Institute, China*
Yan Li, *China Electric Power Research Institute, China*

New Submodule Circuits for Modular Multilevel Current Source Converters with DC Fault Ride through Capability 1468

Xinyu Yu, *Tsinghua University, China*
Yingdong Wei, *Tsinghua University, China*
Qirong Jiang, *Tsinghua University, China*

Voltage and Power Balance Control Strategy for Three-Phase Modular Cascaded Solid Stated Transformer 1475

Zhiyu Zhang, *Zhejiang University, China*
Hengyang Zhao, *Zhejiang University, China*
Shihang Fu, *Zhejiang University, China*
Jianjiang Shi, *Zhejiang University, China*
Xiangning He, *Zhejiang University, China*

Session T30: Multilevel and Matrix Converters for Motor Drives
Location: 102C
March 24, 2016 8:30 - 11:20
Session Chairs: SeonHwan Hwang, *Kyungnam University, Korea*
Xiaohu Liu, *GE*

New Flying-Capacitor-Based Multilevel Converter with Optimized Number of Switches and Capacitors Controlled with a New Logic-Form-Equation based Active Voltage Balancing Technique 1481

Vahid Dargahi, *Clemson University, United States*
Arash Khoshkbar Sadigh, *Extron Electronics, United States*
Keith Corzine, *Clemson University, United States*

New Low-Cost Five-Level Active Neutral-Point Clamped Converter 1489

Hongliang Wang, *Queen's University, Canada*
Lei Kou, *Queen's University, Canada*
Yan-Fei Liu, *Queen's University, Canada*
Paresh C. Sen, *Queen's University, Canada*
Sucheng Liu, *Anhui University of Technology, China*

Medium Voltage (≥ 2.3 kV) High Frequency Three-Phase Two-Level Converter Design and Demonstration using 10 kV SiC MOSFETs for High Speed Motor Drive Applications .. 1497

Sachin Madhusoodhanan, *North Carolina State University, United States*
Krishna Mainali, *North Carolina State University, United States*
Awneesh Tripathi, *North Carolina State University, United States*
Kasunaidu Vechalapu, *North Carolina State University, United States*
Subhashish Bhattacharya, *North Carolina State University, United States*

Novel Three Phase Multi-Level Inverter Topology with Symmetrical DC-Voltage Sources . 1505

Ahmed Salem, *Aswan University, Egypt*
Emad M. Ahmed, *Aswan University, Egypt*
Mahrous Ahmed, *Aswan University, Egypt*
Mohamed Orabi, *Aswan University, Egypt*

A 2 kW, Single-Phase, 7-Level, GaN Inverter with an Active Energy Buffer Achieving 216 W/in^3 Power Density and 97.6% Peak Efficiency 1512

Yutian Lei, *University of Illinois at Urbana-Champaign, United States*
Christopher Barth, *University of Illinois at Urbana-Champaign, United States*
Shibin Qin, *University of Illinois at Urbana-Champaign, United States*
Wen-Chuen Liu, *University of Illinois at Urbana-Champaign, United States*
Intae Moon, *University of Illinois at Urbana-Champaign, United States*
Andrew Stillwell, *University of Illinois at Urbana-Champaign, United States*
Derek Chou, *University of Illinois at Urbana-Champaign, United States*
Thomas Foulkes, *University of Illinois at Urbana-Champaign, United States*
Zichao Ye, *University of Illinois at Urbana-Champaign, United States*
Zitao Liao, *University of Illinois at Urbana-Champaign, United States*
Robert C.N. Pilawa-Podgurski, *University of Illinois at Urbana-Champaign, United States*

Indirect Matrix Converter based Open-End Winding AC Drives with Zero Common-Mode Voltage 1520

Saurabh Tewari, *MTS Systems Corporation, United States*
Ranjan K. Gupta, *First Solar, Inc., United States*
Apurva Somani, *Dynapower Company LLC, United States*
Ned Mohan, *University of Minnesota, United States*

Precharging Strategy for Soft Startup Process of Modular Multilevel Converters based on Various SM Circuits 1528

Jiangchao Qin, *Arizona State University, United States*
Suman Debnath, *Oak Ridge National Laboratory, United States*
Maryam Saeedifard, *Georgia Institute of Technology, United States*

Session T31: System Design Techniques for Reduced EMI
Location: 101B
March 24, 2016 8:30 - 11:20
Session Chairs: John Vigars, *Allegro Microsystems*
Doug Hopkins, *North Carolina State University*

Conducted EMI Analysis and Filter Design for MHz Active Clamp Flyback Front-End Converter 1534

Xiucheng Huang, *Virginia Polytechnic Institute and State University, United States*
Junjie Feng, *Virginia Polytechnic Institute and State University, United States*
Fred C. Lee, *Virginia Polytechnic Institute and State University, United States*
Qiang Li, *Virginia Polytechnic Institute and State University, United States*
Yuchen Yang, *Virginia Polytechnic Institute and State University, United States*

EMC Investigation of a Very High Frequency Self-Oscillating Resonant Power Converter 1541
Jeppe A. Pedersen, *Technical University of Denmark, Denmark*
Arnold Knott, *Technical University of Denmark, Denmark*
Michael A.E. Andersen, *Technical University of Denmark, Denmark*

Numerical Optimization of Passive Line Filter Components for Suppression of Electromagnetic Interference (EMI) 1547
Carsten Henkenius, *Universität Paderborn, Germany*
Norbert Fröhleke, *Universität Paderborn, Germany*
Joachim Böcker, *Universität Paderborn, Germany*
Heiko Figge, *Delta Energy Systems GmbH, Germany*

Electromagnetic Noise Coupling and Mitigation for Fast Response On-Die Temperature Sensing in High Power Modules 1554
Chengcheng Yao, *Ohio State University, United States*
Pengzhi Yang, *Ohio State University, United States*
Mingzhi Leng, *Ohio State University, United States*
He Li, *Ohio State University, United States*
Lixing Fu, *Ohio State University, United States*
Jin Wang, *Ohio State University, United States*
Ke Zou, *Ford Motor Company, United States*
Chingchi Chen, *Ford Motor Company, United States*

Ultra-Low Inductance Vertical Phase Leg Design with EMI Noise Propagation Control for Enhancement Mode GaN Transistors 1561
Xuning Zhang, *Virginia Polytechnic Institute and State University, United States*
Zhiyu Shen, *Virginia Polytechnic Institute and State University, United States*
Nidhi Haryani, *Virginia Polytechnic Institute and State University, United States*
Dushan Boroyevich, *Virginia Polytechnic Institute and State University, United States*
Rolando Burgos, *Virginia Polytechnic Institute and State University, United States*

Decoupling of Interaction between WBG Converter and Motor Load for Switching Performance Improvement 1569
Zheyu Zhang, *University of Tennessee, United States*
Fred Wang, *University of Tennessee, United States*
Leon M. Tolbert, *University of Tennessee, United States*
Benjamin J. Blalock, *University of Tennessee, United States*
Daniel J. Costinett, *University of Tennessee, United States*

Control and Characterization of Electromagnetic Emissions in Wide Band Gap based Converter Modules for Ungrounded Grid-Forming Applications 1577
Robert Cuzner, *University of Wisconsin at Milwaukee, United States*
Rasoul Hosseini, *University of Wisconsin at Milwaukee, United States*
Andrew Lemmon, *University of Alabama, United States*
James Gafford, *Mississippi State University, United States*
Michael Mazzola, *Mississippi State University, United States*

Session T32: Modeling of DC Energy Converters and Systems
Location: 102AB
March 24, 2016 8:30 - 11:20
Session Chairs: Santanu Kapat, *IIT Kharagpur*
Sombuddha Chakraborty, *Texas Instruments*

A Practical Switching Time Model for Synchronous Buck Converters 1585
Yuan Rao, *Texas Instruments Inc., United States*
Surinder P. Singh, *Texas Instruments Inc., United States*
Taisuke Kazama, *Texas Instruments Inc., United States*

**Off-Line Identification of Digitally Controlled Power Converters using an Analog
Frequency Response Analyzer** 1591
Marco Meola, *Zentrum Mikroelektronik Dresden AG, Germany*
Anthony Kelly, *Altera Corporation, Ireland*

**Extended Wide-Load Range Model for Multi-Level DC-DC Converters and a Practical
Dual-Mode Digital Controller** 1597
Nenad Vukadinović, *University of Toronto, Canada*
Aleksandar Prodić, *University of Toronto, Canada*
Brett A. Miwa, *Maxim Integrated, United States*
Cory B. Arnold, *Maxim Integrated, United States*
Michael W. Baker, *Maxim Integrated, United States*

Burst Mode Control and Switched-Capacitor Converters Losses 1603
Michael Evzelman, *Utah State University, United States*
Regan Zane, *Utah State University, United States*

Equivalent Circuit Modeling of LLC Resonant Converter 1608
Shuilin Tian, *Virginia Polytechnic Institute and State University, United States*
Fred C. Lee, *Virginia Polytechnic Institute and State University, United States*
Qiang Li, *Virginia Polytechnic Institute and State University, United States*

Small Signal Modeling of the Hysteretic Modulator with a Current Ripple Synthesizer 1616
Yi Huang, *Intersil Corporation, United States*
Chun Cheung, *Intersil Corporation, United States*

A Black-Box Modeling Approach for DC Nanogrids 1624
A. Francés, *Universidad Politécnica de Madrid, Spain*
R. Asensi, *Universidad Politécnica de Madrid, Spain*
O. García, *Universidad Politécnica de Madrid, Spain*
R. Prieto, *Universidad Politécnica de Madrid, Spain*
J. Uceda, *Universidad Politécnica de Madrid, Spain*

Session T33: Gate Drive Techniques

Location: 103C
March 24, 2016 8:30 - 11:20
Session Chairs: Christopher Bridge, *SIMPLIS Technologies*
Martin Ordonez, *University of British Columbia*

Design and Evaluation of Isolated Gate Driver Power Supply for Medium Voltage Converter Applications 1632
Krishna Mainali, *North Carolina State University, United States*
Sachin Madhusoodhanan, *North Carolina State University, United States*
Awneesh Tripathi, *North Carolina State University, United States*
Kasunaidu Vechalapu, *North Carolina State University, United States*
Ankan De, *North Carolina State University, United States*
Subhashish Bhattacharya, *North Carolina State University, United States*

General-Purpose Clocked Gate Driver (CGD) IC with Programmable 63-Level Drivability to Reduce IC Overshoot and Switching Loss of Various Power Transistors 1640
Koutarou Miyazaki, *University of Tokyo, Japan*
Seiya Abe, *Kyushu Institute of Technology, Japan*
Masanori Tsukuda, *Kyushu Institute of Technology, Japan*
Ichiro Omura, *Kyushu Institute of Technology, Japan*
Keiji Wada, *Tokyo Metropolitan University, Japan*
Makoto Takamiya, *University of Tokyo, Japan*
Takayasu Sakurai, *University of Tokyo, Japan*

An Integrated SiC CMOS Gate Driver 1646
Matthew Barlow, *University of Arkansas, United States*
Shamim Ahmed, *University of Arkansas, United States*
H. Alan Mantooth, *University of Arkansas, United States*
A. Matt Francis, *Ozark Integrated Circuits, Inc., United States*

Digital Active Gate Drives using Sequential Optimization 1650
Daniel J. Rogers, *University of Oxford, United Kingdom*
Boris Murmann, *Stanford University, United States*

One Adaptive Turn-Off Method for PFC Converter with Voltage Spike Limitation 1657
Qunfang Wu, *Nanjing University of Aeronautics and Astronautics, China*
Qin Wang, *Nanjing University of Aeronautics and Astronautics, China*
Lan Xiao, *Nanjing University of Aeronautics and Astronautics, China*
Jialin Xu, *Nanjing University of Aeronautics and Astronautics, China*
Hongxu Li, *Nanjing University of Aeronautics and Astronautics, China*

A Digital Implementation for PWM Phase-Frequency Synchronization in SMPS Systems 1663
Luca Bizjak, *Infineon Technologies Austria AG, Austria*
Emanuele Bodano, *Infineon Technologies Austria AG, Austria*
Ante Gotovac, *Infineon Technologies Austria AG, Austria*
Sergii Tkachov, *Infineon Technologies Austria AG, Austria*

A High Accuracy and High Bandwidth Current Sense Circuit for Digitally Controlled DC-DC Buck Converters 1670
David Stack, *Altera Corporation, Ireland*
Anthony Kelly, *Altera Corporation, Ireland*
Thomas Conway, *University of Limerick, Ireland*

Session T34: Energy Storage Systems
Location: 104B
March 24, 2016 8:30 - 11:20
Session Chairs: Wei Qiao, *University of Nebraska Lincoln*
Yilmaz Sozer, *University of Akron*

Modular Multilevel Dual Active Bridge DC-DC Converter with ZVS and Fast DC Fault Recovery for Battery Energy Storage Systems 1675
Yuxiang Shi, *Florida State University, United States*
Rui Li, *Florida State University, United States*
Hui Li, *Florida State University, United States*

An Analytical Framework to Design a Dynamic Frequency Control Scheme for Microgrids using Energy Storage 1682
Ajit A. Renjit, *Ohio State University, United States*
Feng Guo, *NEC Laboratories America, Inc., United States*
Ratnesh Sharma, *NEC Laboratories America, Inc., United States*

Comparative Evaluation of LiFePO$_4$ Cell SOC Estimation Performance with ECM Structure and Noise Model/Data Rejection in the EKF for Transportation Application 1690
Hyun-jun Lee, *Soongsil University, Korea, South*
Joung-hu Park, *Soongsil University, Korea, South*
Jonghoon Kim, *Chosun University, Korea, South*

A Power Sharing Scheme for Series Connected Offshore Wind Turbines in a Medium Voltage DC Collection Grid 1695
Michael T. Daniel, *Texas A&M University, United States*
Prasad N. Enjeti, *Texas A&M University, United States*

Fault Ride-Through Performance Evaluation of an Interleaved Grid-Connected Converter Employing Low Switching Frequency 1702
Lorand Bede, *Aalborg University, Denmark*
Ghanshyamsinh Gohil, *Aalborg University, Denmark*
Mihai Ciobotaru, *University of New South Wales, Australia*
Tamas Kerekes, *Aalborg University, Denmark*
Remus Teodorescu, *Aalborg University, Denmark*
Vassilios G. Agelidis, *University of New South Wales, Australia*

Analysis of Two Charging Modes of Battery Energy Storage System for a Stand-Alone Microgrid 1708
Jongmin Jo, *Chungnam National University, Korea, South*
Hanju Cha, *Chungnam National University, Korea, South*

Proposition and Experimental Verification of a Bi-Directional Isolated DC/DC Converter for Battery Charger-Discharger of Electric Vehicle 1713
Ryota Kondo, *Mitsubishi Electric Corporation, Japan*
Yusuke Higaki, *Mitsubishi Electric Corporation, Japan*
Masaki Yamada, *Mitsubishi Electric Corporation, Japan*

Session T35: Topics on Inductive and Capacitive Wireless Power Transfer
Location: 104C
March 24, 2016 8:30 - 11:20
Session Chairs: Chris Mi, *San Diego State University*
Omer Onar, *Oak Ridge National Laboratory*

A CLLC-Compensated High Power and Large Air-Gap Capacitive Power Transfer System for Electric Vehicle Charging Applications 1721
Fei Lu, *University of Michigan at Ann Arbor, United States*
Hua Zhang, *Northeastern Polytechnical University, China*
Heath Hofmann, *University of Michigan at Ann Arbor, United States*
Chris Mi, *San Diego State University, United States*

A Large Air-Gap Capacitive Power Transfer System with a 4-Plate Capacitive Coupler Structure for Electric Vehicle Charging Applications 1726
Hua Zhang, *Northwestern Polytechnical University, China*
Fei Lu, *University of Michigan at Ann Arbor, United States*
Heath Hofmann, *University of Michigan at Ann Arbor, United States*
Weiguo Liu, *Northwestern Polytechnical University, China*
Chris Mi, *San Diego State University, United States*

Dynamic Wireless Power Transfer System for Electric Vehicles to Simplify Ground Facilities – Power Control and Efficiency Maximization on the Secondary Side – 1731
Katsuhiro Hata, *University of Tokyo, Japan*
Takehiro Imura, *University of Tokyo, Japan*
Yoichi Hori, *University of Tokyo, Japan*

Uniform-Gain Frequency Tracking of Wireless EV Charging for Improving Alignment Flexibility 1737
Yabiao Gao, *University of Georgia, United States*
Antonio Ginart, *University of Georgia / Sonnenbatterie GmbH, United States*
Kathleen Blair Farley, *Southern Company Services, Inc., United States*
Zion Tsz Ho Tse, *University of Georgia, United States*

Design and Optimization of a Multi-Coil System for Inductive Charging with Small Air Gap .. 1741
Christopher Joffe, *Fraunhofer Institute for Integrated Systems and Device Technology, Germany*
Andreas Roßkopf, *Fraunhofer Institute for Integrated Systems and Device Technology, Germany*
Stefan Ehrlich, *Fraunhofer Institute for Integrated Systems and Device Technology, Germany*
Christian Dobmeier, *Fraunhofer Institute for Integrated Systems and Device Technology, Germany*
Martin März, *Fraunhofer Institute for Integrated Systems and Device Technology, Germany*

Core Design for Better Misalignment Tolerance and Higher Range of Wireless Charging for HEV 1748
Mostak Mohammad, *University of Akron, United States*
Sangshin Kwak, *Chung-ang University, Korea, South*
Seungdeog Choi, *University of Akron, United States*

A 25 kW Industrial Prototype Wireless Electric Vehicle Charger 1756
Mariusz Bojarski, *Hevo Power Inc., United States*
Erdem Asa, *Hevo Power Inc. / New York University, United States*
Kerim Colak, *Istanbul Ulasim A.S., Turkey*
Dariusz Czarkowski, *New York University, United States*

Session T36: Wireless Power Transfer

Location: 103AB
March 24, 2016 8:30 - 11:20
Session Chairs: Sriram Jala Reddy, *Ford Motors*
Michael Masquelier, *WAVE*

Full-Bridge Series Resonant Multi-Inverter Featuring New 900-V SiC Devices for Improved Induction Heating Appliances 1762
Mario Pérez-Tarragona, *Universidad de Zaragoza, Spain*
Héctor Sarnago, *Universidad de Zaragoza, Spain*
Óscar Lucía, *Universidad de Zaragoza, Spain*
José M. Burdío, *Universidad de Zaragoza, Spain*

A Novel Phase Control of Single Switch Active Rectifier for Inductive Power Transfer Applications 1767
Kerim Colak, *Istanbul Ulasim A.S., Turkey*
Erdem Asa, *Hevo Power Inc. / New York University, United States*
Dariusz Czarkowski, *New York University, United States*

Optimal Shaped Dipole-Coil Design and Experimental Verification of Inductive Power Transfer System for Home Applications 1773
Duy T. Nguyen, *Korea Advanced Institute of Science and Technology, Korea, South*
Eun S. Lee, *Korea Advanced Institute of Science and Technology, Korea, South*
Byeung G. Choi, *Korea Advanced Institute of Science and Technology, Korea, South*
Chun T. Rim, *Korea Advanced Institute of Science and Technology, Korea, South*

A Novel Time-Sharing Current-Fed ZCS High Frequency Inverter-Applied Resonant DC-DC Converter for Inductive Power Transfer 1780
Kyohei Konishi, *Kobe University, Japan*
Tomokazu Mishima, *Kobe University, Japan*
Mutsuo Nakaoka, *University of Malaya, Malaysia*

Optimization of Coils for Magnetically Coupled Resonant Wireless Power Transfer System based on Maximum Output Power 1788
Dan Jiang, *Nanjing University of Aeronautics and Astronautics, China*
Yong Yang, *Nanjing University of Aeronautics and Astronautics, China*
Fuxin Liu, *Nanjing University of Aeronautics and Astronautics, China*
Xinbo Ruan, *Nanjing University of Aeronautics and Astronautics, China*
Xuling Chen, *Nanjing University of Aeronautics and Astronautics, China*

Online Regulation of Receiver-Side Power and Estimation of Mutual Inductance in Wireless Inductive Link based on Transmitter-Side Electrical Information 1795
Jeff Po Wa Chow, *City University of Hong Kong, Hong Kong*
Henry Shu-Hung Chung, *City University of Hong Kong, Hong Kong*
Chun Sing Cheng, *City University of Hong Kong, Hong Kong*

Dynamic Period Switching of PRS-PWM with Run-Length Limiting Technique for Spurious and Ripple Reduction in Fast Response Wireless Power Transmission 1802
Takahiro Moroto, *Keio University, Japan*
Toru Kawajiri, *Keio University, Japan*
Hiroki Ishikuro, *Keio University, Japan*

Session T37: Single-Phase AC-DC Converters
Location: 102AB
March 24, 2016 14:00 - 17:30
Session Chairs: Dusty Becker, *Emerson Network Power*
Pritam Das, *National University of Singapore*

A Flyback AC/DC Converter using Power Semiconductor Filter for Input Power Factor Correction 1807
Chung-Pui Tung, *City University of Hong Kong, Hong Kong*
Henry Shu-Hung Chung, *City University of Hong Kong, Hong Kong*

Reducing the Variation Range of the Switching Frequency for CRM Boost PFC Converter by Injecting 3rd Harmonic into the Input Current 1815
Yi Wang, *Nanjing University of Science and Technology, China*
Kai Yao, *Nanjing University of Science and Technology, China*

A Sustained Increase of Input Current Distortion in Active Input Current Shapers to Eliminate Electrolytic Capacitor for Designing AC to DC HB-LED Drivers for Retrofit Lamps Applications 1823
D.G. Lamar, *Universidad de Oviedo, Spain*
M. Arias, *Universidad de Oviedo, Spain*
A. Rodriguez, *Universidad de Oviedo, Spain*
J. Sebastian, *Universidad de Oviedo, Spain*
A. Fernandez, *European Space Agency, Netherlands*
J.A. Villarejo, *Universidad de Cartagena, Spain*

Reduced Current Stress Bridgeless Cuk PFC Converter with New Voltage Multiplier Circuit ... 1831
Yi-Hung Liao, *National Penghu University of Science and Technology, Taiwan*

Implementation of Multi-Level Bridgeless PFC Rectifiers for Mid-Power Single Phase Applications 1835
Trong Tue Vu, *Eisergy Ltd., Ireland*
George Young, *Eisergy Ltd., Ireland*

US Mains Stacked Very High Frequency Self-Oscillating Resonant Power Converter with Unified Rectifier 1842
Jeppe A. Pedersen, *Technical University of Denmark, Denmark*
Mickey P. Madsen, *Technical University of Denmark, Denmark*
Jakob D. Mønster, *Technical University of Denmark, Denmark*
Thomas Andersen, *Technical University of Denmark, Denmark*
Arnold Knott, *Technical University of Denmark, Denmark*
Michael A.E. Andersen, *Technical University of Denmark, Denmark*

Digital-Based Interleaving Control for GaN-Based MHz CRM Totem-Pole PFC 1847
Zhengyang Liu, *Virginia Polytechnic Institute and State University, United States*
Zhengrong Huang, *Virginia Polytechnic Institute and State University, United States*
Fred C. Lee, *Virginia Polytechnic Institute and State University, United States*
Qiang Li, *Virginia Polytechnic Institute and State University, United States*

A Novel AC-to-DC Adaptor with Ultra-High Power Density and Efficiency 1853
Yan-Cun Li, *Virginia Polytechnic Institute and State University, United States*
Fred C. Lee, *Virginia Polytechnic Institute and State University, United States*
Qiang Li, *Virginia Polytechnic Institute and State University, United States*
Xiucheng Huang, *Virginia Polytechnic Institute and State University, United States*
Zhengyang Liu, *Virginia Polytechnic Institute and State University, United States*

**A Single-Stage Single-Phase Isolated AC-DC Converter based on LLC Resonant Unit
and T-Type Three-Level Unit for Battery Charging Applications** 1861
Yikai Gao, *University of Texas at Dallas, United States*
Wen Cai, *University of Texas at Dallas, United States*
Fan Yi, *University of Texas at Dallas, United States*

Session T38: Non-Isolated DC-DC Converters
Location: 101A
March 24, 2016 14:00 - 17:30
Session Chairs: Pradeep Shenoy, *Texas Instruments*
Juan Rivas-Davila, *Stanford*

DC-DC Power Converter Controller for SOC Balancing of Paralleled Battery System 1868
Jaber A. Abu Qahouq, *University of Alabama, United States*
Lin Zhang, *University of Alabama, United States*
Yuan Cao, *University of Alabama, United States*
Bharat Balasubramanian, *University of Alabama, United States*

Ultra-Step-Up DC-DC Converter with Integrated Autotransformer and Coupled Inductor ... 1872
Yam P. Siwakoti, *Aalborg University, Denmark*
Frede Blaabjerg, *Aalborg University, Denmark*
Poh Chiang Loh, *Aalborg University, Denmark*

Optimal Dynamic Phase Add/Drop Mechanism in Multiphase DC-DC Buck Converters 1878
Anandha Ruban T T, *Texas Instruments India Pvt. Ltd., India*
Preetam Tadeparthy, *Texas Instruments India Pvt. Ltd., India*
Sankaran Aniruddhan, *Indian Institute of Technology Madras, India*
Vikram Gakhar, *Texas Instruments India Pvt. Ltd., India*
Muthusubramanian Venkateswaran, *Texas Instruments India Pvt. Ltd., India*

**A Universal Self-Calibrating Dynamic Voltage and Frequency Scaling (DVFS) Scheme
with Thermal Compensation for Energy Savings in FPGAs** 1882
Shuze Zhao, *University of Toronto, Canada*
Ibrahim Ahmed, *University of Toronto, Canada*
Carl Lamoureux, *University of Toronto, Canada*
Ashraf Lotfi, *Altera Corporation, United States*
Vaughn Betz, *University of Toronto, Canada*
Olivier Trescases, *University of Toronto, Canada*

**Morphing Switched-Capacitor Step-Down DC-DC Converters with Variable
Conversion Ratio** 1888
Song Xiong, *University of Hong Kong, Hong Kong*
Ying Huang, *University of Hong Kong, Hong Kong*
Siew-Chong Tan, *University of Hong Kong, Hong Kong*
Shu-Yuen Ron Hui, *University of Hong Kong, Hong Kong*

Compact Modular Switched-Capacitor DC/DC Converters with Exponential Voltage Gain 1894
Ying Huang, *University of Hong Kong, Hong Kong*
Song Xiong, *University of Hong Kong, Hong Kong*
Siew-Chong Tan, *University of Hong Kong, Hong Kong*
Shu-Yuen Ron Hui, *University of Hong Kong, Hong Kong*

Study and Implementation of a High Step-Up Voltage DC-DC Converter using Coupled-Inductor and Cascode Techniques 1900
Tsorng-Juu Liang, *National Cheng Kung University, Taiwan*
Yung-Ting Huang, *National Cheng Kung University, Taiwan*
Jian-Hsing Lee, *National Cheng Kung University, Taiwan*
Lo Pang-Yen Ting, *National Cheng Kung University, Taiwan*

20 mV Input, 4.2 V Output Boost Converter with Methodology of Maximum Output Power for Thermoelectric Energy Harvesting 1907
Taichi Ogawa, *Toshiba Corporation, Japan*
Takeshi Ueno, *Toshiba Corporation, Japan*
Takayuki Miyazaki, *Toshiba Corporation, Japan*
Tetsuro Itakura, *Toshiba Corporation, Japan*

Clarification of Relationship between Current Ripple and Power Density in Bidirectional DC-DC Converter 1911
Hoai Nam Le, *Nagaoka University of Technology, Japan*
Koji Orikawa, *Nagaoka University of Technology, Japan*
Jun-Ichi Itoh, *Nagaoka University of Technology, Japan*

Session T39: Inverter Applications and Technologies
Location: 101B
March 24, 2016 14:00 - 17:30
Session Chairs: Ali Khajehoddin, *University of Alberta*
Wen Cai, *University of Texas, Dallas*

Grid-Voltage Feedforward based Control for Grid-Connected LCL-Filtered Inverter with High Robustness and Low Grid Current Distortion in Weak Grid 1919
Jinming Xu, *Nanjing University of Aeronautics and Astronautics, China*
Qiang Qian, *Nanjing University of Aeronautics and Astronautics, China*
Shaojun Xie, *Nanjing University of Aeronautics and Astronautics, China*
Binfeng Zhang, *Nanjing University of Aeronautics and Astronautics, China*

Evaluation of PV Frequency-Watt Function for Fast Frequency Reserves 1926
J. Neely, *Sandia National Laboratories, United States*
J. Johnson, *Sandia National Laboratories, United States*
J. Delhotal, *Sandia National Laboratories, United States*
S. Gonzalez, *Sandia National Laboratories, United States*
M. Lave, *Sandia National Laboratories, United States*

A Systematic Design Method and Verification for a Zero-Ripple Interface for PV/Battery-to-Grid Applications 1934
Suvankar Biswas, *University of Minnesota, United States*
Ned Mohan, *University of Minnesota, United States*
William Robbins, *University of Minnesota, United States*

Grid-Voltage-Feedforward Active Damping for Grid-Connected Inverter with LCL Filter 1941
Minghui Lu, *Aalborg University, Denmark*
Xiongfei Wang, *Aalborg University, Denmark*
Frede Blaabjerg, *Aalborg University, Denmark*
S.M. Muyeen, *Petroleum Institute, U.A.E.*
Ahmed Al-Durra, *Petroleum Institute, U.A.E.*
Siyu Leng, *Petroleum Institute, U.A.E.*

A High Power Density Single-Phase Inverter using Stacked Switched Capacitor Energy Buffer 1947
Colin McHugh, *University of Colorado at Boulder, United States*
Sreyam Sinha, *University of Colorado at Boulder, United States*
Jeffrey Meyer, *University of Colorado at Boulder, United States*
Saad Pervaiz, *University of Colorado at Boulder, United States*
Jie Lu, *University of Colorado at Boulder, United States*
Fan Zhang, *University of Colorado at Boulder, United States*
Hua Chen, *University of Colorado at Boulder, United States*
Hyeokjin Kim, *University of Colorado at Boulder, United States*
Usama Anwar, *University of Colorado at Boulder, United States*
Ashish Kumar, *University of Colorado at Boulder, United States*
Alihossein Sepahvand, *University of Colorado at Boulder, United States*
Scott Jensen, *University of Colorado at Boulder, United States*
Beomseok Choi, *University of Colorado at Boulder, United States*
Daniel Seltzer, *University of Colorado at Boulder, United States*
Robert Erickson, *University of Colorado at Boulder, United States*
Dragan Maksimovic, *University of Colorado at Boulder, United States*
Khurram K. Afridi, *University of Colorado at Boulder, United States*

A Novel Single-Stage Dual-Active Bridge based Isolated DC-AC Converter 1954
Shiladri Chakraborty, *Indian Institute of Technology Kharagpur, India*
Souvik Chattopadhyay, *Indian Institute of Technology Kharagpur, India*

Ultra-Low Ripple Inverters for Distributed Generation Applications 1962
Ang Shen, *Missouri University of Science and Technology, United States*
Pourya Shamsi, *Missouri University of Science and Technology, United States*
Mehdi Ferdowsi, *Missouri University of Science and Technology, United States*

A 15 kV SiC MOSFET Gate Drive with Power Over Fiber based Isolated Power Supply and Comprehensive Protection Functions 1967
Xuan Zhang, *Ohio State University, United States*
He Li, *Ohio State University, United States*
John A. Brothers, *Ohio State University, United States*
Jin Wang, *Ohio State University, United States*
Lixing Fu, *Texas Instruments Inc., United States*
Mico Perales, *MH GoPower Co., Ltd., Taiwan*
John Wu, *MH GoPower Co., Ltd., Taiwan*

A 15-kV Class Intelligent Universal Transformer for Utility Applications 1974
Jih-Sheng Lai, *Virginia Polytechnic Institute and State University, United States*
Wei-Han Lai, *Enertronics, Inc., United States*
Seung-Ryul Moon, *Virginia Polytechnic Institute and State University, United States*
Lanhua Zhang, *Virginia Polytechnic Institute and State University, United States*
Arindam Maitra, *Electric Power Research Institute, United States*

Session T40: Modeling, Modulation and Control of Motor Drive
Location: 102C
March 24, 2016 14:00 - 17:30
Session Chairs: Jin Wang, *The Ohio State University*
River-TinHo Li, *ABB*

Modulation Technique for Common Mode Voltage Reduction in a Matrix Converter Drive Operating with High Voltage Transfer Ratio 1982
Varsha Padhee, *Rockwell Automation, United States*
Ashish Kumar Sahoo, *University of Minnesota, United States*
Ned Mohan, *University of Minnesota, United States*

Soft-Switched Discontinuous Pulse-Width Pulse-Density Modulation Scheme 1989
Arash Rahnamaee, *University of Illinois at Chicago, United States*
Alireza Mojab, *University of Illinois at Chicago, United States*
Hossein Riazmontazer, *University of Illinois at Chicago, United States*
Sudip K. Mazumder, *University of Illinois at Chicago, United States*
Milos Zefran, *University of Illinois at Chicago, United States*

A Novel Flux Estimator based on SOGI with FLL for Induction Machine Drives 1995
Rende Zhao, *China University of Petroleum, China*
Zhen Xin, *Aalborg University, Denmark*
Poh Chiang Loh, *Aalborg University, Denmark*
Frede Blaabjerg, *Aalborg University, Denmark*

Performance Characterization of Random Pulse Width Modulation Algorithms in Industrial and Commercial Adjustable Speed Drives 2003
Kevin Lee, *Eaton Corporation, United States*
Guangtong Shen, *Purdue University, United States*
Wenxi Yao, *Zhejiang University, China*
Zhengyu Lu, *Zhejiang University, China*

Stability Analysis and Controller Synthesis for Digital Single-Loop Voltage-Controlled Inverters 2011
Xiongfei Wang, *Aalborg University, Denmark*
Poh Chiang Loh, *Aalborg University, Denmark*
Frede Blaabjerg, *Aalborg University, Denmark*

High Efficiency, Hybrid Selective Harmonic Elimination Phase-Shift PWM Technique for Cascaded H-Bridge Inverters to Improve Dynamic Response and Operate in Complete Normal Modulation Indices 2019
Amirhossein Moeini, *University of Florida, United States*
Zhao Hui, *University of Florida, United States*
Shuo Wang, *University of Florida, United States*

Implementation and Experimental Validation of Efficiency Improvement in PMSM Drives through Switching Frequency Reduction 2027
Parag Kshirsagar, *United Technologies Research Center, United States*
Krishnan Ramu, *Virginia Polytechnic Institute and State University, United States*

Sensorless Speed Control of Symmetrical Triple-Star Nine-Phase Interior Permanent Magnet Machines 2035
Olorunfemi Ojo, *Tennessee Technological University, United States*
Medhi Ramezani, *Tennessee Technological University, United States*

Mitigation of Common-Mode Noise in Wide Band Gap Device based Motor Drives 2043

Sneha Narasimhan, *Rockwell Automation, United States*
Saurabh Tewari, *MTS Systems Corporation, United States*
Eric Severson, *University of Minnesota, United States*
Rohit Baranwal, *University of Minnesota, United States*
Ned Mohan, *University of Minnesota, United States*

Session T41: Gate Drivers and Integrated Packaging
Location: 103C
March 24, 2016 14:00 - 17:30
Session Chairs: Qiang Li, *Virginia Tech*
Jean-Luc Schanen, *Ecole Nationale Supérieure de l'Energie*

A High-Efficient Driving Isolated Drive-by-Microwave Half-Bridge Gate Driver for a GaN Inverter 2051

Shuichi Nagai, *Panasonic Corporation, Japan*
Yasufumi Kawai, *Panasonic Corporation, Japan*
Osamu Tabata, *Panasonic Corporation, Japan*
Songbaek Choe, *Panasonic Corporation, Japan*
Noboru Negoro, *Panasonic Corporation, Japan*
Tesuzo Ueda, *Panasonic Corporation, Japan*

Sensing Gallium Nitride HEMT Junction Temperature using Gate Drive Output Transient Properties 2055

He Niu, *University of Wisconsin at Madison, United States*
Robert D. Lorenz, *University of Wisconsin at Madison, United States*

Design and Application of a 1200V Ultra-Fast Integrated Silicon Carbide MOSFET Module ... 2063

Suxuan Guo, *North Carolina State University, United States*
Liqi Zhang, *North Carolina State University, United States*
Yang Lei, *North Carolina State University, United States*
Xuan Li, *North Carolina State University, United States*
Wensong Yu, *North Carolina State University, United States*
Alex Q. Huang, *North Carolina State University, United States*

Active Gate Charge Control Strategy for Series-Connected IGBTs 2071

Fan Zhang, *Xi'an Jiaotong University, China*
Xu Yang, *Xi'an Jiaotong University, China*
Yu Ren, *Xi'an Jiaotong University, China*
Ying Chen, *Xi'an Jiaotong University, China*
Ruifeng Gou, *Xi'an XD Power Systems Co., LTD, China*

A MV Intelligent Gate Driver for 15kV SiC IGBT and 10kV SiC MOSFET 2076

Awneesh Tripathi, *North Carolina State University, United States*
Krishna Mainali, *North Carolina State University, United States*
Sachin Madhusoodhanan, *North Carolina State University, United States*
Akshat Yadav, *North Carolina State University, United States*
Kasunaidu Vechalapu, *North Carolina State University, United States*
Subhashish Bhattacharya, *North Carolina State University, United States*

Linear Temperature Sensors in High-Voltage GaN-HEMT Power Devices 2083
Richard Reiner, *Fraunhofer Institute for Applied Solid State Physics, Germany*
Patrick Waltereit, *Fraunhofer Institute for Applied Solid State Physics, Germany*
Beatrix Weiss, *Fraunhofer Institute for Applied Solid State Physics, Germany*
Matthias Wespel, *Fraunhofer Institute for Applied Solid State Physics, Germany*
Dirk Meder, *Fraunhofer Institute for Applied Solid State Physics, Germany*
Michael Mikulla, *Fraunhofer Institute for Applied Solid State Physics, Germany*
Rüdiger Quay, *Fraunhofer Institute for Applied Solid State Physics, Germany*
Oliver Ambacher, *Fraunhofer Institute for Applied Solid State Physics, Germany*

An Innovative Power Module with Power-System-in-Inductor Structure 2087
Laili Wang, *Sumida Corporation, Canada*
Doug Malcolm, *Sumida Corporation, Canada*
Yan-Fei Liu, *Queen's University, Canada*

Thermal Analysis of a Magnetic Packaged Power Module 2095
Laili Wang, *Sumida Corporation, Canada*
Doug Malcolm, *Sumida Corporation, Canada*
Wenbo Liu, *Queen's University, Canada*
Yan-Fei Liu, *Queen's University, Canada*

**Analysis of a Low-Inductance Packaging Layout for Full-SiC Power Module Embedding
Split Damping** 2102
Yu Ren, *Xi'an Jiaotong University, China*
Xu Yang, *Xi'an Jiaotong University, China*
Fan Zhang, *Xi'an Jiaotong University, China*
Linlin Tan, *Xi'an Jiaotong University, China*
Xiangjun Zeng, *Xi'an Jiaotong University, China*

Session T42: Component Modeling
Location: 103AB
March 24, 2016 14:00 - 17:30
Session Chairs: Sheldon Williamson, *University of Ontario Institute of Technology*
Abhijit Pathak, *Infineon/IR*

**Comprehensive Parametric Analyses of Thermally Aged Power MOSFETs for Failure
Precursor Identification and Lifetime Estimation based on Gate Threshold Voltage** 2108
Serkan Dusmez, *University of Texas at Dallas, United States*
Bilal Akin, *University of Texas at Dallas, United States*

Modeling and Design Guidelines of High Density Power Inductor for Battery Power Unit 2114
Zhigang Dang, *University of Alabama, United States*
Jaber A. Abu Qahouq, *University of Alabama, United States*

Degradation of Low Voltage Metal Oxide Varistors in Power Supplies 2122
Dawood Talebi Khanmiri, *Northeastern University, United States*
Roy Ball, *Mersen USA, United States*
Jerry Mosesian, *Mersen USA, United States*
Brad Lehman, *Northeastern University, United States*

Characterization and Modeling of SiC MOSFET Body Diode 2127
Kang Peng, *University of South Carolina, United States*
Soheila Eskandari, *University of South Carolina, United States*
Enrico Santi, *University of South Carolina, United States*

A Simple Behavioral Electro-Thermal Model of GaN FETs for SPICE Circuit Simulation 2136
Liyao Wu, *Georgia Institute of Technology, United States*
Maryam Saeedifard, *Georgia Institute of Technology, United States*

Decomposition and Electro-Physical Model Creation of the CREE 1200V, 50A 3-Ph SiC Module 2141
Adam J. Morgan, *North Carolina State University, United States*
Yang Xu, *North Carolina State University, United States*
Douglas C. Hopkins, *North Carolina State University, United States*
Iqbal Husain, *North Carolina State University, United States*
Wensong Yu, *North Carolina State University, United States*

A Three-Legged MATLAB/Simulink Transformer Model using a Fictitious Delta Winding 2147
Thomas A. Nondahl, *Rockwell Automation, United States*
Jingbo Liu, *Rockwell Automation, United States*
Peter B. Schmidt, *Rockwell Automation, United States*

A Lifetime Prediction Method for LEDs Considering Mission Profiles 2154
Xiaohui Qu, *Southeast University, China*
Huai Wang, *Aalborg University, Denmark*
Xiaoqing Zhan, *City University of Hong Kong, Hong Kong*
Frede Blaabjerg, *Aalborg University, Denmark*
Henry Shu-Hung Chung, *City University of Hong Kong, Hong Kong*

Enhanced Li-Ion Battery Modeling using Recursive Parameters Correction 2161
Jae-Gu Kim, *Sungkyunkwan University, Korea, South*
Jung-Hoon Ahn, *Sungkyunkwan University, Korea, South*
Byoung-Kuk Lee, *Sungkyunkwan University, Korea, South*

Session T43: Grid and Utility Interface
Location: 104A
March 24, 2016 14:00 - 17:30
Session Chairs: Manish Bhardwaj, *Texas Instruments*
Nan Chen, *ABB*

Robust Sensorless Control of Grid Connected Converters with LCL Line Filters using Frequency Adaptive Observers as AC Voltage Estimators 2167
Vlatko Miskovic, *Danfoss Drives, United States*
Vladimir Blasko, *United Technologies Research Center, United States*
Thomas Jahns, *University of Wisconsin at Madison, United States*
Robert Lorenz, *University of Wisconsin at Madison, United States*
Haojiong Zhang, *Danfoss Drives, United States*

Active Stabilization of Direct Matrix Converter Input Side Filter through Grid Current Control 2175
Martin Leubner, *Technische Universität Dresden, Germany*
Nico Remus, *Technische Universität Dresden, Germany*
Marc Stübig, *Technische Universität Dresden, Germany*
Wilfried Hofmann, *Technische Universität Dresden, Germany*

Impedance-Based Stability Analysis of Single-Phase Inverter Connected to Weak Grid with Voltage Feed-Forward Control 2182
Jiangfeng Wang, *Nanjing University of Aeronautics and Astronautics, China*
Jianhui Yao, *Nanjing University of Aeronautics and Astronautics, China*
Haibing Hu, *Nanjing University of Aeronautics and Astronautics, China*
Yan Xing, *Nanjing University of Aeronautics and Astronautics, China*
Xiaobin He, *Shanghai Institute of Space Power-Sources, China*
Kai Sun, *Tsinghua University, China*

New Configuration of Dynamic Voltage Restorer for Medium Voltage Application 2187
Arash Khoshkbar Sadigh, *Extron Electronics, United States*
Vahid Dargahi, *Clemson University, United States*
Keith Corzine, *Clemson University, United States*

Studies on the Clustered Voltage Balancing Mechanism for Cascaded H-Bridge STATCOM ... 2194
Daorong Lu, *Nanjing University of Aeronautics and Astronautics, China*
Haibing Hu, *Nanjing University of Aeronautics and Astronautics, China*
Yan Xing, *Nanjing University of Aeronautics and Astronautics, China*
Xiaobin He, *Shanghai Institute of Space Power-Sources, China*
Kai Sun, *Tsinghua University, China*
Jianhui Yao, *Nanjing University of Aeronautics and Astronautics, China*

Design of a Fast Response Time Single-Phase PLL with DC Offset Rejection Capability ... 2200
Abhijit Kulkarni, *Indian Institute of Science, India*
Vinod John, *Indian Institute of Science, India*

Four New Applications of Second-Order Generalized Integrator Quadrature Signal Generator 2207
Zhen Xin, *Aalborg University, Denmark*
Rende Zhao, *China University of Petroleum, China*
Xiongfei Wang, *Aalborg University, Denmark*
Poh Chiang Loh, *Aalborg University, Denmark*
Frede Blaabjerg, *Aalborg University, Denmark*

Three-Phase Multiple Harmonic Sequence Detection based on Generalized Delayed Signal Superposition 2215
Yong Lu, *Xi'an Jiaotong University, China*
Guochun Xiao, *Xi'an Jiaotong University, China*
Xiongfei Wang, *Aalborg University, Denmark*
Frede Blaabjerg, *Aalborg University, Denmark*

Hybrid Modelling and Control of Single-Phase Grid-Connected NPC Inverters 2223
Xingda Yan, *University of Southampton, United Kingdom*
Zhan Shu, *University of Southampton, United Kingdom*
Suleiman M. Sharkh, *University of Southampton, United Kingdom*

Session T44: Topics in Renewable Energy Systems II
Location: 104B
March 24, 2016 14:00 - 17:30
Session Chairs: Akshay Kumar Rathore, *Concordia University*
Yichao Tang, *Texas Instruments*

Stability Criterion and Controller Parameter Design of Radial-Line Renewable Systems with Multiple Inverters 2229
Wenchao Cao, *University of Tennessee, United States*
Xuan Zhang, *University of Tennessee, United States*
Yiwei Ma, *University of Tennessee, United States*
Fred Wang, *University of Tennessee, United States*

Stability Analysis and Improvement of Solid State Transformer (SST)-Paralleled Inverters System using Negative Impedance Feedback Control 2237
Qing Ye, *Florida State University, United States*
Hui Li, *Florida State University, United States*

Compensator-Less Structures for Droop Control of Single Phase Inverters in a Flexible Microgrid 2245
Onkar Vitthal Kulkarni, *Indian Institute of Technology Bombay, India*
Suryanarayana Doolla, *Indian Institute of Technology Bombay, India*
B.G. Fernandes, *Indian Institute of Technology Bombay, India*

Comparative Evaluation of the Loss and Thermal Performance of Advanced Three Level Inverter Topologies 2252
Alexander Anthon, *Technical University of Denmark, Denmark*
Zhe Zhang, *Technical University of Denmark, Denmark*
Michael A.E. Andersen, *Technical University of Denmark, Denmark*
Grahame Holmes, *RMIT University, Australia*
Brendan McGrath, *RMIT University, Australia*
Carlos Teixeira, *RMIT University, Australia*

Dual Buck Inverter with Series Connected Diodes and Single Inductor 2259
Liwei Zhou, *Shandong University, China*
Feng Gao, *Shandong University, China*

Magnetic Integration of the Harmonic Filter Inductor for Dual-Converter Fed Open-End Transformer Topology 2264
Ghanshyamsinh Gohil, *Aalborg University, Denmark*
Lorand Bede, *Aalborg University, Denmark*
Remus Teodorescu, *Aalborg University, Denmark*
Tamas Kerekes, *Aalborg University, Denmark*
Frede Blaabjerg, *Aalborg University, Denmark*

Mechanism Analysis and Mitigation of Instability in Grid-Connected Voltage Source Inverter with LCL Filters based on Terminal Impedance 2272
Teng Liu, *Xi'an Jiaotong University, China*
Zeng Liu, *Xi'an Jiaotong University, China*
Jinjun Liu, *Xi'an Jiaotong University, China*
Qingyun Dou, *Xi'an Jiaotong University, China*

Seven-Switch Five-Level Active Neutral-Point Clamped Converter and Optimal Modulation Strategy 2278
Hongliang Wang, *Queen's University, Canada*
Lei Kou, *Queen's University, Canada*
Yan-Fei Liu, *Queen's University, Canada*
Paresh C. Sen, *Queen's University, Canada*
Sucheng Liu, *Anhui University of Technology, China*

A Simple Variable Step Size Method for Maximum Power Point Tracking using Commercial Current Mode Control DC-DC Regulators 2286
Su Sheng, *Northeastern University, United States*
Brad Lehman, *Northeastern University, United States*

Session T45: Envelope Tracking and Resonant Conversion
Location: 104C
March 24, 2016 14:00 - 17:30
Session Chairs: Brian Zahnstecher, *PowerRox*
Davide Giacomini, *Infineon*

Envelope Tracking GaN Power Supply for 4G Cell Phone Base Stations 2292
Yuanzhe Zhang, *University of Colorado at Boulder, United States*
Johan Strydom, *Efficient Power Conversion Corporation, United States*
Michael de Rooij, *Efficient Power Conversion Corporation, United States*
Dragan Maksimović, *University of Colorado at Boulder, United States*

Envelope Tracking Power Supply for Volume-Sensitive Low-Power Applications based on a Resonant Switched-Capacitor Converter 2298
Alon Cervera, *Ben-Gurion University of the Negev, Israel*
Mor Mordechai Peretz, *Ben-Gurion University of the Negev, Israel*

A Passive-Impedance-Matching Concept for Multi-Phase Resonant Converter 2304
Hongliang Wang, *Queen's University, Canada*
Yang Chen, *Queen's University, Canada*
Yan-Fei Liu, *Queen's University, Canada*

LLC Converter with Auxiliary Switch for Hold Up Mode Operation 2312
Yang Chen, *Queen's University, Canada*
Hongliang Wang, *Queen's University, Canada*
Yan-Fei Liu, *Queen's University, Canada*
Jahangir Afsharian, *Murata Power Solutions, Canada*
Zhihua Yang, *Queen's University, Canada*

A Common Capacitor Multi-Phase LLC Resonant Converter 2320
Hongliang Wang, *Queen's University, Canada*
Yang Chen, *Queen's University, Canada*
Zhiyuan Hu, *Queen's University, Canada*
Laili Wang, *Queen's University, Canada*
Yajie Qiu, *Queen's University, Canada*
Wenbo Liu, *Queen's University, Canada*
Yan-Fei Liu, *Queen's University, Canada*
Jahangir Afsharian, *Murata Power Solutions, Canada*
Zhihua Yang, *Murata Power Solutions, Canada*

LLC Resonant Converter Design for Bendable Power Converter 2328

Kwun Yuan Godwin Ho, *University of Hong Kong, Hong Kong*
M.H. Bryan Pong, *University of Hong Kong, Hong Kong*
Shu-Yuen Ron Hui, *University of Hong Kong, Hong Kong*

Design Consideration of MHz Active Clamp Flyback Converter with GaN Devices for Low Power Adapter Application 2334

Xiucheng Huang, *Virginia Polytechnic Institute and State University, United States*
Junjie Feng, *Virginia Polytechnic Institute and State University, United States*
Weijing Du, *Virginia Polytechnic Institute and State University, United States*
Fred C. Lee, *Virginia Polytechnic Institute and State University, United States*
Qiang Li, *Virginia Polytechnic Institute and State University, United States*

A New Capacitor Voltage Balancing Control for Hybrid Modular Multilevel Converter with Cascaded Full Bridge 2342

Mahendra B. Ghat, *Indian Institute of Technology Bombay, India*
Anshuman Shukla, *Indian Institute of Technology Bombay, India*
Richa Mishra, *Indian Institute of Technology Bombay, India*

Sensorless Scheduling of the Modular Multilevel Series-Parallel Converter: Enabling a Flexible, Efficient, Modular Battery 2349

Stefan M. Goetz, *Duke University, United States*
Zhongxi Li, *Duke University, United States*
Angel V. Peterchev, *Duke University, United States*
Xinyu Liang, *North Carolina State University, United States*
Chengduo Zhang, *North Carolina State University, United States*
Srdjan M. Lukic, *North Carolina State University, United States*

Session D01: AC-DC Converters
Location: Poster Area
March 24, 2016 11:30 - 14:00
Session Chairs: Nathan Weise, *Marquette*
Daniel Costinett, *University of Tennessee-Knoxville*

An Input Current Calculation Switching Driver for High Power-Factor and Phase-Cut Dimmer Compatibility 2355

Hyunchul Eum, *Fairchild Semiconductor International, Inc., Korea, South*
Youngjong Kim, *Fairchild Semiconductor International, Inc., Korea, South*
Kuohsien Huang, *Fairchild Semiconductor International, Inc., Taiwan*

High Frequency Range Conducted Common-Mode Noise Suppression in SMPS 2360

Jinping Zhou, *Delta Electronics Shanghai Co., Ltd., China*
Yicong Xie, *Delta Electronics Shanghai Co., Ltd., China*
Min Zhou, *Delta Electronics Shanghai Co., Ltd., China*

Improved Medium Voltage AC-DC Rectifier based on 10kV SiC MOSFET for Solid State Transformer (SST) Application 2365

Qianlai Zhu, *North Carolina State University, United States*
Li Wang, *North Carolina State University, United States*
Liqi Zhang, *North Carolina State University, United States*
Wensong Yu, *North Carolina State University, United States*
Alex Q. Huang, *North Carolina State University, United States*

Suppression of Circulating Current in Parallel Operation of Three-Level Converters 2370
Young-Kwang Son, *Seoul National University, Korea, South*
Seung-Jun Chee, *Seoul National University, Korea, South*
Younggi Lee, *Seoul National University, Korea, South*
Seung-Ki Sul, *Seoul National University, Korea, South*
Changjin Lim, *LG Electronics, Korea, South*
Sungjae Huh, *LG Electronics, Korea, South*
Jaeyoon Oh, *LG Electronics, Korea, South*

Hybrid Bridgeless DCM SEPIC Rectifier Integrated with a Modified Switched Capacitor Cell ... 2376
Paulo Junior Silva Costa, *Universidade Federal de Santa Catarina, Brazil*
Telles Brunelli Lazzarin, *Universidade Federal de Santa Catarina, Brazil*
Carlos Henrique Illa Font, *Universidade Tecnológica Federal do Paraná, Brazil*

LCL Filter Design for Three-Phase Two-Level Power Factor Correction using Line Impedance Stabilization Network 2382
Alireza Kouchaki, *University of Southern Denmark, Denmark*
Morten Nymand, *University of Southern Denmark, Denmark*

Sensorless Current Rebuilding Strategy in a Single Phase Bridgeless PFC 2389
Felipe López, *Universidad de Cantabria, Spain*
Paula Lamo, *Universidad de Cantabria, Spain*
Alberto Pigazo, *Universidad de Cantabria, Spain*
F.J. Azcondo, *Universidad de Cantabria, Spain*

A Compact Electrolytic-Free Two-Stage Universal Input Offline LED Driver 2395
Saad Pervaiz, *University of Colorado at Boulder, United States*
Ashish Kumar, *University of Colorado at Boulder, United States*
Khurram K. Afridi, *University of Colorado at Boulder, United States*

Session D02: DC-DC Converters I
Location: Poster Area
March 24, 2016 11:30 - 14:00
Session Chairs: Charles Sullivan, *Dartmouth*
Mahshid Amirabadi, *Northeastern University*

Design Methodology for a High Insulation Voltage Power Transmission Function for IGBT Gate Driver 2401
Sokchea Am, *Grenoble Institute of Technology, France*
Pierre Lefranc, *Grenoble Institute of Technology, France*
David Frey, *Grenoble Institute of Technology, France*
Mahmoud Ibrahim, *Grenoble Institute of Technology, France*

Optimized Design of GaN Switching Capacitor based Envelope Tracking Power Supply for Satellite Applications 2409
Qian Jin, *Nanjing University of Aeronautics and Astronautics, China*
M. Vasić, *Universidad Politécnica de Madrid, Spain*
O. Garcia, *Universidad Politécnica de Madrid, Spain*
P. Alou, *Universidad Politécnica de Madrid, Spain*
J.A. Oliver, *Universidad Politécnica de Madrid, Spain*
J.A. Cobos, *Universidad Politécnica de Madrid, Spain*

An Isolated High Step-Up Converter with Continuous Input Current and LC Snubber 2415
K.I. Hwu, *National Taipei University of Technology, Taiwan*
W.Z. Jiang, *National Taipei University of Technology, Taiwan*
Y.T. Yau, *National Taipei University of Technology, Taiwan*

Output-Inductor-Less Full-Bridge Converter with SiC-MOSFETs for Low Noise and ZVS Operation 2422
Kazuhide Domoto, *Nagasaki University, Japan*
Yoichi Ishizuka, *Nagasaki University, Japan*
Seiya Abe, *Kyushu Institute of Technology, Japan*
Tamotsu Ninomiya, *Green Electronics Research Institute, Kitakyushu, Japan*

Reduction Technique of Leakage Flux Effects on GaN-HEMTs in 5 MHz / 100 W Isolated DC-DC Converters 2430
Akinori Hariya, *Nagasaki University, Japan*
Tomoya Koga, *Nagasaki University, Japan*
Ken Matsuura, *TDK Corporation, Japan*
Hiroshige Yanagi, *TDK-Lambda Corporation, Japan*
Satoshi Tomioka, *TDK-Lambda Corporation, Japan*
Yoichi Ishizuka, *Nagasaki University, Japan*
Tamotsu Ninomiya, *City of Kitakyushu, Japan*

A High-Voltage Level Shifter with Sub-Nano-Second Propagation Delay for Switching Power Converters 2437
Ahmed Abdelmoaty, *Ohio State University, United States*
Mohammad Al-Shyoukh, *TSMC Inc., United States*
Ayman Fayed, *Ohio State University, United States*

Dual-Output, Three-Level GaN-Based DC-DC Converter for Battery Charger Applications 2441
Ren Ren, *Nanjing University of Aeronautics and Astronautics, China*
Bo Liu, *University of Tennessee, United States*
Edward A. Jones, *University of Tennessee, United States*
Fred Wang, *University of Tennessee, United States*
Zheyu Zhang, *University of Tennessee, United States*
Daniel Costinett, *University of Tennessee, United States*

Quadruple Active Bridge DC-DC Converter as the Basic Cell of a Modular Smart Transformer 2449
Levy F. Costa, *Christian-Albrechts-Universität zu Kiel, Germany*
Giampaolo Buticchi, *Christian-Albrechts-Universität zu Kiel, Germany*
Marco Liserre, *Christian-Albrechts-Universität zu Kiel, Germany*

Analytical Model of a Phase-Shift Controlled Three-Level Zero-Voltage Switching Converter 2457
Cas Bakker, *Prodrive Technologies, Netherlands*
Bas Vermulst, *Technische Universiteit Eindhoven, Netherlands*
Anton Driessen, *Prodrive Technologies, Netherlands*

High Efficiency Design for ISOP Converter System with Dual Active Bridge DC-DC Converter 2465
Masaki Sato, *Nagasaki University, Japan*
Kazuhide Domoto, *Nagasaki University, Japan*
Yoichi Ishizuka, *Nagasaki University, Japan*
Masahiro Yamaguchi, *Tohoku University, Japan*
Shinya Manabe, *RICOH Electronic Devices Co., Ltd., Japan*
Hiizu Okubo, *RICOH Electronic Devices Co., Ltd., Japan*
Atsushi Itagaki, *Ryowa Electronics Co., Ltd., Japan*

Wide Input Range Power Converters using a Variable Turns Ratio Transformer 2473
Ziwei Ouyang, *Technical University of Denmark, Denmark*
Michael A.E. Andersen, *Technical University of Denmark, Denmark*

Design Approaches for Fast Supercapacitor Chargers for Applications like SCATMA, SRUPS 2479
Nicoloy Gurusinghe, *University of Waikato, New Zealand*
Nihal Kularatna, *University of Waikato, New Zealand*
W. Howell Round, *University of Waikato, New Zealand*
D. Alistair Steyn-Ross, *University of Waikato, New Zealand*

Stack Multiphase Asymmetrical Half-Bridge Topology Offering Advance Performance and Efficiency 2485
Trong Tue Vu, *Eisergy Ltd., Ireland*
George Young, *Eisergy Ltd., Ireland*

Session D03: DC-DC Converters II
Location: Poster Area
March 24, 2016 11:30 - 14:00
Session Chairs: Jason Stauth, *Dartmouth*
Yan-Fei Liu, *Queens*

Design of a Novel APWM Half-Bridge DC-DC Resonant Converter with Load-Independent Soft-Switching and Reduced Circulating Current 2491
Kawsar Ali, *National University of Singapore, Singapore*
Sandeep Kolluri, *National University of Singapore, Singapore*
Naga Brahmendra Yadav Gorla, *National University of Singapore, Singapore*
Pritam Das, *National University of Singapore, Singapore*
Sanjib Kumar Panda, *National University of Singapore, Singapore*

A Low-Volume Hybrid Step-Down DC-DC Converter based on the Dual use of Flying Capacitor 2497
S.M. Ahsanuzzaman, *University of Toronto, Canada*
Yingxian Ma, *University of Toronto, Canada*
Abrar Ahmed Pathan, *University of Toronto, Canada*
Aleksandar Prodić, *University of Toronto, Canada*

Fractional Pulse Skipping in Digitally Controlled DC-DC Converters for Improved Light-Load Efficiency and Power Spectrum 2504
Bipin Chandra Mandi, *Indian Institute of Technology Kharagpur, India*
Santanu Kapat, *Indian Institute of Technology Kharagpur, India*
Amit Patra, *Indian Institute of Technology Kharagpur, India*

A New Compact and High Efficiency Resonant Converter 2511
Sheng-Yang Yu, *Texas Instruments Inc., United States*

A 10-MHz eGaN FETs based Isolated Class-Φ_2 DCX 2518
Xuewen Zou, *Nanjing University of Aeronautics and Astronautics, China*
Zhiliang Zhang, *Nanjing University of Aeronautics and Astronautics, China*
Zhou Dong, *Nanjing University of Aeronautics and Astronautics, China*
Yuan Zhou, *Nanjing University of Aeronautics and Astronautics, China*
Xiaoyong Ren, *Nanjing University of Aeronautics and Astronautics, China*
Qianhong Chen, *Nanjing University of Aeronautics and Astronautics, China*

Multi-Level Capacitor Clamped DC-DC Multiplier/Divider with Variable and Fractional Voltage Gain – An (n/m)X DC-DC Converter 2525
Deepak Gunasekaran, *Michigan State University, United States*
Liang Qin, *Wuhan University, China*
Ujjwal Karki, *Michigan State University, United States*
Yuan Li, *Sichuan University, China*
Fang Z. Peng, *Michigan State University, United States*

Multi-Mode Quasi-Z-Source Series Resonant DC/DC Converter for Wide Input Voltage Range Applications 2533
Dmitri Vinnikov, *Ubik Solutions LLC, Estonia*
Andrii Chub, *Tallinn University of Technology, Estonia*
Indrek Roasto, *Ubik Solutions LLC, Estonia*
Liisa Liivik, *Tallinn University of Technology, Estonia*

Hybrid Serial-Output Converter for Integrated LED Lighting Applications 2540
T. McRae, *University of Toronto, Canada*
A. Prodić, *University of Toronto, Canada*
G. Lisi, *Texas Instruments Inc., United States*
W. McIntrye, *Texas Instruments Inc., United States*
A. Aguilar, *Texas Instruments Inc., United States*

Analysis and Modeling of a Modular ISOP Full Bridge based Converter with Input Filter ... 2545
P. Zumel, *Universidad Carlos III de Madrid, Spain*
E. Oña, *Universidad Carlos III de Madrid, Spain*
C. Fernandez, *Universidad Carlos III de Madrid, Spain*
M. Sanz, *Universidad Carlos III de Madrid, Spain*
A. Lazaro, *Universidad Carlos III de Madrid, Spain*
A. Barrado, *Universidad Carlos III de Madrid, Spain*
A. Vazquez, *Universidad de Oviedo, Spain*
D.G. Lamar, *Universidad de Oviedo, Spain*

Wide-Input High Power Density Flexible Converter Topology for DC-DC Applications 2553
Parth Jain, *University of Toronto, Canada*
Aleksandar Prodić, *University of Toronto, Canada*
Alexander Gerfer, *Würth Elektronik eiSos GmbH & Co. KG, Germany*

High Efficiency LLC Converter Design for Universal Battery Chargers 2561
Navid Shafiei, *University of British Columbia, Canada*
Ali Arefifar, *University of British Columbia, Canada*
Mohammad Ali Saket, *University of British Columbia, Canada*
Martin Ordonez, *University of British Columbia, Canada*

A New High Power Density Modular Multilevel DC-DC Converter with Localized Voltage Balancing Control for Arbitrary Number of Levels 2567

Ahmed Morsy, *Texas A&M University, United States*
Yong Zhou, *Texas A&M University, United States*
Prasad Enjeti, *Texas A&M University, United States*

Design and Control of a Fault Tolerant Soft Switching DC-DC Converter for High Power High Voltage Applications 2573

Tao Li, *Rensselaer Polytechnic Institute, United States*
Leila Parsa, *Rensselaer Polytechnic Institute, United States*

Accurate Parametric Steady State Analysis and Design Tool for DC-DC Power Converters 2579

Mohammad Daryaei, *University of Alberta, Canada*
Mohammad Ebrahimi, *University of Alberta, Canada*
S. Ali Khajehoddin, *University of Alberta, Canada*

Analysis of Multi-Output Half-Wave Semi-Synchronous Rectifier with a Uniform Magnetic Field Transmitter 2587

Erdem Asa, *Hevo Power Inc. / New York University, United States*
Kerim Colak, *Istanbul Ulasim A.S., Turkey*
Dariusz Czarkowski, *New York University, United States*

High Gain QZS DC/DC Converter with Coupled Inductor 2592

Rafael V. Silva, *Universidade Federal do Ceará, Brazil*
Antônio A.A. Freitas, *Universidade Federal do Ceará, Brazil*
Marcus R. Castro, *Universidade Federal do Ceará, Brazil*
Fernando L.M. Antunes, *Universidade Federal Rural do Semi-Árido, Brazil*
Edilson M. Sá Jr., *Universidade Federal do Ceará, Brazil*

Session D04: Utility Interface
Location: Poster Area
March 24, 2016 11:30 - 14:00
Session Chairs: Ali Khajehoddin, *University of Alberta*
Julia Zhang, *Oregon State University*

A Power Decoupling Method with Small Capacitance Requirement based on Single-Phase Quasi-Z-Source Inverter for DC Microgrid Applications 2599

Dingyi He, *University of Texas at Dallas, United States*
Wen Cai, *University of Texas at Dallas, United States*
Fan Yi, *University of Texas at Dallas, United States*

Operation Analysis of High Efficiency Grid Connected Bi-Directional Power Conversion System for Various Storage Battery Systems with Bi-Directional Switch Circuit Topology 2607

Go Yamada, *Panasonic Corporation, Japan*
Takaaki Norisada, *Panasonic Corporation, Japan*
Fumito Kusama, *Panasonic Corporation, Japan*
Keiji Akamatsu, *Panasonic Corporation, Japan*
Masakazu Michihira, *Kobe City College of Technology, Japan*

Fault Tolerant Control of MMC with Redundant Sub-Modules based on Carrier Phase Shift Modulation 2613

Kai Li, *Tsinghua University, China*
Zhengming Zhao, *Tsinghua University, China*
Liqiang Yuan, *Tsinghua University, China*
Sizhao Lu, *Tsinghua University, China*
Bing Pan, *State Grid Smart Grid Research Institute, China*
Zhengang Lu, *State Grid Smart Grid Research Institute, China*

A New Topology of Multilevel VSC Converter for Hybrid HVDC Transmission System 2620

Jae-Jung Jung, *Seoul National University, Korea, South*
Shenghui Cui, *RWTH Aachen University, Germany*
Seung-Ki Sul, *Seoul National University, Korea, South*

Performance of Solid State Transformers Under Imbalanced Loads in Distribution Systems 2629

Tao Yang, *University College Dublin, Ireland*
Ronan Meere, *University College Dublin, Ireland*
Cathal O'Loughlin, *University College Dublin, Ireland*
Terence O'Donnell, *University College Dublin, Ireland*

Steady-State Analysis of Modular Multilevel Converter (MMC) Under Unbalanced Grid Conditions 2637

Xiaojie Shi, *University of Tennessee, United States*
Yalong Li, *University of Tennessee, United States*
Zhiqiang Wang, *University of Tennessee, United States*
Bo Liu, *University of Tennessee, United States*
Leon M. Tolbert, *University of Tennessee, United States*
Fred Wang, *University of Tennessee, United States*

Design and Control of a Compensated Submodule Testing Scheme for Modular Multilevel Converter 2645

Yuan Tang, *University of Warwick, United Kingdom*
Li Ran, *University of Warwick, United Kingdom*
Olayiwola Alatise, *University of Warwick, United Kingdom*
Philip Mawby, *University of Warwick, United Kingdom*

A Voltage Independent Islanding Detection Method and Low Voltage Ride through of a Two-Stage PV Inverter 2652

Partha Pratim Das, *Indian Institute of Technology Kharagpur, India*
Souvik Chattopadhyay, *Indian Institute of Technology Kharagpur, India*
Shiladri Chakraborty, *Indian Institute of Technology Kharagpur, India*

Low Cost and High Efficiency Topology for Flexible Integration of Multi-PV and Batteries in Resonant-Based Converters 2660

Ali Elrayyah, *Qatar Environment and Energy Research Institute, Qatar*

Real-Time Integrated Model of a Micro-Grid with Distributed Clean Energy Generators and their Power Electronics 2666

Weiqiang Chen, *University of Connecticut, United States*
Ali M. Bazzi, *University of Connecticut, United States*
James Hare, *University of Connecticut, United States*
Shalabh Gupta, *University of Connecticut, United States*

Minimization of Inter-Module Leakage Current in Cascaded H-Bridge Multilevel Inverters for Grid Connected Solar PV Applications 2673
V.V.S. Pradeep Kumar, *Indian Institute of Technology Bombay, India*
B.G. Fernandes, *Indian Institute of Technology Bombay, India*

Effect of Grid Inductance on Grid Current Quality of Parallel Grid-Connected Inverter System with Output LCL Filter and Closed-Loop Control 2679
Wooyoung Choi, *University of Wisconsin at Madison, United States*
Woongkul Lee, *University of Wisconsin at Madison, United States*
Bulent Sarlioglu, *University of Wisconsin at Madison, United States*

Small Signal Modeling and Control of a Grid Tied Converter without a Syncronization Unit 2687
Subhajyoti Mukherjee, *Missouri University of Science and Technology, United States*
Pourya Shamsi, *Missouri University of Science and Technology, United States*
Mehdi Ferdowsi, *Missouri University of Science and Technology, United States*

Bridgeless SEPIC PFC Converter for Low Total Harmonic Distortion and High Power Factor ... 2693
Yasemin Onal, *Bilecik Seyh Edebali University, Turkey*
Yilmaz Sozer, *University of Akron, United States*

Effectiveness of Pareto-Front Analysis Applied to the Design of a Single-Phase PFC Rectifier 2700
Mahmoud Ibrahim, *Eaton Corporation, France*
Luc Gonnet, *Eaton Corporation, France*
Pierre Lefranc, *Grenoble Institute of Technology, France*
David Frey, *Grenoble Institute of Technology, France*
Jean-Paul Ferrieux, *Grenoble Institute of Technology, France*
Sokchea Am, *Grenoble Institute of Technology, France*

State Space Analysis and Duty Cycle Control of a Switched Reactance based Center-Point-Clamped Reactive Power Compensator 2706
Pankaj Kumar Bhowmik, *University of North Carolina at Charlotte, United States*
Somasundaram Essakiappan, *University of North Carolina at Charlotte, United States*
Madhav Manjrekar, *University of North Carolina at Charlotte, United States*

A SiC-Based Power Converter Module for Medium-Voltage Fast Charger for Plug-In Electric Vehicles 2714
Srdjan Srdic, *North Carolina State University, United States*
Chi Zhang, *North Carolina State University, United States*
Xinyu Liang, *North Carolina State University, United States*
Wensong Yu, *North Carolina State University, United States*
Srdjan Lukic, *North Carolina State University, United States*

Shunt Active Power Filter based on Cascaded Transformers Coupled with Three-Phase Bridge Converters 2720
Gregory A. de Almeida Carlos, *Universidade Federal de Campina Grande, Brazil*
Cursino B. Jacobina, *Universidade Federal de Campina Grande, Brazil*
João Paulo R. Méllo, *Universidade Federal de Campina Grande, Brazil*
Euzeli C. dos Santos Jr., *Indiana University - Purdue University, United States*

Independent DC Link Voltage Control of Cascaded Multilevel PV Inverter 2727
Qingyun Huang, *North Carolina State University, United States*
Wensong Yu, *North Carolina State University, United States*
Alex Q. Huang, *North Carolina State University, United States*

New Active Damping Method for LCL Filter Resonance based on Two Feedback System 2735
Mahmoud A. Gaafar, *Kyushu University, Japan*
Gamal M. Dousoky, *Minia University, Egypt*
Masahito Shoyama, *Kyushu University, Japan*

Static Synchronous Generator Model for Investigating Dynamic Behaviors and Stability Issues of Grid-Tied Inverters 2742
Liansong Xiong, *Xi'an Jiaotong University, China*
Xiaokang Liu, *Xi'an Jiaotong University, China*
Feng Wang, *Xi'an Jiaotong University, China*
Fang Zhuo, *Xi'an Jiaotong University, China*

Session D05: Motor Drives and Inverters: Modeling and Control I
Location: Poster Area
March 24, 2016 11:30 - 14:00
Session Chairs: Liming Liu, *ABB Inc.*
Thomas Gietzold, *United Technologies Aerospace Systems*

Initial Orientation and Sensorless Starting Strategy of Wound-Rotor Synchronous Starter/Generator 2748
Jichang Peng, *Northwestern Polytechnical University, China*
Weiguo Liu, *Northwestern Polytechnical University, China*
Jinhao Meng, *Northwestern Polytechnical University, China*
Tao Meng, *Northwestern Polytechnical University, China*
Guangzhao Luo, *Northwestern Polytechnical University, China*

A Novel Method for Polarity Detection of Non-Salient PMSMs in Initial Position Estimation 2754
Bing Liu, *Nanjing University of Aeronautics and Astronautics, China*
Bo Zhou, *Nanjing University of Aeronautics and Astronautics, China*
Jiadan Wei, *Nanjing University of Aeronautics and Astronautics, China*
Long Wang, *Nanjing University of Aeronautics and Astronautics, China*
Tianheng Ni, *Nanjing University of Aeronautics and Astronautics, China*

A Speed Adaptive Sensorless Flux Observer for the Induction Motor Drive using Sylvester Criterion Design 2759
Mihai Comanescu, *Penn State Altoona, United States*

Discontinuous PWM for Low Switching Losses in Indirect Matrix Converter Drives 2764
Yeongsu Bak, *Ajou University, Korea, South*
Kyo-Beum Lee, *Ajou University, Korea, South*

Model Predictive Control for Extended Kalman Filter based Speed Sensorless Induction Motor Drives 2770
Jie Li, *Xi'an University of Technology, China*
Li-Heng Zhang, *Xi'an University of Technology, China*
Ying Niu, *Xi'an University of Technology, China*
Hai-Peng Ren, *Xi'an University of Technology, China*

Research on Excitation Control Methods for the Two-Phase Brushless Exciter of Wound-Rotor Synchronous Starter/Generators in the Starting Mode 2776

Ningfei Jiao, *Northwestern Polytechnical University, China*
Weiguo Liu, *Northwestern Polytechnical University, China*
Tao Meng, *Northwestern Polytechnical University, China*
Jichang Peng, *Northwestern Polytechnical University, China*
Shuai Mao, *Northwestern Polytechnical University, China*

A High Performance Speed Regulator Design for AC Machines 2782

Adil Khurram, *American University of Sharjah, U.A.E.*
Habibur Rehman, *American University of Sharjah, U.A.E.*
Shayok Mukhopadhyay, *American University of Sharjah, U.A.E.*

Zero-Sequence Current Suppression for Open-End Winding Induction Motor Drive with Resonant Controller 2788

Hajime Kubo, *Meidensha Corporation, Japan*
Yasuhiro Yamamoto, *Meidensha Corporation, Japan*
Takeshi Kondo, *Meidensha Corporation, Japan*
Kaushik Rajashekara, *University of Texas at Dallas, United States*
Bohang Zhu, *University of Texas at Dallas, United States*

Optimized Control of High-Performance Servo-Motor Drives in the Field-Weakening Region 2794

Jack Bermingham, *Moog Ireland Ltd, Ireland*
Gerard O'Donovan, *Moog Ireland Ltd, Ireland*
Ray Walsh, *Moog Ireland Ltd, Ireland*
Michael Egan, *University College Cork, Ireland*
Gordon Lightbody, *University College Cork, Ireland*
John G. Hayes, *University College Cork, Ireland*

Motor Current Reference Generation for Reducing Motor Currents in Drive Systems with Single-Phase Diode Rectifier and Small DC-Link Capacitor 2801

Young-Ho Chae, *Seoul National University, Korea, South*
Jung-Ik Ha, *Seoul National University, Korea, South*

A Simple Double Mapping based SVPWM Method for Balancing DC-Link Capacitor Voltages of Five-Level Diode-Clamped Converters 2806

Aparna Saha, *University of Akron, United States*
Ali Elrayyah, *Qatar Environment and Energy Research Institute, Qatar*
Yilmaz Sozer, *University of Akron, United States*

Session D06: Motor Drives and Inverters: Modeling and Control II
Location: Poster Area
March 24, 2016 11:30 - 14:00
Session Chairs: Bulent Sarlioglu, *University of Wisconsin - Madison*
Yichao Tang, *Texas Instruments*

Capacitor-Clamped Inverter based Transient Suppression Method for Azimuth Thruster Drives 2813

Shantha Gamini Jayasinghe, *Australian Maritime College, University of Tasmania, Australia*
Viknash Shagar, *Australian Maritime College, University of Tasmania, Australia*
Hossein Enshaei, *Australian Maritime College, University of Tasmania, Australia*
Danyal Mohammadi, *Boise State University, United States*
Mahinda Vilathgamuwa, *Queensland University of Technology, Australia*

Active Common-Mode Voltage Reduction in a Fault-Tolerant Three-Phase Inverter 2821
Danyal Mohammadi, *Boise State University, United States*
Said Ahmed-Zaid, *Boise State University, United States*

Power Cycling Lifetime Improvement of Three-Level NPC Inverters with an Improved DPWM Method 2826
Jiangbiao He, *Marquette University, United States*
Lixiang Wei, *Rockwell Automation, United States*
Nabeel A.O. Demerdash, *Marquette University, United States*

Synchronous Optimal Pulsewidth Modulation Digital Implementation Concept for Multilevel Converters 2833
Jackson Lago, *Universidade Federal de Santa Catarina, Brazil*
Marcelo Lobo Heldwein, *Universidade Federal de Santa Catarina, Brazil*

Analytical Determination of Conduction Losses for Modified Flying Capacitor Multicell Converters 2840
Vahid Dargahi, *Clemson University, United States*
Arash Khoshkbar Sadigh, *Extron Electronics, United States*
Keith Corzine, *Clemson University, United States*

Comparison of Electrical Losses in an Inverter-Fed Five-Phase and Three-Phase Permanent Magnet Assisted Synchronous Reluctance Motor 2847
Akm Arafat, *University of Akron, United States*
Seungdeog Choi, *University of Akron, United States*

A Hybrid Adaptive Observer for the Speed and Flux Estimation of Induction Motors 2855
Mihai Comanescu, *Penn State Altoona, United States*

Determination of CM Choke Parameters for SiC MOSFET Motor Drive based on Simple Measurements and Frequency Domain Modeling 2861
Di Han, *University of Wisconsin at Madison, United States*
Casey Morris, *University of Wisconsin at Madison, United States*
Woongkul Lee, *University of Wisconsin at Madison, United States*
Bulent Sarlioglu, *University of Wisconsin at Madison, United States*

An Improved Model Predictive Current Control of Permanent Magnet Synchronous Motor Drives 2868
Yongchang Zhang, *North China University of Technology, China*
Sugu Gao, *North China University of Technology, China*
Wei Xu, *Huazhong University of Science and Technology, China*

Analysis of Magnet Defect Faults in Permanent Magnet Synchronous Motors through Fluxgate Sensors 2875
Taner Goktas, *University of Texas at Dallas, United States*
Kun Wang Lee, *University of Texas at Dallas, United States*
Mohsen Zafarani, *University of Texas at Dallas, United States*
Bilal Akin, *University of Texas at Dallas, United States*

Session D07: Motor Drives and Inverters: Topologies
Location: Poster Area
March 24, 2016 11:30 - 14:00
Session Chairs: Amirnaser Yazdani, *Ryerson University*
Babak Nahid-Mobarakeh, *University of Lorraine*

Performance Comparison of Transfer Switch Topologies in Switched-Doubly-Fed Machine Drives 2881
Arijit Banerjee, *Massachusetts Institute of Technology, United States*
Steven B. Leeb, *Massachusetts Institute of Technology, United States*
James L. Kirtley, *Massachusetts Institute of Technology, United States*

Multilevel Converter Topologies for High-Power High-Speed Switched Reluctance Motor: Performance Comparison 2889
Devendra Patil, *University of Texas at Dallas, United States*
Shiliang Wang, *University of Texas at Dallas, United States*
Lei Gu, *University of Texas at Dallas, United States*

Bidirectional Magnetically Coupled T-Source Inverter for Extra Low Voltage Application 2897
Thomas Baier, *Friedrich-Alexander-Universität Erlangen-Nürnberg, Germany*
Bernhard Piepenbreier, *Friedrich-Alexander-Universität Erlangen-Nürnberg, Germany*

Active Virtual Ground: Single Phase Grid-Connected Voltage Source Inverter Topology ... 2905
River Tin-Ho Li, *ABB China Ltd., China*
Carl Ngai-Man Ho, *University of Manitoba, Canada*

Design and Evaluation of 30kVA Inverter using SiC MOSFET for 180°C Ambient Temperature Operation 2912
Feng Qi, *Ohio State University, United States*
Miao Wang, *Ohio State University, United States*
Longya Xu, *Ohio State University, United States*
Bo Zhao, *State Grid Corporation of China, China*
Zhe Zhou, *State Grid Corporation of China, China*
Xizhou Ren, *State Grid Corporation of China, China*

A DC to Three-Phase Boost-Buck Inverter with Stored Energy Modulation and a Tiny DC Link Capacitor 2919
Mahima Gupta, *University of Wisconsin at Madison, United States*
Giri Venkataramanan, *University of Wisconsin at Madison, United States*

Drive Circuits for Ultra-Fast and Reliable Actuation of Thomson Coil Actuators used in Hybrid AC and DC Circuit Breakers 2927
Chang Peng, *North Carolina State University, United States*
Alex Huang, *North Carolina State University, United States*
Iqbal Husain, *North Carolina State University, United States*
Bruno Lequesne, *E-Motors Consulting, LLC, United States*
Roger Briggs, *Energy Efficiency Research, LLC, United States*

Improved Transformerless Dual Buck Inverters with Buffer Inductors 2935
Liwei Zhou, *Shandong University, China*
Feng Gao, *Shandong University, China*

A 99% Efficiency SiC Three-Phase Inverter using Synchronous Rectification 2942
Shan Yin, *Nanyang Technological University, Singapore*
K.J. Tseng, *Nanyang Technological University, Singapore*
C.F. Tong, *Nanyang Technological University, Singapore*
Rejeki Simanjorang, *Rolls-Royce Singapore Pte. Ltd., Singapore*
C.J. Gajanayake, *Rolls-Royce Singapore Pte. Ltd., Singapore*
Amit K. Gupta, *Rolls-Royce Singapore Pte. Ltd., Singapore*

Comparison and Evaluation of Common Mode EMI Filter Topologies for GaN-Based Motor Drive Systems 2950
Casey T. Morris, *University of Wisconsin at Madison, United States*
Di Han, *University of Wisconsin at Madison, United States*
Bulent Sarlioglu, *University of Wisconsin at Madison, United States*

Analysis of Thermal Cycling Stress on Semiconductor Devices of the Modular Multilevel Converter for Drive Applications 2957
Xiangyu Han, *Georgia Institute of Technology, United States*
Qichen Yang, *Georgia Institute of Technology, United States*
Liyao Wu, *Georgia Institute of Technology, United States*
Maryam Saeedifard, *Georgia Institute of Technology, United States*

Fault Tolerant Topologies of Five-Level Active Neutral-Point-Clamped Converters 2963
Jun Li, *ABB Inc., United States*

Session D08: Advanced Components and Devices
Location: Poster Area
March 24, 2016 11:30 - 14:00
Session Chairs: Abhijit Pathak, *Infineon/IR*
Doug Hopkins, *North Carolina State University*

Dynamic Characterization of the Input and Reverse Transfer Capacitances in Power MOSFETs under High Current Conduction 2969
Cristino Salcines, *Universität Stuttgart, Germany*
Ingmar Kallfass, *Universität Stuttgart, Germany*
Hisao Kakitani, *Keysight Technologies International, Japan*
Atsushi Mikata, *Keysight Technologies International, Japan*

Medium Voltage Power Switch based on SiC JFETs 2973
Xueqing Li, *United Silicon Carbide, Inc., United States*
Hao Zhang, *United Silicon Carbide, Inc., United States*
Peter Alexandrov, *United Silicon Carbide, Inc., United States*
Anup Bhalla, *United Silicon Carbide, Inc., United States*

Numerical Model and Experimental Study on Comparison of Semiconductor Pulsed Power Devices 2981
Lin Liang, *Huazhong University of Science and Technology, China*
Changdong Chen, *Huazhong University of Science and Technology, China*
Fang Luo, *Ohio State University, United States*

A Normalization Procedure of DC-Side Stray Inductance for High-Speed Switching Circuit 2986
Masato Ando, *Tokyo Metropolitan University, Japan*
Keiji Wada, *Tokyo Metropolitan University, Japan*

Thermal Network Parameter Identification of IGBT Module based on the Cooling Curve of Junction Temperature ... 2992

Xiong Du, *Chongqing University, China*
Tengfei Li, *Chongqing University, China*
Jun Zhang, *Chongqing University, China*
Heng-Ming Tai, *University of Tulsa, United States*
Pengju Sun, *Chongqing University, China*
Luowei Zhou, *Chongqing University, China*

Design and Evaluation of High Current PCB Embedded Inductor for High Frequency Inverters ... 2998

Mehrdad Biglarbegian, *University of North Carolina at Charlotte, United States*
Neel Shah, *University of North Carolina at Charlotte, United States*
Iman Mazhari, *University of North Carolina at Charlotte, United States*
Johan Enslin, *University of North Carolina at Charlotte, United States*
Babak Parkhideh, *University of North Carolina at Charlotte, United States*

Prognosis of Wire Bond Lift-Off Fault of an IGBT based on Multisensory Approach ... 3004

Moinul Shahidul Haque, *University of Akron, United States*
Jeihoon Baek, *Korean Rail Research Institute, Korea, South*
Joseph Herbert, *University of Akron, United States*
Seungdeog Choi, *University of Akron, United States*

Electrical Parasitics and Thermal Modeling for Optimized Layout Design of High Power SiC Modules ... 3012

Amir Sajjad Bahman, *Aalborg University, Denmark*
Frede Blaabjerg, *Aalborg University, Denmark*
Atanu Dutta, *University of Arkansas, United States*
Alan Mantooth, *University of Arkansas, United States*

Calculation of Losses in PCB Windings for Multi-Coil Contactless Charging Systems ... 3020

J. Serrano, *Universidad de Zaragoza, Spain*
J. Acero, *Universidad de Zaragoza, Spain*
I. Lope, *BSH Home Appliances Group, Spain*
C. Carretero, *Universidad de Zaragoza, Spain*
J.M. Burdío, *Universidad de Zaragoza, Spain*
R. Alonso, *Universidad de Zaragoza, Spain*

Design of Efficient Loads for Domestic Induction Heating Applications by Means of Non-Magnetic Thin Metallic Layers ... 3026

Jesús Acero, *Universidad de Zaragoza, Spain*
Claudio Carretero, *Universidad de Zaragoza, Spain*
Rafael Alonso, *Universidad de Zaragoza, Spain*
José Miguel Burdío, *Universidad de Zaragoza, Spain*

A New Evaluation Circuit with a Low-Voltage Inverter Intended for Capacitors used in a High-Power Three-Phase Inverter ... 3032

Kazunori Hasegawa, *Kyushu Institute of Technology, Japan*
Ichiro Omura, *Kyushu Institute of Technology, Japan*
Shin-Ichi Nishizawa, *Kyushu Institute of Technology / National Institute of Advanced Industrial Science and Technology, Japan*

Energy Absorption Capability of Low Voltage Metal Oxide Varistors in AC and Impulse Currents 3038
Dawood Talebi Khanmiri, *Northeastern University, United States*
Roy Ball, *Mersen USA, United States*
Craig McKenzie, *Mersen USA, United States*
Brad Lehman, *Northeastern University, United States*

Optimization and Experimental Validation of Medium-Frequency High Power Transformers in Solid-State Transformer Applications 3043
M.A. Bahmani, *Chalmers University of Technology, Sweden*
T. Thiringer, *Chalmers University of Technology, Sweden*
M. Kharezy, *SP Technical Research Institute of Sweden, Sweden*

Evaluation of Core Loss in Magnetic Materials Employed in Utility Grid AC Filters 3051
Remus Beres, *Aalborg University, Denmark*
Xiongfei Wang, *Aalborg University, Denmark*
Frede Blaabjerg, *Aalborg University, Denmark*
Claus Leth Bak, *Aalborg University, Denmark*
Hiroaki Matsumori, *Tokyo Metropolitan University, Japan*
Toshihisa Shimizu, *Tokyo Metropolitan University, Japan*

A Novel Gate Assisted Circuit to Reduce Switching Loss and Eliminate Shoot-Through in SiC Half Bridge Configuration 3058
Shan Yin, *Nanyang Technological University, Singapore*
K.J. Tseng, *Nanyang Technological University, Singapore*
C.F. Tong, *Nanyang Technological University, Singapore*
Rejeki Simanjorang, *Rolls-Royce Singapore Pte. Ltd., Singapore*
C.J. Gajanayake, *Rolls-Royce Singapore Pte. Ltd., Singapore*
Amit K. Gupta, *Rolls-Royce Singapore Pte. Ltd., Singapore*

Session D09: System Design Considerations for Power Electronics
Location: Poster Area
March 24, 2016 11:30 - 14:00
Session Chairs: John Vigars, *Allegro Microsystems*
Ernie Parker, *Crane Aerospace & Electronics*

Methods to Enhance the Thermal Performance of a 3D Power Package 3065
Jonathan Noquil, *Texas Instruments Inc., United States*
Ozzie Lopez, *Texas Instruments Inc., United States*
Tianyi Luo, *Lehigh University, United States*

Highly Reliable and Cost Effective Thick Film Substrates for Power LEDs 3069
Paul Gundel, *Heraeus Deutschland GmbH & Co. KG, Germany*
Ryan Persons, *Heraeus Deutschland GmbH & Co. KG, Germany*
Melanie Bawohl, *Heraeus Deutschland GmbH & Co. KG, Germany*
Mark Challingsworth, *Heraeus Deutschland GmbH & Co. KG, Germany*
Christoph Czwickla, *Heraeus Deutschland GmbH & Co. KG, Germany*
Virginia Garcia, *Heraeus Deutschland GmbH & Co. KG, Germany*
Christina Modes, *Heraeus Deutschland GmbH & Co. KG, Germany*
Ilias Nikolaidis, *Heraeus Deutschland GmbH & Co. KG, Germany*
Jessica Reitz, *Heraeus Deutschland GmbH & Co. KG, Germany*
Caitlin Shahbazi, *Heraeus Deutschland GmbH & Co. KG, Germany*
Torsten Nowak, *Fraunhofer-Institut für Zuverlässigkeit und Mikrointegration, Germany*

Design and Evaluation of SiC-Based High Power Density Inverter, 70kW/Liter, 50kW/kg ... 3075
Koji Yamaguchi, *IHI Corporation, Japan*

An Improved Automatic Layout Method for Planar Power Module ... 3080
Puqi Ning, *Chinese Academy of Sciences, China*
Xuhui Wen, *Chinese Academy of Sciences, China*
Yaohua Li, *Chinese Academy of Sciences, China*
Xiongxuan Ge, *Chinese Academy of Sciences, China*

Practical Implementation Schemes of Motor Speed Measurement by Magnetic Encoder on Electric Power Steering Applications ... 3086
Jae-Hyun Lee, *Hyundai Mobis, Korea, South*

Low-Cost Input Impedance Estimator of DC-to-DC Converters for Designing the Control Loop in Cascaded Converters ... 3090
M. Sanz, *Universidad Carlos III de Madrid, Spain*
A. Lázaro, *Universidad Carlos III de Madrid, Spain*
M. Bermejo, *Universidad Carlos III de Madrid, Spain*
D. López del Moral, *Universidad Carlos III de Madrid, Spain*
P. Zumel, *Universidad Carlos III de Madrid, Spain*
C. Fernández, *Universidad Carlos III de Madrid, Spain*
A. Barrado, *Universidad Carlos III de Madrid, Spain*

On-Chip High Performance Magnetics for Point-of-Load High-Frequency DC-DC Converters ... 3097
Dragan Dinulovic, *Würth Elektronik eiSos GmbH & Co. KG, Germany*
Mahmoud Shousha, *Würth Elektronik eiSos GmbH & Co. KG, Germany*
Martin Haug, *Würth Elektronik eiSos GmbH & Co. KG, Germany*
Alexander Gerfer, *Würth Elektronik eiSos GmbH & Co. KG, Germany*
Mike Wens, *MinDCet NV, Belgium*
Jef Thone, *MinDCet NV, Belgium*

Effects of Auxiliary Source Connections in Multichip Power Module ... 3101
Helong Li, *Aalborg University, Denmark*
Stig Munk-Nielsen, *Aalborg University, Denmark*
Szymon Bęczkowski, *Aalborg University, Denmark*
Xiongfei Wang, *Aalborg University, Denmark*
Emanuel-Petre Eni, *Aalborg University, Denmark*

Session D10: Modeling and Simulation
Location: Poster Area
March 24, 2016 11:30 - 14:00
Session Chairs: Marco Meola, *ZMD AG*
Mehdi Ferdowsi, *Missouri University of Science & Technology*

Modelling Technique Utilizing Modified Sigmoid Functions for Describing Power Transistor Device Capacitances Applied on GaN HEMT and Silicon MOSFET ... 3107
H.L. Yeo, *Nanyang Technological University, Singapore*
K.J. Tseng, *Nanyang Technological University, Singapore*

Design and Precise Modeling of a Novel Digital Active EMI Filter ... 3115
Junpeng Ji, *Xi'an Jiaotong University, China*
Wenjie Chen, *Xi'an Jiaotong University, China*
Xu Yang, *Xi'an Jiaotong University, China*

Development of a Hybrid Emulation Platform based on RTDS and Reconfigurable Power Converter-Based Testbed 3121
Shuoting Zhang, *University of Tennessee, United States*
Yiwei Ma, *University of Tennessee, United States*
Liu Yang, *University of Tennessee, United States*
Fred Wang, *University of Tennessee, United States*
Leon M. Tolbert, *University of Tennessee, United States*

Online Temperature Estimation for Phase Change Composite – 18650 Lithium Ion Cells based Battery Pack 3128
Mohamad Salameh, *Illinois Institute of Technology, United States*
Ben Schweitzer, *AllCell Technologies, United States*
Peter Sveum, *AllCell Technologies, United States*
Said Al-Hallaj, *AllCell Technologies, United States*
Mahesh Krishnamurthy, *Illinois Institute of Technology, United States*

Modeling and Fault Diagnosis of Inter-Turn Short Circuit for Five-Phase PMSM based on Particle Swarm Optimization 3134
Jianwei Yang, *Northwestern Polytechnical University, China*
Manfeng Dou, *Northwestern Polytechnical University, China*
Zhiyong Dai, *Northwestern Polytechnical University, China*
Dongdong Zhao, *Northwestern Polytechnical University, China*
Zhen Zhang, *Northwestern Polytechnical University, China*

Comprehensive Modeling, Testing, and Experimental Validation of Ultracapacitor Open Circuit Voltage Characteristics 3140
Amandeep Singh, *University of Ontario Institute of Technology, Canada*
Najath Abdul Azeez, *University of Ontario Institute of Technology, Canada*
Sheldon S. Williamson, *University of Ontario Institute of Technology, Canada*

Novel SPICE Model for Common Mode Choke Including Complex Permeability 3146
Katsuya Nomura, *Toyota Central R&D Labs., Inc., Japan*
Naoto Kikuchi, *Toyota Central R&D Labs., Inc., Japan*
Yoshitoshi Watanabe, *Toyota Central R&D Labs., Inc., Japan*
Shuntaro Inoue, *Toyota Central R&D Labs., Inc., Japan*
Yoshiyuki Hattori, *Toyota Central R&D Labs., Inc., Japan*

Session D11: Control I
Location: Poster Area
March 24, 2016 11:30 - 14:00
Session Chairs: Bilal Akin, *Univeristy of Texas, Dallas*
Brian Zahnstecher, *PowerRox LLC*

Analysis and Design of Capacitive Power Transmission System Employing Out-of-Band Wireless Feedback Link 3153
Sung-Jin Choi, *University of Ulsan, Korea, South*
Hee-Su Choi, *University of Ulsan, Korea, South*

Introducing Fourier-Based Modeling and Control of Active-Bridge Converters 3158
B.J.D. Vermulst, *Technische Universiteit Eindhoven, Netherlands*
J.L. Duarte, *Technische Universiteit Eindhoven, Netherlands*
C.G.E. Wijnands, *Technische Universiteit Eindhoven, Netherlands*
E.A. Lomonova, *Technische Universiteit Eindhoven, Netherlands*

A Stability Analysis and Efficiency Improvement of Synchronverter 3165
Prasanna Piya, *Mississippi State University, United States*
Masoud Karimi-Ghartemani, *Mississippi State University, United States*

Compensation of Switching Dead-Time Effects in Voltage-Fed PWM Inverters using FPGA-Based Current Oversampling 3172
Bastian Weber, *Leibniz Universität Hannover, Germany*
Tobias Brandt, *Leibniz Universität Hannover, Germany*
Axel Mertens, *Leibniz Universität Hannover, Germany*

Control Strategy of High Power Converters with Synchronous Generator Characteristics for PMSG-Based Wind Power Application 3180
Yuzhi Zhang, *University of Arkansas, United States*
Haoyan Liu, *University of Arkansas, United States*
H. Alan Mantooth, *University of Arkansas, United States*

Phase Compensation, ZVS Operation of Wireless Power Transfer System based on SOGI-PLL 3185
Pingan Tan, *Xiangtan University, China*
Haibing He, *Xiangtan University, China*
Xieping Gao, *Xiangtan University, China*

A Novel Low-Cost Online State of Charge Estimation Method for Reconfigurable Battery Pack 3189
Ni Lin, *University of Nebraska at Lincoln, United States*
Song Ci, *University of Nebraska at Lincoln, United States*
Dalei Wu, *University of Tennessee at Chattanooga, United States*

Effect of Decoupling Terms on the Performance of PR Current Controllers Implemented in Stationary Reference Frame 3193
Sizhan Zhou, *Xi'an Jiaotong University, China*
Jinjun Liu, *Xi'an Jiaotong University, China*

Fuzzy Predictive DTC of Induction Machines with Reduced Torque Ripple and High Performance Operation 3200
Alberto Berzoy, *Florida International University, United States*
Osama Mohammed, *Florida International University, United States*
Johnny Rengifo, *Universidad Simon Bolivar, Venezuela*

Session D12: Control II
Location: Poster Area
March 24, 2016 11:30 - 14:00
Session Chairs: Martin Ordonez, *University of British Columbia*
Jiangbiao He, *GE Global Research*

Fixed-Frequency Generalized Peak Current Control (GPCC) for Inverters 3207
Mohammad Ebrahimi, *University of Alberta, Canada*
S. Ali Khajehoddin, *University of Alberta, Canada*

Improved Control Strategy of 1 MHz LLC Converter for High Frequency Resolution 3213
Hwa-Pyeong Park, *Ulsan National Institute of Science and Technology, Korea, South*
Jee-Hoon Jung, *Ulsan National Institute of Science and Technology, Korea, South*

Bumpless Control for Reduced THD in Power Factor Correction Circuits 3219
Joel Steenis, *Microchip Technology, United States*
Alex Dumais, *Microchip Technology, United States*

Mixed-Signal Hysteretic Internal Model Control of Buck Converters for Ultra-Fast Envelope Tracking 3224
V. Inder Kumar, *Indian Institute of Technology Kharagpur, India*
Santanu Kapat, *Indian Institute of Technology Kharagpur, India*

A Continuous Actor-Critic Maximum Power Point Tracker Applied to Low Power Wind Turbine Systems 3231
J.L. Wattes, *Universidade Federal do Ceará, Brazil*
A.J.S. Dias Jr., *Universidade Federal do Ceará, Brazil*
A.P.S. Braga, *Universidade Federal do Ceará, Brazil*
P.P. Praça, *Universidade Federal do Ceará, Brazil*
A.U. Barbosa, *Universidade Federal do Ceará, Brazil*
D.S. de Souza Oliveira Jr., *Universidade Federal do Ceará, Brazil*

Multi-Band Mixed-Signal Hysteresis Current Control for EMI Reduction in Switch-Mode Power Supplies 3237
Arindam Mandal, *Indian Institute of Technology Kharagpur, India*
V. Inder Kumar, *Indian Institute of Technology Kharagpur, India*
Santanu Kapat, *Indian Institute of Technology Kharagpur, India*

A Parabolic Current Control based Digital Current Control Strategy for High Switching Frequency Voltage Source Inverters 3243
Lanhua Zhang, *Virginia Polytechnic Institute and State University, United States*
Rachael Born, *Virginia Polytechnic Institute and State University, United States*
Xiaonan Zhao, *Virginia Polytechnic Institute and State University, United States*
Jih-Sheng Jason Lai, *Virginia Polytechnic Institute and State University, United States*
Hongbo Ma, *Southwest Jiaotong University, China*

Finite Control Set Model Predictive Control of Dual-Output Four-Leg Indirect Matrix Converter Under Unbalanced Load and Supply Conditions 3248
Ozan Gulbudak, *University of South Carolina, United States*
Enrico Santi, *University of South Carolina, United States*

A Silicon Carbide Integrated Circuit Implementing Nonlinear-Carrier Control for Boost Converter Applications 3255
Richard Kyle Harris, *University of Tennessee, United States*
Benjamin M. McCue, *University of Tennessee, United States*
Benjamin D. Roehrs, *University of Tennessee, United States*
Charles Roberts II, *University of Tennessee, United States*
Benjamin J. Blalock, *University of Tennessee, United States*
Daniel J. Costinett, *University of Tennessee, United States*
Kouros Sariri, *Frequency Management International, United States*
George Megyei, *Frequency Management International, United States*
Cheng-Po Chen, *GE Global Research, United States*
Avinash Kashyap, *GE Global Research, United States*
Reza Ghandi, *GE Global Research, United States*

A New Current Mode Constant on Time Control with Ultrafast Load Transient Response 3259
Syed Bari, *Virginia Polytechnic Institute and State University, United States*
Qiang Li, *Virginia Polytechnic Institute THD State Universy, United States*
Fred C. Lee, *Virginia Polytechnic Institute and State University, United States*

A Web-Based Tool for Compensation Design of Power Converters using Hybrid Optimization 3266

Srikanth Pam, *Texas Instruments Inc., India*
Yudhister Satija, *Texas Instruments Inc., India*
Pradeep Chawda, *Texas Instruments Inc., United States*
Makram Mansour, *Texas Instruments Inc., United States*
Robert Hanrahan, *Texas Instruments Inc., United States*
Jeff Perry, *Texas Instruments Inc., United States*

Second Order Sliding Mode Controlled Point of Load Power Supply 3273

Prasanta K. Achanta, *University of Colorado at Boulder, United States*
David C. Jones, *University of Colorado at Boulder, United States*
Dragan Maksimovic, *University of Colorado at Boulder, United States*
Serhii M. Zhak, *Linear Technology Corporation, United States*
Brett Miwa, *Maxim Integrated, United States*
Cory Arnold, *Maxim Integrated; United States*

Vibration and Torque Ripple Reduction of Switched Reluctance Motors through Current Profile Optimization 3279

Cong Ma, *University of Nebraska at Lincoln, United States*
Liyan Qu, *University of Nebraska at Lincoln, United States*
Rakesh Mitra, *Nexteer Automotive, United States*
Prerit Pramod, *Nexteer Automotive, United States*
Rakib Islam, *Nexteer Automotive, United States*

Modified Predictive Current Control of Neutral-Point Clamped Converter with Reduced Switching Frequency 3286

Dinto Mathew, *Indian Institute of Technology Bombay, India*
Anshuman Shukla, *Indian Institute of Technology Bombay, India*
Santanu Bandyopadhyay, *Indian Institute of Technology Bombay, India*

Implicit Finite Control Set Model Predictive Current Control for Modular Multilevel Converter based on IPA-SQP Algorithm 3291

Hamed Nademi, *ABB AS, Norway*
Lars Einar Norum, *Norwegian University of Science and Technology, Norway*

Resolution Requirements to Avoid Limit Cycling in LLC Resonant Converter 3297

Shadi Dashmiz, *University of Toronto, Canada*
Behzad Mahdavikhah, *University of Toronto, Canada*
Aleksandar Prodić, *University of Toronto, Canada*
Brent McDonald, *Texas Instruments Inc., United States*

Session D13: Renewable Energy Systems I
Location: Poster Area
March 24, 2016 11:30 - 14:00
Session Chairs: Akshay Kumar Rathore, *Concordia University*
Xiaoqiang Guo, *Yanshan University, China*

Reduction of Storage Capacity in DC Microgrids using PV-Embedded Series DC Electric Springs 3302

Ming-Hao Wang, *University of Hong Kong, Hong Kong*
Siew-Chong Tan, *University of Hong Kong, Hong Kong*
Shu-Yuen Ron Hui, *University of Hong Kong, Hong Kong*

A Vector Control Strategy of Grid-Connected Brushless Doubly Fed Induction Generator based on the Vector Control of Doubly Fed Induction Generator 3310
Sheng Hu, *Wuhan University of Technology, China*
Guorong Zhu, *Wuhan University of Technology, China*

An Energy Router based on Multi-Winding High-Frequency Transformer 3317
Xianzhuo Liu, *Tsinghua University, China*
Zedong Zheng, *Tsinghua University, China*
Kui Wang, *Tsinghua University, China*
Yongdong Li, *Tsinghua University, China*

Noise Suppression of the DWT-Based MRA on Mother Wavelet and Decomposition Level Optimization for a Robust Adaptive SOC Estimator in Multi-Cell Battery String 3322
Jonghoon Kim, *Chosun University, Korea, South*
Chang Yoon Chun, *Seoul National University, Korea, South*
Woonki Na, *California State University, Fresno, United States*

A Feedforward Control based Power Decoupling Scheme for Voltage-Controlled Grid-Tied Inverters 3328
Baojin Liu, *Xi'an Jiaotong University, China*
Zeng Liu, *Xi'an Jiaotong University, China*
Jinjun Liu, *Xi'an Jiaotong University, China*
Teng Wu, *Xi'an Jiaotong University, China*
Shike Wang, *Xi'an Jiaotong University, China*

Light Load Efficiency Improvement of Solar Farms Three-Phase Two-Stage Module Integrated Converter 3333
Ahmadreza Amirahmadi, *University of Central Florida, United States*
Utsav Somani, *University of Central Florida, United States*
Mahmood Alharbi, *University of Central Florida, United States*
Charlie Jourdan, *University of Central Florida, United States*
Issa Batarseh, *University of Central Florida, United States*

Switching System Stability Analysis of DC Microgrids with DBS Control 3338
Na Zhi, *Xi'an University of Technology, China*
Hui Zhang, *Xi'an University of Technology, China*
Xi Xiao, *Tsinghua University, China*

A Grid-Connected WECS with Power Limiting Control 3346
Jéssica Santos Guimarães, *Universidade Federal do Ceará, Brazil*
Demercil de Souza Oliveira Jr., *Universidade Federal do Ceará, Brazil*
Juliano de Oliveira Pacheco, *Universidade Federal do Ceará, Brazil*
Paulo P. Peixoto, *Universidade Federal do Ceará, Brazil*

Overshoot Control of the Electromagnetic Torque during Fault Recovery for an SCIG with a STATCOM 3353
Zahra Mahmoodzadeh, *Washington State University, United States*
Mehrdad Yazdanian, *Washington State University, United States*
Hooman Ghaffarzadeh, *Washington State University, United States*
Ali Mehrizi-Sani, *Washington State University, United States*

A Self-Adaptive Power Balance Control Strategy for PV Inverters in Islanded Microgrids 3358
Zhenxiong Wang, *Xi'an Jiaotong University, China*
Hao Yi, *Xi'an Jiaotong University, China*
Fang Zhuo, *Xi'an Jiaotong University, China*
Zhigang Zhang, *Xi'an Jiaotong University, China*

High Performance ZVT with Bus Clamping Modulation Technique for Single Phase Full Bridge Inverters 3364
Yinglai Xia, *Arizona State University, United States*
Raja Ayyanar, *Arizona State University, United States*

Small AC Signal Droop based Secondary Control for Microgrids 3370
Teng Wu, *Xi'an Jiaotong University, China*
Zeng Liu, *Xi'an Jiaotong University, China*
Jinjun Liu, *Xi'an Jiaotong University, China*
Baojin Liu, *Xi'an Jiaotong University, China*
Shike Wang, *Xi'an Jiaotong University, China*

Mode Transition Control Strategy for Multiple Inverter based Distributed Generators Operating in Grid-Connected and Stand-Alone Mode 3376
Onkar Vitthal Kulkarni, *Indian Institute of Technology Bombay, India*
Suryanarayana Doolla, *Indian Institute of Technology Bombay, India*
B.G. Fernandes, *Indian Institute of Technology Bombay, India*

An Autonomous Power Management Strategy based on DC Bus Signaling for Solid-State Transformer Interfaced PMSG Wind Energy Conversion System 3383
Rui Gao, *North Carolina State University, United States*
Iqbal Husain, *North Carolina State University, United States*
Alex Q. Huang, *North Carolina State University, United States*

An Isolated Buck-Boost Type High-Frequency Link Photovoltaic Microinverter 3389
Shiladri Chakraborty, *Indian Institute of Technology Kharagpur, India*
Souvik Chattopadhyay, *Indian Institute of Technology Kharagpur, India*

Energy Management and Stabilization of a Hybrid DC Microgrid for Transportation Applications 3397
Mehdi Karbalaye Zadeh, *Norwegian University of Science and Technology, Norway*
Louis-Marie Saublet, *Université de Lorraine, France*
Roghayeh Gavagsaz-Ghoachani, *Université de Lorraine, France*
Babak Nahid-Mobarakeh, *Université de Lorraine, France*
Serge Pierfederici, *Université de Lorraine, France*
Marta Molinas, *Norwegian University of Science and Technology, Norway*

A Low-Cost Solar Micro-Inverter with Soft-Switching Capability Utilizing Circulating Current 3403
Xiaohu Liu, *GE Global Research, United States*
Mohammed Agamy, *GE Global Research, United States*
Dong Dong, *GE Global Research, United States*
Maja Harfman-Todorovic, *GE Global Research, United States*
Luis Garces, *GE Global Research, United States*

Session D14: Renewable Energy Systems II
Location: Poster Area
March 24, 2016 11:30 - 14:00
Session Chairs: Haoyu Wang, *Shanghai Tech University*
Robert Pilawa-Podgurski, *University of Illinois at Urbana-Champaign*

Design and Stability Analysis for an Autonomous DC Microgrid with Constant Power Load ... 3409
Qianwen Xu, *Nanyang Technological University, Singapore*
Xiaolei Hu, *Nanyang Technological University, Singapore*
Peng Wang, *Nanyang Technological University, Singapore*
Jianfang Xiao, *Nanyang Technological University, Singapore*
Leonardy Setyawan, *Nanyang Technological University, Singapore*
Changyun Wen, *Nanyang Technological University, Singapore*
Lee Meng Yeong, *Rolls-Royce Singapore Pte. Ltd., Singapore*

MPC-SVM Method for Vienna Rectifier with PMSG used in Wind Turbine Systems 3416
June-Seok Lee, *Korea Railroad Research Institute, Korea, South*
Yeongsu Bak, *Ajou University, Korea, South*
Kyo-Beum Lee, *Ajou University, Korea, South*
Frede Blaabjerg, *Aalborg University, Denmark*

An Equivalent Circuit Model for State of Energy Estimation of Lithium-Ion Battery 3422
Kaiyuan Li, *Nanyang Technological University, Singapore*
King Jet Tseng, *Nanyang Technological University, Singapore*

Distributed Optimal Control of Reactive Power and Voltage in Islanded Microgrids 3431
Yanbo Wang, *Aalborg University, Denmark*
Xiongfei Wang, *Aalborg University, Denmark*
Zhe Chen, *Aalborg University, Denmark*
Frede Blaabjerg, *Aalborg University, Denmark*

New Start-Up Scheme for HF Transformer Link Photovoltaic Inverter 3439
Abhijit Kulkarni, *Indian Institute of Science, India*
Vinod John, *Indian Institute of Science, India*

Analysis and Improvement of Harmonic Quasi Resonant Control for LCL-Filtered Grid-Connected Inverters in Weak Grid 3446
Qiang Qian, *Nanjing University of Aeronautics and Astronautics, China*
Jinming Xu, *Nanjing University of Aeronautics and Astronautics, China*
Shaojun Xie, *Nanjing University of Aeronautics and Astronautics, China*
Lin Ji, *Nanjing University of Aeronautics and Astronautics, China*

Model Predictive Control Method to Reduce Common-Mode Voltage and Balance the Neutral-Point Voltage in Three-Level T-Type Inverter 3453
Xiangyang Xing, *Shandong University, China*
Alian Chen, *Shandong University, China*
Zicheng Zhang, *Shandong University, China*
Jie Chen, *Shandong University, China*
Chenghui Zhang, *Shandong University, China*

Convergence Analysis of Distributed Control for Operation Cost Minimization of Droop Controlled DC Microgrid based on Multiagent 3459
Chendan Li, *Aalborg University, Denmark*
Juan C. Vásquez, *Aalborg University, Denmark*
Josep M. Guerrero, *Aalborg University, Denmark*

A Novel Model Predictive Control Algorithm to Suppress the Zero-Sequence Circulating Currents for Parallel Three-Phase Voltage Source Inverters 3465
Zicheng Zhang, *Shandong University, China*
Alian Chen, *Shandong University, China*
Xiangyang Xing, *Shandong University, China*
Chenghui Zhang, *Shandong University, China*

Design of Dynamic Voltage Restorer and Active Power Filter for Wind Power Systems Subject to Unbalanced and Harmonic Distorted Grid 3471
Woei-Luen Chen, *Chang Gung University, Taiwan*
Meng-Jie Wang, *Chang Gung University, Taiwan*

Dynamic Variable Coupling Analysis and Modeling of Proton Exchange Membrane Fuel Cells for Water and Thermal Management 3476
Daming Zhou, *Université de Technologie de Belfort-Montbéliard, France*
Elena Breaz, *Université de Technologie de Belfort-Montbéliard, France*
Alexandre Ravey, *Université de Technologie de Belfort-Montbéliard, France*
Fei Gao, *Université de Technologie de Belfort-Montbéliard, France*
Abdellatif Miraoui, *Université de Technologie de Belfort-Montbéliard, France*
Ke Zhang, *Northwestern Polytechnical University, China*

Voltage and Frequency Control of Electric Spring based Smart Loads 3481
Yun Yang, *University of Hong Kong, Hong Kong*
Siew-Chong Tan, *University of Hong Kong, Hong Kong*
Shu-Yuen Ron Hui, *University of Hong Kong, Hong Kong*

Second Harmonic Current Compensator with Improved One-Cycle-Control 3488
Li Zhang, *Nanjing University of Aeronautics and Astronautics, China*
Xinbo Ruan, *Nanjing University of Aeronautics and Astronautics, China*
Xiaoyong Ren, *Nanjing University of Aeronautics and Astronautics, China*

Frequency Adaptive Control of a Smart Transformer-Fed Distribution Grid 3493
Zhi-Xiang Zou, *Christian-Albrechts-Universität zu Kiel, Germany*
Giovanni De Carne, *Christian-Albrechts-Universität zu Kiel, Germany*
Giampaolo Buticchi, *Christian-Albrechts-Universität zu Kiel, Germany*
Marco Liserre, *Christian-Albrechts-Universität zu Kiel, Germany*

A Synchronization Scheme for Single-Phase Grid-Tied Inverters under Harmonic Distortion and Grid Disturbances 3500
Lenos Hadjidemetriou, *University of Cyprus, Cyprus*
Elias Kyriakides, *University of Cyprus, Cyprus*
Yongheng Yang, *Aalborg University, Denmark*
Frede Blaabjerg, *Aalborg University, Denmark*

Series-Parallel Connection of Low-Voltage Sources for Integration of Galvanically Isolated Energy Storage Systems 3508
Ramy Georgious, *Universidad de Oviedo, Spain*
Jorge Garcia, *Universidad de Oviedo, Spain*
Angel Navarro, *Universidad de Oviedo, Spain*
Sarah Saeed, *Universidad de Oviedo, Spain*
Pablo Garcia, *Universidad de Oviedo, Spain*

Saturation Controller-Based Direct Power Control for Doubly-Fed Induction Generator 3514
Chun Wei, *University of Nebraska at Lincoln, United States*
Zhe Zhang, *Nexteer Automotive, United States*
Wei Qiao, *University of Nebraska at Lincoln, United States*
Liyan Qu, *University of Nebraska at Lincoln, United States*

Inductance-Simulating Control for DFIG-Based Wind Turbine to Ride-Through Grid Faults 3521
Donghai Zhu, *Huazhong University of Science and Technology, China*
Xudong Zou, *Huazhong University of Science and Technology, China*
Yong Kang, *Huazhong University of Science and Technology, China*
Lu Deng, *Wuhan NARI Limited Company of State Grid Electric Power Research Institute, China*
Qingjun Huang, *State Key Laboratory of Disaster Prevention & Reduction for Power Grid Transmission and Distribution Equipment, China*

Session D15: Transportation Power Electronics
Location: Poster Area
March 24, 2016 11:30 - 14:00
Session Chairs: Ted Bohn, *Argonne National Labs*
Khurram Afridi, *University of Colorado, Boulder*

Misalignment Effect on Efficiency of Wireless Power Transfer for Electric Vehicles 3526
Yabiao Gao, *University of Georgia, United States*
Antonio Ginart, *University of Georgia / Sonnenbatterie GmbH, United States*
Kathleen Blair Farley, *Southern Company Services, Inc., United States*
Zion Tsz Ho Tse, *University of Georgia, United States*

Genetic Algorithm Design of a 3D Printed Heat Sink 3529
Tong Wu, *University of Tennessee, United States*
Burak Ozpineci, *Oak Ridge National Laboratory, United States*
Curtis Ayers, *Oak Ridge National Laboratory, United States*

Evaluation of Power Flow Control for an All-Electric Warship Power System with Pulsed Load Applications 3537
J. Neely, *Sandia National Laboratories, United States*
L. Rashkin, *Sandia National Laboratories, United States*
M. Cook, *Sandia National Laboratories, United States*
D. Wilson, *Sandia National Laboratories, United States*
S. Glover, *Sandia National Laboratories, United States*

Reduced Active Switch AC to DC Rectifier with High Frequency Isolation for Electric Vehicle Chargers 3545
José Juan Sandoval, *Texas A&M University, United States*
Taeyong Kang, *Texas A&M University, United States*
Prasad Enjeti, *Texas A&M University, United States*

A Wide Bandgap Device based Multilevel Switched-Capacitor Converter 3553
Diogo Cesar Santos de Moura, *North Dakota State University, United States*
Boris Curuvija, *North Dakota State University, United States*
Dong Cao, *North Dakota State University, United States*

Session D16: Power Topologies, Distribution, and Control
Location: Poster Area
March 24, 2016 11:30 - 14:00
Session Chairs: Tiefu Zhao, *Eaton*
Xiaonan Lu, *Argonne National Laboratory*

Novel Circulating Current Suppression Strategy for MMC based on Quasi-PR Controller 3560
Shengbao Geng, *Shanghai Jiao Tong University, China*
Yiliang Gan, *Shanghai Jiao Tong University, China*
Yungui Li, *Shanghai Jiao Tong University, China*
Lijun Hang, *Shanghai Jiao Tong University, China*
Guojie Li, *Shanghai Jiao Tong University, China*

Assymmetric Duty-Cycle Phase-Shift Modulation for Power Management in Double Half-Bridge Inverter with Partly Coupled Inductive Loads 3566
C. Carretero, *Universidad de Zaragoza, Spain*
H. Sarnago, *Universidad de Zaragoza, Spain*
O. Lucia, *Universidad de Zaragoza, Spain*
J. Acero, *Universidad de Zaragoza, Spain*
J.M. Burdío, *Universidad de Zaragoza, Spain*

Control Implementation for a Wide Voltage Range High Efficiency Power Supply Utilizing Low Voltage MOSFETs 3570
Werner Konrad, *Technische Universität Graz, Austria*
Gerald Deboy, *Infineon Technologies AG, Austria*
Annette Muetze, *Technische Universität Graz, Austria*

A Single-Phase Dual Frequency Inverter based on Multi-Frequency Selective Harmonic Elimination 3577
Chongwen Zhao, *University of Tennessee, United States*
Daniel Costinett, *University of Tennessee, United States*
Brad Trento, *University of Tennessee, United States*
Daniel Friedrichs, *Medtronic, United States*

Grid Connected DC Distribution Network Deploying High Power Density Rectifier for DC Voltage Stabilization 3585
Danillo B. Rodrigues, *Universidade Federal do Triângulo Mineiro, Brazil*
Paulo R. Silva, *Universidade Federal de Uberlândia, Brazil*
Gustavo B. Lima, *Universidade Federal do Triângulo Mineiro, Brazil*
Ernane A.A. Coelho, *Universidade Federal de Uberlândia, Brazil*
Luiz C.G. Freitas, *Universidade Federal de Uberlândia, Brazil*

Even-Harmonic Repetitive Control for Circulating Current Suppression in Modular Multilevel Converters 3591

Shunfeng Yang, *Nanyang Technological University, Singapore*
Peng Wang, *Nanyang Technological University, Singapore*
Yi Tang, *Nanyang Technological University, Singapore*
Michael Zagrodnik, *Rolls-Royce Singapore Pte. Ltd., Singapore*
Xiaolei Hu, *Nanyang Technological University, Singapore*
King Jet Tseng, *Nanyang Technological University, Singapore*

A New DSC-PLL using Recursive Discrete Fourier Transform for Robustness to Frequency Variation 3598

Jaedo Lee, *Korea Institute of Nuclear Safety, Korea, South*
Hanju Cha, *Chungnam National University, Korea, South*

A Four-Quadrant Modulation Technique for Cascaded Multilevel Inverters to Extend Solution Range for Selective Harmonic Elimination/Compensation 3603

Hui Zhao, *University of Florida, United States*
Shuo Wang, *University of Florida, United States*

Online Battery Impedance Spectrum Measurement Method 3611

Jaber A. Abu Qahouq, *University of Alabama, United States*

Analysis and Control of a Reduced Switch Converter for Active Magnetic Bearings 3616

Dong Jiang, *Huazhong University of Science and Technology, China*
Parag Kshirsagar, *United Technologies Research Center, United States*

A Novel Balanced Winding Topology to Mitigate EMI without the Need for a Y-Capacitor 3623

Yongjiang Bai, *Xi'an Jiaotong University, China*
Xu Yang, *Xi'an Jiaotong University, China*
Xinlei Li, *Silergy Corp., China*
Dan Zhang, *Silergy Corp., China*
Wenjie Chen, *Xi'an Jiaotong University, China*

Topology and Control Strategy for Accelerated Lifetime Test Setup of DC-Link Capacitor of Wind Turbine Converter 3629

Youngjong Ko, *Christian-Albrechts-Universität zu Kiel, Germany*
Holger Jedtberg, *Christian-Albrechts-Universität zu Kiel, Germany*
Giampaolo Buticchi, *Christian-Albrechts-Universität zu Kiel, Germany*
Marco Liserre, *Christian-Albrechts-Universität zu Kiel, Germany*

Voltage Droop Compensation based on Resonant Circuit for Generalized High Voltage Solid-State Marx Modulator 3637

Hiren Canacsinh, *Instituto Superior de Engenharia de Lisboa, Portugal*
Luís M. Redondo, *Instituto Superior de Engenharia de Lisboa, Portugal*
J. Fernando Silva, *Instituto Superior Técnico, Portugal*
Beatriz Borges, *Instituto Superior Técnico, Portugal*

Four H-Bridge based Shunt Active Power Filter for Three-Phase Four Wire System 3641

Edgard L.L. Fabricio, *Universidade Federal da Paraíba, Brazil*
Cursino B. Jacobina, *Universidade Federal de Campina Grande, Brazil*
Gregory A.A. Carlos, *Universidade Federal de Campina Grande, Brazil*
Maurício B.R. Correa, *Universidade Federal de Campina Grande, Brazil*

High-Frequency AC Distributed Power Delivery System 3648
Mengqi Wang, *University of Michigan at Dearborn, United States*
Qingyun Huang, *North Carolina State University, United States*
Wensong Yu, *North Carolina State University, United States*
Alex Q. Huang, *North Carolina State University, United States*

**Effect of the Capacitance Distribution on the Output Impedance of the Half-Wave
Cockcroft-Walton Voltage Multiplier** 3655
Liran Katzir, *Tel Aviv University, Israel*
Doron Shmilovitz, *Tel Aviv University, Israel*

Session D17: Emerging and Renewable Power
Location: Poster Area
March 24, 2016 11:30 - 14:00
Session Chairs: Katherine Kim, *Ulsan NIST*
Dimitri Torregrossa, *EPFL*

A Cost Effective High Performance LED Driver Powered by Electronic Ballasts 3659
Jianwen Shao, *STMicroelectronics, United States*
Thomas Stamm, *STMicroelectronics, United States*

**Model Predictive Control of Z-Source Four-Leg Inverter for Standalone Photovoltaic
System with Unbalanced Load** 3663
Sertac Bayhan, *Gazi University, Turkey*
Mohamed Trabelsi, *Texas A&M University at Qatar, Qatar*
Haitham Abu-Rub, *Texas A&M University at Qatar, Qatar*

**Efficiency Optimization of an Integrated Wireless Power Transfer System by a
Genetic Algorithm** 3669
Rosario Pagano, *Integrated Device Technology Inc., United States*
Siamak Abedinpour, *Integrated Device Technology Inc., United States*
Angelo Raciti, *Università degli Studi di Catania, Italy*
Salvatore Musumeci, *Università degli Studi di Catania, Italy*

**Loss Analysis of a High Efficiency GaN and Si Device Mixed Isolated Bidirectional
DC-DC Converter** 3677
Fei Xue, *North Carolina State University, United States*
Ruiyang Yu, *North Carolina State University, United States*
Alex Q. Huang, *North Carolina State University, United States*

**Dynamic Efficiency Tracking Controller for Reconfigurable Four-Coil Wireless Power
Transfer System** 3684
Yuan Cao, *University of Alabama, United States*
Zhigang Dang, *University of Alabama, United States*
Jaber A. Abu Qahouq, *University of Alabama, United States*
Evan Phillips, *University of Alabama, United States*

Wireless Power and Data Transfer System for Smart Bridge Sensors 3690
Yujin Jang, *Korea Advanced Institute of Science and Technology, Korea, South*
Jung Kyu Han, *Korea Advanced Institute of Science and Technology, Korea, South*
Shin Young Cho, *Korea Advanced Institute of Science and Technology, Korea, South*
Gun-Woo Moon, *Korea Advanced Institute of Science and Technology, Korea, South*
Ji-Min Kim, *Korea Advanced Institute of Science and Technology, Korea, South*
Hoon Sohn, *Korea Advanced Institute of Science and Technology, Korea, South*

Inrush Transient Current Analysis and Suppression of Photovoltaic Grid-Connected Inverters during Voltage Sag 3697

Zhongyu Li, *China University of Petroleum, China*
Rende Zhao, *China University of Petroleum, China*
Zhen Xin, *Aalborg University, Denmark*
Josep M. Guerrero, *Aalborg University, Denmark*
Mehdi Savaghebi, *Aalborg University, Denmark*
Peide Li, *Shandong Jinan Power Equipment Factory Co., LTD, China*

A Highly Reliable Single-Stage Converter for Electric Vehicle Applications 3704

S.A.Kh. Mozaffari Niapour, *Northeastern University, United States*
Mahshid Amirabadi, *Northeastern University, United States*

Simple and Efficient Low Power Photovoltaic Emulator for Evaluation of Power Conditioning Systems 3712

Jesus Gonzalez-Llorente, *Universidad Sergio Arboleda, Colombia*
Andres Rambal-Vecino, *Universidad Sergio Arboleda, Colombia*
Luciano A. Garcia-Rodriguez, *University of Arkansas, United States*
Juan C. Balda, *University of Arkansas, United States*
Eduardo I. Ortiz-Rivera, *University of Puerto Rico at Mayaguez, Puerto Rico*

Data Transmission Method without Additional Circuits in Bidirectional Wireless Power Transfer System 3717

Yeongrack Son, *Seoul National University, Korea, South*
Jung-Ik Ha, *Seoul National University, Korea, South*

Improved Impedance Source Inverter for Hybrid/Electric Vehicle Application with Continuous Conduction Operation 3722

Thilak Senanayake, *University of Tsukuba, Japan*
Ryuji Iijima, *University of Tsukuba, Japan*
Takanori Isobe, *University of Tsukuba, Japan*
Hiroshi Tadano, *University of Tsukuba, Japan*

Foreword

31st Annual IEEE Applied Power Electronics Conference and Exposition
March 20-24, 2016
Long Beach Convention & Entertainment Center, Long Beach, California

It is my utmost pleasure to welcome you to the 2016 IEEE Applied Power Electronics Conference and Exposition (APEC 2016), at the Long Beach Convention Center, in Long Beach, California.

As the *Premier Event in Applied Power Electronics*, APEC provides a unique opportunity to power electronics professionals from academia, national laboratories, and industry for exchange of technical knowledge, networking, and exposure to the vibrant indigenous culture.

The APEC 2016 organizing committee has been working wholeheartedly to compose this excellent technical conference for you. Thanks to their dedication and countless hours of work as well as APEC's sponsors: IEEE Industry Applications Society (IAS), IEEE Power Electronics Society (PELS), and Power Sources Manufacturers Association (PSMA).

Like its predecessors, APEC 2016 offers a unique technical program for power electronics community. We have a record breaking and unique conference and exposition planned for you to experience, with a comprehensive program remarkably attractive to the academic researchers, students, educators, industry, government agencies, and general public. The technical presentation papers are selected from an all-time record high 1212 digests submitted from 45 countries from across the globe. The exposition hits its record high participation with 263 exhibitors and 398 booths. The exhibitors will showcase their state-of-the-art technologies, products, and solutions on applied power electronics. Furthermore, this year in our progressively popular industry sessions we have 83 accepted presentations in 18 sessions.

The Professional Educational Seminars, offered by internationally renowned experts, start on Sunday, March 20th. Each of the 21 three-and-a-half hour educational seminars, selected from the record-high 52 submissions, provides an in-depth discussion of important and complex power electronics topics and combines practical application with theory. The Plenary Session, on Monday afternoon, consists of distinguished world-class speakers from industry and academia covering the key power electronics technologies, components, and innovations affecting our industry and the society.

This year, the 30th micro mouse competition will include teams from Japan, Taiwan, Singapore, China, United Kingdom and the United States. The increasingly popular rap sessions include three moderated debates on Future of Semiconductor Technology Development, Power Electronics for Internet of Things, and Advanced Refueling Technologies for Electric Vehicles. This year APEC sponsors have provided 43 travel grants to assist students' participation in this unique conference. In addition, the APEC Mobile App provides access to an interactive directory and map of the exhibitors on their mobile device.

The Wednesday night social event "Surfin' Safari" will be at the Pacific Ballroom – Long Beach Convention Center. The APEC social event provides you with the opportunity to let loose, enjoy great food and network with your colleagues. This year we are thrilled to have live entertainment from California's premiere Beach Boy's cover band, The Beach Toys!

With three convenient local airports, average high temperature of 68°F (20 °C) in March, an enjoyable voyage away from Catalina Island and a convenient trip from Disneyland and Universal Studios Hollywood, Long Beach is a prime location for APEC. In addition to conference outstanding program, I hope you and your families will enjoy your stay at Long Beach and its superlative attractions such as the Ports O' Call Village, Point Vicente Lighthouse, Regal Queen Mary, the Shoreline Village, the Aquarium of the Pacific, its extensive museums, and many elite dining destinations.

I would like to take this opportunity to reiterate my appreciation to APEC attendees, exhibitors, reviewers, volunteers, sponsors, organizing committee members, and steering committee members. We are passionately looking forward to meeting you at APEC 2016, wish you a heartfelt welcome to APEC, and hope that you will have a memorable experience.

Warmest Regards,

Alireza Khaligh
General Chair
2016 IEEE Applied Power Electronics Conference and Exposition

APEC History

Year	Site	Dates	General Chair	Program Chair
1986	Fairmont Hotel New Orleans, Louisiana	April 28 – May 1	John G. Kassakian	R. David Middlebrook
1987	Town and Country Hotel San Diego, California	March 2 – 6	John G. Kassakian	R. David Middlebrook
1988	Fairmont Hotel New Orleans, Louisiana	February 1 – 5	William W. Burns, III	William W. Burns, III
1989	Baltimore Convention Center Baltimore, Maryland	March 13 – 17	William W. Burns, III	Robert V. White
1990	Biltmore Hotel Los Angeles, California	March 11 – 16	Robert V. White	Charles Harm
1991	Hyatt Regency Reunion Hotel Dallas, Texas	March 10 – 15	Charles Harm	Thomas M. Jahns
1992	Weston Copley Plaza Hotel Boston, Massachusetts	February 23 – 27	Thomas M. Jahns	Kevin J. Fellhoelter
1993	Town and Country Hotel San Diego, California	March 7 – 11	Kevin J. Fellhoelter	Douglas McIlvoy
1994	Disney Contemporary Hotel Orlando, Florida	February 13 – 17	Douglas McIlvoy	Thomas Latos
1995	Hyatt Regency Reunion Hotel Dallas, Texas	March 5 – 9	Thomas Latos	Charles E. Mullett
1996	Fairmont Hotel San Jose, California	March 3 – 7	Charles E. Mullett	Thomas G. Wilson, Jr.
1997	Weston Peachtree Hotel Atlanta, Georgia	February 23 – 27	Thomas G. Wilson, Jr.	David Torrey
1998	Disneyland Hotel Anaheim, California	February 15 – 19	David Torrey	F. Dong Tan
1999	Adams' Mark Hotel Dallas, Texas	March 14 – 18	F. Dong Tan	Robert V. White
2000	Fairmont Hotel New Orleans, Louisiana	February 6 – 10	Robert V. White	R. Mark Nelms
2001	Disneyland Hotel Anaheim, California	March 4 – 8	R. Mark Nelms	V. Joseph Thottuvelil
2002	Adams' Mark Hotel Dallas, Texas	March 10 – 14	V. Joseph Thottuvelil	Bruce Miller
2003	Fontainebleau Hotel Miami Beach, Florida	February 9 – 13	Bruce Miller	Jim Kokernak
2004	Disneyland Hotel Anaheim, California	February 22 – 26	Jim Kokernak	Jason Lai
2005	Hilton Austin Austin, Texas	March 6 – 10	Jason Lai	Van Niemela

Year	Site	Dates	General Chair	Program Chair
2006	Hyatt Regency Hotel Dallas, Texas	March 19 – 23	Van Niemela	Russ Spyker
2007	Disneyland Hotel Anaheim, California	February 25 – March 1	Russ Spyker	Steve Pekarek
2008	Austin Convention Center Austin, Texas	February 24 – 28	Steve Pekarek	Kevin Parmenter
2009	Marriott Wardman Park Hotel Washington, District of Columbia	February 15 – 19	Kevin Parmenter	Babak Fahimi
2010	Palm Springs Convention Center Palm Springs, California	February 21 – 25	Babak Fahimi	Patrick Chapman
2011	Fort Worth Convention Center Fort Worth, Texas	March 6 – 10	Patrick Chapman	Frank Cirolia
2012	Coronado Springs in Disney World Orlando, Florida	February 5 – 9	Frank Cirolia	Siamak Abedinpour
2013	Long Beach Convention Center Long Beach, California	March 17 – 21	Siamak Abedinpour	Haidong Yu
2014	Fort Worth Convention Center Fort Worth, Texas	March 16 - 20	Haidong Yu	Aung Thet Tu
2015	Charlotte Convention Center Charlotte, North Carolina	March 15 - 19	Aung Thet Tu	Alireza Khaligh

The APEC Conference Committee

General Chair
Alireza Khaligh, University of Maryland at College Park

Program Chair
Jonathan Kimball, Missouri S&T

Assistant Program Chair
Eric Persson, International Rectifier

Finance Co-Chairs
Mark Nelms, Auburn University
John Vigars, Allegro MicroSystems

Seminar Co-Chairs
Eric Persson, International Rectifier
Jin Wang, Ohio State University

Industry Session Co-Chairs
Tony O'Gorman, PESC Inc.
Conor Quinn, Artesyn Embedded Technologies

Exposition Co-Chairs
Van Niemela, Inventronics
Jose Cobos, Universidad Politécnica de Madrid

Publicity Chair
Greg Evans, Welcomm, Inc.

Rap Sessions Co-Chairs
Omer C. Onar, Oak Ridge National Laboratory
Berker Bilgin, McMaster University

Social Media Chair
Frank Cirolia, Artesyn Embedded Technologies

Special Projects and Local Chair
Doug Hopkins, North Carolina State University

Grants and Awards Chair
Bilal Akin, University of Texas, Dallas

OEM Initiative Chair
Ada Cheng, AdaClock

Publications Chair
Siamak Abedinpour, Integrated Device Technology, Inc.

MicroMouse Chair
David Otten, Massachusetts Institute of Technology

Spousal Hospitality Chair
Jane Wilson

Past General Chair
Aung Thet Tu, Fairchild Semiconductor

Members at Large
Sheldon Williamson, University of Ontario Institute of Technology
Morgan Kiani, Texas Christian University
Maryam Saeedifard, Georgia Tech

Conference Management

Donna Johnson, Tonya Stanback, Bobbie Praske, Michael Nercesian, and Ashley Kesack, Courtesy Associates

Tom Wehner, ePapers.org

Tia Fulmer, The Printing House Inc.

The Sponsors

Power Sources Manufacturers Association

Ernie Parker
Chairman of the Board

Stephen Oliver
Vice President

Eric Persson
President

Michel Grenon
Secretary/Treasurer

IEEE Power Electronics Society

Braham Ferreira
President

Frede Blaabjerg
Vice President – Products

Alan Mantooth
Vice President – Operations

Jian Sun
Treasurer

Mario Pacas
Vice President – Meetings

IEEE Industry Applications Society

David B. Durocher
President

Georges Zissis
Vice President

Tomy Sebastian
President-Elect

Corinne Fields
Treasurer

APEC Steering Committee Members with Sponsoring Organizations

Siamak Abedinpour
Integrated Device Technology
IAS

Frank Cirolia
Artesyn Embedded
Technologies
PSMA

Jose Cobos
U. Politechnica de Madrid
PELS

Babak Fahimi
University of Texas at Dallas
PELS

Alireza Khaligh
University of Maryland
IAS

Kevin Parmenter
Excelsys Technologies
PSMA

Jonathan Kimball
Missouri S&T
PELS

Russell Spyker
Wright-Patterson AFB
IAS

Aung Thet Tu
Fairchild Semiconductor
PSMA

APEC 2016 Program Committee Track Chairs

AC-DC Converters
Gerry Moschopoulos
Dustin Becker

DC-DC Converters
Alexis Kwasinski
Mahshid Amirabadi
Olivier Trescases
Pradeep Shenoy

Power Electronics for Utility Interface
Ali Khajehoddin
Julia Zhang
Babak Nahid Mobarakeh

Motor Drives and Inverters
Ali Bazzi
Bulent Sarlioglu
Maryam Saeedifard
Liming Liu

Devices and Components
Jean-Luc Schanen
Doug Hopkins
Qiang Li

System Integration
Ernie Parker
John Vigars

Modeling and Simulation
Sheldon Williamson
Ali Davoudi
Jaber Abu Qahouq
Marco Meola

Control
Chris Bridge
Martin Ordonez
Bilal Akin

Manufacturing, Quality, and Business Issues
Jim Marinos

Renewable Energy Systems
Khurram Afridi
Akshay Kumar Rathore
Haoyu Wang
Fei Gao

Transportation Power Electronics
Omer Onar
Yingying Kuai

Power Electronics Applications
Deepak Gautam
Juan Manuel Rivas
Robert Pilawa
Zhong Nie

APEC 2016 Reviewers

Abdel-Khalik, Ayman
Abdul Azeez, Najath
Abe, Seiya
Abnavi, Somaye
Abolhassani, Mehdi
Abramov, Eli
Acero, Jesus
Adhikari, Jeevan
Adil, Khurram
Aditya, Kunwar
Aditya, Kunwar
Ahmadi, Mohsen
Ahmadi, Reza
Ahmed, Md Rishad
Alaai, Ramiar
Alam, Muntasir
Alfares, Abdulgafor
Ali, Kawsar
Ali, Syed Qaseem
Aliprantis, D
Allard, Bruno
Anand, Sandeep
Andersen, Michael
Andrade, António
Anthon, Alexander
Anun, Matias
Anwar, Saeed
Anwar, Usama
Athalye, Praneet
Avenas, Yvan
Avestruz, Al-Thaddeus
Ayachit, Agasthya
Azcondo, Francisco
Baek, Jeihoon
Baker, Gary
Bal, Satarupa
Banerjee, Arijit
Barreto, Luiz Henrique
Barth, Chris
Becker, Dustin
Bergveld, Henk Jan
Bernacchia, Giuseppe
Bhardwaj, Manish
Bhattacharya, Subhadeep
Bilgin, Berker
Blaabjerg, Frede
Bobba, Dheeraj
Botting, Chris
Bramerdorfer, Gerd
Brandao, Danilo
C, Hao
Cai, Wen
Caliskan, Vahe
Callanan, Shane
Candan, Enver
Cannon, Michael
Cao, Dong
Cao, Wenchao
Carlos, Gregory
Castro Alvarez, Ignacio

Cengelci, Ekrem
Cervera, Alon
Ceylan, Murat
Chakraborty, Debjani
Chakraborty, Shiladri
Chandar, Subash
Channegowda, Janamejaya
Chazal, Herve
Chen, Alian
Chen, Baifeng
Chen, Baoxing
Chen, Chen
Chen, Fang
Chen, Hao
Chen, Hsin-Chih
Chen, Hung-Chi
Chen, Liqun
Chen, Mengxing
Chen, Minjie
Chen, Nan
Chen, Qingri
Chen, Weiqiang
Chen, Wenjie
Cheng, Ada
Cheng, Lin
Chew, Benjamin
Chinski, Paul
Chiwaridzo, Pride
Choi, Jungwon
Choudhury, Abhijit
Choudhury, Shamim
Chowdhury, Mahmud-Ul-
   Tarik
Chub, Andrii
Chun, Changyoon
Chung, Steven
Cirolia, Frank
Clements, Neal
Colak, Kerim
Comanescu, Mihai
Corradini, Luca
Cui, Shenghui
Cui, Yutian
Cupertino, Francesco
Cuzner, Robert
D, Nicholas
Dahan, Nadav
Dai, Weiran
Dalal, Dhaval
Damodharan, Arun
Danilovic, Milisav
Dargahi, Vahid
Dasika, Jayadeepti
Davari, Pooya
Davidkovich, Vladislav
De, Dipankar
de Bosio, Federico
de Palma, Jean-Francois
Debnath, Suman
Delaforge, Timothe

Delhotal, Jarod
Deng, Junjun
Deng, Yi
Deshpande, Yateendra
Diaz Reigosa, Paula
Dimarino, Christina
Dinavahi, Venkata
Diong, Bill
Dong, Zerui
Dong, Zhou
Doshi, Montu
Doshpande, Yateendra
Du, Chunshui
Du, Xiong
Dube, Sunil
Dubey, Subhash Chander
Dusmez, Serkan
Ebrahimi, Mohammad
Edpuganti, Amarendra
Edwards, Hal
Ekneligoda, Nishantha
Elhami Khorasani, Arash
Erturk, Feyzullah
Erturk, Mete
Ezra, Ofer
Fabricio, Edgard
Faley, Brian
Fall, Mbaye
Farag, Hany
Fedison, Jeff
Fei, Chao
Feng, Wei
Fernandez, Pablo
Fishbune, Richard
Frey, David
Friedrichs, Daniel
Fu, Lixing
Fu, Ruiyun
G, Deepak
G. Lamar, Diego
Gachovska, Tanya
Gaillard, Arnaud
Galiano Zurbriggen, Ignacio
Galigekere, Veda Prakash
Gamini, Shantha
Gao, Fei
Gao, Mingzhi
Gao, Rui
Gao, Yabiao
Garcia, Jorge
Garcia, Pablo
Gauchia, Lucia
Ge, Ting
Ge, Xuefeng
Ghods, Amirhossein
Ghoshal, Anirban
Ghule, Aditya
Ginart, Antonio
Glaser, John

Gohil, Ghanshyamsinh
   Vijays
Golbon, Navid
Gong, Xianzhi
Goward, John
Grezaud, Romain
Gu, Lei
Gu, Yunjie
Guemez, Jorge Guzman
Gulbudak, Ozan
Gunter, Samantha
Guo, Ben
Guo, Feng
Guo, Suxuan
Guo, Xiaoqiang
Gupta, Rahul
Gurpinar, Emre
H, Hamid
Hafez, Bahaa
Hagan, Tobin
Hagiwara, Makoto
Halivni, Bar
Han, Di
Han, Xiangyu
Haniyur, Shey
Hanson, Alex
Harke, Michael
Harrison, Michael
Hartnett, Kevin
Hasan, Iftekhar
Hassanpoor, Arman
He, Liqun
He, Siyu
He, Yiou
Heeger, Derek
Herbert, Edward
Holguin, Fermin
Hopkins, Douglas
Hopkins, Thomas
Hou, Dongbin
Hou, Jun
Hou, Ruoyu
Hoyo, Jose
Hsiao, Felix
Hsu, Ping
Huang, Min
Huang, Xing
Huang, Xiucheng
Huang, Yi
Huangfu, Yigeng
Hussein, Ala A.
Hwang, Seonhwan
Hwu, K. I.
Iannuzzo, Francesco
Im, Wonsang
Irving, Brian
Ishizuka, Yoichi
islam, rakib
IV, Prasanna
J, Sriram

Jain, Sachin
Janabi, Ameer
Janik, Raymond
Jeannin, Pierre-Olivier
Jeon, Woochul
Jeong, In Wha
Ji, Shiqi
Jiang, Li
Jiao, Ningfei
K, Girish
K, Rahul
K, Ramprakash
K, Sandeep
Kadavelugu, Arun
Kampl, Severin
Kan, Tianze
Kanapady, Ramdev
Kang, Yonghan
Kapat, Santanu
Karimi, Masoud
Karki, Ujjwal
Keller, Richard
Kelly, Anthony
Kennel, Ralph
Keogh, Bernard
Keshri, Ritesh
Khajehod, Ali
Khoshkbar Sadigh, Arash
Kim, Byungjin
Kim, Dong-Hee
Kim, Jonghoon
Kim, Ju Hyung
Kim, Katherine
Kim, Yun-Sung
Kirshenboim, Or
Kobravi, Keyhan
Konstantinou, Georgios
Koushki, Behnam
Krishnamoorthy, Harish
Krishnamoorthy, Radha
    Sree
Krishnamurthy, Mahesh
Ksiazek, Peter
Kulkarni, Abhijit
Kumar, Anuj
Kumar, Ashish
Kumar, Dinesh
Kumar, Misha
Kwarteng, Ntiamoah
Kwasinski, Alexis
Kwon, Junbum
Kwon, Yong-Cheol
L, Kevin
Lakshmanan, Padmavathi
Laldin, Omar
Lam, John
Lazaro, Orlando
Lee, Bumkil
Lee, Dong-Choon
Lee, Hyunji
Lee, June-Seok
Lee, Matt

Lee, Yong-Duk
Lefevre, Guillaume
Lei, Yutian
Leong, Kennith
Li, Cong
Li, Dan
Li, Danielle Dan
Li, Fei
Li, Gang
Li, He
Li, Helong
Li, Silong
Li, Tao
Li, Yalong
Li, Yanchao
Li, Ye
Li, Yingjie
Li, Yongjun
Li, Zhiqing
Liang, Lin
Liang, Tsorng-Juu
Liang, Wei
Lim, Seungbum
Liu, Haoyan
Liu, Liming
Liu, Nan
Liu, Wen Chuen
Liu, Xiaosen
Liu, Yiqi
Liu, Yiqi
Liu, Zeng
Liu, Zhengyang
Liu, Zhidong
Loo, Ka Hong
Lu, Fei
Lu, Jie
Lu, Minghui
Lu, Xiaonan
Lyu, Xiaofeng
M, Fariborz
M, Raed
M. Dousoky, Gamal
Madhusoodhanan, Sachin
Mai Xuan, Hung
Mainali, Krishna
Maksimovic, Dragan
Mandic, Goran
Mannarino, Frank
Mao, Saijun
Mao, Yongle
Martin, Kevin
Martinez Martinez, Wilmar
    Hernan
Masquelier, Michael
Massoud, Ahmed
Mazhari, Iman
Mekhilef, Saad
Mirafzal, Behrooz
Miskovic, Vlatko
Mo, Wai Keung
Modepalli, Kumar
Moeini, Amirhossein

Mohammadi, Mehdi
Mohammadi, Mehdi
Mohammadpour, Ali
Mojab, Alireza
Moosavi, Morteza
Morgan, Adam
Morris, Casey
Mosa, Mostafa
Mousavian, Hossein
Mozaffari Niapour,
    Seyyedabdolkhalegh
Mozipo, Aurelien
Mu, Mingkai
Muetze, Annette
Mulkern, Joe
Muni, Bishnu Prasad
Munoz, Alfredo
Na, Woonki
Nabavi-Niaki, Ali
Nademi, Hamed
Naderi, Roozbeh
Nahid, Babak
Nahid-Mobarakeh, Babak
Nakao, Hiroshi
Nan, Chenhao
Narveson, Brian
Nazir, Mohammad Nawaf
Neely, Jason
Nene, Hrishikesh
Nguyen, Luu
Nilles, Jeff
Ning, Puqi
Niu, Geng
Noge, Yuichi
O'Connell, Brian
O'Connell, Tim
Oggier, German
Okuma, Yasuhiro
Oliveira, Rafael
Onal, Yasemin
Onar, Omer
Onwuchekwa, Chimaobi
Orabi, Mohamed
Orikawa, Koji
Orr, Ray
Pagano, Rosario
Pala, Vipindas
Palavicino, Pablo
Pan, Di
Pantic, Zeljko
Parker, Ernie
Parmenter, Kevin
Pathipati, Vamsi
Patil, Devendra
Payami, Saifullah
Pekarek, Steven
Peltoniemi, Pasi
Pena Alzola, Rafael
Peng, Chang
Peng, Kang
Perreault, David
Pervaiz, Saad

Peterson, Bill
Pevere, Alessandro
Pierquet, Brandon
Pieschel, Martin
Pike, Bryan
Pilawa-Podgurski, Robert
Pillonnet, Gael
Pollock, Jenna
Popovic, Jelena
Poshtkouhi, Shahab
Qi, Feng
Qi, Yuan
Qin, Shibin
Raciti, Angelo
Radhakrishnan, Rahul
Radwan, Amr
Rajagopal, Prasanna
Ramachandran, Rakesh
Rao, Appa
Rao, Srinivasa
Rashidi Mehrabadi, Niloofar
Rashkin, Lee
Rathore, Akshay
Raymond, Luke
Rekola, Jenni
Ren, Ren
Ren, Xiaoyong
Renedo Anglada, Jaime
Reusch, David
Rezaei, Mohammad Ali
Rodriguez, Juan
Rodriguez Rogina, Maria
Roig, Jaume
Rouger, Nicolas
S, Ali
S, Aravinth
S, Jason
S, Jayram
S, Navid
S K, Gnana
Sadeghi, Sana
Saeedifard, Maryam
Safaee, Alireza
Saghaleini, Mahdi
Saha, Aparna
Saini, Dalvir
Sajadi, Amirhossein
Saket Tokaldani,
    Mohammad Ali
Salih, Ali
Salmon, John
Samanta, Suvendu
Sandoval, Jose
Sangsefidi, Younes
Santi, Enrico
Santiago, Juan
Scalia, Pietro
Schanen, Jean-Luc
Seltzer, Daniel
Senol, Murat
Seo, Gab-Su
Shadmand, Mohammad B.

Shamsi, Pourya
Shang, Ming
Shao, Jianwen
Shen, Ke
Shen, Zhiyu
Shenoy, Pradeep S.
Shi, Yuxiang
Sinan, Li
Singh, Amandeep
Singh, Amit
Singh, Siddhartha
Siwakoti, Yam
Solis, Carlos J
Song, Xiaoqing
Soni, Jayantika
Souza Júnior, Antonio
    Barbosa de
Spangler, Jim
Spyker, Russell
Sridharan, Srikanthan
Srinivasan, Gamesh Kumar
Stauth, Jason
Stein, Aaron
Stillwell, Andrew
Subbiah, Anandakumar
    Subbiah
Sullivan, Charles
Sun, Kai
Sung, Woongje
Surakitbovorn, Kawin
Surapaneni, Ravi Kiran
Sweet, Mark
Syed, Mudassir
Tabesh, Ahmadreza
Taeed, Fazel
Tahmasebi, Hossein
Tai, Heng-Ming

Takeshita, Takaharu
Tan, Kai
Tang, Lixin
Tang, Tianhao
Tang, Yichao
Tang, Zhuangyao
Tavakoli, Atrin
Tengfei, Qiu
Thomas, Brian
Thong, Weng
Thummala, Prasanth
Tian, Shuilin
Toliyat, Hamid
Town, Graham
Tsai, Kaichien
Vahedi, Hani
Vazquez, Aitor
Vekslender, Timur
Vining, Jennifer
Vk, Kanakesh
Vu, Trong Tue
Vytla, Rajeev Krishna
Wang, Chao
Wang, Cheng
Wang, Fei
Wang, Gangyao
Wang, Haoyu
Wang, Huai
Wang, Li
Wang, Mengqi
Wang, Xianwei
Wang, Xiongfei
Wang, Zhiqiang
Watson, Robert
Wei, Chun
Wei, Kang
Weise, Nathan

Wen, Yue
Whistler, Richard
White, Bob
Williamson, Sheldon
Winterhalter, Craig
Wolf, Christian
Wu, Hao
Wu, Liyao
Wu, Rui
Wu, Weimin
Xia, Bing
Xiong, Song
Xu, Tao
Xu, Yang
Xue, Fei
Xue, Jing
Xue, Lingxiao
Xuewei, Pan
Yadav, Gorla Naga B
Yang, Nanfang
Yang, Qichen
Yang, Xiaofeng
Yang, Xu
Yang, Yongheng
Yang, Yuchen
Yao, Wenli
Yao, Xiu
Ye, Jin
Ye, Zichao
Yelaverthi, Dorai
Yilmaz, Kadir
Yilmaz, Murat
Yin, Congqi
Yu, Liang
Yu, Sheng-Yang
Yu, Wensong
Yu, Xuehong

Yu, Zhehan
Zahid, Zaka Ullah
Zahnstecher, Brian
Zakis, Janis
Zelaya, Hector
Zeltser, Ilya
Zgheib, Rawad
Zhang, Ch
Zhang, Chenmeng
Zhang, Haiyu
Zhang, Hua
Zhang, Julia
Zhang, Lanhua
Zhang, Li
Zhang, Xing
Zhang, Xuan
Zhang, Xuning
Zhang, Yuzhi
Zhang, Zhe
Zhang, Zhemin
Zhao, Chen
Zhao, Chongwen
Zhao, Dongdong
Zhao, Shuze
Zhao, wan
Zhao, Xiaonan
Zhou, Daming
Zhou, Dao
Zhou, Kan
Zhou, Liwei
Zhou, Qi
Zhou, Xin
Zhu, Bohang
Ziaeinejad, Saleh

APEC 2016 Partnerships

DIAMOND Partners

PLATINUM Partners

 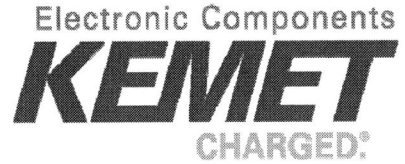

GOLD Partners

SILVER Partners

APEC 2016 Exhibitors

At the time the Proceedings went to print, the companies below were planning to exhibit at APEC 2016. The actual exhibitors at APEC 2016 may differ slightly from this list.

5S Components
Aavid Thermalloy
ABC Trading Beijing Co. LTD.
ACME Electronics Corporation
Acopian Power Supplies
Adaptive Power Systems
Adelser
Agile Magnetics
Agile Switch
Allstar Magnetics
Alpha
Alpha & Omega Semiconductor
Alps Electric Co. Ltd.
Altera Corporation
Ametherm, Inc.
Amogreentech
Amphenol Interconnect Products
Analog Devices
Anpec Electronics
ANSYS, Inc.
APEC
Apex Microtechnology
Ascatron
Athena Energy Corp.
Auxel FTG
AVX
Baknor Thermal & Packaging
BH Electronics, Inc.
Bicron Electronics Co.
Bomatec International Corp.
CalRamic Technologies, LLC
Caton Connector Corporation
Central Semiconductor Corp.
Chroma Systems Solutions, Inc.

CogniPower
Coil Winding Specialist, Inc.
Coilcraft
Coiltron, Inc.
Component Distributors, Inc. (CDI)
Core Technology Group, Inc.
Cornell Dubilier/Illinois Capacitors
Cosmo Ferrites Ltd
CPS Technologies
Cramer Coil & Transformer Co. Inc.
CUI Inc.
Daco Semiconductor Co., Ltd.
Danfoss Silicon Power GmbH
Datatronics
Dau Thermal Solutions North America
Dean Technology, Inc.
Dearborn Electronics Inc.
Dewetron
Dexter Magnetic Technologies
Dialog Semiconductor
Digi-Key Electronics
Dino-Lite Scopes (BigC)
Eaton
EBG Resistors
ECI
EFC/WESCO
Efficient Power Conversion Corporation
    (EPC)
Egston System Electronics Eggenburg
Electro Technik
Electrocube, Inc.
Electronic Concepts, Inc.
Electronic Systems Packaging
Elna Magnetics

Exar Corporation
Fairchild Semiconductor
Fair-Rite Products Corp.
Faratronic Co., Ltd.
Ferroxcube USA, Inc.
FTCAP
Fuji Electric Corp. of America
GAN Systems
Global Choice International LLC
Global Foundries
Global Power Technologies Group
GMW Associates
Gowanda Electronics
GRAPES - NSF I/UCRC
H & H Magnetics
HEFEI ECRIEE-TAMURA Electric Co.
    Ltd.
Hengdian Group DMEGC Magnetics Co.,
    LTD
Heraeus Electronic Materials Division
Hesse Mechatronics, Inc.
Himag Planar Magnetics, Ltd.
Hitachi Metals
Hitachi Semiconductors - AmePower
Holy Stone International
HVR Advanced Power Components, Inc.
I.C.T. Power/ Innovation Plus
IAS
ICE Components, Inc.
Illinois Capacitor Inc.
Indium Corporation
Infineon Technologies Americas Corp.
Infolytica Corporation
INSTEK America
Intepro Systems
Inter Outstanding Electronics, Inc.
Intersil Corporation
Intertape Polymer Group
Isotek Corporation, Subsidiary of
    Isabellenhutte

Itelcond SRL
ITG Electronics
Iwatsu Test Instruments
JARO Thermal
JFE Steel Corporation
Jianghai Capacitor Co. LTD
Johanson Dielectrics, Inc.
John Deere Electronic Solutions
Kanthal Globar, Sandvik Heating
    Technology USA
Kaschke Components GMBH
KDM Zhejiang NBTM Keda
    Magnetoelectricity Co. Ltd.
KEMET
Kendeil srl
KEPCO, Inc.
Keysight Technologies
KITAGAWA INDUSTRIES America, Inc.
Knowles Capacitors
Lee Yuen Electrical Mfy Limited
LEM USA, Inc.
Lenco Electronics, Inc.
LHV Power Corporation
Linear Technology Corporation
Lodestone Pacific
LTEC
Mag Layers USA
MAGDEV Ltd.
MagnaChip Semiconductor
Magna-Power Electronics
Magnetec (Guangzhou) Magnetic Device
    Co. Ltd.
Magnetic Metals Corp
Magnetics
Magsoft Corporation
Magtech & Power Conversion Inc.
Malico Inc.
Marathon Power
MaxQ Technology
Mentor Graphics

Mersen
Mesago PCIM GmbH
Methode Power Solutions Group
MH&W
Microchip Technology Inc.
Micrometals, Inc.
Milplex Circuit (Canada) Inc.
MK Magnetics Inc
Monolith Semiconductor Inc.
Monolithic Power Systems, Inc.
MORNSUN America LLC
Mouser Electronics, Inc.
MPS Industries, Inc.
MS Power Semiconductor Co., Ltd
MTL Distribution
National Magnetics Group/Ceramic
    Magnetics, Inc.
NEC TOKIN America Inc.
New England Wire Technologies
NH Research, Inc.
Nichicon
NORWE, Inc
NXP Semiconductors
Ohmite MFG
ON Semiconductor
Opal-RT
Pacific Sowa Corporation; C/O Epson
    Atmix Corporat
Panasonic
Parker Overseas
Payton America
Pearson Electronics, Inc.
PELS
PINK GmbH Thermosysteme
Plexim
PMK
Power Electronic Measurements Ltd.
Power Integrations
Power Solutions Inc.
PowerELab Ltd.

Powerex, Inc.
Powersim Inc.
POWRMOD DC to DC Converters
Precision Inc.
Prize Stage
Prodrive Technologies
PSMA
Qualtek
Renco Electronics Inc.
Renesas Electronics
Richardson Electronics, Ltd.
Richardson RFPD
Ridley Engineering, Inc.
Rogers Corporation
ROHM
Rubadue Wire Company, Inc.
Rubycon Corporation
RWP Electronic Sales
Samwha USA Inc.
SanRex Corporation
SBE, Inc.
Schaffner Trenco LLC
Schunk Hoffmann Carbon Technology
Schurter, Inc.
Scientific Test, Inc.
Semikron, Inc.
Semtech
ShengYe Electrical Co. Ltd
Shenzhen Poco Magnetic Co., Ltd.
Shenzhen Zeasset Electronic Technology
    Co., Ltd
Sidelinesoft, LLC
Silicon Frontline Technology, Inc.
Simplis Technologies
SMC Diode Solutions
Software Cradle Co., Ltd.
Solantro Semiconductor Corporation
Sonoscan, Inc
SP CONTROL TECHNOLOGIES
Standex-Meder Electronics

Stapla Ultrasonics Corp.
Stellar Industries Corp.
STMicroelectronics, Inc.
Storm Power Components
Sumida America Components Inc.
Synopsys, Inc.
Syrma Technology
Taiwan Semiconductor Inc.
Taiyo Kogyo Co., LTD
Tamura Corporation
TDK Corporation
Tektronix Inc.
Teledyne LeCroy
Texas Instruments
The Allpower Source (Div. of Technology Dynamics)
The Bergquist Company (a Henkel Company)
Thermik Corporation
Toshiba America Electronic Components, Inc.
TowerJazz
Transim Technology
Transphorm
Triad Magnetics
TSC Ferrite International
TT electronics
Typhoon HIL, Inc.
United Chemi-Con
United Silicon Carbide, Inc.
University of Texas - Dallas
VAC Sales USA LLC
Venable Instruments, Inc.
Versatile Power
Viking Tech America Corporation
Vincotech GmbH
Vishay Intertechnology, Inc.
Voltage Multipliers
Wakefield-Vette Thermal Solutions
West Coast Magnetics

Wolfspeed, A Cree Company
Wolverine Tube Inc. - MicroCool Division
Wurth Electronics Midcom Inc.
Wurth Elektronik Wireless Charging Cafe
X-FAB Semiconductor Foundries
Xitron Technologies Inc.
Yokogawa Corporation of America
Yole D,veloppement / System Plus Consurlting
ZES Zimmer
Zhuzhou CSR Times Electric Co., Ltd
Zipalog

APEC 2016 PROFESSIONAL EDUCATION SEMINARS
TABLE OF CONTENTS

At the time the proceedings went to print, the Professional Education Seminars listed below were scheduled for presentation at APEC 2016. The actual seminars presented may have differed slightly from this list.

Professional Education Seminars, Session One
Sunday, March 20, 9:30 am – 1:00 pm

S01 **Exceeding 99% Efficiency for PFC and Isolated DC-DC Converters, GaN Versus Silicon**
Ionel Dan Jitaru, *Rompower Energy Systems Inc.*

S02 **The Invisible Schematic: Non-Idealities in Circuit Elements and System Components**
Ernest H. Wittenbreder, Jr., *Technical Witts, Inc.*

S03 **Getting from 48 V to Load Voltage: Improving Low Voltage DC-DC Converter Performance with GaN Transistors**
Alex Lidow, *Efficient Power Conversion Corporation*
David Reusch, *Efficient Power Conversion Corporation*
John Glaser, *Efficient Power Conversion Corporation*

S04 **A Comprehensive Introduction to Implementing a Fully Digital Power Factor Correction Boost Converter**
Alex Dumais, *Microchip Technology, Inc.*
Joel Steenis, *Microchip Technology, Inc.*

S05 **Basic Switching Power Supply Design**
Marty Brown, *Sierra Energy Management Systems, LLC*

S06 **Solid-State Transformers – Key Design Challenges, Applicability, and Future Concepts**
Johann W. Kolar, *Power Electronic Systems Laboratory*
Jonas E. Huber, *Power Electronic Systems Laboratory*

S07 **Photovoltaic Modeling and Why It Matters for Power Electronics**
Katherine A. Kim, *Ulsan National Institute of Science and Technology*
Jeehoon Jung, *Ulsan National Institute of Science and Technology*

Professional Education Seminars, Session Two
Sunday, March 20, 2:30 pm – 6:00 pm

S08   **Stability and Damping of Grid-Connected Voltage-Source Converters**
Frede Blaabjerg, *Aalborg University*
Xiongfei Wang, *Aalborg University*

S09   **PMBus™: Review and New Capabilities**
Robert V. White, *Embedded Power Labs*

S10   **Wide Bandgap Device Characterization**
Fred Wang, *University of Tennessee*
Zheyu Zhang, *University of Tennessee*
Edward A. Jones, *University of Tennessee*

S11   **High Performance Digital Control**
Hamish Laird, *ELMG Digital Power*

S12   **Non-Linear Thermal Topics in Semiconductors and Electronics**
Roger Stout, *ON Semiconductor*

S13   **Power Architectures, Protection and Control of DC Microgrids**
Tomislav Dragičević, *Aalborg University*
Josep M. Guerrero, *Aalborg University*
Lexuan Meng, *Aalborg University*
Xiaonan Lu, *Argonne National Laboratory*
Juan C. Vasquez, *Virginia Polytechnic Institute and State University / CPES*

S14   **Soft Switching Three-Phase Converters or Inverters**
Mark Dehong Xu, *Zhejiang University*
Rui Li, *Shanghai Jiaotong University*

Professional Education Seminars, Session Three
Monday, March 21, 8:30 am – Noon

S15   **Introduction to Fast Analytical Techniques: Application to Small-Signal Modeling**
Christophe Basso, *ON Semiconductor*

S16   **Reliability of Power Electronic Systems**
Frede Blaabjerg, *Aalborg University*
Francesco Iannuzzo, *Aalborg University*
Huai Wang, *Aalborg University*
Ke Ma, *Aalborg University*

S17   **Addressing Challenges in High Power and High Voltage Designs with IGBTs**
Vittorio Crisafulli, *ON Semiconductor*
Dhaval Dalal, *ON Semiconductor*
Tomas Krecek, *ON Semiconductor*
Dominic Li, *ON Semiconductor*

S18   **A State-Space Design Approach to Digital Feedback Control of DC/DC Converters**
Dorin O. Neacsu, *Gheorghe Asachi Technical University of Iasi*

S19   **How to go from Si to SiC Components in the Design of Converters including Safety and EMC**
Supratim Basu, *Bose Research*
Tore M. Undeland, *Norwegian University of Science and Technology*

S20   **Principles and Practices of Digital Current Regulation for AC Systems**
Grahame Holmes, *Royal Melbourne Institute of Technology*
Brendan McGrath, *Royal Melbourne Institute of Technology*

S21   **Latest Technologies of LLC Converters for High Current, Fast Response, and Wide Input Voltage Range Applications**
Yan-Fei Liu, *Queen's University*

INDUSTRY PRESENTATIONS

Session IS01: Aiding Design Excellence
Location: 201A
March 22, 2016 8:30 - 11:55
Session Chairs: Ada Cheng, *AdaClock*
Paul Greenland, *eIQ Energy*

The Role of Patents in Great Designs: Maximize Rewards to Assignee and Inventor
Laios Burgyan, *LTEC*
Yuji Kakizaki, *LTEC*

Developing a Power Engineering Career
Marty Brown, *Sierra Energy Management Systems*

Fundamentals of Electrical Power Measurements
Yusuf Chitalwala, *Yokogawa Corporation of America*

Today's Power Conversion Devices Require a New Generation of Test and Measurement Technology
Tom Neville, *Tektronix*

Online Power Design Tools: Past, Present and Future
Surinder P. Singh, *Texas Instruments*
Vinay Jayaram, *Texas Instruments*
Jeff Perry, *Texas Instruments*

Part Selection Efficiency & Optimization Aiding Design Excellence
Randall Restle, *Digi-Key Electronics*

Manufacturing and Reliability Perspective for Design Excellence
Aurora Craciun, *Celestica*

Session IS02: 3D Power Packaging
Location: 201B
March 22, 2016 8:30 - 11:55
Session Chairs: Brian Narveson, *Narveson Innovative Consulting*
Ernie Parker, *Crane Aerospace & Electronics*

Additive Manufacturing Technology for Power Electronics Applications
Madhu Chinthavali, *Oakridge National Laboratory*

Some Progress in Cooling and 3D Packaging for EV/HEV Inverters
Yunqi Zheng, *iPowerPak*

Embedded Power from POL to Off-Line Applications
Fred C. Lee, *CPES, Virgina Tech*

Unmet Challenges of Embedded Components for 3D Packaging
Arnold Aldernman, *Anagenesis, Inc.*

Interconnect Reliability – Considerations in Dense Power Packages
Rick Fishbune, *IBM*

Integrated Magnetics for PwrSiP and PwrSoc
Paul McCloskey, *Tyndall National Institute*

Liquid Cooled Transformer Based Power Converters with 3D Printed Micro-Channel Heat Sink
Ernie Parker, *Crane Aerospace & Electronics*
Frank Fan Wang, *Crane Aerospace & Electronics*

Session IS03: Smart Products for the Smart Grid
Location: 202AB
March 22, 2016 8:30 - 11:55
Session Chairs: Edward Herbert, *Independent Consultant*
Dusty Becker, *Independent Consultant*

Virtual Power Plants (VPP)
Alexis Kwasinski, *University of Pittsburgh*

IIoT in Multi-Utility Smart Grid for Community & Smart City
Bharat Shah, *NeoSilica*
Satyam Bheemarasetti, *NeoSilica*
Ravi Prasad Patruni, *NeoSilica*

SmartMeters – Beyond Billing
Marshall Parsons, *Southern California Edison*

DC Line Interactive Uninterruptible Power Supply (UPS) with Load Leveling
Robert Cuzner, *University of Wisconsin, Milwaukee*
Ahmad Hamidi, *University of Wisconsin, Milwaukee*
Adel Nasiri, *University of Wisconsin, Milwaukee*

Advanced Control of PV Grid Connected Converters through the Implementation of the Synchronous Power Controller Concept
Pedro Rodriguez, *Abengoa*

Voltage and VAR Regulation
John Berdner, *Enphase Energy*

Fault Tolerance and Healing
Alexis Kwasinski, *University of Pittsburgh*

Session IS04: Wide Bandgap Semiconductors
Location: 203AB
March 22, 2016 8:30 - 11:55
Session Chairs: Dennis Stephens, *Continental Automotive Systems*
Odile Ronat, *International Rectifier HiRel*

SiC Solution for Industrial Auxiliary Power Supplies
Mitch Van Ochten, *ROHM Semiconductor*

An Industry First: Silicon Carbide Based Intelligent Power Module
Nitesh Satheesh, *AgileSwitch, LLC*
Adam Fender, *AgileSwitch, LLC*
Albert Charpentier, *AgileSwitch, LLC*

Ultra-Wide-Bandgap Semiconductors for Generation-After-Next Power Electronics
Robert Kaplar, *Sandia National Labs*
Andrew Armstrong, *Sandia National Labs*
Arthur Fischer, *Sandia National Labs*
Albert Baca, *Sandia National Labs*
Andrew Allerman, *Sandia National Labs*
Daniel Mauch, *Sandia National Labs*
Fred Zutavern, *Sandia National Labs*
Michael King, *Sandia National Labs*
Jack Flicker, *Sandia National Labs*
Robert Brocato, *Sandia National Labs*
Lee Rashkin, *Sandia National Labs*
Jarod Delhotal, *Sandia National Labs*
Lu Fang, *Sandia National Labs*
Isik Kizilyalli, *Avogy Inc.*
Ozgur Aktas, *Avogy Inc.*
Jason Neely, *Sandia National Labs*

Application-Relevant Qualification of Emerging Semiconductor Power Devices
Sandeep Bahl, *Texas Instruments*
Grant Smith, *Texas Instruments*

Introducing eGaN® IC Targeting Highly Resonant Wireless Power
Michael de Rooij, *Efficient Power Conversion*

Scaling Power Electronic Converter SWaP based on WBG and UWBG Device Characteristics
Jason Neely, *Sandia National Labs*
Jarod Delhotal, *Sandia National Labs*
Robert Kaplar, *Sandia National Labs*
Jack Flicker, *Sandia National Labs*
Lee Rashkin, *Sandia National Labs*

GaN Takes Server Power Supplies' Power Density to New Heights
Jason Cuadra, *Transphorm, Inc.*

Session IS05: Thermal Management
Location: 201A
March 23, 2016 8:30 - 10:10
Session Chairs: Peter Resca, *Advanced Thermal Solutions, Inc.*

Thermal Challenges and Solutions for Industrial Solid State Lighting Applications
Peter Resca, *Advanced Thermal Solutions*

Thermally Managing High Power Devices using Heat Pipe Assemblies
Abdul Samad Jawed, *Mersen*
Cliff Weasner, *Mersen*
Ahmed Zaghlol, *Mersen*

Using Web based Tools for the Thermal Design of a Power Converter
Ahmed Zaghlol, *Mersen*
Jeremy Howes, *Tesla Energy*
David Levett, *Infineon*
Greg Schendel, *Parker SSD Drives*

Reflowable Thermal Devices: Protecting High-Power Automotive Electronics
Barry Brents, *TE Connectivity*

Session IS06: Modeling and Simulation
Location: 201B
March 23, 2016 8:30 - 10:10
Session Chairs: Cahit Gezgin, *Infineon Technologies*
Brian Thomas, *Independent Consultant*

System-Level Crosstalk-Induced Efficiency Impact of DCDC Converter: Simulation to Measurement Correlation
Joerg Goller, *Texas Instruments, Inc.*
Jie Chen, *Texas Instruments, Inc.*
Rajen Murugan, *Texas Instruments, Inc.*

Switching Voltage Regulator Modeling Methodology for Simulation based Power Delivery Design
Wei Xu, *Intel Corporation*
Jiangqi He, *Intel Corporation*
David Figueroa, *Intel Corporation*

CoolSPICE: A New Electrical and Thermal Circuit Simulator for Power Circuit Design with New Wide Bandgap Device Capabilities
Akin Akturk, *CoolCAD Electronics LLC*
Neil Goldsman, *CoolCAD Electronics LLC*
Zeynep Dilli, *CoolCAD Electronics LLC*
Simon Peggs, *CoolCAD Electronics LLC*

Power Converter System Stress and Mechanical Analysis within an Integrated Design Environment
Rehan Iqbal, *Mentor Graphics*
Carl Bycraft, *Mentor Graphics*

Session IS07: Very Low Power Applications
Location: 202AB
March 23, 2016 8:30 - 10:10
Session Chairs: Edward Stanford, *Power Deliver Consultants*
Nick Gruendler, *Celestica*

Energy Harvesting Is Not Fiction Anymore
Lorandt Fölkel, *Würth Elektronik eiSos GmbH*

A New Way to Power the World with High Efficiencies
Michael H. Freeman, *Semitrex*

System Architecture that Extends Battery Life
Matthew Tyler, *ON Semiconductor*

Primary Side Regulation in Flyback Converters Delivers Low Cost, High Reliability and Energy Efficiency
Ramanan Natarajan, *Texas Instruments*
Bing Lu, *Texas Instruments*
Brent McDonald, *Texas Instruments*
Vaibhav Desai, *Texas Instruments*
Peter Fundaro, *Texas Instruments*

Session IS08: Alternative Energy in High Penetration Areas
Location: 203AB
March 23, 2016 8:30 - 10:10
Session Chairs: Michael Harrison, *Enphase Energy*
Bharat Shah, *Independent Consultant*

The Growth of Renewable Energy in California
David Hochschild, *California Energy Commission*

Shine and Drive: The Symbiotic Relationship between Renewables, Electric Vehicles, and the Grid
Carla J. Peterman, *California Public Utilities Commission*

Modernizing the Grid and Enabling Distributed Energy Resources
Heather Sanders, *Southern California Edison*

Promoting Renewable Energy Technologies through Research, Testing and Standards
Ken Boyce, *Underwriters Laboratories*

Session IS09: High Frequency Magnetics; Black Magic, Art or Science?
Location: 201A
March 23, 2016 14:00 - 17:25
Session Chairs: Edward Herbert, *Independent Consultant*
Stephen Carlsen, *Raytheon*

High Frequency Magnetics: Black Magic, Art or Science? Magnetics Core Loss
Ray Ridley, *Ridley Engineering Inc.*

Selecting Magnetics for High Frequency Converters Practical Hints and Suggestions for Getting Started
Len Crane, *Coilcraft*

The Future for SMPS Magnetics
Weyman Lundquist, *West Coast Magnetics*

Accurate Estimation of Losses of Power Inductor in Power Electronics Applications
Ranjith Bramanpalli, *Würth Elektronik eiSos GmbH*

Litz Wire: A Practical Discussion of its Uses and Limitations in High Frequency Transformers
Kyle D. Jensen, *Rubadue Wire Company, Inc.*

Powder Core Materials for Magnetic Components in GaN and SiC Power Devices
Christopher G. Oliver, *Micrometals, Inc.*

EMI Conducted and Radiated Emissions
Mark Rine, *Vacuumschmelze GmbH & Co.*

Session IS10: From the Board to the Datacenter
Location: 201B
March 23, 2016 14:00 - 17:25
Session Chairs: Brian Zahnstecher, *PowerRox LLC*
Wisam Moussa, *Infineon Technologies*

The Technology Behind the World's Smallest 12V, 10A Voltage Regulator
Pradeep Shenoy, *Texas Instruments*

Noise Characterization of Switching Buck Regulators for EMI Analysis
Chunlei Guo, *Intel Corporation*
Jiangqi He, *Intel Corporation*
Yaxiao Qin, *Monolithic Power Systems*
Huaifeng Wang, *Monolithic Power Systems*
Eric Braun, *Monolithic Power Systems*
Jinghai Zhou, *Monolithic Power Systems*

48V Power Delivery to Grantley Reference Board
Donghwi Kim, *Intel*
Jiangqi He, *Intel*
David G. Figueroa, *Intel*

PMBus on Linux: PMBus Support Options for the Linux Platform
Michael Jones, *Linear Technology*

System Power Simplification Utilizing PMBus™ Zone Capabilities
Travis Summerlin, *Texas Instruments*

Power-Defined Software in the Data Center
Brian Zahnstecher, *PowerRox*

Data Center Market and Technology Trends in Power Electronics
Mattin Grao Txapartegi, *Yole Développement*
Pierric Gueguen, *Yole Développement*

Session IS11: Medium Voltage Applications
Location: 202AB
March 23, 2016 14:00 - 17:25
Session Chairs: River Tinh-Ho Li, *ABB (China) Limited*
Alex Craig, *Fairchild Semiconductor*
Nan Chen, *ABB*

Power Semiconductors for Medium Voltage Drives
Uwe Jansen, *Infineon Technologies AG*

Medium Voltage Drives: Design Considerations in Demanding and Special Applications
Wim van der Merwe, *ABB Ltd.*

Advanced Multilevel STATCOM for Flicker-Mitigation in MV Installations
Martin Pieschel, *Siemens AG*

Technologies for Efficient Simulation of Complex MV Power Converters
Min Luo, *Plexim GmbH*

Practical Considerations in Measuring Power and Efficiency on PWM and Distorted Waveforms during Dynamic Operating Conditions
Ken Johnson, *Teledyne LeCroy*

HVDC Transmission Lines – Market, Technology and Geographical Trends
Mattin Grao Txapartegi, *Yole Développement*
Pierric Gueguen, *Yole Développement*

High Power Low Inductance Module Building Blocks for Three-Level Inverters
John Donlon, *Powerex, Inc.*
Marco Honsberg, *Mitsubishi Electric Europe*

Session IS12: Transportation Power Electronics
Location: 203AB
March 23, 2016 14:00 - 17:25
Session Chairs: Ralph Taylor, *Delphi Electronics & Safety*
Fred Weber, *FTW LLC*

More Electric and Electric Aircraft
Kaushik Rajashekara, *University of Texas at Dallas*

Overview of the Unique Requirements and Challenges for Power Electronics in Mining Equipment
Dustin Selvey, *Caterpillar, Inc.*

Medium and Heavy-Duty Vehicle Duty Cycles for Electric Powertrains
Kenneth Kelly, *National Renewable Energy Laboratory*
Kevin Bennion, *National Renewable Energy Laboratory*
Eric Miller, *National Renewable Energy Laboratory*
Bob Prohaska, *National Renewable Energy Laboratory*

Fuel Cells for Material Handling Systems
Fernando Corral, *Plug Power*

Design and Implementation of a LLC-ZCS Converter for Hybrid/Electric Vehicles
Davide Giacomini, *Infineon*
Cesare Bocchiola, *Infineon*

EV-Grid Integration (EVGI) Control and System Implementation – Research Overview
Mithat Kisacikoglu, *NREL*

Assessing the North American Supply Chain for Traction Drive Motors
Steven Boyd, *Department of Energy, Vehicle Technologies Office*
Christopher Whaling, *Synthesis Partners, LLC*

Session IS13: Safety and Compliance
Location: 201A
March 24, 2016 8:30 - 11:30
Session Chairs: Kevin Parmenter, *Excelsys*
Jim Spangler, *Independent Consultant*

Surges and Transients Can't Read Specifications! How to Meet Specifications and Protect against Real-World Threats
Tim Patel, *Littelfuse, Inc.*

Component Level Safety Certification in Systems-IEC60747-17/UL1577
Mark Cantrell, *Analog Devices*

Introduction to EMC and EMC Standards
Ghery S. Pettit, *Pettit EMC Consulting LLC*

Type Testing Primer for Power Converters and Transformers
Brian O'Connell, *Tamura Corp of America*

IEC 60601-1-2, 4th Edition
Darryl P. Ray, *Darryl Ray EMC Consulting, LLC*

Sources for Regulatory Information and Related Search Techniques – How to Find Safety and Compliance Information
Kevin Parmenter, *Excelsys Technologies Ltd.*
Jim Spangler, *Excelsys Technologies Ltd.*

Session IS14a: Topics in Power Integration
Location: 201B
March 24, 2016 8:30 - 10:10
Session Chairs: Dave Hurst, *NextEnergy*

Floating High Voltage Switches Integrated with Standard Logic
Tom Simmonds, *TLSI*

AC/DC to UHV: How Application Requirements Drive IC Technology Requirements
Don Disney, *GLOBALFOUNDRIES*

Multi-Chip Power Module Layout & Design using Q3D Extractor and PowerSynth
Andalib Nizam, *University of Arkansas*
Atanu Dutta, *University of Arkansas*
Tom Vrotsos, *University of Arkansas*
Alan Mantooth, *University of Arkansas*
Steven G. Pytel Jr., *ANSYS, Inc.*

PowerSoC & PowerSiP Markets are Preparing. Are You?
Alex Avron, *Point The Gap*

Session IS14b: Power Electronics Industry in North America
Location: 201B
March 24, 2016 10:40 - 11:30
Session Chairs: Dave Hurst, *NextEnergy*

Driving Collaboration for Power Electronics: Technology Roadmapping Industry Survey
Dave Hurst, *NextEnergy*

Driving Collaboration for Power Electronics: Technology Roadmapping Review
Swad Komanduri, *NextEnergy*
Dave Hurst, *NextEnergy*
Roland Kibler, *NextEnergy*
Dan Radomski, *NextEnergy*

Session IS15: Power Electronics Applications
Location: 202AB
March 24, 2016 8:30 - 11:30
Session Chairs: Bill Peterson, *E&M Power*

Direct Current Emulator: A Wideband Load / Source for Power System Simulation and Testing
Bill Peterson, *E&M Power*

Digital Power Conversion
Herman Vaneijkelenburg, *Adaptive Power Systems*

Latest Solutions to Meet Power Conversion Needs on the "More Electric Aircraft"
Kaz Furmanczyk, *Crane Aerospace & Electronics*

Drive System Loss Reduction by Allpole Sine Filters
Dennis Kampen, *BLOCK Transformatoren-Elektronik GmbH*

Peak Current Controlled ZVS Full Bridge Converter with Digital Slope Compensation
Sabarish Kalyanaraman, *Microchip Technology (India) Pvt. Ltd.*
Ramesh Kankanala, *Microchip Technology (India) Pvt. Ltd.*

Protection of Wide Band Gap Semiconductor Devices used in High Power/High Voltage Applications
Barry Kirkorian, *Mersen USA*

Session IS16: Power Semiconductors Enabling Next Generation Applications
Location: 203AB
March 24, 2016 8:30 - 11:30
Session Chairs: Carl Blake, *Independent Consultant*
John Palmour, *Wolfspeed*

Outlook for Semiconductors & Power Discretes & Modules
Dale Ford, *IHS*

Silicon Carbide Devices for Energy Efficient Infrastructure
Ranbir Singh, *GeneSiC Semiconductor Inc.*

GaN vs. Silicon – Overcoming Barriers to the Rise of GaN
Alex Lindow, *EPC*
Joe Engle, *EPC*

Unlocking the Power of GaN
Dan Kinzer, *Navitas Semiconductor*

GaN in a Silicon World: Competition or Coexistence
Tim McDonald, *Infineon*

Advanced High Power-Density Thermal Packages & Mother-Boards Enable Ultimate Power GaN and SiC Performance & Efficiency
Courtney R. Furnival, *Semiconductor Packaging Solutions*
Arnold Alderman, *Semiconductor Packaging Solutions*

Session IS17: Market Analysis
Location: 201A
March 24, 2016 14:00 - 17:25
Session Chairs: Chris Jones, *Artesyn Embedded Technologies*
Greg Evans, *WelComm*

The New Competitive Environment for Power Semiconductors
Victoria Fodale, *IHS Technology*
Michael Markides, *IHS Technology*

Discrete vs. Integrated Power Solutions IHS Technology
Jonathan Liao, *IHS Global Inc.*
Richard Eden, *IHS Global Inc.*
Michael Markides, *IHS Global Inc.*

GaN on Si HEMT vs SJ Mosfet: Technology and Cost Comparison of Next Generation 600/650V Power Devices
Elena Barbarini, *System Plus Consulting*

Market Forecasts for Silicon Carbide & Gallium Nitride Power Semiconductors
Jonathan Liao, *IHS Technology*
Richard Eden, *IHS Technology*
Michael Markides, *IHS Technology*

Si IGBT and SiC: Which Repartition for Power Devices?
Pierric Gueguen, *Yole Développement*

How Will the Battery Market Build on EV/HEV Adoption?
Pierric Gueguen, *Yole Développement*

Session IS18: LED Lighting
Location: 201B
March 24, 2016 14:00 - 17:25
Session Chairs: Aung Thet Tu, *Independent Consultant*
Brian Johnson, *Texas Instruments*

Damping Circuit for Dimmable Retrofit LED Lamps
Nagaraja Chikkegowda, *OSRAM*

Smart Lighting and the Future of Illumination Markets
Robert F. Karlicek Jr., *Rensselaer Polytechnic Institute,*

The Challenges (and Surprises) of Closed-Loop LED Color and Color Temperature Control
Cary Eskow, *Avnet Electronics*

Session IS19a: ElectroMagnetic Compatibility
Location: 202AB
March 24, 2016 14:00 - 15:40
Session Chairs: Kevin Parmenter, *Excelsys*
Jim Spangler, *Independent Consultant*

The Behavior of Electro-Magnetic Radiation of Storage Inductor in DC-DC Converters
Ranjith Bramanpalli, *Würth Elektronik eiSos GmbH*

EMC Filter Solutions for Switch Mode Power Supplies
Nikila Kareesan, *SCHURTER,Inc.*

Inductor Noise in the Buck Converter GPU Circuit
Zoltan Puskas, *ITG Electronics, Inc.*
David Yu, *ITG Electronics, Inc.*

Session IS19b: Capacitors for Power Applications
Location: 202AB
March 24, 2016 16:10 - 17:25
Session Chairs: Kevin Parmenter, *Excelsys*
Jim Spangler, *Independent Consultant*

New Component Technologies Enable More Robust and Reliable Power System Design
Chris Reynolds, *AVX*

Non-Traditional Supercapacitor Topologies for Traditional Circuit Issues
Nihal Kularatna, *The University of Waikato*

Proposal of Precise SPICE Model of Conductive Polymer Aluminum Solid Capacitors
Shun Koyama, *Nippon Chemi-Con Corporation*
Tomoyuki Goutsu, *Nippon Chemi-Con Corporation*

Session IS20: Active Devices
Location: 203AB
March 24, 2016 14:00 - 17:25
Session Chairs: Sal Akram, *Fairchild Semiconductor*
Kumar Gandharva, *Infineon*

Advantages and Optimized Control of Reverse Conducting Diode Controlled RCDC IGBT's
David Levett, *Infineon*
Tim Frank, *Infineon*

If it Ain't Broke Why Fix It? Design Improvements to the PrimePACK™ IGBT Module for Commercial, Construction and Agricultural Vehicle (CAV) Traction Drives
David Levett, *Infineon*
Tim Frank, *Infineon*

Hybrid Si-SiC High Power Modules for Cost Effective High Voltage, High Current, High Frequency Switching
Eric Motto, *Powerex Inc.*
John Donlon, *Powerex Inc.*
Mike Rogers, *Powerex Inc.*

A New High Power IGBT Module Package
Tim Frank, *Infineon*
David Levett, *Infineon*

Test Setup for Accelerated Lifetime Determination of IGBT Modules
Bram Geene, *Prodrive Technologies*

HybridMOS: Product Development Contributing to Improved Energy Savings
Hiroyoki Ogurisu, *ROHM Semiconductor*
Mitch Van Ochten, *ROHM Semiconductor*

32nd Annual IEEE Applied Power Electronics Conference and Exposition
March 26th – 30th, 2017 at the Tampa Convention Center, Tampa, FL, USA

Announcement and Call for Papers

APEC 2017 continues the long-standing tradition of addressing issues of immediate and long-term interest to the practicing power electronic engineer. Outstanding technical content is provided at one of the lowest registration costs of any IEEE conference. APEC 2017 will provide a) the best power electronics exposition, b) professional development courses taught by world-class experts, c) presentations of peer-reviewed technical papers covering a wide range of topics, and d) venue to network and enjoy the company of fellow power electronics professionals in a beautiful setting. Activities and attractions for guests, spouses, and families are abundant in the Tampa area.

Topics of Interest:

1. **AC-DC Converters:**
   a. Single-Phase and Three-Phase Input
   b. Power Factor Correction, CCM, DCM, CRM/BCM Control, Bridgeless
   c. Embedded AC-DC Power Supplies
   d. External AC-DC Adapters

2. **DC-DC Converters:**
   a. Hard- and Soft-Switched
   b. Resonant Converters
   c. Point-of-Load (PoL) and Multi-Phase Converters
   d. Voltage Regulator Modules (VRM)

3. **Power Electronics for Utility Interface:**
   a. Power Generation, Transmission and Distribution
   b. Power Quality, UPS, Filters
   c. Distributed Energy Systems
   d. SmartGrid
   e. UPS
   f. Solid-State Transformers
   g. Metering

4. **Motor Drives and Inverters:**
   a. AC, DC, BLDC Motor Drives
   b. Single- and Multi-Phase Inverters
   c. Sensor Integration
   d. Actuators
   e. High Performance Drives

5. **Devices and Components:**
   a. Power Silicon MOSFETs, BJTs, IGBTs
   b. GaN HEMTs,
   c. SiC MOSFETs and BJTs
   d. Fast Recovery Diodes
   e. Magnetic Materials and Components
   f. Capacitors, Supercapacitors
   g. Interconnects and Fuses

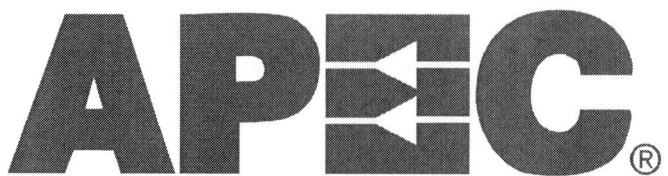

6. **System Integration:**
    a. Power Electronics Packaging
    b. Power Modules
    c. High Power Density Design
    d. Thermal Management
    e. EMI and EMC
    f. Material Science and Nanotechnology
    g. Sensors for Power Electronics

7. **Modeling and Simulation:**
    a. Circuits and Systems
    b. Device and Component Modeling
    c. Parasitics
    d. Software Tools
    e. Rapid Prototyping

8. **Control:**
    a. Control of Power Electronic Converters
    b. Current-Mode and Voltage-Mode Control
    c. Digital Control
    d. Sensor and Sensor-less Control
    e. Gate Drive Circuits
    f. Control ICs
    g. MCUs, DSPs, FPGAs, ASICs

9. **Manufacturing, Quality and Business Issues:**
    a. Quality and System Reliability
    b. Design for Manufacturability
    c. Fault-Tolerant Systems and Lifetime Predictions
    d. Life Cycle Cost Analysis
    e. Material Procurement
    f. Supplier Qualification
    g. Standards
    h. Production Processes

10. **Renewable Energy Systems:**
    a. Photovoltaic (PV) Inverters and Micro Inverters
    b. Maximum power point tracking (MPPT)
    c. Wind Energy Conversion Systems
    d. Fuel Cells
    e. Grid-Tied Systems
    f. Bi-directional Power Converters
    g. Microgrid Systems
    h. Energy Storage Systems

11. **Transportation Power Electronics:**
    a. Vehicular Power Electronic Circuits and Systems
    b. Power Electronics for Hybrid and Electric Cars
    c. Power Electronics for Aerospace
    d. Charging Systems

12. **Power Electronics Applications:**
    a. Lamp Ballasts and LED Lighting
    b. Network and Telecommunication Power Electronics
    c. Defense and Military Power Electronics
    d. AC-DC-AC Applications and Matrix Converters
    e. Portable Power
    f. Energy Harvesting
    g. Wireless Charging (Non-Transportation Applications)

Please note the following time frames (subject to change and posted at www.apec-conf.org/speakers/):

June 1, 2016	Start date to submit paper
July 20, 2016	Deadline for submission of digests
October 5, 2016	Notification that a paper was accepted or declined
November 15, 2016	Final papers and author registrations are due

Submission Requirements: Prospective authors are asked to submit a digest explaining the problem that will be addressed by the paper, the major results, and how this is different from the closest existing literature. Papers presented at APEC must be original material and not have been previously presented or published. The principal criteria in selecting digests will be the usefulness of the work to the practicing power electronic professional. Reviewers value evidence of completed experimental work. Authors should obtain any necessary company and governmental clearance prior to submission of digests. Please visit www.apec-conf.org/speakers/ for all details on digest and final manuscript format and to submit your paper.

If a digest is accepted, authors must submit a final manuscript before the deadline or the manuscript cannot be published in the Proceedings or presented at the conference. Final manuscripts may be subject to charges if their papers are over the page or file-size limit. **At least one of the authors listed on a paper must be registered for either a Full Registration or for the Technical Sessions Only registration, per paper.**

Become an APEC paper Reviewer: APEC relies upon a peer review process to ensure the quality of the technical content. To help maintain the high quality of the program, please contribute a few hours to review digests in your area of expertise by registering at www.apec-conf.org (under "Speakers" and "Paper Reviewer Sign-up").

Calls for Industry Sessions, Professional Education Seminars, and Exhibitor Seminars will be posted at www.apec-conf.org.

Website: www.apec-conf.org	**APEC**	**APEC Sponsors**
Email: apec@courtesyassoc.com	**2025 M Street**	Power Sources Manufacturers Association
Phone: +1-202-973-8664	**Suite 800**	IEEE Industry Applications Society
Facsimile: +1-202-331-0111	**Washington, DC 20036**	IEEE Power Electronics Society

Hardware Implementation and Characterization of SiC-Based Hybrid Three-Phase Rectifier Employing Third Harmonic Injection

M. Makoschitz*, M. Hartmann[†] and H. Ertl*

* Institute of Energy Systems and Electrical Drives, Power Electronics Section
University of Technology Vienna, Austria; Email: markus.makoschitz@tuwien.ac.at

[†] Schneider Electric Power Drives, Section Drives, Power Conversion
Vienna, Austria; Email: michael.hartmann@ieee.org

Abstract—**A recently proposed third harmonic injection rectifier (two half-bridge legs) – serving as optional solution for passive three-phase diode bridge systems with LC output filter – is characterized by unity power factor ($\lambda \approx 1$) and low harmonic input currents (THD $< 5\,\%$). The rectifier at hand, furthermore sticks out of the crowd due to a very low number of active and passive switches compared to similar power electronic circuits based on the same (third harmonic injection) principle. Additionally, the more and more establishing and reliable SiC-MOSFET allows an optimized design for higher switching frequencies without seriously impairing the efficiency of the total system. A proper design of active and passive components of the optional circuit is therefore required and several operating modes and optimization possibilities are existing. Besides basic considerations of control structures and topology limitations, also various implementation issues are discussed. Finally, experimental results of a laboratory prototype based on a 10 kW/72 kHz rectifier system using SiC-MOSFETs verifies the proper behaviour of the proposed upgrade.**

Index Terms—**Three-Phase AC-DC Conversion, Third-Harmonic Injection, SiC-MOSFET**

I. Introduction

Although, numerous promising AC-to-DC rectification circuits evolved during the last decades, recent developments and breakthroughs in the fields of wide band-gap devices heated up debates and discussions about already existing topologies. Rather unconventional rectifier circuits can also come handy again. Besides the very well known bidirectional three-phase six switch boost type (cf., [1]) or unidirectional VIENNA rectifier [2], one very promising solution to guarantee low harmonic input currents is based on the third harmonic principle (as described in [3] and [4]). This specific technique generally consists of two different circuits:

- current shaping circuit
- current injection stage.

Both circuits can be implemented as either passive or active setup. Converter systems based on the third harmonic injection principle can hence be classified into

- passive current shaping/passive current injection [5]–[8]
- active current shaping/passive current injection [9]–[12]
- passive current shaping/active current injection [13]
- active current shaping/active current injection [14]–[20].

Topologies which apply passive current injection and passive current shaping are generally limited in their minimum input current distortions by values close to 5 % (cf., [6]). This limitation is mainly evoked due to the passive implementation of both networks, as only sinusoidal wave shapes with multiples of the fundamental frequency can be generated. These currents are depending on the circuits resonance frequency which is defined due to capacitive and inductive components.

Circuits based on passive current injection and active current shaping are in principle limited in their performance due to similar issues. B6 output currents are indeed now controllable, the injection current is, however, still dependent on the LC injection network or appropriate transformer (YD, zig-zag, etc.). Topologies as the MINNESOTA rectifier therefore similarly suffer from limited input current distortions of approximately 4 %-5 % depending on characteristics of mains input impedance. Nonetheless, the output voltage of the circuit can be regulated, due to the active circuit directly connected to the adjacent passive diode bridge (dependent on the utilized DC/DC converter stage: buck-type, boost-type [9], buck+boost-type [11]).

One very promising solution for active injection and passive shaping is proposed in [13]. Series connected capacitors and a resistive delta or star connected network is used to facilitate properly shaped currents. Capacitors are used for blocking the DC components within the resistive network. The system can be designed to result in a THD$_i$ of 0.6 %. The circuit then, however, suffers from a very poor efficiency (e.g. 90 % for the specified nominal operating point) due to the implemented resistors which form more than 80 % of the total system losses. If improved efficiency of the current shaping network is required, all resistors can hence be replaced by switching resistance emulators, which however will increase the complexity of the enhanced circuit.

Some rectifier circuits where both, the injection and shaping network are active solutions are listed in [14]–[19]. The SWISS rectifier for example offers sinusoidal input currents. Major drawbacks of this circuit are that (i) (at least) two active semiconductors have to process the full amount of output power and (ii) the structure cannot be used to extend an existing passive three-phase rectifier to a low harmonic input stage.

A solution as proposed in [19] would allow such an upgrade for passive three-phase rectifiers. Due to the required 3-level NPC circuits a rather high number of active and passive switches is required. Therefore, a topology with two half-bridges connected in series (less number of switches) is discussed in this paper. If the third harmonic injection circuit is, however, going to serve as option/upgrade for a passive three-phase rectifier with LC output filter some major restrictions are applying in order to fulfill IEEE519/IEC61000-3-2 standards [21], [22]. Current shaping issues are hence going to be discussed in Section II.

II. Operating Principle

A. Basic Considerations

Fig. 1(a) depicts the passive three-phase diode bridge rectifier with LC output filter, upgraded by an additional circuit (active current shaping network and current injection device) based on the third harmonic injection principle. The current shaping network is realized by two half-bridge legs connected in series, whereas the output of the upper converter stage is connected to the positive bus bar and the output of the lower bridge leg linked to the negative output of the B6 (via two chokes L_{cp} and L_{cn}, respectively). The interconnection between AC-side mains and midpoint M of the shaping network is formed via three bidirectional switches $S_{i,ab}$, which are in this specific case implemented as back-to-back IGBTs or SiC-MOSFETs. Also different solutions as reverse blocking IGBTs etc. are applicable. The output voltage V_o of the rectifier is defined due to the mains situation and calculates to

$$V_o = \frac{3\sqrt{3}\hat{V}_N}{\pi}. \qquad (1)$$

The DC-side smoothing inductance current i_L is characterized according to B6 output voltage v_{rec} and V_o and calculates to

$$i_L(\varphi_N) = I_o - \frac{3\sqrt{3}\hat{V}_N}{\pi\omega_N L_{DC}} \sum_k^\infty \frac{2}{(6k)^3 - 6k} \sin(6k\varphi_N). \qquad (2)$$

The current ripple of i_L is mainly defined by a 300 Hz spectral component, evoked due to the 300 Hz passive three-phase rectifier voltage v_{rec}. The half-bridge leg connected to the DC capacitor C_{cp}, has to facilitate a compensational current i_{cp}, which completely suppresses the 300 Hz current ripple and guarantees sinusoidal wave shapes e.g.

$$i_{pos} = \hat{I}_N \cdot \cos(\varphi_N) \quad \text{for} \quad \varphi_N \in \left[-\frac{\pi}{3} \ldots \frac{\pi}{3}\right]. \qquad (3)$$

The current of the positive bus-bar i_{pos} is hence constituted by fractional parts of $\cos(\varphi_N)$, $\cos\left(\varphi_N - \frac{2\pi}{3}\right)$ and $\cos\left(\varphi_N - \frac{4\pi}{3}\right)$. i_{pos} is therefore now defined by $\max(i^*_{N1}, i^*_{N2}, i^*_{N3})$ and hence results in

$$i_{pos} = G_e \frac{3\sqrt{3}\hat{V}_N}{\pi} \left(\frac{1}{2} + \sum_k^\infty \frac{(-1)^{k+1}}{(3k)^2 - 1} \cos(3k\varphi_N)\right). \qquad (4)$$

Same assumptions are applicable for injection current i_{cn} and generated bus-bar current i_{neg}. The third harmonic current i_{h3} is defined according to $i_{cp} - i_{cn}$ and similarly (as i_{pos} and i_{neg})

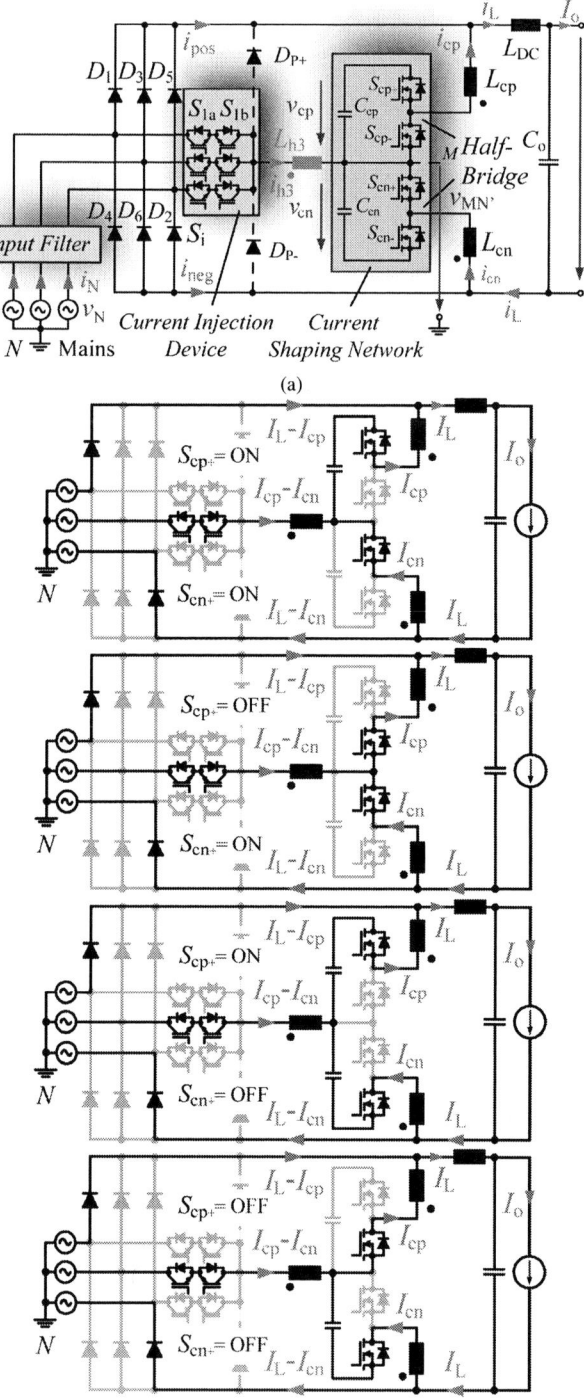

Fig. 1: (a) Passive three-phase rectifier with adjacent LC-output filter equipped by an optional third harmonic injection circuit with two half-bridge stages, three bidirectional switches in back-to-back arrangement and two/three (optional) injection inductors as proposed in [23]. (b) Feasible switching states of third harmonic injection based rectifier for the sector $\varphi_N \in [0 \ldots \frac{\pi}{3}]$.

results in sinusoidal wave shapes. This third harmonic current i_{h3} is going to be injected into the appropriates mains phase which is instantaneously not conducting current. Sinusoidal mains input current for e.g. phase 1 is therefore assembled

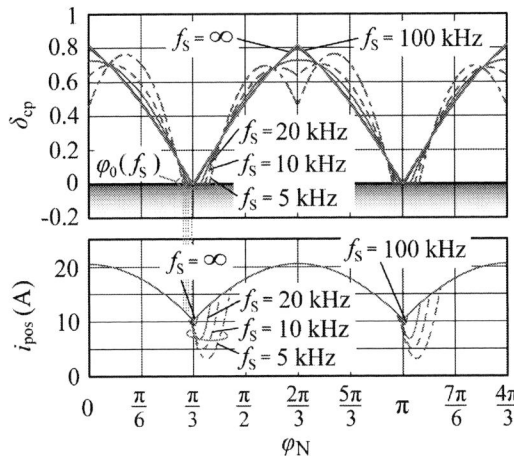

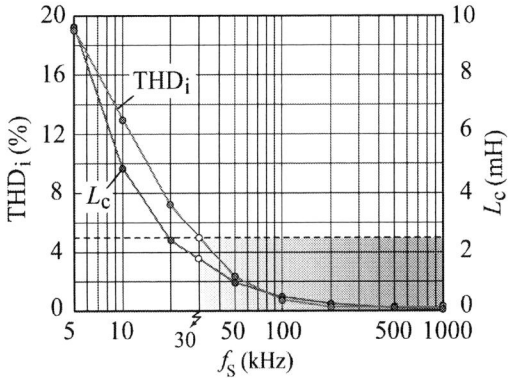

Fig. 3: Dependency of mains input current distortion THD_i on injection chokes $L_{c\frac{p}{n}} = L_c$ and (hence) switching frequency f_s (assumption: two inductors $L_{c\frac{p}{n}}$ and synchronized PWM).

Fig. 2: (Top) – Ideal (red) and real duty cycles δ_{cp} considering voltage drop $v_{L,cp}$, for different switching frequencies (resulting in different inductance values – $f_s = 5\,\text{kHz} \,\hat{=}\, L_{cp} = 9.6\,\text{mH}$, $f_s = 100\,\text{kHz} \,\hat{=}\, L_{cp} = 480\,\mu\text{H}$), if two inductors and synchronized pulse width modulation signals are used. (Bottom) – Appropriate calculated current distortions (no voltage/current controller considered).

according to the following sequence:

- $i_{N1} = i_{pos}$ for $\varphi_N \in [-\pi/3 \ldots \pi/3]$
- $i_{N1} = i_{h3}$ for $\varphi_N \in [\pi/3 \ldots 2\pi/3]$
- $i_{N1} = i_{neg}$ for $\varphi_N \in [2\pi/3 \ldots 4\pi/3]$
- $i_{N1} = i_{h3}$ for $\varphi_N \in [4\pi/3 \ldots 5\pi/3]$

Not only the sequence at hand but also **Fig. 1(b)** indicate that diodes of the passive rectifier are only switched with $f_N(=50\,\text{Hz})$ and the bidirectional switches ($S_{i,ab}$) of the current injection device with $2f_N(=100\,\text{Hz})$. Consequently, only the two half-bridge legs are operating with switching frequency f_s. All possible occurring switching states for one sector e.g. $\varphi_N \in [0 \ldots \frac{\pi}{3}]$, available for current shaping of i_{pos} and i_{neg}, are therefore depicted in **Fig. 1(b)**.

Finally, it has to be mentioned that due to the low switching operation of the passive rectifier bridge, no high-frequency CM voltage appears at the output V_o.

B. Fundamental Issues

As the main operating principle of the proposed third harmonic injection circuit has been discussed, some major operating boundaries are now going to be specified.

The required duty cycles for proper current shaping of the system are defined by

$$\delta_{cp}(\varphi_N) = \frac{v_{pos}(\varphi_N) - v_{h3}(\varphi_N)}{V_c} \tag{5}$$

$$\delta_{cn}(\varphi_N) = 1 + \frac{v_{neg}(\varphi_N) - v_{h3}(\varphi_N)}{V_c} \tag{6}$$

These functions are applying for an ideal consideration of L_{cp} and L_{cn}. If δ_{cp} is now drawn for further investigation, the minimum of the function ($\delta_{cp,min} = 0$) can be found at multiples of $(2k-1)\pi/3$ and the maximum ($\delta_{cp,max} = M = \frac{3\hat{V}_N}{2V_c} = 0.8125$) at multiples of $2k\pi/3$ for $k \in \mathbb{Z}$ (cf., **Fig. 2** - (top) red waveform). For a non-ideal consideration of L_{cp}, an additional voltage drop has to be incorporated (depicted in **Fig. 2** - (top)

blue waveforms for different values of L_{cp}) which results in an adapted minimum of

$$\delta_{cp,min} = -\omega_N L_{cp} \frac{\frac{3\sqrt{3}}{\pi \omega_N L_{DC}}\left(1 - \frac{\pi}{2\sqrt{3}}\right)\hat{V}_N + \frac{\sqrt{3}}{2}\hat{I}_N}{V_c} \tag{7}$$

As $\left(1 - \frac{\pi}{2\sqrt{3}}\right)$ is > 0, $\delta_{cp,min}$ is always < 0 for $L_{cp} > 0$. Each half-bridge, however, is only able to facilitate a duty cycle range of $[0 \ldots 1]$, whereas "0" means upper switch (S_{cp+}) opened and lower switch (S_{cp-}) closed and vice versa. A duty cycle < 0 will hence inevitably lead to current distortions of positive and negative busbar currents i_{pos} and i_{neg}, which is illustrated in **Fig. 2** (top - calculated duty cycles and bottom - resulting current distortions of i_{pos} for different values of L_{cp} (f_s)). It is hence highly preferable to implement an injection choke whose inductance is specified as small as possible which can be achieved due to increased switching frequencies f_s if a specific maximum current ripple of e.g. 30 % of the mains maximum peak current is allowed. According to **Fig. 3**, a $\text{THD}_\text{i} < 5\,\%$ can be achieved for a switching frequency of $\approx 30\,\text{kHz}$. As already discussed in [23] the inductance value of L_{cp} and L_{cn} can be further reduced if a third inductor L_{h3} is used. While the total volume of a three inductor implementation (if three independent chokes are utilized) hardly shrinks, the inductance value of L_{cp} and L_{cn} can be tremendously reduced by $\approx 70\,\%$ which can be expressed by

$$L_{c\frac{p}{n},h3} = \chi_L \cdot L_{c\frac{p}{n}}, \quad \chi_L = \frac{1}{2\sqrt{3}\left(\sqrt{3} - M\right)}. \tag{8}$$

This reduction mainly results due to the injection cell voltage v_{MN} (characterized by voltage levels 0 and $\pm\frac{V_c}{3}$), which is no longer directly clamped to the AC-side mains which would result in $v_{MN} = v_3$. If the optimized inductors are considered in the design of the injection cell the minimum switching frequency, which is required to guarantee a $\text{THD}_\text{i} < 5\,\%$, can hence be improved to values in the range of $\approx 15 - 20\,\text{kHz}$.

It has however be mentioned, that a proper design of the injection inductors is not the only stringent requirement for proper operation of the injection cell. Imbalance of mains input voltages is also evolving as noticeable issue for this type

978-1-4673-9551-9/16 $31.00 © 2016 IEEE

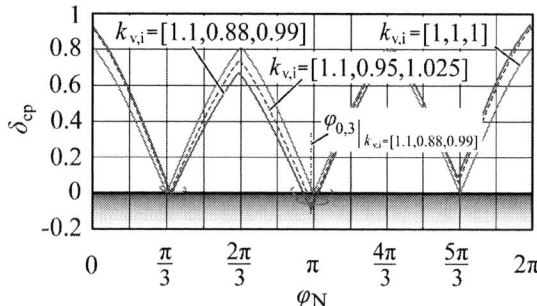

Fig. 4: (Duty cycles δ_{cp} (L_{cp},$L_{cn} = 0$) considering unbalanced (e.g. $k_{v,1} = 1.1$ means $V_{N,rms} = 230\,V + 10\,\%$) mains voltages (red – balanced mains voltages $k_{v,i} = 1$).

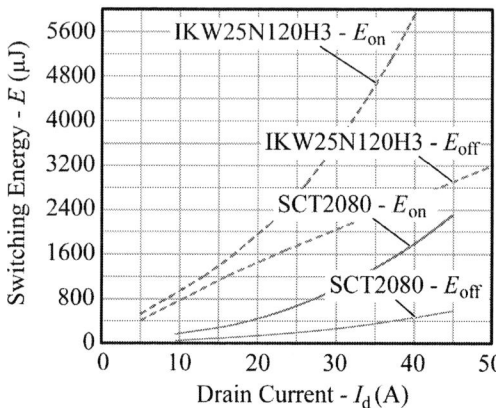

Fig. 5: Comparison of switching losses of an $1200\,V/28\,A$ (@$100\,°C$) SiC-MOSFET (SCT2080KE - losses recorded @ $600\,V$ DC voltage, $R_G = 0\,\Omega$, $V_{GS} = 18/0\,V$, $L = 500\,\mu H$) and a $1200\,V/25\,A$ (@$100\,°C$) IGBT (IKW25N120H3 (high speed switching series) - losses recorded @ $600\,V$ DC voltage, $R_G = 23\,\Omega$, $V_{GS} = 15/0\,V$, $L =$ n/a).

of converter stage. For this investigation an ideal duty cycle is assumed, ignoring injection injection chokes ($L_{cp} \to 0$ ($f_s \to \infty$)). The variable $k_{v,i}$ is going to be introduced which includes the information of relative mains unbalance from the nominal $230\,V_{rms}$ (e.g. $k_{v,1} = 1.1$ which results in a mains input voltage of $253\,V_{rms}$ and hence an increase of $10\,\%$). This is, however, directly affecting the duty cycle of both half-bridges (δ_{cp} and δ_{cn}). The minimum of δ_{cp} is no longer characterized by 0. Considering the modified commutation time instants of the B6 the minimum of the duty cycle can be approximated by

$$\delta_{cp,min}(\varphi_{0,3}) \approx \frac{2}{3} M (k_{v,2} - k_{v,1}) \tag{9}$$

for the sector $\varphi_N \in [\varphi_{0,2} \ldots \varphi_{0,3}]$ which correlates with the sector $\varphi_N \in \left[\frac{2\pi}{3} \ldots \pi\right]$ for balanced mains voltages ($k_{v,i} = 1$, also depicted in **Fig. 4** - red waveform). $\varphi_{0,3}$ is the altered commutation time instant next to π and is defined by

$$\varphi_{0,3} = \frac{2\pi}{3} + \arctan\left(\frac{2k_{v2} + k_{v3}}{\sqrt{3}k_{v3}}\right). \tag{10}$$

Fig. 4 depicts different duty cycles for varying imbalance of mains voltages with $k_v = [1.1, 0.88, 0.99]$ and $k_v = [1.1, 0.95, 1.025]$. It can be seen that for the appropriate sector $\varphi_N \in [\varphi_{0,2} \ldots \varphi_{0,3}]$, the minimum duty cycle is of negative value if $k_{v,1} > k_{v,2}$. Again, it should be pointed out that each half-bridge is only able to facilitate a duty cycle between $[0 \ldots 1]$. Negative duty cycles, hence, lead to increased input current distortions and hence worsened THD_i.

C. SiC-MOSFET or IGBT

The SiC-MOSFET is recently available for blocking voltages $\geq 900\,V$. This allows the implementation of SiC-MOSFETs for half-bridge legs with DC voltage levels $\geq 600\,V$ and offers an operation with increased switching frequency with lower switching losses compared to a conventional IGBT. This is also depicted in **Fig. 5** which shows a comparison of turn-on and turn-off energy losses of a SCT2080KE (SiC-MOSFET) and a IKW25T120H3 (reverse blocking IGBT) with very similar basic characteristics ($V_{max} = 1200\,V$, $I_{d,max@100°C} \approx 25\,A - 28\,A$). It can be observed that both, turn-on and turn-off switching energies of the SiC-MOSFET are significantly smaller than those of an IGBT. The SiC-MOSFET, furthermore, shows ohmic characteristics which will result in improved conduction losses

(compared to an Si-IGBT). The implementation of a SiC-MOSFET hence leads to smaller and less expensive cooling system/volume (if SiC conduction and switching losses are smaller than that of an IGBT). The improved switching loss behaviour will hence lead to an increased efficiency of the total system compared to a conventional IGBT with approximately equal switching frequency. Similar to low-voltage MOSFETs the SiC-MOSFET is characterized by a parasitic body diode due to its physical implementation. As, however, this body diode is only conducting during the adjusted dead time of the half-bridge stage, no external diode is required and the SiC-MOSFET seems to be very well applicable for the chosen topology. The bidirectional switches can also advantageously be implemented by SiC-MOSFETs in order to reduce losses of the current injection stage. It has to be mentioned that the emerging GaN technology could also serve as promising solution for the current shaping network, however, is yet not commercially available for DC voltage levels greater than $650\,V$.

III. CURRENT CONTROLLER

In this section, current control of the proposed injection cell is briefly discussed. The third harmonic injection circuit consists of two half-bridge legs (cf., **Fig. 1(a)**). The half-bridge connected to C_{cp} is therefore responsible for proper control of i_{cp} and the converter stage linked to C_{cn} has to properly adjust i_{cn}, respectively. Thus, two independent current controllers have to be designed.

If a simplified model is applied which is neglecting parasitic effects and components, the basic equations result in

$$\begin{aligned} L_{cp}\frac{di_{cp}}{dt} &= \delta_{cp}v_{cp} + v_{h3} - v_{pos} \\ L_{cn}\frac{di_{cn}}{dt} &= (1 - \delta_{cn})v_{cn} - v_{h3} + v_{neg}. \end{aligned} \tag{11}$$

If it is now assumed that the two DC voltages (v_{cp} and v_{cn}) of the injection cell are regulated such – by the superimposed voltage controller with reduced dynamic – to remain fixed at the same voltage level $V_{cp} = V_{cn} = V_c$, the equations above

978-1-4673-9551-9/16 $31.00 © 2016 IEEE

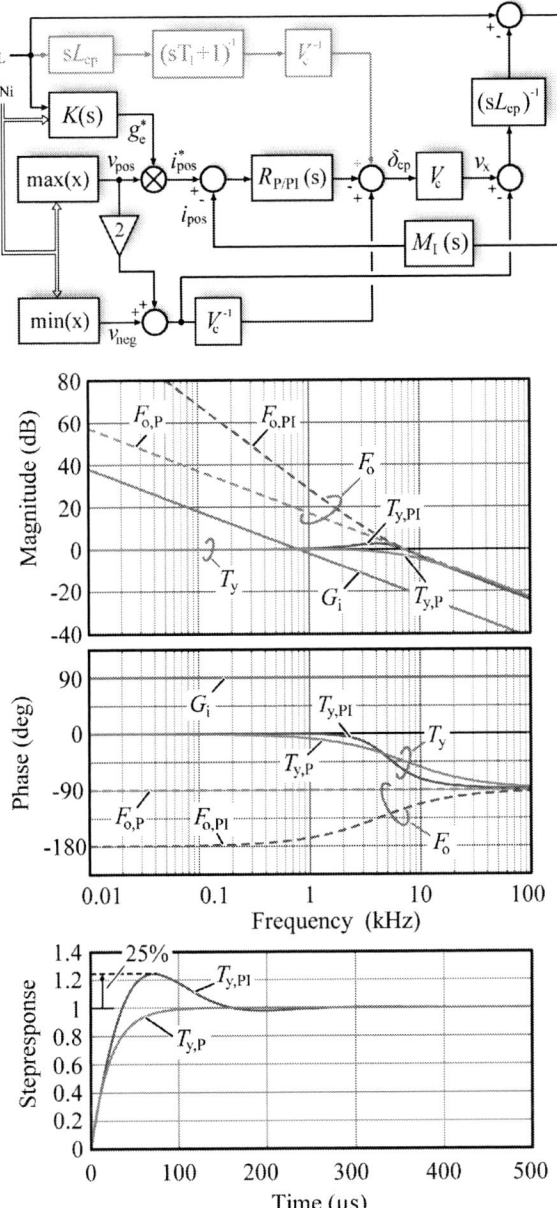

Fig. 6: Basic structure of current control circuit (top). Bode plot (middle) and step-response (bottom) of the designed current controller for a switching frequency of 72 kHz and an inductance value L_{cp} of 200 µH and a gain crossover frequency of 7 kHz), for either P- or PI-type with maximum overshoot of 25 % and minimum phase margin of 60 °).

can be simplified by applying Laplace Transformation which finally results in

$$sL_{cp}i_{pos} = -\delta_{cp}V_c + v_{pos} - v_{h3} + sL_{cp}i_L. \quad (12)$$

If v_{N1}, v_{N2}, v_{N3} (which are required for the generation of an equivalent conductance value, which represents the power demand of the circuit) and i_L are measured, v_{pos}, v_{h3} and $sL_{cp}i_L$ can serve as feedforward signals of the respective current control structure. The hereby resulting model can be described

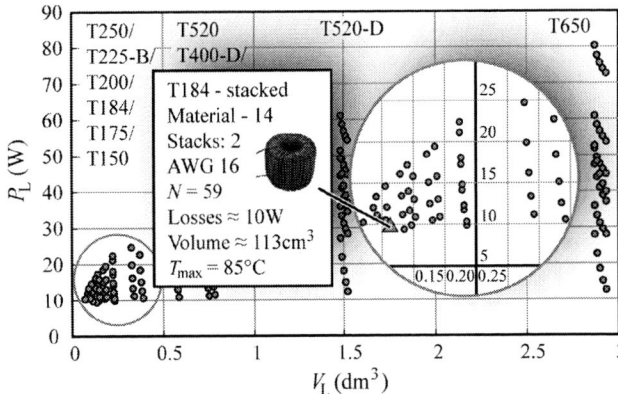

Fig. 7: Evaluation of optimized choke for different toroidal-stacked cores T20-T650, iron powder core materials 14-52 and solid conductors AWG16-AWG20 according to GeckoMAGNETICS and implemented inductor (boxed).

by

$$\delta_{cp} = -\frac{sL_{cp}}{V_c}i_{pos} + \delta_{ff} \quad (13)$$

with the feedforward signal δ_{ff}

$$\delta_{ff} = \frac{1}{V_c}(v_{pos} - v_{h3}) + \frac{sL_{cp}}{(sT_1 + 1)V_c}i_L. \quad (14)$$

The additional damping part $1/(sT_1 + 1)$ is required to guarantee suppression of high frequency noise and/or disturbance. Furthermore, it has to be noticed that the feedforward part which includes i_L could also be omitted for very small values of L_c if it is preferred to reduce the number of current sensors. In **Fig. 6** (top) the basic structure of the proposed current controller is depicted. Furthermore, bode plot (middle) and step-response (bottom) of the designed controller is given (for a switching frequency of 72 kHz and an inductance value L_c of 200 µH and a gain crossover frequency f_c of 7 kHz), for either P- or PI-type (maximum overshoot of 25 %, minimum phase margin 60°) implementation. F_o denotes the open loop characteristic and T_y the closed loop transfer function of controlled current i_{pos} which is defined by

$$T_{y,P/PI}(s) = \frac{R_{P/PI}(s)\,G_i(s)}{1 + M_I(s)\,R_{P/PI}(s)\,G_i(s)}. \quad (15)$$

The reference currents are often generated by a superimposed voltage controller. This voltage controller indirectly determines the power demand by measuring the voltage of the dc-link capacitor. However, voltage control of the injection cell DC voltages is not one of the major topics in this paper and hence not further discussed.

IV. INDUCTOR DESIGN

A stacked toroidal iron powder core (implementation of $L_c = L_{c\underline{p}} = L_{h3}$) finally has been selected which basically allows reduced losses and rather high current saturation limits (compared to ferrite material) which may be necessary due to unbalanced grid voltages and load step behaviour of the total system (discussed in [24] and [25]). In order to minimize input current distortions a maximum inductance of 200 µH has been

TABLE I: Design Specifications of the Built Three-Phase Rectifier.

Mains voltage:	$V_{LL} = 400\,\mathrm{V_{rms}}$
Mains frequency:	$f_N = 50\,\mathrm{Hz}$
Switching frequency:	$f_s = 72\,\mathrm{kHz}$.
Cells DC-link voltage:	$V_{cp} = V_{cn} = 600\,\mathrm{V}$
Output power:	$P_o = 10\,\mathrm{kW}$

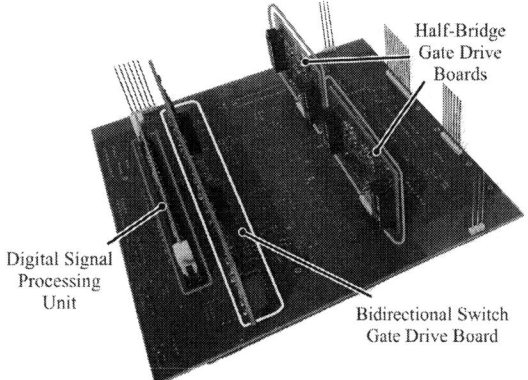

Fig. 8: Controlboard (bottom view) with plug-in DSP board (red), half-bridge gate drive boards (blue) and bidirectional switch boards (yellow).

chosen. Design guidelines for injection inductance values are listed in [23]. The number of turns N for one choke can be calculated by

$$N = \sqrt{\frac{V_c M}{3\sqrt{3} f_s k_i \Delta i_{N,max} A_{L,2\times T184}}} \tag{16}$$

where V_c denotes the DC voltage of the cell (600 V), M the modulation index (0.8125) and $A_{L,2\times T184}$ the core nominal inductance ($2 \cdot 28\,\mathrm{nH}$). It has to be noted that the basic equation to evaluate the required number of turns is given according to

$$L_c = N^2 \cdot A_L. \tag{17}$$

Fig. 7 illustrates an inductor optimization which was performed with GeckoMAGNETICS for different toroidal-stacked cores T20-T650, iron powder core materials 14-52 and solid conductors AWG16-AWG20. The red triangle marks the implemented choke which is depicted in **Fig. 7**.

V. EXPERIMENTAL RESULTS

A 10 kW/72 kHz laboratory prototype was implemented in order to verify proper operation of the proposed circuit. Design specifications of the built three-phase rectifier are given in TABLE I and TABLE II. The system should be constructed for 400 V_{LL}/50 Hz mains voltages, which results in 600 V DC voltage levels (if the modulation index $M = 0.8125$) for the injection cell capacitors C_{cp} and C_{cn}.

ROHMs SCT2080KE SiC-MOSFETs are used for implementing the active switches of the half-bridges connected to C_{cp} and C_{cn}. The bidirectional switches $S_{i,ab}$ are operating in back-to-back arrangement using IKW40T120 IGBTs. It has to be noted that also 1200 V or 900 V SiC-MOSFETs would be applicable and could further reduce losses of the injection

TABLE II: Power Devices Selected for Implementation of the passive rectifier and Third Harmonic Injection Cell.

$S_{ia,b}$	1200 V/40 A IGBT, IKW40T120, Infineon
$S_{c\frac{p}{n}\pm}$	600 V/20 A SiC-MOSFET, SCT2080, ROHM
$C_{c\frac{p}{n}}$	220 μF/400 V, EPCOS B43508-type
$L_{cp}=L_{cn}=L_{h3}$	Iron Powder Core 2 x T184-14, N = 59 turns 200 μH
C_F, C_S	1 μF/275 V_{AC}, MKP X2, Arcotronics
C_o	2.2 mF/400 V, Felsic CO 39 A728848
L_{DC}	2.25 mH, Iron core 2 x UI60a
$D_1 - D_6$	35 A/1600 V, 36MT160, Vishay

Fig. 9: Powerboard consisting of current sensors, inductors, half-bridge stage, bidirectional switches and input filter.

device by $\approx 30\,\%$, which is however only $2 - 3\,\%$ of the total system losses, which does not outweigh the unfavorable additional expense of increasing cost per active component.

The passive three-phase rectifier has to be configured such to allow a maximum total harmonic distortion of input currents of approximately 45 % at nominal load (10 kW) (B6-THD$_i$ - standalone operation). The required DC-side smoothing inductance (L_{DC}) therefore results in 2.25 mH. As no high-frequency common mode at the output of the system is expected a relatively large output capacitor of 2x2.2 mF in series is chosen.

The current injection laboratory prototype mainly consists of two different boards – controller- and power board. The controller board (cf., **Fig. 8**) contains measurement circuits, auxiliary power supply, DSP control unit (TI 320F2808), additional hardware as e.g. zero crossing detection of mains line-to-line voltages and bidirectional switch controller (Lattice CPLD - MachXO 2280). Except for the digital signal processor (which is located at the bottom of the control board) all remaining parts are mainly placed top-side of the board. Gate drives of SiC-MOSFETs and IGBTs are spotted on separate plug-in boards (one board for each half-bridge stage and one additional board for bidirectional switches $S_{i,ab}$). The circuits are located in between control and power board in order to minimize the total volume of the cell and reduce parasitic inductance between gate drive circuit, appropriate switches and DC-link electrolytic capacitor bank (C_{cp} and C_{cn}). The power board, which is depicted in **Fig. 9**, includes main power components as forced

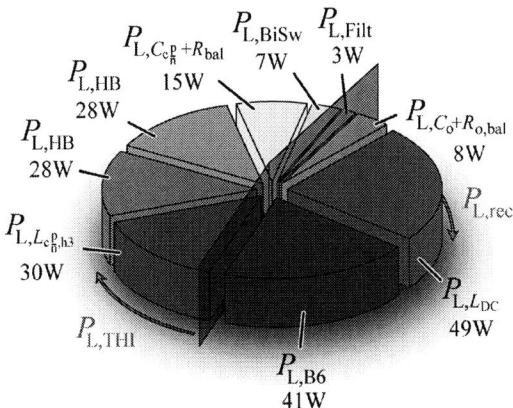

Fig. 10: Losses of the three-phase rectifier system for $230\,\mathrm{V_{rms}}/50\,\mathrm{Hz}$ a switching frequency of $72\,\mathrm{kHz}$ and $10\,\mathrm{kW}$ output power. (a) Apportioned losses by stage and/or passive components. (b) Pie chart which illustrates loss-distribution of passive system and active upgrade.

convection cooling system, IGBTs, SiC-MOSFETs, input filter, current sensors, inductors, electrolytic capacitors and additional hardware which is required for start-up operation of the system. Calculated system losses at nominal load ($P_{\mathrm{N}} = 10\,\mathrm{kW}$) are depicted in **Fig. 10**. As can be seen, the highest losses of all included parts are drawn by the DC-side choke ($\approx 50\,\mathrm{W}$) and the passive three-phase rectifier ($\approx 40\,\mathrm{W}$), which have to transfer the main part of the active power. The injection cell only has to process $\approx 6\,\%P_{\mathrm{o}}$ which obviously results in very low losses of the optional configuration related to the nominal power P_{N}. As can be seen in **Fig. 10**, losses are evenly distributed over both circuits (passive circuit/active circuit - $\approx 50\,\%/50\,\%$, respectively). The total system losses are therefore, calculated to $209\,\mathrm{W}$ which results in an expected system efficiency η_{calc} of $97.95\,\%$.

The performed loss calculation, allows the design of the forced convection (air) cooling system. The chosen air cooled heatsink should serve as loss dissipating surface for SiC-MOSFETs (half-bridges) and IGBTs (bidirectional switches). According to [26] a cooling system performance index (CSPI)

$$\mathrm{CSPI} = \frac{1}{R_{\mathrm{th,hs}} V_{\mathrm{hs}}} \tag{18}$$

has been introduced in order to generally allow a comparison of different cooling system technologies. A forced convection air cooled heatsink typically results in a CSPI between $5 - 12\,\frac{\mathrm{W}}{\mathrm{K \cdot dm^3}}$. The CSPI of the chosen LAM4K (150 mm) air cooled system, as depicted in **Fig. 9**, has been evaluated to $9\,\frac{\mathrm{W}}{\mathrm{K \cdot dm^3}}$. Considering active switch component losses as illustrated in **Fig. 10**, the maximum heatsink temperature T_{hs} results in

$$T_{\mathrm{hs}} = T_{\mathrm{amb}} + \frac{P_{\mathrm{sys}}\left(\eta_{\mathrm{HB+BiSw}}^{-1} - 1\right)}{V_{\mathrm{hs}} \cdot \mathrm{CSPI}} \approx 54^\circ \mathrm{C}. \tag{19}$$

Therefore, the cooling system would still allow an additional decrease of volume for the given active rectifier topology.

The fully assembled prototype (controller board, power board and gate drive boards) is shown in **Fig. 11(a)**. Measurement

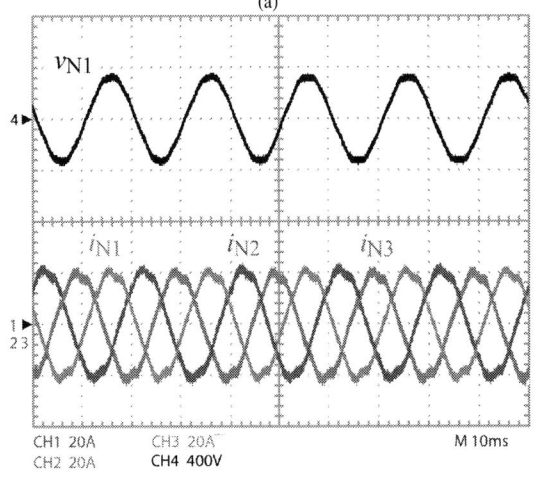

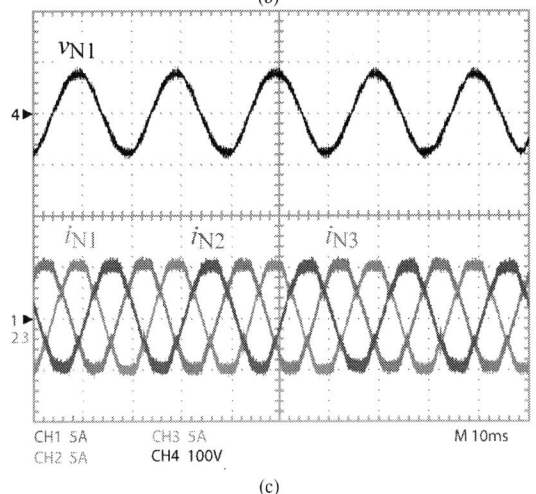

Fig. 11: (a) Assembled $10\,\mathrm{kW}$ laboratory prototype. (b) Mains currents i_{N1} - i_{N3} and mains voltage v_{N1} for $400\,\mathrm{V_{LL,rms}}$ input voltage with implemented P-type controller without feedforward currents results in $\mathrm{THD_i} = 3.6\,\%$ and $\lambda = 0.999$. (b) Mains currents i_{N1} - i_{N3} and mains voltage v_{N1} for $95\,\mathrm{V_{LL,rms}}$ input voltage with implemented PI-type controller without feedforward currents results in $\mathrm{THD_i} = 1.8\,\%$, $\lambda = 0.999$.

results of the $10\,\mathrm{kW}/72\,\mathrm{kHz}$ laboratory prototype are depicted in **Fig. 11(b)** and **Fig. 11(c)**. **Fig. 11(b)** shows input currents i_{N1} - i_{N3} and mains voltage v_{N1} for $230\,\mathrm{V_{LL,rms}}$ input voltage with an implemented P-type controller without feedforward currents i_L and i_{h3}. Nevertheless, input current distortions of $\mathrm{THD_i} = 3.9\,\%$ could be observed. The power factor λ was measured by 0.999. Furthermore, a system efficiency η_{meas} of $97.7\,\%$ has been measured. Heatsink temperature after $20\,\mathrm{min}$ of operation at $10\,\mathrm{kW}$ has resulted in $56\,^\circ\mathrm{C}$ and injection choke inductance temperature of $87\,^\circ\mathrm{C}$ has been measured.

In order to, verify if input current waveforms could be further improved (by utilizing a PI-type controller) a PI-type controller has been implemented for test purpose for $55\,\mathrm{V_{LL,rms}}$ mains voltages (cf., **Fig. 11(c)**) and resulted in a $\mathrm{THD_i} = 1.8\,\%$. Therefore, a PI controller should also be considered and implemented for $230\,\mathrm{V_{LL,rms}}$ mains to improve $\mathrm{THD_i}$ of mains currents.

VI. CONCLUSION

Upgrading a simple three-phase passive rectifier by an optional cell (employing two half-bridge stages) based on the third harmonic injection principle has been basically discussed and analyzed in this work. Drawbacks (sensitive to unbalanced mains voltages, minimum switching frequency required etc.) and advantages (optional upgrade, no high-frequency CM voltage, low harmonic input currents, high power factor, improved efficiency, etc.) are mentioned, illustrated and demonstrated by an implemented $10\,\mathrm{kW}/72\,\mathrm{kHz}$ laboratory prototype. It is shown, that due to the implementation of a simple P-type controller and neglecting feedforward currents i_L and i_{h3}, a $\mathrm{THD_i}$ of $< 4\,\%$ can be achieved (values $< 2\,\%$ can be expected if a PI-type controller is implemented). A power factor λ of 0.999 could be observed, and despite an applied switching frequency of $72\,\mathrm{kHz}$, an efficiency of the total system (B6 + injection cell) close to $98\,\%$ ($\eta_{meas} = 97.7\,\%$) could be measured, which is very well fitting with performed loss calculations.

REFERENCES

[1] R. Bosshard and J.W. Kolar, *Fundamentals and Multi-Objective Design of Inductive Power Transfer Systems*, Tutorial at the 30th Applied Power Electronics Conference and Exposition (APEC 2015), Charlotte, NC, USA, March 15-19, 1st edition, 2005.

[2] J.W. Kolar and F.C Zach, "A Novel Three-Phase Three-Switch Three-Level Pwm Rectifier," in *Proceedings of the Conference for Power Electronics, Intelligent Motion, Power Quality (PCIM)*, Nuernberg, Germany, June 28-30 1994, pp. 125–138.

[3] B.M. Bird, J.F. Marsh, and P.R. McLellan, "Harmonic Reduction in Multiplex Convertors by Triple-Frequency Current Injection," *Proceedings of the Institution of Electrical Engineers*, vol. 116, no. 10, pp. 1730–1734, 1969.

[4] J.F. Baird and J. Arrilaga, "Harmonic Reduction in DC-Ripple Reinjection," *Proceedings of the IEE Generation, Transmission and Distribution*, vol. 127, no. 5, pp. 294–303, 1980.

[5] W.B. Lawrance and W. Mielczarski, "Harmonic Current Reduction in a Three-Phase Diode Bridge Rectifier," *IEEE Transactions on Industrial Electronics*, vol. 39, no. 6, pp. 571–576, 1992.

[6] S. Kim, P. Enjeti, P. Packebush, and I. Pitel, "A New Approach to Improve Power Factor and Reduce Harmonics in a Three Phase Diode Rectifier Type Utility Interface," in *Conference Record of the IEEE Industry Applications Society Annual Meeting*, 1993, pp. 993–1000.

[7] P. Pejovic and Z. Janda, "An Improved Current Injection Network for Three-Phase High-Power-Factor Rectifiers that Apply the Third Harmonic Current Injection," *IEEE Transactions on Industrial Electronics*, vol. 47, no. 2, pp. 497–499, 2000.

[8] P. Pejovic, P. Bozovic, and I. Pavlovic, "A Novel Magnetic Device for Current Injection Based Three-Phase Low-Harmonic Rectifiers that Integrates the Current Injection Device and the Inductor," in *Proceedings of the third IET International Conference on Power Electronics, Machines and Drives*, 2006, pp. 80–84.

[9] N. Mohan, M. Rastogi, and R. Naik, "Analysis of a New Power Electronics Interface with Approximately Sinusoidal 3-Phase Utility Currents and a Regulated DC Output," *IEEE Transactions on Power Delivery*, vol. 8, no. 2, pp. 540–546, 1993.

[10] S. Kim, P. Enjeti, D. Rendusara, and I.J. Pitel, "A New Method to Improve THD and Reduce Harmonics Generated by a Three-Phase Diode Rectifier Type Utility Interface," in *Conference Record of the IEEE Industry Applications Society Annual Meeting*, 1994, pp. 1071–1077.

[11] L.R. Chaar, N. Mohan, and Christopher P. Henze, "Sinusoidal Current Rectification in a Very Wide Range Three-Phase AC Input to a Regulated DC Output," in *Proceedings of the Conference Record of the 1995 Thirtieth IEEE Industry Applications Conference (IAS) Annual Meeting*, 1995, vol. 3, pp. 2341–2347.

[12] M. Rastogi, N. Mohan, and Christopher P. Henze, "Three-Phase Sinusoidal Current Rectifier with Zero-Current Switching," *IEEE Transactions on Power Electronics*, vol. 10, no. 6, pp. 753–759, 1995.

[13] P. Pejovic, "A Novel Low-Harmonic Three-Phase Rectifier," *IEEE Transactions on Circuits and Systems*, vol. 49, no. 7, pp. 955–965, 2002.

[14] J.C. Salmon, "Operating a Three-Phase Diode Rectifier with a Low-Input Current Distortion Using a Series-Connected Dual Boost Converter," *IEEE Transactions on Power Electronics*, vol. 11, no. 4, pp. 592–603, 1996.

[15] M. Hartmann and R. Fehringer, "Active Three-Phase Rectifier System Using a "Flying" Converter Cell," in *Proceedings of the International Energy Conference and Exhibition (ENERGYCON)*, 2012, pp. 82–89.

[16] T.B. Soeiro, T. Friedli, and J.W. Kolar, "Swiss Rectifier — A Novel Three-Phase Buck-Type PFC Topology for Electric Vehicle Battery Charging," in *Proceedings of the Twenty-Seventh Annual IEEE Applied Power Electronics Conference and Exposition (APEC)*, 2012, pp. 2617–2624.

[17] P. Cortes, M.F. Vancu, and J.W. Kolar, "Swiss Rectifier Output Voltage Control with Inner Loop Power Flow Programming (PFP)," in *Proceedings of the 14th Workshop on Control and Modeling for Power Electronics (COMPEL)*, 2013, pp. 1–8.

[18] X. Du, L. Zhou, H. Lu, and H.-M. Tai, "DC Link Active Power Filter for Three-Phase Diode Rectifier," *IEEE Transactions on Industrial Electronics*, vol. 59, no. 3, pp. 1430–1442, 2012.

[19] M. Makoschitz, M. Hartmann, H. Ertl, and R. Fehringer, "Analysis of a New Third Harmonic Injection Active Rectifier Topology Based on an NPC Three-Level Converter Cell," in *Proceedings of the Conference for Power Electronics, Intelligent Motion, Power Quality (PCIM)*, Nuernberg, Germany, May 14-16 2013, pp. 1133–1140.

[20] L. Schrittwieser, J.W. Kolar, and T.B. Soeiro, "Novel Modulation Concept of the SWISS Rectifier Preventing Input Current Distortions at Sector Boundaries," in *in Proceedings of the Sixteenth IEEE Workshop on Control and Modeling for Power Electronics (COMPEL)*, 2015, pp. 1–8.

[21] IEEE Std 519-1992, *IEEE Recommended Practices and Requirements for Harmonic Control in Electrical Power Systems*, New York, NY: IEEE, 1992.

[22] EC61000-3-2, *Electromagnetic compatibility (EMC) Part 3-2: Limits for harmonic current emissions (equipment up to and including 16 A per phase)*, published by the International Electrotechnical Commission (IEC), 2000.

[23] M. Makoschitz, H. Ertl, and M. Hartmann, "A Passive Three-Phase Rectifier Enhanced by a DC-Side High Switching Frequency Add-On SiC-Converter Stage for Unity Power Factor Applications," in *Proceedings of the 17th European Conference on Power Electronics and Applications (EPE'15 ECCE-Europe)*, 2015, pp. 1–8.

[24] M. Makoschitz, M. Hartmann, H. Ertl, and R. Fehringer, "DC Voltage Balancing of Flying Converter Cell," in *Proceedings of the Energy Conversion Congress and Exposition (ECCE)*, 2014, pp. 4071–4078.

[25] M. Makoschitz, M. Hartmanny, and H. Ertl, "Effects of Unbalanced Mains Voltage Conditions on Three-Phase Hybrid Rectifiers Employing Third Harmonic Injection," in *Proceedings of the IEEE International Symposium on Smart Electric Distribution Systems and Technologies (EDST)*, 2015, pp. 417–424.

[26] U. Drofenik and J. W. Kolar, "Thermal Power Density Barriers of Converter Systems," in *Proceedings of the 5th International Conference on Integrated Power Systems (CIPS)*, 2008, pp. 1–5.

Voltage Oriented Control of the Three-level Vienna Rectifier Using Vector Control Method

Jeevan Adhikari, *Student Member, IEEE*, Prasanna I V, *Student Member, IEEE*, S K Panda, *Senior Member, IEEE*,
Department of Electrical and Computer Engineering, National University of Singapore, Singapore

Abstract—This paper proposes a simplified vector control method with output capacitor voltage balance for voltage oriented control of the three-level three-switch Vienna rectifier. The proposed control method makes use of a single carrier based pulse width modulated (PWM) switching technique instead of complex space vector method for the Vienna rectifier. The vector control method is comprehensively compared with the existing hysteresis control method in terms of transient response and input current total harmonic distortion (THD). This method is implemented to control the converter for wide range of load and unbalanced grid supply voltage conditions. A 1 kW prototype of the converter is built and tested in the laboratory environment using the proposed vector control method and compared with the conventional hysteresis control method. The proposed control method provides unity power factor operation with sinusoidal input currents at the input side and balanced output capacitor voltages at the output side.

Index Terms—Three-level rectifier, Vienna rectifier, Vector control, Hysteresis control

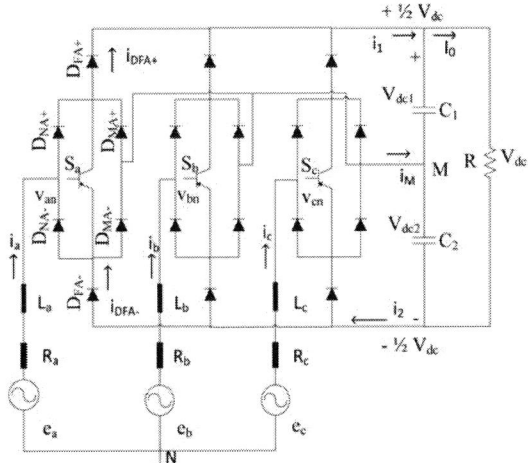

Fig. 1: Schematic of the Vienna rectifier

I. INTRODUCTION

For high voltage rectification process, three-level operation is preferred due to reduced switch stress. Three-level NPC rectifier (bidirectional) consists of twelve semiconductor switches for rectification and thereby increase the losses of the converter. Unidirectional rectification process can be carried out by using a three switch three-level Vienna rectifier. A Vienna rectifier has less number of active semiconductor devices, low current total harmonic distortion (THD), and high power density [1], [2]. Therefore, it is best suited for high voltage and high power unidirectional rectification application. The permanent magnet synchronous generator (PMSG) based wind power generation systems, front-end rectifier for telecommunication applications, medium voltage drive systems (where voltage stress and power density are important) etc can make use of the Vienna rectifier [1]–[8] for front-end rectification process.

Most of the control methods implemented for controlling the Vienna rectifier [1], [2], [4]–[6] makes use of hysteresis current control. This control method regulates the output voltage and the total harmonic distortion (THD) of the source side current. However, this method operates at variable switching frequency that leads to over-design of passive components, higher current THD, and varying switching losses [1], [2], [4]. Different control techniques for the close loop control operation of the Vienna rectifier for various grid conditions are explained in [7], [9]–[12]. The proposed algorithms in [8]–[13] make use of complex non-linear control techniques and space-vector switching technique which is difficult to compute for three switch rectifier.

In this paper, a simplified vector control method is proposed that regulates the input current THD and the output voltage. The proposed algorithm includes voltage balance across two output capacitors by adding zero sequence component generated by a simple PI control. The proposed control method is implemented for wide range of grid voltage and load variations. Easily implementable single carrier based PWM switching method is implemented for switching the semiconductor devices. The proposed method is compared experimentally with the hysteresis control technique in terms of transient response and input current THD and the efficacy of the proposed vector control is demonstrated.

The operations of the Vienna rectifier for different switching conditions are explained in Section II. The modelling and control of the Vienna rectifier is presented in Section III. In Section IV, MATLAB based simulation results are provided. The simulation results are verified experimentally in Section V.

II. OPERATIONS OF THE VIENNA RECTIFIER

In Fig 1, overall schematic diagram of the Vienna rectifier is shown that consists of three switches and eighteen diodes. Active control of the three semiconductor devices ensures sinusoidal input current, desired output voltage, and balanced capacitor voltages. The polarity of the phase current and switching state of the switches determine the operating condition of the Vienna rectifier.

The flow of line currents when S_b is turned on, i_a is positive, i_b and i_c are negative is shown in Fig 2. During this stage, the terminal voltage, v_{an} is $\frac{V_{dc}}{2}$, v_{bn} is zero, and v_{cn} is $-\frac{V_{dc}}{2}$.

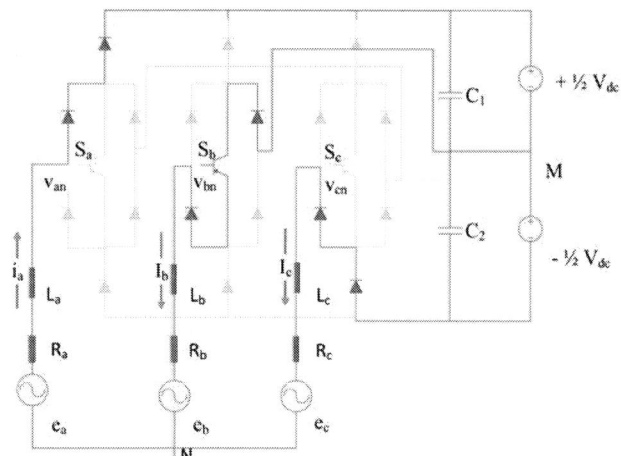

Fig. 2: Operation of the Vienna rectifier when, $i_a i_b i_c = + - -$ and $S_a S_b S_c = 010$

The path of three line currents are presented in Fig 3 when S_b is turned on, i_a and i_c are positive, and i_b is negative. During this stage v_{an} and v_{cn} are clamped to $-\frac{V_{dc}}{2}$ and V_{bn} is zero.

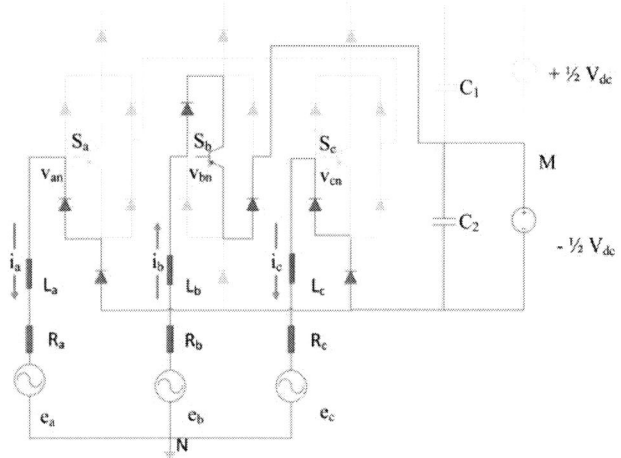

Fig. 3: Operation of the Vienna rectifier when, $i_a i_b i_c = - + -$ and $S_a S_b S_c = 010$

When all three semiconductor switches are turned-off and i_a and i_c are positive, and i_b is negative, the flow of line side currents is illustrated in Fig 4. During this stage v_{an} and v_{cn} are $-\frac{V_{dc}}{2}$ and v_{bn} is $\frac{V_{dc}}{2}$. Therefore, the terminal voltages vary from $-\frac{V_{dc}}{2}$ to zero and zero to $\frac{V_{dc}}{2}$ during all switching conditions. For any switching conditions and current polarity, voltage stress on diodes and switch varies from $-\frac{V_{dc}}{2}$ to $\frac{V_{dc}}{2}$, reducing the voltage stress to half of the DC link voltage.

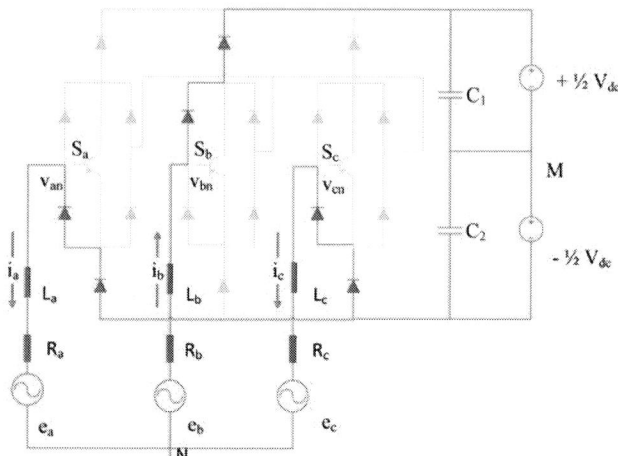

Fig. 4: Operation of the Vienna rectifier when, $i_a i_b i_c = - + -$ and $S_a S_b S_c = 000$

The generation side voltages can be expressed as:

$$
\begin{aligned}
e_a &= E sin(w_s t) \\
e_b &= E sin(w_s t - 120) \\
e_c &= E sin(w_s t + 120)
\end{aligned}
\tag{1}
$$

where e_a, e_b, and e_c are three phase generation voltages. E and w_s are amplitude and angular frequency of the generation voltage.

The state-space equations for the input side of the Vienna rectifier are expressed as:

$$
\begin{aligned}
e_a &= Ri_a + L\frac{di_a}{dt} + v_{an} \\
e_b &= Ri_b + L\frac{di_b}{dt} + v_{bn} \\
e_c &= Ri_a + L\frac{di_c}{dt} + v_{cn}
\end{aligned}
\tag{2}
$$

where v_{an}, v_{bn}, and v_{cn} are the terminal voltages of the Vienna rectifier.

Switching states of the switches and polarity of the current decide the terminal voltage of the Vienna rectifier. Terminal voltage as a function of current polarity and switching states can be expressed as [7], [10]–[12]:

$$
\begin{aligned}
v_{an} &= \frac{V_{dc}}{2}sgn(i_a)(1 - S_a) \\
v_{bn} &= \frac{V_{dc}}{2}sgn(i_b)(1 - S_b) \\
v_{cn} &= \frac{V_{dc}}{2}sgn(i_c)(1 - S_c)
\end{aligned}
\tag{3}
$$

where S_a, S_b, and S_c are switching states of the switches ($S_a, S_b, S_c = 1$ when switches are on and $S_a, S_b, S_c = 0$ when switches are off).

III. PROPOSED CONTROL STRATEGY FOR THE VIENNA RECTIFIER

In [1], [2], [4], the complete hysteresis control diagram for the Vienna rectifier is presented. The outer loop controls the output voltage by adjusting the amplitude of the reference

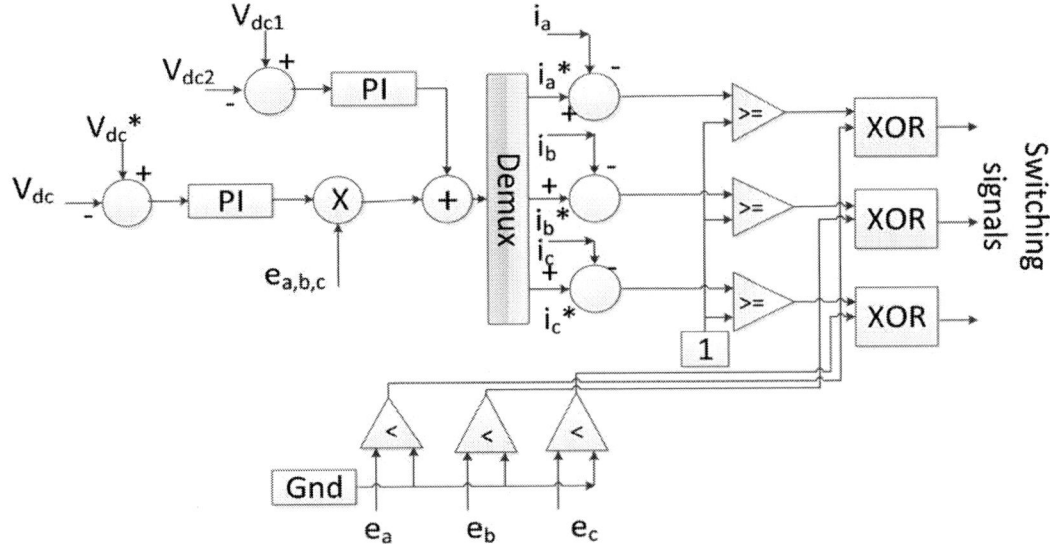

Fig. 5: Existing hysteresis control diagram of the Vienna rectifier

current generated using the three phase voltages. The inner zero-current component is calculated by the inner PI loop and added in the reference current. Three comparators are used to compare the measured currents and the reference currents. The complete block diagram of the hysteresis control is shown in Fig 5.

This paper proposes a simple PI control based vector control technique that controls and balances in the output voltages. The space vector technique is replaced with a simple single carrier based switching method. The $d-q$ model of the rectifier can be expressed as [9]:

$$v_d = e_d - \left(Ri_d + \frac{Ldi_d}{dt}\right) + w_s Li_q \qquad (4)$$

$$v_q = e_q - \left(Ri_q + \frac{Ldi_q}{dt}\right) - w_s Li_d \qquad (5)$$

where e_d and e_q are the $d-$axis and $q-$axis voltages, respectively.

The schematic of the vector control technique for the Vienna rectifier is shown in Fig 6. The outer loop is designed to control the output voltage of the Vienna rectifier. The voltage control loop gives the reference current, i_d. The reference current, i_q is set zero for unity power factor operation. The measured line currents are converted into $d-q$ frame and compared with the respective reference values and the error signals are fed to PI controllers. The compensation terms are added to the output of the PI controller as highlighted by eqn.(4). Then, zero sequence component is added for the output voltage balance.

A. Balancing of the Output Voltages

The output voltage of the Vienna rectifier, V_{dc} is divided into two partial voltages as V_{dc1} and V_{dc2}. The controller should ensure the balance of the partial voltages to reduce the switch voltage stress to half of the DC Link voltage.

The mathematical model for the DC link side can be expressed as:

$$C_1 \frac{V_{dc1}}{dt} + C_2 \frac{V_{dc2}}{dt} = i_1 + i_2 - 2i_0 = \frac{P_{rated}}{V_{dc}} - 2i_0 \qquad (6)$$

$$C_1 \frac{V_{dc1}}{dt} - C_2 \frac{V_{dc2}}{dt} = i_1 - i_2 = i_M \qquad (7)$$

The current through positive side of the DC bus, i_1 and the current through negative side of the DC bus, i_2 are the addition of the respective side (top and bottom) diode currents and can be expressed as:

$$i_1 = i_{DFA+} + i_{DFB+} + i_{DFC+} \qquad (8)$$

$$i_2 = i_{DFA-} + i_{DFB-} + i_{DFC-} \qquad (9)$$

The currents flowing through the top and bottom diodes depend on input current polarity and the switching states of the Vienna rectifier and are expressed in eqns. (10-11). The current flow from the top diodes during negative current polarity and the current through the bottom diodes during positive current polarity are zero [14].

$$i_{DF(A,B,C)+} = (1 - S_{a,b,c})i_{a,b,c} \qquad i_a > 0 \qquad (10)$$

$$i_{DF(A,B,C)-} = -(1 - S_{a,b,c})i_{a,b,c} \qquad i_a < 0 \qquad (11)$$

Finally, the sum of the DC link currents and mid-point current are derived as:

$$i_1 + i_2 = \sum_{x=a,b,c} (1 - S_x)sgn(i_x)i_x \qquad (12)$$

$$i_M = \sum_{x=a,b,c} (i_x S_x) \qquad (13)$$

The average duty cycle, $(d^*_{a,b,c})$ per phase can be expressed as:

$$d^*_{a,b,c} = \sum_{x=a,b,c} (1 - S_x)sgn(i_x)i_x = d'_{a,b,c} + d'_0 \qquad (14)$$

978-1-4673-9551-9/16 $31.00 © 2016 IEEE

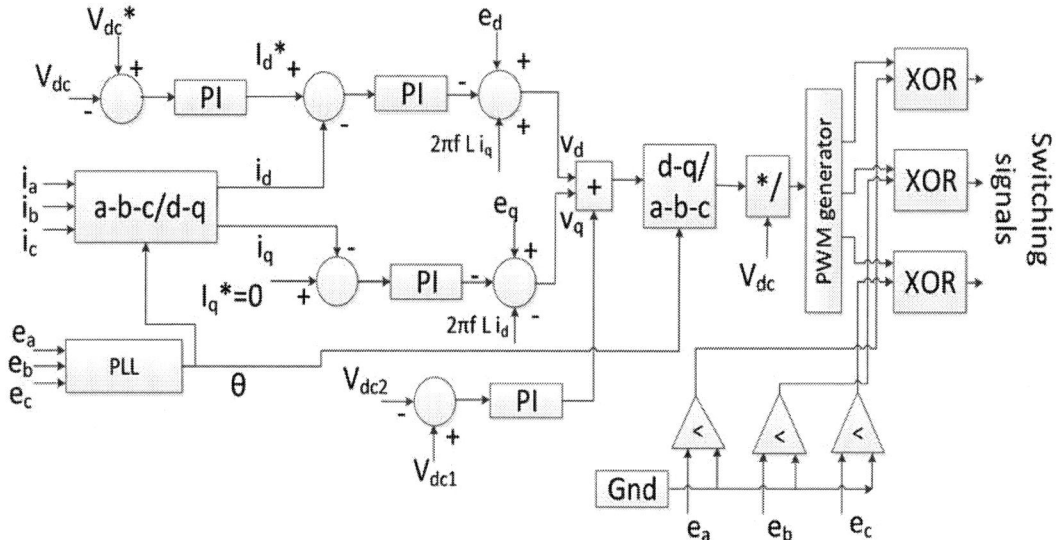

Fig. 6: Proposed vector control diagram of the Vienna rectifier

where $d'_{a,b,c}$ is sinusoidal component to generate the duty cycle and $d'_0 = (d_0 + \Delta d_0)$ is zero-sequence component.

In order to balance the output voltages, the value of capacitances are taken to be equal ($C_1 = C_2 = C$). For maintaining the DC Link voltages, the mid-point current should be as minimum as possible. Therefore, the difference between the output capacitor voltages can be simplified as [14]:

$$\frac{dV_{dc}}{dt} = \frac{i_M}{C} = \frac{6i_M \Delta d_0}{\pi} \quad (15)$$

where dV_{dc} is the difference in the partial voltages.

The centre point current depends on the switching states of the switches and the magnitude of central point current decides the balancing of the capacitor voltages. The converter switching states are calculated from the difference between the reference and measured currents. The zero sequence component is added to v_d and v_q to guarantee zero current injection to the neutral point. The proposed method uses simple PI control for the generation of zero sequence component. Carrier based PWM technique is proposed in the paper. The three reference control signals are compared with the sawtooth/triangular waveform. The output of the comparator is XORed with the phase of the line currents as shown in Fig 6. The mathematical calculations required for determining different sectors in SVM is eliminated in the proposed carrier based technique. The offset for zero sequence component can be added in the modulation signals after converting the signal into $a - b - c$ frame before comparing them with the carrier waveform.

IV. SIMULATION RESULTS

The Vienna rectifier is simulated using simulation software MATLAB. Various simulation parameters are listed in Table I.

Three phase balanced input voltages are applied to the Vienna rectifier. The vector control method manages to reduce

TABLE I: Simulation parameters for the Vienna rectifier

Parameter	Value
Phase Voltage	45 V
Rated Power	1 kW
Output voltage	160 V
Line Frequency	60 Hz
Switching frequency	40 kHz
Line inductance	5 mH
Line resistance	0.4 ohm
Source frequency	50 Hz
Output resistance	25 ohm
Output capacitance	340 uF

total harmonic distortion (THD) below 5%. The source side current waveforms showing unity power factor operation is illustrated in Fig 7. The Vienna rectifier is a three-level converter which is justified by the terminal voltage waveform as shown in Fig 8. Three-level operation reduces the voltage stress of the active/passive devices to half of the DC link voltage. The reduction in voltage stress allows to use low rating devices which eventually reduces switching and conduction losses of the rectifier.

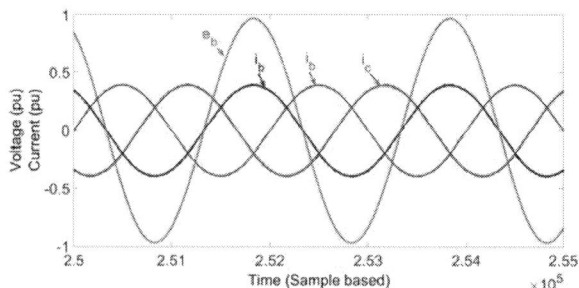

Fig. 7: Input side currents (scaled down by 2) and voltage showing unity power factor operation using vector control

The output side transient response of the converter is

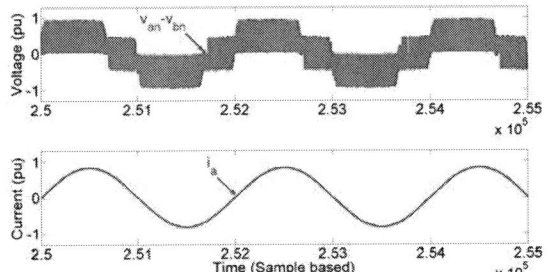

Fig. 8: Input side current and terminal voltage of the Vienna rectifier

presented in Fig 9. The converter tracks the reference voltage balancing the output voltages. The error between the output voltages is observed around 3-4 V. The use of various computational blocks in vector control method increases the settling time of the output parameters but this method regulates the magnitude of overshoot and undershoot during transient and reduces jitter during light load condition.

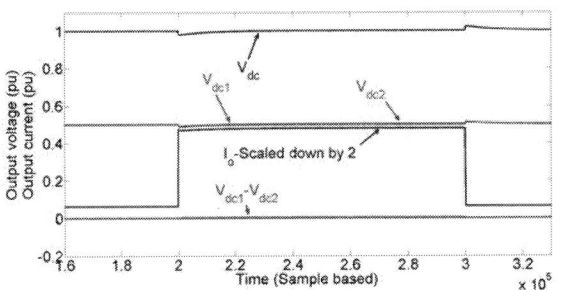

Fig. 9: Transient response of the Vienna rectifier using vector control

The performance of the proposed control system is also evaluated under supply imbalanced conditions. The magnitude of the supply is varied from 1.2 pu to 0.8 pu as exhibited in Fig 11. However, this control method maintains current THD below 5% and balances output DC voltages. Two phase currents and voltages with unity power factor operation during voltage imbalance is shown in Fig 11.

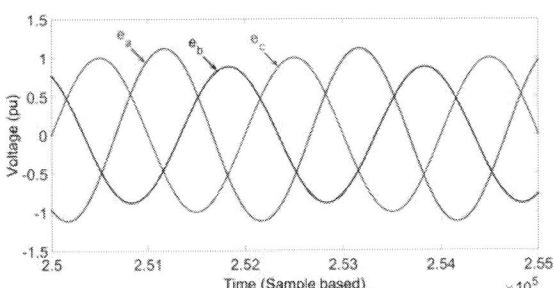

Fig. 10: Input voltages to the rectifier showing amplitude imbalance

The Vienna rectifier transforms 45 V into 160 V and output capacitors share 80 V each. The proposed control method maintains sinusoidal current in the input side and balances the voltages in the output side.

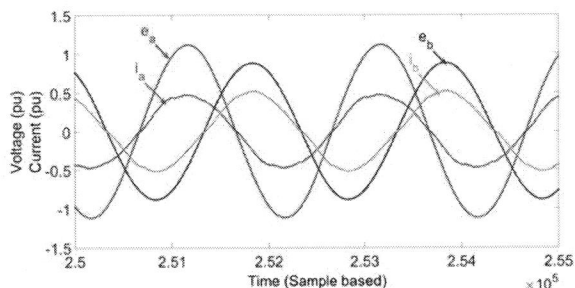

Fig. 11: Input side currents (scaled down by 2) and voltages during imbalance showing unity power factor operation using vector control

V. EXPERIMENTAL RESULTS

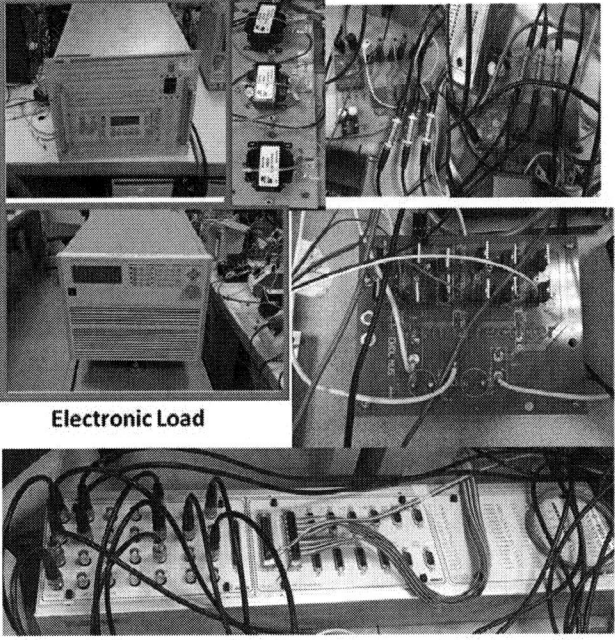

Fig. 12: A laboratory prototype of the Vienna rectifier

A 1 kW laboratory prototype of the Vienna rectifier is designed and tested inside the laboratory environment as shown in Fig 12. The overall ratings of the hardware prototype are listed in Table II. The tuned control parameters are provided in Table III.

Input voltages of 45 V (L-N) is applied to the converter and the reference DC Link voltage of 160 V is set at the output side. The converter is tested with the conventional hysteresis control and the proposed vector control to observe the comparative performance of the proposed controller method.

In Fig 13 and Fig 14 input side currents along with one input voltage are shown for vector control and hysteresis control, respectively. The current waveforms for the proposed vector control method are sinusoidal with less current THD . The current THD is around 4% and 6% for vector control

978-1-4673-9551-9/16 $31.00 © 2016 IEEE   13

TABLE II: Ratings and the list of the components used in designing the hardware prototype of the Vienna rectifier

Major components	Rating/Size	Quantity
Power	1 kW	
Input Voltage	45 V	
Output Voltage	160 V	
Input filter inductors	5 mH	3
IGBT, SPW20N60C3	650V, 20A	3
SiC Diode, CVFD20065A	650V, 20A	18
Filter Capacitors	340 μF	2

TABLE III: Control parameters for hardware experiment

	Control paramters	
	Kp	Ki
Outer voltage loop	1	8
Current loop for I_d	120	20
Current loop for I_q	100	10
Voltage balance loop	0.08	0

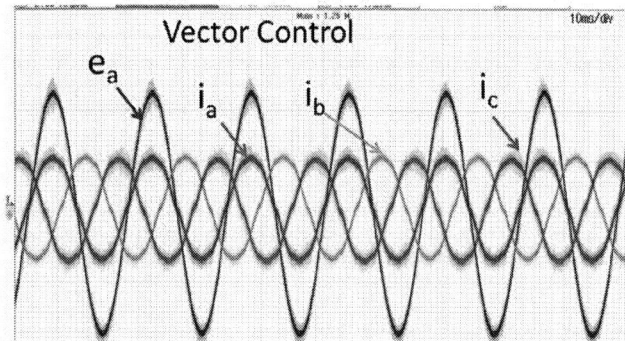

Fig. 13: Input side currents and voltage showing unity power factor operation using vector control, i_a (10 A/div), i_b (10 A/div), i_v (10 A/div), e_a (40 Volts/div)

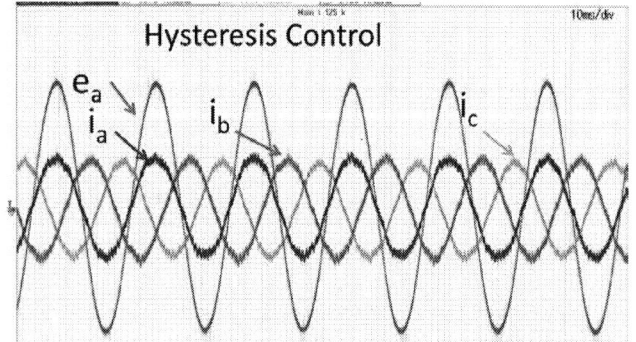

Fig. 14: Input side currents and voltage showing unity power factor operation using hysteresis control, i_a (10 A/div), i_b (10 A/div), i_v (10 A/div), e_a (40 Volts/div)

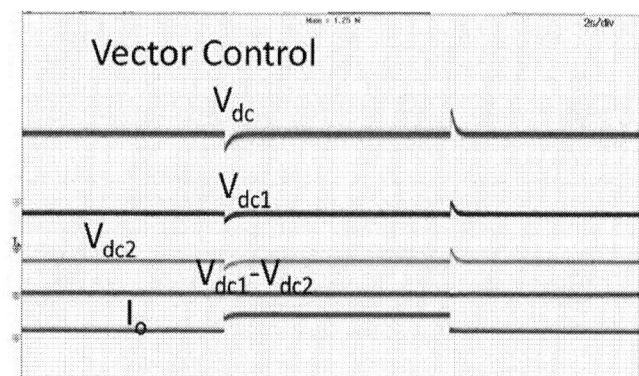

Fig. 15: Transient responses of the rectifier using vector control, V_{dc} (120 V/div), V_{dc1} (120 V/div), V_{dc2} (120 V/div), I_o (10 Volts/div)

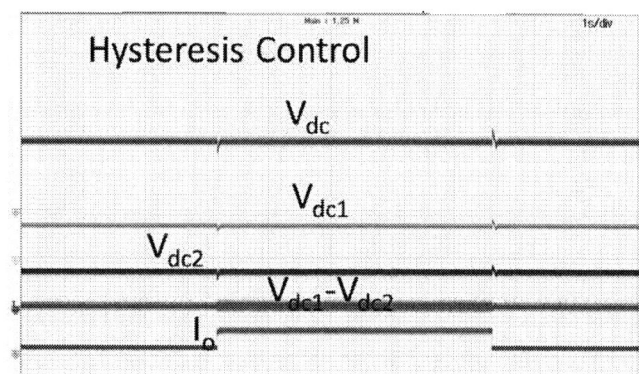

Fig. 16: Transient responses of the rectifier using hysteresis control, V_{dc} (120 V/div), V_{dc1} (120 V/div), V_{dc2} (120 V/div), I_o (10 Volts/div)

and hysteresis control, respectively. The output side load transients are presented in Fig 15 (for vector control) and Fig 16 (for hysteresis control). The proposed control scheme uses the current/voltage transformation blocks which consume time for mathematical computation. Hence, the response of the proposed control method is slightly slower than that of the existing hysteresis control method. However, jitter in the output voltage (at lower loads) is lower in case of the vector control method than that of the conventional method. For lower load, the proposed vector control method maintains the sinusoidal current with better THD than that of hysteresis control as exhibited in Fig 17 and Fig 18.

The proposed control method uses simplified voltage balancing loop to equalize the capacitor voltages. This balance in the capacitor voltages reduces the device stress to half the DC link voltage. The terminal voltage shown in Fig 19 confirms the three-level operating and voltage balance between the capacitor voltages.

The converter is tested with imbalance supply. The magnitude of the supply voltage signals are modified using programmable power source. The unbalanced voltage signals are presented in Fig 20. The input side currents along with unbalanced voltages are exhibited in Fig 21. Despite the imbalance in the input voltage, three phase currents are balanced and the mid-point current is minimal that balances the output voltages. Similarly, the converter is operated with

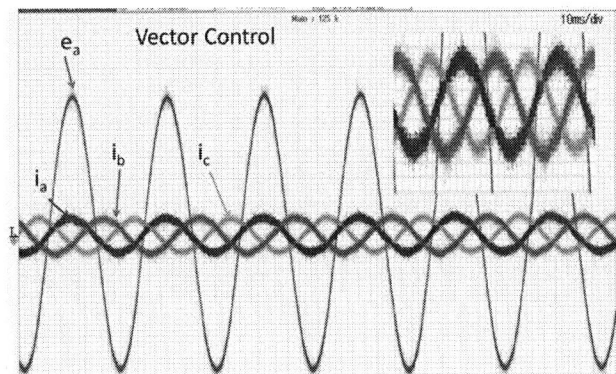

Fig. 17: Input side currents and voltage showing for proposed vector control for lighter load, i_a (10 A/div), i_b (10 A/div), i_c (10 A/div), e_a (70 Volts/div)

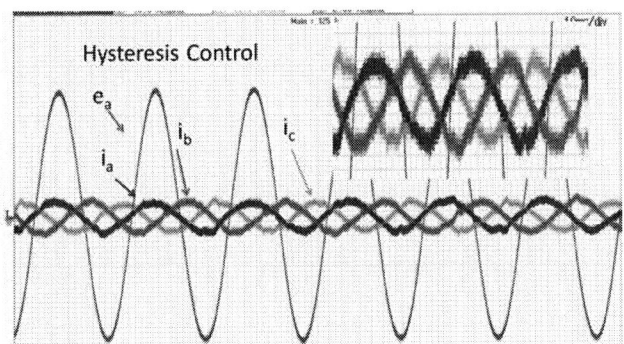

Fig. 18: Input side currents and voltage showing for hysteresis control for lighter load, i_a (10 A/div), i_b (10 A/div), i_c (10 A/div), e_a (70 Volts/div)

distorted grid where third and fifth harmonics are injected in the generation side voltages. The controller manages to control the current harmonic below 6% as shown in Fig 22.

For the rated condition, the converter transforms three phase AC (45 V) into DC voltage of 160 V. The variations of the current THD with respect to loads for the proposed method and the conventional method are shown in Fig 23. The proposed method exhibits better current harmonics for wide range of load with finely tuned filter (small size). The same experiment is repeated with SVM switching technique. The input side

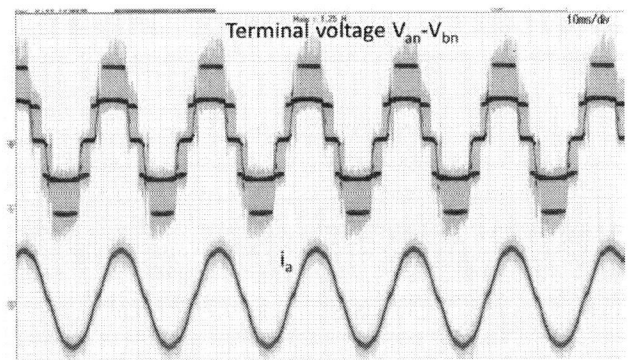

Fig. 19: Terminal voltage and line of the Vienna rectifier using vector control method, $v_{ab} = v_{an} - v_{bn}$ (80 Volts/div), i_a (10 A/div)

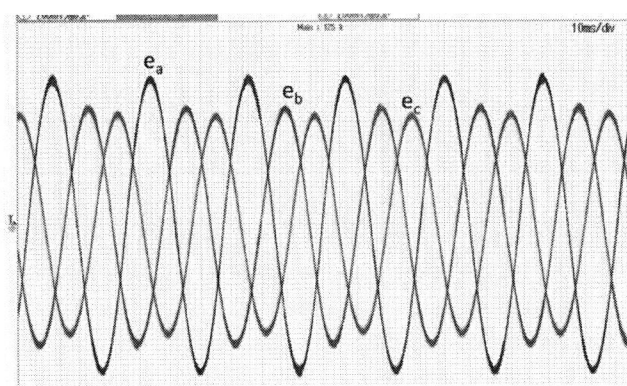

Fig. 20: Imbalance voltages in the input side of the converter, e_a (70 V/div), e_b (70 Volts/div), e_c (70 V/div)

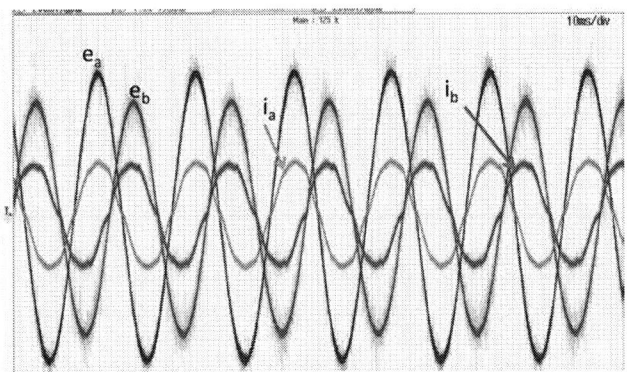

Fig. 21: Input side currents and voltages showing during imbalance condition, i_a (10 A/div), i_b (10 A/div), e_a (70 V/div), e_b (70 Volts/div)

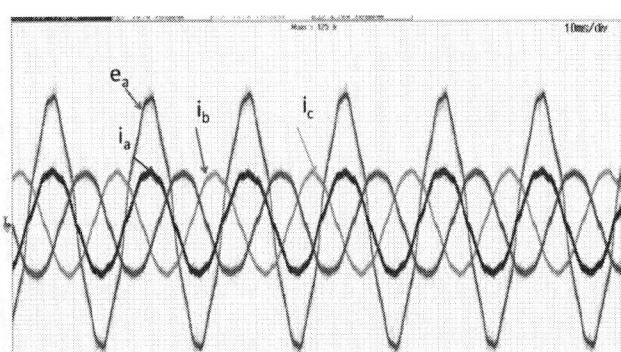

Fig. 22: Input side currents and voltage showing for proposed vector control with third and fifth harmonics injected into the voltage, Input side currents and voltage showing for hysteresis control for lighter load, i_a (10 A/div), i_b (10 A/div), i_c (10 A/div), e_a (70 Volts/div)

current THD for SVM and carrier based technique is almost equal ranging from 4.5% to 6.5% for the complete load profile.

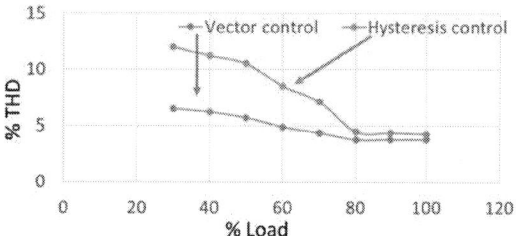

Fig. 23: Current THDs of the input side currents for the Vienna rectifier

VI. CONCLUSION

A simplified vector control method for controlling the Vienna rectifier is proposed in this paper. Mathematical modeling of the proposed control strategy is presented in the paper. The detailed mathematical analysis of the Vienna rectifier is presented. The proposed method provides better current THD and steady voltage transients with less jitter than that of the conventional hysteresis control method. The complex SVM method is replaced with simplified single carrier based PWM technique. The carrier based technique provides slightly better result than SVM switching method. The proposed method is experimentally validated for wide range of load and input voltage variations. The current THD for different loads is found to be within industrial standards. For a rated power of 1 kW, 45 V of phase voltage is converted into 160 V DC link voltage using the Vienna rectifier. The measured current THD at rated load is 3.7% .

REFERENCES

[1] J. Minibock, F. Stogerer, and J. Kolar, "A novel concept for mains voltage proportional input current shaping of a vienna rectifier eliminating controller multipliers. i. basic theoretical considerations and experimental verification," in *Applied Power Electronics Conference and Exposition, 2001. APEC 2001. Sixteenth Annual IEEE*, vol. 1, 2001, pp. 582–586 vol.1.

[2] T. Soeiro and J. Kolar, "Analysis of high-efficiency three-phase two- and three-level unidirectional hybrid rectifiers," *Industrial Electronics, IEEE Transactions on*, vol. 60, no. 9, pp. 3589–3601, Sept 2013.

[3] J. Adhikari, Prasanna, G. Ponraj, and S. Panda, "Power conversion system for low power high altitude wind power generating system," in *Power Electronics and ECCE Asia (ICPE-ECCE Asia), 2015 9th International Conference on*, June 2015, pp. 637–644.

[4] J. Kolar and F. C. Zach, "A novel three-phase utility interface minimizing line current harmonics of high-power telecommunications rectifier modules," *Industrial Electronics, IEEE Transactions on*, vol. 44, no. 4, pp. 456–467, Aug 1997.

[5] J. Kolar and T. Friedli, "The essence of three-phase pfc rectifier systems 2014;part i," *Power Electronics, IEEE Transactions on*, vol. 28, no. 1, pp. 176–198, Jan 2013.

[6] T. Friedli, M. Hartmann, and J. Kolar, "The essence of three-phase pfc rectifier systems 2014;part ii," *Power Electronics, IEEE Transactions on*, vol. 29, no. 2, pp. 543–560, Feb 2014.

[7] T. Viitanen and H. Tuusa, "Space vector modulation and control of a unidirectional three-phase/level/switch vienna i rectifier with lcl-type ac filter," in *Power Electronics Specialist Conference, 2003. PESC '03. 2003 IEEE 34th Annual*, vol. 3, June 2003, pp. 1063–1068 vol.3.

[8] N. Bel Haj Youssef, K. Al-Haddad, and H. Kanaan, "Large-signal modeling and steady-state analysis of a 1.5-kw three-phase/switch/level (vienna) rectifier with experimental validation," *Industrial Electronics, IEEE Transactions on*, vol. 55, no. 3, pp. 1213–1224, March 2008.

[9] L. Hang, H. Zhang, S. Liu, X. Xie, C. Zhao, and S. Liu, "A novel control strategy based on natural frame for vienna-type rectifier under light unbalanced-grid conditions," *Industrial Electronics, IEEE Transactions on*, vol. 62, no. 3, pp. 1353–1362, March 2015.

[10] L. Hang and M. Zhang, "Constant power control-based strategy for vienna-type rectifiers to expand operating area under severe unbalanced grid," *Power Electronics, IET*, vol. 7, no. 1, pp. 41–49, January 2014.

[11] H. Teshnizi, A. Moallem, M.-R. Zolghadri, and M. Ferdowsi, "A dual-frame hybrid vector control of vector modulated vienna i rectifier for unity power factor operation under unbalanced mains condition," in *Applied Power Electronics Conference and Exposition, 2008. APEC 2008. Twenty-Third Annual IEEE*, Feb 2008, pp. 1402–1408.

[12] N. Bel Haj Youssef, K. Al-Haddad, and H. Kanaan, "Implementation of a new linear control technique based on experimentally validated small-signal model of three-phase three-level boost-type vienna rectifier," *Industrial Electronics, IEEE Transactions on*, vol. 55, no. 4, pp. 1666–1676, April 2008.

[13] P. Ide, F. Schafmeister, N. Frohleke, and H. Grotstollen, "Enhanced control scheme for three-phase three-level rectifiers at partial load," *Industrial Electronics, IEEE Transactions on*, vol. 52, no. 3, pp. 719–726, June 2005.

[14] R. Lai, F. Wang, R. Burgos, D. Boroyevich, D. Jiang, and D. Zhang, "Average modeling and control design for vienna-type rectifiers considering the dc-link voltage balance," *Power Electronics, IEEE Transactions on*, vol. 24, no. 11, pp. 2509–2522, Nov 2009.

Compensation of Neutral Point Deviation in 3-level NPC Converter under Unbalanced Grid Conditions

Kyungsub Jung and Yongsug Suh

Dept. of Elec. Eng., Smart Grid Research Center, Chonbuk Nat'l Univ., Jeonju, Korea

Abstract— This paper presents a neutral point deviation compensating control algorithm applied to a 3-level NPC converter. The neutral point deviation is analyzed with a focus on the current flowing out of or into the neutral point of the dc-link. Based on the zero sequence components of the reference voltages, this paper analyzes the neutral point deviation and balancing control for 3-level NPC converter. An analytical method is proposed to calculate the injected zero sequence voltage for neutral point balancing based on average neutral current. This paper also proposes a control scheme compensating for the neutral point deviation under generalized unbalanced grid operating conditions. The positive and negative sequence components of the pole voltages and ac input currents are employed to accurately explain the behavior of 3-level NPC converter. Simulation and experimental results for a test set up of 30kW are shown to verify the validity of the proposed algorithm.

I. INTRODUCTION

Multi-level converters are widely used in high-power applications such as motor drives, utility applications, and, most recently, in wind generation systems. Extensive research has been carried out/on multilevel topologies, modulation, and control strategies [1]. Multilevel converters can provide more than two voltage levels at the output. As a result, the voltage and current waveforms generated have lower Total Harmonic Distortion (THD). Consequently, high voltages can be handled on both the dc and ac sides of the converter [2].

The multilevel converter topology that is most extensively applied at present is the Neutral-Point-Clamped (NPC) converter, which is a 3-level NPC Voltage Source Converter (VSC) in Fig. 1. One of the essential problems of the 3-level NPC converters is that how to keep the voltage of dc-link capacitors balanced, in other words, keep the Neutral-Point (NP) potential stable and suppress the ripple. If the NP potential is not controlled effectively, the output voltage of the converter would deviate from the reference value; moreover, the devices and equipment might be damaged [3]. In practical operations of 3-level NPC VSC, the NP potential variation, in other words, the unbalanced dc-link voltages often lead to a frequent trip of converters due to the over-voltage of either upper dc-link capacitor or lower dc-link capacitor. The control strategies of NP potential that have appeared in the literatures can be grouped according to the Pulse Width Modulation (PWM) algorithm utilized.

If Space Vector PWM (SVPWM) is used, the voltage vectors can be classified into four categories by their magnitude: zero, small, middle, and large vectors. Then, the relationship between NP potential and each switching state vector can be analyzed. It is known that the zero vectors and large vectors have no effect on NP potential, but the middle vectors and small vectors can have an influence on it. It is noticed that there are two switching states (positive and negative) that have reverse action (charging or discharging) on NP potential for one small vector.

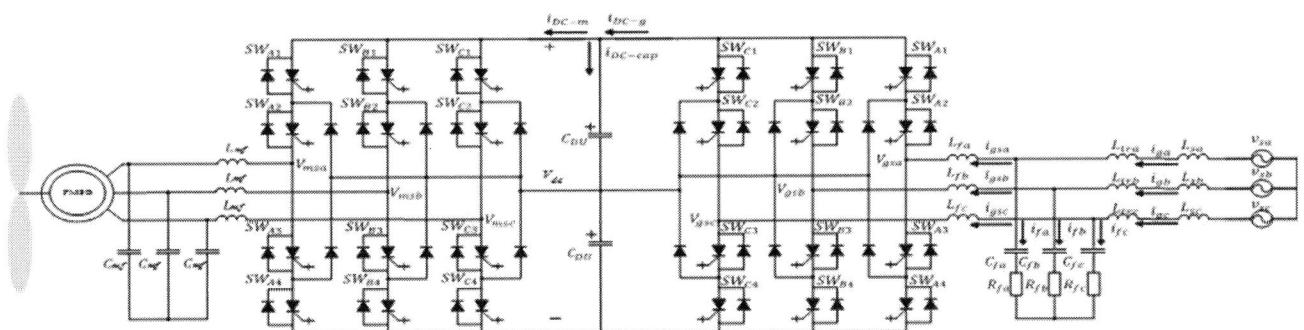

Fig. 1. PMSG wind turbine with a back-to-back 3-Level NPC VSC

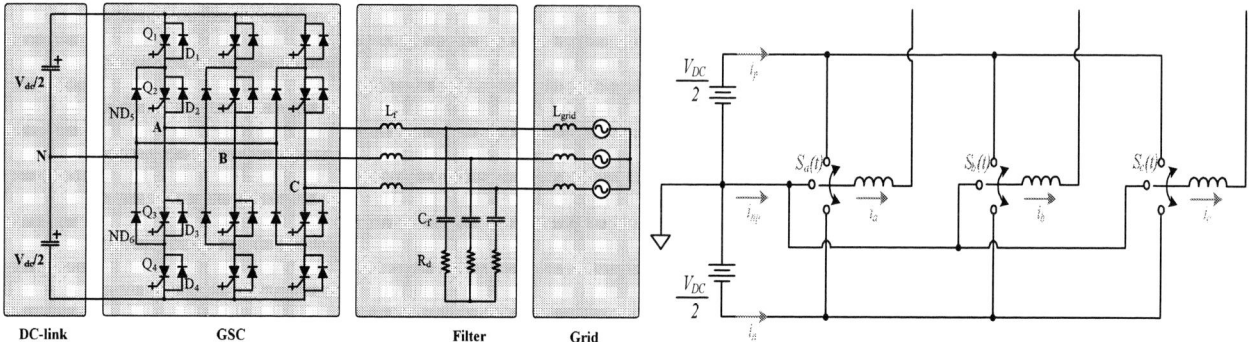

Fig. 2. 3-level NPC voltage source converter and simplified topology using SPTT (Single-pole Triple-Throw)

Therefore, the main task is to adjust the dwell time between the duplicate switching states of small vectors [4]–[9]. In many solutions of the SVPWM strategies for the 3-level NPC converter, one or two switching sequences are strictly assigned to specific subsectors [10], [11]. The control strategies of the dc-link voltage balance are based on the change of switching sequences depending on the unbalanced dc-link voltage [12]-[15]. When Carrier-Based PWM (CBPWM) is used, the control of the NP potential can be considered as the problem of identifying the zero sequence voltage. The zero sequence voltage added to the reference voltages does not change the output line voltages, but influences the switching states and of course the NP potential. The NP potential variation caused by the injected zero sequence voltage has been studied, and some algorithms to keep the NP potential balancing by injecting the appropriate zero sequence voltage were presented in [16]-[20].

Among many possible causes of NP deviation in 3-level NPC converters, unbalanced grid supply can generate NP deviation of significant level. The impact of unbalanced grid input on the NP deviation and suitable compensating control strategies have been paid less attention considering its importance in a practical operation. Also, none of previous works have deeply analyzed the relationship between unbalanced grid condition and NP deviation under the wide range of unbalanced condition within a 3-level NPC converter.

This paper proposes a CB-PWM strategy for a 3-level NPC converter with a zero sequence voltage injection. In this paper, the variation of the neutral point potential is analyzed on the basis of an average current flowing out of or into the neutral point. It is shown that the zero sequence voltage of the NPC-VSI output has an important influence upon the neutral potential variation. The principle of the neutral point potential control by adding a suitable zero sequence voltage is also described. Moreover, this paper proposes NP deviation control scheme for the 3-level NPC converter under generalized unbalanced grid operating conditions. The positive and negative sequence components of the pole voltages and ac input currents are employed to accurately explain the behavior of 3-level NPC converter and its impact on NP deviation.

This paper is structured as follows. In Section 2, the model of neutral point deviation and neutral current are constructed. Section 3 performs an analysis of neutral point deviation under generalized unbalanced grid input conditions. Simulation result verifying the proposed model under unbalanced conditions is explained in Section 4. Finally, Section 5 provides the experimental result to validate the proposed model and control algorithm. The proposed CB-PWM compensating strategy is verified by the real-time simulator of Typhoon HILS. This test equipment provides a real time operating environment of DSP (TI TMS320F28335) so that the effective verification of the proposed compensating algorithm is made possible without performing a time-consuming real hardware experiment.

II. RELATIONSHIP BETWEEN NEUTRAL POINT DEAVIATION AND AVERAGE NEUTRAL CURRENT

In this section, the simplified model of neutral current and zero sequence voltage are implemented. Figure 2 illustrates the simplified 3-level NPC converter using Single-Pole Triple-Throw (SPTT) switches. The operation of this simplified 3-level NPC converter using SPTT switches can be better described by employing the corresponding switching functions as defined in (1). In general, converter output voltage with respect to the mid-point of dc-link can be represented by switching functions as in (2).

$$S_x = [-1,0,1] \,\&\, x = a,b,c \tag{1}$$

$$v_x = S_x(t) \times \frac{V_{dc}}{2}, x = a,b,c \tag{2}$$

Neutral current (i_{np}) can be correlated with phase currents and switching functions as shown in (3). Therefore, by applying (2) into (3), neutral current can be explained by converter output voltages and phase currents as in (4).

$$
\begin{aligned}
i_{np}(t) &= [1-|S_a|] \cdot i_a + [1-|S_b|] \cdot i_b + [1-|S_c|] \cdot i_c \\
&= -(|S_a| \cdot i_a + |S_b| \cdot i_b + |S_c| \cdot i_c)
\end{aligned} \tag{3}
$$

$$i_{np} = -\frac{2}{V_{dc}}(|v_a| \cdot i_a + |v_b| \cdot i_b + |v_c| \cdot i_c) \tag{4}$$

978-1-4673-9551-9/16 $31.00 © 2016 IEEE

In general, converter output voltages can be approximated as PWM waveforms having three levels of $V_{dc}/2$, 0, and $-V_{dc}/2$. These converter output voltages of PWM waveforms can be effectively broken down into three components; fundamental frequency component, harmonic frequency component, and zero sequence component of non-fundamental frequency. These three components are described in (5). The zero sequence component of non-fundamental frequency component is the one which is added to compensate for the neutral point deviation in the proposed control scheme. In this paper, the harmonic frequency components are neglected in the derivation of neutral current model as shown in (6), i.e. localized average model. This assumption does not lead to a significant error in controlling the neutral point deviation since the higher-order harmonic components are effectively attenuated and do not contribute to the neutral point deviation in most of application cases.

$$v_x = v_{x\_fund} + v_{x\_harmonics} + v_{comp}, x = a, b, c \qquad (5)$$

$$v_{x\_harmonics} = 0, x = a, b, c \qquad (6)$$

In Fig. 2, it is readily understood that as long as the neutral current (i_{np}) is kept zero the voltages of upper and lower dc-link capacitors become equal to each other, i.e. zero neutral point deviation, because upper and lower dc-link capacitors charge or discharge at the same current level. In a typical modulation technique, it is inevitable to have non-zero neutral current since some switching vector states of 3-level NPC converter cause the SPTT switches to be connected to the neutral point. However, even in the existence of these switching vectors having SPTT switches connected to the neutral point, the localized average model of neutral current can still be controlled to be zero. This is made possible since the higher-order harmonic components of neutral current is successively neglected in the proposed model as described in (6). As a result, the neutral current being zero is a necessary condition to maintain the voltage balance in upper and lower dc-link capacitors.

It should be also pointed out that this zero neutral current is not a sufficient condition to achieve the dc-link voltage balance. In other words, even under the condition of zero neutral current, the upper and lower dc-link voltage may become unbalanced due to secondary effects such as mismatched upper and lower dc-link capacitance values, unbalanced initial charging condition of capacitor voltages, etc. In this paper these secondary effects are not considered in the modeling of neutral point deviation in order to simplify the dc-link imbalance problem and further focus on the influence of ac grid imbalance upon the neutral point deviation at the dc-link.

Applying (5) and (6) into (4) and also employing the condition of zero neutral current leads to (7). The zero sequence component of non-fundamental frequency component (V_{comp}) obtained from (7) can be utilized in a switching modulation scheme in order to maintain the

condition of zero neutral current, i.e. zero neutral point deviation. In (7), $sgn(V_x)$ refers to a signum function. This is a function that extracts the sign of a real number V_x, yielding -1 if V_x is negative, +1 if V_x is positive, or 0 if V_x is zero. It is noted from (7) that this information of converter output voltages plays an important role in determining the compensating voltage of zero sequence. Additional steps are required to extract the sign information of converter output voltages and solve for the compensating voltage of zero sequence in an iterative way.

III. ANALYSIS OF NEUTRAL POINT DEVIATION UNDER GENERALIZED UNBALANCED 3-PHASE GRID CONDITION

Under unbalanced ac grid conditions, the three-phase input currents flowing through the input filter stage of the grid-side converter also become unbalanced if any proper compensating control measures are not employed. This means that the ac input currents start to contain the negative sequence component under unbalanced grid input. This negative sequence component of ac input current further deteriorates the neutral point deviation of 3-level NPC converter on top of typical causes of dc-link imbalance such as the mismatch of upper and lower dc-link capacitance, switching dead time, asymmetric modulation effects, etc. Therefore, in order to correctly explain the behavior of neutral point deviation and neutral point current under the ac grid imbalance, the negative sequence components of ac input current as well as converter output voltage at the pole of converter should be incorporated into the description of neutral current in (4).

The symmetric components of ac input current and converter output voltage are defined as shown in (8). In (8) subscript of p, n, and o represent the positive, negative, and zero sequence components, respectively. Three-wire configuration of grid side converter naturally assumes the nullified zero sequence component of ac input current in (8). Applying (8) into the description of neutral current of (4) leads to (9). In (9), the higher-order harmonic components are neglected due to the localized average model considered in this paper. In addition, the compensating voltage of zero sequence (V_{comp}) is also neglected in order to analyze the behavior of neutral current under the condition of unbalanced ac grid.

In this paper, the amplitude and phase angle of each symmetric component of converter output voltage and ac input current are defined according to (10) and (11), respectively. After integrating (10) and (11) into (9), the description of neutral current becomes as in (12). The coefficients employed for the simpler expression of (12) are defined as in (13), (14), and (15). It is noted in (12) that the description of neutral current depends on the sign of converter output voltages. Therefore, further development of (12) requires the information of sign of each converter output voltage. In general, there can be eight different combinations of sign values for the converter output voltages. These eight cases are described in Table I for the sake of readers' convenience. It can be readily understood that these sign combinations are closely related with the variable power factor operation of grid-side converter. Under the particular condition in Table I, the description of neutral current in (12) can be further simplified as shown in (16). In (16), the values of coefficients (X, Y, Z) depend on the particular sign combination in Table I. In this paper, the values of coefficients are given for the case of 2 in Table I and summarized in (17). The description of neutral current given in (16) provides useful information in understanding the relationship between the ac grid imbalance condition and neutral potential deviation of 3-level NPC converter.

TABLE I
SIGN VALUES OF CONVERTER OUTPUT VOLTAGE

Case	$\mathrm{sgn}(v_{a\_fund})$	$\mathrm{sgn}(v_{b\_fund})$	$\mathrm{sgn}(v_{c\_fund})$
1	+	+	+
2	+	+	-
3	+	-	+
4	-	+	+
5	+	-	-
6	-	+	-
7	-	-	+
8	-	-	-

Unbalanced depth of grid is quantitatively described by employing a newly defined *IF* (*Imbalance Factor*) in this paper. This IF is defined according to (18) which physically means the ratio of negative sequence component against the total amplitude of grid voltage. In general, unbalanced ac grid type is classified into four different cases as shown in Fig. 3. The newly defined IF can effectively describe the unbalanced depth in all four cases. As an example, the correlation of newly defined IF and the per unit value of single-phase sag voltage of Type B is given in Table II.

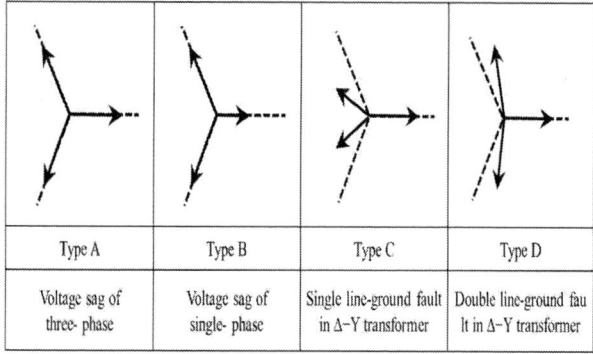

Type A	Type B	Type C	Type D
Voltage sag of three- phase	Voltage sag of single- phase	Single line-ground fault in Δ–Y transformer	Double line-ground fau lt in Δ–Y transformer

$$\text{Imbalance Factor} = \sqrt{\frac{E_n^2}{E_p^2 + E_n^2 + E_o^2}} \quad (18)$$

TABLE II
IF ACCORDING TO UNBALANCED DEGREE OF TYPE B

Voltage sag (pu)	1	0.95	0.9	0.8
IF(%)	0	1.6	3.4	7.1

Using the model of neutral current as described in (12)-(17), the effect of unbalanced ac grid on the neutral point deviation can be analyzed. In this paper, the proposed model and control scheme is verified for the example case of PMSG wind turbine of 2.7MW as illustrated in Fig. 1.

$$v_{comp} = -\frac{\mathrm{sgn}(v_a)\cdot v_{a\_fund}\cdot i_a + \mathrm{sgn}(v_b)\cdot v_{b\_fund}\cdot i_b + \mathrm{sgn}(v_c)\cdot v_{c\_fund}\cdot i_c}{\mathrm{sgn}(v_a)\cdot i_a + \mathrm{sgn}(v_b)\cdot i_b + \mathrm{sgn}(v_c)\cdot i_c}, \quad \text{when } i_{np} = 0 \quad (7)$$

$$\begin{cases} v_{a\_fund} = v_{ap} + v_{an} + v_o, \quad v_{b\_fund} = v_{bp} + v_{bn} + v_o, \quad v_{c\_fund} = v_{cp} + v_{cn} + v_o, \\ i_a = i_{ap} + i_{an}, \quad i_b = i_{bp} + i_{bn}, \quad i_c = i_{cp} + i_{cn} \end{cases} \quad (8)$$

$$i_{np}(t) = -\frac{2}{V_{dc}}\begin{Bmatrix} \mathrm{sgn}(v_{a\_fund})\cdot(v_{ap} + v_{an} + v_o)\cdot(i_{ap} + i_{an}) + \mathrm{sgn}(v_{b\_fund})\cdot(v_{bp} + v_{bn} + v_o)\cdot(i_{bp} + i_{bn}) \\ + \mathrm{sgn}(v_{c\_fund})\cdot(v_{cp} + v_{cn} + v_o)\cdot(i_{cp} + i_{cn}) \end{Bmatrix}, \quad \text{when } v_{comp} = 0 \quad (9)$$

$$\begin{cases} v_{ap} = V_p \cos(\omega t + \theta_p), \quad v_{bp} = V_p \cos(\omega t - \frac{2}{3}\pi + \theta_p), \quad v_{cp} = V_p \cos(\omega t + \frac{2}{3}\pi + \theta_p) \\ v_{an} = V_n \cos(\omega t + \theta_n), \quad v_{bn} = V_n \cos(\omega t + \frac{2}{3}\pi + \theta_n), \quad v_{cn} = V_n \cos(\omega t - \frac{2}{3}\pi + \theta_n), \quad v_o = V_o \cos(\omega t + \theta_o) \end{cases} \quad (10)$$

$$\begin{cases} i_{ap} = I_p \cos(\omega t + \theta_p + \delta_p), \quad i_{bp} = I_p \cos(\omega t - \frac{2}{3}\pi + \theta_p + \delta_p), \quad i_{cp} = I_p \cos(\omega t + \frac{2}{3}\pi + \theta_p + \delta_p) \\ i_{an} = I_n \cos(\omega t + \theta_n + \delta_n), \quad i_{bn} = I_n \cos(\omega t + \frac{2}{3}\pi + \theta_n + \delta_n), \quad i_{cn} = I_n \cos(\omega t - \frac{2}{3}\pi + \theta_n + \delta_n) \end{cases} \quad (11)$$

$$i_{np}(t) = -\frac{2}{V_{dc}} \begin{cases} \mathrm{sgn}(v_{a\_fund}) \cdot \{(A \cdot \cos \omega t - B \cdot \sin \omega t) \times (C \cdot \cos \omega t - D \cdot \sin \omega t)\} \\ + \mathrm{sgn}(v_{b\_fund}) \cdot \{(A' \cdot \cos \omega t - B' \cdot \sin \omega t) \times (C' \cdot \cos \omega t - D' \cdot \sin \omega t)\} \\ + \mathrm{sgn}(v_{c\_fund}) \cdot \{(A'' \cdot \cos \omega t - B'' \cdot \sin \omega t) \times (C'' \cdot \cos \omega t - D'' \cdot \sin \omega t)\} \end{cases} \qquad (12)$$

$$\begin{cases} A = V_p \cos \theta_p + V_n \cos \theta_n + V_o \cos \theta_o, \quad B = V_p \sin \theta_p + V_n \sin \theta_n + V_o \sin \theta_o \\ C = I_P \cos(\theta_p + \delta_p) + I_n \cos(\theta_n + \delta_n), \quad D = I_P \sin(\theta_p + \delta_p) + I_n \sin(\theta_n + \delta_n) \end{cases} \qquad (13)$$

$$\begin{cases} A' = V_p \cos(\theta_p - \frac{2}{3}\pi) + V_n \cos(\theta_n + \frac{2}{3}\pi) + V_o \cos \theta_o, \quad B' = V_p \sin(\theta_p - \frac{2}{3}\pi) + V_n \sin(\theta_n + \frac{2}{3}\pi) + V_o \sin \theta_o \\ C' = I_P \cos(\theta_p - \frac{2}{3}\pi + \delta_p) + I_n \cos(\theta_n + \frac{2}{3}\pi + \delta_n), \quad D' = I_P \sin(\theta_p - \frac{2}{3}\pi + \delta_p) + I_n \sin(\theta_n + \frac{2}{3}\pi + \delta_n) \end{cases} \qquad (14)$$

$$\begin{cases} A'' = V_p \cos(\theta_p + \frac{2}{3}\pi) + V_n \cos(\theta_n - \frac{2}{3}\pi) + V_o \cos \theta_o, \quad B'' = V_p \sin(\theta_p + \frac{2}{3}\pi) + V_n \sin(\theta_n - \frac{2}{3}\pi) + V_o \sin \theta_o \\ C'' = I_P \cos(\theta_p + \frac{2}{3}\pi + \delta_p) + I_n \cos(\theta_n - \frac{2}{3}\pi + \delta_n), \quad D'' = I_P \sin(\theta_p + \frac{2}{3}\pi + \delta_p) + I_n \sin(\theta_n - \frac{2}{3}\pi + \delta_n) \end{cases} \qquad (15)$$

$$i_{np}(t) = -\frac{1}{V_{dc}} \{(X+Z) + (X-Z)\cos 2\omega t + Y \sin 2\omega t\} \qquad (16)$$

$$\text{Case 2}: X = AC + A'C' - A''C'', \, Y = -AD - BC - A'D' - B'C' + A''D'' + B''C'', \, Z = BD + B'D' - B''D'' \qquad (17)$$

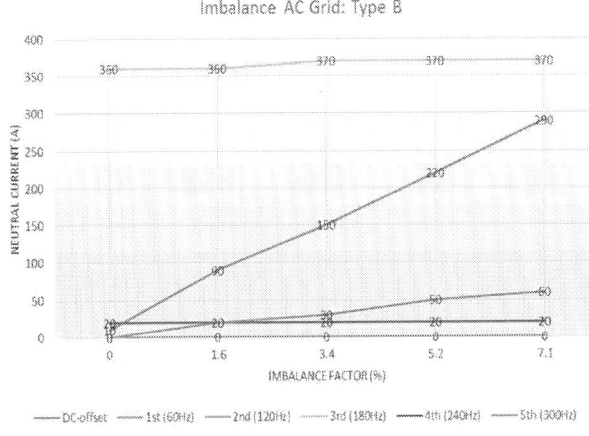

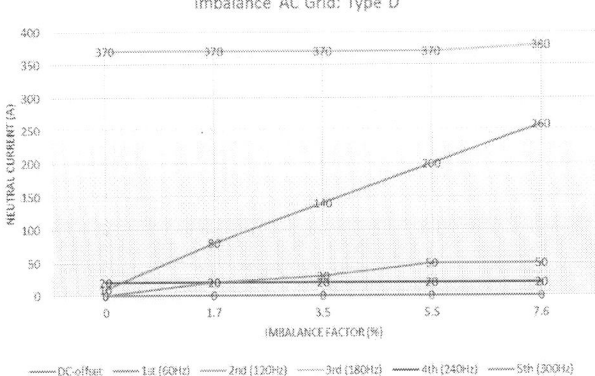

Fig. 4. Neutral point current under unbalanced grid condition
(from the top: Type B and Type D)

Table III describes the circuit parameters and operating conditions of this particular wind turbine system. Figure 4 presents the frequency spectrum of neutral current under two different types of unbalanced ac grid; Type B and Type D. The dominant low-order harmonic components up to 5th-order as well as the dc offset and fundamental component are plotted with respect to various Imbalance Factor values. It is noted from Fig. 4 that under normal balanced grid input condition the neutral current is rich in 3rd-order harmonic component. This observation is consistent with the model of (4). In (4), the product terms of the absolute value of converter output voltage and ac input current result in the neutral current of 3rd-order harmonic. As the depth of unbalanced ac grid becomes severe, i.e. increasing IF, the amplitude of fundamental frequency component increases almost linearly while having relatively constant 3rd-order harmonic component as shown in Fig. 4. Both Type B and D exhibit a similar pattern of increased fundamental component vs. increased IF. This generation of fundamental component in neutral current is attributed to the negative sequence component of ac input current in the model of (12)-(17).

IV. SIMULATION RESULTS OF PROPOSED COMPENSATING METHOD

The compensation control algorithm which is proposed in this paper is verified through the simulation. The compensation control algorithm is implemented through (7). The zero sequence voltage component (V_{comp}) according to (7) is utilized in the switching pwm technique. Circuit simulation is performed based on the circuit parameters and operating conditions summarized in Table III.

TABLE III
CIRCUIT PARAMETERS AND OPERATING CONDITIONS

Parameters	Values
Rated power (P_{rated})	2.7 MW
Rated line voltage ($V_{llrated}$)	3300 V
Rated ac input current (I_{rated})	520 A
Frequency (f_{in})	60 Hz
DC link voltage (V_{DC})	5200 V
DC link capacitance (C_{DC})	6 mF
Converter switching frequency (f_{sw})	1020 Hz
Grid side line inductance (L_s)	1.07 mH (0.1 pu)
Transformer leakage inductance (L_{tr})	0.54 mH (0.05 pu)
Filter inductance (L_f)	1.2 mH (0.112 pu)
Filter capacitance (C_f)	0.24 mF (0.365 pu)
Filter resistance (R_f)	0.3 Ω (0.07 pu)

As for the case without employing the compensation algorithm, the CB-PWM modulation signals are illustrated in Fig. 5. Waveforms in Fig. 5 and 6 are obtained under the condition without dc-link compensating algorithm being employed, i.e. the conventional current regulator of ac input current only. In Fig. 6, the upper/lower half of dc-link voltage, full dc-link voltage, and localized average neutral current are shown. At the time of 1.0 sec, an asymmetric resistive load is applied to the lower half dc-link capacitor only. This scenario effectively models dc-link unbalance conditions such as asymmetric switching modulation and dead time. It is noted from Fig. 6 that, since the application of an asymmetric resistive load at the time of 1.0 sec, the upper and lower half of dc-link voltage start to deviate from each other resulting in the NP voltage deviation. This NP voltage deviation is accompanied by the increase of neutral current which is shown at the bottom of Fig. 6. This is consistent with the argument made in Section II that the neutral current being zero is a necessary condition to maintain the voltage balance in upper and lower dc-link capacitors.

The compensation of unbalanced voltage at dc-link NP is made possible by the injection of compensation voltage to the modulation signals. As for the case employing the proposed compensation algorithm, the CB-PWM modulation signals with compensation voltage are illustrated in Fig. 7. Waveforms in Fig. 8 are obtained under the condition with the proposed dc-link compensating algorithm being employed, i.e. the conventional current regulator of ac input current and its compensation voltage injected. At the time of 1.0 sec, an asymmetric resistive load is applied to the lower half dc-link capacitor in the same manner as done in Fig. 6. It is noted from Fig. 8 that the amplitude of neutral current is decreased to 10A as compared to 15A in Fig. 6, approximately. This smaller neutral current leads to the smaller NP voltage deviation of 200V as compared to 300V of Fig. 6, approximately. This simulation results confirm the fact that the proposed compensation method can actively compensate for the neutral point deviation in 3-level NPC converter.

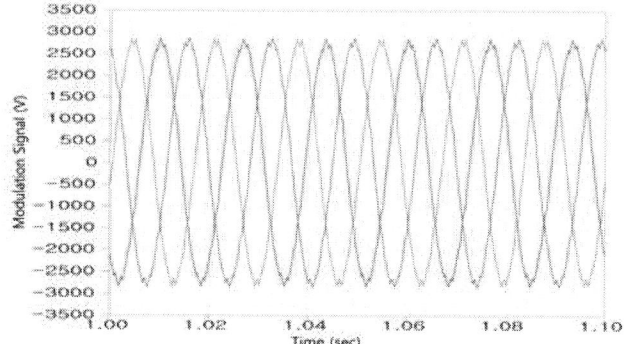

Fig. 5. Modulation signals without compensating algorithm ($v_{grid} = 60Hz$)

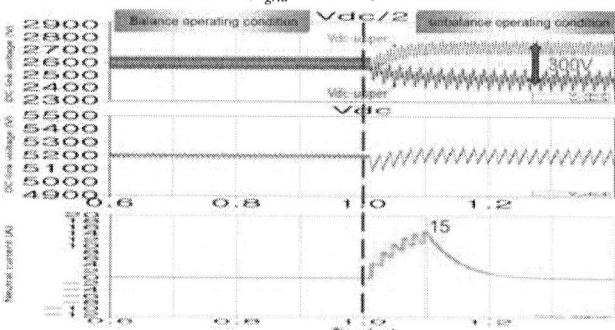

Fig. 6. Simulation waveforms without compensating algorithm (From the top: v_{dc_upper}, v_{dc_lower}, v_{dc}, and i_{np})

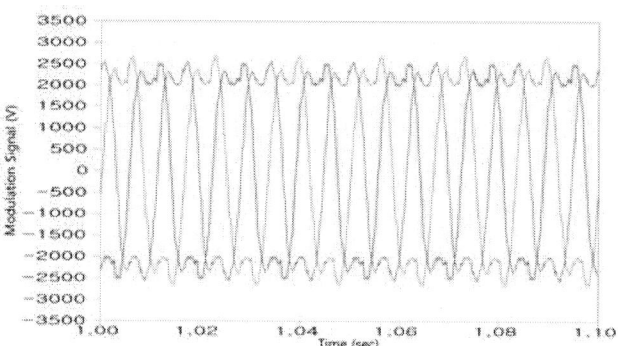

Fig. 7. Modulation signals with compensating algorithm ($v_{fund} = 60Hz$, and $v_{comp} = 180Hz$)

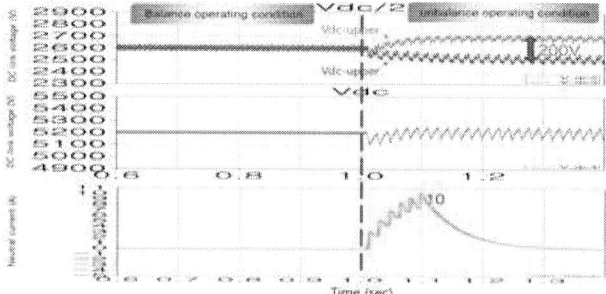

Fig. 8. Simulation waveforms without compensating algorithm (From the top: v_{dc_upper}, v_{dc_lower}, v_{dc}, and i_{np})

V. EXPERIMENTAL RESULT OF PROPOSED COMPENSATION METHOD

The proposed model and control method are verified through the real-time simulator of Typhoon HILS. The parameters of simulator setup and its operating condition are summarized in Table IV. Real-time simulator setup to validate the proposed compensation method is illustrated in Fig. 9. It shows the configuration of HILS and DSP board. Figure 10 describes the circuit model of 3-level NPC converter used in Typhoon HILS.

TABLE IV
CIRCUIT PARAMETERS AND OPERATING CONDITIONS USED FOR EXPERIMENT

Parameters	Values
Rated power (P_{rated})	36 kW
Rated line voltage ($V_{llrated}$)	460 V
Rated ac input current (I_{rated})	46 A
Frequency (f_{in})	60 Hz
DC link voltage (V_{DC})	858 V
DC link capacitance (C_{DC})	1.23 mF
Converter switching frequency (f_{sw})	10 kHz
Grid side line inductance (L_s)	2 mH (0.128 pu)
Transformer leakage inductance (L_{tr})	2 mH (0.128 pu)

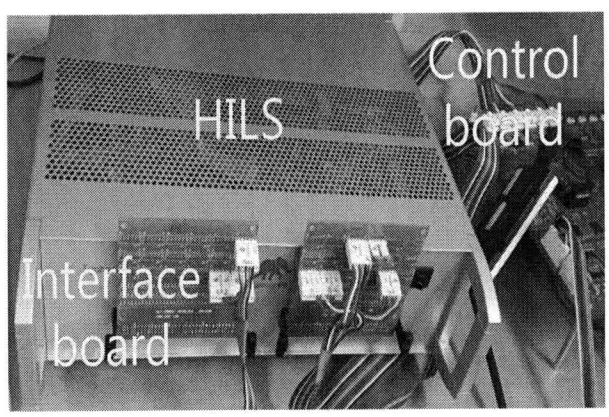

Fig. 9. System configuration of HIL and DSP

Figure 11 presents the real-time simulation result under balanced operating condition as a benchmarking case. It includes the upper/lower half dc-link voltage, full dc-link voltage, and ac grid input currents. The real-time simulation result under unbalanced operating condition is given in Fig. 12. It includes the same kinds of waveforms as in Fig. 11. Figure 13 describes reference voltage of current regulator output and modulation signal for PWM operation under balanced operating condition. The similar types of signals are shown in Fig. 14 for the case of unbalanced operating condition. It is noted from Fig. 14 that the compensation voltage of zero sequence is added to the reference voltage. Experimental result given in Fig. 11 – 14 partly verifies the proposed model and control method in the paper. More detailed result will be presented in future publications.

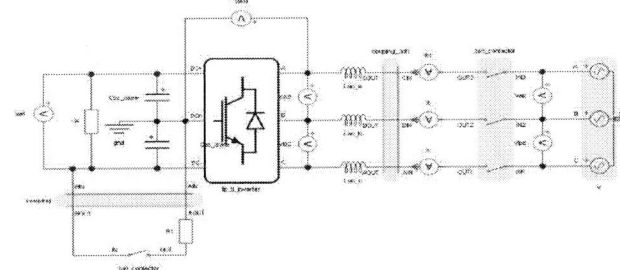

Fig. 10. Circuit model of 3-level NPC converter

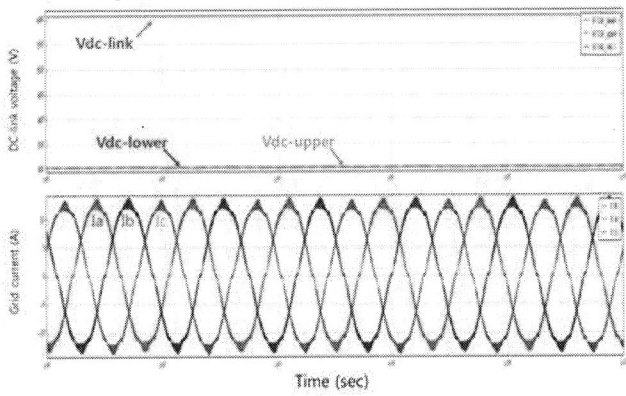

Fig. 11. Waveforms of dc-link voltage and grid current under balanced condition
(From the top : $v_{dc\_link} = 610V$, $v_{dc\_upper} = 305V$, $v_{dc\_lower} = 305V$, and $i_{abs} = 13A$)

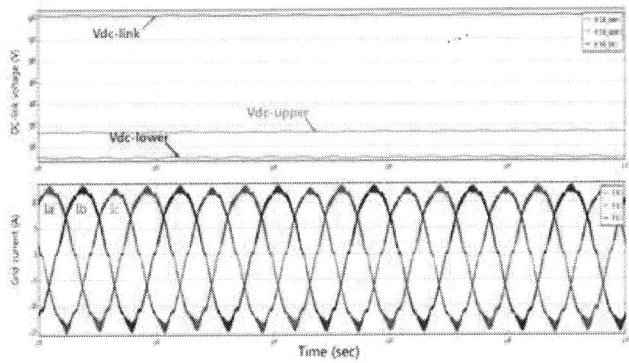

Fig. 12. Waveforms of dc-link voltage and grid current under unbalanced condition
(From the top : $v_{dc\_link} = 610V$, $v_{dc\_upper} = 335V$, $v_{dc\_lower} = 275V$, and $i_{abs} = 13A$)

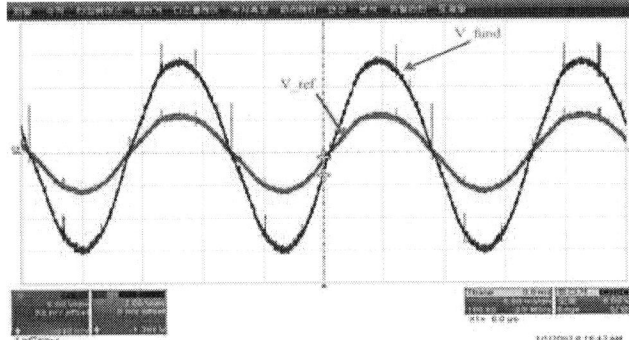

Fig. 13. Waveforms of reference voltage and modulation signal
($v_{fund} = 5.5V$ (2V/div), and $v_{rer} = 5.5V$ (5V/div), $f_{grid} = 16.7ms$ (5ms/div))

978-1-4673-9551-9/16 $31.00 © 2016 IEEE

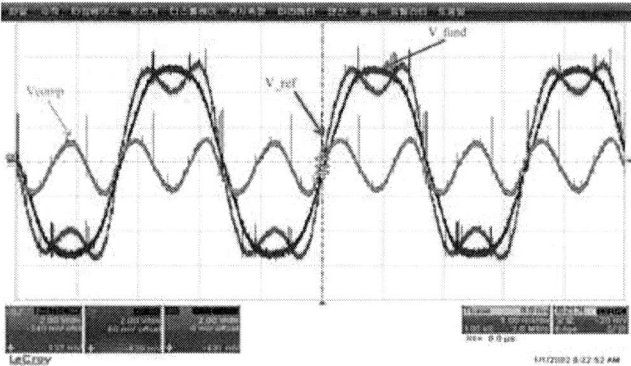

Fig. 14. Waveforms of reference voltage, modulation signal and calculated zero sequence voltage

$(v_{fund} = 5.5V(2V/\text{div}), v_{fund} = 1.7V(2V/\text{div}), v_{ref} = 5.5V(5V/\text{div}), f_{grid} = 16.7ms(5ms/\text{div}))$

VI. CONCLUSION

This paper presents the analysis of the neutral point potential variation of 3-level NPC converter. An analysis is carried out based on the average current flowing at the neutral point of the dc-link. A control scheme to keep the dc-link voltages balanced is also proposed. The potential variation can be eliminated or reduced by controlling the zero sequence voltage. In this paper, control scheme for compensating a neutral point deviation under unbalanced ac grid conditions is also investigated. This control scheme can be effectively analyzed by using both positive and negative sequence components of converter output voltages and ac input currents. Typhoon HILS test environment confirms that the proposed control scheme makes it possible to reduce the neutral point potential variation.

ACKNOWLEDGMENT

This work was supported by the National Research Foundation of Korea (NRF) grant funded by the Korea government (MSIP) (No. 2010-0028509) & (No. 2014R1A2A1A11053678)

REFERENCES

[1] J. Zarogoza, J. Pou, S. Ceballos, E. Robles, and C. Jaen, "Voltage-balance compensator for a carrier-based modulation in the neutral-point-clamped converter," *IEEE Trans. Ind. Electron.*, vol. 56, no. 2, pp. 305-314, Feb. 2009.

[2] J. Zarogoza, J. Pou, S. Ceballos, E. Robles, P. Ibanez, and J.L. Villate, "A Comprehensive Study of a Hybrid Modulation Technique for the Neutral-Point-Clamped Converter," *IEEE Trans. Ind. Electron.*, vol. 56, no. 2, pp. 294-304, Feb. 2009.

[3] C. Wang, and Y. Li, "Analysis and calculation of zero-sequence voltage considering neutral-point potential balancing in three-level NPC converter," *IEEE Trans. Ind. Electron.*, vol. 57, no. 7, pp. 2262-2271, Jul. 2010.P.

[4] C. Wang, and Y. Li, "Analysis and calculation of zero-sequence voltage considering neutral-point potential balancing in three-level NPC converter," *IEEE Trans. Ind. Electron.*, vol. 57, no. 7, pp. 2262-2271, Jul. 2010.

[5] W. Lixiang , W. Yuliang , L. Chongjian , W. Huiqing , L. Shixiang and L. Fahai, "A novel space vector control of three-level PWM converter," *Proc. IEEE Power Electron. Drive Syst.*, pp.745-750, 1999.

[6] N. Celanovic and D. Boroyevich, "A comprehensive study of neutral-point voltage balancing problem in three-level neutral-point-clamped voltage source PWM inverters," *IEEE Trans. Power Electron.*, vol. 15, no. 2, pp.242-249, 2000.

[7] K. Yamanaka, A. M. Hava, H. Kirino, Y. Tanaka, N. Koga, and T. Kume, "A novel neutral point potential stabilization technique using the information of output current polarities and voltage vector," in *Conf. Rec. IEEE IAS Annu. Meeting*, 2001, pp. 851–858.

[8] S. Busquets-Monge, J. D. Ortega, J. Bordonau, J. A. Beristain, and J. Rocabert, "Closed-loop control of a three-phase neutral-point-clamped inverter using an optimized virtual-vector-based pulsewidth modulation," *IEEE Trans. Ind. Electron.*, vol. 55, no. 5, pp. 2061–2071, May 2008.

[9] B. Abdul Rahiman, G. Narayanan, and V. T. Ranganathan, "Modified SVPWM algorithm for three level VSI with synchronized and symmetrical waveforms," *IEEE Trans. Ind. Electron.*, vol. 54, no. 1, pp. 486–494, Feb. 2007.

[10] Y. Lai, Y. Chou, and S. Pai, "Simple PWM technique of capacitor voltage balance for three-level inverter with dc-link voltage sensor only," in *Proc. 33rd Annu. Conf. IEEE IECON*, 2007, pp. 1749–1754.

[11] K. H. Bhalodi and P. Agrawal, "Space vector modulation with dc-link voltage balancing control for three-level inverters," in *Proc. Int. Conf. PEDES*, 2006, pp. 1–6.

[12] J. Suh, C. Choi, and D. Hyun, "A new simplified space-vector PWM method for three-level inverters," in *Proc. 14th Annu. APEC*, 1999, pp. 515–520.

[13] J. Holtz and N. Oikonomou, "Neutral point potential balancing algorithm at low modulation index for three-level inverter medium voltage drives," in *Conf. Rec. IEEE IAS Annu. Meeting*, 2005, pp. 1246–1252.

[14] J. Holtz and N. Oikonomou, "Neutral point potential balancing algorithm at low modulation index for three-level inverter medium voltage drives," *IEEE Trans. Ind. Electron.*, vol. 43, no. 3, pp. 761–768, May/Jun. 2007.

[15] M. Malinowski, S. Stynski, W. Kolomyjski, and M. P. Kazmierkowski, "Control of three-level PWM converter applied to variable-speed-type turbines," *IEEE Trans. Ind. Electron.*, vol. 56, no. 1, pp. 69–77, Jan. 2009.

[16] S. Ogasawara, and H. Akagi, "Analysis of variation of neutral point potential in neutral-point-clamped voltage source PWM inverter," in *Conf. Rec. IEEE IAS Annu. Meeting*, 1993, pp. 965–970.

[17] S. Qiang, L. Wenhua, Y. Qingguang, X. Xiaorong, and W. Zhonghong, "A neutral-point potential balancing algorithm for three-level NPC inverters using analytically injected zero-sequence voltage," in *Proc. IEEE Appl. Power. Electron. Conf.*, 2003, pp. 228–233.

[18] L. Jun, Q. H. Alex, Q. Zhaoming, and Z. Huijie, "A novel carrier-based PWM method for 3-level NPC inverter utilizing control freedom degree," in *Proc. IEEE Power Electron. Spec. Conf.*, 2007, pp. 1899–1904.

[19] A. Videt, P. Le Moigne, N. Idir, P. Baudesson, and X. Cimetiere, "A new carrier-based PWM providing common-mode-current reduction and dc-bus balancing for three-level inverters," *IEEE Trans. Ind. Electron.*, vol. 54, no. 6, pp. 3001–3011, Dec. 2007.

978-1-4673-9551-9/16 $31.00 © 2016 IEEE

High Power Factor Modular Polyphase AC/DC Converters with Galvanic Isolation Based on Resistor Emulators

Javier Sebastián, Ignacio Castro, Diego G. Lamar, Aitor Vázquez and Kevin Martín
Departamento de Ingeniería Eléctrica, Electrónica, de Computadores y Sistemas
University of Oviedo.
Gijón 33204, Spain
e-mail: sebas@uniovi.es

Abstract—This paper deals with a modular, isolated-output, polyphase, AC/DC converter based on the use of Resistor Emulators (REs). A RE is a DC/DC converter that behaves as a resistor at its input port. The value of this resistor (input impedance) is controlled by the converter duty cycle. In the proposed topology, all the REs are controlled to have the same input impedance, whose value is determined by the output-voltage feedback loop. Also the power processed by each RE is the same. As a consequence, the total power is distributed (and also the power losses), thus allowing us to build a modular system. Moreover, the behavior of these DC/DC converters as REs also allows their connection in series and/or in parallel with perfect sharing of voltage (when connected in series) and current (when connected in parallel). This fact makes possible to extend the proposed solution to high power applications.

Keywords—Polyphase AC/DC converters, Power Factor Correction, Modular Converters.

I. INTRODUCTION

Resistor Emulators (REs) [1-7] are DC/DC converters that behave like a resistor at their input port. This behavior is represented schematically by the dependent power source symbol shown in Fig. 1a. This behavior is obtained by choosing the proper converter control. Thus, the control based on an analog-multiplier and an input current feedback loop (see Fig. 2a) is the most popular one when the converter is operating in the Continuous Conduction Mode (CCM) [4, 5]. This type of control is called Multiplier-Based Control (MBC). However, some converters behave as "natural" REs when they have been designed to always operate in the Discontinuous Conduction Mode (DCM). In this case, they are controlled by the so called Voltage-Follower Control (VFC) (Fig. 2b). Moreover, behavior as "almost perfect" RE can also be achieved using One-Cycle Control (OCC) [8-12] or Voltage-Controlled Compensation Ramp Control (VCCRC) [9, 13-14]

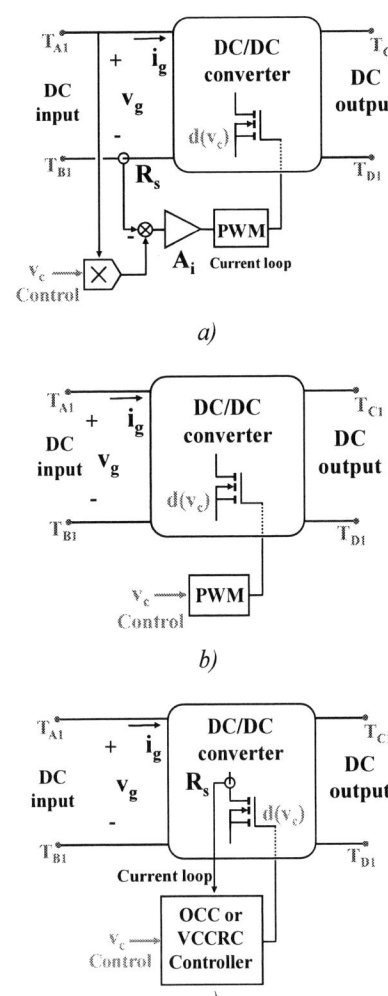

a)

b)

c)

Fig. 2. Resistor Emulator implementations: a) Multiplier-Based Control. b) Voltage-Follower Control. c) One-Cycle Control or Voltage-Controlled Compensation Ramp Control.

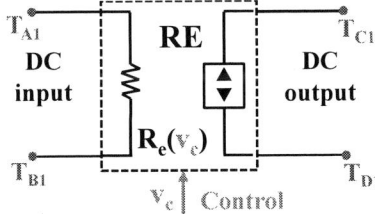

Fig. 1. Resistor Emulator concept.

This work has been supported by the *Spanish Government* under Project MINECO-13-DPI2013-47176-C2-2-R and the *Principality of Asturias* under the grants *"Severo Ochoa"* BP14-140 and BP14-85 and by the Project FC-15-GRUPIN14-143 and by European Regional Development Fund (ERDF) grants.

978-1-4673-9551-9/16 $31.00 © 2016 IEEE

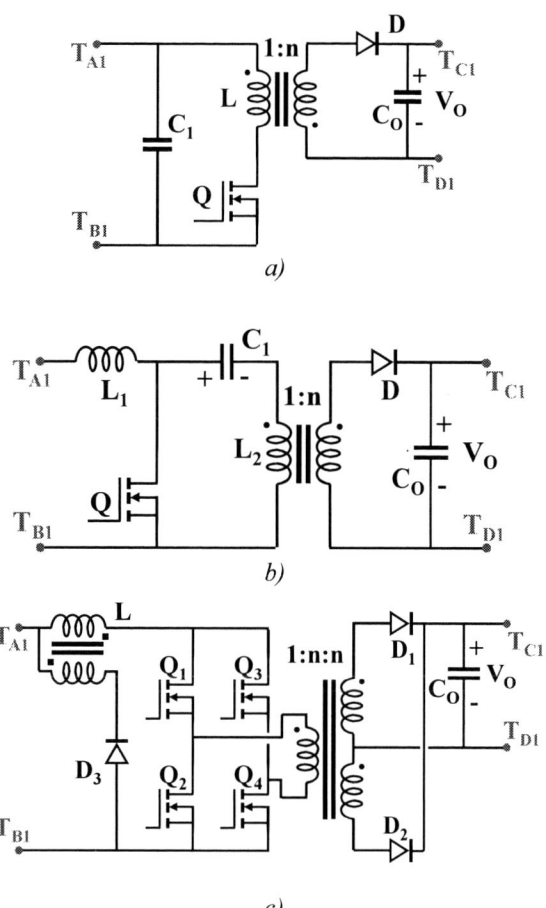

Fig. 3. Examples of DC/DC converters suitable to be used as REs in the case of wide range of input voltage values: a) Flyback. b) SEPIC. c) Current-Fed Full Bridge.

(see Fig. 2c).

If a RE has to operate from very low values of input voltage, then a step-up converter must be selected as practical implementation of the RE power topology. Some examples of step-up converters suitable to be used as RE are the Boost [4], Flyback [5, 6] (see Fig. 3a), SEPIC [15, 16] (see Fig. 3b), Ćuk [16-17] and Zeta converters.

In the case of the Boost converter, ideal operation as RE is achieved by controlling the converter according to different strategies. Thus, if the converter is working in the CCM, ideal behavior as RE is obtained by using MBC. Operation as ideal RE is also achieved if the converter is operating just in the Boundary Conduction Mode (BCM), which means that the converter is always in the limit between CCM and DCM. On the other hand, operation near to ideal RE is obtained by controlling the converter according to the VFC, when the converter is working in the DCM [18-19]. The use of OCC allows operation as "almost perfect" RE, especially if the ripple of the current passing through the input inductor is negligible.

In the case of the Flyback, SEPIC, Ćuk and Zeta converters, ideal RE behavior is obtained when these converters are working in the DCM [6, 15-17] and they are controlled with VFC. As in the case of any DC/DC converter, the use of MBC also allows perfect RE behavior (in the case of these converters even for very low input voltages). The modification of the OCC presented in [13, 14] also allows "almost perfect" RE without using a multiplier when these converters are operating in the CCM.

Flyback, SEPIC, Ćuk and Zeta converters present the important advantage of having galvanic isolation and over-current control possibility (when an overload takes place), which is not possible in the case of the Boost converter. However, these converters are used in relatively low power applications. More complex step-up converters, such as the Current-Fed Push-Pull [20] and the Current-Fed Full Bridge (Fig. 3c) combine relatively high power capability and galvanic isolation.

The main application of the RE is to perform Power factor Correction (PFC) in single-phase, front-end, AC/DC converters (Fig. 4) [4-6, 9, 11-19]. In this case, the cascade connection of a full-wave rectifier (D_{1P}-D_{2N}), a RE and a bulk capacitor C_B (typically an electrolytic one) allows us to obtain an almost perfect sinusoidal line current and a DC output voltage. It should be noted that the size of the bulk capacitor C_B is relatively large due to the instantaneous power handled by the RE, which is pulsating at twice the line frequency.

The high-quality line current obtained using this very well-known arrangement is due to the fact that diodes D_{1P} and D_{2N} remain conducting during the entire positive line half-cycle, whereas D_{2P} and D_{1N} conduct the negative one. Unfortunately, the situation is very different when a full-wave rectifier is connected to a three-phase (or higher order) grid. In this case, the diodes cannot conduct 180°, but a lower angle. For example, in the case of the cascade connection of a three-phase, full-wave rectifier (D_{1P}-D_{3N}), a RE and an output capacitor (see Fig. 5a), the line current angle is 120°, as shown in this figure. The situation is even worse with a higher order rectifier (the line waveform with a six-phase, full-wave rectifier

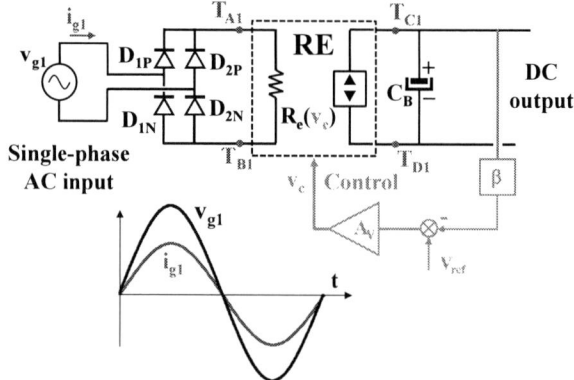

Fig. 4. Use of a RE for single-phase Power Factor Correction.

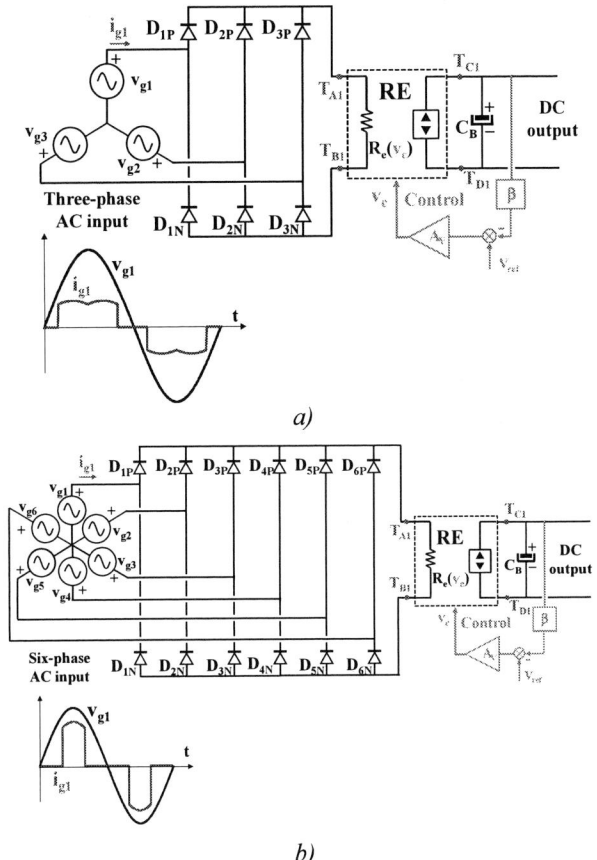

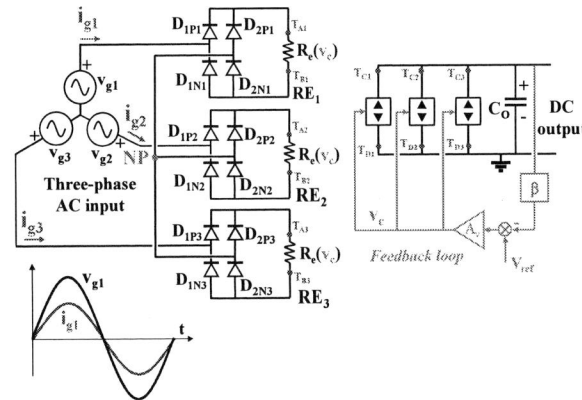

Fig. 6. The three-phase, RE based, AC/DC converter with galvanic isolation proposed in [22].

Fig. 5. Use of a RE at the output of a full-wave, polyphase, rectifier. a) Three-phase case. b) Six-phase case.

is given in Fig. 5b). Due to this, the cascade connection of a full-wave rectifier and a RE is not the right solution to obtain perfect line current in the case of polyphase grids.

Many different power topologies have been proposed in the last years to overcome this problem [21], many of them without galvanic isolation. Among them, the one proposed in [22] achieves perfect sinusoidal line currents and galvanic isolation by using several REs instead of only one (see Fig. 6). In this case, three full-wave, single-phase rectifiers are connected in such a way that one of the input terminals of each single-phase rectifier is connected to each grid phase, whereas the other input terminals are connected together. For each full-wave, single-phase rectifier x, the pair of diodes D_{1Px}-D_{2Nx} conducts for the positive half cycle of v_{gx}, whereas the pair of diodes D_{2Px}-D_{1Nx} conducts for the negative one. Thus, a pair of diodes of each single-phase rectifier is always conducting and, therefore, a balanced load made up of three resistors (the three REs) is seen by the three-phase grid. As a consequence, the three grid currents are proportional to the corresponding grid voltages and unity power factor is achieved.

A different solution, also based on the use of REs, is proposed in this paper. Each RE is considered as a component of the overall converter, which allows the power conversion

process to be spread among them, thus also spreading their power losses. As in [22, 23], each RE is controlled to exhibit the same equivalent input resistance $R_e(v_c)$, whose value is determined by the control voltage v_c, which is the output voltage of an output-voltage feedback loop. Furthermore, all the REs must have galvanic isolation between their input and output ports because the output ports must be connected each other. This connection can be in parallel (see Fig. 7a) or in series (see Fig. 7b). Also as in [22, 23], the power processed by all the REs is delivered to the same point. Therefore, the total power delivered is not pulsating in the case of polyphase grids. This fact allows the use of low-value capacitors (instead of large and bulky electrolytic ones) at the output of the overall converter, thus allowing its faster response.

II. THE PROPOSED MODULAR POLYPHASE AC/DC CONVERTER WITH HIGH POWER FACTOR AND GALVANIC ISOLATION

The proposed converter is based on the connection of the input of a RE in series with each rectifier diode in a full-wave

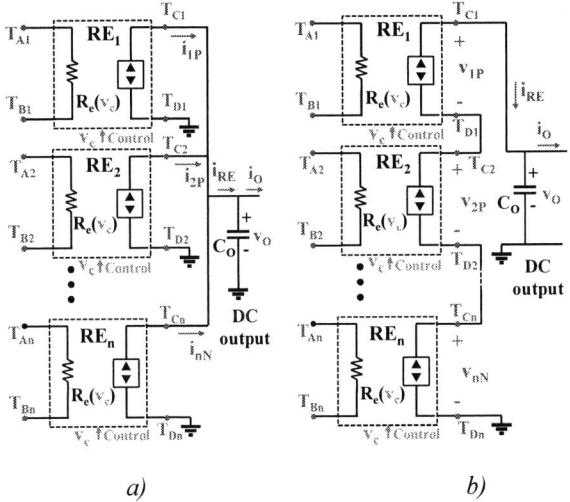

Fig. 7. Connection of RE outputs: a) Parallel connection. b) Series connection.

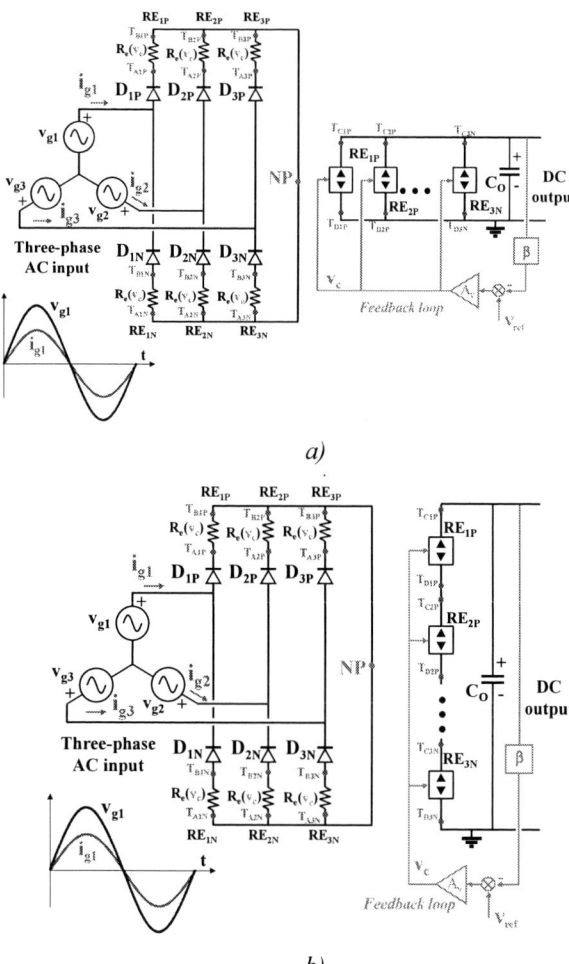

a)

b)

Fig. 8. Proposed modular, isolated output, polyphase, AC/DC converter based on REs (three-phase implementation): a) Case of outputs in parallel connection. b) Case of outputs in series connection.

rectifier (see Fig. 8). All the RE outputs are connected in parallel (see Fig. 8a) or in series (see Fig. 8b) with each other, this connection being the converter output. The full-wave rectifier output is short-circuited in the neutral point NP. The power transfer is achieved from the RE inputs to their outputs, (which is the overall converter output) and it takes place every line half cycle. Thus, the total number of REs needed is $n=2 \cdot p$, where p is the number of phases. It should be noted that each line current has to pass through 1 RE and 1 rectifier diode when circulating from each input voltage source (phase voltage) v_{gx} to the neutral point NP. In the case of the converters presented in [22-23], the total number of RE needed was $n=p$, which is a number of converters twice lower. However, each line current has to pass through 1 RE and 2 rectifier diodes when circulating from each input voltage source to the neutral point NP. Therefore, for a given efficiency in the REs and for a given power processing capability, the proposed converter will exhibit better overall efficiency and

will be capable of handling higher power levels than the ones proposed in [22-23].

The proposed solution can be used in any polyphase grid. However, for the sake of simplicity, the analysis is presented for a three-phase grid. The phase voltages and currents will be defined as (see Fig. 8):

$$v_{gx} = V_g \sin\left[\omega_g t - \tfrac{2\pi}{3}(x-1)\right] \quad (1)$$

$$i_{gx} = \frac{v_{gx}}{R_e(v_C)} \quad (x = 0, 1, 2), \quad (2)$$

where V_g is the peak value of the phase voltage, ω_g is the grid frequency, $R_e(v_c)$ is the resistance emulated by the REs (i.e., the input impedance of all the REs at frequencies as low as the line frequency), v_c is the control voltage that determines the value of resistance emulated by the REs and i_{gx} is the current passing through the phase x. As a consequence, the power handled by each RE can be expressed as:

- if $v_{gx} > 0$, then:

$$p_{xP} = \frac{v_{gx}^2}{R_e(v_C)} =$$

$$= \frac{v_g^2}{2R_e(v_C)}\left\{1 - \cos\left[2\omega_g t - \tfrac{4\pi}{3}(x-1)\right]\right\} \quad (3)$$

$$p_{xN} = 0, \quad (4)$$

- if $v_{gx} < 0$, then:

$$p_{xN} = \frac{v_g^2}{2R_e(v_C)}\left\{1 - \cos\left[2\omega_g t - \tfrac{4\pi}{3}(x-1)\right]\right\} \quad (5)$$

$$p_{xP} = 0, \quad (6)$$

where p_{xP} and p_{xN} are the values of the power handled by the resistor emulator RE_{xP} (those connected to diodes D_{xP}) and RE_{xN} (those connected to diodes D_{xN}), respectively. The waveforms corresponding to the instantaneous power handled by each RE is given in Fig. 9.

The outputs of the REs behave as power sources [1-3] and, therefore, the values of the output voltages and currents depend

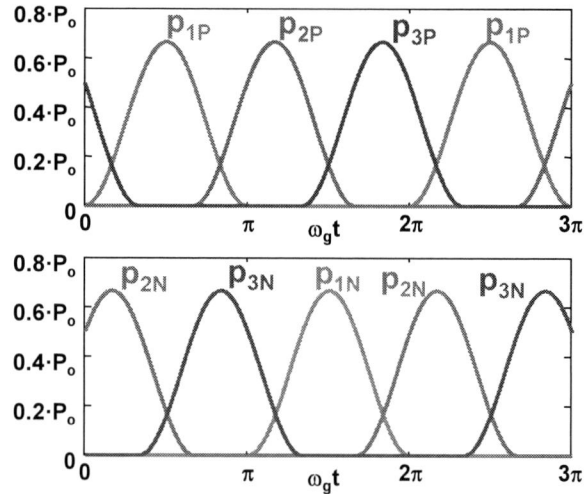

Fig. 9. Power handled for each RE. P_o is the converter total power.

on how these outputs have been connected. Thus, if the outputs are connected in parallel (see Fig. 7a), then all the REs have the same output voltage v_o. In this case, the current supplied by the REs is:

- if $v_{gx} > 0$, then:

$$i_{xP} = \frac{V_g^2}{2v_o R_e(v_C)}\left\{1 - \cos\left[2\omega_g t - \frac{4\pi}{3}(x-1)\right]\right\} \quad (7)$$

$$i_{xN} = 0, \quad (8)$$

- if $v_{gx} < 0$, then:

$$i_{xN} = \frac{V_g^2}{2v_o R_e(v_C)}\left\{1 - \cos\left[2\omega_g t - \frac{4\pi}{3}(x-1)\right]\right\} \quad (9)$$

$$i_{xP} = 0, \quad (10)$$

where i_{xP} and i_{xN} are the values of the current supplied by the resistor emulators RE_{xP} and RE_{xN}, respectively. Fig. 10 shows the waveforms corresponding to these currents. The value of the total current injected into the parallel connection of the output capacitor C_o and the load (not represented in Fig. 7a), i_{RE}, is:

$$i_{RE} = \sum_{x=1}^{3}(i_{xP} + i_{xN}) = \frac{3V_g^2}{2v_o R_e(v_C)}. \quad (11)$$

This expression shows that AC components of twice the grid frequency are completely cancelled out and i_{RE} is just a DC current. As a consequence, the output filter capacitor has to remove only the switching frequency components of the REs output current. As no power pulsating at the grid frequency has to be stored in the output capacitor (only power pulsating at the switching frequency), then the capacitor C_o can be relatively small and it is free of voltage ripple of twice the grid frequency. Due to this, the compensator A_v has not to remove any low frequency component and, therefore, the voltage loop can be relatively fast, clearly faster than in the case of a single-phase AC/DC converter [24-26].

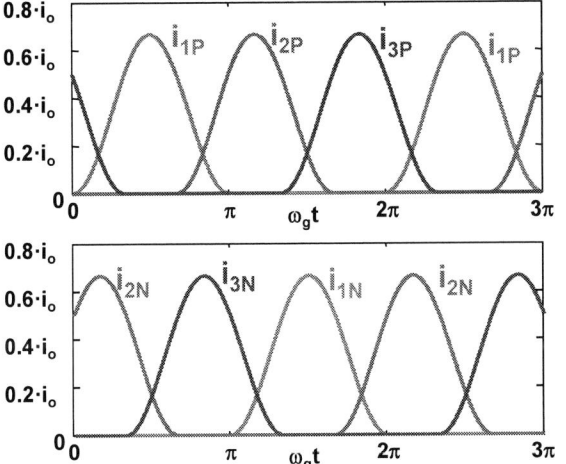

Fig. 10. Current passing through the output of each RE when the outputs have been connected in parallel (see Fig.7a and Fig. 8a).

If the outputs are connected in series (Fig. 7b), then all the REs have the same output current i_{RE}. In this case, the voltage at the output of each RE is:

- if $v_{gx} > 0$, then:

$$v_{xP} = \frac{V_g^2}{2i_{RE} R_e(v_C)}\left\{1 - \cos\left[2\omega_g t - \frac{4\pi}{3}(x-1)\right]\right\} \quad (12)$$

$$v_{xN} = 0, \quad (13)$$

- if $v_{gx} < 0$, then:

$$v_{xN} = \frac{V_g^2}{2i_{RE} R_e(v_C)}\left\{1 - \cos\left[2\omega_g t - \frac{4\pi}{3}(x-1)\right]\right\} \quad (14)$$

$$v_{xP} = 0, \quad (15)$$

where v_{xP} and v_{xN} are the values of the voltage across the outputs of the resistor emulators RE_{xP} and RE_{xN}, respectively. The value of the voltage across the output capacitor C_o and the load (not represented in Fig. 7b), v_o, is:

$$v_o = \sum_{x=1}^{3}(v_{xP} + v_{xN}) = \frac{3V_g^2}{2i_{RE} R_e(v_C)}. \quad (16)$$

As in the previous case, the AC components of twice the grid frequency are completely cancelled out and v_o is just a DC voltage. The same considerations about the size of the output capacitor C_o and the fast response of the converter output can be applied in this case. The waveforms in this case are as the ones shown in Fig. 10, but replacing i_o, i_{xP} and i_{xN} with v_o, v_{xP} and v_{xN}, respectively.

III. POWER TOPOLOGIES FOR THE RESISTOR EMULATORS

The total power supplied by the overall AC/DC converter in the case of a three-phase grid is the addition of the power handled by the six REs:

$$P_o = \sum_{x=1}^{3}(p_{xP} + p_{xN}) = \frac{3V_g^2}{2R_e(v_C)}. \quad (17)$$

Therefore, the average power handled by each RE is:

$$P_{1RE} = \frac{P_o}{6} = \frac{V_g^2}{4R_e(v_C)}. \quad (18)$$

In the case of p-phase grid, (19) becomes:

$$P_{1RE} = \frac{P_o}{2p}. \quad (19)$$

P_{1RE} determines the power to be handled by each RE and therefore, the type of power topology suitable for each specific case. Flyback's family of converters (SEPIC, Ćuk and Zeta) is adequate to implement REs up to 200 W (considering Si power semiconductor devices). Galvanic isolated versions of the Boost converter (e.g., the Current-Fed Full Bridge converter shown in Fig. 3c) are suitable for REs in the range of a few kW. Also, the cascade connection of a Boost converter and a "DC transformer" (a DC/DC converter working at constant duty cycle to achieve soft-switching and, therefore, high efficiency) is an attractive alternative (see Fig. 11).

978-1-4673-9551-9/16 $31.00 © 2016 IEEE

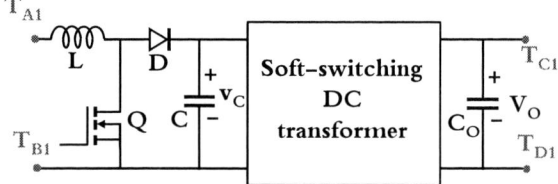

Fig. 11. Implementation of a RE with galvanic isolation to achieve a medium power, high efficiency, cell.

Moreover, the power processing capability can be increased by replacing each RE with the connection of several of them in series and/or parallel (see Fig. 12). It should be noted that ideal current and voltage sharing is achieved if the REs are properly controlled to have the same input impedance $R_e(v_c)$. Finally, the topology can be extended to be used not only in three-phase grids, but also in polyphase grids.

IV. CONTROL OF THE POWER TOPOLOGIES SUITABLE OF BEING USED AS RESISTOR EMULATORS

Regarding the control of REs, several control strategies can be used to achieve RE behavior. The most useful are the following ones:

Multiplier-Based Control, MBC (see Fig.2a).

This control method is very well known because many single-phase, power factor correctors use controllers based on this method. Assuming that the gain of error amplifier of the current loop A_i is high enough at grid frequencies, the average value (averaged in a switching period) of the input current i_g is determined by the multiplier as follows:

$$i_g = \frac{K_M}{R_s} v_g v_c, \qquad (20)$$

where K_M is a constant determined by the multiplier and R_s is the gain of the current sensor. Therefore, the input impedance of the RE is:

$$R_e(v_c) = \frac{v_g}{ig} = \frac{R_s}{K_M} \cdot \frac{1}{v_c}. \qquad (21)$$

This control method can be used when the converter is working in both CCM and DCM.

Voltage-Follower Control, VFC (see Fig. 2b).

This control method is only possible if the converter is

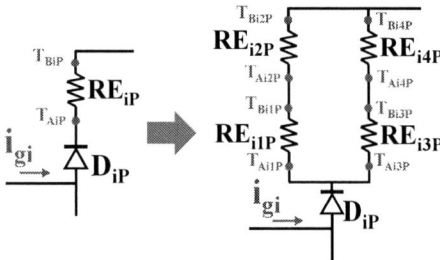

Fig. 12. Use of several REs (instead of just one) to increase the converter power capability.

working in the DCM (for converters derived from the Flyback converter, such as the SEPIC, Ćuk and Zeta) or just in the BCM (for the Boost converter). In the case of converters belonging to the Flyback's family, the average value (averaged in a switching period) of the input current i_g is determined by Faraday's law as follows:

$$i_g = \frac{T_s d^2}{2L} v_g, \qquad (22)$$

where T_s is the switching period, d is the converter duty cycle and L is the inductance of the primary side of the Flyback transformer. In the case of SEPIC, Ćuk and Zeta converters, L is the equivalent inductance of the converter inductors in parallel and referring their values to the transformer primary side. From (22), the value of R_e is:

$$R_e(d) = \frac{v_g}{ig} = \frac{2L}{T_s} \cdot \frac{1}{d^2}. \qquad (23)$$

If the duty cycle d is obtained from a sawtooth waveform in a PWM controller, then its value is given by:

$$d = \frac{v_c}{V_{PV}}, \qquad (24)$$

where V_{PV} amplitude of the sawtooth waveform. Therefore, (23) becomes:

$$R_e(v_c) = \frac{2LV_{PV}^2}{T_s} \cdot \frac{1}{v_c^2}. \qquad (25)$$

In the case of a Boost converter operating in the BCM, the average value (averaged in a switching period) of i_g is: determined by Faraday's law as follows:

$$i_g = \frac{T_s d}{2L} v_g. \qquad (26)$$

From (24) and (26), the value of $R_e(v_c)$ is easily obtained:

$$R_e(v_c) = \frac{2LV_{PV}}{T_s} \cdot \frac{1}{v_c}. \qquad (27)$$

One-Cycle Control, OCC and Voltage-Controlled Compensation Ramp Control, VCCRC (see Fig. 2b).

In the case of the Boost converter [9-13], the grid current is given by:

$$i_g = \frac{v_g}{v_o} \left[\frac{v_c}{R_s} - \frac{(v_o - v_g)T_s}{2L} \right]. \qquad (28)$$

This last equation shows that if the quotient L/T_s satisfies the relationship:

$$\frac{L}{T_s} \gg \frac{(v_o - v_g)R_s}{2v_c}, \qquad (29)$$

then i_g and v_g will be proportional, performing ideal RE behavior (this relationship means that the inductor current ripple is relatively small). In this case, the input impedance of RE is:

$$R_e(v_c) = R_s v_o \cdot \frac{1}{v_c}. \qquad (30)$$

In the case of the converter belonging to the Flyback family, the direct application of the OCC as given in [11] is quite complex and an alternative solution was proposed in [13, 14].

In this case, an exponential ramp is used instead of a linear one. This control is called VCCRC. The time constant of the exponential ramp τ defines the parameter μ as follows:

$$\mu = \frac{T_s}{\tau}. \tag{31}$$

The actual value of the input current is [13]:

$$i_g = \frac{v_o}{v_o + n v_g}\left[\frac{v_c\left(e^{-\frac{v_o}{v_o+nv_g}\mu}-e^{-\mu}\right)}{R_s\left(1-e^{-\mu}\right)} - \frac{v_o v_g T_s}{2L(v_o+nv_g)}\right]. \tag{32}$$

The values of μ can be selected in such a way that the THD of this current when v_g is changing sinusoidaly is minimized for each value of $M = v_o/nV_g$ (V_g being the peak value of v_g) and for each value of the parameter α defined as:

$$\alpha = \frac{L}{L_{crit\_\pi}}, \tag{33}$$

where L_{crit_π} is the value of L that guarantees operation in the CCM during the entire line period for a given value of v_c. It should be noted that (32) is only valid if the converter is working in the CCM [13] during the line period, what implies a value of L higher than L_{crit_π} (i.e., $\alpha > 1$).

After minimizing the THD for each value of M and α, (32) becomes:

$$i_g = \frac{\gamma(M,\alpha)v_g v_c}{R_s V_g}, \tag{34}$$

the values of $\gamma(M, \alpha)$ being represented in Fig. 13. From (34), it can be easily obtained:

$$R_e(v_c) = \frac{R_s V_g}{\gamma(M,\alpha)} \cdot \frac{1}{v_c}. \tag{35}$$

V. EXPERIMENTAL RESULTS

The experimental validation has been carried out by using Flyback converters as REs. These Flybacks have been designed to work in the DCM in all the range of power. Due to this, VFC has been used to control each RE, which is a very simple control method. Based on the use of six of these Flyback converters, a three-phase prototype of AC/DC converter (see Fig. 8a) has been designed and built. In this converter, the RE outputs have been connected in parallel. The overall converter power is 250 W (Output: 48V/5.16 A). The converter has been designed for the European voltage range (400 V line voltage),

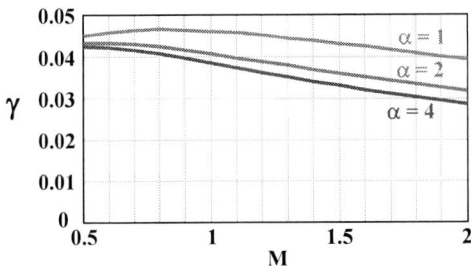

Fig. 13. Values of γ.

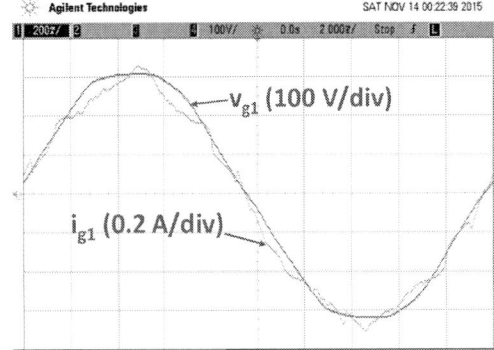

Fig. 14. Grid current and voltage in a phase.

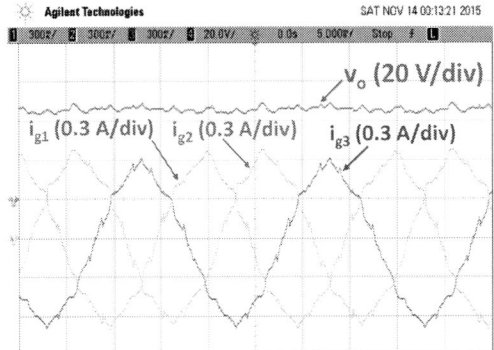

Fig. 15. Grid currents and output voltage.

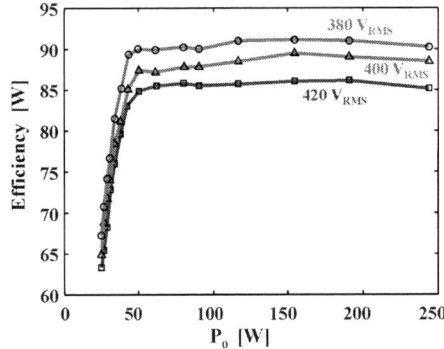

Fig. 16. Converter efficiency versus output power and grid voltage.

achieving 99.68% power factor and 6.5% THD. As Fig. 14 shows, grid current and voltage are in phase.

The measurements have been done with the converter connected to the real power grid, and using a resistive load at the overall converter output. The waveforms have been extracted as data from the oscilloscope and analyzed with the use of MATLAB®. The grid currents (i_{g1}, i_{g2} and i_{g3}) are shown in Fig. 15, as well as, the output voltage obtained with a 10 µF film capacitor at the overall converter output.

It should be noted that, as a three-phase balanced system, it must comply with the IEC 1000-3-2 regulations in Class A [27-29]. The measured grid currents have been analyzed and compared with the limits imposed by those regulations, the measured harmonics being well below these limits.

Finally, the efficiency of the converter is roughly 88% at maximum load (see Fig. 16)

VI. CONCLUSIONS

A polyphase AC/DC converter with galvanic isolation and very high power factor (ideally 1) is presented in this paper. The converter is made up of several (twice the number of phases) REs, which are DC/DC converters controlled to exhibit the same input impedance. The value of this input impedance is determined by an output-voltage feedback loop, which is the feedback loop of the overall AC/DC converter.

The use of REs as building blocks of the overall AC/DC converter allows us not only to achieve ideal power factor and THD, but also to distribute the total power among identical DC/DC converters, that can be considered as independent devices. Moreover, as the instantaneous input power is not time-depending, the instantaneous output power is not pulsating at twice the grid frequency and, therefore, the converter output capacitor is in charge of removing only the components of the switching frequency. As a consequence, this capacitor is relatively small and the output voltage feedback loop can be relatively fast.

Many different power topologies can be used to implement REs and they can be controlled according to several control strategies, which are briefly described in this paper. Moreover, the use of REs as building blocks of converters makes easy to scale up the power handled by the overall converter due to the ideal share of voltage and current obtained by connecting REs in series and in parallel.

Finally, a small prototype of the proposed converter has been built and tested. This prototype is based on six Flyback converters working in the DCM to achieve RE behavior using VFC.

VII. REFERENCES

[1] S. Singer, "Realization of loss-free resistive elements", *IEEE Trans. on Circuits and Systems*, vol. 37, no. 1, pp. 54-60, Jan. 1990.

[2] S. Singer and R.W. Erickson, "Power-source element and its properties", *IEE Proc.-Circuits Devices and Syst.*, vol. 141, no. 3, pp. 220-226, Jun. 1994.

[3] S. Singer and R.W. Erickson, "Canonical modeling of power processing circuits based on the POPI concept", *IEEE Trans. on Power Electronics*, vol. 7, no. 1, pp. 37-43, Jan. 1992.

[4] L. H. Dixon, "High power factor preregulators for off-line power supplies", *Unitrode Power Supply Design Seminar*, 1988, pp. 6.1-6.16.

[5] M. J. Kocher and R. L. Steigerwald, "An ac-to-dc converter with high quality input waveforms", *IEEE Trans. on Industry Applications*, vol. 19, no. 4, pp. 586-599, Jul/Aug. 1983.

[6] R. Erickson, M. Madigan and S. Singer, "Design of a simple high-power factor rectifier based on the flyback converter", *Proc. IEEE APEC 1990*, pp. 792-801.

[7] R. Erickson and D. Maksimovic, "Fundamentals of Power Electronics" *(Second Edition). Kluwer Academic Publishers.*

[8] K. M. Smedley and S. Cuk, "One-cycle control of switching converters", *IEEE Trans. on Power Electronics*, vol. 10, no. 6, pp. 625–633, Nov. 1995.

[9] J. P. Gegner and C. Q. Lee, "Linear peak current mode control: A simple active power factor correction control technique", *Proc. IEEE PESC 1996*, pp. 196–202.

[10] Z. Lai and K. M. Smedley, "A general constant-frequency pulsewidth modulator and its applications," *IEEE Trans. on Circuits and Systems I*, vol. 45, no. 4, pp. 386–396, Apr. 1998.

[11] Z. Lai and K. M. Smedley, "A family of continuous-conduction-mode power-factor-correction controllers based on the general pulse-width modulator". *IEEE Trans. on Power Electronics*, vol. 13, no. 3, pp. 501-510, May 1998.

[12] R. Brown and M. Soldano, "One cycle control IC simplifies PFC designs", *Proc. IEEE APEC* 2005, pp. 825-829.

[13] J. Sebastián, D. G. Lamar, M. Arias, M. Rodríguez and A. Fernández, "The Voltage-Controlled Compensation Ramp: A waveshaping technique for power factor correctors", *IEEE Trans. on Industry Applications*, vol. 45, n° 3, pp. 1016-1027, May/Jun. 2009.

[14] D. G. Lamar, M. Arias, A. Rodríguez, A. Fernández, M. M. Hernando and J. Sebastián. "Design-oriented analysis and performance evaluation of a low-cost high-brightness LED driver based on flyback power factor corrector", *IEEE Transactions on Industrial Electronics*, vol. 60, no. 7, pp. 2614-2626, Jul. 2013.

[15] J. Sebastián, J. Uceda, J. A. Cobos, J. Arau, and F. Aldana, "Improving power factor correction in distributed power supply systems using PWM and ZCS-QR SEPIC topologies". *Proc. IEEE PESC* 1991, pp. 780-791.

[16] D. S. L. Simonetti, J. Sebastián and J. Uceda, "The discontinuous conduction mode SEPIC and Cuk power factor preregulators: analysis and design", *IEEE Trans. on Industrial Electronics*, vol. 44, no. 5, pp. 630-637, Oct. 1997.

[17] M. Brkovic and S. Cuk, "Input current shaper using Cuk converter", *Proc. INTELEC 1992*, pp. 532-539.

[18] J. Lazar and S. Cuk, "Open loop control of a unity power factor, discontinuous conduction mode boost rectifier", *Proc. INTELEC 1995*, pp. 671-677.

[19] K. H. Liu and Y. L. Lin, "Current waveform distortion in power factor correction circuits employing discontinuous-mode boost converter", *Proc. IEEE PESC 1989*, pp. 825–829.

[20] V. J. Thottuvelil, T. G. Wilson and H. A. Owen, "Analysis and design of a Push-Pull current-fed converter", *Proc. IEEE PESC* 1981, pp. 192-203.

[21] B. Singh, B. N. Singh, A. Chandra, K. Al-Haddad, A. Pandey and D. P. Kothari, "A review of three-phase improved power quality AC-DC converters", *IEEE Trans. on Industrial Electronics*, vol. 51, no. 3, pp. 641-660, Jun. 2004.

[22] S. Singer and A. Fuchs, "Multiphase AC-DC conversion by means of loss-free resistive networks", *IEE Proc.-Circuits Devices and Syst.*, vol. 143, no. 4, pp. 233-240, Aug. 1996.

[23] U. Kamnarn and V. Chunkag, "Analysis and design of a modular three-phase AC-to-DC converter using Cuk rectifier module with nearly unity power factor and fast dynamic response", *IEEE Trans. on Power Electronics*, vol. 24, no. 8, pp. 2000-2012, Aug. 2009.

[24] J. Sebastián, D. G. Lamar, M. M. Hernando, A. Rodríguez, and A. Fernández, "Steady-state analysis and modeling of power factor correctors with appreciable voltage ripple in the output-voltage feedback loop to achieve fast transient response" *IEEE Trans. on Power Electronics*, vol. 24, no. 11, pp. 2555-2566, Nov. 2009.

[25] J. Sebastián, D. G. Lamar, A. Rodríguez, M .Pérez, and A. Fernández, "On the maximum bandwidth attainable by power factor correctors with a standard compensator" *IEEE Trans. on Industry Applications*, vol. 46, no. 4, pp. 1485-1497, Jul/Aug. 2010.

[26] A. Fernández, J. Sebastián, P. Villegas, M. M. Hernando and D. G. Lamar, "Dynamic limits of a power-factor preregulator", *IEEE Trans. on Industrial Electronics*, vol. 52, no. 1, pp. 77-87, Feb. 2005.

[27] Electromagnetic Compatibility (EMC)-Part 3: Limits-Section 2: Limits for Harmonic Current Emissions (Equipment Input current < 16 A per Phase), IEC1000-3-2, 1995.

[28] Draft of the Proposed CLC Common Modification to IEC 61000-3-2 Document, 2006.

[29] Draft of the Proposed CLC Common Modification to IEC 61000-3-2/A2 Document, 2010.

978-1-4673-9551-9/16 $31.00 © 2016 IEEE

Reduced Duty-Cycle Loss and Output Inductor Current Ripple in A ZVS Switched Three-Phase Isolated PWM Rectifier

Jahangir Afsharian, Dewei (David) Xu, Tao Zhao
Electrical and Computer Engineering Ryerson University
Toronto, Canada
E-mail: jafshari@ryerson.ca, dxu@ryerson.ca,
zhtao1206@163.com

Bing Gong, Zhihua Yang
Murata Power Solution, Front-end AC/DC Power Module
Toronto, Canada
E-mail:bgong@murata.com, zyang@murata.com

Abstract— **In this paper a PWM method for zero voltage switched three-phase isolated rectifier is presented. The presented PWM scheme features lower maximum output inductor current ripple and lower duty-cycle loss by more than half. Therefore, a smaller output inductor can be used and the ZVS operation can be extended at wider load range for MOSFET devices by increasing the leakage inductance of the transformer. The steady state analysis of duty-cycle loss and inductor current ripple are addressed and verified by the simulation and experimental results for two different PWM schemes.**

Keywords— duty loss, inductor current ripple, SVM, ZVS, PWM, isolated rectifier, Three-phase

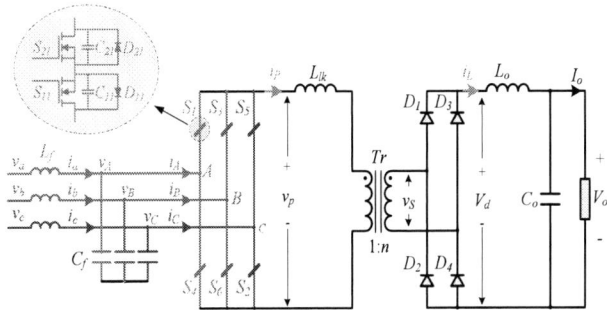

Fig. 1. ZVS three-phase PWM rectifier [3].

I. INTRODUCTION

When reduction of size and weight of converter is mandatory, the electrical isolation has to be performed at high-frequency [1]. Typically, the single-stage power conversion can be realized with a direct matrix-type PFC rectifier that directly converts the mains-frequency AC voltage into a high-frequency AC voltage which is supplied to a high-frequency isolation transformer and whose secondary voltage is then rectified to the desired DC output voltage [2]. A novel three-phase, single-stage, isolated high frequency PWM rectifier proposed in [3, 4] as shown in Fig. 1, capable of power factor correction (unity power factor), low harmonic current distortion, and in the same time realizing zero-voltage switching for all power semiconductor devices. Within any 60° interval, the proposed converter is analyzed in [3] as two full-bridge phase-shifted (FB-PS) converter sub-topologies, leg A and B forms "converter *x*" and leg A and C forms "converter *y*" which operating alternatively within the switching cycle. Same as FB-PS converter, this converter also utilize the transformer leakage inductance to achieve ZVS but at a price of reduced effective duty cycle. The duty-cycle loss increases the conduction losses caused by the circulating current and limited the converter operation at higher switching frequency result in decrease the conversion efficiency and power density. Unlike the FB-PS converter where the input DC bus is constant, the DC bus for "converter *x*" and "converter *y*"

vary during every 60° interval since the DC bus is one of the three line-to-line voltages. In addition, "converter *x*" and "converter *y*" also need to vary their duty cycle to synthesize the grid current to achieve unit power factor. All these variable operating conditions contribute to a large variation on output inductor current ripple. Then a large inductance may be required to limit the ripple current, which will result in reduced power density of the power supply. The aforementioned drawbacks can be improved by recently proposed PWM scheme in [5]. With the proposed PWM scheme, the three-phase converter is operated as one full bridge dc-dc converters by properly arrange the operating sequence of "bridge *x*" and "bridge *y*" within every 60° interval as shown in Fig. 2 and Fig. 3 of [5]. Therefore, the ZVS can be implemented in the similar way as in the FB-PS converter in [6]. In this paper, the duty cycle loss and output current ripple of two different PWM schemes are further studied and discussed. The presented PWM scheme features, lower output current ripples and lower duty cycle losses (reduced by more than half compare to PWM in [3]) which means smaller output inductor can be used and the ZVS operation can be extended at wider load range for MOSFET devices. The rest of this paper is organized as follows. Steady state operation and analysis on duty cycle loss and output inductor current are extensively described in Section II. Simulation and experimental results are presented in Section III, followed by the conclusion and future work in Section IV.

978-1-4673-9551-9/16 $31.00 © 2016 IEEE

II. STEADY-STATE OPERATION AND ANALYSIS

A. Duty Cycle Loss Analysis

Transition time from zero vectors to active vectors is finite depending on the value of L_{lk} and the primary side voltage v_P as shown in Fig. 2 (a) and (b) in the shaded area ΔD_x and ΔD_y. During these intervals, primary current i_P swing from one direction to another direction and there is no energy transfer from AC side to DC side.

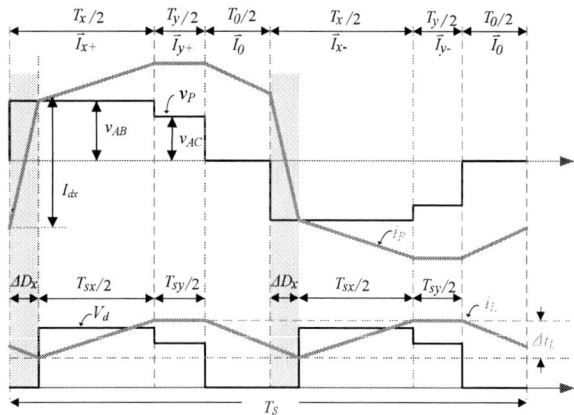

(a) the proposed PWM.

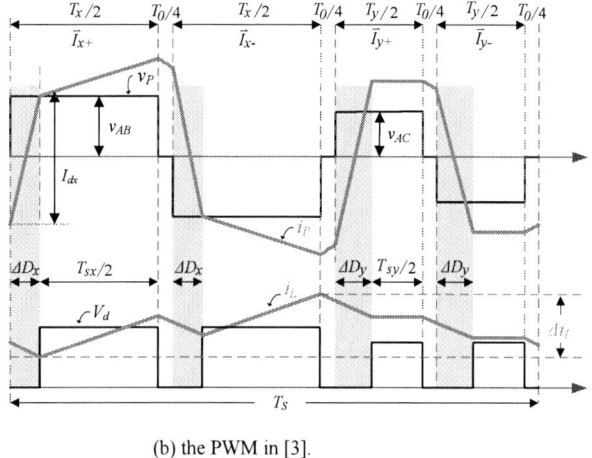

(b) the PWM in [3].

Fig. 2. Duty cycle losses comparison when $-\pi/6 < \theta < 0$ in :

Therefore, the effective duty cycle which is the duty cycle of the transformer secondary side rectifier voltage V_d is less than the duty cycle of the primary side voltage v_P. During the interval of the duty loss, transformer primary current i_P is ramping in a linear fashion from negative to positive or from positive to negative. I_{dx} is the total variation of i_P which is determined by the load current. It is assumed that the output current ripple is small compared to the load current, which is a realistic assumption at full load where the loss of duty cycle is maximal [3].

$$I_{dx} \cong 2nI_L = 2nI_o \qquad (1)$$

Where I_L is the output inductor current, I_o is the load current, n is transformer ratio.

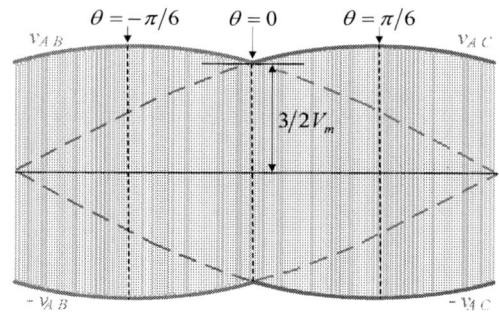

Fig. 3. Envelop of transformer primary voltage $v_P(\theta)$.

The total duty cycle loss of presented PWM can be derived as:

$$\Delta D_{total} = 2\Delta D_x = \frac{4nI_o L_{lk}}{v_P(\theta)T_s} \qquad (2)$$

Where $v_P(\theta)$ is the voltage across the leakage inductance L_{lk} during interval of duty loss ΔD_x, and the angle θ is the angle displacement between the current reference vector $\vec{I}_{ref}$ and angle of α-axis of the α-β plane [5]. Transformer voltage $v_P(\theta)$ is one of the three line-to-line voltages depending on the angle θ as shown in Fig. 3. The maximum duty cycle loss is obtained when the magnitude of $v_P(\theta)$ is minimum at:

$$\theta = 0, \quad v_P(\theta) = \frac{3}{2}V_m,$$

$$\Delta D_{total\_max} = \frac{8}{3}\frac{nI_o L_{lk}}{V_m T_s} \qquad (3)$$

where V_m is the peak value of grid side phase voltage. Since the output voltage can be expressed as [5]:

$$V_o = \frac{3}{2}nm_a V_m \qquad (4)$$

where m_a is modulation index. Substituting (4) into (3), the maximum total duty cycle loss in term of output load is given by

$$\Delta D_{total\_max} = \frac{4n^2 m_a L_{lk}}{R_o T_s} \qquad (5)$$

where $R_o = V_o/I_o$ is the load resistance. In addition to two intervals of ΔD_x, total duty cycle loss of PWM in [3] contains another two intervals of ΔD_y as shown in Fig. 2(b). The interval of ΔD_y is longer than intervals of ΔD_x since ΔD_y is associated to the narrower pulse where magnitude of $v_P(\theta)$ is lower. At the extreme case, when $\theta = \pm\pi/6$, interval of ΔD_y is twice of interval of ΔD_x as shown in Fig. 5(b) since magnitude of $v_P(\theta)$ during interval of ΔD_y is only half of that during interval of ΔD_x. In spite of the two missing pulses for $\vec{I}_y$ on secondary side at $\theta = \pm\pi/6$, two pulses for $\vec{I}_y$ on primary side always exist in order to compensate the duty cycle losses. Therefore, these two pulses contribute no power to the output, but create the losses. The total duty cycle losses of presented PWM and PWM in [3] are compared at different angle θ as shown in Table I. In order to make an easier comparison, all

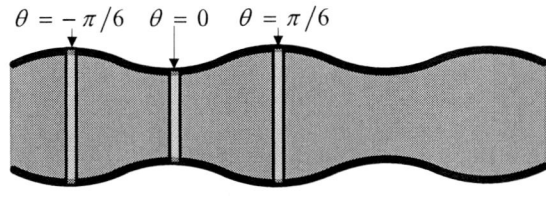

(a) proposed PWM.

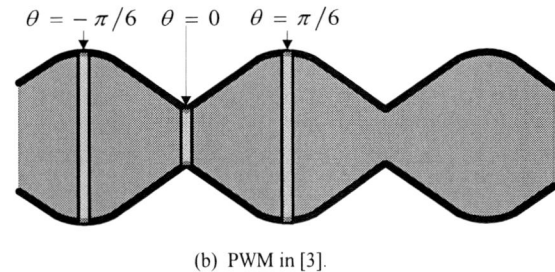

(b) PWM in [3].

Fig. 4. Envelope of output inductor current ripple:

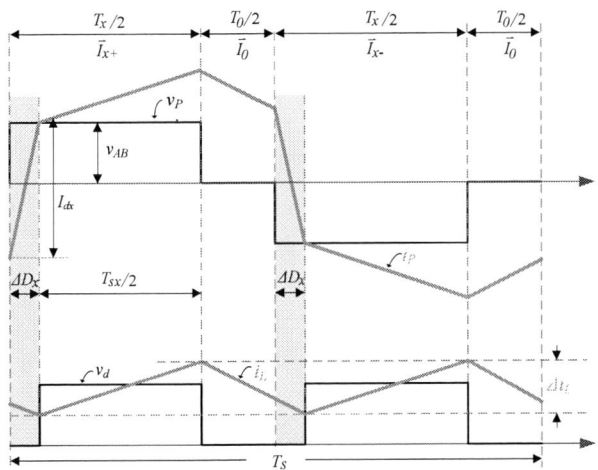

(a) with proposed PWM at $\theta = -\pi/6$.

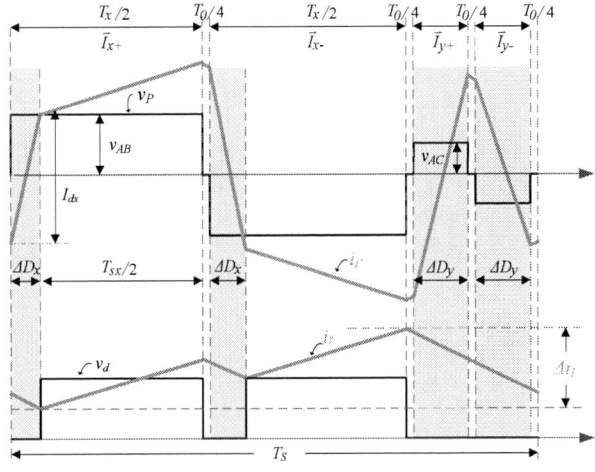

(b) with PWM in [3] at $\theta = -\pi/6$.

Fig. 5. Zoom in maximum current ripple of the current envelop in Fig. 4:

the total duty cycle losses are normalized by using the total duty cycle loss of presented PWM at $\theta = 0$ as a base. It is interesting to note that the total duty cycle loss of PWM in [3] is 2 to 3 times higher than the presented PWM. In summary, compared with PWM scheme in [3], the duty cycle loss of presented PWM is smaller if the same value of L_{lk} is used. In other words, if both designs have the same duty cycle losses, the presented PWM scheme can achieve ZVS with wider load range by using larger value of L_{lk}. This significantly reduces the turn on switching losses at lighter load. Reduced duty cycle losses also enable the power supply operating at higher m_a and reduce the power losses caused by the circulating current during the interval of duty cycle losses.

TABLE I. COMPARISON OF THE NORMALIZED TOTAL DUTY LOSSES.

	ΔD_{total} in Proposed PWM	ΔD_{total} PWM in [3]
$\theta = 0$	1	2
$-\pi/6 < \theta < 0$, $0 < \theta < \pi/6$	$\sqrt{3}/2 < \Delta D_{total} < 1$	$2 < \Delta D_{total} < 3\sqrt{3}/2$
$\theta = \pm\pi/6$	$\Delta D_{total} = \sqrt{3}/2$	$\Delta D_{total} = 3\sqrt{3}/2$

B. Output Inductor Current Ripple

At steady state, the output inductor current ripple varies with phase angle θ as shown in Fig. 5 (a) and (b), at $\theta = -\pi/6$ or $\theta = \pi/6$, the current ripple of presented PWM scheme

reaches maximum and can be derived as:

$$\Delta I_{max} = \frac{V_o(1 - \frac{\sqrt{3}}{2}m_a)T_s}{2L_o} \quad (6)$$

At $\theta = 0$, the current ripple is minimum and can be derived as:

$$\Delta I_{min} = \frac{V_o(1 - m_a)T_s}{2L_o} \quad (7)$$

As shown in Fig. 4 (a) and (b), the maximum envelope of current ripple (at $\theta = \pm\pi/6$) in the presented PWM is lower than the PWM in [3], while the minimum envelope of current ripple in the presented PWM (at $\theta = 0$) is higher than the PWM in [3]. For the PWM in [3], the dwell time of $\vec{I}_0$ is equally divided by four and inserted between $\vec{I}_x$ and $\vec{I}_y$ as shown in Fig. 2 (b). At $\theta = -\pi/6$, the two pulses for vector $\vec{I}_y$ on secondary side disappear so that the rest of two pulses are not evenly distributed, resulting in larger current ripple as shown in Fig. 5 (b). The same situation happens at $\theta = \pi/6$ (not drawn), the two pulses for vector $\vec{I}_x$ on secondary side disappear and the rest of two pulses are not evenly distributed, resulting in larger current ripple as well. It is important to observe that, the frequency across the transformer with PWM in [3] is double of the presented PWM, given that both PWM schemes have the same T_s. Then, lower core losses of the transformer can be expected with PWM in [3]. However, PWM scheme in [3] doesn't offer a benefit in transformer design if the transformer is designed based on the maximum flux density which is normally the most important constraints for transformer design. At the vicinity of $\theta = \pm \pi/6$, one of the two active vectors attained it's maximum magnitude and the other vector reach to minimum. Therefore, the volt-sec on the transformer attained its maximum value at $\theta = \pm \pi/6$ for both PWM scheme. As a result, the maximum flux density on transformer is same for both PWM schemes, although the transformer operating frequency of the presented PWM is only half of [3].

III. SIMULATION AND EXPERIMENTAL RESULTS

The simulation model and the experimental prototype are setup at the rated power of 2.7 KW and $m_a = 0.8$ as shown in Table. II. Fig. 6 and Fig. 7 show that the simulation results match with the experimental results very well. The duty losses can be identified as the intervals between the rising edges of v_P

and v_S. In one cycle, the duty loss in [3] contains four intervals (two associated with wide pulses and two associated with narrow pulses) as shown in Fig. 6 (a) and (b). The duty loss of presented PWM only contains two intervals in one cycle as shown in Fig. 7 (a) and (b). In [3], the two intervals associated with the two wide pulses have the same duration as that of presented PWM, while the other two intervals associated with the two narrow pulses have longer duration because the voltage across the transformer is lower and i_P has longer ramping time. Therefore, the total duty loss in [3] is more than twice of the total duty loss of presented PWM. Although the operating frequency of the presented PWM is only half of [3], the maximum output current ripple in presented PWM is lower than the PWM in [3] as shown in Fig. 6 (c), (d) and Fig. 7 (c), (d).

TABLE II.   SIMULATIN AND EXPERIMENTAL PARAMETERS.

L_f	90 µH	R_o	33.3Ω
C_f	5 µF	V_o	300 V
f_{TR} in propsed PWM	50 KHz	f_{TR} in [3]	100 KHz
f_{grid}	60 Hz	n	2
$V_{LL,rms}$	160 V	L_{lk}	5.7µH
L_o	450 µH	$C_{winding}+C_{diode}$	1300 pF

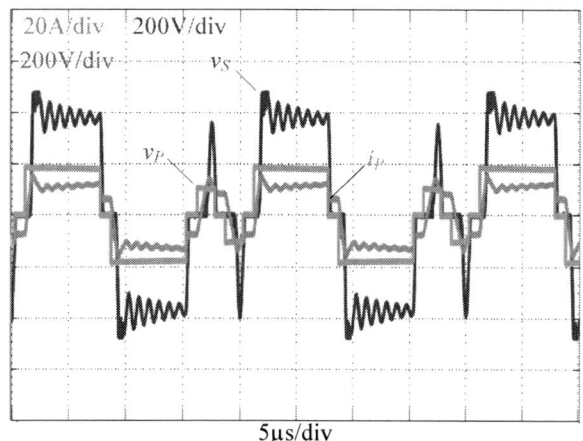

(a) Transformer primary voltage v_P and current i_P and secondary v_S

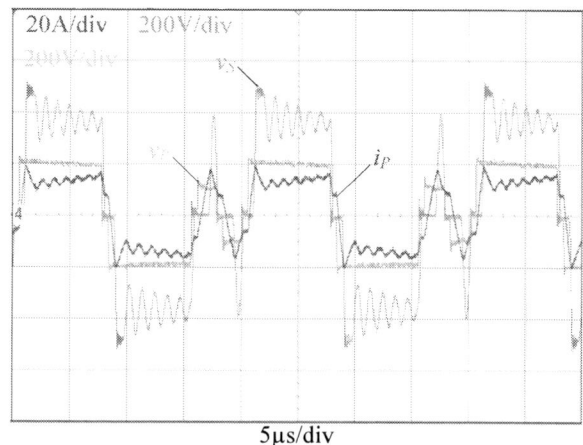

(b) Transformer primary voltage v_P and current i_P and secondary v_S

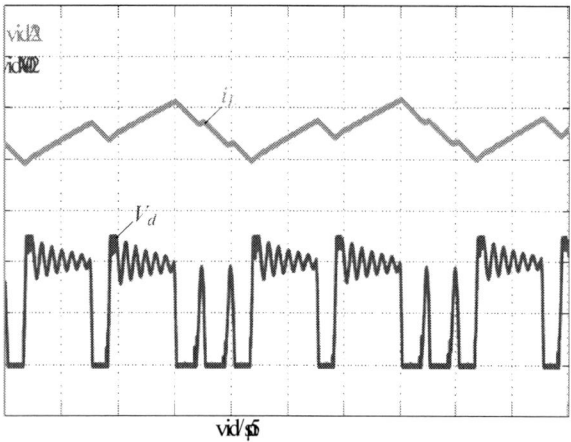

(c) Inductor current i_L and voltage across diode rectifier V_d

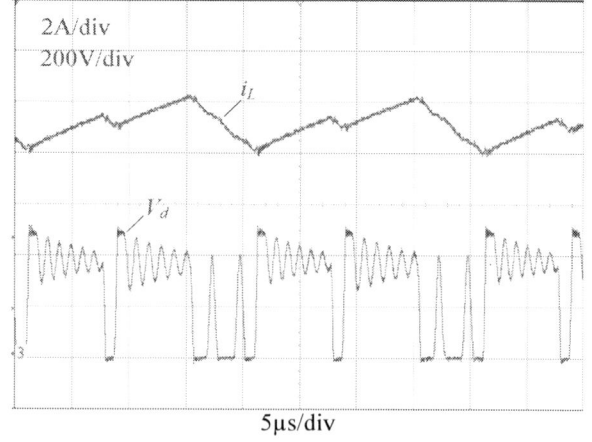

(d) Inductor current i_L and voltage across diode rectifier V_d

Fig. 6. PWM in [3] at f_{TR}=100 KHz (a), (c) simulation and (b), (d) experimental results.

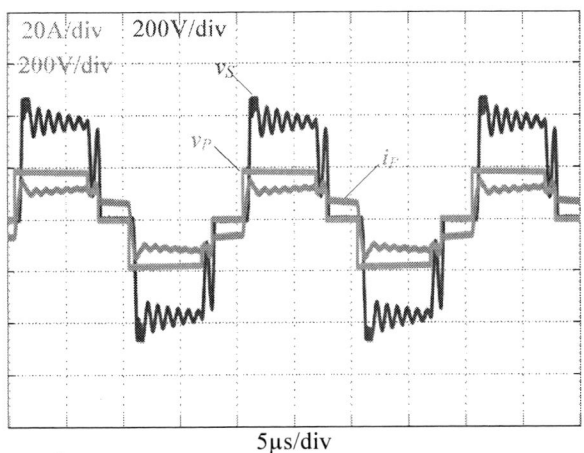

(a) Transformer primary voltage v_P and current i_P and secondary v_S.

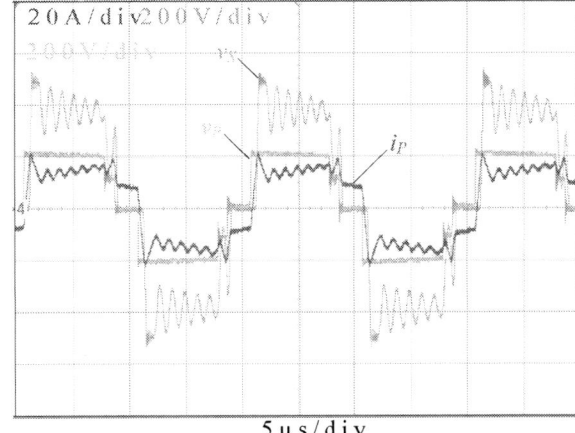

(b) Transformer primary voltage v_P and current i_P and secondary v_S.

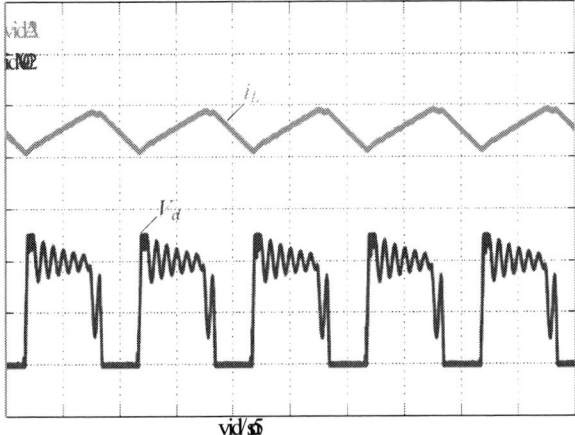

(c) Inductor current i_L and voltage across diode rectifier V_d

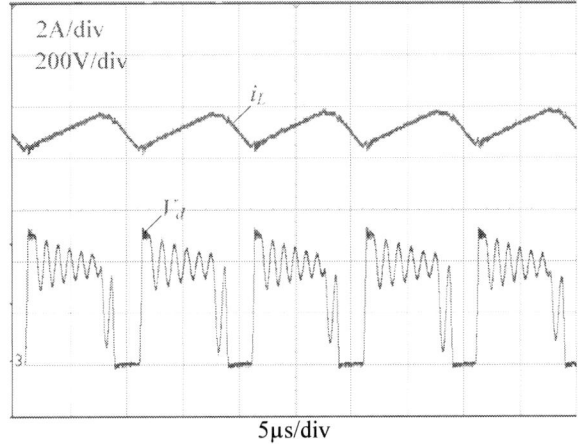

(d) Inductor current i_L and voltage across diode rectifier V_d

Fig. 7. Proposed PWM at f_{TR}=50 KHz (a), (c) simulation and (b), (d) experimental results.

IV. CONCLUSION AND FUTURE WORK

In this paper, the merits of a PWM scheme based on SVM for three-phase isolated high frequency PWM rectifier are presented. Loss of duty cycle and output inductor current ripple is discussed in detail. Compared with the PWM scheme in [3], the presented PWM scheme can significantly reduce the loss of duty cycle with the same ZVS operating range. The maximum output inductor current ripple is also lower even at an half of the switching frequency on transformer compare to [3] which result in smaller inductor size. In addition, the maximum volt-sec on the transformer (at $\theta = \pm \pi/6$) is the same as [3] so that the same design can be applied on transformer. The output inductor current ripple has less variation within each sector due to the fact that the volt-sec on transformer has less variation. A higher core loss on transformer is expected, which can be considered as a drawback compare to [3]. Detail comparison and discussion on the input current THD will be presented in the future publication.

REFERENCES

[1] A. Stupar, T. Friedli, J. Minibock, and J. W. Kolar, "Towards a 99% efficient three-phase buck-type PFC rectifier for 400 V DC distribution systems [J]," *IEEE Trans. Power Electron.*, vol. 27, no. 4, pp.1732–1744, 2012.

[2] R. García-Gil, J. M. Espí, F.Voelker, E. J. Dede, and J. Castelló, "A novel four-quadrant power supply for low-energy correction magnets," *Nucl. Instrum. Methods Phys. Res. A*, vol. 510, pp. 357–361, 2003.

[3] V. Vlatkovic, D. Borojevic, X. Zhuang, and F. C. Lee, "Analysis and design of a zero-voltage switched, three-phase PWM rectifier with power factor correction," Proc. IEEE PESC *Conf,* 1992, pp. 1352-1360.

[4] V. Vlatkovic, D. Borojevic, and F. C. Lee, "A Zero-Voltage Switched, Three-Phase Isolated PWM Buck Rectifier". IEEE Transaction on Power Electronics, vol. 10, n. 2, Mar. 1995, pp. 148 - 157.

[5] J. Afsharian, D. Xu, B. Gong, and Z. Yang "Space Vector Demonstration and Analysis of Zero-Voltage Switching Transitions in Three-Phase Isolated PWM Rectifier". in IEEE Energy Conversion Congress and Exposition (ECCE) Conf, Sep. 2015, pp. 2477 - 2484.

[6] J.A. Sabate, V. Vlatkovic, R.B. Ridley, F.C. Lee, and B.H. Cho, "Design Considerations for High-Voltage High-Power Full-Bridge Zero-Voltage-Switched PWM Converter." in IEEE Applied Power Electronics Conf. '90 Proc., 1990, pp. 275-284.

Analysis, Design, and Evaluation of Three-Phase Three-Wire Isolated AC-DC Converter Implemented with Three Single-Phase Converter Modules

Laszlo Huber, Misha Kumar, and Milan M. Jovanović
Delta Products Corporation
P.O. Box 12173
5101 Davis Drive
Research Triangle Park, NC 27709, USA

Dinggang Ping and Gang Liu
Shanghai Design Center,
Delta Electronics (Shanghai) Company Ltd,
Shanghai 201209,
People's Republic of China

Abstract – In this paper, a control method for three single-phase isolated ac-dc converter modules employed to implement an isolated three-phase, three-wire ac-dc converter is presented and its detailed design procedure provided. In the described control the input voltages of the Y–connected single-phase modules with floating common Y point are kept within a safe range by equalizing their input admittances. It is shown that for obtaining equal input admittances of the three PFC front stages, the outputs of the PFC voltage controllers should be equal. This is achieved by adjusting the reference currents of the current controllers in the output dc-dc stages. The performance of the presented control with both balanced and unbalanced source voltages, as well as with a phase fault (open or short circuit) is evaluated by Matlab/Simulink simulations and verified experimentally on a 6-kW prototype.

I. INTRODUCTION

Three-phase isolated ac-dc converters can be implemented either with a direct three-phase PFC rectifier front end such as the Vienna rectifier or the six-switch PFC boost rectifier followed by an isolated dc-dc converter, or with three single-phase isolated ac-dc converters. Generally, the major advantage of the modular implementation with three single-phase converters is its ease of power expandability. To be able to employ single-phase modules designed for 220/277-Vrms phase-to-neutral voltage in three-phase power systems with nominal phase-to-phase voltage of 380/480 Vrms, the three single-phase modules must be connected in star (Y) configuration. The delta (Δ) configuration cannot be used since it would require that single-phase modules be connected across two phases, i.e., to a voltage exceeding their rating.

In applications where the neutral point of the three-phase voltage source is available, the common point (Y point) of the single-phase converter modules is connected to the source neutral point and the three single-phase converters operate independently from each other with their input voltages equal to the respective phase-to-neutral voltages of the source. However, in applications where the source neutral point is not provided, such as, for example, in standard telecom power supplies, any unbalance in the three-phase source phase voltages and/or in the three single-phase modules will create a potential difference between the Y point of the single-phase modules and the source neutral point, resulting in oscillations and significant variations of the input voltages of the single-phase converters [1]. For a stable and reliable steady-state operation, i.e., operation where the input voltage always stays within a specified range, a balancing control of the three single-phase modules is necessary. The stable and reliable

operation has also to be guaranteed at startup, as well as in case of phase failure (open and short circuit) and load transients.

The balancing control between the three single-phase modules can be achieved with additional passive components used to create a virtual (artificial) neutral point and by using balancing control methods [1]-[8], or without additional passive components by using only balancing control methods [9]. In [1]-[3], three auxiliary transformers with Y-connected primaries and Δ-connected secondaries are used to create an artificial neutral point (ANP). The Y point of the modules is connected to the ANP. With the ANP, the three single-phase modules operate similarly as in the case when the source neutral point is available. When the system is unbalanced, a zero-sequence current flows in the auxiliary transformers that can result in significant losses. To suppress the zero-sequence current, a balancing control method can be employed, where the output reference currents of the single-phase modules, which for a balanced system are equal, are adjusted based on the phase angle of the zero-sequence current [1]. The drawback of using auxiliary transformers is the increased size and cost. Typically, the VA rating of the additional magnetic components is 5% of the VA rating of the rectifier [1].

In [4]-[8], a virtual neutral (VN) point is created by a star connection of three equal resistors. When the three-phase voltage source is balanced, the potential of the virtual neutral point v_{VN} is equal to the potential of the source neutral point v_0 and, therefore, the voltage across a star resistor is equal to the corresponding phase voltage of the three-phase source. Generally, the potential difference between the virtual neutral point and the source neutral point is

$$v_{VN,0} = v_{VN} - v_0 = \frac{v_{a0} + v_{b0} + v_{c0}}{3} \quad , \qquad (1)$$

where, v_{a0}, v_{b0}, and v_{c0} are the three-phase source phase-to-neutral voltages. Therefore, if the phase voltages of the three-phase source are unbalanced and contain a zero-sequence component v_{ZS}, the zero-sequence component v_{ZS} appears as voltage $v_{VN,0}$, i.e., $v_{ZS}=v_{VN,0}$, and, consequently, the voltage across the star resistors will not contain the zero-sequence component. The three-phase source phase voltages can be reconstructed from the voltages across the star resistors following the method presented in [11]. It should be noted that in these implementations the Y point of the single-phase modules is not connected to the virtual neutral point.

978-1-4673-9551-9/16 $31.00 © 2016 IEEE

In [4]-[8], the balancing control between the three single-phase modules is achieved by adjusting the output reference current of the single-phase modules. In [4] and [5], the output reference current of the single-phase modules is adjusted based on the potential difference between the Y point of the modules and the virtual neutral point, $v_{Y,VN}$, whereas in [6]-[8], the output reference current of the single-phase modules is adjusted based on the voltages across the star resistors. In addition, in [6]-[8], the amplitude of an input phase reference current (regulated by average current control) is obtained from the sum of three components: 1) output of a PI-type average-voltage controller that regulates the average value of the output voltages of the three single-phase PFC front ends, 2) an output-power feedforward term (used to improve the control dynamics for load transients), and 3) output of a P-type individual-voltage controller that balances the unequal output voltages of the single-phase PFC front ends. In [4]-[8], the phase currents are controlled to follow the waveform of the voltages across the star resistors, i.e., the source phase potentials referenced to the virtual neutral point.

In [9], the balancing control between the three single-phase modules is achieved by adjusting the output reference current of the single-phase modules based on the output voltages of the voltage controllers of the single-phase PFC stages (employing average current control). Unlike in [4]-[8], in [9] the phase currents are controlled to follow the input voltages of the single-phase modules, i.e. the source potentials referenced to the star point of the modules. Therefore, by using the balancing control method in [9] compared to those in [4]-[8], standard single-phase converter modules can be easier modified for the three-wire three-phase systems. Unfortunately, no control-oriented analysis is provided in [9].

Finally, in [12] a balancing control method that does not require any load side balancing is described. In this method, only two output voltages of the three single-phase PFC front ends are balanced at the same time, which avoids the coupling between the three single-phase modules. However, it should be noted that in [12] the availability of the source neutral point is required, but the Y point of the single-phase modules is not connected to the source neutral point.

In this paper, principles of the balancing control method introduced in [9] are explained and a detailed design

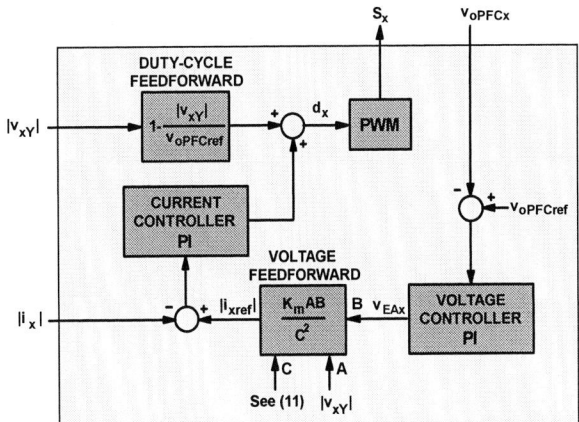

Fig. 2 Block diagram of PFC control circuit.

procedure of the balancing controllers is provided. In addition, the operation with unbalanced source voltages is analyzed with respect to the power distribution between the three single-phase modules, amplitude of the input phase currents, and power factor of the modules. Finally, operation and performance of the balancing control when one phase voltage is reduced and when one phase is disconnected are illustrated with Simulink simulations. Experimental results obtained on a 6-kW prototype are also provided.

II. PRINCIPLES OF BALANCING CONTROL

The block diagram of the three single-phase converter modules with Y connection at the input and parallel connection at the output is shown in Fig. 1. The PFC stage of the modules is implemented with average current control, as shown in Fig. 2. The dc-dc stages are also implemented with average current control, where the output-voltage controller is common for all three dc-dc stages as shown in Fig. 3.

The three single-phase converter modules in Fig. 1 can be equivalently represented with their input admittances Y_k, $k \in \{a,b,c\}$, as shown in Fig. 4 [10]. Using Kirchhoff's voltage and current laws in phasor notation, the potential difference between the Y point of the single-phase modules and the source neutral point (neutral displacement voltage) can be obtained as

$$\underline{V}_{Y0} = \frac{\underline{V}_{a0}\underline{Y}_a + \underline{V}_{b0}\underline{Y}_b + \underline{V}_{c0}\underline{Y}_c}{\underline{Y}_a + \underline{Y}_b + \underline{Y}_c} . \quad (2)$$

If by the balancing control, the input admittances are kept equal, i.e., $\underline{Y}_a = \underline{Y}_b = \underline{Y}_c$, the neutral displacement voltage in

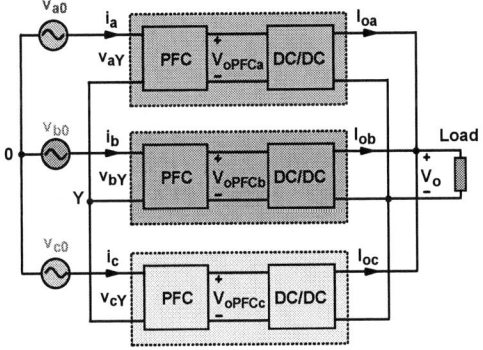

Fig. 1 Block diagram of three single-phase rectifier modules with Y connection at input and parallel connection at output.

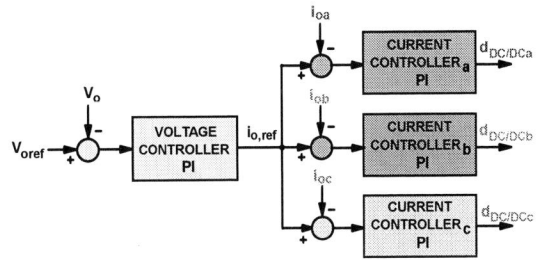

Fig. 3 Block diagram of dc-dc control circuit.

978-1-4673-9551-9/16 $31.00 © 2016 IEEE   39

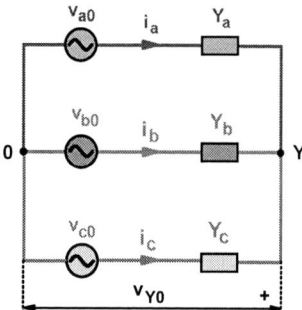

Fig. 4 Simplified equivalent circuit of three single-phase converter modules.

(2) is

$$\underline{V}_{Y0} = \frac{\underline{V}_{a0} + \underline{V}_{b0} + \underline{V}_{c0}}{3} \quad . \qquad (3)$$

Since for balanced source voltages $\underline{V}_{Y0} = 0$, the input voltage of each single-phase PFC module is equal to the corresponding source voltage.

For unbalanced source voltages where the amplitude of one phase voltage is smaller than the amplitude of the other two phase voltages, i.e., $V_{a0} = mV_m$, $m < 1$, and $V_{b0} = V_{c0} = V_m$,

$$\underline{V}_{Y0} = -\frac{1-m}{3} \cdot \underline{V}_{a0} , \qquad (4)$$

i.e., $\underline{V}_{Y0}$ is collinear with $\underline{V}_{a0}$, but it has opposite direction, as shown in the phasor diagram in Fig. 5. It follows from Fig. 5 that the amplitude of the input voltage of all three single phase PFC modules is smaller than V_m, i.e., $V_{kY} < V_m$, $k \in \{a,b,c\}$.

The condition for equal input admittances is derived by recognizing that according to Fig. 2, the reference current of the modules is given as

$$\left| i_{k,ref} \right| = K_m \frac{A_k \cdot B_k}{C_k^2} = K_m \frac{|v_{kY}| \cdot v_{EAk}}{C_k^2}, \ k \in \{a,b,c\}, \quad (5)$$

where, K_m is the multiplier constant and C_k is the voltage feedforward term. By the average current control, the current controllers in the PFC stages enforce the input phase currents to follow the reference currents, i.e., $i_k = i_{k,ref}$. Therefore, the

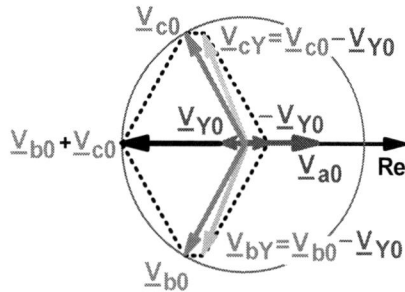

Fig. 5 Phasor diagram of input voltages of single-phase PFC modules when amplitude of phase "a" source voltage is smaller than amplitude of phase "b" and "c" source voltages, i.e., for $V_{a0} = mV_m$, $m < 1$, and $V_{b0} = V_{c0} = V_m$.

input admittance Y_k can be obtained as

$$Y_k = \frac{|i_k|}{|v_{kY}|} = \frac{|i_{k,ref}|}{|v_{kY}|} = K_m \frac{v_{EAk}}{C_k^2}, \ k \in \{a,b,c\} . \quad (6)$$

It follows from (6) that for equal input admittances

$$Y_a = Y_b = Y_c = Y_{in} \ , \qquad (7)$$

the outputs of the voltage controllers as well as the voltage-feedforward terms in the PFC control circuits should be equal, i.e.,

$$v_{EAa} = v_{EAb} = v_{EAc} = v_{EA}, \qquad (8)$$

and

$$C_a = C_b = C_c = C . \qquad (9)$$

The voltage feedforward term C is obtained by considering the total input power

$$P_{in,tot} = \sum_{k=a,b,c} I_{k,RMS} \cdot V_{kY,RMS} = Y_{in} \cdot \sum_{k=a,b,c} V_{kY,RMS}^2 . (10)$$

Then, it follows from (6)-(10), that if

$$C^2 = \sum_{k=a,b,c} V_{kY,RMS}^2 , \qquad (11)$$

the output voltage of the voltage controllers in the PFC control circuit is obtained as

$$v_{EA} = \frac{P_{in,tot}}{K_m}, \qquad (12)$$

i.e., v_{EA} does not depend on the input voltages, which is the goal of the voltage feedforward control method. In the case, no voltage feedforward is implemented, $C = 1$.

With equal input admittances, for balanced source voltages, the total input power is equally distributed between the three single-phase modules. However, for unbalanced source voltages, to achieve equal input admittances, the total input power cannot be equally distributed between the three single-phase modules. In fact, a module connected to a reduced source voltage will draw less input current and, therefore, it will operate at reduced input power. The desired distribution of the total power between the single-phase modules that results in equal input admittances can be achieved by implementing the balancing-control circuit as shown in Fig. 6. In this balancing circuit, reference currents $i_{ok,ref}$, $k \in \{a,b,c\}$ of the current controllers in the dc-dc stages are adjusted by currents $i_{Adj,k}$, $k \in \{a,b,c\}$ which are obtained at the outputs of the balancing controllers that regulate the difference between the average output-voltage of PFC controllers

$$v_{EAavg} = \frac{\sum_k v_{EAk}}{3} , k \in \{a,b,c\} , \qquad (13)$$

978-1-4673-9551-9/16 $31.00 © 2016 IEEE

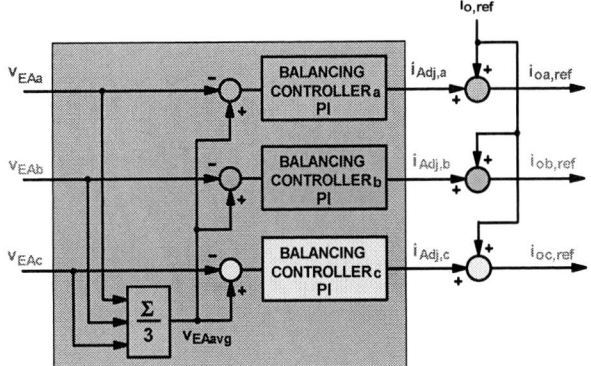

Fig. 6 Block diagram of balancing control circuit.

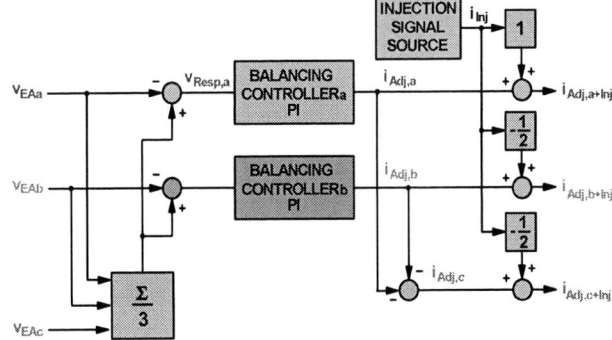

Fig. 8 Block diagram of balancing control circuit with injection signal source used for measuring balancing loop transfer functions.

and corresponding module PFC control voltage v_{EAk}.

The design of the balancing controllers is performed through simulations in Simulink and SIMetrix. In order to reduce the simulation time, in the simulation circuit the dc-dc output stages are replaced with current sources $I_{Load,k}$, $k \in \{a,b,c\}$, which represent the load currents of the PFC stages, i.e., the input currents of the dc-dc stages for balanced source voltages, and with controlled current sources $i_{Adj,k}$, $k \in \{a,b,c\}$, which are controlled by the output signals of the balancing controllers, as shown in Fig. 7. Instead of adjusting the reference currents of the current controllers in the dc-dc stages, the load currents of the PFC stages are adjusted. However, to keep the total load current constant, the total adjustment current must be zero. Therefore, the balancing control circuit in Fig. 7 is modified so that the adjustment current in phase "c" is generated as the negative sum of the adjustment currents in phases "b" and "c", i.e.,

$$i_{Adj,c} = -(i_{Adj,a} + i_{Adj,b}). \tag{14}$$

To measure the balancing loop transfer function in phase "a", a perturbation signal is injected at the output of the "a" balancing controller as shown in Fig. 8. This injection current, i_{Inj}, is added to adjustment current $i_{Adj,a}$,

$$i_{Adj,a+Inj} = i_{Adj,a} + i_{Inj} . \tag{15}$$

To keep the total load current constant, injection signal i_{Inj} is also injected in phases "b" and "c" with opposite sign and with half of its value. The plant transfer function is defined as

$$T_{PL}(s) = \frac{V_{Resp,a}(s)}{I_{Adj,a+Inj}(s)}, \tag{16}$$

where, $V_{Resp,a}(s)$ is the response signal at the input of the balancing controller "a". Finally, the balancing loop gain is defined as

$$T_{LG}(s) = T_{PL}(s) \cdot T_{BC}(s), \tag{17}$$

where, $T_{BC}(s)$ is the transfer function of the balancing controller.

The Bode plots of the plant transfer function, balancing controller transfer function and balancing loop gain are presented in Fig. 9, which were obtained at the zero crossing of source voltage v_{a0}. The balancing controller is designed as a PI controller. The crossover frequency of the balancing loop gain is selected at 3.3 Hz to avoid interaction with the PFC voltage loop whose bandwidth is 10 Hz. At the crossover frequency, the magnitude of the balancing controller transfer function is

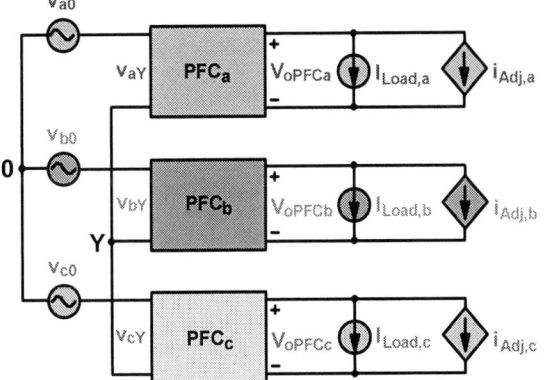

Fig. 7 Block diagram of three single-phase PFC modules used in simulations.

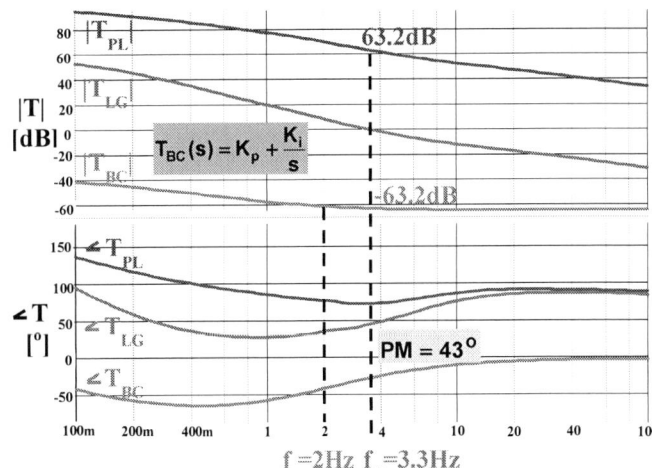

Fig. 9 Bode plots of plant transfer function, balancing controller transfer function, and balancing loop gain.

978-1-4673-9551-9/16 $31.00 © 2016 IEEE   41

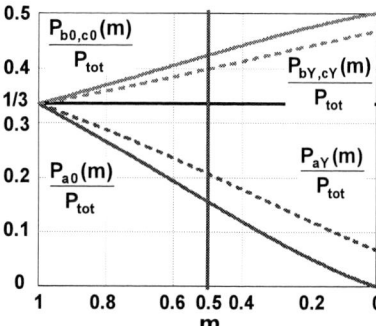

Fig. 10 Calculated power distribution between three single-phase modules.

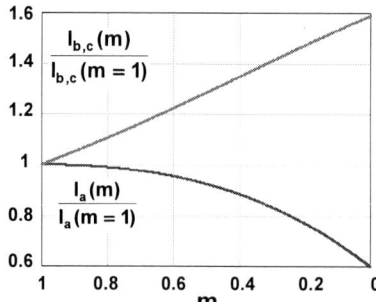

Fig. 11 Calculated amplitude of phase currents.

$$|T_{BC}(\omega_c)|[dB] = 20\log\left(K_p\sqrt{1+\left(\frac{\omega_z}{\omega_c}\right)^2}\right) = |T_{PL}(\omega_c)|[dB] = 63.2\,dB\ . \quad (18)$$

Selecting $f_z = 2$ Hz, it follows from (18) that $K_p = 5.91 \cdot 10^{-4}$. Finally, from

$$\omega_z = \frac{K_i}{K_p}, \quad (19)$$

it is obtained that $K_i = 7.42 \cdot 10^{-3}$.

III. ANALYSIS OF OPERATION FOR UNBALANCED SOURCE VOLTAGES

The operation for unbalanced source voltages is analyzed for the case when the amplitude of one phase voltage is smaller than the amplitude of the other two phase voltages i.e., for $V_{a0} = mV_m$, $m < 1$, and $V_{b0} = V_{c0} = V_m$. The analysis includes the power distribution between the three single-phase modules, the amplitude of the phase currents, and the power factor of the modules.

The calculated power distribution between the three single-phase modules is shown in Fig. 10. The dashed lines present the input power of the modules, P_{kY}, $k \in \{a,b,c\}$, whereas, the solid lines present the power delivered by the phase sources, P_{k0}, $k \in \{a,b,c\}$. As can be seen in Fig. 10, the input power of module "a", whose source voltage is reduced, is provided by its own source "a", but also by sources "b" and "c". For example, for $m = 0.5$, module "a" consumes 20.5%, while modules "b" and "c" each consume 39.7% of the total power.

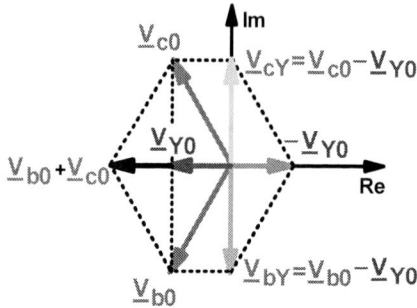

Fig. 12 Phasor diagram of input voltages of PFC modules when phase "a" is disconnected.

However, source "a" delivers only 15.4%, while sources "b" and "c" each deliver 42.3% of the total power.

The calculated amplitude of the phase currents is shown in Fig. 11. It should be noted in Fig. 11 that the amplitude of phase current "a" decreases by only 1% when the amplitude of phase voltage "a" decreases by 20%. It should also be noted in Fig. 11 that at $m = 0$, the amplitude of phase currents "b" and "c" increases by almost 60%. However, if phase "a" is disconnected, the amplitude of phase currents "b" and "c" increases by 73.2%. In fact, when phase "a" is disconnected and the input admittances of phases "b" and "c" are equal, the neutral displacement voltage V_{Y0} is

$$\underline{V}_{Y0} = \frac{\underline{V}_{b0} + \underline{V}_{c0}}{2}\ . \quad (20)$$

As shown by the phasor diagram in Fig. 12, when phase "a" is disconnected, input voltages $\underline{V}_{bY}$ and $\underline{V}_{cY}$, and, consequently, input phase currents $\underline{I}_b = Y \cdot \underline{V}_{bY}$ and $\underline{I}_c = Y \cdot \underline{V}_{cY}$ are equal with opposite signs and they are phase shifted by $30°$ versus the respective input source voltages $\underline{V}_{b0}$ and $\underline{V}_{c0}$. If the total power is maintained before and after phase "a" is disconnected, i.e.,

$$P_{tot} = 3 \cdot \frac{V_m I_m^{3\phi}}{2} = 2 \cdot \frac{V_m I_m^{2\phi}}{2}\cos(30°) = \frac{\sqrt{3}}{2}V_m I_m^{2\phi}\ , \quad (21)$$

it is finally obtained that

$$I_m^{2\phi} = \sqrt{3} \cdot I_m^{3\phi} = 1.732 \cdot I_m^{3\phi}\ . \quad (22)$$

The calculated power factor of modules "b" and "c" when the amplitude of phase voltage "a" decreases is shown in Fig. 13. It can be seen in Fig. 13 that the power factor of modules "b" and "c" is greater than 0.99 for $m > 0.55$. The power factor of module "a" does not decrease when the amplitude of phase voltage "a" decreases as it can be concluded from Fig. 5.

IV. SIMULATION AND EXPERIMENTAL RESULTS

Simulink simulation results obtained for 230-Vrms source phase voltages, 400-V output voltage of the PFC stages, and 6-kW output power are presented in Fig. 14. Figure 14 illustrates the operation with balancing control before and after the amplitude of phase voltage "a" drops by 50%. It

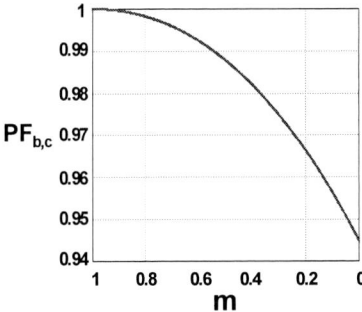

Fig. 13 Calculated power factor of single-phase modules.

should be noted in Fig. 14 that after the amplitude of phase voltage "a" drops by 50%, it takes approximately 0.2 sec for the new steady state to be established. It should also be noted in Fig. 14 that in the new steady state, the adjustment currents are $I_{Adj,a} \approx -2A$, and $I_{Adj,b} \approx I_{Adj,c} \approx 1A$. Before the drop of the amplitude of phase voltage "a", the load current of each PFC stage was 2kW/400V=5A. Therefore, in the new steady state, the adjusted load currents of the PFC stages are $I_{Load,a,adj} \approx 5-2 = 3A$ and $I_{Load,b,adj} \approx I_{Load,c,adj} \approx 5+1 = 6A$, i.e., the output power of the PFC stages is adjusted as $P_{Load,a,adj} \approx 1.2$ kW and $P_{Load,b,adj} \approx P_{Load,c,adj} \approx 2.4$ kW, which means that the output power of PFC stage "a" is reduced to 20% of the total output power, whereas, the output power of each of the PFC stages "b" and "c" is increased to 40% of the total output power. These results are in good agreement with the analytical results for $m = 0.5$ presented in Fig. 10.

Simulation waveforms that illustrate the operation when one phase is disconnected and then reconnected are presented in Fig. 15. The waveforms in Fig. 15 are obtained for 230-Vrms source phase voltages, 400-V output voltage of the PFC stages, and 6-kW output power when all three phases are connected, and at reduced output power of 76% (determined by a maximum overcurrent of 32%, i.e., 1.732·0.76 = 1.32, for the single-phase modules) when one phase is disconnected. It should be noted that when one phase is disconnected, the reference voltage of the balancing controllers is obtained as the averaged value of the output voltages of the PFC voltage controllers of the two active modules. In addition, when one phase is disconnected, the balancing loop bandwidth is increased by increasing balancing coefficient K_p five times. As can be seen in Fig. 15, in both cases (after disconnecting and after reconnecting phase "a"), the new steady-state is established in approximately 0.1 sec. The waveforms of the simulated input voltages v_{kY} and input currents i_k, $k\in\{a,b,c\}$, are in good agreement with the analytical results in Fig. 12.

Experimental results obtained on a 6-kW prototype at full load are presented in Figs. 16(a)-(c). Figure 16(a) shows the source phase voltages and input currents for balanced source voltages $V_{a0}=V_{b0}=V_{c0}=230$ Vrms. Figure 16(b) shows the same waveforms when the RMS value of source voltage "a" is decreased to $V_{a0}=180$ Vrms, i.e., to 78%, whereas, Fig. 16(c) shows the waveforms when the RMS value of source voltage "a" is further decreased to $V_{a0}=150$ Vrms, i.e., to 65%. It should be noted in Figs. 16(a)-(c) that as the RMS

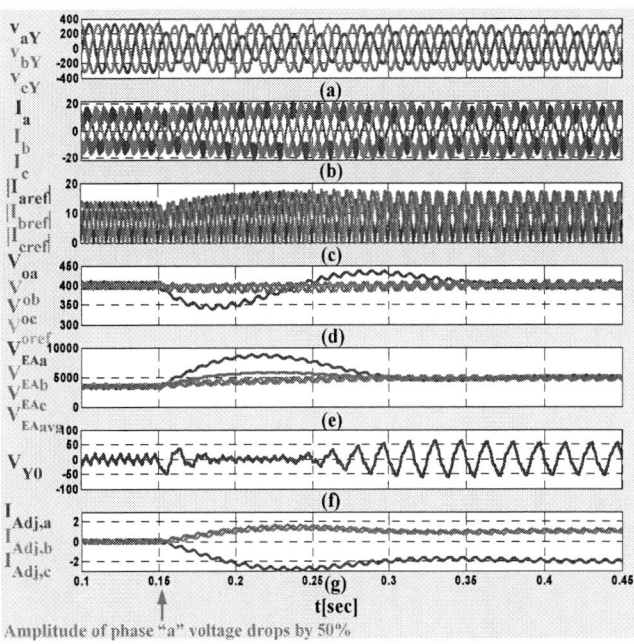

Fig. 14 Key simulated waveforms for 230Vrms source phase voltages and 6-kW load before and after amplitude of phase "a" source voltage drops by 50%: (a) low-pass filtered input phase voltages with respect to Y point [V], (b) input phase currents [A], (c) rectified input phase current references [A], (d) PFC output voltages [V], (e) PFC voltage controller outputs [digital value in Q12 format], (f) low-pass filtered neutral displacement voltage V_{Y0} [V], and (g) adjustment currents [A].

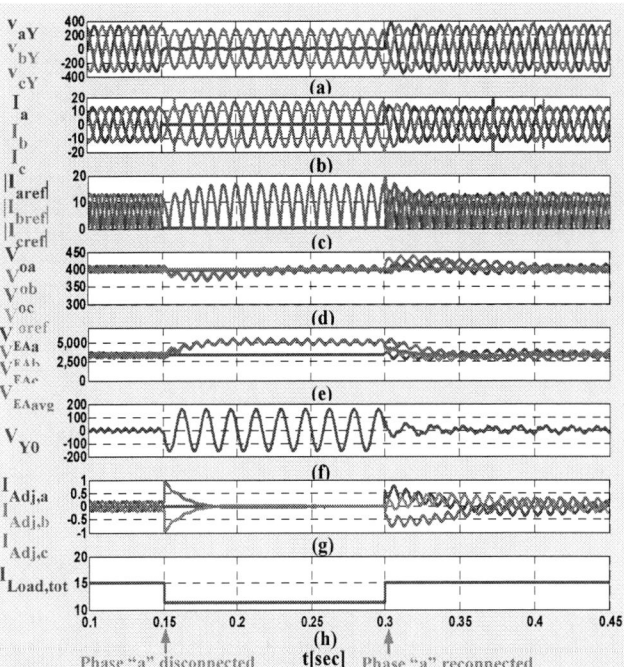

Fig. 15 Key simulated waveforms for 230Vrms source phase voltages and 6-kW load when phase "a" is disconnected and reconnected: (a) low-pass filtered input phase voltages with respect to Y point [V], (b) input phase currents [A], (c) rectified input phase current references [A], (d) PFC output voltages [V], (e) PFC voltage controller outputs [digital value in Q12 format], (f) low-pass filtered neutral displacement voltage V_{Y0} [V], (g) adjustment currents [A], and (h) total load current [A].

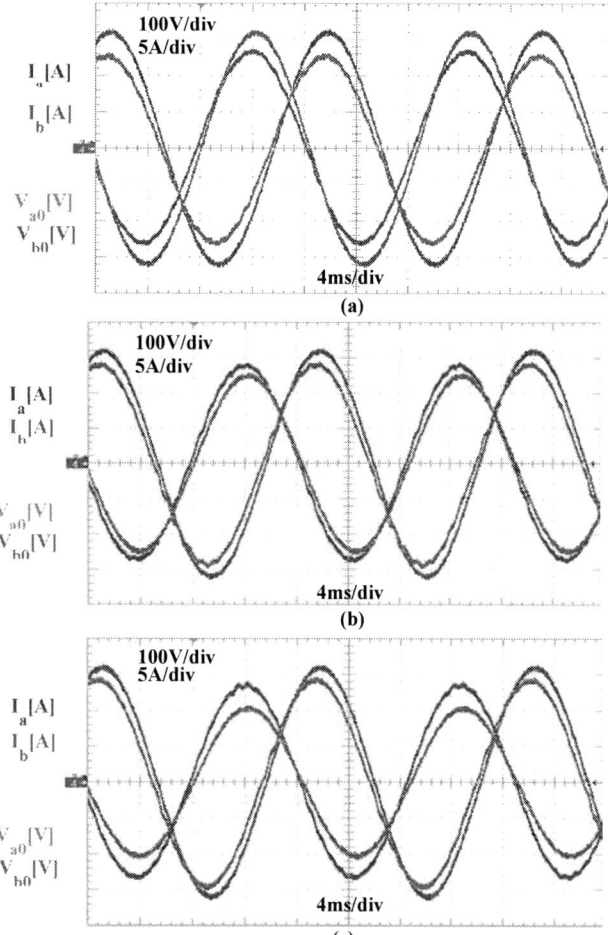

Fig. 16 Experimental waveforms of source phase voltages and input currents obtained on a 6-kW prototype at full load with: (a) balanced source voltages $V_{a0} = V_{b0} = V_{c0} = 230$ Vrms, (b) unbalanced source voltages $V_{a0} = 180$ Vrms, $V_{b0} = V_{c0} = 230$ Vrms, and (c) unbalanced source voltages $V_{a0} = 150$ Vrms, $V_{b0} = V_{c0} = 230$ Vrms.

value of source voltage V_{a0} decreases to 78% and to 65%, the amplitude of the input current of phase "a" is almost constant, whereas, the amplitude of input current of phase "b" increases by approximately 10% and 18%, respectively, which is in good agreement with the analytical results presented in Fig. 11.

V. SUMMARY

In this paper, analysis of the balancing control method for three single-phase isolated ac-dc converter modules employed to implement an isolated three-phase, three-wire ac-dc converter, introduced in [9], is presented. The balancing control between the three single-phase modules is achieved by equalizing the input admittances of the PFC front stages, which operate with average current control. It is shown that for equal input admittances, the outputs of the voltage controllers as well as the voltage feedforward terms in the PFC control circuit should be equal.

With equal input admittances, for balanced source voltages, the total input power is equally distributed between the three single-phase modules. However, for unbalanced source voltages, to achieve equal input admittances, a module

connected to a reduced source voltage draws less input current and, therefore, it operates at reduced input power. The desired distribution of the total power between the single-phase modules is achieved by adjusting the reference current of the current controllers in the dc-dc stages. The adjustment currents are obtained at the outputs of the balancing controllers, which balance the output voltages of the PFC voltage controllers.

A detailed design procedure of the balancing controllers performed through simulations in Simulink and SIMetrix is provided.

The operation with unbalanced source voltages is analyzed with respect to the power distribution between the three single-phase modules, amplitude of the input phase currents, and power factor of the modules. Operation and performance of the balancing control when one phase voltage is reduced and when one phase is disconnected and reconnected are illustrated with Simulink simulations.

Experimental results obtained on a 6-kW prototype are also provided.

REFERENCES

[1] M. Karlsson, C. Thoren, and T. Wolpert, "A novel approach to the design of three-phase AC/DC power converters with unity power factor," *Proc. International Telecommunications Energy Conf. (INTELEC)*, paper 5-1, Jun. 1999.

[2] M. Karlsson, C. Thoren, and T. Wolpert, "Practical considerations concerning a novel 6kW three-phase AC/DC power converter with unity power factor," *Proc. International Telecommunications Energy Conf. (INTELEC)*, pp. 28-33, Sep. 2000.

[3] M. Karlsson and T. Wolpert, "Stable artificial neutral point in a three phase network of single phase rectifiers," U.S. Patent 6,466,466, Oct. 15, 2002.

[4] D. Chapman, D. James, and C.J. Tuck, "A high density 48V 200A rectifier with power factor correction – An engineering overview," *Proc. International Telecommunications Energy Conf. (INTELEC)*, pp. 118-125, Sep. 1993.

[5] C.J. Tuck, D.A. James, and D.A. Chapman, "Power converter with star configured modules," U.S. Patent 5,757,637, May 26, 1998.

[6] R. Greul, S.D. Round, and J.W. Kolar, "Analysis and control of a three-phase unity power factor Y-rectifier," *IEEE Trans. Power Electronics*, vol. 22, no 5, pp. 1900-1911, Sep. 2007.

[7] R. Greul, U. Drofenik, and J.W. Kolar, "Analysis and comparative evaluation of a three-phase three-level unity power factor Y-rectifier," *Proc. International Telecommunications Energy Conf. (INTELEC)*, pp. 421-428, Oct. 2003.

[8] R. Greul, U. Drofenik, and J.W. Kolar, "A novel concept for balancing of the phase modules of a three-phase unity power factor Y-rectifier," *Proc. Power Electronics Specialists Conf. (PESC)*, pp. 3787-3793, Jun. 2004.

[9] R. Girod and D. Weida, "High efficiency true 3-phase compact switch-mode rectifier module for telecom power solutions," *Proc. International Telecommunications Energy Conf. (INTELEC)*, pp. 658-663, Oct. 2013.

[10] M.L. Heldwein A.F. Souza, and I. Barbi, "A simple control strategy applied to three-phase rectifier units for telecommunication applications using single-phase rectifier modules," *Proc. Power Electronics Specialists Conf. (PESC)*, pp. 795-800, Jun. 1999.

[11] T. Atsushi, Y. Itsuo, and O. Tsuyoshi, "The control method of 3-phase PWM converter for 3-phase 3-wired imbalanced ac voltages," *Proc. International Telecommunications Energy Conf. (INTELEC)*, pp. 1-8, Oct. 2011.

[12] J. Biela, U. Drofenik, F. Krenn, J. Miniboeck, and J.W. Kolar, "Three-phase Y-rectifier cyclic 2 out of 3 dc output voltage balancing control method," *IEEE Trans. Power Electronics*, vol. 24, no 1, pp. 34-44, Jan. 2009.

Startup Procedure for Three-Phase Three-Wire Isolated AC-DC Converter Implemented with Three Single-Phase Converter Modules

Misha Kumar, Laszlo Huber, and Milan M. Jovanović
Delta Products Corporation
P.O. Box 12173
5101 Davis Drive
Research Triangle Park, NC 27709, USA

Dinggang Ping and Gang Liu
Shanghai Design Center,
Delta Electronics (Shanghai) Company Ltd,
Shanghai 201209,
People's Republic of China

Abstract – Three-phase isolated ac-dc converters can be implemented either with a direct three-phase PFC rectifier front end followed by an isolated dc-dc converter, or with three single-phase isolated ac-dc converters. Generally, the major advantage of the modular implementation with three single-phase converters is its ease of power expandability. In implementations with Y-connected single-phase modules in three-wire power systems, the major challenge is the design of a reliable control that limits the potential difference between the Y point of the single-phase modules and the neutral point of the three-phase source, V_{Y0}, so that the input voltage of the modules always stays within a specified range. Although a vast amount of literature on the balancing control is available, none of them addresses the startup of the modules, which is an indispensable part of control design of any practical circuit. To fill this gap, in this paper, a detailed startup procedure is described. This two-step startup procedure ensures that even with unbalanced source voltages, V_{Y0} is properly controlled so that the input voltage of the modules stays within a specified limit, while PFC front-end output voltages increase monotonically and their inrush currents are below a specified level. The proposed startup procedure is illustrated with MATLAB/Simulink simulations and also experimentally verified.

I. INTRODUCTION

Three-phase isolated ac-dc converters can be implemented either with a direct three-phase PFC rectifier front end such as the Vienna rectifier or the six-switch PFC boost rectifier followed by an isolated dc-dc converter, or with three single-phase isolated ac-dc converters. Generally, the major advantage of the modular implementation with three single-phase converters is its ease of power expandability. To be able to employ single-phase modules designed for 220/277-Vrms phase-to-neutral voltage in three-phase power systems with nominal phase-to-phase voltage of 380/480 Vrms, the three single-phase modules must be connected in star (Y) configuration. The delta (Δ) configuration cannot be used since it would require that single-phase modules be connected across two phases, i.e., to a voltage exceeding their rating.

In applications where the neutral point of the three-phase voltage source is available, the common point (Y point) of the single-phase converter modules is connected to the source neutral point and the three single-phase converters operate independently from each other with their input voltages equal to the respective phase-to-neutral voltages of the source. However, in applications where the source neutral point is not provided, such as, for example, in standard telecom power supplies, any unbalance in the three-phase source phase voltages and/or in the three single-phase modules will create

a potential difference between the Y point of the single-phase modules and the source neutral point (neutral displacement voltage V_{Y0}), resulting in oscillations and significant variations of the input voltages of the single-phase converters [1]. For a stable and reliable operation, i.e., operation where the input voltage always stays within a specified range, a balancing control of the three single-phase modules is necessary. The stable and reliable operation also has to be guaranteed at startup, as well as in case of phase failure (open and short circuit) and load transients. The balancing control between the three single-phase modules can be achieved with additional passive components used to create a virtual (artificial) neutral point and by employing various balancing control methods [1]-[8], or by using only balancing control without additional passive components [9], [10].

Regardless of the balancing-control method used, the startup of the three single-phase modules should be controlled so that the PFC input currents and the PFC output voltages do not experience abrupt changes, such as significant current spikes and voltage overshoots. Until now, startup control methods of the Y-connected single-phase modules have not been addressed in the literature. Although the general principles of the single-phase ac-dc converter startup control are applied, such as monotonically increasing bulk voltage and limiting input current stress, the startup procedure of the Y-connected single-phase modules is much more complex and far from straightforward because of the coupling between the three single-phase modules.

In this paper, a detailed two-step startup procedure for the Y-connected single-phase modules is described. In the first step, during the precharging of the PFC output capacitors, the inrush currents are limited by employing silicon controlled rectifiers (SCRs). In the second step, the PFC output voltages are increased to their nominal value monotonically, with only minor overshoots, by modifying the conventional average-current control of the PFC stages. The complete startup procedure at both balanced and unbalanced input phase voltages is illustrated with MATLAB/Simulink simulation waveforms. Experimental results are also provided.

II. POWER STAGE AND CONTROL CIRCUIT

The simplified circuit diagram of the three single-phase modules with Y connection at the input and parallel connection at the output used to explain and verify the proposed startup-up procedure is shown in Fig. 1. For the circuit in Fig. 1, the input phase-to-phase voltage range is 320-530 Vrms, the line frequency range is 45-65 Hz, and the

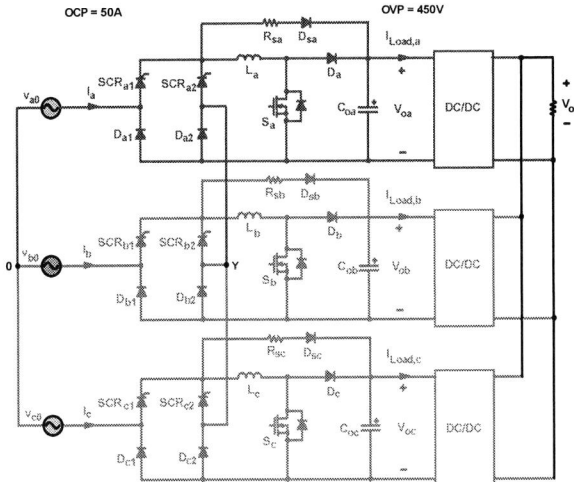

Fig. 1 Simplified circuit diagram of three single-phase modules with Y connection at input and parallel connection at output.

nominal output voltage of the PFC stages is 400 V. The boost inductors and PFC output filter capacitors are $L_a = L_b = L_c = 90~\mu\text{H}$ and $C_{oa} = C_{ob} = C_{oc} = 810~\mu\text{F}$, respectively. The PFC stages operate with average current control as shown in Fig. 2. The switching frequency of the PFC stages is 100 kHz, whereas the current loop and voltage loop bandwidth is 8 kHz and 10 Hz, respectively. The PFC input current OCP is 50 A and the PFC output voltage OVP is 450 V. The dc-dc stages are LLC converters with current-mode control, where the output voltage controller is common for all three dc-dc stages [11]. The maximum output power is 6 kW. The circuit shown in Fig. 1 can be applied for different applications. For example, in telecom power supplies, the nominal output voltage of the dc-dc stages is 48V, whereas, in electric-vehicle chargers the dc-dc output voltage varies from 270 V to 430 V.

The balancing control between the three single-phase modules is achieved by balancing the input admittances of the PFC stages. To maintain equal input admittances, the outputs of the PFC voltage controllers must be equal [10]. With equal input admittances, for balanced source voltages, the total input power is equally distributed between the three single-phase modules. However, for unbalanced source voltages, to achieve equal input admittances, the total input

power cannot be equally distributed between the three single-phase modules. In fact, a module connected to a reduced source voltage will draw less input current and, therefore, it will operate at reduced input power. The desired distribution of the total power between the single-phase modules that results in equal input admittances can be achieved by implementing the balancing-control circuit as introduced in [9] and explained in [10]. In this balancing circuit, the reference currents of the current controllers in the dc-dc stages are adjusted by balancing controllers that regulate the difference between the average output-voltage of the PFC controllers and the output voltage of an individual PFC controller. In the circuit in Fig. 1, the balancing loop bandwidth is 3 Hz.

It should be noted in Fig. 1 that the upper rectifiers in the input full-bridge rectifiers are replaced with silicon-controlled rectifiers (SCRs), which are used to limit the inrush currents during the startup; therefore, eliminating the startup resistors.

III. PROPOSED STARTUP PROCEDURE

Key waveforms during the proposed startup procedure for balanced source voltages and nominal phase voltage of 230 Vrms are shown in Fig. 3. The proposed startup procedure can be divided in two steps. In the first step, during time interval $[T_0\text{-}T_1]$, the PFC output capacitors are precharged to one half of the peak value of the source phase-to-phase voltages ($230 \times \sqrt{2} \times \sqrt{3} / 2 = 282$ V) by controlling the conduction angle of the SCRs, while the boost switches are turned off. During $[T_0\text{-}T_1]$, the dc-dc stages are also turned off, which means that the load current of the PFC stages is zero. In the second step, during time interval $[T_1\text{-}T_2]$, the PFC output capacitors continue to charge at no load by the boost operation of the PFC switches until one output-capacitor voltage increases to the nominal value of 400 V and the corresponding dc-dc stage turns on carrying the total initial output load. During time interval $[T_2\text{-}T_3]$, the voltage of the other two output capacitors sequentially increases to the nominal value of 400 V and the corresponding dc-dc stages turn on. At T_3, after all the three modules start carrying the initial output load, the balancing controller is turned on. Finally, during time interval $[T_3\text{-}T_4]$, the transients of the load current between the three modules are settled and the system reaches a steady-state operation.

The conduction-angle control of the SCRs during the precharging of the PFC output capacitors is illustrated in Fig. 4. It can be seen in Fig. 4 that the control of SCRs is synchronized to the source phase-to-phase voltages. At the beginning of interval T_α before the zero crossing of a phase-to-phase voltage (e.g., $v_{ab} > 0$), the corresponding pair of SCRs (SCR_{a1} and SCR_{b2}) is turned on and two PFC output capacitors connected in series (C_{oa} and C_{ob}) are equally charged through two small series resistors (R_{sa} and R_{sb}) and two series diodes (D_{sa} and D_{sb}) for a short time until the current through the SCRs decreases to zero and the SCRs turn off. By properly selecting the conduction angle of the SCRs, the inrush current pulses are limited below the OCP level. If for balanced source voltages of 230Vrms and $R_{sk} = 0.5~\Omega$,

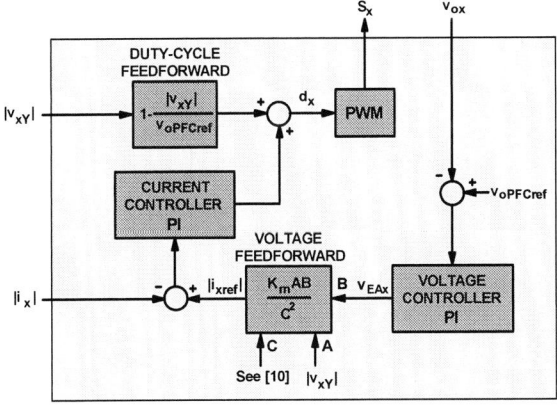

Fig. 2 Block diagram of PFC control circuit.

978-1-4673-9551-9/16 $31.00 © 2016 IEEE

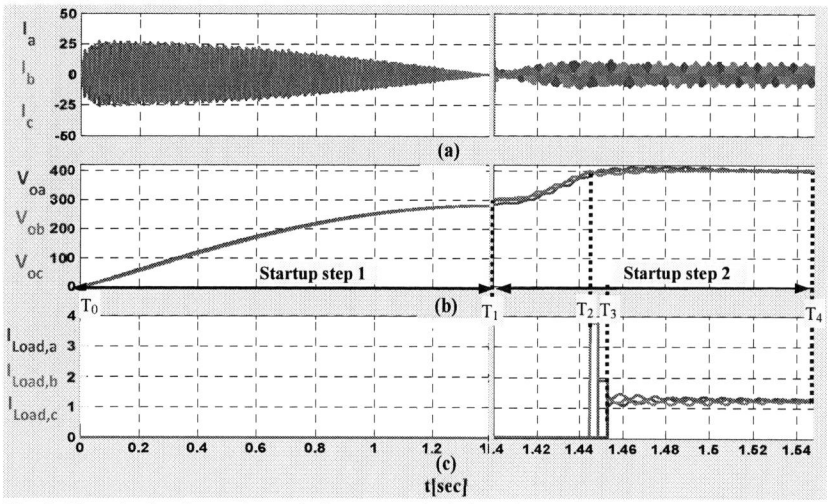

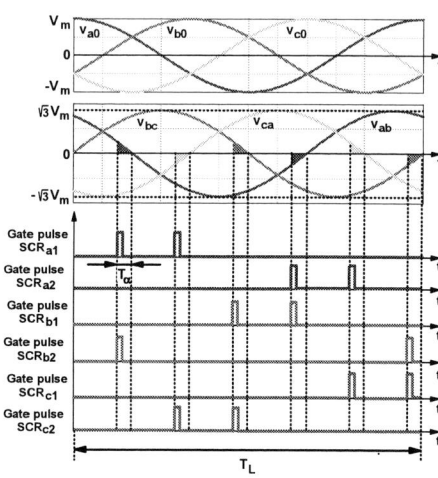

Fig. 3 Key simulated waveforms during startup for balanced source voltages and nominal phase voltage of 230 Vrms: (a) input phase currents [A], (b) PFC output voltages [V], (c) PFC load currents [A].

Fig. 4 Control of SCRs during precharging of PFC output capacitors.

$k \in \{a,b,c\}$, T_α is increased by 50 μs in every line cycle T_L until T_α increases to $T_L/4$, the inrush current pulses are limited below 25 A, which is well below the OCP level (50 A). With limited inrush currents, each PFC output capacitor gradually increases to one half of the peak value of the source phase-to-phase voltages. After T_α increases to $T_L/4$, the gate pulses of the SCRs are permanently set high.

Key simulated waveforms during the precharging of the PFC output capacitors for unbalanced source phase voltages $V_{a0} = V_{b0} = 161$ Vrms and $V_{c0} = 230$ Vrms, as an example, are presented in Fig. 5. It should be noted that waveforms in

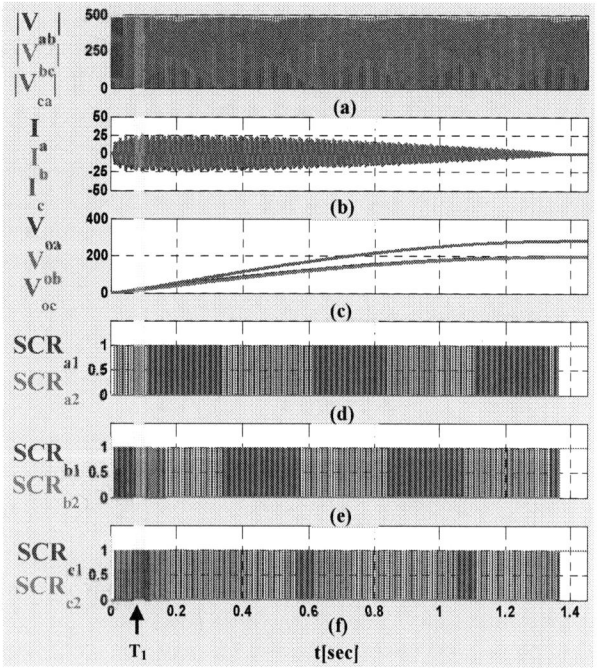

Fig. 5 Key simulated waveforms during precharging of PFC output capacitors for unbalanced input phase voltages ($V_{a0}=V_{b0}= 161$ Vrms, $V_{c0}= 230$ Vrms) and no load: (a) source phase-phase voltages [V], (b) input phase currents [A], (c) PFC output voltages [V], (d), (e), and (f) control pulses for SCR_{x1} and SCR_{x2}, x∈ {a,b,c}.

Fig. 5 are obtained by increasing T_α by 50 μs in every line cycle similarly as in the case of the balanced input voltages in Fig. 3. It can be seen in Fig. 5 that the inrush current pulses are limited below 25A and that the PFC output capacitors are precharged to different voltage levels ($V_{oa} = V_{ob} = 197$ V and $V_{oc} = 284$ V). In fact, the PFC output capacitors are charged so that the total voltage on each pair of PFC output capacitors is equal to the peak value of the corresponding phase-to-phase voltage. It can be easily derived that for unbalanced source phase voltages $V_{a0} = V_{b0} = 161$ Vrms and $V_{c0} = 230$ Vrms, the phase-to-phase voltages are $V_{ab} = 279$ Vrms and $V_{bc} = V_{ca} = 340$ Vrms. Therefore,

$$V_{oa} = V_{ob} = \frac{\sqrt{2} \cdot V_{ab,rms}}{2} = \frac{\sqrt{2} \cdot 279}{2} = 197 \text{ V} \quad (1)$$

and

$$V_{oc} = \sqrt{2} \cdot V_{bc,rms} - V_{ob} = \sqrt{2} \cdot 340 - 197 = 284 \text{ V} \quad . (2)$$

To better illustrate the rise of the PFC output capacitor voltages during the precharging interval, the waveforms in Fig. 5 around instant T_1 are zoomed in in Fig. 6. It can be seen in Fig. 6 that before the zero crossing of a particular phase-to-phase voltage, the corresponding pair of SCRs is turned on and the corresponding two capacitors connected in series are equally charged. However, as the peak values of the phase-to-phase voltages are unequal ($V_{ab,pk} < V_{bc,pk}$, $V_{bc,pk} = V_{ca,pk}$), PFC output capacitor voltages V_{oa} and V_{ob} increase with a smaller step before the zero crossing of phase-to-phase voltage V_{ab} compared to the increase of pairs of PFC output capacitor voltages V_{ob}, V_{oc} and V_{oa}, V_{oc} before the zero crossing of phase-to-phase voltages V_{bc} and V_{ca}, respectively. As voltage V_{oc} always increases with a larger step, its final value at the end of the precharging interval is larger than the final value of voltages V_{oa} and V_{ob}.

In the second step of the proposed startup procedure, the input currents are limited below the OCP level by the soft start of the reference output voltage.

978-1-4673-9551-9/16 $31.00 © 2016 IEEE   48

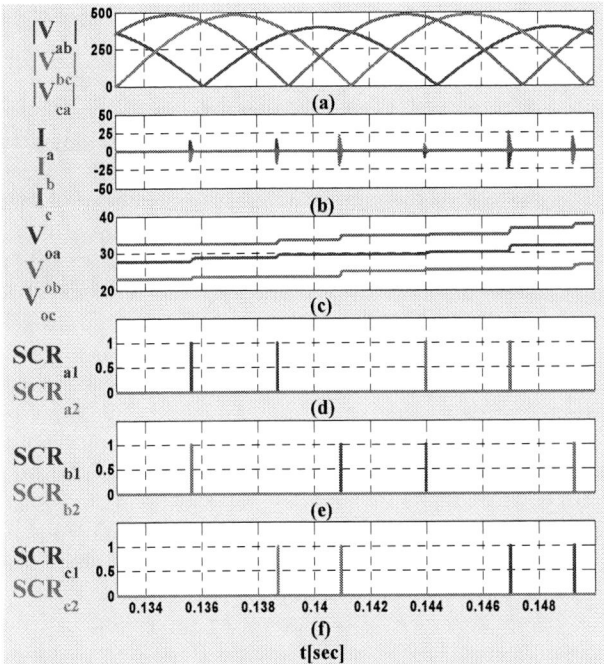

Fig. 6 Zoomed-in waveform around instant T_1 in Fig. 5: (a) source phase-phase voltages [V], (b) input phase currents [A], (c) PFC output voltages [V], (d), (e), and (f) control pulses for SCR$_{x1}$ and SCR$_{x2}$, x∈ {a,b,c}.

It should be noted that in order to reduce the simulation time, in the simulation circuit the dc-dc output stages are replaced with current sources which represent the load currents of the PFC stages, i.e., the input currents of the dc-dc stages for balanced source voltages, and with controlled

current sources which are controlled by the output signals of the balancing controllers [10]. Instead of adjusting the reference currents of the current controllers in the dc-dc stages, the load currents of the PFC stages are adjusted.

IV. MODIFIED AVERAGE-CURRENT CONTROL

During time interval $[T_1\text{-}T_3]$ in Fig. 3, the balancing controller is turned off and, therefore, the neutral displacement voltage V_{Y0} can potentially increase, resulting in significant unbalances in the input voltages of the PFC stages and, eventually, one PFC output voltage could increase to the OVP level. In order to limit the increase of voltage V_{Y0}, during $[T_1\text{-}T_3]$ the conventional average-current control of the PFC stages is modified so that instead of generating a reference input current to be proportional to its voltage-controller output, each reference input current is generated to be proportional to the average value of the outputs of all three voltage controllers. The difference in operation between the conventional and modified average-current control during startup is illustrated with simulated waveforms for balanced source phase voltages (230 Vrms) and no load in Figs. 7 and 8. All simulated results in this paper are obtained in MATLAB/Simulink.

In Fig. 7, with conventional average-current control, PFC output voltages V_{ob} and V_{oc} increase above 450 V, which would trigger OVP, whereas, V_{oa} stays constant at 410 V. After PFC output voltages increase above their nominal value of 400 V, PFC voltage controller outputs V_{EAk}, $k∈\{a,b,c\}$, become unequal ($V_{EAa} > V_{EAb}$, $V_{EAb} ≈ V_{EAc}$) making input admittances Y_k unequal ($Y_a > Y_b$, $Y_b ≈ Y_c$), which results in an increased amplitude of voltage V_{Y0} that is in phase with input

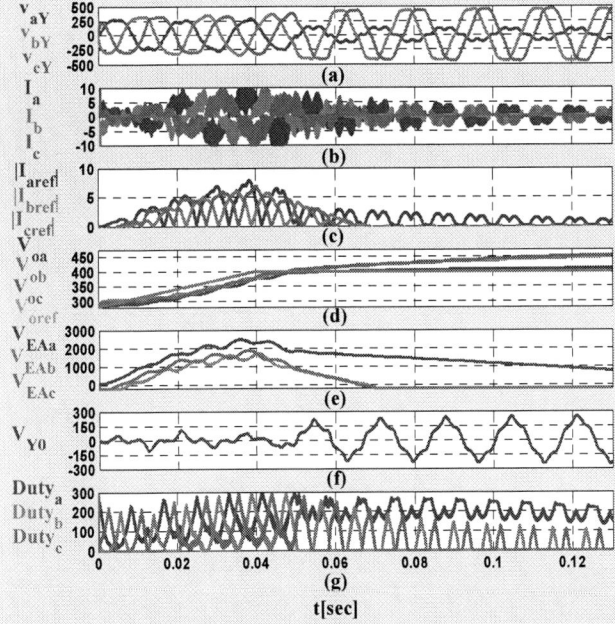

Fig. 7 Key simulated waveforms during startup with conventional average-current control for balanced source phase voltages (230Vrms) and no load: (a) low-pass filtered input phase voltages with respect to Y point [V], (b) input phase currents [A], (c) rectified phase current references [A], (d) PFC output voltages [V], (e) PFC voltage controller outputs [digital value in Q12 format], (f) low-pass filtered V_{Y0} [V], and (g) duty cycle of boost switches [digital values with respect to C_{pk}=300].

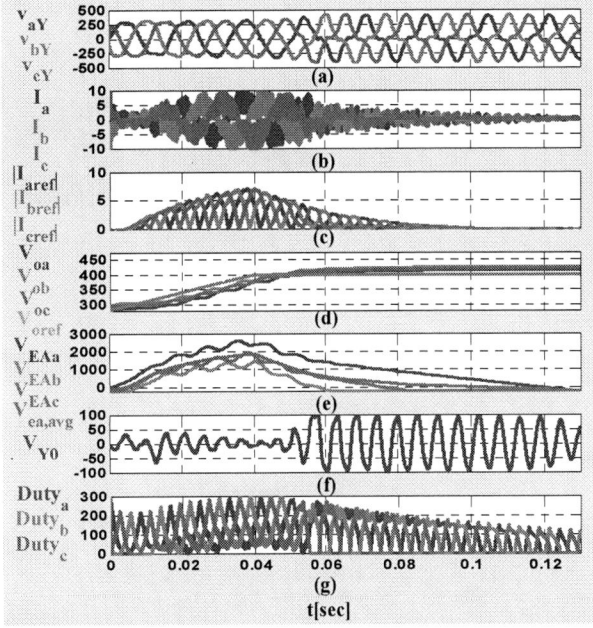

Fig. 8 Key simulated waveforms during startup with modified average-current control for balanced source phase voltages (230Vrms) and no load: (a) low-pass filtered input phase voltages with respect to Y point [V], (b) input phase currents [A], (c) rectified phase current references [A], (d) PFC output voltages [V], (e) PFC voltage controller outputs [digital value in Q12 format], (f) low-pass filtered V_{Y0} [V], and (g) duty cycle of boost switches [digital values with respect to C_{pk}=300].

978-1-4673-9551-9/16 $31.00 © 2016 IEEE

voltage of module "a", V_{aY}, as shown in Fig. 7. Therefore, amplitude of $V_{aY} = V_{a0} - V_{Y0}$ is decreased, whereas, amplitude of V_{bY} and V_{cY} is increased. Due to the increased input voltages of modules "b" and "c", their output voltages V_{ob} and V_{oc} increase above 450 V.

In Fig. 8, with modified average-current control, the three PFC output voltages are approximately equal and well below the OVP limit. After the PFC output voltages increase above their nominal value of 400 V, despite the unequal outputs of their PFC voltage controllers, input admittances Y_k are still approximately equal because the amplitudes of the reference input currents are equal since they are generated from the average value of the outputs of the three PFC voltage controllers. As a result, the amplitude of voltage V_{Y0} is significantly smaller compared to that in Fig. 7. Therefore, the amplitude of input voltages V_{aY}, V_{bY}, and V_{cY} is approximately the same. Also, the amplitude of V_{bY} and V_{cY} is smaller than that in Fig. 7. Consequently, PFC output voltages V_{oa}, V_{ob}, and V_{oc} are approximately equal and they increase to a lower level (≈ 420 V) compared to that in Fig. 7.

It follows from Figs. 7 and 8 that during startup with no load, the PFC output voltages cannot be regulated. To achieve regulation, an initial load should be applied to each module when its PFC output voltage increases to the nominal level of 400 V, as explained in Section III. Key simulated waveforms during the second step of the proposed startup procedure for balanced source voltages (230 Vrms) and 600-W load are presented in Fig. 9. During time interval $[T_1\text{-}T_2]$, the PFC output voltages increase following the reference voltage ramp whose initial value is equal to the precharged value of the PFC output capacitor voltages. At instant T_2, PFC output voltage V_{ob} increases to 400 V and the total 600-W load (i.e., 1.5 A load current) is applied to the output of

PFC stage "b". Shortly after T_2, V_{oc} increases to 400 V and the 1.5-A load current will be equally distributed between modules "b" and "c". At instant T_3, PFC output voltage V_{oa} also increases to 400 V and from that time on the 1.5-A load current will be equally shared between the three modules. It should be noted that during $[T_1\text{-}T_3]$, the PFC stages operate with modified average-current control, where the reference input currents are proportional to the average value of the outputs of the three voltage controllers, V_{EAavg}. As a result, during $[T_1\text{-}T_3]$ amplitude of voltage V_{Y0} is small, as can be seen in Fig. 9. At T_3, when all three PFC stages are loaded, the balancing controller is turned on and the average current control is changed from modified to conventional, where each input reference current is proportional to the output of its own voltage controller. This transition results in increased amplitude of voltage V_{Y0}. At T_4, the transients of the load current between the three modules are settled and the system reaches a steady-state operation with a reduced amplitude of voltage V_{Y0}. It should be noted in Fig. 9 that the overshoot of the PFC output voltages is only 10 V.

Key simulated waveforms during the second step of the proposed startup procedure for unbalanced source phase voltages, $V_{a0} = V_{b0} = 161$ Vrms and $V_{c0} = 230$ Vrms, and 600-W load are presented in Fig. 10. It can be seen in Fig. 10 that PFC output voltage V_{oc} increases above the OVP level ($V_{oc} = 457$V), while V_{oa} increases almost to the OVP level ($V_{oa} = 447$V). It can also be seen in Fig. 10 that compared to Fig. 9, during $[T_1\text{-}T_3]$ the outputs of the PFC voltage controllers deviate more from their average value. As a result, at T_3, when balancing controller is turned on, the amplitude of adjustment currents is larger compared to that in Fig. 9. Also, after T_3, when average current control is changed from modified to conventional, the amplitude of

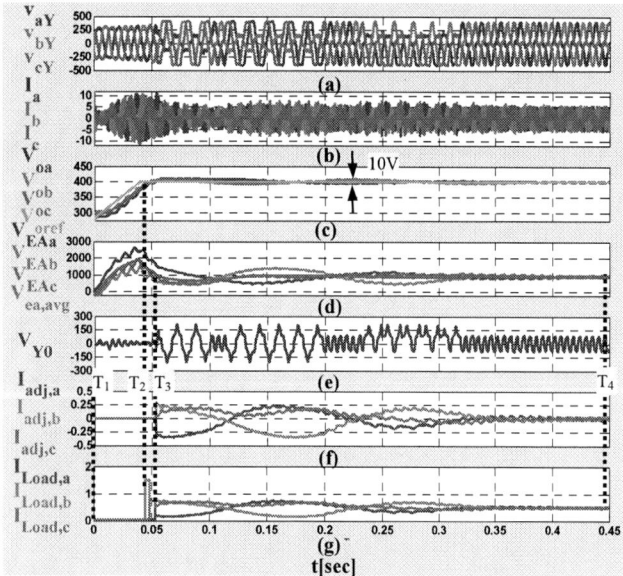

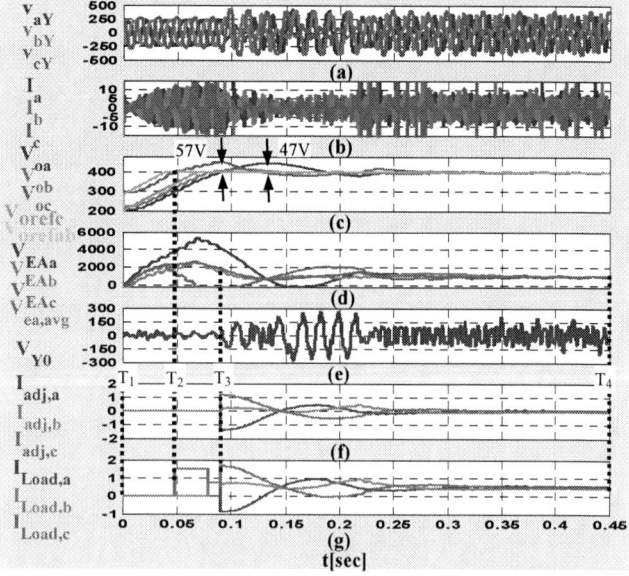

Fig. 9 Key simulated waveforms during startup with modified average-current control for balanced source phase voltages (230Vrms) and 600-W load: (a) low-pass filtered input phase voltages with respect to Y point [V], (b) input phase currents [A], (c) PFC output voltages [V], (d) PFC voltage controller outputs [digital value in Q12 format], (e) low-pass filtered V_{Y0} [V], (f) adjustment currents [A], and (g) PFC load currents [A].

Fig. 10 Key simulated waveforms during startup with modified average-current control for unbalanced source phase voltages ($V_{a0} = V_{b0} = 161$Vrms, $V_{c0} = 230$Vrms) and 600W load: (a) low-pass filtered input phase voltages with respect to Y point [V], (b) input phase currents [A], (c) PFC output voltages [V], (d) PFC voltage controller outputs [digital value in Q12 format], (e) low-pass filtered V_{Y0} [V], (f) adjustment currents [A], and (g) PFC load currents [A].

voltage V_{Y0} increases to a larger value compared to that in Fig. 9.

In order to limit the overshoot of the PFC output voltages below 450V for unbalanced source voltages, two approaches can be used depending on applications. Generally, in applications where the startup can be performed at any load, an increased initial load should be used. However, in applications where the startup cannot be performed at higher loads, the PFC voltage-controller bandwidth should be increased.

To illustrate the startup with an increased initial load, key simulated waveforms during the second step of the proposed startup procedure for unbalanced source phase voltages, $V_{a0} = V_{b0} = 161$ Vrms and $V_{c0} = 230$ Vrms, and 1500-W load are presented in Fig. 11. It can be seen in Fig. 11 that the overshoot of the PFC output voltages is 20 V, i.e., it is well below the OVP level. It can also be seen in Fig. 11 that compared to Fig. 10, during $[T_1\text{-}T_3]$ the outputs of the PFC voltage controllers deviate less from their average value. As a result, at T_3, when balancing controller is turned on and the average current control changes from modified to conventional, the amplitude of the adjustment currents as well the amplitude of voltage V_{Y0} are smaller compared to those in Fig. 10. Finally, at T_4, when the system reaches a steady state, the amplitude of V_{Y0} is significantly smaller than that in Fig. 10.

To illustrate the startup with an increased PFC voltage-controller bandwidth, key simulated waveforms during the second step of the proposed startup procedure for unbalanced source phase voltages, $V_{a0} = V_{b0} = 161$ Vrms and $V_{c0} = 230$ Vrms, 600-W load, and increased voltage-

controller bandwidth from 10 Hz in steady state to 100 Hz are presented in Fig. 12. It can be seen in Fig. 12 that the overshoot of the PFC output voltages is 44 V, i.e., it is slightly below the OVP level. It can also be seen in Fig. 12 that compared to Fig. 10, during $[T_1\text{-}T_3]$ the outputs of the PFC voltage controllers deviate more from their average value. However, at T_3, when the average current control is changed from modified to conventional, the increased bandwidth of the PFC voltage controllers helps to achieve steady state values of the PFC voltage-controller outputs faster. This results in lower amplitude of V_{Y0} compared to that in Fig. 10. Also, after T_3, when balancing controller is turned on, the adjustment currents reach their steady state values faster than in Fig. 10. It should be noted in Fig. 12 that after the adjustment currents reach their steady state values and the load current transients are settled, the PFC voltage-controller bandwidth is changed back to 10Hz from 100Hz. It should also be noted in Fig. 12 that when PFC output capacitor voltage V_{oc} increases above 400V, the output V_{EAc} of the PFC voltage controller "c" becomes saturated. However, during $[T_1\text{-}T_3]$, phase "c" current reference is proportional to V_{EAavg} instead of V_{EAc}. As V_{EAavg} is greater than V_{EAc}, the amplitude of current reference of phase "c" is greater than the required value and, consequently, V_{oc} exhibits an overshoot.

In order to reduce the overshoot of V_{oc} in Fig. 12, the startup procedure is further modified as illustrated in Fig. 13. In fact, at T_2, when V_{oc} increases to 400V, the average current control of phase "c" is changed from modified to conventional, where phase "c" current reference is proportional to the output of its own voltage controller. At

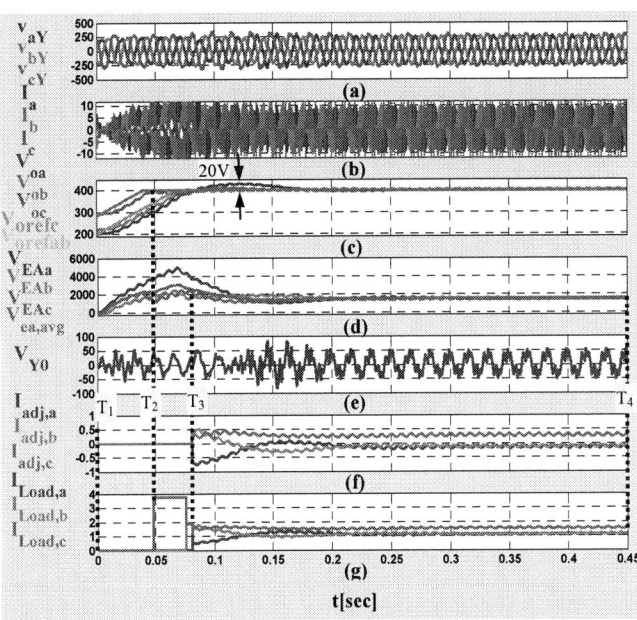

Fig. 11 Key simulated waveforms during startup with modified average-current control for unbalanced source phase voltages ($V_{a0}=V_{b0}=$ 161Vrms, $V_{c0}=$ 230Vrms) and 1500W load: (a) low-pass filtered input phase voltages with respect to Y point [V], (b) input phase currents [A], (c) PFC output voltages [V], (d) PFC voltage controller outputs [digital value in Q12 format], (e) low-pass filtered V_{Y0} [V], (f) adjustment currents [A], and (g) PFC load currents [A].

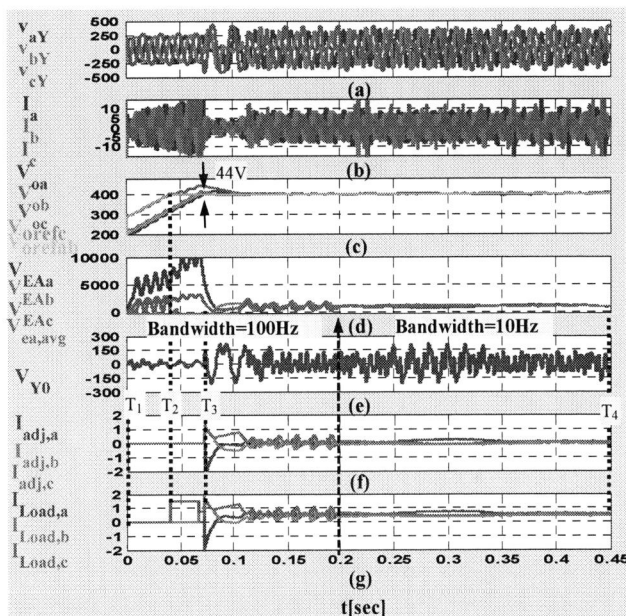

Fig. 12 Key simulated waveforms during startup with modified average-current control with ten times increased PFC voltage controller bandwidth (100Hz) for unbalanced source phase voltages ($V_{a0}=V_{b0}=$ 161Vrms, $V_{c0}=$ 230Vrms) and 600W load: (a) low-pass filtered input phase voltages with respect to Y point [V], (b) input phase currents [A], (c) PFC output voltages [V], (d) PFC voltage controller outputs [digital value in Q12 format], (e) low-pass filtered V_{Y0} [V], (f) adjustment currents [A], and (g) PFC load currents [A].

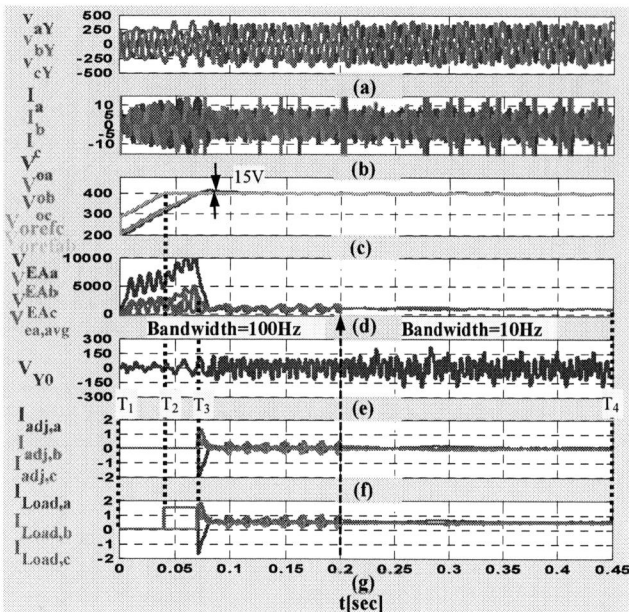

Fig. 13 Key simulated waveforms during startup with further modified average-current control with ten times increased PFC voltage controller bandwidth (100Hz) for unbalanced source phase voltages ($V_{a0}=V_{b0}=$ 161Vrms,$V_{c0}=$ 230Vrms) and 600W load: (a) low-pass filtered input phase voltages with respect to Y point [V], (b) input phase currents [A], (c) PFC output voltages [V], (d) PFC voltage controller outputs [digital value in Q12 format], (e) low-pass filtered V_{Y0} [V], (f) adjustment currents [A], and (g) PFC load currents [A].

the same time, the current references of phases "a" and "b" are still proportional to the average value of the outputs of all three voltage controllers. After T_2, when V_{ob} increases to 400V, the average current control of phase "b" is changed from modified to conventional, while the current reference of phase "a" is still proportional to the average value of the outputs of all three voltage controllers. Finally, at T_3, when V_{oa} increases to 400V, the average current control of phase "a" is changed from modified to conventional. At the same time, the balancing controllers are turned on. It can be seen in Fig. 13 that the overshoot of the PFC output voltages is only 15 V, i.e., it is well below the OVP level. It should be noted in Fig. 13 that the output V_{EAc} of the PFC voltage controller "c" is not saturated.

V. EXPERIMENTAL RESULTS

Experimental results obtained on an 11-kW prototype for balanced (230 Vrms) and unbalanced (207 Vrms, 230 Vrms, and 253 Vrms) source phase voltages and 5.5-kW load are presented in Figs. 14 and 15, respectively. It can be seen in Figs. 14 and 15, that the maximum overshoot of the PFC output voltages is 20 V, which is well below the OVP level of 450 V and that the maximum inrush current spike is approximately 90 A which is below the OCP level of 100 A specified for the 11-kW prototype. It should be noted that the limited overshoot of the PFC output voltages is achieved by implementing the proposed startup procedure at an increased initial load. It should also be noted that during the precharging of the PFC output capacitors, the inrush current spikes are limited by increasing interval T_a before the zero crossing of a relevant phase-to-phase voltage by 20μs in

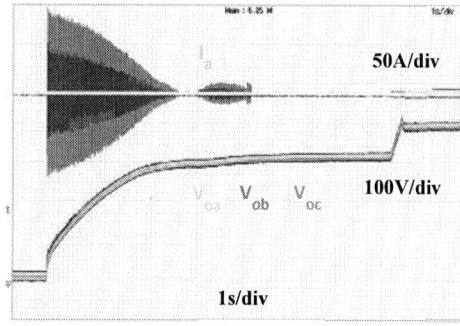

Fig. 14 Experimental waveforms of PFC output voltages V_{oa}, V_{ob}, V_{oc}, and input current I_a during startup for balanced input phase voltages (230Vrms) and 5.5-kW load obtained on a 11-kW prototype.

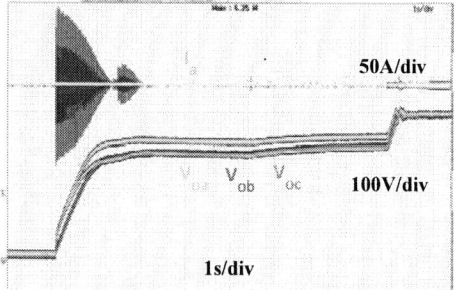

Fig. 15 Experimental waveforms of PFC output voltages V_{oa}, V_{ob}, V_{oc}, and input current I_a during startup for unbalanced input phase voltages (V_{a0}=207Vrms, V_{b0}=230Vrms, V_{c0}=253Vrms) and 5.5-kW load obtained on a 11-kW prototype.

every line cycle T_L until T_a is increased to $T_L/4$. Finally, it should be noted that in both Figs. 13 and 14, that unlike in the simulation results, a second group of current spikes exists near the end of the precharging interval of the PFC output capacitors. These additional current spikes are caused by inaccuracies in determining the line frequency and by setting the gate pulses of the SCRs permanently to high when T_a is still smaller than $T_L/4$.

VI. SUMMARY

In this paper, a detailed two-step startup procedure for the Y-connected single-phase modules is described. The proposed startup procedure ensures that even at unbalanced source voltages, neutral displacement voltage V_{Y0} is properly controlled so that the input voltage of the single-phase modules stays within a specified limit, while PFC front end output voltages increase monotonically and their inrush currents are below a specified level.

During the first step of the proposed startup procedure, when all boost switches and the dc-dc stages are off, the PFC output capacitors are monotonically precharged to one half of the peak value of the source phase-to-phase voltages. The inrush currents are limited by employing conduction-angle control of silicon controlled rectifiers (SCRs). The control of the SCRs is synchronized to the zero crossing of the source phase-to-phase voltages.

During the second step of the proposed startup procedure, the PFC output voltages are increased almost monotonically to their nominal value by modifying the conventional average-current control of the PFC stages. The conventional

978-1-4673-9551-9/16 $31.00 © 2016 IEEE

average-current control of the PFC stages is modified so that instead of generating a reference input current to be proportional to its voltage-controller output, each reference input current is generated to be proportional to the average value of the outputs of all three voltage controllers. With this modified control, the PFC output voltages sequentially increase to the nominal value of 400 V with the dc-dc stages turned off. After a PFC voltage increases to 400 V, its dc-dc stage with some initial load is turned on. After all three dc-dc stages are turned on and the three modules share the load current, the average current control is changed from modified to conventional and the balancing controllers are turned on. At the end of the second step, the transients of the load current between the three modules are settled and system reaches a steady-state operation.

The complete startup procedure at both balanced and unbalanced source phase voltages is illustrated with MATLAB/Simulink simulation waveforms. Experimental results are also provided.

REFERENCES

[1] M. Karlsson, C. Thoren, and T. Wolpert, "A novel approach to the design of three-phase AC/DC power converters with unity power factor," *Proc. International Telecommunications Energy Conf. (INTELEC)*, paper 5-1, Jun. 1999.

[2] M. Karlsson, C. Thoren, and T. Wolpert, "Practical considerations concerning a novel 6kW three-phase AC/DC power converter with unity power factor," *Proc. International Telecommunications Energy Conf. (INTELEC)*, pp. 28-33, Sep. 2000.

[3] M. Karlsson and T. Wolpert, "Stable artificial neutral point in a three phase network of single phase rectifiers," U.S. Patent 6,466,466, Oct. 15, 2002.

[4] D. Chapman, D. James, and C.J. Tuck, "A high density 48V 200A rectifier with power factor correction – An engineering overview," *Proc. International Telecommunications Energy Conf. (INTELEC)*, pp. 118-125, Sep. 1993.

[5] C.J. Tuck, D.A. James, and D.A. Chapman, "Power converter with star configured modules," U.S. Patent 5,757,637, May 26, 1998.

[6] R. Greul, S.D. Round, and J.W. Kolar, "Analysis and control of a three-phase unity power factor Y-rectifier," *IEEE Trans. Power Electronics*, vol. 22, no 5, pp. 1900-1911, Sep. 2007.

[7] R. Greul, U. Drofenik, and J.W. Kolar, "Analysis and comparative evaluation of a three-phase three-level unity power factor Y-rectifier," *Proc. International Telecommunications Energy Conf. (INTELEC)*, pp. 421-428, Oct. 2003.

[8] R. Greul, U. Drofenik, and J.W. Kolar, "A novel concept for balancing of the phase modules of a three-phase unity power factor Y-rectifier," *Proc. Power Electronics Specialists Conf. (PESC)*, pp. 3787-3793, Jun. 2004.

[9] R. Girod and D. Weida, "High efficiency true 3-phase compact switch-mode rectifier module for telecom power solutions," *Proc. International Telecommunications Energy Conf. (INTELEC)*, pp. 658-663, Oct. 2013.

[10] L. Huber, M. Kumar, and M. M. Jovanović, "Analysis, design, and evaluation of a balancing control method for three-phase three-wire PFC rectifiers implemented with three single-phase PFC rectifier modules," *Proc. Applied Power Electronics Conf. (APEC)*, Mar. 2016.

[11] M.L. Heldwein A.F. Souza, and I. Barbi, "A simple control strategy applied to three-phase rectifier units for telecommunication applications using single-phase rectifier modules," *Proc. Power Electronics Specialists Conf. (PESC)*, pp. 795-800, Jun. 1999.

Control of a Single-Stage Three-Phase Boost Power Factor Correction Rectifier

Ayan Mallik[1,2], *Student Member, IEEE*, Bryan Faulkner[3], *Student Member, IEEE,* and
Alireza Khaligh[1,2], *Senior Member, IEEE*

Maryland Power Electronics Laboratory (MPEL)

[1]Electrical and Computer Engineering Department; [2]Institute for Systems Research; University of Maryland, College Park;
[3]Bradley Department of Electrical and Computer Engineering, Virginia Polytechnic Institute and State University

Email: khaligh@ece.umd.edu; URL: http://khaligh.ece.umd.edu

Abstract - **Advances in power electronics are enabling More Electric Aircrafts (MEAs) to replace pneumatic systems with electrical systems. Active power factor correction (PFC) rectifiers are used in MEAs to rectify the output voltage of the three-phase AC-DC boost converter, while maintaining a unity input power factor. Many existing control strategies implement PI compensators, with slow response times, in their voltage and current loops. Alternatively, computationally expensive nonlinear controllers can be chosen to generate input currents with high power factor and low total harmonic distortion (THD), but they may need to be operated at high switching frequencies due to relatively slower execution of control loop. In this work, a novel control strategy is proposed for a three-phase, single-stage boost-type rectifier that is capable of tight and fast regulation of the output voltage, while simultaneously achieving unity input power factor, without constraining the operating switching frequency. The proposed control strategy is implemented, using one voltage-loop PI controller and a linearized transfer function of duty-ratio to input current, for inner loop current control. A 1.5 kW three-phase boost PFC prototype is designed and developed to validate the proposed control algorithm. The experimental results show that an input power factor of 0.992 and a tightly regulated DC link voltage with 3% ripple can be achieved.**

I. INTRODUCTION

Traditional three-phase variable voltage and variable frequency AC/DC rectification topologies in airplane generators utilize passive diode bridges and large DC link capacitors. Passive diode-bridge based rectifiers generate higher harmonics in the input current, have poor input power factor, create input voltage source disturbances, and lack output voltage regulation [1]. To alleviate these problems, recent progresses in high-speed, power semiconductor devices have facilitated the development of active switched-mode AC/DC converters that are controlled by pulse width modulation (PWM) techniques. The dominant topologies for active, single-stage PWM-based AC/DC conversion are boost-type [1-4], buck-type [5-6] and buck-boost type rectifiers [7-8]. Three-phase, boost-type power factor correction (PFC) converters have received attention due to their simple structure with less number of power semiconductor devices, less passive components and more importantly capability of continuous current conduction mode operation. Besides, three-phase buck-boost-type and buck-type AC-DC converter have drawbacks such as operation with discontinuous current conduction mode, higher amount of power semiconductor devices, and lower conversion efficiencies [7-9].

In addition to the fact that these aforementioned topological constraints limit the performance of several power converters, there is a significant and effective role, played by novelty of control methodology in enhancing the performance of any AC-DC converter. Implemented control algorithm acts as a crucial factor in governing power quality of the input current by limiting harmonics content of the PFC stage and also enhancing the conversion efficiency. A novel and noteworthy strategy, proposes a vectorial sliding mode input current control technique for a three-phase AC-DC buck-boost-type PFC [10], which is also applicable for a boost-type AC-DC rectifier with the objective of improving its power quality. Although the simplified vectorial control method is fast and offers robust control; however, considering non-idealities would lead to a computationally time expensive technique, which would limit the operational feasibility of the converter in high-end switching frequencies.

To improve upon existing strategies, and to offer a novel solution without any of the previously mentioned drawbacks, this paper proposes a new control strategy utilizing the input currents and output voltage of the converter. The main objective of the control strategy is to make the input current controller as fast and as robust as possible; to produce high quality input currents (low THD percentage and unity power factor). Instead of using a conventional control loop, that performs Park Transforms from the three-phase (abc) reference frame to the dq0 reference frame [11], our research thrust proposes an input current control structure that manipulates the reference duty ratio of each switch, in order to maintain an appropriate/desired input current shape. The final duty ratio value is derived from a weighted, cross-coupled sum of required change in duty ratios, which are obtained from both active and reactive power controller outputs. This control structure excels in two separate areas: (1) obtaining a fast and robust input current response (with high power factor quality); and (2) achieving a steady state response in a significantly less amount of settling time, under a step change in load or reference output voltage, as compared to conventional PI current compensators. Simply put, the

978-1-4673-9551-9/16 $31.00 © 2016 IEEE

control strategy put forth in this research thrust is simple, fast, and reliable – and is perfectly suited for implementation in the active three-phase boost rectifiers.

This paper is organized as follows. Section II introduces the three-phase AC-DC boost rectifier topology and explains the details of the proposed control strategy. Device selection details and their specifications are provided in Section III. In Section IV, the simulation and experimental results of the designed converter, which is controlled with proposed control strategy, considering some of the specifications for regulated transformer rectifier unit (RTRU) applications in MEAs, are presented and analyzed. Section V evaluates the power loss and efficiency calculation of the converter. Finally, Section VI puts forward conclusions with relevant discussions.

II. TOPOLOGY AND CONTROL OF A THREE-PHASE ACTIVE BOOST RECTIFIER

A. Topology

The overall structure of a three-phase active boost-type rectifier is shown in Fig. 1. In this topology, there are three inductances in series with the AC source. These inductances help to boost the input AC voltage and filter the input current, thus reducing the harmonic contents. The top and bottom set of MOSFETs are switched in a complementary fashion with a fixed deadband. In order to reduce the forward conduction losses, three pairs of diodes are placed in anti-parallel combination with the power MOSFETS. These three external diodes are chosen to be of Schottky type, with less ON-state resistance than existing internal body diodes of the MOSFETs. Thus, the effective path resistance while forward conduction reduces.

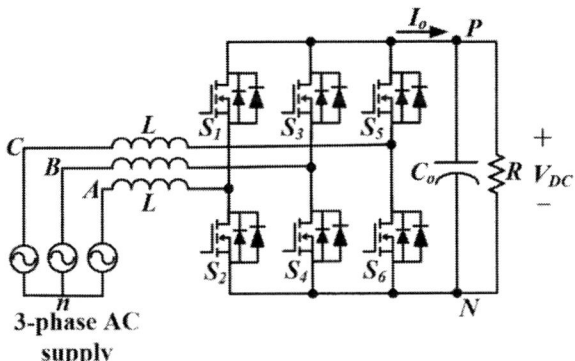

Fig. 1. Three phase boost-type rectifier.

B. Control Strategy

The nomenclatures used in this paper are summarized in Table I.

TABLE I: List of nomenclatures

Symbol	Description
D	Duty ratio
C_o	Output capacitor value
V_{DC}	Output voltage
I	RMS current through inductor L
V_{DC_min}	Minimum possible output DC link voltage in this topology
i_d	Direct current
i_q	Reactive current
$i_m{}^*$ or $v_m{}^*$	Reference current or reference voltage for m^{th} term; (i_a, i_b, i_c) are the input phase 'a', 'b', 'c' currents, respectively

This paper proposes a linear control technique, which uses small signal transfer functions of the converter, derived from the state-space averaging techniques, applied on different modes of operations. State-space averaging techniques allow for reduced computational efforts, and linearize systems around certain operating points. The state-space averaging method is chosen to control the rectification process, due to the fact that the voltage and current transients reach steady state with less oscillation than other methods; zero-order approximations, as an example, are more oscillatory [13]. Using state-space averaging, the small signal transfer function of the three-phase boost-type rectifier [1, 10], shown in Fig. 1, can be derived as shown in eq. (1).

$$T(s) = G_{vd}(s) = \frac{v_o(s)}{d(s)} = \frac{(ILR)s - R(1-D)(V_{DC} - V_{DC\_min})}{RLC_o s^2 + Ls + R(1-D)^2} \quad (1)$$

The transfer function in eq. (1) has a zero, shown in Eq. (2).

$$s = \frac{V_{DC}(1-D) + V_{DC\_min}(D-1)}{LI} \quad (2)$$

Due to the existence of a right half zero, this AC/DC converter acts as a non-minimum phase system [1]. The zero is located in the right half plane because $V_{DC} > V_{DC\_min}$, where V_{DC_min} is the output voltage of the uncontrolled diode-bridge configuration of three-phase active boost rectifier. If any system contains a right-half zero, the dynamic response in output voltage and input current is significantly slower, in comparison to a system with a left-half-plane zero and equivalent gain response [8]. Due to the characteristics of the zero, tuning the control circuit of the rectifier becomes difficult and is best suited for operation only in a particular region. However, since filter components of the converter must be designed to allow for faster input current dynamics, rather than output voltage dynamics, it is obligatory to segregate the dynamics of the input current from the output voltage. Moreover, a cascaded control system can be implemented on a system with separated dynamics [10]. Considering these facts, a suitable control structure is proposed in this paper.

The control goal for the three-phase boost rectifier is to generate sinusoidal input currents in phase with the input voltages, and regulate the DC output voltage. To operate with unity power factor, the reference reactive power should be set as 0 and active power should be set as the total nominal power of the converter. It also implies that in dq control method, the projection of input line current should be 0 on the quadrature axis and its entire projection should lie on direct axis. Thus, the real power and reactive power can be regulated by controlling direct current (i_d) and quadrature current (i_q), respectively. Utilizing Park transforms, as part of our control effort, allows the input line currents, $\{i_a, i_b, i_c\}$ to be transformed into the dq-domain currents, $\{i_d, i_q\}$, as demonstrated in eq. (3) and eq. (4).

$$i_d = -\frac{2}{3}[i_a \cos(\omega t) + i_b \cos(\omega t - \tfrac{2\pi}{3}) + i_c \cos(\omega t + \tfrac{2\pi}{3})] \quad (3)$$

$$i_q = -\frac{2}{3}[i_a \sin(\omega t) + i_b \sin(\omega t - \tfrac{2\pi}{3}) + i_c \sin(\omega t + \tfrac{2\pi}{3})] \quad (4)$$

Applying derivative on both sides of eq. (3) and (4), we get:

$$\partial i_d = -\partial i_a \cos(\omega t) + \frac{1}{\sqrt{3}}(\partial i_a + 2\partial i_b)\sin(\omega t)] \quad (5)$$

$$\partial i_q = -\partial i_a \sin(\omega t) + \frac{1}{\sqrt{3}}(\partial i_a + 2\partial i_b)\cos(\omega t)] \quad (6)$$

In order to achieve unity input power factor, reference quadrature component of line current, i_q, is set to zero. The DC-bus voltage is regulated by controlling the real power through controlling the direct component of line current, i_d. The instantaneous output power of the converter can be presented by eq. (7).

$$P = v_a i_a + v_b i_b + v_c i_c \quad (7)$$

Assuming the converter to be ideal, we can derive eq. (8) using the input and output power balance, considering a resistive output load.

$$v_o^{\,2} = R(v_a i_a + v_b i_b + v_c i_c) \quad (8)$$

Considering a balanced 3-phase system, i.e. $v_a + v_b + v_c = 0$, eq. (8) can be written as:

$$v_o^{\,2} = R \cdot [i_a(v_b + 2v_a) + i_b(v_a + 2v_b)] \quad (9)$$

Differentiating both sides of the equality in eq. (9) with respect to time, leads to the following relationship in eq. (10).

$$2v_o \frac{\partial v_o}{\partial t} = R[\frac{\partial i_a}{\partial t} \times (v_b + 2v_a) + i_a \times (\frac{\partial v_b}{\partial t} + 2\frac{\partial v_a}{\partial t})$$
$$+ \frac{\partial i_b}{\partial t} \times (v_a + 2v_b) + i_b \times (\frac{\partial v_a}{\partial t} + 2\frac{\partial v_b}{\partial t})] \quad (10)$$

Replacing $v_a = k_1 i_a$, $v_b = k_2 i_b$, and $v_c = k_3 i_c$; assuming PFC operation of the converter (where k_1, k_2, k_3 are zero-phase constants), eq. (11) and eq. (12) can be derived.

$$\frac{\partial i_a}{\partial d} = \frac{\partial v_o / \partial d}{\partial v_o / \partial i_a} = \frac{T(s)}{R \cdot [4k_1 i_a + (k_1 + k_2)i_b]} \quad (11)$$

$$\frac{\partial i_b}{\partial d} = \frac{\partial v_o / \partial d}{\partial v_o / \partial i_b} = \frac{T(s)}{R \cdot [4k_2 i_b + (k_1 + k_2)i_a]} \quad (12)$$

The output of the voltage compensator generates a direct current reference, i_d corresponding to active power. The direct component error, of the line current, followed by an equivalent linearized transfer function of $\delta d / \delta i_d$, generates the required change of duty ratio to control the active power, Δd_d. The error in the quadrature component of the line current, followed by an equivalent linearized transfer function of $\delta d / \delta i_q$, generates the required change of duty ratio to control the reactive power, Δd_q.

$$\Delta d = \frac{\partial d}{\partial i_d}(\Delta i_d) + \frac{\partial d}{\partial i_q}(\Delta i_q) \quad (13)$$

The two obtained perturbations in the duty ratios, Δd_d and Δd_q, followed by two gains G_1 (d-loop) and G_2 (q-loop), respectively, add together and generate the total change in the reference duty ratio, Δd, to achieve the reference input

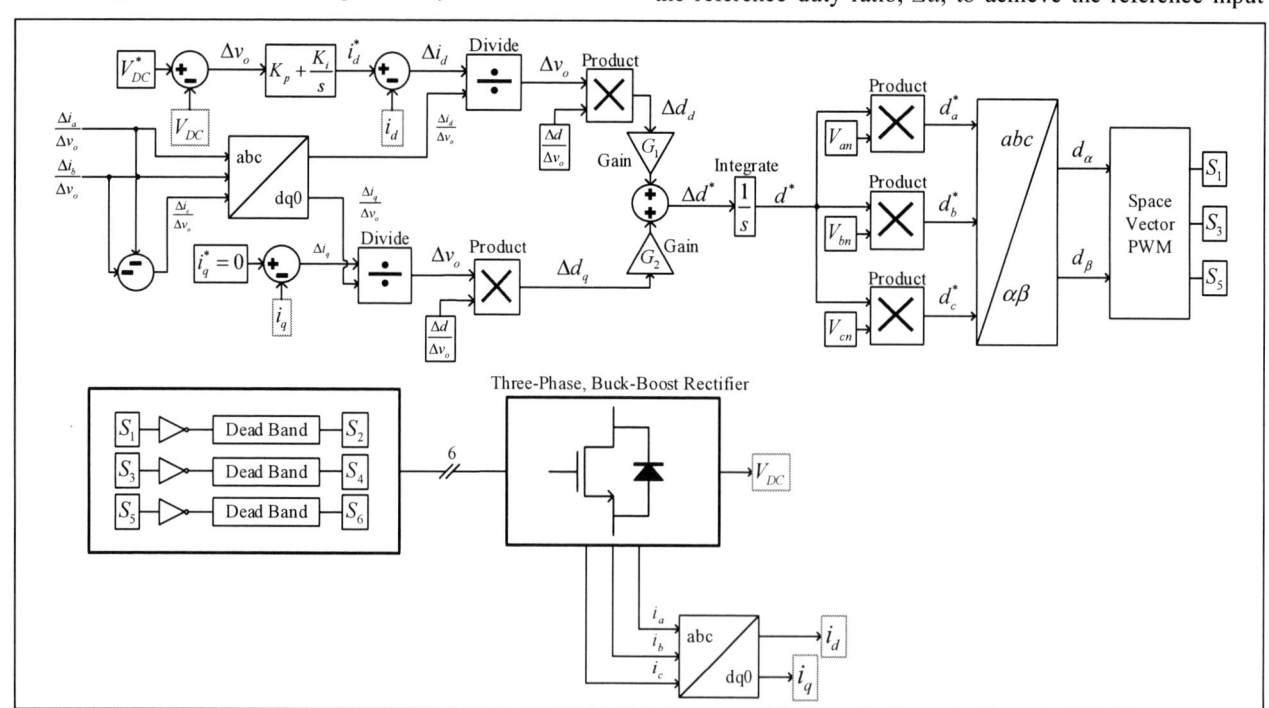

Fig. 2. Control Structure of the Single-Stage, Three-Phase Boost AC/DC Converter.

currents. Taking the integral of the change in the reference duty ratio, Δd, followed by a saturation block (0 to 1), gives the amplitude of the operating duty ratio, from which the phase duty ratios are derived. The individual phase duty ratios are followed by an ABC to $\alpha\beta$ transformation. The duty ratios in the $\alpha\beta$ reference frame, $\{d_\alpha, d_\beta\}$, are then fed to a Space Vector PWM block, which generates switching pulses for the converter.

III. DEVICE SELECTION

The selection of MOSFETs and power diodes depends on input voltages, output DC link voltage, and the converter power rating. Considering 208V RMS (line-line), 400Hz input AC supply, the derived output voltage boundary, between buck and boost mode operations for the converter, is the line-to-line peak voltage; i.e. $208\sqrt{2} = 295V$. Hence, the minimum possible DC link voltage generated by this boost PFC topology is 295V, using SVPWM switching technique. The application in this paper mainly focuses on a RTRU unit inside a MEA, which typically has the second stage of an isolated DC-DC resonant converter. With a higher DC link voltage, the second stage isolated converter would need a transformer with high step-down ratio and hence, more number of turns, which could potentially increase the size of the transformer and the magnetic core loss. In addition, at a higher DC link voltage, the voltage stresses across switches would be higher and result in degrading the efficiency due to more switching losses. Therefore, DC link voltage is regulated at 400V, which leads to a modulation index of 0.74. The reference value of DC link voltage would appear across the parallel combination of a MOSFET and a diode pair. Hence, the maximum voltage stress (V_{SW}) across each MOSFET is 400V.

Under specified DC link voltage ripple requirement, output capacitance (C_o) can be obtained from the fact that capacitor current is the difference between sum of top-leg phase currents and load current, which is provided in eq. (14):

$$\sum_{\substack{i,j=A,B,C \\ i \neq j}} (I_i + I_j) - \frac{V_o}{R} = \frac{\Delta V o}{\Delta T} C_o \qquad (14)$$

where, ΔT is ON period of a lower leg MOSFET of any phase in a switching cycle and I_i, I_j represent currents of the phases, whose upper leg switches are conducting at any time. An output capacitance of 100 µF ensures an output voltage ripple of less than 0.8% (peak-peak). But an output capacitor less than 100 µF results in large undershoot (>15%) during a load step up transient of 100% at the output.

The absolute maximum voltage rating of the power MOSFETs have been chosen based on the fact that it must never be exceeded during operation, irrespective of any fluctuation in input voltage or load transient. A sudden drop in load current could potentially create an overshoot as high as twice of DC link voltage. Therefore, typically the rating

should be chosen at least twice of DC link reference voltage. Though 800V Si-based MOSFET satisfies our requirements, its power loss is more than 1.2 kV SiC MOSFETs at high-end switching frequency operation, which is required for improving the power quality of the converter. Hence, APT40SM120B (1.2 kV, SiC MOSFET) has been selected. Furthermore, 1.2 kV SiC Schottky diode C2D05120 has been chosen, because it satisfies the current and voltage stress requirements with low forward voltage drop and also incurs lower reverse recovery loss. On a nutshell, all the discussed key parameters and power devices of the prototype are listed in Table II.

TABLE II: Component details of boost PFC design

Symbol	Device	Part number
S_1-S_6	SiC Power MOSFET	APT40SM120B
D_1-D_6	SiC Power Schottky diodes	C2D05120A
L	Input inductors	AS225-125A toroid core
C_o	DC link capacitor	DCP4P055009J (polypropylene film capacitor)

IV. SIMULATION AND EXPERIMENTAL RESULTS

The converter is simulated with an input AC voltage of 120V (phase-neutral RMS) at 400 Hz for avionics applications; these specifications are chosen, as they are consistent with the specifications of some of the RTRUs in future MEAs. The simulations are performed in MATLAB-Simulink. The rated power of the simulated converter is 1.5 kW. Fig. 3 provides the simulation results of the converter in boost mode, and also validates the PFC operation with voltage regulation of the converter. The design parameters of the simulated converter include input inductor (L) of 0.3 mH, output capacitor (C_o) of 100 µF and switching frequency of 150 kHz. During the simulation, at 20 ms, the output load power is increased from 1.5kW to 2.25kW. The 2% settling time of the output DC voltage is 7 ms with proposed control, which is faster as compared to a PI compensator that takes 15 ms to reach 2% settling band. The proposed control is able to achieve a unity PFC operation, and 2.3% input current THD.

An experimental prototype of 1.5 kW, with control logic being implemented in DSP (TMS320F28335), is built as a proof-of-concept verification of theoretical analyses. Programmable DC electronic load (BK Precision 8512) has been used for load emulation in this PFC experiment. Photos of the three-phase-boost PFC board and entire experimental setup are shown in Fig. 4.
Table III represents the experimental specification details of the converter in terms of input voltage, output voltage, and load power.

TABLE III: Rating specifications of the PFC

Parameters	Specifications
Input voltage	3-phase 120V RMS (phase-neutral), 400Hz
DC link reference	400V
Load power	1.5kW

978-1-4673-9551-9/16 $31.00 © 2016 IEEE

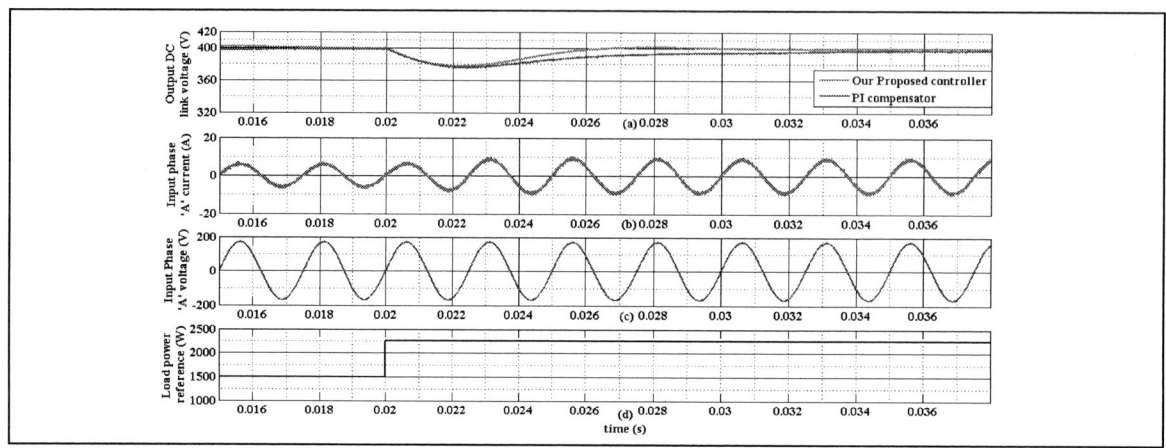

Fig. 3. (a) DC link voltage (V) with our proposed control and PI compensator; (b) Phase 'A' current (A) with our proposed control and PI compensator; (c) Phase 'A' input voltage (V) (d) Output power reference (W).

Fig. 4. The image of the experimental setup of three-phase active boost rectifier.

As Fig. 3 demonstrates, the Fig. 5 shows the boost mode operation of the PFC converter with proposed control logic and represents the stable and settled down output DC link voltage of the PFC. The DC link voltage is regulated at 400V with a ripple of ±2.5% (10V), as implied by Fig. 5(a). As shown in Fig. 5(b) and Fig. 5(c), the input phase current is in phase with the corresponding phase voltage and thus, demonstrates its PFC operation with an input power factor more than 0.99 and a THD below 5%. The output voltage spikes are caused by the acquisition inaccuracy of the voltage probe Tektronix-MDO3014. At 1.5 kW operation, THD and conversion efficiency are measured at 4.8% and 97.5%, respectively.

V. LOSS AND EFFICIENCY CALCULATION

Total power loss in the closed loop PFC system arises from switching, MOSFET conduction and diode conduction voltage drop. The switching loss (P_{SW}) and diode loss (P_{diode_loss}) [1-2] are given by the equations (15) – (17):

$$P_{SW} = P_{SW_{ON}} + P_{SW_{OFF}} = \frac{I_D V_{DS} f_S (t_{on} + t_{off})}{6} \tag{15}$$

$$P_{diode\_loss} = V_f I_{d\_avg} \tag{16}$$

$$I_{d\_avg} = I_a \frac{6 + \sqrt{3}\pi M}{12\pi} \tag{17}$$

The switch conduction loss (P_{cond}) [2] is governed by the eq. (18) and eq. (19).

$$P_{cond} = I_{RMS}^{2} R_{ON} \tag{18}$$

$$I_{RMS} = I_a \sqrt{\frac{4\pi + \sqrt{3}(3 + 4M)}{24\pi}} \tag{19}$$

In the above equations, t_{on} and t_{off} represent the turn-on and turn-off times for the MOSFET and 'M' is the modulation index; I_a is the RMS current of phase 'A'.

Total switching, conduction and diode loss are 15W, 7.5W, 3W respectively, as obtained from theoretical calculation. It leads to an expected efficiency of 98.3%. From the experimental data, total power loss is 40W and hence, obtained efficiency is 97.5%, which closely matches theoretical estimation. Thus, it proves an accurate modelling and precise implementation of our control logic.

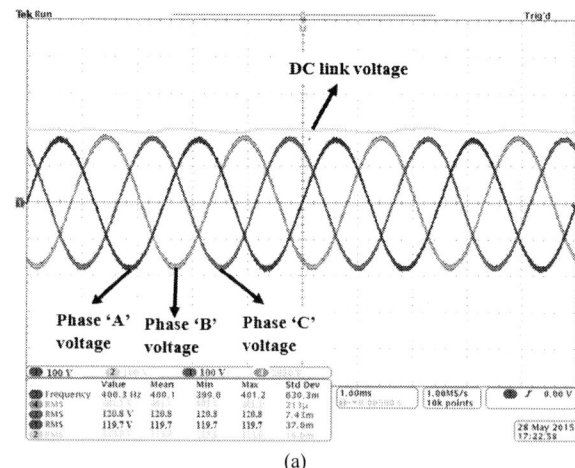

(a)

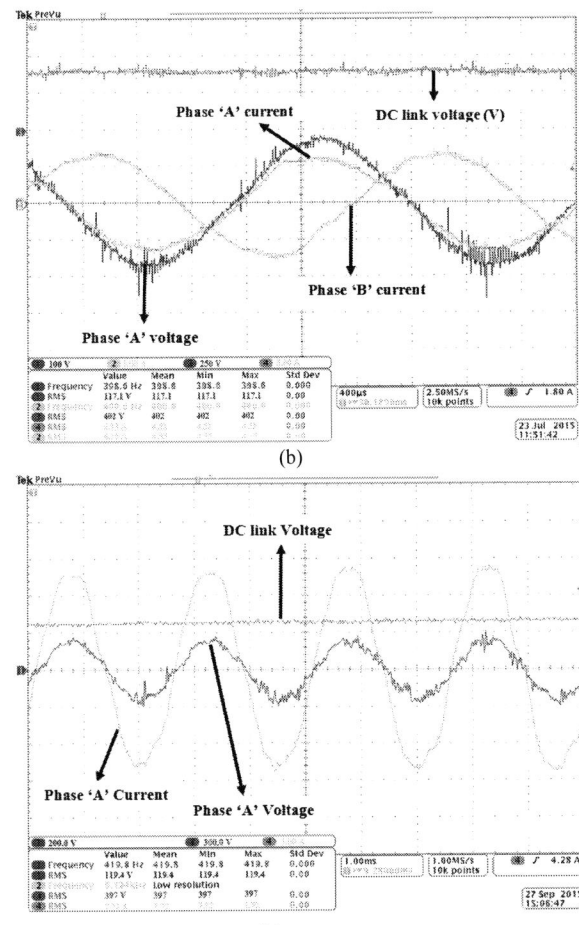

(b)

(c)

Fig. 5. Experimental waveforms of the converter (a) 3-phase input voltage and DC link voltage; (b) Phase 'A' voltage and current, Phase 'B' current, DC link voltage; (c) Phase 'A' voltage & current, DC link voltage

VI. CONCLUSION

In this paper, a novel control methodology, which sets forth a linearized transfer function of duty-ratio to input current, for inner loop control of a three-phase active boost rectifier, has been proposed, analyzed and developed. This strategy has the advantage of faster transient response and also is of less computational complexity than most of the conventional linear control algorithms. The feasibility, suitability and advantages of the proposed control method are discussed, and the design guidelines are provided through the theoretical analyses for a three-phase boost-type rectifier. As a case study, design considerations for a 1.5 kW PFC hardware prototype, which converts 70-120V, 400Hz AC to 400V DC, are provided.

The obtained experimental results demonstrate and verify that an efficiency of 97.5% at 150 kHz switching frequency, an input power factor of 0.993, a THD as low as 4.8% and a tight regulation of output DC link voltage of the converter within ±3% band can be achieved with the proposed control logic without constraining the operating switching frequency. Furthermore, the proposed control ensures the system dynamics and settling to be faster than conventional PI compensator, while undergoing a step change in load power. Hence, the proposed control strategy shows promising performance for 3-phase boost PFC and can be extended to be applied on other single and three phase AC-DC converters with similar plant characteristics for high-switching frequency applications.

ACKNOWLEDGEMENT

This work is sponsored by the Boeing Company and partially by the National Science Foundation grant number EEC 1263063, which are gratefully acknowledged.

REFERENCES

[1] J. W. Kolar, M. Hartmann, and T. Friedli, "Three-phase PFC rectifier and AC-AC converter systems," Tutorial at the *26th Annual IEEE Applied Power Electronics Conference and Exposition*, Fort Worth, TX, USA, Mar. 2011.

[2] J. W. Kolar and T. Friedli, "The essence of three-phase PFC rectifier systems," in *Proc. IEEE International Telecommunications Energy Conference (INTELEC)*, Amsterdam, Oct.2011.

[3] A. Mallik and A. Khaligh, "Comparative study of three-phase buck, boost and buck-boost rectifier topologies for regulated transformer rectifier units," in *Proc. IEEE Transportation Electrification Conference and Exposition*, Dearborn, MI, Jun. 2015.

[4] S. L. Sanjuan, "Voltage oriented control of three - phase boost PWM converters," Master of Science Thesis in Electric Power Engineering, Chalmers University of Technology, Sweden, May 2010.

[5] F. Xu, B. Guo, L.M. Tolbert, F. Wang, and B. J. Blalock, "Evaluation of SiC MOSFETs for a high efficiency three-phase buck rectifier," in *Proc. IEEE Applied Power Electronics Conference and Exposition* , Orlando, FL, Feb. 2012, pp.1762-1769.

[6] A. Stupar, T. Friedli, J. Minibo□ck, M. Schweizer, and J.W. Kolar, "Towards a 99% efficient three-phase buck-type PFC rectifier for 400 V DC distribution systems," in *Proc. IEEE Applied Power Electronics Conference and Exposition*, Fort Worth, TX, Mar. 2011, pp.505-512.

[7] C.T. Pan and T.C. Chen, "Step-up/down three-phase AC to DC converter with sinusoidal input current and unity power factor," *Proc. IEEE*, vol. 141, pp. 77-84, Mar. 1994.

[8] Y. Nishida, A. Maeda, and H. Tomita, "A new instantaneous-current controller for three-phase buck-boost and buck converters with PFC operation," in *Proc. IEEE Applied Power Electronics Conference and Exposition*, Dallas, TX, 1995, Mar. 1995, pp. 875-883.

[9] Y. Nishida and A. Maeda, "A simplified discontinuous-switching-modulation for three-phase current-fed PFC-converters and experimental study for the effects," in *Proc. IEEE Applied Power Electronics Conference and Exposition*, San Jose, CA, Mar. 1996, pp. 552-558.

[10] V. F. Pires, and J.F.A. Silva, "Single-stage three-phase buck-boost type AC-DC converter with high power factor," *IEEE Transactions on. Power Electron.*, vol.16, no.6, pp. 784-793, Nov. 2001.

[11] Z. Shen, M. Jaksic, B. Zhou, P. Mattavelli, D. Boroyevich, J. Verhulst, and M. Belkhayat, "Analysis of Phase Locked Loop (PLL) influence on DQ impedance measurement in three-phase AC systems," in *Proc. IEEE Applied Power Electronics Conference and Exposition*, Long Beach, CA, Mar. 2013, pp. 939-945.

[12] H. Wei and I. Batarseh, "Comparison of basic converter topologies for power factor correction," in *Proc. IEEE Southeastcon*, Orlando, FL, Apr. 1998, pp.348-353.

[13] J. Mahdavi, A. Emaadi, M. D. Bellar, and M. Ehsani, "Analysis of power electronic converters using the generalized state-space averaging approach," *IEEE Transactions Circuits Syst. I, Reg. Papers*, vol.44, no.8, pp.767-770, Aug. 1997.

[14] Y. Lee, A. Khaligh, and A. Emadi, "A compensation technique for smooth-transitions in non-inverting buck-boost converter," *IEEE Transactions on Power Electronics*, vol. 24, no. 4, pp. 1002-1015, Apr. 2009.

A BIDIRECTIONAL SINGLE-STAGE THREE-PHASE RECTIFIER WITH HIGH-FREQUENCY ISOLATION AND POWER FACTOR CORRECTION

Bruno Ricardo de Almeida[1], Demercil de Souza Oliveira Jr.[1], Paulo P. Praça[1]

Federal University of Ceará, Department of Electrical Engineering, Group of Energy Processing and Control
PO Box 6001, P. Code 60455-760, Tel.: +55 (85) 3366-9586
Fortaleza, Ceará - Brazil
allmeida@gmail.com, demercil@dee.ufc.br

Abstract— This paper proposes a single-stage three-phase rectifier with high-frequency isolation, power factor correction, and bidirectional power flow. The presented topology is adequate for dc grids (or smart-grids), telecommunications (telecom) power supplies, and more recent applications such as electric vehicles. The converter is based on the three-phase version of the dual active bridge (DAB) associated with the three-state-switching cell (3SSC), whose power flow between the primary and secondary sides is controlled by the phase-shift angle. A theoretical analysis is presented and validated through simulation and experimental.

Keywords — *PFC; Phase-shift control; Three-state switching cell; Three-phase ac-dc converter;*

I. INTRODUCTION

The use of equipment supplied by dc voltages rated within the range of 380-400 V is currently found in industry, particularly in telecommunications [1]. Besides, the study of dc transmission system has been intensified in the recent literature [2], mainly due to the widespread use of small electrical grids and the intensive exploitation renewable energy sources such as photovoltaic systems and wind energy conversion systems (WECSs), which rely on dc voltages [3]-[5].

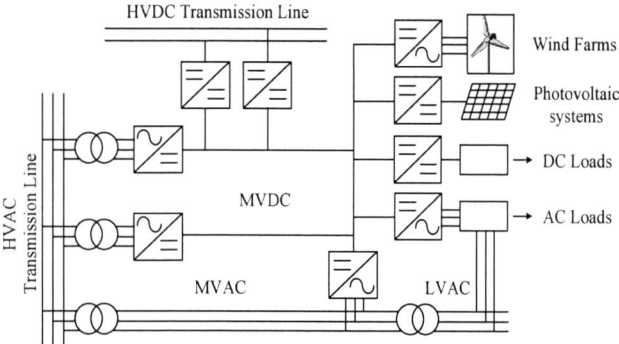

Fig. 1. Typical distribution system composed of HVAC (High voltage alternated current); MVAC (Medium voltage alternated current); LVAC (Low voltage alternated current); HVDC (High voltage direct current); MVDC (Medium voltage direct current); and loads.

With the expansion of the distribution systems in recent decades and the introduction of the smart grid concept, studies focused on topologies with bidirectional capability to manage power flow between the different energy sources (such as wind turbines, photovoltaic modules, fuel cells, among others) and storage systems [6] have been increasingly carried out. Figure 1 presents a typical distribution system composed of high voltage lines and loads supplied by either ac or dc medium and low voltages, while distinct primary sources also exist.

In order meet standards regarding the power factor and harmonic distortion (IEC 61000-3-2) [7], an ac-dc converter must be able to perform the PFC. Besides, high power density associated to high efficiency is a must. Within this context, this work proposes the study and development of a single-stage three-phase ac-dc converter with bidirectional power flow and high frequency isolation adequate to medium power dc distribution systems or telecom power supplies.

After a brief review on basic structures regarding single-stage isolated rectifier systems as presented in [8], a single-stage high-frequency isolated three-phase PWM (pulse width modulation) topology is proposed based on the VIENNA rectifier. The work in [9] proposes a single-stage single-phase bidirectional ac-dc converter, while the work in [10] presents a single-stage three-phase dc-dc converter using dual phase-shift control. By analyzing the aforementioned works and using the DAB converter [11] associated with the 3SSC [12], this work presents a single-stage three-phase bidirectional ac-dc converter with PFC.

II. PROPOSED TOPOLOGY

The proposed converter is shown in Figure 2, which is based on the DAB converter, since the primary side is composed by a full-bridge converter employing the 3SSC. The use of the 3SSC allows good distribution of losses among the semiconductors and also reduction of high-frequency harmonic content for both voltages and currents [13]. In order to obtain reduced reactive power flow through the transformer windings, the secondary is composed by a similar arrangement using full-bridge converters.

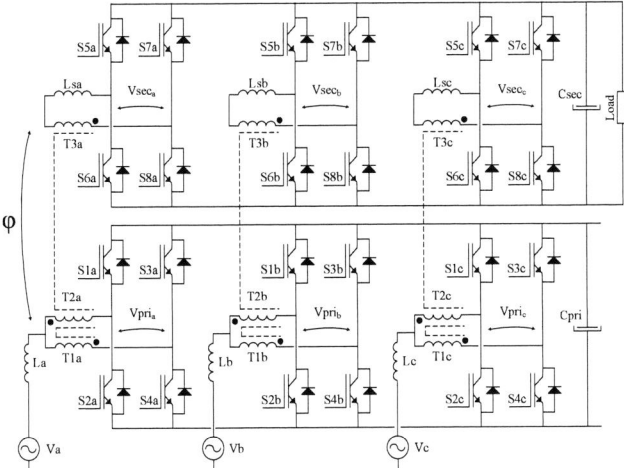

Fig. 2. Proposed three-phase rectifier.

A. Control strategy

The control strategy is shown in Figure 3. The control system of the primary currents is based on the pq theory [14-17]. In order to achieve unity power factor, reference iq is placed at zero and reference id comes from the voltage compensator C3(s) in the primary side. Thus, the inverse Park transform (dq – abc) provides the modulating signals (ma; mb; and mc), which are applied to the modulator in order to generate the drive signals of the primary switches. This strategy ensures PFC and the voltage regulation across the primary bus.

The power flow between the primary and secondary sides can be controlled by the phase-shift angle φ between the voltages across primary windings (Vpri$_a$, Vpri$_b$, and Vpri$_c$) and the voltages across the secondary windings (Vsec$_a$, Vsec$_b$, and Vsec$_c$). Aiming to minimize the reactive content, the secondary side modulator uses the same modulating signals of the primary (m$_a$; m$_b$; and m$_c$) and angle φ as calculated by secondary voltage compensator C4(s), thus generating the drive signals of the secondary switches. Therefore, the switching pattern in this case is similar to the ones in the primary side, but phase-shifted by angle φ, what enables active power flow control.

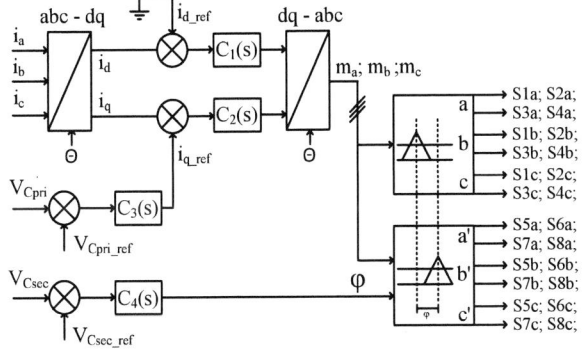

Fig. 3. Proposed control.

The active control of the magnetizing current is supposed to avoid the transformer saturation as presented in Figure 4.

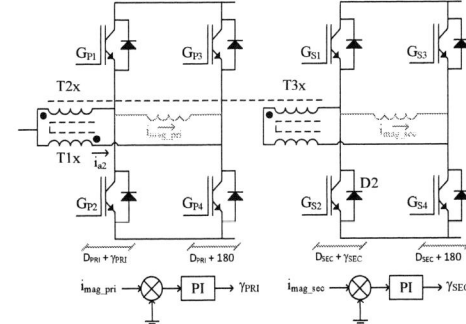

Fig. 4. Illustration of the active control concept for the magnetizing current.

By external inductors, the magnetizing currents are measured indirectly and using a digital PI controller is possible to keep this current as close to zero. The controller generates a signal that is summed with the modulator, causing the PWM pulse of the one leg become more or less wide. Thus they are used to control the magnetizing current six digital controllers, one for each full-bridge cell.

B. Mathematical Analysis

By observing the voltage waveform applied to the transformer (considering the sinusoidal variation of the duty cycle) and expanding it in terms of its respective Fourier series, signal represented by (1) can be obtained.

$$v(t) = [V + v_m \cos(\omega_m \cdot t)]\cos(\omega_c t) \qquad (1)$$

where:

V - peak input voltage;

v_m - peak value of the voltage first harmonic component;

ω_m - angular frequency ($2 \cdot \pi \cdot 120$ Hz);

ω_c - angular frequency of the modulating signal ($2 \cdot \pi \cdot 50$ kHz).

By manipulating (1), expression (2) can be obtained to explicit the fundamental and the harmonic components as:

$$v_{mod}(t) = V\cos(\omega_c t) + \frac{v_m}{2}\cos((\omega_c + \omega_m)t) + \frac{v_m}{2}\cos((\omega_c - \omega_m)t \qquad (2)$$

Considering the use of a low switching frequency in Figure 5, expressions (1) and (2) can be plotted. One can see that the voltage waveform across the transformer presents a pattern similar to modulation AM DSB-FC (amplitude modulation double-sideband full carrier).

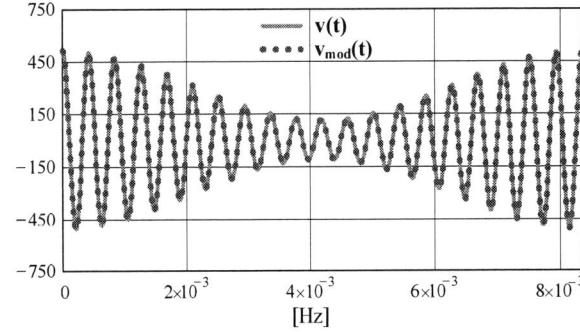

Fig. 5. Transformer voltage waveform (AM DSB-FC).

The representation of the waveform across the secondary side of the transformer is given by (3), where one can see that angle φ is responsible for the phase-shift between the primary and secondary voltages.

$$v_{mod\_sec}(t) = V\cos(\omega_c\, t + \varphi) + \frac{v_m}{2}\cos((\omega_c + \omega_m)t) + \frac{v_m}{2}\cos((\omega_c - \omega_m)t) \quad (3)$$

The basic circuit that represents the fundamental model of the converter is shown in Figure 6.

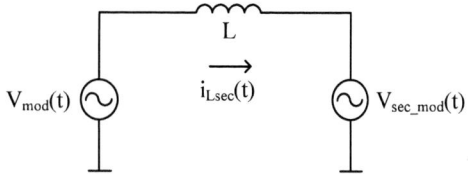

Fig. 6. Fundamental model of DAB.

The equation representing the circuit fundamental model is given by (4).

$$i_{sec}(t) = \frac{v_{mod} - v_{mod\_sec}}{L} \quad (4)$$

Thus, the aforementioned equations allow obtaining the average and apparent power processed by the converter as described in (5) and (6).

$$P_{sec}(t) = \frac{\omega_m}{2 \cdot \pi} \cdot \int (i_{sec}(t)) \cdot \left(v_{sec\_mod}(t)\right) \cdot dt \quad (5)$$

$$S_{sec} = i_{sec\_ef} \cdot v_{sec\_ef} \quad (6)$$

where:

$$i_{sec\_ef} = \sqrt{\frac{\omega_m}{2 \cdot \pi} \cdot \int (i_{sec}(t))^2 \cdot dt}$$

$$v_{sec\_ef} = \sqrt{\frac{\omega_m}{2 \cdot \pi} \cdot \int (v_{sec\_mod}(t))^2 \cdot dt}$$

By varying the phase shift angle, one can see the behavior of the active and apparent powers produced by both fundamental. That powers and the power factor (PF) are present in Figure 7 varying angle φ. For a phase-shift angle of 40°, an average power of 2 kW per phase results and power factor is higher than 0.8

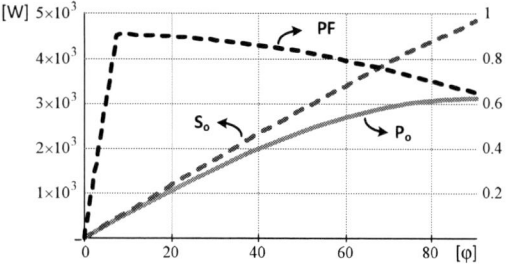

Fig. 7. Behavior of active power, apparent power and power factor, as a function of the phase-shift angle (φ).

C. *Loss analysis*

The design specifications of the proposed converter are presented in Table I.

TABLE I.   DESIGN SPECIFICATIONS

Output Power	6 kW
Input rms voltage	380 Vac
Input voltage frequency	60 Hz
Primary bus voltage	700 Vcc
Secondary bus voltage (output)	400 Vcc
Switching frequency	25-100 kHz
Estimated efficiency	97%

The loss profile was then obtained for different types of semiconductors e.g. MOSFETs (metal oxide semiconductor field effect transistors), IGBTs (insulated gate bipolar transistors) and SiC (silicon carbide) transistors. Figure 8 shows the efficiency as a function of the switching frequency,

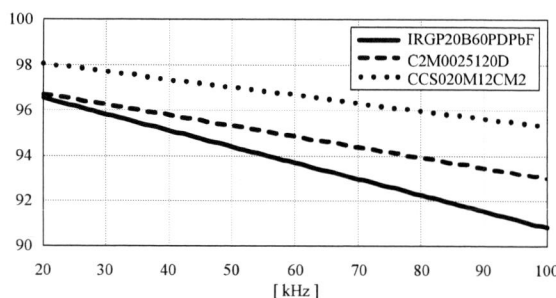

Fig. 8. Efficiency as a function of the switching frequency, for different types of semiconductors.

It can be seen that switching losses increase as the switching frequency does. In order to use compact hardware to achieve improved efficiency, SiC Module CCS0220M12CM2 was chosen. The switching frequency has been defined as 50 kHz, since good tradeoffs can be made between efficiency and dimension of magnetics (transformers and inductors).

III. SIMULATION RESULTS

Simulation results are presented considering the switching frequency of 50 kHz, three-phase rms voltages of 380 V/60 Hz, primary dc link of 700 V, secondary dc link voltage equal to 400 V, and output power of 6 kW. Figure 9 shows the three-phase input current, while THD=4.8% and PF=0.994 are obtained.

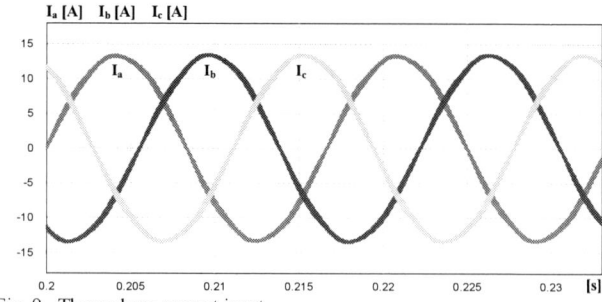

Fig. 9. Three-phase current input.

Figure 10 presents the step response considering the inversion of power flow from +100% to −100%. Fig. 10 (a) presents the input voltage Van and phase current Ia, thus demonstrating good dynamic response. The control system of the primary and secondary dc link maintains the voltages regulated at 700 V and 400 V, respectively with acceptable overshoot. Angle φ is about 40° according to the design specifications given in Table I.

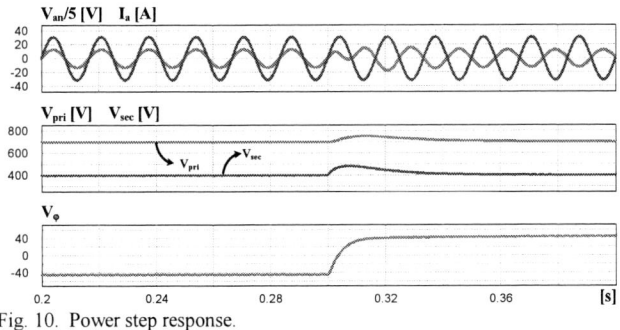

Fig. 10. Power step response.

Figure 11 shows the high-frequency waveforms in detail. Fig. 11 (a) presents voltages V_{Tpri} and V_{Tsec} and the phase-shift angle φ between them. Fig. 11 (b) shows the current through secondary winding T3A.

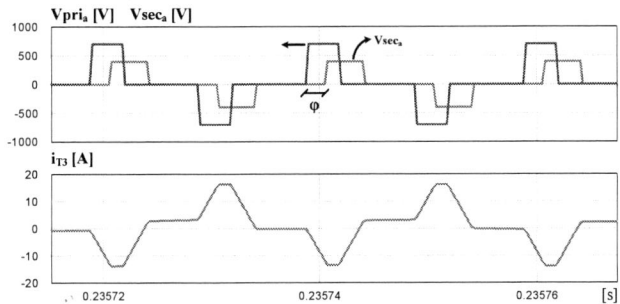

Fig. 11. Voltages across the primary and secondary windings and current iT3.

A short-circuit test is shown in Figure 12. Although the output current is not controlled, it is naturally maintained at acceptable values.

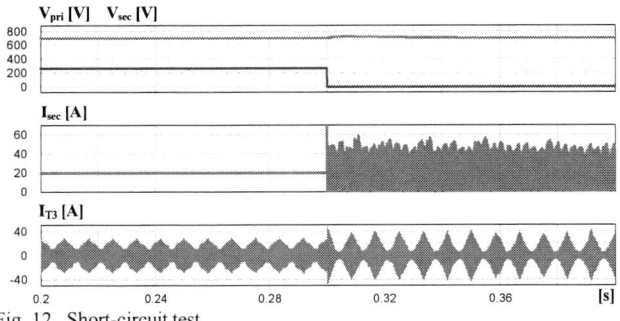

Fig. 12. Short-circuit test.

Figure 13 shows that good sharing is obtained between the currents through primary windings T1A and T2A, which constitute the autotransformer of the 3SSC, as well as the current through secondary winding T3A. Reduced low-frequency ripple exists in the aforementioned currents, which is due to asymmetry of gating pulses applied to the primary- and secondary-side switches, while some reactive power flows through the transformer.

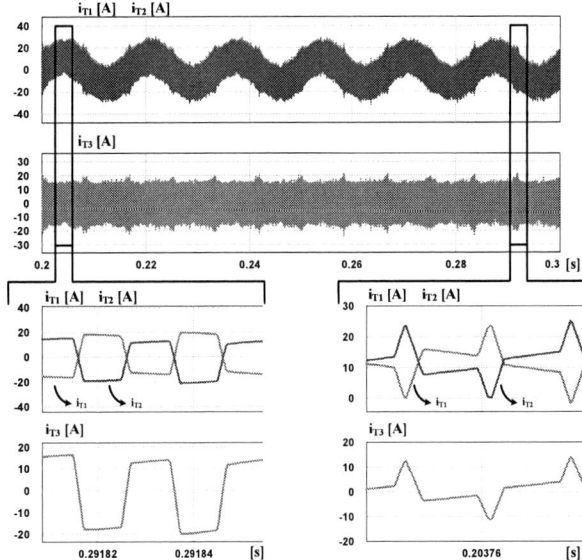

Fig. 13. Transformer currents (i_{T1}, i_{T2} and i_{T3}).

IV. EXPERIMENTAL PROTOTYPE

The 6kW proposed prototype has been implemented in laboratory and can be seen in Figure 14. Was used four three-phase SiC transistor modules (Cree - CCS0220M12CM2 1200V 20A) and four isolated gate drivers (Cree - CGD15FB45P).

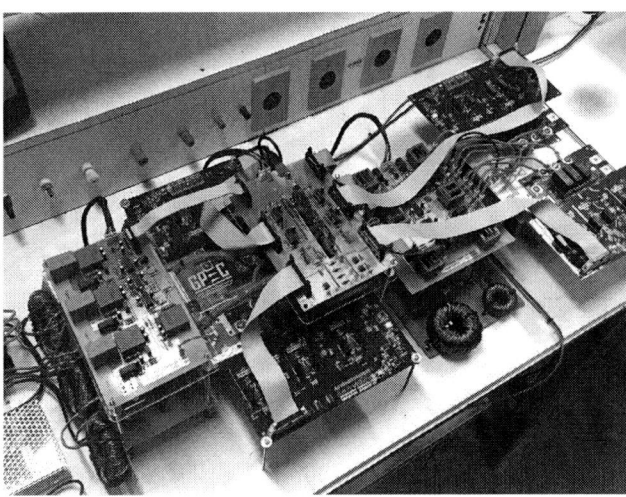

Fig. 14. Proposed 6kW converter prototype.

To controller the system was used the DSP (digital signal processor) model TMS320F28377D by Texas Instruments. This DSP operates at 200 MHz and features 16 channels ADs and 24 PWMs. The following are the preliminary results corresponding to the primary side of the proposed converter. Operating as rectifier, Figure 15 shows the three phase input current waveforms (ILa, ILb and ILc), were the THD obtained is about 2.7%.

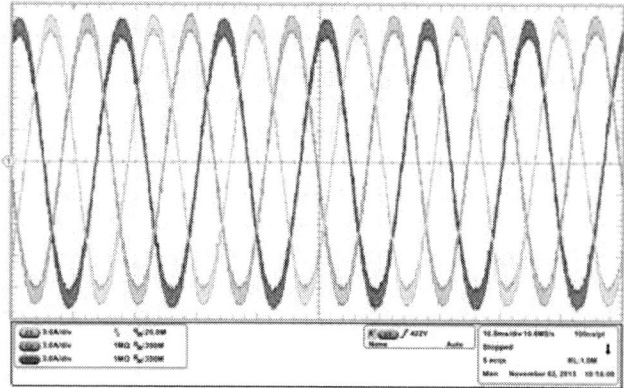

Fig. 15. Three phase input current in rectifier mode.

Figure 16 present the input voltage and current. As it can be seen, the digital q-PLL and input current controller operates successfully and the current are in phase with the input voltage, with a power factor (PF) of 0.997.

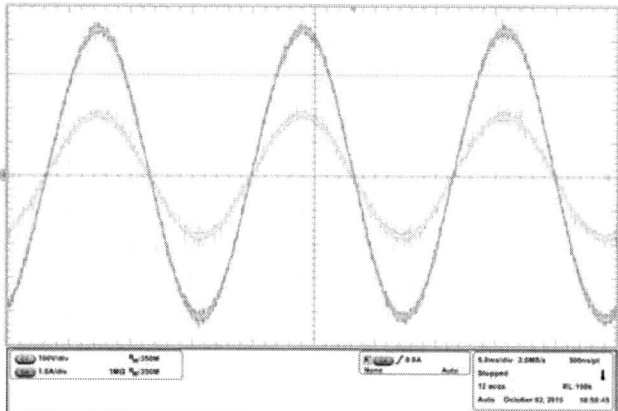

Fig. 16. Input voltage and current (PF = 0.997).

In order to test the dynamic response of the digital controller, in Figure 17 is showed the bus voltage behavior for a load step from 1.2kW to 2.4kW in the primary side of the converter. As it can be seen, the control system operates successfully and the bus voltage keeps on 400 volts, with acceptable overshoot.

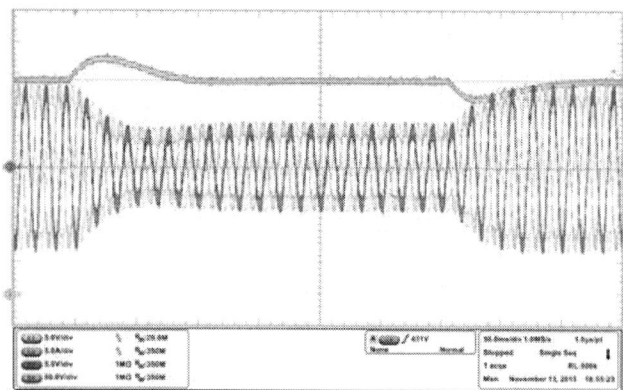

Fig. 17. DC bus voltage at load step from 1.2kW to 2.4kW.

Figure 18 show the soft startup of the input currents. This routine was implemented to reduce possible voltage drops in the grid when in the maximum load.

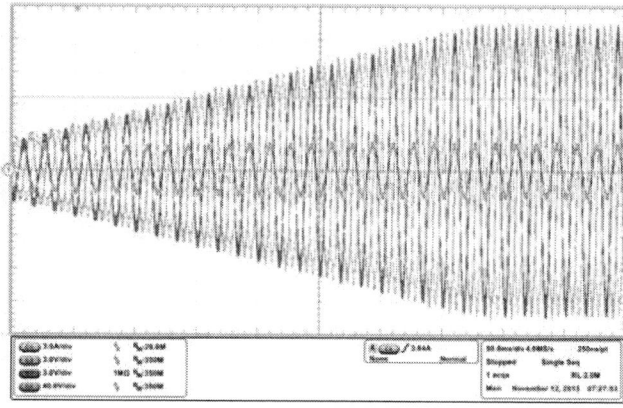

Fig. 18. Soft startup of the input currents.

Shown in Figure 19 the currents in the windings of the auto-transformer of the three states switching cell (3SSC). When the control of the magnetizing current are OFF, note that there is an imbalance between the currents. In the Figure 20 the control are ON, and currents have a good balance. Due to the fast turn off of the SiC semiconductor, some measurements have a certain noise level, as is the case of current delivered.

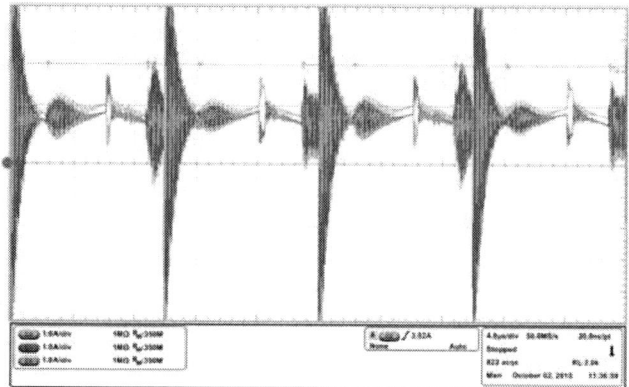

Fig. 19. Control of the magnetizing current OFF.

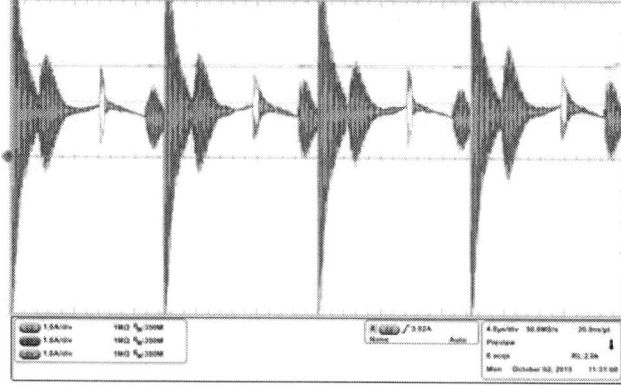

Fig. 20. Control of the magnetizing current ON.

V. CONCLUSIONS

This work has presented a single-stage three-phase ac-dc converter with high-frequency isolation, bidirectional power flow, and power factor correction characteristics. The topology has been validated by simulation, while the main waveforms regarding the control of primary and secondary bus voltages have been presented. It has been shown that varying the phase-shift angle allows controlling power flow direction and respective magnitude, while the duty cycle variation allows power factor correction. Such features make the proposed converter feasible to applications such as battery charging, telecommunications power supplies, electric vehicles, and also low-voltage smart grids.

The theoretical analysis has been validated by simulation. The controller operates satisfactorily, thus maintaining the voltages across the dc links in the primary and secondary sides constant, as the input current presents low THD. The experimental implementation of a prototype is currently in progress, and the results so far have been satisfactory. The input current has a power factor of 0997 and 2.7% of THD. The converter features an efficiency of 96.5% for power 3.5kW.

Secondary side of converter is currently in testing and further results will be presented in future works.

ACKNOWLEDGMENT

The authors acknowledge the Energy Processing and Control Group (GPEC) and the research agencies CNPq and CAPES for the financial support to this work..

REFERENCES

[1] D. E. Burke. (2014) "Reflections on INTELEC 2014". Electronics in Motion and Conversion (Bodo's Power Systems). 22.

[2] A. Pratt, P. Kumar, T. V. Aldridge, "Evaluation of 400V DC distribution in Telco and Data centers to Improve Energy Efficiency", Proc. 29th International Telecommunications Energy Conference, pp. 32–39, 2007.

[3] Z. Weichao, L. Haifeng, B. Zhou, L. Wei, G. Ran, "Review of DC technology in future smart distribution grid," in Innovative Smart Grid Technologies - Asia (ISGT Asia), pp. 1-4, 2012.

[4] S. Grillo, V. Musolino, L. Piegari, E. Tironi, and C. Tornelli, "DC Islands in AC Smart Grids," Power Electronics, IEEE Transactions on, vol. 29, pp. 89-98, 2014.

[5] W. Baochao, M. Sechilariu, F. Locment, "Intelligent DC microgrid with smart grid communications: Control strategy consideration and design," in Power and Energy Society General Meeting (PES), pp. 1-1, 2013.

[6] H. M. de Oliveira Filho, D. S. Oliveira, C. E. de A e Silva, F. L. Tofoli, "ZVS bidirectional isolated three-phase DC-DC converter with dual phase-shift and variable duty cycle," in Industry Applications (INDUSCON), 10th IEEE/IAS International Conference on, pp. 1-8, 2012.

[7] IEC, "IEC 61000-3-2: Electromagnetic Compatibility (EMC) – Part 3: Limits – Section 2: Limits for Harmonic Current Emissions (Equipment input current < 16 A per phase)," vol. Emenda A14, International Electrotechnical Commission, Ed., ed, 2001.

[8] D. S. Oliveira, M. I. V. Batista, L. H. S. C. Barreto, P. P. Praca, "A bidirectional single stage AC-DC converter with high frequency isolation feasible to DC distributed power systems," in Industry Applications (INDUSCON), 10th IEEE/IAS International Conference on, pp. 1-7, 2012.

[9] H. M. de Oliveira Filho, D. S. Oliveira, C. E. de A e Silva, F. L. Tofoli, "ZVS bidirectional isolated three-phase DC-DC converter with dual phase-shift and variable duty cycle," in Industry Applications (INDUSCON), 10th IEEE/IAS International Conference on, pp. 1-8, 2012.

[10] J. W. Kolar, U. Drofenik, F. C. Zach, "VIENNA rectifier II-a novel single-stage high-frequency isolated three-phase PWM rectifier system," Industrial Electronics, IEEE Transactions on, vol. 46, pp. 674-691, 1999.

[11] W. M. dos Santos; D. C. Martins, "Introdução ao conversor DAB monofásico," in Eletrônica de Potência, 2014, vol. 19, no. 1, pp. 36-46, February 2014.

[12] G. V. T. Bascope, I. Barbi, "Generation of a family of non-isolated DC-DC PWM converters using new three-state switching cells," in Power Electronics Specialists Conference PESC. IEEE 31st Annual, pp. 858-863 vol.2, 2000.

[13] D. S. Oliveira, D. de A Honorio, L. H. S. C. Barreto, P. P. Praca, A. Kunzea, and S. Carvalho, "A two-stage AC/DC SST based on modular multilevel converterfeasible to AC railway systems," in Applied Power Electronics Conference and Exposition (APEC), Twenty-Ninth Annual IEEE, pp. 1894-1901, 2014.

[14] E. H. Watanabe, R. M. Stephan, M. Aredes, "New concepts of instantaneous active and reactive powers in electrical systems with generic loads," Power Delivery, IEEE Transactions on, vol. 8, pp. 697-703, 1993.

[15] M. Aredes, H. Akagi, E. H. Watanabe, E. Vergara Salgado, L. F. Encarnacao, "Comparisons Between the p--q and p--q--r Theories in Three-Phase Four-Wire Systems," Power Electronics, IEEE Transactions on, vol. 24, pp. 924-933, 2009.

[16] Sasso, E. M., Sotelo, G., Ferreira, A., Watanabe, E. H., Aredes, M., Barbosa, P. G. "Investigação dos Modelos de Circuitos de Sincronismo Trifásicos Baseados na Teoria das Potências Real e Imaginária Instantâneas (p-PLL e q-PLL)", in Proceedings of the 14th Brazilian Automatic Control Conference, pp. 480-485 (in Portuguese), 2002.

[17] H. Guan-Chyun, J. C. Hung, "Phase-Locked Loop Techniques. A Survey", IEEE Transactions on Industrial Electronics, vol. 43, pp. 609–615, 1996.

A 5 MHz, 12 V, 10 A, Monolithically Integrated Two-Phase Series Capacitor Buck Converter

Pradeep S. Shenoy, Orlando Lazaro, Ramanathan Ramani, Mike Amaro, Wlodek Wiktor, Joseph Khayat, and Brian Lynch

Texas Instruments, Dallas, Texas, USA
pshenoy@ti.com

Abstract—This paper presents the first monolithically integrated two-phase series capacitor buck converter. This converter achieves over 60 A/cm³ current density. The series capacitor buck converter topology enables high frequency (HF) operation up to 5 MHz per phase without special magnetics or compound semiconductors. An adaptive constant on-time controller provides fast load transient response and fixed frequency operation in steady state. The series capacitor is pre-charged before switching and monitored to protect against faults. Experimental results for a 12 V input, 1.2V/10A output application demonstrate higher peak efficiency than a conventional buck converter while operating at 4 times higher switching frequency.

Keywords—voltage regulator, switched capacitor converter, current density, series capacitor buck converter, high frequency dc-dc converter

I. INTRODUCTION

Increasing switching frequency can lead to reduced physical size of power converters. This is especially useful in point-of-load voltage regulators because they often use up considerable circuit board space. Lower values of inductance and capacitance are needed to meet the converter design parameters. Converter transient response is improved with higher switching frequencies since inductor current slew rates increase with lower inductance. There is also the added potential benefit of reducing bill of materials (BOM) costs.

The buck converter is the most commonly used converter topology for voltage regulators. When attempting to increase switching frequency into the HF range (3-30 MHz), switching losses typically becomes prohibitively large for most applications. Additionally, the on-time of the high side switch is very short (e.g. less than 30 ns) for HF, high conversion ratio (>5:1) converters. The HF buck converters on the market today are typically designed with a low input voltage (5 V or less) and low output current (1 A or less). The series capacitor buck converter, shown in Fig. 1, overcomes many challenges faced in high step-down ratio, HF buck converters [1].

Combining a switched capacitor circuit and a buck converter has several advantages. Voltage conversion can be accomplished by the switched capacitor circuit and output regulation is achieved through the buck stage. This approach plays to the strengths of each circuit. Switched capacitor circuits are typically designed for integer ratio voltage conversion (e.g. 2:1, 3:1, etc.). High voltage conversion ratios are attained with relative ease. However, output voltage regulation is realized through adjusting losses. On the other

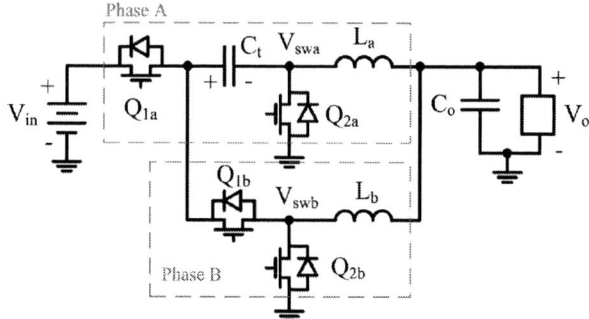

Fig. 1. The two-phase series capacitor buck converter.

hand, a buck converter can adjust its duty ratio efficiently and with high precision. The hybrid approach can be implemented in two stages consisting of a switched capacitor circuit followed by a buck converter [2], [3]. The three level buck converter combines both stages into one with soft charging of the energy transfer capacitor [4]-[7]. The merged capacitive attenuator uses a similar technique [8]-[9]. Downsides of these approaches include more components, current conduction through more than one switch at all times, and potentially a reduction in converter efficiency.

This paper introduces the first monolithically integrated series capacitor buck converter. This converter is designed to operate with up to 5 MHz switching frequency and is targeted at 12 V input and 10 A output applications. Its adaptive on-time controller provides fast transient response and fixed frequency operation in steady state. Design considerations and experimental results are included in the following sections.

II. SERIES CAPACITOR BUCK CONVERTER

The two-phase series capacitor buck converter combines a 2:1 switched capacitor converter and a two-phase buck converter. This topology is also called the double step-down buck [10] or the extended duty ratio converter [11]-[12]. It adds one capacitor (C_t) to the normal two-phase buck converter and rearranges a few connections. The series capacitor is inserted between the phase A high side and low side switches, as shown in Fig. 1. The drain of the phase B high side switch is connected to series capacitor instead of the input supply. Three or more phase versions and coupled inductor implementations have also been explored [13]-[15]. The series capacitor technique has also been applied to the tapped inductor buck converter [16].

The converter configurations and waveforms are very similar to an interleaved, two-phase buck converter. (The

978-1-4673-9551-9/16 $31.00 © 2016 IEEE

reader is referred to the references for detailed analysis [1], [10], [17].) Some key differences are that the duty ratio of the high side switches is doubled, switching occurs with half the drain-to-source voltage experienced by switches in a buck converter, and inductor current balancing is automatic. Reduced voltage switching enables efficient, high frequency energy conversion. Automatic current sharing removes the need for current sensing and current sharing control loops [17]. This offers considerable benefit at high frequency.

The major limitation of the series capacitor buck converter is the voltage conversion ratio. Because the high side switches' on-times cannot overlap (i.e. 50% duty ratio limit) and because there is an inherent 2:1 step down due to the series capacitor, the minimum input voltage must be four times greater than the output voltage, in theory. When converter losses are taken in to account, the practical minimum input voltage may be five times the output voltage. There are point-of-load applications where this limitation is not acceptable; other high conversion ratio applications may not have any problem with this constraint. Another downside to the series capacitor buck converter is that phase shedding and adding is not feasible. Other light load modes using discontinuous conduction mode and pulse frequency modulation can accommodate the need for efficient light load operation [18].

III. INTEGRATED CIRCUIT OVERVIEW

A monolithic integrated circuit (IC) implementation of the series capacitor buck converter was developed for low voltage point-of-load applications (e.g. less than 2 V) powered by a 12 V supply rail. An overview of the IC is shown in Fig. 2. The controller, gate drivers, power MOSFETs, internal regulators, and supporting circuitry are included. Passive components like input and output capacitors, inductors, feedback resistors, etc. are located externally. The chip includes typical protection circuits such as input under/overvoltage lockout, output under/overvoltage protection and overcurrent protection. New features were included to manage the series capacitor voltage. The converter was designed to operate with 2 MHz to 5 MHz per phase switching frequency, 8 V to 14 V input, and up to 10 A output. Several important typical converter parameters are listed in Table 1 for a 2 MHz and a 5 MHz design. The case

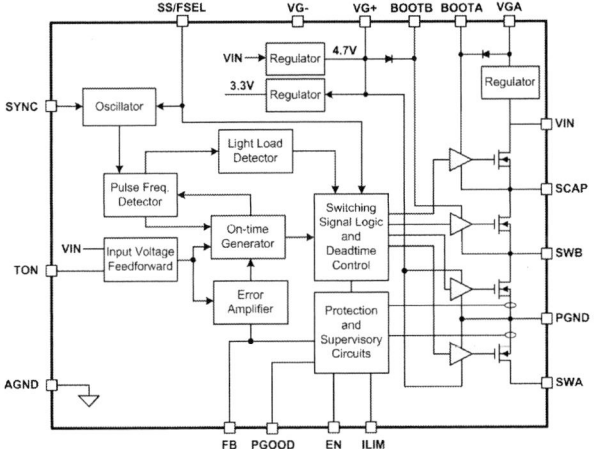

Fig. 2. Integrated series capacitor buck converter block diagram.

TABLE 1. TYPICAL CONVERTER PARAMETERS

Component	Parameter	
Switching frequency (f_s)	2 MHz	5 MHz
Inductance (L)	220 nH (1210)	100 nH (1210)
Inductor DCR	10 mΩ	5 mΩ
Series capacitance (C_t)	2.2 µF (1206)	1 µF (1206)
Output capacitance (C_o)	147 µF	91 µF
$R_{ds,on}$: Q_{1a}, Q_{1b}	27 mΩ	
$R_{ds,on}$: Q_{2a}, Q_{2b}	6.8 mΩ, 9.3 mΩ	

size for the inductors and series capacitor is included using the imperial code.

A. Gate Drive Scheme

The gate driver design is an important practical consideration. Typical, ground referenced low side gate drivers powered from the VG+/- pins were used for the low side MOSFETs. The gate drive supply rail (VG+/-) is supplied by an internal linear regulator or can be supplied externally. The phase B high side MOSFET driver utilized a bootstrap capacitor to bias its floating gate drive circuit.

A unique approach was required to drive the phase A high side MOSFET because its source is floating on top of the series capacitor. A linear regulator generates a supply rail on the VGA pin. The target voltage is dynamically adjusted based on the VG+ voltage and series capacitor voltage. The linear regulator is fairly efficient because the VGA voltage is a small amount lower than the input supply (e.g. 11 V from a 12 V input). The phase A bootstrap capacitor is charged from the VGA rail when the phase A low side MOSFET is on.

B. Protection Circuits

The IC includes several protection circuits. Inductor currents, which indirectly represent the output current, are monitored using the low side MOSFETs. If the measured current exceeds predefined thresholds for three switching cycles in a row, an overcurrent fault is flagged and the converter shuts down. After a wait time of at least 10 ms, the converter restarts automatically in a scheme commonly referred to as a "hiccup" restart.

The input voltage and output voltage is also monitored. If the input voltage falls below 7.65 V or rises above 15.4 V, the converter will shut down and restart when the input voltage is back in bounds. The output voltage is protected in a similar manner. If the output is greater than ±10% away from the target voltage, the converter will shut down and restart after the hiccup wait time has expired.

IV. SERIES CAPACITOR DESIGN AND MANAGEMENT

The series capacitor is a critical component of the converter. Proper capacitor selection and management during operation is important.

A. Capacitor Selection

A major function of the series capacitor is energy transfer. This is a different role from input and output capacitors used in buck converters where decoupling is the primary function. In many ways, the series capacitor is similar to the capacitor used for energy transfer in SEPIC converters and can be designed accordingly. A design objective may be to ensure the series capacitor voltage ripple does not exceed 5% of the nominal voltage under the worst case conditions. The equation that describes the series capacitor voltage ripple is

$$\Delta V_{Ct} = \frac{DTI_{Ct}}{C_t} = \frac{\left(\frac{2V_o}{V_{in}}\right)\left(\frac{I_o}{2}\right)}{C_t f_{sw}} \quad (1)$$

where D is the duty ratio, T is the switching period, I_{Ct} is the series capacitor current, C_t is the series capacitance, V_o is the output voltage, V_{in} is the input voltage, I_o is the output current, and f_{sw} is the switching frequency. Equation (1) can be rearranged to provide the design equation for series capacitor selection which is

$$C_t \geq \frac{\left(\frac{2V_o}{V_{in}}\right)\left(\frac{I_o}{2}\right)}{k_{Ct}\left(\frac{V_{in}}{2}\right) f_{sw}} \quad (2)$$

where k_{Ct} represents the voltage ripple percentage and $V_{in}/2$ is the nominal series capacitor voltage. If the ripple target is 5%, the value for k_{Ct} is 0.05. The largest voltage ripple occurs at full load current (highest I_o), highest duty ratio (lowest input voltage/highest output voltage), and lowest frequency.

Another aspect to consider is capacitor RMS current rating. This impacts the temperature rise of the capacitor. If the temperature rise is too large for a single capacitor, multiple capacitors may be placed in parallel. The series capacitor has the same current profile as the high side MOSFETs. The RMS current squared can be expressed as

$$I_{Ct,RMS}^2 = 2DI_{L,RMS}^2 \quad (3)$$

where $I_{L,RMS}$ is the RMS inductor current of either inductor. The series capacitor RMS current can be expressed as

$$I_{Ct,RMS} = \sqrt{2\left(\frac{2V_o}{V_{in}}\right)\left[\left(\frac{I_o}{2}\right)^2 + \left(\frac{\Delta I_L}{12}\right)^2\right]} \quad (4)$$

where ΔI_L is the inductor current ripple. The largest RMS current occurs at the highest load current and highest duty ratio.

Multilayer ceramic capacitors (MLCC) are well suited for operating as the series capacitor. The equivalent series resistance (ESR) is relatively low (e.g. 5 mΩ to 10 mΩ) which helps to reduce power loss and self heating. The equivalent series inductance (ESL) is fairly low which results in a high self resonant frequency (SRF). There are a few key items that should be considered when designing. First, the effective capacitance decreases with dc bias. This means that the capacitor should be selected based on its capacitance with the nominal voltage of $V_{in}/2$ applied. Temperature variation also reduces effective capacitance. For this reason, X7R capacitors with up to 125°C operating temperature range are

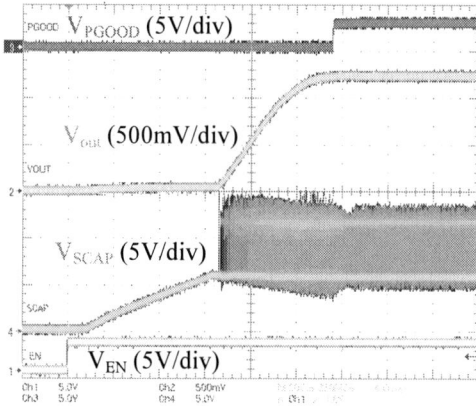

Fig. 3. Startup begins with series capacitor pre-charge (200 μs/div).

recommended. If capacitors are not properly selected, cracking or other failure modes may result.

B. Capacitor Pre-Charge

The converter pre-charges the series capacitor to half the input voltage before switching begins. This ensures predictable, smooth startup and automatic, even current sharing [6], [17]. Since the series capacitor voltage is nominally $V_{in}/2$ during normal operation, that target voltage was selected for the pre-charge circuit. A 10 mA current source charges the series capacitor. The startup delay due to pre-charging the series capacitor can be calculated to be

$$t_{pc} = \frac{C_t\left(\frac{V_{in}}{2}\right)}{I_{pc}} \quad (5)$$

where t_{pc} is the series capacitor pre-charge time and I_{pc} is the pre-charge current (10 mA in this converter). An experimental result demonstrating the series capacitor pre-charge is shown in Fig. 3 for a 1 μF series capacitance. The pre-charge time is approximately 600 μs. This delay can complicate startup tracking functionality. It also highlights a tradeoff with series capacitor selection. A larger series capacitance will result in a longer pre-charge delay.

C. Voltage Monitoring

A unique function included in the IC is series capacitor voltage monitoring. The differential voltage across the series capacitor is continuously measured. A fault is triggered if the voltage strays away from the nominal voltage by more than ±30%. If the voltage is too low, the converter shuts down, waits for a hiccup restart time, pre-charges the series capacitor to half the input voltage, and initiates soft start. If the voltage is too high, the same response occurs except an internal bleed resistor is applied to reduce the series capacitor voltage. This functionality is included to prevent any damage that could be caused by series capacitor voltage going out of bounds.

V. ADAPTIVE ON-TIME CONTROL

An adaptive on-time control scheme is implemented as shown in Fig. 4. Its core operates in much the same way that constant on-time controllers work. The output feedback is compared to an internal voltage reference to trigger a high side switch on-time. This approach provides fast response to load

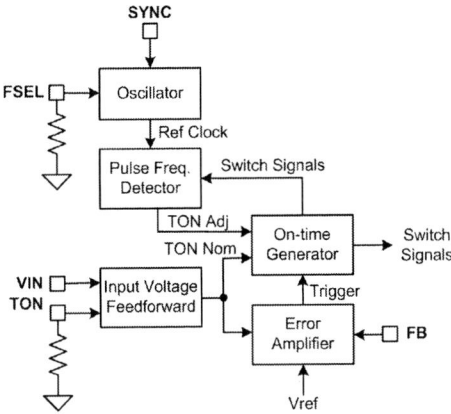

Fig. 4. Adaptive on-time controller block diagram.

transients. When the load steps up, the controller quickly fires successive on-time pulses to increase inductor current. When the load steps down, the high side switches remain off, and the inductor currents fall.

A distinguishing feature of this controller is that the on-time is adapted for fixed frequency operation in steady state. This applies over all variations in the input voltage, load current, and temperature. The nominal on-time is set by a programming resistor and an input voltage feedforward circuit. This is similar to other constant on-time controllers where the on-time is adjusted based on the input voltage. In this controller, as shown in Fig. 4, the on-time is adapted by feeding in a correction term that slowly increases or decreases the on-time based on the output of the pulse frequency detector. The pulse frequency detector compares the on-time pulse train to a reference clock derived from either an internal oscillator or an external synchronization clock signal. This forms a phase lock loop (PLL).

The PLL aligns the frequency and phase of the on-time pulses with the reference clock. When the external synchronization clock is applied to the SYNC pin, it overrides the internal clock. When the external clock is removed, the internal clock will take over after a few switching cycles. Both internal and external clocks are two times the per phase switching frequency because the controller is operating at twice the switching frequency.

The PLL bandwidth is about ten times lower in frequency than the converter closed-loop crossover frequency. This enables the converter to quickly respond to fast, large signal load transients and operate with fixed frequency in steady-state. It keeps the frequency from sliding into unwanted frequency bands. It also allows multiple converters operating from the same input bus to have their on-times interleaved to reduce voltage ripple on the input bus. This is not possible with conventional constant on-time controllers.

VI. EXPERIMENTAL RESULTS

The series capacitor buck converter was tested over various operating conditions. The typical converter parameters are listed in Table 1. A picture of the converter prototype used in testing is shown in Fig. 5. A state-of-the-art 10 A point-of-load buck converter [19] is shown in Fig. 6 for comparison. The

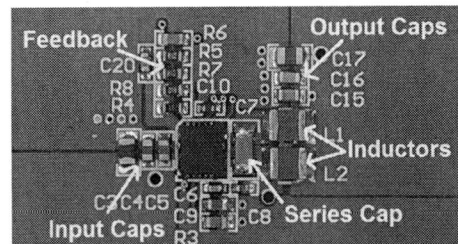

Fig. 5. 10 A, 2-5 MHz per phase series capacitor buck converter prototype.

Fig. 6. 10 A, 500 kHz buck converter reference design.

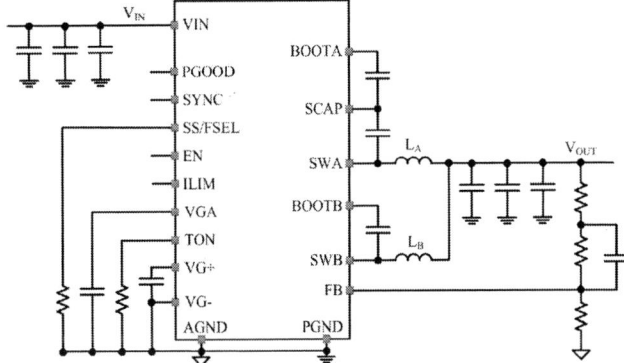

Fig. 7. A simplified schematic of the series capacitor buck converter prototype.

current density of the series capacitor buck converter is improved by use of small inductors that are 3.2 x 2.5 x 1.2 mm as compared with the 500 kHz, 10 A buck converter's inductor that is 7 x 7 x 4 mm. A major benefit is height reduction. Because it is only 1.2 mm tall, the 10 A converter can be placed in low profile locations such as on the back side of the main circuit board or on a plug in card (e.g. DIMM, PCIe). A schematic of the series capacitor buck converter prototype is shown in Fig. 7. Some additional board components used for testing the converter are not shown for simplicity. An additional 50 Ω resistor is included in the feedback network to aid with injecting a perturbation signal for closed loop measurements. A capacitor is placed in parallel with the upper feedback resistor to provide extra phase boost at the crossover frequency. The prototype was designed with a four layer PCB for ease of routing signals and to have sufficient ground planes for ground return current as well as heat removal. The IC uses flip chip, HotRod QFN packaging.

978-1-4673-9551-9/16 $31.00 © 2016 IEEE

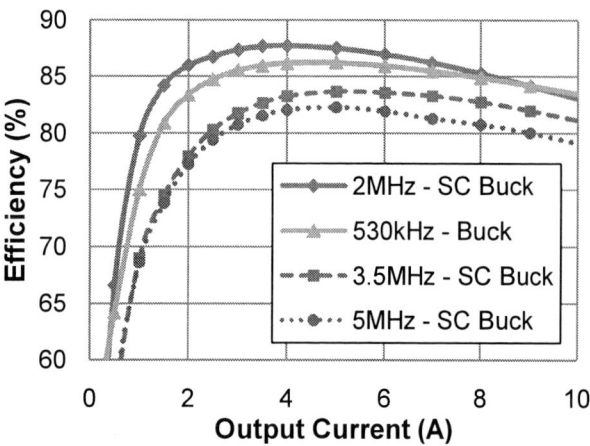

Fig. 8. The peak efficiency of the series cap buck at 2 MHz is higher than a conventional buck at 530 kHz for a 12 V input, 1.2 V output application.

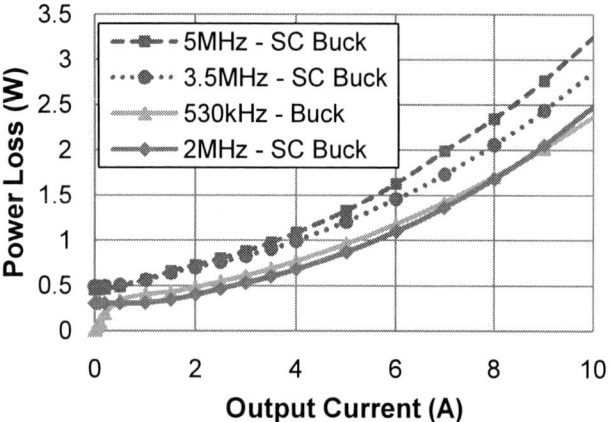

Fig. 9. Power loss comparison of the series cap buck and a standard buck for 12 V input, 1.2 V output application.

A. Efficiency and Power Loss

The measured efficiency and power loss for 12 V input, 1.2 V output, at room temperature are shown in Figs. 8 and 9. The feedback resistors and frequency setting resistor were adjusted for the buck converter reference design [19] to set its output to 1.2 V and switching frequency to approximately 500 kHz. The efficiency data collected includes all power stage, inductors, gate drive, controller, and layout losses. The only additional loss term the buck converter had was an internal linear regulator that provided power from the 12 V input to the 5 V gate driver. An external supply provided the gate drive power for the series capacitor buck converter.

The series capacitor buck converter demonstrates higher peak efficiency (87.7%) at 2 MHz per phase switching frequency than a buck converter operating at 530 kHz switching frequency (86.2%). The lower full load efficiency is attributed primarily to the higher inductor DCR (10mΩ in the series cap buck vs. 3.2mΩ in the buck converter). As expected, the 3.5 MHz and 5 MHz efficiency of the series capacitor buck converter is lower but still above 80% for a large portion of the load range.

The power loss results in Fig. 9 are for the same conditions as the efficiency results in Fig. 8. The series capacitor buck converter was operated in forced continuous conduction mode

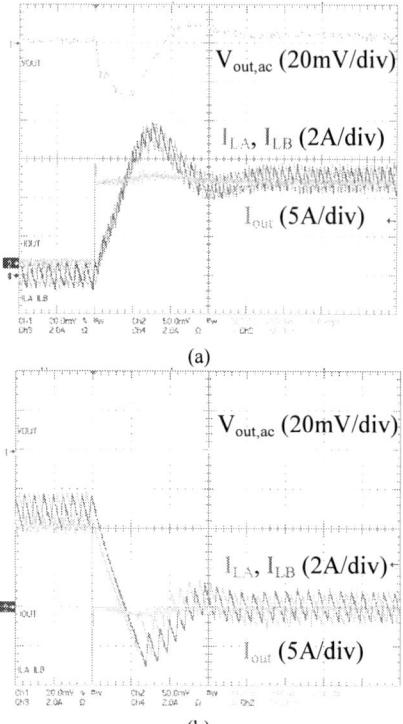

(a)

(b)

Fig. 10. Response to 500 A/μs (a) full load (0 A to 10 A) increase and (b) full load (10 A to 0 A) decrease (2 μs/div).

(FCCM) over the entire load range. The buck converter operated with pulse skipping at light loads that helped to reduce power consumption below 0.5 A output. It is also worth noting that more effort was taken to select the optimal inductor for use at 2 MHz in the series capacitor buck converter; the 3.5 MHz and 5 MHz cases used a different inductor vendor.

B. Large Signal Load Transient Response

The 2 MHz converter response to a 500 A/μs full load increase and decrease is shown in Fig. 10. The output recovery time is 4 μs with peak voltage excursion of approximately ±25mV (i.e. ±2%). The output capacitance was increased for this test to approximately 350 μF. The fast transient load was implemented with an on-board MOSFET load. The inductor currents are well balanced even during the large, fast transient. The inductors currents are slightly imbalanced at the end of the load step down due to the controller response but quickly come back to even current sharing. Faster load transient response is possible when operating at 5 MHz per phase.

C. Closed Loop Response

The bode plot measured for the 5 MHz series capacitor buck converter is shown in Fig. 11 for two different values of output capacitance. The converter demonstrates a 219 kHz crossover frequency with 62.6 degrees of phase margin with 138 μF of output capacitance. The crossover frequency increases to 320 kHz with 53.8 degrees of phase margin when the output capacitance is reduced to 91 μF. Both scenarios were tested for 12 V input and 1.2 V output with a power resistor load drawing approximately 2.5 A. The results demonstrate stability of the system with high closed loop bandwidth. Most buck converters designed for similar

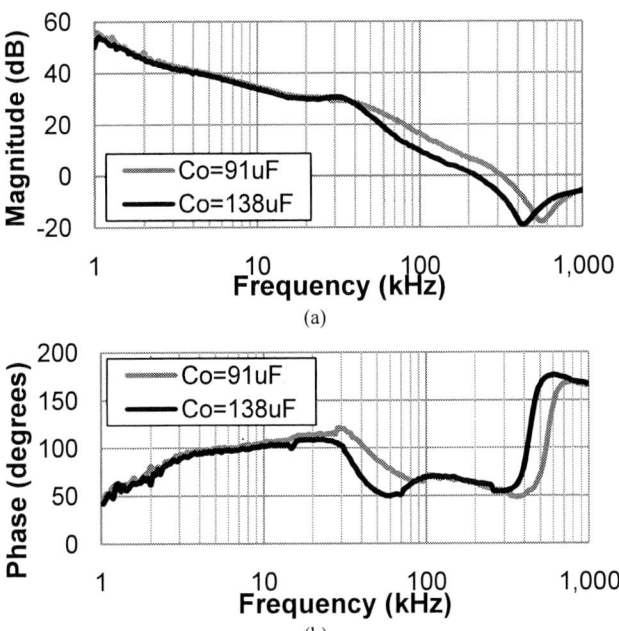

(a)

(b)

Fig. 11. Bode plot showing (a) magnitude and (b) phase for 12 V in, 1.2 V/2.5 A out at 5 MHz for two output capacitance values.

applications have closed loop bandwidths in the tens of kHz range. Extra poles and zeroes can be inferred from the gain and phase plots in the 300 kHz to 700 kHz range. The "wiggling" in the plot is attributed to higher order parasitics in the converter layout and components.

D. Thermal Image

A thermal image of the converter prototype operating at 2 MHz per phase with 12 V input, 1.2V/10A output is shown in Fig. 12. The bar on the right side of the image displays the thermal mapping with temperature measured in degrees Celsius. No air flow was applied and the ambient condition was room temperature. The result indicates that the IC and the series capacitor are the warmest components, and the hottest part of the image reaches 67°C. Most of the power loss is in the IC (more than half the total power loss) so it is understandable that it would be warm. The series capacitor has relatively low power loss but its heat sinking is not very good. The body of the series capacitor is red indicating a high temperature but the ends of the capacitor that are soldered to the PCB are green indicating a much cooler temperature.

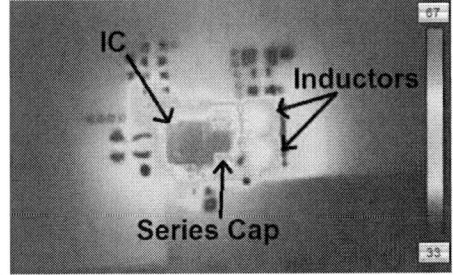

Fig. 12. Thermal image of converter prototype operating at 12 V input, 1.2V/10 A output, 2 MHz per phase switching frequency.

The inductors are fairly cool by comparison. This is noteworthy because most of the overall converter size reduction comes from using smaller inductors. There does not

Fig. 13. Practical, 10 A series capacitor buck converter layout with small size.

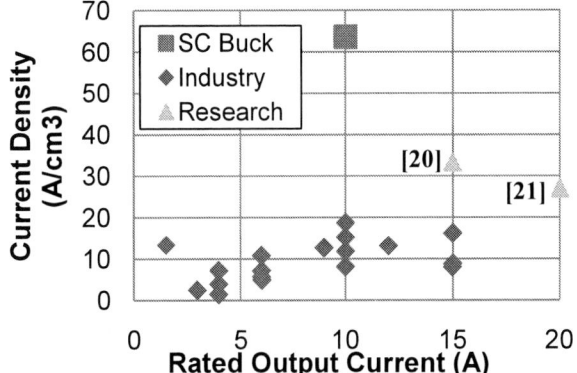

Fig. 14. The integrated series cap buck converter demonstrates several times higher current density than industry state-of-the-art and other research.

appear to be any extra thermal dissipation burden or hot spot when using the small inductors.

VII. SIZE COMPARISON

The current density of the series capacitor buck converter is compared to state-of-the-art buck converters available in industry today as well as a couple research prototypes [20], [21]. A practical, compact layout of the series capacitor buck converter without the extra components used for testing is shown in Fig. 13. The converter footprint is 131 mm^2 and the volume, not including the printed circuit board (PCB) thickness, is 157 mm^3. The resulting current density is 63.6 A/cm^3.

All the converters compared in Fig. 14 are designed for 12 V input, point-of-load applications. The total solution size is estimated including input and output capacitors, inductors, feedback network, etc. The PCB thickness is also included for the research prototypes [20], [21] since the inductor is embedded in the PCB substrate. It is clear from the figure that the series capacitor buck converter has significantly higher current density (over 60 A/cm^3) than any of the conventional buck converters.

Current density is a more appropriate metric for comparing the size of point-of-load dc-dc converters than power density. Power density of a converter is often calculated at the maximum output voltage and output current. This is not useful information for point-of-load converters where the output voltage in a typical application is much lower (e.g. 1.2 V) than the maximum output voltage (e.g. 90% or more of the input voltage). Power density at a specific output voltage of interest could be calculated; however, this ends up being a scaled version of current density.

978-1-4673-9551-9/16 $31.00 © 2016 IEEE

VIII. Conclusions

This paper presents a 2 MHz to 5 MHz, 12 V input, 10 A output, integrated series capacitor buck converter. The converter benefits from lower switching loss, automatic current balancing, and on-time doubling inherent to the series cap buck topology. The IC implementation includes the controller, gate drivers, power MOSFETs, and support circuitry. The series capacitor pre-charge circuit and voltage monitoring are new functions integrated in this converter. The series capacitor design equations are included for reference.

Experimental results for a 12 V to 1.2 V application demonstrate the effectiveness of the series capacitor buck converter. The converter has higher efficiency (87.7% peak) operating at 2 MHz per phase than a conventional buck converter (86.2% peak) operating at 530 kHz. An adaptive constant on-time control scheme is implemented with fast transient response achieving 4 μs recovery time to full load steps (0 A to 10 A) slewing at 500 A/μs. Fixed frequency operation in steady state is also achieved. High closed loop crossover frequency of over 300 kHz was measured with over 50 degrees of phase margin. This converter including all passive components in a practical layout has the highest known current density for a 12 V input, 10 A output point-of-load dc-dc converter.

References

[1] P.S. Shenoy, M. Amaro, D. Freeman, J. Morroni, "Comparison of a 12V, 10A, 3MHz, buck converter and a series capacitor buck converter," in *Proc. IEEE Applied Power Electron. Conf.*, 2015, pp. 461-468.

[2] R.C.N. Pilawa-Podgurski, D.M. Giuliano, and D.J. Perreault, "Merged two-stage power converter architecture with soft charging switched-capacitor energy transfer," in *Proc. IEEE Power Electron. Spec. Conf.*, 2008, pp. 4008-4015.

[3] R.C. N. Pilawa-Podgurski and D.J. Perreault, "Merged two-stage power converter with soft charging switched-capacitor stage in 180 nm CMOS," *IEEE J. Solid-State Circuits*, vol. 47, no. 7, pp. 1557-1567, July 2012.

[4] T.A. Meynard and H. Foch, "Multi-level conversion: high voltage choppers and voltage-source inverters," in *Proc. IEEE Power Electron. Spec. Conf.*, 1992, pp. 397-403.

[5] V. Yousefzadeh, E. Alarcon, and D. Maksimovic, "Three-level buck converter for envelope tracking applications," *IEEE Trans. Power Electron.*, vol. 21, no. 2, pp. 549-552, Mar. 2006.

[6] D. Reusch, F.C. Lee, and M. Xu, "Three level buck converter with control and soft startup," in *Proc. IEEE Energy Conversion Congr. Expo.*, 2009, pp. 31-35.

[7] B. Choi and D. Maksimovic, "Loss modeling and optimization for monolithic implementation of the three-level buck converter," in *Proc. IEEE Energy Conversion Congr. Expo.*, 2013, pp. 5574-5579.

[8] A. Radić and A. Prodić, "Buck converter with merged active charge-controlled capacitive attenuation," *IEEE Trans. Power Electron.*, vol. 27, no. 3, pp. 1049-1054, Mar. 2012.

[9] B. Mahdavikhah, P. Jain, and A. Prodić, "Digitally controlled multi-phase buck-converter with merged capacitive attenuator," in *Proc. IEEE Applied Power Electron. Conf.*, 2012, pp. 1083-1087.

[10] K. Nishijima, K. Harada, T. Nakano, T. Nabeshima, and T. Sato, "Analysis of double step-down two-phase buck converter for VRM," in *Proc. IEEE Telecommun. Energy Conf.*, 2005, pp. 497-502.

[11] J. Yungtaek, M.M. Jovanovic, and Y. Panov, "Multiphase buck converters with extended duty cycle," in *Proc. IEEE Applied Power Electron. Conf.*, 2006, pp. 38-44.

[12] B. Oraw, and R. Ayyanar, "Small signal modeling and control design for new extended duty ratio, interleaved multiphase synchronous buck converter," in *Proc. IEEE Telecommun. Energy Conf.*, 2006.

[13] K. Abe, K. Nishijima, K. Harada, T. Nakano, T. Nabeshima, and T. Sato, "A novel three-phase buck converter with bootstrap driver circuit," in *Proc. IEEE Power Electron. Spec. Conf.*, 2007, pp. 1864-1871.

[14] K. Matsumoto, K. Nishijima, T. Sato, and T. Nabeshima, "A two-phase high step down coupled-inductor converter for next generation low voltage CPU," in *Proc. IEEE Int. Conf. Power Electron. ECCE Asia*, 2011, pp. 2813-2818.

[15] B.S. Oraw and R. Ayyanar, "Voltage regulator optimization using multiwinding coupled inductors and extended duty ratio mechanisms," *IEEE Trans. Power Electron.*, vol. 24, no. 6, pp.1494-1505, June 2009.

[16] M. Chen, P.S. Shenoy, and J. Morroni, "A series-capacitor tapped buck (SC-TaB) converter for regulated high voltage conversion ratio DC-DC applications," in *Proc. IEEE Energy Conversion Congr. Expo.*, 2014, pp. 3650-3657.

[17] P.S. Shenoy, O. Lazaro, M. Amaro, R. Ramani, W. Wiktor, B. Lynch, and J. Khayat, "Automatic current sharing mechanism in the series capacitor buck converter," in *Proc. IEEE Energy Conversion Congr. Expo.*, 2015, pp. 2003-2009.

[18] P.S. Shenoy and M. Amaro, "Improving light load efficiency in a series capacitor buck converter by uneven phase interleaving," in *Proc. IEEE Applied Power Electron. Conf.*, 2015, pp. 2784-2789.

[19] TI reference design, PMP9194, "4.5-V to 17-V Input, 10A Synchronous Buck Converter Optimized for Small Size and Low Output Voltage," Texas Instruments. [Online]. Available: http://www.ti.com/tool/PMP9194.

[20] D. Reusch, F.C. Lee, D. Gilham, and Y. Su, "Optimization of a high density gallium nitride based non-isolated point of load module," in *Proc. IEEE Energy Conversion Congr. Expo.*, Sept. 2012. pp. 2914-2920.

[21] Y, Su, Q. Li, and F.C. Lee, "Design and evaluation of a high-frequency LTCC inductor substrate for a three-dimensional integrated DC/DC converter," *IEEE Trans. Power Electron.*, vol. 28, no. 9, pp. 4354-4364, Sept. 2013.

A 10-MHz Isolated Class-Φ_2 Synchronous Resonant DC-DC Converter*

Yuan Zhou, Zhiliang Zhang, Xue-Wen Zou, Zhou Dong and Xiaoyong Ren

Jiangsu Key Laboratory of New Energy Generation and Power Conversion
Nanjing University of Aeronautics and Astronautics, Nanjing, Jiangsu, P. R. China
{zoezy, zlzhang, zouxuewen, bigzhou, renxy}@nuaa.edu.cn

Abstract—In the multi-MHz low voltage, high current applications, Synchronous Rectification (SR) is strongly needed due to the forward recovery and the high conduction loss of the rectifier diodes. This paper applies the SR technique to a 10-MHz isolated class-Φ_2 resonant converter and proposes a self-driven level-shifted Resonant Gate Driver (RGD) for the SR FET. The proposed RGD can reduce the average on-state resistance and the associated conduction loss of the MOSFET. It also provides precise switching timing for the SR so that the body diode conduction time of the SR FET can be minimized. A 10-MHz prototype with 18 V input, 5 V/2 A output was built to verify the advantage of the SR with the proposed RGD. At full load of 2 A, the SR with the proposed RGD improves the converter efficiency from 80.2% using the SR with the conventional RGD to 82% (an improvement of 1.8%). Compared to the efficiency of 77.3% using the diode rectification, the efficiency improvement is 4.7%.

Keywords—*multi-MHz; synchronous rectification; resonant gate driver; level-shift; isolated resonant converter*

I. INTRODUCTION

Recently, the switching frequency of the power supply has been pushed up to tens of MHz to achieve extremely high power density, which is most limited by the passive components, especially the bulky inductance and high capacitance [1]-[2]. In the multi-MHz resonant converters, the Si Schottky diodes are normally used [3]-[4]. However, the performance of the diodes degrades seriously at multi-MHz because of the forward recovery phenomenon [5]-[6]. The loss dissipated in the rectifier diodes accounts for at least 30% ~ 35% of the total power loss [7]-[8]. Therefore, the SR technique is strongly desired in multi-MHz resonant converters, especially at high current applications.

The challenge coming along with the SR is the high gate drive loss of the MOSFET rectifiers. Since the conventional hard-switched gate drivers are far too inefficient at multi-MHz in low power applications, the RGDs are normally needed. It provides reliable drive voltage and reduces the high frequency gate loss by storing the drive energy in the resonant tank in each switching cycle [9]-[10]. However, because the on-state resistance $R_{DS(on)}$ of the MOSFET decreases with the gate-to-source voltage, the average $R_{DS(on)}$ of the RGD is much higher than that provided in the datasheet, causing high conduction loss in the MOSFET. Moreover, the duty ratio of the switch can hardly be adjusted flexibly, which will also induce additional body diode conduction time in the SR FET, and

cause high conduction loss at multi-MHz.

The objective of this paper is to apply the SR technique to the multi-MHz converter. A self-driven level-shifted RGD is proposed for the implementation of the SR FET in a 10-MHz isolated class-Φ_2 resonant converter. The proposed RGD manages to drive the SR FET with lower $R_{DS(on)}$ and provides precise switching timing for the SR FET.

II. DRIVER CHALLENGES FOR THE SR IN MULTI-MHz RESONANT CONVERTERS

A. Multi-MHz Isolated Class-Φ_2 Resonant Converter with SR

Fig. 1 gives the schematic of the isolated multi-MHz resonant converter and Table I lists the converter specifications. The converter comprises a class-Φ_2 inverter to reduce the voltage stress across the control FET and a class E rectifier. Since the forward recovery causes high conduction loss at multi-MHz in the diodes, especially at high output current, the SR technique is employed to the converter. Therefore, the gate driver for the SR FET in the multi-MHz resonant converter need to be studied to provide efficient gate drive.

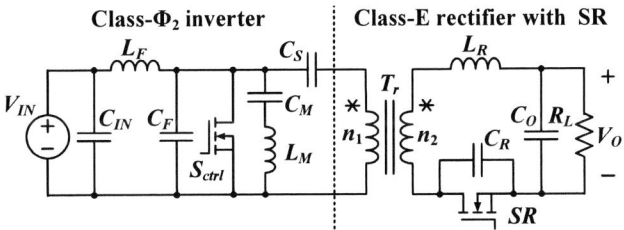

Fig. 1.   Isolated class-Φ_2 synchronous resonant converter

Table I   CONVERTER SPECIFICATIONS

V_{IN}	V_O	I_O	P_O	f_s
18 V	5 V	2 A	10 W	10 MHz

B. Conventional RGD for Multi-MHz Resonant Converters

Fig. 2 gives the typical circuit of the conventional RGD in the multi-MHz resonant converters, where R_G and C_{iss} are the gate resistance and the input capacitance of the MOSFET respectively. The resonant tank comprised of a series inductance L_S and a shunt branch L_P, C_P is connected between the drive signal and the MOSFET to recover the driving energy and reduce the gate loss. However, two main drawbacks of the conventional RGD are as follows.

---

*This work is supported by Natural Science Foundation of China (51377077) and the Fundamental Research Funds for the Central Universities (NUAA), No. NE2014101, No. 3082014NP2014402.

978-1-4673-9551-9/16 $31.00 © 2016 IEEE

1) Fig. 3 presents the $R_{DS(on)}$ versus the gate-to-source voltage v_{gs} based on a Si MOSFET from Vishay (SiS862DN). It is observed that when v_{gs} just exceeds the threshold voltage (V_{th}) or declines close to V_{th}, especially at the turn-on and turn-off instant, the $R_{DS(on)}$ is much high. Conversely, when v_{gs} exceeds the critical voltage V_{cv} (usually 5 ~ 6 V), the $R_{DS(on)}$ is much low and changes smoothly.

Fig. 4 gives the gate voltage of the conventional RGD. Based on the relationship between $R_{DS(on)}$ and v_{gs} in Fig. 3, the conduction time of the MOSFET in one period is divided into three intervals: the conduction time with low $R_{DS(on)}$, $T_{on,2}$, and the conduction time with high $R_{DS(on)}$, $T_{on,1}$ and $T_{on,3}$, as shown in Fig. 4. It is observed that $T_{on,2}$ is quite short (less than 50% of the conduction time T_{on}), while $T_{on,1}$ and $T_{on,3}$ is relatively long (more than 50% of T_{on}). Hence the average $R_{DS(on)}$ of the MOSFET during the conduction time is much higher than the rated value in the datasheet.

2) The conduction time of the MOSFET cannot be adjusted flexibly to meet different duty ratio requirements. The duty ratio is always less than 50%. When applied to the SR FET, the difficulty of tuning the duty ratio will induce the conduction time of the body diode, causing high forward voltage drop and high power loss at multi-MHz.

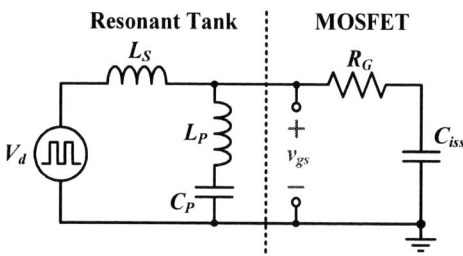

Fig. 2.   Typical circuit of the conventional RGD

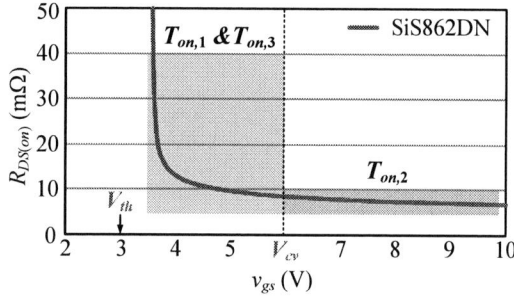

Fig. 3.   On-state resistance $R_{DS(on)}$ vs. gate drive voltage v_{gs}

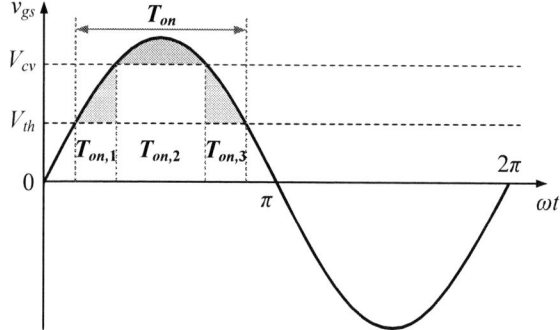

Fig. 4.   Gate voltage of the conv. RGD and three conduction intervals

III. PROPOSED LEVEL-SHIFTED RGD FOR THE SR FET

A. Concept of the Level-shifted RGD

To solve the above problems of the SR with the conventional RGD, a self-driven level-shifted RGD with a tunable DC bias voltage V_B is proposed in Fig. 5 for the SR FET in the multi-MHz isolated converter. The basic idea is to introduce a level-shift circuit between the drive signal and the resonant tank to generate a DC bias voltage to compensate the gate voltage. Fig. 6 gives the drive voltage comparison between the SR with the conventional and the level-shifted RGDs. With the additional DC bias voltage, the benefits of the level-shift RGD are summarized as follows.

1) It increases the peak gate voltage and reduces the average $R_{DS(on)}$ of the MOSFET. As the shaded area shown in Fig. 6, with the DC bias voltage, the conduction time with a low $R_{DS(on)}$, $T'_{on,2}$, is extended significantly compared to the conventional RGD. And the conduction time with a high $R_{DS(on)}$, $T'_{on,1}$ and $T'_{on,3}$, is much shortened. Therefore, the average $R_{DS(on)}$ and the associated conduction loss in the SR FET can be reduced significantly.

2) It provides better design flexibility to realize wide duty ratio of the MOSFET. The total conduction time in one switching cycle can be adjusted in wide range by tuning the bias voltage V_B, shown as T'_{on} in Fig. 6. When used in the SR, the proposed RGD can provide precise switching timing for the SR FET by tuning the DC bias voltage and the phase-shift angle of the resonant tank to minimize the conduction loss and the reverse recovery loss of the body diode.

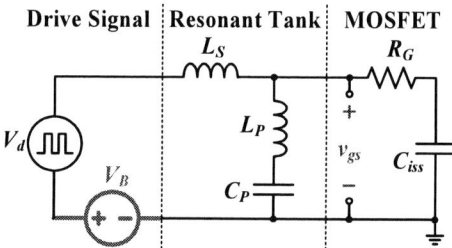

Fig. 5.   Proposed level-shifted RGD with the DC bias voltage for the SR

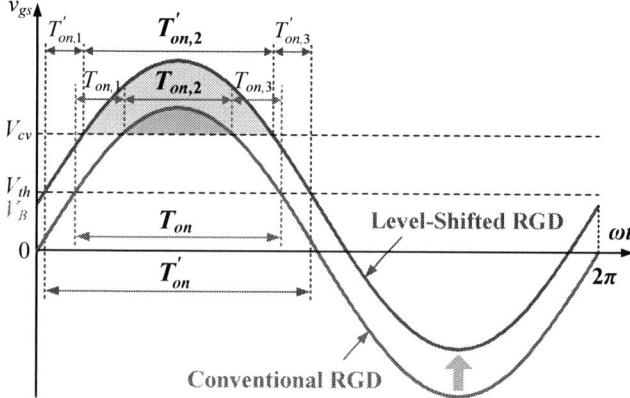

Fig. 6.   Gate voltage comparison between the SR with the conventional and the level-shifted RGDs

B. Circuit of the SR FET with the Level-Shifted RGD

Fig. 7 shows the detailed circuit of the proposed self-driven level-shifted RGD applied to drive the SR FET of the 10-MHz isolated resonant converter. The auxiliary winding n_3 provides the drive signal for the RGD. The resonant tank provides desired voltage gain and phase-shift from the auxiliary winding n_3 to the gate of the SR FET. With the breakdown current provided by the output voltage V_O, the Zener diode D_Z in the level-shift circuit provides the desired bias voltage V_B for the gate voltage. A control stage comprised of a shutdown branch, D_1 and S_{AUX1}, and an auxiliary switch S_{AUX2} is added to ensure fast shutdown and startup of the level-shifted gate voltage so that the converter can work under on/off close-loop control.

Fig. 8 shows the key waveforms of the SR. The voltage across the auxiliary winding, v_3, is transferred to the ac component of the gate voltage $v_{GS,SR}$, which is a sine wave. The voltage across D_Z, v_{DZ}, is kept constant at its Zener voltage V_B. So the bias voltage of the gate voltage is also V_B. At t_0, the drain voltage $v_{DS,SR}$ resonates to zero, and the gate voltage $v_{GS,SR}$ increases in a resonant manner and just exceeds the threshold voltage V_{th} so the SR FET turns on immediately. At t_1, $v_{GS,SR}$ decreases below V_{th} so the SR FET turns off and $v_{DS,SR}$ begins to rise in a resonant manner. After t_1, $v_{GS,SR}$ resonates always below V_{th} until the next switching cycle.

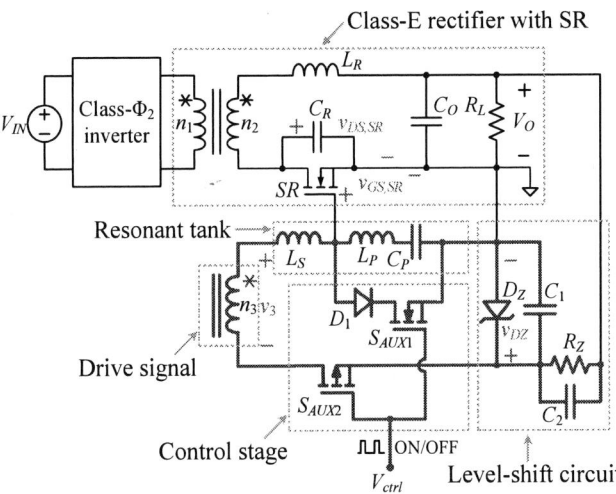

Fig. 7.   Proposed self-driven level-shifted RGD for the SR FET

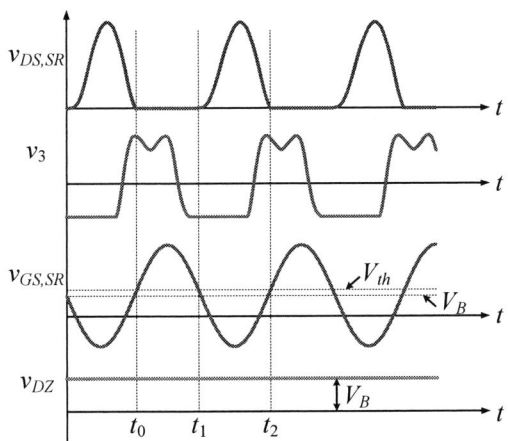

Fig. 8.   Key waveforms of the SR with the level-shifted RGD

IV.   DESIGN OF THE LEVEL-SHIFTED RGD

A. The Auxiliary Winding

In Fig. 7, the drive signal is provided by the auxiliary winding n_3. Since the output of the class-Φ_2 inverter is approximately a square wave swinging between V_{IN} and -V_{IN}, the drive signal v_3 swings between $-n_3 V_{IN}/n_1$ and $n_3 V_{IN}/n_1$. A step-down voltage is preferred so that it is convenient to design the resonant tank to generate the desired phase-shift and voltage gain. Since the turns ratio of the primary and the secondary winding n_1:n_2 is 4:1 to balance the voltage and current stress on the device, there are three possible options for n_1:n_3. The final design of n_1:n_3 is 4:1 because if it is set to 4:2 or 4:3, the peak gate voltage would easily be high or even exceed the limits of the device (±20 V), which results in the damage of the semiconductor devices.

B. The Resonant Tank

The capacitance C_P in the resonant tank blocks the level-shifted DC voltage. The inductances, L_P and L_S, set the winding to gate (ac component) transfer function $H(s)$, which is

$$H(s)=\frac{V_{GS,ac}(s)}{V_3(s)}=\frac{sR_G L_P C_{iss}+L_P}{s^2 L_P L_S C_{iss}+sR_G C_{iss}(L_P+L_S)+L_P+L_S} \quad (1)$$

So the frequency characteristics are

$$H(j\omega)=\frac{j\omega R_G L_P C_{iss}+L_P}{j\omega R_G C_{iss}(L_P+L_S)+L_P+L_S-\omega^2 L_P L_S C_{iss}} \quad (2)$$

$$A(\omega)=|H(j\omega)| \quad (3)$$

$$\varphi(\omega)=\arctan\left(\frac{\mathrm{Im}(H(j\omega))}{\mathrm{Re}(H(j\omega))}\right) \quad (4)$$

The inductances need to be designed to obtain the desired voltage gain and phase-shift. The phase-shift of the drive signal v_3 to the gate-to-source voltage v_{gs}, φ, can be obtained by the simulation of the power train. Substituting φ into (4), the relation between L_P and L_S can be expressed as

$$L_S=f(\varphi,L_P,\omega) \quad (5)$$

The voltage gain $A(\omega)=V_{GS,ac}/V_3$ is supposed to ensure a proper amplitude for the the drive voltage, which is

$$V_{th}<V_{GS,ac}<V_{GS,\max} \quad (6)$$

where $V_{GS,\max}$ is the maximum gate voltage of the device.

And the voltage across n_3 is

$$V_3=\frac{n_3}{n_1}V_{IN} \quad (7)$$

Therefore,

$$\frac{n_1 V_{th}}{n_3 V_{IN,\min}}<A(\omega)<\frac{n_1 V_{GS,\max}}{n_3 V_{IN,\max}} \quad (8)$$

Combining (3), (5) and (8), the values of L_P and L_S can be determined. Some trial-and-error tuning is still required for these reactive components to make the final design.

C. The Level-shift Circuit

As for the level-shift stage, it provides the DC bias voltage V_B for the RGD. The value of V_B is determined by the duty ratio and the threshold voltage of the SR FET. The duty ratio is set to 0.5 to balance the voltage stress and current stress on the device. Hence the DC bias voltage V_B is supposed to be around the threshold voltage. Since the switch chosen for the SR is SiS862DN from Vishay and the threshold voltage is around 2 V, the Zener diode TZS4679 from Vishay with a Zener voltage of 2 V is selected as D_Z.

V. SIMULATION RESULTS

A 10-MHz isolated resonant converter rectified by the SR with the self-driven level-shifted RGD is simulated with LTspice. The simulation parameters are listed in Table II.

Fig. 9 gives the simulated SR FET waveform comparison between the conventional RGD (without the level-shift circuit) and the proposed RGD. The drive voltage of the conventional RGD is a sine wave of 5 V amplitude while the drive voltage of the proposed RGD has a DC bias voltage of 2 V and the amplitude of the ac component is 5 V. With both RGDs, the duty ratio of the SR FET is about 0.48 and the peak drain voltage is 20 V.

From Fig. 9, it is observed that with the conventional RGD, the conduction time of the body diode is 11 ns, more than 20% of the total conduction time in one switching cycle. On the contrary, the proposed RGD provides precise switching timing for the SR FET and the conduction time of the body diode is zero. Another important thing is that, as mentioned in Section II, when the gate voltage $v_{GS,SR}$ exceeds the Critical Voltage V_{cv} of 5.5 V in this case, the MOSFET can conduct with low $R_{DS(on)}$ and reduce the conduction loss. When comparing the gate voltage waveforms in Fig. 9 (the pink lines), the peak gate voltage of the proposed RGD is higher than the conventional RGD owing to the bias voltage of 2 V. With the conventional RGD, the conduction time with low $R_{DS(on)}$ is almost zero while with the level-shifted RGD, the conduction time with low $R_{DS(on)}$ is as much as 26 ns (more than 50% of the total conduction time). So the average $R_{DS(on)}$ of the SR with the level-shifted RGD is reduced significantly.

VI. EXPERIMENTAL VERIFICATION

To verify the proposed level-shifted RGD in the isolated SR resonant converter, an experimental prototype operating at 10 MHz with 18 V input and 5 V/2 A output was built. Fig. 10 gives the photograph of the prototype. Table III gives the component values and part numbers in the power stage and the proposed level-shifted RGD. Two commercial Si N-channel MOSFETs in the PowerPAK package from Vishay are selected as the control and SR FETs. The control FET is Si7898DP (150 V N-channel, $R_{DS(on)}$=76 mΩ @ V_{GS}=6 V, C_{oss}=100 pF @ V_{DS}=45 V) and the SR FET is SiS862DN (60 V N-channel, $R_{DS(on)}$=8.5 mΩ @ V_{GS}=6 V, C_{oss}=545 pF @ V_{DS}=30 V). The air-core inductors from Coilcraft are used in the power stage as the resonant inductance. The values of C_F and C_R listed in Table III are the external capacitances in parallel with the output capacitance of the control and SR FETs.

Table II SIMULATION PARAMETERS OF THE LEVEL-SHIFTED RGD

L_F	220 nH	C_F	420 pF
L_M	220 nH	C_M	220 pF
L_R	18 nH	C_R	3600 pF
$n_1{:}n_2{:}n_3$	4:1:1	C_B	4 μF
L_S	560 nH	R_Z	470 Ω
L_P	150 nH	C_1, C_2	1 μF
C_P	100 nF	D_Z	TZS4679 (2 V)

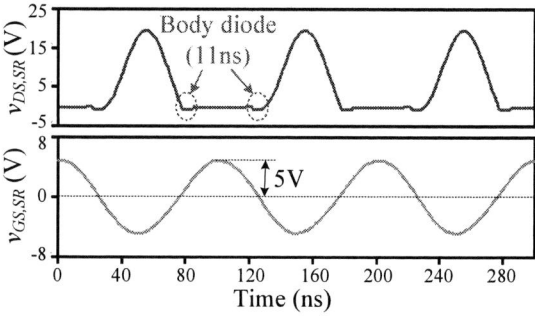

(a) SR with Conventional RGD

(b) SR with Level-shifted RGD

Fig. 9. Simulated waveforms comparison between the SR with the conventional and the level-shifted RGD

(a) Top view

(b) Bottom view

Fig. 10. Photograph of the prototype

978-1-4673-9551-9/16 $31.00 © 2016 IEEE

Table III COMPONENT VALUES AND PART NUMBERS

S_{ctrl}	Si7898DP (Vishay)	SR	SiS862DN (Vishay)
L_F, L_M	2222SQ-221 (Coilcraft)	L_R	A04T (Coilcraft)
C_F	268 pF	C_R	2940 pF
C_M	247 pF	C_B	15 nF
$n_1{:}n_2{:}n_3$	4:1:1	$Core$	ER 14.5/3/7 (TDG TP5)
L_S	680 nH	D_Z	TZS4679 (Vishay)
L_P	150 nH	D_1	BAS170WS (Vishay)
C_P	0.1 µF	S_{AUX1}	FDV303N (Fairchild)
R_Z	470 Ω	S_{AUX2}	NTA4151P (ON Semi)
C_1, C_2	0.22 µF		

Fig. 11 shows the measured drain and gate voltage waveforms of the control FET. It is observed that the drain voltage waveform of the control FET is approximately half-wave symmetric. The peak drain voltage of the control FET is 47 V at 18 V input (2.61 times of the input voltage). Moreover, it can be seen that the converter provides good ZVS achievement for the control FET.

Fig. 12 shows the measured waveforms of the SR FET. It is observed that the proposed RGD provides a sinusoidal drive voltage with a DC offset of 2 V and the amplitude of the ac sine component is 5.8 V, which agrees well with the theoretical analysis and simulation results. The drain and gate voltage oscillation is caused by the parasitic package inductance and the nonlinearity of the parasitic capacitance of the device. The duty ratio of the SR FET is 0.5 and the peak drain voltage is 20 V.

To verify the conduction loss reduction benefit of the SR with the proposed RGD, two additional prototypes rectified by the SR with the conventional RGD and the diodes were built respectively. For fair comparison, other parameters in the power stage are the same. In the prototype with the diode rectification, the Schottky diode PMEG4020EVP (40 V, 2 A, NXP Semiconductors) is selected as the rectifier diode. Two of them are paralleled to reduce the high-frequency conduction loss. The SR FET in the prototype rectified by the SR with the conventional RGD is also SiS862DN.

Fig. 13 gives the thermal imaging comparison between three prototypes operating at steady state at full load under the condition of nature air cooling. In Fig. 13 (a), the temperature of the rectifier diodes is 59.4 ℃ (the other one is on the bottom layer). In Fig. 13 (b), the temperature of the SR FET with the conventional RGD is 61.4 ℃. Although the temperature of the SR FET is almost the same as the diodes, the footprint of the SR FET is only half of the diodes, which means the power loss in the SR FET is much reduced compared to the rectifier diodes. In Fig. 13 (c), the temperature of the SR FET with the proposed RGD is only 45 ℃. The temperature rise is 17 ℃, a reduction of 47% compared to the temperature rise of the SR with the conventional RGD (32 ℃). Since the SR FETs (SiS862DN) have the same thermal resistance, the power loss of the SR with the proposed RGD is reduced by 47% compared to the conventional RGD.

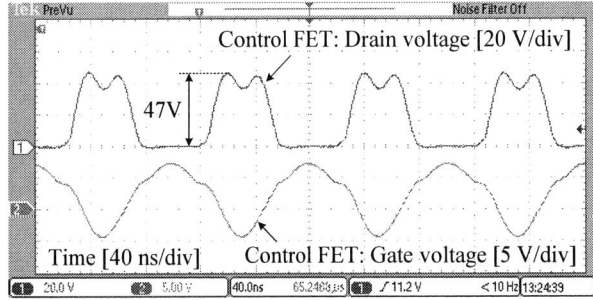

Fig. 11.   Waveforms of the control FET: V_{IN}=18 V, V_O=5 V and f_s=10 MHz

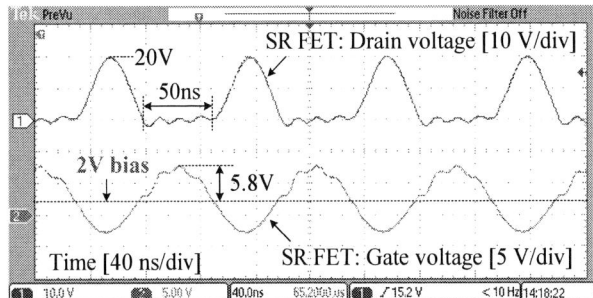

Fig. 12.   Waveforms of the SR FET: V_{IN}=18 V, V_O=5 V and f_s=10 MHz

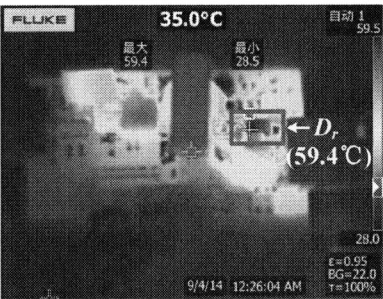

(a) Diode rectification

(b) SR with the conventional RGD

(c) SR with the proposed RGD

Fig. 13.   Thermal imaging comparison: V_{IN}=18 V, V_O=5 V and P_O=10 W

Fig. 14 gives the loss breakdown of the converter with different rectification schemes at 18 V input and 5 V/2 A output. It is observed that the major loss difference among the converters is in the rectifier device. Because of the reduction in the conduction loss, the power loss rectified by the SR with the proposed RGD is 1.89 W, a reduction of 0.23 W compared to the SR with the conventional RGD (2.12 W) and a reduction of 0.73 W compared to the diode rectification (2.62 W).

Fig. 15 gives the close-loop efficiency comparison among three different rectification schemes at 18 V input and 5 V output. It is observed that compared to SR with the conventional RGD and the rectifier diodes, the SR with the proposed RGD achieves significant efficiency improvement in wide load range. At full load of 2 A, the proposed RGD improves the converter efficiency from 80.2% using the conventional RGD to 82% (an improvement of 1.8%). Compared to the efficiency of 77.3% using the diode rectification, the efficiency improvement is 4.7% at full load.

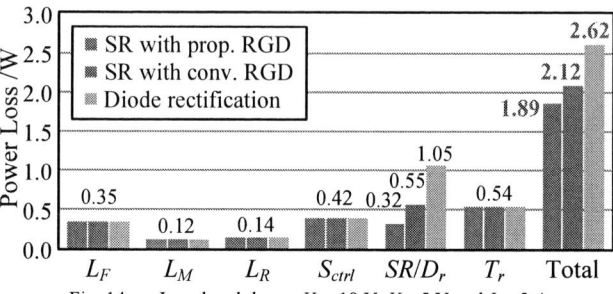

Fig. 14.   Loss breakdown: V_{IN}=18 V, V_O=5 V and I_O =2 A

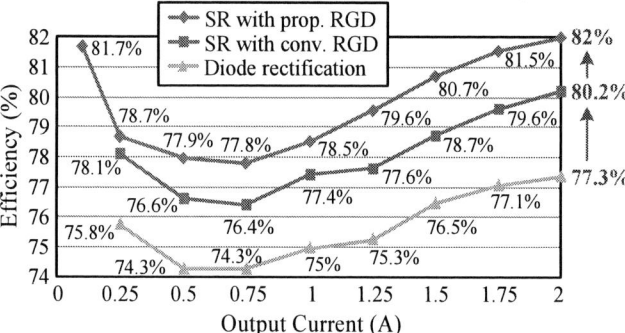

Fig. 15.   Close-loop efficiency comparison (RGD drive loss included): V_{IN}=18 V, V_O=5 V and f_s=10 MHz

VII. CONCLUSION

The forward recovery phenomenon of the diodes causes high conduction loss in the multi-MHz resonant converters. So the SR technique is strongly desired at multi-MHz, especially at high current applications. This paper applies the SR technique to a 10-MHz isolated class-Φ_2 resonant converter. A self-driven level-shifted RGD is proposed to drive the SR FET. The proposed RGD reduces the average on-state resistance of the MOSFET and provides precise drive timing for the SR FET. The advantage of the SR with the proposed RGD is verified by a 10-MHz prototype with 18 V input and 5 V/2 A output. Significant efficiency improvement was achieved in wide load range compared to the SR with conventional RGD and the diode rectification.

REFERENCES

[1] A. Knott, T. M. Andersen, P. Kamby, J. A. Pedersen, M. P. Madsen, M. Kovacevic and M. A. E. Andersen, "Evolution of very high frequency power supplies," *IEEE Journal of Emerging and Selected Topics in Power Electronics*, Vol. 2, No. 3, pp. 386-394, Sept. 2014.

[2] D. J. Perreault, J. Hu, J. M. Rivas, Y. Han, O. Leitermann, R. C. N. Pilawa-Podgurski, A. Sagneri and C. R. Sullivan, "Opportunities and challenges in very high frequency power conversion," in *Proc. IEEE APEC.*, pp. 1-14, Feb. 2009.

[3] M. Madsen, A. Knott and M. A. E. Andersen, "Low power very high frequency switch-mode power supply with 50 V input and 5 V output," *IEEE Trans. on Power Electron.*, Vol. 29, No. 12, pp. 6569-6580, Dec. 2014.

[4] J. M. Rivas, O. Leitermann, Y. Han and D. J. Perreault, "A very high frequency DC-DC converter based on a class-Φ_2 resonant inverter," *IEEE Trans. on Power Electron.*, Vol. 26, No. 10, pp. 2980-2992, Oct. 2011.

[5] J. A. Santiago-Gonzalez, K. M. Elbaggari, K. K. Afridi and D. J. Perreault, "Design of class E resonant rectifiers and diode evaluation for VHF power conversion," *IEEE Trans. on Power Electron.*, vol. 30, no. 9, pp. 4960-4972, 2015.

[6] L. C. Raymond, W. Liang and J. M. Rivas-Davila, "Performance evaluation of diodes in 27.12MHz Class-D resonant rectifiers under high voltage and high slew rate conditions," in *2014 Workshop on Computers and Modeling in Power Electronics (COMPEL 2014)*, Jun. 2014.

[7] T. M. Andersen, S. K. Christensen, A. Knott and M. A. E. Andersen, "A VHF class E dc-dc converter with self-oscillating gate driver," in *Proc. IEEE APEC*, pp. 885-891, Mar. 2011.

[8] A. D. Sagneri, D. I. Anderson and D. J. Perreault, "Transformer synthesis for VHF converters," in *Proc. IEEE IPEC*, pp. 2347-2353, Mar. 2010.

[9] J. Hu, A. D. Sagneri, J. M. Rivas, Y. Han, S. M. Davis and D. J. Perreault, "High frequency resonant SEPIC converter with wide input and output voltage ranges," *IEEE Trans. on Power Electron.*, Vol. 27, No. 1, pp. 189-200, Jan. 2012.

[10] J. Lin, Y. Zhou, Z. Zhang, X. Ruan and Y. Liu, "Analysis and design of a 30 MHz resonant SEPIC converter," in *Proc. IEEE APEC*, pp. 455-460, Mar. 2014.

865 MHz Switching-Speed Step-Down DC-DC Power Converter for Envelope Tracking

Vivek Mehrotra, Andrea Arias, Joshua Bergman, Charles Neft, Miguel Urteaga and Berinder Brar

Teledyne Scientific & Imaging
Thousand Oaks, California

Abstract—Envelope tracking (ET) is an appealing alternative to the widely used Doherty power amplifier (PA) due to its potential to increase efficiency, particularly for high data rate transmissions. In this work, we demonstrate a 28 V, 865 MHz switching speed step-down converter that can be applied to envelope tracking. The converter's ET capability is demonstrated at up to 865 MHz PWM carrier frequency. An integrated RF GaN HEMT and 100 V GaN power switch process is reported, which enables fabrication of a power switch and gate driver on the same IC, necessary for achieving the 865 MHz switching speed. We report a slew rate of the main power switch of 152 V/ns at 50 V, which enables ET with up to 20% less power than that required without ET. To the authors' best knowledge, this is the first demonstration of a GaN MMIC process that integrates a GaN HEMT suitable for X through Ka-band power amplifiers and a high voltage power switch on the same IC.

Keywords—GaN switch; RF GaN HEMT; HF buck converter; DC-DC power converter; envelope tracking; RF power amplifier

I. INTRODUCTION

Highly-efficient, high-frequency power conversion enables an overall efficiency increase of RF transmitters by using envelope tracking (ET) to supply the RF power amplifier (PA) with a drain voltage just slightly in excess of the voltage required for transmission. Unlike Doherty PAs, ET implementation is decoupled from the RF matching [1]. As an added benefit, it has been demonstrated that a close variant of traditional ET can also largely improve the PA's linearity by significantly lowering third-order intermodulation (IM3) distortion while maintaining low fifth-order intermodulation (IM5) [2]. To realize dynamic tracking of drain voltage at frequencies > ~100 MHz, the on-wafer integration of power switches and gate driver is crucial, and, because of GaN's superior combination of high breakdown voltage and bandwidth, GaN was selected as the best-suited material system for this purpose. The GaN switch must be devoid of traps that affect the drain response and significantly increase its R_{on} under switching conditions and the co-integrated RF HEMT as a gate driver must be capable of driving the GaN switch. Previously published results have demonstrated 10 to 40 MHz switching converters up to 60 V input [3-8]. Using a discrete high voltage MOSFET switch, a 2.63 MHz switching resonant converter capable of 120 V input was demonstrated [9]. Buck designs utilizing discrete GaN HEMTs have also been previously reported [10-11], however this approach has limited the converter's maximum switching frequency to a few tens of MHz. A discrete CREE 60 W GaN HEMT was used in [11], wherein a switching frequency of 67 MHz was shown, achieving 25 W of output power and 91.6% of peak efficiency. Using 0.7 μm GaN HEMTs, a driver and power circuit chip was fabricated and tested up to 200 MHz and 20 V bus with almost 90% conversion efficiency [12]. Similarly, using an integrated GaN power switch and gate driver, a synchronous 20 V converter was demonstrated operating at up to 100 MHz switching speed [13]. An integrated GaAs based 100 MHz, 20 V, switch-mode converter has also been reported [14]. Up to 300 MHz DC-DC converters realized in 0.13 μm CMOS have also been demonstrated, but these approaches have been restricted to a maximum input voltage of 2.6 V due to the low voltage nature of high speed sub-micron CMOS devices [15-18]. In addition, a 1 GHz GaAs converter was demonstrated but was limited to 20 V and its output power was less than 1 W [19]. In this work, we have demonstrated a GaN-based power converter capable of switching at 865 MHz at >28 V bus with over 7 W of output power. A 1 GHz converter is capable of tracking up to a 100 MHz envelope and, with a parallel linear stage, the tracking range can be extended to higher frequencies [5]. Considering the need for at least a 10:1 ratio of switching frequency to tracking frequency, this converter could be the basis for ET up to ~86 MHz. Alternatively, an envelope tracker could be implemented as a hybrid converter, using a switching stage in parallel with an analog stage, with the analog stage only being required to supply the higher frequency components beyond the bandwidth of the switching stage. With the hybrid approach, the ET frequency could be extended up to ~500 MHz and still maintain good efficiency.

This work was sponsored by DARPA-MTO under contract number FA865011C7182 - Microscale Power Conversion.

SWITCH
PULSED PERFORMANCE

Vgs from -0.75 to 1.5V in 0.75 V steps
Lg = 0.5 μm, Wg = 6x80 μm
Pulse_Length = 0.2 μs
Pulse_Separation = 1 ms

Red line is quasi-static sweep. Black lines are pulsed IV's at Vgs,q= -4 V and Vds,q= 0 to 50 V in 10 V steps.

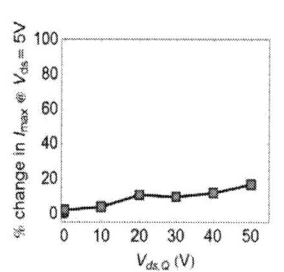

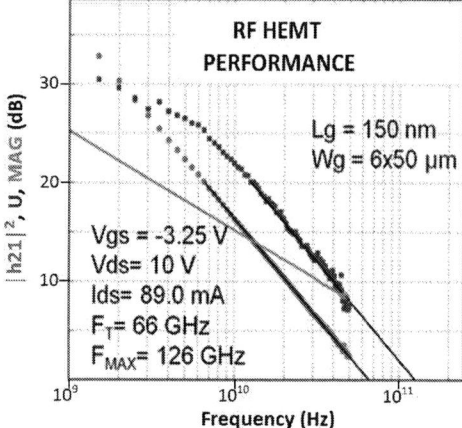

Figure 1: On-wafer Switching and RF performance of GaN Devices

II. GaN MMIC TECHNOLOGY – DESIGN AND FABRICATION

In order to realize a ~GHz switching power converter, a MMIC technology that integrates the converter's commutation path on-chip is essential. The commutation path includes the power switch, Schottky diode, gate driver and input capacitor. The GaN MMIC technology used for the step-down converter fabrication demo is a 2-level metal process fabricated on AlGaN/GaN on SiC. The AlGaN/GaN epitaxy is grown by MOCVD and is composed of a specially designed buffer layer optimized for low DC-to-RF dispersion.

The ohmic contacts for GaN devices are formed through high temperature alloying, which yields a contact resistance of about 0.4 Ω-mm. Ion implantation is used for device isolation. The MMIC process also integrates >80 V breakdown MIM capacitors and thin-film NiCr resistors. The active devices include a 0.5 μm HEMT power switch, a 0.5 μm Schottky diode for the converter circuit and an RF HEMT (0.15 μm T-gate) for the gate driver. The Schottky diode breakdown exceeds 100 V and its zero bias capacitance is 1.6 pF. The off-state gate to drain breakdown voltage of the power switch is also >100 V, its threshold voltage is -1.5 V and its I_{max} is >1 A/mm. The switch dynamic Ron is < 2.7 Ω-mm at 50 V pulsing, and its total gate charge is < 30 pC. The gate driver RF HEMT has an I_{max} of 1.4 A/mm, and can be operated up to 20 V.

The robustness of the epitaxy design for high voltage switching and for gate driver is demonstrated by the low DC-to-RF dispersion of the switch along with the small signal FOMs of the RF HEMT device. Figure 1 shows the on-wafer pulsed-IV measurement of the switch transistor and the small signal gains of the gate driver RF device at Vds= 10 V (RF transistor unit cell). Under the pulse condition of Vds,q= 50 V,

it can be seen that the switch HEMT knee-current reduction from the quasi-static value is < 20%.

From the small signal gains of the RF HEMT, it is seen that the maximum available gain (MAG) of the RF transistor at 30 GHz is > 10 dB. A micrograph of the fabricated integrated circuit that contains the GaN gate driver, power switch, Schottky diode and input capacitor is shown in Figure 2.

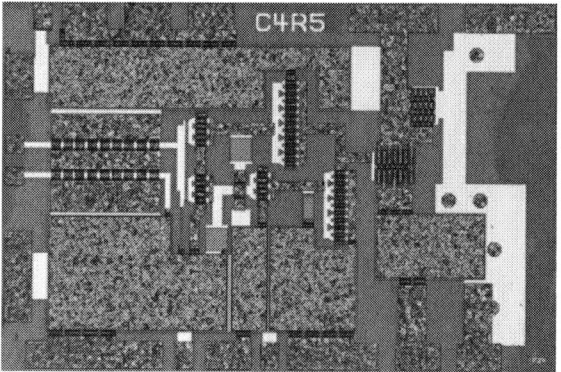

Figure 2: Micrograph of Integrated Gate Driver, Switch, Schottky Diode and Input Capacitor (Chip Size is 3.2 mm × 2.2 mm)

III. GaN GATE DRIVER CIRCUIT

The on-wafer integrated gate driver was designed for a maximum switching frequency of 1 GHz and a voltage swing of 3 V at the switch gate. It is necessary to switch the main device on or off in 100 ps or less for 1 GHz operation in order to ensure adequate performance and efficiency. For an estimated maximum switch transistor gate capacitance of 12.5 pF and desired voltage swing of 3 V, the current into the main device gate - the gate drive output current – must be 0.38 A during the switching interval.

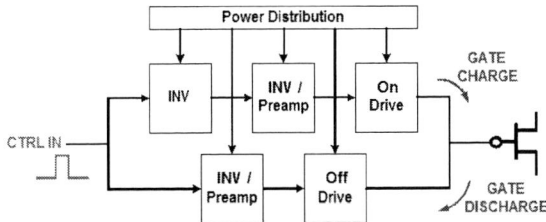

Figure 3: Gate Driver Circuit Block Diagram

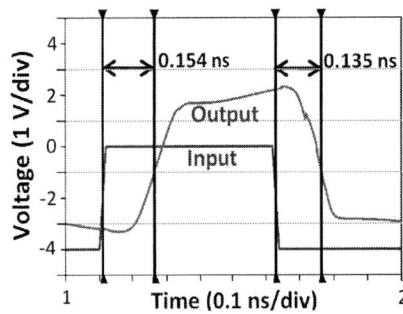

Figure 4: Gate Driver Transport Lag

The gate driver circuit must provide enough amplification to achieve this output current. Also, the gate driver must not introduce any significant transport lag to avoid possible distortion of the power converter output waveform. Two stages of amplification were used in the driver in order to achieve the necessary output current. A block diagram of the gate driver circuit is shown in Figure 3.

The driver output stage consists of two T-gate GaN HEMTs in a totem pole configuration. The upper HEMT is turned on to provide charge to the main switch's gate to achieve turn-on of the main switch, while the lower HEMT is turned on to remove the gate charge for turn-off. Although each HEMT must be capable of delivering 0.38 A when on, the main GaN switch gate current is essentially zero when it is not switching, that is when it is either fully on or off. Thus each of the output FETs undergo a duty cycle of 0.1 and consequently the average current for each is 0.038 A. These two currents define the design space for the two output RF HEMTs of the driver circuit. The three HEMTs in the first and intermediate stages have less gate periphery than the output HEMTs, and provide the first stage of amplification. A total of four input power rails were used to provide flexibility to the device designer in establishing the threshold voltage of the main GaN FET, although in a final switch configuration this would be greatly simplified. Each input rail is referenced to the main device source terminal and supported by an on-chip 100 pF capacitor.

Figure 4 shows the simulated gate driver input and output waveforms. The transport lag as determined from the simulation is 0.154 ns and 0.135 ns for turn-on and turn-off respectively. The difference between the two delays is 19 ps, less than 0.02% of 1 ns fundamental period, which should result in minimal distortion of the power converter output.

IV. STEP-DOWN CONVERTER CIRCUIT

The step-down converter was designed using the on-chip GaN switch, gate driver, integrated Schottky diode and input capacitor. Figure 5 depicts the power circuit schematic.

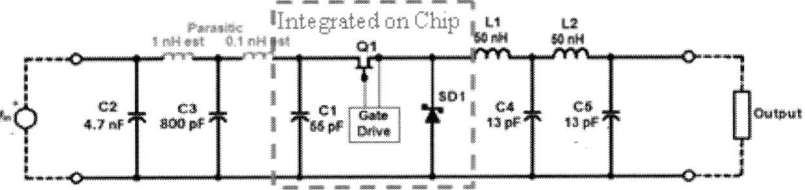

Figure 5: Step-Down Converter Schematic

The output filter is a two-stage LC filter consisting of 50 nH inductors and 13 pF, 100 V capacitors. The natural resonance frequency is ~200 MHz, which was selected to provide sufficient attenuation of the ~1 GHz PWM carrier frequency while still allowing for adequate envelope tracking. The converter circuit, including package parasitics, was fully analyzed by means of E&M simulations, which highlighted the importance of the PWM signal path and its fundamental effect on the functionality of the converter. A top view of a completed step-down converter assembly is shown in Figure 6.

The unit is assembled on a brass carrier with a footprint of 20 mm x 24 mm. In addition to providing a convenient base for the circuit, the carrier also serves to conduct heat away. The integrated part of the circuit, consisting of C1, Q1, SD1 and the gate driver from Figure 5 is on a silicon-carbide substrate (IC detail shown in Figure 2) and is directly bonded to the brass carrier in the same orientation as shown in Figure 6.

The step-down converter assembly is divided into two sections, each with a different common. The main system common is on the right (labeled 'System

978-1-4673-9551-9/16 $31.00 © 2016 IEEE   81

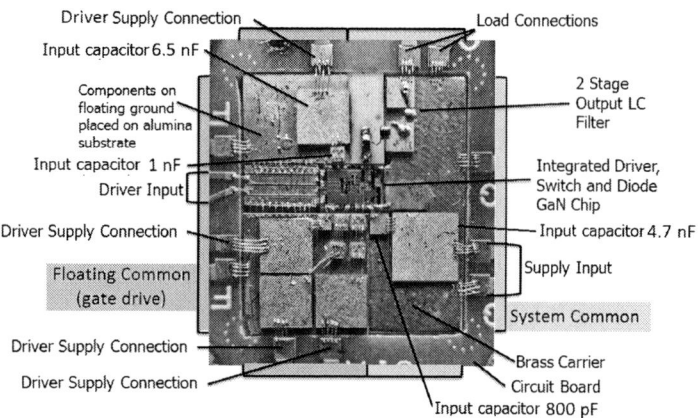

Figure 6: Fabricated Step-Down Converter Top-View

Common' in Figure 6) and is common for the input supply and load. Components connected to this common are mounted directly to the brass carrier.

The floating common is on the left (labeled 'Floating Common' in Figure 6), connects to the GaN FET source terminal, and is the common for the gate driver. An alumina substrate insulates this common from the brass carrier.

The two-stage output filter is located in the upper right part of the assembly. Supply input capacitors are located in the lower right. Gate drive supply capacitors are located in the upper and lower left sections. All of the smaller capacitors, 800 pF for the input and 1 nF for the gate drive are located adjacent to their respective pads on the driver plus switch integrated circuit, with the larger ones located nearby. The gate driver signal input is at the middle left, feeding directly into the drive.

The step-down converter main components described above are attached to a circuit board of which only the inner portion is shown in Figure 6. The PC board has a rectangular cutout through which the circuit elements can be accessed. Pads plated with bondable gold surround this cutout. All electrical connections in the circuit from the board are wire bonded with 1 mil diameter gold wire. For the high current paths, multiple wire bonds are used to

Loss Breakdown By Component - Pout = 10 W

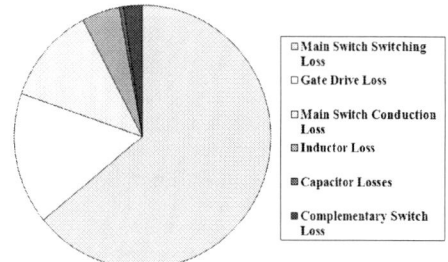

Figure 7: Converter Loss Breakdown by Component

increase current capability and reduce lead inductance. A key feature of the brass carrier is that the top surface is not flat, but rather consists of plateaus of differing elevations. These different elevations accommodate the variations in the thickness of each of the components; after mounting, the tops of all the components to be wire bonded, including the circuit board pads, are at the same elevation thus keeping wire bond lengths and parasitic inductances at a minimum. Off-chip parasitic capacitance and inductance were estimated at 1 nF and 1 nH, which were accounted for in the E&M simulation.

Figure 7 contains the analysis of the overall converter loss breakdown by component. It is worth highlighting that switching losses largely dominate the overall losses of current state-of-the-art ~ 1 GHz converter designs [19]. Figure 8 contains the simulated converter efficiency vs. output power curve.

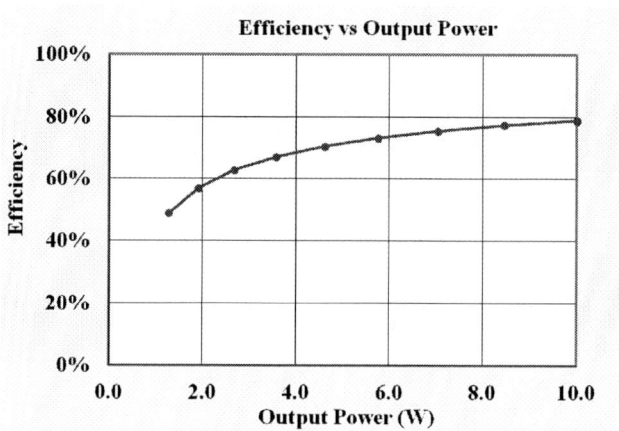

Figure 8: Simulated Converter Efficiency vs. Output Power

V. TEST RESULTS

The gate driver was characterized by on-wafer probing. The test set up included a 20 GHz synthesizer, a 2.6 GHz data generator, an 18 GHz oscilloscope and 40 GHz attenuators. The data generator produced a square output of a maximum amplitude of 2 V peak-to-peak (Vpp), which was used in differential form (by using two 2 V Vpp data generator outputs) as an input to the gate driver. The driver IC was 50 Ω terminated for this test. Figure 9 shows the measured rise and fall times of < 95 ps at 1 GHz switching.

The output slew rate (SR) of the switch was measured at 150 MHz and 50% duty cycle using

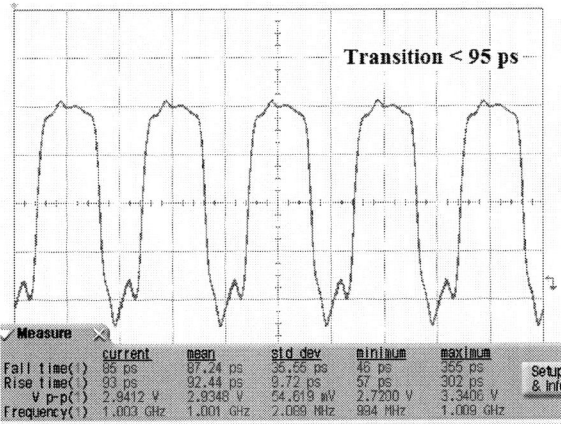

Figure 9: Gate Driver Measured Transition (On-Wafer Measurement)

the integrated gate driver fabricated on the same IC as the switch. The gate driver was biased so as to provide a gate voltage swing of 1.5 V below switch pinch-off (Vp) to +1.5 V gate voltage above pinch-off. The resulting switch test waveform is shown in Figure 10. The trace displayed is the drain voltage, and a switch output slew rate of 152 V/ns was extrapolated from this measurement. The measured switch slew rate (SR) satisfies the buck converter requirement for ~1 GHz switching.

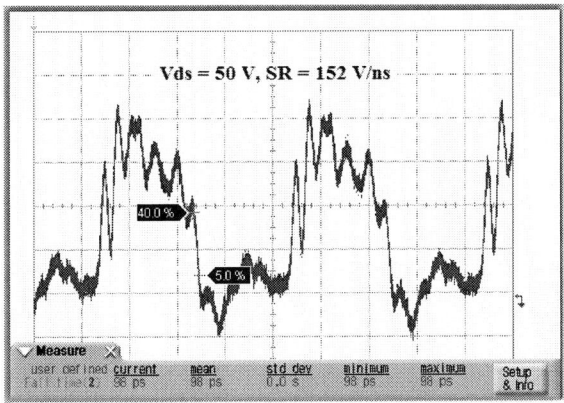

Figure 10: Switch Output Slew Rate Characterization

The step-down converter envelope tracking (ET) capability was tested by varying the duty cycle of driver input at each period and measuring the converter's output. For clarity, the expected converter output voltage ('Converter Vout' trace) is plotted alongside the input signal ('PWM' trace) in Figure 11.

The ET test described above was performed at 500 MHz and 865 MHz. The oscilloscope captures below (Figure 12) demonstrate the buck converter successfully varying its output voltage (envelope) in accordance with the duty cycle variation at the driver

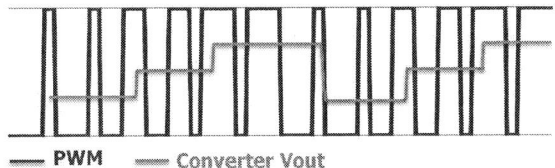

Figure 11: Illustration of Envelope Tracking Test: Driver Input PWM (25%, 50% and 75% Duty Cycles – 2 periods each) and Converter Output Voltage

input (driver input signal was not captured so as to avoid introducing undesired parasitics in the high frequency path). It should be noted that 3 duty cycle variations at 500 MHz (25-50-75%) and two variations at 865 MHz (40-60%) is the data generator's limitation. A fixed load of 50 Ω was used so as to maximize the converter output power obtained.

500 MHz ET (25-50-75%) Duty Cycle

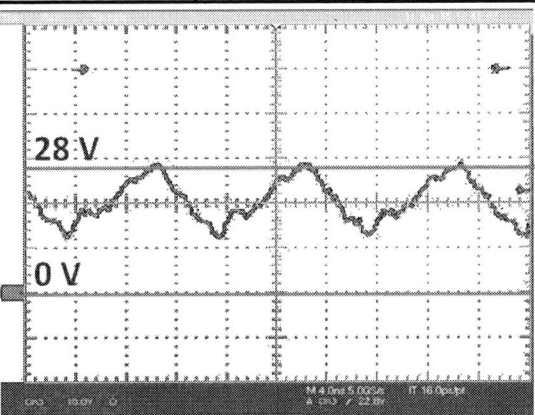

865 MHz ET (40-60%) Duty Cycle

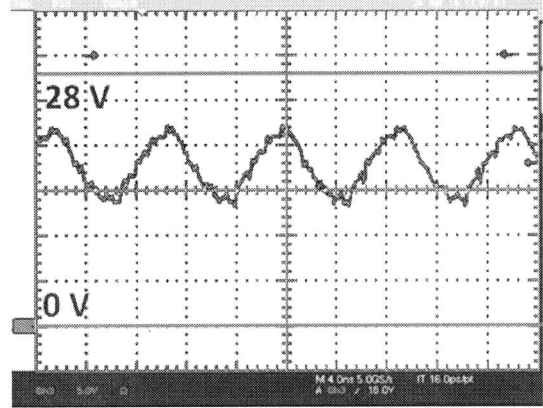

Figure 12: Envelope Tracking Test Results at 500 MHz and 865 MHz

Figure 13 contains the measured efficiency as a function of the converter output power, for 3 different duty cycles at 650 MHz PWM. A comparison is also established between this result (Step-Down Converter w/ ET) and what the output power and efficiency

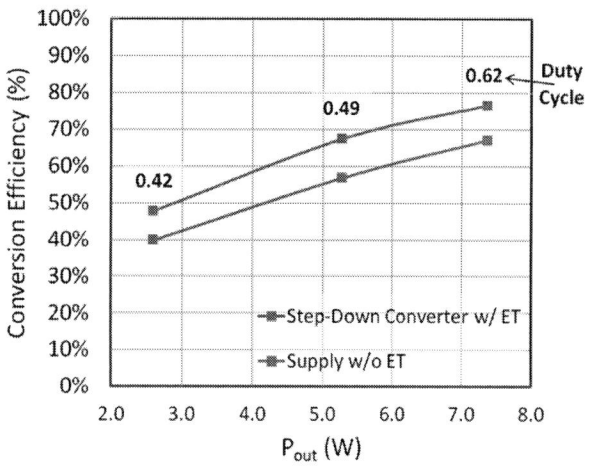

Figure 13: Step-Down Converter Pout vs. Conversion Efficiency at 650 MHz

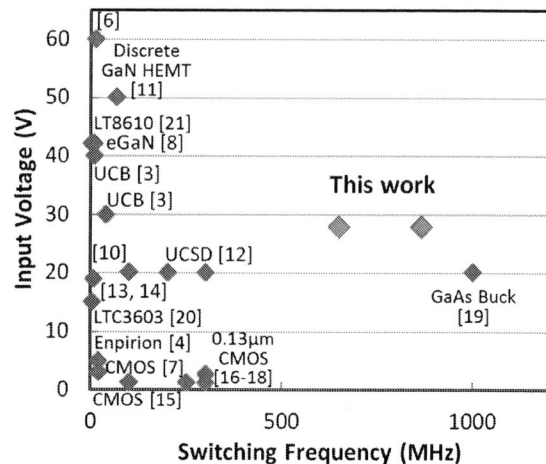

Figure 14: Comparison of this Work to State-of-the-Art

would have been had the converter been purely linear (i.e., no supply modulation). The linear supply calculations (supply w/o ET) show that up to 20% less input power is required with ET.

VI. CONCLUSIONS

In this work, a step-down converter that utilized an integrated gate driver and switch to enable PA supply modulation at ~1 GHz switching speeds was demonstrated. The process technology, which allows the on-wafer integration of RF transistors for gate driver and 100 V GaN power switches, was used to fabricate a switch test IC. The main switch slew rate was measured at 152 V/ns. A comparison with existing products and state-of-the-art research results is summarized in Figure 14.

To the authors' best knowledge, this work contains the combination of highest power delivered and switching frequency within the down-converter work published to date.

ACKNOWLEDGMENTS

The authors gratefully acknowledge the support and guidance of DARPA program managers Dr. John Albrecht and Dr. Daniel Green. The opinions, views and findings contained in this manuscript are those of the authors only and should not be taken as official views of the Defense Advanced Research Projects Agency or the Department of Defense. The authors also wish to thank Dr. Richard Eden for his insight, support and many discussions throughout the course of this effort. Finally, we wish to acknowledge the contributions of our colleagues Dr. Chanh Nguyen, Dr. Aiden Higgins and Mr. Alan Sailer plus the

Teledyne Fabrication Operations team for fabrication of the MMICs.

REFERENCES

[1] S. C. Cripps, *RF Power Amplifiers for Wireless Communications*, Artech House Inch, Norwood, 2006.

[2] Z. Yusoff, J. Lees, J. Benedikt, P. J. Tasker and S. C. Cripps, "Linearity improvement in RF power amplifier system using integrated Auxiliary Envelope Tracking system" *IEEE MTT-S International Microwave Symposium Digest (MTT), 2011*, pp. 1 - 4, 5-10 June 2011

[3] M. Rodriguez, Yuanzhe Zhang, and D. Maksimovic, "High-Frequency PWM Buck Converters Using GaN-on-SiC HEMTs," *IEEE Transactions on Power Electronics*, vol. 29, no. 5, pp. 2462-2473, May 2014.

[4] W. Zhang, M. W. Wilkowski, J. Weld and A. Lofti, "A 20 MHz Monolithic DC-DC Converter Manufactured with the First Commercially Viable Silicon Magnetics Technology" *IEEE 7th International Power Electronics and Motion Control Conference (ECCE Asia), 2012 7th Asian*, pp. 705 - 712, 2-5 Jun. 2012

[5] R. Bondade, Y. Zhang and D. Ma "A Linear-Assisted DC-DC Hybrid Power Converter for Envelope Tracking RF Power Amplifiers," *IEEE Energy Conversion Congress and Exposition (ECCE), 2014*, 14-18 Sept. 2014.

[6] J. Hannon, R. Foley, J. Griffiths, D. O'Sullivan, K. G. McCarthy and M. G. Egan, "A 20 MHz 200-500 mA Monolithic Buck Converter for RF Applications," *Applied Power Electronics Conference and Exposition (APEC), 2009*, 15-19 Feb. 2009.

[7] W. A. Tabisz, P. M. Gradzki and F. C. Y. Lee, "Zero-voltage-switched quasi-resonant buck and flyback converters-experimental results at 10 MHz," *IEEE Transactions on Power Electronics*, vol. 4, no. 2, pp. 194-204, April 1989.

[8] J. Strydom and D. Reusch, "Design and evaluation of a 10 MHz gallium nitride based 42 V DC-DC converter," *Applied Power Electronics Conference and Exposition (APEC), 2014*, 16-20 March 2014

[9] H. B. Kotte, R. Ambatipudi and K. Bertilsson, "High-Speed (MHz) Series Resonant Converter (SRC) Using Multilayered Coreless Printed Circuit Board (PCB) Step-Down Power Transformer," *IEEE Transactions on Power Electronics*, vol. 28, no. 3, pp. 1253-1264, March 2013.

[10] P.F. Miaja, A. Rodriguez and J. Sebastian, "Buck-Derived Converters Based on Gallium Nitride Devices for Envelope Tracking Applications," *IEEE Transactions on Power Electronics*, vol. 30, no. 4, pp. 2084-2095, April 2015.

[11] F. Leroy, N. Le Gallou, C. Delepaut, O. Deblecker and F. Dualibe "A Very High Frequency Step-Down DC/DC Converter for Spaceborne Envelope-Tracking SSPA," *IEEE MTT-S International Microwave Symposium (IMS), 2014,* 1-6 June 2014.

[12] S. Shinjo, Y. Hong, H. Gheidi, D. Kimball and P. Asbeck, "High Speed, High Analog bandwidth buck converter using GaN HEMTs for envelope tracking power amplifier applications," *IEEE Topical Conference on Wireless Sensors and Sensor Networks (WiSNet), 2013,* 20-23 Jan. 2013.

[13] Y. Zhang, M. Rodriguez and D. Maksimovic, "100 MHz, 20 V, 90% efficient synchronous buck converter with integrated gate driver," *IEEE Energy Conversion Congress and Exposition (ECCE), 2014,* 14-18 Sept. 2014.

[14] G. van der Bent, P. de Hek, S. Geurts, A. Telli, H. Brouzes, M. Besselink and F. van Vliet, "A 10 Watt S-Band MMIC Power Amplifier With Integrated 100 MHz Switch-Mode Power Supply and Control Circuitry for Active Electronically Scanned Arrays," *IEEE Journal of Solid-State Circuits*, vol. 48, no. 10, pp. 2285-2295, Oct. 2013.

[15] C. Huang and P. K. T. Mok, "An 84.7% Efficiency 100-MHz Package Bondwire-Based Fully Integrated Buck Converter With Precise DCM Operation and Enhanced Light-Load Efficiency," *IEEE Journal of Solid-State Circuits*, vol. 48, no. 11, pp. 2595-2607, Nov. 2013.

[16] M. Wens and M. Steyaert, "A fully integrated 130 nm CMOS DC-DC step-down converter, regulated by a constant on/off-time control system," *Proc. IEEE Eur. Solid-State Circuits Conf,* pp.62–65, Sep. 2008

[17] S. S. Kudva and R. Harjani, "Fully integrated on-chip DC-DC converter with a 450X output range," *IEEE Journal of Solid-State Circuits*, vol. 46, no. 8, pp. 1940–1951, Aug. 2011.

[18] X. Gong, J. Ni, Z. Hong, and B. Liu, "An 80% peak efficiency, 0.84 mW sleep power consumption, fully integrated DC-DC converter with buck/LDO mode control," *Proc. IEEE Custom Integr. Circuits Conf.,* pp. 1–,. Sep. 2011.

[19] E. Busking, P. de Hek, and F. van Vliet, "1 GHz GaAs Buck Converter for High Power Amplifier Modulation Applications," *Microwave Integrated Circuits Conference (EuMIC), 2012 7th European*, pp. 353 - 356, 29-30 Oct. 2012.

[20] LTC3603 - 2.5 A, 15 V Monolithic Synchronous Step-Down Regulator, Linear Technology LTC3603 Datasheet, (http://cds.linear.com/docs/en/datasheet/3603fc.pdf)

[21] LT8610 – 42 V, 2.5 A Synchronous Step-Down Regulator with 2.5µA Quiescent Current, Linear Technology LT8610 Datasheet, (http://cds.linear.com/docs/en/datasheet/8610fa.pdf)

Current Parking Regulator for Zero Droop/Overshoot Load Transient Response

Sudhir S. Kudva[*], William J. Dally[†], Thomas H. Greer III[*], and C. Thomas Gray [*]

[1]Email: {skudva, bdally, tgreer, tgray}@nvidia.com

[*]NVIDIA Corporation, Durham, NC, 27713

[†]NVIDIA Corporation, Santa Clara CA, 95051

Abstract—Supply voltage integrity during a load transition is a critical problem. Droop on the supply may lead to logic failures, and overshoot can reduce reliability. Voltage deviations from the nominal forces designers to apply margins in design to ensure correct operation. This paper addresses two main causes of droop/overshoot on the supply line namely the sluggish converter response and the parasitics between the converter and the load. We present current parking regulator (CPR), a voltage down-converter with almost zero droop or overshoot during a load transient along with implementation techniques to nullify the effect of the parasitics. The underlying principle involves avoiding the inductor slewing time by parking sufficient excess current in the inductor, which is available to use immediately when the need arises. The system design involves on-die, package, and PCB co-design to minimize the impact of parasitics. Measurement results show negligible droop/overshoot when the load current transitions from 0.8A to 7.5A and vice versa in 2ns with only 200nF of load capacitance on the regulated output voltage node. The design was fabricated in a 28nm CMOS technology and integrated in the same package as the load using an 8 layer substrate. It achieves a peak efficiency of 83% at 8A load current.

Index Terms—current parking, droop mitigation, overshoot mitigation, package integrated

I. Introduction

Maintaining a constant power supply voltage is critical to the performance of processors and other digital systems. A primary cause of supply noise is load transients due to changing workloads, cache stalls, interrupts, etc. [1] Increasing processor speeds and higher circuit density has exacerbated the problem by burdening the power supply network with large instantaneous current [2]. Large on-die decoupling capacitors can reduce the droop/overshoot. However, the available area for these capacitors is typically limited, and they therefore provide only limited droop/overshoot protection. Currently, huge margins are assigned for the supply voltage [3] to account for the droop when the load transition occurs. However, with reducing supply voltage and increasing load current steps [4], it is becoming harder to sustain these design margins and meet design targets.

From a power dissipation stand point, reducing the minimum operating voltage ($V_{op,min}$) of processors is an important goal and design efforts are made to get $V_{op,min}$ as close as possible to the minimum voltage allowed by the process technology ($V_{process,min}$). However, the droop on the supply line during load transients (ΔV_{droop}) dictates the margin that is applied on $V_{process,min}$ such that the operational voltage never drops below $V_{process,min}$ ($V_{op,min} = V_{process,min} + \Delta V_{droop}$). The margin applied results in additional power dissipation compared to the case where the processor is

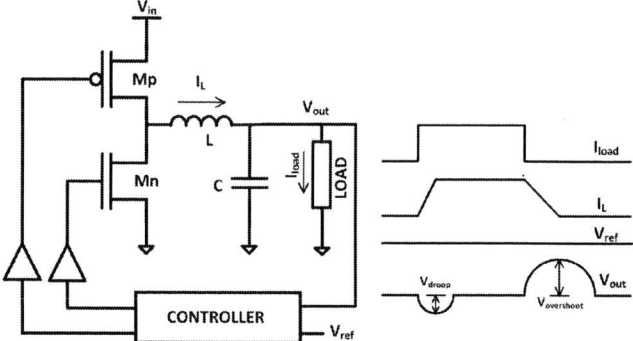

Fig. 1: Conventional buck regulator with current and voltage waveform for load transient

operated at $V_{process,min}$ which can be a significant fraction of the minimum power consumption of the processor.

Similarly, the maximum voltage at which processor is allowed to operate is also restricted to a value lower than the maximum allowed by the process technology because of the reliability concerns caused by the overshoot of the supply voltage when the load current transitions from high to low value. This margin for overshoot limits the maximum performance achievable from the processor. The droop/overshoot margins on the supply voltage can be eliminated if we can ensure a noise free supply voltage.

Figure 1 shows a typical buck converter [5] with current and voltage waveforms at different nodes. The inductor current closely matches the load current (with steady state current ripple) when the load is in a steady state condition. During a load transient when the load current either increases or decreases, the inductor current rises or falls at a rate determined by the difference between the input and output voltage and by the size of the inductor. Ripple considerations dictate a large inductance, but this causes a significant delay as the inductor current lags the load current. The difference in current between the load current and inductor current must be supplied/sinked by the decoupling capacitor. To the extent this decoupling capacitor cannot supply/sink the required current, the supply voltage droops/overshoots.

There are many techniques published in the literature which attack the problem of noise on the supply line due to load transients. A linear regulator to reduce the effective supply impedance is presented in [6]. However this technique requires a separate additional supply and ground network. A band-limited damping circuit to change the current profile is presented in [7] which requires 1mA of damping current per

978-1-4673-9551-9/16 $31.00 © 2016 IEEE

3.67mA of load current. An auxiliary buck converter with a smaller inductor in parallel with the main buck converter [8] speeds up the response at the cost of additional components and board area. [9] uses a transformer in place of the inductor to construct a DC-DC inductance stepping voltage regulator where during a load transient the effective inductance is only leakage inductance of the transformer. However, during a load transient, this technique takes multiple transitions on output voltage before stabilizing to a steady state value. A circuit technique called shortstop is proposed in [10] where an additional dirty supply is used to quickly boost the supply voltage of core under transition and isolates other cores from noise during this transition. This technique requires an additional supply rail and only addresses the droop in the supply voltage and not the overshoot. An auxiliary circuit to charge the inductor [11] results in fast transient response but requires early indication from load, which may not be always feasible.

Our work accomplishes the goal without the drawbacks of [6] - [11] namely the use of additional supply lines or early warning signals. Additionally, CPR not only attains droops free response but also is effective in preventing overshoot on the supply line. The dual frequency SIMO buck presented in [12] has a converter structure similar to CPR regulator. However, our target application demands much higher power and the the regulator cannot be integrated with the load presenting us with the additional problem of parasitics between the regulator and the load which has huge impact due the speed of response of the regulator. In our work, we have demonstrated a practical implementation of the CPR for a processor application where we have comprehensively dealt with the problem of parasitics using an integrated package solution and also shown that the system can be scaled for even higher load currents.

The paper is organized as follows. In the next section we discuss the operation of the CPR and the different parts of the CPR. Different challenges faced and implementation details are explained in Section III. Measurement results are presented in Section IV followed by some conclusions in Section V.

II. Current Parking regulator

We follow a two pronged strategy to generate a droop/overshoot-free supply voltage in the presence of large load transients. First, we have developed a current-parking regulator (CPR) which has an ultra-fast load transient response and then we implement the regulator along with the load in such a way as to reduce the impact of parasitics that exists in a conventional system.

Figure 2 shows the schematic of the CPR as implemented in our test system. Our CPR is divided into three parts:
1) Upstream implemented on a PCB using discrete components with an FPGA for upstream control,
2) Downstream consisting of the CPR switches M3/M4 and downstream control implemented on the CPR die
3) Digital load along with the on-die decoupling capacitors implemented in the load die.

CPR and load die are integrated in the same package in this implementation for the reasons discussed in Section III-D.

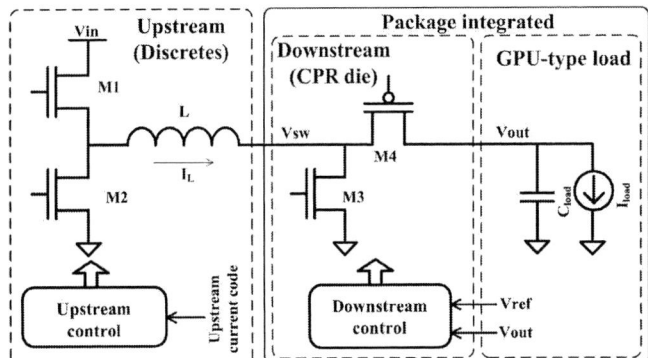

Fig. 2: CPR schematic with dotted lines demarcating the upstream, downstream and load regions

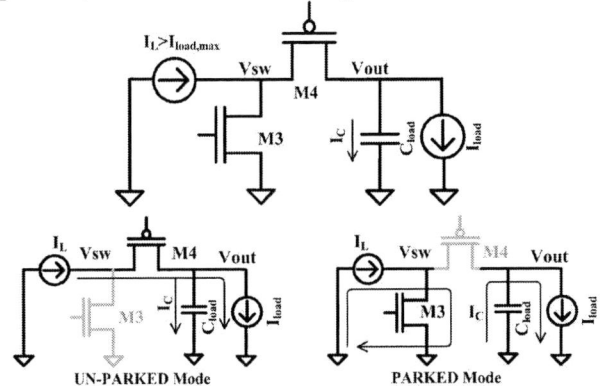

Fig. 3: CPR operating principle showing the parked and un-parked modes

The basic idea of CPR is to avoid inductor slewing when there is a load transient and then use high speed planar CMOS switches to direct the inductor current to the load when the requirement arises thereby speeding the response time of the regulator. To achieve this, the inductor needs to have sufficient current to support any current demanded by the load. This concept of storing excess current in the inductor is referred to as current parking. This makes the upstream behave like a constant current source (neglecting the small ripple on the inductor current) with inductor current $I_L > I_{load,max}$ as shown in Figure 3 (top). The main task of the upstream controller is to sense the current in the inductor and adjust the ON/OFF time of M1 and M2 as shown in Figure 2 to maintain required current. The task of the downstream controller is to appropriately direct the current so that voltage regulation is achieved. The downstream controller transitions between the un-parked and parked modes as shown in Figure 3 (bottom) at high frequency. In the un-parked mode, the inductor current is directed to the load and supports the load current as well as charges the on-die decoupling capacitors. The time that the regulator spends in the unparked mode is given by Equation 1.

$$T_{unpark} = \frac{C_{load} * \Delta V_{ripple}}{I_L - I_{load}} \quad (1)$$

where C_{load} is the on-die decoupling capacitor, ΔV_{ripple} is the voltage ripple on the supply voltage and I_{load} the load current.

In the parked mode, the inductor current is circulated back to the inductor i.e. parked back in the inductor. During the parked mode, the load current is entirely supported by the on-die decoupling capacitors for a time period given by Equation 2. Using Equations 1 and 2, the switching frequency of the downstream part of the regulator can be calculated as in Equation 3.

$$T_{park} = \frac{C_{load} * \Delta V_{ripple}}{I_{load}} \tag{2}$$

$$F_{SW,down} = \frac{1}{C_{load} * \Delta V_{ripple}} * \frac{I_L}{(I_L - I_{load})I_{load}} \tag{3}$$

As can be seen from Equations 2 and 3, the downstream controller varies both the duty-cycle as well as the frequency of switching of the downstream switches to regulate the voltage to desired value. From Equation 3, the maximum downstream switching frequency is attained when $I_{load} = \frac{I_L}{2}$.

The downstream control responsible for voltage regulation is bang-bang type and turns ON M4 and turns OFF M3 when V_{out} dips below V_{ref} (un-parked mode). The downstream enters the parked mode when V_{out} rises above V_{ref}. In this mode, the load current is entirely provided by the on-die decoupling capacitors, $I_C = I_{load}$. The downstream control regulates V_{out} to V_{ref} by toggling between the parked and un-parked mode. The frequency of this toggling varies from tens of MHz to hundreds of MHz depending on the load current.

Inductor current slewing is eliminated by parking excess current in the inductor. Parking current in the inductor involves circuluating the current in a loop formed by M2, L and M3. Ideally, zero loop resistance will result in no loss in parking excess current. However, in actual case the goal is to reduce the loop resistance to minimum (30mΩ in this particular implementation), resulting in small energy loss due to parking. The amount of loss is dependent on the amount of excess current that is parked in the inductor and the resistance of the loop. In exchange for this small loss, we achieve droop/overshoot-less transient response.

III. IMPLEMENTATION

A. Upstream

The upstream portion of the CPR consists of the power FETs M1, M2, inductor L and the upstream control implemented in a FPGA along with the current sensing. The task of the upstream part is to maintain a current in the inductor which is larger than max load current. As inductor slewing is avoided, a large inductor can be used (4.7μH in this implementation) and yet achieve fast response. A large inductor enables low switching frequency in the range of few hundred KHz range in the upstream part which in turn allows for usage of discrete components and reduces the switching losses. A constant-ON time based control is used to control the ON/OFF times of the upstream transistors M1 and M2. The high side transistor M1 is turned ON for constant time interval that is decided by the input voltage (V_{in}) such that the current ripple on the inductor current (ΔI_L) is 0.4A. M1 is turned OFF and M2 is turned ON and the current is monitored using a $5m\Omega$ sense resistor,

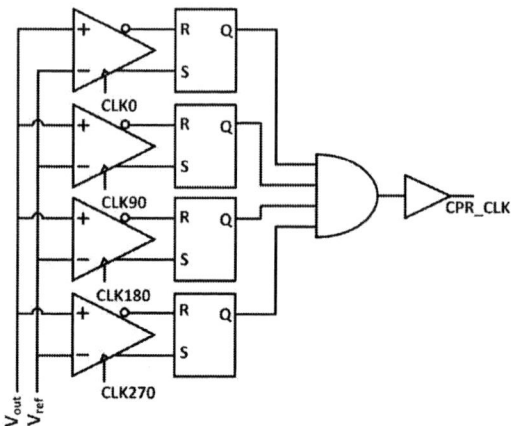

Fig. 4: Interleaved comparator used in downstream control

instrumentation amplifier with a gain of 90, and a flash ADC clocked by the FPGA at a frequency of 12.5MHz. When the current dips below the minimum value set by the max load current specification, M2 is turned OFF and M1 is turned ON. The sensing of the current starts 500ns after M2 is turned ON in order to prevent error in current measurements induced by the noise due to the switching of the power devices. Gallium Nitride transistors have been used for M1 and M2 as these have better characteristics compared to Silicon power FETs with similar ratings [13].

B. Downstream

The downstream is responsible for the ultra-fast load transient response of the CPR as well as the regulation of the output voltage. The downstream switches are planar CMOS transistors implemented on-die along with their control and switch at several hundred MHz frequency. The CPR switches M3 and M4 are sized 60mm/28nm and 80mm/28nm respectively to reduce conduction losses when supporting 2A/CPR die. They are distributed over the entire die area along with the load as will be described in further detail in Section III-C.

The downstream controller consists of a 4-phase interleaved clocked comparator with PMOS input that compares V_{out} and V_{ref} as shown in Figure 4. The different phases of the clock are generated from a local ring oscillator. The clocked comparator is followed by a RS-latch to hold the value when the comparator enters the precharge phase. An AND-gate is used to combine the 4-phases of the interleaved comparator. Therefore, a single phase detecting $V_{out} < V_{ref}$ will make CPR_CLK low to turning ON M4 and turning OFF M3. However, for an opposite transition, comparators in all the phases have to detect the transition for CPR_CLK to go high.

C. Load and Decoupling capacitors

The load and decoupling capacitors are implemented in the same die as the CPR downstream part as shown in Figure 5. The load consists of different types of gates connected in series stimulated with a clock input as a representative processor type switching load. Each load die is capable of consuming a maximum of 2A current at 0.9V supply and has 25nF of on-die bypass capacitance. During the testing process, some

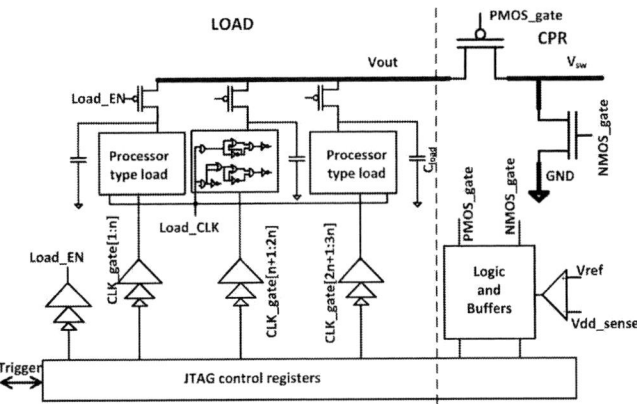

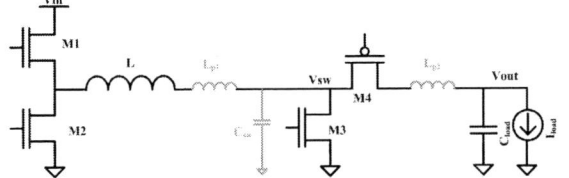

Fig. 5: Load and CPR implementation in the single die

Fig. 6: Parasitics in the design of the CPR based system

dice are configured as load and some as CPR downstream part. In the die configured as load, the transistors M3 and M4 are both turned OFF. A PMOS header is used to isolate the decoupling capacitor in the dice used as CPR which is always turned ON in a load die. The load consists of 32 blocks of identical switching load. In order to vary the load current, a gating signal is used to to gate OFF the load clock to appropriate number of blocks. The length of the chain of the gates can be changed using a 2-bit JTAG code to further vary the load current. V_{out} and GND bumps are shared between both the load and the CPR. In order to uniformly distribute the current through all the available bumps, both the load and CPR switches are spread throughout the die area of 2mm × 2mm.

D. Parasitics

There are three main parasitics associated with a CPR based system as shown in the Figure 6. The parasitic capacitance C_{sw} is predominantly due to the routing capacitance associated with node V_{sw} and the drain capacitance of M3 and M4 which are large transistors to minimize conduction losses. This parasitic capacitor switches at the CPR_CLK frequency resulting in loss of power. We can reduce the routing capacitance portion by placing the inductor L close to the CPR die. Inductance L_{p1} is the parasitic routing inductance between the inductor L and the CPR die. As the CPR operation eliminates the slewing of the inductor current, this parasitic inductance does not impact the operation of the CPR. However, parasitic inductor L_{p2} representing the routing inductance between the CPR and load is critical for the high speed response of the CPR. This inductance sees high frequency switching current when the CPR transitions between the parked and unparked mode. Hence, the placement of the CPR switches with respect to the load die becomes critical for achieving steady V_{out}. The

three possible locations for the placement of the CPR die are shown in Figure 7.

The first option placing the CPR on the PCB as shown in Figure 7a results in a large parasitic inductance L_{p2} between the CPR switches and the load which is not desirable. The second option is to integrate the CPR die in the same package on same side of package substrate as the processor die as shown in Figure 7b. However, with large processor die sizes [14], [15], the lateral inductance from CPR die to the load on the processor die results in large inductance L_{p2}. Finally, the third option places the CPR die on the opposite side of the package substrate right underneath the processor die. Now the inductance L_{p2} is just the vertical inductance between the CPR die on the bottom and processor on the top. Technological advancement in packaging technology [16] have resulted in package substrates which are a few hundred μm thick resulting in very low inductance between the CPR switches and the load. In our implementation, we have used this option with a $200\mu m$ core package substrate with 8 layers.

E. System implementation

The schematic of the entire system as in this implementation is as shown in the Figure 8. In order to support a maximum of 8A load current, four CPR dice are connected in parallel each with a load current capacity of 2A. Each CPR die has its own separate upstream part, localized sense node from the nearest load die (Vout_0B, Vout_1B, Vout_2B and Vout_3B shown in Figure 8) on top, and independent control with no communication with other CPR dice. The downstream and upstream switching events across the four CPR branches are independent of each other and are uncorrelated. The upstream portion of the CPR operating at low frequencies is implemented using discrete components on a PCB along with the FPGA to execute the controller. The downstream (CPR switches + control), the load, and on-die bypass capacitors are implemented in a 28nm CMOS test site measuring 2mm × 2mm. The test chip contains both the CPR switches and the load with CPR switches occupying an area of $0.75mm^2$. The remaining area is consumed by load circuits and the decoupling capacitors distributed across the die. The load and CPR dies are both integrated in the same package with eight load dice on top and four CPR dice on the bottom. The package is mounted on a PCB using an LGA1156 socket as shown in Figure 9. The sense nodes from the four load dice which are not used in CPR for sensing are used to observe (Vout_0A, Vout_1A, Vout_2A and Vout_3A as shown in Figure 8) the V_{out} voltage on each die with each of the lines being routed both on package substrate and PCB with 50Ω impedance to avoid any reflections at high frequency. The package substrate has eight layers with V_{out} and GND planes routed in the form of mesh in each layer of the package substrate. The V_{out} node from each of the CPR dice are shorted on the package substrate. Similarly, the GND nodes are connected together. In order to investigate the scalability of the system another variant where the package containing just two load dice and one CPR die was also fabricated and tested.

978-1-4673-9551-9/16 $31.00 © 2016 IEEE

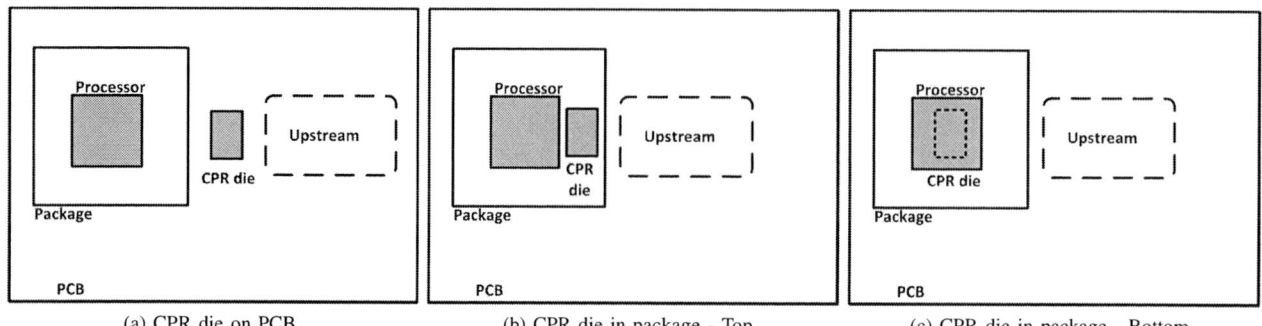

(a) CPR die on PCB   (b) CPR die in package - Top   (c) CPR die in package - Bottom

Fig. 7: Different possible placement of the CPR die

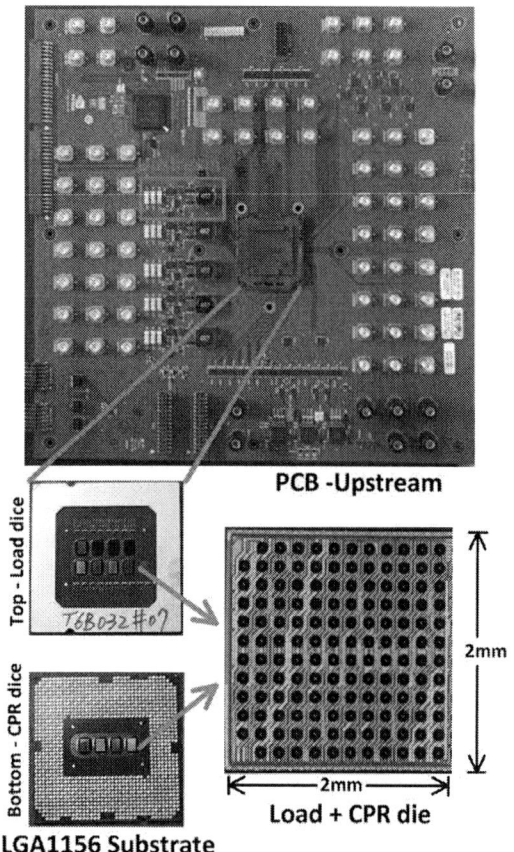

Fig. 9: PCB, package and on-die implementation

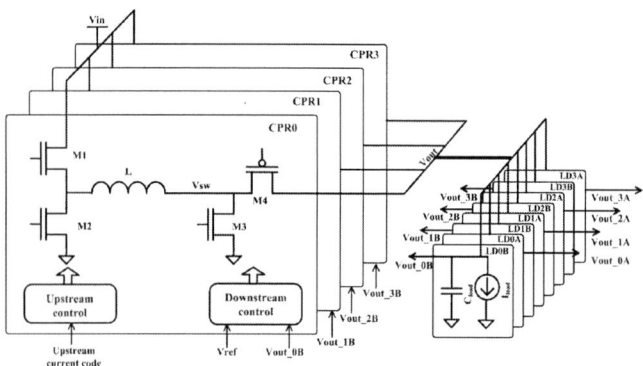

Fig. 8: System implementation schematic

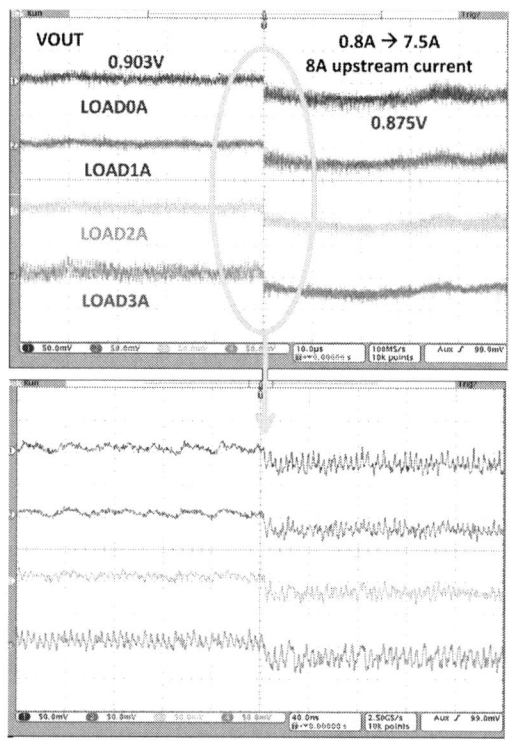

Fig. 10: Load transient response, I_{load} transition from 0.8A to 7.5A and Vout = 0.9V

IV. MEASUREMENT RESULTS

Figure 10 shows the supply voltage on four different load dice located across the package when the load current is changed from 0.8A to 7.5A in 2ns. All load dice consume the same load current and make the transition at the same time instant. The **ONLY** capacitance on the output voltage node is the on-die bypass capacitance of ≈200nF. In spite of the small capacitance on the output voltage node and the large load transient, there is essentially no droop. The CPR has an output impedance of 4.2mΩ and DC output voltage of 0.903V and 0.875V for the low and high load currents respectively. As in the earlier case, there is no voltage V_{out} overshoot when the load current transitions from 7.5A to 0.8A (shown in Figure 11). CPR is effective at maintaining a stable supply voltage for both type of load transitions. Additionally, CPR achieved this response without any pre-warning from the

978-1-4673-9551-9/16 $31.00 © 2016 IEEE   90

load or requiring any additional supply which are some of the main drawbacks of earlier techniques.

Figure 12 shows the steady state output voltage and the CPR clock waveforms when a single CPR die is powering two load dice. CPR is in the unparked state for 10ns and parked state for 2ns when the load is 2A with 2A upstream current. However, when the load current is 0.16A for the same upstream current, CPR is in parked state for 14ns and unparked state for 1.9ns. The downstream controller regulates the output voltage to the desired level by changing the frequency and duty-cycle of the CPR_CLK as shown in Equations 1 and 2. The 500MHz noise on V_{out} is due to the switching of the digital load at this frequency which is more pronounced when load current is 2A as all the load blocks are switching.

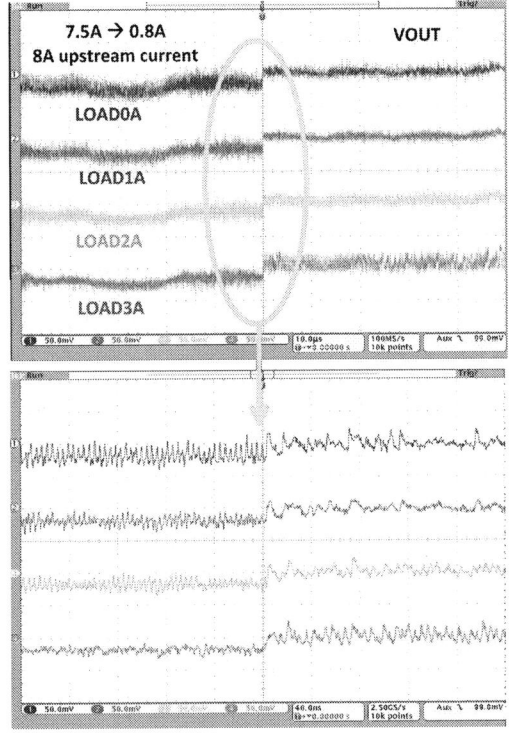

Fig. 11: Load transient response, I_{load} transition from 7.5A to 0.8A and Vout = 0.9V

Figure 13a shows the overall efficiency (V_{in} to V_{out}) of the converter as a function of load current. The overall efficiency is at a maximum at maximum load current i.e. when the load current is close to the upstream inductor current. Compared to a conventional buck converter, the CPR has two additional sources of loss: circulating current losses and downstream switching losses. The upstream inductor current circulates in

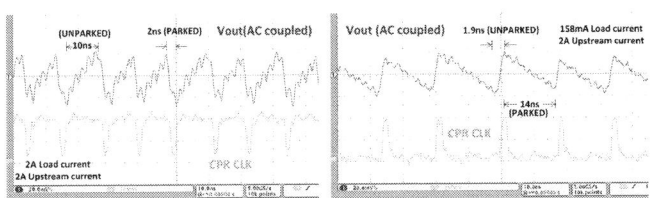

Fig. 12: Steady state output voltage waveforms

the parked loop as described in Section II even when the load current is low in order to be ready for a load current change. The circulating current leads to losses due to the resistance of M2, the inductor, and M3. This loss is independent of load current as the upstream inductor current is constant at 2A/CPR die. This loss is a large fraction of the load power at small load currents resulting in lower efficiency. Efficiency at low load currents can be improved by reducing the amount of current parked in the upstream inductor.

The efficiency with $V_{in} = 12V$ is slightly lower than efficiency at $V_{in} = 5V$. The lower efficiency at higher input voltages is mainly due to two reason. Firstly, the hard switching losses in the upstream switches M1 and M2 are higher at higher voltage given the same inductor current. Secondly, the constant ON time for which the switch M1 is turned ON is smaller for higher input voltage for the same inductor current ripple ΔI_L, resulting higher upstream switching frequency at higher voltage. The downstream efficiency remains unchanged by the input voltage and is only dependent on the inductor current and the load current. We have tested our implementation up to $V_{in} = 12V$. Even higher V_{in} can be used and is limited only by the discrete transistor M1 and M2. No stacking is necessary for the planar transistors M3 and M4 as the node V_{SW} only swings between V_{out} and GND.

Figure 13b shows the efficiency for a case where there are two load and one CPR dice. The max load current in this case is 2A. The efficiency for the system with 8A maximum load current and 2A maximum load current is very similar. Hence, the system can be scaled by having appropriate number of CPR dice in parallel to support the maximum load current.

Figure 14 shows the power dissipated in the downstream portion of the CPR. This includes the power consumed by the controller and buffer chain which are used to switch the distributed M3 and M4 transistor. The downstream switching power peaks when the load current around 4.5A. This is approximately half of the current in the inductor I_L. This is in accordance with the Equation 3 which showed that the downstream switching frequency peaked when $I_{load} = \frac{I_L}{2}$ and hence peaking of power consumed.

Figure 15a and 15b shows the efficiency and ripple respectively for different V_{out} when load current is varied, keeping V_{in} fixed at 1.5V. The kink in the efficiency curve of Figure 15a at $I_{load} = \frac{I_L}{2}$ can be explained due to the variation in switching frequency with load current where the frequency and hence downstream switching power dissipation reaches maximum value.

A. Power-save mode

As can be seen from efficiency plots in Figures 13 and 15 efficiency drops off as the load current decreases which as explained earlier is due to the fixed overhead of maintaining excess current parked in the inductor. We can improve the efficiency at lower load currents by operating in a power-save mode where the amount of current that is parked in the inductor is reduced based on the load current. However, this mode does not support fast load transients with increasing

978-1-4673-9551-9/16 $31.00 © 2016 IEEE

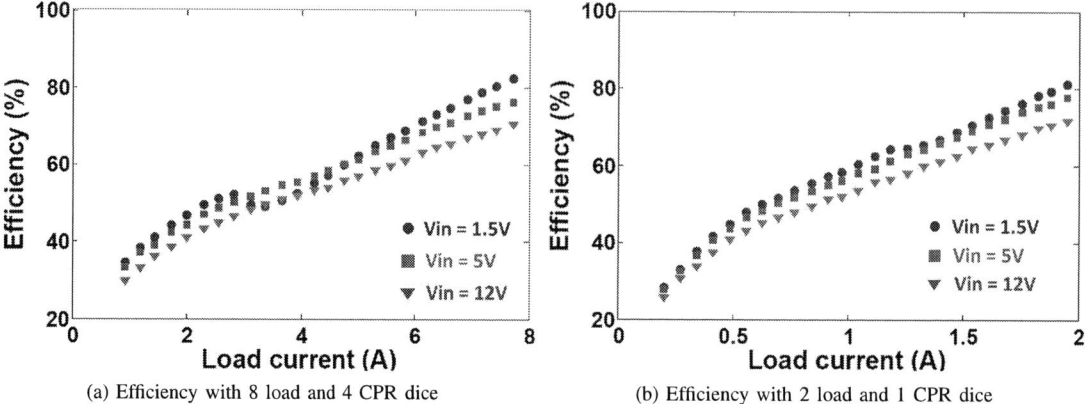

(a) Efficiency with 8 load and 4 CPR dice   (b) Efficiency with 2 load and 1 CPR dice

Fig. 13: Efficiency at different input voltages when load current is varied

current and can be used only when we know that load current will be steady for a some time interval. This mode becomes important when the processor enters the sleep mode or low activity mode where the large load transients are rare and reducing power dissipation is the main aim.

Implementation of the power-save mode requires the detection of the load current and subsequently uses this information to alter the current parked. One of the ways to estimate the load current is by looking at the time for which low-side switch (M2) is turned ON. The high-side switch M1 is turned ON for a constant time $T_{ON,const}$ when the inductor current reaches $I_{L,min}$. This increases the inductor current by ΔI_L. Now M2 is turned ON until I_L again dips below $I_{L,min}$. When M2 is ON and downstream is in the un-parked mode, inductor current reduces at a rate which equals $\frac{V_{out}}{L}$ and remains constant when the downstream is in parked mode as the voltage across the inductor is zero as shown in Figure 16. Total time required for the inductor current to decrease by ΔI_L is given by Equation 4.

$$T_{\Delta I_L} = L\frac{\Delta I_L}{V_L} = L\frac{\Delta I_L}{V_{out}} \qquad (4)$$

The downstream has to be in un-parked mode for N cycles to achieve the total time $T_{\Delta I_L}$ where T_{unpark} is given by Equation 1.

$$N = \frac{T_{\Delta I_L}}{T_{unpark}} = \frac{L\Delta I_L}{V_{out}T_{unpark}} \qquad (5)$$

$$T_{ON,low} = N(T_{unpark} + T_{park})$$
$$= \frac{L\Delta I_L}{V_{out}T_{unpark}}(T_{unpark} + T_{park}) \qquad (6)$$

The total ON time of the low side switch is just N cycles of the downstream cycles where in a single cycle it toggles between parked and unparked mode as shown in Equation 6. Substituting for values of T_{unpark} and T_{park} from Equations 1 and 2 respectively, $T_{ON,low}$ is obtained in terms of I_L and I_{load} as shown in Equation 7.

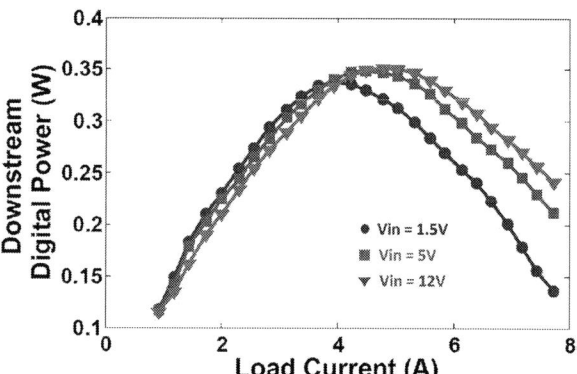

Fig. 14: Power dissipated in downstream control and drivers switching M3 and M4

$$T_{ON,low} = \frac{L\Delta I_L}{V_{out}}\frac{I_L}{I_{load}} \qquad (7)$$

The ON time of the low side switch is proportional to $\frac{I_L}{I_{load}}$. When the load current changes, the current in the inductor can be modified to achieve a constant $T_{ON,low}$ which is programmed in the upstream controller. The onset of the power-save mode is decided by this set value of $T_{ON,low}$ which, as shown in Figure 17, results in entering power-save mode at 5.5A load current giving the same efficiency curve for both the power-save mode and normal mode above 5.5A load current. The efficiency of the CPR in power-save mode is significantly increased compared to the normal mode where the inductor current remains fixed. The components used in this particular implementation prevent the current in each inductor from dropping below 0.5A (total I_L of 2A) hence power-save mode is not effectively applied for load current below 2A. However, with appropriate component selection, power-save mode can be applied to the entire load current range.

V. CONCLUSIONS

In this paper, we have presented the design and implementation of a regulator architecture that parks excess current in the inductor to facilitate ultra-fast response to load transients.

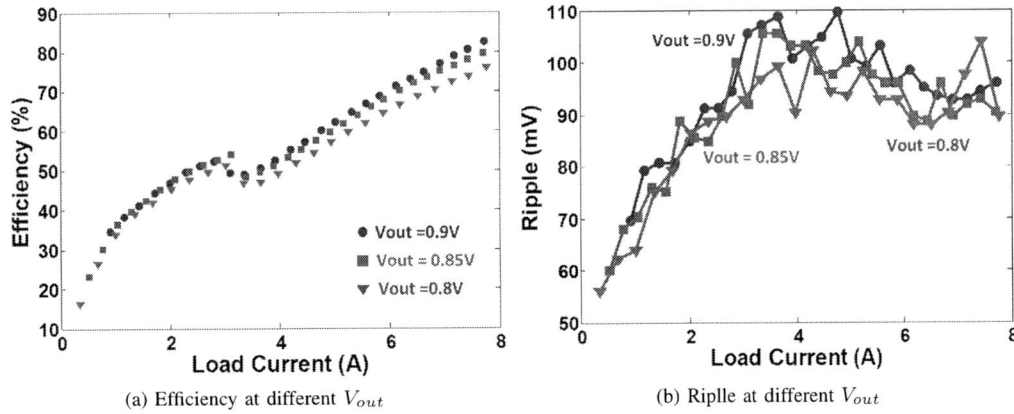

(a) Efficiency at different V_{out}

(b) Riplle at different V_{out}

Fig. 15: Efficiency at different output voltages when load current is varied and $V_{in} = 1.5V$

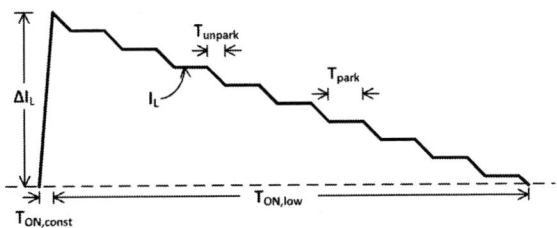

Fig. 16: Inductor current waveform

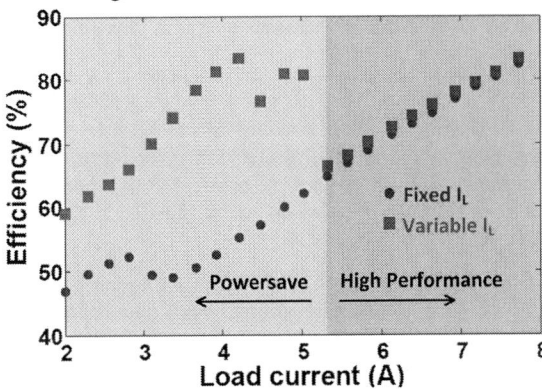

Fig. 17: Efficiency of the CPR in power save mode for varying load current

The regulator is implemented using discrete components as well as standard on-die transistors along with package level integration to reduce the parasitics to a minimum. The regulator, with its downstream portion implemented in a 28nm CMOS technology, achieves a peak overall efficiency of 83%. We have experimentally shown the scaling potential of the system to support even higher load currents and a technique to increase the overall efficiency of the system. The output voltage displays negligible droop or overshoot when the load current transitions between 0.8A and 7.5A in 2ns with only 200nF on-die capacitance on the output node. To the best of our knowledge, this is the best load transient response reported to date using only on-die capacitance.

Acknowledgements: The authors would like to thank the M. Fojtik and S. Tell for IC layout and testing help and H. Law and S. Huff for PCB design and layout help.

REFERENCES

[1] K. Bowman *et al.*, "Dynamic variation monitor for measuring the impact of voltage droops on microprocessor clock frequency," in *IEEE Custom Integrated Circuits Conference*, 2006, pp. 1 – 4.

[2] H. Sanchez *et al.*, "Increasing microprocessor speed by massive application of on-die High-K MIM decoupling capacitors," in *ISSCC*, 2008, pp. 2190 –2191.

[3] N. James *et al.*, "Comparison of split-versus connected-core supplies in power6 microprocessor," in *ISSCC*, 2007, pp. 298–299,604.

[4] *International Technology Roadmap for Semiconductors(ITRS) - System Drivers*, 2013, http://www.itrs.net.

[5] R. Erickson *et al.*, *Fundementals of Power Electronics*. Germany: Springer, 2001.

[6] E. Alon *et al.*, "Integrated regulation for energy-efficienct circuits," *IEEE Journal of Solid-State Circuits*, pp. 1795 – 1807, August 2008.

[7] J. Xu *et al.*, "A band-limited active damping circuit with 13dB power supply resonance reduction," *IEEE Journal of Solid-State Circuits*, pp. 61 – 68, May 2008.

[8] A. Barrado *et al.*, "The fast response double buck dcdc converter (FRDB): Operation and output filter influence," *IEEE Transactions of Power Electronics*, pp. 1261–1270, November 2005.

[9] N. Poon *et al.*, "A low cost dc-dc stepping inductance voltage regulator with fast transient loading response," in *Applied Power Electronics Conference and Exposition, 2001. APEC 2001. Sixteenth Annual IEEE*, vol. 1. IEEE, 2001, pp. 268–272.

[10] N. Pinckney *et al.*, "Shortstop: An on-chip fast supply boosting technique," in *IEEE Symposium on VLSI Circuits (VLSIC)*, 2013, pp. 290 – 291.

[11] Z. Shan *et al.*, "Pre-energized auxiliary circuits for very fast transient loads: Coping with load-informed power management for computer loads," *IEEE Transactions on Circuits and SystemsI: Regular Papers,*, February 2014.

[12] C.-W. Chen *et al.*, "A low-power dual-frequency simo buck converter topology with fully-integrated outputs and fast dynamic operation in 45nm cmos," *IEEE Journal of Solid-State Circuits*, pp. 1 – 13, May 2015.

[13] K. Shah *et al.*, "Simple and accurate circuit simulation model for gallium nitride power transistors," *IEEE Transactions of Electron Devices*, pp. 2735 – 2741, October 2012.

[14] E. Lindholm *et al.*, "Nvidia tesla: A unified graphics and computing architecture," *IEEE micro*, no. 2, pp. 39–55, 2008.

[15] "NVIDIA GeForce GTX 980 Featuring Maxwell, The Most Advanced GPU Ever Made," Nvidia Corporation, Tech. Rep. [Online]. Available: http://international.download.nvidia.com/geforce-com/international/pdfs/GeForce_GTX_980_Whitepaper_FINAL.PDF

[16] L. Zhang *et al.*, "Low cost high performance bare die pop with embedded trace coreless technology and coreless cored build up substrate manufacture process," in *Electronic Components and Technology Conference (ECTC), 2015 IEEE 65th*. IEEE, 2015, pp. 882–887.

978-1-4673-9551-9/16 $31.00 © 2016 IEEE

A 5MHz, 24V-to-1.2V, AO^2T Current Mode Buck Converter with One-Cycle Transient Response and Sensorless Current Detection for Medical Meters

Xugang Ke, Joseph Sankman* and Dongsheng Ma

Department of Electrical Engineering, The University of Texas at Dallas, Richardson, TX 75080, USA
*Battery Management Solutions, Texas Instruments Incorporated, Dallas, TX 75243, USA
Email: {xxk140030, d.ma}@utdallas.edu

Abstract—With a stringent input-to-output conversion ratio (CR) of 20 (24V input and 1.2V output) for a DC-DC converter, two stage cascaded architectures are easy to implement, but suffer from poor efficiency and doubled number of power components. This paper presents a single-stage solution with a proposed Adaptive ON-OFF Time (AO^2T) control. In steady state, the control works as an adaptive ON-time valley current mode control to accommodate the large CR. During load transient periods, both ON- and OFF-time are adaptively adjustable to instantaneous load change, in order to accomplish fast load transient response within one switching cycle. To facilitate the high speed current mode control, a sensorless current detection circuit is also proposed. Operating at 5MHz, the converter achieves a maximum efficiency of 89.8% at 700mA and an efficiency of 85% at 2A full load. During a load current slew rate of 1.8A/200ns, the undershoot/overshoot voltages at V_O are 23mV and 37mV respectively.

Keywords—*Adaptive ON-OFF Time, high conversion ratio (CR), sensorless current detection*

I. INTRODUCTION

With highly integration of microprocessors and memory cells in medical devices, there have been ever-increasing demands on efficient and fast transient on-chip power supplies. These medical devices, such as medical meters, are usually powered by 4 to 5 lithium-ion battery cells in series to extend run time [1], as shown in Fig. 1. However, with up to a 24V input voltage, many functional blocks are still powered at 1.2V, creating challenging step-down conversion ratio (CR) for circuit implementation. Multiple DC-DC converter stages can be employed for the high CR, but it would compromise speed, efficiency and system volume. Hence, a single-stage, step-down converter is highly desirable in this scenario.

While high efficiency is necessary to extend battery life, fast transient performance is essential to maintain a clean and accurate supply voltage. In contrast to fixed-frequency PWM controls, adaptive ON-time (AOT) controls exhibit faster transient responses [2-3]. However, if a load change takes place during T_{ON}, because T_{ON} is fixed, the converter cannot respond immediately, resulting in large output voltage drooping. Several solutions with adjustable T_{ON} have been reported [4-5], however, the extended T_{ON} causes large

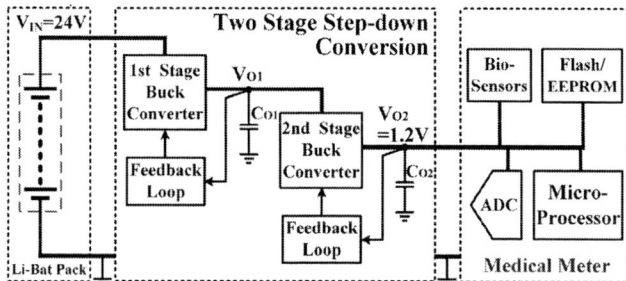

Figure 1. A typical power management system for medical meters.

inductor current ripple and hence overcharged V_O. In addition, because T_{ON} can only be increased, it cannot alleviate output overshoot when the load current drops. Furthermore, if the converter is designed with a linear error amplifier based compensator, the transient response and output voltage drooping would become even worse due to further limited bandwidth.

On a different aspect, as switching frequency is pushed higher for fast transient response and small form factor, inductor current sensing becomes very difficult to achieve, especially with a high step-down CR. Conventional series resistor sensing is accurate, but incurs unacceptable power losses. To minimize unnecessary power loss, inductor DC resistor (DCR) sensing can be employed, but it suffers from low sensing accuracy due to the DCR variations [6]. The sense-FET sensing method is achieved by connecting a much smaller mirror transistor in parallel with power transistor, and forcing their DC biasing voltages equal [7]. However, the complexity of sensing circuits and high bandwidth requirements of the op-amp limit its use at high frequency or with high CR.

In this paper, an adaptive ON-OFF time (AO^2T) current mode control scheme is proposed, in which both T_{ON} and T_{OFF} can be adjusted dynamically during load transient period to achieve one-cycle transient response. For circuit implemented, a fast and accurate sensorless current detection circuit is developed, which overcomes the limitations of DCR variation and avoids complex sense FET circuits. The rest of paper is organized as follows. In Section II, system architecture and operation scheme of this work are introduced, followed by the detailed circuit implementation in Section III. The performance

978-1-4673-9551-9/16 $31.00 © 2016 IEEE

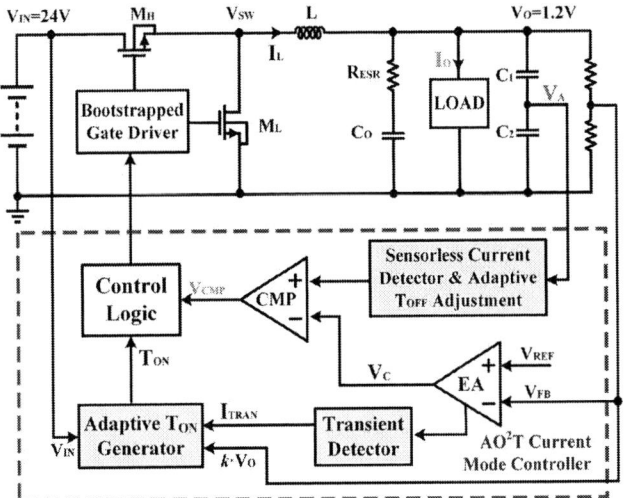

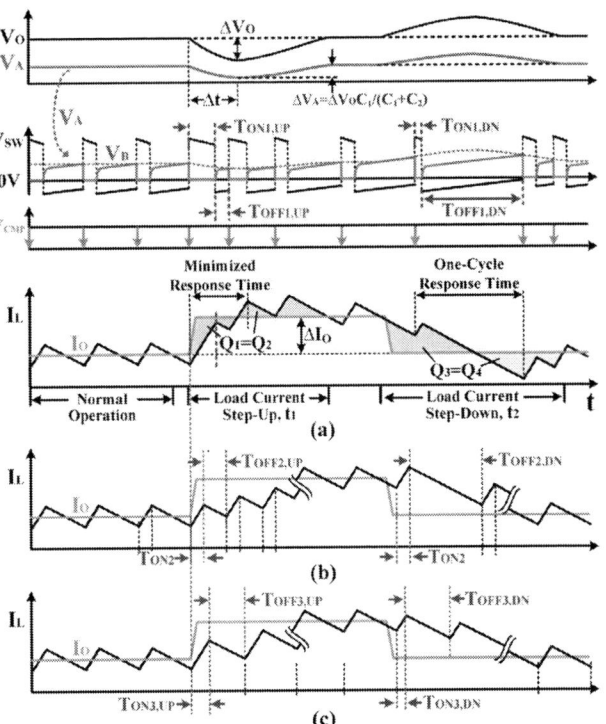

Figure 2. Block diagram of the proposed AO²T current mode buck converter.

of the proposed converter is then verified in Section IV. Finally, the paper is concluded in Section V.

II. SYSTEM ARCHITECTURE & OPERATION SCHEME

Fig. 2 illustrates the system block diagram of the proposed converter. It consists of a step-down high CR power stage and a feedback AO²T current mode controller. The controller itself mainly includes an error amplifier, a transient detector with adaptive ON-time generator and a sensorless current detector with adaptive OFF-time adjustment circuit. In the steady state, the converter is primarily regulated by the adaptive ON-time (AOT) control. For stable V_{IN} and V_{OUT}, T_{ON} remains constant, in which the inductor current is energized, as shown in Fig. 3. After T_{ON} expires, M_L is turned on and the inductor is discharged. The inductor current is measured using a sensorless current detector. The OFF-time ends when V_B surpasses V_A in Fig. 3, triggering the next ON-time to start.

In order to achieve fast response during load transient periods, both T_{ON} and T_{OFF} become adjustable. When the load current rises at t_1 in Fig.3, the output voltage droops down by ΔV_O. The transient detection circuit detects this droop voltage and converts it to a current, I_{TRAN}, which is used to instantly increase the ON-time (T_{ON}). Meanwhile, the capacitive voltage scalar, formed by C_1 and C_2, detects the voltage droop through V_A, causing early trigger of the comparator as well as smaller T_{OFF}. Compared to fixed frequency peak current mode and conventional AOT valley current mode, the T_{ON} and T_{OFF} adjustments during step-up load transient, are defined as

$$T_{ON1,UP} > T_{ON3,UP} > T_{ON2}, \tag{1}$$

$$T_{OFF1,UP} < T_{OFF2,UP} < T_{OFF3,UP}. \tag{2}$$

The maximized T_{ON} and minimized T_{OFF} ensure the inductor current boost to the level of the increased load current at the shortest response time. Similarly, when the load current steps down during t_2, the control scheme functions conversely, with T_{ON} minimized and T_{OFF} maximized to follow the load change, shown as

Figure 3. Transient response behaviors of (a) the proposed AO²T control, (b) conventional AOT valley current mode control, and (c) fixed frequency peak current mode control.

$$T_{ON1,DN} < T_{ON3,DN} < T_{ON2}, \tag{3}$$

$$T_{OFF1,DN} > T_{OFF2,DN} > T_{OFF3,DN}. \tag{4}$$

III. CIRCUIT IMPLEMENTATION

Fig. 4 illustrates the proposed transient detection and T_{ON} adjustment circuits. In regular operation, when V_{FB} is very close to V_{REF}, the transient detection block is inactive and $I_{TRAN}=0$.

When V_O drops significantly and the output current of transient detection amplifier ($=g_m \times (V_{REF}-V_{FB})$) is larger than a threshold current I_{OS1}, the difference current goes to M_4, and M_5 thusly drains more current. Since $I_{TOT}= (I_{CHG}-I_{TRAN})$, the total charging current I_{TOT} decreases and hence T_{ON} increases since the current into C is decreased, which increases the ON-time until the voltage on C reaches V_{DIV}.

If V_O increases significantly, and $g_m \times (V_{FB}-V_{REF}) > I_{OS2}$, the difference current goes to M_7, and M_8 sources current to M_9, increasing I_{TOT} and hence T_{ON} decreases. Thus, during a load transient, T_{ON} time can be adjusted within one cycle if the output voltage runs out of the boundaries set by I_{OS1} and I_{OS2}. Furthermore, V_{IN} and V_O are sensed to maintain a quasi-fixed switching frequency in different V_{IN} and V_O conditions in steady state, as shown in Fig. 4. On the other hand, T_{OFF} can be adjusted during load transient. The value of C_1 and C_2, as shown in Fig. 5, need to be well designed to optimize transient response. When load changes by ΔI_O, the voltage change at V_O is $\Delta V_O=\Delta I_O R_{ESR}+\Delta I_O \Delta t/C_O$ and $\Delta V_O \times C_1/(C_1+C_2)$ at V_A. The voltage change of V_A generates a change in the inductor valley

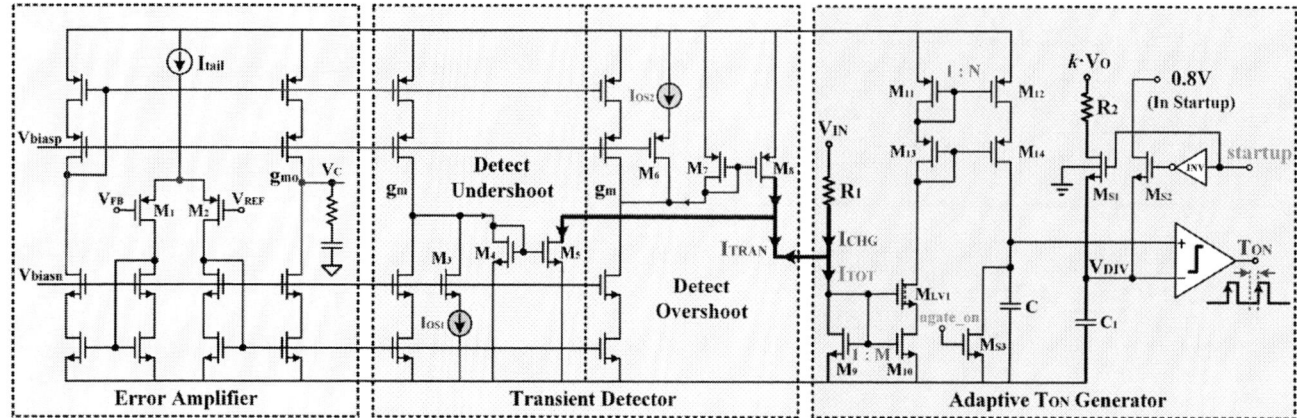

Figure 4. Transient detection and adaptive T_{ON} adjustment circuitry.

current ($I_{L,valley}$), $\Delta I_{L,valley}=\Delta V_A/R_{DSON,ML}$. To minimize the voltage overshoot or undershoot, the change of $\Delta I_{L,valley}$ must match the load change ΔI_O at the output, requires $(C_1+C_2)/C_1 =(R_{ESR}+\Delta t/C_O)/R_{DSON,ML}$. When the load changes from heavy to light, the inductor current can be adequately ramped up within one cycle thanks to unlimited T_{OFF}. When the load increases, T_{OFF} is made smaller to achieve fast tracking of I_L with the load change.

For inductor current detection, the proposed sensorless current detection circuit samples the inductor current during OFF-time, as shown in Fig. 5. This detected voltage, V_{SW}, expressed as $V_{SW} = (-I_L \times R_{DSON,ML})$, is linearly proportional to I_L. Since V_{SW} is negative, it is first level-shifted by a voltage of $(I_C \times R_{DS,MR2} + I_{bias} \times R_{DS,MR2})$ and then compared to V_A, which is set by $(I_{bias} \times R_{DS,MR1})$. I_C is converted from the output V_C of the error amplifier, the switching cycle ends when $(V_{SW} + I_C \times R_{DS,MR2})$ hits 0. Thus, the valley inductor current can be expressed as $I_{L,valley} =(I_C \times R_{DS,MR2})/R_{DSON,ML}$. M_{R1} and M_{R2} operate in linear region, and are well matched to M_L in the layout.

IV. PERFORMANCE VERIFICATION

The proposed design is implemented using a 0.35-μm BCD process. It employs a 0.68μH inductor and an 18μF output capacitor with an ESR of 5mΩ. Operating at 5MHz, it accomplishes a maximum CR from 24V to 1.2V. The design

has been verified by fully transistor level simulations. Fig. 6 measures the step-up and step-down load transient with T_{ON} and T_{OFF} adjustment enabled. When the load current steps up and down between 0.2A and 2A with the transit time of 200ns, the undershoot/overshoot voltages at V_O are 23mV and 37mV, respectively. The response time for the step-up load change is 0.3μs, and for step-down load change is 0.45μs. Both are achieved within one switching cycle. Compared to a conventional AOT valley current mode converter with the same design specifications, the proposed design reduces undershoot and overshoot by 69.6% and 45.6%, respectively, as shown in the waveforms of Fig. 7. It achieves a reduction of undershoot/overshoot by 56.6%/31.5%, compared to the best of prior arts in [2-5]. Fig. 8 shows the efficiency of the proposed design when $V_{IN}=24V$ and $V_O=1.2V$. The efficiency peaks at 89.8% with a load current of 700mA and stays at 85% at the full load of 2A.

V. CONCLUSION

An integrated, 5MHz, 24V-to-1.2V, AO²T current mode buck converter with fast transient response for medical meter applications is demonstrated on a 0.35-μm HV BCD process. With fast, accurate sensorless current detection, the proposed

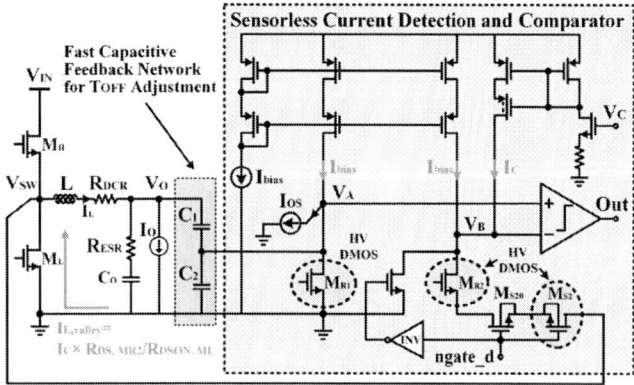

Figure 5. Sensorless current detection and adaptive T_{OFF} adjustment circuitry.

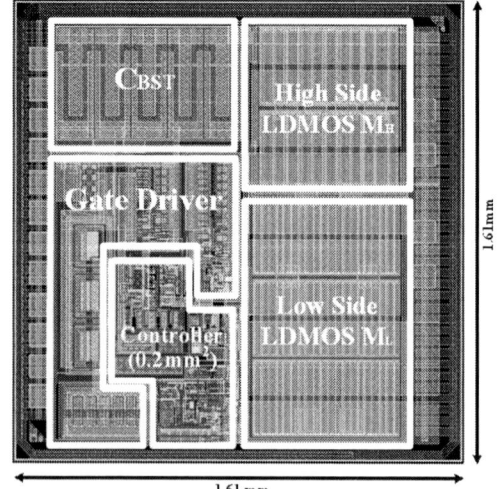

Figure 6. Chip layout.

978-1-4673-9551-9/16 $31.00 © 2016 IEEE

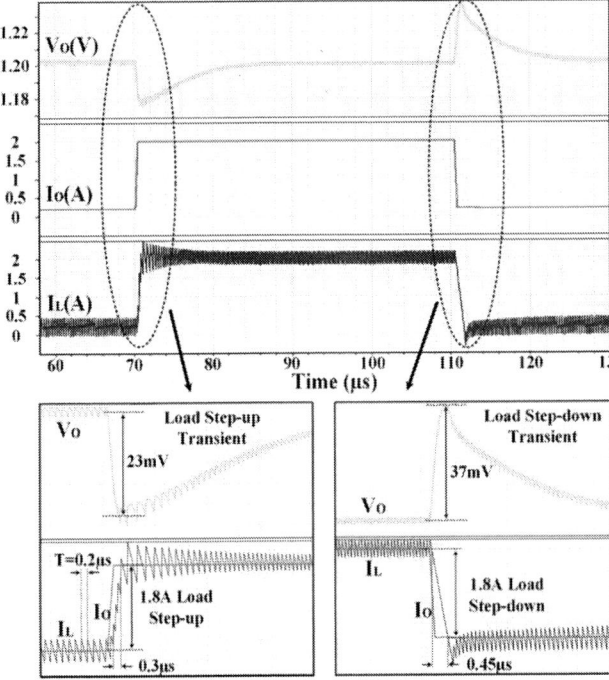

Figure 7. Load transient response.

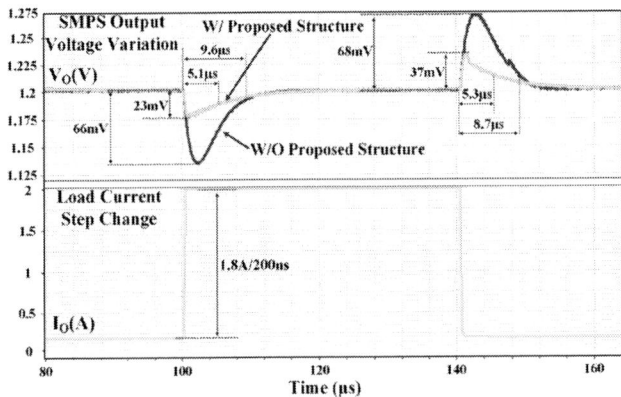

Figure 8. Load transient performance comparison with and without proposed structure.

converter can operate at 5MHz switching frequency and with the conversion ratio of 24V to 1.2V. Thanks to the proposed T_{ON} and T_{OFF} adjustment scheme, the converter achieves within-one-cycle transient response with minimized drooping voltages. In addition, the efficiency for this 24V-to-1.2V voltage conversion is significantly improved due to the single stage architecture. Table I demonstrates performance summary of the proposed AO^2T current mode buck converter.

ACKNOWLEDGMENT

The authors would also like to thank Rais Miftakhutdinov and Tuli Dake of Texas Instruments for reviewing the paper and giving valuable inputs.

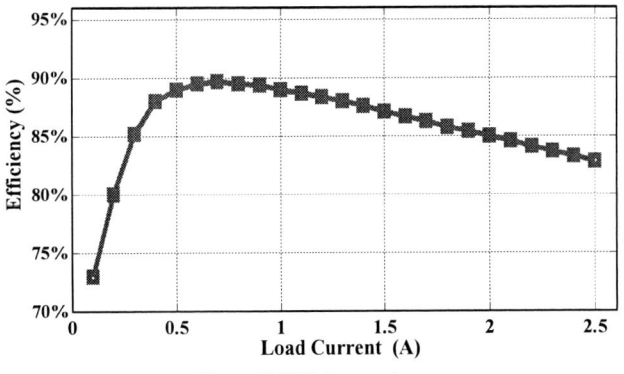

Figure 9. Efficiency plot.

TABLE I PERFORMANCE SUMMARY

Technology	0.35-μm HV BCD Process
V_{IN}	24V
V_{OUT}	1.2V
Control	AO^2T Current Mode
Switching Frequency	5 MHz
Output Capacitor	$C_{OUT} = 18\mu F$
ESR	5mΩ
Inductance	0.68μH
V_{OUT} Undershoot/Overshoot / Load Step (A/ns)	23mV/37mV / 1.8A/200ns
Response Time	0.3μs/0.45μs
Output Ripple	3.4mV
Peak Efficiency	89.8%
Efficiency at I_{MAX}	85%

REFERENCES

[1] Texas Instruments Application Note: "Consumer Medical Applications Guide," www.ti.com, 2010.

[2] F. F. Ma, W. Z. Chen, and J. C. Wu, "A monolithic current-mode buck converter with advanced control and protection circuits," IEEE Trans. Power Electron., vol. 22, no. 5, pp. 1836–1846, Sep. 2007.

[3] X. C. Jing, P. K. T. Mok, and M. C. Lee, "Current-slope-controlled adaptive-on-time DC-DC converter with fixed frequency and fast transient response," in Proc. IEEE Symp. Circuits and Systems, pp. 1908–1911, May 2011.

[4] C.-H. Tsai, S.-M. Lin, and C.-S. Huang, "A fast-transient quasi-V2 switching buck regulator using AOT control with a load current correction (LCC) technique," IEEE Trans. Power Electron., vol. 28, no. 8, pp. 3949–3957, Aug. 2013.

[5] H. C. Lin, B. C. Fung, and T. Y. Chang, "A current mode adaptive on-time control scheme for fast transient DC-DC converters," in Proc. IEEE Symp. Circuits and Systems, pp. 2602–2605, May 2008.

[6] H. P. Forghani-zadeh and G. A. Rincon-Mora, "An accurate, continuous, and lossless self-learning CMOS current-sensing scheme for inductor-based DC-DC converters," IEEE J. Solid-State Circuits, vol. 42, pp. 665–679, 2007.

[7] W. H. Ki, "Current sensing technique using MOS transistors scaling with matched bipolar current sources", U.S. Patent 5,757,174, May 26, 1998.

Capacitively-Aided Switching Technique for High-Frequency Isolated Bus Converters

Seungbum Lim, Alex J. Hanson, Juan A. Santiago-González, and David J. Perreault

Department of Electrical Engineering and Computer Science
Massachusetts Institute of Technology
Cambridge, Massachusetts 02139-4307
Email: sblim@mit.edu

Abstract—This paper presents a new capacitively-aided zero voltage switching (ZVS) technique for isolated bus converters. The proposed technique helps to achieve ZVS conditions for both inverter and rectifier switches with the aid of magnetizing current of the transformer and capacitors interconnecting primary and secondary side switching nodes. With the proposed topology, the ZVS conditions for inverter and rectifier switches occur at the same time, and (ideally) maintain constant required dead time independent of the load level of the converter. These features are particularly advantageous especially for designs operating at high frequencies where timing control becomes critical. The proposed approach is demonstrated in a 1.4 MHz prototype converter which operates from 36 Vdc and supplies a 12 Vdc output load (i.e., fixed input voltage and fixed voltage transformation ratio, 3 : 1) at a 36 W rating. Experimental results show ZVS operation of all devices independent of the load level, and the prototype converter achieves 94 % peak efficiency with 300 W/in³ power density.

I. INTRODUCTION

Advances in power distribution systems (e.g., for telecommunications, data centers, and dc distribution systems) have been driven by a demand for higher efficiency, smaller volume, and lower cost. The intermediate-bus architecture, which incorporates a "bus converter" stage (or "dc transformer") that provides isolation and voltage transformation but without the regulation capability, has become popular due to its advantages (e.g., system efficiency and power density) [1]–[4]. Moreover, there are many other power conversion architectures and applications in which the isolation and transformation function of the power electronics can be separated from any required regulation capability [2], [5]–[10]. Practical applications of such systems depend upon bus converters having small size, low loss, and high performance.

Recent advances in gallium-nitride (GaN) transistor technology facilitate higher frequency operation with improved transistor figures of merit (e.g., $R_{on} \times C_{oss}$ [11] or $\sqrt{R_{on} \times Q_g}$ [12], [13]), such that low-loss soft switching can be realized at high frequency [2], [14]–[19]. Furthermore, as shown in the study [20], increasing operating frequencies into the multi-MHz range can yield high performance in terms of the capabilities of available magnetic materials. However, leveraging the achievable performance of these devices and materials requires topologies and control techniques to address the challenging constraints of control timing (e.g., for ZVS switching and dead time) as frequency increases. This paper explores a topology and switching technique suitable for bus converters at multi-MHz frequencies.

There are many resonant converter topologies processing (quasi) sinusoidal waveforms that may be useful for isolated bus conversion (e.g., series-resonant [21], parallel-resonant [22], and LLC [23]–[27]). Differences among these converters are based on how the resonant tank and transformer are configured and utilized. While many of these converters can be designed with diode-based rectifiers, synchronous rectifiers are often used to reduce diode conduction loss and achieve higher efficiency (e.g., [1], [28]). Control and driver circuitry are then needed to accurately adjust gate-source signals of the secondary-side (output side) transistors. There are many commercial control / driver ICs which are specifically designed for controlling the secondary-side transistors to behave like diodes using only information available on the secondary. These integrated circuits usually sense the drain-source voltage of the secondary switch and compare it with threshold voltages to control the switch gate-source voltage. However, these techniques tend not to work well at frequencies above 1 MHz, because parasitics (e.g., parasitic capacitance/inductance from PCB and switch) make it difficult to achieve the necessary switch timing using drain-source voltage measurements. In addition, fast (low propagation delay) comparators consume a lot of control power and there are limits on the achievable response speed of such comparators, especially at low overdrive of the comparator inputs. Consequently, there is significant difficulty in making this self-sensing synchronous rectification approach work well in the multi-MHz regime. Direct control of the secondary side switches (e.g., from the primary side) likewise becomes challenging for these topologies at high frequencies. Control timing (and its variation with operating condition) thus becomes an increasingly challenging problem as frequency increases for many conventional resonant converter topologies employing synchronous rectification.

An alternative to resonant topologies in the bus converter (or dc transformer) application is the dual-active-bridge (DAB) converter [28]. In this topology there is active control of the timing of the primary switches relative to the secondary switches (such that the secondary switches are not operated like diodes, but do achieve ZVS). In addition to ZVS operation of all of the switches, this topology has the advantage that, for fixed conversion ratio applications. One achieves

978-1-4673-9551-9/16 $31.00 © 2016 IEEE

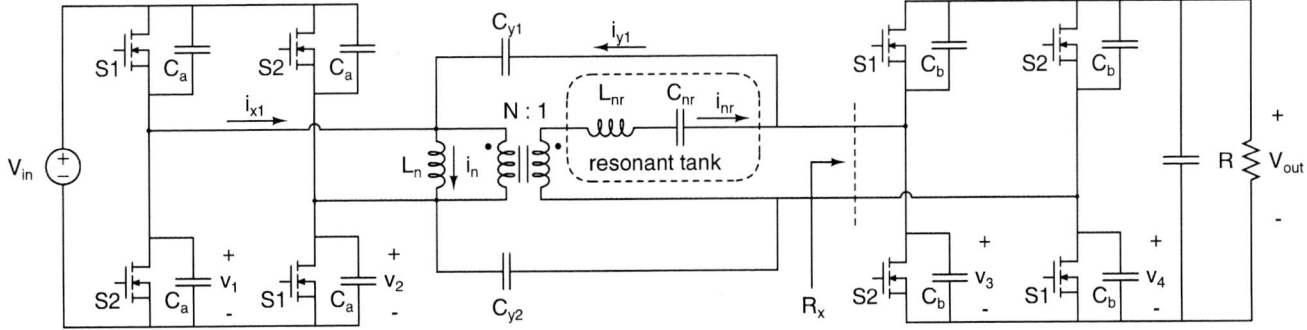

Fig. 1. The schematic shows the proposed isolated converter topology. It is designed with full-bridge inverter, transformer, resonant tank, full-bridge synchronous rectifier, and Y-rated capacitors interconnecting corresponding inverter and rectifier switch sets.

quasi-trapezoidal current waveforms that provide very low rms currents in the devices and transformer, yielding low conduction dissipation [29]. However, the driving signals need to be precisely controlled for all the switches, such that the DAB converter requires complex and accurate control circuitry with variable dead time, duty ratio, and phase shift (between inverter stage and rectifier stage) for various loads and input voltages. This complex control becomes increasingly difficult at high frequency and imposes heavy constraints on the achievable frequency, especially considering the timing delay and timing variability among signals crossing the isolation barrier. Overall, there is a difficulty in operating many conventional isolated converters with synchronous rectification at high frequency because of the complexities of achieving the necessary control timing among all of the switches when working across the isolation barrier. This highlights the need for new topologies and control approaches more suitable to this frequency range.

Here we propose a new ZVS technique applicable for unregulated isolated converters which addresses the challenge of maintaining ZVS operation with accurate switch timing in the MHz frequency range. Section II of the paper introduces the new bus converter topology and illustrates its operation, design considerations, and advantages. Section III of the paper demonstrates an implementation of the converter operating at 1.4 MHz and presents experimental results. Finally, in Section IV we conclude that the proposed topology can maintain high efficiency with simple control as operating frequency is increased into the high frequency regime.

II. PROPOSED BUS CONVERTER TOPOLOGY

The proposed converter is illustrated in Fig. 1. It is designed to act as a "dc transformer" in which the voltage conversion ratio is relatively fixed and no regulation capability is required. In our intended application the converter is used as a bus converter, in which input voltage is fixed in addition to conversion ratio, though it can operate well at variable input voltages within limitations constrained by device capacitance nonlinearities. The converter comprises a full-bridge inverter, transformer, resonant tank, full-bridge synchronous rectifier, and capacitors interconnecting primary and secondary side

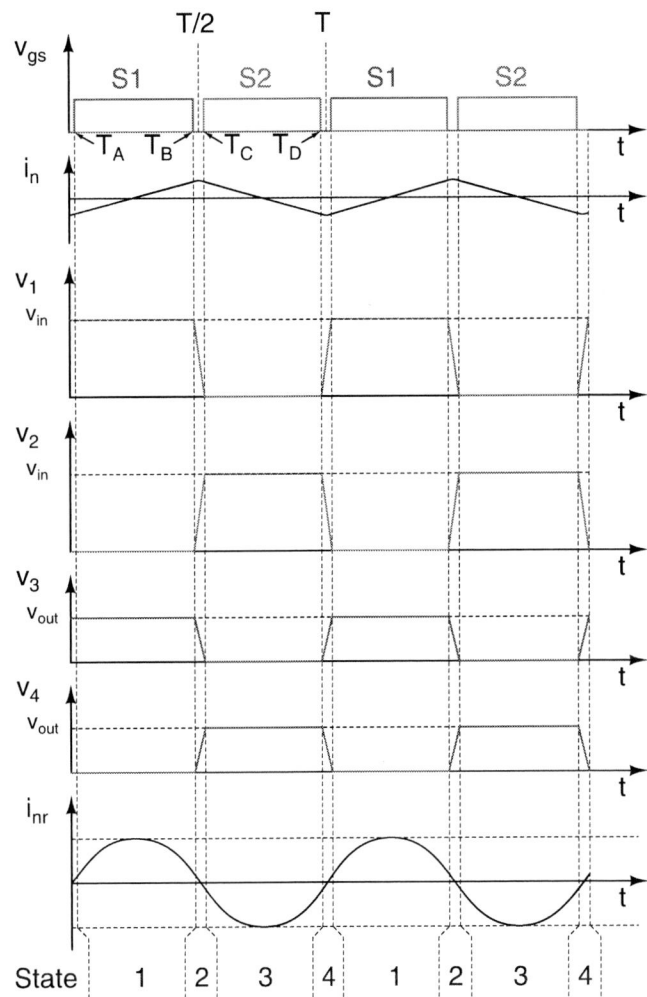

Fig. 2. The waveform describes the voltage and current of the proposed isolated converter topology shown in Fig. 1.

switching nodes. The capacitors C_{y1} and C_{y2} are implemented with Y-rated capacitors to meet isolation requirements.

The operation of the proposed converter topology naturally absorbs the leakage and magnetizing inductances of the trans-

978-1-4673-9551-9/16 $31.00 © 2016 IEEE

former as well as the device and transformer capacitances, making it suitable for high-frequency operation. In considering the transformer parasitics, we utilize the cantilever model of the transformer, such that we only consider a single "leakage inductance" in addition to a "magnetizing inductance", and a "turns ratio" to characterize the transformer; these model components thus do not necessarily reflect the exact physical values (e.g., of the turns ratio), though we can directly map between this model and physical parameters [30].

As shown in Fig. 1, C_a and C_b are the output capacitances of the inverter and rectifier switches (plus any added capacitance), respectively, and L_n is the magnetizing inductance of the N:1 turns ratio transformer. For the selected operating frequency, the inductor L_{nr} and capacitor C_{nr} form a resonant tank, such that (ideally) only fundamental (first harmonic) frequency current flows through the resonant tank (i.e., $i_{nr} = I_{nr} \, sin(2\pi f_{sw} t)$). In addition, all the switches at primary/secondary side are synchronized and controlled by either S1 or S2 gate signals as shown in Fig. 1.

A. Operation

Fig. 2 illustrates the ideal switch timing and operating waveforms of the proposed isolated bus converter topology. The resonant tank (composed of inductor L_{nr} and capacitor C_{nr}) is tuned to the selected operating frequency, such that the current i_{nr} approximates a sinusoidal current as shown in Fig. 2. The converter operates at fixed frequency and all the switches are operated to have the same duty ratio with a fixed dead time as shown in Fig. 1 and Fig. 2.

To simplify the analysis, we assume 100% efficiency and small dead time between S1 and S2 signals such that no significant power is transferred through the interconnecting capacitors. With these assumptions, the input power (1) and output power (1) of the transformer are equal (2). The current flowing through the magnetizing inductor does not receive/send power through the transformer on average, so it drops out of the integral as shown in (3). Solving equations (1)–(3) show that the output voltage is simply V_{in} / N.

$$P_{in} = \frac{2}{T} \int_0^{\frac{T}{2}} V_{in} \left(\frac{I_{nr}}{N} \, sin\,(\omega t) + I_n(t) \right) dt$$
$$= \frac{2V_{in}}{T} \int_0^{\frac{T}{2}} \frac{I_{nr}}{N} \, sin\,(\omega t) \, dt \qquad (1)$$

$$P_{out} = \frac{2V_{out}}{T} \int I_{nr} \, sin\,(\omega t) \, dt \qquad (2)$$

$$P_{in} = P_{out} \quad \Rightarrow \quad V_{out} = \frac{V_{in}}{N} \qquad (3)$$

With the above assumptions, the converter cycles through four states as described below, with the polarity of the voltages and currents as illustrated in Fig. 1.

State 1 : $T_A < t < T_B$ — In this state, the S1 switches are turned on and S2 switches are turned off. V_{in} is applied across the transformer primary such that magnetizing current increases linearly. Voltages v_1, v_2, v_3, and v_4 are clamped to V_{in}, 0, V_{out}, and 0, respectively. The capacitors C_{y1} and C_{y2} do not conduct current because the voltages across them remain fixed.

State 2 : $T_B < t < T_C$ — State 2 starts as the S1 switches turn off under ZVS conditions using their output capacitances to snub the switch turn offs. During this dead time, the current through the resonant tank (approximately sinusoidal current) is in phase with the voltage applied to the transformer, so that current i_{nr} is close to zero for the short dead time (or, even considering non-zero instantaneous current, the average current of i_{nr} during the dead time is zero, so that i_{nr} current does not affect to voltage on v_3 and v_4 during the dead time). After the S1 switches turn off, the magnetizing current starts to decrease the voltage v_1, and also v_3 through C_{y1}. Likewise, the magnetizing current charges the voltage of v_2 and v_4 via C_{y2}. By selecting correct capacitances C_{y1} and C_{y2} with respect to the secondary-side switch output capacitances, the required dead times to achieve ZVS for the primary switches and secondary switches can be matched (this will be further described in section II-B). When the drain-source voltage of the S2 switches are discharged (close enough) to zero, the S2 switches are turned on with ZVS condition and the system enters state 3.

State 3 : $T_C < t < T_D$ — In this state, the S2 switches are turned on and S1 switches are turned off. Therefore, $-V_{in}$ is applied across the primary magnetizing inductance of the transformer, such that magnetizing current decreases monotonically.

State 4 : $T_D < t < T + T_A$ — All the S2 switches turn off with ZVS conditions with the aid of the transistor output capacitances snubbing the turn off. Then, similar to state 2 but with opposite polarity, utilizing the magnetizing current and capacitors C_{y1} and C_{y2}, the drain-source voltages of the S1 switches decrease and create an opportunity for ZVS turn-on of the S1 devices. When the S1 switches turn on, state 1 is entered and the cycle repeats.

The equations for magnetizing and resonant tank currents are as follows (dead time between S1 and S2 is ignored here for simplicity).

$$i_n = \begin{cases} \dfrac{V_{in}}{L_n} \left(t - \dfrac{T}{4} \right) & 0 < t < \dfrac{T}{2} \\[2mm] -\dfrac{V_{in}}{L_n} \left(t - \dfrac{3T}{4} \right) & \dfrac{T}{2} < t < T \end{cases} \qquad (4)$$

$$i_{nr} = I_{nr} \, sin\,(2\pi f_{sw} t) = \frac{\pi}{2} \frac{V_{out}}{R} \, sin\,(2\pi f_{sw} t) \qquad (5)$$

B. Design parameters

For the desired operation as illustrated above, the Y-rated capacitors need to be selected properly to achieve ZVS conditions for the secondary-side switches. During dead time (when switches S1 and S2 are all off), the current through the switches and resonant tank is close to zero. Then, the magnetizing current i_n flows through inverter side output capacitors (C_a), Y-rated capacitors (C_{y1} and C_{y2}), and rectifier side output capacitors (C_b). The current and voltage relations during the dead time (state 2 and 4) are as follows:

$$i_n(t) = i_{x1}(t) + i_{y1}(t) \qquad (6)$$

$$i_{x1}(t) = -2C_a \frac{dv_1}{dt} \qquad (7)$$

$$i_{y1}(t) = C_{y1} \frac{d(v_3 - v_1)}{dt} = -2C_b \frac{dv_3}{dt} \qquad (8)$$

To achieve ZVS conditions for inverter switches and rectifier switches at the same time, the value of capacitors C_{y1} and C_{y2} can be selected based on the relations in (8). If the converter achieves ZVS for both primary and secondary switches during the dead time, the voltages of v_1 and v_3 at the beginning and the end of the dead time at state 2 are as shown in Fig. 2 and equations (9) and (10).

$$v_1(T_B) = V_{in}, \quad v_3(T_B) = V_{out} = \frac{V_{in}}{N} \qquad (9)$$

$$v_1(T_C) = 0, \qquad v_3(T_C) = 0 \qquad (10)$$

Integrating eq. (8) over the dead time (i.e., from T_B to T_C), we can calculate the desired capacitance for C_{y1} and C_{y2} as illustrated in (11)–(13).

$$\int_{T_B}^{T_C} i_{y1}(t)dt = C_{y1}(v_3(t) - v_1(t)) \, |_{T_B}^{T_C} \qquad (11)$$

$$= -2C_b(v_3(t)) \, |_{T_B}^{T_C} \qquad (12)$$

$$\Rightarrow \quad C_{y1} = \frac{2C_b}{N-1} \qquad (13)$$

This illustrates that the values of the interconnecting capacitors C_y should be selected based on the effective capacitance C_b and turns ratio of the transformer. Then the ratio between C_y and C_b makes the required ZVS dead time for inverter switches (v_1) and rectifier switches (v_3) the same. It should be noted that if the switch capacitors are significantly nonlinear, we may need to select an effective value C_b. Also, the range of input voltages (and scaled output voltages) that the converter can accommodate with ZVS may be limited by how much effective capacitance Cb changes with variations in output voltage.

Furthermore, we can estimate the dead time for the converter from the equations above. If we treat the magnetizing current as almost constant during the relatively short dead time, then we can regard the magnetizing current during dead time (i.e., states 2 and 4) as follows:

$$i_n(T_B) \simeq i_n(T_C) \simeq i_n(T/2) \simeq \frac{V_{in}T}{4L_n} \qquad (14)$$

Then, the required dead time can be calculated from (6)-(8), (13), and (14) as follows.

$$\int_{T_B}^{T_C} i_n(t)dt = -2C_a v_1(t) \, |_{T_B}^{T_C} - 2C_b v_3(t) \, |_{T_B}^{T_C}$$

$$= 2C_a V_{in} + 2C_b \frac{V_{in}}{N} \qquad (15)$$

$$\Rightarrow \quad \frac{V_{in}}{L_n} \frac{T}{4}(T_C - T_B) = 2C_a V_{in} + \frac{2C_b V_{in}}{N}$$

$$\Rightarrow \quad T_C - T_B = DeadTime = \frac{8L_n(C_a + C_b/N)}{T} \qquad (16)$$

The dead time between high and low side switches can therefore be easily calculated for the selected converter designs and it is independent of load level. However, it should be noted that both L_n and T should be selected together to keep the magnetizing current in a reasonable range. That is, the magnetizing current needs to be large enough to fully discharge/charge capacitors (C_a and C_b) for ZVS with a modest dead time duration, but is also constrained not to be too high to reduce conduction loss and to get high efficiency.

Another design consideration is selecting the quality factor (Q) of the resonant tank of the converter. For the series-resonant tank structure used on secondary side, the Q is defined as shown in (17), where R_x is the effective impedance looking into the synchronous rectifier from the resonant tank as illustrated in Fig. 1 and defined by (18) [21].

$$Q = \frac{Z_{Lr}}{R_x} = \frac{Z_{Cr}}{R_x} = \frac{\omega L_r}{R_x} = \frac{1}{\omega C_r R_x} \qquad (17)$$

$$R_x = \frac{8}{\pi^2} R \qquad (18)$$

With a selected Q the voltage across the resonant capacitor is as follows.

$$v_{Cr} = \frac{1}{C_{nr}} \int i_{nr}(t) \, dt = \frac{1}{C_{nr}} \int \frac{\pi}{2} \frac{V_{out}}{R} sin(\omega_{sw}t) \, dt$$

$$= -\frac{\pi}{2} \frac{V_{out}}{C_{nr} \, \omega_{sw} \, R} cos(\omega_{sw} \, t)$$

$$= -\frac{4}{\pi} Q V_{out} \, cos(\omega_{sw} \, t) \qquad (19)$$

Larger Q values give ideal sinusoidal currents (at fundamental frequency), but require larger resonant inductances (L_{nr}) and larger voltage rating of the resonant capacitors (C_{nr}) as described in (17) and (19), and consequently higher loss.

To keep the resonant components physically small, especially L_{nr} so that it can be replaced by the leakage inductance of the transformer, we prefer a low-Q resonant tank if possible. It turns out that since the voltage applied to the transformer is an odd waveform, even a very low quality factor resonant tank does not alter the basic operation of the converter as long as the resonant tank current is small and crossing zero during the dead time. From numerous simulations of the proposed approach, it is found that the converter can maintain good operation with Q as low as 0.1 at full load (and so is effective

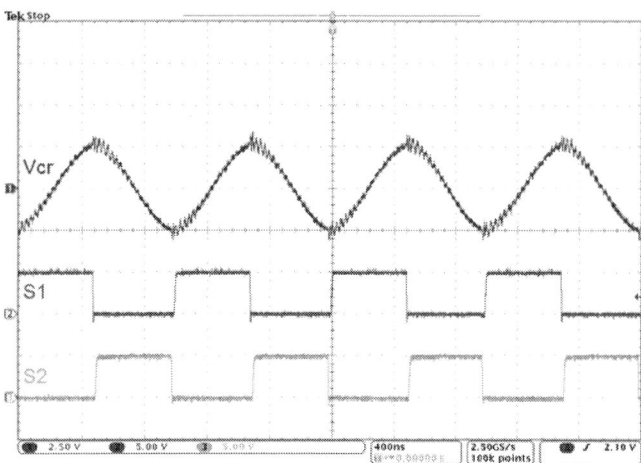

Fig. 3. The figure shows the waveform of the prototype converter when the resonant tank is tuned at operating frequency. ch1 shows the voltage across the resonant capacitor (2.5 V/div), ch2 and ch3 show the gate-source signal for S1 and S2 switches (5 V/div). The resonant tank can be simply tuned by measuring the phase of the resonant capacitors referenced to the control signals.

even at light loads). In our designs, we chose Q values between 0.1 and 1, and tuned the resonant tank as described in section III-A.

C. Advantage

The proposed isolated bus converter has several advantages over conventional topologies at high frequency. The first advantage is reliably achieving ZVS for the rectifier switches, even at high frequency. This made possible by the inclusion of the Y-rated capacitors interconnecting primary and secondary switches. Compared to many conventional resonant converters (which do not obtain ZVS operation for the rectifier stage), the ZVS capabilities for the proposed topology offers high efficiency at high frequency operation.

Secondly, the switch control circuitry and timing are greatly simplified because the ZVS conditions happen simultaneously for both inverter switches and rectifier switches independent of the load level. This advantage originates from the characteristic of the proposed converter which completely splits the function of different current components: The first component (resonant tank current, i_{nr}) is used to deliver power through the transformer, while a second component (magnetizing current, i_n) is for achieving ZVS.

Because the resonant tank is tuned at the operating frequency and thus flows sinusoidal current in phase with the control signals, the power delivering current (i_{nr}) becomes close to zero during the dead time, such that it ideally does not affect the zero-voltage switching conditions of the converter; instead, when the load level changes, only the amplitude of the power delivering current varies. On the other hand, magnetizing current does not deliver power from input voltage to output voltage, instead it can be used to achieve ZVS for primary and secondary switches. For a given input voltage at a certain operating frequency, the magnetizing current waveform

TABLE I
SPECIFICATIONS AND COMPONENTS OF THE PROTOTYPE CONVERTER

Specification	
Input Voltage	36 Vdc
Output Voltage	12 Vdc
Output Power	up to 36 W
Switch and Driver IC	
Inverter	GaN switch EPC 2014C Coss $\simeq$ 150 pF
Rectifier	GaN switch EPC 2015C Coss $\simeq$ 700 pF
Driver IC	LM5113 half-bridge driver, TI
Y-rated capacitors	
C_{y1}, C_{y2}	680 pF Y3 X7R 250 Vac (rating) and 1500 Vac (withstand), Johanson Dielectrics
Transformer	
Core	Ferroxcube 3F45 EQ13 with 0.28 mm gap
Number of Turns	6 : 2 (i.e, N=3)
Winding	Planar structure with 62 mil thickness PCB 12 layers of 2 oz copper
L_N	5.8 μH (measured from primary side)
L_{nr}	60 nH (estimated)
C_{nr}	0.22 μF C0G/NPO 50V, TDK
Control	
Microcontroller	TMS320F28035, TI
Switching Frequency	1.4 MHz
Dead time	16.66 ns

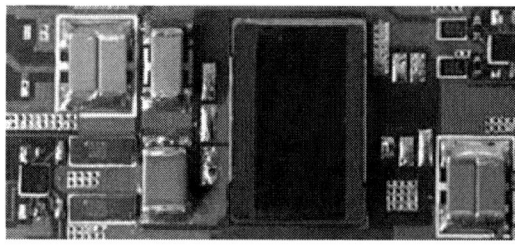

Fig. 4. The figure shows the top and bottom sides of implemented prototype converter. The specifications and components are presented in Table I. The main power stage converter has a size of 1.2 in $\times$ 0.6 in $\times$ 0.17 in, yielding a power density of 300 W/in^3.

is always the same independent of the load current, such that the same dead time is expected (and the dead time is also maintained as voltage levels vary as long as the capacitors are linear). Therefore, even though the load level changes, the converter always achieves ZVS conditions.

Once the switch components, transformer, and specifications of the converter are decided, the switch driving signals can be specified to have dead time as shown in (16). The same switch driving signals, then, can be used for various load levels while

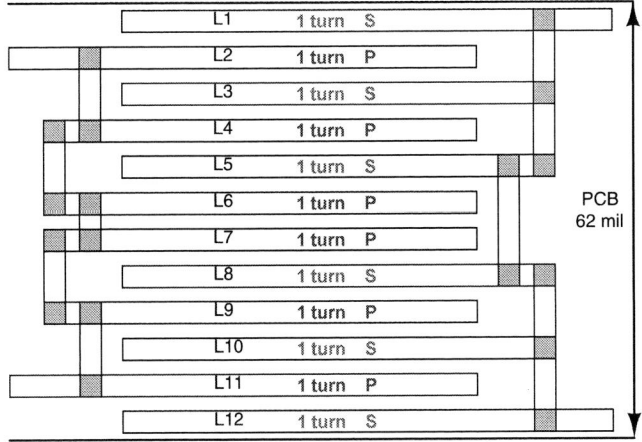

Fig. 5. The figure shows the layer configuration of the planar transformer of prototype converter as shown in Fig. 4. Primary side is wound 6 turns around core by connecting 6 layers of 1 turn/layer copper traces, and secondary side is wound 2 turns by connecting two of 3 parallel copper trace layers. The copper layers of planar transformer is then fully interleaved to reduce proximity effect.

maintaining ZVS operation, such that the converter can operate efficiently for wide load range at high frequency without adding any extra control complexity.

III. IMPLEMENTATION AND EXPERIMENTAL RESULTS

In this section, we present an implementation of the proposed capacitively-aided isolated bus converter, and show the experimental results. The prototype converter is designed and tested from 36 Vdc input voltage to the 12 Vdc load supplying up to 36 W. (These specifications are drawn from the application in which the proposed bus converter will be used, but it should be emphasized that the topology's potential application is not limited to this voltage range, step-down ratio, or power level.)

A. Converter design and tuning

To operate the prototype converter as described in section II, the resonant tank of the converter needs to be tuned at the operating frequency. For the inductance (L_{nr}) of the resonant tank, one can use (the cantilever model) leakage inductance of the transformer or add external inductors for a higher inductance level, but since the leakage inductance is relatively small (i.e., up to 100 nH) compared to the magnetizing inductance, it is hard to measure the exact leakage inductance. As a consequence, it is hard to predict the exact resonant frequency for a selected resonant capacitance (C_{nr}). Thus, iterative tuning steps are necessary to match the resonant frequency and operating frequency in a given design.

Since measuring high frequency current is significantly hard, one simple strategy instead for this iterative tuning is using the voltage across the resonant capacitor. If we assume that sinusoidal current flows through the resonant tank, the voltage across the capacitor is sinusoidal as well, with 90 degree phase shift as shown in (5) and (19). For that reason, we can simply check the resonant tank by measuring the resonant capacitance

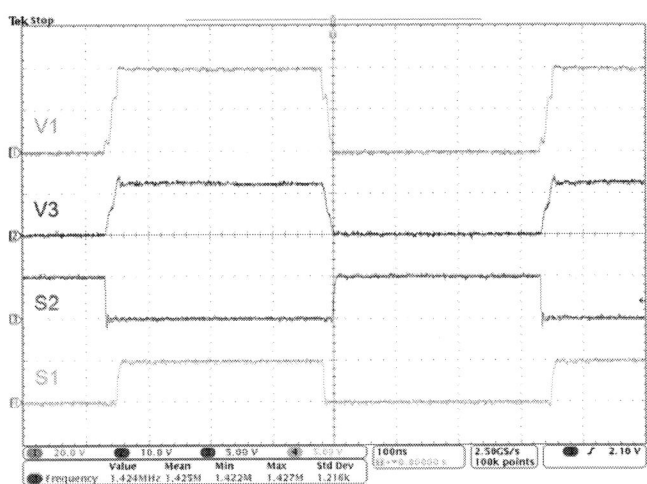

Fig. 6. The figure shows the waveform of the prototype converter when the converter operates from 36 Vdc to 12 Vdc supplying 36 W power. ch1 shows drain-source voltage for primary switch switch (v_1) (20 V/div), ch2 shows drain-source voltage for secondary switch (v_3) (10 V/div), and ch3 / ch4 show gate-source signal voltage for S2 and S1 switches respectively, (5 V/div). As can be seen, the primary and secondary switches achieve zero voltage switching at the same time using same control signals.

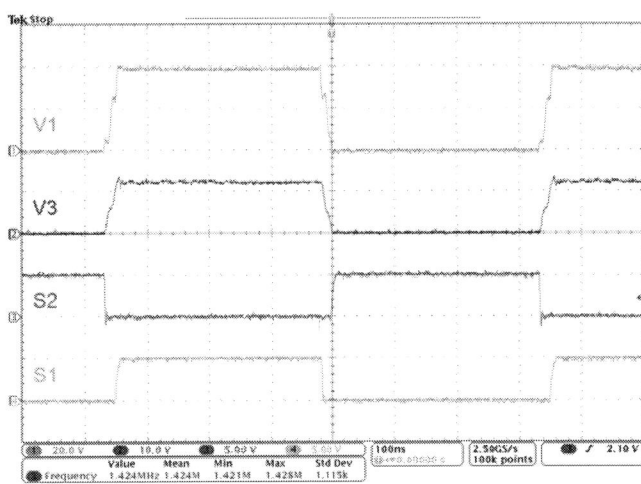

Fig. 7. The figure shows the waveform of the prototype converter when the converter operates from 36 Vdc to 12 Vdc supplying 18 W power. ch1 shows drain-source voltage for primary switch switch (v_1) (20 V/div), ch2 shows drain-source voltage for secondary switch (v_3) (10 V/div), ch3 / ch4 show gate-source signal voltage for S2 and S1 switches respectively, (5 V/div). As can be seen, the primary and secondary switches still keep zero voltage switching condition at different power level even though the same control signals are used.

voltage and checking the phase information compared to the gate-source signals. Then we can either tune the operating frequency of the converter or the resonant capacitance to get the right phase shift between voltage across resonant capacitor and gate-source voltages. To illustrate, Fig. 3 shows example measured waveforms of voltage across the resonant capacitor (ch1 – v_{cr}) and gate-source signals (ch2 – S1 and ch3 – S2) when the resonant tank is adjusted correctly.

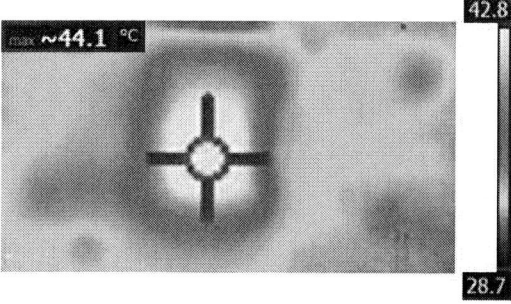

Fig. 8. This figure shows the captured image from infrared temperature camera when the converter operates from 36 Vdc to 12 Vdc at full power, 36 W. Comparing the temperature information to the prototype converter shown in Fig. 4, low loss dissipation is expected from the switch by achieving zero-voltage switching.

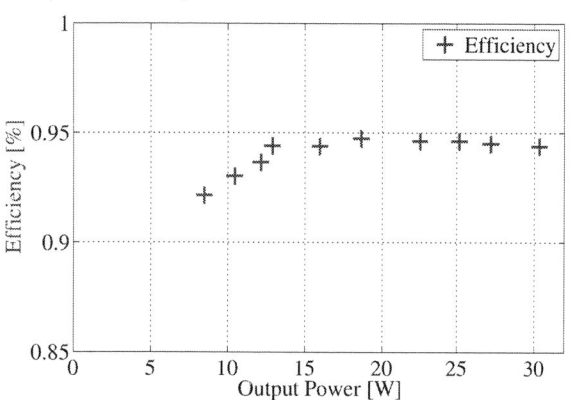

Fig. 9. The figure shows the measured efficiency of the prototype converter across the power range when the converter operates from 36 Vdc to 12 Vdc. The converter shows high efficiency up to 94 % for wide load range because of maintaining zero-voltage switching conditions independent of the load.

B. Prototype Converter Implementation

The selected components for the prototype converter are described in Table I. We used gallium-nitride (GaN) high-electron-mobility transistors (HEMT) driven by LM5113 half-bridge driver ICs. A microcontroller produces two switching signals (S1 and S2) at 1.4 MHz frequency with a 16.66 ns dead time between them, and controls the primary and secondary sides switches with these signals. We selected 680 pF, Y-rated capacitors as these are commercially available and close to the capacitance calculated from (13). (One could also add additional capacitance across the switches or transformer to get a precise match.)

We used 3F45 core material with an EQ13 core structure and distributed 0.28 mm gap (i.e., center post and side legs are all gapped by 0.28 mm), and designed the planar transformer with 5.8 μH magnetizing inductance utilizing 12 layers of 2 oz copper PCB traces. Six turns of copper trace (one turn per layer, and six layers connected in series) are wound on primary side, and two turns of trace coppers (one turn per layer, with two series-connected sets of three parallel copper layers) are fabricated on the 62 mil thickness PCB as illustrated in Fig.

5. As can be seen, primary and secondary windings are fully interleaved to reduce the ac resistance and conduction loss from proximity effect.

We used the leakage inductance of the transformer (approximately 60 nH) as the resonant inductance, and selected a 0.22 μF resonant tank capacitance through the iterative process described above (using a C0G/NPO capacitor with low dissipation factor). From the tuned resonant capacitance and the selected operating frequency, Q-value was set to be around 0.16 at full load. Fig. 4 shows the front and bottom side of the implemented prototype converter with the components described above.

C. Experimental Results

Fig. 6 illustrates the measured converter waveforms when the converter operates from 36 Vdc to 12 Vdc at 36 W load (1.4 MHz switching frequency). Channel 1 and 2 show the drain-source voltage for primary and secondary switches (i.e., v_1 and v_3 respectively), and channel 3 and 4 show the measured gate-source signal waveform for S2 and S1 switches. As can be seen from the measured waveform, the prototype converter presents nice ZVS operation for both primary and secondary switches at the same time. Likewise, Fig. 7 shows the measured waveforms when the converter operates at same input / output voltage but delivers 18 W power with same gate-source signals at the same switching frequency and same dead time. As can be seen, the converter maintains zero-voltage switching conditions independent of the load level as described in section II-C.

Fig. 8 shows a captured image from FLIR E6 infrared camera when the converter operates from 36 Vdc to 12 Vdc at 36 W load. Comparing the temperature information to the photograph of the prototype converter shown in Fig. 4, low temperature rise can be seen near the switches, which is expected since ZVS operation dissipates very little power in the switches. With the aid of zero-voltage switching operation and maintaining ZVS independent of the load level, the prototype converter shows up to 94 % efficiency over a wide load range as illustrated in Fig. 9, yielding a high power density of 300 W/in^3.

IV. CONCLUSION

A new capacitively-aided zero voltage switching (ZVS) technique for isolated bus converters is presented. The proposed technique has significant advantages for achieving ZVS condition and precise control timing for both the inverter and rectifier switches through the use of the Y-rated capacitors interconnecting the inverter and rectifier switch nodes. With properly selected interconnecting capacitors, the ZVS conditions for inverter and rectifier switches occur at the same time, and furthermore (ideally) the required dead time for ZVS operation becomes independent of the load of the converter. These advantages allow to use simple control circuit design, especially when the converter operates at high frequency. The proposed approach is implemented with a prototype converter which operates from 36 Vdc to 12 Vdc, up to 36 W load (i.e., a fixed input voltage and a fixed voltage transformation ratio of $3:1$). Experimental results show ZVS soft-switching and excellent efficiency for a wide load range, and the prototype converter achieves up to 94 % efficiency with a high 300 W/in^3 power density.

REFERENCES

[1] Y. Ren, M. Xu, J. Sun, and F. Lee, "A family of high power density unregulated bus converters," *Power Electronics, IEEE Transactions on*, vol. 20, no. 5, pp. 1045–1054, Sept 2005.

[2] S. Lim, J. Ranson, D. Otten, and D. Perreault, "Two-stage power conversion architecture suitable for wide range input voltage," *Power Electronics, IEEE Transactions on*, vol. 30, no. 2, pp. 805–816, Feb 2015.

[3] R. Miftakhutdinov, "Power distribution architecture for tele- and data communication system based on new generation intermediate bus converter," in *Telecommunications Energy Conference, 2008. INTELEC 2008. IEEE 30th International*, Sept 2008, pp. 1–8.

[4] L. Mweene, C. Wright, and M. Schlecht, "A 1 kw 500 khz front-end converter for a distributed power supply system," *Power Electronics, IEEE Transactions on*, vol. 6, no. 3, pp. 398–407, Jul 1991.

[5] M. Chen, K. K. Afridi, S. Chakraborty, and D. J. Perreault, "A high-power-density wide-input-voltage-range isolated dc-dc converter having a multitrack architecture," in *Energy Conversion Congress and Exposition (ECCE), 2015 IEEE*, Sept 2015, pp. 2017–2026.

[6] P. Yeaman and E. Oliveira, "A high efficiency high density voltage regulator design providing vr 12.0 compliant power to a microprocessor directly from a 48v input," in *Applied Power Electronics Conference and Exposition (APEC), 2013 Twenty-Eighth Annual IEEE*, March 2013, pp. 410–414.

[7] M. Salato and U. Ghisla, "Optimal power electronic architectures for dc distribution in datacenters," in *DC Microgrids (ICDCM), 2015 IEEE First International Conference on*, June 2015, pp. 245–250.

[8] M. Salato, "Datacenter power architecture: Iba versus fpa," in *Telecommunications Energy Conference (INTELEC), 2011 IEEE 33rd International*, Oct 2011, pp. 1–4.

[9] R. Farrington and M. Schlecht, "Intermediate bus architecture with a quasi-regulated bus converter," US Patent 7,787,261, 2010.

[10] M. Schlecht, "High efficiency power converter," US Patent 7,269,034, 2007.

[11] J. Rivas, R. Wahby, J. Shafran, and D. Perreault, "New architectures for radio-frequency dc-dc power conversion," *Power Electronics, IEEE Transactions on*, vol. 21, no. 2, pp. 380–393, March 2006.

[12] B. Baliga, "Power semiconductor device figure of merit for high-frequency applications," *Electron Device Letters, IEEE*, vol. 10, no. 10, pp. 455–457, Oct 1989.

[13] A. Huang, "New unipolar switching power device figures of merit," *Electron Device Letters, IEEE*, vol. 25, no. 5, pp. 298–301, May 2004.

[14] S. Lim, J. Ranson, D. M. Otten, and D. J. Perreault, "Two-stage power conversion architecture for an LED driver circuit," in *Applied Power Electronics Conference and Exposition (APEC), 2013 Twenty-Eighth Annual IEEE*, Mar. 2013, pp. 854–861.

[15] S. Lim, D. Otten, and D. Perreault, "Power conversion architecture for grid interface at high switching frequency," in *Applied Power Electronics Conference and Exposition (APEC), 2014 Twenty-Ninth Annual IEEE*, Mar. 2014, pp. 1838–1845.

[16] D. Perreault, J. Hu, J. Rivas, Y. Han, O. Leitermann, R. Pilawa-Podgurski, A. Sagneri, and C. Sullivan, "Opportunities and challenges in very high frequency power conversion," in *Applied Power Electronics Conference and Exposition, 2009. APEC 2009. Twenty-Fourth Annual IEEE*, Feb 2009, pp. 1–14.

[17] M. Rodriguez, Y. Zhang, and D. Maksimovic, "High-frequency pwm buck converters using gan-on-sic hemts," *Power Electronics, IEEE Transactions on*, vol. 29, no. 5, pp. 2462–2473, May 2014.

[18] L. Xue, Z. Shen, D. Boroyevich, and P. Mattavelli, "Gan-based high frequency totem-pole bridgeless pfc design with digital implementation," in *Applied Power Electronics Conference and Exposition (APEC), 2015 IEEE*, March 2015, pp. 759–766.

[19] B. Macy, Y. Lei, and R. Pilawa-Podgurski, "A 1.2 mhz, 25 v to 100 v gan-based resonant dickson switched-capacitor converter with 1011 w/in3 (61.7 kw/l) power density," in *Applied Power Electronics Conference and Exposition (APEC), 2015 IEEE*, March 2015, pp. 1472–1478.

[20] A. J. Hanson, J. A. Belk, S. Lim, D. J. Perreault, and C. R. Sullivan, "Measurements and performance factor comparisons of magnetic materials at high frequency," in *Energy Conversion Congress and Exposition (ECCE), 2015 IEEE*, Sept 2015, pp. 5657–5666.

[21] R. Steigerwald, "A comparison of half-bridge resonant converter topologies," *Power Electronics, IEEE Transactions on*, vol. 3, no. 2, pp. 174–182, Apr 1988.

[22] V. Vorperian, "Approximate small-signal analysis of the series and the parallel resonant converters," *Power Electronics, IEEE Transactions on*, vol. 4, no. 1, pp. 15–24, Jan 1989.

[23] B. Yang, F. Lee, A. Zhang, and G. Huang, "Llc resonant converter for front end dc/dc conversion," in *Applied Power Electronics Conference and Exposition, 2002. APEC 2002. Seventeenth Annual IEEE*, vol. 2, 2002, pp. 1108–1112 vol.2.

[24] D. Fu, Y. Liu, F. Lee, and M. Xu, "A novel driving scheme for synchronous rectifiers in llc resonant converters," *Power Electronics, IEEE Transactions on*, vol. 24, no. 5, pp. 1321–1329, May 2009.

[25] X. Wu, G. Hua, J. Zhang, and Z. Qian, "A new current-driven synchronous rectifier for series-parallel resonant (llc) dc-dc converter," *Industrial Electronics, IEEE Transactions on*, vol. 58, no. 1, pp. 289–297, Jan 2011.

[26] J. Zhang, J. Wang, G. Zhang, and Z. Qian, "A hybrid driving scheme for full-bridge synchronous rectifier in llc resonant converter," *Power Electronics, IEEE Transactions on*, vol. 27, no. 11, pp. 4549–4561, Nov 2012.

[27] W. Feng, P. Mattavelli, and F. C. Lee, "Pulsewidth locked loop (pwll) for automatic resonant frequency tracking in llc dc-dc transformer (llc -dcx)," *Power Electronics, IEEE Transactions on*, vol. 28, no. 4, pp. 1862–1869, April 2013.

[28] D. Costinett, D. Maksimovic, and R. Zane, "Design and control for high efficiency in high step-down dual active bridge converters operating at high switching frequency," *Power Electronics, IEEE Transactions on*, vol. 28, no. 8, pp. 3931–3940, Aug 2013.

[29] R. De Doncker, D. Divan, and M. Kheraluwala, "A three-phase soft-switched high-power-density dc/dc converter for high-power applications," *Industry Applications, IEEE Transactions on*, vol. 27, no. 1, pp. 63–73, Jan 1991.

[30] R. Erickson and D. Maksimovic, "A multiple-winding magnetics model having directly measurable parameters," in *Power Electronics Specialists Conference, 1998. PESC 98 Record. 29th Annual IEEE*, vol. 2, May 1998, pp. 1472–1478 vol.2.

978-1-4673-9551-9/16 $31.00 © 2016 IEEE

A 10 MHz, 48-to-5 V Synchronous Converter with Dead Time Enabled 125 ps Resolution Zero-Voltage Switching

Alexander Barner[1], Juergen Wittmann[2], Thoralf Rosahl[1] and Bernhard Wicht[2]

[1]Robert Bosch GmbH, Reutlingen, Germany

[2]Robert Bosch Center for Power Electronics, Reutlingen University, Reutlingen, Germany

Email: juergen.wittmann@reutlingen-university.de

Abstract—An integrated synchronous buck converter with a high resolution dead time control for input voltages up to 48 V and 10 MHz switching frequency is presented. The benefit of an enhanced dead time control at light loads to enable zero voltage switching at both the high-side and low-side switch at low output load is studied. This way, compact multi-MHz DCDC converters can be implemented at high efficiency over a wide load current range. The concept also eliminates body diode forward conduction losses and minimizes reverse recovery losses. A dead time resolution of 125 ps is realized by an 8-bit differential delay chain. A further efficiency enhancement by soft switching at the high-side switch at light load is achieved with a voltage boost of the switching node by dead time control in forced continuous conduction mode. The monolithic converter is implemented in an 180 nm high-voltage BiCMOS technology. At $V_{IN} = 48$ V, $V_{OUT} = 5$ V, 50 mA load, 10 MHz switching frequency and 500 nH output inductance, the efficiency is measured to be increased by 14.4 % compared to a conventional predictive dead time control. A peak efficiency of 80.9 % is achieved at 12 V input.

I. INTRODUCTION

At high volume and low power applications below 10 W, e.g. in server or automotive applications, converters are often locally implemented at the point of load and require a high level of integration. Many applications benefit from highly integrated power management ICs. Fully integrated synchronous converters, Fig. 1, save board space and reduce ringing as parasitic capacitance and stray inductance are small. Ringing is an increasing concern for fast switching power stages as it negatively impacts EMC and efficiency and puts additional overvoltage stress onto various devices. DCDC converters with fully integrated power transistors can operate at high switching frequencies [1], [2]. This enables downsizing of passives. However, it also results in higher frequency depending losses [3], [4]. These losses increase even more for applications with larger input voltages of up to 48 V. However, the synchronous converter suffers from additional losses caused by a non-optimal dead time between the turn off and turn on instant of the low-side or high-side switch. Main loss components are (1) cross conduction between high-side and low-side, (2) diode forward conduction, (3) reverse recovery at the low-side body diode and (4) non-zero voltage switching. Compared to converters with asynchronous freewheeling, in which the low-side switch is usually replaced by a Schottky diode, synchronous

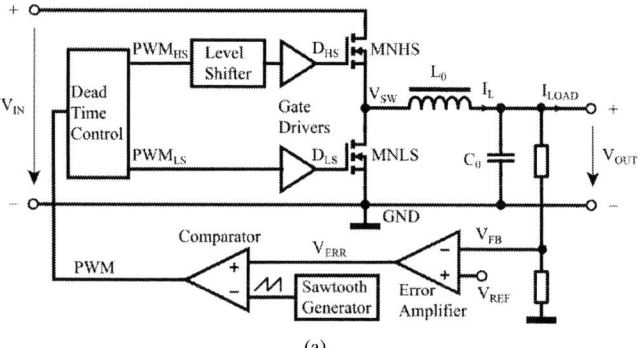

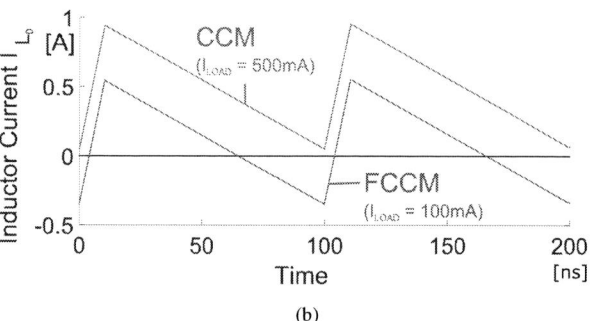

Fig. 1: a) Implemented buck converter with b) inductor current over time for two different load currents of 100mA and 500mA at $V_{IN} = 48$ V and $V_{OUT} = 5$ V.

buck converters achieve a better on-state resistance at the low-side switch at the cost of a higher gate charge losses [1].

In conclusion, fast switching synchronous buck converters are only superior in efficiency if the dead time can be adjusted to the optimal switching point. As the dead time highly depends on the load current and the input voltage, especially at turn off, optimal efficiency requires a dead time regulation. As multi-MHz (>10 MHz) converters require fast switching slopes at very short switching periods, a dead time resolution in the picosecond range is required. Advanced techniques are applied to reduce the dead time depending losses in continuous conduction mode (CCM), e.g. predictive dead time control [5], [6], [7], [8], maximum efficiency point tracking (MEPT) [9]

978-1-4673-9551-9/16 $31.00 © 2016 IEEE

and adaptive dead time control [6]. A further reduction of the size of the main coil in the converter results in higher current ripple. Thus, at low output current, the inductor current gets negative for a portion of the switching period (Fig. 1b). In this case, conventional converters go into discontinuous conduction mode (DCM) with complex loop compensation [10] or they operate in forced continuous conduction mode (FCCM) with excessive losses due to hard switching of the high-side switch. Soft-switching techniques for this case were presented, which require to operate in critical or boundary conduction mode (CRM, BCM) [11]–[13]. In CRM or BCM the converters operate exactly at the boundary between DCM and CCM, i.e. the low-side switch or the freewheeling diode turn off when the inductor current is zero. The parasitic capacitances and the inductor create a voltage ringing at the switching node while the high-side switch is turned on at the maximum amplitude with reduced drain-source voltage and thus reduced switching losses. As the amplitude of the resonating switching node can achieve a maximum of twice the output voltage, soft-switching is only possible for low input voltages or requires more complex tapped inductors. Moreover, CRM can only be achieved with a variable switching frequency. In other concepts, known as Zero-Voltage-Switching Quasi-Square-Wave (ZVS-QSW) converters [14]–[16], the inductor current is significantly forced negative by the low-side switch. When the low side is turned off, the switching node is pulled up by the negative inductor current, the body diode of the high-side switch gets conducting and the high-side switch can be turned on with zero-voltage-switching (ZVS). The disadvantage is that a very high peak current in the inductor is required, and additional losses in the forward conducting body diode of the high-side switch occur.

In addition to a high-resolution dead time control for higher load currents in CCM, this paper proposes a technique to improve the efficiency in a synchronous buck converter at light loads, by utilizing an adjustable high-resolution dead time at switching frequencies as high as 10 MHz, input voltages of up to 48 V and an output voltage of 5 V. At light loads, where the efficiency is significantly reduced in conventional converters, negative inductor current boosts the switching node up to the input voltage, while the dead time controls the negative peak current of the inductor and thus the amplitude of the switching node. The amplitude reaches exactly the input voltage and the high-side switch can be turned on with ZVS without body diode conduction. High converter efficiency can be obtained due to the soft-switching over a wide range of input voltage and load current, even at a fixed switching frequency. At higher load, the converter operates in continuous conduction mode (CCM) with predictive dead time control as presented in [5]. In this work, the benefit of high-resolution dead time control for light load conditions is investigated and compared to the dead time control used for CCM over the full load range.

In Section II, the losses in a synchronous converter are discussed and the effect of the dead time to these losses are explained. The proposed concept is presented in Section III. Experimental results are shown in Section IV.

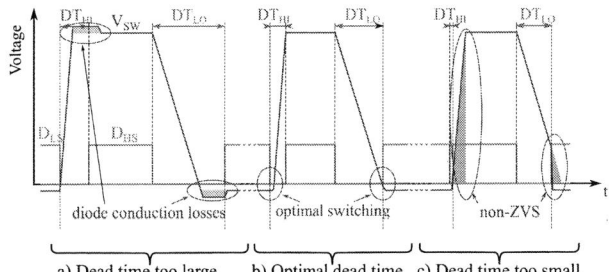

Fig. 2: Switching states at the switching node of the dead time controller.

II. DEAD TIME DEPENDING LOSSES

A. The Buck Converter

In the synchronous DCDC buck converter of Fig. 1, a high-side power switch and an inversely controlled low-side switch (MNHS and MNLS), generate a pulsed voltage at the switching node V_{SW}, which is filtered by L_0 and C_0 to obtain a DC output voltage V_{OUT}. Error amplifier, sawtooth generator and comparator generate the pulse-width modulated signal PWM. PWM is the input signal for the proposed dead time control, in which the low-side and high-side control signals PWM_{LS}, PWM_{HS} are derived. To operate at high input voltages in the multi-MHz range, high-speed level shifter and gate driver concepts are used to achieve gate control pulses of a few nano-seconds and thus, high-conversion ratios [1], [2].

B. Losses Influenced by Dead Time Control

Two dead times have to be considered, the dead time DT_{HI} between the low-side switch turn off and high-side switch turn on (rising edge of V_{SW}), as well as the dead time DT_{LO} between the high-side switch turn off and lowside switch turn on (falling edge of V_{SW}). Fig. 2 shows the switching behavior for three cases in FCCM for light load conditions. Both dead times can be either too large (Fig. 2a), at optimum (Fig. 2b) or too short (Fig. 2c). Is the dead time too large, the body diode of the high-side and low-side switch gets forward biased due to the forced inductor current of L0, which is negative in FCCM at light load when the low-side switch turns off, and positive when the high side switch turns off. Thus, V_{SW} is pulled high or low, respectively, to a positive or negative diode forward voltage V_F. This leads to excessive losses depending on the output current I_{Load}, the switching frequency f_{SW}, and the respective dead time $DT_{LO/HI}$. At the end of DT_{HI}, when the high-side switch turns on, the current commutates from the body diode to the high-side or low-side switch, respectively. The optimal switching point (Fig 2b) is achieved, when the high-side or low-side switch, respectively, turns on a very short instant before the body diode of the low-side or high-side switch starts conducting. In this case, the losses are minimal as the low-side and high-side switch are turned on with ZVS. This can be achieved, if the parasitic capacitances at the switching node are fully charged and discharged by the inductor current while the low-side or high-side switch,

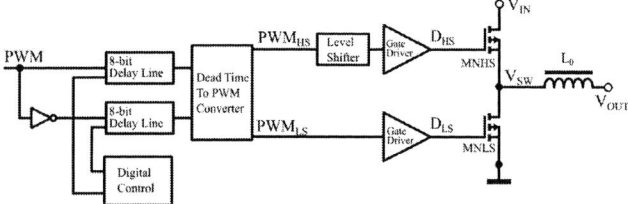

Fig. 3: Proposed dead time generation at the power stage of a synchronous buck converter.

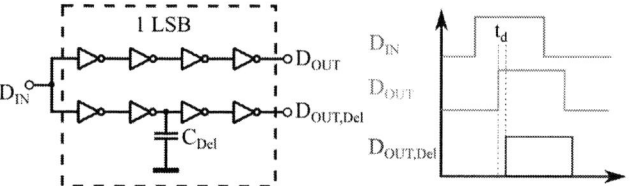

Fig. 4: Proposed dead time generation to overcome the technology limiting parameters.

respectively, is still turned off. Dead times, which are too small or even negative (Fig. 2c), lead to high power losses, as the low-side switch is not turned on with ZVS (hard switching), or cross-conduction occurs as both switches are turned on at the same time. While Fig. 2a represents a typical operating point of ZVS-QSW converters, the extended dead time control for light load conditions, analyzed in this work, aims to achieve the optimal operating point shown in Fig. 2b.

III. PROPOSED CONCEPT

A. General Description

Fig. 3 shows the power stage of the synchronous buck including the proposed dead time generation. The dead times are realized using two digitally adjustable delay lines for each the turn on (DT_{HI}) and the turn off dead time (DT_{LO}). 8-bit delay lines are adjusted by a digital control circuit.

Dead time control with high resolution in the picosecond range is required because of fast rising and falling transitions at the switching node, as the parasitic capacitances, charged and discharged by the inductor current, are low in integrated power stages. In addition, picosecond resolution keeps the influence of the dead time on the duty cycle as small as possible, as adjusting the dead times causes a change in the duty cycle. In multi-MHz converters, this influence becomes dominant. At 10 MHz switching frequency, a dead time adjustment of 1 ns results in nearly a duty cycle change of 1 %. A continuously adjustment or regulation of the duty cycle would cause a perceivable increase of the output voltage ripple at V_{OUT}. Conventional concepts of ZVS-QSW converters with fixed dead time or poor dead time resolution use a larger dead time to assure that the body diodes of the low-side and high-side switch always conducts before the switch is turned on. State-of-the-art integrated applications achieve a dead time resolution ≥ 1 ns [6], [7].

In this paper, the dead time is generated by 8-bit digitally adjustable delay elements with a dead time resolution of 125 ps. The high resolution is achieved by a differential dead time generation, shown in Fig. 4 [5]. Two parallel inverter chains are installed. A delay capacitance C_{Del} in one of the two branches increases the propagation delay slightly with respect to the other branch. The difference of the propagation delay t_d (Fig. 4) between the two branches of the differential delay chain is used to generate the dead time. This way, the adjustable dead time can be smaller than the technology related minimum inverter propagation delay. The time difference

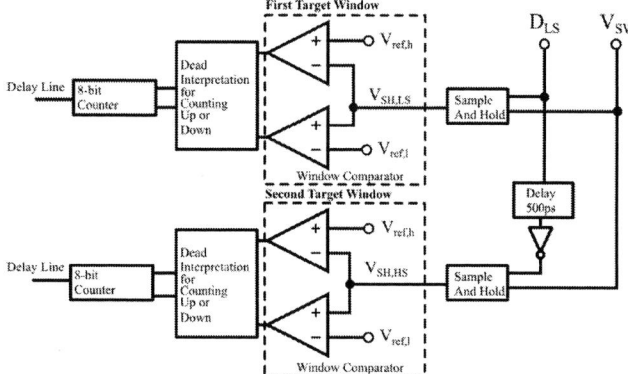

Fig. 5: Proposed dead control for continuous conduction mode.

between D_{OUT} and $D_{OUT,Del}$ is small, but each of the signals has a large absolute propagation delay compared to the PWM input D_{IN}. This delay is still small enough to not influence the stability of the output voltage regulation loop.

In the buck converter, presented in this paper, two operating modes are implemented, further explained in Section III-B and C below: (1) the delay line bits can be provided externally by a serial interface to adjust an optimal soft-switching (Fig. 2b) and to study the implementation of an extended dead time control for light loads when negative inductor currents are possible; (2) the digital circuit represents the conventional dead time control for higher load conditions suitable for CCM [5].

B. Dead Time Control for Light Load Conditions

For negative inductor currents in FCCM (Fig. 1b), the power stage can be considered as a boost converter, formed by the low-side switch and the inductor, while V_{OUT} acts as the boost converter's input and V_{SW} as its output. After turning off the low-side switch, the inductor current charges the switching node up to V_{IN}. The key idea is to control the dead time to switch off the low-side when the current is at its optimal negative value such that V_{SW} reaches V_{IN} exactly at the instant when the high-side switch is turned on. Compared to CCM and DCM, losses can be reduced by soft switching at turn on of the high-side switch [10], [17]. The operating mode of the converter with dead time control for light-load is similar to an operating mode in forced continuous conduction mode, which simplifies the loop compensation compared to an operation in discontinuous mode.

C. Dead Time Control for CCM

To achieve an optimal efficiency over a wide load current range, the predictive mixed signal dead time control presented in [5] is used for CCM (positive inductor currents). For the dead time adjustment of low-side ZVS, the voltage V_{SW} at the falling transition of the switching node is sampled when the low side switch is turned on. The sampled voltage is analyzed by a window comparator to be within a defined target window (first target window, see Fig. 5). If the analyzed voltage is above the target window, the low-side switch is turned on with a positive drain-source voltage (no ZVS). The 8-bit counter value and in consequence DT_{LO} is increased. If the sampled voltage is too low, the turn on of the low-side switch happens close or during body diode conduction (Fig. 2a). At the next switching cycle, the dead time is increased and the turn on event, and thus the sampled voltage, should be closer to the target window. A steady-state is achieved, when the sampled voltage at the switching node is within the target window, which has a width from zero to 500 mV. Turning on the high-side switch in CCM is controlled accordingly, using a second sampling stage (Fig. 5). The voltage at the switching node is sampled 500 ps after the low side switch is turned off, while the sampled voltage is compared by another window comparator to a second target window. The delay lines are adjusted such that the sampled voltage is regulated to be close to zero volts in the second target window (Fig. 5). Is the sampled voltage too low, the high-side switch is turned on too late and the low-side body diode conducts. Is the sampled voltage too high, the high-side switch is turned on too early and cross conduction occurs (Fig. 2c).

For the enhanced dead time control for light load conditions (Section III-B), this control technique could be expanded to implement a maximum point detection circuit of the switching node V_{SW} [18], which controls the counter for DT_{HI} in such a way that the high-side switch is turned on close to the maximum of V_{SW} and thus with a minimum drain-source voltage.

IV. EXPERIMENTAL RESULTS

The dead time generation and the synchronous power stage are implemented in a 180 nm high-voltage BiCMOS technology [5]. The die (Fig. 6a) with an active area of around 2.5 mm² is directly bonded to the PCB, Fig. 6b. The dead times at both switching events are adjusted by the implemented dead time generator. Measurements are performed with $L_0 = 500$ nH and $C_0 = 10\,\mu$F. Fig. 7 shows the switching node voltage V_{SW} and the gate drive signals D_{HS} and D_{LS} for an optimal dead time for 150 mA, where ZVS is achieved at both switching events. An appropriate DT_{HI} enables to charge the switching node up to V_{IN} for ZVS. Fig. 8 shows the efficiency of the proposed synchronous buck converter design in dependence of load current at 48 V input, 5 V output and 10 MHz switching frequency. A peak efficiency of 64.3 % is achieved. Fig. 8 also shows the efficiency for the converter operating with conventional dead time control (Section III-C) (i.e. for minimized cross-conduction and reverse recovery losses), with ZVS at

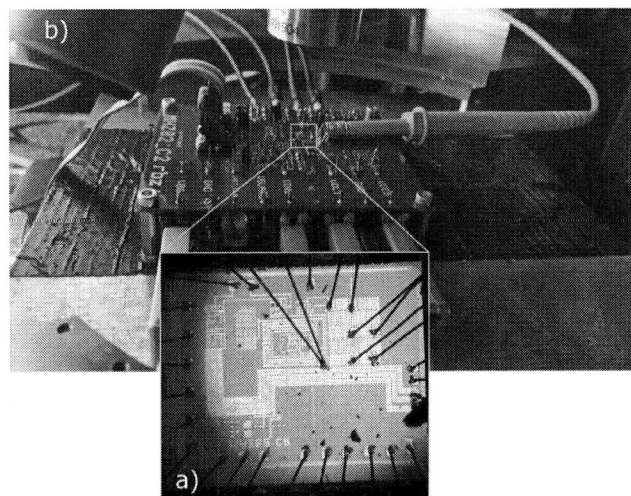

Fig. 6: Microphotograph of the testchip, directly bonded to the PCB.

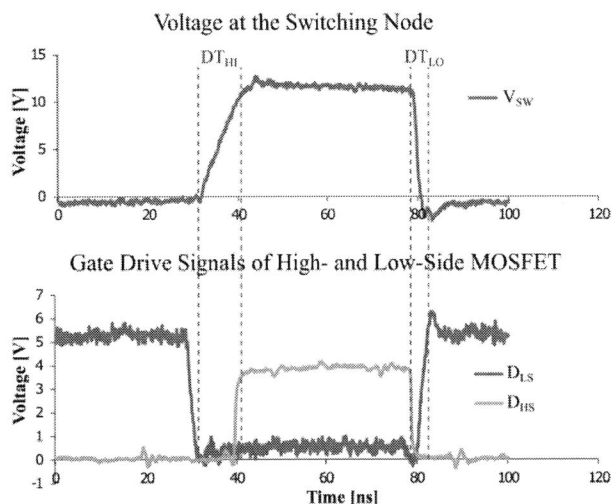

Fig. 7: Measured voltages at the switching node and the corresponding gate drive signals at light load with enhanced dead time control achieving ZVS at both switching events.

the low-side switch and hard switching with minimum dead time at the high-side switch [5]. The efficiency improves by up to 14.4 % for low output current. For a conversion from 12 V input to 5 V output, Fig. 9, soft switching at both switches leads to a high converter peak efficiency of 80.9 %, with an improvement to the conventional dead time control [5] of up to 4.7 %. The proposed dead time concept with soft switching at both switches is most effective at large input voltages demonstrated for 48 V with high switching frequencies above 10 MHz and large load range up to 500 mA.

V. CONCLUSION

A technique to reduce switching losses at low output loads by using dead time control with a resolution of 125 ps

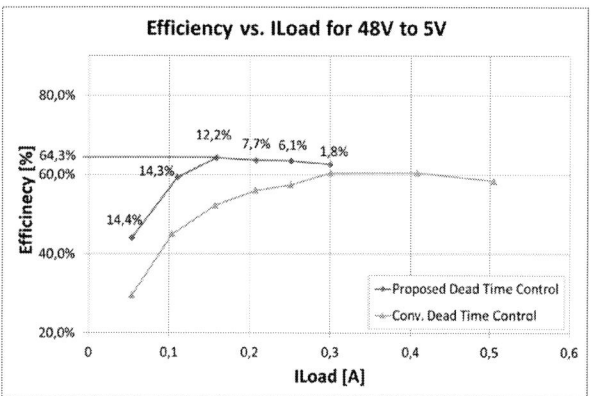

Fig. 8: Efficiency measurements from 48 V to 5 V with soft switching in FCCM and minimal dead time by a predictive dead time control.

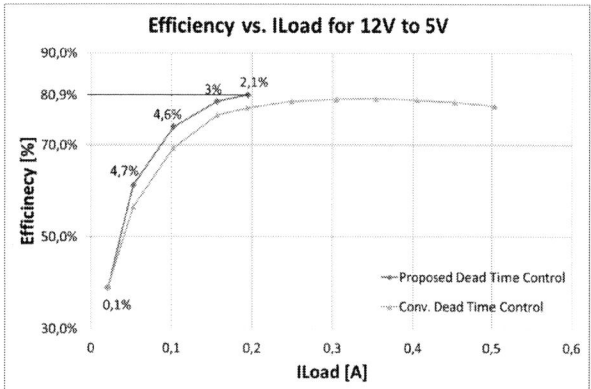

Fig. 9: Efficiency measurements from 12 V to 5 V with soft switching in FCCM and minimal dead time by a predictive dead time control.

is presented. The dead time control is used to charge the switching node up to V_{IN} before the high-side switch turns on. This enables ZVS for the high-side switch. Similarly, ZVS at the low-side switch is accomplished. The dead time control is implemented in a test chip with a fully integrated synchronous buck converter in an 180 nm high-voltage BiCMOS technology, suitable for 10 MHz operation. The dead time resolution is achieved by offsetting the delay in one branch of an 8-bit differential delay chain. This way, the proposed circuit implementation is well suitable to reduce switching losses in multi-MHz DCDC converters. The efficiency gain due to ZVS is predominant at high V_{IN}. Experimental results confirm an efficiency increase compared to conventional dead time control of up to 14.4 % for a 48 V to 5 V conversion at 10 MHz switching with a main inductor of 500 nH. At 12 V input, a peak efficiency of 80.9 % is achieved The efficiency increases by up to 4.7 %. Besides loss reduction, the proposed soft-switching concept significantly reduces ringing [19] due to parasitic inductances.

REFERENCES

[1] J. Wittmann and B. Wicht, "Mhz-converter design for high conversion ratio," in *Power Semiconductor Devices and ICs (ISPSD), 2013 25th Int. Symp. on*, May 2013, pp. 127–130.

[2] J. Wittmann, T. Rosahl, and B. Wicht, "A 50v high-speed level shifter with high dv/dt immunity for multi-mhz dcdc converters," in *European Solid State Circuits Conference (ESSCIRC), ESSCIRC 2014 - 40th*, Sept 2014, pp. 151–154.

[3] T. Y. Man, P. K. T. Mok, and M. Chan, "Analysis of switching-loss-reduction methods for mhz-switching buck converters," in *IEEE Int. Conf. on Electron. Devices and Solid-State Circuits*, Dec 2007, pp. 1035–1038.

[4] X. Wang, J. Park, E. Van Brunt, and A. Huang, "Switching Losses Analysis in MHz Integrated Synchronous Buck Converter to Support Optimal Power Stage Width Segmentation in CMOS Technology," in *Energy Conversion Congress and Expos. (ECCE), 2010 IEEE*, Sept. 2010, pp. 2718 –2724.

[5] J. Wittmann, A. Barner, T. Rosahl, and B. Wicht, "A 12v 10mhz buck converter with dead time control based on a 125 ps differential delay chain," in *European Solid State Circuits Conference (ESSCIRC), ESSCIRC 2014 - 40th*, Sep 2015, p. in Press.

[6] S. Mappus, *Predictive Gate Drive^{TM} Boosts Synchronous DC/DC Power Converter Efficiency*, ser. Texas Instruments Application Report SLUA281, Apr 2003.

[7] G. Maderbacher, T. Jackum, W. Pribyl, M. Wassermann, A. Petschar, and C. Sandner, "Automatic dead time optimization in a high frequency dc-dc buck converter in 65 nm cmos," in *ESSCIRC (ESSCIRC), 2011 Proceedings of the*, Sept 2011, pp. 1919–1922.

[8] W. Yan, C. Pi, W. Li, and R. Liu, "Dynamic dead-time controller for synchronous buck dc-dc converters," in *Electronics Letters (Volume:46 , Issue: 2)*, Jan 2010, pp. 164 – 165.

[9] J. Abu-Qahouq, H. Mao, H. Al-Atrash, and I. Batarseh, "Maximum efficiency point tracking (mept) method and digital dead time control implementation," in *IEEE Transactions on Power Electronics, VOL. 21, NO. 5*, Sept 2006, pp. 1273–1281.

[10] S. Cuk and R. Middlebmok, "A general unified approach to modelling switching dc-to-dc converters in discontinuous conduction mode," in *IEEE Power Electronics Specialists Conference (PESC'77)*, Jun 1977, pp. 36–57.

[11] J.-H. Park and B.-H. Cho, "The zero voltage switching (zvs) critical conduction mode (crm) buck converter with tapped-inductor," *Power Electronics, IEEE Transactions on*, vol. 20, no. 4, pp. 762–774, July 2005.

[12] J. Wang, F. Zhang, J. Xie, S. Zhang, and S. Liu, "Analysis and design of high efficiency quasi-resonant buck converter," in *Electronics and Application Conference and Exposition (PEAC), 2014 International*, Nov 2014, pp. 1486–1489.

[13] C. Y. Chinag and C. L. Chen, "Zero-voltage-switching control for a pwm buck converter under dcm/ccm boundary," in *IEEE Trans. Power Electronics*, Sep 2009, pp. 2120–2126.

[14] S. Chen, O. Trescases, and W. T. Ng, "Fast dead-time locked loops for a high-efficiency microprocessor-load zvs-qsw dc/dc converter," in *Electron Devices and Solid-State Circuits, 2003 IEEE Conference on*, Dec 2003, pp. 391–394.

[15] X. Zhou, P.-L. Wong, P. Xu, F. Lee, and A. Huang, "Investigation of candidate vrm topologies for future microprocessors," *Power Electronics, IEEE Transactions on*, vol. 15, no. 6, pp. 1172–1182, Nov 2000.

[16] V. Vorperian, "Quasi-square-wave converters: Topologies and analysis," in *IEEE Trans. Power Electronics*, Apr 1988, pp. 183–191.

[17] T. Y. Man, P. K. T. Mok, and M. Chan, "A 0.9-v input discontinuous-conduction-mode boost converter with cmos-control rectifier," in *IEEE J. Solid-State Circuits*, Sep 2008, pp. 2036–2046.

[18] T. Funk, J. Wittmann, T. Rosahl, and B. Wicht, "A 20 v, 8 mhz resonant dcdc converter with predictive 1 ns resolution zero voltage switching," in *Circuits and Systems (ISCAS), 2015 IEEE International Symposium on*, May 2015, pp. 1742 – 1745.

[19] H. Chung, S. Hui, and K. Tse, "Reduction of power converter em1 emission using soft-switching techniques," in *Electromagnetic Compatibility, IEEE Transactions on (Volume:40 , Issue: 3)*, Aug 1998, pp. 282 – 287.

Plug-and-Play Electronic Capacitor for VRM Applications

Or Kirshenboim, *Student Member, IEEE*, Alon Cervera, *Student Member, IEEE*, Bar Halivni, Eli Abramov, *Student Member, IEEE*, and Mor Mordechai Peretz, *Member, IEEE*

The Center for Power Electronics and Mixed-Signal IC
Department of Electrical and Computer Engineering
Ben-Gurion University of the Negev
P.O. Box 653, Beer-Sheva, 8410501 Israel

orkir@post.bgu.ac.il   cervera@bgu.ac.il   barhalivni@gmail.com   eliab@post.bgu.ac.il   morp@ee.bgu.ac.il
http://www.ee.bgu.ac.il/~pemic

Abstract - **This paper introduces a new plug-and-play transient suppression unit (TSU) to enhance the performance and reduce the overall volume of voltage regulator modules (VRMs). The TSU acts as an electronic capacitor that is realized by switched-capacitor technology, mimics an increased capacitance during load transients. The unit connects in parallel to any existing tightly-regulated power supply without affecting the performance, and does not require any changes or interference in the design of the VRM. The resultant dynamic performance for load transients is significantly improved while the steady-state precision of the original design is intact. Furthermore, the unit is fully independent and is connected at the load-side of the converter, and as a result does not affect the input filter. The operation of the electronic capacitor is verified on a 30W, 12V-to-5V commercial buck converter evaluation module, demonstrating a near-ideal transient recovery with reduced output voltage deviation and settling time.**

I. INTRODUCTION

A target feature in present-day VRMs is the ability to maintain a well-regulated, virtually constant, output voltage under wide range of load changes while maximizing power density. A key consideration to achieve this goal is the size of the passive components that prohibits full integration of the solution. Many modern applications raise the switching frequency and employ multi-phase converters to enhance the transient response that allow integration of the inductor [1]-[3]. On the other hand, sizing of the output capacitor in VRM applications primarily depends on the load transient magnitude and rate, and therefore consumes a significant portion of the PCB area [4].

To minimize the effect of load transient, several approaches to enhance the control bandwidth that result in saturation of the duty ratio have been described. Methods such as current-programmed mode control and its derivatives [5]-[10], time-optimal and minimum-deviation control [11]-[18], have shown transient response with virtually the smallest possible voltage deviation, restricted only by the inductor current slew-rate. The main limitation of these methods is the weak regulation during unloading transient due to the high input-to-output conversion ratio.

State-of-the-art solutions that exceed the performance of the time-optimal control method, especially for unloading transients, propose several circuit extensions in order to increase the inductor current slew-rate, either by internal changes to the topology [4], [19], [20], addition of a fast auxiliary circuits in parallel to the main converter [21]-[25], or by connecting an auxiliary unit at the load side [26], [27]. These solutions often require specially-tailored controller (sometimes combined with a digital design) or multi-mode compensation schemes.

The additional layers of complexity are the prominent reason for the lack of absorbance, of such promising technology, in commercial VRM applications. As evident, the majority – if not all, VRM solutions rely on the well-established analog compensators to guarantee reliability, performance and above all reduced complexity and cost. It would be extremely advantageous, and potentially better absorbed by the industry, if the auxiliary transient suppression unit (TSU) could be integrated as an add-on unit to the VRM without the need to interfere, replace or modify the original design.

The objective of this paper is therefore to introduce a plug-and-play TSU for VRM applications *that trades the output capacitance by a silicon-based solution without affecting the steady-state operation, the originally designed compensation network and the input filter*. As detailed in Fig. 1, the TSU comprises a bi-directional current source that is realized by a gyrator resonant switched-capacitor converter (GRSCC) [28] that connects in parallel to the output capacitor, and a transient response accelerator that connects in parallel to the output of the error-amplifier. Since the new Electronic Capacitor is active only during load transients, the steady-state precision is not jeopardized and the design procedure for the buck converter is intact. In addition, the GRSCC which implements the TSU does not require a magnetic element, making it ideal for integration, simple and cost-effective.

The rest of the paper is organized as follows: Section II describes the transient suppression concept and details the operation. Section III briefly reviews the operation of the GRSCC in the context of an electronic capacitor and provides a simulation case study. Experimental verification is presented in Section IV. Section V concludes the paper.

II. TRANSIENT SUPPRESSION CONCEPT

A key factor for assisting the recovery of the main converter from a load transient is the capability of the auxiliary circuit to rapidly sink or source the current mismatch between the new load current and the main inductor current. To analyze the required behavior and control mechanism of the auxiliary TSU, an idealized bi-directional current source that is connected to the output terminals of the buck converter can be assumed as depicted in Fig. 2.

A. Principle of Operation

The description is aided by Fig. 3 which shows the waveforms for consecutive loading and unloading transients with a magnitude of ΔI_{out} and the flowchart of Fig. 4. Transient operation is initiated upon its detection by the upper or lower comparators (with reference voltage assignment of $V_{ref,H}$ and $V_{ref,L}$, respectively), indicating a charge mismatch in the output capacitor. Upon detection of a transient, two actions are simultaneously performed. The duty ratio is saturated to either maximum or minimum, depending on the transient type, and the current source is enabled and sinks or sources with a constant magnitude of I_{max} (the converter's nominal current).

Since i_{aux} is higher than the current mismatch between i_{buck} and i_{load}, the output voltage returns to the steady-state value. This is detected by an additional comparator with voltage reference set to $V_{ref,M}$. At this point, the auxiliary current source is halted while the duty ratio continues to be saturated. In case that a current mismatch still exists, the output voltage moves away from the steady-state value, crossing the comparator threshold again, re-triggering the auxiliary circuit. This procedure continues until the steady-state comparator (with threshold $V_{ref,M}$) is triggered twice (or triggered and remains in the new state), which indicates that charge balance is achieved, i.e. $i_{buck} \approx i_{load}$; $v_{out} = V_{ref,M}$, and the duty ratio saturation is discontinued.

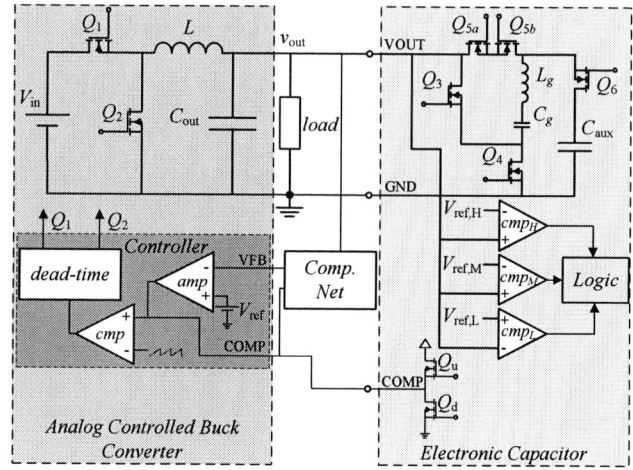

Fig. 1. Electronic capacitor circuit connected to a buck converter controlled by an analog controller.

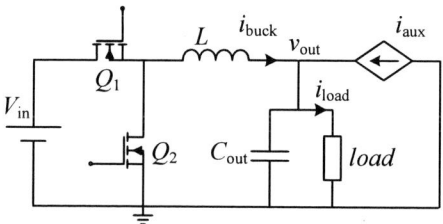

Fig. 2. Simplified circuit with the auxiliary circuit modelled as a controlled current source, demonstrating the current relationships towards the load.

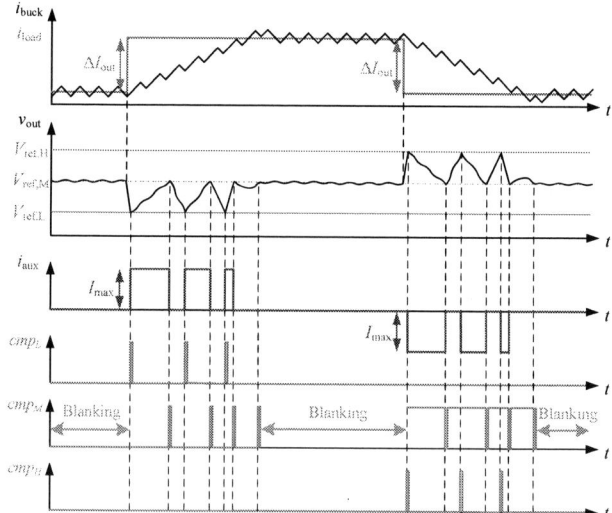

Fig. 3. Typical waveforms of loading and unloading transients with the electronic capacitor.

B. Transient Response Accelerator

To successfully recover from a load transient, both the output voltage and the inductor current must move to the new steady-state operating point. This typical feature presents a challenge for so-called perfect transient response where virtually zero output voltage deviation is evident. In this study, since the auxiliary TSU acts as an infinite capacitor and is connected *in-situ* to a tightly-compensated voltage regulator, the error signal at the error amplifier (E/A) terminals is zero. As a result, the inductor current would not ramp up or down to the new steady-state point as depicted in Fig. 5.

To overcome this challenge, a third port of the electronic capacitor with a response accelerator unit is added as shown in Fig. 1. It consists of two complementary transistors, for pull-up and pull-down, and connects to the output of the analog controller's E/A (This port is readily available in most external-compensation designs). By activating the pull-up and pull-down transistors during load transients, the desired duty ratio saturation is obtained. Since zero error signal is maintained at the E/A terminals during this operation, the E/A can be momentarily bypassed by the transient response accelerator and then restored without any concerns for integrator windup or compensation reset [29]. It should be further noted that this approach does not interfere with the compensation loop and avoids the need to redesign the network - an advantage over many dual-mode applications.

C. Comparators Thresholds Settings

As described earlier, the thresholds values dictate the worst-case voltage deviation that is allowed. Since the operation of the electronic capacitor is enabled during transient events, it is necessary to set the thresholds so that the steady-state voltage ripple avoids false transient triggering. As extra measure to avoid false detection is to assure large margins between the thresholds to accommodate the voltage ripple that is caused by current sinking or sourcing of the auxiliary circuit. The larger value of the two is caused by the latter since the current source is designed to sink or source the nominal current.

The largest voltage swing by auxiliary current source occurs when the current mismatch is small ($i_{buck} \approx i_{load}$), that is:

$$V_{ref,H} - V_{ref,L} \geq Q_g / C_{out} = 8 V_{out} C_g / C_{out} , \quad (1)$$

where Q_g is the charge delivered from the auxiliary circuit during a single discharge cycle, C_g is the GRSCC resonant tank capacitor and $V_{ref,M}$ is the steady-state value:

$$V_{ref,M} = \frac{V_{ref,H} + V_{ref,L}}{2} . \quad (2)$$

The implementation of the detection circuit can be realized as shown in Fig. 6. It comprises two voltage divider ladders, one for the reference voltage setting and the other for the output voltage measurement. Using this configuration, the reference voltages can be designed by:

$$V_{ref,H} = \frac{R_2 + 2R_3}{R_1 + R_2 + 2R_3} V_{out},$$

$$V_{ref,M} = \frac{R_2 + R_3}{R_1 + R_2 + 2R_3} V_{out}, \quad (3)$$

$$V_{ref,L} = \frac{R_2}{R_1 + R_2 + 2R_3} V_{out}.$$

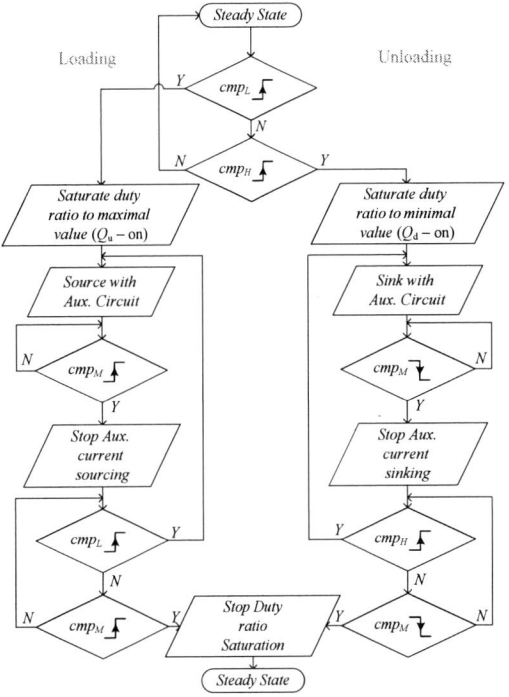

Fig. 4.   Flowchart of the electronic capacitor circuit operation algorithm.

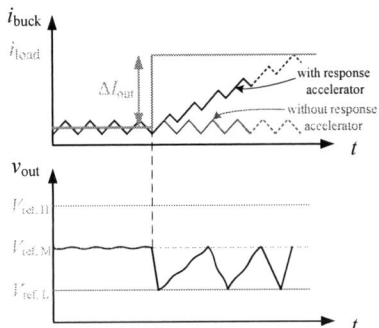

Fig. 5.   Inductor current response with and without the response accelerator.

III. AUXILIARY CURRENT SOURCE REALIZATION

A. Gyrator Resonant Switched Capacitor Converter

An auxiliary current source can be realized by either linear-mode regulators [24], switched-inductor converters [22], [25], [26] or by switched-capacitor converters [27]. The latter is adopted in this study in the form of the GRSCC [28] which has been found the most suitable for the application. It does not require a magnetic element, can be operated at high frequencies with soft-switching and maintains high efficiency over a wide and continuous step-up/down conversion ratio. Furthermore, it has a bi-directional current sourcing behavior and is able to react *immediately* to create current step response with bandwidth of up to half its maximal switching frequency [30].

A voltage doubling variation of the GRSCC has been implemented in this study and is shown as the auxiliary circuit in the electronic capacitor of Fig. 1. It is structured relying on a classical voltage multiplying resonant switched capacitor converter topology, shifting the GRSCC optimal efficiency point from V_{out} to $V_{aux} = 2V_{out}$. The main reason for the selection of this topology is to increase the power density of the auxiliary storage capacitor C_{aux} by increasing its rated voltage, but without adding voltage stress to the transistors. Another advantage of the doubling realization is that the desired current, i.e. I_{max}, can be obtained by a higher characteristic impedance of the resonant network. This implies that higher target efficiency of the GRSCC can be obtained for a given loop resistance.

The GRSCC is resonant in nature and can be completely halted at zero-current after each cycle, as can be observed in Fig. 7. As a result, the nominal current can be resumed within one cycle. In the context of this study, this zero-order current-stepping capability enables the GRSCC to be used as the auxiliary current source unit. Moreover, there is no limitation to scalability as the resonant tank values can be determined for any desired V_{out} and operating frequency. The bridge configuration also guarantees that the maximum stress on any given switch will be around V_{out}, which translates into small area requirements of the power transistors.

B. Simulation Case Study

A simulation of the GRSCC as an auxiliary current source assisting a buck converter to handle a loading transient is depicted in Fig. 8. The buck is a 12V to 5V converter, controlled by an analog controller with a type III compensation network as depicted in Fig. 9. The thresholds of the electronic capacitor's comparators are set to be ±50mV off the output voltage steady-state value and the GRSCC is designed to source a current of 5A to the output terminals of the buck converter.

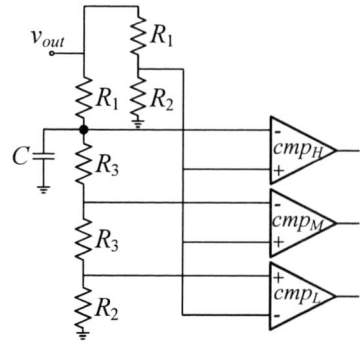

Fig. 6. Comparators references voltage generating circuit.

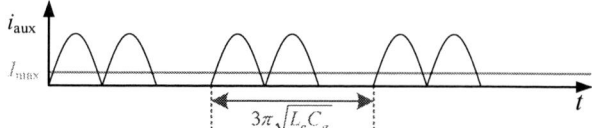

Fig. 7. GRSCC output current waveform.

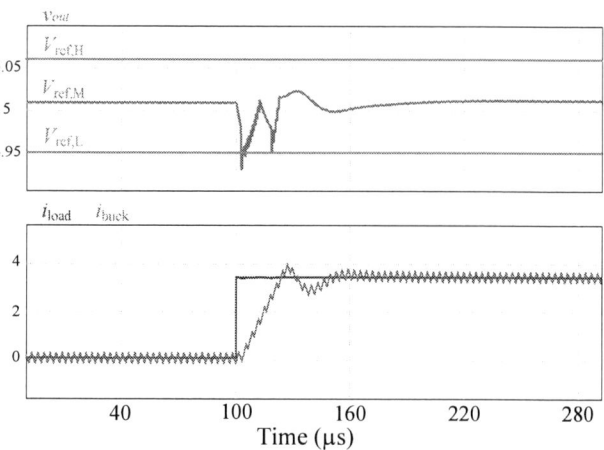

Fig. 8. Simulated loading transient response of a buck converter assisted by the electronic capacitor circuit.

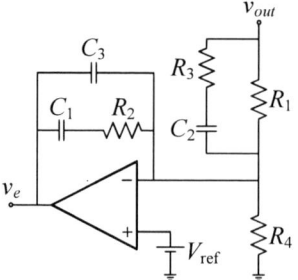

Fig. 9. Type III compensation network.

A loading transient from 0A to 3.5A causes the output voltage to drop and cross $V_{ref,L}$, triggering cmp_L and a loading transient is detected by the electronic capacitor. The transient response accelerator is activated, the duty ratio is saturated to the maximal value, and the GRSCC sources 5A. Since $i_{aux} + i_{buck} > i_{load}$, the output voltage rises and crosses $V_{ref,M}$, halting the GRSCC current sourcing. At this point, the load current is still higher than the inductor current and the output voltage drops again, crossing $V_{ref,L}$ once more, re-triggering the GRSCC. After the output voltage has risen and when it crossed $V_{ref,M}$ for the second time, the GRSCC operation is halted again, the inductor current is higher than the load current and charge balance is achieved. At this point, the end of transient is detected, the transient response accelerator is deactivated and steady-state operation is resumed without any need for compensator reset or update.

IV. EXPERIMENTAL VERIFICATION

In order to validate the operation of the electronic capacitor concept and to demonstrate the plug-and-play feature of the solution, an off-shelf evaluation module (EVM) of a 30W 12-to-5V analog-controlled synchronous buck converter from Texas Instruments (TPS40055) was selected to serve as the already compensated (type III scheme) and optimized voltage regulator. The electronic capacitor module developed in this study was connected as an add-on circuit to the EVM reference design, as described in Fig. 1. The auxiliary circuit was realized by a GRSCC with sinking and sourcing current capability of 6A, as described in Section III. The transient suppression unit's state-machine is implemented on an Altera Cyclone IV FPGA [31]. Table I lists the components values and parameters of the experimental prototype and the comparator's threshold voltages setting. The load step signal is also generated by the FPGA, independently, without synchronization to the controller.

TABLE I – EXPERIMENTAL PROTOTYPE VALUES

Component	Value / Type
Buck converter – Evaluation module	TI - TPS40055
Input voltage V_{in}	12 V
Output voltage V_{out}	5 V
Switching frequency f_s	300 kHz
Output capacitor C_{out}	330 µF
Inductor L	22 µH
MOSFETs	Si4946BEY, 41 mΩ
Comparator upper threshold $V_{ref,H}$	2.515 V
Comparator middle threshold $V_{ref,M}$	2.499 V
Comparator lower threshold $V_{ref,L}$	2.487 V
GRSCC switching frequency. f_g	1.66 MHz
GRSCC resonant tank capacitor C_g	200 nF
GRSCC resonant tank inductor L_g	20 nH (stray inductance)
Auxiliary capacitor C_{aux}	20 µF

It should be further emphasized that voltage regulator has been assigned as prescribed by the reference design, including the exact bill of materials. The three ports of the electronic capacitor were connected to the output voltage terminal (VOUT), the output of the analog controller's E/A (COMP) and to GND.

A loading transient of 6A, depicted in Fig. 10, is generated in order to compare the buck converter's response without (see Fig. 10(a)) and with (see Fig. 10(b)) the assistance of the electronic capacitor. As can be observed, without the electronic capacitor the output voltage undershoot is 500mV and the response with the assistance of the electronic capacitor exhibits an output voltage undershoot of 25mV. Fig. 11 presents a 6A unloading transient response for the same cases. The output voltage overshoot without the electronic capacitor is now 240mV (see Fig. 11(a)), whereas the output voltage overshoot with the electronic capacitor sums to be 30mV (see Fig. 11(b)). To get a full view of the system performance and automated TSU operation, a consecutive 6A loading-unloading transient response was measured with and without the electronic capacitor, as depicted in Fig. 12. Using the electronic capacitor,

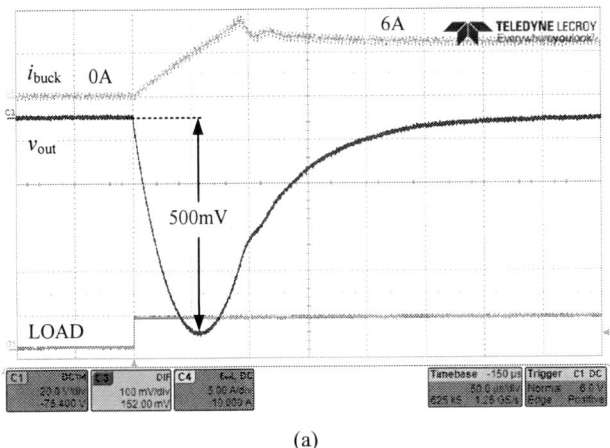

(a)

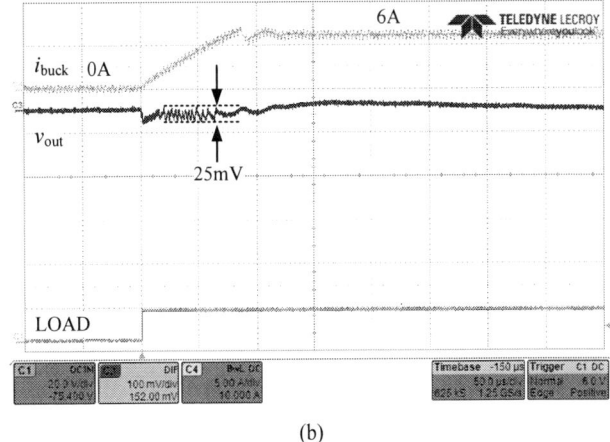

(b)

Fig. 10. Experimental results of a 6A loading transient response: (a) without the assistance of the electronic capacitor, (b) with the assistance of the electronic capacitor. Inductor Current (top – green) 5A/div, output voltage (middle - blue) 100mV/div, time scale 50µs/div, CH1 - load step signal.

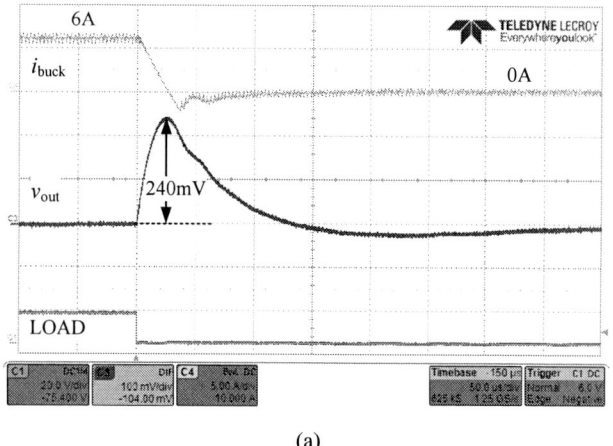
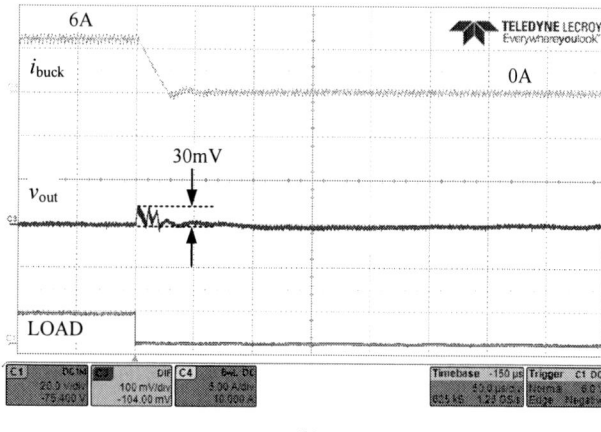

(a)   (b)

Fig. 11. Experimental results of a 6A unloading transient response: (a) without the assistance of the electronic capacitor, (b) with the assistance of the electronic capacitor. Inductor Current (top – green) 5A/div, output voltage (middle - blue) 100mV/div, time scale 50µs/div, CH1 - load step signal.

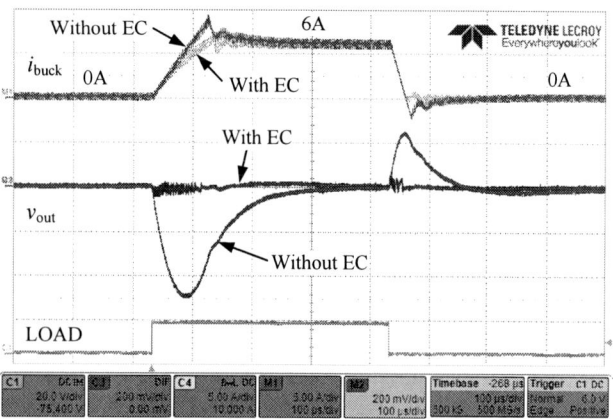
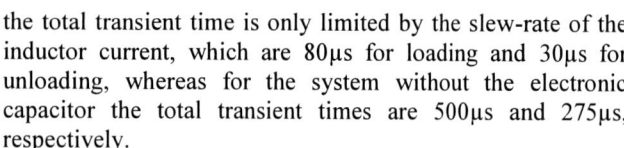

Fig. 12. Experimental results of consecutive 6A loading-unloading transients and the response of the system with and without the electronic capacitor. Inductor Current (top) 5A/div, output voltage (middle) 200mV/div, time scale 100µs/div, CH1 - load step signal.

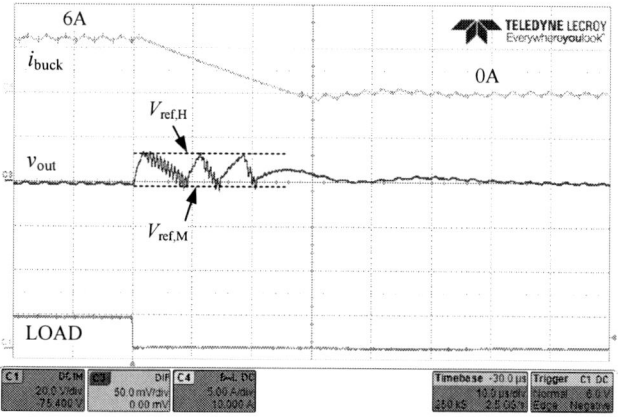

Fig. 13. Experimental results of a 6A unloading transient response showing the effective reference voltages based operation of the electronic capacitor. Inductor Current (top) 5A/div, output voltage (middle) 50mV/div, time scale 10µs/div, CH1 - load step signal.

the total transient time is only limited by the slew-rate of the inductor current, which are 80µs for loading and 30µs for unloading, whereas for the system without the electronic capacitor the total transient times are 500µs and 275µs, respectively.

Fig. 13 zooms in to the electronic capacitor operation which is based on the reference voltages $V_{ref,H}$ and $V_{ref,M}$ for an unloading event. As can be observed, the electronic capacitor maintains the output voltage between the two thresholds, sinks current when the output voltage crosses $V_{ref,H}$ and halts the operation when the output voltage reaches $V_{ref,M}$. This leads to the conclusion that the output voltage deviation for a system with the electronic capacitor is now determined by the comparator's thresholds, and that minimizing the difference between these thresholds is a function of: (1), the steady-state

voltage ripple and the noise in the system. As a result, the output capacitance can be significantly reduced and sized to the steady-state specifications of the output voltage ripple, as opposed to sizing by the requirements of load transients.

V. CONCLUSION

A plug-and-play electronic capacitor for improved loading and unloading transient response of voltage regulator modules has been presented. The improvement has been achieved by the addition of a load-side transient suppression unit, implemented by a recently presented GRSCC topology. The electronic capacitor circuit has three ports and can be connected as an add-on unit to any closed-loop power supply with external compensation network, or access to the PWM generator. The output capacitance can be significantly reduced at the cost of small additional semiconductors and few capacitors, with no ferromagnetic elements, and therefore has the potential to be space conserving and cost-effective when integrated on-chip.

The experimental results demonstrated the performance and benefits of the new transient suppression approach for both loading and unloading transients when applied on an off-shelf commercial buck converter specified by its reference design. Using the electronic capacitor, the output voltage deviation is reduced by up to 20 times compared to the original design along with settling time that is up to 9 times shorter, providing a strong evidence for a significant volume reduction capability.

ACKNOWLEDGEMENTS

This research is supported by Vishay Ltd., Siliconix division.

REFERENCES

[1] E. A. Burton, G. Schrom, F. Paillet, J. Douglas, W. J. Lambert, K. Radhakrishnan, and M. J. Hill, "FIVR — Fully integrated voltage regulators on 4th generation Intel® Core™ SoCs," in *Proc. IEEE Appl. Power Electron. Conf. Expo. 2014*, pp. 432-439, Mar. 2014.

[2] S. Abedinpour, B. Bakkaloglu, and S. Kiaei, "A multistage interleaved synchronous buck converter with integrated output filter in 0.18μm SiGe process," *IEEE Trans. Power Electron.*, vol. 22, no. 6, pp. 2164–2175, Nov. 2007.

[3] G. Schrom, "A 100 MHz eight-phase buck converter delivering 12 A in 25 mm^2 using air-core inductors," in *Proc. IEEE Appl. Power Electron. Conf. Expo. 2007*, pp. 727-730, Mar. 2007.

[4] S. Ahsanuzzaman, A. Parayandeh, A. Prodic, and D. Maksimovic, "Loadinteractive steered-inductor dc–dc converter with minimized output filter capacitance," in *Proc. IEEE Appl. Power Electron. Conf. Expo. 2010*, pp. 980–985, Feb. 2010.

[5] M. del Viejo, P. Alou, J. A. Oliver, O. Garcia, and J. A. Cobos, "Fast control technique based on peak current mode control of the output capacitor current," in *Proc. IEEE Energy Convers. Congr. Expo.*, pp. 3396–3402, Sep. 2010.

[6] M. del Viejo, P. Alou, J. A. Oliver, O. Garcia, and J. A. Cobos, "V2IC control: A novel control technique with very fast response under load and voltage steps," in *Proc. IEEE Appl. Power Electron. Conf. Expo. 2011*, pp. 231–237, Mar. 2011.

[7] J. Chen, A. Prodić, R. Erickson, and D. Maksimović, "Predictive digital current programmed control," *IEEE Trans. Power Electron.*, vol. 18, no. 1, pp. 411–419, Jan. 2003.

[8] S. Chattopadhyay and S. Das, "A digital current-mode control technique for dc-dc converters," *IEEE Trans. Power Electron.*, vol. 21, no. 6, pp. 1718–1726, Nov. 2006.

[9] Y. Qiu, X. Chen, and H. Liu, "Digital average current-mode control using current estimation and capacitor charge balance principle for dc-dc converters operating in DCM," *IEEE Trans. Power Electron.*, vol. 25, no. 6, pp. 1537–1545, Jun. 2010.

[10] P. Midya, P. T. Krein, and M. F. Greuel, "Sensorless current mode control-an observer-based technique for DC-DC converters," *IEEE Trans. Power Electron.*, vol. 16, pp. 522 – 526. 2001.

[11] A. Babazadeh and D. Maksimović, "Hybrid digital adaptive control for fast transient response in synchronous buck DC–DC converters," *IEEE Trans. Power Electron.*, vol. 24, no. 11, pp. 2625 – 2638, Nov. 2009.

[12] L. Corradini, A. Costabeber, P. Mattavelli, and S. Saggini, "Parameter-independent time-optimal digital control for point-of-load converters," *IEEE Trans. Power Electron.*, vol. 24, no. 10, pp. 2235–2248, Oct. 2009.

[13] A. Babazadeh, L. Corradini, and D. Maksimović, "Near time-optimal transient response in DC-DC buck converters taking into account the inductor current limit," in *Proc. IEEE Energy Convers. Conf. Expo.*, Sep. 2009, pp. 3328–3335.

[14] V. Yousefzedah, A. Babazadeh, B. Ramachandran, E. Alarcon, L. Pao, and D. Maksimović, "Proximate time-optimal control for synchronous buck DC–DC converters," *IEEE Trans. Power Electron.*, vol. 23, no. 4, pp. 2018–2026, Jul. 2008

[15] L. Corradini, A. Babazadeh, A. Bjeletić, and D. Maksimović, "Current-limited time-optimal response in digitally controlled dc–dc converters," *IEEE Trans. Power Electron.*, vol. 25, no. 11, pp. 2869–2880, Nov. 2010

[16] G. E. Pitel and P. T. Krein, "Minimum-time transient recovery for DC-DC converters using raster control surfaces," *IEEE Trans. Power Electron.*, vol. 24, no. 12, pp. 2692 - 2703, Dec. 2009.

[17] E. Meyer, Z. Zhang, and Y. F. Liu, "An optimal control method for buck converters using a practical capacitor charge balance technique", *IEEE Trans. Power Electron.*, vol. 23, no. 4, pp. 1802-1812, Jul. 2008.

[18] A. Radić, Z. Lukić, A. Prodić, and R. de Nie, "Minimum deviation digital controller IC for DC-DC switch-mode power supplies," *IEEE Trans. Power Electron*, vol. 28, no. 9, pp. 4281-4298, Sep. 2013.

[19] A. Stupar, Z. Lukić, and A. Prodić, "Digitally-controlled steered-inductor buck converter for improving heavy-to-light load transient response," in *Proc. IEEE Power Electron. Spec. Conf. 2008*, pp. 3950–3954, Jun. 2008.

[20] W. Jing, A. Prodić, and W. T. Ng , "Mixed-signal-controlled flyback-transformer-based buck converter with improved dynamic performance and transient energy recycling," *IEEE Trans. Power Electron* , vol. 28, no. 2, pp. 970-984, Feb. 2013.

[21] P.S. Shenoy, P.T. Krein, and S. Kapat, "Beyond time-optimality: Energy-based control of augmented buck converters for near ideal load transient response," in *Proc. IEEE Appl. Power Electron. Conf. Expo. 2011*, pp. 916-922, Mar. 2011

[22] Y. Wen and O. Trescases, "DC-DC converter with digital adaptive slope control in auxiliary phase for optimal transient response and improved efficiency," *IEEE Trans. Power Electron.*, vol. 27, no. 3, pp. 1314–1326, Mar. 2012.

[23] E. Meyer, Z. Zhang, and Y. F. Liu, "Controlled auxiliary circuit to improve unloading transient response of buck converters," *IEEE Trans. Power Electron.*, vol. 25, no. 4, pp. 806–819, Apr. 2010.

[24] V. Svikovic, J. A. Oliver, P. Alou, O. García, and J. A. Cobos, "Synchronous buck converter with output impedance correction circuit," *IEEE Trans. Power Electron.*, vol. 28, no. 7, pp. 3415–3427, Jul. 2013.

[25] V. Svikovic, J. J. Cortes, P. Alou, J. A. Oliver, O. Garcia, and J. A. Cobos, "Multiphase current-controlled buck converter with energy recycling output impedance correction circuit (OICC)," *IEEE Trans. Power Electron.*, vol. 30, no. 9, pp. 5207-5222, Sep. 2015

[26] Z. Shan, S. C. Tan, and C. K. Tse, "Transient mitigation of dc-dc converters for high output current slew rate applications," *IEEE Trans. Power Electron.*, vol. 28, no. 5, pp. 2377–2388, May 2013.

[27] O. Kirshenboim, A. Cervera, and M. M. Peretz, "Improving loading and unloading transient response of a voltage regulator module using a load-side auxiliary gyrator circuit," in *Proc. IEEE Appl. Power Electron. Conf. Expo. 2015*, pp. 913-920, Mar. 2015.

[28] A. Cervera, M. Evzelman, M.M. Peretz, and S. Ben-Yaakov, "A high efficiency resonant switched capacitor converter with continuous conversion ratio," *IEEE Trans. Power Electron.*, vol. 30, no. 3, pp. 1373-1382, Mar. 2015.

[29] B. C. Kuo, *Automatic Control Systems*, Englewood Cliffs, NJ, Prentice-Hall, 1982.

[30] A. Cervera and M. M. Peretz, "Resonant switched-capacitor voltage regulator with ideal transient response," *IEEE Trans. Power Electron.*, vol. 30, no. 9, pp. 4943-4951, Sep. 2015.

[31] DE2 Development and Education Board user manual, Altera Corporation, 2006.

Adaptive Voltage Positioning (AVP) Design of Multi-Phase Constant on-Time I^2 Control for Voltage Regulators with Ramp Compensations

Kuang-Yao (Brian) Cheng and Yipeng Su

Controller and Digital Solution (CDS)
Texas Instruments Inc.
Manchester, NH 03101 USA

Abstract—This paper proposes a multi-phase constant on-time I^2 control architecture with adaptive voltage positioning (AVP) for voltage regulators (VRs). By including both fast inner current loop and the slow outer current loop, fast load transient performance and accurate current control can be achieved. In addition, in order to overcome the current ripple cancellation effects over wide duty ratios for future microprocessors, external ramp compensation is added to the proposed control architecture. The small-signal model of the proposed architecture is derived to provide the AVP design guideline, and to understand the design constraints about ramp compensations designs for different operating phases. The Simplis simulation and experimental results are provided to show the good correlations of the derived small-signal model and the effectiveness of the proposed control architecture. The parasitic of the typical output capacitor bank for VRs has also been included in the Simplis simulation bench to predict the load transient performance precisely with different load frequencies.

Index Terms—Constant on time control, voltage regulators, multi-phase buck converters, and ramp compensations.

I. INTRODUCTION

With growing markets on the mobile devices, the infrastructures, such as servers, storages, or telecommunication equipment, to support the increasing demands on the cloud computing need to be upgraded with better computing power, better efficiency, and higher power densities. The microprocessors inside the equipment determine the computing power, and the voltage regulators (VRs) [1]-[3] for powering these microprocessors are the key to improve its dynamic performance, reliability, and efficiency with less costs and higher power densities. High-performance microprocessors require low-voltage and high-current VRs with fast load transient response. Besides that, the adaptive voltage positioning (AVP) design is very important for VRMs to fulfill the strict load transient requirements [4]-[5]. Current-mode control schemes have been widely used in multiphase VRs due to its simplicity of the compensation and the AVP designs [4]-[6]. However, the conventional current mode control has accuracy limitations for the AVP design, and additional offset cancellation circuits have been proposed in for improving the voltage regulation

accuracy for both peak-current mode [7] and constant on-time (COT) current mode control schemes [8].

In order to achieve fast transient response and accurate current control, a COT I^2 average current mode (ACM) control has been proposed in [9] for single-phase buck converters. However, due to the high load current with extremely high slew-rate requirements for microprocessors, multi-phase converters are required for VRs to reduce the size of filter components, to distribute heat dissipation, and to increase the equivalent switching frequency and bandwidth [10]-[11]. In addition, in order to support later than 2013 Intel microprocessor platforms, a wide operating voltage range of 0.5V to 2.3V is required, which may cause jitter issues of COT current-mode control due to the ripple cancellation effect of the multi-phase operations [11]. Fig. 1(a) shows the normalized output current ripple with different duty ratios and operating phase numbers, and Fig. 1(b) shows the individual phase current ripple and summed inductor current ripple of a four-phase converter with 10% and 20% duty ratios. As shown in Fig. 1, the summed current becomes smaller with increasing the operating phase numbers and the duty ratios, which will cause challenges for the typical COT current mode control. External ramp compensation has been proposed in [12] to improve the jitter performance for COT

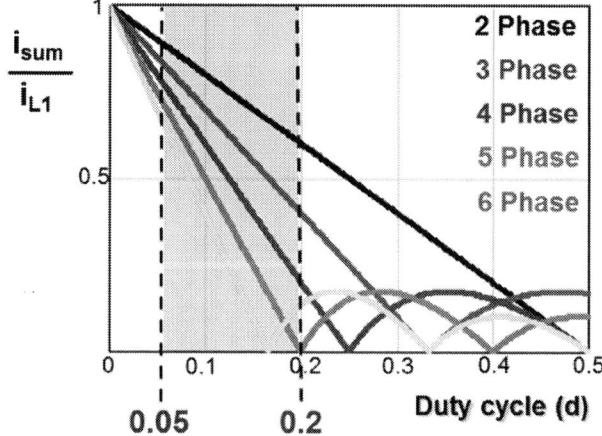

(a) normalized output current ripple with different duty ratios and operating phases number.

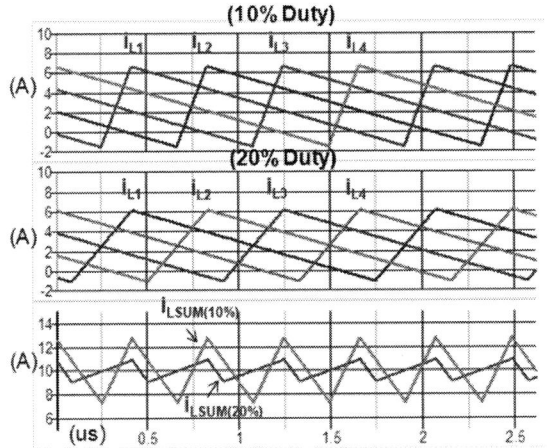

(b) 4-phase current ripple waveforms with 10% and 20% duty ratios.
Fig.1. Ripple cancellation effect.

current-mode control multi-phase VRs. Fig. 2 shows the comparison of the jitter performance with and without external compensation. As shown in Fig. 2 that larger jitter will lead to larger voltage ripple, which will cause worse load transient performance. However, the AVP design cannot be achieved in [12] without having an additional accurate current control path.

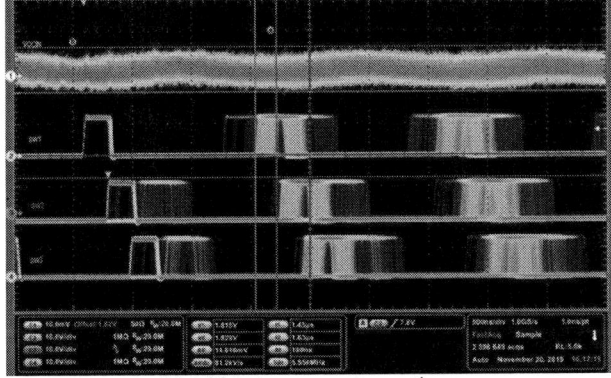

(a) without ramp compensations;

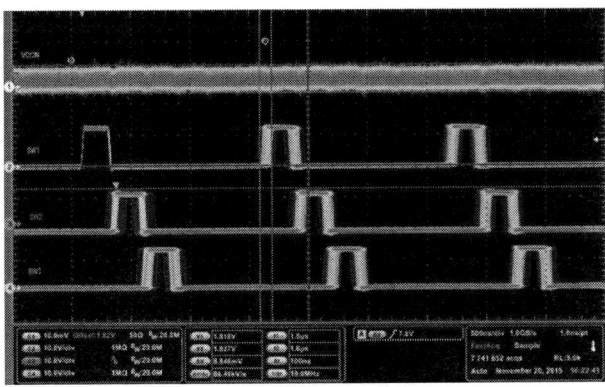

(b) with ramp compensations

Fig. 2. Jitter performance of a 6-phase ripple-based constant on-time VR (Vin=12V, Vout=1.8V, D=0.15):

II. PROPOSED COT I^2 CONTROL FOR VRs WITH RAMP COMPENSATION

This paper proposes a multi-phase COT I^2 control architecture with external ramp compensation to meet the accurate AVP design requirements for VRs while keeping the fast load transient characteristics and good jitter performance under different operating phase numbers. Fig. 3 shows the block diagram and the illustrated operational waveforms of the proposed architecture. A two-phase operational waveform is used as an example, which can be easily added with more phases. The architecture contains of 3 different loops. The most inner loop is to connect the summed inductor current to the modulator to provide fast and direct current feedbacks. An external ramp is also generated to the modulator to improve the jitter performance. The 2nd loop is to have an average current feedback from the summed current to provide accurate AVP design. The most outer loop is the voltage loop to have accurate voltage regulation performance.

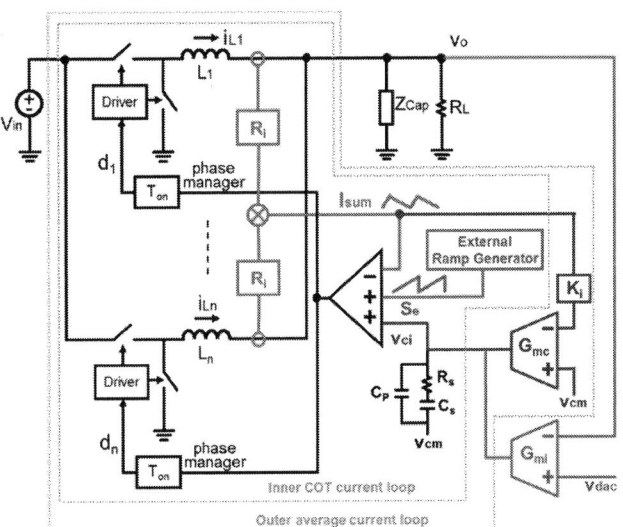

(a) block diagram for N-phase VR

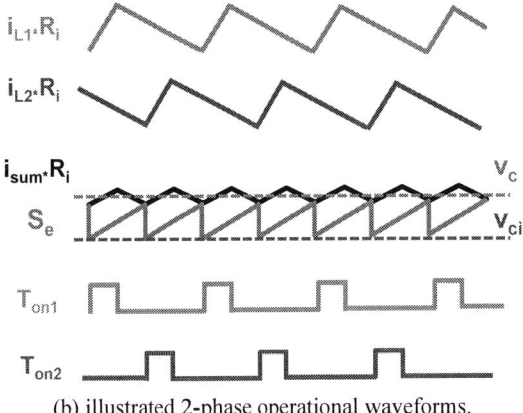

(b) illustrated 2-phase operational waveforms.

Fig. 3. Proposed Constant on-Time (COT) I^2 control with external ramp compensation.

In order to understand the characteristics of the architecture and the design guideline of the ramp and loop compensations, a small-signal model is derived in this paper. By treating the entire inner COT current loop with ramp compensation as a single entity for model derivations [13], the control-to-inductor current transfer function of the inner current loop of the proposed architecture can be derived based on the describing functions as follows [12]:

$$G_{c2iL}(s) = \frac{i_{SUM}(s)}{v_{ci}(s)} \approx \frac{n}{R_i} \cdot \frac{1}{\left(1 + \frac{s}{Q_1 \omega_1} + \frac{s^2}{\omega_1^2}\right)} \cdot \frac{1 + \frac{T_{SW}}{n}s}{1 + \left(\frac{s_e}{s_f} + \frac{1}{2}\right)\frac{T_{SW}}{n}s} \quad (1)$$

where $Q_1 = 2/\pi$, $\omega_1 = n\pi/T_{on}$, T_{on} is the on-time, T_{sw} is the switching period of each phase, R_i is the current sensing gain, s_e is the external ramp slope, s_f is the down slope of the phase inductor current, n is the operating phase number of the multi-phase converter. By closing the outer average current loop, the new control-to-inductor current transfer function can be derived as follows:

$$G_{c2iL2}(s) = \frac{i_{SUM}(s)}{v_c(s)} = \frac{G_{c2iL}(s)}{1 + G_{c2iL}(s) \cdot R_{droop} \cdot H_i(s)} \quad (2)$$

$$R_{droop} = K_i \cdot R_i \quad (3)$$

$$H_i(s) = GM_i \cdot Z_{\cdot comp}(s) \quad (4)$$

where R_{droop} is desired droop resistance of AVP design, which can be adjusted by K_i, and $H_i(s)$ is the outer current loop compensator to provide accurate current control. The loop gain of the proposed architecture can be derived as follows based on (2):

$$G_{loop}(s) = H_v(s) \cdot G_{c2iL2}(s) \cdot Z_{cap} \quad (4)$$

where $H_v(s)$ is the outer voltage loop compensator and Z_{cap} is the equivalent output impedance, which consist of bulk caps, cavity caps, and the socket impedance for VRs as shown in Fig. 4.

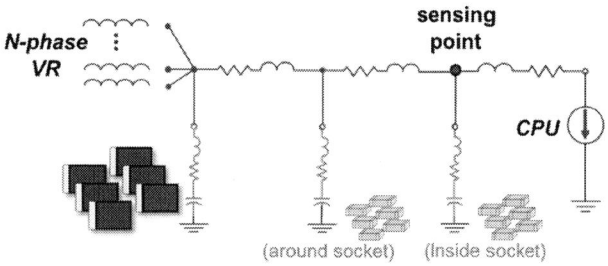

Fig. 4. The output stage configuration of the multi-phase VR including parasitic resistance and inductance.

III. Construction of SIMPLIS Simulation Model and Prototypical Hardware

To verify the effectiveness of the proposed control architecture and the derived analytical small signal model in section II, the SIMPLIS simulation model and the experimental prototype are constructed. Fig. 5 illustrates the SIMPLIS simulation bench, including the models of controller, multi-phase power stages, loads, interposer, EVM board, socket and output capacitors. As we know in the real motherboard of the microprocessor, the output capacitors are distributed in such a manner that the bulk capacitors are located close to the VR, while the ceramic capacitors are placed close to or inside the socket, which has been illustrated in Fig. 4. The parasitics in the output stage (from the output inductor to the remote sensing point) of the multi-phase VR has significant impact on the system performance, which has to be considered in the SIMPLIS simulation. Therefore, the parasitics between bulk capacitors and ceramic capacitors introduced by PCB copper trace is modeled by the "EVM board" block, the parasitics between cavity ceramic capacitors and the microprocessor is modeled by "Interposer" and "Socket" blocks. The PCB parasitics values are extracted by Cadence Sigrity PowerDC according to the PCB layout of the real motherboard shown as Fig. 6. The parasitics of interposer and socket are provided by Intel.

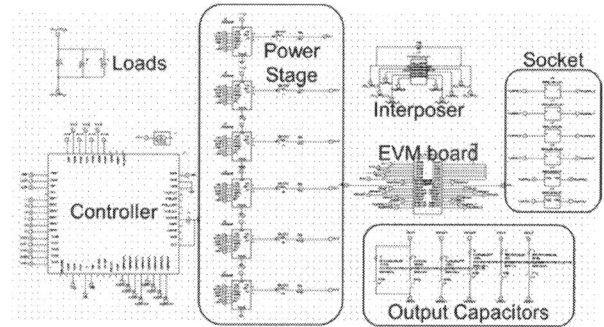

Fig. 5. SIMPLIS simulation bench

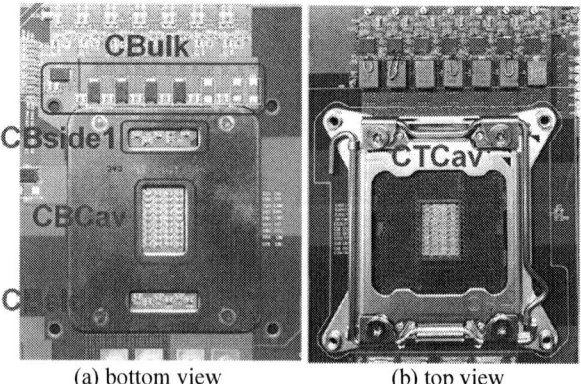

(a) bottom view   (b) top view

Fig. 6. Experimental platform for 6-phase VR with proposed COT I^2 control architecture.

IV. CORRELATION OF THE ANALYTICAL, SIMULATED AND EXPERIMENTAL DATA

In this section, the AC loop gain obtained from the analytical small signal model, SIMPLIS simulation bench and experimental platform are correlated in order to validate the analytical and simulation models. The first studied case is 6-phase VR with V_{in}=12V, V_{out}=1.8V, f_s=650kHz. The type two compensation network is designed to be R_s=6kΩ, C_s=200pF and C_p=10pF. The amplitude of the external ramp is chosen as 260mV and the DC loadline (R_{droop}) is 1mΩ. The loop gain at no load condition from analytical model, simulation and measurement are plotted and compared in Fig. 7. In the second studied case, the DC loadline is decreased from 1mΩ to 0.5mΩ, and the other parameters are unchanged. The loop gain correlation is shown in Fig. 8. In the third case, the DC loadline and the ramp height are kept the same as the first case (i.e. 1mΩ and 260mV), while the phase number is reduced from 6-phase to single-phase. The loop gains for the single-phase VRs with the same external ramp as 6-phase VR are correlated in Fig. 9.

Form the correlations for all the studied cases shown in Fig. 7 to Fig. 9, it can be seen that the analytical small signal model and SIMPLIS simulation model can match each other very well, and they both have good agreement with the bench measurement. Besides that, the comparisons between different studied cases show that (a) reducing DC loadline can further boost the gain and push the bandwidth, however,

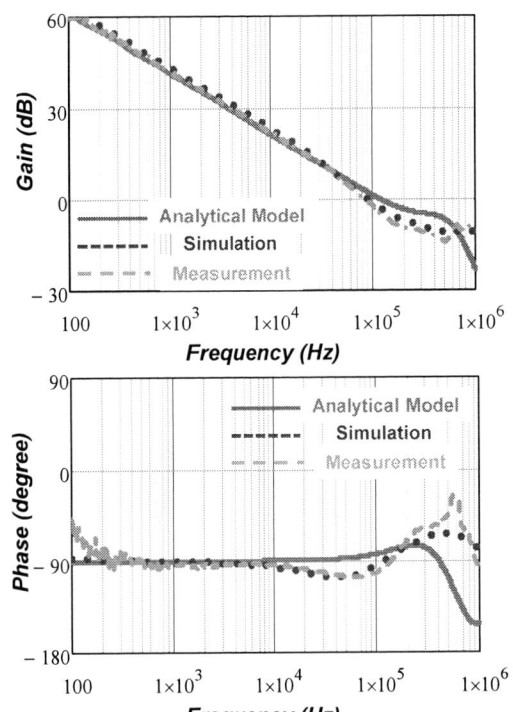

Fig. 8. Correlation of the loop gain data for **6-phase** VR with 0.5mΩ DC loadline and 260mV ramp.

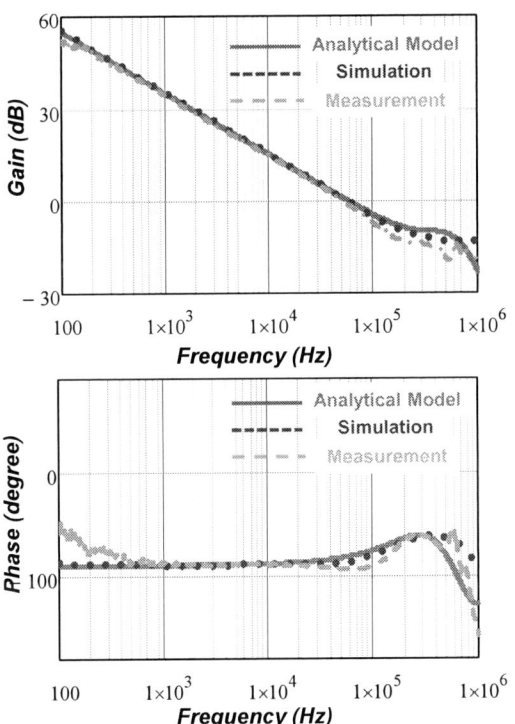

Fig. 7. Correlation of the loop gain data for **6-phase** VR with 1mΩ DC loadline and 260mV ramp.

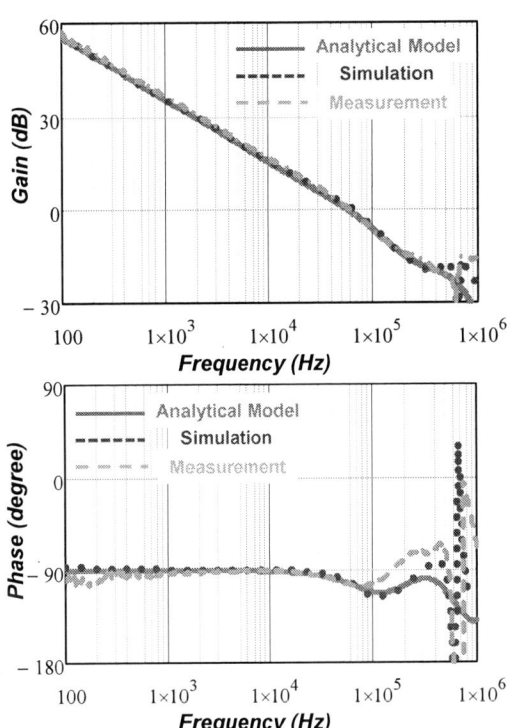

Fig. 9. Correlation of the loop gain data for **single-phase** VR with 1mΩ DC loadline and 260mV ramp.

the phase margin drops; (b) if the phase is changed from 6-phase to single-phase while the ramp height is still kept as the same, although there is almost no difference for the gain curve comparing Fig. 7 and Fig. 9, a prominent phase drop can be found at the crossover frequency.

Actually, when the VR operation is changed from 6-phase to single-phase, the down slope of the current ripple (s_f), with which the on-time is generated, becomes much smaller than the external ramp slope (s_e). The current mode COT control becomes more like the voltage mode COT control. The phase margin will drop significantly if the type tow compensation is still used. A detailed analysis on the impact of the external ramp on the system performance will be presented in section V, along with a design guideline of the external ramp.

The output voltage waveforms of the 6-phase VR in time-domain are obtained from both simulation and measurement. The 6-phase VR is operated at the same settings as the first studied case in the loop gain correlation. The load steps up and down between 50A and 208A with 632A/μs slew rate. The simulated waveforms are compared with the measured results in Fig. 10 under four different load repeating-rate frequencies, (a) 1 kHz low rep-rate; (b) 50kHz and (c) 100kHz close to the cross-over frequency; (d) 400kHz close to the switching frequency. The good agreement between the simulated and measured output voltage demonstrates that the SIMPLIS model can achieve very accurate results to predict the transient performance of the multi-phase VR, if all the parasitics in the output stage has been well extracted and modeled. Only very little discrepancy is that the measured waveforms have slightly larger ripple, which is contributed by the noise coupled from the probe. Even if there is no signal input, a small voltage ripple around 10mV can be observed on the oscilloscope. Therefore, the developed SIMPLIS model can be used as a powerful tool to design and optimize the motherboard layout and the output capacitors, before the hardware is implemented.

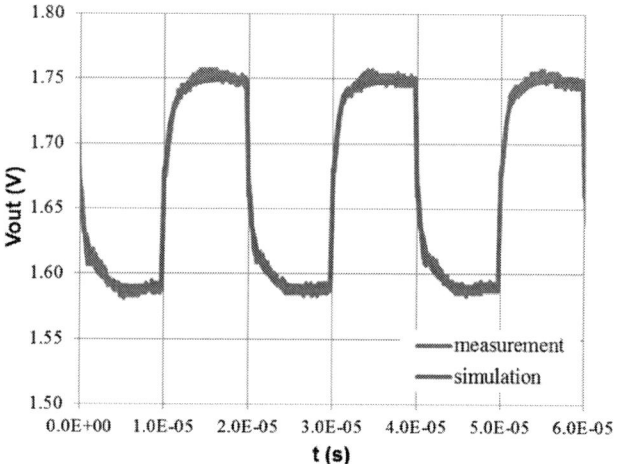

(b) 50kHz load repeating-rate frequency

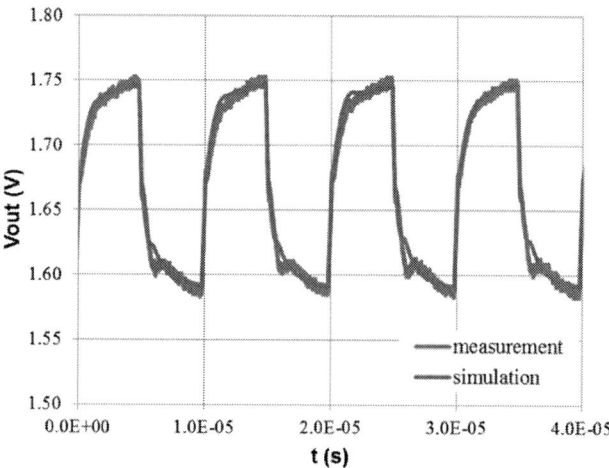

(c) 100kHz load repeating-rate frequency

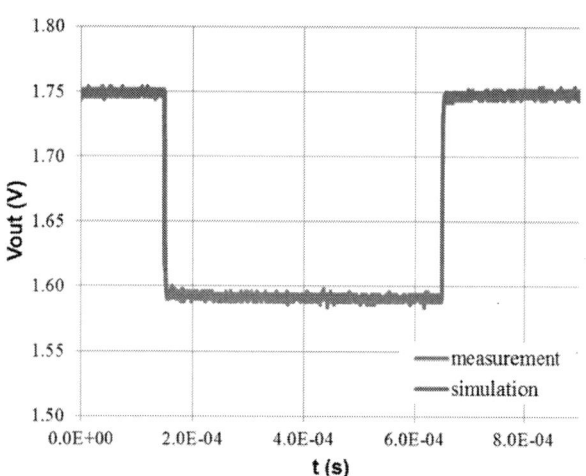

(a) 1kHz load repeating-rate frequency

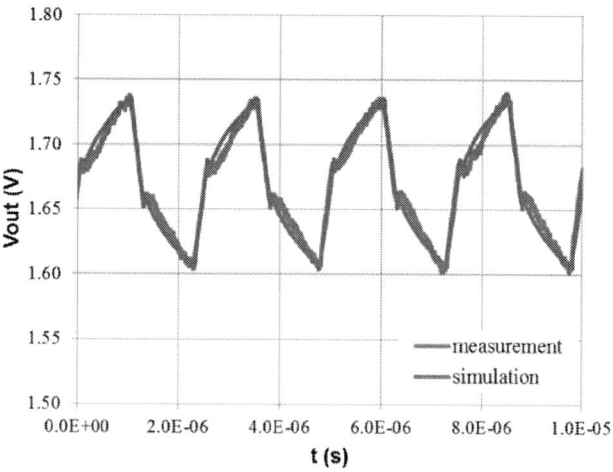

(d) 400kHz load repeating-rate frequency

Fig. 10. Comparison of the 6-phase VR output voltage between simulation and measurement under different load frequency.

V. AVP AND RAMP COMPENSATION DESIGN GUIDELINE

Based on the correlation results presented in the previous section, the derived analytical small-signal model can provide accurate predictions of the system performance, and can be used to provide the design guidelines for different VR applications. Fig. 11 shows the loop gain comparisons with different loadline settings for AVP design of a 6-phase VR. As shown in Fig. 11, the lower loadline settings will make higher bandwidth of the system as the control architecture becoming more like the COT current control schemes. Different microprocessors may require different loadline settings, and also the AVP design can also improve the load transient performance even if the loadline is not required in some applications. Hence the proposed control architecture and the design guideline can be used for any VR application.

Fig. 12 shows the loop gain comparisons with different ramp compensation of a 1-phase VR. As shown in Fig. 12, the phase margin can be significantly affected by the ramp compensation. Fundamentally, since there is no ripple cancellation effect in 1-phase operations, external ramp compensation may not be required, so the ramp compensation can be adjusted dynamically based on the operating phase numbers to improve the overall performance of the proposed control architecture

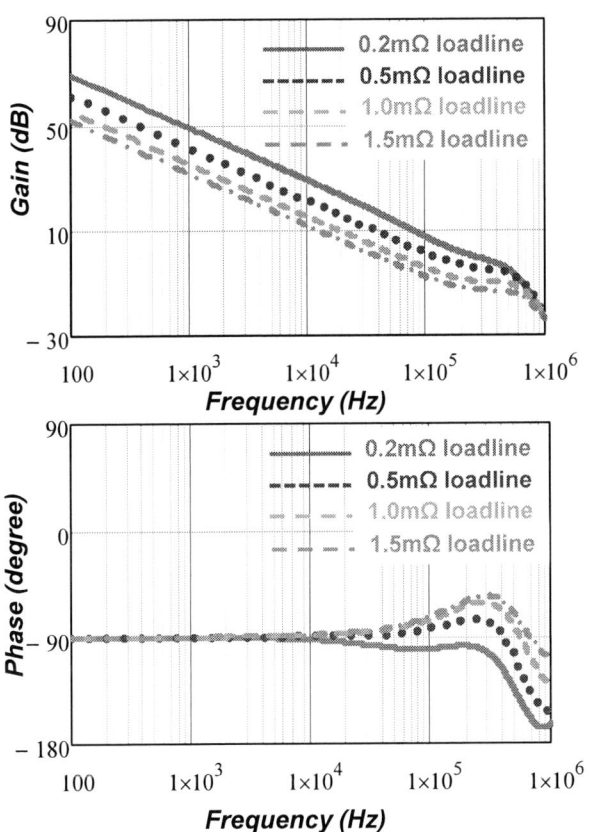

Fig. 11. Loop gain comparisons with different loadline settings for AVP designs of a 6-phase VR.

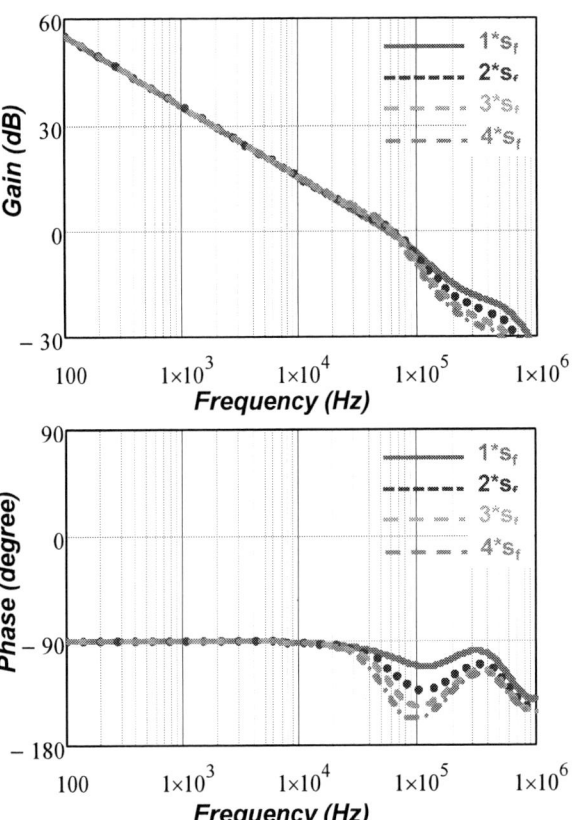

Fig. 12. Loop gain comparisons with different ramp compensation of a 1-phase VR.

VI. CONCLUSIONS

The current-mode control is attractive for multi-phase VRs due to its accurate current control and the independency of the resonant double pole formed by the output inductor and the output capacitance. The proposed COT I^2 control provides fast transient response and accurate AVP design for multi-phase VRs. The derived small-signal model of the proposed architecture is verified by the SIMPLIS simulation and the experimental measurement, which can provide an optimal design guideline for the external ramp and the loop compensation design for both voltage and current loops. The parasitics in the output stage of the multi-phase VRs are well extracted and modeled in the SIMPLIS simulation, according to the layout of the real motherboard, socket and interposer, this important step leads to a good agreement between the simulated and measured output voltage. Therefore, the developed SIMPLIS model can be used as a powerful tool to design and optimize the motherboard layout and the output capacitors, before the hardware is implemented.

REFERENCES

[1] K. Yao, Y. Ren, and F. C. Lee, "Critical bandwidth for the load transient response of voltage regulator modules," *IEEE Trans. Power Electron.*, vol. 19, no. 6, pp. 1454-1461, Nov. 2004.

[2] K. Y. Cheng, F. Yu, S. L. Tian, P. Mattavelli, and F. C. Lee, "Digital hybrid ripple-based constant on-time control for voltage regulator modules," *IEEE Trans. Power Electron.*, vol. 29, no. 6, pp. 3132-3144, June 2014.

[3] VR12.5 Pulse Width Modulation Specification, Intel Corporation, 2012.

[4] K. Yao, Y. Ren, J. Sun, K. Lee, M. Xu, J. Zhou, and F. C. Lee, "Adaptive voltage position design for voltage regulators", in *Proc. IEEE APEC'04 Conf.*, 2004, pp. 272-278.

[5] K. Yao, M. Xu, Y. Meng, and F. C. Lee, "Design considerations for VRM transient response based on the output impedance," *IEEE Trans. Power Electron.*. vol. 18, no. 6, pp. 1270-1277, Nov. 2003.

[6] M. Lee, D. Chen, K. Huang, C. W. Liu, and B. Tai, "Modeling and design for a Novel Adaptive Voltage Positioning (AVP) Scheme for Multiphase VRMs," *IEEE Trans. Power Electron.*, vol. 23, no. 4, pp. 1733-1742, Jul. 2008.

[7] C. J. Chen, D. Chen, C. S. Huang, M. Lee and K. L. Tseng, "Modeling and Design Considerations of a Novel High-Gain Peak Current Control Scheme to Achieve Adaptive Voltage Positioning for DC Power - Converters," *IEEE Trans. Power Electron.*, vol. 24, no. 12, pp. 2942-2950, Dec. 2009.

[8] Y. J. Chen, Dan Chen, Y. C. Lin, C. J. Chen, and C. H. Wang, "A Novel Constant On-Time Current-Mode Control Scheme to Achieve Adaptive Voltage Positioning for DC Power Converters," in *Proc. IEEE IECON'12 Conf.*, 2012, pp. 104-109.

[9] Y. Yan, F. C. Lee, P. Mattavelli, and P. H. Liu, "I^2 average current mode control for switching converters," *IEEE Trans. Power Electron.*, vol. 29, no. 4, pp. 2027-2036, Apr. 2014.

[10] K. Lee, K. Yao, X. Zhang, Y. Qiu, and F. C. Lee, "A novel control method for multiphase voltage regulators," in *Proc. IEEE APEC'03 Conf.*, 2003, pp. 738-743.

[11] K. Y. Cheng, F. Yu, Y. Yan, F. C. Lee, P. Mattavelli, and W. Wu, "Analysis of multi-phase hybrid ripple-based adaptive on-time control for voltage regulator modules," in *Proc. IEEE APEC'12 Conf.*, 2012, pp. 1088-1095.

[12] S. Tian, F. C. Lee, J. Li, and Q. Li, "Equivalent Circuit Model of Constant On-time Current Mode Control with External Ramp Compensation," in *Proc. IEEE ECCE'14 Conf.*, 2014, pp. 3747-3754.

[13] J. Li and F. C. Lee, "New modeling approach and equivalent circuit representation for current-mode control," *IEEE Trans. Power Electron.*, vol. 25, no. 5, pp. 1218-1230, May 2010.

Reactive Power Support Capabilities of Nonsynchronous Interconnection Systems in Microgrid Applications

Yong-Duk Lee and Sung-Yeul Park
University of Connecticut
Department of Electrical and Computer Engineering
Storrs, CT USA
yongduk@engr.uconn.edu and supark@engr.uconn.edu

Abstract— This paper presents the capabilities and benefits of reactive power support of the nonsynchronous interconnection system, "GridLink", in microgrid applications. This equipment provides grid interconnection using a back to back converter. Particularly, it also provides ancillary services including reactive power, spinning reserve for frequency regulation, fault mitigation, and aids in black start for generators. Since GridLink facilitates bidirectional power flow, it can support ancillary services for microgrid side generators and loads, and utility grid side. In addition, the dynamic response of GridLink is faster than that of a synchronous generator, so that it has fewer constraints with respect to operation conditions. In this paper, the benefits of the GridLink, are described. These benefits are energy efficiency and procurement of reactive power capacity of microgrid. Cost savings result from these benefits in terms of variable operation and maintenance (O&M) cost. Furthermore, voltage stability of microgrid will be enhanced by supplying reactive power via GridLink. The reactive power support capabilities of GridLink, are verified using MATLAB/Simulink simulation.

Keywords—GridLink, Reactive Power, Microgrid

I. INTRODUCTION

Due to load growth in the expanding distribution power network, utility companies, regional transmission organizations (RTO) and independent system operators (ISO) have common concerns with respect to reactive power support and financial analysis related to their capital expense. By providing sufficient reactive power, active power can be efficiently transmitted from the generation site to the customer. Conversely, insufficient reactive power can cause problems including: 1) voltage collapse, 2) limited line capacity, 3) system losses and 4) additional expense for reactive power support [1-4].

The first issue is voltage collapse. Insufficient reactive power induces voltage collapse that causes the problems of voltage stability and outage [1-2]. [1] reported an outage issue due to insufficient reactive power which was the Northeast

blackout on August 14, 2003. Thus, sufficient reactive power supply should be ensured to avoid issues such as nuisance trips and shutdown due to overheating of generators [1].

The second issue is the limited capacity of transmission line. From an analytical standpoint, reactive power is working for transmission line and transformers are worked by magnetic field excitation. Thus, active power can be limited by proportion of reactive power. Insufficient active power causes a frequency deviation due to large reactive power demand [1, 5].

The third issue is related to losses of generators and transmission lines. Even though the load power consumption is the same, large reactive power demand causes an increased line current, which increases transmission and generator losses [1, 5]. Usually, in order to prevent these losses apparent power capacity needs to be increased and thus capital cost increases.

The fourth issue is reactive power cost. Reactive power support is one type of an ancillary services and is not separately charged to end-users. However, most utility companies include reactive power cost in their transmission or delivery fees. For commercial and industrial consumers, utility companies may impose additional reactive power cost or power factor penalties [1-5]. In addition, indirect cost, which is lost opportunity cost (LOC), can be considered for evaluating reactive power cost. In the case of independent distribution power generation, they can control reactive power individually. Thus, in normal operation, end-users of this independent power generation don't need to pay additional reactive power cost. However, it is still incurred in operation and maintenance (O&M) costs for the generator.

Thus, in a growing power network, reactive power support is challenging to improve the energy efficiency and voltage stability, because it directly reflects the cost with respect to the electricity charge cost, operation, maintenance, and fuel consumption.

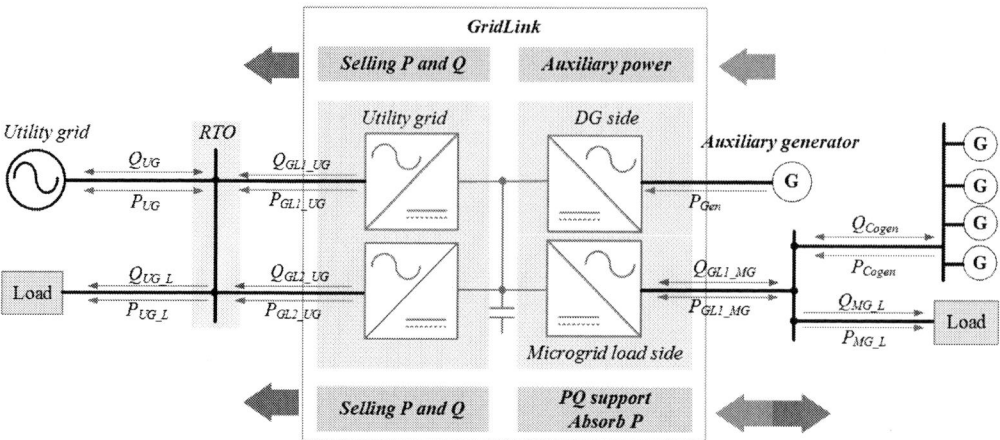

Fig.1. Proposed operation of GridLink.

There are many existing solutions which support reactive power, including synchronous generators, capacitors, grid-connected inverters with renewable energy sources, static compensators, unified power quality conditioners and active power filters. Amongst these solutions, the most representative method of reactive power support comes in the form of shunt capacitors, which relieve the burden of producing all of the necessary reactive power at the generation site [1, 6-8]. However, although the capital costs of capacitors are much lower than other devices, they are not flexible enough in environments that operate under a diverse set of conditions, because they can't adapt to rapid changes during system contingencies. In addition, inrush current due to switching and an expense to protect the capacitor are disadvantages and the factor to weaken reliability. For this reason, it is clear that inverter-based reactive power supplies are preferable in microgrid environments due to their fast response, available capacity, and flexibility [1, 9]. Typically, this flexibility facilitates additional ancillary services including voltage regulation, peak shaving, generator black start, fault ride through and contingency reserve. Recently, inverter-based interconnection technology was introduced, which is GridLink shown in Fig.1. It includes all flexibilities and has the advantage with respect to reactive power support. GridLink is a non-synchronous interconnection device that resides between the utility grid and a microgrid. This technology enables the feature of an indefinitely islanded microgrid [10].

In this paper, reactive power support is presented that makes maximal utilization of GridLink resources, because this configuration is very useful for reactive power support as an ancillary service on both sides due to the back to back converter topology. The proposed operations of GridLink are as follows:

Microgrid side:

✓ Improving power factor → Efficiency improvement of generator → Reduction of annual O&M cost
✓ Voltage support → Improving reliability of power network → Avoiding LOC

Utility grid side:

✓ In normal operation, reactive power procurement with GridLink capacity → Selling reactive power
✓ No payment w.r.t reactive power to ISO or RTO

In addition, cost savings are analyzed with respect to reactive power support in terms of annual O&M cost. Voltage stability on microgrid is verified by supplying reactive power via GridLink.

II. INSUFFICIENT REACTIVE POWER IN COGENERATION PLANT

Cogeneration plant is typical microgrid and supplies power generated by a synchronous generator for end-user. It is necessary to analyze the impact of insufficient reactive power, in order to evaluate the benefit of the proposed operation in cogeneration plant as follows:

✓ In standalone mode, the system voltage decreases.
✓ Efficiency of generator is reduced.

In a microgrid operating in standalone mode, a synchronous generator works as a PV bus, which is an automatic voltage regulator which controls the generator terminal voltage by adjusting the field current. It is related to

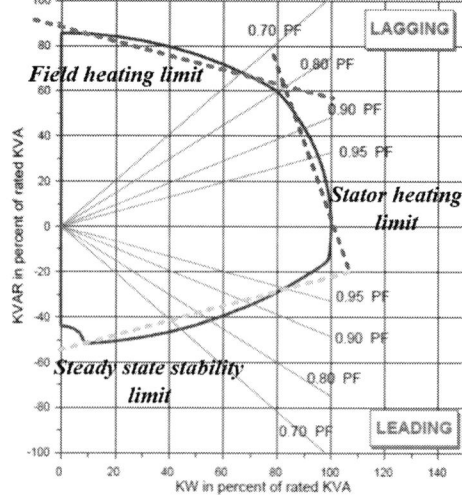

Fig.2. D-curve of generator [11].

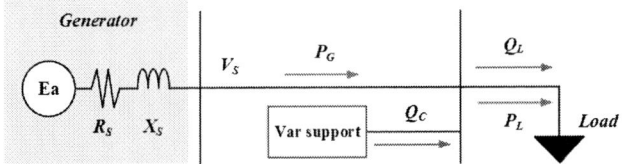

Fig.3. Active and reactive Power flow of the generator.

reactive power output of the synchronous generator. Typically, this control can be defined by the reactive power capability curve, "D-curve" shown in Fig.2. This curve describes the constraints of a synchronous generator in terms of the power vector diagram and thermal losses. Typically, the top line of the D-curve from the vertical axis presents the field heating limit. Thus, reactive power is limited by field current. The field heating limit and reactive power determines the system output voltage. In order to raise the synchronous generator's output voltage, the magnetic field force needs to be increased in the field windings by increasing the current. It can be expressed mathematically a follows[12-13]:

$$\frac{\partial Q}{\partial V} = \frac{V}{X_S^2}\left(\frac{E_m^2}{\sqrt{(E_m V / X_S)^2 - P^2}} - 2X_S\right) \qquad (1)$$

where V is the output voltage of synchronous generator, X_s is the synchronous reactance, E_m is the internal EMF, Q is the produced reactive power from generator and P is the output real power.

In accordance with the internal EMF limit, the output voltage can be changed [12-13], as shown in Table.I. Note that V_s is the system voltage.

TABLE.I SYSTEM VOLTAGE AND REACTIVE POWER

Field current is limited → Voltage is fixed.
$V = V_S, \dfrac{\partial Q}{\partial V} = 0$
Generator can actively support the system voltage.
$V > V_S, \dfrac{\partial Q}{\partial V} < 0$
Output voltage decreases.
$V < V_S, \dfrac{\partial Q}{\partial V} > 0$

A generator's primary function is to convert energy from fuel to electrical power. The copper loss of a generator is I^2R and one of efficiency factors. If the output current is increased, the output efficiency can be reduced. Fig.3 shows the active and reactive power flow of the generator. The efficiency of a generator worsens due to large reactive power demand.

In this case, even though the power consumption of the load is not changed, large reactive power demand increases the power losses of the generator. By compensating reactive power in normal operation, the power factor of the generator side increases and the generator losses can be reduced. The line current for generator can be calculated as follows:

-The line current without compensation is

$$I_G = \sqrt{P_{cogen}^2 + Q_{Load}^2}\bigg/ \sqrt{3}V_S \qquad (2)$$

-The line current with compensation is

$$I_G = \sqrt{P_G^2 + (Q_{Load} - Q_{GridLink})^2}\bigg/ \sqrt{3}V_S \qquad (3)$$

where V_s is the system voltage with 13.8kV.

This line current directly affects the losses with respect to the copper losses. The power losses are as follows:

$$P_{loss\_copper} = I_G^2 R_S \qquad (4)$$

where R_s is the stator resistance.

Typically, in mechanical power losses, fuel is directly related to output electrical power losses. Thus, by decreasing reactive power, O&M cost can be reduced.

III. GRIDLINK OPERATION AND REACTIVE POWER COST

A. GridLink Operation

Basically, GridLink can support ancillary services as follows:

✓ Black start

✓ Emergency control: Fault ride through, voltage support and frequency regulation.

✓ *Normal operation*: Improving power factor, selling active and reactive power to utility grid

This paper takes into account the reactive power flow of normal operation. Fig.4 shows a new power vector diagram of Cogeneration plant with GridLink. In normal operation of Cogeneration plant side, synchronous generators cover active power demand and reactive power demand of auxiliary

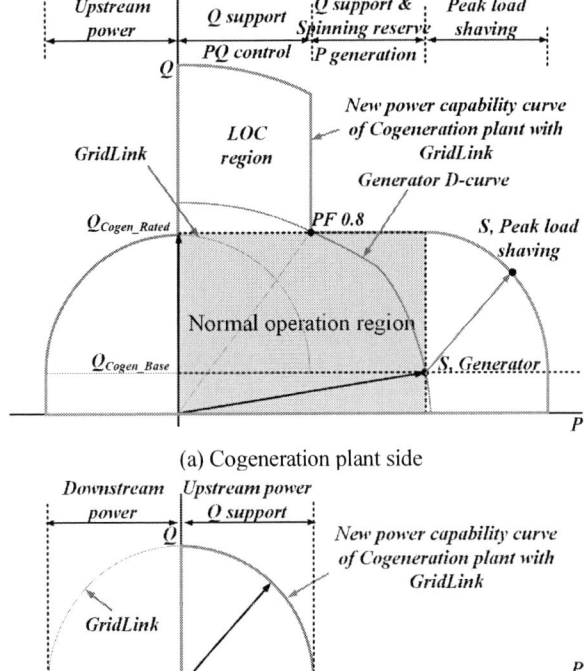

(a) Cogeneration plant side

(b) Utility grid interconnection side

Fig.4. Power diagram of Cogeneration with GridLink:
(a) Cogeneration plant side and (b) Utility grid interconnection side.

TABLE.II PROPOSED OPERATION OF GRIDLINK

Mode	Precondition	Sources		Capacity	Operation Reference
Normal Operation	Precondition $P_{Cogen\_Max} \geq P_{MG\_L}$ $Q_{GL1\_MG\_Max} \geq Q_{MG\_L}$	Generator_50Hz		5MW	$P_{Gen50Hz} = 0, Q_{Gen50Hz} = 0$
		Parallel Gas turbines		14.6MW	$P_{Cogen} = P_{MG\_L}, Q_{Cogen} = Q_{Cogen\_Base}$
		Microgrid	GridLink MG1	5MVA	$P_{GL1\_MG} = 0, Q_{GL1\_MG} = Q_{MG\_L}$
			GridLink MG2	5MVA	$P_{GL2\_MG} = 0, Q_{GL2\_MG} = 0$
		RTO	GridLink UG1	5MVA	$P_{GL1\&2\_UG} = 0$
			GridLink UG1	5MVA	$Q_{GL1\&2\_UG} = \begin{cases} Q_{UG\_L} & if\ Q_{CP} > Q_{UG\_L} \\ Q_{Contract} & if\ Q_{CP} < Q_{UG\_L} \end{cases}$

equipment. GridLink supplies reactive power demand from the load. At utility grid side, GridLink supplies contracted reactive power demand to RTO. This operation is determined by the power management system of GridLink that receives observation data of Cogeneration plant. Based on this data and GridLink capability, the mode is determined shown in Table.II.

B. Reactive Power Cost

Cogeneration plant does not purchase total electricity from the utility grid. This power plant receives the lack of active and reactive power from the utility grid. Thus, Cogeneration plant owner can pay the charge with respect to demand charge and the PF penalty case by case. In order to avoid these charges, the Cogeneration plant operator controls power factor and power flow using their multiple generators. Typically, this operation can be limited as follows:

✓ Initial investment

✓ O&M and Fuel cost

Thus, reactive power cost is also included in this cost.

In order to define GridLink operation cost, reactive power cost can be categorized into two sections as follows:

✓ Cogeneration plant support: saving cost (fuel cost

and loss cost) and voltage support

✓ Utility grid: selling reactive power based on utility grid policy.

Cost saving is obtained by improving efficiency of generator and voltage stability. Additional earnings can be obtained by selling reactive power to RTO. The cost saving and additional benefit is used to return the investment cost of GridLink.

In normal operation of Cogeneration plant side, the reactive power cost regions are as follows:

✓ $0 \sim Q_{Cogen\_base}$: Auxiliary equipment.

✓ $Q_{Cogen\_base} \sim Q_{Cogen\_Rated}$: Losses cost → Efficiency

✓ $Q_{Cogen\_rated} <$: Losses cost and LOC.

Usually, losses cost with respect to generator efficiency improvement is dominant. The event cost is not considered over Q_{cogen_rated}. At utility grid side, the selling of reactive power can be determined by RTO and ISO.

IV. SIMULATION OF GRIDLINK OPERATION

Fig. 5 shows the GridLink configuration. This power network is separated by GridLink and the 4×7.5MW gas turbines are connected with non-synchronous interconnection systems with the utility grid. In order to verify the

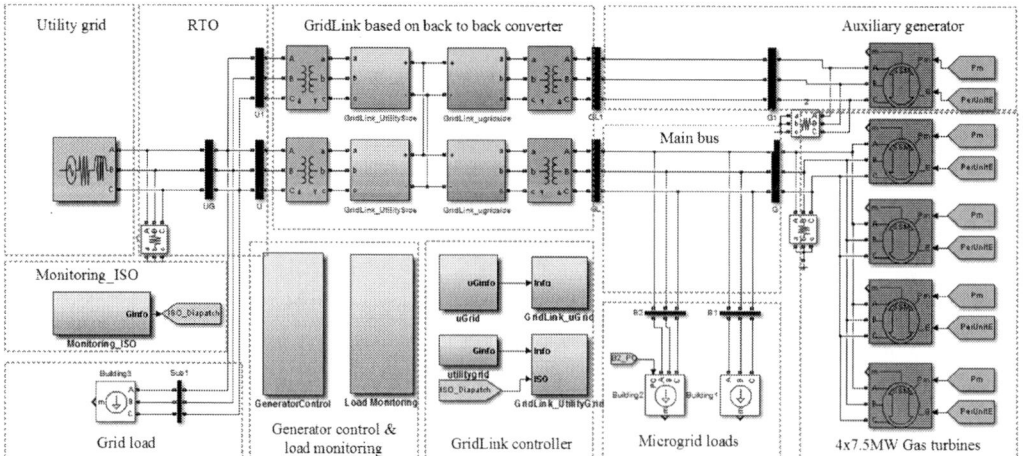

Fig.5. Simulation model of the cogeneration plant with GridLink.

978-1-4673-9551-9/16 $31.00 © 2016 IEEE

effectiveness of reactive power support as an ancillary service, both sides were simulated under reactive power compensation conditions. Note that simulation time is not normalized. Thus, system response is faster than the original design. The simulation conditions in the matlab/simulink are simplified in order to show the effectiveness of the power control. However, actual profile, practical parameters, and increased simulation duration will be necessary in the real time simulation platform.

A. Cogen plant support

Fig.6 shows the capability of reactive power support from GridLink. In Fig 6 (a), GridLink's output power, the microgrid load's power consumption and the synchronous generator's output power are described. Before 1.5sec, Gridlink does not support reactive power and the generator transmits all of the demanded power. After this, GridLink enables reactive power support and the amount of reactive power produced by generator is reduced. As a result, the total KVA demand is reduced.

In order to verify the relationship between the system voltage and reactive power, the system bus voltage is measured. The predefined system voltage is 13.8 kV. Before 1.5sec, the bus voltage is 12.87 kV which is a 7% drop compared to the predefined voltage level shown in Fig 6 (b). After reactive power compensation, the bus voltage recovers to the predefined voltage level.

B. Utility Grid Support to Sell Reactive Power

Fig. 7 shows the effectiveness of reactive power compensation in improving the power factor of the utility grid and the performance of both microgrid and utility sides. At 2.75sec, GridLink enables reactive power support, shown in Fig. 7 (a). Before this point, the grid power factor was below 0.9. After compensation, the grid power factor is increased to unity. Generally, in the case that the ISO does not requisition GridLink as a reactive power compensator, GridLink does not have a responsibility to support this power. However, if they request this service, the GridLink owner can sell this capacity dynamically. This revenue depends on the contract. Fig. 7 (b) shows the performance of both sides during reactive power compensation. At 1.25sec, GridLink supports the reactive

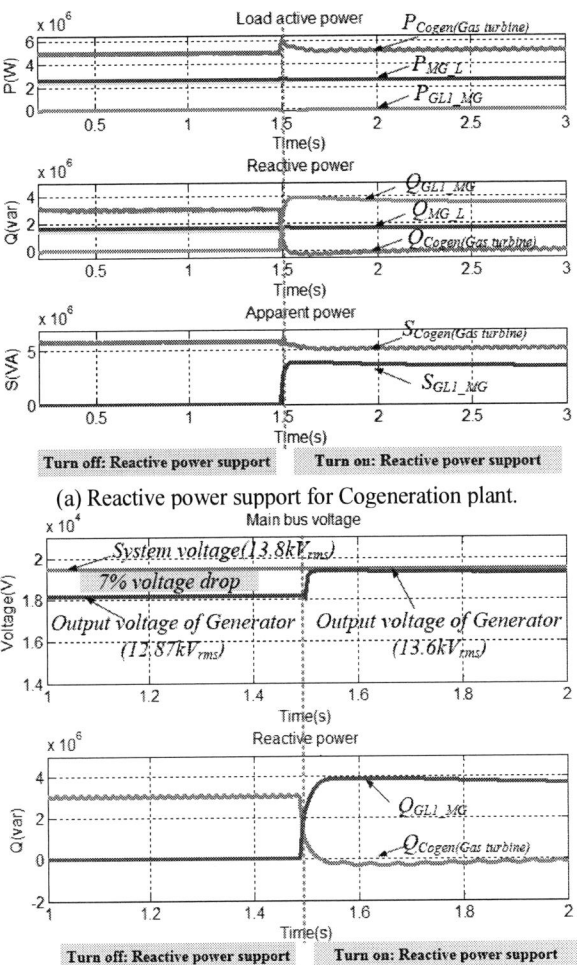

(a) Reactive power support for Cogeneration plant.

(b) Stability improvement of bus voltage.

Fig.6. Cogeneration plant support of GridLink: (a) KVA demand reduction of generator and (b) relationship between reactive power and the bus voltage.

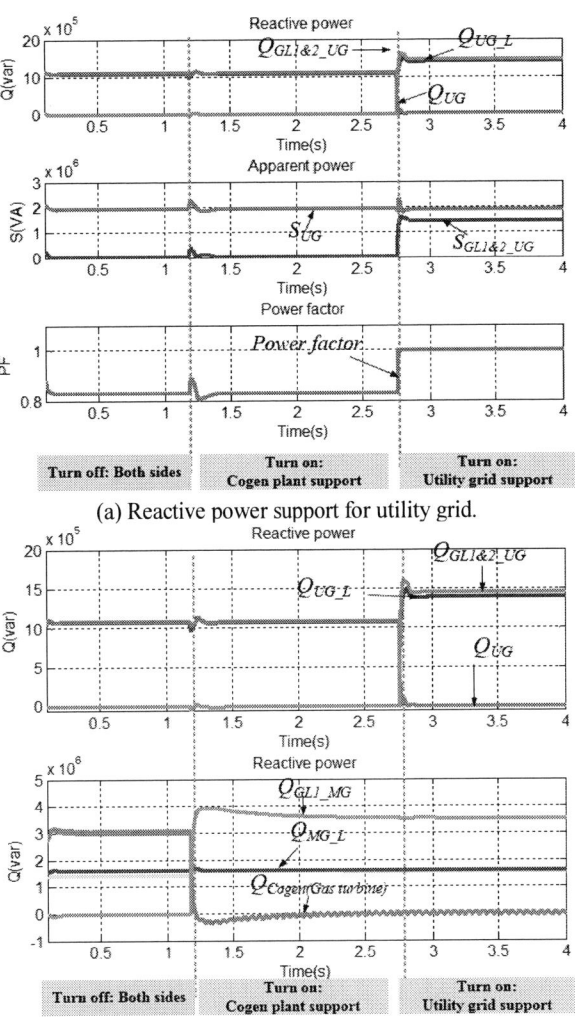

(a) Reactive power support for utility grid.

(b) Reactive power capability of GridLink.

Fig.7. Utility grid support of GridLink: (a) Utility grid power factor compensated by GridLink and (b) Verification of both reactive power support operation for utility grid and microgrid.

power in the microgrid. At 2.75sec, GridLink begins reactive power compensation for the utility grid. As a result, GridLink can support both sides simultaneously and continuously.

V. CASE STUDY OF O&M AND FUEL COST REDUCTION

In order to calculate dynamic saving cost, a one day load profile is used shown in Fig.8. The saving cost is calculated by using variable O&M cost as follows:

$$C_{saving\,cost} = \lambda P_{out}(\eta_1 - \eta_2) \qquad (5)$$

Where λ is O&M cost, η_1 is efficiency without GridLink and η_2 is the efficiency with GridLink.

Fig.8 (a) describes mechanical power difference of generator in reactive power support of GridLink and no support. The generator's mechanical power is reduced by compensating reactive power of GridLink. The reduced power can reduce cost losses, because the efficiency of generator increases as shown in Fig.8 (b).

A Variable O&M cost of 94.6$/MWh is used. By reducing reactive power output from the generator, 254 $/day is saved compared to conventional connection.

VI. CONCLUSION

This paper presents the reactive power support based on GridLink. Reactive power is required for operation in an AC power system. In order to efficiently produce reactive power, it is necessary to overcome many constraints and cost issues. To supplement O&M cost and efficiently operate a synchronous generator, reactive power control of GridLink is proposed. Based on this feature, the compensation method and benefit of reactive power are analyzed with a case study. Based on scaled O&M cost, the saving cost is analyzed. By reducing reactive power output from the generator, 254 $/day is saved compared to conventional connection. GridLink is very helpful in the United States, because the environment of the U.S power network is already microgrid consisting of about 10,000 synchronous generators. The performed simulation results effectively shows the benefit of reactive power support for the short duration. Furthermore, a longer duration of simulation will be considered for future work.

Acknowledgement

The authors thank Pareto Energy for their enormous technical and financial support as well as discussion.

REFERENCES

[1] F. Fran Li John Kueck Tom Rizy Tom King Oak Ridge National Laboratory, "A Preliminary Analysis of the Economics of Using Distributed Energy as a Source of Reactive Power Supply," The U.S. Department of Energy, ORNL/TM-2006

[2] Feng Dong; Chowdhury, B.H.; Crow, M.L.; Acar, Levent, "Improving voltage stability by reactive power reserve management," *Power Systems, IEEE Transactions on* , vol.20, `no.1, pp.338,345, Feb. 2005

[3] Jin Zhong; Bhattacharya, K., "Toward a competitive market for reactive power," *Power Systems, IEEE Transactions on* , vol.17, no.4, pp.1206,1215, Nov 2002

[4] Gross, G.; Shu Tao; Bompard, E.; Chicco, G., "Unbundled reactive support service: key characteristics and dominant cost component," in *Power Systems, IEEE Transactions on* , vol.17, no.2, pp.283-289, May 2002

[5] Principles for efficient and reliable reactive power and consumption, Federal Energy Regulatory Commission, No.Ad05-1-000, Feb 2005

[6] Dixon, J.; Moran, L.; Rodriguez, J.; Domke, R., "Reactive Power Compensation Technologies: State-of-the-ArtReview," *Proceedings of the IEEE* , vol.93, no.12, pp.2144,2164, Dec. 2005

[7] Das, J.C., "Passive filters - potentialities and limitations," *Industry Applications, IEEE Transactions on* , vol.40, no.1, pp.232,241, Jan.-Feb. 2004

[8] A. Edris, " Low Cost Emergency VAR Compensator," EPRI, Technical Progress, Nov 2000

[9] Majumder, R., "Aspect of voltage stability and reactive power support in active distribution," *Generation, Transmission & Distribution, IET* , vol.8, no.3, pp.442,450, March 2014

[10] http://www.paretoenergy.com/

[11] Ideal Electric Company, "Horizontal Synchronous Brushless Generator Operation and Maintenance Manual"

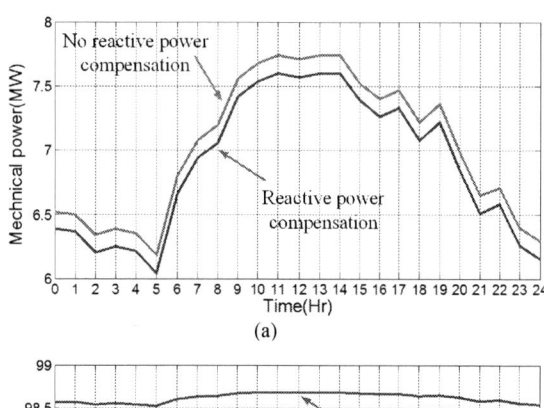

(a)

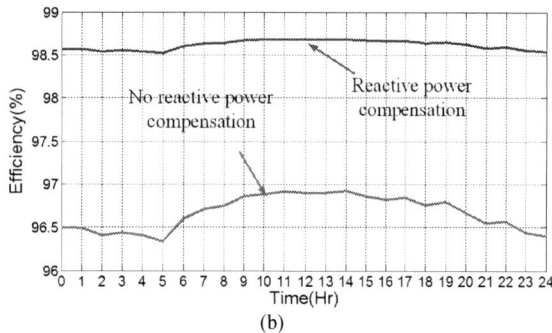

(b)

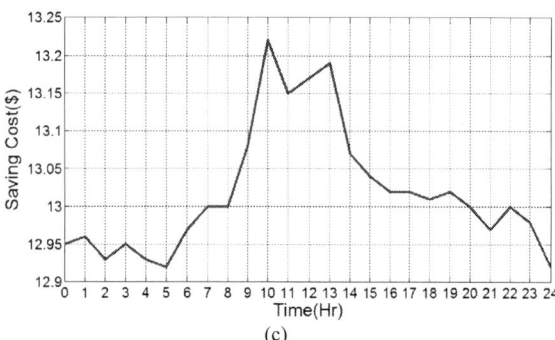

(c)

Fig.8. One day profile of load: (a) mechanical power of generator and (b) efficiency of generator and (c) saving cost.

[12] Efthymiadis, A.E.; Guo, Y.-H., "Generator reactive power limits and voltage stability," *Power System Control and Management, Fourth International Conference on (Conf. Publ. No. 421)* , vol., no., pp.196,199, 16-18 Apr 1996

[13] Dobson, I.; Lu, L., "Voltage collapse precipitated by the immediate change in stability when generator reactive power limits are encountered," *Circuits and Systems I: Fundamental Theory and Applications, IEEE Transactions on* , vol.39, no.9, pp.762,766, Sep 1992

Zero standby power high efficiency hot plugging outlet for 380VDC power delivery system

Kai Tan, Chang Peng, Pengkun Liu, Xiaoqing Song and Alex Q. Huang

FREEDM Systems Center, Department of Electrical and Computer Engineering

North Carolina State University

Raleigh, NC

Abstract— The DC power delivery system is becoming an appealing research topic and real world solution due to its higher energy efficiency compared with AC delivery system. It has already been applied in data centers, commercial buildings, electrical vehicle charge stations and micro grid systems, etc. However, the electrical arc and related potential for fire and human injury is a main safety concern for the DC system. Besides, the inrush current caused by the load capacitance when hot plugged in to the DC system is also a major issue that must be considered when designing DC system. In this paper, a smart hot plugging outlet is proposed and developed by embedding solid state device into the DC outlet. Besides, over temperature and dual threshold over current protections have been also integrated to realize current limiting and trip function for each outlet. Analysis and experimental results based on 380V DC, which is a common voltage level adopted in data centers and DC micro grids, are discussed and presented.

Keywords—LVDC system; 380V outlet; Hot plugging; Saturation; Current limiting.

I. INTRODUCTION

Today, 120V (or 220V in some countries) AC power systems are widely used worldwide. However, low voltage DC (LVDC) distribution is becoming more and more utilized in electrical systems thanks to the development of power electronics technology. For power distribution, more and more DC loads appear in our daily life, such as LED lights, computers, communications and numerous types of batteries used in consumer electronics. Currently, these loads need a power rectifier stage to transfer AC power to DC power which increases the cost and power loss. For massive electronics DC load system, such as a DC data centers, DC buildings, electrical vehicle charge stations, etc.[1]–[3], the power loss will become significant which makes a direct DC distribution system an attractive high efficiency option[4]. To support plug-and-play of these applications, the DC hot plugging outlets need to be developed.

In 2012, ABB and Green® launched 380V DC data center employing HP servers with ±190V DC bus in Zurich, Switzerland [5]. Many other manufactures and researchers also devoted a lot of efforts to the 380V DC and a draft standard has also been developed [6]–[11].

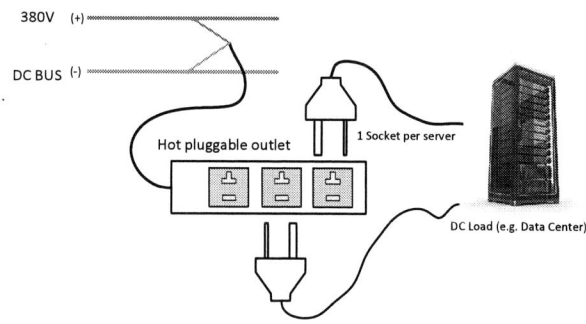

Figure 1. DC hot pluggable outlet

The typical power ratings of 2U and 4U servers in data centers are about 800W to 1400W which helps to define the current rating requirement for the 380V DC outlet. The state-of-the-art TO-247 packaged Si/GaN/SiC power devices can easily cover this power range with an acceptable steady state thermal stress.

Considering all major requirements for a 380V DC distribution system, a new zero standby power hot-plugging outlet is proposed, developed and tested which is programmable by a micro-controller to work under various load conditions. Analysis and experimental results have been demonstrated on a 380V/5A prototype.

II. MAJOR CHALLENGES AND EXSISTING SOLUTIONS

For 380V rating DC distribution system, there are mainly three issues for designing a reliable hot plugging outlet.

A. Major challenges of plug and outlet in 380V DC system

1) Arc at the moment of unplug

The electric arc in 380V DC is much severe than traditional AC system due to the lack of a zero crossing point in voltage and current waveforms. The arc discharge will vaporize the contacts material and cause contamination or corrosion on the contact surface and increase the contact resistance. The arc may cause damage to contacts such as melting and welding, destruction of insulation, and fire. An arc flash also presents a hazard to humans [12].

2) Inrush current at the moment of plug in

Computer servers and other electronic loads behave like a capacitive load when they are plugged into the 380V DC bus. A large inrush current will flow through the contacts to charge capacitor voltage to 380V. Therefore it performs like a short circuit current due to the low impedance of the initially un-energized capacitors. The inrush current is higher for larger capacitive loads. The peak value of this inrush current can be 150A to 450A for 0.68uF to 6.8uF capacitors [13] which induces a large current stress on conductors and will damage the weak joint point on the current loop. On the other hand, a large inrush current may also cause a voltage sag on the 380V DC bus due to the transient low impedance. This voltage sag may cause under voltage lock out on other equipment connected on the bus.

3) The large and fast discharge current with faults

Once a fault occurs, the fault current will also be large and di/dt rate is very high due to the discharge of the energized capacitive components on the DC bus. If a low impedance fault occurs on one load, the input capacitor will be discharged by the fault and further generates a low impedance for other loads in neighborhood. This current can be several hundred amperes with milliseconds rise speed [14]. The large and fast discharge current will affect other users until the low impedance fault is cleared.

B. Pros and Cons of Exisiting Solutions

Currently, there are several solutions and products designed for 380V or similar DC voltage rating DC system. The Saf-D-Grid® plug and receptacle of Anderson Power Products® [15] is designed for 400V/40A application with Hi Temp Nylon and Polycarbonate housings to contain the arc whether connectors are mated or unmated while under load to minimize the risk to operators. However, the design is still based on mechanical optimization so that the arc and sparks still exist. 250 cycles is the lifetime of this plug and receptacle under 400V and 20A, according to the datasheet.

A 430V/10A power distribution units developed by FUJITSU® has enhanced the lifetime with arc by designing a sealed arc chamber with built-in permanent magnet to extinguish the arc [16]. This increases the life time to 5000 cycles. However, the inrush current is not limited and it only has 500 cycles lifetime with 300A inrush current.

TE Connectivity® proposed and developed a DC plug with polymer positive temperature coefficient (PPTC) materials [17]. The resistance of the PPTC increases with temperature. With such characteristic, it is designed to be parallel with main current path but with first mate, last break connection. The current in PPTC heat itself to have current limiting capability. During unplugging, the PPTC reduce the final turn off current to shorten the arc between contacts into a safe region. One issue in this design is that the current reduction in the PPTC depends on the temperature and time, therefore both the speed of unplug and the value of load current will have an influence on the result. Another issue lies in that although the high resistance can reduce the current value, the current still exists

rather than zero. Consequently, the arc may still exist and will damage the contact.

In 2011, ref [18] proposed a DC plug and socket in its DC Nano grid system. The basic idea is to treat the circuit as a buck converter. Extra power pin is necessary for driving circuit. For turn on, increasing the duty cycle makes the output voltage increase from 0 to Vin. For turn off, decreasing the duty cycle makes the output voltage decrease from Vin to 0. The problem of this proposal is the complicated control with extra pins. And the inrush current is not considered with SOA of solid state devices. No experimental data is provided.

III. PROPOSED HOT PLUGGING OUTLET

Considering all important aspects mentioned above, a zero standby power, smart and hot plugging outlet is proposed, developed and tested. The novel outlet with one N channel MOSFET device, a DC/DC power supply circuit, a controller, a driver and other passive components is shown in Figure 2. The negative bus is blocked with embedded N channel MOSFET. This MOSFET is also used to limit the inrush current, conduct load current, extinguish arc and limit the large discharge current. One input capacitor and output freewheeling diode form the snubber circuit are inserted to protect the MOSFET with minimum loop inductance. A DC/DC converter connected with Energized Pad is designed to generate power for the controller and drive circuit. The controller, sensing and drive circuit is designed for close loop control. A unique feature of this design is the positive pole of the plug can be utilized as a path to connect the 380V with the energized pad without any other extra pins. The "last mate, first break" for the energized pad and the embedded MOSFET are naturally achieved.

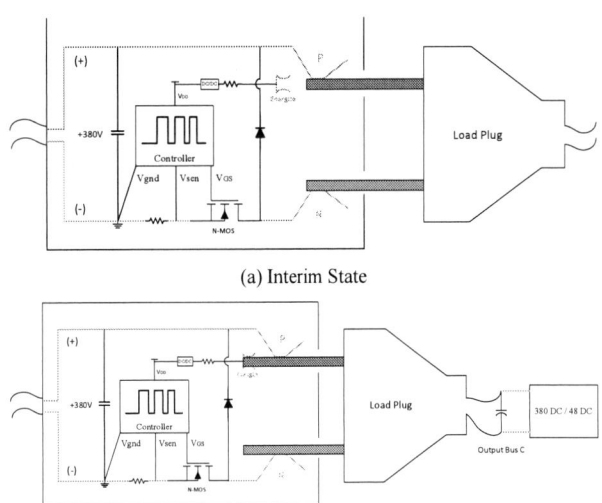

(a) Interim State

(b) Normally On State

Figure 2. Topology and two states of designed hot plugging outlet

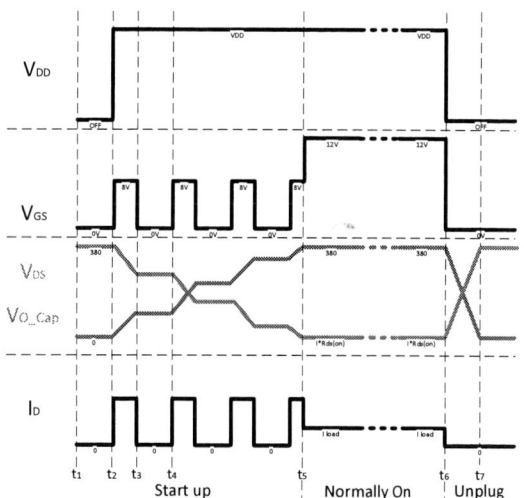

Figure 3. Operation modes of proposed outlet

The earth wire is a default and is not shown in the diagram. The operation modes are demonstrated in Figure 3. They are classified as standby mode, startup mode, normally on mode and unplug mode. The saturation characteristics of the solid state devices helps the designed outlet to limit the inrush and discharge current within programmed values.

A. Standby mode

Before t_1, the whole circuit in outlet is not powered since the energized pad is not connected. The NMOS is naturally 'off' because the gate to source voltage is zero. The power consumption of whole circuit is zero.

B. Startup mode:

At the moment of t1, the plug is in interim state as (a) in Figure 2. The drain to source voltage on NMOS is 380V because the bus capacitor voltage is zero. The NMOS is still in off state with only leakage current. When the plug keeps inserting in, at the moment of t_2, the red male pin of the plug touches the energized pad at the bottom of outlet (Fig.2.b) which energizes the control circuit. First the DC/DC power supply circuit starts to work and generates V_{DD} for the control circuit. The control circuit then starts the designed hiccup startup and generates a VGS to ensure the NMOS device works in the saturation region to limit the inrush current with short pulses (t_2-t_3). The output capacitor is charged in a train of pulses like this and its voltage is increased. The drain to source voltage on the NMOS device is still large which makes power stress large during the pulses because it operates in saturation. Therefore the pulse width and the value of limited current need to be designed within the safe operation area (SOA) of the NMOS device to avoid the device failure due to the excessive power dissipation. After each short pulse, the device is off again from t_3-t_4 to control the average power stress and cool the device down. The voltage on the device and capacitor remain the same in this period.

The current on the drain is sensed with close loop control to a given value with adaptive V_{GS}. The on-off hiccup process continues until the voltage of the output capacitor is close to the bus voltage. Then the control circuit fully turns on the NMOS device with 12V VGS at the moment of t_5. Because the drain to source voltage is already close to zero, there is no inrush current, the normal load current flows continuously.

C. Normally on mode:

The plug remains in position (b) in Fig.2. And the NMOS device keeps fully turn on to keep the lowest $R_{DS(ON)}$ so the power loss on the solid state device is minimized.

D. Unplug mode:

The plug is pulled out and separated from the energized pad and the V_{DD} becomes 0 at t_6. Then NMOS device is tuned off immediately. The 380 V DC voltage is applied to the NMOS device which is in off state in series with the output capacitor of the load converter. The voltage of the output capacitor at the load keeps decreasing because of the converter load until under voltage lock out (UVLO) occurs. The drain to source voltage on the NMOS will increase. At the moment of t7, the plug is fully unplugged with no arc because NMOS is off. There is also no inrush and arc for the energized pad because the total control circuit power is only about 0.4W which is about 1mA for 380V system.

E. Dual Threshold Over Current Protection

Besides normal plug and unplug features similar to traditional AC outlet, the developed DC outlet also has over current protection capability which helps this outlet act as a current limiter/breaker during low and high impedance faults.

A dual threshold over current protection strategy is implemented by the controller to monitor over current or short fault. These two thresholds I_{Limit} and I_{SC} are adopted to trip the real over current fault and filter nuisance transient noises.

- I_{Limit} – Fault timer runs if $I_{Load} > I_{Limit}$ and the circuit breaker trips when timer runs out of set up $T_{(fault)}$ which is usually been used to filter nuisance transient noises.

- I_{SC} – Short circuit threshold (typically 1.3 to 2.5 times of I_{Limit}). I_{SC} is programmed as a current limiting value to control the MOSFET back to saturation before final preset trip time to limit the capacitor discharge current and isolate the low impedance fault with affecting others.

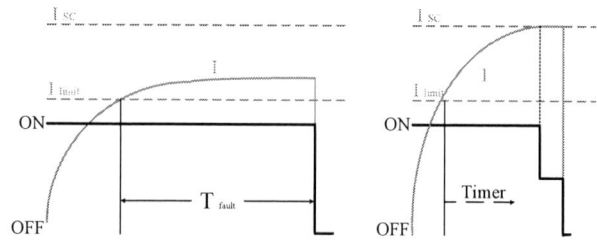

Figure 4. Dual thresholds off strategy for over current protection

978-1-4673-9551-9/16 $31.00 © 2016 IEEE

F. Over Temperature Protection

An over temperature protection is also designed to limit the inrush hiccup cycles. The threshold is set with margin to 125°C as the transient junction temperature is not accurately monitored in hot swappable outlet.

IV. HARDWARE AND EXPERIMENTAL RESULTS

A. Hardware of designed DC hot plugging outlet

Fig.5 shows the developed prototype. The external view of DC hot plugging outlet can have similar structure with traditional AC outlet. The only difference is the positive and negative bus need to be identified more clearly. The embedded circuit is about 1.5 inch square with one TO-247 MOSFET. Figure 5 (b) shows the internal circuit connection. The red circle indicates the energized pad and the corresponding positive pole. The black circle indicates the TO-247 device with simple heatsink and temperature sensor. Only one device is applied to conduct 5A in steady state. The white connectors are used as the input and output ports of the outlet.

(a) External view of the developed DC hot plugging outlet

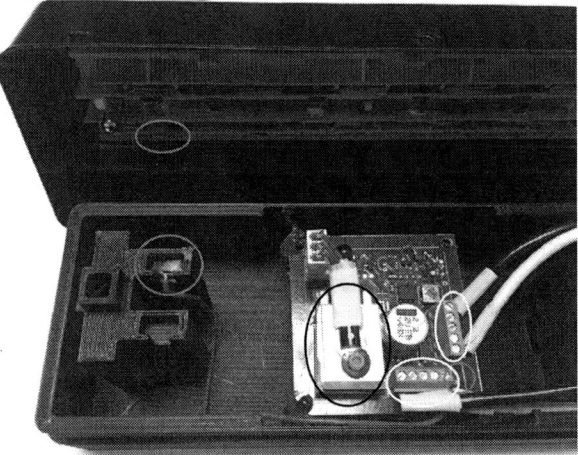

(b) Internal circuit connection
Figure 5. Hardware of proposed outlet

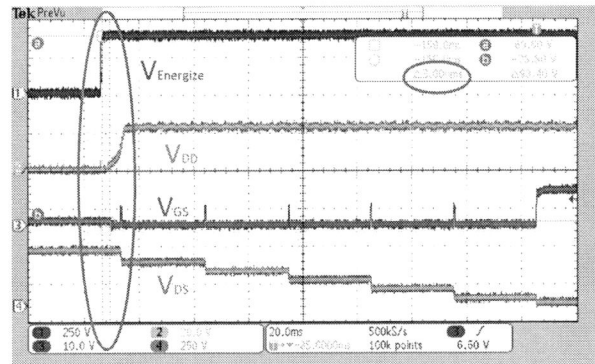

Figure 6. Plug in process waveforms of DC hot pluggable outlet

B. Modes of circuit with energized pad

The hot plugging outlet is tested with 47uF capacitor load on the output side with a 380V input voltage. Figure 6 shows the waveforms of $V_{Energize}$, V_{DD}, V_{GS} and V_{DS} from t_1 to t_5. The circuit spends about 3ms (circled) to start the startup mode. After V_{DD} generating with a threshold value, the controller and driver circuit start working to generate V_{GS} to limit the inrush current within the saturation mode. Figure 7 shows the full waveforms of the hiccup startup mode with V_{GS}, V_{DS}, VC_{AP} and I_D. The inrush current is limited to about 9A. The 47uF capacitor on the load side takes about 70ms to be charged to 380V. This value can be programmed as long as the power stress of MOSFET is with its SOA.

Figure 8 shows the waveforms of $V_{Energize}$, V_{DD}, V_{GS} and V_{DS} in unplug mode. After disconnecting the energized pad, the input voltage of DC/DC buck converter starts to decrease. It takes about 16ms to lose the V_{DD} from breaking $V_{Energize}$ and fully power off the MOSFET. The delay can be further reduced by decreasing the capacitance in the power supply circuit.

Figure 9 presents the normally on mode, the steady state thermal image at 380V/5A with 650V/33mΩ Si N channel MOSFET. The stable hot spot is only 60.5 □ C with natural convection. The lower $R_{DS(ON)}$ device like latest CoolMOS or

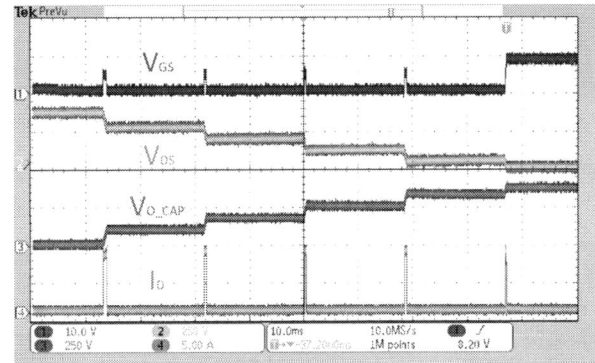

Figure 7. Hiccup startup process waveforms of DC hot pluggable outlet

978-1-4673-9551-9/16 $31.00 © 2016 IEEE

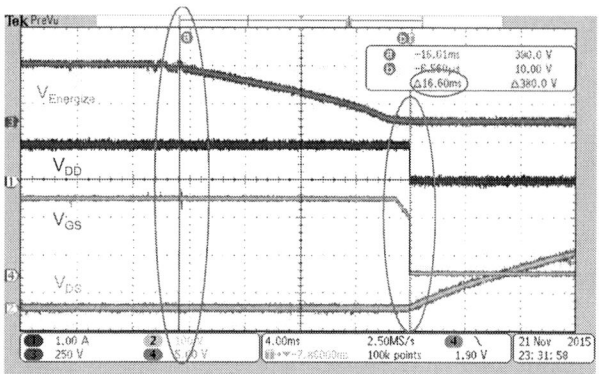

Figure 8. Unplug process waveforms of DC hot pluggable outlet

SiC MOSFET can handle even larger load. The total power loss is about 1W on the MOSFET and 0.4W on the power, sensing and drive circuit. Total 1.4W power loss and 0 stand by power loss makes the efficiency of the designed DC hot plugging outlet very high compared with the 2kW DC load.

C. Dual Threshold Over Current Portection

Figure 10 shows the waveforms of dual threshold over current protection. From waveforms we can find that once the sensed current is over $I_{threshold}$ (=5.4A), the controller will trip the circuit after the preset t_{fault} (=680us) timer. Once the controller decides to trip, the circuit only spends about 1us to finish this process like an instantaneous trip.

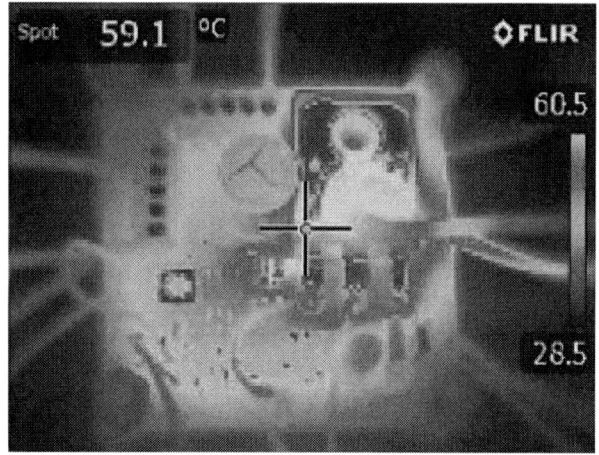

Figure 9. Normally on Steady state thermal image

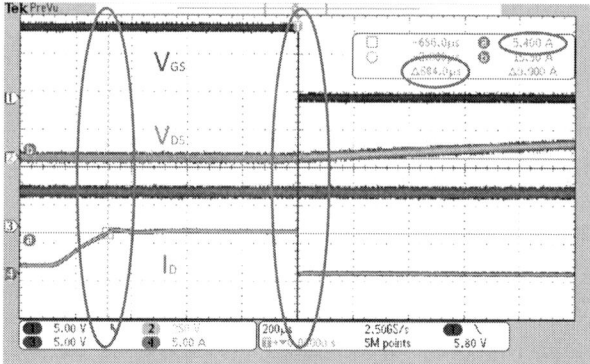

(a) I_{limit} OCP with programmed 680us timer

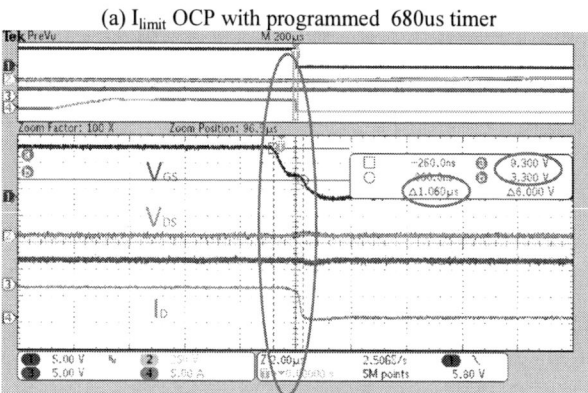

(b) The transient trip speed after present timer
Figure 10. Lower threshold I_{limit} with timer over current protection waveforms of DC hot pluggable outlet

Figure 11. Over current protection waveforms for Short circuit low impedance fault

Figure 11 shows the waveforms of I_{SC} over current protection. Once the current is larger than setup value 9A, the designed circuit starts to limit the current to I_{SC} with adaptive V_{GS} to make the MOSFET work in saturation mode. The V_{DS} starts to increase because the low impedance fault pulled down the voltage on output capacitor. The increase is not that much because the output capacitor is 47uF. With smaller output capacitor, the V_{DS} increases much faster to a high voltage and the power stress on MOSFET needs to be considered with setup trip time ensure the safe operation area with MOSFET.

This feature is very important to protect the bus voltage from being affected when a low impedance fault occurs on one single outlet and isolate the faulted load with other healthy loads in the neighborhood.

D. Over Temperature Protection

Figure 12 shows the over temperature protection of the MOSFET. From the waveforms we can find that once the sensed temperature reaches 100 C (1.23V from Vo of sensor), the circuit will be tripped by the controller in about 1.68 us.

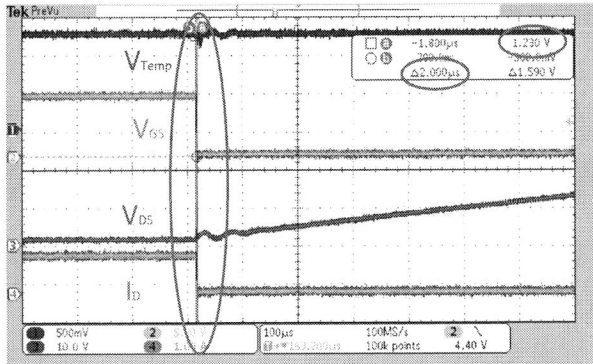

Figure 12. Over temperature protection waveforms of DC hot pluggable outlet

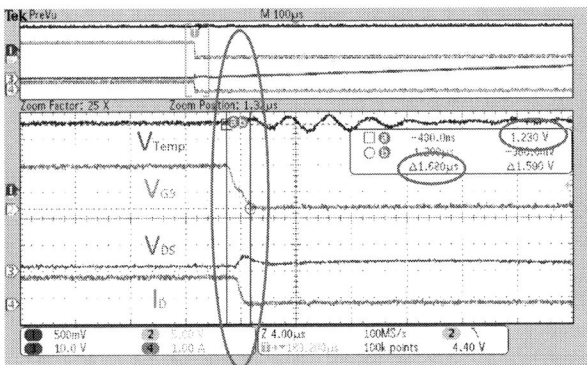

Figure 13. Over temperature protection waveforms (transient)

V. CONCLUSIONS

A novel hot plugging outlet is proposed and developed for the 380V DC distribution system with the following important features:

1. The standby power of the outlet is zero when no load is plugged in.

2. The novel outlet design achieves natural "the last energize, first de-energize" for the downstream equipment.

3. The embedded solid state device helps to limit the inrush current and eliminate the arc with low power consumption. The total power loss from is about 1.4W under 2 kW (380V, 5A) load condition.

4. Dual threshold current trip is realized to protect the upstream equipment against high or low impedance faults.

6. Over temperature protection is designed and embedded for protecting the solid state device from exceeding the thermal limit.

REFERENCES

[1] D. J. Becker and B. J. Sonnenberg, "DC microgrids in buildings and data centers," *2011 IEEE 33rd Int. Telecommun. Energy Conf.*, pp. 1–7, 2011.

[2] Y. Zhang, J. Umuhoza, H. Liu, F. Hossain, C. Farnell, and H. A. Mantooth, "Realizing an Integrated System for Residential Energy Harvesting and Management," pp. 3240–3244, 2015.

[3] Y. Liu, C. Farnell, J. C. Balda, and H. A. Mantooth, "Laboratory Test Bed," pp. 697–702, 2015.

[4] M. Ton and B. Fortenbery, "DC Power for Improved Data Center Efficiency," no. March, pp. 1–74, 2008.

[5] G. Allee and W. Tschudi, "Edison redux: 380 Vdc brings reliability and efficiency to sustainable data centers," *IEEE Power Energy Mag.*, vol. 10, no. 6, pp. 50–59, 2012.

[6] E. Alliance, "380 Vdc Architectures for the Modern Data Center," pp. 1–29, 2013.

[7] D. E. Geary, "380V DC POWER FOR DATA CENTERS (AN ENGINEERING PERSPECTIVE)."

[8] D. E. Geary, U. Electric, C. Starline, and D. C. Solutions, "380V DC Eco-system Development Present Status and Future Challenges Reported 380VDC Distribution Advantages," no. October, 2013.

[9] T. Lai and P. Manager, "Using 380vDC Power Feeds for Data Centers."

[10] D. P. Symanski, "380VDC Data Center At Duke Energy From Dept of Energy Secretary Steven Chu," *Power*, 2010.

[11] R. Mehl and P. Meckler, "Comparison of advantages and disadvantages of electronic and mechanical Protection systems for higher Voltage DC 400 V," no. October, pp. 236–242, 2013.

[12] S. Baek, T. Yuba, K. Kiryu, A. Nakamura, H. Miyazawa, M. Noritake, and K. Hirose, "Development of plug and socket-outlet for 400 volts direct current distribution system," *8th Int. Conf. Power Electron. - ECCE Asia "Green World with Power Electron. ICPE 2011-ECCE Asia*, pp. 218–222, 2011.

[13] B. Davies, "Analysis of inrush currents for DC powered IT equipment," *2011 IEEE 33rd Int. Telecommun. Energy Conf.*, pp. 1–4, 2011.

[14] R. M. Cuzner and G. Venkataramanan, "The status of DC micro-grid protection," *Conf. Rec. - IAS Annu. Meet. (IEEE Ind. Appl. Soc.*, pp. 1–8, 2008.

[15] A. P. Products, "Safe Connections for Higher Voltage Power Distribution Systems."

[16] K. Features, "FUJITSU Component Power Distribution Units 10A-430V DC Outlet Units , Socket-outlets and Plugs for DC Power Distribution Systems," 1956.

[17] I. B. M. Power and T. Symposium, "Outline of Presentation Reliable DC Interconnections."

[18] I. Cvetkovic, "Modeling , Analysis and Design of Renewable Energy Nanogrid Systems," pp. 1–118, 2010.

Design of Control System for Smooth Mode Transfer in Smart Microgrid Application

Mingzhi Gao, Canhui Zhang, Maohang Qiu, Min Chen
Department of Applied Electronics
Zhejiang University
Hangzhou, China
Xiaosan433@zju.edu.cn

Aron Levy
Technology Dynamics Inc.
Bergenfield, NJ, United States
aron.levy@theallpower.net

Abstract—**A complete control system of smart microgrid is proposed in this paper, in order to realize the smooth mode transfer of the grid-connected mode (GTM) and islanding mode (ILM). In the system, every DG unit will work as the voltage source and use the same control method of output voltage, whatever in GTM or ILM. The only difference of GTM and ILM is how to regulate of the reference voltage of the voltage control. Meanwhile, a grid-tracking control loop is proposed in order to synchronize the AC bus of microgrid with the grid before microgrid switches from ILM to GTM. Finally, the experiment results are presented, which validates the performance of the smooth mode transfer.**

Keywords—Smooth mode transfer; Voltage source; Microgrid

I.   INTRODUCTION

Traditionally, distributed generations (DG) of microgrid usually work as the current sources to realize the accurate power-flow control in the grid-tied mode (GTM). In the islanding mode (ILM), at least one DG should work as the voltage source to sustain local voltage [1]. Meanwhile, if there are more than two voltage-source DGs, the control strategy of voltage-source parallel operation should be introduced to realize the accurate power sharing and eliminate the circulating current [2].

When the operation modes switch between GTM and ILM, the outputs of the control strategy of voltage source and current source may not be equal during the transfer instant, which will cause the current or voltage spikes during the switching process [3]. On the other hand, if the DG units can work as the voltage source in GTM by adopting a control method which is analogous to that in ILM, it will be much easier to realize the smooth mode transfer than traditional strategy of mode transfer [4].

Based on this assumption, this paper designs a complete control system for the whole microgrid, including the control strategies of GTM, ILM and mode transfer. In this control system, the most important characteristic is that every DG unit will work as the voltage source and use the same control method of output voltage, whatever in GTM or ILM. The difference of GTM and ILM is how to regulate of the reference voltage V_{ref} of the output voltage control.

This work was supported by National Natural Science Foundation of China (Project 51477153).

In this paper, GTM adopts an improved grid-tied control strategy of voltage-source inverter. ILM adopts an improved parallel-operation control strategy, which is proposed in [5]. Moreover, a grid-tracking control loop is proposed in this paper, which is used to synchronize the AC bus voltage U_{bus} of microgrid with the grid voltage V_{grid} before microgird switches from ILM to GTM. Subsequently, the experiment results are presented in section VI, which validates the performance of the smooth mode transfer. Finally, some conclusions of this paper are given in section VII.

II.   STRUCTURE OF MICROGRID SYSTEM

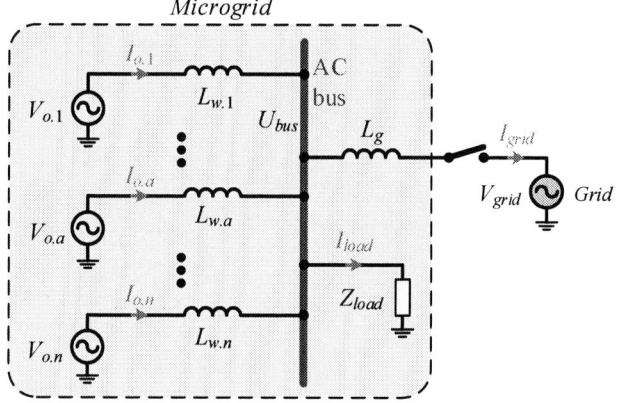

Fig. 1.   Diagram of microgrid system

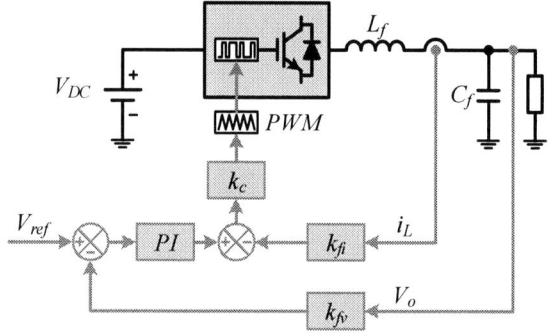

Fig. 2.   Control method of output voltage

Fig. 1 expresses the diagram of microgrid system, which consists of *n* voltages-source inverters. The letter "*a*" represents any inverter in the system. $V_{o.a}$ and $I_{o.a}$ is the output voltage and current of any inverter respectively, $L_{w.a}$ is the wire inductor between the inverter and AC bus, which comply with the relation of (1).

$$k_a \cdot L_{w.a} = L_e \quad (a = 1...n) \tag{1}$$

In this equation, k_a is the weighted coefficient of every inverter, which is related with the rated output power $S_{R.a}$ of every inverter, as shown in (2).

$$k_a = \frac{S_{R.a}}{\sum_{j=1}^{n} S_{R.j}} \tag{2}$$

In the microgrid, every VSI inverter will work as a robust voltage source and use the same control method of output voltage, whatever in GTM or ILM. The voltage control method is presented as shown in 0. The only difference of GTM and ILM is how to regulate the reference voltage V_{ref} of every inverter.

III. CONTROL STRATEGY IN GRID-TIED MODE

The principle of grid-tied control strategy of VSI in GTM can be described as Fig. 3. And the control equation of every inverter is presented as (3).

$$\begin{cases} \omega_{a.k} = \omega_{a.k-1} - m_{\omega.a}(P_{o.a.k-1} - P_{ref.a.k-1}) \\ V_{a.k} = V_{a.k-1} - n_{v.a}(Q_{o.a.k-1} - Q_{ref.a.k-1}) \end{cases} \tag{3}$$

In this equation, the letter k and $k-1$ represents the kth and $(k-1)^{th}$ control cycle, ω_a and V_a is the angular frequency and amplitude of the inverter's voltage. Meanwhile, $\omega_{a.0}$ and $V_{a.0}$ is the initial value of ω_a and V_a when k equals to 0, and the initial value is defined as the rated angular frequency ω_r and the rated amplitude V_r.

Moreover, $P_{o.a}$ and $Q_{o.a}$ is the actual active and reactive output power of VSI, $m_{\omega.a}$ and $n_{v.a}$ is the droop control coefficient of $P_{o.a}$ and $Q_{o.a}$ which complies with the relation of (4). $P_{ref.a}$ and $Q_{ref.a}$ is the assigned active and reactive power of VSI injected to the grid which is determined by itself.

$$\begin{cases} m_{\omega.a} \cdot k_a = m_{\omega.e} \\ n_{v.a} \cdot k_a = n_{v.e} \end{cases} \quad (a = 1...n) \tag{4}$$

The FFT module is used to analyze the harmonic components of grid voltage. Based on the results of FFT module, the harmonic compensation module will calculate the harmonic reference voltage $V_{ref.H}$ to simulate the harmonics of grid, which is used to eliminate the harmonic current injected into the grid. In the harmonic compensation, only the low-frequency odd harmonics will be analyzed, such as 3rd - 9th odd harmonics, and the high-frequency harmonics will be ignored.

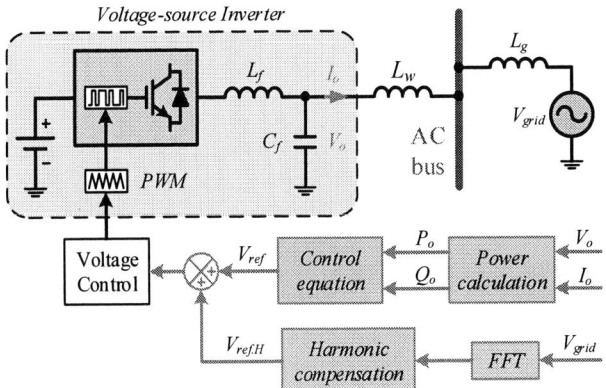

Fig. 3. Control strategy of VSI in GTM

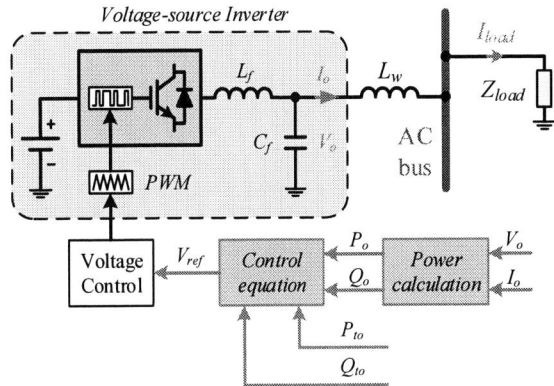

Fig. 4. Control strategy of VSI in ILM

IV. CONTROL STRATEGY OF ISLANDING MODE

The principle of islanding control strategy can be described as Fig. 4, which is proposed in [5]. The control equation of every inverter is presented as (5), which is analogous to (3). In this equation, $m_{\omega.a}$ and $n_{v.a}$ also complies with the relation as shown in (4).

$$\begin{cases} \omega_{a.k} = \omega_r - m_{\omega.a}(P_{o.a.k-1} - P_{ref.a.k-1}) \\ V_{a.k} = V_{a.k-1} - n_{v.a}(Q_{o.a.k-1} - Q_{ref.a.k-1}) \end{cases} \tag{5}$$

The major difference of (3) and (5) is, (5) replace $\omega_{a.k-1}$ with the rated angular frequency ω_r. The reason is that, the AC bus voltage U_{bus} is controlled by grid in GTM, and controlled by VSIs in ILM. Therefore, every VSI should track the frequency of grid In GTM, which has a certain variation range. While in ILM, every VSI should have a fixed frequency reference to sustain U_{bus} stable.

Moreover, $P_{ref.a}$ and $Q_{ref.a}$ in ILM is determined by the load sharing of microgrid, which is calculated by (6). Therefore the realization of (6) needs communication among VSIs to analysis the active and reactive power of the whole microgrid system.

$$\begin{cases} P_{ref.a} = k_a \cdot P_{to} = k_a \sum_{j=1}^{n} P_{o.j} \\ Q_{ref.a} = k_a \cdot Q_{to} = k_a \sum_{j=1}^{n} Q_{o.j} \end{cases} \quad (6)$$

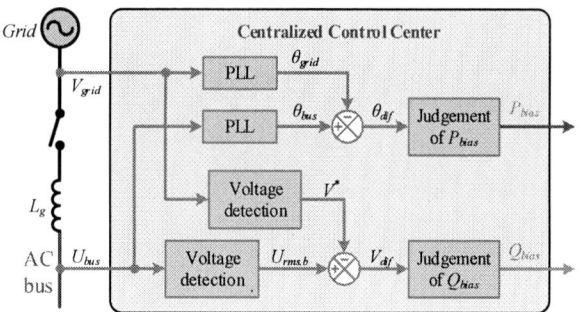

Fig. 5. Grid-tracking control strategy

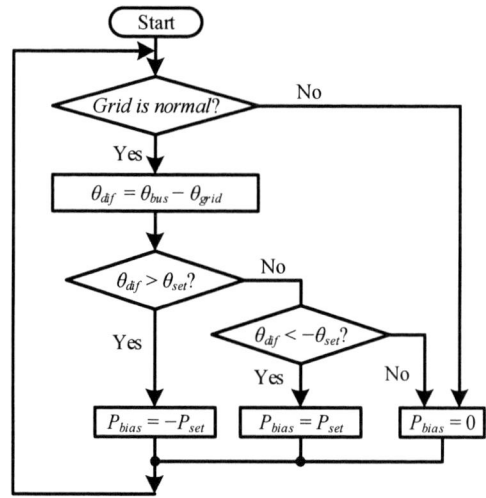

Fig. 6. Control method of phase tracking

V. GRID-TRACKING CONTROL STRATEGY

The AC bus voltage U_{bus} of microgrid should be synchronized with the grid voltage V_{grid} before microgird switches from ILM to GTM. In this paper, a grid-tracking control strategy is proposed to solve this problem. This method works as an additional loop of the control strategy of ILM to ensure U_{bus} has the same amplitude, phase and frequency with V_{grid}, and it will be deactivated when the grid is abnormal or unavailable.

The proposed grid-tracking method is realized by introducing the bias power P_{bias} and Q_{bias} in (5), which is shown as (7). And the diagram of this method can be presented as shown in Fig. 5.

$$\begin{cases} \omega_{a.k} = \omega_r - m_{\omega.a}(P_{o.a.k-1} - P_{ref.a.k-1}) - m_{\omega.a}k_a \cdot P_{bias.k-1} \\ V_{a.k} = V_{a.k-1} - n_{v.a}(Q_{o.a.k-1} - Q_{ref.a.k-1}) - n_{v.a}k_a \cdot Q_{bias.k-1} \end{cases} \quad (7)$$

In this method, a centralized control center (3C) is used to measure and analyze the voltages of AC bus and grid, and send the same P_{bias} and Q_{bias} to every VSI. The principle of phase tracking and voltage tracking can be presented by Fig. 6 and Fig. 7.

As shown in the two figures, when the grid is normal, 3C will calculate the phase difference θ_{dif} and amplitude difference V_{dif} of U_{bus} and V_{grid}. Subsequently, P_{bias} and Q_{bias} will be assigned a fixed value respectively according to the polarity of θ_{dif} and V_{dif}. The fixed value should guarantee U_{bus} can synchronize with V_{grid} during a given period, and avoid the voltage jump of U_{bus}.

When the grid is abnormal or unavailable, 3C will assign 0 to P_{bias}, so every VSI can control freely its own frequency around the rated angular frequency ω_r. Meanwhile, 3C will calculate the amplitude difference of U_{bus} and the rated value V_r, and guarantee the amplitude of U_{bus} in a reasonable range by Q_{bias}.

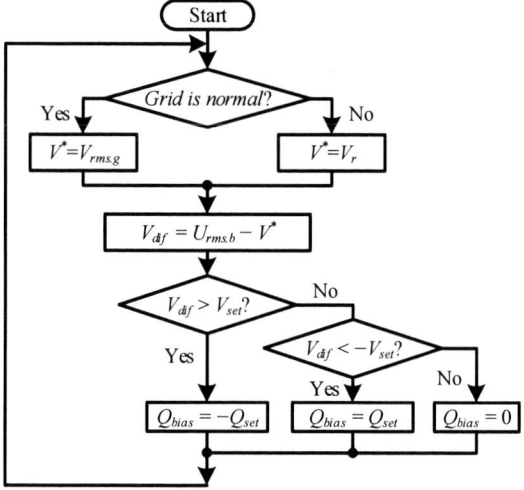

Fig. 7. Control method of amplitude tracking

VI. EXPERIMENT RESULTS

A prototype system of microgrid is implemented in order to verify the control system proposed in this paper. The prototype system is on the basis of the power stage of *Technology Dynamics Inc.*, which consists of a control center and two *3kVA* single-phase inverters. The diagram of the system can be expressed as Fig. 1.

In this system, every inverter consists of a single-phase IGBT full-bridge and an *LC* output filter, and is connected to AC bus with an inductor L_w. The RMS value and frequency of grid voltage is *115V* and *50Hz*, and the grid is connected to AC bus with an inductor L_g.

Parameters of the system are shown in TABLE I. , and parameters' values of the inverters are the same. And the experiment results of GTM, grid-tracking in ILM and mode transfers are shown as the following figures.

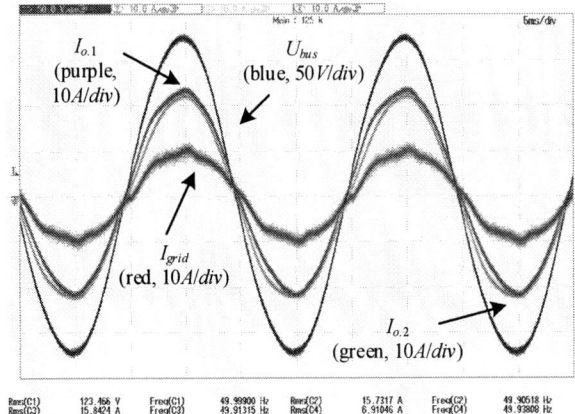

Fig. 8. Experiment results of GTM

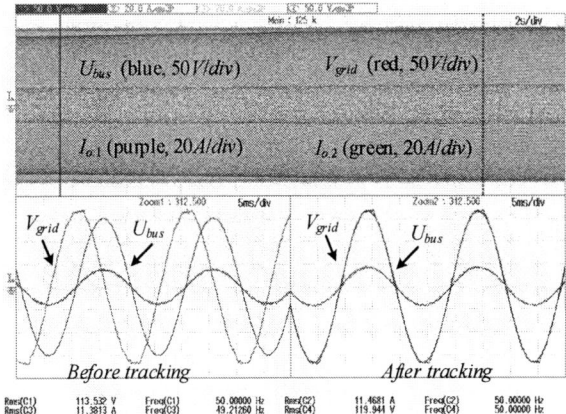

Fig. 9. Experiment results of grid-tracking in ILM

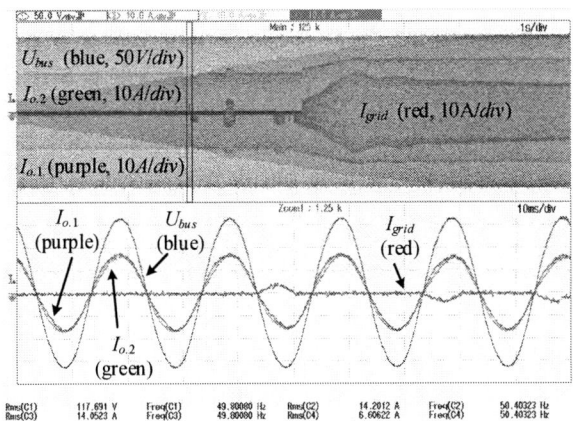

Fig. 10. Experiment results of mode transfer from ILM to GTM

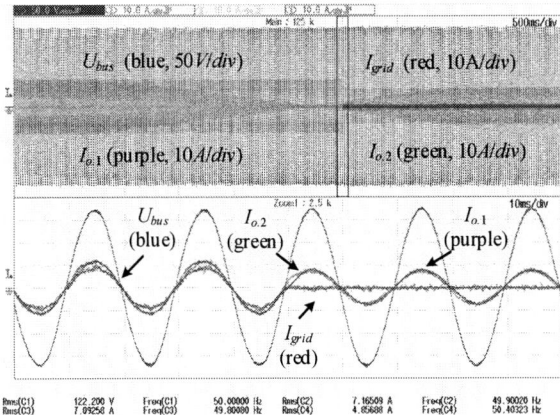

Fig. 11. Experiment of mode transfer from GTM to ILM

TABLE I.   PARAMETERS OF THE PROTOTYPE SYSTEM

Parameter	Values
Inductance of output filter (L_f)	$600\mu H$
Capacitance of output filter (C_f)	$20\mu F$
Wire inductance (L_w)	$250\mu H$
Grid inductance (L_g)	$2mH$
Droop coefficient of P_o (m)	$0.0001Hz/W$
Droop coefficient of Q_o (n)	$0.001V/Var$
Switching frequency (f_s)	$20kHz$

Fig. 8 shows the experiment results of the prototype system under the control strategy in GTM, when the P_{ref} and Q_{ref} of every inverter are assigned as $2000W$ and $90VA$. A linear load (5.1Ω) is connected with the AC bus, so about $850VA$ power is injected into the grid.

Fig. 9 shows the experiment results of the proposed grid-tracking method under the linear load (5.1Ω) in ILM. According to this figure, both the large phase difference and amplitude difference of U_{bus} and V_{grid} are eliminated by the grid-tracking control in about 14 seconds

Fig. 10 shows the experiment results of mode transfer from ILM to GTM, when the linear load is 5.1Ω. Meanwhile, P_{ref} and Q_{ref} of every inverter will be assigned as $2000W$ and $90VA$ after the mode transfers to GTM.

Fig. 11 shows experiment results of mode transfer from GTM to ILM, when the linear load is 11.1Ω. Meanwhile, P_{ref} and Q_{ref} of every inverter are assigned as $1000W$ and $20VA$ in GTM.

According to Fig. 10 and Fig. 11, the waveform of AC bus voltage has little vibration and distortion at the transient state of mode transfer, which prove the excellent smooth mode transfer performance of the proposed control system in this paper.

VII. CONCLUSIONS

A complete control system of smart microgrid is proposed in this paper, in order to realize the smooth mode transfer of the GTM and ILM.

In the system, every DG unit will work as the voltage source and use the same control method of output voltage, whatever in GTM or ILM. The only difference of GTM and ILM is how to regulate of the reference voltage of the voltage-source control. Therefore, this control system can easily realize the smooth mode transfer of the control methods between the GTM and ILM, and avoids the vibration and distortion of AC bus voltage simultaneously.

Moreover, a grid-tracking control loop is proposed for the control method in ILM, in order to synchronize the AC bus of microgrid with the grid before microgrid switches from ILM to GTM.

Finally, a prototype system of the microgrid is build, and the experiment results of GTM, grid-tracking in ILM and mode transfers are presented, which validates the excellent smooth mode transfer performance of the proposed control system in this paper.

ACKNOWLEDGMENT

The authors would like to thank Technology Dynamics Inc., for the support of the inverter power stage during the project. The authors also would like to thank PLEXIM Inc., for the support of the powerful simulation tools PLECS.

REFERENCES

[1] Peng, F.Z., Yun Wei Li, Tolbert, L.M., "Control and protection of power electronics interfaced distributed generation systems in a customer-driven microgrid," Power & Energy Society General Meeting, 2009. PES '09. IEEE , pp.1-8, 26-30 July 2009

[2] J.M. Guerrero, Loh P C, Tzung-Lin Lee, Mukul Chandorkar, "Advanced control architectures for intelligent microgrids—Part II: Power quality, energy storage, and AC/DC microgrids," Industrial Electronics, IEEE Transactions on, 2013, 60(4): 1263-1270.

[3] Zhilei Yao, Lan Xiao, Yangguang Yan, "Seamless Transfer of Single-Phase Grid-Interactive Inverters Between Grid-Connected and Stand-Alone Modes," Power Electronics, IEEE Transactions on , vol.25, no.6, pp.1597-1603, June 2010

[4] Yunwei Li, D.M. Vilathgamuwa, Poh Chiang Loh, "Design, analysis, and real-time testing of a controller for multibus microgrid system," Power Electronics, IEEE Transactions on , vol.19, no.5, pp.1195-1204, Sept. 2004

[5] Mingzhi Gao, Canhui Zhang, Maohang Qiu, Weiheng Li, Min Chen, Zhaoming Qian, "An accurate power-sharing control method based on circulating-current power model for voltage-source-inverter parallel system," 2015 IEEE APEC, pp.1815-1821, 15-19 March 2015

Resonance Propagation Modeling and Analysis of AC Filters in a Large-Scale Microgrid

Yusi Liu[1], Chris Farnell, H. Alan Mantooth[2], Juan Carlos Balda[3], Roy A. McCann[4], Cheng Deng
National Center for Reliable Electric Power Transmission, Department of Electrical Engineering
University of Arkansas
Fayetteville, AR, USA
[1]yusiliu@uark.edu, [2]mantooth@uark.edu, [3]jbalda@uark.edu, [4]rmccann@uark.edu,

Abstract— Low-pass ac filters are frequently adopted in microgrid power electronic interfaces that convert distributed generators' dc power to ac power because most of today's distribution grids have ac voltages. Compared to a simple L filter, higher-order filters, such as LC or LCL filters, are preferred due to their more effective reduction of switching-frequency harmonics and smaller sizes. As the converter power ratings increase, resonance problems caused by the higher-order filters challenge microgrid stability. This paper investigates a microgrid resonance propagation circuit model, which includes paralleling multiple converters with their filters. Each converter has a power rating in the MW range. The converter control design is illustrated and hardware experimental results validate the design and modeling.

Index terms— *microgrid, voltage source converter, high power, resonance propagation*

I. INTRODUCTION

The microgrid having the ability to operate in both grid-connected and islanded modes, may be one of the best concepts for enabling integration of distributed generation (DG) such as photovoltaic (PV), battery, fuel cell and combined heat and power (CHP) [1]-[3], while enhancing the power reliability for critical loads. Moreover, it is able to provide uninterruptable power from DGs when the main power grid is unavailable. Power electronic converters using pulse width modulation (PWM) methods are well adapted to convert power between the grid and DGs in the applications mentioned above [1]-[3].

Low-pass ac filters are extensively studied and successfully applied for reducing high-frequency PWM harmonics [7]. LCL and LC filters are preferred in industrial applications because they have higher damping capability (up to -60 *dB/dec* at the PWM switching frequency) compared to L filters (up to -20 *dB/dec*) [8]. Therefore, the volume of high-order filters could be smaller which makes them a cost-effective solution for the mitigation of the harmonic currents and voltage problems of voltage source converter (VSC) applications in microgrids.

Most literature refers to microgrids which have DG's power ranges in kW or tens of kW. Recently, high-power converters, such as central solar inverters, high-power fuel cells or battery stations, receive increasing attention because they may achieve better overall efficiency, easier management/maintenance and lower cost over power ratios ($/kW). When a single DG's power rating is in the MW range, it becomes more demanding to design higher-order ac filters because the filter sizes are

becoming larger due to higher current ratings and their costs become a major portion of the entire system's cost.

The DG's ac filter inductor values are usually designed to be larger than the grid short-circuit equivalent impedance in a low-power application, so the filter designer may neglect the grid impedance affects. However, in the case of designing high-power filters (in the MW range) and considering it operates within a weak grid (or islanded mode), the filter inductor values could be on the same order (or even smaller) when compared to the grid short-circuit impedance. A change of grid equivalent impedance, which happens quite often in a microgrid, could greatly affect the impedance "seen" by a high-power DG. Furthermore, the PWM switching frequencies of a MW power converter (considering most off-the-shelf silicon products) are usually limited to a few kHz due to the heat stresses. These constraints make high-power microgrid converter design more challenging.

The remainder of the paper is organized as follows: a brief overview of a microgrid test bed (MGTB) prototype and its high power converters is given in Section II. The modeling of high power converter with its controller is presented in Section III. The resonance propagation issue caused by paralleling multiple converters is discussed in Section IV. Experimental results are illustrated in Section V. The conclusions and future work are given in Section VI.

II. OVERVIEW OF THE MICROGRID TEST BED

The concept of the high power microgrid is sketched in Fig. 1. It represents a microgrid that spans a few miles and has medium voltage power distribution lines for reducing the low voltage losses. It mimics microgrid applications of a small town in a remote area, an industrial park, a military base, etc. A high-power medium voltage MGTB is under construction in the authors' lab [9] for high-power microgrid research. It includes: DG emulators (DGE), power electronic interfaces, medium-voltage (MV) transformers, critical/noncritical loads, power lines (overhead lines or cables) and a microgrid central controller (communications with each DG).

The one-line diagram of MGTB prototype is shown in Fig. 2. A three-phase, 1.5 MVA utility transformer (UT) connects the test facility to a 12.47-kV distribution line. The main service bus1 (MSB1) is a 480-V ac bus that feeds several low-voltage circuit breakers (LVCBs) and connects to the microgrid

---

This project is supported by NSF I/UCRC GRid-connected Advanced Power Electronics Systems (*GRAPES*).

978-1-4673-9551-9/16 $31.00 © 2016 IEEE

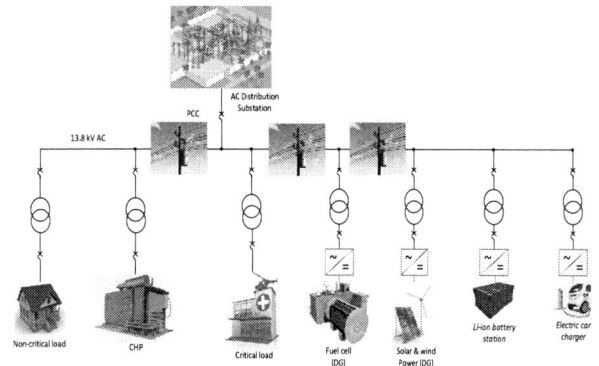

Fig. 1. Concept of a large-scale microgrid.

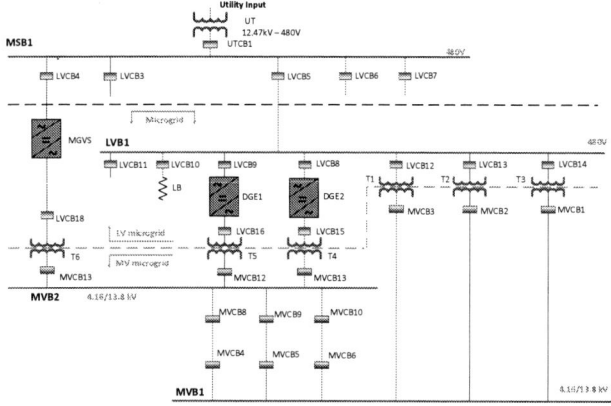

Fig. 2. One-line diagram of the microgrid test bed.

voltage source (MGVS) converter, through LVCB4. The MGVS is able to provide power to the rest of the microgrid, as an MG forming inverter [10] in islanded mode, through transformer T_6 and MV bus MVB2. T_1 though T_6 are 0.48Δ/4.16×13.8Y kV 2.5-MVA transformers. The MV components are used to emulate the microgrid MV distribution line. All converters have the same topologies of a back-to-back (B2B) three-phase, two-level circuit as shown in Fig. 3 (corresponding to the red blocks in Fig. 2).

The voltage source converter (VSC) is able to emulate the characteristics of many electric systems and equipment [11]-[13]. Two distributed generator emulators DGE1 and DGE2 have been built as shown in Fig. 2. Each DGE consists of B2B ac-dc converters and is able to work only in the grid-connected mode [10]. One advantage of the B2B converter is that it is able to recirculate/regenerate the real power during high-power tests [6]. The black dashed line indicates the virtual microgrid.

Each VSC is controlled by a Texas Instruments F2812, a 32-bit fixed-point digital signal processor (DSP). Xilinx Field Programmable Gate Arrays (FPGAs) are also installed to provide redundant and faster overcurrent and over temperature protection; they are able to detect fault signals faster than the DSP PWM interrupt and turn off all IGBTs before the next DSP switching period happens (PWM switching signals are sent to FPGA from the DSP, then bypassed to IGBT gate drivers during normal operation).

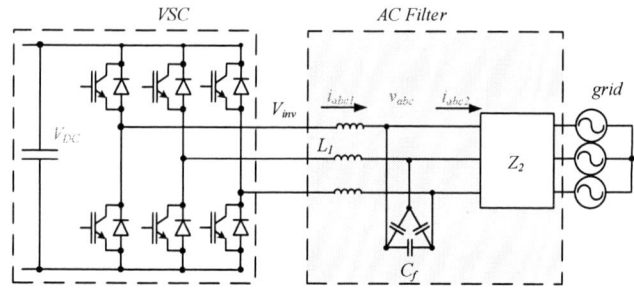

Fig. 3. Three-phase grid-connected converter with ac filter.

TABLE I.   VSC PARAMETERS

Parameter		Nominal Value
MGVS rated power	S_{MGVS}	0.75 MVA
DGE1 rated power	S_{DGE}	2 MVA
MGVS IGBT swiching frequency	$f_{sw\text{-}MGVS}$	8 kHz
DGE1 IGBT swiching frequency	$f_{sw\text{-}DGE}$	4 kHz
MGVS ac inductor	$L_{ac\text{-}MGVS}$	110 μH (0.135 p.u.)
DGE1 ac inductor	$L_{ac\text{-}DGE}$	20 μH (0.065 p.u.)
ESR of *DGE1* ac inductor	$R_{L\text{-}DGE}$	0.8 mΩ (0.007 p.u.)
MGVS ac capacitor	$C_{f\text{-}MGVS}$	Δ3× 1920 μF
DGE1 ac capacitor	$C_{f\text{-}DGE}$	Δ3× 480 μF
ESR of *DGE1* ac capacitor	$R_{cf\text{-}DGE}$	0.2 mΩ
DC link capacitor	C_{DC}	46.2 mF
Rated ac voltage	v_{ac}	480 V

TABLE II.   SYSTEM P.U. BASE VALUES OF 2 MVA VSC

Parameter	Formula	Nominal Value	
Power	S_B	-	2.0	MVA
Voltage	V_B	-	480	V
Frequency	f_B	-	60	Hz
Current	I_B $\dfrac{P_B}{\sqrt{3}V_B}$	2400	A
Radian frequency	ω_B $2\pi f_B$	377	Rads/s
Base impendance	Z_B $\dfrac{V_B^2}{P_B}$	0.115	Ω
Base Inductance	L_B $\dfrac{z_B}{\omega_B}$	300	μH
Base capacitor	C_B $\dfrac{1}{z_B\omega_B}$	23	mF

MGVS and DGE parameters are shown in Table I. Each VSC has a LC filter connected to its PWM output V_{inv}. More details about the test bed were illustrated in [9]. It is well-accepted that designing filters is based on the converter per unit value [1]. Table II illustrates the corresponding base values of the high-power converter for the readers' convenience.

III. MODELING OF HIGH POWER CONVERTERS

Fig. 3 shows the basic topology and feedback signals of a DGE converter. If the impedance Z_2 between the VSC LC filter output and the infinite bus could be simplified to inductor L_2, the VSC connects to the infinite bus through an LCL filter, which is composed of the converter-side inductor L_1 ($L_{ac\text{-}DGE}$ in Table I), delta-connected filter capacitor C_f, and grid-side inductor L_2. The L_2 is the sum of inductance values of step-up transformer leakage inductor, distribution line, and the equivalent grid short-circuit inductor. Sometimes the VSC

978-1-4673-9551-9/16 $31.00 © 2016 IEEE   144

output impedance is more complicated than a simple LCL filter due to the changes of Z_2 which is discussed later in Section IV.

Two types of ac current feedback are adopted among designers: converter-side current i_{abc1} and grid-side current i_{abc2} as indicated in Fig. 3. Neglecting all equivalent series resistances (ESR) of passive components for straightforward inspection, two transfer functions from converter PWM voltage V_{inv} to ac output current i_{abc1} and i_{abc2} are given as

$$G_{V \to I1}(s) = \frac{I_1(s)}{V_{inv}(s)} = \frac{s^2 + \omega_0^2}{L_1 s(s^2 + \omega_{Res}^2)} \tag{1}$$

$$G_{V \to I2}(s) = \frac{I_2(s)}{V_{inv}(s)} = \frac{1}{L_1 L_2 C_f s(s^2 + \omega_{Res}^2)} \tag{2}$$

where $\omega_{Res} = 2\pi f_{Res} = \sqrt{(L_1 + L_2)/(L_1 L_2 C_f)}$ and $\omega_0 = 2\pi f_0 = \sqrt{1/(L_1 C_f)}$. The main difference between (1) and (2) is that (1) has a pair of zeros. Since $\omega_{Res}/\omega_0 = \sqrt{(L_1 + L_2)/(L_2)}$ is always greater than 1, firstly the zeros of (1) introduce a negative magnitude peak at ω_0 and raise the phase angle of the transfer function (1) $+180°$, then the resonant poles decrease the phase angle $-180°$ at the resonant frequency ω_{Res}. (2) has a theoretical -60 dB/dec noise damping rate while (1) has -20 dB/dec after ω_{Res}. Obviously, i_{abc2} has less current ripple compared to i_{abc1}. Most DGs are controlled as current sources which inject/extract real and reactive power to/from the grid, so grid-side currents are the intuitive choice for the feedback signal.

However, it is important that the converter is able to interrupt all IGBTs as fast as possible when there is a fault. i_{abc1} as a feedback current is a better choice in this case since it is physically closer to the IGBTs. Another advantage of i_{abc1} as the feedback signal is that (1) has an inherent damping term in its transfer function, compared to (2), as is explained in [14]. Considering these trade-offs, the converter-side current i_{abc1} is chosen as the current feedback signal to the controller in this high power application.

The inner current control loop has higher bandwidth than other outer control loops (such as power or voltage loops) so the current loop mostly determines the system dynamic response and stability. Only the current control loop design is described in this paper and its block diagram is illustrated in Fig. 4 [9]. A phase-locked loop (PLL) synchronizes the converter phase angle to the grid. The classic Park's transformation (abc-dq) is applied to achieve zero steady-state error of sinusoidal current tracking by proportional-integral (PI) controller. The converter controls its grid current i_{abc2} indirectly by controlling i_{abc1}. If the unity power factor at the point of common coupling (PCC) is preferred, the reactive current reference I_q^* should be set to $\omega_{60} C_f V_d$ to compensate for the reactive power consumed by the filter capacitor C_f, where ω_{60} is the grid fundamental angular frequency and V_d is the d-axis voltage in the d-q reference frame.

All state feedback signals are scaled to per unit values by either voltage gain H_v or current gain H_i as shown in Fig. 4, which are more convenient in the control design and necessary in a fixed-point DSP application (the F2812 DSP is not able to

directly process the nominal values of current and voltage in such high power applications based on its decimal point setting). The impedance decoupling term ωL is also converted to a per unit value using the base impedance Z_B.

For more realistic analyses, ESRs of the passive components in Fig. 3 are included in a later discussion, for example, Z_1 is the s-domain impedance of the converter-side inductor $Z_1 = sL_1 + R_{L1}$. Similarly, Z_2 and Z_3 are grid-side impedance and ac filter capacitor impedance, respectively. A per-phase equivalent ac filter model is sketched in Fig. 5.

Including ESR2, more general transfer functions of (1) and (2) are given by

$$G_{V \to I1}(s) = \frac{I_1(s)}{V_{inv}(s)} = \frac{Z_2 + Z_3}{Z_1 Z_2 + Z_1 Z_3 + Z_3 Z_2} \tag{3}$$

$$G_{V \to I2}(s) = \frac{I_2(s)}{V_{inv}(s)} = \frac{Z_3}{Z_1 Z_2 + Z_1 Z_3 + Z_3 Z_2} \tag{4}$$

Combining Figs. 4 and 5, the converter current closed-loop controller and filter block is simplified to Fig. 6.

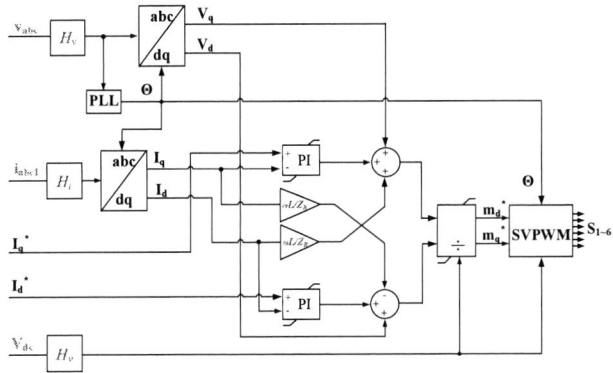

Fig. 4. Current loop control block.

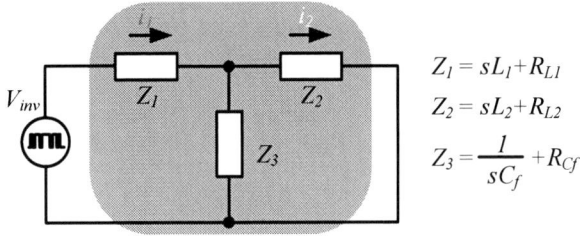

Fig. 5. Per-phase LCL filter model.

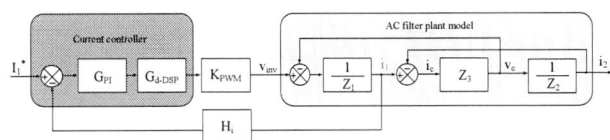

Fig. 6. Current control model.

The well-accepted PI compensator is applied here as G_{PI} and is given by

$$G_{PI}(s) = K_{i\_p} + \frac{K_{i\_i}}{s} \tag{5}$$

The power stage of a three-phase half-bridge can be modeled as a linear amplifier which has a gain $K_{PWM} = V_{dc}/2$. A practical DSP implementation uses a finite amount of time to complete its calculations after sampling the measured currents (it is finished in one DSP interrupt period T_{sw} here). A fair approximation of the DSP transport and sampling delay in this case is $G_{d-DSP} = e^{-1.5sT_{sw}}$. Without accounting for the DSP current controller, the current forward path transfer functions are given as

$$T_{1ol} = H_i K_{PWM} G_{V \to I1} \tag{6}$$

$$T_{2ol} = H_i K_{PWM} G_{V \to I2} \tag{7}$$

Figs. 7 (a) and (b) show the open-loop Bode diagrams for (7) and (6), respectively. Different L_2 values are considered for illustrating different microgrid situations. Z_1, Z_3, and all ESRs values are kept the same as shown in Table I. Defining the numbers of positive crossings of the phase -180° line is N$^+$, while the numbers of negative crossings of the same line is N$^-$. The Nyquist stability criterion concludes: in the frequency range with gains above 0 dB of the open-loop Bode diagram, the subtraction of (N$^+$ - N$^-$) must equal to zero for a stable system [15]. As shown in Fig. 7 (a), there is a negative crossing of -180° line but no positive crossing which indicates that the use of the grid-side current as a feedback signal may introduce system instability. This is another reason that the converter-side current is selected to be the feedback signal in this paper.

The converter-side current forward path transfer function including the DSP controller is given as

$$T_{1cl} = H_i G_{PI} G_{d-DSP} K_{PWM} G_{V \to I1} \tag{8}$$

The PI controller is able to shift the bandwidth (cross-over frequency $\omega_c = 2\pi f_c$) of the current loop to a desired value. In order to avoid the phase margin (PM) decreasing at f_c, the corner frequency of the PI controller $f_{crn} = K_{i\_p}/2\pi K_{i\_i}$ is usually tuned sufficiently lower than f_c. The magnitude of $G_{PI}(s)$ at f_c is almost the same as K_{i_p}. The current bandwidth is chosen at 350 Hz in this paper as a trade-off between a high magnitude gain at low frequencies and high damping capability around the PWM switching frequency. Based on the parameters shown in Fig. 7 (b), the filter gain $G_{V \to I1}$ is dominated by the sum of two inductors $s(L_1+L_2)$ at the cross-over frequency f_c. Substituting $G_{PI}(s) = K_{i\_p}$ into (8) and considering the loop gain is unity at f_c yields

$$1 = H_i K_{i\_p} K_{PWM} \omega_c (L_1 + L_2) \tag{9}$$

Using (9), the K_{i_p} value is able to be selected. Around the PI controller corner frequency f_{crn}, the magnitude slope of (8) changes from -40 to -20 dB/dec, and the phase rises from -180° to -90°, ideally. A small phase margin at f_c translates as a big overshoot in the time-domain response, so a small f_{crn} is preferred. Alternatively, a big f_{crn} is demanded to achieve a high gain in (8) at low frequencies for fast dynamic response and rejection of grid voltage disturbances. The integral gain is selected as a trade-off between two considerations mentioned above. As shown in the Fig. 7 (b), the phase response reaches -90° asymptote after f_{RES}. However, a delay term $G_{d\_DSP,}$ which is inevitable in any real application, pushes the phase response to -∞° as the frequency increases. Fig. 8 showed one design example of the Bode diagram of (8) It is clear that when L_2 increases, the system tends to be unstable (the phase crosses the -180° line while it has positive gain in the case of $L_2 = 100 \mu H$, around 2 kHz). This proves that if the microgrid becomes weak, the high power converter may become unstable.

It is noteworthy that the loop transfer functions of the high-power converter current are more vulnerable to grid impedance changes compared to low-power converters because the high-power filter inductors may be sized on a similar order to that of the grid impedance. For example, when L_2 increases

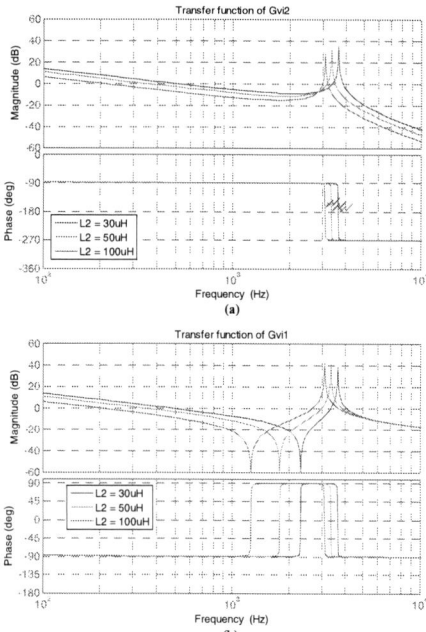

Fig. 7. Open-loop current forward path Bode diagram without considering DSP controller.

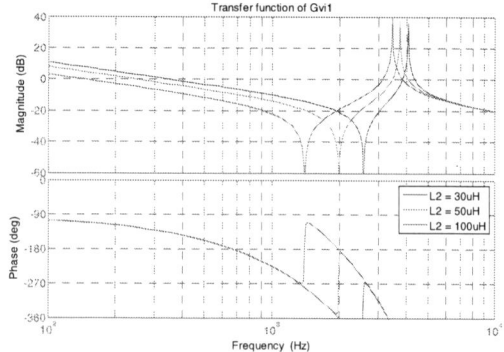

Fig. 8. Open-loop current forward path Bode diagram.

from 30 μH to 100 μH as shown in Fig. 8, (8) approaches to its stability margin. However, a grid-side inductor of low-power LCL filter could be designed in the range of mH [1]. The total L_2 only changes slightly because the grid impedance is only a small fraction of L_2, which benefits converter robustness.

IV. RESONANCE PROPAGATION

A future high-power microgrid may be supported by multiple ac-dc converters with different power ratings as shown in Fig. 1. Assuming that the DG VSCs connect to the ac grid at LV ports (three-phase 208 or 480 V ac in the U.S.), the high-power converters have lower output impedances compared to low-power converters as discussed previously. This section focus on paralleling multiple high-power VSCs since they are more important and more vulnerable during the microgrid dynamic transition.

There are totally four 2-MW ac-dc/dc-ac VSCs in DGE1 and DGE2 as shown in Fig. 2. All four converters have the identical LC filters whose parameters are shown in Table I. The ac-dc active front end (AFE) of DGE1 has been selected as the target under study in this section. Fig. 9 (a) shows the equivalent filter circuit of DGE1 in grid-connected mode from the VSC of DGE1 AFE, according to Fig. 2. Fig. 9 (b) shows the equivalent filter circuit of both DGE1 and DGE2. The dominant component of each impedance is depicted inside the impedance symbol block and all ESRs are considered. Table III briefly describes the physical components in Fig. 9.

Compared to Fig. 5 where the grid-side impedance Z_2 is mainly an inductor, Z_2 as indicated in the dashed blocks of Figs. 9 (a) and (b) are more complicated; they are higher-order circuits. Deriving a mathematical expression of the equivalent impedance Z_2 is cumbersome. Computing tools, such as MATLAB, are used here to calculate the impedance Z_2 and

TABLE III. PASSIVE COMPONENT DESCRIPTIONS

Symbol	Physical Component
Z_{2-1}	Cable
Z_{2-2}	Grid equivalent impedance
Z_{2-3}	MV transformer T_1
Z_{2-4}	MV transformer T_5
Z_{2-5}	DGE1 inverter filter capacitor
Z_{2-6}	MV transformer T_2
Z_{2-7}	MV transformer T_4
Z_{2-8}	DGE2 inverter filter capacitor
Z_{2-9}	DGE2 AFE filter capacitor

show the frequency responses of these parallel multiple converter circuits. Fig. 10 shows the Bode diagrams of the two examples given in Figs. 9 (a) and (b). The transfer function in (8) is used again, but G_{V_{VSC}/I_l} has now a more complex impedance term Z_2.

In the case of running one B2B DGE1 (two converters in parallel), the phase angle has the first negative crossing of the -180° line around 700 Hz (indicated by the blue line), then rises roughly 180° due to the zeros which are mainly caused by the DGE1 AFE filter capacitor around 1.8 kHz (the second is positive crossing of the -180° line). Thanks to the delay term G_{d_DSP}, the phase angle has the third negative crossing of the -180° line at around 2 kHz which is before the first resonant frequency (around 2.3 kHz). According to the Nyquist stability criterion, running DGE1 in the microgrid is stable.

However, running DGE1 and DGE2 together introduces a system instability illustrated in Fig. 10 (green line). Due to the additional capacitor of the DGE2 AFE LC filter, the zeros shift the phase angle 180° to around 1.5 kHz. The first negative peak frequency ω_0 is smaller compared to the case of DGE1. The phase angle has the third negative crossing of the -180° line at a frequency of around 2 kHz, while the gain has a value greater than 0. The first resonant gain reduces to 0 at a frequency around 2.2 kHz. Therefore, the Nyquist stability criterion is not satisfied as highlighted by the red arrow in Fig. 10. The resonance propagation is caused by paralleling multiple high-power converters in the microgrid.

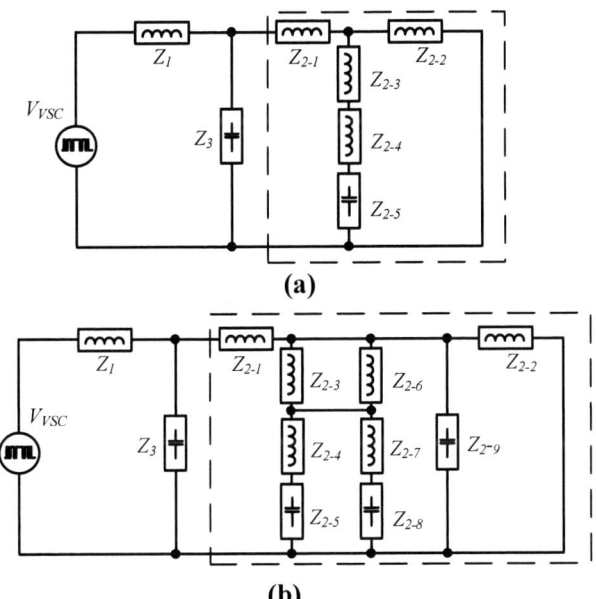

(a)

(b)

Fig. 9. Filter equivalent circuits of paralleled converters.

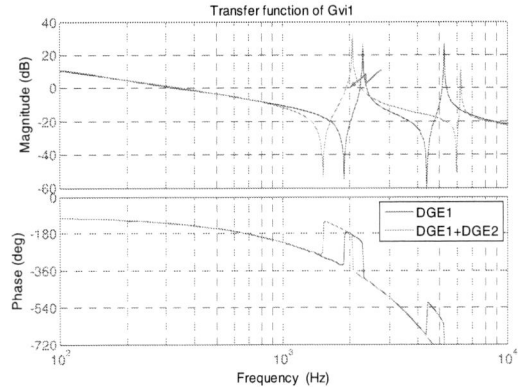

Fig. 10. Bode diagram of paralleling converters.

978-1-4673-9551-9/16 $31.00 © 2016 IEEE

V. EXPERIMENTAL RESULTS

The high power MGTB, which is shown in Section II, is used to verify the modeling proposed in the previous section. The LVCB5 is closed so all DGEs are in the grid-connected mode. The grid-side phase A currents of DGE1 AFE and inverter (as indicated as i_{abc2} in Fig. 3) are measured by Rogowski coils, respectively. Again, the grid-side currents are controlled indirectly since the current feedback is the converter-side current in this paper (as indicated as i_{abc1} in Fig. 3).

The current waveforms while running DGE1, which is basically recycling power by one set of B2B converter, are shown in Fig. 11 (a). The inverter active current reference is set to 0.5 p.u. and the system is stable. The result agrees with the Bode diagram of Fig. 10 (blue line).

The current waveforms of running DGE1 and DGE2 simultaneously, which is recycling power by two sets of B2B converters, are shown in Fig. 11 (b). It is clear that the AFE grid-side current has switching noise components. This coincides with the instability frequency which is obtained in Fig. 10. The inverter active current reference is only 0.2 p.u. in this case because the authors concern that further increasing the reference current may trigger the overcurrent protection.

VI. CONCLUSIONS AND FUTURE WORKS

The microgrid ac filter resonance propagation problem caused by paralleling VSCs in the MVA range was presented and analyzed in this paper. The design of closed-loop current control and its response were illustrated in detail. When

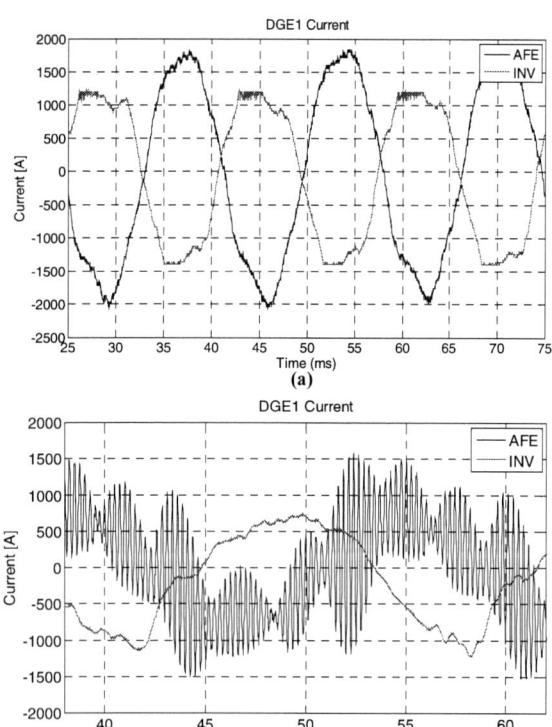

Fig. 11. Experimental current waveforms for the grid-connected mode.

multiple high power converters with their LC/LCL filters operate in a microgrid, the system stability does not only rely on a single converter, but multiple converter interactions.

High-power experimental results from paralleling different numbers of 2 MW converters validated the proposed modeling.

Future work for solving the resonance propagation (running DGE1 and DGE2 together) include: 1) install grid-side inductors at AFE to form LCL filters instead of LC filters; 2) digital filter or passive damping for reducing resonant peaks; 3) active damping by feeding back more state variables (such as filter capacitor current).

ACKNOWLEDGMENT

The authors are grateful for the financial support from the National Science Foundation Industry/University Cooperative Research Center on Grid-connected Advanced Power Electronics Systems (GRAPES).

REFERENCES

[1] M. Liserre, F. Blaabjerg, S. Hansen, "Design and control of an LCL-filter-based three-phase active rectifier," *IEEE Trans. Ind. Appl.*, vol. 41, no. 5, pp. 1281-1291, Sep.-Oct. 2005.

[2] R. H. Lasseter, J. H. Eto, B. Schenkman, J. Stevens, H. Vollkommer, D. Klapp, E. Linton, H. Hurtado, J. Roy, "CERTS Microgrid Laboratory Test Bed," *IEEE Trans. Power Del.*, vol. 26, no. 1, pp. 325-332, Jan. 2011.

[3] H. A. Mohammadpour, Y. Shin, E. Santi, "SSR analysis of a DFIG-based wind farm interfaced with a gate-controlled series capacitor," in *Proc. IEEE Applied Power Electronics Conference and Exposition (APEC)*, pp. 3110-3117, Mar. 2014.

[4] J. He, Y. Li, D. Bosnjak, B. Harris, "Investigation and Active Damping of Multiple Resonances in a Parallel-Inverter-Based Microgrid," *IEEE Trans. Power Electron.*, vol. 28, no. 1, pp. 234-246, Jan. 2013.

[5] L. Che, M. Shahidehpour, A. Alabdulwahab, Y. Al-Turki, "Hierarchical Coordination of a Community Microgrid With AC and DC Microgrids," *IEEE Trans. Smart Grid*, vol. PP, no. 99, pp. 1, Mar. 2015.

[6] K. Tan, R. Yu, S. Guo, A. Q. Huang, "Optimal design methodology of bidirectional LLC resonant DC/DC converter for solid state transformer application," in *Proc. IEEE IECON Annual Conference*, pp. 1657-1664, Oct. 2014.

[7] G. Shen, D. Xu, L. Cao, X. Zhu, "An Improved Control Strategy for Grid-Connected Voltage Source Inverters With an LCL Filter," *IEEE Trans. Power Electron.*, vol. 23, no. 4, pp. 1899-1906, Jul. 2008.

[8] M. Liserre, A.D. Aquila, F. Blaabjerg, "Genetic algorithm-based design of the active damping for an LCL-filter three-phase active rectifier," *IEEE Trans. Power Electron.*, vol. 19, no. 1, pp. 76-86, Jan. 2004.

[9] Y. Liu, C. Farnell, J. C. Balda, and H. A. Mantooth, "A 13.8-kV 4.75-MVA microgrid laboratory test bed," ," in *Proc. IEEE Applied Power Electronics Conference and Exposition (APEC)*, pp. 697-702, Mar. 2014.

[10] J. Rocabert, A. Luna, F. Blaabjerg, P. Rodríguez, "Control of Power Converters in AC Microgrids," *IEEE Trans. Power Electron.*, vol. 27, no. 11, pp. 4734-4749, Nov. 2012.

[11] J. Wang, Y. Ma, L. Yang, L. M. Tolbert, F. Wang, "Power converter-based three-phase induction motor load emulator," in *Proc. IEEE Applied Power Electronics Conference and Exposition (APEC)*, pp. 3270-3274, Mar. 2013.

[12] W. Cao, Y. Ma, J. Wang, F. Wang, "Virtual series impedance emulation control for remote PV or wind farms," in *Proc. IEEE Applied Power Electronics Conference and Exposition (APEC)*, pp. 411-418, Mar. 2014.

[13] Y. Ma, L. Yang, J. Wang, F. Wang, L. M. Tolbert, "Emulating full-converter wind turbine by a single converter in a multiple converter

based emulation system," in *Proc. IEEE Applied Power Electronics Conference and Exposition (APEC)*, pp. 3042-3047, Mar. 2014.

[14] Y. Tang, P. C. Loh, P. Wang, F. H. Choo, F. Gao, "Exploring Inherent Damping Characteristic of LCL-Filters for Three-Phase Grid-Connected Voltage Source Inverters," *IEEE Trans. Power Electron.*, vol. 27, no. 3, pp. 1433-1443, Mar. 2012.

[15] J. Dannehl, M. Liserre, F.W. Fuchs, "Filter-Based Active Damping of Voltage Source Converters With Filter," *IEEE Trans. Ind. Electron.*, vol. 58, no. 8, pp. 3623-3633, Aug. 2011.

A New Bidirectional DC-DC Converter for Fuel Cell, Solar Cell and Battery Systems

Ankur Patel

Applications Engineering
Vicor Corporation
Andover, MA, USA
apatel@vicorpower.com

Abstract—This paper describes a new zero voltage switching (ZVS) bidirectional DC-DC converter (BDC) module. Compared to existing bidirectional DC-DC converter (BDC) modules for the same application, the new BDC module has the advantage of high efficiency, high power density and isolation. These advantages make the new BDC promising for medium and high power fuel cell, solar cell and battery applications where high power density, high efficiency, high reliability and lightweight power converters are needed. A real world implementation of 384 V to 48 V and 48 V to 384 V BDC is described, implemented and simulated using existing old components and proposed new power components for power level up to 1.65 kW. A new BDC provides a +13.5% increase in efficiency at 10% load, a +3.4% increase at 50% load and a +1.5% increase at 100% load in both forward and backward mode and more than double power density.

Keywords—Bidirectional DC-DC converter (BDC), power conversion modules, power components, power electronic building blocks, high efficiency, high power density, galvanic isolation, lightweight, high reliability, 380 V DC distribution

I. INTRODUCTION

A. The Need

The bidirectional DC-DC converter (BDC) along with energy storage has become a promising option for many high power systems, including battery applications in hybrid and electric vehicle, ups, telecom and grid, fuel cell and solar cell applications in renewable energy and so forth. All these high power systems call for high efficiency from light load to full load, high power density and lightweight.

B. Background of existing solution

Most existing ZVS bidirectional [1] and unidirectional [2] DC-DC converters are low power and their narrow input voltage range makes operation difficult to source and load in transient and steady state conditions. They often are paralleled to meet the high power requirements in the above applications as shown in Fig. 1. Parallel arrays are very efficient under high load conditions, but can suffer from inefficiency under light load or no load operation. Many techniques have also been developed to limit the no load power dissipation and improve the light load efficiency. However, these techniques are not effective in dealing with the fast load current increase. Moreover, AC distribution does not benefit renewable energy

Vicor Corporation sponsored this paper.

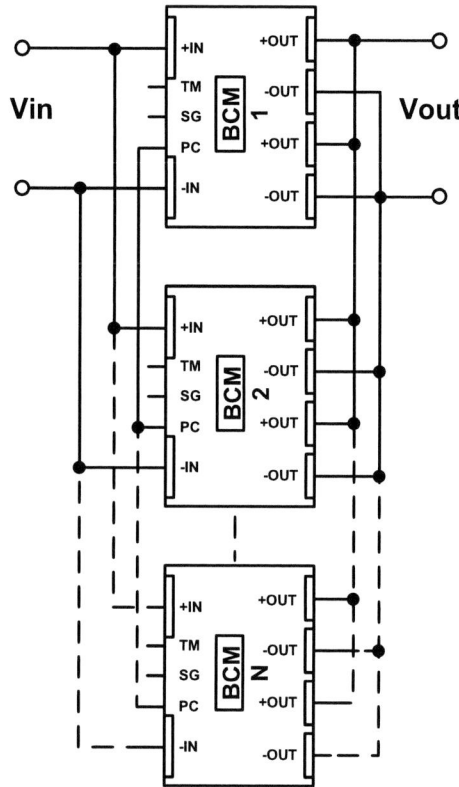

Fig. 1.   Existing Bidirectional DC-DC Converter for high power applications

generation that generates DC power, such as solar cells, fuel cells and wind turbine due to an increased number of energy conversion steps. Furthermore, using an AC line frequency transformer for isolation and voltage matching in the AC grid applications makes the bidirectional energy storage system heavy and bulky.

The proposed new designed BDC solves the problems of old and existing BDCs by utilizing its own technical attributes such as high efficiency, high power density and lightweight. It benefits the solar cell, fuel cell and battery systems in various application markets by reducing the number of conversion steps.

II. PROPOSED NEW BIDIRECTIONAL DC-DC CONVERTER

A. Theory of operation

This new BDC as shown in Fig. 2 is a fixed ratio DC-DC converter. It supports a wide input voltage range and high power [3], [4]. It is one of the important power electronic building block of the 384 V DC distribution. It uses unique Sine Amplitude Converter (SAC) Topology [5] with open loop control. Its primary and secondary circuit is transformer coupled to provide high frequency galvanic isolation. Primary circuit is stacked half bridge and secondary is center tap with synchronous rectification. Its voltage and current ratios are defined by following equation in ideal condition.

$$\frac{V_{OUT}}{V_{IN}} = \frac{I_{IN}}{I_{OUT}} = K \qquad (1)$$

Where V_{IN} = Input Voltage, V_{OUT} = Output Voltage, I_{IN} =Input Current, I_{OUT} = Output Current and K = Turns ratio or Transformation factor of fixed ratio DC-DC converter.

This new BDC works like a DC-DC transformer which uses a ZVS and ZCS soft switching techniques and operates at 1.1 MHz fixed switching frequency to provide low noise output voltage, high efficiency and high power density. All four input capacitors (C_{IN}) have the same characteristics to balance the voltage in stacked half bridge primary circuit. Voltage balancing helps to store the same energy in the inductor of both primary windings. The output voltage is proportional to the input voltage, minus the voltage drop due to the resistance. The resistance term is due to PCB resistance, MOSFET's R_{DS-ON} and transformer's winding resistance. Because of the resistance term and open loop control, output voltage has a natural droop in output voltage vs. output current graph.

So in order to realize the benefits and needs, the new BDC is designed with the following technical attributes.

1. Primary (input) side stacked half bridge to obtain low voltage MOSFET to achieve reduced conduction losses

2. ZVS/ZCS technology to achieve low switching losses with high frequency operation across the entire load range from light load to full load

3. Planar transformer design to achieve low transformer losses as well as solid galvanic isolation and therefore enhanced reliability

4. Resonant sine amplitude converter (SAC) topology control to achieve smooth bidirectional power flow control

B. Circuit implementation

This new BDC operates in two modes to provide bidirectional power flow as depicted in Fig. 2. The first is forward mode and the second is backward mode. It provides step-down DC-DC conversion when operating in a forward mode direction from input to output terminals. Forward mode is implemented by connecting a high voltage source to the

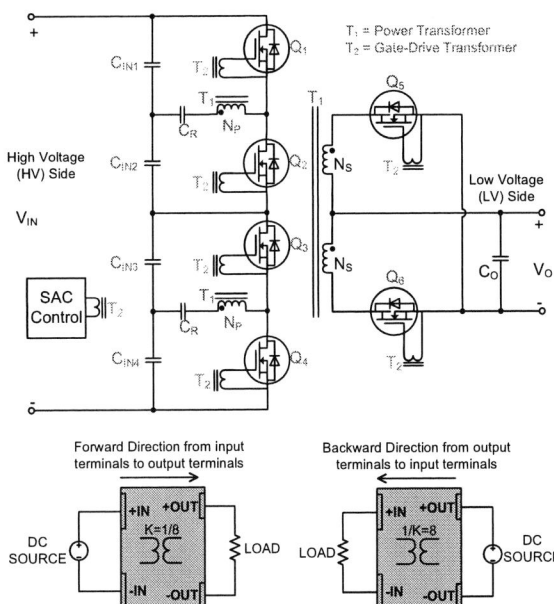

Fig. 2. Proposed New Bidirectional DC-DC Converter for high power applications

input terminals and low voltage load to the output terminals. The BDC turns on in a forward direction by applying the voltage to its input terminals between low line and high line. To enable backward mode, the BDC first needs to be turned on in forward mode because SAC control is on the input side. This is done by applying the low line voltage to the input terminals. When the voltage applied to output terminals exceed the turns ratio times the input voltage then the BDC starts processing the current in backward direction. It provides step-up DC-DC conversion when operating in a backward mode direction from output to input terminals. Backward mode is implemented by connecting a low voltage source to the output terminals and high voltage load to the input terminals. So in the backward mode, output is the input and input is the output.

In order to enable the backward power flow from the low voltage side, a flyback topology based start up circuit may be designed. The startup circuit receives its input from the low voltage side. It powers the high voltage side to provide the bias voltage and current to SAC control. The low line of BDC on the high voltage side is 260 V. There is an undervoltage turn on point below the low line where the BDC turns on. The startup circuit is designed to provide current of about 50 to 100 mA and a voltage of about 260 V.

The new BDCs are interfaced between high voltage bus (e.g. 384 V DC) and low voltage energy sources like solar cells, fuel cell, wind turbines and 12 V, 24 V and 48 V batteries. The BDC with transformation factor (K) of 1/8 is used to implement 384 V to 48 V bidirectional conversions.

III. SIMULATION

Simulation of existing and new BDC is performed using Vicor's power-bench whiteboard tool. Efficiency and power loss are analyzed for both BDCs at 384 V DC input, 50% load and 25 °C operating temperature as shown in Fig. 3 and Fig. 4.

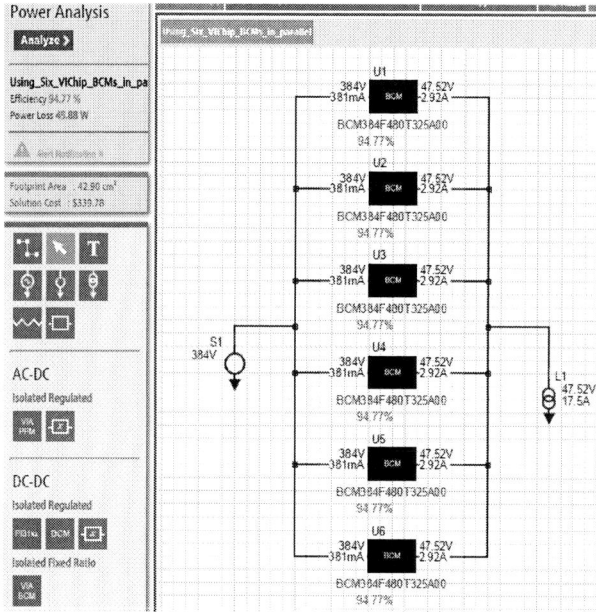

Fig. 3. Power analysis of existing bidirectional DC-DC converter using Vicor's six BCM384x480T325A00 modules

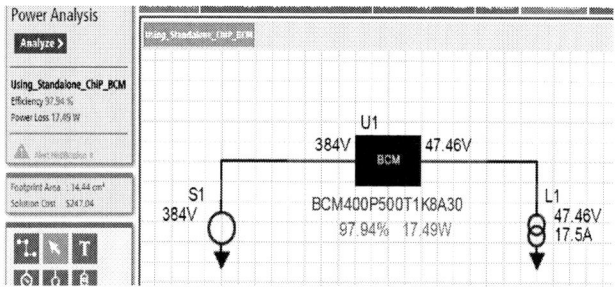

Fig. 4. Power analysis of proposed new bidirectional DC-DC converter using Vicor's single BCM400P500T1K8A30 module

The performance of both BDCs is noted in Table I based on power analysis simulation.

TABLE I. SIMULATION PERFORMANCE

Parameters	Existing BDC	New BDC
Efficiency (%)	94.77	97.94
Power loss (W)	45.88	17.49
Footprint Area (cm^2)	42.90	14.44
Solution Cost ($)	339.78	247.04

IV. SYSTEM IMPLEMENTATION AND EXPERIMENTAL RESULTS

Two PCB prototypes have been designed as shown in Fig. 5 and Fig. 6 to implement the existing BDC and new proposed BDC. This section also presents the experimental results of the existing and new BDC. The input voltage is 384 V in the forward mode direction and 48 V in backward mode direction. The load current range is 0 to 35 A in forward direction and 0 to 4.5 A in backward direction.

Fig. 5. System implementation of existing bidirectional DC-DC converter using six such boards

Fig. 6. System implementation of proposed new bidirectional DC-DC converter using single such board

The new BDC operates at 1.1 MHz switching frequency, compared to existing BDC at 1.75 MHz. This results in 37% lower switching frequency, which helps to lower the core losses and ultimately no load losses in new BDC.

A comparison in terms of measured efficiency between the proposed new and existing BDC is shown in Fig. 7 and Fig. 8. As can be seen, the proposed new BDC can achieve more than 96% efficiency from 15% load to 100% load in both directions, whereas the efficiency of the existing BDC approaches 96% at 100% load. In forward mode, at 384 V input and 25 °C temperature, the efficiency of the new BDC is maximum when its no load losses (9 W) are equal to resistive losses ($I^2_{OUT}*26.5$ mΩ). So the output current at peak efficiency point for a given condition is 18.4 A (52.6% load). The measured peak efficiency of the new BDC is 98% at 17.5 A (50% load) output current, which matches closely with analysis.

A comparison in terms of measured power loss between the proposed new and existing BDC is shown in Fig. 9 and Fig. 10. As can be seen, the new BDC has 30 W lower no load power dissipation than existing BDC in both directions. Lower no load power dissipation improves its efficiency in light load conditions. The power dissipation of new BDC is not only lower at no load, but also at other load conditions. The new BDC has 27 W lower power dissipation at 50% load and 19 W lower power dissipation at 100% load as compared to existing BDC.

A comparison in terms of power density between proposed new BDC, existing old BDC and old unidirectional DC-DC converter is shown in Fig. 11. Power density is measured at 1.65 kW power levels. As can be seen, the proposed new BDC provides the highest power density of 157.4 w/cm^3, which is

978-1-4673-9551-9/16 $31.00 © 2016 IEEE

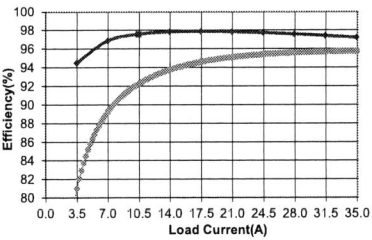

Fig. 7. Measured efficiency in forward mode at 384 V input

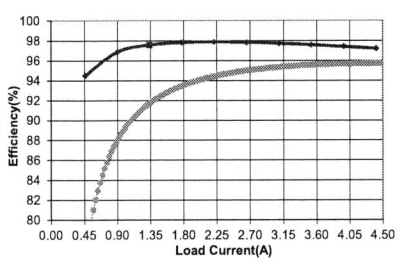

Fig. 8. Measured efficiency in backward mode at 48 V input

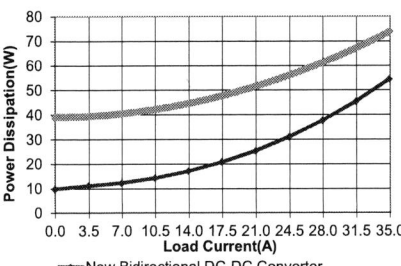

Fig. 9. Measured power dissipation in forward mode at 384 V input

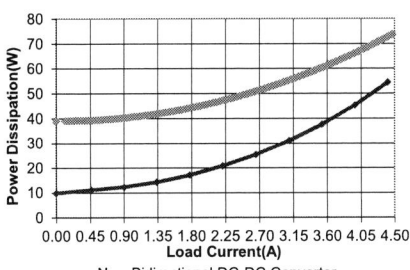

Fig. 10. Measured power dissipation in backward mode at 48 V input

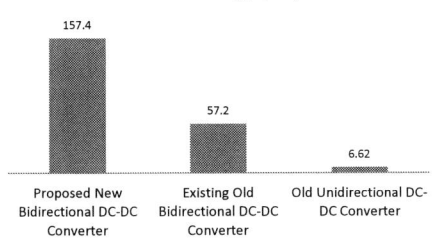

Fig. 11. Power density comparison between proposed new BDC, existing BDC and old unidirectional DC-DC converters

more than double the existing BDCs and old unidirectional DC-DC converters.

V. BENEFITS

The proposed new ZVS/ZCS bidirectional DC-DC converter has the following benefits for 384 V to 48 V bidirectional conversions.

A. Higher power capability

Maximum output current rating of existing BDC is limited to 7 A. Where as a proposed new BDC is rated up to 35 A. This higher output current rating provides higher output power.

B. Less number of converters

Existing BDC is rated for low power level up to 325 W. Six such BDCs are paralleled to achieve the power level of 1.65 kW where as a new BDC processes up to 1.65 kW of power in a single converter housed in a package.

C. Higher efficiency over extended load range

The total no load power dissipation of existing BDC is 39 W for 1.65 kW output power. The proposed single new BDC has lower no load power dissipation (fixed loss) and lower resistance (variable loss I^2R). Therefore, the efficiency of new BDC is higher than the existing BDC. High efficiency minimizes the need for cooling to remove heat generated by the converter.

D. Higher power density

The total volume of existing BDC is 28.9 cm^3, whereas the volume of new BDC is 10.48 cm^3. The new standalone BDC provides high power density by processing more power in a smaller form-factor.

Following Table II notes the performance gains of 384 V to 48 V new BDC for power level up to 1.65 kW.

TABLE II. PERFORMANCE GAINS OF NEW BDC

Parameters	Existing BDC	New BDC	Gains of New BDC
Number of converters	6	1	Less number of converters
Input Voltage Range (V)	360 to 400	260 to 410	Wider input voltage range
Efficiency at 10% load (%)	81	94.5	13.5% better
Efficiency at 50% load (%)	94.4	97.8	3.4% better
Efficiency at 100% load (%)	95.7	97.2	1.5% better
Output resistance (mΩ)	28.3	22.6	Lower resistive losses
No load power dissipation (W)	39	10	Lower no load losses
Volume (cm^3)	28.86	10.48	Occupy less PCB space
Power density (W/cm^3)	57.2	157.4	2.75 times more
Weight (g)	84	41	Weight is half

VI. APPLICATIONS

A. The future DC home model

The DC micro-grid [6], [7], depicted in Fig. 12, has two major sides: the high-voltage DC (HVDC) and the low-voltage DC (LVDC). The high voltage side is to be established at 384 V and the low voltage side is to be established at 48 V, 24 V or 12 V. The external AC Grid is expected to be supplying the majority of energy to the household at 230 V AC source. The rectifier will convert the AC voltage to DC and enter through the HV Bus and then proposed new BDCs in forward mode will convert the 384 V to 48 V, 24 V or 12 V for the electrical and electronic devices that are to be operated at 48 V, 24 V or 12 V. The ratings of rectifiers and BDCs will be based on the household loads.

In order to power the DC home, the AC voltage from the grid has to be converted into DC. A PFC boost rectifier is utilized to convert 230 V AC to 384 V DC. This is a centralized component, so any home electric and electronic devices utilizing the high voltage side of the micro grid will not require the rectification at each point of use, allowing for maximum efficiency by reducing the number of energy conversions. Also, by utilizing centralized rectification, the grid rectifiers are to be designed for higher load and capacity, which improves efficiency as well.

The proposed new BDC can also be easily interfaced between 384 V Bus and renewable energy sources such as solar cells, fuel cells and wind turbines for step-up DC-DC conversion. If renewable energy is utilized in the micro grid, then a 48 V, 24 V or 12 V battery should be attached as well in order to maximize the energy usage by storing unused energy from renewable sources. Once again, the proposed new BDC can be easily interfaced between a 48 V, 24 V or 12 V battery and 384 V Bus. The grid rectifier implemented is a bidirectional rectifier and capable of inversion, allowing for the grid to receive power from the home.

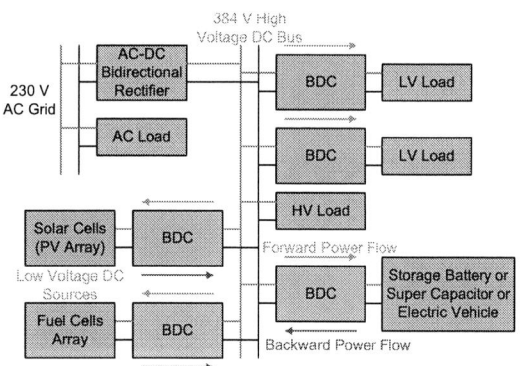

Fig. 12. Typical power distribution application in microgrid renewable energy system

B. The smart office buildings and commercial facilities

The above DC home model can also be utilized in smart office buildings or commercial facilities. Smart buildings are also part of the green energy trend. In the USA, for example, some supermarkets are capturing solar energy on their rooftops, distributing it at 380 V DC and converting to a bus voltage that is typically 48 V, 24 V or 12 V. The proposed new BDC can provide significant gains in energy efficiency throughout the facility as described in [6], [7].

C. The hybrid and electric vehicle (HEV and EV)

In HEVs and EVs, there are two or more different voltage buses for vehicle operation in different conditions. There are needs of galvanically isolated BDCs to link different DC voltage buses and transfer energy back and forth. The high efficiency, compact size, lightweight and high reliability of new BDC is a key technology for several automotive manufacturers [8].

D. The AC/DC grid system

As described in [9], [10] using a 50 or 60 Hz line frequency transformer in an AC grid system of Fig. 13 provides voltage matching and isolation which makes it heavy and bulky. This problem can be overcome by using a proposed new high frequency transformer as an isolation stage as shown in Fig 14. In this way, the isolation barrier moves from the low frequency bulky transformer to the high frequency transformer integrated into the new BDC. Using high frequency transformers lead to more compact and flexible systems. Furthermore, this configuration provides more flexibility in terms of selection of DC voltage amplitude in different stages for optimized operation.

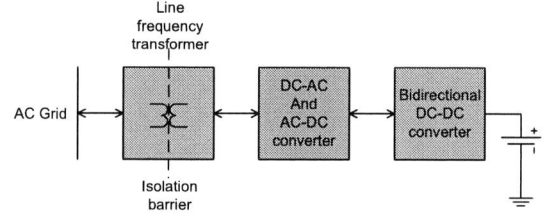

Fig. 13. Using line frequency transformer in the AC grid system for isolation

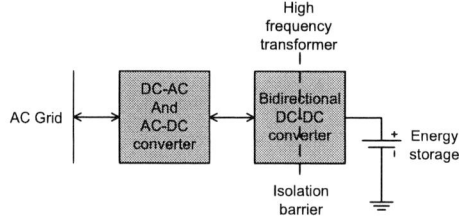

Fig. 14. Using New Bidirectional DC-DC Converter in the AC grid system

E. Battery backup system (BBS)

The proposed new BDC is the promising option for bidirectional power processing in energy storage systems. In such systems it would provide current to charge a battery bank in the forward direction and then provide energy from the battery bank to hold up a bus voltage in the backward direction. Similar to BBS as described in [11] for the 48 V DC bus, one new BDC with two post regulator modules (PRMs)

[12] can form bidirectional BBS as shown in Fig. 15 for the 384 V DC bus. New BDC runs bidirectionally while PRM A and PRM B are in opposite directions. In such system, BDC provides fixed ratio conversion where as PRM provides regulation against line and load and it is implemented using Vicor's ZVS buck-boost regulator modules. In one direction, 384 V bus charges a battery when PRM A is ON and PRM B is OFF. In the other direction, battery supports 384 V bus when PRM A is OFF and PRM B is ON.

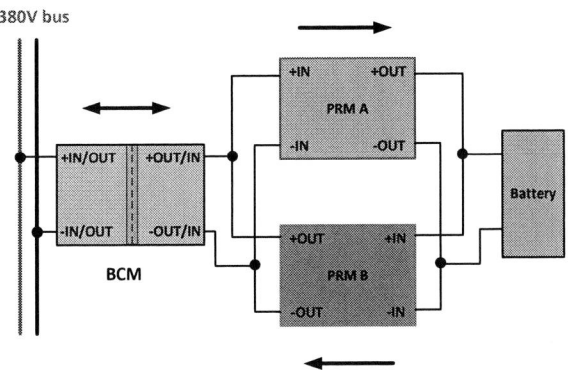

Fig. 15.   New BCD and PRM topology for 384 V Bus

In all above applications, new BDC instead of existing BDC with transformation factor (K) of 1/8 can be utilized to implement 384 V to 48 V bidirectional conversions as needed. Besides K of 1/8 BDC, there are BDCs with K of 1/16 and 1/32 available and can also be utilized based on the line and load requirements in the end applications and systems.

VII.   CONCLUSION

This paper demonstrates a new ZVS and ZCS DC-DC converter for 384 V to 48 V and 48 V to 384 V bidirectional DC-DC power systems. This new BDC is built on an inherently bidirectional topology. It can allow current to flow in both directions and stepping the voltage up or down as required. Using new BDC as a bidirectional converter is much simpler and more compact than having two separate unidirectional converters.

A new BDC provides a +13.5% increase in efficiency at 10% load, a +3.4% increase at 50% load and a +1.5% increase at 100% load in both forward and backward mode and more than double power density. It has the clear advantage of higher efficiency, higher power density, lightweight and smaller form factor. In many real world applications of battery, fuel cell and solar cell systems where bidirectional power transfer is needed, the existing 384 V to 48 V, 24 V or 12 V converters and AC line transformers can be easily transformed by proposed new bidirectional DC-DC converter to achieve better system performance in terms of system efficiency, power density and light weight.

ACKNOWLEDGMENT

The author would like to thank Maurizio Salato and colleagues for giving an opportunity to write this paper and

also like to thank the APEC reviewers for reviewing the digest for this paper.

REFERENCES

[1] VI Chip BCM Bus Converter Module Data Sheet, [online] Rev1.9, 04/2011, http://cdn.vicorpower.com/documents/datasheets/BCM384_480_325A0 0.pdf.

[2] Maxi DC-DC Converter Data Sheet [online] Rev9.2, 09/2014, http://cdn.vicorpower.com/documents/datasheets/ds_375vin-maxi-family.pdf.

[3] Chip BCM Bus Converter Module Data Sheet [online] Rev1.3, 05/2015, http://www.vicorpower.com/documents/datasheets/ds_BCM400P500T1 K8A31.pdf.

[4] VIA BCM High Voltage Bus Converter Module Data Sheet [online] Rev1.1   06/2015 http://www.vicorpower.com/documents/datasheets/ds_BCM4914xD1E5 135yzz.pdf

[5] Vicor technical marketing and application engineering, "Vicor Factorized Power Architecture and VI Chips," Vicor White Paper http://www.vicorpower.com/documents/whitepapers/fpa101.pdf

[6] http://www.us.tdk-lambda.com/ftp/brochures/TDK_TJ023_DC-DC_E_1105.pdf

[7] Webb, Victor-Juan, "Design of a 380 V/24 V DC micro-grid for residential DC distribution" (2013). *Theses and Dissertations*. Paper 231.

[8] Sonya Gargies, Hongjie Wu and Chris Mi, "Isolated Bidirectional DC-DC Converter for Hybrid Electric Vehicle Application" (2006)

[9] Fan Haifeng, "Adavnaced Medium-Voltage Bidirectional DC-DC Conversion Systems for Future Electric Energy Delivery and Management Systems" (2011) Electronic Theses, Treatises and Dissertations. Page 4507.

[10] Hamid R. Karshenas, Hamid Daneshpajooh, Alireza Safaee, Praveen Jain and Alireza Bakhshai, "Bidirectional DC-DC Converters for Energy Storage Systems, Energy Storage in the Emerging Era of Smart Grids, Prof. Rosario Carbone (Ed.), ISBN: 978-953-307-269-2, InTech, Available from: http://www.intechopen.com/books/energy-storage-in-the-emerging-era-of-smart-grids/bidirectional-dc-dcconverters-for-energy-storage-systems

[11] Xiaoyan Yu and Paul Yeaman, "A new high efficiency isolated bi-directional DC-DC converter for DC-bus and battery bank interface," in 2014 Applied Power Electronics Conference

[12] VI Chip PRM Pre Regulator Module Data Sheet, [online] Rev1.2, 08/2013 http://cdn.vicorpower.com/documents/datasheets/PRM48BF480T500A0 0_ds.pdf

A Multiport Isolated DC-DC Converter

Yan-Kim Tran, Drazen Dujic

Power Electronics Laboratory - PEL
École Polytechnique Fédérale de Lausanne - EPFL
Lausanne CH-1015, Switzerland
yan-kim.tran@epfl.ch, drazen.dujic@epfl.ch

Abstract—**This paper presents a multi-port isolated DC-DC converter for DC applications. A three-port structure is presented, characterized with full bidirectional power flow and simple control. Galvanic isolation is achieved by means of a multi-winding medium frequency transformer which is a part of a resonant LLC converter. To provide controllable power exchange with external DC ports, two out of three ports are equipped with additional bidirectional buck/boost stages. They serve to provide active power flow control, while the inner resonant stage provides galvanic isolation, operates in the open-loop and adapts its mode of operation based on the actual power flow. Both switched mode and rea-time hardware-in-the-loop simulations, with control algorithm deployed on the digital signal processor, are used to verify and demonstrate various operational modes.**

I. INTRODUCTION

Most of the newly installed energy sources (e.g. PV) or storage technologies (e.g. batteries) are DC by nature, resulting in an increased interest in the development of small or large scale DC grids. Flexibility and simplicity of interfacing power electronics as well as suitable protection technologies are the key to enable new grid architectures. Considering energy production, consumption and storage, multi-port power electronic converters are viable solution to bring together all three aspects and meet demands of micro-grids, while being fully flexible for the integration of different technologies on each side.

So far, only a few investigations related to high power multi-port converters have been reported. The work of [1] deals with converters based on interconnection through a common DC bus. Despite some advantages due to the simplicity of the control, the voltage range is limited and defined by choice of the bus voltage. This limitation can be overcome by means of electromagnetic coupling through medium or high frequency transformers. In [2],[3] and [4] authors have presented solutions derived from the Dual Active Bridge (DAB) or Dual Half Bridge (DHB) with inclusion of additional ports. Limited soft switching operating range of DAB, resulted in considerations towards the resonant converters thanks to their advantages in reduced switching losses. A multi-port LLC converter has been presented in [5] and consists of a three winding medium frequency transformer (MFT) with distributed resonant capacitors tuned to multiple leakage inductors, in order to create resonant tank. The authors have presented configuration where one port act as a source, while two other ports are acting as sink (load). Experiments from a low power setup (300W), demonstrate good load regulation but gave no consideration to different operating modes. On the other hand, work of [6]

shows a similar topology where two ports are configured as sources and are equipped with resonant capacitors, while the third port has only a rectifier as it is behaving as the load. Authors have demonstrated controllable power flow and load sharing between sources, using phase shift control. While, several works have been dealing already with LLC or multi-port converters, the focus has been mainly on a low power rated converters and unidirectional power flow.

This paper is an extension of the previous work [7] and it presents a novel multi-port topology and its operation in section II. Design considerations and sizing of main components are presented in section III, while the control scheme is provided in section IV. PLECS simulations as well as hardware-in-the-Loop simulations, with control algorithm deployed on the digital signal processor (DSP) are given in section V. A summary and conclusion are provided in section VI.

II. TOPOLOGY AND OPERATING MODES

The topology considered in this paper is based on a three winding medium frequency transformer as shown in Fig. 1. It is fully bidirectional, even with one of the three ports deactivated. The converter basic stage is composed of a multi-port LLC resonant circuit equipped with two additional buck/boost stages. The ports may have different power ratings ($P_1 \neq P_2 \neq P_3$) and arbitrary output voltages ($V_{DC_1} \neq V_{DC_2} \neq V_{DC_3}$).

A. Multiport LLC Resonant Stage

The inner part is made by a three winding MFT with turn ratio n_i according to the desired output voltages and equipped with three half bridges. In order to achieve high efficiency, a resonant LLC topology is used and operated in the open-loop mode below resonant frequency [8]. Since the full bi-directionality of the converter implies certain symmetry requirements, the resonant tank is not located on a single port, like in [5], but is split between all three ports. Each resonant tank is composed of a capacitor C_i and an inductor L_i which is the combination of the stray inductor of the MFT and, if needed, an additional inductor. The design of the resonant tank requires proper sizing of all distributed L_i and C_i in a way that guaranties soft switching conditions for any operating point within the power specifications. The resonant part is intended to be operated as a DC transformer without any complex control scheme. It is only controlled by activating or deactivating PWM on $S_{1,2,3}$, depending on the power flow direction. Only the ports that are providing power

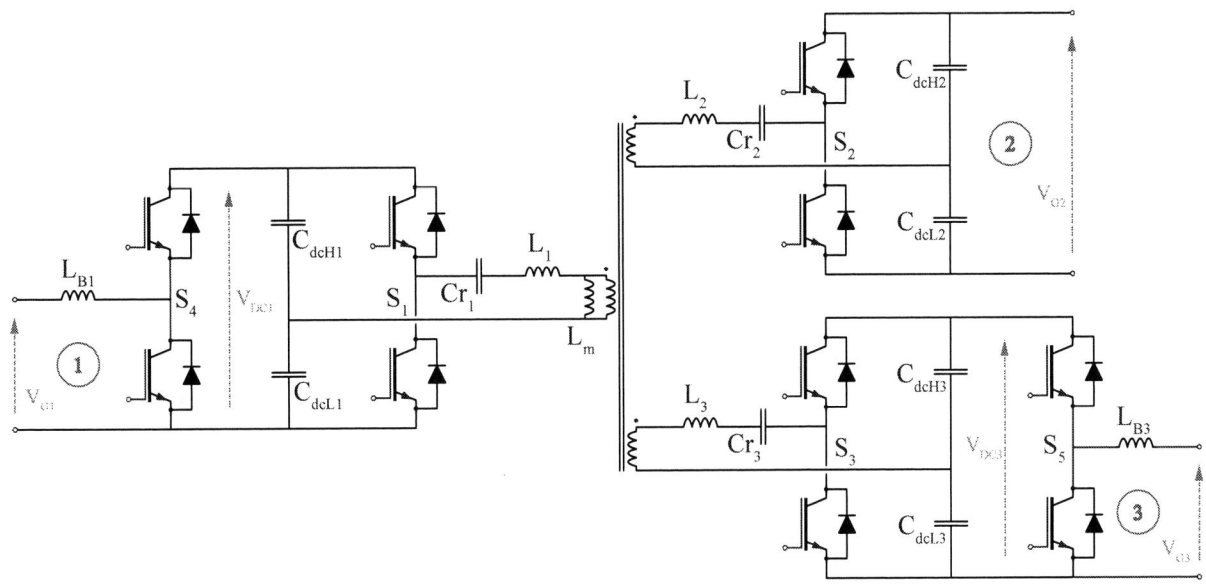

Fig. 1. Topology of a Multiport Isolated DC-DC Converter

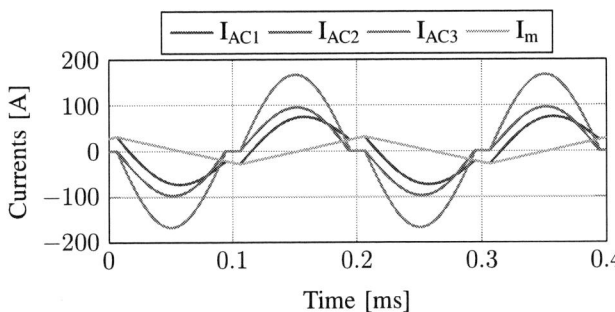

Fig. 2. Source port: 1 ($P_1 = 90kW$); Sink ports: 2 ($P_2 = -50kW$) and 3 ($P_3 = -40kW$)

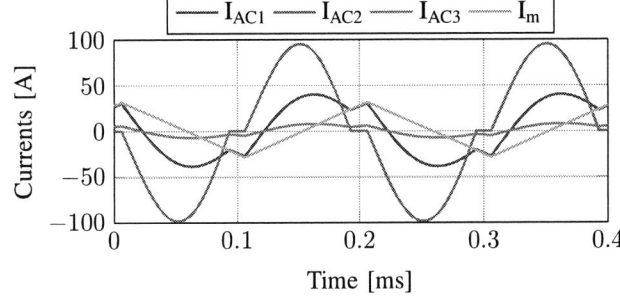

Fig. 3. Source ports: 1 ($P_1 = 50kW$) and 3 ($P_3 = 10kW$); Sink port: 2 ($P_2 = -60kW$)

are actively switched. Load ports are turned off and their free-wheeling diodes are used as passive rectifiers (voltage doubler configuration as shown in Fig. 1). When a port is active, its high and low side switches are operated with opposite polarity at a fixed frequency f_{sw} and a constant duty cycle of approximately 50% (considering dead-time). Typical current waveforms in such a structure are shown in Fig. 2 and Fig. 3.

B. Buck/Boost Stages

Since no active control is applied on the inner resonant stage, the power flow must be regulated by two additional stages (S_4 and S_5) on the ports 1 and 3. These stages are based on a bidirectional buck/boost converter and thus can be operated either as boost or as buck depending on the actual mode of operation, as it will be discussed shortly.

C. Operating Modes

There are several operating modes of this converter that influence the overall control scheme, as summarized in Table I. The resonant stage modes, related to switching cells S_1, S_2 and S_3, simply describe if the respective switching cells are active (PWM on) or passive (PWM off). However, the operation of S_4 and S_5 define if the stage is used as boost or buck converter. In this paper, ports 1 and 2 are the main ports rated for the full power of the converter and in charge of the power flow. Bidirectional operation implies that power flow from the port 1 to the port 2 (modes A1, B1 and C1) or vice versa (modes A2, B2 and C2). In both cases, the voltage on port 2 is directly controlled through port 1 by the buck/boost stage S_4. Port 3 is considered as an additional port that can source (mode B1 and B2) or sink (mode A1 and A2) power with reduced ratings, regardless of the power direction between the ports 1 and 2. Its buck/boost stage S_5, regulates the current I_{b3}, compensating

978-1-4673-9551-9/16 $31.00 © 2016 IEEE   157

for voltage variations due to cross-coupling with ports 1 and 2. In the modes C1 and C2, no energy is exchanged with the port 3 and S_5 is deactivated.

Previous description assumes that multiport converter controls its voltage on port 2, however in case of presence of additional converter, (e.g. AC grid connected inverter), overall role of different ports may change, as it is usually the grid side converter that controls its DC link voltage (e.g. port 2 voltage).

TABLE I
OPERATING MODES

Mode	S_1	S_2	S_3	S_4	S_5
A1	Active	Passive	Passive	Boost	Buck
A2	Passive	Active	Passive	Buck	Buck
B1	Active	Passive	Active	Boost	Boost
B2	Passive	Active	Active	Buck	Boost
C1	Active	Passive	Passive	Boost	Off
C2	Passive	Active	Passive	Buck	Off

III. DESIGN METHODOLOGY

For this case study, power and voltage ratings are chosen somewhat arbitrarily and are summarized in Table II.

TABLE II
POWER AND VOLTAGE RATINGS USED FOR THE CASE STUDY

Port	P_i	V_{DC_i}	V_{G_i}
1	125 kW	3600 V	3000 V
2	125 kW	1500 V	1500 V
3	25 kW	750 V	400 V

A. LLC Tank Design

The design of the resonant part is based on a methodology presented in [7]. The aim is to determine the resonant parameters in such a way that ensures the ZVS of the primary switches (active port), soft commutation of rectifier diodes (passive port) and minimize the amount of circulating energy (conduction losses). For a given ratio f_n, of the switching and the resonant frequency and for a given operating point represented by an AC load R_{AC_i}, the maximum value of L_{r_i} required to stay in the inductive region is given by:

$$L_{max_i} = \frac{f_n R_{AC_i} \pi^2}{8\sqrt{2}\omega_0} \quad (1)$$

The quality factor Q_{opt_i} is defined as the ratio between the load and the tank impedance, under the maximum load conditions, for which the tank RMS current is the smallest. For the operating point given by R_{AC_i}, the resonant inductor which provides the optimal quality factor is defined, according to [9] and [10], as:

$$L_{opt_i} = \frac{R_{AC_i}^2}{\omega_0^2 L_{m_i}} \quad (2)$$

$$Q_{opt_i} = \frac{R_{AC_i}}{\omega_0 L_{m_i}} \quad (3)$$

We can then obtain the optimal Q factor for each port. The distributed resonant inductors are calculated from the equations linking the three power quality factors to the three resonant inductors:

$$L_i \leq Q_{opt_i} R_{AC_i} - \left(\frac{1}{(n_i/n_j)^2 L_j} + \frac{1}{(n_i/n_k)^2 L_k} \right)^{-1} \quad (4)$$

$$C_i = \frac{1}{L_i(2\pi f_{res})^2}; i \neq j \neq k \quad (5)$$

Resulting parameters for the L_i and the C_i are shown in Table III.

In order to benefit from the ZVS, the magnetizing inductor has to be sized accordingly. Its value will define the turn-off current of the active ports and thus turn-off switching losses. Since this turn-off current corresponds to the negative turn-on current of the free-wheeling diode of the complementary switch, its value cannot be selected too low. To preserve ZVS, the resonant tank current should not change polarity within the dead-time. The duration of diode conduction can be approximated by division of the turn-off current by the slope of the resonant tank current at this point. To maintain ZVS within the dead-time T_{dt}, L_m must satisfy:

$$I_{off_{min}} \approx T_{dt} I_{DC_n} \frac{\pi}{2} \quad (6)$$

$$L_{m_i} \leq \frac{V_{DC_i} T_{sw}}{8 I_{off_{min}}} \quad (7)$$

Considering that the quality factor Q is linked to the magnetizing inductor through (3), it has to be chosen such to keep the optimal L_{opt}, smaller than L_{max}:

$$L_{m_i} \geq \frac{R_{AC_i} 8\sqrt{2}}{\omega_0 f_n \pi^2} \quad (8)$$

The magnetizing inductors seen from each port, L_{m_i} must fulfill the condition of (7) and (8) for all three ports:

$$L_{m_1} = L_m \qquad L_{m_2} = L_m \frac{n_2^2}{n_1^2} \qquad L_{m_3} = L_m \frac{n_3^2}{n_1^2} \quad (9)$$

Resulting parameters of the distributed resonant tank, for the ratings used in this paper, are summarized in Table III.

B. DC Link Capacitors Sizing

DC link capacitors are sized according to voltage ripple specifications and resulting values are shown in Table IV.

$$C_{DC_i} = \frac{V_{DC_i} T_{sw}}{2 R_{dc} \Delta V_i} \quad (10)$$

$$C_{dcH_i} = C_{dcL_i} = 2 C_{DC_i} \quad (11)$$

Since DC capacitors are in series with the resonant tank, resonant capacitors are adjusted to match the value calculated in (5).

$$C_{r_i} = \frac{C_i 4 C_{DC_i}}{4 C_{DC_i} - C_i} \quad (12)$$

TABLE III
DISTRIBUTED RESONANT TANK PARAMETERS

L_m	10.0 mH	$f_{sw_{llc}}$	5 kHz
L_1	13.7 μH	C_1	66.6 μF
L_2	2.6 μH	C_2	354.1 μF
L_3	29.8 μH	C_3	30.7 μF

TABLE IV
DC LINK CAPACITORS FOR EACH PORT

$Port_i$	$\Delta V_i/V_{dc_i}$	ΔV_i	C_{DC_i}	C_{r_i}
1	0.01	36 V	96.5 μF	80.5 μF
2	0.01	15 V	555.6 μF	421.2 μF
3	0.01	7.5 V	444.5 μF	31.23 μF

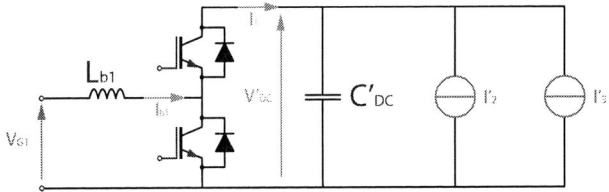

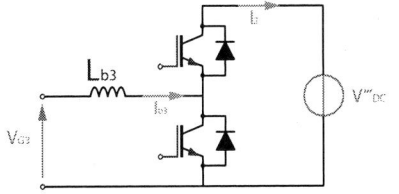

Fig. 4. Simplified model of the converter assuming unity transfer function of the resonant stage

C. Buck/Boost Stage Design

The switching frequencies for the two buck/boost stages are chosen in such a way that switching frequency of the port 1 is lower than the LLC switching frequency, while the switching frequency of the port 3 is higher. It should be noted that present values are not result of any optimization, since actual hardware design is not the main scope of this paper. Chosen values are shown in Table V. The buck/boost inductors are sized according to the ripple specifications.

$$L_{B_i} = \frac{V_{G_i} T_{sw_{boosti}}}{2\Delta I_{b_i}} \tag{13}$$

TABLE V
BUCK/BOOST STAGES PARAMETERS

F_{B_1}	2.5 kHz	$\Delta Ib1$	4 A	L_{B_1}	40 mH
F_{B_3}	10 kHz	$\Delta Ib3$	4 A	L_{B_3}	5 mH

IV. CONTROL SYSTEM DESIGN

Simplified control scheme is presented here, considering different ratings for the ports. The output voltage of the port 2 is regulated from the port 1 buck/boost S_4 stage while the power through port 3 is locally regulated by its own buck/boost stage S_5. In this way a large degree of freedom exists for integration of different energy storage technologies on port 3.

A. Plant model

Considering that the resonant stage is operated in an open-loop with a fixed switching frequency, its DC transfer function (gain) is considered constant and close to unity. For that reason it is neglected in the analysis and the DC link capacitors can be considered in parallel, taking into account the turn ratio of the transformer. Thus the circuit can be modeled as shown in Fig. 4. The voltage of the port 2, proportional to the voltage $V'_{C_{DC}}$ on C'_{DC} (14), has to be regulated and the control input is the duty-cycle D_1 of the buck/boost stage S_4. The perturbations to this circuit are the voltage on the buck/boost S_4 input V_{G1} and the currents in ports 2 and 3, I'_2 (15) and I'_3 (16), respectively (all referred to port 1 side). The buck/boost stage S_5 current (I_{b3}) has to be regulated and the control input is its duty-cycle D_3. The perturbation to this part is the voltage on C_{DC}

referred to the third winding. The average model of two sub-circuits from Fig. 4 is represented in Fig. 5, where $G_{I_1}(s)$ (18), $G_{I_3}(s)$ (20) are the transfer functions of the boost inductors L_{B_1}, L_{B_3} and $G_{Uc}(s)$ (19) is the transfer function of the combined DC capacitors C_{DC}.

$$C'_{DC} = C_{DC_1} + \frac{n_2^2}{n_1^2}C_{DC_2} + \frac{n_3^2}{n_1^2}C_{DC_3} \tag{14}$$

$$I'_2 = \frac{n_2}{n_1}I_2 \tag{15}$$

$$I'_3 = \frac{n_3}{n_1}I_3 \tag{16}$$

$$V'''_{DC} = \frac{n_3}{n_1}V'_{DC} \tag{17}$$

$$G_{I_1}(s) = \frac{1/R_{l_1}}{1 + s(L_{b_1}/R_{l_1})} = \frac{K_{I_1}}{1 + sT_{I_1}} \tag{18}$$

$$G_{Uc}(s) = \frac{1}{sC'_{DC}} = \frac{1}{sT_U} \tag{19}$$

$$G_{I_3}(s) = \frac{1/R_{l_3}}{1 + s(L_{b_3}/R_{l_3})} = \frac{K_{I_3}}{1 + sT_{I_3}} \tag{20}$$

B. Control system

Rather simple control approach has been followed and the voltage on the port 2 is regulated through two cascaded PI loops. The inner loop controls the inductor current I_{B_1} with the PI regulator G_{RI_1} while the outer loop controls the voltage V_{DC_2} with the PI regulator G_{RU}. The current in the port 3 is regulated with a single PI loop and the regulator G_{RI_3}. Please note that in case of a connection of energy storage (e.g. battery, ultra-capacitors) there would be a need for outer loop for the purpose of the energy management. This is however not the scope of this paper and the used control structure is presented in Fig. 6, with the parameters of the regulators provided in the Table VI. These values are obtained according to description provided in the Appendix.

The control structure shown in Fig. 6 does not include all implementation details associated with the changes of the modes of operation. To avoid large inrush currents during the

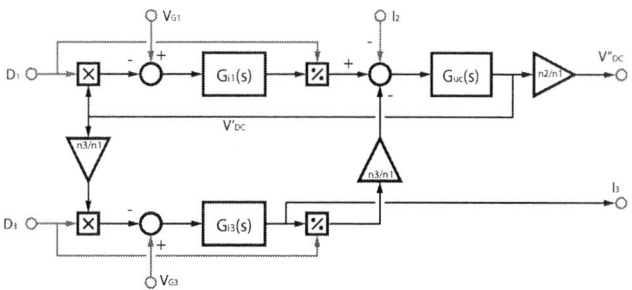

Fig. 5. Average model of the circuit from Fig. 4

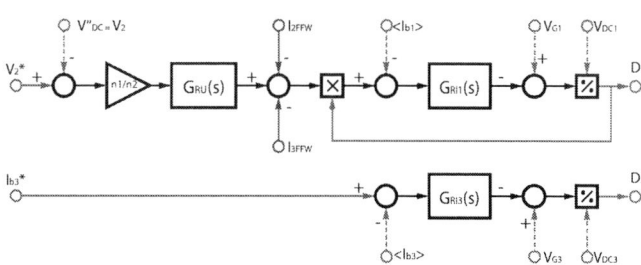

Fig. 6. Control structure used in conjunction with the model from Fig. 5

TABLE VI
PI REGULATOR'S PARAMETERS

$T_{n_{I1}}$	0.8	$T_{i_{I1}}$	0.012
$T_{n_{uc}}$	0.0032	$T_{i_{uc}}$	0.0241
$T_{n_{I3}}$	0.1	$T_{i_{I3}}$	0.018

startup of S_1, S_2 or S_3, the duty cycle is ramp-up to final 50%. During this transition sequence, the ZVS is not necessarily achieved.

V. SIMULATION RESULTS

To verify design of the distributed resonant tank and the performance of the control scheme in the different modes of operation, different simulations have been performed. The closed-loop switched model has been implemented in PLECS as well as on the real-time power electronics hardware-in-the-loop (HIL) simulator controlled with digital signal processor (DSP).

A. PLECS Simulations

Simulation results with several mode transitions are shown in Fig. 7 and 8. The unfiltered DC voltages of the resonant stage are shown in the upper plots in per unit values, while the RMS and average values of relevant currents, respectively, are shown in the bottom plots. Detailed description of the simulation scenarios is provided in the captions of Fig. 7 and 8. As it can be seen, the output voltage V_{DC_2} has been effectively regulated to its rated value, in all modes of operation.

B. HIL Simulations

As a second step, the complete converter has been modeled and deployed on the Typhoon HIL 400 real-time power

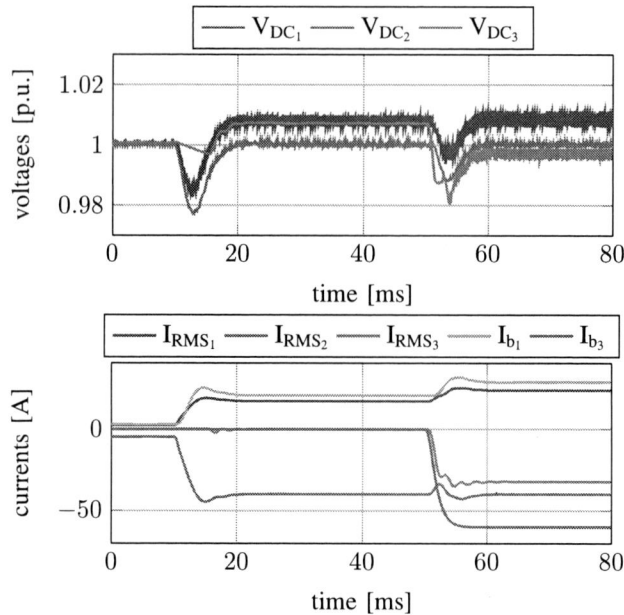

Fig. 7. PLECS simulations: At t=10ms, the load on the port 2 is increased from 0 to -40A. The current in the boost S_4 tracks this current step resulting in an increase of V_{DC_1}. After the transient, the voltage V_{DC_2} returns to its reference (1 p.u.). At t=50ms, the load on the port 3 is increased from 0 to -60A resulting in a decrease of V_{DC_3}. After the transient, I_{RMS_2} remains the same and I_{RMS_1} is slightly increased by its contribution in I_{RMS_3}. V_{DC_2} is regulated back to 1 [p.u.]

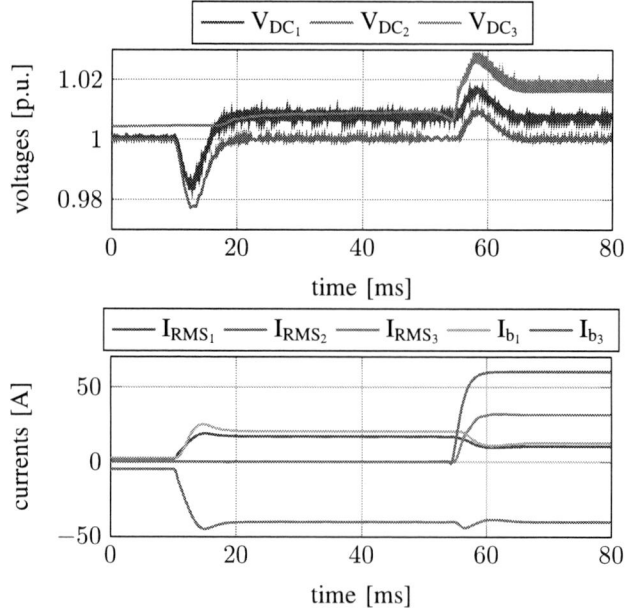

Fig. 8. PLECS simulations: At t=10ms, the load on the port 2 is increased from 0 to -40A. I_{B_1} follows this current step resulting in an increase of V_{DC_1}. Since there is no power exchange through the port 3, S_5 is deactivated and V_{DC_3} is following V_{DC_1}. At t=50ms, S_3 is turned on and starts with a modulation ramp. At t=55ms the ramp is finished and the buck/boost stage S_5 of port 3 starts to regulate the current I_{b_3} to 60A.

electronics simulator, while the control algorithm has been

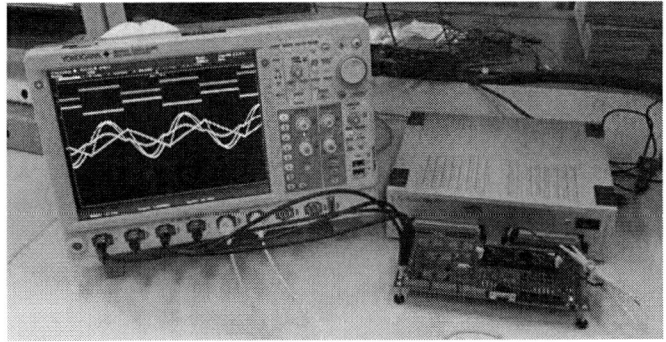

Fig. 9. HIL test setup with TI C2000 DSP

implemented on the TI TMS320F28335 DSP. This allows to carry out the simulations of numerous test cases and verify performances of the digital controller in the real-time. The complete test setup is shown in Fig. 9 (actual resonant current waveforms observed during testing are visible on the scope), while the captions of Fig. 10 and 11 provide description of the modes of the operation. Change of loading conditions on the port 3 has been tested and closed loop response of the control algorithm responsible for the main ports has been successfully verified.

VI. CONCLUSIONS

The topology of a bidirectional multiport resonant DC-DC converter has been presented in this paper. It is based on the combination of an open-loop operated multiport resonant converter with distributed resonant tank, and additional closed-loop operated boost/buck stages on two ports. The operating modes of the converter have been described and suitable control scheme is presented. The inner resonant stage operates as DC transformer, and its main function is to provide galvanic isolation between different ports. Auxiliary third port has reduced ratings and is intended for connection of electrical energy storage elements, as support for the main ports.

Simulation results demonstrate operating principles, while further verification has been performed with digital controller deployed on the DSP and tested in the real-time with hardware-in-the-loop simulator. Topology, of the presented multiport converter, allow for further extensions for the higher operating voltages. Several converter stages from Fig. 1 can be stacked in input-series (port 1) output-parallel (port 2) fashion, with multiple ports 3 either left floating or tied together. In this way, connection to the higher DC voltages on the port 1 side is possible (insulation requirements for the MFT are naturally increased). This, together with the hardware design is the subject of the ongoing work, and will be reported separately.

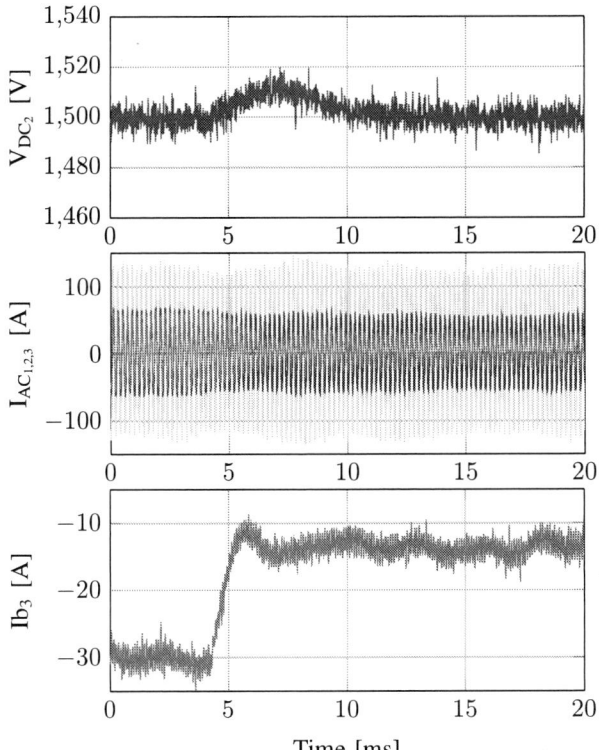

Fig. 10. HIL simulations: At t=4ms, the load on the port 3, I_{b_3} is decreased from -30 to -15 A. In response, the voltage on the port 2 V_{DC_2} is slightly increased, but immediately regulated by buck/boost stage S_4 of the port 1. All three resonant tank currents are shown in the middle plot, from where their envelope values can be observed.

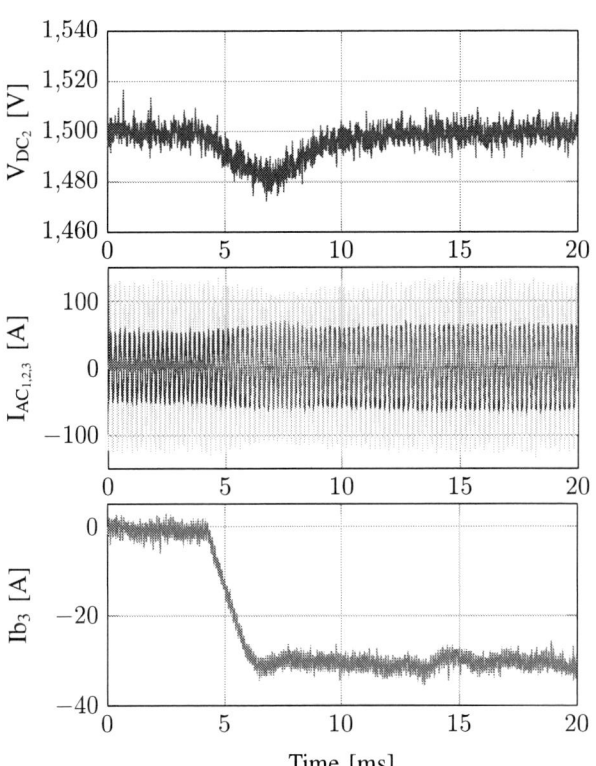

Fig. 11. HIL simulations: At t=4ms, the load on the port 3, I_{b_3} is increased from 0 to -30 A. In response, the voltage on the port 2 V_{DC_2} is slightly decreased, but immediately regulated by the buck/boost stage S_4 of the port 1. All three resonant tank currents are shown in the middle plot, from where their envelope values can be observed.

978-1-4673-9551-9/16 $31.00 © 2016 IEEE

ACKNOWLEDGEMENT

This research project is part of the National Research Programme "Energy Turnaround" (NRP70) of the Swiss National Science Foundation (SNSF). Further information on the National Research Programme can be found at www.nrp70.ch.

APPENDIX

The parameters of the regulators G_{RI_1} and G_{RI_3} are determined according to the principle of the symmetric optimum while the regulator G_{RU} is tuned according to the principle of the magnitude optimum.

$$T_{PE_1} = 0.5T_{sample} + T_{sw_{boost1}} \tag{21}$$

$$T_{n_{I1}} = T_{I_1} \tag{22}$$

$$T_{i_{I1}} = 2K_{I_1}T_{PE_1} \tag{23}$$

$$T_{CL} = 0.5T_{sample} + 2T_{PE_1} \tag{24}$$

$$T_{n_U} = 4T_{CL} \tag{25}$$

$$T_{i_U} = 8T_{CL}^2/T_U \tag{26}$$

$$T_{PE_3} = 0.5T_{sample} + T_{sw_{boost3}} \tag{27}$$

$$T_{n_{I3}} = T_{I_3} \tag{28}$$

$$T_{i_{I3}} = 2K_{I_3}T_{PE_3} \tag{29}$$

Please note that for the simulation studies, the sampling time T_{sample} is equal to the switching period of the resonant stage T_{sw}. Resulting parameters of the regulators are given in Table VI.

REFERENCES

[1] H. Tao, A. Kotsopoulos, J. L. Duarte, and M. A. M. Hendrix, "Family of multiport bidirectional dc-dc converters," *IEE Proceedings - Electric Power Applications*, vol. 153, no. 3, pp. 451–458, 2006.

[2] H. Tao, A. Kotsopoulos, J. L. Duarte, and M. A. M. Hendrix, "Triple-half-bridge bidirectional converter controlled by phase shift and pwm," in *Proceedings of the 21st Annual IEEE Applied Power Electronics Conference and Exposition - APEC*, 2006, pp. 1256–1262.

[3] H. Tao, A. Kotsopoulos, J. L. Duarte, and M. A. M. Hendrix, "Transformer-coupled multiport zvs bidirectional dc/dc converter with wide input range," *Power Electronics, IEEE Transactions on*, vol. 23, no. 2, pp. 771–781, 2008.

[4] H. Tao, A. Kotsopoulos, J. L. Duarte, and M. A. M. Hendrix, "A soft-switched three-port bidirectional converter for fuel cell and supercapacitor applications," in *Proceedings of the 36th Annual IEEE Power Electronics Specialists Conference - PESC*, 2005, pp. 2487–2493.

[5] Z. Pavlovic, J. A. Oliver, P. Alou, O. Garcia, and J. A. Cobos, "Bidirectional multiple port dc/dc transformer based on a series resonant converter," in *28th Annual IEEE Applied Power Electronics Conference and Exposition - APEC*, 2013, pp. 1075–1082.

[6] H. Krishnaswami and N. Mohan, "Constant switching frequency series resonant three-port bi-directional dc-dc converter," in *39th Annual IEEE Power Electronics Specialists Conference - PESC*, 2008, pp. 1640–1645.

[7] Y.-K. Tran, D. Dujic, and P. Barrade, "Multiport Resonant DC-DC Converter," in *Proceedings of the 41st Annual Conference of the IEEE Industrial Electronics Society - IECON*, 2015, pp. 3839–3844.

[8] D. Dujic, G. Steinke, E. Bianda, S. Lewdeni-Schmid, C. Zhao, J. K. Steinke, and F. Canales, "Soft switching characterization of a 6.5kV IGBT for high power LLC resonant DC-DC converter," in *Proceedings of International Power Conversion and Intelligent Motion Conference - PCIM*, 2012, pp. 625–631.

[9] I.-O. Lee and G.-W. Moon, "The k-q analysis for an llc series resonant converter," *Power Electronics, IEEE Transactions on*, vol. 29, no. 1, pp. 13–16, 2014.

[10] R.-L. Lin and C.-W. Lin, "Design criteria for resonant tank of llc dc-dc resonant converter," in *Proceedings of the 36th Annual Conference of IEEE Industrial Electronics Society - IECON*, 2010, pp. 427–432.

A Seamless Transfer Control Method with High Load Sharing Performance for Modular ESS

Jung-Hoon Ahn, Won-Yong Sung, Chang-Yeol Oh
and Byoung-Kuk Lee [†]

Department of Electrical and Computer Engineering
Sungkyunkwan University
Suwon, Gyeonggi-do, Korea
E-mail: bkleeskku@skku.edu

Yun-Sung Kim

Research & Development Center
Dongahelecomm Corporation
Yongin, Gyeonggi-do, Korea

Abstract— **This paper describes a theoretical and experimental study on a seamless transfer control strategy for the parallel operation of three-phase voltage source inverters (VSIs) for energy storage systems (ESSs). Moreover, phase-locked loop, seamless transfer control, and parallel control algorithm are optimized for simplicity in order to implement in a unified-controller. Finally, the proposed algorithm is validated through informative simulation waveforms and experimental results**

Keywords—seamless transfer control, load sharing control

I. INTRODUCTION

For the efficient and reliable usage of electric energy, various power conditioning systems (PCSs) have been lively developed for various applications, such as energy storage systems (ESSs), uninterrupted power supplies (UPSs), active power filters (APFs), and etc.. And their markets have grown independently to each other to date. However, the development of each technique has blurred the boundaries between the markets because there exists a common ground on these techniques: usage of voltage source inverters (VSIs). A representative example of these technology is the unique featured ESS, which has an additional function to supply uninterrupted and continuous power to critical loads [1-2].

The current research trends in such ESSs can be classified into two categories. The first is the parallel operation of VSIs, which allows the processed load power to be distributed among VSIs, creating a redundant system and making the power expansion flexible [3-14]. And the second is a seamless transfer control. Once a grid-fault event occurs, conventional controllers should be changed from the current controlled mode (grid-connection) to the voltage controlled mode (stand-alone). A main function of the seamless transfer control is to stabilize this transient-state across the critical load [15-21]. Contrary to these trends, however, there has been little research on a unified-control algorithm which functions as the seamless transfer control and the parallel control of VSIs, simultaneously.

In this paper, a seamless transfer control method with high load sharing performance for a modular ESS is proposed. And by illustrating a series of development processes from improving each function to merging all functions into the unified-controller, logicality of the proposed control method is presented with informative figures. Finally, the feasibility of the proposed controller is verified by simulation results and an actual prototype. As discussed in details by the following sections, a novel seamless transfer control for parallel VSIs is presented.

II. REQUIREMENTS OF THE TARGET ENERGY STORAGE SYSTEM (ESS)

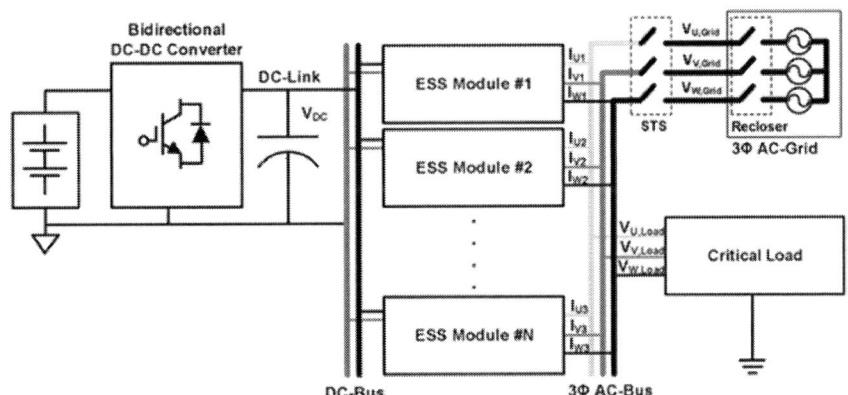

Fig. 1. Configuration of target energy storage system (ESS).

Table I. Specifications of ESS

Item	Specification
V_{LL}	380 V_{RMS}
V_P	220 V_{RMS}
f_{Grid}	60 Hz
$V_{DC\text{-}Link}$	650 V_{DC}
P_{rated}	10 kW
f_{sw}	20 kHz
L_i	800 uH
L_g	300 uH
C_f	4.7 uF

Fig. 1 shows the configuration of the target ESS which consists of a bidirectional DC/DC converter for regulating DC-Link to 650V regardless of battery terminal voltage, static transfer switches (STSs) for the safe separation with faulty-grid, critical loads, and two power-rates of VSI Modules (detailed in Table I). The target ESS should provide uninterrupted and continuous power to critical loads even if any fault occurs in the grid. Therefore, VSIs have to operate in both grid-connection and stand-alone modes.

The entire control sequence of the target ESS can be divided into the four modes as indicated in Fig.2. Among these modes, Mode 2 is the most difficult to control because any VSIs do not perceive the change of the grid-state. Thus, VSIs cannot adjust their control logic in respond to the grid-state, resulting in the uncontrollable output voltage which is determined only by the reference for the current controller and an unknown load condition. Furthermore, it is more difficult to utilize the parallel VSIs under the risk of grid failure since they must achieve a high load sharing performance with tight output regulation at the same time to solve the above problems.

Therefore, the control method for the target system must satisfy the following fundamental requirements: 1) a highly reliable seamless transfer control; 2) a proper load sharing

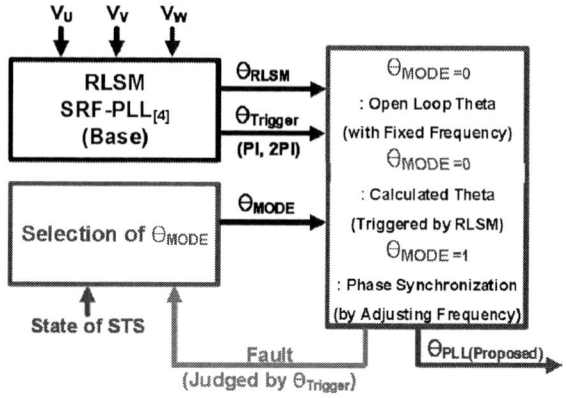

(a) Logic diagram of proposed PLL

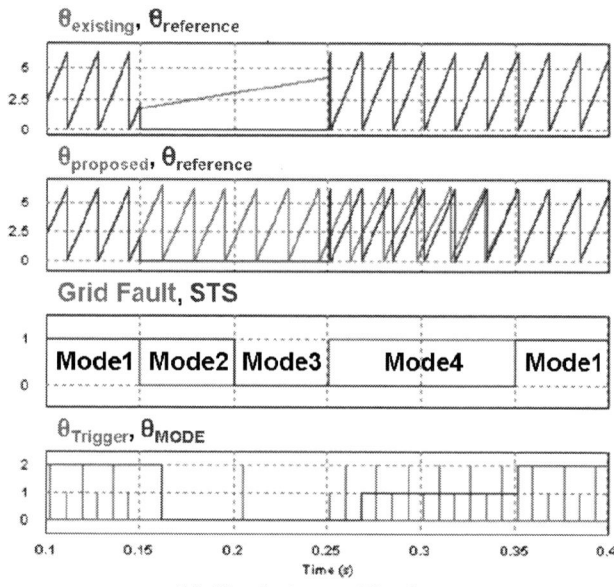

(b) Simulation verification

Fig. 3. Proposed phase-locked loop technique.

control in any case; 3) their compatibility with each other. In other words, for the proposed unified control algorithm, both of the logics should be improved to achieve a high performance and simplicity at once.

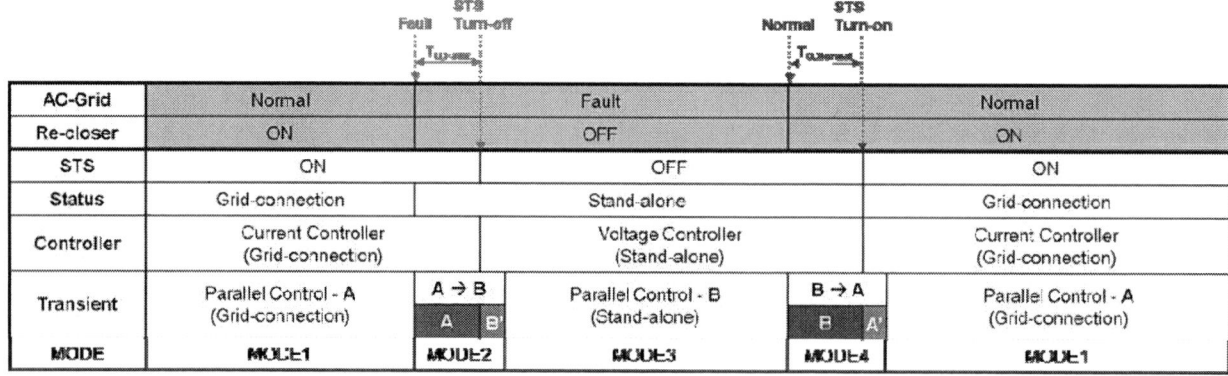

Fig. 2. Entire control sequence according to a grid-fault event.

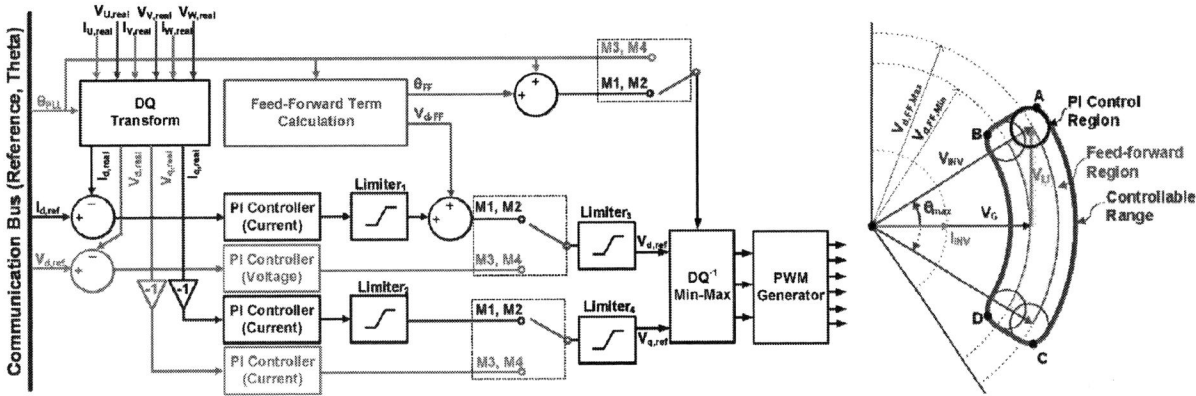

(a) Block diagram of the proposed seamless control algorithm (M1~4)　　　(b) Control principle (M1, M2)

Fig. 4. Proposed seamless transfer control algorithm for the ESS consisting of single VSI module.

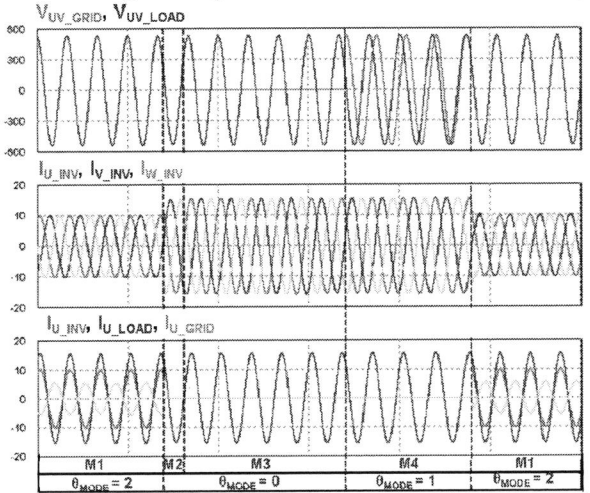

Fig. 5. Simulation verification (M1~4) of proposed seamless transfer control algorithm.

III. PROPOSED SEAMLESS TRANSFER CONTRLOL ALGORITHM WITH ENHANCED PLL

Firstly, a unique phase-locked loop (PLL) technique, which is essential for the entire logic, is required for securing the reliable theta against any change of grid-state. The proposed PLL is based on RLSM SRF-PLL [22-24] robust to the imbalanced and/or unstable-grid, and it facilitates the seamless

transfer through the additional functions: the open-loop operation for Mode 2 (θ_{MODE} = 0) and the phase-synchronization for Mode 4 (θ_{MODE} = 1) as shown in Fig. 3.

Most literature [15-18] on seamless transfer has focused only on the transient-state from the current controlled mode to the voltage controlled mode (referred as B' in Fig. 2). On the other hand, a few researches [19-21] considering the entire Mode 2 needs additional sensors or the unusual LCL filter with complicated logics. However, the proposed controller of Figs. 4 and 5 can achieve high seamless transfer performance in all the modes as verified in Fig. 5 without any additional components and alterations of LCL filter. The key point of the proposed controller is a unique control region, which is small enough to prevent the unexpected divergence and large enough to cover the entire control object. The control region is accomplished easily by utilizing a simple combination: a strong feed-forward term calculated by Eq. (1-2) and a weak PI controller as shown in Fig. 4.

$$V_{d,FF} = \sqrt{\left(i_{d,inv} \times \omega\left(L_i + L_g\right)\right)^2 + (V_G)^2} \qquad (1)$$

$$V_{d,FF,Max} = \sqrt{\left(i_{d,inv,Max} \times \omega\left(L_i + L_g\right)\right)^2 + (V_G)^2} \qquad (2)$$

$$\theta_{FF} = \tan^{-1}\left(i_{d,inv} \times \omega\left(L_i + L_g\right) + V_G\right) \qquad (3)$$

$$\theta_{Max} = \tan^{-1}\left(i_{d,inv,Max} \times \omega\left(L_i + L_g\right) + V_G\right) \qquad (4)$$

TABLE II. Classification and comparison of conventional parallel control methods

	Redundancy	Performance	Simplicity	Compatibility with STC*
Master-Slave [3-4]/ Central Mode [5-6]	X	O	O	△
Distributed Controls [7-11]	△	O	△	O
Without Control Interconnection [12-14]	◎	X	X	△
Proposed algorithm	O	O	O	O

(* STC means Seamless Transfer Control)

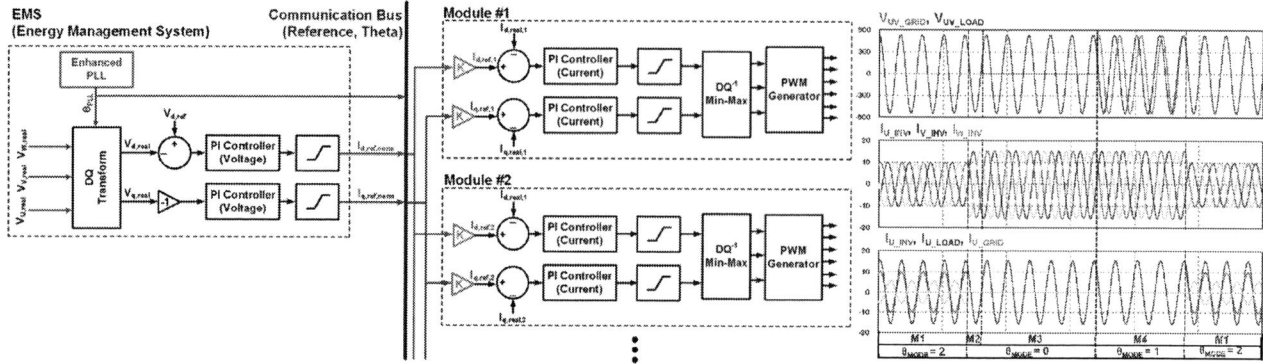

(a) Block diagram of the proposed parallel control algorithm (M3 and M4)   (b) Simulation verification (M3 and M4)

Fig. 6. Proposed parallel control algorithm.

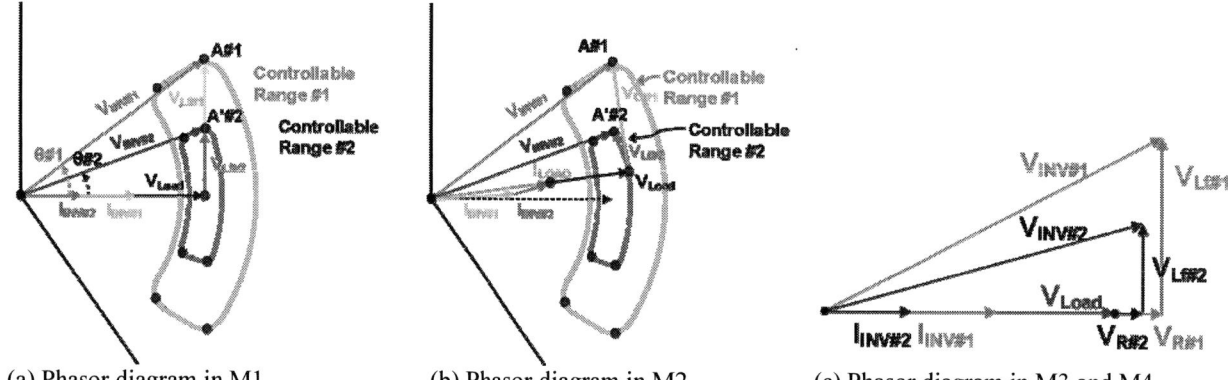

(a) Phasor diagram in M1   (b) Phasor diagram in M2   (c) Phasor diagram in M3 and M4

Fig. 7. Control Principles of the proposed parallel/seamless transfer control algorithm.

IV. PROPOSED PARALLEL CONTROL ALGORITHM COMPATIBLE WITH SEAMLESS TRANSFER

Table II indicates the comparison results of conventional parallel controllers [3-14]. Without Control Interconnection [12-14] has the highest redundancy property due to needlessness for any communication around VSIs. However, the communication bus is an essential element for the energy management system (EMS) so that it is ineffective to implement this logic for an ESS. And other controllers [3-6], in which the controllers of VSIs should be different to each other, can complicate the seamless transfer control and have a low redundancy property.

The proposed parallel controller (Fig. 6) can be classified as Distributed Control. The voltage controller located in EMS provides the reference of the nominalized current and by multiplying K-factor (proportional to their rated-power) to this, the final current references for VSIs are derived. Hereby, each VSI can achieve an excellent load sharing performance without knowledge of their rated-powers and how many VSIs are connected as verified in Fig. 6 (b).

This simple configuration also has an excellent compatibility with the seamless transfer control because VSIs

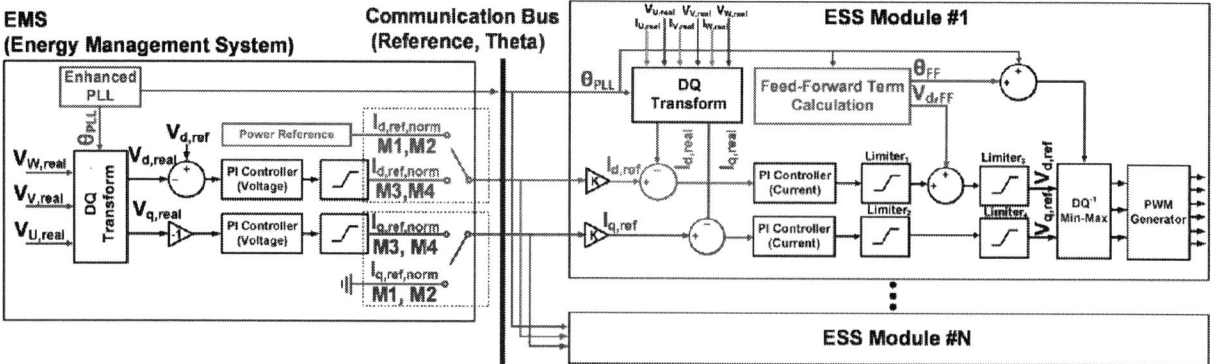

Fig. 8. Block diagram of the proposed parallel/seamless transfer control algorithm (M1~4).

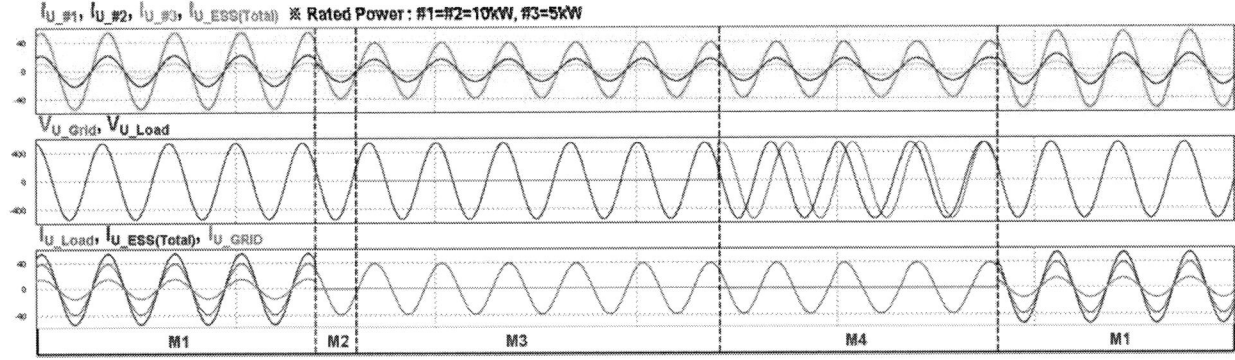

Fig. 9. Simulation verification of the proposed parallel/seamless transfer controller (M1~4).

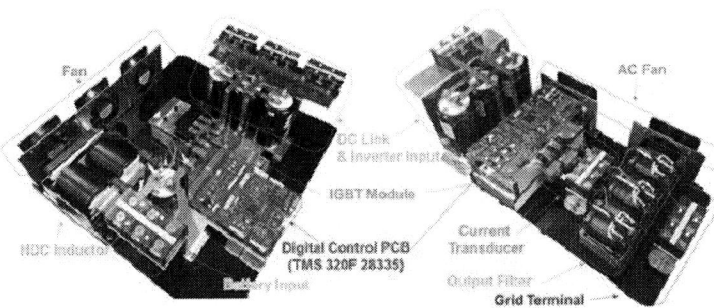

(a) Bidirectional DC/DC converter   (b) VSI module

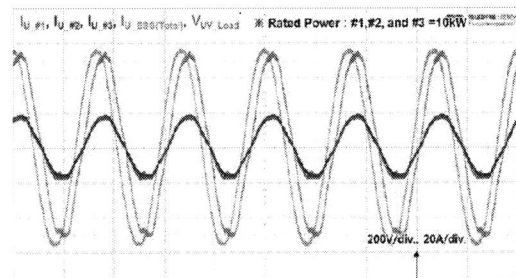

(c) Experimental waveforms in M3

Fig. 10. Hardware and experimental result.

are always operated in the current control mode. In other words, it is possible that unchanged-unified controller of VSIs responds to all the cases. As a result, the proposed seamless transfer controller for the parallel VSIs is completed by merging the two controllers as the way of Fig. 8. And simulation results of Fig. 9 verify that its control logic (indicated in Fig. 7) is reasonable. (A series of experiments has been actively carried out with help of the hardware of Fig. 10.)

V.   CONCLUSIONS

In the digest manuscript, the seamless transfer control of the parallel VSIs is proposed for a new type of energy storage system (ESS). And the operating principles of the proposed controller are described with various visual aids and simulation results. Through a series of development processes, the feasibility of the proposed controller is verified with competitive merits with respect of redundancy, reliability, performance, and simplicity to employ.

REFERENCES

[1] H. T. Le, S. Santoso, and   T. Q. Nguyen, "Augmenting Wind Power Penetration and Grid Voltage Stability Limits Using ESS: Application Design, Sizing, and a Case Study," *IEEE Trans. Power System*, vol. 27, no. 1, pp. 161–171, 2012.

[2] O. Erdinc, N.G. Paterakis, T.D.P. Mendes, A.G. Bakirtzis, J.P.S. Catalao, "Smart Household Operation Considering Bi-Directional EV and ESS Utilization by Real-Time Pricing-Based DR," *IEEE Trans. Smart Gird*, vol. 6, no. 3, pp. 1281–1291, 2015.

[3] C. Jiann-Fuh and C. Ching-Lung, "Combination voltage-controlled and current-controlled PWM inverters for UPS parallel operation," *IEEE Trans. Power Electron.*, vol. 10, pp. 547–558, Sep. 1995.

[4] N. Ainsworth and J. Murphree, "Paralleling of 3-phase 4-wire dc-ac inverters using repetitive control," in *Proc. IEEE Appl. Power Electron. Conf. Expo.*, 2009, vol. 1, pp. 116–120.

[5] S. Xiao, L. Yim-Shu, and X. Dehong, "Modeling, analysis, and implementation of parallel multi-inverter systems with instantaneous average current-sharing scheme," *IEEE Trans. Power Electron.*, vol. 18, no. 3, pp. 844–856, May 2003.

[6] M. Pascual, G. Garcera, E. Figueres, and F. Gonzalez-Espin, "Robust model-following control of parallel UPS single-phase inverters," *IEEE Trans. Ind. Electron.*, vol. 55, no. 8, pp. 2870–2883, Aug. 2008.

[7] T. Jingtao, L. Hua, Z. Jun, and Y. Jianping, "A novel load sharing control technique for paralleled inverters," in *Proc. Power Electron. Spec. Conf.*, 2003, vol. 3, pp. 1432–1437.

[8] C. Yeong Jia and E. K. K. Sng, "A novel communication strategy for decentralized control of paralleled multi-inverter systems," *IEEE Trans. Power Electron.*, vol. 21, no. 1, pp. 148–156, Jan. 2006.

[9] H. Zhongyi and X. Yan, "Distributed control for UPS modules in parallel operation with RMS voltage regulation," *IEEE Trans. Ind. Electron.*, vol. 55, no. 8, pp. 2860–2869, Aug. 2008.

[10] J. M. Guerrero, L. Hang, and J. Uceda, "Control of distributed uninterruptible power supply systems," *IEEE Trans. Ind. Electron.*, vol. 55, no. 8, pp. 2845–2859, Aug. 2008.

[11] H. Ming, H. Haibing, X. Yan, and H. Zhongyi, "Distributed control for ac motor drive inverters in parallel operation," *IEEE Trans. Ind. Electron.*, vol. 58, no. 12, pp. 5361–5370, Dec. 2011.

[12] H. Ming, H. Haibing, X. Yan, and H. Zhongyi, "Distributed control for ac motor drive inverters in parallel operation," *IEEE Trans. Ind. Electron.*, vol. 58, no. 12, pp. 5361–5370, Dec. 2011.

[13] R. Turner, S. Walton, and R. Duke, "Stability and bandwidth implications of digitally controlled grid-connected parallel inverters," *IEEE Trans. Ind. Electron.*, vol. 57, no. 11, pp. 3685–3694, Nov. 2010.

[14] Z. Yao and M. Hao, "Theoretical and experimental investigation of networked control for parallel operation of inverters," *IEEE Trans. Ind. Electron.*, vol. 59, no. 4, pp. 1961–1970, Apr. 2012.

978-1-4673-9551-9/16 $31.00 © 2016 IEEE

[15] Z. Xiaotian and J. W. Spencer, "Linear voltage-control scheme with duty ratio feed-forward for digitally controlled parallel inverters," *IEEE Trans. Power Electron.*, vol. 26, no. 12, pp. 3642–3652, Dec. 2011.

[16] R. Tirumala, N. Mohan, and C. Henze, "Seamless transfer of grid-connected PWM inverters between utility-interactive and stand-alone modes," in *Proc IEEE Appl. Power. Electron. Conf. Expo.*, 2002, pp. 1081–1086.

[17] S. Jung, Y. Bae, S. Choi, and H. Kim, "A low-cost utility interactive inverter for residential fuel cell generation," *IEEE Trans. Power Electron.*, vol. 22, no. 6, pp. 2293–2298, Nov. 2007.

[18] G. Q. Shen, D. H. Xu, and X. M. Yuan, "A novel seamless transfer control strategy based on voltage amplitude regulation for utility-interconnected fuel cell inverters with an LCL-filter," in *Proc. IEEE Power Electron. Spec. Conf.*, 2006, pp. 1–6.

[19] T. Hwang, K. Kim, and B. Kwon, "Control strategy of 600kW E-BOP for molten carbonate fuel cell generation system," in *Proc Int. Conf. Elect. Mach. Syst.*, 2008, pp. 2366–2371.

[20] Z. Yao, L. Xiao, and Y. Yan, "Seamless transfer of single-phase grid-interactive inverters between grid-connected and stand-alone modes," *IEEE Trans. Power Electron.*, vol. 25, no. 6, pp. 1597–1603, Jun. 2010.

[21] H. Kim, T. Yu, and S. Choi, "Indirect current control algorithm for utility interactive inverters in distributed generation systems," *IEEE Trans. Power Electron.*, vol. 23, no. 3, pp. 1342–1347, May 2008.

[22] J. Kwon, S. Yoon, and S. Choi, "Indirect Current Control for Seamless Transfer of Three-Phase Utility Interactive Inverters," *IEEE Trans. Power Electron.*, vol. 27, no. 2, pp. 773–781, Feb. 2012.

[23] C. H. da Silva, R. R. Pereira, L. E. B da Silva, G. Lambert-Torres, Bimal K. Bose, and S. U. Ahn, "A Digital PLL Scheme for Three-Phase System Using Modified Synchronous Reference Frame", *IEEE Transactions on Industrial Electronics*, vol. 57, no. 11, pp. 3814-3821, November 2010.

[24] R. M. Santos Filho, P. F. Seixas, P. C. Cortizo, Leonardo A. B. Torres, and André F. Souza, "Comparison of Three Single-Phase PLL Algorithms for UPS Applications", *IEEE Transactions on Industrial Electronics*, vol. 55, no. 8, pp. 2923-2932, August 2008.

[25] L. G. Barbosa Rolim, D. R. da Costa, Jr., and Maurício Aredes, "Analysis and Software Implementation of a Robust Synchronizing PLL Circuit Based on the pq Theory", *IEEE Transactions on Industrial Electronics*, vol. 53, no. 6, pp. 1919-1926, December 2006

A Plug-and-Play Ripple Mitigation Approach for DC-Links in Hybrid Systems

Sinan Li[1], Albert T. L. Lee[1], Siew-Chong-Tan[1] and S. Y. (Ron) Hui[1,2]
Email: snli@eee.hku.hk, tlalee@eee.hku.hk, sctan@eee.hku.hk, ronhui@eee.hku.hk
[1]Department of Electrical and Electronic Engineering, The University of Hong Kong, Hong Kong, China
[2]Department of Electrical and Electronic Engineering, Imperial College London, U.K.

Abstract—In this paper, a plug-and-play ripple mitigation technique is proposed. It requires only the sensing of the DC-link voltage and can operate fully independently to remove the low-frequency voltage ripple. The proposed technique is non-intrusive to the existing hardware and enables hot-swap operation without disrupting the normal functionality of the existing power system. It is user-friendly, modular and suitable for plug-and-play operation. The experimental results demonstrate the effectiveness of the ripple-mitigation capability of the proposed device. The DC-link voltage ripple in a 110 W miniature hybrid system comprising an AC/DC converter and two resistive loads is shown to be significantly reduced from 61 V to only 3.3 V. Moreover, it is shown that with the proposed device, the system reliability has been improved by alleviating the components' thermal stresses.

Keywords—Plug-and-play ripple mitigation, ripple pacifier, DC-link, AC/DC power system.

I. INTRODUCTION

In recent years, with the rapid penetration of distributed renewable energy sources (such as solar photovoltaic, wind power) into the traditional AC power network, it is envisaged that a mixture of AC power grids and emerging DC power grid will emerge as a future form of power network. A typical infrastructure of a hybrid power grid is shown Fig. 1(a), where power conversions such as AC/DC, DC/AC and DC/DC link the various AC and DC power sources/loads together. In the system, the electrical energy is transmitted and distributed through an intermediate DC voltage link. A stable and reliable operation of DC-link is of vital importance because a large variation of the DC-link voltage can lead to efficiency and performance degradation of its upstream/downstream converters, increased voltage stresses of the system and coupled interference between the DC and AC utilities [1], [2]. For specific applications, the DC-link voltage fluctuation will generate flickers in the LED lightings [3], [4], shorten the life expectancy of a battery in electric vehicle applications [5], [6], and reduce the power efficiency in the PV panels [7], [8] and fuel cells [9]. Large voltage ripple across the electrolytic capacitors (E-caps) also leads to significant capacitor current ripple and thus internal resistive loss and temperature rise inside the E-Cap.

With the growing use of low-power-rating single-phase AC/DC converters, such as the plug-in AC modules for PV

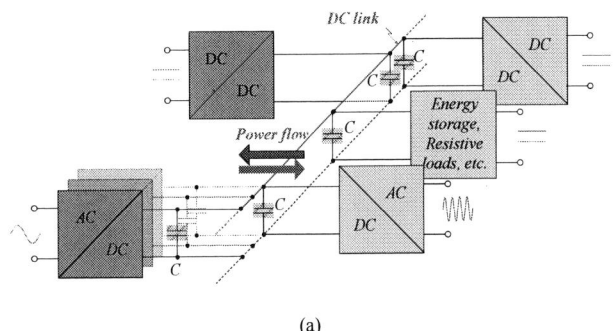

(a)

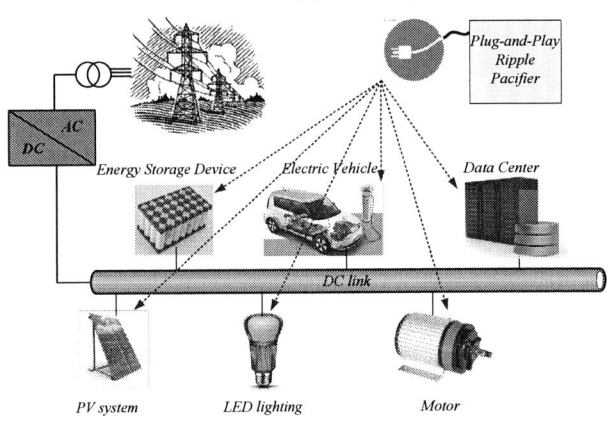

(b)

Fig. 1. (a) Typical power stage infrastructure of power electronic systems with an intermediate DC-link; (b) overview of the proposed plug-and-play ripple mitigation scheme for the DC-link in a hybrid power systems.

applications [10], one critical challenge with the DC-link in an AC-DC hybrid system is that the it suffers from a steady-state oscillation with a double-line frequency component (i.e. 100 Hz or 120 Hz for a 50 Hz or 60 Hz grid, respectively). The oscillation is caused by the instantaneous power difference between the AC and DC side in single-phase AC/DC converters [1], [2]: at the AC side, power will be varying at double-line frequency around a DC offset; while at the DC side, a constant power is desired. To buffer the power imbalance and stabilize the DC-link voltage, large and bulky capacitors must be installed in parallel with the DC-link either dispersedly over the transmission and distribution line or locally near the source/load converters (as the ones

978-1-4673-9551-9/16 $31.00 © 2016 IEEE

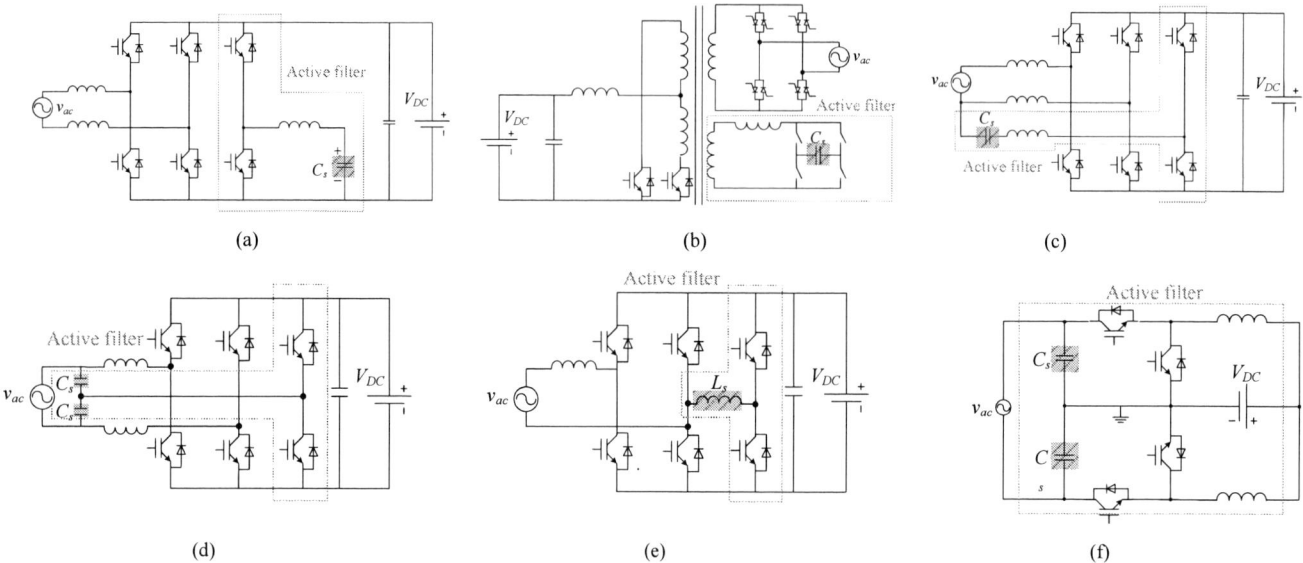

Fig. 2. Various circuit diagrams for implementing an active-filter with an AC/DC converter.

shown in Fig. 1(a) and labeled as C). Typically, E-caps are common selections of DC-link capacitors due to their space-saving and cost-effective nature. However, their high failure rates substantially undermine the system's reliability [5]–[8], [11], [12]. Their limited lifetime is also highly incompatible with that of existing renewable technologies. It is reported that a typical E-cap has a lifetime of merely 1000–7000 hours at 105 $^\circ C$, and the number is halved with each additional 10 $^\circ C$ rise in the junction temperature [13]–[15]. In contrast, PV is reported to have a life expectancy of more than 20 years [16], [17]. A recent trend to tackle the reliability issue is to eliminate the use of E-caps with active-power-filter-based approaches. The principles of operation are to store the ripple power in a separate energy storage component and to allow a large voltage/current fluctuation across it. In this way, small but long lifetime film capacitors can be used to stabilize the DC-link voltage with enhanced system reliability. Another benefit of this approach is that the volume of the switching converters in the host system can be significantly shrunk with the removal of the relatively large E-caps [18].

Despite the advantages of the active-filter-based approach, the installation and maintenance of an active filter can be a tedious task. As will be reviewed in Section II, the prior arts solutions typically require modifications and re-design of the existing host system before an active filter can be used. This inevitably increases the cost of the overall systems at both power supplies' and demands' ends. In this paper, a new concept called plug-and-play ripple mitigation is proposed. It is physically realized as a two terminal device, whose major objective is to achieve DC-link voltage stabilization based on the active-filter concept, but it differentiates itself from prior arts with a distinctive feature of hot-swap operation, i.e., device can be easily attached and detached on-line from the DC-link. Therefore, the device requires no modification of its host DC system and can operate as a stand-alone equipment. The proposed approach

includes a patent-pending device known as the plug-and-play ripple pacifier (RP), as shown in Fig. 1(b) [28]. It can simply be plugged into the DC-link distributedly and/or locally. For example, it can be installed locally on the PV side of an AC inverter module for the purpose of power decoupling. In the event that the proposed device breaks down and fails to operate, it can be swapped out with a new one in a plug-and-play manner. The success of the device largely relies on (i) the selection of a proper power electronics topology and (ii) the proposed DC-ripple-based control techniques. Importantly, by performing the small-signal analysis, the device is shown to be non-intrusive to the normal operation of the DC power system. The effectiveness of the ripple-mitigating function, the non-intrusive property of the RP and viability of hot-swapping operation are experimentally verified in a 110 W miniature AC-DC hybrid system.

II. EXISTING ACTIVE-FILTER-BASED APPROACHES

A. Hardware

The circuit configurations of an active filter with the existing host systems can be manipulated in several ways. It can be (i) a complete switching converter that can work independently (Fig. 2(a) and (b)), (ii) a semi-complete switching converter which can only operate normally by sharing some of the components with the host systems (Fig. 2(c)–(e)), and (iii) a fully integrated switching converter that shares all the components with the host systems (Fig. 2(f)). The energy storage components in these active filters can be a capacitor (Fig. 2 (a)–(d), (f)), an inductor (Fig. 2(e)) or batteries, with bipolar or unipolar operating waveforms.

Of these active filter configurations, the integrated solution (Fig. 2(f)) achieves active filtering by modifying the operation waveforms in the original host converters which potentially increases the current/voltage stress. Therefore, the original hardware, including its controllers, must be re-designed. For the combined solution (e.g. Fig. 2(c), (e) and

978-1-4673-9551-9/16 $31.00 © 2016 IEEE

(f)), no modification of the existing system is necessary. However, these active filters require the access of several terminals of the host system, e.g. the AC side, the DC side and/or the phase leg switches, and thus a hot-swap installation of such an active filter is difficult. For the solution which employs a complete switching converter, there might still be problems with the installation. For instance, in Fig. 2(b), a complete re-design of the high-frequency transformer is required. This is because the active filter is magnetically coupled to the host system, and the energy storage requirement of the high-frequency transformer can be different from that in the original designs.

B. Sensing and Control

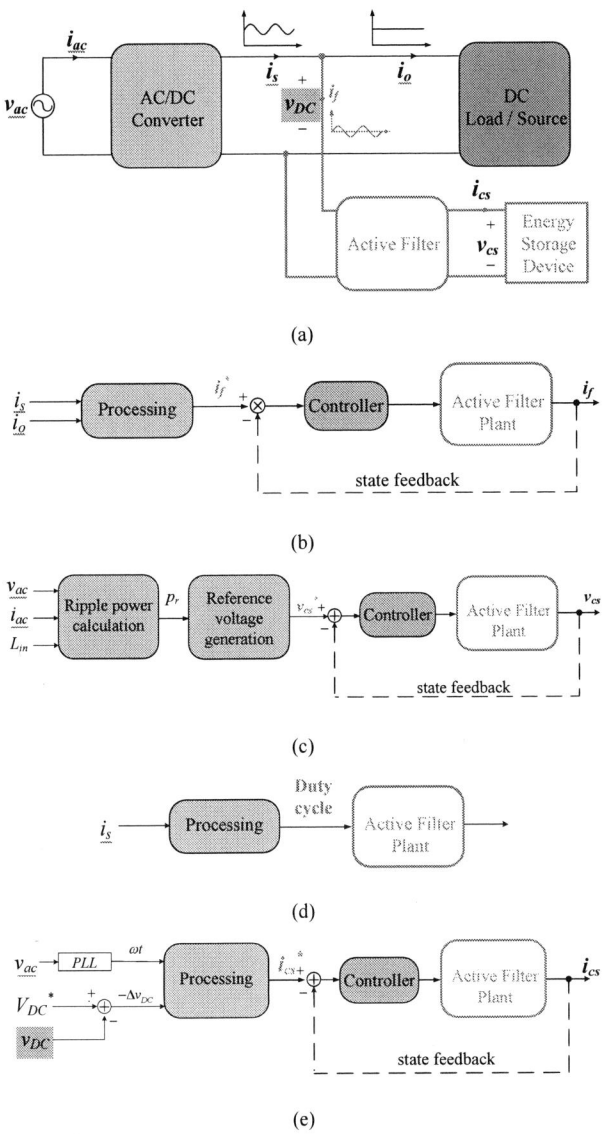

(a)

(b)

(c)

(d)

(e)

Fig. 3. (a) A simplified schematics of a DC system with active filter and (b–e) existing control methods for an active filter used in a DC system.

The implementation of sensing and control in an active filter can also prohibit a true plug-and-play operation of the ripple mitigation. This can be explained using a simplified

schematic in Fig. 3(a), where a single-phase AC/DC converter, a DC load/ source (converter) and a shunt active filter are included. It should be noted that all the configurations in Fig. 2 can be equivalently translated into Fig. 3(a). Assuming a constant DC-link voltage v_{DC}, the DC-link current from the AC/DC converter i_s will contain a DC and a double-line frequency content. By introducing a compensation current i_f into the DC-link, the shunt device can eliminate the low-frequency oscillations in the DC-link, leading to an stable DC current i_o for the DC load/source. The compensation current i_f from the shunt active filter can be controlled directly or indirectly. To date, there are four basic methods to control the active filter, as illustrated in Fig. 3(b)–(e), and they are described in more detail as follows.

(i) Direct control of i_f in a closed-loop such that it follows a reference signal i_f^*, as shown in Fig. 3(b). Since i_f should only compensate the AC component of the DC-link current i_s, the derivation of i_f^* often involves the measurement of i_s. For instance, i_f^* can be obtained by calculating $i_f^* = i_s - I_o$, where I_o is the DC portion of i_o [19]. In [11], [20]–[23], a high-pass filter, a neural filter and a virtual-capacitor-based control is applied to i_s to exact the AC component .

(ii) Direct control of the instantaneous power in the energy storage device in a closed-loop, such that it matches the ripple power generated by the AC/DC converter, as shown in Fig. 3(c). Such a method involves an estimation of the ripple power from the AC/DC stage and a reference generation algorithm for the energy storage device to track. To estimate the ripple power, a measurement of v_{ac} and i_{ac} is needed. To obtain a more accurate result, the values of the internal inductors, capacitors and power loss in the host systems must be known [7], [24].

(iii) Direct calculation of the required duty cycle based on the AC component in the DC-link current i_s, as shown in Fig. 3(c) [18], [25].

(iv) Direct regulation of the instantaneous DC-link voltage v_{DC} in a closed-loop, as shown in Fig. 3(e). However, the measurement of v_{ac} is still required to provide phase information for the active filter reference generation [26].

With reference to the four control methods, it is worth noting that direct measurements of variables such as the DC-link current i_s, the DC load current i_o or the AC side current/ voltage/ phase information is required. Since implementing a current sensor requires opening the AC line or the DC-link, and the AC voltage sensing is typically achieved far away from the local point of connection of an active filter, the requirements of these measurements render these methods unsuitable for plug-and-play operation. In a more practical system, where multiple AC and DC sources/loads are interconnected, collecting all the necessary information to formulate a proper reference command for the active filter will be a costly and inconvenient task.

III. REQUIREMENTS FOR BEING A TRUE PLUG-AND-PLAY DEVICE

Based on the review in Section II, to enable the true plug-and-playable feature, two requirements must be satisfied.

First, the hardware of the device must be a complete switching converter and connected directly with the DC-link without intermediate coupling or connections, as shown in Fig. 4(a). Second, the controller must also be DC-link based which means that the only measurement required is the local DC-link voltage v_{DC} at the point of connection, as depicted in Fig. 4(b). Unlike previous methods, the proposed control scheme does *not* need to measure any voltage or current information (v_{ac}, i_{ac}, i_s) from the AC side or the current information (i_o) from the DC side. In addition, it does not require a priori knowledge of the exact system parameters used in the existing system. Moreover, the method is local-information-based which does *not* require any communications between the various AC and DC sources/loads. In this paper, an active filter with a plug-and-play ripple mitigation feature is referred to as a plug-and-play ripple pacifier (RP).

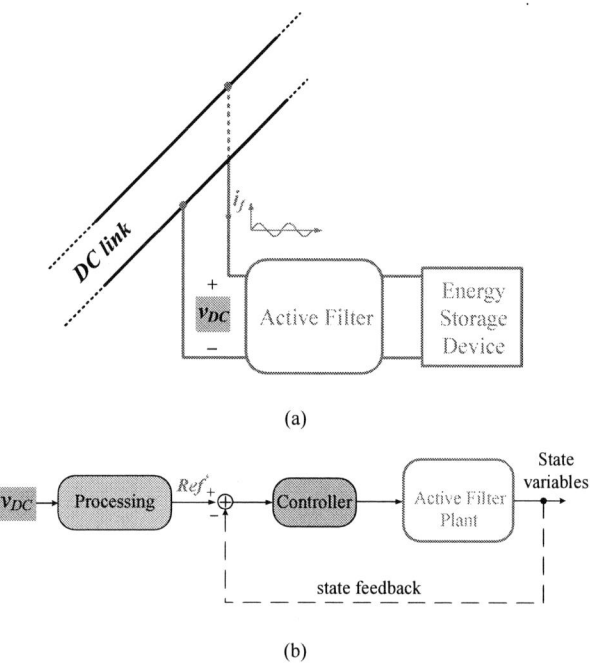

(a)

(b)

Fig. 4. Requirements for plug-and-play operation, namely (a) a shunt connection must exist on the DC port of the PFC rectifier and (b) the DC-link voltage v_{DC} is the only information required for the controller.

IV. IMPLEMENTATIONS OF A PLUG-AND-PLAY RIPPLE PACIFIER

A. Hardware configuration

An RP deals with AC ripple power only, and thus the switching converter must support bidirectional power conversion. Since the input of the RP is the DC-link voltage, while the output is connected to an energy storage device, whose voltage waveform can either be DC or AC, there is a myriad of DC-voltage-sourced topologies for selection, including at least step-down (forward) converter, step-up (boost) converter, step-up/down (flyback) converter and their combinations. Some of these converters can be constructed with a half-bridge or a full-bridge configuration. Switched capacitor converters are also possible candidates [27]. For

very high voltage or power application, multilevel power converter can also be used if necessary.

With respect to the voltage level of the target DC-link voltage, there are mainly three rules which governs the circuit topology selection of an RP as follows.

a) The energy utilization rate of the energy storage should be large.

b) The size and cost of the energy storage device should be small.

c) High voltage stresses should be avoided.

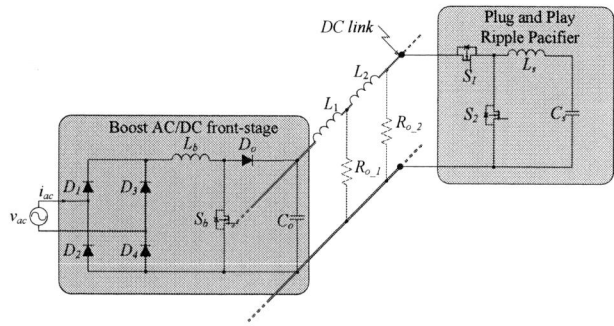

Fig. 5. Ideal circuit model of a boost-type PFC rectifier with a buck-type RP.

This paper focuses on a hybrid system with a high voltage DC-link as shown in Fig. 5, where a boost PFC rectifier is employed as the front-end stage of the DC-link with an average output voltage of V_{DC} = 400 V, and two resistive loads R_{o1} and R_{o2} are used to emulate the practical loading conditions with intermediate transmission impedance L_1 and L_2. A bi-directional buck-type RP with a capacitive energy storage C_s is selected and plugged into the DC-voltage link at the point of common coupling (PCC) near the load R_{o2}. Also, there is a pre-installed DC-link capacitor C_o, whose effective capacitance is deliberately chosen to be small enough to mimic the effect of aging and heating of the liquidated chemical contained in the E-Caps .

B. DC-Ripple-Based Control Techniques

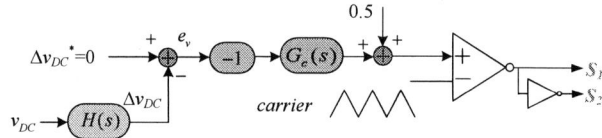

Fig. 6. Schematic diagram of the proposed ripple-voltage-based control method for the plug-and-play RP.

Fig. 6 illustrates the proposed DC-ripple-based control scheme, where the only requisite information is v_{DC}. The motivation is that the DC-link voltage ripple is caused by the unbalanced input and output power that flows into and out of the DC-link capacitor C_o, and that the DC-link voltage ripple alone should contain sufficient information to predict the unbalanced ripple power p_r or the ripple current in i_s. In Fig. 6, the ripple voltage Δv_{DC} is initially obtained from v_{DC} through $H(s)$, e.g. a high-pass filter, and then tightly regulated following a reference $\Delta v_{DC}^* = 0$ through the closed-loop feedback control, thereby realizing zero voltage ripple on the DC-link. A feed-forward term of 0.5 is added to the final modulation signal such that the voltage across C_s will contain a $V_{DC}/2$ offset. It should be emphasized that albeit the proposed DC-ripple-based control appears to be simple and straightforward, it is the first control method reported that is suitable for a true plug and play operation.

C. Small-Signal Analysis

To facilitate the small-signal analysis of the RP with the proposed control schemes, an equivalent circuit diagram of Fig. 5 is obtained in Fig. 7(a). The transmission inductances have been neglected to ease the analysis and demonstrate the proof-of-concept. The current source i_s containing a DC and a double-line frequency component represents the AC/DC converter; the resistor R_o represents the total load on the DC voltage link, and r_p is the equivalent series resistance (ESR) in L_s and C_s. By performing state-space averaging and linearization of the model in Fig. 7(a), the small-signal control block diagram of the RP system can be obtained in Fig. 7(b), where $G_{im}(s)$ is the RP plant transfer function from the duty cycle m to the average input current i_{S1}, as

$$G_{im}(s) = \frac{\widetilde{\langle i_{S1} \rangle}(s)}{\widetilde{m}(s)} = \frac{sC_sMV_{DC}}{s^2C_sL_s + sC_sr_p + 1}, \qquad (1)$$

where $\langle \cdot \rangle$ is an averaging operator over a switching period, and the symbol $\sim$ denotes small-signal perturbations. M is the steady-state value of the duty cycle m.

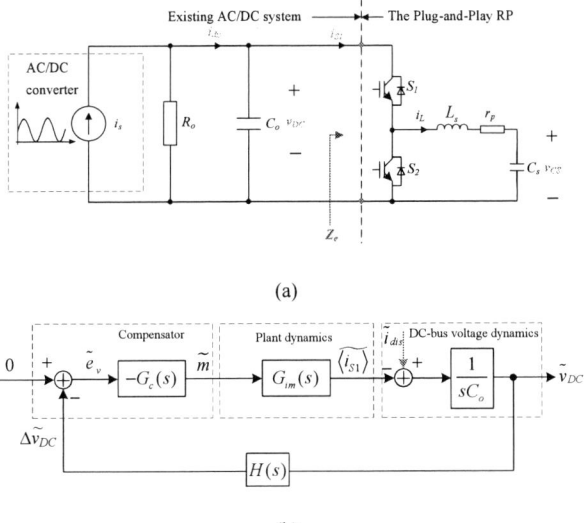

(a)

(b)

Fig. 7. (a) Equivalent circuit diagram of Fig. 5 and (b) The small-signal control block diagram of the RP system based on the linearized model.

From Fig. 7(b), the total open-loop gain $\ell(s)$ and the closed-loop $\tilde{v}_{DC}$ are given by (2) and (3) respectively as

$$\ell(s) = \frac{1}{sC_o}G_c(s)G_{im}(s)H(s). \qquad (2)$$

$$\tilde{v}_{DC}(s) = \frac{1}{sC_o\left[\ell(s)+1\right]}\tilde{i}_{dis}(s), \qquad (3)$$

In other words, to suppress the ripples in v_{DC}, $\ell(s)$ should be designed to have a high enough gain at frequencies with respect to $\tilde{i}_{dis}(s)$. A proportional-resonant (PR) controller whose resonant poles are designed precisely at the double-line frequency can be selected for the compensator $G_c(s)$. To further complement the ripple mitigating performance, multi-RP controllers or repetitive controllers can be used for reducing high order ripples.

It is interesting to examine the impact of the RP to the host system by examining the input impedance Z_e of the RP, as defined in Fig. 7(a). Based on Fig. 7(b), Z_e can be derived in (4) whose Bode plots are shown in Fig. 8.

$$Z_e(s) = \frac{\tilde{v}_{DC}(s)}{\widetilde{\langle i_{S1} \rangle}(s)} = \frac{1}{H(s)G_c(s)G_{im}(s)} = \frac{1}{s\left[C_o\ell(s)\right]} \qquad (4)$$

The zero impedance of the RP at 100 Hz exemplifies the ripple-mitigating capability of the RP from a filter point of view: any 100 Hz ripple power will be directly shorted to ground without going to the DC load. Meanwhile, the infinitely large impedance at low frequency (including DC) implies that the RP behaves like an open-circuit, so that adding the RP in parallel with the DC-link has negligible, if not zero, influence on the DC voltage regulation in the host AC/DC converter(s). Therefore, it can be concluded that applying an RP on the DC-link will only suppress the designated low-frequency ripple, and will not interfere with the existing AC/DC system.

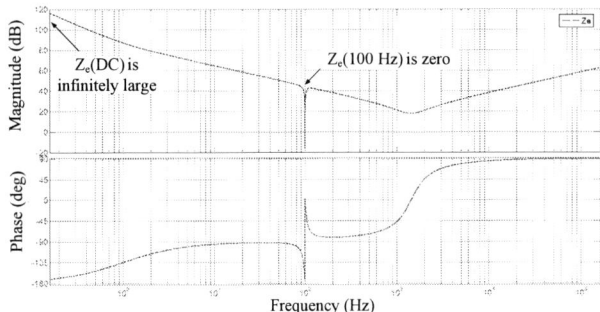

Fig. 8. Bode plots of the input impedance Z_e of the RP with a proportional-resonant controller whose resonant poles are set at 100 Hz.

V. EXPERIMENT VERIFICATIONS

The effectiveness of the proposed plug-and-play ripple reduction scheme is experimentally verified in a system as that given in Fig. 5. Table I summarizes the key circuit parameters used for the experimental setup. Note that the pre-installed DC-link capacitor C_o has a small capacitance of merely 5 μF. Based on the power ratings of the loads which is 100 W, the energy storage capacitor C_s can be chosen as 5 μF to minimize the required energy storage. In a system with medium power ratings, a larger C_s should be utilized. For a high power system, multiple RPs with relative small C_s may

TABLE I.   KEY CIRCUIT PARAMETERS FOR EXPERIMENTAL SETUP

Plug-and-Play RP Parameters	Values	Host System Parameters	Values
Input voltage V_{DC} (V)	400	Nominal power P_{DC} (W)	100
Energy storage capacitor C_s (μF)	5	AC voltage v_{ac} (V)	220
Inductor L_s (mH)	2.5	Output voltage V_{DC} (V)	400
Switching frequency f_{s1} (kHz)	25	Dc-link capacitor C_o (μF)	5
		ESR of C_o @ 400 V (Ω)	3.51
		Boost inductor L_b (μH)	390
		Load 1: R_{o_1} (Ω)	8000
		Load 2: R_{o_2} (Ω)	2000

be connected to the system in a distributive manner to operate simultaneously.

A. Steady-State Ripple Mitigation Performance

Fig. 9 show the measured steady-state DC-link voltage waveforms before and after the proposed plug-and-play ripple mitigation device is applied to the DC-link.

Fig. 9(a) shows that before the RP is applied, there is a significant 100 Hz ripple in the v_{DC}. The measured peak-to-peak DC-link voltage ripple is 61 V, which accounts for 15.25% of the average DC-link voltage. The large ripple is mainly due to the a small capacitor C_o used on the DC-link. When the plug-and-play RP is activated and compensating at 100 Hz (through a single PR controller with a pair of resonant poles at 100 Hz), the DC-link voltage ripple is significantly mitigated from 61 V to 6.3 V, as shown in Fig. 9(b). At the same time, the voltage across C_s is pulsating significantly at 100 Hz between zero and v_{DC}. Since only a single PR controller is used, the capacitor voltage v_{cs} has a smooth and sinusoidal waveform. To further mitigate the high frequency ripples, multiple PR controllers (with resonant poles at 100 Hz, 200 Hz and 300 Hz) are used in a third experiment and the plug-and-play device is able to compensate ripples at 100 Hz, 200 Hz and 300 Hz. The measured v_{DC} is shown in Fig. 9(c) with a ripple of merely 3.3 V. With a multiple PR controller, a much sharper waveform is observed around the valley of v_{cs}, as compared with that in Fig. 9(b). The results in Fig. 9 successfully confirms the ripple mitigation capability with the proposed DC-ripple-based control techniques.

B. Non-Invasive Property During Load Transient and Steady-State

To demonstrate the non-invasive property of the RP, transient performance of the RP are examined, with four sets of step load change experiments (from full load to 20% load and from 20% load to full load) with and without the RP. The change of load power is realized by switching on and off the load R_{o_2} (80 W). The transient waveforms for the DC-

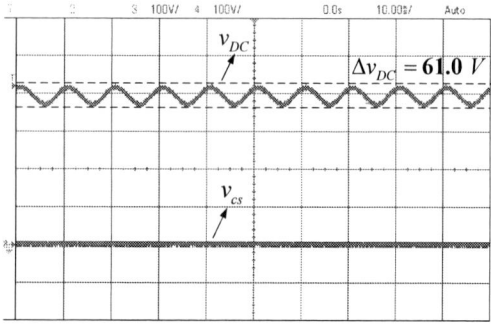

(a)

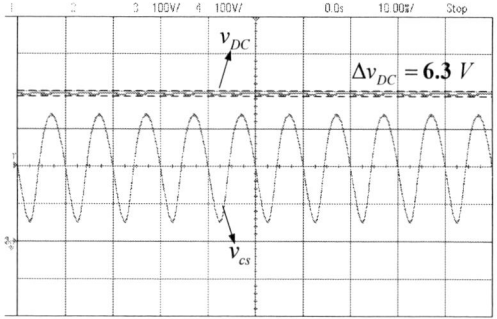

(b)

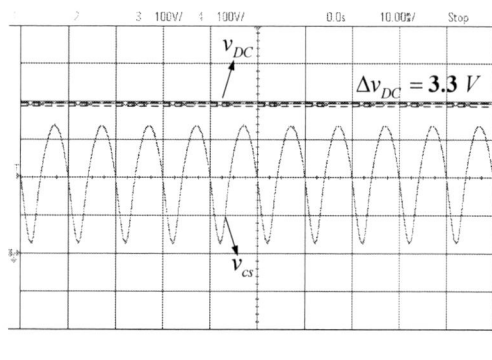

(c)

Fig. 9. Steady-state voltage waveforms of the DC-link and the energy storage capacitor with and without the plug-and-play ripple pacifier. (a) Without the ripple pacifier; (b) With the ripple pacifier (using a single PR controller resonant at 100 Hz); (c) With the ripple pacifier (using multiple PR controllers resonant at 100 Hz, 200 Hz, and 300 Hz).

link and the energy storage capacitor are shown in Fig. 10 and Fig. 11.

In both sets of experiments, the DC-link voltage with and without a plug-and-play RP has similar dynamics in terms of the overshoot/undershoot ratio and settling time during the load change transient. Meanwhile, during steady-state, the DC-link voltage is unchanged before and after the load transient. These observations verify the non-invasive property of the proposed RP to the host system in regulating the average DC-bus voltage.

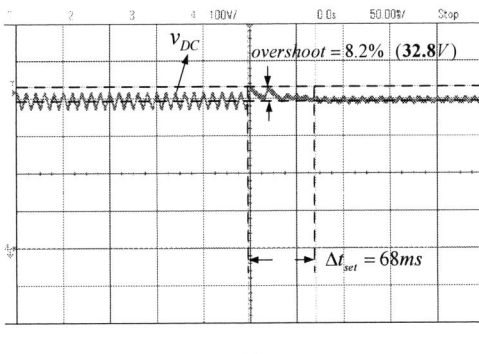

(a)

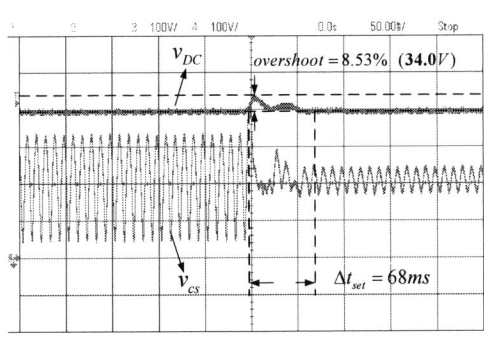

(b)

Fig. 10. Transient voltage waveforms from full load to 20% load (a) without the plug-and-play RP (b) with the plug-and-play RP.

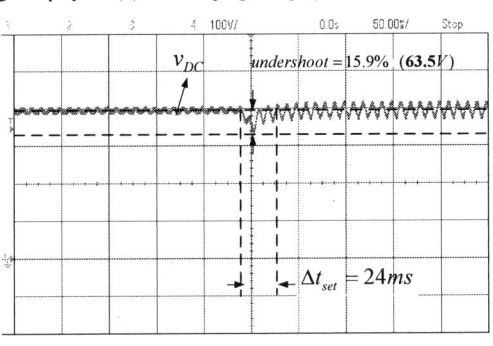

(a)

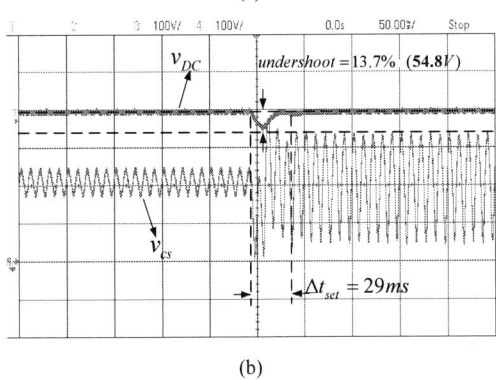

(b)

Fig. 11. Transient voltage waveforms from 20% load to full load (a) without the plug-and-play RP (b) with the plug-and-play RP.

C. Hot-Swapping with Soft-Start

To avoid undesired transient caused by the inrush current when the RP is activated or hot-swapped into the DC-link, a soft start function must be incorporated into the RP with a proper powering sequence. Fig. 12 illustrates the waveforms of v_{DC} and v_{cs} during the plug-in interval of an RP to the DC-link. At the plugged-in instant, the RP detects the connection with the DC-link and gradually increases the duty cycle of from zero to 50 %. In this way, v_{cs} is softly charged to half of the DC-link voltage, which is also the steady-state voltage of v_{cs}. No severe overshoot/undershoot or oscillations is observed in v_{DC}. The controller in the RP will be activated as soon as the soft-start ends. An immediate reduction of Δv_{DC} will then be obtained.

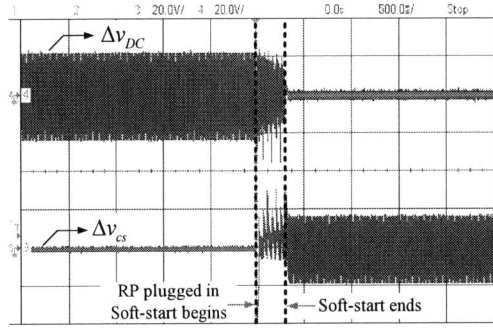

Fig. 12. Voltage waveforms of the DC-link ripple Δv_{DC} and energy storage ripple Δv_{cs} during hot-swapping interval of an RP.

D. Theraml and Reliability Improvement of the Hybrid System

By eliminating the ripple current inside the DC-link capacitor and ensuring a stable DC-link voltage, the junction temperature of C_o as well as that in the power devices and magnetics in the host AC/DC converters are expected to decrease.

TABLE II. MEASURED TEMPERATURE WITH AND WITHOUT RP

Measured Surface	Without RP (°C)	With RP (°C)	Temperature reduction ΔT (°C)
Plastic cover of E-Cap C_o	42.5	35.0	7.5
Aluminum top of E-Cap C_o	42.2	34.1	8.1
Output diode D_o	50.7	41.4	9.3
Power MOSFET S_b	42.4	33.9	8.5
Input diode bridge $D_1 \sim D_4$	60.0	52.3	7.7
Magnetic core of L_b	64.7	54.3	10.4

Table II shows the experimentally measured surface temperatures of the boost PFC rectifier front stage taken after three hours of continuous operation under free convection at the ambient temperature of 25.4 °C with and without the use of RP. The measurement is conducted using thermal couplers and a FLUKE data acquisition unit. As expected, with the use of RP, all the power devices, magnetics, and the output E-cap have much lower temperatures than those without RP. The surface (junction) temperature of all the components has reduced by more than 7.5 ℃ which improves the operating lifetime of components and hence the reliability of the whole system. For instance, according to [13], a 8.1 °C reduction of the surface temperature of an E-cap could extend its lifetime by 1.75 times.

VI. CONCLUSIONS

In this paper, a plug-and-play ripple mitigation technique for stabilizing the DC-link voltage is proposed. The major benefits of the proposed solution are that it is simple to use, cost-effective, modular in structure, and non-invasive to the existing AC/DC power system. Also, the capacitor on the DC bus can be made significantly smaller than that required in conventional AC/DC systems without the proposed auxiliary ripple energy storage. Longer lifetime film capacitors with lower equivalent series resistance (ESR) can therefore be used to replace the short-lifetime E-caps with high ESR. The experimental results confirm the effectiveness of the ripple-reduction capability of the proposed plug-and-play RP device on a boost-type PFC rectification system.

ACKNOWLEDGMENT

This work is supported by the Hong Kong Research Grant Council under the Theme-based project T22-715/12N.

REFERENCES

[1] H. Wang and F. Blaabjerg, "Reliability of capacitors for dc-link applications in power electronic converters—an overview," *IEEE Trans. Ind. Appl.*, vol. 50, no. 5, pp. 3569–3578, Sep. 2014.

[2] L. Gu, X. Ruan, M. Xu, and K. Yao, "Means of eliminating electrolytic capacitor in AC/DC power supplies for LED lightings," *IEEE Trans. Power Electron.*, vol. 24, no. 5, pp. 1399–1408, May 2009.

[3] B. Lehman, A. Wilkins, S. Berman, M. Poplawski, and N. Johnson Miller, "Proposing measures of flicker in the low frequencies for lighting applications," in *2011 IEEE Energy Conversion Congress and Exposition*, 2011, pp. 2865–2872.

[4] S. Li, S. C. Tan, C. K. Lee, C. K. Tse, E. Waffenschmidt, S. Y. R. Hui, "A Survey, Classification and Critical Review of Light-Emitting Diode Drivers", *IEEE Transactions on Power Electronics*, vol. 31, no. 2, pp. 1503–1516, Mar. 2015.

[5] T. Shimizu, T. Fujita, G. Kimura, and J. Hirose, "A unity power factor PWM rectifier with DC ripple compensation," *IEEE Trans. Ind. Electron.*, vol. 44, no. 4, pp. 447–455, 1997.

[6] T. Shimizu, Y. Jin, and G. Kimura, "DC ripple current reduction on a single-phase PWM voltage-source rectifier," *IEEE Trans. Ind. Appl.*, vol. 36, no. 5, pp. 1419–1429, 2000.

[7] P. T. Krein, R. S. Balog, and M. Mirjafari, "Minimum energy and capacitance requirements for single-phase inverters and rectifiers using a ripple port," *IEEE Trans. Power Electron.*, vol. 27, no. 11, pp. 4690–4698, Nov. 2012.

[8] S.-H. Lee, T.-P. An, and H.-J. Cha, "Mitigation of low frequency AC ripple in single-phase photovoltaic power conditioning systems," *J. power Electron.*, vol. 10, no. 3, pp. 328–333, 2010.

[9] R. S. Gemmen, "Analysis for the effect of inverter ripple current on fuel cell operating condition," *J. Fluids Eng.*, vol. 125, no. 3, p. 576, May 2003.

[10] S. B. Kjaer, J. K. Pedersen, and F. Blaabjerg, "A review of single-phase grid-connected inverters for photovoltaic modules," *IEEE Trans. Ind. Appl.*, vol. 41, no. 5, pp. 1292–1306, Sep. 2005.

[11] S. Wang, X. Ruan, K. Yao, S.-C. Tan, Y. Yang, and Z. Ye, "A flicker-free electrolytic capacitor-less AC–DC LED driver," *IEEE Trans. Power Electron.*, vol. 27, no. 11, pp. 4540–4548, Nov. 2012.

[12] S. Li, G.-R. Zhu, S. C. Tan, and S. Y. Hui, "Direct AC/DC rectifier with mitigated low-frequency ripple through waveform control," *IEEE Transactions on Power Electronics*, vol. 30, no. 8, pp. 4336–4348, Sep. 2014.

[13] S. G. Parler, "Application guide, aluminum electrolytic capacitors." [Online]. Available: www.cornell-dubilier.com.

[14] S. K. Maddula and J. C. Balda, "Lifetime of electrolytic capacitors in regenerative induction motor drives," in *IEEE 36th Conference on Power Electronics Specialists*, 2005, pp. 153–159.

[15] H. Hu, S. Harb, N. Kutkut, I. Batarseh, and Z. J. Shen, "A review of power decoupling techniques for microinverters with three different decoupling capacitor locations in PV systems," *IEEE Trans. Power Electron.*, vol. 28, no. 6, pp. 2711–2726, Jun. 2013.

[16] The Eco Experts, "The life expectancy of solar panels." [Online]. Available: http://www.theecoexperts.co.uk/life-expectancy-solar-panels.

[17] E. D. Dunlop, "Lifetime performance of crystalline silicon PV modules," in *Photovoltaic Energy Conversion, 2003. Proceedings of 3rd World Conference on*, 2003, pp. 2927–2930.

[18] R. Wang, F. Wang, D. Boroyevich, R. Burgos, R. Lai, P. Ning, and K. Rajashekara, "A high power density single-phase PWM rectifier with active ripple energy storage," *IEEE Trans. Power Electron.*, vol. 26, no. 5, pp. 1430–1443, May 2011.

[19] S. Dusmez and A. Khaligh, "Generalized Technique of Compensating Low-Frequency Component of Load Current With a Parallel Bidirectional DC/DC Converter," *IEEE Trans. Power Electron.*, vol. 29, no. 11, pp. 5892–5904, Nov. 2014.

[20] L. Palma, "An active power filter for low frequency ripple current reduction in fuel cell applications," *Speedam 2010*, pp. 1308–1313, Jun. 2010.

[21] R.-J. Wai and C.-Y. Lin, "Development of active low-frequency current ripple control for clean-energy power conditioner," in *2010 5th IEEE Conference on Industrial Electronics and Applications*, 2010, pp. 695–700.

[22] R.-J. Wai and C.-Y. Lin, "Active low-frequency ripple control for clean-energy power-conditioning mechanism," *IEEE Trans. Ind. Electron.*, vol. 57, no. 11, pp. 3780–3792, Nov. 2010.

[23] W. Cai, B. Liu, S. Duan, and L. Jiang, "An Active Low-Frequency Ripple Control Method Based on the Virtual Capacitor Concept for BIPV Systems," *IEEE Trans. Power Electron.*, vol. 29, no. 4, pp. 1733–1745, Apr. 2014.

[24] Y. Tang, F. Blaabjerg, P. C. Loh, C. Jin, and P. Wang, "Decoupling of fluctuating power in single-phase systems through a symmetrical half-bridge circuit," *IEEE Trans. Power Electron.*, vol. 30, no. 4, pp. 1855–1865, Apr. 2015.

[25] Y. Yang, X. Ruan, L. Zhang, J. He, and Z. Ye, "Feed-forward scheme for an electrolytic capacitor-less AC/DC LED driver to reduce output current ripple," *IEEE Trans. Power Electron.*, vol. 29, no. 10, pp. 5508–5517, Oct. 2014.

[26] Y. Tang, Z. Qin, F. Blaabjerg, and P. C. Loh, "A dual voltage control strategy for single-phase PWM converters with power decoupling function," *IEEE Trans. Power Electron.*, vol. 30, no. 12, pp. 7060–7071, Dec. 2014.

[27] M. Chen, Y. Ni, C. M. Serrano, B. J. Montgomery, D. J. Perreault, and K. K. Afridi, "An electrolytic-free offline LED driver with a ceramic-capacitor-based compact SSC energy buffer," in *Proceedings of the IEEE Energy Conversion Congress and Exposition (ECCE)*, 2014.

[28] S. Li, Albert Lee, S. C. Tan and S. Y. R. Hui, "A Plug and Play Ripple Pacifier", PCT/CN2015/092649.

Active Control of Low Frequency Common Mode Voltage to Connect AC Utility and 380 V DC Grid

Fang Chen, Rolando Burgos, Dushan Boroyevich, Xuning Zhang
Center for Power Electronics Systems (CPES)
Virginia Polytechnic Institute & State University
Blacksburg, VA, US
fangchen@vt.edu

Abstract—DC system is promising in data centers and future homes. It can be interconnected with ac system through power converters. One typical case is connecting the 380 V dc grid to the single-phase ac utility through a transformerless two-stage ac/dc converter. In such configuration, the ac and dc system common mode (CM) quantities are coupled through the common ground. While the high frequency noise is filtered by passives, the asymmetric dc and low frequency CM voltage need to be controlled actively for the connection of bipolar dc systems. In this paper, a passive floating filter is used to contain the high frequency noise within the converter. An active CM duty cycle injection method is proposed to suppress the low frequency (e.g. double line frequency) CM voltage ripple and generate symmetric dc bus voltage to the ground. The operation range is identified and the impact from bus voltage and grounding scheme is analyzed. Detailed CM circuit is modeled. The CM transfer function is derived and verified by measurement to complete the control loop design. Design trade-off is provided. Experimental results validate the performance and stability of the CM voltage control loop. The control method is also generalized to three-phase cascaded converters.

Keywords—dc microgrid; utility interface; renewables; grounding; common mode voltage

I. INTRODUCTION

Nowadays, renewable energy sources like solar and wind contribute more and more in the electricity generation. Energy storages is used to buffer the intermittent energy. To integrate different renewable sources and energy storage, the dc bus is an attractive solution because of the higher efficiency and reliability [1]–[5]. The small dc grids can then be connected to the ac utility through interface converters and exchange extra or insufficient energy. The converter provides bidirectional power flow, regulates the ac current and dc voltage, and decouples the dynamic of interconnected systems.

To connect ac and dc systems, the grounding scheme is critical. Fig. 1 and Fig. 2 present some typical interfaces and grounding schemes of ac and dc distribution systems. In Fig. 1, three typical single-phase and three-phase ac configurations are drawn. In North America., the most common residential power interface is the 120 V split-phase system as shown in Fig. 1(a). The phase to neutral voltage is 120 Vrms and feeds lighting and plug loads. The line to line voltage is 240 V and feeds heavy loads such as heater, electric range and air conditioner. Fig. 1(b)

This work is sponsored by the Renewable Energy and Nano-grid (REN) mini-consortium of the Center for Power Electronics System (CPES) at Virginia Tech.

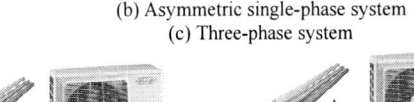

Fig. 1. Different residential ac distribution systems.
(a) Split-phase single-phase system
(b) Asymmetric single-phase system
(c) Three-phase system

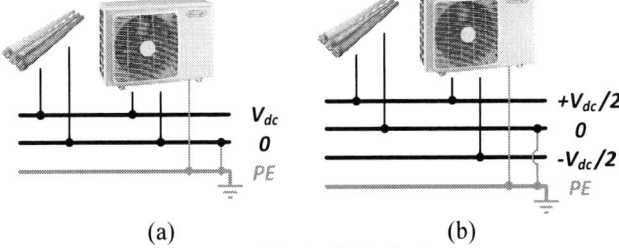

Fig. 2. Unipolar (a) and bipolar (b) dc distribution systems.

is an asymmetric single-phase system. It can be drawn from one phase of the three-phase transformer and provides electricity to several families, which is the current practice in China. It can be also a single wire earth return system that is used for rural area to save one transmission line. The three-phase system in Fig. 1(c) is broadly used in European residential buildings. It is also used in higher power applications, like commercial buildings around the world.

On the dc side, there exist different voltage levels and grounding schemes. The most common voltage levels are 380 V and 48 V for home appliances and telecommunication equipment respectively. The grounding can be unipolar or bipolar as shown in Fig. 2. In the unipolar configuration, one power line is grounded. The voltage across the active line and the ground is the dc bus voltage. The unipolar method only requires two lines for the power transmission. But from the safety point of view, the full dc bus voltage will be applied if contacting the active line. Compared with unipolar, the bipolar uses a third middle line for grounding. So the voltage on the positive and negative dc bus is only half of the total dc bus voltage, which reduces the hazard. Also, the system provides

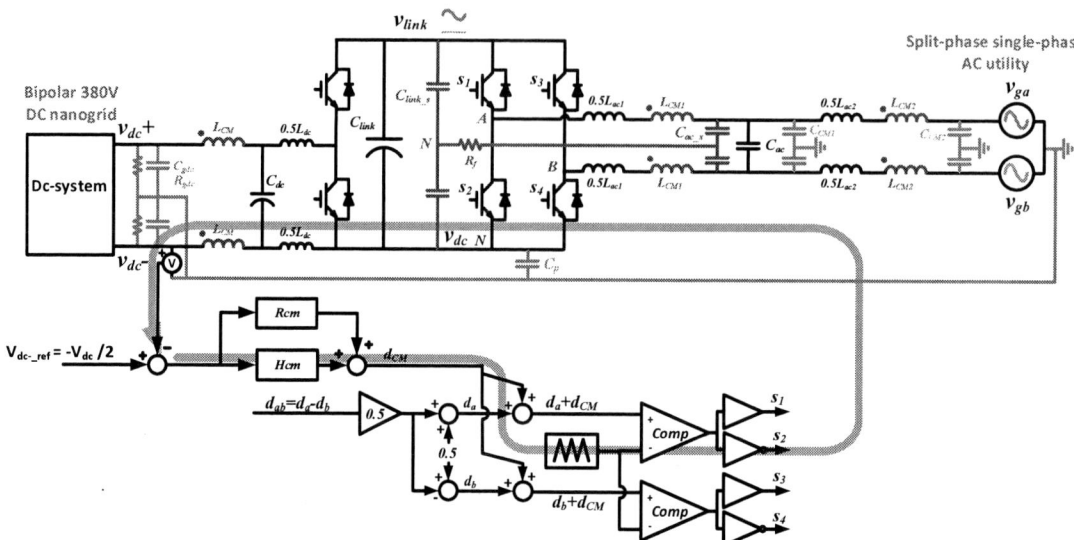

Fig. 3. Structure of floating filter and CM voltage control loop.

both full dc bus voltage and half dc bus voltage to loads. Small appliances can be connected between active and neutral line while heavy loads connected between the positive and negative line [6].

In [7]–[9], a transformerless two-stage bidirectional ac/dc converter was proposed to connect the 380 V residential dc nano-grid to single-phase ac utility. Compared with isolated topologies, the non-isolated topology is usually simpler and more efficient. One main concern is the circulating CM current, which is also called the leakage current. It flows between the ac and dc system through the common ground. The leakage current introduces extra loss, accelerates parts ageing and jeopardizes human safety.

The common mode issue can be considered in high and low frequency range separately. In the high frequency range, because the parasitic impedance to the ground is low, the voltage excitation generates noticeable leakage current. This value is limited by safety and EMI standards. This problem is especially severe in photovoltaic applications where the parasitic capacitance from the cells to the frame and ground is large [10]–[12]. In motor drive applications, the leakage current flows through the stator and rotor of the machine, reduces the life time of bearings [13], [14]. A lot of research has been done to mitigate the high frequency CM problem. The improvement can be about topology, modulation and filter design. These methods reduce the noise source and increase the impedance of transmission path. So the CM current is reduced.

Compared with high frequency noise, the research on low frequency CM voltage control is not much. The parasitic impedance at low frequency is usually high so the leakage current is not so severe. But this issue is important when two grounded systems are connected, especially through a non-isolated power converter. If the low frequency CM voltage is not properly controlled, continuous dc or low frequency current can circulate between the two systems through the ground. Also, the bus voltage needs to be symmetric to the ground for the bipolar dc bus system as shown in Fig. 2(b).

In this paper, a floating filter is used to contain the high frequency CM noise within the converter. An active CM duty cycle injection mothed is proposed in Section II to reduce the low frequency voltage ripple and generate symmetric dc bus voltage. In Section III, the operation range is identified. The impact from ac voltage, dc-link voltage and ac grounding is analyzed. In Section IV, a detailed CM circuit model is derived using a single-phase example to get the transfer function from CM duty cycle to the dc bus voltage. The control loop is then designed and verified by measurement. The control method is generalized to three-phase ac/dc converters in Section V. The experimental waveforms and designed trade-off is concluded in Section VI.

II. COMMON MODE DUTY CYCLE INJECTION TO CONTROL DC AND LOW FREQUENCY CM VOLTAGE

To demonstrate the effectiveness of the proposed low frequency CM voltage control, a two-stage bidirectional single-phase ac/dc converter with both ac and dc side filters is used as an example and is shown in Fig. 3. On the ac side, the converter is connected to the U.S. residential utility (split-phase 120 Vrms ac with the midpoint grounded). On the dc side, the 380 V dc system with high resistance midpoint grounding (HRMG) is adopted for the sake of efficiency and human safety [15]–[17].

The power stage topology is a full bridge converter cascaded with a bidirectional buck converter. By allowing large voltage ripple on the intermediate dc-link, the required capacitance to store the double line frequency ripple power is significantly reduced. This topology can also limit the short circuit current if ac or dc port is shorted. To reduce the volume of ac inductor, unipolar modulation is used.

The ac/dc stage uses a floating CM filter, which consists of CM choke L_{CM1} and split capacitors C_{ac_s}. Different from traditional filter, the midpoint of the CM capacitors C_{ac_s} is not connected to the ground but to the midpoint of dc-link (point N) through a damping resistor R_f. Because of the impedance mismatch ($Z_{Cac\_s} \ll Z_{CM2}$), most of the high frequency noise is contained within the floating filter loop. The floating filter not

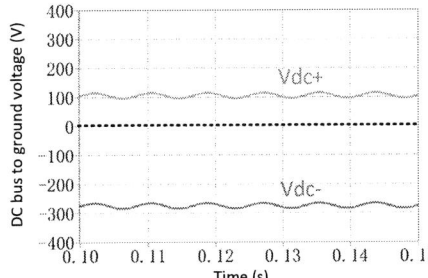

Fig. 4. DC bus to the ground voltage is asymmetric and has ripple.

only avoids the interaction between ac and dc CM filters but also reduces the filter volume by allowing larger Z_{Cac_s}. The detailed filter design can be found in [7]. The DM inductors are put in a symmetric way to reduce the coupling between DM and CM quantities. C_p is the equivalent parasitic capacitance between the power device and the ground.

In Fig. 4, it shows the dc bus voltage to the ground with the floating filter. All the high frequency components are filtered, but the low frequency ripple still exists. The positive and negative bus voltage is also not symmetric to the ground because only the positive dc-link is modulated by the dc/dc stage. The asymmetric bus cannot be used for the bipolar dc bus as shown in Fig. 2(b).

To generate a symmetric dc bus voltage without low frequency voltage ripple, a CM voltage control loop is added to the ac/dc stage (see Fig. 3). The feedback loop controls the dc bus CM voltage by changing the CM voltage of the ac/dc stage. When the negative dc bus is controlled to be half dc bus voltage below ground, the positive bus voltage will be half of the dc bus voltage above ground. In other words, the positive and negative bus voltage is symmetric. Since this is a closed loop system, it can also suppress the low frequency voltage ripple that exists in single-phase applications.

In the feedback loop, the error between the sensed negative bus voltage and its reference is passed through a CM voltage controller. The controller is a multi-pole multi-zero controller (H_{CM}) paralleled with a resonant controller (R_{CM}). Different from the Proportional Resonant (PR) controller, this controller also has an integrator in H_{CM} to accurately control the dc offset of the bus voltage. The resonant controller at 120 Hz is to suppress the double line frequency voltage ripple on the dc bus. The controller generates the required d_{CM} and add it to the ac/dc stage duty cycles d_a and d_b, which are from the ac current control loop. Since both of the two duty cycles are added a same value, the differential mode (DM) voltage between the two phase legs does not change. The control of the ac current is not affected while the CM voltage is controlled by the added loop.

III. OPERATION RANGE OF THE CM DUTY CYCLE INJECTION

Because the proposed method uses the extra control freedom of ac/dc phase legs to control the CM voltage, the margin for the injection of CM duty cycle is critical. The operation range is identified in this section.

Since the target of the duty cycle injection is to only control the low frequency (dc and multiple line-frequency) CM voltage,

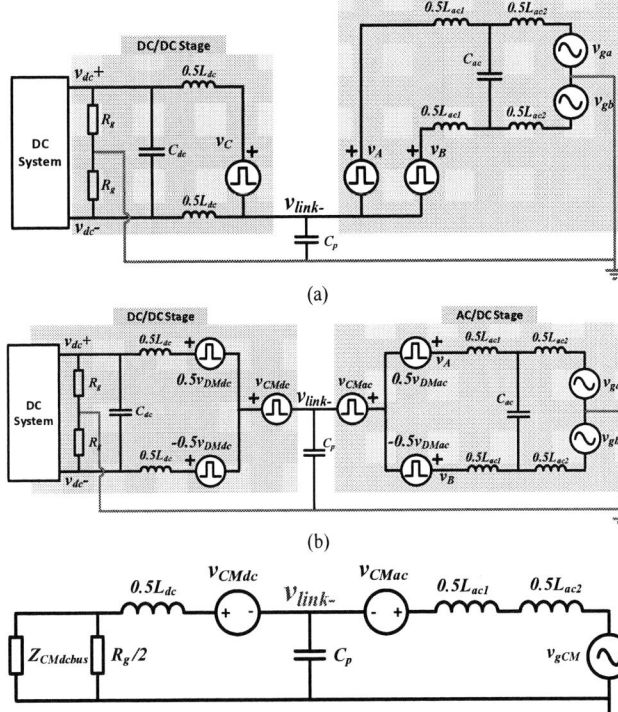

(a)

(b)

(c)

Fig. 5. Derivation of CM circuit for low frequency discussion
(a) Power stage circuit with controlled phase leg voltage sources.
(b) Separation of DM and CM sources
(c) CM equivalent circuit for operation range analysis

the operation range can be identified by looking at the low frequency model of the power circuit as shown in Fig. 5(a). The equivalent phase leg voltage sources are defined as

$$v_A = d_a v_{link}, v_B = d_b v_{link} \qquad (1)$$

The high frequency CM filter is neglected because it is only for the reduction of high frequency noise and has little impact on the operation range of the low frequency control. The complete circuit model and the filter's impact on control loop design will be discussed in the following sections.

Replace the phase legs with separated DM and CM sources

$$v_{DMac} = v_A - v_B = (d_a - d_b) v_{link} \qquad (2)$$

$$v_{CMac} = 0.5 \times (v_A + v_B) = 0.5 \times (d_a + d_b) v_{link} \qquad (3)$$

$$v_{DMdc} = v_C, v_{CMdc} = 0.5 \times v_C \qquad (4)$$

$$v_{gCM} = 0.5 \times (v_{ga} + v_{gb}) \qquad (5)$$

The circuit can be redrawn in the form of Fig. 5(b). The common mode equivalent circuit Fig. 5(c) can then be derived by only considering the CM variables. It needs to be clarified that the CM sources are defined with respect to the negative dc-link. $Z_{CMdcbus}$ is the total CM impedance from appliances on the dc bus. If the circuit is not symmetric, the DM variables will also impact the CM quantities. This is called the mixed mode. The detailed analysis of this effect can be found in [18], [19].

At very low frequency, the inductor can be treated as short and capacitor as open. If the ac input is symmetric, the CM voltage v_{gCM} from ac grid is zero. Using superposition to calculate the effect of the two CM sources, the negative dc-link voltage, which is the voltage across C_p can be expressed in (6). It has the same low frequency value as the negative dc bus voltage.

$$v_{bus-} = v_{link-} = -v_{CMac} \qquad (6)$$

Equation (6) shows that we can control the negative dc bus voltage by controlling the ac/dc stage CM voltage, which is a function of the ac/dc stage duty cycles. It also shows the dc/dc stage does not influence the negative dc bus voltage.

Based on the control target of the DM and CM voltage, we have the following two expressions.

$$v_{DMac} \approx v_{ac} \qquad (7)$$

$$v_{bus-} = -v_{CMac} = -\frac{v_{dc}}{2} \qquad (8)$$

Equation (7) means the ac DM voltage will roughly track the ac grid voltage to generate sinusoidal ac current. Equation (8) shows the CM control target it to regulate the negative dc bus voltage to the desired value.

If the original duty cycles from ac current loop is defined as

$$d_a = 0.5 + d_{DM} \qquad (9)$$

$$d_b = 0.5 - d_{DM} \qquad (10)$$

Then after adding d_{CM} to both d_a and d_b,

$$d_a = 0.5 + d_{DM} + d_{CM} \qquad (11)$$

$$d_b = 0.5 - d_{DM} + d_{CM} \qquad (12)$$

Then we can define DM and CM duty cycles as

$$d_{DM} = \frac{1}{2}\left(d_a - d_b\right) = \frac{v_{ac}}{2v_{link}} \qquad (13)$$

$$d_{CM} = \frac{1}{2}\left(d_a + d_b\right) - 0.5 = \frac{v_{dc}}{2v_{link}} - 0.5 \qquad (14)$$

Equation (13) implies the required DM duty cycle to control the ac current. Equation (14) implies the necessary CM duty cycle to control the dc bus voltage symmetric to the ground. The summation of DM and CM duty cycle is limited between 0 and 1. In other words, it should always satisfy

$$\left|d_{DM}\right| + \left|d_{CM}\right| < 0.5 \qquad (15)$$

Substituting (13) and (14) into (15), we have

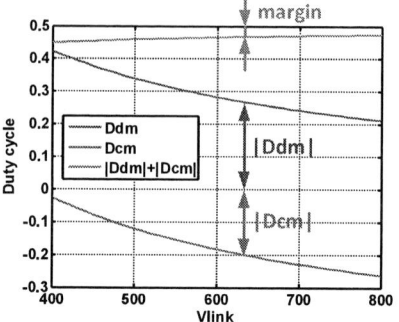

Fig. 6. Impact from dc-link voltage on CM control margin

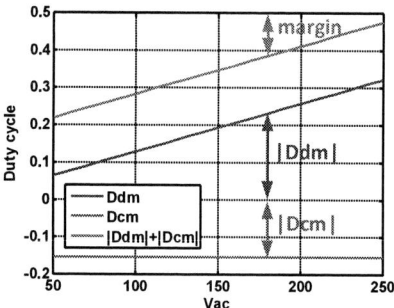

Fig. 7. Impact from ac voltage on CM control margin

$$\left|\frac{v_{ac\_max}}{2v_{link}}\right| + \left|\frac{v_{dc}}{2v_{link}} - 0.5\right| < 0.5 \qquad (16)$$

$$\left|v_{ac\_max}\right| < v_{dc}$$
$$v_{dc} + \left|v_{ac\_max}\right| < 2v_{link} \qquad (17)$$

Considering the topology of the example converter, the dc-link voltage is always higher than the ac and dc terminals voltage. Equation (17) can be further simplified as

$$\left|v_{ac\_max}\right| < v_{dc} < v_{link} \qquad (18)$$

In fact, (18) is a very weak requirement. When the input ac/dc stage is a boost type converter, the dc-link voltage is always higher than the peak ac input. If the dc/dc stage is a buck type converter that steps down the dc-link voltage, the requirement that dc voltage less than dc-link voltage is also satisfied. The only condition needs to be considered is the ac peak voltage should be smaller than the dc bus voltage. This is satisfied in this example. The ac peak voltage is 340V ($240\sqrt{2}$) and the dc bus voltage is 380 V. This shows the example has sufficient margin to inject the CM duty cycle and control dc bus voltage symmetrically.

A. The impact from dc-link voltage and ac voltage

The impact from dc-link voltage and ac voltage on the CM duty cycle injection margin is drawn in Fig. 6 and Fig. 7 by evaluating (13) and (14). It can be observed from Fig. 6, the dc-link voltage does not influence the control margin much. Though increasing the dc-link voltage reduces the required DM duty cycle, the injected CM duty cycle is increased. The sum of

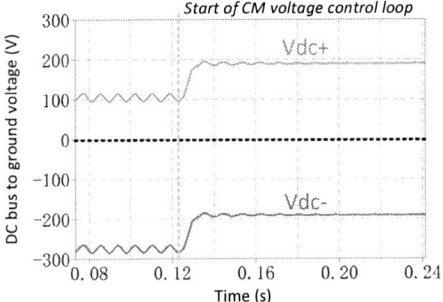

Fig. 8. DC bus voltage before and after CM control with symmetric ac

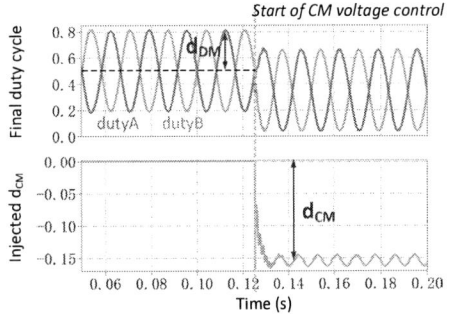

Fig. 9. Total and injected duty cycle with symmetric ac

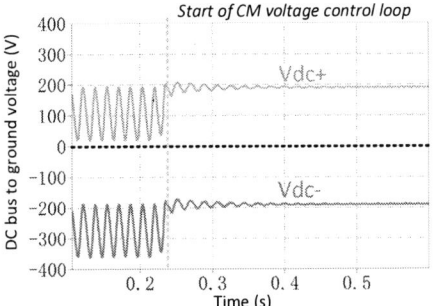

Fig. 10. DC bus voltage before and after CM control with asymmetric ac

Fig. 11. Total and injected CM duty cycle with asymmetric ac

these two does not change much with different dc-link voltage. On the contrary, ac input voltage has big impact on the control margin. Lower ac input voltage gives larger control margin. Fig. 8 shows the time domain simulation with 550 V dc-link voltage and 240 V split-phase ac input. After enabling the low frequency CM voltage control, the dc bus is controlled symmetric to the ground. Fig. 9 shows the duty cycle before and after enabling the CM control. A constant CM duty cycle with small ac ripple is injected.

B. The impact from asymmetric ac grounding

As Fig. 1(b) shows the ac grounding can be asymmetric. The equivalent CM circuit in Fig. 5 is still valid except the CM grid voltage is no longer zero and need to be compensated by the control.

$$v_{gCM} = \frac{1}{2}(v_{ac} + 0) = \frac{1}{2}v_{ac} \qquad (19)$$

The corresponding control requirement and operation range are changed to

$$v_{link-} = -v_{CMac} + \frac{v_{ac}}{2} = -\frac{v_{dc}}{2} \qquad (20)$$

$$\frac{|v_{ac}|}{2v_{link}} + \left| \frac{v_{dc} + v_{ac}}{2v_{link}} - 0.5 \right| < 0.5 \qquad (21)$$

Fig. 10 and Fig. 11 shows the time domain simulation with 120 V asymmetric ac input and 380 V dc output. Without the CM voltage control, the dc bus voltage has around 170 V CM voltage ripple. The injected CM duty cycle is no longer dc but sinusoidal. The shape of the duty cycle for the two phase legs is completely different. Since the d_{CM} has large ripple, the

theoretical analysis can only provide a rough estimation and needs to be verified by simulation.

IV. COMMON MODE CIRCUIT SMALL SIGNAL MODELING AND CONTROL LOOP DESIGN

Within the operation range, a closed loop compensator can be designed to control the dc bus to ground voltage. A CM circuit small signal model is derived in this section for the compensator design.

Based on the definition of CM quantities, the complete CM equivalent circuit with the floating CM filter is drawn in Fig. 12. The physical meaning of each component can be found in Fig. 3. A line stabilization network (LISN) is placed between the converter and ac grid to bypass the uncertainty of the line impedance from ac side transformer and utility. The component in red is the important ones that influence the transfer function from CM duty cycle to dc bus CM voltage. C_{p1} and C_{p2} are the parasitic capacitance from the device to the ground.

To simplify the derivation of small signal transfer functions, the impedances (22) - (28) are defined. The paralleled R_{CMi} is the leakage resistance of each choke and needs to be considered to evaluate the damping effect.

$$Z_1 = sL_{CM1} \parallel R_{CM1} \qquad (22)$$

$$Z_2 = sL_{CM2} \parallel R_{CM2} \qquad (23)$$

$$Z_0 = sL_{CM} \parallel R_{CM} \qquad (24)$$

$$Z_{gdc} = R_{gdc} \parallel \frac{1}{sC_{gdc}} \qquad (25)$$

978-1-4673-9551-9/16 $31.00 © 2016 IEEE   181

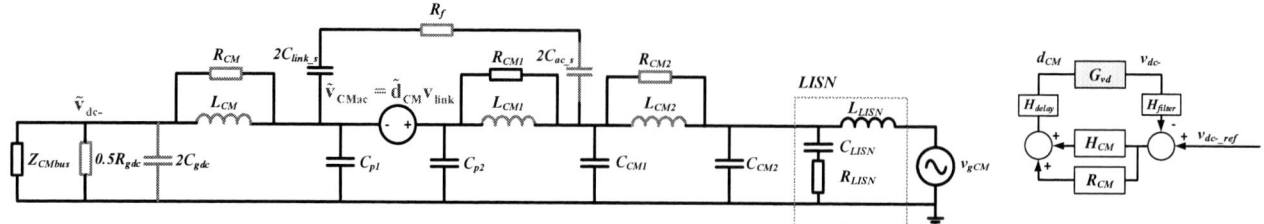

Fig. 12. Complete CM equivalent circuit and CM voltage control loop

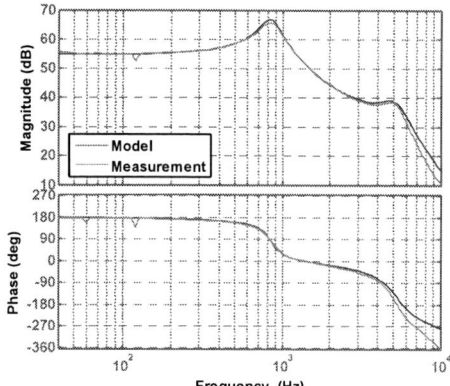

Fig. 13. Power stage transfer function G_{vd}

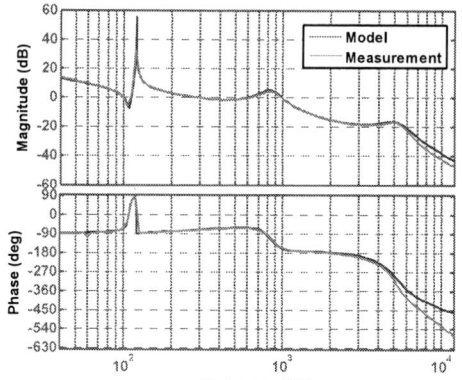

Fig. 14. Loop gain with resonant controller

$$Z_f = R_f + \frac{1}{sC_{ac\_s}} + \frac{1}{sC_{link\_s}} \quad (26)$$

$$Z_{LISN} = \left(\frac{1}{sC_{LISN}} + R_{LISN} \right) \| sL_{LISN} \quad (27)$$

$$Z_{total} = Z_2 + Z_{LISN} + Z_{gdc} + Z_0 \quad (28)$$

By solving the circuit shown in Fig. 12, the transfer function from injected CM duty cycle to the negative dc bus voltage can be solved as shown in (29).

$$\tilde{v}_{dc-} = -\tilde{d}_{CM} v_{link} \frac{Z_f \| Z_{total}}{Z_1 + Z_f \| Z_{total}} \frac{Z_{gdc}}{Z_{total}} \quad (29)$$

To validate the model, an online transfer function measurement is executed on a hardware prototype with digital processor. The switching frequency in the experiment is 20 kHz. The passive parameters are listed in Table I. The DSP injects a series of perturbation with different frequency to the power stage and measures the response. The magnitude and phase information of the perturbation and response is sent back to the host computer for further processing and graph. The measured power stage transfer function from CM duty cycle to negative dc bus voltage is compared with the modeled result in Fig. 13. They match well up to half of the switching frequency.

From the power stage transfer function, we can observe two resonant peaks at 850 Hz and 5 kHz. They are from the LC resonance of the CM filters. The peak at 850 Hz is caused by the floating filter, which consists of L_{CM1}, C_{ac_s} and C_{link_s}. It causes 180 degree phase drop and is near the crossover frequency. This needs to be compensated by the control loop zeros. There is also a high frequency peak at 5 kHz. This peak is caused by the

TABLE I
FILTER PARAMETERS

	Symbol	Value
AC filters	L_{CM1}	3.54 mH
	R_{CM1}	10 kΩ
	L_{CM2}	17 mH
	R_{CM2}	8 kΩ
	C_{CM1}	1 nF
	C_{CM2}	1 nF
	LISN	50 uH, 50 Ω, 0.1 uF
DC filters	L_{CM}	12.5 mH
	R_{CM}	10.5 kΩ
	C_{gdc}	10 nF
	R_{gdc}	100 kΩ
Floating filter	C_{ac_s}	5.6 uF
	R_f	5 Ω
	C_{link_s}	50 uF

resonance of ac and dc side CM filters together and appears in this system because the ac and dc sides are coupled by the ground. AC side choke L_{CM2}, in series with the dc side choke L_{CM} resonant with the dc side grounding capacitor C_{gdc}. Though the grounding capacitance is only several nano farad, the total CM inductance is big. This frequency is above the crossover frequency but also needs to be considered. During the resonant peak, the gain curve increases and can cross the 0 dB line again to cause instability. In the control loop design, poles need to be placed before the second resonant frequency to damp the resonant peak from going back to the 0 dB line.

With the small signal model, the compensator can be designed based on the gain and phase margin requirement. The final compensator parameters are shown in (30) and (31). The closed loop gain has 14 dB gain margin and 35 degree phase margin. The control bandwidth is around 1 kHz. Again the model and measurement results are compared in Fig. 14. The peak at 120 Hz is from the resonant controller to suppress the

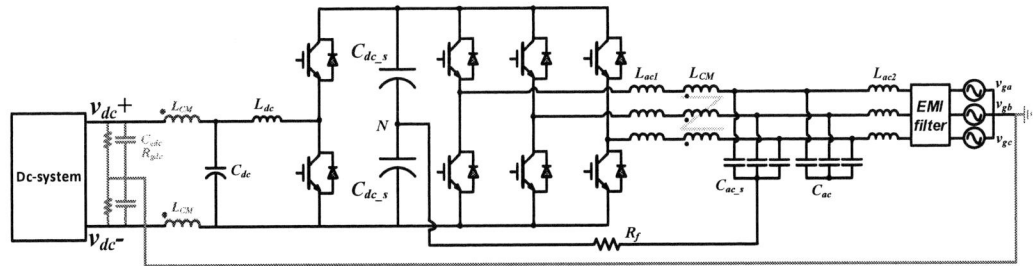

Fig. 15. Two-stage ac/dc converter with three-phase front-end

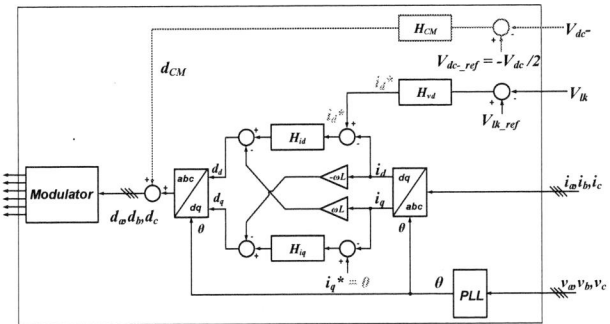

Fig. 16. CM voltage control with SPWM

Fig. 17. DC bus to ground voltage with SPWM

double line frequency ripple. The sag before the peak is from the parallel of integrator and resonant controller.

$$H_{CM} = -0.00242 \times \frac{\left(s + 4.37 \times 10^3\right)\left(s + 6.32 \times 10^3\right)}{s\left(s + 3.17 \times 10^4\right)} \quad (30)$$

$$R_{CM} = -0.5 \times \frac{s}{s^2 + 0.5s + \left(2\pi \times 120\right)^2} \quad (31)$$

V. GENERALIZATION TO THREE-PHASE AC INPUT

In high power cases, the front-end ac/dc is usually fulfilled by three-phase structure. Fig. 15 shows a non-isolated three-phase cascaded ac/dc converter. The ac/dc stage regulates the dc-link voltage and ac current. The dc/dc stage steps down the dc-link voltage to the required dc bus voltage. In this two-stage structure, the front-end ac/dc part can be also used to control the dc bus CM voltage.

The modulation scheme for three-phase converters can be carrier-based SPWM and space vector modulation (SVM). SVM has the benefits of better dc-link voltage utilization and possibility to reduce switching loss. But it generates a low frequency CM voltage on the dc-link, e.g. third order harmonic, which is not good for the connection of dc grid.

SPWM shows a symmetric and constant CM voltage on the dc-link. Like the single-phase case, a CM duty cycle is added to all the three phase leg duty cycles, so the CM voltage can be adjusted without impacting the DM voltages and currents. Fig. 16 shows the control scheme in DQ frame. The d axis current reference is from the dc-link voltage control loop. The q axis current reference is zero. The CM voltage control loop generates the required d_{CM} by comparing the negative dc bus voltage with reference. The injected duty cycle d_{CM} is added to the duty cycles d_a, d_b and d_c. The final dc bus voltage is

symmetric to the ground as shown in Fig. 17. In the three-phase case, there is no low frequency voltage ripple on the dc bus and the resonant controller is unnecessary.

VI. EXPERIMENTAL RESULTS AND TRADE-OFF

To verify the proposed low frequency CM voltage control method, a 10-kW single-phase bidirectional ac/dc converter is built. The CM filter parameters are shown in Table I.

Fig. 18 to Fig. 20 show the steady state and transient experiment results. Fig. 18 shows the steady state ac voltage, ac current, positive and negative dc bus voltage, with and without the CM voltage control. After enabling the control loop, the dc bus voltage is adjusted to be symmetric to the ground. The low frequency ripple is also suppressed. Fig. 19 shows the start-up process after enabling the CM control loop. The CM voltage smoothly changes to the desired value with reduced ripple. Fig. 20 shows the load step-up test. The voltage spike on dc bus DM voltage and voltage to the ground are measured. It shows the load transient will not influence the operation of the CM voltage control.

It is worth noting that, the CM duty cycle injection do have impact on the grid side ac current. The amplitude of switching frequency current ripple increases a little. It can be observed by comparing the ac current in Fig. 18(a) and (b). The reason can be explained by looking at the distribution of the power devices' on and off time intervals within each switching period. For single-phase full bridge converter, the unipolar modulation has small current ripple because it not only doubles the equivalent switching frequency but also evenly distributes the on and off time. When the CM duty cycle is injected, though the switching frequency is still doubled, the distribution of these time intervals is changed. The on and off time of switches is not as even as before. So the peak to peak current ripple increases. This issue can be solved by properly design the ac side DM filter.

978-1-4673-9551-9/16 $31.00 © 2016 IEEE
183

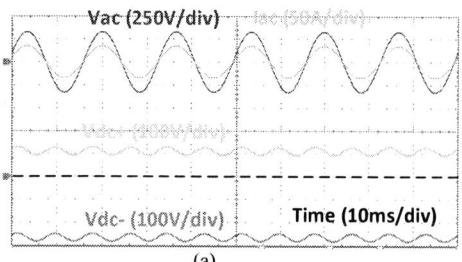
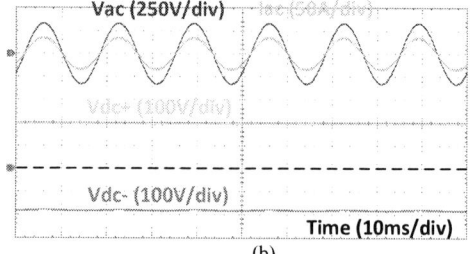

|(a)|(b)|

Fig. 18. DC bus CM voltage before (a) and after (b) enabling CM control

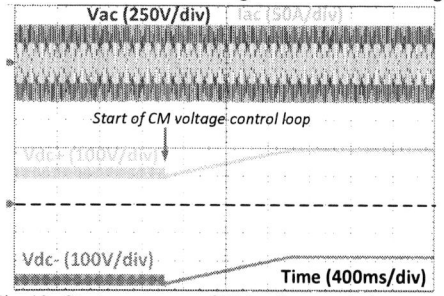

Fig. 19. Start-up process of CM voltage control

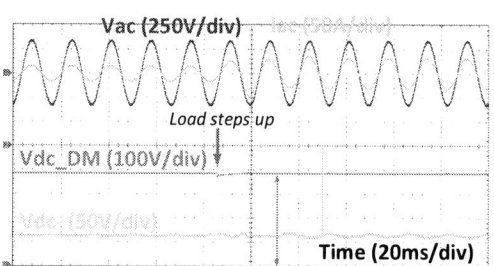

Fig. 20. Load step-up test for DM and CM bus voltage

VII. CONCLUSION

In this paper, a transformerless two-stage ac/dc converter was used as an example to interconnect dc and ac grid. But it generates asymmetric dc bus voltage and has low frequency ripple. A CM duty cycle injection method was proposed to actively control the dc bus to ground voltage. As a result, the dc bus voltage is symmetric and the low frequency voltage ripple is suppressed. The operation range of the proposed method was identified and proved to be easy to satisfy. The impact from different voltage level and asymmetric ac grounding was analyzed. The complete CM circuit model was derived and verified by hardware, based on which the closed loop controller was designed. The experimental result verifies the control method in steady state and transient. The trade-off was clarified. The control method was also generalized to three-phase ac scenario.

REFERENCES

[1] D. Boroyevich, I. Cvetkovic, R. Burgos, and D. Dong, "Intergrid: A Future Electronic Energy Network?," *IEEE J. Emerg. Sel. Top. Power Electron.*, vol. 1, no. 3, pp. 127–138, 2013.

[2] A. Kwasinski, "Quantitative evaluation of DC microgrids availability: Effects of system architecture and converter topology design choices," *IEEE Trans. Power Electron.*, vol. 26, no. 3, pp. 835–851, 2011.

[3] D. Salomonsson and A. Sannino, "Low-voltage DC distribution system for commercial power systems with sensitive electronic loads," *IEEE Trans. Power Deliv.*, vol. 22, no. 3, pp. 1620–1627, 2007.

[4] A. Sannino, G. Postiglione, and M. H. J. Bollen, "Feasibility of a DC network for commercial facilities," *IEEE Trans. Ind. Appl.*, vol. 39, no. 5, pp. 1499–1507, 2003.

[5] M. G. Simoes, R. Roche, E. Kyriakides, S. Suryanarayanan, B. Blunier, K. D. McBee, P. H. Nguyen, P. F. Ribeiro, and A. Miraoui, "A comparison of smart grid technologies and progresses in Europe and the U.S.," *IEEE Trans. Ind. Appl.*, vol. 48, no. 4, pp. 1154–1162, 2012.

[6] H. Kakigano, Y. Miura, and T. Ise, "Low-voltage bipolar-type dc microgrid for super high quality distribution," *IEEE Trans. Power Electron.*, vol. 25, no. 12, pp. 3066–3075, 2010.

[7] D. Dong, F. Luo, X. Zhang, D. Boroyevich, and P. Mattavelli, "Grid-interface bidirectional converter for residential DC distribution systems - Part 2: AC and DC interface design with passive components

minimization," *IEEE Trans. Power Electron.*, vol. 28, no. 4, pp. 1667–1679, 2013.

[8] Dong, I. Cvetkovic, D. Boroyevich, W. Zhang, R. Wang, and P. Mattavelli, "Grid-interface bidirectional converter for residential DC distribution systems - Part one: High-density two-stage topology," *IEEE Trans. Power Electron.*, vol. 28, no. 4, pp. 1655–1666, 2013.

[9] F. Chen, R. Burgos, D. Boroyevich, and D. Dong, "Control Loop Design of a Two-stage Bidirectional AC / DC Converter for Renewable Energy Systems," *2014 IEEE Appl. Power Electron. Conf. Expo. - APEC 2014*, pp. 2177–2183, 2014.

[10] D. Barater, G. Buticchi, E. Lorenzani, and C. Concari, "Active common-mode filter for ground leakage current reduction in grid-connected PV converters operating with arbitrary power factor," *IEEE Trans. Ind. Electron.*, vol. 61, no. 8, pp. 3940–3950, 2014.

[11] M. C. Cavalcanti, K. C. De Oliveira, A. M. De Farias, F. a S. Neves, G. M. S. Azevedo, and F. C. Camboim, "Modulation techniques to eliminate leakage currents in transformerless three-phase photovoltaic systems," *IEEE Trans. Ind. Electron.*, vol. 57, no. 4, pp. 1360–1368, 2010.

[12] N. Vazquez, M. Rosas, C. Hernandez, E. Vazquez, and F. J. Perez-Pinal, "A New Common-Mode Transformerless Photovoltaic Inverter," *IEEE Trans. Ind. Electron.*, vol. 62, no. 10, pp. 6381–6391, Oct. 2015.

[13] H. Akagi and T. Doumoto, "An approach to eliminating high-frequency shaft voltage and leakage current from an inverter-driven motor," *IEEE Trans. Ind. Appl.*, vol. 1, no. 4, pp. 1162–1169, 2003.

[14] H. Akagi and S. Tamura, "A passive EMI filter for eliminating both bearing current and ground leakage current from an inverter-driven motor," *IEEE Trans. Ind. Appl.*, vol. 21, no. 5, pp. 1459–1468, 2006.

[15] J. C. Das and R. H. Osman, "Grounding of AC and DC low-voltage and medium-voltage drive systems," *IEEE Trans. Ind. Appl.*, vol. 34, no. 1, pp. 205–216, 1998.

[16] T. Dragicevic, X. Lu, J. Vasquez, and J. Guerrero, "DC Microgrids-Part II: A Review of Power Architectures, Applications and Standardization Issues," *IEEE Trans. Power Electron.*, vol. 8993, no. c, pp. 1–1, 2015.

[17] T. Dragicevic, X. Lu, J. Vasquez, and J. Guerrero, "DC Microgrids-Part I: A Review of Control Strategies and Stabilization Techniques," *IEEE Trans. Power Electron.*, vol. 8993, no. c, pp. 1–1, 2015.

[18] M. Jin and M. Weiming, "A New Technique for Modeling and Analysis of Mixed-Mode Conducted EMI Noise," *IEEE Trans. Power Electron.*, vol. 19, no. 6, pp. 1679–1687, Nov. 2004.

[19] S. Qu and D. Y. Chen, "Mixed-mode EMI noise and its implications to filter design in offline switching power supplies," in *APEC 2000. Fifteenth Annual IEEE Applied Power Electronics Conference and Exposition*, 2000, vol. 2, no. 4, pp. 707–713.

A Three-level Space Vector Modulation Scheme for Paralleled Two Converters to Reduce Zero-sequence Circulating Current and Common Mode Voltage

Zhongyi Quan, Yunwei Li
Department of Electrical & Computer Engineering
University of Alberta
Edmonton, Canada
zquan@ualberta.ca, yunwei.li@ualberta.ca

Abstract— **Circulating current has been the major concern for the implementing of paralleled converters. This paper proposes a three-level space vector modulation (SVM) scheme for the system constructed by two paralleled voltage source converters (VSCs) and common mode inductor (CMI). The proposed scheme aims to reduce the zero-sequence circulating current (ZSCC) and the magnitude of common mode voltage (CMV) of the system. The ZSCC pattern with respect to modulation schemes are first analyzed to provide a clear understanding of the generation of ZSCC. And then the proposed scheme is introduced. Analysis regarding the ZSCC peak value, impact on the common mode current (CMC), and switching losses are made and compared with the existing methods. The proposed method has been verified in simulation and experiment.**

Keywords— Parallel VSCs; space vector modulation (SVM); zero-sequence circulating current (ZSCC); common mode voltage (CMV).

I. INTRODUCTION

Three-phase VSCs have been very popular in many power conversion applications such as active-front-end (AFE) and motor drive inverter. There is an increasing interest in using VSCs in paralleled manner, as shown in Fig. 1, to meet the increasing requirement of power rating through current sharing among the VSCs [1]. Paralleling converters also provide benefits like improved reliability by enabling implementation of flexible fault-tolerant techniques [2]. An important concern for implementing parallel VSCs is the ZSCC. Due to parameter and control asymmetries, the terminal voltages of the paralleled phase legs may be different, generating ZSCC among the VSCs [3]. If not dealt properly, such ZSCC brings adverse effects to the operation of the system like increasing the current stresses and conduction losses of the switching devices [4]. On the other hand, interleaving the VSCs has been increasingly attractive due to its capability of multilevel output, line current harmonics reduction, and passive component size reduction [5]. In the case of interleaving, ZSCC can be enlarged since the voltage differences among the paralleled phase legs are always created intentionally. Therefore ZSCC suppression methods for paralleled VSCs have been intensively studied.

Basically, the methods can be classified into three categories, passive methods, control methods and modulation methods. Passive methods increase the size and cost of the system significantly. And the size and cost of passive components like coupled inductor (CI) [6] and common mode inductor (CMI) [7] are strongly dependent on the pattern of ZSCC, which can only be altered by active methods. Active ZSCC control methods utilize a zero-sequence controller to regulate ZSCC [8]. But only low-frequency ZSCC can be controlled due to the limitation of the control bandwidth. However, the fundamental component of ZSCC is at the switching frequency, which can only be regulated by modulation methods. Different Pulse Width Modulation (PWM) strategies have also been investigated, including Selective Harmonic Elimination (SHE) based method [9], interleaved PWM methods [10], and the multilevel modulation methods [11]. But SHE method suffers from low dynamic performances. Interleaved PWM method proposed in [10] introduces additional switching action in one sampling period and there are times that two phases are switched on and off simultaneously. While the existing multilevel modulation schemes are designed for system using CI as filter. These schemes cannot be directly used for system with CMI.

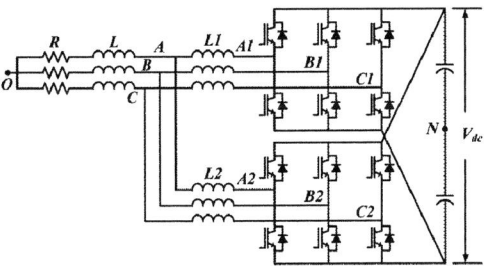

Fig. 1 Two parallel connected VSCs with common DC link

One common drawback of above modulation methods is that the reduction of CMV of the converter system is not considered. For applications like grid connected PV inverters, there is a requirement on the level of CMV. And high CMV peak will also bring high value of CMC, which in turn increases the size of CMI. For paralleled VSCs, the maximum value of CMV (Vdc/2) happens when the same null-vector is applied in the VSCs. Therefore the interleaved modulation methods introduced above also have the peak CMV of Vdc/2 since they suppress ZSCC by intentionally apply the same null-vector at the VSCs.

978-1-4673-9551-9/16 $31.00 © 2016 IEEE

To reduce the peak value of ZSCC and CMV at the same time, while still maintain the high quality output of the interleaved VSCs, an optimized SVM scheme is designed for the system with two paralleled VSCs in this paper, as shown in Fig. 1. The filter of each converter can be single phase inductor or CMI in this case. The peak value of ZSCC is suppressed to the same level as that in [10]. And CMV of the system is reduced from Vdc/2 to Vdc/6.

II. ZSCC AND CMV OF PARALLELED VSCs

Considering a system with two paralleled two-level VSCs, each VSC has eight switching states. The states are presented as: V0 [0 0 0], V1 [1 0 0], V2 [1 1 0], V3 [0 1 0], V4 [0 1 1], V5 [0 0 1], V6 [1 0 1], V7 [1 1 1]. When different switch states are applied to the two converters, ZSCC and CMV can have different patterns.

A. ZSCC pattern

Seen from Fig. 1, when the terminal voltages of a paralleled phase leg are different, circulating current will be generated in this phase leg. For instance, when the voltages of A1 and A2 are different, there will be a voltage applied across the inductance of L1+L2, then current will flow through the inductance and circulating within phase A, as common DC link is used. The circulating current in a single phase can be expressed as

$$\frac{di_{X,0}}{dt} = \frac{v_{X1N} - v_{X2N}}{L_1 + L_2} \tag{1}$$

where X is A, B, or C; N is the neutral point of the DC link; $i_{X,0}$ stands for the circulating current of phase X; L_1 and L_2 are the filter inductance of the two converters. $L_1 + L_2$ can be assumed equal to Lcm. For each converter, the ZSCC can be given by

$$I_{01} = I_{A1} + I_{B1} + I_{C1} = -I_{02} \tag{2}$$

where I_{01} and I_{02} are the ZSCC of VSC1 and VSC2 respectively; IX1, X=A, B, C, are the three-phase currents of VSC1. When using CMI as ZSCC filter, the size of CMI is influenced by the peak value of I_{01} and I_{02}. Lower peak value of ZSCC will give smaller CMI size [10]. Hence the goal is to minimize the ZSCC peak value.

The ZSCC pattern can be analyzed using a case by case method. Basically, the combinations of switching vectors that can generate ZSCC can be classified into five cases, as shown in Fig. 2. The conditions that these five cases represent are summarized in Table I.

TABLE I. THE FIVE CONDITIONS THAT GENERATE ZSCC

Case	Conditon
Case 1	Two active vectors applied, and only one phase has circulating current
Case 2	Two active vectors applied, and two phases have circulating current
Case 3	One active vector and one active vector applied, and two phases have circulating current
Case 4	Two active vectors applied, and three phases have circulating current
Case 5	Different null-vectors applied, and three phases have circulating current

In case 1, the circulating current only exists in phase B, from VSC2 to VSC1. Based on equation (1) and (2), the dynamic of circulating current for phase B and VSC1 in this case can be respectively given by equation (3) and (4).

$$\frac{di_{B,0}}{dt} = -\frac{V_{dc}}{L_1 + L_2} \tag{3}$$

$$\frac{di_{01}}{dt} = \frac{di_{A,0}}{dt} + \frac{di_{B,0}}{dt} + \frac{di_{C,0}}{dt} = \frac{V_{dc}}{L_1 + L_2} \tag{4}$$

In case 2, although both phase A and phase B have circulating current, the overall ZSCC remains same as the directions of the circulating currents in phase A and phase B are opposite. While the overall ZSCC in case 3 will vary since the two phases have circulating currents with same direction. In case 4 and case 5, all three phases have circulating currents. The difference is that the directions of the three phase circulating currents are different in case 4 while same in case 5. Similar to case 1, the dynamics of ZSCC for VSC1 in case 2 to case 5 can be written as

$$\frac{di_{01}}{dt} = \frac{V_{dc} - V_{dc}}{L_1 + L_2} = 0 \tag{5}$$

$$\frac{di_{01}}{dt} = \frac{2V_{dc}}{L_1 + L_2} \tag{6}$$

$$\frac{di_{01}}{dt} = \frac{V_{dc}}{L_1 + L_2} \tag{7}$$

$$\frac{di_{01}}{dt} = \frac{3V_{dc}}{L_1 + L_2} \tag{8}$$

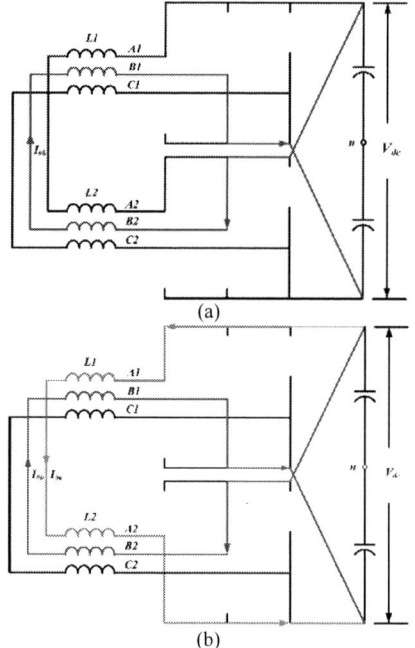

(a)

(b)

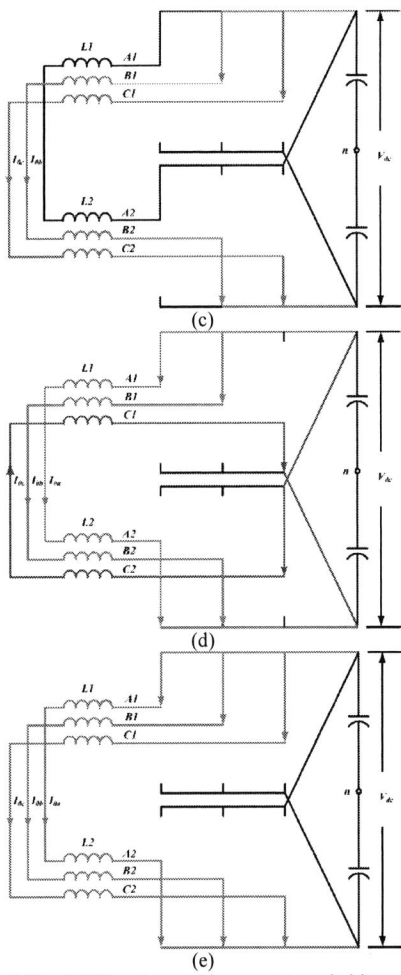

(c)

(d)

(e)

Fig. 2 The ZSCC patterns with respect to switching states

From equations (4) to (8), one can see that under different modulation conditions, the varying speed of the overall ZSCC are different. And case 5 is what has been called pure zero-sequence circulating condition. Given the same amount of duration, the peak value of ZSCC will be different under these cases. For modulation strategy design, case 1, case 2 and case 4 are more preferred while case 3 and case 5 which represent fast dynamic of ZSCC should be avoided.

Summarizing all the switching state combinations, the ZSCC dynamics with respect to switching states can be obtained as shone in Table II, where 0 means the ZSCC keeps constant; 1 means that the ZSCC varies by excitation of Vdc; 2 means ZSCC varies by excitation of 2Vdc; and 3 means the ZSCC varies by excitation of 3Vdc. The signs '+' and '-' denote the varying direction of ZSCC from the view point of VSC2. The analysis result is identical with that presented in [13]. But the analysis in this paper provides a better understanding of the generation of ZSCC.

TABLE II. THE ZSCC DYNAMICS WITH RESPECT TO DIFFERENT SWITCHING STATE COMBINATIONS

		VSC2							
		V0	V1	V2	V3	V4	V5	V6	V7
VSC1	V0	0	-1	-2	-1	-2	-1	-2	-3
	V1	+1	0	-1	0	-1	0	-1	-2
	V2	+2	+1	0	+1	0	+1	0	-1
	V3	+1	0	-1	0	-1	0	-1	-2
	V4	+2	+1	0	+1	0	+1	0	-1
	V5	+1	0	-1	0	-1	0	-1	-2
	V6	+2	+1	0	+1	0	+1	0	-1
	V7	+3	+2	+1	+2	+1	+2	+1	0

B. CMV pattern

The relationship between CMV and the switching states of a VSC is expressed in (9).

$$V_{CM} = \frac{S_A + S_B + S_C}{3} \cdot V_{dc} - \frac{V_{dc}}{2} \tag{9}$$

where V_{CM} is the value of CMV; S_X is the switching actions of the three phases, X=A, B, C.

From (9) one can see that the use of null vector corresponds to the highest CMV peak, i.e. Vdc/2. While other states only result in CMV peak of Vdc/6. Thus the CMV reduction SVM strategies avoid using null vectors to reduce CMV peak. For paralleled VSCs, the CMV pattern can be obtained by adding the CMV of the two VSCs. Table III lists the CMV values with respect to different switching state combinations for a system with two paralleled VSCs.

TABLE III. RESULTED CMV OF DIFFERENT VECTOR COMBINATIONS

		VSC2							
		V0	V1	V2	V3	V4	V5	V6	V7
VSC1	V0	-Vdc/2	-Vdc/3	-Vdc/6	-Vdc/3	-Vdc/6	-Vdc/3	-Vdc/6	0
	V1	-Vdc/3	-Vdc/6	0	-Vdc/6	0	-Vdc/6	0	Vdc/6
	V2	-Vdc/6	0	Vdc/6	0	Vdc/6	0	Vdc/6	Vdc/3
	V3	-Vdc/3	-Vdc/6	0	-Vdc/6	0	-Vdc/6	0	Vdc/6
	V4	-Vdc/6	0	Vdc/6	0	Vdc/6	0	Vdc/6	Vdc/3
	V5	-Vdc/3	-Vdc/6	0	-Vdc/6	0	-Vdc/6	0	Vdc/6
	V6	-Vdc/6	0	Vdc/6	0	Vdc/6	0	Vdc/6	Vdc/3
	V7	0	Vdc/6	Vdc/3	Vdc/6	Vdc/3	Vdc/6	Vdc/3	Vdc/2

III. PROPOSED SVM SCHEME

From both Table II and Table III, one can see that the null vectors are the major reason of both large ZSCC variation and high CMV peak. If null vectors are not used in modulation, the ZSCC variation speed and CMV peak are naturally reduced. This principle is similar to the 'RCMV SVM' method for single two-level VSCs. But the difference is, for single VSC, the null vector is replaced by active vectors, thus zero voltage vectors are completely eliminated during the modulation process. The output quality will be negatively influenced. While for the paralleled VSCs, although null vectors for single VSC are not used, zero voltage vectors for multilevel output can still be realized. The voltage vectors for three-level output are shown in Table IV. It is shown that there are redundancies for voltage vector V7 to V18, thus the switching sequence design can be very flexible to regulate the waveform of ZSCC and CMV.

TABLE IV. COMPLETE VOLTAGE VECTORS FOR THREE-LEVEL OUTPUT

		VSC2					
		V1	V2	V3	V4	V5	V6
VSC1	V1	1	7	14	0	18	12
	V2	7	2	8	15	0	13
	V3	14	8	3	9	16	0
	V4	0	15	9	4	10	17
	V5	18	0	16	10	5	11
	V6	12	13	0	17	11	6

A. Designed switching scheme

In the proposed three-level SVM scheme, each sector is divided into four sub-sectors. The locations of the voltage space vectors and the division of subsectors are shown in Fig. 3. The voltage vectors can be classified by magnitude, i.e., zero vector (V0), large vectors (V1-V6), medium vectors (V7-V12), and small vectors (V13-V18).

Taking sector 1 as an example, the designed switching sequences within subsectors I and II are listed in Table V and Table VI respectively. Subsector III and IV have similar sequences.

TABLE V. SWITCHING SEQUENCE IN SUBSECTOR I OF SECTOR 1

Vector	V13	V7	V1	V7	V13	V7	V1	V7	V13
VSC1	2	2	1	1	6	1	1	2	2
VSC2	6	1	1	2	2	2	1	1	6

TABLE VI. SWITCHING SEQUENCE IN SUBSECTOR II OF SECTOR 1

Vector	V0	V13	V7	V13	V0	V13	V7	V13	V0
VSC1	3	2	2	1	6	6	6	1	2	2	3
VSC2	6	6	1	2	2	3	2	2	1	6	6

The ZSCC and CMV variation pattern can be illustrated in Fig. 4, taking sub-sector I and II as examples. Sub-sector III

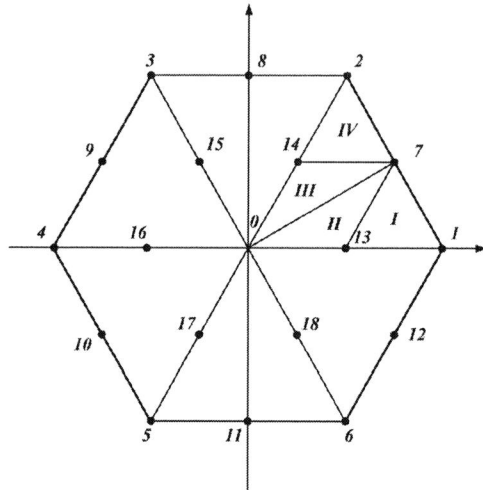

Fig. 3 The division of sectors and locations of vectors

and IV have the same pattern. Seen from Fig. 4, the peak value of ZSCC is limited by quarterly dividing the dwell time of voltage vectors and reversing the direction of ZSCC by redundant switching vectors. And the average ZSCC value is controlled to 0 in one sampling cycle.

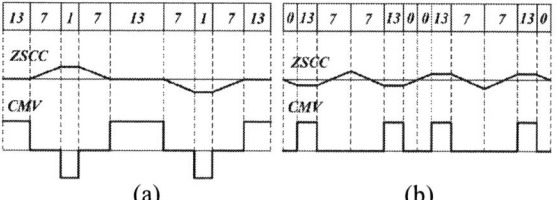

(a)   (b)
Fig. 4 ZSCC and CMV pattern in sub-sector I and II of sector 1. (a) sub-sector I; (b) sub-sector II.

Moreover, in the proposed SVM scheme, each VSC only has one switching at a time, which means the problem of instantaneous line-to-line voltage reversal is avoided. This feature makes the proposed scheme practical for real application [14]. Also the switching frequency is limited within the sampling frequency.

The dwell time calculation of the voltage vectors is given in Table VII, where T0, Ta, Tb, and Tc represent the dwell time of zero vector, small vectors, medium vectors, and large vectors; Ts represents the sample period; θ stands for the reference voltage vector angle within the sector; and M stands for the modulation index (range from 0 to 1) which is defined in (10).

$$M = \frac{\sqrt{3}\,|V_{ref}|}{V_{dc}} \qquad (10)$$

TABLE VII.   THE DWELL TIME CALCULATION OF THE PROPOSED SCHEME

	T0	Ta	Tb	Tc
I		$T_s\left[2-2M\cdot\sin(\frac{\pi}{3}+\theta)\right]$	$2M\cdot\sin(\theta)T_s$	$T_s\left[2M\cdot\sin(\frac{\pi}{3}-\theta)-1\right]$
II	$T_s\left[1-2M\cdot\sin(\frac{\pi}{3}-\theta)\right]$	$2\sqrt{3}M\cdot\sin(\frac{\pi}{6}-\theta)T_s$	$2M\cdot\sin(\theta)T_s$	
III	$T_s\left[1-2M\cdot\sin(\theta)\right]$	$2\sqrt{3}M\cdot\sin(\theta-\frac{\pi}{6})T_s$	$2M\cdot\sin(\frac{\pi}{3}-\theta)T_s$	
IV		$T_s\left[2-2M\cdot\sin(\frac{\pi}{3}+\theta)\right]$	$2M\cdot\sin(\frac{\pi}{3}-\theta)T_s$	$T_s\left[2M\cdot\sin(\theta)-1\right]$

B.   Relationship of ZSCC peak value and modulation index

The relationship of ZSCC peak value and modulation index can be theoretically calculated. For a certain modulation index, the ZSCC reaches its peak value when θ is 30°. At this point, only medium vectors and zero vector are applied (when M≠1), hence the peak value of ZSCC is determined by the dwell time of these two vectors. The ZSCC patterns with respect to the modulation index range are shown in Fig. 5, taking sector 1 as example. When M = 1, only medium vector V7 is applied. Thus the ZSCC peak is determined by quarter of the dwell time of V7. When M is between 0.5 and 1, the dwell time of V0 is smaller than that of V7. As the switching sequence is designed such that the ZSCC direction reverses when V0 and V7 are alternated, the peak value of ZSCC will be lower than that of M = 1 condition. When M = 0.5, the application time of V0 and V7 are equal. When M is between 0 and 0.5, the peak value of ZSCC is solely determined by the quarter of the dwell time of V0, as shown in Fig. 5.

The above analysis can be mathematically expressed by the dwell time calculation equations listed in Table VII. The lowest ZSCC peak value occurs when the dwell time of V7 equals to two times the dwell time of V0, that is when M = 2/3. After that, the ZSCC peak value is only determined by the dwell time of V0 since (T_b-T_0) is smaller than T_0. Considering that the sampling frequency, DC link voltage and ZSCC filter inductance are constant, the ZSCC peak value is solely determined by the application time of V0 and V7, thus the ZSCC peak value, normalized with respect to $V_{dc}\cdot T_s/L_{cm}$, under different modulation index range can be given by

$$I_{0,peak} = \left|\frac{1}{4}T_b - \frac{1}{4}T_0\right| = \left|\frac{1}{2}M - \frac{1}{4}\right| \quad \frac{2}{3} < M \le 1 \quad (11)$$

$$I_{0,peak} = \left|\frac{1}{4}T_0\right| = \frac{1}{4}|1-M| \quad 0 \le M \le \frac{2}{3} \quad (12)$$

For the method proposed in [10], the normalized ZSCC peak value can be expressed by

$$I_{0,peak} = \frac{1}{4}M \quad 0 \le M \le 1 \quad (13)$$

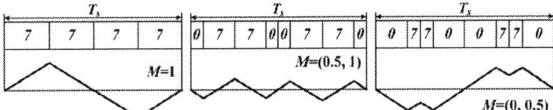

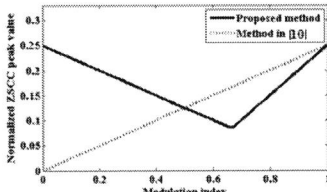

Fig. 5 The ZSCC pattern under different modulation index range when reference voltage angle is 30°

Fig. 6 ZSCC peak value with respect to the modulation index

From Fig. 6, one can see that the proposed method and the method in [10] have the same maximum ZSCC peak value, which means the two methods can have the same size of CMI. And within the modulation index range of (0.5, 1), the ZSCC peak value of the proposed method is smaller than that of the method in [10]. This can effectively reduce the semiconductor losses and current stress of the converters. But when the modulation index is between 0 and 0.5, the ZSCC peak of the proposed method will increase.

C.   Common mode current

The CMI is also influenced by the value of CMC. For both conventional SVM and DSVM modulated inverters, the value of CMC usually becomes higher when modulation index is low. This is because when M is small, the dwell time of zero vector will be long. During this time, the magnitude of CMV is Vdc/2, resulting in high CMC peaks. This problem also goes to parallel VSC application if same null-vectors are applied simultaneously to suppress ZSCC. But in the proposed scheme, the CMV magnitude is limited to Vdc/6, and at the same time, the zero vector V0 does not generate CMV, only small vectors contribute to the generation of CMV. Therefore the lower modulation index is, the smaller dwell time of small vectors become, thus the lower value of CMC.

Compared with the method in [10], the proposed method has lower CMC under low modulation index range, though the ZSCC is higher. But for CMI sizing, both ZSCC and CMC contribute to the flux linkage of the CMI. Thus the comparison of the two methods under low modulation index range should be based on specific parameters.

D. Switching losses

When modulation index is 1, the proposed scheme becomes DSVM. Hence the switching frequency is of 2/3 of the sampling frequency. The specific switching frequency of the proposed scheme cannot be directly calculated as it is affected by the modulation index and carrier-ratio. But the general relationship between switching frequency f_{sw} and modulation index M can be obtained. When the modulation index drops from 1 to $\sqrt{3}/3$, the reference voltage vector starts entering sub-sector II and III, then the switching frequency increases. When modulation index drops below $\sqrt{3}/3$, the reference voltage vector only passes sub-sector II and III, then the proposed scheme becomes continuous SVM (CSVM) and the switching frequency is equal to the sampling frequency. Hence the switching loss is higher in low modulation index range than that of high modulation index range. But the overall switching loss of the proposed scheme is between the conventional SVM and DSVM.

Another issue regarding the switching losses is the switching loss with respect to the power factor angle. When the M is smaller than $\sqrt{3}/3$, the switching losses are independent of the load condition. But when M is higher than $\sqrt{3}/3$, one phase is clamped to up or low rail for a certain degrees. Hence the switching losses are determined by the load conditions. The switching losses can be evaluated by the switching loss function (SLF) proposed in [15]. The modulation signal when M above $\sqrt{3}/3$ is shown in Fig. 7. The clamping period is represented by δ, which is 60° when M = 1. And when M = 1, the SLF is the same as that of DPWM1, which is written as

$$
SLF_{M=1} = \begin{cases} 1 - \dfrac{1}{2}\cos(\varphi) & 0 \leq \varphi \leq \dfrac{\pi}{3} \\[2mm] \dfrac{\sqrt{3}}{2}\sin(\varphi) & \dfrac{\pi}{3} \leq \varphi \leq \dfrac{\pi}{2} \end{cases} \quad (14)
$$

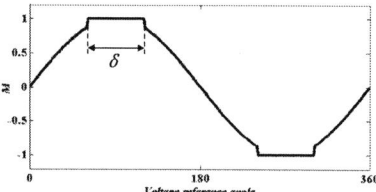

Fig. 7 The modulating signal of the proposed method when $M > \sqrt{3}/3$

When M drops but is larger than $\sqrt{3}/3$, the clamping period of each phase is shorten. Then δ enters the SLF, as expressed in (15).

$$
SLF_{\frac{\sqrt{3}}{3}<M<1} = \begin{cases} 1 - \sin\left(\dfrac{\delta}{2}\right)\cos(\varphi) & 0 \leq \varphi \leq \dfrac{\pi}{2} - \dfrac{\delta}{2} \\[2mm] \cos\left(\dfrac{\delta}{2}\right)\sin(\varphi) & \dfrac{\pi}{2} - \dfrac{\delta}{2} \leq \varphi \leq \dfrac{\pi}{2} \end{cases} \quad (15)
$$

The relationship between δ and M can be obtained by

$$
\delta = \frac{2\pi}{3} - 2\arcsin\left(\frac{1}{2M}\right) \qquad 0 < \delta < \frac{\pi}{3} \quad (16)
$$

Substituting (16) into (15), the SLF of the proposed method with respect to modulation index and power factor angle can be obtained. The SLF is plotted in Fig. 8.

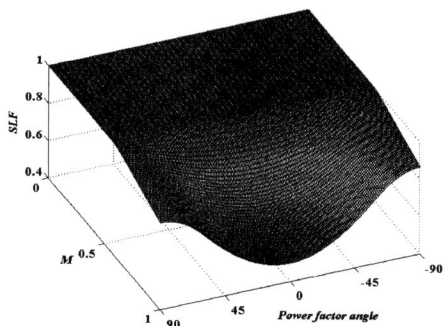

Fig. 8 SLF of the proposed method with respect to modulation index and power factor angle

From Fig. 8 we can see that when M is larger than $\sqrt{3}/3$, the switching losses are always lower than that of conventional SVM over all power factor angle range. Compared to that of the method in [10], the lower switching losses make the proposed method suitable for various kinds of applications, not only for reactive power compensation.

IV. SIMULATION

Simulation verification of the proposed method is conducted using Simulink/Matlab. The simulation parameters are listed in Table VIII. The simulated waveforms of phase A current (I_A), ZSCC, and CMV when M is 1, 0.7, 0.5, and 0.3, are shown in Fig. 9 to Fig. 12 respectively. The ZSCC and CMV waveforms under these three conditions suggest that the simulation results are in agreement with the analysis presented in Section III. Low frequency fluctuation of ZSCC is eliminated by the designed switching sequence. And magnitude of CMV is limited within Vdc/6 (100V). The simulation results of ZSCC peak value with respect to modulation index is shown in Fig. 13, which is identical with the analytical result shown in Fig. 6. The THD curve of phase A current of the proposed method is presented in Table IX, compared with the THD of interleaved SVM and interleaved DSVM1. It shows that the output quality of the proposed method is better than that of interleaved SVM but worse than that of the interleaved DSVM1 when modulation index is low. But at higher modulation index range the proposed method shows best THD performance.

TABLE VIII.   THE SIMULATION PARAMETERS.

Parameter	Value
DC link voltage, V_{dc}	600 V
Sampling frequency, f_s	3000 Hz
ZSCC filter, L_{cm}	7 mH
Load resistance, R	10 Ohm

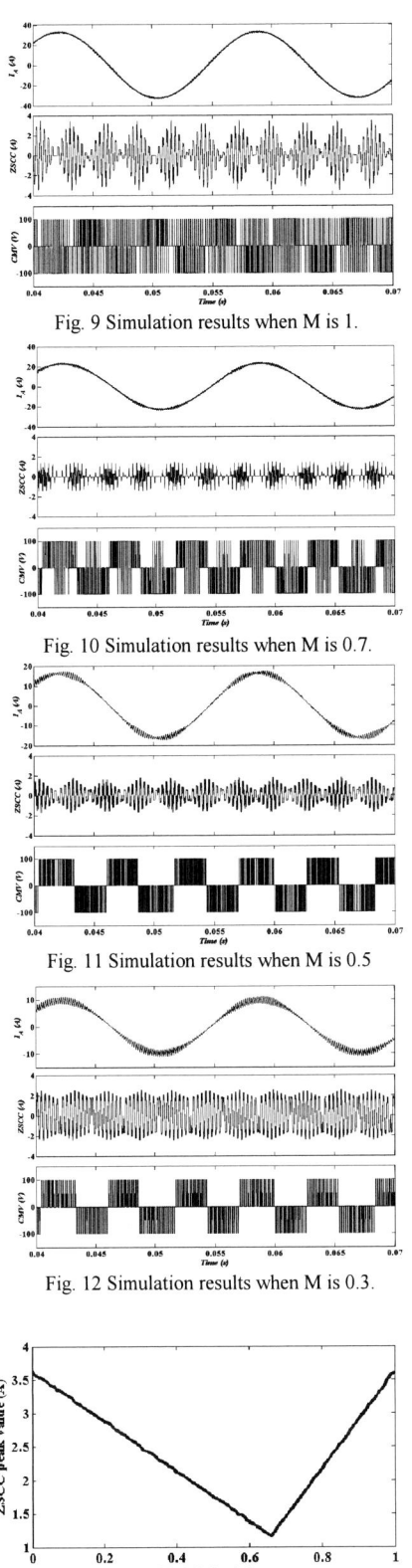

Fig. 9 Simulation results when M is 1.

Fig. 10 Simulation results when M is 0.7.

Fig. 11 Simulation results when M is 0.5

Fig. 12 Simulation results when M is 0.3.

Fig. 13 The simulation results of ZSCC peak value with respect to modulation index

TABLE IX.   THD PERFORMANCE COMPARISON

M	Proposed method	Interleaved SVM	Interleaved DSVM1
0.1	12.66%	13.20%	11.92%
0.2	10.80%	11.84%	9.34%
0.3	8.97%	10.48%	6.86%
0.4	7.20%	9.13%	4.65%
0.5	5.53%	7.79%	3.36%
0.6	4.09%	6.47%	3.35%
0.7	3.17%	5.17%	3.27%
0.8	2.57%	3.91%	3.08%
0.9	2.03%	2.73%	2.69%
1.0	1.67%	1.84%	2.49%

V.   EXPERIMENTAL VERIFICATION

The proposed method is implemented on DSP TMS320F2812 and verified experimentally. The line inductor of 2.5mH is used. A resistor load is used and the load resistance is 10 Ohm. The DC link voltage is set to 200V. The sampling frequency is 2880Hz. The experimental results when M is 0.9, 0.7, and 0.3 are presented in Fig. 14 to Fig. 16. In the figures, the line current of phase A is illustrated in green, ZSCC is in red, and CMV is in blue. The ZSCC is measured by summating the three-phase current of one converter, as presented in equation (2). As shown from the experimental results, the waveform of ZSCC is identical with the analysis. But due to the existence of hardware asymmetric, the waveform cannot be as perfect as in simulation. But no obvious low-frequency ZSCC is observed. The ZSCC peak values in full modulation range are measured, as shown in Fig. 17. The relationship between ZSCC and modulation index can still be confirmed. The experiment results also show that the magnitude of CMV is also limited to about 33V, which is identical with the analytical value of Vdc/6.

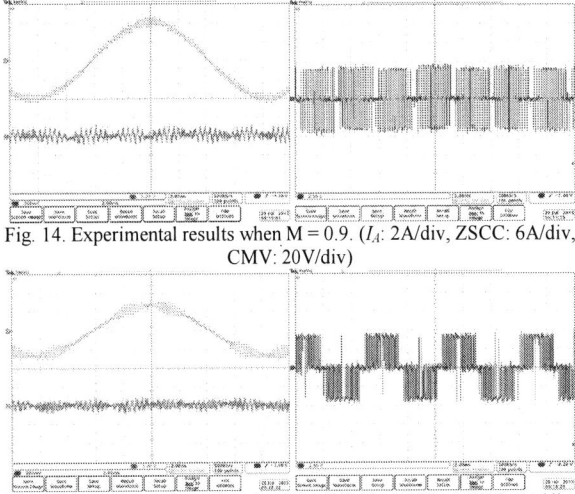

Fig. 14. Experimental results when M = 0.9. (I_A: 2A/div, ZSCC: 6A/div, CMV: 20V/div)

Fig. 15. Experimental results when M = 0.6. (I_A: 2A/div, ZSCC: 6A/div, CMV: 20V/div)

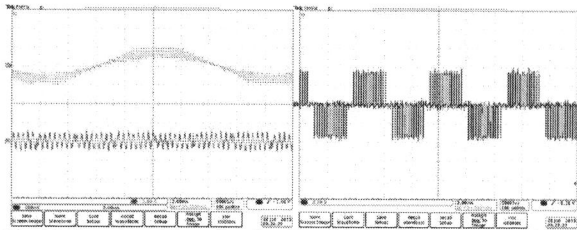

Fig. 16. Experimental results when M = 0.3. (I_A: 2A/div, ZSCC: 6A/div, CMV: 20V/div)

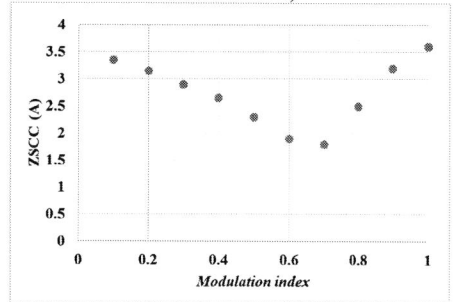

Fig. 17. Experimental results of ZSCC peak values with respect to modulation index

VI. CONCLUSION

A SVM scheme is proposed in this paper for the parallel VSCs system with single phase inductor or CMI as filter. The proposed scheme aims to reduce the peak value of ZSCC and CMV yet provide high quality output. In the proposed scheme, the null-vectors for single converter are not used. But for three-level output, zero-vector can be realized by some combinations of active vectors of the two converters. And through proper design of switching sequence, the magnitude of ZSCC and CMV can be reduced. Compared with other modulation methods, the proposed method has the lowest ZSCC peak value when modulation index is between 0.5 and 1. When modulation index drops below 0.5, the ZSCC in the proposed method will be increased but the maximum value is still the same as that when modulation index is 1, which means the proposed method can achieve the same CMI size reduction effect as the method in [10]. On the other hand, when modulation index is low, the CMC of the proposed method should be lower than in other methods. This may bring additional CMI size reduction since the magnitude of CMV is reduced from Vdc/2 to Vdc/6 during entire modulation index range. Furthermore, the proposed method is suitable for practical application since only there is only one switching action at a time. When modulation index is higher than $\sqrt{3}/3$, the switching losses are lower than that of conventional SVM for full power factor angle range. Otherwise the switching losses are the same as that of conventional SVM. The THD performance of the proposed method is verified through simulation study, suggesting that the proposed method has a good THD performance for full modulation index range. The ZSCC and CMV suppression performance of the proposed scheme are also verified through experimental results. Therefore the proposed method is particularly suitable for grid

connected applications where CMI is used and modulation index is usually high.

REFERENCES

[1] Chien-Liang Chen; Yubin Wang; Jih-Sheng Lai; Yuang-Shung Lee; Martin, D., "Design of Parallel Inverters for Smooth Mode Transfer Microgrid Applications," Power Electronics, IEEE Transactions on , vol.25, no.1, pp.6,15, Jan. 2010

[2] Xiaoxiao Yu; Khambadkone, A.M., "Reliability Analysis and Cost Optimization of Parallel-Inverter System," Industrial Electronics, IEEE Transactions on , vol.59, no.10, pp.3881,3889, Oct. 2012

[3] Chen, T.-P., "Circulating zero-sequence current control of parallel three-phase inverters," IEE Proceedings - Electric Power Applications, vol.153, no.2, pp.282-288, 2 March 2006

[4] Ching-Tsai Pan; Yi-Hung Liao, "Modeling and Coordinate Control of Circulating Currents in Parallel Three-Phase Boost Rectifiers," Industrial Electronics, IEEE Transactions on , vol.54, no.2, pp.825,838, April 2007

[5] Di Zhang; Wang, F.; Burgos, R.; Rixin Lai; Boroyevich, D., "DC-Link Ripple Current Reduction for Paralleled Three-Phase Voltage-Source Converters With Interleaving," Power Electronics, IEEE Transactions on , vol.26, no.6, pp.1741,1753, June 2011

[6] Capella, G.J.; Pou, J.; Ceballos, S.; Zaragoza, J.; Agelidis, V.G., "Current-Balancing Technique for Interleaved Voltage Source Inverters With Magnetically Coupled Legs Connected in Parallel," Industrial Electronics, IEEE Transactions on , vol.62, no.3, pp.1335,1344, March 2015

[7] Asiminoaei, L.; Aeloiza, E.; Enjeti, P.N.; Blaabjerg, F., "Shunt Active-Power-Filter Topology Based on Parallel Interleaved Inverters," Industrial Electronics, IEEE Transactions on , vol.55, no.3, pp.1175,1189, March 2008

[8] Zhihong Ye; Boroyevich, D.; Jae-Young Choi; Lee, F.C., "Control of circulating current in two parallel three-phase boost rectifiers," Power Electronics, IEEE Transactions on , vol.17, no.5, pp.609,615, Sep 2002

[9] Narimani, M.; Moschopoulos, G., "Three-Phase Multimodule VSIs Using SHE-PWM to Reduce Zero-Sequence Circulating Current," Industrial Electronics, IEEE Transactions on , vol.61, no.4, pp.1659,1668, April 2014

[10] Gohil, G.; Maheshwari, R.; Bede, L.; Kerekes, T.; Teodorescu, R.; Liserre, M.; Blaabjerg, F., "Modified Discontinuous PWM for Size Reduction of the Circulating Current Filter in Parallel Interleaved Converters," Power Electronics, IEEE Transactions on , vol.30, no.7, pp.3457,3470, July 2015

[11] Cougo, B.; Gateau, G.; Meynard, T.; Bobrowska-Rafal, M.; Cousineau, M., "PD Modulation Scheme for Three-Phase Parallel Multilevel Inverters," IEEE Trans. Ind. Electron., vol.59, no.2, pp.690-700, Feb. 2012

[12] Tallam, R.M.; Leggate, D.; Kirschnik, D.W.; Lukaszewski, R.A., "Reducing Common-Mode Current: A Modified Space Vector Pulsewidth Modulation Scheme," Industry Applications Magazine, IEEE , vol.20, no.6, pp.24,32, Nov.-Dec. 2014

[13] Ogasawara, S.; Takagaki, J.; Akagi, H.; Nabae, A., "A novel control scheme of a parallel current-controlled PWM inverter," Industry Applications, IEEE Transactions on , vol.28, no.5, pp.1023,1030, Sep/Oct 1992

[14] Hava, A.M.; Un, E., "Performance Analysis of Reduced Common-Mode Voltage PWM Methods and Comparison With Standard PWM Methods for Three-Phase Voltage-Source Inverters," Power Electronics, IEEE Transactions on , vol.24, no.1, pp.241,252, Jan. 2009

[15] Hava, A.M.; Kerkman, R.J.; Lipo, T.A., "Simple analytical and graphical methods for carrier-based PWM-VSI drives," Power Electronics, IEEE Transactions on , vol.14, no.1, pp.49,61, Jan 1999

Nonlinearity Analysis and Linear Modulation Method for Two Level Voltage Source Inverter with Low Switching to Operating Frequency Ratio

Yongjae Lee and Jung-Ik Ha

Department of Electrical and Computer Engineering
Seoul National University
Seoul, Korea
E-mail: {yongjaelee, jungikha}@snu.ac.kr

Abstract—**Recently, high speed motor drive is widely studied thanks to its advantages such as high reliability and power density. In these applications, the ratio of switching to operating frequency, F_{ratio}, becomes very low due to the limitation of the switching frequency of switching devices. Such a low F_{ratio} drive is also discussed in traction and high power applications due to the efficiency issue. Among the various issues regarding the system with low F_{ratio}, this paper analyzes the nonlinearity in modulation caused by vector rotating direction of space vector modulation. This paper also presents the correcting method for linear modulation. Simulations and experiments with reactor are accomplished to verify the analysis and compensation method for inverter with low F_{ratio}.**

Keywords—*DC-AC power converters; nonlinear distortion; pulse width modulation; three-phase electric power*

I. INTRODUCTION

It is widely known that switching to operating frequency ratio, F_{ratio} ($=f_{samp}/f_e$), should be over 30 to reduce low order harmonics for double sampling case [1]. The low F_{ratio} operation, however, is inevitable in particular applications for some reasons, high speed drive and limited switching frequency. Firstly, operating frequency of the high speed motor drive applications such as drill and turbo compressor are above 1 kHz, and even reach to 5 kHz. Because the physical limitation of IGBT's switching frequency is 20 kHz in maximum up to now, F_{ratio} becomes extremely low for these application [2]. Secondly, F_{ratio} is constrained in mid and high power applications such as fan, pump, and various grid converters. Although the operating frequency of these applications are not extremely high, F_{ratio} is very low because the maximum switching frequencies of the switching devices for high power applications, GTO and IGCT, are limited to 1kHz due to the thermal limitation. Even with the fast devices, low switching frequency operation is sometimes preferred in several applications including traction for the higher efficiency.

According to the above merits and demands for the low F_{ratio} drive, researches for the low F_{ratio} drive are accomplished and several control problems are reported by many literatures [4]-[10]. In [4], control delay caused by digital implementation is analyzed and $1.5T_{samp}$ compensation method is proposed. In [5], [6], current sampling error in low F_{ratio} system is reported. The beat phenomenon is also pointed out as the stability hazards in

This work was supported by the Brain Korea 21 Plus Project in 2015.

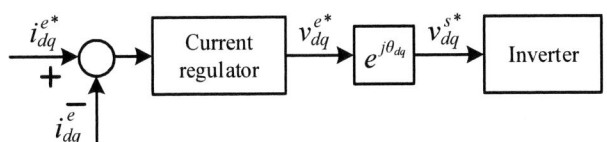

Fig. 1. Control block diagram with conventional modulation method.

low F_{ratio} drive system [7], [8]. Because the position sensors for the high speed machines cannot withstand the mechanical stresses, sensorless algorithms for the ultrahigh speed machines are also studied [9]. The current control methods for low F_{ratio} drive systems are also studied in [10].

Due to the low order harmonics, various voltage modulation methods are studied for the low F_{ratio} drive system including V/F control, optimal PWM, predictive control method, and synchronized PWM instead of asynchronous space vector (SV) PWM. These voltage modulation methods have their own merits and demerits in the sense of performance criteria such as control performance, dynamic performance, and calculation burden. However, most of the studies are concentrated on the harmonics of the low F_{ratio} drive system and only few papers report about inverter nonlinearity such as amplitude and phase distortion of the fundamental components [11]-[13].

In this paper, the nonlinearity of the low F_{ratio} drive system is analyzed and relieved through the modification of PWM duty. The second section presents the nonlinearity of the general SVPWM methods in low F_{ratio} drive system. The third section gives the linear modulation method for SVPWM methods with double sampling method, peak and valley sampling. The fourth section presents the simulation and experimental results regarding the nonlinearity in low F_{ratio} drive and the proposed modification method. The conclusion is provided in the sixth section.

II. NONLINEARITY IN VOLTAGE MODULATION

Fig. 1 shows the fundamental control block diagram of inverter with conventional modulation method. Here, v_{dq}^{s*} is directly modulated from v_{dq}^{e*}, for the single point with angle θ_{dq}. It is basic idea of vector control in synchronous reference frame that the DC value in the synchronous reference frame is converted to the circle or arc in the stationary reference frame.

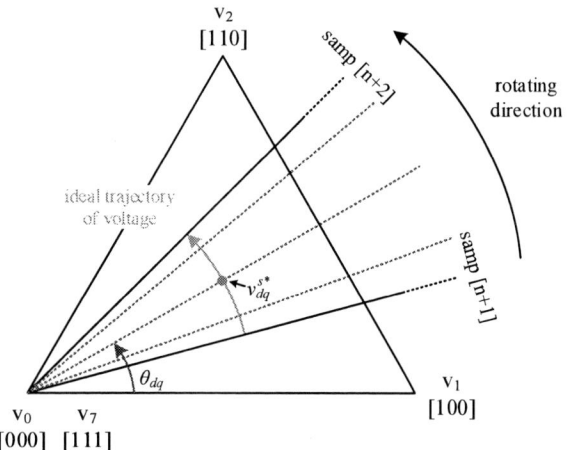

Fig. 2. Voltage reference of conventional modulation method and ideal trajectory.

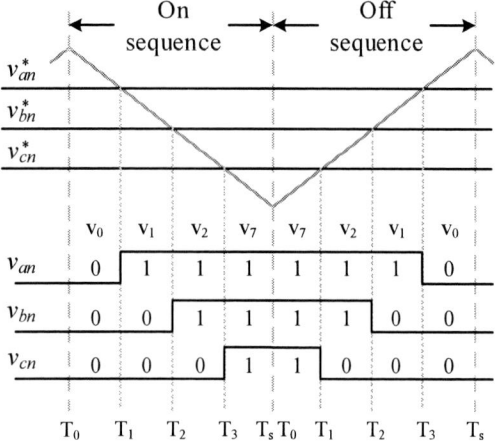

Fig. 3. Voltage references and gate signals.

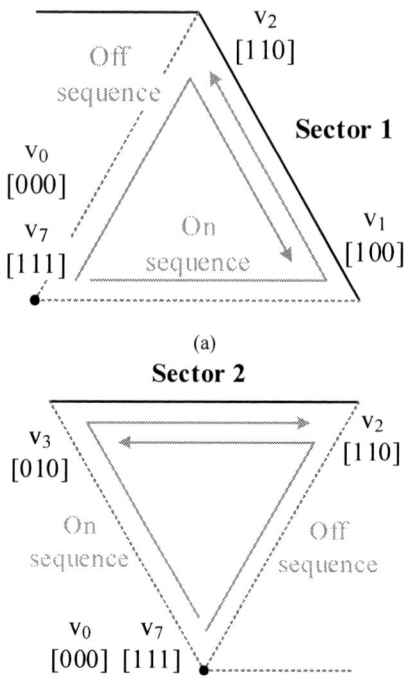

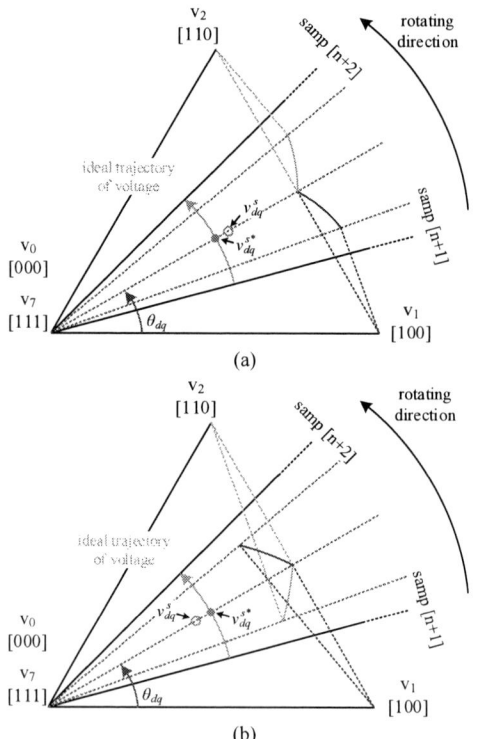

Fig. 4. Vector rotating sequence of SVPWM. (a) Sector 1. (b) Sector 2.

Thus, the desirable trajectory of output voltage can be represented as an arc as shown in Fig. 2. The PWM, however, cannot satisfy the desired trajectory due to the discontinuity in modulation and digital sampling. For this reason, it is inevitable to modulate certain point, v_{dq}^{s*}, during one sampling on behalf of the continuous trajectory. Especially, as the length of the arc increases as F_{ratio} reduces, so the nonlinearity increases when v_{dq}^{s*} contains the nonlinearity. Thus, calculation of accurate v_{dq}^{e*} which can represent v_{dq}^{s*} is important for precise modulation.

Thanks to the $1.5 T_{samp}$ compensation method [4], the phase delay comes from the digital sampling can be remarkably reduced. Although Bae and Sul also take the amplitude variation into account, only averaging effect between the ideal voltage trajectory and v_{dq}^{s*} is considered. In this paper, nonlinearity caused by PWM is analyzed based on SVPWM with double sampling method.

Fig. 3 shows the PWM carrier, phase voltage references, and corresponding switching sequence and voltage vectors for sector 1 where $v_{an}^* > v_{bn}^* > v_{cn}^*$. As shown in the figure, switches are

Fig. 5. Example of projected switching vectors. (a) Forward direction. (b) Reverse direction.

turned on sequentially when carrier count decreases and vice versa. This sequential turn on and off is important feature of the

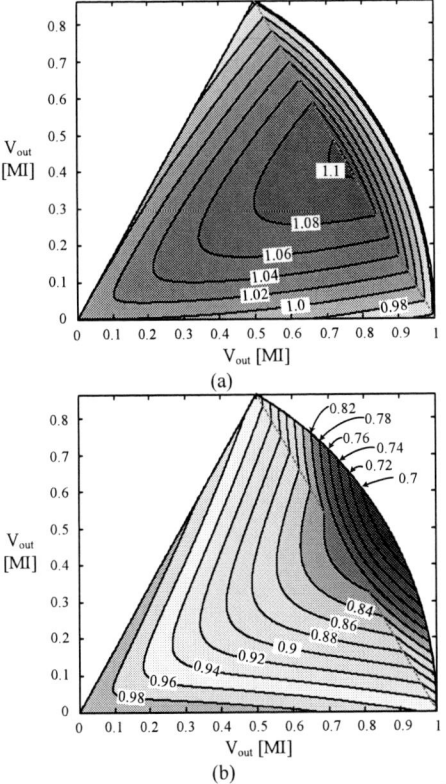

Fig. 6. Amplitude variation ratio for sector 1 without correction when F_{ratio}=6. (a) Forward direction. (b) Reverse direction.

space vector PWM and can be represented as rotation of voltage vector as shown in Fig. 4. As presented in Fig. 4(a), voltage vector rotates counterclockwise, same with rotating direction for the on sequence. However, it becomes clockwise for the off sequence. It becomes opposite for the sector 2 as shown in Fig. 4(b). In the similar manner, the voltage vector rotates in other 4 sectors.

This voltage vector rotation affect the voltage modulation performance and cause the nonlinearity. The effect of vector rotating is negligible in high F_{ratio} drive system. It, however, becomes considerable in low F_{ratio} drive system. Fig. 5 illustrates the effect of voltage vector rotation on the modulated amplitude by projecting the voltage vectors to the rotating voltage reference vector. The blue and green solid lines show the

projected voltage vector from v_1 and v_2, respectively. Because the amplitudes of v_0 and v_7 are zero, these are not represented in the figure. The average amplitude of the modulated vectors can be calculated from the average radius of projected line. In the similar way, average quadrature axis voltage distortion can be calculated by averaging the quadrature component of vectors to the rotating voltage reference frame. The resultant direct and quadrature axis components regarding the voltage reference frame for the both rotating direction, $v_{d,f}^v$, $v_{d,r}^v$, $v_{q,f}^v$, and $v_{q,r}^v$ can be formulated as (1) and (2).

The calculated voltages after modulation $v_{d,f}^v$, $v_{d,r}^v$, $v_{q,f}^v$, and $v_{q,r}^v$ can be marked as open circles in Fig. 5. As shown in Fig. 5, the amplitude of the average output voltage is larger than the reference when the rotating direction of the reference and switching sequence are coincide, forward direction, and vice versa. In other words, the output voltage from the inverter, v_{dq}^s, does not coincide with the voltage reference v_{dq}^{s*}.

Fig. 6 shows the amplitude variation ratio (v_d^v/v_d^{v*}) due to the voltage vector rotating effect for sector 1 when F_{ratio}=6. As shown in Fig. 6, the nonlinearity occurs from -17 to 10% even inside the hexagon. The nonlinearity increases as modulation index increases and it is maximized along the perpendicular line from the origin to the side of the hexagon. Although the nonlinearity is reduced as F_{ratio} increases, it still shows error of few percent when F_{ratio} is under 20.

III. LINEAR MODULATION METHOD

Aforementioned nonlinearity can cause current harmonics in asynchronous PWM operation, degrade the sensorless performance by incorrect output voltage feedback, and deteriorate the current control performance. This section provides linear modulation methods to improve control performance for low F_{ratio} operation.

For the precise output modulation, output amplitude should match v_{dq}^{e*}, and quadrature distortioni should be eliminated. This requirements can be formulated as

$$\begin{cases} v_d^v = v_{dq}^{e*} \\ v_q^v = 0 \end{cases} \tag{3}$$

There are two conditions in (3) and three switching timing

$$\begin{cases} v_{d,f}^v = \dfrac{2v_{dc}}{3T_s}\left[\int_{T_1}^{T_2} \cos\left\{ \theta_{dq} + \omega_e\left(\tau - \dfrac{1}{2}T_s\right) \right\} d\tau + \int_{T_2}^{T_3} \cos\left\{ \dfrac{\pi}{3} - \theta_{dq} - \omega_e\left(\tau - \dfrac{1}{2}T_s\right) \right\} d\tau \right] \\ v_{d,r}^v = \dfrac{2v_{dc}}{3T_s}\left[\int_{T_1}^{T_2} \cos\left\{ \dfrac{\pi}{3} - \theta_{dq} - \omega_e\left(\tau - \dfrac{1}{2}T_s\right) \right\} d\tau + \int_{T_2}^{T_3} \cos\left\{ \theta_{dq} + \omega_e\left(\tau - \dfrac{1}{2}T_s\right) \right\} d\tau \right] \end{cases} \tag{1}$$

$$\begin{cases} v_{q,f}^v = \dfrac{2v_{dc}}{3T_s}\left[\int_{T_1}^{T_2} -\sin\left\{ \theta_{dq} + \omega_e\left(\tau - \dfrac{1}{2}T_s\right) \right\} d\tau + \int_{T_2}^{T_3} \sin\left\{ \dfrac{\pi}{3} - \theta_{dq} - \omega_e\left(\tau - \dfrac{1}{2}T_s\right) \right\} d\tau \right] \\ v_{q,r}^v = \dfrac{2v_{dc}}{3T_s}\left[\int_{T_1}^{T_2} \sin\left\{ \dfrac{\pi}{3} - \theta_{dq} - \omega_e\left(\tau - \dfrac{1}{2}T_s\right) \right\} d\tau + \int_{T_2}^{T_3} -\sin\left\{ \theta_{dq} + \omega_e\left(\tau - \dfrac{1}{2}T_s\right) \right\} d\tau \right] \end{cases} \tag{2}$$

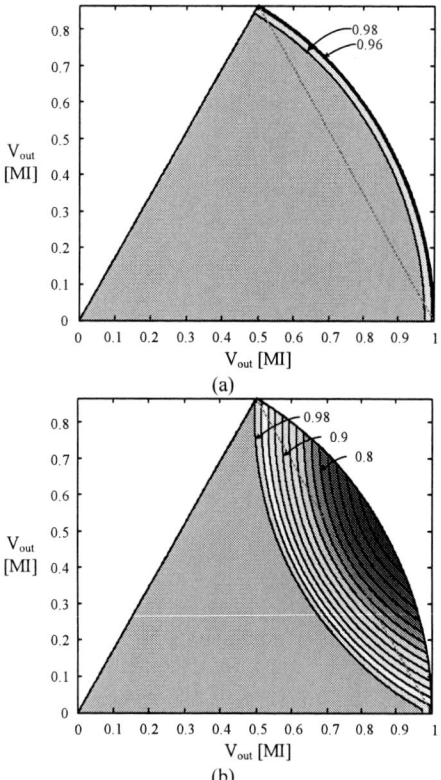

(a)

(b)

Fig. 7. Amplitude variation ratio for sector 1 with proposed method when F_{ratio}=6. (a) Forward direction. (b) Reverse direction.

which should be calculated. Thus, there should be one more constraint for calculation. In this paper, the duty of zero vectors, v_0 and v_7, are treated as equal to sustain the features of the SVPWM. This assumption can be formulated as

$$T_3 = T_s - T_1 \tag{4}$$

From these requirements and above equations, the corrected switching angle can be derived as (5) and (6).

$$\theta_{2,f} = \theta_x - \frac{\pi}{3} + \cos^{-1}\left(\frac{3\omega_e T_s v_{dq}^{e*}}{2v_{dc}}\sin\theta_x\right), \tag{5-1}$$

$$\theta_{1,f} = \theta_x + \sin^{-1}\left\{\frac{\sin\left(\theta_{2,f} + \pi/3\right)}{2\cos\theta_x} - \frac{3\omega_e T_s v_{dq}^{e*}}{4\cos\theta_x v_{dc}}\right\}, \tag{5-2}$$

$$\theta_{3,f} = 2\theta_{dq} - \theta_{1,f}. \tag{5-3}$$

$$\theta_{2,r} = \theta_x + \frac{2\pi}{3} - \cos^{-1}\left(\frac{3\omega_e T_s v_{dq}^{e*}}{2v_{dc}}\sin\theta_x\right), \tag{6-1}$$

$$\theta_{1,r} = \theta_x + \frac{\pi}{3} - \sin^{-1}\left\{\frac{\sin\left(\theta_{2,r} + \pi/3\right)}{2\cos\theta_x} + \frac{3\omega_e T_s v_{dq}^{e*}}{4\cos\theta_x v_{dc}}\right\}, \tag{6-2}$$

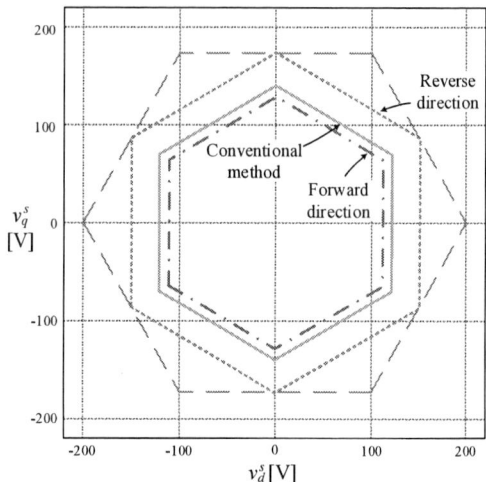

Fig. 8. Voltage references according to the vector rotating direction when F_{ratio}=6.

$$\theta_{3,r} = 2\theta_{dq} - \theta_{1,r}. \tag{6-3}$$

where $\theta_x = \theta_{dq} - (n-1)\pi/3 - \pi/6$, and $\theta_k = \theta_{dq} + \omega_e(T_k - 0.5T_s)$, where n indicates the nth sector and k is 1, 2, and 3. The switching timing T_k can be calculated from θ_k. Fig. 7 shows the modulation results with the proposed method in the same condition with Fig. 6. As shown in the figure, output voltage is linearized to its physical limitation.

IV. SIMULATION AND EXPERIMENTAL RESULTS

To verify the analysis results and the effectiveness of the proposed modulation method, simulations are conducted by PLECS. In the simulation, DC link voltage of the inverter is 300V. v_{dq}^{e*} of the inverter is 140V and modulated by conventional and proposed modulation methods and different vector rotating sequences. Voltage is modulated with three representative F_{ratio}, 6, 12, and 18.

Fig. 8 shows the modulated voltage references, v_{dq}^{s*}, according to the vector rotating direction when F_{ratio}=6. The three Lissajous curve show the output vector location of conventional method and proposed method with forward and reverse directions for same voltage reference, 140V. Due to the nonlinearity, the amplitude of output voltage with conventional methods are 152.6 and 119V for forward and reverse direction, respectively. With the proposed modulation method, however, output voltage matches the voltage reference. With the proposed method, voltage references are corrected from green solid line to blue dot-dashed line and red dashed line according to the rotating sequence, respectively. For the forward sequence, voltage reference is reduced to compensate the voltage increasing effect from the voltage vector rotation. On the other hand, voltage reference is enlarged to compensate the decreasing effect.

Fig. 9 shows the Lissajous curve for the conventional and proposed method when F_{ratio}=12. When F_{ratio}=6, the rotating direction is not changed because on/off sequence alternates with sector simultaneously. In this case, however, the rotating direction is alternating in one sector. As shown in the Fig. 9, the

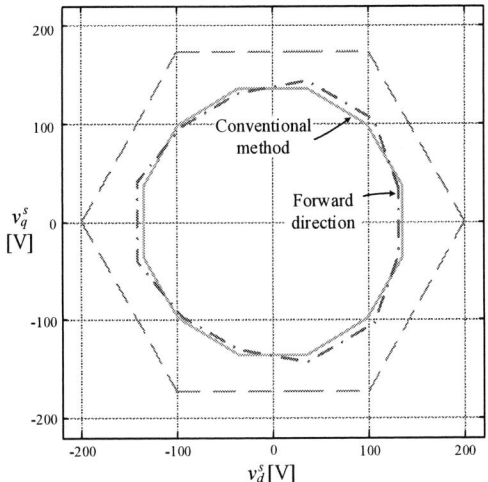

Fig. 9. Voltage references when F_{ratio}=12.

TABLE I. INVERTER NONLINEARITY IN SIMULATION

Modulation method	Initial rotating direction	F_{ratio}		
		6	12	18
Conventional method	Forward	1.09	0.994	0.996
	Reverse	0.85	0.994	0.996
Proposed method	Forward	1.00	1.00	1.00
	Reverse	1.00	1.00	1.00

output vector with the proposed method goes inside and outside of that with the conventional method to compensate the nonlinearity.

Table I shows the nonlinearity of the modulated voltage in simulation when v_{dq}^{e*} is 140V. As shown in the table, nonlinearity is maximized when F_{ratio} is 6. Nonlinearity is relatively low when F_{ratio} is 12 and 18, because the increasing and decreasing effect cancel out each other. As depicted in table, proposed method perpectly compensates the nonlinearity and genereates the required voltage.

Experiments are accomplished to verify the analysis. In the experiments, switching frequency is 6kHz, v_{dc}=300V, and v_{dq}^{e*} is 120V. Voltage amplitudes are measured by power analyzer PM6000. Fig. 10 shows the line to line voltage of the PWM output voltage when F_{ratio} is 6. As shown in the figure, line to line voltages show difference waveforms due to the rotating sequence. While active voltages are close together in forward direction, they are more far apart from each other than that of the forward direction. This difference makes nonlinearity in PWM with low F_{ratio}.

Fig. 11 shows the experimentally achieved voltage references according to the modulation methods. As same with the simulation results, voltage reference is decreased for forward operation and increased for reverse operation. Table II shows the summarized experimental results. Line-to-line voltage for the given voltage reference is 207.85V and the generated output to reference voltage ratios are listed in the table. As shown in the table, proposed method generates wanted voltage amplitude as

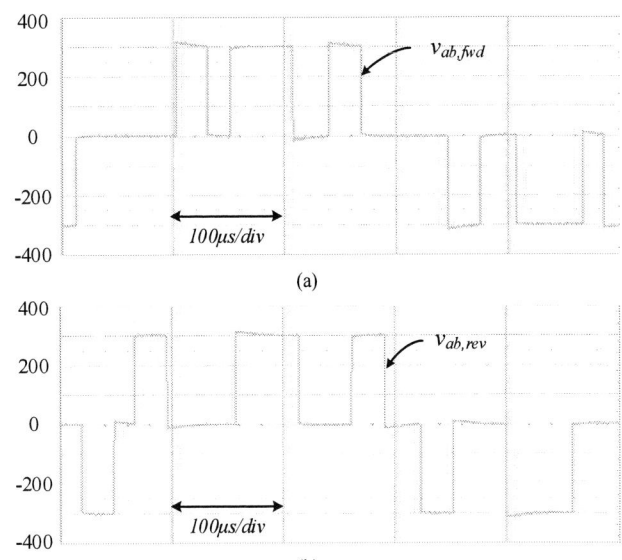

Fig. 10. Line-to-line voltage of inverter output when F_{ratio}=6. (a) Forward direction. (b) Reverse direction.

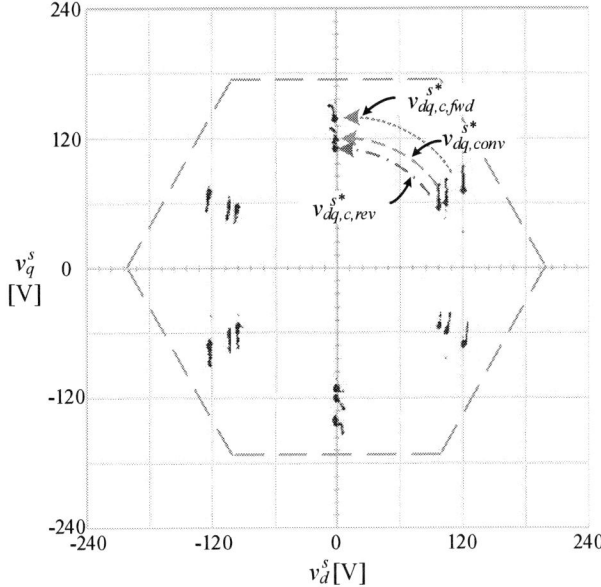

Fig. 11. Lissajous figures of voltage reference with and without the compensation.

TABLE II. INVERTER NONLINEARITY IN EXPERIMENT

Modulation method	Initial rotating direction	F_{ratio}	
		6	12
Conventional method	Forward	1.083	0.992
	Reverse	0.871	0.992
Proposed method	Forward	1.001	1.002
	Reverse	0.997	1.000

same with the simulation results. Without the proposed method, however, generated output voltages are quite different with the voltage reference.

V. CONCLUSION

This paper presents the analysis results regarding the nonlinearity of the PWM in low F_{ratio} drive system. The linear modulation method is also proposed to eliminate the nonlinearity. The linearization of the output voltage is expected to improve the current control and sensorless performance. The simulation and experimental results prove the nonlinear operation of the conventional modulation method and the linear modulation performance of the proposed method.

REFERENCES

[1] J. Holtz, "Pulsewidth modulation-a survey," *IEEE Trans. Ind. Electron.*, vol. 39, no. 5, pp. 410–420, Dec. 1992.

[2] C. Zwyssig, S. D. Round, and J. W. Kolar, "An ultrahigh-speed, low power electrical drive system," *IEEE Trans. Ind. Electron.*, vol. 55, no. 2, pp. 577–585, Feb. 2008.

[3] T. A. Burress, C. L. Coomer, S. L. Campbell, L. E. Seiber, L. D. Marlino, R. H. Staunton, and J. P. Cunningham, "Evaluation of the 2007 Toyota Camry Hybrid Synergy Drive System," Oak Ridge National Laboratory (ORNL), Oak Ridge, TN, USA; Pacific Northwest National Laboratory (PNNL), Richland, WA, USA, Tech. Rep. ORNL/TM- 2007/190, [Online]. Available: http://www.osti.gov/scitech/servlets/purl/928684.

[4] B.-H. Bae and S.-K. Sul, "A compensation method for time delay of full-digital synchronous frame current regulator of PWM AC drives," *IEEE Trans. Ind. Appl.*, vol. 39, no. 3, pp. 802–810, May/Jun. 2003.

[5] J.-S. Yim, S.-K. Sul, B.-H. Bae, N. R. Patel, and S. Hiti, "Modified current control schemes for high-performance permanent-magnet AC drives with

low sampling to operating frequency ratio," *IEEE Trans. Ind. Appl.*, vol. 45, no. 2, pp. 763–771, Mar./Apr. 2009.

[6] J. Holtz and N. Oikonomou, "Estimation of the fundamental current in low-switching-frequency high dynamic medium-voltage drives," *IEEE. Trans. Ind. Appl.*, vol. 44, no. 5, pp. 1597–1605, Sep./Oct. 2008.

[7] Y. Iwaji, T. Sukegawa, T. Okuyama, T. Ikimi, M. Shigyo, and M. Tobise, "A new PWM method to reduce beat phenomenon in large-capacity inverters with low switching frequency," *IEEE Trans. Ind. Appl.*, vol. 35, no. 3, pp. 606–612, May/Jun. 1999.

[8] P. Stumpf, R. K. Jardan, and I. Nagy, "Subharmonics generated by space vector modulation in ultrahigh speed drives," *IEEE Trans. Ind. Electron.*, vol. 59, no. 2, pp. 1029–1037, Feb. 2012.

[9] P. Kshirsagar and R. Krishnan, "Sensorless control of permanent magnet motors operating at low switching frequency for climate control systems," in *Proc. Symp. SLED*, Sep. 2012, pp. 1–8.

[10] B. P. McGrath, S. G. Parker, and D. G. Holmes, "High-performance current regulation for low-pulse-ratio inverters," *IEEE Trans. Ind. Appl.*, vol. 49, no. 1, pp. 149–158, Jan./Feb. 2013.

[11] H. Badizadeh, N. Farokhniah, H. Toodeji, A. Kavousi, "Formulation of line-to-line voltage total harmonic distortion of two-level inverter with low switching frequency," *IET Power Electron.*, vol. 6, no. 3, pp. 561–571, 2013.

[12] K. Yang, Q. Zhang, R. Yuan, W. Yu, J. Yuan, and J. Wang, "Selective harmonic elimination with groebner bases and symmetric polynomials," *IEEE Trans. Power Electron.*, [Online]. Available: DOI: 10.1109/TPEL.2015.2447555.

[13] T. A. Sakharuk, A. M. Stanković, G. Tadmor, and G. Eirea, "Modeling of PWM inverter-supplied AC drives at low switching frequencies," *IEEE Trans. Circuit Syst.*, vol. 49, no. 5, pp. 621–631, May 2002.

Synchronization Strategies in Cascaded H-Bridge Multi Level Inverters for Carrier Based Sinusoidal PWM Techniques

Saroj Kumar Sahoo, *Student Member IEEE*
Department of Electrical Engineering
IIT Kharagpur
Kharagpur, India-721302
sarojsahoo.mdpe@gmail.com

Tanmoy Bhattacharya, *Member IEEE*
Department of Electrical Engineering
IIT Kharagpur
Kharagpur, India-721302
btanmoy@ee.iitkgp.ernet.in

Abstract— This paper proposes synchronization strategies for cascaded H-Bridge multi level inverter (CHBMLI) topologies with carrier based sinusoidal pulse width modulation (PWM) techniques. It is presented how synchronous carriers are generated in desired phase and frequency from the instantaneous voltage references to maintain 3-Φ, half wave and quarter wave symmetries among inverter pole voltage outputs for carrier based sinusoidal PWM of CHBMLI. To achieve dynamic synchronization, the triangular carriers are generated from the instantaneous voltage references in a phase locked manner. The scheme is experimentally tested with phase shifted PWM technique for both open loop Vbyf and closed loop vector control of squirrel cage induction motor drive supplied from a 3-Φ 5 level CHBMLI and the results are presented.

Index Terms—cascaded H-Bridge multilevel inverter, sinusoidal PWM technique, LSPWM technique, PSPWM technique, 3-Φ symmetry, half wave symmetry, quarter wave symmetry.

I. INTRODUCTION

With the increasing demand of medium voltage and high power drives, the use of multilevel inverters is gradually becoming more and more important. This is due to reduced voltage stress on semiconductor devices, reduced harmonics in inverter output voltage and lesser electromagnetic interferences. For medium voltage and high power applications the switching frequency and device ratings are limited by the existing semiconductor technology. Increasing the power rating by minimizing switching frequency, while still maintaining reasonable power quality is an important requirement and a persistent challenge. Hence, the use of multilevel inverters (MLIs) become relevant as it suitably distributes the stress among the semi conductor switches. Among the different MLI topologies, the cascaded H-Bridge (CHB) topology is a preferred choice for medium voltage drives for its modularity and this is also the target converter for the proposed PWM technique in this paper.

In high power and medium voltage applications, the power converters are operated at low switching frequency. This is very well defined in the literature as low pulse ratio operation of the converters. The applications consist of traction drives

(both VSI and CSI fed drives), grid applications (as bidirectional converters, active power filters) etc. As the pulse ratio is less in these power converters, lower order harmonics including sub-harmonics are introduced in line currents resulting in higher total harmonic distortion (THD). Hence, synchronization among PWM voltages is necessary. Along with synchronization, the PWM voltage should maintain half wave, quarter wave and 3-Φ symmetries [1]. This is valid for both two level and MLIs. Reference [2] proposes an optimal PWM scheme for two level inverters to reduce current harmonics with offline calculations of switching pattern. This optimized PWM strategy can also be extended to multi level inverters. Reference [3] shows the application of synchronous optimal PWM (SOPWM) technique for a cascaded 9 level inverter. Reference [4] compares the performances of 5 level and 7 level neutral point clamped (NPC) inverters with SOPWM technique. But, the SOPWM technique is an offline calculation based optimization strategy. The switching angles are pre-calculated assuming steady state condition and this requires storage of large data for better accuracy. Reference [5] proposes the model predictive pulse pattern control for a 5 level NPC inverter with optimized PWM technique.

Selective harmonic elimination and selective harmonic mitigation PWM techniques are the other alternatives. Reference [6] shows the application of SHEPWM technique for cascaded MLIs, whereas [7] shows the use of SHMPWM technique. But they are also offline calculation based PWM technique. Hence, along with model predictive control they are used for drive applications.

All the above MLI PWM techniques are offline calculation based technique. The calculation complexity increases with increase in number of voltage levels or increase in the number of commutation angles at lower modulation indexes. Even if the offline calculation is completely ignored, the real time implementation requires commutation angles to be stored in look-up tables which in turn require huge memory storage capacity of the controller platform. The carrier based PWM techniques are independent of motor parameters and do not require commutation angles to be stored in look-up tables. Hence, carrier based PWM technique is well suited to MLIs although the harmonic content of the inverter output voltage is

978-1-4673-9551-9/16 $31.00 © 2016 IEEE

not optimized. Two carrier based PWM techniques are available in literature for MLIs. They are:- (i) Level shifted PWM technique (LSPWM technique) and (ii) Phase shifted PWM technique (PSPWM technique). But, the major challenge for carrier based PWM techniques for MLIs is to position the zero crossings of the carriers w.r.t. the zero crossings of the voltage references so that different symmetries of the pole voltage waveforms can be maintained.

For a two level inverter, the carrier should be synchronized with the voltage references i.e. the zero crossings of the carrier should match with the zero crossings of the voltage references and the ratio ($p=f_c/f_s$) should be an odd integer (multiple of 3). This condition will maintain 3-Φ symmetry, half wave symmetry and quarter wave symmetry of inverter pole voltage waveform. The carrier synchronization with the voltage references is sufficient for the LSPWM technique for CHBMLIs, as only one synchronous carrier is sufficient for implementation of different voltage levels. Hence, the pole voltage maintains all the basic properties of an ideal synchronous PWM technique. But the power distribution and average device switching frequencies of different HBs are different.

But this scenario is different for PSPWM technique, as multiple phase shifted synchronous carriers are used for different HBs. Hence, the positions of the voltage references w.r.t. to the carriers play an important role for maintaining different basic properties of an ideal synchronous PWM, as stated in the previous paragraph. Hence, for maintaining symmetry among inverter pole voltage patterns, PSPWM technique is analytically studied in this paper.

The carriers are generated from the instantaneous voltage references, as in [8] and always maintain an odd integer ratio p. The carrier frequency f_c changes as $500Hz$-$21f_s$-$15f_s$-$9f_s$-$3f_s$ (where f_s=fundamental frequency) with the increase in fundamental frequency f_s. This paper experimentally verifies the validity of the proposed PSPWM technique for both open loop Vbyf and closed loop vector control of a squirrel cage induction motor drive supplied from a 3-Φ 5 level CHBMLI.

This paper is arranged as follows. Section II deals with the explanation of LSPWM technique to maintain 3-Φ symmetry, HWS and QWS of the CHBMLI pole voltage waveform. Section III deals with the analytical explanation for PSPWM technique to maintain 3-Φ symmetry, HWS and QWS of the CHBMLI pole voltage waveform. Experimental results are given in section IV and the paper is concluded in Section V.

II. LEVEL SHIFTED PWM STRATEGIES IN CHBMLI TOPOLOGIES

The main goal of any carrier based PWM technique for low pulse ratio inverters is to maintain HW, QW and 3-Φ symmetries among the inverter pole voltage waveforms. The LSPWM technique can be implemented with multiple level shifted co-phasor carriers or a single carrier with modified voltage references for each level. The approach of using a single carrier with modified voltage references is adopted in this work. Hence, synchronous carrier generation method of [8] can directly be used. The single triangular carrier is generated from the instantaneous voltage references in a phase

locked manner. The voltage references are modified for each level. Fig.1.(b) shows one leg of 3-Φ 5 level CHBMLI. Fig.2 shows LSPWM technique for one leg of 3-Φ 5 level CHBMLI. Fig.2.(a) shows the modified voltage references for HB1 and HB2 respectively along with carriers C_{HB1} and C_{HB2} (where $C_{HB1}=C_{HB2}$ i.e. same carrier is used for both bridges). Fig.2.(b) shows HB1, HB2 and resultant output voltages respectively. From the figures it can be observed that the individual bridge voltages and the resultant output voltage maintain HW and QW symmetries.

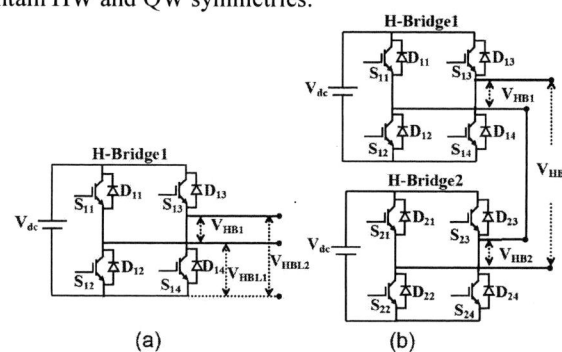

Fig. 1. (a) Single H-Bridge, (b) Double cascaded H-Bridge.

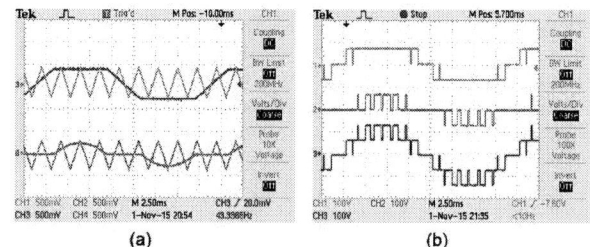

Fig. 2. (a) Ch.1: Carrier C_{HB1}, Ch.2: Carrier C_{HB2}, Ch.3: Reference R_1, Ch.4: Reference R_2 (b) Ch.1: V_{HB1}, Ch.2: V_{HB2}, Ch.3: V_{HB} for modulation index=0.9 and p=9.

III. PHASE SHIFTED PWM STRATEGIES IN CHBMLI TOPOLOGIES

A. Verification of 3-Φ symmetry and HWS

Circuit schematic of a 5 level CHBMLI is shown in Fig.1.(b). Unipolar switching strategy is employed for each H-Bridge inverter. With synchronous carriers having 3n times the fundamental frequency (n being an odd integer) being used for the PSPWM technique, 3-Φ Symmetry is always maintained among individual HB output pole voltages. Also, from Fig.3 it can be observed that, the region from π rad to 2π rad is equivalent to a mirror image of the region from 0 rad to π rad with respect to x-axis. Hence, the intersection points C_1 to C_6 are the mirror images of points from C_7 to C_{12} respectively. Hence, the pole voltage patterns in the positive portion of the voltage reference are the exact replica of pole voltage patterns in the negative portion of the voltage reference. Hence, the condition for HWS at points C_1 and C_7 satisfies (1).

$$\theta_7 = \pi + \theta_1 \qquad (1)$$

Similarly, the angle relations at other intersection points can also be stated for HWS. Though, the example is shown for p=3, the HWS is satisfied for any carrier having its frequency equal to odd integer multiple of the fundamental frequency. Hence, it can be concluded that the inverter pole voltage maintains 3-Φ symmetry for carriers having 3n times the fundamental frequency (n being any integer) and HWS for carriers having m times the fundamental frequency (m being any odd integer). With carriers having 3n times the fundamental frequency (n being any integer), both 3-Φ and HW symmetries are ensured. It is hence only necessary to determine the conditions for QWS in the inverter pole voltage.

B. Verification of QWS

1) Single H-Bridge

The QWS among pole voltage patterns of an HB (Fig.1.(a)) can be maintained, if the zero crossings of the voltage reference coincide with the zero crossings of the carrier. The carrier can be of two types. They are: - (i) In phase carrier (positive zero crossing of positive voltage reference coincides with the positive zero crossing of the carrier) (ii) Out of phase carrier (positive zero crossing of the positive voltage reference coincides with the negative zero crossing of the carrier). It is also possible to maintain QWS among single HB output pole voltage V_{HB1}, when the carrier zero crossings do not match with voltage reference zero crossings. The phase relation between the reference and the carrier for QWS of the HB output voltage is analytically derived in the following section.

Fig.3. shows the zero crossings of carrier C_{HB} having p=3, is Φ rad lagging w.r.t. the zero crossings voltage references R_1 and R_2. For maintaining QWS among pole voltage patterns V_{HB1}, the condition to be satisfied among voltage reference and carrier intersection points C_1 to C_6 is given by (2).

$$\theta_{7-k} = \pi - \theta_k \quad \text{for k = 1, 2 and 3} \tag{2}$$

For k=1, the values of θ_1 and θ_6, at points C_1 and C_6 can be found out by equating the equations of voltage references and carrier and can be written as (3) and (4) respectively.

$$-m\sin\theta_1 = \left(\frac{6}{\pi}\right)(\theta_1 - \phi) \tag{3}$$

$$m\sin\theta_6 = \left(\frac{6}{\pi}\right)\left(\theta_6 - \frac{2\pi}{3} - \phi\right) \tag{4}$$

Equation (4) can be modified as (5) by putting the condition of QWS (2).

$$m\sin\theta_1 = \left(\frac{6}{\pi}\right)\left(\frac{\pi}{3} - \theta_1 - \phi\right) \tag{5}$$

By adding (3) and (5), the value of ϕ can be found as (6).

$$\phi = \frac{\pi}{6} rad \tag{6}$$

From (6) it can be observed that, the QWS among pole voltage V_{HB1} can be maintained if the carrier zero crossings are lagging the voltage reference zero crossings by π/6 rad. Equation (6) gives the value of Φ in terms of fundamental period of voltage reference. If the value of Φ is expressed in terms of carrier period, then Φ should be equal to π/2 rad for p=3. In a similar way the conditions for QWS at other

intersection points can also be stated. Similarly, when the carrier zero crossings lead the voltage reference zero crossings by an angle Φ=π/2 rad, QWS among pole voltage patterns V_{HB1} in a single HB can be maintained.

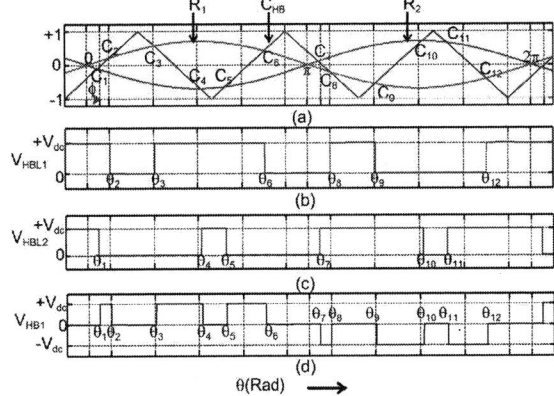

Fig. 3. (a) Carrier C_{HB} lagging the voltage references R_1 and R_2 by Φ rad (f_c/f_s=3) (b) Leg 1 voltage V_{HBL1} (c) Leg2 voltage V_{HBL2} (d) Bridge pole voltage V_{HB1}

Equation (6) shows the condition for QWS among pole voltages V_{HB1}, when p=3. For carriers having frequency p=3n (where n=1,3,5,7,9,11,......etc.) times the fundamental frequency with their zero crossings lagging the fundamental voltage reference zero crossings by Φ rad, the condition for QWS among pole voltage patterns V_{HB1} can be found by using equations (7)-(11). Here, for maintaining QWS, the condition is given by (7).

$$\theta_{6n+1-k} = \pi - \theta_k \quad \text{for k = 1, 2, ...3n} \tag{7}$$

For k=1

$$-m\sin\theta_1 = \left(\frac{6n}{\pi}\right)(\theta_1 - \phi) \tag{8}$$

$$m\sin\theta_{6n} = \left(\frac{6n}{\pi}\right)\left\{\theta_{6n} - \phi - \frac{(3n-1)}{3n}\pi\right\} \tag{9}$$

By putting the condition of QWS $\theta_{6n} = \pi - \theta_1$ in (9), (9) can be simplified as (10).

$$m\sin\theta_1 = \left(\frac{6n}{\pi}\right)\left(\frac{\pi}{3n} - \theta_1 - \phi\right) \tag{10}$$

By adding (8) and (10), the value of ϕ can be found as (11).

$$\phi = \left(\frac{1}{3n}\right)\left(\frac{\pi}{2}\right)rad \tag{11}$$

From (11) it can be observed that, QWS among pole voltage patterns V_{HB1} can be maintained if the zero crossings of 3n carriers lag the zero crossings of voltage references by $\Phi=(1/3n)(\pi/2)$ rad. If the value of Φ is expressed in terms of carrier period, then Φ should be equal to $\pi/2$ rad. Similarly, for maintaining QWS among pole voltage patterns V_{HB1} the value of Φ becomes $\pi/2$ rad, when the zero crossings of 3n carriers lead the zero crossings of the voltage references by an angle Φ rad.

Hence, the conditions for maintaining QWS among pole voltage patterns V_{HB1} in a single HB can be summarised as below:-

➢ The zero crossings of the voltage references should coincide with the zero crossings of the carrier.

➢ The zero crossings of the carrier should be placed at $\pm\pi/2$ rad w.r.t. the zero crossings of the voltage references where 2π rad denotes one carrier period.

2) Double H-Bridge

For two cascaded HBs (Fig.1.(b)), the QWS among individual pole voltages V_{HB1} and V_{HB2} can be maintained by using two carriers. They are:-(i)The zero crossing of the carrier C_{HB1} in phase with the zero crossings voltage references and (ii)The zero crossing of carrier C_{HB2} at $\pm\pi/2$ rad w.r.t. the carrier C_{HB1}. The above two conditions maintain QWS among individual pole voltages V_{HB1} and V_{HB2} along with resultant pole voltage V_{HB}. It is also possible to maintain QWS of the resultant pole voltage V_{HB}, without maintaining QWS of the individual bridge voltages V_{HB1} and V_{HB2}. Two approaches are possible as shown below. Both the approaches are analytically derived in the coming sections.

➢ Zero crossings of carriers C_{HB1} and C_{HB2} are equidistantly placed on both sides of the carrier C^{ref1} (where C^{ref1} is a fictitious carrier whose zero crossings are in phase with the zero crossings of the voltage references R_1 and R_2) zero crossings.

➢ Zero crossings of carriers C_{HB1} and C_{HB2} are equidistantly placed on both sides of the carrier C^{ref2} (where C^{ref2} is a fictitious carrier whose positive or negative peak are placed at zero crossings of the voltage references R_1 and R_2) zero crossings.

a) Approach I

Fig.4 shows the case, where the zero crossings of carrier C_{HB1} lag the zero crossings of carrier C^{ref1} by Φ_1 rad, whereas the zero crossings of carrier C_{HB2} lead the zero crossings of carrier C^{ref1} by Φ_2 rad for p=3. Fig.4.(b), (c) and (d) show that the individual bridge pole voltages V_{HB1}, V_{HB2} and resultant pole voltage V_{HB} do not maintain QWS. For maintaining QWS among resultant pole voltage patterns V_{HB}, the condition to be satisfied among voltage reference and carrier intersection points C_1 to C_{12} is given by (12).

$$\theta_{13-k}=\pi-\theta_k \quad \text{for k=1, 2, 3, 4, 5 and 6} \quad (12)$$

For k=1, the values of θ_1 and θ_{12}, at points C_1 and C_{12} can be found out by equating the equations of voltage references and carriers and can be written as (13) and (14) respectively.

$$-m\sin\theta_1=\left(\frac{6}{\pi}\right)(\theta_1-\phi_1) \quad (13)$$

$$-m\sin\theta_{12}=-\left(\frac{6}{\pi}\right)(\theta_{12}-\pi+\phi_2) \quad (14)$$

Equation (14) can be modified as (15) by putting the condition of QWS (12).

$$m\sin\theta_1=\left(\frac{6}{\pi}\right)(\phi_2-\theta_1) \quad (15)$$

By adding (13) and (15), the condition for QWS can be found out as (16).

$$\phi_1=\phi_2 \quad (16)$$

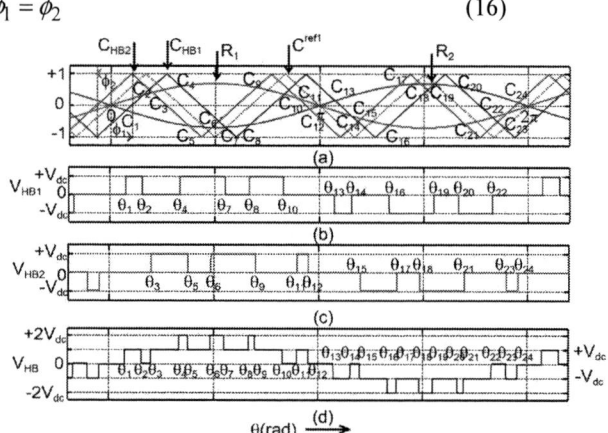

Fig. 4. References R_1 and R_2 and Carriers C_{HB1} and C_{HB2} when C^{ref1} is in phase with voltage references (p=3) (b) H-Bridge1 pole voltage V_{HB1} (c) H-Bridge2 pole voltage V_{HB2} (d) Resultant pole voltage V_{HB}.

From (16) it can be observed that, the QWS among resultant pole voltage patterns V_{HB} can be maintained if $\Phi_1=\Phi_2$, i.e. the zero crossings of carriers C_{HB1} and C_{HB2} are placed equidistantly from the zero crossings of carrier C^{ref1}. In a similar way the conditions for QWS at other intersection points can also be derived and every pair of intersection points will result in the condition of (16). This condition is also valid for carriers having frequency equal to 3n times the fundamental frequency.

b) Approach II

Fig.5 shows the zero crossings of carrier C_{HB1} lead the zero crossings of carrier C^{ref2} by Φ_1 rad, whereas the zero crossings of carrier C_{HB2} lag the zero crossings of carrier C^{ref2} by Φ_2 rad for p=3. Fig.5.(b), (c) and (d) show that the individual bridge pole voltages V_{HB1} and V_{HB2} and resultant pole voltage V_{HB} do not maintain QWS. For maintaining QWS among resultant pole voltage patterns V_{HB}, the condition to be satisfied among voltage reference and carrier intersection points C_1 to C_{12} is given by (17).

$$\theta_{13-k}=\pi-\theta_k \quad \text{for k=1, 2, 3, 4, 5 and 6} \quad (17)$$

For k=1, the values of θ_1 and θ_{12}, at points C_1 and C_{12} can be found out by equating the equations of voltage references and carriers and can be written as (18) and (19) respectively.

$$-m\sin\theta_1=\left(\frac{6}{\pi}\right)\left(\theta_1-\frac{\pi}{6}+\phi_1\right) \quad (18)$$

978-1-4673-9551-9/16 $31.00 © 2016 IEEE

$$m \sin \theta_{12} = \left(\frac{6}{\pi}\right)\left(\theta_{12} - \frac{5\pi}{6} - \phi_2\right) \quad (19)$$

By putting the condition of QWS (17) in (19), (19) can be simplified as (20).

$$m \sin \theta_1 = \left(\frac{6}{\pi}\right)\left(\frac{\pi}{6} - \theta_1 - \phi_2\right) \quad (20)$$

By adding (18) and (20), the condition for QWS can be found out as (21).

$$\phi_1 = \phi_2 \quad (21)$$

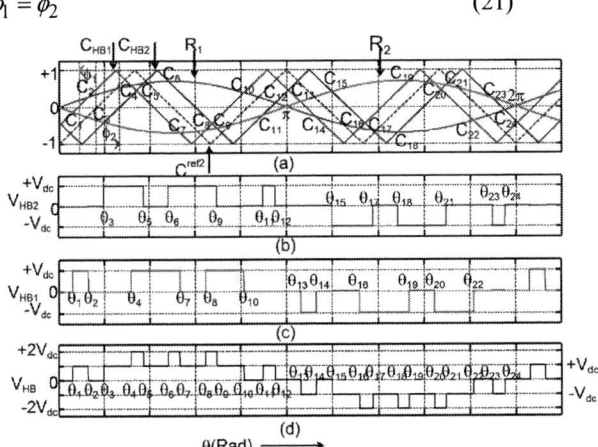

Fig. 5. References R_1 and R_2 and Carriers C_{HB1} and C_{HB2} when C^{ref2} is $\pi/2$ rad lagging w.r.t. voltage references (p=3)) (b) H-Bridge1 pole voltage V_{HB1} (c) H-Bridge2 pole voltage V_{HB2} (d) Resultant pole voltage V_{HB}.

From (21) it can be observed that, the QWS among resultant pole voltage patterns V_{HB} can be maintained if $\Phi_1 = \Phi_2$, i.e. the zero crossing of carriers C_{HB1} and C_{HB2} are equidistant from the zero crossings of carrier C^{ref2}. The condition of QWS can also be shown as $\Phi_1 = \Phi_2$, if the zero crossing of carrier C^{ref2} lead $\pi/2$ rad from the zero crossings of the voltage references. This condition is valid for carriers having frequency equal to 3n times the fundamental frequency.

3) Possible Carrier Positions for Maintaining QWS

For maintaining QWS among resultant pole voltage patterns V_{HB} in two cascaded HBs, several placements of carrier zero crossings w.r.t. voltage reference zero crossings are possible. For two cascaded HBs, the harmonic profile of resultant pole voltage V_{HB} is better when the carrier zero crossings are placed at $\pi/2$ rad apart from each other. Hence, for maintaining QWS, only those carriers are considered whose zero crossings are placed $\pi/2$ rad apart.

Table-I shows the various possible placements of zero crossings of carriers C_{HB1} and C_{HB2} w.r.t. the zero crossings of voltage references for maintaining QWS among resultant pole voltage V_{HB} in two cascaded HBs. Fig.6 shows the cases for p=9. Hence, the conditions for maintaining QWS among resultant pole voltage patterns V_{HB} in two cascaded HBs, where two carriers are placed $\pi/2$ rad apart, can be summarised as :-

➤ The zero crossings of the voltage references should coincide with the zero crossings of any carrier.

➤ The zero crossings of the voltage references should be placed exactly at the mid-point of the carriers zero crossings.

TABLE-I (5 Level CHBMLI)

Case	Position of C_{HB1}	Position of C_{HB2}
1	0 rad	$\pm\pi/2$ rad
2	$\pm\pi/4$ rad	$\mp\pi/4$ rad
3	$\pm\pi/4$ rad	$\pm 3\pi/4$ rad

Positive zero crossing of voltage reference is placed at 0 rad

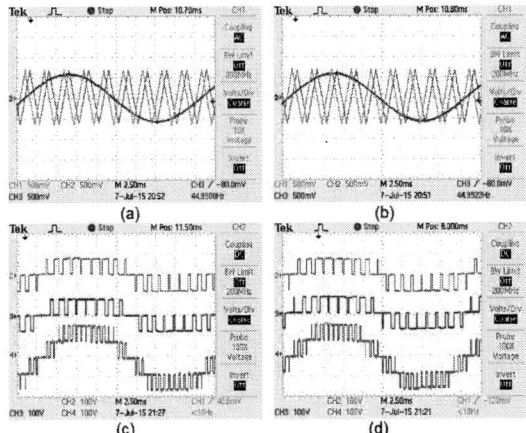

Fig. 6. (a) and (b) Ch.1: Carrier C_{HB1}, Ch.2: Carrier C_{HB2}, Ch.3: Reference R_1 (c) and (d) Ch.2: V_{HB1}, Ch.3: V_{HB2}, Ch.4: V_{HB} for modulation index=0.9 and p=9.

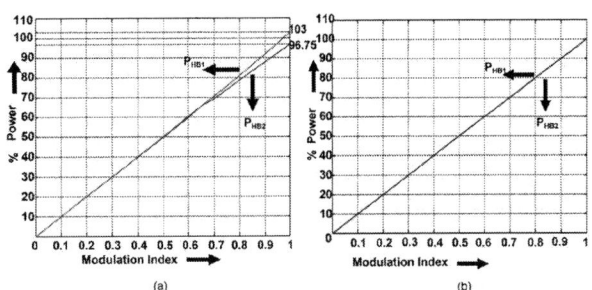

Fig. 7. Power distribution between the HBs (a) Zero crossings of reference R_1 coincide with the zero crossings of C_{HB2} and (b) Zero crossings of reference R_1 are placed at $\pm\pi/4$ rad w.r.t. the crossings of C_{HB1} and C_{HB2} for p=3.

But, the FFT of the different bridge voltages and the resultant leg voltage show that, as the ratio p decreases, the individual cells handle unequal power when the positive zero crossing of the voltage reference coincides with positive zero crossing of C_{HB2} after a modulation index of m_i=0.65, as shown in Fig.7. Hence, in this paper, the condition where the positive zero crossing of the voltage reference is placed in between the positive zero crossings of C_{HB1} and C_{HB2} is used for experimental verification.

Similarly, for three CHBs the conditions for maintaining QWS among pole voltages are tabulated in Table-II. Hence, in general the QWS in three cascaded H-Bridges can be maintained if the following conditions are valid.

➢ Any of the carriers is synchronized with the voltage references.

➢ The voltage references are placed at the mid-point between any two adjacent carriers.

TABLE-II (7 Level CHBMLI)

Case	Position of C_{HB1}	Position of C_{HB2}	Position of C_{HB3}
1	$\mp \pi/3$ rad	0 rad	$\pm \pi/3$ rad
2	$\mp \pi/6$ rad	$\pm \pi/6$ rad	$\pm \pi/2$ rad
3	$\mp \pi/6$ rad	$\mp \pi/2$ rad	$\mp 5\pi/6$ rad
4	0 rad	$\mp \pi/3$ rad	$\mp 2\pi/3$ rad

Positive zero crossing of voltage reference is placed at 0 rad

4) Generalized Conditions for maintaining QWS among the resultant pole voltage patterns of an n number of CHBs

In previous sections the conditions for maintaining QWS among resultant pole voltage V_{HB} in two CHBs are discussed analytically. The analytical discussions can be extended to n (where n may be odd/even) number of CHBs, where separate carriers are used for different HBs. The zero crossings of the carriers are separated from each other by a phase of π/n rad. Three cases are considered below for showing QWS among resultant pole voltage patterns for odd number (n=odd number) of HBs.

Case:I (Equal numbers of carriers are present around the carrier placed at 0th position)

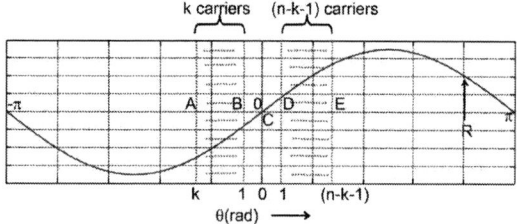

Fig. 8. Positive zero crossing of normalized pole voltage reference R coincides with the positive zero crossing of 0th carrier.

Fig.8. shows the positions of positive zero crossings of carriers w.r.t. the positive zero crossing of normalized pole voltage reference R for n (where n=1,3,5,7,9,....etc.) number of HBs. Positive zero crossing of R is synchronized with the positive zero crossing of the 0th carrier. It can be observed that k numbers of carriers are placed on left side and (n-k-1) numbers of carriers are placed on right side of positive zero crossing of 0th carrier.

The position of kth carrier should satisfy the constraint (22).

$$k\frac{\pi}{n}rad \le \frac{\pi}{2}rad \Rightarrow k \le \frac{n}{2} \qquad (22)$$

As n=odd, $k \ne n/2$ (k^{th} carrier cannot be present at $\pi/2$ rad). Hence k= (n-1)/2.

$$\left.\begin{array}{l} L.H.S = k = \dfrac{n-1}{2} \\[2mm] R.H.S. = n-k-1 = \dfrac{n-1}{2} \end{array}\right\} \qquad (23)$$

It can be observed from (23) that equal numbers of carriers are present on both sides of 0th carrier. Hence, the resultant pole voltage patterns maintain QWS.

Case:II (If k number of carriers are present on both sides of the carrier placed at 0th position, then the carriers present at $(k+1)^{th}$ and $(n-k)^{th}$ positions are placed equidistantly from the fictitious carrier placed at $\pi/2$ rad)

Fig.9. shows the positive zero crossing of normalized pole voltage reference R for n (where n=1,3,5,7,9,....etc.) number of HBs coincides with the positive zero crossing of 0th carrier. On both sides of the 0th carrier k numbers of carriers are placed between points A and B and D and E. On the right hand side of 0th carrier between points F and H (n-2k-1) numbers of carriers are present. For maintaining QWS among resultant pole voltage of (2n+1)level cascaded HB it can be shown that the $(n-2k-2)^{th}$ and $(n-2k-1)^{th}$ carriers are placed symmetrically w.r.t. $\pi/2$ rad (where $\pi/2$ rad is calculated w.r.t. carrier period). The distances GF and GH are calculated as $\pi/2n$ by putting k=(n-1)/2 in (24).

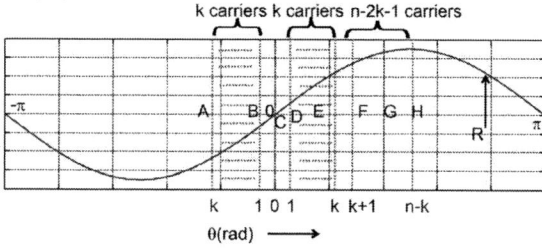

Fig. 9. Positive zero crossing of normalized pole voltage reference R coincides with the positive zero crossing of 0th carrier.

$$\left.\begin{array}{l} GF = \dfrac{\pi}{2} - \left\{(n-2k-2)\dfrac{\pi}{n} + k\dfrac{\pi}{n} + \dfrac{\pi}{n}\right\} = \dfrac{\pi}{2n} \\[3mm] GH = \left\{(n-2k-1)\dfrac{\pi}{n} + k\dfrac{\pi}{n} + \dfrac{\pi}{n}\right\} - \dfrac{\pi}{2} = \dfrac{\pi}{2n} \end{array}\right\} \qquad (24)$$

Case:III (If k numbers of carriers are present around the carrier placed in between 0th and 1st position, then the carrier at $(n-k)^{th}$ position coincides with the fictitious carrier placed at $\pi/2$ rad).The zero crossings of the 0th and 1st carrier are placed equidistantly on both sides of the zero crossing of the normalized reference.

Fig.10. shows the positive zero crossing of normalized voltage reference R is placed between the positive zero crossings B and C of 0th and 1st carriers respectively. On both sides of the positive zero crossing of R, k numbers of carriers are present between points A to B and C to D. On right side of positive zero crossing of R (n-2k) number of carriers are present between points E and F. For maintaining QWS among resultant pole voltage patterns it can be shown that the $(n-k)^{th}$ carrier is present at $\pi/2$ rad (where $\pi/2$ rad is calculated w.r.t. carrier period).

As the voltage reference positive zero crossing is placed in between the positive zero crossings of 1st and 0th carrier, it can be stated that k=(n-1)/2. Hence, the distance GF is calculated as given in (25). Hence, the $(n-k)^{th}$ carrier is placed at $\pi/2$ rad.

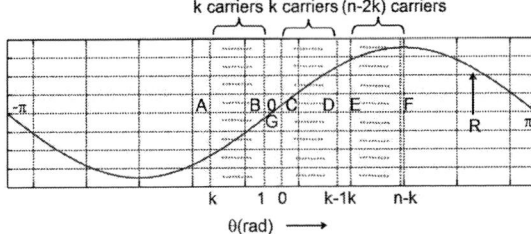

Fig. 10. Positive zero crossing of normalized pole voltage reference R is placed between the positive zero crossings of 0th and 1st carrier.

$$GF = \left\{ (n-k-1)\frac{\pi}{n} + \frac{\pi}{2n} \right\} = \frac{\pi}{2} \qquad (25)$$

All the above proofs are done for odd number of HBs. Similarly the proofs for even number of HBs can also be done. The conditions are:-

➤ Equal numbers of carriers are present around the 0th position.

➤ If k numbers of carriers are present on both sides of the carrier placed at 0th position, then the carrier present at $(n-k)^{th}$ position is placed at $\pi/2$ rad.

➤ If k numbers of carriers are present around the carrier placed in between 0th and 1st position, then the carriers present at $(n-k)^{th}$ and k^{th} positions are placed equidistantly from $\pi/2$ rad.

Hence, the generalized statements for maintaining QWS among pole voltage patterns for an n (where n may be even or odd) number of CHBs using PSPWM technique are summarized in Table-III.

TABLE-III

	CONDITIONS FOR QWS
1	Zero crossings of voltage references in phase with the zero crossing of any carrier
2	Zero crossings of voltage references are placed at the mid-point between any two adjacent carrier pair zero crossings

IV. EXPERIMENTAL VERIFICATION

The proposed PWM technique is experimentally verified with the help of a 3-Φ five level CHBMLI supplying a squirrel cage induction motor drive both in open loop Vbyf as well as closed loop vector control. The field weakening control similar to [8] is performed in order to validate the PWM scheme at higher fundamental frequencies. Fig.11 shows the variation of switching frequency f_c along with fundamental frequency f_s. The switching frequency varies as $500Hz$-$21f_s$-$15f_s$-$9f_s$-$3f_s$ as the fundamental frequency increases. Fig.12 shows the synchronous carrier generation scheme from the instantaneous voltage references. The transitions between the synchronous carriers occur at the zero crossings of a carrier having frequency $3f_s$.

Fig.13 shows the positions of carriers C_{HB1} and C_{HB2}, voltage reference R_1 and carrier C ($3f_s$) during the transition from p=9 to p=3. Fig.14 shows the response of R-phase current i_R during the transitions from (a) p=15 to p=9 and (b) p=9 to p=3 respectively, which signifies smooth transition

between carriers during open loop Vbyf control of a squirrel cage induction motor drive (f_s is changing from 0 to 55Hz). Fig.15 shows the steady state pole voltages V_{RO} and V_{YO} along with the R and Y-phase currents for (a) p=9 and (b) p=3 respectively.

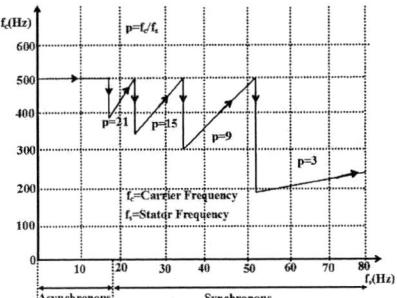

Fig. 11. Variation of f_c Vs. f_s.

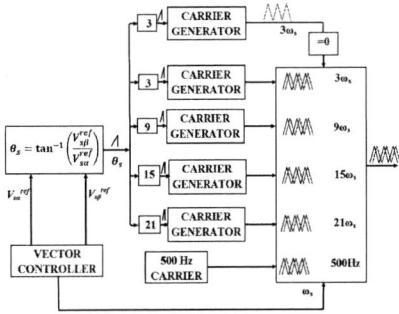

Fig. 12. Synchronous carrier generation.

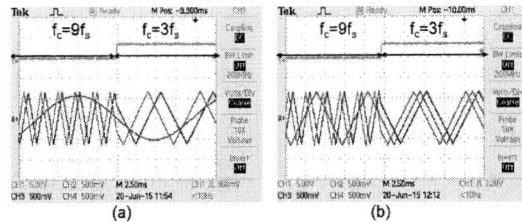

Fig. 13. (a) Ch.1: Transition signal, Ch.2: C_{HB1}, Ch.3: C_{HB2}, Ch.4:R_1 and (b) Ch.1: Transition signal, Ch.2: C_{HB1}, Ch.3: C_{HB2}, Ch.4:C ($3f_s$) for transition from p=9 to p=3 during open loop Vbyf control.

The field weakening scheme is implemented with GENESYS VERTEX-5 based FPGA Platform. The above proposed synchronized PSPWM technique is used to control the 5 level CHBMLI. Harmonic current estimation scheme similar to [8] is also implemented along with synchronous PWM scheme in FPGA platform. The experiments are performed on a 5H.P, 400V (L-L), 7.7Amp squirrel cage induction motor drive. The motor is a 6 pole squirrel cage induction motor having parameters:-$L_s=L_r=198.8mH$, $L_m=190mH$, $R_s=0.9\Omega$, $R_r=0.78\Omega$. The full load rated speed of the motor is 960R.P.M. For the sake of experimental verification and safety purpose, the rated motor voltage is assumed to be 150V r.m.s. line to line and having a phase current of 3.5Amp r.m.s. The dc link voltage of each bridge is kept at 75V during experiment. Fig.16.(a) shows the responses of actual speed(estimated speed) w.r.t. speed reference, torque and flux currents for a step speed reference of 1.5 p.u. Fig.16.(b) shows the variation of effective device switching frequency along with the modulation index for the above case. Fig.16.(c) shows the R-phase pole voltage along with R-phase current at a steady state speed of 1.5 p.u. Fig.16.(d) shows the responses of actual speed(estimated speed) w.r.t. speed reference, torque and flux currents for the speed reference changing as 0p.u-0.8p.u-1.5p.u.

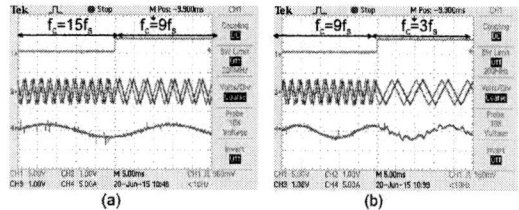

Fig. 14. Ch.1: Transition signal, Ch.2: C_{HB1}, Ch.3: C_{HB2}, Ch.4:i_R for transition from (a) p=15 to p=9 (b) p=9 to p=3 during open loop Vbyf control.

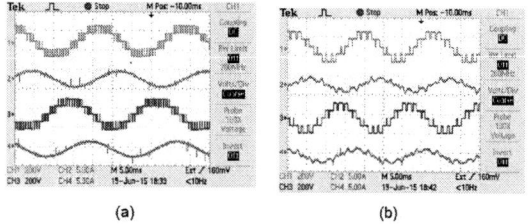

Fig. 15. Ch.1: V_{RO}, Ch.2: i_R, Ch.3: V_{YO}, Ch.4:i_Y for (a) p=9 (f_s=45Hz) (b) p=3 (f_s=55Hz) for modulation index 0.9 during open loop Vbyf control.

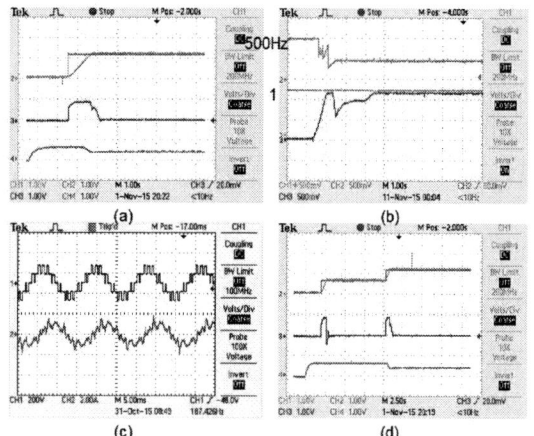

Fig. 16. (a) Ch.1:ω_m,Ch.2:ω_m^*,Ch.3:i_{sq},Ch.4:i_{mr} (b) Ch.2: Switching frequency variation, Ch.3:1, Ch.4: modulation index (c) steady state R-phase pole voltage V_{RO} and R-phase current i_R for a speed reference of 1.5 p.u (d) Ch.1:ω_m,Ch.2:ω_m^*,Ch.3:i_{sq},Ch.4:i_{mr} for speed reference changing as 0p.u-0.8 p.u-1.5 p.u.

V. CONCLUSION

This paper analytically shows the possible positions of zero crossings of the carriers with respect to the zero crossings of voltage references for the CHBMLIs using PSPWM technique for maintaining 3-Φ symmetry, HWS and QWS. 3-Φ and HW symmetries are maintained among HB pole voltages for any position of zero crossing of carrier w.r.t. the zero crossing of the voltage references, as synchronous carriers having frequency 3n (where n=1,3,5,7,9,11,......etc.) times the fundamental frequency are used for any HB. But the positions of zero crossings of the carriers w.r.t. to the zero crossings of voltage references are important for maintaining QWS among pole voltages. This is analytically studied in this paper for single and two cascaded HBs and generalized for any number of HBs. The study is experimentally verified with the help of a 3-Φ 5 level CHBMLI laboratory prototype for both the open loop Vbyf and closed loop vector control of a squirrel cage induction motor drive and the results are presented.

VI. REFERENCES

[1] G. Narayanan and V.T. Ranganathan, "Two novel synchronized bus-clamping PWM strategies based on space vector approach for high power drives," *IEEE Trans. Power.Electron.*, vol.17, no.1, pp.84-93,Jan-2002.

[2] Guiseppe S. Buja and Goivanni B. Indri, "Optimal Pulsewidth Modulation for Feeding AC Motors", *IEEE Trans.Ind.Appl.*,vol.IA-13,no.1,pp.38-44,Jan./Feb.1977.

[3] A.Edpuganti and A.K.Rathore, "Optimal Low-Switching Frequency Pulsewidth Modulation of Medium Voltage Seven-Level Cascade-5/3H Inverter," *IEEE Trans.Power.Electron* ,vol.30, no.1, pp.496-503, Jan.2015.

[4] A.K.Rathore, J.Holtz and T.Boller, "Synchronous Optimal Pulsewidth Modulation for Low-Switching-Frequency Control of Medium-Voltage Multilevel Inverters," *IEEE Transactions on Industrial Electronics,*, vol.57, no.7, pp.2374-2381, July 2010.

[5] T.Geyer, N.Oikonomou, G.Papafotiou and F.D.Kieferndorf, "Model Predictive Pulse Pattern Control," *IEEE Transactions on Industry Applications* , vol.48, no.2, pp.663-676, March-April 2012.

[6] Law Kah Haw, M.S.A.Dahidah and Haider.A.F.A , "SHE- PWM Cascaded Multilevel Inverter With Adjustable DC Voltage Levels Control for STATCOM Applications," *IEEE Trans.Power.Electron* , vol.29, no.12, pp.6433-6444, Dec.2014.

[7] J.Napoles, J.I.Leon, R.Portillo, L.G. Franquelo and M.A.Aguirre, "Selective Harmonic Mitigation Technique for High-Power Converters,"*IEEE Trans.Ind.Appl.*, vol.57, no.7, pp.2315-2323, July - 2010.

[8] S.K.Sahoo, T.Bhattacharya and M. Aravind, "A synchronized sinusoidal PWM based rotor flux oriented controlled induction motor drive for traction application," *in Proc. Twenty-Eighth Annual IEEE Applied Power Electronics Conference and Exposition (APEC), 2013* , vol., no., pp.797-804, 17-21 March-2013.

Design and Implementation of a Sinusoidal Flux Controller for Core Loss Measurements

Burak Tekgun   Ali R. Boynuegri   Md Asif Mahmood Chowdhury   Yilmaz Sozer

Department of Electrical and Computer Engineering
The University of Akron
Akron, Ohio, USA

Abstract—In this paper, design and implementation of a sinusoidal flux controller has been proposed for a core loss tester for eliminating the higher order harmonics from the flux passing through the core. The core loss test is performed with a toroidal transformer that consists of a main and sense windings wound on a toroidal core. The flux through the core is calculated with the numerical integration of the sense coil voltage. The controller commands the voltage applied to the main winding to keep the sense coil voltage thus the flux waveform sinusoidal. A single-phase SiC inverter operating at 150 kHz has been developed to generate the waveforms, an LC filter is used between the converter and the toroidal transformer to smooth the pulsating inverter output voltages. To be able to achieve zero steady state error and fast tracking as well as being robust to the periodic errors, a new controller is proposed. The proposed controller includes the periodic, conventional feedback and feedforward controllers and coordinates them. The designed controller has been simulated and it has been found that the flux waveform tracks the reference fast enough with a satisfactory steady state error. The system has been tested through experiments and experimental results are in good agreement with the simulation results.

Keywords—flux control, SiC inverter, core loss, periodic control.

I. INTRODUCTION

Achieving higher efficiencies while meeting design specifications is one of the most critical purpose for motor designers. Over the years, there have been plenty of researchers working on the loss segregation of electrical machines [1-3]. Some of the loss components such as core losses associated with the magnetic field distortion, rotor AC fields, rotating magnetic fields are hard to determine due to the complexity of their nature [3-6]. These losses might add up to 2% of the total output power in some cases. The analysis, prediction and consequent reduction of the mentioned losses is extremely important for designing high efficiency electric machines [4].

Many researchers have spent significant effort to make the accurate measurement of the core losses [7,8] and proposed variety of methods. One of the methods in measuring these losses is the thermal approach, which uses the isolated chamber to measure the temperature difference between inlet and outlet coolant [9]. This method is considered as universal; however, segregation of copper loss is hard and time consuming.

Another method was proposed for high frequency core loss measurements that use the resonance of the series connected capacitor and the inductor wound on a core under the test [10]. Hence, the power taken from the source has only the active components, which are copper and core losses. However, the method is only valid for sinusoidal voltages. Four-wire method is also very commonly used core loss measurement technique [11]. Primary winding is used for the excitation and secondary winding is used for sensing the induced voltage. The core loss can be calculated by integration of the product of the sensed voltage and the current passing through the primary winding [8, 12,13]. The method provides very fast and accurate BH loop determination through the sense coil voltage and the primary winding current. On the other hand, the method has also some drawbacks such as excluding the winding loss from the measured loss and the phase discrepancy due to the physical limitations of the current measurement devices. It is also very much sensitive to phase discrepancy due to 90☐ phase difference between primary winding current and secondary winding voltage [14, 15]. Couple of methods are also proposed to get rid of this phase discrepancy issue through the calibration of the measurements accordingly [8,16]. Finite element analysis (FEA) tools are also used to calculate the flux in the electric machines in determining the core losses [5,17-19]. Calculation of the core losses is typically carried out based on Steinmetz equations and parameters [20].

In core loss measurement, it is important to apply the flux waveform at the specified frequency without additional harmonics. For a sinusoidal voltage excitation, the flux passing through the magnetic material becomes distorted and higher order harmonics are added to the flux waveform due to saturation of the core. Analyzing core losses using saturated flux waveforms require core loss calculation for each harmonic frequency component [17-19]. Even though the method sounds straightforward, the calculations are very sensitive to the phase angles and magnitudes of the harmonics, small parameter errors cause high errors in the calculations [12].

We propose a core loss test system and the associated controller, which can directly control the flux passing through the magnetic core. The desired flux waveform at the specific frequency can be generated without having additional harmonics through the proposed system. Therefore, the core loss analysis can be done in each frequency at a time without worrying about the additional harmonic components.

978-1-4673-9551-9/16 $31.00 © 2016 IEEE

In Section II, the modeling and design of the proposed flux controller is provided. Section III provides the simulation results, the experimental verifications are presented in Section IV. The summary of the work is provided in Section V.

II. Modeling and Design of The Sinusoidal Flux Controller

The proposed system is mainly designed for the BH loop core loss test [12], and it consists of a single phase high frequency SiC based inverter, an *LC* filter and a toroid which has evenly distributed two windings (see Fig. 1)

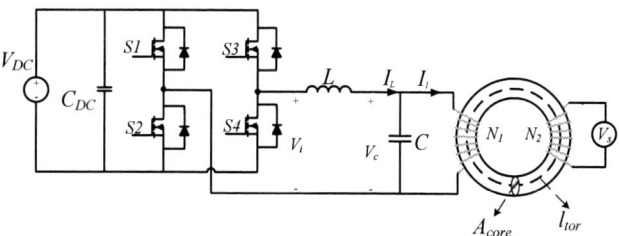

Figure 1. BH loop core loss test setup

The H-bridge SiC inverter operating at 150 kHz to synthesize flux waveforms at a range of fundamental frequencies up to 10 kHz. Pulsed PWM voltages generated at the output of the inverter are filtered through an *LC* filter.

The flux linkage λ is estimated through the integration of the voltage, sensed from the sense coil, and then flux density *B* can be calculated as:

$$B = \frac{1}{N_2 A_{core}} \underbrace{\int V_s(t)\,dt}_{\lambda} \qquad (1)$$

where, N_2 is the number of turns on the sense coil, A_{core} is the cross sectional area of the core, V_s is the induced voltage on the sense winding.

Due to the nonlinear behavior of the core, sinusoidal output voltage itself will not be able to generate sinusoidal flux when the core is saturated. In most cases the core is saturated to reach required flux density levels beyond 2 T for some cases [21]. Controlling only the voltage of the main winding will not result in a sinusoidal flux through the core. Direct flux measurement would provide better control, but still requires higher performance from the controller especially during the saturation.

Due to the poor performance to the periodic disturbances, closed loop controllers are not enough to compensate such systems [22]. Hence, a repetitive controller, which is based on internal model principle [23], is added to the system to eliminate the periodic distortions. Fig. 2 presents the block diagram of the proposed repetitive flux control structure. The digital control law, which is composed of three terms, is given as:

$$u(k) = u_{ff}(k) + u_{fb}(k) + u_{per}(k) \qquad (2)$$

The feedforward controller gain is selected based on the toroid transformer turns ratio. The feedback controller is a conventional feedback controller which can be selected as PI, deadbeat, state feedback, etc. which utilized to improve the dynamic response of the system while improving the stability margins.

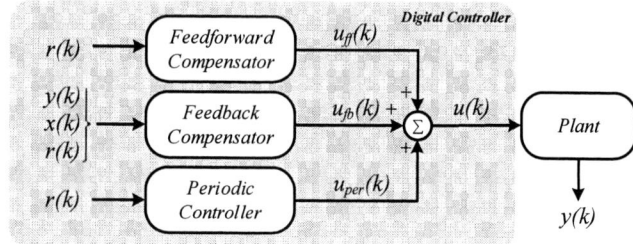

Figure 2. Simplified control structure

A. Modeling

In order to calculate the state feedback gains, state space model of the system needs to be developed. Capacitor voltage, filter inductor current and sense coil voltage have been selected as state variables and plant model is driven accordingly. Fig. 3 shows the equivalent circuit model of the core loss test system.

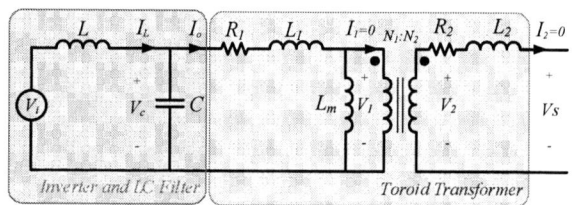

Figure 3. Equivalent circuit model

Here, *L* and *C* are the output filter inductor and capacitor, R_1 and L_1 are primary winding resistance, and leakage inductance, L_m is magnetizing inductance, R_2 and L_2 are secondary winding resistance and inductance.

The core loss test does not require any loading on the secondary winding; therefore, the secondary winding parameters are not included in the model and $V_s = V_2$. Hence, first two state equations will be as following:

$$\dot{i}_L = -\frac{1}{L}V_c + \frac{1}{L}V_i \qquad (3)$$

$$\dot{V}_c = \frac{1}{C}i_L - \frac{1}{R_1 C}V_c + \left(\frac{N_2}{N_1 R_1 C}\frac{L_1 + L_m}{L_m}\right)V_s \qquad (4)$$

The primary winding leakage inductance is much smaller (approx. 30 times) than the magnetizing inductance, and as there is no load on the secondary, we can combine L_m and L_1. The simplified circuit representation is shown in Fig. 4.

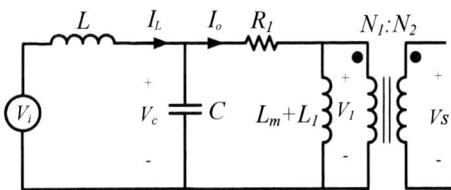

Figure 4. Simplified equivalent circuit model

Following equation is derived using the simplified equivalent circuit:

$$\dot{V}_s = \frac{N_2}{N_1 C} i_L - \frac{N_2}{N_1 R_1 C} V_C + \left(\frac{L_1 + L_m}{L_m + R_1 C} - \frac{R_1}{L_m + L_1} \right) V_s \quad (5)$$

Using the equations given above, system can be defined as following:

$$\dot{\mathbf{X}} = \mathbf{AX} + \mathbf{BU} \quad (6)$$
$$\mathbf{Y} = \mathbf{CX}$$

where,

$$\mathbf{X} = \begin{bmatrix} i_L & V_c & V_s \end{bmatrix}^T \quad (7)$$

$$\mathbf{U} = V_i \quad (8)$$

$$\mathbf{Y} = V_s \quad (9)$$

$$\mathbf{A} = \begin{bmatrix} 0 & -\dfrac{1}{L} & 0 \\[2mm] \dfrac{1}{C} & -\dfrac{1}{R_1 C} & \left(\dfrac{N_2}{N_1 R_1 C} \dfrac{L_1 + L_m}{L_m} \right) \\[2mm] \dfrac{N_2}{N_1 C} & \dfrac{N_2}{N_1 R_1 C} & \left(\dfrac{L_1 + L_m}{L_m + R_1 C} - \dfrac{R_1}{L_m + L_1} \right) \end{bmatrix} \quad (10)$$

$$\mathbf{B} = \begin{bmatrix} \dfrac{1}{L} & 0 & 0 \end{bmatrix}^T \quad (11)$$

$$\mathbf{C} = \begin{bmatrix} 0 & 0 & 1 \end{bmatrix} \quad (12)$$

The corresponding discrete time model can be driven as:

$$\mathbf{x}(k+1) = \mathbf{A_d}\mathbf{x}(k) + \mathbf{B_d}\mathbf{u}(k) \quad (13)$$

$$\mathbf{y}(k) = \mathbf{Cx}(k) \quad (14)$$

$$\mathbf{A_d} = e^{\mathbf{A}T_s} \quad (15)$$

$$\mathbf{B_d} = \mathbf{A}^{-1} \left(e^{\mathbf{A}T_s} - \mathbf{I} \right) \mathbf{B} \quad (16)$$

$$\mathbf{x}(k) = \begin{bmatrix} i_L(k) & V_c(k) & V_s(k) \end{bmatrix}^T \quad (17)$$

where, k being the sample number and T_s is the sampling time.

B. State Feedback Control Design

The control input signal is the voltage applied from the converter defined as:

$$\mathbf{u}(k) = V_i(k) \quad (18)$$

The control can be expressed as:

$$\mathbf{u}(k) = \mathbf{KX} \quad (19)$$

Here, K is the optimal control coefficients vector, and it can be calculated as linear quadratic optimal control.

$$\mathbf{K} = \begin{bmatrix} K_{p2,1} & K_{p2,2} & K_{p2,3} \end{bmatrix} \quad (20)$$

The cost function that will be used to find the optimal control can be expressed as

$$J = \int_0^\infty \left(\mathbf{x}^T \mathbf{Q} \mathbf{x} + \mathbf{u}^T \mathbf{R} \mathbf{u} \right) dt \quad (21)$$

where Q and R are positive semi definite and positive definite matrices, respectively. Correspondingly, the linear time invariant (LTI) static state feedback control can be calculated as:

$$\mathbf{K} = -\mathbf{R}^{-1} \mathbf{B}^T \mathbf{M} \quad (22)$$

where $M \geq 0$ is the solution of the following algebraic Riccati equation

$$\mathbf{A}^T \mathbf{M} + \mathbf{M} \mathbf{A} - \mathbf{M} \mathbf{B} \mathbf{R}^{-1} \mathbf{B}^T \mathbf{M} + \mathbf{Q} = 0 \quad (23)$$

C. Periodic Control Design

The periodic controller is based on the internal model principle and which guarantees that the plant output tracks the reference signal without having a steady state error as long as the closed loop system that generates the periodic signal is stable. In this work, periodic disturbances consist of more than one harmonic component. Hence, the periodic controller should be able to include the model of these signals [22]. Following periodic controller transfer function can generate periodic signals with harmonic components of the fundamental frequency.

$$\frac{u_{per}(z)}{e(z)} = \frac{K_{p1} z^{N-a}}{1 - Q(z) z^{-a}} \quad (24)$$

where, $e(z)$ is the z-transform of the error $e(k) = r(k) - y(k)$, K_{p1} is the gain of the periodic controller, N is the time advance step size, a is the number of delayed samples caused by the low-pass filter, and $Q(z)$ is the transfer function of a low pass filter. The low pass filter has to be included into the system to avoid the accumulation of the high frequency measurement noise, that way overall robustness of the system is also improved. The cutoff frequency of the low pass filter needs to be selected as high as possible to improve the sense voltage THD. If the cutoff frequency is low, the controller will not be able to act for the harmonic components which has higher frequencies than the cutoff frequency. The gain of the periodic controller effects the convergence time, higher the gain is faster the convergence. However, increasing the gain makes the system more vulnerable to the high frequency measurement noise and may cause instability.

The detailed structure of the controller presented in Fig. 5. It is designed such that the induced flux waveform will be sinusoidal. The reference voltage of the sense coil is calculated using (1). The error between the reference and the measured sense coil voltages is fed through the periodic compensator. The periodic controller brings the system closer to the point where the error is minimal. The state feedback is taken from the states compensating the non-periodic disturbances, providing a better tracking performance. The feedforward path is added to the controller for faster response during transitions.

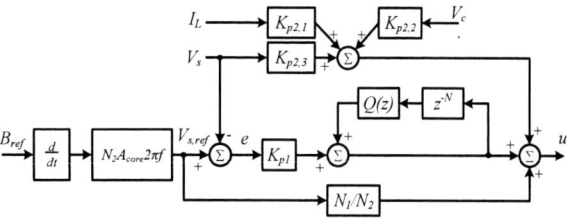

Figure 5. Proposed sinusoidal flux controller

III. SIMULATION RESULTS

To verify the effectiveness of the proposed control system, a set of simulations have been performed under periodic disturbances caused by the magnetic saturation in Matlab/Simulink environment.

At 200 Hz, the flux density in the core while the reference amplitude is 2.25 T is shown in Fig. 4. The error convergence of the sense coil voltage is provided in Fig. 5 where the sense coil voltage reaches to sinusoidal shape progressively. The THD of the resulting waveform after fourth cycle is 1.4%. Generated flux waveform is quite close to a pure sine wave.

After the progression of the cycles where the error gets minimum, the commanded voltage, primary winding voltage and current are presented in Fig. 6. As shown in this figure, although the secondary voltage is sinusoidal, inverter output quantities are highly distorted.

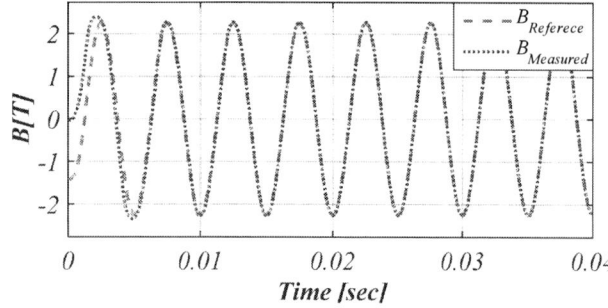

Figure 4. Flux density waveform under periodic distortion

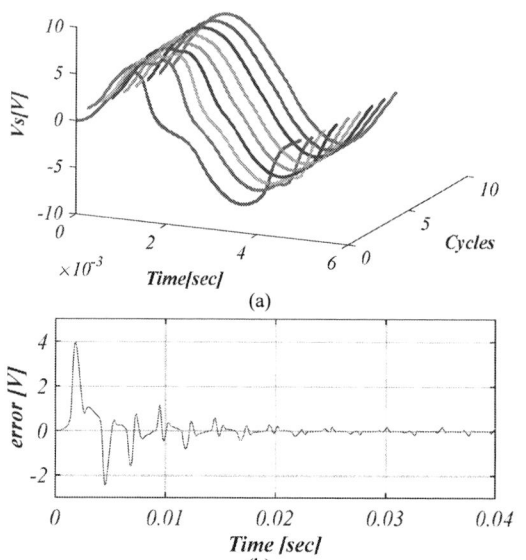

(a)

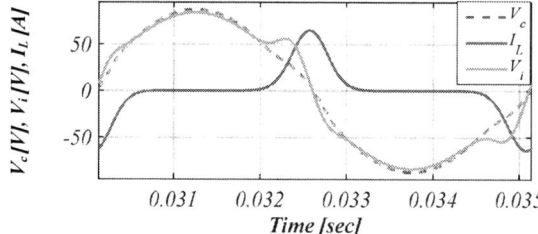

(b)

Figure 5. (a)Sense coil error voltage convergence in every cycle, (b) instantaneous error variation

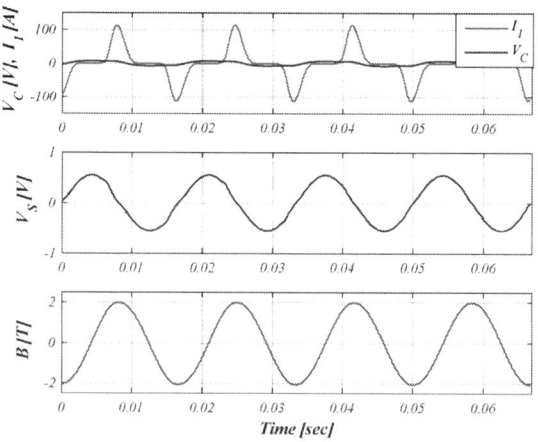

Figure 6. Output waveforms.

The control operation is tested at various output frequencies such as 60 Hz, 250 Hz, 400 Hz and 1000 Hz. Fig. 7 through Fig. 10 shows the primary winding voltage and current, sense coil voltage and magnetic flux densities at each given frequency. As shown from these figures, controller is able to generate sinusoidal flux waveforms through the core at various range of frequencies.

Figure 7. Primary winding current and primary winding voltage, sense coil voltage and calculated flux density waveform at 60 Hz

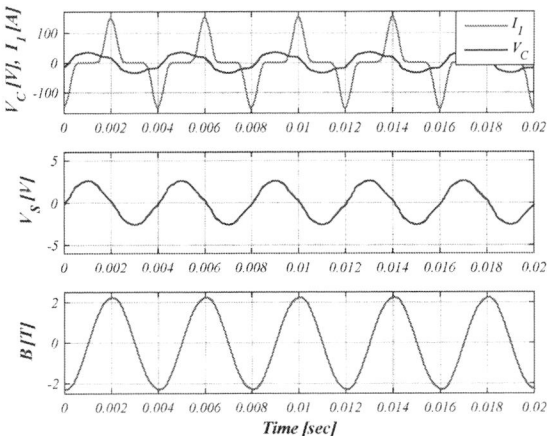

Figure 8. Primary winding current and primary winding voltage, sense coil voltage and calculated flux density waveform at 250 Hz

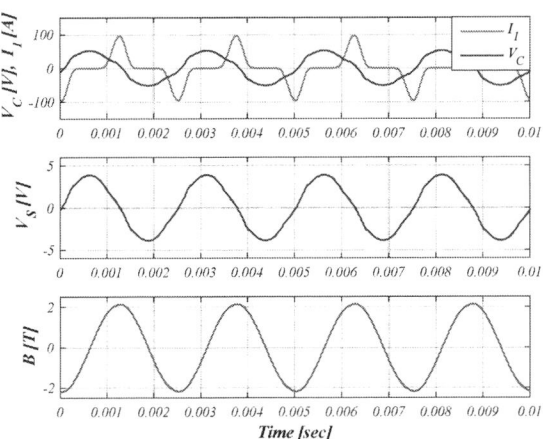

Figure 9. Primary winding current and primary winding voltage, sense coil voltage and calculated flux density waveform at 400 Hz

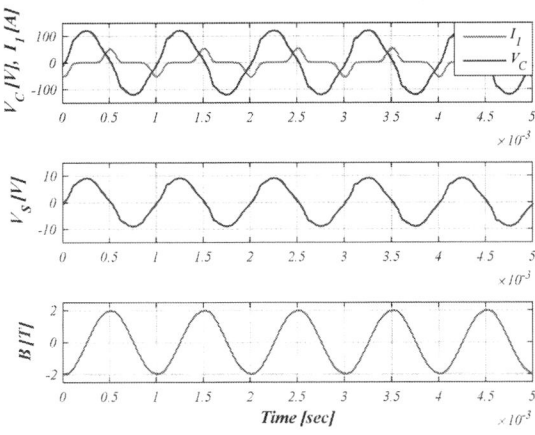

Figure 10. Primary winding current and primary winding voltage, sense coil voltage and calculated flux density waveform at 1000 Hz

IV. EXPERIMENTAL VERIFICATIONS

Experimental design includes hardware design, software coding and interfacing between them. The single-phase inverter is an H-bridge inverter operating at 150 kHz. Since the inverter operates at very high switching frequencies at high power levels, it is very important to reduce the parasitic inductance between the DC bus capacitors and mosfet modules to decrease the ringing effect. A printed circuit board (PCB) has been used as laminated bus bar for eliminating the parasitic inductance between the DC bus capacitors and mosfet modules. Distributed capacitors are connected through PCB between DC supply and switches ensure steady DC bus. In addition, variety of capacitors at different capacitance range have been placed close to the mosfet modules for further elimination of the parasitic inductance, ringing and overshoots.

Gate driver circuits have been designed for each switch individually; commercially available gate driver chips from IXYS have been used in the application. The gate driver has 3000V galvanic isolation and capable of providing 30 A peak currents for small amount of time. Protection circuits are also available to trip the system in case of any short circuit incidents and exceeding current limits. The mosfets were switched with an external 10 Ω resistor to ensure the fast turn on and turn off. Fig. 11 exhibits the experimental setup of the inverter system.

(a)

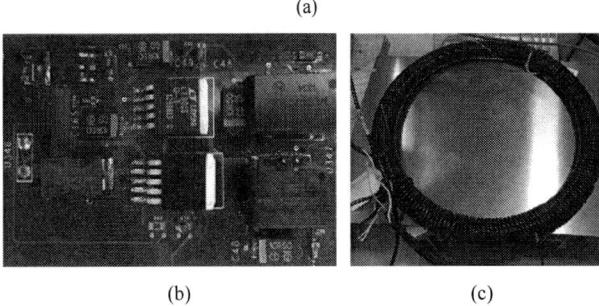

(b)   (c)

Figure 11. (a) Experimental setup, (b) gate driver circuit, (c) nonlinear inductive load

Texas Instruments floating point DSP model TMS320F28335 is used for control with a sampling time $T_s = 10$ μs ($f_s = 100$ kHz). The measured calculation time is less than 8 μs.

The core used in the experiments is laminated annealed Nucor type 6 steel, it has evenly distributed 400 turns on the primary and 30 turns on the secondary windings.

Fig. 12. (a) and (b) show the steady state experimental result taken at 60 Hz when the command flux amplitude is 2.0 T. The calculated flux density is given in Fig. 12.c which has a THD of 1.58%.

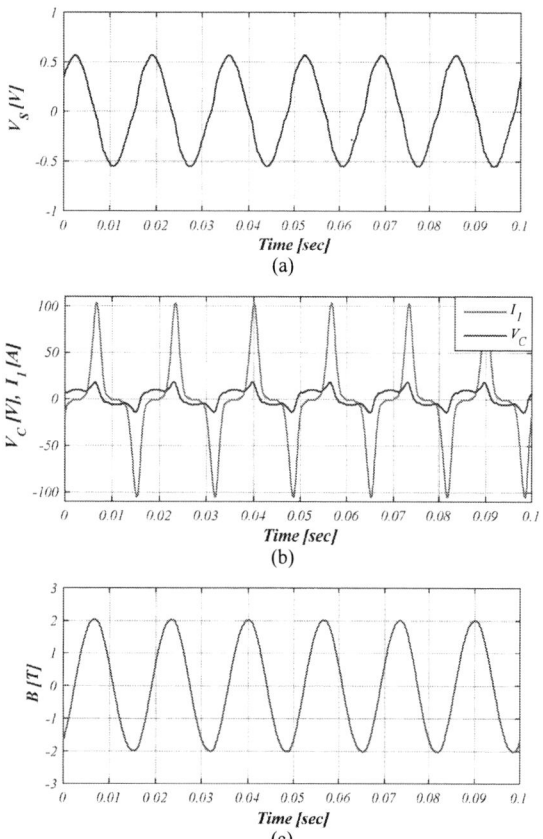

Figure 12 (a) Measured sense coil voltage and (b) corresponding primary winding current and primary winding voltage and (c) calculated flux density waveform at 60 Hz

Fig. 13 presents the steady state experimental results at 250 Hz. Fig. 13.a shows the sense coil voltage, Fig.13. (b) and (c) show the corresponding non-sinusoidal primary winding voltage and current waveforms, and calculated flux density waveform. Calculated THD is 1.01%. The peak amplitude of the current reaches to 135 A.

Fig. 14 exhibits the steady state results obtained at 400 Hz. In The sense coil voltage is given in Fig. 14 (a), Fig.14 (b) and (c) show the corresponding non-sinusoidal primary winding voltage and current waveforms, and flux density waveform. The THD of the flux density is calculated as 1.24%.

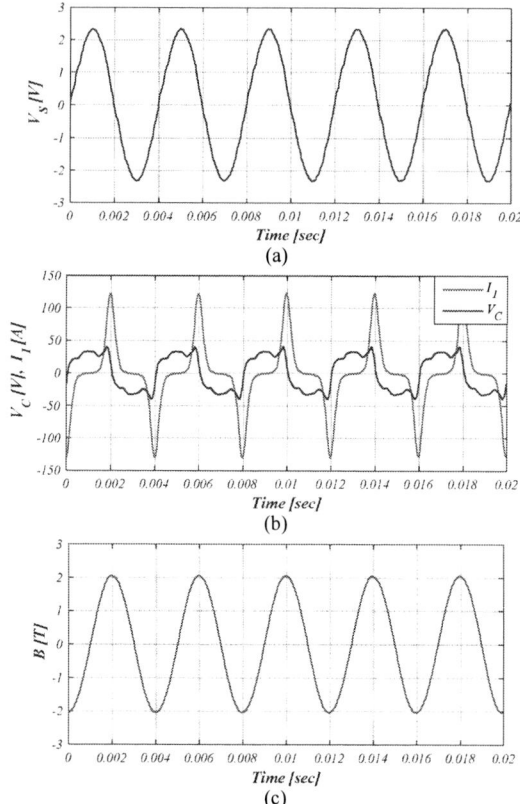

Figure 13 (a) Measured sense coil voltage and (b) corresponding primary winding current and primary winding voltage and (c) calculated flux density waveform at 250 Hz

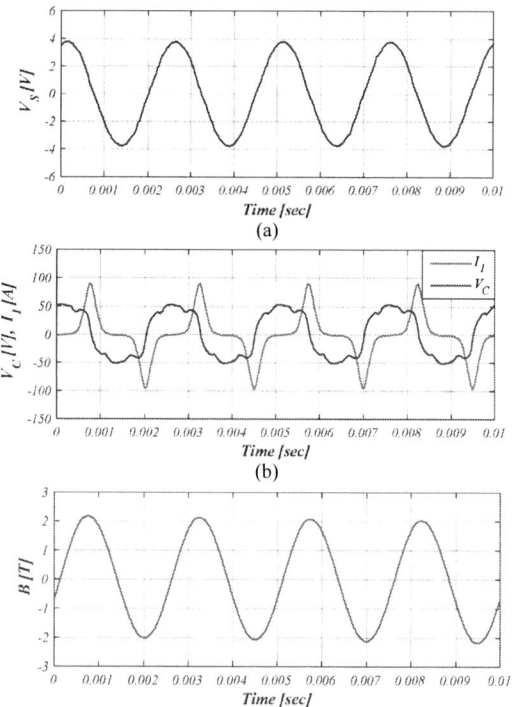

Figure 14 (a) Measured sense coil voltage and (b) corresponding primary winding current and primary winding voltage and (c) calculated flux density waveform at 400 Hz

The steady state results obtained at 1 kHz shown in Fig.15. The sense coil voltage provided in Fig. 15 (a), and the Fig. 15 (b) and (c) show the corresponding non-sinusoidal primary winding voltage and current waveforms, as well as the flux density waveform. The THD of the flux density is calculated 0.83%.

As it can be observed from the waveforms, the simulation and the experimental results are matching closely having a very low THD.

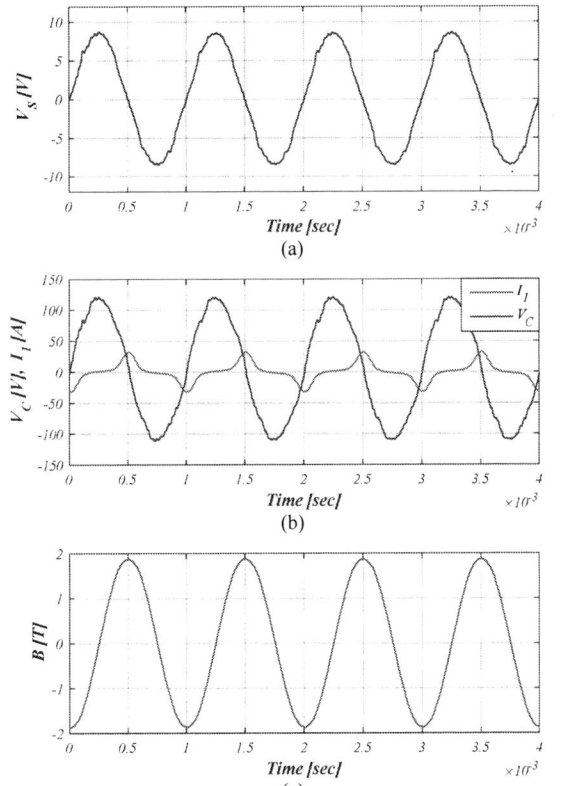

Figure 15 (a) Measured sense coil voltage and (b) corresponding primary winding current and primary winding voltage and (c) calculated flux density waveform at 1000 Hz

V. CONCLUSION

This paper has presented a flux control technique in order to ensure distortion free robust sinusoidal flux through the core under test, which is essential for measuring core loss precisely. The proposed technique keeps the core loss computation procedure simple while preserving plausible accuracy level. Proposed flux control scheme incorporates both feedback and feed-forward compensation along with the periodic controller. It exhibits satisfactory result in tracking the reference flux signal, moreover it is capable of acting fast enough to be implemented in SiC 20 kVA H-bridge inverter system operating at 150 kHz switching frequency. Measured THD levels of the flux density waveforms are not exceeding 1.58%. Therefore, generated signals are quite close to the pure sine wave.

The importance of having sinusoidal flux waveform in the electrical steel core gives us the freedom of performing the

core loss measurement and analysis in single frequency even in deep saturation regions. As a result, each harmonic frequency components of the flux waveforms at various parts of the electric machine can be generated, core losses can be measured. Then, these losses can be combined considering the volume and the mass. Therefore the total loss of the electric machine can be estimated with a good accuracy.

ACKNOWLEDGMENT

This research was supported by fellowships from ABB Corporate Research. The authors are also grateful to Baldor for providing steel cores for the experimental work.

REFERENCES

[1] J. Pippuri, and A. Arkkio, "Challenges in the segregation of losses in cage induction machines," *18th International Conf. on Electrical Machines,* pp. 1-5, 6-9 Sept. 2008.

[2] C.A. Hernandez-Aramburo, T.C. Green, and A.C. Smith, "Estimating rotational iron losses in an induction machine," *IEEE Trans. Magnetics,* vol. 39, no. 6, pp. 3527-3533, Nov. 2003.

[3] A. Boglietti, A. Cavagnino, L. Ferraris, M. Lazzari, "Induction Motor Equivalent Circuit Including the Stray Load Losses in the Machine Power Balance," *IEEE Trans. on Energy Conv.,* vol. 23, no. 3, pp.796-803, Sept. 2008.

[4] B. Tekgun, Y. Sozer, and I. Tsukerman, "Measurement of core losses in electrical steel in the saturation region under DC bias conditions," *IEEE, Applied Power Electron. Conf. and Expo. (APEC),* pp. 276-282, March 2015.

[5] N. Alatawneh, and P. Pillay, "Design of a Novel Test Fixture to Measure Rotational Core Losses in Machine Laminations," *IEEE Trans. on Ind. Appl.,* vol. 48, no. 5, pp. 1467-1477, Sept.-Oct. 2012.

[6] D. Son, J.D. Sievert, and Y. Cho, "Core loss measurements including higher harmonics of magnetic induction in electrical steel," *Journal of Magnetism and Magnetic Materials,* vol. 160, pp. 65-67, July 1996.

[7] X. Chucheng, C. Gang, and W. G. H. Odendaal, "Overview of Power Loss Measurement Techniques in Power Electron. Systems," *IEEE Trans. on Ind. Appl.,* vol. 43, pp. 657-664, 2007.

[8] M. Mingkai, F. C. Lee, L. Qiang, D. Gilham, and K. D. T. Ngo, "A high frequency core loss measurement method for arbitrary excitations," *IEEE, Applied Power Electron. Conf. and Exposition (APEC),* pp. 157-162, 2011.

[9] R. Linkous, A. W. Kelley, and K. C. Armstrong, "An improved calorimeter for measuring the core loss of magnetic materials," *IEEE Applied Power Electron. Conf. and Exposition (APEC),* pp. 633-639 vol. 2, 2000.

[10] H. Yehui, G. Cheung, L. An, C. R. Sullivan, and D. J. Perreault, "Evaluation of magnetic materials for very high frequency power applications," *IEEE Power Electron. Specialists Conf.,* pp. 4270-4276, 2008.

[11] V. J. Thottuvelil, T. G. Wilson, and H. A. Owen, Jr., "High-frequency measurement techniques for magnetic cores," *IEEE Trans. Power Electron.,* vol. 5, pp. 41-53, 1990.

[12] H. Kosai, Z. Turgut, and J. Scofield, "Experimental Investigation of DC-Bias Related Core Losses in a Boost Inductor," *IEEE Trans. Magnetics,* vol. 49, pp. 4168-4171, 2013.

[13] C. P. Steinmetz, "On the Law of Hysteresis," *Trans. of the American Institute of Electrical Engineers,* vol. IX, pp. 1-64, 1984.

[14] W. A. Roshen, "A Practical, Accurate and Very General Core Loss Model for Nonsinusoidal Waveforms," *IEEE Trans. on Power Electron.,* vol. 22, pp. 30-40, 2007.

[15] Valchev, V. Cekov, and A. Van den Bossche, *Inductors and transformers for power electronics,* CRC press, 2005.

[16] C. A. Baguley, U. K. Madawala, and B. Carsten. "A new technique for measuring ferrite core loss under DC bias conditions." *IEEE Trans. on Magnetics,* vol. 44, pp. 4127-4130, 2008.

[17] D. M. Ionel, M. Popescu, S. J. Dellinger, T. J. E. Miller, R. J. Heideman, and M. I. McGilp, "Factors Affecting the Accurate Prediction of Iron

Losses in Electrical Machines," *IEEE International Conf. Electric Machines and Drives*, pp. 1625-1632, May 2005.

[18] N. Alatawneh, and P. Pillay, "Design of a Novel Test Fixture to Measure Rotational Core Losses in Machine Laminations," *IEEE Trans. on Ind. Appl.*, vol. 48, no. 5, pp.1467-1477, Sept.-Oct. 2012.

[19] M. Popescu, D.M. Ionel, A. Boglietti, A. Cavagnino, C. Cossar, and M.I. McGilp, "A General Model for Estimating the Laminated Steel Losses Under PWM Voltage Supply*," IEEE Trans. on Ind. Appl.*, vol. 46, no. 4, pp.1389-1396, July-Aug. 2010.

[20] J. Muhlethaler, J. Biela, J. W. Kolar, and A. Ecklebe, "Core Losses Under the DC Bias Condition Based on Steinmetz Parameters," *IEEE Trans. on Power Electron.*, vol. 27, pp. 953-963, 2012.

[21] E. Barbisio, F. Fiorillo, and C. Ragusa, "Predicting loss in magnetic steels under arbitrary induction waveform and with minor hysteresis loops," *IEEE Trans. on Magnetics*, vol. 40, pp. 1810-1819, 2004.

[22] C. Rech, and J.R. Pinheiro, "New repetitive control system of PWM inverters with improved dynamic performance under nonperiodic disturbances," *IEEE Power Electron. Specialists Conf.*, vol. 1, no., pp. 54-60 vol. 1, 20-25 June 2004.

[23] B. A. Francis, and W. M. Wonham, 'The internal model principle of control theory," *Automotica*, 12, pp. 457-465, May 1976.

Implementation of Deadbeat-Direct Torque and Flux Control for Synchronous Reluctance Machines to Minimize Loss Each Switching Period

Michael Saur, Francisco Ramos, Aday Perez, Dieter Gerling
Universität der Bundeswehr München
Neubiberg, Germany
Michael.Saur@unibw.de

Robert D. Lorenz
University of Wisconsin-Madison, WEMPEC
Madison, WI, USA
rdlorenz@wisc.edu

Abstract- **Deadbeat-direct torque and flux control (DB-DTFC) has been emerged as a high performance digital control law for induction machines and permanent magnet machines. Its key attributes are excellent dynamic performance, the use of only one single control law over a wide torque/speed range and direct applicability to overmodulation strategies. This paper extends prior research by presenting the implementation of DB-DTFC for synchronous reluctance machines (SynRMs). In addition, a control law for minimizing SynRM losses each switching period is proposed. Real-time implementation on a 5.5kW test bench is demonstrated.**

I. INTRODUCTION

Loss minimizing control (LMC) of SynRMs without considerable dynamic performance degradation is challenging [1]. The high saliency leads inherently to a slow direct-axis time constant. To reduce flux linkage and iron losses at higher speeds the direct-axis time constant can influence the dynamic performance considerably. Furthermore, at high speeds the voltage that can be applied in order to change the flux is limited by the voltage source inverter which physically limits the dynamic performance. In addition, the very strong magnetic saturation and cross-coupling phenomenon in SynRMs leads to eigenvalue migration and affects the tuning of current regulators. In [2] the current regulator gains were adapted with increasing saturation to improve the dynamic performance. Nevertheless, the bandwidth of current regulators is inherently limited. In the flux weakening region, saturation of current regulators has to be avoided and a modified vector control algorithm is needed. No general-purpose flux weakening algorithm has been proposed yet.

Conventional direct torque control (DTC) methods have been proposed for SynRMs [3]. However, the resulting high harmonic content in the phase currents and acoustic noise are essential drawbacks. Thus, DTC methods with space vector modulation (SVM) have been emerged and investigated [4,5,6]. Unfortunately, these methods do not decouple the manipulated inputs in the rotor reference frame which generally leads to dynamic performance degradation.

Deadbeat-direct torque and flux control (DB-DTFC) is a digital control law that has been proposed for induction machines [7] and for permanent magnet machines [8]. DB-DTFC's direct control of the rate of change of torque by

applying the torque differential equation leads to excellent dynamic performance in torque and flux simultaneously. Because DB-DTFC decouples the control of flux and torque it inherently enables loss minimizing control each switching period. DB-DTFC for SynRMs has been discussed briefly in [9], but the focus of that work was on experimental validation of loss models and steady state loss minimization capabilities.

In this paper, the implementation of DB-DTFC for SynRMs is presented and a control algorithm to minimize losses each switching period is proposed and evaluated.

The paper is structured as follows: After introducing DB-DTFC for SynRMs, the methodology to minimize losses each switching period is derived mathematically. The real-time LMC algorithm as well as the inherent full inverter utilization within DB-DTFC is explained. In the following section, dynamic flux linkage-trajectories of the well-known control strategies maximum torque-per-ampere (MTPA), maximum torque-per-voltage (MTPV) and maximum torque-per-losses (MTPL) are compared. In section V, the LMC algorithm is evaluated via simulation. Finally, in section VI experimental results are presented: the performance of the proposed control law is shown by a torque command response of measured current as well as estimated torque and flux.

II. DEADBEAT-DIRECT TORQUE AND FLUX CONTROL FOR SYNCHRONOUS RELUCTANCE MACHINES

The torque of SynRMs is calculated as the product of number of pole pairs p, the currents i_d and i_q and the apparent inductances $L_d(i_d,i_q)$, $L_q(i_d,i_q)$ as depicted in (1).

$$T = \frac{3}{2} p \left(L_d - L_q \right) i_d i_q \qquad (1)$$

The currents i_d and i_q can be expressed as $\Psi_d/L_d(i_d,i_q)$ and $\Psi_q/L_q(i_d,i_q)$ respectively. This leads to a flux linkage-based torque equation (2).

$$T = \frac{3}{2} p \frac{\left(L_d - L_q \right)}{L_d L_q} \psi_d \psi_q \qquad (2)$$

One of the main ideas of DB-DTFC is to control the rate-of-change of torque instead of only controlling torque magnitude. In order to do so, saturation in one switching period is assumed to be small for high switching frequencies and equation (2) is derived with respect to time.

$$\frac{dT}{dt} = \frac{3}{2} p \frac{(L_d - L_q)}{L_d L_q} \left(\dot{\psi}_d \psi_q + \psi_d \dot{\psi}_q \right) \tag{3}$$

The use of this differential torque equation is a clear distinction between DB-DTFC and conventional DTC methods.

The derivatives of the flux linkages Ψ_d and Ψ_q can be obtained by rearranging the general voltage equations of SynRMs into (4) and (5).

$$\dot{\psi}_d = \frac{\partial \psi_d}{\partial t} = u_{dt} = u_d - R \cdot i_d + \omega \cdot \psi_q \tag{4}$$

$$\dot{\psi}_q = \frac{\partial \psi_q}{\partial t} = u_{qt} = u_q - R \cdot i_q - \omega \cdot \psi_d \tag{5}$$

The voltages which change the flux are u_{dt} and u_{qt} and will be used throughout the remainder of this paper.

Inserting the expressions (4) and (5) into (3) leads to (6).

$$\frac{dT}{dt} = C \cdot \psi_q \left(u_d - R \cdot \frac{\psi_d}{L_d} + \omega \cdot \psi_q \right) + \\ C \cdot \psi_d \left(u_q - R \cdot \frac{\psi_q}{L_q} - \omega \cdot \psi_d \right) \tag{6}$$

Where C is given as:

$$C = \frac{3}{2} p \cdot \frac{(L_d - L_q)}{L_d \cdot L_q} $$

Equation (6) shows explicitly that the rate-of-change of torque can be written as a function of voltage. Since the voltage source inverter manipulates the voltage fed to the machine, this can be used to select an optimal voltage vector each switching period, i.e. the optimal Volt-seconds.

III. MINIMIZING LOSSES EACH SWITCHING PERIOD VIA DB-DTFC

In order to minimize losses each switching period, a real-time analytical loss computation methodology is developed and presented in this section.

A. Real-Time Determination of the Loss-Minimizing Flux Linkage

Machine losses are modelled as copper and iron losses. To obtain minimal computation time iron losses are described with the compact ordinary Steinmetz equation (OSE). Total machine losses can be written as (7):

$$P_{Losses} = \frac{3}{2} \cdot R \cdot I^2 + k_i \cdot \psi^{\beta} \cdot f^{\alpha} \tag{7}$$

The coefficients k_i=0.86, α=1.19 and β=2.17 of the Steinmetz equation are experimentally identified. Fig. 1 shows the agreement of measurement and model:

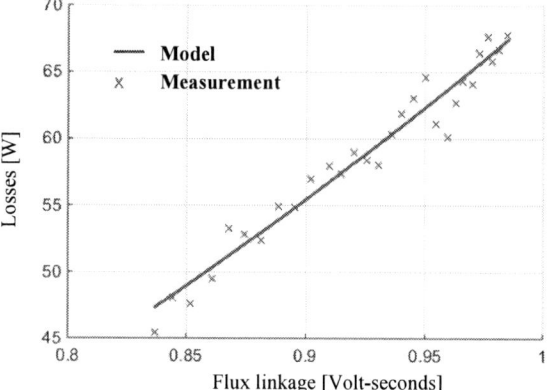

Fig. 1. Experimental identification of iron loss coefficients k_i=0.86, α=1.19 and β= 2.17 at n=1600rpm and T=17.5Nm

Since the coefficients vary over a wide torque speed range they are modelled as variables and can be adapted using a lookup-table or fitted polynomial.

DB-DTFC simultaneously and independently manipulates torque and flux. Thus, total machine losses have to be modelled based on flux linkage. Total current can be expressed as (8) and total flux as (9).

$$I^2 = \left(\frac{\psi_d}{L_d} \right)^2 + \left(\frac{\psi_q}{L_q} \right)^2 \tag{8}$$

$$\psi^2 = \psi_d^2 + \psi_q^2 \tag{9}$$

Rearranging the torque equation (2) and the total flux expression (9) and inserting into (8) total current can be written as a function of flux and torque as shown in (10).

$$I^2 = \frac{C \cdot \psi^2 + \sqrt{C^2 \cdot \psi^4 - 4 \cdot T^2}}{2 \cdot C \cdot L_q^2} \\ + \frac{2 \cdot T^2}{C \cdot L_d^2 \cdot \left(C \cdot \psi^2 + \sqrt{C^2 \cdot \psi^4 - 4 \cdot T^2} \right)} \tag{10}$$

Once the current is obtained as a function of the flux linkage, the total machine losses can be written as follows in (11).

$$P_{Losses} = \frac{3 \cdot R \cdot C \cdot \psi^2 + \sqrt{C^2 \cdot \psi^4 - 4 \cdot T^2}}{4 \cdot C \cdot L_q^2} \\ + \frac{3 \cdot R \cdot T^2}{C \cdot L_d^2 \cdot \left(C \cdot \psi^2 + \sqrt{C^2 \cdot \psi^4 - 4 \cdot T^2} \right)} \\ + k_i \cdot \psi^{\beta} \cdot f^{\alpha} \tag{11}$$

The loss minimizing flux linkage can be calculated by taking the derivative of (11) with respect to Ψ and setting it equal to zero as depicted in (12).

$$\frac{\partial P_{Losses}}{\partial \psi} = 0 \tag{12}$$

Solving (12), one obtains several solutions. One of them is in the positive half-plane and corresponds to the minimum total losses. It can be calculated with (13).

$$\psi = \sqrt{\frac{T \cdot \sqrt{\sigma_1} \cdot (3 \cdot L_d^2 \cdot R + 3 \cdot L_q^2 \cdot R + 4 \cdot k_i \cdot L_d^2 \cdot L_q^2 \cdot f^\alpha)}{C \cdot L_d \cdot L_q \cdot \sigma_1}}$$

Where σ_1 is given as: $\tag{13}$

$$\sigma_1 = \left(3 \cdot R + 2 \cdot k_i \cdot L_d^2 \cdot f^\alpha\right) \cdot \left(3 \cdot R + 2 \cdot k_i \cdot L_q^2 \cdot f^\alpha\right)$$

B. Solving the Inverse SynRM Model to Achieve Deadbeat Response Each Switching Period

In order to achieve the desired torque and flux in one switching period T_S, the inverse model of the SynRM has been solved in the discrete time domain. Equation (14) and (15) describe the discrete time flux linkage and differential torque.

$$\psi^2(k+1) = \left(\psi_d(k) + u_{dt} \cdot Ts\right)^2 + \left(\psi_q(k) + u_{qt} \cdot Ts\right)^2 \tag{14}$$

$$\frac{\Delta T}{Ts} = C \cdot u_{dt} \cdot \left(\psi_q(k) + \frac{u_{qt} \cdot Ts}{2}\right) + C \cdot u_{qt} \cdot \left(\psi_d(k) + \frac{u_{dt} \cdot Ts}{2}\right) \tag{15}$$

Inserting the loss minimizing flux linkage (13) and the discrete time flux linkage equation (14) into (15) and rearranging one obtains (16).

$$u_{dt} = -\frac{2 \cdot \psi_d(k) - \sqrt{C \cdot \psi^2(k+1) + \sqrt{\sigma_2}} \cdot \sqrt{\frac{2}{C}}}{2 \cdot Ts}$$

Where σ_2 is given as: $\tag{16}$

$$\sigma_2 = C^2 \cdot \psi^4(k+1) - 4 \cdot C^2 \cdot \psi_d^2(k) \cdot \psi_q^2(k)$$
$$- 8 \cdot C \cdot \Delta T \cdot \psi_d(k) \cdot \psi_q(k) - 4 \cdot \Delta T^2$$

After obtaining u_{dt} with (16), u_{qt} can be calculated using (17) depending on the sign of the desired torque $T(k+1)$ in the next step.

$$u_{qt} = -\frac{\psi_q(k) \pm \sqrt{\psi(k+1) + \psi_d(k) + Ts \cdot u_{dt}}}{Ts} \cdot \frac{\sqrt{\psi(k+1) - \psi_d(k) - Ts \cdot u_{dt}}}{Ts} \tag{17}$$

C. Maximizing Inverter Utilization and Dynamic Performance Utilizing Full DC-Bus Voltage

For any specific ΔT command, there exists an infinite number of voltage vectors which each achieve the desired change of torque but inherently lead to a different total flux in the machine. Fig. 2 shows an example of the loss distribution versus stator flux linkage.

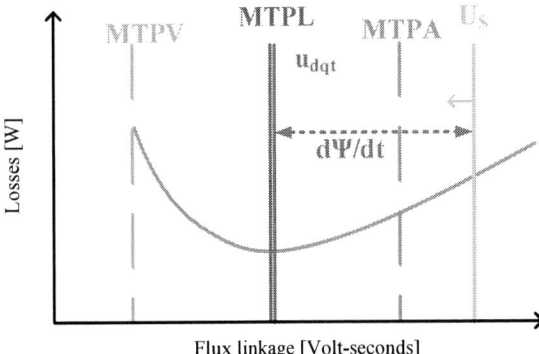

Fig. 2. Loss properties versus flux linkage for different LMC methods

MTPV (pale blue) achieves the desired torque with minimum flux linkage and therefore minimal iron losses. Current and copper losses are higher as compared to the MTPA solution (purple) which in contrast, applies a higher flux linkage. In between both strategies is the MTPL solution (brown) which causes minimum total losses.

The steady state voltage border U_s (orange) for this particular operating point corresponds to a specific flux linkage slightly higher than the MTPA flux linkage.

Depending on the amplitude of the torque and flux command, a change of flux and thus transient voltage u_{dqt} is needed. If the desired torque and flux change is not feasible in one switching period, a simple one step optimization method is applied: the torque command $(\Delta T / dT_S)$ is reduced to such an extent that the total voltage border $U_s + u_{dqt}$ (red) and the loss optimal MTPL flux coincide. This direct consideration of the inverter limits enables full inverter utilization and an improvement in the dynamic performance.

IV. DYNAMIC FLUX LINKAGE-TRAJECTORIES FOR MTPV, MTPA AND MTPL

Depending on the loss minimizing control strategy and on the operating point, different flux linkages are calculated. Fig. 3 shows a comparison of flux linkage trajectories for an applied torque command of 17.5Nm for MTPV (blue), MTPA (green) and MTPL (red).

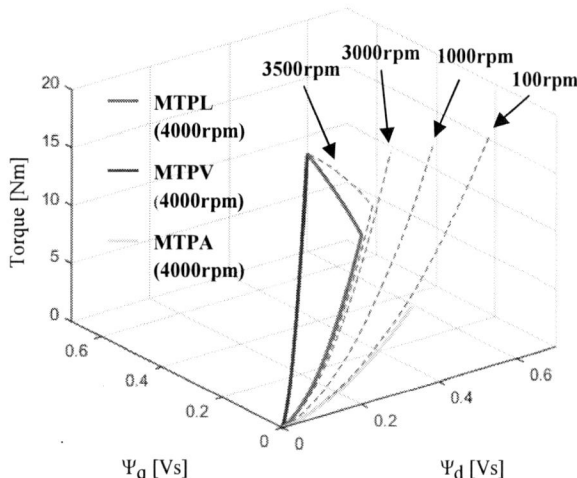

Fig. 3. Flux linkage-trajectories for MTPV (blue), MTPA (green) and MTPL (red) at n=4000rpm in the flux weakening region and MTPL (red dashed) for lower speeds

For the MTPV solution, d- and q-axis flux linkages are approximately equal. MTPA applies a higher d-axis and total flux and it is not possible to achieve the desired torque in the flux weakening region at 4000rpm. The MTPL flux linkage trajectory is located between MTPA and MTPV and moves towards the MTPV solution when approaching the inverter voltage limit. Furthermore it is shown that the MTPL solution gets closer to the MTPA solution when speed is reduced and iron losses decrease (red dashed lines).

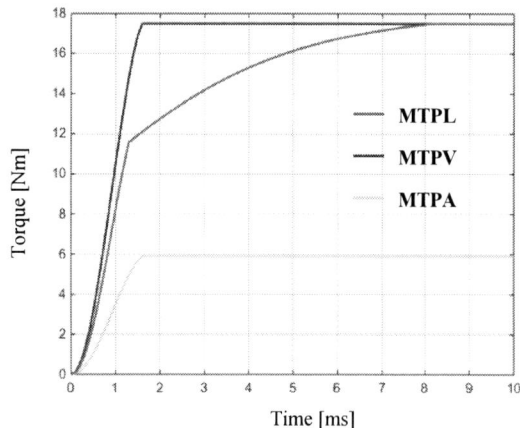

Fig. 4. Comparison of dynamic performance for MTPV (blue), MTPA (green) and MTPL (red) at n=4000rpm for a torque command of T_{rated}=17.5Nm

Fig. 4 shows the comparison of the dynamic response of the three control strategies at n=4000 rpm. The dynamic performance of MTPL is inferior to MTPV since a higher flux linkage is required. Approaching the inverter voltage limit, this effect leads to perceptible dynamic performance deterioration.

V. SIMULATION RESULTS

In this section, simulation results are presented to show the behavior of the proposed loss minimizing control law under various operating conditions. A torque command sequence is applied to investigate the dynamic response of torque, flux and current at two different speeds. The simulation model emulates typical test bench effects such as noise in the current and position measurements as well as parameter estimation errors. Fig. 5 shows the torque command sequence (green) and the torque response for n_{rated}=3000rpm (blue) and n=1000rpm (red).

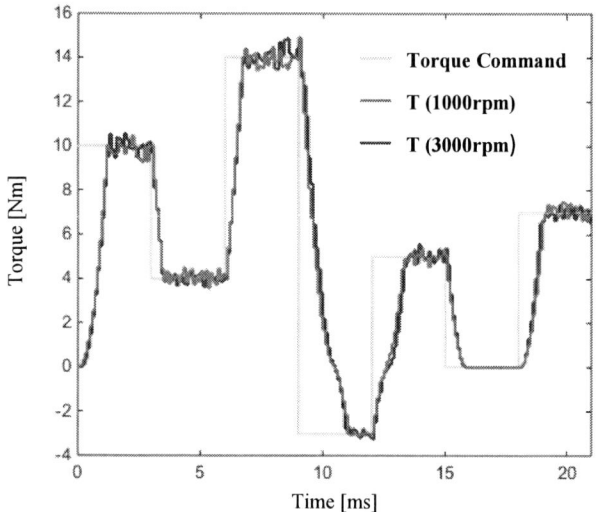

Fig. 5. Torque response for torque command sequence (green) of DB-DTFC MTPL at n=1000rpm (red) and n_{rated}=3000rpm (blue)

The dynamic performance for both speeds is approximately equal. At n=1000rpm the required loss minimizing flux linkages are higher but more transient voltage is available. Total fluxes for both speeds are shown in Fig. 6.

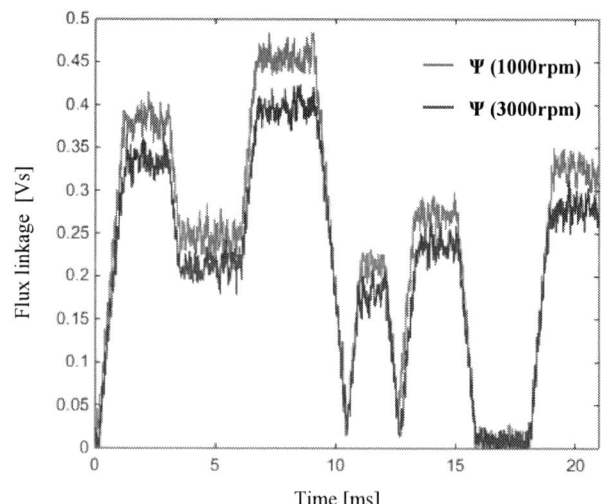

Fig. 6. Flux linkage response for torque command sequence of DB-DTFC MTPL at n=1000rpm (red) and n_{rated}=3000rpm (blue)

In Fig. 7 the d- and q-axis currents are depicted. It can be seen that for 1000rpm, the currents are closer to a MTPA solution.

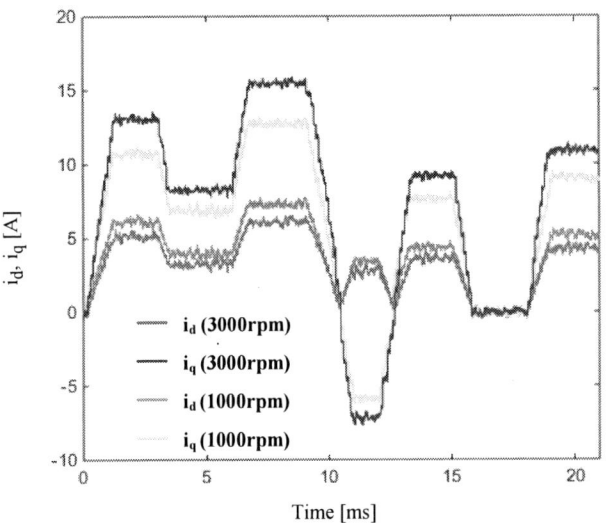

Fig. 7. Current response i_d, i_q for n=3000rpm (red, blue) and n_{rated}=1000rpm (magenta, pale blue)

VI. EXPERIMENTAL EVALUATION

A. Test Bench Configuration

The test bench setup shown in Fig. 8 consists of a 5.5kW SynRM test machine controlled with a dSpace MicroLabBox. The proposed LMC is implemented in real-time on a freescale QorIQ P5020 64-bit dual-core processor with 2 GHz CPU clock. Space vector modulation and current sampling are implemented on a Xilinx Kintex-7 FPGA. The algorithm is computed at a switching frequency of 10kHz.

The test machine is fed by a GustavKlein DC power supply and a GVA two level inverter. The DC load machine is fed by a separated DC power supply. Current is measured with LEM current sensors. Torque is measured at the shaft with a Magtrol torque transducer.

The dynamic responses of the measured current and estimated torque and flux linkage are recorded using the oscilloscope functions of Control Desk 5.4.

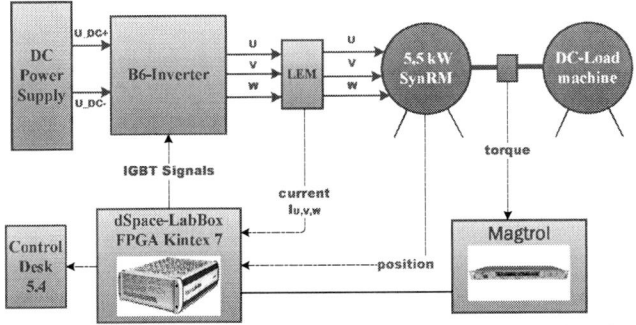

Fig. 8. Test bench setup consisting of a DB-DTFC controlled 5.5kW SynRM and a DC load machine

The SynRM parameters and mechanical characteristics are given in Table I:

TABLE I
PARAMETERS OF THE INVESTIGATED SYNCHRONOUS RELUCTANCE MACHINE

R_{S90}	0.357 Ω (at ϑ_0=20°C)
J_p	0.019 kgm^2
T_S	100 μs
Pole Pairs p	2
Rated Voltage	220 V
Rated Power	5.5 kW
Rated Torque	17.5 Nm
Rated Speed	3000 rpm

The inductances have been identified in [9] and are shown in Fig. 9:

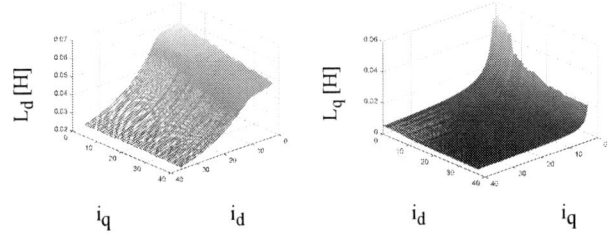

Fig. 9. Experimentally identified inductances $L_d(i_d,i_q)$ and $L_q(i_d,i_q)$ [9]

B. Experimental Results

Fig. 10 shows the estimated torque, three phase currents I_U, I_V, I_W as well as the currents i_d (red), i_q (green) and flux linkages Ψ_d (purple), Ψ_q (blue) in the rotary reference frame. A torque command step from 1Nm to 9Nm at a speed of 300rpm is applied and the responses are recorded for one second. DC-bus voltage has been reduced to U_{DC} =150V.

In Fig. 10b it is clearly demonstrated that a SynRM controlled by DB-DTFC can have regular sinusoidal steady state currents which are virtually the same as those with Current Vector Control (CVC). Not having a current regulator does not lead to an increase in current ripple or acoustic noise.

Fig. 10c shows the currents transformed to the dq-rotor reference frame. It can be seen that no current regulator is necessary to get a smooth and fast current response. The inherent decoupling of d- and q-axis applying the deadbeat control law is demonstrated.

Fig.10d depicts the flux linkage responses using a discrete time Gopinath-style flux linkage observer. Ψ_d and Ψ_q are the actual controlled states of DB-DTFC. For low speeds, the current model of the Gopinath-style flux linkage observer is parameter sensitive to the estimated inductances L_d and L_q. The saturation effects illustrated in Fig. 9 are considered with the Langevin function as discussed in [9]. Thus, even for low speeds such as n=0.1·n_{rated}=300rpm, smooth flux-linkages can be achieved.

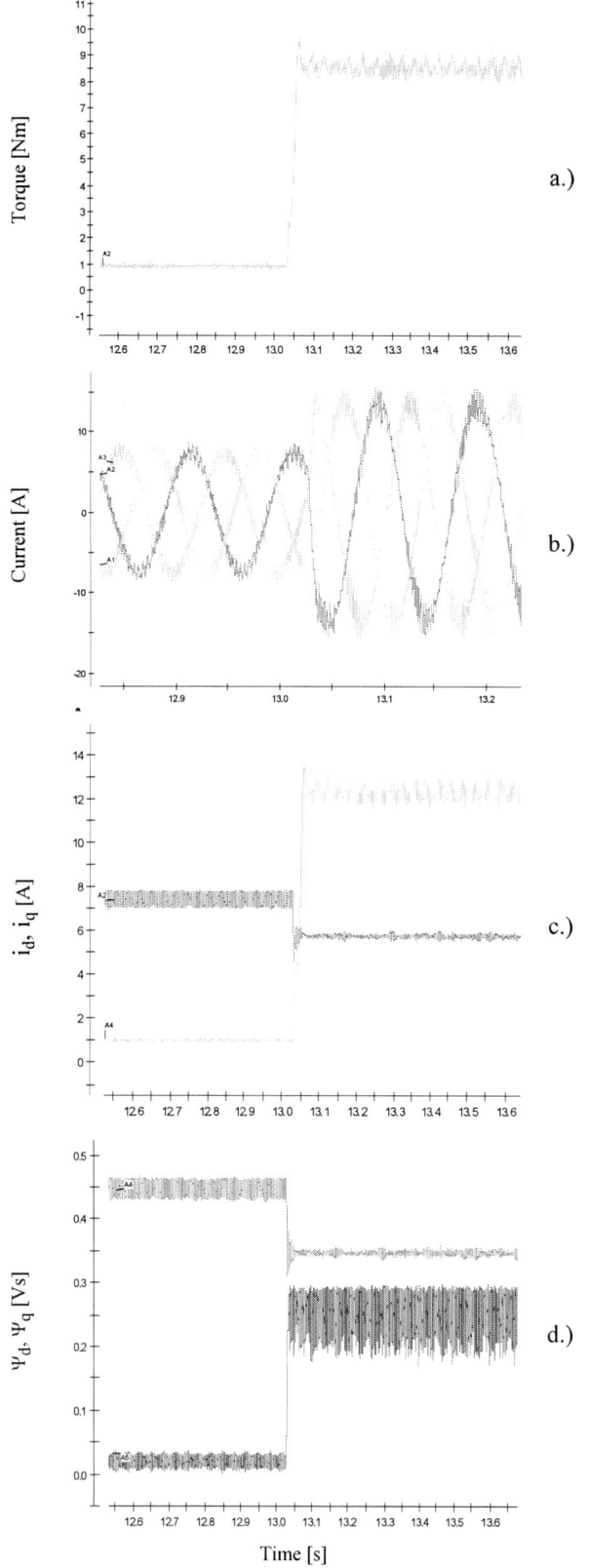

Fig. 10. Experimental results for estimated torque, measured three phase currents I_U, I_V, I_W, dq-reference frame currents i_d, i_q and flux linkages Ψ_d, Ψ_q for an applied torque command step from 1Nm to 9Nm at n=300rpm.

VII. CONCLUSIONS

This paper extends prior research by investigating deadbeat-direct torque and flux control for synchronous reluctance machines. The key conclusions of this work are:

- The application of DB-DTFC to SynRM drives including the utilization of the differential torque equation leads to a smooth and fast torque and flux response. Simulation and experiment validate excellent dynamic behavior.

- Experimental results show regular sinusoidal three phase currents at steady state. Current ripple is small compared to conventional DTC methods since SVM is used. Therefore, there is no acoustic noise increase as compared to CVC.

- A control algorithm for loss minimization each switching period is derived mathematically.

- The proposed algorithm has been implemented in real-time on a SynRM dynamometer test bench. Loss minimizing flux linkage computation together with the selection of the corresponding voltage vectors each switching period is demonstrated for T_S=100μs.

REFERENCES

[1] Boldea, I, "Reluctance Synchronous Machines and Drives", London, U.K.: Oxford Univ. Press, 1996

[2] Kilthau, A.; Pacas, J.M., "Parameter-measurement and control of the synchronous reluctance machine including cross saturation," Industry Applications Conference, 2001. Thirty-Sixth IAS Annual Meeting. Conference Record of the 2001 IEEE , vol.4, no., pp.2302,2309 vol.4, Sept. 30 2001-Oct. 4 2001

[3] Lagerquist, R.; Boldea, I.; Miller, T.J.E., "Sensorless-control of the synchronous reluctance motor," in Industry Applications, IEEE Transactions on , vol.30, no.3, pp.673-682, May/Jun 1994

[4] Morales-Caporal, R.; Pacas, M., "Encoderless Predictive Direct Torque Control for Synchronous Reluctance Machines at Very Low and Zero Speed," in Industrial Electronics, IEEE Transactions on , vol.55, no.12, pp.4408-4416, Dec. 2008

[5] Longya Xu; Xingyi Xu; Lipo, T.A.; Novotny, D.W., "Vector control of a synchronous reluctance motor including saturation and iron loss," in Industry Applications, IEEE Transactions on , vol.27, no.5, pp.977-985, Sep/Oct 1991

[6] Inoue, Y.; Morimoto, S.; Sanada, M., "A Novel Control Scheme for Maximum Power Operation of Synchronous Reluctance Motors Including Maximum Torque Per Flux Control," in *Industry Applications, IEEE Transactions on* , vol.47, no.1, pp.115-121, Jan.-Feb. 2011

[7] Kenny, B.H.; Lorenz, R.D., "Stator- and rotor-flux-based deadbeat direct torque control of induction machines," Industry Applications, IEEE Transactions on , vol.39, no.4, pp.1093,1101, July-Aug. 2003

[8] Jae Suk Lee, Chan-Hee Choi, Jul-Ki Seok and R.D. Lorenz, "Deadbeat-direct torque and flux control of interior permanent magnet synchronous machines with discrete time stator current and stator flux linkage observer," Industry Applications, IEEE Transactions on, vol. 47, no. 4, pp. 1749-1758 2011.

[9] Saur, M.; Lehner, B.; Hentschel, F.; Gerling, D.; Lorenz, R.D.: "DB-DTFC as Loss Minimizing Control for Synchronous Reluctance Drives"; 41st Annual Conference of the IEEE Industrial Electronics Society (IECON-2015), page 1412-1417, Nov. 2015

Addressing the Unbalance Loading Issue in Multi-Drive Systems with A DC-Link Modulation Scheme for Harmonic Reduction

Yongheng Yang[†], *IEEE Member*, Pooya Davari[†], *IEEE Member*, Firuz Zare[‡], *IEEE Senior Member*, and Frede Blaabjerg[†], *IEEE Fellow*

[†]Department of Energy Technology, Aalborg University, Aalborg 9220, Denmark
[‡]Department of EMC and Harmonics, Global R& D Center, Danfoss Power Electronics A/S, Gråsten 6300, Denmark
yoy@et.aau.dk; pda@et.aau.dk; fza@danfoss.dk; fbl@et.aau.dk

Abstract— Concerning cost, volume, and efficiency, Adjustable Speed Drive (ASD) systems are commonly employed with diode rectifiers or Silicon Controlled Rectifiers (SCR) as the front-ends (i.e., ac-dc converters). Apart from low cost, small volume, and high reliability, harmonic currents are significantly produced at the grid side by the rectification apparatus (i.e., diode rectifiers or SCR). As a simple strategy, the phase-shifted current control can be applied to the SCR-fed drive systems, where the input currents for the SCR are phase-shifted in such a way to cancel out the harmonics of interest (e.g., the 5th order harmonic). However, this solution is effective only when same amounts of currents are drawn by the rectifiers. Unfortunately, in practice, the loading is different among the parallel drive systems, leading to degradation in the harmonic mitigation. In this paper, unequal loading conditions in multi-drive systems are thus addressed. A load-adaptive scheme by means of varying the dc-link voltage is proposed, where a unified dc-link current modulation scheme is also employed in the dc-link. The proposed load-adaptive scheme can ensure that the rectified currents (i.e., rectifier output currents) are equal, and thus the harmonic reduction enabled by the phase-shifted current control is enhanced. The principle of the proposed method is demonstrated on a two-drive system consisting of a diode rectifier and a SCR. Experimental tests have validated the effectiveness of the proposed scheme in terms of harmonic mitigation for multiple parallel ASD systems.

I. INTRODUCTION

Currently, around 65% of the industrial electrical energy is consumed by motors, which thus is calling for energy-efficient motor drives [1], [2]. Substantial energy savings are enabled by means of variable speed drive systems [3], [4], where Diode Rectifiers (DR) or Silicon-Controlled Rectifiers (SCR) employed as the front-ends (i.e., ac-dc converters) are still popular [5]–[7]. This is mainly due to low cost, simple control, small volume, and high reliability during operation in contrast to their counterparts, e.g., active front-end based drive systems [8]–[11] and multi-pulse transformer based drive units [11]–[16]. However, beyond the above benefits, either the DR- or the SCR-fed drive systems bring significant distorted currents to the grid that is connected to [6], [17]. If the harmonic issue is not properly addressed, the overall efficiency will be affected, violating the harmonic regulations or guidelines [18], [19] and potentially inducing resonance in the entire system.

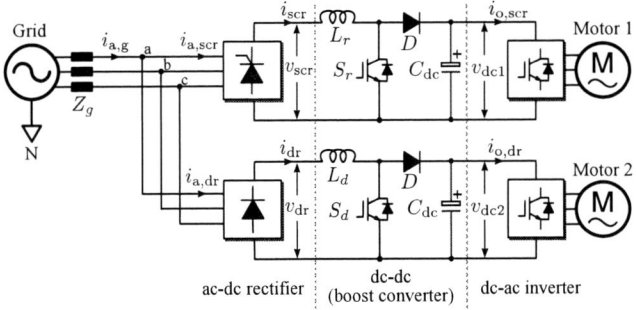

Fig. 1. Hardware schematic of a two-drive system consisting of a diode rectifier and a silicon-controlled rectifier, where dc-dc boost converters are adopted in the dc-links.

Typically, in the variable-frequency drive applications, the ac grid voltage is firstly rectified into a dc voltage, as it is exemplified in Fig. 1. It is observed that a dc-dc converter is adopted and placed into the dc-link in order to increase the control flexibility of the "uncontrolled" (DR) and the "half-controlled" (SCR) rectifiers, being a Power Factor Correction (PFC) circuit. This ac-dc configuration (i.e., ac-dc rectifier and dc-dc converter) offers one possibility to do proper modulations for the rectified currents (i.e., i_{scr} and i_{dr} in Fig. 1) in such a manner that the currents (e.g., $i_{a,scr}$ and $i_{a,dr}$ in Fig. 1) drawn from the grid by the rectifiers can be "modulated" (controlled), as it has been discussed in [2], [20] and [21]. At the same time, owing to the PFC system, it is also possible to control the dc-link voltage (i.e., $v_{dc1,2}$ in Fig. 1) to be constant (e.g., 700 V), which is independent of the actual grid voltage [5]. Nevertheless, with the PFC configuration demonstrated in Fig. 1, the total grid currents (e.g., $i_{a,g}$ in Fig. 1) can be modulated as multi-level, where certain targeted harmonics (e.g., the 5th order harmonic) can be mitigated in theory. This will contribute to improved quality of the grid currents in terms of a lower Total Harmonic Distortion (THD), which is demanded in relevant standards [18].

In addition, in the case of multi-drive systems consisting of parallel SCR-fed drives, by shifting the SCR currents, the total grid current quality can be further enhanced [17], being a phase-shifted current control [21]. Taking the two-drive system shown in Fig. 1 as an example, in particular, when the rectified currents (e.g., i_{scr} and i_{dr} in Fig. 1) are controlled as dc currents at the same level through dc-dc converters (e.g., a boost converter), the total grid current becomes multi-level. As a consequence, an even better THD of the grid currents is achieved. However, the effectiveness of this harmonic mitigation strategy is affected by the loading conditions of the parallel drive systems. That is to say, if the currents drawn by multiple parallel drive units are not equal, the harmonic cancellation enabled by the phase-shifted current control cannot be accomplished completely [21]. Moreover, in practice, all the drive units are rarely operating at the same loading condition. As a result, the harmonic mitigation enabled by the phase-shifted current control even with the dc-link modulation scheme will be degraded.

In order to address the above issues, this paper proposes a load-adaptive phase-shifted control scheme, which can ensure that each drive unit draws the same amount of currents from the grid. The principle of the load-adaptive scheme is demonstrated on a two-drive system referring to Fig. 1. Firstly, a unified dc-link modulation scheme aiming at selective harmonic cancellations is introduced in § II, followed by the load-adaptive scheme. Experimental tests have been conducted, and the results are presented in § IV, which validate the effectiveness of the load adaptive phase-shifted current control in terms of harmonic cancellations in multi-drive systems. Finally, § V draws the conclusions.

II. UNIFIED DC-LINK MODULATION SCHEME

As mentioned, for the drive system with a dc-dc converter in the dc-link, it is possible to apply advanced modulation schemes, which makes the system behave as a PFC system [5]. In that case, the dc-link voltage can also be adjusted to a constant level (e.g., 700 V). In order to implement the dc-link current modulation scheme, a desired current pattern should be designed and pre-programmed, as it is illustrated in Fig. 2, where it shows how the unified dc-link current modulation scheme is synthesized. Specifically, referring to Fig. 1, when assuming that the grid voltages (e.g., the line-line voltages v_{ab}, v_{bc}, and v_{ca}) are balance, the rectified output voltage (i.e., v_{scr} and v_{dr} in Fig. 1) will contain six identical segments with a conduction angle of 60° for each [22], [23]. As a consequence, it is feasible to synthesize the reference for the rectified output current (i.e., i_{scr} and i_{dr} in Fig. 1) by summing up the absolute three-phase desired current signals, as it is highlighted in Fig. 2. This is the principle of the unified current modulation scheme for the dc-link with a dc-dc converter. As long as a specific desired current signal is designed and implemented according to Fig. 2, the phase currents drawn by the rectifiers (e.g., $i_{a,scr}$ and $i_{a,dr}$ in Fig. 1) will follow the shape of the designed current signal. It explains the possibility to cancel out certain harmonics by designing a proper desired current signal. The following gives two modulation signals, which makes the grid currents "three-level" and "five-level", respectively.

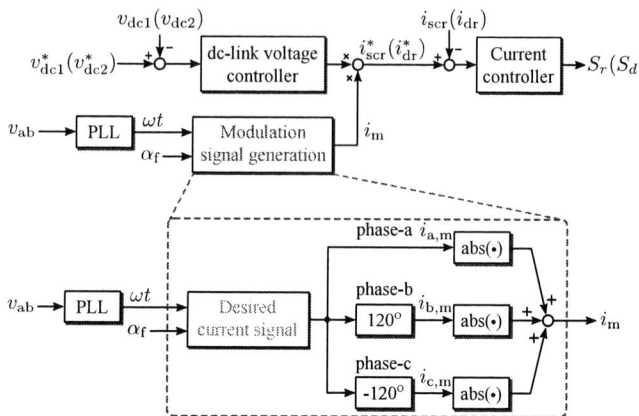

Fig. 2. Control block diagrams of the dc-dc boost converter as the dc-link in a SCR-fed or a DR-fed drive system (Fig. 1) with the unified dc-link current modulation scheme (PLL - Phase Locked Loop), where v_{ab} is the grid line-line voltage and α_f is the SCR firing angle (in the case of a DR-fed drive system, $\alpha_f = 0$).

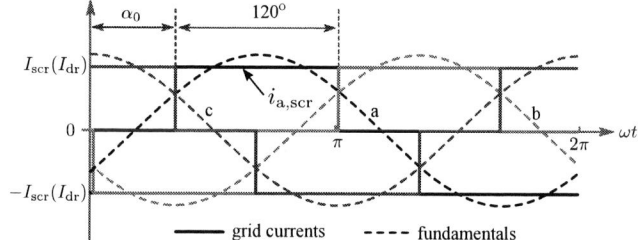

Fig. 3. Typical mains current waveforms of a SCR-fed or a DR-fed drive system when the rectified currents are controlled as purely dc, in which $\alpha_0 = \pi/6 + \alpha_f$ (in the case of a DR-fed drive system, $\alpha_f = 0$).

A. Square Modulation Signal

For the rectifiers shown in Fig. 1, if the rectified currents (e.g., i_{scr} and i_{dr}) are controlled as purely dc currents (denoted as I_{scr} and I_{dr}, respectively), the mains current will be of a square waveform [21], [22], as it is shown in Fig. 3. Hence, a simple modulation signal can be generated by taking the rectangular waveform of phase-a shown in Fig. 3 as the desired current signal that should be implemented according to Fig. 2. By doing so, the grid currents will be rectangular waveforms (see Fig. 3), but it brings significant distortions to the grid, which is further illustrated by means of a Fourier analysis.

According to Fig. 3, applying the Fourier analysis to a specific phase current of the SCR unit (e.g., $i_{a,scr}$) yields

$$i_{a,scr}(t) = \sum_{h}^{\infty} i_{a,scr}^{h}(t) = \sum_{h}^{\infty} \left[a^h \cos(h\omega t) + b^h \sin(h\omega t) \right]$$

(1)

in which $i_{a,scr}^{h}(t)$ is the hth order harmonic of the grid current $i_{a,scr}(t)$ drawn by the SCR, and a^h, b^h are the Fourier coefficients that can be calculated by

$$\begin{cases} a^h = \dfrac{2}{\pi} \displaystyle\int_0^\pi i_{a,scr}(t) \cos(h\omega t)\, d(\omega t) \\ b^h = \dfrac{2}{\pi} \displaystyle\int_0^\pi i_{a,scr}(t) \sin(h\omega t)\, d(\omega t) \end{cases}$$

(2)

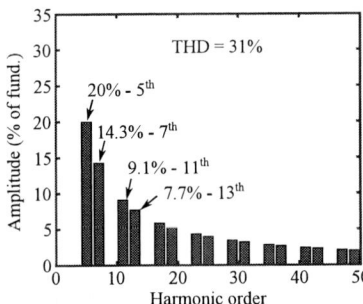

Fig. 4. Harmonic distributions of the rectangular grid currents shown in Fig. 3 when the dc-link modulation scheme is implemented according to Fig. 2.

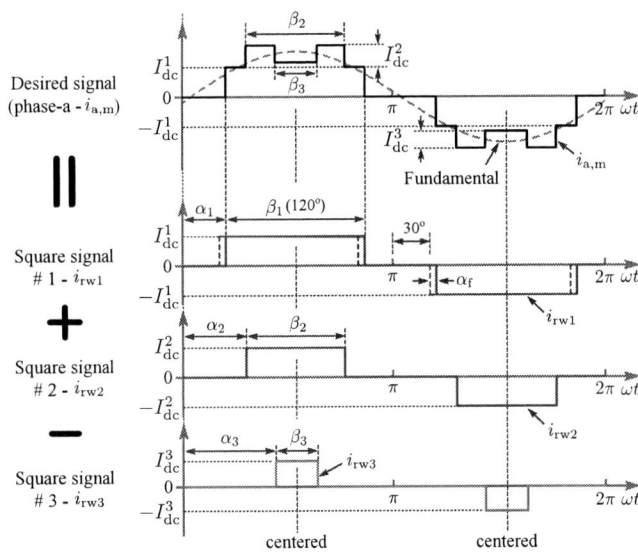

Fig. 5. Generation of a five-level modulation signal (phase-a) for the unified dc-link modulation scheme with the purpose of canceling out certain low-order harmonics using three rectangular signals (i.e., i_{rw1}, i_{rw2}, and i_{rw3}).

with $h = 1, 3, 5, \ldots$ being the harmonic order and ω being the angular grid frequency. Subsequently, the Root-Mean-Square (RMS) value of the hth order harmonic in the grid current can be obtained as

$$I_{a,\,scr}^h = \frac{\sqrt{2}}{2}\left[(a^h)^2 + (b^h)^2\right]^{1/2} \tag{3}$$

where $I_{a,\,scr}^h$ is the RMS magnitude of the hth harmonic current.

Accordingly, the harmonic distributions of the resultant rectangular grid currents discussed above can be obtained as demonstrated in Fig. 4. It can be observed in Fig. 4 that the square modulation signal implemented according to the unified dc-link current modulation scheme (see Fig. 2) will contribute to a poor current quality with the THD of 31%, although the grid currents become three-level. In particular, the low-order harmonics (e.g., the 5th, the 7th, the 11th, and the 13th) are significant, which however are not desired in the ASD applications. Hence, advanced schemes to mitigate these harmonics should be developed in order to achieve a satisfied power quality.

B. Five-Level Modulation Signal

In order to cancel out certain harmonics of the square waveforms in Fig. 3, a five-level modulation signal has been designed for the unified dc-link modulation scheme. Fig. 5 shows the generation process of the desired five-level modulation signal for phase-a in a SCR system, i.e., $i_{a,m}$. It can be identified in Fig. 5 that the five-level desired current signal is composed of three rectangular waveforms (i.e., $i_{a,m} = i_{rw1} + i_{rw2} - i_{rw3}$). Furthermore, it is shown that these square signals (i.e., #1 − i_{rw1}, #2 − i_{rw2}, and #3 − i_{rw3}) have a conduction angle of β_1 (120°), β_2, and β_3, a phase-shift of α_1, α_2, and α_3, and an amplitude of I_{dc}^1, I_{dc}^2, and I_{dc}^3, correspondingly. Additionally, for the DR-fed dc-dc converter, a five-level modulation signal can also be generated by setting $\alpha_f = 0$. That is to say, all the waveforms shown in Fig. 5 will be shifted back by a degree of α_f (i.e., $\alpha_1 = 30°$). It should be noted further that, in order to avoid triplen harmonics, the square signals (i_{rw2} and i_{rw3}) should be centered in respect to the square signal #1 − i_{rw1}, as annotated in Fig. 5. Moreover, the levels (i.e., $I_{dc}^1 + I_{dc}^2$) of the desired signal should be symmetrical, and thus

$$\beta_1 = \beta_2 + \beta_3 = 120° \tag{4}$$

which should be considered during the design phase as well as optimization of the five-level modulation signal.

According to Fig. 5, the hth order harmonic component of each rectangular waveform used to generate the five-level modulation signal can be identified as

$$i_{rwk}^h(t) = a_k^h \cos(h\omega t) + b_k^h \sin(h\omega t) \tag{5}$$

with $k = 1, 2, 3$ being the signal index, a_k^h and b_k^h being the corresponding Fourier coefficients. Substituting the k-th square signal (i.e., $i_{rwk}(t)$) into (2) to replace $i_{a,scr}(t)$ gives the Fourier coefficients as

$$\begin{cases} a_k^h = \dfrac{2I_{dc}^k}{h\pi}\left[-\sin(h\alpha_k) + \sin(h\alpha_k + h\beta_k)\right] \\ b_k^h = \dfrac{2I_{dc}^k}{h\pi}\left[\cos(h\alpha_k) - \cos(h\alpha_k + h\beta_k)\right] \end{cases} \tag{6}$$

where h is the harmonic order as defined previously.

As aforementioned, when the five-level modulation signal is implemented according to Fig. 2, the currents drawn from the grid (i.e., $i_{a,scr}$) will follow the desired five-level current signal. According to the superposition principle, Fig. 5, and (5), the harmonic component of the resultant grid phase-a current (i.e., $i_{a,scr}(t)$) will follow

$$i_{a,m}^h(t) = (a_0^h + a_1^h - a_2^h)\cos(h\omega t) + (b_0^h + b_1^h - b_2^h)\sin(h\omega t) \tag{7}$$

and its RMS magnitude can then be obtained as

$$I_{a,m}^h = \frac{\sqrt{2}}{2}\left[\left(a_0^h + a_1^h - a_2^h\right)^2 + \left(b_0^h + b_1^h - b_2^h\right)^2\right]^{1/2} \tag{8}$$

As a consequence, in order to eliminate certain harmonics with the desired five-level modulation scheme, the corresponding harmonic amplitude should be zero. Specifically, if $I_{a,m}^h = 0$ with $h \neq 1$ is solved, the hth order harmonic will be completely canceled in theory.

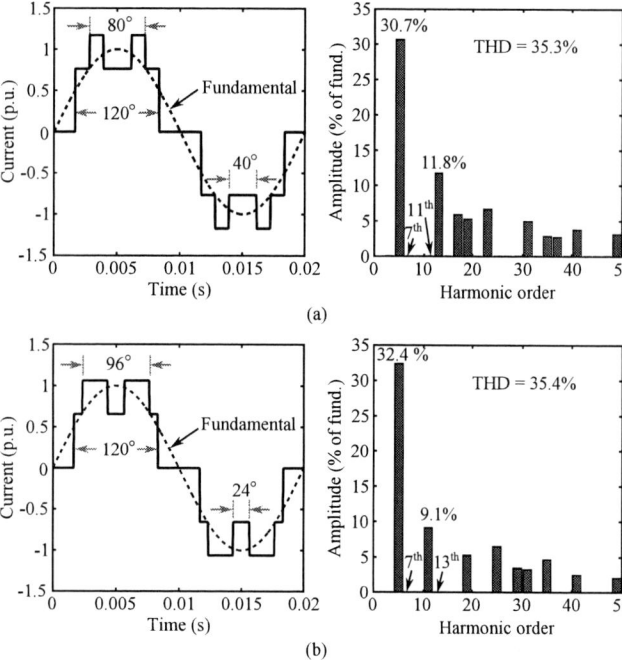

Fig. 6. Examples of five-level modulation signals that can be implemented into the unified dc-link current modulation scheme for selectively mitigating: (a) the 7th and the 11th harmonics with $I_{dc}^1 = 0.765$ p.u., $I_{dc}^2 = I_{dc}^3 = 0.408$ p.u., $\beta_2 = 80°$ and (b) the 7th and the 13th harmonics with $I_{dc}^1 = 0.656$ p.u., $I_{dc}^2 = I_{dc}^3 = 0.406$ p.u., $\beta_2 = 96°$, where $\alpha_f = 0$.

Fig. 6 demonstrates two examples of the five-level modulation signal designed according to Fig. 5, where a Global Search Algorithm [24] has been adopted to solve the equations (i.e., (4) through (8)), noted that other software (e.g., the EES - Engineering Equation Solver from F-Chart Software, LLC) is also applicable. It can be observed in Fig. 6 that, with a five-level modulation signal, selective harmonic cancellations are enabled, when it is implemented according to the unified dc-link modulation scheme shown in Fig. 2. However, due to the inherent relationships among the harmonics, it is impossible to cancel out the 5th and the 7th harmonics at the same time. Moreover, also because of this cross-impact, when certain harmonics are completely mitigated (e.g., the 7th and the 11th), some of the rest harmonics within the spectrum will be exaggerated in contrast to those shown in Fig. 4, leading to a high THD of the resultant currents. In order to improve the current quality, a phase-shifted current control scheme is introduced in the following, where the unequal loading impact is also taken into account.

III. LOAD-ADAPTIVE PHASE-SHIFTED CURRENT CONTROL SCHEME

A. Principle of Phase-Shifted Control

The above has demonstrated the unified dc-link current modulation scheme for three-phase rectifiers, which enables the flexibility to control the currents drawn from the grid. However, in order to ensure the performance (i.e., the rectified currents follow the desired modulation signals as exemplified in Fig. 5), the current controllers as shown in Fig. 2 should be taken care of. In particular, the current dynamics have to be fast enough in such a way that the grid currents will be as close to the designed modulation signal as possible. In light of this, hysteresis controllers have been employed, and thus guaranteeing the control effectiveness of the grid currents.

Additionally, as exemplified in Figs. 4 and 6, the unified dc-link current modulation scheme can only achieve a selective harmonic mitigation in a cost- and size-effective way, rather than an improved current quality. In practice, multiple ASD systems may operate in parallel, and be connected to the point of common coupling (e.g., Fig. 1). This initiates the phase-shifted current control to further improve the current quality, where SCR-fed drives should be utilized. To illustrate, the two-drive system is taken as an example. According to Fig. 3 and (1), the corresponding h^{th} harmonic components of the SCR and the DR can be represented as phasors,

$$\mathbf{I}_{a,scr}^h = \sqrt{2} I_{a,scr}^h e^{j\phi_{a,scr}^h} \text{ and } \mathbf{I}_{a,dr}^h = \sqrt{2} I_{a,dr}^h e^{j\phi_{dr}^h} \quad (9)$$

with $I_{a,scr}^h$ and $I_{a,dr}^h$ being the RMS magnitudes, $\phi_{a,scr}^h$ and $\phi_{a,dr}^h$ being the phases of the hth order harmonic of the SCR and the DR, respectively. The magnitudes and the phases of an individual harmonic can be calculated using the corresponding Fourier coefficients [22].

According to the superposition principle and (9), the phasor of the hth order harmonic component appearing in the grid (i.e., $i_{a,g}$ in Fig. 1) can be obtained as

$$\mathbf{I}_{a,g}^h = \mathbf{I}_{a,scr}^h + \mathbf{I}_{a,dr}^h = \sqrt{2} I_{a,scr}^h e^{j\phi_{a,scr}^h} + \sqrt{2} I_{a,dr}^h e^{j\phi_{a,dr}^h} \quad (10)$$

which indicates that the hth order grid current harmonic can be fully mitigated, only when

$$I_{a,scr}^h = I_{a,dr}^h \text{ and } \phi_{a,scr}^h = \phi_{a,dr}^h - \pi \quad (11)$$

where the latter criterion can be fulfilled by properly introducing a phase shift to the SCR of Fig. 1 (i.e., $\alpha_f = 180°/h$). However, in practice, the magnitudes of the currents drawn by the rectifiers (or the rectified current amplitudes) cannot be always the same (i.e., $I_{a,scr}^h \neq I_{a,dr}^h$), as discussed previously and elaborated in Fig. 7. It is demonstrated in Fig. 7 that incomplete cancellation of the 5th order harmonic by introducing a phase shift of 180°/5 (i.e., $\alpha_f = 36°$) occurs, when the magnitudes of the currents are not equal. As a consequence of the unbalanced loading condition, the performance of the phase-shifted current control is degraded, and the 5th order harmonic appears again in the grid current. Thus, load-adaptive schemes should be developed to ensure that conditions in (11) are always satisfied during operation.

Nevertheless, the phase-shifted current control combined with the five-level modulation schemes (see Fig. 6) can significantly enhance the grid current quality. Fig. 8 gives an example of the resultant grid current, assuming that the two drives are drawing the same amount of currents from the grid. As exemplified, the grid current become even multi-level, and a low THD of 12% is reached. Furthermore, since the firing angle for the SCR has been set as $\alpha_f = 36°$, the 5th order and

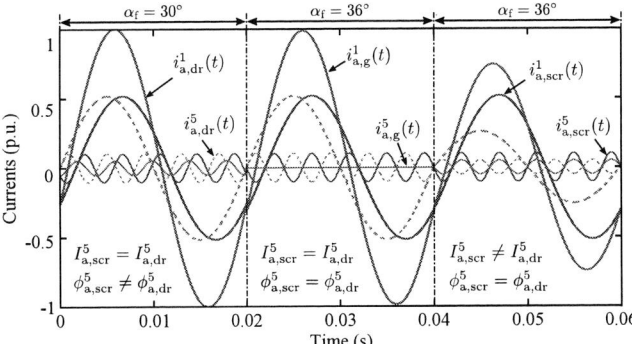

Fig. 7. Illustration of the unbalanced loading impact on the performance of phase-shifted current control in harmonic cancellations, where only the fundamental (denoted by the subscript - 1) currents and the 5th order (denoted by the subscript - 5) harmonic currents are shown.

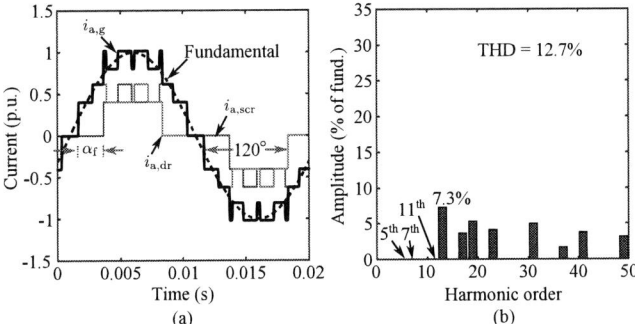

Fig. 8. Grid current harmonic characteristics of the two-drive system with the phase-shifted current control ($\alpha_f = 36°$), where the five-level modulation signal (Fig. 6(a)) has been applied to both rectifiers according to the unified dc-link modulation scheme: (a) grid current waveform and (b) harmonic distributions of the grid current.

the harmonics of fivefold the grid fundamental frequency are all eliminated by the phase-shifted control, in contrast to these harmonic contents shown in Fig. 6(a).

B. Load-Adaptive Control Scheme

As it is shown in Fig. 7, unequal loading will result in an incomplete cancellation of the harmonics of interest (e.g., the 5th order harmonic). Hence, a load-adaptive control scheme is proposed in the following. It is clear to all that the firing angle α_f controls the SCR input current phase (e.g., the phase-a current drawn from the grid as $i_{a,scr}$ in Fig. 1), and also the average rectified voltage that is given as

$$\bar{v}_{scr} = \bar{v}_{dr} \cos \alpha_f = \frac{3}{\pi} V_{LL} \cos \alpha_f \quad (12)$$

where $\bar{v}_{scr}$, $\bar{v}_{dr}$ are the average rectified voltages of the SCR and the DR, respectively, and V_{LL} is the amplitude of the line-to-line voltages (e.g., v_{ab}) of a balanced grid. On the condition that communication is available in the multi-drive system shown in Fig. 1, the loading information can then be obtained, where a power ratio γ and a load current ratio λ are defined as

$$\gamma = \frac{P_{scr}}{P_{dr}} \text{ and } \lambda = \frac{\bar{i}_{o,scr}}{\bar{i}_{o,dr}} \quad (13)$$

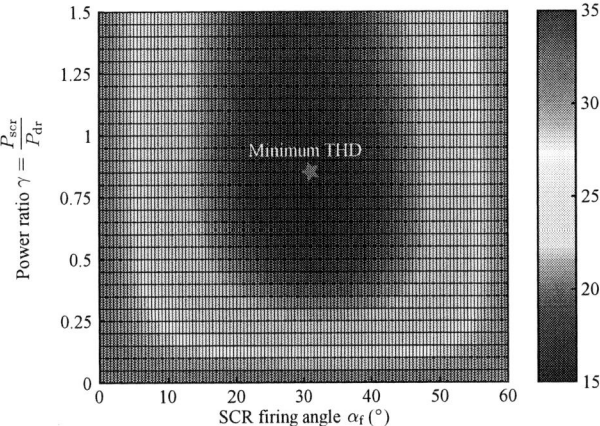

Fig. 9. THD distribution of the grid currents (e.g., phase-a $i_{a,g}$) in the two-drive system with the phase-shifted control (different firing angles for the SCR), where the square modulation signal (see § II.A) has been employed for both rectifiers.

with $P_{scr} = \bar{v}_{dc1} \cdot \bar{i}_{o,scr}$ and $P_{dr} = \bar{v}_{dc2} \cdot \bar{i}_{o,dr}$ being the boost output powers (i.e., the input powers for the inverters), where $\bar{i}_{o,scr}$ and $\bar{i}_{o,dr}$ are the average load currents, and $\bar{v}_{dc1}$ and $\bar{v}_{dc2}$ are the average dc-link voltages, respectively.

Ignoring the power losses on the boost converters gives $P_{scr} \approx \bar{v}_{scr} \cdot I_{scr}$ and $P_{dr} \approx \bar{v}_{dr} \cdot I_{dr}$. Thus, the following can be obtained:

$$\gamma = \frac{P_{scr}}{P_{dr}} = \frac{\bar{v}_{dc1} \cdot \bar{i}_{o,scr}}{\bar{v}_{dc2} \cdot \bar{i}_{o,dr}} \approx \frac{\bar{v}_{scr} \cdot I_{scr}}{\bar{v}_{dr} \cdot I_{dr}} \quad (14)$$

where I_{scr} and I_{dr} are averages of the rectified currents (controlled as dc) of the SCR and the DR, respectively. Substituting (12) into (14) yields

$$\gamma = \frac{P_{scr}}{P_{dr}} \approx \frac{I_{scr}}{I_{dr}} \cos \alpha_f \quad (15)$$

Then, with the knowledge of the loading information, it is possible to obtain the resultant grid current quality of the two-drive system (see Fig. 1), as shown in Fig. 9, where the square modulation signal discussed in § II.A has been adopted for the two rectifiers. It is further demonstrated that, under unbalanced loading between the two rectifier-fed drives, the grid current quality is affected. This impact is related to the power ratio (unbalance condition) as well as the phase shift angle. More important, it can be observed in Fig. 9 that a minimum THD of 16% (red star) is achieved at around $\gamma = 0.848$ and $\alpha_f = 32°$, corresponding to $I_{scr} = I_{dr}$.

Consequently, in order to ensure the performance of the phase-shifted current control, the load-adaptive control scheme should always maintain $I_{scr} = I_{dr}$, which accordingly results in $\gamma = \cos \alpha_f$. To accomplish this, the dc-link reference voltages for the two rectifier systems should be set in accordance to

$$\frac{v^*_{dc1}}{v^*_{dc2}} = \frac{\cos \alpha_f}{\lambda} \quad (16)$$

where v^*_{dc1} and v^*_{dc2} are the dc-link reference voltages for the SCR and the DR rectifier system, respectively. It should be noted that in order to ensure proper operation of the boost

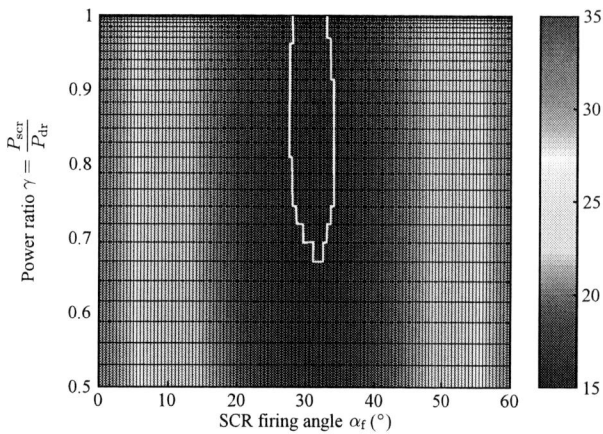

Fig. 10. THD distribution of the grid currents (e.g., phase-a $i_{a,g}$) in the two-drive system with the load-adaptive phase-shifted control and the square modulation signal, when the SCR is operating at partial loading condition.

TABLE I
PARAMETERS OF THE MULTI-DRIVE SYSTEM (FIG. 1).

Parameter	Symbol	Value
DC-link inductor	L_r, L_d	2 mH
DC-link capacitor	C_{dc}	470 μF
SCR output voltage reference	v_{dc1}^*	650 V
Grid frequency	f_g	50 Hz
Grid phase voltage (RMS)	$v_{abc, N}$	220 V
Grid impedance	$Z_g (L_g, R_g)$	0.18 mH, 0.1 Ω
PI controller	k_p, k_i	0.01, 0.1
Hysteresis band	-	2 A

converters, $v_{dc1}^* \geq v_{scr}$ and $v_{dc2}^* \geq v_{dr}$. If (16) is included in the control scheme shown in Fig. 2, the amounts of currents drawn by both rectifier systems will be equal, leading to an effective harmonic cancellation enabled by the phase-shifted current control. In that case, the SCR should operate in partial loading, i.e., $P_{scr} = P_{dr} \cos \alpha_f$, as shown in Fig. 10. It is highlighted in Fig. 10 that the SCR should be operated with 75% to 100% loading of the DR system and the firing angle of 28° to 34° (i.e., the white line zone), where the THD of the grid current will be 16% to 18%. Additionally, when the voltage references are set according to (16), the unified dc-link modulation scheme associated with the phase-shifted current control becomes load-adaptive no matter what modulation signals are used, and thus benefiting for harmonic reduction in multi-drive applications.

IV. EXPERIMENTAL RESULTS

In order to verify the discussion, experiments have been carried out on a two-drive system, referring to Figs. 1 and 2. The system parameters are given in Table I. The control algorithms are implemented in digital signal processors. A hysteresis controller is adopted to control the rectified currents (i.e., i_{scr} and i_{dr}). The output dc-link voltages (i.e., v_{dc1} and v_{dc2}) are controlled through a Proportional Integral (PI) controller for each drive, and the PI transfer function in the z-domain is given as

$$G_{PI}(z) = k_p + \frac{k_i T_s}{2} \cdot \frac{1 + z^{-1}}{1 - z^{-1}} \tag{17}$$

with k_p and k_i are the gains of the PI controller, and T_s is the sampling period. All the control parameters are listed in Table I. It should be noted that in the experimental tests resistive loads have been used. A second order generalized integrator based PLL system has been employed to synchronize, and its design can be found in [25].

Two cases are studied to validate the unequal loading impact on the harmonic cancellation. In the first case, the load resistors are the same for both rectifiers, and the experimental

results are shown in Fig. 11, where the square modulation signal has been implemented according to the unified dc-link modulation scheme (see Fig. 2) and $\alpha_f = 32°$, $v_{dc2}^* = v_{dc1}^* = 650$ V. It can be seen in Fig. 11(a) that, with the phase-shifted current control, the THD of the grid current in the two-drive system has been brought to 16.3%, which is close to the theoretical value (16% in Fig. 9). The difference is induced by the unequal currents that are drawn by the rectifiers (i.e., $I_{scr} \neq I_{dr}$), although the output powers are almost the same. Subsequently, the load-adaptive scheme is applied to the drive system, where the dc-link output voltage for the DR has been changed to $v_{dc2}^* = 705$ V. As it is shown in Fig. 11(b), the grid current THD is lowered to 16% as the theoretical one. Furthermore, it is observed that, by changing the dc-link voltage reference, the load-adaptive phase-shifted current scheme can ensure the input currents of both rectifiers are at the same level, and the SCR is operating at partial loading condition in respect to the power of the DR (i.e., $P_{scr} \approx P_{dr} \cos \alpha_f$).

In the second study case, the loading of the SCR is around 80% of the DR (i.e., $\lambda \approx 0.8$ and $\bar{i}_{o,scr} \approx 0.8 \bar{i}_{o,dr}$), if the dc-link voltage references are the same (i.e., $v_{dc1}^* = v_{dc2}^*$). This will lead to unequal input currents for the rectifiers, and thus it will deteriorate the phase-shifted current control in terms of higher THD of the grid currents. Hence, in order to enhance the harmonic cancellation performance, the dc-link voltage reference for the DR-fed boost converter is reduced to $v_{dc2}^* = 630$ V according to (16) with $\alpha_f = 32°$. As it is shown in Fig. 12, the load-adaptive phase-shifted current control can maintain an almost-equal-current drawing from the grid for the rectifiers, and thus the grid current quality is enhanced. The slight current difference (i.e., $I_{scr} \approx I_{dr}$ in Fig. 12) is induced by the resistance variations in the resistive loads (e.g., temperature rises during operation). Nevertheless, the above cases have demonstrated the effectiveness of the load-adaptive phase-shifted current control scheme in multi-drive applications, where thus the unbalanced loading issues are addressed.

In addition to the above cases with the square modulation signal, more tests have been carried out on the same setup, where the five-level modulation signal shown in Fig. 6(a) has

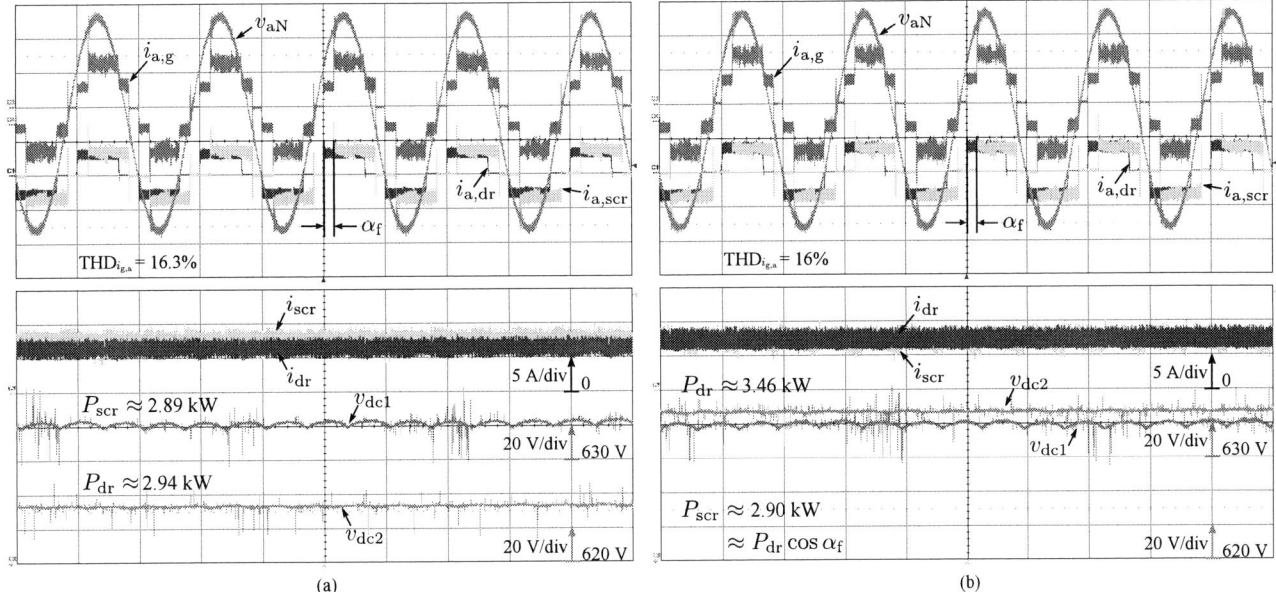

Fig. 11. Experimental results of the two-drive system with the unified dc-link current modulation scheme (square modulation signal), where the phase-shift for the SCR is $\alpha_f = 32°$ (top: grid voltage v_{aN} [100 V/div], grid current $i_{a,g}$ [10 A/div], SCR input current $i_{a,scr}$ [10 A/div], DR input current $i_{a,dr}$ [10 A/div], time [10 ms/div]; bottom: SCR rectified current i_{scr}, DR rectified current i_{dr}, SCR dc-link voltage v_{dc1}, DR dc-link voltage v_{dc2}, time [5 ms/div]): (a) without the load-adaptive scheme (power factor: 0.94) and (b) with the load-adaptive control scheme (power factor: 0.94).

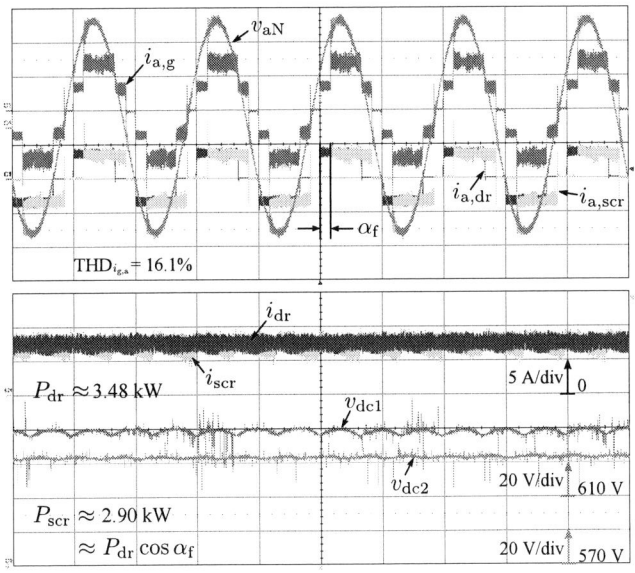

Fig. 12. Experimental results of the two-drive system with the unified dc-link current modulation scheme (square modulation signal) and the load-adaptive control scheme, where $\alpha_f = 32°$ and the power factor is 0.94. (top: grid voltage v_{aN} [100 V/div], grid current $i_{a,g}$ [10 A/div], SCR input current $i_{a,scr}$ [10 A/div], DR input current $i_{a,dr}$ [10 A/div], time [10 ms/div]; bottom: SCR rectified current i_{scr}, DR rectified current i_{dr}, SCR dc-link voltage v_{dc1}, DR dc-link voltage v_{dc2}, time [5 ms/div]).

been implemented into both rectifier systems. In this case study, the resistors are the same, and a phase-shift of 36° has been introduced to the SCR in order to mitigate the harmonics of fivefold the fundamental grid frequency. According to the discussions in § III and the previous experimental tests,

the dc-link voltage references should be set considering the constraints of (16), in such a way that the averages of the rectified currents will be equal, leading to improved grid current quality. Thus, $v_{dc1}^* = 650$ V, while $v_{dc2}^* = 720$ V. The experimental results are presented in Fig. 13.

It can be seen from Fig. 13 that, with the unified dc-link current modulation scheme, the currents drawn by the rectifiers can be flexibly shaped or controlled (e.g., as the five-level waveforms shown in Fig. 6(a)). As a consequence of the load-adaptive phase-shifted current control, the resultant grid current becomes even multi-level, similar to the shape of Fig. 8(a). Hence, the grid current THD can be lowered to 11.6%, as shown in Fig. 13. Additionally, in contrast to the theoretical harmonic distribution shown in Fig. 8(b), the low-order harmonics (i.e., the 5th, the 7th, and the 11th) are not completely eliminated, but are also relatively low with the unified dc-link modulation scheme. In all, the experimental tests have demonstrated that the load-adaptive phase-shifted control employed with the unified dc-link current modulation scheme can contribute to significant improvements of the grid current quality in motor drive applications. This lies in: 1) the flexibility to design the desired modulation signals for selective harmonic mitigation and 2) the enhancement of harmonic cancellation enabled by the load-adaptive phase-shifted current control scheme.

V. CONCLUSION

In this paper, the unequal loading condition induced harmonic mitigation degradation of a phase-shifted current control scheme has been addressed in the DR and the SCR based three-phase motor drive applications, where dc-dc boost

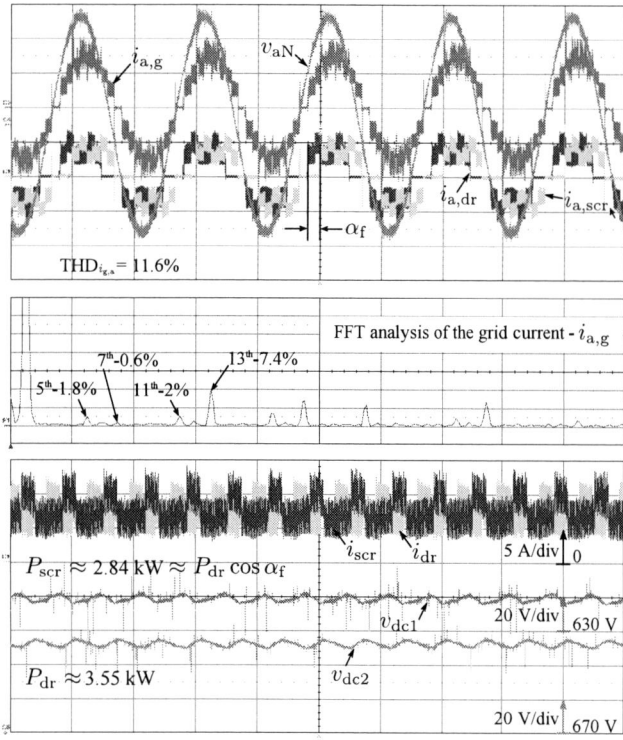

Fig. 13. Experimental results of the two-drive system with the unified dc-link current modulation scheme (five-level modulation signal shown in Fig. 6(a)) and the load-adaptive control scheme, where $\alpha_f = 36°$ and the power factor is 0.93. (top: grid voltage v_{aN} [100 V/div], grid current $i_{a,g}$ [10 A/div], SCR input current $i_{a,scr}$ [10 A/div], DR input current $i_{a,dr}$ [10 A/div], time [10 ms/div]; middle: Fast Fourier Transform - FFT of the grid phase-a current [500 mA/div], frequency [200 Hz/div]; bottom: SCR rectified current i_{scr}, DR rectified current i_{dr}, SCR dc-link voltage v_{dc1}, DR dc-link voltage v_{dc2}, time [5 ms/div]).

converters are employed in the dc-link. A unified dc-link current modulation scheme is also applied to the rectifiers in such a way to shape the currents drawn from the grid, and thus to further mitigate the harmonics of interest. Together with the phase-shifted current control, the grid current quality can be improved a lot. However, the performance of this harmonic mitigation strategy (i.e., the phase-shifted current control) is affected by the loading among the paralleled drive units. Accordingly, a load-adaptive scheme is proposed in this paper by changing the dc-link output voltage dynamically, which in return can always ensure that the currents drawn by the rectifiers are equal, and thus a good quality of the total grid current is maintained. Experiments have verified the effectiveness of the proposal in terms of harmonic mitigation in a size- and cost-effective way for motor drive systems.

REFERENCES

[1] P. K. Steimer, "High power electronics innovation," *presented at ICPE - ECCE Asia*, pp. 1–37, Jun. 2015.

[2] P. Davari, Y. Yang, F. Zare, and F. Blaabjerg, "A multi-pulse pattern modulation scheme for harmonic mitigation in three-phase multi-motor drives," *IEEE J. Emerg. Sel. Top. Power Electron.*, vol. PP, no. 99, pp. 1–11, in press, 2016.

[3] P. Barbosa, C. Haederli, P. Wikstroem, M. Kauhanen, J. Tolvanen, and A. Savolainen, "Impact of motor drive on energy efficiency," in *Proc. of PCIM*, pp. 1-6 2007.

[4] P. Waide and C. U. Brunner, "Energy-efficiency policy opportunities for electric motor-driven systems," International Energy Agency, Tech. Rep., 2011.

[5] J. W. Kolar and T. Friedli, "The essence of three-phase PFC rectifier systems: Part I," *IEEE Trans. Power Electron.*, vol. 28, no. 1, pp. 176–198, Jan. 2013.

[6] D. Kumar and F. Zare, "Harmonic analysis of grid connected power electronic systems in low voltage distribution networks," *IEEE J. Emerg. Sel. Top. Power Electron.*, vol. PP, no. 99, pp. 1–10, in press, 2016.

[7] J. Holtz and X. Qi, "Optimal control of medium-voltage drives - an overview," *IEEE Trans. Ind. Electron.*, vol. 60, no. 12, pp. 5472–5481, Dec 2013.

[8] H. Akagi, "Active harmonic filters," *Proceedings of the IEEE*, vol. 93, no. 12, pp. 2128–2141, Dec. 2005.

[9] W.-J. Lee, Y. Son, and J.-I. Ha, "Single-phase active power filtering method using diode-rectifier-fed motor drive," *IEEE Trans. Ind. Appl.*, vol. 51, no. 3, pp. 2227–2236, 2015.

[10] X. Du, L. Zhou, H. Lu, and H.-M. Tai, "DC link active power filter for three-phase diode rectifier," *IEEE Trans. Ind. Electron.*, vol. 59, no. 3, pp. 1430–1442, Mar. 2012.

[11] H. Akagi and K. Isozaki, "A hybrid active filter for a three-phase 12-pulse diode rectifier used as the front end of a medium-voltage motor drive," *IEEE Trans. Power Electron.*, vol. 27, no. 1, pp. 69–77, Jan. 2012.

[12] M. L. Zhang, B. Wu, Y. Xiao, F. A. Dewinter, and R. Sotudeh, "A multilevel buck converter based rectifier with sinusoidal inputs and unity power factor for medium voltage (4160-7200 V) applications," *IEEE Trans. Power Electron.*, vol. 17, no. 6, pp. 853–863, Nov. 2002.

[13] S. Choi, P. N. Enjeti, and I. J. Pitel, "Polyphase transformer arrangements with reduced kVA capacities for harmonic current reduction in rectifier-type utility interface," *IEEE Trans. Power Electron.*, vol. 11, no. 5, pp. 680–690, Sept. 1996.

[14] F. Meng, W. Yang, Y. Zhu, L. Gao, and S. Yang, "Load adaptability of active harmonic reduction for 12-pulse diode bridge rectifier with active inter-phase reactor," *IEEE Trans. Power Electron.*, vol. 30, no. 12, pp. 7170–7180, Dec. 2015.

[15] B. Singh, V. Garg, and G. Bhuvaneswari, "A novel T-connected autotransformer-based 18-pulse AC-DC converter for harmonic mitigation in adjustable-speed induction-motor drives," *IEEE Trans. Ind. Electron.*, vol. 54, no. 5, pp. 2500–2511, Oct. 2007.

[16] M. M. Swamy, "An electronically isolated 12-pulse autotransformer rectification scheme to improve input power factor and lower harmonic distortion in variable-frequency drives," *IEEE Trans. Ind. Appl.*, vol. 51, no. 5, pp. 3986–3994, Sept 2015.

[17] S. Hansen, P. Nielsen, and F. Blaabjerg, "Harmonic cancellation by mixing nonlinear single-phase and three-phase loads," *IEEE Trans. Ind. Appl.*, vol. 36, no. 1, pp. 152–159, Jan./Feb. 2000.

[18] IEC, "Electromagnetic compatibility (EMC) - part 3-2: Limits - limits for harmonic current emissions (equipment input current $\leq$ 16 A per phase)," *IEC/EN 61000-3-2*, 2006.

[19] J. A. Pomilio and G. Spiazzi, "A low-inductance line-frequency commutated rectifier complying with EN 61000-3-2 standards," *IEEE Trans. Power Electron.*, vol. 17, no. 6, pp. 963–970, Nov. 2002.

[20] F. Zare, "A novel harmonic elimination method for a three-phase diode rectifier with controlled DC link current," in *Proc. of PEMC*, pp. 985-989, 21-24 Sept. 2014.

[21] Y. Yang, P. Davari, F. Zare, and F. Blaabjerg, "A DC-link modulation scheme with phase-shifted current control for harmonic cancellations in multi-drive applications," *IEEE Trans. Power Electron.*, vol. 31, no. 3, pp. 1837–1840, Mar. 2016.

[22] N. Mohan, T. M. Undeland, and W. P. Robbins, *Power electronics: converters, applications, and design*, 3rd ed. John Wiley & Sons, Inc., Chapter 6 (pp. 138-147), 2007.

[23] P. Davari, F. Zare, and F. Blaabjerg, "Pulse pattern modulated strategy for harmonic current components reduction in three-phase AC-DC converters," in *Proc. of ECCE*, pp. 5968-5975, Sept. 2015.

[24] *MATLAB – Global Optimization Toolbox - User's Guide*, R2015b ed. The MathWorks, Inc, Chapter 11 (pp. 11-44-11-48), 2015.

[25] M. Ciobotaru, R. Teodorescu, and F. Blaabjerg, "A new single-phase PLL structure based on second order generalized integrator," in *Proc. of PESC*, pp. 1-6, Jun. 2006.

Input current interharmonics in Adjustable Speed Drives caused by fixed-frequency modulation techniques

Hamid Soltani, Pooya Davari, Poh Chiang Loh, and Frede Blaabjerg
Department of Energy Technology
Aalborg University
Aalborg East, Denmark
hso@et.aau.dk, pda@et.aau.dk, pcl@et.aau.dk, fbl@et.aau.dk

Firuz Zare
Danfoss Drives A/S
Global Research & Development Centre
Ulsnaes 1, DK-6300 Graasten, Denmark
fza@danfoss.com

Abstract—**Adjustable Speed Drives (ASDs) based on double-stage conversion systems may inject interharmonics distortion into the grid, other than the well-known characteristic harmonic components. The problems created by interharmonics make it necessary to find their precise sources, and, to adopt an appropriate strategy for minimizing their effects. This paper investigates the ASD's input current interharmonic sources caused by applying symmetrical regularly sampled fixed-frequency modulation techniques on the inverter. The interharmonics generation process is precisely formulated and comparative results of different fixed-frequency modulation techniques are presented to evaluate the drive input current interharmonic components. The theoretical analysis and simulation studies are validated with experimental results on a 2.2 kW motor drive system.**

I. INTRODUCTION

Adjustable speed drives based on double-stage conversion systems are considered as one of the major sources of interharmonics in the grid, other than the well-known characteristic harmonics [1]. Fig. 1 illustrates the typical block diagram of a double-stage voltage source inverter-fed ASD, where a three-phase front-end diode rectifier is connected back-to-back to a rear-end inverter sharing a common DC link. With connecting two systems running at separate frequencies, the ASD operation can result in interharmonic distortions at both terminals via interactions between the associated harmonics.

Interharmonics are spectral components of voltages or currents, which are not multiple integer of the fundamental supply frequency [2]. Although small in magnitude, interharmonics may cause their own unique problems as well as those in common with the harmonic distortions [3], [4]. The problems mainly have roots in the interharmonic characteristics, where they can spread at a wide range in the frequency spectrum. Light flicker, sideband torques on the motor/generator shaft, interference with control and protection signals, dormant resonance excitations are some of the most significant negative effects caused by interharmonic distortions [5]–[8].

Over the years, many investigations have been dedicated to interharmonics issue, mainly by considering their origins, negative effects on the power supply, accurate detections and identification methods [9]–[19]. Moreover, the accurate

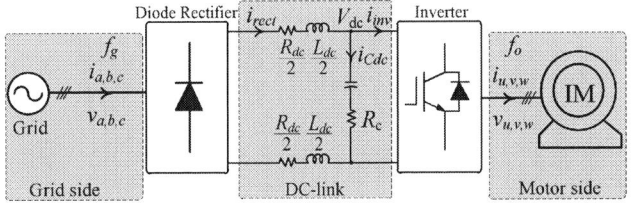

Fig. 1. Equivalent circuit diagram of a typical adjustable-speed drive for system analysis with an Induction Motor (IM).

modeling of the ASD for interaction analysis of interharmonics has initiated several studies [20]–[22]. The ASD interharmonic analyses at the presence of the output currents imbalance, and also, the unbalanced supply side voltages have been investigated in [8], [23]. In addition to the above-mentioned studies, the authors in [24], [25] proposed active compensation methods in order to reduce the drive input current interharmonics at the presence of load currents imbalance.

Interharmonics are usually smaller than the characteristic harmonics and this is why less attention has been devoted to them compared with harmonics issue. But with the rapid growth of power electronic applications, and consequently increasing the interharmonics distortion in the grid, special concerns have been raised to define limitations for interharmonics in respect to their frequencies and amplitudes.

Besides several investigations done so far related to interharmonics, a precise mapping of their origins in ASD applications has always been a challenging issue. This mapping should cover all the operating points of the ASD. Prediction of the drive input current interharmonic frequencies corresponding with a specific operating condition of ASD can prevent undesirable interference with control and protection signals in the power system. With knowing the interharmonics frequencies, the potential dormant resonance excitations can also be avoided.

The authors in [26] have evaluated the interharmonics generation process in ASDs, where a naturally sampled Sinusoidal Pulse Width Modulation (SPWM) technique with

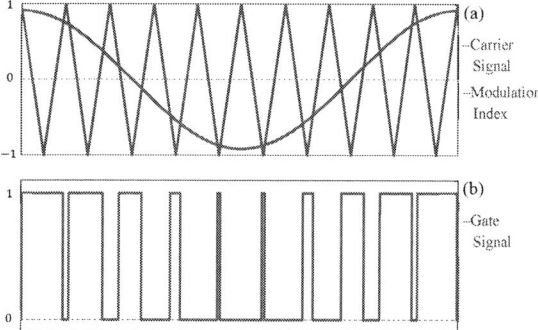

Fig. 2. (a) Symmetrical naturally sampling modulation technique, (b) Switching pulse train for inverter.

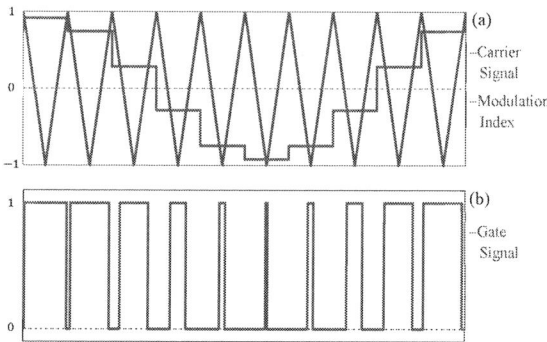

Fig. 3. (a) Symmetrical regular sampling modulation technique, (b) Switching pulse train for inverter.

low switching frequency has been considered for the inverter operation. Introducing proper definitions for the input side, the DC link and the output side harmonics and interharmonics can make it useful for understanding the harmonics interactions and consequently the associated interharmonics.

In this paper, a general analysis is proposed to find the ASD's input current interharmonic components caused by applying symmetrical regularly sampled fixed-frequency modulation techniques on the inverter. The Space Vector Modulation (SVM) and Discontinuous Pulse Width Modulation-30 degree lagging clamp (DPWM2) are investigated as the most commonly used modulation techniques in motor drive applications. The reference is made to ideal supply conditions in order to only evaluate those interharmonics generated by the interaction between the front-end diode rectifier and the rear-end inverter.

First, the harmonics transfer from the inverter output side to the rectifier input side, which leads to the drive input current interharmonic components is comprehensively formulated in Section II. As a result, the drive input current interharmonic frequencies can precisely be obtained. Thereafter in Section III, a comparative evaluation addresses the results obtained for the symmetrical regularly sampled SVM and DPWM2 techniques. Although the interharmonics amplitude estimation is so complicated due to uncertainties in the system, the obtained results indicate that the DPWM2 may not be suitable choice of modulation from an interharmonics point of view. Finally, the theoretical analysis and simulation studies are validated with experimental results performed on a 2.2 kW ASD.

II. INTERHARMONICS GENERATION PROCESS

In a double-stage voltage source inverter fed ASD, shown in Fig. 1, the input three-phase AC voltages are rectified first by the diode rectifier to provide a DC voltage for powering the inverter. The DC link is made up of a capacitor and two inductors to make a smoother voltage and current at the DC stage. The inverter then converts the DC voltage to a set of three-phase balanced voltages for driving the motor. The fundamental frequency of the inverter is determined by the

demanded motor speed, which typically is not the same as the grid frequency.

A. Harmonic Transfer at Inverter Level

Referring to Fig. 1, in a two-level pulse width modulation strategy, the inverter output pulse trains are generated by comparing two independent periodic signals; a low-frequency reference signal against a high frequency carrier signal. Fig. 2 illustrates the symmetrical naturally sampled PWM technique, where the intersections between the reference sinusoid and the triangle carrier determine the switching pulse edges. The resultant pulsating voltages (which are periodic with respect to the modulated and the modulating signals) harmonic components can be obtained by using the double Fourier integral approach. As a result of this evaluation, the pulsating output voltages v_x ($x = u, v, w$) are composed of a DC offset, baseband harmonics (simple harmonics of the fundamental output frequency f_o), harmonics of the carrier frequency f_c, and carrier sidebands located around the carrier components [27], and they can generally be represented as

$$
v_x(t) = \frac{A_{00}}{2} + \sum_{n=1}^{\infty} \left[A_{0n} \cos(n[w_o t - p\frac{2\pi}{3}]) + B_{0n} \sin(n[w_o t - p\frac{2\pi}{3}]) \right] \\
+ \sum_{m=1}^{\infty} \sum_{n=-\infty}^{\infty} \left[A_{mn} \cos(m w_c t + n[w_o t - p\frac{2\pi}{3}]) + B_{mn} \sin(m w_c t + n[w_o t - p\frac{2\pi}{3}]) \right]
\tag{1}
$$

where the carrier index and the baseband index variables are notated as m and n, respectively. The fundamental and carrier angular frequencies are also presented as w_o and w_c, respectively. The p values are 0, 1 and -1 with respect to the output phases u, v and w. The harmonic coefficients A_{mn} and B_{mn}, which are defined by a double Fourier integral, can be evaluated in accordance with the selected modulation strategy. The hth harmonic component is also defined in terms of m and n, and it is given as

$$
h = m(\frac{w_c}{w_o}) + n
\tag{2}
$$

In practical applications, it is difficult to implement the naturally sampled PWM in a digital modulation system. Consequently, the regularly sampled modulation technique has

978-1-4673-9551-9/16 $31.00 © 2016 IEEE

been widely applied instead of the previous method, due to the simplicity of its implementation. In accordance to the regularly sampled PWM strategy, the reference waveform is sampled and kept constant during each carrier interval, and then, it is compared with the triangle carrier waveform, as shown in Fig. 3.

Regarding the Induction Motor (IM) as a highly inductive load, the three-phase AC output currents i_x ($x = u, v, w$) are assumed to be sinusoidal and they are given as

$$i_u(t) = I_o \cos(w_o t + \theta) \tag{3}$$

$$i_v(t) = I_o \cos(w_o t + \theta - \frac{2\pi}{3}) \tag{4}$$

$$i_w(t) = I_o \cos(w_o t + \theta + \frac{2\pi}{3}) \tag{5}$$

with I_o and θ stated as the amplitude of the output currents and a displacement factor, respectively. Assuming that the inverter does not dissipate or generate power, under balanced load condition, the DC-link inverter-side current can be written as

$$i_{inv}(t) = \frac{1}{V_{dc}}[v_u(t) \cdot i_u(t) + v_v(t) \cdot i_v(t) + v_w(t) \cdot i_w(t)] \tag{6}$$

where V_{dc} represents the DC value of the DC-link voltage. The time-domain expression given in (6) can be applied to obtain the DC-link harmonic components caused by the rear-end inverter operation. It should be noted that the v_x/V_{dc} ($x = u, v, w$) terms in (6) are usually referred to as inverter switching functions.

Here, in order to identify the interharmonics sources, the effects of baseband and carrier group harmonics have been studied separately. Substituting the baseband harmonic components (first summation in (1)) and (3)-(5) into (6), the corresponding DC-link oscillations i_{inv-b} can be written as (7). It is worth to mention that, in respect to the applied symmetrical regularly sampled modulation techniques, the baseband harmonics coefficients A_{0n} and B_{0n} are potentially existing for all values of n. However, further investigations of

(7) show that only the triple multiples of the output frequency w_o will be transferred to the DC link under balanced condition. Thus, the contribution of baseband harmonic components on the DC-link oscillation f_{dc-b}^h can generally be presented as

$$f_{dc-b}^h = 3 \cdot k \cdot f_o, \qquad k = 1, 2, 3, ... \tag{8}$$

In respect to (8) it should be mentioned that the *odd* triple multiples of the output frequency (i.e. $3f_o$, $9f_o$, ...) are caused by the corresponding *even* order baseband harmonics of the output frequency (i.e. $\{2^{th}, 4^{th}\}$, $\{8^{th}, 10^{th}\}$, ...), and, the *even* triple multiples (i.e. $6f_o$, $12f_o$, ...) are generated by the associated *odd* order baseband harmonics of the output frequency(i.e. $\{5^{th}, 7^{th}\}$, $\{11^{th}, 13^{th}\}$, ...).

The effects of the carrier sideband harmonic components of the pulsating output voltage on the DC-link oscillations can be evaluated by the same procedure. With substitution of the carrier sideband harmonics (double summation in (1)) and (3)-(5) into (6), the associated DC-link oscillations i_{inv-s} can be obtained as (9). With a precise inspection of (9), it can be concluded that although the output pulsating voltage contains all sideband carrier harmonics corresponding with the values of m and n, the DC-link inverter-side current only inherits oscillations with the frequencies of the carrier harmonic components, and their differences from the triple multiple of the fundamental frequency w_o. Consequently, the contributions of the carrier sideband harmonics on the DC-link oscillation f_{dc-s}^h can be given as

$$f_{dc-s}^h = \{(m \cdot f_c), (m \cdot f_c \pm 3 \cdot k \cdot f_o)\}, \qquad k = 1, 2, 3, ... \tag{10}$$

Regarding the DC-link oscillation frequencies in (10), it is worthwhile noting that the first carrier sidebands produce the carrier frequency oscillations at the DC-link. Moreover, the *odd* triplen sidebands of f_{dc-s}^h (i.e. $(m \cdot f_c)-3f_o$, $(m \cdot f_c)-9f_o$, ...) are caused by the corresponding *even* order sidebands of the output voltages (with the sets of n as $\{-2, -4\}$, $\{-8, -10\}$, ...), and, the *even* triplen sidebands (i.e. $(m \cdot f_c) - 6f_o$, $(m \cdot$

$$
\begin{aligned}
i_{inv-b}(t) = \sum_{n=1}^{\infty} \big[& \tfrac{A_{0n}I_o}{2V_{dc}}[\cos((n+1)w_o t + \theta) + \cos((n-1)w_o t - \theta) \\
& + \cos((n+1)w_o t + \theta - (n+1)\tfrac{2\pi}{3}) + \cos((n-1)w_o t - \theta - (n-1)\tfrac{2\pi}{3}) \\
& + \cos((n+1)w_o t + \theta + (n+1)\tfrac{2\pi}{3}) + \cos((n-1)w_o t - \theta + (n-1)\tfrac{2\pi}{3})] \\
+ & \tfrac{B_{0n}I_o}{2V_{dc}}[\sin((n+1)w_o t + \theta) + \sin((n-1)w_o t - \theta) \\
& + \sin((n+1)w_o t + \theta - (n+1)\tfrac{2\pi}{3}) + \sin((n-1)w_o t - \theta - (n-1)\tfrac{2\pi}{3}) \\
& + \sin((n+1)w_o t + \theta + (n+1)\tfrac{2\pi}{3}) + \sin((n-1)w_o t - \theta + (n-1)\tfrac{2\pi}{3})]]
\end{aligned}
\tag{7}
$$

$$
\begin{aligned}
i_{inv-s}(t) = \sum_{m=1}^{\infty} \sum_{n=-\infty}^{\infty} & \tfrac{A_{mn}I_o}{2V_{dc}}[\cos(mw_c t + (n+1)w_o t + \theta) + \cos(mw_c t + (n-1)w_o t - \theta) \\
& + \cos(mw_c t + (n+1)w_o t + \theta - (n+1)\tfrac{2\pi}{3}) + \cos(mw_c t + (n-1)w_o t - \theta - (n-1)\tfrac{2\pi}{3}) \\
& + \cos(mw_c t + (n+1)w_o t + \theta + (n+1)\tfrac{2\pi}{3}) + \cos(mw_c t + (n-1)w_o t - \theta + (n-1)\tfrac{2\pi}{3})] \\
+ & \tfrac{B_{mn}I_o}{2V_{dc}}[\sin(mw_c t + (n+1)w_o t + \theta) + \sin(mw_c t + (n-1)w_o t - \theta) \\
& + \sin(mw_c t + (n+1)w_o t + \theta - (n+1)\tfrac{2\pi}{3}) + \sin(mw_c t + (n-1)w_o t - \theta - (n-1)\tfrac{2\pi}{3}) \\
& + \sin(mw_c t + (n+1)w_o t + \theta + (n+1)\tfrac{2\pi}{3}) + \sin(mw_c t + (n-1)w_o t - \theta + (n-1)\tfrac{2\pi}{3})]
\end{aligned}
\tag{9}
$$

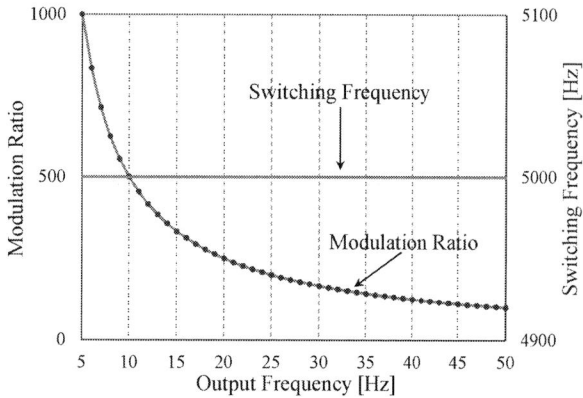

Fig. 4. The inverter switching frequency and the corresponding modulation ratio at different output frequency f_o variation range.

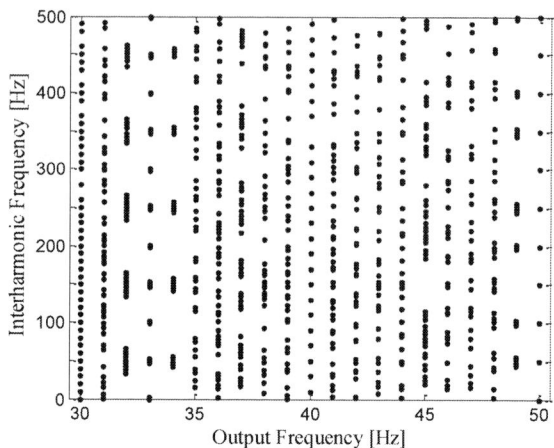

Fig. 5. ASD input current i_a interharmonics frequency location with respect to output frequency f_0 variations, by using (15).

$f_c) - 12f_o$, ...) are produced by the related *odd* order sideband carriers of the output voltages (with the sets of n as $\{-5, -7\}$, $\{-11, -13\}$, ...).

The general expression for the DC-link inverter-side current oscillations, inherited from the switching operation of the inverter can then be obtained with combining (8), and (10) and it can be written as

$$f_{dc}^h = \{f_{dc-b}^h, f_{dc-s}^h\} \qquad (11)$$

B. Harmonic Transfer at Rectifier Level

The DC-link inverter-side current oscillations, after flowing through the DC link stage, will be multiplied by the well-known six-pulse diode rectifier switching functions $\{S_a(t), S_b(t), S_c(t)\}$, defined in (12)-(14),

$$
\begin{aligned}
S_a(t) = 2\sqrt{3}/\pi[&\cos(w_g t) - 1/5 \, \cos(5w_g t) \\
&+ 1/7 \, \cos(7w_g t) - 1/11 \, \cos(11w_g t) + ...]
\end{aligned} \qquad (12)
$$

$$S_b(t) = S_a(t - T/3) \qquad (13)$$

$$S_c(t) = S_a(t + T/3) \qquad (14)$$

where the grid voltage fundamental period and the angular frequency are notated as T and w_g.

The modulation between the DC-link harmonics, caused by inverter operation, and the rectifier switching functions results in the input current interharmonic frequencies, and can be expressed as

$$f_{ih} = \left|[6 \cdot (v-1) \pm 1] \cdot f_g \pm f_{dc}^h\right| \quad v = 1, \, 2, \, 3, ... \quad (15)$$

with f_g stated as the input voltage fundamental frequency.

Fig. 4 illustrates the selected modulation strategy for the inverter, where the switching frequency is considered a constant value of 5 kHz during the output frequency f_o variations from 5 Hz to 50 Hz. The drive input current interharmonic locations, obtained using (15), are plotted in Fig. 5, only for the output frequency f_o range of 30 Hz to 50 Hz. It should be noted that, for plotting Fig. 5, the interactions between the three significant AC side harmonics (with $v = 1, 2$ in (15)) and the following DC-link harmonics have been considered:

- the DC-link harmonics below 400 Hz caused by output-side baseband harmonics
- the DC-link harmonics below 300 Hz caused by the first carrier group sidebands
- the DC-link harmonics below 300 Hz caused by the second carrier group sidebands

As it can be observed in Fig. 5, the drive input current interharmonics change their locations with respect to output frequency variations. Most interharmonics overlaps occur at the output frequencies of 33 Hz and 50 Hz, where the interharmonics accommodate at the input side harmonic frequencies and very close to them.

III. SIMULATION AND LABORATORY TEST RESULTS

The calculation presented in Section II was a general evaluation of the drive input-side current interharmonics with respect to output frequency variations, when a symmetrical regularly sampled modulation strategy is adopted. This strategy is normally implemented in a fixed-frequency modulation technique. To validate the accuracy of the theoretical analysis, a set of simulation and experimental tests were considered based on the adjustable speed drive system shown in Fig. 1 with the parameter values listed in Table I. In this investigation, the most common modulation techniques in ASDs (i.e. SVM and DPWM2) were evaluated in terms of the drive input current interharmonics.

TABLE I
Simulation and Experimental Parameters Values.

Symbol	Parameter	Value
$v_{a,b,c}$	Grid phase voltage	225 V_{rms}
f_g	Grid frequency	50 Hz
L_{dc}, R_{dc}	DC link inductor & resistor	8 mH, 360 mΩ
C_{dc}, R_c	DC Link Capacitor & Resistor	125 μF & 500 mΩ
f_{sw}	Inverter switching frequency	5 kHz
v_{LL}	Induction motor rated voltage	380 V_{rms}
P_{IM}	Induction motor rated power	2.2 kW

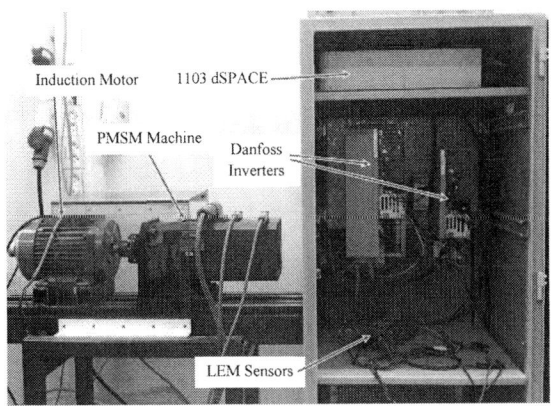

Fig. 6. Laboratory setup for experimental verification.

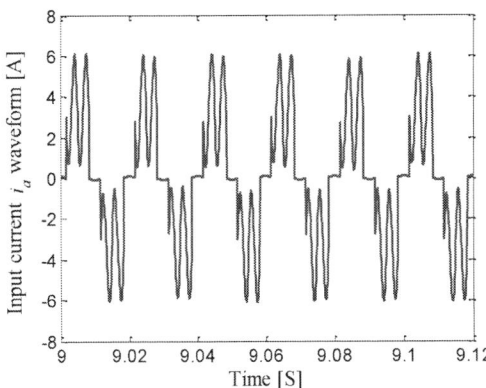

Fig. 8. The simulated ASD input current i_a, when the induction motor is operating at the output frequency $f_o = 40$ Hz and with a load torque value of 12 Nm using SVM.

Fig. 6 shows the employed experimental setup using two Danfoss inverters rated at 2.2 kW and 10 kW. Moreover, a California MX30 three-phase grid simulator was used to remove the potential grid background distortion. The induction motor is controlled with a constant Voltage-to-Frequency (V/F) strategy using a 2.2 kW inverter, and, the load torque was implemented by controlling a Permanent Magnet Synchronous Machine (PMSM) coupled with the induction motor via a 10 kW inverter. The control algorithm was executed on a dSPACE1103 real-time platform.

A. SVM Modulation Technique

The space vector modulation technique is an advanced, computation-intensive and arguably the best among all PWM strategies for adjustable speed drive applications, where the neutral point of the load is normally isolated. This method considers the interaction of the inverter output phases and optimizes the harmonic content of the three-phase induction motor. In this respect, the ASDs interharmonic components, when a SVM modulation technique is implemented on the inverter, can be subjected to further investigation.

The phase-u reference waveform for applying the space vector modulation is drawn in Fig. 7. By applying a (V/F) control strategy and also considering the drive operating condition, the

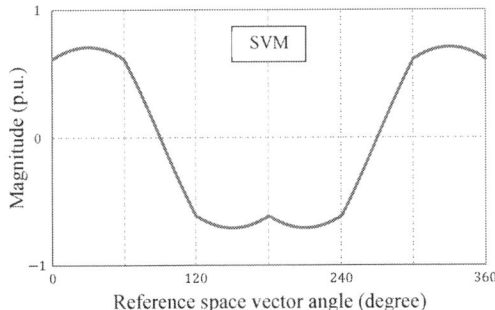

Fig. 7. Phase-u reference waveform for SVM scheme with modulation index M=0.818. The induction motor is controlled with V/F technique at the output frequency f_o= 40 Hz.

modulation index M will be close to 0.818, when the motor is working at the output frequency of 40 Hz. The simulated drive input current i_a waveform is illustrated in Fig. 8, where the SVM modulation technique has been selected as the rear-end inverter switching strategy. The input current i_a interharmonic components obtained using the MATLAB simulation and experimental results are shown in Fig. 9. The interharmonics frequency mapping, as already shown in Fig. 5, has been rescaled in Fig. 9(c) for further clarification. An intersection between the horizontal line at the output frequency f_o of 40 Hz with the plotted black points in Fig. 9(c) results in the corresponding drive input current interharmonic frequencies. As it can be observed, the interharmonics locations obtained using the theoretical analysis in (15) are in good agreement with the MATLAB simulation and experimental results, as illustrated in Figs. 9(a) and (b).

B. DPWM2 Modulation Technique

The discontinuous pulse width modulation-30 degree lagging clamp (DPWM2) scheme, which clamps the inverter pole terminals to the positive and negative terminals of the DC link for a 60° interval each in a fundamental period, is shown in Fig. 10. Since it usually is preferred to avoid the switching when the current through the devices is near its peak value, applying the DPWM2 technique can be found suitable for the applications where the power factor pf is close to 0.866 lagging such as the motor drive application.

Fig. 11 shows the adjustable speed drive input current i_a interharmonics, when the motor is operating at the output frequency f_o of 40 Hz and having a load torque value of 12 Nm, and, a DPWM2 modulation technique is selected for the inverter operation. Like the SVM modulation technique, the reference waveform modulation index M for the inverter operation is close to 0.818 to provide the desired output voltage fundamental frequency based on the selected (V/F) control strategy. As it can be seen in Fig. 11, there is a good agreement between the simulation and experimental results, and the plotted interharmonic frequencies. It is worth to note that the presence of two distinctive interharmonic components at 70 Hz and 170 Hz is due to higher amplitude of the DC-link third harmonic (of the output frequency) oscillation. As already discussed in Section II, the DC-link third harmonic (of the output frequency) oscillation is basically generated by the second and the fourth order frequency of output voltages.

Generally, addressing the drive input current interharmonics amplitude is a complicated topic, which needs a very precise model of interharmonics interactions, when the ASD is working at different operating conditions. This issue will be more sophisticated knowing that the interharmonics may usually have some overlaps at specific frequencies. In the overlap condition they could easily cancel or intensify each others. But theoretically, the inverter modulation index M, the switching frequency f_{sw} and the passive filter components of the drive

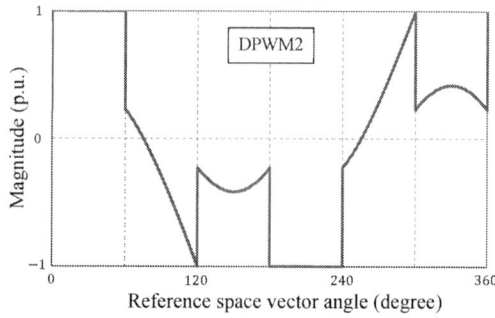

Fig. 10. Phase-u reference waveform for DPWM2 scheme with a modulation index M=0.818. The induction motor is controlled with V/F technique at the output frequency f_o= 40 Hz.

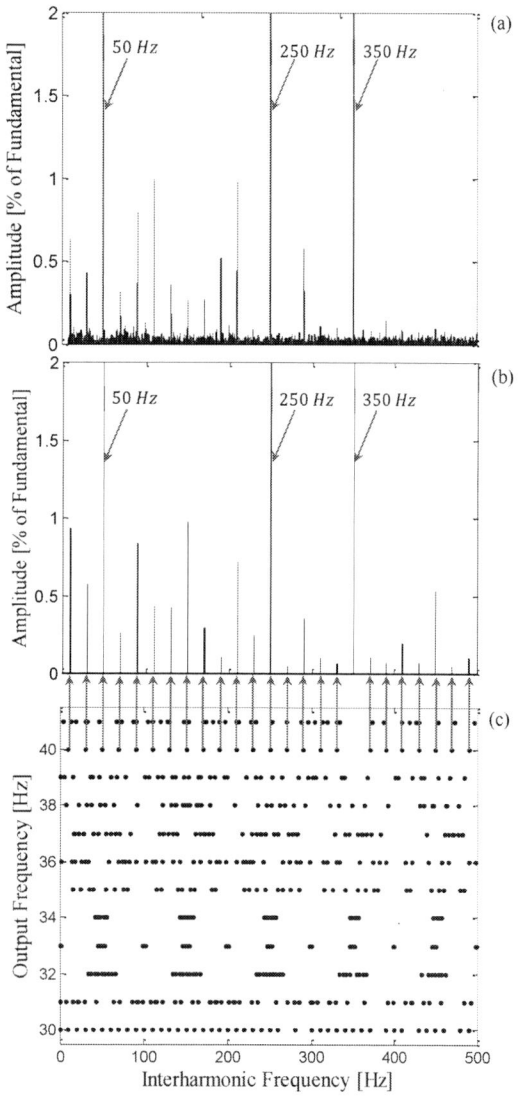

Fig. 9. Drive input current i_a interharmonics at the output frequency f_o= 40 Hz and load torque value of 12 Nm with applying SVM modulation technique: (a) Experimental result, (b) Simulation result, and (c) Output frequency versus estimated interharmonic frequencies using (15).

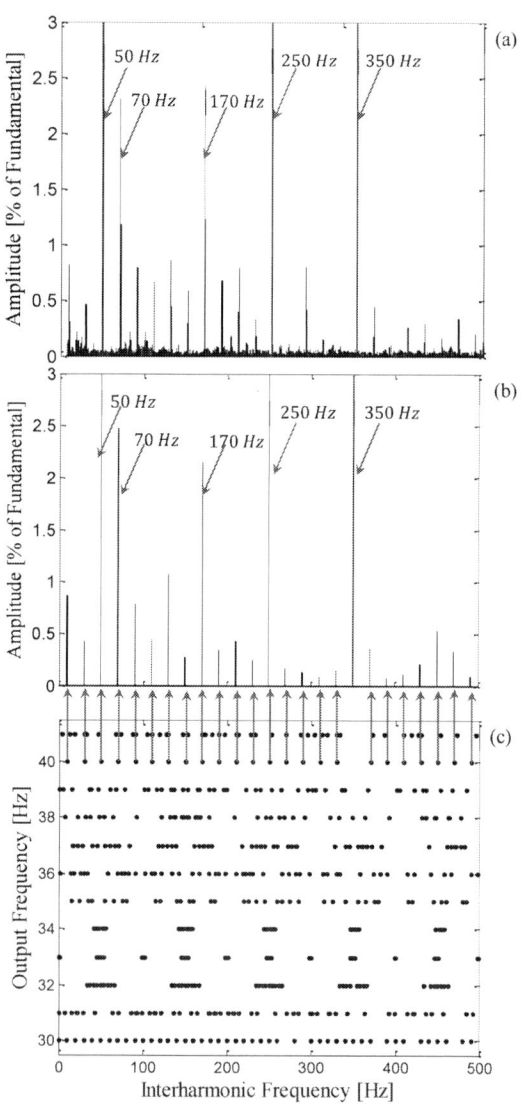

Fig. 11. Drive input current i_a interharmonics at the output frequency f_o= 40 Hz and load torque value of 12 Nm with applying DPWM2 modulation technique: (a) Experimental result, (b) Simulation result, and (c) Output frequency versus estimated interharmonic frequencies using (15).

will determine which modulation strategy may result in lower amplitude input current interharmonics in ASDs.

IV. CONCLUSION

In this paper, the adjustable speed drive input current interharmonics generated by double-edge symmetrical regularly sampled modulation strategies have comprehensively been analyzed. The investigation has been done at two different fixed-frequency pulse width modulation techniques (i.e. SVM, DPWM2). The harmonics transfer from the output side of the rear-end inverter to the input side of the front-end diode rectifier, which causes the interharmonics distortion, has been analyzed with respect to the baseband harmonics and the carrier sideband harmonics separately. Then, the drive input current interharmonic frequencies have been plotted using the proposed analysis. Finally, the results obtained by MATLAB simulation and experimental tests demonstrate accuracy of the analytical calculations. The investigations provide a precise benchmark for estimating the ASD′s input current interharmonic frequencies, which help to choose the correct frequency resolution for the DFT spectrum analysis.

With respect to the interharmonics amplitudes, it has been shown that applying the DPWM2 modulation technique may give rise to higher interharmonic amplitudes compared with the SVM modulation technique, when the ASD is operating at high switching frequency.

REFERENCES

[1] P. Davari, Y. Yang, F. Zare, and F. Blaabjerg, "A multi-pulse pattern modulation scheme for harmonic mitigation in three-phase multi-motor drives," *IEEE J. Emerg. Sel. Top. Power Electron.*, vol. PP, no. 99, pp. 1–12, Jul. 2015.

[2] A. Testa, M. Akram, R. Burch, G. Carpinelli, G. Chang, V. Dinavahi, C. Hatziadoniu, W. Grady, E. Gunther, M. Halpin *et al.*, "Interharmonics: theory and modeling," *IEEE Trans. Power Del.*, vol. 22, no. 4, pp. 2335–2348, Oct. 2007.

[3] D. Gallo, R. Langella, A. Testa, and A. Emanuel, "On the effects of voltage subharmonics on power transformers: a preliminary study," in *Proc. 11th ICHQP*, Lake Placid, NY, 2004, pp. 501–506.

[4] J. de Abreu and A. Emanuel, "Induction motor thermal aging caused by voltage distortion and imbalance: loss of useful life and its estimated cost," *IEEE Trans. Ind. Appl.*, vol. 38, no. 1, pp. 12–20, Jan. 2002.

[5] D. Gallo, C. Landi, R. Langella, and A. Testa, "IEC flickermeter response to interharmonic pollution," in *Proc. 11th ICHQP*, Lake Placid, NY, 2004, pp. 489–494.

[6] M. Hernes and B. Gustavsen, "Simulation of shaft vibrations due to interaction between turbine-generator train and power electronic converters at the visund oil platform," in *Proc. IEEE Power Convers. Conf. (PCC)*, 2002, pp. 1381–1386.

[7] C. Bowler, "Proposed steady-state limits for turbine-generator torsional response," in *Proc. IEEE Summer Power Meeting*, 2002.

[8] D. Basic, "Input current interharmonics of variable-speed drives due to motor current imbalance," *IEEE Trans. Power Del.*, vol. 25, no. 4, pp. 2797–2806, Oct. 2010.

[9] R. Yacamini, "Power system harmonics. Part 4: Interharmonics," *Power Eng. J.*, vol. 10, no. 4, pp. 185–193, Aug. 1996.

[10] E. W. Gunther, "Interharmonics in power systems," in *Proc. IEEE Power Eng. Soc. Summer Meeting*, 2001, pp. 813–817.

[11] D. Gallo, R. Langella, and A. Testa, "On the processing of harmonics and interharmonics in electrical power systems," in *Proc. IEEE Power Eng. Soc. Winter Meeting*, 2000, pp. 1581–1586.

[12] M. Rifai, T. H. Ortmeyer, and W. J. McQuillan, "Evaluation of current interharmonics from AC drives," *IEEE Trans. Power Del.*, vol. 15, no. 3, pp. 1094–1098, Jul. 2000.

[13] G. Chang, C. Chen, Y. Liu, and M. Wu, "Measuring power system harmonics and interharmonics by an improved fast Fourier transform-based algorithm," *IET Gener. Transm. Distrib.*, vol. 2, no. 2, pp. 193–201, Mar. 2008.

[14] C.-I. Chen and Y.-C. Chen, "Comparative study of harmonic and inter-harmonic estimation methods for stationary and time-varying signals," *IEEE Trans. Ind. Electron.*, vol. 61, no. 1, pp. 397–404, Jan. 2014.

[15] F. Cupertino, E. Lavopa, P. Zanchetta, M. Sumner, and L. Salvatore, "Running DFT-based PLL algorithm for frequency, phase, and amplitude tracking in aircraft electrical systems," *IEEE Trans. Ind. Electron.*, vol. 58, no. 3, pp. 1027–1035, Mar. 2011.

[16] I. Y.-H. Gu and M. H. Bollen, "Estimating interharmonics by using sliding-window ESPRIT," *IEEE Trans. Power Del.*, vol. 23, no. 1, pp. 13–23, Jan. 2008.

[17] G. W. Chang and C.-I. Chen, "An accurate time-domain procedure for harmonics and interharmonics detection," *IEEE Trans. Power Del.*, vol. 25, no. 3, pp. 1787–1795, Jul. 2010.

[18] D. Gallo, R. Langella, and A. Testa, "A self-tuning harmonic and interharmonic processing technique," *Eur. Trans. Elect. Power*, vol. 12, no. 1, pp. 25–31, Jan./Feb. 2002.

[19] C.-I. Chen and G. W. Chang, "An efficient Prony-based solution procedure for tracking of power system voltage variations," *IEEE Trans. Ind. Electron.*, vol. 60, no. 7, pp. 2681–2688, Jul. 2013.

[20] W. Xu, H. W. Dommel, M. B. Hughes, G. W. Chang, and L. Tan, "Modelling of adjustable speed drives for power system harmonic analysis," *IEEE Trans. Power Del.*, vol. 14, no. 2, pp. 595–601, Apr. 1999.

[21] R. Carbone, F. De Rosa, R. Langella, and A. Testa, "A new approach for the computation of harmonics and interharmonics produced by line-commutated AC/DC/AC converters," *IEEE Trans. Power Del.*, vol. 20, no. 3, pp. 2227–2234, Jul. 2005.

[22] G. W. Chang and S. K. Chen, "An analytical approach for characterizing harmonic and interharmonic currents generated by VSI-fed adjustable speed drives," *IEEE Trans. Power Del.*, vol. 20, no. 4, pp. 2585–2593, 2005.

[23] G. W. Chang, S. K. Chen, H. J. Su, and P. K. Wang, "Accurate assessment of harmonic and interharmonic currents generated by VSI-fed drives under unbalanced supply voltages," *IEEE Trans. Power Del.*, vol. 26, no. 2, pp. 1083–1091, Apr. 2011.

[24] H. Soltani, P. C. Loh, F. Blaabjerg, and F. Zare, "Interharmonic analysis and mitigation in adjustable speed drives," in *Conf. Rec. IECON 40th Annu. Meeting*, 2014, pp. 1556–1561.

[25] H. Soltani, P. Loh, F. Blaabjerg, and F. Zare, "Interharmonic mitigation of adjustable speed drives using an active DC-link capacitor," in *Proc. ICPE-ECCE Asia*, 2015, pp. 2018–2024.

[26] F. De Rosa, R. Langella, A. Sollazzo, and A. Testa, "On the interharmonic components generated by adjustable speed drives," *IEEE Trans. Power Del.*, vol. 20, no. 4, pp. 2535–2543, Oct. 2005.

[27] D. G. Holmes and T. A. Lipo, *Pulse width modulation for power converters: principles and practice.* John Wiley & Sons, 2003, vol. 18.

978-1-4673-9551-9/16 $31.00 © 2016 IEEE

Low-Frequency Voltage Ripples in the Flying Capacitors of the Nested Neutral-Point-Clamped Converter

Amer M. Y. M. Ghias[1], Josep Pou[2], Salvador Ceballos[3], and Vassilios G. Agelidis[2]

[1]University of Sharjah, Sharjah 27272, United Arab Emirates.

[2]Australian Energy Research Institute & School of Electrical Engineering and Telecommunications, The University of New South Wales, UNSW Sydney, NSW 2052, Australia.

[3] Tecnalia Research and Innovation, Energy and Environment Division, 48160 Derio, Spain.

Abstract—The flying capacitors (FCs) of the nested neutral-point-clamped (NNPC) converter show an inherent voltage ripple at fundamental frequency. This ripple can be significantly large under some operating conditions of the converter. In this paper, the amplitudes of the low-frequency voltage ripples in the FCs are determined. An averaged model of the NNPC converter is introduced and used in the analysis. The amplitudes of the capacitor voltage ripples are provided using normalized variables so that this information can be used to size the FCs of the converter in different applications. The results of the analysis are validated experimentally in a laboratory prototype.

I. Introduction

Multilevel converters are well-suited for medium- and high-power applications [1]–[10]. This is because multilevel converter topologies produce output voltages with lower total harmonic distortion (THD), provide lower power losses and allow higher voltage/power ratings than the conventional two-level converter counterpart. The most popular multilevel topologies are the cascaded multilevel converter [11], the modular multilevel converter (MMC) [12], the neutral-point-clamped (NPC) converter [13], and the flying capacitor (FC) converter [14]. Most of these topologies are well established within industry nowadays.

Recently, hybrid multilevel converter topologies have been introduced and are considered competitive solutions when compared with the popular multilevel topologies, since they require less energy storage and offer other interesting features [15], [16]. The four-level nested neutral-point-clamped (NNPC) converter was proposed in [17]. The NNPC converter is a hybrid topology from a FC converter [14] and a NPC converter [13]. When compared with the four-level NPC converter [18], the NNPC topologies requires fewer clamping diodes and capacitor voltage balance can be achieved very easily.

Some active capacitor voltage balancing methods applied to the NNPC converter have been proposed [17], [19], [20]. Although voltage balance can always be achieved, a fundamental-frequency voltage ripple appears in the FCs. Depending on the capacitor values and the operating conditions of the converter, such ripple can take significant amplitude. As a result, the output voltages contain low-frequency distortion and the maximum voltages applied to the capacitors and the

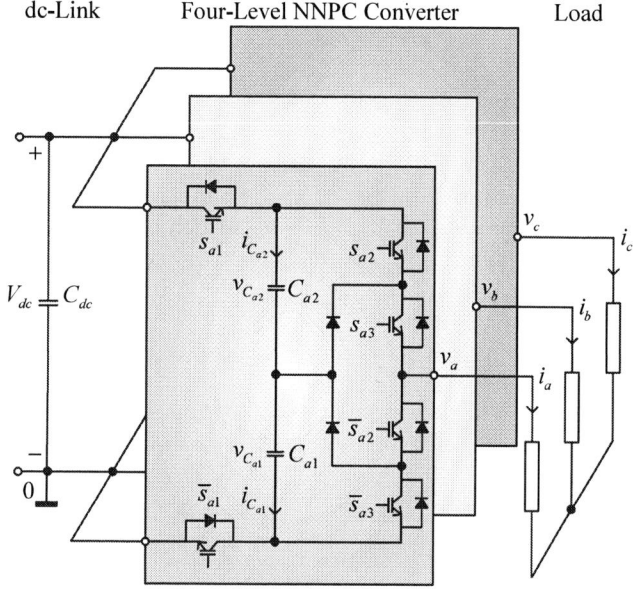

Fig. 1. Three-phase four-level NNPC converter.

power semiconductors of the converter increase. Although the low-frequency ripples may have a significant impact on the operation of the converter, none of the current publications have evaluated the amplitude of such ripples.

The aim of this paper is to evaluate the low-frequency voltage ripples in the FCs of the NNPC converter. An averaged model of the converter is developed by using mathematical equations and is used to determine the amplitudes of the low-frequency voltage ripples. The information obtained from this model is used to size the FCs of the converter. The theoretical analysis is validated experimentally.

The paper is organized as follows. Section II describes the fundamentals of the NNPC converter and an averaged model of the converter is developed. The model is used to determine the low-frequency ripples in the FCs. Section III presents experimental results obtained from a three-phase four-level NNPC converter. Finally, the conclusions of this work are summarized in Section IV.

TABLE I
FOUR-LEVEL NNPC CONVERTER: VOLTAGE LEVELS, SWITCHING STATES, FC CURRENTS, AND EFFECTS ON THE FC VOLTAGES

Output Voltage Level (v_{x0})			Switching States			FC Currents		FC Voltages	
			s_{x1}	s_{x2}	s_{x3}	i_{Cx1}	i_{Cx2}	v_{Cx1}	v_{Cx2}
4	l_4	V_{dc}	1	1	1	0	0	NC	NC
3	$l_{3,2}$	$V_{dc} - v_{C1} = \frac{2V_{dc}}{3}$	1	0	1	0	i_x	NC	↑
	$l_{3,1}$	$v_{C1} + v_{C2} = \frac{2V_{dc}}{3}$	0	1	1	$-i_x$	$-i_x$	↓	↓
2	$l_{2,2}$	$V_{dc} - v_{C1} - v_{C2} = \frac{V_{dc}}{3}$	1	0	0	i_x	i_x	↑	↑
	$l_{2,1}$	$v_{C2} = \frac{V_{dc}}{3}$	0	0	1	$-i_x$	0	↓	NC
1	l_1	0	0	0	0	0	0	NC	NC

Note: The charging/discharging effects in the FC are given assuming that i_x is positive ($i_x > 0$) with the following notation:

↑ Capacitor voltage increases

↓ Capacitor voltage decreases

NC No change in the capacitor voltage

II. NNPC CONVERTER

A. Topology and Fundamentals

Fig. 1 shows the circuit diagram of a three-phase four-level NNPC converter. It consist of two FCs per phase, C_{x1} and C_{x2} with $x \in \{a, b, c\}$, with a neutral point connection. For proper operation of the converter, each capacitor voltage is regulated to one third of the dc-link voltage ($V_{dc}/3$). The converter can provide up to four voltage levels to each output, i.e., $v_{x0} \in \{0, V_{dc}/3, 2V_{dc}/3, V_{dc}\}$. The switch control functions s_{xy} for $y \in \{1, ..., 3\}$, indicate the switch number corresponding to a particular cell in the phase-leg x of the converter. The switch control functions can take two values $s_{xy} \in \{0, 1\}$, with 0 and 1 corresponding to the switch in the off and on states, respectively. The switch pairs in each phase-leg (s_{xyz} and $\overline{s}_{xyz}$) operate in a complementary manner.

B. Averaged Mathematical Model

In order to evaluate the amplitudes of the low-frequency capacitor voltage ripples, an averaged NNPC converter model has been developed. Assuming a positive output current ($i_x > 0$) and according to the information shown in Table I, the output voltage of the converter with respect to the negative dc-link and the FCs currents can be expressed as:

$$v_{x0} = s_{x1}V_{dc} + (s_{x2} - s_{x1})v_{Cx2} + (s_{x3} - s_{x1})v_{Cx1}, \quad (1)$$

$$i_{Cx1} = -(s_{x3} - s_{x1})i_x \quad \text{and} \quad (2)$$

$$i_{Cx2} = -(s_{x2} - s_{x1})i_x. \quad (3)$$

The locally averaged representation of the output voltage (1) and FC currents (2) and (3) calculated over a switching period are:

$$\bar{v}_{x0} = d_{x1}V_{dc} + (d_{x2} - d_{x1})\bar{v}_{Cx2} + (d_{x3} - d_{x1})\bar{v}_{Cx1}, \quad (4)$$

$$\bar{i}_{Cx1} = -(d_{x3} - d_{x1})\bar{i}_x \quad \text{and} \quad (5)$$

$$\bar{i}_{Cx2} = -(d_{x2} - d_{x1})\bar{i}_x, \quad (6)$$

where $\bar{v}_{Cx1}$ and $\bar{v}_{Cx2}$ are the locally-averaged voltages of the capacitors C_{x1} and C_{x2}, respectively. $\bar{i}_{Cx1}$ and $\bar{i}_{Cx2}$ are the locally-averaged currents of the capacitors C_{x1} and C_{x2}, respectively. $\bar{i}_x$ is the locally-averaged output current, and d_{x1}, d_{x2} and d_{x3} are the duty cycles of the switch control functions s_{x1}, s_{x2} and s_{x3}, respectively. The average voltage of the FCs can be expressed as:

$$\bar{v}_{Cx1} = \frac{1}{C_{x1}} \int_0^t \bar{i}_{Cx1} dt + V_{C10} \quad (7)$$

$$\bar{v}_{Cx2} = \frac{1}{C_{x2}} \int_0^t \bar{i}_{Cx2} dt + V_{C20}, \quad (8)$$

where V_{C10} and V_{C20} are the initial voltages of the capacitors C_{x1} and C_{x2}, respectively. The locally averaged output voltage and FCs current in (4) to (6) can be expressed in terms of the duty cycle of the level as:

$$\bar{v}_{x0} = V_{dc}(l_{2,2}d_{xl2} + l_{3,2}d_{xl3} + d_{xl4}) + \bar{v}_{Cx1}[-l_{2,2}d_{xl2} +$$
$$(l_{3,1} - l_{3,2})d_{xl3}] + \bar{v}_{Cx2}[(l_{2,1} - l_{2,2})d_{xl2} + l_{3,1}d_{xl3}],$$
(9)

$$\bar{i}_{Cx1} = \bar{i}_x[(-l_{2,1} + l_{2,2})d_{xl2} - l_{3,1}d_{xl3}] \quad \text{and} \quad (10)$$

$$\bar{i}_{Cx2} = \bar{i}_x[l_{2,2}d_{xl2} + (-l_{3,1} + l_{3,2})d_{xl3}], \quad (11)$$

where $l_{2,1}$, $l_{2,2}$, $l_{3,1}$ and $l_{3,2}$ are the control functions of the redundant states shown in Table I. They take the value 1 when the particular state is activated and 0 otherwise. The particular redundant state to be applied will be selected by the voltage balancing algorithm. The duty cycles of the levels 1 to 4 are defined by the variables d_{xl1}, d_{xl2}, d_{xl3} and d_{xl4}, respectively. In the case of applying phase-disposition PWM (PD-PWM), the duty cycles of the output voltage levels can be obtained as follows:

$$\text{for} \quad \frac{2}{3}k - 1 \leq v_{xref} \leq \frac{2k-1}{3}:$$
$$d_{xlk} = k - \frac{3v_{xref} + 1}{2}, \quad (12)$$

$$\text{and for} \quad \frac{2k-5}{3} \leq v_{xref} \leq \frac{2}{3}k - 1:$$
$$d_{xlk} = \frac{3v_{xref} + 5}{2} - k, \quad (13)$$

$$\text{otherwise} \quad d_{xlk} = 0, \quad (14)$$

where v_{xref} is the modulation signal that ranges in the interval $[-1, 1]$ under linear operation mode, k is the particular voltage level $k \in \{1, ..., 4\}$, and d_{xlk} is its duty cycle.

Equations (7) to (14) represent the averaged model of the NNPC converter.

C. Voltage Balancing Method

The optimal states of the levels with redundant switching states available (l_2 and l_3) are selected for each switching period. The selection is performed according to the minimization of a cost function [17], [22], given as follows:

$$J_x = \frac{1}{2}(C_{x1}\Delta\bar{v}_{Cx1}^2 + C_{x2}\Delta\bar{v}_{Cx2}^2)$$
$$\Delta\bar{v}_{Cx1} = \bar{v}_{Cx1} - V_{dc}/3 \quad (15)$$
$$\Delta\bar{v}_{Cx2} = \bar{v}_{Cx2} - V_{dc}/3.$$

The cost function (15) is positively defined and it becomes zero if both FC voltages equal their reference value ($V_{dc}/3$). Therefore, it has to be minimized for each switching period to attain capacitor voltage balancing. One of the methods for minimizing the cost function is through differentiation, as follows:

$$\frac{d}{dt}J_x = \frac{d}{dt}\frac{1}{2}(C_{x1}\Delta\bar{v}_{Cx1}^2 + C_{x2}\Delta\bar{v}_{Cx2}^2)$$
$$= \Delta\bar{v}_{Cx1}\bar{i}_{Cx1} + \Delta\bar{v}_{Cx2}\bar{i}_{Cx2} \leq 0, \quad (16)$$

where $\bar{i}_{Cx1}$ and $\bar{i}_{Cx2}$ are the currents in the capacitors defined by (10)-(14), which depend on the redundant switching states and the load current. At any switching period, the cost function

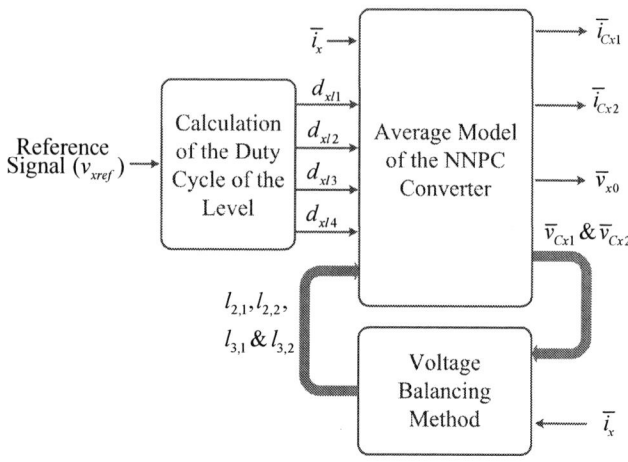

Fig. 2. Block diagram of the average model with embedded capacitor voltage balance.

is evaluated for all redundant state. Based on the calculated values, the redundant switching states that provides the minimum value to the cost function:

$$\min\left(\Delta\bar{v}_{Cx1}\bar{i}_{Cx1} + \Delta\bar{v}_{Cx2}\bar{i}_{Cx2}\right) \quad (17)$$

are the ones selected and used to obtain $\bar{v}_{x0}$, $\bar{i}_{Cx1}$ and $\bar{i}_{Cx2}$ from the mathematical model.

Fig. 2 shows the block diagram of the proposed average model for the four-level NNPC converter, which is used to evaluate the amplitudes of the low-frequency voltage ripples in the FCs.

D. Low-Frequency Capacitor Voltage Ripples

In this subsection, the amplitudes of the low-frequency capacitor voltage ripples are obtained. The voltage amplitudes are represented using a normalized magnitude in order to present general results that are useful for different applications and operating conditions, as follows:

$$\frac{\Delta V_{FCn}}{2} = \frac{\Delta V_{FC}/2}{I_{rms}/fC} \quad (18)$$

where ΔV_{FC} is the peak-to-peak low-frequency FC voltage ripple, f is the fundamental output frequency, I_{rms} is the root mean square (rms) output current, and C is the value of the FCs.

Assuming sinusoidal output currents, Fig. 3 shows the amplitudes of the FC voltage ripples for all the operating conditions of the converter, i.e., modulation indices and phase current angles. This information is very helpful to size the FCs in a practical application. The maximum amplitude is produced when operating with a modulation index $m = 0.4$ and the output current angle is $\theta = 20°$ or $\theta = -160°$. Under such conditions, the maximum normalized ripple is:

$$\frac{\Delta V_{FCn}}{2} = \frac{\frac{\Delta V_{FC}}{2}}{\frac{I_{rms}}{fC}} = 0.08. \quad (19)$$

If the values of the parameters in a particular application were, for instance, $I_{rms} = 20A$, $f = 50Hz$, and $C = 1000\mu F$, the

978-1-4673-9551-9/16 $31.00 © 2016 IEEE

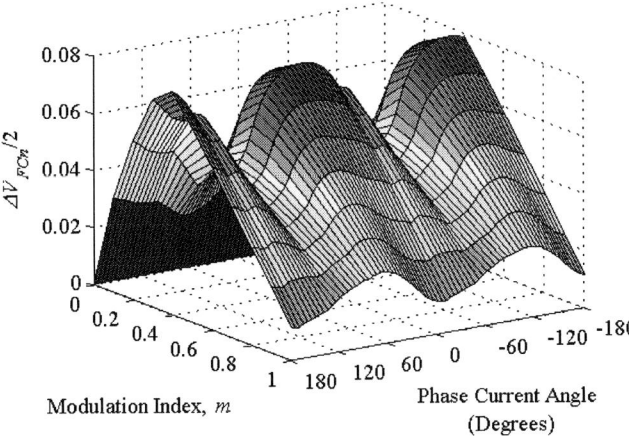

Fig. 3. Normalized amplitudes of the FC voltage ripples.

maximum amplitude of the FC voltage ripples would be:

$$\frac{\Delta V_{FC}}{2} = \frac{\Delta V_{FCn}}{2}\frac{I_{rms}}{fC} = 0.08\frac{20}{50 \cdot 0.001} = 32V. \quad (20)$$

Other parameters and operating conditions can be considered in (20) to estimate the FC voltage ripples. Therefore, the information provided in Fig. 3 is extremely useful to size the FCs in the NNPC converter.

III. EXPERIMENTAL RESULTS

Experimental results are obtained from a low-power three-phase NNPC converter to validate the results of the theoretical analysis. For the control of the converter, a DSPACE 1006 controller with integrated DS 5203 FPGA board is used [23], [24]. In this experiment, the dc voltage is $V_{dc} = 100V$ and a Wye-connected linear RL load ($R = 10\Omega$, $L = 4mH$) is used. The value of the FCs is 440μF. The fundamental and the carrier frequencies are $f = 50Hz$ and $f_s = 5kHz$, respectively.

Fig. 1 shows experimental results when the converter operates with a modulation index of 0.4. The waveforms presented

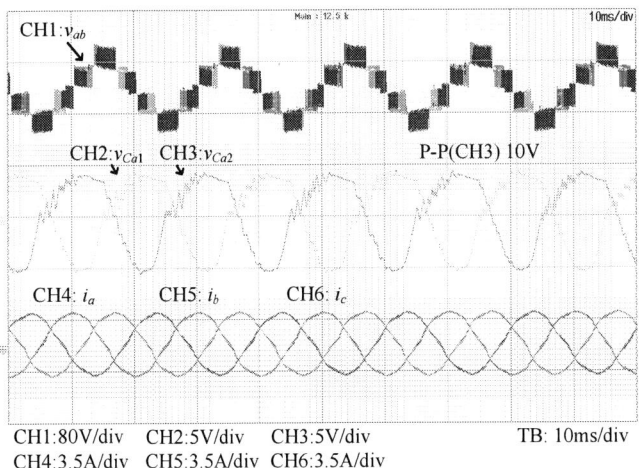

CH1:80V/div   CH2:5V/div   CH3:5V/div   TB: 10ms/div
CH4:3.5A/div CH5:3.5A/div CH6:3.5A/div

Fig. 4. Experimental results showing a line-to-line voltage, FC voltages of phase a, and the output currents.

in this figure are the line-to-line voltage (v_{ab}), the FC voltages (v_{Ca1} and v_{Ca2}), and the load currents (i_a, i_b and i_c). The rms output current at the operating point is $I_{rms} = 1.45A$, with an angle of 2.29°. In such conditions, the expected amplitude of the FC voltage ripple according to the theoretical analysis would be:

$$\frac{\Delta V_{FC}}{2} = \frac{\Delta V_{FCn}}{2}\frac{I_{rms}}{fC} = 0.078\frac{1.45}{50 \cdot 0.00044} \approx 5.1V, \quad (21)$$

which matches the amplitude obtained in the experiment.

Due to the voltage ripple, the capacitors and the power devices must withstand higher voltages. The reference voltage for the FCs is $Vdc/3 = 33.3V$, however the maximum voltage in the capacitors under this operating condition is:

$$V_{FC\,Max} = \frac{V_{dc}}{3} + \frac{\Delta V_{FC}}{2} \approx 38.4V, \quad (22)$$

which is 15.2% larger.

Furthermore, the capacitor voltage ripples produce low-frequency distortion to the output voltages generated by the converter. This effect will eventually distort the output currents, as it can be appreciated in Fig. 4.

In this analysis, the switching-frequency ripple in the FC is neglected because it is significantly smaller compared to the fundamental-frequency ripple. This is generally acceptable and can be assumed in many applications. However, if the switching-frequency ripples were more significant, they should be taken into consideration and added to the calculated value of the maximum voltage (22).

The results of the theoretical analysis performed in this paper can also be used to size the FCs given the specification of maximum voltage ripple. This will provide an initial estimation of the capacitances based on which the converter can be designed to meet other operating criteria.

IV. CONCLUSION

In this paper, the low-frequency voltage ripples that appear in the FCs of the NNPC converter have been determined. An averaged model of the converter has been developed and used to obtain the amplitudes of the capacitor voltage ripples. The results of the analysis are presented using a nondimensional variable and therefore they are general and useful for different applications of the converter. The results can also be used to size the FCs given an specification of maximum voltage ripple. A low-power laboratory prototype has been used to confirm the theoretical results.

REFERENCES

[1] S. Kouro, M. Malinowski, K. Gopakumar, J. Pou, L. G. Franquelo, B. Wu, J. Rodriguez, M. A. Perez, and J. I. Leon, "Recent advances and industrial applications of multilevel converters," *IEEE Trans. Ind. Electron.*, vol. 57, no. 8, pp. 2553–2580, Aug. 2010.

[2] W. Jinn-Chang and C. Chia-Wei, "A solar power generation system with a seven-level inverter," *IEEE Trans. Power Electron.*, vol. 29, no. 7, pp. 3454–3462, Jul. 2014.

[3] V. Yaramasu, B. Wu, M. Rivera, and J. Rodriguez, "A new power conversion system for megawatt PMSG wind turbines using four-level converters and a simple control scheme based on two-step model predictive strategy-Part I: Modeling and theoretical analysis," *IEEE J. Emerg. and Sel. Topics Power Electron.*, vol. 2, no. 1, pp. 3–13, Mar. 2014.

[4] V. Yaramasu, B. Wu, M. Rivera, and J. Rodriguez, "A new power conversion system for megawatt PMSG wind turbines using four-level converters and a simple control scheme based on two-step model predictive strategy-Part II: Simulation and experimental analysis," *IEEE J. Emerg. and Sel. Topics Power Electron.*, vol. 2, no. 1, pp. 14–25, Mar. 2014.

[5] V. Yaramasu and B. Wu, "Model predictive decoupled active and reactive power control for high-power grid-connected four-level diode-clamped inverters," *IEEE Trans. Ind. Electron.*, vol. 61, no. 7, pp. 3407–3416, Jul. 2014.

[6] M. A. Parker, L. Ran, and S. J. Finney, "Distributed control of a fault-tolerant modular multilevel inverter for direct-drive wind turbine grid interfacing," *IEEE Trans. Ind. Electron.*, vol. 60, no. 2, pp. 509–522, Feb. 2013.

[7] E. Solas, G. Abad, and J. A. Barrena, S. Aurtenetxea, A. Carcar, and L. Zajac, "Modular multilevel converter with different submodule concepts Part II: experimental validation and comparison for HVDC application," *IEEE Trans. Ind. Electron.*, vol. 60, no. 10, pp. 4536–4545, Oct. 2013.

[8] G. Bergna, E. Berne, P. Egrot, P. Lefranc, A. Arzande, J. C. Vannier, and M. Molinas, "An energy-based controller for HVDC modular multilevel converter in decoupled double synchronous reference frame for voltage oscillation reduction," *IEEE Trans. Ind. Electron.*, vol. 60, no. 6, pp. 2360–2371, Jun. 2013.

[9] G. Buticchi, D. Barater, E. Lorenzani, C. Concari, and G. Franceschini, "A nine-level grid-connected converter topology for single-phase transformerless PV systems," *IEEE Trans. Ind. Electron.*, vol. 61, no. 8, pp. 3951–3960, Aug. 2014.

[10] C. D. Townsend, T. J. Summers, and R. E. Betz, "Impact of practical issues on the harmonic performance of phase-shifted modulation strategies for a cascaded H-bridge statcom," *IEEE Trans. Ind. Electron.*, vol. 61, no. 6, pp. 2655–2664, Jun. 2014.

[11] F. Z. Peng, J. S. Lai, J. W. McKeever, and J. VanCoevering, "A multilevel voltage-source inverter with separate dc sources for static VAr generation," *IEEE Trans. Ind. Appl.*, vol. 32, no. 5, pp. 1130–1138, Sep./Oct. 1996.

[12] A. Lesnicar and R. Marquardt, "An innovative modular multilevel converter topology suitable for a wide power range," in *Proc. IEEE Power Tech Conference*, 23–26 Jun. 2003, vol. 3, pp. 6.

[13] A. Nabae, I. Takahashi, and H. Akagi, "A new neutral-point-clamped PWM inverter," *IEEE Trans. Ind. Appl.*, vol. IA-17, no. 5, pp. 518–523, Sep./Oct. 1981.

[14] T. A. Meynard and H. Foch, "Multi-level conversion: High voltage choppers and voltage-source inverters," in *Proc. IEEE Power Electronics Specialists Conference (PESC)*, 29 Jun.-3 Jul. 1992, vol. 1, pp. 397–403.

[15] T. A. Meynard, H. Foch, F. Forest, C. Turpin, F. Richardeau, L. Delmas, G. Gateau, and E. Lefeuvre, "Multicell converters: Derived topologies," *IEEE Trans. Ind. Electron.*, vol. 49, no. 5, pp. 978–987, Oct 2002.

[16] G. Gateau, T. A. Meynard, and H. Foch, "Stacked multicell converter (SMC): Properties and design," in *Proc. IEEE Power Electronics Specialists Conference (PESC)*, 2001, vol. 3, pp. 1583–1588.

[17] M. Narimani, B. Wu, Z. Cheng and N. R. Zargari, "A novel and simple single-phase modulator for the nested neutral-point clamped (NNPC) converter," *IEEE Trans. Power Electron.*, vol. 30, no. 8, pp. 4069–4078, Aug. 2015.

[18] J. Pou, R. Pindado, and D. Boroyevich, "Voltage-balance limits in four-level diode-clamped converters with passive front ends," *IEEE Trans. Ind. Electron.*, vol. 52, no. 1, pp. 190–196, Oct. 2005.

[19] M. Narimani, B. Wu, Z. Cheng, and N. R. Zargari, "A new nested neutral point-clamped (NNPC) converter for medium-voltage (MV) power conversion," *IEEE Trans. Power Electron.*, vol. 29, no. 12, pp. 6375–6382, Dec. 2014.

[20] M. Narimani, B. Wu, and N. R. Zargari, "A novel five-level voltage source inverter with sinusoidal pulse width modulator for medium-voltage applications," *IEEE Trans. Power Electron.*, vol. 31, no. 3, pp. 1959–1967, Mar. 2016.

[21] J. Pou, D. Boroyevich, and R. Pindado "New feedforward space-vector PWM method to obtain balanced AC output voltages in a three-level neutral-point-clamped converter," *IEEE Trans. Ind. Electron.*, vol. 49, no. 5, pp. 1026–1034, Oct. 2002.

[22] A. M. Y. M. Ghias, J. Pou, V. G. Agelidis, and M. Ciobotaru, "Voltage balancing method for a flying capacitor multilevel converter using phase disposition PWM," *IEEE Trans. Ind. Electron.*, vol. 61, no. 12, pp. 6538–6546, Dec. 2014.

[23] MATLAB/SIMULINK software package, version R2011b, The Math-Works. [Online]. Available: http://www.mathworks.com

[24] DSPACE 1006, Solutions for Control. [Online]. Available: http://www.dspace.com

DC Bus Capacitor Discharge of Permanent Magnet Synchronous Machine Drive Systems for Hybrid Electric Vehicles

Ziwei Ke, Julia Zhang
School of Electrical Engineering and Computer Science
Oregon State University
Corvallis, OR, USA
zhangjul@eecs.oregonstate.edu

Michael W. Degner
Research and Innovation Center
Ford Motor Company
Dearborn, MI, USA
mdegner@ford.com

Abstract—This paper investigates control methods to quickly and safely discharge the high voltage DC bus capacitor for the permanent magnet synchronous machine (PMSM) drive systems of hybrid electric vehicles (HEVs) during emergency situations such as a crash event. The published DC bus capacitor discharge strategies either need impractically heavy and bulky hardware or are only effective for low machine speeds. Applying the existing methods under high speeds will lead to loss of the current control and a system shutdown, followed by the bus voltage recharging again. The proposed new control method applies a modulation index controller together with a bus voltage regulator to quickly discharge the bus voltage and avoid any bus recharge regardless of the electric machine speed. Simulation and experimental results are presented to verify the effectiveness of the proposed control algorithm.

Keywords: active discharge, permanent magnet synchronous machines, bus regulation, modulation index regulation, hybrid electric vehicles.

I. INTRODUCTION

A typical topology for the PMSM drive systems in HEVs is shown in Fig. 1 [1-3]. The high voltage batteries, also called traction batteries, in the HEV system generally operate between 250 V and 650 V [2]. The battery pack is connected to the power electronic converters through a contactor, which is closed during the normal operation of the vehicle. The contactor will be disconnected in response to either a passive or active discharge request. The passive discharge is used for the normal key-off events that do not need a rapid discharge of the high voltage bus V_{dc}. A high resistance value resistor parallel to the bus capacitor to bleed the bus energy can typically meet the requirements of normal key-off events. For emergency situations such as a crash event, the vehicle system requires the DC bus voltage V_{dc} to drop below the safe voltage level, i.e., 60 V DC or 30 V AC, within 5 seconds to avoid any electrical shock hazards according to the United Nation Vehicle Regulation ECE R94 [4].

One general practice to actively discharge the DC bus is to parallel a dynamic brake circuit with the bus capacitor as shown in Fig. 2 [5, 6]. The dynamic break circuit consists of a low-resistance, high-power bleeding resistor in series with a power switch as shown. The size and weight of the dynamic

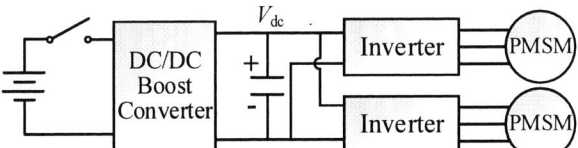

Fig. 1. A typical topology for the traction machine drive systems of hybrid electric vehicles.

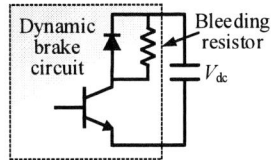

Fig. 2. Dynamic brake circuit to discharge high voltage DC bus.

brake circuit will compromise the compactness and cost of the electric machine drive systems for HEVs.

Another commonly-used active discharge method is to use the electric machine windings to dissipate the bus capacitor energy by applying a non-zero *d*-axis current and zero *q*-axis current to the PM machine [7, 8]. In this case, no additional hardware is required to implement the discharge function. An active discharge event can occur at any electric machine speed in the real world. This method is only effective when the electric machine speed is lower than the threshold speed where the machine back electromotive force (back-emf) is equal to the safe voltage. The waiting time, determined by the machine speed when the discharge is requested and the moment of inertia of the machine, can be very long (multiple minutes in duration) since the wheels may be freely spinning. Section II will explain the issues of this method in detail.

The contribution of this work is to investigate and implement a new control algorithm that can rapidly discharge the high voltage bus down to a safe range regardless of the electric machine speed whenever an active discharge event is requested. Although the existing methods meet the current vehicle standards, the proposed method provides improved dynamic results as a DC bus recharge problem is avoided, and the discharging time is significantly reduced from seconds or minutes to a few hundred milliseconds compared with the existing algorithms. Both simulation and experimental test

978-1-4673-9551-9/16 $31.00 © 2016 IEEE

results are presented to verify the proposed active discharge control algorithm.

II. TRADITIONAL HIGH VOLTAGE DC BUS DISCHARGE ALGORITHM USING MACHINE WINDINGS

This section reviews the traditional high voltage DC bus active discharge algorithms using electric machine windings and related issues. Experiments were conducted to demonstrate the issues.

A) Active Discharge Starts Immediately upon an Active Discharge Request

Upon an active discharge request, immediately applying a large *d*-axis current and zero *q*-axis current for the electric machine first seems to be a reasonable solution to quickly reduce the DC bus voltage. Fig. 3 demonstrates the series of events that actually happen.

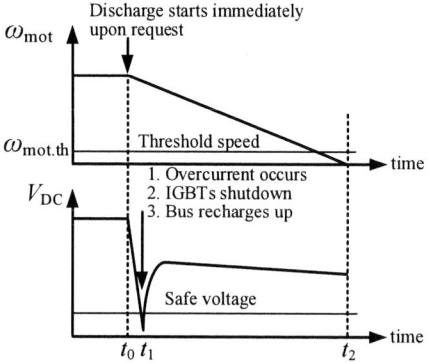

Fig. 3. Series of events when the discharge starts immediately upon an active discharge request.

The back-emf of a PM machine is proportional to its speed. The threshold speed $\omega_{mot.th}$ is defined as the speed at which the back-emf of the PM machine is equal to the safe voltage 60 V. The discharge starts at t_0 right after the discharge is requested, the bus voltage will first decrease quickly due to the large copper loss in the machine windings. When the bus voltage is discharged to close to zero, the current regulators will not be able to operate appropriately due to low voltage. Under extremely low voltage, the steady state *d*- and *q*-axis currents of a PM machine will approach the machine characteristic current, λ_m/L_d, and zero, where λ_m is the flux linkage of the permanent magnets, and L_d is the machine *d*-axis inductance. The transient current will trigger the overcurrent protection at t_1 followed by the power switches shutdown. Since the machine is still running at a higher speed than the threshold speed, the machine will begin running as a generator to charge up the bus voltage again.

Experimental tests were performed using an interior permanent magnet (IPM) synchronous machine to demonstrate the above bus recharge problem if the discharge begins above the threshold speed at which the machine back-emf is higher than the safe voltage. The test results are shown in Fig. 4. All the analyses, simulations, and experimental tests of this study are based on this IPM machine. The major parameters of the machine can be found in Table I in Section

IV. The threshold speed of this machine is about 1350 rpm. The active discharge is requested at 0.052 sec when the machine speed is 2,500 rpm, which is above the threshold speed. The discharge starts immediately using -100 A *d*-axis current. The bus voltage declines from 250 V to about 20 V when an over current occurs. All power electronics are then shut down and the machine begins to run as a generator and recharges the bus above 100 V. Since the power electronics and machine are not consuming any power after the shutdown, the bus voltage drops extremely slowly. In an HEV system, the bus energy will be dissipated slowly by the bleeding resistor across the bus capacitor after the power converter experiences a shutdown.

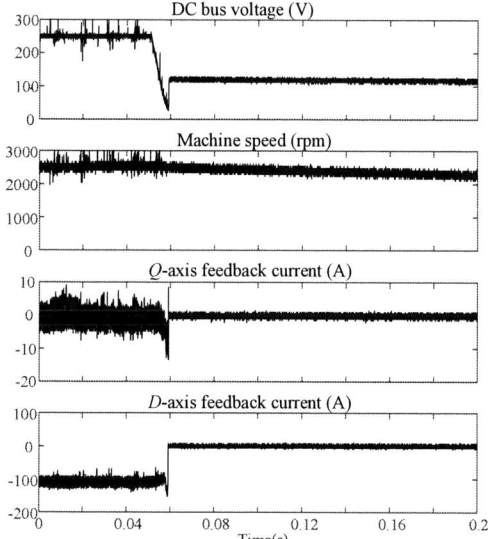

Fig. 4. Experimental test results of discharge immediately upon an active discharge request.

B) Active Discharge Starts after Machine Speed Drops below Threshold Value

To avoid the bus recharge problem under high machine speed, the actual discharge will not begin until the machine speed drops below the threshold value at which the machine back-emf is equal to the safe voltage, as demonstrated in Fig. 5.

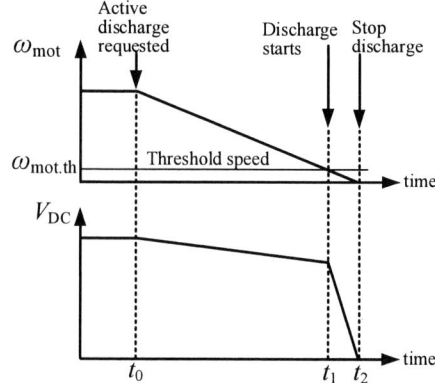

Fig. 5. Active discharge starts after the machine speed falls below the threshold value.

The machine is operating in the normal condition before t_0. At t_0, an active discharge is requested. As the machine speed is above the threshold speed, a zero current command is given to the machine. Due to the mechanical friction, the machine is decelerating and the bus voltage is decreasing. When the machine speed drops below the threshold speed, a large d-axis current is applied to the drive system that will quickly dissipate the bus capacitor energy. However, once the bus voltage reduces to a very low level close to zero, the similar bus recharge problem described in the previous section will happen as well.

Experiments were conducted to test the above algorithm and the results are shown in Fig. 6. The active discharge is requested at 0.1 sec while the machine speed is higher than the threshold value. Both the d- and q-axis currents are controlled to be zero and the machine is decelerating due to the mechanical friction. At 0.6 sec, the machine speed falls below the threshold speed and a large d-axis current, -80 A, is applied and the bus voltage immediately drops from approximately 80 V to 20 V. As the bus voltage reduces to close to zero (around 0.69 sec), the machine current increases quickly and triggers the over current protection at 0.7 sec. All power switches are then disabled and the current drops to zero in a short period of time. The back-emf and energy stored in the winding inductance recharge the DC bus capacitor back to 80 V again. This test was conducted under 2500 rpm. For higher speeds, the bus can easily be recharged back up to a few hundred volts.

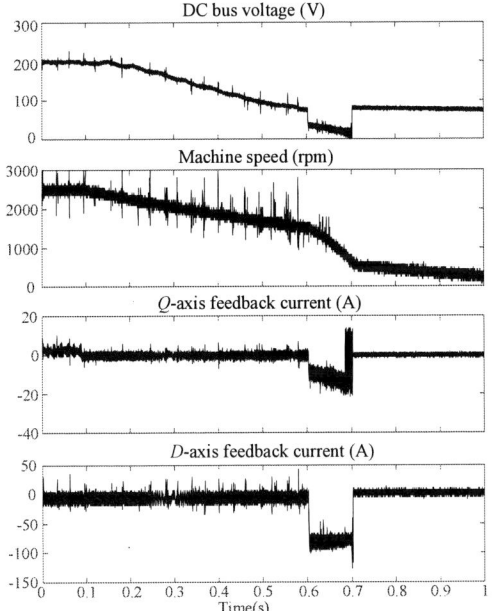

Fig. 6. Experimental test results of active discharge that starts after the machine speed drops below the threshold value.

III. PRINCIPLES OF PROPOSED HIGH VOLTAGE DC BUS DISCHARGE ALGORITHM

This section discusses the desired active discharge performance and proposes a new active discharge algorithm

that can quickly and safely discharge the bus to the safe voltage regardless of the machine speed at which the active discharge is requested. A single-machine system is used to explain the control algorithm details. The control algorithm is then generalized to accommodate the multiple-machine vehicle systems.

A) Active Discharge for a Single-machine Vehicle System

Fig. 7 illustrates the desired high voltage DC bus capacitor discharge performance. The electric machine speed is above the threshold speed when the active discharge is requested. The entire discharge process can be divided into the following three phases:

1) Phase 1 (t_1 to t_2): the fast discharge phase. Once the active discharge event is requested at t_1, the highest d-axis current that will not cause magnet demagnetization is applied to the electric machine to bleed the DC bus voltage to the safe level $V_{DC \cdot safe}$ (60 V). Simultaneously, the q-axis current is kept at zero to prevent any torque production.

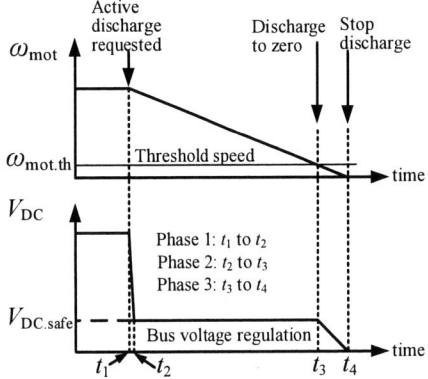

Fig. 7. Desired discharge performance using one machine for example.

2) Phase 2 (t_2 to t_3): the bus voltage regulation phase. From t_2, the DC bus voltage regulation algorithm, shown in Fig. 7, is activated to maintain the bus voltage slightly lower than the safe voltage until the machine speed drops below the threshold, i.e., the machine back-emf is lower than the safe voltage. The DC bus voltage regulator produces a q-axis current command to provide enough energy to keep the DC bus voltage from completely discharging. The modulation index feedback control is used to produce the d-axis current command so that the machine operating point will not exceed the control limit where loss of control of the currents could result in uncontrolled generation and the DC bus voltage quickly recharging.

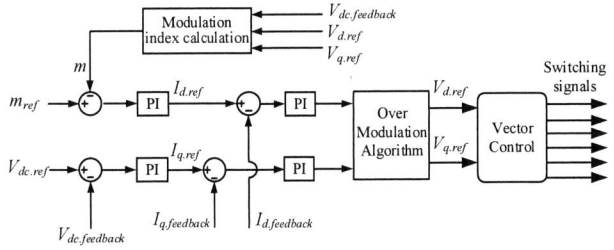

Fig. 8. Proposed active discharge algorithm using bus voltage regulation and modulation index regulation.

3) Phase 3 (t_3 to t_4): the active discharge wind down phase. After the machine decelerates to a speed lower than the threshold speed $\omega_{mot.th}$ at which the machine back-emf is lower than the safe voltage, the discharge currents (both d-axis and q-axis) will be ramped down to zero. Turning them off slowly instead of immediately turning off all power electronics in a very short period of time prevents the bus voltage from recharging again due to the stored energy in the machine's inductance. The current ramp rate was selected so that the power loss generated by the discharge current should be higher than the regeneration power during discharge.

Fig. 9 shows three voltage limit ellipses of an IPM machine in the i_d-i_q plane for three different levels of DC bus voltage under a constant machine speed of 3,000 rpm. The possible operating area of the IPM machine is constrained by both the circuit limit circle and the voltage limit ellipse. During the fast discharge phase, the voltage limit ellipse is shrinking as the bus voltage decreases rapidly. The closer the machine operating point is to the voltage limit ellipse, the higher the modulation index is. In the simulations and experiments presented later, the reference modulation index is selected to be 0.9. The modulation index in this study is defined as:

$$m = \frac{\sqrt{v_d^2 + v_q^2}}{V_{dc}/\sqrt{3}} \tag{1}$$

where v_d and v_q are the command d- and q-axis voltages, and $V_{dc}/\sqrt{3}$ is the maximum phase voltage considering that the space vector PWM and linear modulation are used. Using linear modulation, the maximum modulation index is 1 for the above definition.

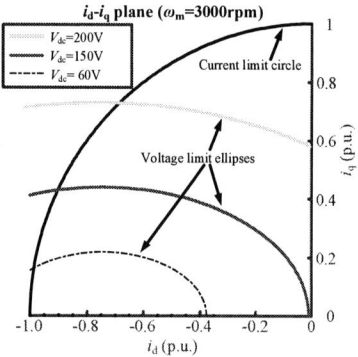

Fig. 9. Voltage limit ellipses under various DC bus voltage levels.

B) Active Discharge for a Multiple-machine HEV System

For a multiple-machine HEV system, the discharge also consists of the fast discharge, bus regulation, and discharge wind down phases. At least one or more machines can be used to discharge the bus energy depending on the operating speeds of all machines when an active discharge is requested and how fast the system requires the voltage to reduce below the safe voltage.

The machine(s) running above the threshold speed(s) should be involved in the discharge process. The threshold speed may differ from machine to machine. The machine(s) running below the threshold speed(s) do not have to be used to discharge the bus voltage if the other machine(s) are sufficient enough to reduce the bus voltage to the safe voltage in the required time. The machine(s) operating below the threshold speed(s) can also be used to discharge the bus voltage if the goal is to reduce the voltage as quickly as possible.

IV. SIMULATION AND EXPERIMENTAL VERIFICATIONS

Simulations and experimental tests were conducted on an IPM synchronous machine drive system to verify the active discharge algorithm. Key machine parameters and ratings are presented in Table I. These machine parameters were also used in the simulations. The back-emf of this machine exceeds the safe voltage above 1350 rpm approximately, i.e., the threshold speed as defined in the previous sections, $\omega_{mot.th}$.

Table I. Key machine parameters and ratings

Machine parameters	Values
Number of poles	8
d-axis inductance	0.15 mH
q-axis inductance	0.25 mH
Phase winding DC resistance	0.01 Ω
Switching frequency	5 kHz
PM flux linkage	0.045 V-s
Maximum current	500 A

A) Simulation Results and Analysis

In the simulations, a circuit breaker was added between the DC power supply and the high voltage DC bus of the three-phase inverter for the IPM machine. This circuit breaker was used to simulate the disconnection of the battery from the inverter during a fault condition in the vehicle. Fig. 10 shows the simulation results of an active discharge test. Before the discharge, the IPM machine is running at 3,000 rpm and the DC bus voltage is 250 V. The active discharge request occurs at 0.1 sec. The q-axis current is then set to be zero to prevent any torque production and the d-axis current is kept at -160 A to quickly bleed the bus voltage. Meanwhile, the machine is decelerating. Since the mechanical transient is much slower than the electrical transient, the mechanical speed of the machine only has a small change during the fast discharge phase. It is necessary to point out that a higher d-axis current can be used to discharge the bus voltage and the higher the d-axis current is, the quicker the bus will be discharged. Here -160 A was applied as the discharge current for this current was sufficient enough to discharge the bus voltage below the safe voltage in the required time.

Fig. 11 shows that it takes about 34 ms for the bus voltage to drop from 250 V to 60 V. The only loss considered in the simulation system was the copper loss of the electric machine windings. The fast discharge will take much shorter time in the experimental tests due to other loss sources such as the inverter switching and conduction losses, machine iron losses, and mechanical friction loss.

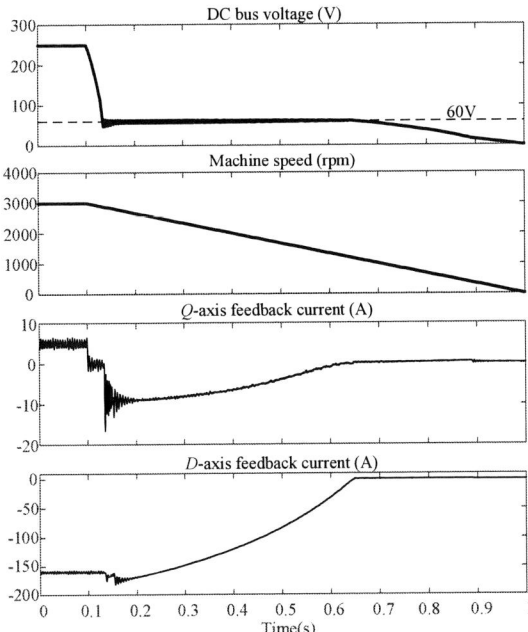

Fig. 10. Simulation results of proposed active discharge algorithm using a single PM machine.

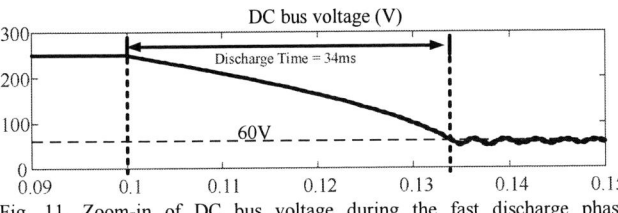

Fig. 11. Zoom-in of DC bus voltage during the fast discharge phase (simulation results).

The bus regulation phase starts at 0.0134 sec when the bus voltage drops below the safe voltage, 60V. From 0.0134 sec to 0.65 sec, the machine is running in the bus regulation phase as the back-emf generated by the machine is still higher than the safe voltage. During this phase, the bus voltage command is set to be 60 V and the modulation index command 0.9. It can be observed that the feedback DC bus voltage follows the command well. While the machine is decelerating, the discharge current is decreasing. At 0.65 sec, the machine speed already falls below the threshold speed, the current ramps down to zero. Afterward, the bus voltage gradually reduces to zero as the machine is slowing down. No bus recharge problem occurs.

B) Experimental Verification

In the experimental tests, a relay was installed between the DC power supply and the high voltage bus capacitor of the three-phase inverter to simulate the battery contactor disconnection during a fault condition in the vehicle. Once an active discharge is requested, a command is sent through an I/O port of the microcontroller to disconnect the relay and activate the active discharge control algorithm. Fig. 12 shows a picture of the experimental test system.

Fig. 12. Experimental test system.

Fig. 13 demonstrates the experimental test results including the bus voltage, machine mechanical speed, d- and q-axis currents. Before the active discharge, the IPM machine is running at 3,000 rpm under no load condition and the bus voltage is 250 V.

The active discharge command occurs at approximately 0.077 sec. The zoom-in of the DC bus voltage in Fig. 14 shows that it takes about 13 ms for the bus voltage to discharge from 250 V to 60 V, which is much faster compared to the simulation results. As mentioned earlier, the power electronics losses, machine iron losses, and mechanical loss are all ignored in simulations, which results in a longer discharge time.

The bus voltage is discharged to the safe voltage at around 0.091 sec when the bus regulation algorithm begins to work. During the DC bus regulation phase, the modulation index command is set to be 0.9 and the DC bus voltage regulator generates a small negative q-axis current command to maintain the bus voltage. The torque produced by this q-axis current is negligible and contributes slightly to slowing down the machine faster. As the machine is slowing down, the voltage limit ellipse is expanding, i.e., the right intersection point between the voltage limit ellipse and the i_d-axis is moving right toward the origin. Therefore, the required d-axis current is decreasing to maintain the constant modulation index 0.9.

At 0.38 sec, the machine speed drops below the threshold value and the machine enters the discharge wind down phase, i.e., the current is ramping down to zero. At even lower speeds, the back-emf of the machine is smaller than the safe voltage. Therefore, the bus voltage will gradually drain as the machine is decelerating.

Both simulation and test results show that the proposed active discharge algorithm has improved dynamic performance compared to the traditional bus discharge algorithm, i.e., the bus recharge problem is avoided, and the bus discharge can start immediately after the active discharge event is requested without waiting for the machine speed to fall below the safe threshold, which can take tens of seconds or even minutes due to the moment of inertia of the machine, and freely spinning of wheels.

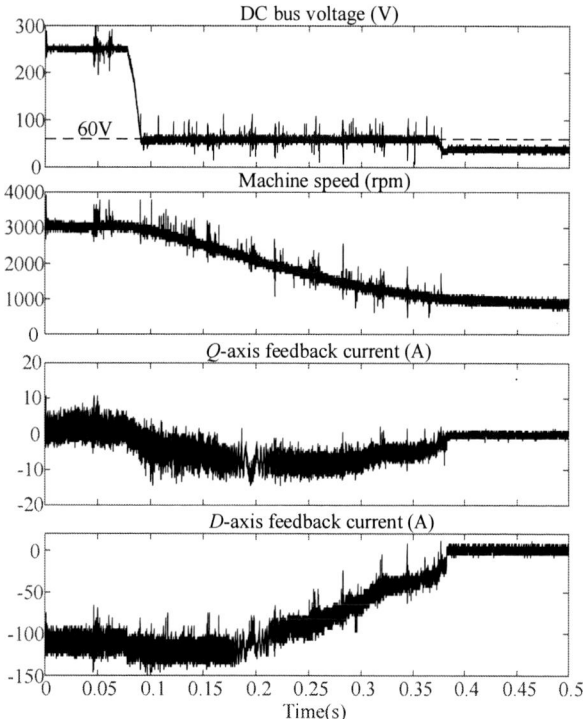

Fig. 13. Experimental test results of the proposed active discharge algorithm using a single electric machine.

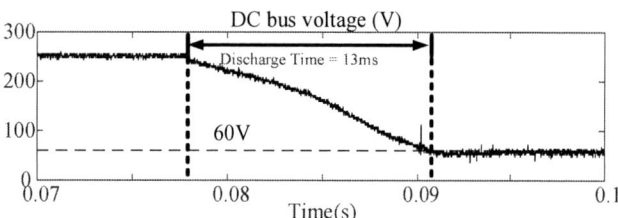

Fig. 14. Zoom-in of DC bus voltage during the fast discharge phase (test results).

V. CONCLUSION

This paper discusses the published high voltage DC bus discharge algorithms for the electric machine drive systems of HEVs. Published methods suffer not being enable to start the discharge function until the machine speed falls below the safe threshold due to the machine back-emf.

A new control method was proposed to quickly and safely discharge the DC bus capacitor. This method allows immediate discharge in response to any active discharge events regardless of the machine speed. For high machine speeds, the large d-axis current generates high copper loss in the electric machine to rapidly discharge the bus voltage down to the safe voltage. A bus voltage regulator together with a modulation index regulator are developed to maintain the bus voltage slightly lower than the safe voltage until the machine speed drops below certain threshold. Both simulation and experimental results are presented to verify the proposed bus discharge algorithm.

REFERENCES

[1] S. Hirose, Power conversion device, method of controlling power conversion device, and vehicle with the same mounted thereon, U.S. Patent, US20120039100 A1, 2012.
[2] T. A. Burress, S.L. Campbell, C. L. Coomer, C. W. Ayers, A. A. Wereszczak, J. P. Cunningham, L. D. Marlino, L. E. Seiber, and H. T. Lin, Evaluation of the 2010 Toyota Prius hybrid synergy drive system, Oak Ridge National Laboratory Report for U.S. Department of Energy, 2011.
[3] M. Degner, V. Sankaran, and C. Chen, Power converter topologies for better traction drive packaging, U.S. Patent, US7339345 B2, 2008.
[4] United Nation Economic Commission for Europe Vehicle Regulation No. 94 (ECE R94), Unifrom provisions concerning the approval of vehicles with regard to the protection of the occupants in the event of a frontal collision, Rev. 2, Annex 11, Aug. 2013.
[5] K. Benson, "Regeneration for AC drive systems," Fiber and Film Industry Technical Conference, pp. 1-3, Greenville, SC, May 1994.
[6] A.K. Kaviani, B. Hadley, and B. Mirafzal, "A time-coordination approach for regenerative energy saving in multiaxis motor-drive systems," *IEEE Transactions on Power Electronics*, vol. 27, no.2, pp. 931-941, Feb. 2012.
[7] T. Goldammer, T. Le, J. Miller, and J. Wai, Active high voltage bus bleed down, U.S. Patent, US 20120161679 A1, Jun. 2012.
[8] S. Ashida, K. Yamada, M. Nakamura, T. Shimana, and T. Soma, Electric vehicle, and control apparatus and control method for electric vehicle," U.S. Patent, US 8631894 B2, Jan. 2014.

C_{OSS} Hysteresis in Advanced Superjunction MOSFETs

J.B. Fedison, M.J. Harrison
Enphase Energy, Inc.
Petaluma, CA, USA

Abstract— In this work, a Sawyer-Tower circuit is employed to characterize the output capacitance (C_{OSS}) of advanced superjunction MOSFETs. It is shown that some of the most advanced superjunction MOSFETs exhibit significant hysteresis in their output capacitance which leads to unrecoverable power loss. This work shows that the conventional impedance analyzer method can only measure C_{OSS} accurately when hysteresis is not present while measurement of C_{OSS} with a Sawyer-Tower circuit gives accurate results regardless of whether hysteresis is present or not. Accurate measurement of C_{OSS} with a Sawyer-Tower circuit not only enables designers to more accurately calculate and predict power loss but even more importantly allows the power semiconductor industry to more effectively advance future generations of superjunction MOSFETs for optimum efficiency, especially for use in resonant converter applications.

Keywords—superjunction MOSFET, output capacitance (C_{OSS}), hysteresis, zero-voltage switching (ZVS)

I. INTRODUCTION

In a previous paper the authors discussed an anomalous behavior in state-of-the-art superjunction MOSFETs where significant energy dissipation is observed in the process of charging and discharging the parasitic output capacitance (C_{OSS}) of the MOSFET while the gate is shorted to the source [1]. This observation of unusually large C_{OSS} energy dissipation challenges the long standing assumption that the dominant source of dissipation in the output capacitance is drift layer resistance. In one studied device, C_{OSS} was found to dissipate more energy than the apparent stored energy based on calculation from the small-signal C_{OSS} versus V_{DS} curve from the datasheet. In the author's original study, calorimetric loss measurement techniques were used and this allowed detailed measurement of the C_{OSS} energy dissipation but did not allow effective characterization of the charge-discharge process. This report investigates the same group of five advanced superjunction MOSFETs as in the original study but utilizes a different measurement technique that allows the direct measurement of the charge versus voltage characteristics of the parasitic output capacitance during both the charging and discharging phases using entirely current driven commutations. This work follows the convention that small-signal quantities are denoted with lower-case and large signal quantities are denoted with upper-case.

II. MOTIVATION

With the increasing use of resonant topologies in power converters, there is a heightened need to ensure that the semiconductor switches utilized do not contribute extra losses during the zero-voltage-switching (ZVS) transition. The latest superjunction devices exhibit area specific on-resistance that is many times lower than the unipolar limit for silicon, hence there is intense interest in utilizing these devices to minimize conduction loss while maintaining small device footprint, low input capacitance and low cost. For switches in a resonant converter, the main contributors to power loss are the on-resistance ($P_{loss} = I^2_{rms} R_{DSON}$) and gate dive loss ($P_{loss} = \frac{1}{2} C_{ISS} V^2_{GS}$). The output capacitance has traditionally not been considered as a source of power loss in resonant converters because it has always been assumed that C_{OSS} exhibits negligible energy dissipation. This study reveals the detailed behavior of the C_{OSS} charge and discharge processes showing, in agreement with the findings of reference [1] that for the more advanced superjunction MOSFETs C_{OSS} energy dissipation is not negligible. This work also describes an easily implemented setup that allows accurate characterization of the charge and discharge behavior of the nonlinear output capacitance of a MOSFET even in the presence of hysteresis. An example of the measured voltage versus charge curves for one of the studied devices is shown in Fig 1.

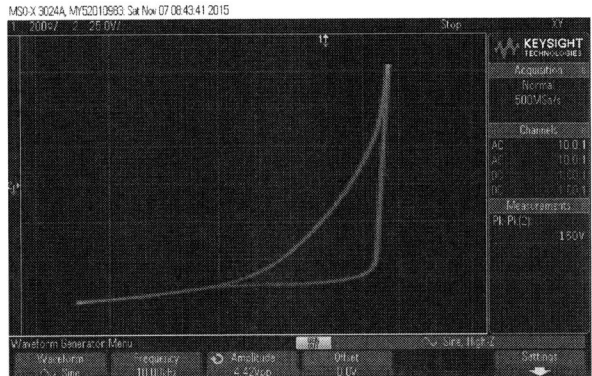

Fig. 1. Oscilloscope capture for Device A being measured in the Sawyer-Tower test setup with the oscilloscope in X-Y mode where X represents the output charge and Y represents the drain-to-source voltage of the MOSFET under test.

978-1-4673-9551-9/16 $31.00 © 2016 IEEE

III. EXPERIMENTAL

Various techniques can be employed to measure the capacitance of a component. The industry standard method used to characterize parasitic capacitance in power MOSFETs employs an impedance meter such as the Agilent 4284A Precision LCR Meter where a low amplitude voltage or current stimulus is applied to the device under test and the phase shift in the current relative to the voltage is used to determine the impedance at the chosen test frequency and DC bias point. This technique, while convenient, has the limitation that it gives only a single value for the impedance while only measuring the small-signal value and therefore is not useful for measuring a capacitance that exhibits hysteresis.

The parasitic output capacitance of the MOSFETs examined in this study exhibit varying degrees of hysteresis. This hysteresis is of interest because it gives a direct measure of the energy loss over the full charge-discharge cycle. The fact that the conventional impedance analyzer based method used to measure output capacitance cannot correctly measure a capacitance that has hysteresis has previously been recognized [2]. The present study employs the Sawyer-Tower circuit shown in Fig. 2 which is able to measure hysteresis. This measurement technique was originally used to study ferroelectric dielectric materials for their potential use in making compact, large-value capacitors [3]. More recently the same technique has been used to study multi-layered ceramic capacitors made from class 2 dielectric materials [4]. The usefulness of the Sawyer-Tower circuit stems from its ability to characterize a capacitor with hysteresis by measuring separately the large-signal charging and the discharging characteristics of a non-linear capacitor. In addition to the charge integration function, in this case C_{Ref} also serves to provide a dc bias function to ensure that the body diode of the DUT MOSFET is reverse biased.

The setup shown in Fig. 2 consists of a signal generator feeding a power amplifier whose output is applied to the series combination of the MOSFET under test (DUT) and a reference capacitor. It is important that the reference capacitor be linear, exhibit low-loss and have a voltage rating equal to the peak-to-peak stimulus voltage (V_{pp}). Most film capacitors fit this requirement and a polypropylene film capacitor is used here. Specific details of the setup are listed in Table 1. The MOSFET is purposely held in the off state by shorting the gate to the source. Charge versus voltage measurements are taken after steady-state conditions are reached when the DC bias voltage across the reference capacitor is charged to one-half V_{pp} and the body diode of the MOSFET is reverse biased. A digital oscilloscope configured in X-Y mode measures the total voltage across the reference capacitor, C_{Ref}, and the MOSFET,

TABLE I. EQUIPMENT USED IN SAWYER-TOWER CIRCUIT

Description	Supplier	Model Number	Specification
Power Amplifier	Pintek	HA-205	Vpp,max: 170V Bandwidth: 3MHz Slew Rate: 2500V/us
Digital Oscilloscope	Keysight	MSO-X 3024A	Bandwidth: 200MHz Sample Rate: 4GSa/s With internal function generator option.
Oscilloscope Probes	Keysight	N2890A	10:1 Passive Probe Bandwidth: 500MHz Impedance: 10MΩ//11pF Max Voltage: 300Vrms
Reference Capacitor "Cref"	EPCOS	B32671P6104K	100nF, 630VDC, Polypropylene Film Capacitor

V_Y, and the voltage across just the reference capacitor, V_X. A Python script controls the oscilloscope and signal generator and collects the measured data. To enable further research this test script has been made available as an open source Python script [5]. At each sample point the output charge of the MOSFET (Q_{OSS}) is computed as

$$Q_{OSS} = C_{Ref} V_X. \qquad (1)$$

The MOSFET drain-to-source voltage, V_{DS}, is computed as

$$V_{DS} = V_Y - V_X. \qquad (2)$$

The V_{DS} versus Q_{OSS} curves for each MOSFET tested are shown in Fig. 3 and will be discussed further in the next section.

IV. RESULTS AND DISCUSSION

The V_{DS} versus Q_{OSS} measurements shown in Fig. 3 illustrate the range of charge-discharge behavior of the superjunction MOSFETs examined. The various devices show different degrees of hysteresis where device A clearly exhibits the most hysteresis while device D shows minimal hysteresis. It should be noted that the particular hysteresis shown in Fig. 3 for each device is consistent with ferroelectric behavior. The conditions for ferroelectric behavior are met when the V versus Q curve shows hysteresis where there is a region of saturation and the V versus Q curve is convex [6].

All the results shown in Fig. 3 are for an excitation frequency of 10 kHz and a peak-to-peak voltage of 160V. To examine the frequency dependence of the the hysteresis related power loss, the excitation frequency is varied in the range of 1kHz to 100kHz and the V_{DS} versus Q_{OSS} curves show negligible change with frequency. The voltage dependence of the power loss is also examined in the range of 30V to 160V peak-to-peak. In this study, the maximum peak-to-peak voltage is limited to 160V due to limitations of the power amplifier. From the V_{DS} versus Q_{OSS} curves the C_{OSS} hysteresis energy loss, $E_{Hysteresis}$, is calculated as the area between the charging and discharging V_{DS} versus Q_{OSS} curves and the hysteresis power loss is defined as

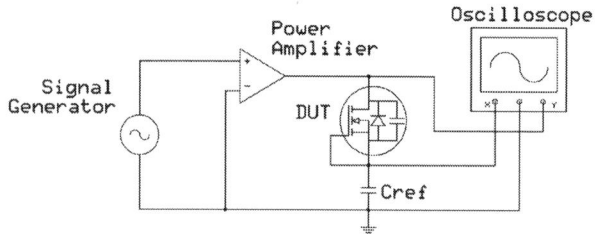

Fig. 2. Sawyer-Tower circuit used to measure output capacitance, C_{OSS} of the DUT.

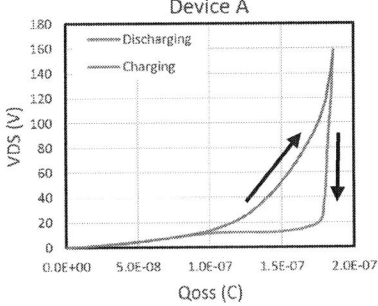

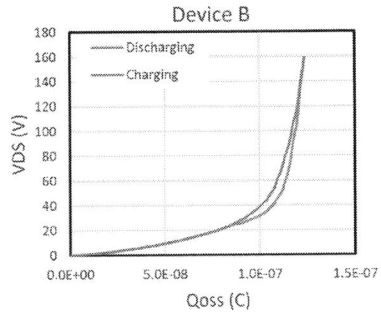

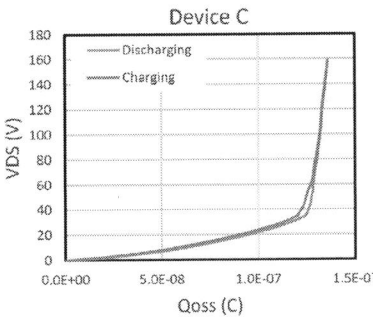

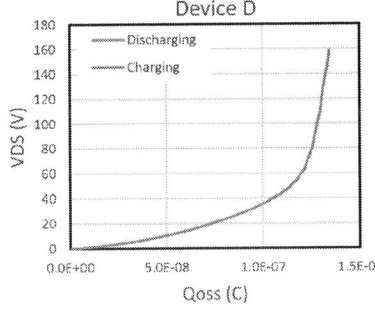

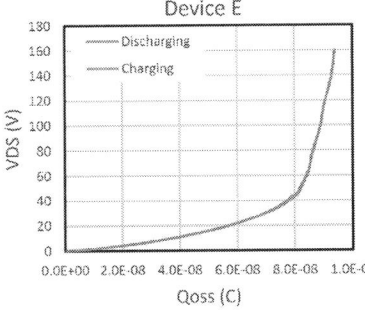

Fig 3. V_{DS} versus Q_{OSS} curves measured with the Sawyer-Tower circuit for devices A – E. Sweep frequency = 10kHz and peak-to-peak voltage = 160V.

$$P_{Hysteresis} = f\, E_{Hysteresis} \qquad (3)$$

where f is excitation frequency. $P_{Hysteresis}$ measurements for device A as a function of excitation frequency and peak-to-peak voltage are summarized in Fig. 4 in the form of Steinmetz curves. These curves are a fit of the hysteresis power loss to the Steinmetz equation [7] such that

$$P_{Hysteresis} = k\, f^{\alpha}\, V^{\beta} \qquad (4)$$

where k is a constant, f is excitation frequency, V is the peak-to-peak voltage across the DUT and the exponents describe the frequency dependence of the power loss, α, and amplitude dependence, β. Curves fitted to these data for device A indicate an α value of 1.04 which indicates that energy loss per charge-discharge cycle due to hysteresis is virtually independent of frequency while β has a value of 2 at low voltage ($V_{DS} < 100V$) and 0.5 at higher voltage ($V_{DS} > 100V$). The reduced amplitude dependence at higher voltage indicates that the hysteresis occurring in the output capacitance is confined mainly to lower voltages. This behavior suggests that the hysteresis is related to the depletion of the superjunction pillars present in the device structure. At drain-to-source voltages below 100V the superjunction pillars within the device cause a mix of both vertical and lateral depletion while above 100V the regions between the superjunction pillars remain fully depleted so that any further depletion occurs in the vertical direction only. The fact that hysteresis power loss shows strong voltage dependence only at low voltage suggests that the hysteresis effect is related to three-dimensional depletion between the superjunction pillars occurring below 100V. Evidence of a pillar related origin to the hysteresis loss has recently been shown based on TCAD simulations where charge between the pillars becomes stranded during the discharging phase but not the charging phase [8].

The utility of the measurements made with the Sawyer-Tower circuit is two-fold. Firstly, this technique gives a direct measure of Q_{OSS} over a selected voltage range which is directly relevant to the conditions present in a power converter. Secondly, the Sawyer-Tower circuit allows the measurement of both the charge and discharge curves for a complete charge-discharge cycle giving the ability to detect hysteresis. When hysteresis is present, the dissipation energy over the complete

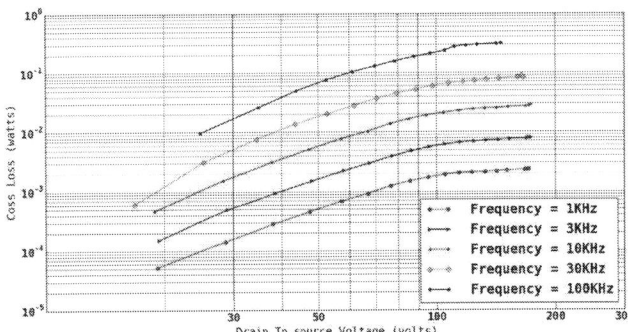

Fig. 4. Measured C_{OSS} power dissipation data (points) and Steinmetz curve fit (lines) for device A.

TABLE II.   SPECIFICATIONS AND MEASURED E_{OSS} RELATED ENERGIES FOR SUPERJUNCTION MOSFETS EXAMINED

Device ID	VDSS at 25C [V]	RDSON Max at 25C [mΩ]	Die Area Specific RDSON, Typ. at 25C [mΩ cm2]	E_{OSS}[1] [μJ]	$E_{OSS,Charging}$[2] [μJ]	$E_{OSS,Discharging}$[2] [μJ]	$E_{Hysteresis}$[3] [μJ]	$E_{Hysteresis}$[4] [μJ]
A	650	148	19	1.16	4.90	-2.25	2.65	4.05
B	600	160	27	2.36	2.98	-2.81	0.17	0.57
C	600	178	26	2.28	2.62	-2.54	0.08	0.24
D	600	158	30	3.68	3.49	-3.72	<0.01	0.14
E	600	160	33	2.11	2.29	-2.40	<0.01	0.08

[1] From the datasheet at V_{DS}=160V.

[2] Using Sawyer-Tower circuit with $V_{DS,max}$ = 160V.

[3] $E_{Hysteresis}$ = $E_{OSS,charging}$ + $E_{OSS,discharging}$, with $V_{DS,max}$ = 160V.

[4] Calorimetry based measurement with $V_{DS,max}$=200V, taken from ref [1].

charge-discharge cycle ($E_{Hysteresis}$) is determined by computing the area between the V_{DS} versus Q_{OSS} charge and discharge curves as will be shown mathematically below.

Table 2 provides a summary of the specifications of the five state-of-the-art superjunction MOSFETs studied as well as the various energies extracted from V_{DS} versus Q_{OSS} measurements using the Sawyer-Tower circuit. The E_{OSS} value shown in column 5 is read directly from the datasheet at V_{DS}=160V. E_{OSS} can also be calculated from the V_{DS} versus Q_{OSS} curves measured using the Sawyer-Tower circuit. As such the charging energy is computed as

$$E_{OSS,Charging} = \int_{0}^{Qmax} V_{DS}(Q_{OSS})\, dQ_{OSS} \tag{5}$$

and the discharging energy is computed as

$$E_{OSS,Discharging} = \int_{Qmax}^{0} V_{DS}(Q_{OSS})\, dQ_{OSS} \tag{6}$$

The above energies are listed in columns 6 and 7 of the table at a peak-to-peak voltage of 160V. Listed in column 8 is the energy loss due to hysteresis given as

$$E_{Hysteresis} = E_{OSS,Charging} + E_{OSS,Discharging} \tag{7}$$

also at a peak-to-peak voltage of 160V. The last column of Table 2 shows the hysteresis energy measured by calorimetry at V_{DS} of 200V [1].

The results in Table 2 show that for the devices with negligible hysteresis loss, E_{OSS} from the datasheet is in reasonable agreement with $E_{OSS,charging}$ measured with the Sawyer-Tower circuit. In particular, devices D and E show agreement within 10 percent, however, for devices A - C, the agreement diminishes as $E_{Hysteresis}$ increases. The deviation in $E_{OSS,charging}$ measured using the Sawyer-Tower circuit relative to E_{OSS} from the datasheet is shown in Fig. 5 as a function of the hysteresis energy. It is not surprising that the deviation between these two measures of E_{OSS} increases the larger the hysteresis energy since the E_{OSS} measurement reported on the datasheet is obtained through the small-signal C_{oss} measurements that are known to be inaccurate in the presence

of hysteresis. These results suggest the importance of including large-signal Sawyer-Tower measurements as part of the datasheet especially when a device exhibits large hysteresis such as Device A. In the presence of large hysteresis, the E_{OSS} value based on small-signal C_{oss} measurements gives an erroneously low E_{OSS} reading whereas measurements made using the Sawyer-Tower method give results that are in better agreement with calorimetry as discussed below.

The small-signal C_{OSS} versus V_{DS} curves for the five MOSFETs examined can be calculated from the measured V_{DS} versus Q_{OSS} data by the relation

$$C_{oss} = \frac{dQ_{OSS}}{dV_{DS}} \tag{8}$$

The corresponding small-signal C_{oss} curves are shown in figure 6 for each MOSFET examined. For comparison the small-signal capacitance from the datasheet is also shown for each MOSFET. It should be pointed out that the small-signal C_{oss} charging curves appear to have an unusual spike where the capacitance appears to be non-monotonic, most noticeable for Device A. The large signal capacitance for Device A is also shown in Fig. 7 where, as expected, the large-signal capacitance, defined here as C_{OSS} = Q_{OSS} / V_{DS} is consistently

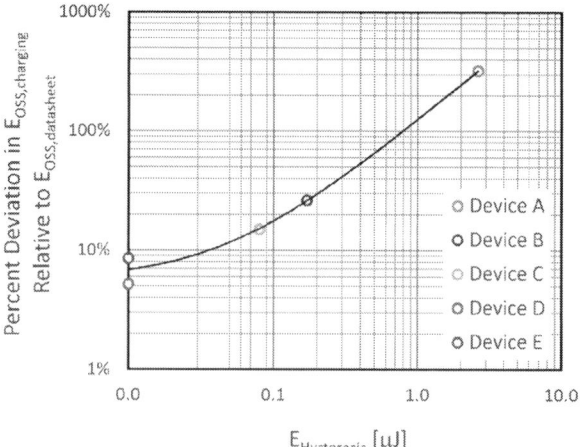

Fig. 5. Deviation in $E_{OSS,charging}$ as measured using the Sawyer-Tower circuit relative to E_{OSS} reported on the datasheet to at VDS=160V.

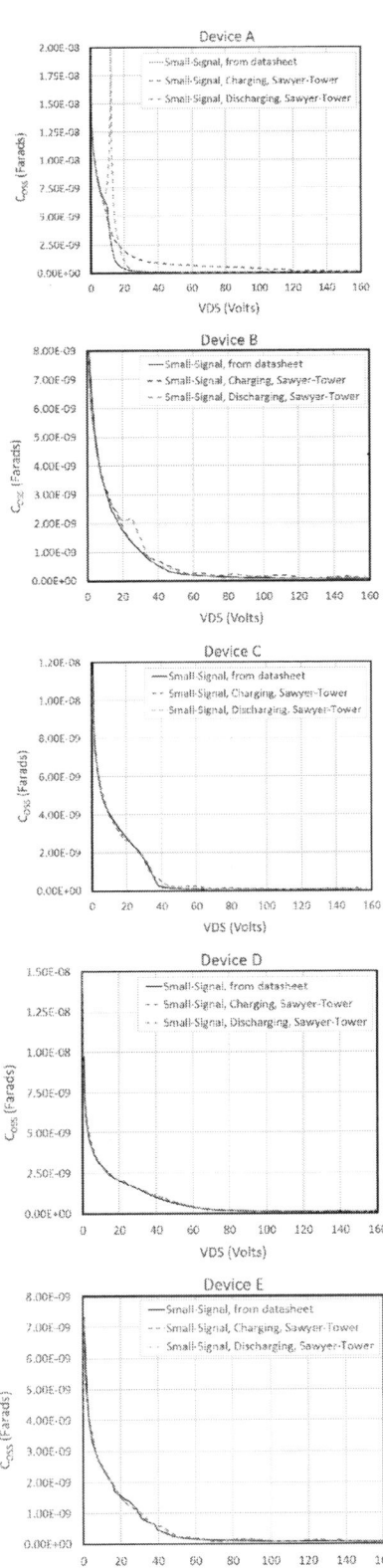

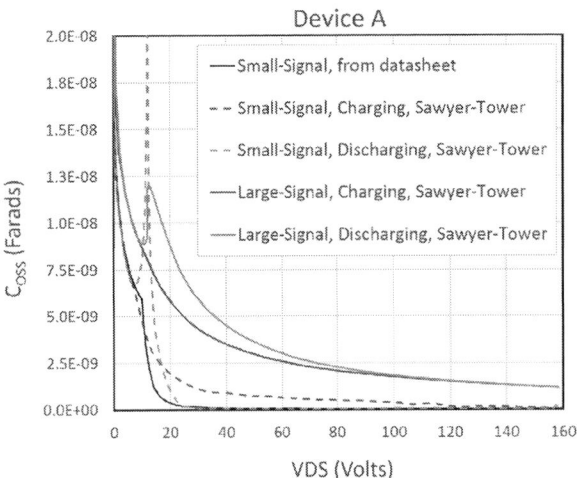

Fig. 7. Large-signal and small-signal output capacitance for device A.

Fig. 6. Small-signal output capacitance for devices A-E as reported on the datasheet and as measured in the Sawyer–Tower circuit at a frequency of 10kHz and a peak to peak voltage of 160V.

higher in value than the small-signal capacitance due to the concave shape of the Q_{OSS} versus V_{DS} curve. The spike in C_{OSS} is most likely not related to a physical capacitance within the device structure but likely is an artifact of the internal loss mechanism causing hysteresis related energy dissipation.

As discussed in our earlier paper, to account for the hysteresis we observe in C_{OSS} (and Q_{OSS}), there must be a non-reciprocal charge transfer process that affects charging C_{OSS} differently from discharging C_{OSS}. One possible explanation for this could be the fact that when C_{OSS} is charging, the n and p pillars initially have mobile carriers while, when C_{OSS} is discharging, the pillars are completely depleted of mobile carriers. If, during the charging phase, some of the mobile carriers get islanded where non-uniform spread of the depletion region causes pockets of undepleted regions along both p and n pillars, this could lead to the non-reciprocal process that ultimately leads to hysteresis and thereby dissipation of some of the energy in C_{OSS}.

It is instructive to compare the dissipation energy measured with the Sawyer-Tower circuit in this work to that obtained by calorimetry as described in our previous work [1]. Figure 8 shows the comparison for devices A and B where it can be seen that results from the two methods are in reasonable agreement.

V. RECOMMENDATIONS TO MOSFET VENDORS

Based on the findings of this body of work, we make the following recommendations to the manufacturers of superjunction MOSFETs:

1) Ensure that the C_{OSS} and E_{OSS} related data on the datasheet is relevant for the intended application of the MOSFETs, in particular this data should be relevant for large-signal environments like those encountered in power electronics.

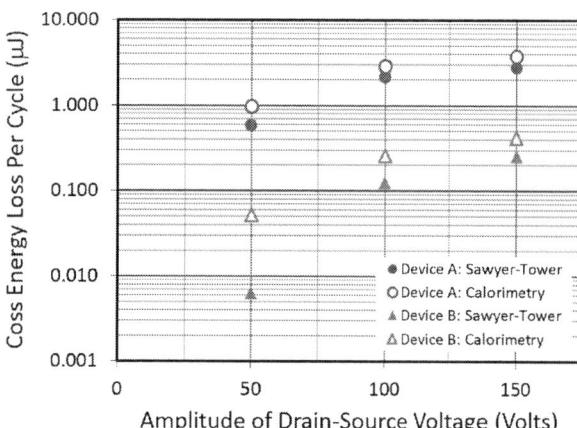

Fig. 8. Comparison of C_{OSS} energy loss per charge-discharge cycle measured using the Sawyer-Tower method presented in this work and the calorimetry method reported in [1].

2) Adopt the Sawyer-Tower method for characterizing C_{OSS} of superjunction MOSFETs.

3) Publish the Sawyer-Tower derived Q_{OSS} and C_{OSS} curves along with a E_{OSS} loss figure per switching cycle on the data sheets of superjunction MOSFETs.

4) Base any Figure of Merit on the E_{OSS} measured using the Sawyer-Tower method.

5) Explain that the traditional C_{oss} curve is a characterization of the linear mode performance of the MOSFET that is measured using small-signal capacitance methods. In addition explain that the new C_{oss} curve is a characterization of the switching performance of the MOSFET that is measured using the Sawyer-Tower method. Alternatively consider simply removing the traditional C_{oss} curve to avoid confusion.

VI. SUMMARY

This work has illustrated that there can be very large differences in the C_{OSS} versus voltage curves depending on the measurement method. In particular, E_{OSS} calculated from industry standard small-signal C_{OSS} measurements appear to significantly underestimate E_{OSS} determined by large-signal (Sawyer-Tower) and calorimetry based measurements. The

Sawyer-Tower circuit employed in this work directly gives large-signal V_{DS} versus Q_{OSS} curves for both the charging and discharging phases and these curves are more relevant to predicting performance in switch-mode power converters. The power semiconductor industry should consider incorporating large-signal V_{DS} versus Q_{OSS} curves as a standard part of the datasheet and the Sawyer-Tower circuit is suggested as a suitable measurement method. This work has also shown that the more advanced the superjunction MOSFET is, the more severe is the hysteresis in Q_{OSS} and C_{OSS} and this hysteresis is a direct measure of the energy loss that will occur in resonant converters, which have traditionally been assumed to be immune from C_{OSS} related loss. Finally, the C_{OSS} dissipation energy measured with the Sawyer-Tower circuit in this work are in good agreement with measurements made by calorimetry in our previous work [1].

ACKNOWLEDGEMENTS

The authors would like to acknowledge Krunal Patel (previously with Santa Clara University, now with IXYS Corporation) and Erik Weyker (Enphase Energy, Inc.) for their contributions in the development of the test platform and Python test script that enabled the measurements in the work.

REFERENCES

[1] J.B. Fedison, M. Fornage, M.J. Harrison, and D.R. Zimmanck, "Coss Related Energy Loss in Power MOSFETs Used in Zero-Voltage-Switched Applications," APEC 2014 Proceedings, pp. 150-156, 16-20 March 2014.

[2] L.E. Mosley, "Capacitor Impedance Needs for Future Microprocessors," CARTS 2006.

[3] C.B Sawyer, and C.H. Tower, "Rochelle Salt as a Dielectric," Physical Review, vol. 35, pp. 269-273, 1930.

[4] L.E. Mosley, and J.S. Schrader, "Hysteresis Measurements of Multi-Layer Ceramic Capacitors Using a Sawyer-Tower Circuit," CARTS USA 2007 Proceedings, pp. 309-319, Albuquerque, NM, 26-29 March 2007.

[5] https://github.com/SawyerTower

[6] J.F. Scott, "Ferroelectrics Go Bananas," Journal of Physics: Condensed Matter, vol. 20, issue 2, pp. 021001-021002, 2008.

[7] C.P. Steinmetz, "On the Law of Hysteresis," AIEE Transactions, vol. 9, pp. 3-64, 1892.

[8] J. Roig, F. Bauwens, "Origin of Anomalous COSS Hysteresis in Resonant Converters With Superjunction FETs," Electron Devices, IEEE Transactions on, vol.62, no.9, pp.3092-3094, Sept. 2015.

Compact Electrothermal Models for Unbalanced Parallel Conducting Si-IGBTs

Roozbeh Bonyadi, Olayiwola Alatise, Ji Hu, Zarina Davletzhanova, Yeganeh Bonyadi, Jose Ortiz-Gonzalez, Li Ran, Phil Mawby

School of Engineering
University of Warwick
Coventry, United Kingdom
r.bonyadi@warwick.ac.uk, o.alatise@warwick.ac.uk, ji.hu@warwick.ac.uk, z.davletzhanova@warwick.ac.uk,
y.bonyadi@warwick.ac.uk, j.a.ortiz-gonzalez@warwick.ac.uk, l.ran@warwick.ac.uk, p.a.mawby@warwick.ac.uk.

Abstract—For high current applications, silicon IGBTs are normally connected in parallel to deliver the required current ratings. The devices are normally designed to have identical electrothermal parameters for equal current and power sharing. However, over the mission profile of the device, non-uniform degradation of the electro-thermal properties like solder de-lamination or gate contact resistance as well as unequal heat extraction from the heat sink, can cause the parallel connected IGBTs to have different electrothermal properties. In this paper, a compact and accurate electro-thermal model for parallel connected IGBTs has been developed and validated by experimental measurements.

Keywords—Parallel IGBT, Current Sharing, Electrothermal, Model

I. INTRODUCTION

IGBT based power modules are normally comprised of several dies connected in parallel to deliver a defined current rating [1-3]. Fig. 1. shows the power module of a (a) Nissan Leaf EV and (b) Infineon wind energy power electronic converter where several parallel dies can be seen. Furthermore, the Tesla Model S power inverter uses 14 parallel IGBTs in TO-247 packages which are mounted on a PCB board [4]. This is unlike other hybrid electric vehicles (HEV), plug-in hybrid electric vehicles (PHEV) or electric vehicles which use the packaged power module as shown in Fig. 1. Consequently, ensuring synchronized electrical switching and balanced electrothermal parameters between these parallel devices becomes more crucial for high power rated inverters with high current capability. Balanced power dissipation between the parallel connected devices is required for optimal temperature distribution [5, 6]. Current sharing between the devices depends on the device parameters as well as the circuit parameters such as the gate inductance and gate resistance. Hence, in order to have balanced current sharing between the devices, they all need to switch ON and OFF at the same rate i.e. switching needs to be synchronized. While this is usually achieved by power module designers, however, over the mission profile of the module, it is possible that non-uniform

This work was supported by EPSRC funding in collaboration with Jaguar Land Rover Automotive PLC and EPSRC funding through the underpinning power electronics devices theme research project (EP/L007010/1) and the components theme research project (EP/K034804/1).

degradation of the device electrothermal parameters can cause electro-thermal imbalance between the devices.

Common failure modes like solder/die attach voiding and delamination can cause a non-uniform thermal resistance across the die or DBC substrate, meaning that the parallel devices may be subject to different junction temperatures. Other parameters that may vary between the devices include the internal and external gate resistances due to increased gate wire-bond contact resistance due to thermo-mechanical stress cycling. These variations in the gate resistance will introduce variations in the switching rate and switching energy. Hence, the two principal parameters under investigation in this paper are electrical switching rates and the thermal resistance of the devices. The electrical switching rate is set by varying the gate resistance of the parallel connected IGBTs while the junction temperature of the devices is varied using a hotplates connected to the base of the device.

It is important that accurate compact models are developed with the capability of probing the effect of these electrothermal variations between parallel connected devices. These compact models should be computationally efficient and be physics-based. Finite element methods are computationally expensive and time consuming. Furthermore, the electrical switching time constants are on the order of microseconds while thermal response time constants range from milliseconds to seconds. Hence, finite element methods will be a cumbersome approach. SPICE based IGBT models on the other hand do not capture some physics-based thermal effects that will impact the parallel operation of the IGBTs. Hence, in this paper, a physics based compact modelling approach is used to capture the internal physics of the IGBT and couple it with circuit equations. The Ambipolar-Diffusion-Equation (ADE) that describes the carrier distribution profile in the IGBT is solved using the Fourier series approach developed elsewhere [6-15]. Look-up tables are used to de-couple the electrical time constants from the thermal time constants, thereby ensuring a fully coupled electrothermal model of parallel connected IGBTs. Using compact models to predict impact of electrothermal variation on temperature imbalance between the parallel connected IGBTs is a useful tool for reliability analysis.

(a)

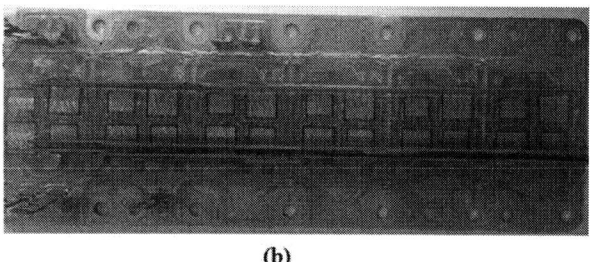

(b)

Fig. 1. (a) Nissan Leaf power inverter with 4 parallel IGBTs for each top and bottom switches in three phase voltage source converter, (b) Infineon power module (FF1000R17IE4) with 1.7 kV/1 kA power rating showing 6 parallel IGBTs with a very long gate path and unbalanced gate resistance and inductance.

II. IGBT MODELLING

The IGBT device model has been explained in detail in several publications [6-15]. This model takes account for the excessive carrier concentration in the drift region of the IGBT by solving the ADE. This equation explains the transient electron and hole distribution in the charge storage region of the IGBT and it depends on time and distance in the drift region. The device model takes physical parameters of the device such as die size, device thickness, channel size, doping, carrier mobility and etc. into account and solves the drift-diffusion equations of the device. The model uses quasi-static temperature as an input and it considers the effect of this temperature on the temperature-dependent parameters of the device such as threshold voltage, intrinsic carrier concentration, electron/hole mobility, diffusivity, minority carrier-lifetime in the drift region, electron/hole recombination rate, saturation velocity, MOS-transconductance etc. Hence, accurate transient current and voltage switching waveforms can be obtained from the model. This is very important as small differences in temperature or gate resistance between two parallel devices lead to unbalanced current sharing between the devices. Consequently, in order to understand how the current is shared between the devices it is very important to be able to model the behavior of two parallel IGBTs working at different temperature or having different switching rates. This IGBT model along with a physics-based PiN diode model [15-18] are used in a clamped inductive switching circuit with two parallel IGBTs. The models can accurately simulate current sharing

between the parallel devices. A clamped inductive switching circuit with two parallel IGBTs is shown in Fig. 2. The equations below are derived by applying Kirchhoff's current and voltage laws (KCL and KVL respectively) on the circuit shown in Fig. 2.

$$V_{UP} = I_s R_s + {}^1\!/_{C_s} \int I_s dt \tag{1}$$

$$I_s = I_{AK} + I'_{C1} + I'_{C2} - I_L \tag{2}$$

$$I_{AK} = {}^1\!/_{L_D} \int (-V_{UP} - V_{AK} - V'_{CE}) dt \tag{3}$$

$$I'_{C1} = {}^1\!/_{L_{s1}} \int (V_{DC} - V_{UP} - V'_{CE1}) dt \tag{4}$$

$$I'_{C2} = {}^1\!/_{L_{s2}} \int (V_{DC} - V_{UP} - V'_{CE2}) dt \tag{5}$$

$$I'_{G1} = {}^1\!/_{L_{G1}} \int (V_{gg1} - I'_{G1} R_{G1} - V'_{GE1}) dt \tag{6}$$

$$I'_{G2} = {}^1\!/_{L_{G2}} \int (V_{gg2} - I'_{G2} R_{G2} - V'_{GE2}) dt \tag{7}$$

$$I_{G1} = I'_{G1} + (V_{CE1} - V_{GE1})\left(\frac{sC_{FB1}R_{FB1}}{1+sC_{FB1}R_{FB1}}\right)\left(\frac{1}{1+s\tau_{lim}}\right) \tag{8}$$

$$I_{G2} = I'_{G2} + (V_{CE2} - V_{GE2})\left(\frac{sC_{FB2}R_{FB2}}{1+sC_{FB2}R_{FB2}}\right)\left(\frac{1}{1+s\tau_{lim}}\right) \tag{9}$$

$$V'_{GE1} = V_{GE1} + L_{E1}(I_{C1} + I_{G1})\left(\frac{s}{1+s\tau_{lim}}\right) \tag{10}$$

$$V'_{GE2} = V_{GE2} + L_{E2}(I_{C2} + I_{G2})\left(\frac{s}{1+s\tau_{lim}}\right) \tag{11}$$

$$R_{dson1} = {}^{V_{CE1}}\!/_{I_{C1}} \tag{12}$$

$$R_{dson2} = {}^{V_{CE2}}\!/_{I_{C2}} \tag{13}$$

The schematic shown in Fig. 2 is a half-bridge configuration which is the basis of common inverters. In this circuit, two pulses are given to the gate of bottom IGBTs. During the first pulse, the inductive load is charged and during the second pulse the switching behavior of the IGBT and the diode is captured.

To achieve an accurate full electro-thermally coupled model of the parallel connecting Si-IGBTs, it is necessary to de-couple the electrical switching time constants which occur on the microsecond scale from the thermal time constants which occur on the millisecond to second scale. To achieve this, an electrothermal model that incorporates the physics of the IGBT is used for modeling the instantaneous power over microseconds to generate the look-up table of losses. Afterwards, a thermal model fed from the look-up table is developed for longer timescale simulations capable of capturing electrothermal imbalance between the parallel devices.

Fig. 3 shows how the model works. The electrothermal model is comprised of a Cauer-network that calculates the transient junction and case temperatures of the devices due to conduction and switching losses. The thermal resistances and thermal capacitances of the Cauer-network are derived based on the dimensions of the TO-247 IGBT package and the thermal properties of the material for each layer. These losses are calculated from the voltage and current waveforms obtained from the Fourier series solution of the ADE in the IGBT and diode drift regions. The losses have also been

978-1-4673-9551-9/16 $31.00 © 2016 IEEE

determined for different switching rates which have been set by using different gate resistances on the low side IGBTs. Hence, electrothermal imbalance between the parallel IGBTs is simulated simply by using different look-up-tables for each IGBT in the thermal model. Once the look-up-table is complete for different gate resistances and junction temperatures, the purely thermal model takes over and calculates the junction temperature of the parallel IGBTs using the Cauer Network and look-up-table. The purely thermal model is able to complete simulations lasting several minutes in a computationally effective manner because it does not solve the detailed physics based ADE equations that have already been pre-solved by the electrothermal model. Hence, the model is both accurate and computationally efficient. When, there is

electrothermal imbalance between the parallel IGBTs, the respective Cauer Networks take account of the different junction temperatures and updates the temperature sensitive electrical parameters like the threshold voltage and on-state resistance accordingly.

For the electro-thermal simulation, the block diagram shown in Fig. 3 is used. The repetitive switching was carried out using fixed frequency pulses with a duty cycle of 10%. A typical power loss transient profile for one period of repetitive clamped inductive switching is shown in Fig. 4. In this graph the voltage was set to 50 V similar to the experimental conditions and the current passing through the device was approximately 6 A (12 A in total). The thermal simulation block shown on the right hand-side of the Fig. 3 is used to

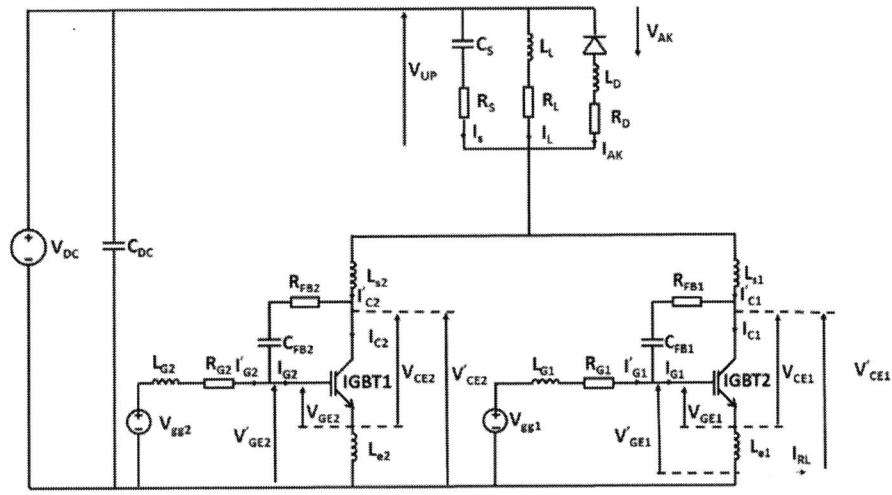

Fig. 2.   Schematic of the clamped inductive switching circuit with all the parasitic inductances.

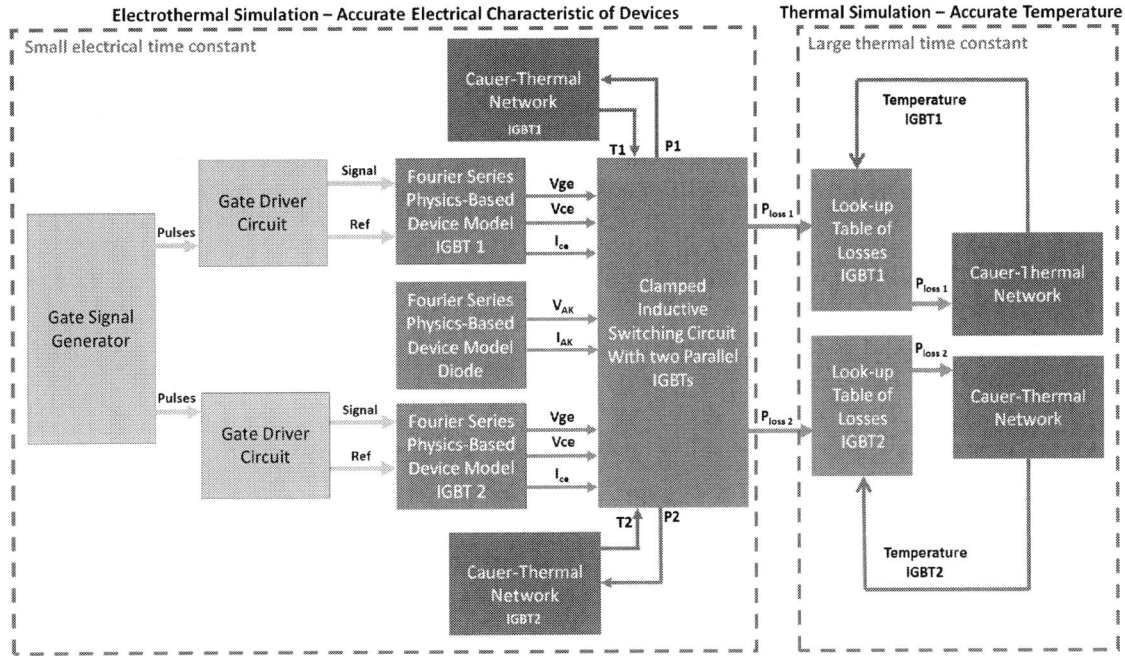

Fig. 3.   Block diagram of the electro-thermal simulation.

978-1-4673-9551-9/16 $31.00 © 2016 IEEE   255

accurately model the temperature rise in the IGBT due to unbalanced current sharing using the look-up table of losses and two independent Cauer-thermal networks for each switching device when the simulation time is significantly larger than the minimum simulation time-step of the electro-thermal model. This increases the performance of the simulation and reduces the computational load.

As can be seen in Fig. 4 the input power to the Cauer-thermal network is a pulse shape that has two overshoots and undershoots at the both edges of the pulse. From the graph it can be observed that the turn-off loss for the IGBT is significantly larger than the turn-on loss. This is due to the excess amount of charge stored in the drift region of the IGBT which brings about the current tail of the IGBT. For simplicity, the integration of the power loss during a period, which is the energy loss in one period, was calculated and used in the look-up table of losses. In the thermal simulation, it is assumed that the overshoot and undershoot does not exist and the energy loss consists the sum of turn-on losses, conduction losses and turn-off losses. This assumption is valid as the thermal time constants ($R_{th} \times C_{th}$) is significantly larger than the electrical time constant (the time that it takes for the device to turn on and off). At the end of each simulation step, the junction temperature of each device was calculated and fed as an input to the look-up table of losses and a new value of losses was obtained based on the new temperature.

Next, the energy loss corresponding to the junction

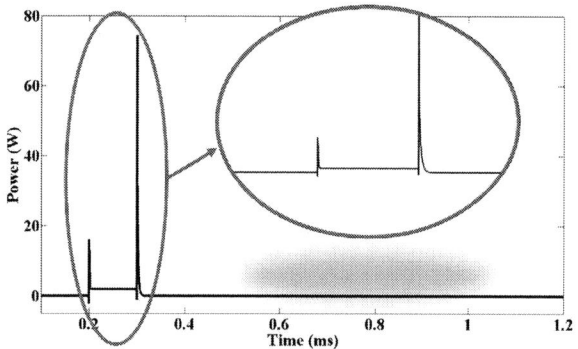

Fig. 4. A typical power loss during the repetitive clamped inductive switching for one period with duty cycle of 10%.

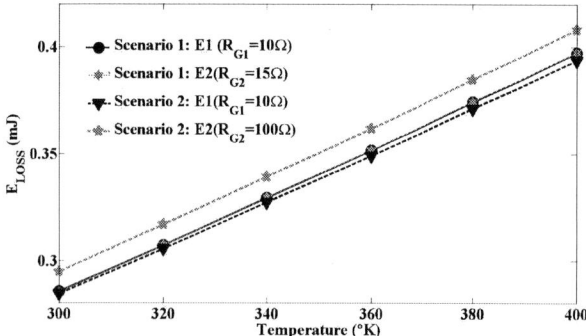

Fig. 5. Energy losses for two scenarios - scenario 1: R_{G1}=10Ω and R_{G2}=15Ω and scenario 2: R_{G1}=10Ω and R_{G2}=100Ω

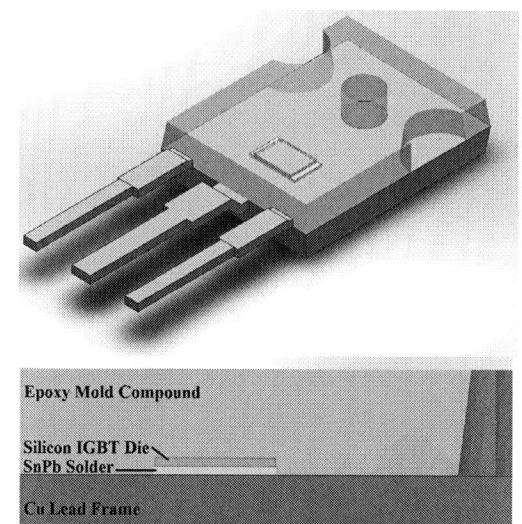

Fig. 6. CAD design for the descrete TO247 IGBT package and the cross-section view of the material stack.

Table 1   Thermal parameters for the material used in power electronics packaging

Material	Thermal Conductivity	Thermal Capacity	Density
	W/mK	J/kgK	kg/m³
Epoxy Mold Compound	0.72	794	2020
Silicon (at 25˚C)	148	712	2328.9
SnPb Solder	50	150	8500
Cu Lead Frame	360	380	8890

Table 2   Calculated 5 level Cauer-thermal network for TO247 Package

Silicon IGBT Die	R1	0.010135	C1	0.0249
SnPb Solder Layer	R2	0.0277	C2	0.0367
Cu Lead Frame	R3	0.021515	C3	2.7260
Epoxy Mold Compound	R4	10.7575	C4	1.2942
Cu Heatsink	R5	0.0222	C5	3.7564

temperature at the certain gate resistance was divided by the time of the pulse (duty cycle multiplied in the period) and used as the input power to the Cauer-thermal network. As the temperature rises, the amplitude of the pulse also increases and consequently the input power to the thermal network rises. The graph showing energy loss versus temperature for two scenarios for two IGBTs switching at different gate resistances (R_{G1}=10Ω and R_{G2}=15Ω for scenario 1 and R_{G1}=10Ω and R_{G2}=100Ω for scenario 2) is shown in Fig. 5. As can be seen, the energy loss for scenario 1 are very close and the device switching at R_{G2}=15Ω shows slightly higher losses. However, for the second scenario the energy losses are significantly different. Moreover, comparing scenario 1 and scenario 2, the device with $R_{G1}=10\Omega$ has less losses in scenario 2 than the first scenario. This is further investigated in the result section.

IGBT devices used during the experiments are International Rectifier IGBTs with datasheet reference IRG4PH20KPBF rated at 1.2 kV / 11 A. This IGBT does not have an antiparallel diode, hence all the heat generated in the device is due to the switching and conduction losses of the IGBT die. The dimensions of this device was taken from the device datasheet and the CAD design of the TO247 package was drawn in Solidworks. Fig. 6 shows the CAD design of this transistor and the cross-section view of the device package. Equation (14) and (15) were used to derive the thermal resistance and the thermal capacitance of each layer for the Cauer-thermal network.

$$R_{th} = \frac{l}{K_{th}A} \qquad (14)$$

$$C_{th} = V\rho C_p \qquad (15)$$

In these equations, R_{th} and C_{th} are the thermal resistance and capacitance respectively. K_{th} is the thermal conductivity of the material, A is the area of each layer and l represent the thickness of the layer. V represents the volume of each layer and ρ is the density of the material and C_p is the specific heat of the material. Table 1 shows the parameters used to calculate the thermal conductance and capacitance of each layer of the device. The stack of material used in this type of package and the cross-sectional view of this type of package are summarized in Fig. 6. Table 2 shows the thermal parameters calculated for each layer of this graph. These calculated parameters are used in the Cauer-thermal network of the electro-thermal and independent thermal models. In the thermal model, a resistor is added in the last stage of the Cauer-thermal network which determines the heat exchange between the ambient temperature (voltage source with the value of 300°K) and the backside Cu lead frame.

III. Experimental Setup

The experimental setup used to capture the current and

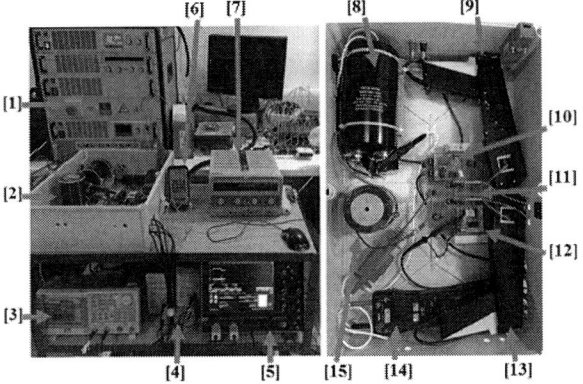

Fig. 7. Clamped inductive switching test rig with two parallel IGBTs and all the circuit components.

voltage waveforms of the parallel IGBTs is illustrated in Fig. 7. The device under test were discrete IGBTs in TO247 packages. They were driven using two separate gate drives. Consequently, they could be driven using different gate resistances. As explained earlier, non-uniform degradation of the electro-thermal properties like solder de-lamination or gate contact resistance, can cause the parallel connected IGBTs to have different temperatures or different gate resistances. Two heating elements were mounted on the back side of the two IGBTs and the devices were tested under the condition that they had different junction temperatures. Fig. 2 shows the schematic of the test circuit which is also used in the model. Different components used in this circuit are as below:

1) The high voltage power supply

2) The test rig enclosure

3) The function generator to generate double pulse

4) The current probes amplifiers

5) Teledyne oscilloscope

6) Temperature logger

7) Logic power supply for the driver board

8) DC link capacitor

9 and 13) Current probes

10 and 12) gate driver boards

11) Heating elements at the backside of TO247 IGBTs

14) Differential voltage probes

15) Inductive load

IV. Results and Discussion

During the experiments, two IGBTs were connected in parallel at the bottom side of the half-bridge and double pulse test was carried out on both devices. Fig. 8 shows the experimental results and the simulation results obtained from the physics-based model when one of the devices was working at higher junction temperature than the other i.e. the cooler device was at 25 °C while the hotter device was at 110 °C. As can be seen, the device with higher case temperature, takes less current than the cooler device. This is due to the positive temperature coefficient of the IGBT's on-state resistance and the current divider rule which means that more current flows through the less resistive IGBT.

In another case scenario, two devices were driven with different gate resistances i.e. the slower switching device was switched with $R_G=47\Omega$ while the faster switching device was switched with $R_G=10\Omega$. The experimental and simulation results showing turn-on and turn-off waveforms for the two devices are shown in Fig. 9 and Fig. 10. Fig. 9 shows that the faster switching device takes on more load current since the slower switching device turns ON later. Fig. 10 shows that during turn OFF, the slower switching device takes on more of the load current. This can be used in condition monitoring of power IGBTs where one device is degraded at higher rate than the other IGBT and shows higher gate resistance due to gate wire bond de-lamination. As can be seen in Fig. 10, the device with higher gate resistance switches off slower than the device with lower gate resistance. This results in a peak current in this device as the faster device is switched-off and consequently puts more load current on the slower device. As can be seen, the model can accurately predict the device behavior.

Fig. 11 and Fig. 12 show the turn-on and turn-off switching current waveforms of two parallel IGBTs when working under different case temperatures. The cooler device is working at

room temperature (25°C) while the other device is heated to (55°C). The results indicate that when the junction temperature of devices increase, the device with lower junction temperature takes higher current and it switches at different switching speed (dI/dt). As can be seen, the model can accurately predict this behavior. Moreover, Fig. 8 shows the hotter device is working at 110°C. Comparing these results with the results at 55°C in Fig. 8, indicates that the current sharing between two devices becomes even more unbalanced. Consequently, the amount of current which passes through the warmer device reduces more

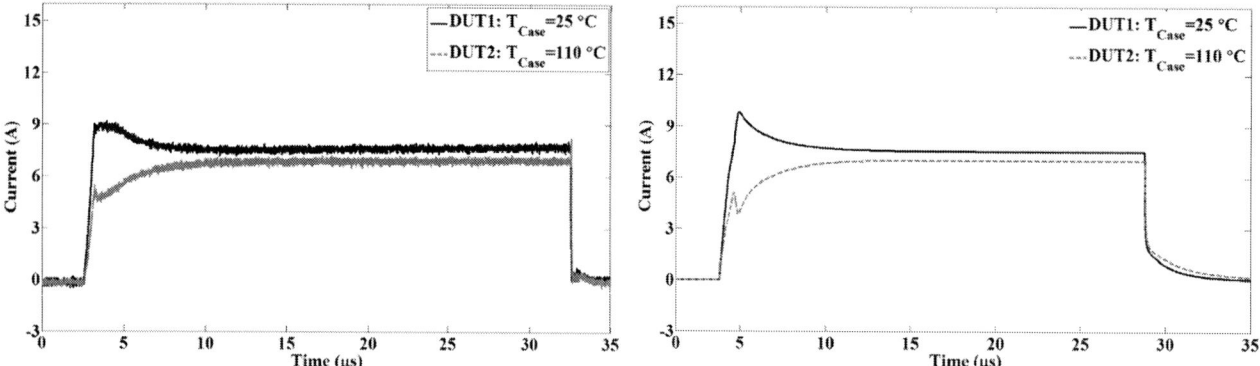

Fig. 8. Experimental results (Left) and simulation results (Right) of turn-on and turn-off switching current waveforms of two parallel IGBTs working at different junction temperatures.

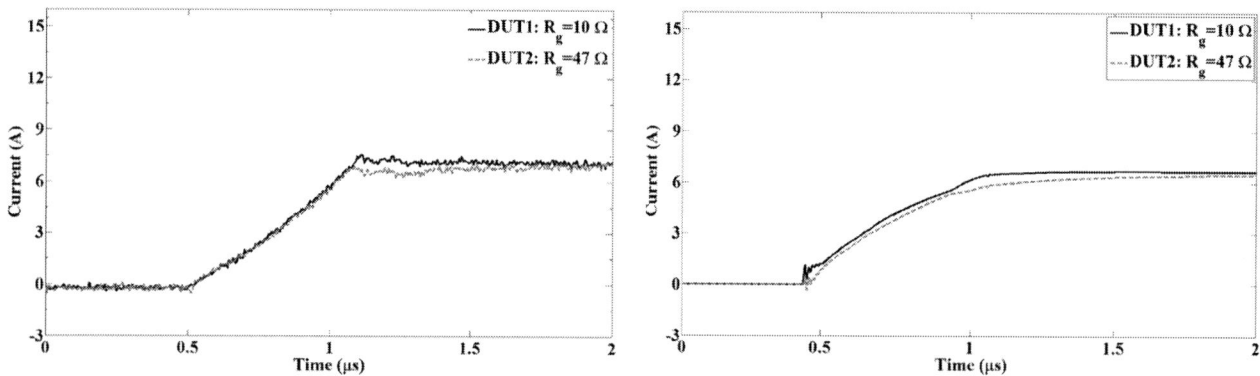

Fig. 9. Experimental results (Left) and simulation results (Right) of turn-on switching current waveforms of two parallel IGBTs working with two different gate resistances.

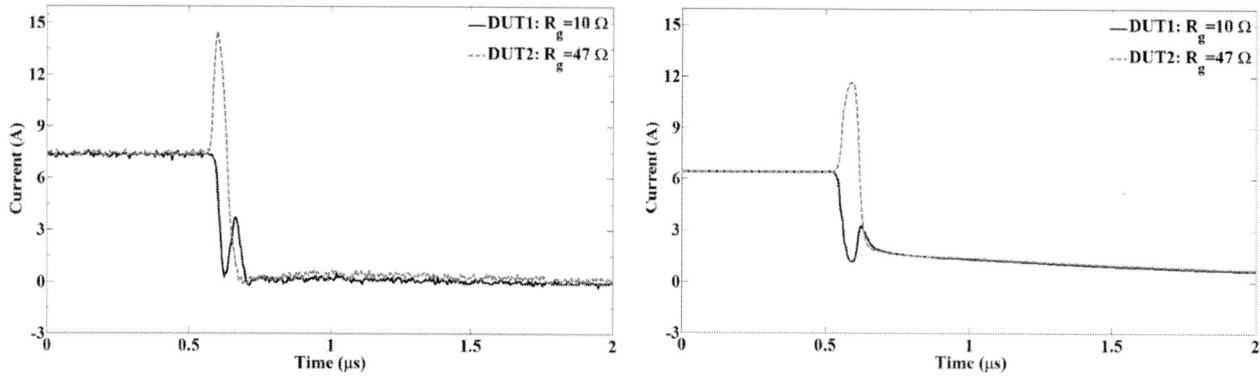

Fig. 10. Experimental results (Left) and simulation results (Right) of turn-off switching current waveforms of two parallel IGBTs working with two different gate resistances.

978-1-4673-9551-9/16 $31.00 © 2016 IEEE

significantly.

During the experiment, in order to evaluate the temperature rise in the unbalanced parallel IGBTs, the devices were switched under repetitive clamped inductive switching using different gate resistors. The temperature rise on the case of the devices was logged using a thermocouple which was mounted at the backside of each IGBT. The frequency of pulses were set to 1 kHz and the duty cycle of the pulses were set to 10%. Due to the unbalanced switching between the parallel devices, the device with the higher gate resistance showed a higher case temperature rise after 600 seconds of repetitive switching. Fig. 13 shows the experimental result and the simulation result obtained from the Cauer-thermal network for two parallel IGBT under repetitive clamped inductive switching using two different gate resistances. As can be seen, the thermal network can predict the temperature rise in the devices due to the unbalanced current sharing between the two IGBTs. The device with a larger gate resistance switches slower than the device with a smaller gate resistance, hence more current

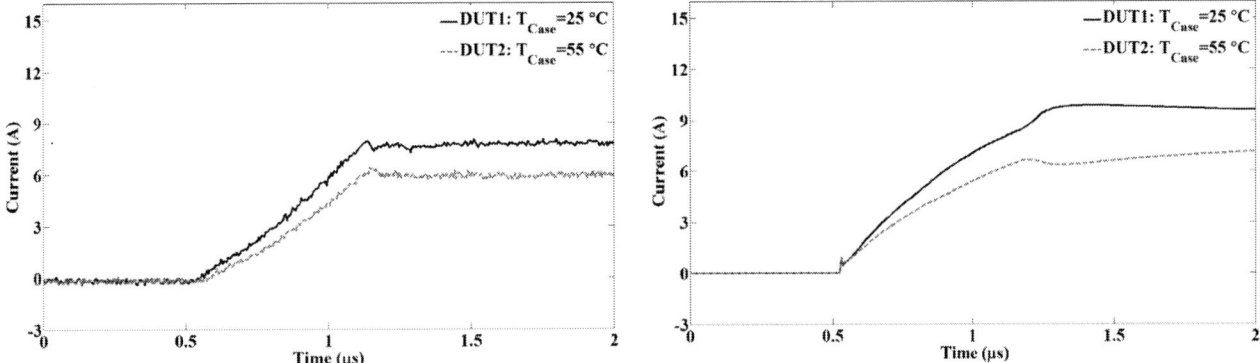

Fig. 11. Experimental results (Left) and simulation result (Right) of turn-on switching current waveform of two parallel IGBTs working at different case temperature (25°C and 55°C).

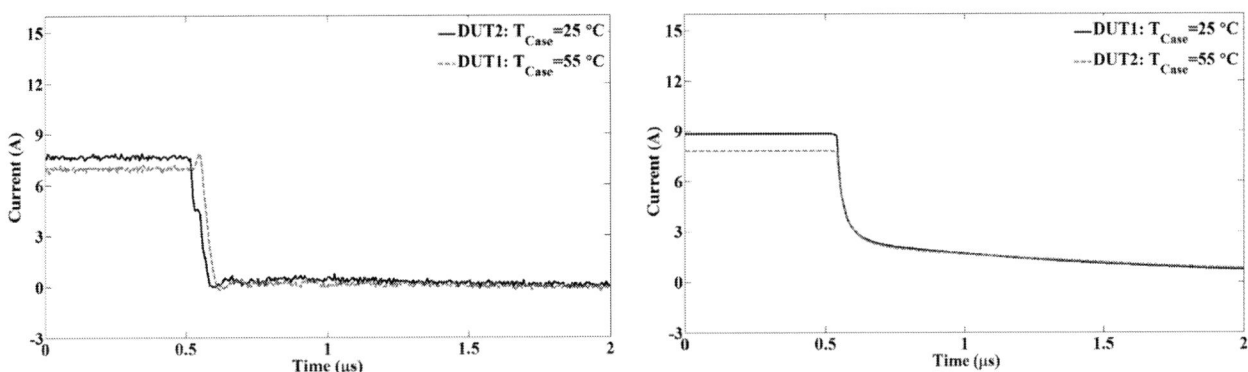

Fig. 12. Experimental results (Left) and simulation results (Right) of turn-off switching current waveform of two parallel IGBTs working at different case temperatures (25°C and 55°C).

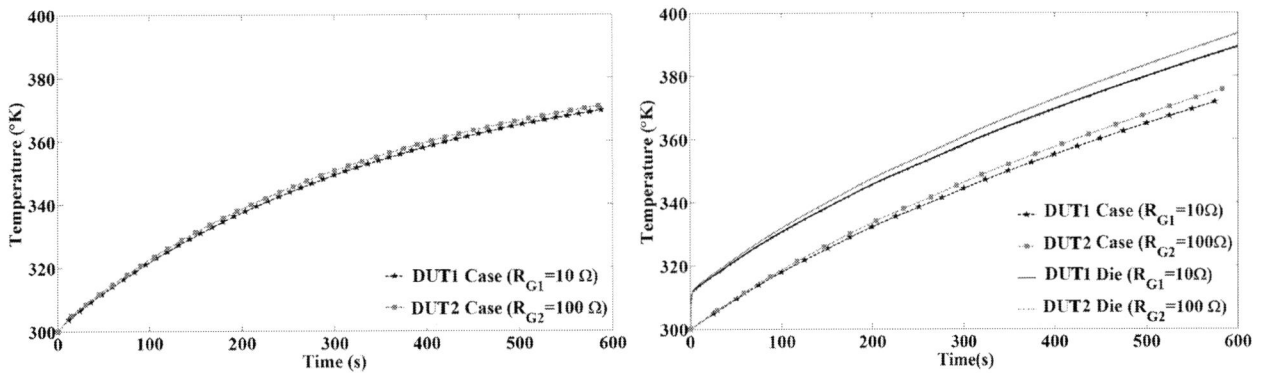

Fig. 13. Experimental results (Left) and simulation results (Right) of temperature rise within two parallel IGBTs with different gate resistances under repetitive clamped inductive switching.

passes through this device and consequently the temperature of the device rises more rapidly than the device with a smaller gate resistance. As the temperature of the device increases, the carrier lifetime and the diffusivity of minority carriers increases and consequently the tail current of the device increases which brings about higher losses. Moreover, the threshold voltage of the device reduces as the temperature rises which brings about even slower turn-off rate. Turn-off losses are relatively larger than turn-on losses or conduction losses for IGBT and consequently the effect of turn-off losses is more dominant in temperature rise within the device when the current sharing is unbalanced.

Because the model does not take heat radiation into consideration, the final temperature obtained in the simulation is slightly higher than the experimental results. Moreover, the simulation results show that the temperature of dies are around 20°C higher than the case temperature and the temperature difference between two dies operating using different gate resistances is around 7°C. It is worthwhile to mention that this becomes more significant in the power modules with higher current rating and working at the higher voltages.

V. CONCLUSIONS

A physics-based temperature-dependent IGBT and diode models were used to model the current sharing between two parallel IGBTs working under different electrothermal conditions. In one scenario, the gate resistance of each device was varied and in another case scenario the junction temperature of devices were varied. The devices were tested in a clamped inductive switching test rig and the current passing through devices were measured. Using the model, the switching current waveforms were predicted accurately. This model can be used by power electronic engineers to model non-uniform degradation of devices or reliability issues arising from solder de-lamination or gate wire bond degradation in individual devices in a module. The electrothermal model is able to simulate the electrical switching characteristics of the IGBT very accurately on the microsecond scale while also being able to simulate long thermal transients on the millisecond to second scale. This has been achieved by using 2 Cauer networks and a look-up table to decouple the electrical and thermal time constants. Hence, the effects of electrothermal imbalance between parallel connected IGBTs can be captured accurately in a computationally inexpensive methodology.

REFERENCES

[1] F. Blaabjerg, M. Liserre, and M. Ke, "Power Electronics Converters for Wind Turbine Systems," *Industry Applications, IEEE Transactions on*, vol. 48, pp. 708-719, 2012.

[2] J. Colmenares, D. Peftitsis, H. P. Nee, and J. Rabkowski, "Switching performance of parallel-connected power modules with SiC MOSFETs," in *Power Electronics Conference (IPEC-Hiroshima 2014 - ECCE-ASIA)*, 2014 International, 2014, pp. 3712-3717.

[3] D. P. Sadik, J. Colmenares, D. Peftitsis, L. Jang-Kwon, J. Rabkowski, and H. P. Nee, "Experimental investigations of static and transient current sharing of parallel-connected silicon carbide MOSFETs," in

Power Electronics and Applications (EPE), 2013 15th European Conference on, 2013, pp. 1-10.

[4] A Avron, 'In a Tesla model S, there is no IGBT packaging trick' Sep. 15, 2015. [online]. Available: http://www.pointthepower.com/on-tesla-electric-vehicles-semiconductor-packaging. [Accessed: 10- Nov- 2015].

[5] G. Breglio, A. Irace, E. Napoli, M. Riccio, and P. Spirito, "Experimental Detection and Numerical Validation of Different Failure Mechanisms in IGBTs During Unclamped Inductive Switching," *Electron Devices, IEEE Transactions on*, vol. 60, pp. 563-570, 2013.

[6] Y. Shaoyong, A. Bryant, P. Mawby, X. Dawei, L. Ran, and P. Tavner, "An Industry-Based Survey of Reliability in Power Electronic Converters," *Industry Applications, IEEE Transactions on*, vol. 47, pp. 1441-1451, 2011.

[7] R. Bonyadi, O. Alatise, S. Jahdi, H. Ji, J. A. O. Gonzalez, R. Li, and P. A. Mawby, "Compact Electrothermal Reliability Modeling and Experimental Characterization of Bipolar Latchup in SiC and CoolMOS Power MOSFETs," *Power Electronics, IEEE Transactions on*, vol. 30, pp. 6978-6992, 2015.

[8] A. Bryant, "Simulation and Optimisation of Diode and IGBT Interaction in a Chopper Cell," Doctor of Philosophy, Queens' College, University of Cambridge, Cambridge, 2005.

[9] R. Bonyadi, O. Alatise, S. Jahdi, J. Hu, L. Evans, and P. A. Mawby, "Investigating the reliability of SiC MOSFET body diodes using Fourier series modelling," in *Energy Conversion Congress and Exposition (ECCE)*, 2014 IEEE, 2014, pp. 443-448.

[10] A. T. Bryant, G. J. Roberts, A. Walker, and P. A. Mawby, "Fast Inverter Loss Simulation and Silicon Carbide Device Evaluation for Hybrid Electric Vehicle Drives," in *Power Conversion Conference - Nagoya, 2007. PCC '07*, 2007, pp. 1017-1024.

[11] R. Bonyadi, O. Alatise, S. Jahdi, J. O. Gonzalez, L. Ran, and P. A. Mawby, "Modeling of temperature dependent parasitic gate turn-on in silicon IGBTs," in *Power Electronics and ECCE Asia (ICPE-ECCE Asia), 2015 9th International Conference on*, 2015, pp. 560-566.

[12] L. Lu, A. Bryant, E. Santi, J. L. Hudgins, and P. R. Palmer, "Physical Modeling and Parameter Extraction Procedure for p-i-n Diodes with Lifetime Control," in *Industry Applications Conference, 2006. 41st IAS Annual Meeting. Conference Record of the 2006 IEEE*, 2006, pp. 1450-1456.

[13] R. Bonyadi, O. Alatise, S. Jahdi, J. Ortiz-Gonzalez, Z. Davletzhanova, L. Ran, et al., "Physics-based modelling and experimental characterisation of parasitic turn-on in IGBTs," in *Power Electronics and Applications (EPE'15 ECCE-Europe), 2015 17th European Conference on*, 2015, pp. 1-9.

[14] P. A. Mawby, A. T. Bryant, P. R. Palmer, E. Santi, and J. L. Hudgins, "High Speed Electro-Thermal Models for Inverter Simulations," in *Microelectronics, 2006 25th International Conference on*, 2006, pp. 166-173.

[15] A. T. Bryant, K. Xiaosong, E. Santi, P. R. Palmer, and J. L. Hudgins, "Two-step parameter extraction procedure with formal optimization for physics-based circuit simulator IGBT and p-i-n diode models," *Power Electronics, IEEE Transactions on*, vol. 21, pp. 295-309, 2006.

[16] A. Bryant, N. A. Parker-Allotey, D. Hamilton, I. Swan, P. A. Mawby, T. Ueta, et al., "A Fast Loss and Temperature Simulation Method for Power Converters, Part I: Electrothermal Modeling and Validation," *Power Electronics, IEEE Transactions on*, vol. 27, pp. 248-257, 2012.

[17] A. T. Bryant, P. R. Palmer, E. Santi, and J. L. Hudgins, "A Compact Diode Model for the Simulation of Fast Power Diodes including the Effects of Avalanche and Carrier Lifetime Zoning," in *Power Electronics Specialists Conference, 2005. PESC '05. IEEE 36th*, 2005, pp. 2042-2048.

[18] A. T. Bryant, L. Liqing, E. Santi, P. R. Palmer, and J. L. Hudgins, "Physical Modeling of Fast p-i-n Diodes With Carrier Lifetime Zoning, Part I: Device Model," *Power Electronics, IEEE Transactions on*, vol. 23, pp. 189-197, 2008.

General 3D Lumped Thermal Model with Various Boundary Conditions for High Power IGBT Modules

Amir Sajjad Bahman, Ke Ma, Frede Blaabjerg

Center of Reliable Power Electronics (CORPE)
Department of Energy Technology, Aalborg University, 9220 Aalborg, Denmark
asb@et.aau.dk, kema@et.aau.dk, fbl@et.aau.dk

Abstract— Accurate thermal dynamics modeling of high power Insulated Gate Bipolar Transistor (IGBT) modules is important information for the reliability analysis and thermal design of power electronic systems. However, the existing thermal models have their limits to correctly predict these complicated thermal behaviors in the IGBTs. In this paper, a new three-dimensional (3D) lumped thermal model is proposed, which can easily be characterized from Finite Element Methods (FEM) based simulation and acquire the thermal distribution in critical points. Meanwhile the boundary conditions including the cooling system and power losses are modeled in the 3D thermal model, which can be adapted to different real field applications of power electronic converters. The accuracy of the proposed thermal model is verified by experimental results.

Keywords—Insulated gate bipolar transistors, Thermal modeling, Boundary conditions, Finite element method, Reliability, Power electronic converters.

I. INTRODUCTION

Insulated Gate Bipolar Transistor (IGBT) modules are widely applied in power electronic conversion systems especially in high power application like renewable energy systems, traction industries and HVDC [1]. Since industries demand for higher power densities, more integrated packaging, and cost saving, the risk of failures and reliability of the power electronics become more crucial. Consequently, thermal management of power semiconductors finds more importance due to increased heat generation inside the devices [2]. The first step in thermal management of the power semiconductor devices is to identify accurate and detailed temperature information in critical locations by using compact thermal models [3]. However, accurate thermal modeling of high power IGBT modules is of great challenges due to several physical and operational factors. The physical factors are related to geometries – e.g. size, thickness and positon of the semiconductor chips – as well as thermal properties of materials used in different layers of the IGBT module. These factors lead to uneven thermal distribution among the chips as well as the sub-layers inside the IGBT module due to thermal coupling effects [4]. On the other hand, operational factors include those related to long-term mission profiles and thermal dynamics, which are crucial for lifetime calculation and thermal design/management of high power module. In many lifetime models, temperature cycles in critical locations such as junction and solder layers are important factors to be identified for a reliable design of IGBT modules [5].

Currently, various compact thermal models have been introduced. The first group of thermal models is based on one dimensional lumped RC networks, e.g. Cauer or Foster type models. Typically the equivalent thermal circuit consists of several RC pairs to identify the dynamics of the device temperatures in respect to the power losses injected to the semiconductor chips [6]. Conventionally, Cauer or Foster thermal networks are given by the manufacturer in the IGBT module datasheet. These thermal models are all based on a one dimensional (1D) modeling approach of the heat conduction and can be used for a rough and fast calculation of junction temperature. However, they cannot be used for studying the three dimensional (3D) heat spreading effects to come up with accurate temperatures in different locations. The other group of thermal models is based on analytical solutions to the heat equation. Due to the complexity of the thermal system, Finite Element Method (FEM) or Finite Difference Method (FDM) simulations have been used to calculate the 3D thermal behaviors in power module [7]-[9]. However, these methods are not efficient for long-term mission profile-based analysis of IGBT modules since they demand large computational cost and may also lead to divergence for high dynamic operation.

Apart from thermal dynamics, variation of boundary conditions is neglected in the thermal models, which are inevitable on the thermal analysis of IGBT modules. The boundary conditions in term means a set of conditions that is required to be satisfied at all or one part of the boundaries of a design geometry in which a set of differential equations is to be solved [10]. In an IGBT module, boundary conditions consist of heat sources i.e. power losses in the semiconductor chips and heat sink i.e. cooling system [11]. On the other hand, in a reliable design of high power IGBT modules, it is important to evaluate the dynamic thermal behavior with real load profiles associated with e.g. renewable or automotive applications. For this reason, a circuit simulator has the benefits of fast simulation for given load profiles.

In the this paper, a method to transform the boundary conditions from the FEM environment to a circuit simulator is given, which has benefits of FEM's accuracy and circuit simulator speed. A generic 3D thermal model will be introduced with flexible RC elements to be used for different heating and cooling conditions, and is able to calculate temperatures at different locations and layers of the IGBT. The introduced thermal model is verified by thermography measurements from a power cycling test setup.

978-1-4673-9551-9/16 $31.00 © 2016 IEEE

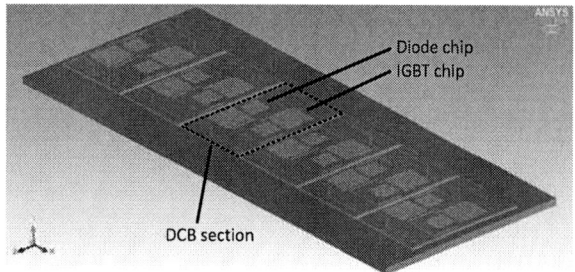

Fig. 1. Schematic of high power IGBT module modeled in ANSYS Icepak for FEM analysis.

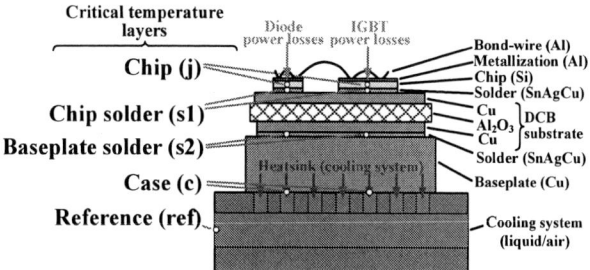

Fig. 2. IGBT module layers and boundary conditions of high power IGBT module modeled in ANSYS Icepak for FEM analysis.

II. THE PROPOSED THREE-DIMENSIONAL THERMAL NETWORK

A high power IGBT module consisting 6 full-bridge Direct Copper Bonded (DCB) sections connected in parallel is shown in Fig. 1. The materials of IGBT module and boundary conditions for thermal analysis are shown in Fig. 2. There are two boundary conditions in this case study, one is at the chip junction as the heat source (power losses), and the other one at the bottom of the baseplate as the cooling capability (heat sink/cooling system). The thermal characteristics for the materials used in the IGBT module under study are given in Table I. It is noted that the conductivity of some materials is set to be temperature dependent according to [12]. For simplicity of analysis it is assumed that the IGBT module is adiabatic from the top and the lateral surfaces and therefore all generated heat is dissipated in the cooling system.

When the model is ready to be used in the FEM environment, specific indexes should be defined to identify the boundary condition related effects on transient thermal behavior of IGBT module. Therefore, the transient thermal impedance curves are defined in the design geometry of IGBT module. In dynamic operation, the temperature difference between every two points is calculated using the transient thermal impedance curve, $Z_{th}(t)$. The transient thermal impedance between two points is explained by

$$Z_{th(a-b)}(t) = \frac{T_a(t) - T_b(t)}{P} \qquad (1)$$

where $T_a(t)$ and $T_{b(t)}$ are transient temperatures in two points and P is the power dissipation, which is generated in the

TABLE I. IGBT MODULE MATERIAL THERMAL PROPERTIES

Material	Density kg/m^3	Specific heat $J/(kg \cdot K)$	Conductivity $W/(m \cdot K)$	
			Temp. (°C)	Conductivity
Silicon (Si)	2330	705	0.0	168
			100.0	112
			200.0	82
Copper (Cu)	8954	384	0	401
			100	391
			200	389
Al₂O₃	3890	880	all	35
SnAgCu	7370	220	all	57

device. To derive the thermal impedances, a single square power pulse with amplitude P is applied to a heat source in the power module (IGBT chip or diode chip) until the junction temperature reaches the steady state. Then, by dividing the temperature difference between each two neighboring region to power loss, a transient thermal impedance curve is obtained. To derive the transient thermal impedances, a step response analysis is implemented in FEM simulation [13]. The curves are fitted into a finite number of exponential terms by

$$Z_{th(a-b)}(t) = \sum_{i=1}^{n} R_{thi} \cdot (1 - e^{-t/\tau_{thi}}) \qquad (2)$$

where R_{thi} is thermal resistance, τ_{thi} is time constant which equals to $R_{thi} * C_{thi}$ and n is the number of exponential terms. Commonly, four exponential terms are enough to fit $Z_{th(a-b)}(t)$ with enough accuracy for the intended application. The number of exponential terms determines the number of RC pairs in the RC thermal network. The details of the extraction process have been explained in [13]. The thermal network, which is used in this work is a Foster network and is widely used by industry due to the simplicity of extraction of parameters.

Most of the IGBT module failures occur due to bond wire lift-off or solder crack, so the temperature profiles in these locations are critical [14]. Thus, the IGBT module is divided vertically into four sections and the thermal impedance curves are derived between layers to study the transient thermal behavior of IGBT module. These sections are shown in Fig. 2. Moreover, to study the thermal stress on bond wire heel locations, temperatures at different points on the chip surface are required to be identified. Based on the described critical temperature locations, a 3D thermal network is extracted, which includes all the mentioned nodes, different temperature locations on material layers, heat sources, heat sink and thermal coupling effects from other heat sources. The schematic of the 3D thermal network is shown in Fig. 3.

III. CHARACTERIZATION AND MODELING OF BOUNDARY CONDITIONS FOR THE THERMAL ANALYSIS

In order to understand the importance of the boundary condition effects in thermal impedance of IGBT module, the cooling system variations are modeled by FEM simulations and temperature responses are extracted in the corresponding points in the 3D thermal network. To represent the capability

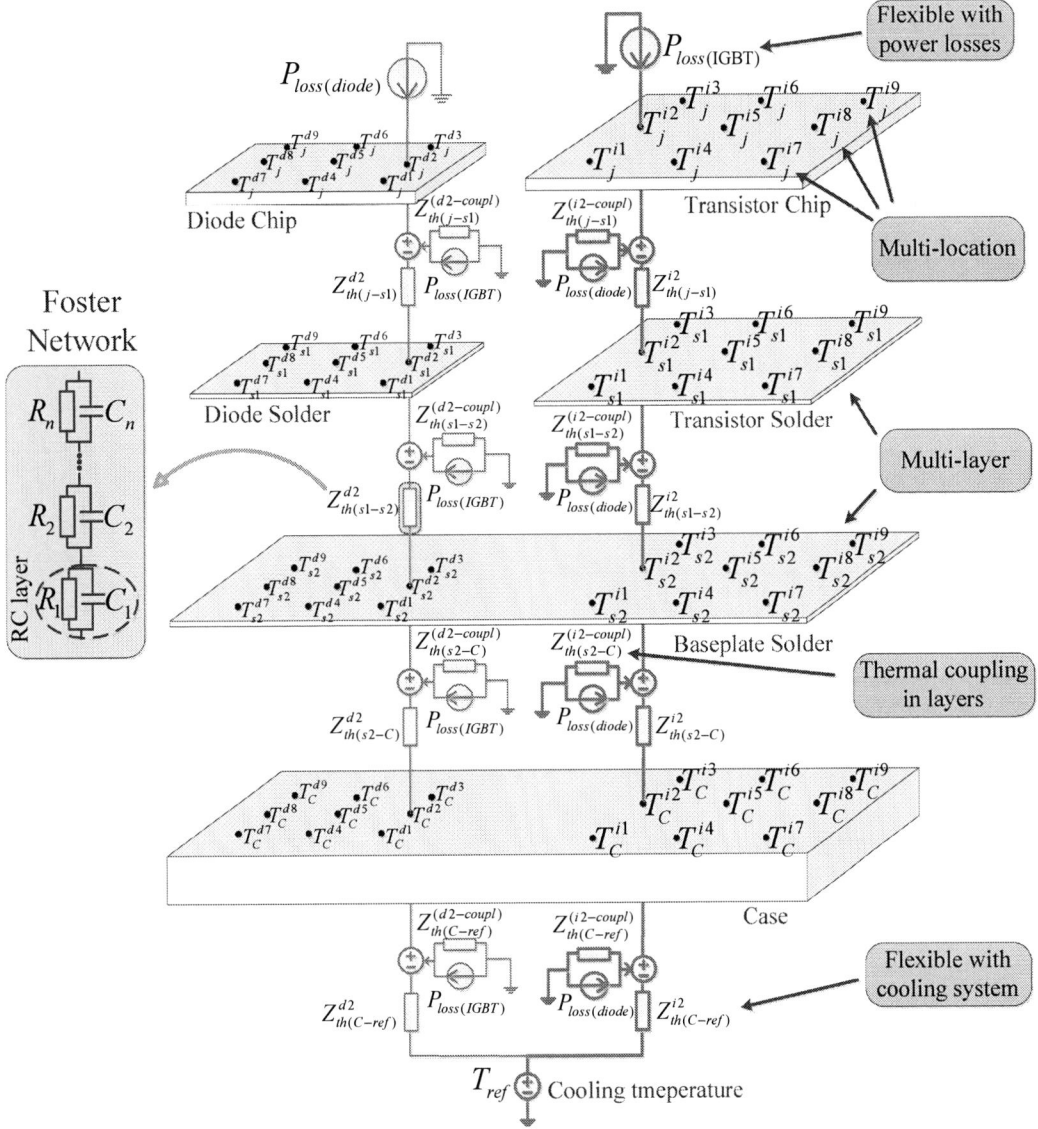

Fig. 3. 3D thermal network from chip (junction) to reference (cooling temperature).

of the fluid cooling systems, different cooling mechanisms are considered in the heatsink. For each cooling mechanism, the equivalent heat transfer coefficient, *htc*, of the cooling system is extracted and modelled as a thick plate beneath the baseplate. The equivalent *htc* is a measure, which stands for the amount of heat, which is transferred by convection between a solid and a fluid [12]. The *htc* in concept is the proportionally coefficient between the heat flux and the temperature difference between the solid and fluid:

$$htc = \frac{q}{\Delta T} \ [W \,/\, m^2 \cdot K] \tag{3}$$

where *q* is the amount of heat, which is transferred between two materials (heat flux) and ΔT is the temperature difference between the solid surface and surrounding fluid area. The heat flux, *q*, in turn is defined as the thermal power (or power losses in the IGBT module) per unit area:

$$q = \frac{d\dot{Q}}{dA} \ [W \,/\, m^2] \tag{4}$$

where *A* is defined as the effective area for heat dissipation of the heatsink. Moreover, the thermal resistance between the IGBT module and the heatsink, $R_{th(c\text{-}ref)}$, can be defined based on the definition of heat transfer coefficient

$$R_{th(c-ref)} = \frac{1}{htc \cdot A} \ [K \,/\, W] \tag{5}$$

From eq. (5), it can be seen that a higher *htc* leads to a smaller R_{th}. With a *htc*, heat flux in the power module is more localized beneath the IGBT chips that lead to a smaller heat spreading specially in the baseplate. This will reduce the effectiveness of the baseplate area in spreading the heat dissipation; so the temperature difference between the junction

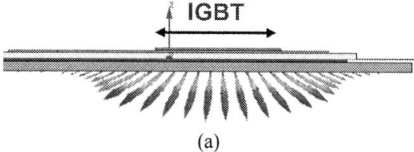

(a)

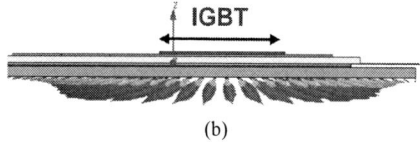

(b)

Fig. 4. Heat flux distribution in a power module for heat transfer coefficient: (a) 10000 W/m²·K, (b) 1000 W/m²·K.

TABLE II. HEAT TRANSFER COEFFICIENTS FOR SOME COMMON FLUIDS (W/m²·K)

Free convection-Air	5-25
Free convection-Water	20-100
Forced convection-Air	10-200
Forced convection-Water	50-10000
Boiling water	3000-100000
Condensing water vapor	5000-10000

and case will be increased. This phenomenon is shown in Fig. 4. As listed in [12], the equivalent *htcs* can vary from *10 W/m²·K* for natural convection systems to *10⁵ W/m²·K* for phase change cooling systems. Typical values for those cooling systems are listed in Table II. With the method described in section II, the transient thermal impedance curves are extracted for different *htcs*. The *htcs* within the ranges of *3000<htc<100000 W/m²·K* are used in this paper, which represents reasonable cooling conditions for the IGBT module under study. The transient thermal impedances under different *htcs* are shown in Fig. 5. As it is shown, the most influenced thermal impedance is the section from case to reference (cooling fluid temperature) due to its closer distance to the heatsink.

On the other hand, the effect of a hot plate (fixed case temperature) variation beneath the baseplate on thermal impedance of IGBT module is studied. To model the fixed case temperature, in FEM environment, a thick plate is placed under the baseplate and boundary condition between the baseplate and wall is set to a very high *htc* (close to infinite). The reference temperature at the back of thick plate is then changed to have the same case temperature as the hot plate should perform. In this study, the case temperature is varied in the range of 20°C to 120°C. The results are shown in Fig. 6. As it is seen the most affected section is the junction to chip solder. The reason originates from thermal blocking behavior of the case surface of the device, which prevents the heat to be dissipated in the heatsink. So, the heat generated in the chip does not propagate to the lower layers and tends to be remained in the upper layers.

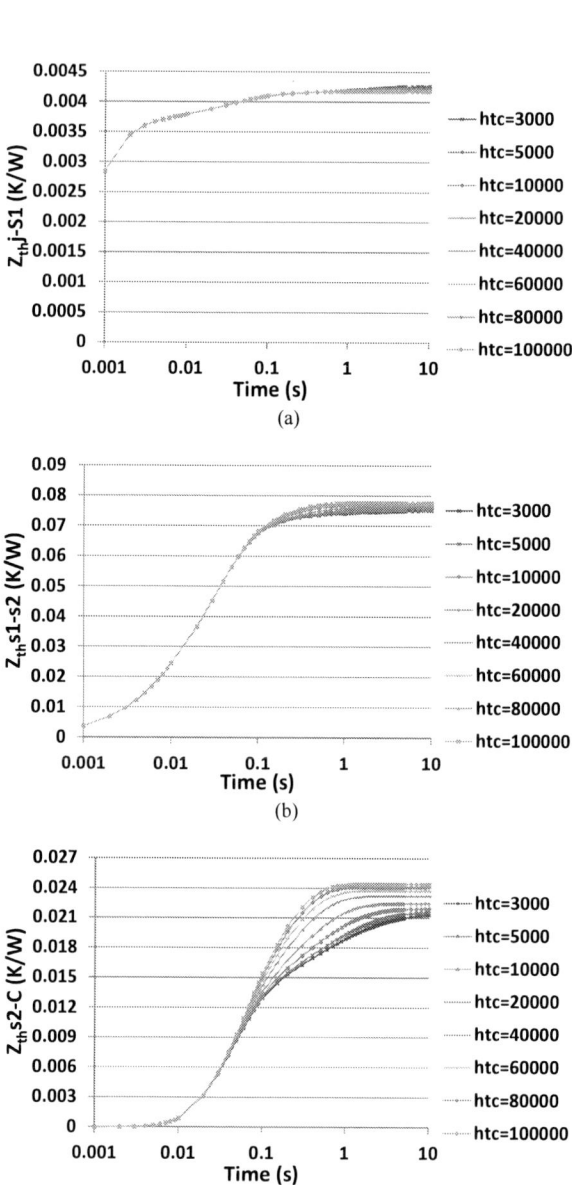

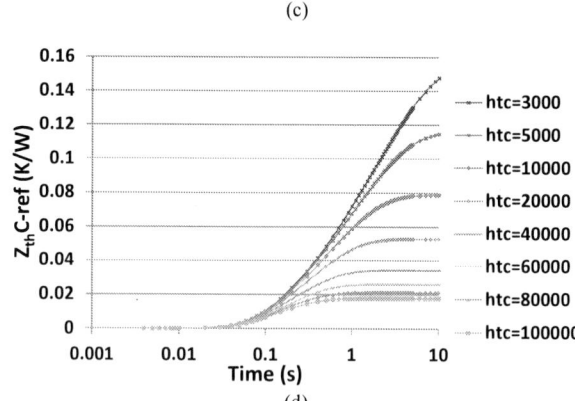

Fig. 5. Transient thermal impedances in different layers for various **cooling systems (*htc* in W/m²·K)**. (a) junction to chip solder, (b) chip solder to baseplate solder, (c) baseplate solder to case, (d) case to reference.

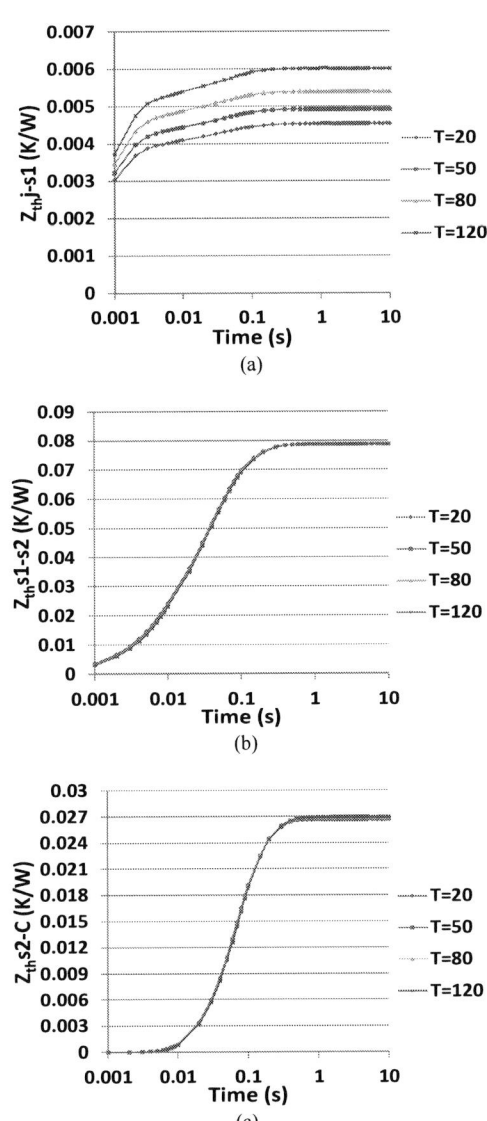

Fig. 6. Transient thermal impedances in different layers for various **hotplates** (**T in °C**). (a) junction to chip solder, (b) chip solder to baseplate solder, (c) baseplate solder to case, (d) case to reference.

IV. TRANSFORMATION OF BOUNDARY CONDITIONS FROM FEM MODEL TO LUMPED RC NETWORK

It was discussed in section I that FEM simulations can be used in the case of short-term load profiles. For longer load profiles (e.g. 1 day or 1 year for a complete mission profile), FEM simulation will be too time-consuming and demands for high computational facilities. Therefore, simplified thermal models are needed to be used in circuit simulators. But, in circuit simulators it is very hard to model the effect of boundary conditions in the compact thermal model. The boundary conditions need to be translated from FEM to circuit simulator in order to develop a more general thermal model. This is possible to implement by using a step response analysis for different boundary conditions in FEM. By using a

step response analysis, the transient thermal impedance curves are extracted and mathematically fitted to a 3D thermal network. As shown in Fig.3, Foster networks in the 3D thermal network can vary from one RC layer to multiple RC layers depending on the accuracy of the curve-fitting. For simplicity of modeling of the boundary conditions, one RC layer is used in this work.

The RC element values in respect to the different cooling systems are shown in Fig. 7. For a higher accuracy of the thermal model, the thermal coupling branches are connected to the main branch as controlled voltage sources. As described in section II, the highly affected regions are from baseplate solder to case and from case to reference. The variation is mathematically curve fitted to find the generic model for various cooling mechanisms. In the given curves, the horizontal axis shows different *htc*s (different cooling conditions) and the vertical axis shows the respected thermal resistance and thermal capacitance values. The curve fitted linear mathematical model and respected R-squared values are also shown beside the curves. For all cases, the R-squared values are at least 0.9 for a better accuracy of the curve-fitting [15].

The generic thermal models for variation of case temperatures and power losses follow the same approach. The schematic of one branch of 3D thermal network (highlighted in red in Fig. 3) with a variation of cooling system is also shown in Fig. 8. As it is seen, the RC elements in the regions which are not varied by the cooling system are shown as constant values. By variation of the RC elements in the 3D thermal network, a flexible thermal network is developed in which RC elements are dependent on the boundary conditions. In other words, by the presented approach, thermal model of IGBT module can get feedback from the boundary conditions in transient operation, and calculate accurately the temperatures at different locations with a very high simulation speed. The parameters needed to construct this network are the geometries and materials of the IGBT module, the equivalent *htc* that can be extracted by Computational Fluid Dynamics (CFD) simulations of the real cooling system or rough values given in heat transfer handbooks, the power loss level that module will be used, the ambient temperature and the case temperature. Of course, the 3D thermal network can be constructed by the parameters given in one condition; however, to extract a generic thermal model, RC elements can be modeled as variable parameters for a few possible operating conditions.

V. EXPERIMENTAL VERIFICATION

The boundary-dependent thermal model is tested in a real power cycling operation. For this purpose, an experimental setup is established. The IGBT module is loaded with a three-phase DC-AC two-level Voltage Source Converter (2L-VSC). The detailed converter specifications are listed in Table III. The fundamental frequency of the converter is set to 6 Hz, which is usual in reliability power cycling tests [16]. A black painted, opened IGBT module is being monitored by an IR camera (Fig. 9). The IGBT module is mounted on a direct liquid cooling system where the liquid cooling temperature and flow rate can be controlled for each experiment.

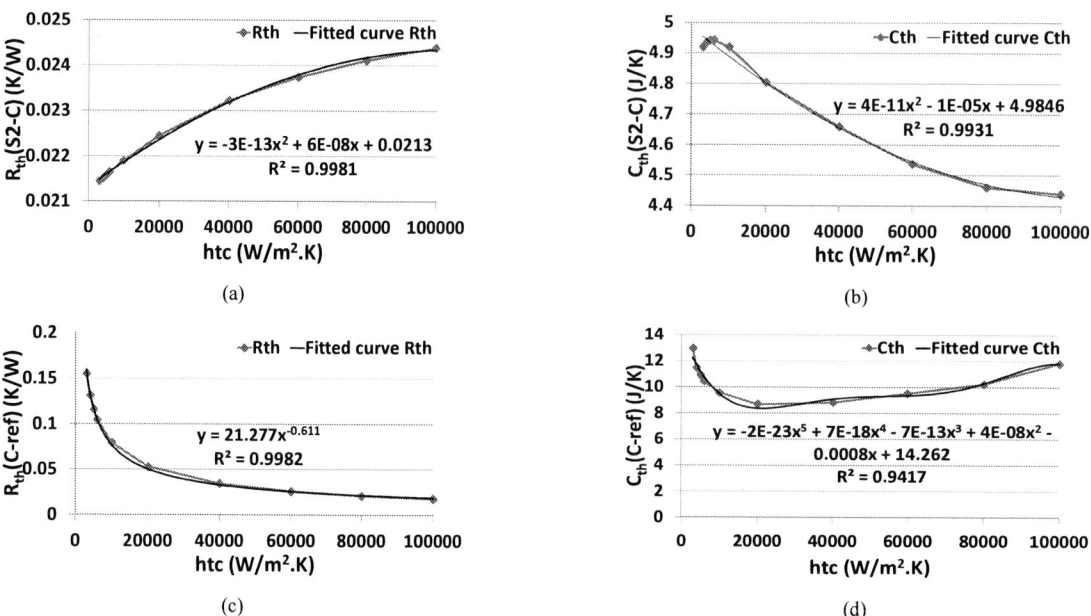

Fig. 7. Curve fitted thermal resistance and thermal capacitance for various **cooling systems**. (a) baseplate solder to case thermal resistance, (b) baseplate solder to case thermal capacitance, (c) case to reference thermal resistance, (d) case to reference thermal capacitance.

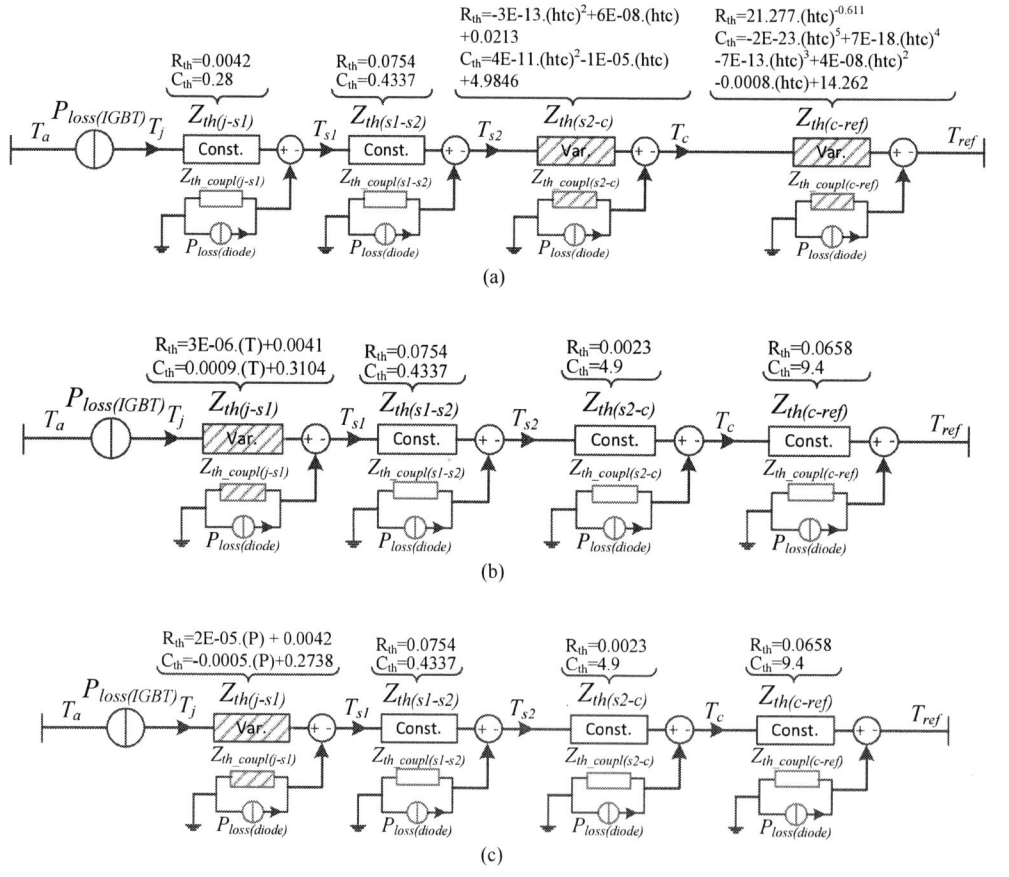

Fig. 8. Schematics of one branch of 3D thermal network (highlighted in red in Fig. 3) with different boundary conditions. (a) variation of fluid cooling system, (b) variation of hotplate, (c) variation of power losses.

TABLE III. PARAMETERS OF THE TWO LEVEL VOLTAGE SOURCE CONVERTER

DC bus voltage V_{dc}	450 V
Rated load current I_{load}	variable up to 900 A (peak)
Fundamental frequency f_o	6 Hz
Switching frequency f_{sw}	2.5 kHz
Filter inductor L_l	350 µH
IGBT module	1700V/1000A

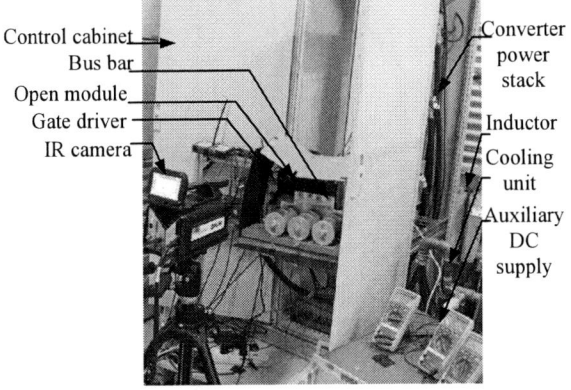

Fig. 9. Test setup featured with the infra-red camera.

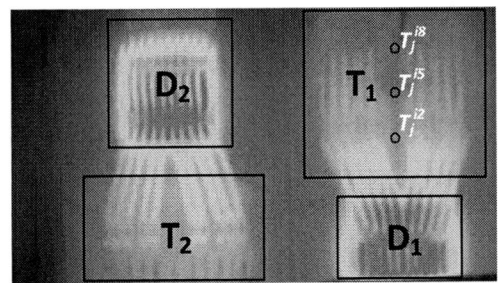

Fig. 10. Thermographical picture of one DCB section at the surface of IGBT module.

The cooling system is able to dissipate the heat from the IGBT module homogenously [17]. The infrared thermal image of one DCB section of the IGBT module is shown in Fig. 10. To validate the presented model, similar monitoring points as in the 3D thermal network are considered on the surface of the IGBT chip and the diode chip. The monitoring points should be considered between the bond-wires and be on the surface of the chips to prevent false temperature monitoring of the bond-wires.

In the simulation environment (e.g. PLECS), the same converter topology is established and the power losses are applied into the IGBT chips and diode chips. The peak of the load current, $I_{load(peak)}$, is fixed to *500 A*, cooling liquid flow rate, $\dot{V}$, is set to *5 m^3/hr* and cooling liquid temperature, $T_{cooling}$, is set to *45°C*. For the cooling system, the equivalent *htc* is set to 7000 $W/m^2 \cdot K$ as suggested by the manufacturer of

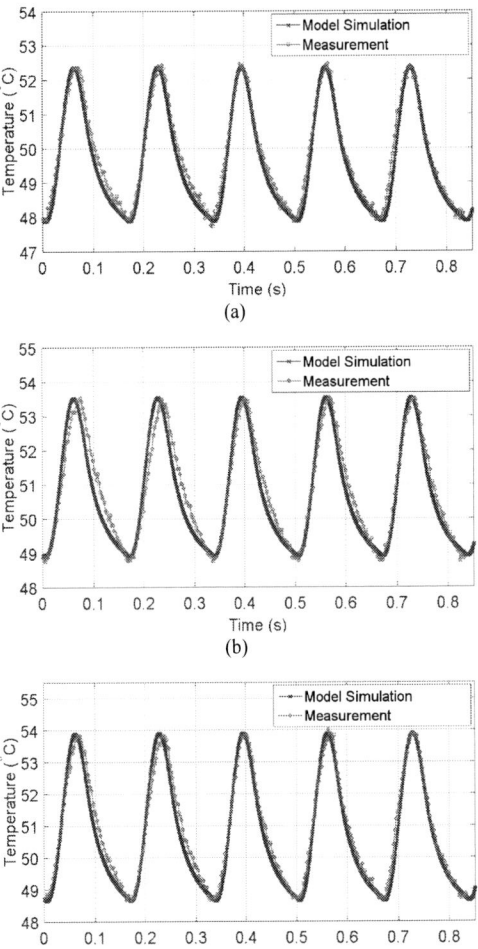

Fig. 11. Comparison of junction temperatures by the thermal model simulation and measurement in $I_{load(peak)}=500A$, $\dot{V}=5m^3/hr$, and $T_{cooling}=45°C$. (a) i_2, (b) i_5, and (c) i_8 (see Fig. 3).

the cooling system for the mentioned flow rate. The equivalent *htc* of the cooling system is dependent on the liquid flow rate and geometries of cooling system and can be calculated by CFD simulations. Since it is difficult to access the accurate power losses in the experimental setup, the power losses used in the thermal model are calculated based on the IGBT module datasheet, and power losses applied to the thermal model were adjusted in such a way to achieve the same case temperature as the experimental setup. However, the loss is only adjusted within ±10%, which is a reasonable range for the loss estimation error by datasheet. The junction and case temperatures are extracted for three different points: i_2, i_5 and i_8 (see Fig. 3). The results are shown in Fig. 11. The boundary-dependent 3D thermal model accurately calculates the same junction temperatures as seen in the experimental results. It should be mentioned that the main feature in the presented thermal model is its high accuracy in the calculation of steady-state peak-to-peak temperature, ΔT, and maximum

978-1-4673-9551-9/16 $31.00 © 2016 IEEE

temperature, T_{max}, of junction temperature, which are the main factors in life-time models for IGBT modules. For both parameters, the thermal model shows less than 2% error in steady-state compared to the experimental results. To be ensured about the validity of the thermal model on the other boundary conditions, the next validation is implemented for these conditions: $I_{load(peak)}=700$ A, $\dot{V}=1$ m^3/hr $(htc=3000$ $W/m^2{\cdot}K)$, and $T_{cooling}=34\,°C$. Results are shown in Fig. 12 for temperature monitoring points: i_2, i_5 and i_8. Similarly to the previous case, thermal model results are consistent with the experimental results in both steady-state ΔT and T_{max}.

VI. CONCLUSIONS

In this paper, a simplified boundary-dependent thermal model for high power IGBT modules has been presented. The boundary conditions, which were applied in the model, are the heat source (power losses) and the heatsink (cooling system). The presented thermal model is a generic RC lumped network model, which is controlled by the variation of boundary conditions. It has been proved that by changing in the cooling system, the lower layers closer to the heatsink are more affected. So, the baseplate solder layer is more stressed and is face to reliability issues. By translation of the FEM thermal model to a circuit simulator a boundary-dependent 3D thermal network has been extracted, which can estimate detailed and accurate temperatures of the power module in different locations and layers. This thermal model has the benefits of FEM accuracy and circuit simulator speed and it can be used for accurate and detailed temperature estimation in real operating conditions. The simulated temperature profiles can be used for accurate life-time estimation of the IGBT module for long-term mission profiles.

REFERENCES

[1] S. Yantao, and W. Bingsen, "Survey on Reliability of power electronic systems," *IEEE Trans. Power Electron.*, vol. 28, no. 1, pp. 591-604, Jan. 2013.

[2] H. Wang, M. Liserre and F. Blaabjerg, "Toward reliable power electronics: challenges, design tools, and opportunities," *IEEE Ind. Electron. Mag.*, vol. 7, no. 2, pp.17-26, Jun. 2013.

[3] G. Moreno, S. Narumanchi, K. Bennion, S. Waye, and D. DeVoto, "Gaining Traction: Thermal Management and Reliability of Automotive Electric Traction-Drive Systems," *IEEE Electrification Mag.*, vol.2, no.2, pp.42-49, June 2014.

[4] A. S. Bahman, K. Ma and F. Blaabjerg, "Thermal mpedance model of high power IGBT modules considering heat coupling effects," in *Proc. Electron. App. Conf. Expo. (PEAC)*, 2014, pp.1382-1387.

[5] H. Wang, M. Liserre, F. Blaabjerg, P. de Place Rimmen, J.B. Jacobsen, T. Kvisgaard, J. Landkildehus, "Transitioning to Physics-of-Failure as a Reliability Driver in Power Electronics," *IEEE J. Emerg. Sel. Topics Power Electron.* , vol.2, no.1, pp.97-114, March 2014.

[6] ABB, *Thermal design and temperature ratings of IGBT modules*, App. Note 5SYA 2093-00, pp. 5-6.

[7] D. Cottet, U. Drofenik, and J.-M. Meyer, "A systematic design approach to thermal-electrical power electronics integration," in *Proc. Electron. Syst.-Integr. Technol. Conf.*, 2008, pp. 219–224.

[8] T. Kojima, Y. Yamada, Y. Nishibe, and K. Torii, "Novel RC compact thermal model of HV inverter module for electro-thermal coupling simulation," in *Proc. Power Convers. Conf. (PCC)*, 2007, pp. 1025–1029.

[9] S. Carubelli, and Z. Khatir, "Experimental validation of a thermal modelling method dictated to multichip power modules in operating conditions," *Microelectron. J.*, vol. 34, pp. 1143–1151, Jun. 2003.

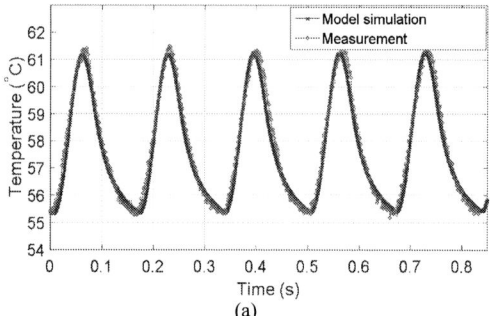

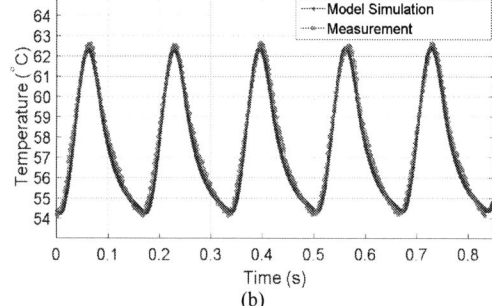

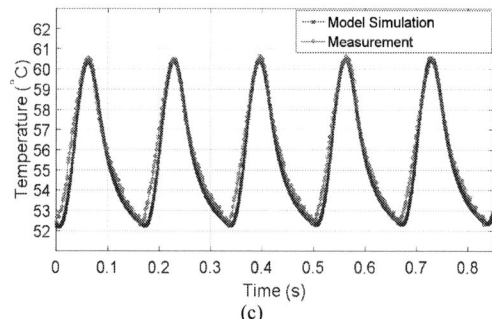

Fig. 12. Comparison of junction temperatures by the thermal model simulation and measurement in $I_{load(peak)}=700A$, $\dot{V}=1m^3/hr$, and $T_{cooling}=34°C$. (a) i_2, (b) i_5, and (c) i_8 (see Fig. 3).

[10] D. V. Hutton, *Fundamentals of Finite Element Analysis*, Freiburg, Germany: Mcgraw-Hill, 2003.

[11] M. Ibrahim, S. Bhopte, B. Sammakia, B. Murray, M. Iyengar, and R. Schmidt, "Effect of Transient Boundary Conditions and Detailed Thermal Modeling of Data Center Rooms," *IEEE Trans. Compon. Packag. Manuf. Technol.*, vol.2, no.2, pp.300-310, Feb. 2012.

[12] J. H. Lienhard IV, and J. H. Lienhard V, *A Heat Transfer Textbook*, 4th ed. Englewood Cliffs, New Jersey: Prentice-Hall Inc., 2013.

[13] A. S. Bahman, K. Ma, P. Ghimire, F. Iannuzzo, and F. Blaabjerg, "A 3D Lumped Thermal Network Model for Long-term Load Profiles Analysis in High Power IGBT Modules," *IEEE J. Emerg. Sel. Topics Power Electron.*, to be published.

[14] N.Y.A. Shammas, "Present problems of power module packaging technology," *Microelectron. Rel.*, vol. 43, pp. 519-527, April 2003.

[15] MATLAB version 8.1.0.604, The MathoWorks Inc., 2013.

[16] P. Ghimire, A. R. de Vega, S. Beczkowski, B. Rannestad, S. M. -Nielsen, and P. Thogersen, "Improving Power Converter Reliability: Online Monitoring of High-Power IGBT Modules," *IEEE Ind. Electron. Mag.*, vol.8, no.3, pp.40-50, Sept. 2014.

[17] K. Olesen, R. Bredtmann, and R. Eisele, "ShowerPower® New Cooling Concept", in *Proc. PCIM'2004*, 2004. pp.1-9.

Improved 6.5kV FREEMD-Pair Based on SiC JFET and Si IGBT

Xiaoqing Song, *Student Member, IEEE,* Alex Q. Huang, *Fellow, IEEE,* Chang Peng, Liqi Zhang

NSF FREEDM System Center
North Carolina State University
Raleigh, North Carolina 27695, USA
xsong8@ncsu.edu

Abstract— The newly proposed FREEDM-Pair is an ideal and economical solution to address high cost issue in high power SiC power devices. The FREEDM-Pair, in which a Si IGBT and a SiC JFET are connected in parallel, combines the advantages of SiC JFET's low switching losses and Si IGBT's superior forward conduction characteristics. One issue of the JFET based FREEDM-Pair is the incompatible gate drive voltage for the SiC JFET and Si IGBT which complicates the gate driver design and increases the total cost. Also, the high voltage SiC JBS reverse diode in FREEDM-Pair is indispensable for the reverse current conduction, leading to higher cost and larger package size. To address these issues, an improved FREEDM-Pair is proposed in this paper, in which the SiC JBS diode is eliminated and the normally-off SiC JFET is operated in cascode configuration to unify the gate driver voltage level and speed up the switching of the JFET. The design and operation of improved FREEDM-Pair is elaborated and experimental results verified its advantages. Also, the affordable cost demonstrates that this promising concept is an ideal step to introduce high voltage SiC power devices.

Keywords—Silicon Carbide; JFET; IGBT; High Power; Hybrid; Cascode

I. INTRODUCTION

6.5kV power devices like IGBT are widely used in medium voltage motor drives, solid state transformer (SST), HVDC, FACTS and traction applications because of its simple gate drive interface and high current carrying capability due to its conductivity modulation contributes [1-4]. However, its high switching loss, especially higher turn off losses caused by the current tail during turn off period limit its applications to less than 1kHz switching frequency converters. The development of high voltage wide bandgap (WBG) semiconductors such as SiC, has attracted great attentions due to its inherent material advantages over Si [5-7]. Unipolar SiC power switches like JFET or MOSFET show significantly better characteristics over Si IGBT in terms of significantly reduced switching losses [8-10]. A major issue facing large scale adoption of SiC power devices is still the much higher material and fabrication cost. Also, one inherent drawback of the unipolar semiconductor power devices is still the high conduction loss, even for SiC JFETs or MOSFETs as shown in Fig.2. This is especially true at high voltages and high temperatures. Hybrid device solutions [11-12] are proposed to address this issue. A 6.5 kV hybrid was published in [13-14] which integrates a high voltage SiC JFET, a Si IGBT and a SiC JBS diode connecting in parallel as shown in Fig.1. The authors named the high

voltage Si/SiC hybrid power module FREEDM-Pair. In the FREEDM-Pair, the SiC JFET and Si IGBT are turned on simultaneously, and the SiC JFET is turned off T_d time delay after the Si IGBT as shown in Fig. 1(a). In this way, the low forward conduction voltage drop is achieved by the FREEDM-Pair, taking advantage of IGBT's lower conduction loss capability. The switching loss of the IGBT will be significantly reduced due to the ultra-low turn-off loss of the SiC JFET. The physics behind the FREEDM-Pair is also discussed in [13-14].

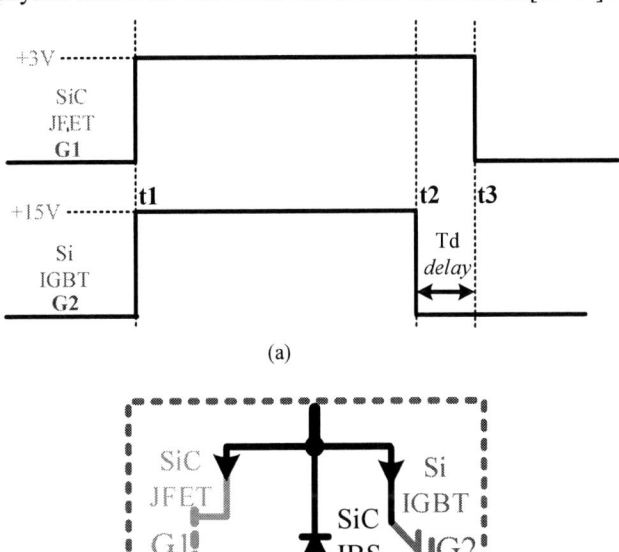

Fig. 1: (a) FREEDM-Pair gate driver signal, (b) SiC JFET based FREEDM-Pair configuration.

Fig. 2 shows the forward conduction characteristics of the FREEDM-Pair compared with the Si IGBT and SiC JFET. It can be found that the FREEDM-Pair combines the advantages of the Si IGBT at high current and the SiC MOSFET at lower current, which means at lower current, most current will go through the SiC JFET while at higher currents, Si IGBT will conduct most current resulting in lower overall forward voltage drop. The forward conduction characteristic of the FREEDM-Pair is even better than Si IGBT, because the SiC JFET also conducts a portion of the total current.

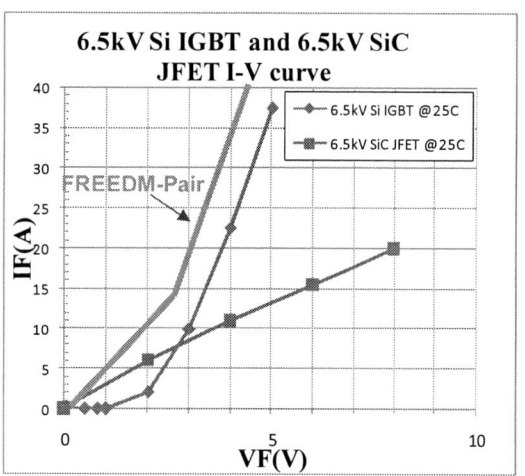

Fig. 2: Forward conduction comparison of Si IGBT, SiC JFET and FREEDM-Pair.

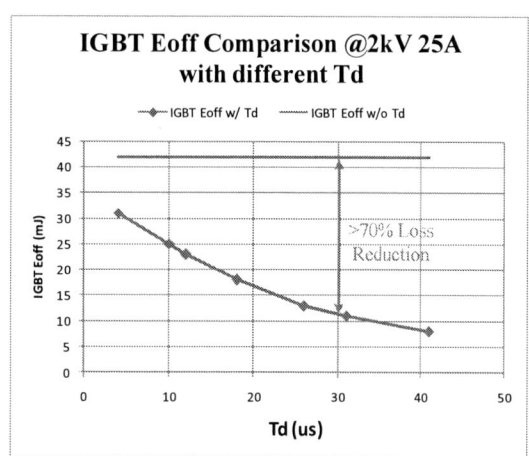

Fig. 3: IGBT turn-off loss reduction in FREEDM-Pair with different delay time T_d.

Fig. 3 shows the IGBT turn-off loss reduction potential at 2kV/25A with the help of the SiC JFET. With 30us delay time, the IGBT turn-off loss could be >70% less. The delay time may limit the allowable switching frequency, but it should be mentioned that the 6.5kV Si IGBT are usually operated at <1kHz switching frequency due to its large switching loss [1], so the tens of milliseconds delay time are totally acceptable. In [13-14], as a tradeoff between the switching frequency and turn-off loss reduction, the delay time is set at 10us.

Also, it is experimentally verified that with the simultaneous turn-on of SiC JFET and Si IGBT, the turn-on loss could also be reduced, and a ~30% turn-on loss reduction

compared to Si IGBT or SiC JFET turn-on individually was obtained experimentally.

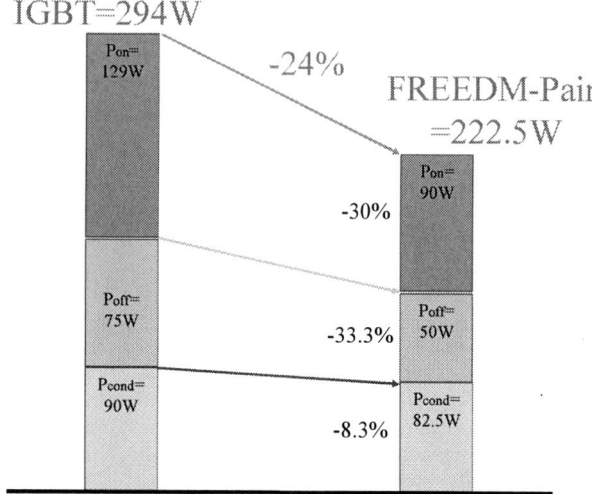

Fig. 4: Converter loss comparison between IGBT based and FREEDM-Pair based converters. (f=500 Hz, D=50%, I=30A, T_{jmax}=125C)

Fig. 4 gives the converter loss comparison of Si IGBT based converter and FREEDM-Pair based converter. The converter is operated at 120kW, 500Hz switching frequency, 50% duty cycle. In the FREEDM-Pair based converter, the conduction loss, turn-off loss and turn-on loss can be 8.3%, 33.3% and 30% lower, respectively than the Si IGBT based converter. A total 24% loss reduction could be achieved in the FREEDM-Pair based converter. This loss reduction could be even larger if operating at higher switching frequency.

However, there are still some issues with the proposed FREEDM-Pair. One of the issues is the incompatible gate driver voltage level for SiC JFET and Si IGBT, which complicates the gate driver design. Also, the integrated high voltage SiC JBS diode in FREEDM-Pair for the reverse current conduction increases the total FREEDM-Pair cost. To address these issues, an improved FREEDM-Pair is proposed in this paper, in which the SiC JBS diode is eliminated and the SiC JFET is operated in cascode configuration to solve the different gate driver voltage level issue as well as to speed up the switching of SiC JFET. In the following parts, the configuration and operation of improved FREEDM-Pair will be discussed and the advantages are elaborated. The experimental results verified its superior performances over SiC JFET and Si IGBT.

II. DESIGN AND OPERATION OF THE IMPROVED FREEDM-PAIR

A. Configuration of the improved FREEDM-Pair

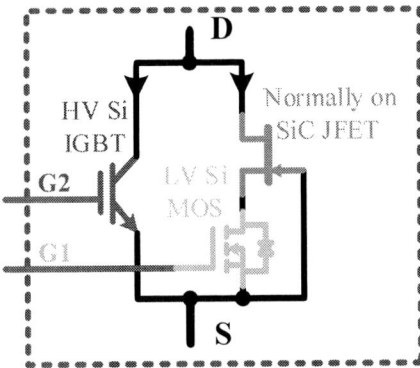

Fig. 5: Improved FREEDM-Pair based on Normally-on SiC JFET

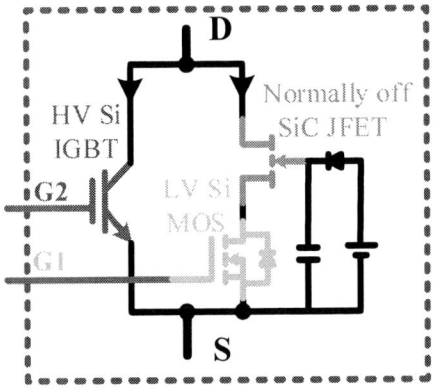

Fig. 6: Improved FREEDM-Pair based on Normally-off SiC JFET

Fig. 5 and Fig. 6 show the improved FREEDM-Pair based on normally-on SiC JFET and normally-off SiC JFET in cascode configuration, respectively. Due to the lack of 6.5kV normally-on SiC JFET, the improved FREEDM-Pair prototype in this paper is developed based on the 6.5kV/15A normally off SiC JFET from USCi. According to the datasheet, Si IGBT's gate driver voltage for turn on is 15V while for the normally off SiC JFET is only 3V. With the assistance of low voltage MOSFET in cascode, the gate driver voltage level for Si IGBT and SiC JFET are unified to 15V. Also, in Fig. 6, the capacitor at the gate of SiC JFET can help to speed up the turn-on of the SiC JFET. The experimental validation is given in Part III.

The improved FREEDM-Pair's gate driver signal is shown in Fig. 1(a). The Si IGBT is turned off first, and then all the current will commutate to Si MOSFET and SiC JFET. After a carefully designed delay time T_d, the Si MOSFET is turned off and the increased drain-source voltage over Si MOSFET will automatically turn off the SiC JFET. In this way, the low forward conduction voltage drop is achieved by FREEDM-Pair taking advantage of IGBT's low conduction loss capability. The switching loss will be significantly reduced due to the ultra-low turn-off loss of the SiC JFET.

One main challenge of the FREEDM-Pair is to determine the appropriate Si MOSFET turn off delay time T_d. The relationship of FREEDM-Pair's turn-off loss with the delay time T_d is depicted in Fig. 7. The turn-off loss of FREEDM-Pair is mainly consisted of two parts: the Si IGBT's turn off loss at t_3 and the conduction loss of SiC JFET during t_2 and t_3. It can be found that with the increase of delay time T_d, the conduction loss of SiC JFET is linearly increasing while $E_{off\_IGBT@t3}$ is exponentially decreasing. So there is an optimal delay time T_{d_opt}, with which the turn-off loss of FREEDM-Pair will reach a minimum value.

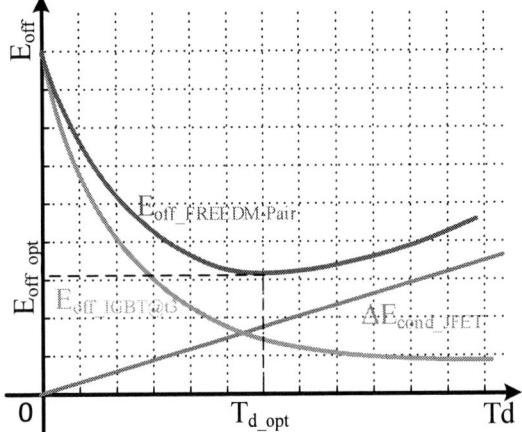

Fig. 7: Relationship of $E_{off\_FREEDM-Pair}$ with T_d

Fig. 8 shows the single pulse test waveform of the FREEDM-Pair with a 10us delay time. At first, the SiC JFET and Si IGBT are turned on at the same time. Due to the lower voltage drop of the SiC JFET at lower current, almost all the current go through the SiC JFET. With the increase of the total current, a larger portion of the total current goes through the Si IGBT. It can be seen that at high current, most of the current goes through the Si IGBT and the current through SiC JFET I_{ds_JFET} almost remains constant. During turning off state, the SiC JFET is turned off 10us after Si IGBT.

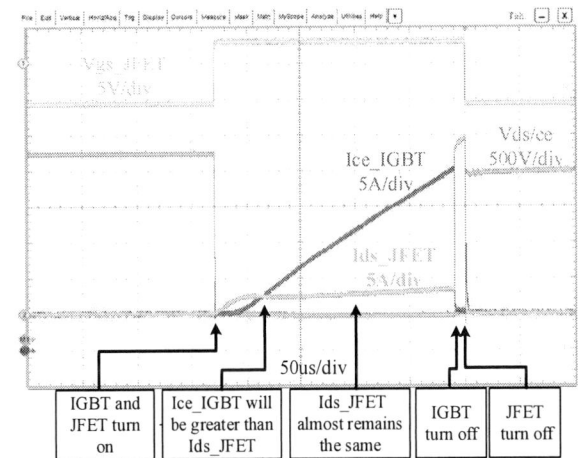

Fig. 8: Single pulse waveforms of FREEDM-Pair with clamped inductive load

In the following part, some considerations when designing the FREEDM-Pair are given.

B. Selection of Low Voltage Si MOSFET

There are two basic criteria when selecting the low voltage Si MOSFET: one is the conduction resistance $R_{ds(on)}$, and the other one is the reverse recovery charge (RRC) of the body diode. A large current will go through the Si MOSFET during conduction, so a smaller $R_{ds(on)}$ MOSFET is preferred. Usually, the $R_{ds(on)}$ the Si MOSFET could be about 20% of the SiC JFET[15]. Also the RRC of the Si MOSFET's body diode determines the RRC of the FREEDM-Pair, so a Si MOSFET with smaller body diode RRC will greatly reduce the switching loss. Based on the above considerations, a 60V/120A Si MOSFET with 2.4mΩ Rds(on) and 40nC RRC is selected.

C. Gate Driver Design

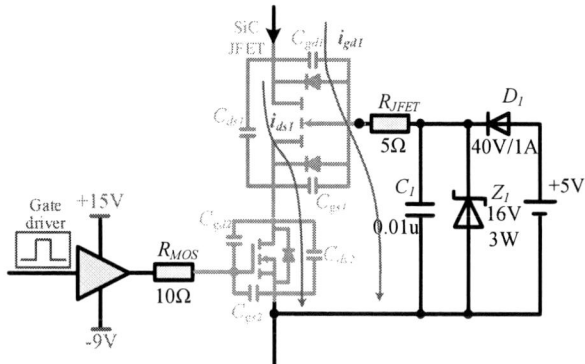

Fig. 9: Parasitic capacitances charging path during turn-off of the SiC JFET in cascode.

To drive the normally-off SiC JFET in cascode configuration, a 5V voltage supply is applied to the gate of the SiC JFET through a diode (shown in Fig. 9), so SiC JFET is transformed from a normally-off to a normally-on state. The gate capacitor C_1 connected between the SiC JFET's gate and Si MOSFET's source plays an important role on the SiC JFET switching speed. During SiC JFET turn-off, C_1 and C_{gd1} is charged by the same current i_{gd1} as shown in Eq. 1 and Eq. 2. Based on Eq. 1 and Eq. 2, the voltage increase over the gate capacitor C_1 is determined by its capacitance, the parasitic capacitance C_{gd1} and the drain-source voltage of SiC JFET, as expressed in Eq. 3.

$$i_{gd1} = C_{gd1} \frac{dV_{ds}}{dt} \tag{1}$$

$$i_{gd1} = C_1 \frac{dV_{C_1}}{dt} \tag{2}$$

$$\frac{dV_{C_1}}{dV_{ds}} = \frac{C_{gd1}}{C_1} \tag{3}$$

If the SiC JFET is switched at 2kV and C_{gd1} is estimated as 50pF according to the datasheet, a gate capacitance with 0.01uF is selected, and then the voltage increase over C_1 will be around 10V. Experiment results shown in Fig. 11 to Fig. 14 verify that there is a 10V voltage increase on the capacitor C_1. During turn-on, the extra charge in the gate capacitor C_1 will quickly discharge to the input capacitance of the SiC JFET, so

the turn-on speed of the SiC JFET will be faster due to the large discharging current.

There are two gate resistors in the gate driver of the SiC JFET in cascode. One is the gate resistor R_{MOS} for the low voltage Si MOSFET, and the other one is the R_{JFET} for the SiC JFET. R_{JFET} can be as small as possible to speed up the SiC JFET's switching. In the prototype, R_{JFET} is selected as 5Ω to damp the oscillation caused by the parasitic inductances in the JFET gate loop. With further reduction of the parasitic in the JFET gate loop, the value of the R_{JFET} could be zero. The effect of R_{MOS} on the switching speed is also studied through a double pulse tester as shown in Fig. 10, in which the voltage over the SiC JFET in cascode V_{ds_JFET}, the voltage over the Si MOSFET V_{ds_MOSFET}, the gate driver voltage of the Si MOSFET V_{g_MOSFET} as well as the current through the SiC JFET in cascode I_{ds} are monitored.

To compare the effect of the Si MOSFET's gate resistor on the switching speed, the turn-on and turn-off waveforms at 2kV/10A are presented in Fig. 11 to Fig. 14 with R_{JFET} equal to 10Ω and 100Ω, respectively.

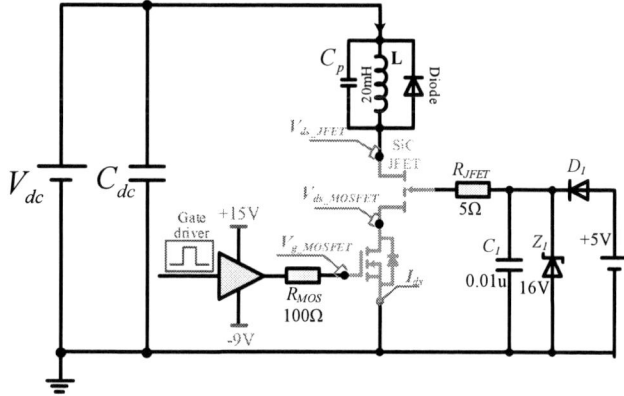

Fig. 10: Double pulse tester with clamped inductive load for the switching test of the SiC JFET in cascode.

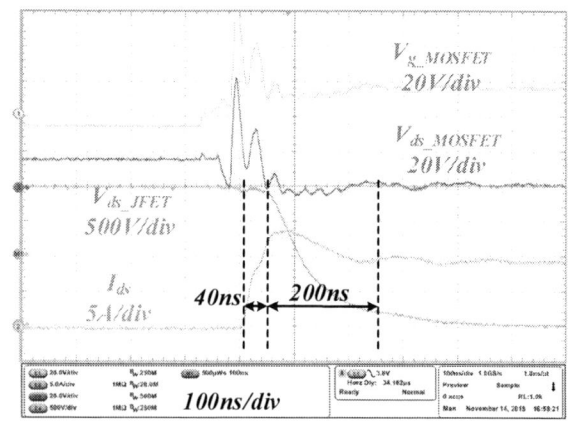

Fig. 11: Turn on waveforms of SiC JFET in cascode at 2kV/10A with R_{MOS} equal to 3Ω. (Eon=2.7mJ)

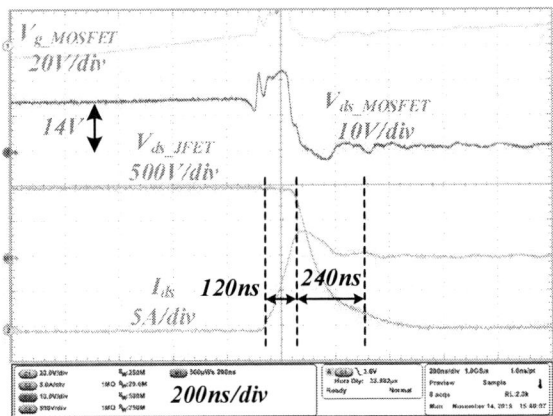

Fig. 12: Turn on waveforms of SiC JFET in cascode at 2kV/10A with R_{MOS} equal to 100Ω. (Eon=3.93mJ)

Fig. 11 and Fig. 12 compared the turn-on of the SiC JFET in cascode with different gate resistor for Si MOSFET. With 3 Ω gate resistor, the current rise time can be reduced from 120ns to 40ns, while voltage fall times drops from 240ns to 200ns. So the smaller gate resistor for the Si MOSFET can greatly affect the di/dt and the turn-on loss is reduce from 3.93mJ to 2.7mJ, correspondingly.

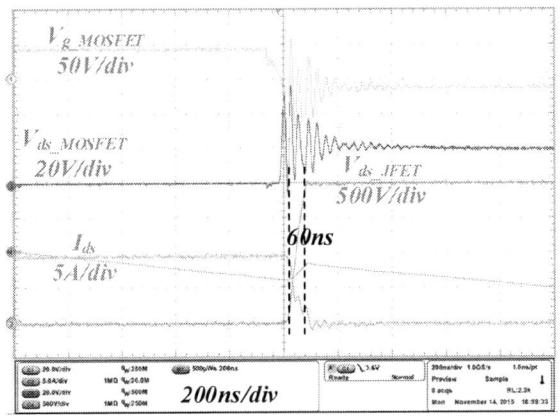

Fig. 13: Turn off waveforms of SiC JFET in cascode at 2kV/10A with R_{MOS} equal to 3Ω. (Eoff=0.3mJ)

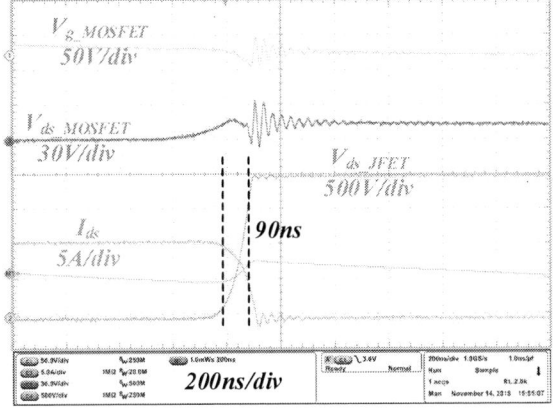

Fig. 14: Turn off waveforms of SiC JFET in cascode at 2kV/10A with R_{MOS} equal to 100Ω. (Eoff=3mJ)

Fig. 13 displays the 2kV/10A turn-off waveforms of the SiC JFET in cascode with 3Ω R_{MOS} whiles Fig. 14 shows the turn-off waveforms with 100Ω R_{MOS}. Both the di/dt and dv/dt are increased with lower gate resistor R_{MOS}. Also the turn-off loss for the 3Ω $R_{MOS\ case}$ is about 10 times lower, from 3mJ to 0.3mJ. But it is also obvious that due to higher dv/dt and di/dt, the oscillation introduced on the gate driver voltage and drain-source voltage of Si MOSFET V_{ds_MOSFET} is much larger, which could even trigger the false operation of the SiC JFET. Within the acceptable EMI, the gate resistor of the low voltage Si MOSFET can be as small as possible.

III. ADVANGTAGES OF THE IMPROVED FREEDM-PAIR

The advantages of the improved FREEDM-Pair could be summarized as following:

A. Simplified Gated Driver Design

The gate driver voltage for the SiC JFET and Si IGBT are different. So this incompatibility not only increases the complexity of the gate driver design, but also increases the cost. In the improved FREEDM-Pair, the SiC JFET is turned on and off with the assistance of Si MOSFET, whose driver voltage is the same with Si IGBT. In this way, the gate driver voltage level is unified and the gate driver design is greatly simplified. Also, for a normally-on SiC JFET, it is difficult for it to parallel with the Si IGB so the cascode configuration is needed. Fig. 15 shows the complete FREEDM-Pair configuration based on the normally off SiC JFET. There is an unified gate driver voltage level for the Si IGBT and SiC JFET. The delay circuit will automatically produce a gate driver signal for Si IGBT and a delay turn-off signal for the SiC JFET.

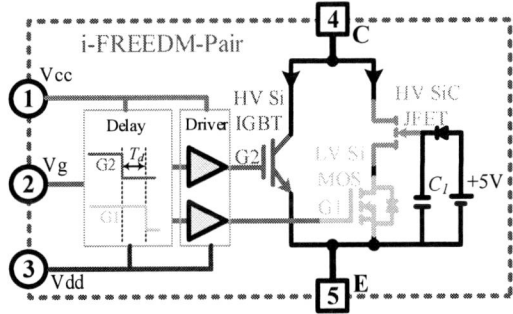

Fig. 15: 6.5kV Normally-off SiC JFET based FREEDM-Pair configuration

B. Reduced Reverse Recovery Charge

Due to the reverse conduction characteristics of the SiC JFET, no additional high voltage reverse diode is needed in the improved FREEDM-Pair, and the reverse current conducts through the body diode of the Si MOSFET and the SiC JFET channel. Generally the low voltage Si MOSFET's RRC is much smaller than that of a high voltage Si diode, so one advantage of the improved FREEDM-Pair is the greatly reduced RRC. In addition, the switching losses can be further reduced due to the smaller RRC.

The RRC is tested in a double pulse tester with the DUT as the clamped diode in parallel with the inductor. A 10kV SiC

MOSFET is selected as the lower active switch. The SiC MOSFET is switched at 2kV/10A with a 10Ω gate driver resistor. Then the current through the SiC MOSFET I_{ds} during turn-on is monitored to calculate the RRC. Fig. 16 shows the RRC comparison of three devices: 6.5kV/15A SiC Schottky barrier diode (SBD) from USCi (green line), 6.5kV SiC JFET in cascode (blue line) and a 6.5kV/25A Si diode from ABB (red line). The current above 10A is the actual reverse recovery current. It can be found that the SiC SBD and SiC JFET has similar reverse recovery current. Actually, the reverse recovery current is introduced mostly due to the parasitic capacitance of the DUT. On the other hand, the 6.5kV Si diode has much larger RRC, which is almost 30 times larger than the SiC SBD or SiC JFET.

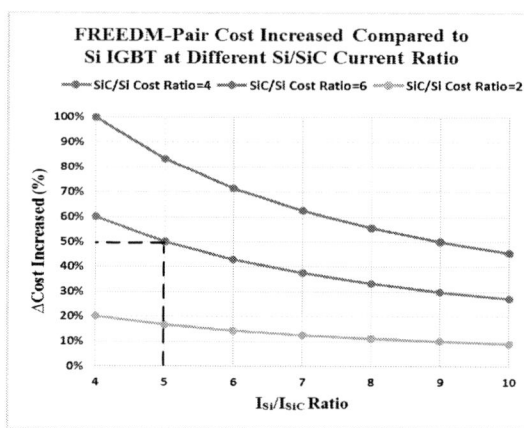

Fig. 17: Cost analyses of FREEDM-Pair with different Si/SiC current ration and cost ration.

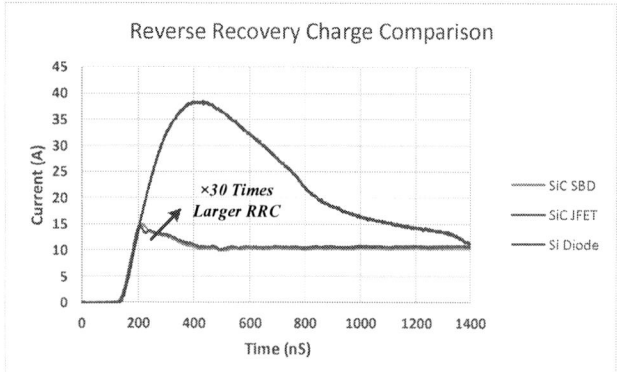

Fig. 16: RRC comparison of the 6.5kV SiC SBD, SiC JFET and Si diode.

C. Further Reduced Cost and Package Size

The cost of FREEDM-Pair is determined by the cost ratio of SiC devices to Si devices and the current distribution between the Si IGBT and SiC MOSFET, as show in Fig.17. For example, when the cost per Ampere of SiC devices is 4 times as much as of Si devices (SiC/Si cost ratio is 4) and the rated current for Si IGBT is 5 times as much as the rated current for SiC MOSFET (Si/SiC current ratio is 5), only 50% extra cost is needed to realize up to 70% turn off loss reduction compared to Si IGBT. The cost analysis suggests a very moderate increase in the cost while achieving significant loss reduction. The improved FREEDM-Pair has an even better cost per Ampere performance due the elimination of an extra reverse diode. Although a Si MOSFET is added, the cost and chip size of the low voltage Si MOSFET is negligible compared to the high voltage reverse diode.

D. Faster Switching Speed

Thanks to the cascode configuration, the switching speed of the SiC JFET is also faster than if the JFET is directly driven. Fig. 18 and Fig. 19 show the switching waveforms of the SiC JFET with a 5Ω gate driver resistor.

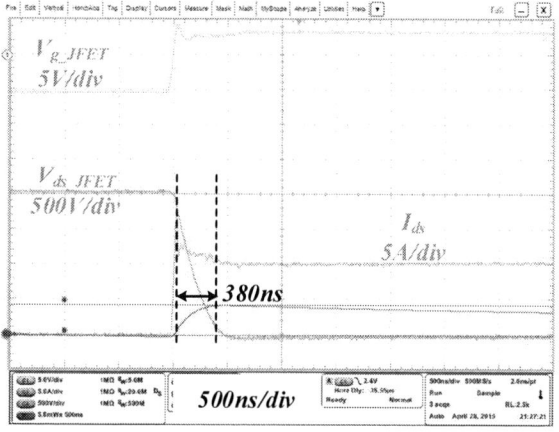

Fig. 18: Normally-off SiC JFET turn-on waveforms with +3V/-5V gate driver voltage and 5Ω gate resistor.

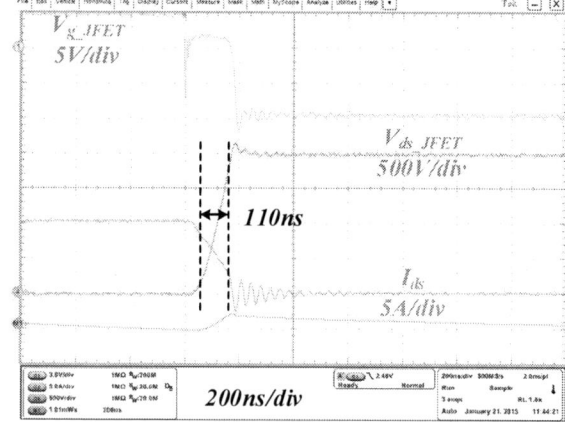

Fig. 19: Normally-off SiC JFET turn-off waveforms with +3V/-5V gate driver voltage and 5Ω gate resistor.

TABLE I. SWITCHING PERFORMANCE COMPARISON BETWEEN SIC JFET IN CASCODE AND DIRECT DRIVE

	Turn-on			Turn-off		
	dv/dt (kV/us)	di/dt (A/us)	E_{on} (mJ)	dv/dt (kV/us)	di/dt (A/us)	E_{off} (mJ)
Cascode	10	325	2.7	33.3	125	0.3
Direct Drive	5.3	400	4.4	18.2	83	2.5

Table I. compared the switching performances of the SiC JFET between cascode configuration and directly driven. The green color number means better performances and the red number means slower speed or higher loss. It can be found that during turn-on or turn-off, the dv/dt is almost twice of the SiC direct driven JFET. Both the turn-on loss and turn-off loss of the SiC JFET in cascode is smaller than the direct driven one, especially the turn-off loss. Only the di/dt of the directly driven SiC JFET is slight better than the cascode configuration, which may be due to the large parasitic inductance caused by the series connected Si MOSFET. But this can be improved with better packaging design in the FREEDM-Pair.

IV. CONCLUSIONS

In this paper, an economical and efficient FREEDM-Pair is proposed to address the incompatible gate driver voltage issue and to further drive down the cost by eliminating the SiC JBS diode. The design considerations of the FREEDM-Pair are illuminated. The improved performance is demonstrated. Even lower cost is also predicted for the new FREEDM-Pair making it a promising and economical concept to accelerate high voltage SiC power device adoption.

ACKNOWLEDGMENT

The authors also would like to thank USCi and ABB for the supply of 6.5kV normally off SiC JFET and Si IGBT respectively.

REFERENCES

[1] Gangyao Wang; Xing Huang; Jun Wang; Tiefu Zhao; Bhattacharya, S.; Huang, A.Q., "Comparisons of 6.5kV 25A Si IGBT and 10-kV SiC MOSFET in Solid-State Transformer application,"Energy Conversion Congress and Exposition (ECCE), 2010 IEEE , vol., no., pp.100,104, 12-16 Sept. 2010

[2] Kai Tan; Ruiyang Yu; Suxuan Guo; Huang, A.Q., "Optimal design methodology of bidirectional LLC resonant DC/DC converter for solid

state transformer application," Industrial Electronics Society, IECON 2014 - 40th Annual Conference of the IEEE , vol., no., pp.1657,1664, Oct. 29 2014-Nov. 1 2014

[3] Chang Peng; Xiaoqing Song; Rezaei, M.A.; Xing Huang; Widener, C.; Huang, A.Q.; Steurer, M., "Development of medium voltage solid-state fault isolation devices for ultra-fast protection of distribution systems," Industrial Electronics Society, IECON 2014 - 40th Annual Conference of the IEEE , vol., no., pp.5169,5176, Oct. 29 2014-Nov. 1 2014

[4] Chang Peng; Huang, A.Q.; Xiaoqing Song, "Current commutation in a medium voltage hybrid DC circuit breaker using 15 kV vacuum switch and SiC devices," in Applied Power Electronics Conference and Exposition (APEC), 2015 IEEE , vol., no., pp.2244-2250, 15-19 March 2015

[5] Wang, Gangyao; Huang, Alex Q.; Wang, Fei; Song, Xiaoqing; Ni, Xijun; Ryu, Sei-Hyung; Grider, David; Schupbach, Marcelo; Palmour, John, "Static and dynamic performance characterization and comparison of 15 kV SiC MOSFET and 15 kV SiC n-IGBTs," Power Semiconductor Devices & IC's (ISPSD), 2015 IEEE 27th International Symposium on , vol., no., pp.229,232, 10-14 May 2015 .

[6] Xiaoqing Song; Huang, A.Q.; Mengchia Lee; Chang Peng; Lin Cheng; O'Brien, H.; Ogunniyi, A.; Scozzie, C.; Palmour, J., "22 kV SiC Emitter turn-off (ETO) thyristor and its dynamic performance including SOA," Power Semiconductor Devices & IC's (ISPSD), 2015 IEEE 27th International Symposium on , vol., no., pp.277,280, 10-14 May 2015

[7] Xiaoqing Song, Alex Q. Huang, Xijun Ni, Liqi Zhang, "Comparative Evaluation of 6kV Si and SiC Power Devices for Medium Voltage Power Electronics Applications", WiPDA 2015

[8] Hostetler, J.L.; Alexandrov, P.; Li, X.; Fursin, L.; Bhalla, A., "6.5 kV SiC normally-off JFETs — Technology status," Wide Bandgap Power Devices and Applications (WiPDA), 2014 IEEE Workshop on , vol., no., pp.143,146, 13-15 Oct. 2014

[9] Ni, Xijun; Gao, Rui; Song, Xiaoqing; Huang, Alex Q.; Yu, Wensong, "Development of 6kV SiC hybrid power switch based on 1200V SiC JFET and MOSFET," in Energy Conversion Congress and Exposition (ECCE), 2015 IEEE , vol., no., pp.4113-4118, 20-24 Sept. 2015

[10] Suxuan Guo, Liqi Zhang, Yang Lei, Wensong Yu, Alex Q. Huang, "3.38 Mhz Operation of 1.2kV SiC MOSFET With Integrated Ultra-Fast Gate Drive," In 2015 IEEE Workshop on Wide Bandgap Power Devices and Applications (WiPDA)

[11] Ortiz, G.; Gammeter, C.; Kolar, J.W.; Apeldoorn, O., "Mixed MOSFET-IGBT bridge for high-efficient Medium-Frequency Dual-Active-Bridge converter in Solid State Transformers," Control and Modeling for Power Electronics (COMPEL), 2013 IEEE 14th Workshop on , vol., no., pp.1,8, 23-26 June 2013.

[12] Rahimo, M.; Canales, F.; Minamisawa, R.A.; Papadopoulos, C.; Vemulapati, U.; Mihaila, A.; Kicin, S.; Drofenik, U., "Characterization of a Silicon IGBT and Silicon Carbide MOSFET Cross-Switch Hybrid," Power Electronics, IEEE Transactions on , vol.30, no.9, pp.4638,4642, Sept. 2015

[13] Song, Xiaoqing; Huang, Alex Q., "6.5kV FREEDM-Pair: Ideal high power switch capitalizing on Si and SiC," in Power Electronics and Applications (EPE'15 ECCE-Europe), 2015 17th European Conference on , vol., no., pp.1-9, 8-10 Sept. 2015

[14] Huang, A.Q.; Xiaoqing Song; Liqi Zhang, "6.5 kV Si/SiC hybrid power module: An ideal next step?," in Integrated Power Packaging (IWIPP), 2015 IEEE International Workshop on , vol., no., pp.64-67, 3-6 May 2015

[15] http://www.unitedsic.com/littleboxchallenge/pdf/usci_introduction_to_u sci_cascode_devices.pdf

On the comparative assessment of 1.7 kV, 300 A full SiC-MOSFET and Si-IGBT power modules

Muhammad Nawaz, Kalle Ilves

ABB Corporate Research
Forskargränd 7, SE-721 78 Västerås, Sweden
muhammad.nawaz@se.abb.com,   kalle.ilves@se.abb.com

Abstract—This paper deals with comparative assessment through static and dynamic measurements performed for full SiC-based MOSFET and Si-based IGBT power modules. The full SiC-MOSFET and Si-IGBT based power modules all have an identical voltage rating of 1.7 kV and a current rating of 300 A. Full SiC-MOSFETs presents a lowest on-resistance (R_{ON}) of 10.0 mΩ, blocking voltage of 1800 V and a threshold voltage around 2.5 – 3.0 V at 300 K, for most of the transistor samples, as promised by the manufacturer. Dynamic tests using a single pulse test setup have been performed with commercial gate drive unit from Cree that is especially compatible to full SiC- and Si-IGBT power modules footprint. A turn on (turn off) energy loss of full SiC-MOSFET module at 1000 V (230 A) and 1200 V (275 A) was 12 (28) mJ and 19 (41) mJ, respectively using 300 µH load inductance at 25 °C. Similarly, a turn on (turn off) energy loss of hybrid power module at 1000 V (235 A) and 1200 V (280 A) was 68 (86) mJ and 85 (115) mJ, respectively using 300 µH load inductance at 25 °C. Compared to Si-IGBT modules, no reverse recovery is observed for full SiC power modules. Temperature independent switching loss performance is also observed.

Keywords—Power Semiconductor Modules, Silicon Carbide, MOSFET, 4H-SiC MOSFETs, Power Devices.

I. Introduction

Gradual increase in the energy demand as a result of worldwide population and business growth requires highly efficient and compact power conversion system. For a number of high power electronic applications such as PV inverters, motor drives, traction, HVDC, and FACTS, wide bandgap semiconductor devices such as silicon carbide (SiC) based MOSFETs/IGBTs/GTOs [1 - 4] provide interesting replacement options to more conventional but leading horse silicon (Si) based counterpart. This is quite understandable, primarily due to larger bandgap values of SiC material (i.e., 3x than that of Si), higher critical field (i.e., 10x than that of Si), higher saturation velocity (i.e., 2x than that of Si), larger thermal conductivity values (i.e., 3x than that of Si), and higher radiation resistance compared to Si material. Hence, these interesting physical properties of SiC directly lead to a thinner device with lower on-resistance (i.e., lower conduction and switching losses), more efficient heat extraction, more radiation resistant, and well-suited for high temperature environments. Presently, various device manufacturers [5 - 15] have launched their products in the commercial market (i.e., discrete package

and high power modules) with SiC based Schottky diodes and SiC based MOSFETs of voltage rating 1.2 – 1.7 kV and current rating ranging from 100 – 800 A. On the other side, 15 kV and 27 kV n-IGBT [16, 17] have recently been presented making them further compatible for energy transmission and distribution network. At 10 kV bus voltage with 5 A and 10 A load current, n- IGBT can be operated up to 6.2 and 3.9 kHz respectively [16], where the junction temperature is limited to 150 °C with liquid cooling. On the other side, the n-IGBT can be switched at 5.1 and 3.2 kHz at 10 kV for a load current of 5 A and 10 A respectively, with air cooling. Low voltage (1.2 kV, 120 – 800 A) SiC-MOSFET power modules have recently been evaluated [18, 19, 21, 22] with static and dynamic tests and also compared with Si-IGBTs in [20].

Main objective of this paper is to give a first comparative evaluation through static and dynamic results obtained for full SiC-MOSFET and Si-IGBT power modules [5 - 6] with identical current rating of 300 A per phase leg and voltage rating of 1700 V at 25 °C. Note that these are the first engineering samples in the market being assessed for such a high current rating at 1.7 kV by putting several dies in parallel. Note that each full SiC-MOSFET module has an extra Schottky barrier diode across each MOSFET die, besides internal body diode of the MOSFET, which allows conduction in the reverse direction contrary to IGBTs. The overall main objective of these power modules [5 - 6] is to fairly balance the requirement for larger current handling capability with a smaller form factor suitable for a number of high power applications such as power converters for welding, switched mode power supplies, uninterruptible power supplies and motor control systems, as claimed by the manufacturer. SiC power MOSFETs further offer the lowest On-resistance (R_{ON}) with high level of device integration and using AlN as a substrate material for efficient thermal performance. This further allows stable temperature behavior, very rugged performance using direct mounting to heatsink with possible low junction to case thermal resistance.

II. Results and Discussion

Fig. 1 illustrates the physical footprint and electrical connectivity of the SiC based MOSFET and Si based IGBT power modules. The module is capable to be used for a

978-1-4673-9551-9/16 $31.00 © 2016 IEEE

maximum junction temperature of 150 °C as recommended by the manufacturer. Absolute maximum rating of the power

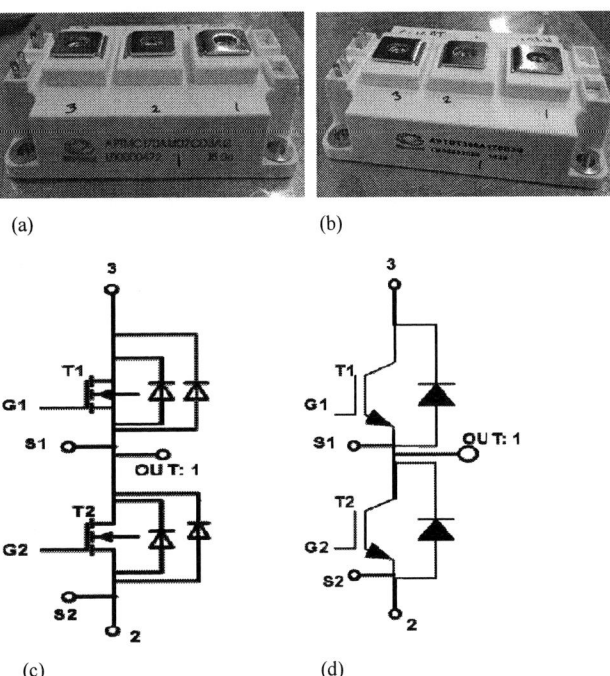

(a)　　　　　　　(b)

(c)　　　　　　　(d)

Fig. 1. Physical footprint of Silicon Carbide (SiC) MOSFET power module: APTMC170AM07CD3AG [5] (a) and Si-IGBT based power module: APTGT300A170D3G [6] (b) from Microsemi. Also shown is the electrical configuration for SiC-MOSFET module (c) and Si-IGBT module (d).

modules offers continuous drain current of 300 A at 25 °C. A maximum pulsed drain current of 600 A is allowed with maximum gate-source voltage swing of -10/25 V and power dissipation of 1470 W for Si-IGBT and 2080 W for SiC-MOSFET modules. Since both modules have the same housing with similar voltage and current rating, comparable stray inductances are expected. Note that the SiC-MOSFETs have an internal gate resistance of 1.1 Ω besides external variable gate resistance on the driver circuit board. The unipolar Schottky barrier diode (SBD) in the SiC-MOSFET module stays in parallel to the internal body diode of the SiC-MOSFET, which is a bipolar emerged from the pn-junction formation diode. A junction to case thermal resistance of 0.085 and 0.075 °C/W is listed for Si-IGBT [6] and SiC-MOSFET [5] modules, respectively.

All fully functional power modules (i.e., a total of 5 power modules with two switches each) have been tested using Tek371A curve tracer with a maximum voltage measurement range of 3000 V and current range of 300 A. Threshold voltage defined at drain bias of 2.0 V lies around 3.5 – 4.0 V for almost all SiC-MOSFET devices. Fig. 2 shows the typical transistor characteristics (i.e., I_{DS} versus V_{DS} at various gate biases (a), transfer curves at drain bias of 2 V (b), On-resistance at different gate voltages (c), and blocking voltage curves (d)) at different temperatures. A decrease in the threshold voltage and drop in the peak drain current (and hence maximum transconductance) of the SiC-MOSFET device with temperature is noticed for all devices. A maximum shift in

threshold voltage of 1.8 V was observed with varying temperature from 300 to 425 K. On the other side,

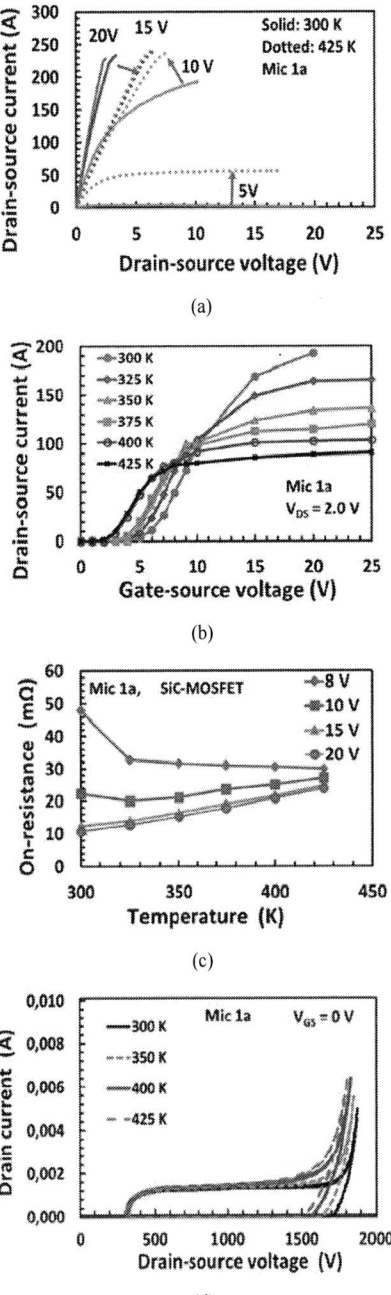

(a)

(b)

(c)

(d)

Fig. 2. Static current-voltage characteristics (a, b) of full SiC MOSFET module at different temperature. Also shows is the extracted on-resistance (c) and blocking voltage characteristics (d) at different temperatures.

ON-resistance (R_{ON}) as illustrated in Fig. 2c, decreases initially at lower gate voltages (e.g. 8 V) with the increase of temperature and then slightly increases with temperature at higher gate biases (e.g., 15, 20 V) once the MOS channel is fully opened. This is due to the fact of various competing processes in the MOS device, i.e., increase of effective channel

978-1-4673-9551-9/16 $31.00 © 2016 IEEE　　　　　277

mobility with temperature (i.e., due to reduced Coulomb scattering with temperature) at lower gate biases (i.e., at low field) and decrease of overall effective channel mobility and

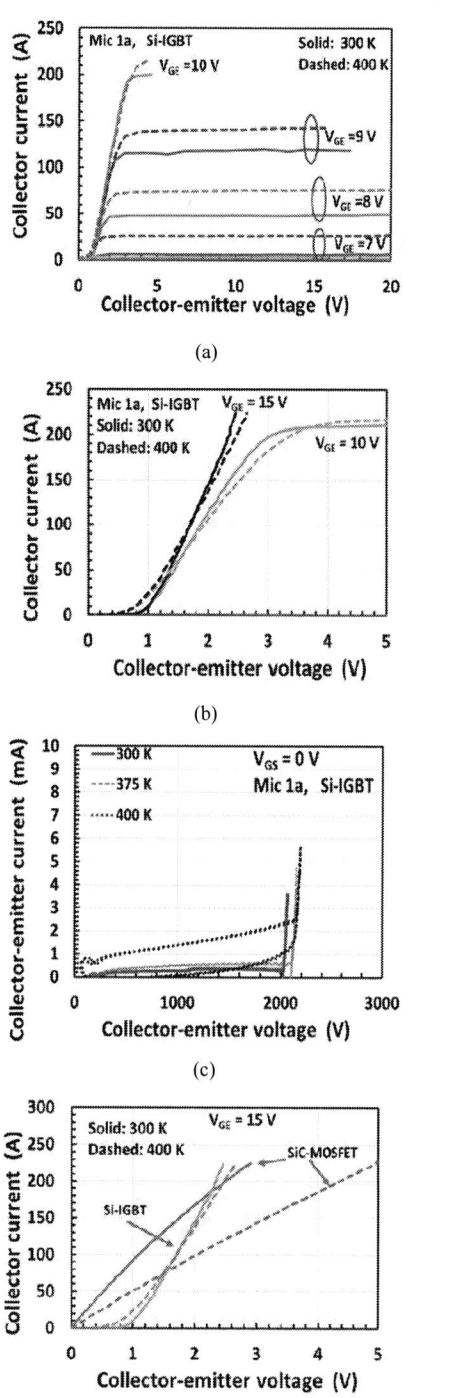

Fig. 3. Static current-voltage characteristics (a, b) of Si IGBT module at different temperature. Also shows is the blocking voltage characteristics (c) at different temperatures and both full SiC –MOSFET and Si-IGBT are shown at 300 K and 400 K.

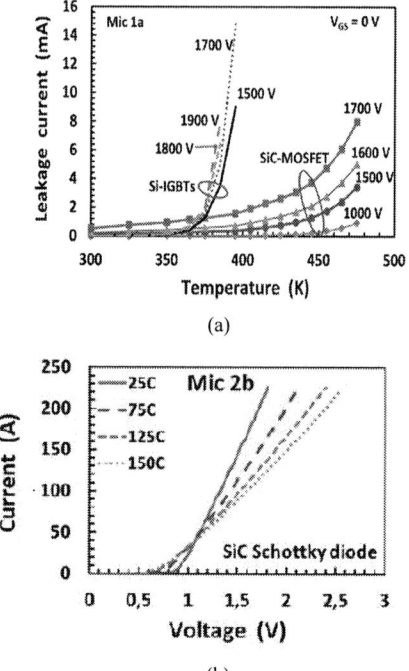

Fig. 4. Leakage current in the off state of full SiC-MOSFET and Si-IGBT module (a). Also shown is the temperature dependent SiC forward diode characteristics (b).

bulk drift mobility with temperature (i.e., due to increased bulk phonon scattering) in the drift layer at higher gate biases (i.e., at high field). This increase of R_{ON} with temperature (i.e., positive temperature coefficient of resistance at V_{GS} of 15.0 and 20.0V) thus facilitates paralleling of several devices for higher power ratings. Blocking voltage of all purchased devices stay around 1700 – 1750 V (defined at 1.0 mA current).

Static characteristics of the Si-IGBT modules is illustrated in Fig. 3 at different temperatures. A threshold voltage of 6.0 - 7.0 V has been observed for the Si-IGBT modules. A blocking voltage of 2000 V is observed for all Si-IGBT modules that is 300 V higher than stated rated blocking voltage (i.e., 1700 V) of devices and hence gives a fair margin for turn off voltage overshoot. Note that at lower current values, the SiC-MOSFETs show lower forward voltage drop compared to the Si-IGBTs (see Fig. 3d). At 190 A and 25 °C the forward voltage drop of both modules is approximately the same. Leakage current of SiC-MOSFET and Si-IGBT power module (see Fig. 4a) is very much sensitive to the temperature variation and this current increases significantly around 375 – 400 K for a given collector-emitter voltage close to the breakdown voltage of the device. Note that the leakage current of Si-IGBT is approaching to 15 mA at only 400 K for a given collector-emitter voltage of 1700 V (V_{BR} =2.0 kV for Si-IGBTs) which is much larger than that of SiC-MOSFETs (i.e., < 2- 3 mA at 450 K for 1700 V, as shown in Fig. 4a). Overall, the SiC-MOSFET package is more resistant to temperature variation when measured up to prescribed junction temperature, $T_j = 150$ °C. A forward diode characteristics of SiC Schottky barrier diode is illustrated in Fig. 4b that shows positive temperature coefficient as expected. It is interesting to note that

978-1-4673-9551-9/16 $31.00 © 2016 IEEE

both modules could not reproduced exact I-V characteristics given in the datasheet, hence the extracted On-resistance (R_{ON}) values is slightly inferior to that given in the datasheet of power modules.

Top view of experimental setup for dynamic tests is shown in Fig. 5 along with its electrical configuration of the test setup. A DC link capacitor of 520 µF and nominal load inductance of 300 µH is used in the experiments. However, the load inductance is varied to get required load current (i.e., L_{load} decreases to get higher load current). Both power modules have been tested using same setup and with same gate drive units namely GDU-Ka and GDU-Cr with gate voltage driving swing of -8/20 V. Note that both gate drive units have nominal gate resistance of 10 Ω. Fig. 6 illustrates the double pulse transient for a full SiC-MOSFET module using GDU-Ka and GDU-Cr at 1000 V. Both drivers under nominal design conditions (i.e., same gate resistance) performed equally well with almost similar dv/dt and di/dt. Extracted rise time (90 %) and fall time (90 %) at 1000 (1200 V) during turn on and turn off was approximately 60 ns (80 ns) and 40 ns (60 ns), respectively. A dv/dt and di/dt during turn on phase of -10.5 (-14) kV/µs and 3.18 (3.48) kA/µs has been observed at supply voltage of 1200 V (1000 V). Similarly, a dv/dt and di/dt during turn off phase of 12.55 (11) kV/µs and -5.92 (-4.6) kA/µs has been obtained at supply voltage of 1200 V (1000 V).

Double pulse transient for Si-IGBT and SiC-MOSFET is compared in Fig. 7 at supply voltage of 1200 V. Compared to

Si-IGBT, a small gate charge is needed to switch the device and the relative low transconductance value for the SiC-MOSFET eliminate the gate-source Miller plateau which is contrarily observed for Si-IGBT power module.

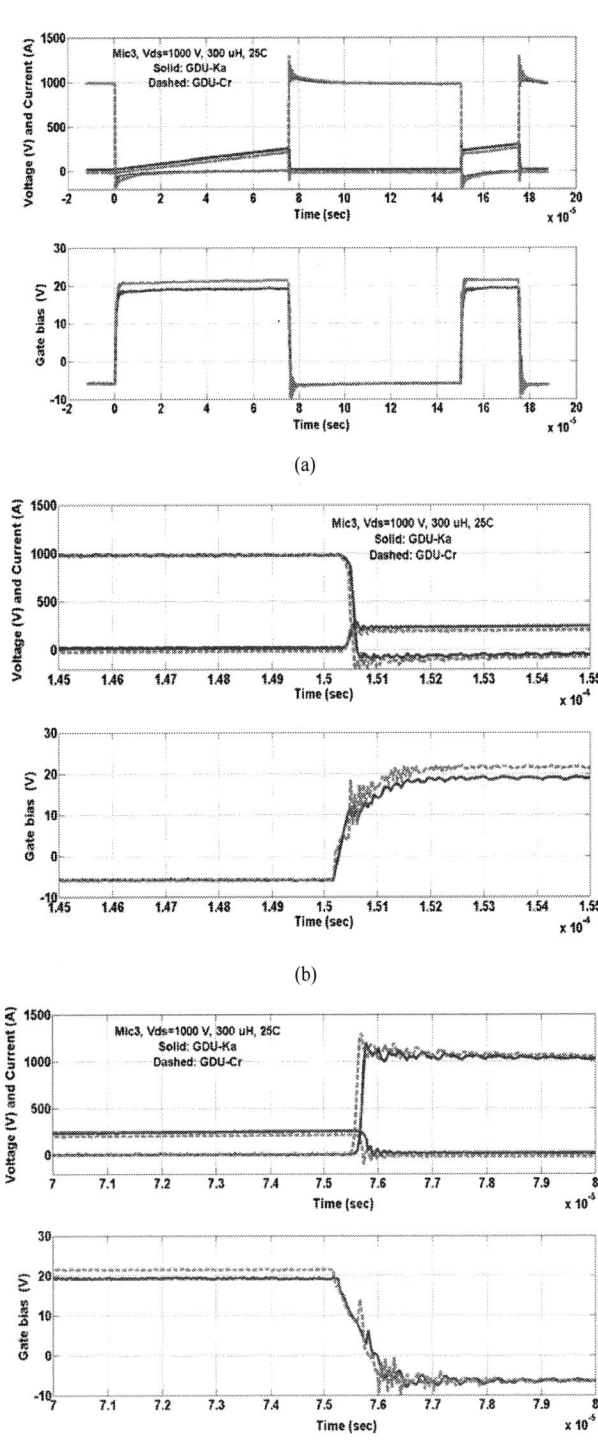

Fig. 6. Full double pulse transient for GDU-Ka (solid) and GDU-Cr (dashed) (a) is shown along with zoomed in view of turn on (b) and turn off behavior (c) at 25 °C for supply voltage of 1000 V.

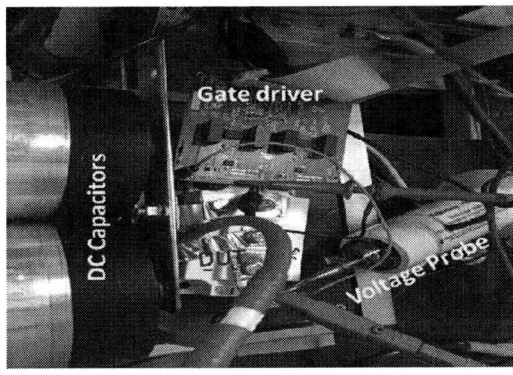

(a)

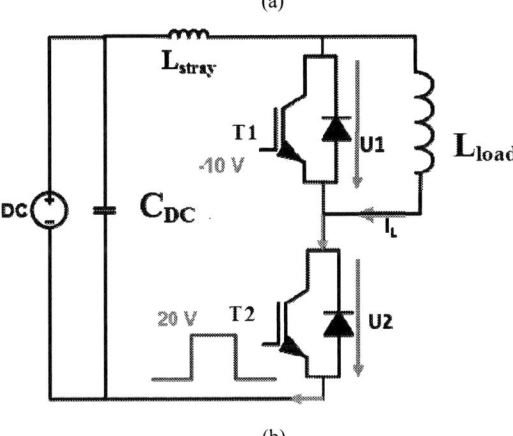

(b)

Fig. 5. Top view of experimental setup for double pulse dynamic tests (a) and its electrical configuration (b) using C_{DC} of 520 µF and L_{load} = 300 µH.

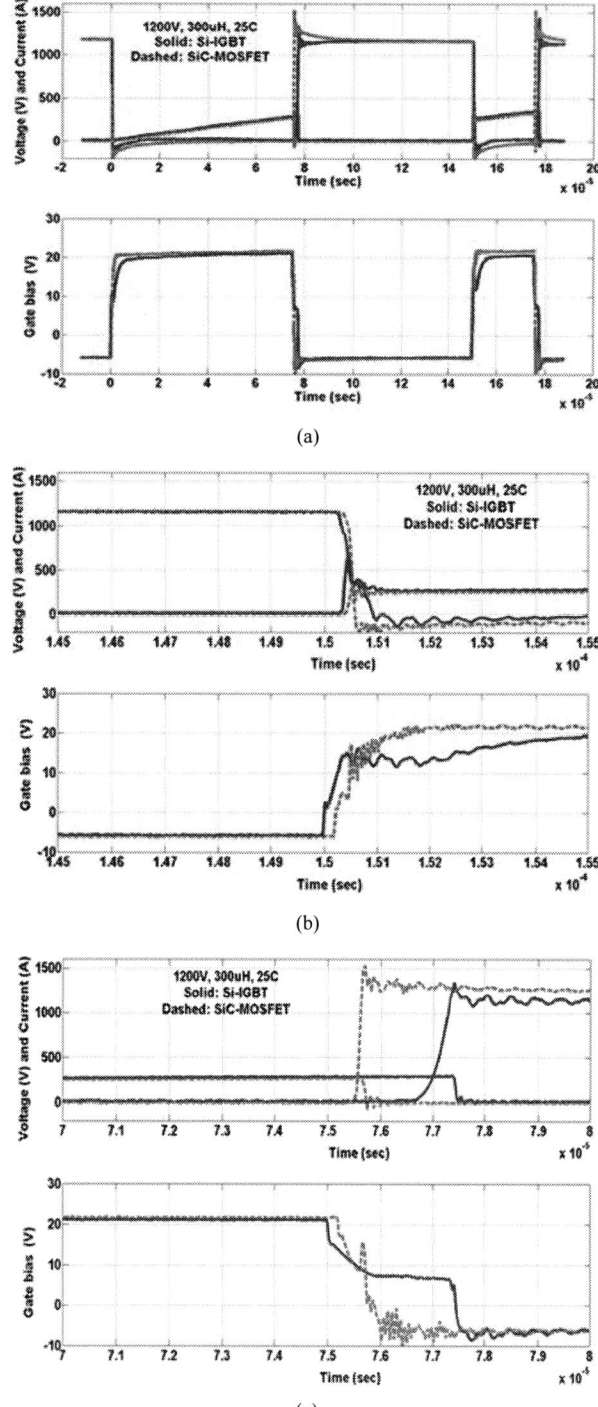

Fig. 7. Full double pulse transient for Si-IGBTs (solid) and SiC MOSFETs (dashed) modules (a) is shown along with zoomed in view of turn on (b) and turn off behavior (c) at 25 °C for supply voltage of 1200 V.

Extracted energy losses are summarized in Table I for Si-IGBT and in Table II for SiC-MOSFET. Both GDU units reflect almost the same amount of energy losses for the SiC-MOSFET as is visible in Table II, where the results for GDU-Cr are shown in brackets. A slight extra turn off loss for GDU-Ka at 1200 V is due to some oscillations in the gate turn off phase.

This is, however, a driver related issue and is not caused by the device itself.

TABLE I: EXTRACTED ENERGY LOSSES FOR Si-IGBT MODULE.

Parameter	400 V 100A	600 V 145 A	800 V 190 A	1000 V 235 A	1200 V 280 A
Eon (mJ)	12	25	42	68	85
Eoff (mJ)	16	34	60	86	115
E_{total} (mJ)	28	59	102	154	200

TABLE II: EXTRACTED ENERGY LOSSES FOR SiC-MOSFET MODULE WITH GDU-Ka AND GDU-Cr (INSIDE BRACKET).

Parameter	400 V 100A	600 V 145 A	800 V 190 A	1000 V 235 A	1200 V 280 A
Eon (mJ)	2 (1)	5 (3)	11 (7)	18 (12)	29 (19)
Eoff (mJ)	3 (4)	8 (8)	17 (16)	26 (28)	79 (41)
E_{total} (mJ)	5 (5)	13 (11)	28 (23)	44 (40)	108(60)

A steep turn off voltage and current behavior (i.e., higher dv/dt and -di/dt) and larger turn off voltage overshoot and turn on current overshoot for Si-IGBT modules have resulted in higher energy losses than that of SiC-MOSFETs with similar load current and supply voltage. As expected, extracted energy losses increase almost linearly with the increase of load current for fixed supply voltage as illustrated in figure 8. Overall, turn on energy loss for Si-IGBT module is 5.6x (4.5x) higher than that of SiC-MOSFET module at 1000 V (1200 V). Similarly, turn off energy loss for Si-IGBT module is 3.0x (2.8x) higher than that of SiC-MOSFET module at 1000 V (1200 V). Note that turn on and turn off energy loss of Si-IGBT increases with the increase of temperature contrary to SiC MOSFET power module that shows independent switching behavior as witnessed in the Table III. Generally speaking, all dynamic measurements for both power modules have been performed using nominal settings of the gate drive unit (i.e., Rg of 10.0 Ω), meaning that there is fair room to further decrease the energy losses with lower gate resistance (i.e., say 2.5 Ω) and with more compact design leading to lower overall stray inductance of the test setup. Energy loss dependence on the gate resistance is presented in [19] where on average 3 times lower losses were recorded with gate resistance variation from 10 to 2.5 Ω.

Note that the voltage overshoot of the SiC MOSFET merely depends on the device switching speed and the stray inductance of the circuit commutation loop, whereas the voltage overshoot of the Si-IGBT is partly composed of due to stray inductance in the circuit loop and partly due to the peak forward voltage of the freewheeling silicon diode [20]. Generally speaking, with far lower switching losses for SiC-MOSFETs than for comparable Si-based IGBTs, SiC devices can operate at switching frequencies two to five times greater than present Si-counterparts and hence leads to overall system design benefits from smaller, compact and lighter passive components. On the other side, very low leakage current of SiC MOSFET may boost system reliability and consistency even when subject to elevated reverse voltages or higher ambient temperatures. SiC-MOSFET is a majority

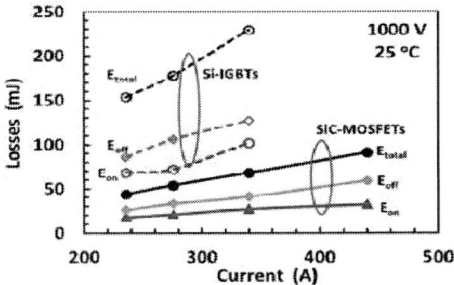

Fig. 8. Extracted energy losses as a function of load current at 1000V.

carrier device that does not offer any tail current and hence the turn off energy loss for SiC-MOSFET is primarily due to possible overlap of drain-source voltage and drain current during voltage rise and current fall time. As observed in our measurements, turn off losses are higher than that that of turn on losses which can further be lowered down separately by reducing gate resistance during turn off phase to drain more current form the gate terminal.

A reverse recovery behavior is presented in Fig. 9 at 1000 V and 300 A for both power modules. A reverse recovery loss increases with temperature for Si-hybrid power module as expected. Contrary to Si-IGBT module, negligible reverse recovery loss of full SiC-MOSFET is observed in our findings. Finally, a short circuit test under hard switch fault (HSF) condition is performed (Fig. 10) to investigate the survivability of the SiC-MOSFET power module. A gate pulse of 4.0 μs was applied at the gate terminal of device under test while upper switch was shorted by ordinary cable. Short circuit was carried out at supply voltage of 1000 V at 25 °C. A short circuit current reaches to a maximum value of 4000 A before it safely turns off. An increase in the gate signal beyond its maximum on voltage limit (i.e., 20 V) is observed. This could be caused either by the experimental setup due to common ground of ordinary voltage probe, gate drive unit and the SiC-MOSFET device under test, or extra voltage drop in the emitter leg of the device as a result of stray inductance associated to it. More recently, various SiC-MOSFETs power modules with 1.2 kV and 120 – 800 A rating have been assessed through short circuit test under HSF condition [18]. A short circuit survivability time of 10 μs has been reported for these devices although with slow gate drivability and a bit larger overall loop inductance in the test setup.

TABLE III: ENERGY LOSSES AT VARIOUS TEMPERATURES FOR SiC-MOSFETS AND Si-IGBTS (INSIDE BRACKET).

Parameter	1000 V 230A, 25°C	1000 V 230A, 75°C	1000 V 230A,125°C
Eon (mJ)	12 (68)	9 (89)	8 (--)
Eoff (mJ)	28 (86)	26 (99)	27 (--)
E_{total} (mJ)	40 (154)	35 (188)	35 (--)

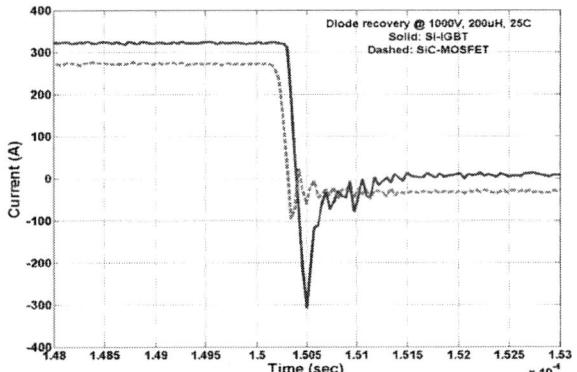

Fig. 9. Reverse recovery behavior of Si-IGBT and SiC-MOSFET power module at 25 °C for supply voltage of 1000 V.

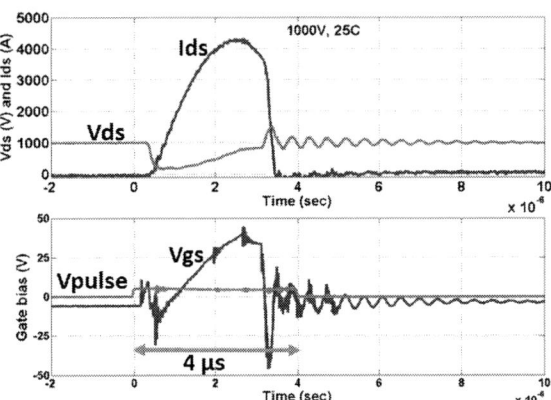

Fig. 10. Short circuit behavior under HSF condition at 1000 V supply voltage for SiC MOSFET at 25 °C.

III. CONCLUSIONS

A comparative assessment of first commercial SiC based MOSFETs and Si based IGBT power modules with current rating of 300 A (several dies in parallel) and voltage rating of 1700 V is presented in this work. Note that this is the first set of commercial power modules with very high current rating and with voltage rating of 1700 V being evaluated here to support and facilitate the converter design. Overall, MOSFET devices show ON-resistance values of 10.0 mΩ for a given gate bias of 20 V and may meet the requirement of power applications for reliable switch performance. Dynamic tests have been performed at different temperatures. Contrary to Si-IGBTs, no tail current was noticed. Since SiC-MOSFET devices switch at a very high speed, a reliable gate driver is needed that should have some on-board functionality to control the oscillations and simultaneously tailored the turn on/off behavior for different applications. Turn on/off energy losses are quite reasonable and slightly exceeding the datasheet values (partly due to higher gate resistance (10 Ω is used here) and partly due to larger overall loop inductance used in our setup). A threshold voltage shift and decay in the peak trans-conductance is observed as expected with the increase of temperature. Overall, energy losses remain approximately unchanged with variation in device temperature for SiC-MOSFETs. Compared to Si-IGBTs, SiC-MOSFET power modules offer lower energy losses. SiC-MOSFETs modules tested in the diode configuration did not show any reverse recovery behavior neither due to change in the DC voltage supply nor due to change in the temperature while a significantly higher reverse recovery has been witnessed for Si-IGBT hybrid power modules. A short circuit test is presented here and show SiC-MOSFET survivability up to 4.0 μs with 13 times higher current than nominal rated current of 300 A for the power module.

REFERENCES

[1] V. Pala, Edward V. Brunt, L. Cheng, M. O'Loughlin, J. Richmond, A. Burk, S. T. Allen, D. Grider, J. W. Palmour, "10 kV and 15 kV Silicon Carbide Power MOSFETs for Next-Generation Energy Conversion and Transmission Systems," IEEE Energy Conversion Congress and Exposition (ECCE), 2014, pp. 449-454.

[2] J. Millan, P. Godignon, X. Perpina, A. Perez-Tomas and J. Rebollo "A survey of wide bandgap power semiconductor devices", IEEE Trans. Power Electronics., vol. 29, no. 5, 2014, pp. 2155 -2163.

[3] E. Cilio, J. Hornberger, B. McPherson, R. Schupbach, A. Lostetter, and J. Garrett, "A Novel High Density 100 kW Three-Phase Silicon Carbide (SIC) Multichip PowerModule (MCPM) Inverter," in Proc. APEC '07, 2007, pp. 666 - 672.

[4] T. Funaki, J.C. Balda, J. Junghans, A.S. Kashyap, H.A. Mantooth, F. Barlow, T.Kimoto, and T. Hikihara, "Power Conversion With SiC Devices at Extremely High Ambient Temperatures," IEEE Transactions on Power Electronics, Vol. 22, no. 4, 2007, pp. 1321 – 1329.

[5] Datasheet for SiC-MOSFETs module, "APTMC170AM07CD3AG-Rev 0", 2014.

[6] Datasheet for hybrid Si-IGBT module, "APTGT300A170D3G-Rev 2", 2014.

[7] Datasheet Rohm SiC MOSFET power module (BSM120D12P2C005), 2014.

[8] ROHM Application Note, "SiC Power Devices and Modules; Application Note", Rev. 1, 13103EAY01, June 2013.

[9] Microsemi power products group, Datasheet of 1.2 kV and 714 A MOSFETs, "APTMC120DUM08D3AG", 2013.

[10] Mitsubishi power products, Datasheet of 1.2 kV and 800 A MOSFETs, "CMF800-24-S001", 2014.

[11] Powerex power module, Data sheet of 1.2 kV and 1.2 kA MOSFETs, "FMF1200DX1-24A", 2014.

[12] Semikron power modules, Datasheet of 1.2 kV and 541 A MOSFETs, "SKM500MB120SC", 2014.

[13] Genesic power module, Datasheet of 6.5 kV and 40 A asymmetric thyristor "GA040TH65", 2014.

[14] Cree power modules, Datasheet of 1.2 kV and 404 A MOSFETs, "CAS300M12BM2", 2014.

[15] Infineon power module, Datasheet of 1.2 kV and 25 A JFETs in TO-247 Package, "IJW120R070T1", 2013.

[16] A. Kadavelugu, S. Bhattacharya, Sei-H. Ryu, E. V. Brunt and D. Grider, "Experimental Switching Frequency Limits of 15 kV SiC N-IGBT Module", The 2014 International Power Electronics Conference, 2014, pp. 3726 – 3733.

[17] E. Van Brunt, L. Cheng, M. O'Loughlin, J. Richmond, V. Pala, J. W. Palmour, C. W. Tipton and C. Scozzie, "27 kV, 20 A 4H-SiC n-IGBTs", Materials Science Forum, Vol. 821-823, 2015, pp 847-850.

[18] M. Nawaz, N. Chen, "Static and dynamic analysis of SiC Based commercial MOSFET power modules", ISBN: 9789075815238, EPE'15 ECCE Europe, 2015.

[19] F. Chimento, M. Nawaz, "On the short circuit robustness evaluation of silicon carbide high power modules", pp. 920 – 926, IEEE ECCE-2015.

[20] A. März, R. Horff, M. Helsper, Mark-M. Bakran, "Requirements to change from IGBT to Full SiC modules in an on-board railway power supply", ISBN: 9789075815238, EPE'15 ECCE Europe, 2015.

[21] M. Nawaz, F. Chimento, K. Ilves, "Static and dynamic performance assessment of commercial SiC MOSFET power modules", IEEE ECCE-15, pp. 4899 – 4906, 2015.

[22] M. Nawaz and N. Chen, "Evaluating 4H-SiC Based Commercial MOSFETs Power Modules", IEEE PEDS-2015, pp. 462 – 466, 2015.

978-1-4673-9551-9/16 $31.00 © 2016 IEEE

Suppression of reverse recovery ringing 3.3kV/450A Si/SiC hybrid in low internal inductance package

next High Power Density Dual; nHPD^2

Katsuaki Saito

Power Device Division,
Hitachi Europe Ltd.
Maidenhead Berkshire, United Kingdom
Katsuaki.saito@hitachi-eu.com

Daisuke Kawase, Masamitsu Inaba, Keiichi
Yamamoto, Katsunori Azuma and Seiichi Hayakawa
Hitachi Power Semiconductor Ltd.
Hitachi, Japan

Abstract— **Suppression of reverse recovery ringing from 3.3kV Si-IGBT SiC Schottky barrier diode hybrid was verified. Reducing loop inductance, consisting of internal module, busbar connection and capacitor, is effective. To realize low inductance within the module and external connection with busbar, the gap between p and n terminal is minimized whilst maintaining the creepage and clearance required by regulation standards. Functional isolation is adopted instead of basic isolation. We achieved an inductance value 9nH for a 3.3kV, 450A dual IGBT, that leads to the extinction of reverse recovery oscillation. Details of switching characteristics of Si + SiC hybrid module is compared with that of Si IGBT module with low inductance.**

Keywords— *3.3kV; IGBT; SiC; hybrid; low inductance; isolation coodination*

I. INTRODUCTION

Wide band gap semiconductor are desired switching devices because of the low switching loss characteristics [1][8]. However, because of no tail current during the recovery and turn-off periods, switching oscillation may create an EMI problem. In Fig.1, Turn-on and reverse recovery waveforms of 3.3kV / 1200A devices are compared. On the left hand side: Si IGBT and Si diode module; on the right hand side Si IGBT and SiC SBD (Schottky barrier diode) hybrid module, are shown. Si-SiC hybrid decreases reverse recovery energy by more than 99% and turn-on energy by 50%. In this case, simple Rg(on/off) gate resistances were used. Si-Si module does not show oscillation in both turn-on or reverse recovery waveforms. To solve oscillation problems, active gate control is proposed [12]. Using this technology, the direction of diode current can be detected and after diode current crosses zero current, turn on speed is actively slowed down to avoid switching oscillation without increasing turn on loss.

But as long as using a simple Rg(on/off) gate driver, reducing loop inductance seems to be the principal way to avoid the oscillation. Assuming high voltage devices of 3.3kV, using conventional package types, it is impossible to reduce the inductances drastically to the value at which switching oscillation can be suppressed. Slower speed in turn-on and turn-off may be adopted, but that will destroy the fundamental benefit of WBG devices. Adding snubber circuitry is also

proposed [11]. Several low inductance packages were proposed targeting potential high voltage SiC devices [2], [3], [4], [9].

In this study, we have set the target inductance value by evaluation of SIC-MOS and SBD at chip level. Target value for 3.3kV / 450A SiC devices was set at a level where oscillation can be suppressed. We developed 3.3kV/ 450A Si-IGBT + SiC SiC (Schottky barrier diode) hybrid module and verified the suppression of switching oscillation.

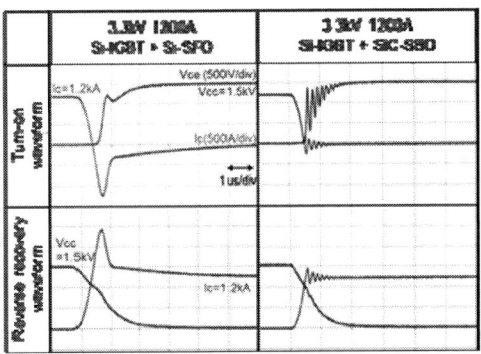

Fig. 1. Turn-on and reverse recovery waveforms of 3.3kV, 1200A IGBT module (MBN1200E33E) and 3.3kV, 1200A Si+SiC hybrid module (MBN1200E33E-C). *(Conditions; Vcc=1500V, Ic=1200A, Tj=125℃, Ls=100nH, RG=3.9/3.9Ω)*

II. TARGET OF THIS STUDY

A. How oscillation condition can be avoided.

In this section, using a simple LCR model, the direction to suppress the oscillation is explained. Fig.2 (a) shows the circuit diagram of switching measurement. This circuit can be converted into Fig 2. (b). IGBT turn on is equivalent to a resistor whose resistance changes from constant high value to low value. Bottom side IGBT and FWD (Free Wheel Diode) are equivalent to the parallel connection of a resister and capacitor. The resistance changes from low value to high value corresponding to tail current. The capacitor is a constant capacitance during switching conditions. The differential equation to solve this circuit can be deployed.

$$Ls \cdot \frac{di_{Ls}}{dt} + R_p \cdot i_{Ls} + \frac{1}{C_n} \int i_{Cn} \cdot dt = 0 \qquad (1)$$

$$\frac{1}{C_n} \int i_{Cn} \cdot dt = R_n \cdot i_{Rn} \qquad (2)$$

$$i_{Ls} = i_{Cn} + i_{Rn} \qquad (3)$$

When considering Si devices, IGBT turn-on time is much faster than the diode recovery. In such a case, R1 can be neglected and the equivalent circuit can be Fig.2 (c). To avoid oscillation, the imaginary part of the solution of this differential equation has to be 0.

$$R_n \leq 2\sqrt{\frac{L_s}{C_n}} \qquad (4)$$

In handling SiC SBD, Rn changes much faster than Rp. Therefore, the equivalent circuit can be converted to Fig.2 (d). The solution where oscillation can be avoided is as follows:

$$R_p \geq 2\sqrt{\frac{L_s}{C_n}} \qquad (5)$$

To realize this condition only 3 parameters are valuable. Rp has to be a large value which means slower turn-on speed and leads to large turn on energy. Cn is already approximately ten times larger than Si diode because of a narrower space charge region. To increase it further, adding a snubber capacitance is one idea, but it is not a desirable solution. Decreasing Ls is the only viable solution to realize eq. (5).

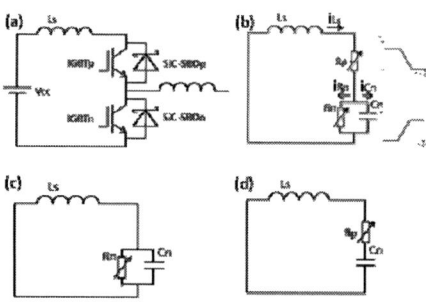

Fig. 2. (a) Measuremnet circuit, (b) Equivalent circuit during reverse recovery period, (c) Converted equivalent circuit during reverse recovery of Si-SFD (Soft and fast recovery diode), and (d) Converted equivalent circuit for reverse recovery of SiC-SBD (Schotkey barrier diode) hybrid.

B. Setting of target inductance experimentally

Before developing high current modules, inductance target values are set. Dependence of turn on energy on loop inductance and gate resistance were investigated

experimentally. A 25A rated SiC-MOS was used for switching device and 25A SiC SBD was used as the free wheel diode. Results are shown in Fig.3. Turn on energy decreases by increasing the internal inductance and increases by increasing Rg. On the other hand, the amplitude of reverse recovery oscillation becomes smaller by an increase of Rg and decreasing loop inductance. Eventually oscillation can be suppressed. There is a border for oscillation on the inductance and turn-on energy surface. There is no oscillation above the border and reverse recovery oscillation below the border. To realize both low turn on energy and no oscillation, we could set the target at 40nH for the 450A rated current which corresponds to 700nH for a 25A chip. Of the 40nH, 10nH was assigned to module internal inductance. Another 10nH was assigned to the busbar, including module connection. The balance 20nH was assigned to the capacitor, including the connection between capacitor and busbar.

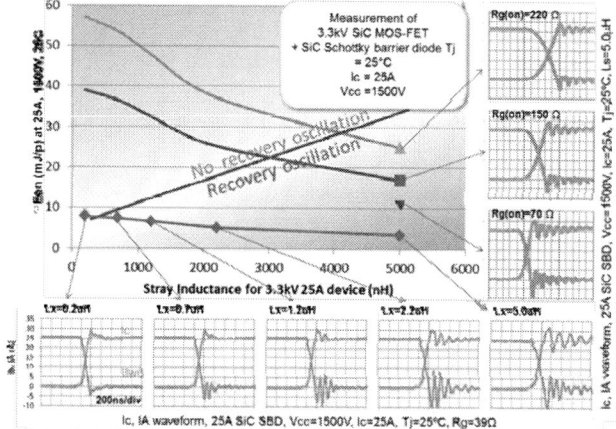

Fig. 3. Dependence of turn-on energy on stray inductance with the parameter of turn-on gate resistance(Rg) and respective turn-on (pink) and reverse recovery current of SiC-SBD hybrid (light blue). In the graph, green area shows no reverese recovery oscilation and red area shows reverserecovery oscillation.

III. PREPERATION OF LOW INDUCTANCE 3.3KV PACKAGE

In this study, two package types were investigated, (Fig. 4). Both have the same foot print size of 140mm*100mm, which allows high power density rated 3.3kV-450A in a dual switch configuration. In Fig. 4 (a), p and n terminals are aligned side by side. It has comparable clearance and creepage distances to conventional 3.3kV IGBT modules. Fig. 4 (b) is based on the 1.7kV package proposed in ref. [2]. Due to the high current of 1700V, twice the number of terminals are aligned to handle twice the current. The gap between p and n terminals is shorter than conventional 3.3kV modules.

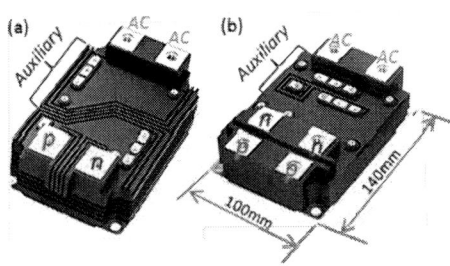

Fig.4. Pictures of modules investigated in this study

Electromagnetic analysis was conducted to optimize the internal and external structure. In Fig. 5, the model of electromagnetic simulation (b) and example of the simulated results (b) are shown. One of the most influencial parameters of internal inductance is the gap between p and n terminals. In Fig. 5, the dependence of inductance, with and without busbar, on the p-n terminals is shown. Intermural inductance increases linearly versus the gap between p and n terminals with an offset of 5nH, which corresponds to the internal inductance inside the package housing. Inductances between p-n busbars also increase linearly. The difference between the inductance at the busbar edge and module terminals corresponds to the inductance at the connection of the busbar to modules. This value also increases proportionally to the gap between p and n terminal. According to these results, the design should make the gap as small as possible.

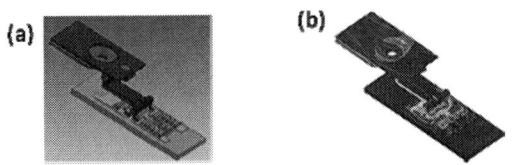

Fig.5. (a) Model for electromagnetic simulation and (b) simulated results.

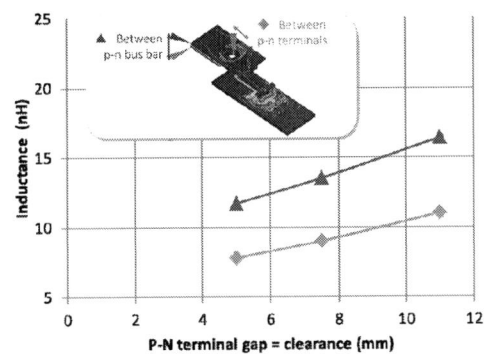

Fig.6. Dependence of inductance with and without busbar on the gap between P-N terminals.

But a minimum clearance must be met according to the requirement of isolation coordination. Conventional packages have met these requirements. Despite device isolation voltage from terminals to base plate of 6.0kVrms, the allowable maximum voltage between the terminals is 3.3kVpeak. Otherwise over voltage may deteriorate the blocking characteristics. These specifications should be coordinated. Adding a safety factor, a 4kV impulse voltage is considered to define functional isolation, instead of basic isolation (Table 1). By adopting functional isolation, the minimum clearance, case (b), can expect 9nH. In this paper, the module is called nHPD2 because it is a dual package enabling high power density [2]

TABLE I.   TABLE STYLES

#	Item		Dimension	Condition		Remark
1	Clearance	terminal -heatsink	15.0mm	Over voltage category :OV2 Impulse voltage :10kV Pollution degree :PD3		Basic Insulation
2		terminal -terminal	7.5mm	Over voltage category :OV2 Impulse voltage :4kV Pollution degree :PD3		Functional insulation
3	creepage distance	terminal -heatsink	27.5mm	Material group Working voltage	:I(CTI>600) :2120kVrms (1500Vrmsx)	
4		terminal -terminal	27.5mm	Pollution degree :PD3		

In Fig 7 the cross sectional view around the main terminal connection with a bus bar is shown. Both clearance and creepage comply with the specification of isolation distances. Pink lines in Fig. 7 (b) show clearances. The shortest path is the directly from p terminal to n terminals with a distance of 7.5mm. Green lines in Fig.7 (c) show the creepage. The shortest path is the path between p n terminals via the groove with wall inside.

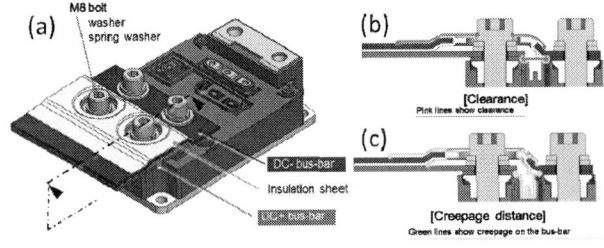

Fig.7. (a) Bird view of low inductance module, (b) clearance paths and (c) creepage paths.

IV.   MODULE CHARACTERISTICS

In this section, electric characteristics of the completed modules with low inductance bus bar are demonstrated. Fig.8, shows the evaluation system. Modules are mounted on the heatsink controlled by a heater. Current is measured with a current transformer.

978-1-4673-9551-9/16 $31.00 © 2016 IEEE

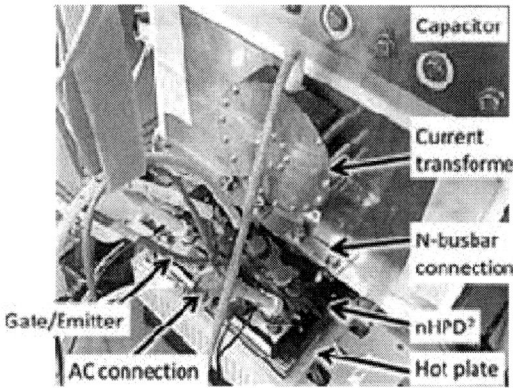

Fig.8. Picture of the evaluation system

As a first step, measurements of module internal inductances were conducted. We set target inductances of 10nH in the first section. The measured value is 9nH. The total allowable inductance of 40nH can be easily realized and minimum value of this system is 20nH.

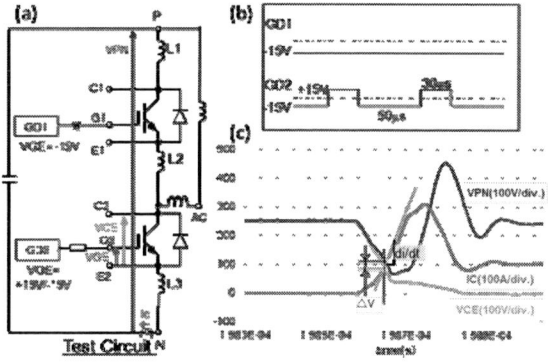

Fig.9. Measurement of nHPD². Measurement circuit, (b)IGBT gate pulse during measurement and (c) Measurement waveforms

Differences of switching characteristics were investigated using the 3.3kV chip generation. Conventional high voltage package and nHPD² were compared. Conventional type was MBN1800F33F [5], which has an inductance of 7nH for a rating of 1800A. Total inductance is 80nH including 7nH for upper arm and 7nH bottom arm module internal inductance. In the case of nHPD², total inductance is 40nH including 9nH for module internal inductance. The products of total inductance and rated current are 144μA·H and 18μA·H respectively for MBN1800F33F and nHPD². nHPD² decreases loop inductance by 88%.

Turn off waveforms were compared. Spike voltage overshoot from Vcc are 500V and 250V respectively. A reduction of spike voltage by 50% was confirmed. But, considering 88% decrease in loop inductances, 50% decrease appears to be a small improvement. Reason for this phenomenon is handled in the next paragraph.

Comparing turn-on waveforms, the voltage kink during positive di/dt shows the difference of the lower internal inductances.

The knee voltage in reverse recovery of MBN1800F33F can be seen, but no knee voltage in nHPD2 appears. This leads to a lower reverse recovery dv/dt (4.8kV/μs to 2.5kV/μs).

Using the same low inductance package (nHPD²) IGBT chip and circuit configurations, Si diode and SiC SBD are compared. Firstly, despite the same loop inductance, the spike voltage shows a difference. Spike voltage decreases from 250V to 60V. The spike voltage is the sum of the product of stray inductance and di/dt and forward recovery voltage. Si device does not change to a complete condition state. After current starts to flow inside of n(-) region, the modulated carrier gradually increases to a fully modulated state. During this period the forward voltage drop of diode is much higher than final on-state voltage. However the SiC SBD does not show forward recovery voltage because it is unipolar device and the conductance modulation effect is not used.

Fig. 8 shows the comparison of forward recovery waveforms. Measurements were done in the same system. In the case of Si, forward recovery voltage (Vfr) is 240V, whereas for SiC SBD the forward recovery voltage peak is not observed. Therefore, a lower spike voltage during turn off but higher di/dt can be achieved; and that leads to lower turn off energy by 11%. As for the turn-on waveform, due to no superimposed reverse recovery current, turn-on energy decreases drastically by 52%.

Final comparison is reverse recovery waveform. Almost no reverse recovery loss was generated, which decreases by 99.6%. The reverse recovery switching oscillation previously observed in the conventional package (Fig.1 right hand side) can be extinguished completely.

V. MODULE CHARACTERISTICS

Suppression of reverse recovery ringing from 3.3kV Si-IGBT SiC Schottky barrier diode hybrid was verified. Reducing the loop inductance consisting of internal module, busbar connection and capacitor is effective. To realize low inductance internally within module and external busbar connection, the gap between p and n terminal is minimized whilst maintaining the creepage and clearance required by regulation standards. Functional isolation is adopted in place of basic isolation. We achieved an inductance value of 9nH for a 3.3kV / 450A dual IGBT that leads to the extinction of reverse recovery oscillation.

Details of the switching characteristics of Si + SiC hybrid module were compared with that of Si IGBT module with low inductance. SiC SBD lowers turn off spike voltage by eliminating forward recovery voltage generated by bipolar devices. All turn on, turn off and recovery switching energy could be reduced by applying SiC SBD.

ACKNOWLEDGMENT

Authors appreciate Dr. Kurosu, Mr. Koike, Mr. Takayanagi, Mr. Yoshino Dr. Mori Dr. Horiuchi and Mr.Markham for their valuable discussion and strong back up for this work.

	MBN1800F33F	MBN450FS33F	MBM450FS33F-C
Device	Si IGBT Si diode Conventional high voltage package 140x190mm²	Si IGBT Si diode nHPD² This study 100x190mm²	Si IGBT SiC SBD nHPD² This study 100x190mm²
Conditions	Vcc=1800V, Ic=1800A, Tj=150°C, Ls=80nH Vg=15V/15V, Rg(on/off)=4.7Ω/5.6Ω	Vcc=1800V, Ic=450A, Tj=150°C, Ls=40nH Vg=15V/15V, Rg(on/off)=12Ω/12Ω	Vcc=1800V, Ic=450A, Tj=150°C, Ls=40nH Vg=15V/15V, Rg(on/off)=12Ω/12Ω
Turn-off	Eoff=3300mJ/p(=820mJ/p*4) ΔVce=500V	Eoff=640mJ/p ΔVce=250V di/dt=1.0kA/μs	Eoff=570mJ/p ΔVce=60V di/dt=1.4kA/μs
Turn-on	Eoff=3700mJ/p (=920mJ/p*4) ΔVce=750V	Eoff=750mJ/p ΔVce=70V	Eoff=360mJ/p
Reverse recovery	4.8kV/μs Eoff=2400mJ/p (=600mJ/p*4)	2.4kV/μs Eoff=650mJ/p	Eoff=3mJ/p

Fig.10. Switching waveforms of (a) conventional IGBT module Si IGBT and Si diode in conventional high voltage package, (b)Si IGBT and diode in nHPD² and (c) Si IGBT and SiC SBD in hybrid nHPD²

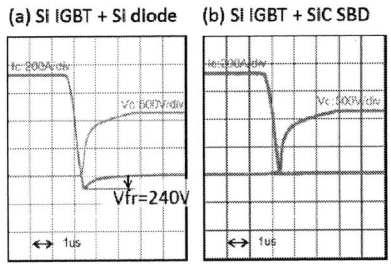

Fig.11. Forward recovery waveforms of (a) Si diode and (b) SiC-SBD

REFERENCES

[1] K. Ogawa, K. Ishikawa, N. Kameshiro, H. Onose, and M. Nagasu, "Traction inverter that applied SiC hybrid module"; PCIM2011.

[2] D. Kawase, M. Inaba, K. Horiuchi and K. Saito, "High voltage module with low internal inductance for next chip generation - next High Power Density Dual (nHPD2) –", PCIM2015.

[3] R. Schnell, S. Hartmann, D. Trüssel, F. Fischer and A. Baschnagel, "LinPak, a new low induct active phase-leg IGBT module with easy paralleling for high power density converter designs", PCIM2015.

[4] S. Buchholz, M. Wissen and T. Schütze, "Electric performance of a low inductive 3.3kV half bridge IGBT module"

[5] T. Kushima, K. Azuma, Y. Nemoto, K. Saito and Y. Koike, "3.3kV/1800A IGBT module using advanced trench HiGT structure and module design optimization", PCIM2014

[6] R. Horff, A. März and M. Bakran, "Analysis of reverse recovery behavior of SiC MOSFET body diode-regarding dead time", PCIM2015.

[7] C. Neeb, J. Teichrib, R. Doncker, L. Boettcher, and A. Ostmann, "A 50 kW IGBT power module for automotive applications with extremely low DC-link inductance", EPE2014.

[8] T. Ishigaki, H. Kageyama, A. Shima, D. Hisamoto, K. Tomiyama, Y. Yasaki and S. Ibori, "Free wheel diode less SiC-inverter with fast short circuit protection for industrial applications", PCIM2015.

[9] E. Kraft, B. Laska, A. Nageland J. Weigel, "A New Standard IGBT Housing for High-Power Converters", EPE2015.

[10] E. Velander, A. Löfgren, K. Kretschmar and H. Nee, "Novel Solutions for suppressing Parasitic Turn-on behaviour on Lateral Vertical JFETs", EPE2014.

[11] M. Joko, A. Goto, M. Hasegawa, S. Miyashita and H. Murakami, "Snubber circuit to suppress the voltage ringingfor SiCdevice.", PCIM2015.

[12] K Onda, A Konno and J Sakano, "New concept high-voltage IGBT gate driver with self-adjusting active gate control function for SiCSBD hybrid module", Proceedings of The 25th International Symposium on Power Semiconductor Devices & ICs, Kanazawa, (2013)

New Layout Concepts in MW-Scale IGBT Modules for Higher Robustness during Normal and Abnormal Operations

Paula Diaz Reigosa, Francesco Iannuzzo, Stig Munk-Nielsen, Frede Blaabjerg
Department of Energy Technology
Aalborg University
Pontoppidanstraede 101, Aalborg East DK-9220, Denmark
pdr@et.aau.dk, fia@et.aau.dk, smn@et.aau.dk, fbl@et.aau.dk

Abstract—This paper presents six different layout designs for MW-level Insulated-Gate Bipolar Transistor (IGBT) modules. The aim is to optimize the normal switching and abnormal short circuit performances of IGBT modules in terms of stray inductance reduction and inductance coupling cancellation. Using the Finite-Element-Method AnSYS Q3D Extractor, electromagnetic simulations are conducted to extract the self and mutual inductance from the six different layouts. PSpice simulations are used to reveal that the stray parameters inside the module play an important role under normal and abnormal operations, and thus the best performing layout strategies can be selected. The prototypes are fabricated to validate the results of the Q3D simulation analysis, in which the self-inductance is measured by using a high precision impedance analyzer.

I. Introduction

Insulated-Gate Bipolar Transistor (IGBT) modules are the most prevalent packaging in modularization and integration of power electronic systems from medium to high power applications. The increasing demand for fast switching devices has reached a point, where the traditional semiconductor packages constrain the full performance of the IGBT modules due to the internal parasitic RLC networks. To make use of the full benefit of the fast switching power devices, novel packaging and layout designs have to be designed for achieving low-inductive modules.

In practical applications, many failures of the power semiconductor devices are caused by parasitic effects [1]–[4]. For example, when the IGBT turns off, the collector-emitter over voltage is a function of the stray inductance of the power loop and the *di/dt* of the IGBT collector current. With both high switching speed and high power handling requirements, the reduction of the stray inductance becomes very important in order to reduce the voltage and current stresses. As a consequence, the device will be faster and the switching power losses will be reduced [5], [6]. If the layout is not effectively optimized, the voltage spikes during IGBT turn off in normal and abnormal operations can be beyond the maximum voltage rating of the IGBT (i.e., the breakdown voltage of the device). This means that an IGBT chip with higher voltage rating must be selected, which leads to increased IGBT chip sizes and lower power density of the overall power module.

On the other hand, reducing the inductance is not always an improvement. For instance, the increased IGBT turn-on losses and the increased ElectroMagnetic Interference (EMI) caused by the reverse recovery behaviour of the free-wheeling diode is a main drawback for increased turn-on speed [7]. As a consequence, a compromise between turn-on and turn-off speed must be considered during normal operation. In contrast, during abnormal operations, an ultra-low inductance is always the main target since the free-wheeling diode does not contribute to the short-circuit event.

In the past, the stray inductance influence on the normal switching behaviour has been discussed in terms of switching loss reduction and switching oscillations, leading to very robust technologies [7]–[9]. Nevertheless, the short circuit performance may be compromised if the targeted inductance for switching conditions is not sufficiently small. In other words, the expected IGBT module robustness during switching operations may be constrained due to a weak performance during short circuit operations (e.g., critical oscillations), as discussed in [10]. The present study tackles the problem by proposing six different layout solutions aiming for a better compromise among normal and abnormal operations.

This paper is organized as follows: Section II describes the proposed layouts from a theoretical point of view to understand the impact of parasitic inductances in power modules. Section III includes the self and mutual inductance simulations done by AnSYS Q3D Extractor, which evidence the concepts proposed in Section II. Section IV gives a fundamental insight into the influence of parasitic inductances on both normal and abnormal operations. Section V shows the fabricated prototypes together with the measurement results. Finally, Section VI gives the conclusions of the paper.

II. Proposed Layout Designs

This section shows a comparison between six different layouts with the aim of minimizing both the self and mutual parasitic inductances for a specific MW-level IGBT module. The parasitic inductances found inside the power module are mainly coming from: i) the power terminals, ii) the Direct Bonded Copper (DBC) substrates, and iii) the bond wires. The

978-1-4673-9551-9/16 $31.00 © 2016 IEEE

study in this paper is focused on the optimization of the DBC substrate to achieve a low-inductive power module. Even if the DBC substrate takes the smallest contribution to the total module inductance, it becomes relevant when paralleling several power devices, which is the most prevalent methodology for increasing the current capability in high power applications. A reduced DBC inductance can be achieved by the following methods: i) shorten the trace length, ii) widen the trace width, and iii) cancelling the magnetic field by bringing the return current close to the forward current. The selected designs were chosen bearing in mind one of the following reasons: they were determined through simulations to be near optimal designs or they were chosen for comparison purposes. For the sake of comparison, the six DBC substrates are similar in terms of substrate size and position of the active chips and power and control terminals.

A. Layout A

To make the study more coherent with the industry standard package, the layout outline of a commercial 1.7 kV/ 1 kA IGBT module was selected [11]. As can be seen in Fig. 1, the module has 5 power terminals, 4 control terminals and 6 identical DBC substrates arranged in parallel. Each section contains two IGBT chips and two free-wheeling diodes, which are configured as a half-bridge. The following study will be focused on the improvement of the DBC substrate (i.e., one section of the module). In general, it is expensive and time-consuming to fabricate DBCs, therefore, a more economical and faster solution based on PCB technology has been chosen to verify the proposed designs. In order to extract the self and mutual stray inductances of the new layout concepts, the software tool AnSYS Q3D Extractor has been used [12]. Prior to the simulations, Solidworks [13] has been used to develop the geometries and define the material properties. The detailed dimensions used for the simulations are listed in Table I, which are common for all the layouts.

Fig. 2a represents the traditional DBC substrate layout of the commercial 1.7 kV/ 1 kA IGBT module. The current path from points 1-2 and 4-5 are the gate loops of the high side and low side respectively. The current paths from points 2-3 and 5-6 are the gate return loops of the high side and low side, respectively. It can be observed that the upper side IGBT and diode devices are located at one side, while the lower side IGBT and diode devices are located on the other side. The loop from point 7-8 and 8-9 form the power loop from the positive and negative DC terminal to the output phase terminal.

TABLE I

DIMENSIONS OF THE PCB SUBSTRATES

Component	Value
PCB area	39 x 53 mm^2
PCB thickness	0.105 mm (Cu) + 1.6 mm (FR4) + 0.105 mm (Cu)
IGBT area	13.58 x 12.58 mm^2
Diode area	9.58 x 9.58 mm^2
Bond wire	0.25 mm diameter (Al)

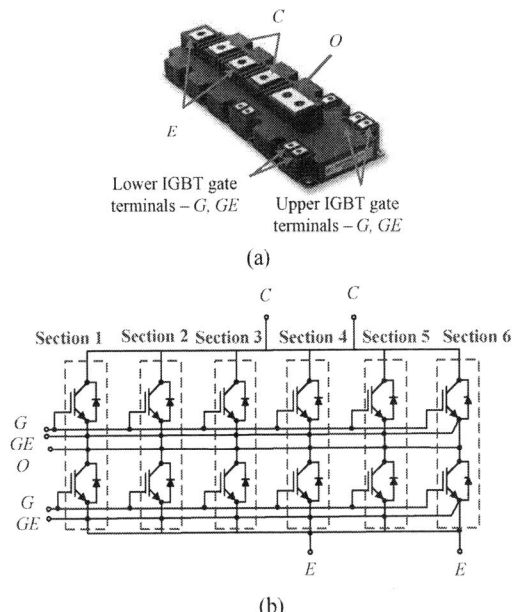

Fig. 1. The 1.7 kV/ 1 kA IGBT module including its terminals: (a) a picture of the package, and (b) its half-bridge structure.

B. Layout B

Fig. 2b shows layout B. According to layout A in Fig. 3a, the inductance of the emitter bond wires and DBC trace forming the power loop shares a common path with the gate return trace. This shared inductance, L_σ, prevents fast current changes by voltage feedback on the gate loop. As a result, a lower over voltage across the IGBT device is expected during the turn-off. The trade off is that the gate-return inductance increases, where a long copper trace around the IGBT and diode chips has been adopted. Fig. 3b points out how to minimize the gate return path inductance by separating the gate drive loop from the power loop with an additional bond wire from the IGBT emitter to the gate return trace. This solution helps to shorten the gate return path and to widen the trace width of the power path as pointed out in Fig. 3b. It is also expected that layout B has an additional coupling term between the gate bond wire and the gate-return bond wire, which minimizes the overall inductance.

Some interesting considerations can be made from the cancellation of the parasitic shared inductance, L_σ. One of the advantages is that the device will turn-off much faster but with the limitation that a higher over voltage is expected during the turn off transient.

C. Layout C

The designs in Figs. 4a and 4b aim at compare layout A and C, respectively. Layout C is adopted for minimizing the power loop inductance from the DC positive to the DC negative terminals. As can be seen in Fig. 4b, the upper diode bond wires of layout C have been rotated 90 degrees clockwise with the aim of shortening the loop. This improvement would be

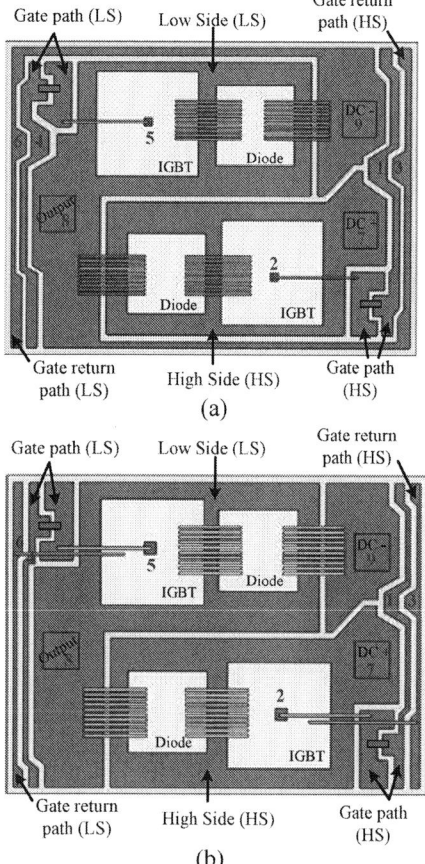

Fig. 2. Single DBC design section for the IGBT power module (a) layout A, and (b) layout B.

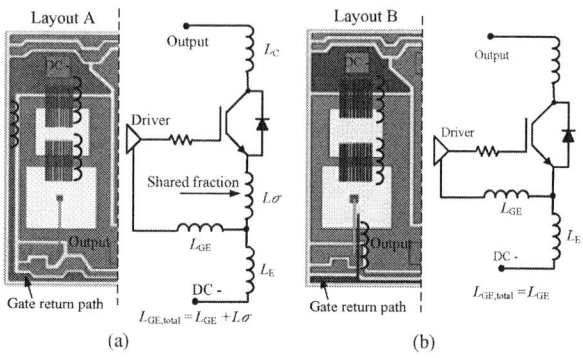

Fig. 3. Equivalent circuit of the low-side IGBT power module together with the layout configuration: (a) layout A, and (b) layout B.

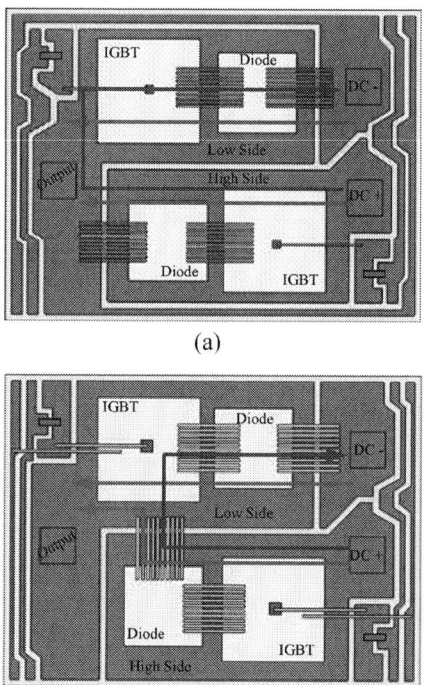

Fig. 4. Short circuit current path and switching current path: (a) layout A, and (b) layout C.

beneficial in case that a short circuit occurs when both high side and low side IGBTs are turned on at the same time (shoot-through) [6]. For example, in Fig. 4b the blue line which starts from the DC positive terminal, goes through the upper IGBT and its bond wires, and then reaches the DC negative terminal. This is much shorter than the same loop in layout A, as shown in Fig. 4a. This configurations has a longer current loop during normal operations, however the possible effect is negligible.

D. Layout D

Layout B demonstrated that the best solution for minimizing the gate-return path inductance is by separating the gate path from the power loop path. Another way to have inductance minimized is the proposed layout D in Fig. Fig. 5a. This configuration places a bond wire from the emitter IGBT to the gate return trace, keeping minimizing the effect of the shared inductance, L_σ. The emitter forms a loop which can be analysed as a triangle of inductances as pointed out in Fig. 5a: 1) from the IGBT emitter to the gate-return path emitter, 2) from the IGBT emitter, through the bond wires to the output phase terminal, and 3) from the output phase terminal to the gate return bond wire. With a basic Δ - Y transformation,

the stray inductance values can be calculated for analysis in circuit simulators, such as PSpice.

It is worth to note that the layout is symmetrical for both low and high sides, resulting in equal distribution of the stray inductances and equal coupling terms during switching conditions. This improvement is crucial for abnormal operations, where the symmetry among high and low side represents a great advantage.

E. Layout E

Previous experiments on MW-level IGBT power modules have revealed critical gate voltage oscillations during short circuit events [14]. One of the methods to suppress the

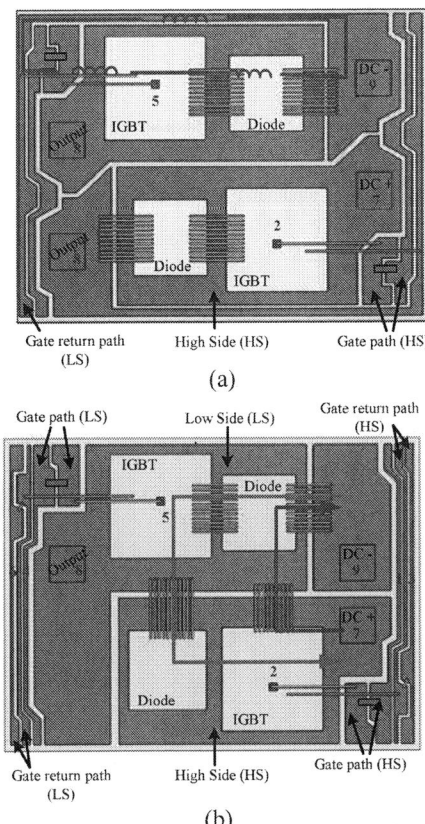

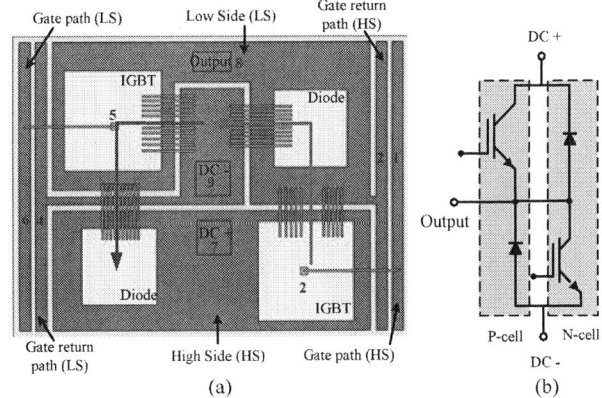

Fig. 6. Single DBC design section for the IGBT power module (a) layout F, and (b) equivalent circuit pointing out the P-cell and N-cell.

Fig. 5. Single DBC design section for the IGBT power module (a) layout D, and (b) layout E.

This improvement would be interesting for both normal and abnormal operations, since the power loops are remarkably smaller in contrast with layouts A, B, C, D and E. Fig. 6 shows the layout F by applying the P-cell and N-cell strategy. It is worth to note that the DC positive, DC negative and output phase terminals have been positioned in the center of the DBC substrate area, differently from layouts A, B, C, D and E. Nevertheless, the DC positive and DC negative busbars are just as close as the previous layouts. The output phase is located near to the DC+ and DC- terminals, but no influence effect is expected. Referring to the gate-return path, the gate loop and control loop are not separated, but it could easily be modified in order to apply the concepts of layouts B or D. Last but not least, layout F has symmetrical low and high sides parasitic inductances, which guarantees equally distributed inductance inside the package.

III. PARASITIC INDUCTANCE EXTRACTION

The AnSYS Q3D parasitic extraction tool is applied for the inductance estimation of the six proposed layouts. Q3D performs electromagnetic field simulations using a combination of the Finite Element Methods (FEM) and the Method of Moments (MoM). This tool is able to perform a DC and AC analysis. The results from the AC analysis are considered since the skin effect is taken into account. For the inductance estimation, the information about the geometry is designed through Solidworks and then exported to Q3D. Electromagnetic simulations are conducted at high frequencies, specifically at 100 kHz, ensuring that the skin effect becomes well developed so that the current flows only on the conductor surfaces and magnetic field effects control the path the current flows through. In the inductance matrix, the off-diagonal entries represent the mutual inductance and the diagonal entries represent the self-inductance of the loop. The sign of the mutual inductance depends on the direction of the currents. The mutual inductance between the components of the paths in which the currents flow in opposite directions is negative. If the currents flow in the same direction, the mutual inductance is positive.

oscillations is to optimize the layout by reducing the emitter inductance, as discussed in [15]–[17]. Thus, the layout E presented in Fig. 5b has been designed with two gate-return paths in parallel for minimizing the gate return path inductance. Furthermore, the high side IGBT device has been arranged closer to the lower side diode in order to reduce the loop between the positive and negative terminals, as highlighted in Fig. 5b with the blue line. Similarly, the red line indicates the loop when the low side IGBT and high side diode are conducting, that is the loop from the DC negative terminal to the DC positive terminal. In this case a compromise between the performances during the positive half cycle compared with the negative half cycle is needed. It may be useful in the case that it is necessary to counteract the possible dissimilarities of the external layout (i.e., DC-link and/or gate driver).

F. Layout F

The previous layout solutions are based on the traditional anti-parallel phase-leg inverters, which means that the high-side IGBT and high side diode are connected together and independently of the lower side IGBT and diode. As discussed in [7] and [18], P-cells and N-cells can form a phase-leg where the high side IGBT and low side diode (P-cell) and low side IGBT and high side diode (N-cell) are closer to each other.

978-1-4673-9551-9/16 $31.00 © 2016 IEEE

For the sake of conciseness, only the results of the high side are presented here. The simulations have been conducted in the following way: when calculating the gate inductance loop, that is points 1-2-3, two sources are placed in points 1 and 3 and one sink is placed in point 2 (source and sink are the measurement points). For the power loop inductance between DC+ / output / DC-, that is points 7-8-9, two sources are placed in points 7 and 9 and one sink in point 8. The path from point 7-9 is the commutation loop from the positive bus to the negative bus, the source has been placed in point 7 and the sink has been placed in point 9. For the special case of layout F, the materials of the conducting devices are set to copper and the materials of the non-conducting devices are set to silicon, implying that only the high side conducts. The bus bars are not included in the simulation, neither the IGBT and diode chips for the obvious reason that all the DBC layouts share the same dimensions.

Fig. 7a presents the self-inductance values of the above mentioned loops and Fig. 7b presents the results of the self-inductance including the effect of the mutual coupling terms. The loop from points 1-2 and 2-3 represent the gate loop, in which the layout A is the worst case among the six possibilities. On the other hand, layout F is the best solution, where the inductance has been reduced by 2 times for the gate path and by 4 times for the gate-return path. For the commutation loop (points 7-8-9), the layout F reveals once again the best solution for minimizing the inductance. In contrast with the worst solutions (layouts A, B, D), layout F effectively minimizes the inductance by a factor of 3. For short circuit operations, the results from points 7-9 in Fig. 7b can be analysed, implying that layout D is the worst case for abnormal operations followed by layouts A and B, in contrast with layouts F and E which are more suitable solutions.

From these results, the question is how these parasitics influence the maximum over voltage during normal and abnormal operations (i.e., effect of shared inductance between gate and power loop and/or effect of power inductance loop). Furthermore, another consideration arises from the maximum possible slope across the IGBT, which depends on the resonant circuit formed by the total power loop inductance and output capacitance (miller capacitance and gate-emitter capacitance). The package designer must be aware of the above limitations when selecting a given layout.

IV. PERFORMANCE EVALUATION OF THE PROPOSED LAYOUTS

PSpice simulations have been performed in this study to further investigate the effects of the stray inductances distribution among the proposed layouts. A physics-based lumped-charge IGBT Spice model [19] is used, which has been demonstrated a remarkable accuracy for high voltage IGBTs, with respect to the embedded PSpice IGBT models. Based on the datasheet information, the IGBT model has been calibrated at different conditions for the IGBT study case rated at 1.7 kV/ 1 kA. With the calibrated IGBT model and the extracted stray parameters, PSpice simulations are performed under switching and short

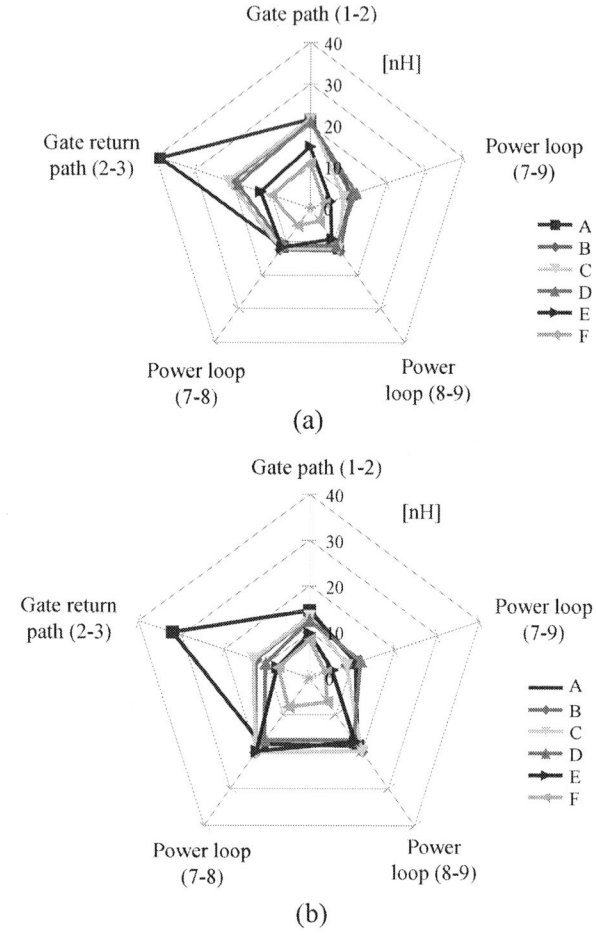

Fig. 7. Parasitic inductance extracted from the six layouts by Q3D at 100 kHz: (a) self-inductance, and (b) self- inductance and mutual inductance.

circuit conditions. The procedure is the following: firstly, a typical Double Pulse Test (DPT) is used to characterize the normal operation behaviour, secondly, a typical short circuit type I test is used to characterize the abnormal operation behavior.

A. Simulations under normal operations

In order to characterize the switching behaviour, the PSpice simulation circuit plotted in Fig. 9a is used. $L_{DC-link}$ represents the stray inductance of the DC-link loop. L_{load} is the load stray inductance. L_C and L_E represent the stray inductance of the power loop, respectively, the collector and emitter stray inductances, which are extracted from the Q3D simulations. L_g and L_{ge} represent the stray inductance of the control loop, respectively, the gate stray inductance and the gate return path stray inductance. Finally, R_{on} and R_{off} represent the gate resistances for the turn-on and turn-off. The simulations are performed under the following conditions: 1000 V DC voltage, 150 μH load inductance and 20 nH DC-link inductance.

In Fig. 8a, the results of the DPT simulations of the

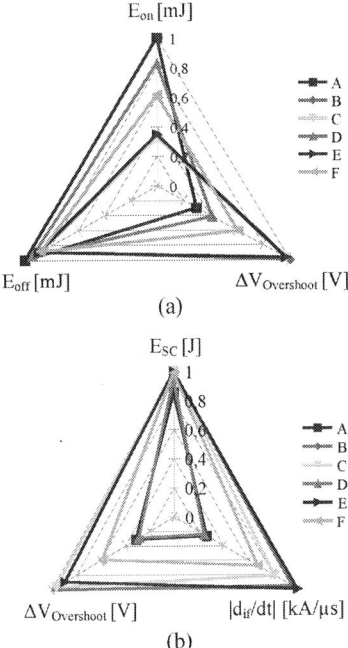

(a)

(b)

Fig. 8. Comparison of the six PCB layouts by means of PSpice simulations - E_{on} is the turn-on energy loss, E_{off} is the turn-off energy loss, $\Delta V_{overshoot}$ is the overshoot voltage during turn-off, E_{SC} is the short circuit energy, d_{if}/dt is the turn-off current falling rate: (a) switching normal conditions, (b) short circuit conditions.

six layouts are compared in terms of turn-on energy loss, E_{on}, turn-off energy loss, E_{off}, and turn-off over voltage, $\Delta V_{overshoot}$. The results have been normalized from 0 to 1 for comparison purposes. From this result, it can be understood the existent compromises among the six proposed layouts during normal operations. The aim during normal operations is to minimize E_{on}, E_{off} and $\Delta V_{overshoot}$. In overall, layout F seems to be the most convenient layout.

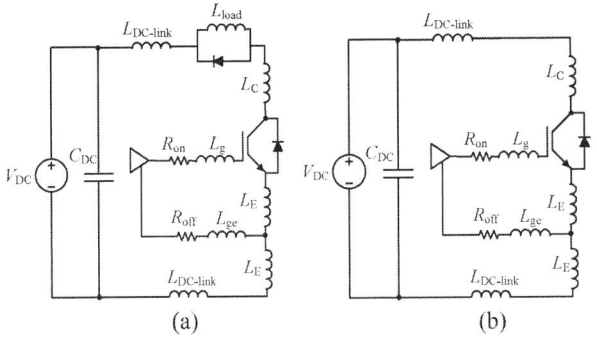

(a)   (b)

Fig. 9. The PSpice circuits for simulating the six proposed layouts: (a) double pulse test for normal operation, (b) short circuit type I for abnormal operation.

B. Simulations under abnormal operations

In order to characterize the short-circuit behaviour, the PSpice simulation circuit plotted in Fig. 9b is used. The

simulations are performed under the following conditions: 1000 V DC voltage, 10 μs short circuit time and 20 nH DC-link inductance.

In Fig. 8b, the short circuit results of the six layouts are compared in terms of short circuit energy, E_{SC}, turn-off over voltage, $\Delta V_{overshoot}$ and turn-off current falling rate, d_{if}/dt. Once again, the results have been normalized from 0 to 1. It can be understood from the results that compromises are needed among the six proposed layouts during abnormal operations. The aim during short circuit operations is to minimize the short circuit energy which should be smaller than the critical energy. Also, the turn-off overshoot voltage must be lower than the breakdown voltage of the IGBT, which can be minimized by limiting the di/dt during turn off. On the other hand, the increase of the di/dt falling rate is very relevant for IGBTs. The IGBT can fail a few microseconds after its turn-off due to heating above the maximum heating dissipation capability of the device, mainly caused by high tail currents. Taken into consideration the above compromises, layouts D and E guarantee lower turn-off voltage spikes in contrast with a very slow turn-off commutation. Thus, it seems reasonable to accept the compromise of layout F, whose performance is the most optimal one.

V. EXPERIMENTAL EVALUATION AND TESTING

For each of the six design configurations evaluated in the previous section by Q3D simulations, prototypes were constructed and measured in the laboratory. The layouts are built without the IGBT and diode chips, similarly to what has been done in the Q3D simulations, in which the wires are bonded directly to the PCB. The bonding wire process has been done thanks to the wire-bonding machine available in the laboratory of CORPE (Center of Reliable Power Electronics) at Aalborg University, Denmark, as can be seen in Fig. 10a. A picture of the six fabricated layouts can be seen in Fig. 11.

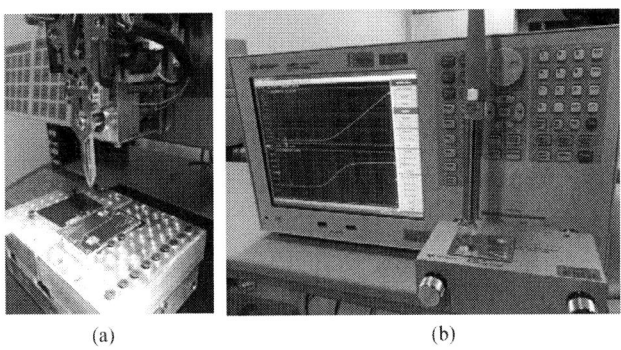

(a)   (b)

Fig. 10. Experimental test setup: (a) wire-bonding machine, and (b) measurement done by means of the impedance analyzer and probe fixture.

The stray inductance has been measured by means of an Agilent precision impedance analyzer E2499A within a frequency range from 10 kHz up to 120 MHz, as shown in Fig. 10b. The probe fixture is crucial for the measurement of small inductances, thus a pin probe is adopted (e.g. accessory 42941A),

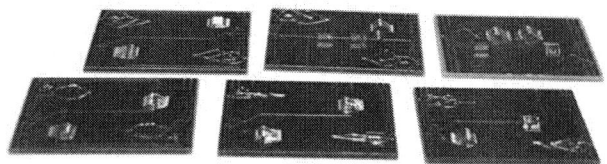

Fig. 11. Fabricated PCB substrates of the proposed layout designs. From lower-left right and up: layouts A, B, C, D, E and F.

TABLE II

COMPARISON OF THE MEASUREMENT AND SIMULATION AT 100 KHZ

		Gate loop (1-2) [nH]	Gate loop (2-3) [nH]	Power loop (7-9) [nH]
A	Simulation	21.54	38.93	10.85 +1.13
	Measurement	28.49 ± 0.67	37.30 ± 0.68	10.30 +1.13
B	Simulation	21.54	20.59	10.67
	Measurement	27.19 ± 0.39	23.60 ± 0.68	10.30+1.13
C	Simulation	21.51	21.21	8.75
	Measurement)	27.19 ± 0.39	23.60 ± 0.68	9.14+1.02
D	Simulation	20.92	18.70	12
	Measurement	22.15 ± 0.18	20.24 ± 0.12	--
E	Simulation	14.92	12.76	5.35
	Measurement	23.15 ± 0.39	11.16 ± 0.20	6.45 + 0.13
F	Simulation	10.99	9.97	3.74
	Measurement	17.73 ± 0.74	10.37 ± 0.14	4.94 + 0.30

which can also be observed in Fig. 10b. However, when using this type of probe, the maximum allowable distance between two measuring points is fixed, thus the measurement between two points far from each other is not possible. In order to measure the parasitic inductance, the impedance analyzer applies an alternating current between the two measurement points and acquires the voltage. The resistance, capacitance and inductance can be read from the output graphs. The measurements have been repeated 10 times in order to give a more accurate result in terms of mean and standard deviation. Only the high side has been measured for consistency with the simulation results. The results can be seen in Table II for different commutation loops and compared with the simulation ones. Simulations and experiments are in good agreements with an error below 15 %, whose feeble differences may be due to the uncertainties during the fabrication process and measurement. It is worth to note that the measurement results show the same trends as in the simulation results.

VI. CONCLUSION

In order to minimize the parasitic effects in a state-of-the-art MW-level IGBT power module, different layout concepts have been proposed. The improvements by the proposed layouts are compared and studied through layout design, stray inductance extraction, PSpice simulation and experimental measurement. The results demonstrate that the best candidate is the layout F for both normal and abnormal operations. The remaining layouts need to be critically judged depending on the application limits and trades-off between normal and abnormal operations. For instance, layout B and C represent a good trade-off for

both normal and abnormal operations with the compromise of higher turn-off overshoot voltage, which must be lower than the breakdown voltage of the device. On the other hand, the layouts that have a lower turn-off overshoot voltage, such as layout D, represent a good solution for normal operations but, it shows a poor performance during abnormal ones.

REFERENCES

[1] K. Xing, F. Lee, and D. Boroyevich, "Extraction of parasitics within wire-bond IGBT modules," in *Proc. of the thirteenth Annual Applied Power Electronics Conference and Exposition, 1998. APEC '98l*, vol. 1, Feb 1998, pp. 497–503 vol.1.

[2] P. Zhang, X. Wen, and Y. Zhong, "Parasitics consideration of layout design within IGBT module," in *Proc. of the International Conference on Electrical Machines and Systems (ICEMS)*, Aug 2011, pp. 1–4.

[3] E. Hoene, A. Ostmann, and C. Marczok, "Packaging very fast switching semiconductors," in *Proc. of the 8th International Conference on Integrated Power Systems (CIPS)*, Feb 2014, pp. 1–7.

[4] S. Hartmann, V. Sivasubramaniam, D. Guillon, D. E. Hajas, R. Schuetz, D. Truessel, and C. Papadopoulos, "Packaging technology platform for next generation high power IGBT modules," in *Proc. of the International Exhibition and Conference for Power Electronics, Intelligent Motion, Renewable Energy and Energy Management (PCIM)*, May 2014, pp. 1–7.

[5] A. Wintrich, U. Nicolai, W. Tursky, and T. Reimann, "Application manual power semiconductors, "Semikron International GmbH"," 2011.

[6] A. Volke and M. Hornkamp, *IGBT Modules*. Infineon Technologies AG, Second edition, 2012.

[7] S. Li, L. Tolbert, F. Wang, and F. Z. Peng, "Stray inductance reduciton of commutation loop in the P-cell and N-cell based IGBT phase leg module," *IEEE Proceedings on Circuits, Devices and Systems*, vol. 29, no. 7, pp. 3616 – 3624, July 2014.

[8] K. Mochizuki and Y. Tomomatsu, "Igbt module," Patent US 20 050 194 660, Sept. 08, 2005.

[9] M. Meisser, M. Schmenger, and T. Blank, "Parasitics in power electronic modules: How parasitic inductance influences switching and how it can be minimized," in *Proc. of the International Exhibition and Conference for Power Electronics, Intelligent Motion, Renewable Energy and Energy Management, PCIM 2015*, May 2015, pp. 1–8.

[10] R. Wu, P. Diaz Reigosa, F. Iannuzzo, L. Smirnova, H. Wang, and F. Blaabjerg, "Study on oscillations during short circuit of MW-scale IGBT power modules by means of a 6-kA/1.1-kV nondestructive testing system," *IEEE Journal of Emerging and Selected Topics in Power Electronics,*, vol. 3, no. 3, pp. 756–765, Sept 2015.

[11] FF1000R17IE4 datasheet. [Online]. Available: http://www.infineon.com

[12] "Ansoft Q3D extractor manual, ansoft inc." 2004.

[13] Solidworks. [Online]. Available: http://www.solidworks.com/

[14] P. Reigosa, R. Wu, F. Iannuzzo, and F. Blaabjerg, "Robustness of MW-level IGBT modules against gate oscillations under short circuit events," *Journal of Microelectronics Reliability*, vol. 55, pp. 1950–1955, August-September 2015.

[15] T. Basler, J. Lutz, R. Jakob, and T. Bruckner, "The influence of asymmetries on the parallel connection of igbt chips under short-circuit condition," in *Proc. of the 14th European Conference on Power Electronics and Applications (EPE 2011)*, Aug 2011, pp. 1–8.

[16] P. Palmer and J. Joyce, "Circuit analysis of active mode parasitic oscillations in IGBT modules," *IEEE Proceedings on Circuits, Devices and Systems*, vol. 150, no. 2, pp. 85–91, Apr 2003.

[17] M. Takei, Y. Minoya, N. Kumagai, and K. Sakurai, "Analysis of IPFM current oscillation under short circuit condition," in *Proc. of the 10th International Symposium on Power Semiconductor Devices and ICs, 1998. ISPSD 98.*, Jun 1998, pp. 89–93.

[18] J. Chen, L. Yang, D. Boroyevich, and W. Odendaal, "Modeling and measurements of parasitic parameters for integrated power electronics modules," in *Proc. of the Nineteenth Annual IEEE Applied Power Electronics Conference and Exposition*, vol. 1, 2004, pp. 522–525 Vol.1.

[19] F. Iannuzzo and G. Busatto, "Physical CAD model for high-voltage IGBTs based on lumped-charge approach," *IEEE Transactions on Power Electronics*, vol. 19, no. 4, pp. 885–893, July 2004.

Design, Package, and Hardware Verification of a High Voltage Current Switch

Ankan De, Adam Morgan, Vishnu Mahadeva Iyer, Haotao Ke, Xin Zhao, Kasunaidu Vechalapu,
Dr. Subhashish Bhattacharya, Dr. Douglas C. Hopkins
North Carolina State University
Raleigh, NC 27606
Emails: ade@ncsu.edu , ajmorga4@ncsu.edu , vmahade@ncsu.edu
hke@ncsu.edu , xzhao20@ncsu.edu , kvechal@ncsu.edu , sbhatta4@ncsu.edu , dchopkins@ncsu.edu

Abstract—In this paper, an attempt has been made to demonstrate various package design considerations to accommodate series connection of high voltage Si-IGBT (6500V/25A die) and SiC-Diode (6500V/25A die). The effects of connecting the cathode of the series diode to the collector of the IGBT versus connecting the emitter of the IGBT to the anode of the series diode have been analyzed in regards to parasitic line inductance of the structure. Various simulation results have then been used to redesign and justify the optimized package structure for the final current switch design. The package is fabricated using the optimized parameters. A double pulse test-circuit has been assembled. Initial hardware results have been shown to verify functioning. The main motivation of this work is to enumerate detailed design considerations for packing a high voltage current switch package.

Keywords— Wide-Bandgap, Current Switch, High Voltage, Series IGBT and Diode, Silicon Carbide, Packaging.

I. INTRODUCTION

The current switch (series connected switch and diode) has found its application in various current source based converters [1]-[5]. The main advantages of using current-source based topologies can be linked to the fact that these converters usually use less number of active switches, have a more rugged natural protection (owing to the series diode), are well suited for zero-current and zero-voltage based soft switching, etc [6]-[7]. These converters are usually made using thyristors with external commutation circuits, albeit at low switching frequency. The frequency of operation can be pushed to significantly higher values by using faster devices, like IGBTs. With the advent of Silicon-Carbide Devices, singular power electronic devices can now operate beyond 10kV-15kV levels [8]-[12]. This effectively reduces the number of series connected components in the circuit, but drastically increases the dielectric and thermal stresses that develop within the power module. The non-sinusoidal nature of this stress which is punctuated with high dV/dt and dI/dt, leads to several steep constraints in terms of parasitic series inductance, shunt stray capacitance, capacitive coupling, thermal conduction, etc [13].
As previously mentioned, the current switch has an additional diode in series with the regular active switch. Packaging this structure comes with an inherent problem of dealing with peak positive (owing to the switch) and negative (owing to the diode) voltage stress within a span of less than 100µs.

This is considerably larger than regular switches for the same terminal voltage rating. In order to understand fully how these non-conventional stresses affect the current switch packaging, and therefore the overall switch operation, multi-physics simulations have been performed for various layouts of this series diode and regular switch topology based on a material selection that is capable of the required dielectric strength. Physics based accurate device models have been studied with estimated parasitic parameters in SIMPLIS-SIMETRIX. Various cross simulation results have then been used in an iterative fashion to redesign and analyze the optimized package structure for the final current switch design. The paper is divided into several subparts. First, the basic structure and common testing practices of current switch are enumerated. Two different approaches of fabrication are discussed. MAXWELL and Q3D based simulations depicting the voltage/current distribution in the module and extraction of various parasitic parameters are then studied. 3D Printed Package housing and initial double pulse results have been shown. Finally, the paper is concluded depicting the major critical design parameters and proposed design considerations.

II. PRINCIPLE STRESS AND OPERATIONS OF CURRENT SWITCH

As in most power electronic circuits, the switches usually undergo hard-switching states. In hard switching transitions, the device undergoes a state when both its voltage and current are non-negligible. This state can last for as low as sub 100ns to as high as above 100us. This leads to increased amount of power loss in the device and the overall converter. It is therefore essential to study the ill effects of this stress in the package. The main motivation of this part of the work is to demonstrate the typical stress in terms of turn-on and turn-off voltage and current as seen in the device during such hard switching conditions. Fig. 1 shows the test circuit schematic. In this test, Vin is set to 1000V. At first S1 is turned on and consequently, the inductor (Lo) current rises linearly. At some point of time S2 is turned on. As the series connected diode (D2) is reverse biased, S1 continues to conduct. At a following point in time, S1 is turned off and the non-zero inductor current is forced to flow through S2 and D2. A similar gating-pulse is again passed through S1 and S2. This forms a complete double pulse test. Overall, this results in hard turn-on and turn-off of the switch S1. Fig. 2 and 3 show the test circuit with parasitic inductance and capacitance

978-1-4673-9551-9/16 $31.00 © 2016 IEEE

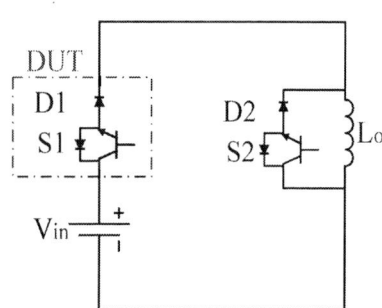

Figure 1: Double Pulse Test Circuit Schematic.

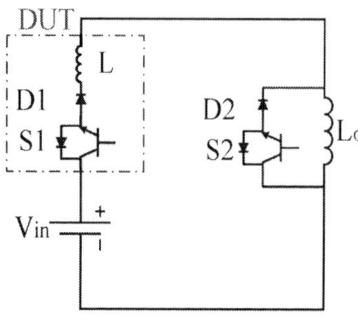

Figure 2: Test Circuit Schematic with parasitic inductance.

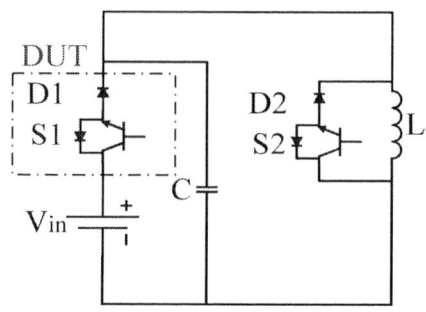

Figure 3: Test Circuit Schematic with parasitic capacitance.

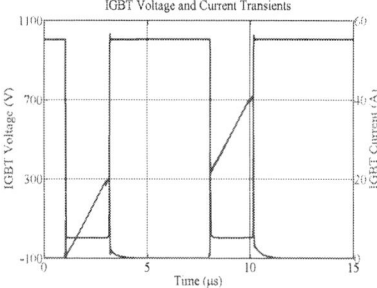

Figure 4: Double Pulse Test Results corresponding to the circuit mentioned in Fig. 1.

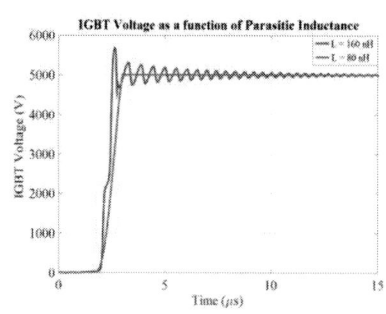

Figure 5: Double Pulse Test Results corresponding to the circuit mentioned in Fig. 2.

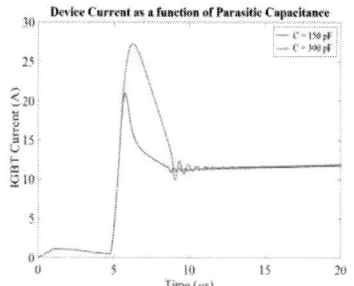

Figure 6: Double Pulse Test Results corresponding to the circuit mentioned in Fig. 3.

included, respectively. It should be noted that even though both parasitic inductance and capacitance exists at all times in the understudied setup, it has been studied separately to emphasize on contribution of individual parasitic component. The inclusion of circuit parasitics aid the understanding of the maximum allowable non-idealities the circuit can tolerate before the circuit is unable to operate within a safe operating area.

Fig. 4, 5 and 6 show the simulation results of the double pulse test of three different configurations previously mentioned. As expected, during the intervals switch S1 is turned on, the inductor current rises linearly. During the turned-off state, the inductor current free-wheels via S2 and D2. A detailed analysis of low voltage hardware testing for current switch and various hard and soft switching stresses has been presented [14]-[16]. The packaging of these devices usually adds some additional parasitic components which may lead to unwanted voltage/current overshoot. This increases device stress, losses, EMI, etc. and might lead to a dysfunctional system. The dominating series parasitic inductance (as shown in Fig. 2) is due to bond wires used to form interconnects between the contact pads of the semiconductor devices and the Cu pads of the Direct Bond Copper (DBC) substrate owing to their slender structure. Fig. 5 shows a typical increase in overshoot voltage as the parasitic inductance parameter is increased. It should be noted that this increased peak voltage could go beyond the rated blocking voltage of the device. Continuous over voltage operation can lead to eventual device malfunction. The values of parasitic inductances used are 80nH and 160nH. These values are chosen specifically to emulate a practical scenario as mentioned in later sections.

Similarly, the additional parasitic capacitance of the module can adversely affect the peak switching current seen by the device. As shown in Fig. 6, the overshoot current at C=300pF is larger than the rated current of the understudied device. Continuous operation of the device under similar stress can lead to premature failure and reduction of efficiency.

The anti-parallel diode across the IGBT is used to provide a low impedance path to the reverse recovery current of the series diode. If this path is not provided, the negative current will be forced to go through the IGBT. This can lead to unwanted behavior and eventual failure. The stress gets severe when a Si-PiN Diode is used as the series diode instead of SiC JBS Diode. During forward conduction however, this diode does not participate.

III. PRIMARY TOPOLOGIES OF CURRENT SWITCH

The arrangements of series connection of IGBT and Diode can be done in two broad ways. The Cathode of the diode can be connected to the collector of the IGBT or the emitter of the IGBT can be connected to the anode of the diode. As shown in the prior section, the series parasitic inductance in this package usually leads to unwanted stress, thereby leading to unreliable operation. It should be noted that the cathode-collector connected design leads to two distinct bond wire bridges in the conduction path, thereby considerably increasing the parasitic inductance (~160 nH) and resistance of the design. This would also lead to increased probability of module failure in the case of current magnitudes greater than or equal to the fusing currents necessary to burn open the bond wires. Fig. 7 shows

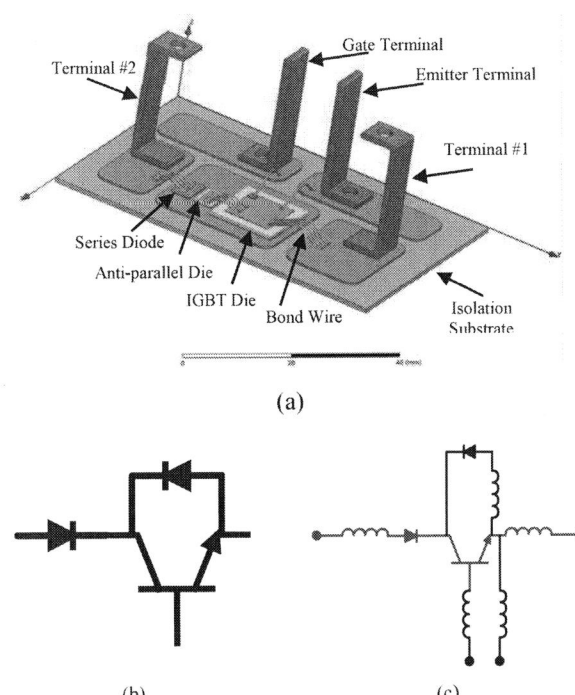

(a)

(b)   (c)

Figure 7: (a) 3D Structure of Cathode-Collector Connected Current Switch (left), (b) Schematic of the ideal circuit and (c) Circuit schematic of the circuit with parasitic inductance.

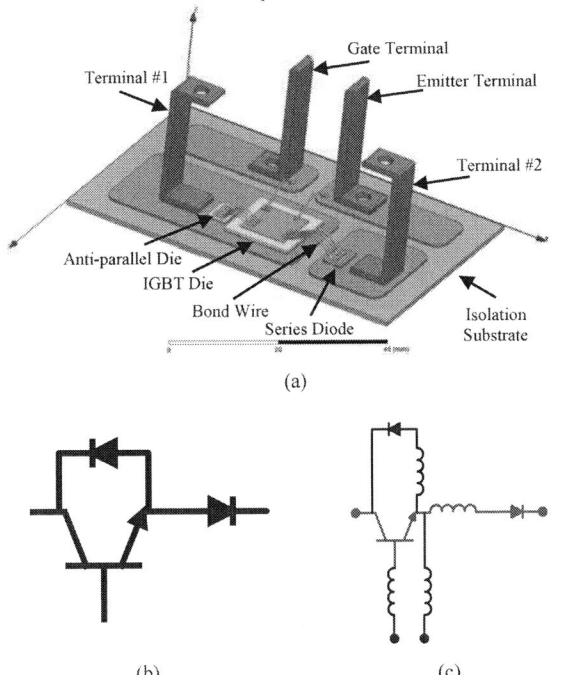

(a)

(b)   (c)

Figure 8: (a) 3D Structure of Emitter-Anode Connected Current Switch (left), (b) Schematic of the ideal circuit and (c) Circuit schematic of the circuit with parasitic inductance.

the fundamental circuit schematic, 3D Model and schematic with extracted parasitic.

Emitter-Anode connected designs on the other hand leads to a more compact design with just one series connected bond Wire

Bridge. This considerably reduces the parasitic parameters (~80nH) and has safer operating conditions as compared to the previous design. Fig. 8 shows the 3D Model depicting the preferred locations of the dies and bond wires for this configuration. During the reverse blocking turn off transition, the emitter-anode voltage changes rapidly. To make sure this doesn't affect the gating signals, the parasitic capacitance of gate and emitter signaling terminal with respect to ground should be as small as possible. The gate and emitter terminal pad should be made as optimally small as possible to reduce this capacitance. This way spurious gating transitions can be avoided. From next section onwards, only the configuration shown in Fig. 8 is considered.

IV. CURRENT DISTRIBUTION AND THERMAL STUDY

In order to obtain a basic understanding of initial thermal constraints of the package design, a Heat Transfer in Solids (conduction) study was done using COMSOL Multi-physics software. The thermal simulation analyzed the total power loss of the series current switch package based on the summation of the typical, rated conduction loss and switching loss of both the Si-IGBT and SiC-Diode and I²R losses of the DBC, as portrayed by Tab. 3.

Based on the total power dissipation conditions and the planar cross-sectional areas of the Si- IGBT and series SiC-Diode, appropriate heat source boundary conditions were established for the power die within the package. The Si-IGBT die has a planar cross-sectional area of 13.56mm x 13.56mm, and a die thickness of 0.670mm. The series SiC-Diode die has a cross-sectional area of 3.94mm x 5.70 mm, and a die thickness of 0.387mm. Therefore, specific heat flow rates were calculated for these die dimensions depending on the operating point of study as a function of switching frequency. The frequency dependent heat flow rates for each die were calculated based on the loss data of Table 3 and Equations 1 and 2 below.

Si-IGBT Loss (f_{sw}) = (Total Rated Switching Loss + Conduction Power Loss)/(Si-IGBT die area) = (0.125J * f_{sw} + 105 W)/(13.56mm x 13.56mm) [W/m^2]   (1)

SiC-Diode Loss (f_{sw}) = (Total Rated Switching Loss + Conduction Power Loss)/(SiC-Diode die area) = (0.01J * f_{sw} + 40 W)/(3.94mm x 5.70mm) [W/m^2]   (2)

It was assumed that the ambient temperature on the back-side of the DBC, where a cold-plate would be mounted, has a safe operating temperature of 45℃. During the package fabrication process, a silicone gel will be used as an encapsulate material to cover the top-side of the DBC substrate along with the power semiconductor die, the bottom third of the terminals, and the wire bonds. The silicone gel used in the simulation (Silopren* Gel 6209) has a thermal conductivity of 0.17 W/m-K and a thickness of 2.0 cm in order to electrically isolate the devices and interconnects up past the height of the wire bond loops. The thermal resistance of the silicone gel was calculated to be 0.09375 m²K/W. This resistive boundary layer was taken into consideration within the thermal simulation as well, and was applied to all top-side surfaces within the package that comes into contact with the encapsulate material. Simulations were also done without the silicone gel, simply using the thermal conductivity of air for all exposed top-side surface boundary conditions, and a negligible difference in thermal performance was observed. It should also be noted that the wire bond interconnects were removed from the 3D CAD model within COMSOL due to their negligible effect on the heat transfer compared to the entirety of the package.

V. MAXIMUM ATTAINABLE FREQUENCY

The thermal simulation is swept as a function of the overall loss in Si-IGBT and SiC-Diode at different frequencies for different isolation substrate materials. This will enable the maximum attainable frequency constraint for which the loss is same as that which leads to a peak internal temperature of 150 ℃.

According to the data sheets, typical loss data as a function of rated current and voltage is shown in Tab. 1. Using this table, the overall loss of IGBT and Diode is estimated as a function of frequency, as previously described in (1) and (2). All calculations have been done at V = 3600V and I_c = 20A. These loss data at different frequencies are used as boundary condition inputs to the COMSOL thermal model of the design. The maximum temperature for each simulation is noted, and the maximum frequency for which the peak temperature is about 150 °C is tabulated. The isolation substrates studied were Alumina, AlN, and BeO. Tab. 2 shows the material properties of these substrates

Fig. 9-11 show the simulation results of the package at the maximum permissible frequency for Alumina, AlN and BeO substrates, respectively. As expected, the package with BeO as the insulation substrate limited the peak temperature to about 150 ℃ at a reasonably higher frequency compared to the Alumina and AlN substrates.

Tab. 3 shows the maximum attainable switching frequencies of the structure using these materials for which the maximum device temperature is held to approximately 150 ℃. Fig. 12 shows the variation of Peak Temperature as a function of switching frequency. This data can be further used for an optimum converter design. The size and efficiency of an overall converter comprising of power modules, magnetic components like inductors and transformers, and capacitors varies as a function of frequency. Higher frequency operation is preferred as it reduces the size of passive components. The data presented in this paper can hence be used to model a high density power converter of desirable efficiency.

Table 1: Typical loss data as a function of rated Current and Voltage at 150 °C

Parameters	Value
Conduction Power Loss of Si-IGBT	105 W
Total Switching Loss of Si-IGBT	1.5e-4*I_c^2+1.8e-3*I_c+0.03 J
Total Switching Loss of Si-IGBT (@ V=3600V and I_c=20A)	0.125J
Total Switching Loss of SiC-Diode (@ V=3600V and I_c=20A)	0.01J
Conduction Power Loss of SiC-Diode	40 W
I^2R Losses (using Aluminum Bond Wires)	36 W

Table 2: Material properties of the studied substrates [17]

Material	Thermal Conductivity [W/m*K]	Heat Capacity [J/kg*K]	Density [kg/m^3]
Alumina	27	765	3970
AlN	165	745	3260
BeO	270	1047	3000

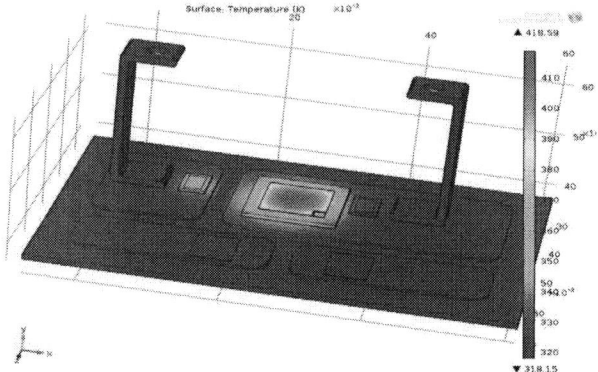

Figure 9: Temperature Distribution in the package with Alumina Substrate

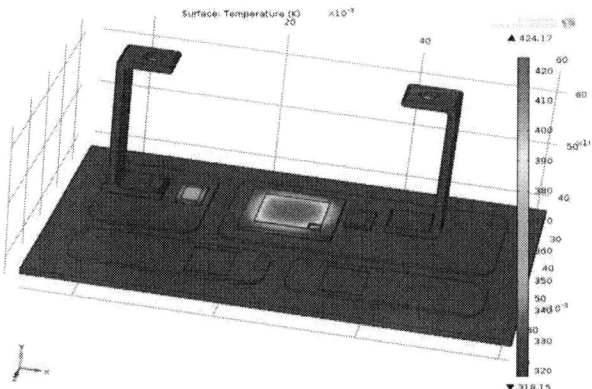

Figure 10: Temperature Distribution in the package with AlN Substrate

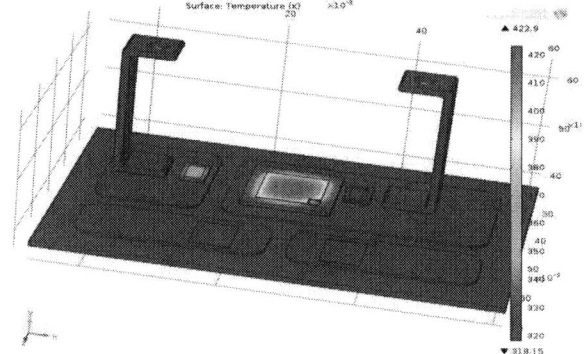

Figure 11: Temperature Distribution in the package with BeO Substrate

Table 3: Maximum attainable Switching Frequency

Material	Maximum Frequency (Hz)
Alumina	4000
AlN	10500
BeO	12000

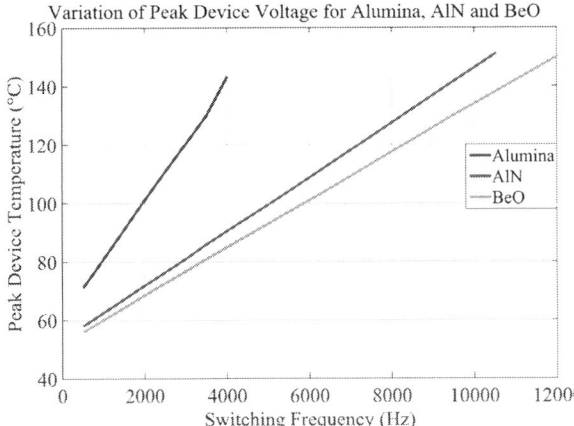

Figure 12: Peak Package Temperature as a funtion of Overall Loss of IGBT+Diode

VI. MODULE COMPONENTS AND FABRICATION PROCESSES

Fig. 13 shows various packaging components of a standard power module which will be used in the following sections for reference. The internal structure of the power stage of the package is shown in Fig. 14. The module housing is fabricated using a Stratasys PolyJet Connex350 3D printer with VeroWhitePlus RGD836 material. Curamik Alumina 12/25/12mil direct bonded copper (DBC) substrate was patterned, etched, and finally nickel-plated. The power semiconductor die were an ABB 6.5kV Si-IGBT, a CREE 6.5kV SiC-JBS diode, and an ABB 6.5kV Si-PiN diode. The Si-IGBT and Si-PiN diode have a planar cross-sectional area of 13.56mm x 13.56mm, and a die thickness of 0.670mm. The series SiC-Diode die has a cross-sectional area of 3.94mm x 5.70 mm, and a die thickness of 0.387mm. The dies were mounted down to the DBC using a Nordson EFD solder paste (Sn63/Pb37, Flux NC-D501).

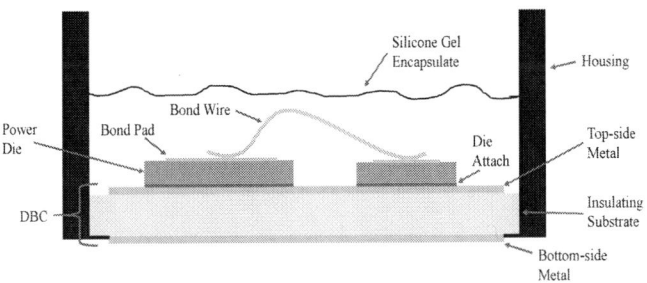

Figure 13: General Structure and components of a typical package

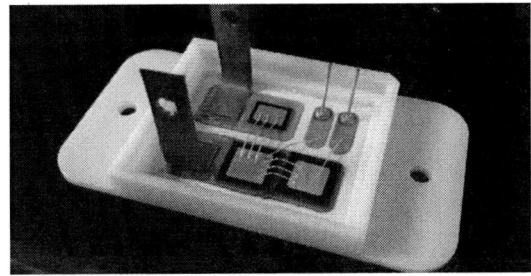

Figure 14: Custom, fabricated series current switch module

During the package fabrication process, a Hesse-Mechatronics BJ939 wire bonder was used to form the wire bonds using Heraeus 15mil Al wire. The power stage was attached inside the module housing using a 3M ceramic adhesive 18042, and lastly a Silopren* Gel 6209 silicone gel was used as an encapsulate material to cover the top-side of the DBC substrate along with the power semiconductor die, the bottom third of the copper terminals, and the wire bonds.

VII. DOUBLE PULSE TEST

The fabricated modules have been tested under low (<800V) and high (<4000V) voltages. To comprehensively characterize the modules, a double pulse test circuit has been built. Special care has been taken to make sure the loop inductances in the setup is as low as possible. Fig. 15-17 show the double pulse characteristics at 200V, 500V and 800V. This has been tested to check the basic functionality of the gate drivers and the modules in general. As expected the modules performed with minimal current (~5% @ 35A) and voltage overshoot (~7% @ 800V). The turn on and off losses were verified to closely match the ones provided in the datasheet of internal devices.

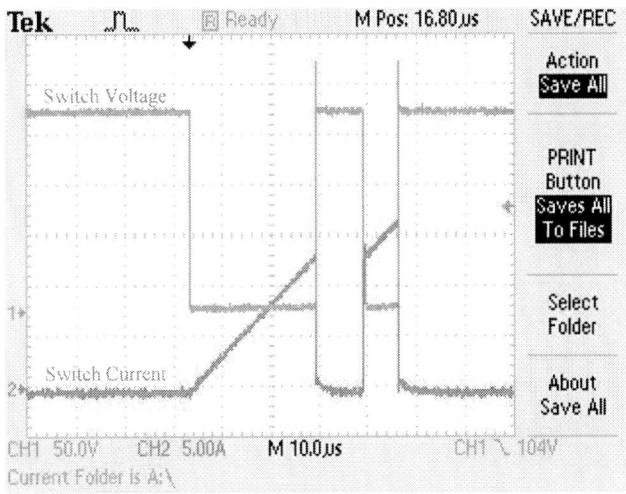

Figure 15: Double Pulse Test at 200V and peak current of 16A

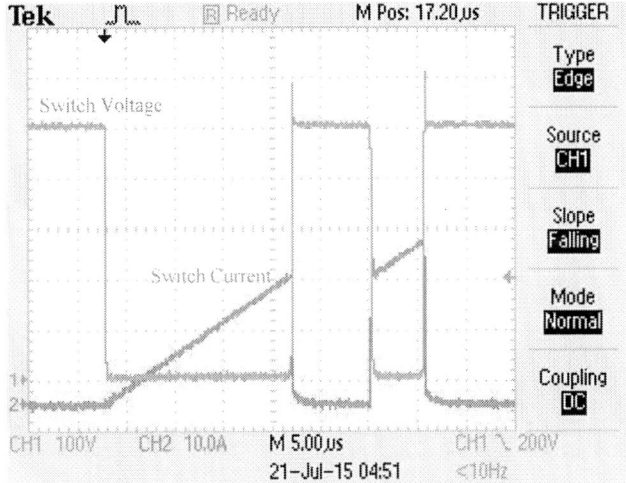

Figure 16: Double Pulse Test at 500V and peak current of 32A

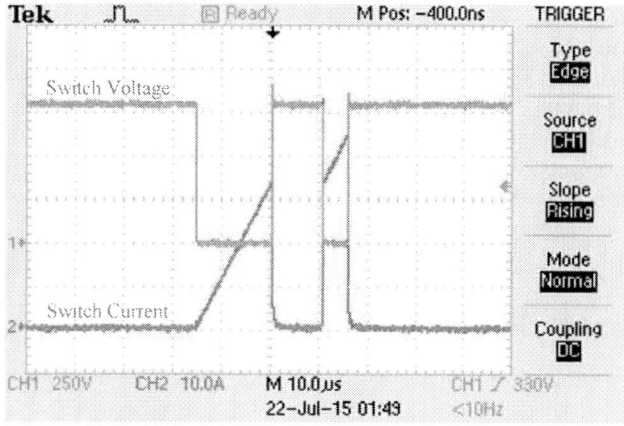

Figure 17: Double Pulse Test at 800V and peak current of 45A

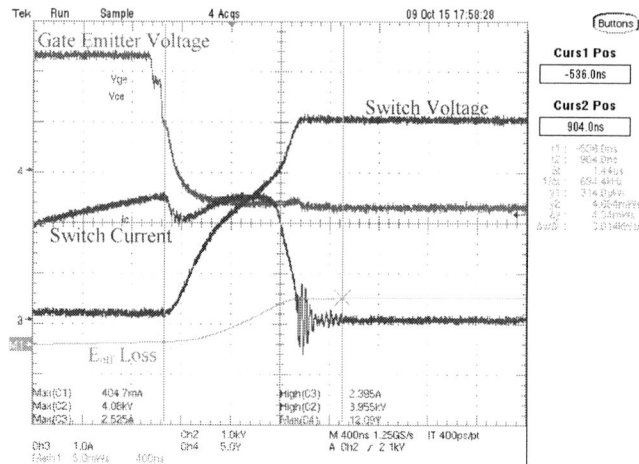

Figure 20: Turn off Test at 4000V

The modules were then tested at higher voltage to verify functionality and stability. Fig. 18-20 show the turn off transitions at 1000V, 2500V and 4000V. The losses bore a close resemblance to the ones enumerated in the device datasheet.

Fig. 21-23 show the turn-on transitions of the module at 1000V, 2000V and 2500V. The module performed as expected during both turn on and off transitions.

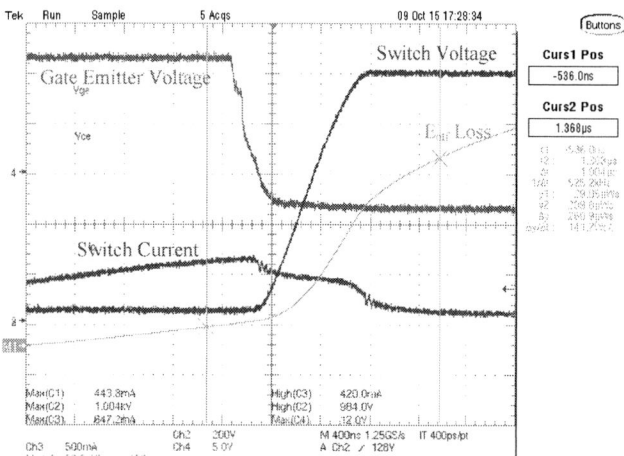

Figure 18: Turn off Test at 1000V

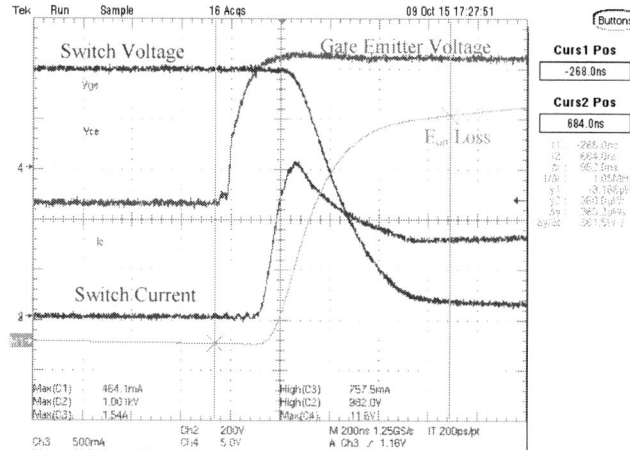

Figure 21: Turn off Test at 1000V

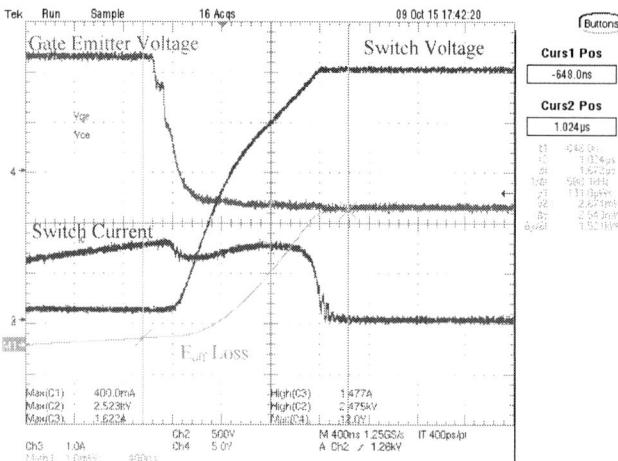

Figure 19: Turn off Test at 2500V

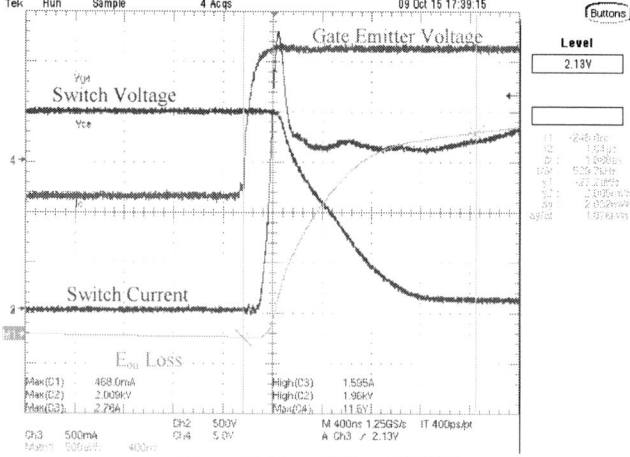

Figure 22: Turn off Test at 2000V

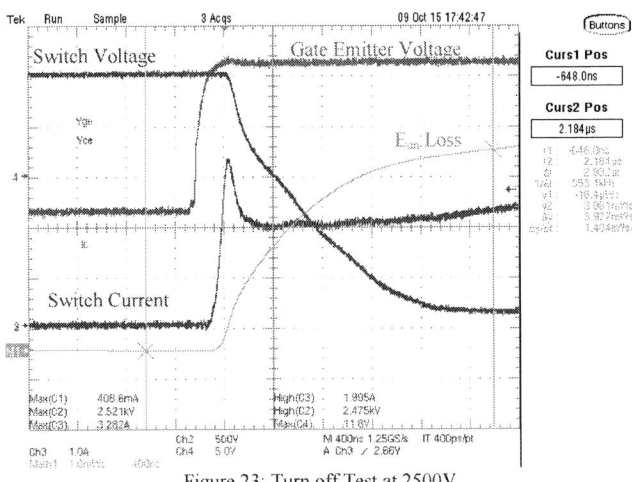

Figure 23: Turn off Test at 2500V

Fig. 24 shows the high voltage hardware setup used for this test.

Figure 24: High-V double pulse test setup of the fabricated module (A: Fabricated module, B: 10kV/10A SiC JBS diode, C: 8mH inductor, D: 125μF capacitor bank, E: gate driver, F: high-V probe, G: high-BW current probe)

VIII. CONTINUOUS PULSE TEST

The thermal behavior of the modules have been studied in a continuous pulse based converter (buck-boost) test. The main motivation is to verify if the temperature of the module remained under stable norms while performing at low voltage and rated current. Fig. 25-27 show the hardware results.

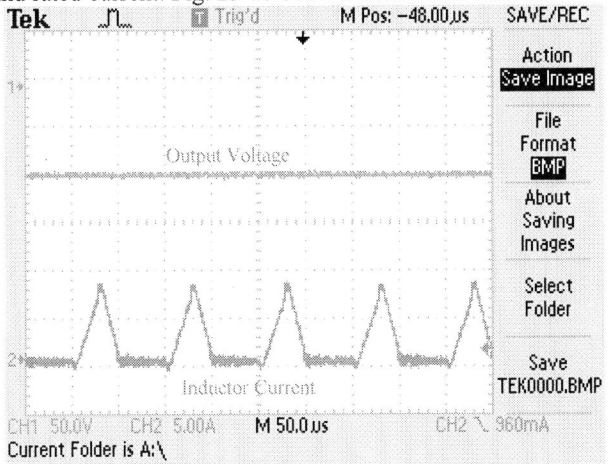

Figure 25: Converter Test at Vout = 100V, Vpeak ~200V and Ipeak ~ 9A

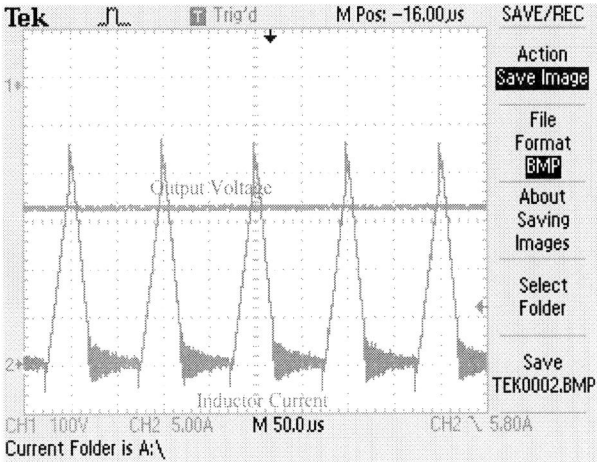

Figure 26: Converter Test at Vout = 250V, Vpeak ~500V and Ipeak ~ 23A

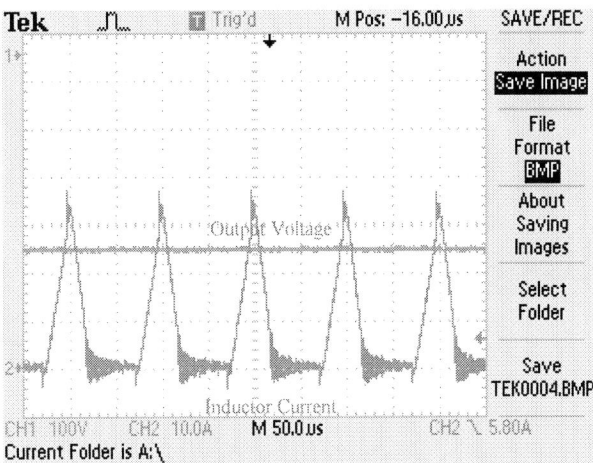

Figure 27: Converter Test at Vout = 400V, Vpeak ~800V and Ipeak ~ 36A

It should be noted that the modules have been tested under low voltage (~800V) condition. A more thorough high voltage evaluation needs to be done to make an all-round assessment of the fabricated modules. Fig. 28 shows the low-voltage hardware setup.

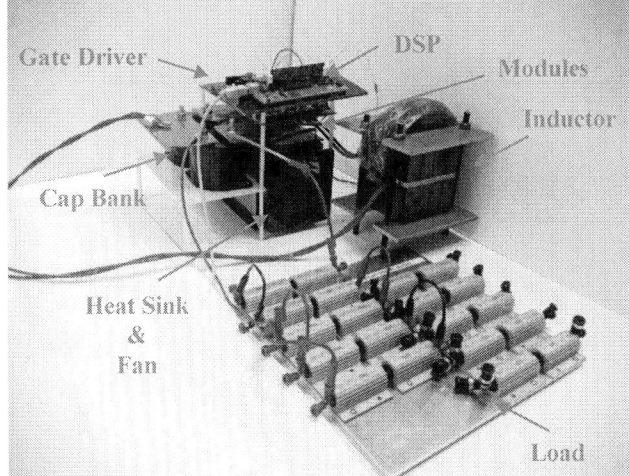

Figure 28: Buck Boost Converter Hardware Test Setup

IX. CONCLUSION

The main motivation of the paper is to demonstrate various package design considerations to accommodate series connection of high voltage Si-IGBT (6500V/25A Die) and SiC-Diode (6500V/25A die). Fundamental hard switching simulation results have been enumerated. Various ill effects of parasitic parameters have been listed. It has been shown that the series parasitic inductance leads to increased over-shoot voltage. The stray shunt capacitance on the other hand leads to increased current spikes. The advantages of connecting the emitter to the anode, as opposed to cathode to collector, are shown in terms of effective series parasitic inductance. FEA simulations were used to understand the current distribution within the package along with the thermal performance over a select switching frequency range. Custom current switch modules have been fabricated and tested under static and dynamic conditions. The test results show lower losses and device stress (voltage and current overshoot) for transients of 4kV. The continuous pulse test showed stable thermal characteristics.

REFERENCES

[1] L. Palma, "Current source converter topology selection for low frequency ripple current reduction in PEM fuel cell applications," on IEEE Rec. Conf. IECON 2013. pp.1577-1582, 10-13 Nov. 2013.

[2] B. Wu, J. Pontt, J. Rodriguez, S. Bernet, and S. Kouro, "Current-source converter and cycloconverter topologies for industrial medium-voltage drives," IEEE Trans. Ind. Electron., vol. 55, no. 7, pp. 2786–2797, Jul. 2008.

[3] T. Noguchi and S. Suroso, "Review of novel multilevel current-source inverters with H-bridge and common-emitter based topologies," presented at the IEEE Energy Convers. Congress Expo., Atlanta, GA, USA, Sep. 2010.

[4] B.R. Pelley, "Thyristor Phase-Controlled Converters and Cycloconverters ", New York: Willey, 1971.

[5] C.L. Neft, et al, "Theory and design of a 30-hp matrix converter", IEEE Trans. on Ind App, Vol.28, No.3, pp. 546-55 I, 1992.

[6] Patent Publication # US20130201733 A1, Aug 8, 2013. "Isolated dynamic current converters", by Deepakraj M. Divan, Anish Prasai, Hao Chen.

[7] T. Nussbaumer, M. Baumann, and J. W. Kolar, "Comprehensive design of a three-phase three-switch buck-type PWM rectifier," IEEE Trans. on Power Electronics, vol. 22, no. 2, pp. 551-562, Mar. 2007.

[8] K. Hatua, S. Dutta, A. Tripathi, S. Baek, G. Karimi, and S. Bhattacharya, "Transformer less Intelligent Power Substation design with 15kV SiC IGBT for grid interconnection", in proc. 2011 IEEE Energy Conversion Congress and Exposition, Phoenix, AZ, 2011, pp. 4225-4232.

[9] A. Kadavelugu, S. Bhattacharya, S. H. Ryu, E. V. Brunt, D. Grider, A. Agarwal, and S. Leslie, "Characterization of 15 kV SiC n- IGBT and its application considerations for high power converters", in proc. 2013 IEEE Energy Conversion Congress and Exposition, Denver, CO, pp. 2528-2535.

[10] D. C. Patel, A. Kadavelugu, S. Madhusoodhanan, K. Hatua, S. Leslie, S. H. Ryu, D. Grider, A. Agarwal, and S. Bhattacharya, "15 kV SiC IGBT Based Three-Phase Three-Level Modular-Leg Power Converter", in proc. 2013 IEEE Energy Conversion Congress and Exposition, Denver, CO, pp. 3291-3298.

[11] S. Madhusoodhanan, K. Hatua, and S. Bhattacharya, "Control technique for 15 kV SiC IGBT based active front end converter of a 13. 8 kV grid tied 100 kVA Transformerless Intelligent Power Substation", in proc. 2013 IEEE Energy Conversion Congress and Exposition, Denver, CO, pp. 4697-4704.

[12] L.Y. Yang, T.F. Zhao, J. Wang and A.Q. Huang, "Design of analysis of a 270kW five-level DC/DC converters for solid state transformer using 10kV SiC power devices," Proc. IEEE 38th Power Electronics Specialists Conference, June 2007, pp. 245 – 251.

[13] A. Kadavelugu, Bhattacharya, S. ; Leslie, S. ; Sei-Hyung Ryu ; Grider, D. ; Hatua, K., "Understanding dv/dt of 15 kV SiC N-IGBT and its control using active gate driver", in proc. 2014 IEEE Energy Conversion Congress and Exposition, Pittsburgh, PA, pp. 2213-2220.

[14] De, A.; Roy, S.; Bhattacharya, S.; Divan, D. M., "Performance Analysis and Characterization of Current Switch under Reverse Voltage Commutation, Overlap Voltage Bump and Zero Current Switching", Proceedings of the 28th Applied Power Electronics Conference and Exposition (APEC 2013), Long Beach, California, USA, March 17-21, 2013.

[15] De, A.; Roy, S.; Bhattacharya, S.; Divan, D. M., "Characterization and performance comparison of reverse blocking SiC and Si based switch", IEEE Workshop on Wide Bandgap Power Devices and Applications (WiPDA 2013), Columbus, Ohio, USA, Oct. 27-29, 2013.

[16] B. J. Baliga, Fundamentals of Power Semiconductor Devices: Springer, 2008, Chap. 9.

[17] W. W. Sheng; R. P. Colino, Power Electronic Modules Design and Manufacture: CRC Press, 2004, Chap. 3.

Investigation of short circuit in a IGBT power module with Three-Level Neutral Point Clamped Type 2 (NPC2, T-NPC, mixed voltage) topology

Kevin Lenz, Vladan Jerinic and Reiner Hinken
Danfoss Silicon Power
Flensburg, Germany
kevin.lenz@danfoss.com, vladan.jerinic@danfoss.com

Abstract— during standard robustness evaluations of a NPC type 2 IGBT power module a non-described effect in a Three Level NPC 2 topology was observed. A full description of that effect can help to protect IGBT power modules in this topology against blow ups and can help in a post mortem analysis to understand the reason for this. With a full understanding of that effect the lifetime of the inverter can be increased and the time and costs for qualification reduced.

Keywords—Three Level; NPC 2; Neutral Point Clamped; Short Circuit; IGBT; Power Module; desaturation; commutation;

I.   INTRODUCTION

Three level topologies are implemented in an increasing number of applications. Especially Solar, UPS and Active Filter [1] use the benefit of low switching losses for a high efficient inverter or use higher switching frequencies to reduce the filter size. Motor drive and Wind turbine applications further benefit of lower dv/dt and bearing currents what can extend the lifetime of the motor or generator [2]. Better shape of sinusoidal current, smaller common mode voltage and less dv/dt can help saving money in filter design for all applications. In literature [1], [3] two types of neutral point clamped (NPC) three-level topologies are described.

II.   THREE LEVEL NPC-2 TOPOLOGY

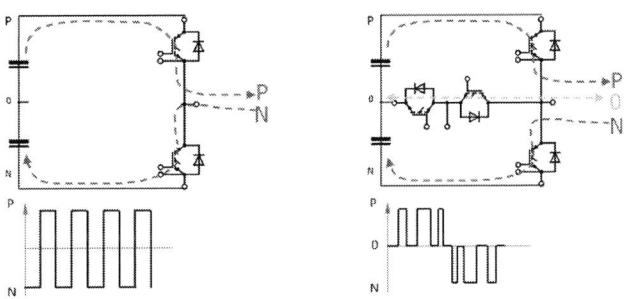

Fig. 1. topologies with current flow and principal flow of output voltage; (left) Two Level Halfbridge; (right) Three Level NPC2

In the NPC type 2 topology (also known as NPC2, T-NPC, mixed voltage NPC or MNPC) a bidirectional switch connects the neutral point (0) to the output. This switch is realized with two IGBTs including FWD in series. The collectors (common collector) or the emitters (common emitter) of the IGBTs need to be connected.

The IGBTs and diodes in the bidirectional switch can have a lower blocking voltage than the halfbridge, for example 650V at 1200V halfbridge or 1200V at 1700V halfbridge. This helps further to keep switching and conduction losses of the system low and increases the efficiency.

III.   IGBT POWER MODULE CONCEPT P3L

The Danfoss P3L IGBT power module package is known as a standard in multi-level power application. A full neutral point clamped Type 2 (NPC2) topology with low stray inductance commutation paths offers the three level benefits for high power applications. The same P3L package is also able to accommodate Type 1 (NPC1) topologies [4].

Fig. 2. Danfoss Silicon Power's P3L module

The internal module topology was optimized to create an improved NPC2 topology. To minimize inductive loop in the switches with the highest switching rate the topology is split into a high side bridge (T1, D1, T2, D2) and a low side bridge (T4, D4, T3, D3). To address the reverse recovery current of the diodes D2 and D3 protection diodes D5 and D6 are required. They have to handle only the reverse recovery current of D2 respectively D3 and can therefore be very small compared to the other semiconductors, e.g. 90A instead of 900A.

978-1-4673-9551-9/16 $31.00 © 2016 IEEE

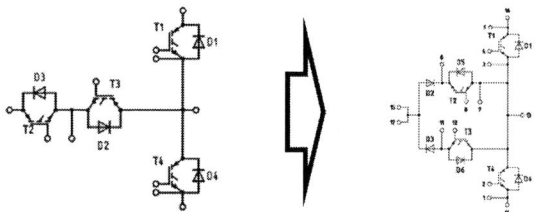

Fig. 3. for higher power density optimized NPC2 topology

The module is available in different current/voltage versions:

IGBT current	900A	700A
T1/T4	1200V	1700V
T2/T3	650V	1200V
name	DP900N1200TU104204	DP700N1700TU104202

Fig. 4. available P3L modules in NPC2 topology

IV. SHORT CIRCUIT MODES

To validate the robustness of the module several short circuit tests have been performed. Short circuit tests are very important and common in design evaluation because they simulate high currents in application.

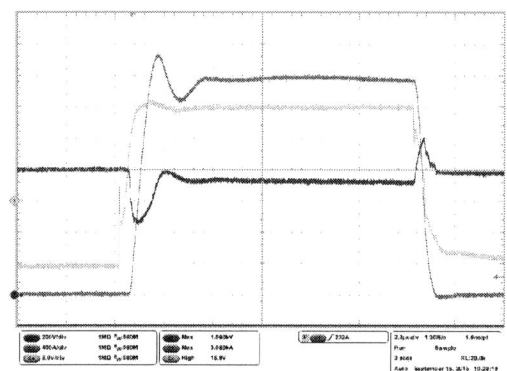

Fig. 5. Short Circuit type 1 (SC1) of T1, Ch2 (blue line): V_{ceT1}, Ch3 (red line): I_{T1}, Ch4 (green line): V_{geT1}

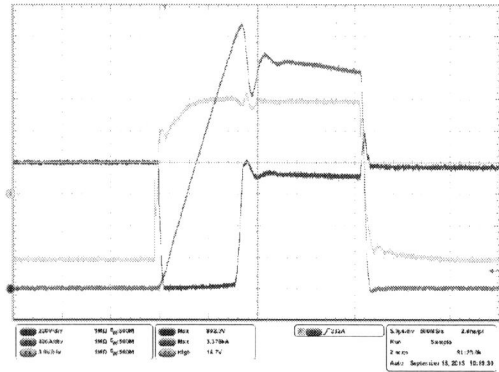

Fig. 6. Short Circuit type 2 (SC2) of T1, Ch2 (blue line): V_{ceT1}, Ch3 (red line): I_{T1}, Ch4 (green line): V_{geT1}

There are mainly two different types of short circuit (SC) modes for an IGBT - they are named as SC1 and SC2 [5]. SC1 (Fig. 5) is the case when an IGBT turns on to an existing short circuit. This case is simulated with a short cable (a few nH) between AC and DC terminals. SC2 (Fig. 6) is the case when an IGBT is already turned on before the short circuit occurs. This case is simulated with a cable (a few hundred µH) between AC and DC terminals.

V. TEST SETUP

Following components were used in the test:

- DC capacitor bank C1 = C2 = 2x400uF/1100V, film capacitors (see Fig. 7)

- Gate driver evaluation board CONCEPT 2SC0435T2D0-17 (see Fig. 7)

- Short-circuit wire (L=1uH)

- Danfoss Silicon Power (DSP) IGBT module DP900N1200TU104204 (DUT)

The Danfoss Silicon Power IGBT power module (DUT = Device Under Test) was connected to the DC capacitor bank. Current transducers were placed around the positive terminal of the busbar (I_{DC+}, Fig. 7) to measure the T1 current and the short-circuit wire (I_L, Fig. 7) for measuring the load. The gate driver board was mounted on top of the DUT and connected via galvanic isolation NI6602 to the pulse generator. The DUT output terminal 13, was shorted with a wire to the busbar minus terminal.

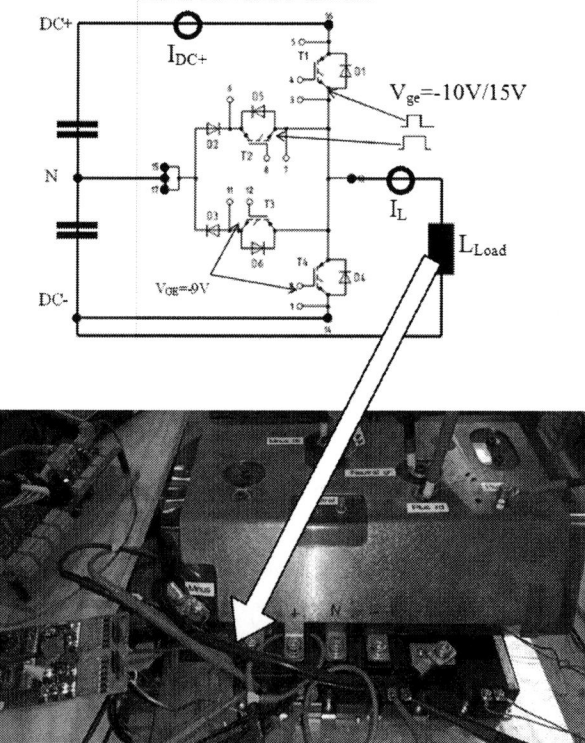

Fig. 7. test setup initial test

VI. INITIAL TEST RESULTS

Different short circuit test have been performed to evaluate the robustness of the module. Most of the test showed expected behavior. Fig. 5 and Fig. 6 show exemplary test results where T1 is switched onto a SC1 and SC2. Here the IGBT T1 was switched on until desaturation. After turning off the IGBT the current commutates into the freewheeling diode D4. For a typical IGBT power module characterization this test has to be done for all IGBTs. T2 has to commutate to D4, T3 and T4 to D1.

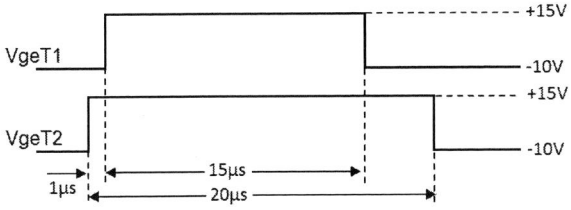

Fig. 8.   Pulse pattern T1 and T2

In regards to a typical three level PWM pattern a test with two turned on devices was done too. In this mode T2 and T1 were conducting at the same time (Fig. 8). This simulates a typical pulse pattern of a three level topology [6]. The IGBT T2 was switched on for 1µs and not in short circuit mode when T1 turned on.

As expected, after a few micro seconds T1 went into desaturation. The current through T1 was not rising further, it remains constant. The load current I_{load} however continued to rise.

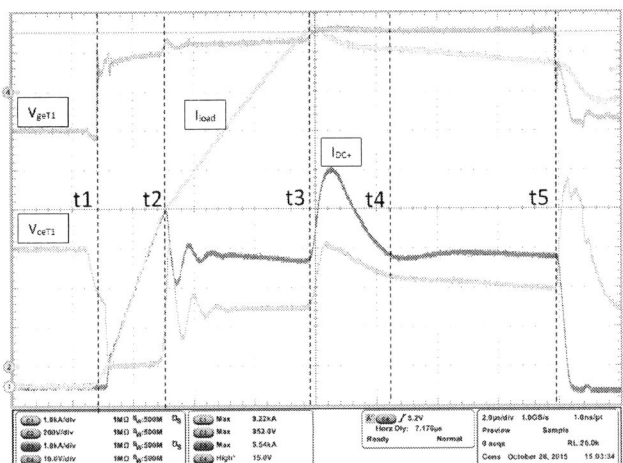

Fig. 9.   initial test result, Ch1 (yellow line): Loadcurrent I_{load}, Ch2 (blue line): V_{ceT1}, Ch3 (lila line): $I_{dc+}=I_{T1}$, Ch4 (green line): V_{geT1}, $R_{gonT2}=5R$, $R_{goffT2}=15R$, $R_{gonT1}=0R56$, $R_{goffT1}=3R3$, $V_{DC}=600V$, $T_{Junction}=25°C$

t1: Gate T1 switch on

t2: T1 in desaturation. But the load current is rising further until t3. V_{ceT1} is $V_{dc}/2=300V$ in that period.

t3: load current I_{load} shows a desaturation effect. A current peak in T1 is measured. V_{ceT1} rises for a short time up to $V_{dc}=600V$.

t4: current through T1 (I_{dc+}) falls to the short circuit level of $V_{dc}/2$

t5: V_{geT1} is turned off; I_{dc+} falls to zero amps.

It was not expected that the load current rise further after T1 is desaturated. Furthermore the current peak between moment t3 and t4 were not expected. The test was repeated at junction temperature of 150°C. Due to the semiconductor behavior the value of the short circuit current was less [7]. But both effects, the rising of I_{load} after desaturation of I_{T1} and the current peak after t3, were observed too. To discuss the reason for that effect more values have been measured. The results are shown in this paper.

VII. INVESTIGATION

To identify the reason for the described effect the test setup (Fig. 7) was used but additional signals were measured. In the investigation all measurements were done at 25°C. Additional to the current in T1 and in the load the currents through Neutral Point (T2, D2) and D4 were measured to get a picture of the current sharing in the IGBT power module. Additional to the Collector Emitter Voltage of T1 the voltage of D2, T2 and D4 was measured.

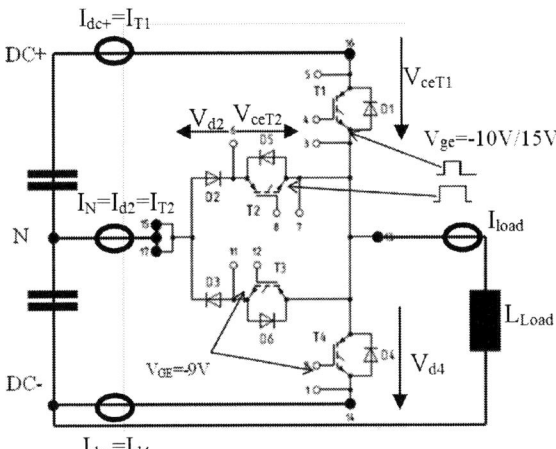

Fig. 10. Test setup investigation

The measurements (Fig. 11) have shown that at the moment when T1 desaturates the IGBT T2 turns on too. Now both T1 and T2 conduct current and the load current rise further. At the moment where T2 desaturates the diode D4 starts to conduct current in the same time I_{T2} decrease.

Further measurements show that this happens also in a standard two level IGBT halfbridge. In a halfbridge the current can commutate into the freewheeling diode at the moment where the IGBT desaturates. In a three level topology this commutation current can commutate additionally into the opened IGBT (in this case T1). Therefore current is flowing in T1, T2 and D4 at the same time. This explains the current-dip in Fig. 9 between t3 and t4.

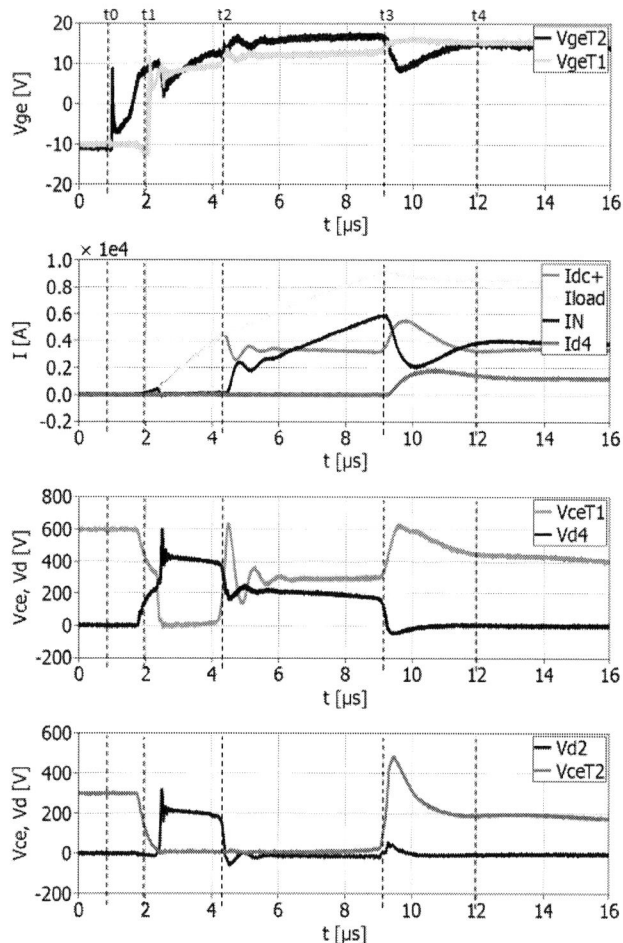

Fig. 11. investigation result, pane 1: V_{geT1} (green line), V_{geT2} (black line), pane 2: I_{dc+} (lila line), Iload (yellow line), I_N (black line), I_{dc-} (red line), pane 3: V_{ceT1} (blue line), V_{d4} (black line), pane 4: V_{ceT2} (red line), V_{d2} (black line); R_{gonT2}=5R, R_{goffT2}=15R, R_{gonT1}=0R56, R_{goffT1}=3R3, V_{DC}=600V, $T_{Junction}$=25°C, no gate clamping

t0: Gate T2 switch on.

t1: Gate T1 switch on

t2: T1 in desaturation, as expected the current I_{T2} doesn't rise further on. But the load current is rising further until t3. V_{ceT1} is $V_{dc}/2$=300V in that period. V_{ceT2} rise until $V_{dc}/2$. Now T2 can take over current. The current doesn't commutate from T1 in T2 – both IGBTs conducting right now and the load current is the sum of I_{T1} and I_{T2}.

t3: T2 desaturates. I_{T2} and I_{load} don't rise further on. The current in T2 falls and commutate in T1 and D4.

t4: I_{T1} and I_{T2} are in desaturation with each $V_{dc}/2$. The diode D4 conducts further on.

The behavior of the system after the time t4 depends on the turn off sequence of the IGBTs. The IGBT which turns off first commutates the current in the other IGBT and the diode D4. When both IGBTs are turned off the diode will take over the current.

Further measurements on other NPC2 IGBT power modules have shown that it the described effect appears in those IGBT power modules too. So it's not a bug or a feature only of the so called P3L module.

VIII. SUMMARY

It is always important to know the worst case scenario of different parameters like the maximum current in a system. The investigation has shown that the parallel short circuit of T1 and T2 leads to very high currents in a NPC 2 IGBT power module, probably the highest possible in the system. It is also shown that the diode D4 will take over the sum of the short circuit currents of T1 and T2. The knowledge of the described effect can help to design robust and reliable NPC 2 IGBT power modules and inverter.

On the IGBT power module side for example there are approaches to reduce the size of the diode D1 and D4 in applications where the reactive power is low or even zero. If there is low or no reactive power in the system the diodes D1 and D4 are not stressed hard thermal wise. Theoretical the die size could be smaller than the other semiconductors to safe costs and to get a higher power density in the power module. But with the knowledge that the maximum current in that diodes will occur during the short circuit of T1 and T2 a shrink of these diodes should be verified.

On the inverter side cables, DC link and current sensors of the inverter have to withstand the sum of short circuit currents of T1 and T2. The influence of the gate driver on that effect has to be investigated. Clamping [8] for example can influence the behavior of the IGBTs in the short circuit mode.

If this effect is not considered in the qualification loop the knowledge of the behavior of three level NPC 2 topology in different short circuit mode can help to investigate IGBT modules after a blow up. For example if T1, T2 and D4 show over-current marks the described short circuit behavior could be the reason. It is recommended to test this mode in all NPC 2 applications in the module and the inverter qualification.

As expected the increased number of semiconductors in a three level topology compared to a two level halfbridge makes life not easier. There are more commutation paths which make the full understanding of behavior in all cases more sophisticated. This paper gives a better understanding for one of these cases.

ACKNOWLEDGMENT

Thanks to our colleagues Marco Bäßler and Timo Stuber. They discussed with us the effects and measurements from other IGBT modules.

REFERENCES

[1] T. B. Soeiro, M. Schweizer, J. Linner, P. Ranstad, W. Kolar, Comparison of 2- and 3-level Active Filters with Enhanced Bridge-Leg Loss Distribution

[2] Yaskawa Product Application Note, Motor Bearing Current Phenomen and 3-Level Inverter Technology

[3] A. Nagae, I. Takahashi, H. Akagi, A new neutral-point-clamped PWM inverter, IEEE 1981

[4] K. Lenz, J. Rudzki, F. Osterwald, U. Pandey, M. Poech, New IGBT Power Module concept in NPC Topology with Extended Reliability, PCIM 2015

[5] J. Lutz, H. Schlangenotto, U. Scheuermann, R. De Doncker, Semiconductor Power Devices, ISBN 978-3-642-11125-9

[6] I. Staudt, Semikron Application Note AN-11001

[7] V. Bolloju, J. Yang, Influence of Short Circuit conditions on IGBT Short circuit current in motor drives, Whitepaper

[8] O. Garcia, J. Thalheim, N. Meili, Safe Driving of Multi-Level Converters Using Sophisticated Gate Driver Technology, PCIM 2013

Closed-Loop Design and Time-Optimal Control for a Series-Capacitor Buck Converter

Timur Vekslender, *Student Member, IEEE*, Ofer Ezra, *Student Member, IEEE*, Yevgeny Bezdenezhnykh, *Student Member,* and Mor Mordechai Peretz, *Member, IEEE*

The Center for Power Electronics and Mixed-Signal IC, Department of Electrical and Computer Engineering
Ben-Gurion University of the Negev, P.O. Box 653, Beer-Sheva, 8410501 Israel
timurv@post.bgu.ac.il, oferez@post.bgu.ac.il, yevgenyb@post.bgu.ac.il, and morp@ee.bgu.ac.il
http://www.ee.bgu.ac.il/~pemic

Abstract— This paper explores the large-signal and small-signal dynamics of a series-capacitor (SC) buck-type converter and introduces an optimal closed-loop control scheme to accommodate both the steady-state and transient modes. As opposed to a conventional buck converter, where time-optimal control is realized by a single on-off cycle, in the SC-buck topology there is a need to distribute the switching phases to satisfy the charge-balance of the flying capacitor. The new control method hybrids a voltage-mode small-signal controller for steady-state operation and a non-linear, state-plane based transient-mode control scheme for load transients. A detailed principle of operation of the SC-buck converter is provided and explained through an average behavioral model and state-plane analysis. The operation of the controller is experimentally verified on a 12W 12V-to-1.5V converter, demonstrating voltage-mode control operation as well as time-optimal response for load transients.

Keywords— Time-optimal control, state-space control, dc-dc converters, voltage regulation

I. INTRODUCTION

In recent years, a significant effort is made to enhance the performance of voltage regulator modules (VRMs) for high-performance ICs that operates with low supply voltage and high current. Tighter voltage regulation, high efficiency, and accommodating load transient are key factors in the design of the switch-mode power supplies (SMPS), in particular for high step-down conversion ratio applications. Several converter topologies and circuit extensions have been discussed in the literature to minimize the size of passives and improve the dynamics of the VRM. One direction of VRM implementation is based on multi-phase interleaved converters, allowing high frequency operation and size reduction at the cost of complex control for current sharing [1]-[2]. Another approach is by multi-level converters where the lower component stress allow better sizing of the components and efficiency improvements [3].

The series capacitor (SC) buck converter, also known as a double step-down two-phase buck converter, originally presented in [4] and recently revised in [5]-[6], merges an interleaved buck converter with a switched-capacitor front-end and by doing so allows high-frequency operation in the MHz range and better system dynamics with reduced stress on the components. Additional attractive features of the SC-buck converter are natural current balancing between phases and effectively doubles the pulse width, which make it suitable for large conversion ratio applications.

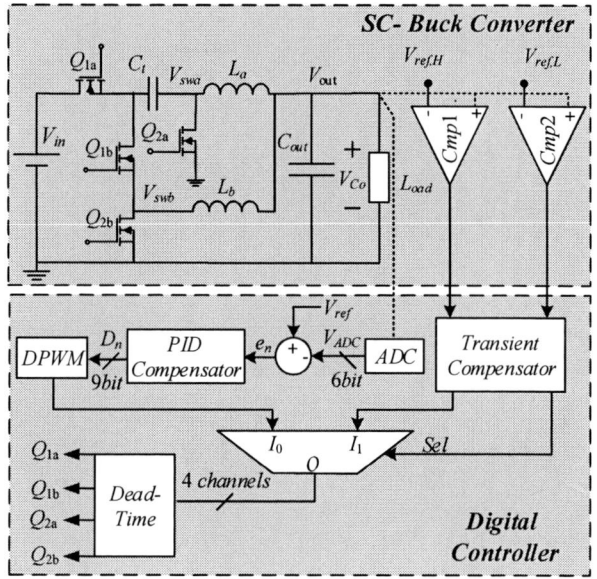

Fig. 1. SC-buck converter and a hybrid controller

Recent studies have quantified the attributes of the SC-buck topology at high frequency [7] and demonstrated improved light load efficiency when operating in DCM [8]. Further extensions presented a two-phase, four-inductor, converter which emphasizes its current balancing feature when the power is distributed between multiple phases [9]. In spite of all the major benefits for this topology, neither closed-loop operation nor dynamic analysis for controller design have been investigated to-date. It would be further advantageous to examine the converter suitability for time-optimal controller assignment in order to be considered attractive for VRM applications.

The objective of this study is therefore to investigate the dynamic features of a SC-buck converter and *to introduce an optimal closed-loop controller that hybrids a small-signal controller for steady-state operation and a time-optimal control scheme for load transient,* as detailed in Fig. 1. In this study, two modeling approaches are presented, the first by an average-behavioral model to examine the control-to-output response and design a small-signal voltage-mode controller. The second technique is to obtain a state-space representation which will be the basis for the design of a non-linear time-optimal controller for loading transient. The latter is

978-1-4673-9551-9/16 $31.00 © 2016 IEEE

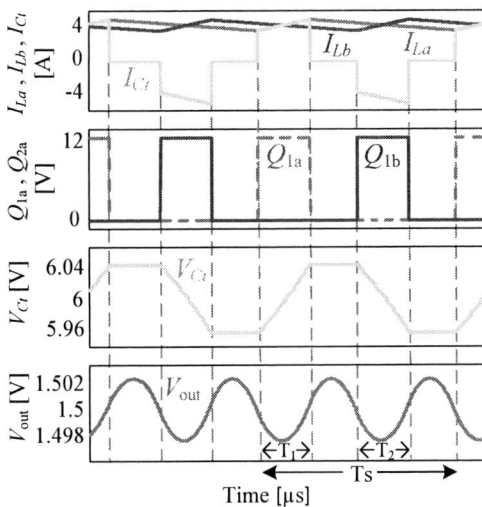

Fig. 2. Typical waveforms of a SC-buck converter. Case study of 12V-1.5V.

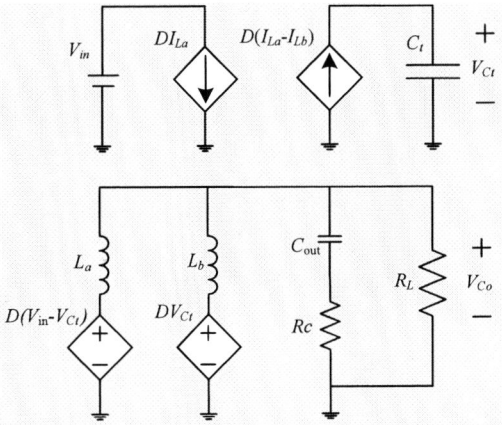

Fig. 3. Average-behavioral model of the SC-buck converter

significantly more challenging in the context of the SC-buck converter since it involves combined operation of two pseudo-balanced converters.

The rest of the paper is organized as follows: Section II describes the steady-state operation of the converter, extracts an average-behavioral model representation, and provides a design procedure of a small-signal voltage-mode compensation scheme. Section III provides a state-space analysis and a time-optimal control of the SC-buck converter. Experimental validation is presented in Section IV. Section V concludes the paper.

II. Steady-State Operation and Voltage-Mode Controller

The description of the SC-buck converter is assisted by topology and typical waveforms, as shown in Fig. 1 and 2, respectively. The steady-state operation is similar to that of an interleaved, two-phase buck converter with a slight difference that the back-end converter is fed by a flying capacitor C_t that is charged to approximately $V_{in}/2$. The duty ratio for both phases (a and b) is identical and each phase is time-interleaved with a 180-degree phase delay. As a result, four switching-

states are identified. State-1, Q_{1a} and Q_{2b} are on, resulting in V_{swa} equals $V_{in}/2$ and the inductor current I_{La} ramps up with a slope of $(V_{in}/2 - V_{Co})/L_a$ and I_{Lb} ramps down with a slope of $-V_{Co}/L_b$. In state-2, Q_{1a} is turned off, Q_{2a}, Q_{2b} are on, and the operation resembles a conventional buck converter in off state. In state-3 Q_{2a} and Q_{1b} are on, the flying capacitor C_t, acts as the source, and the inductor current I_{Lb} ramps up with a slope of $(V_{in}/2 - V_{Co})/L_b$ while I_{La} ramps down with a slope of $-V_{Co}/L_a$. State-4 is identical to state-2. Charge balance of C_t is naturally maintained by this operation allowing both charge and discharge action per cycle [4], this naturally stabilizes V_{Ct} to half the input voltage.

Following the switching sequence and assuming CCM operation, the behavioral operation of the converter is obtained by averaging [10]-[11]. The average voltage across the inductors, $\langle v_{La} \rangle$, $\langle v_{Lb} \rangle$ and the average capacitor current $\langle i_c \rangle$ can be expressed as:

$$\langle i_c \rangle = D_1 i_{La} - D_2 i_{Lb}, \qquad (1)$$

$$\langle v_{La} \rangle = D_1(V_{in} - V_{Ct}) - V_{out}, \qquad (2)$$

$$\langle v_{Lb} \rangle = D_2 V_{C_t} - V_{out}, \qquad (3)$$

where $D_1 = T_1/T_s$ and $D_2 = T_2/T_s$ are the duty ratios related Q_{1a} and Q_{2a} conduction time, respectively. V_{out} is the output voltage.

Fig. 3 shows a graphical representation for an average-behavioral model as described by (1)-(3). Assuming that $D_1 = D_2$ and applying small-signal linearization, the full control-to-output dynamic expression can be obtained. To simplify the expressions, V_{Ct} is assumed constant by small-ripple approximation [12], resulting in a control-to-output expression of the form:

$$\frac{v_{out}}{d}(s) = \frac{\dfrac{V_{in}}{2}\left(sC_o R_c + 1\right)}{s^2 \dfrac{C_o\left(L_a \| L_b\right)}{R_L}\left(R_L + R_c\right) + s\left(C_o R_c + \dfrac{L_a \| L_b}{R_L}\right) + 1}. \qquad (4)$$

As can be observed from (4), this expression is similar to the control-to-output response of a classical 2-phase buck converter. The frequency response of (4) is presented in Fig. 4 along with the required compensator that its design is detailed in the next sub-section.

Closed-Loop Discrete-Time Compensator Design

To satisfy the requirements for loop-gain stability and high bandwidth, the crossover frequency f_c of the closed-loop system is chosen to be greater than the double pole frequency by approximately 50% while the target phase margin is set above 45°. Based on the control-to-output behavior, the setting of the target parameters in this way guarantees suitability for PID compensation scheme [13]. The extraction of the PID coefficients (a,b,c) is based on the methodology that has been presented in [14] with minor adjustments to a frequency-domain design. The design procedure is as follows:

978-1-4673-9551-9/16 $31.00 © 2016 IEEE

1) Specify the crossover frequency f_c and the phase margin of the desired closed-loop $A_{CL}(s)$ frequency response based on a knowledge of the control-to-output response $A(z)$.

2) Obtain the denominator of $A_{CL}(z)$ by a pole-zero matching s-to-z transformation.

3) Derive the numerator of $A_{CL}(z)$ such that the closed-loop response is of second order system [14].

4) Derive the transfer function of an ideal compensator $B_{ideal}(z)$ that yields the desired closed-loop response.

5) Obtain the response of a template PID compensator $B_{PID}(z)$ from the first three samples of the ideal compensator $B_{ideal}(z)$ by evaluation of difference equations.

The designed PID compensator has been validated through Matlab simulations as a full closed-loop system with a 12-to-1.5 V SC-buck converter, operating at 1.25 MHz ($L_a = L_b = 0.6$ µH; $C_f = 10$ µF; $C_o = 50$ µF). The target closed-loop parameters were crossover frequency of 60 KHz and phase margin of 45°. Fig. 4 shows the frequency response of the converter (blue), the $1/B$ of the PID compensator (red) and the loop-gain (green). It should be noted that higher bandwidth could be obtained by higher gain settings of the compensator or by different target specifications. However, since large transients are accommodated by an optimal controller, a conservative crossover frequency of 50% beyond the double pole location has been satisfactory for steady-state regulation.

III. STATE-SPACE REPRESENTATION AND TIME-OPTIMAL CONTROL

To enhance the load transient performance of the SC-buck converter, it is essential to obtain the information of the possible state-trajectories that are available for the converter to recover to the new steady-state operating point after a load transient. Since in this study, a small-signal compensator is assumed for steady-state operation, the objective of the transient controller is to minimize the recovery time of the converter and as a result minimizing the output voltage deviation, i.e. time-optimal control [15]-[18]. Unlike a conventional two-phase buck converter, in the SC-buck converter case, charge balance of the flying capacitor must be satisfied during the transient time to allow smooth transition back to the steady-state operation. This implies that the 'simple' on-off time-optimal cycle as carried out by many applications would not hold in this case and, as a matter of fact, would worsen the overall performance.

To realize the required switching sequence, the first task is to map the behavior of the state-variables with respect to the new loading conditions [19]-[20], then the required switching sequence can be derived from the possible trajectories of the state-variable. The state equations for state-1 can be expressed as:

$$\frac{dV_{C_o}}{dt} = \frac{1}{C_o}\left(I_{L_a} + I_{L_b} - \frac{V_{C_o}}{R_{out}}\right), \tag{5}$$

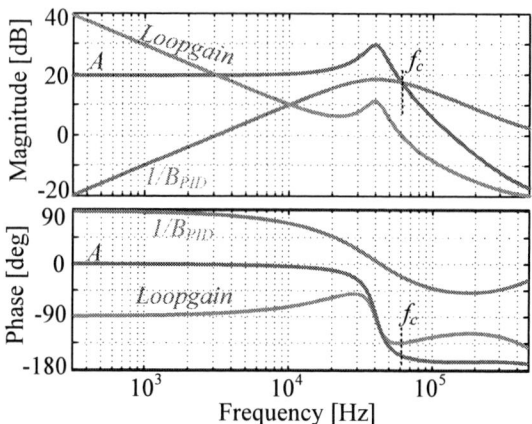

Fig. 4. Frequency responses of: control-to-output A, inverse compensator $1/B_{PID}$, and the Loop-Gain. Crossover frequency is marked f_c.

$$\frac{dI_{L_b}}{dt} = -\frac{V_{C_o}}{L_b}, \tag{6}$$

$$I_{L_a} = I_{L_b}(0) - \frac{L_a}{L_b}\frac{V_{in}/2 - V_{c_o}}{V_{c_o}}\left(I_{L_b} - I_{L_b}(0)\right), \tag{7}$$

where $I_{La}(0)$ and $I_{Lb}(0)$ are the inductors currents at the beginning of state-1. As can be seen in (7), the current difference of $I_{La} - I_{La}(0)$ depends on the current difference of $I_{Lb} - I_{Lb}(0)$. Substituting (7) and (6) into (5) yields:

$$\frac{C_o}{3}V_{C_o}{}^3 - L_b\frac{I_{L_b}}{R_{out}}V_{C_o}{}^2 + \left(I_{L_b}{}^2 L_b + I_{L_a}(0)I_{L_b}L_b - I_{L_b}(0)I_{L_b}L_b\right)V_{C_o} + C_{onstant} + L_bV_{in}\left[\frac{I_{L_b}(0)I_{L_b}}{2} - \frac{I_{L_b}{}^2}{4}\right] = 0 \tag{8}$$

where $C_{onstant}$ is defined by the initial values of I_{Lb} and V_{Co}. $I_{La}(0)$ and $I_{Lb}(0)$ are the inductors currents at the beginning of state-1. The first solution of this function yields the state-1 trajectories of the converter in the form of $V_{Co}=f(I_{Lb})$ with three initial conditions: $I_{La}(0)$, $I_{Lb}(0)$ and $V_{Co}(0)$.

By symmetry in the operation of state-3 to state-1 and proper variable assignment, the state-trajectories are derived from (8). The variables are assigned as: $V_{Cf}=V_{in}/2$, I_{La} swaps with I_{Lb}, and L_a swaps with L_b. States 2 and 4 are the off states and identical. The states' equations can be expressed as:

$$\frac{dV_{C_o}}{dt} = \frac{1}{C_o}\left(I_{L_a} + I_{L_b} - \frac{V_{C_o}}{R_{out}}\right), \tag{9}$$

$$\frac{dI_{L_a}}{dt} = -\frac{V_{C_o}}{L_a}, \frac{dI_{L_b}}{dt} = -\frac{V_{C_o}}{L_b}, \tag{10}$$

$$I_{L_a} = I_{L_a}(0) + \frac{L_a}{L_b}(I_{L_b} - I_{L_b}(0)), \tag{11}$$

and the state trajectories for each state are obtained by the variable assignment as described earlier.

For example I_{La} can be expressed as:

$$V_c^2 \frac{C_o}{2} - V_c \frac{L_a}{R_{out}} + C_{onstant} + L_a(I_{L_a}^2 + (I_{L_b}(0) - I_{L_a}(0))I_{L_a}) = 0 \quad (12)$$

Solving (12) yields the trajectories of the converter $V_{Co} = f(I_{La})$ for states 2 and 4.

To summarize, following the above derivations, the state trajectories for the SC-buck converter are defined as a conventional buck converter with two expansions. First, the converter includes two on-states 1 and 3 and two off states 2 and 4. The second expansion is that there are three initials conditions instead of two. As a result, the procedure to obtain the graphical state-space map as presented in Fig. 5 is as follows:

- Horizontal axis variable for all states is the output capacitor voltage V_{Co}.

- For states 1 and 2, the state vertical axis variable is I_{Lb}; for states 3 and 4 use I_{La}. i.e., the state variable is an inductor current in the off state and the progress direction of all the trajectories is down toward the vertical axis.

- Transition on the map between states 1 to 3 is not continuous, but depends on the actual value of the inductors current. As a result of this so-called singularity, it is possible to view the climb-up of the inductor current from a lower point to a higher one.

The above procedure enables to draw a state-space map with two trajectories instead of four, when states 1 and 3 share one trajectory (on) and states 2 and 4 share the other (off). This is facilitated by duplication of the vertical axis such that it represents both I_{La} and I_{Lb} as shown in Fig. 5. The blue trajectories represent an on state; state-1 is monitored by I_{Lb} and state-3 by I_{La}. The red trajectories represent an off state; state-2 monitored by I_{Lb} and state-4 by I_{La}. For an easier view, a single trajectory is depicted in Fig. 6 for a case of loading transient. (full movement along the trajectory is detailed in the next sub-section).

Transient Controller

Observation of the resulting state-space map for the SC-buck converter reveals one of the main differences of this converter topology with respect to a multi-phase buck. As exemplified by Fig. 7 (a), during on state, while one current ramps up, and may satisfy the required charge balance to the output, the other phase's current ramps down and may result in unstable convergence around the new steady-state point. In addition, since only one phase carries the load, the minimum possible deviation is not obtained. It should be noted that for demonstration purposes, the load transient convergence in Fig. 7(a) has been obtained using an extremely overly-sized flying capacitor (1 mF) to hold the charge during the exceeding long on time. The situation worsens when C_t is sized to the steady-state requirements (50 µF) as depicted in Fig. 7(b).

Based on the behavior of the converter and by observing the trajectories map, better results are obtained by distribution of the on periods between the phases. As presented in Fig. 6,

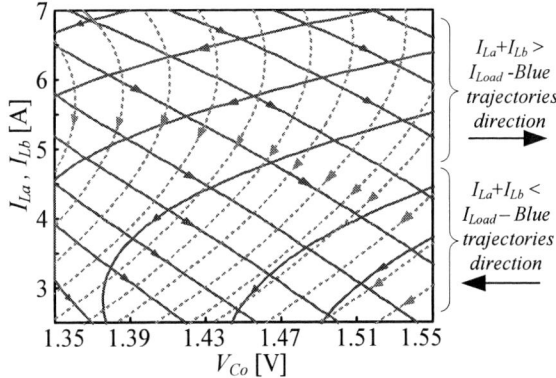

Fig. 5. State-space map for the SC-buck converter. On (states (a) and (c)-solid-blue) and off (states (b) and (d)-doted-red) trajectories.

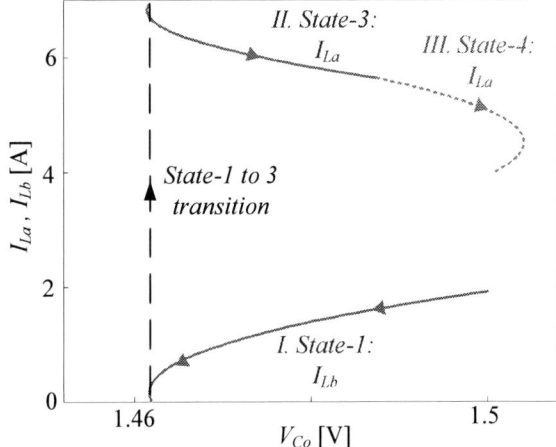

Fig. 6. Optimal trajectories for loading transient: I. First on period of phase a (state-1). II. Second on period of the alternating phase (state-3). III. Off period state-2 or 4. Note the singular transitioning point – monitoring different currents using the same plot.

switching between one on sequence to another, and then applying the off phase, results in a smaller voltage drop down to the minimum deviation of V_{Co} and a time-optimal-like behavior.

A time-optimal switching sequence for the SC-buck converter is as follows (described for loading transient):

- At the detection of a load change, the controller switches to one of the on phases (1 or 3).

- A second on period of the alternating phase is initiated when sum of the currents equals to the new load current. This point is detected by the output voltage minimum [21].

- Third, off state (state-4 or 2) can be initiated based on the charge balance of the output capacitor, which can be achieved by transient time calculation. Q_{charge} and $Q_{discharge}$, which represents the value of the capacitor charge and discharge, must be equal, as shown in Fig. 9. By assuming that the sum of the inductors current $I_{La} + I_{Lb}$ ramps up with constant slope, $V_{in}/2 - 2V_{Co}/L$, and ramps down with a slope of $-2V_{Co}/L_b$, the second state ends when $T_3 = T_1 D^{0.5}$ [16]. This assures the desired equilibrium, and can be implemented using counters.

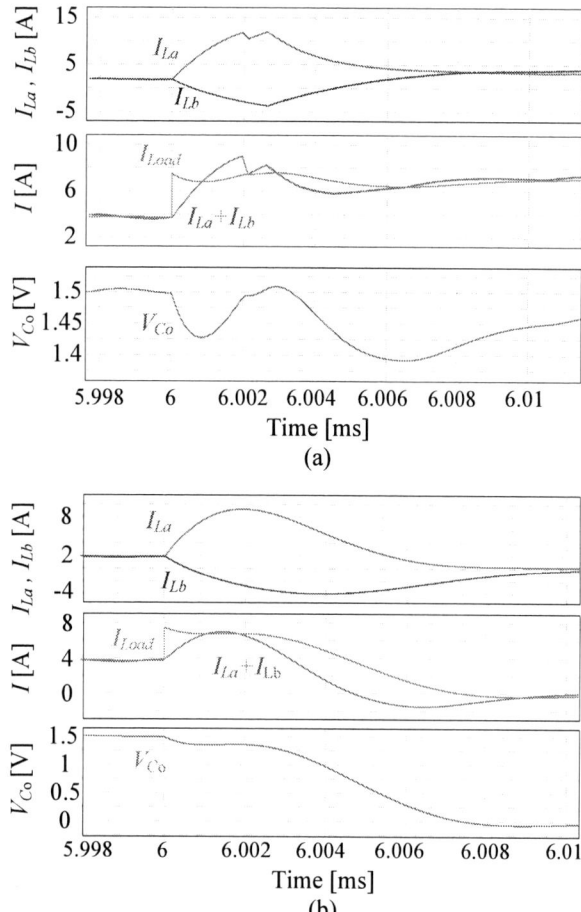

(a)

(b)

Fig. 7. Attempts of time-optimal recovery from load transient with uneven distribution of the on phases: (a) large flying capacitor (1 mF) and (b) 50μF.

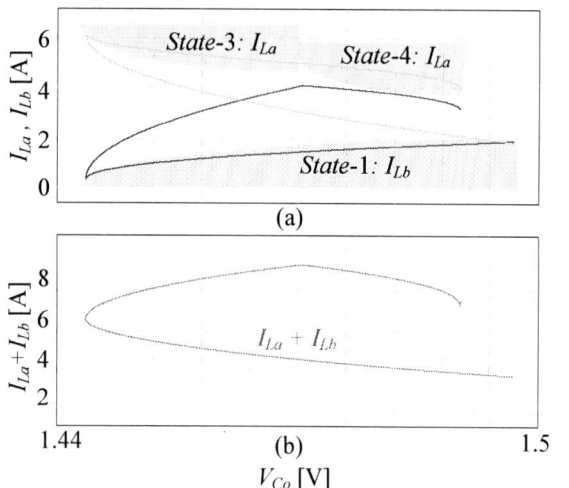

(a)

(b)

Fig. 8. Simulation results for load step heavy-to-light on the state domain: (a) the inductor currents and (b) for the sum of the inductor currents

To facilitate fast transient detection and end-of-transient phase, the first is assisted by two auxiliary comparators with two thresholds, well below the maximum allowed voltage deviation. This assists in the detection of both loading and unloading events.

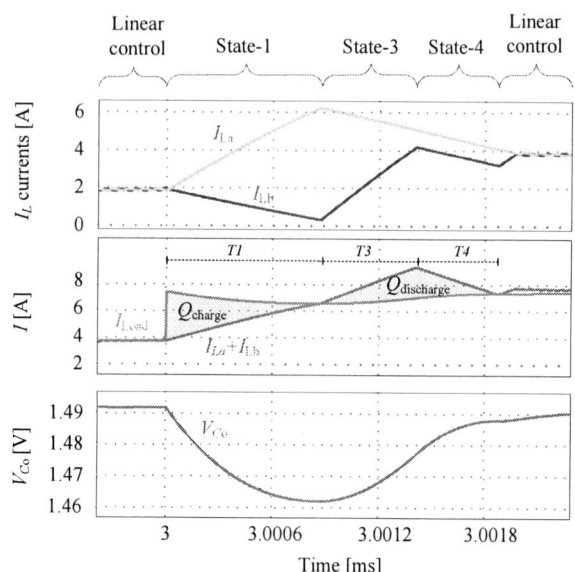

Fig. 9. Time-optimal recovery sequence for loading transient.

Figs. 8 and 9 are used to demonstrate the full sequence procedure for a loading transient, showing the individual inductor currents, sum of the currents and the output capacitor voltage for both in the time-domain and in the state-space. As can be seen, the resulting recovery trajectory of the sum of currents exactly matches a time-optimal behavior for a step load of single-phase buck converter.

To establish that time-optimal response is facilitated, the movement of the state-variables along the trajectories is examined, in the context of load transient, by observation of the output capacitor voltage and the sum of the inductors current, i.e. $I_{La}+I_{Lb}$. As a consequence, the resultant state-plane map (Fig. 8) resembles one of a conventional buck converter. As can be observed in Figs. 8 and 9, the trajectory is the one where the minimal output voltage deviation is obtained, i.e. time-optimal control [22].

IV. EXPERIMENTAL RESULTS

To validate the operation of the small-signal compensator and the time-optimal controller, two 12-to-1.5 V SC-buck converters prototype have been built and tested. One converter has been designed to operate at 200 kHz and the second converter at 1.25 MHz. The main components of both units are listed in Table I. The digital controller comprises the steady-state voltage-mode compensator and the transient mode controller as shown in Fig. 1. The controller has been realized

TABLE I. EXPERIMENTAL COMPONENT SYMBOLS AND VALUES

Hardware components parameters		
Parameter	200 kHz	1.25 MHz
Inductors (L_a, L_b)	5 uH	0.6 uH
Series capacitor (C_t)	20 uF	10 uF
Output capacitor (C_o)	100 uF	50 uF
Output capacitor resistance	10 mΩ	5 mΩ
Input capacitor (C_{in})	100 uF	100 uF

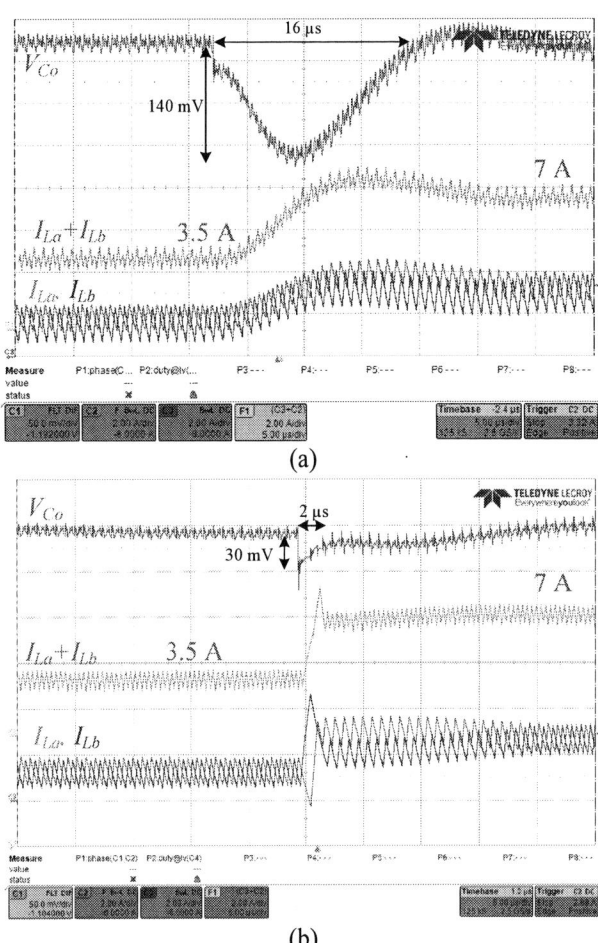

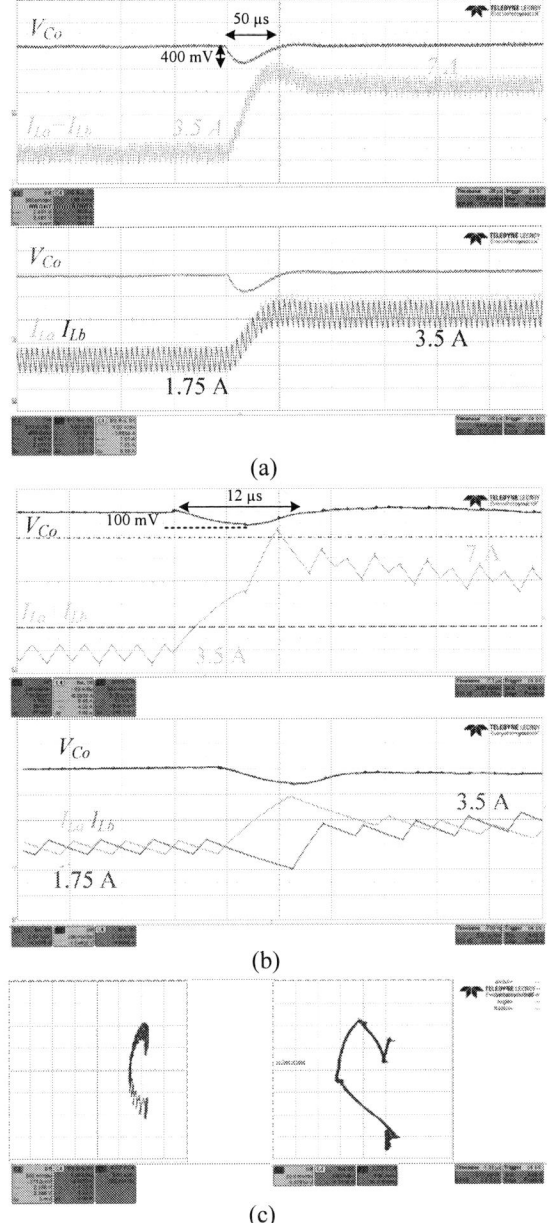

Fig. 11. 1.25 MHz experimental results for transient response from 3.5A to 7A by voltage-mode control (a) and time-optimal control (b). (50mV/div, 2A/div. Time scale is 5µs/div)

Fig. 10. 200 kHz experimental results for loading transient response from 3.5A to 7A by voltage-mode control (a) (500mV/div, 1A/div. Time scale is 50µs/div) and time-optimal control (b) (200mV/div, 1A/div for the sum of the currents and 2A/div for I_{La} and I_{Lb} . Time scale is 5µs/div), and (c) the state trajectories.

entirely on Altera Cyclone IV FPGA [23], including custom design of all related peripherals with an all-digital delay-line ADC and digital PWM as described in [24]-[25].

Fig. 10 shows the recovery of the 200kHz SC-buck converter from a loading transient of 50%, comparing the operation and performance of a small-signal voltage-mode compensation to time-optimal control scheme that has been developed. Also shown in Fig. 10 are the state-plane trajectories that are obtained using each control method, validating the theoretical analysis. Fig. 11 shows recovery results for the 1.25MHz prototype. A significant improvement

of the transient recovery using the time-optimal approach can be observed for both designs. Fig. 10 demonstrates output voltage undershoot of the TOC is 100 mV and settling time of 12 µs, compared to 400 mV and 50µs obtained by small-signal voltage-mode control. In Fig. 11 the undershoot has been trimmed from 140 to 30 mV while the settling time has been reduced from 12 µs to 2 µs.

V. CONCLUSIONS

In this study, an optimal closed-loop control scheme for a SC buck converters has been presented. The controller hybrids a voltage-mode small-signal compensator for steady-state operation and a transient-mode time-optimal controller for load transients. In the theoretical analysis, an average-behavioral model as well as state-space representation of the converter have been derived, then an optimal controller has been developed. The analysis revealed that a key factor for recovery of the converter from load transients is the capability of the controller to satisfy the charge-balance of the flying capacitor at all times (including during load transients). As a result, the time-optimal controller developed distributes the on-time period between the phases and by doing so, a smooth transient recovery is obtained. The experimental validation of the controller operations has been found to be in very good agreement with the theoretical predictions. Also shown are the differences between small-signal voltage mode and time-optimal controls, demonstrating the superiority of the state-variable based approach.

REFERENCES

[1] X. Zhou, X. Zhang, J. Liu, P. Wong, J. Chen, H. Wu, L. Amoroso, F. C. Lee, and D. Chen, "Investigation of candidate VRM topologies for future microprocessors," in *Proc. IEEE Applied Power Electronics Conf.*, pp. 145-150, Feb. 1998.

[2] Y. Panov and M. Jovanović, "Design considerations for 12-V/1.5-V, 50-A voltage regulator modules," in *Proc. IEEE Applied Power Electronics Conf.*, pp. 39-46, Feb. 2000.

[3] Y. Ren, M. Xu, K. Yao, Y. Meng, and F. C. Lee, "Two-stage approach for 12-V VR," *IEEE Trans. Power Electron.*, vol. 19, no. 6, pp. 1498–1506, Nov. 2004.

[4] K. Nishijima, K. Harada, T. Nakano, T. Nabeshima, and T. Sato, "Analysis of double step-down two-phase buck converter for VRM," in *Proc. IEEE Telecommun. Energy Conf.*, pp. 497-502, Sept. 2005.

[5] J. Yungtaek, M. M. Jovanovic, and Y. Panov, "Multiphase buck converters with extended duty cycle," in *Proc. IEEE Applied Power Electron. Conf.*, pp. 38-44, March. 2006.

[6] B. Oraw and R. Ayyanar, "Small signal modeling and control design for new extended duty ratio, interleaved multiphase synchronous buck converter," in *Proc. IEEE Telecommun. Energy Conf.*, Sept. 2006.

[7] P.S. Shenoy, M. Amaro, D. Freeman, and J. Morroni, "Comparison of a 12V, 10A, 3MHz buck converter and a series capacitor buck converter," in *Proc. IEEE Applied Power Electron. Conf.*, March. 2015.

[8] P.S. Shenoy and M. Amaro, "Improving light load efficiency in a series capacitor buck converter by uneven phase interleaving," in *Proc. IEEE Applied Power Electron. Conf.*, March. 2015.

[9] K. Matsumoto, K. Nishijima, T. Sato, and T. Nabeshima, "A two-phase high step down coupled inductor converter for next generation low voltage CPU," in *Power Electronics and ECCE Asia (ICPE & ECCE), IEEE 8th International Conference on* pp. 2813-2818, May. 2011.

[10] S. Ben-Yaakov, "Average simulation of PWM converters by direct implementation of behavioral relationships," *International journal of electronics*, vol. 77, no. 5, pp. 731-746, 1994.

[11] J. Sun, D. M. Mitchell, M. F. Greuel, and R. M. Bass, "Averaged modeling of PWM converters in discontinuous conduction mode," *IEEE Trans. Power Electron.*, vol. 16, pp. 482–492, July 2001.

[12] R. W. Erickson and D. Maksimović, *Fundamentals of Power Electronics*, 2nd ed. Boston, MA: Kluwer, 2000.

[13] M. M. Peretz and S. Ben-Yaakov, "Revisiting the closed loop response of PWM converters controlled by voltage feedback," in *Proc. Applied Power Electron. Conf. and Expo*, pp. 28-64, Feb. 2008.

[14] M. M. Peretz and S. Ben-Yaakov, "Time-domain design of digital compensators for PWM DC-DC converters," *IEEE Trans. Power Electron.*, vol. 27, no. 1, pp. 284-293, Jan. 2012.

[15] M. Ordonez, M. T. Iqbal, and J. E. Quaicoe, "Selection of a curved switching surface for buck converters," *IEEE Trans. Power Electron.*, vol. 21, no. 4, pp. 1148–1153, Jul. 2006.

[16] G. Feng, E. Meyer, and Y-F. Liu, "A new digital control algorithm to achieve optimal dynamic performance in DC-to-DC converters," *IEEE Trans. Power Electron*, vol. 22, no. 4, pp. 1489–1498, 2007.

[17] V. Yousefzadeh, A. Babazadeh, B. Ramachandran, E. Alarcon, L. Pao, and D. Maksimović, "Proximate time-optimal digital control for synchronous buck DC-DC converters," *IEEE Trans. Power Electron*, vol. 23, no. 4, pp. 2018–2026, Jul. 2008.

[18] A. Babazadeh and D. Maksimović, "Hybrid digital adaptive control for fast transient response in synchronous buck DC–DC converters," *IEEE Trans. Power Electron*, vol. 24, no. 11, pp. 2625–2638, 2009.

[19] W. W. Burns and T. G. Wilson, "State trajectories used to observe and control DC-to-DC converter," *IEEE Trans. Aerosp. Electron. Syst.*, vol. 12, no. 6, pp. 706–717, Nov. 1976.

[20] W. W. Burns and T. G. Wilson, "Analytical derivation and evaluation of a state trajectory control law for DC-to-DC converters," in *Proc. Power Electron. Specialists Conf.*, pp. 70–85, Jun. 1977.

[21] A. Radic, Z. Lukic, A. Prodić, and R. Nie, "Minimum deviation digital controller IC for single and two phase DC-DC switch-mode power supplies," *IEEE Applied Power Electronics Conference and Exposition (APEC)*, pp.1-6, Feb. 2010.

[22] E. Meyer, Z. Zhang, and Y-F. Liu, "An optimal control method for buck converters using a practical capacitor charge balance technique," *IEEE Transactions on Power Electronics*, vol. 23, no. 4, pp. 1802-1812, Jul. 2008.

[23] DE2 development and education board user manual, Altera Corporation, 2006.

[24] Y. Halihal, Y. Bezdenezhnykh, I. Ozana, and M. M. Peretz, "Full FPGA-based design of a PWM/CPM controller with integrated high-resolution fast ADC and DPWM peripherals," *IEEE Workshop on Control and Modeling for Power Electronics (COMPEL)*, Jun. 2014.

[25] Y. Bezdenezhnykh, T. Vekslender, E. Abramov, A. Cervera, and M. M. Peretz, "Design and IC implementation of a fully digital power management delay-line ADC," In *Electrical & Electronics Engineers in Israel (IEEEI)*, pp. 1-5, Dec. 2014.

Unified Constant On/Off-time Hybrid Compensation for Fast Recovery in Digitally Current-mode Controlled Point-of-Load Converters

K Hariharan, *Student Member, IEEE*, Santanu Kapat *Member, IEEE*, and
Siddhartha Mukhopadhyay, *Member, IEEE*
Embedded Power Management Lab
Department of Electrical Engineering
Indian Institute of Technology Kharagpur, West Bengal, India
Email: harikarsum@gmail.com, santanu.kapat@ieee.org,
and siddhartha.mukhopadhyay@gmail.com

Abstract—Constant on-time control offers fast recovery during a step-down transient; however, the step-up recovery gets penalized, which often requires extra arrangements for real-time on-time adaptation. On the other hand, constant off-time control can achieve fast step-up (transient) recovery; however, the step-down recovery is generally degraded. This paper proposes a unified constant on/off-time hybrid compensation technique in a digitally current-mode controlled (CMC) synchronous buck converter with the inductor current in the analog domain. This utilizes a single mono-shot timer and a single digital voltage controller G_c , which can be configured to either constant on-time or off-time control with a simple modification. Thus, improved step-up/down transient performance is achieved through a real-time configuration with an anti-windup controller transition because of sharing of a common voltage controller. An asynchronous error voltage sampling can achieve unconditional current-loop stability along with robust voltage-loop stability. The mono-shot counter can be updated in real-time using a time-to-digital converter for fixed-frequency operation, without the need for input voltage sensing. A buck converter prototype is made, and the proposed control is implemented using an FPGA device.

I. INTRODUCTION

Ripple-based control [1], [2], such as constant on-time and off-time control techniques find widespread applications in point-of-load converters, because of their simplicity and the ability to achieve fast transient response along with an improvement in the light-load efficiency [3], [4]. This can inherently improve the resolution of the digital pulse width modulator (DPWM) [5], [6], while a fixed-frequency DPWM requires a very high frequency clock [7], [8]. However, the former suffers from a varying switching frequency at steady-state [4], and often requires an input voltage sensing for a quasi-fixed frequency operation [9]. Also constant on-time control suffers from poor step-up transient because of fixed on-time, while constant off-time control suffers from poor step-down transient because of fixed off-time, which often requires adaptation in on- or off-time set values [10]- [12] to improve the performance further. A hybrid combination can drastically improve the performance [13]; however, this requires separate timers and voltage controllers along with an additional anti-windup arrangement. This paper proposes a hybrid control strategy by unifying constant on/off-time DPWM architecture in a synchronous buck converter. The proposed scheme adapts the constant off-time modulation during a step-up transient, and the constant on-time modulation during a step-down transient. This inherently improves the transient performance. The former uses the rising edge of the high side gate signal u_H to sample the error voltage, while the later uses the falling edge of u_H. This achieves inherent current loop stability and robust voltage-loop stability.

II. THE PROPOSED CONTROL SCHEME

A. Unified constant on/off- time hybrid compensation

Figure 1 shows the mixed signal current mode control implementation [14], [15] in a synchronous buck converter with the proposed constant on/off-time modulation. This considers a digital voltage controller G_c , in which the error voltage v_e is sampled at the rising edge of the high-side gate signal u_H for constant off-time modulation, while a falling edge of u_H is used for constant on-time modulation. A digital-to-analog converter converts the controller output into an analog voltage v_A, which is directly compared with the sensed inductor current using a high speed analog comparator. The comparator output u_c is used to generate high-side and low-side gate signals using the mono-shot timer followed by the dead-time circuit as shown in Fig. 2. The select line F_{con}, which carries the information of the load step, configures the proposed control either to a constant on-time modulator for $F_{con} = 0$ or to a constant off-time modulator for $F_{con} = 1$ as shown in Fig. 3. Subsequently, F_{con} considers either a falling edge or a rising edge of u_H to sample v_e based on modulator configuration, and the controller G_c is computed in synchronism with u_H.

978-1-4673-9551-9/16 $31.00 © 2016 IEEE

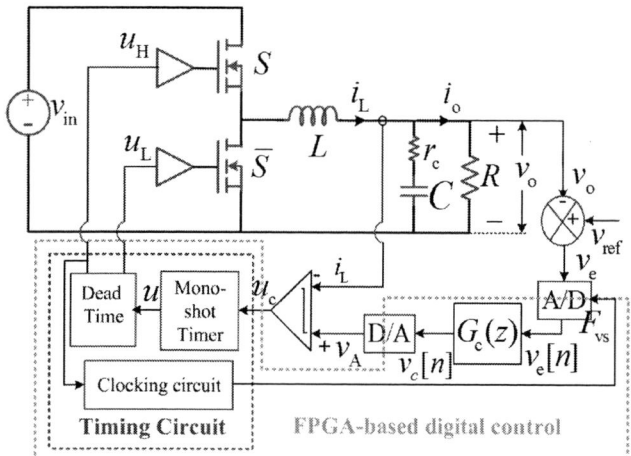

Fig. 1. Synchronous Buck converter with the proposed constant on/off-time hybrid compensation

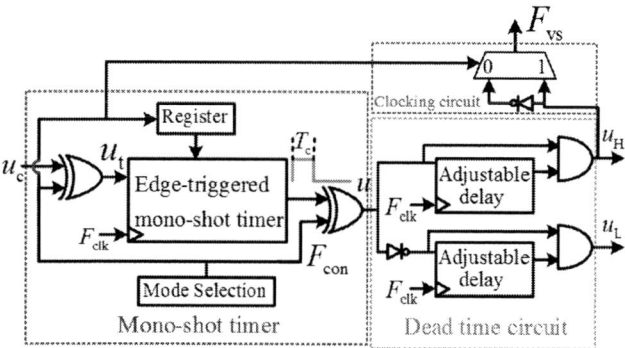

Fig. 2. Timing circuit of Fig. 1

B. General open-loop modeling of the buck converter

The state space model of the synchronous buck converter (in Fig. 1) is given by

$$\dot{\mathbf{x}} = A\mathbf{x} + Bv_{\text{in}}$$
$$y = Q\mathbf{x} \tag{1}$$

where

$$A = \begin{bmatrix} -(R_1 + r_c R_2)/L & -R_2/L \\ R_2/C & -R_2/RC \end{bmatrix}; B = \begin{bmatrix} u/L \\ 0 \end{bmatrix};$$

$Q = [r_c R_2 \ R_2]$ where $R_1 = r_{\text{on}} + r_L$, $R_2 = R/(R + r_c)$, and $\mathbf{x} = [i_L \ v_{\text{cap}}]^T$ where i_L is the inductor current, v_{cap} is the capacitor voltage, y is the output voltage v_o. If the control signal $u = 1$, the high side MOSFET is ON; otherwise, OFF. The average model is

$$\dot{\bar{\mathbf{x}}} = A\bar{\mathbf{x}} + [B|_{u=1}d + B|_{u=0}(1-d)] \tag{2}$$

where the effective duty ratio $d = t_{\text{on}}/(t_{\text{on}} + t_{\text{off}})$, where t_{on} and t_{off} are the on-time and the off-time of the converter. The

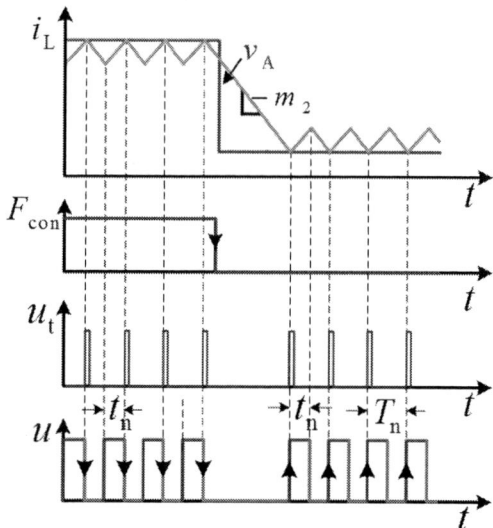

Fig. 3. Timing diagram of the proposed control scheme

perturbed model of (2) is given by [17]

$$\dot{\hat{\mathbf{x}}} = A\hat{\mathbf{x}} + \left(\frac{A\mathbf{X} + B|_{u=1} V_{\text{in}}}{T_{\text{on}} + T_{\text{off}}}\right)\hat{t}_{\text{on}}$$
$$+ \left(\frac{A\mathbf{X} + B|_{u=0} V_{\text{in}}}{T_{\text{on}} + T_{\text{off}}}\right)\hat{t}_{\text{off}} + B|_{u=1} D\hat{v}_{\text{in}} \tag{3}$$

where $\mathbf{X} = -A^{-1} B|_{u=1} DV_{\text{in}}$, and $D = T_{\text{on}}/(T_{\text{on}} + T_{\text{off}})$.

C. Control law for the proposed scheme

1) Constant on-time CMC: The constant on-time current mode control is analogous to valley current mode control. Figure 4(a) shows the control waveform of the proposed constant on-time current mode control scheme. Referring to Fig. 4(a), the average inductor current is given by

$$\bar{i}_L = \left[\left(v_c + \frac{m_1 T_{\text{on}}}{2}\right)d + \left(v_c + \frac{m_2 t_{\text{off}}}{2}\right)(1-d)\right] \tag{4}$$

where v_c acts as the current reference. The perturbed small signal model of the current programmed control law is given by

$$\hat{i}_L = \hat{v}_c + \left(\frac{T_{\text{on}} D}{2L}\right)\hat{v}_{\text{in}} + \left(\frac{T_{\text{off}} - T_{\text{on}}}{2L}\right)\hat{v}_o$$
$$+ \left(M_2(1-D) - \frac{M_1 D}{2}\right)\hat{t}_{\text{off}} \tag{5}$$

where $M_1 = (V_{\text{in}} - V_o)/L$, and $M_2 = V_o/L$ are the up-slope and the down-slope of the inductor current, respectively.

2) Constant off-time CMC: The constant off-time current mode control is analogous to peak current mode control. Figure 4(b) shows the control waveform of the proposed constant off-time current mode control scheme. Referring to Fig. 4(b), the average inductor current is given by

$$\bar{i}_L = \left[\left(v_c - \frac{m_1 t_{\text{on}}}{2}\right)d + \left(v_c - \frac{m_2 T_{\text{off}}}{2}\right)(1-d)\right] \tag{6}$$

978-1-4673-9551-9/16 $31.00 © 2016 IEEE   316

The perturbed small signal model of the current programmed control law is given by

$$
\hat{i}_{\mathrm{L}} = \hat{v}_{\mathrm{c}} - \left(\frac{T_{\mathrm{on}}D}{2L}\right)\hat{v}_{\mathrm{in}} + \left(\frac{T_{\mathrm{on}} - T_{\mathrm{off}}}{2L}\right)\hat{v}_{\mathrm{o}}
$$
$$
+ \left(\frac{M_2(1-D)}{2} - M_1 D\right)\hat{t}_{\mathrm{on}} \qquad (7)
$$

D. Control-to-output transfer function of constant on-time CMC

The control-to-output transfer function with the outer voltage loop open is given by

$$
G_{\mathrm{vv_c}} = \left.\frac{\hat{v}_{\mathrm{o}}}{\hat{v}_{\mathrm{c}}}\right|_{\hat{v}_{\mathrm{in}}=0;\hat{t}_{\mathrm{on}}=0} \qquad (8)
$$

where

$$
\frac{\hat{v}_{\mathrm{o}}}{\hat{v}_{\mathrm{c}}} = \left.\left(\frac{\hat{v}_{\mathrm{o}}}{\hat{t}_{\mathrm{off}}} * \frac{\hat{t}_{\mathrm{off}}}{\hat{v}_{\mathrm{c}}}\right)\right|_{\hat{v}_{\mathrm{in}}=0;\hat{t}_{\mathrm{on}}=0} \qquad (9)
$$

where $\hat{v}_{\mathrm{c}}$ is the perturbed signal of the reference of the inductor current. From (3), the open loop control-to-output transfer function of the constant on-time modulation is given by

$$
G_{\mathrm{vt_{off}}} = \left.\frac{\hat{v}_{\mathrm{o}}}{\hat{t}_{\mathrm{off}}}\right|_{\hat{v}_{\mathrm{in}}=0;\hat{t}_{\mathrm{on}}=0}
$$
$$
= \frac{1}{\Delta}\left(-\frac{DV_{\mathrm{in}}R_2 r_{\mathrm{c}}}{L(T_{\mathrm{on}} + T_{\mathrm{off}})}\right)\left(s + \frac{1}{Cr_{\mathrm{c}}}\right) \qquad (10)
$$

where

$$
\Delta = s^2 + \left(\frac{R_2}{RC} + \frac{R_1 + R_2 r_{\mathrm{c}}}{L}\right)s + \frac{R_2}{RLC}(R_1 + R)
$$

From (5),

$$
\frac{\hat{t}_{\mathrm{off}}}{\hat{v}_{\mathrm{c}}} = -\frac{1}{\left(M_2(1-D) - \frac{M_1 D}{2}\right)}
$$
$$
+ \frac{1}{2L}\left(\frac{T_{\mathrm{on}} - T_{\mathrm{off}}}{M_2(1-D) - \frac{M_1 D}{2}}\right)\frac{\hat{v}_{\mathrm{o}}}{\hat{v}_{\mathrm{c}}} \qquad (11)
$$

Using (10), (11) in (9), the closed loop control-to-output transfer with the outer voltage loop open is obtained as

$$
G_{\mathrm{vv_c}} = \frac{K_1\left(s + \frac{1}{Cr_{\mathrm{c}}}\right)}{s^2 + P_1 s + P_2} \qquad (12)
$$

where

$$
K_1 = \frac{2DV_{\mathrm{in}}R_2 r_{\mathrm{c}}}{(T_{\mathrm{on}} - T_{\mathrm{off}})(V_{\mathrm{o}}(2-D) - V_{\mathrm{in}}D)}
$$
$$
P_1 = \frac{R_2}{RC} + \frac{R_1 + R_2 r_{\mathrm{c}}}{L} + \frac{DV_{\mathrm{in}}R_2 r_{\mathrm{c}}(2D-1)}{L(V_{\mathrm{o}}(2-D) - V_{\mathrm{in}}D)}
$$
$$
P_2 = \frac{R_2(R_1 + R)}{RLC} + \frac{V_{\mathrm{in}}D(2D-1)R_2}{LC(V_{\mathrm{o}}(2-D) - V_{\mathrm{in}}D)}
$$

E. Control-to-output transfer function of constant off-time CMC

The control-to-output transfer function with the outer voltage loop open is given by

$$
G_{\mathrm{vv_c}} = \left.\frac{\hat{v}_{\mathrm{o}}}{\hat{v}_{\mathrm{c}}}\right|_{\hat{v}_{\mathrm{in}}=0;\hat{t}_{\mathrm{off}}=0} \qquad (13)
$$

where

$$
\frac{\hat{v}_{\mathrm{o}}}{\hat{v}_{\mathrm{c}}} = \left.\left(\frac{\hat{v}_{\mathrm{o}}}{\hat{t}_{\mathrm{on}}} * \frac{\hat{t}_{\mathrm{on}}}{\hat{v}_{\mathrm{c}}}\right)\right|_{\hat{v}_{\mathrm{in}}=0;\hat{t}_{\mathrm{off}}=0} \qquad (14)
$$

From (3), the open loop control-to-output transfer function of the constant off-time modulation is given by

$$
G_{\mathrm{vt_{on}}} = \left.\frac{\hat{v}_{\mathrm{o}}}{\hat{t}_{\mathrm{on}}}\right|_{\hat{v}_{\mathrm{in}}=0;\hat{t}_{\mathrm{off}}=0}.
$$
$$
= \frac{1}{\Delta}\left(\frac{V_{\mathrm{in}}(1-D)R_2 r_{\mathrm{c}}}{L(T_{\mathrm{on}} + T_{\mathrm{off}})}\right)\left(s + \frac{1}{Cr_{\mathrm{c}}}\right) \qquad (15)
$$

where

$$
\Delta = s^2 + \left(\frac{R_2}{RC} + \frac{R_1 + R_2 r_{\mathrm{c}}}{L}\right)s + \frac{R_2}{RLC}(R_1 + R)
$$

From (7),

$$
\frac{\hat{t}_{\mathrm{on}}}{\hat{v}_{\mathrm{c}}} = -\frac{1}{\left(\frac{M_2(1-D)}{2} - M_1 D\right)}
$$
$$
- \frac{1}{2L}\left(\frac{T_{\mathrm{on}} - T_{\mathrm{off}}}{\frac{M_2(1-D)}{2} - M_1 D}\right)\frac{\hat{v}_{\mathrm{o}}}{\hat{v}_{\mathrm{c}}} \qquad (16)
$$

Using (15), (16) in (14), the closed loop control-to-output transfer with the outer voltage loop open is obtained as

$$
G_{\mathrm{vv_c}} = \frac{K_2\left(s + \frac{1}{Cr_{\mathrm{c}}}\right)}{s^2 + P_3 s + P_4} \qquad (17)
$$

where

$$
K_2 = -\frac{2V_{\mathrm{in}}(1-D)R_2 r_{\mathrm{c}}}{(T_{\mathrm{on}} + T_{\mathrm{off}})(V_{\mathrm{o}}(1+D) - 2V_{\mathrm{in}}D)}
$$
$$
P_3 = \frac{R_2}{RC} + \frac{R_1 + R_2 r_{\mathrm{c}}}{L} + \frac{V_{\mathrm{in}}(1-D)R_2 r_{\mathrm{c}}(2D-1)}{L(V_{\mathrm{o}}(1+D) - 2V_{\mathrm{in}}D)}
$$
$$
P_4 = \frac{R_2(R_1 + R)}{RLC} + \frac{V_{\mathrm{in}}(1-D)(2D-1)R_2}{LC(V_{\mathrm{o}}(1+D) - 2V_{\mathrm{in}}D)}
$$

III. DESIGN AND ANALYSIS OF THE PROPOSED SCHEME

A. Compensation design

The feedback controller for the constant on/off-time modulation technique is designed using small signal based approach. Typical feedback loop design consideration includes steady state error less than 0.1% for a unit step input, the phase margin of at least 50°, and the crossover frequency f_{c} of $1/8^{th}$ of the switching frequency, F_{s}; here $F_{\mathrm{s}} = 200$ kHz corresponds to crossover frequency $f_{\mathrm{c}} = 25$ kHz.

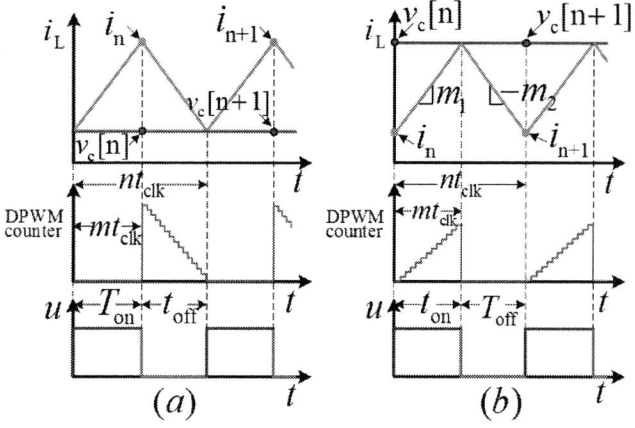

Fig. 4. Control waveforms (a) constant on-time modulation, (b) constant off-time modulation

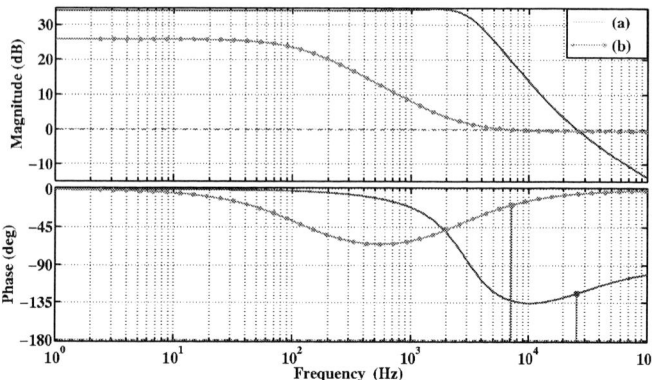

Fig. 5. Frequency response of the proposed control scheme under constant on-time modulation (a) control-to-output transfer function $G_{\mathrm{vt_{off}}} = \hat{v}_o/\hat{t}_{off}$ (b) a feedback compensator $G_{\mathrm{cT_{on}}}$

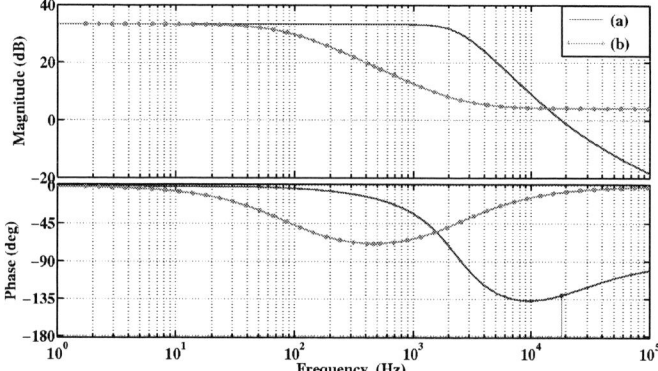

Fig. 6. Frequency response of the proposed control scheme under constant off-time modulation (a) control-to-output transfer function $G_{\mathrm{vt_{on}}} = \hat{v}_o/\hat{t}_{on}$ (b) a feedback compensator $G_{\mathrm{cT_{off}}}$

1) Constant on-time control: Figure 5(a) shows the frequency response of control-to-output transfer function of the constant on-time modulation with outer voltage loop open. From Fig. 5(a), the phase margin of the system is 55°. So, a lag compensator is designed for achieving the desired steady state response without much altering the transient performance, which is given by

$$G_{\mathrm{cT_{on}}} = \frac{0.001249s + 19.61}{0.001324s + 1} \qquad (18)$$

Figure 5(b) shows the frequency response of (18). The compensated system has a phase margin of 50.5^o with the crossover frequency of 25 kHz; thus the desired specifications are achieved.

2) Constant off-time control: Figure 6(a) shows the frequency response of control-to-output transfer function of the constant off-time modulation with outer voltage loop open. From Fig. 6(a), the phase margin of the system is 50° with the crossover frequency of 18 kHz. A lag compensator is designed for achieving the desired specification, which is given by

$$G_{\mathrm{cT_{off}}} = \frac{0.00297s + 46.63}{0.00185s + 1} \qquad (19)$$

Figure 6(b) shows the frequency response of (19). The compensated system has a phase margin of 50.1^o with the crossover frequency of 25 kHz, which meet the desired specifications.

B. Stability Analysis

Figure 4 shows the control waveforms under the proposed control scheme. For constant on-time modulation, the error voltage is sampled at every falling edge of high side gate signal u_H as shown in Fig. 4(a). Let i_n be the peak current during the nth cycle, and $v_c[n]$ is the digital controller output. The peak current during the $(n+1)^{\mathrm{th}}$ cycle is

$$i_{\mathrm{n+1}} = v_c[n] + m_1 T_{\mathrm{on}} \qquad (20)$$

where m_1 is the rising slope of inductor current, and T_{on} is the fixed on-time. For the constant off-time modulation, $v_e[n]$

is sampled at every rising edge of u_H as shown in Fig. 4(b). The valley current during $(n+1)^{\mathrm{th}}$ cycle is

$$i_{\mathrm{n+1}} = v_c[n] - m_2 T_{\mathrm{off}} \qquad (21)$$

where m_2 is the falling slope of inductor current, and T_{off} is the fixed off-time. For both the cases, the final value of inductor current $i_{\mathrm{n+1}}$ does not depend on the initial current i_n; thus the current-loop has unconditional zero-input stability. The closed-loop stability completely depends on the voltage loop; thus it is important to investigate the voltage loop stability, since the closed-loop stability completely depends on it. For the analytical study, a simple proportional controller with a gain K is considered.

1) Constant on-time: The output voltage at $(n+1)^{th}$ is

$$v_o[n+1] = v_o[n] - \left[R_2 r_c \left(1 + \frac{T_{\mathrm{on}}}{2r_c C} \right) \right] (m_2 t_{\mathrm{off}} - m_1 T_{\mathrm{on}}) \qquad (22)$$

From Fig. 4(a),

$$t_{\mathrm{off}} = \frac{i_n - v_c[n]}{m_2} \qquad (23)$$

where $i_\text{n} = v_\text{c}[n-1] + m_1 T_\text{on}$.

From (22) and (23),

$$\Delta v_\text{o}[n+1] = -K R_2 r_\text{c} \left(1 + \frac{T_\text{on}}{2 r_\text{c} C}\right) \Delta v_\text{o}[n] \qquad (24)$$

where

$$\Delta v_\text{o}[n+1] = v_\text{o}[n+1] - v_\text{o}[n]$$
$$v_\text{c}[n] - v_\text{c}[n-1] = -K \Delta v_\text{o}[n]$$

Thus, the voltage loop will be stable for

$$K < \frac{1}{R_2 r_\text{c}} \left(1 + \frac{T_\text{on}}{2 r_\text{c} C}\right)^{-1} \qquad (25)$$

which can be approximated as

$$K < \frac{1}{R_2 r_\text{c}} \left(1 - \frac{T_\text{on}}{2 r_\text{c} C}\right) \qquad (26)$$

where $r_c C > \dfrac{T_\text{on}}{2}$ [4].

2) Constant off-time: The output voltage at $(n+1)^{th}$ is

$$v_\text{o}[n+1] = v_\text{o}[n] + \left[R_2 r_\text{c} \left(1 + \frac{T_\text{off}}{2 r_\text{c} C}\right)\right] (m_1 t_\text{on} - m_2 T_\text{off}) \qquad (27)$$

From Fig. 4(b),

$$t_\text{on} = \frac{v_\text{n}[n] - i_\text{n}}{m_1} \qquad (28)$$

where $i_\text{n} = v_\text{c}[n-1] - m_2 T_\text{off}$.

From (27), and (28)

$$\Delta v_\text{o}[n+1] = -K R_2 r_\text{c} \left(1 + \frac{T_\text{off}}{2 r_\text{c} C}\right) \Delta v_\text{o}[n] \qquad (29)$$

Thus, the voltage loop will be stable for

$$K < \frac{1}{R_2 r_\text{c}} \left(1 + \frac{T_\text{off}}{2 r_\text{c} C}\right)^{-1} \approx \frac{1}{R_2 r_\text{c}} \left(1 - \frac{T_\text{off}}{2 r_\text{c} C}\right) \qquad (30)$$

provided that $r_c C > \dfrac{T_\text{off}}{2}$.

Although the stability analysis of voltage loop considers only a proportional voltage controller, this method can be extended to any other compensators such as PI, lag-/lead- compensators, etc.

C. Real-time on/off-time adaptation

Steady-state switching frequency under constant on/off modulation varies with the input voltage, which requires an input voltage feed-forward [9] or a PLL-based synchronization [18] for frequency regulation. However, this requires considerable hardware resources. The proposed modulator uses a time-to-digital converter (TDC), and updates the timer parameter in one step to regulate the steady-state switching frequency. This eliminates extra sensing circuits and expedites the frequency regulation process in one iteration.

IV. HARDWARE RESULTS

A. Implementation of the proposed scheme

A buck converter prototype has been made, and the proposed technique is implemented using an FPGA device.

The power circuit parameters are as follows: the input voltage $V_\text{in} = 6V$, the output inductance and capacitance $L = 10\mu H$ and $C = 470\mu F$, the load current $I_\text{o} \in (2,\ 6)\ A$, the reference voltage $V_\text{ref} = 3.3V$, and the nominal switching frequency $F_\text{s} = 200kHz$. The time period of an FPGA clock is $10ns$. A lag compensator is implemented using

$$v_\text{c}[n] = K_1 v_\text{c}[n-1] + K_2 v_\text{e}[n] + K_3 v_\text{e}[n-1] \qquad (31)$$

where K_1, K_2, and K_3 are the gains of the lag compensator. The optimized gain parameters can be easily updated in real time [16] for performance improvement over a wide operating range. Also the timing parameters can be updated based on the modulation scheme. The proposed scheme minimizes the hardware resources by utilizing a single controller and a single timer, yet achieves fast recovery by updating the parameters in real-time based on operating point information, such as load current and input voltage, which are now-a-days available from PMBUS™ in smart digital devices. For LED drivers, it is readily available from the LED gate signals. Then F_con can suitably configure the modulator for fast load transient recovery.

B. Discussion

Figures 7 and 8 shows the load transient performance in a buck converter using the constant on-time modulator throughout, which results in $450\mu S$ settling time and $150mV$ peak voltage undershoot for a step-up transient [shown in Fig. 7]. However, the same controller can speed-up a step-down recovery [shown in Fig. 8] with $280\mu s$ settling time and $150mV$ peak (voltage) overshoot. On the other hand, Figs. 9 and 10 shows that the constant off-time modulator can achieve fast step-up recovery with $24\mu s$ settling time and $130mV$ peak undershoot, while the step-down performance is significantly degraded with $490\mu s$ settling time and $170mV$ peak overshoot. It is clear from Figs. 8 and 9 that constant on-time modulator offers fast step-down recovery, whereas constant off-time modulator offers fast step-up recovery. The proposed scheme utilizes a hybrid combination to make use of their best features and improved performance is achieved for both step-up and step-down transients using constant off- and on-time modulators respectively as shown in Figs. 11 and 12. The voltage controller structure, however, remains the same for both the cases. A further improvement is possible by adaptively varying the timing parameters for an active modulator. This provides opportunities for implementing high performance control algorithms with minimized hardware resources.

V. CONCLUSION

A hybrid constant on/off-time modulator was proposed in digitally current-mode controlled point-of-load converters. This utilizes the best features of the individual modulators and achieves improved load transient performance over a wide operating range. Either of the modulators uses a common digital voltage controller while remains active; thus an anti-windup controller transition is inherently achieved

978-1-4673-9551-9/16 $31.00 © 2016 IEEE

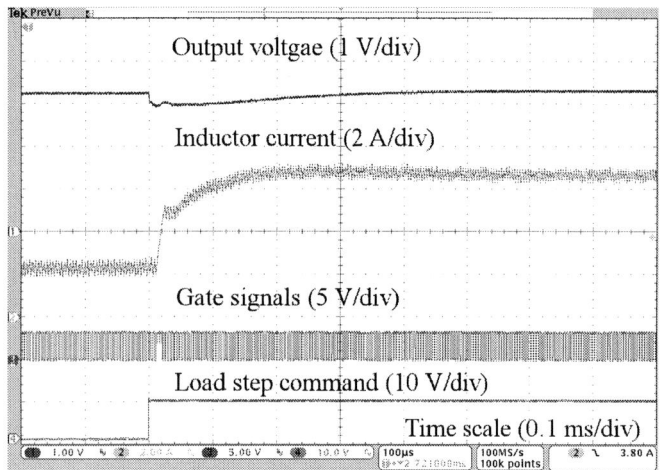

Fig. 7. Load transient response of a buck converter at 6 V input using constant on-time modulation technique for the load step-up from 2.15 A to 7 A

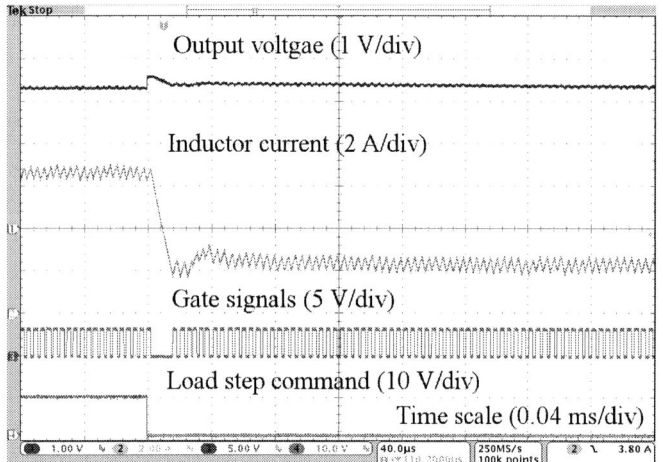

Fig. 8. Load transient response of a buck converter at 6 V input using constant on-time modulation technique for the load step-down from 7 A to 2.15 A

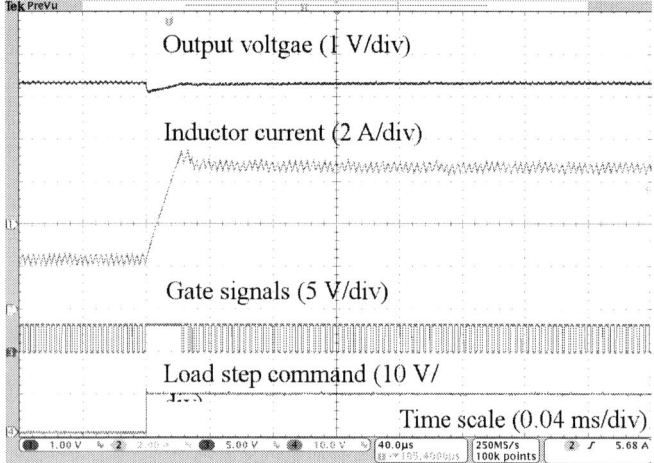

Fig. 9. Load transient response of a buck converter at 6 V input using constant off-time modulation technique for the load step-up from 2.15 A to 7 A

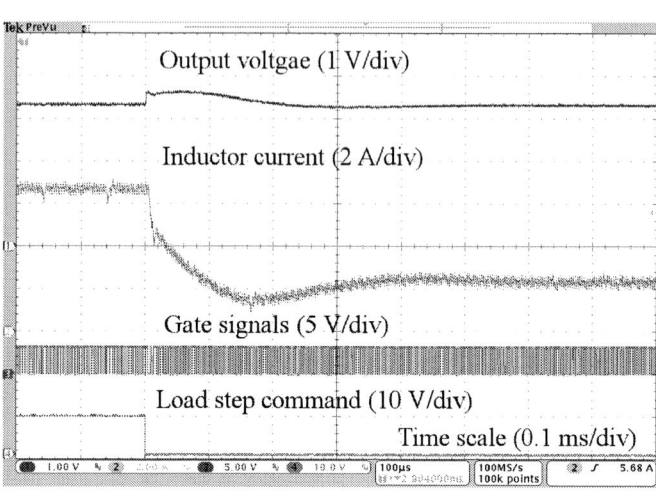

Fig. 10. Load transient response of a buck converter at 6 V input using constant off-time modulation technique for the load step-down from 7 A to 2.15 A

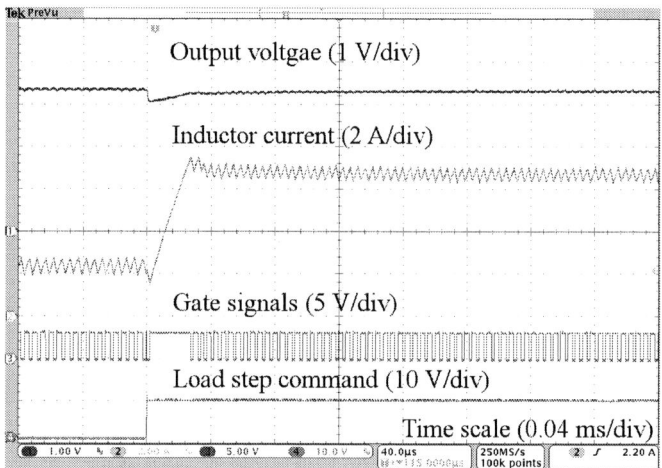

Fig. 11. Load transient response of a buck converter at 6 V input using constant on/off-time hybrid modulation technique for the load step-up from 2.15 A to 7 A

and hardware resource is also minimized. This achieves robust closed-loop stability and a fixed frequency operation at steady-state without sensing the input voltage. The proposed scheme was implemented using an FPGA device and tested in a buck converter. The proposed scheme is simple to implement and provides opportunities to implement high performance control.

ACKNOWLEDGMENT

The authors would like to thank Amit Kumar Singha for his help in experimentation.

REFERENCES

[1] K. Y. Cheng, S. Tian, F. Yu, F. C. Lee, and P. Mattavelli,"Digital hybrid ripple-based constant on-time control for voltage regulator modules," *IEEE Trans. Power Electron.*, vol. 29, no. 6, pp. 3132–3144, Jun. 2014.

[2] N. Kong, D. S. Ha, J. Li, and F. C. Lee,"Off-time prediction in digital constant on-time modulation for DC-DC converters," *in Proc. IEEE ISCAS*, pp. 3270–3273, May 2008.

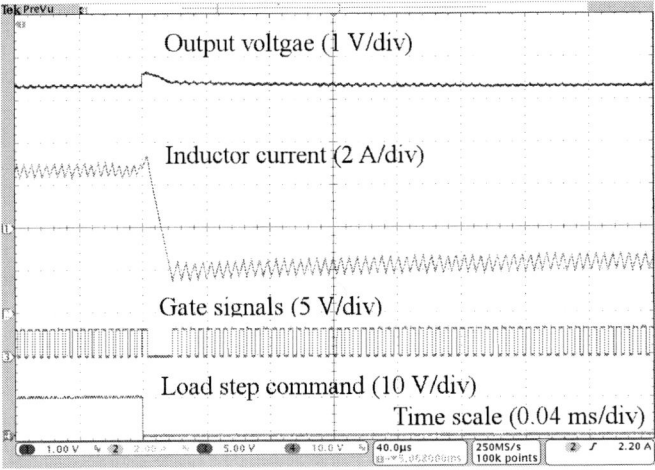

Fig. 12. Load transient response of a buck converter at 6 V input using constant on/off-time hybrid modulation technique for the load step-down from 7 A to 2.15 A

[3] J. Sun,"Characterization and performance comparison of ripple based control for voltage regulator modules," *IEEE Trans. Power Electron.*, vol. 21, no. 2, pp. 346–353, Mar. 2006.

[4] W. R. Redl and S. Jian, "Riple-based control of switching regulators–An overview," *IEEE Trans. Power Electron.*, vol. 24, no. 12, pp. 2669–2680, Dec. 2009.

[5] J. Li, Y. Qiu, Y. Sun, B. Huang, M. Xu, D. S. Ha, and F. C. Lee,"High resolution digital duty cycle modulation schemes for voltage regulators," *in Proc. IEEE APEC*, pp. 871–876, Mar. 2007.

[6] Y. Yan, F. C. Lee, and P. Mattavelli,"Comparison of small signal characteristics in current mode control schemes for point-of-load buck converter applications," *IEEE Trans. Power Electron.*, vol. 28, no. 7, pp. 3504–3414, Jul. 2013.

[7] H. Peng, D. Maksimovic, A. Prodic A, and E. Alarcon, "Modeling of quantization effects in digitally controlled DC-DC converters," *IEEE Trans. Power Electron.*, vol. 22, no. 1, pp. 208–215, Jan. 2007.

[8] A. V. Peterchev and S. R. Sanders, "Quantization resolution and limit cycling in digitally controlled PWM converters," *IEEE Trans. Power Electron.*, vol. 18, no. 1, pp. 301–308, Jan 2003.

[9] W. Fu, S. T. Tan, M. Radhakrishnan, R. Byrd, and A. A. Fayed,"A DCM-only buck regulator with hysteretic assisted adaptive minimum-on-time control for low-power microcontrollers," *IEEE Trans. Power Electron.*, Early access, 2015.

[10] S. Bari, Q. Li, and F. C. Lee, "Fast adaptive on time control for transient performance improvement,"*in Proc. IEEE APEC*, pp. 397–403, Mar. 2015.

[11] B. Sahu and G. A. Rincon-Mora, "An accurate, low-voltage, CMOS switching power supply with adaptive on-time pulse-frequency modulation (PFM) control," *IEEE Trans. Cir. Syst.-I*, vol. 54, no. 2, pp. 312–321, Feb. 2007.

[12] H. C. Lin, B. C. Fung, and T. Y. Chang,"A current mode adaptive on-time control scheme for fast transient DC-DC converters," *in Proc. IEEE ISCAS*, pp. 2602–2605, May 2008.

[13] C. A. Yeh and Y. S. Lai, "Digital pulsewidth modulation technique for a synchronous buck DC/DC converter to reduce switching frequency," *IEEE Trans. Ind. Electron.*, vol. 59, no. 1, pp. 550–561, Jan. 2012.

[14] S. Saggini, M. Ghioni, and A. Geraci,"An innovative digital control architecture for low-voltage high-current DC–DC converters with tight load regulation," *IEEE Trans. Power Electron.*, vol. 19, no. 1, pp. 210–218, Jan. 2004.

[15] O. Trescases, A. Prodic, and W. T. Ng,"Digitally controlled current-mode DC-DC converter IC," *IEEE Trans. Cir. Syst.-I*, vol. 58, no. 1, pp. 219–231, Jan. 2011.

[16] D. Maksimovic, R. Zane, and R. Erickson, "Impact of digital control in power electronics," *in Proc. IEEE ISPSD*, pp. 13–22, May 2004.

[17] R. Priewasser, M. Agostinelli, C. Unterrieder, S. Marsili, and M. Huemer, "Modeling, control, and implementation of DC-DC converters for variable frequency operation," *IEEE Trans. Power Electron.*, vol. 29, no. 1, pp. 287–301, Jan. 2014.

[18] P. Li, D. Bhatia, L. Xue, and R. Bashirullah, "A 90–240 MHz hysteretic controlled DC-DC buck converter with digital phase locked loop synchronization," *IEEE J. Solid-State Cir.*, vol. 46, no. 9, pp. 2108–2119, Sept. 2011.

Digital Implementation of Adaptive Synchronous Rectifier (SR) Driving Scheme for LLC Resonant Converters

Chao Fei, Student Member, IEEE, Fred C. Lee, Fellow, IEEE, Qiang Li, Member, IEEE

Center for Power Electronics Systems
Virginia Tech, Blacksburg, VA 24061 USA
feichao@vt.edu

Abstract—In this paper, an adaptive synchronous rectifier (SR) driving scheme for the LLC resonant converters using the ripple counter concept is proposed, along with two methods of implementation. With the proposed scheme, the SR drain to source voltage is sensed to detect the body diode conduction, based on which the SR on-time can be well tuned to eliminate the body diode conduction. One proposed implementation tunes the SR on-time every switching cycle based on the ripple detection; another proposed implementation tunes the SR on-time every n^{th} switching cycle (n = 1, 2, 3 ...) based on the ripple counter, which is suitable for the high frequency LLC converters. The proposed SR driving scheme has the simple implementation, requires only low-cost digital controllers and occupies very few controller resources. More importantly, since the digital controllers have already been widely adopted in the control of the LLC converters, the proposed adaptive SR driving method can be embedded into these digital controllers with little extra cost. Furthermore, how to integrate the proposed SR driving method with closed-loop control is explained in details. Experimental results are demonstrated on a 130kHz LLC converter with 100MHz microcontroller (MCU) and a 500kHz LLC converter with a 60MHz MCU and a ripple counter.

Keywords—LLC resonant converter, high frequency converters, synchronous rectification, digital control.

I. INTRODUCTION

The LLC resonant converter has been widely used as a DC-DC converter due to its high efficiency and hold-up capability [1][2]. With the fast development of the information technology, the demand for higher efficiency keeps growing. Since most of the IT applications require the DC-DC converters to provide a low-voltage high-current output, the diode rectifiers will induce very large conduction loss. The synchronous rectifiers (SR) are critical for the LLC converters to improve the efficiency by tremendously reducing the conduction loss of the diode rectifiers.

However, the SR driving scheme is quite challenging due to the discrepancy between the primary driving signal and the SR

This work is supported by the Power Management Consortium of Center for Power Electronics Systems (CPES) Industry Partnership.

driving signal. The topology of the LLC converters is shown in Fig. 1, and the corresponding SR on-time under different switching frequency is shown in Fig. 2. If the switching frequency is below the resonant frequency, the SR on-time is smaller than the primary switch on-time; if the switching frequency is above the resonant frequency, the SR on-time is larger than the primary switch on-time; if the switching frequency is equal to the resonant frequency, the SR on-time is equal to the primary switch on-time. Besides, the SR on-time is also dependent on the load condition.

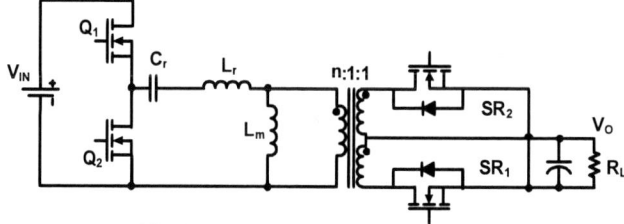

Fig. 1. Topology of the LLC converters

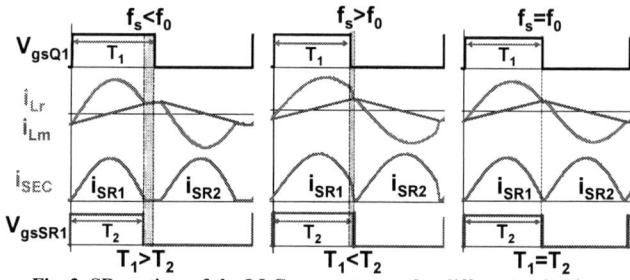

Fig. 2. SR on-time of the LLC converters under different switching frequency

Several SR driving methods have been proposed, most of which can be sorted into three categories. The first category of the SR driving methods is using the current sensing. One solution [3] is to use the current transformer to sense the SR current, and generate the corresponding driving signal, which is accurate but induces large loss due to the large current in the SRs. Another solution [4] using the CLL topology senses the primary current to generate the SR driving signal, but it requires an additional inductor with large inductance. Another SR driving method by sensing the primary resonant current was proposed in [5], but it requires the complex current-compensating winding in the current transformer to cancel out the magnetizing current and generate the suitable driving

978-1-4673-9551-9/16 $31.00 © 2016 IEEE

signals for the SRs. The second category of the SR driving methods is to use the independent driving circuit based on the SR drain to source voltage V_{ds_SR}, which works in the following principle: at the beginning, the SR is in the off-state; when there is the body diode conduction, it would result in a large forward voltage drop; then the controller would compare the V_{ds_SR} with the turn-on threshold voltage; if the V_{ds_SR} is larger than the threshold, the controller would turn on the SR; in the LLC converters, during the SR on-time, the current in SR will first increase and then decrease to zero; as the current approaches zero, the V_{ds_SR} also becomes very small, which is compared with the turn-off threshold voltage to determine when to turn off the SR. Most of the solutions in the second category of the SR driving methods are the smart ICs [6][7][8]. However, the accuracy of the SR driving methods in the second category is highly affected by the SR package. The actual SR on-time is shorter than the expected value due to the inevitable package inductance of the SRs. This problem is extremely severe in the high frequency application. A compensation network can be connected to the sensed terminals to solve this problem [9]. The third category of the SR driving methods is the adaptive SR driving scheme. The drain to source voltage of SR V_{ds_SR} is detected to tune the gate driving signal. If there is the body diode conduction, the large forward voltage drop in V_{ds_SR} will be sensed and the SR on-time will be increased accordingly. One solution [10] proposes to use a compensator to generate the SR on-time. Another solution [11] proposes to detect the body diode conduction just at the SR turn-off moment, and tune the SR on-time step-by-step until it reaches around the optimal point.

The digital controllers are gradually taking the place of the analog controllers in the control of the LLC converters. Among the digital controllers, the cost-effective microcontrollers (MCU) are preferred in the industrial applications. The high frequency LLC converters can reduce the total cost due to its high power density and integrated magnetics [12][13]. And with the fast development of the wideband gap devices and the novel magnetic materials [14][15][16][17], the trend of pushing switching frequency higher continues.

Recently, a lot of efforts have been made to achieve the low-cost digital implementation of Simplified Optimal Trajectory Control (SOTC) for the LLC converters. Compared with the conventional control methods for the LLC converters, the SOTC has the benefits of the fast transient response and relatively simple implementation [18]. For the first time, the SOTC has been implemented by a low-cost MCU and demonstrated on a 130kHz LLC converter with 100MHz MCU [19]. The whole control system integrates the fast load transient response, the soft start-up with minimum stresses and the burst mode for the light load efficiency improvement, whose control scheme is shown in Fig. 3.

Furthermore, a lot of research have been conducted to analysis how to control the high frequency LLC converters with the low-cost MCUs based on the SOTC concept, including: detailed analysis of the optimal trajectory control for the soft start-up of the high frequency LLC converters [20], the digital implementation of the SOTC for the fast load transient response of the high frequency LLC converters [21], and the light load efficiency improvement for the high frequency LLC converters based on the SOTC [22]. The results in these papers are demonstrated on a 500kHz LLC converter with a 60MHz MCU.

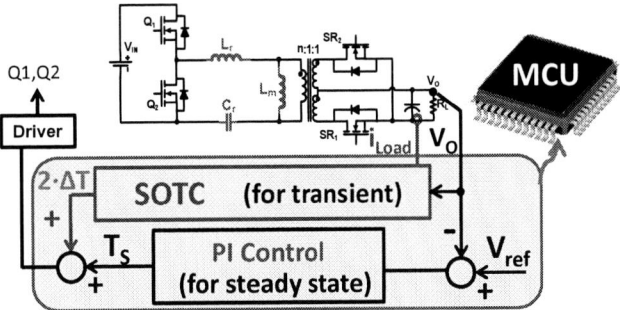

Fig. 3. Control scheme of SOTC with MCU implementation

To sum up, it is of great meaning to integrate the SR driving within the cost-effective digital controllers and combine with closed-loop control while at the same time utilizing very few controller resources. Furthermore, it is even more challenging to integrate the SR driving within the digital controller for the high frequency LLC converters with the minimum additional cost. Compared with the first and the second categories of the SR driving methods, which are independent solutions, the third category, which is adaptive SR driving, is more suitable to be integrated within the digital controllers.

In this paper, Section II investigates the limitation of the previous adaptive SR driving methods. Section III proposes the adaptive SR driving scheme based on the ripple detection, and analysis how to combine the proposed SR driving method with the closed-loop control. Section IV extends the proposed adaptive SR driving scheme to the high frequency LLC converters by adding a ripple counter. Section V presents the experimental results on a 130kHz LLC converter with a 100MHz MCU and a 500kHz LLC converter with a 60MHz MCU. The conclusions are given in Section VI.

II. THE LIMITATIONS OF THE PREVIOUS ADAPTIVE SR DRIVING METHODS

The control scheme and waveforms of the adaptive SR driving method using linear compensator [10] are shown in Fig. 4. The SR body diode forward voltage drop is detected and compared with a threshold voltage. The output of the comparator is connected to the input of a linear compensator. The control signal of the linear compensator is connected to the positive input of the PWM generator, and a triangular waveform generated from the primary driving signal is connected to the negative input of the PWM generator. If the body diode conduction is detected, the compensator will increase the SR on-time gradually until the steady state. If no body diode conduction is detected, the compensator will decrease the SR on-time gradually to the new steady state.

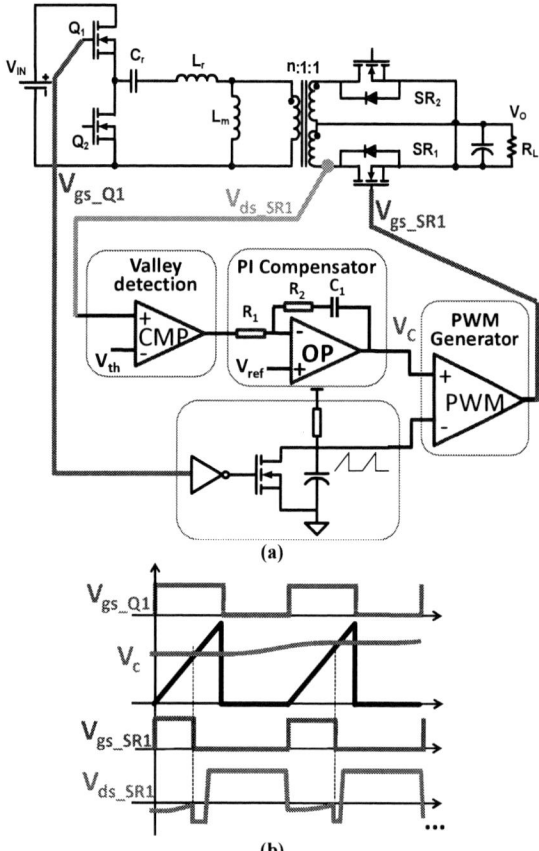

(a)

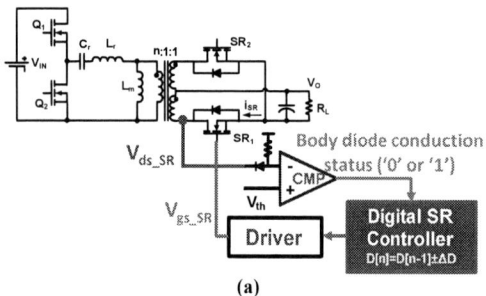

(b)

Fig. 4. Adaptive SR driving method using the linear compensator: (a) control scheme; (b) waveform

There would be two main limitations if this SR driving method is integrated within the cost-effective digital controllers. The first limitation is that it requires a linear compensator, which would either require the additional circuit or occupy a lot of CPU resources. The second limitation is that since the negative input of the PWM generator is from the primary driving signal, the maximum SR on-time cannot be large than the on-time of the primary switches, which is not suitable when the switching frequency is above the resonant frequency.

The control scheme and the control flowchart of the adaptive SR driving method based on the digital tuning [11] are shown in Fig. 5. The SR body diode forward voltage drop is detected and compared with a threshold voltage. The turn-on of the SRs in the LLC converters is synchronized with the primary switches. The on-time of the SRs is tuned step by step to be around the optimal point.

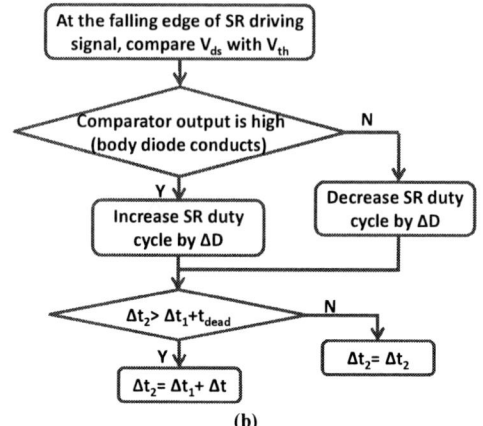

(a)

Fig. 5. Adaptive SR driving method based on the digital tuning: (a) control scheme; (b) control flowchart

The limitation with this SR driving method is that it requires comparing the V_{ds_SR} with the threshold voltage (which means sensing the output of the comparator) just at the falling edge of the SR driving signal. The implementation of this function requires the additional logic circuit or an FPGA controller as demonstrated in [11], which cannot be integrated within the cost-effective digital controllers.

III. THE PROPOSED ADAPTIVE SR DRIVING SCHEME BASED ON THE RIPPLE DETECTION

To integrate the adaptive SR driving within the cost-efficiency digital controllers, an adaptive SR driving method based on the ripple detection is proposed, whose control scheme is shown in Fig. 6(a). The V_{ds_SR} is sensed and compared with the threshold voltage and the output of the comparator is connected to the ripple detection function of the digital controller, i.e. the external interrupt of the microcontrollers. Since the proposed adaptive SR driving method is integrated within the digital controller, the primary driving signal can be used to control the enable/disable of the ripple detection.

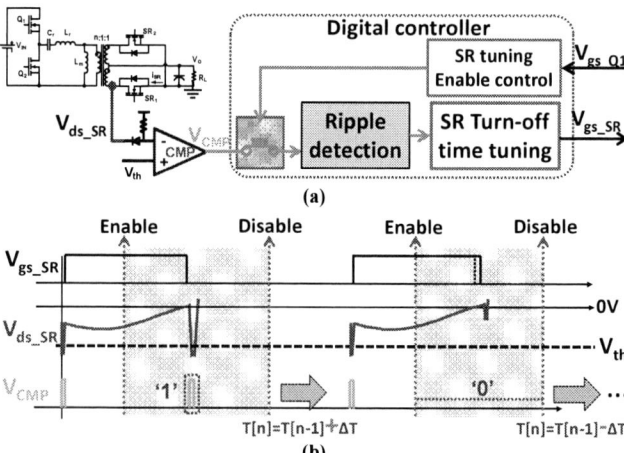

Fig. 6. The proposed adaptive SR driving based on the ripple detection: (a) control scheme; (b) waveform around the steady state

The control principle is explained in the followings. The turn-on time of the SR is synchronized with the turn-on of the

978-1-4673-9551-9/16 $31.00 © 2016 IEEE   324

primary switch, and there is a very small duration of the body diode conduction. After that, the ripple detection is enabled. And before the next SR turn-on moment, the ripple detection is disabled. If there is the body diode conduction, the controller would detect the ripple at the output of the comparator, and will increase the SR on-time by ΔT (ΔT is a very small period of time) for the next switching cycle. If there is no body diode conduction, the controller cannot detect the ripple, and will decrease the SR on-time by ΔT for the next switching cycle. The tuning process is the same as the previous adaptive SR driving method. And Fig. 6(b) shows the waveforms around the steady state.

To integrate the proposed adaptive SR driving method based on the ripple detection and the closed-loop control within the digital controller, execution of the closed-loop control and the adaptive SR driving must be properly allocated to the CPU of the digital controller. As shown in Fig. 7, within one switching cycle, the execution of the adaptive SR driving (red shade area) is divided into 3 parts and inserted into the closed-loop control (yellow shaded area). The Part 1 is to enable the ripple detection; the Part 2 is to store the result of the ripple detection; and the Part 3 is to disable the ripple detection and update the SR on-time for the next switching cycle.

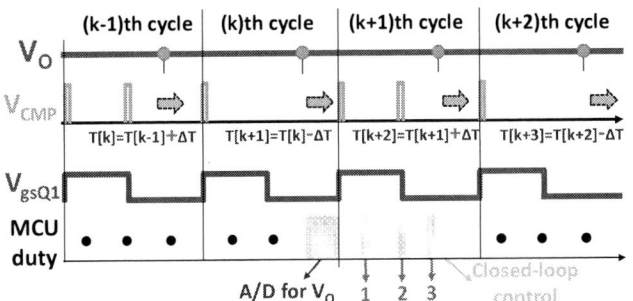

Fig. 7. Allocation of MCU duty for the closed-loop control and the adaptive SR driving within one switching cycle

Furthermore, the SR on-time must be changed accordingly during the transient to guarantee the proper operation when the proposed adaptive SR driving works together with the closed-loop control, especially for the fast load transient response proposed in [18][19].

When the switching frequency suddenly decreases as shown in Fig. 8(a), which means the pulse width of the primary switches suddenly increases from T_S to $T_S + \Delta T_{UP}$, in such case, the current in the SR may reach zero earlier than the corresponding primary switch since magnetizing inductor will take part in the resonance when the switching frequency is below the resonant frequency. So the SR on-time is kept the same as the previous switching cycle when the switching frequency suddenly increases.

When the switching frequency suddenly increases as shown in Fig. 8(b), which means the pulse width of the primary switches suddenly decreases from T_S to $T_S - \Delta T_{DOWN}$, in such case, the current in the SR may reach zero later than the corresponding primary switch. To guarantee there is no shoot-

through in the SRs, the SR on-time is also reduced by ΔT_{DOWN}. Since both the closed-loop control and the adaptive SR driving are integrated in the digital controller, the SR on-time can always be updated accordingly as mentioned above to guarantee the optimal transient and no shoot-through.

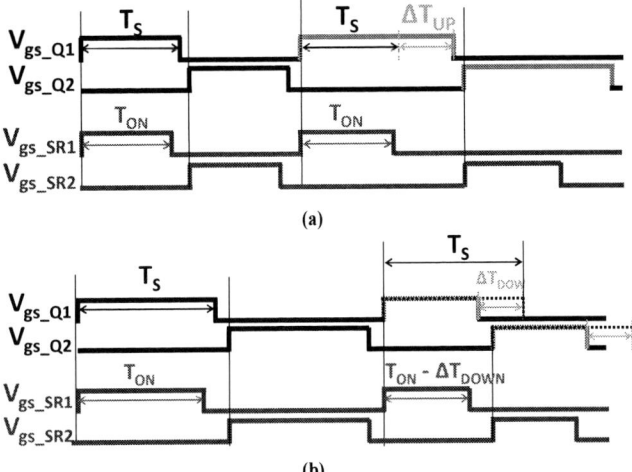

Fig. 8. Change the SR on-time accordingly during transient: (a) switching period suddenly increases; (b) switching period suddenly decreases

Compared with the other adaptive SR driving methods, the proposed adaptive SR driving method can be easily integrated within the digital controller. It only requires a comparator in addition to the digital controller. And it takes very few controller resources, so that the execution of the closed-loop control functions is not impacted. For a 130kHz LLC converter with a 100MHz digital controller (TMS320F2808), the execution of the proposed adaptive SR driving method requires a maximum of 50 CPU cycles at each switching cycle, which corresponds to only 6.5% CPU utilization.

IV. THE EXTENSION OF THE PROPOSED ADAPTIVE SR DRIVING METHOD FOR THE HIGH FREQUENCY LLC CONVERTERS USING THE RIPPLE COUNTER

The proposed adaptive SR driving method based on the ripple detection is suitable for the conventional LLC converter (resonant frequency is below 150kHz) with the cost-effective digital controllers. However, when this method is applied to the high frequency LLC converters, there would be a problem caused by the dramatically increased CPU utilization. For a given 60MHz digital controller (TMS320F28027), the execution of the proposed adaptive SR driving method requires a maximum of 50 CPU cycles; if this is applied to a 500kHz LLC converter, it would take 42% CPU utilization, which means the controller has little spare CPU time for the closed-loop control function.

To solve the challenge of the SR driving for the high frequency LLC converters, the proposed adaptive SR driving method is extended to the high frequency LLC converters by adding a ripple counter. The control scheme and the waveforms of the extended proposed method are shown in Fig. 9. Compared with the previous adaptive SR driving method based on the ripple detection, the digital controller does not need to provide the ripple detection function, instead, a ripple counter is required to be added to the control scheme.

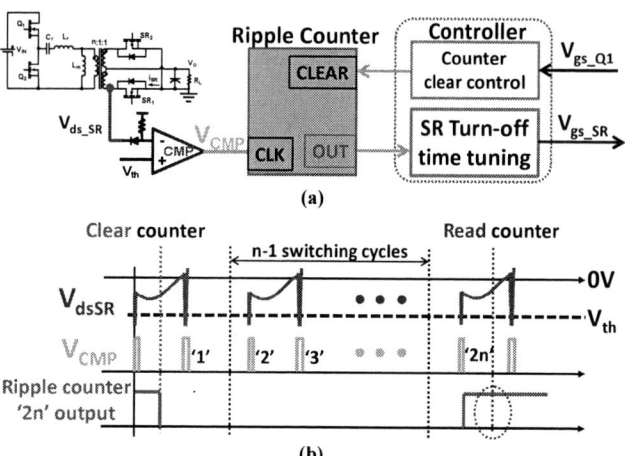

(a)

(b)

Fig. 9. The extension of the proposed adaptive SR driving scheme by adding a ripple counter: (a) control scheme; (b) waveforms

By using the ripple counter, the extended proposed method can tune the SR on-time every $(n+1)^{th}$ switching cycle ($n = 1, 2, 3…$), as shown in Fig. 9(b). The turn-on time of the SRs is still synchronized with the primary switches, so there is always a very small time of the body diode conduction at the turn-on moment. The extended proposed method counts the ripples at the output of comparator to determine if there is extra body diode conduction after the SR turn-off. The controller clears the ripple counter after the SR turn-on of the first switching cycle; then after the SR turn-on of the $(n+1)^{th}$ switching cycle ($n = 1, 2, 3 …$), the controller reads the ripple counter. If the output is '2n', it means that there is the body diode conduction after the SR turn-off. Otherwise, there is no body diode conduction.

The tuning process is shown in Fig. 10. The SR tuning cycle is $n+1$ switching cycles ($n = 1, 2, 3…$). At the beginning, there is a large body diode conduction after the SR turn-off, and the ripple counter indicates $2n$ ripples. So the controller keeps increasing the SR on-time. Every $n+1$ switching cycles, the SR on-time is increased by ΔT, which continues until when the ripple counter indicates that there is only n ripples. Then the controller decreases SR on-time by ΔT. In the next $n+1$ switching cycles, there are $2n$ ripples again. Thus, the SR on-time is tuned step-by-step to eliminate the body diode conduction, and finally it's around the optimal point.

For the extension of the proposed adaptive SR driving scheme by adding a ripple counter, the ripple counter could be cleared at the kth switching cycle in the SR tuning cycle and read at the mth switching cycle in one SR tuning cycle ($n+1$ switching cycles), as long as $k < m < n+1$, and in such case, '$2 \cdot (m-k)$' ripples mean the extra body diode conduction after the SR turn-off.

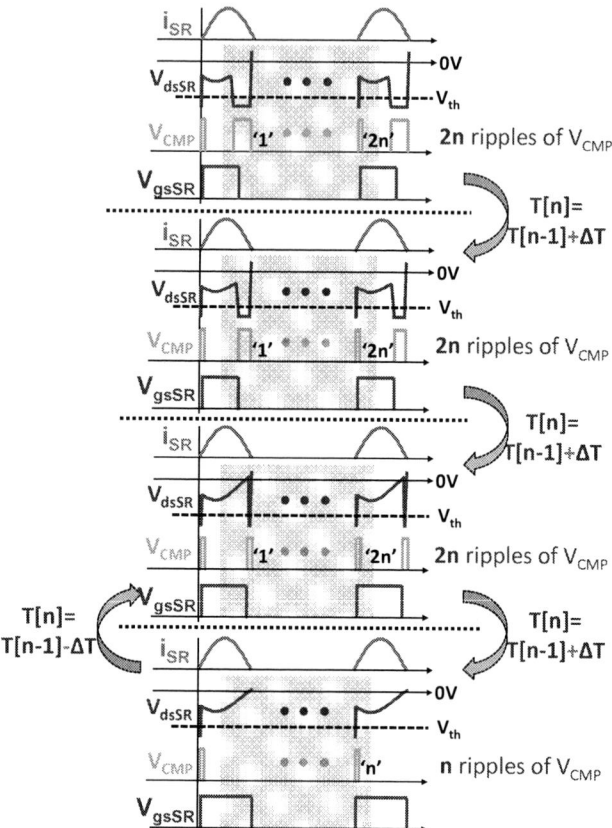

Fig. 10.Tuning process of the extended proposed adaptive SR driving scheme using the ripple counter

For most digital control system in the high frequency applications, the control loop are executed every several switching cycles since the digital delay is longer than the switching period, as mentioned in [21]. For these applications, the extended proposed adaptive SR driving method can be easily combined with the closed-loop control within the digital controller. Fig. 11 is an example of the control loop executed every third switching cycle. Within one control cycle (3 switching cycles), the execution of the adaptive SR driving (red shade area) is divided into 2 parts and inserted into the closed-loop control (yellow shaded area). The Part 1 is to clear the ripple counter; the Part 2 is to read the ripple counter and change T_{ON} accordingly. All the PWM signals are updated together at the end of the third switching cycle.

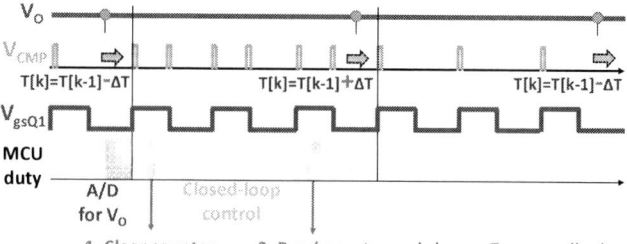

Fig. 11. Allocation of MCU duty for the closed-loop control and the adaptive SR driving within 3 switching cycles

There are three major benefits of the extended proposed method. The first benefit is that it does not require the digital controller to provide the ripple detection function, which means less resources occupation and less CPU utilization of the digital controllers. It takes less than 20 CPU cycles with TI's 60MHz MCU TMS320F28027 to execute the proposed method once. The second benefit is that the SR tuning can be selected to be executed every $(n+1)^{th}$ switching cycle to further reduce the CPU utilization. For a 500kHz LLC converter with a 60MHz MCU, if the SR tuning is selected to be executed every third switching cycle, the CPU utilization is significantly reduced to 5%. The third benefit is that it can easily cooperate with the closed-loop control if the control cycle is the same as the closed-loop control.

V. EXPERIMENTAL RESULTS

The proposed adaptive SR driving method based on the ripple detection is verified on a 130kHz 300W 380V/12V LLC converter as shown in Fig. 12, which is TI's demo board TMDSHVRESLLCKIT with the following circuit parameters: L_r = 55uH, C_r = 24nF, L_m = 280uH, C_O = 1.32mF. The controller is the 100MHz MCU TMS320F2808.

Fig. 12. The 130kHz LLC converter hardware

The experimental waveform of the proposed adaptive SR driving method based on the ripple detection is shown in Fig. 13. It is shown clearly that the body diode conduction is minimized in the steady state.

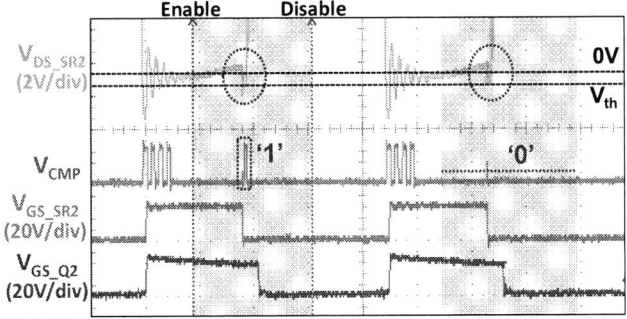

Fig. 13. Steady state waveform of the proposed adaptive SR driving method based on ripple detection on the 130kHz LLC converter

The extended proposed adaptive SR driving method by adding a ripple counter is verified on a 500kHz LLC converter as shown in Fig. 14. The 500kHz LLC converter is designed based on the matrix transformers for the LLC converters [17], which was originally a 1MHz 1kW 390V/12V unregulated DCX with GaN devices. Here the LLC converter is redesigned

as a 500kHz 1kW 390V/12V regulated DC/DC converter with Si devices. The parameters of the 500kHz LLC converter is: Lr = 4.5uH, Cr = 22nF, Lm = 22uH. The controller is the 60MHz MCU TMS320F28027.

Fig. 14. The 500kHz LLC converter hardware

The tuning process of the extended proposed adaptive SR driving method on the 500kHz LLC converter is shown in Fig. 15. The initial SR on-time is relatively small, which means the body diode conduction time is very large. The proposed adaptive SR driving would tune the SR on-time step-by-step until finally the body diode conduction is around the minimum point.

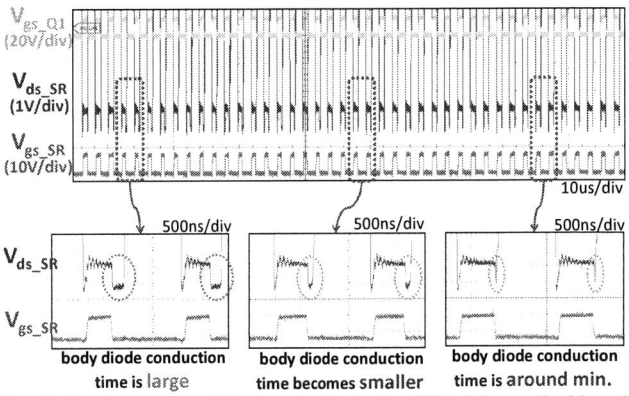

Fig. 15. Tuning process of the proposed adaptive SR driving method based on ripple counter on the 500kHz LLC converter

The efficiency curve of the 500kHz LLC converter with the extended proposed adaptive SR driving method is shown in Fig. 16 with a peak efficiency of above 95% and a full load efficiency of above 94%.

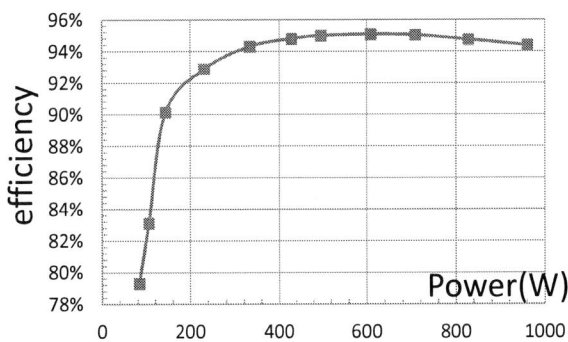

978-1-4673-9551-9/16 $31.00 © 2016 IEEE

Fig. 16. Efficiency of the 500kHz LLC converter with the proposed adaptive SR driving method based on the ripple counter

VI. CONCLUSIONS

In this paper, the adaptive SR driving for the LLC converters is investigated and its limitation with the digital implementation is explained in details. The previous adaptive SR driving methods require either the additional circuit or the high-performance digital controllers.

To solve this challenge, the adaptive SR driving method based on the ripple detection and its extension for the high frequency LLC converters are proposed. The original proposed method is suitable for the conventional LLC converters (resonant frequency is below 150kHz). It requires very little CPU utilization, thus, can be implemented by the low-cost digital controllers. The extension of the proposed method is suitable for the high frequency LLC converters. With the help of a ripple counter and by tuning the SR on-time every $(n+1)^{th}$ switching cycle ($n = 1, 2, 3…$), it can also be implemented by the low-cost digital controllers with the minimum CPU utilization.

The original proposed method is verified on a 130kHz LLC converter with a 100MHz microcontroller. It's very simple, requiring only a comparator. However, the CPU utilization for the SR driving will increase as the switching frequency increases, which means it is not quite suitable for the high frequency applications.

The extension of the proposed method is verified on a 500kHz LLC converter with a 60MHz microcontroller and a ripple counter. Compared with the original proposed method, an additional ripple counter is needed, however, the CPU utilization for the SR driving is reduced significantly.

Both methods require very little CPU utilization and can approach around the optimal point at the steady state.

REFERENCES

[1] B. Yang, F. C. Lee, A. J. Zhang, and G. Huang. "LLC resonant converter for front end DC/DC conversion." In *Applied Power Electronics Conference and Exposition, 2002. APEC 2002. Seventeenth Annual IEEE*, vol. 2, pp. 1108-1112. IEEE, 2002.

[2] B. Lu, W. Liu, Y. Liang, F. C. Lee, and J. D. Van Wyk. "Optimal design methodology for LLC resonant converter." In *Applied Power Electronics Conference and Exposition, 2006. APEC'06. Twenty-First Annual IEEE*, pp. 533-538. IEEE, 2006.

[3] X. Xie, J. Liu, F. N. K. Poon, and M. Pong, "A novel high frequency current-driven SR applicable to most switching topologies," *IEEE Trans. Power Electron.*, vol. 16, no. 5, pp. 635–648, Sep. 2001.

[4] D. Huang, D. Fu, and F. C. Lee, "High switching frequency, high efficiency CLL resonant converter with synchronous rectifier," In *Proc. IEEE Energy Convers. Congr. Expo.*, 2009, pp. 804–809.

[5] X. Wu, G. Hua, J. Zhang, and Z. Qian, "A new current-driven synchronous rectifier for series–parallel resonant (LLC) DC-DC converter," *Industrial Electronics, IEEE Transactions on*, vol. 58, no. 1, pp. 289–297, Jan. 2011.

[6] NXP Semiconductors, "TEA1795T: GreenChip synchronous rectifier controller" (Nov. 2010). [Online]. Available: http://www.nxp.com/documents/data_sheet/TEA1795T.pdf

[7] International Rectifier, "IR11682S: DUAL SmartRectifier DRIVER IC" (Jul. 2011). [Online]. Available: http://www.irf.com/product-info/datasheets/data/ir11682spbf.pdf

[8] STMicroelectronics, "Synchronous rectifier smart driver for LLC resonant converters" (Aug. 2013). [Online]. Available: http://www.st.com/st-web-ui/static/active/en/resource/technical/document/datasheet/CD00282226.pdf

[9] D. Fu, Y. Liu, F. C. Lee, and M. Xu, "A novel driving scheme for synchronous rectifiers in LLC resonant converters," *Power Electronics, IEEE Transactions on*, vol. 24, no. 5, pp. 1321–1329, May 2009.

[10] L. Cheng, T. Liu, H. Gan, and J. Ying, "Adaptive synchronous rectification control circuit and method thereof," U.S. Patent 7.495.934, Feb. 24, 2009

[11] W. Feng, F. C. Lee, P. Mattavelli, and D. Huang, "A universal adaptive driving scheme for synchronous rectification in LLC resonant converters," *Power Electronics, IEEE Transactions on*, vol. 27, no. 8, pp.3775-3781, Aug. 2012

[12] B. Yang, R. Chen, and F. C. Lee. "Integrated magnetic for LLC resonant converter." In *Applied Power Electronics Conference and Exposition, 2002. APEC 2002. Seventeenth Annual IEEE*, vol. 1, pp. 346-351. IEEE, 2002.

[13] D. Fu, B. Lu, and F. C. Lee. "1MHz high efficiency LLC resonant converters with synchronous rectifier." In *Power Electronics Specialists Conference*, 2007. PESC 2007. IEEE, pp. 2404-2410. IEEE, 2007.

[14] D. Reusch, F. C. Lee, D. Gilham, and Y. Su. "Optimization of a high density gallium nitride based non-isolated point of load module." In *Energy Conversion Congress and Exposition (ECCE)*, 2012 IEEE, pp. 2914-2920. IEEE, 2012.

[15] X. Huang, Z. Liu, Q. Li, and F. C. Lee. "Evaluation and application of 600V GaN HEMT in cascode structure." In *Applied Power Electronics Conference and Exposition (APEC)*, 2013 Twenty-Eighth Annual IEEE, pp. 1279-1286. IEEE, 2013.

[16] S. Ji, D. Reusch, and F. C. Lee. "High-frequency high power density 3-D integrated gallium-nitride-based point of load module design." *Power Electronics, IEEE Transactions* on 28, no. 9 (2013): 4216-4226.

[17] D. Huang, S. Ji, and F. C. Lee. "LLC resonant converter with matrix transformer." In *Applied Power Electronics Conference and Exposition (APEC)*, 2014 Twenty-Ninth Annual IEEE, pp. 1118-1125. IEEE, 2014.

[18] W. Feng, F. C. Lee, and P. Mattavelli. "Simplified optimal trajectory control (SOTC) for LLC resonant converters." *Power Electronics, IEEE Transactions on* 28, no. 5 (2013): 2415-2426.

[19] C. Fei, W. Feng, F. C. Lee, and Q. Li. "State-trajectory control of LLC converter implemented by microcontroller." In *Applied Power Electronics Conference and Exposition (APEC), 2014 Twenty-Ninth Annual IEEE*, pp. 1045-1052. IEEE, 2014.

[20] C. Fei, F. C. Lee, and Q. Li. "Soft start-up for high frequency LLC resonant converter with optimal trajectory control." In *Applied Power Electronics Conference and Exposition (APEC), 2015 IEEE*, pp. 609-615. IEEE, 2015.

[21] C. Fei, F. C. Lee, and Q. Li. "Multi-step Simplified Optimal Trajectory Control (SOTC) for fast transient response of high frequency LLC converters." In *Energy Conversion Congress and Exposition (ECCE), 2015 IEEE*, pp. 2064-2071. IEEE, 2015.

[22] C. Fei, F. C. Lee, and Q. Li. "Light load efficiency improvement for high frequency LLC converters with Simplified Optimal Trajectory Control (SOTC)." In *Energy Conversion Congress and Exposition (ECCE), 2015 IEEE*, pp. 1653-1659. IEEE, 2015.

DIGITAL SYNCHRONOUS RECTIFICATION CONTROLLER FOR LLC RESONANT CONVERTERS

Maryam. S.Amouzandeh, Behzad Mahdavikhah,
Aleksandar Prodic,

Laboratory for Power Management and Integrated Switch-Mode Power Supplies
University of Toronto, Toronto, ON, Canada

Brent McDonald
Texas Instruments Incorporated
Dallas, TX, United States

Abstract—This paper introduces a robust, hardware efficient, mixed-signal control loop that governs the switching actions of the secondary-side synchronous rectifier (SR) switches of LLC resonant converters. The new SR control method minimizes switching and conduction losses of the SR switches through online optimization of their on-off timing using an auto-tuning process that takes into account the effect of parasitic elements, mainly leakage inductances. In this controller, the information from body diode conduction detection circuits across the SR switches and switching frequency available from digital controller are utilized by a digitally implemented auto-tuning algorithm to determine optimal switching times to achieve zero current switching. In comparison with the existing SR controllers the introduced solution has more precise driving scheme. Moreover, the controller requires simple hardware implementation

Experimental results obtained with a 350 W, 400 V to 12 V, isolated LLC experimental resonant converter, verify the operation of the introduced SR driving scheme and show efficiency improvement over the whole operating range resulting in up to 9% reduction of converter power losses during light load operation.

Keywords—LLC resonant converters, auto-tuned SR controller.

I. INTRODUCTION

The LLC resonant converter is becoming more and more popular in numerous applications [1-2], due to its ability to achieve zero-voltage switching (ZVS) for the primary side switches and zero-current switching (ZCS) for the secondary side diode rectifier [1]. For improving efficiency the diode rectifier can be replaced with a synchronous rectifier (SR), which can significantly reduce conduction losses. However, a major difficulty in achieving this is creation of the SR gate driving scheme [3-10], which needs to be adjusted for different operating condition.

Fig.1 shows an LLC resonant converter with SR on the secondary side. In previous publications [3-10] number of SR driving schemes have been proposed. The analog solutions [3-7] are mostly based on the SR current sensing, to generate SR driving signal by use of current transformers, which reduces the efficiency due to extra transformer winding resistance. Digital solutions utilizing computational power, rather than direct sensing, have also been presented [8]-[10].

This work of Laboratory for Power Management and Integrated SMPS is supported by Texas Instruments.

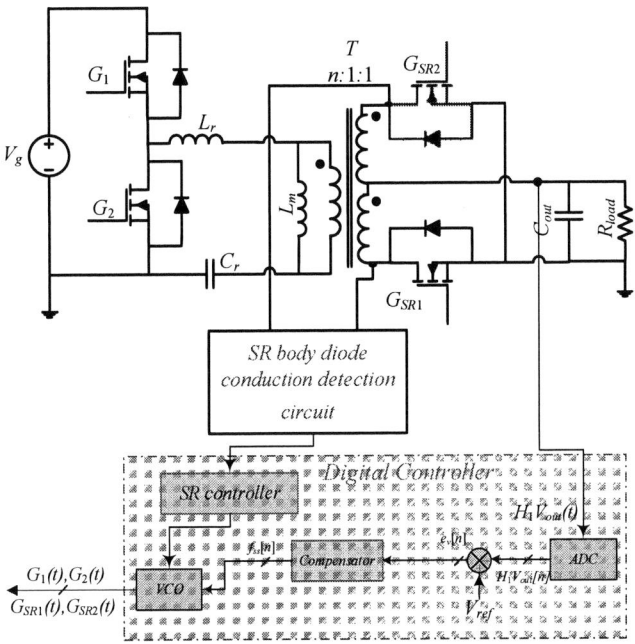

Fig. 1. LLC resonant converter with SR on the secondary side and digital controller

Among the most effective digital solutions is [8], in which the SR is turned on synchronously with the main switches and the turn off process is auto tuned. However, as demonstrated in [11], for a realistic converter, the optimal SR turn on time instant varies with respect the main switch turn-on time, depending on the operating conditions. The slow auto-tuned turn-off time adjustment introduced in that paper is precise, improving robustness and reducing susceptibility to the noise associated with the body diode conduction detection circuits [8].

This paper introduces a new hardware-efficient control method for driving the SR switches that can minimize losses associated with secondary side switches regardless of the converter's operating condition. The new SR controller utilizes the readily available switching frequency information from the digital controller and the body diode detection circuit (Fig.1). Based on the obtained information a simple algorithm adjusts both turn-on and turn-off timing instants, through a digital auto-tuning process. The paper is organized as follows: Section II studies LLC resonant converter operation and discusses the desired driving scheme. In Section III the principle of operation and the practical implementation of the introduced method is described. Section IV presents the experimental results and the conclusions are given in the last section.

II. LLC RESONANT OPERATION

To show importance of variable turn-on time of the SR switches, with respect to turn on timing of the primary switches, in the following subsection LLC waveforms in the presence of parasticics are analyzed. The desired SR gate driving signals for the LLC resonant converter are reviewed for two modes of operation, for the operations below and above the resonance, while taking into account the transformer leakage inductance and parasitic capacitors. The analysis presented in [11] and simulation results shown in the following figures demonstrate that, in order to achieve ZCS for secondary switches during operation below resonance, the SR must be turned on before turning on the primary side switch. On the other side, when operating above resonance [11] the SR switch should be turned on after the primary side switch commutation.

In order to achieve ZCS for the secondary side, the SR gating signals should be applied for the same time duration that the SR body diodes are conducting. Therefore, in order to find out the desired SR signals, an LLC resonant converter with a diode rectifier is simulated. Afterwards the diode rectifier is replaced with synchronous rectifier and the effect of driving the SR synchronously with the primary side is examined.

Fig.2 shows the simulation results for the LLC converter while operating below resonance and the corresponding equivalent circuits are shown in Fig. 3. A half cycle, i.e. period from t_0 to t_4, can be divided into four subintervals. In the subinterval t_0 to t_1, the equivalent circuit is shown in Fig.3.a, the output capacitor of the switch Q_1 is being discharged while the output capacitance of the switch Q_2 is charging. At the end of this subinterval the current through Q_1 is still negative, so the next subinterval (t_1 to t_2) starts and the body diode of Q_1 starts conducting along with SR diode, D_{SR1}. In order to achieve ZVS for the primary side switches, the switch Q_1 needs to be turned on during this subinterval. Fig.3.b shows the t_1 to t_2 subinterval, in which Q_1 and D_{SR1} are on. This subinterval ends when the magnetizing indictor current (I_{LM}) reaches the series inductor current (I_{Lr}), and the SR current reaches zero. During the third subinterval (t_2 to t_3) both diodes on secondary sides are off and magnetizing inductor makes a resonant circuit with resonance capacitor and inductor, C_r and L_r. This subinterval ends when the main switch turns off. The second half cycle is similar to first half cycle.

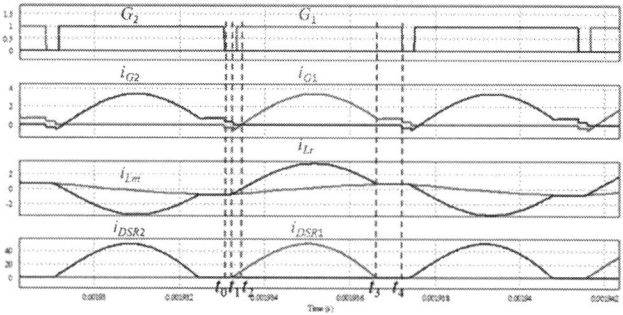

Fig. 2. Simulation results of LLC operating below resonance; G_1 and G_2 are main switches driving signal; I_{G1} and I_{G2} are main switches currents; I_{Lm} and I_{Lr} are magnetizing and tank inductor currents respectively; I_{DSR1} and I_{DSR2} are rectifier body diode currents.

A. LLC Resonant Converter With Diode Rectifier

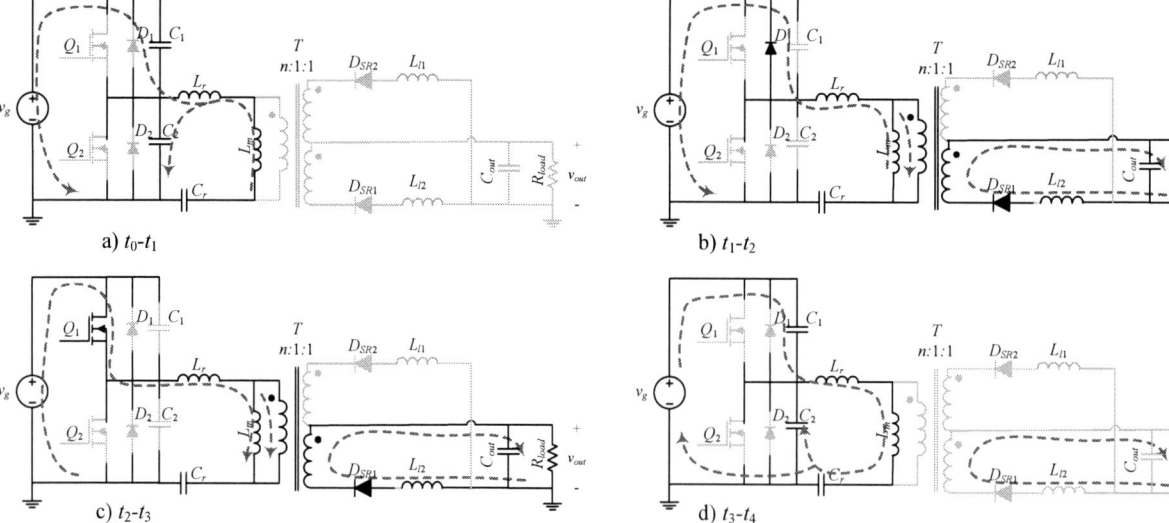

Fig. 3. LLC resonant converter equivalent circuits operating below resonance.

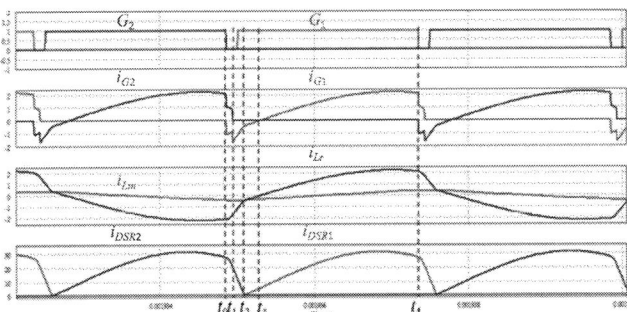

Fig. 4. Simulation results of LLC operating above resonance; G_1 and G_2 are main switches driving signal; I_{G1} and I_{G2} are main switches currents; I_{Lm} and I_{Lr} are magnetizing and tank inductor currents respectively; I_{DSR1} and I_{DSR2} are rectifier body diode currents.

Similarly, Fig.4 shows the simulation results for the LLC converter while operating above the resonant frequency. A half cycle can be divided into four modes of operation. The equivalent circuits are shown in Fig.5.

As shown in Figs. 2 and 4 the secondary side current starts growing slightly before the turn-on of the main switches while converter is operating below the resonant frequency ($f_s < f_0$),

and slightly after that when converter is operating above the resonant frequency ($f_s > f_0$). SR turn-on duration also depends on the converter's operating condition. Therefore, if the diode rectifier is replaced with a SR, the SR turn-on and turn-off time instants have to be executed precisely, in order to avoid circulating energy or the body diode conduction loss.

B. LLC Resonant Converter With Synchronous Rectifier

To examine the effect of inappropriate SR turn on, the simulation results of a LLC resonant converter with a SR on the secondary side has been presented here.

Fig.6.a shows the simulation results while operating below resonance and the turn on of the SR happens synchronously with the main switch. During the ΔT_{BR} the body diode conducts while the SR switch turns on after this time. This delay increases the conduction losses, which can be avoided/minimized by a proper SR switching scheme. Fig.6.b shows the same simulation results while operating above resonance. Again the SR turn on is in synch with the main switch. Simulation results show a reverse current through the SR in ΔT_{AR} time. This reverse current will cause circulating energy and decrease the efficiency.

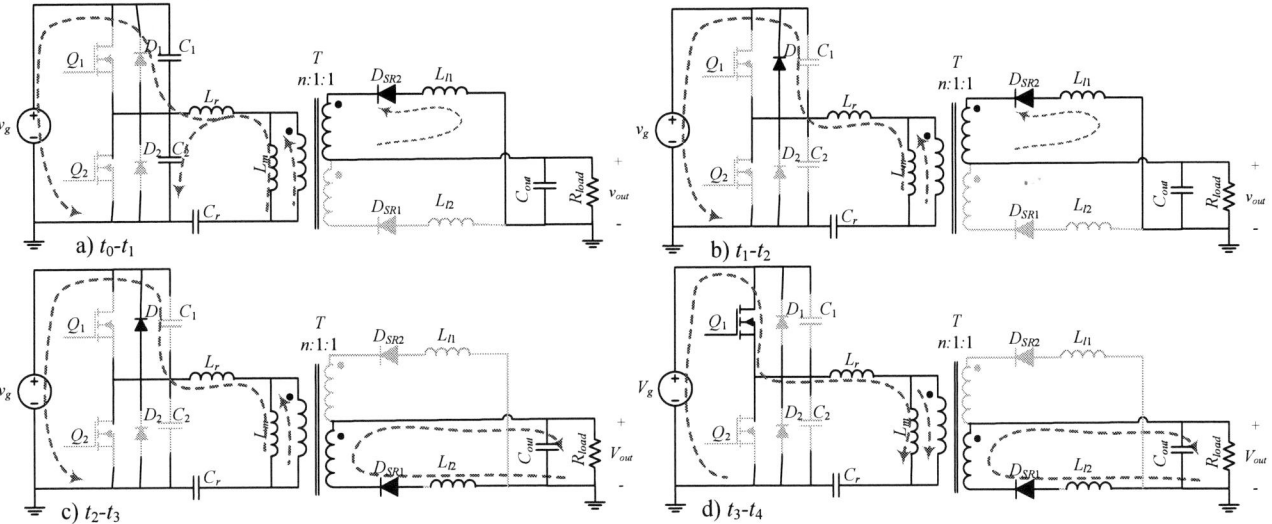

Fig. 5. LLC resonant converter equivalent circuits operating above resonance.

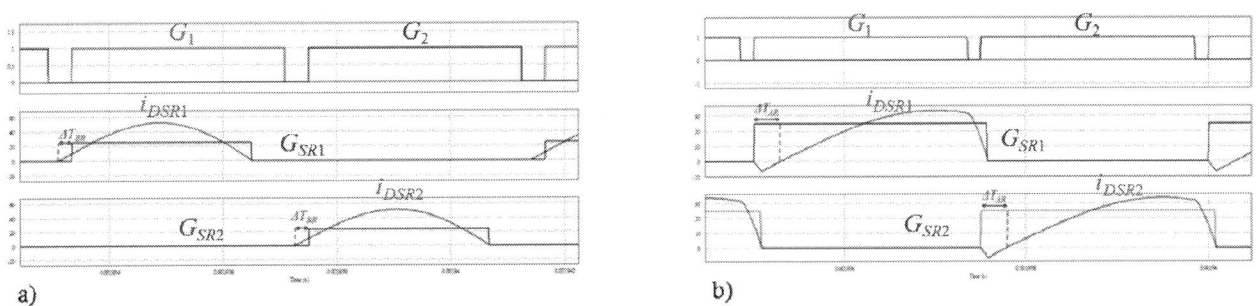

Fig. 6. Simulation results of LLC resonant converter with SR while SR synchronously turns on with the main switch. a) Operating below resonance. b) Operating above resonance; G_1 and G_2 are main switches driving signal; I_{DSR} and G_{DSR1} are the SR current and driving signals.

978-1-4673-9551-9/16 $31.00 © 2016 IEEE

III. PRINCIPLE OF OPERATION OF THE SR CONTROLLER AND PRACTICAL IMPLEMENTATION

Here, a practical and fast SR control scheme that optimizes switching sequence timing is introduced. It takes into account all the details discussed in the previous section. From the previous discussion it can be seen that it is highly desirable to turn on the SR switch right before the diode would be starting its conduction and to turn it off right after the diode would stop conducting. Fig.7 shows the controller flowchart that explains how the introduced SR driving/control scheme achieves that goal. At the moment the body diode conduction is sensed, by a detection circuit, the SR turns on and that time instant, in respect to the switching cycle is captured in register. In the next cycle, if the switching frequency remains the same, the controller takes into account the registered value, and SR turns on ΔT_{on} time earlier than in the previous cycle. Therefore except for the first cycle, in each operating point all other cycles start without body diode conduction. This process is activated each time a change of frequency is detected and, if the frequency remains the same, the algorithm is restarted every 20 cycles, to prevent possible errors.

A similar algorithm is used to find the SR turn-off time instant. For the first time in a particular switching cycle, the turn off time is determined based on the maximum allowable switching frequency (minimum time period T_m).

When the SR turns off the body diode starts conducting. The extra time that SR needs to remain in the on state, is the time duration which the body diode conducts or the duration of the active pulse at the output of the body diode detection circuit.

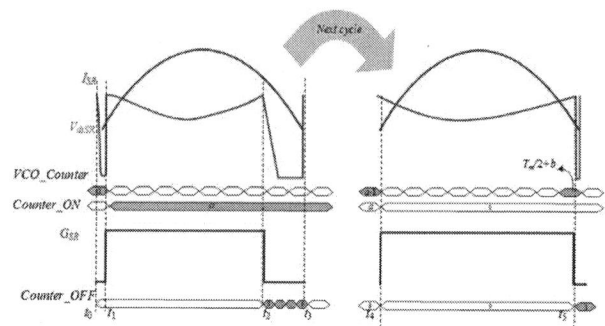

Fig. 8. Timing diagram, I_{SR} is the SR current, V_{dsSR} is the SR switch drain to source voltage, G_{SR} is the SR gating signal.

The remaining time (ΔT_{off}) is measured (by a counter and register) and added to the initial value of the turn on pulse width ($T_m/2$). In this way the turn off time instant is determined as $T_{off} = T_m/2 + \Delta T_{off}$. This algorithm restarts every time switching frequency changes, so it guarantees to work in all operating frequency regions. The implementation of the fast two step SR controller is shown in Fig.8. At t_0 the body diode conduction circuit detects conduction. SR switch turns on and the value of $VCO\_counter$ is registered as a $Counter\_ON$ variable at t_1. At t_2 the on time period reaches the maximum allowable time period and G_{SR} turns off. At this point body diode starts conducting and $Counter\_off$ starts measuring the time until body diode stops conducting at t_3. During the next cycle, when the $VCO\_Counter$ reaches $Counter\_ON$-1 at t_4, the G_{SR} turns on. At t_5, which is the time that $VCO\_counter$ reaches $Counter\_off+T_m/2$, G_{SR} turns off.

IV. EXPERIMENTAL RESULTS

To validate the performance and functionality of the introduced SR controller a modified 400 V to 12 V, 350 W half bridge LLC resonant converter evaluation board [12] is used. The existing controller of the board is replaced with a custom built FPGA controller, based on the diagrams of Figs.1 and 7. Fig.9.a shows the experimental result while the SR turns on synchronously with the primary switch. As mentioned in the previous section and as can be seen from the experimental results, the body diode start conducting for some time in the beginning of the cycle. Fig.9.b shows the operation with modified SR switch timing. It can be seen that with applied algorithm, resulting in proper on and off time the body diode conduction is avoided.

The efficiency results of both introduced and conventional method are shown in Fig.10. The introduced method improves efficiency by 1% at light loads (reduces losses by 9%) while converter is operating above resonance. As the load increases, the efficiency gains are minimized mainly due to the domination of conduction losses at higher power levels and partially because the converter operating frequency starts approaching resonant frequency and the optimum SR turn on time approaches the primary switch turn on time.

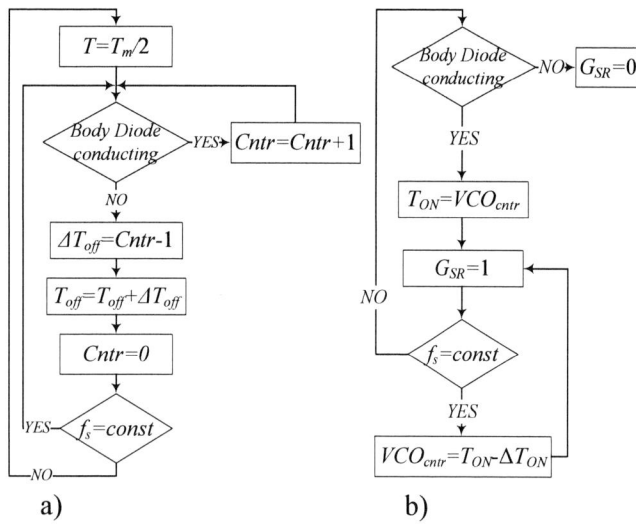

Fig. 7. a) SR turn off controller flowchart; b) SR turn on controller flowchart.

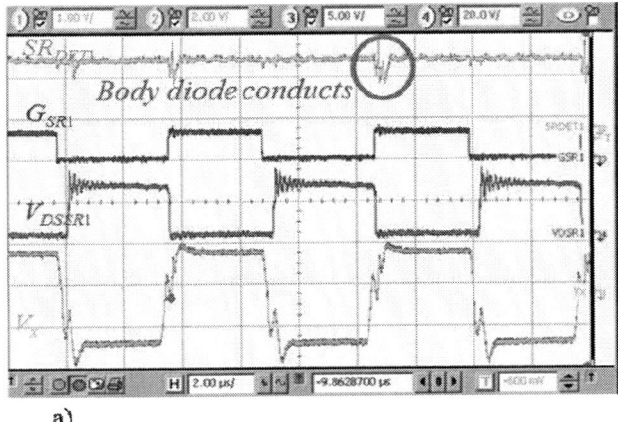

a)

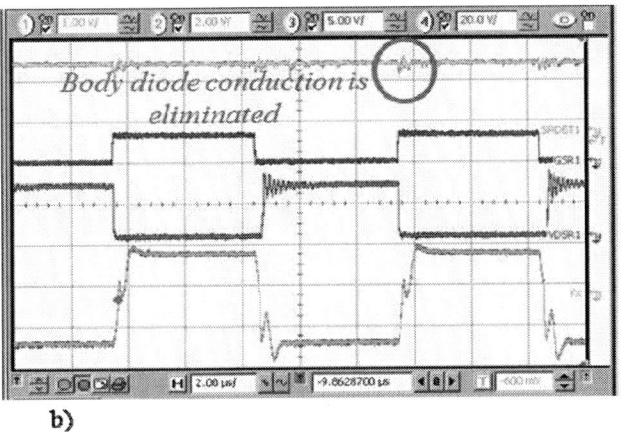

b)

Fig. 9. Experimental results, a) SR driving signals in synchronous with the primary side switches; b) SR driving signals with modified on time. SRDET1 is the QSR1 body diode conduction detection signal; GSR1 is the QSR1 gating signal; VDSSR1 is the QSR1 drain to source voltage 20V/div.; Vx primary side witching node voltage. Time is 2us/div

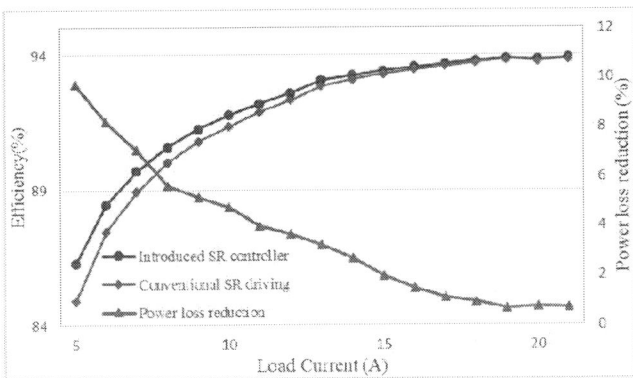

Fig. 10. Efficiency and loss reduction comaprison of the introduced and conventional method.

V. CONCLUSIONS

A simple and fast auto-tuned digital controller for optimization of the synchronous rectifiers' conduction times in LLC converters is introduced. The introduced solution eliminates the body diode conduction and avoids the reverse SR current, therefore reduces the conduction losses. Moreover the controller prevents reverse current build-up, therefore it limits the circulating energy and improves the efficiency. The driving scheme consists of two fast auto tuned processes, for turn-on and turn-off, which tune the SR driving signal in one cycle. The advantages of this method, which could potentially be used in other applications, over other existing methods are its precise driving signals and fast auto tuning, resulting in lower power losses. The effectiveness of the introduced method is verified experimentally, by comparison with a conventional SR driving scheme and improvements in efficiency are demonstrated.

VI. REFERENCES

[1] R. D. Middlebrook and S. Cuk, "A general unified approach to modeling switching-converter power stages," in Proc. IEEE Power Electron. Spec. Conf., 1976, pp. 36–57.

[2] Bo, Yang, and F.C. Lee, "LLC resonant converter for front end DC/DC conversion, " in Seventeenth Annual IEEE Applied Power Electronics Conference and Exposition, APEC 2002, 2002. vol.2: pp. 1 108- 1 1 12.

[3] X. Xie, J. Liu, F. N. K. Poon, and M. Pong, "A novel high frequency current-driven SR applicable to most switching topologies," IEEE Trans. Power Electron., vol. 16, no. 5, pp. 635–648, Sep. 2001.

[4] D.Huang, D. Fu, and F. C. Lee, "High switching frequency, high efficiency CLL resonant converter with synchronous rectifier," in Proc. IEEE Energy Convers. Congr. Expo., 2009, pp. 804–809.

[5] D. Fu, Y. Liu, F. C. Lee, and M. Xu, "A novel driving scheme for synchronous rectifiers in LLC resonant converters," IEEE Trans. Power Electron., vol. 24, no. 5, pp. 1321–1329, May 2009.

[6] International Rectifier, "IR11672AS: Advanced smart rectifier control IC," (Jul. 2011). [Online]. Available: http://www.irf.com/product-info/ datasheets/data/ir11672aspbf.pdf

[7] X.Wu, G. Hua, J. Zhang, and Z. Qian, "A new current-driven synchronous rectifier for series–parallel resonant (LLC) DC-DC converter," IEEE Trans. Ind. Electron., vol. 58, no. 1, pp. 289–297, Jan. 2011.

[8] Weiyi Feng; Lee, F.C.; Mattavelli, P.; Daocheng Huang, "A Universal Adaptive Driving Scheme for Synchronous Rectification in LLC Resonant Converters," Power Electronics, IEEE Transactions on , vol.27, no.8, pp.3775,3781, Aug. 2012

[9] Abe, Seiya; Yang, Sihun; Shoyama, Masahito; Zaitsu, Toshiyuki; Yamamoto, Junichi; Ueda, Shinji; Ninomiya, Tamotsu, "Adaptive driving of synchronous rectifier for LLC converter without signal sensing," Applied Power Electronics Conference and Exposition (APEC), 2013 Twenty-Eighth Annual IEEE , vol., no., pp.1370,1375, 17-21 March 2013

[10] Yu-Shan Cheng; Jing-hsiao Chen; Yi-hua Liu; Kuo-Liang Huang; Zong-zhen Yang, "Design of a digitally-controlled LLC resonant converter with synchronous rectification," Future Energy Electronics Conference (IFEEC), 2013 1st International , vol., no., pp.772,776, 3-6 Nov. 2013

[11] Ki-Bum Park; Byoung-Hee Lee; Gun-Woo Moon; Myung-Joong Youn, "Analysis on Center-Tap Rectifier Voltage Oscillation of LLC Resonant Converter," Power Electronics, IEEE Transactions on , vol.27, no.6, pp.2684,2689, June 2012

[12] (Jul. 2013). Texas Instrument, UCD3138LLCEVM-028: Digitally Controlled LLC Resonant Half-Bridge DC-DC Converter [Online]. Available: http://www.ti.com/lit/ug/sluu979a/sluu979a.

A Novel Adaptive Synchronous Rectification Method for Digitally Controlled LLC Converters

Fan Wang, Brent A. McDonald, Jeff Langham and Bo Fan

High Voltage Power Solutions, Texas Instruments Inc.

Dallas, TX, United States

Email: {fan-wang, b-mcdonald, jlangham, ball-fan}@ti.com

Abstract— In this paper, a novel adaptive synchronous rectification method for digitally controlled LLC converters is proposed. By sensing the synchronous rectifier (SR) body diode forward drop, both the SR turn-on and turn-off edges are optimized for efficiency. Negative current prevention is utilized to improve the system robustness and is enhanced by simple digital control capabilities. Compared with a conventional analog SR control approach, this method achieves higher system efficiency and flexibility. This control method has been implemented in a Texas Instruments digital power controller UCD3138A and a companion gate driver UCD7138 [1][2].

Keywords—synchronous rectification; LLC resonant converters; digital control

I. INTRODUCTION

LLC resonant converters are becoming increasingly popular for their ability to achieve high efficiency. For an even higher improvement in efficiency, the secondary side rectification diodes are replaced with synchronous MOSFETs to reduce the conduction losses. However, the control of SRs is non-trivial as both the turn-on and turn-off edges should be exactly at the SR current zero crossing point, which will move during operation depending on input voltage and load current. In addition, turning the SR on/off too early or too late will result in lower efficiency, negative current, or high drain-to-source stresses. Substantial research and commercial products have been focusing on providing an improved SR control method for LLC converters [3-8]. Fig. 1 shows a half bridge LLC resonant converter with SR MOSFETs.

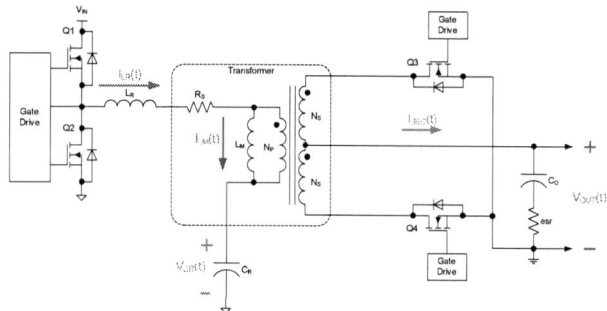

Fig. 1 Half Bridge LLC Resonant Converter with SR MOSFETs

State-of-the-art LLC resonant converter SR driving strategies fall into two categories: 1) MOSFET turn-on resistance (R_{dson}) voltage drop sensing methodology [4][5][6];

2) SR pulse width clamp methodology [7][8]. In the first method, the MOSFET is turned off once the voltage drop at the MOSFET R_{dson} is detected above a small negative voltage threshold. Turn-on and turn-off blanking times are required to avoid false tripping. One drawback of this method is that the voltage drop on R_{dson} is often too small to detect, and varies with PCB layout and the type of MOSFETs being used. In high current applications where there are several MOSFETs in parallel, the R_{dson} is so small that the MOSFETs may be turned off when the current is still large.

In the second method, at frequencies equal to or above the resonant frequency, the SR turn-off edge varies with switching frequency. At below resonant frequency, the SR pulse-width is clamped to half of the resonant period. This method is simple to implement with digital power controllers; however, it requires resonant tank information to program the SR pulse width clamp value. The calibration process of the resonant tank generally takes too much time for a factory application, which drives production costs higher [7][8].

Table I is a summary of the two categories of state of the art LLC SR control methods.

TABLE I. STATE OF THE ART LLC SR CONTROL METHODS

Method	Advantages	Disadvantages
Rdson voltage drop sensing	- Does not require gate drive signal input	- Not suitable for high current application with several MOSFETs in parallel - Layout parasitic challenges
SR pulse width clamp	- Digital controller friendly - Simply system configuration	- Requires calibration in production - No SR turn on edge optimization

A digital adaptive scheme for LLC SRs is discussed in [3], however, there are two issues with this method, which makes the load/line transient operation less robust: (1) The primary and SR turn-on edges are time aligned, while the SR turn off edge can extend beyond the primary side turn-off edge. This may result in a shoot through or drain-to-source over stress condition. (2) At the SR turn-off edge, there is no special body diode conduction detection region. When the SR on time is too long and there is negative current, the body diode may

978-1-4673-9551-9/16 $31.00 © 2016 IEEE

conduct as well. Without specific detection region, the SR on-time may adjust in a direction which decreases efficiency and over stresses the MOSFETs.

To overcome the previously mentioned drawbacks of the existing solutions, a novel body diode conduction time (DCT) sensing adaptive SR control method is proposed. Compared with conventional solutions, the benefits of the proposed method are:

- Achieve high efficiency over a wider load range.

- Large signal detection, easy layout, no parasitic concern.

- No minimum on/off time or blanking time constraints.

- Better noise immunity compared with competitive solutions.

- Good performance for both low current applications and high current applications (i.e. with MOSFETs in parallel).

- Robust transient operation.

The second section will describe the operation principles in detail. In the third section, a hardware implementation method is outlined. In the fourth section, experimental results are presented and discussed. The last section concludes the paper and summarizes the benefits.

II. BODY DIODE CONDUCTION SENSING BASED ADAPTIVE SR CONTROL

A. Desired SR Operation

Without optimization, the SR on time can either be too long or too short. This can be observed on the SR drain-to-source voltage (V_{DS}) waveforms, as shown in Fig. 2 first row. The desired waveforms are shown in the second row.

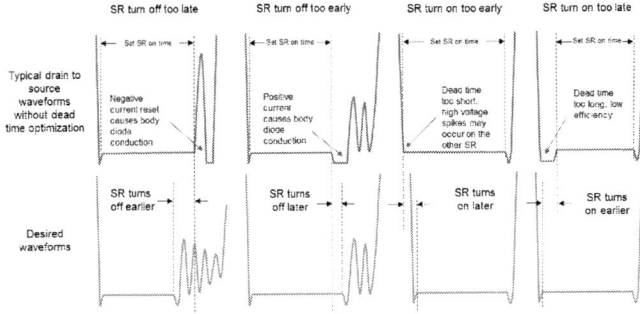

Fig 2: Desired SR V_{DS} Waveforms

As seen in Fig. 2, the turn-on edge can be optimized by turning on the SR only after a brief body diode conduction period is detected; the turn-off edge optimization adds difficulty because in both cases the body diode conducts.

The reason for SR body diode conduction when SR on-time is too long is that the negative current charges up the SR MOSFET drain-to-source capacitance. Fig. 3 shows the SR current and the V_{DS} waveform and demonstrates the circuit behavior when SR on-time is too long.

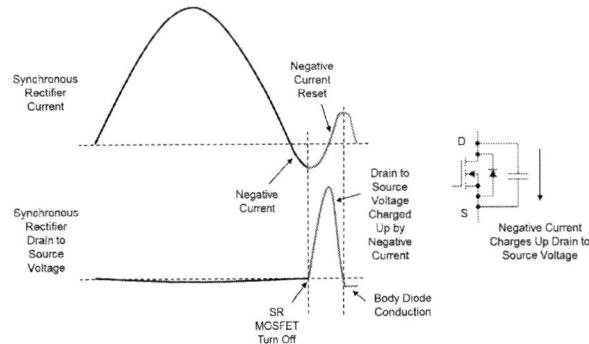

Fig.3 LLC Converter V_{DS} Shoot Up When Negative Current Flows

By way of monitoring the body diode conduction from V_{DS} waveforms, the SR turn off conditions, too early or too late, can be extracted. In order to isolate the conditions, the sensing circuit needs to be active at specific times.

B. Body Diode Conduction Detection at SR Turn-Off Edge

The body diode conduction detector is a comparator with a −150 mV threshold and outputs a logic low when body diode conducts. Fig.4 shows the two cases of SR body diode conduction as described in the previous section:

1) MOSFET turns off too early, positive current flows; body diode conducts right after SR turning off;

2) MOSFET turns off too late, negative current flows; drain to source voltage shoot up first, and then body diode conducts.

In each case, the SR on time should be adjusted in opposite directions; ergo a detection window is utilized to distinguish between the two cases. The detection window begins immediately after the SR gate drive turns off, and the length of the detection window is configured based on the delay in the circuit. If an extended period of the comparator is low, this indicates the SR turned off too early. If the comparator is low for a short timeframe, this indicates the SR turned off too late. Based on this information, the digital controller adjusts the SR on time of the next switching cycle to increase efficiency and prevent damage to the SR.

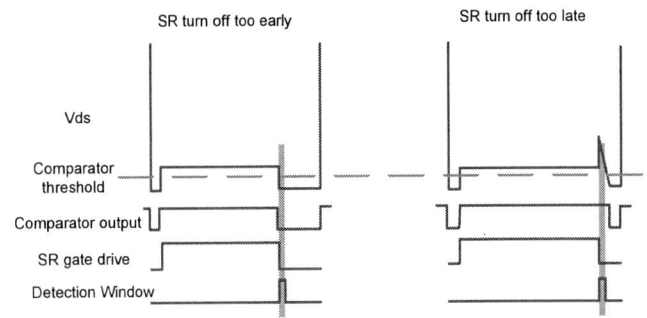

Fig. 4 Body Diode Conduction Detection

C. SR Turn-On and Turn-Off Edge Optimization

In the proposed method, the digital controller will output SR gate drive commands based on the SR clamp method as

978-1-4673-9551-9/16 $31.00 © 2016 IEEE

mentioned in the first section of this paper. This SR gate drive command is called IN. It has a rising edge with a fixed dead time to the primary side gate drive signal. The turn-off edge of IN transitions together with the primary side gate drive turn-off edge at above or equal to resonant frequency. At lower than resonant frequency, the turn-off edge of IN is fixed at half of the resonant period.

The output of the DCT comparator is low when the body diode conducts.

The actual SR gate signal is determined by both the digital controller (IN), and DCT. The OUT signal can only be high when IN is high. If at IN rising edge, DCT is already low, turn on the gate driver output immediately; if at IN rising edge, DCT is still high, turn on the gate driver output as soon as the DCT falling edge is received.

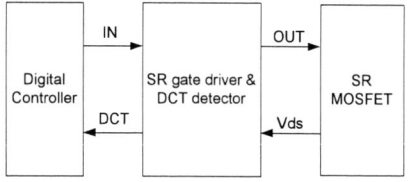

Fig. 5 Block Diagram of the Proposed Method

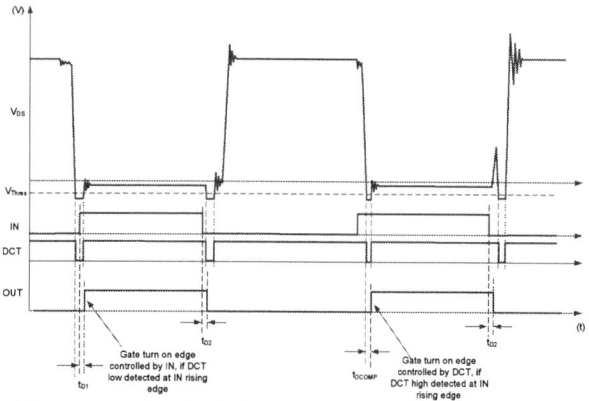

Fig. 6 Turn-On Edge Optimization

On the SR gate driver and DCT detector side, the falling edge of OUT is determined by IN only. The gate is turned off immediately at the IN falling edge. The optimization of the SR turn off edge is done by the digital controller by utilizing a high resolution digital counter to determine how long DCT is low during the detection window. If during the detection window the DCT low time is too short, the falling edge of IN will be moved backward by 4ns in the next cycle. Conversely, if the DCT low time is too long, the falling edge of IN will be moved forward by 4ns.

Fig. 5 shows the block diagram of the proposed method, and the critical signals mentioned in this section. The turn on edge optimization is illustrated in Fig. 6. The turn off edge optimization can be explained by Fig. 4.

D. Negative Current Prevention

When the SR pulse is on for too long, the drain to source voltage will peak momentarily due to negative current, as shown in Fig. 2 and Fig. 3. There are two possible ways to detect this situation:

1) to set a positive threshold to detect the V_{DS} peak;

2) or use the same detection window and body diode conduction detector as describe in the previous section – when there is no body diode conduction during the detection window, negative current may have occurred and the digital controller will take action to protect the system from damage.

To avoid adding another comparator, the method implemented in this paper is Method 2). The action is to reduce SR on-time by a pre-programmed amount.

III. HARDWARE IMPLEMENTATION

To prove the proposed method, a digital controller is implemented in silicon to control a 360W LLC resonant converter and matched with a gate driver IC with the DCT detection comparator.

The advanced dead time control module in the digital controller contains a high resolution timer capture module to measure the DCT pulse width received from the DCT detector, a dead time adjustment accumulator to set the SR turn off edge offset from the original position, and a fault action module. Table II shows the key registers and detailed functions implemented in the digital controller. There are clamps implemented on the dead time adjustment accumulator to avoid over adjustment.

Fig. 7 shows how the DCT pulse width is measured and how the digital controller subsequently takes action to adjust the SR on time. This figure shows that a DCT detection window is determined by a DCT blanking time register and a DCT detection window length register.

TABLE II.   HARDWARE IMPLEMENATION INTERFACE REGISTERS

DCT0[6:0]	SR0 diode conduction time, updated every switching cycle
DCT1[6:0]	SR1 diode conduction time, updated every switching cycle
DCT0DELTA_FW[9:0]	SR0 firmware write value
DCT1DELTA_FW[9:0]	SR1 firmware write value
MEAS_WINDOW[7:0]	Measurement window pulse width
MEAS_WINDOW_BLANK[6:0]	Measurement window blanking time
DCT_TARGET[6:0]	Body diode conduction time regulation target
DCT_TARGET_HYSTERESIS[1:0]	Body diode conduction time regulation target hysteresis, positive offset.
SR_NL_CTRL_THRES[3:0]	SR pulse width big step back off threshold, LSB is 4ns. Set to 0 to disable big step back off
SR_NL_STEP[2:0]	SR pulse with big back off step size
SR_ADJ_EN	Enable or disable SR increase/decrease by 4ns function, default is disabled.
SR_MIN_CLAMP[9:0]	Min DEADTIME_ADJ clamp.
SR_MAX_CLAMP[9:0]	Max DEADTIME_ADJ clamp.
SR_INTERRUPT_EN	Interrupt enable bit.
SR_MAX_FAULT_COUNT[1:0]	SR sequential fault counter configuration register

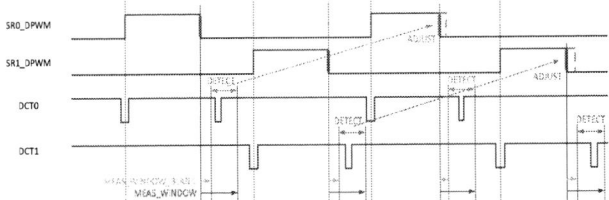

Fig. 7 Hardware Implementation – Digital Controller

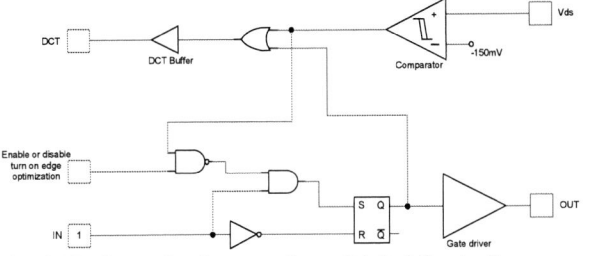

Fig. 8 Hardware Implementation – Digital Controller

As shown in Fig. 7, the DCT signal is only detected during the narrow detection window. Any signal outside of the detection window is ignored, which achieves good noise immunity when compared with the conventional method mentioned in the initial section.

In Fig. 8, the system block diagram of the SR gate driver together with the DCT detector is illustrated. As can be seen from the figure, the turn on edge optimization can be implemented in a few basic digital building blocks.

IV. EXPERIMENTAL RESULTS

A. Cases of Body Diode Conduction

Fig. 9 shows the different cases of body diode conduction measured on a 360W 400V input, 12V output half bridge LLC application.

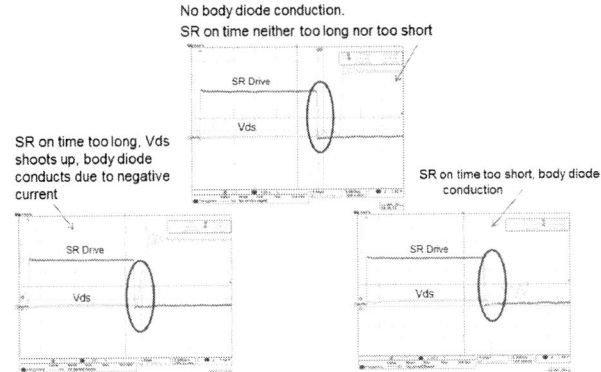

Fig. 9 Cases of Body Diode Conduction Waveforms Measured on a Typical Half Bridge LLC Application

B. Body Diode Conduction Detector Output and V_{DS} Signal

Fig. 10 shows the turn-off edge optimization waveforms. The upper left figure shows that when the SR on time is too long. The lower left figure shows that when the SR on time is too short. After optimization, both conditions get corrected

and a very small body diode conduction time (control target) is maintained.

Some blanking time is added to the detection window so that the noise spikes on V_{DS} waveform won't affect the results.

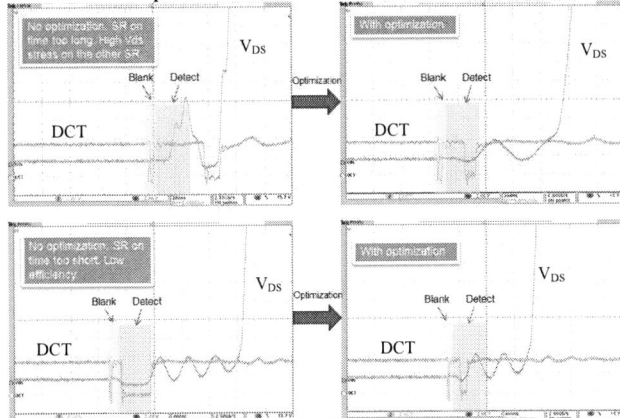

Fig. 10 Turn-Off Edge Optimization

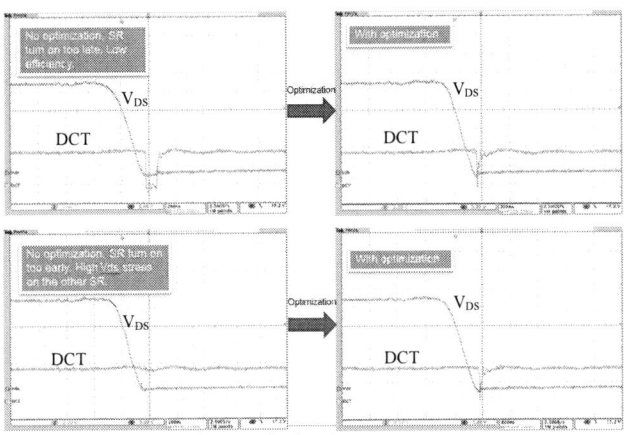

Fig. 11 Turn-On Edge Optimization

Fig. 11 shows the turn-on edge optimization waveforms. The upper left figure shows when the SR is turned on too late. The lower left figure shows when the SR is turned on too early. After optimization, both conditions are corrected. A very small dead time can be maintained at the gate driver turn on edge.

C. Transient Response

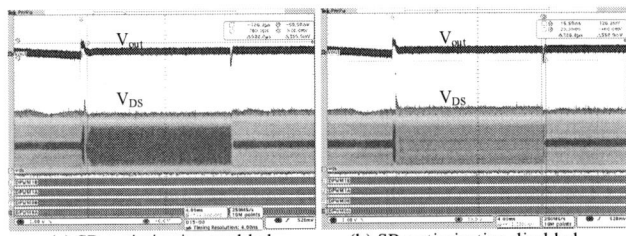

(a) SR optimization enabled   (b) SR optimization disabled

Fig. 12 0A~30A load transient, 1A/us

Fig. 12 shows the load transient response of the 360W LLC converter. From the figure, with or without SR optimization

978-1-4673-9551-9/16 $31.00 © 2016 IEEE

enabled, the V_{DS} stress has no big difference. The V_{DS} stress with SR optimization enabled is slightly lower. This is due to the fast negative current protection scheme.

D. Transient Response

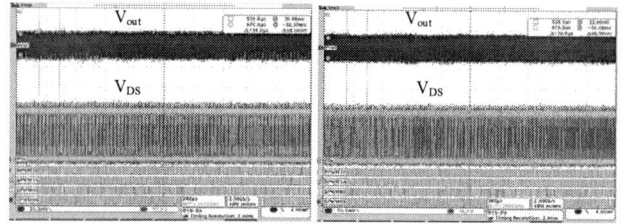

| (a) SR optimization enabled | (b) SR optimization disabled |

Fig. 13 Steady State Operation with V_{IN} = 380V, I_{OUT} = 15A

From Fig. 13, if dead times are properly tuned, the output voltage ripple and V_{DS} stress have no difference in steady state operation with or without SR optimization. The reason is that the SR optimization in steady state is done in 4-ns adjustment step size, which is fine enough to prevent extra jitter. With SR optimization enabled, dead time fine tuning can be greatly simplified.

E. Efficiency Improvement

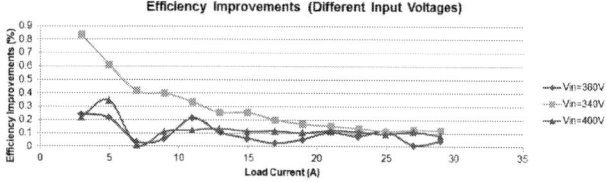

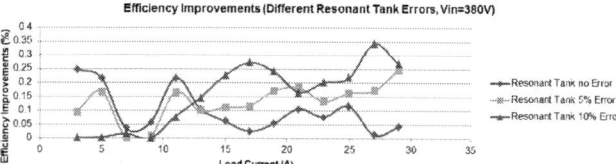

Fig. 14 Efficiency Improvements

Compared with conventional digital control, the proposed method can improve the system efficiency by up to 0.8% with the experiment setup. The upper diagram in Fig. 14 shows the efficiency improvements in different input voltages. The lower diagram in Fig. 14 shows three test conditions all with 380V input voltage, but with different percentage error in resonant tank values. The proposed method helps more when the resonant tank error goes bigger.

V. CONCLUSION

The proposed adaptive synchronous rectification method overcomes the drawbacks of various conventional analog and digital SR control method, resulting in higher system efficiency and improved robustness. With body diode conduction sensing, both the turn on edge and turn off edge of the SRs are optimized. Compared with the R_{dson} sensing method [4][5][6], the proposed method senses body diode forward drop, which is a large signal and is minimally affected by layout and load current. This makes the hardware design easier to implement, and results in a more robust solution. Comparing to the SR pulse width clamp method [7][8], the proposed method eliminates the need of resonant tank calibration in production. Comparing with previously proposed digital SR adaptive control method [3], the proposed method achieves robust transient operation and provides a practical solution.

REFERENCES

[1] UCD3138A datasheet: http://www.ti.com/lit/ds/slusc66b/slusc66b.pdf, Texas Instruments

[2] UCD7138 datasheet: http://www.ti.com/lit/ds/slvscs1b/slvscs1b.pdf, Texas Instruments

[3] W. Feng, FC. Lee, P. Mattavelli, and D. Huang, "A Universal Adaptive Driving Scheme for Synchronous Rectification in LLC Resonant Converters," IEEE Transactions on Power Electronics, vol. 27, No.. 8, August 2012.

[4] IR11672 datasheet: http://www.irf.com/product-info/datasheets/data/ir11672aspbf.pdf , International Rectifier

[5] NCP4303 datasheet: http://www.onsemi.com/PowerSolutions/product.do?id=NCP4303 , ON Semiconductor

[6] MP6922 datasheet: http://www.monolithicpower.com/DesktopModules/DocumentManage/API/Document/getDocument?id=136 , Monolithic Power Systems

[7] UCD3138 datasheet: http://www.ti.com/lit/ds/symlink/ucd3138.pdf, Texas Instruments

[8] CM6901 datasheet: http://www.championmicro.com.tw/datasheet/Analog%20Device/CM6901.pdf, Champion Microelectronic Corporation

[9] UCD3138A LLC Evaluation Module: http://www.ti.com/tool/ucd3138allcevm150, Texas Instruments

Influence of the ADC Zero Bin on the Performance of an Integrated DC-DC Converter

S. Vesti, M. Agostinelli, H. Koltsov, and S. Marsili
Infineon Technologies Austria AG
Villach, Austria

Abstract— **The purpose of this paper is to provide a detailed description of a technique that modifies the characteristics of analog-to-digital converter (ADC). The overall design objective for any power converter is to reduce the size, cost and power losses. In a digitally controlled system-on-chip (SoC) application, this results in a tradeoff between the performance and resolution of the ADC and digital PWM and required silicon footprint, which translates to additional cost. Therefore, the main functional blocks of the digital control loop, the ADC and digital PWM are often realized with limited resolution which might introduce static inaccuracy. Therefore, in order to obtain tight regulation while relaxing the resolution requirements, a zero bin technique is applied. This method is a simple post-processing technique to avoid the "dead zone" of the ADC. However, it violates the conventional "no limit cycling" conditions but maintains the output voltage ripple to an acceptable level.**

I. INTRODUCTION

The main design target of low power dc-dc converter for a SoC application is to obtain inexpensive integrated circuit (IC) with small silicon area and high efficiency. In addition, requirements for static accuracy, dynamic regulation and system monitoring need to be fulfilled. Application of digital control provides many advantages over analog schemes, such as increased flexibility, possibility to implement more advanced control methods and special features [1-5]. Therefore, digital control scheme is widely used for power electronics converters in various applications [1, 6-8] and it has also gained popularity in the control of integrated low power dc-dc converters [1, 8-10]. However, digital control introduces various challenges within the overall system design.

A well-known disadvantage in the digital control implementation is limit cycling [11, 12], which is due to the quantization errors introduced by the ADC and DPWM blocks. In order to avoid it as well as achieve tight regulation, the conditions given in [11, 12] must be satisfied. These conditions set strict requirements to the quantization resolutions of the analog-to-digital converter and the digital PWM: the effective DPWM resolution must be higher than in the ADC. Therefore, in order to obtain good static accuracy both blocks are required to have high resolution thus increasing the design complexity as well as the power losses and silicon area.

Various methods exist to improve the overall performance of the digital control loop without significantly increasing the area and power consumption of the controller. In [13, 14] different ways to increase the effective DPWM resolution without augmenting the clock frequency are presented. For a better dynamic response, the ADC characteristics can be also modified [6, 15]. In [6] dead zone of the ADC is used to eliminate the output capacitor ripple in the feedback-loop of a low harmonic rectifier whereas in [15] a nonuniform ADC gain is introduced, resulting improved transient responses in comparison to a uniform A/D quantizer. The starting point for applying both of these modification methods is to comply with the no-limit cycling conditions.

The design and implementation of the digital control loop, especially the DPWM block significantly contributes to the area and power consumption of the overall IC [14,16,17]. Thus in order to obtain high efficiency as well as small size, ADC and DPWM blocks are realized with limited speed and resolution. However, this might lead to static inaccuracy which is typically not acceptable in regulator applications. In order to obtain tight output voltage regulation while relaxing the requirements for the ADC and DPWM, zero bin technique can be applied in order to avoid the ADC "dead zone". In [8] this method is briefly discussed, however, without providing theoretical description of the method or insight on the impact to the overall system.

The purpose of this paper is to introduce the zero bin technique in detail and its influence on the overall converter performance. This method violates the conventional "no limit-cycling" condition, but allows to relax the requirements set for the ADC and thus for the DPWM resolution while maintaining the output voltage ripple to an acceptable level. The advantage of the zero bin technique is that it modifies the ADC characteristics without the need for any additional hardware thus enabling optimized digital controller design for low power SoC applications. The rest of the paper is structured as follows: second chapter introduces the comprehensive theoretical background of the zero bin technique discussing its influence on the dc-dc converter performance. In Chapter III more detailed system description is provided and experimental results are shown in Chapter IV in time domain as well as in frequency domain illustrating the zero bin influence.

II. ZERO BIN TECHNIQUE

In order to avoid the problem of limit cycling and to achieve tight regulation, the conditions outlined in [11] are typically satisfied during the design of a digital controller for a dc-dc converter. However, given the limitations on the area and power consumption in the IC design, the resolution of the ADC and DPWM are often relaxed. Nevertheless, high static accuracy and good dynamic performance are still required. In order to achieve all of these goals, a post-processing of the ADC codes is proposed in this paper.

A. Principle

The main principle of the proposed solution is to change the ADC gain in the two first error bins around the reference voltage. This method is illustrated in Fig. 1, where the upper diagram shows the A/D quantizer characteristics and the lower diagram the output of the re-encoded/post-processed ADC. The main idea is to place the reference voltage (V_{ref}) of the feedback loop exactly at the threshold of two consecutive ADC codes e.g. 4 and 5, as demonstrated in the upper part of Fig. 1. The output of the ADC is then re-encoded as shown in the bottom part, where the two "zero bins" around the reference voltage are denoted as ZB.

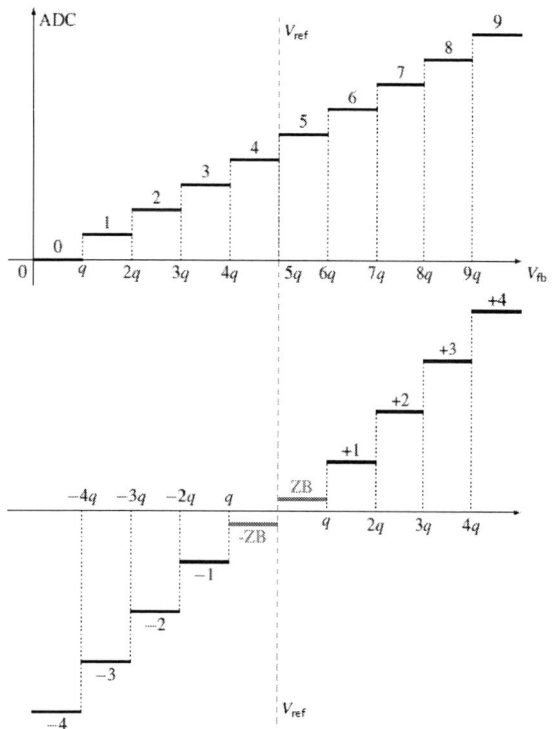

Fig. 1. A/D quantizer characteristics and post-processing of the ADC codes.

The zero bins are programmable and in the proposed concept, it is possible to set the ZB either to 0, LSB/8, LSB/4, LSB/2. Thus the ADC gain is effectively changed in the two error bins around the reference voltage. In Fig. 1, the zero bin setting corresponds to LSB/4 whereas setting the ZB to zero results in a dead zone of 2 LSBs around the reference voltage. The zero bin technique can be considered as an application of the nonuniform gain [15] in the two central error bins.

The describing function of the ADC quantizer [12] is plotted in Fig. 3, with the aforementioned ZB values. It is worth noting that changing the ZB setting results in different gains for small variations around the reference voltage. Therefore, depending on the programmed ZB value the loop-

gain magnitude changes while the phase remains unaltered. Therefore, the bandwidth and the phase margin of the system might be affected.

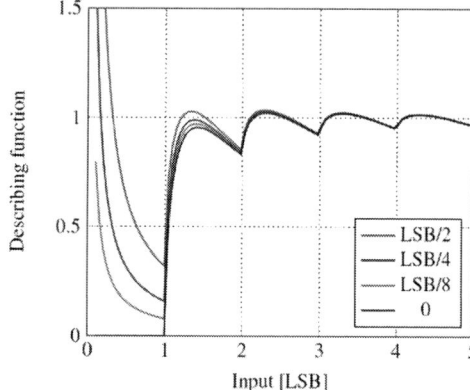

Fig. 2 Describing function of the ADC quantizer with different zero bin (ZB) gains.

B. Limit cycling

The condition to guarantee no-limit cycling requires that the DPWM resolution is higher than for the ADC. However, applying the ZB technique enables the system design without complying with this condition. While the ZB is set to a value other than 0, the system is forced to enter a high-frequency limit cycling condition since the output of the ADC (i.e. the calculated error) never equals 0. In fact, the duty cycle computed by the digital PID controller repeatedly changes at each switching period.

The net effect is that the output of the DPWM is dithered and then inherently filtered by the low-pass nature of the dc-dc converter itself, resulting in a low output voltage ripple. The precision of the regulation is improved due to the fact that any small deviation from the reference value causes an update of the ADC. Conversely in the conventional approach, where the reference voltage is placed in the middle of the LSB, the ADC output is not updated unless the error is larger than at least LSB/2. Therefore, the zero bin technique allows to significantly relax the requirements set for the ADC and DPWM, due to the constraints on the area and power consumption.

III. System Description

The proposed concept of the zero bin technique is applied to an integrated boost converter and the system structure is described in Fig. 3. The power management IC is emphasized with dashed lines and the digital blocks are illustrated with color. The main objective of the power management IC is to adjust the reference for the control loops, program the controller parameters as well as system monitoring [18,19].

The controller is implemented as full-custom digital and no external microcontrollers or FPGA's are used. The design objective is to reduce the silicon area as well as the power losses while complying with the performance requirements. As previously discussed, the DPWM is the main contributor on the digital power losses as well as in the size: the higher the required DPWM resolution, the higher the losses and the area [14,16,17].

The application of the ZB technique relaxes the resolution requirements for the DPWM and ADC while guaranteeing sufficient static accuracy because the zero bin around the reference voltage can be modified. Thus the application of this concept, allows a simple counter-based DPWM implementation in the actual system. The final digital controller of Fig. 3 includes additionally a $\Sigma\Delta$-modulator to further increase the effective DPWM resolution.

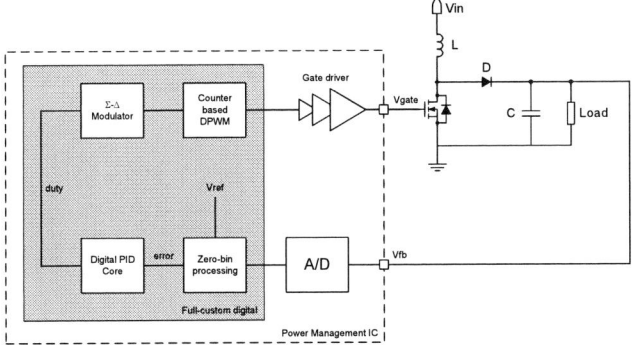

Fig. 3. Digital controller implementation of integrated boost converter.

The output voltage is sampled by the ADC and subsequent to post-processing the resulting error is fed to the controller, which is a conventional digital PID. While the ZB is set to a value other than 0, the calculated error never equals 0. Therefore, the duty cycle computed by the controller, changes repeatedly at each switching period. The PID core output is then fed into a second order $\Delta\Sigma$- modulator to increase the effective resolution of the DPWM. However, the usage of this modulator is not necessary in order to obtain sufficient static accuracy [8]. The implementation of the ADC post-processing as well as the $\Sigma\Delta$-modulator require additional digital blocks but they have negligible contribution on the overall area and power consumption.

IV. EXPERIMENTAL RESULTS

The influence of the zero bin technique is first illustrated in time domain for the integrated boost converter described in Fig. 3. The converter operates from input voltage of 6V-20V and regulates its output to 23.8V with maximum current of 270mA. The passive components are C =330µF, L = 47µH, and the switching frequency is 400kHz. The ADC samples the output voltage once per switching period and has a resolution of 100mV. For the following experimental results, the $\Sigma\Delta$-modulator is disabled in order to fully observe the influence of the ZB setting on the system performance.

A. Time domain analysis

The steady state output voltage is measured under different settings of the ZB and shown in Fig. 4 (time scale: 2µs/div) together with the switching node while the ZB is set to zero. Fig. 5 presents the same measurement condition but with

different time scale (5ms/div). From these figures low frequency limit cycling, is clearly observable. This is caused by the fact that the DPWM resolution is lower than the one of the ADC (100mV). While applying the ZB settings, the limit cycling can be suppressed to higher frequencies and filtered by the converter power stage. The measured steady state output voltage with the ZB setting of LSB/8 is shown in Fig. 6 using the same time scale than in Fig. 5. Based on these figures, it can be clearly observed that the ADC zero bin setting provides sufficient static accuracy. Additionally, the $\Sigma\Delta$-modulator can be applied to further increase the effective resolution of the DPWM.

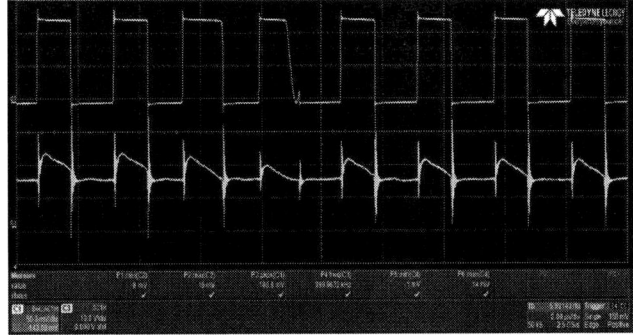

Fig.4. Boost converter output voltage, (yellow, 50mV/div) and the switching node (blue, 10V/div) under ZB setting 0 with time scale of 2µs/div.

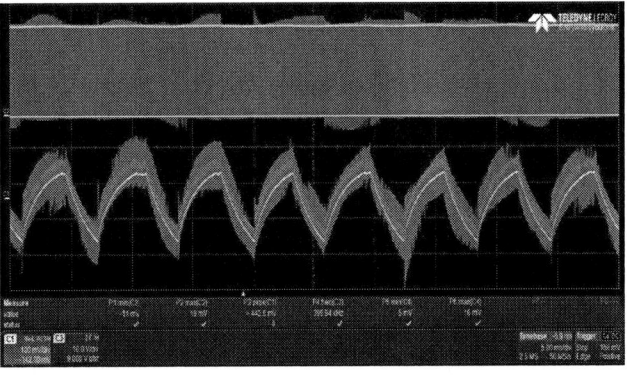

Fig. 5. Boost converter output voltage (yellow, 100mV/div) and the switching node (blue, 10V/div) under ZB setting 0 with time scale of 5ms/div.

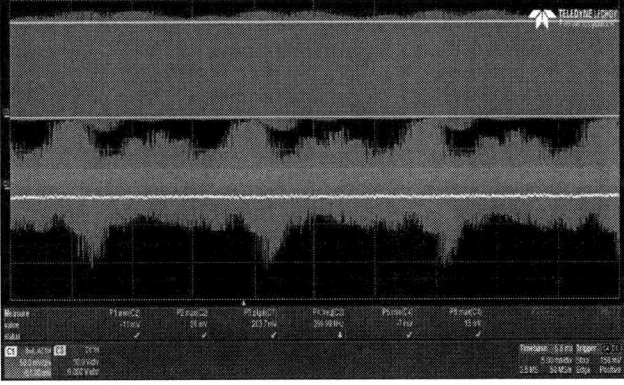

Fig. 6. Boost converter output voltage (yellow, 50mV/div) and the switching node (blue, 10V/div) under ZB setting LSB/8 with time scale of 5ms/div.

B. Frequency domain analysis

The measured loop-gains of the converter with different zero bin settings are shown in Fig. 7 with the input 7V and while the load current is 75mA. In the figure, the settings $ZB_{LSB/2}$ (solid line), $ZB_{LSB/4}$ (dashed line) and $ZB_{LSB/8}$ (dotted line) corresponds to the ADC offsets illustrated in Figs. 1 and 2. The frequency domain measurement of the loop gain while the ZB is 0 is not possible due to the steady state limit cycling. Based on the measured loop gains, it can be clearly observed that depending on the ZB setting, the gain is altered while the phase behavior remains the same thus validating the theoretical conclusions.

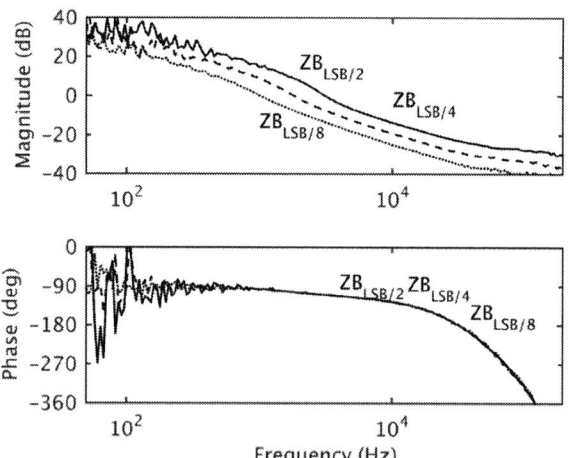

Fig. 7 Measured loop gains of the boost converter under different ZB settings.

V. CONCLUSIONS

In this paper a zero bin technique was introduced as a method to relax the resolution requirements for the quantizers in the digital control loop, while maintaining high static accuracy. This simple method modifies the ADC characteristics without the need of any additional hardware. This method was described in detail as well as its influence on the converter performance was analyzed. The presented method violates the traditional no-limit cycling conditions. Nevertheless, it enables the optimized controller design while maintaining sufficient static accuracy of the output voltage. This technique is especially useful in low power SoC applications where silicon area and power consumption need to be minimized. Experimental results were provided for a power management IC with full custom digital implementation in time and frequency domain. The results clearly validate the presented theoretical analysis.

REFERENCES

[1] A. Berger, M. Agostinelli, R. Priewasser, S. Marsili and M. Huemer, "Unified digital sliding mode control with inductor current ripple reconstruction for DC-DC converters," in *Proc. IEEE ISCAS,* 2015 pp.213-216.

[2] M. Shirazi, R. Zane, D. Maksimovic, "An Autotuning Digital Controller for DC–DC Power Converters Based on Online Frequency-Response Measurement," *IEEE Trans. Power Electron.,* vol. 24, no.11, pp.2578-2588, Nov. 2009.

[3] A. Costabeber, P. Mattavelli, S. Saggini, A. Bianco, "Digital autotuning of dc-dc converters based on model reference impulse response," in *Proc. IEEE APEC Conf.,* 2010, pp.1287-1294.

[4] S. Kapat, "Voltage-mode digital pulse train control for light load DC-DC converters with spread spectrum," in *Proc. IEEE APEC Conf.,* 2015 pp. 966-971.

[5] H. Siyu, R. M. Nelms, "Digital I^2 average current mode control for swith-mode power supplies," in *Proc. IEEE APEC Conf.,* 2015 pp.628-634.

[6] A. Prodic, D. Maksimovic, R. W. Erickson, "Dead-zone digital controllers for improved dynamic response of low harmonic rectifiers," *IEEE Trans. Power Electron.,* vol. 21, no.1, pp.173-181, Jan. 2006.

[7] M. Pahlevani, A. Bakhshai, P. Jain, "A novel digital peak-current-mode self-sustained oscillating control (PCM-SSOC) technique for a Dual-Active Bridge DC/DC converter," in *Proc. IEEE APEC Conf.,* 2015, pp. 3150-3156.

[8] F. Kuttner, H. Habibovic, T. Hartig, M. Fulde, G. Babin, A. Santner, P. Bogner, C. Kropf, H. Riesslegger, U. Hodel, "A digitally controlled DC-DC converter for SoC in 28nm CMOS," in Proc. *IEEE Int. Solid-State Circuits Conf.,* Feb. 2011 pp.384-385.

[9] M. Meola, A. Cinti, A. Kelly, "Controller scalability methods for digital Point Of Load converters," in *Proc. IEEE APEC Conf.,* 2015, pp.430-436.

[10] B. Patella, A. Prodic, A. Zirger, D. Maksimovic, "High-frequency digital PWM controller IC for DC-DC converters," *IEEE Trans. Power Electron.,* vol.18, no.1, pp.438-446, Jan. 2003.

[11] A. Peterchev, S. Sanders, "Quantization resolution and limit cycling in digitally controlled PWM converters," *IEEE Trans. Power Electron.,* vol.18, no.1, pp.301-308, Jan. 2003.

[12] P. Hao, A. Prodic, E. Alarcon, D. Maksimovic, "Modeling of quantization effects in digitally controlled DC–DC converters," *IEEE Trans. Power Electron.,* vol.22, no.1, pp.208-215, Jan. 2007.

[13] J. Mooney, M. Halton, P. Iordanov, V. O'Brien, "Dithered multi-bit sigma-delta modulator based DPWM for DC-DC converters," *Proc. IEEE APEC Conf.,* 2015 pp. 2835-2839.

[14] Z. Lukic, W. Kun A. Prodic, "High-frequency digital controller for dc-dc converters based on multi-bit Σ-Δ pulse-width modulation," in *Proc. IEEE APEC Conf.* 2005, pp. 35-40.

[15] H. Haitao, V. Yousefzadeh, D. Maksimovic, "Nonuniform A/D quantization for improved dynamic responses of digitally controlled DC–DC converters," *IEEE Trans. Power Electron.,* vol. 23, no.4, pp.1998-2005, Jul. 2008.

[16] L. Jian, F. C. Lee, Q. Yang, "New Digital Control Architecture Eliminating the Need for High Resolution DPWM," in *Proc. IEEE PESC,* 2007, pp.814-819.

[17] A. Syed, E. Ahmed, D. Maksimovic, E. Alarcon, "Digital pulse width modulator architectures," in *Proc. IEEE PESC,* pp.4689-4695.

[18] B. A. Mather, D. Maksimovic, I. Cohen, "Input Power Measurement Techniques for Single-Phase Digitally Controlled PFC Rectifiers," in *Proc. IEEE APEC Conf.,* 2009, pp.767-773.

[19] J. Morroni, A. Dolgov, M. Shirazi, R. Zane, D. Maksimovic, "Online Health Monitoring in Digitally Controlled Power Converters in *Proc. IEEE PESC,* 2007, pp.112-118.

Improved Current-Mode Control with Single-Cycle Load Transient

Virginia Li, Peihsin Liu, Qiang Li, Fred C. Lee
Center for Power Electronics Systems (CPES)
Virginia Polytechnic Institute and State University
Blacksburg, VA 24060 USA

Abstract—Non-linear single-cycle load transient improvement structures are proposed for current-mode controlled converters. Two popular variable-frequency current-mode controls are Constant On-Time (COT) and Ramp Pulse Modulation (RPM) due to their fast transient response. However, conventional COT and RPM both need multiple steps to increase the inductor current during load step-up transient. The fastest transient response is a single-cycle response. In this paper, two methods are proposed to achieve single-cycle load transient for current-mode control by improving the present control structures. For both methods of improvement, the operation mechanism of the control remains unchanged during steady-state. The proposed improvements can be used for RPM and COT in single-phase and multiphase.

Keywords—*Constant On-Time, Ramp Pulse Modulation, Control, VRM, Single-Cycle, Transient Improvement*

I. INTRODUCTION

For DC-DC converters, variable-frequency controls are widely used to improve the light-load efficiency, increase transient-response speed, and reduce the amount of output capacitors by utilizing high-bandwidth designs. One popular variable-frequency current-mode control is Constant On-Time control (COT), shown in Fig. 1 [1]-[3]. COT control achieves variable frequency by utilizing a fixed on-time (T_{on}). As the input and output of the system varies during operation, the frequency changes automatically to meet the load demand. However, during transient, the speed of the response is limited by the fixed T_{on}.

Prior methods have been proposed to improve the performance of COT during load step-up transient. One method is the Quick Response feature proposed by Richtek [4], [5]. In this method, a threshold-based undershoot-detection is utilized. Once undershoot is detected, a pre-determined longer T_{on} is used. However, with this method, the pre-defined threshold and pulse-width may create a ring-back issue.

The transient limitations of COT is improved by another variable-frequency current-mode control, Ramp Pulse Modulation (RPM), shown in Fig. 2 [6]-[9]. Unlike COT,

This work is supported by the Power Management Consortium in CPES, Virginia Tech.

This work was conducted with the use of SIMPLIS donated in kind by Simplis Technologies of the CPES Industry Consortium Program.

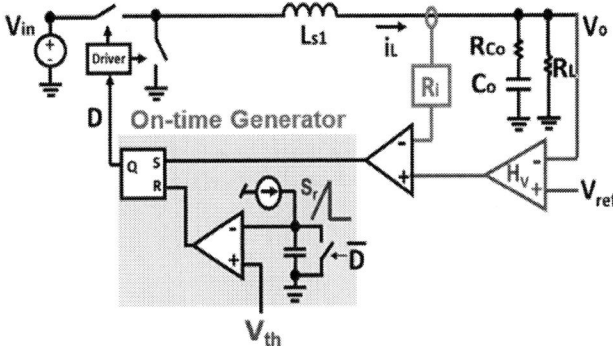
Fig. 1. General control scheme of COT.

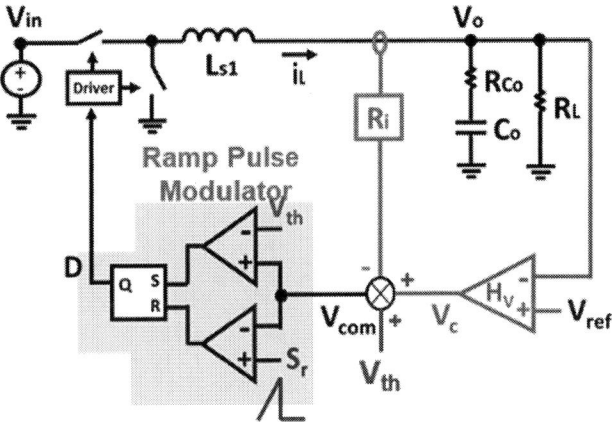
Fig. 2. General control scheme of RPM.

RPM operates with variable frequency and variable T_{on}. As a result, the transient performance of RPM is better than COT [10]. To fully understand these two control schemes, the steady-state and transient performances of the controls are examined.

II. PERFORMANCE OF COT AND RPM

A. Steady-State Performance

For a general control scheme of COT, the steady-state waveforms are shown in Fig. 3. From Fig. 3, the beginning of T_{on} is determined by the sensed inductor-current ($i_L R_i$) compared with the compensator voltage (V_c), and the end of T_{on} is determined by a threshold voltage (V_{th}) compared with the ramp (S_r). For a general control scheme of RPM, the steady-state waveforms are shown in Fig. 4. From Fig. 4, the beginning of T_{on} is determined by V_{th} compared with a V_{com} signal, which is the difference of V_c and $i_L R_i$, and the end of T_{on} is determined by V_{com} compared with S_r. For the two control schemes, the steady-state performances are similar, but the T_{on} for RPM is not fixed like COT. To fully see the difference, the transient behavior of the two controls are examined.

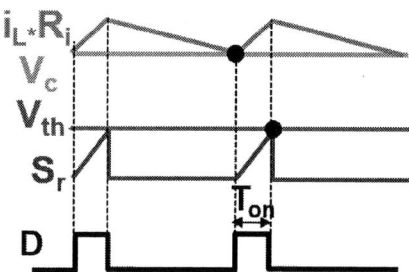

Fig. 3. Steady-state performance of COT.

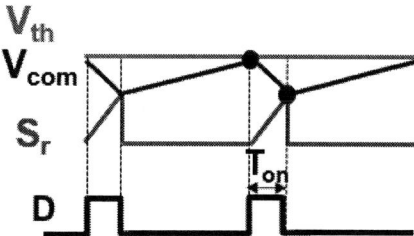

Fig. 4. Steady-state performance of RPM.

B. Load Transient Performance

From the transient waveforms, the limitations of COT during transient can be more clearly seen. The load step-up transient operation of COT is shown in Fig. 5(a). From Fig. 5(a), when the duty cycle saturates, the transient speed is limited by the fixed T_{on} and T_{off_min}. When this occurs, the energy transfer to the output is limited, causing an undershoot to be exhibited in the output voltage (V_o). The load step-down transient operation of COT is shown in Fig. 5(b). From Fig. 5(b), transient speed is also limited by the fixed T_{on}. When this occurs, the energy continues to be transferred to the output, causing an overshoot to be exhibited in V_o.

The load step-up transient operation of RPM is shown in Fig. 6(a). From Fig. 6(a), when the duty cycle saturates, T_{on} is automatically extended during transient due to V_{com} naturally operating higher than the steady-state range. Since the end of T_{on} is determined by V_{com} compared with S_r, the increase in V_{com} will extend T_{on}. The load step-down transient operation of RPM is shown in Fig. 6(b). From Fig. 6(b), T_{on} is automatically truncated during transient due to V_{com} naturally operating lower than the steady-state range. Due to the automatic T_{on} extension during load step-up and reduction during load step-down, RPM has better transient performance than COT. However, like COT, RPM may also require multiple cycles to reach steady-state after a load step-up transient occurs.

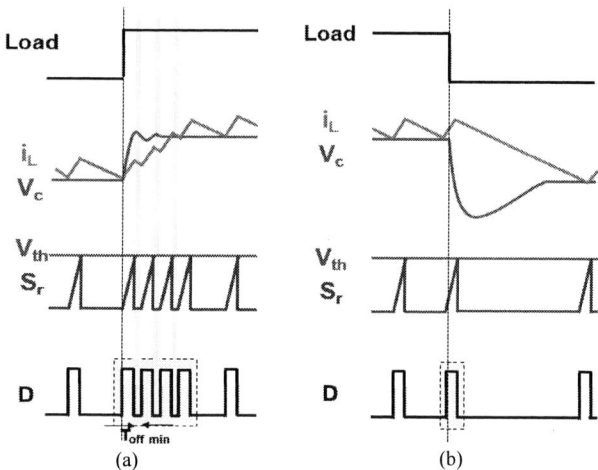

Fig. 5. (a) Load step-up, and (b) load step-down operation of COT.

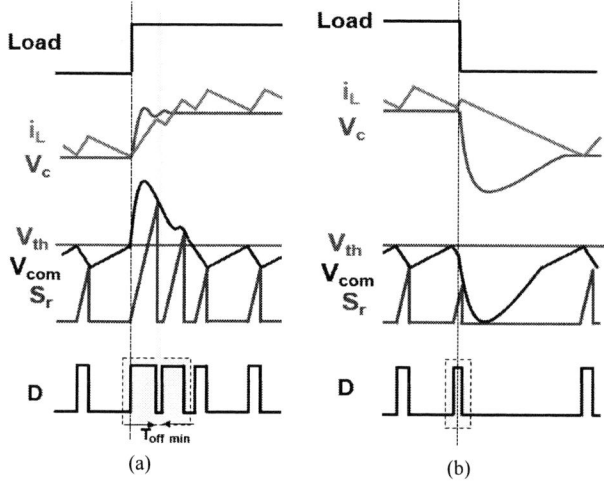

Fig. 6. (a) Load step-up, and (b) load step-down operation of RPM.

III. SINGLE-CYCLE TRANSIENT DETECTION

The fastest achievable transient is a single-cycle transient. In order to achieve a single-cycle response, the beginning and end of the transient must be determined. Using RPM as the bases, the concept of a single-cycle transient response is shown in Fig. 7. In Fig. 7, once transient is detected, the T_{on} pulse is extended for the duration it takes for the inductor current to achieve its next steady-state peak value.

By observing the V_{com} waveform, the two requirements can be achieved. Because V_{com} is the difference between V_c and $i_L R_i$, at steady-state, it operates within a steady-state band mainly determined by the inductor current ripple. During transient when the inductor is unable to follow the change in the load, V_{com} naturally increases higher than the steady-state band. As such, the beginning of transient can be detected using a threshold higher than the maximum steady-state operation of V_{com}. During steady-state operation, when the inductor current reaches its peak, V_{com} reaches its minimum (V_{com_min}). As such, steady-state can be detected by using a threshold at V_{com_min}.

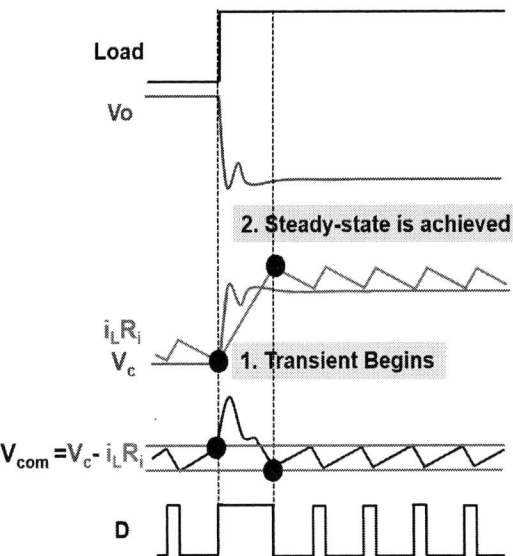

Fig. 7. Concept of single-cycle load step-up transient.

IV. A METHOD TO ACHIEVE SINGLE-CYCLE LOAD TRANSIENT

A. For RPM

The schematic of the proposed method, using RPM as an example, is shown in Fig. 8 and the transient waveform is shown in Fig. 9. During the steady-state operation, the upper bound of V_{com} is bounded by V_{th}. When load step-up transient occurs, V_{com} increases higher than V_{th} until the next steady-state is achieved. Due to this behavior, the beginning of transient can be detected using a V_{top} threshold slightly higher than V_{th}. When V_{com} is higher than V_{top}, transient is detected and the turn-off mechanism is modified such that the S_r is compared with a maximum voltage threshold (V_{max}) to turn off T_{on}. When V_{com} reaches V_{bot}, located at V_{com_min}, the turn-off mechanism goes back to normal operation of S_r compared with V_{com}. When this occurs, the change from V_{max} to V_{com} will intersect with S_r and turn off T_{on} to achieve single-cycle transient response.

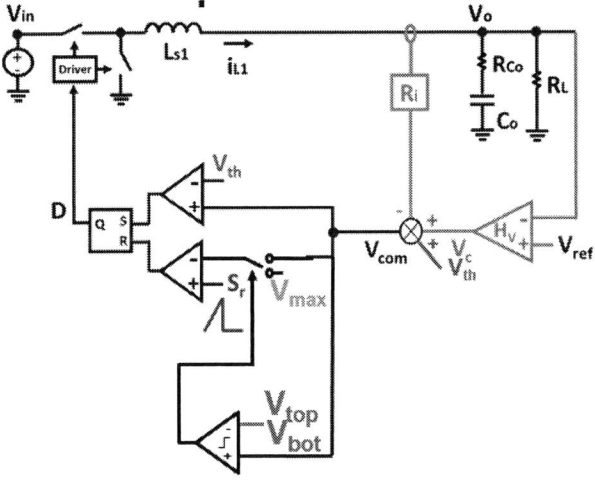

Fig. 8. RPM with proposed method to achieve single-cycle load transient.

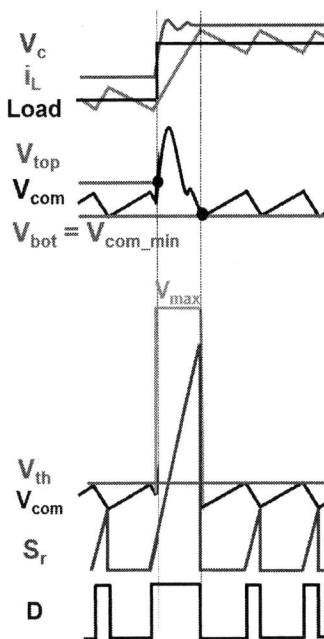

Fig. 9. Transient waveforms of RPM with proposed method.

The original RPM and proposed single-cycle RPM are simulated with V_{in}=8.4V, V_o=1.8V, T_{off_min}=200ns, L=280nH, f_s=800kHz, load step-up=22A, and di/dt=320A/us. The simulated transient waveforms are shown in Fig. 10 and Fig. 11. From the waveforms, single-cycle response is achieved using the proposed method and no undershoot is observed under worst-case scenario. Another case with small load-step change, load step-up=4A, is shown in Fig. 12. From Fig. 12, single-cycle response is also achieved with no ring-back issues with small load-step change. Taking a closer look at V_{com}, the extent and duration of which V_{com} operates higher than the steady-state operation range decreases as the load-step decreases. As a result, the T_{on} extension provided by the proposed method can be automatically adjusted with various load-step changes.

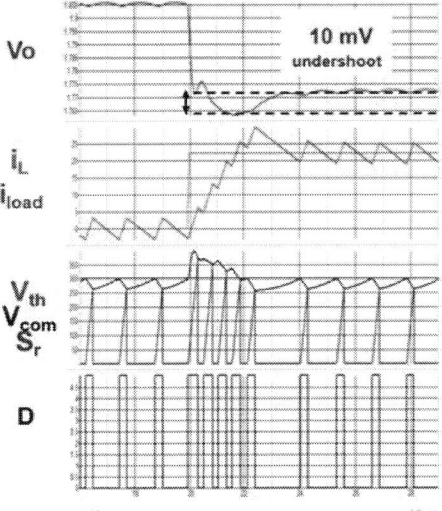

Fig. 10. Simulation results of the original RPM with V_{in}=8.4V, V_o=1.8V, T_{off_min}=200ns, L=280nH, f_s=800kHz, load step-up=22A, and di/dt=320A/us.

978-1-4673-9551-9/16 $31.00 © 2016 IEEE

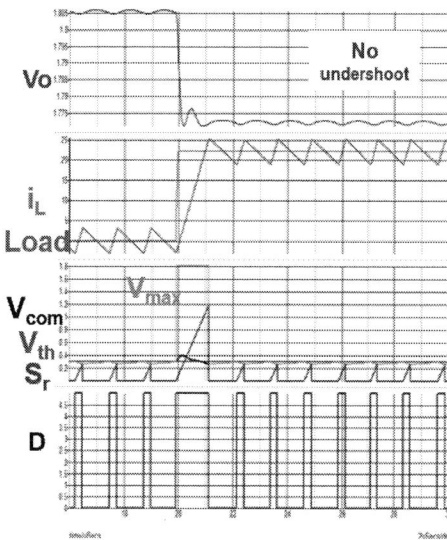

Fig. 11. Simulation results of the single-cycle RPM with V_{in}=8.4V, V_o=1.8V, T_{off_min}=200ns, L=280nH, f_s=800kHz, load step-up=22A, and di/dt=320A/us.

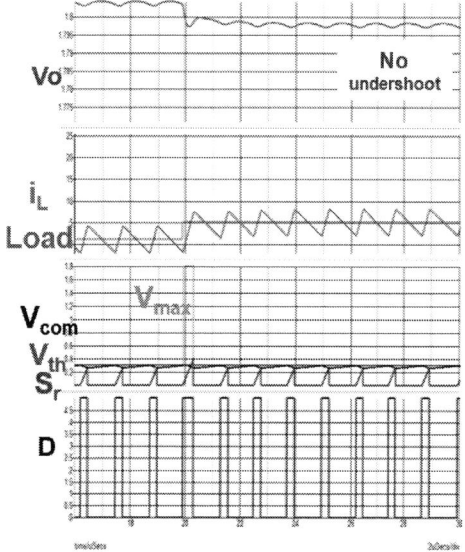

Fig. 12. Simulation results of the single-cycle RPM with V_{in}=8.4V, V_o=1.8V, T_{off_min}=200ns, L=280nH, f_s=800kHz, load step-up=4A, and di/dt=320A/us.

B. For COT

This single-cycle method can be extended to COT. The schematic of the proposed method, using COT as an example, is shown in Fig. 13 and the transient waveform is shown in Fig. 14. Unlike RPM, COT does not use V_{com} to operate. However, the information necessary to create V_{com}, V_c and $i_L R_i$, are used for control. As such, V_{com} can be created using these two signal and a DC offset (V_x) for transient detection. During the steady-state operation, the upper bound of V_{com} is bounded by V_x. When load step-up transient occurs, V_{com} increases higher than V_x until the next steady-state is achieved. Due to this behavior, the beginning of transient can be detected using a V_{top} threshold slightly higher than V_x. When V_{com} is higher than V_{top}, transient is detected and the

turn-off mechanism is modified such that the S_r is compared with a maximum voltage threshold, V_{max}, to turn off T_{on}. When V_{com} reaches V_{bot}, located at V_{com_min}, the turn-off mechanism goes back to normal operation of S_r compared with V_{th}. When this occurs, the change from V_{max} to V_{th} will intersect with S_r and turn off T_{on} to achieve single-cycle transient response.

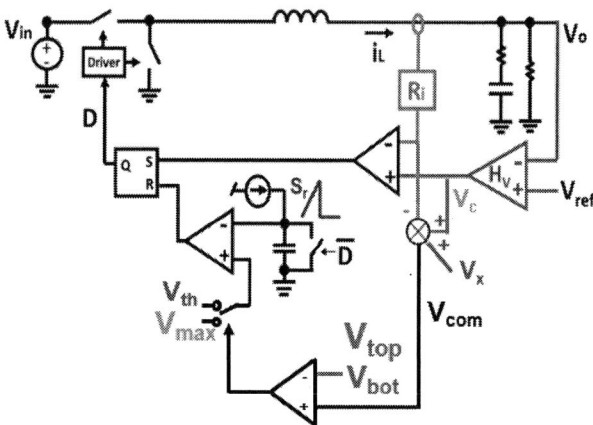

Fig. 13. COT with proposed method to achieve single-cycle load transient.

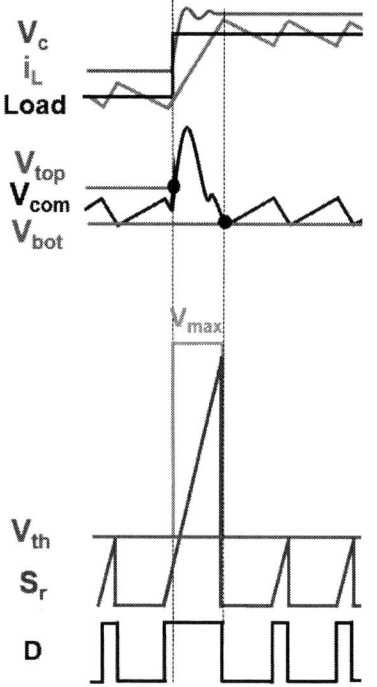

Fig. 14. Transient waveforms of COT with proposed method.

C. For Two-Phase COT

Since the single-cycle method was easily extended to COT, extension into multiphase is also examined. Using a two-phase COT as an example, the schematic of the proposed method is shown in Fig. 15. Unlike the single phase operation of COT, in multiphase, the inductor current information is summed. As a result, the V_{com} created to detect transient will control both phases at once during transient.

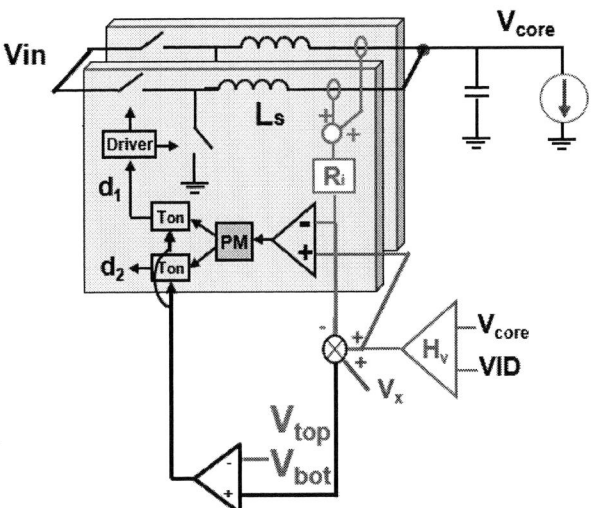

Fig. 15. Two-phase COT with proposed method.

The two-phase operation is simulated with V_{in}=8.4V, V_o=1.8V, T_{off_min}=200ns, L=120nH, f_s=800kHz, load step-up=55A, and di/dt=300A/us. The simulated transient waveforms are shown in Fig. 16. From Fig. 16, it can be observed that V_{com} with the summed inductor current terminates the T_{on} extension for both phase at the same time. It can also be observed that interleaving is maintained and the system transitions easily between the steady-state operation and the single-cycle modifications.

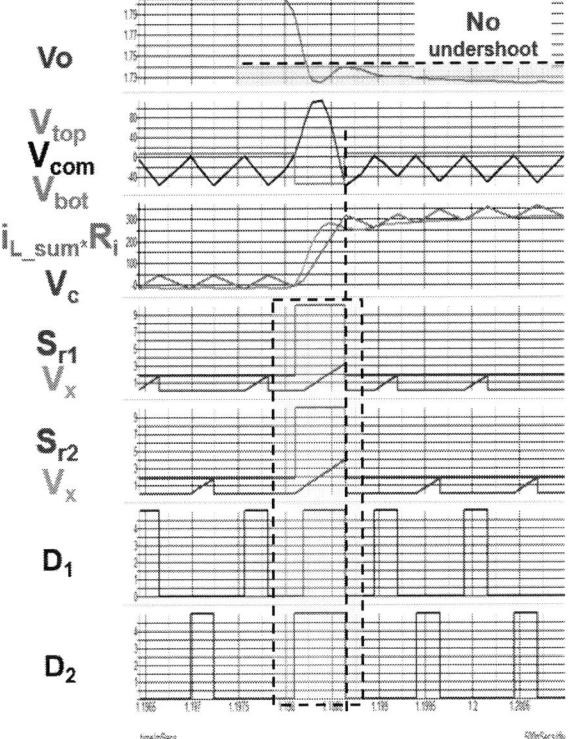

Fig. 16. Simulation results of the 2-phase COT with single-cycle modifications with V_{in}=8.4V, V_o=1.8V, T_{off_min}=200ns, L=120nH, f_s=800kHz, load step-up=55A, and di/dt=300A/us.

D. Transient Detection Design

For this method, the location of V_{top} and V_{bot} are crucial to the performance of single-cycle control. The location of V_{top} determines how fast transient is detected. If the location of V_{top} is too far away from V_{th}, it may not detect small transient load-steps as the magnitude of V_{com} may never reach V_{top}. If the location of V_{top} is too close to V_{th}, false transient detections may be detected due to noise.

The location of V_{bot} ensure the T_{on} extension turns off at the desired optimal location. If the system utilizes V_{in} feedforward to the S_r and V_o feedback to V_{th} to pseudo-fix frequency, the location of V_{bot} can also be fixed. If not, the location of V_{bot} must follow the change in location of V_{com_min} to ensure single-cycle transient. Otherwise, a multi-cycle transient with T_{on} extension or single-cycle with ring-back issue may occur.

V. A SIMPLIFIED METHOD TO ACHIEVE SINGLE-CYCLE LOAD TRANSIENT

A simplified version of the previous method is proposed. In the simplified method to achieve single-cycle load transient, the design process is simplified by eliminating the need of V_{bot}. Unlike the previous proposed method where the transition between the modified and original turn-off mechanism also turns off T_{on}, the simplified method relies only on the original turn-off mechanism to turn off T_{on} to achieve single-cycle response.

A. For RPM

The schematic of the simplified method, using RPM as an example, is shown in Fig. 17 and the transient waveform is shown in Fig. 18. During the steady-state operation, the upper bound of V_{com} is bounded by V_{th}. When load step-up transient occurs, V_{com} increases higher than V_{th} until the next steady-state is achieved. Due to this behavior, the beginning of transient can be detected using a V_{top} threshold slightly higher than V_{th}. When V_{com} is higher than V_{top}, transient is detected and the turn-off mechanism is modified such that the S_r is clamped at zero. When V_{com} reaches V_{th}, the turn-off mechanism goes back to normal operation of S_r rising to compare with V_{com}. The first cycle of the normal operation

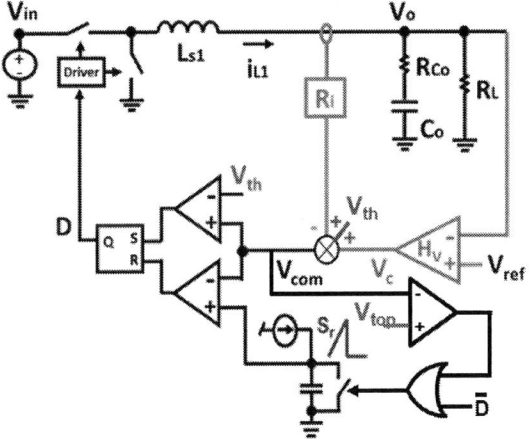

Fig. 17. RPM with proposed simplified method to achieve single-cycle load transient.

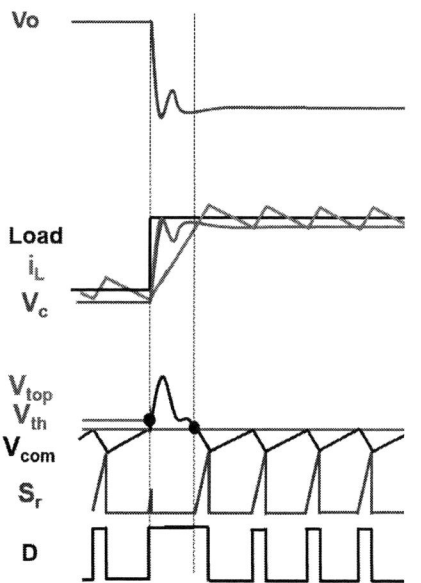

Fig. 18. Transient waveforms of RPM with simplified method.

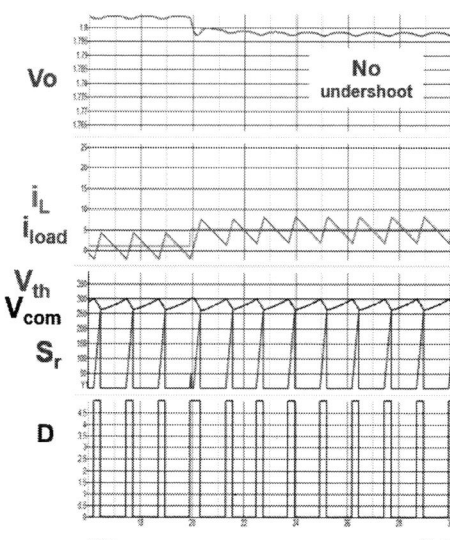

Fig. 20. Simulation results of the simplified single-cycle RPM with V_{in}=8.4V, V_o=1.8V, T_{off_min}=200ns, L=280nH, f_s=800kHz, load step-up=4A, and di/dt=320A/us.

turn-off mechanism will turn off T_{on} to achieve single-cycle transient response.

The simplified single-cycle RPM is simulated with V_{in}=8.4V, V_o=1.8V, T_{off_min}=200ns, L=280nH, f_s=800kHz, load step-up=22A, and di/dt=320A/us. The simulated transient waveforms are shown Fig. 19. From the waveforms, single-cycle response is achieved using the simplified method and no undershoot is observed under worst-case scenario. Another case with small load-step change, load step-up=4A, is shown in Fig. 20. From Fig. 20, single-cycle response is also achieved with no ring-back issues with small load-step change. Like the previous method, the T_{on} extension provided by the simplified method can be automatically adjusted with various load-step changes.

B. For COT

This simplified single-cycle method can also be extended to COT. The schematic of the simplified method, using COT as an example, is shown in Fig. 21 and the transient waveform is shown in Fig. 22. Again, the V_{com} signal will be created the same way for transient detection purposes. During the steady-state operation, the upper bound of V_{com} is bounded by V_x. When load step-up transient occurs, V_{com} increases higher than V_x until the next steady-state is achieved. Due to this behavior, the beginning of transient can be detected using a V_{top} threshold slightly higher than V_x. When V_{com} is higher than V_{top}, transient is detected and the turn-off mechanism is modified such that the S_r is clamped at zero. When V_{com} reaches V_x, the turn-off mechanism goes back to normal operation of S_r rising to compare with V_{th}. The first cycle of the normal operation turn-off mechanism will turn off T_{on} to achieve single-cycle transient response. Like the previous method, the simplified method can also be extended to two-phase COT.

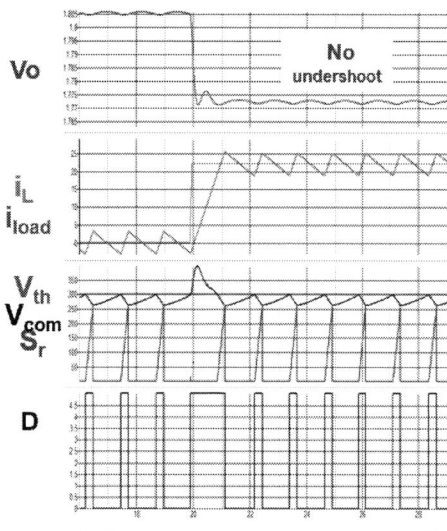

Fig. 19. Simulation results of the simplified single-cycle RPM with V_{in}=8.4V, V_o=1.8V, T_{off_min}=200ns, L=280nH, f_s=800kHz, load step-up=22A, and di/dt=320A/us.

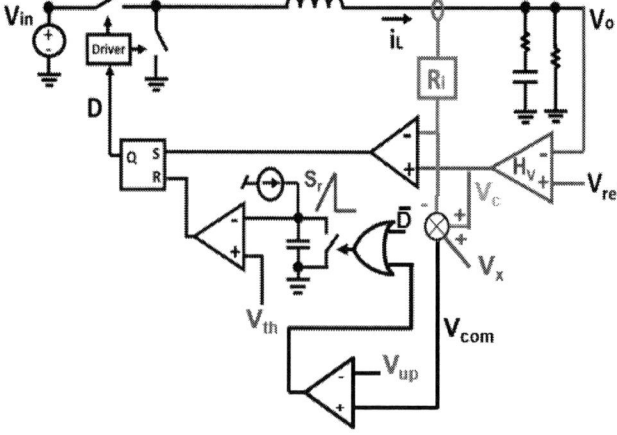

Fig. 21. COT with simplified method to achieve single-cycle load transient.

978-1-4673-9551-9/16 $31.00 © 2016 IEEE   348

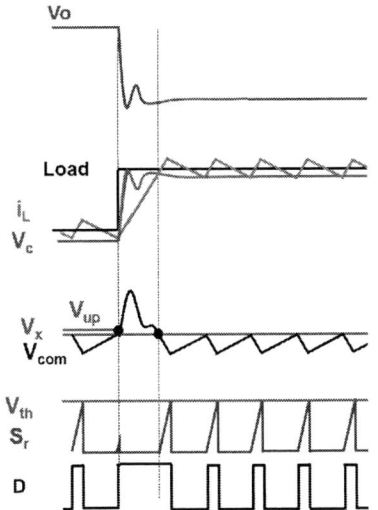

Fig. 22. Transient waveforms of COT with simplified method.

VI. EXPERIMENTAL RESULTS USING EXISTING COT CONTROLLER

Experimental results for the simplified single-cycle method for COT is obtained by making external modifications to the TPS5960 evaluation board by Texas Instruments after three 470uF bulk output capacitors are removed. For single-phase operation with V_{in}=12V, V_{ref}=1.2V, and using the on-board 32A dynamic load, the performance with the original COT control is shown in Fig. 23, and COT with simplified single-cycle modifications is shown in Fig. 24.

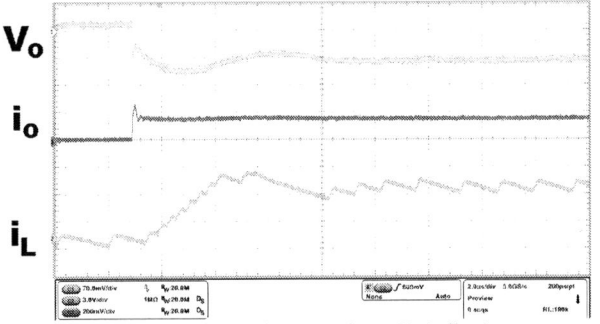

Fig. 23. Measured waveforms of output voltage, V_o (yellow), output current (pink), and inductor current (blue) using COT.

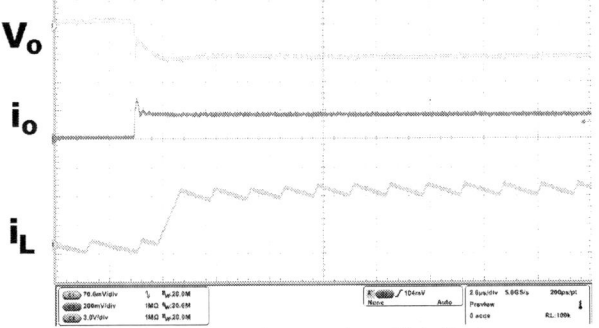

Fig. 24. Measured waveforms of output voltage, V_o (yellow), output current (pink), and inductor current (green) using COT with single-cycle improvement.

After the load step-up transient is detected, COT with simplified single-cycle is able to go from one steady-state to another steady-state much faster than the conventional COT. As a result, an output-voltage undershoot of 28mV is observed for conventional COT while a 3mV undershoot is observed for COT with simplified single-cycle modifications.

VII. CONCLUSION

In this paper, two simple and smart methods to achieve single-cycle load transients are proposed. For both methods, single-cycle load transient can be achieved by monitoring the behavior of V_{com}. Once transient is detected, a single-cycle response is realized using the proposed modifications. Some benefits of the proposed methods include T_{on} extension that are automatically adjusted with load changes, and easy modifications for existing control schemes, such as COT and RPM, to operate at the fastest transient available. The methods can also be extended to use in multiphase operation. Due to the nature of the proposed methods, the modifications only improve the transient behavior and does not impact the steady-state operation of the original control schemes. Experimental results of the simplified single-cycle method using COT is presented. Hardware verifications for the proposed single-cycle method using RPM and COT will be presented as part of the future works.

REFERENCES

[1] Texas Instrument, "Adaptive Constant On-Time (D-CAP) Control Study in Notebook Applications," www.ti.com, Dec. 2007.

[2] Y. Chen, D. Chen, Y. Lin, and C. Chen, "A Novel Constant On-time Current Mode Control Scheme to Achieve Adaptive Voltage Positioning for DC Power Converters," in Proc. IEEE IECON, pp. 104-109, Oct. 2012.

[3] Analog Devices, "ADP1882/ADP1883: Synchronous Current-Mode with Constant On-Time, PWM Buck Controller," www.analog.com, 2010.

[4] US 20130314060 A1, "Load Transient of a COT Mode Power Supply" by Richtek Corp. Pub: Nov 2013.

[5] Richtek, "Datasheet of RT8889C - 3-Phase Controller with Triple Integrated Driver for VR12.5 Mobile CPU Core Power Supply," 2014.

[6] T. F. Schiff, "Switching Power Supply Control," U.S. Patent, no. US 7,595,617, Sept. 2009.

[7] On Semiconductor, "ADP3211: 7-Bit, Programmable Single-Phase Synchronous Buck Controller," www.onsemi.com, Nov. 2009.

[8] Y.C. Wan, J.R. Huang, C.H. Wang, C.F. Chung, "Voltage Regulator and Pulse Width Modulation Signal Generation Method Thereof," U.S. Patent, no.20120146608, Jun. 2012.

[9] K. Lee, "Method for Generating a Signal and Structure Therefor," U.S. Patent, no. US 8,643,355, Feb. 2014.

[10] K. Lee, and H. Zou, "Comparison between Ramp Pulse Modulation (RPM) and Constant Frequency Modulation for the Beat Frequency Oscillation in Voltage Regulators," in Proc. IEEE ECCE'10, pp. 3101-3106, Sept. 2010.

A Mixed-Signal Ripple-Based Controller for a 16 V, 10 MHz Integrated Buck Converter

Sergii Tkachov, Matteo Agostinelli
Infineon Technologies Austria AG
9500 Villach, Austria
Email: sergii.tkachov@infineon.com,
matteo.agostinelli@infineon.com

Abstract—A high-performance mixed-signal controller, based on the V^2 architecture, has been designed to regulate a high-frequency (up to $10\,\mathrm{MHz}$) $16\,\mathrm{V}$-input synchronous buck converter with integrated driver and power stage in a $130\,\mathrm{nm}$ **Bipolar-CMOS-DMOS (BCD) technology. The inner loop has been implemented fully in the analog domain to support high switching frequencies, while the outer loop has been realized digitally for maximum programmability and flexibility. The controller can optionally employ the inductor or capacitor current to further improve stability and performance in the case of low-ESR output capacitors. A hardware prototype has been developed and experimental results are provided in order to confirm the performance of the controller.**

I. INTRODUCTION

A trend leading to an increase of the switching frequency of DC-DC converters can be observed in many applications, ranging from point-of-load converters [1] to automotive microcontroller's power supplies [2]. One of the main motivations behind this trend is the possibility to reduce the size of bulky external components or even to fully integrate the passives inside the package (Power System-In-Package, PwrSiP) or on-chip (PwrSoC) [3], [4].

Regulating a high-frequency DC-DC converter with a high-bandiwdth digital ripple-based controller presents many challenges and poses tough requirements for the Analog-to-Digital and Digital-to-Analog Converters (ADC and DAC, respectively). In this paper, a mixed-signal implementation of a ripple-based controller, based on the constant-frequency V^2 architecture [5], is presented. The controller has been successfully employed to regulate the output voltage of an integrated high-voltage (up to $16\,\mathrm{V}$ input voltage, which requires high voltage devices in an integrated solution), high-frequency (up to $10\,\mathrm{MHz}$ switching frequency) buck converter. The output voltage can be regulated in the range from

$3.3\,\mathrm{V}$ to $5\,\mathrm{V}$. The controller can be configured to use either the inductor current sense or capacitor current sense, yielding a $V^2 I_L$ or $V^2 I_C$ [6], [7] architecture, to further improve stability in low-ESR applications [8]. The internal Fast Feedback (FFB) loop, optionally including current information, has been implemented in the analog domain, in order to guarantee fast transient response at high switching frequencies, while the outer Slow Feedback (SFB) loop has been realized digitally. Therefore, this approach exploits both the simplicity and performance of the analog domain and the flexibility and programmability of the digital implementation.

A hardware prototype has been developed, where the power stage and driver have been designed in a $130\,\mathrm{nm}$ Bipolar-CMOS-DMOS (BCD) technology and integrated on the same silicon die. In order to validate the controller concept, the analog FFB loop has been realized with discrete components, while the digital SFB has been implemented by means of an industrial microcontroller [9]. The long-term goal is to integrate all circuitries and components (including the passive filter) together with the power stage and driver to realize a PwrSiP or PwrSoC solution.

II. CONTROLLER ARCHITECTURE AND IMPLEMENTATION

A schematic representation of the overall system is illustrated in Fig. 1. The architecture is based on a conventional V^2 controller with feed-forward of the inductor or capacitor current. The current sensing has been implemented with a shunt resistor of $15\,\mathrm{m\Omega}$ in order to avoid excessive penalization of the overall efficiency, but at the same time guarantee an accurate reading. The current information is then processed by an instrumentation amplifier topology. In a future integrated design, the current sense could be implemented with a lossless technique, e.g. by sensing the current through the

978-1-4673-9551-9/16 $31.00 © 2016 IEEE

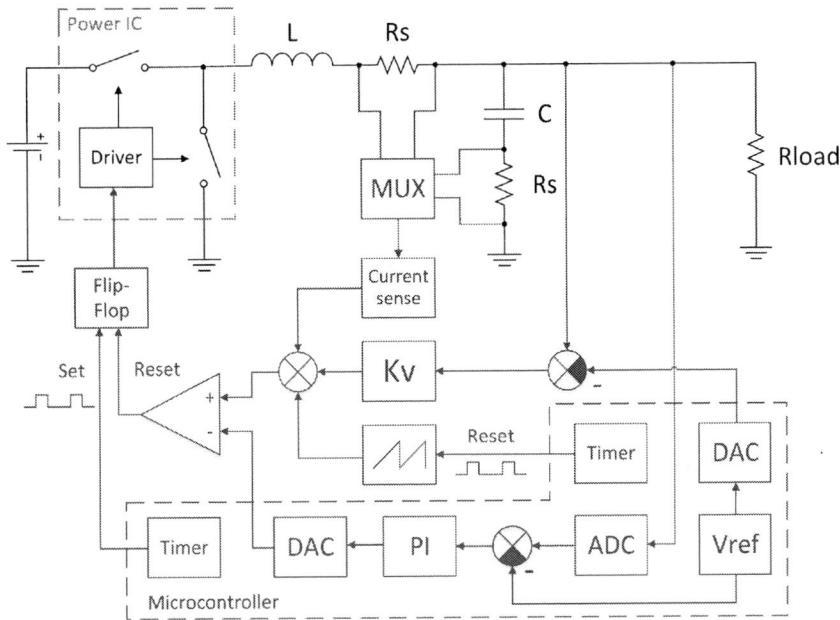

Fig. 1: Schematic representation of the controller architecture. The parts implemented in the analog (digital) domain are represented in blue (red) color.

high-side switch with a SenseFET-based approach [10]. Alternatively, the capacitor current could be estimated by means of a trans-impedance amplifier, as proposed in [11].

The fast feedback loop (FFB, depicted in blue in Fig. 1) has been implemented in the analog domain with discrete components, in order to guarantee fast operation, while the slow feedback loop (SFB, realized by the ADC-PI-DAC path inside the dashed red box in Fig. 1) has been realized by means of an industrial microcontroller unit, for maximum programmability and flexibility. The output voltage of the buck converter is passed to the analog subtractor unit, based on a high speed amplifier. The reference signal V_{ref} is set by a DAC of the microcontroller. The reference voltage can be programmed in the range from 3.3 V to 5 V and can be updated online, thanks to the digital implementation.

In order to avoid subharmonic oscillations, a ramp generator, operating at the switching frequency f_{sw} of the converter, is used for slope compensation. The ramp is realized by charging a capacitor with a constant current and a fast discharge phase, that takes up to 8 ns. The timing of the ramp generator is managed by digital signals generated from the microcontroller. A non-inverting summing amplifier adds the current sense, output voltage error and ramp signals together, weighted by appropriately designed coefficients, which are defined by resistors.

The SFB consists of a digital proportional-integral (PI) regulator. The ADC samples the output voltage with a resolution of 12 bits. The DAC has also a resolution of 12 bits. The computations are performed by the Floating Point Unit (FPU) of the microcontroller, which is based on the ARM Cortex-M4 architecture. The output of the SFB is updated with a rate of approximately 250 kHz. Furthermore, the microcontroller can control the driver of the buck converter through an SPI interface, in order to implement protection features and to regulate various parameters, such as the dead-time to avoid shoot-through currents. The microcontroller also dictates various timings of the system. The PWM modulator is a conventional constant-frequency peak control modulator [12], with a nominal switching frequency $f_{\mathrm{sw}} = 10\,\mathrm{MHz}$, which is set by the microcontroller.

III. SIMULATION RESULTS

The controller architecture presented in Sec. II has been simulated with a mixed-signal circuit simulator (SIMetrix) and SIMPLIS. In all the results presented in this section, capacitor current feedback is enabled, therefore yielding a $V^2 I_C$ controller. The dynamic performance of the controller during load transients has been verified, as shown in Fig. 2. In the figure, a response to a 200 mA load step can be observed, with $V_{\mathrm{in}} = 8\,\mathrm{V}$ and $V_{\mathrm{ref}} = 3.3\,\mathrm{V}$. The spikes occuring at each switching

978-1-4673-9551-9/16 $31.00 © 2016 IEEE

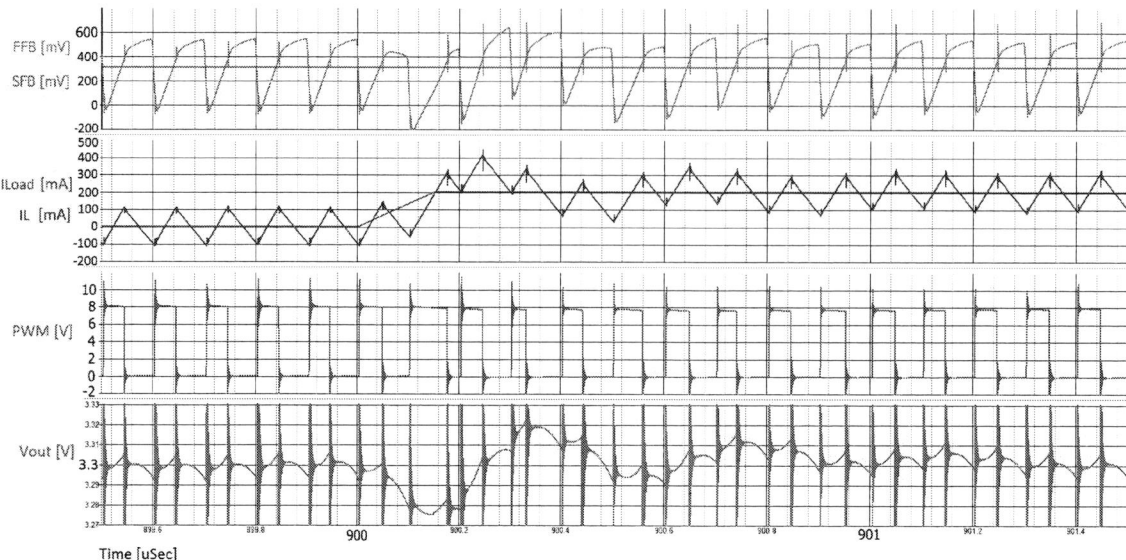

Fig. 2: Transient simulation showing the response of the system to a load jump from 0 to 200 mA in 150 n sec. From top to bottom, the waveforms represent the SFB and FFB signals, the inductor and load current, the switching node, and the output voltage.

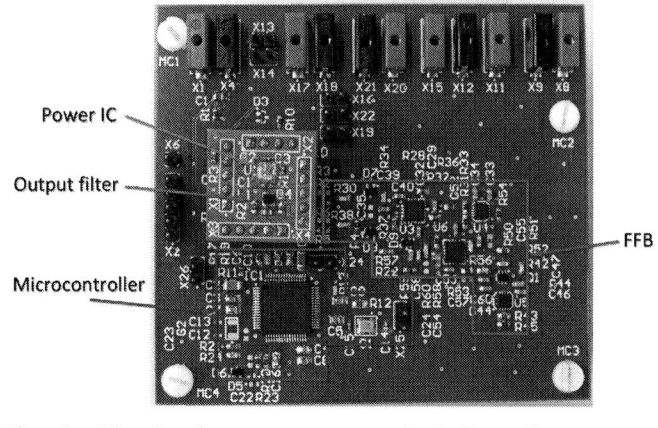

Fig. 3: The hardware prototype, including the power stage and driver IC with the external passive components, the microcontroller and the controller realized with discrete components.

event stem from the parasitic inductance that has been considered in the simulations.

IV. EXPERIMENTAL RESULTS

A picture of the developed hardware prototype is shown in Fig. 3. The driver and power stage integrated circuit (IC) is visible in the top left side, together with the external passive components of the output filter. The capacitance of the output filter is $C = 470$ nF, while the inductance is $L = 1\,\mu$H. The microcontroller can be

seen in the bottom left part, while the analog part of the controller (FFB) occupies the remaining area. The size of the board is 85 mm×75 mm. The area of the power IC is 1 mm^2, while the output filter occupies approximately 6 mm^2, not including sense resistors.

The same load variation shown in Sec. III has been verified experimentally, through the aforementioned hardware prototype. The measurement is reported in Fig. 4, where the top waveform corresponds to the output voltage, the middle one is the switching node and finally the bottom curve represents the load current. The undershoot caused by a 200 mA load jump equals approximately 85 mV. The overshoot resulting from a load drop from 200 mA to 0 (not shown in the figure) equals approximately 60 mV.

The frequency-domain response of the prototype has also been validated experimentally. The control-to-output transfer function is plotted in Fig. 5 for different values of the slope of the compensation ramp. The obtained results are in good agreement with the model proposed by [5]. Current feedback is disabled in this measurement.

The same transfer function has been measured for two different output capacitors, one with high and one with low ESR. The results can be observed in Fig. 6 and, again, are in line with the model proposed by [5].

978-1-4673-9551-9/16 $31.00 © 2016 IEEE   352

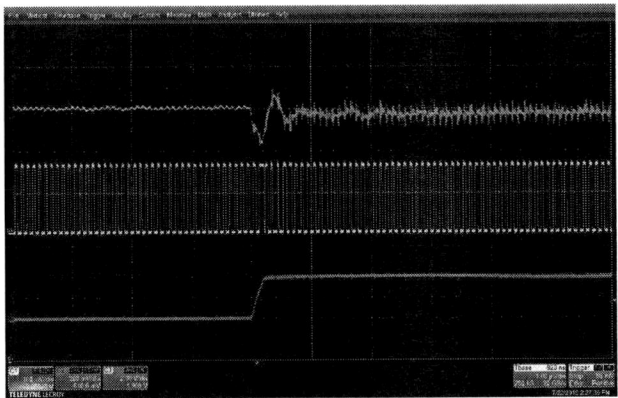

Fig. 4: Measurement of a load transient, with the load current jumping from 0 to 200 mA, $V_{\mathrm{in}} = 8\,\mathrm{V}$ and $V_{\mathrm{out}} = 3.3\,\mathrm{V}$. From top to bottom, the waveforms correspond to the output voltage, the switching node, and the load current.

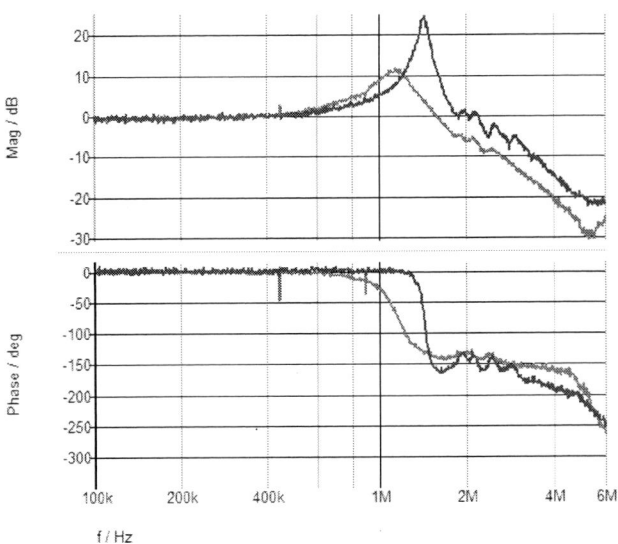

Fig. 6: Measurements of the control-to-output transfer function for different output capacitor. The red curve corresponds to an output capacitor with ESR of $100\,\mathrm{m\Omega}$ while the blue curve corresponds to a low-ESR ($10\,\mathrm{m\Omega}$) output capacitor.

V. CONCLUSIONS

A mixed-signal implementation of a V^2 or $V^2 I_C$ ($V^2 I_L$) controller is presented in this paper. The capacitor (or inductor) current feedback is particularly important in order to improve stability and performance in the case of low-ESR capacitors. The fast part of the feedback loop has been implemented in the analog domain, in order to support high switching frequencies. The digital part, implemented with an industrial microcontroller, gives additional flexibility and programmability.

The performance of the controller has been simulated and validated experimentally by means of a hardware prototype, which includes a $10\,\mathrm{MHz}$ $16\,\mathrm{V}$-input buck converter, implemented in a $130\,\mathrm{nm}$ BCD technology. Excellent dynamic performance during load transients has been proven.

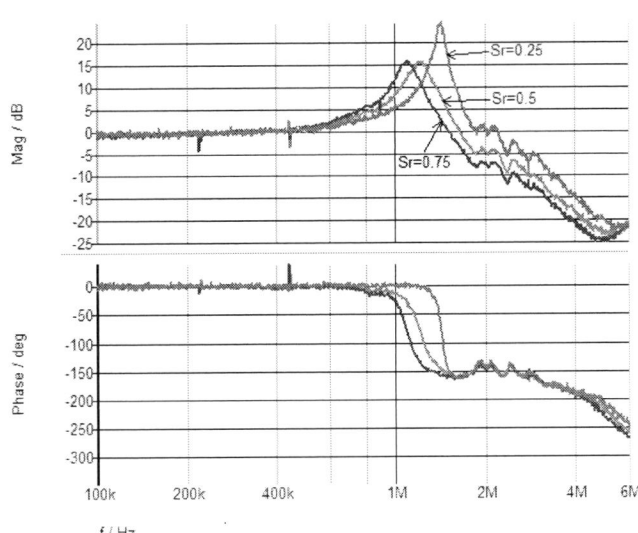

Fig. 5: Measurements of the control-to-output transfer function for different values of the slope of the compensation ramp. The red curve corresponds to a ramp slope coefficient $S_r = 0.25$, the green one to $S_r = 0.5$ and the blue one to $S_r = 0.75$. The slope of the compensation ramp is given by $S_r \cdot 22\,\mathrm{mV\,ns^{-1}}$.

ACKNOWLEDGEMENT

This work is partly funded by the EU under FP7-ICT-2011-8 – PowerSWIPE – Project no.: 318529.

REFERENCES

[1] E. Burton, G. Schrom, F. Paillet, J. Douglas, W. J. Lambert, K. Radhakrishnan, and M. J. Hill, "FIVR - Fully integrated voltage regulators on 4th generation Intel® Core™ SoCs," in *Proc. IEEE Appl. Power Electron. Conf. Expo.*, 2014, pp. 432–439.

[2] C. Sandner, "Towards a PowerSoC Solution for Automotive Microcontroller Applications," in *International Power Supply on Chip (PwrSoC) Workshop*, 2014.

[3] C. Ó Mathúna, N. Wang, S. Kulkarni, and S. Roy, "Review of integrated magnetics for power supply on chip (PwrSoC)," *IEEE Trans. Power Electron.*, vol. 27, no. 11, pp. 4799–4816, Nov. 2012.

[4] Y.-F. Liu and C. Ó Mathúna, "Editorial: Special Issue on Power Supply on Chip, 2013," *IEEE Trans. Power Electron.*, vol. 28, no. 9, pp. 4125–4126, 2013.

[5] S. Tian, F. C. Lee, and P. Mattavelli, "Small-Signal Analysis and Optimal Design of Constant Frequency V^2 Control," *IEEE Trans. Power Electron.*, vol. 30, no. 3, pp. 1724–1733, 2015.

[6] M. Del Viejo, P. Alou, J. A. Oliver, O. Garcia, and J. A. Cobos, "V^2I_C Control :a Novel Control Technique with Very Fast Response Under Load and Voltage Steps," in *Proc. IEEE Appl. Power Electron. Conf. Expo.*, 2011, pp. 231–237.

[7] J. Cortes, V. Svikovic, and P. Alou, "Comparison of the behavior of voltage mode, V^2 and V^2I_c control of a buck converter for a very fast and robust dynamic response," in *Proc. IEEE Appl. Power Electron. Conf. Expo.*, 2014, pp. 2888–2894.

[8] P. H. Liu, Y. Yan, P. Mattavelli, and F. C. Lee, "Digital V^2 control with fast-acting capacitor current estimator," in *Proc. IEEE Energy Convers. Congr. Expo.*, 2012, pp. 1833–1840.

[9] "32-bit XMC4000 industrial microcontroller ARM® Cortex®-M4," http://www.infineon.com/xmc4000, accessed: 2015-07-07.

[10] C. Lee and P. Mok, "A monolithic current-mode CMOS DC-DC converter with on-chip current-sensing technique," *IEEE J. Solid-State Circuits*, no. 39, pp. 3–14, 2004.

[11] S. Huerta, P. Alou, J. Oliver, O. Garcia, J. Cobos, and A. Abou-Alfotouh, "Design methodology of a non-invasive sensor to measure the current of the output capacitor for a very fast non-linear control," in *Proc. IEEE Appl. Power Electron. Conf. Expo.*, 2009, pp. 806–811.

[12] R. Redl and N. O. Sokal, "Current-mode control, five different types, used with the three basic classes of power converters - Small-signal ac and large-signal dc characterization, stability requirements, and implementation of practical circuits," in *Proc. IEEE Power Electron. Specialists Conf.*, 1985, pp. 771–785.

New Control Concept for Soft-Switching Flyback Converters with Very High Switching Frequency

A.M. Connaughton[†], K.Krischan[†], K.K Leong[‡], and A. Muetze[†]

[†]Electric Drives and Machines Institute, TU Graz, Austria
[‡]Infineon Technologies Austria AG, Villach, Austria

Abstract—**A conventional Flyback converter's synchronous rectification switch is controlled from the primary side via opto-isolators; introducing propagation delay. A new control concept is presented here that eliminates the need for opto-isolators, communicating instead through the coupled inductor to assist more precise soft-switching control at much higher frequencies. As a precursor to a 1 MHz GaN demonstrator, an initial 100 kHz prototype has been built to prove the feasibility of the concept.**

Keywords—Flyback, secondary side control.

I. INTRODUCTION

The grid-connected Flyback converter is a commonly used topology in switched-mode-power-supplies (SMPS) due to its low component count and inherent galvanic isolation [1]: see Fig. 1. Conventionally, the basic Flyback converter is controlled from the primary side and is driven in either hard-switching mode (with a diode in place of S2), or in quasi- or full-resonant mode with the synchronous rectification controlled via opto-isolators; maintaining galvanic isolation [2-7].

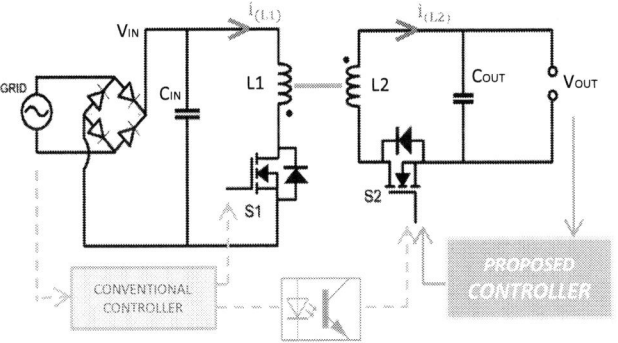

Fig. 1 - Principal schematic of Flyback converter with comparison of conventional and proposed controllers.

Pushing the Flyback to a higher switching frequency (f_s) in the MHz range allows the use of a smaller inductor core and significant improvement to the eventual power density [8]. However, the isolated opto-coupling necessary with the conventional controller can be expensive and add propagation delay up to 150 ns (significant at 1 MHz) to secondary gate signals and to the output voltage measurement; limiting the achievable switching frequency in resonant applications [9].

By moving the main controller to the secondary side and communicating to the primary switch (S1) through the power flow, we eliminate the need for opto-isolators, and gain direct access to the output voltage. This not only reduces the component count but also removes significant propagation delay between primary and secondary side; assisting precise high-frequency control.

II. NEW CONTROL CONCEPT

A. Steady-State Behaviour

The central improvement to the state-of-the-art is that the primary drain-source voltage ($V_{ds(S1)}$) provides the signal to turn ON the primary switch, S1, rather than auxiliary digital signals. Fig. 2 presents various waveforms of the Secondary Side controlled Flyback (SSF) during steady-state operation.

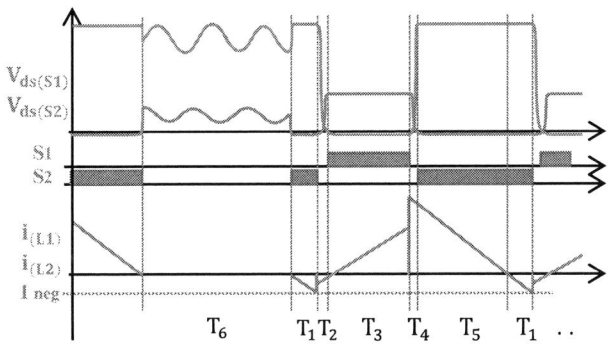

Fig. 2 - Ideal primary (green) and secondary (pink) V_{ds}, logic signals, and inductor currents during steady state.

The controller uses an adapted form of pulse-skipping control [10], along with triangular current mode (TCM) operation to generate the primary side turn ON "signals" and to simultaneously achieve full zero voltage switching (ZVS) [11]. By energizing the secondary side inductor to a predetermined negative current (I_{neg}), it is possible to forcibly discharge the non-linear parasitic output capacitance of S1 ($C_{oss(S1)}$) prior to turn-ON, and cause $V_{ds(S1)} \leq 0$ V. I_{neg} is a constant defined as the minimum negative current required on the secondary side to induce the zero crossing of $V_{ds(S1)}$ that commands S1 to turn ON. During T_6, both switches are OFF

978-1-4673-9551-9/16 $31.00 © 2016 IEEE

and the controller waits for the output voltage (V_{out}) to fall below its desired threshold value. The secondary switch ($S2$) is then turned ON for a time T_1 to energize the inductor to negative current I_{neg} and induce ZVS across $S1$ (T_2).

$$T_1 = (I_{neg}L_2)/V_{OUT} \qquad (1)$$

Independently, the primary side acts as a slave to the secondary side controller. It constantly waits for $V_{ds(S1)} \leq 0$ V and once it occurs, will turn ON $S1$ for a fixed time T_3 – the time required to transfer rated power plus the energy for primary ZVS.

$$P_{neg} = \frac{1}{2} I_{neg} V_{OUT} T_1 f_s \qquad (2)$$

$$T_3 = \sqrt{(2L_1(P_{rated} + P_{neg}))/(V_{IN}{}^2 f_s)} \qquad (3)$$

Once T_3 has elapsed, $S1$ turns OFF and begins waiting again for $V_{ds(S1)}$ to fall below zero. At this point $S2$ acts as a synchronous rectifier: turning ON at $V_{ds(S2)} < 0$ V (at the end of T_4). $S2$ is ON during T_5 and remains ON even after $i_{(L2)}$ falls to zero in order to generate a negative current pulse for the next switching cycle (T_1). Once the desired output voltage is reached, the secondary side refrains from sending a negative current pulse at the end of T_5 entering the pulse-skipping wait period T_6. The circuit should be dimensioned such that at rated power, operation is continuous with no pulse skipping.

B. High- and Low-Line Input Voltage

From (1) and (3), it is evident that the control timings will change depending on whether the circuit is operating from high-line (240 Vrms) or low-line (120 Vrms) input voltage. For low-line, T_1 will be shorter due to the reduced energy stored in $C_{oss(S1)}$, but T_3 will be longer due to the smaller V_{in} across $L1$. Therefore, both primary and secondary side controls must be aware of the input voltage. The secondary side can infer V_{in} from the reflected blocking voltage across $S2$.

C. Start-Up Routine

The SSF control concept requires the output capacitor (C_{OUT}) to be sufficiently charged to achieve the negative current pulses for primary side ZVS. There must therefore be a suitable method of charging C_{OUT} without primary side measurement or computation. Briefly stated, the routine is as follows: the primary side sends bursts of low-power pulses via $S2$ body diode until the secondary side "wakes up". Secondary side then sends a negative current pulse, putting the primary side into "slave" mode whereby proposed SSF control begins.

More specifically, once C_{OUT} is charged beyond a threshold (e.g. 50% of rated V_{out}), the secondary side controller enters "charge-mode" and takes responsibility for charging C_{OUT}. In this quasi steady-state, the controller samples V_{out} to determine the required time (T_1) to achieve I_{neg} and after a set number of switching cycles, operation halts (T_6) and V_{out} is resampled. This cycle repeats until V_{out} approaches rated V_{out}. Full steady-state operation with constant T_1 then begins.

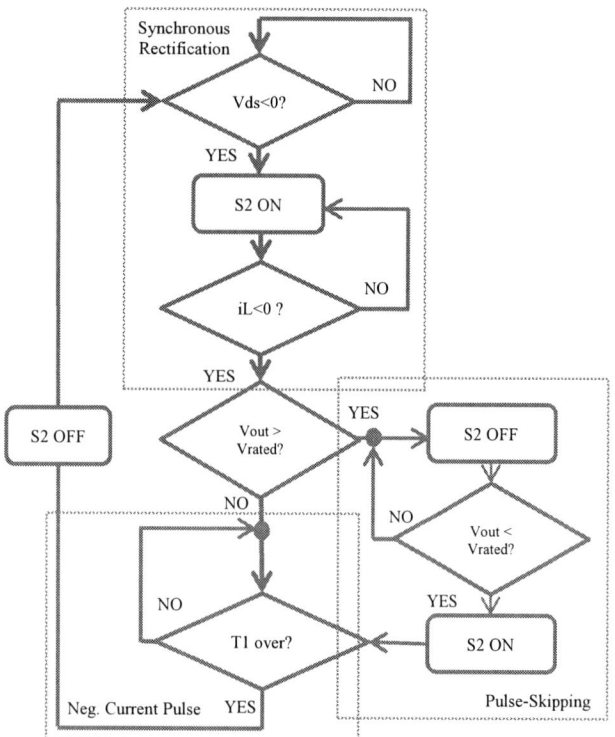

Fig. 3 - Simplified flow chart of secondary side steady-state control.

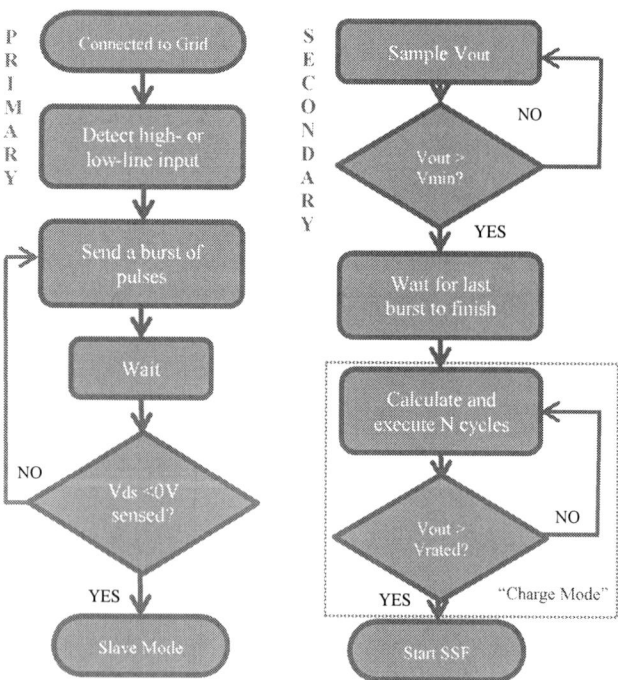

Fig. 4 - Simplified flow chart of start-up routine.

III. DRAIN-SOURCE VOLTAGE SENSING

The zero voltage crossing of both switches' V_{ds} is sensed by high speed comparators with their positive input nodes connected to local ground.

A. Primary Side Switch

The primary side comparator is supplemented with a high impedance (n-channel) MOSFET (Q) in a source-follower arrangement. Fig. 5 shows the principal schematic of the primary side V_{ds} sensing sub-circuit. The source-follower uses this high voltage "small-signal-transistor" to block the primary input voltage when S1 is OFF. A constant voltage (5 V) between its gate and ground acts as a cut-off voltage: when $V_{ds(S1)}$ rises above this cut-off voltage, the transistor autonomously turns OFF - protecting the comparator from the high input voltage. Once Q is conducting and $V_{ds(S1)}$ falls below 0 V, the comparator will be triggered. The rising edge of the output is used to initiate the gate signal for S1.

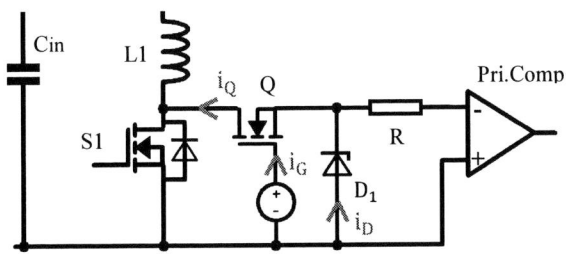

Fig. 5 - Primary V_{ds} sensing - comparator with source-follower.

A major benefit of the source-follower circuit is that it can distinguish between intentional zero voltage crossings (caused by the energy delivered from the secondary side) from undesired crossings that may occur during low-line input conditions while both S1 and S2 are OFF (during T_6).

In the first case, the steep dV/dt across S1 prior to primary turn-ON is also seen by Q, discharging its parasitic capacitance ($C_{oss(Q)}$) via D_1. If the current through the parasitic capacitance of Q ($C_{oss(Q)}$) is greater than its maximum gate current ($i_{G(max)}$), the parasitic capacitance of D_1 (charged only to the cut-off voltage) will also discharge.

- The zener diode's parasitic capacitance will discharge before S1's - triggering the comparator.
- Once i_q falls below $i_{G(max)}$, D_1's parasitic capacitance recharges, disabling comparator output.
- Once $V_{ds(S1)}$ is lower than the cut-off voltage, Q turns-ON and the voltage across D_1 follows $V_{ds(S1)}$.

Once $V_{ds(S1)}$ is below zero, the comparator is triggered again.

In the second case, the low frequency oscillations during T_6 incur a smaller dV/dt that is not enough to cause $i_Q > i_{G(max)}$. The parasitic capacitance of D_1 never fully discharges and the comparator is thus only triggered once (at $V_{ds(S1)} < 0$ V).

By adapting the primary control accordingly, the first comparator output pulse can be acknowledged, but only a second pulse occurring within a given time window counts as a genuine signal to turn ON S1. A measured implementation of this technique is shown in Fig. 6.

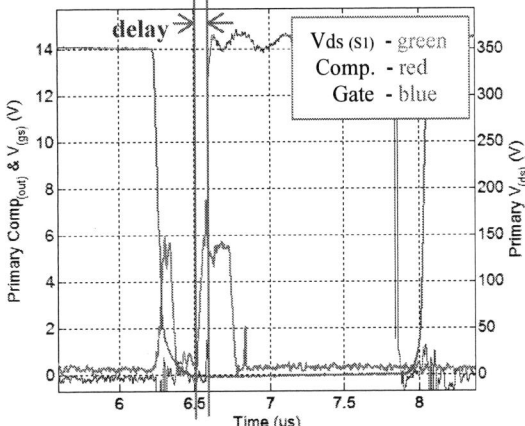

Fig. 6 - Measurement of source-follower circuit in operation.

Using a series gate resistance, the maximum gate current of Q (i_G) can be used to set the sensitivity of the source-follower's blanking capability. It should fulfill the following criteria:

$$\omega_0 \widehat{V}_{ds(S1)} C_{oss(Q)} < i_G < C_{oss(Q)} \frac{dV_{ds(S1)}}{dt} \quad (4)$$

where $\widehat{V}_{ds(S1)}$ is the peak primary drain-source voltage, and ω_0 is the resonant frequency between $C_{oss(S1)}$ and L1. The upper-limit ensures that the parasitic capacitance of D_1 is forced to discharge during steep dV/dt, the lower limit ensures that it does not discharge during Flyback oscillations. Setting maximum i_G in the middle of these limits will provide room for parameter drift and tolerances.

B. Secondary Side Switch

The secondary side V_{ds} sensing must provide the controller with two signals: when the V_{ds} falls to zero for turn-ON, and when the current falls to zero for turn-OFF. Fig. 7 shows the basic comparator sub-circuit used to achieve this:

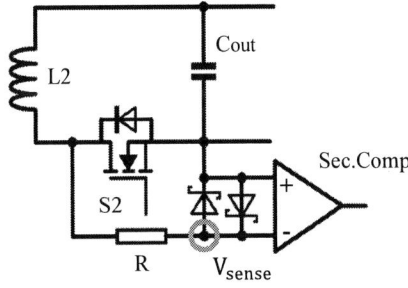

Fig. 7 - Secondary V_{ds} sensing sub-circuit.

Once the body diode of S2 conducts, the potential V_{sense} becomes negative, triggering the comparator. This signal indicates that the synchronous rectification phase has begun and that S2 can be turned ON. This state remains until L2 is fully discharged, at which point the potential across the switch flips direction, causing the comparator output to go low. This indicates that the synchronous rectification phase has ended.

IV. 100 kHz PROTOTYPE

As a precursor to a 1 MHz demonstrator, a 100 kHz prototype was built to demonstrate the SSF control concept. It was designed to operate at 65 W with a 20 Vdc output.

A. Coupled Inductor Design

The coupled inductor should avoid saturation, have low leakage and incur minimal losses. It is also preferable that switching frequency at low- and high-line is similar. A large turns-ratio (N) allows use of secondary side MOSFETs with a lower voltage rating (lower device losses), but causes high-line switching frequency to rise further away from low-line:

$$\frac{1}{f_s} = T_s = L_1 \left(\hat{\imath}_{L1} + \frac{I_{neg}}{N} \right) \left(\frac{1}{V_{IN}} + \frac{1}{NV_{OUT}} \right) \quad (5)$$

where $\hat{\imath}_{L1}$ is the peak primary inductor current.

A turns ratio of N = 5 was found to offer a good balance between Δf_s and losses. Using the inductances required for ~100 kHz operation and determining peak currents and flux densities, an appropriate core shape/size could be found.

Pri. Inductance	L1	232 µH
Sec. Inductance	L2	9.3 µH
Turns	pri:sec	35:7 (N=5)
Low-line Frequency	fs low	~90 kHz
High-line Frequency	fs high	~130 kHz
Low-line peak currents	$\hat{\imath}_{L1} : \hat{\imath}_{L2}$	2.5 A , 12.5 A
High-line peak currents	$\hat{\imath}_{L1} : \hat{\imath}_{L2}$	1.7 A , 8.5 A
Core	-	RM8i (3C95)
Air gap	lg	300 µm
Winding Type	pri:sec	Litz : Foil
Estimated Max. Loss	core:winding	1.2 W, 0.4 W

Tab. 1 - Key parameters of the coupled inductor.

B. Control Boards

To easily experiment with primary ON-times, this prototype used an FPGA to control the primary switch (future demonstrators will use simple dedicated on-board analog logic). Both controllers were based on external boards.

PRIMARY SIDE	LATTICE iCE40-HX8K FPGA
SECONDARY SIDE	INFINEON XMC4500

Tab. 2 - External control devices used.

The FPGA was operated with a 27 MHz clock, introducing 37ns additional delay to primary side turn-ON. The secondary controller had a turn-ON delay of 81 ns – though this is relatively inconsequential since current can conduct through S2's body diode during this time.

C. Power Board

A board with the following specifications was built:

Input Capacitance	C_IN	110 µF
Output Capacitance	C_OUT	400 µF
Primary MOSFET	S1	CoolMOS P6 255mΩ
Secondary MOSFET	S2	OptiMOS 5 2.8 mΩ
Source Follower	Q	SigMOS BSS127
Gate Driver	-	Infineon Eice
Comparator	-	ADCMP600

Tab. 3 - Key parameters and components of the power board.

Figs. 8 & 9 show the assembled board. The resulting dimensions are 3.0 x 6.5 x 2.3cm yielding an approximate power density of 23W/inch³ (however no EMI filters were included).

Fig. 8 - Realized 65 W 100 kHz prototype - top side.

To minimise the loop inductance, the inductor current measurement loops can be short circuited when not needed.

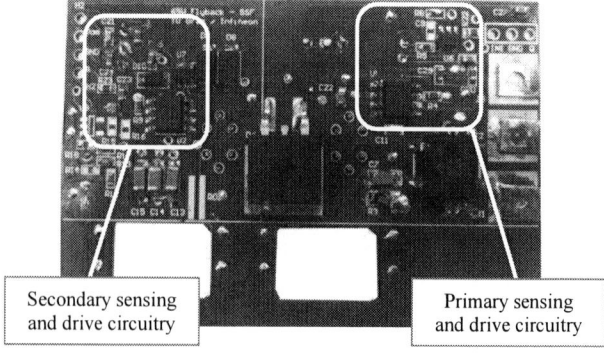

Fig. 9 - Realized 65 W 100 kHz prototype - bottom side.

V. EXPERIMENTAL RESULTS

The following measurements are presented in this paper:

- High- and low-line steady-state operation,
- Primary side turn-ON delay,
- Autonomous zero crossing blanking.

A. Steady-State Operation

1) High-Line Input Voltage

High-line steady state operation was tested using 325 Vdc input voltage and with the fixed primary ON time (T_3) and negative current pulse time (T_1) shown in Table 4.

Secondary Neg. Current Time	T_1	2.00 µs
Primary ON-Time	T_3	1.65 µs

Tab. 4 - Fixed switching times for steady-state high-line operation.

Fig. 10 shows drain-source voltages for both S1 and S2 during steady-state operation at high-line. At t=76 µs, the output voltage exceeds the 20 V threshold and the secondary controller starts pulse-skipping, refraining from sending the subsequent negative pulse.

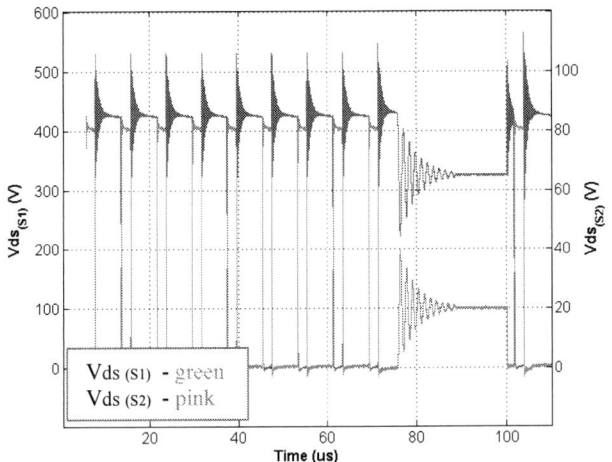

Fig. 10 - Pri. & Sec. i_L & V_{ds} in high-line steady state operation.

At t=100 µs, the end of the pulse-skipping phase can be observed whereby S2 turns-ON in order send a negative current pulse and reinitiate steady-state operation.

1) Low-Line Input Voltage

Low line steady state tests used 170 Vdc input voltage.

Secondary Neg. Current Time	T_1	1.70 µs
Primary ON-Time	T_3	3.85 µs

Tab. 5 - Fixed switching times for steady-state low-line operation.

Fig. 11 shows drain-source voltages along with the secondary side's comparator and gate-drive signals.

The measurement was taken immediately after T_6 - when the secondary controller reinitiates operation. Between t=5.5µs and t=10.5µs comparator output is high; indicating that the current through S2 is positive and that synchronous rectification is on-going. Afterwards, the comparator output goes low indicating that current is now negative - S2 is then kept ON for the fixed time (T_1) required to achieve I_{neg}.

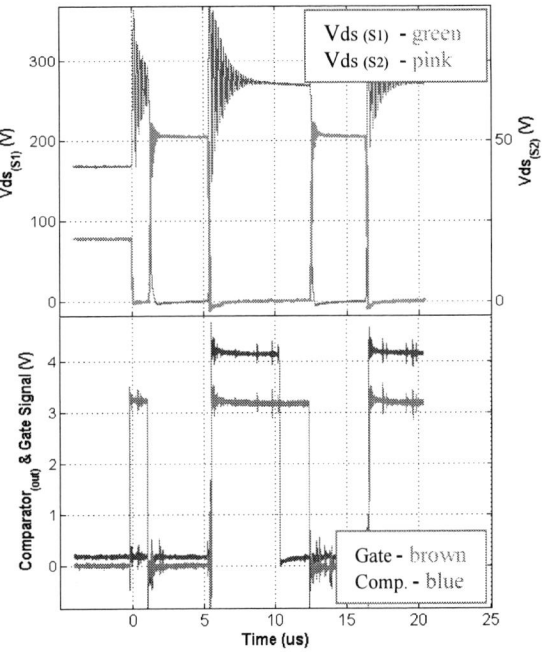

Fig. 11 - Pri. & Sec. side V_{ds} with Sec. comparator & gate-drive signals during steady-state low-line operation.

B. Source Follower Behaviour

1) Primary Turn-ON Delay

As seen in Fig. 12 & 13, the total signal delay between the drain-source zero voltage crossing of S1 and the subsequent gate signal into the driver was measured to be 71ns at high-line, and 75 ns at low-line input voltage.

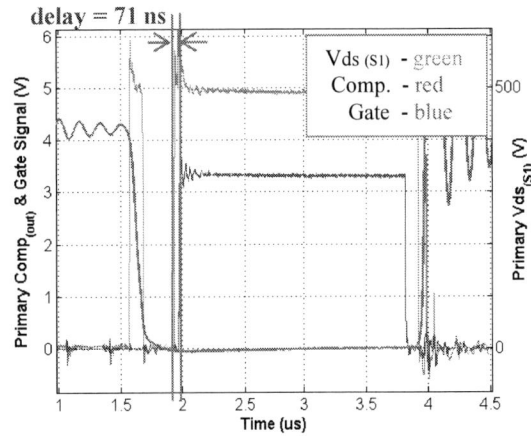

Fig. 12 - Primary side V_{ds} sensing at high-line input voltage.

978-1-4673-9551-9/16 $31.00 © 2016 IEEE

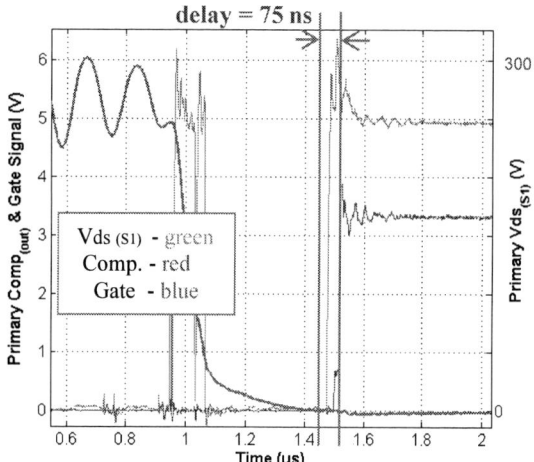

Fig. 13 - Primary side V_{ds} sensing at low-line input voltage.

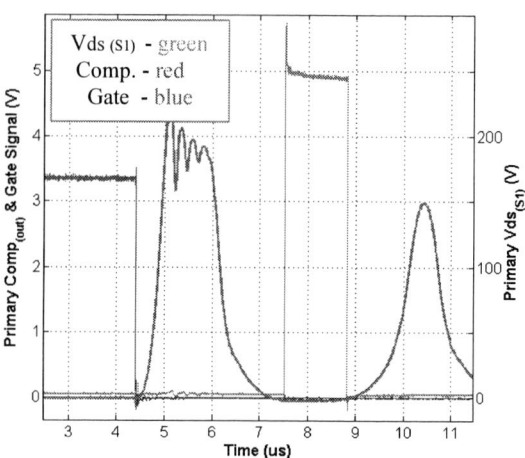

Fig. 14 - Demonstration of ignored zero voltage crossings.

Importantly, the FPGA clock is responsible for 37 ns of this delay, meaning that the actual circuit delay is 34 ns and 38 ns for high- and low-line respectively.

To compete with the conventional control approach, the total signal delay should be less than the propagation delay of a modern standard opto-isolator (70-150 ns) and compete with ultra-fast digital opto-isolators (25-50 ns) [12] [13].

These measurements indicate that the SSF approach is indeed able to outperform standard opto-isolators. In comparison to ultra-fast opto-isolators, an inferior total signal delay was observed due to the FPGA clock used. A faster clock or dedicated analog logic would bring the SSF prototype in line with the fastest opto-isolators. If the gate-driver propogation delay and MOSFET turn-ON time are also considered, this prototype achieved an overall delay time of approximately 112 ns which is itself as fast as standard opto-isolators alone. Table 4 lists a breakdown of the primary turn-ON delay.

Source	Delay Time (ns)	
	Low Line	High Line
Source Follower	34	30
Comparator	4	4
Total Circuit Delay	**38**	**34**
FPGA	37	37
Total Signal Delay	**75**	**71**
Driver Propagation	20	20
Turn-ON Time	18	20
TOTAL	**113**	**111**

Tab. 6 – Measured primary side sensing & turn-ON delays

2) Flyback Voltage Blanking

To test the autonomous filtering of false turn-ON signals, circuit parameters were altered in order to force the primary Flyback voltage to cross 0 V after turn-OFF. Fig. 14 shows the primary control's response in such conditions.

No gate signal is generated at t=7.5 us despite the definite zero crossings of $V_{ds(S1)}$ observed. The dV/dt across S1 prior to the zero crossing is insufficient to trigger the comparator and thus the primary control only receives a single pulse, no gate signal is generated and S1 successfully remains OFF.

VI. CONCLUSIONS

The secondary side Flyback control concept has been proven in hardware with all "communication" between primary and secondary sides occurring without opto-isolator based signals, rather through the coupled inductor via zero crossings of the primary side MOSFET's drain-source voltage.

Steady state operation with pulse-skipping triangular current mode operation has been demonstrated at high- and low-line input voltage with ~100 kHz switching frequency. In such conditions, the primary side sensing delay with high-line input voltage is 34 ns and is 38 ns at low line; much faster than the typical propagation delay of a standard opto-isolator.

The primary-side V_{ds} sensing was executed with a source-follower based sub-circuit. Whilst also protecting the comparator from the input voltage, it has been proven that this arrangement can be used to autonomously distinguish zero voltage crossings due to oscillations in wait periods, from genuine primary turn-ON requests from the secondary side. This helps keep the primary side control relatively simple.

In future publications, the operating efficiency of such a secondary side controlled Flyback converter will be presented along with a reliable start-up routine. The possibility of using a positive non-zero (e.g. 40V) reference for primary turn-ON will also be investigated. The control scheme will then be extended to a 1 MHz GaN based demonstrator. In such an application the advantages of the concept (no opto-isolator propagation delay and direct access to the output voltage) can be exploited more fully to achieve precise soft-switching operation at very high switching frequencies.

REFERENCES

[1] B.R.Lin, "Implementation of the ZVS Converter with Synchronous Rectifier," in *Electric Power Applications*, 2006, pp. 361-368.

[2] R. Erickson, "Design of a Simple High-Power-Factor Rectifier Based on the Flyback Converter," in *Applied Power Electronics Conference and Exposition*, 1990, pp. 792-801.

[3] K.-H. Liu, "Zero-Voltage Switching Technique in DC/DC Converters," in *IEEE Transactions on Power Electronics*, 1990, pp. 293-304.

[4] H. Dong, "A Variable-frequency One-cycle Control for BCM ," in *Industrial Electronics Society, IECON 2014*, 2014, pp. 1161-1166.

[5] J. Park, "A CCM/DCM Dual-Mode Synchronous Rectification Controller for a High-Efficiency Flyback Converter ;" in *IEEE Transactions on Power Electronics*, 2013, pp. 768-774.

[6] W. Yuan, "A Novel Soft Switching Flyback Converter with Synchronous Rectification," in *Power Electronics and Motion Control Conference*, 2009, pp. 551-555.

[7] X. Huang, "A Novel Variable Frequency soft switching Method ," in *Applied Power Electronics Conference and Exposition*, 2010, pp. 1392-1396.

[8] X. Huang, "Evaluation and Application of 600V GaN HEMT in Cascode Structure," in *Applied Power Electronics Conference and Exposition 2013* , 2013, pp. 1279-1286.

[9] B. Mahato, "Hardware Design and Implementation of Unity Power Factor Rectifiers using Microcontrollers," in *Power Electronics (IICPE)*, 2014, pp. 1-5.

[10] H. Hu, "Efficiency improvement of grid-tied inverters at low input power using pulse skipping control strategy," in *Applied Power Electronics Conference and Exposition* , 2010, pp. 627-633.

[11] C. Marxgut, "Ultraflat Interleaved Triangular Current Mode (TCM) Single-Phase PFC Rectifier," in *IEEE Transactions on Power Electronics* , 2013, pp. 873-882.

[12] PLC, Optek TT Electronics. (2015, Sep.) High Speed Opto-Isolator OPTI1268S. http://optekinc.com/datasheets/opi1268s.pdf.

[13] T. I. Inc. (2008, Jun.) 5V High Speed Digital Isolators. http://www.ti.com/lit/ds/symlink/iso721m-ep.pdf.

Analysis, Modeling and Control of an Interleaved Isolated Boost Series Resonant Converter for Microinverter Applications

Luciano A. Garcia-Rodriguez[1], Cheng Deng[2],
Juan Carlos Balda[3],
Department of Electrical Engineering
University of Arkansas
Fayetteville, AR, 72701, USA
[1]lgarciar@uark.edu, [2]cdeng@uark.edu, [3]jbalda@uark.edu

Andrés Escobar-Mejía,
Department of Electrical Engineering
Universidad Tecnológica de Pereira
Pereira, Colombia
andreses1@utp.edu.co

Abstract — The analysis, modeling and control of an interleaved boost series resonant converter (IBSRC) is presented in this paper. The converter consists of an interleaved boost stage, a high-frequency transformer (HF-XFMR) and a voltage doubler rectifier. The HF-XFMR's leakage inductance and the voltage doubler bridge rectifier's capacitors form a resonant tank that allows zero-voltage switching (ZVS) of the main semiconductor devices and zero-current switching (ZCS) of the rectifier's diodes. This capability enables the IBSRC to operate at high switching frequencies leading to a compact design while keeping efficiency high. This makes the IBSRC suitable for the dc-dc converter stage of a dual-stage microinverter. The IBSRC dynamics are extensively investigated using the extended describing functions (EDF) methodology, whereas the converter small-signal model is derived to design the closed-loop control strategy. Experimental results on a 350-W prototype are presented to demonstrate the feasibility of the proposed model and the performance of the controller during input and output variations.

Keywords — extended describing functions, microinverter, series resonant converter, small-signal modeling, interleaved synchronous boost converter

I. INTRODUCTION

Large-scale solar photovoltaic facilities usually require connecting several PV modules to the grid using high-power central or string inverters. For medium-power facilities the string system is recommended, while microinverters are more attractive for small power applications, e.g., commercial and residential. Other considerations such as cost, efficiency and application are critical to selecting the grid connection method. In the case of microinverters, variables such as frequency and volume dominate the design. Increasing the frequency to achieve a compact converter increases the total losses, thus reducing the system efficiency.

Several topologies for a microinverter have been proposed to interface a PV system with the grid [1]–[4]. These can be classified depending on whether or not the conversion process involves galvanic isolation [5]. Topologies such as the PWM-based converter with high-frequency link [6] and the resonant converter with isolated high-frequency link [7] comprise of one

or two stages for the energy conversion process. Even though transformerless topologies have several advantages over transformer-based topologies, they are not recommended since they do not provide enough voltage gain to step up the PV module voltage [5]. Conventionally, the well-known flyback topology is used for a single PV module. This topology includes a high-frequency link with a high leakage inductance that limits the energy transfer. This makes the flyback topology an unappealing solution in high-power applications (>200 W) [8]. Other concerns, such as reducing switching losses, air gap length and better design for higher PV panel ratings, need to be addressed to make this topology suitable for widespread use [8], [9].

Resonant converter schemes, such as the IBSRC, are attractive for microinverters since they allow high switching frequencies to achieve compact converters while keeping efficiencies high [10]. As shown in Fig. 1, the IBSRC consists of two synchronous boost converters connected to a HF-XFMR and operating with a phase shift of 180°. The converters' modulation schemes impress three-level square waveforms that resonate with the tank circuit formed by the transformer's leakage inductance (L_k) and the voltage doubler rectifier's capacitors (C_3, C_4).

This paper analyzes the small-signal modeling of the IBSRC to present the converter behavior during input voltage (v_{in}), switching frequency (f_{sw}) and duty cycle (d) changes as well as the control scheme, which offers satisfactory closed-loop performance for the specified operating region. The paper is organized as follows: The topology description, modes of operation and the main design equations are presented in Section II. The non-linear state equations obtained combining the ones from the different operation modes are presented in Section III. The harmonic approximation and the EDF are covered in Section IV. The harmonic balance and steady-state solution is shown in Section V. Then, the space-state small-signal model is described in Section VI. Finally, the experimental results and conclusions are shown in Sections VII and VIII.

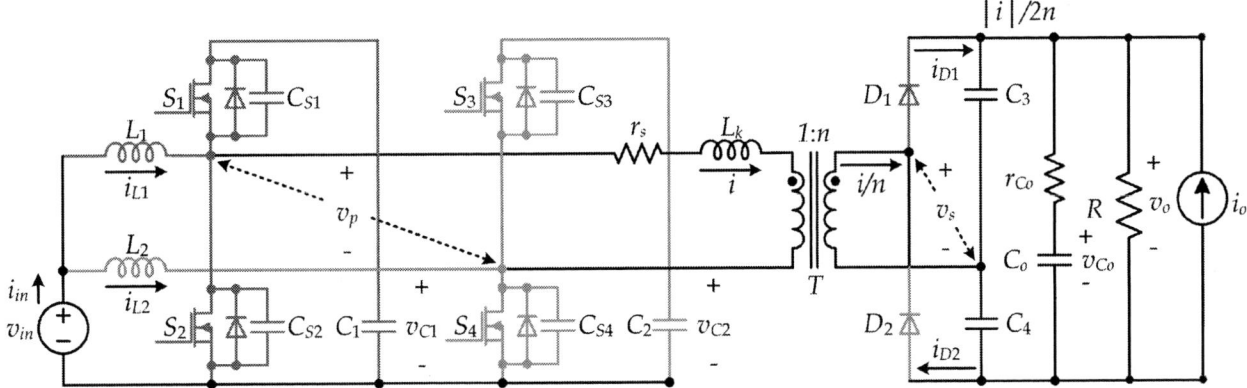

Fig. 1. IBSRC circuit; the current source i_o represents the perturbation at the output.

II. TOPOLOGY DESCRIPTION, MODES OF OPERATION AND MAIN DESIGN CONSIDERATIONS

Per Fig. 1, the IBSRC is composed of two synchronous interleaved boost converters, each one connected to one terminal of the primary side of the HF-XFMR T. The interleaved synchronous boost converters operate 180 degrees out of phase with respect to each other.

In a synchronous boost converter, the inductor current will only operate in the continuous conduction mode (CCM). Therefore, the steady-state voltage at the capacitors C_1 and C_2 can be calculated as:

$$V_{C_1} = V_{C_2} = \frac{v_{in}}{1 - d/2}. \tag{1}$$

The equivalent discontinuous conduction mode (DCM) of a boost converter translates for the synchronous boost converter case into CCM, with the exception that the inductor current has a negative value. The polarity change in the inductor current can be used to provide zero voltage soft-switching (ZVS) turn on for some of the devices. The use of an interleaved boost stage accounts for the increment in the current ripple of each boost inductor.

The secondary side of the HF-XFMR is connected to a voltage doubler rectifier, which reduces the required transformer turns ratio by a factor of two. The steady-state voltage of capacitors C_3 and C_4 can be obtained as:

$$v_{C_3} = v_{C_4} = \frac{v_o}{2}, \tag{2}$$

where v_o is the output votage of the converter.

The voltage at the primary side of the transformer, v_p, alternates between the values v_{c_1}, zero, and $-v_{c_2}$ depending on the position of the switches.

When S_1 and S_3 are turned on, capacitors C_1 and C_2 are charged by the boost inductances L_1 and L_2, and ideally have the same voltage. As a result, v_p equals zero, and thus the primary current is zero, and diodes D_1 and D_2 become reversed biased. The equivalent circuit of this mode is shown in Fig. 2 (a). The following equations describe the situation when v_p has a value of zero:

$$v_{in} - v_{C_1} = L_1 \frac{di_{L1}}{dt}, \tag{3}$$

$$v_{in} - v_{C_2} = L_2 \frac{di_{L2}}{dt}, \tag{4}$$

$$v_p = v_{C_1} - v_{C_2} \approx 0, \tag{5}$$

$$C_1 \frac{dv_{C_1}}{dt} = i_{C_1} = i_{L_1}. \tag{6}$$

$$C_2 \frac{dv_{C_2}}{dt} = i_{C_2} = i_{L_2}. \tag{7}$$

When S_1 and S_4 turn on at the same time, inductor L_2 is short-circuited to the input voltage source, causing it to charge, while inductor L_1 is discharged through capacitor C_1. The voltage applied to the transformer is equal to v_{C_1}. The voltage applied to the transformer is reflected to the secondary side, turning on D_1. The transformer leakage inductance L_k starts resonating with the output capacitors C_3 and C_4 of the voltage doubler rectifier. It can be seen in the equivalent circuit of Fig. 2(b) that capacitors C_4 and C_o are connected in series. Due to the fact that $C_4 \ll C_o$, the series connection of C_4 and C_o is modeled as C_4, with an initial condition equal to $V_o/2$. The equivalent circuit with the parallel connection of C_3 and C_4 referred to the primary side is given in Fig. 2(b). A voltage equal to $v_{C1} - v_o/2n$ is applied to the resonant tank ($L_k, n^2(C_3 + C_4)$).

The following are the differential equations that represent when a positive voltage is applied to the primary side of the HF-XFMR and the resonant tank is active:

$$v_p = v_{C_1} = ir_s + L_k \frac{di}{dt} + v + \frac{v_o}{2n}, \tag{8}$$

$$n^2(C_3 + C_4) \frac{dv}{dt} = i, \tag{9}$$

$$C_1 \frac{dv_{C_1}}{dt} = i_{L_1} - i, \tag{10}$$

$$C_2 \frac{dv_{C_2}}{dt} = 0, \tag{11}$$

$$v_{in} - v_{C_1} = L_1 \frac{di_{L_1}}{dt}, \tag{12}$$

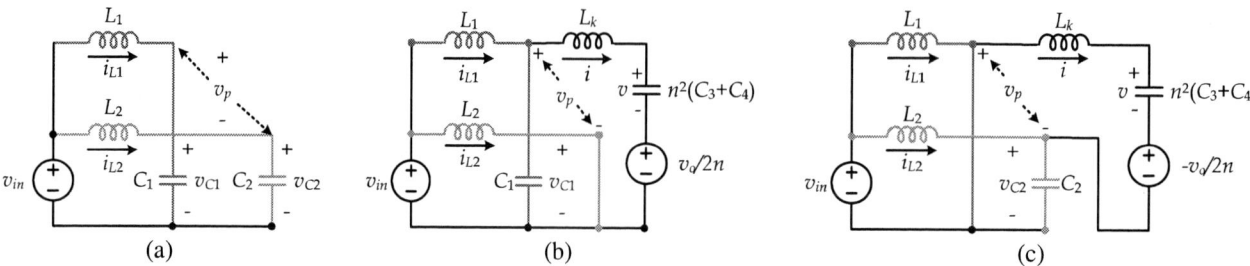

Fig. 2. IBSRC equivalent circuits referred to the HF-XFMR's primary side.

$$v_{in} = L_2 \frac{di_{L_2}}{dt}. \tag{13}$$

When S_2 and S_3 turn on at the same time, the voltage applied to the transformer is now $-v_{C_2}$ and the rectifier diode D_2 becomes forward biased. Inductor L_1 is charged by the input voltage source while inductor L_2 is discharged through capacitor C_2. The leakage inductance of the transformer L_k resonates with the output capacitors C_3 and C_4 of the voltage doubler rectifier. It can be seen that capacitors C_3 and C_o are series connected. However, since $C_3 \ll C_o$, the series connection of C_3 and C_o is modeled as C_3 with an initial condition equal to $-V_o/2n$.

The equivalent circuit with the parallel connection of C_3 and C_4 referred to the primary side is illustrated by Fig. 2(c). It is shown that a voltage approximately equal to $-v_{C2} + V_o/2n$ gets applied to the resonant tank $(L_k, n^2(C_3 + C_4))$. The following equations model the equivalent circuit of Fig. 2(c).

$$v_p = -v_{C_2} = i r_s + L_k \frac{di}{dt} + v - \frac{v_o}{2n}, \tag{14}$$

$$n^2 (C_3 + C_4) \frac{dv}{dt} = i, \tag{15}$$

$$v_{in} = L_1 \frac{di_{L_1}}{dt}, \tag{16}$$

$$v_{in} - v_{C_2} = L_2 \frac{di_{L_2}}{dt}, \tag{17}$$

$$C_1 \frac{dv_{C_1}}{dt} = 0, \tag{18}$$

$$C_2 \frac{dv_{C_2}}{dt} = i_{L_2} + i_p. \tag{19}$$

The resonant current will have a period that can be calculated using the following equation:

$$T_o = 2\pi \sqrt{L_k n^2 (C_3 + C_4)}. \tag{20}$$

There exists a tradeoff when selecting the values of the inductors for the synchronous interleaved boost converter. As mentioned earlier, if the current changes polarity through the inductor, the bottom switches of the boost stage also switch on at ZVS. Therefore, the switching losses can be reduced. However, the RMS currents through the switches are going to be larger than for a case where the inductor current ripple is smaller. In [11], it is proposed to design the minimum negative peak of the boost inductor currents to allow ZVS when the output power of the converter is 75% of the rated value. The

latter means that ZVS will not be achieved for an output power higher than 75%. The following equations can be used to select the required boost inductors based on the minimum values for the boost inductor currents necessary to provide ZVS:

$$L_1 i_{L1pk}^2 = L_2 i_{L2pk}^2 > 2C_{oss} V_{C1} = 2C_{oss} V_{C2}, \tag{21}$$

$$i_{L_1pk} = \frac{P_o}{2V_{in}} - \frac{D T_{sw} v_{in}}{2L_1}. \tag{22}$$

To achieve ZCS of the output diodes it is necessary to allow the resonant current to decrease down to a value of zero. This is accomplished when the duty cycle d times the switching period T_{sw} is larger than the resonant period T_o.

III. NON-LINEAR STATE EQUATIONS

The goal of this section is to obtain a single set of nonlinear equations that represent all the possible operating modes. The ideal operating waveforms of the IBSRC for the case where the operating switching frequency is close to the resonant frequency f_o are presented in Fig. 3. Under these conditions, the resonant current is zero when the primary voltage v_p is also zero. Equations (5), (8) and (14), can be combined to obtain (23) using the describing function $sgn(i)$:

$$L_k \frac{di}{dt} + i r_s + v + sgn(i) \frac{v_o}{2n} = v_p. \tag{23}$$

Similarly, equations (3), (12) and (16) can be represented by (24) using the describing function $(sgn(di_{L1}/dt) - 1)/2$:

$$L_1 \frac{di_{L1}}{dt} = v_{in} + v_{C_1} \frac{sgn\left(\frac{di_{L_1}}{dt}\right) - 1}{2}, \tag{24}$$

and equations (4), (13) and (17) are expressed by (25) using the function $(sgn(di_{L_2}/dt) - 1)/2$:

$$L_2 \frac{di_{L2}}{dt} = v_{in} + v_{C_2} \frac{sgn\left(\frac{di_{L_2}}{dt}\right) - 1}{2}. \tag{25}$$

Equations (6), (10) and (18) modeling the current through the capacitor C_1 are replaced by (26):

$$C_1 \frac{dv_{C_1}}{dt} = (i - i_{L_1}) \frac{sgn\left(\frac{di_{L_1}}{dt}\right) - 1}{2}. \tag{26}$$

Similarly, the current through the capacitor C_2 expressed by (7), (11), and (19) is modelled by (27):

$$C_2 \frac{dv_{C_2}}{dt} = -(i_{L_2} + i) \frac{sgn\left(\frac{di_{L_2}}{dt}\right) - 1}{2}. \tag{27}$$

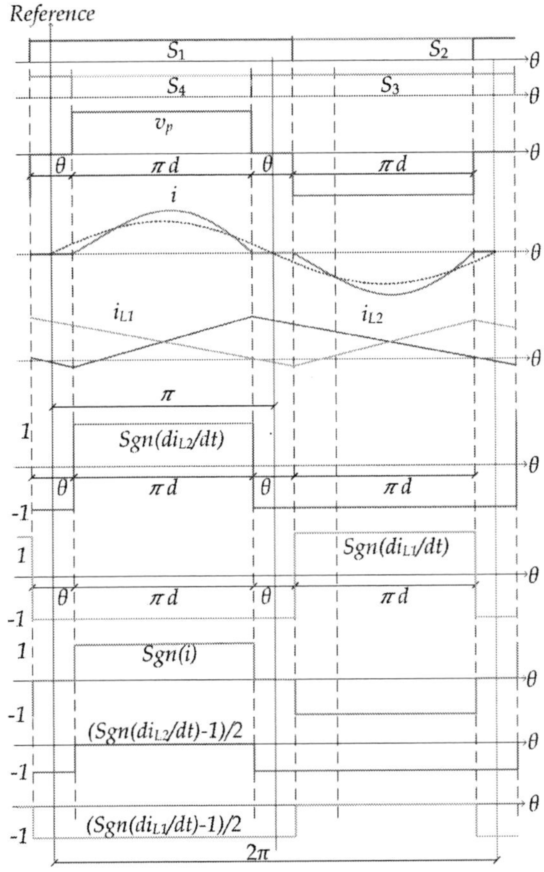

Fig. 3. IBSRC theoretical waveforms.

Equation (28) models the resonant tank current and (29) is obtained by applying Kirchhoff's current law to the load side node:

$$n^2(C_3 + C_4)\frac{dv}{dt} = i, \tag{28}$$

$$\frac{v_{C_o}}{R} + C_o \frac{dv_{C_o}}{dt}\left(1 + \frac{r_{C_o}}{R}\right) = i_o + \frac{|i|}{2n}. \tag{29}$$

The variables of interest are the average input current i_{in} and the output voltage v_o which are calculated as:

$$\langle i_{in} \rangle = \frac{1}{T_s} \int_0^{T_s} i \frac{v_p}{v_{in}} dt, \tag{30}$$

$$\langle v_o \rangle = \left(i_o + \frac{|i|}{2n}\right) r'_{C_o} + v_{C_o} \frac{r'_{C_o}}{r_{C_o}}, \tag{31}$$

where $r'_{C_o} = R \| r_{C_o}$.

IV. HARMONIC APPROXIMATION AND EXTENDED DESCRIBING FUNCTION

In order to obtain an approximate model, only the first harmonic terms of the resonant current i and voltage v are considered [12]. Therefore, they are expressed as:

$$i = i_s \sin(\omega_s t) + i_c \cos(\omega_s t), \tag{32}$$

$$v = v_s \sin(\omega_s t) + v_c \cos(\omega_s t). \tag{33}$$

The derivatives of i and v are given by:

$$\frac{di}{dt} = \left(\frac{di_s}{dt} - \omega_s i_c\right)\sin(\omega_s t) + \left(\frac{di_c}{dt} + \omega_s i_s\right)\cos(\omega_s t), \tag{34}$$

$$\frac{dv}{dt} = \left(\frac{dv_s}{dt} - \omega_s v_c\right)\sin(\omega_s t) + \left(\frac{dv_c}{dt} + \omega_s v_s\right)\cos(\omega_s t). \tag{35}$$

The output capacitor voltage v_{C_o} and the input current i_{in} are approximated to their dc components. The same approximation is applied to the current though the boost inductors L_1 and L_2 and the voltage across the capacitors C_1 and C_2. Furthermore, in (24) to (27) it is required to approximate the "sgn" functions to their dc components. Assuming a phase shift of 180° between the two synchronous boost converters, the dc components of the "sgn" functions, related to the inductors L_1 and L_2, are:

$$\frac{\left(sgn\left(\frac{di_{L_1}}{dt}\right) - 1\right)}{2} = \frac{\left(sgn\left(\frac{di_{L_2}}{dt}\right) - 1\right)}{2}$$

$$= \frac{1}{2\pi} \int_0^{2\pi} \frac{\left(sgn\left(\frac{di_{L_1}}{dt}\right) - 1\right)}{2} d\omega_s t$$

$$= \frac{1}{2\pi} \int_0^{\pi+\frac{\theta}{2}} - d\omega_s t + \frac{1}{2\pi} \int_{2\pi-\frac{\theta}{2}}^{2\pi} - d\omega_s t$$

$$= -\frac{1}{2\pi}(\pi + \theta) = -1 + \frac{d}{2}. \tag{36}$$

As indicated, the zero state of the primary voltage θ is expressed in terms of the duty cycle d as $\theta = \pi - \pi d$.

From (26), the resonant current i flowing through capacitor C_1 can be approximated as the average of the resonant current during the time interval when the derivative of the inductor current i_{L1} is negative $(\theta + \pi)$.

$$i_{C1} = \frac{1}{\theta + \pi}\left[\int_0^{\pi+\frac{\theta}{2}} i_p \sin(\omega_s t + \alpha)\, d\omega_s t + \int_{2\pi-\frac{\theta}{2}}^{2\pi} i_p \sin(\omega_s t + \alpha)\, d\omega_s t\right],$$

$$i_{C1} = \frac{2i_s}{\pi(2-d)}\sin\left(\frac{\pi d}{2}\right). \tag{37}$$

A similar analysis can be done for the resonant current that flows through capacitor C_2.

The dc approximation for the magnitude of the resonant current $|i|$ which is part of (29) and (31) is calculated as:

$$\frac{1}{2\pi}\int_0^{2\pi} |i| d\omega_s t = \frac{1}{\pi}\int_{-\alpha}^{\pi-\alpha} |i_p \sin(\omega_s t + \alpha)| d\omega_s t,$$

$$= \frac{2}{\pi}i_p = \frac{2}{\pi}\sqrt{i_s^2 + i_c^2}. \tag{38}$$

The sign of the resonant current $sgn(i)$ present in (23) is approximated by its first harmonic as:

$$C_1 = \frac{1}{2}(a_1 - jb_1) = \frac{1}{2\pi}\int_0^{2\pi} sgn(i)e^{-j\omega_s t}\, d\omega_s t,$$

$$= \frac{1}{2\pi}\int_0^{2\pi} sgn(i_p \sin(\omega_s t + \alpha))e^{-j\omega_s t}\, d\omega_s t,$$

$$= -\frac{2j}{\pi}(\cos(\alpha) + j\sin(\alpha)).$$

Making $\sin(\alpha) = i_c/i_p$ and $\cos(\alpha) = i_s/i_p$, the $sgn(i)$ is given by:

$$sgn(i) \approx \frac{4}{\pi} \frac{i_s}{i_p} \sin(\omega_s t) + \frac{4}{\pi} \frac{i_c}{i_p} \cos(\omega_s t). \tag{39}$$

The transformer primary side voltage v_p is also approximated to its first harmonic component. Due to the odd symmetry, only sine terms are considered in the Fourier series:

$$a_1 = 0, \quad b_1 = \frac{1}{\pi} \int_{-\pi}^{\pi} v_p \sin(\omega_s t)\, d\omega_s t,$$

$$b_1 = \frac{1}{\pi} \int_{-\pi+\frac{\theta}{2}}^{-\frac{\theta}{2}} -V_C \sin(\omega_s t)\, d\omega_s t + \frac{1}{\pi} \int_{\frac{\theta}{2}}^{\pi-\frac{\theta}{2}} V_C \sin(\omega_s t)\, d\omega_s t.$$

The zero state θ can be expressed in terms of d ($\theta = \pi(1-d)$), allowing the primary voltage v_p to be approximated as:

$$v_p \approx \frac{4V_C}{\pi} \sin\left(\frac{\pi d}{2}\right) \sin(\omega_s t). \tag{40}$$

To solve the integral in (30), the resonant current i is first replaced using (32), then the primary side voltage v_p is replaced using (40) and, finally, the boost capacitor voltage V_C is replaced by $2v_{in}/(2-d)$ resulting in the following expression:

$$i_{in} = \frac{1}{T_s} \int_0^{T_s} i \frac{v_p}{v_{in}}\, dt = \frac{4}{\pi} \frac{\sin\left(\frac{\pi d}{2}\right)}{(2-d)} i_s. \tag{41}$$

V. Harmonic Balance and Steady-State Solution

The extended describing functions derived in section IV are replaced with the non-linear state equations presented in section III. Furthermore, the resonant current and voltage, and their derivatives, are substituted by (32) though (35). Then, the dc components, and the sine and cosine terms are grouped together to get:

$$\frac{di_s}{dt} = \frac{1}{L_k}\left(\frac{4V_C}{\pi}\sin\left(\frac{\pi d}{2}\right) - \frac{2}{\pi}\frac{i_s}{i_p}\frac{v_o}{n} - v_s - r_s i_s\right) + \omega_s i_c, \tag{42}$$

$$\frac{di_c}{dt} = \frac{1}{L_k}\left(-\frac{2}{\pi}\frac{i_c}{i_p}\frac{v_o}{n} - v_c - i_c r_s - \omega_s i_s\right), \tag{43}$$

$$\frac{di_{L1}}{dt} = \frac{1}{L_1}\left(v_{in} + v_{C_1}\left(-1+\frac{d}{2}\right)\right), \tag{44}$$

$$\frac{di_{L2}}{dt} = \frac{1}{L_2}\left(v_{in} + v_{C_2}\left(-1+\frac{d}{2}\right)\right), \tag{45}$$

$$\frac{dv_{C_1}}{dt} = \frac{1}{C_1}\left(i_{L_1} - \frac{2i_s}{\pi(2-d)}\sin\left(\frac{d\pi}{2}\right)\right)\left(1-\frac{d}{2}\right), \tag{46}$$

$$\frac{dv_{C_2}}{dt} = \frac{1}{C_2}\left(i_{L_2} - \frac{2i_s}{\pi(2-d)}\sin\left(\frac{d\pi}{2}\right)\right)\left(1-\frac{d}{2}\right), \tag{47}$$

$$\frac{dv_s}{dt} = \frac{1}{n^2(C_3+C_4)}i_s + \omega_s v_c, \tag{48}$$

$$\frac{dv_c}{dt} = \frac{1}{n^2(C_3+C_4)}i_c - \omega_s v_s, \tag{49}$$

$$i_{in} = \frac{4}{\pi}\frac{1}{2-d}i_s \sin\left(\frac{\pi}{2}d\right), \tag{50}$$

$$\frac{dv_{C_o}}{dt} = \frac{1}{C_o\left(1+\frac{r_{C_o}}{R}\right)}\left(i_o + \frac{1}{\pi n}\sqrt{i_s^2+i_c^2} - \frac{v_{C_o}}{R}\right). \tag{51}$$

This set of equations represents the non-linear large signal model of the IBSRC. The resonant current and voltage sine and cosine terms plus the capacitor voltages and inductor currents constitute the state variables vector $x = \{i_s, i_c, v_s, v_c, v_{c_o}, i_{L1}, i_{L2}, v_{C1}, v_{C2}\}$. In order to obtain the steady-state solution, the derivative terms of the large signal model are set to zero and evaluated around the operation point: $\{V_{IN}, D, \Omega_s, I_o, R\}$. In steady-state conditions, the notation of the state variables vector is represented as: $X = \{I_s, I_c, V_s, V_c, V_{c_o}, I_{L1}, I_{L2}, V_{C1}, V_{C2}\}$. The steady-state values of the state vector are given by:

$$I_s = n^2(C_3+C_4)\Omega_s V_e \frac{\beta}{\alpha^2+\beta^2}, \tag{52}$$

$$I_c = n^2(C_3+C_4)\Omega_s V_e \frac{\alpha}{\alpha^2+\beta^2}, \tag{53}$$

$$V_s = \frac{V_e \alpha}{\alpha^2+\beta^2}, \tag{54}$$

$$V_c = -\frac{V_e \beta}{\alpha^2+\beta^2}, \tag{55}$$

$$V_{C_o} = \frac{R}{\pi n}\frac{n^2(C_3+C_4)\Omega_s}{\sqrt{\alpha^2+\beta^2}}V_e, \tag{56}$$

$$I_{L_1} = I_{L_2} = \frac{2n^2(C_3+C_4)\Omega_s V_e \frac{\beta}{\alpha^2+\beta^2}}{\pi(2-D)}\sin\left(\frac{\pi D}{2}\right), \tag{57}$$

$$V_{C_1} = V_{C_2} = \frac{V_{in}}{1-\frac{d}{2}}, \tag{58}$$

$$\alpha = 1 - n^2(C_3+C_4)\Omega_s^2 L_k, \tag{59}$$

$$\beta = \left(r_s + \frac{2R}{(\pi n)^2}\right)n^2(C_3+C_4)\Omega_s. \tag{60}$$

Considering r_s and r_{co} equal to zero and D equal to one, the voltage conversion ratio for the IBSRC can be expressed as:

$$\frac{V_o}{V_{in}} = \frac{\frac{8}{\pi^2}nR(C_3+C_4)\Omega_s}{\sqrt{\left(1-n^2\Omega_s^2 L_K(C_3+C_4)\right)^2 + \left((C_3+C_4)\Omega_s\frac{2R}{\pi^2}\right)^2}}. \tag{61}$$

VI. IBSRC Space-State Small-Signal Model

In order to obtain the small-signal model of the converter, the nonlinear terms of the large-signal model equations are first approximated by a first-order Taylor expansion as listed in Table I, where $C_{eq} = n^2(C_3+C_4)$. After replacing the linear terms of Table I in (42) to (51), the large-signal model is perturbed around the equilibrium point $\{v_{in} = V_{in} + \tilde{v}_{in},\ d = D + \tilde{d},\ \omega_s = \Omega_s + \tilde{\omega}_s, i_o = 0 + \tilde{i}_0\}$. Additionally, the state variables are replaced by a dc term plus a small signal perturbation: $\{i_s = I_s + \tilde{i_s},\ i_c = I_c + \tilde{i_c}, v_s = V_s + \tilde{v_s}, v_c = V_c + \tilde{v_c}, i_{L_1} = I_{L_1} + \tilde{i_{L_1}},\ i_{L_2} = I_{L_2} + \tilde{i_{L_2}}, v_{C_1} = V_{C_1} + \tilde{v_{C1}}, v_{C_2} = V_{C_2} + \tilde{v_{C2}}, v_{c_o} = V_{C_o} + \tilde{v_{C_o}}\}$. Similarly, the output

TABLE I. LINEARIZED IBSRC LARGE-SIGNAL MODEL TERMS

$$f(x,y) \approx f(a,b) + \frac{\partial f}{\partial x}(a,b)(x-a) + \frac{\partial f}{\partial y}(a,b)(y-b)$$

$$\frac{i_c}{\sqrt{i_s^2 + i_c^2}} \approx \frac{1}{\sqrt{\alpha^2+\beta^2}}\left(\alpha + \frac{\beta^2}{C_{eq}\Omega_s V_e}(i_c - I_c) - \frac{\alpha\beta}{C_{eq}\Omega_s V_e}(i_s - I_s)\right)$$

$$\frac{i_s}{\sqrt{i_s^2 + i_c^2}} \approx \frac{1}{\sqrt{\alpha^2+\beta^2}}\left(\beta + \frac{\alpha^2}{C_{eq}\Omega_s V_e}(i_s - I_s) - \frac{\alpha\beta}{C_{eq}\Omega_s V_e}(i_c - I_c)\right)$$

$$\sqrt{i_s^2 + i_c^2} \approx \frac{1}{\sqrt{\alpha^2+\beta^2}}\left(C_{eq}\Omega_s V_e + \beta(i_s - I_s) + \alpha(i_c - I_c)\right)$$

$$\frac{1}{\pi}i_s \sin\left(\frac{\pi}{2}d\right) \approx \frac{1}{\pi}I_s \sin\left(\frac{\pi}{2}D\right) + \frac{1}{\pi}\sin\left(\frac{\pi}{2}D\right)(i_s - I_s) + \frac{1}{2}I_s \cos\left(\frac{\pi}{2}D\right)(d - D)$$

$$\frac{1}{2-d}i_s \sin\left(\frac{\pi}{2}d\right) \approx \frac{1}{2-D}\Big(I_s \sin\left(\frac{\pi}{2}D\right) + \sin\left(\frac{\pi}{2}D\right)(i_s - I_s) + \left(\frac{1}{2-D}I_s \sin\left(\frac{\pi}{2}D\right) + \frac{\pi}{2}I_s \cos\left(\frac{\pi}{2}D\right)\right)(d - D)\Big)$$

$$\frac{1}{2-d}v_{in} \sin\left(\frac{\pi}{2}d\right) \approx \frac{1}{2-D}\Big(V_{in} \sin\left(\frac{\pi}{2}D\right) + \sin\left(\frac{\pi}{2}D\right)(v_{in} - V_{in}) + \left(\frac{1}{2-D}V_{in} \sin\left(\frac{\pi}{2}D\right) + \frac{\pi}{2}V_{in} \cos\left(\frac{\pi}{2}D\right)\right)(d - D)\Big)$$

$$\frac{4}{\pi}\sin\left(\frac{\pi}{2}d\right)v_C \approx \frac{4}{\pi}\sin\left(\frac{\pi}{2}D\right)V_C + \frac{4}{\pi}\sin\left(\frac{\pi}{2}D\right)(v_C - V_C) + 2V_C \cos\left(\frac{\pi}{2}D\right)(d - D)$$

variables are replaced by a dc component plus a perturbation term $\{v_o = V_o + \tilde{v_o},\ i_{in} = I_{in} + \tilde{i_{in}}\}$. By solving equations (42) to (51), three types of terms can be found: dc terms, first-order ac terms and second-order ac terms. Only the first-order ac terms are considered, which are the ones composed by the product of a single perturbation variable and a dc term. The resulting linear equations of the IBSRC small-signal model are presented in the state-space form $\dot{\tilde{x}} = A\tilde{x} + B\tilde{u}, y = C\tilde{x} + E\tilde{u}$, where $\tilde{x} = (\tilde{i}_s, \tilde{i}_c, \tilde{v}_s, \tilde{v}_c, \tilde{i}_{L1}, \tilde{i}_{L2}, \tilde{v}_{C1}, \tilde{v}_{C2}, \tilde{v}_{co})^T$, $\tilde{u} = (\tilde{v}_{in}, \tilde{d}, \tilde{\omega}_{SN}, \tilde{i}_o)^T$ and $\tilde{y} = (\tilde{v}_o, \tilde{i}_{in})^T$. The state-space matrixes A, B, C and E are shown in Table II.

The state-space model makes it possible to obtain the transfer function between the output variables and control variables, so the required compensator network can be designed. For example, the input current i_{in} and output voltage v_{on} of the IBSRC can be controlled using either the duty cycle d or the switching frequency f_{sw}.

$$\begin{bmatrix} G_{v_o d}(s) & G_{v_o \omega_s}(s) \\ G_{i_{in} d}(s) & G_{i_{in} \omega_s}(s) \end{bmatrix} = C[sI - A]^{-1}B + E. \qquad (62)$$

VII. EXPERIMENTAL RESULTS

In order to demonstrate the accuracy of the small-signal model and the feasibility of the developed controller, a 350-W prototype presented in Fig. 4 is built and tested under the conditions listed in Table III. The complete description of the topology's components is listed in Table IV. Both inductor currents, the transformer primary-side current and primary voltage are shown in Fig. 5.

A wide bandgap modular realization is built for all active devices. GaN MOSFETs from GaN Systems are used to

TABLE II. IBSRC STATE-SPACE MODEL MATRIXES

$$A = \begin{bmatrix}
-\left[\frac{2}{\pi}\frac{V_{co}}{n}\frac{\alpha^2}{C_{eq}\Omega_s V_e\sqrt{(\alpha^2+\beta^2)}} + r_s\right]\frac{1}{L_{lk}} & \frac{1}{L_{lk}}\frac{2}{n}\frac{V_{co}}{\pi}\frac{\alpha\beta}{C_{eq}\Omega_s V_e\sqrt{\alpha^2+\beta^2}} + \Omega_s & -\frac{1}{L_{lk}} & 0 & 0 & 0 & 0 & 0 & -\frac{1}{L_{lk}}\frac{2}{n\pi}\frac{\beta}{\sqrt{\alpha^2+\beta^2}} \\
\frac{1}{L_{lk}}\frac{V_{co}}{n}\frac{2}{\pi}\frac{\alpha\beta}{C_{eq}\Omega_s V_e\sqrt{\alpha^2+\beta^2}} - \Omega_s & -\left[\frac{1}{L_{lk}}\frac{V_{co}}{n}\frac{2}{\pi}\frac{\beta^2}{C_{eq}\Omega_s V_e\sqrt{\alpha^2+\beta^2}} + \frac{1}{L_{lk}}r_s\right] & 0 & -\frac{1}{L_{lk}} & 0 & 0 & 0 & 0 & -\frac{1}{L_{lk}}\frac{2}{n\pi}\frac{\alpha}{\sqrt{\alpha^2+\beta^2}} \\
\frac{1}{C_{eq}} & 0 & 0 & \Omega_s & 0 & 0 & 0 & 0 & 0 \\
0 & \frac{1}{C_{eq}} & -\Omega_s & 0 & 0 & 0 & 0 & 0 & 0 \\
0 & 0 & 0 & 0 & 0 & 0 & \frac{1}{L_1}\left(\frac{D}{2}-1\right) & 0 & 0 \\
0 & 0 & 0 & 0 & 0 & 0 & 0 & \frac{1}{L_2}\left(\frac{D}{2}-1\right) & 0 \\
-\frac{1}{\pi C_1}\sin\left(\frac{\pi}{2}D\right) & 0 & 0 & 0 & \frac{1}{C_1}\left(1-\frac{D}{2}\right) & 0 & 0 & 0 & 0 \\
-\frac{1}{\pi C_1}\sin\left(\frac{\pi}{2}D\right) & 0 & 0 & 0 & \frac{1}{C_2}\left(1-\frac{D}{2}\right) & 0 & 0 & 0 & 0 \\
\frac{1}{\left(1+\frac{r_{Co}}{R}\right)C_o}\frac{1}{\pi n}\frac{\beta}{\sqrt{\alpha^2+\beta^2}} & \frac{1}{\left(1+\frac{r_{Co}}{R}\right)C_o}\frac{1}{\pi n}\frac{\alpha}{\sqrt{\alpha^2+\beta^2}} & 0 & 0 & 0 & 0 & 0 & 0 & -\frac{1}{\left(1+\frac{r_{Co}}{R}\right)C_o}\frac{1}{R}
\end{bmatrix}$$

$$B = \begin{bmatrix}
\frac{1}{L_{lk}}\frac{8}{\pi}\frac{1}{2-D}\sin\left(\frac{\pi}{2}D\right) & \left[\frac{1}{2-D}\sin\left(\frac{\pi}{2}D\right) + \frac{\pi}{2}\cos\left(\frac{\pi}{2}D\right)\right]\frac{8}{\pi}\frac{1}{(2-D)}V_{in}\frac{1}{L_{lk}} & I_c & 0 \\
0 & 0 & -I_s & 0 \\
0 & 0 & V_c & 0 \\
0 & 0 & -V_s & 0 \\
1/L_1 & \frac{V_{C_1}}{2L_1} & 0 & 0 \\
1/L_2 & \frac{V_{C_2}}{2L_2} & 0 & 0 \\
0 & -\frac{1}{2C_1}I_s\cos\left(\frac{\pi}{2}D\right) & 0 & 0 \\
0 & -\frac{1}{2C_2}I_s\cos\left(\frac{\pi}{2}D\right) & 0 & 0 \\
0 & 0 & 0 & \frac{1}{\left(1+\frac{r_{Co}}{R}\right)C_o}
\end{bmatrix}$$

$$C = \begin{bmatrix}
\frac{1}{\pi n}r'_{co}\frac{\beta}{\sqrt{\alpha^2+\beta^2}} & \frac{1}{\pi n}r'_{co}\frac{\alpha}{\sqrt{\alpha^2+\beta^2}} & 0 & 0 & 0 & 0 & 0 & 0 & \frac{r'_{co}}{r_{co}} \\
\frac{4}{\pi}\frac{1}{2-D}\sin\left(\frac{\pi}{2}D\right) & 0 & 0 & 0 & 0 & 0 & 0 & 0 & 0
\end{bmatrix}$$

$$E = \begin{bmatrix}
0 & 0 & 0 & r'_{co} \\
0 & \frac{4}{\pi}\frac{1}{(2-D)^2}I_s\sin\left(\frac{\pi}{2}D\right) + \frac{2}{(2-D)}I_s\cos\left(\frac{\pi}{2}D\right) & 0 & 0
\end{bmatrix}$$

TABLE III. IBSRC MICROINVERTER WORKING CONDITIONS

Output Power, P_O	350 W
DC-Link Output Voltage, V_O	400 V
AC Output Voltage	240 Vac
Input Voltage Range, V_{IN}	19 ~ 70 V
Switching frequency Range, f_{Sw}	100 ~ 200 kHz

TABLE IV. IBSRC MICROINVERTER PROTOTYPE PARAMETERS

Boost Converter Inductors $L_1 \sim L_2$	9 μH
Boost Converter Switches $S_1 \sim S_4$	100 V/ 90 A E-GaN GS61008P
Bare Die Diodes $D_1 \sim D_2$	1200 V/ 10 A CPW4-1200-S010B
Full Bridge Bare Die Switches	1200 V/ 11 A CPM2-1200-0160B
Boost Converter Capacitors $C_1 \sim C_2$	33 μF
Resonant Capacitors $C_3 \sim C_4$	22 nF
DC-Link Capacitor C_O	3x68 μF
Transformer Turn Ratio n_p/n_s	7/19
Transformer Leakage Inductance L_k	5.6 μH

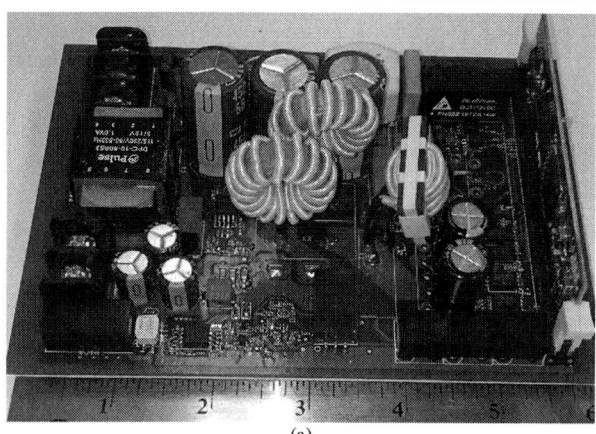

(a)

(b)

Fig. 4(a). IBSRC microinverter prototype rated at 350 W and (b) modularized realization of the power stage (synchronous boost converter GaN switches at the left and full bridge inverter SiC switches and voltage doubler rectifier diodes at the right).

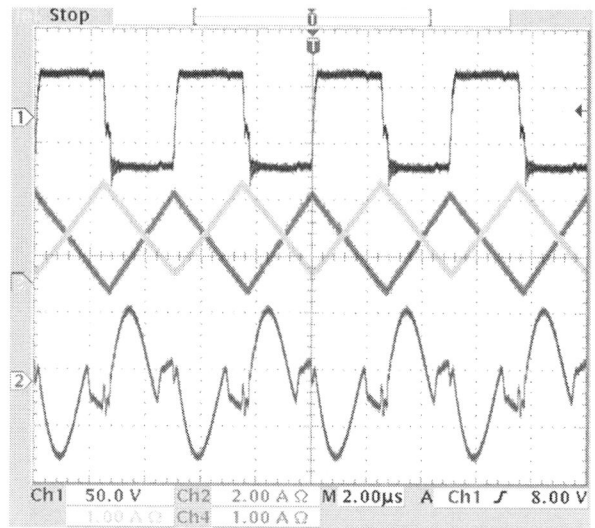

Fig. 5. IBSRC experimental waveforms. Transformer input voltage (upper), current (lower), and inductor currents (middle).

implement the synchronous boost converter devices. The voltage doubler rectifier diodes are implemented with SiC bare dies diodes from Cree. The inverter full bridge stage is built using SiC bare dies MOSFETs from Cree. The resonant switching frequency is approximately equal to 120 kHz and the input voltage is set to 20 V. The resonant tank shapes the current waveform flowing through the primary of the transformer to eliminate the reverse recovery current in the rectifier diodes at the output stage when the operating switching frequency is close to the resonant frequency.

If the boost inductors are designed according to (21) and (22), the devices S_2 and S_4 operate under ZVS, and with an appropriate time delay, S_1 and S_3 operate at ZVS. The converter can be controlled by frequency and duty cycle at the same time. In [11], the duty cycle is chosen as the control variable, but the switching frequency is adjusted in order to allow the resonant current period to be completed. Hence, the switching operating frequency must be less than the resonant frequency. This allows ZCS for the voltage doubler rectifier diodes. However, the controller can be designed to regulate the output variables by increasing the switching frequency from the resonant point and either leaving constant or modifying the duty cycle.

VIII. CONCLUSIONS

The small-signal model for an interleaved boost series resonant converter using the extended describing functions method was presented in this paper. Furthermore, the procedure as well as the analysis to obtain the small signal model was documented. The presented linear model made possible the design of a controller to meet the required gain and phase margins needed for a specific application. Additionally, the presented model made possible the design of a controller to achieve a fast rejection of disturbances with a high bandwidth response. Experimental results on a 350 W prototype are presented. The prototype was built with a modularize structure implemented using GaN and SiC bare dies. In comparison with a discrete prototype, a volume reduction of about 45% can be achieved.

ACKNOWLEDGMENTS

The authors are grateful to Mr. Brett Schauwecker and Mr. Jordan Gamble for their help to conclude this paper.

The authors are grateful for the financial support from the NSF-EPSCoR Vertically Integrated Center for Transformative Energy Research (www.victercenter.com). Dr. Andrés Escobar-Mejía is grateful for the financial support as assistant professor of the Universidad Tecnológica de Pereira (Colombia). Dr. Cheng Deng is grateful for the financial support of the National Natural Science Foundation of China (51577161, 51277156), Natural Science Foundation of Hunan Province, China (2015JJ2135).

REFERENCES

[1] Y. Fang, X. Ma, "A novel PV microinverter with coupled inductors and double-boost topology," *IEEE Transactions on Power Electronics*, vol. 25, no.12, pp. 3139–3147, December, 2010.

[2] H. Haibing, S. Harb, N.H. Kutkut, Z.J. Shen, I. Batarseh, "A single-stage microinverter without using eletrolytic capacitors," *IEEE Transactions on Power Electronics*, vol. 28, no. 6, pp. 2677–2687, June, 2013.

[3] H. Haibing, S. Harb, N.H. Kutkut, I. Batarseh, Z.J. Shen "A review of power decoupling techniques for microinverters with three different decoupling capacitor locations in PV systems," *IEEE Transactions on Power Electronics*, vol. 28, no. 6, pp. 2711–2726, June, 2013.

[4] C. Lin, A. Amirahmadi, Z. Qian, N. Kutkut, I. Batarseh, "Design and implementation of three-phase two-stage grid-connected module integrated converter," *IEEE Transactions on Power Electronics*, vol. 29, no. 8, pp. 3881–3892, August, 2014.

[5] W.J., Cha, Y.W. Cho, J.M. Kwon, B.H. Kwon, "Highly efficient microinverter with soft-switching step-up converter and single-switch-modulation inverter," *IEEE Transactions on Industrial Electronics*, vol. 62, no. 6, June, 2015.

[6] K. de Souza, M. de Castro, F. Antunes, "A dc/ac converter for single-phase grid-connected photovoltaic systems," in *Proceedings of the 28th Annual Conference of the Industrial Electronics Society*, IECON 2002, pp. 3268–3273, November, 2002.

[7] H. Krishnaswami, "Photovoltaic microinverter using single-stage isolated high-frequency link series resonant topology," in *Proceedings of the IEEE Energy Conversion Congress and Exposition*, ECCE 2011, pp. 495–500, September, 2011.

[8] B. Tamyurek, B. Kirimer, "An interleaved high-power flyback inverter for photovoltaci applications," *IEEE Transactions on Power Electronics*, vol. 30, no. 6, June, 2015.

[9] N. Sukesh, M. Pahlevaninezhad, P.K. Jain, "Analysis and implementation of a single-stage flyback PV microinverter with soft switching," IEEE Transactions on Industrial Electronics, vol. 64. No. 4, April, 2014.

[10] B. York, Y. Wensong, L. Jih-Sheng, "An integrated boost resonant converter for photovoltaic applications," *IEEE Transactions on Power Electronics*, vol. 28, no. 3, pp. 1199–1207, March, 2013.

[11] J.B. York, Jr., "An Isolated Micro-Converter for Next-Generation Photovoltaic Infrastructure," Ph.D Dissertation, Virginia Polytechnic Institute and State University, 2013.

[12] E.X. Yang, F.C. Lee, M.N. Jovanovic, "Small-signal modeling of series and parallel resonant converters," in *Proceedings of the 7th Applied Power Electronics Conference and Exposition*, APEC 1992, pp. 785–792, February, 1992.

Benchmarking of Constant Power Generation Strategies for Single-Phase Grid-Connected Photovoltaic Systems

Ariya Sangwongwanich[1], Yongheng Yang[2], *IEEE Member*, Frede Blaabjerg[3], *IEEE Follow*,
and Huai Wang[4], *IEEE Member*

Department of Energy Technology
Aalborg University
Pontoppidanstraede 101, Aalborg, DK-9220 Denmark
ars@et.aau.dk[1], yoy@et.aau.dk[2], fbl@et.aau.dk[3], hwa@et.aau.dk[4]

Abstract—With a still increase of grid-connected Photovoltaic (PV) systems, challenges have been imposed on the grid due to the continuous injection of a large amount of fluctuating PV power, like overloading the grid infrastructure (e.g., transformers) during peak power production periods. Hence, advanced active power control methods are required. As a cost-effective solution to avoid overloading, a Constant Power Generation (CPG) control scheme by limiting the feed-in power has been introduced into the currently active grid regulations. In order to achieve a CPG operation, this paper proposes three CPG strategies based on: 1) a power control (P-CPG), 2) a current limit method (I-CPG) and 3) the Perturb and Observe algorithm (P&O-CPG). However, the operational mode changes (e.g., from the maximum power point tracking to a CPG operation) will affect the entire system performance. Thus, a benchmarking of the proposed CPG strategies is also conducted on a 3-kW single-phase grid-connected PV system. Comparisons reveal that either the P-CPG or I-CPG strategies can achieve fast dynamics and satisfactory steady-state performance. In contrast, the P&O-CPG algorithm is the most suitable solution in terms of high robustness, but it presents poor dynamic performance.

Index Terms—Active power control, constant power control, maximum power point tracking, PV systems, power converters.

I. Introduction

Photovoltaic (PV) systems have a high growth rate during the last several years, and will play an even more significant role in the future mixed power grid [1]–[3]. Currently, a Maximum Power Point Tracking (MPPT) is mandatory for the PV systems in most active grid codes and also to ensure the maximum energy yield from the sun power [4]. At a high penetration level of PV systems in the near future, the grid may face a challenge of overloading during peak power generation periods through a day if the power capacity of the grid remains the same. For instance, it was reported by BBC that parts of the Northern Ireland's grid were overloaded by the increased number of grid-connected PV systems in a sunny and clear day [5]. In order to enable more PV installations and address such issues, the control algorithms have to be feasible to flexibly regulate the active power generated by PV systems [4], [6]–[8]. For instance, limiting the feed-in power of PV systems to a certain level has been found as a cost-effective approach to overcome overloading, and thus it is currently required in Germany through the grid codes [9]. Actually, this active power control strategy corresponds to an absolute power

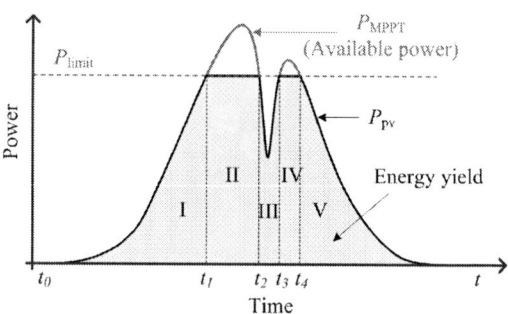

Fig. 1. Constant Power Generation (CPG) concept for PV systems: 1) MPPT mode during I, III, V, and 2) CPG mode during II, IV [13].

constraint defined in the Danish grid code [10], and is also referred to as a Constant Power Generation (CPG) control in prior-art work [11], [12].

According to [11], [13], the most intuitive and effective way to achieve the CPG control is through the modification of the MPPT algorithm at the PV inverter level. Specifically, as long as the PV output power P_{pv} is below the setting-point P_{limit}, the PV system continues operating in the MPPT mode with injection of the maximum power. However, when the output power reaches the level of P_{limit}, the PV system will inject a constant active power, i.e., $P_{\mathrm{pv}} = P_{\mathrm{limit}}$, by regulating the PV output power at the so-called Constant Power Point (CPP). The operational principle of the CPG scheme can be illustrated in Fig. 1 and

$$P_{\mathrm{pv}} = \begin{cases} P_{\mathrm{MPPT}}, & \text{when} \quad P_{\mathrm{pv}} \leq P_{\mathrm{limit}} \\ P_{\mathrm{limit}}, & \text{when} \quad P_{\mathrm{pv}} > P_{\mathrm{limit}} \end{cases} \quad (1)$$

where P_{pv} is the PV output power, P_{MPPT} is the maximum available power (according to the MPPT operation), and P_{limit} is the power limit, which is the setting-point.

In the prior-art work, several CPG control strategies have been introduced. For example, in [14], a P&O based CPG algorithm has been used in single-stage three-phase PV systems. However, its operating region is limited due to the single-stage configuration. A conditioning switch to change the operating modes has been employed in [15] and [16], which requires the initialization of the controllers during the operational mode changes, while a compensation to stabilize the dc-link voltage

978-1-4673-9551-9/16 $31.00 © 2016 IEEE

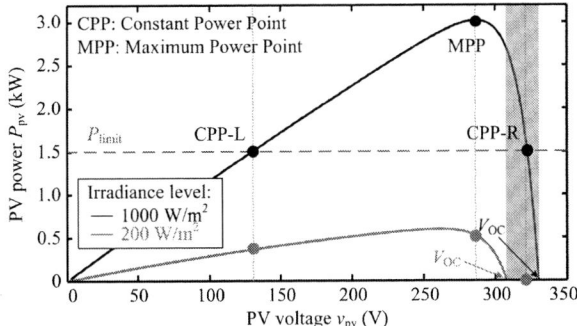

Fig. 2. Stability issues of the conventional CPG algorithms, when the operating point is normally located at the right side of the MPP for a PV panel system [12].

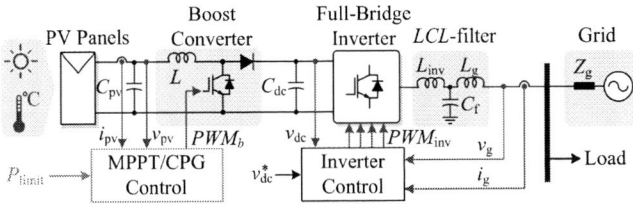

Fig. 3. Hardware schematics and overall control structure of a two-stage single-phase grid-connected PV system.

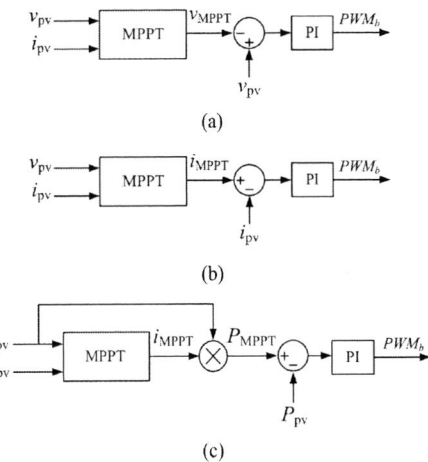

Fig. 4. Implementation of different MPPT controllers: (a) PV output voltage, (b) PV output current, and (c) PV output power, where PI represents a proportional-integral controller.

TABLE I
PARAMETERS OF THE TWO-STAGE SINGLE-PHASE PV SYSTEM (FIG. 3).

PV rated power	3 kW
Boost converter inductor	$L = 1.8$ mH
PV-side capacitor	$C_{pv} = 1000\ \mu F$
DC-link capacitor	$C_{dc} = 1100\ \mu F$
LCL-filter	$L_{inv} = 4.8$ mH, $L_g = 4$ mH, $C_f = 4.3\ \mu F$
Switching frequency	Boost converter: $f_b = 16$ kHz, Full-Bridge inverter: $f_{inv} = 8$ kHz
DC-link voltage	$V_{dc} = 450$ V
Grid nominal voltage (RMS)	$V_g = 230$ V
Grid nominal frequency	$\omega_0 = 2\pi \times 50$ rad/s

is needed in [17], and increasing the overall complexity. Additionally, most of the state-of-the-art CPG methods [14]–[17] cannot always ensure a stable operation (e.g., during a fast change in the irradiance), since the operating region is restricted to the right side of the Maximum Power Point (MPP) in the power-voltage (P-V) curve (i.e., at the CPP-R) shown in Fig. 2. In that region, the CPG operation can potentially introduce instability, since the operating point may go to the open-circuit condition when the PV systems experience a fast decrease of the irradiance condition [12].

In the light of the above issues, this paper proposes three CPG control methods for two-stage single-phase PV systems. The performances under both dynamic and steady-state conditions are benchmarked experimentally on a 3-kW two-stage single-phase PV system. Finally, conclusions are drawn on the comparison.

II. CONTROL STRUCTURE OF TWO-STAGE SINGLE-PHASE GRID-CONNECTED PV SYSTEMS

A. System Configuration

The system configuration and its control structure are shown in Fig. 3, where a two-stage single-phase grid-connected PV system is adopted. The system parameters are given in Table I. The PV arrays are connected to a boost converter, allowing a wide-range operation during both MPPT and CPG operations [18]. In other words, with the use of the boost converter, the PV system can operate at a lower PV voltage v_{pv} (e.g., at the left side of the MPP in the case of the CPG operation), since the PV output voltage v_{pv} can be stepped up to match the required dc-link voltage (e.g., 450 V) for the PV inverter [19].

This may not be possible in the single-stage configuration, where the PV output voltage v_{pv} is directly fed to the PV inverter (i.e., $v_{pv} = v_{dc}$ with v_{dc} being the dc-link voltage). Practically, the v_{dc} is required to be higher than the grid voltage level (e.g., 325 V) to ensure the power delivery [20].

In the boost converter stage, either the MPPT or CPG control can be implemented in order to control the power extraction from the PV arrays. Then, the extracted power is delivered to the ac grid through the control of the full-bridge inverter. In this case, the control of the full-bridge inverter keeps the dc-link voltage to be constant through the control of the injected grid current [21].

B. Boost Converter Controller

As aforementioned, the boost converter plays a major role to control the power extraction from the PV arrays. Usually, the MPPT control (i.e., P&O MPPT algorithm) is implemented in the boost converter, which can be achieved by regulating the PV output voltage v_{pv} according to the reference voltage v_{MPPT} from the MPPT algorithm, as it is shown in Fig. 4(a). Actually, it is also possible to control the boost converter through the PV output current i_{pv} or the power P_{pv} [22] (e.g., Figs. 4(b) and (c)), which are of less robustness [23]. This is due to the

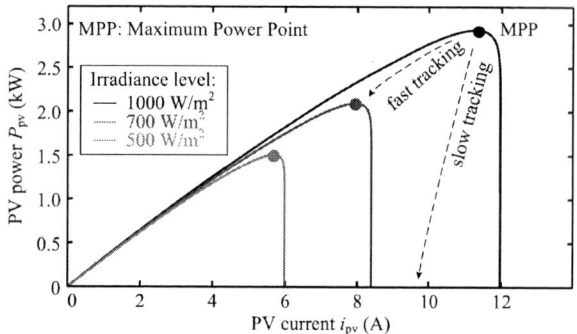

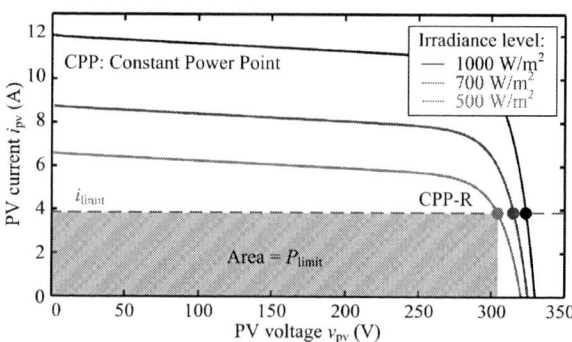

Fig. 5. Stability issues of the MPPT controller based on the PV output current due to the high slope (dP_{pv}/di_{pv}) at the right side of the MPP [23].

Fig. 7. Operational principle of the Constant Power Generation (CPG) scheme based on a current limit (I-CPG).

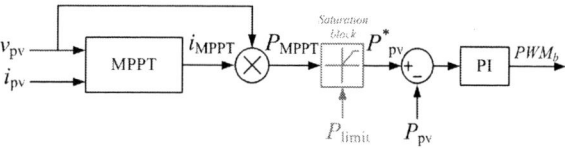

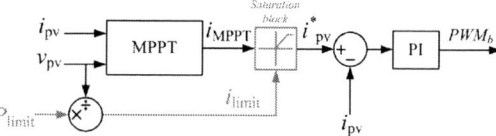

Fig. 6. Control structure of the Constant Power Generation (CPG) scheme based on a power control (P-CPG).

Fig. 8. Control structure of the Constant Power Generation (CPG) scheme based on a current limit (I-CPG).

very steep slope (i.e., large dP_{pv}/di_{pv}) on the right side of the MPP in the power-current (P-I) curve of the PV arrays, as it is shown in Fig. 5. The operating point of the PV system may go into the short-circuit condition under a sudden decrease of the irradiance condition (if the MPPT algorithm cannot track fast enough), when the PV output current is controlled [23].

III. PROPOSED CONSTANT POWER GENERATION STRATEGIES

From the P-V characteristic curve of the PV arrays shown in Fig. 2, there are two possible operating points – CPP-L and CPP-R for the CPG mode at a certain power level (i.e., P_{limit}). Generally, the demands for the CPG control schemes are

- In the steady-state CPG operation, the CPG strategies should keep the PV systems operating at one of the CPPs with a minimum deviation, in order to minimize the power losses yield in the steady-state.
- Under a changing irradiance condition (e.g., in a cloudy day), the CPG control scheme should be able to track either the MPP or the CPP, depending on the operating mode, and at the same time ensure a stable transition.

Accordingly, three CPG strategies are proposed in the following based on: 1) a power control (P-CPG), 2) a current limit (I-CPG), and 3) the Perturb and Observe algorithm (P&O-CPG), where the above demands are taken as the benchmarking criteria.

A. CPG based on a Power Control (P-CPG)

As shown in Fig. 4(c), it is possible to directly control the PV output power P_{pv} by multiplying the reference current i_{MPPT} from the MPPT algorithm with the PV voltage v_{pv}. In order to achieve a CPG operation, the power reference P_{pv}^* is limited by using a saturation block, as it is shown in Fig.

6. Namely, when P_{MPPT} reaches the power limit P_{limit}, the power reference will be kept as a constant, i.e., $P_{pv}^* = P_{limit}$ and the PV system enters into the CPG mode. Otherwise, the PV system will operate in the MPPT mode with a maximum power injection (i.e., $P_{pv}^* = P_{MPPT}$). The operational principle can be further summarized as

$$P_{pv}^* = \begin{cases} P_{MPPT}, & \text{when} \quad P_{MPPT} \leq P_{limit} \\ P_{limit}, & \text{when} \quad P_{MPPT} > P_{limit} \end{cases} \quad (2)$$

where P_{MPPT} is the maximum available power (according to the MPPT operation), and P_{limit} is the power limit, as defined previously. Note that the P-CPG controller will regulate P_{pv} at the CPP-R, where the PV voltage v_{pv} is almost constant.

B. CPG based on a Current Limit (I-CPG)

Since the PV voltage v_{pv} is almost constant at the right side of the MPP (at the CPP-R), as it is shown in Fig. 7, the PV power P_{pv} can effectively be controlled through the PV current i_{pv} in this region. Thus, it is possible to achieve a CPG operation by limiting the reference current i_{MPPT} from the MPPT algorithm according to $i_{limit} = P_{limit}/v_{pv}$ [16], [17], as it is shown in Fig. 8. The power limit P_{limit} corresponds to the rectangular area under the CPP-R in Fig. 7.

According to the CPG concept in (1), the performance of the controller during the MPPT operation should not be diminished by the current limit. This can be ensured when considering

$$\frac{P_{MPPT}}{v_{pv}} \leq \frac{P_{limit}}{v_{pv}}$$

and thus,

$$i_{MPPT} \leq i_{limit}$$

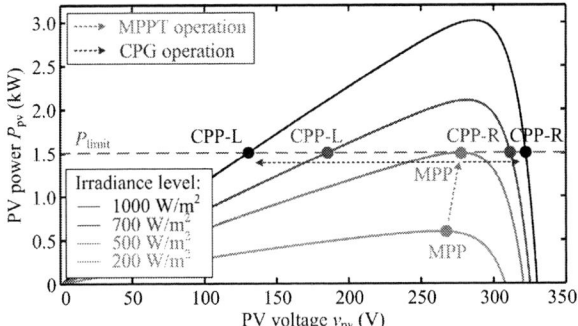

Fig. 9. Operational principle of the Constant Power Generation (CPG) scheme based on the P&O algorithm (P&O-CPG).

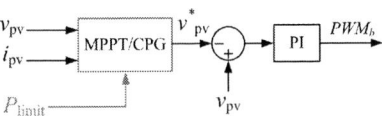

Fig. 10. Control structure of the Constant Power Generation (CPG) scheme based on the P&O algorithm (P&O-CPG).

where it can be seen that the current limit will not be activated as long as $P_{\mathrm{MPPT}} \leq P_{\mathrm{limit}}$.

C. CPG based on the P&O Algorithm (P&O-CPG)

A CPG operation can also be realized by means of a Perturb and Observe (P&O) algorithm. During the MPPT operation, the reference PV voltage v_{pv}^* is determined from the MPPT algorithm. However, in the case of the CPG operation, the PV voltage v_{pv} is continuously perturbed towards one CPP, i.e., $P_{\mathrm{pv}} = P_{\mathrm{limit}}$, as illustrated in Fig. 9. After a number of iterations, the operating point will be reached and oscillate around the corresponding CPP. Notably, the PV system with the P&O-CPG control can operate at either the CPP-L or the CPP-R, depending on the perturbation direction. However, the power oscillation in the steady-state is larger at the CPP-R compared to that at the CPP-L due to the high slope of the P-V curve on the right side of the MPP (i.e., large $dP_{\mathrm{pv}}/dv_{\mathrm{pv}}$). The control structure of the algorithm is shown in Fig. 10, where v_{pv}^* can be expressed as

$$v_{\mathrm{pv}}^* = \begin{cases} v_{\mathrm{MPPT}}, & \text{when} \quad P_{\mathrm{pv}} \leq P_{\mathrm{limit}} \\ v_{\mathrm{pv}} - v_{\mathrm{step}}, & \text{when} \quad P_{\mathrm{pv}} > P_{\mathrm{limit}} \end{cases} \quad (3)$$

if the PV system operates at the CPP-L, or

$$v_{\mathrm{pv}}^* = \begin{cases} v_{\mathrm{MPPT}}, & \text{when} \quad P_{\mathrm{pv}} \leq P_{\mathrm{limit}} \\ v_{\mathrm{pv}} + v_{\mathrm{step}}, & \text{when} \quad P_{\mathrm{pv}} > P_{\mathrm{limit}} \end{cases} \quad (4)$$

if the PV system operates at the CPP-R, where v_{MPPT} is the reference voltage from the MPPT algorithm (i.e., the P&O MPPT algorithm) and v_{step} is the perturbation step size.

IV. BENCHMARKING OF CONSTANT POWER GENERATION (CPG) STRATEGIES

In order to benchmark the discussed CPG control strategies, experiments have been carried out referring to Fig. 3, where

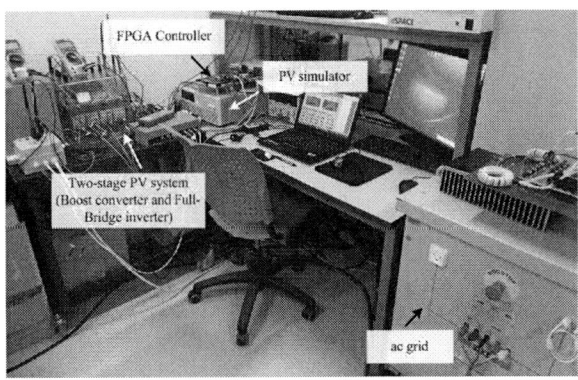

Fig. 11. Experimental setup of the two-stage single-phase grid-connected PV system.

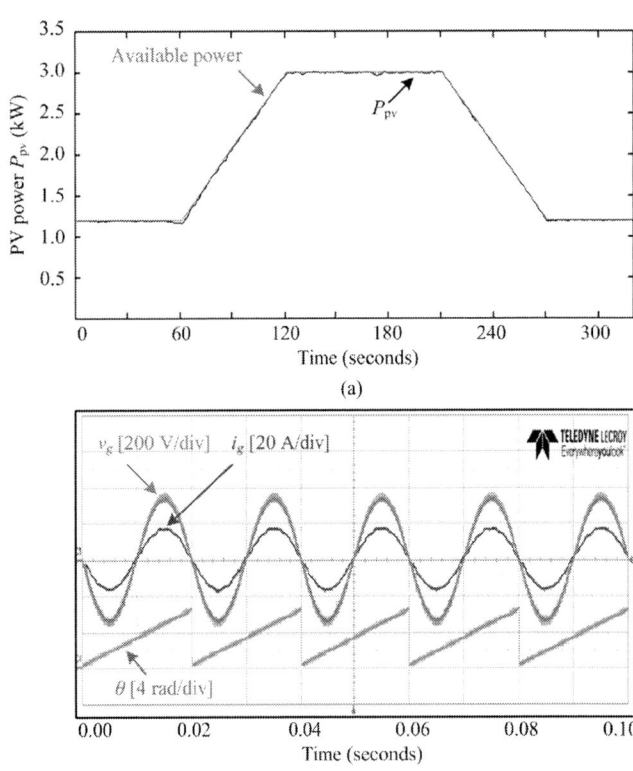

Fig. 12. Performance of the two-stage single-phase grid-connected PV system: (a) the PV power extraction during the MPPT operation, and (b) the grid voltage v_g, grid current i_g and the phase angle θ during the steady-state MPPT operation (3 kW).

the experimental test-rig is shown in Fig. 11. The performance of the two-stage single-phase PV system during the MPPT operation are demonstrated in Fig. 12(a). Here, the sampling frequency of the MPPT (and also CPG) algorithms is chosen as 10 Hz. For the PV inverter controller, the dc-link voltage v_{dc} is regulated at 450 ± 5 V and the extracted power is delivered to a single-phase 50-Hz ac grid with a peak voltage of 325 V, as it can be seen from Fig. 12(b).

In the experiments, a 3-kW PV simulator has been adopted,

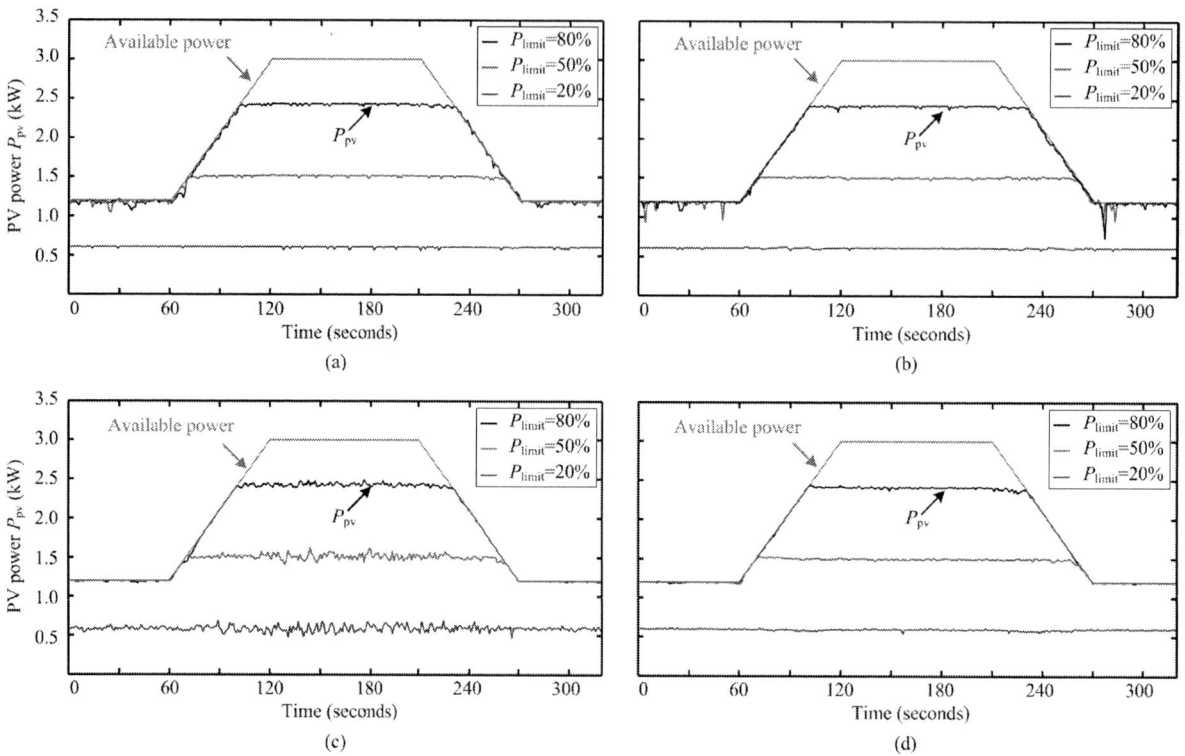

Fig. 13. Experimental results of the Constant Power Generation (CPG) scheme based on: (a) the power control, (b) the current limit, (c) the P&O at the right side of the MPP, and (d) the P&O at the left side of the MPP under a slow changing irradiance condition.

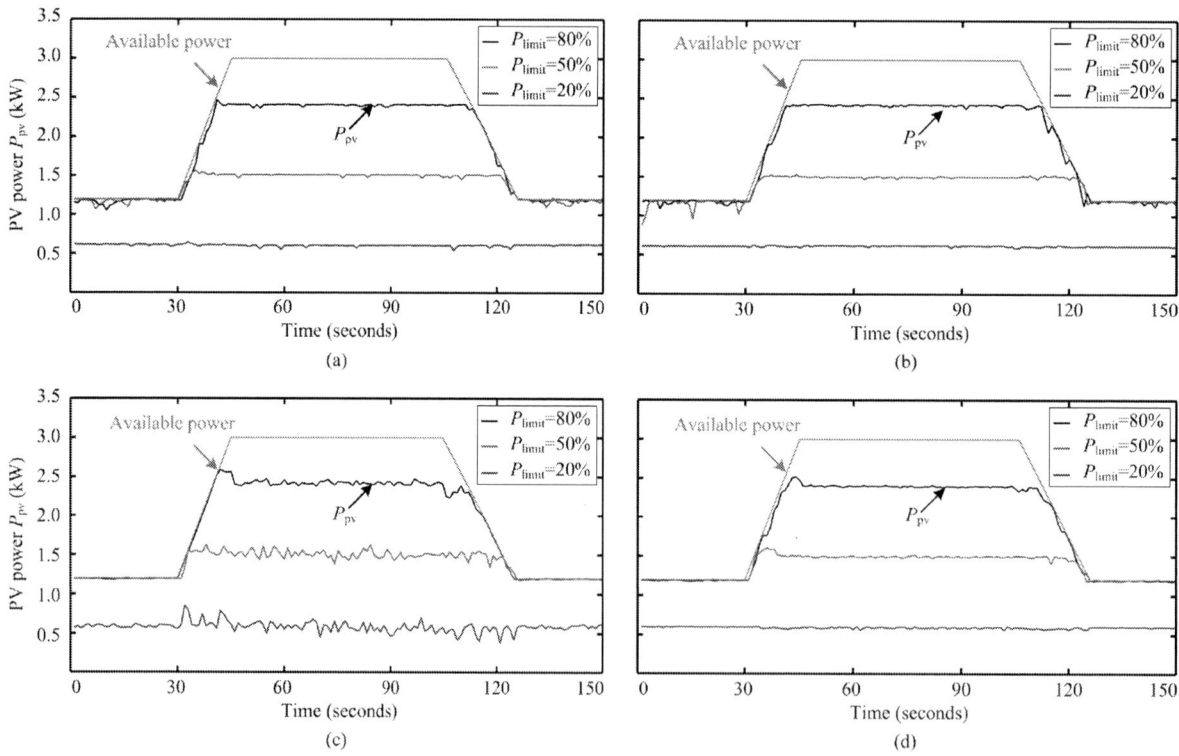

Fig. 14. Experimental results of the Constant Power Generation (CPG) scheme based on: (a) the power control, (b) the current limit, (c) the P&O at the right side of the MPP, and (d) the P&O at the left side of the MPP under a fast changing irradiance condition.

978-1-4673-9551-9/16 $31.00 © 2016 IEEE

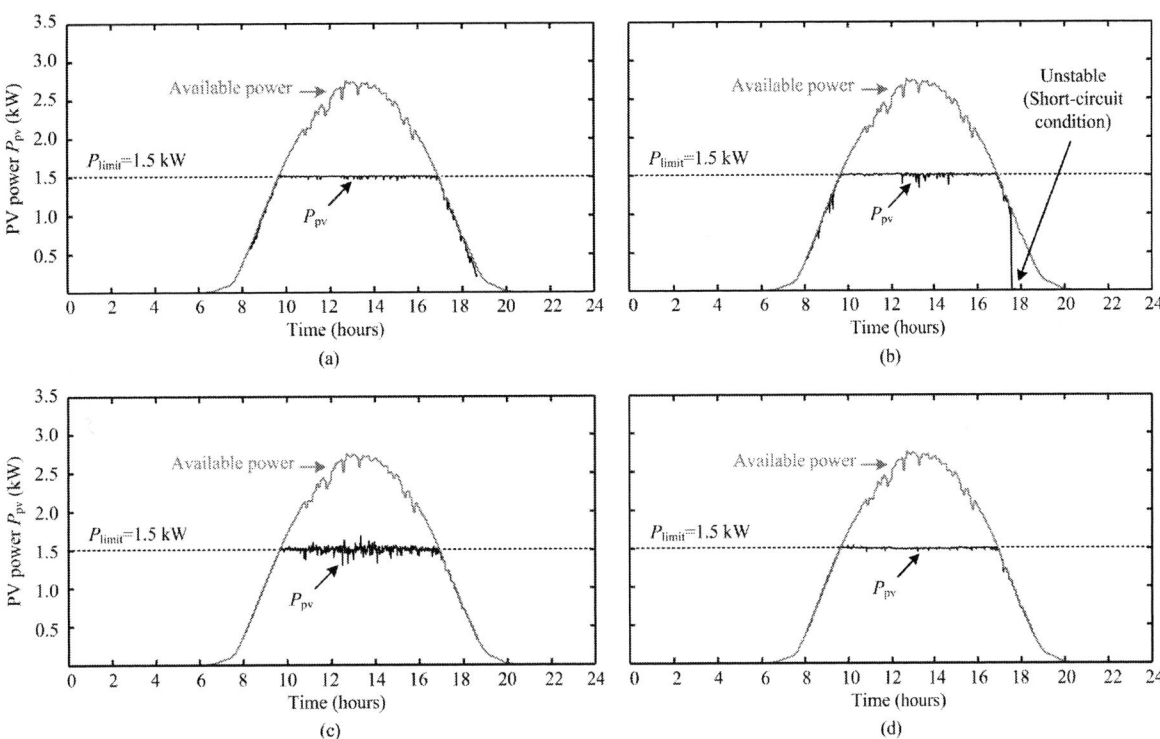

Fig. 15. Experimental results of the Constant Power Generation (CPG) scheme based on: (a) the power control, (b) the current limit, (c) the P&O at the right side of the MPP, and (d) the P&O at the left side of the MPP under a clear day condition.

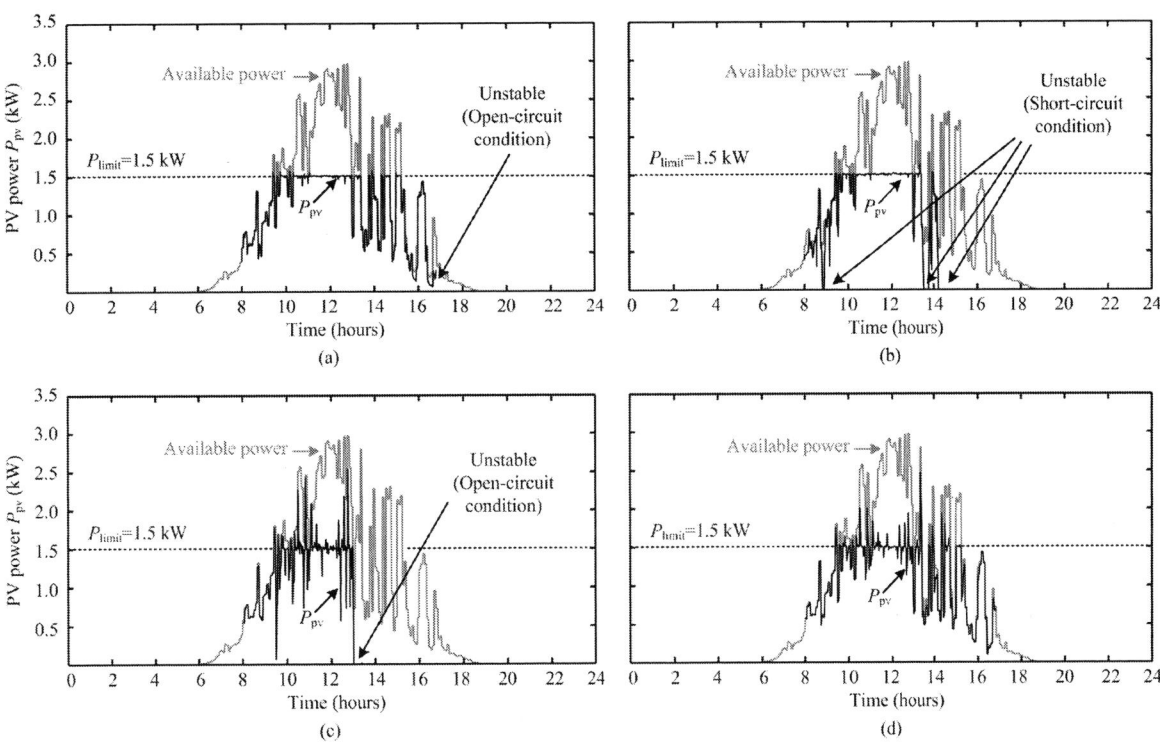

Fig. 16. Experimental results of the Constant Power Generation (CPG) scheme based on: (a) the power control, (b) the current limit, (c) the P&O at the right side of the MPP, and (d) the P&O at the left side of the MPP under a cloudy day condition.

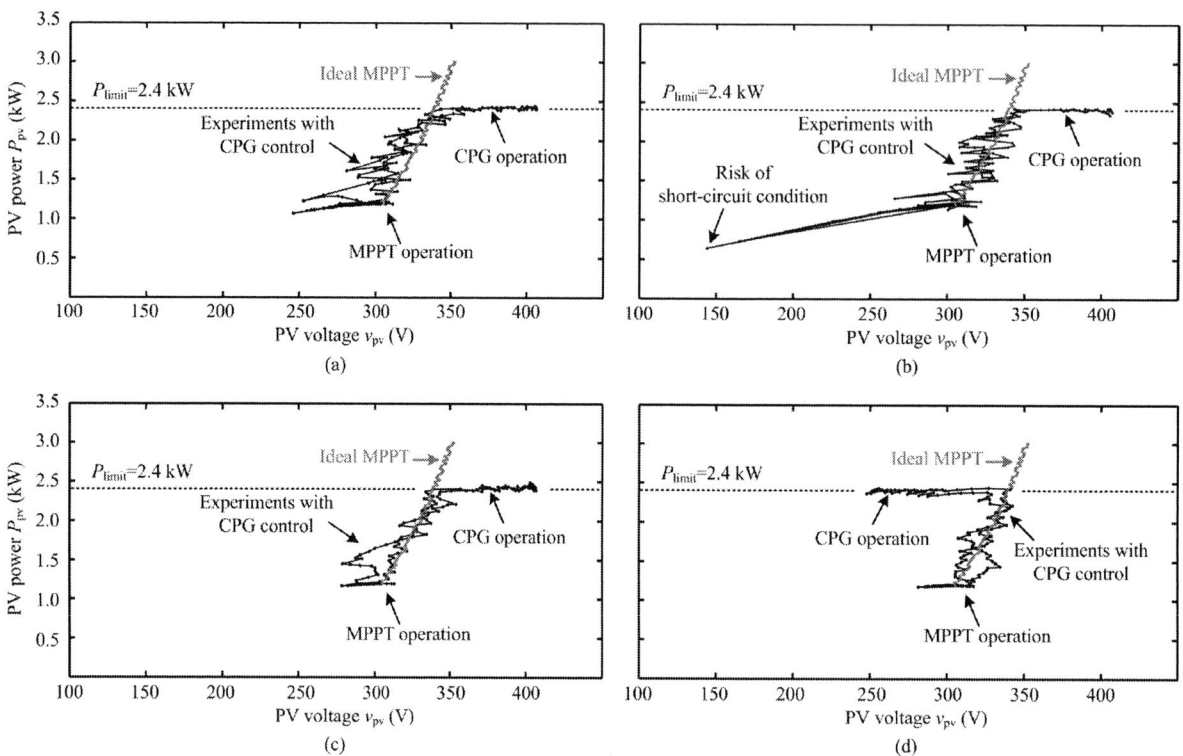

Fig. 17. Trajectory of the operating point of the Constant Power Generation (CPG) scheme based on: (a) the power control, (b) the current limit, (c) the P&O at the right side of the MPP, and (d) the P&O at the left side of the MPP under a slow changing irradiance condition (Fig. 13), when $P_{\mathrm{limit}} = 2.4$ kW.

TABLE II
BENCHMARKING OF THE CONSTANT POWER GENERATION ALGORITHMS.

CPG Strategy	Dynamic Responses		Steady-state Responses	Stability	Complexity
	MPPT→CPG	CPG→MPPT			
Power control (P-CPG)	+ +	+	+ +	-	-
Current limit (I-CPG)	+	- -	+	- -	+ +
P&O-CPG at CPP-R	- -	-	- -	-	-
P&O-CPG at CPP-L	- -	-	+ +	+ +	-

Note: the more +, the better stability and less complexity.

where two trapezoidal solar irradiance profiles are programmed in order to emulate a slow changing (i.e., Fig. 13) and a fast changing (i.e., Fig. 14) irradiance conditions. Here, three different values of power limit P_{limit} (i.e., 20, 50, and 80 % of the rated power) are used to confirm the feasibility of the CPG strategies. Furthermore, two real-field solar irradiance and ambient temperature profiles are also programmed in order to observe the performance of the CPG algorithms in the real operation, as it is shown in Figs. 15 and 16, where $P_{\mathrm{limit}} = 1.5$ kW (i.e., 50 % of the rated power). An example of the operating trajectory of the CPG strategies are also illustrated in Fig. 17, where the irradiance condition in Fig. 13 is used.

A. Dynamic responses

The dynamic responses can be observed during the CPG to MPPT transition and vice versa. In Figs. 13 and 15, all the CPG strategies have a smooth transition, since the irradiance

changes relatively slow. However, in the case of fast changing solar irradiance, the P&O-CPG scheme presents large overshoots during the MPPT to CPG transition, as it is shown in Figs. 14 (c) and (d). Similar power overshoots also appear in the P&O-CPG algorithm under a cloudy day irradiance condition in Figs. 16 (c) and (d). In contrast, it is observed in Figs. 14 and 16 that the P- and I-CPG algorithms have a very fast dynamic response almost without any overshoots during the CPG transients.

B. Steady-state responses

In the steady-state, the CPG algorithm should regulate the PV power P_{pv} with minimum deviations, as discussed in § III. Most of the CPG algorithms have a satisfactory steady-state performance (see Figs. 13 and 15). However, when the P&O-CPG algorithm is employed to regulate the PV power at the right side of the MPP (i.e., the CPP-R), large power

978-1-4673-9551-9/16 $31.00 © 2016 IEEE

oscillations appeared as shown in Figs. 13(c) and 15(c). This is due to the large dP_{pv}/dv_{pv} at the CPP-R (see Fig. 2).

C. Stability

Stability is another important aspect for the CPG control schemes. Thus, the proposed CPG strategies are also benchmarked in terms of stability. Instability can occur in the case of the P- and P&O-CPG algorithms when the operating point is chosen at the CPP-R. The operating point may go to the open-circuit condition if the PV power is regulated too far at the right side of the MPP, since the open-circuit voltage in the P-V curve decreases as the irradiance level drops. Figs. 16 (a) and (c) verify that the P-CPG or the P&O-CPG at the right side of the MPP can go into instability during transients. Furthermore, the I-CPG algorithm can also introduce instability to the PV system under a decreasing irradiance condition as it is shown in Figs. 15(b) and 16(b). However, in this case, it is due to the less robust MPPT schemes, which may result in a short-circuit condition, as it is explained in Fig. 5. In fact, it can be seen in Figs. 15 and 16 that the P&O-CPG algorithm can always ensure a stable operation regardless of the irradiance conditions, only when the PV system operating point is regulated at the CPP-L.

D. Complexity

When comparing all the above CPG strategies, it is found that the I-CPG algorithm has the simplest control structure, where only one additional current limiter needs to be added to the original MPPT controller in Fig. 4(b). Besides, the calculation of the i_{limit} is also simple by dividing P_{limit} by the measured PV voltage v_{pv}. The control structure of the P-CPG algorithm is more complicated, basically due to the MPPT controller in Fig. 4(c). In the case of the P&O-CPG algorithm, the modification needs to be done at the MPPT algorithm level as it can be seen from Fig. 10. This makes the design of a P&O-CPG controller more complicated than the other two CPG algorithms.

Table II further summarizes a comparison of the results of the proposed CPG control schemes, in terms of dynamic and steady-state performances, stability, and complexity. The benchmarking results have validated the effectiveness of the proposed CPG strategies, and that the P&O-CPG algorithm (when operating at the CPP-L) is the most suitable approach to realize the CPG control practically due to its robustness and feasible to be used for the future grid codes.

V. CONCLUSION

In this paper, three Constant Power Generation (CPG) control solutions for PV systems have been presented. A benchmarking of the three CPG control methods has also been conducted in terms of dynamic and steady-state performances, stability, and complexity. Comparisons have revealed that the CPG strategy based on a current limit (I-CPG) has the simplest control structure. Additionally, the power control based CPG scheme (P-CPG) has fast dynamics and good steady-state responses. However, instability may occur in both I-CPG and

P-CPG methods during the operational mode transition, e.g., in the case of a fast change in the solar irradiance. It can be concluded that the CPG based on the P&O algorithm (P&O-CPG) is the best one in terms of high robustness among the three CPG control strategies once the PV system is operating at the left side of the maximum power point.

REFERENCES

[1] REN21, "Renewables 2015: Global Status Report (GRS)," 2015. [Online]. Available: http://www.ren21.net/.

[2] Fraunhofer ISE, "Recent Facts about Photovoltaics in Germany," May 19, 2015. [Online]. Available: http://www.pv-fakten.de/.

[3] Solar Power Europe, "Global Market Outlook For Solar Power 2015 - 2019," 2015. [Online]. Available: http://www.solarpowereurope.org/.

[4] Y. Yang, P. Enjeti, F. Blaabjerg, and H. Wang, "Wide-scale adoption of photovoltaic energy: Grid code modifications are explored in the distribution grid," *IEEE Ind. Appl. Mag.*, vol. 21, no. 5, pp. 21–31, Sep. 2015.

[5] D. Maxwell, "Parts of Northern Ireland's electricity grid overloaded," *BBC News NI*, 2013. [Online]. Available: http://www.bbc.com/.

[6] T. Stetz, F. Marten, and M. Braun, "Improved low voltage grid-integration of photovoltaic systems in Germany," *IEEE Trans. Sustain. Energy*, vol. 4, no. 2, pp. 534–542, Apr. 2013.

[7] A. Ahmed, L. Ran, S. Moon, and J.-H. Park, "A fast PV power tracking control algorithm with reduced power mode," *IEEE Trans. Energy Convers.*, vol. 28, no. 3, pp. 565–575, Sep. 2013.

[8] T. Caldognetto, S. Buso, P. Tenti, and D. Brandao, "Power-based control of low-voltage microgrids," *IEEE Trans. Emerg. Sel. Topics Power Electron.*, vol. 3, no. 4, pp. 1056–1066, Dec. 2015.

[9] *German Federal Law: Renewable Energy Sources Act (Gesetz fur den Vorrang Erneuerbarer Energien) BGBl*, Std., Jul. 2014.

[10] Energinet.dk, "Technical regulation 3.2.2 for PV power plants with a power output above 11 kW," Tech. Rep., 2015.

[11] Y. Yang, H. Wang, F. Blaabjerg, and T. Kerekes, "A hybrid power control concept for PV inverters with reduced thermal loading," *IEEE Trans. Power Electron.*, vol. 29, no. 12, pp. 6271–6275, Dec. 2014.

[12] A. Sangwongwanich, Y. Yang, and F. Blaabjerg, "High-performance constant power generation in grid-connected PV systems," *IEEE Trans. Power Electron.*, vol. 31, no. 3, pp. 1822–1825, Mar. 2016.

[13] Y. Yang, F. Blaabjerg, and H. Wang, "Constant power generation of photovoltaic systems considering the distributed grid capacity," in *Proc. of APEC*, pp. 379-385, Mar. 2014.

[14] R.G. Wandhare and V. Agarwal, "Precise active and reactive power control of the PV-DGS integrated with weak grid to increase PV penetration," in *Proc. of PVSC*, pp. 3150-3155, Jun. 2014.

[15] W. Cao, Y. Ma, J. Wang, L. Yang, J. Wang, F. Wang, and L. Tolbert, "Two-stage PV inverter system emulator in converter based power grid emulation system," in *Proc. of ECCE*, pp. 4518-4525, Sep. 2013.

[16] A. Urtasun, P. Sanchis, and L. Marroyo, "Limiting the power generated by a photovoltaic system," in *Proc. of SSD*, pp. 1-6, Mar. 2013.

[17] Y. Chen, C. Tang, and Y. Chen, "PV power system with multi-mode operation and low-voltage ride-through capability," *IEEE Trans. Ind. Electron.*, vol. 62, no. 12, pp. 7524–7533, Dec. 2015.

[18] H. Ghoddami and A. Yazdani, "A bipolar two-stage photovoltaic system based on three-level neutral-point clamped converter," in *Proc. IEEE Power Energy Soc. Gen. Meet*, pp. 1–8, Jul. 2012.

[19] S.B. Kjaer, J.K. Pedersen, and F. Blaabjerg, "A review of single-phase grid-connected inverters for photovoltaic modules," *IEEE Trans. Ind. Appl.*, vol. 41, no. 5, pp. 1292–1306, Sep. 2005.

[20] B. Yang, W. Li, Y. Zhao, and X. He, "Design and analysis of a grid-connected photovoltaic power system," *IEEE Trans. Power Electron.*, vol. 25, no. 4, pp. 992–1000, Apr. 2010.

[21] F. Blaabjerg, R. Teodorescu, M. Liserre, and A.V. Timbus, "Overview of control and grid synchronization for distributed power generation systems," *IEEE Trans. Ind. Electron.*, vol. 53, no. 5, pp. 1398–1409, Oct. 2006.

[22] C. Rosa, D. Vinikov, E. Romero-Cadaval, V. Pires, and J. Martins, "Low-power home PV systems with MPPT and PC control modes," in *Proc. of CPE*, pp. 58–62, Jun. 2013.

[23] N. Femia, G. Petrone, G. Spagnuolo, and M. Vitelli, *Power electronics and control techniques for maximum energy harvesting in photovoltaic systems*. CRC press, 2012.

Advanced Slip Mode Frequency Shift Islanding Detection Method for Single Phase Grid Connected PV Inverters

Bahador
Mohammadpour
ECE, Queen's University
Kingston, Canada
bahador.mohammadpour
@queensu.ca

Majid Pahlevani
ECE, Queen's University
Kingston, Canada
mp60@queensu.ca

Sajjad Makhdoomi
Kaviri
ECE, Queen's University
Kingston, Canada
12ssmk@queensu.ca

Praveen Jain
ECE, Queen's University
Kingston, Canada
praveen.jain@queensu.ca

Abstract— **This paper presents an active islanding detection method which has zero non-detection zone. This method is the advanced version of the conventional slip mode frequency shift islanding detection method (SMS IDM). In this method the parameters of SMS IDM change based on the local load impedance value. The parameters value are calibrated periodically by measuring the impedance of the local load. The local load impedance can be calculated by injecting harmonic currents to the point of common coupling (PCC) and measuring its voltage during the calibration period. Simulation and experimental results show the accuracy of the proposed method in detection of islanding in any loading conditions.**

Keywords—islanding detection method; slip mode frequency shift; single phase grid connected inverter;

I. INTRODUCTION

Due to growing concerns with fossil fuels, emphasis has been placed on clean and sustainable energy generation [1]-[2]. This has resulted in the increase in renewable energy sources and energy storage being integrated into the utility system [3]-[6]. The integration of renewable energy sources and energy storage have raised some concerns for utility power systems. One of the main concerns is islanding detection that is enforced by grid interconnection regulatory standards such as UL1741, IEEE 1547, etc. [7]-[8]. According to these standards, the islanding is "a condition in which a portion of the utility system that contains both load and distributed resources remains energized while isolated from the remainder of the utility system". Therefore, the islanding detection is an integral part of any grid –connected power converters.

Numerous methods for islanding detection have been introduced in the literature [9]-[10]. They can be categorized into local methods and remote methods. Remote methods, also referred to as communication based methods, have a small non-detection zone (NDZ) but the implementation cost is higher and more complicated than the implementation of local methods [11]. The local methods are categorically divided into passive and active methods [12]-[13]. Passive methods detect islanding by measuring the point of common coupling (PCC) voltage and the inverter output current [14]. These methods are inexpensive and simple to implement but they fail to detect islanding when the local load power consumption closely matches the inverter's output power generation [15]. The active methods inject intentional disturbances to the utility grid and measure the grid's signals to detect whether islanding has occurred [16]. Active methods generally have smaller NDZs but the injecting disturbances will slightly degrade the power quality and reliability of the power system. Slip Mode Frequency Shift Islanding Detection Method (SMS IDM) is an active method which uses positive feedback for islanding detection. In this method, the phase angle of the converter is controlled to have a sinusoidal function of the deviation of the PCC voltage frequency from the nominal grid frequency. This method has some non-detection zone which means it fails to detect islanding for some specific local load conditions.

Advanced Slip Mode Frequency Shift Islanding Detection Method (Advanced SMS IDM) which has been introduced in this paper eliminates the non-detection zone of the SMS IDM. In this method the parameters of SMS IDM change based on the local load impedance value. The parameters value are calibrated periodically by measuring the impedance of the local load. The local load impedance can be calculated by injecting harmonic currents to the point of common coupling (PCC) and measuring its voltage. In section II, a method for estimation of power system frequency is proposed. In section III, slip mode frequency shift islanding detection method is explained and the loading conditions which SMS IDM fails to detect islanding are specified. In section IV, a method for load impedance calculation is proposed. In section V, the Advanced SMS IDM is explained in detail. Finally, in section VI, the simulation results and experimental results of the islanding detection methods have been provided.

II. ESTIMATION OF POWER SYSTEM FREQUENCY

An estimation of the power system frequency is essential for the implementation of the proposed islanding detection method

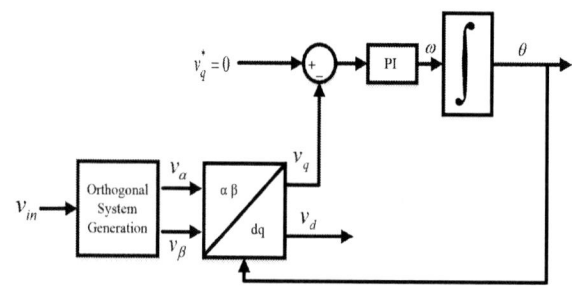

Fig. 1. Orthogonal system generation based PLL

and also for the synchronization of the inverter with the power system. In [17], a brief review of the synchronization techniques for single-phase converter-interfaced distributed generation (DG) systems has been presented. The capability of tracking frequency variation of the grid signal has been compared for these techniques. Orthogonal System Generation (OSG) based PLL is one of the PLL techniques that has good speed for following the system frequency. The OSG based PLL block has been shown in Fig. 1. As evident, the OSG based PLL has an orthogonal signal generation block, which is followed by an αβ-dq transformation block. Park transformation is used for reorienting the signals from αβ to dq frames, as given in Equation (1).

$$\begin{bmatrix} v_d \\ v_q \end{bmatrix} = \begin{bmatrix} sin\theta & cos\theta \\ -cos\theta & sin\theta \end{bmatrix} \begin{bmatrix} v_\alpha \\ v_\beta \end{bmatrix} \tag{1}$$

As it can be seen from the PLL block diagram, the implementation of the PLL requires a block for orthogonal signal generation, which gives a 90 degree phase shift. In [18], some methods for shifting the single phase signal and making two orthogonal signals have been explored. Phase Delay Block, Differentiator Block and All Pass Filter (APF) Block are the methods that could be used for this purpose. Using an all pass filter block is the best method for orthogonal signal generation. One of the main advantages of this method is its fast, dynamic performance. Moreover, this method is more noise-resistant than other methods and its implementation is easy and inexpensive.

The transfer function of the filter is given in Equation (2) and a bode plot of it is shown in Fig. 2. As evident from the bode plot of the filter, the filter does not change the magnitude of the signal and the phase shift of the filter is a function of the signal frequency. As the operating frequency changes in this application, the center frequency of the filter (ω) should be provided to the filter from the PLL such that the APF always has a 90 degree phase shift for all operating frequencies. Since the all pass filter should be implemented digitally, the differential equation of the filter is derived in Equations (3), (4), (5), (6) and (7). Bilinear Transform has been used here for the transformation of the filter transfer function from the s domain to the z domain.

$$\frac{v_\beta(s)}{v_\alpha(s)} = \frac{s - \omega_f}{s + \omega_f} \tag{2}$$

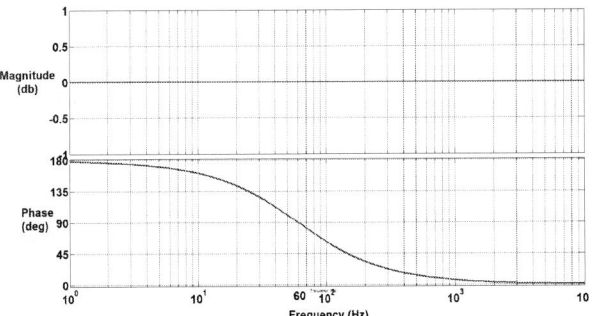

Fig. 2. Bode plot of the all pass filter (APF) block

$$H(z) = H(s)\big|_{s \rightarrow \frac{2}{T_s}\frac{z-1}{z+1}} = H(\frac{2}{T_s}\frac{z-1}{z+1}) \tag{3}$$

$$y[n] = k_3.y[n] + k_1.y[n-1] + k_1.x[n] + k_2.x[n-1] \tag{4}$$

$$k_1 = \frac{2 - \omega_f T_s}{2} \tag{5}$$

$$k_2 = \frac{-2 - \omega_f T_s}{2} \tag{6}$$

$$k_3 = \frac{-\omega_f T_s}{2} \tag{7}$$

III. SLIP MODE FREQUENCY SHIFT ISLANDING DETECTION METHOD (SMS IDM)

All grid-connected inverters have some sort of passive islanding detection. All inverters are equipped with Over/Under Voltage Relays (OVR/UVR) and Over/Under Frequency Relays (OFR/UFR). When the islanding occurs, the inverter is the only source providing power to the load. Therefore, the voltage and frequency of the PCC may change. This is due to the active and reactive power mismatch between the local load power consumption and the inverter output power generation, as shown in Equations (8) and (9).

$$P_{load} = \frac{V_{pcc}^2}{R} \tag{8}$$

$$Q_{Load} = V_{pcc}^2 (\frac{1}{2.\pi.f.L} - 2.\pi.f.C) \tag{9}$$

For some loading conditions, the small power mismatch between load power and inverter power results in the potential for the voltage and frequency of the PCC to be in the normal range of operation after the islanding occurs. In this case, the protective relays fail to detect islanding. As such, other types of islanding detection methods are needed to make sure the disconnection of the inverter from the utility grid is detected within the required time interval, which is 150 milliseconds based on IEEE standard.

SMS IDM method is one of the local active methods which uses positive feedback for islanding detection [19]; it applies positive feedback to the phase angle of the inverter. SMS IDM is implemented through the PLL implementation. Consider there is no SMS in the inverter control system and phase angle of the current, and voltage is controlled to be zero at all times. In this case, if the frequency of the islanded part is perturbed downward, a negative phase error would be detected by the PLL. Therefore, the PLL would reduce its frequency to make the output current and voltage in phase. Now consider the inverter output current phase angle follows the SMS IDM phase angle characteristic. When the frequency of the islanded part increases, the SMS IDM characteristic causes a positive phase error to the PLL and the PLL increases its frequency. When the frequency increases, the phase error increases and makes the PLL further increase the frequency. Therefore, the PLL control acts in the wrong direction to correct the error.

This condition will continue until the phase angle of the inverter output current and phase angle of the local load become equal. The inverter output current can be described by Equation (10).

$$i_{inv} = I.\sin(2.\pi.f.t + \theta_{SMS}) \qquad (10)$$

The phase angle of the inverter current is controlled to have a sinusoidal function of the deviation of the PCC voltage frequency from the nominal operating frequency of the utility system. The SMS IDM phase angle characteristic is described by Equation (11),

$$\theta_{SMS} = \frac{2.\pi}{360}\theta_m \sin(\frac{\pi}{2}\frac{f-f_g}{f_m}) \qquad (11)$$

where θ_m and f_m are the parameters of the SMS IDM characteristic, θ_m (degree) is the maximum phase shift, f_m (Hz) is the frequency at which the maximum phase shift occurs, f is the system operating frequency, and f_g is the utility normal operating frequency of 60 Hz or 50 Hz. When the inverter is connected to the utility, the solid frequency of the PCC voltage provides a zero phase angle for the inverter, but when the islanding occurs, the frequency of the PCC varies upward or downward based on the direction of the perturbation. The steady state value of the frequency can be calculated as the frequency at which the phase angle of the inverter and the phase angle of the local load are the same.

$$\theta_{Load} + \theta_{SMS} = 0 \qquad (12)$$

$$\theta_{Load} = -\tan^{-1}(R.(2.\pi.f.C - \frac{1}{2.\pi.f.L})) \qquad (13)$$

Graphically, the steady state point is at the intersection of the SMS IDM phase response curve and load phase response curve, as shown in Fig. 3. The intersection point can be either stable or unstable. When the point is unstable, it moves either

upward or downward based on the direction of perturbation and settles in a new stable operating point. If the final stable operating point is in the allowed range of frequency, which is between 59.3 Hz and 60.5 Hz, then the SMS IDM fails to detect islanding and the inverter will continue injecting power to the islanded part of the grid. Therefore, the conventional SMS IDM should be improved to make sure that islanding can be detected for any loading conditions. Advanced SMS IDM has zero non detection zone, which means by changing the parameters of SMS IDM based on the value of local load impedance, islanding detection is guaranteed for all loading conditions.

IV. LOAD IMPEDANCE CALCULATION

The IEEE test bench system model which is shown in Fig. 4 is employed to investigate islanding detection. At the initializing stage of a grid connected inverter, which is the time interval before connecting the inverter to the grid, the impedance of the local load can be calculated. The process of local load impedance calculation is repeated periodically after connecting the inverter to the grid. This process is done during the periodic shut downs of the inverter which last only a few cycles. It is possible to calculate the impedance of the local load at a harmonic frequency by injecting the harmonic current by the PV inverter into the PCC and measuring the voltage response [20]. When the PV is not connected to the grid the impedance transfer function and equivalent local load impedance can be expressed as in Equation (14),(15) and (16).

$$H(j\omega_h) = \frac{V_{pcc}(j\omega_h)}{I_{inv}(j\omega_h)} = R_{h,eq} + j.X_{h,eq} \qquad (14)$$

$$R_{h,eq} = \frac{V_{rms}}{I_{rms}}\cos(\angle V - \angle I) \qquad (15)$$

$$X_{h,eq} = \frac{V_{rms}}{I_{rms}}\sin(\angle V - \angle I) \qquad (16)$$

The calculated value of inductance is the equivalent inductance of the parallel L and C. Since the inductance is calculated in a harmonic frequency, the equivalent inductance will be approximately equal to the capacitor inductance.

$$X_{h,eq} = X_{h,c} \parallel X_{h,L} \cong X_{h,c} = \frac{1}{5}X_c \qquad (17)$$

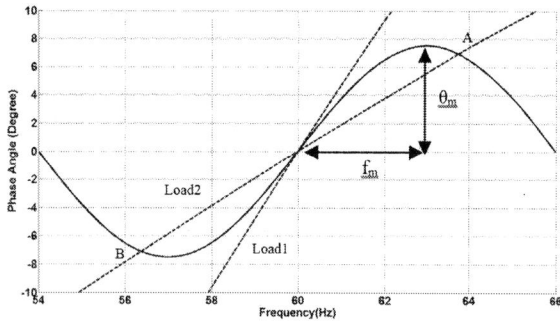

Fig. 3. SMS IDM and load phase angle response curves (Load1: load with Qf=2.5 which has stable point at the fundamental frequency, Load2: load with Qf=1.5 which has unstable point at the fundamental frequency and final stable point at point A or B)

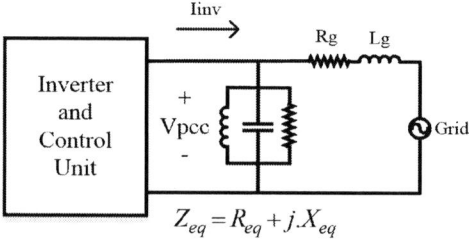

Fig. 4. IEEE test bench system model for islanding detection

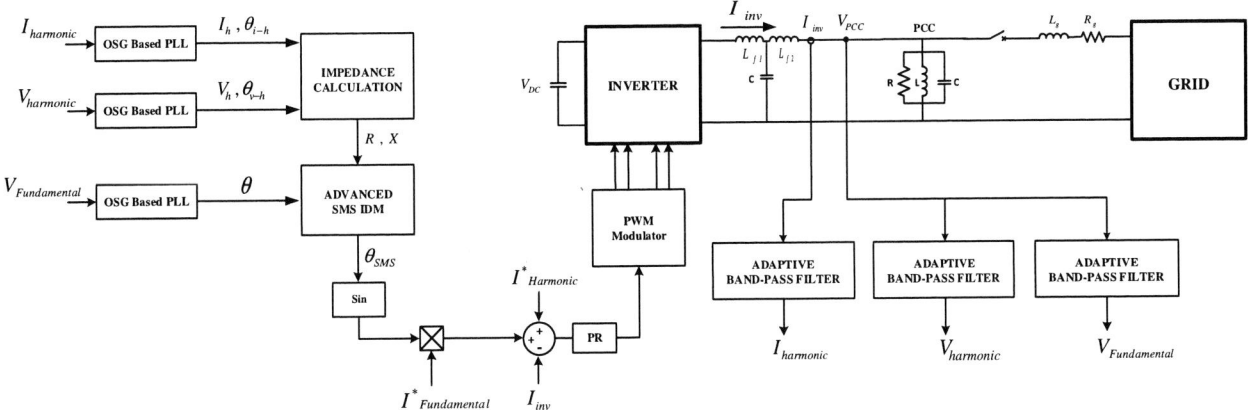

Fig. 6. Advanced SMS block and inverter's control block diagram

The amplitude and phase angle of the voltage and current are required for measuring the transfer function. An adaptive band pass filter has been proposed for extracting the harmonic signal from the main signal. The filter is a second order band pass filter, the transfer function is given in Equation (18) and bode plot of the filter is shown in Fig 5. The filter is always tuned to the harmonic frequency because the center frequency of the band pass filter comes from the PLL. This is so when the PCC voltage's frequency changes, the center frequency of the filter changes as well. The filter is implemented digitally in the inverter control system. The differential equation of the filter is derived in Equations (19), (20), (21) and (22). Orthogonal system generation (OSG) based PLL is proposed to extract the amplitude and phase angle of the extracted harmonic signal. The block diagram of load impedance calculation and other control blocks of the inverter are shown in Fig. 6.

$$\frac{y(s)}{x(s)} = \frac{B.s}{s^2 + B.s + \omega_0^2} \tag{18}$$

$$y[n] = k_1.x[n] + k_2.x[n-2] + k_3.y[n-1] + k_4.y[n-2] \tag{19}$$

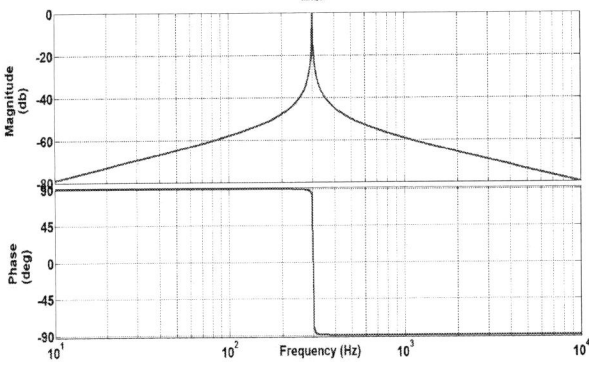

Fig. 5. Bode plot of second order band pass filter

$$k_1 = -k_2 = \frac{2.B.T_s}{4 + 2.B.T_s + \omega_0^2.T_s^2} \tag{20}$$

$$k_3 = \frac{8 - 2.\omega_0^2.T_s^2}{4 + 2.B.T_s + \omega_0^2.T_s^2} \tag{21}$$

$$k_4 = \frac{-4 + 2.B.T_s - \omega_0^2.T_s^2}{4 + 2.B.T_s + \omega_0^2.T_s^2} \tag{22}$$

V. ADVANCED SMS IDM

In section III, it was mentioned that the steady state point is at the intersection of the SMS IDM phase response curve and load phase response curve. The intersection point is stable if the slope of the load phase response curve at the intersection point is greater than the slope of the SMS IDM phase response curve at that point. On the other hand, the intersection point is unstable if the slope of the load phase response is smaller than the slope of the SMS IDM phase response curve. Fig. 3 shows two different types of intersection points. When an intersection point is unstable it moves upward or downward, based on the direction of disturbance, and it settles in a new stable operation point. As one can see from the Fig. 3, it is possible that the stable operation point lies within the allowed range of operating frequency. So for these loading conditions, conventional SMS IDM fails to detect islanding. Advanced SMS IDM is a modified version of conventional SMS IDM, which utilizes an algorithm to set the value of f_m and θ_m at the initializing stage of a grid connected inverter to ensure that islanding will be detected.

The algorithm is shown in Fig. 7. The first step in the Advanced SMS IDM algorithm is to set initial values for parameters f_m and θ_m. In the next step of the algorithm, the slope of the load phase response is calculated. Equation (23), (24) and (25) show how to calculate the slope of the load line at the fundamental frequency.

$$Q_f = \frac{R}{2.\pi.f_0.L} = 2.\pi.f_0.R.C \tag{23}$$

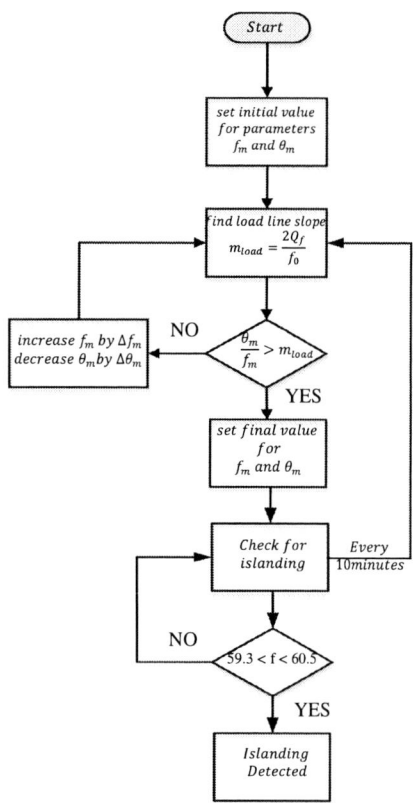

Fig. 7. Advanced SMS IDM algorithm for adjusting the parameters values and detection of islanding

$$\theta_{Load} = -\tan^{-1}(R(2.\pi.f.C - \frac{1}{2.\pi.f.L})) =$$

$$= -\tan^{-1}(Q_f(\frac{f_0}{f} - \frac{f}{f_0})) \quad (24)$$

$$\frac{d\theta_{load}}{df} = \frac{Q_f(-\frac{f_0}{f^2} - \frac{1}{f_0})}{1 + Q_f^2(\frac{f_0}{f} - \frac{f}{f_0})^2}\Bigg|_{f=f_0} = \frac{2.Q_f}{f_0} \quad (25)$$

In Equation (25), f_0 is the fundamental frequency and the Q_f can be calculated from the harmonic inductance, which was calculated earlier in section IV.

$$Q_f = 2.\pi.f_0.R.C = R.X_c = R.h.X_{h,c} \quad (26)$$

The next step in Advanced SMS IDM algorithm is to check whether the intersection point is a stable point which lies outside of the allowed range of operating frequency. The value of θ_m/f_m has been used here as an approximation for the value of slope of the SMS IDM phase response curve at fundamental frequency. If the value of θ_m/f_m is greater than the load line slope at fundamental frequency, it means the initial intersection point is an unstable point. The point will slide to one side of the curve

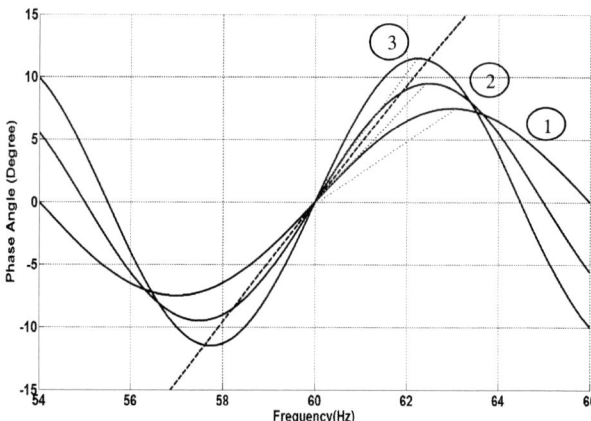

Fig. 8. Graphical display of Advanced SMS IDM algorithm for adjusting the parameters values (Qf=1.5, value of f_m decreases and θ_m increases from curve 1 to 3)

and settle at a point which is further than f_m, as shown for Load 2 in Fig. 3. So in this case, when islanding occurs the final stable frequency point will slip out of the allowed range of frequency operation. Therefore, the values of f_m and θ_m are set as final values for Advanced SMS IDM characteristic parameters. But if the value of θ_m/f_m is smaller than the load line slope at fundamental frequency, the intersection point will be a stable point in the allowed range of frequency operation which is not desirable for us, like the load 1 which is shown in Fig. 3. So in this case, the algorithm increase the slope by decreasing the value of f_m by Δf_m and increasing the value of θ_m by $\Delta\theta_m$ and checks the condition again. The values of f_m and θ_m change until the condition for an unstable point becomes true. After this, the values of f_m and θ_m are set as final values for Advanced SMS IDM characteristic parameters. The operation of the algorithm for setting the parameters is shown graphically in Fig. 8.

After setting the final values of f_m and θ_m, the inverter is ready to be connected to the grid. It is known that the value of the local load might change during the inverter operation. Therefore, periodic calibration of the parameters is needed to ensure the NDZ of this method is always zero and that islanding will be detected in the required time frame. As it has been explained in section IV, the local load impedance value can be calculated periodically. The calculated value of the impedance has been employed in the algorithm to calibrate the parameters. The parameter calibration is done every 10 minutes.

The last block in the algorithm is the block for comparing the frequency value with the lower limit and upper limit of the allowed frequency range. If the frequency at any time exceeds this limit, the islanding is detected and the inverter shuts down.

VI. EXPERIMENTAL RESULTS

In order to illustrate the design feasibility of the proposed Advanced SMS IDM, a PSIM simulation model for the grid-connected converter system is developed to perform a digital simulation. Moreover, the method is implemented on a 1 kW

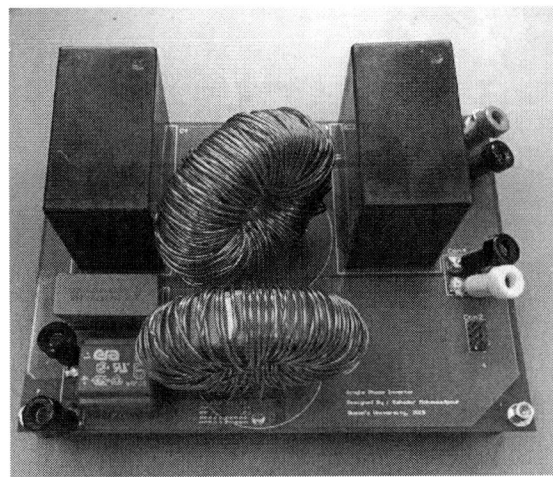

Fig. 9. Prototype of the inverter

grid connected inverter in the lab which is shown in Fig. 9. The specifications of the inverter are given in Table 1.

Table I. Specifications of the prototype inverter

Parameter	Notation	Value
Rated output power	P	1 Kw
Grid voltage	V_g	120 Vrms
DC bus voltage	V_{dc}	400 V
Switching Freq.	f_{sw}	20 kHz
Filter Parameters	L_{f1} L_{f2} C_f	2.0 mH 0.65 mH 2.2 uF

The implemented grid-connected inverter includes a full-bridge inverter and an LCL-filter in order to attenuate switching harmonics at the output of the inverter [21]-[23]

The inverter system control and Advanced SMS IDM are digitally implemented with a F28335 DSP. The following figures show the PCC voltage and frequency of the inverter for islanding detection tests. Fig. 10 and Fig. 11 show the PCC voltage, inverter output current, and frequency before and after islanding. As one can see for this loading condition, SMS IDM cannot detect islanding, and the frequency and voltage of the PCC stay the same while the inverter is disconnected from the utility grid. Experimental results shown in Fig. 12 and Fig. 13 verifies the simulation results. The experimental result for the frequency of inverter is the digital value of the frequency which is been shown using Code Composer Studio graph utility. Fig. 14 and Fig. 15 show the PCC voltage, inverter current, and frequency for the case which Advanced SMS IDM has been implemented. As it can be seen, Advanced SMS IDM can slip the frequency out of the allowed range and can detect islanding in less than 6 cycles (100 milliseconds) after islanding occurs. Experimental results shown in Fig. 16 and Fig. 17 verify the ability of this method in detection of islanding in the required time frame.

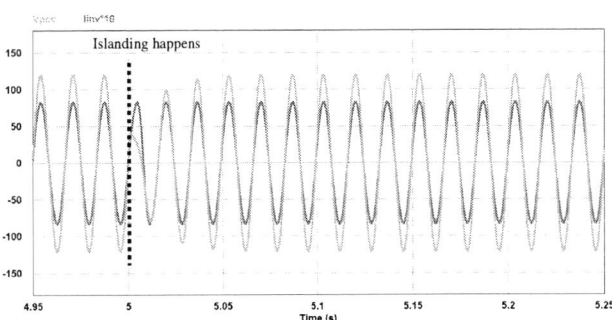

Fig. 10. Voltage and current of the inverter for load with Q=2.5(islanding happens at t=5s and SMS IDM fails to detect the islanding)

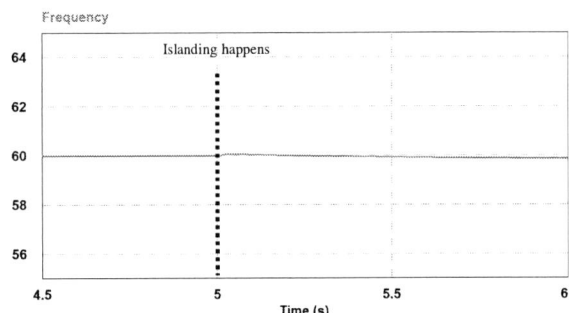

Fig. 11. Frequency of the inverter for load with Q=2.5(islanding happens at t=5s and SMS IDM fails to detect the islanding)

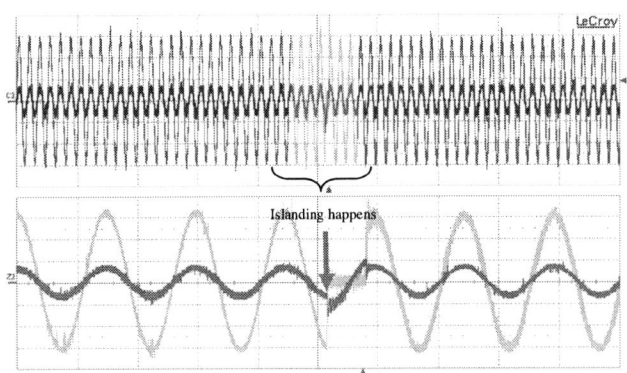

Fig. 12. Experimental waveforms of voltage and current of the inverter for load with Q=2.5(islanding happens at t=5s and SMS IDM fails to detect the islanding)

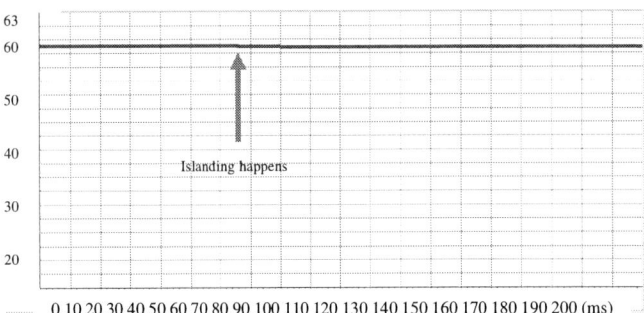

Fig. 13. Frequency of the inverter for Q=2.5, digital output of the DSP (islanding happens at t= 90 ms)

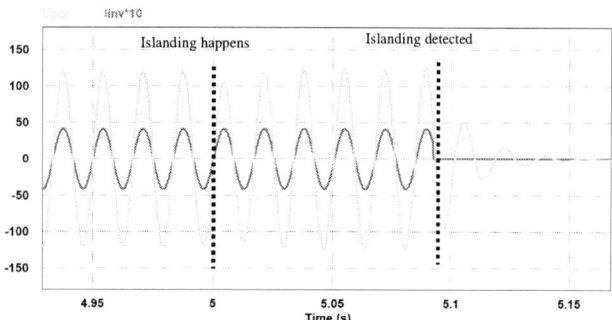

Fig. 14. Voltage and current of the inverter for load with Q=2.5(islanding happens at t=5s and Advanced SMS IDM detects the islanding in 100 ms)

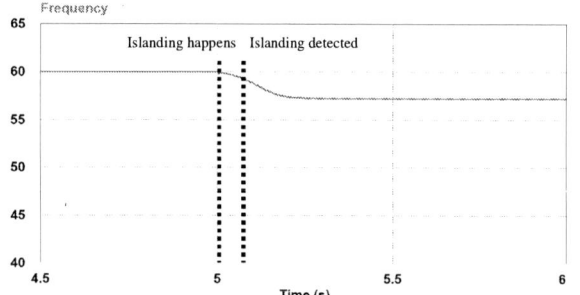

Fig. 15. Frequency of the inverter for load with Q=2.5(islanding happens at t=5s and Advanced SMS IDM detects the islanding in 100 ms)

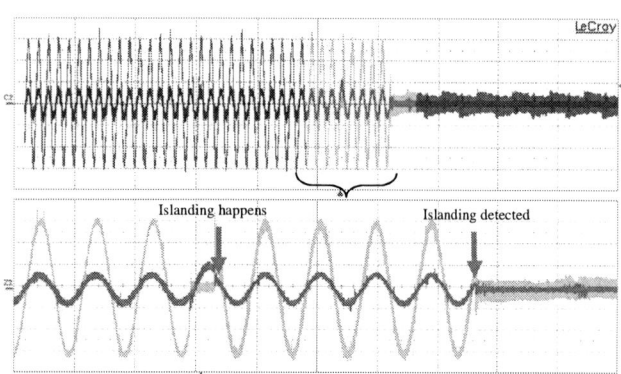

Fig. 16. Experimental waveforms voltage and current of the inverter for load with Q=2.5(islanding happens at t=5s and Advanced SMS IDM detects the islanding in 100 ms)

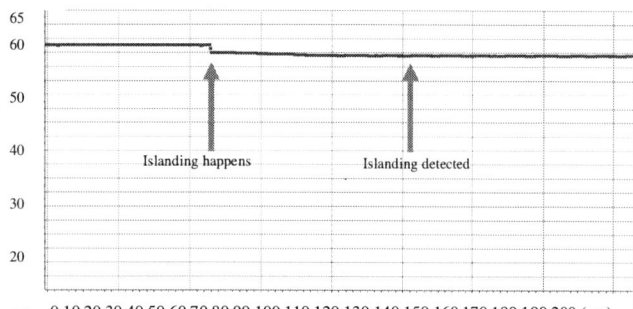

Fig. 17. Frequency of the inverter for load with Q=2.5 (islanding happened at t= 70 ms and detected at t=150 ms)

VII. CONCLUSION

An algorithm for automatically setting the parameters of the SMS IDM is introduced. In this algorithm by injecting a harmonic current by the PV inverter and processing the PCC voltage signal the local load impedance is calculated. An adaptive band pass filter and an OSG based PLL is proposed for extracting the harmonic signals from the ambient signals. It has been shown that the non-detection zone of the conventional SMS IDM method has been completely eliminated and the PV inverter can detect islanding in any loading conditions. The proposed algorithm is experimentally verified using a DSP-based platform.

REFERENCES

[1] Guerrero, J.M.; Berbel, N.; Matas, J.; Sosa, J.L.; de Vicuna, L.G., "Droop Control Method with Virtual Output Impedance for Parallel Operation of Uninterruptible Power Supply Systems in a Microgrid," in Applied Power Electronics Conference, APEC 2007 - Twenty Second Annual IEEE , vol., no., pp.1126-1132, Feb. 25 2007-March 1 2007.

[2] Pahlevaninezhad, M.; Eren, S.; Guerrero, J.M.; Jain, P., "A Hybrid Estimator for Active/Reactive Power Control of Single-Phase Distributed Generation Systems with Energy Storage," in Power Electronics, IEEE Transactions on , vol.PP, no.99, pp.1-1, 2016.

[3] Sukesh, N.; Pahlevaninezhad, M.; Jain, P.K., "Analysis and Implementation of a Single-Stage Flyback PV Microinverter With Soft Switching," in Industrial Electronics, IEEE Transactions on , vol.61, no.4, pp.1819-1833, April 2014.

[4] Castilla, M.; Miret, J.; Matas, J.; Garcia de Vicuna, L.; Guerrero, J.M., "Control Design Guidelines for Single-Phase Grid-Connected Photovoltaic Inverters With Damped Resonant Harmonic Compensators," in Industrial Electronics, IEEE Transactions on , vol.56, no.11, pp.4492-4501, Nov. 2009.

[5] Pahlevaninezhad, M.; Das, P.; Drobnik, J.; Jain, P.K.; Bakhshai, A., "A Novel ZVZCS Full-Bridge DC/DC Converter Used for Electric Vehicles," in Power Electronics, IEEE Transactions on , vol.27, no.6, pp.2752-2769, June 2012.

[6] Pahlevaninezhad, M.; Das, P.; Drobnik, J.; Jain, P.K.; Bakhshai, A., "A ZVS Interleaved Boost AC/DC Converter Used in Plug-in Electric Vehicles," in Power Electronics, IEEE Transactions on , vol.27, no.8, pp.3513-3529, Aug. 2012.

[7] UL1741 Standard, "Inverters, Converters, Controllers and Interconnection System Equipment for Use With Distributed Energy Resources" Jan. 2010.

[8] IEEE 1547 Standard for Interconnecting Distributed Resources with Electric Power Systems, July, 2003.

[9] W. Bower, M. Ropp, "Evaluation Of Islanding Detection Methods For Photovoltaic Utility-Interactive Power Systems," 2002.

[10] A. Khamis, H. Shareef, E. Bizkevelci, T. Khatib, "A review of islanding detection techniques for renewable distributed generation systems," Renewable and Sustainable Energy Reviews, December 2013, Volume 28, Pages 483–493.

[11] A. Timbus, A. Oudalov, and C. N. M. Ho, "Islanding detection in smart grids," in 2010 IEEE Energy Conversion Congress and Exposition, 2010, pp. 3631-3637.

[12] F. De Mango, M. Liserre, A. D. Aquila, and A. Pigazo, "Overview of Anti-Islanding Algorithms for PV Systems. Part I: Passive Methods, " in 2006 12th International Power Electronics and Motion Control Conference, 2006, pp. 1878-1883.

[13] F. De Mango , M. Liserre and A. D. Aquila , "Overview of anti-islanding algorithms for PV systems. Part II: Active methods, " Proc. EPE-PEMC, pp.1884 -1889, 2006.

[14] F. De Mango, M. Liserre, A. D. Aquila, and A. Pigazo, "Overview of Anti-Islanding Algorithms for PV Systems. Part I: Passive Methods," in 2006 12th International Power Electronics and Motion Control Conference, 2006, pp. 1878-1883.

[15] Z. Ye, A. Kolwalkar, Y. Zhang, P. Du and R. Walling, "Evaluation of Anti-Islanding Schemes Based on Nondetection Zone Concept, " IEEE Transactions on Power Electronics, Vol. 19, No. 5, September 2004, pp. 1171-1176.

[16] F. De Mango, M. Liserre and A. D. Aquila, "Overview of anti-islanding algorithms for PV systems. Part II: Active methods, " Proc. EPE-PEMC, pp.1884 -1889, 2006.

[17] D. Yazdani, M. Pahlevaninezhad, A. Bakhshai, "Single-Phase grid-synchronization algorithms for converter interfaced distributed generation system, " Canadian Conference on Electrical and Computer Engineering,CCECE'09, 3-6 May 2009, pp. 127-131.

[18] B. Mohammadpour, M. Pahlevaninezhad, S. Makhdoomi, P. Jain, "Islanding Detection for a Single Phase Bidirectional Converter in Telecommunication Power Systems, " IEEE International Telecommunication Energy Conference (INTELEC), 2015.

[19] G. Hung, C. Chang, and C. Chen, "Automatic phase-shift method for islanding detection of grid-connected photovoltaic inverter," IEEE Trans. Energy Conversion, vol. 18, no. 1, pp. 169-173, Mar. 2003.

[20] A. Moallem, D. Yazdani, A. Bakhshai, and P. Jain, "Frequency domain identification of the utility grid parameters for distributed power generation systems," in Applied Power Electronics Conference and Exposition (APEC), 2011 Twenty-Sixth Annual IEEE, 2011, pp. 965-969.

[21] Eren, S.; Pahlevani, M.; Bakhshai, A.; Jain, P., "An Adaptive Droop DC-Bus Voltage Controller for a Grid-Connected Voltage Source Inverter With LCL Filter," in Power Electronics, IEEE Transactions on , vol.30, no.2, pp.547-560, Feb. 2015.

[22] Eren, S.; Pahlevaninezhad, M.; Bakhshai, A.; Jain, P.K., "Composite Nonlinear Feedback Control and Stability Analysis of a Grid-Connected Voltage Source Inverter With LCL Filter," in Industrial Electronics, IEEE Transactions on , vol.60, no.11, pp.5059-5074, Nov. 2013.

[23] Eren, S.; Bakhshai, A.; Jain, P., "Control of grid-connected voltage source inverter with LCL filter," in Applied Power Electronics Conference and Exposition (APEC), 2012 Twenty-Seventh Annual IEEE , vol., no., pp.1516-1520, 5-9 Feb. 2012.

Direct MPPT Control of PWM Converters for Extreme Transient PV Applications

Ignacio Galiano Zurbriggen, Francisco Paz, and Martin Ordonez
Electrical and Computer Engineering
The University of British Columbia
Vancouver, BC, Canada
Email: {igaliano, franciscopaz, mordonez}@ieee.org

Abstract— **Solar panels require of Maximum Power Point Tracking algorithms in order to ensure the amount of power extracted is maximized. The control of the power stage connected to the photovoltaic panel, and the MPPT algorithm are traditionally implemented as two independent building blocks, which derives in a relatively slow response. A novel approach is introduced in this work by embedding the MPPT algorithm into the power converter control, allowing faster dynamics, smaller reactive components and reduced switching losses. The scheme being introduced combines MPPT concepts with large-signal geometric control to achieve a high-performance, reliable solution. The theoretical analysis is supported by detailed mathematical procedures and validated by simulations and experimental results.**

I. INTRODUCTION

Photovoltaic (PV) cells have been one of the most popular alternative energy sources during the past decade, with a Compound Annual Growth Rate (CAGR) of 44% from 2000 to 2013 [1], and an accumulative capacity expected to exceed 230GW for 2015 [2]. Battery chargers and grid-tie applications are amongst the most common targets for PV applications. Nevertheless, the popularity of alternative applications such as vehicle rooftop PV [3, 4] and wearable technology [5] is rising. Maximum Power Point Tracking (MPPT) algorithms must be included in the system architecture in order to maximize the power transfer from the solar panel to the load. The MPPT algorithm must be implemented at the first stage of the energy conversion chain, where DC-DC converters are undoubtedly the most popular choice.

Many MPPT techniques have been presented in the literature with diverse degrees of accuracy and complexity [6, 7]. Hill-climbing techniques, such as Incremental Conductance (IC) and Perturb and Observe (P&O), are one of the most popular family of algorithms due to their flexibility and balance between complexity and accuracy. However, traditional implementation of these algorithms are specially prone to

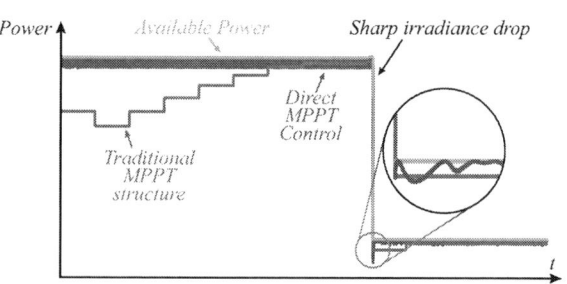

Fig. 1. Fast irradiance changes require very fast tracking in order to maximize the power extracted from rooftop solar panels for vehicle applications.

errors during rapid changing conditions [8, 9], typical of applications such as rooftop PV and wearable technology. An example of changing environmental conditions can be observed in Fig. 1, where the solar panel on top of the vehicle moves through shadows that cause rapid changes in irradiance. The plot in Fig. 1 illustrates tracking issues typical of traditional MPPT algorithms under of rapid changes in the environmental conditions. As illustrated in Fig. 2, MPPT algorithms are typically implemented in a stand-alone fashion by setting the reference voltage (or current) to the DC-DC converter [8–13]. The MPPT algorithm must wait for the converter to reach the indicated operating point prior

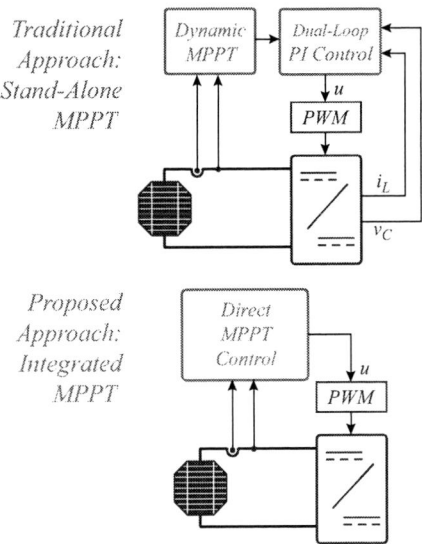

Fig. 2. Traditional and proposed Solar MPPT/control structures

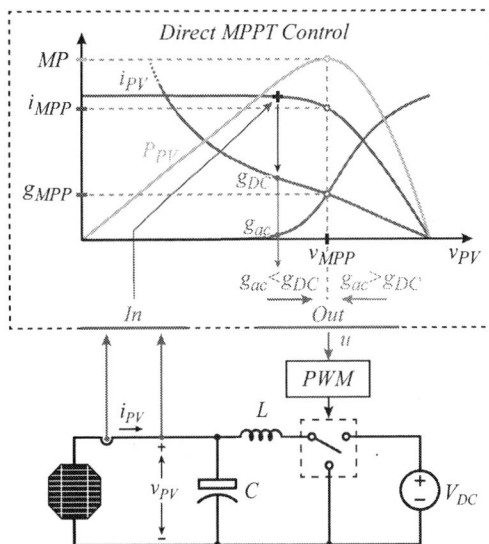

Fig. 3. Incremental Conductance basis employed to define the control law

taking any other action, which leads to a relatively slow performance even when the algorithm is optimized [14–16].

In contrast with traditional approaches, this work proposes a direct MPPT-duty-cycle control structure that controls the power converter based on the MPPT laws. The MPPT algorithm is embedded into the power converter controller by combining Incremental Conductance principles with geometric modelling of the power stage. In this way, no fixed set-point is defined and the control law always leads the operating point towards the MPP. This work proposes a dynamic solution to solve the dynamic MPPT problem. The geometric modelling of the basic PWM-based DC-DC topologies has been studied in [17–19], showing fast and predictable transient response, large-signal reliable operation, and reduced reactive components size. Employing geometric concepts as cornerstone, the Direct-MPPT control enables substantial improvements in dynamic performance, efficiency (by reducing switching losses), and power density. The theoretical findings are supported by detailed mathematical procedures and validated by simulations and experimental results.

II. DYNAMIC MPPT-BASED DIRECT CONTROL BASIS

A direct MPPT-based control law that rules the dynamic behaviour of solar powered converters is derived in this section. Since the converter is desired to operate at the maximum power point, the control law being proposed must be derived from MPPT algorithms. Due to the unknown nature of the maximum power point, it is not possible to determine a desired steady state voltage or current. Instead, the control law must indicate the direction towards which the operating point must move in order to reach the maximum power point.

Several known MPPT algorithms could be employed as basis to determine the control law. Due simplicity and suitability for the application, a variation of the Incremental Conductance (IC) method is adopted. The power delivered by the PV cell is given by:

$$P_{PV} = i_{PV} \ v_{PV} \qquad (1)$$

Therefore, at the maximum power point:

$$\frac{dP_{PV}}{dv}(MPP) = v_{MPP}\frac{di_{PV}}{dv_{PV}}(MPP) + i_{MPP} = 0 \quad (2)$$

Rearranging terms to the form of conductances, and taking only positive values:

$$g_{MPP} = \frac{|di_{PV}|}{|dv_{PV}|}(MPP) = \frac{i_{MPP}}{v_{MPP}} \qquad (3)$$

Due to the characteristic behaviour of the PV cells, shown in Fig. 3, the i_{PV} to v_{PV} ratio decreases with the PV cell voltage v_{PV}, while the incremental conductance g_{ac} increases along the entire range. Since (3) is only satisfied at the Maximum Power Point, the left and right sides of the MPP voltage v_{MPP} are given by:

$$v_{PV} < v_{MPP} \Rightarrow g_{ac} = \frac{|di_{PV}|}{|dv_{PV}|} < \frac{i_{PV}}{v_{PV}} = g_{DC}; \text{ and}$$
$$v_{PV} > v_{MPP} \Rightarrow g_{ac} = \frac{|di_{PV}|}{|dv_{PV}|} > \frac{i_{PV}}{v_{PV}} = g_{DC} \qquad (4)$$

These simple mathematical identities provide the basis to perform MPPT. As illustrated in Fig. 3, the expressions (4) provide information sufficient to determine whether the PV cell voltage must increase or decrease in order to reach the MPP. The main control law is found by rearranging (4) into simpler expressions with no divisions:

$$\text{if: } |\Delta i_{PV}| v_{PV} \leq |\Delta v_{PV}| i_{PV} \Rightarrow \textit{Move to the right}$$
$$\textit{else if: } |\Delta i_{PV}| v_{PV} > |\Delta v_{PV}| i_{PV} \Rightarrow \textit{Move to the left} \quad (5)$$

Traditional IC methods are based on the same mathematical expressions but the information provided by them is usually employed to define new steady state points the converter must reach instead of indicating the desired dynamic behaviour. This work focuses on embedding the MPPT laws into the converter control. Since the dynamic behaviour of the selected topology plays an important role in achieving this objective, it is analysed in the following section.

III. INVERTED BUCK CONVERTER BEHAVIOUR

The *input-voltage controlled boost* converter, or simply *inverted-buck* topology, shown in Fig. 4 is a very common alternative for implementing the DC-DC stage in PV harvesting systems. The behaviour of the converter is described by the following averaged differential equations:

$$L \frac{di_L}{dt} = v_{PV} - u \, V_{DC} \quad (6)$$

$$C \frac{dv_{PV}}{dt} = i_{PV} - i_L \quad (7)$$

In order to gain generality, this work takes the analysis to a normalized domain, where: $t_n = \frac{t}{T_0}$, $Z_{xn} = \frac{Z_x}{Z_0}$, $v_{xn} = \frac{v_x}{v_{norm}}$, and $i_{xn} = \frac{i_x}{i_{norm}}$. The employed base quantities are the filter characteristic impedance $Z_0 = \sqrt{L/C}$, and resonance period $T_0 = 2\pi\sqrt{LC}$, as well as the reference voltage $v_{norm} = V_{DC}$ and current $i_{norm} = \frac{V_{DC}}{Z_0}$.

The resulting normalized expressions are independent from the L and C parameters as well as voltage levels:

$$\frac{1}{2\pi} \frac{di_{Ln}}{dt_n} = v_{PVn} - u \quad (8)$$

$$\frac{1}{2\pi} \frac{dv_{PVn}}{dt_n} = i_{PVn} - i_{Ln} \quad (9)$$

The equilibrium (traditional steady state) conditions are given when the derivatives of the state variables are null, and therefore: $u = v_{PVn}$, and $i_{Ln} = i_{PVn}$.

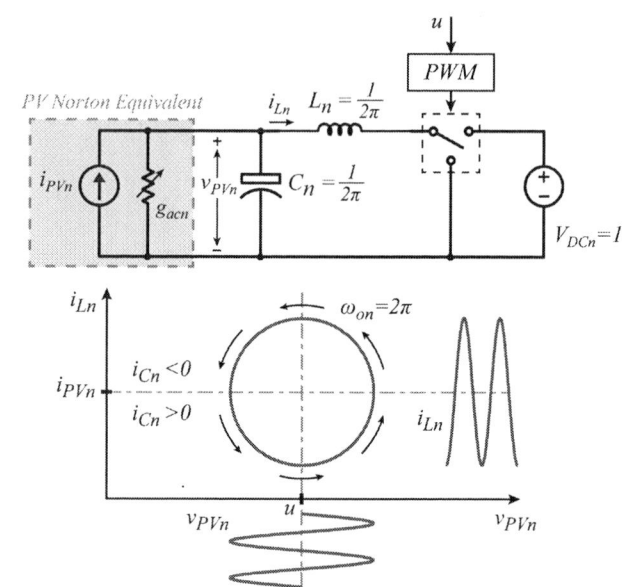

Fig. 4. Inverted buck converter dynamic geometric behaviour: circular trajectories with centre at (u, i_{PVn}), and angular speed $\omega_{on} = 2\pi$

Time-domain expressions that describe the evolution of the state variables can be obtained by solving (8) and (9):

$$v_{PVn}(t_n) = u + [v_{PVn(0)} - u]\, cos(\theta_0 + 2\pi\, t_n)$$
$$- [i_{Ln(0)} - i_{PVn}]\, sin(\theta_0 + 2\pi\, t_n) \quad (10)$$

$$i_{Ln}(t_n) = i_{PVn} + [i_{Ln(0)} - i_{PVn}]\, cos(\theta_0 + 2\pi\, t_n)$$
$$+ [v_{PVn(0)} - u]\, sin(\theta_0 + 2\pi\, t_n) \quad (11)$$

Merging these two expressions by eliminating the time variable t_n, the behaviour of the converter can be represented by a geometric shape in the state-plane:

$$\lambda_u : (v_{PVn} - u)^2 + (i_{Ln} - i_{PVn})^2$$
$$= (v_{Cn(0)} - u)^2 + (i_{Ln(0)} - i_{PVn})^2 \quad (12)$$

The obtained equation indicates the operating point moves counter-clock-wise along a circular trajectory for constant values of i_{PVn} and u, as illustrated in Fig. 4. It is worthwhile mentioning that when the converter is connected to a PV panel, the panel current i_{PVn} will vary along with the voltage v_{PVn}, showing an dynamic resistance effect that causes the circle to turn into an ellipsoidal spiral. Nevertheless, the circle equation (12) provides an accurate description of the worst condition from the control point of view, where the oscillations are not damped by any dissipative component.

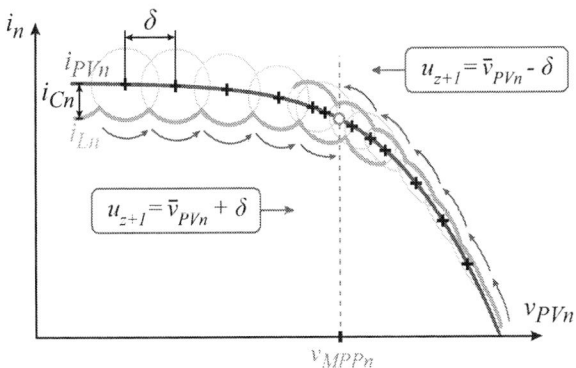

Fig. 5. Dynamic MPPT-based direct control scheme caracteristic behaviour.

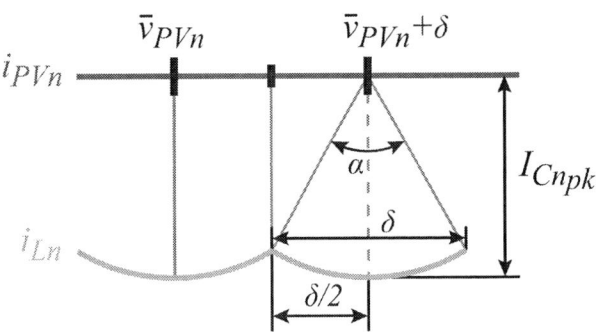

Fig. 6. Dynamic MPPT-based direct control scheme critical parameters.

IV. MPPT-Based Control Algorithm

The proposed control scheme is introduced in this section by combining the basis for MPPT set in section II with the geometric model of the converter developed in section III. Control laws are defined in order to track the MPP and the control actions are taken according to the geometric behaviour of the converter. In this way, the proposed controller not only ensures large-signal reliable operation but also accomplishes the task of operating at the MPP.

As illustrated in Fig. 4, the converter's operating point moves counter-clock-wise with angular speed of $\omega_{on} = 2\pi$ along a circular trajectory with centre coordinates at the PWM duty cycle value u and normalized panel current i_{PVn}. If the inductor current i_{Ln} is lower than the panel current i_{PVn}, the capacitor current $i_{Cn} = i_{PVn} - i_{Ln}$ is positive, and the operating point moves to the right (PV voltage increases). In a similar way, when the inductor current is larger than the current provided by the panel, the operating point moves to the left (PV voltage decreases).

The control laws being proposed are based on the idea of placing the updated center of the circle (given by the PWM duty cycle) slightly away from the equilibrium condition corresponding to the measured voltage ($u_{z+1} = v_{PVn} \pm \delta$) in order to make the operating point move in the desired direction. Inserting this control actions into (5), the converter's control rules are determined:

$$if: \ |\Delta i_{PV}| \, v_{PV} \leq |\Delta v_{PV}| \, i_{PV} \Rightarrow u_{z+1} = \bar{v}_{PVn} + \delta$$
$$else \ if: \ |\Delta i_{PV}| \, v_{PV} > |\Delta v_{PV}| \, i_{PV} \Rightarrow u_{z+1} = \bar{v}_{PVn} - \delta$$
$$(13)$$

Where $\bar{v}_{PVn}$ is the average PV voltage measured during the previous fraction of circle and $\delta << 1$ maintains the converter slightly off from the equilibrium condition. The

operating point never settles and the proposed technique shows a dynamic solution for the dynamic problem of MPPT. Characteristic trajectories obtained by employing the stated control laws are illustrated in Fig. 5. The analysis of the critical parameters that must be adjusted for the proposed technique to perform properly is provided in the following sub-sections.

A. Duty Cycle Update Frequency:

Due to the simple algorithm (voltage sensing and averaging, plus one addition) the duty cylce can be updated as fast as every switching cycle. In order to ensure the PV voltage only moves in the desired direction, the duty cycle value must be updated fast enough so the angle $\alpha < \pi$, as shown in 6. Since the angular speed is determined by the LC circuit resonance frequency F_0, the duty cycle must be updated at least twice per resonance period:

$$\frac{f_U}{F_0} = f_{Un} \geq 2 \qquad (14)$$

B. Center Displacement δ:

The capacitor current is proportional to the PV voltage derivative and it defines the speed at which the operating voltage changes and the voltage oscillations around the desired operating point. A large value in this current will derive in large voltage oscillations and a drop in the average power absorbed as consequence. Since the PV current cannot be deliberately controlled (it only depends on the PV voltage), the capacitor current can only be controlled by modifying the inductor current. As illustrated in Fig. 6, the capacitor current peak (maximum difference between PV and inductor currents) is determined by the duty cycle increment δ and the angle α. The capacitor current peak, and therefore the

TABLE I
SIMULATION PLATFORM PARAMETERS

PARAMETER	VALUE
V_{DC}	50 V
L	55 μH
C	6.5 μF
F_0	8kHz
$f_{Sw} = f_U$	50 kHz
δ	3%

TABLE II
EXPERIMENTAL PLATFORM PARAMETERS

PARAMETER	VALUE
V_{DC}	19 V
L	68 μH
C	14 μF
F_0	5.16kHz
f_{Sw}	100 kHz
f_U	33 kHz
δ	5%

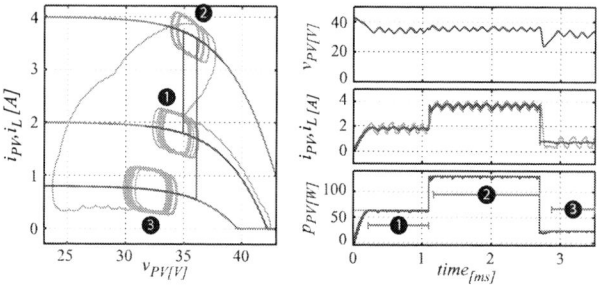

Fig. 7. Simulation results of the proposed scheme response during start-up and sharp irradiance changes: transients solved in fractions of ms

voltage oscillations, can be controlled by properly defining the value of the centre displacement δ. The capacitor peak current is found performing geometric analysis of Fig. 6:

$$i_{Cn_{pk}} = \frac{\delta}{2\,sin\left(\frac{\pi}{f_{Un}}\right)} \quad (15)$$

V. SIMULATION AND EXPERIMENTAL RESULTS

The parameter values employed for the simulation are detailed in Table I. The switching frequency is only 6.25 times higher than the LC resonance frequency, in order to illustrate how the proposed technique allows reducing the size/cost of reactive components. The duty cycle is updated every switching cycle, leading to $F_U = 6.25$, which complies with (14). Simulation results of start-up, and irradiance steps up/down are shown in Fig. 7 in both geometric and time domains.

A hardware platform was built using the prototype detailed in Table II connected to a BP350U solar panel. Experimental results are shown in Figs. 8, 9, and 10, in both geometric and time domains. The start-up transient is shown in Fig. 8, the MPP is reached in 1.75 ms, highlighting the extremely fast transient response achieved. The XY plot (PV power vs voltage) shows how the operating point oscillates around the MPP. A zoomed view of the steady state behaviour at the MPP is shown in Fig. 9. A very stable behaviour is shown, with a tracking efficiency of

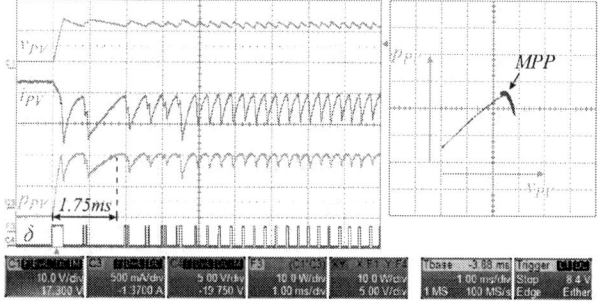

Fig. 8. Experimental results of the proposed scheme response under start-up transient: MPP found in 1.75ms

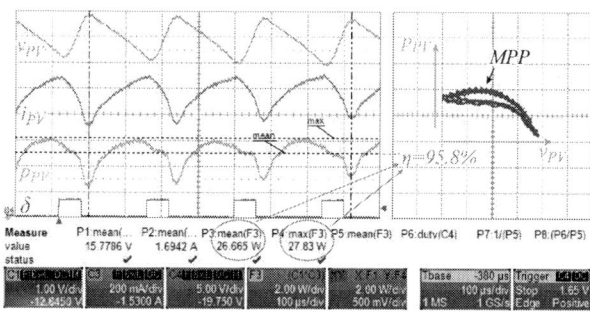

Fig. 9. Experimental results of the proposed scheme operating in steady state: reliable operation and 95.8% tracking efficiency.

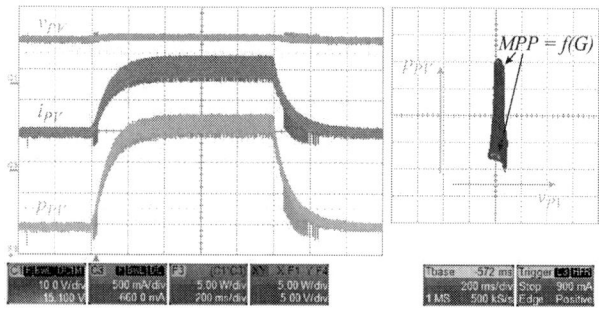

Fig. 10. Experimental results of the proposed scheme response under irradiance changes

95.8%. The XY plot shows a zoomed view of the oscillatory behaviour around the MPP. The experimental capture in Fig. 10 shows the response of the proposed technique under changes of irradiance. The step speed is limited due to thermal inertia of the lamps employed to generate the irradiance transients. Nevertheless, the proposed algorithm shows reliable behaviour over the whole range of irradiance values and good tracking of the MPP. It is worthwhile observing how the effect of parasitic capacitance at the panel and inductance at the cables is manifested as two different trajectories followed when the operating point moves in opposite directions.

The experimental results provided in this section highlight the contribution of this work to the applied field. Enhanced dynamic behaviour is achieved while the size of the reactive components and switching losses are reduced due to the reduced switching-resonance frequency ratio.

VI. CONCLUSIONS

A novel approach for photovoltaic energy harvesting was introduced in this work by embedding the MPPT into the PWM DC/DC converter controller. This structural change enables performing the MPPT algorithm as fast as the switching frequency, enhancing the dynamic tracking capabilities of the converter by orders of magnitude. Furthermore the switching losses and the reactive components size can be reduced due to the low switching-to-resonance frequencies ratio required. The controller was designed combining concepts of traditional MPPT algorithms with geometric modelling of power converters. Detailed mathematical procedures were provided to define the system components and parameters. The analytical expressions were validated by simulations and experimental results.

REFERENCES

[1] Fraunhofer Institute for solar energy systems ISE, "Photovoltaics report," Tech. Rep., Oct. 2014. [Online]. Available: www.ise.fraunhofer.de/de/downloads/pdf-files/aktuelles/photovoltaics-report-in-englischer-sprache.pdf

[2] B. I. Gatan masson, Sinead Orlandi, "Global market outlook 2015-2019," SolarPower Europ, Tech. Rep., 2015.

[3] M. Abdelhamid, R. Singh, A. Qattawi, M. Omar, and I. Haque, "Evaluation of on-board photovoltaic modules options for electric vehicles," *Photovoltaics, IEEE Journal of*, vol. 4, no. 6, pp. 1576–1584, Nov 2014.

[4] P. Armstrong, R. Armstrong, R. Kang, R. Camilleri, D. Howey, and M. McCulloch, "A reconfigurable pv array scheme integrated into an electric vehicle," in *Hybrid and Electric Vehicles Conference 2013 (HEVC 2013), IET*, Nov 2013, pp. 1–7.

[5] Q. Brogan, T. O'Connor, and D. S. Ha, "Solar and thermal energy harvesting with a wearable jacket," in *Circuits and Systems (ISCAS), 2014 IEEE International Symposium on*, June 2014, pp. 1412–1415.

[6] M. de Brito, L. Galotto, L. Sampaio, G. de Azevedo e Melo, and C. Canesin, "Evaluation of the main mppt techniques for photovoltaic applications," *IEEE Trans. Ind. Electron.*, vol. 60, no. 3, pp. 1156 – 1167, Mar. 2013.

[7] T. Esram and P. Chapman, "Comparison of photovoltaic array maximum power point tracking techniques," *IEEE Trans. Energy Convers.*, vol. 22, no. 2, pp. 439–449, Jun. 2007.

[8] M. Elgendy, B. Zahawi, and D. Atkinson, "Assessment of perturb and observe mppt algorithm implementation techniques for pv pumping applications," *IEEE Trans. Sustain. Energy*, vol. 3, no. 1, pp. 21 –33, jan. 2012.

[9] ——, "Assessment of the incremental conductance maximum power point tracking algorithm," *IEEE Trans. Sustain. Energy*, vol. 4, no. 1, pp. 108 –117, jan. 2013.

[10] N. Femia, M. Fortunato, G. Petrone, G. Spagnuolo, and M. Vitelli, "Dynamic model of a grid-connected photovoltaic inverter with one cycle control," in *35th IEEE Annu. Conf. Industrial Electronics (IECON)*, Porto, Portugal, Nov. 2009, pp. 4561–4565.

[11] E. Koutroulis, K. Kalaitzakis, and N. Voulgaris, "Development of a microcontroller-based, photovoltaic maximum power point tracking control system," *IEEE Trans. Power Electron*, vol. 16, no. 1, pp. 46 –54, jan. 2001.

[12] D. Sera, R. Teodorescu, J. Hantschel, and M. Knoll, "Optimized maximum power point tracker for fast-changing environmental conditions," *IEEE Trans. Ind. Electron.*, vol. 55, no. 7, pp. 2629 –2637, Jul. 2008.

[13] F. Paz and M. Ordonez, "Zero oscillation and irradiance slope tracking for photovoltaic mppt," *IEEE Trans. Ind. Electron.*, vol. 61, no. 11, pp. 6138 – 6147, Nov. 2014.

[14] A. Latham, R. Pilawa-Podgurski, K. Odame, and C. Sullivan, "Analysis and optimization of maximum power point tracking algorithms in the presence of noise," *IEEE Trans. Power Electron*, vol. 28, no. 7, pp. 3479–3494, Jul. 2013.

[15] N. Femia, G. Petrone, G. Spagnuolo, and M. Vitelli, "Optimization of perturb and observe maximum power point tracking method," *IEEE Trans. Power Electron*, vol. 20, no. 4, pp. 963 – 973, Jul. 2005.

[16] S. Zhang, Z. Xu, Y. Li, and Y. Ni, "Optimization of mppt step size in stand-alone solar pumping systems," in *Power Engineering Society General Meeting, 2006. IEEE*, 2006, pp. 6 pp.–.

[17] I. Galiano Zurbriggen, M. Ordonez, and M. Anun, "PWM-geometric modelling and centric control of basic DC-DC topologies for sleek and reliable large-signal response," *IEEE Trans. Ind. Electron.*, vol. 62, no. 4, pp. 2297–2308, April 2015.

[18] I. Galiano Zurbriggen and M. Ordonez, "Average natural trajectories (ANTs) for boost converters: Centric-based control," in *Proc. IEEE Appl. Power Electron. Conf. and Expo.*, March 2014, pp. 1190–1197.

[19] I. Galiano Zurbriggen, M. Ordonez, and M. Anun, "Average natural trajectories (ANTs) for buck converters: Centric-based control," in *Appl. Power Electron. Conf. and Exp. (APEC), 2013 Twenty-Eighth Annual IEEE*, 2013, pp. 1346–1351.

Feeding Partial Power into Line Capacitors for Low Cost and Efficient MPPT of Photovoltaic Strings

Ali Elrayyah

Qatar Environment and Energy Research
Institute (QEERI)
Doha, Qatar

aelrayyah@qf.org.qa

Mohammed Badawey

Electrical and Computer Engineering
Department
The University of Akron
Akron-OH, USA
mob4@zips.uakron.edu

Yilmaz Sozer

Electrical and Computer Engineering
Department
The University of Akron
Akron-OH, USA
ys@uakron.edu

Abstract—In PV strings, a number of PV modules are connected in series where the same current passes over all of them. To allow the maximum power points of the PV modules to be tracked individually, DC/DC converter known as optimizers could be used to interface each PV module with string. Processing the entire PV power by the optimizers increases the cost and degrades the system efficiency. In this paper, a method is proposed to have partial processing of the PV power using DC/DC converters which feed their output into capacitors connected in the same string of the PV modules. To further decrease the system cost and improve the efficiency, more than one PV modules can feed their partial power into the same capacitor. As the number of PV modules that share the same capacitor increases, the system cost decreases further. The validity of the proposed method is verified through simulation studies and hardware experimentation.

Keywords—PV strings; partial power processing; maximum power point tracking

I. INTRODUCTION

Recently, a significant growth is witnessed in the energy supplied to utility grids from renewable energy sources (RESs). This growth is motivated by various economic, environmental and technical benefits provided by deploying distributed energy sources (DESs) and RESs. Among RESs, solar PV is receiving a lot of interest and as its cost continues to drop; its installations are increasing all over the world. Power electronics converters are used to perform the required conversion for the DC power produced by the PV modules to the AC form that suits the voltage and current in the grid power lines. It is always desirable to keep the system cost low and one of the approaches to achieve that is by interfacing a number of PV modules with the grid trough the same power electronics converter. This approach is effective as several PV modules are usually interfaced with utility grids at the same location. The common practice is to connect the modules in series forming what is called PV strings before it get attached to the DC/AC power electronics inverter. This approach helps also in raising up the DC voltage to a level that allows smooth interfacing with the grid.

Factors such as manufacturing variability, shading and accumulation of dirt/dust may lead serially connected PV modules to have different maximum power point (MPP) characteristics [1-2]. Accordingly, the current that passes over each module needs to be controlled separately to ensure that the MPPs of the individual PV modules are tracked.

Among available technologies, two kinds of systems are used to track the MPPs of individual PV modules which are the microinverters and the DC power optimizers. Microinverters convert the DC power produced by a single PV module to an AC power that is supplied to the grid. DC power optimizers use DC/DC converters to interface the PV modules to the string. But as stated before, DC power optimizers could be more preferred for utility scale application as they reduce the cost and support the high voltage required for utility-tied inverters [3].

Figure 1 shows the use of DC power optimizers to interface PV modules with input side of a DC/AC inverter. As the PV modules have different MPP currents, the DC power optimizers, which are DC/DC converters, process these current to match the string current at their outputs.

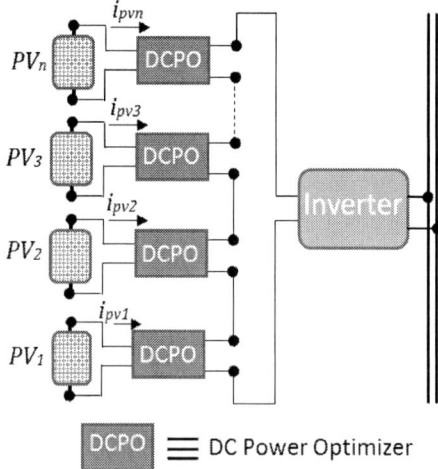

Figure 1. Using DC power optimizers to interface PV module to a single inverter

---

This research is sponsored by Qatar Environment and Energy Research Institute for the project # GC5004 .

978-1-4673-9551-9/16 $31.00 © 2016 IEEE

In DC power optimizers, the PV produced power is processed entirely by DC/DC converters which increases the converters cost and losses. Many researchers proposed technologies for partial PV power processing (PPP) such that only a small amount of the PV power is processed by DC/DC converters. These methods are based on using PV strings that carry a high share of the produced power and then to use converters with low power ratings to manipulate the differences in MPP currents of the individual modules [4-5]. In this way the current i_{str} shown in Fig. 2 passes over all modules such that most of the power is extracted through this current. The currents, I_{d1}, I_{d2}, ... I_{dN} when combined with i_{str} produce the MPP currents of PV_1, PV_2,...PV_N, respectively. The use of PPP improves the conversion efficiency and reduces the system cost [6].

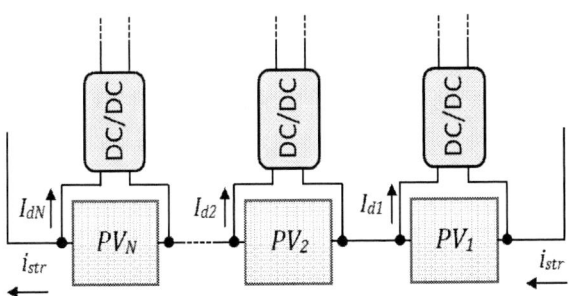

Figure 2. Partial Power processing in PV strings

The real challenge in PPP is to maintain the system simplicity and modularity at the same time. The simplicity in the hardware and in the control algorithms is needed for cost saving and implementation smoothness. The modularity feature is also important since it allows every PV module to operate as an independent unit unaffected by status and the number of connected module in the string and it, therefore, supports the scalability feature.

The main difference among various methods for PPP is how to process the partial power extracted by the converters. Some of the techniques exchange the partial power with full bus of the entire string [7-9], but in this case high voltage stress is experiences in the converter output side [3]. Another technique is to connect the outputs of all converters across an isolated low voltage DC bus [8] where the PV modules with relatively high power supply power to the bus while those with low power consume power from the bus [3]. The main problem of this method is the lack of modularity as it needs centralized controller and complex control algorithms. A similar concept is to use PV-to-PV power processing [10-11].

In this paper, the objective is to develop a low cost system for PPP that supports the modularity of system structure and the control algorithm. The method presented in this paper for PPP can use non-isolated DC/DC converters that process 1/3 of the PV produced power. Moreover, further reduction in the converters cost could be achieved if isolated DC/DC converters are used while applying low current stresses in its output components. The simplicity of the needed control algorithms

makes the proposed controller very attractive to be adopted in commercial products.

II. STRUCTURE OF THE PROPOSED PV-TO-LINE CAPACITOR PPP

Figure 3 shows one version of proposed structure for PPP where a capacitor (C_{lin}) is placed between the modules # $2n-1$ and $2n$, n= 1, 2 ... $N_p/2$ where N_p is the total number of the modules in the string. The capacitor is fed by currents from the converters con_{2n-1} and con_{2n}.

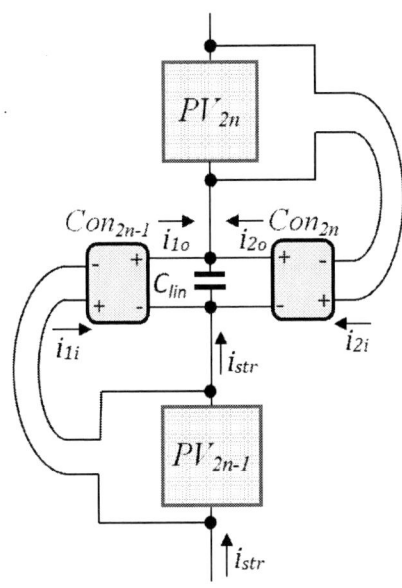

Figure 3. Proposed structure for PPP system with only two-module

The operation concept of the proposed topology is to feed most of the power through the main string while a portion of the PV produced power is to be fed into the line capacitor C_{lin}. The two converters feed currents into C_{lin} while the string line current discharges the current out of the capacitor. The appropriate operation could then be obtained by making the value of the string current around 2/3 of the PV currents. Each of the PVs will then feeds 1/3 of its current into the capacitor through its converter and the sum of the two components matches the string current. Accordingly, each one of the converters con_{2n-1} and con_{2n} processes around $1/3^{rd}$ of the modules power into C_{lin}. When compared with DC power optimizer, the components ratings of the converters in the proposed method is one third of the equivalent optimizers which helps in cost reduction. Moreover, the efficiency of the proposed system is better than the optimizer case. Consider a case of an optimizer whose efficiency is η. Then the total loss in the DC power processing part as a percentage of the total power is given by $1- η$. On the other hand, if efficiency of the converters in Fig. 3 is also η, then the total percentage loss will be given by $(1- η)/3$ which is one third of the optimizers' case.

978-1-4673-9551-9/16 $31.00 © 2016 IEEE

To verify this point, consider the case of when the two modules in Fig. 3 have different in the MPP currents. Let i_{2n-1} and i_{2n} be the MPP current of the $2n$-1th and the $2n^{th}$ modules, respectively, and consider the difference between them be Δ, where

$$i_{2n} = i_{2n-1} + \Delta \qquad (1)$$

The current of every PV module is a combination of two parts. One part comes from the main string current i_{str} while the other part is the one processed by the DC/DC converter connected across its terminals. To determine the amount of currents processed by the converter, consider the case when the capacitor voltage is close to the MPP voltage of the two modules. This assumption is quite reasonable, as for number of PV connected at the same location their MPP voltages are very close in their values [12]. Therefore, the gain of the converters can easily be adjusted around the unity value. From Fig. 3, the by equating the input and output currents of the capacitors it's found that:

$$i_{str} = i_{1o} + i_{2o} \qquad (2)$$

However, as the converters have gains close to unity, then

$$i_{1o} \approx i_{1i} \qquad (3)$$

$$i_{2o} \approx i_{2i} \qquad (4)$$

By analyzing the current of PV_{2n}, it can be shown that:

$$i_{2n} = i_{str} + i_{2i} \approx i_{str} + i_{2o} \qquad (5)$$

Similarly:

$$i_{2n-1} \approx i_{str} + i_{1o} \qquad (6)$$

By adding Eqn. 5 to Eqn. 6 it can be seen than:

$$i_{2n} + i_{2n-1} \approx 3i_{str} \rightarrow 2i_{2n-1} + \Delta \approx 3i_{str} \qquad (7)$$

Accordingly, the relation between the PV modules currents and the main string current can be derived as:

$$i_{str} = \frac{2}{3}i_{2n-1} + \frac{1}{3}\Delta = \frac{2}{3}i_{2n} - \frac{1}{3}\Delta \qquad (8)$$

Equation 8 shows when the value of Δ is zero, the line current becomes 2/3 of the PV modules currents as stated before. Moreover, the effect of difference in MPP current is reduced. In this example, the current processed by con_{2n} is given by:

$$i_{2i} = i_{2n} - i_{str} = \frac{1}{3}i_{2n} + \frac{1}{3}\Delta \qquad (9)$$

Accordingly, even when there is a significant difference between i_{2n} and $i_{2n-1,}$ the power processed by the converter is only one third of that difference.

Isolated converters can help in further reducing the required converter rating in the proposed structure by allowing more converters to share the same capacitor. Fig. 4 shows the case when four converters are used. In this case, each module can supply 20% of its current through the converters to be supplied to the capacitor C_{lin}. Accordingly, the value of i_{str} would be 80% of the module current. Since the capacitor C_{lin} receives around 20% power from each converter, the power combined

form the four module provides the required value of the 80% of the module current.

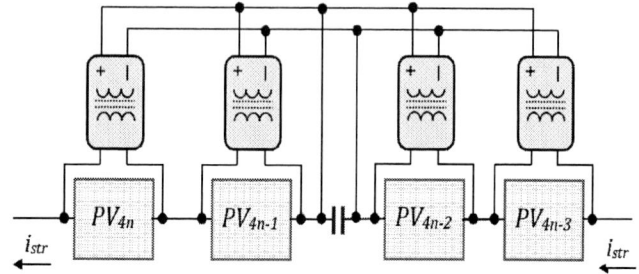

Figure 4. Sharing the same line capacitor by four PV modules

Further reduction in converters rating could be achieved through the sharing of the same output capacitor by more converters, however, the complexity in the modules connection could be a limiting factor in this case. Sharing the same capacitor by four modules can easily by achieved by arranging the modules as shown in Fig. 5. In this case, rather than having one capacitor attached to the line, the capacitor C_{lin} is formed by the paralleling four capacitors each one of them is connected at the output of one converter. Clearly, this enhances the reliability as a failure of any converter makes allows the other three to compensate. Moreover, a short circuit fault in any capacitor cancel the effect of the partial power processing system and thus in its worst case performance it will provide the same amount of power of a conventional PV strings.

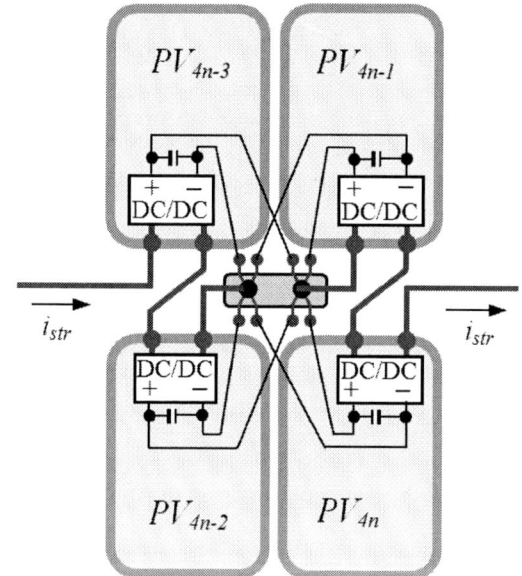

Figure 5. Field installation for the structure in Fig. 4

III. SIMULATION RESULTS

The system shown in Fig. 6 is considered for simulation. For simplicity, the string is considered to have two modules only and they feed part of their power through DC/DC converters into the line capacitor. The DC/DC converter is of buck-boost type, but other types like CUK converters could be used as well.

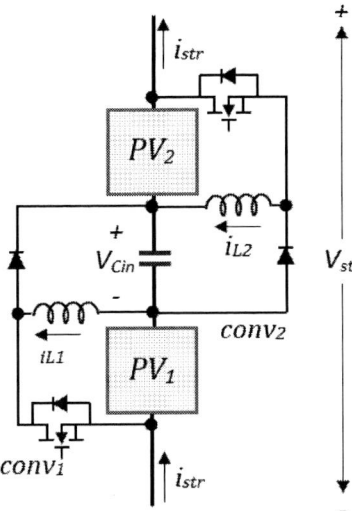

Figure 6. Simulated PV string

The assumed MPP currents of the two modules are listed in Table 1 where the modules produce different current at MPP operation. The controllers of the DC/DC converters are driven to track the MPP of the two modules. On the other hand, the string voltage is regulated by adjusting i_{str} through the inverter controller. The inverter sets a reference value for the voltage of the entire string; based on that voltage the value of i_{str} is adjusted. The reference value of the line voltage is set to make the voltage across the line capacitor (V_{Cin}) comparable with the MPP voltage of the PV modules such that the power extracted from this capacitor becomes around one third of the total power. The MPP voltage of the two modules is considered to be around 30 V, therefore, the reference value of the string voltage is taken as 90V. Clearly, the initial value for the DC bus is considered by rough estimation and the converter actions modify this value till it reach the actual MPP of the PV modules.

Table 1. MPP currents of the two modules during different operation intervals

Time	$0 \leq t < 0.1$	$0.1 \leq t < 0.2$	$0.2 \leq t$
PV_1 current (A)	12	12	9
PV_2 current (A)	10.5	12.5	12.5

Figure 7 shows the currents that flow from the two modules. By comparing Fig. 7 with Table 1, it can be seen clearly the perfect tracking of the modules MPPs. As stated before, the controller of the DC/DC converters adjust their duty ratios until the MPP operation is ensured.

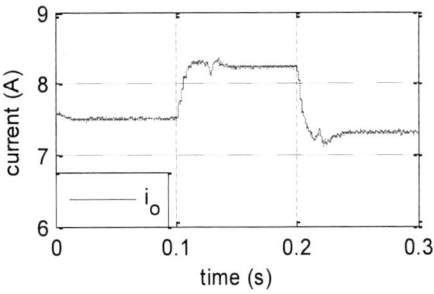

Figure 7. Output currents of the two PV modules in the string shown in Fig. 6

The string current (i_{str}) is shown in Fig. 8. The string current takes the values close to the 2/3 of the PV modules and that makes 2/3 of the total produced power to flow directly into the string leaving only 1/3 of the total power to be processed by the DC/DC converters. Note that the line current increases as the PV output power increases to maintain a regulated value for the string voltage.

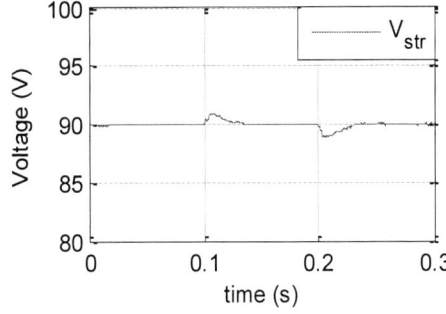

Figure 8. String currents I_{str} in the string shown in Fig. 6

The voltage of the entire string (V_{str}) and the voltage V_{Cin} are shown in Figs. 9 and 10, respectively. The string voltage is always maintained at the required value of 90 V and that makes the voltage V_{Cin} be adjusted around 30 V.

Figure 9. String line Voltage for the system in Fig. 6

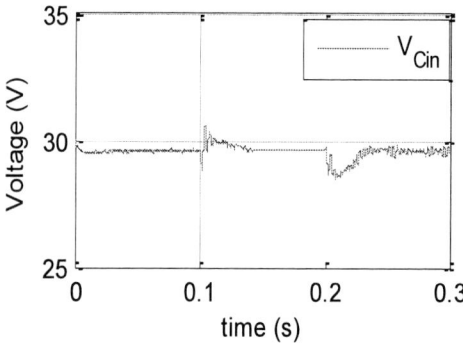

Figure 10. Line capacitor Voltage for the system in Fig. 6

IV. EXPERIMENTAL RESULTS

To verify the validity of the proposed method experimentally, the setup shown in Fig. 11 is used. Two programmable DC sources are used to represent the PV modules. The MPP currents of the two sources are assumed to have different values which vary with time. The MPP currents of the two PV sources are listed in Table 2. The structure shown in Fig. 6 is used to implement the converters and the used value for the converters inductors is 3mH while the value of C_{lin} is 3.3 mF.

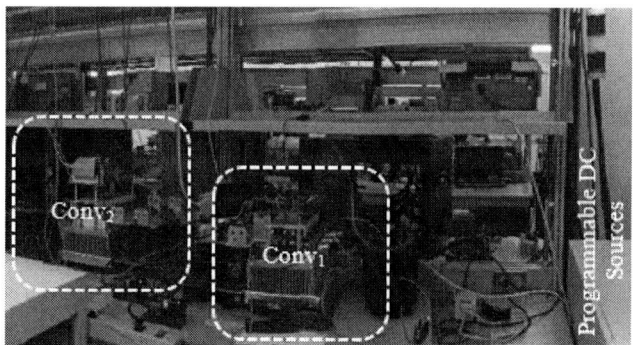

Figure 11. Experimental Setup

Table 2. MPP currents of the two PV sources during different operation intervals in the experimental testing

Time	$0 \leq t < 9.6$	$9.6 \leq t < 25.2$	$25.2 \leq t$
PV_1 current (A)	4.4	2	2
PV_2 current (A)	3.4	3.4	4.4

Similar to the simulation case, the MPP voltages of the two sources is around 30 V and therefore, the reference value for the string voltage is taken as 90 V. The inverter controls the value of the DC bus to adjust the value of the string current. Figure 12 shows the output current of the two PV sources where the MPP currents are precisely tracked by the controllers of the two modules.

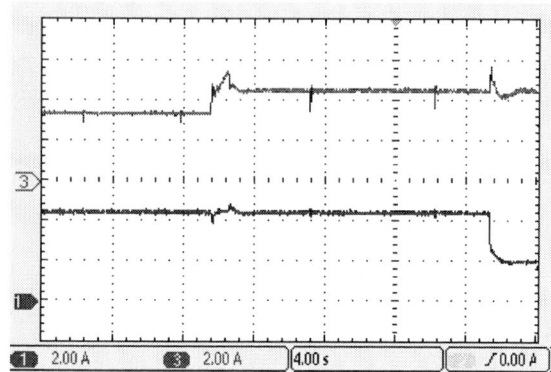

Figure 12. Output currents of the two PV modules in the string tested in the experimental setup

The string and the line capacitor voltages are shown in Figs. 13 and 14, respectively. As expected, the string voltage is always regulated at the required value of 90 V. Also, the line capacitor voltage is maintained at around 30 V indicating the processing of only one third of the string power by the DC/DC converters.

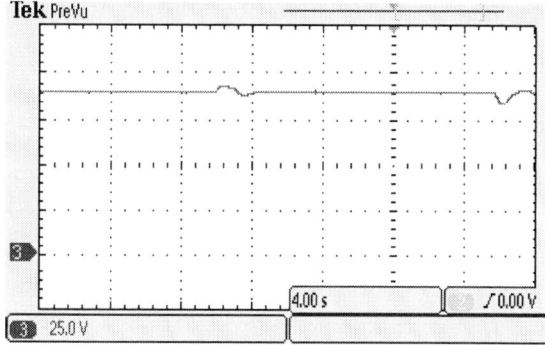

Figure 13. String line Voltage for the system in the experimental setup

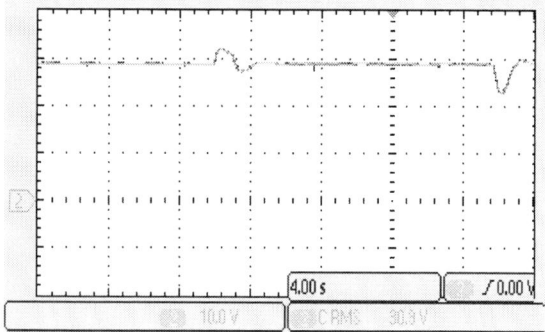

Figure 14. Line capacitor Voltage for the system in the experimental setup

V. CONCLUSION

In this paper, a low cost and highly efficient method is proposed to perform partial power processing to the track the maximum power points of PV modules connected in strings. Unlike DC power optimizers, the DC/DC converters in the proposed method process only a fraction of the PV produced power. In the proposed method, a number of PV modules supply current to a capacitor connected in the same string of those modules. As many PV modules supply current to the same capacitor, the power provided by every module becomes only a percentage of the module total power. The simplest case of two modules sharing one capacitor is studied where each module is found to supply around one third of its power through the converter to the capacitor. As the number of the sharing modules increases, the power processed by the converters decreases. E. g., in the case of four PV modules sharing the capacitor, only one fifth of the PV power is required to be processed by the converters. The proposed method is modular and distributed as there is no need for central controllers of inter-controllers communication. This fact enhances the system scalability and simplifies its design and deployment. The effectiveness of the proposed method in tracing the MPPs of the PV modules by partial power processing is demonstrated by simulation studies and hardware experiments.

REFERENCES

[1] C. Olalla, C. Deline, and D. Maksimovic, "Performance of Mismatched PV Systems with Submodule Integrated Converters," Proc. IEEE Photo. Spec. Conf., pp. 1-9, 2013.

[2] K. A. Kim, P. S. Shenoy and P. T. Krein, "Photovoltaic Differential Power Converter Trade-offs as a Consequence of Panel Variation", Cont. Mod. Power Electron., 2012.

[3] K. A. Kim, P. S. Shenoy, P. T. Krein, "Converter Rating Analysis for Photovoltaic Differential Power Processing Systems," IEEE Trans. Power Electron., vol. 30, no. 4, pp.1987-1997, 2015.

[4] T. Shimizu, M. Hirakata, T. Kamezawa, and H. Watanabe, "Generation Control Circuit for Photovoltaic Modules," EEE Trans. Power Electron., vol. 16, pp. 293-300, 2001.

[5] P. S. Shenoy, K. A. Kim and P. T. Krein, "Comparative Analysis of Differential Power Conversion Architectures and Controls for Solar Photovoltaics," IEEE Workshop Control Model. Power Electron., 2012.

[6] P. S. Shenoy, K. A. Kim, P. T. Krein, P. L. Chapman, "Differential Power Processing for Efficiency and Performance Leaps in Utility-scale Photovoltaics," IEEE Photo. Spec. Conf., pp. 1357-1361, 2012.

[7] Y. Nimni and D. Shmilovitz, "A Returned Energy Architecture for Improved Photovoltaic Systems Efficiency", in Proc. IEEE Int. Symp. Circuits Sys., pp. 2191-2194, 2010.

[8] C. Olalla, M. Rodriguez, D. Clement and D. Maksimovic, "Architectures and Control of Submodule Integrated DC-DC Converters for Photovoltaic Applications", IEEE Trans. Power Electron., vol. 28, no. 6, pp. 2980-2997, 2013.

[9] S. Poshtkouhi, A. Biswas and O. Trescases, "DC-DC Converter for High Granularity, Sub-string MPPT in Photovoltaic Applications Using a Virtual-parallel Connection", Proc. IEEE Appl. Power Electron. Conf. Expo., pp. 86-92, 2012.

[10] P. S. Shenoy, B. Johnson and P. T. Krein, "Differential Power Processing Architecture for Increased Energy Production and Reliability of Photovoltaic Systems," IEEE Appl. Power Electron. Conf., pp. 1987 -1994, 2012.

[11] H. J. Bergveld, D. Buthker, C. Castello, T. Doorn, A. de Jong, R. van Otten, and K. de Waal, "Module-level DC/DC Conversion for Photovoltaic Systems," Int. Teleco. Energy Conf., pp. 1-9, 2011.

[12] Z H. Zhou, J. Zhao, and Y. Han, "PV Balancers: Concept, Architectures, and Realization," IEEE Trans. Power Electron, vol. 30, no. 7, pp. 3479-3487, 2015.

Single Phase Cascaded H5 Inverter with Leakage Current Elimination for Transformerless Photovoltaic System

Xiaoqiang Guo, Xiayu Jia, Zhigang Lu
Department of Electrical Engineering
Yanshan University, Qinhuangdao, 066004, China
Email: gxq@ysu.edu.cn

Josep M. Guerrero
Institute of Energy Technology
Aalborg University, 9220 Aalborg East, Denmark
Email: joz@et.aau.dk

Abstract—Leakage current reduction is one of the important issues for the transformelress PV systems. In this paper, the transformerless single-phase cascaded H-bridge PV inverter is investigated. The common mode model for the cascaded H4 inverter is analyzed. And the reason why the conventional cascade H4 inverter fails to reduce the leakage current is clarified. In order to solve the problem, a new cascaded H5 inverter is proposed to solve the leakage current issue. Finally, the experimental results are presented to verify the effectiveness of the proposed topology with the leakage current reduction for the single-phase transformerless PV systems.

I. INTRODUCTION

The transformerless photovoltaic (PV) inverters have the advantages of low cost, small size, light weight and high efficiency [1]−[3]. However, the leakage current will arise due to lack of galvanic isolation. The undesirable leakage current may lead to electromagnetic inferences, current harmonic distortion and safety concerns. Therefore, it is crucial to eliminate the leakage current in transformerless PV systems.

Many interesting single-phase topologies have been reported such as Heric, H5, H6, and so on. But they are limited to three-level inverters [4]−[8]. On the other hand, the multilevel inverters can decrease the voltage stress of dv/dt on switches and increase the output waveform quality [9]−[11]. However, few papers have been reported regarding eliminating the leakage current for the single-phase cascaded multilevel inverters. A significant contribution by Zhou and Li is the filter-based leakage current suppression solution for the single-phase cascaded multilevel PV inverter [12]. But the topology-based solution is rarely discussed in literature, and needs further investigation.

The main contribution of this paper is to present a new cascaded H5 inverter to achieve the leakage current reduction. The experimental results are presented to verify the effectiveness of the proposed topology with the leakage current reduction for the single-phase transformerless PV systems.

II. SINGLE-PHASE CASCADED H4 INVERTER

The schematic diagram of the single-phase cascaded H4 inverter is shown in Fig.1. Since each H-bridge unit requires an independent dc source, it is necessary to consider the stray capacitance of each PV panel to the ground. The common mode voltage (CMV) and differential-mode voltage (DMV) for the upper and lower H-bridge unit are defined as follows:

$$U_{cma} = (U_{AN} + U_{BN})/2 \qquad (1)$$

$$U_{dma} = U_{AN} - U_{BN} \qquad (2)$$

$$U_{cmb} = (U_{A'N'} + U_{B'N'})/2 \qquad (3)$$

$$U_{dmb} = U_{A'N'} - U_{B'N'} \qquad (4)$$

Fig. 2 shows the common-mode model for single-phase cascaded H4 PV system, where $C_{pva}=C_{pva1}//C_{pva2}$, $C_{pvb}=C_{pvb1}//C_{pvb2}$. Based on the common-mode model, the leakage current for the cascaded H4 inverter can be calculated as (5).

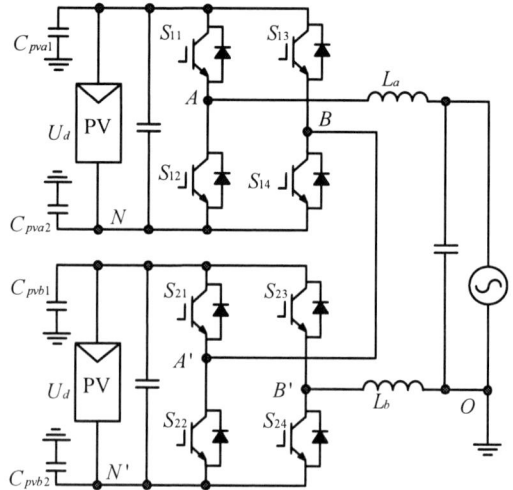

Fig. 1. Single-phase cascaded H4 inverter

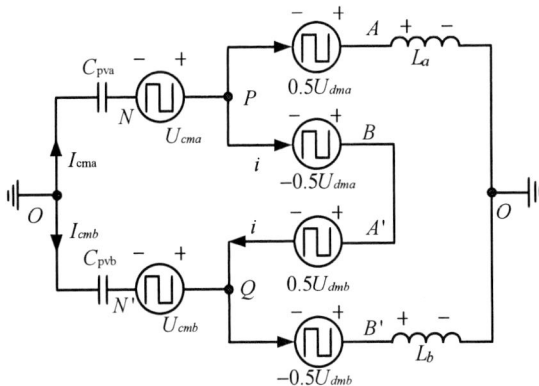

Fig. 2.   Common-mode model of cascaded H4 inverter

The following equations can be derived from Fig. 2.

$$\begin{cases} -\dfrac{U_{PO}-U_{cma}}{Z_{cpva}} = i + \dfrac{U_{PO}+0.5U_{dma}}{Z_{La}} \\[2mm] -\dfrac{U_{QO}-U_{cmb}}{Z_{cpvb}} = -i + \dfrac{U_{QO}-0.5U_{dmb}}{Z_{Lb}} \\[2mm] U_{PQ} = 0.5U_{dma}+0.5U_{dmb} \qquad (5) \\[2mm] I_{cma} = \dfrac{U_{OP}+U_{cma}}{Z_{cpva}} \\[2mm] I_{cmb} = \dfrac{U_{OQ}+U_{cmb}}{Z_{cpvb}} \end{cases}$$

where $Z_{cpva}=1/(sC_{pva})$, $Z_{cpvb}=1/(sC_{pvb})$.

$$\begin{cases} I_{cma} = \dfrac{AU_{cma}+BU_{dma}+CU_{cmb}+DU_{dmb}}{Z_{cpva}[(Z_{cpva}+Z_{cpvb})Z_{La}Z_{Lb}+(Z_{La}+Z_{Lb})Z_{cpva}Z_{cpvb}]} \\[4mm] I_{cmb} = \dfrac{A'U_{cma}+B'U_{dma}+C'U_{cmb}+D'U_{dmb}}{Z_{cpvb}[(Z_{cpva}+Z_{cpvb})Z_{La}Z_{Lb}+(Z_{La}+Z_{Lb})Z_{cpva}Z_{cpvb}]} \end{cases}$$

$$(6)$$

where $Z_{La}=sL_a$, $Z_{Lb}=sL_b$,

$A = Z_{cpva}\left(Z_{La}Z_{Lb}+Z_{cpvb}Z_{Lb}+Z_{cpvb}Z_{La}\right)$,

$B = -0.5Z_{cpva}\left(Z_{La}Z_{Lb}-Z_{cpvb}Z_{Lb}+Z_{cpvb}Z_{La}\right)$,

$C = -Z_{cpva}Z_{La}Z_{Lb}$, $D = -0.5Z_{cpva}Z_{La}(Z_{Lb}+Z_{cpvb})$,

$A' = -Z_{cpvb}Z_{a}Z_{Lb}$, $B' = Z_{cpvb}Z_{Lb}(0.5Z_{a}+Z_{cpva})$,

$C' = Z_{cpvb}(Z_{a}Z_{Lb}+Z_{cpva}Z_{Lb}+Z_{cpva}Z_{La})$,

$D' = 0.5Z_{cpvb}(Z_{a}Z_{Lb}+Z_{cpva}Z_{Lb}-Z_{cpva}Z_{La})$

For (6), it can be concluded that the leakage current is dependent on many factors, e.g. the CMV and DMV of each H-bridge unit. Therefore, it is difficult to eliminate the leakage current of single-phase cascaded H4 inverter in an effective way.

III.   PROPOSED CASCADED H5 INVERTER

In order to solve the abovementioned problem, a new single-phase cascaded H5 inverter is proposed, as shown in Fig. 3. The common-mode model of cascaded H5 inverter is shown in Fig. 4, from which the following equation can be derived.

$$\begin{cases} I_{cma} = \dfrac{U_{cma}+U_{dm\_a}}{Z_{cpva}+Z_{Lab}} \\[3mm] I_{cmb} = \dfrac{U_{cmb}+U_{dm\_b}}{Z_{cpvb}+Z_{La'b'}} \end{cases} \qquad (7)$$

Considering that the filter inductors are generally designed as the same value, the differential mode voltages $U_{dm\_a}=U_{dm\_b}=0$. So Eq. (7) can be rewritten as

$$\begin{cases} I_{cma} = \dfrac{U_{cma}}{Z_{cpva}+Z_{Lab}} \\[3mm] I_{cmb} = \dfrac{U_{cmb}}{Z_{cpvb}+Z_{La'b'}} \end{cases} \qquad (8)$$

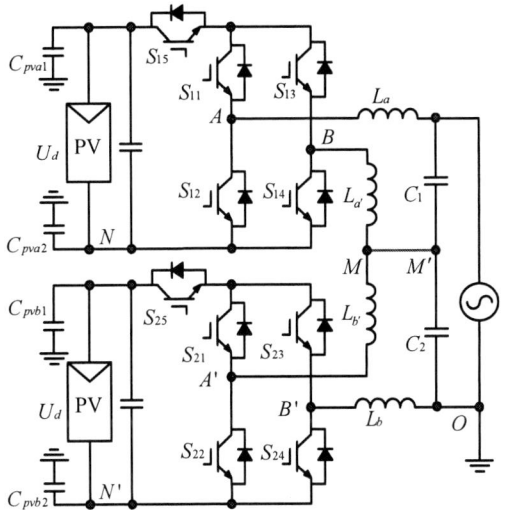

Fig. 3.   Single-phase cascaded H5 inverter

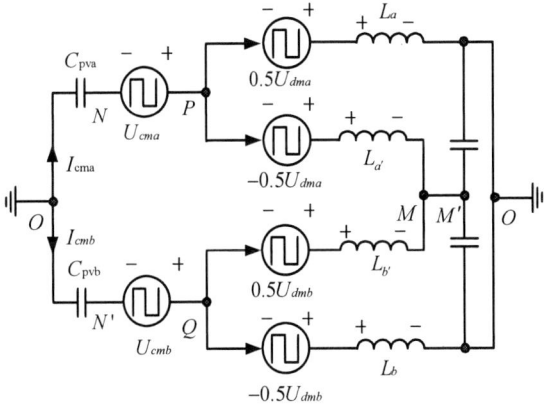

Fig. 4.   Common-mode model of cascaded H5 inverter

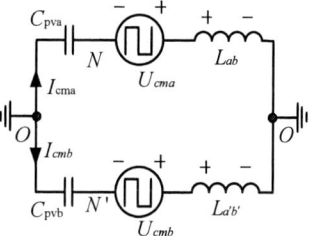

Fig. 5. Simplified common-mode model of cascaded H5 inverter

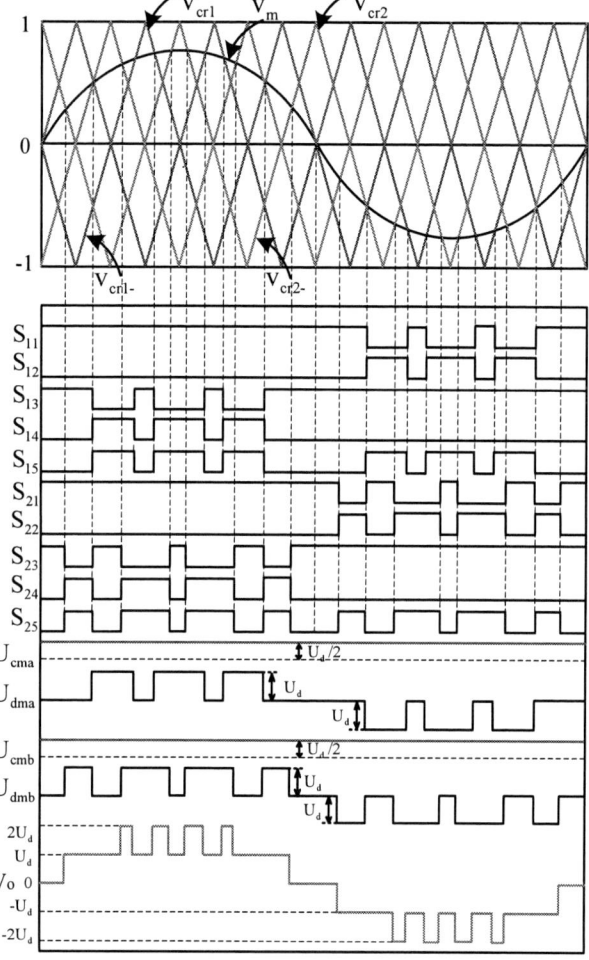

Fig. 6. Proposed modulation strategy

From (8), it can be concluded that the leakage current can be eliminated if the common mode voltage of each cascaded unit is constant. Fig. 6 shows the proposed modulation strategy for the cascaded H5 inverter. It can be observed that the proposed method can achieve both the constant common mode voltage and five-level output voltage. That is, the high quality output waveforms with the leakage current reduction can be achieved.

IV. EXPERIMENTAL RESULTS

The experimental prototype of the proposed cascaded H5 topology is controlled with TMS320F28335 DSP and XC3S400 FPGA. The experimental parameters are listed

as follows. The dc link voltage of each cascaded unit is 120V, switching frequency is 10 kHz. The filter inductor is 5mH. The output filter capacitor is 9.4μF. The parasitic capacitance is 150nF.

The experimental results are shown in Fig. 6 and Fig. 7, from which, it can be observed that the both cascaded H4 and H5 topology can achieve five-level output waveforms. However, for the cascaded H4 topology, the voltage across the stray capacitor is polluted with the high frequency components, which result in very high leakage currents.

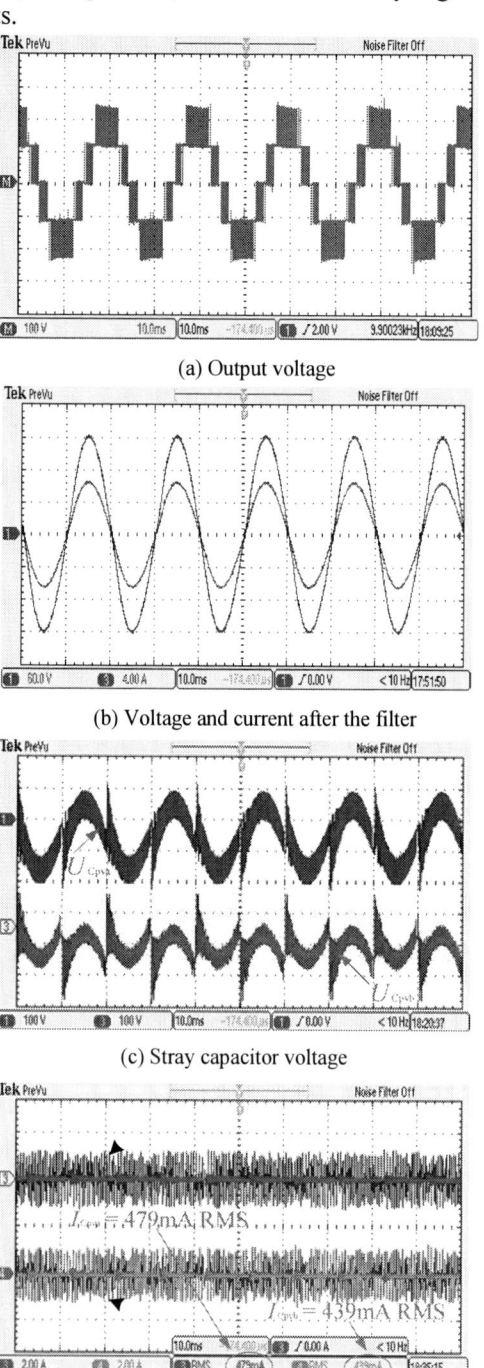

(a) Output voltage

(b) Voltage and current after the filter

(c) Stray capacitor voltage

(d) Leakage current

Fig. 7. Experimental results of cascaded H4 topology

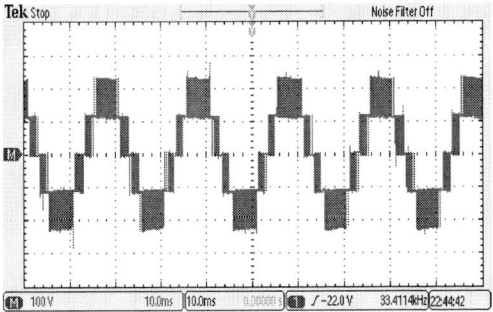

(a) Output voltage

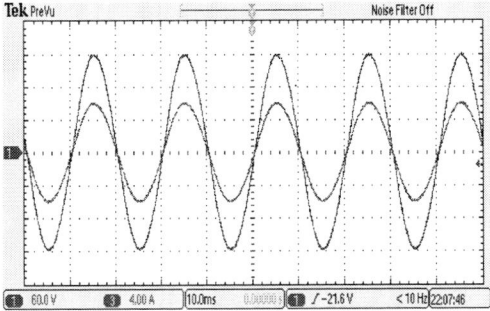

(b) Voltage and current after the filter

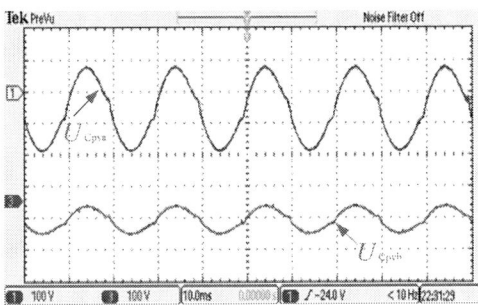

(c) Stray capacitor voltage

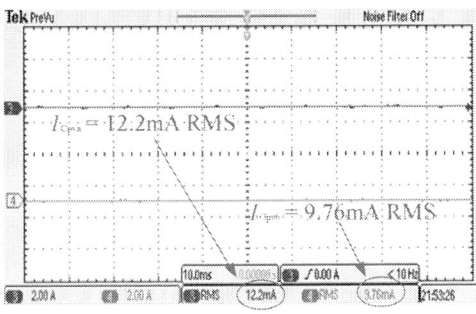

(d) Leakage current

Fig. 8. Experimental results of proposed topology

As shown in Fig. 8, for the cascaded H5 topology, the voltage across the stray capacitor is free of any high frequency components. Consequently, the high frequency leakage current is significantly reduced, which is well below 300mA, as specified in VDE 0126-1-1.

V. CONCLUSIONS

This paper has presented the theoretical analysis and experimental verification of the leakage current suppression capability of single-phase cascaded H-bridge topologies for the transformerless PV systems. The experimental results indicate that the conventional single-phase cascaded H-bridge topology fail to reduce the leakage current. On the other hand, the proposed topology and new modulation strategy can ensure that the stray capacitor voltage is free of any high-frequency components, and the leakage current can be effectively reduced. Therefore, it is attractive for single-phase transformerless PV systems.

ACKNOWLEDGMENT

This work was supported by National Natural Science Foundation of China (51307149).

REFERENCES

[1] M. C. Cavalcanti, K. C. de Oliveira, A. M. de Farias, F. A. S. Neves, G. M. S. Azevedo, and F. Camboim, "Modulation techniques to eliminate leakage currents in transformerless three-phase photovoltaic systems," *IEEE Trans. Power Electron.*, vol. 57, no. 4, pp. 1360-1367, Apr. 2010.

[2] O. Lopez, F. D. Freijedo, A. G. Yepes, P. Fernandez-Comesaa, J. Malvar, R. Teodorescu, and J. Doval-Gandoy, "Eliminating ground current in a transformerless photovoltaic application," *IEEE Trans. Energy Conver.*, vol. 25, no. 1, pp. 140-147, Mar. 2010.

[3] M. C. Cavalcanti, A. M. Farias, K. C. Oliveira, F. A. S. Neves, and J. L. Afonso, "Eliminating leakage currents in neutral point clamped inverters for photovoltaic systems," *IEEE Trans. Ind. Electron.*, vol. 59, no. 1, pp. 435–443, Jan. 2012.

[4] Wuhua Li, Yunjie Gu, Haoze Luo, Wenfeng Cui, Xiangning He, and Changliang Xia, "Topology review and derivation methodology of single phase transformerless photovoltaic inverters for leakage current suppression," *IEEE Trans. Ind. Electron.*, vol. 62, no. 72, pp. 4537-4551, Jul. 2015.

[5] Huafeng Xiao, Shaojun Xie, Yang Chen, and Ruhai Huang, "An optimized transformerless photovoltaic grid connected inverter," *IEEE Trans. Ind. Electron.*, vol. 58, no. 5, pp. 1887-1895, May. 2011.

[6] L. Zhang, K. Sun, Y. Xing, and M. Xing, "H6 transformerless full-bridge PV grid-tied inverters," *IEEE Trans. Power Electron.*, vol. 29, no. 3, pp. 1229–1238, Mar. 2014.

[7] S. V. Araujo, P. Zacharias, and R. Mallwitz, "Highly efficient single phase transformerless inverters for grid-connected photovoltaic systems," *IEEE Trans. Ind. Electron.*, vol. 57, no. 9, pp. 3118–3128, Sep. 2010.

[8] T. K. S. Freddy, N. A. Rahim, W. P. Hew, and H. S. Che, "Comparison and analysis of single-phase transformerless grid-connected PV inverters," *IEEE Trans. Power Electron.*, vol. 29, no. 10, pp. 5358–5369, Oct. 2014

[9] Ebrahim Babaei, Sara Laali, and Zahra Bayat, "A single-phase cascaded multilevel inverter based on a new basic unit with reduced number of power switches," *IEEE Trans. Ind. Electron.*, vol. 62, no. 2, pp. 922–929, Feb. 2015.

[10] E. Babaei, S. Alilu, and S. Laali, "A new general topology for cascaded multilevel inverters with reduced number of components based on developed H-bridge," *IEEE Trans. Ind. Electron.*, vol. 61, no. 8, pp. 3932–3939, Aug. 2014.

[11] Xiaotian Zhang, and Spencer, J.W., "Study of multisampled multilevel inverters to improve control performance," *IEEE Trans. Power Electron.*, vol. 27, no. 11, pp. 4409–4416, Nov. 2012.

[12] Y. Zhou and H. Li, "Analysis and suppression of leakage current in cascaded-multilevel-inverter-based PV systems," *IEEE Trans. Power Electron.*, vol. 29, no. 10, pp. 5265–5277, Oct. 2014.

Optimal Low Switching Frequency Pulse Width Modulation of Current-Fed Three-Level Inverter for Solar Integration

Gnana Sambandam K*, *Student Member, IEEE*, Akshay K. Rathore*, *Senior Member, IEEE*,
Amarendra Edpuganti, *Student Member, IEEE*, and Dipti Srinivasan*, *Senior Member, IEEE*
Department of Electrical and Computer Engineering
*Solar Energy Research Institute of Singapore
National University of Singapore, 117583, Singapore
gnana@u.nus.edu; eleakr@nus.edu.sg; amarendra@u.nus.edu; dipti@nus.edu.sg;

Abstract—In high power conversion, low device switching frequency operation is preferred in order to satisfy the thermal constraints of the semiconductor devices and also to improve efficiency. However, low device switching frequency operation leads to higher total harmonic distortion (THD) of the converter output currents. Synchronous optimal pulse-width modulation (SOP) technique is an emerging low device frequency modulation technique that maintains minimal THD on converter output currents. The state-of-the-art generalized SOP has only been applied to voltage source inverters so far. Hence, the aim of this paper is to propose a modified SOP technique for current source inverter (CSI). In addition, a simple conversion technique has been introduced in the modulation stage to incorporate additional switching constraints of CSI. The experimental results show effectiveness of the proposed modified SOP technique on CSI by achieving better total harmonic distortion for range of modulation indexes without compromising on device switching frequency.

Index Terms—Multilevel inverters, current source inverter, solar power integration, synchronous optimal pulse-width modulation

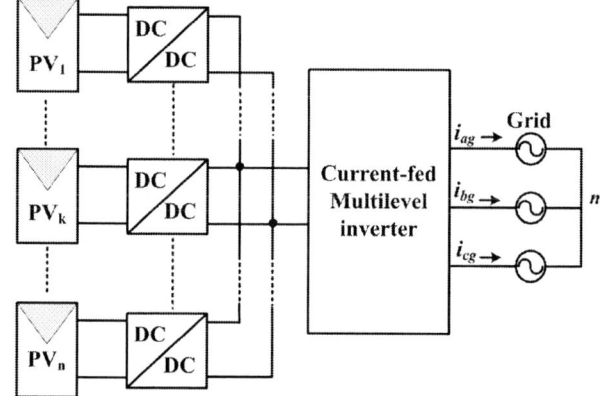
Fig. 1: Multi-string configuration for large scale photo-voltaic power generation.

I. INTRODUCTION

Generation of electricity from solar energy is rapidly growing in recent years due to continual decline in the cost of photo-voltaic (PV) cells. Large scale photo-voltaic (LSPV) power plants are installed across the world by growing group of companies [1]. The power conditioning stage is the major component of LSPV system. It involves collection of power from various photo-voltaic modules using suitable configuration and integrating the total power into the grid. Centralized and multi-string configurations are two widely researched configuration in the solar-grid integration [2]. In this paper, multi-string configuration has been chosen as it extracts maximum power efficiently compared to centralized configuration. This configuration has dedicated dc-dc converter for each string and one central grid-tied inverter as shown in Fig. 1. The dc-dc converter handles only fraction of power whereas the dc-ac inverter handles the total power of solar energy conversion system.

High power inversion stage can be achieved using voltage source inverter (VSI) or current source inverter (CSI). In high power applications such as megawatt-motor-drive, the

CSI has been commercially utilized due to their advantages like transformer-less operation, constant input current, direct output current control and reliable over-current/short-circuit protection [3]. However, high power drive applications utilize current source rectifier along with large dc chokes to achieve constant input current. The large dc chokes leads to poor dynamic performance. In case of LSPV power plant with multi-string configuration, the constant dc current is provided by the dc-dc converters [4], hence the large dc-chokes can be avoided. Therefore, CSI is a suitable high power inverter for LSPV power plants.

For high power applications, thermal constraints of semiconductor devices and requirement on higher efficiency of the inverter limits the device switching frequency to a lower value (< 1 kHz). However, low device switching frequency operation leads to higher harmonic distortion of output currents. Therefore, the challenge is to minimize the harmonic distortion of output current at low device switching frequency operation. Classical modulation techniques such as sinusoidal pulse-width modulation (SPWM) and space vector modulation (SVM) techniques require higher device switching frequency to achieve better quality of output current waveforms [5]–[7]. Some of the notable low device switching frequency modulation techniques include selective harmonic elimination (SHE) technique, staircase modulation technique, synchronous

978-1-4673-9551-9/16 $31.00 © 2016 IEEE

optimal pulse-width modulation technique (SOP) and model predictive control (MPC) technique. Among them, SOP and MPC techniques have better steady-state as well as dynamic performance [8]. However, key issue of MPC technique is higher computational burden.

SOP is an emerging low device switching frequency modulation technique that has been successfully implemented in commercial medium voltage (MV) drives [9]. It is based on an off-line optimization technique to compute switching patterns that minimize harmonic distortion of inverter output currents. Till now, implementation of SOP technique has been limited to voltage-fed multilevel converter (MLC) topologies [6]. The state-of-the-art SOP technique requires some modifications to modulate current source topologies. Because, the operation of CSI imposes constraints on switching commutation. The goal of our research is to propose a modified SOP technique to include the operational constraints of CSI. The proposed modulation technique is implemented for voltage boost current source operation and total harmonic distortion (THD) of output current is achieved within grid integration standards, while limiting the device switching frequency to 650 Hz.

The contents of the paper are organized as follows: Circuit topology and operation of three-level (3L) CSI is discussed in Section II; Modified SOP technique for current source operation is proposed in Section III; Experimental results are demonstrated in Section IV to validate the proposed technique.

II. TOPOLOGY AND OPERATION OF 3L CSI

The topology of 3L CSI is shown in Fig. 2. It consists of three phase legs and each leg has one top, S_{xp} and one bottom S_{xn} semiconductor device with series power diodes D_{xp}, and D_{xn} ($x \in a, b, c$). The series diode and semiconductor device combination provides a reverse voltage blocking capability and unidirectional input-current flow of the inverter. When the inverter is supplied with a constant dc current I_{dc}, it can produce output current i_x with three levels I_{dc}, 0 and $-I_{dc}$. In order to assist smooth commutation of power semiconductor devices, CSI requires three phase capacitors C_{af}, C_{bf}, and C_{cf}. For instance, when power semiconductor device S_{ap} is turned off, the inverter output current i_a falls to zero for a very short period of time. Then, the capacitor provides a current path for the energy trapped in the grid side inductance.

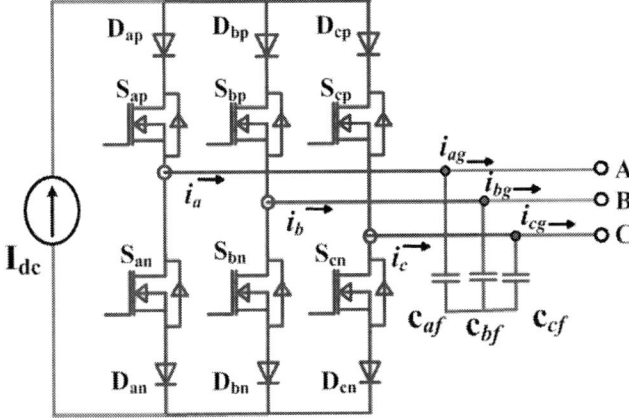

Fig. 2: 3L CSI topology.

TABLE I
SWITCHING STATES AND OUTPUT CURRENT OF 3L CSI

State	I_a	I_b	I_c	S_{ap}	S_{an}	S_{bp}	S_{bn}	S_{cp}	S_{cn}
1	0	I_{dc}	$-I_{dc}$	0	0	1	0	0	1
2	I_{dc}	$-I_{dc}$	0	1	0	0	1	0	0
3	I_{dc}	0	$-I_{dc}$	1	0	0	0	0	1
4	$-I_{dc}$	0	I_{dc}	0	1	0	0	1	0
5	$-I_{dc}$	I_{dc}	0	0	1	1	0	0	0
6	0	$-I_{dc}$	I_{dc}	0	0	0	1	1	0
7	0	0	0	1	1	0	0	0	0
8	0	0	0	0	0	1	1	0	0
9	0	0	0	0	0	0	0	1	1

Otherwise, a high-voltage spike will appear across the semi-conductor devices which damages the devices. In addition, capacitor also acts as a harmonic filter, improving the grid current waveforms.

The operation of CSI imposes following constraints [3] :
1) continuous path for input DC current i_{dc}
2) inverter output current i_x should be defined

These constraints imply that at any instant of time, there are only two semiconductor devices conducting, i.e., one of the top devices S_{xp} and one of the bottom devices S_{xn}. Owing to these constraints, the number of possible switching states with six semiconductor devices is reduced from 27 to 9. The switching states and corresponding output currents for the CSI are given in Table I. It should be observed from the Table I that the summation of three phase current levels is zero for each switching state.

III. MODULATION AND CONTROL

The modulation and control of the CSI involve identifying optimal reference signal and realizing it by assigning gating pulse for each semiconductor device. The optimal reference signal is obtained through SOP method and a conversion method is utilized to include constraints mentioned in Section II. Conversion method and SOP optimization technique are given next.

A. Conversion method

The purpose of the conversion method is to transform a reference signal into realizable waveform that satisfies the operation constraints mentioned in Section II, while maintaining the objective of the reference signal. It should be noticed from Table I that, at any instant of time, the sum of three phase output current levels should be zero. For example, the inverter needs to be modulated in order to produce three-phase output current waveforms, $i_{a,ref}$, $i_{b,ref}$, and $i_{c,ref}$, as shown in Fig. 3. The equation that governs the constraint is given by,

$$i_{a,ref} + i_{b,ref} + i_{c,ref} = 0 \qquad (1)$$

where, $i_{a,ref}$, $i_{b,ref}$, and $i_{c,ref}$ are reference currents for three phases with 120° rotation. For a small time interval Δt, the sum of these three waveforms gives a non-zero value 1. Therefore, these references cannot be utilized for operating 3L CSI.

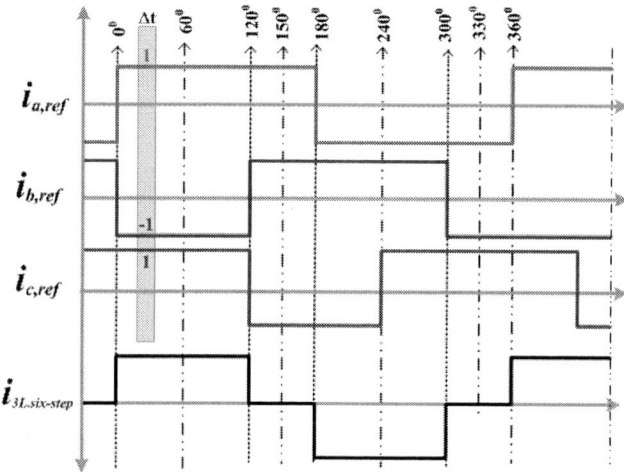

Fig. 3: Reference waveforms considered for analysing conversion method.

Fourier-series analysis demonstrates that the sum of three waveforms with 120° phase shift will have only third order harmonic components $(3, 6, 9, 12, ..)$. Hence, in order to satisfy (1), the current reference waveform should not contain third order harmonics. Elimination of third order harmonic component on any reference signal can be achieved by delaying it to 120^0 and subtracting the delayed waveform from actual waveform. The signal $i_{3L,six-step}$ ($i_{a,ref}$-$i_{b,ref}$) can be utilized as a constraint reference signal for operating 3L CSI. The suffix $six - step$ refers to the highest possible 3L waveform obtained from conversion of two-level (2L) signal (i.e.) amplitude modulation index, m=1. It should be noted that conversion of 2L reference signals produces 3L waveforms. The fundamental component magnitude of 3L signal is $\sqrt{3}$ times that of 2L signal.

B. Modified SOP for current source operation

SOP technique is the combination of synchronous PWM and optimization. Initially the technique has been proposed for 3L VSI topology [10], and later it has been generalized for voltage-fed MLC with any number of voltage levels [6]. SOP is applied for low device switching frequency applications where device switching frequency f_s and sinusoidal control signal frequency f are synced with each other, i.e. $\frac{f_s}{f}$ is an integer. The synchronization is utilized to eliminate subharmonic frequencies, which are undesirable in many applications. The first step in SOP technique is to determine the number of switching angles for each steady state operating point and then optimization is performed to obtain switching angles that minimize the harmonic distortion of inverter output currents. In the last step, optimal switching angles are assigned to each semiconductor device to realize optimal current waveforms based on a systematic procedure.

The conventional SOP technique requires some modifications in order to modulate 3L CSI. It can be achieved by two methods: direct and in-direct. In the direct method, the optimization algorithm should be modified to directly obtain 3L optimal switching angles that satisfies operational constraints of 3L CSI topology. In the in-direct method, the last step of SOP technique need to be modified to convert 2L optimal switching angles into 3L optimal switching angles

[11], and assign them to each power semiconductor device . In this paper, the second approach has been implemented for SOP modulation of 3L current-fed MLC topology.

1) Mathematical analysis: Consider 2L reference current waveform i_{2L} shown in Fig. 4 with switching angles at α_1 to α_6 in a quarter period. To eliminate all even order harmonics half-wave and quarter-wave symmetries is introduced in the switching pattern. Using Fourier series analysis, harmonic components of i_{2L} can be obtained as,

$$i_{k,2L} = \frac{I_{peak}}{k\pi}\left(1 + \sum_{i=1}^{N_\alpha} s(i)cos(k\alpha_i)\right) \quad (2)$$

where, I_{peak} represents the peak value of reference current waveform, k is the harmonic order (k=1,3,5,7. . .), $i_{k,2L}$ is the amplitude of kth harmonic current component of i_{2L}, N_α is the number of switching angles for 2L waveform in a quarter period, $s(i)$ represents the slopes of switching transients at switching angles α_i, $s(i)=\pm2$ when switching to a positive or to a negative current, respectively. The total harmonic rms current of i_{2L} is given by,

$$i_{h,2L} = \sum_k \sqrt{i_{k,2L}^2} \quad (3)$$

$$i_{h,2L} = \frac{I_{peak}}{\pi}\sqrt{\sum_k \frac{1}{k^2}\left(1 + \sum_{i=1}^{N_\alpha} s(i)cos(k\alpha_i)\right)^2} \quad (4)$$

The switching angles are optimized to reduce the harmonic distortion. In order to eliminate the system parameters in optimization function, distortion factor d is obtained as follows [10],

$$d = \frac{i_{h,2L}}{i_{h,2L,six-step}} \quad (5)$$

where $i_{h,2L}$ represents the total harmonic content of 2L reference current at a given operating point and $i_{h,2L,six-step}$

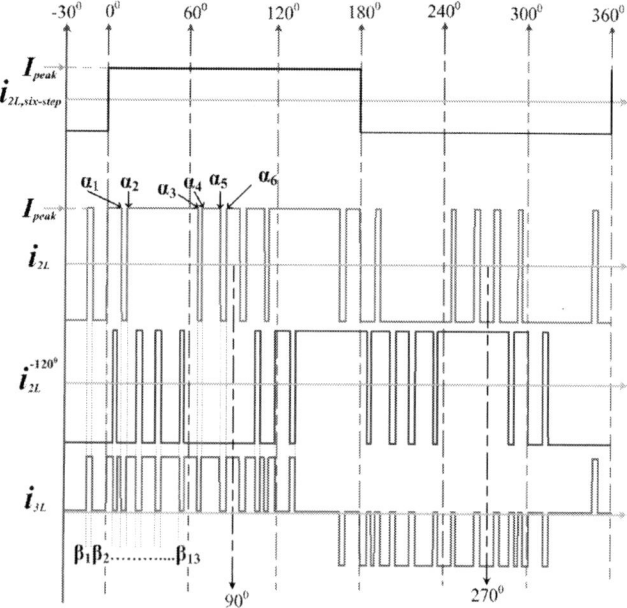

Fig. 4: 2L and 3L waveforms with optimal switching angles.

978-1-4673-9551-9/16 $31.00 © 2016 IEEE

represents the total harmonic content of 2L six-step wave-form $i_{2L,six-step}$ in Fig. 4. The total harmonic contents of $i_{2L,six-step}$ is given by,

$$i_{h,2L,six-step} = \frac{I_{peak}}{\pi}\sqrt{\sum_k \frac{1}{k^2}} \qquad (6)$$

After simplifying (5) and (6), the final expression for d is obtained as,

$$d = \frac{\sqrt{\sum_k \frac{1}{k^2}\left(1 + \sum_{i=1}^{N_\alpha} s(i)cos(k\alpha_i)\right)^2}}{\sqrt{\sum_k \frac{1}{k^2}}} \qquad (7)$$

From (7), it should be noted that, d is dependent only on k, $s(i)$ and α_i. SOP technique utilizes optimization algorithm for determining 2L switching angles α_i to minimize the value of d in (7). However these 2L switching angles can not be utilized for operating CSI. Hence these 2L angles need to be converted into 3L angles by utilizing the conversion method explained in Section III-A. For example, consider the 2L angles α_1, α_2, ...α_6, of waveform i_{2L} in Fig. 4 are the optimal switching angles obtained from SOP technique. On applying the conversion method, these six angles are converted into thirteen 3L angles β_1, β_2,....β_{13}. The 3L waveform i_{3L} with the resultant thirteen angles is shown in Fig. 4.

In general, the conversion of N_α number of 2L angles returns/obtains N_β number of 3L angles as $N_\beta = (2*N_\alpha+1)$.

The relation between device switching frequency f_s and selection of pulse number N_α is given by

$$N_\alpha = \frac{\left[\text{floor}(\frac{f_s}{f}) - 1\right]}{2} \qquad (8)$$

The total harmonic content of i_{3L} is obtained as, $i_{h,3L} = \sqrt{3}i_{h,2L}$. The maximum value of fundamental component on inverter output current can be obtained if 3L CSI is modulated with reference signal as 3L six-step waveform $i_{3L,six-step}$ in Fig. 3, i.e. $m=1$. The total harmonic content of $i_{3L,six-step}$ is given by $i_{3L,six-step} = \sqrt{3}i_{h,2L,six-step}$. The distortion factor d with respect to 3L waveform is given by,

$$d = \frac{i_{h,3L}}{i_{h,3L,six-step}} = \frac{i_{h,2L}}{i_{h,2L,six-step}} \qquad (9)$$

Hence, the value of distortion factor is unchanged with the conversion operation.

SOP technique requires optimal switching patterns to be calculated offline for all steady-state operating points. The modulation index m is defined as,

$$m = \frac{i_{1,2L}}{i_{1,2L,six-step}} \qquad (10)$$

where $i_{1,2L,six-step}$ is the amplitude of fundamental component of $i_{2L,six-step}$ and $i_{1,2L}$ is the fundamental amplitude of i_{2L} at the given operating point.

To obtain the desired fundamental amplitude of inverter output current, the switching angles should satisfy the following equality constraint.

$$m = \frac{i_{1,2L}}{i_{1,2L,six-step}} = \left(1 + \sum_{i=1}^{N_\alpha} s(i)cos(k\alpha_i)\right) \qquad (11)$$

2) *Inverter Control:* A detailed signal flow graph that explains control algorithm is shown in Fig. 5. Angle-selection in the control flow utilizes the magnitude of the reference current vector i_{ref} to select optimal 3L pattern $P(m, N_\alpha)$ that consists of optimized switching angles along with switching transitions $s(i)$. The optimal 3L pattern $P(m, N_\beta)$, phase angle of reference current vector and fundamental frequency f, are given as input to modulator which generates 3L switching state vector $i_k^{(3L)}$. The state-selector utilizes the redundant states in Table I to minimize the overall switching transition and produce the gating signals for each semiconductor device of CSI. The mechanism of redundant state selection can be referred to [12].

3) *Optimization Algorithm:* The optimized angles are generated for $0 < m < 1$ by modified SOP optimization algorithm in order to minimize d. The flowchart of modified SOP optimization algorithm is shown in Fig. 6. The constraints of optimization for CSI are as follows:

a) Sufficient gap (10 µs) between consecutive switching angles to allow for minimum ON times and OFF times of the power semiconductor devices;

b) To maintain current modulation index value that satisfies the relation (11).

The step-by-step procedure of proposed modified SOP algorithm are as follows:

i) Calculate number of 2L pulses required for desired device switching frequency f_s using (8),

ii) For each modulation index m, MATLAB function "randn" is used to generate the initial values of switching angles for further optimization while satisfying the relation (11).

iii) The gradient method 'FMINCON', a built-in MATLAB function is used for obtaining optimized switching patterns for each modulation indexes. For distortion-factor calculations, only harmonic components up to 100 are considered.

iv) The optimization loop runs for modulation index range $(0 < m < 1)$.

v) If switching angles for consecutive modulation index values differ by more than 5 degrees, post-optimization is performed starting with optimized switching angles as initial values. Due to post-optimization, transients in output currents are reduced but the distortion factor d is slightly compromised.

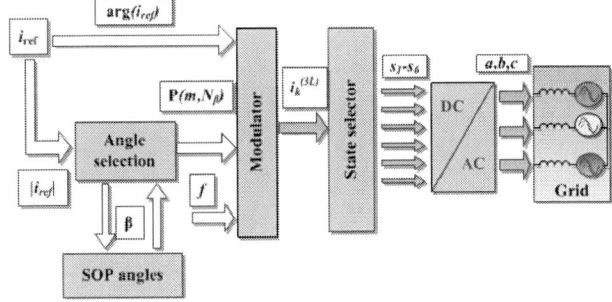

Fig. 5: Modified SOP control Flow for 3L CSI.

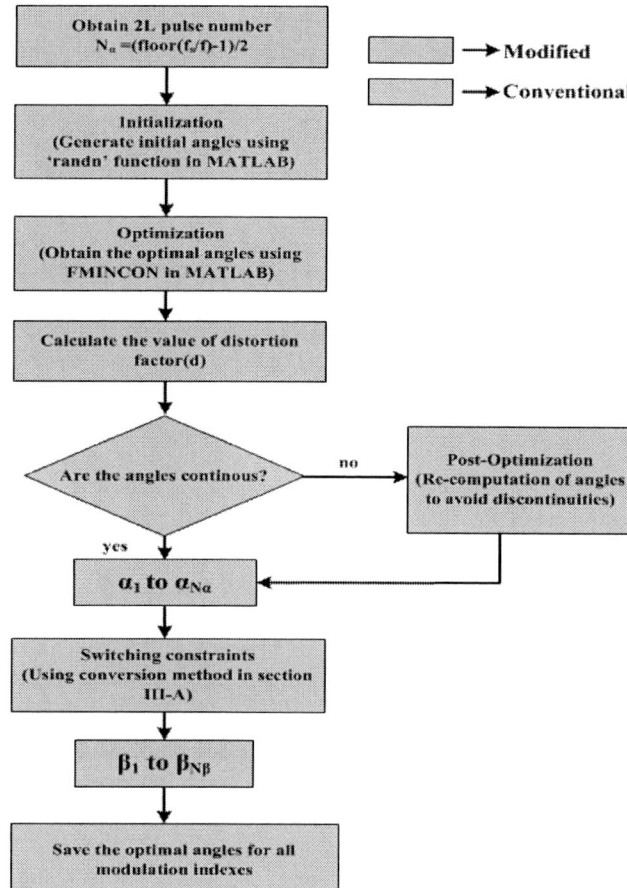

Fig. 6: Modified SOP optimization method.

vi) The final optimal switching angles are stored as complete patterns $P(m, N_\beta)$ in a DSP and they are retrieved during real time operation depending on the output current to be delivered to the grid.

4) Optimization Results: The distortion factor for 3L CSI for $0 < m < 1$ is shown in Fig. 7. It should be observed that distortion factor approached unity at $m = 1$. This is because inverter output will be similar to six-step waveform $i_{3L,six-step}$ shown in Fig. 3.

In addition, due to reduced harmonic distortion, the proposed modified SOP technique reduces output filter size needed for CSI in Fig. 1. In case CSI is modulated with classical techniques such as SPWM, the harmonic content of

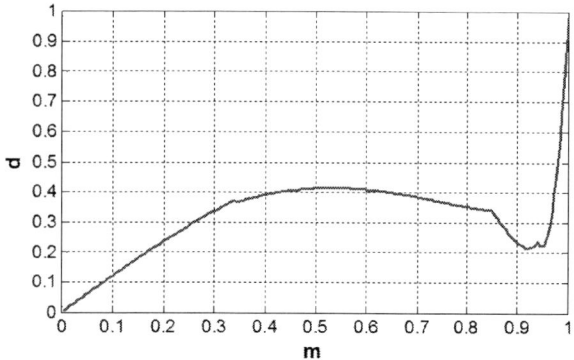

Fig. 7: Distortion factor d versus modulation index m for 3L CSI.

TABLE II
DEVICES AND PARAMETERS

Device/ Parameter	Part number/ Value
Converter power	500 W
Input DC Current	4 A
Switching frequency	650 Hz
AC output Frequency	50 Hz
Load Resistance	15.6 Ω
Load Inductance	10 mH
Power Diode	STTH30R04-Y
	I_f 30 A, V_{RRM} 400 V
MOSFET device	IPI60R199CP
	V_{DS} 650 V, I_D 16 A
Six pack driver	SKHI 61R

TABLE III
2L SOP SWITCHING ANGLES

m	α_1	α_2	α_3	α_4	α_5	α_6
0.949	3.82	6.84	11.65	14.06	21.13	22.86
0.502	12.60	15.12	63.86	70.63	79.08	86.57
0.3098	13.86	15.33	62.50	72.48	77.55	87.84

3L CSI output current i_a appears at multiples of switching frequency f_s, hence the capacitor value is decided by the switching frequency. However, in SOP technique the harmonic components are shifted to frequencies higher than the switching frequency as shown in Fig. 8. Hence, the requirement on output filter capacitor value with modified SOP technique is reduced.

IV. EXPERIMENTAL RESULTS

The proposed SOP technique was implemented for controlling a 3L CSI supplied with 500 W constant input current source using Chroma solar array simulator 62020H-150S. The experimental setup of 3L CSI with RL load is shown in Fig. 9. Each leg of the CSI was implemented by using discrete MOSFET device from Infineon and a six-pack driver SKHI 61R from Semikron was used as a driver. The switching signals were programmed on a Texas Instrument TMS320F28335 and a sampling frequency of 20 kHz is utilized. Table II shows the list of major components along with their parameters.

Proposed modified SOP technique is utilized for generating optimal switching angles at three different operating points: ($m = 0.949$, $N_\alpha = 6$), ($m = 0.502$, $N_\alpha = 6$), and (m = 0.3098, $N_\alpha = 6$). The 2L and 3L angles for these three operating points are shown in Table III and IV, respectively.

The pulse number for the operating points are maintained at N_β=13, so the device switching frequency of all semiconductor devices should be equal to 650 Hz (8). The gating signals for all the six semiconductor devices of CSI for three operating points are shown in Fig. 10 (a)-(c), and it is clear from the waveforms that each semiconductor devices is turned ON and OFF for 13 times within one fundamental cycle, i.e. switching frequency is 650 Hz.

The waveforms of (1) inverter output current and its FFT spectrum, (2) filtered output current for 3L current source

978-1-4673-9551-9/16 $31.00 © 2016 IEEE

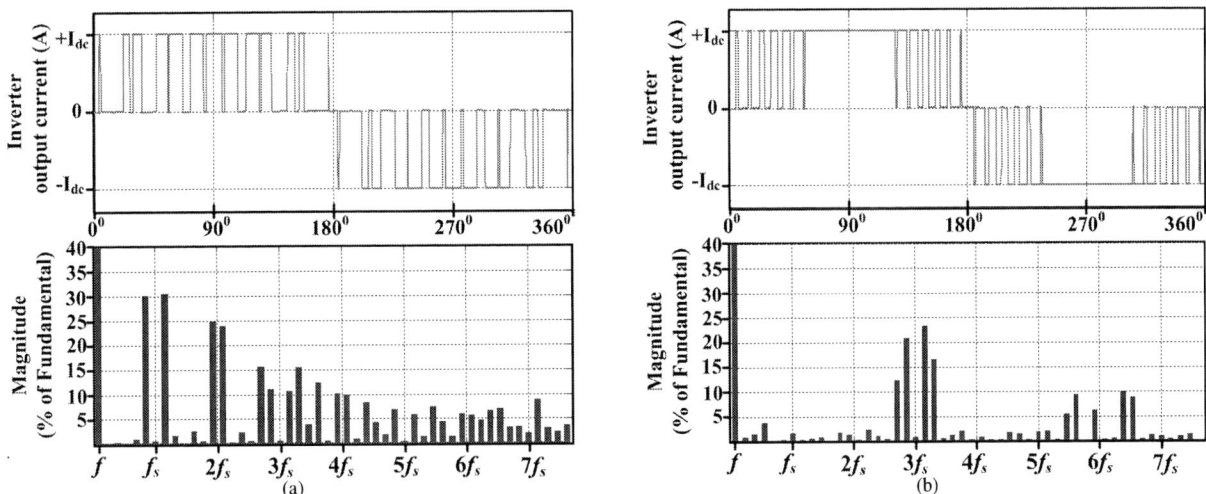

Fig. 8: Modulation technique comparison (a) Inverter output current and FFT spectrum using SPWM modulation. (b) Inverter output current and FFT spectrum using SOP technique.

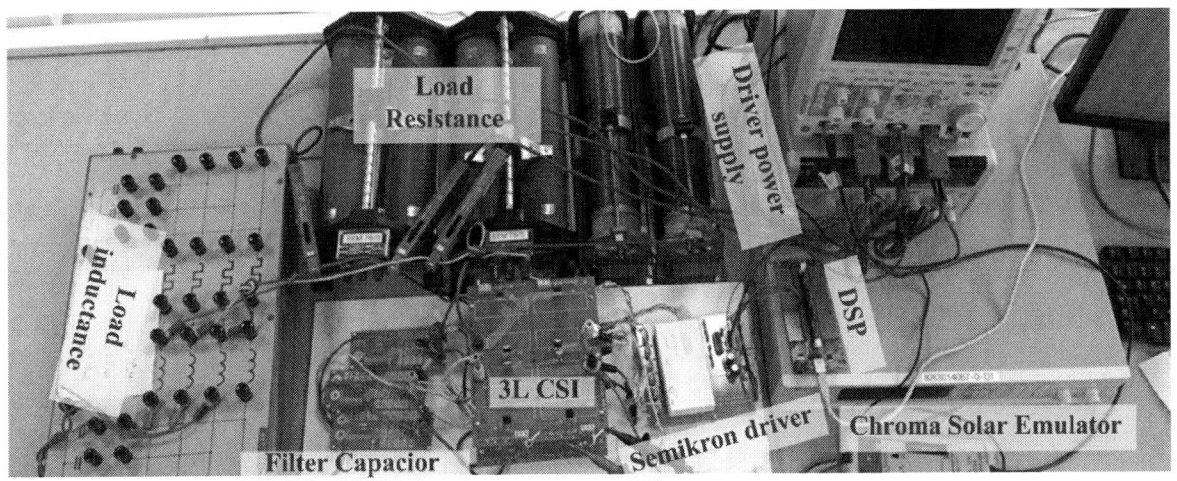

Fig. 9: Experimental setup of 3L CSI.

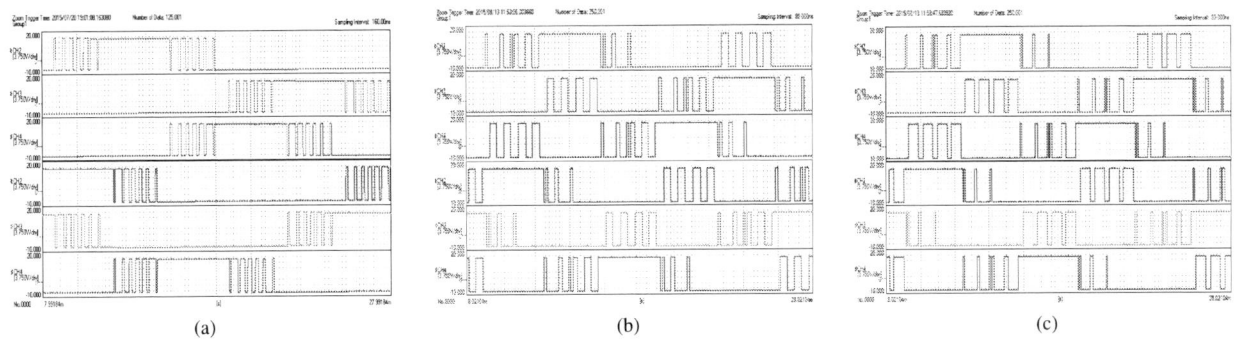

Fig. 10: Inverter gating pulses (Y-axis: 3.75 V/div; X-axis: 2 ms/div) (a) $m = 0.949$. (b) $m = 0.502$. (c) $m = 0.3098$.

TABLE IV
3L SOP SWITCHING ANGLES

m	β_1	β_2	β_3	β_4	β_5	β_6	β_7	β_8	β_9	β_{10}	β_{11}	β_{12}	β_{13}
0.949	7.14	8.87	15.94	18.34	23.16	26.18	30.00	33.82	36.84	41.66	44.06	51.13	52.86
0.502	14.87	17.40	30.00	33.86	40.63	42.60	45.12	49.08	56.57	63.43	70.92	79.37	86.14
0.3098	14.67	16.14	30.00	32.50	42.48	43.86	45.33	47.55	57.84	62.16	72.45	77.52	87.50

978-1-4673-9551-9/16 $31.00 © 2016 IEEE

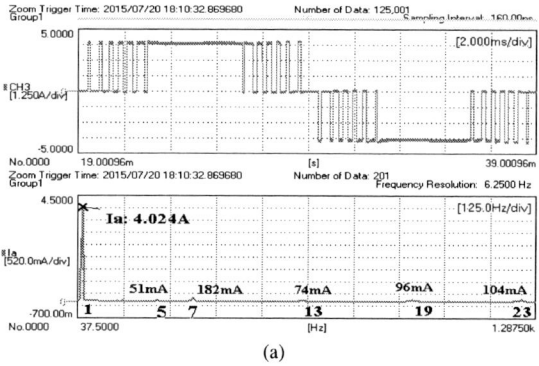 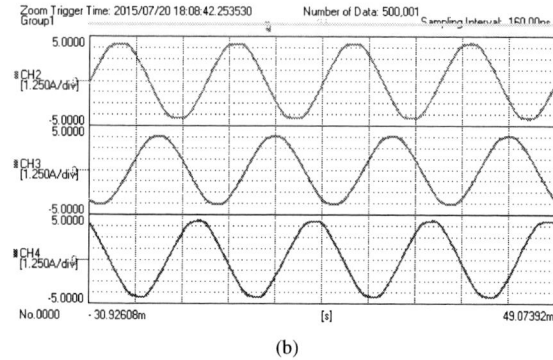

(a)   (b)

Fig. 11: Experimental results for $m = 0.949$ (a) Inverter output current, i_a (Y-axis: 1.25 A/div, X-axis: 2 ms/div) and its FFT spectrum (Y-axis: 0.52 A/div, X-axis: 125 Hz/div). (b) Filtered currents, i_{ag}, i_{bg}, i_{cg} (Y-axis: 1.25 A/div, X-axis: 8 ms/div).

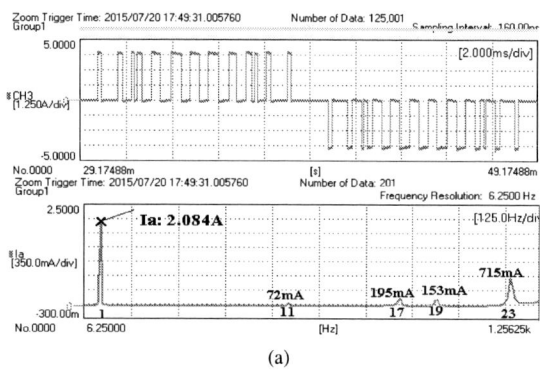 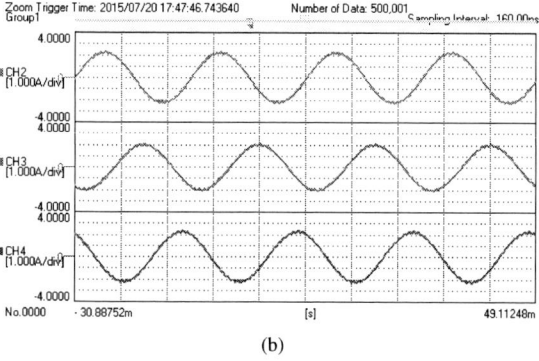

(a)   (b)

Fig. 12: Experimental results for $m = 0.502$ (a) Inverter output current, i_a (Y-axis: 1.25 A/div, X-axis: 2 ms/div) and its FFT spectrum (Y-axis: 0.35 A/div, X-axis: 125 Hz/div). (b) Filtered currents, i_{ag}, i_{bg}, i_{cg} (Y-axis: 1 A/div, X-axis: 8 ms/div).

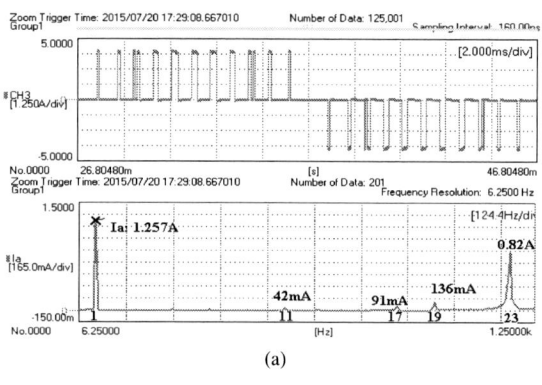 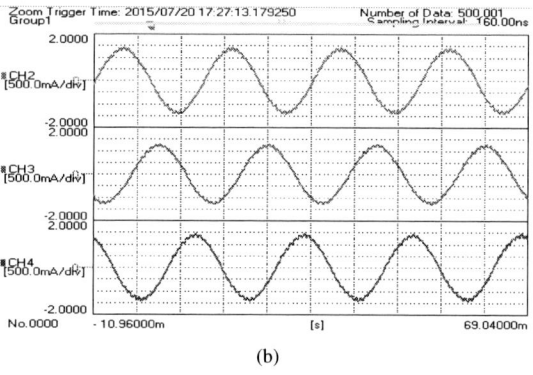

(a)   (b)

Fig. 13: Experimental results for $m = 0.3098$ (a) Inverter output current, i_a (Y-axis: 1.25 A/div, X-axis: 2 ms/div) and its FFT spectrum (Y-axis: 0.16 A/div, X-axis: 125 Hz/div). (b) Filtered currents, i_{ag}, i_{bg}, i_{cg} (Y-axis: 0.5 A/div, X-axis: 8 ms/div).

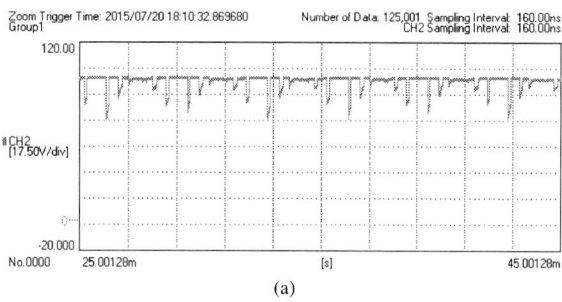 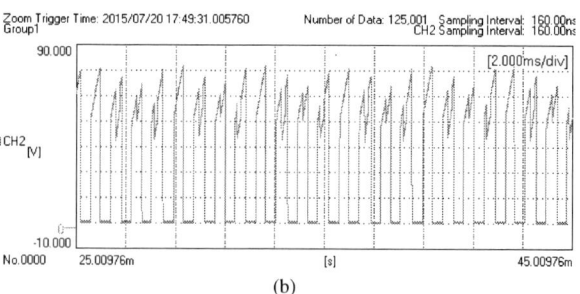

(a)   (b)

Fig. 14: Input voltage of 3L CSI (X-axis: 2 ms/div) for (a). $m = 0.949$ (Y-axis : 17.5 V/div). (b) $m = 0.502$ (Y-axis : 12.5 V/div).

inverter pertaining to operating point ($m = 0.949$, $N_\alpha = 6$, $f = 50Hz$) are shown in Fig.11 (a)-(b), respectively. It could be noticed from the FFT spectrum that the lower order harmonic components such as 5th, 7th, 13th, 19th and 23rd of the inverter output current are infinitesimal compared to the amplitude of fundamental. The magnitude of these harmonic components are also shown in Fig. 11 (a). It should also be noticed that the even order harmonic and third order harmonic components are eliminated. The filtered line current i_{ag} of 3L CSI is nearly sinusoidal although the device switching frequency is reduced to 650 Hz. The THD of the filtered current is obtained as 2.4%.

Similar observations about inverter output current, FFT spectrum and filtered currents for operating points $m = 0.502$, $m = 0.3098$ are made from Fig. 12 to Fig. 13. It should be noticed that lower order harmonic components (<1 kHz) of inverter output current are infinitesimal for these operating points. At lower modulation index $m = 0.3098$, the magnitude of 23rd harmonic component is comparatively higher that of other three operating points. However, with the same capacitive filter the grid current is maintained to near sinusoidal. The THD of output filtered currents for these two operating points are obtained as, 3.1%, and 3.6%, respectively. These values are well below ($< 5\%$) the PV grid integration standard [13].

The waveforms of input voltages for operating points $m = 0.949$ and $m = 0.502$, are shown in Fig. 14. It should be noticed that there are no voltage spikes during switching transitions. This has been achieved by providing sufficient overlap (1µs) between switching transition. It should also be noticed that, input voltage for operating point $m = 0.502$ is zero for short intervals of time, which is due to turning on both top (S_{xp}) and bottom (S_{xn}) semiconductor devices of one phase leg.

V. CONCLUSION

For higher power applications like large scale solar to grid integration, thermal constraint and efficient operation call for low switching frequency operation of semiconductor devices. In this paper, a modified synchronous optimal pulse-width modulation technique has been proposed, analysed and implemented for operating three-level current-fed multilevel converter topology at low device switching frequency. A laboratory prototype has been designed, developed and tested at 500 W to verify the proposed technique. Experimental results demonstrated the effectiveness of the proposed method and from the experimental results, it should be noticed that the inverter output current is nearly sinusoidal. The THD of line current has been maintained below 5% at all operating points of power flow.

REFERENCES

[1] (2014, August) Top 400 Solar Contractors. [Online]. Available: http://www.solarpowerworldonline.com/2014-top-400-solar-contractors/

[2] S. Kouro, J. Leon, D. Vinnikov, and L. Franquelo, "Grid-Connected Photovoltaic Systems: An Overview of Recent Research and Emerging PV Converter Technology," *IEEE Ind. Electron. Mag.*, vol. 9, no. 1, pp. 47–61, March 2015.

[3] B. Wu, J. Pontt, J. Rodriguez, S. Bernet, and S. Kouro, "Current-Source Converter and Cycloconverter Topologies for Industrial Medium-Voltage Drives," *IEEE Trans. Ind. Electron.*, vol. 55, no. 7, pp. 2786–2797, July 2008.

[4] V. Vekhande and B. Fernandes, "Central multilevel current-fed inverter with module integrated DC-DC converters for grid-connected PV plant," in *IEEE Energy Convers. Cong. and Expo. (ECCE '13)*, Sept 2013, pp. 1933–1940.

[5] J. Holtz and X. Qi, "Optimal Control of Medium-Voltage Drives-An Overview," *IEEE Trans. Ind. Electron.*, vol. 60, no. 12, pp. 5472–5481, Dec 2013.

[6] A. K. Rathore, J. Holtz, and T. Boller, "Generalized Optimal Pulsewidth Modulation of Multilevel Inverters for Low-Switching-Frequency Control of Medium-Voltage High-Power Industrial AC Drives," *IEEE Trans. Ind. Electron.*, vol. 60, no. 10, pp. 4215–4224, Oct 2013.

[7] T. Boller, J. Holtz, and A. Rathore, "Optimal Pulsewidth Modulation of a Dual Three-Level Inverter System Operated From a Single DC Link," *IEEE Trans. Ind. Appl.*, vol. 48, no. 5, pp. 1610–1615, Sept 2012.

[8] A. Edpuganti and A. Rathore, "A survey of low-switching frequency modulation techniques for medium-voltage multilevel converters," in *IEEE Ind. Appl. Soc. Annu. Meeting (IAS '14)*, Oct 2014, pp. 1–8.

[9] T. Boller, J. Holtz, and A. Rathore, "Neutral-Point Potential Balancing Using Synchronous Optimal Pulsewidth Modulation of Multilevel Inverters in Medium-Voltage High-Power AC Drives," *IEEE Trans. Ind. Appl.*, vol. 50, no. 1, pp. 549–557, Jan 2014.

[10] J. Holtz, "Pulsewidth modulation for electronic power conversion," *Proc. IEEE*, vol. 82, no. 8, pp. 1194–1214, Aug 1994.

[11] X. Wang and B.-T. Ooi, "Unity PF current-source rectifier based on dynamic trilogic PWM," *IEEE Trans. Ind. Electron.*, vol. 8, no. 3, pp. 288–294, Jul 1993.

[12] M. Aguirre, L. Calvino, and M. Valla, "Multilevel Current-Source Inverter With FPGA Control," *IEEE Trans. Ind. Electron.*, vol. 60, no. 1, pp. 3–10, Jan 2013.

[13] "IEEE Application Guide for IEEE Std 1547(TM), IEEE Standard for Interconnecting Distributed Resources with Electric Power Systems," *IEEE Std 1547.2-2008*, pp. 1–217, April 2009.

Low Leakage Current Single-Phase PV Inverters with Universal Neutral-Point-Clamping Method

Liwei Zhou, Feng Gao
School of Electrical Engineering
Shandong University
Jinan, China
18769785783@163.com

Abstract—**The transformerless inverters have the advantages of low cost and high efficiency, which attract more attentions in the field of photovoltaic (PV) power generation system. However, the transformerless grid connected system has the main challenge for keeping low leakage current due to the non-isolation configuration. To better attenuate the leakage current in single-phase system, a few of neutral point clamped (NPC) topologies have been proposed. This paper further proposes a kind of universal neutral point clamping method to build the single-phase NPC PV inverters, which only employs one additional switch to connect the midpoint of dc link and the circulating circuit. Since only the leakage current flows through the additional switch, the single-phase NPC inverter could still maintain the high operational efficiency with the significantly attenuated leakage current. The working principles of the proposed method and the corresponding NPC inverters are analyzed in detail. The simulation and experimental results verified the theoretical findings.**

Keywords—common-mode voltage; neutral point clamping circuit; photovoltaic inverter; leakage current

I. INTRODUCTION

The transformerless grid-tied single-phase PV system has a lot of advantages, such as high efficiency, small volume and light weight [1]. However, without the galvanic isolation, the high-frequency fluctuation of common-mode voltage will excite a high level of common-mode leakage current [2]. The leakage current through the stray capacitance of the solar panel threatens the safety of the PV system. In order to eliminate the common-mode leakage current, the most effective method is to make the common-mode voltage constant as half of the DC bus voltage. The bipolar SPWM modulation is a kind of solution. However, the bipolar SPWM has some obvious disadvantages compared to the unipolar SPWM modulation, such as large current ripple and high switching losses. Another solution to attenuate the common-mode current when using the unipolar SPWM method is to construct the extra freewheeling paths in the inverter circuitry. Several single-phase topologies have been proposed to make the common-mode voltage constant, for example, H5, HERIC, and H6 topologies [3-6]. These topologies have the capability of making the common-mode voltage half of the DC bus voltage in the freewheeling modes theoretically. However, the common-mode voltage is actually variable during the freewheeling modes because of the potential fluctuation induced by the charging and discharging of switch junction capacitance [7]. Therefore, a few of neutral-point-clamped circuits have been proposed to clamp the

common-mode voltage to half of DC bus voltage more effectively [7-10] as shown in Fig. 1.

Being different, this paper proposes a universal neutral point clamping method to build two families of single-phase NPC PV inverters by only assuming one additional switch, whose working principles are analyzed in detail. The simulation and experimental results have verified the leakage current attenuation performance of the proposed circuits.

II. A CONCEPT OF UNIVERSAL NPC METHOD

The common-mode circuit of single-phase H-bridge PV inverter is shown in Fig. 2, where the four switches operate at high frequency and the corresponding common-mode voltage can be expressed as [11-14]:

$$u_{cm} = \frac{u_{ao} + u_{bo}}{2} \qquad (1)$$

In the above equation, u_{ao} and u_{bo} are the pulsating voltages between the midpoint of bridges and the negative DC rail, respectively. The common-mode current i_{cm}, which will flow through switches, filter inductor, grid and stray capacitor of PV panel as indicated by the red lines in Fig. 1, is mainly induced by the fluctuation of u_{cm} as expressed in (2) [15-16].

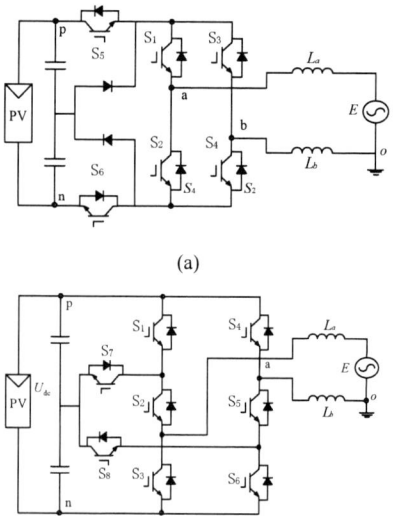

(a)

(b)

Fig. 1. Traditional neutral point clamped topologies proposed in (a) [9] and (b) [10].

978-1-4673-9551-9/16 $31.00 © 2016 IEEE

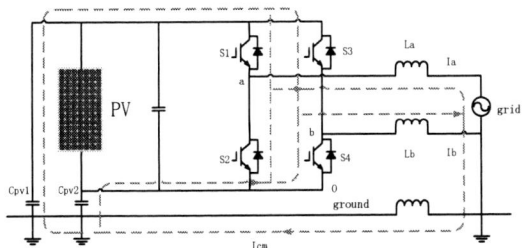

Fig. 2. The common-mode circuit of single-phase H-bridge PV inverter.

$$i_{cm} = C_{PV} \frac{du_{cm}}{dt} \qquad (2)$$

Without the isolation transformer and other attenuation methods, the leakage current may exceed the permissible levels defined by the mandatory standards, e.g. DIN VDE 0126-1-1, and will cause the safety problems and distort the output current [17-18].

Several modified single-phase inverters have the capability of attenuating the common-mode current, such as Heric, HB-ZVR and H6 topologies. These inverters can use the auxiliary switches and diodes to provide the additional freewheeling path to make the common-mode voltage constant during the whole fundamental period in theory. However, because of the charging of switch junction capacitance during the freewheeling operation modes, the potentials of output voltages vary during the freewheeling period and then the common-mode voltage is indeed not constant [7]. Hence, it is necessary to generate an additional neutral-point-clamped path to keep the common-mode voltage constant more effectively.

The design of neutral-point-clamped circuit for a single-phase inverter should follow some principles. Firstly, the clamping circuit should be bidirectional to adjust the potential of the freewheeling paths. Secondly, the split DC capacitances should not be shoot-through by the clamping circuit. Finally,

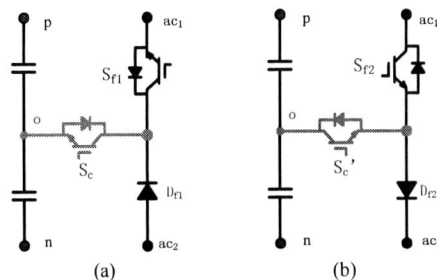

Fig. 3. The neutral point clamping circuits with (a) common collectors or (b) common emitters connection.

the clamping circuits should be connected to the midpoint of the DC link and form the additional freewheeling paths.

Following the principles above, a concept of universal NPC method is proposed to maintain the common-mode voltage strictly constant as half of the DC voltage. The proposed connecting method of the NPC circuits is shown in Fig. 3. The novel NPC circuits consist of two parts: (1) part of the traditional freewheeling circuits used in H6 inverters or HB-ZVR inverter, (2) the extra clamping switch, S_c. It is noted that the clamping switch can be connected to different positions of the freewheeling circuits. Based on the connecting directions of the clamping switches, the NPC circuits can be divided into two types, which are shown in Fig. 3(a) and (b), respectively. To avoid the DC capacitor being shorted, the collectors of S_c, S_{f1} and the cathode of the freewheeling diode, D_{f1}, are connected together as shown in Fig. 3(a), where the clamping switch, S_c, has been highlighted in red. Alternatively, the emitters of S_c', S_{f2} and the anode of the freewheeling diode, D_{f2}, can be connected together as shown in Fig. 3(b), where the clamping switch, S_c', has been highlighted in blue. In the following sections, the NPC circuit in Fig. 3(a) is named as the common-collector NPC circuit while the NPC circuit in Fig. 3(b) is named as the common-emitter NPC circuit. Based on

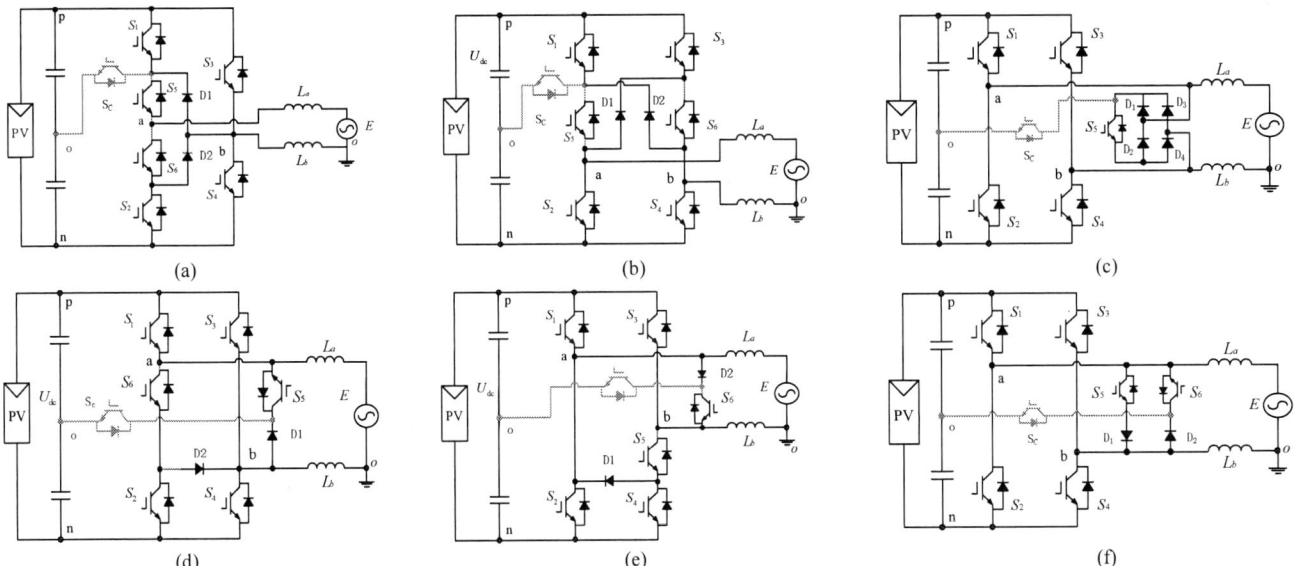

Fig. 4. A family of single-phase inverters with common-collector NPC circuit.

978-1-4673-9551-9/16 $31.00 © 2016 IEEE

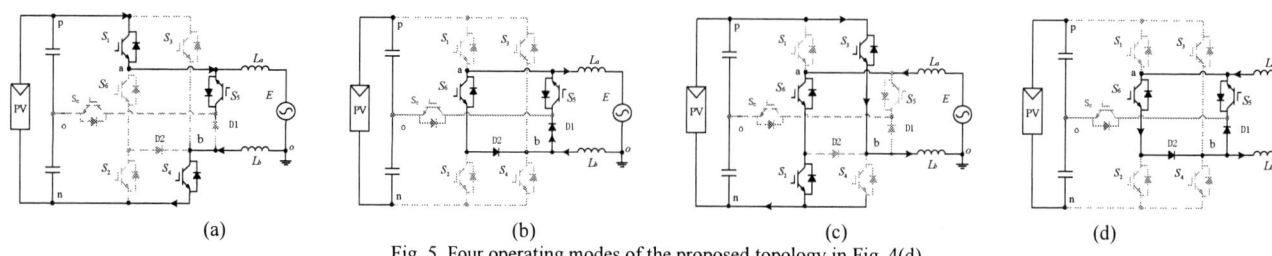

(a)   (b)   (c)   (d)

Fig. 5. Four operating modes of the proposed topology in Fig. 4(d).

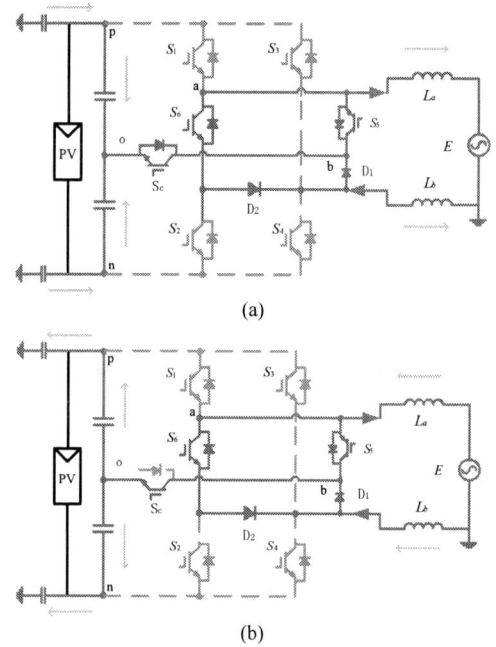

(a)

(b)

Fig. 6. The common-mode current flow paths of the NPC inverter in Fig 4(d) when the potential of the freewheeling circuit is (a) lower or (b) higher than the middle value of DC link voltage.

the proposed universal neutral point connecting method, two families of novel single-phase neutral-point-clamping inverters are proposed in the next Section.

III. FAMILIES OF NOVEL INVERTERS WITH COMMON-COLLECTOR NPC CIRCUITS

This section presents two families of single-phase inverters using the neutral-point-clamped circuits in Fig. 3(a) and (b), respectively, whose circuitry details and operational principles will be fully elaborated below.

A. Single-Phase NPC Inverters with Common Collector Connection

Fig. 4 shows a family of inverters based on the common-collector NPC circuit, where the clamping circuit can be inserted in different positions of the traditional single-phase inverters to form a variation of topologies. It is noted that Fig. 4 are the direct modification for various H6 and HB-ZVR topologies to improve the common-mode voltage behavior.

To generally illustrate the operational principle of the proposed single-phase inverters with the common-collector clamping circuit, the topology of Fig. 4(d) is assumed as an example, which has four operation modes. Fig. 5 shows the

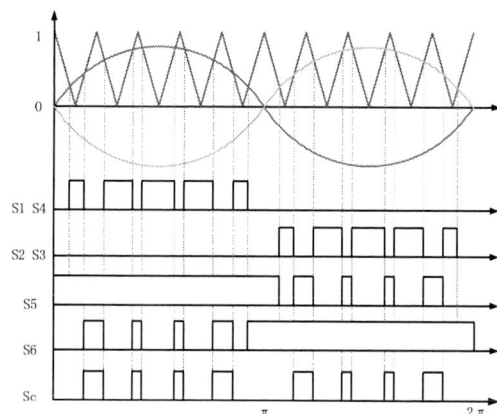

Fig. 7. Modulation illustration for single-phase NPC inverter.

specific current flowing paths during the energy transferring modes and the freewheeling modes.

Mode 1: During the positive half period of modulation reference, S_1 and S_4 are modulated in high frequency, while S_5 is always ON and S_6 and S_c are commanded complementarily to S_1 and S_4. When S_1 and S_4 are on, the current flows through S_1, grid and S_4 successively. The common-mode voltage is

$$u_{cm} = \frac{u_{an}+u_{bn}}{2} = \frac{u_{dc}+0}{2} = \frac{u_{dc}}{2} \qquad (3)$$

Mode 2: When S_1 and S_4 are off, the current flows through D_1, S_5 and grid successively. The common-mode voltage is

$$u_{cm} = \frac{u_{an}+u_{bn}}{2} = \frac{u_{dc}/2+u_{dc}/2}{2} = \frac{u_{dc}}{2} \qquad (4)$$

Mode 3: During the negative half period, S_2 and S_3 are modulated in high frequency, while S_6 is always ON and S_5 and S_c are commanded complementarily to S_2 and S_3. When S_2 and S_3 are on, the current flows through S_3, grid, S_6 and S_2 successively. The common-mode voltage is:

$$u_{cm} = \frac{u_{an}+u_{bn}}{2} = \frac{0+u_{dc}}{2} = \frac{u_{dc}}{2} \qquad (5)$$

Mode 4: When S_2 and S_3 are off, the current flows through S_6, D_2 and grid successively. The common-mode voltage is

$$u_{cm} = \frac{u_{an}+u_{bn}}{2} = \frac{u_{dc}/2+u_{dc}/2}{2} = \frac{u_{dc}}{2} \qquad (6)$$

Besides, when the potential of points a and b falls to lower than that of the midpoint of DC link voltage during freewheeling modes, the common-mode current flows through S_c and its body diode, thus it will balance the potential of freewheeling path as shown in Fig. 6(a). On the other hand, when the potential of points a and b rises to be higher than that

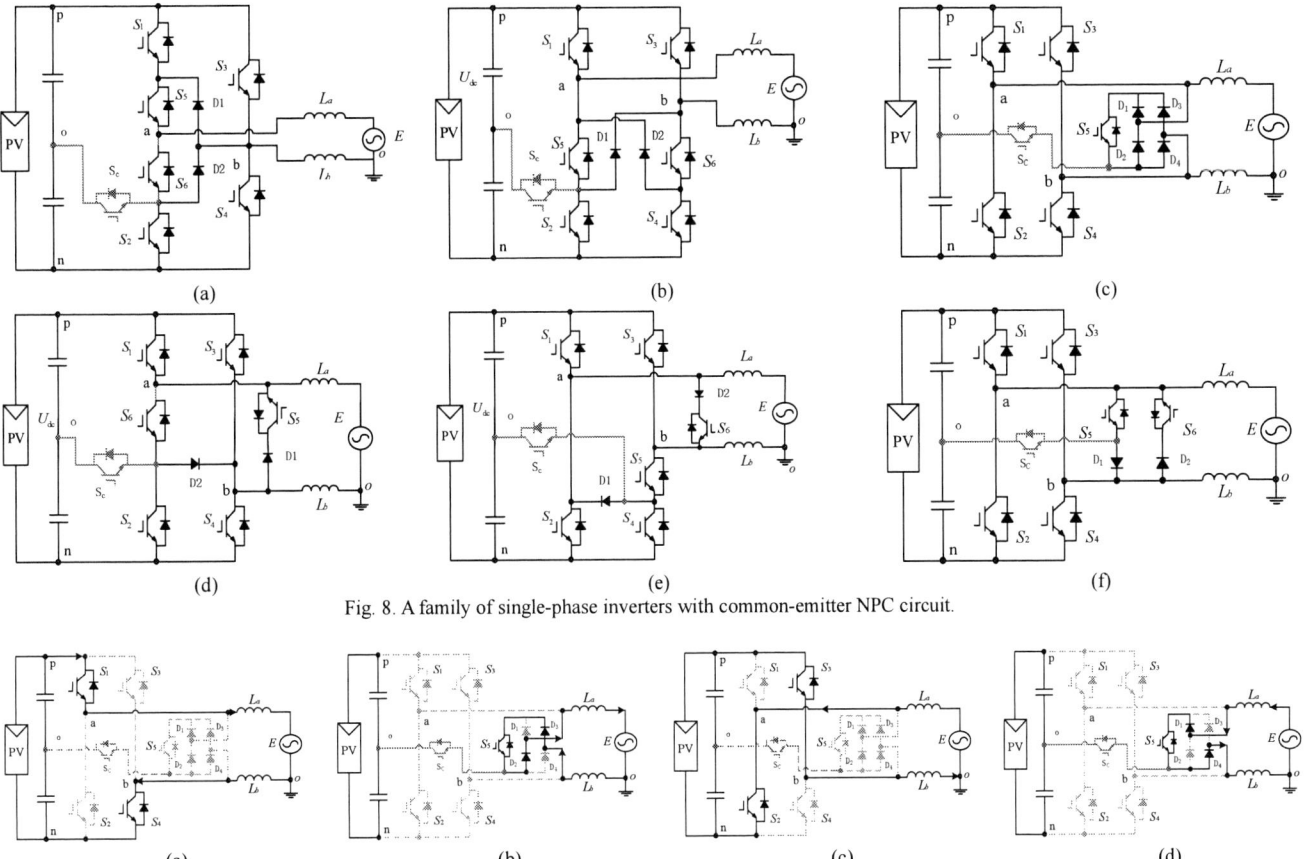

Fig. 8. A family of single-phase inverters with common-emitter NPC circuit.

Fig. 9. (a) The transferring mode and (b) the freewheeling mode during the positive half period; (c) the transferring mode and (d) the freewheeling mode during the negative half period.

of the midpoint of DC link voltage during freewheeling modes, the common-mode current will flow through S_c and balance the potential of two bridges as illustrated in Fig. 6(b). Therefore, the common-mode voltage can keep constant during the whole fundamental period.

The corresponding modulation strategy is shown in Fig. 7, where S_1-S_6 and S_c are the gating signals of the proposed inverter in Fig. 4(d). It is noted that S_c has the same gating commands as S_6 during the first half fundamental period and as S_5 during the second half fundamental period, respectively. Furthermore, the same modulation method is indeed suitable for all proposed single-phase NPC inverters except of the inverter in Fig. 4(c). In Fig. 4(c), S_c and S_5 have the same gating signals, which are complimentary to S_1 and S_4 in the positive modulation period and S_2 and S_3 in the negative modulation period, respectively.

B. Single-Phase NPC Inverters with Common Emitter Connection

This subsection presents another family of the proposed inverters using the neutral-point-clamped circuit in Fig. 3(b). Being similarly, a family of single-phase inverters based on the common-emitter NPC circuit is presented in Fig. 8, whose operational principle can be demonstrated by specifically assuming the topology of Fig. 8(c) as an example. In detail, Fig.

9 shows the current flowing paths during the energy transferring modes and the freewheeling modes of Fig. 8(c). Being similarly, when the potential of points a and b falls to lower than that of the midpoint of DC link in the freewheeling modes, the common-mode current flows through S_c, thus it will balance the potential of freewheeling path. On the other hand, when the potential of points a and b rises to higher than that of the midpoint of DC link, the common-mode current will flow through S_c and its body diode and balance the potential of two bridges either. Thus, the common-mode voltage can keep constant during the whole fundamental period. The same modulation method as shown in Fig. 7 can be employed to control the single-phase NPC inverters with the common-emitter NPC circuit.

It is noted that the proposed topologies shown in Fig. 4(c)-Fig. 4(f) and Fig. 8(c)-Fig. 8(f) will have smaller conduction losses when comparing to the traditional NPC inverters shown in Fig. 1 because less switches will be involved in the current flow path during both energy transferring modes and freewheeling modes. Besides, the additional clamping switch only conducts the leakage current during operation, which means its losses can be neglected.

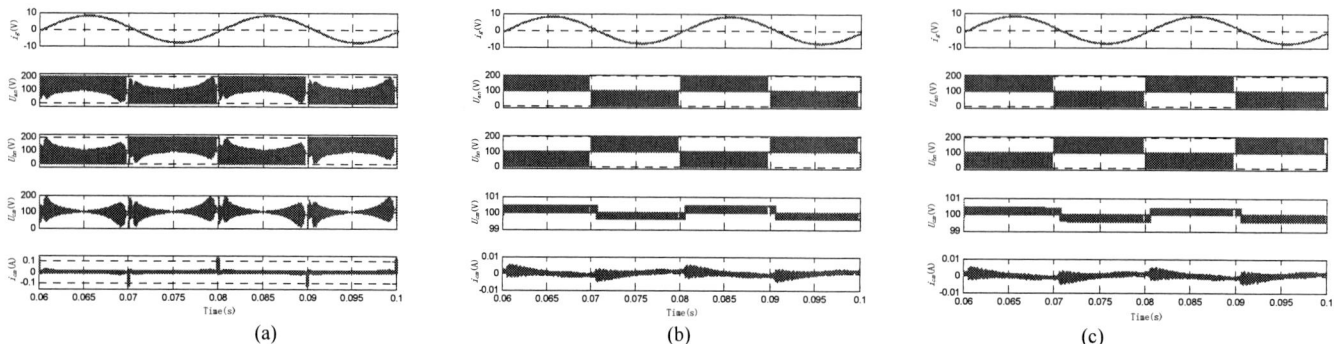

(a)   (b)   (c)

Fig. 10. Comparison of (a) the traditional H6 inverter (b) the proposed inverter in Fig. 4 and (c) the proposed inverter in Fig. 8.

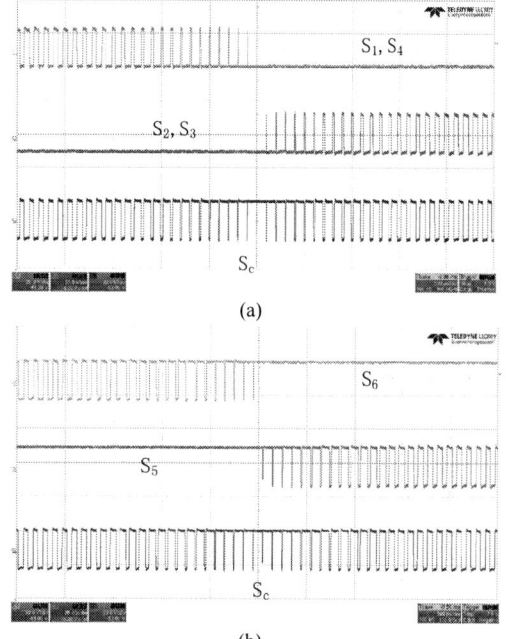

(a)

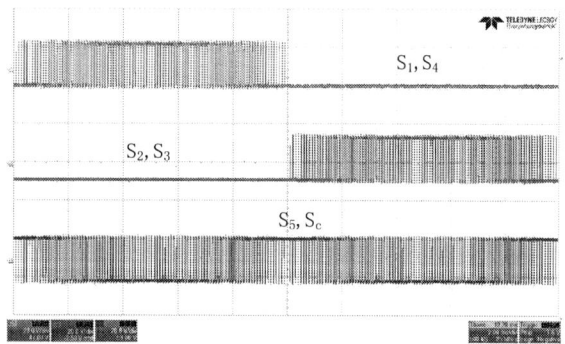

(b)

Fig. 11. The gating signals of (a) S_1-S_4 and S_c, (b) S_5-S_6 and S_c of the proposed inverter in Fig. 4(d).

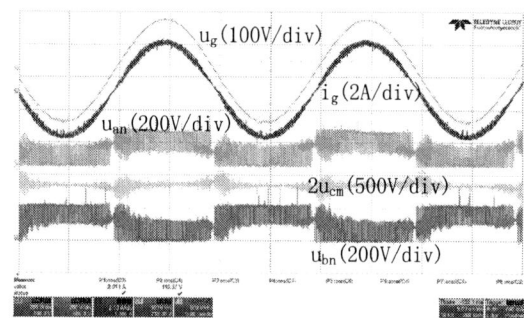

Fig. 13. The experimental waveforms of the traditional H6 inverter with common mode voltage captured.

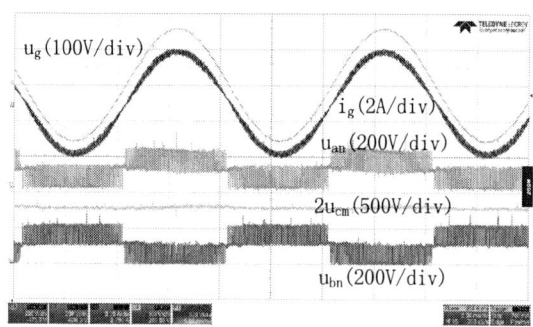

Fig. 14. The experimental waveforms of the proposed NPC inverter with common mode voltage captured.

Fig. 12. The gating signals of the proposed inverter in Fig. 8 (c).

IV. SIMULATION AND EXPERIMENTAL RESULTS

The simulation and experimental results are shown in this section. The proposed inverters in Fig. 4(d) and Fig. 8(c) and the traditional H6 inverter in [14] were simulated in Matlab/Simulink. The DC voltage is 200V, and the grid voltage is 110V/50Hz. The switching frequency is 10kHz. The output inductors are both 2mH. The stray capacitance between DC source and ground is 470nF. The grid current is controlled by a conventional PR controller. Fig. 10 shows the comparison of the common-mode characteristics of traditional H6 inverter and the proposed inverters with the universal clamping circuits. The proposed topologies can achieve a more constant common-mode voltage and lower leakage current as expected.

In order to further validate the proposed topologies and the modulation strategies, the 1kW experimental prototypes were built with the same parameters as the simulation model. The experimental gating signals of the proposed inverters in Fig. 4(d) and Fig. 8(c) are shown in Fig. 11 and Fig. 12 respectively. Also, Fig. 13 and Fig. 14 show the comparison of the output

978-1-4673-9551-9/16 $31.00 © 2016 IEEE

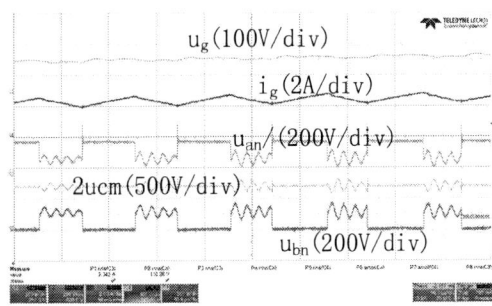

Fig. 15. Zoomed view of Fig. 13.

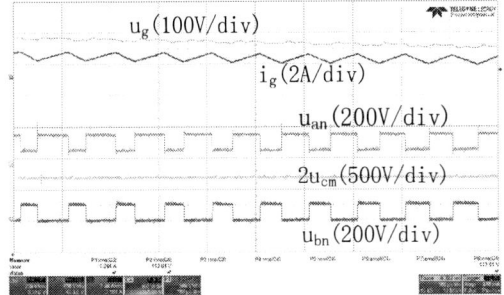

Fig. 16. Zoomed view of Fig. 14.

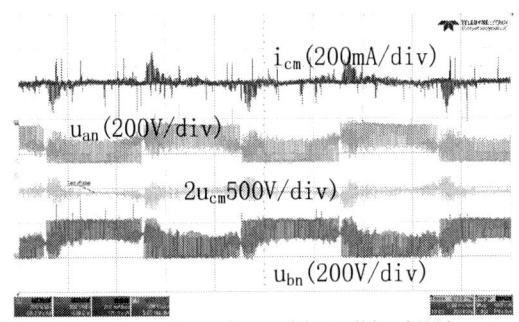

Fig. 17. Experimental waveforms of the traditional H6 inverter: (from TOP to BOTTOM) leakage current, bridge A voltage, common mode voltage, bridge B voltage.

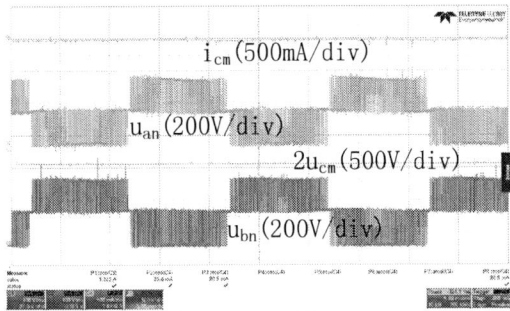

Fig. 18. Experimental waveforms of the proposed H6 inverter: (from TOP to BOTTOM) leakage current, bridge A voltage, common mode voltage, bridge B voltage.

waveforms between the traditional H6 inverters and the proposed NPC topologies. It is especially noted that the proposed inverter demonstrates a more constant common mode

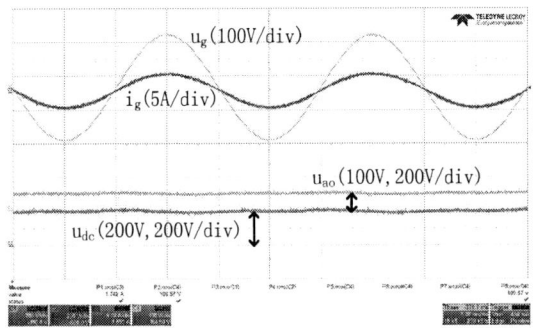

Fig. 19. The captured experimental waveforms of the proposed NPC inverter.

voltage as expected. Fig. 15 and Fig. 16 show the zoomed view of captured waveforms in Fig. 13 and Fig. 14, respectively. Obviously, the proposed inverters show a better common mode behavior. Besides, Fig. 17 and Fig. 18 capture the leakage currents, switched phase voltage and common mode voltage, where the proposed inverter achieves a lower leakage current. The neutral point voltage of the NPC inverter was captured in Fig. 19, where it is half of the DC link voltage.

V. CONCLUSION

This paper proposes a concept of universal NPC method and consequently presents two families of single-phase neutral-point-clamped inverters, where the common mode voltage can be effectively maintained constant as half of dc-link voltage by only assuming one additional clamping switch, which only conducts the leakage current during operation. The simulation and experimental results verified the performance of the proposed inverters.

REFERENCES

[1] O. Lopez, F. D. Freijedo, A. G. Yepes, P. Fernandez-Comesana, J. Malvar, R. Teodorescu, and J. Doval-Gandoy, "Eliminating ground current in a transformerless photovoltaic application," *IEEE Trans. Energy Convers.*, vol. 25, no. 1, pp. 140–147, Mar. 2010.

[2] T. Kerekes, R. Teodorescu, and M. Liserre. "Common mode voltage in case of transformerless PV inverters connected to the grid," IEEE International Symposium on Industrial Electronics, 2008, pp. 2390-2395.

[3] M. Victor, G. Kaufungen, and B. Alheim, et al., "Method of converting a direct current voltage of a source of direct current voltage, more specifically of a photovoltaic source of direct current voltage, into an alternating current voltage," U.S Patent 7411 802 B2, 2008.

[4] T. Kerekes, R. Teodorescu, P. Rodriguez, G. Vazquez, E. Aldabas, "A new high-efficiency single-phase transformerless PV inverter topology," *IEEE Trans. Ind. Electron.*, vol. 58, no. 1, pp. 184-191, Jan. 2011.

[5] S. Heribert, S. Christoph, and K. Jurgen, "Inverter for transforming a DC voltage into an AC current of an AC voltage," Europe Patent 1 369 985 (A2), December 2003.

[6] W. S. Yu, J.-S. Lai, H. Qian, C. Hutchens, J. H. Zhang, G. Lisi, A. Djabbari, G. Smith, and T. Hegarty, "High-efficiency inverter with H6-type configuration for photovoltaic non-isolated AC module applications," in Proc. IEEE Appl. Power Electron. Conf. Expo., 2010, pp. 1056–1061.

[7] H. Xiao, S. Xie, Y. Chen, and R. Huang, "An optimized transformerless photovoltaic grid-connected inverter," IEEE Trans. Ind. Electron., vol. 58, no. 5, pp. 1887–1895, May 2011.

[8] S. J. Hu, W. F. Cui and X. N. He, "A High-Efficiency Single-Phase Inverter for Transformerless Photovoltaic Grid-Connection," in Proc. IEEE ECCE 2014, 2014, pp. 4232-4236.

[9] R. Gonzalez, J. Lopez, P. Sanchis, and L. Marroyo, "Transformerless inverter for single-phase photovoltaic systems," *IEEE Trans. Power Electron.*, vol. 22, no. 2, pp. 693–697, Mar. 2007.

[10] L. Zhang, K. Sun, L. Feng, H. Wu, and Y. Xing, "A family of neutral point clamped full-bridge topologies for transformerless photovoltaic grid-tied inverters," *IEEE Trans. Power Electron.*, vol. 28, no. 2, pp. 730–739, Feb.2013.

[11] F. Blaabjerg, Z. Chen, and S. B. Kjaer, "Power electronics as efficient interface in dispersed power generation systems," *IEEE Trans. Power Electron.*, vol. 19, no. 5, pp. 1184–1194, Sep. 2004.

[12] S. B. Kjaer, J. K. Pedersen, and F. Blaabjerg, "A review of single-phase grid-connected inverters for photovoltaic modules," *IEEE Trans. Ind. Appl.*, vol. 41, no. 5, p. 1292, Sep. 2005.

[13] Q. Li and P. Wolfs, "A review of the single phase photovoltaic module integrated converter topologies with three different dc link configurations," *IEEE Trans. Power Electron.*, vol. 23, no. 3, pp. 1320–1333, May 2008.

[14] W. Yu, J. S. Lai, H. Qian, and C. Hutchens, "High-efficiency MOSFET inverter with H6-type configuration for photovoltaic non-isolated AC-module applications," *IEEE Trans. Power Electron.*, vol. 56, no. 4, pp. 1253-1260, Apr. 2011.

[15] H. Xiao and S. Xie, "Transformerless split-inductor neutral point clamped three-level PV grid-connected inverter," *IEEE Trans. Power Electron.*, vol. 27, no. 4, pp. 1799–1808, Apr. 2012.

[16] T. Kerekes, R. Teodorescu, M. Liserre, C. Klumpner, and M. Sumner, "Evaluation of three-phase transformerless photovoltaic inverter topologies," *IEEE Trans. Power Electron.*, vol. 24, no. 9, pp. 2202–2211, Sep. 2009.

[17] B. Yang, W. Li, Y. Gu, W. Cui, and X. He, "Improved transformerless inverter with common-mode leakage current elimination for photovoltaic grid-connected power system," *IEEE Trans. Power Electron.*, vol. 27, no. 2, pp. 752-762, Feb. 2012.

[18] S. V. Araujo, P. Zacharias, and R. Mallwitz, "Highly efficient single-phase transformer-less inverters for grid-connected photovoltaic systems," *IEEE Trans. Power Electron.*, vol. 57, no. 9, pp. 3118–3128, Sep. 2010.

Modular Subpanel Photovoltaic Converter System: Analysis and Control

Yuan Li[1,2], Yue Zheng[2], Su Sheng[2], Brad Scandrett[3], Brad Lehman[2]

Email: yli@scu.edu.cn bscan@powerfilmsolar.com lehman@ece.neu.edu

1. Department of Electrical Engineering and Information, Sichuan University, Chengdu, Sichuan, China
2. Department of Electrical and Computer Engineering, Northeastern University, Boston, Massachusetts, USA
3. PowerFilm, Inc, Ames, IA, USA

Abstract—A modular plug-and-play photovoltaic (PV) converter system for portable PV applications is proposed. Small subpanel of PV cells have individual dc/dc converters performing distributed maximum power point tracking. The dc/dc submodules are connected in parallel to a DC bus. Challenges for the proposed PV system to accommodate diverse types of loads are addressed. Based on scenario analysis, the operating mode and the operating range of the PV system are illustrated. A peer-to-peer control strategy is developed to maintain stable power supply automatically with PV power and/or load power changing. The DC bus is semi-regulated in accordance with different operating scenarios. The presented system features advantages of 1) compatibility of various types of PV subpanels (different chemistry, power levels, and operating characteristics); 2) capability of drawing higher power from mismatched PV subpanels; 3) high reliability due to its parallel structure and peer-to-peer operating mechanism. A prototype of the system has been developed in the lab, and experimental results are provided to verify the method.

Keywords—photovoltaic; converter; modular; peer-to-peer control

I. INTRODUCTION

High efficiency thin film solar cell technology [1]-[3], e. g. amorphous silicon, thin film Gallium Arsenide (GaAs), etc., have made it feasible to package and directly connect solar energy with medium power level outdoor devices. Fig. 1(a), for example, illustrates a typical foldable, flexible (120W) solar panel with amorphous silicon subpanels. Portable panels are utilized by hobbyists, recreationalist or soldiers-on-the-move to charge or power directly their portable electronic devices. Often the panels are connected to a single, centralized DC-DC converter with maximum power point tracker (MPPT) [4]. Since the MPPT is centralized, then partial shading, submodule failure, or PV characteristics mismatch, will cause reduction of PV power output on the other PV submodules [5]-[6]. To account for this reduction, recent research investigates both distributed MPPT [7]-[11] as well as differential power processing [12]-[14] to improve system efficiency. This paper focuses on the distributed MPPT technique where each PV submodule has its own maximum power point tracker so that mismatch between smaller PV units can be decoupled. The outputs of the converters are subsequently connected in parallel to reach desired current level.

A difficulty with individualized distributed MPPT in applications, such as the portable PV, is that the output of the

This work was partially funded through grants by PowerFilm and DOD contract W56KGU-14-C0014.

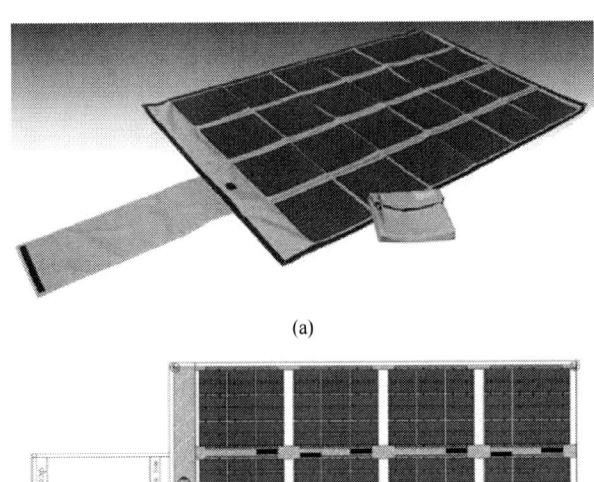

(a)

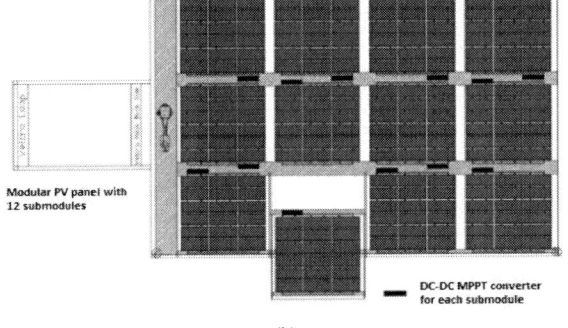

(b)

Fig. 1. (a) Existing commercial (PowerFilm-120W) foldable, flexible PV panel with submodules sewn into fabric. (b) Proposed modular PV system with each plug-and play module containing distributed MPPT.

MPPT is often directly connected to the load. Unlike grid-connected applications, many loads are power limited or with constant power, so that it is not always possible to force each submodule to operate at its MPPT. Coordination and control is needed for power management in the proposed PV system. Recently, several research works have addressed similar demands, for instance, in standalone DC microgrids [15]-[26]. Load shedding [20]-[21], curtailing of the renewable power generation [22]-[23], droop control [24]-[26] are general approaches that are utilized in different purpose, e.g., to regulate the DC bus voltage, to share the power in proportion to generator ratings, or to increase the life-span of batteries. Differently, for modular subpanel PV converter systems presented in this paper, the DC bus voltage is not strictly required to be regulated to precise levels because of load flexibility. Instead, individual MPPT has a higher priority to increase PV power harvesting for portable applications.

Specifically, this paper presents the plug-and-play, peer-to-peer submodule architecture for a foldable PV solar system as illustrated in Fig. 1(b). Challenges of power balancing/shedding between PV source and diverse types of loads are addressed. Based on load scenario analysis, the criterion for bus voltage regulation is illustrated. The DC bus is semi-regulated with the ability to deliver the maximum PV/load power in accordance with different operating scenarios. The presented system features advantages of 1) flexible modular configuration; 2) individual maximum power point tracking for higher PV power harvesting capability; 3) diverse load accommodation; 4) dynamic operating mode assignment for each PV submodule; and 5) high overall system conversion efficiency.

II. SYSTEM STRUCTURE

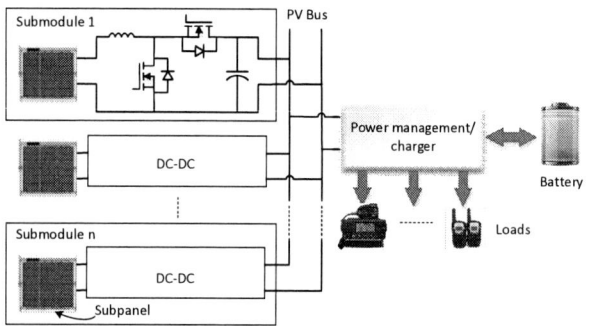

Fig. 2 Presented subpanel PV converter system.

As shown in Fig. 1(b), each subpanel of the solar PV slides in and out of a solar blanket using a plug-and-play configuration. Herein, subpanel refers to multi-PV-cell integration, but the power rating is usually 5% ~ 10% of a commercial PV panel. The subpanel can be removed or replaced while the solar blanket PV maintains operation. Each subpanel has an individual power converter/submodule connecting the output bus. I²C is utilized as communication protocol to coordinate power control of each converter, but other protocols could be implemented as well, such as CAN bus. So the connector of each submodule is interfacing both power delivery and communication.

Fig. 2 shows the block diagram of the subpanel PV converter system structure. In the experimental prototype, a boost DC-DC converter is used as basic power conversion module because of the specific voltage gain demand. However, buck or buck-boost converter is still selectable and can be in effect in other cases. The outputs of the DC-DC converters are then connected in parallel. For the specific prototype, each PV subpanel voltage ranges from 4 V to 9 V, depending on PV chemistry, irradiance levels, and shading configuration. The PV bus voltage could be in a wide range from the minimum output voltage of boost converter to the maximum input voltage of loads, i.e. 10 V ~ 25 V. As for portable application, a typical load might be a battery charger, centralized MPPT converter, a simple resistive load (for heating), or a DC appliance.

Based on this structure, the presented PV system achieves an extremely flexible operating manner with the ability to draw the maximum power from each PV subpanel. While this would not be possible for centralized converter structure because the operating point of each PV subpanel is clamped to one another by series or parallel connection. If the I-V characteristics are not identical, operating voltage and current would deviate from the maximum power point. Because of the same reason, it is inherently robust to partial shading and subpanel/submodule failure, where mismatch of characteristics is decoupled as well.

III. ANALYSIS ON SYSTEM OPERATING MODE

In this section, challenges of the proposed PV system with converters to diverse load are addressed. Possible working scenarios are first investigated in order to derive operating mode from the system's viewpoint. Secondly, discussion about overall conversion efficiency related to system operating mode is provided. In the following analysis, P_{MPPT} refers to the maximum power that all PV subpanels can produce, P_{PV} refers to the actual power that all PV subpanels are producing, and P_{load_max} refers to the maximum power that load can consume. Assume V_H and V_L are the upper and lower operating voltage limits of the PV bus (see Fig. 2), where V_L is the minimum output voltage of boost converter, and V_H is either the upper limit of load voltage or the boost converter output voltage, whichever is smaller.

A. Operating mode analysis

1) Scenario I: Resistive load, $P_{MPPT} \leq P_{load\_max}$

For resistive load, P_{load_max} is actually determined by V_H, where

$$P_{load\_max} = V_H^2 / R_{load}. \qquad (1)$$

In this case, all PV subpanels are possible to operate at individual MPPT mode because the PV bus voltage would not exceed V_H. In fact, in order to harvest as much PV power as possible, it is necessary for all PV subpanels to operate at their individual maximum power point. Thus $P_{PV} = P_{MPPT}$. Fig. 3 shows the operating range in this case. The operating point would be the intersection (O) of load curve R_{load} and power contour P_{MPPT}. In the case when P_{MPPT} changes due to the environmental changes (irradiance, temperature, etc.), the operating point moves between O_L and O_H along the red straight line (solid), where under/over voltage will be triggered. Similarly, if the load changes, the operating point moves between O'_L and O'_H along the blue curve.

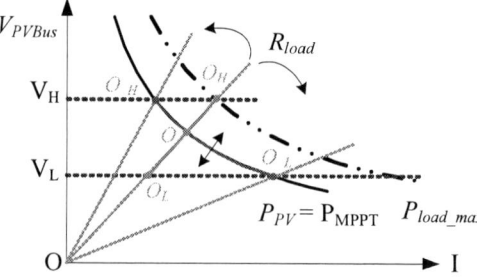

Fig. 3 Operating range for resistive load when $P_{MPPT} \leq P_{load\_max}$.

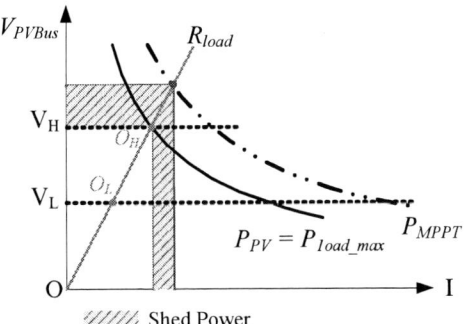

Fig. 4 Operating range for resistive load when $P_{MPPT} > P_{load\_max}$.

2) Scenario II: Resistive load, $P_{MPPT} > P_{load\_max}$

In this case, power shedding should be conducted, i.e. $P_{PV} < P_{MPPT}$, otherwise output voltage will raise up over V_H. There is only one valid operating point O_H (see Fig. 4, the intersection of R_{load} and V_H) that harvests the maximum PV power from subpanels. At this point $P_{PV} = P_{load\_max}$. It should be noted that system can operate at other points along R_{load} line in between O_L and O_H. However, in this case P_{PV} would be less than P_{load_max}, which implies PV power is not fully used. The area shaded in Fig. 4 represents the amount of shed power in this case.

3) Scenario III: Constant power load, $P_{MPPT} \leq P_{load\_max}$

For constant power load, P_{load_max} has real meaning: if a battery charger is the load, P_{load_max} refers to battery voltage times charging current limitation; if a centralized MPPT converter is used, P_{load_max} is actually the same as P_{MPPT}.

In this case, individual MPPT is expected to be applied in order to increase energy harvesting capability of the PV system. However, the operating range of the PV system is not as apparent as in Scenario I or II. The formidable challenge is that there is a second stage of power conversion as in the load, e.g., a buck converter to charge battery stacks or centralized DC-DC converter conducting MPPT. This implies PV bus voltage is determined by both operating point of modular submodules in this paper and the load converter. Unfortunately, the operating point of the load is totally independent and often not controllable in the PV system.

The purpose of the following analysis is two-fold: 1) to illustrate the feasibility that individual MPPT can be implemented in this scenario; 2) to derive the operating range of the PV system. For simplicity of explanation, assume that a buck converter charging battery is the secondary power conversion stage in this analysis process.

For system voltage, we have

$$V_{PVBus} = V_{in1} / (1 - d_1), \text{ and } V_{Bat} = d_{Buck} \cdot V_{PVBus}, \qquad (2)$$

where V_{in} and d_1 are input voltage and duty ratio of a PV submodule and its DC-DC converter, respectively; and V_{Bat} and d_{Buck} are the output voltage and duty ratio of the secondary buck converter, respectively. Thus the voltage gain of the two stages becomes

$$V_{Bat} / V_{in1} = d_{Buck} / (1 - d_1). \qquad (3)$$

Assuming the operating range of duty ratio for each DC-DC converter is 0.2 to 0.8, i.e. $0.2 < d_1 < 0.8$; $0.2 < d_{Buck} < 0.8$, one can get voltage gain according to (3), as shown in Fig. 5.

In a specific application, for example, $V_{Bat} = 5$ V, and assuming PV submodule 1 is tuned to work on its MPP where $V_{MPP} = V_{in1} = 7$ V, thus the voltage gain should be $V_{Bat} / V_{in1} = 0.71$. According to (3), infinite pairs of d_1 and d_{Buck} satisfies the demanded voltage gain. This implies that with any d_{Buck} given by the secondary converter, a proper d_1 can be settled by convention MPPT algorithm. However, it should be pointed out that the PV bus voltage V_{PVBus} is variable even with the same voltage gain. Fig. 6 shows the range of d_1 and d_{Buck} to achieve MPPT, where each single curve represents a constant voltage gain; it also shows the PV bus voltage accordingly. To draw the curves, the upper and lower voltage bus limits were constrained to be 10 V $\leq V_{PV\_Bus} \leq$ 15 V with input voltage limits 3 V $\leq V_{in} \leq$ 9 V. And other parameters are $V_{Bat} = 5$ V;

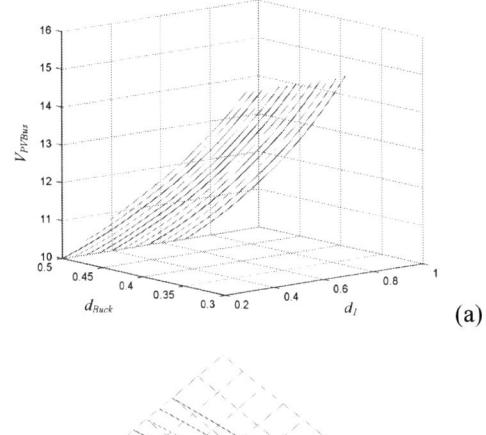

(a)

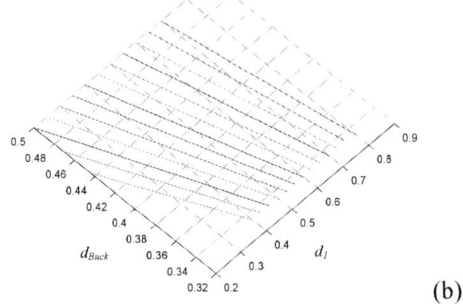

(b)

Fig. 6 PV bus voltage is variable with D and d_1 (a) front view, (b) top view

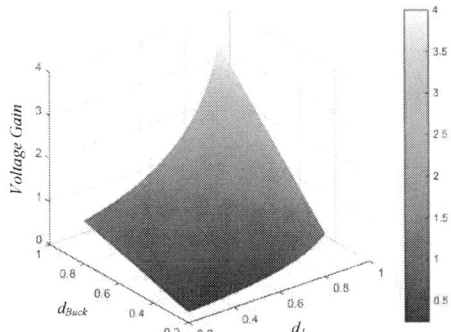

Fig. 5 Voltage gain of the two stage power processing system

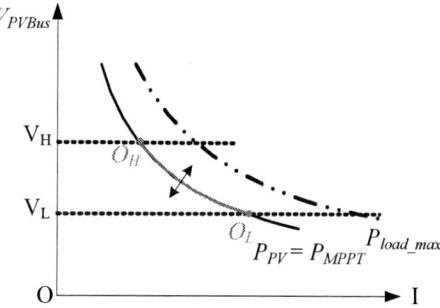

Fig. 7 Operating range for constant power load when $P_{MPPT} \leq P_{load\_max}$.

$0.2 < d_l < 0.8$; $0.2 < d_{Buck} < 0.8$. It can be confirmed by these curves that 1) if there is a second power conversion stage, individual MPPT still can be realized by conventional algorithms, such as P & O, that deliberately search for a maximum power producing duty cycle; 2) during this approach, PV bus voltage is semi-regulated by both individual MPPT and the secondary power conversion stage, which is variable even with the same voltage gain and power delivered.

According to the above analysis, the operating range for constant power load is marked by the red curve in Fig. 7. There is no similar load-line intersection point as in Fig. 3 or Fig. 4. Instead, the operating point would be any point from O_L to O_H on the power contour of P_{MPPT}, where individual MPPT is applied to every submodule. Thus, $P_{PV} = P_{MPPT}$. If P_{MPPT} is changing with circumstance, operating point would still be settled on a new power contour after individual MPPT process, within V_L and V_H.

4) Scenario IV: Constant Power load, $P_{MPPT} > P_{load\_max}$

This case often happens when a battery charger is used as the load. Because of current limitations of the load (battery), not all the PV power can be absorbed.

The analysis on this situation is similar as Scenario III. The operating point would be from O_L to O_H, lying on the power contour of P_{load_max} as shown by the red curve in Fig. 8. The area shaded represents the amount of shed power in this case. However, it is important to regulate the PV bus voltage to V_H in this case. Because when operating on other points except O_H, it is difficult for controller to decide whether more power should be extracted from the PV source.

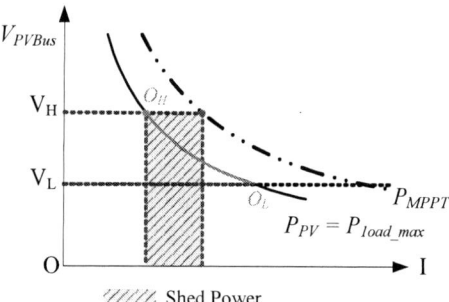

Fig. 8 Operating range for constant power load when $P_{MPPT} > P_{load\_max}$.

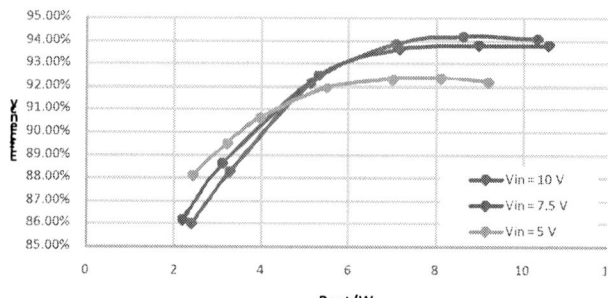

Fig. 9 Measured efficiency curve of DC-DC prototype in the lab

In summary, the PV bus voltage of the converter system is a vital criterion to decide the operating mode from the system's viewpoint: if $V_{PVBus} > V_H$, power shedding is necessary; if $V_{PVBus} < V_H$, individual MPPT can be conducted. And this criterion is applicable to both resistive and constant power load.

B. Overall Conversion Efficiency

There are two intuitive ways to shed PV power when necessary: 1) shut down PV subpanel/submodule; 2) reduce output of every PV subpanel/submodule proportionally. The second method can be realized by droop control and the power can be controlled decreasing gradually. However, this paper presents a combination approach to shed PV power: 1) coarse shedding: to turn on/off a converter and 2) fine shedding: to use one submodule as a voltage regulator.

For example, assume there is a load consuming 80 W power out from 20 PV submodules, where the maximum power can be produced by every submodule is 10 W. If we have every submodule work at 4 W output, according to the efficiency curve measured from each DC-DC submodule prototype shown in Fig. 9, the overall system efficiency is calculated as

$$80 / (\frac{4}{90.5\%} \times 20) = 90.5\%, \qquad (4)$$

given that every submodule has an individual efficiency of 90.5% at 4 W output. On the other hand side, if there were 8 submodules operating at MPP (assume $P_{MPP} = 9.5$ W), one operating as voltage regulator which produces 4 W, the overall system efficiency can be estimated as

$$80 / [(\frac{9.5}{93.8\%} \times 8) + \frac{4}{90.5\%}] = 93.6\%. \qquad (5)$$

Apparently, the system overall efficiency is higher if the combination approach is applied.

IV. PEER-TO-PEER CONTROL STRATEGY

According to the above analysis, a peer-to-peer control strategy is presented. For the case that no power shedding is needed, each individual DC-DC converter will conduct MPPT independently. Voltage of the PV bus is not constant during the MPPT procedure. Instead, it will fluctuate within a range from V_L to V_H. For the case that power shedding is necessary, the combination approach can be used.

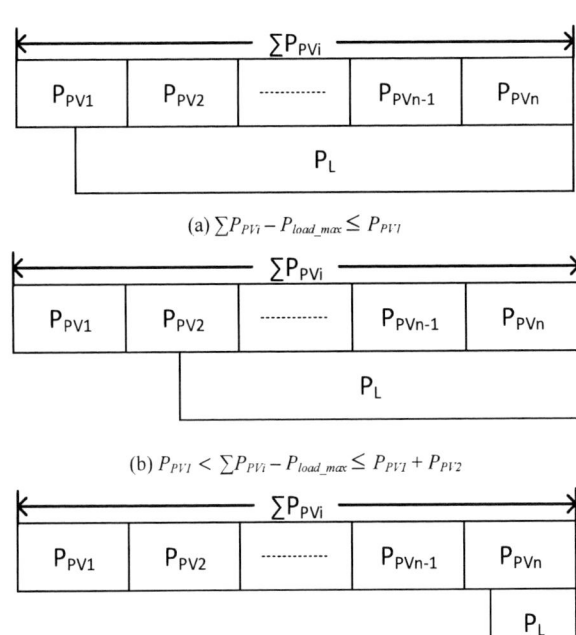

(a) $\sum P_{PVi} - P_{load\_max} \leq P_{PV1}$

(b) $P_{PV1} < \sum P_{PVi} - P_{load\_max} \leq P_{PV1} + P_{PV2}$

(c) $P_{PV1} + P_{PV2} + \ldots + P_{PVn-1} < \sum P_{PVi} - P_{load\_max} \leq \sum P_{PVi}$

Fig. 10 Unbalanced power between PV and load

Power shedding can further fall into different categories, as shown in Fig. 10. For category (a), one DC-DC converter is used to create a constant output voltage to balance the PV bus voltage V_H. This converter is not operating at MPP. The other converters, however, operate at MPPT independently. For category (b), one DC-DC converter conducts output voltage control, one is shut down and the remaining converters perform individual MPPT. This process of shutting down converters, as necessary, to reduce power is continued until case (c), where one DC-DC converter performs constant voltage regulation, and all the other DC-DC converters are shut down.

Based on the analysis, the flow chart for master MCU and the criterion to choose different control methods are shown in Fig. 11. PV bus voltage is compared to V_H to decide whether to increase or decrease input PV power. The master MCU receives measured information of individual P_{PVi}. It then

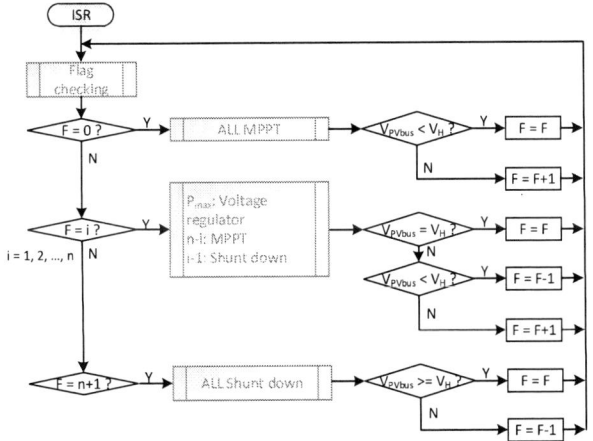

Fig. 11 Flow chart of master MCU.

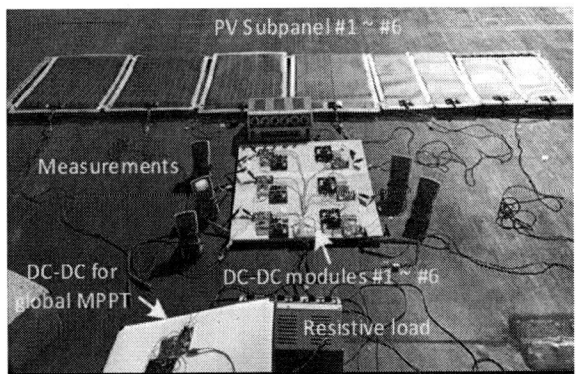

Fig. 12 Experimental setup for PV subpanels converter system.

commands the DC-DC module with the maximum output power to perform any necessary voltage regulation, and lets the DC-DC module with the minimum output power be shut down. With this dynamic operating mode assignment method, minimum switching times between different operating manners of an individual DC-DC module can be achieved. It is worth noting that it is, also, feasible for every DC-DC to operate without been dropped out. However, the proposed method will achieve a higher overall system conversion efficiency.

V. EXPERIMENTAL RESULTS

Fig. 12 shows the experimental setup for PV converter system composed of up to 6 amorphous silicon subpanels, each with distributed MPP and outputs connected in parallel.

Experiment 1-Individual MPPT with resistive Load. In the first test condition, the designed system is connected to a resistive load, with 4 of the 6 subpanels connected. The test result is listed in Table I. It needs to be pointed out that the total output power for individual MPPT and centralized MPPT has not been compared directly. This is because constant irradiance and temperature could not be guaranteed in the outdoor tests on different days when these two tests were conducted. However, since non-shedding and shedding tests were conducted one immediately after another, the percentage of output power can be compared fairly. Table I shows that when subpanels are intentionally shaded, 72.5% of PV power (18.7 W) remained. As a comparison, with a centralized MPPT, the shaded PV submodule behaved as a load, which draws current from other parallel connected submodules. It can be calculated that when one subpanel is shaded in the same way as individual MPPT, only 62.9% of the non-shading power remains, which is a significant reduction compared to individual MPPT strategy.

Table I Comparison of power captured by two MPPT approaches

Output power of DC-DC	Individual MPPT	Centralized MPPT
Nonshading	25.7 W	19.3 W
Shading	18.7 W	12.1 W
Power remains %	72.5%	62.9%

Experiment 2 -Power shedding and the dynamic of V_{PVBus} with load step change. In order to demonstrate the proposed control strategy, experiments are conducted under the following condition: two submodules work following MCU commands in accordance of V_{PVBus}. In the experiment, the maximum power of #1 and #2 subpanel is 9.1 W and 8.3 W, respectively. The PV bus voltage is semi-regulated with a hysteresis band of 15 V $\pm$ 2 V, i.e. $V_{PVBus} > 17$ V is regarded as $V_{PVBus} > V_H$ (see Fig. 11) and $V_{PVBus} < 13$ V is regarded as $V_{PVBus} < V_H$. Resistive load is used for step change, where 10 ohms, 27 ohms, and 47 ohms correspond to three operating modes: Operating Mode I (OP-I) – two submodules are performing MPPT (MPPT $\times$ 2); OP-II – one submodule is performing MPPT and one is performing voltage regulation (MPPT + VR); and OP-III – one submodule is performing voltage regulation and the other one is shut down (VR + SD), respectively. The results shown in Fig. 13 indicate the following accomplishments of the proposed control method: 1) necessary power shedding when PV bus voltage is higher than reference ; 2) a stable response of the dynamic load change; and 3) individual MPPT for each submodule leading to PV bus voltage lower than regulated voltage reference.

VI. CONCLUSIONS

A peer-to-peer control strategy for PV system is proposed to accommodate diverse power of loads. The control strategy maintains a stable power supply in the subpanel PV converter system. The presented system is designed and experimentally verified for portable applications, which is able to extract more energy than the centralized MPPT approach. With the proposed control method, this system achieves self-power shedding and can work under diverse types of load conditions. Experimental results also demonstrate dynamic operating mode assignment capability with load step change. A stable operation mode shift can be observed.

REFERENCES

[1] R. Windisch, B. Dutta, M. Kuijk, A. Knobloch, S. Meinlschmidt, S. Schoberth, P. Kiesel, G. Borghs, G. H. Dohler, and P. Heremans, "40% efficient thin-film surface-textured light-emitting diodes by optimization of natural lithography," *IEEE Transactions on Electron Devices*, vol. 47, no. 7, pp. 1492 - 1498, July. 2000.

[2] K. A. Munzer, K. T. Holdermann, R. E. Schlosser, and S. Sterk, "Thin monocrystalline silicon solar cells," *IEEE Transactions on Electron Devices*, vol. 46, no. 10, pp. 2055 - 2061, Oct. 1999.

[3] B. M. Kayes, N. Hui, R. Twist, S. G. Spruytte, F. Reinhardt, I. C. Kizilyalli, and G. S. Higashi, "27.6% Conversion efficiency, a new record for single-junction solar cells under 1 sun illumination," *37th IEEE Photovoltaic Specialists Conference (PVSC)*, 2011, pp. 4 – 8.

[4] K. H. Hussein, I. Muta, T. Hoshino, and M. Osakada, "Photovoltaic power tracking: an algorithm for rapidly changing atmospheric conditions," *IEE Proceedings-Generation, Transmission and Distribution*, vol. 142, no. 1, pp. 59 - 64, 1995.

[5] A. Mäki, ; and S. Valkealahti, "Power Losses in Long String and Parallel-Connected Short Strings of Series-Connected Silicon-Based Photovoltaic Modules Due to Partial Shading Conditions," *IEEE Trans. Energy Conversion*, vol. 27, no. 1, pp. 173 - 183, March. 2012.

[6] P. Bakas, A. Marinopoulos, and B. Stridh, "Impact of PV module mismatch on the PV array energy yield and comparison of module, string and central MPPT," *Photovoltaic Specialists Conference (PVSC)*, 2012 38th IEEE, pp. 1393–1398.

[7] S. M. MacAlpine, R. W. Erickson, and M. J. Brandemuehl, "Characterization of power optimizer potential to increase energy capture in photovoltaic systems operating under nonuniform

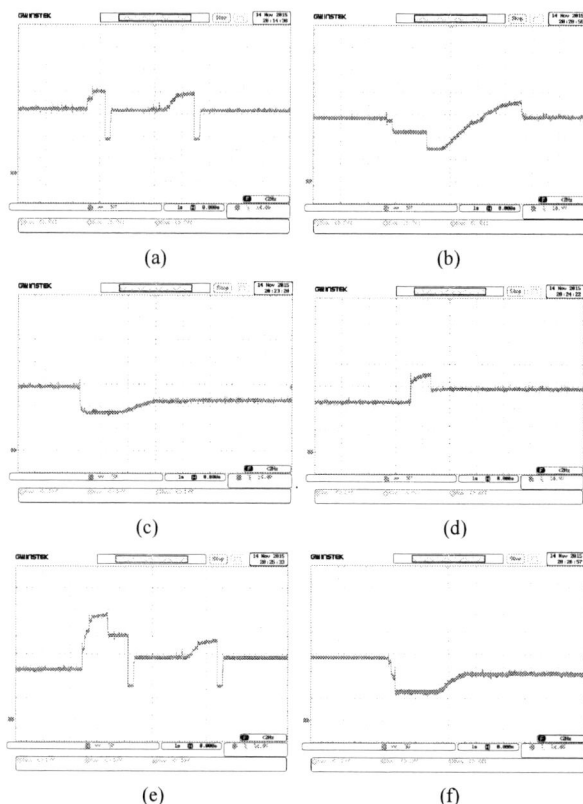

Fig. 13 Dynamic response of V_{PVBus} with load step change, where R_{load} changes (a) from 27 ohms (VR +MPPT) to 47 ohms(VR + SD); (b) from 47 ohms (VR + SD) to 27 ohms (VR +MPPT); (c) from 10 ohms (MPPT $\times$2) to 27 ohms (VR +MPPT); (d) from 27 ohms (VR +MPPT) to 10 ohms (MPPT $\times$2); (e) from 10 ohms (MPPT $\times$2) to 47 ohms(VR + SD); and (f) from 47 ohms (VR + SD)to 10 ohms (MPPT $\times$2).

conditions," *IEEE Trans. Power Electron.*, vol. 28, no. 6, pp. 2936–2945, Jun. 2013.

[8] F. Wang, X. K. Wu, F. C. Lee, Z. J. Wang, P. J. Kong and F. Zhuo, "Analysis of Unified Output MPPT Control in Subpanel PV Converter System," *IEEE Trans. Power Electron.*, vol. 29, no. 3, pp. 1275–1284, Mar. 2014.

[9] L. Linares, R. W. Erickson, S. MacAlpine, and M. Brandemuehl, "Improved energy capture in series string photovoltaics via smart distributed power electronics," *in Proc. IEEE 24th Annu.Appl. Power Electron. Conf. Expo.*, 2009, pp. 904–910.

[10] N. Femia, G. Lisi, G. Petrone, G. Spagnuolo, and M. Vitelli, "Distributed maximum power point tracking of photovoltaic arrays: novel approach and system analysis," *IEEE Trans. Ind. Electron.*, vol. 55, no. 7, pp. 2610– 2621, Jul. 2008.

[11] B. L. Xiao, L. J. Hang, J. Mei, C. Riley, L. M. Tolbert, and B. Ozpineci, "Modular Cascaded H-Bridge Multilevel PV Inverter with Distributed MPPT for Grid-Connected Applications," *IEEE Trans. on Industry Applications*, vol. 51, no. 2, pp. 1722–1731, Mar./Apr. 2015.

[12] P. S. Shenoy, K. A. Kim, B. B. Johnson, and P. T. Krein, "Differential Power Processing for Increased Energy Production and Reliability of Photovoltaic Systems," *IEEE Trans. Power Electron.*, vol. 28, no. 6, pp. 2968–2979, Jun. 2013.

[13] P. Sharma, and V. Agarwal, "Maximum Power Extraction From a Partially Shaded PV Array Using Shunt-Series Compensation," *IEEE Journal of Photovoltaics*, vol. 4, no. 4, pp. 1128–1137, Jul. 2014.

[14] J. D. Bastidas-Rodriguez, E. Franco, G. Petrone, C. Andrés Ramos-Paja, and G. Spagnuolo, "Maximum power point tracking architectures for photovoltaic systems in mismatching conditions: a review," *IET Power Electronics*, vol. 7, no. 6, pp. 1396–1413, 2014.

[15] A. M. Dizqah, A. Maheri, K. Busawon, and A. Kamjoo, "A Multivariable Optimal Energy Management Strategy for Standalone DC Microgrids," *IEEE Trans. Power System*, vol. 30, no. 5, pp. 2278-2287, 2015.

[16] F. Nejabatkhah, and Y. W. Li, "Overview of Power Management Strategies of Hybrid AC/DC Microgrid," *IEEE Trans. Power Electron.*, vol. 30, no. 12, pp. 7072 – 7089, 2015.

[17] Y. J. Gu; W. H. Li; and X. N. He, "Frequency-Coordinating Virtual Impedance for Autonomous Power Management of DC Microgrid," *IEEE Trans. Power Electron.*, vol. 30, no. 4, pp. 2328 – 2337, 2015.

[18] S. Anand, B. G. Fernandes, and M. Guerrero, "Distributed control to ensure proportional load sharing and improve voltage regulation in low-voltage DC microgrids," *IEEE Trans. Power Electro.*, vol. 28, no. 4, pp. 1900–1913, 2013.

[19] V. Nasirian, S. Moayedi, A. Davoudi, and F. L. Lewis, "Distributed Cooperative Control of DC Microgrids," *IEEE Trans. Power Electron.*, vol. 30, no. 4, pp. 2288 – 2303, 2015.

[20] R. S. Balog, W. W. Weaver, and P. T. Krein, "The load as an energy asset in a distributed DC smartgrid architecture," *IEEE Trans. Smart Grid*, vol. 3, no. 1, pp. 253–260, 2012.

[21] L. Xu and D. Chen, "Control and operation of a DC microgrid with variable generation and energy storage," *IEEE Trans. Power Del.*, vol. 26, no. 4, pp. 2513–2522, Oct. 2011.

[22] H. Kanchev, D. Lu, F. Colas, V. Lazarov, and B. Francois, "Energy management and operational planning of a microgrid with a PV-based active gen. for smart grid applications," *IEEE Trans. Ind. Electron.*, vol. 58, no. 10, pp. 4583–4592, 2011.

[23] C. T. Hsu, B. Lehman and T. Qian, "A new power stage architecture and control scheme to optimize maximum power point tracking for photovoltaic systems," *in Proc. IEEE 29th Annu.Appl. Power Electron. Conf. Expo.*, 2014, pp. 684-690.

[24] A. P. Nobrega Tahim, D. J. Pagano, E. Lenz, and V. Stramosk, "Modeling and Stability Analysis of Islanded DC Microgrids Under Droop Control," *IEEE Trans. Power Electron.*, vol. 30, no. 8, pp. 4597 – 4607, 2015.

[25] T. Dragicevic, J. M. Guerrero, J. C. Vasquez, and D. Skrlec, "Supervisory Control of an Adaptive-Droop Regulated DC Microgrid With Battery Management Capability," *IEEE Trans. Power Electron.*, vol. 29, no. 2, pp. 695 – 706, 2014.

[26] X. N. Lu; J. M. Guerrero, K. Sun, and J. C. Vasquez, "An Improved Droop Control Method for DC Microgrids Based on Low Bandwidth Communication With DC Bus Voltage Restoration and Enhanced Current Sharing Accuracy," *IEEE Trans. Power Electron.*, vol. 29, no. 4, pp. 1800 – 1812, 2014.

Integrated DC-DC Converter Design for Electric Vehicle Powertrains

Saeed Anwar, Weimin Zhang, Fred Wang, and Daniel J. Costinett
Center for Ultra-wide-Area Resilient Electric Energy Transmission Networks (CURENT)
Department of Electrical Engineering and Computer Science
The University of Tennessee, Knoxville, TN, 37996, USA
sanwar@vols.utk.edu

Abstract—In this paper, an integrated, reconfigurable DC-DC converter for plugin and hybrid Electric Vehicles (EV) is proposed. The converter integrates functionality for both EV powertrain and charging operation into a single unit. During charging, the proposed converter functions as a DAB converter, providing galvanic isolation. For powertrain operation, the converter functions as an interleaved boost converter. During light load powertrain operation, the efficiency of the converter can be further improved by employing the integrated DAB. The proposed integrated converter does not require any extra relays or contactors for charging and powertrain operation. By using such integration, the overall volume and weight of the power electronics circuits, passives and associated cooling system can be improved. In addition, the power flow efficiency from EV battery to the high voltage DC bus for the motor inverter can be improved. The experimental results of the prototype are presented to verify the functionality of the proposed converter.

Keywords—*Electric vehicle, SiC device, integrated magnetics, powertrain, on-board charger, integrated converter.*

I. INTRODUCTION

Electric vehicles (EVs) are gaining popularity due to growing environmental awareness, as well as the price volatility and security of the fossil fuel supply [1]. The design targets of power electronics for plug-in or hybrid EVs include efficiency improvement, volume and weight reduction of the passive components, achieving high power density, and overall cost reduction of the power converter. An integrated converter, combining and re-using some charging components and powertrain components together, can reduce converter volume and weight with higher power density.

For powertrain operation, the interleaved boost converter has been widely studied for high power transfer with reduced EMI and lower current ripple at the HV bus [2-4]. For isolated battery chargers, the DAB, LLC resonant converters [5-7], and phase shift full bridge converters [8] are commonly employed. The DAB exhibits favorable characteristics due to fixed frequency operation, lower stress on the switches, zero voltage switching (ZVS) operation and bidirectional power flow capability. In [9, 10], a high efficiency DAB for charging application is presented where the leakage inductance of the transformer is utilized with significant improvement of efficiency and energy density. Different integrated power electronic converters for EVs are studied for reduced number of components without galvanic isolation although isolation is preferred in automotive industry for user safety [9, 10].

In this paper, an integrated DC-DC converter design is proposed which contains both a drivetrain boost and isolated DAB on-board charger. In the proposed converter, the boost inductor is realized by the magnetizing inductance of the DAB transformer. The DAB secondary side transistors are shared with the boost converter, reducing the total number and semiconductor area of power devices. No additional reconfiguration switches are required for the converter operation in different modes, saving power dissipation, cost, and size.

As an additional benefit, the integration of the isolated charger into the drivetrain boost enables the converter to selectively process power through either topology during traction operation. The design of the boost can be optimized for high power operation where frequent high current is required, while the lower-power DAB can be optimized for low current operating points, increasing power conversion efficiency at a subset of the most frequent operating points of the drivetrain.

The paper is organized as follows. The proposed integrated converter topology and the converter operation is presented in Section II. The magnetics design integrating the transformer and inductor is presented in Section III. The experimental prototype and experimental results are presented in Section IV. Conclusions and future work are stated in Section V.

II. PROPOSED TOPOLOGY

In conventional EV power conversion architecture with an on-board charger (OBC), shown in Fig. 1, the charging and powertrain circuits are separate. During the charging operation, the battery contactors ($S_{11} - S_{12}$) are connected and the EV is

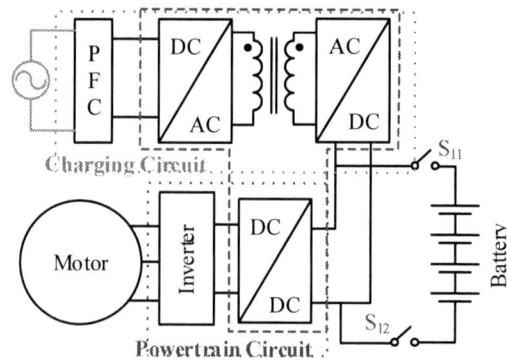

Fig. 1. Conventional EV power conversion architecture.

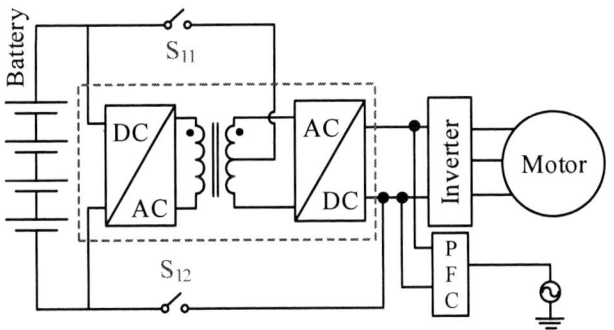

Fig. 2. Proposed EV power conversion architecture.

electronics. In the proposed topology, shown in Fig. 2, the charging and powertrain circuits are integrated into a single converter indicated by the dashed line. Such integration enables the magnetics, cooling system, and select devices to be re-used for both the OBC and traction DC-DC. The same battery contactors ($S_{11} - S_{12}$) are used for reconfiguration between the two DC-DC converter topologies and, when opened, isolation between the battery and remaining vehicle systems is maintained.

To implement the integrated converter, the magnetics design approach of the conventional DAB converter is altered to accommodate both DAB and boost operation in a single magnetic element. With proper reconfiguration of the converter topology, the efficiency of traction operation can be further improved at light load, relative to the non-integrated topology. Thus, the advantages of this integration are the reduction of weight and volume of power electronics, lower component count, and higher efficiency.

The design specifications of the proposed integrated DC-DC converter are shown in Table 1. The experimental prototype, detailed in Section IV, is a scaled-down implementation of a 50 kW powertrain boost converter and 6.6 kW Level 2 on-board charger [14, 15], where voltages are maintained but power levels are reduced by a factor of ten.

charged from the AC grid through the power factor correction circuit (PFC) and isolation transformer. The isolation transformer is used to isolate the AC power ground from the EV ground. Separately, a DC-DC converter is commonly used to boost from the high voltage battery to a DC bus used as the input to the motor drive inverter. The voltage of this bus may be a fixed value, or vary with the operating point of the traction drive [11-13]. This discrete implementation of each converter requires separate magnetic devices for the isolation transformer of the DAB and inductors of the interleaved boost converter. These magnetics increase weight and volume of the EV which impacts the overall vehicle performance and consumes usable space. Additionally, each converter requires separate cooling systems.

The integration of the OBC and powertrain boost converter can significantly reduce the size and weight of on-board power

TABLE I. DESIGN SPECIFICATIONS

Converter	Boost Converter	DAB Converter
Battery Voltage (V_B)	200 – 250 V	200 – 250 V
HV Bus Voltage (V_{HV})	300 – 500 V	400 – 500 V
Power	5 kW	660 W

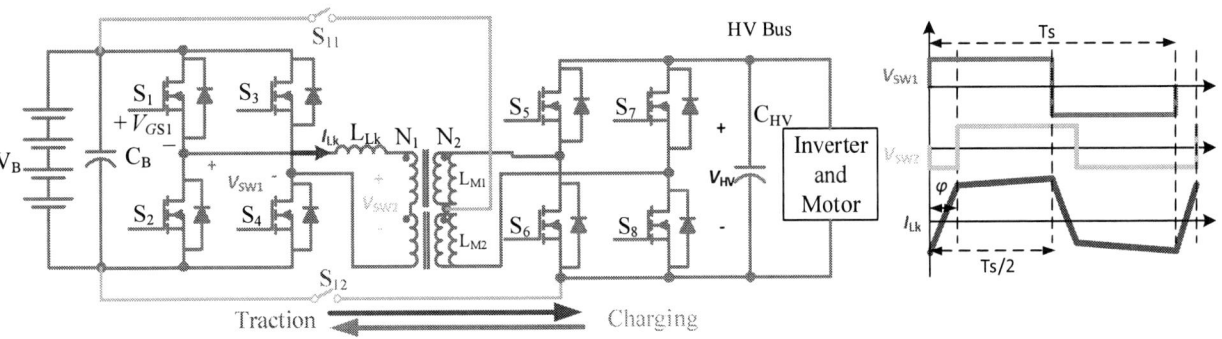

Fig. 3. Proposed integrated EV DC-DC converter working as DAB converter.

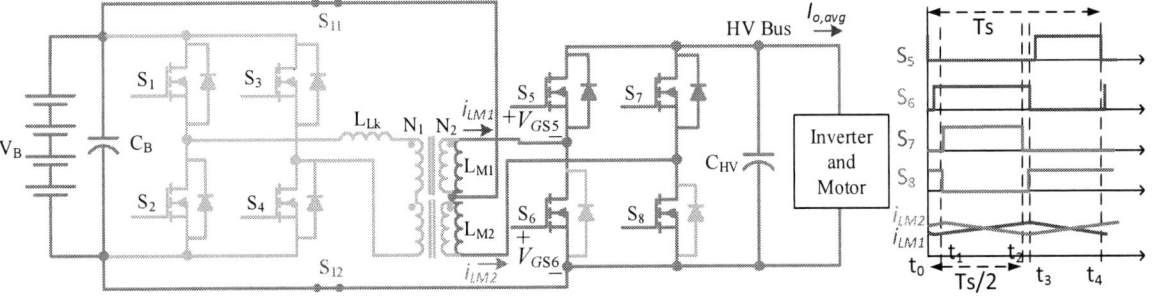

Fig. 4. Proposed integrated EV DC-DC converter working as boost configuration.

The circuit diagram for the proposed topology is shown in Figs. 3-4. In the topology, eight active switches ($S_1 - S_8$) and the two battery contactors ($S_{11} - S_{12}$) are used. The converter can work as interleaved boost converter or DAB converter based on the connection of the contactors, and supplied modulation signals to each switch. During EV powertrain operation, the converter works as an interleaved boost converter by connecting S_{11} and S_{12} switches as shown in Fig. 3. In this mode, four switches (S_5 to S_8) are modulated and other four switches (S_1 to S_4) are turned-off.

During EV charging operation, shown in Fig. 4, the converter transfers power from the HV bus to the battery while maintaining galvanic isolation by disconnecting the S_{11} and S_{12}. All eight switches ($S_1 - S_8$) are modulated, and the converter functions as a DAB.

During powertrain operation the efficiency of the interleaved boost converter decreases at low power operation and high conversion ratio due to the large switching loss and high rms currents present in a high step-up boost. In order to improve the efficiency further, the DAB operating mode can instead be used to increase the efficiency, so long as the output power is within the achievable range of the DAB converter. The configuration in traction mode of the DAB converter is shown in Fig. 4.

III. CONVERTER DESIGN

The main challenge in designing the proposed topology is the design of the integrated magnetics. The isolation transformer used in the proposed converter is designed to serve as required magnetic element for different operating modes. For the powertrain boost mode, the magnetizing inductance of the transformer is used as boost inductor. For the DAB operation, both the primary and secondary sides of the isolation transformer are used for power flow. This design interdependency leads to inherent tradeoffs in performance between the two operating modes. As an example, the portion of the magnetics winding area dedicated to the boost, $K_{u,\text{boost}}$, will have direct relationship with the boost efficiency but inverse relationship with the DAB efficiency. As such, the definition of an optimum design is dependent not only on traditional tradeoffs between efficiency and power density, but also on tradeoffs between performances of each operating mode.

A flowchart of the employed design process for this combined converter is given in Fig. 7. The boost converter is designed first, and an optimal is selected based on design specifications on efficiency η and power density α. Often these specifications are the subject of designer preference, and not quantitatively defined. The DAB is then designed in an optimization process which is more direct, as the core, secondary devices, and window area are pre-defined by the boost design. The process is then iterated as necessary, where DAB performance can be improved by either increasing the winding area dedicated to DAB operation (decreasing $K_{u,\text{boost}}$), or altering the selected design, e.g. making compromises in α to increase core size and allow greater area for the additional DAB windings.

The remainder of this section details the analysis employed in the design flowchart.

A. Boost converter design

Based on the specifications provided for the boost converter, the boost inductance is expressed as a function of the boost converter switching frequency using fixed inductor current ripple approximation. From the specified maximum power operation and HV bus voltage, the average current is calculated first. The required magnetization inductance for the boost converter can be expressed as

$$L_{M1,2} \geq \frac{V_{HV} - V_B}{\Delta I_M f_s V_{HV}} \tag{1}$$

where $L_{M1,2}$, V_{HV}, V_B, f_s and ΔI_M represents the magnetization inductance of the boost converter, HV bus voltage, battery voltage, switching frequency, and allowable boost current ripple, respectively. The worst case scenario may be defined for lowest battery voltage and HV bus voltage assumed constant for initial design.

To design the inductor, a core database is constructed for different commercially-available core geometries and core material characteristics. Initially, a first guess fill factor for the boost windings, $0 < K_{u,\text{boost}} < 1$, is assumed. The maximum magnetic flux density is also set for the boost converter based on the saturation flux density and B-H curve characteristics [16]. Since the boost converter is designed for high power operation, a larger portion of the window area is generally dedicated to it to accommodate higher AWG wires. The remaining portion of the window area, $1-K_{u,\text{boost}}$, will be used for the primary winding construction of the DAB later.

For a designer-specified range of switching frequencies, the number of turns N_2, required air gap l_g, and maximum wire gauge A_{w2} are calculated to achieve the required inductance. The result is a range of boost inductor design with varying switching frequency, core geometry, material, and winding design. A more detailed analysis of magnetics design for varying materials is given in [16].

For each of the resulting designs, analytical estimates of power loss are calculated. Core and copper loss are considered for the inductor; switching and conduction loss are considered for the semiconductors. Core loss is approximated using the traditional Steinmetz equation and copper losses are considered for DC resistance only, assuming Litz wire or other means of limiting AC losses are employed.

In order to obtain switching energy loss associated with the power switches, double pulse tests (DPT) are performed at different loads in a single phase leg. For different power devices the switching energy loss database is created. For the selected set of frequencies, the switching power loss is evaluated. From the datasheet of the device, the on-resistance is used to evaluate the conduction loss for the specified operating condition. The conduction loss and switching losses can be calculated as

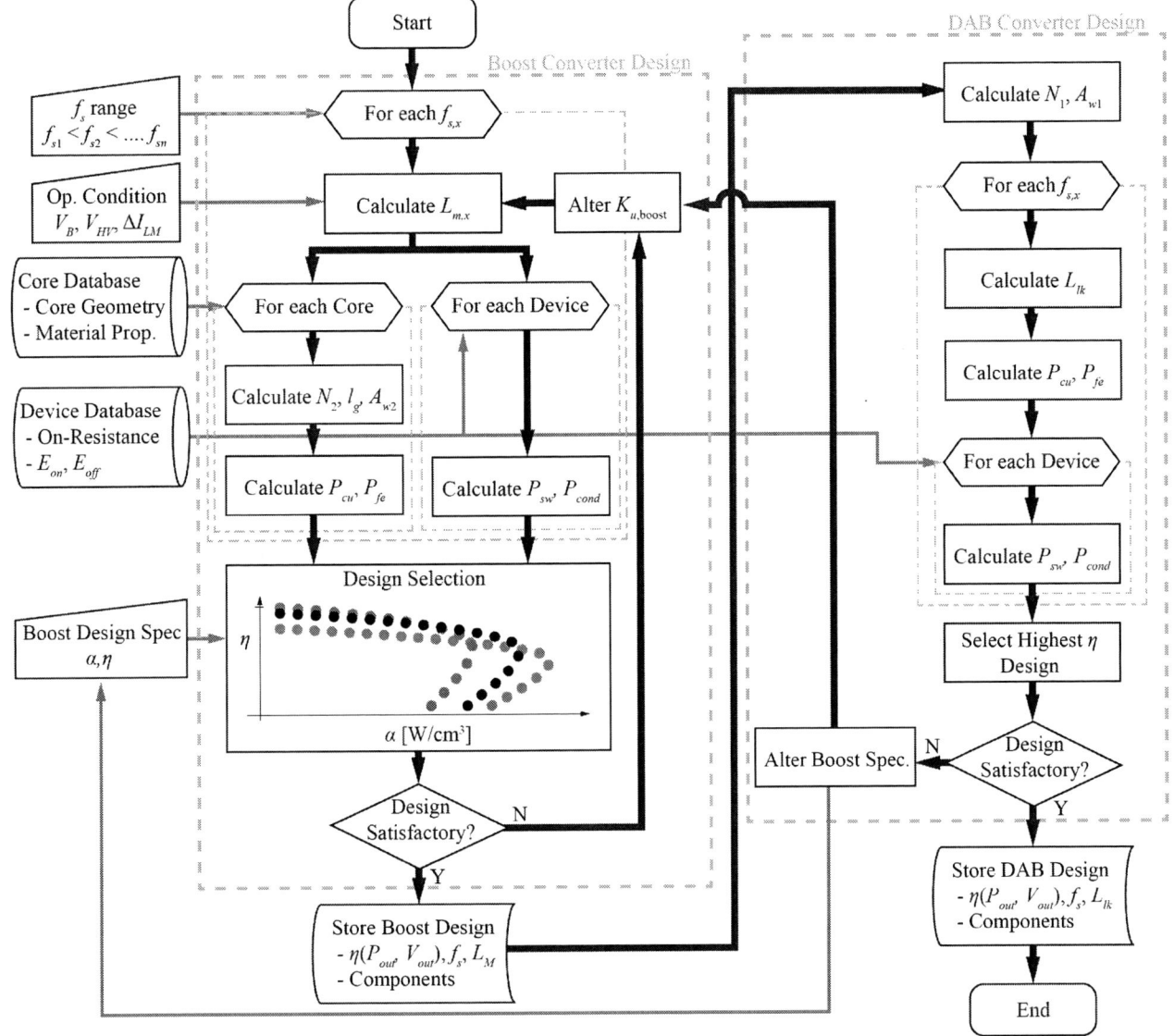

Fig. 5. Design flow chart for the integrated converter operating for traction and charging operation.

$$P_{COND_{main}} = I_{LM}^2 D \left(1 + \frac{1}{3} \left(\frac{\Delta I_{LM}}{I_{LM}} \right)^2 \right) R_{on_{main}} \quad (2)$$

$$P_{SW} = f_s E_{loss} \quad (3)$$

where I_{LM} and E_{loss} represent the average current through the inductor and the switching (E_{on} and E_{off}) energy loss of the semiconductor devices, respectively. For the boost converter, the main switch causes both turn-on and turn-off switching loss. The synchronous switch causes only turn-off switching loss due to the soft turn-on characteristics of a CCM boost converter.

From the device losses and the transformer losses, the efficiency and power density of the boost converter is evaluated. The optimum design considering the efficiency and power density is selected. The converter performance is then evaluated for all operating conditions. If the converter design is deemed unsatisfactory, given the broad range of designs considered, design choices are to increase $K_{u,boost}$, or alter the external inputs to the design process (e.g. expand the core database, or altering design specifications). If the converter specifications are met, the DAB converter design is initiated.

B. DAB converter design

Once the boost converter design is set, a limited set of freedoms in design parameters are available for the DAB design. As the core size, number of secondary windings, and secondary devices have already been determined, only the primary devices and DAB switching frequency are unconstrained.

Though the number or primary turns used in the DAB transformer is also available, it is well established that the

DAB converter achieves highest efficiency, under phase shift modulation, with turns ratio equal to the converter conversion ratio [17]. Thus, the turns ratio is fixed as

$$n_t = \frac{N_2}{N_1} = \frac{V_{HV}}{V_b} \qquad (4)$$

and wire size is selected to completely fill the core window area not occupied by windings already used in the boost.

The power flow for a DAB under phase shift modulation can be expressed as

$$P_{DAB} = \frac{n V_{HV} V_b \phi (\pi - |\phi|)}{2\pi^2 f_{s\,DAB} L_{LK}} \qquad (5)$$

where $f_{s\,DAB}$, ϕ, and L_{LK} represents switching frequency, phase shift of the DAB converter, and leakage inductance of the isolation transformer respectively. The DAB switching frequency selection and corresponding leakage inductance selection is the main challenge. With a high core utilization, ΔB of the core is much larger during DAB operation than boost operation.

However, core loss also increases with higher switching frequency. So, the design tradeoffs are switching frequency, leakage inductance and core loss. At higher switching frequency, the requirement of the leakage inductance is reduced. The advantage of having lower leakage inductance requirement is that the transformer leakage may be sufficient to satisfy the minimum leakage requirement to achieve ZVS. Otherwise, an extra inductor is need to extend the ZVS range, which eventually will increase the converter weight, volume, and loss. The minimum required leakage inductance constraint can be expressed as

$$\frac{1}{2} L_{Lk} I_{LkP}^2 > \frac{1}{2} C_{eq} (2V_B)^2 \qquad (6)$$

where L_{Lk}, I_{LkP}, and C_{eq} represents the leakage inductance of the transformer, the peak leakage current and equivalent switch node capacitance respectively. The condition for the ZVS boundary is plotted in Fig. 6. The average current for the converter operation at the rated power is also shown. The leakage inductance can be approximated by analytical methods provided in [18, 19]. The estimated leakage inductance can be verified by simulation. Since the leakage inductance is more depends on the winding configuration, gap between the layers; the leakage inductance is measured after the construction of the isolation transformer. According to the measured leakage inductance, the designer may either update the analytical model, or alter winding configuration and/or supplement with external inductance as required. In this work, after satisfying the boost magnetization inductance requirement, the DAB winding is constructed with calculated number of turns and AWG wire for the selected switching frequency. Since the minimum leakage inductance requirement is satisfied, the constructed DAB was used for the experiment.

The expected efficiency of the combined converter during the powertrain operation from analytical calculation is shown in Fig. 7, using analytical loss modeling of the converter and the magnetics. The model shows that the DAB converter can significantly improve the efficiency in the high output voltage,

low power operating region, where the vehicle frequently operates in normal city driving.

IV. EXPERIMENTAL RESULTS

The experimental prototype for the proposed converter is constructed and tested to validate the operation and performance of the proposed converter. CREE C2M0080120D SiC MOSFETs are used to implement the eight switches (S_1-S_8). Film capacitors with 200μF used for C_B and C_{HV} which are located at the battery terminal and high voltage terminal of the integrated converter respectively. The integrated transformers are built using two Nanocrystalline cores. The isolation transformer parameters are provided in Table II. The switching frequency of the DAB operation was set initially to 100 kHz, but due to higher core loss generated by the isolation transformer, the switching frequency is reduced to 50 kHz.

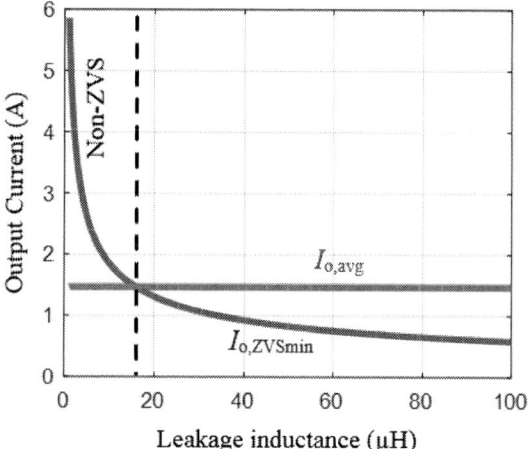

Fig. 6. ZVS boundary for the DAB operation.

TABLE II. ISOLATION TRANSFORMER PARAMETERS

PARAMETER	VALUES
Core type	C
Core model	F3CC06.3
Primary number of Turns, N_1	20
Primary equivalent AWG	16.5
Secondary number of Turns, N_2	45
Secondary equivalent AWG	14
Leakage inductance (Primary)	26.13 μH
Magnetizing inductance (Primary)	86.88 μH
Leakage inductance (Secondary)	135.57 μH
Magnetizing inductance (Secondary)	457.49 μH
Magnetizing coil Resistance (Sec.)	1.1 Ω
Saturation current	19 A
Boost Switching frequency	50 kHz
DAB Switching frequency	50 kHz

Double pulse tests are performed to evaluate the switching loss at light loads at required voltage level for the CREE C2M0080120D SiC MOSFETs in a single phase leg. The waveform of the drain to source voltage (V_{DS}), gate to source voltage (V_{GS}) and drain current (I_D) during the switching turn-off and turn-off operation are shown in Fig. 8 for 200 V bus voltage and 5 A load current. The obtained switching energy loss data is shown in Fig. 9.

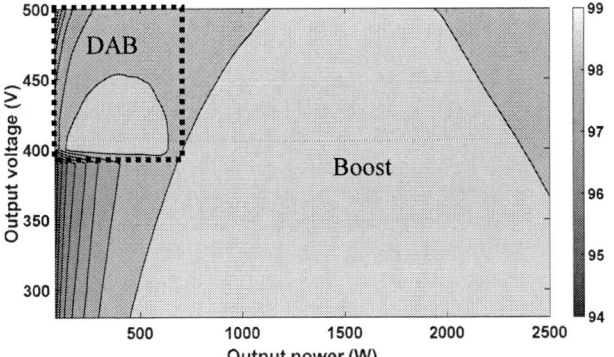

Fig. 7. Calculated efficiency for the integrated converter.

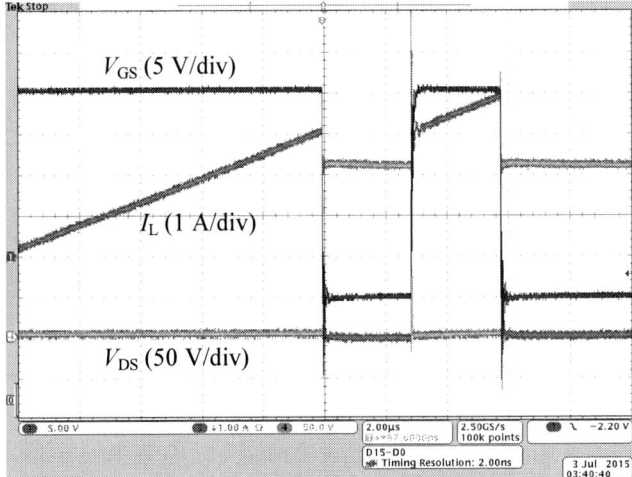

Fig. 8. Double pulse test waveform to evaluate switching loss for DC bus voltage at 200 V and 5 A load current.

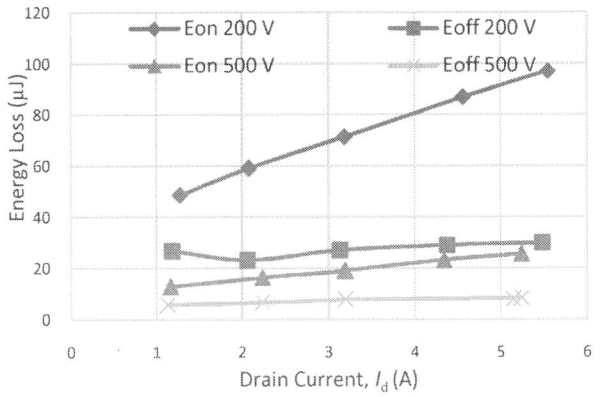

Fig. 9. Turn-on and turn-off energy loss from double pulse test at 500 V and 200 V DC bus voltage at different drain current.

The proposed integrated converter is tested in steady-state at a range of operating conditions. The converter in boost mode is tested at 5 kW power with input voltage 200 V and HV bus voltage 450 V. The waveforms of the inductor current for each phase (I_{LM1} and I_{LM2}), switch node voltage (V_{DS6}), and the gate drive signal applied to the MOSFET S_6 (V_{GS6}) are shown in

Fig. 10. The current is also evenly distributed between the two inductors. The power level is swept from 160 W to 5.2 kW at 200 V battery input voltage and 500 V HV bus voltage, and the measured efficiency is shown in Fig. 11. The interleaved boost converter can provide high efficiency from 700 W to 5 kW operation. However, when the load is lower than 700 W, the efficiency falls below 98%. At maximum load of 5 kW, the total loss is approximately 75 W. All devices reached thermal equilibrium with converter temperatures below 70 °C.

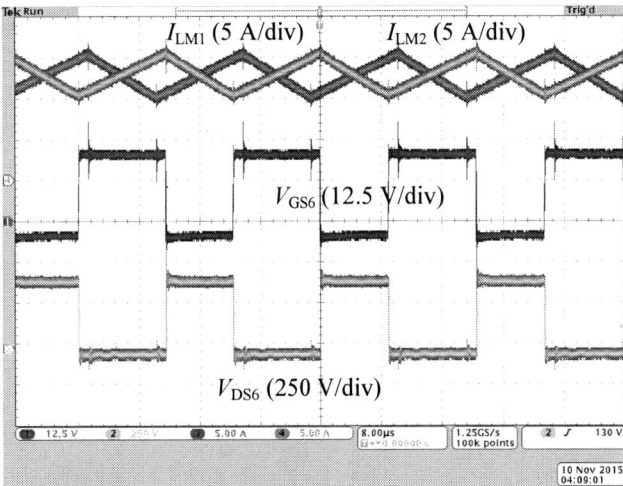

Fig. 10. Waveform of the integrated converter at interleaved boost mode at 5 kW.

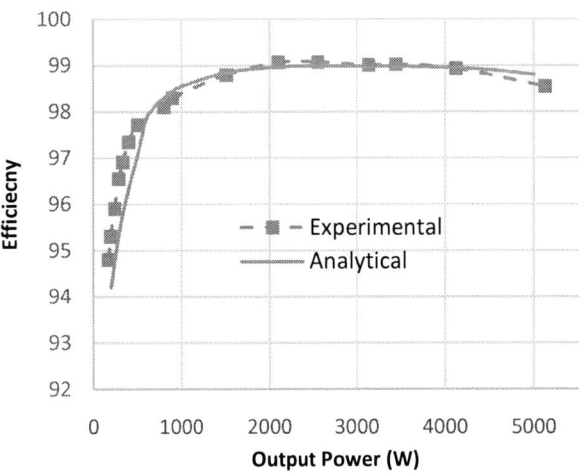

Fig. 11. Efficiency of the interleaved boost converter.

In order to improve the light load efficiency of the interleaved boost converter, the proposed converter is instead operated in DAB mode. The gate to source voltage of S_1 (V_{GS1}), transformer leakage current (I_{LK}), HV bus voltage (V_{HV}) and voltage across the transformer (V_{SW2}) waveforms in DAB mode at 660 W with 200 V input and 450 V output are shown in Fig. 12. The converter operation is tested from 170 W to 660 W. The efficiency of the DAB and interleaved boost mode are compared in Fig. 13. The boost converter suffers lower efficiency at light load because the switching loss dominates at the light load region. Since, DAB can overcome the switching

978-1-4673-9551-9/16 $31.00 © 2016 IEEE   429

loss with ZVS at light load condition, it can provide higher efficiency.

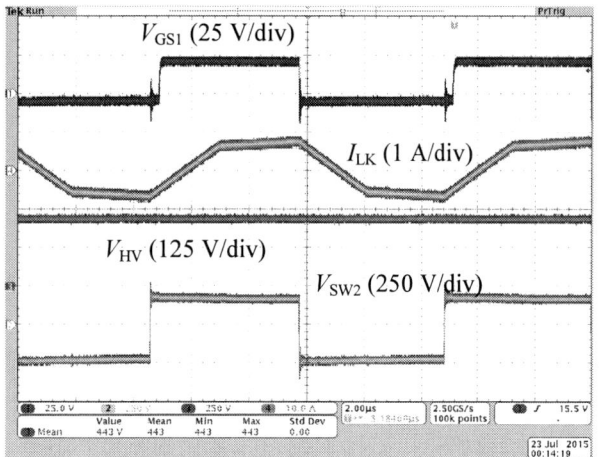

Fig. 12. Waveforms in DAB powertrain mode at 660 W.

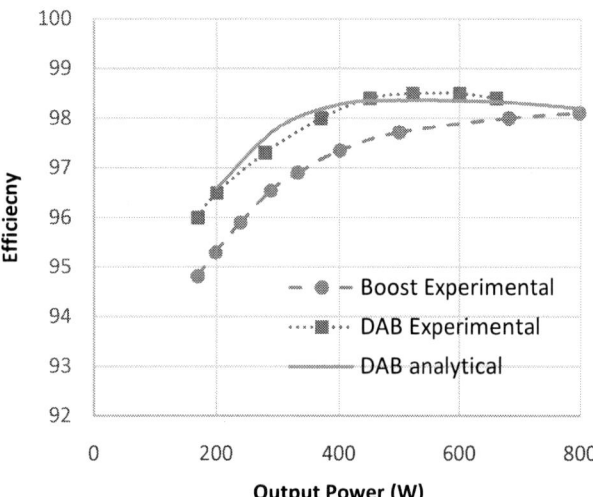

Fig. 13. Efficiency comparison of interleaved boost converter and DAB converter at light load.

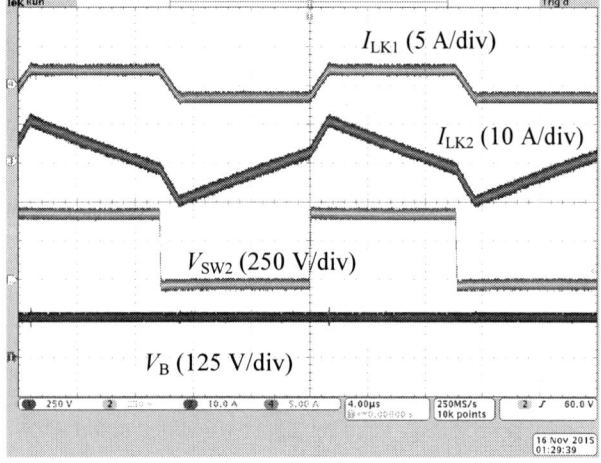

Fig. 14. Waveforms in DAB in charging mode at 660 W.

For charging operation, the integrated converter is tested in DAB mode for the power flow from HV bus to the battery. The gate to source voltage of S_1 (V_{GS1}), transformer leakage current (I_{LK2}), transformer leakage current (I_{LK2}) and voltage across the transformer (V_{SW2}) waveforms in DAB mode at 660 W with 450 V input and 250 V output is shown in Fig. 14. The efficiency of the DAB converter in charging mode is shown in Fig. 15. The efficiency of the DAB converter is lower than the operation of DAB converter in powertrain mode. The main reason is, with the deviation from the unity conversion ratio operation, the ZVS range also decreases for the DAB converter. However, the design can also be optimized for higher charging efficiency having the benefit of shared magnetics.

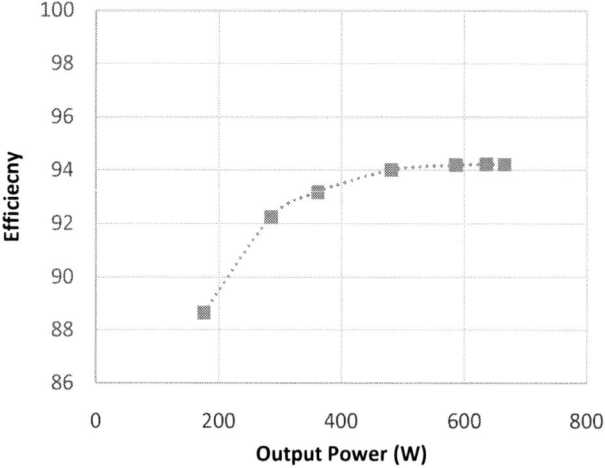

Fig. 15. Efficiency of the converter in DAB mode during charging at 660 W.

V. CONCLUSIONS AND FUTURE WORK

In this paper, an integrated bidirectional converter topology for the EV powertrain is proposed and implemented which can save volume and weight of conventional EV. The reduced weight will help to increase the overall efficiency. The design steps of the integrated converter are presented which can be used to optimize the design for the targeted application.

The proposed topology exhibits key benefits relative to a non-integrated architecture. The number of components in the EV is reduced through sharing of power devices, gate drives, sensing, and control circuitry, and through the sharing of magnetic components between the OBC and drivetrain DC-DC. Further, converter cooling systems are shared. Additionally, the presence integration of the OBC into the drivetrain boost allows a different mode of operation to be employed, which increases system efficiency at light load operating points, where an EV will frequently operate during city driving profiles.

A scaled-down experimental prototype is built and tested to verify the operation of the proposed topology. Results confirm the merit of the proposed topology. In future, the integrated converter model will be used to design and test a full power integrated converter to further showcase the benefits of the system.

978-1-4673-9551-9/16 $31.00 © 2016 IEEE

ACKNOWLEDGMENT

This work made use of the Engineering Research Center Shared Facilities supported by the Engineering Research Center Program of the National Science Foundation and DOE under NSF Award Number EEC-1041877 and the CURENT Industry Partnership Program.

REFERENCES

[1] A. Poullikkas, "Sustainable options for electric vehicle technologies," *Renewable and Sustainable Energy Reviews,* vol. 41, no. 0, pp. 1277-1287, 2015.

[2] O. Hegazy, J. Van Mierlo, and P. Lataire, "Analysis, Modeling, and Implementation of a Multidevice Interleaved DC/DC Converter for Fuel Cell Hybrid Electric Vehicles," *IEEE Transactions on Power Electronics,* vol. 27, no. 11, pp. 4445-4458, 2012.

[3] O. Hegazy, R. Barrero, J. Van Mierlo, P. Lataire, N. Omar, and T. Coosemans, "An Advanced Power Electronics Interface for Electric Vehicles Applications," *IEEE Transactions on Power Electronics,* vol. 28, no. 12, pp. 5508-5521, 2013.

[4] D. Yu, Z. Xiaohu, B. Sanzhong, S. Lukic, and A. Huang, "Review of non-isolated bi-directional DC-DC converters for plug-in hybrid electric vehicle charge station application at municipal parking decks," in *Applied Power Electronics Conference and Exposition (APEC), 2010 Twenty-Fifth Annual IEEE,* pp. 1145-1151, 2010.

[5] W. Haoyu, S. Dusmez, and A. Khaligh, "Design and Analysis of a Full-Bridge LLC-Based PEV Charger Optimized for Wide Battery Voltage Range," *IEEE Transactions on Vehicular Technology,* vol. 63, no. 4, pp. 1603-1613, 2014.

[6] D. Junjun, L. Siqi, H. Sideng, C. C. Mi, and M. Ruiqing, "Design Methodology of LLC Resonant Converters for Electric Vehicle Battery Chargers," *IEEE Transactions on Vehicular Technology,* vol. 63, no. 4, pp. 1581-1592, 2014.

[7] A. Hillers, D. Christen, and J. Biela, "Design of a Highly efficient bidirectional isolated LLC resonant converter," in *Power Electronics and Motion Control Conference (EPE/PEMC), 2012 15th International,* pp. DS2b.13-1-DS2b.13-8, 2012.

[8] B. Whitaker, A. Barkley, Z. Cole, B. Passmore, D. Martin, T. R. McNutt, A. B. Lostetter, L. Jae Seung, and K. Shiozaki, "A High-Density, High-Efficiency, Isolated On-Board Vehicle Battery Charger Utilizing Silicon Carbide Power Devices," *IEEE Transactions on Power Electronics,* vol. 29, no. 5, pp. 2606-2617, 2014.

[9] S. Dusmez and A. Khaligh, "A Compact and Integrated Multifunctional Power Electronic Interface for Plug-in Electric Vehicles," *IEEE Transactions on Power Electronics,* vol. 28, no. 12, pp. 5690-5701, 2013.

[10] L. Young-Joo, A. Khaligh, and A. Emadi, "Advanced Integrated Bidirectional AC/DC and DC/DC Converter for Plug-In Hybrid Electric Vehicles," *IEEE Transactions on Vehicular Technology,* vol. 58, no. 8, pp. 3970-3980, 2009.

[11] T. Schoenen, M. S. Kunter, M. D. Hennen, and R. W. De Doncker, "Advantages of a variable DC-link voltage by using a DC-DC converter in hybrid-electric vehicles," in *IEEE Vehicle Power and Propulsion Conference (VPPC)* pp. 1-5, 2010.

[12] R. Karimi, T. Koeneke, D. Kaczorowski, T. Werner, and A. Mertens, "Low voltage and high power DC-AC inverter topologies for electric vehicles," in *IEEE Energy Conversion Congress and Exposition,* pp. 2805-2812, 2013.

[13] S. Tenner, S. Gunther, and W. Hofmann, "Loss minimization of electric drive systems using a DC/DC converter and an optimized battery voltage in automotive applications," in *Vehicle Power and Propulsion Conference (VPPC), 2011 IEEE,* pp. 1-7, 2011.

[14] M. Yilmaz and P. T. Krein, "Review of Battery Charger Topologies, Charging Power Levels, and Infrastructure for Plug-In Electric and Hybrid Vehicles," *IEEE Transactions on Power Electronics,* vol. 28, no. 5, pp. 2151-2169, 2013.

[15] *SAE Electric Vehicle and Plug-in Hybrid Electric Vehicle Conductive Charge Coupler,* SAE International Standard J1772, Jan. 2010.

[16] W. Zhang, S. Anwar, D. Costinett, and F. Wang, "Investigation of Cost-effective SiC Based Hybrid Switch and Improved Inductor Design Procedure for Boost Converter in Electrical Vehicles Application," in *SAE Technical Paper* 2015.

[17] D. Costinett, D. Maksimovic, and R. Zane, "Design and Control for High Efficiency in High Step-Down Dual Active Bridge Converters Operating at High Switching Frequency," *IEEE Transactions on Power Electronics,* vol. 28, no. 8, pp. 3931-3940, 2013.

[18] P. Gomez, L. de, x, and F. n, "Accurate and Efficient Computation of the Inductance Matrix of Transformer Windings for the Simulation of Very Fast Transients," *IEEE Transactions on Power Delivery,* vol. 26, no. 3, pp. 1423-1431, 2011.

[19] M. A. Bahmani and T. Thiringer, "Accurate Evaluation of Leakage Inductance in High-Frequency Transformers Using an Improved Frequency-Dependent Expression," *IEEE Transactions on Power Electronics,* vol. 30, no. 10, pp. 5738-5745, 2015.

A 1 MHz Bi-directional Soft-switching DC-DC Converter with Planar Coupled Inductor for Dual Voltage Automotive Systems

Chenhao Nan and Raja Ayyanar
School of Electrical, Computer and Energy Engineering
Arizona State University
Tempe, Arizona 85281
Email: cnan2@asu.edu

Abstract—The 48V – 14V automotive power system is gaining acceptance - due to the increasing number and power rating of electrical and electronic components in vehicles to support advanced functionalities. Multiphase synchronous bi-directional buck/boost converter is currently employed for connecting two bus voltages. However, it has low efficiency at high frequency operation and high EMI noise due to its hard-switching. A zero-voltage transition bi-directional buck/boost converter with coupled inductor is proposed for this application, which provides ZVS for main switches and ZCS for auxiliary switches, and features wide ZVS range and low loss in auxiliary branch. The operating principles including ZVS/ZCS mechanism, details of circuit design, and experimental results from a 1 MHz and 250 W prototype are presented.

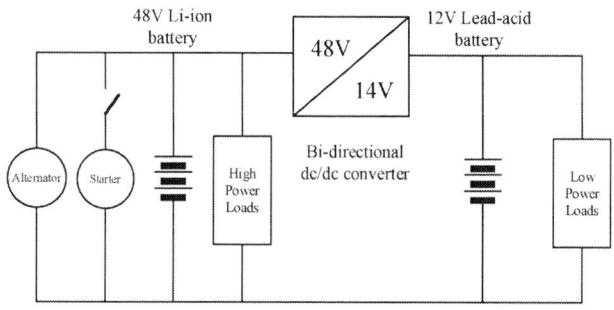

Fig. 1. System layout of dual voltage automotive system.

I. Introduction

As modern automobiles are getting smarter, safer, and more luxurious, more and more electronic and electrical components are being implemented in automotive vehicles. Nowadays, the total electronic and electrical loads on vehicles are around 2 kW and, in the near future, the total electric power required by most internal combustion engine (ICE) vehicles may exceed 5 kW [1]. The conventional 14V PowerNet on most vehicles has reached its limit to support vehicles electrification due to the high current level with 14V bus and hence thicker and lossy wirings. Therefore, 42V PowerNet was proposed in 2000 [2] and recently, 48V PowerNet was introduced as a supplementary power supply for high power applications on vehicles [3], such as Electrical Power Steering (EPS), Electronic Braking Systems (EBS) and HVAC systems. With the 48V voltage, these loads can operate more efficiently and achieve better performance. Moreover, wiring and component weight is reduced leading to reduced fuel consumptions and CO2 emission. Meanwhile, 14V PowerNet is still kept for providing power for low power loads.

Since there are dual voltage systems on board, a bi-directional non-isolated dc-dc converter is required to connect the 14V low voltage and 48V high voltage systems, as shown in Fig. 1. Typically, the power rating of this bus converter is up to 3 kW according to alternator capacity.

Currently, the most popular and accepted solution for this application is multiphase interleaved synchronous buck/boost converter [4]–[7], due to its simplicity, robustness, good thermal management, and scalability for power rating. However, the issues with synchronous buck/boost converter are also obvious: low efficiency at high switching frequency operation and high EMI noise due to its hard-switching and fast slew rate. Thus, most synchronous buck/boost based converter for this application are running with a switching frequency less than 200 kHz and hence have bulky passive components [4]–[7]. In addition, there are very stringent requirements on EMI performance for automotive dc-dc converters, and it is desired to push the switching frequency of automotive converters above 2 MHz, to avoid AM band interference [8], [9] and for compact size and light weight of the converter. Therefore, the previous two problems of synchronous buck/boost converter are significant and imperative to be solved.

Besides bi-directional synchronous buck/boost converter, there are some other topologies have been proposed for this application. Reference [1] and [10] proposed two switched-capacitor based converters; reference [11] proposed a multilevel switching cell based converter with fault tolerant. However, all of them are still with hard-switching. Therefore, the high efficiencies reported in [1], [10] and [11] are only achieved at very low switching frequencies (<100 kHz). A ZVS cuk converter was proposed by [12], but it employs too many components and is not easy for interleaving and hence scaling up the power rating. If a single converter is used to output 1 kW to 3 kW, usually a large heatsink is required resulting in a bulky and heavy converter, which is not desired for automotive applications. Therefore, interleaving several converters is still preferred for this application. For achieving high efficiency and good EMI performance, it is desired to have soft-switching converters to be interleaved. Many resonant or quasi-resonant converters with frequency control or even PWM control are not suitable for interleaving. However, the zero-voltage-transition (ZVT) converters with resonance only occurs in the switching transitions are well compatible with interleaving technique.

There are several non-synchronous ZVT converter proposed previously [13]–[15], but they suffer from one or more of the following drawbacks – limited ZVS range, high voltage rating for auxiliary MOSFETs, and high current magnitude and hence conduction loss in auxiliary branch. In this paper, a multiphase interleaved bi-directional soft-switching (ZVT) dc-dc converter with coupled inductor, is proposed for dual voltage automotive systems to achieve high efficiency and low EMI at high frequency operation. Since each phase is exactly the same, in this paper, only a single phase of the proposed bi-directional ZVT dc-dc converter with coupled inductor is used for analysis and design, and to demonstrate the converter operation and efficiency performance.

II. Theoretical Analysis of the Proposed Topology

Fig. 2 shows the proposed multiphase interleaved bi-directional ZVT converter. Compared with conventional synchronous buck/boost converter, an additional resonant network consisting of auxiliary back-to-back MOSFETs (Saux1 and Saux2) and a coupled inductor is added in each phase. The coupled inductor integrates the buck/boost filter inductor and an extra winding into one magnetic structure. L_f is the primary side inductance employed as the filter inductor and L_l is the leakage of the coupled inductor referred to the auxiliary side, used as the resonant inductor during the operation. The primary to the auxiliary side turns ratio is 1:n. C_r is the resonant capacitor which combines external C_{ds} and output capacitance of the main MOSFETs (S1 and S2).

A. Operating Principles

For simplifying the analysis of operation, as shown in Fig. 3, the coupled inductor is modeled as an ideal transformer in parallel with the filter inductance, and the leakage inductance L_l is in series with the secondary winding. Further, the filter inductance is assumed sufficiently large so that it could be considered as a constant current source I_L. Similarly, the output capacitor and load could also be represented by a voltage source V_o assuming very large output capacitance.

Fig. 4 shows the key waveforms of the bi-directional ZVT dc-dc converter with coupled inductor in buck mode and boost mode, respectively. It should be noted that, in key waveforms of boost mode, the high voltage output side is still labeled as V_{in} and the low voltage input side is labeled as V_o. Since the operations of two modes are dual, only buck mode operating

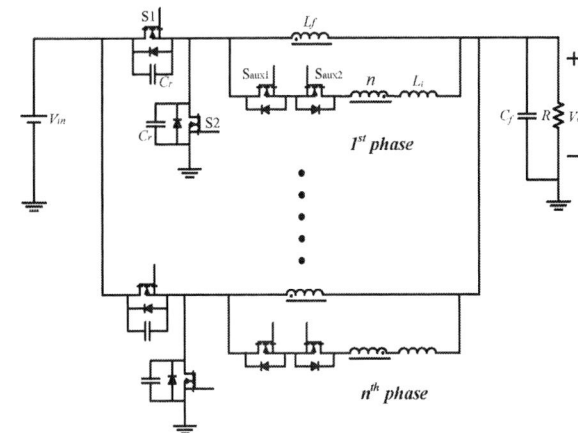

Fig. 2. Proposed multiphase interleaved bi-directional ZVT converter with coupled inductor.

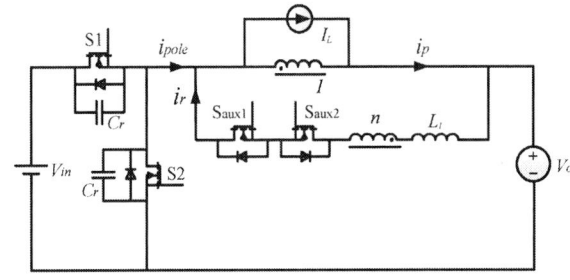

Fig. 3. Simplified circuit diagram for a single phase bi-directional ZVT converter with coupled inductor.

principle is described as following. In steady state operation, a switching cycle could be divided into eight intervals and Fig. 5 shows the topological state in each interval.

Interval 1 (T_0-T_1): high side MOSFET (S1) is off and synchronous rectifier (S2) is conducting the current IL. Auxiliary branch is off and hence no current is in the ideal transformer.

Interval 2 (T_1-T_2): at T1, auxiliary switch Saux2 turns on and the leakage inductance L_l is linearly charged by $(n+1)\cdot V_o$ through Saux2 and the body diode of Saux1. Meanwhile, the current in the ideal transformer primary winding decreases proportionally. During this interval, the switching pole current i_{pole} also changes from positive to negative.

Interval 3 (T_2-T_3): at T2, S2 is turned off, then i_{pole} charges and discharges C_r in a resonant manner till the voltages across S2 and S1 reach V_{in} and zero at T3, respectively.

Interval 4 (T_3-T_4): the body diode of S1 is forward biased during this interval. Leakage inductance L_l is linearly discharged by $(n+1)\cdot(V_{in}-V_o)$, and i_{pole} is still negative. S1 should be turned on within this interval to realize ZVS.

Interval 5 (T_4-T_5): L_l continues being linearly discharged till its current i_r falls to zero and the body diode of Saux1 is turned off with ZCS. At the same time, the ideal transformer primary winding current rises back to zero and the switching pole current i_{pole} increases back to filter inductor current I_L.

978-1-4673-9551-9/16 $31.00 © 2016 IEEE   433

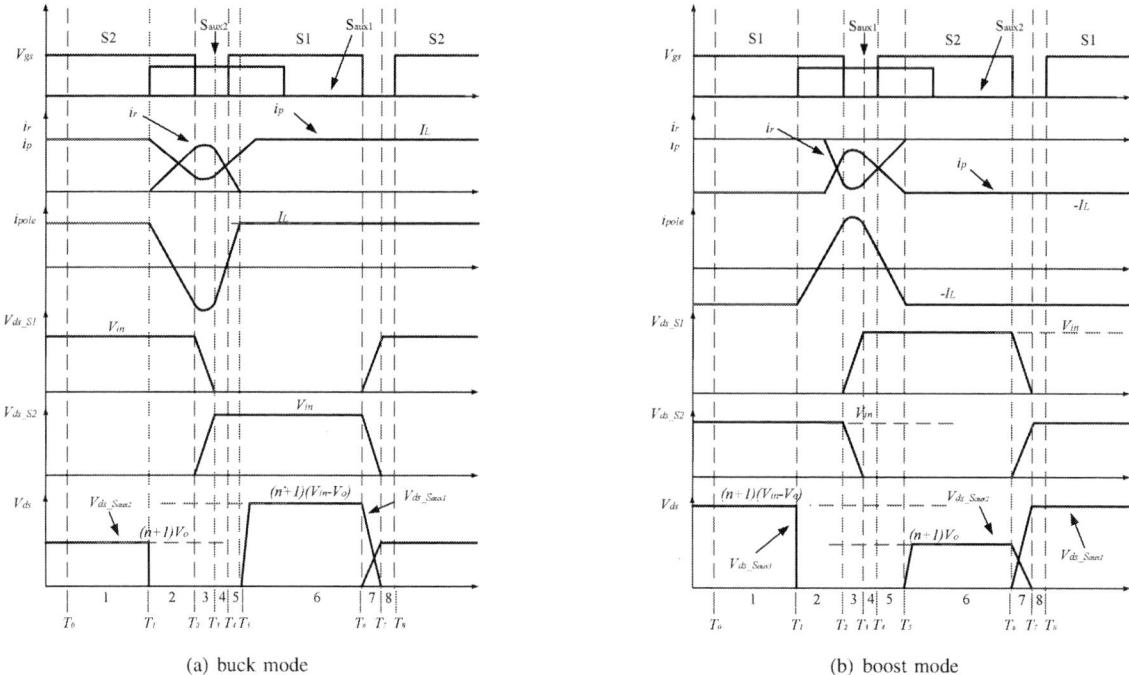

(a) buck mode   (b) boost mode

Fig. 4. Key waveforms in bi-directional ZVT converter with coupled inductor.

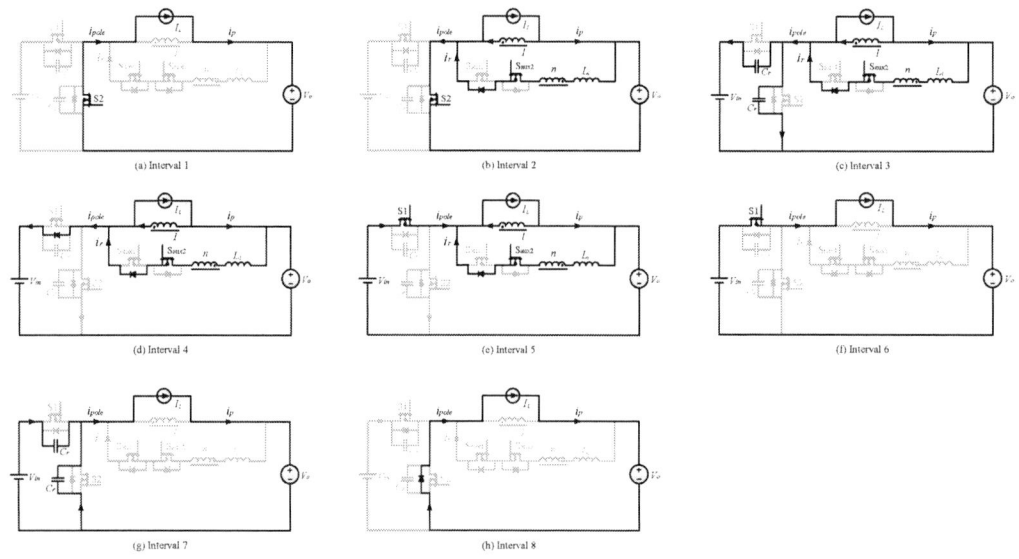

Fig. 5. Topological state in each interval.

At T5, Saux2 is switched off with ZCS.

Interval 6 (T_5-T_6): auxiliary branch is off. S2 is off and S1 is conducting I_L, like in conventional buck converter.

Interval 7 (T_6-T_7): at T6, S1 is turned off. During this interval, current is transferred from S1 to S2 body diode in a soft-switching manner. At T7, S2 body diode is forward biased and stars carrying current. Transition in this interval is the same as that in conventional synchronous buck converter.

Interval 8 (T_7-T_8): during this interval, S2 body diode is conducting and S2 should be turned on within this interval to achieve ZVS. After T8, S1 is off and S2 is on. Circuit state is the same as that in Interval 1.

B. Converter Features

From the analysis of the operation, it could be seen that the proposed bi-directional ZVT converter with coupled inductor has several advantages for dual voltage automotive systems:

1) Soft-switching for all switches: buck/boost converter main switches S1 and S2 are switched with ZVS and auxiliary switches Saux1 and Saux2, and their body diodes, are switched

978-1-4673-9551-9/16 $31.00 © 2016 IEEE   434

with ZCS. Therefore, it has potential to achieve high efficiency at high frequency and low EMI noise.

2) Wide ZVS range: in most other soft-switching converters, ZVS is lost at light load or at high input voltage. Proposed converter can still achieve ZVS at light load and at high voltage only if the leakage current is sufficiently charged and then the energy stored in leakage inductance would be enough to charge and discharge resonant capacitors. Actually, for this ZVT converter, achieving ZVS at light load is easier.

3) Low loss in auxiliary branch: because of the coupled inductor, the peak current required in the auxiliary branch could be much lower than the filter inductor current while the ZVS of S1 (buck mode) or S2 (boost mode) is still realized. Furthermore, with the coupled inductor, the charging and discharging voltage across the leakage inductance during Interval 2 and 4 are at least doubled compared to the charging and discharging voltages with a separate resonant inductor. Then the charging and discharging time is much shortened. Therefore, with the coupled inductor, the average current and hence conduction loss in the auxiliary branch, which is a very lossy part in high current applications like in automotive 48/14V converter, could be much reduced. Another benefit comes from the coupled inductor is the small core loss compared with the separate resonant inductor, which helps to improve the efficiency further.

4) Flexibility for interleaving: since the resonance only occurs during switch transitions, basic buck/boost converter control as well as the interleaving for buck/boost converter are not affected. With the interleaving technique and the high efficiency of each phase, the 1 kW 3 kW automotive 48/14V converter consisting of several phases may even avoid the use of heatsink, reducing size and weight significantly.

5) Compact size: compared with traditional buck converter filter inductor, coupled inductor employed by this ZVT converter only needs a secondary winding which carries small averaging current. Thus the space occupied by the secondary winding could be negligible if a thin wire is employed as the secondary winding. For planar coupled inductor with PCB as windings, only one layer might be enough for the secondary winding. If both windings are in one PCB, the space occupied by the secondary winding is even more negligible. Therefore, the size of the coupled inductor will only be slightly bigger than a single filter inductor. In addition, only two auxiliary MOSFETs and their associated drivers are required so the size of the converter could be compact.

C. Design Considerations

1) Coupled inductor turns ratio selection: there is a trade-off in determining the coupled inductor turns ratio between auxiliary branch current magnitude and the voltage rating of auxiliary MOSFETs. The higher the turns ratio n is, the lower the magnitude of the auxiliary current i_r is required for achieving ZVS for S1 and S2. However, the voltage rating for auxiliary MOSFETs, Saux1 and Saux2, also increases proportionally as n increases. With a turns ratio of n, the

voltage ratings for Saux1 and Saux2 are $(n+1) \cdot (V_{in} - V_o)$ and $(n+1) \cdot V_o$, respectively, for both buck and boost modes.

2) Auxiliary switch timing: auxiliary switch timing is critical for achieving ZVS of S1 in buck mode or S2 in boost mode. Assuming $i_{r,zvs}$ is the desired peak current in auxiliary branch to achieve ZVS, then for ZVS of S1, the interval t_{12} shown in Fig. 4(a) should be $t_{12} = i_{r,zvs}/[(n+1) \cdot V_o/L_l]$ For ZVS of S2, the interval t_{12} shown in Fig. 4(b) should be $t_{12} = i_{r,zvs}/[(n+1) \cdot (V_{in} - V_o)/L_l]$.

III. CONVERTER DESIGN AND IMPLEMENTATION

A. Converter Design

1) Per phase power rating: as the interleaving technique is used, the converter power capability could be scaled up to 3 kW with enough phases. However, the power capability for each phase needs to be determined. Lower power rating for each phase helps to reduce the power loss and better dissipate the heat while it requires more components and hence cost. For a compromise of thermal management and number of components, the power rating for each phase is specified as 250 W. Thus, the design specifications for one phase converter is shown in Table I.

TABLE I
CONVERTER DESIGN SPECIFICATIONS

Rated Power	250 W
Switching Frequency	1 MHz
High Side Voltage	48 V
Low Side Voltage	14 V

2) Turns ratio of coupled inductor: aforementioned analysis shows that higher turns ratio n results in higher voltage rating for auxiliary MOSFETs. In order to take the advantages of 100V MOSFETs for auxiliary switches, considering the potential voltage ringing during the turn-off of auxiliary MOSFETs, a turns ratio of 1:1 is designed for the coupled inductor. Then, the peak current required in auxiliary branch to achieve ZVS of main MOSFETs is reduced by half compared to the value required in ZVT converters with a separate resonant inductor. However, the voltage rating for Saux1 and Saux2 should be at least 68V and 28V, respectively.

3) Simulation results: Table II shows the designed parameters for each phase converter and Fig. 6 shows simulation results at $48V_{in}, 14V_{out}/18A$ in buck mode. It can be seen that the ZVS for both S1 and S2 are realized even when the auxiliary branch peak current is less then the primary winding current.

B. Circuit Design and Implementation

1) Coupled inductor design: Coupled inductor is a key component in the proposed ZVT converter. For achieving a low average current and hence low conduction loss in auxiliary branch, 40 nH leakage inductance is designed for the coupled inductor. For the conventional design with vertical cores and wire-based windings, it is very difficult to achieve such a

TABLE II
DESIGNED PARAMETERS

Primary Inductance L_f	1.5 uH
Leakage Inductance L_l	40 nH
Output Capacitance C_o	80 uF
Resonant Capacitance C_r	1 nF
Turns Ratio 1 : n	1 : 1

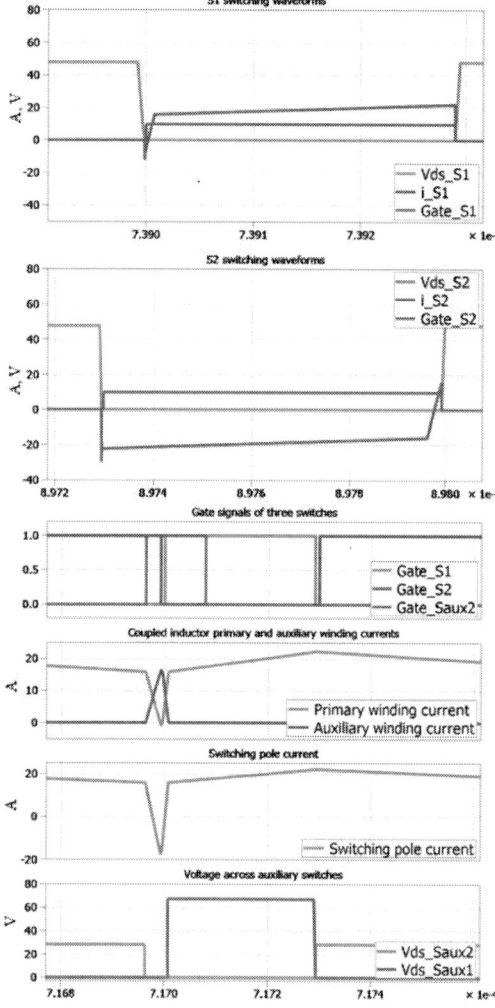

Fig. 6. Simulation results in buck mode operation.

low leakage for coupled inductor. Therefore, a planar coupled inductor with PCB windings is designed since it has lower leakage compared with vertical coupled inductor. Moreover, the fabrication of planar design with PCB windings makes it easier to achieve repeatable and significantly lower parameter tolerance between two coupled inductors, which is critical for a multiphase converter considering the current sharing issue. If the components tolerance causes significant current unbalance among phases, active current-sharing method might be needed [15], [16].

Considering the thermal management and the efficiency requirement, the DCR for the primary winding (high RMS current) is limited to 4 m and the core loss at 1 MHz should be less than 1 W. Design specifications are summarized in Table III. Table IV shows the designed parameters, and Fig. 7 shows the structure and Maxwell 3D modeling of the designed coupled inductor. A ferrite core 3F45 from Ferroxcube suitable for ¡3MHz operation is chosen. Two separate but same 4-layer PCBs are employed as primary and auxiliary windings, respectively. Each layer has only one turn and the configuration is paralleled layer 1 and 3, in series with paralleled layer 2 and 4. For further improving the efficiency, two 6-layer PCBs could be employed to reduce DCR.

TABLE III
DESIGN SPECIFICATIONS FOR COUPLED INDUCTOR

Turns Ratio 1 : n	1 : 1
Primary Inductance	1.5 uH
Leakage Inductance	40 nH
Coupling Coefficient	0.973
DCR	<4 mΩ
Core Loss (at 1 MHz)	<1 W

TABLE IV
DESIGNED PARAMETERS OF COUPLED INDUCTOR

Core Material	3F45
Core Shape	E22/6/16+PLT22/16/2.5
Primary/auxiliary number of turns	2 turns
ΔB_{max}	60 mT
DCR	3.1 mΩ
Air Gap l_g	0.3 mm
Leakage Gap	0.2 mm

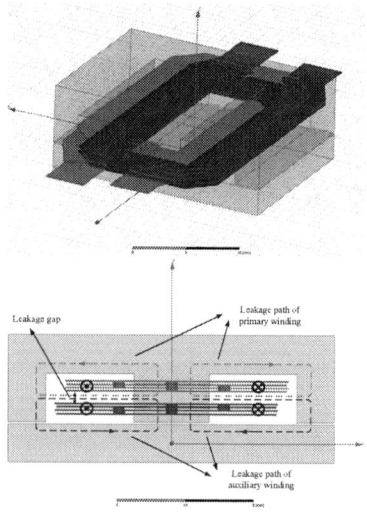

Fig. 7. Maxwell 3D modeling of designed coupled inductor.

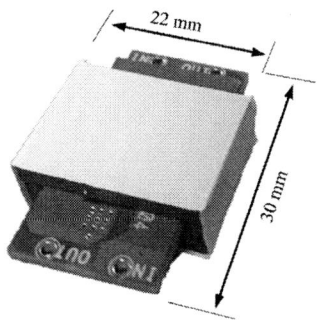

Fig. 8. Constructed planar coupled indcutor.

One important parameter for this coupled inductor design is the leakage inductance. To precisely control the leakage inductance, two separate PCBs are used for windings and leave the leakage gap between them adjustable. The shorter leakage gap increases the reluctance of the leakage path and leads to lower leakage inductance. To obtain a 40 nH leakage inductance, Maxwell 3D simulation shows that the leakage gap should be 0.2 mm. Finally, the constructed inductor, as shown in Fig. 8, gives a total leakage inductance of 30 nH. Considering the PCB trace and interconnection stray inductance, 30 nH is assumed to be enough for the design. In fact, once the leakage gap is finalized by experiment, only one PCB is enough for both primary and secondary windings, with distance between two winding layers the same as the leakage gap. Since the secondary winding carries very small average current, only one layer is enough for secondary winding. Then, the secondary winding almost takes no space, and the coupled inductor size could be kept the same as the filter inductor size in conventional buck converter.

2) Power state design: In order to verify the proposed bi-directional ZVT converter with coupled inductor, a 1 MHz, single phase 250 W prototype is constructed, as shown in Fig. 9. For higher power rating, more phases could be easily interleaved with the built prototype. The design specifications and parameters are shown in Table I and II. For achieving high efficiency, two MOSFETs are paralleled for both S1 and S2. Table V shows the key components selection and parameters.

TABLE V
KEY COMPONENTS SELECTION AND VALUES

Components	Part Number	Parameters
Main MOSFETs	BSZ123N08NS3G	80V, 12.3mΩ, 6.3nC
Auxiliary MOSFETs	BSZ440N10NS3G	100V, 44mΩ, 9.1nC
Main MOSFETs driver	UCC27211	bootstrap
Auxiliary MOSFETs driver	UCC27524	2-channel, low side

IV. EXPERIMENTAL RESULTS

Fig. 10 shows the gate signals of S1, S2 and Saux2 in buck mode operation. The gate signal for Saux1 is constantly off. Fig. 11 and 12 are the switching waveforms of S1 and S2, respectively, at 48V input, 14V output and 15A load.

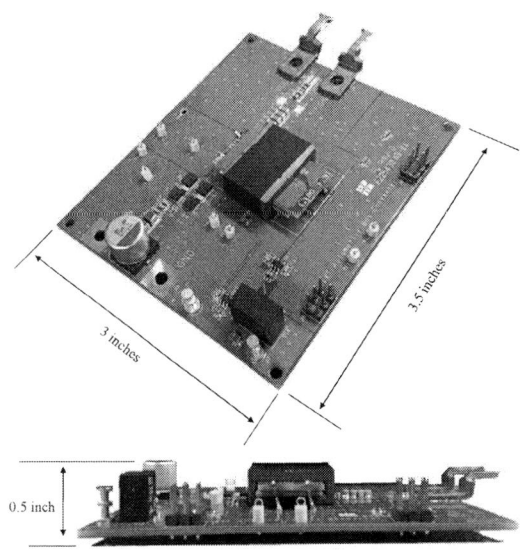

Fig. 9. 1 MHz, 250W prototype of the bi-directional ZVT converter with coupled inductor.

It could be seen that ZVS of both switches are realized. At light load, ZVS realization of S1 is more easily. In addition, since two small capacitors are placed in parallel with S1 and S2 respectively, slowing the voltage rising at turn-off, the turn-off of S1 and S2 is almost ZCS. Fig. 13 shows the voltage waveforms across auxiliary MOSFETs in buck mode operation. There is a voltage ringing during the turn-off of Saux1 due to the resonance between the leakage inductance and the output capacitance of Saux1. However, this voltage could be easily clamped by a diode placed between input voltage and the drain node of Saux1.

The waveforms associated with boost mode operation are shown in Fig. 14 to Fig. 17. Fig. 14 is the gate signals of S1, S2 and Saux1, and Saux2 is constantly off in boost mode. The switching waveforms of S1 and S2, at 14V input, 48V output and 4A load, are shown in Fig. 15 and 16, respectively. Both switches can achieve ZVS turn-on and almost ZCS turn-off. Fig. 17 shows the auxiliary MOSFETs voltage waveforms. Still, Saux1 withstands the higher voltage.

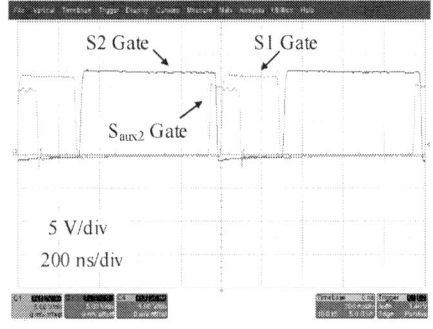

Fig. 10. Gate signals in buck mode.

978-1-4673-9551-9/16 $31.00 © 2016 IEEE

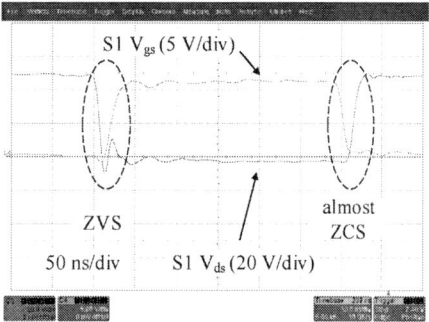

Fig. 11. Switching waveforms of S1 in buck mode.

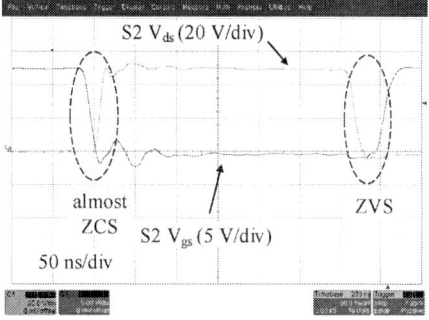

Fig. 12. Switching waveforms of S2 in buck mode.

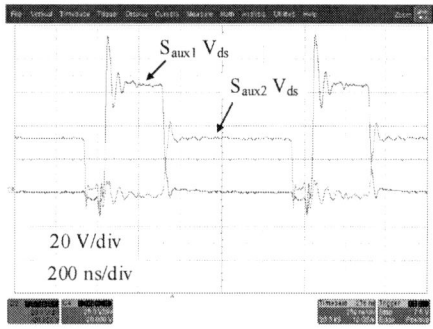

Fig. 13. Voltage waveforms of auxiliary MOSFETs in buck mode.

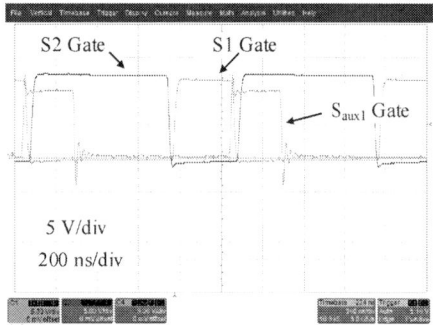

Fig. 14. Gate signals in boost mode.

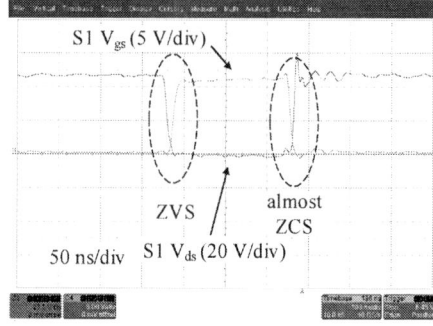

Fig. 15. Switching waveforms of S1 in boost mode.

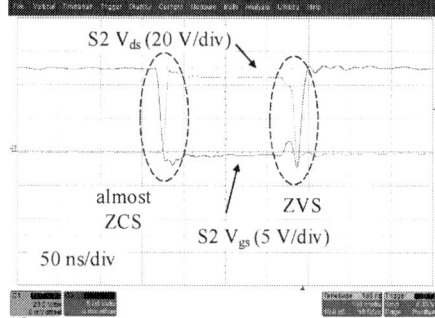

Fig. 16. Switching waveforms of S2 in boost mode.

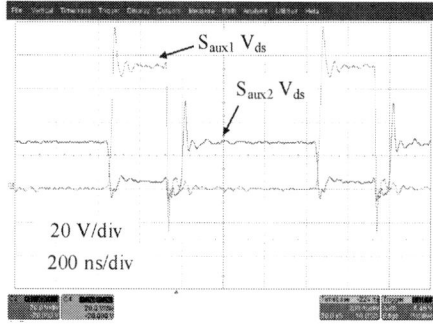

Fig. 17. Voltage waveforms of auxiliary MOSFETs in boost mode.

The conversion efficiency is measured for both buck mode (48V to 14V) and boost mode (14V to 48V), and is shown in Fig. 18. A peak efficiency of 93.98% is achieved at 133 W output in buck mode operation. In boost mode, the peak efficiency is 92.99% at 160 W.

V. CONCLUSION

In this paper, a novel high frequency bi-directional ZVT converter with coupled inductor is proposed for automotive dual voltage systems. With the combination of the ZVT and coupled inductor, this converter features high efficiency, low EMI noise, scalability for high power output, and low profile design. The operating principles, ZVS analysis as well as the details of the converter design and implementation are presented. Comprehensive experimental results from a 1 MHz,

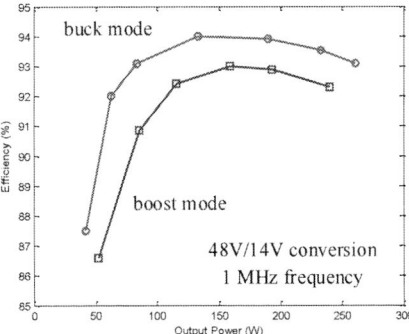

Fig. 18. Efficiency curves.

250 W prototype verified the operation and good performance of the proposed bi-directional ZVT converter with coupled inductor.

REFERENCES

[1] Flores Cortez, D.; Waltrich, G.; Fraigneaud, J.; Miranda, H.; Barbi, I., "DCDC Converter for Dual-Voltage Automotive Systems Based on Bidirectional Hybrid Switched-Capacitor Architectures," in *Industrial Electronics, IEEE Transactions on* , vol.62, no.5, pp.3296,3304, May 2015

[2] Kassakian, J.G., "Automotive electrical systems-the power electronics market of the future," in *Applied Power Electronics Conference and Exposition, 2000. APEC 2000. Fifteenth Annual IEEE* , vol.1, no., pp.3-9 vol.1, 2000

[3] Jan Fischer-Wolfarth, Gereon Meyer, Advanced Microsystems for Automotive Applications 2014: Smart Systems for Safe, Clean and Automated Vehicles, Springer, 2014

[4] Garcia, O.; Zumel, P.; de Castro, A.; Cobos, J.A., "Automotive DC-DC bidirectional converter made with many interleaved buck stages," in *Power Electronics, IEEE Transactions on* , vol.21, no.3, pp.578-586, May 2006

[5] Czogalla, J.; Jieli Li; Sullivan, C.R., "Automotive application of multiphase coupled-inductor DC-DC converter," in *Industry Applications Conference, 2003. 38th IAS Annual Meeting. Conference Record of the* , vol.3, no., pp.1524-1529 vol.3, 12-16 Oct. 2003

[6] Neugebauer, T.C.; Perreault, D.J., "Computer-aided optimization of DC/DC converters for automotive applications," in *Power Electronics, IEEE Transactions on* , vol.18, no.3, pp.775-783, May 2003

[7] Seung-Yo Lee; Pfaelzer, A.G.; van Wyk, J.D., "Comparison of Different Designs of a 42-V/14-V DC/DC Converter Regarding Losses and Thermal Aspects," in *Industry Applications, IEEE Transactions on* , vol.43, no.2, pp.520-530, March-april 2007

[8] John Rice and Sanmukh Patel, Designing an EMC-Compliant Automotive Switching Buck Regulator, Texas Instruments.

[9] Chenhao Nan; Ayyanar, R.; Youhao Xi, "High frequency active-clamp buck converter for low power automotive applications," in *Energy Conversion Congress and Exposition (ECCE), 2014 IEEE*, vol., no., pp.3780-3785, 14-18 Sept. 2014

[10] Fang Zheng Peng; Fan Zhang; Zhaoming Qian, "A magnetic-less DC-DC converter for dual-voltage automotive systems," in *Industry Applications, IEEE Transactions on* , vol.39, no.2, pp.511-518, Mar/Apr 2003

[11] Gleissner, M.; Bakran, M.M., "Design and control of fault-tolerant non-isolated multiphase multilevel dcdc converters for automotive power systems," in *Industry Applications, IEEE Transactions on* , vol.PP, no.99, pp.1-1

[12] Jose, P.; Mohan, N., "A novel ZVS bidirectional Cuk converter for dual voltage systems in automobiles," in *Industrial Electronics Society, 2003. IECON '03. The 29th Annual Conference of the IEEE* , vol.1, no., pp.117-122 vol.1, 2-6 Nov. 2003

[13] Hua, G.; Leu, C.S.; Yimin Jiang; Lee, F.C.Y., "Novel zero-voltage-transition PWM converters," in *Power Electronics, IEEE Transactions on* , vol.9, no.2, pp.213-219, Mar 1994

[14] de Freitas, L.C.; Coelho Comes, P.R., "A high-power high-frequency ZCS-ZVS-PWM buck converter using a feedback resonant circuit," in *Power Electronics, IEEE Transactions on* , vol.10, no.1, pp.19,24, Jan 1995

[15] Smith, K.M., Jr.; Smedley, K.M., "A comparison of voltage-mode soft-switching methods for PWM converters," in *Power Electronics, IEEE Transactions on* , vol.12, no.2, pp.376-386, Mar 1997

[16] Shiguo Luo; Zhihong Ye; Ray-Lee Lin; Lee, F.C., "A classification and evaluation of paralleling methods for power supply modules," in *Power Electronics Specialists Conference, 1999. PESC 99. 30th Annual IEEE* , vol.2, no., pp.901-908 vol.2, 1999

[17] Chenhao Nan; Angkititrakul, S.; Zhixiang Liang, "Optimal design of a redundant high current DC/DC converter," in *Applied Power Electronics Conference and Exposition (APEC), 2015 IEEE* , vol., no., pp.2109-2115, 15-19 March 2015

A Bridgeless Totem-pole Interleaved PFC Converter for Plug-In Electric Vehicles

Yichao Tang[1,2,3], *Member, IEEE*, Weisheng Ding[2,3], *Student Member, IEEE*, and Alireza Khaligh[2,3], *Senior Member, IEEE*

[1]Kilby Labs, Texas Instruments, Santa Clara, CA, USA; [2]Maryland Power Electronics Laboratory; [2]Electrical and Computer Engineering Department; [3]Institute for Systems Research; [2,3]University of Maryland, College Park, MD 20742; Email: khaligh@ece.umd.edu; URL: http://khaligh.umd.edu/

Abstract- This paper proposes a bidirectional bridgeless totem-pole interleaved power-factor-correction (PFC) converter using SiC MOSFETs as the front-end stage of an onboard charger for plug-in electric vehicles (PEVs). The proposed converter provides bidirectional operation enabling both grid-to-vehicle (G2V) charging and vehicle-to-grid (V2G) ancillary services. The converter is suitable for efficient G2V V2G onboard charging due to its superiorities in terms of bidirectional operation, smaller current ripple, lower EMI, lower conduction losses and switching losses. A 3.3kW PFC converter is designed and developed, using Silicon-Carbide MOSFETs with fast recovery body diodes, for validation of and V2G operating modes. Utilizing SiC MOSFETs enables continuous current mode (CCM) operation of the totem-pole PFC converter in high-power applications. The converter is capable of converting 85Vac-265Vac line voltages into a regulated dc voltage in the range of 300V to 600V. The maximum efficiency of converter reaches up to 99.2% with 0.99 power factor.

Keywords: Bridgeless, interleaved, onboard charger, plug-in electric vehicles, power factor correction, sillicon-carbide.

I. INTRODUCTION

In a typical plug-in electric vehicle (PEV), a single-phase power factor correction (PFC) ac-dc converter is used as the front-end stage of an onboard grid-to-vehicle (G2V) charger, followed by an isolated dc-dc converter as the second stage, as illustrated in Fig. 1 [1]-[3]. The most commonly used PFC converter in an onboard charger is a single-phase diode bridge rectifier followed by a boost converter. It converts the 85Vac~265Vac single-phase ac voltages to a regulated dc voltage (typically around 390V) [4]-[6].

Although the conventional PFC boost converter is the most popular topology, its efficiency suffers from the conduction losses of the front-end diode bridge rectifier and it is not bidirectional [7]. Therefore, bridgeless PFC boost converters are investigated to reduce the number of diodes and increase the efficiency [8]-[16]. Moreover, bridgeless topologies can be slightly modified to enable bidirectional operation to provide vehicle-to-grid (V2G) ancillary services. Dual boost bridgeless PFC converter is a commonly used bridgeless topology, but its serious common mode noise requires a large electromagnetic interference (EMI) filter [8]-[10]. Semi-boost bridgeless PFC converter can reduce the common mode noise; however, each of its two bridges has to handle the peak input current, increasing the size and cost of the components [11], [12]. A dual boost bridgeless PFC converter using bidirectional switches is also considered by researchers [13]; however, its major drawbacks include difficult gate drive design and the inefficient post-end diode

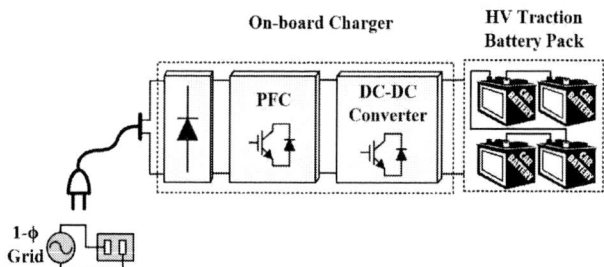

Fig. 1. Typical onboard charger with a front-end PFC stage in a PEV powertrain.

rectifier. Another variant topology of dual boost bridgeless PFC is the pseudo totem-pole bridgeless PFC converter [14]. However, the difficult control and gate drive design limit its practical implementation. Totem-pole bridgeless PFC converter, shown in Fig. 2(a), overcomes the issues of other bridgeless topologies [15], [16]. However, reverse recovery issue of Silicon (Si) MOSFET's intrinsic body diodes causes large reverse recovery current in continuous current mode (CCM) and makes it impractical for high power applications. Fast recovery diodes can be used in anti-parallel, but they increase the size and cost of the converter.

This paper introduces a Silicon-Carbide (SiC) based bridgeless totem-pole interleaved PFC converter as the front-end stage of an onboard charger capable of G2V and V2G operating modes. In comparison to other topologies, the proposed topology has superiorities in terms of high efficiency, small common mode noise, small ac current ripple, small reverse recovery current, less number of components, simple control and simple gate drive design. The low reverse recovery charge of SiC body diode and the low turn-on resistance of SiC MOSFET make the converter an efficient and cost effective solution for bidirectional onboard chargers. The converter is capable of interfacing with 85Vac-265Vac universal single-phase ac-line voltages. Furthermore, due to the high voltage capability of SiC MOSFETs, depending on the input voltage, the dc link voltage can be regulated in the range from 300V to 600V.

II. BRIDGELESS INTERLEAVED PFC CONVERTER

The proposed bidirectional bridgeless totem-pole interleaved CCM PFC converter using six SiC MOSFETs is shown in Fig. 2(b). Two boost interleaved phases (L_1, S_1, S_2 and L_2, S_3, S_4) are driven with 180 degrees phase difference. In each boost phase, the high-side switch and the low-side switch are driven complementarily with a deadband. Two

978-1-4673-9551-9/16 $31.00 © 2016 IEEE

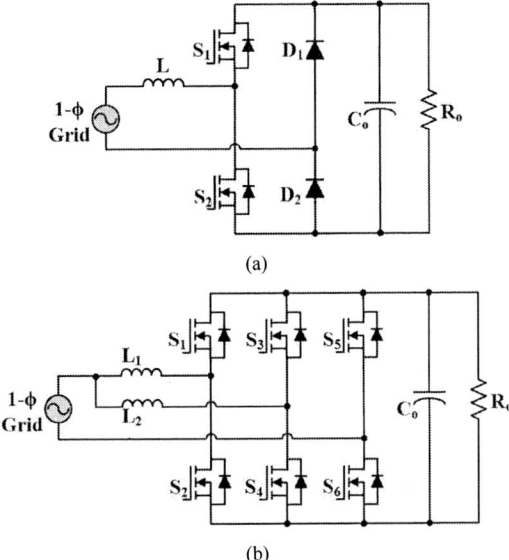

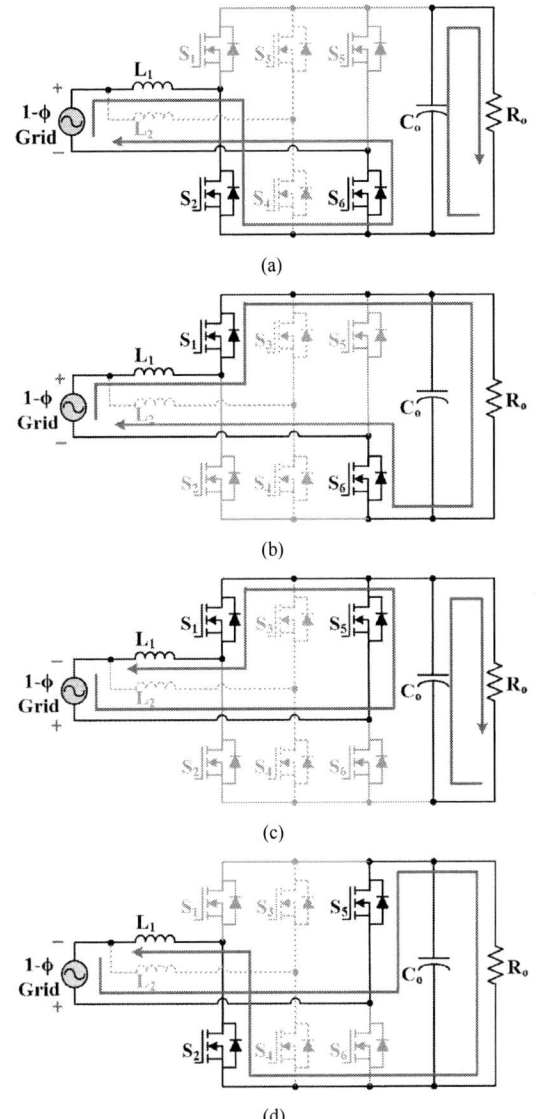

Fig. 2. (a) Unidirectional bridgeless totem-pole DCM PFC converter; (b) proposed bidirectional bridgeless totem-pole interleaved CCM PFC converter.

line rectification diodes are replaced with two SiC MOSFETs (S_5 and S_6) to be used for synchronous rectification.

The operation principles of one leg of totem-pole interleaved PFC converter for G2V charging is depicted in Fig. 3. In the positive half-cycle of ac line, as shown in Fig. 3(a), S_6 is turned on, connecting the low potential of ac line to the dc ground. When S_2 is turned on, the ac source charges the inductor L_1. The current of L_1 increases linearly,

$$\frac{di_{L1}}{dt} = \frac{v_{ac}}{L_1} \qquad (1)$$

When S_2 is turned off, a deadband interval is introduced before S_1 is turned on to avoid overshoot current. After the deadband, S_1 is turned on, as shown in Fig. 3(b), generating a freewheeling path for inductor current and discharging the stored energy in inductor to the dc link. The current of L_1 decreases linearly according to Eq. (2),

$$\frac{di_{L1}}{dt} = \frac{v_{ac} - V_o}{L_1} \qquad (2)$$

The turn-on time of S_2 is determined by the boost duty ratio of pulse-width-modulation (PWM) signal, while the turn-on time of S_1 is complementary to S_2.

In the negative half-cycle of ac line, S_5 is turned on, connecting the high potential of ac line to the dc link. When S_1 is turned on, as shown in Fig. 3(c), the ac source charges the inductor L_1, and the current increases linearly as Eq. (1). S_2 is turned on after S_1 is turned off with a deadband interval, which enables synchronous rectification and generates a freewheeling path for inductor current, as shown in Fig. 3(d). The inductor current decreases linearly as Eq. (2). In the negative half-cycle of ac line, the turn-on time of S_1 is determined by the boost duty ratio.

S_3 and S_4 are used to construct the second interleaved phase, driven with 180 degrees phase difference with respect to S_1 and S_2. The operation of the second interleaved phases is similar to that of the first interleaved phase, increasing the effective switching frequency by two times.

Fig. 3. Operation of one leg of bridgeless totem-pole interleaved CCM PFC converter for G2V operation.

For V2G operation, the switching operation modes are similar to those of G2V. In the positive half-cycle of ac line, as shown in Fig. 4(a), S_6 is turned on, connecting the low potential of ac line to the dc ground. When S_1 is turned on, the dc link charges the inductor L_1. The current of L_1 increases linearly according to Eq. (3),

$$\frac{di_{L1}}{dt} = \frac{V_o - v_{ac}}{L_1} \qquad (3)$$

When S_1 is turned off, S_2 is turned on with a deadband interval in between, as shown in Fig. 4(b). The inductor current freewheels through S_2 and releases the stored energy to the ac line. The current of L_1 decreases linearly,

$$\frac{di_{L1}}{dt} = -\frac{v_{ac}}{L_1} \qquad (4)$$

The turn-on time of S_1 is determined by the buck duty ratio of pulse-width-modulation (PWM) signal, while the turn-on time of S_2 is complementary to S_1. In the negative half-cycle

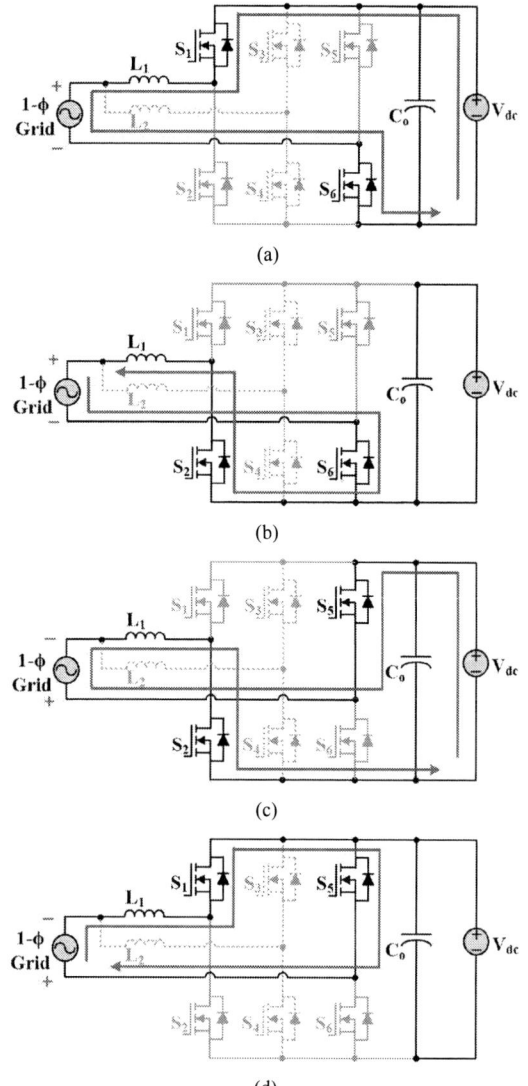

Fig. 4. Operation of one leg of bridgeless totem-pole interleaved CCM PFC converter for V2G operation.

of ac line, S_5 is turned on, connecting the high potential of ac line to the dc link. The dc link charges the inductor L_1 when S_2 is turned on, as shown in Fig. 4(c). S_1 is turned on after S_2 is turned off to enable synchronous rectification, as shown in Fig. 4(d). In the negative half-cycle of ac line, the buck PWM duty ratio determines the turn-on time of S_2.

By introducing a current reference, 180 degree out of phase with respect to the ac line voltage, the turn-on time of switches can be determined by the buck PWM duty ratio. Therefore, the inductor current can be shaped 180 degree out of phase with respect to the ac line voltage.

III. DESIGN GUIDELINE AND CONTROL SCHEME

A. DC Link Voltage Range

For universal input line voltages (85Vac-265Vac), the dc link voltage can be varied in the range of 300V to 600V. The minimum dc link voltage is regulated at 380V when the ac-line voltage is 155Vac-265Vac. When the ac-line voltage

is 85Vac-155Vac, the maximum dc link voltage is regulated at 525V. Therefore, a high-voltage traction battery (200V-420V) can be charged using a second stage dc-dc resonant converter with 0.5-0.8 voltage gain [17]. The dc link voltage variation can alleviate the need for a wide voltage gain range for the second stage, which in turn reduces the design difficulty and increases the efficiency of the second stage of an onboard charger.

B. PFC Interleaved Inductors

The interleaved inductors are designed to meet the requirement of high power factor over a wide range of ac-line voltages. Two non-coupled inductors are used. The inductance value of inductors should be high enough to avoid discontinuous current mode (DCM) operation of each boost leg and acquire high power factor, low total harmonic distortion (THD) and low electromagnetic interference (EMI). As the inductance value of an inductor decreases at higher operating currents, the nominal value of inductance is designed at maximum operating current, corresponding to full load condition (3.3kW) at lower line voltage (85Vac).

The inductance value is determined by the switching frequency, which is set at 100kHz, and the current ripple of each boost leg, which is selected to be equal to 20% (Δ_i) of the maximum ac current, yielding

$$L_1 = L_2 = \frac{V_{ac,\min}^2}{\Delta_i P_{o,\max} f_{sw}}(1 - \frac{\sqrt{2}V_{ac,\min}}{P_{o,\max}}) \tag{5}$$

The maximum current through the inductor can be calculated as

$$I_{L1,\max} = I_{L2,\max} = \frac{\sqrt{2}P_{o,\max}}{2V_{ac,\min}}(1 + \frac{\Delta_i}{2}) \tag{6}$$

In this work, 70µ permeability Kool Mµ toroid cores from Magnetics Inc. are used, which are wound up with 32 turns of 1.5mm litz wire. The inductance value is 340µH under no load and 100µH under full load at low line-voltage. The dc resistance is around 30mΩ.

C. DC-link Capacitor

The capacitance value of dc-link capacitor is determined by the low frequency voltage ripple and the loading hold-up time. The capacitor needs to be large enough to suppress the voltage ripple (as small as 10% (Δ_v) of minimum dc link voltage) caused by low-frequency line voltage.

$$C_o = \frac{P_{o,\max}}{\Delta_v 2\pi f_{ac}V_{o,\min}^2} \tag{7}$$

In addition, the capacitor is required to store enough energy to hold up the dc-link voltage above its minimum value for 20ms (t_{hold}) under full load, yielding

$$C_o = \frac{2P_{o,\max}t_{hold}}{V_{o,\max}^2 - V_{o,\min}^2} \tag{8}$$

Therefore, the capacitance of dc link capacitor is set at 880µF. In this work, two 220µF/400V aluminum capacitors are connected in series to handle 600V. Eight groups of such series-connected capacitors are connected in parallel to create 880µF.

D. SiC MOSFETs

MOSFET selection is considered based on (1) high drain-to-source breakdown voltage; (2) low turn-on

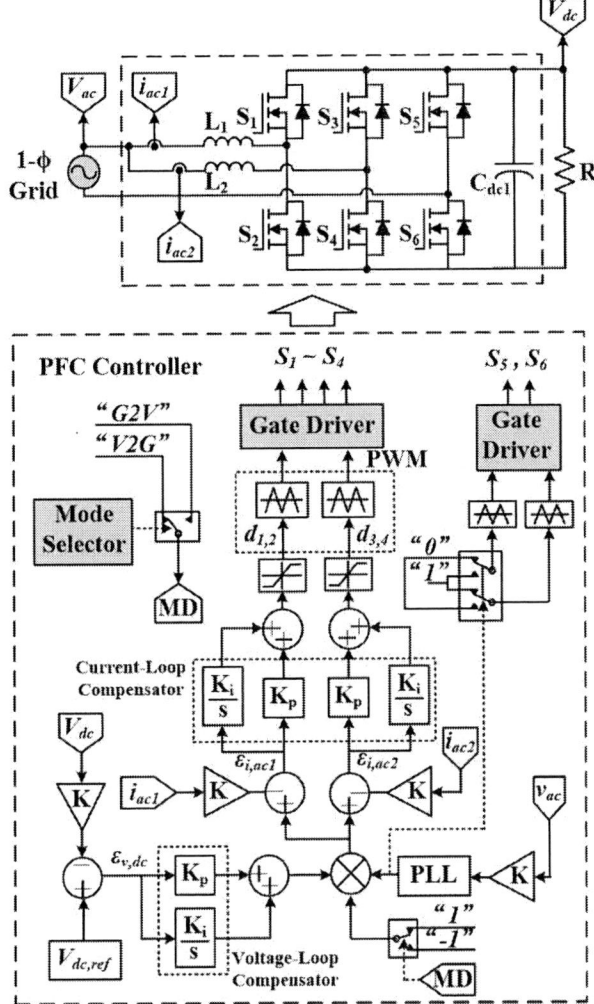

Fig. 6. A 3.3kW SiC-based bidirectional bridgeless totem-pole interleaved PFC converter.

Table I: Specifications of a 3.3kW prototype of the proposed bridgeless interleaved PFC converter.

Components	Value	Quantity
AC-line Voltage (V_{ac})	85Vac ~ 265Vac	/
DC-link Voltage (V_{dc})	300V ~ 600V	/
Rated Power (P_{rated})	3.3kW	/
Switching Frequency (f_{sw})	100kHz	/
First-phase Boost Inductor (L_1)	345µH/20A (Customized)	1
Second-phase Boost Inductor (L_2)	336µH/20A (Customized)	1
MOSFETs ($S_1 \sim S_6$)	SiC FET 1200V/36A (Cree C2M0160120D)	6
DC-link Capacitor (C_o)	220µF/400V	16

Fig. 5. PFC control loop for the bridgeless totem-pole interleaved PFC converter.

resistance; (3) fast switching turn-on and turn-off times; (4) low gate charge; (5) small parasitic output capacitance; and (6) fast reverse recovery body diode. SiC power MOSFETs (C2M0160120D) from CREE are selected due to their superiorities in terms of high voltage capability, low turn-on resistance and low reverse recovery charge of body diode. In this work, six SiC FETs (1200V/36A) are used. They are able to handle voltage spikes twice of maximum dc-link voltage. 80mΩ turn-on resistance reduces the conduction losses. The reverse recovery charge of body diodes is 192nC, in comparison to typical 2µC for a counterpart Si body diode, making it suitable for PFC high power CCM operation.

E. Control Scheme

A dual closed-loop pulse-width-modulation (PWM) control, as shown in Fig. 5, is used to control the ac-dc interleaved PFC stage. During G2V operation, the ac-line voltage v_{ac} is sampled, and a phase-locked-loop (PLL) is used to create a reasonably accurate sinusoidal ac-line current reference. The dc-link voltage (V_{dc}) is sampled and controlled through a voltage-loop compensator to regulate the magnitude of ac-line current reference. By adjusting the

ac-line current magnitude or input power, the dc-link voltage can be regulated at $V_{dc,ref}$. Inductor current of each interleaved phase (i_{ac1} or i_{ac2}) is individually sampled and controlled through a current loop compensator to shape a sinusoidal inductor current. Inductor current, rather than ac-line current, is regulated to ensure equal current sharing in two interleaved phases and achieve fast response.

Two compensated duty cycles ($d_{1,2}$: $d_1=1-d_2$; $d_{3,4}$: $d_3=1-d_4$) are generated by each current-loop compensator. Due to the operation of proposed topology, the duty cycles have abrupt changes between "0" to "1" at ac line zero-crossing. By comparing the duty cycles to triangle carrier signals, four PWM signals are then generated to control $S_1 \sim S_4$. To achieve synchronous rectification, S_5 and S_6 are turned on or turned off complimentarily based on the polarity of ac-line voltage. During V2G operation, a negative feedback is introduced to the ac-line current reference to shape a sinusoidal ac current with 180° phase difference with respect to the ac-line voltage.

IV. EXPERIMENTAL RESULTS

A 3.3kW prototype of the proposed SiC-based bridgeless PFC stage, as illustrated in Fig. 6, is designed and developed to validate the operation of the converter. The converter is capable of converting 86Vac-265Vac universal single-phase ac-line voltages into a regulated dc-link voltage in the rane of

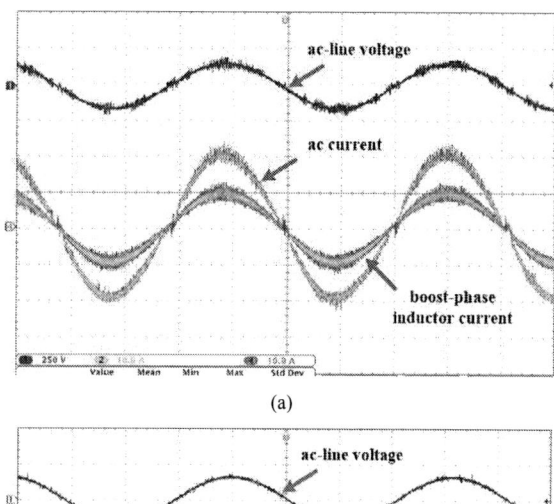

(a)

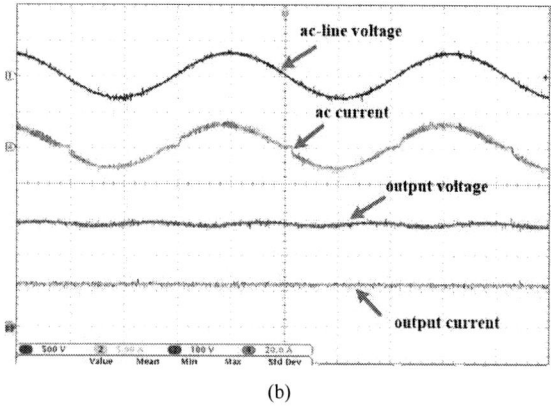

(b)

Fig. 7. (a) The ac voltage, ac current and boost-phase inductor current of the proposed PFC stage during G2V operation at 1.5kW, 110Vac line voltage; (b) The ac voltage, ac current, output voltage and output current of the proposed PFC stage during G2V operation at 2kW, 220Vac line voltage.

300V to 600V. With consideration of efficiency and power density, the switching frequency is selected at 100kHz. The circuit specifications are listed in Table I.

The proposed ac-dc PFC stage is tested for performance evaluation of converting both high-voltage and low-voltage ac-line voltages. Fig. 7(a) illustrates the ac-line voltage, ac current and boost-phase inductor current of the PFC stage at 110Vac/60Hz ac-line voltage during G2V charging at 1.5kW. In order to avoid current spikes, a soft start is generated at every ac-line zero-crossing by setting a number of disabled switching cycles. The dc-link voltage is regulated at 420V with a 20V/120Hz ripple, which is 5% of the dc-link voltage. The PFC stage is then connected with a second stage dc-dc converter. The voltage and current waveforms of the PFC stage at 220Vac/60Hz ac-line voltage, 2kW G2V charging are shown in Fig. 7(b). The dc-link voltage ripple is reduced to 15V, which is 3.5% of dc-link voltage. At 2kW, the G2V efficiency can reach up to 98.8% at 220Vac/60Hz ac-line voltage. The power factor is close to 0.99.

A load step change is set from 1.2kW to 700W to evaluate the dynamic response of the converter. The ac-line current drops from 10A to 6.4A. The dc current of the PFC stage drops from 4A to 2.6A. It takes 0.18 seconds, corresponding to 10 ac-line cycles, to regulate the dc-link voltage. Fig. 8 shows the charger dynamic response of load step change (from 1.2kW to 700W). Fig. 9 demonstrates the efficiency of

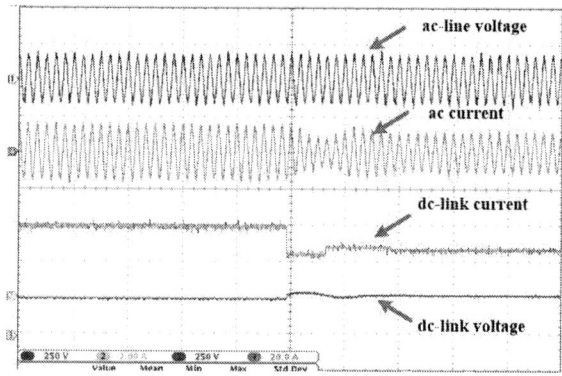

Fig. 8. Dynamic response of the proposed PFC converter to load step change (from 1.2kW to 700W) during G2V operation.

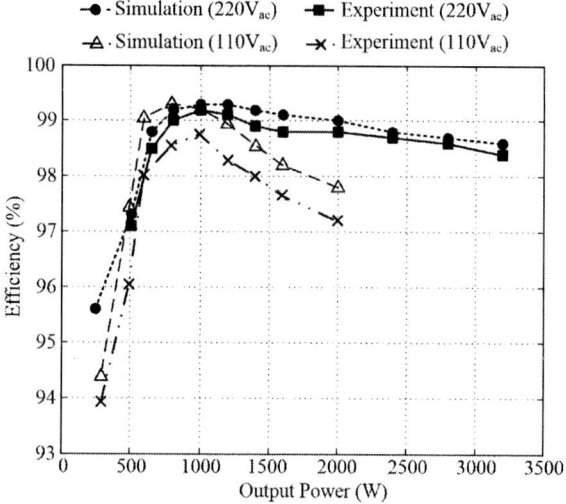

Fig. 9. Simulation and experimental efficiencies of the proposed PFC converter at different ac-line voltages and output power.

the proposed converter for 110Vac and 220Vac input voltages. As shown in this figure, the simulation results adequately coincide with the experimental measurements.

V. CONCLUSIONS

This paper outlines a SiC-based bidirectional bridgeless totem-pole interleaved PFC converter as the front-end stage of an onboard charger for PEVs. Due to its bidirectional operation, the proposed converter is suitable for both G2V charging and providing V2G ancillary services. SiC MOSFETs with fast recovery body diodes significantly reduce the reverse recovery current, making the topology a cost effective solution for CCM operation in high power applications. A 3.3kW PFC converter is designed and developed for validation of onboard charging. Experimental results are carried out to validate its capability of converting 85Vac-265Vac line voltages into a regulated dc voltage in the range from 300V to 600V. The measured efficiency reaches up to 98% at 110Vac line voltage and 1.5kW charging; and 98.8% at 220Vac line voltage and 2kW charging.

978-1-4673-9551-9/16 $31.00 © 2016 IEEE

VI. ACKNOWLEDGEMENT

This work is partially sponsored by a Maryland Innovation Initiative (MII) grant from Maryland Technology Development Corporation (TEDCO), which is gratefully acknowledged.

REFERENCES

[1] A. Emadi, S. S. Williamson, and A. Khaligh, "Power electronics intensive solutions for advanced electric, hybrid electric, and fuel cell vehicular power systems," *IEEE Trans. Power Electron.*, vol. 21, no. 3, pp. 567-577, May 2006.

[2] C. C. Chan, A. Bouscayrol, and K. Chen, "Electric, hybrid, and fuel-cell vehicles: Architectures and modeling," *IEEE Trans. Veh. Technol.*, vol. 59, no. 2, pp. 589-598, Feb. 2010.

[3] A. Khaligh and S. Dusmez, "Comprehensive topological analysis of conductive and inductive charging solutions for plug-in electric vehicles," *IEEE Trans. on Veh. Technol.*, vol. 61, no. 8, pp. 3475-3489, Oct. 2012.

[4] J. G. Kassakian, "Future automotive electrical systems- The power electronics market of the future," *in Proc. IEEE Appl. Power Electron. Conf.*, New Orleans, LA, 2000, pp. 3-9.

[5] S.M. Lukic, J. Cao, R.C. Bansal, F. Rodriguez, and A. Emadi, "Energy Storage Systems for Automotive Applications," *IEEE Trans. on Ind. Electron.*, vol. 55, no. 6, pp. 2258-2267, 2008.

[6] H. Wang, S. Dusmez, and A. Khaligh, "Design and analysis of a full bridge LLC based PEV charger optimized for wide battery voltage range," *IEEE Trans. on Veh. Technol.*, vol. 63, no. 4, pp. 1603-1613, May 2014.

[7] J. P. M. Figuerido, F. L. Tofili, and B. L. A. Silva, "A review of single-phase PFC topologies based on the boost converter," in *Proc. IEEE Int. Conf. Indus. Appl.*, Sao Paulo, Brazil, Nov. 2010, pp. 1-6.

[8] H. Ye, Z. Yang, J. Dai, C. Yan, X. Xin, and J. Ying, "Common mode noise modeling and analysis of dual boost PFC circuit," in *Proc. Int. Telecommun. Energy Conf. (INTELEC)*, Sep. 2004, pp. 575-582.

[9] B. Lu, R. Brown, and M. Soldano, "Bridgeless PFC implementation using one cycle control technique," in *Proc. IEEE App. Power Electron. (APEC) Conf.*, Mar. 2005, pp. 812-817.

[10] P. Kong, S. Wang, and F.C. Lee, "Common mode EMI noise suppression in bridgeless boost PFC converter," in *Proc. CPES Power Electron. Conf.*, Apr. 2006, pp. 65-70.

[11] F. Musavi, W. Eberle, and W. G. Dunford, "A high-performance single-phase bridgeless interleaved PFC converter for plug-in hybrid electric vehicle battery chargers," *IEEE Trans. on Ind. Appl.*, vol. 47, no. 4, pp. 1183-1143, 2011.

[12] F. Musavi, W. Eberle, and W. G. Dunford, "A phase shifted semi-bridgeless boost power factor corrected converter for plug-in hybrid electric vehicle battery chargers," in *Proc. IEEE Appl. Power Electron. Conf. and Expo.*, 2011, pp. 821-828.

[13] D. Tollik and A. Pietkiewicz, "Comparative analysis of 1-phase active power factor correction topologies," in *Proc. Int. Telecommun. Energy Conf. (INTELEC)*, Oct. 1992, pp. 517-523.

[14] L. Huber, Y. Jang, and M. M. Jovanovic, "Performance evaluation of bridgeless PFC boost rectifiers," *IEEE Trans. on Power Electron.*, vol. 23, no. 3, pp. 1381-1390, May 2008.

[15] S. S. Darly, P. V. Ranjan, K. V. Bindu, and B. J. Rabi, "A novel dual boost rectifier for power factor improvement," *Int. Conf. on Elec. Energy Syst.*, 2011, pp. 121-127.

[16] B. Su, J. M. Zhang, and Z.Y. Lu, "Totem-pole boost bridgeless PFC rectifier with simple zero-current detection and full-range ZVS operating at the boundary of DCM/CCM," *IEEE Trans. on Power Electron.*, vol. 26, no. 2, pp. 427-435, Feb. 2011.

Stability analysis of hybrid AC/DC power systems for More Electric Aircraft

Mehdi Karbalaye Zadeh[1], Roghayeh Gavagsaz-Ghoachani[3], Babak Nahid-Mobarakeh[3], Serge Pierfederici[3], and Marta Molinas[2]

[1]Department of Electric Power Engineering, Norwegian University of Science and Technology, Trondheim, Norway
[2]Department of Engineering Cybernetics, Norwegian University of Science and Technology, Trondheim, Norway
[3]GREEN Laboratory, University of Lorraine, Vandœuvre-lès-Nancy, France
Email: mehdi.zadeh@ntnu.no

Abstract—This paper presents the stability analysis of a hybrid AC/DC power distribution system, for embedded applications, particularly for more electric aircraft. In the studied system, a balanced AC source supplies the hybrid system, and a DC distribution system feeds multi-converter controlled loads. The load converters are tightly controlled, behaving as constant power loads with low damped *LC* input filters. An analytical model, taking to account the AC and DC characteristics of the system, is developed to investigate the dynamic behavior of the system. The AC system is modeled in a *d-q* synchronous reference frame. The stable and unstable operating points are identified using the proposed method, and then, the stability pattern of the system is established. The impacts of the constant power loads and interactions between the source and the load converters on the system's stability are studied. Experimental tests, conducted on a laboratory scale microgrid, validate the analytical system analysis.

Keywords— AC/DC microgrid; power electronics; hybrid energy systems; stability analysis; transportation

I. INTRODUCTION

Hybrid AC/DC distribution systems are attractive for use in industrial applications, since they facilitate integrating energy storage devices and modern electric loads [1-4]. The hybrid systems provide optimized system architectures for the next generation transportation systems like electric ships, more electric aircraft (MEA) and advanced automotive systems [5-9]. The MEA, as the trend of the future aircraft, uses the power electronics capabilities toward providing reliable, low emissions, and high-efficiency energy conversion [10-12]. Mechanical actuators in aircrafts are replaced by converter-controlled electric drives, which are supplied by local DC distribution systems [13].

However, the stability of such power-electronics-based system is a major design consideration, mainly due to the converter controlled loads which behave as constant power loads (CPLs) [14-16]. The consequent nonlinear dynamics of these converters create complex system interactions that can lead to instability issues. A typical hybrid system with DC power distribution architecture is shown in Fig. 1.

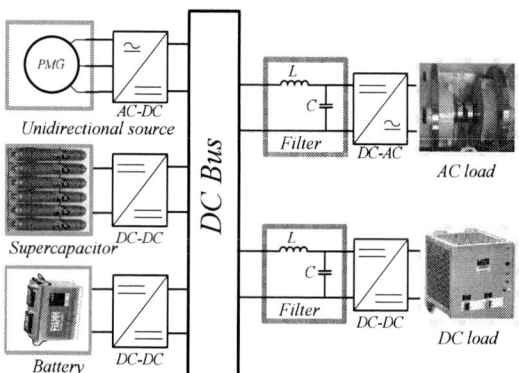

Fig. 1. Schematic of the hybrid system.

In aircraft power system, the main energy source is usually a turbine engine, as a unidirectional source. Therefore, energy storage systems can be employed to store the surplus energy and to maintain the DC bus voltage constant [17, 18]. The load converters transfer the DC bus voltage to tightly regulated power for the AC and DC loads. Since the level of the voltage of the source is not suitable for the load side, DC-DC converters are used to adapt the voltage level [19]. *LC* filters are added to the input of the converter in order to limit the harmonic effects and electromagnetic interferences [20]. The weight, volume, and cost of power components have the highest importance for the aircraft industry [21]. Therefore, these filters are usually poorly damped for reducing losses as well as optimizing the cost of the total system. In such a system, instability is potentially a major issue because of the nonlinear dynamics of power electronics converters, as well as the complex interaction between the source, the low damped *LC* filter, and the constant power characteristics of the load.

This paper, for the first time, presents the small signal stability analysis of the complete hybrid system, including the three-phase AC system, DC distribution system, and constant power loads. The small signal model of the AC system is established in *d-q* frame. The system analysis is as generic as taking into account any such hybrid AC/DC systems.

978-1-4673-9551-9/16 $31.00 © 2016 IEEE

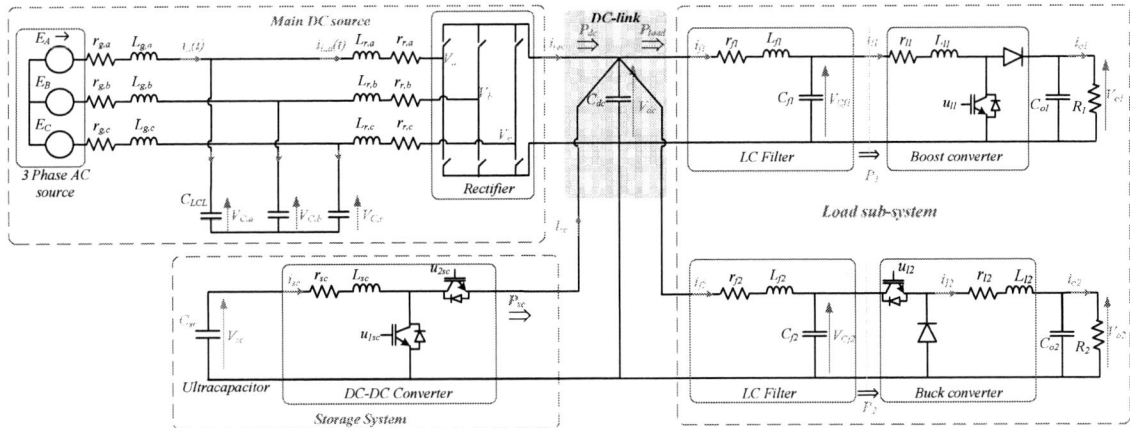

Fig. 2. Electrical diagram of the studied system.

The results of the analytical modeling are used to predict the stable and unstable operating points, and consequently to derive the stability region of the complete system. Furthermore, the experimental results in stable and unstable conditions, validate the effectiveness of the stability analysis method.

II. DYNAMIC MODEL OF THE HYBRID SYSTEM

Fig. 2 presents the electrical diagram of the complete system used for the stability analysis. In this model, the AC source is assumed to be a balanced three phase network. A voltage source converter (VSC), as a boost rectifier, then supplies the DC bus. The VSC is connected to the AC grid with an *LCL* filter. The *LCL* filter is utilized to reject the current harmonics [22]. A supercapacitor (SC) bank is used as the storage unit. The SC bank is controlled by a bidirectional DC-DC converter, which regulates the DC bus voltage, and provides the load power under transient loads and overloads. This converter is called SC converter. The load subsystem is comprised of a boost (load converter 1) and a buck (load converter 2) DC-DC converter, which absorb controlled constant powers, namely P_1 and P_2. The load converters are connected to the DC link through *LC* input filters.

In this circuit diagram: $r_{f1,2}$, $C_{f1,2}$, and $L_{f1,2}$ are the resistance, capacitance, and inductance of the input filters; C_{dc} is the capacitance of the DC bus; $r_{l1,2}$ and $L_{l1,2}$ are the resistance and inductance of the load converters; C_{sc} is the total capacitance of the SC bank; r_{sc} and L_{sc} are resistance and inductance of the bidirectional converter; C_{LCL} is the capacitance of the *LCL* filter; $r_{r_{a,b,c}}$ and $L_{r_{a,b,c}}$ are the *LCL* filter's resistance and inductance, on the rectifier side, and $r_{g_{a,b,c}}$ and $L_{g_{a,b,c}}$ are the filter's parameters on the grid side. The grid side parameters represent either the input filter or parameters of cable. In this study, the filter is balanced and the parameters are equal for the three phases.

In the dynamic and steady-state operation of the system, the load converters regulate the load currents i_{l1} and i_{l2} such that the current's variations are proportional to the variations of the

filter voltages V_{cf1} and V_{cf2} in the constant powers P_1 and P_2: $i_{l1\_ref} = \frac{P_1}{V_{cf1}}$, $i_{l2\_ref} = \frac{P_2}{V_{02}}$. Consequently, the system model has intrinsic nonlinearities because of the load profile of the load converters.

The system model is established in state-space taking to account the physical state variables and the control variables. The electrical model of the AC system is established in a *d-q* synchronous reference frame (SRF). The phase angle of the SRF is calculated by a phase-locked loop (PLL), which measures the phase voltages on the filter capacitor. All the system's characteristics are then synchronized with the resulting rotating frame.

A. Control System

The control algorithm of the hybrid system can be divided into three control levels: energy management or outer loop control; power control; and current control or inner-loop control. The outer loop control regulates the energy stored in the storage unit y_{sc} ($y_{sc} = \frac{1}{2} C_{sc} V_{sc}^2$) and also in the DC bus y_{dc} ($y_{dc} = \frac{1}{2} C_{dc} V_{dc}^2$). Consequently, the DC bus voltage V_{dc} and the voltage of SC bank V_{sc} are regulated at the set-points. The energy management is not the focus of this paper. This controller finally produces the reference power for the VSC and the bidirectional converter (SC converter). As the result, the reference power components of the VSC are calculated, such as the active power $P_{dc_{ref}}$ and reactive power Q_{ref}. Similarly, the reference of the supplied power by the SC converter P_{sc} is determined by the energy control loop. By knowing these values, the *dq*-current references $i_{d_{ref}}$ and $i_{q_{ref}}$ can be calculated as follows:

$$\begin{cases} P_{dc_{ref}} = V_d \cdot i_{d_{ref}} + V_q \cdot i_{q_{ref}} \\ Q_{dc_{ref}} = V_q \cdot i_{d_{ref}} - V_d \cdot i_{q_{ref}} \end{cases} \quad (1)$$

$$\begin{cases} i_{d_{ref}} = \dfrac{P_{dc_{ref}} \cdot V_d + Q_{dc_{ref}} \cdot V_q}{V_d^2 + V_q^2} \\ i_{q_{ref}} = \dfrac{P_{dc_{ref}} \cdot V_q - Q_{dc_{ref}} \cdot V_d}{V_d^2 + V_q^2} \end{cases} \quad (2)$$

978-1-4673-9551-9/16 $31.00 © 2016 IEEE

The current controller of the VSC is based on the decoupling control approach, and uses a proportional integral (PI) compensator. Consequently, the control vector can be expressed by:

$$\begin{pmatrix} V_d \\ V_q \end{pmatrix} = k_p \begin{pmatrix} i_{Ld} - i_{Ld_{ref}} \\ i_{Lq} - i_{Lq_{ref}} \end{pmatrix} - k_i \begin{pmatrix} \int \left(i_{Ld_{ref}} - i_{Ld} \right) \\ \int \left(i_{Lq_{ref}} - i_{Ldq} \right) \end{pmatrix}$$
$$- \begin{pmatrix} \dfrac{-r_r}{L_r} & \omega \\ -\omega & \dfrac{-r_r}{L_r} \end{pmatrix} \begin{pmatrix} i_{Ld} \\ i_{Lq} \end{pmatrix} + \begin{pmatrix} V_{Cd} \\ V_{Cq} \end{pmatrix} \tag{3}$$

After rejecting the coupling term, the controller is tuned based on the pole placement [23].

For the SC converter, the reference current $i_{sc_{ref}}$ is obtained from the power control scheme defined as follows:

$$i_{sc_{ref}} = \frac{2P_{max}}{V_{sc}} \left(1 - \sqrt{1 - \frac{P_{sc_{ref}}}{P_{max}}} \right) \tag{4}$$
$$\text{where: } P_{max} = \frac{V_{sc}^2}{4r_{sc}}$$

In the inner-loop control of the SC converter, the inductor current i_{sc} is regulated using a digital PWM controller based on the equivalent control approach, as detailed in [24]. The resulting duty cycle is defined as follows:

$$D_{sc} = \frac{1}{V_{dc}} \Bigg(-V_{sc} + V_{dc} + r_{sc} \, i_{sc}$$
$$- L_{sc} \Big((K_x + \lambda) \left(i_{sc} - i_{sc_{ref}} \right) \tag{5}$$
$$+ K_x \lambda \int \left(i_{sc} - i_{sc_{ref}} \right) \Big) \Bigg)$$

where K_x And λ are the control coefficients of the inner-loop control.

The system model can then be described with a state-space form as presented in (6).

$$\dot{x} = A \cdot x + B \cdot u \tag{6}$$

In order to develop a complete small-signal model that can capture the dynamics of the hybrid system, the system equations are driven for the three main subsystems: AC sub-system as detailed in II.B; DC sub-system, described in II.C; and load sub-system, given in II.D.

B. AC System Model

The state variables presented in (7) are defined as small-signal variations of the physical variables around the operating point of the system: $x_i = X_i(t) - X_{i,0}$. Thus, the AC system's model is composed of eight state variables: grid-side dq-currents, capacitors dq-voltages, converter-side dq-currents and two additional state variables corresponding to the integral terms of the inner-loop control.

$$x_{AC} = [x_1, x_2, x_3, x_4, x_5, x_6, x_7, x_8]^T$$
$$= \left[i_d, i_q, v_{Cd}, v_{Cq}, i_{Ld}, i_{Lq}, I_d, I_q \right]^T \tag{7}$$
$$\text{where: } I_d = \int \left(i_{Ld_{ref}} - i_{Ld} \right), I_q = \int \left(i_{Lq_{ref}} - i_{Lq} \right)$$

The state variables, δI_d and δI_q, are associated with the integral term of the d and q current components, respectively, which are used in the inner-loop control of the rectifier. The variables of the system are then calculated at the operating point to solve the equations of the system. The system's differential equations can then be written as follows:

$$\begin{cases} \dfrac{d}{dt} i_d = -\dfrac{r_g}{L_g} i_d + \omega i_q - \dfrac{1}{L_g} v_{Cd} + \dfrac{1}{L_g} e_d \\[2mm] \dfrac{d}{dt} i_q = -\dfrac{r_g}{L_g} i_q - \omega i_d - \dfrac{1}{L_g} v_{Cq} + \dfrac{1}{L_g} e_q \\[2mm] \dfrac{d}{dt} v_{Cd} = \dfrac{1}{C_g} i_d - \dfrac{1}{C_g} i_{Ld} + \omega v_{Cd} \\[2mm] \dfrac{d}{dt} v_{Cq} = \dfrac{1}{C_g} i_q - \dfrac{1}{C_g} i_{Lq} - \omega v_{Cq} \\[2mm] \dfrac{d}{dt} i_{Ld} = -\dfrac{k_p}{L_r} i_{Ld} + \dfrac{k_i}{L_r} I_d + \dfrac{k_p}{L_r} i_{Ld_{ref}} \\[2mm] \dfrac{d}{dt} i_{Lq} = -\dfrac{k_p}{L_r} i_{Lq} + \dfrac{k_i}{L_r} I_q + \dfrac{k_p}{L_r} i_{Lq_{ref}} \\[2mm] \dfrac{d}{dt} I_d = -I_d + i_{Ld_{ref}} \\[2mm] \dfrac{d}{dt} I_q = -I_q + i_{Lq_{ref}} \end{cases} \tag{8}$$

Here, the AC network provides a regulated three phase voltage. Therefore, the grid voltage is assumed to be independent of the load variations. In this case, the dq-components of the grid voltage E_d and E_q are supposed to be constant, and hence, their variations e_d and e_q are equal to zero.

C. DC System model

The presented state-space-based modeling approach is to be developed to take into account the storage system and the bidirectional converter. The state vector of the DC sub-system x_{DC} is given in (9). The corresponding small-signal state-space model can then be derived as presented in (10).

$$x_{DC} = [x_9, x_{10}, x_{11}, x_{12}, x_{13}, x_{14}]^T$$
$$= \left[i_{sc}, v_{dc}, v_{sc}, v_{cf}, I_{isc}, I_{y_{dc}}, I_{y_{sc}} \right]^T$$
$$\text{where: } I_{isc} = \int \left(i_{sc_{ref}} - i_{sc} \right); I_{y_{sc}} = \int \left(y_{sc_{ref}} - y_{sc} \right); \tag{9}$$
$$I_{y_{dc}} = \int \left(y_{dc_{ref}} - y_{dc} \right)$$

978-1-4673-9551-9/16 $31.00 © 2016 IEEE

$$\begin{cases}
\dfrac{d}{dt}i_{sc} = \dfrac{1}{L_{sc}}\big[v_{sc} - r_{sc}i_{sc} - (1 - \tilde{d}_{sc})v_{dc}\big]\\[2mm]
\dfrac{d}{dt}v_{dc} = \dfrac{1}{C_{dc}}\big[(1 - d_{sc})i_{sc} + i_{rec} - i_{f1} - i_{f2}\big]\\[2mm]
\dfrac{d}{dt}v_{sc} = \dfrac{1}{C_{sc}}i_{sc}\\[2mm]
\dfrac{d}{dt}I_{isc} = i_{sc_{ref}} - i_{sc}\\[2mm]
\dfrac{d}{dt}I_{ydc} = y_{dc_{ref}} - y_{dc}\\[2mm]
\dfrac{d}{dt}I_{ysc} = y_{sc_{ref}} - y_{sc}
\end{cases} \tag{10}$$

Here, $\tilde{d}_{sc}$ is the small signal model associated with the duty cycle of the SC converter.

D. Load Converters

The load sub-system can be expressed by the state equations (11) and (12):

$$x_l = [x_{15}, x_{16}, x_{17}, x_{18}, x_{19}, x_{20}, x_{21}, x_{22}, x_{23}, x_{24}]^T$$
$$= [i_{f1}, v_{cf1}, i_{f2}, v_{cf2}, i_{l1}, v_{o1}, i_{l2}, v_{o2}, I_{i_{l1}}, I_{i_{l2}}]^T \tag{11}$$

where: $I_{l1} = \int \big(i_{L1_{ref}} - i_{L1}\big),\ I_{l2} = \int \big(i_{L2_{ref}} - i_{L2}\big)$

$$\begin{cases}
\dfrac{d}{dt}i_{f1} = \dfrac{1}{L_{f1}}\big[v_{dc} - r_{f1}i_{f1} - v_{cf1}\big]\\[2mm]
\dfrac{d}{dt}v_{cf1} = \dfrac{1}{C_{f1}}\big[i_{f1} - i_{l1}\big]\\[2mm]
\dfrac{d}{dt}i_{f2} = \dfrac{1}{L_{f2}}\big[v_{dc} - r_{f2}i_{f2} - v_{cf2}\big]\\[2mm]
\dfrac{d}{dt}v_{cf2} = \dfrac{1}{C_{f2}}\big[i_{f2} - i_{l2}\big]\\[2mm]
\dfrac{d}{dt}i_{l1} = \dfrac{1}{L_{l1}}\big[v_{cf1} - r_{l1}i_{l1} - (1 - \tilde{d}_{l1})v_{o1}\big]\\[2mm]
\dfrac{d}{dt}v_{o1} = \dfrac{1}{C_{o1}}\Big[(1 - \tilde{d}_{l1})i_{l1} - \dfrac{v_{o1}}{R_1}\Big]\\[2mm]
\dfrac{d}{dt}i_{l2} = \dfrac{1}{L_{l2}}\big[\tilde{d}_{l2}.v_{cf2} - r_{l2}i_{l2} - v_{o2}\big]\\[2mm]
\dfrac{d}{dt}v_{o2} = \dfrac{1}{C_{o2}}\Big[i_{l2} - \dfrac{v_{o2}}{R_2}\Big]\\[2mm]
\dfrac{d}{dt}I_{l1} = -i_{L1} + i_{L1_{ref}}\\[2mm]
\dfrac{d}{dt}I_{l2} = -i_{L2} + i_{L2_{ref}}
\end{cases} \tag{12}$$

For the stability analysis, the sub-system models of (8), (10) and (12) are put in the state-space form presented in (6). The system equations of the three sub-systems are regrouped in matrices A and B, and u contains the input variables and the nonlineities of the system. As the result of the small-signal modeling, Jacobian matrix of the system is established and system's eigenvalues are calculated. These eigenvalues are then used in Section III to analyze the stability of the complete system.

III. STABILITY ANALYSIS

Asymptotic stability of the hybrid power system, shown in Fig. 2, is investigated using the analytical model of the system. The resulting eigenvalues are used to identify the stable and unstable operating points. The stability pattern of the system is then established using the stable and unstable operating points. This stability pattern was first utilized in [25] to study the stability of a discrete-time model. Here, we use this method for the small-signal analysis of the system. The influence of different control parameters, such as the load powers P_1 and P_2 and the DC bus voltage, on the stability of the system is studied.

In the first study, the system's stability pattern is developed by changing the reference power of the two load converters P_1 and P_2, as shown in Fig. 3. The stable and unstable operating points are shown with black circles and red crosses, respectively. In this case, the DC bus voltage is regulated on $V_{dc} = 120\,V$. In general, by increasing the load power of converter 1 (P_1), the maximum stable power by converter 2 (P_2) is reduced. When $P_1 = 550W$, the maximum stable power through converter 2 is $P_{2max} = 240W$. In a similar way, by increasing P_2, less power can be delivered by converter 2. With $P_2 = 300W$, maximum stable power is equal to $P_{1max} = 400W$.

In the second case, the system analysis is repeated for different DC bus voltages. The reference voltage is changed: $V_{dc} = 80\,V \rightarrow 200V$, and the reference power of the first load converter is changed: $P_1 = 100\,W \rightarrow 1000W$, while the second load power is fixed at $P_2 = 50\,W$. Fig. 5 presents the stability pattern with two variables: load power 1 and voltage V_{dc}. Clearly, the maximum stable load power is increased by increasing the DC voltage. The maximum stable power is also called "critical power". The stability border of the system is depicted with a green curve. It shows that the maximum deliverable power with the DC distribution architecture, is highly dependent on the voltage of the DC link.

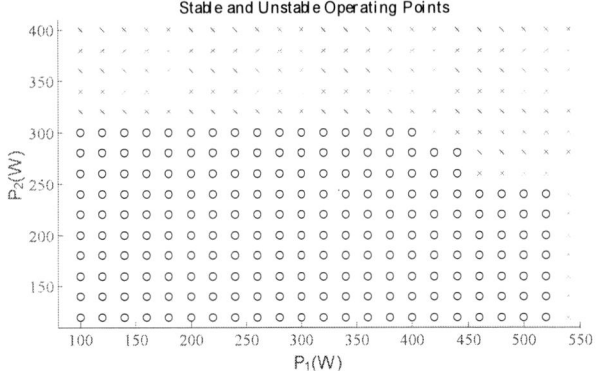

Fig. 3. Stability pattern of CPLs, indicating stable operating points (black circles) and unstable operating points (red crosses).

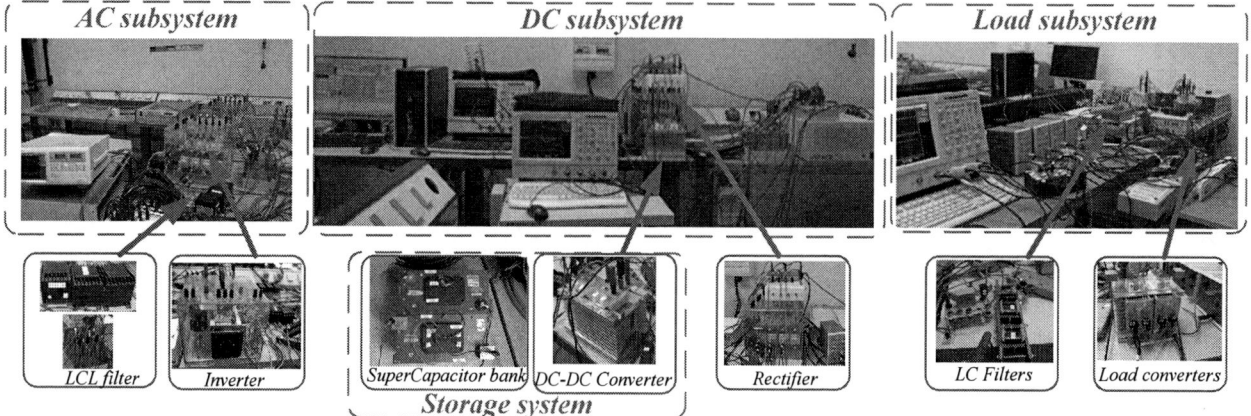

Fig. 4. Laboratory test-bed.

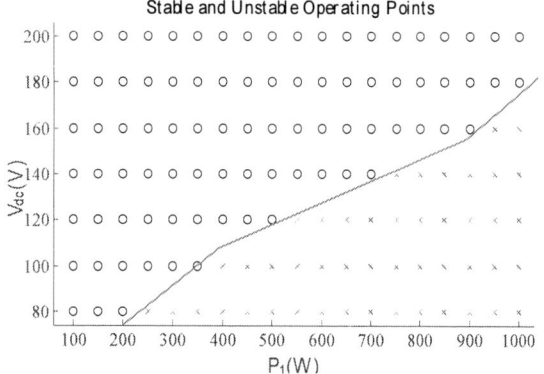

Fig. 5. Stability pattern by changing DC voltages; black circles: stable, red crosses: unstable operating points.

Obviously, the change of the stability margin for different voltages is not uniform, since the system model is nonlinear. This analysis gives the stability region of the hybrid system.

IV. EXPERIMENTAL RESULTS

The efficiency of the proposed method is experimentally validated using a test-bed hardware prototype, which is indicated in Fig. 4. Three parts constitute the experimental platform. The AC network is implemented by a voltage source inverter (VSI), which is connected to the point of common coupling through an *LCL* filter. This inverter regulates the AC voltage, providing the three phase AC source (E_{ABC}). A VSC converts the AC voltage to the controlled power in DC side. It provides the main source of the hybrid system. The storage unit comprises a SC bank, including two ultracapacitor modules in series, which has a total rated voltage of 96 V, and equivalent capacitance of $C_{sc} = 82F$.

The SC converter is power controlled in order to regulate the DC bus voltage and to balance the load power. The load converters are implemented with a boost and a buck DC-DC converter. The load converters are controlled in constant power and are connected to the DC bus through two *LC* filters, corresponding to the system model. The controllers are implemented with three dSPACE real-time control cards: DS1105 for the inverter, DS1005 for the DC subsystem (rectifier and supercapacitor), and DS1104 the load converters.

The experiments are performed by changing the load powers P_1 and P_2. This is performed by changing the reference power of the load converter 1 and 2: boost and buck converter. The stability of the system is then investigated with different load powers. In this test, the DC bus voltage is regulated at $V_{dc} = 120\,V$.

First, the reference power of the converter 1 (buck converter) is set to a low value, and P_1 is changed to investigate the destabilizing effect of the increased power from the boost converter. Fig. 6 presents the experimental waveforms. In Fig. 6 (a), $P_2 = 200W$ and $P_1 = 500W$: the system characteristics are obviously stable. Next, P_1 is increased to $650W$, and the corresponding results are shown in Fig. 6 (b): the system currents and the DC bus voltage get unstable oscillations, as predicted by the theoretical analysis. In Fig. 6 (c), $P_2 = 300W$ and $P_1 = 550W$: an unstable operating point is observed from the current and voltage waveforms. Clearly, the maximum stable power of converter 1 P_{1max} is reduced by increasing P_2. These observations are consistent with the theoretical results presented in Fig. 3.

Secondly, P_1 is kept at a low value, and P_2 is changed to reach the critical power of load 2. Fig. 7 presents the experimental waveforms. Initially, a stable operating point is studied with $P_1 = 250W$ and $P_2 = 300W$. The experimental results are indicated in Fig. 7 (a), which shows the steady-state of the system's characteristics. The load power 2 is then increased to $P_2 = 400W$ without changing P_1. In this case, unstable oscillations occur, as can be seen from Fig. 7 (b). This was foreseen by the analytical method. In Fig. 7 (c), the load power 1 is increased to $P_1 = 300W$, and the load power 2 is reduced to $P_2 = 450W$. Again, the system characteristics are oscillating. Indeed, by increasing P_1, the maximum stable power of load 2 P_{2max} is reduced. This is in line with the theoretical observations.

978-1-4673-9551-9/16 $31.00 © 2016 IEEE   450

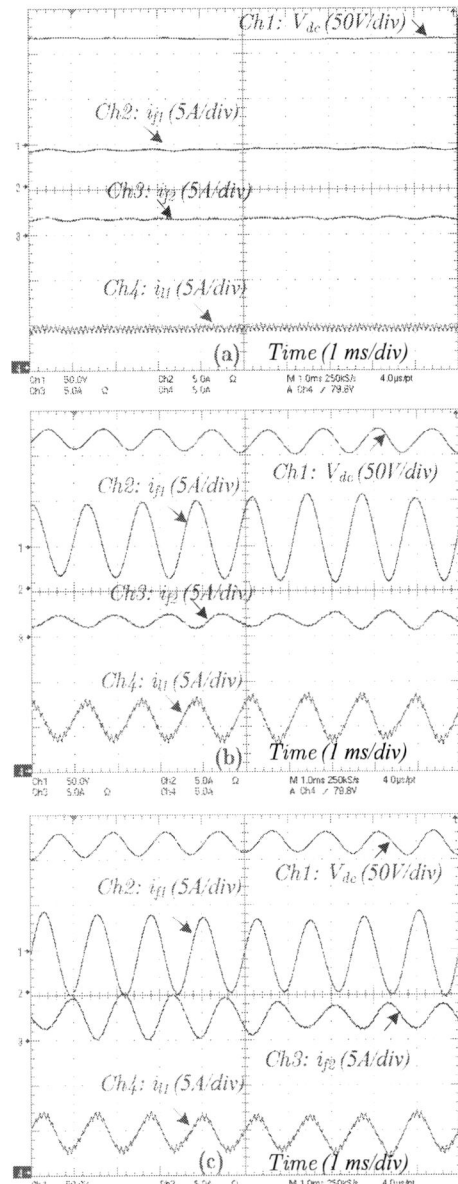

Fig. 6. Experimental waveforms: (a) P2=200W, P1=500W; (b) P2=200W, P1=650W; (c) P2=300W, P1=550W.

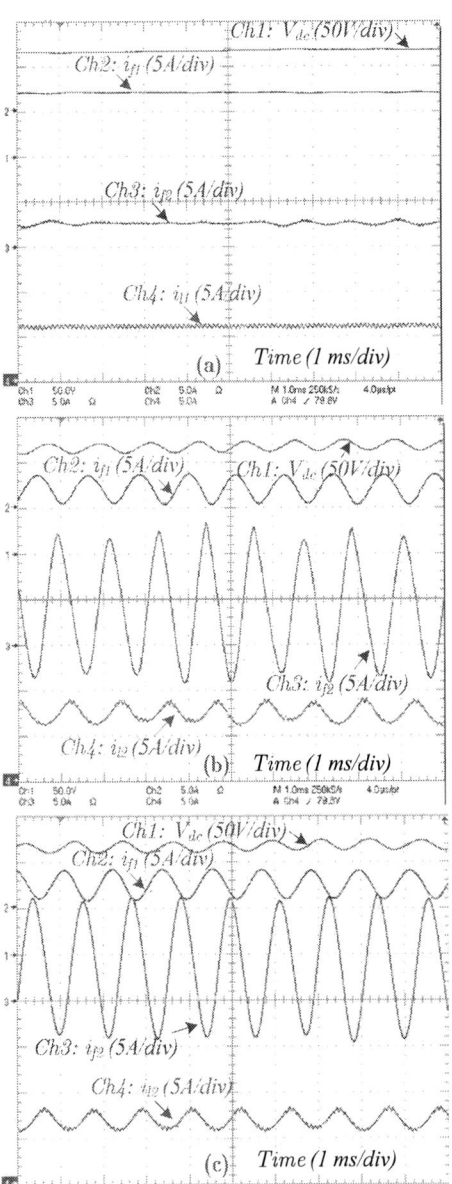

Fig. 7. Experimental waveforms: (a) P1=250W, P2=300W; (b) P1=250W, P2=450W; (c) P1=300W, P2=400W.

The experimental results presented in this section verify the critical powers and the system instabilities, which were predicted by the proposed analytical method. Indeed, the presented stability analysis can be applied to guarantee the stability of the complete system during the design process.

V. CONCLUSION

Stability analysis of hybrid AC/DC distribution systems with multiple constant power loads has been presented in this paper using an analytical model of the complete system. A generalized state-space modeling is developed that can capture the dynamics of the system.

The proposed model is used to study the performance and the stability of the system under load changes, as well as the different voltage levels on the DC side. According to the theoretical analysis, stable and unstable operating points are identified, and are used to establish the stability pattern of the hybrid system. The presented stability analysis has been then validated experimentally, on a test-bed hardware prototype.

REFERENCES

[1] P. C. Loh, D. Li, Y. K. Chai, and F. Blaabjerg, "Autonomous operation of hybrid microgrid with AC and DC subgrids," *IEEE Trans. Power Electron.*, vol. 28, pp. 2214-2223, 2013.

[2] R. S. Balog and P. T. Krein, "Bus selection in multibus DC microgrids," *IEEE Trans. Power Electron.*, vol. 26, pp. 860-867, 2011.

[3] A. Kwasinski and C. N. Onwuchekwa, "Dynamic behavior and stabilization of DC microgrids with instantaneous constant-power loads," *IEEE Trans. Power Electron.*, vol. 26, pp. 822-834, 2011.

[4] M. K. Zadeh, M. Amin, J. A. Suul, M. Molina, and O. B. Fosso, "Small-signal stability study of the Cigré DC grid test system with analysisof participation factors and parameter sensitivity of oscillatory modes " in *Proc. 18th Power Syst. Comput. Conf. (PSCC'14)*, Wroclaw, Poland, 2014.

[5] P. Magne, B. Nahid-Mobarakeh, and S. Pierfederici, "Active stabilization of DC microgrids without remote sensors for More Electric Aircraft," *IEEE Trans. Ind. App.*, vol. 49, pp. 2352-2360, 2013.

[6] M. K. Zadeh, B. Zahedi, M. Molinas, and L. E. Norum, "Centralized stabilizer for marine DC microgrid," in *Proc. 39th Annual Conf. IEEE Ind. Electron. Society (IECON'13)*, 2013, pp. 3359-3363.

[7] A. Khaligh, "Realization of parasitics in stability of DC–DC converters loaded by constant power loads in advanced multiconverter automotive systems," *IEEE Trans. Ind. Electron.*, vol. 55, pp. 2295-2305, 2008.

[8] J. Cao and A. Emadi, "A new battery/ultracapacitor hybrid energy storage system for electric, hybrid, and plug-in hybrid electric vehicles," *IEEE Trans. Power Electron.*, vol. 27, pp. 122-132, 2012.

[9] A. Ferreira, J. A. Pomilio, G. Spiazzi, and L. de Araujo Silva, "Energy management fuzzy logic supervisory for electric vehicle power supplies system," *IEEE Trans. Power Electron.*, vol. 23, pp. 107-115, 2008.

[10] B. Sarlioglu and C. Morris, "More Electric Aircraft; review, challenges and opportunities for commercial transport aircraft," *IEEE Trans. Transport. Electrific.*, vol. PP, pp. 1-1, 2015.

[11] R. Burgos, G. Chen, F. Wang, D. Boroyevich, W. G. Odendaal, and J. D. Van Wyk, "Reliability-oriented design of three-phase power converters for aircraft applications," *IEEE Trans. Aero. Electron. Syst.*, vol. 48, pp. 1249-1263, 2012.

[12] G. Gong, M. L. Heldwein, U. Drofenik, J. Miniböck, K. Mino, and J. W. Kolar, "Comparative evaluation of three-phase high-power-factor AC-DC converter concepts for future More Electric Aircraft," *IEEE Trans. Ind. Electron.*, vol. 52, pp. 727-737, 2005.

[13] M. K. Zadeh, S. Pierfederici, B. Nahid-Mobarakeh, and M. Molinas, "Stability analysis and dynamic performance evaluation of a power electronics-based DC distribution system with active stabilizer," *IEEE J. Emerg. Sel. Topics Power Electron.*, 2015.

[14] A. Emadi, A. Khaligh, C. H. Rivetta, and G. A. Williamson, "Constant power loads and negative impedance instability in automotive systems: definition, modeling, stability, and control of power electronic converters and motor drives," *IEEE Trans. Veh. Tech.*, vol. 55, pp. 1112-1125, 2006.

[15] A. Griffo and J. Wang, "Large signal stability analysis of More Electric Aircraft power systems with constant power loads," *IEEE Trans. Aero. Electron. Syst.*, vol. 48, pp. 477-489, 2012.

[16] K. Areerak, S. Bozhko, G. Asher, L. De Lillo, and D. Thomas, "Stability study for a hybrid ac-dc more-electric aircraft power system," *IEEE Trans. Aero. Electron. Syst.*, vol. 48, pp. 329-347, 2012.

[17] H. Zhang, F. Mollet, C. Saudemont, and B. Robyns, "Experimental validation of energy storage system management strategies for a local dc distribution system of more electric aircraft," *IEEE Trans. Ind. Electron.*, vol. 57, pp. 3905-3916, 2010.

[18] S. N. Motapon, L. Dessaint, and K. Al-Haddad, "A comparative study of energy management schemes for a fuel-cell hybrid emergency power system of more-electric aircraft," *IEEE Trans. Ind. Electron.*, vol. 61, pp. 1320-1334, 2014.

[19] M. Zandi, A. Payman, J.-P. Martin, S. Pierfederici, B. Davat, and F. Meibody-Tabar, "Energy management of a fuel cell/supercapacitor/battery power source for electric vehicular applications," *IEEE Trans. Veh. Tech.*, vol. 60, pp. 433-443, 2011.

[20] P. Magne, B. Nahid-Mobarakeh, and s. Pierfederci, "Dynamic consideration of DC microgrids with constant power loads and active damping system; a design method for fault-tolerant stabilizing system," *IEEE J. Emerg. Sel. Topics Power Electron.*, vol. 2, pp. 562-570, 2014.

[21] R. Gavagsaz-Ghoachani, J. Martin, S. Pierfederici, B. Nahid-Mobarakeh, and B. Davat, "DC power networks with very low capacitances for transportation systems: dynamic behavior analysis," *IEEE Trans. Power Electron.*, vol. 28, pp. 5865-5877, 2013.

[22] Y. Tang, P. C. Loh, P. Wang, F. H. Choo, and F. Gao, "Exploring inherent damping characteristic of LCL-filters for three-phase grid-connected voltage source inverters," *IEEE Trans. Power Electron.*, vol. 27, pp. 1433-1443, 2012.

[23] M. K. Zadeh and M. Molinas, "A controllable distributed energy resource with active filtering capability based on online harmonic detection," in *Proc. IEEE Grenoble PowerTech (POWERTECH'13)*, 2013, pp. 1-6.

[24] M. K. Zadeh, G.-G. Roghayeh, S. Pierfederici, N.-M. Babak, and M. Molinas, "A discrete-time tool to analyze the stability of weakly filtered active front-end PWM converters," in *Proc. IEEE Transportation Electrification Conf. & Expo. (ITEC'14)*, 2014.

[25] M. K. Zadeh, R. Gavagsaz-Ghoachani, J. P. Martin, S. Pierfederici, B. Nahid-Mobarakeh, and M. Molinas, "Discrete-time modelling, stability analysis, and active stabilization of dc distribution systems with constant power loads," in *Proc. IEEE Applied Power Electronics Conf. & Expo. (APEC'15)*, 2015, pp. 323-329.

978-1-4673-9551-9/16 $31.00 © 2016 IEEE

On the Concept of the Multi-Source Inverter

Lea Dorn-Gomba, *Student Member, IEEE,* Pierre Magne, *Member, IEEE,* Clement Barthelmebs,
and Ali Emadi *Fellow, IEEE*
Electrical and Computer Engineering Department
McMaster University, Hamilton, ON, Canada
Email: dorngoml@mcmaster.ca

Abstract—This paper presents an inverter topology, which aims to connect several DC sources to the same AC output using a single stage of conversion. This topology has been given the name of multi-source inverter. This concept has been developed for applications such as hybrid electric powertrains. Specifically, this topology allows different DC-link voltages to drive a traction motor using the battery voltage as one of the voltage levels. Thus, the DC/DC converter used in hybrid powertrain architectures can be downsized by reducing its power rating. In this paper, the multi-source inverter topology is introduced and its different operating modes are determined through an analysis of the inverter circuit. Closed-loop simulations with an interior permanent magnet (IPM) synchronous machine were performed to verify the theoretical principles of operation. A scaled-down prototype was built and successfully tested in open loop control with an RL load. The experimental results verify the effectiveness of the proposed topology and concept.

Index Terms—Electric vehicles, hybrid electric vehicles, hybrid powertrains, motor drives, multi-source inverters, plug-in hybrid electric vehicles, power electronics, pulse-width modulation.

I. INTRODUCTION

Motivated by the need to significantly reduce fuel consumption and harmful emissions in transportation, hybrid electric vehicles (HEVs), plug-in HEVs (PHEVS), and electric vehicles (EVs) have been developed. They currently achieve promising performances in comparison to the internal combustion engine (ICE) vehicles [1]–[3].

Although EVs are a very good candidate to overcome the current environmental issues associated with transportation, they possess drawbacks including the energy-storage limitations of their battery. HEVs and PHEVs, however, have improved range and vehicle performance thanks to the combination of one or two electric motors with the ICE. In a series-parallel architecture (such as in the Toyota Prius), two electric machines are mechanically coupled with the ICE through a planetary gear system and can operate either as motor or generator. This combination of power sources allows for different operating modes at high efficiency [4]. Moreover, a smaller battery pack can be used compared to the one integrated in EVs allowing volume and cost reductions. Power electronic circuits connect the battery pack to the traction drive and are composed of various components with numerous interactions between them. Therefore, even if combining electrical and combustion engines reduces the volume and the cost of the battery pack, it also increases the overall complexity of the system. Power electronic converters and

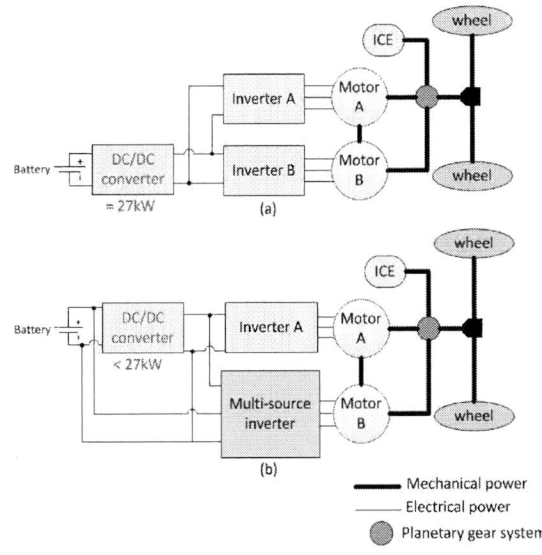

Fig. 1. (a) Toyota Prius architecture (b) proposed architecture with the multi-source inverter.

electric machines used in these system architectures play a major role since they enable higher efficiency and, hence, fuel consumption reductions compared to ICE vehicles [5], [6]. These technologies are also improving power density and cost [7], [8].

A more power-dense and lower-cost alternative is to develop new electrical topologies dedicated to hybrid powertrains [9]–[13]. Several bi-directional inverters for traction drive were suggested in literature. Located between the battery and the electric machines, they highly contribute to the overall system performance. Solution applied on the Toyota Prius consists of using a DC-DC boost converter between the battery and the DC-link of the traction inverters [14], [15] (Fig. 1-a). By doing so, the voltage of the battery is stepped up accordingly to the operating conditions [16]–[19]. This architecture presents several advantages such as extending the constant torque characteristic, reducing current rating, and having a higher inverter power supply voltage without increasing the battery voltage. However, a high-power DC/DC boost converter also offers several drawbacks since the inductor is usually bulky and expensive. Furthermore, as this requires the addition of a power conversion stage, it directly impacts the overall efficiency of the system.

This paper presents an inverter topology called the multi-source inverter [20]. This design allows different DC-link sources to drive a motor while directly using the battery voltage as one of the voltage levels. Thus, the power rating of the DC/DC converter can be decreased allowing size and cost reductions (Fig. 1-b). The following will first present the topology and the operating modes of the multi-source inverter. Then, a modified SVPWM control strategy and the controller structure are detailed. Finally, simulations in closed-loop control with an interior permanent magnet (IPM) synchronous machine and experimental results in open loop control with an RL load are presented to verify the concept of the proposed multi-source inverter.

II. MULTI-SOURCE INVERTER

The multi-source inverter (Fig. 2) presents a good tradeoff between topologies available in literature for hybrid vehicles. Similar to a multilevel inverter, the multi-source inverter can generate several output voltages. For instance, Fig. 2 shows a multi-source inverter with two DC inputs. According to the use of switches, different voltages can be applied to the output (line-to-line):

- Switches $U_{1,2,3}$ and $V_{1,2,3}$ enables V_{dc1} to be applied across V_{AB}, V_{BC}, V_{CA};
- Switches $W_{1,2,3}$ and $V_{1,2,3}$ enables V_{dc2} to be applied across V_{AB}, V_{BC}, V_{CA};
- Switches $U_{1,2,3}$ and $W_{1,2,3}$ enables $V_{dc1} - V_{dc2}$ to be applied across V_{AB}, V_{BC}, V_{CA}.

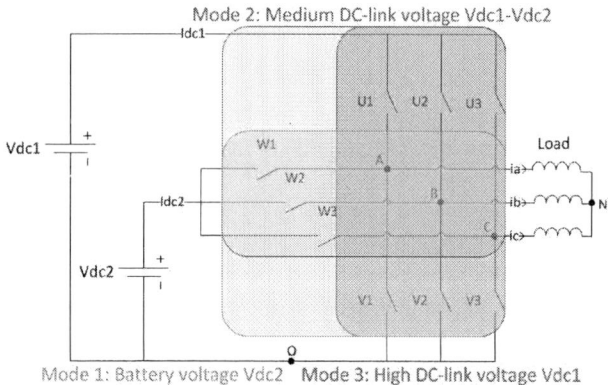

Fig. 2. Multi-source inverter topology with two DC inputs.

As a result, the combination of the two DC sources enables to supply the load with seven different voltages according to the state of the switches (Table I).

With this topology, a low voltage source, such as a battery, can directly drive a motor without stepping up its voltage with a DC/DC boost converter. Thus the power rating of this DC/DC converter can be decreased, enabling weight, volume and cost reductions. Moreover, when only the battery is used to drive the motor, this topology does not add additional conversion stage between the motor and the battery, hence improving the overall efficiency of the system.

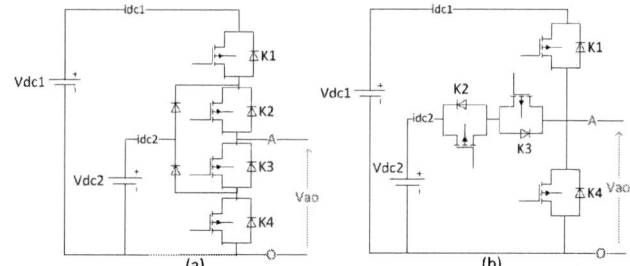

Fig. 3. (a) NPC single-phase leg (b) TNPC single-phase leg.

In each leg of the multi-source inverter, a Neutral Point Clamped (NPC) single-phase or a T-type NPC (TNPC) single phase is used (Fig. 3). These topologies present several advantages such as a low total harmonic distortion and low power losses [21]–[23].

A. Operating modes

In hybrid vehicles, the battery is commonly used to start the motor and for low speeds while the high DC bus provides power for middle and high speeds. According to the state of the switches, the multi-source inverter (Fig. 2) is composed of three standard three-phase inverters. This implies three different operating modes:

- Mode 1: Only the battery (V_{dc2}) supplies the motor ;
- Mode 2: The high DC-link voltage (V_{dc1}) supplies the motor while charging the battery (V_{dc2}). The voltage applied to the output is equal to V_{dc1}-V_{dc2};
- Mode 3: Only the high DC-link (V_{dc1}) voltage supplies the motor. The battery is not used.

According to Fig. 3-a, the line-to-neutral voltages $[V_{AO}, V_{BO}, V_{CO}]$ can be expressed as functions of the gate signal of the switches and the DC-bus voltages :

$$
\begin{aligned}
V_{A0} &= F_{h1}V_{dc1} + F_{m1}V_{dc2} - r.i_a \\
V_{B0} &= F_{h2}V_{dc1} + F_{m2}V_{dc2} - r.i_b \\
V_{C0} &= F_{h3}V_{dc1} + F_{m3}V_{dc2} - r.i_c
\end{aligned} \tag{1}
$$

where

r the phase resistance of the load;

$F_{(h1,h2,h3)} = K_1.K_2$ with $'.'$ the AND sign in Boolean logic;

$F_{(m1,m2,m3)} = K_1 \oplus K_2$ with $'\oplus'$ the XOR sign in Boolean logic.

Considering a Y connection of the load (3 phases AC motor), the DC-side current equations $[i_{dc1}, i_{dc2}]$ can be expressed as follows:

$$
\begin{aligned}
i_{dc1} &= F_{h1}i_a + F_{h2}i_b + F_{h3}i_c \\
i_{dc2} &= F_{m1}i_a + F_{m1}i_b + F_{m1}i_c
\end{aligned} \tag{2}
$$

The phase voltages $[V_{AN}, V_{BN}, V_{CN}]$ are calculated with the well-known relation:

$$
\begin{bmatrix} V_{AN} \\ V_{BN} \\ V_{CN} \end{bmatrix} = \frac{1}{3} \begin{bmatrix} 2 & -1 & -1 \\ -1 & 2 & -1 \\ -1 & -1 & 2 \end{bmatrix} \times \begin{bmatrix} V_{AO} \\ V_{BO} \\ V_{CO} \end{bmatrix} \tag{3}
$$

TABLE I
SWITCHING COMBINATIONS OF THE MULTI-SOURCE INVERTER.

Modes	States of switches									Line-to-line voltages		
	U_1	U_2	U_3	V_1	V_2	V_3	W_1	W_2	W_3	V_{AB}	V_{BC}	V_{CA}
1	Always open			0	1	1	1	0	0	V_{dc2}	0	$-V_{dc2}$
				0	0	1	1	1	0	0	V_{dc2}	$-V_{dc2}$
				1	0	1	0	1	0	$-V_{dc2}$	V_{dc2}	0
				1	0	0	0	1	1	$-V_{dc2}$	0	V_{dc2}
				1	1	0	0	0	1	0	$-V_{dc2}$	V_{dc2}
				0	1	0	1	0	1	V_{dc2}	$-V_{dc2}$	0
2	1	0	0	Always open			0	1	1	$V_{dc1}-V_{dc2}$	0	$-(V_{dc1}-V_{dc2})$
	1	1	0				1	0	0	1	$V_{dc1}-V_{dc2}$	$-(V_{dc1}-V_{dc2})$
	0	1	0				1	0	1	$-(V_{dc1}-V_{dc2})$	$V_{dc1}-V_{dc2}$	0
	0	1	1				1	0	0	$-(V_{dc1}-V_{dc2})$	0	$V_{dc1}-V_{dc2}$
	0	0	1				1	1	0	0	$-(V_{dc1}-V_{dc2})$	$V_{dc1}-V_{dc2}$
	1	0	1				0	1	0	$V_{dc1}-V_{dc2}$	$-(V_{dc1}-V_{dc2})$	0
3	1	0	0	0	1	1	Always open			V_{dc1}	0	$-V_{dc1}$
	1	1	0	0	0	1				0	V_{dc1}	$-V_{dc}$
	0	1	0	1	0	1				$-V_{dc1}$	V_{dc1}	0
	0	1	1	1	0	0				$-V_{dc1}$	0	V_{dc1}
	0	0	1	1	1	0				0	$-V_{dc1}$	V_{dc1}
	1	0	1	0	1	0				V_{dc1}	$-V_{dc1}$	0

The line-to-line voltages $[V_{AB},V_{BC},V_{CA}]$ can be expressed as follows:

$$\begin{aligned} V_{AB} &= V_{AN} - V_{BN} \\ V_{BC} &= V_{BN} - V_{CN} \\ V_{CA} &= V_{CN} - V_{AN} \end{aligned} \qquad (4)$$

From these above-mentioned equations a Simulink model of the multi-source inverter can be built.

B. PWM strategy

Space vector pulse width modulation (SVPWM) is usually preferred in automotive motor drive application because it offers better controllability at high speed in comparison to traditional sinusoidal PWM techniques. Also, it presents better performances in term of total harmonic distortion in control of AC motor and DC bus utilization [24]–[27]. A modified SVPWM model was developed to control switches in the multi-source inverter and simulations were carried out to verify the principle of operation.

Fig. 3-a shows that each leg of the multi-source inverter is very similar to the well-known three-level inverter made with a NPC topology. Thus, the modified SVPWM strategy will be closed to the usual one. As the widely used SVPWM, the new strategy consists of applying the reference voltage by a combination of the two nearest vectors and one zero vector in the alpha-beta reference plane [28]. However in the modified SVPWM, the DC voltage changes according to the multi-source's modes which is not the case in a standard SVPWM.

With a three-level inverters, three switching states -1, 0 and 1 represent the three phase voltage status of each leg (respectively $-V_{dc}/2$, 0, $V_{dc}/2$) (Fig. 4-a). Because the multi-source has two different input voltages, each mode is reachable by only two of the three switching states in each leg (Fig. 4-b and Table II).

In mode 1, the switching states are 0 or -1 which means that for each leg K1 is always opened, K3 is always closed and only K2 and K4 are switching. If the switching states are 0 or 1, the multi-source is in mode 2 where K2 stays closed, K4

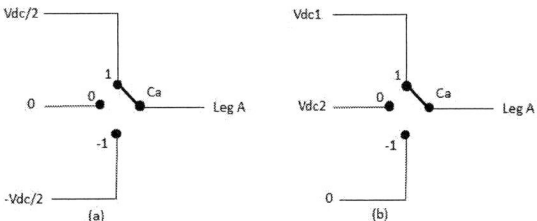

Fig. 4. (a) Switching states in a three-level inverter and (b) in the multi-source inverter.

TABLE II
SWITCHING STATES OF THE MULTI-SOURCE INVERTER.

Switching states	Device switching states	Multi-source's modes
1	K1/K2 on K3/K4 off	2 or 3
0	K2/K3 on K1/K4 off	1 or 2
-1	K3/K4 on K1/K2 off	1 or 3

is always opened and only K1 and K3 are switching. Finally mode 3 is only reachable when the switching states are 1 or -1, implying the switching of K1, K2, K3 and K4.

The spatial representation in alpha-beta diagram of the new SVPWM strategy is composed of three hexagons, whose size varies depending on the DC input voltage values (Fig. 5). Each hexagon is composed of six sectors. Once the mode and the sector are determined, the switching duration for each vector is calculated such as in a standard SVPWM control scheme. Finally, symmetric switching sequences were chosen here in order to reduce the THD and the number of switching, allowing power loss reductions.

III. CONTROLLER

Today's high-performance microprocessors allow the development and implementation of complex control algorithms and

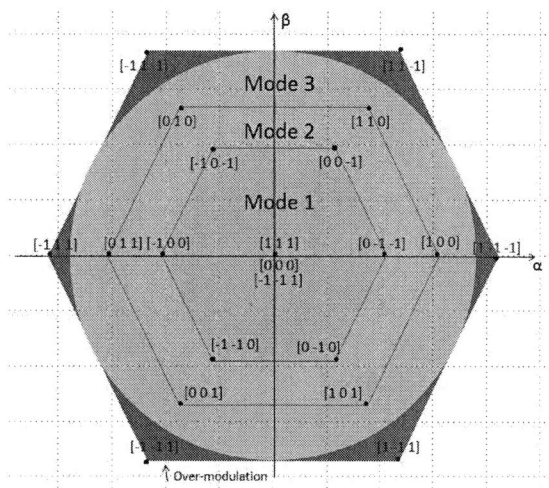

Fig. 5. Space representation in alpha-beta diagram of the modified SVPWM.

the generation of advanced PWM control scheme [29]. DSP algorithms provide many advantages over analog systems such as flexibility and fast prototyping capabilities as the code can be easily updated without hardware modification, reliability, capacity to repeat precisely from one unit to another and ability to manage sophisticated applications which would not be possible with traditional analog techniques [30].

The multi-source inverter requires generating gate signals for the power semiconductor devices. Hence a control algorithm of the modified SVPWM was implemented in Simulink and automatically compiled in language C/C++ (Fig. 6). The DSP used is TMS320F28335 from Texas Instrument [31], [32].

As shown in Fig. 6, reference, battery and high DC-link voltages must be first converted into digital format through the use of Analog-to-Digital converters included in the DSP. Then, calculation of the sectors, the time duration and the switching sequences are similar to those calculated in simulations until the generation of the pulse gate signals. By using a NPC topology in each leg (Fig. 3-a), switches K1 and K3 (and respectively K2 and K4) are complementary, allowing the use of solely six PWM peripherals with their complementary output channels.

An open loop system model was built with Matlab/Simulink and then compiled in C/C++ code [33]. The hardware Code Composer Studio was used to interface between the DSP board and the computer.

IV. SIMULATIONS AND EXPERIMENTAL RESULTS

A. Simulations in closed- loop control

A closed-loop control simulation model of the multi-source inverter driving an IPM was developed in Matlab/Simulink.

TABLE III
SIMULATION PARAMETERS OF THE MULTI-SOURCE INVERTER.

Parameters	V_{dc1}	V_{dc2}	Switching frequency
Units	600V	200V	10kHz

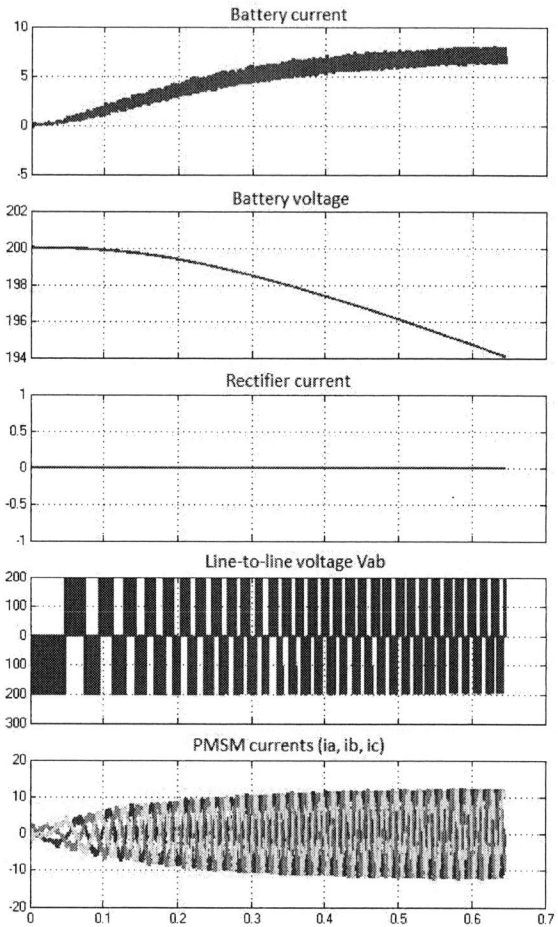

Fig. 7. Battery current and voltage, rectifier current, line-to-line voltage V_{AB}, and IPM currents in mode 1.

Parameters are listed in Table III. Input DC voltages were chosen in the same range of those commonly used in hybrid vehicles. Simulations were performed with three different speed references, showing the behavior of the multi-source inverter in each operating mode.

The simulation waveforms, presented in Figs. 7, 8 and 9, are in line with the theory:

- In mode 1 (Fig. 7), the battery supplies the motor whereas the rectifier voltage is not being used. From Fig. 7 it can be seen that the battery current is positive and its voltage is decreasing while the rectifier current remains null;
- In mode 2 (Fig. 8), a positive current at the output of the rectifier and a negative battery current are drawn. As expected, battery voltage is increasing, confirming its charge from the rectifier while the motor is running;
- In mode 3 (Fig. 9), the battery current is null and its voltage stays constant, showing that the battery is not being used. In this mode the motor is driven by the rectifier output voltage only.

In Figs. 7 to 9, the line-to-line voltage V_{AB} varies according to the mode of the multi-source inverter as predicted by the

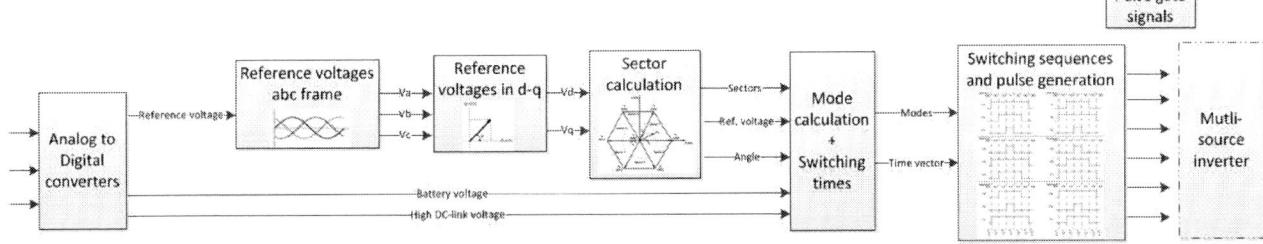

Fig. 6. Control structure of the multi-source inverter (open-loop).

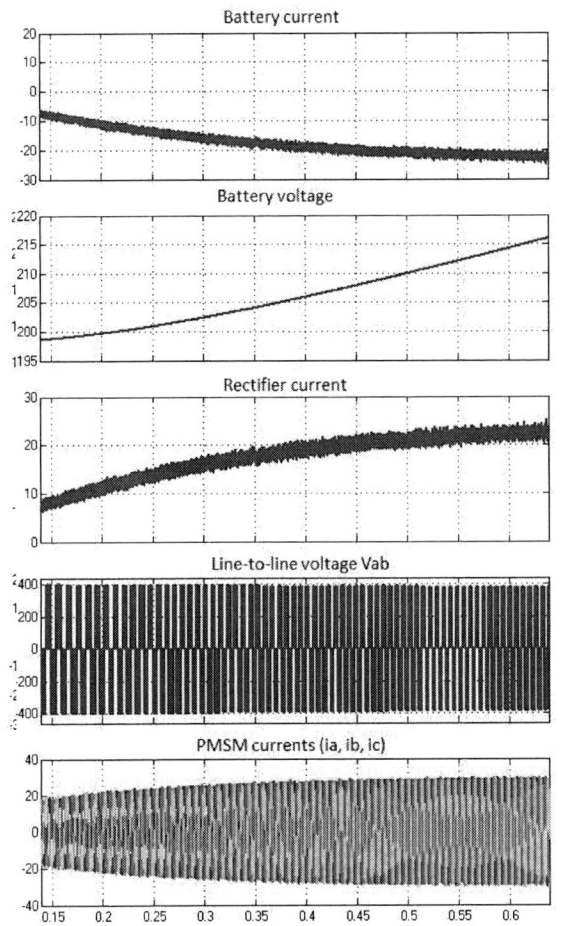

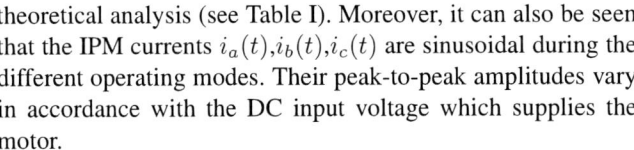

Fig. 8. Battery current and voltage, rectifier current, line-to-line voltage V_{AB}, and IPM currents in mode 2.

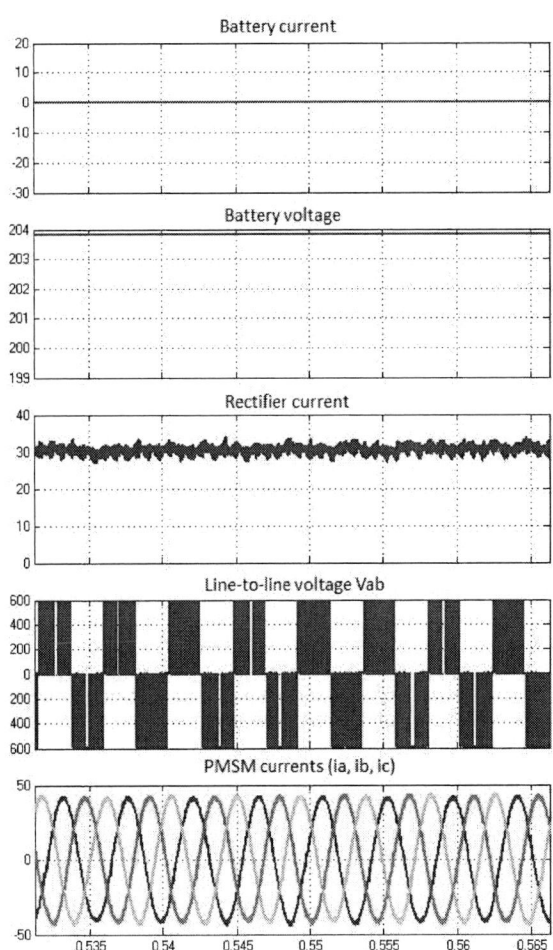

Fig. 9. Battery current and voltage, rectifier current, line-to-line voltage V_{AB}, and IPM currents in mode 3.

theoretical analysis (see Table I). Moreover, it can also be seen that the IPM currents $i_a(t), i_b(t), i_c(t)$ are sinusoidal during the different operating modes. Their peak-to-peak amplitudes vary in accordance with the DC input voltage which supplies the motor.

These results verify the theoretical principle of operation of the multi-source inverter in closed-loop control of an IPM motor drive.

B. Experimental results in open-loop control

A scaled-down prototype was developed (Fig. 10) to validate the effectiveness of the converter. The PCB is composed of three IGBT modules type F3L50R06W1E3-B11 and the control signals are obtained using the DSP TMS320F28335. Experiments were carried out with an RL load (10Ω and 256μH), an input battery voltage of 12.5V, and a high DC-link voltage of 30V provided by a power supply. Results in modes 1, 2, and 3 are presented in Figs. 11, 12 and 13 respectively.

978-1-4673-9551-9/16 $31.00 © 2016 IEEE   457

Fig. 10.   Prototype of the multi-source inverter.

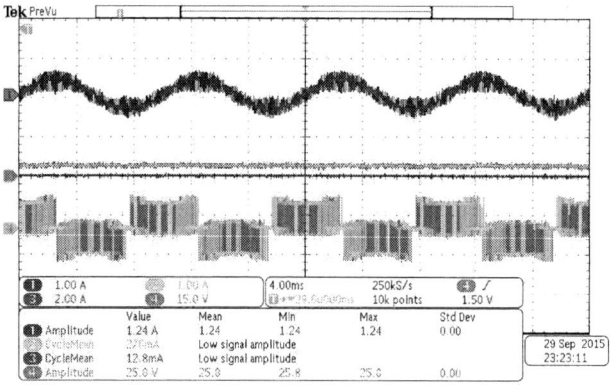

Fig. 11.   Load current in one phase (in blue), battery current (in cyan), high DC-link current (in pink), and line-to-line voltage Vab (in green) during mode 1 with an RL load.

As shown in Fig. 11, the measured load current is sinusoidal and the average battery current is positive while the high DC-link current is almost null. These results confirm the theory where in mode 1 only the battery supplies the load. Moreover, as predicted in Table I, the line-to-line voltage V_{AB} has three different values: V_{dc2}, $-V_{dc2}$ or 0.

In mode 2 (Fig. 12), the load current presents a higher amplitude compared to the one measured in mode 1 and the average battery current is negative. This confirms that the high

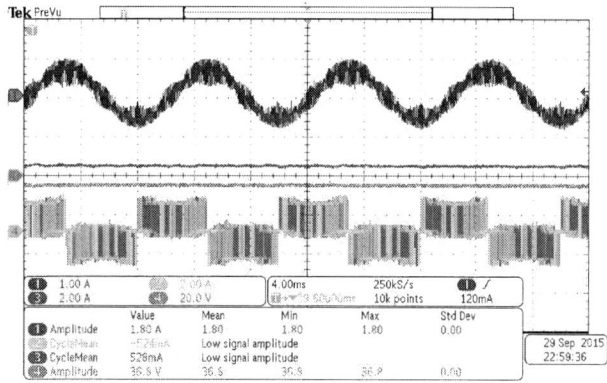

Fig. 12.   Load current in one phase (in blue), battery current (in cyan), high DC-link current (in pink), and line-to-line voltatge Vab (in green) during mode 2 with an RL load.

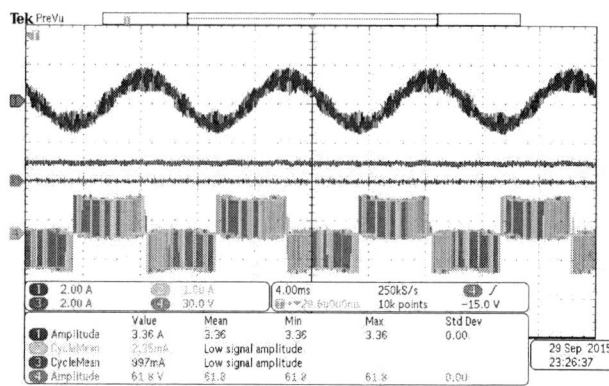

Fig. 13.   Load current in one phase (in blue), battery current (in cyan), high DC-link current (in pink), and line-to-line volatge Vab (in green) during mode 3 with an RL load.

DC-link voltage supplies the load and charges the battery at the same time. The line-to-line voltage V_{AB} matches with the theory and has three values: $V_{dc1} - V_{dc2}$, $-(V_{dc1} - V_{dc2})$ or 0.

From Fig. 13 it can be seen that the battery current displays a value of a few mA, allowing considering the battery as off. The amplitude of the phase load current is the greatest one compared to the other modes, which is coherent since only the high DC-link supply the load in mode 3. Finally, the line-to-line voltage values of V_{AB} are V_{dc1}, $-V_{dc1}$ or 0 which is still consistent with the theory.

According to the theoretical principle of operation and simulations, the experimental results are consistent. Moreover, each mode commutes well according to the voltage reference requested.

V. CONCLUSION

In this paper, an inverter topology suitable for hybrid electric vehicles was proposed. The multi-source inverter allows different DC-link voltages to drive the propulsion motor while directly using the battery voltage as one of the voltage levels. Simulation results verify its theoretical principle of operation for a motor drive. A scaled-down prototype was developed, built, and successfully tested with an RL load. Future experimental work will be conducted to confirm validation of the proposed concept. The multi-source inverter could offer a new option for the electrical propulsion system architecture in electrified vehicles.

ACKNOWLEDGEMENT

This research was undertaken, in part, thanks to funding from the Canada Excellence Research Chairs (CERC) Program.

REFERENCES

[1] B. Bilgin, P. Magne, P. Malysz, Y. Yang, V. Pantelic, M. Preindl, A. Korobkine, W. Jiang, M. Lawford, and A. Emadi, "Making the Case for Electrified Transportation," *IEEE Transactions on Transportation Electrification*, vol. 1, no. 7, pp. 4–17, Jun. 2015.

[2] C. C. Chan, "The State of the Art of Electric, Hybrid, and Fuel Cell Vehicles," *Proceedings of the IEEE*, vol. 95, no. 4, pp. 704–718, Apr. 2007.

[3] C. Ma, M. Song, J. Ji, J. Park, S. Ko, and H. Kim, "Comparative Study on Power Characteristics and Control Ctrategies for Plug-in HEV," in *2011 IEEE Vehicle Power and Propulsion Conference (VPPC)*, Chicago, IL, USA, Sep. 2011, pp. 1–6.

[4] A. Emadi, *Advanced Electric Drive Vehicles*. CRC Press Taylor and Francis Group, Oct. 2014.

[5] A. Emadi, S. S. Williamson, and A. Khaligh, "Power Electronics Intensive Solutions for Advanced Electric, Hybrid Electric, and Fuel Cell Vehicular Power Systems," *IEEE Transactions on Power Electronics*, vol. 21, no. 3, pp. 567–577, May 2006.

[6] J. M. Miller, "Power Electronics in Hybrid Electric Vehicle Applications," in *Eighteenth Annual IEEE Applied Power Electronics Conference and Exposition, 2003. APEC '03*, vol. 1, Miami Beach, FL, USA, Feb. 2003, pp. 23–29 vol.1.

[7] U. D. of Energy. Electrical and Electronics Technical Team Roadmap. [Online]. Available: http://www.eere.energy.gov/

[8] M. Pavlovsky, Y. Tsuruta, and A. Kawamura, "Recent Improvements of Efficiency and Power Density of DC-DC Converters for Automotive Applications," in *Power Electronics Conference (IPEC), 2010 International*, Sapporo, Japan, Jun. 2010, pp. 1866–1873.

[9] H. Ye, Y. Yang, and A. Emadi, "Traction Inverters in Hybrid Electric Vehicles," in *2012 IEEE Transportation Electrification Conference and Expo (ITEC)*, Dearborn, MI, USA, Jun. 2012, pp. 1–6.

[10] A. Battiston, J.-P. Martin, E.-H. Miliani, B. Nahid-Mobarakeh, S. Pierfederici, and F. Meibody-Tabar, "Comparison Criteria for Electric Traction System Using Z-Source/Quasi Z-Source Inverter and Conventional Architectures," *IEEE Journal of Emerging and Selected Topics in Power Electronics*, vol. 2, no. 3, pp. 467–476, Sep. 2014.

[11] B. A. Welchko and J. M. Nagashima, "The Influence of Topology Selection on the Design of EV/HEV Propulsion Systems," *IEEE Power Electronics Letters*, vol. 1, no. 2, pp. 36–40, Jun. 2003.

[12] B. Welchko and J. Nagashima, "A Comparative Evaluation of Motor Drive Topologies for Low-Voltage, High-Power EV/HEV Propulsion Systems," in *2003 IEEE International Symposium on Industrial Electronics, 2003. ISIE '03*, Rio de Janeiro, Brazil, Jun. 2003, pp. 379–384 vol. 1.

[13] O. Hegazy, J. Van Mierlo, and P. Lataire, "Design and Control of Bidirectional DC/AC and DC/DC Converters for Plug-in Hybrid Electric Vehicles," in *2011 International Conference on Power Engineering, Energy and Electrical Drives (POWERENG)*, Malaga, Spain, May 2011, pp. 1–7.

[14] T. A. Burress, S. L. Campbell, C. L. Coomer, C. W. Ayers, A. A. Wereszczak, J. P. Cunningham, L. D. Marlino, L. E. Seiber, and H. T. Lin, "Evaluation of the 2010 Toyota Prius Hybrid Synergy Drive System," Oak Ridge National Laboratory, Tech. Rep., May 2011.

[15] K. Muta, Y. M., and J. Tokieda, "Development of New-generation Hybrid System THS II - Drastic Improvement of Power Performance and Fuel Economy," in *2004 SAE World Congress*, Detroit, MI, Mar 2004.

[16] K. Asano, Y. Inaguma, H. Ohtani, E. Sato, M. Okamura, and S. Sasaki, "High Performance Motor Drive Technologies for Hybrid Vehicles," in *Power Conversion Conference, 2007. PCC '07*, Nagoya, Japan, Apr. 2007, pp. 1584–1589.

[17] H. Bai and C. Mi, "The Impact of Bidirectional DC-DC Converter on the Inverter Operation and Battery Current in Hybrid Electric Vehicles," in *2011 IEEE 8th International Conference on Power Electronics and ECCE Asia (ICPE ECCE)*, Jeju Province, South Korea, May 2011, pp. 1013–1015.

[18] H. Ye, P. Magne, B. Bilgin, S. Wirasingha, and A. Emadi, "A Comprehensive Evaluation of Bidirectional Boost Converter Topologies for Electrified Vehicle Applications," in *IECON 2014 - 40th Annual Conference of the IEEE Industrial Electronics Society*, Dallas, TX, USA, Oct. 2014, pp. 2914–2920.

[19] M. Olszewski, "Boost Converters for Gas Electric and Fuel Cell Hybrid Electric Vehicles," Oak Ridge National Laboratory, Tech. Rep., Jun 2005.

[20] A. Emadi and P. Magne, "Power Converter," Patent U.S. 0 117 770 A1, May 1, 2014.

[21] I. Staudt, "3L NPC and TNPC Topology," SEMIKRON, Tech. Rep., Sep 2012.

[22] M. Schweizer and J. W. Kolar, "Design and Implementation of a Highly Efficient Three-Level T-Type Converter for Low-Voltage Applications," *IEEE Transactions on Power Electronics*, vol. 28, no. 2, pp. 899–907, Feb. 2013.

[23] J. Rodriguez, S. Bernet, P. K. Steimer, and I. E. Lizama, "A Survey on Neutral-Point-Clamped Inverters," *IEEE Transactions on Industrial Electronics*, vol. 57, no. 7, pp. 2219–2230, Jul. 2010.

[24] S. P. Singh and R. K. Tripathi, "Performance Comparison of SPWM and SVPWM Technique in NPC Bidirectional Converter," in *2013 Students Conference on Engineering and Systems (SCES)*, Allahabad , India, Apr. 2013, pp. 1–6.

[25] X. Jing, J. He, and N. A. O. Demerdash, "Loss Balancing SVPWM for Active NPC Converters," in *2014 Twenty-Ninth Annual IEEE Applied Power Electronics Conference and Exposition (APEC)*, Fort Worth, TX, USA, Mar. 2014, pp. 281–288.

[26] O. Dordevic, M. Jones, and E. Levi, "A Comparison of Carrier-Based and Space Vector PWM Techniques for Three-Level Five-Phase Voltage Source Inverters," *IEEE Transactions on Industrial Informatics*, vol. 9, no. 2, pp. 609–619, May 2013.

[27] K. V. Kumar, P. A. Michael, J. J. P., and S. K. S., "Simulation and Comparison of SPWM and SVPWM Control for Three Phase Inverter," *ARPN Journal of Engineering and Applied Sciences*, vol. 5, no. 7, pp. 61–74, Jul 2010.

[28] R. L. Mallikarjuna, G. D. Naik and S. H. Jangamshetti, "Space Vector Modulation Technique for Three Level Diode Clamped Inverter," 2012.

[29] A. Y. E. Lesan, M. Doumbia, and P. Sicard, "DSP-Based Sinusoidal PWM Signal Generation Algorithm for Three Phase Inverters," in *2009 IEEE Electrical Power Energy Conference (EPEC)*, Montreal, QC, Canada, Oct. 2009, pp. 1–6.

[30] S. M. Kuo and L. B. H., *Real-Time Digital Signal Processing*. Wiley, 2001.

[31] D. Ward, I. Husain, C. Castro, A. Volke, and H. M., *Fundamentals of Semiconductors for Hybrid-Electric Powertrain*. Infeneon Technologies North America Corp, 2013.

[32] Z. Maha and B. Faouzi, "Implementation of Space Vector Modulation Using DSP," in *2013 International Conference on Electrical Engineering and Software Applications (ICEESA)*, Hammamet, Tunisia, Mar. 2013, pp. 1–7.

[33] Z. Yu, "Space-vector PWM with TMS320C24x/F24x Using Hardware and Software Determined Switching Patterns," Texas Instrument, Tech. Rep., Mar 1999.

978-1-4673-9551-9/16 $31.00 © 2016 IEEE

Time-Domain Analysis of A Wide-DC-Range Series Resonant Dual-Active-Bridge Bidirectional Converter with a New Passive Auxilliary Circuit

Alireza Safaee, Praveen Jain, Alireza Bakhshai
Queen's University Kingston ON, Canada
E-mail: az_safaee@ieee.org

Abstract— In this paper, an isolated bidirectional series resonant converter is proposed for transportation applications such as mid-size electrified airplanes. Using time-domain analysis closed form formulae for operation in two mutually exclusive and collectively exhaustive regimes are achieved. A modulation scheme is suggested allowing bidirectional power transfer up to the nominal level over the entire wide dc range of terminal voltages under zero voltage or zero current switching. A novel robust and low cost passive auxiliary is introduced to guarantee zero voltage switching for all the voltage and power levels. Effectiveness of the modulation scheme in achieving zero voltage switching was experimentally validated.

Keywords—Series Resonant Converter; Bidirectional Converter; Dual Active Bridge; Soft Switching; Wide range; step-up; step down.

I. INTRODUCTION

Many stationary and mobility systems take advantage of bidirectional converters for example uninterruptible power supplies, energy storage systems, regenerative braking energy retriever units in rail and road electric and hybrid vehicles, and dual-dc-bus aircraft [1-3]. Achievement of high efficiency, power density and reliability is essential for any system onboard vehicles with no exception for bidirectional converter units [4-6].

Some bidirectional converters use a series resonant tank as the impedance between two ac sources [7-10] which brings the same benefits of the unidirectional series resonant converters including: intrinsic low switching losses (depending on the modulation scheme), taking advantage of inevitable parasitic inductances as a part of resonant tank, lower EMI, inductor-free output filer, and natural short circuit protection. Many studies use the fundamental harmonic approximation for the current in the resonant tank due to the band pass nature of the tank with a high quality factor. This necessitates the paper to verify the validity of the approximation.

This paper presents a precise description of bidirectional series resonant converter using time-domain analysis which does not require approximating the resonant current with a sinusoidal waveform. Time-domain analysis is frequently used to study the steady state operation of converters [11-14] and the auxiliary circuits [15,16]. The closed form formulae from the time-domain steady state analysis enable us to propose a modulation scheme ensuring zero current switching (ZCS) for the entire range of dc voltages on both the dc terminals and power levels. Then an adaptive auxiliary circuit is introduced

to maintain zero voltage switching (ZVS) for all the operating points and power levels.

II. ANALYSIS IN TIME-DOMAIN

Fig. 1 sows the dual active bridge (DAB) bidirectional converter with a series resonant network coupling. The inductor L can include the leakage inductor of the transformer.

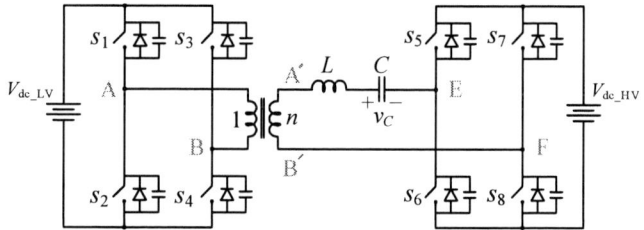

Fig. 1: Dual active bridge bidirectional converter with series resonant coupling network.

A. Assumptions

The assumption used in this text are:

- The converter has settled down to steady state operation.
- All the switches, on both the low voltage (LV) and high voltage (HV) sides, have 50% on and 50% off times. The effect of dead-time is neglected.
- The entire components, passive and active, are ideal, lossless, and with no parasitics.
- Transformer is ideal, linear, with no magnetization branch.
- The switches act instantaneously with no delays.

Due to no significant parallel branch, we have $i_E = -i_A/n$.

B. Key Definitions

The angular switching frequency, ω_{SW}, is higher than the natural resonant frequency of the series resonant tank, ω_0, given by:

$$\omega_0 = 1/\sqrt{LC} \quad , \quad Z_0 = \sqrt{L/C} \tag{1}$$

Z_0 is the natural impedance of the resonant tank. The notation of the referred voltage ratio is defined as:

$$d \triangleq \frac{V_{dc\_HV}}{nV_{dc\_LV}} \tag{2}$$

The voltage ratio d has a major role in determining the soft switching behavior of the converter. As far as the soft switching is concerned only the ratio of the voltages is important and not the actual values.

Fig. 2 shows the basic waveforms and defines the inter-bridge phase-shift (φ), LV side duty-cycle (D_{LV}) and HV side duty-cycle (D_{HV}).

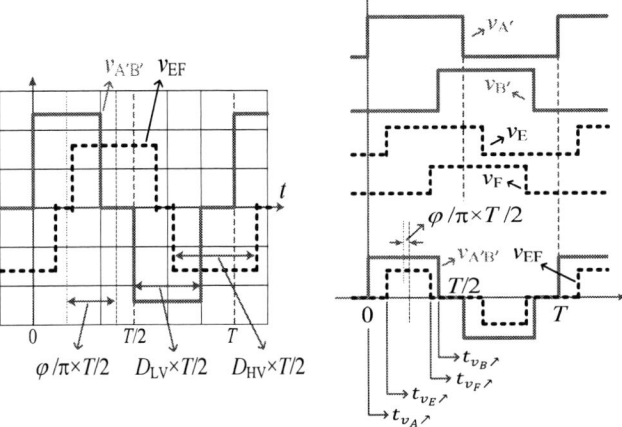

Fig. 2: Major voltage waveforms.

Based on Fig. 2, the parameter ranges are $0 \leq \varphi \leq 2\pi$, and $0 \leq D_{LV}, D_{HV} \leq 1$. For the most general case the rising edge moments for $v_{A'}$ and $v_{B'}$ (referred v_A and v_B to HV side) and for v_E and v_F are shown in Fig. 2, and their relations with φ, D_{LV} and D_{HV} are given in TABLE I.

TABLE I.   RISING EDGE MOMENTS IN FIG. 3

$t_{v_A\nearrow} = 0$	$t_{v_B\nearrow} = (2D_{LV})T/4$
$t_{v_E\nearrow} = \left(\dfrac{\varphi}{\pi/2} + D_{LV} - D_{HV}\right)T/4$	$t_{v_F\nearrow} = \left(\dfrac{\varphi}{\pi/2} + D_{LV} + D_{HV}\right)T/4$

We select the beginning of the switching period with the rising edge of v_A and thus $t_{v_A\nearrow}$ is equal to zero for all the values of φ, D_{LV} and D_{HV}. Also $t_{v_B\nearrow}$ is independent of φ and D_{HV} and only depends on D_{LV} and also switching period T.

C. Description of Regime 1: $d \geq 1$

First the case of $nV_{dc\_LV} \leq V_{dc\_HV}$ or $d \geq 1$ is considered. For each power level it is desirable to have lower conduction losses, so the rms currents should be reduced. Thus, the available ac voltage should be taken full advantage of i.e. the largest duty-cycle should be used for the ac voltage with the lower amplitude. Here $v_{A'B'}$ has a peak of nV_{dc_LV} which is less than peak of v_{EF} equal to V_{dc_HV}. Thus $D_{LV} = 1$ is selected.

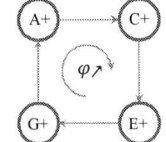

Fig. 3: Four modes is Regime 1.

The four possible modes for Regime 1 are presented in Fig. 3 and TABLE II.

TABLE II.   MODES IN IN REGIME 1
($d \geq 1$ and $D_{LV} = 1$)

Mode	Condition
$A+$	$0 \leq \varphi/(\pi/2) < 1 - D_{HV}$ or $3 + D_{HV} \leq \varphi/(\pi/2) < 4$
$C+$	$1 - D_{HV} \leq \varphi/(\pi/2) < 1 + D_{HV}$
$E+$	$1 + D_{HV} \leq \varphi/(\pi/2) < 3 - D_{HV}$
$G+$	$3 - D_{HV} \leq \varphi/(\pi/2) < 3 + D_{HV}$

The transition between the four modes by increasing φ is shown in Fig. 3 and the waveforms of the modes are illustrated in Fig. 4. Also TABLE III. gives the transition moments.

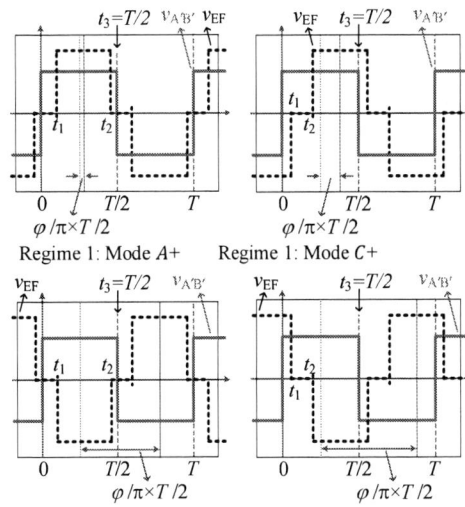

Fig. 4: Voltage waveforms of four modes is Regime 1.

TABLE III.   VOLTAGE TRANSITION MOMENTS IN REGIME 1

MODE	$\dfrac{t_1}{T/4}$	$\dfrac{t_2}{T/4}$	$\dfrac{t_3}{T/4}$
$A+$	$\dfrac{\varphi}{\pi/2} + 1 - D_{HV}$	$\dfrac{\varphi}{\pi/2} + 1 + D_{HV}$	2
$C+$	$\dfrac{\varphi}{\pi/2} - 1 + D_{HV}$	$\dfrac{\varphi}{\pi/2} + 1 - D_{HV}$	2
$E+$	$\dfrac{\varphi}{\pi/2} - 1 - D_{HV}$	$\dfrac{\varphi}{\pi/2} - 1 + D_{HV}$	2
$G+$	$\dfrac{\varphi}{\pi/2} - 3 + D_{HV}$	$\dfrac{\varphi}{\pi/2} - 1 - D_{HV}$	2

The boundary of Modes $A+$ and $C+$ is of special interest.

D. Description of Regime 2: $d \leq 1$

Similarly for the case of $nV_{dc\_LV} \geq V_{dc\_HV}$ ($d \leq 1$) the choice of $D_{HV} = 1$ is made. The four possible modes for Regime 1 are presented in Fig. 5 and TABLE II.

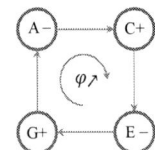

Fig. 5: Four modes is Regime 2.

TABLE IV.   MODES IN IN REGIME 2
(d ≤ 1 and $D_{HV} = 1$)

Mode	Condition
$A-$	$0 \leq \varphi/(\pi/2) < 1 - D_{LV}$ or $3 + D_{LV} \leq \varphi/(\pi/2) < 4$
$C+$	$1 - D_{LV} \leq \varphi/(\pi/2) < 1 + D_{LV}$
$E-$	$1 + D_{LV} \leq \varphi/(\pi/2) < 3 - D_{LV}$
$G+$	$3 - D_{LV} \leq \varphi/(\pi/2) < 3 + D_{LV}$

The transition between the four modes by increasing φ is shown in Fig. 5 and the waveforms of the modes are illustrated in Fig. 6. There is a specific interest in the boundary of Modes $A-$ and $C+$. TABLE V. provides the transition moments.

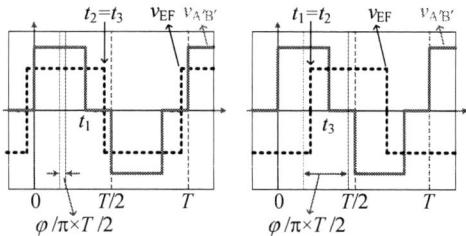

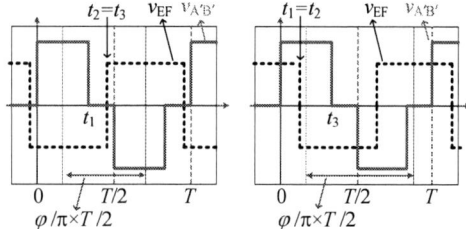

Fig. 6: Voltage waveforms of four modes is Regime 2.

TABLE V.   VOLTAGE TRANSITION MOMENTS IN REGIME 2

MODE	$\dfrac{t_1}{T/4}$	$\dfrac{t_2}{T/4}$	$\dfrac{t_3}{T/4}$
$A-$	$2D_{LV}$	$\dfrac{\varphi}{\pi/2} + D_{LV} + 1$	$\dfrac{\varphi}{\pi/2} + D_{LV} + 1$
$C+$	$\dfrac{\varphi}{\pi/2} + D_{LV} - 1$	$\dfrac{\varphi}{\pi/2} + D_{LV} - 1$	$2D_{LV}$
$E-$	$2D_{LV}$	$\dfrac{\varphi}{\pi/2} + D_{LV} - 1$	$\dfrac{\varphi}{\pi/2} + D_{LV} - 1$
$G+$	$\dfrac{\varphi}{\pi/2} + D_{LV} - 3$	$\dfrac{\varphi}{\pi/2} + D_{LV} - 3$	$2D_{LV}$

The boundary of Modes $A-$ and $C+$ is also important. The boundaries are studied in the next section.

E. Boundary Between Modes $A \pm$ and $C+$

For Regime 1, the value of φ at the boundary of modes $A+$ and $C+$, named as φ_{A+C+}, is:

$$\varphi_{A+C+} = (1 - D_{HV})\frac{\pi}{2} \tag{3}$$

For Regime 2, the φ at the boundary of modes $A-$ and $C+$, named as φ_{A-C+}, is:

$$\varphi_{A-C+} = (1 - D_{LV})\frac{\pi}{2} \tag{4}$$

F. Steady-State Analysis

Here we introduce the frequency ratio as:

$$r = \omega_{SW}/\omega_0 \tag{5}$$

Due to the intention to study the soft switching behavior of the converter, in this text only the equation for the transferred power and the resonant inductor current at the voltage transition moments are provided here. In the full paper the entire set of equations for the steady state behavior of the resonant inductor current, $i_L(t)$, and the resonant capacitor voltage, $v_C(t)$ will be presented.

For Regimes 1 and 2 the power equations are (6) and (7), respectively. In practice the range of $0 \leq \varphi \leq \pi/2$ $(3\pi/2 \leq \varphi \leq 2\pi)$ for transferring power from V_{dc_LV} to V_{dc_HV} $(V_{dc\_HV}$ to $V_{dc\_LV})$. As expected at $\varphi = 0$ and $\pi/2$ the power is zero and the maximum, respectively.

G. Currents at Boundary of Modes $A \pm$ and $C+$

Fig. 7(a) and (b) show the boundaries $A+C+$ and $A-C+$ for Regimes 1 and 2. The currents at the transition moments in the $A+C+$ boundary for Regime 1, named as $I_{x\_v_x \nearrow\_A+C+}$, $(x = A, B, E$ and $F)$ are given in (8) to (10). Also, the transition moment currents in the $A-C+$ boundary for Regime 2, i.e. $I_{x\_v_x \nearrow\_A+C+}$, $(x = A, B, E$ and $F)$ are provided in (11) to (13).

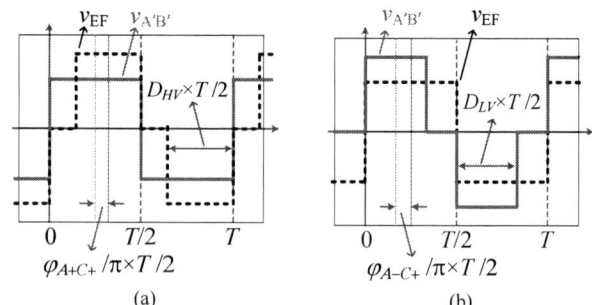

Fig. 7: Voltage waveforms of Boundaries (a) $A+C+$ in Regime 1, and (b) $A-C+$ in Regime 2.

$$P = \frac{2n}{\pi Z_0} V_{dc\_LV} V_{dc\_HV} \frac{r}{\cos\left(\frac{\pi}{2r}\right)} \times \begin{cases} \sin\left(\frac{\pi}{2r} D_{HV}\right) \sin\left(\frac{\varphi}{r}\right) & 0 < \varphi < \varphi_{A+C+} \quad \text{(Mode A +)} \\ \cos\left(\frac{\varphi}{r} - \frac{\pi}{2r}\right) \cos\left((1 - D_{HV})\frac{\pi}{2r}\right) - \cos\left(\frac{\pi}{2r}\right) & \varphi_{A+C+} < \varphi < \pi - \varphi_{A+C+} \quad \text{(Mode C +)} \\ \sin\left(\frac{\pi}{2r} D_{HV}\right) \sin\left(\frac{\pi - \varphi}{r}\right) & \pi - \varphi_{A+C+} < \varphi < \pi + \varphi_{A+C+} \quad \text{(Mode E +)} \\ -\cos\left(\frac{\varphi}{r} - \frac{3\pi}{2r}\right) \cos\left((1 - D_{HV})\frac{\pi}{2r}\right) + \cos\left(\frac{\pi}{2r}\right) & \pi + \varphi_{A+C+} < \varphi < 2\pi - \varphi_{A+C+} \quad \text{(Mode G +)} \\ \sin\left(\frac{\pi}{2r} D_{HV}\right) \sin\left(\frac{\varphi - 2\pi}{r}\right) & 2\pi - \varphi_{A+C+} < \varphi < 2\pi \quad \text{(Mode A +)} \end{cases} \quad (6)$$

$$P = \frac{2n}{\pi Z_0} V_{dc\_LV} V_{dc\_HV} \frac{r}{\cos\left(\frac{\pi}{2r}\right)} \times \begin{cases} \sin\left(\frac{\pi}{2r} D_{LV}\right) \sin\left(\frac{\varphi}{r}\right) & 0 < \varphi < \varphi_{A-C+} \quad \text{(Mode A −)} \\ \cos\left(\frac{\varphi}{r} - \frac{\pi}{2r}\right) \cos\left((1 - D_{LV})\frac{\pi}{2r}\right) - \cos\left(\frac{\pi}{2r}\right) & \varphi_{A-C+} < \varphi < \pi - \varphi_{A-C+} \quad \text{(Mode C +)} \\ \sin\left(\frac{\pi}{2r} D_{LV}\right) \sin\left(\frac{\pi - \varphi}{r}\right) & \pi - \varphi_{A-C+} < \varphi < \pi + \varphi_{A-C+} \quad \text{(Mode E −)} \\ -\cos\left(\frac{\varphi}{r} - \frac{3\pi}{2r}\right) \cos\left((1 - D_{LV})\frac{\pi}{2r}\right) + \cos\left(\frac{\pi}{2r}\right) & \pi + \varphi_{A-C+} < \varphi < 2\pi - \varphi_{A-C+} \quad \text{(Mode G +)} \\ \sin\left(\frac{\pi}{2r} D_{LV}\right) \sin\left(\frac{\varphi - 2\pi}{r}\right) & 2\pi - \varphi_{A-C+} < \varphi < 2\pi \quad \text{(Mode A −)} \end{cases} \quad (7)$$

$$I_{A\_v_A \nearrow\_A+C+} = I_{B\_v_B \nearrow\_A+C+} = \frac{n^2 V_{dc\_LV}}{Z_0 \cos\left(\frac{\pi}{2r}\right)} \left\{ -\sin\left(\frac{\pi}{2r}\right) + d.\sin\left(D_{HV}\frac{\pi}{2r}\right) \cos\left((1 - D_{HV})\frac{\pi}{2r}\right) \right\} \quad (8)$$

$$I_{E\_v_E \nearrow\_A+C+} = -\frac{n V_{dc\_LV}}{Z_0 \cos\left(\frac{\pi}{2r}\right)} \left\{ \sin\left(\frac{\pi}{2r}\right) + d.\sin\left(D_{HV}\frac{\pi}{2r}\right) \cos\left((1 - D_{HV})\frac{\pi}{2r}\right) \right\} \quad (9)$$

$$I_{F\_v_F \nearrow\_A+C+} = \frac{n V_{dc\_LV}}{Z_0 \cos\left(\frac{\pi}{2r}\right)} \left\{ \sin\left(\frac{\pi}{2r}\right) - d.\sin\left(D_{HV}\frac{\pi}{2r}\right) \cos\left((1 - D_{HV})\frac{\pi}{2r}\right) \right\} \quad (10)$$

$$I_{A\_v_A \nearrow\_A-C+} = \frac{n^2 V_{dc\_LV}}{Z_0 \cos\left(\frac{\pi}{2r}\right)} \left\{ -\sin\left(D_{LV}\frac{\pi}{2r}\right) \cos\left((1 - D_{LV})\frac{\pi}{2r}\right) + d.\sin\left(\frac{\pi}{2r}\right) \right\} \quad (11)$$

$$I_{B\_v_B \nearrow\_A-C+} = -\frac{n^2 V_{dc\_LV}}{Z_0 \cos\left(\frac{\pi}{2r}\right)} \left\{ \sin\left(D_{LV}\frac{\pi}{2r}\right) \cos\left((1 - D_{LV})\frac{\pi}{2r}\right) + d.\sin\left((1 - 2D_{LV})\frac{\pi}{2r}\right) \right\} \quad (12)$$

$$I_{E\_v_E \nearrow\_A-C+} = I_{F\_v_F \nearrow\_A-C+} = -\frac{n V_{dc\_LV}}{Z_0 \cos\left(\frac{\pi}{2r}\right)} \left\{ \cos\left((1 - D_{LV})\frac{\pi}{2r}\right) \sin\left(D_{HV}\frac{\pi}{2r}\right) - \sin\left(\frac{\pi}{2r}\right).d \right\} \quad (13)$$

H. Boundary Duty-cycle Selection for Soft Switching

Equations (8) to (10) in Regime 1 reveals the possibility of selecting D_{HV} at the mode boundary such that all three transition moments currents be equal to zero. We name this duty-cycle as $D_{HV\_A+C+\_ZCS}$ (because of ZCS) governed by:

$$D_{HV\_A+C+\_ZCS} = \frac{1}{2} + \frac{r}{\pi} \sin^{-1}\left[\left(\frac{2}{d} - 1\right) \sin\left(\frac{\pi}{2r}\right)\right] \quad (14)$$

and using (3):

$$\varphi_{HV\_A+C+\_ZCS} = \frac{\pi}{4} - \frac{r}{2} \sin^{-1}\left[\left(\frac{2}{d} - 1\right) \sin\left(\frac{\pi}{2r}\right)\right] \quad (15)$$

At $d = 1$ we get $D_{HV\_A+C+\_ZCS} = 1$ as expected. Similarly (11) to (13) calculations for Regime 2 lead to:

$$D_{LV\_A-C+\_ZCS} = \frac{1}{2} + \frac{r}{\pi} \sin^{-1}\left[(2d - 1) \sin\left(\frac{\pi}{2r}\right)\right] \quad (16)$$

$$\varphi_{LV\_A-C+\_ZCS} = \frac{\pi}{4} - \frac{r}{2} \sin^{-1}\left[(2d - 1) \sin\left(\frac{\pi}{2r}\right)\right] \quad (17)$$

Again at $d = 1$ we get $D_{LV\_A-C+\_ZCS} = 1$ as expected. The importance of knowing these quantities is in the possibility of developing a new modulation scheme such that the soft switching is maintained for all the combinations of dc voltages and power levels (in both directions).

I. Modulation Scheme for Soft Switching

In Fig. 8 a modulation scheme is proposed to control the converter as well as to maintain the soft switching property. This is done by introducing the remaining non-unity duty-cycle as a function of the inter-bridge phase-shift, φ. By doing this, the control space becomes one dimensional and it will be possible to apply well-established control methods directly.

The first modulation scheme includes a sequence of steps in each switching cycle as follows:

- The quantities $V_{dc\_LV}, V_{dc\_HV}, I_{dc\_LV}$, and I_{dc_HV} are sampled and fed to the controller unit. The controller considers the proper output variable on the receiving terminal to be controlled (voltage, current, or power) based on the input command for the step up or step down operation mode. Any over current and short circuit events are also detected. The output of the controller unit is the inter-bridge phase-shift φ which is between -90° and 90°.

- The frequency ratio r is determined based on V_{dc_LV} and V_{dc_HV} following the formula (6) or (7) such that the peak power is kept constant. This step is independent from the φ value in the previous step.

- The voltage ratio d is calculated from V_{dc_LV} and V_{dc_HV} using (2). For $d > 1$ ($d < 1$) we select $D_{LV} = 1$ ($D_{HV} = 1$) i.e. operation in Regime 1 (Regime 2).

- Using r and for $d > 1$ ($d < 1$) we calculate $D_{HV\_A+C+\_ZCS}$ and $\varphi_{HV\_A+C+\_ZCS}$ ($D_{LV\_A-C+\_ZCS}$ and $\varphi_{LV\_A-C+\_ZCS}$).

- For $d > 1$ we calculate D_{HV} using φ and $D_{HV\_A+C+\_ZCS}$ such that $D_{HV}(\varphi = \pi/2) = 1$ and $D_{HV}(\varphi = \varphi_{HV\_A+C+\_ZCS}) = D_{HV\_A+C+\_ZCS}$. Similarly, for $d < 1$ we calculate D_{LV} using φ and $D_{LV\_A+C+\_ZCS}$ such that $D_{LV}(\varphi = \pi/2) = 1$ and $D_{LV}(\varphi = \varphi_{LV\_A-C+\_ZCS}) = D_{LV\_A-C+\_ZCS}$.

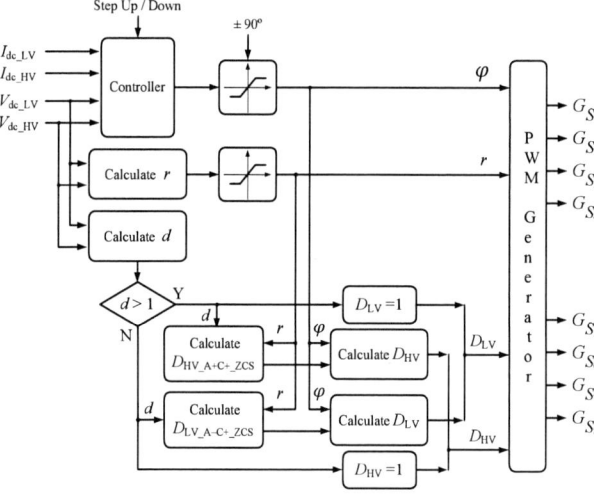

Fig. 8: Modulation scheme.

There are infinite monotonically increasing functions that satisfy the above conditions. It is better to select a simple function with a minimized calculation time. Here the simplest which is a linear function is selected. For the case of Regime 1 ($d > 1$):

$$D_{HV}(\varphi) = (1 - k_{SC1})\frac{\varphi}{\pi/2} + k_{SC1} \tag{18}$$

where:

$$k_{SC1} = 2 - \frac{1}{D_{HV\_A+C+\_ZCS}} \tag{19}$$

Similarly for the case of Regime 2 ($d < 1$):

$$D_{LV}(\varphi) = (1 - k_{SC2})\frac{\varphi}{\pi/2} + k_{SC2} \tag{20}$$

where:

$$k_{SC2} = 2 - \frac{1}{D_{LV\_A-C+\_ZCS}} \tag{21}$$

Note that k_{SC1} and k_{SC2} are positive numbers less than or equal to one (at $d = 1$).

III. PROPOSED PASSIVE AUXILIARY CIRCUIT

The proposed converter topology is depicted in Fig. 9. The auxiliary circuit is indicated inside the dashed lines. There is only one auxiliary circuit, composed of two robust magnetic components, to ensure ZVS of the entire converter which means an even smaller component count.

TABLE VI. and TABLE VII. show the equations of auxiliary currents at the rising edge moments of four ac node voltages in Regimes 1 and 2, respectively. The importance of knowing these currents is that the auxiliary current is negative at all the ac node voltage rising edge moments and can be selected properly to maintain ZVS at any phase-shift for all dc voltages.

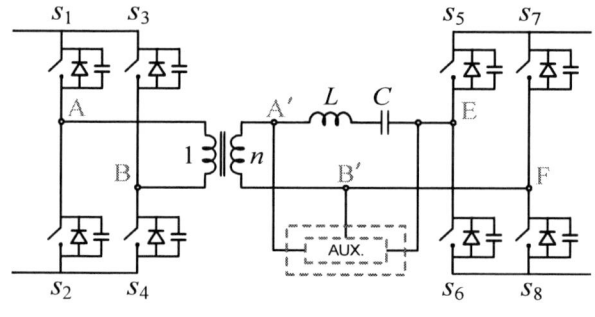

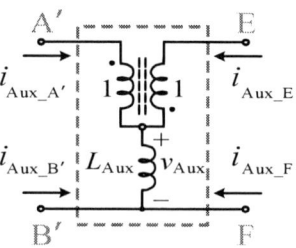

Fig. 9: DAB series resonant bidirectional converter with proposed passive auxiliary circuit.

TABLE VI.   AUXILIARY CURRENT AT AC NODE VOLTAGE RISING EDGE MOMENTS IN REGIME 1 (STEP-UP OPERATION)

Mode	Quantity	Value
A +	$I_{Aux\_A\_v_A \nearrow}$ $I_{Aux\_B\_v_B \nearrow}$	$-\dfrac{nT}{8L_{Aux}}[D_{HV}+d]nV_{dc\_LV}$
	$I_{Aux\_E\_v_E \nearrow}$	$-\dfrac{T}{8L_{Aux}}\left[D_{HV}+\left(D_{HV}-\dfrac{\varphi}{\pi/2}\right)d\right]nV_{dc\_LV}$
	$I_{Aux\_F\_v_F \nearrow}$	$-\dfrac{T}{8L_{Aux}}\left[-D_{HV}-\left(D_{HV}+\dfrac{\varphi}{\pi/2}\right)d\right]nV_{dc\_LV}$
C +	$I_{Aux\_A\_v_A \nearrow}$ $I_{Aux\_B\_v_B \nearrow}$	$-\dfrac{n}{L_{Aux}}\dfrac{T}{8}\left\{D_{HV}+\left(2-\dfrac{\varphi}{\pi/2}-D_{HV}\right)d\right\}nV_{dc\_LV}$
	$I_{Aux\_E\_v_E \nearrow}$	$-\dfrac{1}{L_{Aux}}\dfrac{T}{8}\left\{1-\dfrac{\varphi}{\pi/2}+(2D_{HV}-1)d\right\}nV_{dc\_LV}$
	$I_{Aux\_F\_v_F \nearrow}$	$-\dfrac{1}{L_{Aux}}\dfrac{T}{8}\left\{\dfrac{\varphi}{\pi/2}-1-d\right\}nV_{dc\_LV}$

TABLE VII.   AUXILIARY CURRENT AT AC NODE VOLTAGE RISING EDGE MOMENTS IN REGIME 2 (STEP-UP OPERATION)

Mode	Quantity	Value
A −	$I_{Aux\_A\_v_A \nearrow}$	$-\dfrac{nT}{8L_{Aux}}\left[D_{LV}+\left(D_{LV}+\dfrac{\varphi}{\pi/2}\right)d\right]nV_{dc\_LV}$
	$I_{Aux\_B\_v_B \nearrow}$	$-\dfrac{nT}{8L_{Aux}}\left[-D_{LV}+\left(\dfrac{\varphi}{\pi/2}-D_{LV}\right)d\right]nV_{dc\_LV}$
	$I_{Aux\_E\_v_E \nearrow}$ $I_{Aux\_F\_v_F \nearrow}$	$-\dfrac{T}{8L_{Aux}}[D_{LV}+d]nV_{dc\_LV}$
C +	$I_{Aux\_A\_v_A \nearrow}$	$-\dfrac{nT}{8L_{Aux}}\left[D_{LV}+\left(2-\dfrac{\varphi}{\pi/2}-D_{LV}\right)d\right]nV_{dc\_LV}$
	$I_{Aux\_B\_v_B \nearrow}$	$-\dfrac{nT}{8L_{Aux}}\left[-D_{LV}+\left(\dfrac{\varphi}{\pi/2}-D_{LV}\right)d\right]nV_{dc\_LV}$
	$I_{Aux\_E\_v_E \nearrow}$ $I_{Aux\_F\_v_F \nearrow}$	$-\dfrac{T}{8L_{Aux}}\left[\left(1-\dfrac{\varphi}{\pi/2}\right)+d\right]nV_{dc\_LV}$

IV. EXPERIMENTAL RESULTS

The effectiveness of the proposed converter and the modulation scheme was verified using the prototype converter with the components listed in TABLE VIII. The operating range of converter includes a range of dc voltage variation from 20 to 32 V (240 to 290 V) for LV (HV) terminal with a nominal power of 2kW in both directions. The switching frequency was 200kHz.

TABLE VIII.   EXPERIMENTAL SETUP COMPONENTS

Component	Part Number	Location	Value
MOSFET	IRFP4110	S_1 - S_4	100V, 125A, 3.5mΩ
MOSFET	IPW65R041CFD	S_5 - S_8	650V, 68.5A, 41mΩ
Transformer	Custom	T_1	ETD 49, 3F3, 2:20
Inductor	Custom, 36.6 μH	L	PQ44, 3C85, 12 turns, 1.6mm air gap
Capacitor	2×0.022μF	C	2× Film Capacitor 1600V 0.022μF 5% PCM22.5
Inductor	Custom, 45.2 μH	L_{Aux}	PQ44, 3C85, 15 turns, 1.6mm air gap

Fig. 10 to Fig. 13 illustrate the waveforms of v_{AB}, v_{Aux}, v_{EF} and $i_{L_{Aux}}$ for operation at four combination of V_{dc_LV} of 20V and 32V and V_{dc_HV} of 240V and 290V. ZVS is evident in the scope shots with slow rises and no voltage spikes. The light-load situation is shown due to loss of inherent ZVS in the original converter, as known from literature [1].

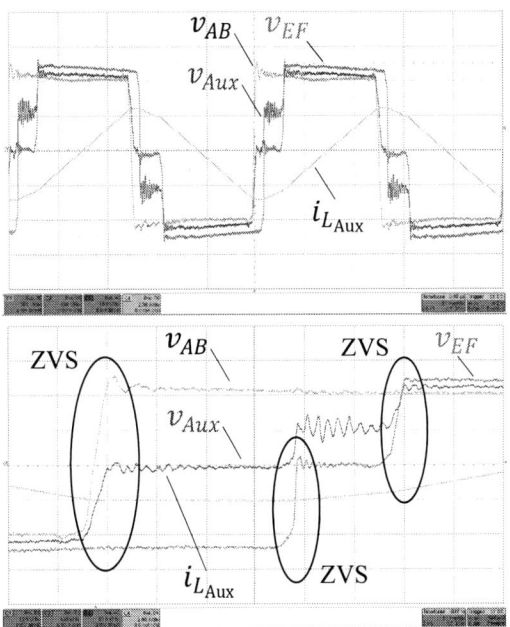

Fig. 10: Waveforms of v_{AB}, v_{Aux}, v_{EF} and $i_{L_{Aux}}$ for $(V_{dc\_LV}, V_{dc\_HV}) = (20,240)$.

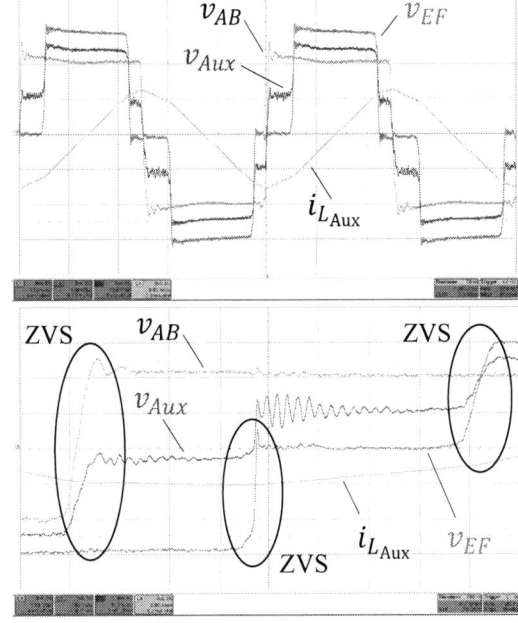

Fig. 11: Waveforms of v_{AB}, v_{Aux}, v_{EF} and $i_{L_{Aux}}$ for $(V_{dc\_LV}, V_{dc\_HV}) = (20,290)$.

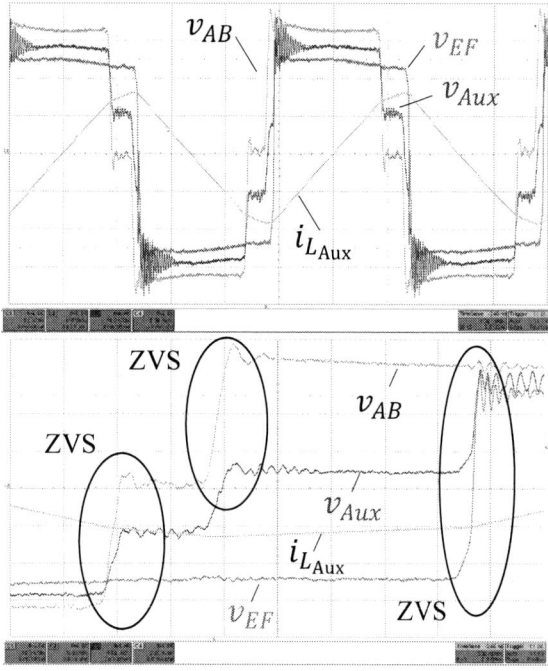

Fig. 12: Waveforms of v_{AB}, v_{Aux}, v_{EF} and $i_{L_{Aux}}$ for $(V_{dc\_LV}, V_{dc\_HV}) = (32,240)$.

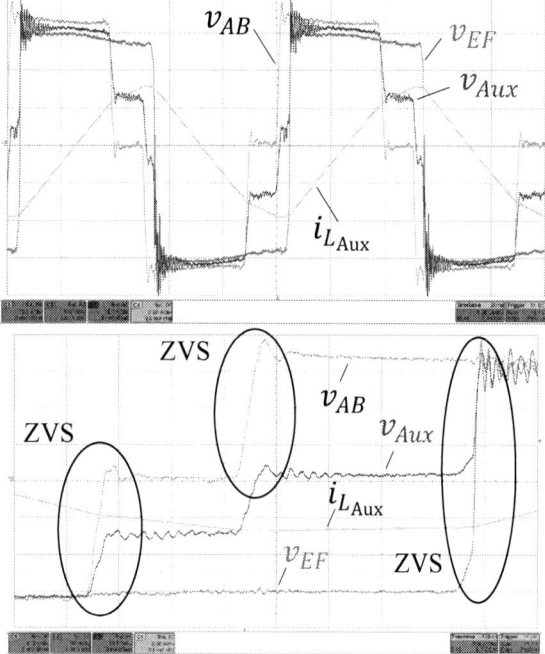

Fig. 13: Waveforms of v_{AB}, v_{Aux}, v_{EF} and $i_{L_{Aux}}$ for $(V_{dc\_LV}, V_{dc\_HV}) = (32,290)$.

Fig. 14 illustrates the experimental efficiency results for both step-up and step-down operation at all the choices of V_{dc_LV} of 20 and 32 V as well as V_{dc_HV} of 240 and 290V without the auxiliary circuit. In all the tests the highest

efficiency is found to be around 50% of nominal power. The efficacy drop above 50% of nominal power is from conduction loss in the switches and the PCB. Note that for $V_{dc\_LV} = 20V$ and 2kW the average of I_{dc_LV} reached to 100A and the switches experienced a huge current stress. The low power efficiency reduction is a result of ZVS loss (although the modulation provides ZCS).

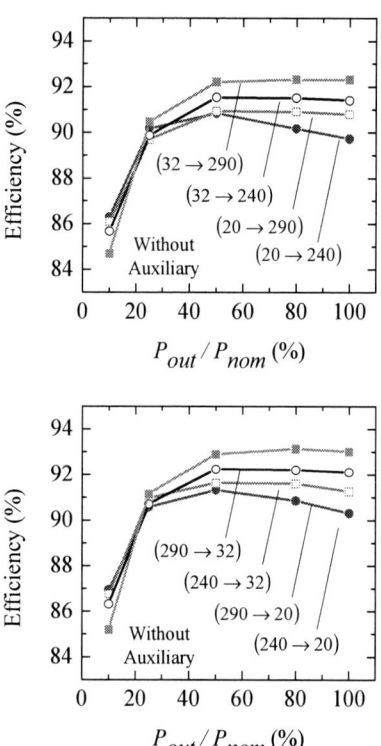

Fig. 14: Efficiency of converter in step-up and step-down operations without auxiliary circuit.

Fig. 15 shows the enhancement of efficiency at lower power levels after adding the auxiliary circuit. The efficiency curves are not completely flat due to having the unavoidable conduction loss caused by the auxiliary current.

V. CONCLUSION

A bidirectional dual active bridge series resonant converter with a novel robust passive auxiliary circuit only two magnetic elements is proposed and analyzed. Time-domain analysis provides insight, eliminates the need for the first harmonic approximation, and allows defining a modulation scheme to ensure zero current switching. The auxiliary circuit maintains the ZVS operation of switches for the entire operational conditions of the resonant dual active bridge converter. Effectiveness of the proposed auxiliary in achieving zero voltage switching at low power operation was experimentally validated.

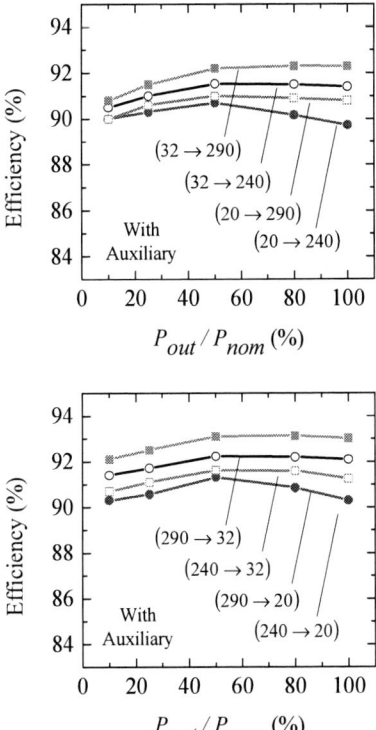

Fig. 16: Efficiency of converter in step-up and step-down operations with auxiliary circuit.

REFERENCES

[1] R. W. De Doncker, D. M. Divan, and M. H. Kheraluwala, "A three-phase softswitched high-power-density DC/DC converter for high-power applications," *IEEE Transactions on Industry Applications*, vol. 27, no. 1, pp. 63-73, Jan./Feb. 1991.

[2] S. Inoue, and H. Akagi, "A Bidirectional DC-DC converter for an Energy storage system With Galvanic Isolation," *IEEE Transactions on Power Electronics*, vol. 22, no. 6, pp. 2299-2306, 2007.

[3] H. Li, and F. Z. Peng, "Modeling of a new ZVS bi-directional dc-dc converter," *IEEE Transactions on Aerospace and Electronic Systems*, vol. 40, no. 1, pp. 272-283, Jan. 2004.

[4] G. G. Oggier, G. O. Garcia, and A. R. Oliva, "Switching Control Strategy to Minimize Dual Active Bridge Converter Losses, " *IEEE Transactions on Power Electronics*, vol. 24, no. 7, pp. 1826-1838, July 2009.

[5] L. Zhu, "A Novel Soft-Commutating Isolated Boost Full-Bridge ZVS-PWM DC–DC Converter for Bidirectional High Power Applications," *IEEE Transactions on Power Electronics*, vol. 21, no. 2, pp. 422-429, March 2006.

[6] T. Tao, J. L. Duarte, and M. A. M. Hendrix, "Three-Port Triple-Half-Bridge Bidirectional Converter With Zero-Voltage Switching", *IEEE Transactions on Power Electronics*, vol. 23, no. 2, pp. 782-792, March 2008.

[7] L. Corradini, D. Seltzer, D. Bloomquist, R. Zane, D. Maksimovic, and B. Jacobson, "Minimum current operation of bidirectional dual-bridge series resonant DC/DC converters," *IEEE Transactions on Power Electronics*, vol. 27, no. 7, pp. 3266-3276, July 2012.

[8] A. Safaee, A. Bakhshai, and P. Jain, "A resonant bidirectional dc-dc converter for aerospace applications," *in IEEE ECCE*, 2011, pp. 3075-3079, 2011.

[9] B. Ray, "Single-cycle Resonant Bidirectional DC/DC Power Conversion," in APEC, 1993, pp. 44-50, 1993.

[10] F. Krismer and J. Kolar, "Accurate small-signal model for an automotive bidirectional Dual Active Bridge converter, " in COMPEL 2008, 2008.

[11] L. Fiorella, C. Di Miceli, T. Raimondi and C. Cutrona, "Analysis of a Series Resonant Converter", *Telecommunications Energy Conference, 1989. INTELEC '89. Conference Proceedings., Eleventh International*, pp. 20.3/1 - 20.3/6 vol. 2.

[12] B. Lee and H. Cha, "Comparative Analysis of Charging Modes of Series-Resonant Converter for an Energy Storage Capacitor," *Plasma Science, IEEE Transactions on*, vol. 41, no. 3, pp. 570-577, Mar. 2013.·

[13] A. Safaee, A. Bakhshai, and P. Jain, "A ZVS Pulse Width Modulation Full Bridge Converter with A Low-RMS-Current Resonant Auxiliary Circuit," *IEEE Transactions on Power Electronics*, August 2015.

[14] A.K.S. Bhat, "A Generalized Steady-State Analysis of Resonant Converters Using Two-Port Model and Fourier-Series Approach," *Power Electronics, IEEE Transactions on*, vol. 13, no. 1, pp. 142-151, Jan. 1998.

[15] A. Safaee, P.K. Jain, and A. Bakhshai, "An Adaptive ZVS Full-Bridge DC–DC Converter With Reduced Conduction Losses and Frequency Variation Range," *Power Electronics, IEEE Transactions on*, vol.30, no.8, pp.4107-4118, Aug. 2015.

[16] A. Safaee, P.K. Jain, and A. Bakhshai, " A ZVS Pulse Width Modulation Full Bridge Converter with A Low-RMS-Current Resonant Auxiliary Circuit," *Power Electronics, IEEE Transactions on*, August 2015.

A New High Capacity Compact Power Modules for high power EV/HEV inverters

Seiichiro Inokuchi, Shoji Saito, Arata Izuka
Yuki Hata and Shinji Hatae
Mitsubishi Electric Corporation
1-1-1 Imajukuhigashi Nishi-ku
Fukuoka JAPAN

Toshiya Nakano, Eric R. Motto
Powerex, Inc.
173 Pavilion Lane
Youngwood, PA 15697 USA

Abstract— This paper presents a new power semiconductor IGBT module family dedicated for Electric Vehicle (EV) and Hybrid Electric Vehicle (HEV) power-train inverter applications, especially for higher power requirements. The new IGBT module family adopts a 6-in-1 circuit configuration with an optimized internal layout, Direct Lead Bond (DLB) structure and an integrated Al fin for direct liquid cooling. As a result, this new power module family has simultaneously achieved high performance, low self-inductance, small size and light weight. Compared to conventional products with similar power capabilities, the adoption of these innovative technologies has led to a 20% improvement in thermal performance, 30% reduction of self-inductance, 50% reduction of the module's footprint, and a 70% reduction of the module's weight.

Keywords—Automotive, EV, HEV, IGBT Module, Diode, Direct Liquid Cooling, High Power, Automotive Power-Train Inverter

I. INTRODUCTION

Since the very successful release of the world's first mass-produced HEV in 1997, the number of HEVs and EVs has been gradually increasing in major cities around the world. Reliable and efficient semiconductor power modules transferring energy between the battery and the motor-generator represent the heart of the electric power-train that realizes the electric-mobility concept. During the early years of EV/HEV inverter development, the power module's design was mainly custom-made or tailored to suit particular layout or certain application requirements.

As more and more EV/HEV designs are increasingly projected by various automakers as well as inverter unit makers, automotive power modules designed to suit a wide range of application requirements are being introduced. A new high capacity compact power module called "High Power J1-Series" has been developed to adapt to essential requirements which are high power, high reliability, compact size and high efficiency from automotive market. The High Power J1-Series has been implemented using 7th Generation CSTBT^{TM} and Relaxed Field of Cathode (RFC)-diode chips, a 6-in-1 circuit configuration, Direct Lead Bond (DLB) structure and direct cooling structure. The optimized combination of these technologies has brought successful improvement of the High Power J1-Series performance dedicated for EV/HEV application. The High Power J1-Series power module's external appearance and circuit configuration are given in Fig.1 and Fig.2. Dimensions and corresponding current/voltage ratings are given in Table 1.

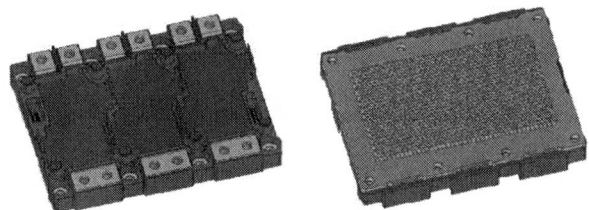

Fig.1.   External appearance of High Power J1-Series

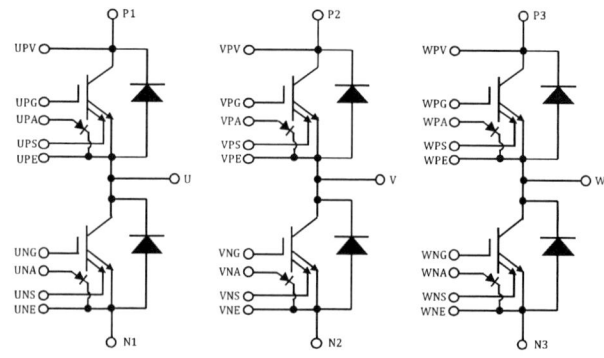

Fig.2.   High Power J1-Series Six-Pack circuit configuration

Table.1 High Power J1-Series Line-up

Model	Ratings (Vces/Ic)	Package Size
CT1000CJ1B060	650V/1000A	163×124.5×33.6mm
CT600CJ1B120	1200V/600A	(including Control terminal and Pin-fin, 6-in-1)

II. PACKAGE TECHNOLOGY

High Power J1-Series's package is characterized by several features including a highly reliable Direct Lead Bonding (DLB) structure, compact size, light weight, and high power handling capability. Especially, the newest member of power modules for EV/HEV applications are the proposed module employing a built-in Aluminum cooling-fin as shown in Fig.3.

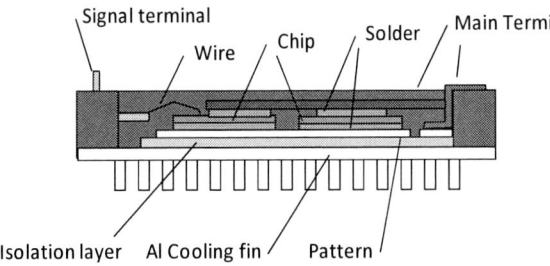

Fig.3.   Internal structure of High Power J1-Series

In comparison with more conventionally packaged products (J-Series T-PM), the new power module family reduces the footprint more than 50% (Fig.4). The reduced size of the High Power J1-Series is the result of combining an optimized aluminum pin-fin structure with high efficiency 7th Generation CSTBT^{TM}/RFC-Diode chips. Despite the fact that Aluminum cooling-fins have lower thermal conductivity compared to copper cooling-fin structures, this selection provides several advantages to EV/HEV applications. Among these advantages the most prominent ones are corrosion resistance when Aluminum is exposed directly to coolants and light weight as much as 70% weight reduction when comparing 6-in-1 power module inverter solutions. Aluminum is not susceptible to galvanic corrosion like copper. If a copper fin is used it becomes necessary to apply thick nickel plating to prevent corrosion. In addition, the light weight of aluminum contributes to the reduction of electricity costs and fuel consumption in the EV/HEV.

Additionally High Power J1-Series has eliminated two layers in the thermal path. One is the solder layer between the substrate and the baseplate, the other is the grease layer between the baseplate and water jacket. As a result a 20% thermal performance improvement is achieved when comparing 6-in-1 power module inverter solutions. At the same time, the reduction of layers contributes to improved thermal cycling capability.

These solutions compared in Fig.4, Fig.5 and Fig.6 are based on equivalent module current and voltage ratings for three-phase EV/HEV motor drives.

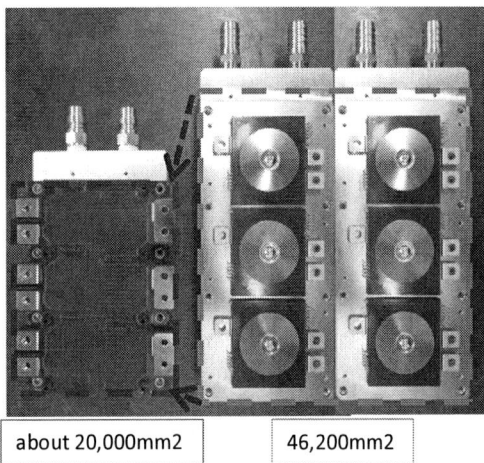

Fig.4. High Power J1-Series footprint comparison with conventional T-PM solution.

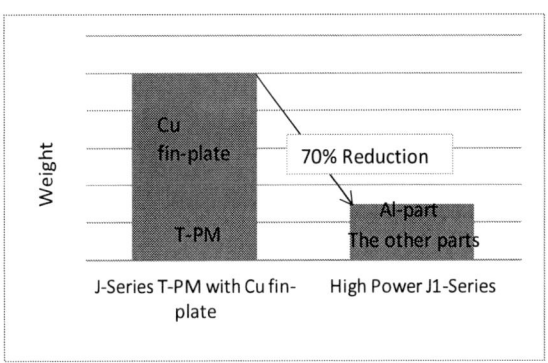

Fig.5. High Power J1-Series weight comparison vs. conventional T-PM with Cu fin.

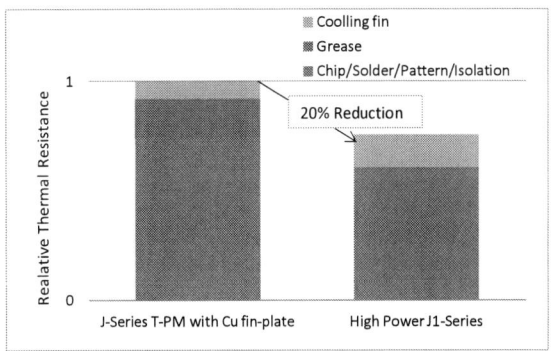

Fig.6. High Power J1-Series thermal resistance comparison vs. conventional T-PM with Cu fin.

High power capacity modules (e.g. 1000A/650V or 600A/1200V) require large internal leads and power terminals which tends to increase the module package size. Moreover, bigger packages tend to have larger self-inductance than smaller ones which is a critical problem for high power applications due to the surge voltage caused by high di/dt conditions. However, by adopting an optimized internal power lead design and chip layout aimed at cancelling the magnetic flux between the PN terminals, the newly developed High Power J1-Series successfully achieved low overall self-inductance. Fig.7 shows the inductance measurement results of the newly developed High Power J1-Series compared with conventional wire bonding design.

In addition, by utilizing the DLB structure, the package's internal contact resistance and self-inductance can be reduced compared to conventional wire-bonding structures. A further important advantage of the increased chip contact area is uniform temperature distribution across the chip surface, hence reducing the peak temperature value and resulting in lower stress for the entire construction. Thus, the DLB structure addresses the power-cycling stress issues usually encountered in conventional wire-bonded packages. This is reflected in the DLB high power-cycling capability compared with conventional wire-bonding structure.

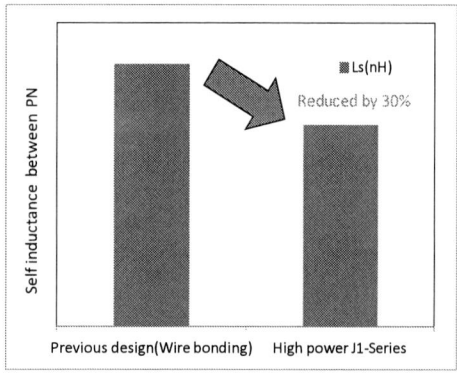

Fig.7.   Self inductance between P and N

III.   EXPERIMENTAL RESULTS WITH LATEST CHIP TECHNOLOGY

The new IGBT series' power handling capability in conjunction with the performance of the thermal interface construction were experimentally verified under the following test conditions, 650V/1000A : Main battery voltage = 450V; PWM switching frequency (fc) = 5kHz, 10kHz; coolant temperature (Tw) = 65°C; coolant flow-rate = 10 l/min, Rth(j-w)Q=max, 1200V/600A : Main battery voltage = 600V; PWM switching frequency (fc) = 5kHz, 10kHz; coolant temperature (Tw) = 65°C; coolant flow-rate = 10 l/min, Rth(j-w)Q=max.

Under these conditions the inverter output current of 650V/1000A can exceed 400Arms (corresponding to more than 120kW output power) at a maximum operation junction temperature less than 150°C. Moreover Inverter output current of 1200V/600A can exceed 400Arms (corresponding to more than 120kW output power) at a maximum operation junction temperature less than 150°C.

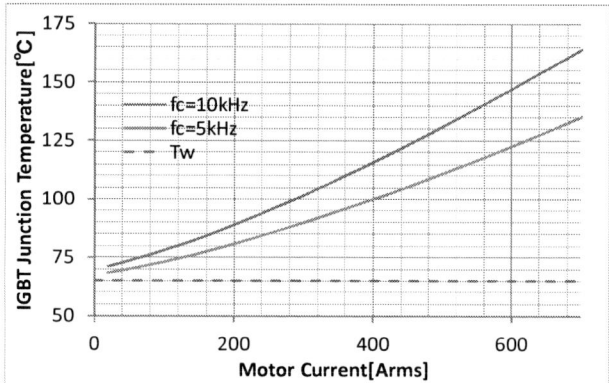

Fig.8.   Experimental performance of the High Power J1-Series(650V/1000A, representative example)

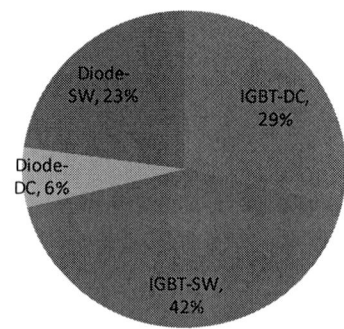

Fig.9 Power loss ratio of of the High Power J1-Series Io=600Arms, Vcc=450V, fc=5kHz, Pf=0.8, Mo=1 Tj=150°C(650V/1000A,representative example)

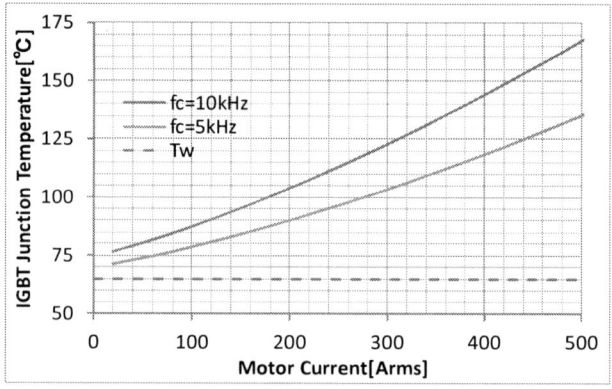

Fig.10.   Experimental performance of the High Power J1-Series(1200V/600A, representative example)

978-1-4673-9551-9/16 $31.00 © 2016 IEEE   470

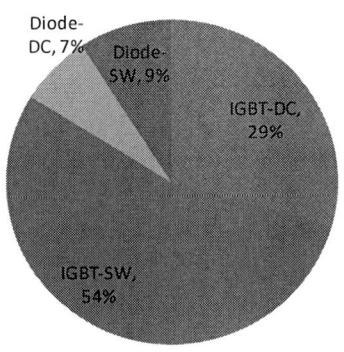

Fig.11 Power loss ratio of of the High Power J1-Series Io=600Arms, Vcc=450V, fc=5kHz, Pf=0.8, Mo=1 Tj=150°C(1200V/600A,representative example)

This extraordinary result has been provided by state-of-the-art 7th Generation CSTBT[TM] and RFC-diode chip technology. Advances in IGBT technology have always been driven by the continuing demand for higher power densities and higher efficiencies as reflected in the progress in IGBT generations with improved internal structures aiming at optimizing the well-known Vce(sat)-vs-Eoff tradeoff characteristic. By adding an extra layer of carriers within the IGBT structure, the CSTBT[TM] concept achieves higher efficiency by reducing both the saturation voltage VCE(sat) and the switching loss EOFF [1~3]. The 7th Generation IGBT further optimizes the CSTBT[TM] VCE(sat) vs. EOFF trade-off characteristic as illustrated in Fig.1 which pictorially summarizes the continuous improvement in IGBT characteristics with advanced generations. Considering the very compact size of the J1-Series innovative package design (power module volume less than 0.68 liters), the very high power density level realized by the utilization of the 7th Generation IGBT is clearly evident.

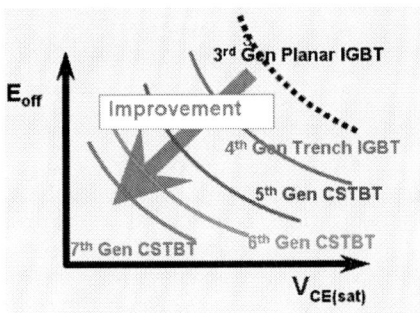

Fig.12 IGBT trade-off characteristics improvement

IV. FUTURE

A family of the next generation automotive power modules, the J1-Series is utilizing the technology described here to realize a wide range line-up.

V. CONCLUSION

New High Capacity Compact Power Modules "High Power J1-Series" has been developed to meet the requirements of the evolving EV/HEV market. The High Power J1-Series achieves high performance, low self-inductance, compact size and light weight. These attractive features were made possible through the combination of optimized package structure technology such as wire-bond-less Direct Lead Bonding (DLB) and direct liquid cooling of an aluminum cooling fin base plate and state-of-the-art chip technology such as 7th Generation CSTBT[TM] and RFC-diode. In conclusion, the High Power J1-Series can realize a wide range inverter operation and accommodate a variety of requests for EV and HEV applications.

REFERENCES

[1] H. Takahashi, E. Haruguchi, H. Hagino, and T. Yamada, "Carrier stored trench-gate bipolar transistor (CSTBT)- a novel power device for high voltage application –", Proc. ISPSD 1996, pp. 349-352.

[2] T. Nishiyama and Y. Miyazaki, "The IGBT module with 6th generation IGBT," PCIM Proceedings 2009.

[3] K. Nakamura, et al., "Advanced wide cell pitch CSTBTs having light punch-through structures," International Symposium on Power Semiconductor Devices and ICs Conference Record 2002, pp. 277-280.

[4] M. Ishihara et al., "New compact-package Power Modules for Electric and Hybrid Vehicles (J1-Series)", PCIM -Europe 2014, pp. 1093-1097

[5] S. Inokuchi et al., "A new versatile high power Intelligent Power Module (IPM)", PCIM -ASIA 2015 pp.205-209

[6] K. Hussein, et al., "IPMs Solving Major Reliability Issues in Automotive Applications", IEEE-ISPSD 2004, Proceedings, pp. 89-92.

[7] T. Ueda, et al., "Simple Compact, Robust and High-performance Power module T-PM (Transfermolded Power Module)," ISPSD 2010, pp. 47-50.

[8] K. Hussein, et al., "New Compact, High Performance 7th Generation IGBT Module with Direct Liquid Cooling for EV/HEV Inverters",APEC 2015, pp1343-1346.

[9] K. Hussein, S. Saito, M. Ishihara, and S. Hatae, "Advanced power modules series for EV/HEV inverters", EVTeC & APE Japan 2014.

Modular PET, Two-Phase Air-Cooled Converter Cell Design and Performance Evaluation with 1.7kV IGBTs for MV Applications

Frederick Kieferndorf, Uwe Drofenik, Francesco Agostini, Francisco Canales

ABB, Ltd.
Corporate Research
Baden, Switzerland
frederick.kieferndorf@ch.abb.com

Abstract—This paper explores the optimization and design of a medium voltage, modular power electronics transformer based on industrial 1.7 kV IGBTs. The design of the medium frequency transformer is outlined. The thermal management using a two-phase, air-cooled thermosyphon is described. The overall performance of one cell is detailed from measurements on a full-scale laboratory demonstrator. From the data, the impact of converter variables on power throughput and efficiency is illustrated. In addition, the waveforms are analyzed in terms of converter parasitics and circuit configuration.

Keywords—PET, modular converter, thermosyphon, medium frequency transformer, magnetic optimization, LLC resonant converter, device characterization.

I. INTRODUCTION

Modular converters are increasingly considered for Medium Voltage (MV) high power applications. In order to be practical, such converters must exhibit high efficiency, high power density, high reliability and low cost. Some suitable applications are data centers, MVDC marine applications and high power photovoltaic parks. One especially interesting application for this type of converter is in traction [1][2][3][4]. In some European networks the electrical supply is provided by a MV single phase catenary, for example in Germany or Switzerland a 15 kV line operating at 16 ⅔ Hz is used. The high voltage and low fundamental frequency results in a relatively large transformer to step the grid voltage down to the level of the power electronics in the rectification stage.

A typical option for such a modular converter is to use MV semiconductors (for example 3.3 kV silicon devices) to build the basic cell, thus minimizing the number of cells needed to reach the required voltage level of the application. In power electronic transformer (PET) converters, the size of the medium frequency transformer (MFT) is closely related to the switching frequency (f_{sw}) of the semiconductor devices. The f_{sw} is quite limited for MV devices, even in soft switching operation, resulting in higher weight and volume of the MFT [1][2][3][4]. An alternative is to use low voltage (LV) semiconductor devices that can switch at a much higher frequency than MV devices. The overall size and weight of the

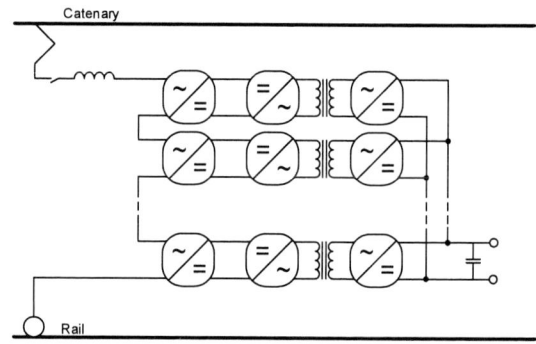

Fig. 1 – Traction rectifier application for a PET

TABLE 1 – SYSTEM NOMINAL SPECIFICATIONS

	System	Cell
Nominal Input voltage (16 ⅔ Hz)	15 kV	600 V
Input Voltage range (16 ⅔ Hz)	11 – 19 kV	440-760 V
Nominal Output power	1.2 MW	50 kW
Overload Output power	1.8 MW	75 kW
Nominal Output voltage (DC)	750 V	1000 V
Number of cells	25	1

MFT in each cell can be greatly reduced because of the high switching frequency attained by LV devices. In addition, the semiconductor cost can be quite low due to the use of standard industrial devices. At the frequencies of interest (around 10 kHz in soft switching operation) the MFT can also be made with inexpensive ferrite cores. As a further benefit, the use of air-cooling with a two-phase cooler, to efficiently remove the semiconductor losses, contributes to the weight reduction. In this paper, the performance of LV IGBTs are investigated as the switching device in a modular converter cell.

The cell topology analyzed in this paper consists of a two-stage AC/DC converter as seen in Fig. 2. The first stage,

978-1-4673-9551-9/16 $31.00 © 2016 IEEE

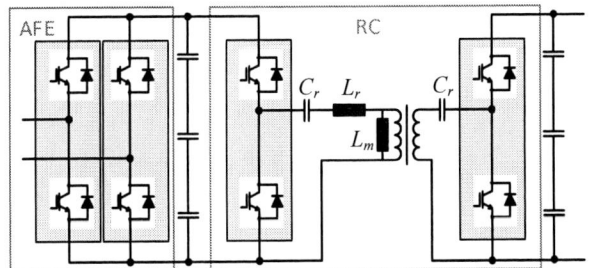

Fig. 2 – Converter and cell topology: AFE – active front end, RC – resonant converter

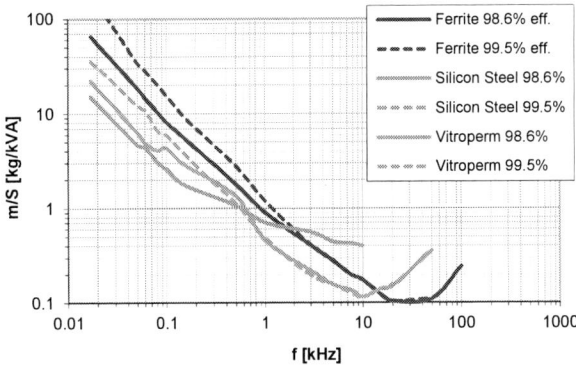

Fig. 3 – Unconstrained optimization of the MFT normalized weight over operating frequency for different cores and different target efficiencies.

an Active Front End (AFE), is used to control the power factor and regulate the DC link voltage. The second stage, a Resonant Converter (RC), operating in open loop, provides galvanic isolation and very high efficiency. The isolated RC consists of a LLC DC/DC converter [5] [9] with the inductive resonant components integrated into the MFT.

An additional consideration in traction is that the strict weight limitation for trains typically results in a penalty on the efficiency. By using an optimized PET type configuration, the efficiency can be significantly improved while also reducing the weight [1]. The system specifications for a traction rectifier in the described network are shown in TABLE 1 and the circuit configuration for a PET converter in a train is shown in Fig. 1. The cell output voltage was modified from the traction specification since 1000 V is a typical industrial specification. In addition, it simplifies the testing procedure by allowing for a 1:1 MFT design.

The design optimizations for the MFT will be described first, followed by the design and performance capabilities of the two-phase cooler. Finally, the experimental electrical performance is analyzed in detail, including critical semiconductors timing parameters for zero voltage switching (ZVS) operation and maximum power handling capability.

TABLE 2 – MFT SPECIFICATIONS

Input voltage (rectangular)	500 V peak	Insulation voltage stress	30 kV
Input power	50 kW	Magnetizing inductance	1.25 mH
Winding ratio	1:1	Stray inductance	12 uH
Ambient temperature	60°C		

II. MFT DESIGN/OPTIMIZATION

With the specifications given in TABLE 2, a MFT was systematically optimized without geometrical constraints based on the analytical models and procedures described in [10]. Fig. 3 shows the minimum weight design over the operating frequency. The following assumptions were applied: different core materials, maximum temperatures below 120°C, and 30 kV insulation between the high-voltage and low-voltage windings. The curves, created by automatic optimization, are not perfectly smooth because of the discrete nature of the winding layer number and the assumption of square and/or round-shaped litz wire cross sections. In similar simulations two overall system optima have been identified,

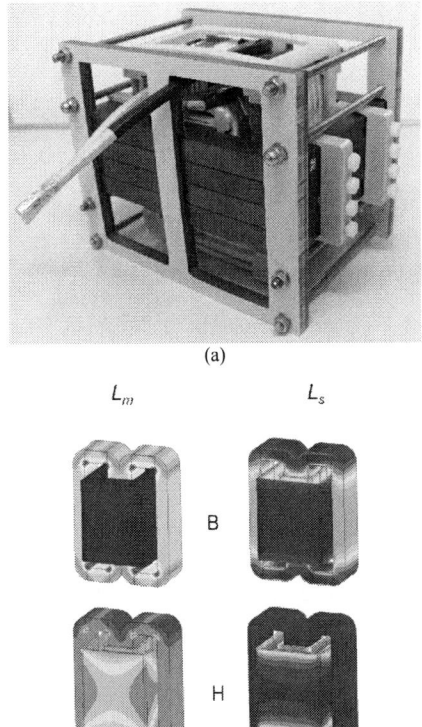

Fig. 4 – (a) MFT with air cooling, air-insulation and ferrite core. (b) B and H-field distributions from 3D FEM simulations: no-load-case to calculate the transformer magnetizing inductance, Lm; full load-case to calculate the transformer stray inductance, Ls.

one with a large number of small cells (25 cells), achieved with LV devices and a comparable one with a small number of larger cells (7 cells), achieved with MV class devices. In this paper, we discuss the system optimum with the smaller LV cells.

The system simulation was performed with nanocrystalline cores, amorphous cores, several types of silicon steel, ferrites and powder core materials. Based on cost-optimization (considering material cost) and core loss minimization, ferrite

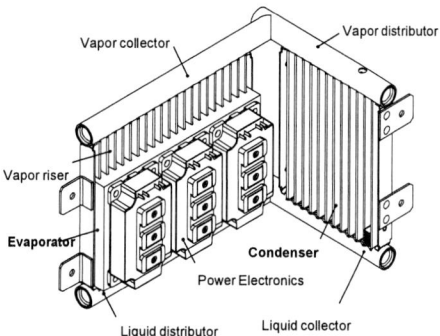

Fig. 5 – Two-phase thermosyphon assembly

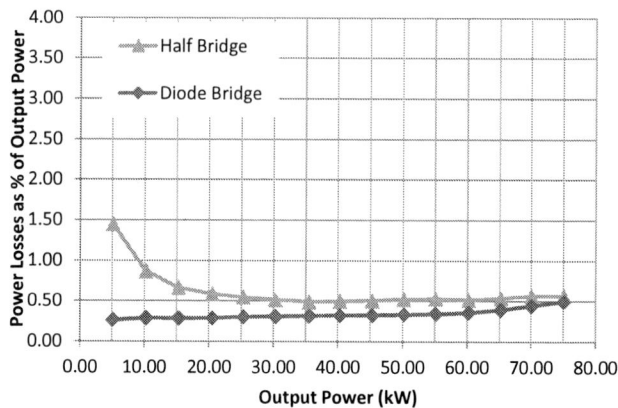

Fig. 6 – Power losses as a function of the total power output

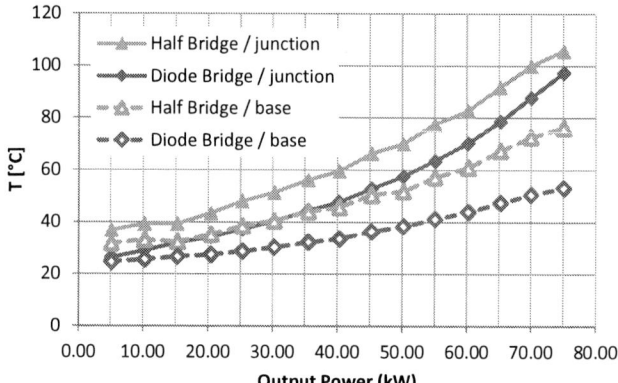

Fig. 7 – Maximum junction and base temperatures as a function of the total power output

was the material with the best performance. Nanocristalline cores allow smaller size at lower frequency, but this comes at much higher cost, losses and core cooling effort.

The final transformer design was optimized based on the largest standard ferrite C-cores available at the time. The resulting weight-optimized transformer (Fig. 4a) with copper-litz windings and a ferrite core is air-cooled via forced convection and air-insulated by providing a 30 mm wide air gap between the high-voltage winding and the low-voltage winding. In clean environments, the electric strength of air between parallel planes is about 3 kV/mm. By setting the insulation distance to 30 mm and requiring 30 kV insulation, the maximum stress in the air is 1 kV/mm, which was experimentally verified by PD testing. Since the core is grounded, the 30 mm insulation distance is also required between high-voltage winding and core.

The outer volume of the transformer prototype without the mechanical clamping structures is 250×200×200 mm^3 and the weight is 15 kg. The efficiency is 99.6%, measured operating temperature at 70 kW is 120°C, and the measured magnetic properties are L_m=1.88 mH and L_σ=23 µH. Due to the small contact gap between the ferrite cores, in the range of several hundred micrometers, it is difficult to design the targeted inductances exactly. There is a strong dependence on the clamping force applied to the cores, and individual core surface inhomogeneity has a non-negligible impact. Therefore, sufficient design margin has to be considered.

III. THERMAL MANAGEMENT WITH 2-PHASE COOLING

The specially designed cooler, shown in Fig. 5, holds three half-bridge modules where one is used as part of the LLC converter and the other two form the input AC boost converter. The second half bridge for the resonant converter is mounted on a separate aluminum heat sink.

The two-phase cooler is a local, compact, L-shaped thermosyphon with refrigerant R245fa. The advantage of the thermosyphon approach compared to a common heat sink is that while keeping the system air-cooled, it provides high cooling performances enabling high power density and an optimal geometrical arrangement with a low pressure loss. The original design, as given in [11], is such that for a volumetric air flow of 400 m^3/h, air inlet temperature of 42°C, maximum outlet temperature of 60°C and maximum junction temperature of 125°C, the achieved maximum cooling power

is 1400 W with a thermal resistance of 45 K/kW. Details on the physical working principle and on the thermal performances are given in [11] where an equivalent device is analyzed.

The distribution of the power losses as a percentage of the total power input is given in Fig. 6, the half bridge power losses decrease from 1.50 to 0.55% going from 5 to 75 kW while for the diode bridge they increase from 0.25 to 0.50%. This shows the effect of constant switching losses even at low power output vs. the conduction losses that increase with the power output. The maximum junction temperature and base temperature for the half bridge and for the diode bridge are represented in Fig. 7. The cooling systems guarantee a safe and reliable operation of the power semiconductors over the investigated power range.

IV. LLC RESONANT TANK CHARACTERISTICS

The key characteristics of a LLC converter [5] are described in this section. The switching waveforms and critical timing parameters for the LLC converter operation are defined in Fig. 8. The primary resonant frequency, dominant during the input/output power transfer period, is determined by the resonant capacitor (C_r) and the transformer integrated resonant inductor (L_r), shown in (1). A second resonant frequency (2), active during the hold-off time, incorporates the transformer magnetizing inductance (L_m).

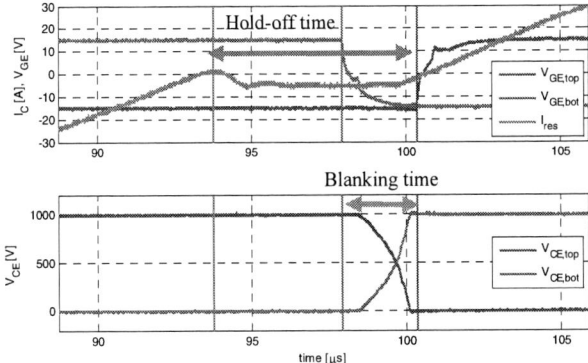

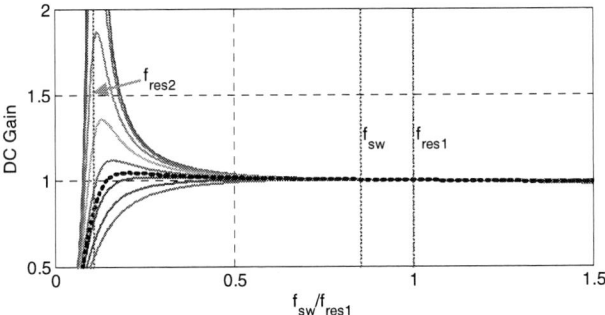

Fig. 8 – Critical timing parameters in the switching waveform of an LLC converter, top and bottom IGBTs where: V_{CE} – collector to emmiter voltage, V_{GE} – gate to emmiter voltage, I_{res} – resonant current

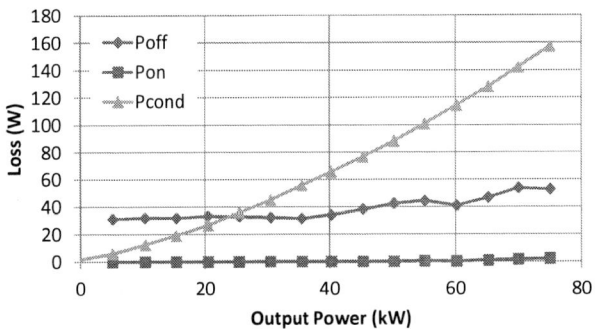

Fig. 10 – Breakdown of measured IGBT losses as a function of output power

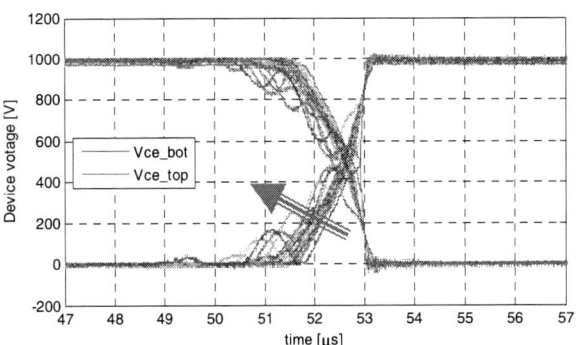

Fig. 9 – DC gain as a function of the normalized switching frequency, with f_{sw} = 0.85 f_{res1}, nominal power curve is shown with black dashes

Fig. 11 – Voltage commutation time showing the increasing time required as output power increases (arrow direction)

$$f_{res1} = \frac{1}{2\pi\sqrt{L_r C_r}} \quad (1)$$

$$f_{res2} = \frac{1}{2\pi\sqrt{(L_r + L_m)C_r}} \quad (2)$$

The location of the resonant frequencies, normalized to f_{res1}, are noted in the ideal DC gain characteristic, based on the fundamental harmonic assumption [7], shown in Fig. 9. Another key parameter to consider is the quality factor:

$$Q = \left(\frac{\pi}{8}\right)^2 \sqrt{\frac{L_r}{C_r}} \frac{P_{out}}{V_{dc}^2} \quad (3)$$

To achieve a load independent gain for the resonant converter, the resonant tank must be designed for a very low Q at the rated power and operated close to resonant frequency f_{res1} [6].

Charge carriers are injected into the conducting device during the sinusoidal resonant pulse of current and they must be removed during the hold-off time so the device can turn-off commutating the current to the diode of the complementary switch. The hold-off time is defined as half the difference between the switching period and the resonant period, when only the magnetizing current is flowing in the primary side circuit as indicated in Fig. 8. In the ZVS mode of operation the switching frequency is below resonance and constant to provide sufficient hold-off time for the stored charge to be removed from the turning-off IGBT. Then the complementary switch can turn-on with ZVS. It is very difficult to estimate the actual hold-off time and magnetizing current level needed

to remove enough of the charge carriers to achieve ZVS since both parameters are determined by the semiconductor properties of the devices, which are not well characterized for turn-off at such low currents with resonant waveforms.

The longer the hold-off time is, the more carriers can be removed before the switching action occurs resulting in lower switching losses, but the larger the peak current is for a given output power. This is because the resonant pulse occupies less of the switching period as the hold-off time increases, leading to higher current stress in the IGBTs. Thus, it is desired to use the minimum possible hold-off time. A good compromise value is f_{sw}=0.85$\cdot f_{res}$. Once the ratio f_{sw}/f_{res}, is fixed, the remaining design parameters are the turn-off current and the blanking time, which can only be determined based on the measured waveforms. To achieve the desired low switching losses in a LLC converter, ZVS must be maintained at turn-on, and very low turn-off losses must be guaranteed by proper design of the MFT magnetizing inductance.

The LV IGBTs under these switching conditions, exhibit losses dominated by the conduction losses at high load and the turn-off losses at no load as shown by measurement data in Fig. 10. In addition, despite the ZCS operation there is a small diode reverse recovery occurring in the output diode bridge, which is clearly seen in in the I_{res} waveform during the hold-off period in Fig. 8. Compared to MV devices the most important effect is the commutation time of the device voltage (called T_{min} in [4]) since it gives an idea of what the blanking time should be to enable ZVS for the complementary switch. A commutation range from 1.1 µs at low power to 4 µs at high power can be seen in Fig. 11. The electrical performance of the 1 kV converter cell will be described next.

978-1-4673-9551-9/16 $31.00 © 2016 IEEE   475

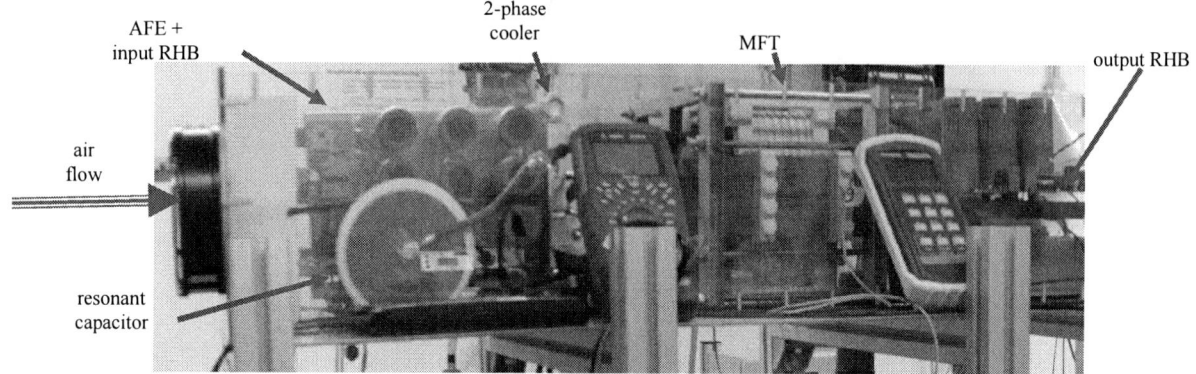

Fig. 12 – Experimental test cell: active front end (AFE), input resonant half bridge (RHB), MFT and output RHB.

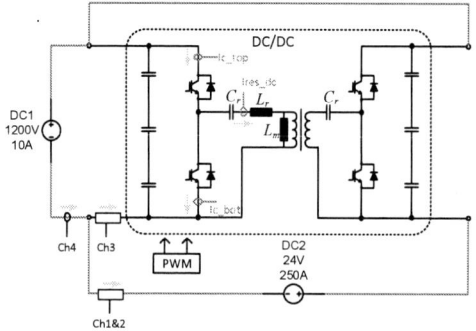

Fig. 13 – Experimental test setup for measurement of LLC converter performance

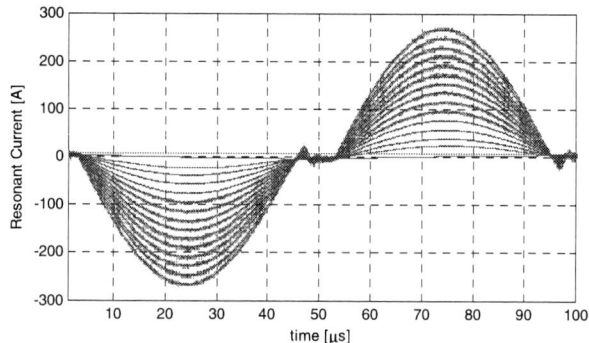

Fig. 14 – DC-DC resonant current at different power levels: red line shows turn-off current level: 7 A for f_{sw} = 10 kHz, $f_{res1.}$ = 11.7 kHz.

V. EXPERIMENTAL SETUP AND MEASUREMENTS

The experimental work was carried out to verify the power limits and identify the required timing parameters for proper operation. The test setup is shown in Fig. 12. The selected semiconductors showed the lowest losses in simulated test studies: Infineon FF300R17KE3, 1.7 kV, 300 A. The arrangement of components in the cell was made to maximize the cooling of the critical components. The input DC link capacitors are nearest to the air inlet since they are the most thermally sensitive components. Next, the condenser of the thermosyphon provides cooling for the AFE+input resonant half-bridge modules. Then the open MFT design allows sufficient airflow for its own cooling as well as for the cooling of the output resonant half bridge on an aluminum heatsink.

Since the MFT was designed with a 1:1 turns ratio the switching behavior and efficiency could be tested using the configuration shown in Fig. 13 [12]. Two DC power supplies are used, one (DC1) providing the desired DC link voltage at the input, in this case 1000 V and the second (DC2) generating the desired input/output voltage required for each desired operating point for the characterization.

The measured resonant current waveforms over the full power range of the tested converter are shown in Fig. 14. As can be seen the peak current varies from about 25 A up to 280 A, by the red line. It is also clear from the figure that despite the ZCS (zero current switching) operation of the

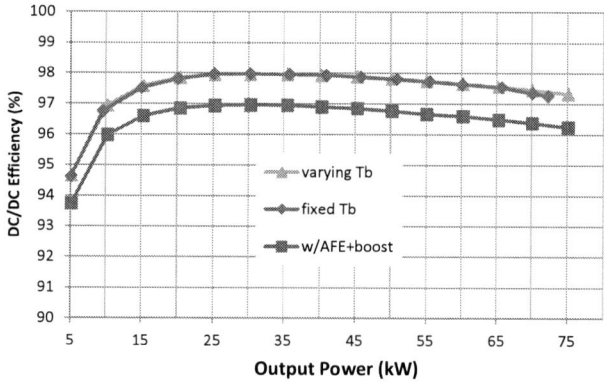

Fig. 15 –Converter cell efficiency curves (T_b – blanking time)

output diodes, there is reverse recovery current during the turn-off. This is evidenced by the current passing through zero near 44μs and then recovering to the magnetizing current before the turn-off. It is important to point it out that as expected, this becomes worst at higher current levels and results in added losses on the rectifier side.

The measured efficiency curve of the DC/DC converter is shown in Fig. 15 including gate drives and all other auxiliary power. A Yokogawa WT3000 Power meter was used to achieve an accuracy of +/- 1%. The input power and the delivered power from each DC power supply was directly

978-1-4673-9551-9/16 $31.00 © 2016 IEEE

measured and used to calculate the total efficiency. The converter was operated at switching frequencies in the design range of 10-12 kHz. It was found that the maximum efficiency could be achieved at the low end of this range and so the f_{sw} is set to 10 kHz for all the analysis. From Fig. 15 it can be seen that in the range from 20 kW to 50 kW the RC efficiency is very close to 98% and only drops below 97% when the power is less than 10 kW. The full efficiency considering the AFE and additional boost inductor is estimated using data sheet values based on the hard switched AFE waveforms in the full application converter with 25 levels.

VI. Switching Performance Analysis

Two typical switching waveforms are shown in Fig. 16, one at low power with a nearly ideal waveform and one at the rated power, both showing ZVS at turn-on. It can be noticed that at rated power, additional oscillations are already seen at turn-off, but the ZVS conditions are still maintained. However, these effects result in slightly higher switching losses. At higher power, this phenomenon becomes exaggerated as will be discussed later. Using the oscilloscope waveforms, analysis software developed in Matlab is used to calculate the switching losses. The algorithm also identifies the switching instants and then multiplies the device voltage and current and integrates the result to obtain the turn-on and turn-off energy. The switching loss calculation gives a relative sense of the distribution of the total losses and the variation in switching losses at different operating conditions. For example, for a load of 30 kW (Fig. 16(a)) the calculated switching losses were 32 W per IGBT device and for a load of 50 kW (Fig. 16(b)) 42 W of losses were calculated, both well within the operating limits.

In Fig. 17, two waveforms at full overload power (75 kW) are shown. At this operating point, the turn-on occurs beyond the limit of ZVS. The blanking time in Fig. 17(a) is 2.4 μs with switching losses of 121 W and the blanking time in Fig. 17b is 3.9 μs with switching losses of 55 W. It is clear from the commutation time of the collector emitter voltages (V_{CE} in the figure) why the switching losses are so different. In the first case, the voltage has dropped by only 23% by the time the in-coming switch turns on, whereas in the second case, the voltage has dropped by nearly 77%, and thus the switching losses are only slightly elevated compared to the 50 kW operating point. The main cause for the extended turn-off time is that the stored charge is higher due to the over load condition and that it is removed more slowly because of the strong oscillations seen during the hold-off time. Only when the current is positive is the charge removed. Thus, the required blanking time is much longer than that at lower power. This is true if the magnetizing current is not reversed by the intrinsic operation of the resonant circuit, i.e. if the voltage across the resonant capacitor is too high.

Under the conditions of fixed switching frequency and fixed resonant frequency, the effect of increasing blanking time adjustment is to begin turning off the resonating IGBT sooner in the switching cycle. The hold-off time is fixed in this case and turning the device off sooner helps to remove more of the stored charge and gives more time for the voltage to rise because of the fact that there is still sufficient energy in

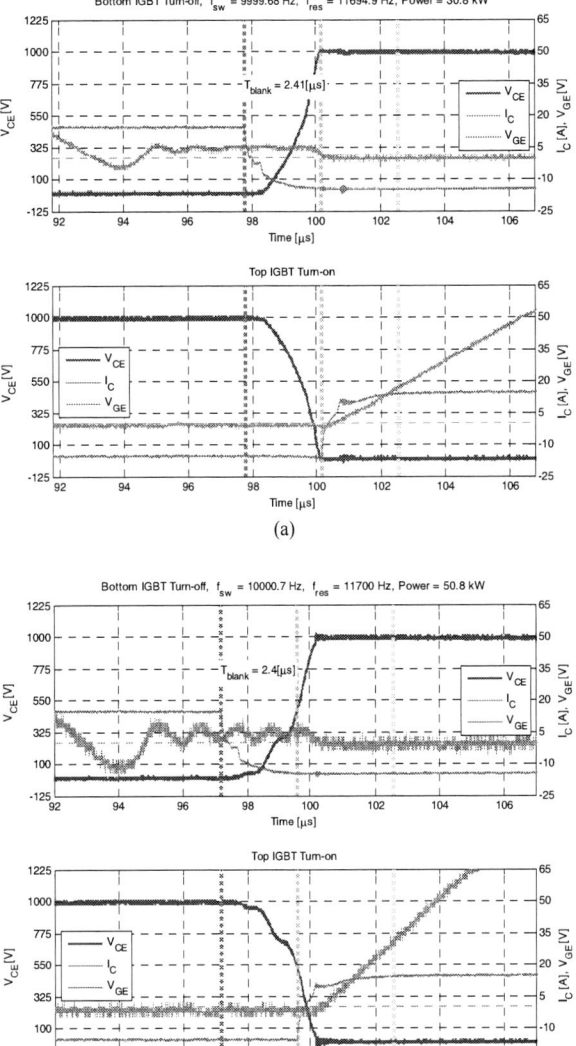

Fig. 16 – Switching waveforms at (a) 30 kW: P_{on} = 0 W, P_{off} = 32 W and (b) 50 kW: P_{on} = 0.13 W, P_{off} = 42 W

the magnetizing current. It is also possible to increase the switching period (i.e. reduce the switching frequency) by exactly the increase in blanking time, thus turning off the device at the same time in the hold-off period. This can take more advantage of the magnetizing current and reduce the turn-off losses even more.

In addition to the current spikes during turn-on, the full load waveforms demonstrate the non-linear behavior of the IGBT devices at full overload power. One feature apparent in both waveforms is the strong oscillation in the current of the turning off device. This effect is due to a resonant interaction with the output diode bridge. The oscillation in the diode voltage and current is seen in Fig. 17(c). The high frequency resonance with the rectifier diode limits the removal of stored charge in the device, which, again, is why a longer blanking time is required.

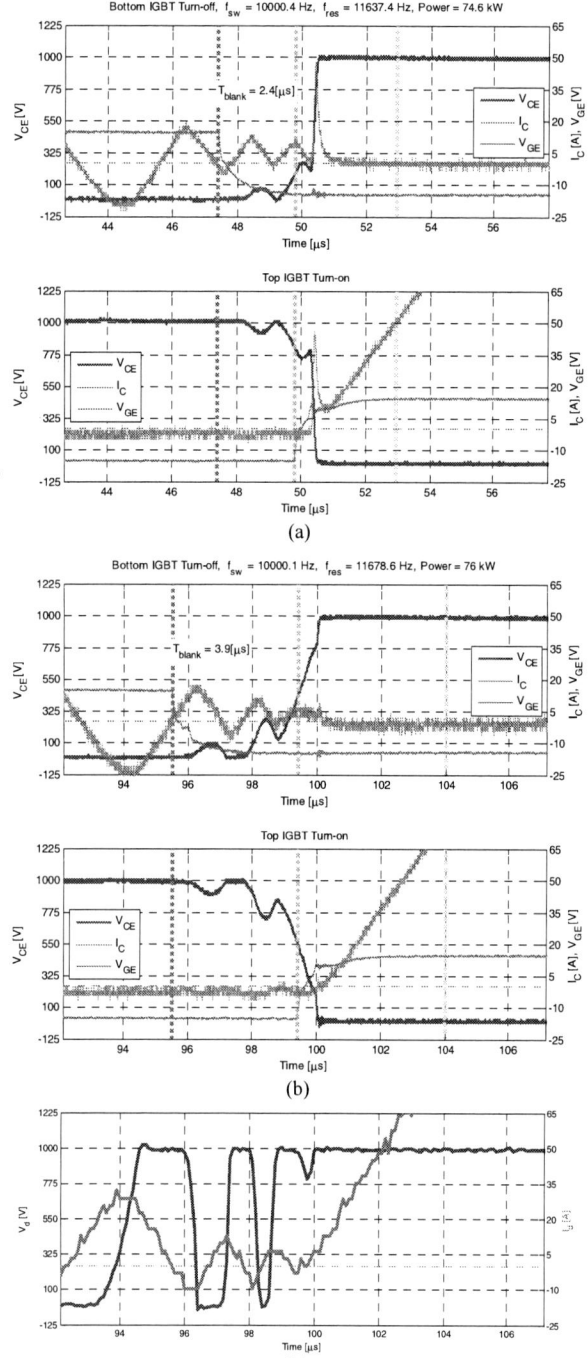

Fig. 17 – Switching waveforms at 75 kW and blanking times of
(a) 2.4 μs: P_{on} = 17 W, P_{off} = 104 W and (b) 3.9 μs: P_{on} = 2.3 W,
P_{off} = 52.7 W (c) RC rectifier side diode voltage and resonant
current showing oscillations

Another way to help remove the charge more quickly is to design the transformer with a smaller magnetizing inductance to increase the turn-off current. It then becomes a tradeoff between switching losses and output power.

To see the effect of a fixed, longer blanking time over the full operating range, we examine the case at low power, where

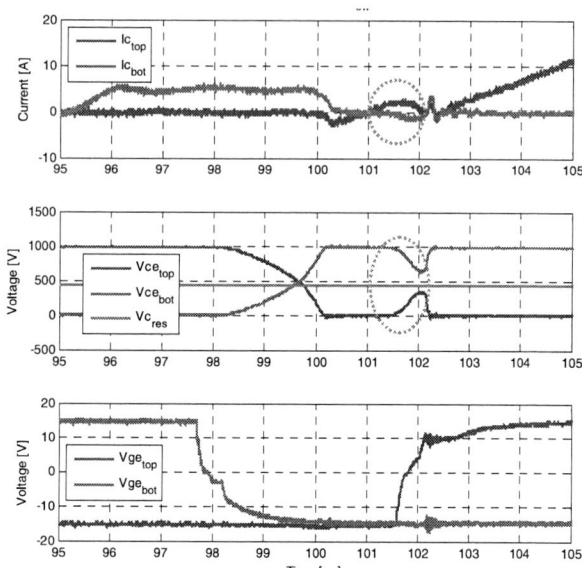

Fig. 18 – Waveforms at low power (15.4kW) showing resonance after
voltage commutation with T_{blnk} = 3.9μs.

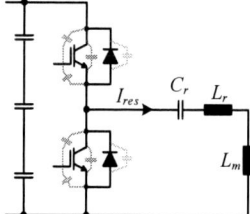

Fig. 19 – Equivalent resonant circuit during the hold-off time when both
devices are off showing parasitic device and package
capacitances.

the blanking time is set to the value needed for full overload power. In the case with f_{sw} = 10 kHz and f_{res} = 11.6 kHz, the required value was shown to be 3.9 μs. At the operating condition of 15.4 kW output power, the waveforms are shown in Fig. 18. The equivalent circuit of the RC for the time when both switches are fully off is shown in Fig. 19.

During this time period the bottom switch has completed the turn-off transition and the top switch diode, which had taken over the magnetizing current, now becomes positive. The diode no longer conducts and both switches are off, so the only place for energy in the resonant tank to go is into charging/discharging the device and package stray capacitances. Thus, the voltage across the bottom switch begins to rise as the upper switch voltage falls. This continues until the top switch is finally turned onm and small pulses of current are seen in both switches. It should be pointed out that there are no snubber capacitors in the converter. Thus, these voltages are developed across the device junction capacitances and the energy is very small. The penalty in losses for this mode of operation is a change in switching energy from 3.26 mJ to 3.7 mJ. The resulting power loss is changed from 32.6 W to 37 W. This could be unacceptable in MV devices as they are much more dominated by switching losses. In the LV case, there are also consequences for the efficiency at low load.

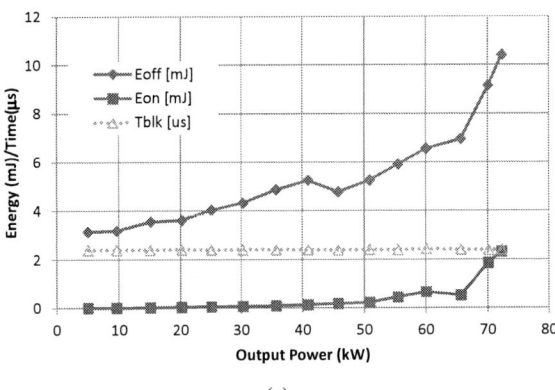

(a)

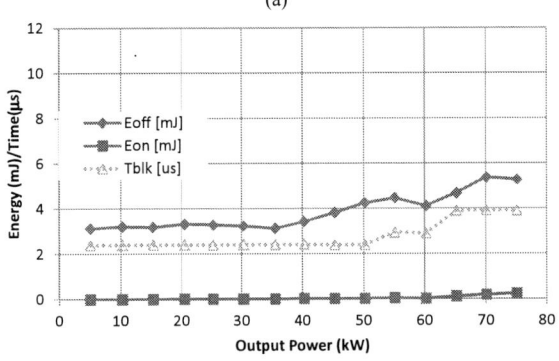

(b)

Fig. 20 – (a)Without blanking time control the converter loses ZVS above 65 kW, (b) with control the full rated overload power of 75 kW is reached with nearly constant turn-off energy.

By proper control of the blanking time, full power can be easily achieved without losing ZVS and the additional losses are quite small even at full overload power. The recommended blanking time adjustment can be done as an online function of the output power, which would require some additional control to be implemented. A simpler alternative is to use a fixed maximum blanking time value required for the full overload power. This can be acceptable if the gate drive unit is robust enough to handle the added common mode voltage. In the tested converter, both the cooling and the gate drive unit were sufficient.

The full operating range switching energy is shown in Fig. 20. The increase in switching energy for both turn-on and turn-off, In Fig. 20(a), increases sharply above 50 kW clearly indicating the loss of ZVS with a constant blanking time insufficient for overload operation. On the other hand, Fig. 20(b) demonstrates that when the blanking time is dynamically adjusted to the proper value, the turn-off losses can be nearly constant in the rated power range with only a small increase in the overload range to allow the excess carriers to recombine before the turn-on of the next device.

As mentioned earlier, this performance can be improved further by extending the switching period by the increase in the blanking time. This will result in a longer hold-off time, thus allowing the magnetizing current to increase further during the device turn-off to sweep out more carriers. The optimum values for the blanking time must be determined before final implementation of the converter control.

VII. CONCLUSIONS

In this paper, the design considerations for the LLC resonant tank components and thermal design are described for a LV IGBT modular cell based on 1.7 kV industrial devices. The critical timing parameters are investigated under soft-switching resonant operation and demonstrated with experimental waveforms. The device properties are investigated to better understand the limitations and requirements for maximum power throughput. With optimal design of the MFT as well as the use of the two-phase forced air-cooling solution, the size, weight and cost of the overall converter cell can be significantly reduced.

VIII. REFERENCES

[1] Zhao, C.; Lewdeni-Schmid, S.; Steinke, J.; K., Weiss, M.; Chaudhuri, T.; Pellerin, M.; Duron, J. & Stefanutti, P. "Design, implementation and performance of a modular power electronic transformer (PET) for railway application", 14th European Conference on Power Electronics and Applications, EPE '11 ECCE Europe, 2011, pp.1-10.

[2] Reinold, H. & Steiner, M. "Characterization of semiconductor losses in series resonant DC-DC converters for high power applications using transformers with low leakage inductance", 8th European Conference on Power Electronics and Applications, EPE '99, 1999, pp.1-10.

[3] Hoffmann, H. & Piepenbreier, B. "High voltage IGBTs and medium frequency transformer in DC-DC converters for railway applications", International Symposium on Power Electronics Electrical Drives Automation and Motion (SPEEDAM), 2010, pp.744-749.

[4] Dujic, D.; Steinke, G.; Bianda, E.; Lewdeni-Schmid, S.; Zhao, C.; Steinke, J. & Canales, F. "Soft switching characterization of a 6.5kV IGBT for high power LLC resonant DC-DC converter", PCIM Europe, 2012, pp.1-7.

[5] Yang, B.; Lee, F.; Zhang, A. & Huang, G. "LLC resonant converter for front end DC/DC conversion", Seventeenth Annual IEEE Applied Power Electronics Conference and Exposition, APEC, 2002, vol.2, pp.1108-1112.

[6] Coccia, A.; Canales, F.; Barbosa, P. & Ponnaluri, S. "Wide input voltage range compensation in DC/DC resonant architectures for on-board traction power supplies", 12th European Conference on Power Electronics and Applications, EPE '07, 2007, pp.1 -10.

[7] Duerbaum, T. "First harmonic approximation including design constraints" Twentieth International Telecommunications Energy Conference, 1998. INTELEC, 1998, pp.321-328.

[8] Ranstad, P. & Nee, H.-P. "On dynamic effects influencing IGBT losses in soft-switching converters", IEEE Transactions on Power Electronics, 2011, vol.26, no.1, pp.260-271.

[9] Steigerwald, R. "A comparison of half-bridge resonant converter topologies", IEEE Transactions on Power Electronics, 1988, vol.3, no.2, pp.174-182.

[10] Drofenik, U. "A 150kW medium frequency transformer optimized for maximum power density", 7th International Conference on Integrated Power Electronics Systems, (CIPS), 2012, pp.1-6.

[11] Agostini, F., Gradinger, T. "L-shaped thermosyphon loop with vertical evaporator for power electronics cooling", 15th International Heat Transfer Conference, IHTC-15, Kyoto, Japan, 2014, pp.1-10.

[12] Nagel, A.; Backhaus, K. & Reinold, H. "A resonant power supply for distributed systems", 6th European Conference on Power Electronics and Applications, EPE '95, 1995, vol.2, pp.833-838.

A Phase Shift Full Bridge Based Reconfigurable PEV Onboard Charger With Extended ZVS Range and Zero Duty Cycle Loss

Haoyu Wang, *Member, IEEE*
School of Information Science and Technology
ShanghaiTech University
Shanghai, China
wanghy@shanghaitech.edu.cn

Abstract—In this paper, an integrated onboard charger architecture is proposed for plug-in electric vehicle (PEV). In this architecture, the phase shift full bridge (PSFB) converter serves as the main high voltage battery charging topology, and the half bridge LLC resonant converter serves as the low voltage battery charging topology. Under light charging mode, the half-bridge LLC is reconfigured to be paralleled with the PSFB topology, to guarantee zero voltage switching (ZVS) of the lagging-leg MOSFETs. Practical design considerations are presented for both the PSFB and the half bridge LLC converters. Switching frequency and the shifted phase angle provide two degrees of freedom to regulate the output voltage/current of both converters. The proposed architecture maintains low cost and high efficiency in this specific application. A 390V input, 420V/2.4A, 14V/21A outputs converter prototype is designed, simulated, and analyzed to verify the proof of concept.

Keywords—*integrated charger; LLC; onboard charging; PEV; PSFB; ZVS;*

I. INTRODUCTION

Both the environmental problems and the energy crisis have been pushing the transition from conventional internal combustion engine vehicles towards more electrified plug-in electric vehicles (PEV) [1], [2]. In order to achieve a longer electric mileage, a high voltage Li-ion battery pack is installed onboard. Thus, an onboard battery charger is mandatary to charge this high voltage battery pack [3], [4]. Plus, a low voltage lead-acid battery (12V/14V) is installed onboard to provide power to the auxiliary loads, such as the air conditioner, head lights, stereo systems [5].

The typical configuration of the power management system in PEV is plotted in Fig. 1. There are three major power modules: a) onboard charger for the high voltage battery, b) propulsion motor drive, and c) low voltage battery charger. It should be noted that both the high voltage charger and the low voltage charger require an isolated DC/DC conversion stage.

The PSFB dc/dc topology as shown in Fig. 2. (a), enjoys the benefits of a) simple circuit structure with reduced components count, b) zero voltage switching of MOSFETs,

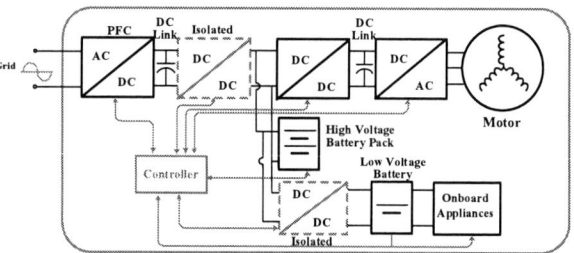

Fig. 1. Power management architecture of a full electric vehicle.

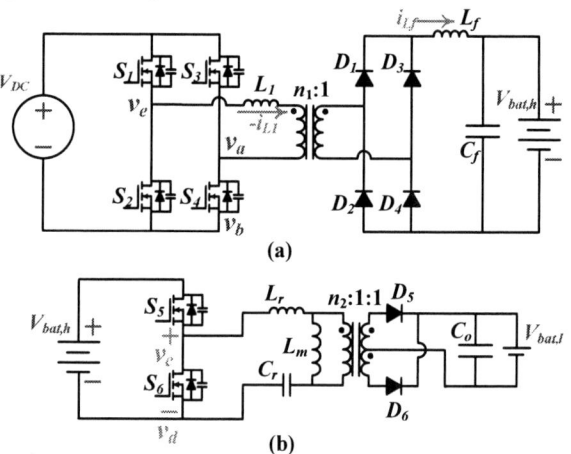

Fig. 2. Conventional onboard charger topologies: a) PSFB based high voltage battery charger, and b) Half bridge LLC based low voltage battery charger.

and c) easy to control with pulse width modulation. Therefore, it has been widely used in the PEV onboard battery chargers [6]–[8]. Regarding to the low voltage battery charger, half bridge LLC topology as shown in Fig. 2. (b), is considered as a good candidate due to its wide voltage gain range and ZVS features.

However, the traditional ZVS PSFB dc–dc converter has two fundamental limitations: a) the lagging leg MOSFETs lose ZVS feature under light load conditions, and b) the duty cycle loss problem. Many research efforts have been made to solve those problems [9]–[13]. Ideas on utilizing the ZVS half bridge

978-1-4673-9551-9/16 $31.00 © 2016 IEEE

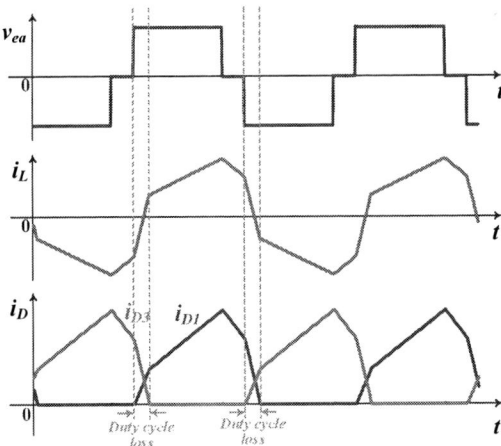

Fig. 3. Duty cycle loss of PSFB converter at heavy load condition.

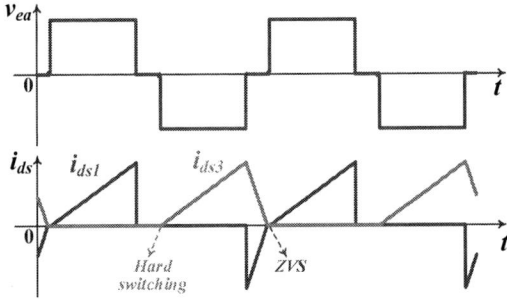

Fig. 4. PSFB MOSFET drain-source I-V waveforms under light load condition.

LLC topology to mitigate the hard switching problems on the PSFB lagging leg have attracted special attention [14]–[17]. In [14], a dual output dc/dc converter combining PSFB and half bridge LLC topologies is proposed to achieve the ZVS of the lagging leg. However, the proposed converter still suffers from duty cycle loss problems and design considerations are not optimized for the PEV specific applications. In [15], a PSFB and LLC integrated topology is proposed with the dual outputs in series; in [16], a PSFB and LLC integrated topology is proposed with the LLC output in series with the PSFB auxiliary capacitor; while in [17], PSFB and LLC outputs are simply in parallel. However, all those converters significantly increases the components count, and the secondary side always suffers from doubled diode conduction losses. Moreover, the parallel topology proposed in [17] suffers from increased control complexity.

In this paper, a self-reconfigurable PEV onboard charger is proposed. In the proposed architecture, an auxiliary circuit is added to the secondary side of the PSFB topology. This modification eliminates the duty cycle loss problem and mitigates the turning off di/dt on the secondary diodes. Moreover, under light load charging mode, the half bridge LLC based low voltage battery charger can be switched to the DC link (AC/DC stage output) of the PSFB converter. Therefore, the proposed architecture can achieve those benefits simultaneously: a) full ZVS range of the lagging leg; b) zero duty cycle loss and reduced circulating currents; c) relative low

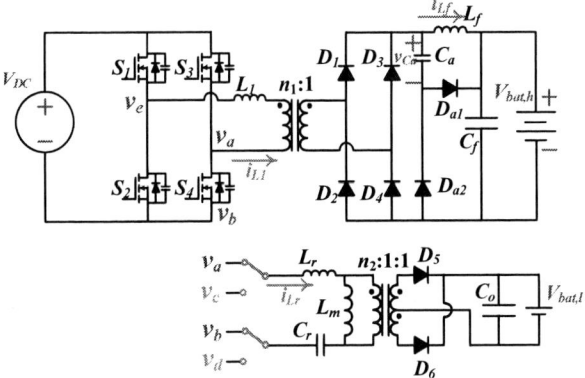

Fig. 5. Proposed self-reconfigured onboard charger topology combination.

circuit components count due to the topology reuse; and d) reduced secondary side diode turning off losses.

II. Issues with Conventional PSFB Topology

PSFB converter is a classic ZVS isolated dc/dc topology and has been well studied. By actively controlling the phase shift amount between the leading phase leg and the phase lagging leg, the root mean square voltage exposed to the transformer primary side can be actively controlled. Therefore, the voltage/current regulation can be achieved by phase shift control. Its main issues are summarized as below.

A. Duty cycle loss at heavy load condition

At heave load condition, PSFB topology suffers from duty cycle loss. Duty cycle loss occurs when the current entering into the transformer primary side (-i_{L1} as defined in Fig. 1) intersect with the secondary side output current (i_{Lf} as defined in Fig. 1). From this moment on, all the diodes (D_1-D_4) on the secondary side are off. This phenomenon is marked in Fig. 3.

During the duty cycle loss mode, the transformer secondary side sees a short circuit. Thus, no power is delivered to the load side, and the current will be circulating on the primary side tank. This squeezes the effective power supply duty cycle and increases the conduction losses. Moreover, in order to deliver the required amount of power to the load, the circuit components need to stand higher current stresses due to the low power supply utilization rate.

B. ZVS feature loss in the lagging leg at light load condition

The ZVS mechanism of power MOSFET is detailed as: before the current conduction through the MOSFET channel, there is a current from the source terminal to the drain terminal of the MOSFET. This current goes through the MOSFET body diode and creates a zero voltage condition for the MOSFET channel. Thus, ZVS is achieved when the current intersects with zero and the conduction is switched from the body diode to the channel. This ZVS feature can be observed among all the power MOSFETs.

However, with the decrease of load power, this source to drain current on the lagging leg MOSFETs decays. When the integral of this current during the dead band cannot fully

978-1-4673-9551-9/16 $31.00 © 2016 IEEE   481

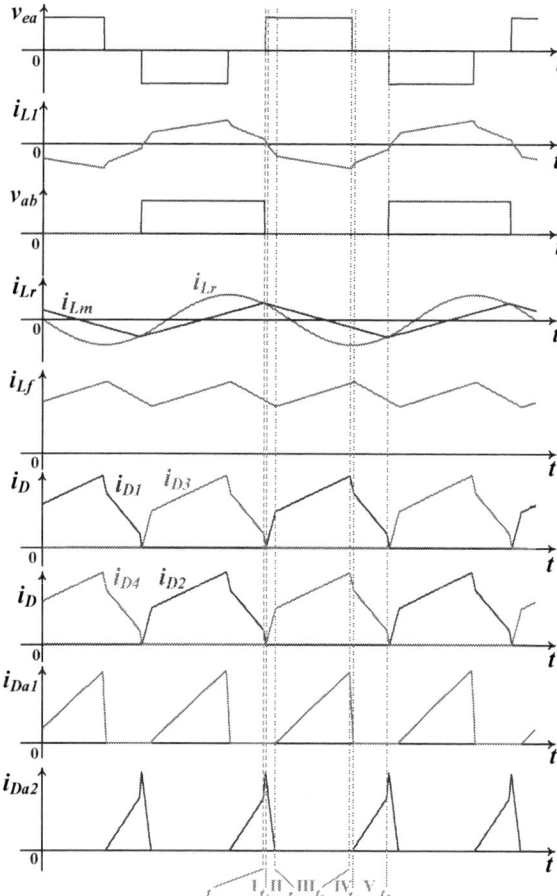

Fig. 6. Key waveforms of the circuit.

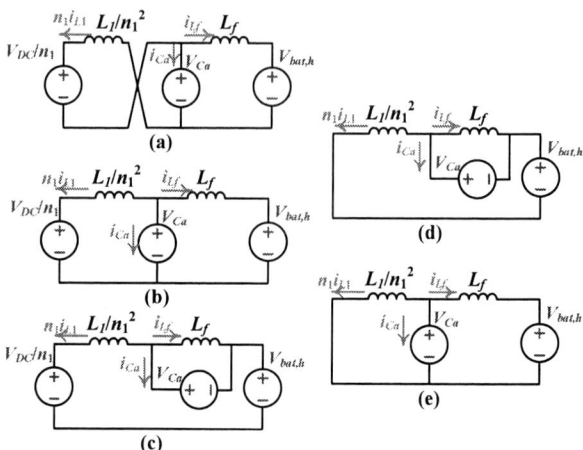

Fig. 7. PSFB equivalent circuits of different operation modes during one half switching cycle. a) Mode I: $t_0 \leq t < t_1$; b) Mode II, $t_1 \leq t < t_2$; c) Mode III, $t_2 \leq t < t_3$; d) Mode IV, $t_3 \leq t < t_4$; e) Mode V, $t_4 \leq t < t_5$.

B. Operation Principle

Key waveforms of the proposed circuit is plotted in Fig. 6. As shown, in each half switching cycle, there are five different operating modes. The next half switching cycle is symmetrical to the first switching cycle. To facilitate the analysis, one specific half switching period, $[t_0, t_5)$ is extracted from Fig. 6. The five operating modes of PSFB stage during this half switching cycle corresponds to five equivalent circuits as shown in Fig. 7. It should be noted that the primary side voltage source and impedance are both equivalent to the secondary side.

The following analysis is based on the assumption: C_a, C_f, and C_o are sufficiently large such that one can ignore their voltage ripples. Thus, those capacitor voltages are considered as dc voltages, V_{Ca}, $V_{bat,h}$, and $V_{bat,l}$, respectively.

Mode I: $[t_0, t_1)$. Mode I starts when S_3 is turned off and S_4 is turned on. The full bridge generates a positive voltage V_{DC}. D_2, D_3 conduct on the secondary side. Therefore, V_{DC} is coupled to be $-V_{DC}/n_1$ on the secondary side. D_{a2} conducts. Thus, negative terminal of C_a is connected to the isolated ground, as demonstrated in Fig. 7(a). Mode I ends when i_{L1} reaches zero. i_{L1} decreases linearly as,

$$\frac{di_{L1}}{dt} = \frac{-V_{DC} - n_1 V_{Ca}}{L_1} \tag{1}$$

i_{Lf} decreases linearly as,

$$\frac{di_{Lf}}{dt} = \frac{-V_{bat,h} + V_{Ca}}{L_f} \tag{2}$$

It should be noted that $i_{Ca} = n_1 i_{L1} - i_{Lf}$ in mode I.

Mode II: $[t_1, t_2)$. D_2, D_3 are off while D_1, D_4 are on. Therefore, V_{DC} is coupled to be V_{DC}/n_1 on the secondary side. D_{a2} keeps conducting. Thus, negative terminal of C_a is connected to the isolated ground, as demonstrated in Fig. 7(b). Mode II ends when i_{Da2} reaches zero and D_{a2} is turned off. In this mode, i_{L1} decreases linearly as,

---

charger and discharge the lagging leg MOSFET output capacitances, the ZVS feature is lost. This can be clearly observed from Fig. 4. As shown on the waveform of i_{ds3} in Fig. 4, before i_{ds3} crosses with zero, there is not a negative current to pre-charge or pre-discharge C_{oss}.

This ZVS loss problem brings severe ringing and EMI problems to the circuit. To reduce the ringing and EMI, typically a lossy snubber and an additional bulky passive filtering tank is required. Thus, both the conversion efficiency and the power density decays.

III. PROPOSED RECONFIGURABLE PEV ONBOARD CHARGER

A. Topology Description

The proposed reconfigurable PEV onboard charger is plotted in Fig. 5. As shown, by adopting two single-pole-double-throw relays, the half bridge LLC topology in the low voltage battery charger can be re-connected to the lagging leg of the PSFB converter. Moreover, a capacitor diode network is added on the secondary side of the PSFB topology. This network can eliminate the duty cycle loss problem of PSFB converter.

$$\frac{di_{L1}}{dt} = \frac{-V_{DC} + n_1 V_{Ca}}{L_1} \tag{3}$$

In Mode II, i_{Lf} still decreases linearly following Eq. (2). While $i_{Ca} = -n_1 i_{L1} - i_{Lf}$ in modes II-V. The initial value of $i_{L1}(t_1) = 0$, which can facilitate solving the time-domain expression for $i_{L1}(t)$.

Mode III: [t_2, t_3). D_1, D_4 are still on and V_{DC} is coupled to be V_{DC}/n_1 on the secondary side. D_{a1} is on. Thus, C_a is paralleled with the filtering inductor L_f, as demonstrated in Fig. 7(c). Mode III ends when S_1 is turned off and S_2 is turned on. In this mode, i_{L1} decreases linearly as,

$$\frac{di_{L1}}{dt} = \frac{n_1 V_{Ca} - V_{DC} + n_1 V_{bat,h}}{L_1} \tag{4}$$

i_{Lf} increases linearly as,

$$\frac{di_{Lf}}{dt} = \frac{V_{Ca}}{L_f} \tag{5}$$

Mode IV: [t_3, t_4). S_2 and S_4 are both on. Thus, the output of the full bridge is zero. D_1, D_4 are still on and zero is coupled to the secondary side. D_{a1} is still on. Thus, C_a is still paralleled with the filtering inductor L_f, as demonstrated in Fig. 7(d). Mode IV ends when i_{Da1} reaches zero and D_{a1} is turned off. In this mode, i_{L1} increases linearly as,

$$\frac{di_{L1}}{dt} = \frac{n_1 V_{Ca} + n_1 V_{bat,h}}{L_1} \tag{6}$$

In Mode III, i_{Lf} still increases linearly following Eq. (5).

Mode V: [t_4, t_5). S_2 and S_4 are both on. Thus, the output of the full bridge is zero. D_1, D_4 are still on and zero is coupled to the secondary side. D_{a2} conducts. Thus, negative terminal of C_a is connected to the isolated ground, as demonstrated in Fig. 7(e). Mode V ends when S_3 is turned on and S_4 is turned off and the circuit operation enters into the second half cycle. In this mode, i_{L1} increases linearly as,

$$\frac{di_{L1}}{dt} = \frac{n_1 V_{Ca}}{L_1} \tag{7}$$

i_{Lf} decreases linearly following Eq. (2). Itt should be noted that,

$$t_3 - t_0 = \frac{T_s}{2\pi}(\pi - \varphi) \tag{8}$$

where, φ is the shifted phase angel between the leading and lagging legs. φ is an actively controlled variable.

Assuming that all the circuit components are ideal, all the input power is delivered to the high voltage battery pack. According to the law of energy conservation,

$$P_{in} = \frac{\int_{t_0}^{t_0 + \frac{T_s}{2\pi}(\pi - \varphi)} V_{DC} i_L(t) dt}{T_s / 2} = V_{bat,h} I_{charge} \tag{9}$$

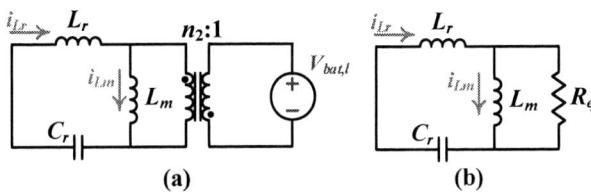

Fig. 8. a) LLC equivalent circuit during one half switching cycle, [t_0-t_5); b) circuit model using FHA.

According to the charge balance of the auxiliary capacitor, C_a,

$$\int_{t_0}^{t_0 + \frac{T_s}{2}} i_C(t) dt = 0 \tag{10}$$

Therefore, if circuit parameters (L_1, L_f, n_1, V_{DC}, I_{charge}, φ) are considered as known variables, the two unknown voltages, V_{Ca}, and $V_{bat,h}$ could be represented by those known variables based on the two equations (9-10). This means that the dc operating point of this converter can be solved numerically.

Regarding to the half bridge LLC converter, during this half switching cycle defined by [t_0-t_5), S_4 is always on. Thus, the input to the LLC resonant tank is always grounded, as shown in Fig. 8(a). D_6 is always on. Thus, the low voltage battery (-$V_{bat,l}$) is connected to the secondary side of the transformer. The magnetizing inductor voltage is clamped to be -$n_2 V_{bat,l}$. Thus, i_{Lm} decrease linearly. i_{Lr} and v_{Cr} resonates sinusoidally. This could be observed from Fig. 6.

The LLC circuit can be modeled based on the first harmonic approximation (FHA) [18]. The secondary side of the transformer is equivalent to an effective resistance,

$$R_{eff} = \frac{8n^2}{\pi^2} \frac{V_{bat,l}}{I_{bat,l}} \tag{11}$$

The LLC circuit model using FHA is plotted in Fig. 8(b). Utilizing this circuit model, the voltage gain can be predicated. It should be noted that this predication demonstrates good accuracy when the switching frequency is close to the resonant frequency between L_r and C_r [19].

IV. DESIGN CONSIDERATIONS

A. ZVS condition of the lagging leg MOSFETs

The ZVS of the lagging leg MOSFETs is achieved by the joint force of i_{L1} and i_{Lr}. The corresponding critical waveforms are plotted in Fig. 9, where Fig. 9(b) illustrates the more detailed waveforms during the dead band (t_{tead}).

Fig. 10 shows the equivalent circuit during the dead band. Both S_3 and S_5 have their channels off. $i_{L1} + i_{Lr}$ jointly charges C_{oss3} from 0V to V_{DC}, and discharges C_{oss4} from V_{DC} to 0V. Therefore, the ZVS condition is defined as,

$$\int_{t_0}^{t_0 + t_{dead}} [i_{L1}(t) + i_{Lr}(t)] dt \geq 2C_{oss} V_{DC} \tag{12}$$

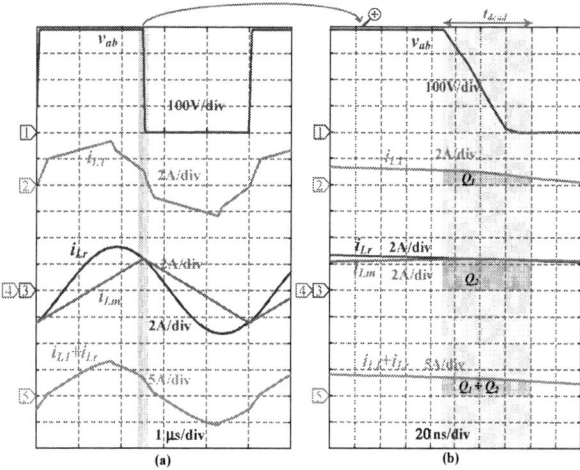

Fig. 9. ZVS waveforms of the lagging leg.

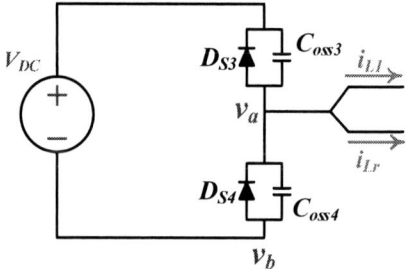

Fig. 10. Charging/discharging MOSFETs output capacitors during the dead band of the lagging leg.

As can be observed from Fig. 9(a), $i_{L1}(t)$ can be considered as a constant current source during this narrow dead band. $i_{L1}(t)$ can be calculated as,

$$i_{Lr}(t)\big|_{t\in[t_0,t_0+t_{dead}]} = i_{Lr}(t_o) = \frac{T_s n_2 V_{bat,l}}{4L_m} \quad (13)$$

It should be noted that $i_{L1}(t)$ can be derived based on the circuit analysis in Section III. Thus, the ZVS condition of lagging leg MOSFETs can be achieved once Eq. (12) is satisfied.

B. LLC tank parameters selection

Regarding to the LLC part, the resonant tank L_r, C_r, L_m are the most important design parameters. To guarantee the optimized operation of the LLC converter, the switching frequency should be equal to the resonant frequency,

$$f_s = \frac{1}{2\pi\sqrt{L_r C_r}} \quad (14)$$

The voltage gain is also determined by the quality factor, Q, which is defined as the ratio between the characteristic impedance and effective load resistance,

$$Q = \frac{\sqrt{L_r / C_r}}{R_{eff}} \quad (15)$$

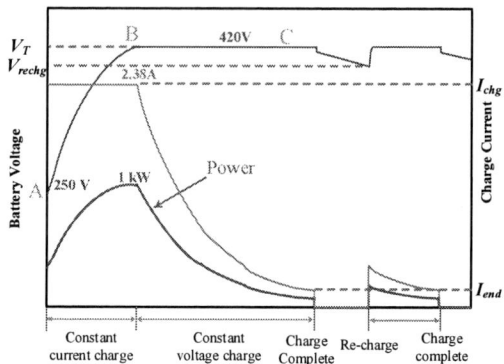

Fig. 11. 1kW charging profile of a 250V-420V battery pack.

Detailed selection of Q can be found in [18]. The basic idea is to make sure that the voltage gain curve at f_r has a sharp slope, such that the f_s has a narrow range and close to the resonant frequency. The gain fluctuation range corresponds to the fluctuation on the DC link input voltage. This fluctuation origins from the grid and the PFC rectification stage. After selecting Q and f_s, the L_r and C_r could be calculated.

Generally, large L_m means reduced circulating current and conduction losses. However, large L_m could results in ZVS loss in the MOSFETs. L_m can be selected by its largest possible value which still guarantees ZVS of the lagging leg MOSFETs.

C. Semiconductor devices selection

MOSFETs S_1-S_4 must have their voltage stress equal to V_{DC}. Typically, a 20% margin should be reserved. Thus, the voltage rating of MOSFETs can be calculated. The current stress of the MOSFETs can be calculated based on the expression of i_{L1} at the rated power of PSFB. It should be noted that at the PSFB rated power, the LLC topology is not activated.

Regarding to power diodes, D_1-D_4 have their voltage stress equal to $V_{bat,h} + V_{Ca}$, D_{a1} and D_{a2} have their voltage stress equal to $V_{bat,h}$, while D_5 and D_6 have their voltage stress equal to $2V_{bat,l}$. Typically, a 20% margin should be reserved. The current stress of the diodes can be calculated based on the expression of secondary side currents at the rated power, which can be derived based on the circuit analysis in Section III.

V. RESULTS

The specifications and design parameters of the proposed integrated charger are summarized in Table I.

Fig. 11 shows the 1kW charging profile of a PEV onboard battery pack. The charging is classified into constant current charging and constant voltage charging. While in the transition between those two charging modes, the charging power reaches its maximum value. Three important operating points (A, B, C) are marked in Fig 11.

Based on the circuit parameters shown in Table I. The normalized voltage gain of LLC converter versus f_s under different load conditions is plotted in Fig. 12. As shown, within the specified dc link voltage range (390V ± 10V), f_s can always be constrained in a narrow range close to f_r.

978-1-4673-9551-9/16 $31.00 © 2016 IEEE   484

TABLE I
SPECIFICATIONS AND DESIGNED PARAMETERS OF THE PROPOSED CHARGER

Symbol	Parameter	Symbol	Parameter
V_{DC}	390V±10V	L_r	35 μH
$V_{bat,h}$	250V-420V	C_r	18 nF
$P_{PSFB,max}$	1 kW	L_m	100 μH
$V_{bat,l}$	14 V	n_2:1:1	40:3:3
f_s	200 kHz	n_1:1	12:20
$P_{LLC,max}$	300 W	L_1	40 μH
Co	100 μF	L_f	560 μH
C_f	20 μF	C_a	1 μF

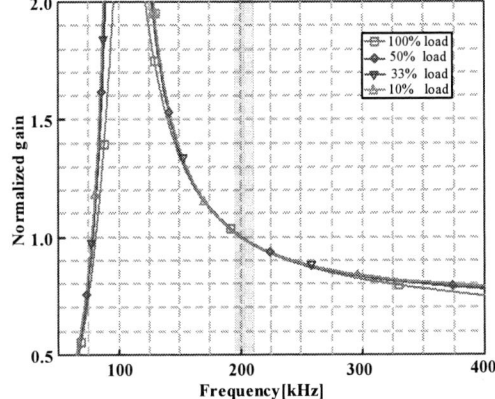

Fig. 12. Normalized gain versus the frequency for the selected design parameters.

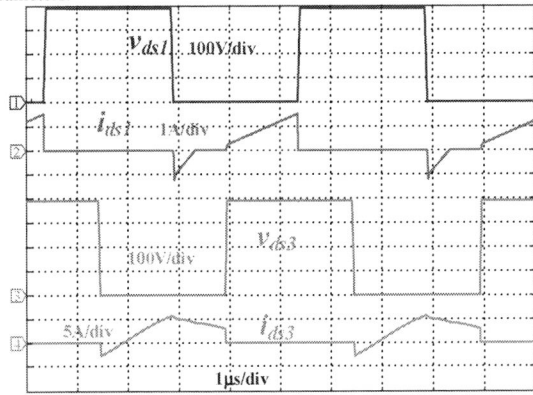

Fig. 13. PSFB circuit waveforms at V_{bat} = 420V, I_{bat} = 0.48A, LLC

Simulation is conducted based on the parameters provided in Table I. MOSFET C_{oss} is set to be 150 pF. Simulation data is presented in figures 13-15. Fig. 13 demonstrates the circuit operation at PSFB light load condition, where the charging power for the high voltage battery is 200W (point C in Fig. 11). At this operating point, the LLC converter is reconfigured and activated. As shown, V_{ds3} drops to zero before the conduction of MOSFET. Therefore, ZVS is achieved on the lagging leg.

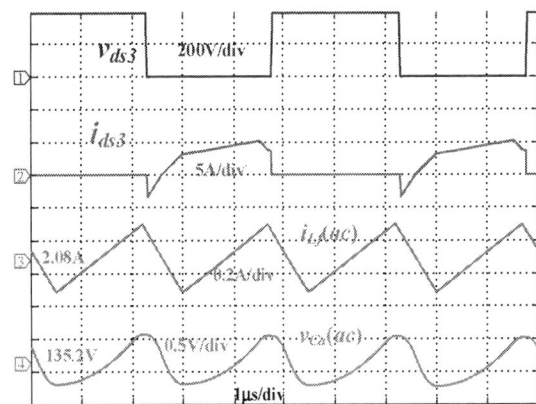

Fig. 14. PSFB circuit waveforms at V_{bat} = 420 V, I_{bat} = 2.38A, LLC not activated.

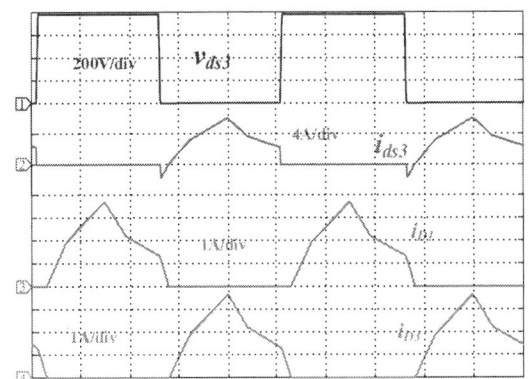

Fig. 15. PSFB circuit waveforms at V_{bat} = 250 V, I_{bat} = 2.38A, LLC not activated.

Fig. 14 shows the circuit operation at the PSFB full load condition, where the charging power for the high voltage battery is 1 kW (point B in Fig. 11). At this operating point, the LLC converter is in idle mode. As can be observed on the curve of i_{ds3} and v_{ds3} The PSFB can achieve ZVS on the lagging leg by itself. Ac coupled filter inductor current (i_{Lf}) and ac coupled auxiliary capacitor voltage (v_{Ca}) waveforms are also captured. i_{Lf} ripple is 0.42A with its dc offset as 2.08A. v_{Ca} ripple is 0.8V with its dc offset as 135.2 V. The ripple voltage is much smaller than its dc offset.

Fig. 15 shows the circuit operation at the beginning of constant current charging mode, where the battery voltage is 250V and the charging power is 595W (point A in Fig. 11). At this operating point, the LLC converter is also in idle mode. The PSFB achieve ZVS on the lagging leg without the assistance of LLC converter. i_{D1} and i_{D3} are captured. The waveforms shows that duty cycle loss is successfully eliminated.

VI. CONCLUSIONS

In this paper, a self-reconfigured PEV onboard charging architecture is proposed. The proposed architecture adopts an auxiliary circuit on the secondary side of the PSFB converter, which help to eliminate the duty cycle loss problem. Moreover, at light charging mode, a half bridge LLC topology is reconfigured to be connected to the DC link. This help the

PSFB converter to realize ZVS on the lagging leg. The proposed converter demonstrates the benefits of a) optimized LLC switching frequency; a) full ZVS range of the lagging leg; b) zero duty cycle loss and reduced circulating currents; c) relative low circuit components count due to the topology reuse; and d) reduced secondary side diode turning off losses. Circuit analysis and design considerations are both conducted in detail. A 1 kW charging prototype is designed and simulated to verify the proof of concept. Future work will be focused on the hardware implementation of this designed charger.

REFERENCES

[1] Z. Li, O. Onar, A. Khaligh, and E. Schaltz, "Design and Control of a Multiple Input DC/DC Converter for Battery/Ultra-capacitor Based Electric Vehicle Power System," in *2009 Twenty-Fourth Annual IEEE Applied Power Electronics Conference and Exposition*, 2009, pp. 591–596.

[2] O. C. Onar, J. Kobayashi, D. C. Erb, and A. Khaligh, "A Bidirectional High-Power-Quality Grid Interface With a Novel Bidirectional Noninverted Buck–Boost Converter for PHEVs," *IEEE Trans. Veh. Technol.*, vol. 61, no. 5, pp. 2018–2032, 2012.

[3] M. Yilmaz and P. T. Krein, "Review of battery charger topologies, charging power levels, and infrastructure for plug-in electric and hybrid vehicles," *IEEE Trans. Power Electron.*, vol. 28, no. 5, pp. 2151–2169, 2013.

[4] A. Khaligh and S. Dusmez, "Comprehensive Topological Analysis of Conductive and Inductive Charging Solutions for Plug-In Electric Vehicles," *IEEE Trans. Veh. Technol.*, vol. 61, no. 8, pp. 3475–3489, Oct. 2012.

[5] H. Wang, A. Hasanzadeh, and A. Khaligh, "Transportation Electrification: Conductive Charging of Electrified Vehicles," *IEEE Electrif. Mag.*, vol. 1, no. 2, pp. 46–58, Dec. 2013.

[6] S. Moisseev, K. Soshin, and M. Nakaoka, "Tapped-inductor filter assisted soft-switching PWM DC-DC power converter," *IEEE Trans. Aerosp. Electron. Syst.*, vol. 41, no. 1, pp. 174–180, 2005.

[7] H. Cha, L. Chen, R. Ding, Q. Tang, and F. Z. Peng, "An alternative energy recovery clamp circuit for full-bridge PWM converters with wide ranges of input voltage," *IEEE Trans. Power Electron.*, vol. 23, no. 6, pp. 2828–2837, 2008.

[8] O. C. Onar, J. M. Miller, S. L. Campbell, C. Coomer, C. P. White, and L. E. Seiber, "A novel wireless power transfer for in-motion EV/PHEV charging," in *2013 Twenty-Eighth Annual IEEE Applied Power Electronics*

Conference and Exposition (APEC), 2013, pp. 3073–3080.

[9] T. T. Song and N. Huang, "A novel zero-voltage and zero-current-switching full-bridge PWM converter," *IEEE Trans. Power Electron.*, vol. 20, no. 2, pp. 286–291, 2005.

[10] B. Gu, J.-S. Lai, N. Kees, and C. Zheng, "Hybrid-Switching Full-Bridge DC–DC Converter With Minimal Voltage Stress of Bridge Rectifier, Reduced Circulating Losses, and Filter Requirement for Electric Vehicle Battery Chargers," *IEEE Trans. Power Electron.*, vol. 28, no. 3, pp. 1132–1144, Mar. 2013.

[11] M. Pahlevaninezhad, P. Das, J. Drobnik, P. K. Jain, and A. Bakhshai, "A novel ZVZCS full-bridge DC/DC converter used for electric vehicles," *IEEE Trans. Power Electron.*, vol. 27, no. 6, pp. 2752–2769, 2012.

[12] Y. Do Kim, C. E. Kim, K. M. Cho, K. B. Park, and G. W. Moon, "ZVS phase shift full bridge converter with controlled leakage inductance of transformer," *INTELEC, Int. Telecommun. Energy Conf.*, pp. 2–6, 2009.

[13] W. Chen, P. Rong, and Z. Lu, "Snubberless bidirectional DC-DC converter with new CLLC resonant tank featuring minimized switching loss," *IEEE Trans. Ind. Electron.*, vol. 57, no. 9, pp. 3075–3086, 2010.

[14] Y. Chen, X. Pei, L. Peng, and Y. Kang, "A high performance dual output DC-DC converter combined the phase shift full bridge and LLC resonant half bridge with the shared lagging leg," *Conf. Proc. - IEEE Appl. Power Electron. Conf. Expo. - APEC*, pp. 1435–1440, 2010.

[15] C. Liu, B. Gu, J.-S. Lai, M. Wang, Y. Ji, G. Cai, Z. Zhao, C.-L. Chen, C. Zheng, and P. Sun, "High-Efficiency Hybrid Full-Bridge–Half-Bridge Converter With Shared ZVS Lagging Leg and Dual Outputs in Series," *IEEE Trans. Power Electron.*, vol. 28, no. 2, pp. 849–861, Feb. 2013.

[16] B. Gu, C.-Y. Y. Lin, B. F. Chen, J. Dominic, and J.-S. S. Lai, "Zero-Voltage-Switching PWM Resonant Full-Bridge Converter With Minimized Circulating Losses and Minimal Voltage Stresses of Bridge Rectifiers for Electric Vehicle Battery Chargers," *IEEE Trans. Power Electron.*, vol. 28, no. 10, pp. 4657–4667, Oct. 2013.

[17] M. Yu, D. Sha, and X. Liao, "Hybrid PS Full Bridge and LLC Half Bridge DC-DC Converter for Low-Voltage and High-Current Output Applications," no. 3132032, pp. 1088–1094, 2014.

[18] H. Wang, S. Dusmez, and A. Khaligh, "Design and Analysis of a Full Bridge LLC Based PEV Charger Optimized for Wide Battery Voltage Range," *IEEE Trans. Veh. Technol.*, vol. 63, no. 4, pp. 1603–1613, 2014.

[19] H. Wang, S. Dusmez, and A. Khaligh, "Design Considerations for a Level-2 On-board PEV Charger Based on Interleaved Boost PFC and LLC Resonant Converters," in *2013 IEEE Transportation Electrification Conference and Expo (ITEC)*, 2013, pp. 1–8.

978-1-4673-9551-9/16 $31.00 © 2016 IEEE

Series Arc Fault Detection Method Based on Statistical Analysis for DC Microgrids

Gab-Su Seo, Jung-Ik Ha, and Bo-Hyung Cho
Department of Electrical and Computer Engineering
Seoul National University
Seoul, South Korea
gabzzu@snu.ac.kr

Kyu-Chan Lee
Smart Power Supply Co., Ltd.
Seoul, South Korea
kyuchan6@empas.com

Abstract—Arc fault detection is required for the promising dc Microgrids because they are susceptible to dc arcing which may cause fire without protection. This paper proposes a new arc fault detection method based on statistical analysis. The statistical analysis provides design insight to establish the arc detection strategy by helping to determine the key parameters and operation principle of the arc fault detector. Since it enables guaranteeing the reliability of a fault detection method, the arc fault detector can be designed to meet the system requirements. The hardware prototype verifies the effectiveness of the method. The prototype arc fault detector can achieve 99.999% of fault detection accuracy in given arcing condition and 0.027% of false detection in given non-arcing condition while avoiding malfunctions from the switching noise of dc Microgrids interfacing power circuits.

Keywords—arc fault detection; dc distribution; fault protection; Microgrids; series arc;

I. INTRODUCTION

DC Microgrid using low voltage dc (LVDC) and medium voltage dc (MVDC) has been taken lots of attention as they are considered to be an effective solution to construct a power network interfacing a variety of and widespread distributed generations (DG) and electric loads. The dc power network is advantageous to achieve higher conductor efficiency in transmission/distribution and higher power conversion efficiency with the same level of voltage than ac ones [1], [2]. In addition, it does not require phase and frequency synchronization between the system components, which highly reduces system complexity. Owing to the advantages, it has been considered to be applied to data centers, commercial buildings, military base, and ships [3].

Despite the advantages, the dc Microgrids is not popular yet because protection measures for current breaking and dc arcing are not fully prepared in practical ways [4], [5]. Among the three faults in power networks: 1) short circuit, 2) ground, and 3) arc fault, the arc fault poses the most urgent threat since it is difficult to not only extinguish but also detect [6]. Especially in case of a series arc fault, the arc fault is not detected by a conventional circuit breaker as its fault current cannot exceed the threshold of the breaker [7]. Also, due to the lack of zero crossing in dc, it is difficult to self-extinguish while the arc fault in ac has higher probability of self-

extinction. On the other hand, parallel arc fault is likely to cause higher fault current than circuit breaker threshold like short circuit which enable higher probability to be detected [6]. Therefore, this paper focuses its discussion on the series arc fault detection in dc Microgrids.

The conventional arc detection methods recognize the presence of arcing through its original features which appear in the fault condition such as harmonic increase [8] and increased baseline noise [9]. However, they have difficulties in guaranteeing their reliability, because the arcing phenomenon is not predictable but chaotic [4]. Therefore, in implementing an arc fault detector, it is inevitable to observe the field test data to achieve high reliability [10], [11].

Pre-recorded data from the field test can be effective in developing an arc fault detector for a simple structured system such as photovoltaic (PV) generation system, which is one of the most popular dc system. The PV system not only has relatively simple structure with a limited number of known components but also is less likely to expand. In that kind of system, one can predict the electrical characteristics of the baseline and arcing and protect the entire system from faults as shown in Fig. 1 [12]. However, dc Microgrids comprise a variety of DGs and loads in which interface power circuits interact with each other depending on the system operating conditions. They are more likely to expand by combining new sources and loads as shown in Fig. 2. As a result, the conventional methods designed to operate in a simple system are inherently subject to malfunction by operation of power interfacing circuits [5]. In conclusion, it is challenging to develop a highly reliable arc fault detector operable in dc Microgrids.

To overcome the limitations of the conventional methods, this paper presents a statistical analysis method used to establish an efficient fault detection strategy and to guarantee its reliability such as probability of fault detection and nuisance tripping. In addition, it proposes an arc fault detection algorithm which suppresses the effect of interfacing unit's operation, power circuits for system components. As the algorithm is able to recognize the operation of power circuits and eliminate their effect in frequency spectrum, the algorithm can achieve high detection reliability regardless of their presence.

This work was supported by Brain Korea 21 Plus Project in 2015.

978-1-4673-9551-9/16 $31.00 © 2016 IEEE

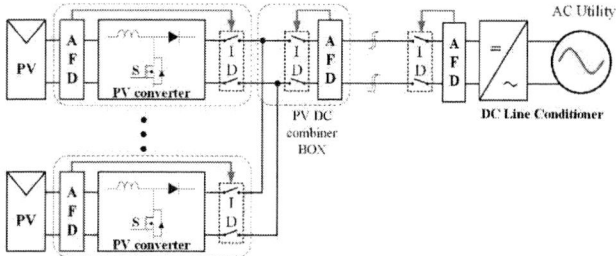

Fig. 1. PV generation system comprising a limited number of components and arc fault detectors.

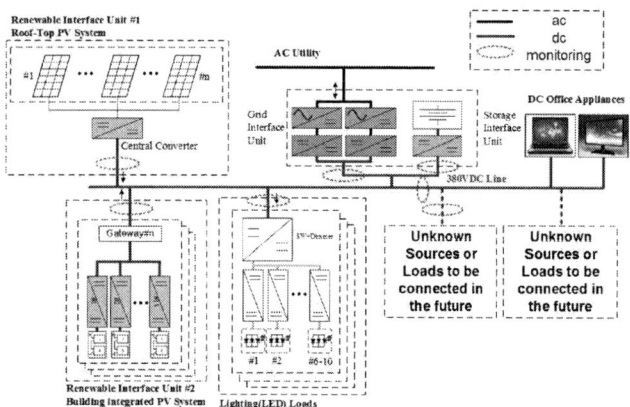

Fig. 2. DC Microgrid incorporating expandable sources and loads.

TABLE I. EXPERIMENTAL CONDITION FOR ARC CHARACTERIZATION

Item	Contents
Electrode	Copper
DC Voltage [V]	380
Load Current [A]	1.25, 2.5, 3.75, 5
Gap Distance [mm]	0.8, 1.6, 2.4, 3.2, 4
Temperature [°C]	27.5
Humidity [%]	57
Sampling Rate [MS/s]	1
Total Recorded Time [s]	10
Analysis Data Block [ms]	16 (16 378 samples)

The contents of this paper are delivered as follows. Section II characterizes the dc arc fault and presents the statistical analysis to derive the arc fault detection method. Based on the analysis, a new detection algorithm with data processing to suppress switching noise of power circuits is proposed in Section III. A hardware prototype is implemented in Section IV, and it verifies the feasibility of the proposed method.

II. ANALYSIS OF DC ARC CHARACTERISTIC

To derive an arc fault detection method for dc Microgrids, characterization of dc arc from the signals in normal and arcing condition is necessary. To acquire the arc fault signals, this paper uses an arc fault generator complying with UL1699B which is employed to evaluate the performance of an arc fault detector for PV system [13]. Using the arc fault generator, this study conducts a set of arc fault experiments with the common

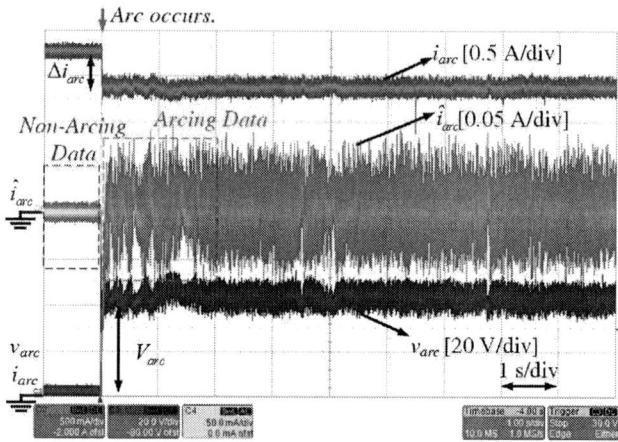

Fig. 3. Experimental waveform of dc arcing (3.75 A, 3.2 mm).

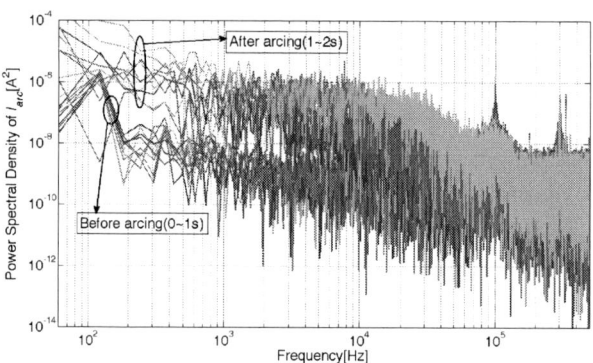

Fig. 4. Frquency spectrum for two seconds from normal to arcing.

operating conditions detailed in Table I, and the arc signals, arc voltage and current, are acquired by an oscilloscope as highlighted in Fig. 3. As the data for the first two seconds of arcing are critical to prevent fire accidents [14], the critical time data are analyzed while one second normal data before arcing is taken as depicted in Fig. 3.

A. Frequency Spectrum Analysis of Normal and Arc Signals

As the analysis focuses its discussion on the frequency domain feature changes by arcing to derive a detection method, the two groups of data are analyzed by fast Fourier transform (FFT). FFT is widely used due to the less computation requirements and ease of application on digital signal processors for ease of implementation rather than digital Fourier transform [15]. Fig. 4 illustrates the change of power spectral density of the current at the monitoring point for the first two seconds with 0.1 second interval under condition of 0.75 ampere load and 0.8 millimeter gap distance. As shown in Fig. 4, occurrence of arc fault leads to change in frequency spectrum, but it is hard to clearly recognize the difference between normal and arcing data. For clarity, each data group is averaged. Fig. 5 shows the averaged data of normal and arcing signals. As shown in Fig. 5, arcing signal shows pink noise

978-1-4673-9551-9/16 $31.00 © 2016 IEEE 488

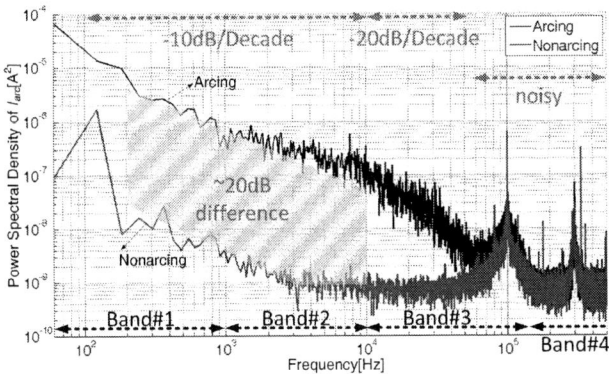

Fig. 5. Power spectral density comparison of normal and arc signal.

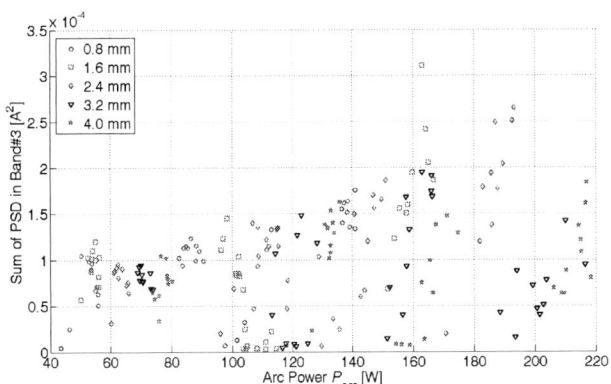

Fig. 6. Power spectral densities of arc signals in band #3.

characteristic [16]. A function of power spectrum density of a specific signal is considered to have a pink noise characteristic if

$$S(f) \propto \frac{1}{f^{\alpha}} \qquad (1)$$

where α ranges from 0 to 2, which means it has -10dB/decade with α equal to 1. As illustrated in Fig. 5, the power spectrum density of the monitored current is significantly increased as the arc fault occurs, but it decays and has similar level of magnitude with that of non-arcing signals after a certain frequency, 100 kHz in this case. The strong signal around 100 kHz results from the switching operation of the grid interface unit regulating the dc link voltage.

This study classifies the frequency band into four to determine the usable frequency range as illustrated in Fig 5. The band #1 ranges from dc to 10^3 Hz. Although the significant signal difference is observed in band #1, it is not useful because it is highly susceptible to the harmonics of 100/120 Hz utility frequency. Besides, the inherent spectral leakage of FFT would cause inaccuracy of its result [15], so it is not recommended for fault detection.

Band #2 ranging from 10^3 to 10^4 Hz would be useful because it shows -20dB steady difference in average. However, the effect of motor drive operating in the band should be taken into account to avoid false detection. Band #3 which spans from 10^4 to $1.5·10^5$ Hz also could be used for fault detection. However, the band should be also affected by power interfacing circuits in dc Microgrids operating in this band.

Band #4 covers the range behind band #3 up to the half of sampling frequency. Even though the spectrum magnitude in the frequency band would be kept under a certain value thanks to the electromagnetic interference (EMI) regulation, pink noise characteristic of arc makes the band less useful causing significant signal attenuation. In addition, it should be noted that the usage of band #4 requires high performance of data acquisition devices and processors which forces the costly implementation; as a result, it deteriorates its practicality.

In conclusion, band #2 and #3 are available for arc fault detection if the interference can be effectively suppressed. If the interference that can cause malfunction of detector appears, a data processing to remove its effect is required to ensure high detection reliability. After achieving a specific band data, the data of interest is summed into a value, and the resultant value is directly used for determination of fault presence.

B. Investigation of Power Spectral Density for Arc Detection

As shown in Fig. 4, the arcing signal clearly increase the power spectral density of the current. If the strength of the signal has a relationship to its power dissipation that is closely connected to probability of fire by the arc, it can be directly utilized to determine the significance of the arc. Fig. 6 plots the signal strength in part of band #3 as a function of the power dissipation by arc. As shown in Fig. 6, the power spectral density rarely depends on the power dissipation. Although it has a tendency that the signal strength increases as the arc power increases, it is difficult exactly to determine the significance of arcing only by this frequency domain data. In addition, it is clear that an arc fault detector would not detect an arc fault as the strength of some arcing signal is relatively approaching to that of non-arcing signal. As a result, it is hard to guarantee reliability of arc fault detection due to chaoticness of arc.

This paper presents a statistical method to address the unpredictable arcing phenomenon and to derive an arc fault detector whose reliability can be designed by experimental data and its statistical analysis. By statistically analyzing the experimental data of different conditions, the detection accuracy and probability of nuisance tripping can be derived. Fig. 7 and Fig. 8 display the probability distribution of non-arcing and arcing signals, respectively. The data come from the same experiments specified in Table I. The signals are assumed to have normality. As predicted and confirmed in these data, the probability distribution of non-arcing signals is far away from that of the fault signals while some of arc signals such as 5-A/4-mm condition extend into the area of non-arcing signals. The extended distribution should affect the reliability of arc fault detection which will be the determining factor for detector design.

978-1-4673-9551-9/16 $31.00 © 2016 IEEE

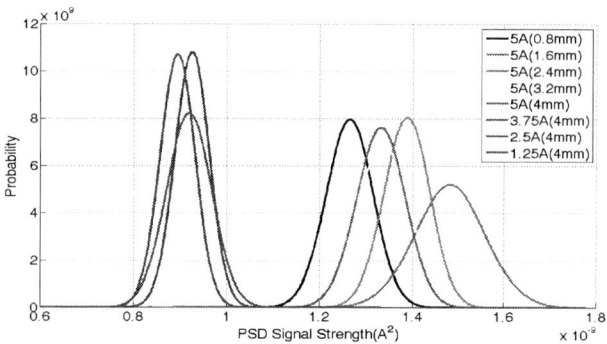

Fig. 7. Probability distribution of non-arcing signals.

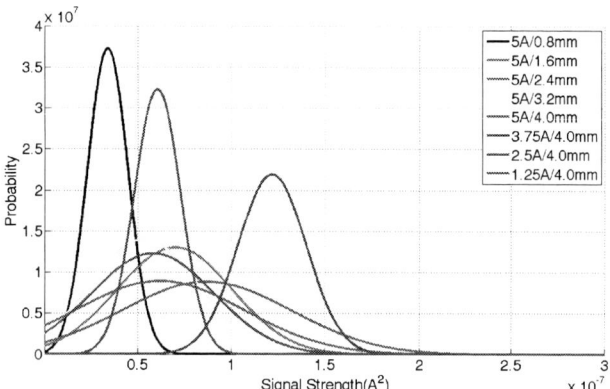

Fig. 8. Probability distribution of arcing signals.

Lots of phenomena in nature show normality, so they can be interpreted using characteristics of normal distribution [17]. And in general, the normality becomes strong as the number of samples increases. However, with an insufficient number of data, the data would not show normality. Because of the chaoticness of arcing and the limited number of samples, most of the arcing signals do not show the normality. Without confirmation of normality, it is not possible to analyze the signal distribution using the characteristic of normal distribution. In this case, normalization methods can be employed. For detector design, three statistical techniques can be utilized as follows: 1) normality test to check whether the normal/arcing data can be considered as normal distribution, 2) Box-Cox or Johnson transform to normalize a specific data set to follow normal distribution if it does not have normality, and 3) t-test to confirm the two groups of data are distinguishable. After the statistical analysis, we can determine the design parameters of an arc fault detector satisfying the system requirement.

III. ARC FAULT DETECTION METHOD BASED ON STATISTICAL ANALYSIS WITH DATA PROCESSING

Based on the statistical analysis, a new arc fault detection method is proposed. Through the aforementioned step-by-step

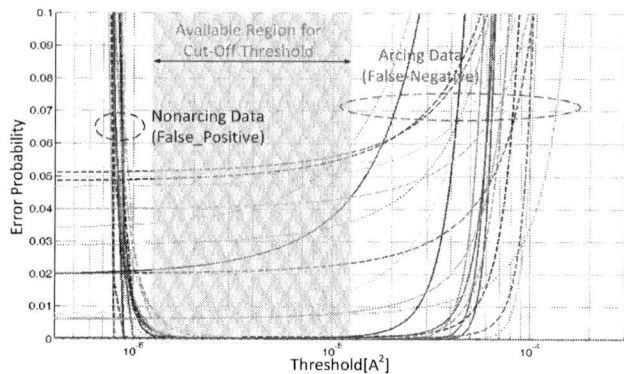

Fig. 9. Graphical method for fault detection threshold determination.

statistical approach, the probability curves of all operating conditions as a function of threshold values can be plotted. With the conditions in Table I, the detection threshold can be graphically determined as shown in Fig. 9. Accordingly, maximum detection accuracy achievable in the given conditions can be recognized. There is a clear trade-off between false-positive and false-negative. In this paper, false-positive is defined as the malfunction of detector that it determines the occurrence of arc in non-arcing condition, while false-negative is defined as the failure of detection in arcing condition. As a result, the probability of false-positive increases as one tries to set lower threshold to achieve higher probability of successful detection. Besides, it can be noted that the achievable accuracy of a detector is limited by the signal distribution; e.g. the false-negative cannot be reduced to lower than 5% in the curve even if the probability of the false positive reaches 100%.

Based on the distribution curves from statistical analysis, the fault detection strategy can be finalized. To overcome the aforementioned inherent limitation in achieving high reliability and optimize the trade-offs between the detection and nuisance tripping, multiple decision makings within the time allowed for detection should be useful. For example, ten decisions are available for 300 W arcing in two seconds according to UL1699B [13], which makes the false negative reduces in square manner. According to the analysis of trade-off between false-negative and false-positive, it is found that to achieve less false-negative (high arc detection accuracy) is more difficult than to achieve less false-positives because the arc data shows higher standard deviation as shown in Fig. 8. As a result, multiple decision makings are indispensable to obtain higher detection accuracy. In addition, noise immunity also should be taken attention in design process as load transients (turning on/off) can cause nuisance tripping. Nuisance tripping from abrupt operating condition change can be addressed by averaging of multiple samples of the power spectrum density.

To attain the robustness of detection to noise from operations of system components in dc Microgrids, the system modeling and analysis should be conducted. This paper employs a data processing method for noise suppression of switching power circuits according to [5]. Through the method, the arc fault detector can recognize the operation of switching

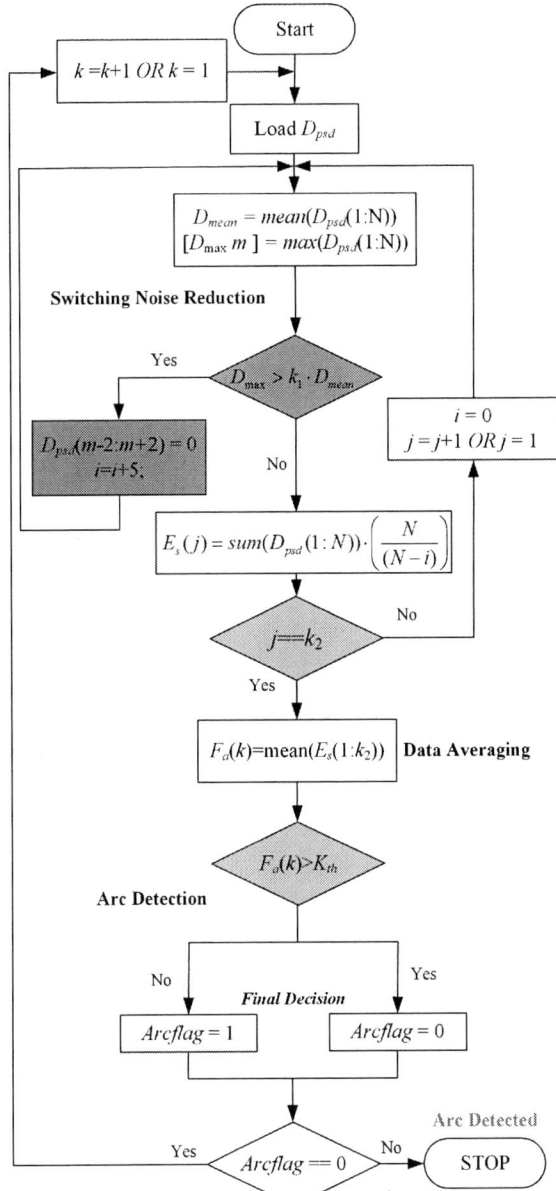

Fig. 10. Flow chard of proposed arc fault detector.

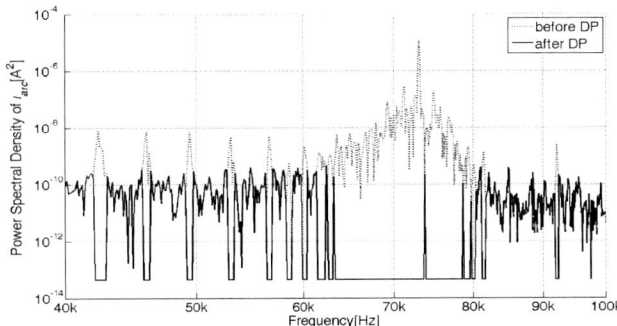

Fig. 11. Power spectral densities of LED driver with data processing (DP).

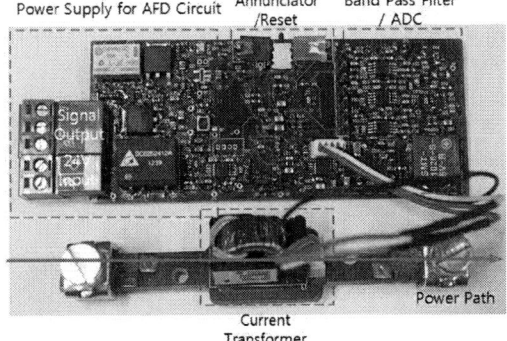

Fig. 12. Arc fault detector prototype hardware.

operations and avoid false detection with improved signal clarity.

Fig. 10 shows the flow chart of the proposed arc fault detection method equipped with the noise reduction technique for power circuits. The noise reduction technique is derived from the modeling of power circuits and electric arc. And the factor k_1 which determines the amount of the data processing (DP) can be designed from the statistical analysis. Fig. 11 shows a representative example of data set processed by the proposed noise reduction technique when k_1 is 5 which confirms its robustness to noise and its applicability. The frequency spectrum is obtained from the current with light

emitting diode (LED) driver operating on the monitoring branch, and due to the strong frequency signature around 70 kHz, it may cause serious nuisance tripping of the detectors with conventional detection algorithm. All the parameters can be designed referring to the system requirement and based on the statistical analysis.

IV. EXPERIMENTAL RESULTS

To verify the proposed analysis method and detection algorithm, a prototype arc fault detector is implemented. Fig. 12 shows the prototype and it employs current transformer for ac current sensing, high speed analog-to-digital converter for data acquisition, and digital signal processor for suppressing switching noise and detecting the arc fault. The design parameters for the prototype is as follows:

- k_1: 20; k_2: 5; N:512;

- f_s (Sampling frequency): 250 kHz;

- T_p (Sampling Period): 20 ms;

- K_{th} (Detection Threshold): $1 \cdot 10^{-5}$;.

The prototype is designed to achieve 99.999% detection accuracy in arcing condition by multiple detection trials which equals to 10 ppm false-positive, while it has 270 ppm false-negative probability with the threshold value $1 \cdot 10^{-5}$ and the other design parameters. To suppress the malfunction from system transients, averaging factor k_2 is set to 5.

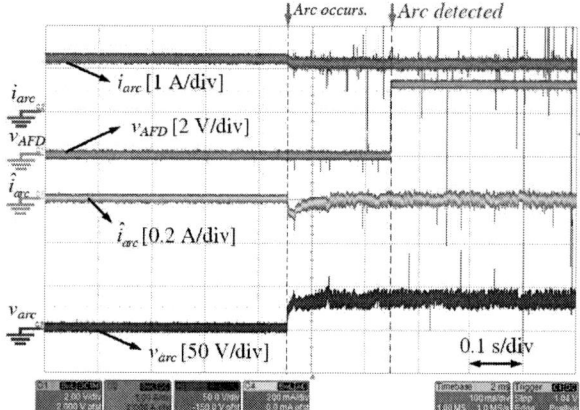

Fig. 13. Experimental waveform of arc fault detection under 1.25-A load and 0.8-mm arc gap distance.

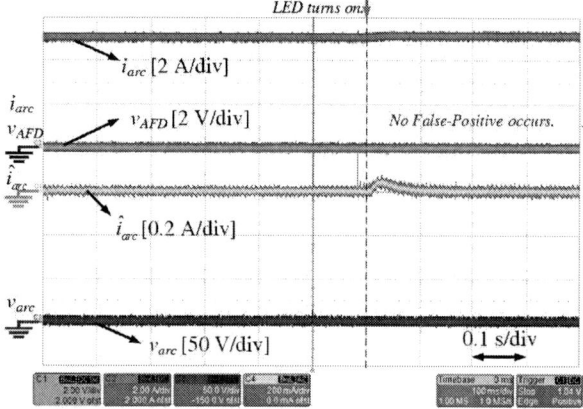

Fig. 14. Experimental waveform of false-positive surppression of prototype under 5-A resistive load and 24-W LED driver operation.

Fig. 13 verifies the detection performance in arcing condition; it detects the arc at the time of 0.183 seconds. The output signal of the prototype can be used as a trigger to control the circuit breaker addressing the monitoring branch. Also, the time required to take a complete action from detection to protection by the branch switch should also be considered to guarantee the system safety in the design process. Fig. 14 verifies the immunity of the prototype to the switching noise of power circuits; LED driver is a representative dc load and presents challenging false-positive issues since it operates in boundary conduction mode which changes its operation frequency depending on the operating conditions.

V. CONCLUSION

Based on the statistical analysis, a new arc fault detection method has been proposed. With the analysis and the noise reduction technique, not only can the reliability of arc fault detector be guaranteed in the design stage, but also false

tripping caused by power circuits of dc Microgrids can be minimized. To verify the feasibility of the proposed arc fault detection method, a prototype is implemented and it can achieve the high detection reliability with suppressed false-positive.

REFERENCES

[1] D. Salomonsson and A. Sannino, "Low-voltage DC distribution system for commercial power systems with sensitive electronic loads," *IEEE Trans. Power Deliv.,* vol. 22, no. 3, pp. 1620-1627, Jul. 2007.

[2] G.-S. Seo, K. C. Lee, and B. H. Cho, "A new dc anti-islanding technique of electrolytic capacitor-less photovoltaic interface in dc distribution systems," *IEEE Trans. Power Electron.,* vol. 28, no. 4, pp. 1632-1641, Apr. 2013.

[3] P. Cairoli, "Fault protection in dc distribution systems via coordinated control of power supply converters and bus tie switches," Ph.D. dissertation, Dept. Elect. Eng., University of South Carolina, Columbia SC, 2013.

[4] X. Yao, L. Herrera, S. Ji, K. Zou, and J. Wang, "Characteristic study and time-domain discrete-wavelet-transform based hybrid detection of series dc arc faults," *IEEE Trans. Power Electron.,* vol. 29, no. 6, pp. 3103-3115, Jun. 2014.

[5] G.-S. Seo, K. A. Kim, K.-C. Lee, K.-J. Lee, and B.-H. Cho, "A new dc arc fault detection method using dc system component modeling and analysis in low frequency range," in *Proc. IEEE Applied Power Electronics Conf. and Expo.,* 2015, pp. 2438-2444.

[6] G. D. Gregory, K. Wong, and R. F. Dvorak, "More about arc-fault circuit interrupters," *IEEE Trans. Ind. Appl.,* vol. 40, no. 4, pp. 1006-1011, Jul./Aug. 2004.

[7] M. K. Alam, F. Khan, J. Johnson, and J. Flicker, "A comprehensive review of catastrophic faults in PV arrays: types, detection, and mitigation techniques," *IEEE J. Photovolt.,* vol. 5, no. 3, pp. 982-997, May 2015.

[8] C. Hong, C. Xiaojuan, L. Fangyun, and W. Cong, "Series arc fault detection and implementation based on the short-time fourier transform," in *Proc. Asia-Pacific Power and Energy Engineering Conf.,* 2010, pp. 1-4.

[9] G. Healy and G. Roemer, "Arc fault detection using Rogowski coils," in *Proc. PCIM Europe,* 2014, pp. 1-5.

[10] B. D. Russell, K. Mehta, and R. P. Chinchali, "An arcing fault detection technique using low frequency current components-performance evaluation using recorded field data," *IEEE Trans. Power Deliv.,* vol. 3, no. 4, pp. 1493-1500, Oct. 1988.

[11] J. Johnson and J. Kang, "Arc-fault detector algorithm evaluation method utilizing prerecorded arcing signatures," in *Proc. IEEE Photovoltaic Specialists Conf.,* 2012, pp. 1378-1382.

[12] G.-S. Seo, H. Bae, B.-H. Cho, and K.-C. Lee, "Arc protection scheme for dc distribution systems with photovoltaic generation," in *Proc. Int. Conf. Renewable Energy Research and Applications,* 2012, pp. 1-5.

[13] *Outline of Investigation for Phovotoltaic DC Arc-Fault Circuit Protection.* UL1699B, 2011.

[14] J. K. Hastings, M. A. Juds, C. J. Luebke, and B. Pahl, "A study of ignition time for materials exposed to dc arcing in PV systems," in *Proc. IEEE Photovoltaic Specialists Conf.,* 2011, pp. 3724-3729.

[15] S. Rapuano and F. J. Harris, "An introduction to FFT and time domain windows," *IEEE Instrum. Meas. Mag.,* vol. 10, no. 6, pp. 32-44, Dec. 2007.

[16] M. Wendl, M. Weiss, and F. Berger, "HF characterization of low current dc arcs at alterable conditions," in *Proc. IEEE Int. Conf. Electrical Contacts,* 2014, pp. 1-6.

[17] D. P. Doane and L. E. Seward, *Applied statistics in business and economics.* New York, NY: McGraw-Hill/Irwin, 2005.

Arc Welding Inverter With Embedded Digital Active EMI Controller

Junpeng Ji*[†], Wenjie Chen* and Xu Yang*

Email: jijunpeng@xaut.edu.cn, {cwj, yangxu}@mail.xjtu.edu.cn

*School of Electrical Engineering, Xi'an Jiaotong University, Xi'an, Shaanxi, China

[†]Department of Electrical Engineering, Xi'an University of Technology, Xi'an, Shaanxi, China

Abstract—Digital high-frequency arc welding inverter is widely used in welding areas because of its many advantages such as high efficiency, small size, easy to control etc. However, with the development of high frequency of the switching, the conducted EMI generated by arc welding inverter becomes increasingly serious, especially common mode conducted EMI. Based on Digital Active EMI Filter (DAEF) controller, this paper proposes a new method which can reduce common mode conduction EMI for arc welding inverter. It is a closed EMI control loop by sensing EMI signal, sampling, digital compensation control, DAC output and injection. The EMI control can be embedded in original controller of arc welding inverter to reduce costs. Moreover, since there is no component in series connect with the power circuit, DAEF solves the problem of traditional EMI filter in filter size and power consumption fundamentally. In this paper, the arc welding inverter system with DAEF is designed, molded and compensated. Meanwhile, by simulation and analyses, the advantages of arc welding inverter with DAEF in system stability and dynamic performance are showed. Finally, experiment results showed that the proposed method has advantages in reducing conducted EMI emission and improving dynamic performance of arc welding inverter.

Keywords—*arc welding inverte; EMI controller; conducted EMI; common mode; digital active; embedded*

I. INTRODUCTION

As welding technology highly improved and welding automation deeply developed, digital high-power high-frequency inverter welding power supply has become a trend of modern welding power source [1, 2]. But the problem of EMI produced by high frequency switching control has grown more serious [3], the Common Mode (CM) conducted EMI is particularly serious. This generates a higher challenge for EMI filtering technologies in arc welding inverters. This conducted EMI not only pollutes the power grid, and affects the normal operation of other equipment, but also affects arc welding inverters own stability and reliability. For conducted EMI problem, the traditional passive EMI filters [4, 5], active EMI filter [6, 7] and hybrid active EMI filter [8-10] are making progress.

On the other hand, with the improvement of digital processing technique, such as high speed and high precision FPGA, Analog-Digital Converter (ADC) and Digital-Analog Converter (DAC), the dream of using digital active EMI filtering (DAEF) technique in arc welding inverters is ready to bring into reality.

Since there is no component in series connect with the power circuit of arc welding inverters, DAEF solves the problem in filter size and power consumption of traditional welding supply. Moreover, DAEF can selectively suppress the CM and DM EMI in accordance with the actual situation of CM and DM EMI in the arc welding inverters. It can also effectively prevent interconversion between CM and DM interference. Meanwhile, due to embedded DAEF, the control system performance of arc welding inverters in the high frequency is improved.

The concept of DAEF was first proposed in 2012 by the Canadian academic Hamza, and its effectiveness had been verified in the photovoltaic grid-connected inverter [11]. In the following year, DAEF technology was applied in DC-DC converters of electric vehicle and other switched-mode power converters [12-14]. However, current DAEF only suppresses EMI to ground in the AC (or DC) ports. It doesn't definitely aim to CM or DM interference. About the relationship between DAEF control loop and switched-mode control loop, current DAEF literature not mentioned.

On considering the above mentioned problems, based on digital active EMI filter control method [12-14], a new method to reduce CM conduced EMI of arc welding inverters is proposed. The goal of this paper is try to find an effective filtering method to eliminate the CM EMI in the arc welding inverter of high power density. Moreover, try to find the relationship of control performance between DAEF control loop and switched-mode control loop. The advantage of the proposed circuit and control method lies in that it can be embedded into digital controller resources of welding inverter power supply to decrease the cost of EMI filter, Meanwhile, it can also realized the digital closed-loop control of conducted EMI signals and improve the stability and dynamic performance of arc welding inverter power supply. To achieve and verify it, the control system of an arc welding power supply embedded DAEF is designed and modeled, and the actual supply prototype is constructed. It form closed loop control by sensing EMI signal, sampling, digital compensation control, DAC output and injection. By this embedded DAEF, not only that conductive EMI is suppressed, but also that the system behavior is improved for arc welding inverters.

This work was supported by the National Natural Science Foundation of China (51277145), Science and technology planning of Beilin District of Xi'an in 2015 (GX1508), and Featured research program of Xi'an University of technology (2014TS010).

II. Arc Welding Power Supply System Design and Modeling

A. Arc Welding Power Supply Control System Design

The topology of arc welding inverter with DAEF is shown in Fig. 1. The main circuit consists of non-controlled rectifier circuit and DC/DC full bridge inverter circuit. Control system is mainly consists of three control circuit, two DAEF control circuit control two power line's conductive EMI, and a current control circuit control the output current, block diagram of the control system is shown in Fig. 2. DAEF control circuit includes six parts such as EMI signal sense circuit, injection circuit, ADC sampling circuit, controller, DAC output and decoupling circuit between the injected point and sensed point. EMI signal sense circuit (High-pass filter) picks up conducted EMI signal V_{LEMI} and V_{NEMI} to ground on the L line or N line. A/D converter acquisition detected conduction EMI signal of arc welding inverter power supply and sends it to EMI controller. According to equation 1 and equation 2, EMI controller can calculate the CM EMI voltage $V_{Lc}(V_{Nc})$ and DM EMI voltage V_D. By controller $G_C(s)$ compensating, DAEF control system outputs EMI current signal I_{LEMI}, which will cancel the sum of CM EMI on L line and DM EMI by injection circuit (low pass filter). As well as DAEF control system outputs EMI current signal I_{NEMI} to cancel the CM EMI on N line. I_{LEMI} and I_{NEMI} signal is converted back to analog signal for injection back to L line or N line. DAC output is EMI cancel signal of 0~20mA which can suppress 120dBμV EMI signal. The capacitor C_{iL} and C_{iN} of injection can prevent overloading to DAC from the power converter. Injection circuit (Low pass filter) can cancel the sum of CM EMI on L line and DM EMI. High impedance RF inductor L_{jL} and L_{jN} is located between the injection point and sense point as the decoupling circuit, which reduced the coupling between the two points to improve the filtering performance. In current control loop current signal I_{out} is detected by hall sensor, and voltage current is detected by resistive subdivision. Current controller sampling and compensation produces driving signal to control DC/DC full-bridge converter [15,16].

$$V_{Lc}(V_{Nc}) = \frac{V_{LEMI} + V_{NEMI}}{2} \tag{1}$$

$$V_D = \frac{V_{LEMI} - V_{NEMI}}{2} \tag{2}$$

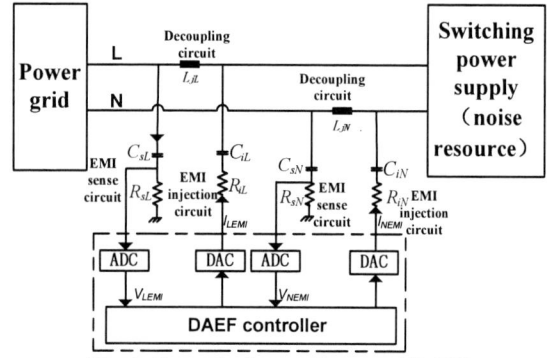

Fig. 1 General scheme of proposed DAEF

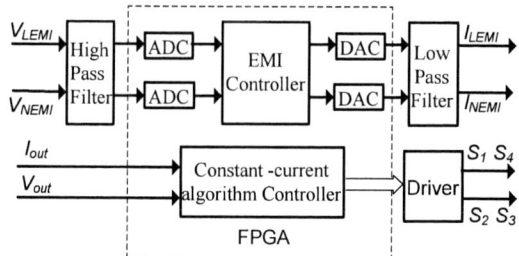

Fig. 2 Control system diagram of arc welding inverter with DAEF

B. Scheme and Principles of Proposed DAEF

DAEF control loop picks up conducted EMI signal to ground on the power line, phase reverse processing by the controller and DA analog output, and at last injects back to power line. The closed-loop block diagram of DAEF is shown in Fig. 3.

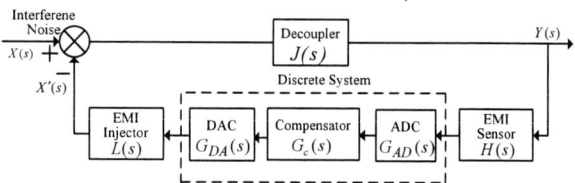

Fig. 3 Closed-loop system diagram of DAEF

In Fig. 3, X(s) is EMI noise signal before filter, Y(s) is EMI noise signal after filter, X'(s) is the injected EMI noise after digital control. In theory, X(s) is equal in magnitude to X'(s), and their phase is opposite. It can achieve full nullification of the EMI noise which is generated by power converter.

EMI sense circuit is a RC high-pass filter, its transfer function $H(s)$ can be expressed as

$$H(s) = \frac{s}{s + \omega_1} \tag{3}$$

where $\omega_1 = 2\pi f_1 = \dfrac{1}{R_S C_S}$ is the corner frequency of high-pass filter. According to GB/T 21419-2013 [17], f_1 should be under 150 kHz to pick EMI signal above 150 kHz.

EMI controller select inverse proportional compensation control, its transfer function $G_c(s)$ can be expressed as

$$G_c(s) = -K \tag{4}$$

D/A convert selects zero-order holder, its transfer function $D_{zoh}(s)$ can be expressed as

$$D_{zoh}(s) = \frac{1 - e^{-sT}}{s} \tag{5}$$

where T is sampling period.

EMI injection circuit is a RC low-pass filter, its transfer function $B(s)$ can be expressed as

$$B(s) = \frac{1}{1 + s / \omega_2} \tag{6}$$

where $\omega_2 = 2\pi f_2 = \dfrac{1}{R_i C_i}$ is the corner frequency of low-pass filter. According to GB/T 21419-2013, f_2 should be about 30MHz to inject necessary EMI signal below 30MHz.

Decoupling inductance select single-turn inductor in order not influence the value of DAEF and the advantages of minimize power consumption, its transfer function $J(s)$ can be expressed as

$$J(s) = L_j s \tag{7}$$

DAEF closed-loop control system transfer function can be expressed as

$$G_{DAEF} = \frac{L_j s^3 + L_j(\omega_1 + \omega_2)s^2 + \omega_1 \omega_2 s}{s^2 + [\omega_1 + \omega_2 + KL_j\omega_2(1 - e^{-sT})]s + \omega_1 \omega_2} \tag{8}$$

C. Control System Modeling of Arc Welding Power Supply

Closed loop control system model of arc welding power supply with DAEF is shown in Fig. 4.

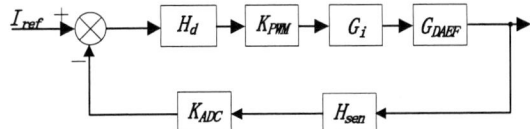

Fig. 4 Control system model of arc welding power supply

In Fig. 4, G_{DAEF} is transfer function of DAEF control system, H_d is transfer function of delay in control system. It can be expressed as

$$H_d = e^{-\tau_d s} \tag{9}$$

where τ is delay time constant of the control system.

K_{ADC} is the gain of A/D converter, its transfer function can be expressed as

$$K_{ADC} = \frac{2^n}{V_{ref}} \times e^{-\tau_{AD} s} \tag{10}$$

where V_{ref} is reference voltage of A/D converter, n is the bit of A/D converter, τ_{AD} is delay time constant of the A/D converter.

$G_i(s)$ is transfer function of current control in DC/DC full-bridge converter, it can be expressed as

$$G_i(s) = \frac{V_{in}(1 + R_o C_f s)}{R_o C_f L_f s^2 + L_f s + R_o} \tag{11}$$

where R_o is load impedance, C_f is output capacitance, L_f is output inductor, V_{in} is DC bus voltage after uncontrolled rectifier.

K_{PWM} is the gain of PWM control, its transfer function can be expressed as

$$K_{PWM} = \frac{2^m}{T_s} \tag{12}$$

where T_s is switching period of the switch tube, m is the bit of counter register.

$H_{sen}(s)$ is the transfer function of current sensor, it consists of sensor gain K_{sen} and a first order RC filter. It can be expressed as

$$H_{sen}(s) = \frac{K_{sen}}{1 + \tau_{RC} s} \tag{13}$$

where sensor gain K_{sen} can be determined by turns ratio and vice side resistance, τ_{RC} is time constant of RC filter.

Open loop transfer function of discrete control model of arc welding inverter power supply system with DAEF can be expressed as

$$T_{uncomop\_i} = H_d \cdot K_{PWM} \cdot G_i \cdot G_{DAEF} \cdot H_{sen} \cdot K_{ADC} \tag{14}$$

III. CONTROL SYSTEM ABILITY ANALYSIS AND COMPENSATOR DESIGN

Analysis the stability of arc welding power system with DAEF, the open-loop bode diagram is the dotted lines as shown in Fig. 5. As can be seen from the figure, magnitude margin of the system is -41.1dB, phase margin is $-16°$, the system is unstable, it needs compensation control.

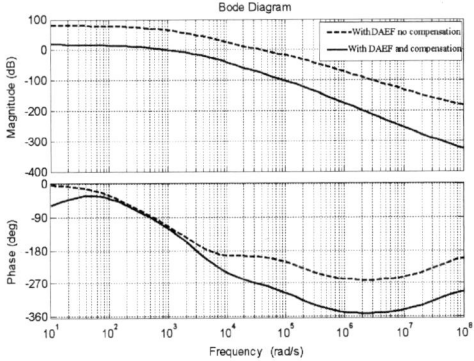

Fig. 5 Open-loop bode diagram of arc welding inverter with DAEF

This paper adopts double pole-double zero PI compensating network system to compensate, its transfer function can be expressed as

$$G_B(s) = \frac{K_p(T_{z1}s + 1)(T_{z2}s + 1)}{s(T_{p1}s + 1)(T_{p2}s + 1)} \tag{15}$$

The compensator can produces two zero points $-1/T_{z1}$ and $-1/T_{z2}$ in the left half of S plane, produces two pole points $-1/T_{p1}$ and $-1/T_{p2}$.

The control model of arc welding power source control system after compensation is shown in Fig. 6.

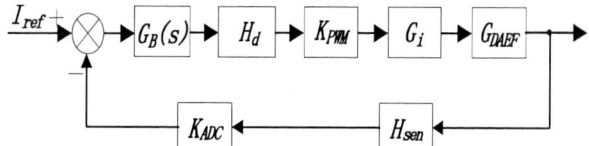

Fig. 6 Control system model of arc welding inverter after compensation

Open loop transfer function of arc welding inverter power supply system with compensation can be expressed as

$$T_{comop\_i} = G_B \cdot H_d \cdot K_{PWM} \cdot G_i \cdot G_{DAEF} \cdot H_{sen} \cdot K_{ADC} \quad (16)$$

Open loop bode diagram after compensation of the arc welding power system is the solid lines as shown in Fig. 5. As can be seen from the figure, after the system compensation, magnitude margin of the system is 20.5dB, phase margin is $64°$, the system is stable. Compared with the bode diagram before compensation in Fig. 5, stability of the system has been increased.

IV. DAEF CONTROL SYSTEM PERFORMANCE ANALYSIS

DAEF control system not only controls the conduction EMI, but also affects the performance of power control system. The open-loop bode diagram of arc welding power system without DAEF is the dotted lines as shown in Fig. 7. The open-loop bode diagram of arc welding power system with DAEF is the solid lines as shown in Fig. 7.

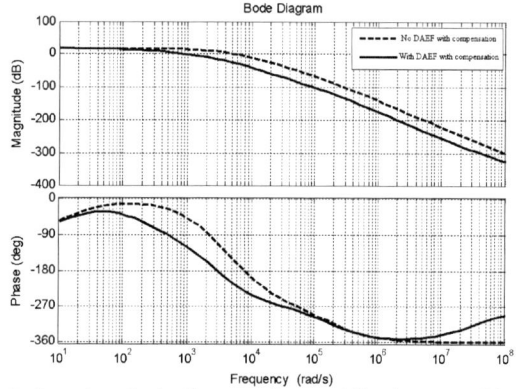

Fig. 7 Open loop bode diagram of arc welding inverter with and without DAEF

As can be seen from the Fig. 7, arc welding power system with DAEF is the dotted lines, its magnitude margin is 7.3dB, phase margin is $20°$. After compensation, magnitude margin improves to 20.5dB, phase margin improves to $64°$. So with DAEF, the stability of the arc welding power supply system has been increased.

Closed loop step response without DAEF of the arc welding power system is the dotted line as shown in Fig. 8. Closed loop step response with DAEF of the arc welding power system is the solid line as shown in Fig. 8.

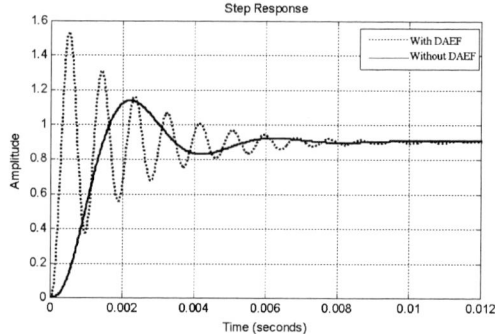

Fig. 8 Close-loop step response of arc welding inverter

As can be seen from the Fig. 8, without DAEF, system overshoot is 53.7%. with DAEF, system overshoot decreased to 14%. Rising time of the system with DAEF becomes longer, but setting time reduced slightly. This shows that dynamic performance of the system with DAEF has been improved as a whole.

It can be seen that adding DAEF control system in arc welding power supply control system not only improves its stability, it can also improves the dynamic performance.

V. EXPERIMENTAL RESULTS

This paper builds a 4.5kW high-frequency arc welding inverter power supply prototype to test the system, its power supply parameters shown in Table I.

Tab. I Parameters of the arc welding inverter

variable	parameters
Output Power	4.5kW
Output Voltage	DC:20V
Output Current	DC:225A
Switch Frequency	20kHz

The main control chip uses FPGA controller to realize control algorithms, high-speed high-precision ADC and DAC module to sampling and conversion EMI signal. the main parameters of DAEF circuit is shown in Table II.

Tab. II Parameters of the DAEF system

circuit	Code name	Parameters
Sampling circuit	ADC	14bit、250MSPS
	DAC	14bit、250MSPS
Sense circuit	R_{sL} / C_{sL}	1kΩ/0.1uF
	R_{sN} / C_{sN}	1kΩ/0.1uF
Rejection circuit	R_{zL} / C_{zL}	30Ω/1nF
	R_{zN} / C_{zN}	30Ω/1nF
Decoupling circuit	L_{zL} / L_{zN}	0.54uH

The complete system test setup is depicted in Fig. 9. Using of ESL3 interference receiver to measurement conducted EMI of the L line and N line.

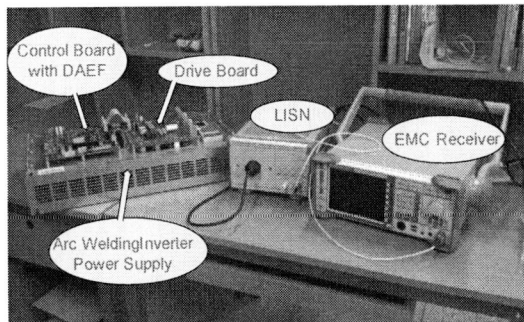

Fig. 9 Test setup of conducted EMI of arc welding inverter with DAEF

A. CM conduction EMI of the system

Common mode conducted EMI in arc welding inverter system is the average conducted EMI of the L line and N line. Then get common mode conducted EMI spectrum of the arc welding inverter system. The spectrum of CM EMI emission without DAEF is shown in Fig. 10. The thick solid line indicates the class I common mode conducted EMI emission limits environment in GB / T 21419-2013.

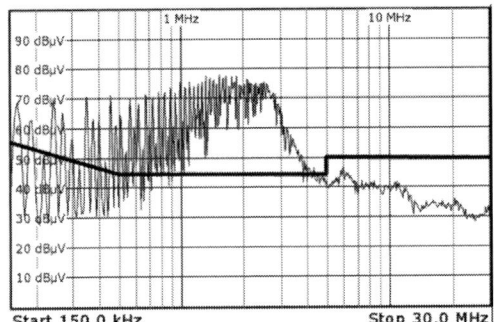

Fig. 10 Spectrum of common mode conducted EMI of arc welding inverter without EMI filter

From Fig. 10, it can be seen that when arc welding inverter power supply without EMI filter, the spectrum of CM conductive EMI is higher than standard limits curve, especially under 5MHz. It needs EMI filter to suppression.

The spectrum of CM EMI emission with proposed DAEF is shown in Fig. 11.

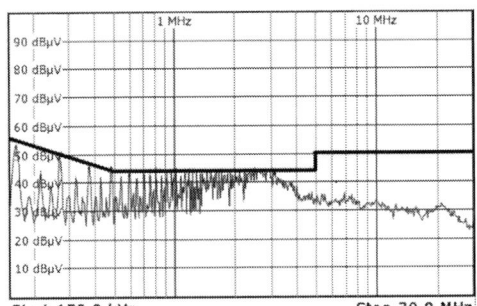

Fig. 11 Spectrum of common mode conducted EMI of arc welding inverter with DAEF

B. System Dynamic Performance

In this paper, dynamic performance of the system is verified by load mutation, the load starts with 75% to 50% and

then rises to 75% from 50%. Without DAEF, the output voltage and switching tube waveform during mutation load are shown in Fig. 12. With DAEF, the output voltage and switching tube waveform during mutation load are shown in Fig. 13.

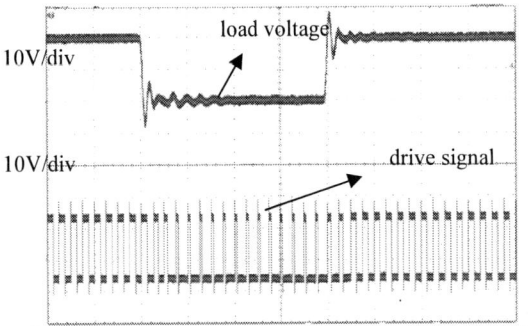

Fig. 12 Transient response to the system without DAEF

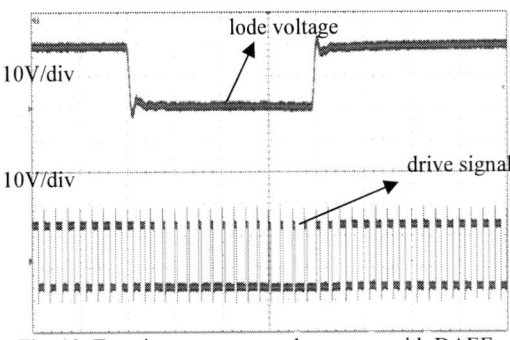

Fig. 13 Transient response to the system with DAEF

From Fig. 12 and Fig. 13, it can be seen that dynamic performance of arc welding inverter power supply with DAEF is better, the overshoot of the system from 46% to 14%, and the setting time is reduced. It shows that after embedded DAEF in arc welding power supply, dynamic performance of the control system has been improved.

VI. CONCLUSION

For arc welding inverters, this paper proposed a digital active EMI filtering method, which can reduces CM conductive EMI and minimizes the size and lowest power consumption, and improves the control system performance in high frequency. The method calculates the EMI signals to ground acquired from L and N lines, separately. It obtains actual CM and DM noise of converters. Then DAEF cancel CM and DM EMI effectively and separately. Compared to the traditional DAEF, the proposed method may adjust the suppression capacity of CM and DM EMI effectively, and improve stability and dynamic performance of the arc welding inverters. Since there is no component in series connect with the power circuit of arc welding inverters, the proposed DAEF solves the problem in filter size and power consumption of traditional welding supply fundamentally. In this paper, the prototype of arc welding inverters is constructed, the suppression effect of CM EMI and improving stability and dynamic performance was verified.

REFERENCES

[1] Datao Nie, Jianping He, Fuxin Wang, et al. "Status and analyze of the digital arc welding power source," Electric Welding Machine, vol. 44, no. 6, pp. 21-25, Jun. 2014.

[2] N. C. Alejandro, M. L. Victor, C. Rosario, et al. "Digital control for an arc welding machine based on resonant converters and synchronous rectification," IEEE Transactions on Industrial Informatics, vol. 9, no. 2, pp. 839-847, May. 2013.

[3] L. Yu-Kang, C. Huang-Jen, S. Tzu-Herng, et al. "A software-based CM and DM measurement system for the conducted EMI," IEEE Transactions on Industrial Electronic, vol. 47, no. 4, pp. 977-978, Aug. 2000.

[4] Shishan Wang, Min Gong, Zhenyang Yu. "Theoretical investigation of mutual transformation between differential mode and common mode noise and its implementation in the design of planar EMI filter," Transactions of China Electrotechnical Society, vol. 29, no. 2, pp. 239-246, Feb. 2014.

[5] Dianbo Fu, Shuo Wang, Pengju Kong, et al. "Novel techniques to suppress the common-mode EMI noise caused by transformer parasitic capacitances in DC-DC converters," IEEE Transactions on industrial electronics, vol. 60, no. 11, pp. 4968-4977, Nov. 2014.

[6] D. Shin, S. Kim, G. Jeong, et al. "Analysis and design guide of active EMI filter in a compact package for reduction of common-mode conducted emissions," IEEE Transactions on Electromagnetic Compatibility, vol. PP, no. 99, pp. 1-12, Feb. 2015.

[7] Xinli Chang, Yuehong Yang, Wenjie Chen, et al. "Implementation of a novel active common-mode filter used in DC-DC converters," in General Assembly and Scientific Symposium, 2014, pp. 1-4.

[8] Shuo Wang, Y. Y. Maillet, Fei Wang, et al. "Investigation of hybrid EMI filters for common-mode EMI suppression in a motor drive system," IEEE Transactions on Power Electronics, vol. 25, no. 4, pp. 1034-1045, Apr. 2010.

[9] Wenjie Chen, Xu Yang, Zhaoan Wang. "A novel hybrid common-mode EMI filter with active impedance multiplication," IEEE Transactions on Industrial Electronics, vol. 58, no. 5, pp. 1826-1834, May. 2011.

[10] Wenjie Chen, Xu Yang, Jing Xue and Fred Wang. "A novel filter topology with active motor CM impedance regulator in PWM ASD system", IEEE Transactions on Industrial Electronics, vol. 61, no. 12, pp.6938-3946, Dec. 2014.

[11] D.Hamza, MeiQiu, P.K.Jain. "Implementation of an EMI active filter in grid-tied PV micro-inverter controller and stability verification," in 2012-38th Annual Conference on IEEE Industrial Electronics Society, 2012, pp. 477-482.

[12] D. Hamza, Mei Qiu. "Digital active EMI control technique for switch mode power converters," IEEE Transactions on Electromagnitic Compatibility, vol. 55, no. 1, pp. 81-88, Feb. 2013.

[13] D. Hamza, MeiQiu, P. K. Jain. "Application and stability analysis of a novel digital active EMI filter used in a grid-tied PV micro-inverter module," IEEE Transactions on Power Electronics, vol. 28, no. 6, pp. 2867-2874, Jun. 2013.

[14] D. Hamza, M. Pahlevaninezhad, P. K. Jain. "Implementation of a novel digital active EMI technique in a DSP-based DC–DC digital controller used in electric vehicle(EV)," IEEE Transactions on Power Electronics, vol. 28, no. 7, pp. 3126-3137, Jul. 2013.

[15] Yihan Chen, Zheng Wei, Chunying Gong. "Study of output impedance and control loop optimization theoretical for average current mode control phase-shift full bridge DC-DC converter," Transactions of China Electrotechnical Society, vol. 28, no. 4, pp. 43-49, Apr. 2013.

[16] Lin Wu, Zhigang Liu, Xiang Hong. "Comparison and analysis of power control characteristic for isolated bidirectional full-bridge DC-DC converter," Transactions of China Electrotechnical Society, vol. 28, no. 10, pp. 179-187, Oct. 2013.

[17] Standardization Administration of the People's Republic of China. GB/T 21419-2013 Safety of transformers, reactors, power supply units and combinations thereof - EMC requirements [S]. Beijing: China Standards Press, 2013.

A Thermo-Sensitive Electrical Parameter with Maximum dI_C/dt during Turn-off for High Power Trench/Field-stop IGBT Modules

Yuxiang Chen, Haoze Luo, Wuhua Li, Xiangning He
College of Electrical Engineering
Zhejiang University
Hangzhou, China
Email: woohualee@zju.edu.cn

Jun Ma, Guodong Chen, Ye Tian, Enxing Yang
Technology Center, Power Transmission & Distribution
Group
Shanghai Electric
Shanghai, China
Email: chengd@shanghai-electric.com

Abstract—Junction temperature monitoring of IGBT modules is crucial for power devices in high power applications. In this paper, a thermo-sensitive electrical parameter based on the maximum collector current falling rate $-dI_C/dt_{max}$ for trench/field-stop IGBT junction temperature extraction is outlined. The inherent monotonic relationship between the maximum collector current falling rate and chip temperature is explored. Fortunately, the intrinsic parasitic inductance L_{eE} of IGBT module can be directly used as the maximum collector current falling rate sensor. Experimental measurements of Trench/Field-Stop IGBT module rated at 1700V/3600A are implemented to show that the dependence between the IGBT chip temperature and maximum collector current falling rate approximates in a linear way. This indicates that the collector current falling rate is a potential thermo-sensitive electrical parameter (TSEP) for high power IGBT modules junction temperature extraction.

Keywords—*Thermo-sensitive electrical parameter; junction temperature extraction; trench/field-stop IGBTs; maximum collector current declining rate*

I. INTRODUCTION

High power IGBT modules are the core components in the voltage-source-converter based high voltage direct current (VSC-HVDC) systems [1]. It is reported that the power device failure is the major failure reason for the power electronic conversion systems and nearly 55% of device failures are thermally induced [2].This means the real-time junction temperature monitor of IGBT modules is of great importance for the health management for VSC-HVDC systems. To avoid the severe IGBT degradation or even destruction, the special attention should be paid to the junction temperature characteristics during operation.

Generally speaking, three main methods are currently used to measure and/or predict the junction temperature of IGBT modules, including the optical-based, physical contact-based and thermo-sensitive electrical parameter (TSEP)-based

approaches [3]. Although the infrared radiation (IR) camera and thermocouple, as typical representations for the optical-based and physical contact-based methods respectively, can be accurate, they need a destructive modification of power modules. And also taking the relatively slow dynamic response of both methods into consideration, the industrial applications of optical-based and physical contact-based methods in practical system are limited [4]. However, the TSEP-based approach is taken as a promising candidate for the IGBT module junction temperature monitor because it employs the power device itself as a temperature sensor with high accuracy and fast response.

Considering the inherent relationship between the IGBT module junction temperature and its electrical characteristics, TSEP methods use a temperature-dependent electrical parameter of the device as a junction temperature indicator. So far, the introduced TESP methods can be categorized into the static-characteristic related and dynamic-characteristic related techniques. The static TSEP methods measure the on-state or off-state characteristics of the power devices to extract the junction temperature, such as the widely used low current injection method and saturation current method [4]. For the low current injection method, it requires an auxiliary current source injection circuit to ensure the current flowing through the device is low enough to avoid the self-heating effects [5]. And for the saturation current method, the gate-emitter voltage v_{ge} is needed to be maintained a slightly higher than the threshold voltage V_{th} of the power devices, which is far from the normal operation value [4]. Clearly, both former static TSEP methods require special device operation which interrupts the normal working mode of the IGBT modules. As a matter of fact, it is common for most of static-characteristic related TSEP approaches that has strong operation-mode dependency. However, the dynamic TSEP methods detect the device's junction temperature during the routine switching states and have no requirement for special operation mode. As a result, the dynamic-characteristic related TSEP approaches make the

---

This work is sponsored by the National Basic Research Program of China (973 Program 2014CB247400) and the National Nature Science Foundation of China (51490682).

IGBT module junction temperature monitor much easier to implement.

In this paper, the dynamic characteristics during the collector current falling transition for Trench/Field-Stop IGBT modules is disclosed in the voltage-source converters with typical inductive loads. A thermo-sensitive electrical parameter with maximum collector current falling rate $-dI_C/dt_{max}$ for the trench/field-stop IGBT junction temperature extraction is outlined. Owing to the intrinsic parasitic inductance L_{eE} between the Kelvin and power emitter of the high power IGBT modules, $-dI_C/dt_{max}$ during the collector current falling transition can be directly acquired by monitoring the induced voltage v_{eE} across L_{eE}, which is readily detectable. And by investigating the device structure properties, the temperature-dependency of the maximum collector current falling rate $-dI_C/dt_{max}$ and its operation condition-dependency are discovered and evaluated experimentally.

The paper is structured as follows. In Section II, the typical collector current falling transition characteristics of the trench/field-stop IGBT is highlighted. In Section III, the temperature-dependency of the maximum collector current falling rate $-dI_C/dt_{max}$ and its operation condition-dependency are detailed explored. In Section IV, the experimental measurements are implemented to validate the effectiveness of the proposed method. Finally, the conclusions drawn from the investigation are provided in Section V.

II. TYPICAL COLLECTOR CURRENT FALLING TRANSITION CHARACTERISTICS OF TRENCH/FIELD-STOP IGBTs

A. Switching Characteristics Analysis of Trench/Field-Stop IGBTs

The IGBT technology improvements have led to higher voltage-blocking capability and lower overall loss. The trench/field-stop structure is the state-of-the-art structure for high power IGBTs, which combines both the trench cell and field-stop concept. A typical trench/field-stop IGBT structure is illustrated in Fig.1.

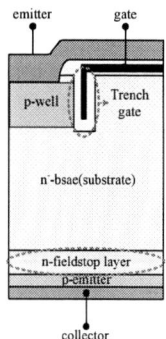

Fig. 1. *Basic trench/field-stop IGBT Structure*

In the case of an inductive load, regardless of the IGBT structures, the collector-emitter voltage rising transition and collector current falling transition are two characteristic transitions of the IGBT turn-off process and dominate the device's dynamic performance. *Fig.2* shows the typical trench/field-stop IGBT inductive turn-off waveforms.

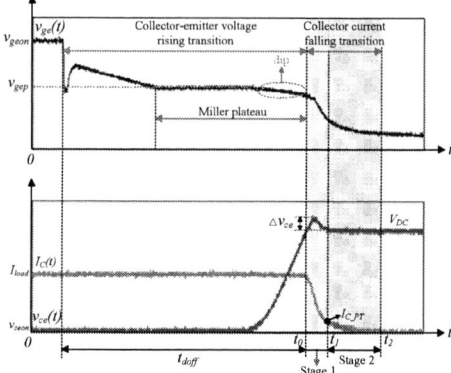

Fig..2. *Typical trench/field-stop IGBT inductive turn-off process*

Benefiting from the advanced trench cell concept, the n‾-base region of a typical trench/field-stop IGBT module is characterized by a more pronounced carrier storage effect during the on-state. As a result, different turn-off behavior of the trench/field-stop IGBT module can be discovered, compared with former generations. The inductive turn-off waveform displays a dip (indicated in *Fig.2*) in the gate voltage v_{ge} at the end of the Miller plateau [6]. In general, for the trench/field-stop IGBTs, it can be considered that the MOS channel is fully pinched off before the falling of collector current I_C. That is to say, during the collector current falling transition, the collector current is thus totally sustained by the storage carrier extraction mechanism in the n‾-base region. As a consequence, the collector current falling slope $-dI_C/dt$ is intrinsically limited rather than controlled by the gate resistance. Based on this assumption, the collector current falling transition can be divided into two stages and described as follow, illustrated in *Fig.2*.

Stage 1 [t_0-t_1]: At t_0, the collector current I_C starts to fall from the load current I_{load} and the collector current falling transition begins. At this stage, there is still some part of stored charges left in the n‾-base region near the field-stop layer. And the collector current I_C drops sharply until to a level $I_{C\ PT}$, which denotes the transient value of the collector current when the left storage charges in the n‾-base region are fully extracted. This process is mainly governed by the remaining storage carrier extraction mechanism in the n‾-base region, which is triggered by the electron injection into the p-emitter.

Stage 2 [t_1-t_2]: At t_1, the collector current I_C reaches $I_{C\ PT}$ and the space charge region (SCR) almost extends through the entire n‾-base region. At this stage, there is nearly no stored charge left in n‾-base region, which still presents in the field-stop layer. This means that the falling characteristics of I_C are mainly dominated by the carrier recombination mechanism in the field-stop layer.

B. Feasibility of TESP Monitor Considering Internal Parasitic Inductance

For the standard high power IGBT modules, the traditional power emitter terminal works as a common part for both the drive loop and power loop of the device. Thus, the parasitic inductance of the power emitter terminal is integrated in both the gate loop and power loop. During switching transitions, the collector current changing rate would result in an induced voltage across the parasitic inductance. Because this induced voltage is opposite to the drive voltage, unintended parasitic turn-off and turn-on processes may occur [7]. In order to overcome this problem, an additional terminal named as Kelvin emitter is provided to the emitter in the new generation of high power IGBT modules. The Kelvin emitter is used as a part of the drive loop to decouple from the power loop. The high power IGBT module rated at 1700V/3600A from Infineon is used as an example to illustrate its package and equivalent circuit, as shown in *Fig.3*. L_{Cc} is the inherent parasitic inductance between the collector and auxiliary collector, L_{eE} is the inherent parasitic inductance between Kelvin and traditional power emitters. Due to the Kelvin emitter, the drive loop and power loop of the IGBT module in *Fig.3 (b)* are completely independent from each other. In addition, C_{ge}, C_{gc}, and C_{ce} represent the intrinsic parasitic capacitors of the IGBT module.

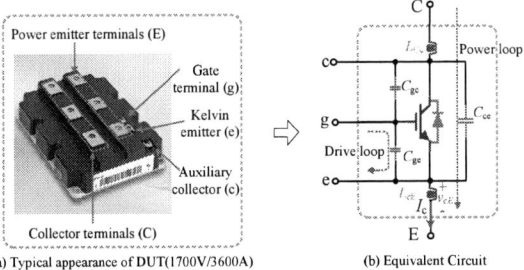

(a) Typical appearance of DUT(1700V/3600A)   (b) Equivalent Circuit

Fig. 3. *High power IGBT module package structure/equivalent circuit in terms of parasitic inductance*

Thanks to the advanced Kelvin emitter concept of high power IGBT modules, an intrinsic parasitic inductance L_{eE} is introduced between Kelvin and traditional power emitters as plotted in *Fig.3 (b)*. Because of just integrating in the device power loop, L_{eE} can be used as a collector current changing rate dI_C/dt sensor during the device switching transition. Thus, dI_C/dt can be directly acquired by monitoring the induced voltage v_{eE} across the parasitic inductance L_{eE}. Moreover, the maximum collector current falling rate $-dI_C/dt_{max}$ can be simply represented by the peak value of the induced voltage v_{eEmax} during the collector current falling transition

$$v_{eE\max} = -L_{eE} \frac{dI_C}{dt}\bigg|_{\max} \quad (1)$$

As a result, extra high current sensor is not required for the maximum collector current falling rate detection and the IGBT module itself can be used as an indicator to extract the maximum collector current falling rate.

III. TEMPERATURE DEPENDENCY CHARACTERIZATION OF MAXIMUM COLLECTOR CURRENT FALLING RATE

The transient collector current falling model should be built to explore the temperature-dependency of the maximum collector current falling rate $-dI_C/dt_{max}$ and its operation condition-dependency.

A. Collector current falling model during stage 1 based on carrier extraction

Due to the thin and relatively low doped p-emitter design, the trench/field-stop IGBT inherits the low p-emitter injection efficiency attribute of the traditional NPT IGBT [8]. Therefore, for the trench/field-stop IGBTs, the electron current supported by the electron backside injection into the p-emitter is the dominant part of the total p-emitter current. Moreover, because the trench/field-stop IGBT is characterized by the fact that the MOS channel pinches off before the start of *Stage 1*. As a result, all electrons, which need to maintain the continuous electron backside injection during *Stage 1*, are supplied by the remaining storage electron extraction from the n⁻-base region. Consequently, the storage carrier extraction behavior of the n⁻-base region is the dominating mechanism to govern the characteristics of *Stage 1*.

The stored carrier extraction behaviors in the n⁻-base region during *Stage 1* are described in *Fig.4*. Where the remaining excess holes in the n⁻-base region are extracted to the emitter side and swept out via the reverse-biased p-well-n⁻-base junction, the stored electrons in the n⁻-base region are extracted to the collector side to maintain the needed electron backside injection, and the electrical field $E(w)$ is built in the space charge region (SCR) in which the stored carriers are exhausted.

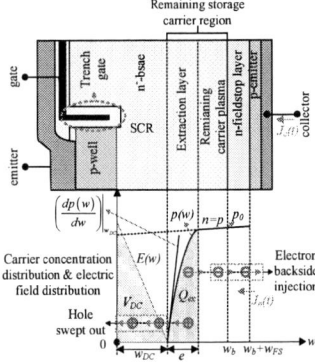

Fig. 4. *Storage carriers Extraction process in Stage 1*

Based on the carrier extraction behavior, during *Stage 1*, the remaining storage carrier region of the n⁻-base can be divided into two parts [9], as indicated in *Fig.4*: one is a carrier-extraction taking place layer denoted as the extraction layer; the other is the remaining carrier plasma which keeps the same carrier distribution as the on-state.

That is to say, actually it is the stored carrier behaviors in the extraction layer that determine the collector current falling characteristics of *Stage 1*. What is more, the extraction layer is defined by three parameters, as illustrated in *Fig.4*. They are the carrier concentration distribution $p(w)$, the start point location of the extraction layer w_{DC}, and the extraction layer width e.

For the extraction layer, considering the current continuity equation and charge control equation, the transient collector current density during *Stage 1* $J_{C1}(t)$ can be derived as

978-1-4673-9551-9/16 $31.00 © 2016 IEEE   501

$$
\begin{cases}
J_{C1}(t) = \dfrac{1+b}{b} q D_a \dfrac{2 p_0}{e} \\[2mm]
e = \sqrt{e_0^{\,2} + \dfrac{40}{7} D_a t} \\[2mm]
e_0 = \dfrac{1+b}{b} q D_a \dfrac{2 p_0}{J_C}
\end{cases}
\tag{2}
$$

where $b = \mu_n/\mu_p$ is the ratio of the mobilities, q is the electronic charge (1.6×10^{-19}C), D_a is the ambipolar diffusivity, p_0 the stored carrier concentration at $w = w_b$, e_0 is the initial extraction layer width of *Stage 1*.

B. Collector current falling model during stage 2 based on carrier recombination

At *Stage 1*, the stored carrier concentration in the n$^-$-base region decreases with the decline in I_C by carrier extraction until the SCR almost extends through the entire n$^-$-base region. And then *Stage 2* starts, as presented in *Fig.5*, where w_{DC_fin} and e_{fin} are the width of the SCR and extraction layer at the beginning of *Stage 2*, and $w_{DC\_fin} + e_{fin} = w_b$, w_b is the width of the n$^-$-base region. It indicates that the extraction layer and remaining carrier plasma completely overlap with each other. In other words, the extraction layer cannot move to the collector side any more, hence, the carrier extraction mechanism is no more applicable. As a result, during *Stage 2*, it's the carrier recombination behaviors of both the n$^-$-base region and field-stop layer that contribute to the remaining storage carrier sweeping out [10].

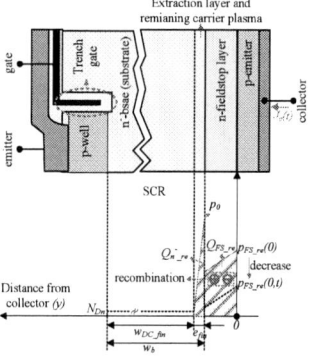

Fig. 5. *Excess carrier profile of the n$^-$-base region and the field-stop layer during stage 2*

Typical, most storage carriers in the n$^-$-base region have already been exhausted during *Stage 1*. Thus, the remaining stored carrier charge left in the n$^-$-base region Q_{n-_re} at the start of *Stage 2* is negligible compared with the remaining stored carrier charge left in the field stop layer Q_{FS_re}, as indicated in *Fig.5*. In general, the excess carrier lifetimes of the n$^-$-base region and field stop layer are also incomparable [10]. The one of the n$^-$-base region is usually much larger. Consequently, the recombination behavior of the remaining storage carriers in the field stop layer plays the dominating role in governing the collector current falling process of *Stage 2*.

For the filed stop layer, considering the carrier charge continuity equation, the transient collector current density during Stage 2 $J_{C2}(t)$ can be expressed as

$$
\begin{cases}
J_{C2}(t) = J_{C\_PT}\, e^{-2t/\tau_{HL\_FS}} \\[3mm]
\sqrt{\dfrac{2\varepsilon_0 \varepsilon_{si} V_{DC}}{q\left(N_{Dn} + \dfrac{J_{C\_PT}}{q v_{psat}}\right)}} + \dfrac{1+b}{b} q D_a \dfrac{2 p_0}{J_{C\_PT}} = w_b
\end{cases}
\tag{3}
$$

where J_{C_PT} is the transient collector current density when the Stage 2 starts, τ_{HL_FS} is high level carrier lifetime in field stop layer, N_{Dn} is the doping concentration of the n$^-$-base region and v_{psat} is the hole saturation velocity, V_{DC} is the bus voltage.

C. Maximum collector current falling rate dependence analysis

With the above-derived model of the collector current falling transition, it is possible to calculate the transient value of the collector current density $J_C(t)$ for the given bus voltage V_{DC} and load current density J_{load}. The governing equations describing the collector current falling characteristics of the both stages are implemented in MATLAB/Simulink, in terms of the typical structure parameters of the 1200V/1700V series trench/field-stop IGBT modules shown in *Table I* [10-11].

TABLE I.   DEVICE STRUCTURE PARAMETER

Parameter	Symbol	Value
Base doping	N_{Dn}	10^{13}-10^{14}cm^{-3}
Base width	W_b	120-140µm
Carrier lifetime in the n$^-$-base region	τ_{pb}	10µs
Field-stop layer doping	N_{DF}	10^{15}-10^{16}cm^{-3}
Field-stop layer width	τ_{np}	7-10µm
Carrier lifetime in the p-emitter	W_{FS}	0.1µs
p-emitter doping	N_{Ap}	10^{16}-10^{17}cm^{-3}

In order to examine the relationship between the device junction temperature T_j and its electrical characteristics during the collector current falling transition, the temperature-dependent factors are also incorporated into the governing equations, given in *Table II* [12].

Details of the temperature and operation condition dependencies of the corresponding parameters within the governing equations for the collector current falling transition is illustrated in *Fig.6*. The effects of the load current density J_{load} are mainly imposed by the static concentration characteristic p_0 of the stored carrier. And the influence of the bus voltage V_{DC} is successfully incorporated into the model through J_{C_PT}. What is more, the relevant temperature-dependent parameters are also clear here, as listed in *Table II*. Based on the parameters in *Table II* and the dependency in *Fig.6*, the simulation with MATLAB/Simulink can be implemented to show the collector current falling transition with typical 1700V trench/field-stop IGBT modules.

TABLE II.   DEVICE STRUCTURE PARAMETER

Parameter	Temperature – dependent expression
Hole saturation velocity ($v_{p,sat}$)	$v_{p,sat}(T_j)=(8.36\times10^6)(300/T_j)^{0.52}$
Electron mobility (μ_n)	$\mu_n(T_j)=1500(300/T_j)^{2.5}$
Hole mobility (μ_p)	$\mu_p(T_j)=450(300/T_j)^{2.5}$
Electron diffusivity (D_n)	$D_n(T_j)=\mu_n(kT_j/q)$
hole diffusivity (D_p)	$D_p(T_j)=\mu_p(kT_j/q)$
Ambipolar diffusivity (D_a)	$D_a(T_j)=2D_nD_p/(D_n+D_p)$
Electron and hole lifetime ($\tau_{n,p}$)	$\tau_{n,p}(T_j)=\tau_{n,p}(300K)(T_j/300)^{1.7}$

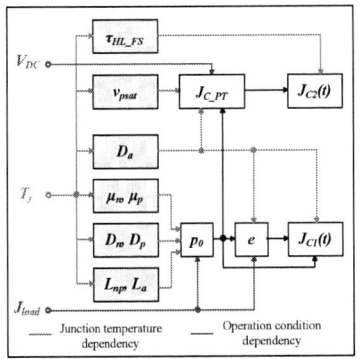

Fig. 6. *Details of dependencies of the collector current falling transition*

Considering the bus voltage of 900V as an example, the $-dJ_C/dt_{max}$ is extracted from the $J_C(t)$ waveforms with varying T_j from 300K to 400K and J_{load} from 20A/cm² to 100A/cm², indicated in *Figs. 7* and *8*.

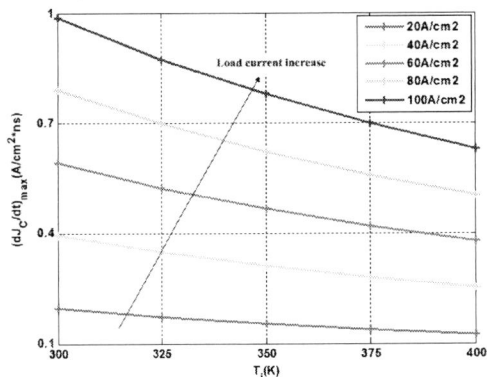

Fig. 7. *$-dJ_C/dt_{max}$ values and linear fitted curves against T_j with changing J_{load}*

In *Fig.7*, for fixed J_{load}, $-dJ_C/dt_{max}$ decreases with increasing T_j with a near linear relationship. And this negative temperature-dependency of $-dJ_C/dt_{max}$ is more pronounced at a high J_{load}. Specifically, the sensitivity ratio changes from -0.713A/μs·K at 20A/cm² to -3.564 A/μs·K at 100A/cm².

In *Fig.8*, not only the same changing trend of $-dJ_C/dt_{max}$ temperature-dependency against J_{load} can be found, the proportional load current density-dependency of $-dJ_C/dt_{max}$ is also clearly illustrated.

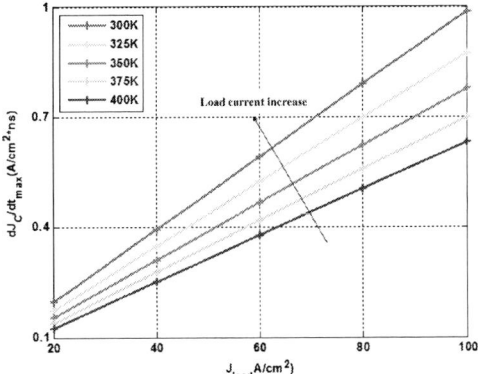

Fig. 8. *$-dJ_C/dt_{max}$ values and linear fitted curves against J_{load} with changing T_j*

Generally, $-dJ_C/dt_{max}$ has a linear and negative temperature-dependency in wide bus voltage and load current density ranges. And during this situation, a high load current density level can enable the temperature-dependency of $-dJ_C/dt_{max}$ to further strengthen. As a result, $-dJ_C/dt_{max}$ is a promising TSEP for trench/field-stop IGBT modules, especially at large current applications.

IV.   EXPERIMENTAL INVESTIGATION

In order to validate temperature-dependency of $-dI_C/dt_{max}$ and its operation condition-dependency during the collector current falling transition, an Infineon trench/field-stop IGBT module rated at 1700V and 3600A is tested under inductive switching conditions. A heater is used to uniformly heat the IGBT and its surroundings through its baseplate in the sealed box over a long period before experimental data is taken.

The measured waveforms of the induced voltage v_{eE} on the inherent parasitic inductance L_{eE} during the collector current falling transition are illustrated in *Fig.9*. For the given bus voltage of 900V and load current of 1100A, the junction temperature of the device under test is set to 25℃.From an estimated L_{eE} of nearly 3nH, a -13.2V peak value of v_{eE} during the collector current falling transition corresponds to $-dI_C/dt_{max}$ of about -4.4A/ns, for the given test conditions.

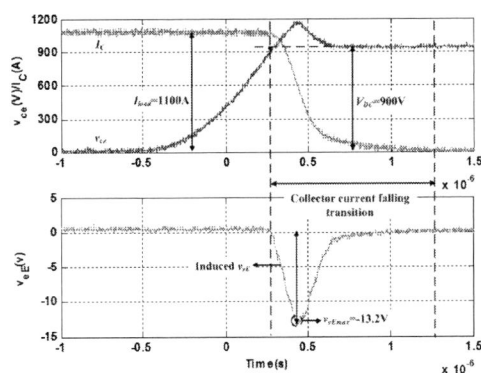

Fig. 9. *Experimental waveforms of turn-off i_c, v_{ce} and v_{eE}*

Fig.10 shows measured v_{eEmax} results against I_{load} of varying T_j, at given V_{DC}. It can be clearly seen from this figure that v_{eEmax} decreases with the increasing T_j for given test conditions.

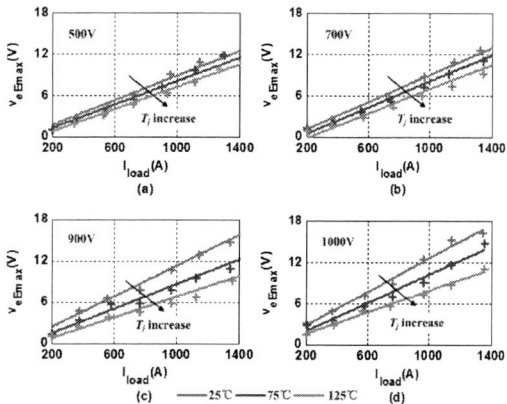

Fig. 10. *Tested v_{eEmax} values and linear fitted curves against I_{load} of varying T_j at given V_{DC}*

In *Fig.10*, a positive bus voltage-dependency of the v_{eEmax} can also be identified. What is more, the temperature sensitivity of the v_{eEmax} increases significantly when the V_{DC} reaches 900V. And within this situation, a large I_{load} enable the v_{eEmax} temperature-dependency to further strengthen.

The negative temperature-dependent characteristics of v_{eEmax} are illustrated with more details in *Fig.11*.

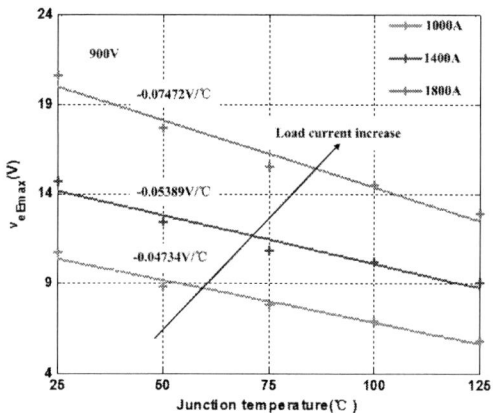

Fig. 11. *Tested v_{eEmax} values and linear fitted curves against T_j of varying I_{load} at given V_{DC}*

As shown in *Fig.11*, a high linearity relationship between v_{eEmax} and T_j can be identified for each given test conditions. And for fixed V_{DC}, the temperature sensitivity of v_{eEmax} has a significant increase with the rising of I_{load}. At the given V_{DC} of 900V, the temperature sensitivity increase from -0.04734V/□ to -0.07472 V/□ when I_{load} changes from 1000A to 1800A

Without interfering the switching dynamic characteristics of the IGBT module, $-dI_C/dt_{max}$ can be directly acquired by monitoring the peak value v_{eEmax} of the induced voltage on the inherent parasitic inductance L_{eE}. It is more convenient for the detect circuit implementation. And with high linearity and sensitivity at the high voltage and large current conditions, $-dI_C/dt_{max}$ is a feasible candidate for extracting the junction temperature of trench/field-stop IGBT modules, especially for high power applications.

V. CONCLUSION

This paper proposes a junction temperature extraction approach for the high power Trench/Field-Stop IGBT modules that utilizes the maximum collector current falling rate - dI_C/dt_{max} as a thermo-sensitive electrical parameter. Benefiting from the advanced package with Kelvin terminal, an inherent parasitic inductance L_{eE} has been introduced which can transfer the high collector current I_C information to the lower induced voltage v_{eE} information. Then $-dI_C/dt_{max}$ can be directly extracted from the peak value v_{eEmax}. As a result, a more cost-effective detection circuit can be implemented. The junction temperature-dependency of $-dI_C/dt_{max}$ and its operation condition-dependency have been disclosed and validated with experimental measurements of a high power IGBT module rated at 1700V/3600A from Infineon. The high linearity relationship between $-dI_C/dt_{max}$ and T_j has been identified, with a relatively high sensitivity for the high bus voltage and large collector current applications. Consequently, $-dI_C/dt_{max}$ is a hopeful TSEP for the real-time junction temperature detecting of high power Trench/Field-Stop IGBT modules, especially in high power applications.

REFERENCES

[1] S. Debnath, Jiangchao Qin, B. Bahrani, and M. Saeedifard, "Operation, control, and applications of the modular multilevel converter: A review," IEEE Trans. Power Electron., vol. 30, no. 5, pp. 37-53, 2014.

[2] J1211, "Handbook for robustness validation of automotive electrical/electronic modules," 2009.

[3] Haoze Luo, Yuxiang Chen, Pengfei Sun, Wuhua Li and Xiangning He, " Junction Temperature Extraction Approach with Turn-off Delay Time for High-voltage High-power IGBT Modules",IEEE Trans. Power Electron., 2015.(online published)

[4] Haoze Luo, Wuhua Li and Xiangning He, "Online High-power P-i-N diode Chip Temperature Extraction and Prediction Method With Maximum Recovery Current di/dt" ,IEEE Trans. Power Electron., vol. 30, no. 5, pp. 2395-2404, 2014.

[5] Dawei Xiang, L. Ran, P. Tavner and Shaoyong Yang, "Condition Monitoring Power Module Solder Fatigue Using Inverter Harmonic Identification," IEEE Trans. Power Electron., vol. 27, no.1, pp. 235-247, 2012.

[6] B. Reinhold, "IGBT-driver circuit for desaturated turn-off with high desaturation level," U. S. Patent 7768337 B2, Aug. 2010.

[7] F. Stueckler and E. Vecino, "CoolMOSTM C7 650V switch in a Kelvin source configuration," Infineon Technologies Austria AG Press, 2013.

[8] R. Kraus and K. Hoffmann, "An analytical model of IGBTs with low emitter efficiency," in Proc. 5th Int. Sym. Power Semiconductor Devices and ICs, Monterey, May 1993, pp. 30-34.

[9] J. Schumann and H. Eckel, "Charge carrier extraction IGBT model for circuit simulators," 15th Int. Conf. Power Electronics and Motion Control, Novi Sad, Sept. 2012, pp. 1-7.

[10] B. Baliga, Advanced High Voltage Power Device Concepts. New York: Springer Verlag, 2012.

[11] A. Volke, M. Baessler, F. Umbach, F. Hille, W. Rusche and M. Hornkamp, "The new power semiconductor generation: 1200V IGBT4 and EmCon4 Diode," in Proc. India Int. Conf. Power Electronics, Chennai, Dec. 2006, pp. 77-82.

[12] K. Vinod, Insulated Gate Bipolar Transistor IGBT Theory and Design. New Jersey: Wiley-IEEE, 2003.

A Software Frequency Response Analysis Method to Monitor Degradation of Power MOSFETs in Basic Single-Switch Converters

[1]Serkan Dusmez, *Student Member, IEEE,* [1,2]Manish Bhardwaj, *Student Member, IEEE,* [1]Lei Sun, *Student Member, IEEE,* [1]Bilal Akin, *Senior Member, IEEE*

[1]Electrical and Computer Science Department, Power Electronics and Drives Laboratory
University of Texas at Dallas, Richardson, TX, USA
[2]Texas Instruments, Houston, TX, USA
serkan.dusmez@utdallas.edu, mbhardwaj@ti.com, lxs141130@utdallas.edu, bilal.akin@utdallas.edu

Abstract— **The efforts on more reliable power conversion systems are gaining momentum in the recent years. Majority of the studies concerning reliability of power switches focus on the package related failures, mainly caused by the thermal stress. The basic failure precursor for this type of stress has been identified as increased on-state resistance for power MOSFETs in recent literature. However, calculation of on-state resistance requires a voltage sensing circuit which can block the high voltage across the switch during off-state not to damage the measurement or control unit. This also limits the implementation as it requires additional hardware. This paper proposes a software frequency response analyzing algorithm to determine the health status of the power MOSFETs through evaluating the variation in the plant model using the same DSP that is used for control purposes. The proposed concept has been analyzed for basic single-switch converters, and experimentally verified on a dc/dc boost converter.**

Keywords—failure diagnosis, power MOSFET, frequency response, health monitoring, power converter.

I. INTRODUCTION

The power switches of power electronics converters undergo both mechanical and environmental stresses which causes wear out over time. The environmental factors such as humidity, ambient temperature, and thermal swing due to power cycling all contribute to increased rate of package related failures [1]-[5]. For instance, the temperature swing amplitude of the traction drive can be up to 80°C degree in electric trains [6], which significantly reduces the reliability of the packaging of the power switches. The failure precursors for both IGBTs and power MOSFETs have been exhaustively researched in recent literature [4], [7]-[11]. It has been shown that the on-state resistance of MOSFETs and saturation voltage of IGBTs are the two basic failure precursors. The measurements of these variables under continuous operation are critical for health monitoring of power switches.

The measurements of these precursors are quite similar, where the switch current has to be known for power MOSFETs in addition to the voltage drop across it. In switch-

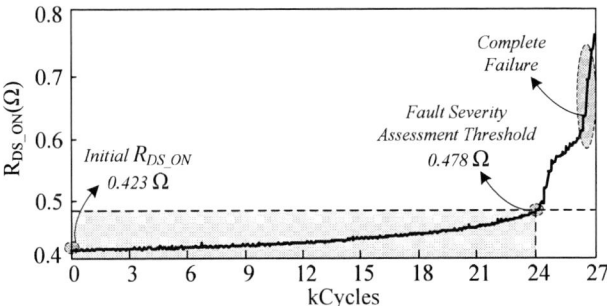

Fig. 1. $R_{ds,on}$ variation of sample #Rb1 versus thermal cycles.

mode power converters, the current of the switch is usually equal to the current of the energy storage unit, most commonly inductor, when the switch is on. Since this state variable is the control objective of interest, the feedback is available in most of the circuits. On the other hand, the voltage drop measurement is quite trivial as the voltage across the switch is very high at blocking state. The direct measurement through voltage dividers or using an isolated voltage sensor could result in low resolution measurements.

To alleviate this problem, an off-state high-voltage blocking circuit can be used; however, this approach is not too practical and cost-effective for many applications. Therefore, it is of interest to detect the variation of the on-state resistance without additional sensors or circuitry, and only using the on-board sensors utilized for control purposes such as average mode current control.

With the recent advances in DSP technology, it is possible to extract the transfer plant models in real time. The changes in the transfer plants can indicate the parameter shifts in the components caused by wear-out. In this paper, the frequency response of the inductor current in basic dc/dc converters with respect to duty cycle perturbation is analyzed to detect the on-state resistance variation. The proposed concept has been experimentally verified on a 315V/250W dc/dc boost converter prototype.

978-1-4673-9551-9/16 $31.00 © 2016 IEEE

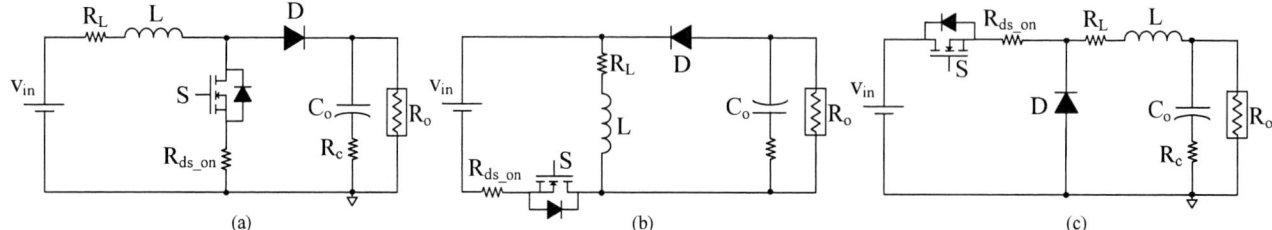

Fig. 2. Analyzed single-switch dc/dc converters; (a) boost, (b) buck/boost, (c) buck.

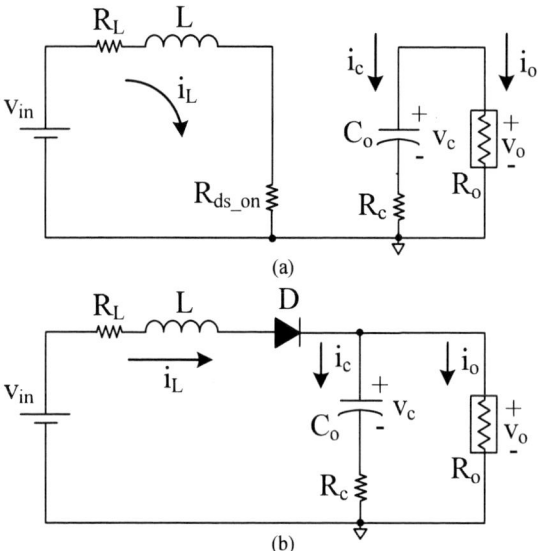

Fig. 3. Boost converter equivalent circuits when the switch is; (a) on, (b) off.

II. FAILURE PRECURSOR ON-STATE RESISTANCE VARIATION

In previously published work, an accelerated thermal aging platform has been designed and IRFP340 11A/400V discrete package MOSFETs have been thermally cycled under different thermal swings [10]. In Fig. 1, the result of on-state resistance variation under swing amplitude of 80°C and a maximum junction temperature of 160°C are presented. Based on the experimentally obtained results, two distinct features are observed. The on-state resistance variation 1) exponentially increases to some value, 2) jumps to a much higher value before complete failure.

The fail mechanism is identified as the loss of gate control. The devices aged to this level can be classified as "faulty", as the characteristics deviate from normal operating conditions. In an earlier study [11], 50mΩ increase from initial $R_{ds,on}$ had been defined as the threshold value after exhaustive tests to estimate the remaining useful lifetime for safety critical applications. However, as seen from Fig. 1, $\Delta R_{ds,on}$ might be much higher when the devices actually fail. The time difference between reaching 50mΩ and actual failure is not significantly large. Thus, $\Delta R_{ds,on}$ of 50mΩ is a reasonable threshold value in lifetime estimation of safety critical systems. Yet, further increase from this threshold can be assessed to monitor the package related wear-out condition of

the switches and send a warning signal to warn the user for protective action.

The measurement of on-state resistance is of critical importance in health monitoring of power converters. Particularly, measuring the voltage drop across the switch can be a challenging task. The direct measurement through voltage dividers or using an isolated voltage sensor could result in low resolution measurements. To alleviate this problem, the off-state high-voltage blocking such as the one proposed in [12] can be used. Basically the high voltage across the switch is blocked by the series connected blocking diode when the switch is turned off. Even though this circuit can solve the problem, it is not too practical and cost-effective for many applications as it introduces additional hardware.

III. PROPOSED DEGRADATION MONITORING ALGORITHM

In order to make health monitoring of power switches more practical, a method that takes advantage of on-board sensors utilized for control purposes is proposed. The proposed switch degradation detection method is based on only sensing inductor current or output voltage. In a simple way, this can be achieved by deriving circuit dynamics and track the changes of the variable of interest. For instance, the increase in the switch resistance in boost converter given in Fig. 2(a) limits the energy stored in the inductor as it causes higher voltage drop and reduces the voltage applied on the inductor. At a given sampling time, the inductor currents can be compared to evaluate the difference in the on-state resistances. However, this difference is considerably small, which requires high resolution analog to digital converters making the measurement trivial. Thus, it would be difficult to clearly observe the difference using the circuit analysis in some of the converters.

On the other hand, it is clear that $R_{ds,on}$ impacts the damping of the resonant tank between L and C//R_o. To observe the effect, an ac signal with a frequency close to that of the double pole frequency should be added on top of the steady-state duty cycle at DC operating point, where damping effect can be clearly seen. The frequency response of the output can be obtained through software frequency response analysis (SFRA) technique in real time, which can extract the magnitude and phase of the output with respect to duty cycle perturbation at different frequencies. Then frequency range of interest can be evaluated to detect the variations in the on-state resistances. This method is still topology dependent, yet it can

978-1-4673-9551-9/16 $31.00 © 2016 IEEE

be applied to other topologies having more switches at ease. The topology dependency of the proposed method is due to the determination of the frequency range at which variations from healthy condition can be observed. The effect of on-state resistance on damping can be identified both from the frequency response of the output voltage or inductor current with respect to duty cycle perturbation.

The circuit analysis for boost dc/dc converter is analyzed here for illustration. In state-space form, the dynamics of a system is expressed as

$$\dot{x} = Ax + Bu$$
$$y = Cx + Eu \qquad (1)$$

When the switch is turned on, the equivalent circuit becomes as given in Fig. 3(a). The coefficients given in Eq. (1) for this interval is expressed as

$$A_1 = \begin{bmatrix} -\dfrac{(R_L + R_{ds,on})}{L} & 0 \\ 0 & -\dfrac{1}{C(R_o + R_c)} \end{bmatrix}, B_1 = \begin{bmatrix} \dfrac{1}{L} & 0 \\ 0 & 0 \end{bmatrix} \qquad (2)$$

where, R_L, R_c, R_o are the resistances of inductor, capacitor and load, respectively. When the switch is turned off, the equivalent circuit becomes as in Fig. 3(b). The coefficients in the off-time interval are

$$A_2 = \begin{bmatrix} -\dfrac{R_L}{L} - \dfrac{R_c R_o}{(R_c + R_o)L} & -\dfrac{R_o}{(R_c + R_o)L} \\ \dfrac{R_o}{C(R_c + R_o)} & -\dfrac{1}{C(R_c + R_o)} \end{bmatrix}, B_2 = \begin{bmatrix} \dfrac{1}{L} & -\dfrac{1}{L} \\ 0 & 0 \end{bmatrix} \qquad (3)$$

Averaging Eq. (2) and Eq. (3) over the switching period yields

$$A = \begin{bmatrix} -\dfrac{R_{ds,on}}{L}d_1 - \dfrac{R_L}{L} - \dfrac{R_c R_o}{(R_c + R_o)L}(1-d_1) & -\dfrac{R_o}{(R_c + R_o)L}(1-d_1) \\ \dfrac{R_o}{C(R_c + R_o)}(1-d_1) & -\dfrac{1}{C(R_c + R_o)} \end{bmatrix},$$

$$B = \begin{bmatrix} \dfrac{1}{L} & -\dfrac{1}{L}(1-d_1) \\ 0 & 0 \end{bmatrix} \qquad (4)$$

The output to input transfer function for CCM mode can be written in the form of

$$H(s) = C[sI - A]^{-1}\left[(A_1 - A_2)X + (B_1 - B_2)U\right] + (C_1 - C_2)X + (F_1 - F_2)U \qquad (5)$$

Using Eq. (5), open loop transfer function of inductor current to the duty cycle is expressed as

$$\frac{\tilde{i}_L(s)}{\tilde{d}(s)} = \frac{\lambda_1\left(Cs(R_c + R_o)+1\right) - R_o I_L \beta_1}{Cs^2(R_c + R_o) - s\left(\alpha_1 C(R_c + R_o)-1\right) - \alpha_1 - (1-D)R_o \beta_1} \qquad (6)$$

where

$$\alpha_1 = -\frac{DR_{ds,on}}{L} - \frac{R_L}{L} - \frac{(1-D)R_c R_o}{L(R_c + R_o)};$$

$$\beta_1 = -\frac{(1-D)R_o}{(R_c + R_o)L}; \qquad (7)$$

$$\lambda_1 = -\frac{R_{ds,on}I_L}{L} + \frac{R_c R_o I_L}{(R_c + R_o)L} + \frac{R_o V_c}{(R_c + R_o)L} + \frac{V_D}{L};$$

Similarly, the transfer functions for buck/boost and buck converters are expressed as in Eq. (8) and Eq. (10).

$$\frac{\tilde{i}_L(s)}{\tilde{d}(s)} = \frac{C\lambda_2(R_c + R_o)s + \lambda_2 + \beta_2 R_o I_L}{Cs^2(R_c + R_o) - s\left(\alpha_1 C(R_c + R_o)-1\right) - \alpha_1 + (1-D)R_o \beta_2} \qquad (8)$$

where

$$\beta_2 = -\beta_1;$$

$$\lambda_2 = -\frac{R_{ds,on}I_L}{L} + \frac{R_c R_o I_L}{(R_c + R_o)L} - \frac{R_o V_c}{(R_c + R_o)L} + \frac{V_D + V_{in}}{L} \qquad (9)$$

and

$$\frac{\tilde{i}_L(s)}{\tilde{d}(s)} = \frac{C\lambda_3(R_c + R_o)s + \lambda_3}{C(R_c + R_o)s^2 - \left(C\alpha_3(R_c + R_o)-1\right)s - \alpha_3 - R_o \beta_3} \qquad (10)$$

where

$$\alpha_3 = -\frac{DR_{ds,on}}{L} - \left(\frac{R_L}{L} + \frac{R_o R_c}{(R_c + R_o)L}\right); \beta_3 = -\frac{R_o}{L(R_c + R_o)};$$

$$\lambda_3 = \frac{V_{in} + V_D - I_L R_{ds,on}}{L}; \qquad (11)$$

The transfer functions given in Eq (6), Eq. (8) and Eq. (10) are plotted for both healthy condition ($R_{ds,on}$=0.3Ω) and after aging ($R_{ds,on}$=0.5Ω, $R_{ds,on}$=0.7Ω, $R_{ds,on}$=0.9Ω) using the following values; L=150μH, C=300μF, R_L=40mΩ, V_D=1.5V. The increased $R_{ds,on}$ represents different wear-out conditions. For the boost converter, the parameters are selected as V_{in}=100V, V_o=315V, D=0.7, R_o=422Ω. The buck/boost converter has the same parameters as of boost converter, resulting in V_o=220V. For the buck converter, the remaining parameters are; V_{in}=318V, V_o=97V, D=0.31, R_o=41Ω.

The bode plots for dc/dc boost converter showing the magnitude and phase of the output is given in Fig. 4. As it can be seen even though there is a gain difference at very low frequencies, it is minimal and not easy to distinguish the difference between healthy and aged devices. On the other hand, the variations on the double pole frequency are apparent. To observe it in detail, the zoom-in profile is given in Fig. 5. It is observed that the gain has been decreased to 52.7dB and 51.2dB and 50dB from 54.4dB when the $\Delta R_{ds,on}$ increases to 0.2Ω, 0.4Ω, and 0.6Ω, respectively. It is also pertinent that double loop frequency remains the same, which makes the measurement and comparison easier.

The phase and magnitude around double pole frequency has

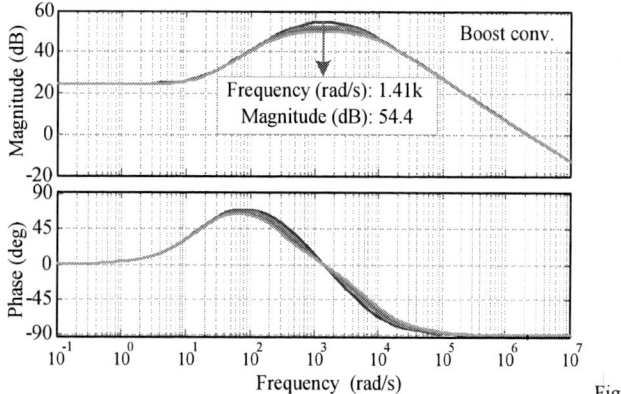

Fig. 4. Plant model transfer function of dc/dc boost converter under healthy ($\Delta R_{ds,on}$=0Ω) and aged ($\Delta R_{ds,on}$=0.6Ω) conditions.

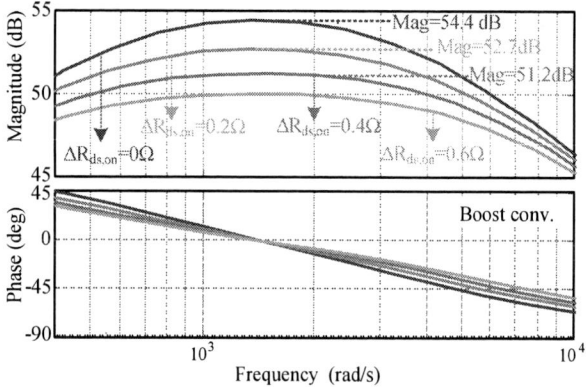

Fig. 5. Zoom into double pole frequency of dc/dc boost converter at healthy ($\Delta R_{ds,on}$=0Ω) and aged ($\Delta R_{ds,on}$=0.2Ω, 0.4Ω, 0.6Ω) conditions.

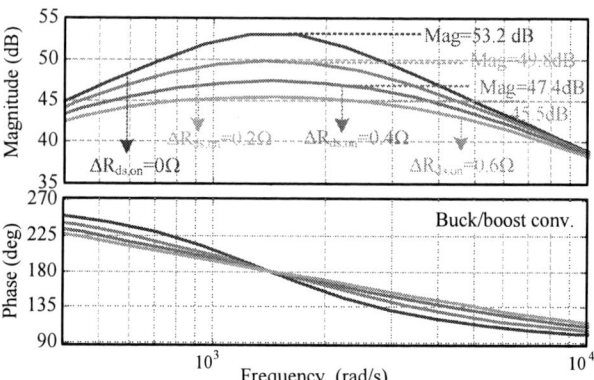

Fig. 6. Plant model transfer function of dc/dc buck/boost converter at healthy ($\Delta R_{ds,on}$=0Ω) and aged ($\Delta R_{ds,on}$=0.2Ω, 0.4Ω, 0.6Ω) conditions.

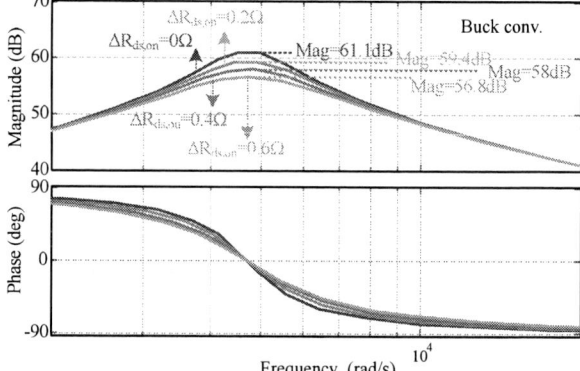

Fig. 7. Plant model transfer function of dc/dc buck converter at healthy ($\Delta R_{ds,on}$=0Ω) and aged ($\Delta R_{ds,on}$=0.2Ω, 0.4Ω, 0.6Ω) conditions.

been plotted for dc/dc buck/boost converter in Fig. 6. Similar to the boost converter, the magnitude decreases from 53.2dB to 49.8dB, 47.4dB and 45.5dB, respectively, when on-state resistance increases. Likewise, bode plots of a buck converter for different on-state resistance values are given in Fig. 7. There is 4.3dB difference between the cases when $\Delta R_{ds,on}$=0 and $\Delta R_{ds,on}$=0.6Ω.

It is pertinent that the increment in the on-state resistance damps the gain at the double pole frequency. Thus, it is possible to find the on-state resistance from the gain measured at the double pole frequency. In this paper, this mathematical relation has been analyzed for boost converter. In order to reduce the computation effort, the relations given in Eq. (7) is simplified as

$$\alpha_1 = -\frac{DR_{ds,on}+R_L+(1-D)R_c}{L}; \beta_1 = -\frac{(1-D)}{L}; \lambda_1 = \frac{V_c}{L} \quad (12)$$

The magnitude and phase responses of the plant can be found by rewriting Eq. (6) in the frequency domain as

$$H(j\omega) = \frac{j\omega C\lambda R_o + \lambda - R_o I_L \beta}{j\omega(1-CR_o\alpha)-(1-D)R_o\beta-\alpha-CR_o\omega^2} \quad (13)$$

where, the phase response with respect to frequency is expressed as

$$\angle H(j\omega) = \text{atan}\left[\frac{C\lambda R_o\omega}{\lambda - I_L R_o I_L \beta}\right] - \text{atan}\left[\frac{(1-CR_o\alpha)\omega}{-(1-D)R_o\beta-\alpha-CR_o\omega^2}\right] \quad (14)$$

The denominator of Eq. (13) can be further simplified as $(1-D)R_o\beta_1 \gg \alpha_1$. Using this simplification, the magnitude response is expressed as

$$|H(j\omega)| = \sqrt{\frac{(C\lambda R_o\omega)^2+(\lambda-R_o I_L\beta)^2}{\left[(1-CR_o\alpha)\omega\right]^2+\left[(1-D)R_o\beta+CR_o\omega^2\right]^2}} \quad (15)$$

The on-state resistance both at healthy and aged conditions can be found by measuring the gain at the double pole frequency and solving Eq. (15). In order to reduce the number of calculations, it is of interest to find the on-state resistance of an aged device, $R_{ds,onx}$, using the measured gain difference. Thus, it is necessary to first find the expression for gain difference.

$$\Delta H = 20\log|H_1|-20\log|H_x| \quad (16)$$

The gain difference for measured two points is expressed as

$$\Delta H = 10\log\frac{\left[(1-CR_o\alpha_1)\omega\right]^2+\left[(1-D)R_o\beta+CR_o\omega^2\right]^2}{\left[(1-CR_o\alpha_x)\omega\right]^2+\left[(1-D)R_o\beta+CR_o\omega^2\right]^2} \quad (17)$$

TABLE I. ACTUAL AND ESTIMATED $R_{DS,ON}$ IN BOOST CONVERTER

Actual $R_{ds,on}(\Omega)$	ΔH(dB)	Estimated $R_{ds,onx}(\Omega)$
0.3	0	0.2983
0.5	-1.73	0.4999
0.7	-3.17	0.7022
0.9	-4.42	0.9051

TABLE II. SFRA CYCLE ON FIXED AND FLOATING POINT PROCESSORS

Math	SFRA Injection	SFRA Collection
Fixed Point	45	81
Floating Point	43	78

where α_x represents the simplification which involves the $R_{ds,onx}$, and is expressed as

$$\alpha_x = -\frac{DR_{ds,onx} + R_L + (1-D)R_c}{L} \qquad (18)$$

From Eq. (17), the relationship between α_x and measured gain difference, ΔH, is extracted as

$$\alpha_x = \frac{1}{CR_o} - \sqrt{\frac{\left[(1-CR_o\alpha_1)\omega\right]^2 + \left(1-10^{\frac{\Delta H}{10}}\right)\left[(1-D)R_o\beta_1 + CR_o\omega^2\right]^2}{(CR_o\omega)^2 10^{\frac{\Delta H}{20}}}} \qquad (19)$$

Using Eq. (18) and (19), the on-state resistance of the aged device is calculated as

$$R_{ds,onx} = -\frac{L\alpha_x + R_L + (1-D)R_c}{D} \qquad (20)$$

The aforementioned simplifications cause small error in the calculation as illustrated in Table I. As discussed earlier, the gain at the double pole frequency should be measured in order to observe the effects of the on-state variation. For boost converter, the value of the double pole frequency is given as

$$\omega_n = \sqrt{\frac{-\alpha_1 - (1-D)R_o\beta_1}{C(R_c + R_o)}} \qquad (21)$$

The determination of ω_n also requires initial $R_{ds,on}$ value. However, this can be an estimate as it does not really impact the result, $(1-D)R_o\beta_1 \gg \alpha_1$.

IV. EXPERIMENTAL RESULTS

In order to experimentally verify the proposed concept, a cycle efficient SFRA has been implemented on one of the legs of the interleaved boost stage of TI experimental kit using TMS320F28035 as shown in Fig. 8. The circuit parameters are the same as the ones used to plot the Bode diagrams but with a current sensing gain. The SFRA is first used to identify the plant model when the switches are at healthy condition, and then compare the variations in the expected frequency range to determine the level of wear-out condition, through injection of small signal disturbance in the control software.

In earlier studies, FRA approach has been demonstrated

Fig. 8. Photo of the test board.

mainly for tuning the control parameters online or monitoring capacitor degradation [13]-[17]. The plant input and output are decomposed using DFT as in Eq. (22) where multiplication of the measured signal with sine and cosine of the interested frequency, gives the real and the imaginary value of frequency response.

$$X_k = \sum_{n=0}^{N-1} x_n \cdot \left(\cos\left(\frac{2\pi kn}{N}\right) - j\sin\left(\frac{2\pi kn}{N}\right)\right) \qquad (22)$$

The number of CPU cycles SFRA takes, plays a critical role in determining the feasibility of its integration in a digital power converter. SFRA implementation on fixed and floating point processors are compared in Table II. Cycles represent time taken by the SFRA routine, including the ADC read, compensation calculation, PWM update, SFRA and software overhead including context save and restore. A simple control loop takes ~197 cycles on a fixed point processor like TMS320F28035. Thus on a 60MHz processor controlling a 100KHz SMPS, with SFRA included, CPU utilization is 33%. Given the trend for high switching frequencies, a multi core approach on the digital controller can be used to alleviate the cycle burden from the CPU that is controlling the power stage, when SFRA is included. For such applications, Control Law Accelerator (CLA) which is a small footprint co-processor available on F28035 can be used to offload the SFRA task [16].

The experimentally obtained frequency response results are given in Fig. 9. Using Eq. (21), the frequency of the pole is found as 1.41 krad/s. From the zoom-in profile given in Fig. 9(b), the gain at healthy condition ($\Delta R_{ds,on}$ is equal to zero) corresponding to 225Hz is 20.85dB. In the case that $\Delta R_{ds,on}$ is increased to 0.3Ω and 0.68Ω, the amplitude is measured as 19.81dB and 18.7dB, respectively. Using the Eq. (20), the estimated $\Delta R_{ds,on}$ values are found as 0.285Ω and 0.697Ω.

V. CONCLUSION

The degradation monitoring of power switches is of critical importance for high reliability systems. In this work, the effect of on-state resistance variations of power MOSFETs on the plant model is first analyzed in basic single-switch dc/dc converters. Based on the analysis, it is shown that the resistance variation affects the damping of the transfer functions differently depending on the topology. The variations in the Bode plots are illustrated for switches under

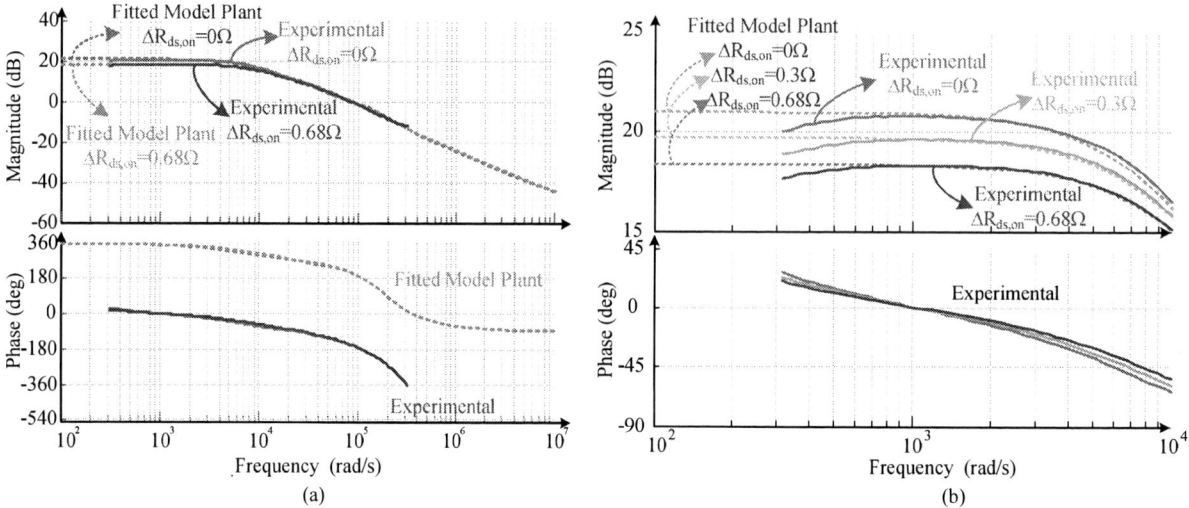

Fig. 9. Experimentally extracted plant model transfer function of dc/dc boost converter under healthy ($\Delta R_{ds,on}$=0Ω) and aged ($\Delta R_{ds,on}$=0.3Ω, 0.6Ω) conditions; (a) full scale Bode plot, (b) ω=100-10000 rad/s.

various aged conditions in accordance with the experimental results obtained from earlier work of the authors. The proposed frequency response analysis technique is implemented online in a low-cost DSP and verified on a boost dc/dc converter. The experimental results successfully matches with the theoretical results, suggesting that the proposed technique can be used to identify the wear-out condition of the switches even in converters having higher number of switches or more control complexity.

VI. ACKNOWLEDGEMENT

This work was supported in part by the Office of Naval Research (ONR) under Award Number N00014-15-1-2325, and TXACE/SRC under TASK 1836.154. Authors would like to thank Priority Labs for providing the failure analysis results which form the basis of this work.

REFERENCES

[1] K. Li, G.Y. Tian, L. Cheng, A. Yin, W. Cao, and S. Crichton, "State Detection of Bond Wires in IGBT Modules Using Eddy Current Pulsed Thermography," *IEEE Trans on Power Electron.*, vol. 29, no. 9, pp. 5000-5009, Sept. 2014.

[2] S. Yang, D. Xiang, A. Bryant, P. Mawby, L. Ran, and P. Tavner, "Condition Monitoring for Device Reliability in Power Electronic Converters: A Review," *IEEE Trans on Power Electron.*, vol. 25, no. 11, pp. 2734-2752, Nov. 2010.

[3] W. Kexin, D. Mingxing, X. Linlin, and L. Jian, "Study of Bonding Wire Failure Effects on External Measurable Signals of IGBT Module," *IEEE Trans on Device and Materials Reliability*, vol. 14, no. 1, pp. 83-89, March 2014.

[4] V. Smet, F. Forest, J,-J. Huselstein, F. Richardeau, Z. Khatir, S. Lefebvre, and M. Berkani, "Ageing and Failure Modes of IGBT Modules in High-Temperature Power Cycling," *IEEE Trans on Ind. Electron.*, vol. 58, no. 10, pp. 4931-4941, Oct. 2011.

[5] S. Yang, D. Xiang, A. Bryant, P. Mawby, L. Ran, and P. Tavner, "Condition Monitoring for Device Reliability in Power Electronic Converters: A Review," *IEEE Trans on Power Electron.*, vol. 25, no. 11, pp. 2734-2752, Nov. 2010.

[6] M. Held, P. Jacob, G. Nicoletti, P. Scacco, and M. H. Poech, "Fast power cycling test of IGBT modules in traction application," in *Proc. Int. Conf. Power Electron. Drive Syst.*, 1997, pp. 425–430.

[7] N. Patil, J. Celaya, D. Das, K. Goebel, and M. Pecht, "Precursor Parameter Identification for Insulated Gate Bipolar Transistor (IGBT) Prognostics," *IEEE Trans on Reliability*, vol. 58, no. 2, pp. 271-276, June 2009.

[8] J. Celaya, A. Saxena, P. Wysocki, S. Saha, and K. Goebel, "Towards Prognostics of Power MOSFETs: Accelerated Aging and Precursors of Failure", in *Proc. Annual Conference of the Prognostics and Health Management Society*, pp. 1-10, 2010.

[9] J. R. Celaya, N. Patil, S. Saha, P. Wysocki, and K. Goebel, "Towards Accelerated Aging Methodologies and Health Management of Power MOSFETs (Technical Brief)", in *Proc. Annual Conference of the Prognostics and Health Management Society*, pp. 1-8, 2009.

[10] S. Dusmez and B. Akin, "An accelerated thermal aging platform to monitor fault precursor on-state resistance", in *Proc. IEEE Int. Electric Machines and Drives Conf.*, pp. 1-6, May 2015.

[11] S. Dusmez, and B. Akin, "Remaining Useful Lifetime Estimation For Degraded Power MOSFETs Under Cyclic Thermal Stress", in *Proc. IEEE Energy Conversion Conference & Expo.*, pp. 3846-3851, 2015.

[12] P. Ghimire, S. Bęczkowski, S. Munk-Nielsen, B. Rannestad, P. B. Thøgersen, "A review on real time physical measurement techniques and their attempt to predict wear-out status of IGBT," in *Proc. IEEE Power Electron and Appl. Conf.*, pp. 1–10, 2013.

[13] G. Buiatti and e. al, "An Online and Noninvasive Technique for the Condition Monitoring of Capacitors in Boost Converters," *IEEE Trans on Instrumentation and Measurement*, vol. 59, no. 8, pp. 2134-2143, 2010.

[14] Y.-M. Chen and e. al, "Online Failure Prediction of the Electrolytic Capacitor for LC Filter of Switching-Mode Power Converters," *IEEE Trans on Ind. Electron.*, vol. 55, no. 1, pp. 400-406, 2008.

[15] J. Morroni, A. Doglov, M. Shirazi, R. Zane, and D. Maksimovic, "Online Health Monitoring in Digitally Controlled Power Converters," in *Proc. Power Electronics Specialist Conference*, pp. 112-118, Jun. 2007.

[16] Akin, B.; Bhardwaj, M.; Choudhury, S., "An Integrated Implementation of Two-Phase Interleaved PFC and Dual Motor Drive Using Single MCU With CLA," *IEEE Trans on Ind. Informatics*, vol. 9, no. 4, pp. 2082-2091, Nov. 2013.

[17] M. Bhardwaj, S. Choudhury, R. Poley, and B. Akin, "Online frequency response analysis: A powerful plug-in tool for compensation design & health assessment of digitally controlled power converters," in *Applied Power Electronics Conference and Expo.*, vol., no., pp. 838-843, Mar. 2014.

A new Capacitance Estimation method of Supercapacitor Bank using a Bank Impedance and Current Injection

Junwon Lee, Hyunsik Jo and Hanju Cha

Department of Electrical Engineering, Chungnam National University, Daejeon, Korea

Abstract—This paper proposes a capacitance estimation method of the supercapacitor bank in the supercapacitor energy storage system (SCESS). To diagnose a deterioration, the capacitance of the supercapacitor bank is estimated by using a current injection. The injected current causes an AC ripple voltage and AC ripple current in the supercapacitor bank, which are extracted by using the proposed signal processing method. The proposed method provides an accurate value of the capacitance for reliability and durability of the supercapacitor energy storage system. Usefulness of the proposed estimation method is verified through simulation and experiment with a prototype of a 10kW SCESS. Experimental results shows that the maximum estimation error rate is less than 3% at both 2.25F and 2.57F capacitance bank.

I. INTRODUCTION

A highly reliable power supply is required for critical loads and this requirement can be fulfilled by a supercapacitor energy storage system (SCESS), where the energy is stored in a supercapacitor bank. Supercapacitor is a key energy storage element for supporting various emerging energy systems and applications such as renewable energy power generation, transportation, power system, and it has a very rapid dynamic response and high power density [1][2]. Lifetime of supercapacitor is usually shorter than that of the other components of an energy storage system, and its capacitance value decreases with aging. Therefore, issues of supercapacitor health and lifetime are important and supercapacitor health diagnosis system should be able to estimate the state of health (SOH) of supercapacitor, or to give prediction about the possibility of failure. Some technical references have focused on evaluating different approaches for health and lifetime estimation of capacitor [3][4][5][6]. In general, the lifetime of supercapacitor is considered to be finished when the capacitance is reduced by more than 20% from the initial value [7][8].

Supercapacitors used in SCESS are installed inside the system, making it difficult to measure the whole capacitance without detaching the supercapacitor, which is especially troublesome for the system. Accordingly, to overcome these difficulties, this paper proposes a capacitance estimation of the supercapacitor bank in the SCESS based on bank impedance extraction method. To diagnose the degree of deterioration, capacitance of the supercapacitor bank is estimated by injecting a current in this paper. The injected current causes an AC ripple voltage and AC ripple current in the supercapacitor bank, which are extracted by using the proposed signal

processing scheme. In other words, capacitance of the supercapacitor bank is estimated by a bank impedance derived from the ratio of AC ripple voltage to AC ripple current.

In this paper, a new capacitance estimation method by using a current injection is proposed, and the usefulness of the proposed capacitance estimation method is verified through simulation and experiment with a prototype of a 10kW SCESS.

II. CAPACITANCE ESTIMATION OF SUPERCAPACITOR BANK

A. Modeling of Supercapacitor Bank

Fig. 1 (a), (b) show a unit supercapacitor and supercapacitor bank, respectively. Capacitance of each supercapacitor is 360F and equivalent series resistor (ESR) of each supercapacitor is 3.2mΩ. Supercapacitor bank is assembled with 160 supercapacitors connected in series and has the voltage range from 280V to 400V. Therefore, total capacitance is reduced to 2.25F and ESR of supercapacitor bank increases to 512mΩ.

In general, equivalent circuit of electrolytic capacitor does not consider ESR because electrolytic capacitor has a very low ESR [9]. However, ESR of supercapacitor bank is larger than one of electrolytic capacitor, thus ESR of supercapacitor bank should be considered. Fig 2 shows an equivalent circuit of supercapacitor bank including ESR and capacitance. When current I_{SC} flows into the equivalent circuit of supercapacitor bank, voltage V_{SC} is calculated by (1) and (2). Then, supercapacitor bank voltage can be written by the Laplace transform in (3).

$$v_c(t) = \frac{1}{C}\int i_{sc}(t) + v_c(t_0) \tag{1}$$

$$v_{sc}(t) = \frac{1}{C}\int i_{sc}(t) + v_c(t_0) + (R \times i_{sc}(t)) \tag{2}$$

$$V_{sc}(s) = (R + \frac{1}{sC}) \times I_{sc}(s) \tag{3}$$

Fig. 1. (a) Unit supercapacitor and (b) supercapacitor bank

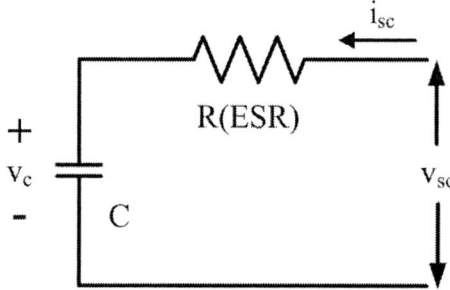

Fig. 2. Equivalent circuit of supercapacitor bank

B. Configuration of SCESS and Control for Estimation

Fig. 3 shows configuration of three-phase SCESS and control block diagram for capacitance estimation of the supercapacitor bank. Three-phase SCESS consists of a three-phase voltage source inverter (VSI), supercapacitor bank, line-frequency transformer, filters and bidirectional thyristor switches. Table I shows the parameters of three-phase SCESS.

In the normal mode, grid voltage is connected to the critical load through bidirectional thyristor switches. VSI operates for maintaining the supercapacitor bank voltage with full-charged state and prepares for an abrupt voltage sag or interruption. When a voltage sag or interruption occurs, bidirectional thyristor switches are blocked to separate the disturbed grid from the critical load and then VSI supplies rated voltage continuously to load from the supercapacitor bank.

To estimate a capacitance of the supercapacitor bank, SCESS operates in the estimation mode, where supercapacitor bank voltage is maintained with a specific value. In the estimation mode, controller consists of outer DC-link voltage control loop and inner current control loop, where both controllers are PI controller type. Grid currents are transformed into d- and q- axis values in the synchronous reference frame. D- axis current is maintained as zero for a unity power factor operation, whereas q- axis current is controlled to maintain the DC voltage with the specific value. It is difficult to obtain any information from supercapacitor bank without variation of DC link voltage in the estimation mode. Then DC link voltage is perturbed by injecting a current and the current injection signal is added to q- axis reference current generated by the outer voltage loop. Current variation in the q- axis causes variation of supercapacitor bank voltage and then capacitance of supercapacitor bank can be estimated by using the relationship between current variation and voltage variation.

Table I System parameters

Rated voltage	Three-phase 220V
Rated frequency	60Hz
Rated power	10kW
DC-link capacitance	2.25F
Transformer	10kVA, 220V:130V
AC Filter, L/C	2mH/ 50uF
Switching frequency	10kHz

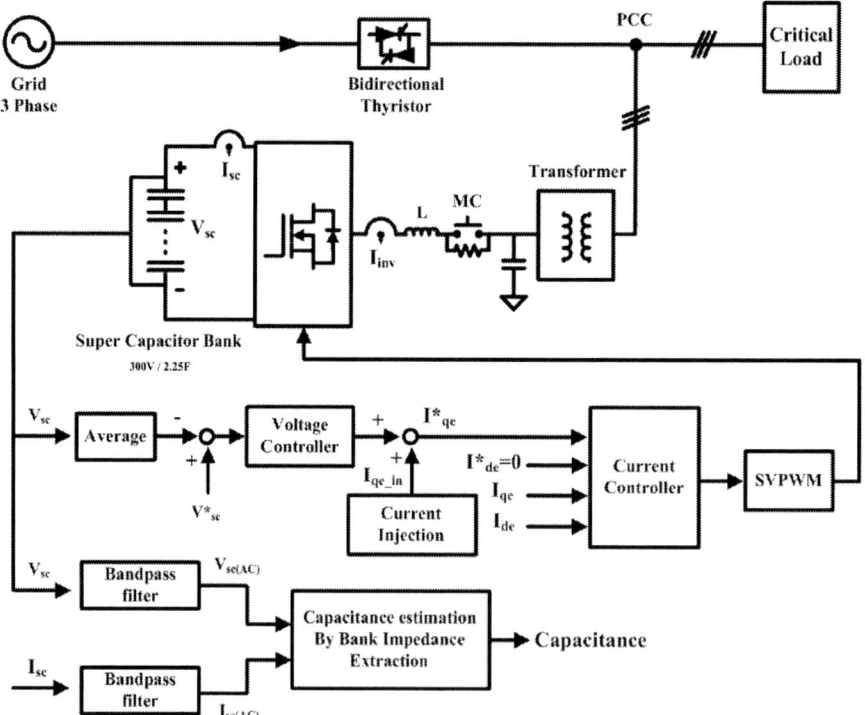

Fig. 3. Configuration and control block diagram for supercapacitor energy storage system

978-1-4673-9551-9/16 $31.00 © 2016 IEEE

C. Current Injection

Injected current reference in the synchronous reference frame is given as

$$I_{qe\_in} = I_o \sin(2\pi \times f_0 t) \tag{4}$$

When I_{qe_in} is well controlled and is transformed into the stationary reference frame, the AC current of the sinusoidal waveform with frequency of f_0 Hz flows to the supercapacitor bank. I_{SC} consists of DC component and AC component by injecting I_{qe_in}. Likewise, V_{SC} has DC component and AC component. To estimate capacitance of supercapacitor bank, both current and voltage AC components are only required in order to calculate impedance of supercapacitor bank.

A band pass filter is employed to extract AC component and equation (5) shows transfer function of the band pass filter, where k is gain, b is bandwidth and ω_c is center frequency. In this paper, gain, bandwidth and center frequency 5, 1rad/s and 0.5Hz, respectively.

$$G(s) = \frac{kbs}{s^2 + bs + \omega_c^2} \tag{5}$$

Voltage and current outputs of the band pass filter are shown in (6), (7).

$$I_{sc(AC)} = \frac{1.5(V_{qe} \times I_{qe\_in})}{Vdc} \tag{6}$$

$$V_{sc(AC)} = (R + \frac{1}{j \times 2\pi \times f_0 \times C})I_{sc(AC)} \tag{7}$$

Through the band pass filter, $V_{sc(AC)}$ and $I_{sc(AC)}$ have only base frequency component

D. Bank Impedance Extraction Method

Fig. 4 shows the block diagram of the proposed estimation method which uses a magnitude and phase relation of $I_{sc(AC)}$ and $V_{sc(AC)}$. In order to obtain the magnitude and phase, single phase PLL is used, and V_m, I_m, V_θ and I_θ are magnitude and phase of $V_{sc(AC)}$, $I_{sc(AC)}$, respectively. Impedance of supercapacitor bank (Z) is derived into (8) and the capacitance is estimated by (9).

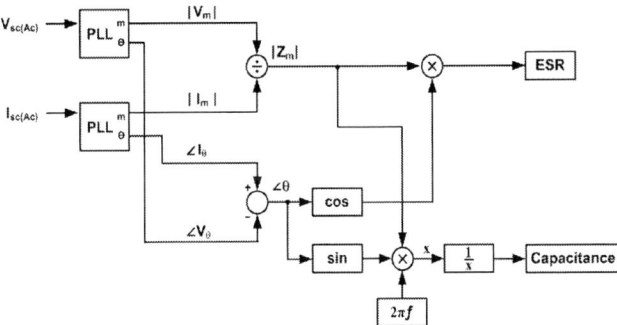

Fig. 4. Block diagram for the proposed capacitance estimation

$$Z = \frac{V_m \angle V_\theta}{I_m \angle I_\theta} = Z_m \angle \theta \tag{8}$$

$$Capacitance = \frac{1}{2\pi \times f_0 \times Z_m \times \sin\theta} \tag{9}$$

III. EXPERIMENT RESULTS

To verify effectiveness of the proposed method, PSIM simulations were carried out with SCESS. Sampling time of the DC-link voltage and the phase current is $100\,\mu$ sec which corresponds to the 10kHz switching frequency of the inverter, and capacitance of the supercapacitor bank is set to 2.25F.

Fig. 5 shows simulation results for the capacitance estimation using the bank impedance extraction method. It shows $I_{sc(AC)}$, $V_{sc(AC)}$, I_m and V_m. I_m and V_m are represented as the magnitude of $I_{sc(AC)}$, $V_{sc(AC)}$, and they has a constant value. Fig. 6 shows the estimated capacitance and phase angle difference between $I_{sc(AC)}$ and $V_{sc(AC)}$. The capacitance is perfectly converged to 2.25F in the simulation.

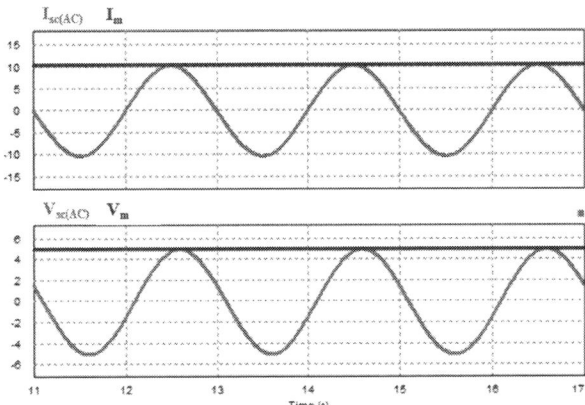

Fig. 5. Waveforms of $I_{sc(AC)}$, $V_{sc(AC)}$, I_m and V_m in the simulation

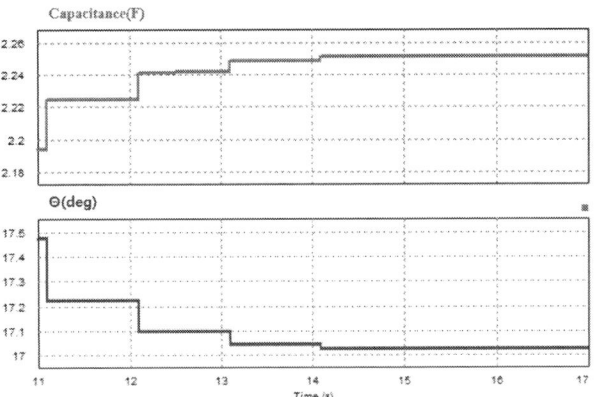

Fig. 6. Waveforms of estimated capacitance and theta in the simulation

10kW SCESS is controlled by a floating-point digital signal processor (DSP), TMS320F28335 and sampling frequency is 10kHz. Fig.7 shows the front and side view of the 10kW SCES, where (a) for the display panel located in front side, (b) for supercapacitor bank. Fig.8 shows the rear and top view of the 10kW SCESS, where (a) for transformer placed in the

978-1-4673-9551-9/16 $31.00 © 2016 IEEE   513

bottom of the stack, (b) for control board and sensor board placed at the top of the system.

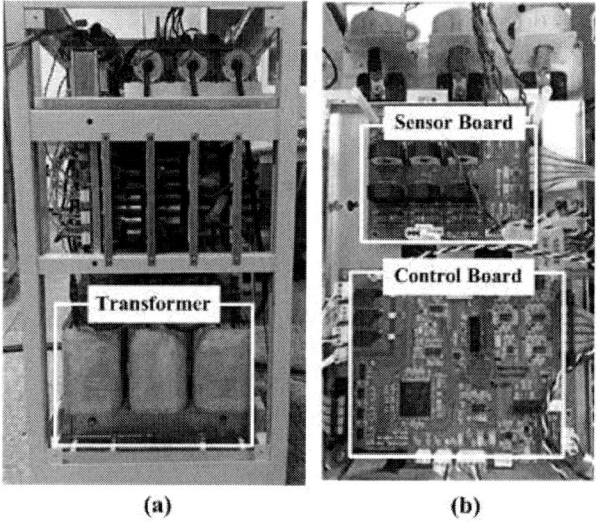

(a)   (b)

Fig. 7. (a) Front (b) Side view of the 10kW SCESS

(a)   (b)

Fig. 8. (a) Rear (b) Top view of the 10kW SCESS

Fig. 9 shows $I_{sc(AC)}$, $V_{sc(AC)}$ waveforms when the I_{qe_in} current is injected. Injected current is 20A peak and 0.5Hz frequency component. So $I_{sc(AC)}$, $V_{sc(AC)}$ are also 0.5Hz component of signals. It is confirmed that $V_{sc(AC)}$ is lagging the $I_{sc(AC)}$ in the supercapacitor bank. Fig. 10 shows experimental results of capacitance estimation by the impedance extraction method, and current injection makes the voltage ripple in the DC-link. Capacitance estimation routine is started after the 2 period (4s) of current injection. The estimated capacitance is stabilized after the 5 period (10s) of that capacitance estimation routine is started. Fig. 11 shows that estimated value of capacitance is 1.97F. It is approximately 15% smaller than 2.25F rated value of capacitance of supercapacitor bank. These result shows that capacitance of supercapacitor which is used in the experiment has 15% smaller than rated value of

capacitance at 0.5Hz. The rate of reduced capacitance is depends on frequency of the injected current.

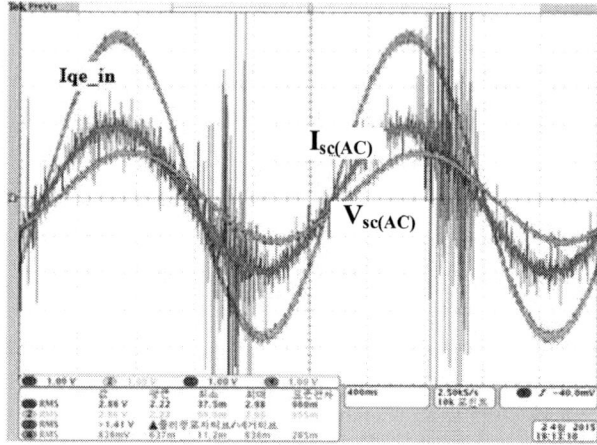

Fig. 9. Waveforms of I_{qe_in}, $I_{sc(AC)}$ and $V_{sc(AC)}$ in the experiment

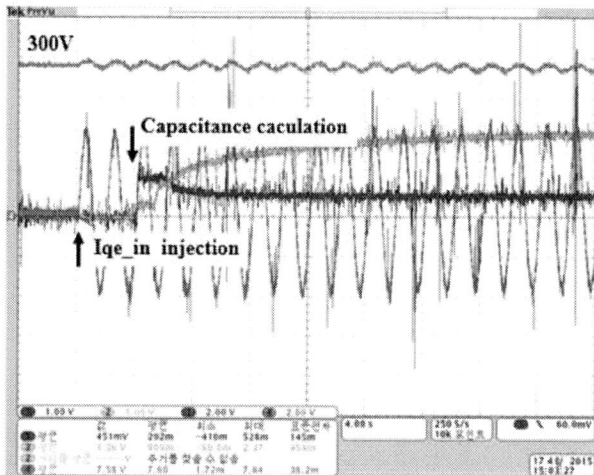

Fig. 10. The experimental results of the capacitance estimation method

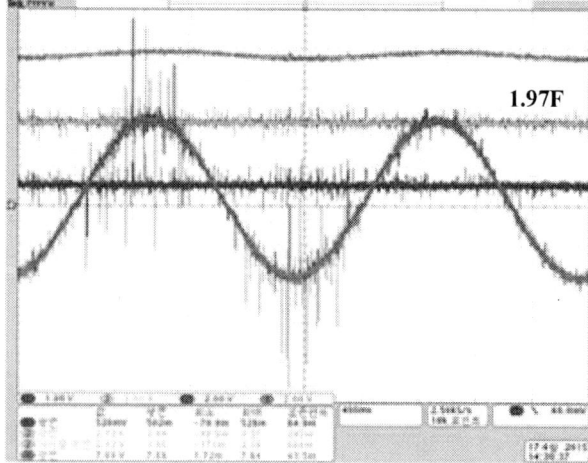

Fig. 11. The estimated capacitance is 1.97F

To verify effectiveness of the capacitance estimation method at 0.5Hz, experiments were also carried out with 2.57F rated value of capacitance of supercapacitor bank. It is assembled with 140 supercapacitors connected in series by bypassing 20 supercapacitors of 160 supercapacitors. Fig. 12 shows experimental results of estimated capacitance at both 2.25F and 2.57F rated value of capacitance bank for 10 days. It is confirmed that 1.97F is estimated at 2.25F Bank, and 2.26F is estimated at 2.57F Bank. The maximum estimation error rate is less than 3% for 10days. These results show that proposed capacitance estimation method has repetitiveness, and the capacitance reduced in the identical ratio (15%) at any capacitance of supercapacitor bank. The state of health (SOH) of supercapacitor bank can be estimated through initial value of capacitance which is 15% reduced from the rated value of capacitance.

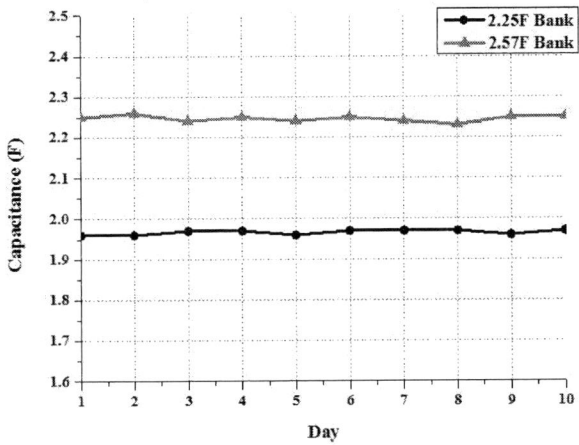

Fig. 12. The estimated capacitance at 2.25F and 2.57F bank for 10days

IV. CONCLUSION

In this paper, a new capacitance estimation method of supercapacitor bank has been proposed. Capacitance of the supercapacitor bank has been estimated by using the current injection. The injected current caused an AC ripple voltage and AC ripple current in the supercapacitor bank, which were extracted by using the proposed signal processing scheme. The proposed method provided an accurate value of the capacitance for reliability and durability of the supercapacitor energy storage system. Usefulness of the proposed capacitance estimation methods has been verified through simulation and experiment with a prototype of a 10kW SCESS. Experimental results show that the maximum estimation error is less than 3% at both 2.25F and 2.57F capacitance bank.

REFERENCES

[1] Hyunsik Jo, Wujong Lee, Hanju Cha "Three-Phase Voltage Sag Compensator For Smart Grid And Infrastructure" IEEE Vehicle Power and Propulsion Conference(VPPC), pp. 1463-1468 2012.10

[2] D. Casadei, G. Grandi, and C. Rossi, "A supercapacitor-based power conditioning system for power quality improvement and uninterruptible power supply," in Proc. IEEE ISIE, 2002, vol. 4, pp. 1247–1252.

[3] C. Lee, J. K. Seok, and J. W. Choi, "Online capacitance estimation of DC-link electrolytic capacitors for three-phase AC/DC/AC PWM converters using recursive least squares method," Proc. Inst. Electr. Eng.—Electr. Power Appl., vol. 152, no. 6, pp. 1503–1508, Nov. 2005.

[4] Oukaour, B. Tala-lghil, M. AI Sakka, H. Gualous, R. Gallay, and B. Boudart, "Calendar ageing and health diagnosis of supercapacitor," Electric Power Systems Research, vol. 95, pp. 330-338, 2013.

[5] M. Uno and K. Tanaka, "Accelerated charge-discharge cycling test and cycle life prediction model for supercapacitors in alternative battery applications," IEEE Trans. Ind. Electron., vol. 59, pp. 4704-4712, 2012.

[6] J.-J. Moon, W.-S. Im, and J.-M. Kim, "Capacitance estimation of DC-link capacitor in brushless DC motor drive systems," in Proc. IEEE ECCE, Jun. 2013, pp. 525–529.

[7] Catelani, M., Ciani, L., Marracci, M.. Tellini, B "Frequency dependent failure region definition for supercapacitors" 2014 IEEE International Instrumentation and Measurement Technology Conference (I2MTC), pp. 1021-1025, 2014.

[8] W. Lajnef, J.-M. Vinassa, O. Briat, H. El Brouji, S. Azzopardi, and E. Woirgard, "Quantification of ageing of ultracapacitors during cycling tests with current profile characteristics of hybrid and electric vehicles applications," IET Electr. Power Appl., vol. 1, no. 5, pp. 683–689, Sep. 2007.

[9] N. Devillers, S. Jemei, M-C. Péra, D. Bienaimé and F. Gustin, "Review of characterization methods for supercapacitor modelling", Journal of Power Sources, vol. 246, pp. 596 – 608, 2014.

Gate Driver Design for 1.7kV SiC MOSFET Module with Rogowski Current Sensor for Shortcircuit Protection

Jun Wang, Zhiyu Shen, Christina DiMarino, Rolando Burgos, Dushan Boroyevich
Center for Power Electronics Systems
Virginia Polytechnic Institute and State University
Blacksburg, VA 24061, USA
junwang@vt.edu

Abstract—This paper shows a gate driver design for 1.7 kV SiC MOSFET module as well a Rogowski-coil based current sensor for effective shortcircuit protection. The design begins with the power architecture selection for better common-mode noise immunity as the driver is subjected to high dv/dt due to the very high switching speed of the SiC MOSFET modules. The selection of the most appropriate gate driver IC is made to ensure the best performance and full functionalities of the driver, followed by the circuitry designs of paralleled external current booster, Soft Turn-Off, and Miller Clamp. In addition to desaturation, a high bandwidth PCB-based Rogowski current sensor is proposed to serve as a more effective method for the shortcircuit protection for the high-cost SiC MOSFET modules.

Keywords—*SiC MOSFET Module; Gate Driver; Rogowski Current Sensor; EMI*

I. INTRODUCTION

SiC MOSFET as a wide-bandgap device has superior performance for its high breakdown electric field, low on-state resistance, fast switching speed and high working temperature [1]. High switching speed enables high switching frequency, which improves the power density of high power converters. The cost of SiC MOSFET is also gradually decreasing due to the growing of usage in industry applications. Therefore, SiC MOSFET is of great potential in medium-voltage (MV) high power applications as a substitution of MV IGBTs. A well-performed gate driver with sufficient protections is critical to ensure excellent performance of a SiC MOSFET module. The driver must be capable of features such as low propagation delay, high driving current, good dv/dt immunity, and effective protections that includes correct detection and fast response. The basic specifications of the gate driver is shown in Table I, where one example SiC MOSFET Module that the gate driver can be applied to is also shown. The target of this paper is to design a driver that meets the specifications.

For the 1.2 kV IGBTs, a number of commercial driver ICs has been well developed with sufficient isolation rating and protection functionalities. Those driver ICs are fast enough to drive 1.2 kV SiC MOSFETs, and some commercial driver boards are designed based on the ICs. When it comes to 1.7 kV or higher rated voltage of devices, the isolation barrier of the driver ICs becomes insufficient. In commercial designs, fiber-optic cables [2] or pulse transformers [3] are used to realize the

TABLE I. SiC MOSFET and Gate Driver Specifications

Example device part number	Cree CAS300M17BM2
Device voltage rating	1700 V
Device current rating	225 A @T$_C$=90°C
Gate charge required	1076 nC
Device Package	62 mm
Maximum switching frequency	100 kHz
Maximum Propagation delays	150 ns
Maximum Propagation distortion	20 ns
Driving voltage	+20 V / −4 V
Maximum gate driving current	24 A
Functionalities	Under-voltage lockout; Shortcircuit; Soft Turn-Off; Active Miller Clamp

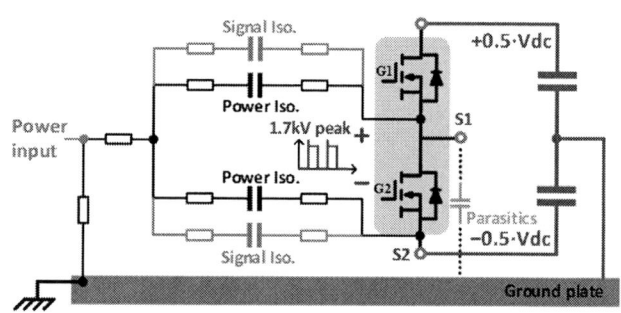

Fig. 1. Common model model of a isolated gate driver circuit

MV isolation for signal channels of the gate driver. Although the isolation requirement is satisfied by using optocouplers, fiber-optics or pulse transformers, the logic signal may still suffer from common-mode noise that is generated by the switching-transient dv/dt. This is analyzed in Section II, and different power/signal architectures are compared based on parasitics model. A best solution is found to minimize the chance of falsely turn-on of device that causes shortcircuit. In Section III, The driver IC is carefully screened and selected among a number of candidates. Detailed circuit design is presented in section IV. A solution of self-balanced paralleled current booster was proposed not only to meet the driving current capability but also to enable the two-level soft turn-off upon shortcircuit. The Soft Turn-Off (STO) mechanism is analyzed, and the parameter is design. A circuit for Active Miller Clamp (AMC) is proposed to solve a threshold issue

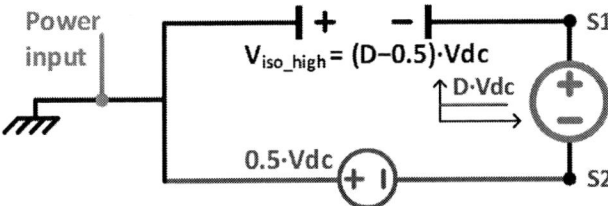

Fig. 2. DC analysis of the gate driver model

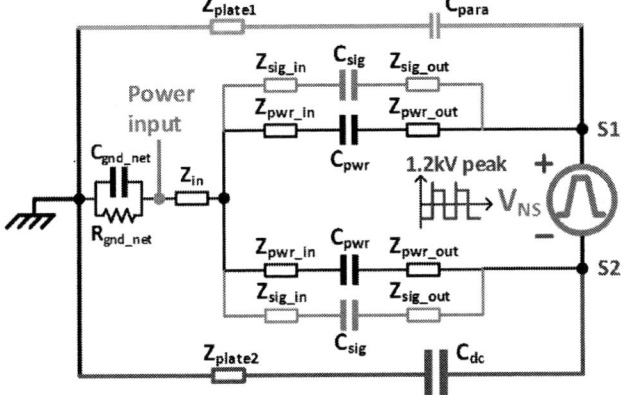

Fig. 3. AC analysis of the CM model of gate driver considering parasitics

limited by the driver IC. Due to the difference in output characteristics between SiC MOSFET and IGBT, the desaturation approach for shortcircuit protection may not be effective anymore. This is analyzed in Section V, and a Rogowski-coil based current sensor is proposed as another protection layer of shortcircuit. Experimental validations are shown together with the description of the designs.

II. POWER ARCHITECTURE DESIGN

According to the specifications, the maximum switching transient can be as high as 1700/50 = 34 ns. Then the gate driver must be able to operate under the noise source below 15 MHz. A simplified model with gate driver parasitics in Fig. 1 represents the common-mode (CM) noise propagation in conventional gate driver configurations. Each trace in the figure indicates the ground path of different isolated voltages. The capacitor in the figure represents the coupling effect at isolation barriers. The rectangles in the figure is the impedances of the circuit traces. Note that the high-side gate driver ground is sitting on the source terminal of the high-side SiC MOSFET, which is jumping between the positive and negative rail at the switching transient. The voltage between S1 and S2 can be modeled as a noise voltage source because it is not impacted by the parasitics of the CM model. The signal isolation barrier is generally achieved by an isolated driver IC/optocoupler for 1.2 kV device, or by fiber-optics/pulse transformer for 1.7 kV device or higher. Any high-frequency noise current flowing through the signal path isolator can induce a voltage drop on the ground trace impedance, which will be added to the differential-mode logic signal and cause a falsely turn-on of the SiC MOSFET. In conventional designs, the power and signal isolation channels share the same ground at the primary and secondary side, respectively.

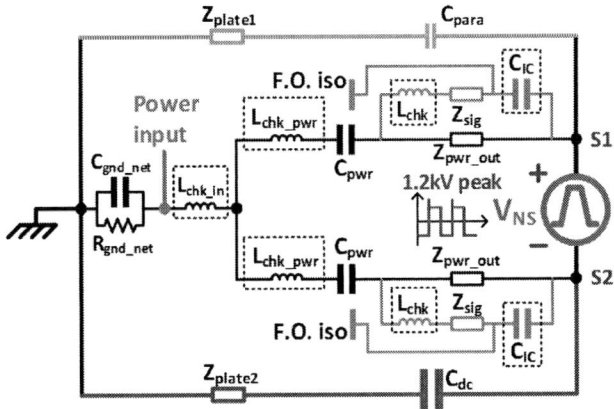

Fig. 4. CM model of the proposed gate driver configurations with high CM noise immunity for the control/logic circuit

A. DC Analysis of the Conventional Design

Fig. 2 shows the DC analysis of the CM model of the gate driver, where the capacitors are open-circuit and all the trace impedances are treated as short-circuit if they are in series with the capacitors. The voltage of one DC-link capacitor C_{dc} is $0.5 \cdot V_{dc}$, and can be treated as a DC voltage source. The DC voltage of the low-side isolation barrier is $0.5 \cdot Vdc$ and the high-side one is $(D-0.5) \cdot V_{dc}$, where D is the averaged duty cycle of the high-side switch. Accordingly, the low side isolation barrier will take most of the DC voltage stress, for both the signal and power isolations in the conventional design.

B. AC Analysis of the Conventional Design

The AC CM model of the gate driver circuit is shown in Fig. 3. The stray inductance of ground plate is estimated as 30 nH and the resistance as 1 mΩ. The same ground impedance assumption is made for the parasitic capacitance C_{para} whose value is around 0.2 nF. A low high-frequency impedance ground network is connected between the power-input terminal of the gate driver and the ground. All the other ground trace impedances are assumed to be 1 nH and 1 mΩ. The frequency-domain analysis of Fig. 5 tells, firstly, the high-side isolation barrier takes almost all of the high frequency voltage stress up to 20 MHz. At frequency high than 20 MHz, the isolation stress of high side is even higher than the noise source as the coupling capacitance C_{pwr} and C_{sig} resonate together with the ground plane impedance Z_{plane1} and Z_{plane2}. This indicates that the high side isolation barrier is subjected to almost all the AC noise which can even be amplified at certain frequency. Plus, all the AC noise stress will be placed on the coupling capacitance of C_{sig} such that large CM noise current is induced in the signal path. The noise in signal path can cause falsely turn-on, malfunction of logic, and even disable the controller if the noise current flow to the controller along with the signal path. Inserting CM chokes at the signal path will help attenuate the noise current, but the leakage inductance of the chokes will introduce delay or resonance to deteriorate the differential-mode signal.

In order to prevent the CM noise from flowing into the controller, the signal barrier is normally strengthened by using the fiber-optical cables instead of optocouplers. However, the

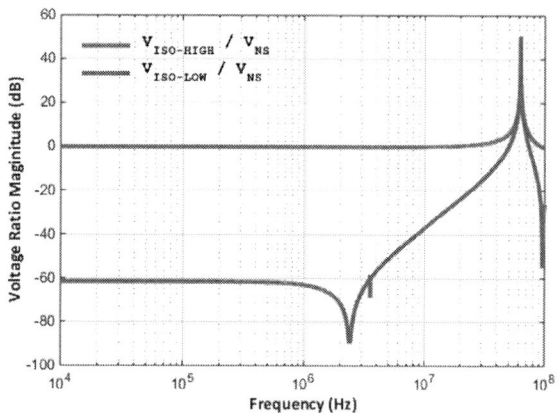

Fig. 5. AC voltage sharing between the high-side and low-side isolations

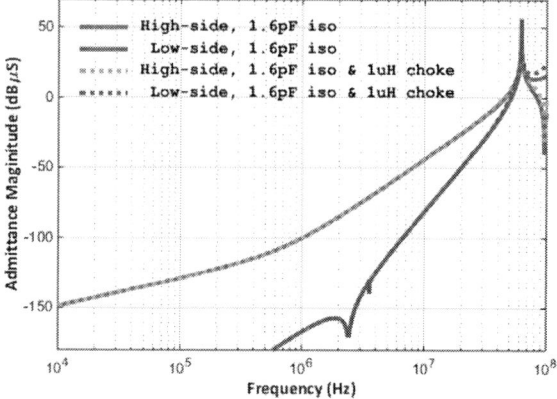

Fig. 7. Comparison to decide if CM choke is needed for signal path

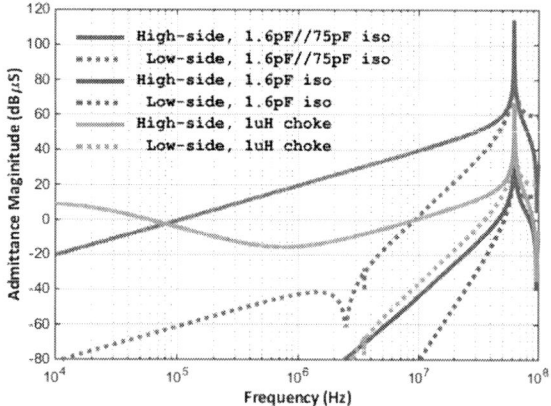

Fig. 6. Comparison between noise-immunity solutions for signal path

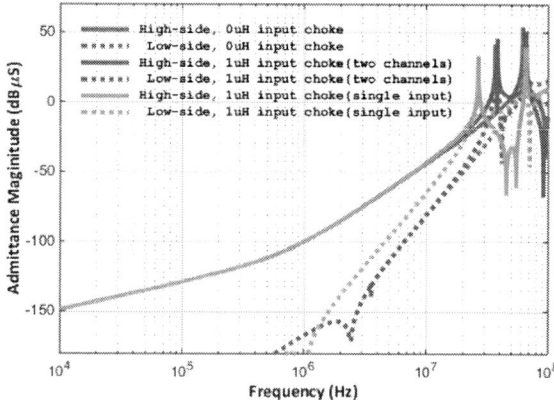

Fig. 8. Comparison to decide if CM choke is needed for power-input path

noise current at the signal path of the driver is still large, as long as the secondary sides of the power/signal isolation barriers share the same ground. The design objective is to increase the impedance mismatch between the signal path Z_{sig_out} and power path Z_{pwr_out}. The improved configuration is shown in Fig. 5. Either adding another isolation barrier or introducing a CM choke for the logic signal power supply will be a helpful solution. For comparison, the admittance from the signal-path noise current to noise voltage source is set as the criteria to represent the performance in noise attenuation. Fig. 6 shows the comparison result of different solutions where the CM choke inductance L_{chk} is selected to be 1 μH and the coupling capacitance of additional isolation barrier C_{IC} is measured to be 1.6 pF. Both of them show significant improvement than the conventional design. The isolation solution has lower admittance than the CM choke one at the wide low-frequency range, and even about 40 dB less at 10 MHz. Plus, as the signal path is paralleled with a low impedance power path, the added isolation barrier will not take any high voltage stress. Accordingly, the additional-isolation solution is selected for the driver architecture design.

To further identify if it brings any benefit by using the choke and the additional-isolation together, a comparison is made in Fig. 7 that shows they have almost the same performance below 50 MHz. Thus, a CM choke is not needed if the isolated-IC is already designed in the configurations.

In order to find out if CM chokes at the power input can further bring down the noise flowing through the signal path, two designs at the power input are made for comparison. In Fig. 8, it shows that one design is to add two 1 μH chokes L_{chk} at the two power channels, respectively, while the other is to add one 1 μH chokes L_{chk_in} at the common input. From the result, the admittance is not reduced significantly, but two more resonant peak is created at around 20~30 MHz, which can possibly be covered by the noise source. Therefore, CM chokes for the power path are not recommended.

It is worth noting that for all the comparison cases, the low-side signal path is always subjected to much less noise current than the high-side. The reason that the grounding network impedance Z_{gnd_net} and the ground plate impedance Z_{plate2} in series with the DC-link capacitor impedance is much lower than the low-side channel at frequency less than 60 MHz. This explains why most of the noise issues in signal path occur at the high-side channel. Based on all the analysis above, the power architecture is designed as shown in Fig. 9. Careful impedance control also need to be taken while designing the PCB layouts.

III. GATE DRIVER IC SELECTION

A. Critical characteristics of a gate driver IC

TABLE II. COMPREHENSIVE FUNCTIONALITIES COMPARISON OF GATE DRIVER IC CANDIDATES

Manufacturer	Part Number	Critical Functionalities				
		Propagation	UVLO	Desaturation	Soft Turn-Off	Active Miller Clamp
Avago	ACPL-332J	(d) 180 ns, 180 ns (r/f) 50 ns, 50 ns	+ 10.3 V	(th) 6.5 V, (r) 300 ns	LRTO	V_{EE} + 2.0 V
Avago	HCPL-316J	(d) 300 ns, 320 ns (r/f) 100 ns, 100 ns	+ 11.1 V	(th) 7.0 V, (r) 300 ns	LRTO	N/A
Fairchild	FOD8318	(d) 300 ns, 250 ns (r/f) 34 ns, 34 ns	+ 10.0 V	(th) 7.0 V, (r) 850 ns	LRTO	V_{EE} + 2.2 V
Toshiba	TLP5214	(d) 85 ns, 90 ns (r/f) 32 ns, 18 ns	+ 10.3 V	(th) 6.5 V, (r) 180 ns	LRTO	V_{EE} + 3.0 V
Infineon	1ED020I12	(d) 170 ns, 165 ns (r/f) 30 ns, 50 ns	+ 11.0 V	(th) 9.0 V, (r) 350 ns	N/A	V_{EE} + 2.1 V
On Semi	MC33153	(d) 80 ns, 120 ns (r/f) 17 ns, 17 ns	+ 11.0 V	(th) 6.5 V, (r) 300 ns	N/A	N/A
ROHM	BM6102FV-C	(d) 150 ns, 150 ns (r/f) 50 ns, 50 ns	+ 11.5 V	(th) 10 V, (r) 3250 ns	2LTO	V_{EE} + 2 V
STMicro	STGAP1S	(d) 100 ns, 100 ns (r/f) 25 ns, 25 ns	+ 13.0 V − 2.0 V	(th) 10 V, (r) 100 ns	2LTO	GND + 2 V

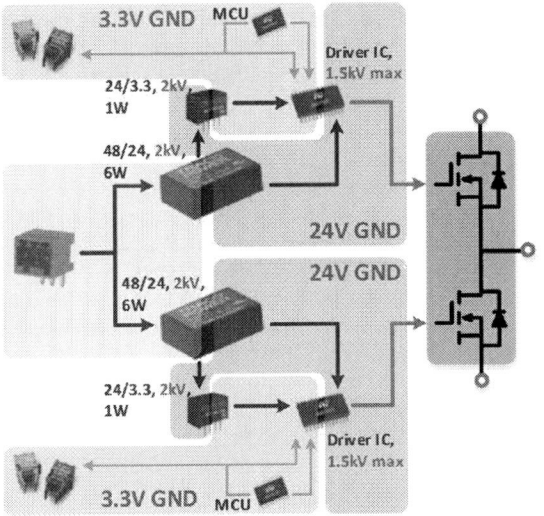

Fig. 9. Designed power architecture for the gate driver

As analyzed that an additional isolation barrier is needed for noise attenuation, it will be preferred to use a gate driver IC instead of an optocoupler, because the IC can provide more functionalities and save space for the driver board. As a gate driver IC for SiC MOSFET, it should propagate the gate signals with low delay, low distortions and low jittering. It should also be able to provide functionalities including undervoltage lockout (UVLO), shortcircuit protection, Soft Turn-Off (STO), and Active Miller Clamp (AMC). When selecting the driver IC, the critical characteristics of the IC should be marked at the first place.

For the SiC MOSFET design to switch at up to 100 kHz, thus the switching period is 10 μs. Typically, it takes 50~100 ns for the given Cree MOSFET module to complete the switching transient. The propagation delay of the driver IC should be as short as possible to ensure the MOSFET can promptly respond to the command from the controller for the control and protection purpose. The mismatch of the driver delays and rise/fall time between the turn-on and turn-off will cause duty-cycle distortions, but it is acceptable if the mismatch are close to delay characteristics of the device itself.

The UVLO is used to monitor the status of the power supply for the driving loop to ensure high switching speed and low conduction loss of the SiC MOSFET. If the power supply cannot maintain the designed voltage any more, the device should shut down and inform the controller. It is preferred to set a higher UVLO threshold, and to monitor the positive and the negative supply voltage at the same time.

The shortcircuit protection is the most critical functionality of the gate driver. As analyzed in Section II, the high-side signal path is usually subjected to high CM noise that possibly brings about falsely turn-on and shortcircuit if the low-side gate signal is also high. The shortcircuit protection can be realized by detecting the device current at its on-state. The detection must respond fast and have high bandwidth as the shortcircuit current can rise at very high di/dt slope. In high current applications, the conventional method to detect IGBT current is to measure the on-state voltage V_{CE} and calculate the collector current according to the device output characteristics. As soon as a very high V_{CE} is detected at on-state, the device is turned off. This is normally named desaturation (DeSat) protection because at shortcircuit the IGBT is deviated from its saturation region. In recent designs, the DeSat is also implemented to protect the SiC MOSFET from shortcircuit. Despite that the DeSat for SiC MOSFET is not as effective as for IGBT, the DeSat functionality is still used in the gate driver of this paper as a second-stage shortcircuit protection. To achieve the DeSat functionality, a driver IC should provide high enough threshold voltage for a given output characteristics of a SiC MOSFET. Also, the response time between detection and turn-off should be as short as possible.

A STO functionality is usually combined together with shortcircuit protection. The turn-off current will be extremely high when the shortcircuit protection is triggered, and a hard turn-off will kill the device due to the induced turn-off voltage spike. Therefore, either a soft turn-off process or a V_{DS} clamp circuit by using TVS diode is necessary for the shortcircuit turn-off. In this paper, the STO is selected because in the latter approach the breakdown voltage of TVS varies in a wide range such that the V_{DS} clamp voltage is difficult to control. In state-of-art designs, two typical STO mechanism are adopted by different driver ICs manufacturers. One is to increase the turn-

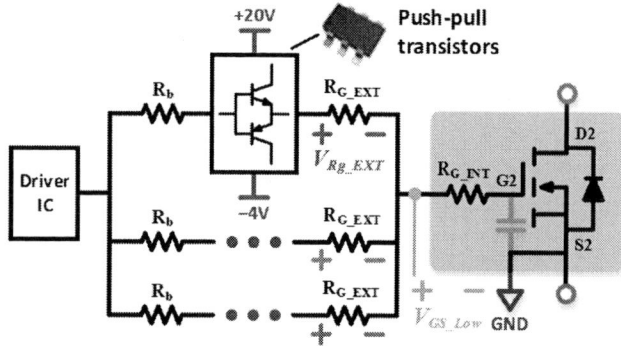

Fig. 10. Schematics of the self-balanced paralleled current boosters

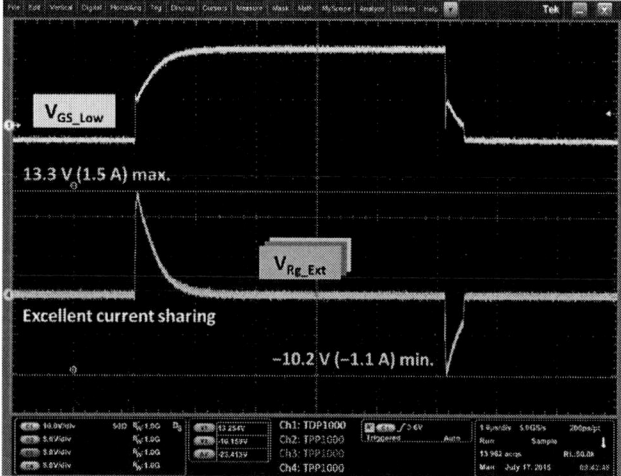

Fig. 11. Current comparison of three paralleled current boosters

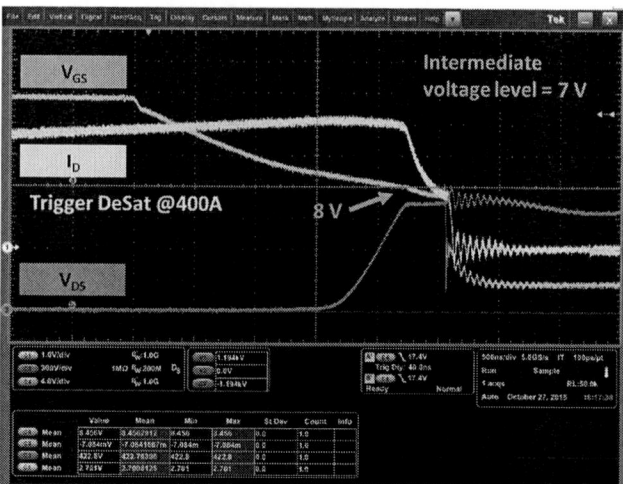

Fig. 12. Test waveform of Soft Turn-Off

off resistance hundreds times higher than normal when soft turn-off is triggered (LRTO), while the other is to use Two-Level Turn-Off (2LTO) voltages. The former approach does not apply to the designs where an external gate current booster is required. The driver IC with 2LTO is preferred, otherwise the 2LTO circuit has to be built by discrete components.

The AMC is the functionality that create a very low-impedance in the gate loop for the device in its off state. In this manner the cross-talk induced noise current will not cast a large noise voltage at the gate voltage V_{GS}. The Miller Clamp detects the gate voltage of the device, and enable the low-impedance path as soon as the gate voltage drop below the threshold. Usually an external bipolar transistor will be used to create the low impedance path, so the on-state resistance of AMC transistor inside the driver IC is not a critical parameter. An important requirement is that the threshold voltage of Miller Clamp should be lower than the device threshold voltage with enough margin, with consideration of the delays brought by the internal gate resistors inside the device package.

B. Gate Driver IC Comparison and Selection

Eight gate driver ICs from seven manufacturers are selected for comparison regarding the critical parameters as mentioned above. In Table II, the values are obtained from public datasheets of the IC manufacturers. In column "Propagation",

"(d)" means delays and the first number indicates the turn-on delay while the second indicates the turn-off. "(r/f)" means rise/fall time and they are corresponding to the two numbers in row, respectively. In column "UVLO", the "+ number" indicates the threshold voltage where the under-voltage lockout will be triggered when the positive supply voltage drops, while the "− number" represents the threshold of the negative one. In column "Desaturation", the first value means the maximum threshold voltage while the second indicates the reaction time. In the column "Soft Turn-Off", two types of soft turn-off mechanisms are shown. In the column "Active Miller Clamp", the thresholds of gate voltage to enable the low-impedance gate loop are presented. Overall, the Driver IC STGAP1S has the best performance in the first four characteristics. The only shortcoming of it is that the AMC threshold voltage is almost the same as the device gate threshold, so the AMC circuit should be designed for this concern. Eventually, STGAP1S is selected to be the driver IC with an isolation barrier.

IV. FUNCTIONAL CIRCUITS DESIGN

The detailed circuits design is carried out based on the selected driver IC. Since the design for propagation and UVLO is determined by the driver IC and the DeSat circuit design can follow [3], this paper focuses only on the design of the current booster, STO, and AMC.

A. Current booster design

As specified in the selected driver IC datasheet [], the source/sink shortcircuit current at the driver output is 5 A, so the SiC MOSFET switching speed may be limited by the driver IC. To meet the 24 A peak driving current in the driver specifications, an external current booster is necessary. Typically, the current booster is a pair of push-pull MOSFETs or bipolar transistors. There is voltage drop on the on-resistance (about 0.3 Ω) of the MOSFET-based booster, which can be as high as 7.2 V when the highest peak current is reached. Therefore, the MOSFET-based booster will take 30% of the total driving loss if the total gate resistance is designed to be 1 Ω, then a very good cooling solution is. As a contrast, the bipolar transistor have a very low voltage drop such that the loss on the current booster will be very small. The

978-1-4673-9551-9/16 $31.00 © 2016 IEEE

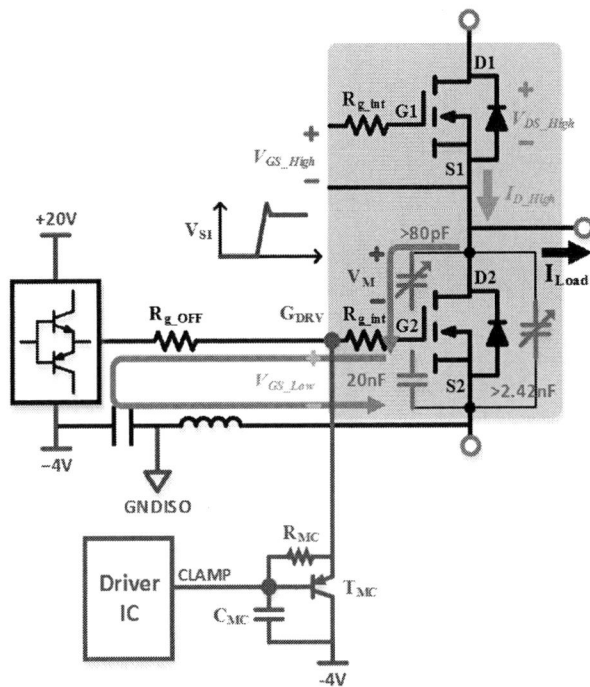

Fig. 13. Schematics of the Active Miller Clamp

disadvantage of the bipolar transistor is its output becomes high-Z if the gate voltage is higher than the positive rail voltage or lower than the negative rail voltage. This will occur when the cross-talk noise current generates a negative voltage spike when the device is at its off-state, as shown in Fig. 15. The recovery time span is determined by the discharging resistance across the device gate and source.

The current booster should be able to work together with the STO. For LRTO, [4] recommended a method by increasing the input capacitance of the current booster but the switching transient will also be delayed by hundreds of nanoseconds even at normal operations, which is not favorable to the SiC MOSFET. For 2LTO, the current booster should be able to deliver the intermediate voltage level following the driver IC output and only the bipolar transistor based current booster can achieve this goal. The highest peak current rating of available commercial transistor-based current boosters is 10 A, so three of the current booster are paralleled as shown in Fig.10. This design has a natural closed loop to balance the output currents of three boosters. When one of the boosters delivers higher current, the voltage drop on R_{G_Ext} will increase so the voltage on the base resistor R_b will decrease. Then the base current will decrease and the transistor emitter current is reduced. Fig. 11 shows the perfect balance of the three paralleled current boosters at switching transient.

B. Soft Turn-Off design

The transfer characteristics (I_D vs. V_{GS}) of the MOSFET determines how the gate voltage controls the drain current. The intermediate voltage level in 2LTO should be designed corresponding to the peak normal-operation current according to the transfer characteristics, as the normal turn-off is allowed at this peak current. The duration of this intermediate voltage

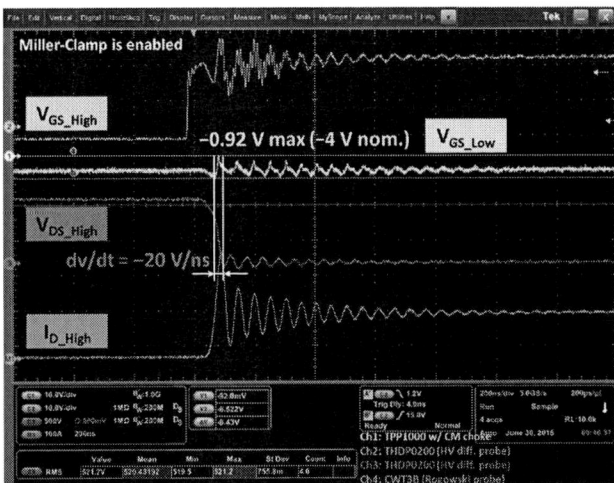

Fig. 14. Test waveform of the cross-talk issue at turn-on

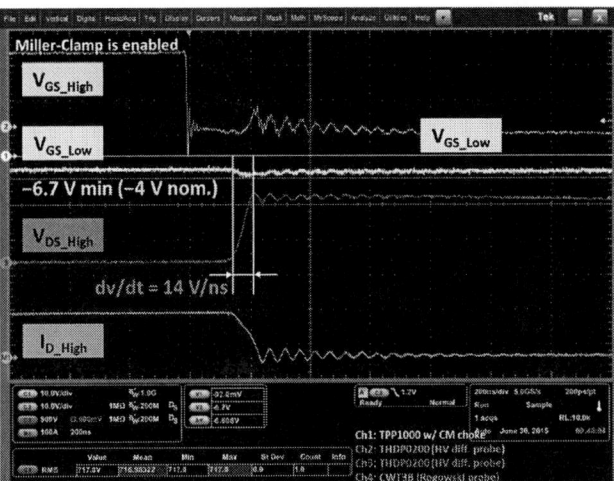

Fig. 15. Test waveform of the cross-talk issue at turn-off

level should be long enough for the drain current drops to the peak normal-operation current. As the date sheet only give the transfer characteristics at one drain-source voltage, but in practical application the drain-source voltage varies with the drain current, the proper intermediate voltage can only be design by tests. Fig.12 shows the STO behavior when the intermediate voltage is designed as 7 V and the time span for the current to drop is about 2 μs.

C. Active Miller Clamp design

The AMC design schematic is shown in Fig. 13. A PNP transistor T_{MC} is added to boost the clamp current capability of the driver IC. A resistor R_{MC} is put across the base and emitter to detect the gate voltage. Because the threshold voltage of AMC of the selected driver IC is very close the device threshold voltage, it is possible that the SiC MOSFET is not fully turned off when the AMC is activated, and the MOSFET is always turned off at zero gate resistance. Therefore, a capacitor C_{MC} is added to delay the detection of gate voltage drop, by which means the real gate voltage has dropped to the cutoff region when the AMC is triggered.

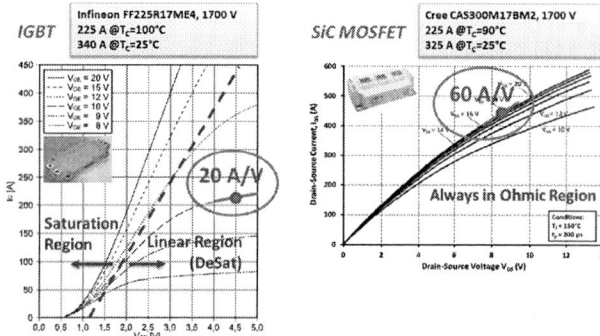

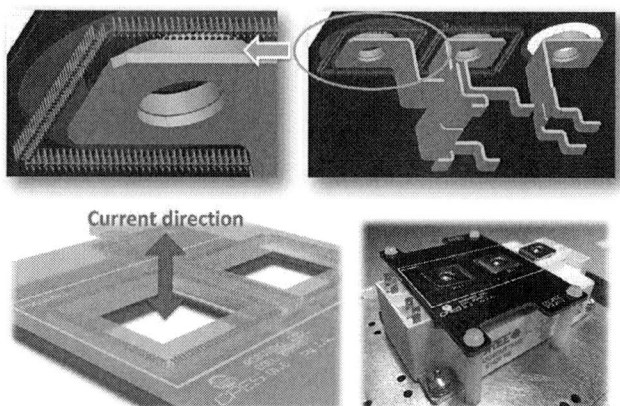

Fig. 16. Device output characeristics comparison, IGBT vs. SiC MOSFET

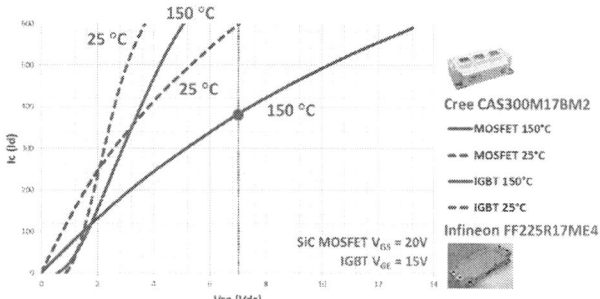

Fig. 17. Device output characeristics comparison against junction temperature

Fig. 18. Rogowski coil 3D model and assembly

As the DeSat protection may not function well for SiC MOSFET, a current sensor is a good additional option to detect the fault current. Reference [6] reviews and compares different current sensing method, including shunt, Hall, current transducer, Rogowski coil, GMR and GMI, concluding that the Rogowski coil has outstanding performances in terms of bandwidth, accuracy, linearity, implementation, profile and cost. Thus, the Rogowski-coil-based current sensor is selected and designed for shortcircuit protection.

V. ROGOWSKI CURRENT SENSOR DESIGN

A. Motivations to design current sensor for gate driver

In recent protection designs for SiC MOSFET, the DeSat method is simply borrowed from the IGBT applications. However, DeSat protection for SiC MOSFET is not as effective as for IGBT because of two main reasons [5].

1) As shown in Fig.17, when shortcircuit occurs in IGBTs, the device drifts away from the saturation region and enters linear region where the current rising slope is getting smaller. Even at steady state of shortcircuit when the IGBT block the entire DC voltage, its collector current is still limited. Thus, the in-time shut down can be achieved easily even with detection delays and sensing errors because the current won't increase very fast. In comparison, however, The SiC MOSFET is still in its omic region where the current rising slope is much faster than IGBTs. At steady state, when the SiC MOSFET blocks the DC voltage, the drain current will rise to an extrmely high value. Therefore, delays or on-stage voltage sensing errors can lead to higher possibility of device damage before it can be shut down in time and safely.

2) Fig.2 shows that the on-state voltage of the SiC MOSFET is more temperature dependent than the IGBT. If a DeSat protection is designed for the nominal junction temperature, then it can never be effective when the device temperature is still low at the startup of the converter. Instead, if the protection is designed for low junction temperature, then it can be falsely triggered at nominal junction temperatures. As a result, the difference between two output characteristics curves in different temperature of the SiC MOSFET make it very difficult to design DeSat threshold voltage.

B. Rogowski coil design

Because of the limited space between the two power terminals of the 62 mm package, the winding width is designed to be 1 mm. In order to minimize the noise from adjacent conductors, the coil almost encloses the terminal busbar, and has a one-turn winding in the opposite direct to compensate the one-turn effect of the coil. The physical assembly of coil is shown in Fig. 16. The turn number is designed to be 176 turns to minimize the noise flux from adjacent conductors, which is the maximum number that can be achieved by regular PCB fabrication techniques. The measured mutual inductance self-inductance L_S is 359 nH and the equivalent paralleled capacitance (EPC) of the winding C_S is 8.13 pF. These two parasitics compose a low-pass filter with the resonant frequency at 93 MHz, which is almost 5 times of the designed bandwidth of 20 MHz. A damping resistor connected to the output of the coil is designed to be $0.5 \cdot (L_S/C_S)^{1/2} = 105\ \Omega$ for critically damped response of the LC resonant circuit.

C. Signal processing circuit design

Signal processing circuit is designed to integrate the di/dt information obtained from the Rogowski coil output. Active integration circuit using operational amplifier is selected instead of RC passive signal circuit to achieve wider sensor bandwidth [7]. The output of the integrator circuit is sent to a low-delay comparator. The output of the comparator is given to high-voltage side of the isolated gate driver IC to switch off the SiC MOSFET. The delay from the device current I_D to the comparator output V_{CMP_OUT} is designed to be less than 20 ns.

The input resistor R_{i1} and integration capacitor C_f determines the transducer gain G_{SENSOR} from device current I_D to comparator input V_{CMP_IN}. Note that the di/dt polarities of the

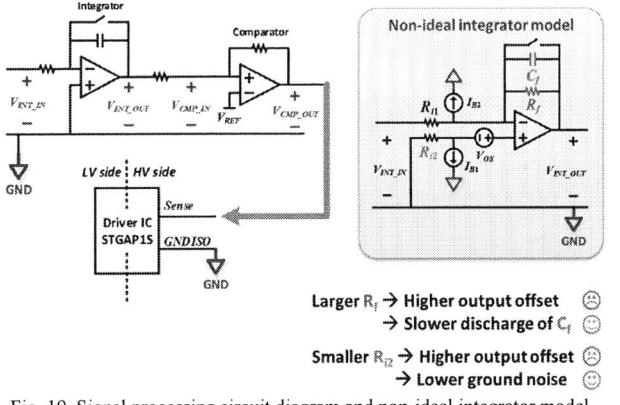

Fig. 19. Signal processing circuit diagram and non-ideal integrator model

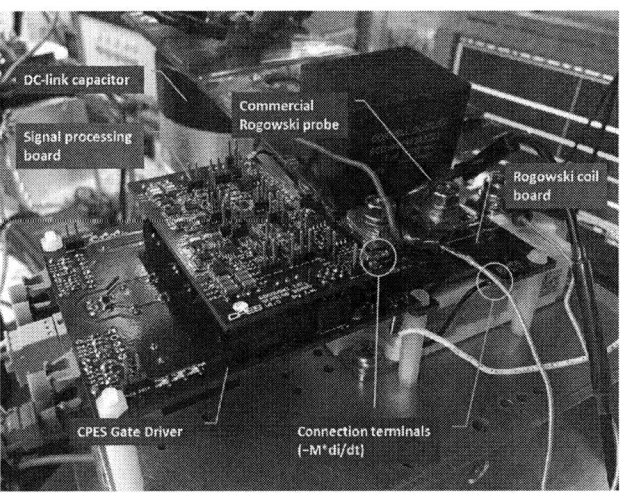

Fig. 20. Test setup and circuit boards

positive and negative busbar are opposite and the values are identical with negligible EPC of the load inductor. Therefore, the equivalent mutual inductance is the different between the mutual inductance from the measured conductor and the adjacent conductor.

$$G_{SENSOR} = (M_1 - M_2) / (R_{i1} \cdot C_f) \qquad (1)$$

Larger C_f is preferred to prevent integration error caused by C_f discharging. It is finally fine-tuned that $R_{i1} = 403 \ \Omega$ and $C_f = 200$ pF to achieve $G_{SENSOR} = 0.02 \ \Omega$. Then the mutual inductance is calculated to be $M_1 - M_2 = 1.61$ nH, which is very close to the simulation result 1.70 nH. The reasonable error is likely caused by the difference between the simulation model and the real assembly.

The general non-idea characteristics of an operational amplifier can be modeled as shown in Fig. 19. The bias current I_{B1} and I_{B2} and offset voltage V_{OS} cause an output voltage offset at V_{OUT} when the input voltage V_{IN} is zero. In conventional designs, the feedback resistor R_f is designed to minimize the effect from offset voltage V_{OS}, and two identical input resistors R_{i1} and R_{i2} is selected to cancel out the effect of from bias current I_{B1} and I_{B2}. The larger R_f leads to higher output offset, but to slower discharge of the C_f such that the sensor gain will not decrease if the device conducting time is long. The smaller R_{i2} leads to higher output offset, but to lower ground noise at the non-inverting input terminal of operational amplifier. The R_f is finally designed to be open-circuit and R_{i2} is zero, in order to maximize the sensing performance. The offset issue can be resolved by an active reset switch. The reset switch is designed to turn on for a very short time when the SiC MOSFET is switched off. The tests setup is in Fig. 20. The result in Fig.21 shows that the Rogowski current sensor has fast enough response to capture the shortcircuit current of SiC MOSFET.

VI. CONCLUSIONS

The most important aspects in designing an industry-oriented gate driver for SiC MOSFET has been presented in this paper. Solutions to resolve critical issues are proposed and supported by experimental validations.

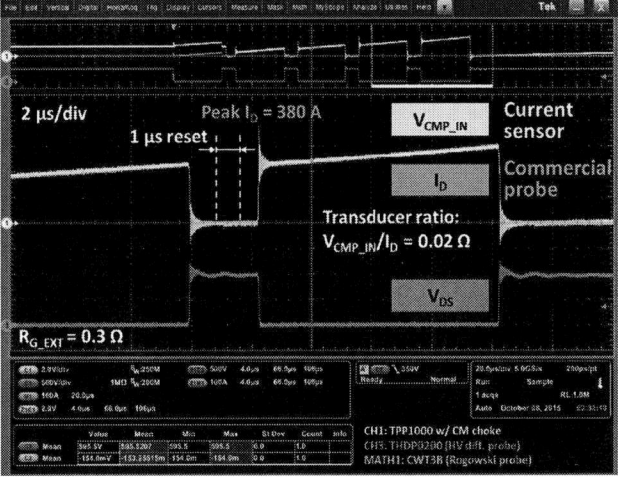

Fig. 21. Switching transient performances of the Rogowski current sensor

REFERENCES

[1] J. Millan, P. Godignon, X. Perpina, A. Perez-Tomas, J. Rebollo, "A survey of wide bandgap power semiconductor devices," IEEE Trans. Power Electron., vol. 29, no. 5, pp. 2155-2163, May, 2014.

[2] "1SP0635 Manual," CONCEPT gate driver manual, 2014.

[3] "gapDRIVE™: galvanically isolated single gate driver," STMicro gate driver manual, 2015.

[4] "Soft Turn-off Features – Application Note 5315," Avago Technologies application notes, 2010.

[5] J. Wang, Z. Shen, R. Burgos, and D. Boroyevich, "Design of a high-bandwidth rogowski current sensor for gate-drive shortcircuit protection of 1.7 kV SiC MOSFET Power Modules," in Proc. IEEE WiPDA, 2016.

[6] C. Xiao, L. Zhao, T. Asada, W. G. Odendaal, and J. D. van Wyk, "An overview of integratable current sensor technologies," in 2003, vol. 2, pp. 1251-1258.

[7] W. F. Ray and C. R. Hewson, "High performance Rogowski current transducers," in Proc. IEEE Ind. Applic. Conf., 2000, vol. 5, pp. 3083-3090.

2 MHz High-Density Integrated Power Supply for Gate Driver in High-Temperature Applications

Remi Perrin, Bruno Allard,
Cyril Buttay and Nicolas Quentin
Univ. Lyon, Ampere, INSA Lyon,
CNRS UMR 5005
21 avenue Jean Capelle,
France-69621 Villeurbanne

Wenli Zhang, Rolando Burgos and
Dushan Boroyevic
CPES - Virginia Tech
1185 Perry Street
Blacksburg, VA 24061
USA

Philipe Preciat and
Donatien Martineau
SAFRAN Labinal Power Systems
Rond Point Ren-Ravaud - BP42
77551 Moissy-Cramayel
France

Abstract—**A PCB embedded transformer for harsh environment (i.e., ambient temperature above $200\,°C$) applications is presented and used in the design of a 2 MHz integrated power-supply prototype for gate driver. The main benefits of using this developed PCB embedding process are the capability to customize the air-gap for the flyback transformer and volume reduction for the converter. The easy modulation of the air-gap distance can be achieved using PCB material in specfic thickness. Moreover, the design of a coplanar-winding transformer structure with very low inter-winding capacitance needs a large winding area and raises the interest for the PCB integration approach. Two-machined ferrite pieces in UI shape were sandwiched into a multi-layer PCB laminate with multiple pressing processes. A converter prototype built with the PCB embedded transformer and other components (GaN transistors, gate driver and passives) mounted above it shows 72% of power efficiency. One thousand thermal cycles between $-55\,°C$ and $200\,°C$ were performed on the PCB embedded transformer without observation of any major defects, such as delamination and cracking. The thermal reliability test validates the compatibility between the selected ferrite core and PCB materials as well as the feasibility of this developed transformer embedding method. A three-time volume reduction is achieved when comparing with a benchmark converter prototype using discrete transformer.**

I. Introduction

The integration of passive components is a key enabler for high-power-density power supply [1], [2]. The low-cost printed circuit board (PCB) is the most widely used substrate material in electronic applications. The PCB material selected as the substrate material to integrate a high power-density converter is mainly because of its capability for high-volume production using standard lamination process. The use of wide-band gap power transistors switched at above 1 MHz with soft-switching topology is favorable to reduce the size of passive components and thus to make it possible to embed passives into PCB substrate [3], [4].

Two methods for fabrication of PCB embedded transformers have been reported in [5]–[7]. The first approach introduces the use of a toroidal-shape core inside a PCB. The conductive vias and traces as winding are wrapped around the core formed by standard etching, drilling and plating processes. However, the reliability of the assembly against temperature variation is compromised due to the large number of winding vias. Additionally, this structure with the external windings generates a significant magnetic field emission, which cannot

fit with the aeronautical standard.

The second method is to pot the soft magnetic composite materials around conductive winding for the transformer fabrication. The transformer is connected by pin connection to a PCB stack with the active layer. But the low operating temperature of this soft magnetic material makes it not suitable for high-temperature ($> 200\,°C$) applications.

A series of studies have been performed on the realization of

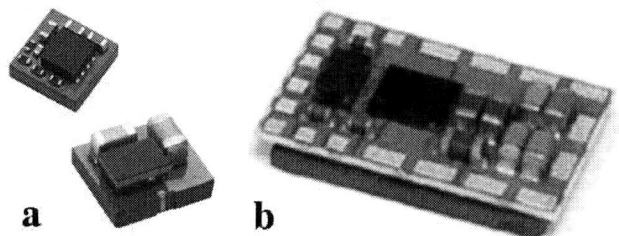

Fig. 1: *Non-isolated LTCC inductor integrated POL converters a) Murata LTCC power supply [8]; b) CPES POL module with LTCC [9]*

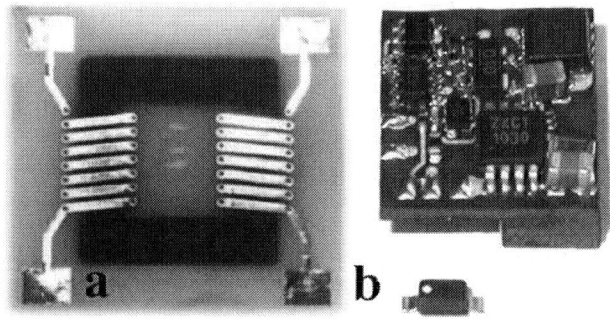

Fig. 2: *a) PCB integrated EMI Filter with coupled inductor [10] ; b) Integrated flyback dc-dc converter built using ferrite-based LTCC materials [11]*

multi-megahertz integrated POL converter with LTCC planar inductor and GaN devices [3], [7], [9]. Fig. 1(a) presents the Murata's Micro DC-DC converts, which use a LTCC inductor with a Flip-Chip IC on the top [12] and Fig. 1(b) shows a LTCC ferrite planar inductor and assembled POL module

---

978-1-4673-9551-9/16 $31.00 © 2016 IEEE   524

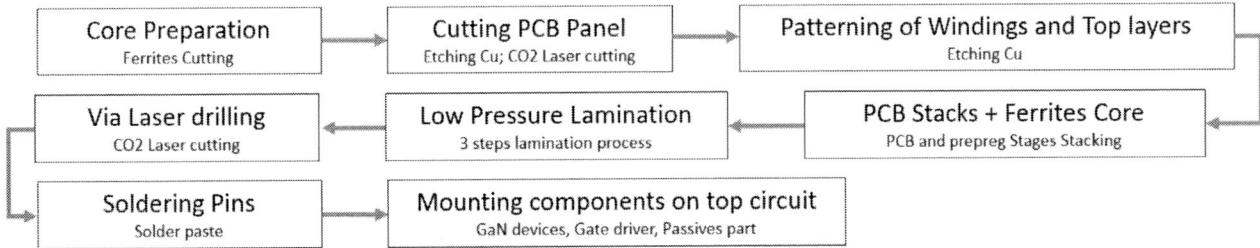

Fig. 3: *Process flow for power supply with PCB embedded transformer*

working at 5 MHz. Nevertheless, a few papers deal with the integration of fully isolated converters. Fig. 2(a) demonstrates a PCB integrated EMI filter with two coupled inductor with a toroidal core [10]. Fig. 2(b) present a LTCC transformer for high-voltage Flyback application, the transformer allows a low frequency switching frequency and the measured isolation capacitance is around 30 pF [11], [13]. [14] presents a PCB-integrated flyback transformer optimization for a 1 mm thin PFC rectifier without any experimental results.

The PCB embedding process presented in this work utilizes traditional Mn-Zn ferrite core and properly selected PCB materials with high glass-transition temperature (Tg) and low co-efficient of thermal expansion (CTE) to develop a transformer with coplanar winding structure for the highly integrated power supply used in harsh environment.

II. PCB MATERIAL SELECTION

Most commercially available PCB materials with high Tg have very high CTE ($> 30 \cdot 10^{-6}/\,^{\circ}$C) along z-direction, as listed in Table I. The embedded Mn-Zn ferrite material has a CTE of around 7 to $10 \cdot 10^{-6}/\,^{\circ}$C. Significant amount of thermal stresses could be applied on both ferrite core and PCB laminate due to this dramatic CTE mismatch especially for high-temperature applications. Delamination on the multilayer PCB assembly and cracking of ferrite core are the two most expected defects in the embedded structure, which would cause failure of the integrated power supply.

The Panasonic R-1515 PCB material is a ceramic-fiber enforced PCB material and presents a low CTE of $22 \cdot 10^{-6}/\,^{\circ}$C in z-axis (Table I) which is more compatible with that of ferrite materials. The reliability of embedded transformer would be improved accordingly. An embedded transformer prototype using traditional FR-4 material (epoxy resin with fiberglass) with higher CTEs ($60 \cdot 10^{-6}/\,^{\circ}$C in z-axis) was also manufactured as benchmark. Thermal cycling test has been used to evaluate the impact of different CTE mismatches on the integrity of the assemblies.

III. PCB LAMINATION PROCESS

Fig. 4 shows a cross-section view of the PCB embedded transformer with different layers. The ferrite core and coplanar winding were embedded into the PCB (Panasonic R-1515W) to implement the transformer. All components including active GaN devices were mounted on the top surface of the PCB substrate. This paper introduces the embedding process of transformer and the full integration process including active

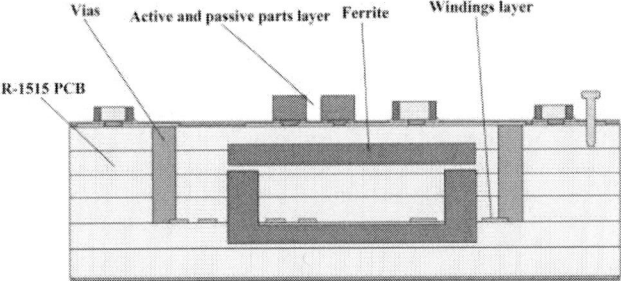

Fig. 4: *Schematic of power supply with PCB embedded transformer*

device. The detailed information about the PCB embedding process of transformer is introduced below:

- Two commercially available ferrite cores (ER 9.5, Ferroxcube) were machined to U- and I-shapes, respectively (Fig. 5). This Mn-Zn ferrite core material was used due to its low core loss density at 1-2 MHz and high Curie temperature (300 °C) [15]. A sectioning machine with diamond blade and a grinding/polishing table were used to reshape the ferrite cores.

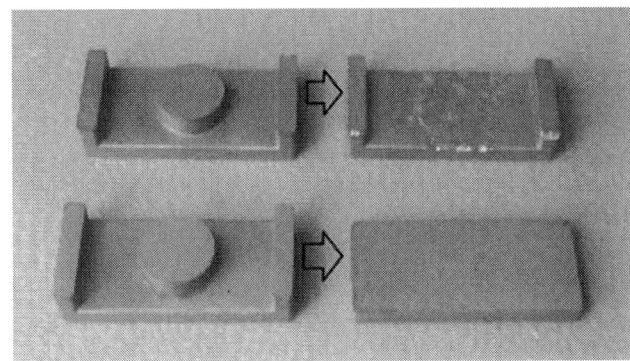

Fig. 5: *Machined UI-shape ferrite core*

- PCB panels were cut using a CO2 laser machine. Unneeded copper (Cu) on PCB panels was removed by ferric chloride etching solution.

- Two R-1515W laminate sheets with 1 oz. Cu were used for the preparation of embedded winding layer

TABLE I: High-Temperature PCB Materials

Manufacturer	PCB	Tg [C]	CTEz[10^{-6}]	CTEx[10^{-6}]	CTEy[10^{-6}]	Material
Arlon	35N	250	51	16	16	Polymide
Arlon	85N	260	50	16	16	Polymide
Panasonic	R1515E	250	22	9	9	N/A
Panasonic	R1515W	250	22	9	9	N/A
Rogers	Duroid 6202R	326	30	15	15	PTFE

and the top active layer. The top Cu surfaces were patterned and etched to form the winding for transformer and circuitry for power supply, respectively.

- The PCB stack consisting of alternate c-stage laminates and b-stage prepregs was aligned with ferrite cores.

- The lamination process for the embedded transformer was divided into three steps: The flat I-shape ferrite core with the top circuitry layer was laminated into one piece at the first. Then, the U-shape core with the internal winding layer was embedded into the PCB multilayer substrate. The last lamination was performed to integrate the two parts made in the previous steps. The customized air-gap distance can be adjusted using different thicknesses and/or numbers of the bonding prepreg layer. Fig. 6 shows the picture of the final assembled passive substrate.

- Vias were drilled by a laser process, and the connections between the internal winding layer and the top circuitry were made with soldered Cu pins.

Fig. 6: *Assembled passive substrate with embedded transformer*

The integration process for the power supply with PCB embedded transformer is illustrated in Fig. 3. The total thickness of the assembled passive substrate is about 3.6 mm. The complete converter prototype was built by mounting other components on top of it.

IV. THERMAL CYCLING OF PCB EMBEDDED TRANSFORMER

The reliability of the PCB with embedded ferrite core transformer was evaluated using thermal cycling method. Due to the difference in CTE between the ferrite and PCB substrate materials, major defects such as delamination at interfaces

and cracking of ferrite core could happen. The cycling profile applied to the PCB samples follows the aeronautic constraint, with a temperature range from $-55\,^\circ$C to $200\,^\circ$C in the heating/cooling rate of $20\,^\circ C/min$ and 20-min holding period at each extreme. In order to make a comparative study, two specific samples were manufactured for this test using Panasonics R-1515W and Isolas FR-406 PCB materials, respectively. After 1000 cycles, no delamination and other major defects were observed on the sample using Panasonics PCB materials, as shown in Fig. 7b. The color changing may be the results of an oxidation and the air-void was created during the abrasive process for the sectional view. Serious delamination and ferrite cracking can be observed in Fig. 7d on the PCB-embedded sample using FR-4 materials after only 150 cycles.

(a) Panasonic R1515 sample before cycling test (b) Panasonic R1515 samples at 1000 temperature cycles

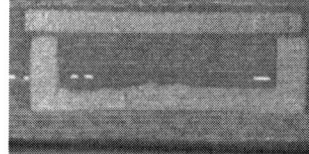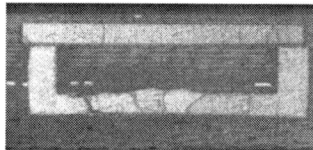

(c) FR-4 samples before cycling test (d) FR-4 samples after 150 temperature cycles

Fig. 7: Temperature cycling test

V. EXPERIMENTAL RESULTS

The circuit diagram of this power-supply prototype with PCB embedded transformer is shown in Fig. 8. This converter uses an active-clamp flyback topology operating at resonant mode [16], [17]. The switching frequency is 2 MHz, 15 V for the input voltage and 6 V for the output voltage. This topology could compensate the high leakage inductance from the transformer using a clamp capacitor. The LC resonant tank was designed to achieve ZVS mode for the transistors and ZCS mode for the diodes in the rectifier. The volume of this fabricated converter is 15 mm x 22 mm x 3.6 mm (Fig. 10).

In Fig. 9, the resonant phase on the orange drain-source voltage (1) and the GaN body-diode conduction during the dead time (2), allow to check the right design of the resonant

978-1-4673-9551-9/16 $31.00 © 2016 IEEE

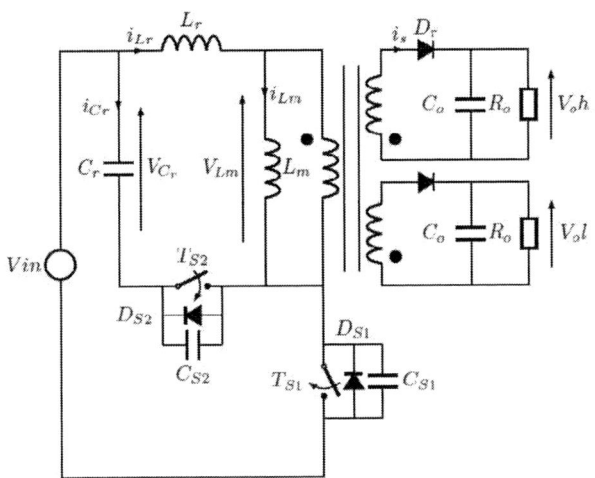

Fig. 8: *Active Clamp Flyback circuit*

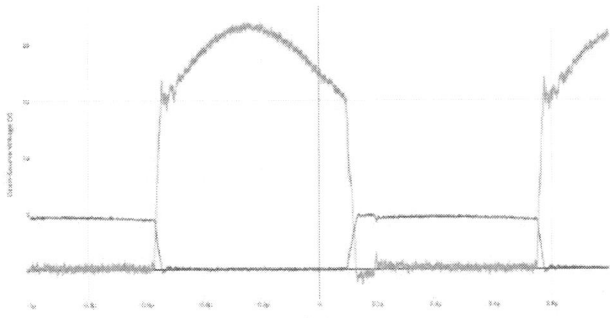

Fig. 9: *Ts1 and Ts2 Drain Voltages*

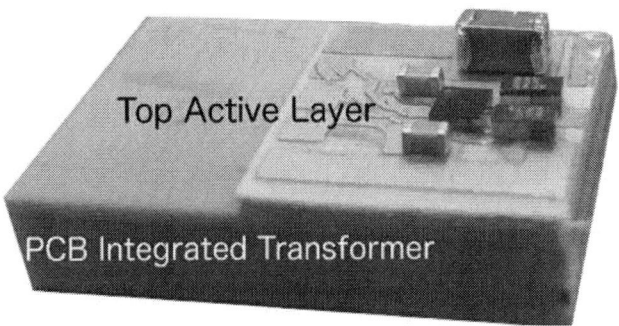

Fig. 10: *First prototype with top components*

point.

Low temperature gate driver and the EPCs GaN transistors (EPC8003) were used in the first prototype. But all passive components were chosen based on their high-temperature capabilities. The clamp capacitor is NPO Kemet ceramic capacitor and the decoupling capacitor comes from the high-temperature (250 °C) Ipedia silicon capacitor process.

The PCB embedded transformer was characterized by measuring the primary and secondary inductances and the variation

between simulation and experimental results does not exceed 4%. The measured efficiency of this integrated converter is about 72% at 2 W output power. In Fig. 11 the efficiency of the

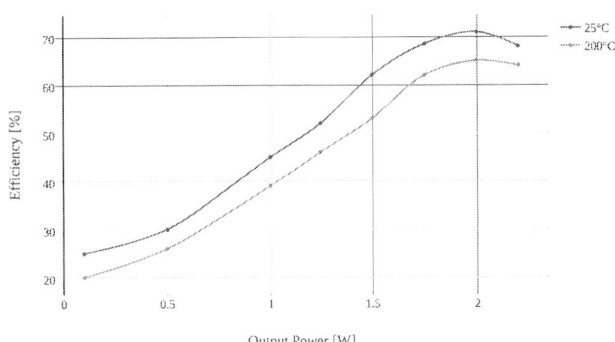

Fig. 11: *Power supply efficiency at ambient temperature and* 200 °C *environment*

power supply is experimentally verified. The decrease of the efficiency is mainly due to the increase of drop-voltage in the rectifier (0.8 to 1.3 V). The rectifier drop voltage is responsible of 55% of the global losse at 200 °C. The transformer losses are quiet constant (25%) because of the low maximum value of the core saturation (0.15 mT). Different experimentation on JBS SiC diode technology are currently performed to reduce the drop-voltage in high-temperature.

VI. CONCLUSION

A new integration process for the PCB embedded transformer has been proposed in this paper. In comparison with the conventional tools and know-how, this developed fabrication process allows the embedding of commercial ferrites with a good control of the air-gap. The high-temperature application requires the use of a PCB material with high Tg and suitable CTE compatible with embedded ferrite core material. The reliability of PCB embedded transformer fabricated in this study has been demonstrated using thermal cycling test. The first 2-W converter prototype shows 72% power efficiency at 2 MHz switching frequency. The final assembly between the top active-layer and the PCB winding layer will be improved in the future work.

ACKNOWLEDGMENT

The authors would like to thank Panasonic for providing the PCB materials used in this study. And The MEGaN project.

REFERENCES

[1] J. Popovic and J. Ferreira, "Converter concepts to increase the integration level," *Power Electronics, IEEE Transactions on*, vol. 20, no. 3, pp. 558–565, May 2005.

[2] ——, "An approach to deal with packaging in power electronics," *Power Electronics, IEEE Transactions on*, vol. 20, no. 3, pp. 550–557, May 2005.

[3] Y. Su, Q. Li, and F. Lee, "Design and evaluation of a high-frequency ltcc inductor substrate for a three-dimensional integrated dc-dc converter," *Power Electronics, IEEE Transactions on*, vol. 28, no. 9, pp. 4354–4364, Sept 2013.

[4] P. Artillan, M. Brunet, D. Bourrier, J.-P. Laur, N. Mauran, L. Bary, M. Dilhan, B. Estibals, C. Alonso, and J. Sanchez, "Integrated lc filter on silicon for dc-dc converter applications," *Power Electronics, IEEE Transactions on*, vol. 26, no. 8, pp. 2319–2325, Aug 2011.

[5] M. Ludwig, M. Duffy, T. Donnell, and C. O Mathuna, "Pcb integrated inductors for low power dc-dc converter," in *Applied Power Electronics Conference and Exposition, 2002. APEC 2002. Seventeenth Annual IEEE*, vol. 1, 2002, pp. 319–325 vol.1.

[6] Q. Chen, Z. Gong, X. Yang, Z. Wang, and L. Zhang, "Design considerations for passive substrate with ferrite materials embedded in printed circuit board (pcb)," in *Power Electronics Specialists Conference, 2007. PESC 2007. IEEE*, June 2007, pp. 1043–1047.

[7] D. Hou, Y. Su, Q. Li, and F. Lee, "Improving the efficiency and dynamics of 3d integrated pol," in *Applied Power Electronics Conference and Exposition (APEC), 2015 IEEE*, March 2015, pp. 140–145.

[8] Murata, "Ltcc murata power supply lxdc55kaaa datasheet."

[9] W. Zhang, Y. Su, M. Mu, D. Gilham, Q. Li, and F. Lee, "High-density integration of high-frequency high-current point-of-load (pol) modules with planar inductors," *Power Electronics, IEEE Transactions on*, vol. 30, no. 3, pp. 1421–1431, March 2015.

[10] M. Ali, E. Laboure, F. Costa, B. Revol, and C. Gautier, "Hybrid integrated emc filter for cm and dm emc suppression in a dc-dc power converter," in *Integrated Power Electronics Systems (CIPS), 2012 7th International Conference on*, March 2012, pp. 1–6.

[11] A. Roesler, J. Schare, and C. Hettler, "Integrated power electronics using a ferrite-based low-temperature co-fired ceramic materials system," in *Electronic Components and Technology Conference (ECTC), 2010 Proceedings 60th*, June 2010, pp. 720–726.

[12] M. Note, "Low temperature co-fired ceramics (ltcc) multi-layer module boards," *http://www.murata.com/ /media/webrenewal/support/library/catalog/products/substrate/ltcc/n20e.ashx*, 01/2014.

[13] A. Roesler, J. Schare, S. Glass, K. Ewsuk, G. Slama, D. Abel, and D. Schofield, "Planar ltcc transformers for high-voltage flyback converters," *Components and Packaging Technologies, IEEE Transactions on*, vol. 33, no. 2, pp. 359–372, June 2010.

[14] C. Marxgut, J. Muhlethaler, F. Krismer, and J. Kolar, "Multi-objective optimization of ultra-flat magnetic components with a pcb-integrated core," in *Power Electronics and ECCE Asia (ICPE ECCE), 2011 IEEE 8th International Conference on*, May 2011, pp. 460–467.

[15] "3f45 datasheet."

[16] R. Watson, F. Lee, and G. Hua, "Utilization of an active-clamp circuit to achieve soft switching in flyback converters," *Power Electronics, IEEE Transactions on*, vol. 11, no. 1, pp. 162–169, Jan 1996.

[17] B.-R. Lin, K. Huang, and D. Wang, "Analysis, design, and implementation of an active clamp forward converter with synchronous rectifier," *Circuits and Systems I: Regular Papers, IEEE Transactions on*, vol. 53, no. 6, pp. 1310–1319, June 2006.

Design Consideration of Gate Driver Circuits and PCB Parasitic Parameters of Paralleled E-mode GaN HEMTs in Zero-Voltage-Switching Applications

Juncheng Lu, Hua(Kevin) Bai
Department of Electrical and Computer Engineering
Kettering University
Flint, MI 48504, USA
hbai@kettering.edu

Alan Brown, Matt McAmmond
Hella Corporate Center USA Inc.
Plymouth Twp, MI 48170, USA
matt.mcammond@hella.com

Di Chen, Julian Styles
GaN Systems Inc.
Ottawa, ON. Canada K2K 3G8
jstyles@gansystems.com

Abstract-**This paper designed the gate driver circuits and optimized the PCB layout in a 7.2kW battery charger using paralleled GaN HEMTs. 650V/60A enhancement mode GaN HEMTs provided by GaN Systems Inc are adopted. To optimize the switching performance of paralleled GaN HEMTs with low loss and high reliability, effects of parasitic inductance and capacitance are modeled and analyzed. Through cancelling the flux in the commutation loop, the power-loop parasitic inductance is reduced to only 0.7nH, which significantly decreases the electrical stress in the switch turn-off process. A diverse-parameter gate driver design has been proposed to achieve the reliable switching off. The Finite-Element-Analysis and Spice simulation show our current design could effectively suppress the voltage overshoot and gate-drive ringing on HEMTs. Experiments were carried out on both double pulse test platform and the 7.2kW charger to verify the proposed design strategy.**

Keywords—**gallium nitride, high electron mobility transistor, zero voltage switching, parasitic, double pulse test, battery charger, electric vehicle**

I. INTRODUCTION

It is believed that wide-bandgap (WBG) devices like Silicon Carbide (SiC) and Gallium Nitride (GaN) represent the next-generation power electronics switches, which exhibit the higher switching frequency, lower switching loss and better thermal capability. As an attempt to utilize such switches, in this paper authors adopted 650V/60A GaN HEMTs with zero-voltage-switching (ZVS) control for a level-2 charger (208VAC/7.2kW) used in electric vehicles (EVs). The control algorithm is similar to [1], and the topology is an indirect matrix converter shown in Fig. 1. Here the front-end rectifier stage (M1~M4) operates at 60Hz, converting the sinusoidal waveform into a 120Hz single-polarity DC voltage. As the result, no back-to-back switches are needed like the conventional matrix converter, which facilitates the system design and control. On the other hand, the switching frequency of the front-end rectifier is 60Hz, indicating the switching loss is negligible. For other switches (P_1~P_4, S_1~S_4), GaN HEMTs are the excellent candidates due to ultra-fast switching transitions and ultra-low gate-drive power consumption compared to the traditional Si MOSFETs.

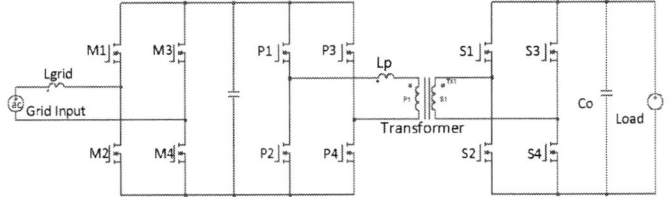

Fig.1 Proposed Indirect Matrix Converter

Among all types of GaN HEMTs available on the market [2], we selected the GS66516T from GaN Systems Inc. Such device is compatible with the regular gate-driver of Si MOSFETs, which significantly facilitates the system design. Meanwhile the top-cooled package makes it easy to dissipate the heat. The switch has much smaller area compared to Si MOSFETs, shown in Fig.2(a) saving much space in the final prototype, or the same space could be utilized to conduct the higher current. Based on our double pulse test (DPT) results shown as Fig. 2(b), a brief loss breakdown analysis of our charger has been carried out as Fig.3(a), in which different solutions are compared at the same switching frequency(153k~500kHz). The conduction loss always dominates. Therefore paralleling more switches will significantly decrease the system loss and increase the system efficiency.

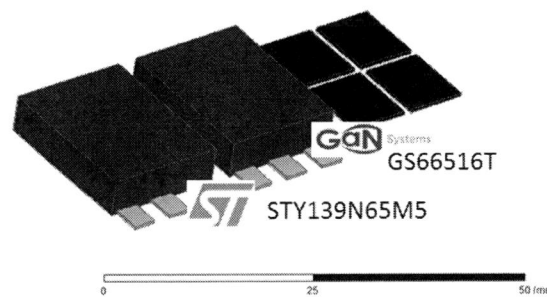

Fig.2 (a): 4 GaN HEMTs GS66516T vs 2 Si MOSFETs

However, the paralleled devices bring other challenges. For example, parasitics in the gate driver and power loops will cause the gate-signal ringing and voltage spikes on

978-1-4673-9551-9/16 $31.00 © 2016 IEEE

switches. Therefore the system layout needs be particularly taken care of [3].

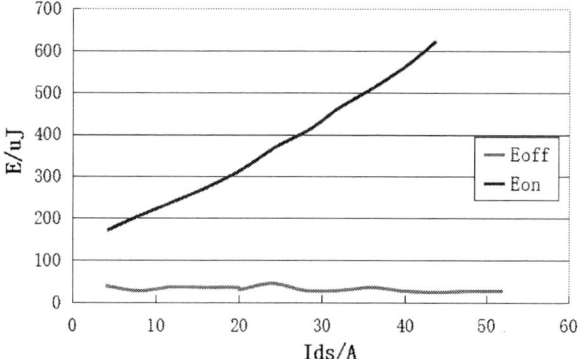

Fig.2 (b): Measured Switching Characteristics of GS66516T

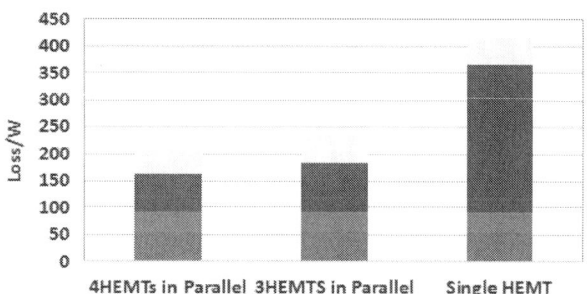

Fig. 3 (a): Loss Breakdown Comparison of different Solutions

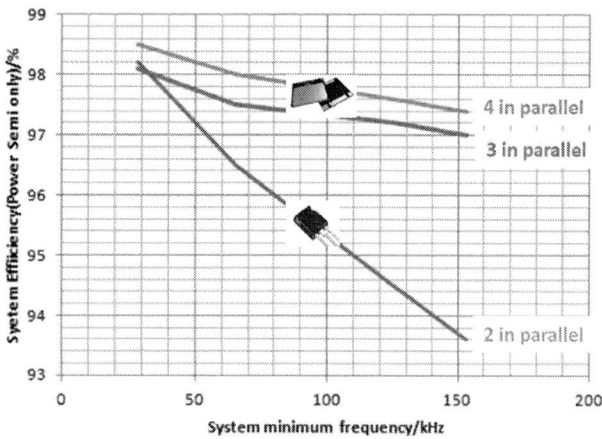

Fig. 3 (b): System efficiency comparison

In this paper, the device gate-drive circuit and PCB design will be emphasized. The control algorithm will not be detailed. Section II proposes a special gate-drive design particularly beneficial for ZVS applications using paralleled switches. Section III presented the system layout to shrink the system parasitics especially the gate-drive and power loop inductance. Section IV is the experimental validation through the double-pulse test. Section V is the conclusion.

II. DESIGN CONSIDERATION OF GATE DRIVER CIRCUITS

A. Assessment of Parasitics and Proposed Diverse Parameter Gate-driver Circuit

The parasitic capacitance of Enhancement-mode GaN HEMTs used in this design is much smaller than traditional Si devices. All the capacitance is at ~pF level. This will secure its faster switching speed, however, requires more critical system design to shrink the parasitic inductance. Take one leg made of two complementary switches as an example. Various parasitic inductance exists in the gate-drive loop and main power loop, as shown in Fig.4. To facilitate the analysis, here we only parallel two switches for the easy illustration. Their side effects are summarized as Table.1. In our ZVS application where only hard switching off exists, reducing the commutation loop inductance could allow driving GaN HEMTs faster thereby reduce the switching-off loss.

To minimize the gate-drive loop inductance and the cross talking between the gate-drive loop and power loop, the Kelvin terminal is usually employed in high-frequency applications. Although the paralleled switches share the same gate driver and have Kelvin connection, the loop-inductance, as highlighted in Fig.5 could potential be problematic due to the existence of the quasi-common-source inductance(L17,L18 and etc). The unbalanced quasi-common source inductance and di/dt will eventually cause an unbalanced voltage across the $V_{GS.}$

Shown in Fig.5, The loop current could be described as Eqn. (1). The induced ringing on the gate voltage of HEMT 1 and HEMT 2 is shown as Eqns (2) and (3), respectively. Here i_{loop} means the current circulating among paralleled switches.

$$Ls1\frac{di_1}{dt} - Ls2\frac{di_2}{dt} = i_{loop} * (Zss1 + Zss2) \tag{1}$$

$$V_{feedback\_1} = Zss1\frac{Ls1\dfrac{di_1}{dt} - Ls2\dfrac{di_2}{dt}}{Zss1 + Zss2} \tag{2}$$

$$V_{feedback\_2} = Zss2\frac{Ls1\dfrac{di_1}{dt} - Ls2\dfrac{di_2}{dt}}{Zss1 + Zss2} \tag{3}$$

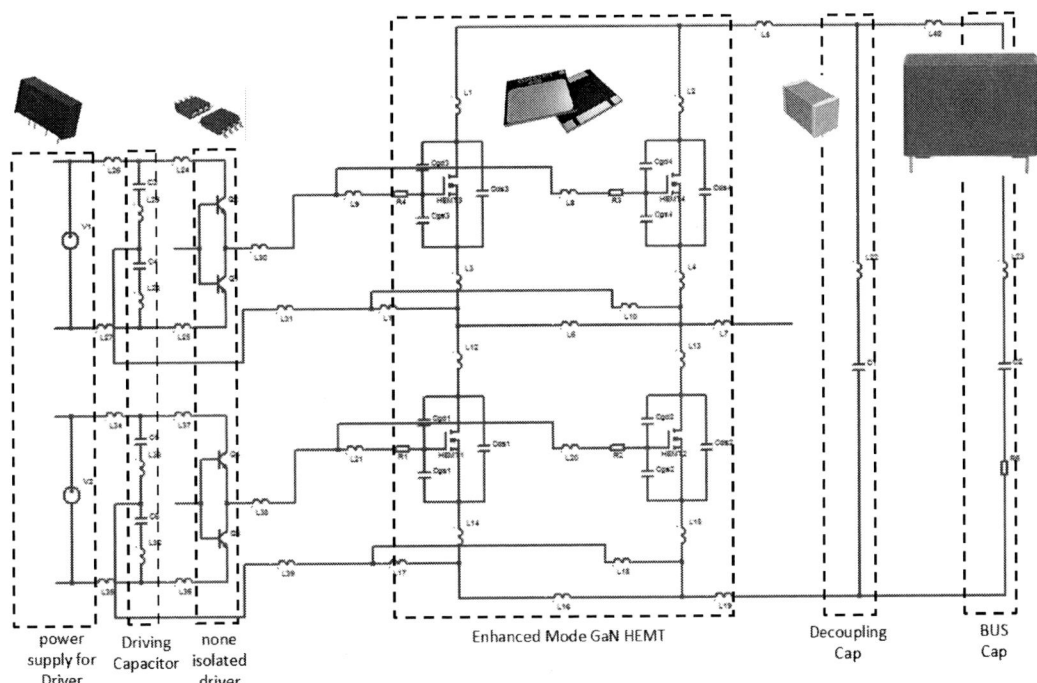

Fig.4: Equivalent Circuit of A Half Bridge with Two HEMTs in Parallel

Table 1: Effect of Parasitic Parameters on the Switching Performance

Parameter	Description	Effect		Priority	Design Consideration
		Switching On	Switching off		
L26 L27 L34 L36	Stray inductance	None	None	Low	The smaller the better
C3~ C6	Driving capacitor	Charging or discharge Ciss		Medium	At least 10 times Larger than Ciss
L24 L29 L33 L37	Switching-on Lg	Increase ringing on Vgs	None	Very High	The smaller the better
L25 L28 L32 L36	Switching-off Lg	None	Increase ringing on Vgs	Very High	
Q1 Q3	Totem pole switches for switching-on	Slow down the switching	None	Very High	
Q2 Q4	Totem pole switches for switching-off	None	Slow down switching	Very High	
L30 L38 L31 L39 L9 L21 L8 L20	Lg	Increase ringing on Vgs		Very High	
R1 R2 R3 R4	Rg	Slow down switching		Very High	Depends on applications
Cgd	Miller Capacitance	Charge or discharge current from gate driver when Vds changes		Very High	Decided by Switches
Ciss	Input capacitance	Bypass miller current Increase switching delay and gate driver loss		High	
Coss	Output capacitance	Decrease dv/dt of Vds during switching		High	
L14 L15 L3 L4	Common source inductance	Feedback voltage on the gate when Ids changes		Extremely High	The smaller the better
L17 L18 L16 L6 L10 L11	Quasi Common source inductance	Same as the common source inductance			
L19 L5 L1 L2 L22	Power loop inductance	Increase voltage spike on Vds			
C1	Decoupling cap	Bypass di/dt to decrease the voltage overshoot		High	At least 10 times Larger than Coss
C2	Bus cap	Stabilize the DC-bus voltage		High	Depends on application
L40 l23	Stray inductance	Increase voltage ringing across decoupling cap		Medium	The smaller the better

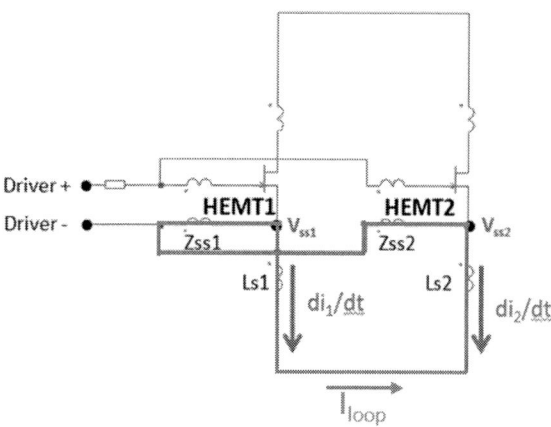

Fig. 5: Equivalent Circuit of A Half Bridge with Two HEMTs in Parallel

Eqn(1) indicates that increasing Zss1 and Zss2 in the gate-loop, e.g., increasing the gate-resistance could limit the loop current I_{loop}. However, such effort will slow down the switching speed and not eliminate the ringing on the gate, shown as Eqns (2)~(3). Shown in Fig.6(a), a 1nH difference between the common-source inductance (Ls1 and Ls2) of two paralleled GaN HEMTs will result in the huge ringing on the gate, which will mis-trigger switches or damage the gate. One possible solution is a diverse-parameter based gate-drive circuit shown in Fig.6(b). Instead of using the same gate-drive resistance, we purposely placed a bigger gate-resistance for one switch, e.g., HEMT 2. When turned on, since they are all ZVS turned on, the different turn-on speed will not affect the switching-on behaviors. When turned off, HEMT1 will first turn off at ZVS condition since HEMT2 is still on. The gate ringing will be fully eliminated, shown in Fig.6(b). Since HEMT 2 eventually will handle all the current when being switched off, when selecting the switches we need leave enough current allowance. Note the simulated waveform is obtained through the Pspice model.

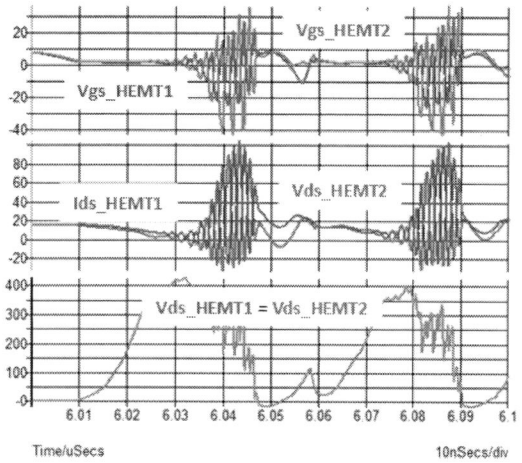

Fig.6(a) Gate-drive Signals Based on the Conventional Gate-Drive Method

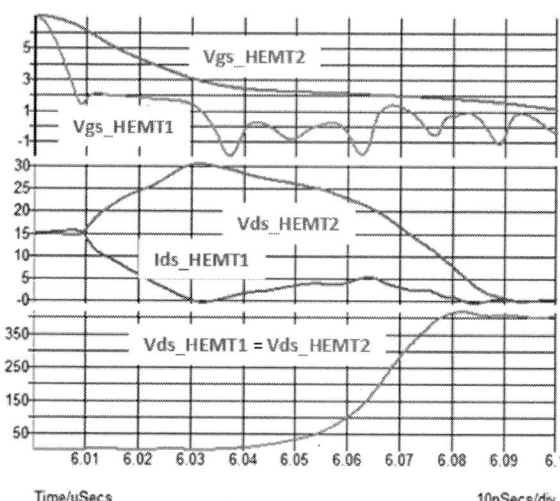

Fig.6(b) Gate-Drive Signals Based on the Proposed Gate-Drive Method

(Top: V_{GS}, Mid: I_D, Bottom: V_{DS})

B. A Novel Gat-Drive Circuit with Different Switching on and off Speed

The gate driver design is very challenging in high-speed GaN HEMT based application because of its fragile gate. The equivalent circuits of the gate driver are shown as Fig.7.

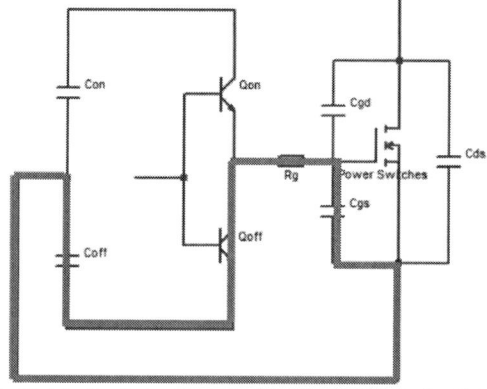

Fig. 7(a) the Equivalent Circuit of the Gate Driver During Switching Off

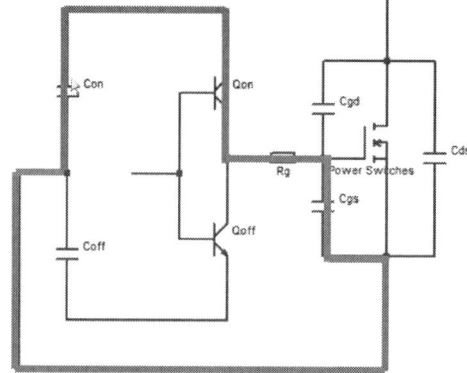

Fig. 7(b) the Equivalent Circuit of Gate Driver During Switching On

To be immune from the miller affect, a miller clamp circuit is proposed as Fig.8(a). When the device is fully

978-1-4673-9551-9/16 $31.00 © 2016 IEEE

switched off, the miller clamping transistor is on to bypass the miller charge. Fig.8(b) is using different gate parameters to achieve different switching on and off speed. The gate drive circuit in our system is shown in Fig.8(c). Changing Ron and Roff will change the switching on and off speed.

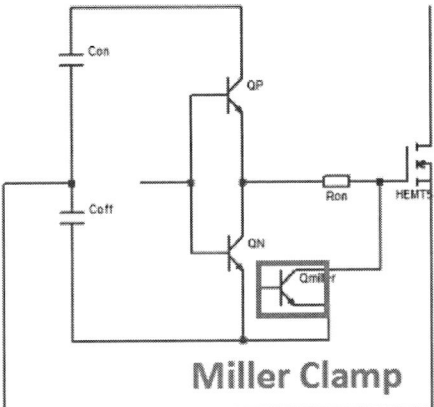

Fig. 8(a) Miller Clamp Circuit

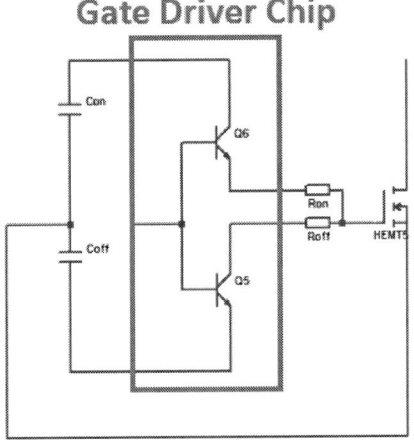

Fig. 8(b) Special Gate Driver Chip with Separate Output

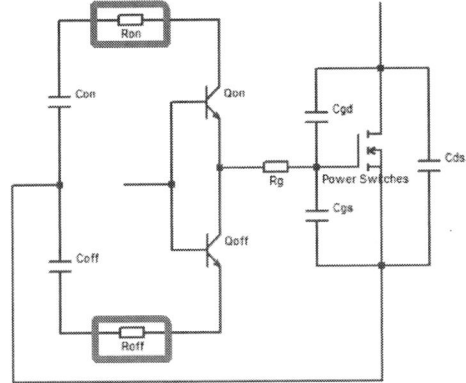

Fig. 8(c) Proposed Gate Driver Circuit

C. Dead-band loss

Fig.9 shows the current distribution in the switching-off process of a half bridge.

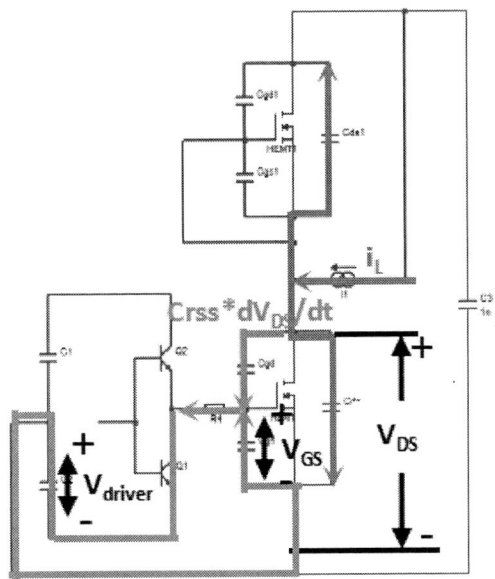

Fig. 9 Switching-off Process

$$\frac{V_{GS} + V_{driver}}{Rg} - Crss * \frac{dV_{DS}}{dt} = -Ciss * \frac{dV_{GS}}{dt} \quad (4)$$

$$\frac{dV_{DS}}{dt} = \frac{i_L}{2 * Coss + Cpcb} \quad (5)$$

Increasing V_{driver} could make the first item of Eqn (4) dominate the gate current so that the device could be immune from the miller effect, which results in a higher gate current thereby accelerates the switching off. The switching-off loss comparison with different V_{GS} is shown as Fig.10. The system with +6.8V/-5.2V V_{GS} has a very flat Eoff-Ids curve compared to +6.8V/0V.

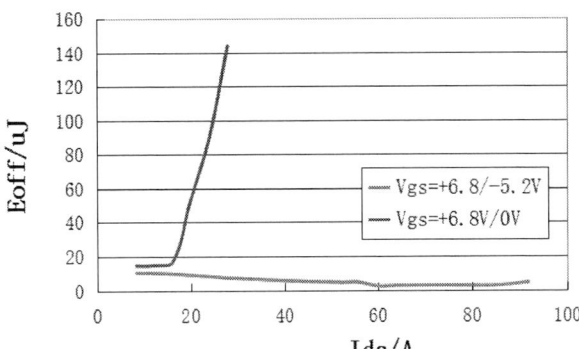

Fig. 10 DPT based Eoff Comparison with Different Vgs

Because of the low turn-on threshold gate voltage of GaN HEMTs, a negative V_{GS} is always preferred to shut off the switch and enhance the system reliability. However, different from Si MOSFET, no body diode exists inside GaN HEMTs. In the reverse conduction mode, the drain will behave as the source and the source will act as the drain. When V_{GD} is higher than the reverse threshold voltage $V_{th\_GD,}$ slightly higher than Vth_GS, the switch will be on to freewheel the current. The voltage drop across the switch is

$$Vsd = Vth\_gd - Vgs\_off + id * Rdson \qquad (6)$$

The more negative the gate turn-off voltage, the more dead-band loss. A systematic comparison with different negative gate voltage will be made in Section IV.

III. PROPOSED LAYOUT STRATEGY

With the magnetic-flux-canceling design, the parasitics of the PCB could be greatly reduced. The direction of the high frequency current between two layers are opposite so that the generated flux could be cancelled. The layout of four HEMTs in parallel is modeled in Ansys Q3D, and the power loop and gate-driver loop inductance are evaluated by Finite Element Analysis in Fig. 11(a). The power-loop inductance is only 0.7nH. The gate-drive loop has 4.3~7.8nH inductance.

Compared to Direct Bonded Copper(DBC), the PCB could easily adopt the multi-layer structure and smaller distance between layers to achieve a better magnetic flux canceling effect. A comparison with previously released power module designs is shown as Fig.11(b). The power loop inductance of the proposed design is only 1/4 of the best E-mode GaN HEMT power module.

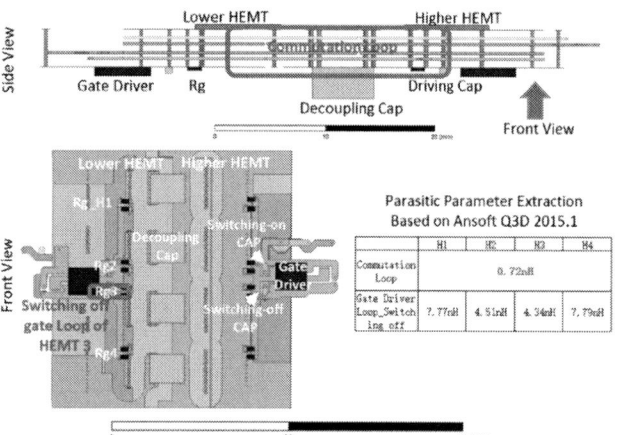

Fig. 11(a) Loop Inductance Evaluation of Four Paralleled GaN HEMTs

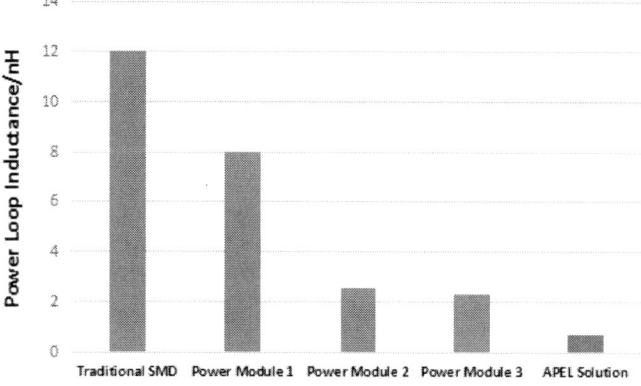

Fig. 11(b) Comparison of Different Layout Methods

IV. EXPERIMENTAL VALIDATION

A DPT platform with four GaN HEMTs and current shunts in parallel is built as Fig.12(a). The switching current waveforms @400V are shown in Fig.14(b)-(d). The paralleled switches could switch off 400V/92A. The voltage spike across the switch is 72V@di/dt=15A/nS. The current sharing among switches is even. Experimental waveforms support the previous theoretical analysis.

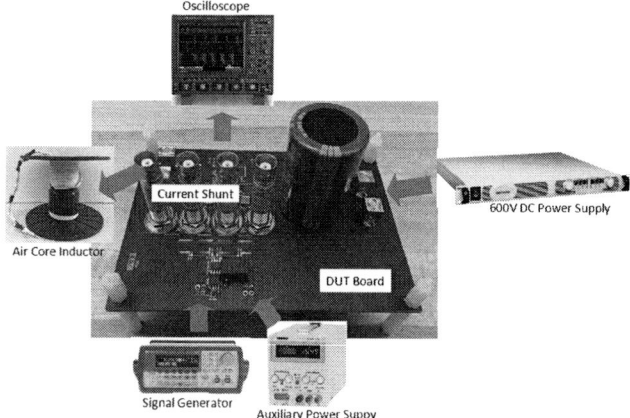

Fig. 12(a)   Double Pulse Test Platform with Current Shunts

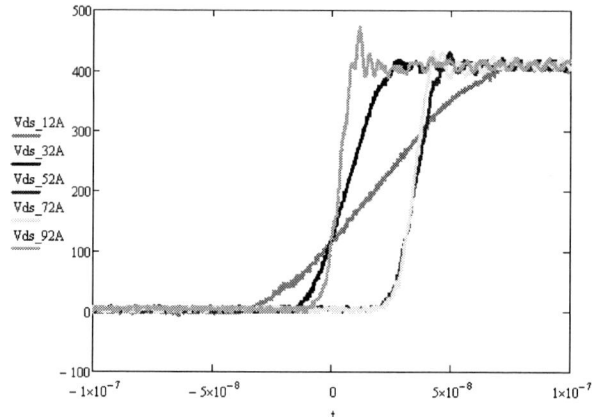

Fig. 12(b) V_DS at the Different Switching Current

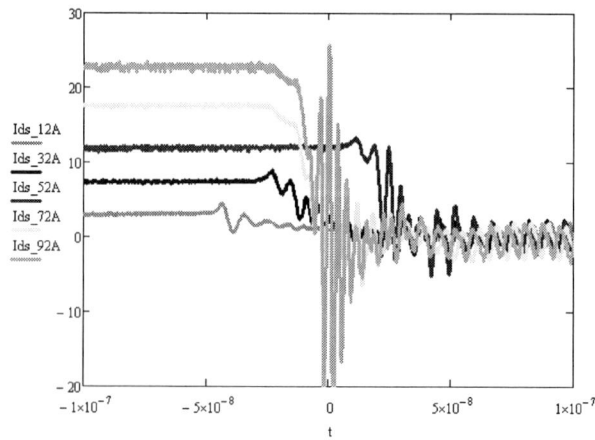

Fig. 12(c) Different Switching-off Current

978-1-4673-9551-9/16 $31.00 © 2016 IEEE

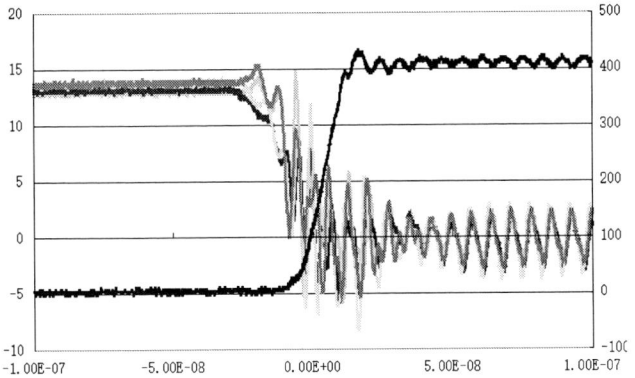

Fig. 12(d) Switching-off Current for Each Switch@400V/54A

Based upon the test above, the indirect matrix converter based 7.2kW on-board charger is prototyped as Fig.13(a). Different negative gate voltage are applied to switch off the devices and compared as Fig 13(b). At this power stage, the dead-band loss dominates the system loss, and the -2.3V solution leads to a higher system efficiency. With the full-power test being carried out in the near future, the optimal negative gate-voltage will be found.

Fig. 13(a)   Indirect Matrix Converter Based Level-2 On-board Charger

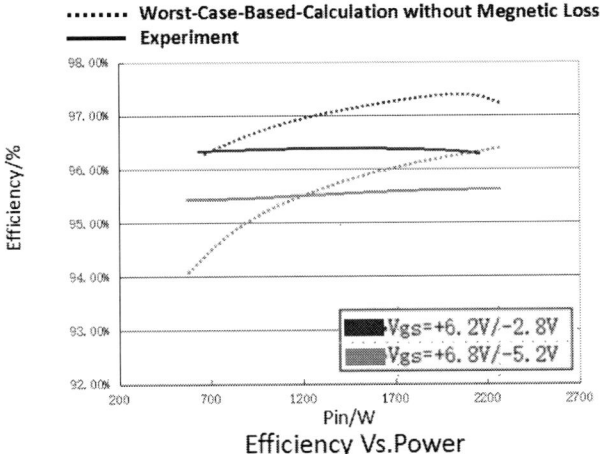

Fig. 13(b) System Efficiency Comparison

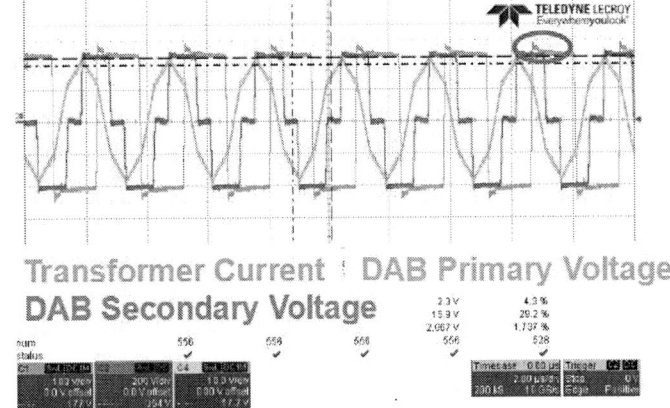

Fig. 13 (c) Test Waveform@Vgs=6.2V/-2.8V

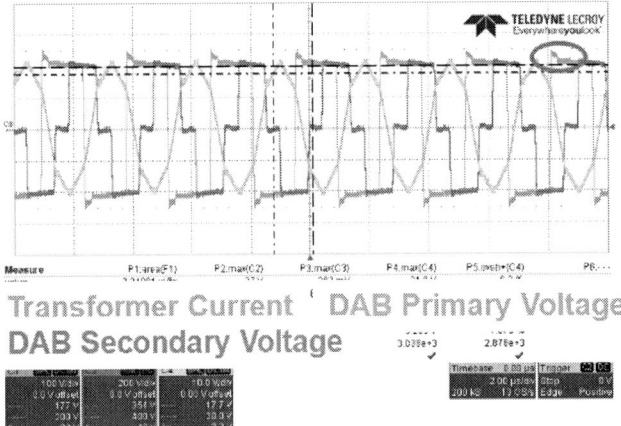

Fig. 13(d) Test Waveform@Vgs=6.8V/-5.2V

V. CONCLUSION

This paper paralleled four GaN HEMTs to form one switch module, which is applied in an indirect matrix converter based 7.2kW charger. To guarantee the uniform switching performance among paralleled switches, PCB layout and a diverse-parameter based gate-drive circuit are proposed. The preliminary test result on the DPT platform and the charger validated its effectiveness. More detailed results will be published once the full-power test is finished.

REFERENCES

[1]   F.   Jauch   and   J.Biela,   "Single-phase single-stage bidirectional isolated ZVS AC-DC converter with PFC", 15th international Power Electronics and Motion Control Conference, 2012, pp. LS5d.1-1 - LS5d.1-8.

[2]   E.A. Jones, F. Wang and B. Ozpineci, "Application-based review of GaN HFETs", IEEE Workshop on Wide-Bandgap Power Devies and Applications, 2014, pp. 24 – 29.

[3]   F.Luo, Z.Chen, L.Xue, P.Mattavelli, D.Boroyevich, B.Hughes, "Design Considerations for GaN HEMT Multichip Half- bridge Module for High-Frequency Power Converters", online

A Gate Driver of SiC MOSFET for Suppressing the Negative Voltage Spikes in a Bridge Circuit

Qi Zhou, Feng Gao
School of Electrical Engineering
Shandong University
Jinan, China
qzhousdu@gmail.com

Abstract—**SiC MOSFET has low on-state resistance and can work on high switching frequency, high voltage and some other tough conditions with less temperature drift, which could provide the significant improvement of power density in power converters. However, for the bridge circuit in an actual converter, high dv/dt during fast switching transient of one MOSFET will amplify the negative influence of parasitic components and produce significant voltage spikes on the complementary MOSFET, which will threaten its safe operation. This paper proposes a new gate driver circuit for SiC MOSFET to attenuate the negative voltage spikes in a bridge circuit. The proposed gate driver adopts a simple voltage dividing circuit to generate a negative gate-source voltage as traditional and a passive triggered transistor with a series capacitor to suppress the negative voltage spikes, which could satisfy the stringent requirements of fast switching SiC MOSFETs under high dc voltage condition with low cost and less complexity. An analysis is presented in this paper based on the simulation and experimental results with the performance comparison evaluated.**

Keywords—SiC MOSFET; gate driver; bridge configuration; auxiliary transistor.

I. INTRODUCTION

Silicon-based technology has to some extent reached its physical limits for power handling and switching frequency capability. To meet the application demands of high power density, high switching frequency and high temperature, wide band gap devices have become important alternatives because of their unique characteristics and better performances [1]. As the promising wide band gap device, SiC MOSFET is an excellent alternative of Si IGBT for high power converters [2]. However, high switching speed may amplify the negative influence of parasitic components and produce the significant gate-source voltage spikes during switching transient [3]. Especially, compared to Si MOSFETs, the Silicon Carbide (SiC) MOSFETs can withstand the smaller negative gate-source voltage [4], which put forward the higher requirements for gate driver's design. Taking the commercial product of CREE's SiC MOSFET for example, the minimum allowable negative gate-source voltage of the first generation CREE's typical 1.2kV SiC MOSFET CMF20120D is only -5V. The second generation CREE's typical 1.2kV SiC MOSFET C2M0040120D has

significant improvement, but its negative gate-source voltage still cannot exceed -10V [5]-[6].

Bridge configuration is a commonly used topology in actual converters, where turning-on and turning-off of one switching device will generate the gate-source voltage spikes on its complementary device especially under the high switching speed condition due to the unavoidable cross talk phenomena. Gate impedance control and gate voltage control are two basic ideas assumed for cross talk suppression [7]. In specific, gate impedance control is to make the gate impedance of one switch become smaller during the switching transient of its complementary. Then most current will pass through the gate loop, which reduces the current that would induce spurious gate voltage. The normal method of gate impedance control is mainly to decrease the switching speed for reducing dv/dt, whose simple way is to add a capacitor between gate and source of SiC MOSFET [8]. This approach is simple and effective. However, it will unavoidably increase switching losses and ignore some switching signals. Reference [9] proposed to use the actively controlled gate resistance to reduce switching spikes, which however cannot fully satisfy the fast switching requirement of SiC MOSFET since it needs to detect gate voltage and make a feedback control.

On the other hand, gate voltage control is to provide a negative gate-source voltage when the positive gate voltage spikes occur. Reference [10] designed a level-shift circuit without using extra voltage source, which could generate the negative gate voltage to lower down the positive gate voltage spikes and speed up the switch OFF operation. However, the negative gate voltage brings the risk for damaging the devices when the negative gate-source spikes exceed the maximum allowable negative biased gate voltage. Reference [11] provided another method to eliminate the positive gate voltage spikes by controlling an auxiliary transistor without slowing down the switching speed of SiC JFET, but it cannot eliminate the negative spikes. So it is suitable for SiC JFET because its minimum negative gate-source voltage is -25V and the maximum drain-source voltage is only 30V [12]. The combined method proposed in reference [13] can eliminate both positive and negative spikes, which makes the gate impedance become small during the switching transient and uses an additional voltage source to provide negative gate voltage. However, the auxiliary actively controlled

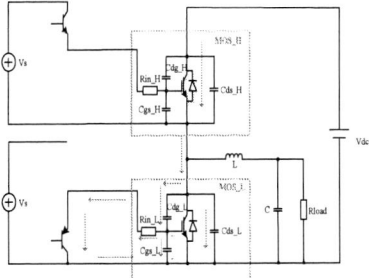

Fig. 1. Turning-on transient of the upper switch in a bridge configuration.

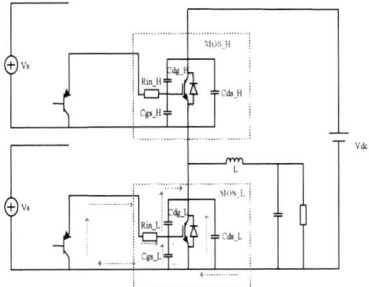

Fig. 2. Turning-off transient of the upper switch in a bridge configuration.

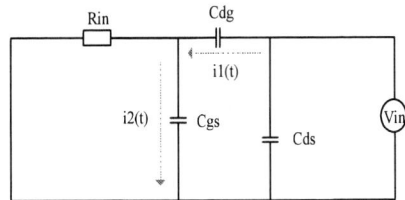

Fig. 3. Equivalent circuit of the lower switch's gate driver during transient process.

transistors and the additional voltage source will unavoidably increase the complexity and cost. Reference [14] assumed the level shift circuit to generate the negative gate-source voltage and added another passively controlled part to eliminate the negative gate-source voltage spikes. However, this circuit is still a little bit complex and has adverse effects on positive gate-source voltage spikes. Some soft switching drivers are also proposed for high switching frequency [15]-[16]. This paper proposes an enhanced gate driver of SiC MOSFET for achieving better performance on eliminating the gate-source voltage spikes, which adopts a traditional voltage dividing circuit to generate a negative gate-source voltage and a passive triggered transistor with a series capacitor suppress the negative voltage spikes. A simulation using LTspice and an experimental prototype were carried out to show the performance of the proposed gate driver under different drain-source voltages with the comparison to other traditional gate circuits.

II. GENERATION MECHANISM OF NEGATIVE GATE-SOURCE VOLTAGE SPIKES

This paper assumes the synchronous buck converter to analyze the generation mechanism of negative gate-source voltage spikes. In specific, when the upper switch is turning

on, the voltage between drain and source of the lower switch is suddenly rising up, which induce a current through the parasitic capacitance C_{dg_L} and it will charge up the gate-source capacitance C_{gs_L}, as shown in Fig. 1. Consequently, the gate voltage is pushed up and generates a spurious triggering pulse. On the other hand, when the upper switch is turning off, most of the load current will flow through the parasitic diode of the lower MOSFET. However, still part of the load current will flow through the gate-source capacitance C_{gs_L} as shown in Fig. 2. The gate voltage is pulled down and generates a negative gate-source voltage spike [17]-[18].

In order to analyze the values of turning-on and turning-off spikes caused by the complementary switch in a bridge configuration, the equivalent circuit of this transient process is given in Fig. 3, where V_{in} represents a time dependent external voltage source and is defined as:

$$V_{in} = k \times t \tag{1}$$

Where, k represents the changing rate of the external voltage. Furthermore, $i_{1(t)}$, $i_{2(t)}$, u_{Cgs} and u_{Cdg} can be derived as:

$$
\begin{cases}
i_1(t) = C_{dg} \dfrac{du_{Cdg}}{dt} \\[2mm]
i_2(t) = C_{gs} \dfrac{du_{Cgs}}{dt} \\[2mm]
u_{Cgs} = R_{in}[i_1(t) - i_2(t)] \\[2mm]
u_{Cgs} + u_{Cdg} = V_{in} = kt
\end{cases}
\tag{2}
$$

Referring to (2), the peak value of spurious gate-source voltage can be calculated as (3) [19]-[20].

$$u_{Cgs(max)} = kR_{dg}C_{dg} - kR_{in}C_{dg}e^{-\frac{V_{in}}{k(C_{dg}+C_{gs})R_{in}}} \tag{3}$$

When k approaches ∞, the peak value of U_{Cgs} is simplified as:

$$u_{Cgs(max)} = \frac{V_{in}}{1 + C_{gs}/C_{dg}} \tag{4}$$

It is noted that the peak value is determined by the equivalent parameters of switching device itself and the maximum value of the external voltage. When considering the parasitic inductance of gate driver, the spurious gate-source voltage is even larger. Further, it is observed that the peak value of U_{Cgs} can be reduced by increasing C_{gs} like those proposed in [8].

III. PROPOSED GATE DRIVER CIRCUIT AND ITS OPERATIONAL PRINCIPLE

Being different, the proposed gate driver neither assumes a large uncontrollable capacitor cross gate and source terminals nor implements an actively controlled circuit to

978-1-4673-9551-9/16 $31.00 © 2016 IEEE

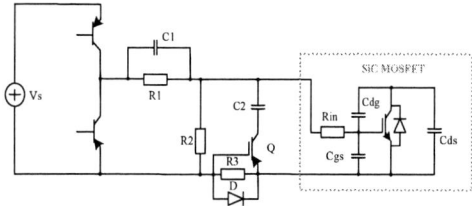

Fig. 4. The proposed gate driver circuit.

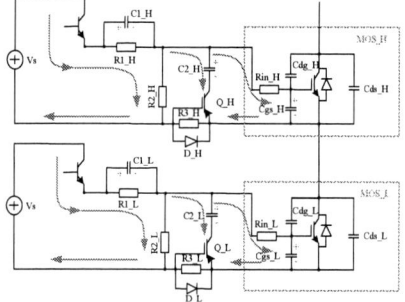

Fig. 5. Pre-charge process of proposed gate circuit.

achieve variable gate capacitance. According to the voltage between base and emitter terminals, the passively triggered auxiliary transistor in the proposed circuit can connect a capacitor to the gate terminal when needed. The detailed gate driver circuit in a bridge configuration is shown in Fig. 4, where the negative gate voltage is generated by using a voltage level shifter as traditional. The turning-on speed will be faster because C_{gs} can be charged by capacitor C2.

In details, this circuit consists of three resistors R1, R2 and R3, two capacitors C1 and C2, one diode D and a PNP transistor Q. R1 and R2 make up a series voltage divider circuit and can generate various negative gate-source voltages based on the values of R1 and R2 without any extra voltage source. The capacitor C1 can make sure to attain steady-state values of negative gate-source voltage. Therefore, C1 should be larger enough to keep the voltage almost be constant. Capacitor C2 series with a PNP transistor Q provide another small impedance loop when the negative gate-source voltage spike occurs. Resistor R3 controls the transistor passively and when the positive spike happens, the diode D can eliminate adverse effects on positive gate-source voltage spikes. The value design for the components in this circuit asks for detailed transient analysis.

A. Pre-charge Processing

The main reason for this process is making sure the voltage across C1 and C2 to attain steady-state values. Before the main circuit begins to work, the gate circuit should be pre-charged by gate voltage source to guarantee that the capacitors obtain enough energy. The value of positive gate voltage is the voltage of C2, which can be written as:

$$V_{gs+} = \frac{R_2}{R_1 + R_2} \times V_{source} \tag{5}$$

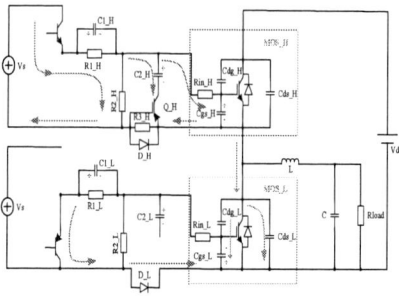

Fig. 6. Turning-on transient of MOS_H in a phase-leg configuration.

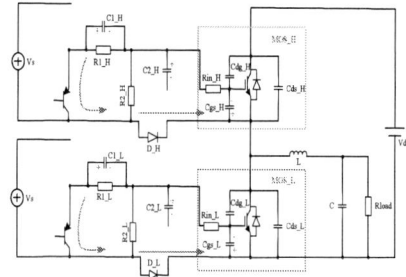

Fig. 7. Turning-off transient of MOS_H in a bridge configuration.

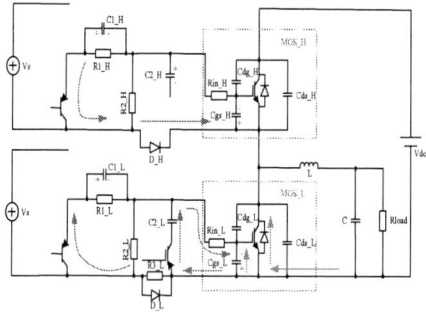

Fig. 8. Principle illustration of suppressing negative spurious gate voltage in a bridge configuration.

And the value of negative gate voltage is the voltage of C1, which can be derived as:

$$V_{gs-} = \frac{R_1}{R_1 + R_2} \times V_{source} \tag{6}$$

The capacitance of C2 should be much larger than C_{gs} so that the input capacitance of MOSFET would not be affected. During this process, part of the current will go through C_{gs} first, which make the voltage between the base and emitter of Q is negative.

$$i_s = C_1 \frac{dv_{R1}}{dt} + \frac{v_{R1}}{R_1} = C_{gs} \frac{dv_{Cgs}}{dt} + \frac{v_{R2}}{R_2} \tag{7}$$

$$v_{R2} = (R_{in} + R_3) \times C_{gs} \times \frac{dv_{Cgs}}{dt}$$

$$v_{eb} = R_3 \times C_{gs} \times \frac{dv_{Cgs}}{dt} \tag{8}$$

R3 should be large enough to make v_{eb} being larger than the threshold voltage. The PNP transistor Q turns on and

TABLE I COMPARISON OF DIFFERENT NOVEL GATE DRIVERS

	Positive spikes	Negative spikes	Complexity	Stability
Proposed gate driver	Eliminating	Eliminating	less	Normal
(a)	Eliminating	Eliminating	Less	Normal
(b)	Eliminating	No	Less	Normal
(c)	Eliminating	Eliminating	Additional control signal	Less
(d)	Eliminating	Increasing	More	Normal

TABLE II MAIN PARAMETERS OF C2M0040120D

Parameters	Values
Gate Source Voltage	-10V/+25V
Gate Threshold Voltage	2.8V
Internal Gate Resistance	1.8Ω
Drain-Source Breakdown Voltage	1200V
Input Capacitance	1893pF
Reverse Transfer Capacitance	10pF
Drain-Source On-State Resistance	40mΩ

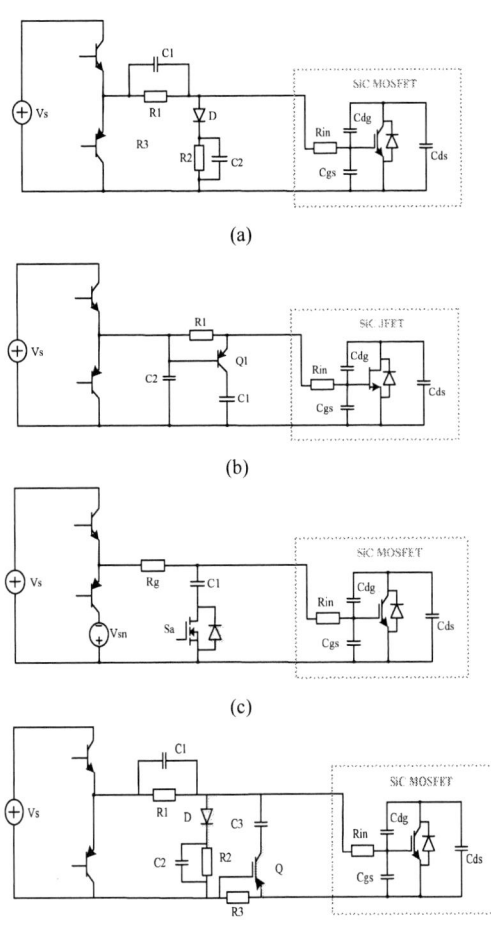

Fig. 9. Topologies of different novel gate drivers.

then the capacitor C2 can be charged to V_{gs+}.

This process is shown in Fig.5. After pre-charge processing, the control signal of one bridge turns to low and both the upper switch and the lower switch turn off. Then, the main circuit begins to work.

B. Turning-on Transient

When the MOS_H turns on, as the pre-charge processing, part of driver current go through R3_H, which makes the voltage between the emitter and base of Q_H positive. As shown in Fig. 6, the PNP transistor Q_H in the off state turns on and the gate-source capacitance C_{gs}_H of MOS_H is charged by the capacitor C2_H. Therefore, the speed of turning on transient will not be slowed down in this circuit. However, part of inductive load current will charge C_{gs}_L and raise the gate voltage. The unwanted gate voltage rise may trigger the MOS_L to shoot through the main circuit. The added negative gate voltage can eliminate this risk and prevents the unwanted shoot-through.

C. Turning-off Transient

When the MOS_H turns off, the voltage across C1 makes C_{gs}_H of MOS_H get a negative voltage through the diode

D_H instead of R3_H. And the capacitance of C1 should be large enough to keep the negative voltage be almost constant during one switching period. The voltage between the base and emitter of Q_H is positive because of the internal resistance of the diode. Therefore the PNP transistor Q_H turns off and C2 does not work. The switching speed won't be affected. On the contrary, this negative voltage accelerates the discharge of gate-source capacitance C_{gs}_H of MOS_H. This process is shown in Fig.7.

D. Negative Gate Voltage Spike Suppression

The lower switch is now used as the device of analysis. Its operational principle is illustrated in Fig. 8. When the upper switch turns off, the inductive current will flow through the parasitic diode of MOS_L. And generally, part of load current will go through the parasitic capacitance C_{gs_L} and generate the negative gate voltage spikes. However, the proposed gate driver circuit will make the part of load current go through the resistor R3_L, which makes the emitter voltage of Q_L higher than the base voltage of Q_L. Therefore, Q_L turns on and C2_L connects to the gate-source terminal of MOS_L. The capacitance of C2_L is much larger than the parasitic capacitance C_{gs_L}, which offers a low impedance loop to the inductive current during the turning-off transient of upper switch. Therefore, the negative gate voltage spike can be suppressed. On the other hand, when the emitter voltage of Q_L is equal to the base voltage of Q_L, the PNP transistor Q_L will be turned off.

E. Summary

SiC power devices have relatively low minimum allowable negative gate-source voltage restricted mainly by the present semiconductor manufacturing technology. Therefore, it is very important to suppress the negative gate voltage spikes to avoid device failure. In order to satisfy this driving requirement, the proposed gate driver circuit adds

TABLE III PARAMETERS OF MAIN PASSIVE COMPONENTS

Parameters	Values
R1	2.5kΩ
R2	9.5kΩ
R3	5Ω
C1	0.2uF
C2	0.1uF

another low impedance loop by using a passive triggered transistor with a series capacitor.

In order to show the advantages of the proposed gate driver, some comparisons with other novel gate drivers are presented in details. Fig. 9 shows some novel gate drivers proposed in Reference [13]-[17]. These gate drivers can improve the performance in some aspects. However, they all have some disadvantages in other aspects more or less. TABLE I presents comparison of different novel gate drivers. The proposed gate driver in this paper has better characteristics comparing to other novel gate drivers.

IV. SIMULATION VERIFICATIONS

A bidirectional buck converter using the C2M0040120D SiC MOSFETs was assumed for simulation and experimental verification. TABLE II lists the main parameters of C2M0040120D when the temperature is 25°C. Cree Inc. offers the Spice model of C2M0040120D when using the LTspice. Therefore, LTspice is assumed to simulate the proposed gate driver.

Compared to Si MOSFETs, SiC MOSFETs have a lower drift region resistance, but a higher channel resistance. At the low Gate-Source voltages ($V_{GS} < 13V$), the channel resistance dominates the total R_{ON} , which has a negative temperature coefficient. Therefore, it is always recommended to turn on SiC MOSFETs with V_{GS} higher than 18V. In order to be consistent with experiment, the input source voltage of gate driver is set to 24V and the calculated positive gate-source voltage and negative gate-source voltage is set to +19V and -5V, respectively.

The DC voltage of main circuit varies from 200V~600V and the loads set to 100Ω. A parasitic inductance 5nH of gate driver is also considered in the simulation. Fig. 10 shows the gate-source voltage spike's waveform of the traditional gate driver in different DC voltages. It is observed that when the DC voltage is above 400V, the negative voltage spike exceeds the maximum allowable negative biased gate voltage of SiC MOSFET, which will degrade the SiC device's reliability. However, the drain-source breakdown voltage of C2M0040120D is 1200V. Therefore, the traditional gate driver is unable to exert the full advantages of SiC devices.

In order to show the advantages of proposed gate driver, the switching speed and the gate voltage spikes are compared among three different gate drivers, which refer to the traditional gate driver, the gate driver tied a capacitor

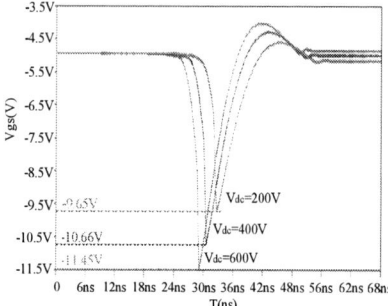

Fig. 10. Gate-source voltage spike's waveform of traditional gate driver in different DC voltages.

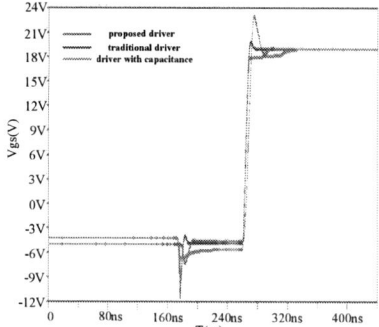

Fig. 11. Turning-on transient of three different gate drivers in a bridge configuration.

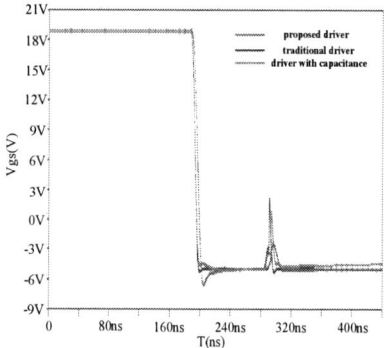

Fig. 12. Turning-off transient of three different gate drivers in a bridge configuration.

between gate and source terminals, and the proposed gate driver, respectively. The DC voltage of all the three gate drivers is set to 600V.

The added capacitance between gate and source terminals of the second gate driver is 5nF. A 5nH parasitic inductor is added between the negative voltage generation section and the voltage spikes elimination section in the third gate driver. According to the calculated positive gate-source voltage and negative gate-source voltage, the values of R1 and R2 are selected as 2.5kΩ and 9.5kΩ. All the values of passive elements in the proposed gate driver are given in TABLE III. The turning-on transient and the turning-off transient of three gate drivers can be seen in Fig. 11 and Fig. 12.

The second gate driver with a sole capacitor between gate and source terminals shows the slowest turning-on and

978-1-4673-9551-9/16 $31.00 © 2016 IEEE   540

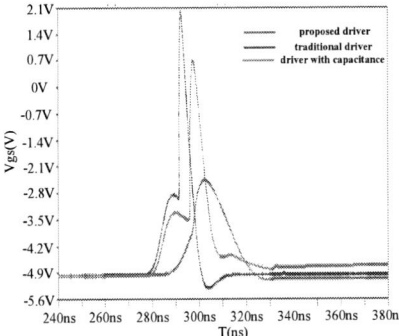

Fig. 13. Positive gate driver spikes of three different gate drivers in a bridge configuration.

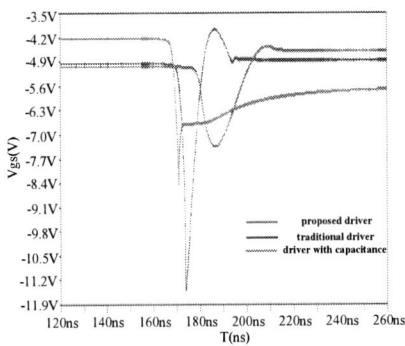

Fig. 14. Negative gate driver spikes of three different gate drivers in a phase-leg configuration.

TABLE IV MAXIMUM POSITIVE GATE-SOURCE VOLTAGE SPIKE OF DIFFERENT GATE DRIVERS

Gate Driver	Maximum positive gate-source voltage spike
traditional	1.99V
With sole capacitor	-2.45V
proposed	0.70V

TABLE V MAXIMUM NEGATIVE GATE-SOURCE VOLTAGE SPIKE OF DIFFERENT GATE DRIVERS

Gate Driver	Maximum positive gate-source voltage spike
traditional	-11.40V
With sole capacitor	-7.34V
proposed	-8.40V

turning-off transient process, which will increase the switching losses and decrease the operation efficiency. And the proposed gate driver has the same turning-on and turning-off transient process with the traditional gate driver. Also, it is noted from Fig. 12 that the negative gate voltage produced by the passive level shifter can fasten the turning-off process and effectively avoid the spurious ON signal appeared around 290ns. What's more, the negative gate-source voltage is a bit larger than -5V. That's mainly because the voltage of capacitance C1 has a little drop during the discharging process.

In specific, Fig. 13 shows the positive gate-source voltage spikes of lower switch when the upper switch is turning on, which is the zoomed view of Fig. 12. TABLE IV

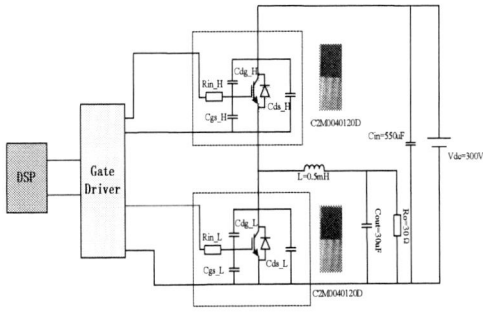

Fig. 15. Schematic of experimental circuit

Fig. 16. Prototype of SiC MOSFET bidirectional buck converter

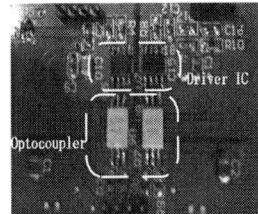

Fig. 17. Picture of gate driver prototype

TABLE VI EXPERIMENT PARAMETERS

Parameters	Values
Carrier frequency	100kHz
DC voltage	300V
Inductor	550uH
Capacitor	550uF
Load	30Ω
Duty cycle	50%
Dead time	0.3us

gives the maximum values of the gate-source voltage spikes appeared in three gate drivers and all these gate drivers are below the gate threshold voltage of 2.8V, which means the negative gate voltage can prevent the unwanted shoot-through in the phase-leg configuration effectively.

V. EXPERIMENTAL VERIFICATIONS

In this paper, the proposed gate driver circuit was also validated by an experimental prototype. Fig. 15 shows the schematic of experimental circuit and Fig. 16 is the prototype of SiC MOSFET bidirectional buck converter. The main parameters of this circuit are listed as TABLE VI.

In order to prevent shooting-through in the test, a

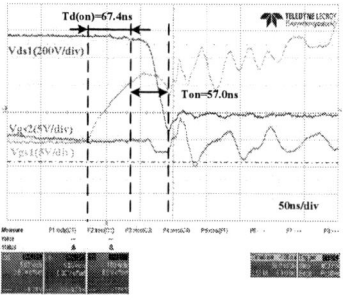

(a) Conventional gate drive

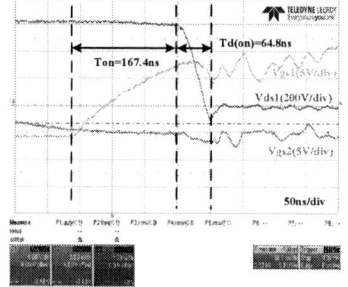

(b) The gate driver with a gate-source capacitor

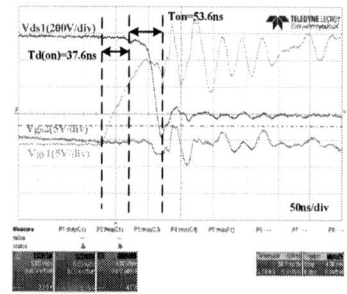

(c) Proposed gate driver

Fig.18 Turning-on process of the upper switch.

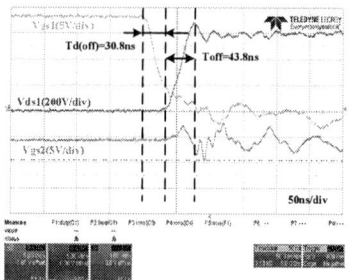

(a) Conventional gate drive

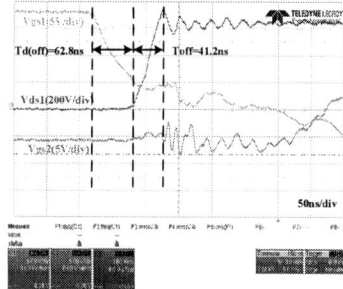

(b) The gate driver with a gate-source capacitor

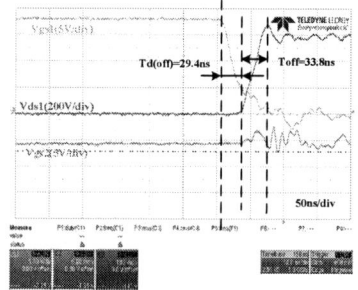

(c) Proposed gate driver

Fig.19 Turning-off process of the upper switch.

TABLE VII. COMPARISON OF DIFFERENT GATE DRIVERS

	Conventional	with sole capacitor	proposed
Turn-on delay time ($t_{d(on)}$)	67.4ns	167.4ns	37.6ns
Turn-on time (t_{on})	57.0ns	64.8ns	53.6ns
Positive spike (V_p)	-0.95V	-3.40V	-2.35V
Turn-off delay time($t_{d(off)}$)	30.8ns	62.8ns	29.4ns
Turn-off time (t_{off})	43.8ns	41.2ns	33.8ns
Negative spike (V_n)	-9.60V	-8.30V	-7.25V

negative gate-source voltage -5V was added in the traditional gate driver and the one tied an additional gate-source capacitor. Therefore, three tested gate drivers all have the positive gate voltage of +19V and the negative gate voltage of -5V. The gate-source voltage is generated by MORNSUN's B0524XT-1WR2 and B0505XT-1WR2. The optocoupler is ACPL-W346 from Avago Technologies and the additional PNP transistor is ZXTP25140BFHTA manufactured by Zetex Semiconductors. The experimental parameters are listed in TABLE V. Fig. 17 shows the captured pictures of proposed gate driver.The switching waveforms of the three different gate driver circuits are shown in Fig. 18 and Fig. 19 and the proposed gate driver worked properly.

From the experimental results, the second gate driver with an additional gate-source capacitor can eliminate both positive and negative gate-source voltage spikes as shown in Fig. 18(b) and Fig. 19(b). However, it slows down the speed of turning-on and turning-off, which will affect the narrow pulse width in a voltage source inverter and increase

switching losses. The proposed gate driver can eliminate the negative gate-source voltage spikes without slowing down the switching speed as shown in Fig. 19(c). The comparison of these gate drivers is listed in TABLE VII. In order to show the advantages of proposed gate driver, the positive and negative spikes of three different gate drivers are presented in Fig. 20 and Fig. 21, respectively. The oscillations are mainly caused by parasitic inductance of

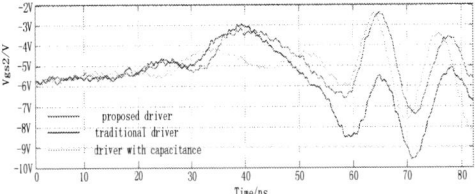

Fig.20 Positive spikes of different gate drivers

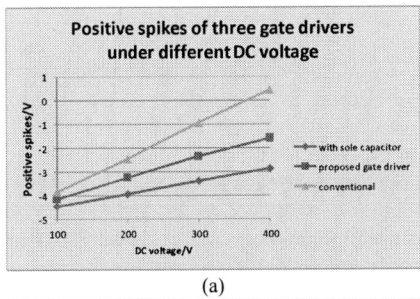

Fig.21 Negative spikes of different gate drivers

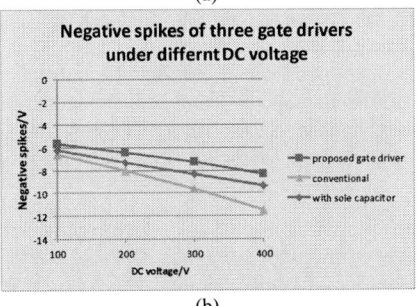

Fig.22 Positive and negative spikes of three gate drivers under different DC voltages.

gate drivers. This paper also compared the positive spikes and negative spikes of three gate drivers under different DC voltages. The results are shown in Fig. 22.

VI. CONCLUSION

This paper proposes a novel gate driver of SiC MOSFET with a passive triggered auxiliary transistor. The auxiliary transistor will connects a series capacitor into the gate circuit to suppress the negative gate voltage spikes upon the complementary switch turns off. Since the auxiliary transistor is controlled passively, the circuit configuration is simple. By employing a RC level shifter, the proposed gate driver can reduce the magnitude of positive gate voltage spikes and then fully satisfy the stringent driving requirements of SiC MOSFETs with high switching speed. The simulation and experimental results also validated the performance of the proposed gate driver.

REFERENCES

[1] R. B. Campbell,"Whatever happened to silicon carbide," IEEE Trans. Ind. Electron., vol. IE-29, no. 1, pp. 124-128, Feb. 1982.

[2] J. Biela, M. Schweizer, S. Wafer, and J. W. Kolar,"SiC versus Si evaluation of potentials for performance improvement of inverter and DC–DC converter systems by SiC power semiconductors," IEEE Trans. Ind. Electron., vol. 58, no. 7, pp. 2872-2882, Jul. 2011.

[3] Z. Wang, X. Shi, Y. Xue, L. M. Tolbert, F. Wang and B. J. Blalock, "Design and performance evaluation of overcurrent protection schemes for silicon carbide (SiC) power MOSFETs," IEEE Trans. Ind. Electron., vol. 61, no. 10, pp. 5570-5581, Oct. 2014.

[4] Singh R, Hefner AR. Reliability in SiC MOS devices. Solid-State Electron 2004; 48: 1717-20

[5] www.cree.com/Power/Landing-pages/MOSFET-products

[6] CREE CPWR AN-08. Application considerations for silicon carbide MOSFETs. [Online]. Available: http://www.cree.com

[7] Z. Zhang, F. Wang, L. M. Tolbert and B. J. Blalock, "A gate assist circuit for cross talk suppression of SiC devices in a phase-leg configuration," in IEEE 2013 Energy Conversion Congress and Exposition (ECCE), 2013, pp. 2536-2543.

[8] Z. Chen, M. Dailovic, D. Boroyevich, Z. Shen, "Modulized design consideration of a general-purpose, high-speed phase-leg PEBB based on SiC MOSFETs," In Proc. Eur. Power Electron. Appl. Conf., Aug. 2011, pp. 1-10.

[9] K. Ishikawa, K. Ogawa, S. Yukutake, N. Kameshiro, and Y Kono, "Traction Inverter that applies compact 3.3kV/1200A SiC hybrid module," The 2014 International Power Electronics Conference, 2140-2144, 2014.

[10] J. Wang and H.S.H. Chung, "A novel RCD level shifter for elimination of spurious turn-on in the bridge-leg configuration," IEEE Trans. Power Electron., vol. 30, no. 5, pp. 976-984, FEB. 2015.

[11] Y. Zushi, S. Sato, K. Matsui, Y. Murakami, S. Tanimoto, "A novel gate assist circuit for quick and stable driving of SiC-JFETs in a 3-phase inverter," in Proc. IEEE Applied power Electronics Conference and Exposition 2012, Feb. 2012, pp. 1734-1739.

[12] http://www.infineon.com/dgdl/lnfineon-BSC030P03NS3G-DS-v02_01-en.pdf?fileId=db3a30431d8a6b3c011d90d084910435

[13] Z. Zhang, F. Wang, L.M. Tolbert and B. J. Blalock, "Active Gate Driver for Crosstalk Suppression of SiC Devices in a Phase-Leg Configuration," IEEE Trans. Power Electron., vol. 29, no. 4, pp. 1986-1997, APR. 2014.

[14] Q. Zhou, F. Gao and T. Jiang, "A gate driver of SiC MOSFET with passive triggered auxiliary transistor in a phase-leg configuration," to be published in IEEE 2015 Energy Conversion Congress and Exposition(ECCE), Sep. 2015.

[15] H. Yu, J. Lai, X. Huang, J. Zhao, J. Zhang, X. Hu, J. Carter and L. Fursin, "A gate driver based soft-switching SiC Bipolar Junction Transistor," in Proc. IEEE Applied power Electronics Conference and Exposition 2003, Feb. 2003, pp. 968-973.

[16] J. Jiang, X. Wu, B. Hu, J. Liao and L. Zhao, "Development and Production of ZCS Soft Switching Converter-based Gate Driver IC," in Proc. IEEE 8th International Conference on ASICON 2009, Oct. 2009, pp. 1058-1061.

[17] Q. Zhao and G. Stojcic, "Characterization of Cdv/dt induced power loss in synchronous buck DC-DC converters," IEEE Trans. Power Electron., vol. 22, no. 4, pp. 1505-1513, Jul. 2007.

[18] M. E. Jacobs, K.J. Timm, and V.J. Thottuvelli, "Appartus and method for generating negative bias for isolated MOSFET gate-driver circuits," EP Patent 0 693 825, Jan. 2, 1996.

[19] Fairchild Semiconductor AN-7019, "Limiting Cross-Conduction Current in Synchronous Buck Converter Designs,"
http://www.fairchildsemi.com/an/AN/AN-7019.pdf#page=1

[20] J. Wang and H. S. H. Chung, "Impact of Parasitic Elements on the Spurious Triggering Pulse in Synchronous Buck Converter," in Energy Conversion Congress and Exposition (ECCE), pp. 480-487, Sep. 2013.

Interleaved Boost Based AC/DC Bidirectional Converter with Four Quadrant Power Control Based on One-Cycle Controller (OCC)

Snehal Bagawade, Praveen Jain
Department of Electrical and Computer Engineering
Queen's University
Kingston, Ontario, Canada
Email: snehal.bagawade@queensu.ca, praveen.jain@queensu.ca

Abstract— **This paper presents a new bidirectional AC-DC interleaved converter topology and its control technique. The proposed control technique, based on One Cycle Controller (OCC), is used to process power in all four quadrants in the V-I plane. A fictitious reactive current is introduced in the control law that decides the PWM switching action. Further, a size reduction of the input side magnetics by the use of coupled inductors is proposed. The simulation and experimental results show a low THD content in the input current and a fast control of power factor, load and switch between modes of operation. The experimental and simulation results for a 2KW lab prototype are included in this paper which show a close match of theoretical and experimental results.**

Keywords—AC/DC; bidirectional; one cycle controller; reactive power control; interleaved, coupled inductor

I. INTRODUCTION

The rising interest in electric vehicles is pushing the technology in this field towards optimization of its various systems and sub-systems, such as the electric drive train system, electric breaking system, battery charging system etc. A high efficiency of operation for all the systems is very important as it directly translates into money saved on energy expenditure. The battery charging system in electric vehicles is a very specialized system. It needs to connect to the utility grid and provide energy suitable to charge a battery. It is the front end AC-DC converter of the battery charging system that converts the power from the grid which is in the form of alternating currents and voltages into DC power. This DC power is then processed further to charge the battery. The present day chargers need to be highly efficient, and must have extremely good transient performance. The AC/DC PFC converters used for the front end of the electric vehicle battery chargers are typically in the range of 2KW-3KW. It is a general practice to use a diode bridge rectifier in series with a boost converter or interleaved boost converter for this application. The main disadvantage of this topology is that the losses in the diode bridge become high at such high power levels [5], [9]. Hence to reduce the losses in this front end, the bridge-less interleaved topology [4] and ZVS boost interleaved PFC topology [6], [7] were proposed. In the first one the diode bridge is eliminated, while in the latter, a passive circuit is used in the interleaved boost converter to reduce the switching losses. To avoid all the problems of losses in diode bridge at high power, the use of extra passive elements and to have good transient performance, an interleaved full bridge topology with coupled inductors is proposed in this paper. A novel four quadrant control of power based on One-Cycle Control (OCC) technique is proposed in this paper. The main objectives of this paper are (1) To introduce the interleaved full bridge topology for boost-based AC-DC bidirectional converter, and (2) Extend the One-Cycle Control technique for reactive power control with full bridge boost based AC-DC bidirectional converters.

The use of full bridge converters in interleaved mode creates problems of current sharing based on the amount of input side inductance is introduced on it. Further the current sharing by different inductors in balanced and unbalanced inductance on both input legs is seen. Both the problems are tackled by controlling the fictitious current that is introduced in the controller.

The technique of introducing a fictitious current in the controller is very powerful. It creates a virtual environment for the controller under which it operates. This technique is used to control the reactive current in the controller. The second order generalized integrator along with PLL is utilized for reactive current reference generation. Based on the amount of reactive current required, the magnitude of reference is changed and added to the active current reference in the controller.

II. INTERLEAVED TOPOLOGY FOR AC-DC BIDIRECTIONAL FULL BRIDGE CONVERTER

A. Proposed topology

The circuit diagram of the proposed interleaved converter topology is shown in Fig.1. This circuit employs eight active switches in two full bridge configurations. The switching pulses of the two bridges are phase shifted by 180 degrees. The interleaved topology used here helps in reducing the current rating of a single cell of the converter. Also it helps reduce the ripple in the source current and hence the THD content is also less.

The operation of converter has eight distinct times where different switching devices work. In this section the operation of the converter is analyzed. It is assumed that the MOSFET

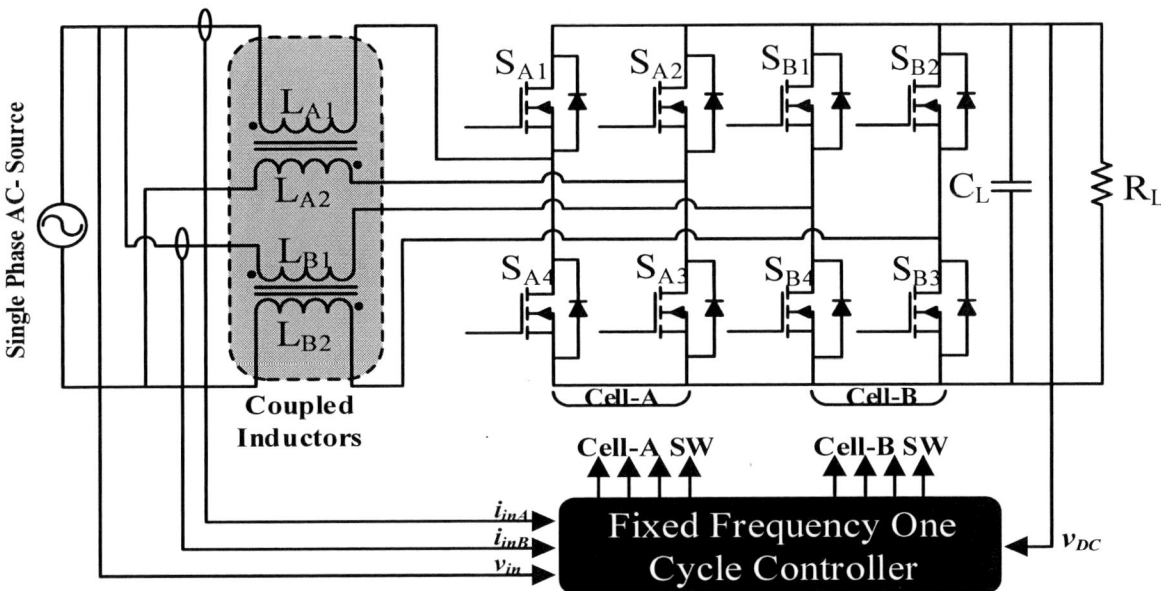

Fig. 1. Circuit Diagram of proposed converter with coupled inductors

switches and diodes are ideal, the coupled inductors used here are balanced, i.e. the inductance in both the input connections is equal and the circuit is considered to operate in steady state, the AC voltage, V_{in} has a frequency much smaller than the switching frequency, so that the AC voltage is assumed constant for one switching period. The analysis is made for the positive half of the AC line voltage. For the negative half of AC line voltage the analysis remains analogous. For the analysis of the circuit operation the current waveforms and switching times are shown in Fig.4 and corresponding current paths in the circuit are shown in Fig.2 by the bold paths. All the components that are shown faded are not conducting at that moment.

The switching cycle begins with the switches S_{A1} and S_{A3} turning off at $t = t_0$ and Cell-A going in the dead band, the circuit diagram of Fig.2.(a) depicts the operation. During the dead band the body diodes of S_{A1} and S_{A3} conduct. At this time the Cell B switches S_{B1} and S_{B3} remain on. This period goes on for the dead time and ends at time $t = t_1$. At this time the switches S_{A2} and S_{A4} are switched on (Fig.2.(b)). The current in the inductors L_{A1} and L_{A2} starts rising. The switches are hard switched at this instant since they have the full DC bus voltage applied across them and current in the diodes from previous interval gets transferred to these switches. Whereas in Cell-B the switches S_{B1} and S_{B3} remain on. This switching interval ends at $t = t_2$. The Cell-A again enters the dead band mode at $t= t_2$, where the body diodes of S_{A1} and S_{A3} conduct again (Fig.2(c)). The current in the inductors L_{A1} and L_{A2} starts falling and keeps a negative slope until next time the switches S_{A2} and S_{A4} are switched on. This interval ends at $t = t_3$ when the switches S_{A1} and S_{A3} are switched on(Fig.2.(d)). Since the body diodes of the same switches were conducting in the interval before this, so the voltage across these switches was zero, and hence these switches are soft switched under the zero voltage condition. The Cell-A remains in this condition for the remaining switching period. Till this point the switches S_{B1} and S_{B3} of the Cell-B remain on. At the end of this interval, at $t = t_4$ the switches S_{B1} and S_{B3} turn off and Cell-B

enters the dead band mode when the body diodes of switches S_{B1} and S_{B3} switch on (Fig.2.(e)). The diodes conduct till $t = t_5$, when the switches S_{B2} and S_{B4} are switched on (Fig.2.(f)). These switches are hard switched because the voltage across the switch at the time of turning on was not zero and current rises to inductor current at that time. The current in the inductors L_{B1} and L_{B2} starts rising. This interval remains on for a time period of DT_s and the interval ends at time $t = t_6$ when the Cell B again enters the dead band and the body diodes of S_{B1} and S_{B3} are switched on due to the direction of current in the inductors L_{B1} and L_{B2} (Fig.2.(g)). From this point on the current in the inductors L_{B1} and L_{B2} starts falling and keeps a negative slope until the switches S_{B2} and S_{B4} are switched on again. This period ends at time $t = t_7$. Finally at this instant the switches S_{B1} and S_{B3} are turned on in a soft manner as their body diodes were conducting in the previous cycle (Fig.2.(d)). This period goes on till the time $t = t_8$, when the new cycle starts and all this repeats again.

The point worth noting is that in each switching cycle two of the switches of each cycle are soft switched and only two switches are hard switched.

B. Coupled Inductors

The interleaved current sourced converter as described above usually has four bulky inductors on the input side. The use of coupled inductors in this application greatly reduces the size of the input side magnetics. For the same size and core material, the A_L value (A_L: specified by the manufacturers) is constant. A comparison of a two-cell interleaved converter with and without coupled inductors shows a number of advantages like reduced copper losses, reduced iron losses (hence an improvement in the efficiency) and a size reduction.

As an example consider a two cell interleaved converter, the number of turns required for an inductor is given by:

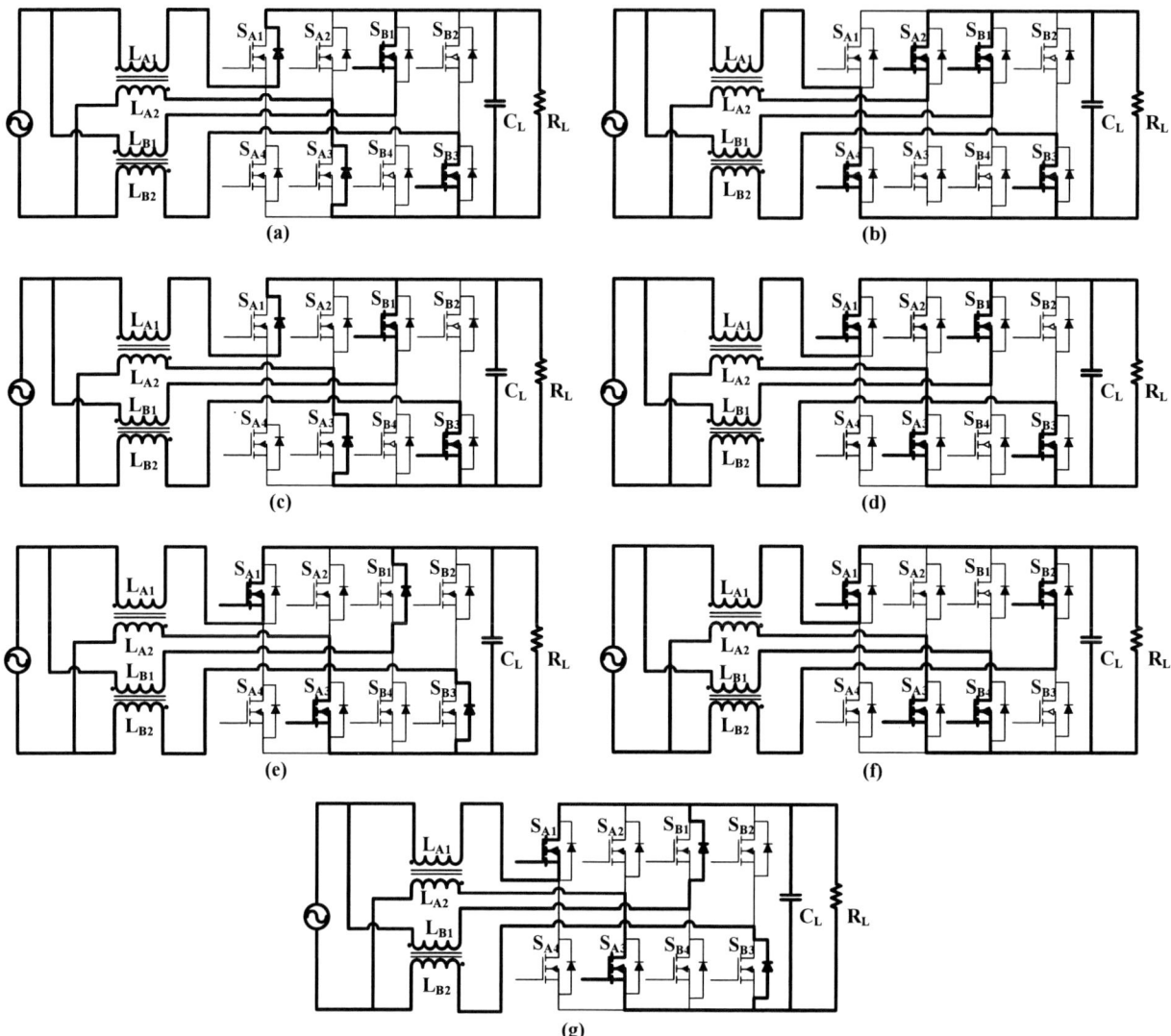

Fig. 2. Circuit Diagram showing current paths in each switching interval

$$N = 1000\sqrt{\frac{L}{A_L}} \qquad (1)$$

Where N is the number of turns, L the inductance value in mH and A_L in mH per 1000 turns squared. For, four inductors of the same value L, the number of turns will be 4N.

For the case of coupled inductors, two inductors of the same cell are wound together as shown in the figure. Hence a value of 2L is required by winding on a single core. The number of turns required for making the inductor for a single cell is:

$$N_{12} = N_{34} = 1000\sqrt{\frac{2L}{A_L}} = 1.414\,N \qquad (2)$$

Where N_{12} and N_{34} are total number of turns on a single coupled inductor. This means that the total number of turns for a two cell converter is 2.828 times the number of turns required for a converter without mutual inductance. Which is around 30% reduction in the number of turns and a 50% reduction in the number of cores. The total number of turns is reduced by 30 percent but to maintain the same fill factor for the magnetic core, a thinner wire may have to be used, which may increase the resistance of the wire and cause more losses, also since we are down from two magnetic cores to one core, the magnetic field density now would be higher and may again lead to a higher losses. In this case a tradeoff may be made as per the application whether a single coupled inductor is more bulky and tougher to manufacture or two separate inductors are bulkier and costlier.

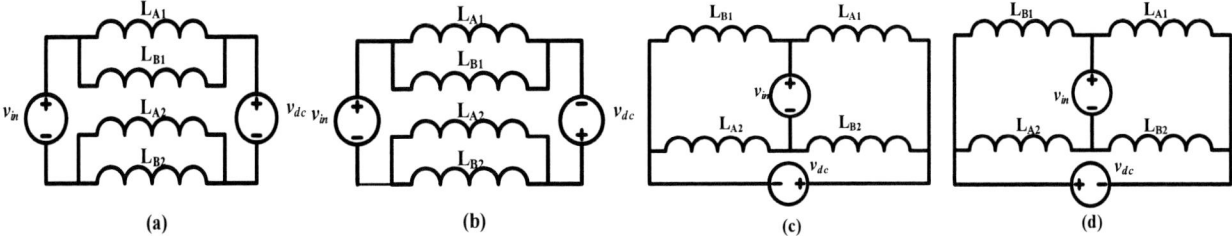

Fig. 3. Equivalent Circuits for current ripple analysis in coupled inductors

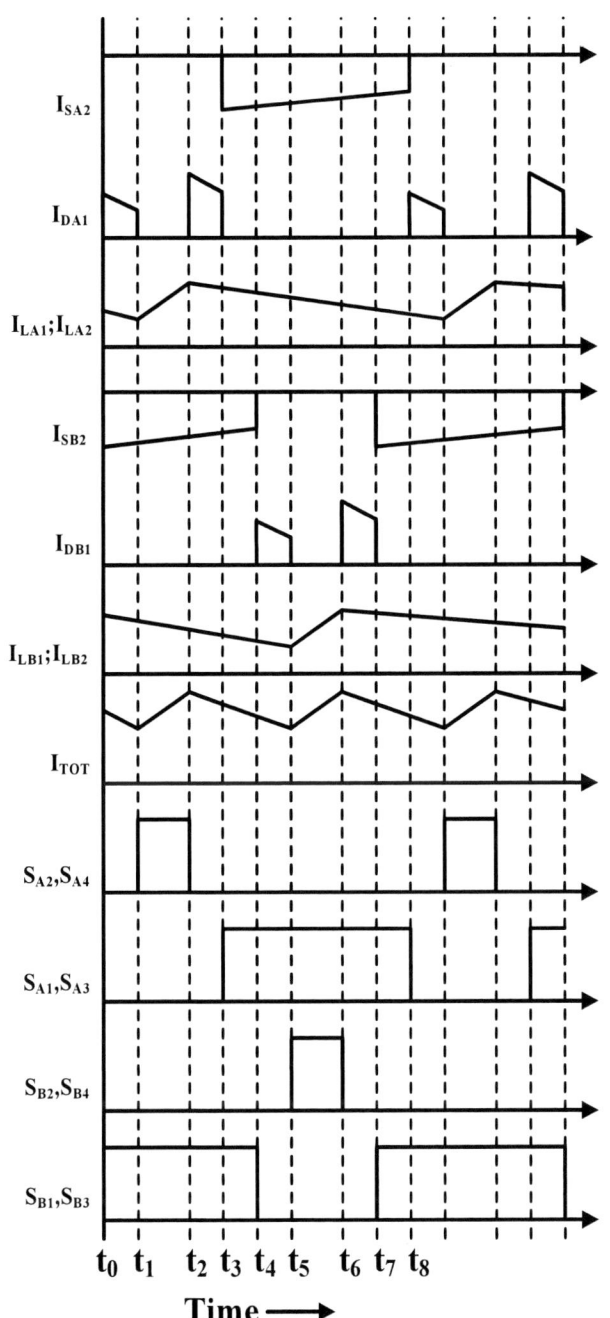

Fig. 4. Timing Diagram for the converter operation

C. Operation of Converter with Balanced Coupled Inductor

The effect of inductors in each incoming line is analyzed in this section using the equivalent circuit diagrams of the circuit in each of the four different switching instants. The analysis presented here is for a two cell converter having four input side inductors, which are realized using two coupled inductors.

For the analysis, it is considered that the change in line voltage for one switching period is negligible, since the switching frequency is much higher than the line frequency. The equivalent circuits presented here are for the positive half of the line cycle and since voltage is considered constant, it is represented as a DC voltage source of same value as that of instantaneous AC voltage. Also, because the output voltage is considered to be a constant DC value, the change in inductor current in one switching period should be zero. So a focus on the change in currents in one switching period will provide us information about current sharing by different inductors.

For the first switching instant, if S_{B1} & S_{B3} are in on state and S_{A2} & S_{A4} are switched on at t = 0. The equivalent circuit is as shown in Fig.3.(d) and the corresponding equations according to superposition theorem:

$$v_{in} = ((L_{A1} \parallel L_{B1}) + (L_{A2} \parallel L_{B2}))\frac{\Delta i_{in1}}{\Delta t} \quad (3)$$

$$v_{dc} = ((L_{A1} \parallel L_{B2}) + (L_{A2} \parallel L_{B1}))\frac{\Delta i_{in2}}{\Delta t} \quad (4)$$

This switching interval finishes at t = DTs. Now the switches SA2 & SA4 are switched off and S_{A1} & S_{A3} are turned on. The switched S_{B1} & S_{B3} are still conducting. At this moment the equivalent circuit is as shown in Fig.3.(a) and the change in current through inductors is given by equations:

$$v_{in} - v_{dc} = ((L_{A1} \parallel L_{B1}) + (L_{A2} \parallel L_{B2}))\frac{\Delta i_{in}}{\Delta t} \quad (5)$$

This switching interval ends at time t = T$_s$/2. At this time the Switches S_{B1} & S_{B3} are turned off and switches S_{B2} & S_{B4} are turned on. The switches S_{A1} & S_{A3} remain on during this transition. The equivalent circuit is shown in the Fig.3(c) and corresponding change in inductor currents are again given by similar equations according to superposition theorem:

$$v_{in} = ((L_{A1} \parallel L_{B2}) + (L_{A2} \parallel L_{B1}))\frac{\Delta i_{in1}}{\Delta t} \qquad (6)$$

$$v_{dc} = ((L_{A1} \parallel L_{B1}) + (L_{A2} \parallel L_{B2}))\frac{\Delta i_{in2}}{\Delta t} \qquad (7)$$

This interval ends at $t = T_s(D + 1/2)$. At this time the switches S_{B2} & S_{B4} are switched off and S_{B1} & S_{B3} are turned on while the switches S_{A1} & S_{A3} remain on. The equivalent circuit for this interval is shown in Fig.3.(a) and (5) can be applied. This interval finishes at $t = T_s$, and a new switching cycle starts where the same process repeats again.

In the case when the coupled inductors are manufactured well, the inductance in both the legs of a single converter would be same in a single cell. This means $L_{A1} = L_{A2} = L_{B1} = L_{B2}$. These four inductances form different parallel combinations at different time intervals. Since all of them are equal, they share voltage across them and hence the current ripple equally. If identical coupled inductors are used in the circuit, the current ripple in all the inductors would be same. In this case the ripple in the source current is minimum and hence the THD is minimum.

D. Operation of Converter with Unbalanced Coupled Inductor

The unbalance in the inductors can be of two types: (1) the coupled inductors of two cells are balanced individually, but both of them have different inductance, i.e. $L_{A1} = L_{A2} \neq L_{B1} = L_{B2}$. (2) The Coupled inductors are not balanced, i.e. $L_{A1} \neq L_{A2} \neq L_{B1} \neq L_{B2}$. In the first case, the unbalance in the inductance in two cells would just lead to different current ripples for equal current sharing by the two cells. In the second case there would be a change in the current slope in each inductor every time either of the cells switch, hence there would be a second harmonic of the switching frequency in all the four inductors. In, both cases there is not much that can be done, the ripple will be different in both inductors but source current will not have any distortion other than the current ripple double the switching frequency.

III. ONE CYCLE CONTROLLER FOR AC-DC CONVERTERS

A. One Cycle Controller Operation

The One Cycle Control scheme [1] is widely used for various applications in power electronic converters such as active power filters, power factor correction converters, etc. Their fast transient response and simplicity of implementation makes them extremely useful in such applications. There have been many modifications proposed to this control scheme to achieve different objectives for both the single phase and three phase systems. One such modification was presented by Ghodke et.al [2], who were able to use it for bidirectional power flow.

The fixed frequency operation of One Cycle controller for a converter as shown in Fig.5 is explained here. At the starting of each switching cycle, the switches S_2 and S_4 are turned on and Switches S_1 and S_3 are turned off. The input current, which is considered constant for one switching cycle, is compared with a (negative slope) ramp, the switching happens when both cross each other. In the next instant the switches S_2 and S_4 are turned off, and switches S_1 and S_3 turned on. The control equation is given by,

$$V_m(1 - 2D) = R_s i_{in} \qquad (8)$$

$$V_m = \frac{R_s}{R_e} V_{dc} \qquad (9)$$

Where, R_e is the equivalent resistive load as seen by the AC source, R_s is current sense resistor, i_{in} is the input source current and V_{dc} is the output load voltage. The details of this control scheme can be found in [1] and [2].

Ghodke et.al. modified the above equation by adding a fictitious current to the right hand side of the control law, which enabled the use of this technique for bidirectional active power flow.

B. Proposed OneCycle Controller

The Reactive power control using the OCC technique provides an advantage of simplicity of implementation, since the reactive power control is naturally independent of active power control there is no need to decouple them explicitly. This property of OCC avoids the use of complex mathematical operations such as Park's transformations for splitting the two components in d-q axes components [8].

The input current for converters controlled by OCC technique used for power factor correction is:

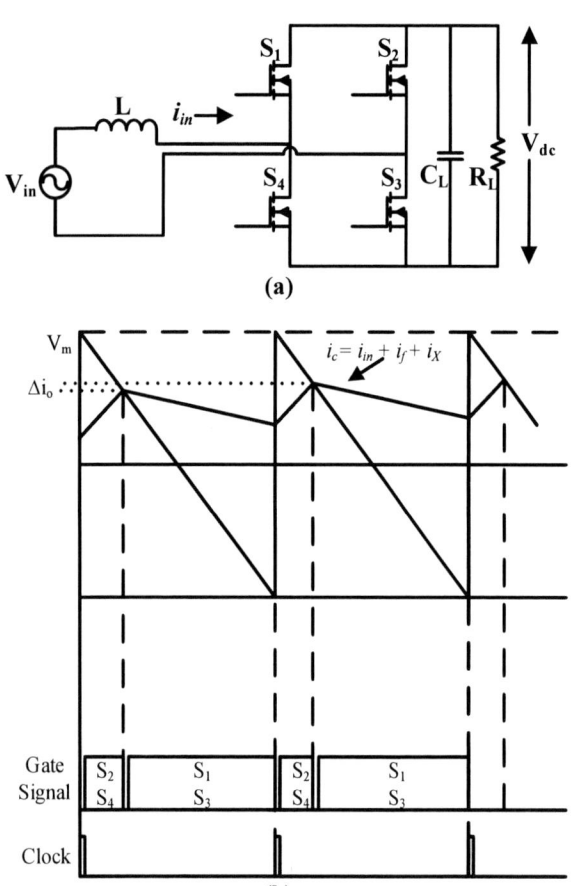

Fig.3. (a) Single cell converter, (b) OCC control scheme

$$i_c = i_{in} = \frac{v_s}{R_e} \qquad (10)$$

Where i_c is the current seen by the controller, i_{in} is the input current to the converter, v_s is the grid voltage and R_e is the equivalent resistance as it appears to the grid. An addition of a component, proportional to the input voltage, to the control law changes the control goal to,

$$i_c = i_{in} + i_f = \frac{v_s}{R_e} + \frac{v_s}{R_f} \qquad (11)$$

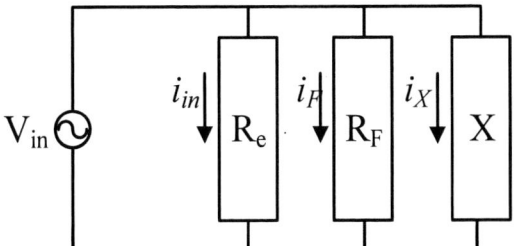

Fig. 4. Equivalent Circuit of the proposed converter control scheme.

The point worth noting in this technique is that the control goal is dealing with only active power and no term involving reactive power appears in any expression. This confirms the property of OCC technique to have a natural capability of independent active and reactive power control. To achieve reactive power control, an orthogonal current component is introduced in the control law. In this paper, an addition of a fictitious reactive current is proposed which is proportional to the derivative of the source voltage, i.e.

$$i_X = X \frac{dv_s}{dt} \qquad (12)$$

Where the sign of X gives us the leading (positive sign) or the lagging current (negative sign).

The new control law changes the control goal to,

$$i_c = i_{in} + i_f + i_X = \frac{v_s}{R_e} + \frac{v_s}{R_f} + X \frac{dv_s}{dt} \qquad (13)$$

The equivalent circuit of the converter as seen by the controller is depicted in Fig.6.

The control block diagram for the single phase full bridge converter is shown in the Fig.7. The orthogonal current reference used here can be generated by the use of any of the available techniques like the differentiation of the input voltage, all pass filters with 90 degree phase shift at the required 60Hz frequency or SOGI based double integration technique. Here SOGI with PLL was used to generate the orthogonal current reference. An appropriate gain is then applied to this reference current and added to the real current to achieve the required power factor for the input current.

C. Stability of the controller under proposed control scheme

The stability of the converter under reactive power control mode depends on the magnitude of effective impedance posed by the converter. Further, the stability of interleaved converter depends on the load sharing by each converter cell. Hence, for an interleaved converter to provide full control of active and reactive power in rectifying and inverting modes, we need to consider both active and reactive impedance.

Stability for OCC for AC-DC systems is explained in [1] and [2]. For stable operation of converter, following condition should be met at all times:

$$-v_{in} < \frac{2L\,V_m}{T_s\,R_s} \qquad (14)$$

According to this equation the negative peak of the input sinusoid poses the most dangerous situation. But this equation is only dependent on input voltage, switching frequency and series inductance. It does not depend on input current, which means that as long as the above equation is satisfied, the power factor angle does not affect the stability of the converter.

Since the control law of the converter changes by incorporating reactive component, so there is a change in V_m.

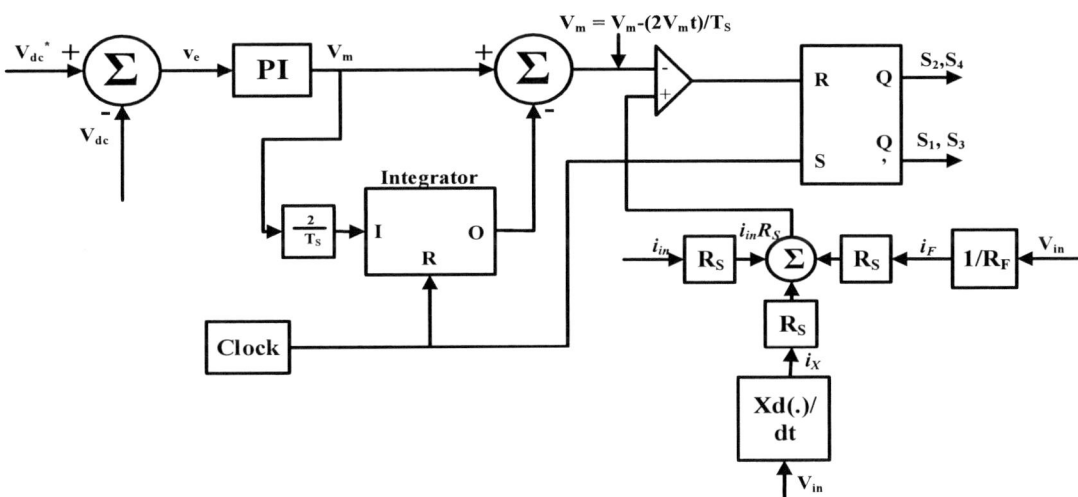

Fig.5. : Control Block Diagram of the proposed four quadrant power flow control

$$V_{dc}(1 - 2D) = v_{in} \qquad (15)$$

$$v_{in} = Z_e i_{in} \qquad (16)$$

Where, Z_e is the equivalent series impedance posed by the converter to the grid. Including current sense resistor, R_s, (16) becomes,

$$V_{dc}R_s(1 - 2D) = Z_e i_{in}R_s \qquad (17)$$

$$V_m(1 - 2D) = i_{in}R_s \qquad (18)$$

Where, $\qquad V_m = \dfrac{V_{dc}R_s}{|Z_e|} \qquad (19)$

The converter remains stable if this new value of V_m satisfies (14).

IV. EXPERIMENTAL RESULTS

The proposed converter topology and its control scheme was simulated on PSIM9.3 64-bit and then a 1KW experimental setup was prepared to verify the theoretical results. The parameters of the lab prototype are shown in Table-I. The control scheme discussed in this paper was implemented using C-2000 Delfino series TI-DSP TMS320F28335. Fig.8 shows the photographs for the experimental setup prepared for the converter. The simulated and experimental waveforms for converter operation are shown in Fig.9 The figure, Fig.9(a) and Fig.9(b) illustrates the difference in current ripple for Cell-A and

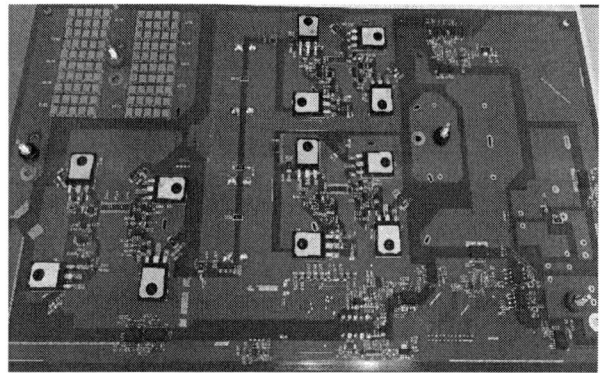

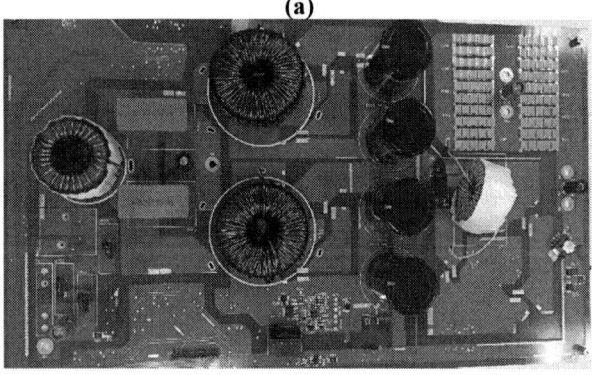

Fig. 8. Lab Prototype for proposed converter, (a) Bottom side of the PCB, (b) Top side of the PCB

Cell-B of converter with balanced and unbalanced coupled inductors. It can be observed from the waveforms that the source

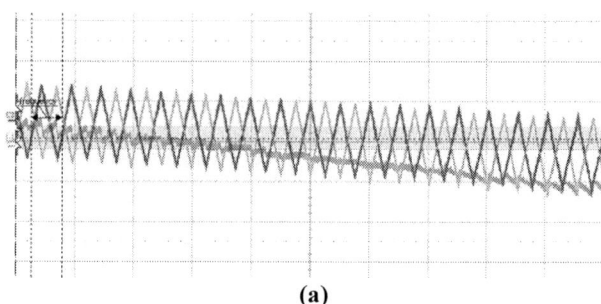

(a)

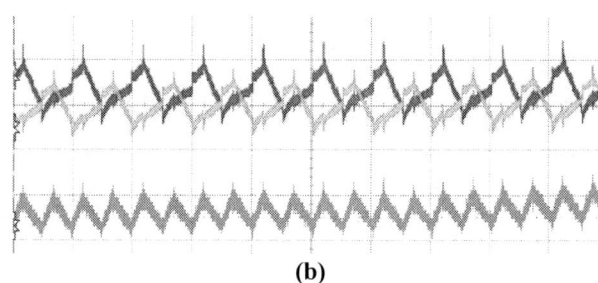

(b)

Fig.9. Current ripple in the two coupled inductors under balanced (a) and unbalanced (b) windings on coupled inductors

current shape in both the cases is similar but the individual currents through the inductors change. Further, the waveforms shown in fig.10 are for the active and reactive power flow in both the rectifying mode and inverting mode of operation.

The current measurement in this application is done using the low cost hall effect sensors ACS-710 from Allegro Microsystems, which provide good enough accuracy to perform control for this application. The input side voltage sensor is a potential divider with isolation amplifier (AMC-1200) from Texas Instruments which again provides a low cost solution as compared to potential transformers for isolated voltage measurement in DSP based applications where analog signals need to be isolated in a circuit.

TABLE I. PARAMETERS OF CONVERTER CIRCUIT

Sl. No.	Passive Component		
	Description	*Value*	Quantity
1.	Input Side Inductors (Coupled Inductor)	1.65 mH	2
2.	DC Link Capacitors	470µF, 450V	4
3.	Output Load	160Ω, 1KW	1
4.	Input AC voltage	240V$_{rms}$	-
5.	DC Bus Voltage	400V	-

V. CONCLUSION

An interleaved boost based AC-DC converter topology is introduced in this paper. Coupled inductors are used for the input boost inductors used for this topology, to reduce the total number of inductors for this application. Also a novel reactive power control technique for use in AC-DC bidirectional converters is proposed in this paper. This control technique is

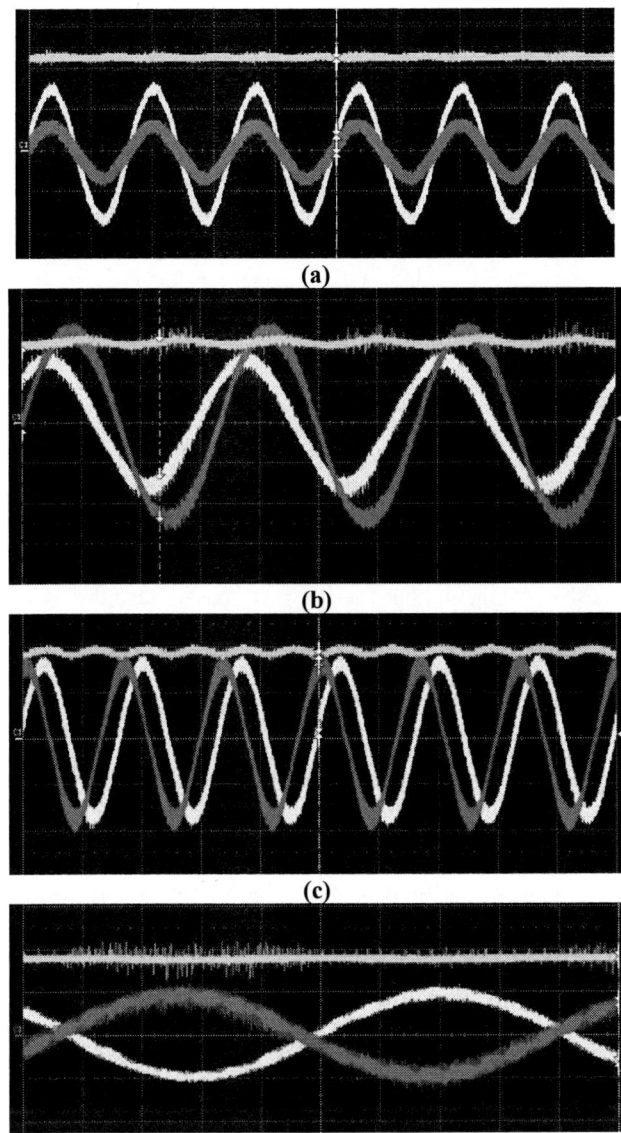

Fig.10. Converter operation in (a) Unity power factor mode, (b) Lagging Power factor mode, (c) Leading Power factor mode, (d) Inverting mode. Yellow curve is the grid voltage, Pink is the line current and blue is the DC bus voltage

based on OCC. The interleaved topology used here helps not only to increase the power handling capacity of the converter but also to reduce the THD. Use of this strategy provides the fast transient response of One Cycle controller for higher power applications (where the current handled by single converter is too high for acceptable size magnetics). The use of coupled helps in the reduction of number of turns by 30% (as compared with the inductors without coupled magnetics) and also improve the efficiency of the converter. The experimental waveforms are presented in this paper, which include application of proposed control technique to the interleaved AC-DC bidirectional converters, including the power factor correction and reactive power flow. The future work would be directed towards implementing unipolar modulation and soft switching techniques for the full bridge switches using auxiliary network of passive components as presented in [7].

REFERENCES

[1] K.M. Smedley, L. Zhou, and C. Qiao, "Unified Constant-Frequency Integration Control of Active Power Filters – Steady – State and Dynamics," IEEE Transactions on Power Electronics, vol. 16, no. 3, pp. 428-436, May 2001.

[2] D.V. Ghodke, K. Chatterjee, and B.G. Fernandes, "Modified One-Cycle Controlled Bidirectional High-Power Factor AC-to-DC Converter," IEEE Transactions on Industrial Electronics, vol. 55, no. 6, pp. 2459-2472, June 2008.

[3] D.V. Ghodke, K. Chatterjee, and B.G. Fernandes, "One-Cycle Controlled Bidirectional AC-to-DC Converter with Constant Power Factor," IEEE Transactions on Industrial Electronics, vol. 56, no. 5, pp. 1499-1510, May 2009.

[4] F. Musavi, W. Eberle, and W.G. Dunford, "High Performance Single Phase Bridgeless Interleaved PFC Converter for Plug-in Hybrid Electric Vehicle Battery Chargers," IEEE Transactions on Industry Applications, vol. 47, no. 4, pp. 1833-1843, July/August 2011.

[5] F. Musavi, M. Edington, W. Eberle, and W.G. Dunford, "Evaluation and Efficiency Comparison of Front End AC-DC Plug-in Hybrid Charger Topologies," IEEE Transactions on Industry Applications, vol. 47, no. 4, pp. 1833-1843, July/August 2011.

[6] D.S. Gautam, F. Musavi, M. Edington, W. Eberle and W.G. Dunford, "An Automotive On-board 3.3KW Battery Charger for PHEV Application," IEEE Transactions on Vehicular Technology, vol. 61, no. 8, pp. 3466-3474, October 2012.

[7] M. Pahlevaninezhad, P. Das, J. Drobnik, P.K. Jain and A. Bakhshai, "A ZVS Interleaved Boost AC/DC Converter Used in Plug-in Electric Vehicles," IEEE Transactions on Power Electronics, vol. 27, no. 8, pp. 3513-3529, August 2012.

[8] N.Akel, M. Pahlevaninezhad, and P. Jain, "A D-Q rotating frame reactive power controller for single-phase bi-directional converters," in Telecommunications Energy Conference (INTELEC), 2014, pp. 1-5.

[9] F. Musavi, W. Eberle, and W.G. Dunford, "Efficiency Evaluation of Single-Phase Solutions for AC-DC PFC Boost Converters for Plug-in-Hybrid Electric Vehicule Battery Chargers," in Vehicle Power and Propulsion Conference (VPPC), 2010 IEEE, pp. 1-6.

A New Control Scheme to Improve Load Transient Response of Single Phase PWM Rectifier with Auxiliary Current Injection Circuit

Naga Brahmendra Yadav Gorla, Sandeep Kolluri, Pritam Das and Sanjib Kumar Panda
Department of Electrical and Computer Engineering
National University of Singapore, Singapore 117580
Email:naga@u.nus.edu

Abstract—**This paper presents a new control scheme for a single-phase PWM rectifier with Auxiliary Current Injection Circuit (ACIC) to improve the load transient response and DC bus voltage controller bandwidth. The PWM rectifier and ACIC are controlled to mitigate the 2nd harmonic ripple in the DC bus under steady-state as well as the peak undershoot and overshoot during load transients. Analytical equations governing the operation of the single phase PWM rectifier with proposed ACIC controller are derived and presented. The proposed controller is designed and simulated using MATLAB Simulink and implemented on a 100 W test setup. The simulation and experimental results are presented to validate the performance of proposed controller.**

I. INTRODUCTION

In single-phase PWM rectifiers, bulky electrolytic capacitors (E-caps) are necessary to filter the 2nd harmonic ripple current generated as a result of the power transfer between the AC source and the DC load. Moreover, E-caps have limitations on ripple current capability and operating life cycles [1]. In order to attenuate the effect of 2nd harmonic ripple in the generated reference current, the DC bus voltage controller has to be designed at low bandwidth (< 20 Hz) [2]. As a result, the DC bus voltage response to the step changes in load becomes very slow.

Several passive [3]- [4] and active [5]- [6] decoupling circuits have been proposed for single-phase PWM rectifiers to reduce the 2nd harmonic ripple in the DC bus. In active decoupling circuits, the active switches are appropriately modulated to store the ripple energy either in an inductor [7]- [8] or in a capacitor [5]- [6], [9]- [10]. Another classification of power decoupling circuits is based on the polarity of ripple on energy storage elements. The circuit with energy storage elements having unipolar ripple are categorised as DC decoupling circuits [6] whereas those with bipolar ripple are known as AC decoupling circuits [11]. A benchmark of AC and DC decoupling circuits has been presented in [12] and it has been shown that DC decoupling circuits outperform the AC decoupling circuits in terms of efficiency and power density. The experimental results presented in literature [5]- [7] show that the active decoupling circuits can minimize the DC bus capacitance requirement and improve the DC bus controller bandwidth. It can be observed from the DC bus voltage waveforms [5]- [7] that the overshoot and undershoot for step change in load is about 25 %.

The decoupling circuits proposed in [5]- [7] behave as current injection circuits that can be controlled actively to mitigate the 2nd harmonic voltage at the DC bus. These Auxiliary Current Injection Circuits (ACIC) have energy storage elements that can be actively controlled to suppress the peak overshoot and undershoot in DC bus voltage during load transients. The scope of this paper is to study the improvement in transient performance of single phase PWM rectifier through proper utilization of energy available in the energy storage elements.

In this paper, a new control scheme is proposed to control the ACIC, connected in parallel with the PWM rectifier, to regulate the DC bus voltage during load transients. The bidirectional converter presented in [6], [9] for compensating 2nd harmonic ripple in the DC bus is used here as the ACIC. The objectives of the proposed control scheme are two-fold: firstly to mitigate the 2nd harmonic ripple in the DC bus and secondly to inject or absorb the required current during load transient to suppress the disturbances at the DC bus voltage [13]- [14]. The paper is organized with Section II explaining the steady-state and transient operation and section III covering the mathematical analysis and design of the single phase PWM rectifier with proposed ACIC control scheme. Simulation and experimental results are presented in section IV. Finally, section V concludes the paper.

II. OPERATION OF A SINGLE PHASE PWM RECTIFIER WITH ACIC

The operation of the single-phase PWM rectifier in parallel with auxiliary boost converter as shown in Fig. 1, to mitigate the 2nd harmonic ripple in the DC bus voltage under steady-state and to obtain better dynamic performance (overshoot, undershoot and settling time) during transients, is explained in this section.

A. Steady-State Operation

The apparent power (S_{ac}) at the input of PWM rectifier as shown in Fig. 1 (assuming Unity Power Factor) is given by

$$
\begin{aligned}
S_{ac} &= v_g i_g \\
&= V_g sin(wt) I_g sin(wt) \\
&= \frac{V_g I_g}{2} - \frac{V_g I_g}{2} cos(2wt) \\
&= P_{av} + P_{rip}
\end{aligned}
$$

978-1-4673-9551-9/16 $31.00 © 2016 IEEE

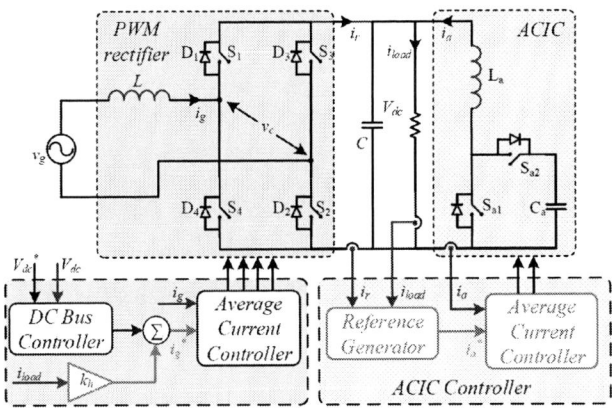

Fig. 1. Single phase PWM rectifier with Auxiliary Current Injection Circuit in parallel to the DC bus.

The apparent power S_{ac} has two components viz., average power (P_{av}) and ripple power (P_{rip}). Under steady-state, load power ($P_o = V_{dc}I_{load}$) is equal to the average power from the ac source (P_{av}). The ripple power (P_{rip}) is reflected as a predominant 2nd harmonic output voltage ripple in the single phase PWM rectifier. DC bus filter capacitor has to be designed to filter-out not only the high frequency switching ripple in the PWM rectifier, but also the 2nd harmonic ripple due to ripple power (P_{rip}). This high capacitance requirement calls for the use of bulky electrolytic capacitor which decreases the power density of the converter. Moreover electrolytic capacitors have lower operating life cycles as compared to film capacitors.

A boost converter when connected to the DC bus as shown in Fig.1 can act like an active filter to absorb 2nd harmonic ripple current from the PWM rectifier. The auxiliary boost converter, when controlled appropriately, can take the 2nd harmonic ripple current, completely bypassing the DC bus capacitor (C). This reduces the DC bus filter capacitance requirement as it has to filter-out the high frequency switching ripple alone. The 2nd harmonic ripple current (i_{r2}) for a given DC bus voltage V_{dc} is given by

$$i_{r2} = \frac{P_{rip}}{V_{dc}}$$
$$= \frac{V_g I_g}{2V_{dc}} cos(2wt)$$

The energy storage elements in the auxiliary boost converter (auxiliary capacitor C_a and inductor L_a) can be designed to have higher ripple as they are not contributing to the power transfer from source to load. The auxiliary converter can be controlled to improve the dynamic performance of the PWM rectifier, the details of which are explained in the next section.

B. Transient operation

The conventional cascaded two loop control structure (viz., DC bus control loop and average current control loop) for a PWM rectifier is shown in Fig. 1 and Fig. 2 (lower part). During transients the load power and the input power will not be balanced instantaneously until the DC bus controller provide necessary current reference. As a result the DC bus capacitor will be discharged or overcharged to maintain the

power balance resulting in severe undershoot or overshoot during transients of the DC bus voltage. A new control scheme is proposed to improve the dynamic response of the single phase PWM rectifier with auxiliary converter during transients. The proposed controller is designed to maintain the power balance by momentarily transferring energy (difference between source and load) from the auxiliary circuit to the DC bus during transients. To realize this, the sensed load current is fed forward to the controller. As the auxiliary circuit is absorbing the 2nd harmonic current during steady-state and injecting an auxiliary current into DC bus during transients, it is more appropriate to name it as Auxiliary Current Injection Circuit (ACIC).

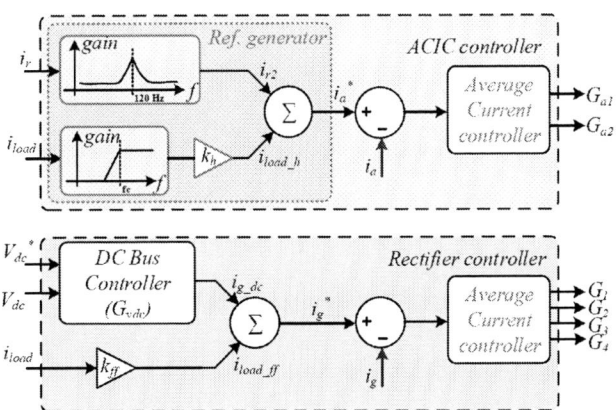

Fig. 2. Control scheme for ACIC and PWM rectifier with load current feedforward.

The control scheme for PWM rectifier with ACIC is shown in Fig.2. The reference generator circuit in the ACIC controller synthesizes the current reference (i_a^*) based on 2nd harmonic component of rectifier current (i_{r2}) and high frequency component of load current (i_{load_h}). The 2nd harmonic component of rectifier current (i_{r2}) is obtained by feeding the rectifier current (i_r) to a resonant filter tuned to 2nd harmonic frequency (i.e., 120 Hz). In case the converter is operating at a fundamental grid frequency of 50 Hz, the resonant filter should be tuned to 100 Hz. The high frequency component of load current is obtained by passing the sensed load current through a high-pass filter. Because of the ACIC controller action the DC bus controller (G_{vdc}) does not generate appropriate grid current reference during transients [13]- [14]. In order to avoid this a load current feedforward loop is added to the PWM rectifier controller as shown in Fig. 2. The controller parameters are tuned to reduce overshoot and undershoot of the DC bus voltage during transients. The load current feedforward element can cause instability if the gain k_{ff} is not designed properly. The design guidelines along with the small signal model of PWM controller with feedforward controller is discussed in the next section.

III. ANALYSIS, MODELLING AND DESIGN OF PWM RECTIFIER WITH THE PROPOSED ACIC CONTROL SCHEME

As discussed earlier, ACIC compensates the 2nd harmonic ripple during steady-state and mitigates the dynamics in the

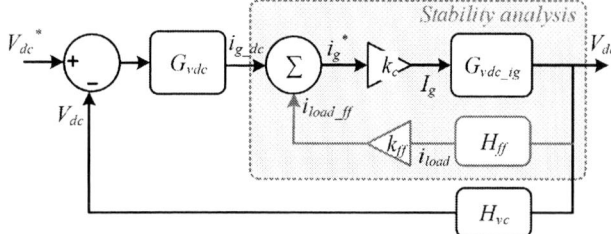

Fig. 3. Closed loop single phase PWM rectifier with load current feedforward.

PWM rectifier output voltage (overshoot and undershoot) during load transients by momentarily supplying a part of energy demanded by the load. In order to realize this, the bandwidth of ACIC controller should be designed much higher (atleast 10 times) than that of the PWM rectifier. Using conventional state space averaging technique, the small signal model equation of the PWM rectifier with ACIC is as follows

$$
\begin{bmatrix} \dfrac{d\hat{i_g}}{dt} \\ \dfrac{d\hat{v_{dc}}}{dt} \end{bmatrix} = \begin{bmatrix} 0 & -\dfrac{m_g}{L} \\ \dfrac{m_g}{C} & -\dfrac{1}{RC} \end{bmatrix} \begin{bmatrix} \hat{i_g} \\ \hat{v_{dc}} \end{bmatrix} + \begin{bmatrix} \dfrac{1}{L} & 0 & -\dfrac{V_{dc}}{L} \\ 0 & \dfrac{1}{C} & \dfrac{I_g}{C} \end{bmatrix} \begin{bmatrix} \hat{v_g} \\ \hat{i_a} \\ \hat{m_g} \end{bmatrix}
$$

where $\hat{i_g}, \hat{v_{dc}}, \hat{v_g}, \hat{i_a}$ are small disturbances in the input current (i_g), output voltage (V_{dc}), input voltage (v_g) and ACIC current (i_a) respectively and $\hat{m_g}$ is the control input. In order to simplify the analysis, the effect of right half plane zero is neglected. The control to input current ($\dfrac{\hat{i_g}}{\hat{m_g}}$), control to output voltage ($\dfrac{\hat{v_{dc}}}{\hat{m_g}}$) and output impedance ($\dfrac{\hat{v_{dc}}}{\hat{i_g}}$) transfer functions are given by (1), (2) and (3) respectively. The single phase PWM rectifier model and the simplified two loop control structure with load current feedforward is as shown in Fig. 3. Since load current is fed as a positive feedback, its gain k_{ff} has to be designed properly to avoid instability.

$$
G_{ig\_mg} = \frac{\hat{i_g}}{\hat{m_g}} = \frac{-\left(s\dfrac{V_{dc}}{L} + \dfrac{V_{dc}}{RLC} + \dfrac{m_g i_g}{LC}\right)}{s^2 + s\dfrac{1}{RC} + \dfrac{m_g^2}{LC}} \tag{1}
$$

$$
G_{vdc\_mg} = \frac{\hat{v_{dc}}}{\hat{m_g}} = \frac{s\dfrac{I_g}{C} - \dfrac{V_{dc}m_g}{LC}}{s^2 + s\dfrac{1}{RC} + \dfrac{m_g^2}{LC}} \tag{2}
$$

$$
\begin{aligned}
G_{vdc\_ig} = \frac{\hat{v_{dc}}}{\hat{i_g}} &= \frac{\left(\dfrac{V_{dc}m_g}{LC}\right)\left(1 - s\dfrac{I_g L}{V_{dc}m_g}\right)}{s\dfrac{V_{dc}}{L} + \dfrac{V_{dc}}{RLC} + \dfrac{m_g i_g}{LC}} \\
&\approx \frac{\dfrac{V_{dc}m_g}{LC}}{s\dfrac{V_{dc}}{L} + \dfrac{V_{dc}}{RLC} + \dfrac{m_g i_g}{LC}}
\end{aligned} \tag{3}
$$

In order to determine stability of the single phase rectifier with load current feedforward element, its current reference

(i_{g_dc}) to DC bus voltage (V_{dc}) transfer function needs to be examined.

$$
\frac{\hat{i_{g\_dc}}}{\hat{v_{dc}}} = \frac{G_{vdc\_ig}k_c}{1 - G_{vdc\_ig}k_{ff}k_c H_{ff}} \tag{4}
$$

where, k_c is the approximated low frequency loop gain of the current loop. In order to ensure the stability of the current loop, feedforward gain k_{ff} is designed such that all the poles of the $\dfrac{\hat{i_{g\_dc}}}{\hat{v_{dc}}}$ transfer function lies on the left half of s-plane. Therefore, the following condition should be satisfied.

$$
k_{ff} < \frac{\dfrac{V_{dc}}{R} + m_g i_g}{i_{load}m_g k_c} \tag{5}
$$

TABLE I. PARAMETERS OF THE TEST SETUP

Parameter name	Symbol	Value
Grid voltage and frequency	V_g and f_s	115 $Vrms$, 60 Hz
Grid side inductor	L	5 mH
DC link capacitor without ACIC	C	1 mF
DC link capacitor with ACIC	C	0.1 mF
DC link voltage	V_{dc}	200 V
Rectifier switching frequency	f_{sr}	20 kHz
ACIC switching frequency	f_{sa}	50 kHz
ACIC inductor	L_a	1 mH

IV. RESULTS AND DISCUSSIONS

The ACIC with proposed control scheme is connected in parallel to the single phase PWM rectifier to study the performance of the controller. This section presents the results obtained from simulation and experiment with the system.

A. Simulation Results

The proposed control scheme is simulated in MATLAB Simulink and DC bus voltage transient response results for 90% step change in load are shown in Fig. 4. Following are some of the inferences from the simulations.

1) Transient response of the PWM rectifier without ACIC is shown in Fig. 4(a). Since the ACIC is not present, the Steady-State Ripple (SSR) in the DC bus voltage at full load is 3.35 % with 1 mF DC bus capacitor. A peak overshoot of 27.6 V (13.8 % of rated DC bus voltage), undershoot of 27.4 V (13.7 % of rated DC bus voltage) and maximum settling time of 123.9 ms are observed for 90 % step change in load.

2) Transient response of the PWM rectifier with ACIC when controlled to mitigate the 2nd harmonic ripple in the DC bus is shown in Fig. 4(b). A DC bus capacitance of 100 μF is used to filter out the switching harmonics of PWM rectifier and ACIC. It can be observed from Fig. 4(b) that the SSR in the DC bus voltage at full load is 0.3 V, which is 0.15 % of rated DC bus voltage. A peak overshoot of 50 V (25 % of rated DC bus voltage) and peak undershoot of 45 V (25 % of rated DC bus voltage) are observed for 90 % step change in load.

3) Transient response of the PWM rectifier with the proposed controller for ACIC is shown in Fig. 4(c). It can be observed from Fig. 4(c) that the maximum SSR in the DC bus voltage is 1.3 V (0.65 % of rated DC bus voltage). The peak overshoot and undershoot are within 5 V for 90 % step change in load. For comparison purpose the

978-1-4673-9551-9/16 $31.00 © 2016 IEEE

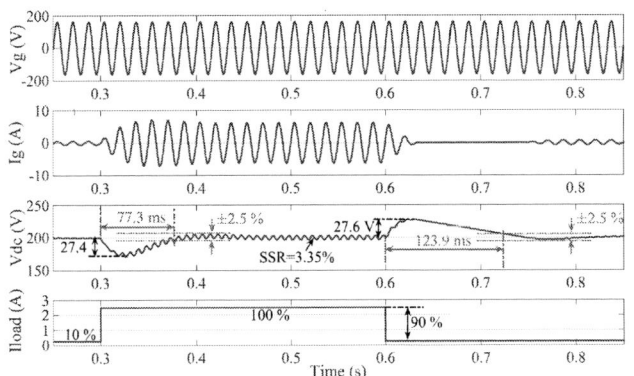
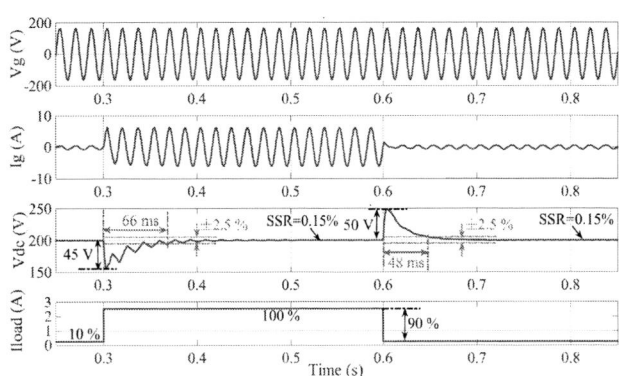

(a) Case (i): Transient response of 500 W single phase PWM rectifier with 1 mF DC bus capacitor and without ACIC.

(b) Case (ii): Transient response of 500 W single phase PWM rectifier with 100 μF DC bus capacitor and with ACIC controlled to mitigate 2nd harmonic ripple in the DC bus.

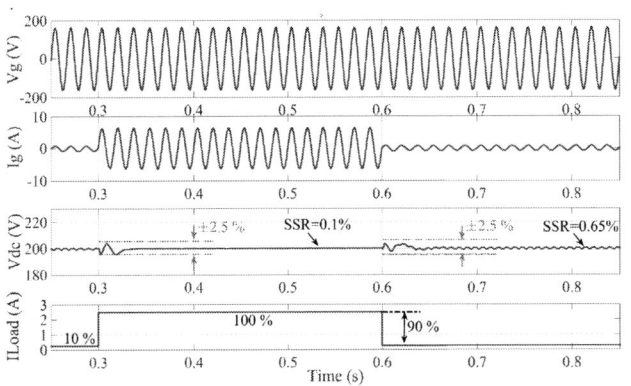
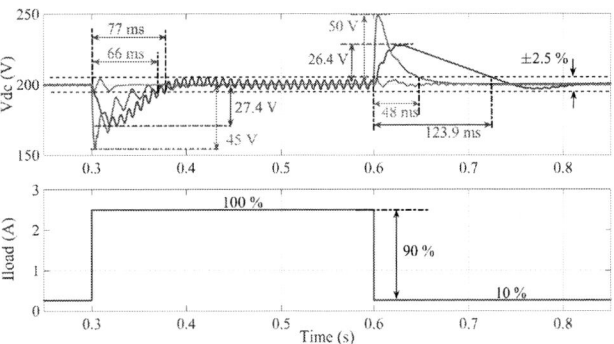

(c) Case (iii): Transient response of 500 W single phase PWM rectifier with 100 μF DC bus capacitor, with proposed control scheme for ACIC.

(d) Transient response of 500 W single phase PWM rectifier (i) without ACIC and 1 mF capacitor in blue, (ii) with ACIC for 2nd harmonic ripple compensation and 100 μF capacitor in green and (iii) with proposed control scheme for ACIC and 100 μF capacitor in red.

Fig. 4. Simulation results of proposed control scheme for ACIC in parallel with single phase PWM rectifier for 90 % step change in load.

transient response of PWM rectifier for all the three cases are shown in Fig. 4(d).

4) It can be seen from the Figs. 4(a), 4(b) and 4(c) that the grid current waveform is sinusoidal because of the following reasons. In the case of PWM rectifier without ACIC the bandwidth of the DC bus controller is designed at low frequency (8 Hz) to generate a sinusoidal reference current. In the second and third cases, ACIC is controlled to mitigate the 2nd harmonic ripple in the DC bus and therefore reference current is sinusoidal even at higher DC bus controller bandwidth. It is evident from the grid current waveforms that the proposed scheme does not affect the performance of current controller and hence it can be designed independently.

B. Experimental Results

The experimental setup of single phase PWM rectifier is operated with a grid voltage (v_g) of 30 V_{rms} at 60 Hz, DC bus voltage (V_{dc}) of 60 V and other parameters as shown in Table. I. The ACIC with and without the proposed control scheme is connected in parallel with the single phase PWM rectifier and tested upto 100 W. The experimental results are shown in Fig. 5. Following are some of the inferences.

1) As discussed earlier, the 2nd harmonic ripple in the DC

bus is responsible for distorted grid current in a single phase PWM rectifier. This problem can be solved by connecting ACIC across the DC bus. The grid current waveforms without and with ACIC being connected to the DC bus are shown in Fig. 5(a) and Fig. 5(b) respectively. A phase plot with measured grid current (i_g) on x-axis and derived imaginary term ($\int i_g$) on Y-axis is plotted in both the cases. It can be inferred from the circular phase plot (refer to Fig. 5(b)) that the measured grid current is sinusoidal without harmonics because of ACIC operation.

2) DC bus voltage (V_{dc}), grid voltage (v_g), grid current (i_g) and ACIC current (i_a) when ACIC is controlled to reduce 2nd harmonic ripple alone are shown in Fig.5(c) & Fig. 5(d).

- It can be seen in Fig. 5(c) & Fig. 5(d) that the 2nd harmonic ripple in the DC bus is minimum because of ACIC control action.
- It can be observed from Fig. 5(c) that for a 90% step decrease in load (from 100 W to 10 W), the peak overshoot is 26.2%.
- Similarly from Fig. 5(d), it can be observed that for a 90% step increase in load (from 10 W to 100 W), the peak undershoot is 22.6%.

3) DC bus voltage (V_{dc}), grid voltage (v_g), grid current (i_g) and ACIC current (i_a) when ACIC is controlled with the

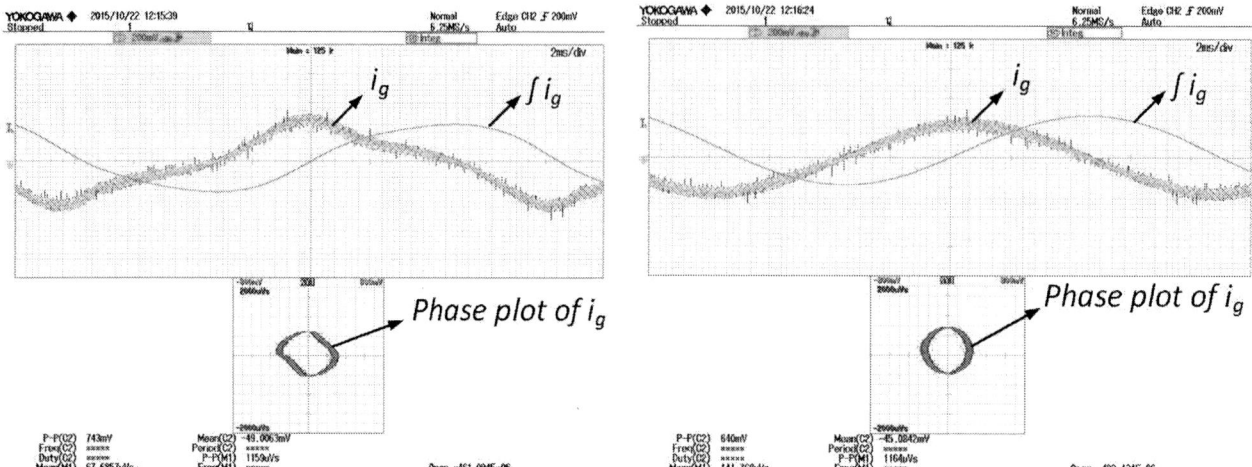

(a) ig: Grid current (2 A/div) and plot with real value of grid current on x axis and derived imaginary component on y axis without ACIC in operation.

(b) ig: Grid current (2 A/div) and phase plot with real value of grid current on x axis and derived imaginary component on y axis with ACIC and proposed controller.

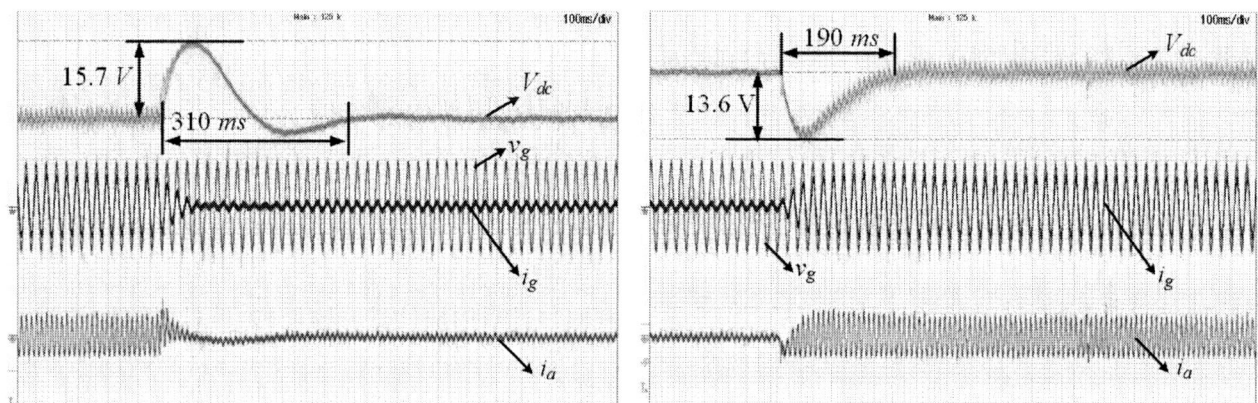

(c) Vdc: DC bus voltage of single phase PWM rectifier (10 V/div), vg: Grid voltage (50 V/div), ig: Grid current (10 A/div) and ia: ACIC current (5 A/div) for 90 % step change in load (100 W to 10 W). ACIC is controlled to reduce 2nd harmonic ripple in the DC bus.

(d) Vdc: DC bus voltage of single phase PWM rectifier (10 V/div), vg: Grid voltage (50 V/div), ig: Grid current (10 A/div) and ia: ACIC current (5 A/div) for 90 % step change in load (10 W to 100 W). ACIC is controlled to reduce 2nd harmonic ripple in the DC bus.

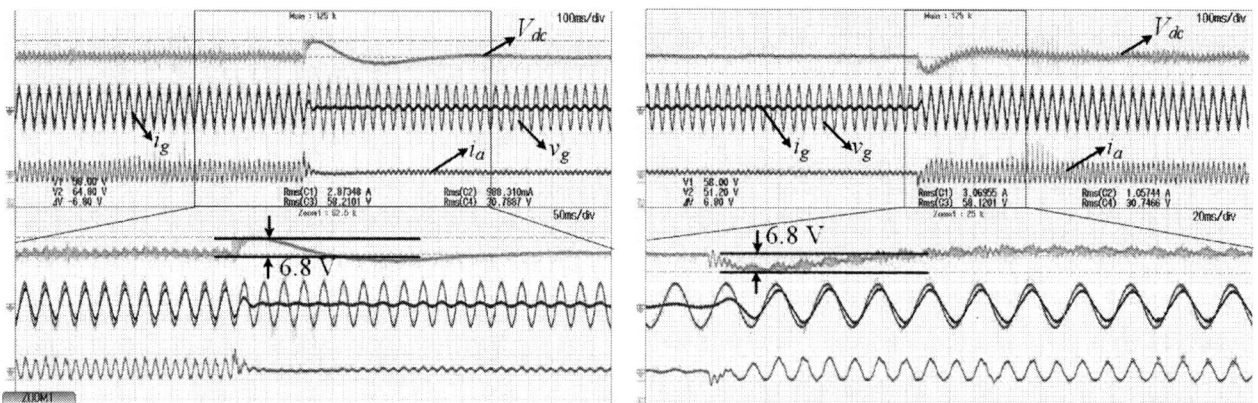

(e) Vdc: DC bus voltage of single phase PWM rectifier (10 V/div), vg: Grid voltage (50 V/div), ig: Grid current (10 A/div) and ia: ACIC current (5 A/div) for 90 % step change in load (100 W to 10 W) with the proposed control scheme.

(f) Vdc: DC bus voltage of single phase PWM rectifier (10 V/div), vg: Grid voltage (50 V/div), ig: Grid current (10 A/div) and ia: ACIC current (5 A/div) for 90 % step change in load (10 W to 100 W) with the proposed control scheme.

Fig. 5.   Experimental results for the proposed control scheme for ACIC connected in parallel with single phase PWM rectifier.

978-1-4673-9551-9/16 $31.00 © 2016 IEEE

proposed control scheme are shown in Fig. 5(e) & Fig. 5(f).

- It can be seen in Fig. 5(e) & Fig. 5(f) that even though the ACIC has current feed-forward and high frequency load current terms, it does not affect the basic operation of the ACIC. The 2nd harmonic ripple in the DC bus voltage (V_{dc}) is minimal.

- It can be observed from Fig. 5(e) that for a 90% step decrease in load (from 100 W to 10 W), the peak overshoot is 11%.

- Similarly, it can be observed from Fig. 5(d) that for a 90% step increase in load (from 10 W to 100 W), the peak undershoot is 11%.

4) Overall there is a 50% (apx.) reduction in peak overshoot and undershoot for 90% step change in the load (for both step-up and step-down cases).

The peak overshoot and undershoot in the DC bus voltage can be further reduced by adaptively changing the k_{ff} and k_h gains with load (this is implemented in simulation to improve the stability), otherwise stability of the system will be affected. The design parameters at any load should not violate the stability condition provided by (5).

V. CONCLUSION

In this paper a new control scheme for ACIC connected in parallel with a single phase PWM rectifier to improve the DC bus voltage dynamic response during load transients is proposed. The ACIC with the proposed control scheme is explained with proper small signal modelling. Simulation and experimental results are presented to validate the operation of ACIC with proposed control. Results show that the proposed control scheme improves the PWM rectifier dynamic response during load transients.

REFERENCES

[1] H. Wang, M. Liserre, and F. Blaabjerg, "Toward reliable power electronics: challenges, design tools, and opportunities," *Industrial Electronics Magazine, IEEE*, vol. 7, no. 2, pp. 17–26, 2013.

[2] R. Wang, F. Wang, R. Lai, P. Ning, R. Burgos, and D. Boroyevich, "Study of energy storage capacitor reduction for single phase pwm rectifier," in *Applied Power Electronics Conference and Exposition, 2009. APEC 2009. Twenty-Fourth Annual IEEE*, pp. 1177–1183, IEEE, 2009.

[3] J. Das, "Passive filters-potentialities and limitations," in *Pulp and Paper Industry Technical Conference, 2003. Conference Record of the 2003 Annual*, pp. 187–197, IEEE, 2003.

[4] P. T. Krein and R. S. Balog, "Cost-effective hundred-year life for single-phase inverters and rectifiers in solar and led lighting applications based on minimum capacitance requirements and a ripple power port," in *Applied Power Electronics Conference and Exposition, 2009. APEC 2009. Twenty-Fourth Annual IEEE*, pp. 620–625, IEEE, 2009.

[5] R. Wang, F. Wang, D. Boroyevich, R. Burgos, R. Lai, P. Ning, and K. Rajashekara, "A high power density single-phase pwm rectifier with active ripple energy storage," *Power Electronics, IEEE Transactions on*, vol. 26, no. 5, pp. 1430–1443, 2011.

[6] Q.-C. Zhong, W.-L. Ming, X. Cao, and M. Krstic, "Reduction of dc-bus voltage ripples and capacitors for single-phase pwm-controlled rectifiers," in *IECON 2012-38th Annual Conference on IEEE Industrial Electronics Society*, pp. 708–713, IEEE, 2012.

[7] Y. Tang, Z. Qin, F. Blaabjerg, and P. C. Loh, "A dual voltage control strategy for single-phase pwm converters with power decoupling function," *IEEE Transactions on Power Electronics*, no. 99, pp. 1–12, 2014.

[8] M. Su, P. Pan, X. Long, Y. Sun, and J. Yang, "An active power-decoupling method for single-phase ac–dc converters," *Industrial Informatics, IEEE Transactions on*, vol. 10, no. 1, pp. 461–468, 2014.

[9] W. Cai, B. Liu, S. Duan, and L. Jiang, "An active low-frequency ripple control method based on the virtual capacitor concept for bipv systems," *Power Electronics, IEEE Transactions on*, vol. 29, no. 4, pp. 1733–1745, 2014.

[10] H. Li, K. Zhang, H. Zhao, S. Fan, and J. Xiong, "Active power decoupling for high-power single-phase pwm rectifiers," *Power Electronics, IEEE Transactions on*, vol. 28, no. 3, pp. 1308–1319, 2013.

[11] S. Fan, Y. Xue, and K. Zhang, "A novel active power decoupling method for single-phase photovoltaic or energy storage applications," in *Energy Conversion Congress and Exposition (ECCE), 2012 IEEE*, pp. 2439–2446, IEEE, 2012.

[12] Z. Qin, Y. Tang, P. Loh, and F. Blaabjerg, "Benchmark of ac and dc active power decoupling circuits for second order harmonic mitigation in single-phase inverters," *Emerging and Selected Topics in Power Electronics, IEEE Journal of*, vol. PP, 2015.

[13] S. Kolluri *et al.*, "A new isolated auxiliary current pump module for load transient mitigation of isolated/non-isolated step-up/step-down dc-dc converters," *Power Electronics, IEEE Transactions on*, vol. 30, 2015.

[14] S. Kolluri and N. Lakshmi Narasamma, "A new auxiliary current injection circuit for improved transient response of step-up/step-down dc-dc converters," in *Industrial Electronics Society, IECON 2013-39th Annual Conference of the IEEE*, pp. 216–221, IEEE, 2013.

Active Capacitor with Ripple-Based Duty Cycle Modulation for AC-DC Applications

Ching-Chieh Yang
Department of Electrical
Engineering,
National Taiwan University
Taipei, Taiwan
f03921025@ntu.edu.tw

Yang-Lin Chen
Department of Electrical
Engineering,
National Taiwan University
Taipei, Taiwan

Yaow-Ming Chen
Department of Electrical
Engineering,
National Taiwan University
Taipei, Taiwan

Abstract − **Active capacitor with long lifetime film capacitor can be adopted to replace the short lifetime electrolytic capacitors to improve the reliability of the power converter. However, the conventional active capacitor suffers from the complicated control method. Based on the derived mathematical equations of the active capacitor, a ripple-based duty cycle modulation (RDCM) method for the active capacitor is proposed. The proposed RDCM is simple and can be easily realized by using analog circuits. Also, it can be applied to all kinds of active capacitors with different circuit topologies. Computer simulations have confirmed the performance of the proposed active capacitor with the RDCM control.**

I. INTRODUCTION

Due to the rapidly developed sold-state electronic technologies, the demand of AC-DC converter has been increased rapidly [1]-[2]. To produce the demanded constant voltage, a large capacitor bank in parallel with the output terminal is essential. The electrolytic capacitor has been used in most of converters to fulfill this task. However, the electrolytic capacitor has shorter lifetime compared with other electronic components and it will reduce the reliability of the converter [3]-[4]. Therefore, the long lifetime film capacitor becomes a proper substitution for the electrolytic capacitor [5]. Unfortunately, under the same capacitance value requirement, the film capacitor brings the issue of high cost and large volume [18]. As a result, the active power decoupling circuit (APDC) has been proposed to reduce the required capacitance [6]-[10].

It is true that the APDC can be adopted to replace the electrolytic capacitor in different power converters. Normally, the APDC is a basic decoupling cell connected in parallel or in series to the AC-DC converter. Also, the operation of APDC is highly depend on the main circuit of AC-DC converter, by sharing the components with the AC-DC converter partially and even fully [6]-[10]. It is inevitable that the control method of the AC-DC converter should be modified to meet the APDC's operation and its complexity will be increased. However, the APDC needs a complex control method and is not easy for implementation. Therefore, the active capacitor has been proposed to replace the use of APDC and electrolytic capacitor [11]-[19].

Theoretically, the active capacitor can be used in all

kinds of AC-DC converters to reduce the output voltage ripple. Both open-loop and closed-loop control methods can be applied to the active capacitor to achieve the desired function [11]-[19]. Open-loop control strategies can validly control the circuit steady but the average voltage of auxiliary capacitor in the active capacitor will be relatively high which also results in the shorter lifetime[11],[12]. To solve this problem, many closed-loop control methods have been proposed. Basically, three circuit topologies, buck-type, buck-boost-type, and boost-type, are commonly adopted to realize the active capacitor with closed-loop control [13]-[19].

The storage energy of the buck-type active capacitor is limited because of the low voltage on the auxiliary capacitor. To increase the storage energy, the buck-boost-type active capacitor usually has its the auxiliary capacitor operated in the high voltage level. Eventually, the boost circuit topology is commonly used to realize the active capacitor. Also, the boost-type active capacitor has continuous compensation current at the DC-link terminal, which is convenient for current control implementation [17]-[19]. However, complex control loop and extra circuitries or components, including a current sensor, are always required.

Therefore, a novel and simple control method for the boost-type active capacitor is proposed in this paper. The relation between the output voltage ripple and the required capacitance for an AC-DC converter will be introduced. Then, the characteristics of the active capacitor will be presented. Based on the mathematical analysis of the active capacitor, the new control method, ripple-based duty cycle modulation (RDCM), is proposed. The proposed RDCM is simple and can be applied to all kinds of active capacitors with different circuit topologies.

II. CAPACITOR FOR AC-DC CONVERTER

Fig.1 shows the conceptual diagram of an AC-DC converter with an output capacitor, C_{bus}, to smooth out the output voltage ripple. Different circuit topologies, such as bridge rectifier with a boost converter, can be adopted to realize the task of converting an AC source into a DC voltage source. Theoretically, the input power of the AC-DC converter is composed of two components, the active power

P_o, which is a constant value, and the reactive power $p_{C_{bus}}(t)$, which has a double-line frequency variation. The constant active power P_o will be transferred to the load while the reactive power $p_{C_{bus}}(t)$ is delivered to the output capacitor C_{bus}. The relation between $p_{C_{bus}}(t)$ and P_o can be expressed as:

$$p_{C_{bus}}(t) = P_o \cos 2\omega t$$
$$= v_o(t)i_{C_{bus}}(t) = V_0 C_{bus}\frac{d}{dt}v_{o,r}(t) \qquad (1)$$

where ω is the angular line frequency, $v_o(t)$ is the instantaneous output voltage, V_o is the average output voltage, $v_{o,r}(t)$ is the output ripple voltage, and $i_{C_{bus}}(t)$ is the output capacitor current. From (1), the output ripple voltage, $v_{o,r}(t)$, can be expressed as:

$$v_{o,r}(t) = \frac{P_o \sin(2\omega t)}{2\omega V_o C_{bus}} = \frac{\Delta V_o \sin(2\omega t)}{2} \qquad (2)$$

where ΔV_o is the peak to peak voltage ripple of the output voltage.

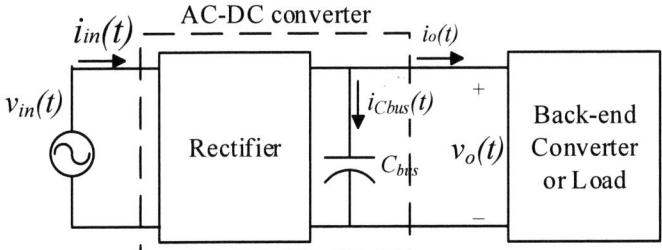

Fig. 1 The AC-DC converter with an output capacitor C_{bus}.

From (2) the required output capacitance C_{bus} for a specified load demand and with a limited ripple voltage ΔV_o can be expressed as:

$$C_{bus} = \frac{P_o}{\omega V_o \Delta V_o} = \frac{I_o}{\omega \Delta V_o} \qquad (3)$$

Usually, the output average power P_o and output average voltage V_o are determined by the specifications of the AC-DC converter. The angular frequency ω is a constant, too. Then, the required capacitance value C_{bus} is inversely proportional to the output ripple voltage ΔV_o, according to (3). In order to suppress the output voltage ripple, the required capacitance is essentially high. Hence, the electrolytic capacitor, which has high capacitance value, is widely used in the AC-DC converter. The concept of active capacitor, which consists of a bi-directional power converter and a small capacitor, has been proposed to imitate the characteristics of a capacitor to eliminate the use of electrolytic capacitor [11]-[19].

III. THE PROPOSED RDCM

The conception diagram of the active capacitor used in an AC-DC converter is shown in Fig. 2. The active capacitor consists of a bidirectional power converter and an auxiliary capacitor C_a, which has an average voltage V_a and a peak to peak voltage ΔV_a. The active capacitor alone with a small output capacitor C_o acts like the equivalent capacitor C_{bus} in an AC-DC converter. Among many circuit topologies, the boost converter have been found to be the best candidate to fulfill the bidirectional power converter inside the active capacitor [17]-[19]. Therefore, the proposed active capacitor adopts the boost converter but a novel control method is proposed.

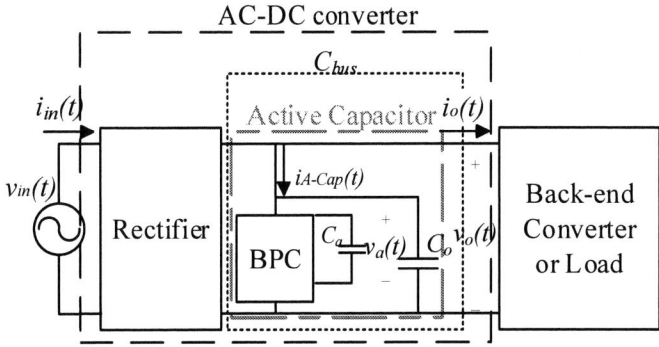

Fig. 2 The AC-DC converter with an active capacitor.

Theoretically, the double line frequency voltage ripple at the output terminal of the AC-DC converter should be absorbed by the active capacitor. The conceptual waveforms of input instantaneous power $p_a(t)$, input current $i_{A-Cap}(t)$, and the voltage across the output auxiliary capacitor $v_a(t)$ of the active capacitor can be plotted in Fig. 3. Since the active capacitor only consume reactive power, its average power is zero. Because of the line frequency ripple voltage at the DC side, which is the input terminal of the active capacitor, the instantaneous power of the active capacitor is a function with the same line frequency. Therefore, the instantaneous power of the auxiliary capacitor C_a should be sinusoidal, too.

According to the waveforms shown in Fig. 3 and equations (1) to (3), the charging energy ΔE_a into the active capacitor can be expressed in (4). On the other hand, the charging energy of the auxiliary capacitor C_a can be expressed in (5). Assuming an ideal power converter with no power loss, the charging energy enter the active capacitor will be equal to the charging energy of the auxiliary capacitor. From (4) and (5), the relationships among C_a, C_o, and C_{bus} can be derived and expressed in (6). It implies that the equivalent capacitance of the active capacitor C_{bus} is related to the auxiliary capacitor's average voltage V_o and ripple voltage ΔV_o.

Based on the derived equation shown in (6), to achieve a high C_{bus}, which leads to a low output voltage ripple ΔV_o, with the same capacitance values of C_a and C_o, both V_a and

ΔV_a should be as high as possible. To achieve this goal, the RDCM method is proposed.

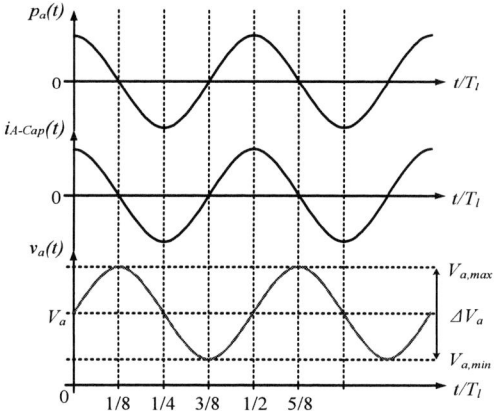

Fig. 3 The conceptual waveforms of the active capacitor.

$$\Delta E_a = \int_{3T/8}^{5T/8} p_a(t)$$

$$= \left(C_{bus} - C_o\right)V_o \left(\frac{V_{rms}I_{rms}}{\omega V_o C_{bus}}\right) \qquad (4)$$

$$= \left(C_{bus} - C_o\right)V_o \Delta V_o$$

$$\Delta E_a = \frac{1}{2}C_a V_{a,max}^2 - \frac{1}{2}C_a V_{a,min}^2$$

$$= C_a \left(\frac{V_{a,max} + V_{a,min}}{2}\right)\left(V_{a,max} - V_{a,min}\right) \qquad (5)$$

$$= C_a V_a \Delta V_a$$

$$\frac{C_{bus} - C_o}{C_a} = \frac{V_a \Delta V_a}{V_o \Delta V_o} \qquad (6)$$

Conventional PWM converter has a near constant duty ratio to obtain a stable output voltage under the steady state operation. The propose RDCM method intentionally induce a ripple voltage into control signal to generate the desired PWM signal to achieve the desired V_a and ΔV_a. The conceptual block diagram and its corresponding waveforms of the proposed RDCM are shown in Fig. 4.

The information of output voltage ripple, v_{ripple}, which can be obtained by subtracting the average output voltage V_o from the output voltage $v_o(t)$, is used to control the ΔV_a. The offset value, v_{offset}, of the control signal, $v_{control}$, is used to regulate V_a. For a larger output voltage ripple, the amplitude variation of $v_{control}$ will be increased so that ΔV_a will be increased, too. Eventually, the equivalent capacitance C_{bus} of the active capacitor will be increased to mitigate the output voltage ripple.

Since all control signals, v_{offset}, v_{ripple}, and $v_{control}$, need to meet the constraint of the PWM carrier signal v_{tri}, mathematical equations among these control signals should be conducted.

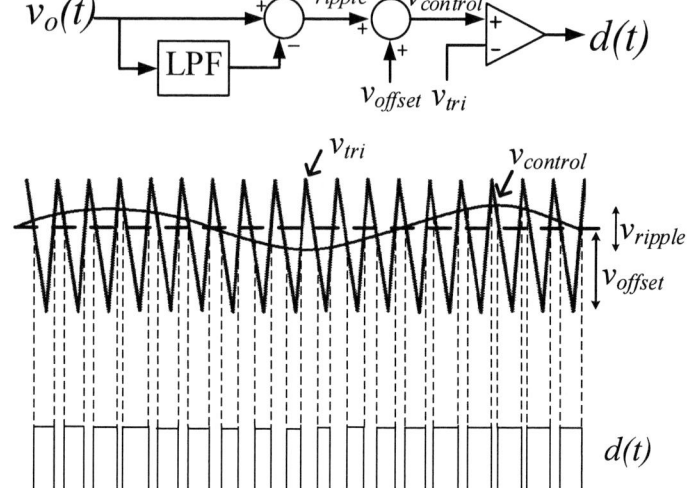

Fig. 4 The conceptual control block diagram and waveforms for the proposed RDCM.

IV. THE IMPLEMENTATION OF ACTIVE CAPACITOR WITH RDCM

The proposed RDCM method is applied to a rectifier with an active capacitor to verify its performance. The implementation of the testing circuit can be separated in two parts, the rectifier and the active capacitor. In this paper, the rectifier is chosen to be the commonly used boost-type power factor correction (PFC) converter, and the active capacitor is realized by a boost-type circuit topology.

A. PFC converter

Fig. 5 shows the circuit diagram of the commonly used boost-type PFC converter which consists of a full-bridge rectifier and a boost DC-DC converter. The full-bridge rectifier converts the input AC voltage into a rectified sinusoidal voltage and the boost DC-DC converter will produce a rectified sinusoidal input current while regulating the DC output voltage. Many PFC control methods have been proposed [20]-[22]. A typical controller which consists of an error amplifier and a PWM generator is shown in Fig. 5 where the output voltage $v_o(t)$ is feedback to the input of the error amplifier to determine the duty ratio level of the PWM generator [20].

Usually, the control system needs a low pass filter in the feedback loop to obtain a stable duty ratio for the PWM generator. However, it will reduce the transient response dramatically. Therefore, a large capacitor bank, C_{bus} is required to reduce the ripple of the feedback output voltage so the low pass filter in the control loop can be eliminated and the high transient response of the PFC converter can be achieved. To realize the large capacitor bank, C_{bus}, the electrolytic capacitors are always needed. In the paper, the active capacitor with a high life-time film capacitor is adopted to realize the large capacitor bank.

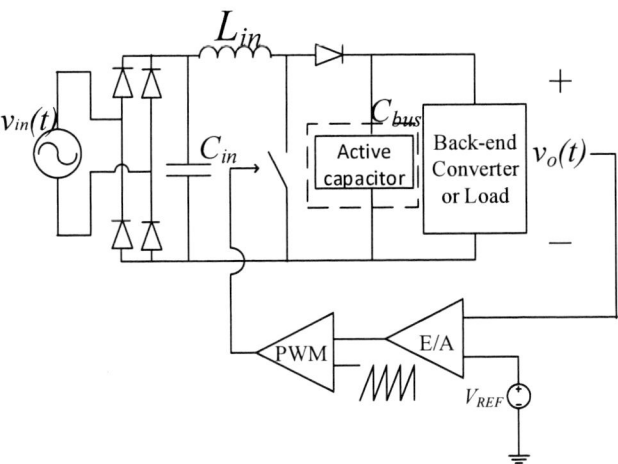

Fig. 5 A typical power stage and control stage for a PFC converter.

B. Active Capacitor with RDCM

Fig. 6 shows the schematic diagram of the active capacitor, which is a function block in Fig. 5, with the proposed RDCM. As shown in Fig. 6, the boost-type circuit topology, which is capable of bi-directional power transferring, is adopted to realize the power stage of the active capacitor. It should be noticed that the output voltage of the PFC converter is the input voltage of the active capacitor. The input capacitor C_o and the auxiliary capacitor C_a require small capacitance and can be realized by film capacitors.

The control stage consists of a low pass filter (LPF), a differential amplifier (Diff.), a signal voltage offset v_{offset}, and a PWM generator. The LPF is needed to obtain the average output voltage V_o. The output voltage ripple v_{ripple} of the PFC converter can be obtained by subtracting the average output voltage V_o from the input voltage. After level shifting by an offset singlet v_{offset}, the control signal $v_{control}$, as shown in Fig. 4 can be obtained. The PWM gate signal $d(t)$ is generated by comparing the $v_{control}$ and v_{tri}. The PWM gate signal $d(t)$ and its inverted one are used to control the two power switches of the active capacitor.

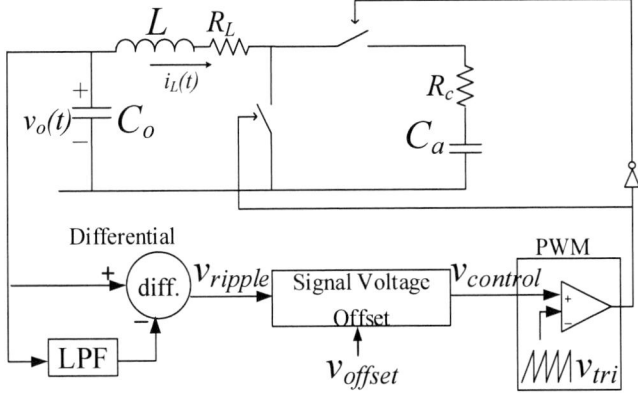

Fig. 6 The schematic diagram of the active capacitor with the proposed RDCM.

V. COMPUTER SIMULATION

An AC-DC converter with the proposed active capacitor is built and tested. The specification of the PFC converter is shown in Table I while the one for the active capacitor is shown in Table II. The computer software SIMPLIS is used to simulate the performance of the active capacitor with the proposed RDCM control.

To compare the performance of the proposed active capacitor, three different types of capacitors with different capacitance are adopted to realize the output capacitors. There are a 30µF film capacitor, a 1.4mF electrolytic capacitor, and the proposed active capacitor. These three capacitors are applied to the same PFC converter with the power stage and the control stage shown in Fig. 5. Input voltage and current waveforms as well as the output DC voltage waveforms are used to compare the performance of the proposed active capacitor.

Fig. 7 shows the output voltage $v_o(t)$ and input voltage and current waveforms of the PFC converter with a 30µF film output capacitor. Since the capacitance of the output capacitor is relative small, the ripple voltage of the output voltage is found to be about 48V, which meets the result from the derived equation shown in (3). The input voltage is sinusoidal while the input current is almost a sinusoidal one with slightly distortion. Since the main purpose of the computer simulation is not to verify the performance of the PFC function, it is objective to compare the performance of the active capacitor with the same PFC converter.

To mitigate the output ripple ΔV_o, a larger capacitance is required. By replacing the output capacitor with a 1.4mF electrolytic one, similar waveforms can be obtained and shown in Fig. 8. It can be observed that the output voltage ripple ΔV_o has been reduced to about 1V, which is the expected value based on the derived equation shown in (3). It should be mentioned that the input current shown in Fig. 8(b) is also improved with less distortion because of the reduction of the DC output voltage ripple. It also implies that a low output voltage ripple is essential to achieve a high power factor.

For the proposed active capacitor with the RDCM control, the similar waveforms are shown in Fig. 9. Using only two film capacitors, 30µF each, the output ripple ΔV_o is about 1V, which is the same with the one shown in Fig. 8 by using a 1.4mF electrolytic capacitor. Also, Fig. 9 (b) shows that the proposed active capacitor will not affect the performance of the original PFC converter and a high power factor and low harmonic distortion current can be achieved. It reveals that the proposed active capacitor has the equivalent capacitance C_{bus} up to 1.4 mF by using two 30µF film capacitors.

By comparing the simulated waveforms shown in Fig. 7 through Fig. 9, it can be verified that the active capacitor with the proposed RDCM control can achieve the desired performance successfully. It can use long life-time film capacitor with small capacitance quantity to achieve an equivalent large capacitance to improve the life-time of the power converter.

TABLE I.   SPECIFICATIONS OF THE PFC CONVERTER.

V_{in}	110Vrms 60Hz	V_o	400Vdc
P_o	200W	L_{in}	100μH
C_o	30μF	f_{s1}	50kHz

TABLE II.   SPECIFICATIONS OF THE ACTIVE CAPACITOR
CIRCUIT.

C_a	30μF	V_a	800Vdc
f_{s2}	200kHz	L	500μH

VI.   CONCLUSION

A novel RDCM control method for the active capacitor is proposed. In this paper, the required output capacitance for the AC-DC converter is introduced and the concept of the active capacitor is presented. One major disadvantage of the active capacitor is its control complexity. The proposed RDCM control method is simple and can be easily realized by analog circuits. Simulation results using different output capacitance values for the same PFC converter are presented. It has been proved that the proposed RDCM method can control the active capacitor successfully and high equivalent capacitance can be achieved to reduce the output voltage ripple. Also, the results also verify that the proposed active capacitor will not affect the original desired performance of the PFC converter.

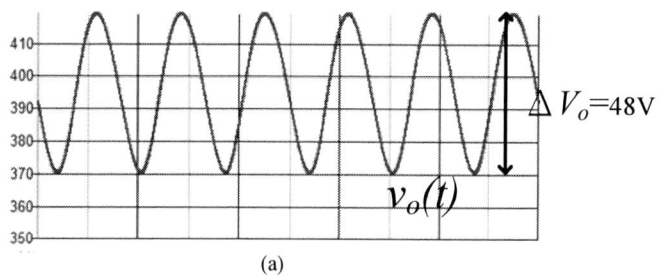

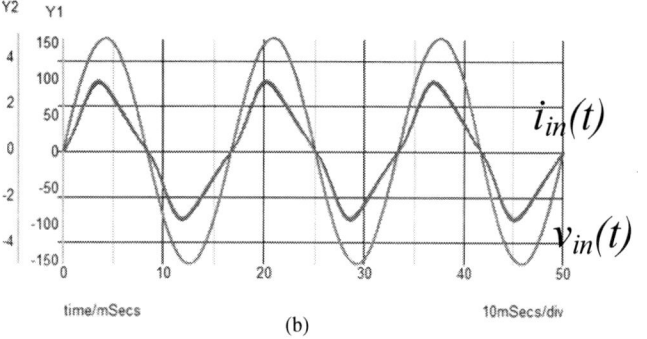

Fig. 7   Waveforms of the PFC converter with a 30μF film output capacitor, (a) the output voltage (b) the input voltage and input current.

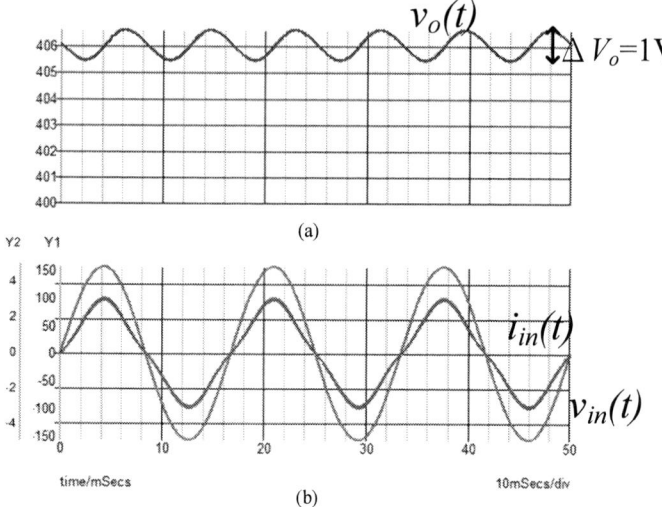

Fig. 8   Waveforms of the PFC converter with a 1.4mF electrolytic output capacitor, (a) the output voltage (b) the input voltage and input current

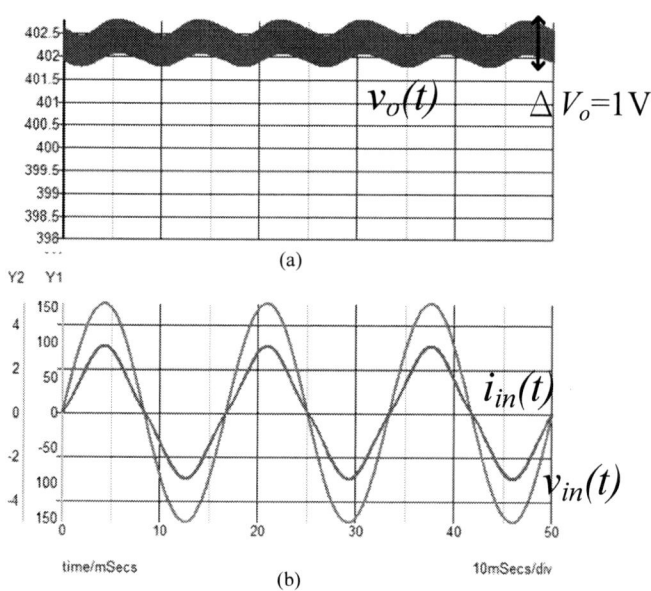

Fig. 9   Waveforms of the PFC converter with the proposed active capacitor, (a) the output voltage (b) the input voltage and input current.

REFERENCE

[1]   Y.-C. Li and C.-L. Chen, "A Novel Single-Stage High-Power-Factor AC-to-DC LED Driving Circuit With Leakage Inductance Energy Recycling," *IEEE Transactions on Industrial Electronics*, vol.59, no.2, pp.793-802, Feb. 2012.

[2]   P. Das, M. Pahlevaninezhad, and G. Moschopoulos, "Analysis and Design of a New AC–DC Single-Stage Full-Bridge PWM Converter With Two Controllers," IEEE Transactions on Industrial Electronics, vol.60, no.11, pp.4930-4946, Nov. 2013.

[3]   S. Yang, A. Bryant, P. Mawby, D. Xiang, L. Ran, and P. Tavner,

"An Industry-Based Survey of Reliability in Power Electronic *IEEE Transactions on Industry Applications*, vol.47, no.3, pp.1441-1451, May-Jun. 2011.

[4] H. Wang, M. Liserre, and F. Blaabjerg, "Toward Reliable Power Electronics: Challenges, Design Tools, and Opportunities," *IEEE Industrial Electronics Magazine*, vol.7, no.2, pp.17-26, Jun. 2013.

[5] P. T. Krein and R. S. Balog, "Cost-Effective Hundred-Year Life for Single-Phase Inverters and Rectifiers in Solar and LED Lighting Applications Based on Minimum Capacitance Requirements and a Ripple Power Port," *APEC 2009*, pp.620-625.

[6] T. Shimizu, Y. Jin, and G. Kimura, "DC ripple current reduction on a single-phase PWM voltage-source rectifier," IEEE Transactions on Industrial Electronics, vol.36, no.5, pp.1419-1429, Sep./Oct. 2000.

[7] H. Li, K. Zhang, H. Zhao, S. Fan, and J. Xiong, "Active Power Decoupling for High-Power Single-Phase PWM Rectifiers," IEEE Transactions on Power Electronics, vol.28, no.3, pp.1308-1319, Mar. 2013.

[8] S. Fan, Y. Xue, and K. Zhang, "Novel Active Power Decoupling Method for Single-Phase Photovoltaic or Energy Storage Applications," in Proc. IEEE ECCE, Raleigh, 2012. pp 2439-2446

[9] S. Liang, X. Lu, R. Chen, Y, Liu"A Solid State Variable Capacitor with Minimum DC Capacitor" in Proc. IEEE APEC, Fort Worth, TX, 2014. pp 3496-3501

[10] R. Chen, Y. Liu, and F. Z. Peng, "DC Capacitor-Less Inverter for Single-Phase Power Conversion With Minimum Voltage and Current Stress," IEEE Trans. Power Electon., vol. pp, no. 99, pp.1, Nov, 2014.

[11] R. Wang, F. Wang, R. Lai, P. Ning, R. Burgos, and D. Boroyevich, "Study of Energy Storage Capacitor Reduction for Single Phase PWM Rectifier," *APEC 2009*, pp.1177-1183

[12] R. Wang, F. Wang, D. Boroyevich, R. Burgos, Rixin Lai, P. Ning, and K. Rajashekara, "A High Power Density Single-Phase PWM Rectifier With Active Ripple Energy Storage," IEEE Transactions on Power Electronics, vol.26, no.5, pp.1430-1443, May 2011..

[13] K.-H. Chao, P.-T. Cheng, and T. Shimizu, "New control methods for single phase PWM regenerative rectifier with power decoupling

function," PEDS 2009, pp.1091-1096

[14] X. Zhang, X. Ruan, H. Kim, and C. K. Tse, "Adaptive Active capacitor for Improving Stability of Cascaded DC Power Supply System," IEEE Transactions on Power Electronics, vol.28, no.4, pp.1807-1816, Apr. 2013.

[15] M. Jang, M. Ciobotaru, and V. G. Agelidis, "A Single-Stage Fuel Cell Energy System Based on a Buck-Boost Inverter with a Backup Energy Storage Unit," IEEE Trans. Power Electron., vol. 27, no 6, pp. 2825-2834, Jun, 2012.

[16] M. Jang, and V. G. Agelidis, "A Minimum Power-Processing Stage Fuel Cell Energy System Based on A Boost-Inverter with A Bi-Directional Back-Up Battery Storage," IEEE Trans. Power Electron., vol. 26, no 5, pp. 1568-1577, May, 2011.

[17] Y. Tang, D. Zhu, C. Jin, P. Wang, and F. Blaabjerg, "A Three-Level Quasi-Two-Stage Single-Phase PFC Converter with Flexible Output Voltage and Improved Conversion Efficiency," IEEE Trans. Power Elecron., vol. 30, no 2, pp. 717-726, Feb, 2015.

[18] S. Y. Lee, Y. L. Chen, Y. M. Chen, and K. H. Liu, "Development of the Active Capacitor for PFC Converter," in Proc. IEEE ECCE, PA, 2014. pp. 1522-1527.

[19] Y. Yang, X. Ruan, L. Zhang, J. He, and Z. Ye, "Feed-Forward Scheme for an Electrolytic Capacitor-Less AC-DC LED Driver to Reduce Output Current Ripple," IEEE Trans. Power Electron., vol. 29, no 10, pp. 5508-5517, Jun, 2014.

[20] J.Sebastian, M. Jaureguizar, and J. Uceda, "An overview of Power Factor Correction In Single-Phase Off-line Power Supply System" *IEEE Industrial Electronics*, vol.3, pp.1688 – 1693, Sep 1994

[21] Wei Ma, Mingyu Wang, Shuxi Liu, Shan Li, and Peng Yu, "Stabilizing the Average-Current-Mode-Controlled Boost PFC Converter via Washout-Filter-Aided Method" IEEE Transactions on circuit and system—II: Express Briefs, vol. 58, no. 9, Sep 2011

[22] Siu-Chung Wong , Tse, C.K., Orabi, M., Ninomiya, T. "The method of double averaging: an approach for modeling power-factor-correction switching converters," IEEE Transactions on circuit and system—I: Regular Papers, pp. 454 – 462, Feb. 2006

Novel Approach to Current-mode Control in DCM/CCM Boundary Boost PFC

Giovanni Gritti, Claudio Adragna

STMicroelectronics s.r.l

Power Conversion BU – I&PC Division

20864 Agrate Brianza (MB), Italy

Abstract— Traditionally, DCM/CCM Boundary Boost PFC Converters with current-mode control require line voltage sensing (via a resistor divider) to define the sinusoidal shape of the reference for the inner current loop and a multiplier to adjust the amplitude of this reference to regulate the output voltage. In contrast, using voltage-mode control neither line voltage sensing nor a multiplier is needed. Lower external part count, lower cost of control and lower power consumption (particularly useful when considering no-load conditions) are the resulting benefits.

This paper presents a novel approach to current-mode control that changes this paradigm. With the proposed technique the sinusoidal reference of the inner current loop is synthesized with no sensing of the line voltage and in a way that does not require a multiplier to adjust its amplitude. In this way Power Supply Designers can enjoy both the performance of current-mode control and the savings offered by voltage-mode control.

Keywords— *Converter control; Power factor correction; Current-mode control; Voltage-mode control; Analog multiplier*

I. INTRODUCTION

There are fundamentally two basic control techniques used in today's DCM/CCM Boundary Boost PFC Converters: voltage-mode control - aka Constant-ON-time (COT) control – and current-mode control [1] - [2]. All known implementations of these techniques can be described as: current-mode control requires input voltage sensing to define the shape of the reference for the inner current loop and a multiplier to adjust the amplitude of this reference (see fig. 1a), whereas voltage-mode control does not (see fig. 1b). As a result, as compared to current-mode, voltage-mode control results in lower external part count, lower cost of control and lower power consumption due to the absence of the resistor divider sensing the line voltage. This last point is particularly significant in those applications that are specified to meet very low no-load input power consumption targets, such as those envisaged in [3].

Unlike CCM Boost PFC Converters, where over the years a number of different techniques ([4] - [7] are just a few of them) have been proposed that simplify the control and reduce the external components by removing the multiplier and the line voltage sensing, to authors' knowledge much less work with this aim has been done on DCM/CCM Boundary Boost PFC Converters with current-mode control. The most likely reason is that voltage-mode control already provides a very simple way to meet this target. Further, the few pieces of work available in the literature [8] - [10] propose solutions that do eliminate the line voltage sensing divider but still use the multiplier.

Yet, in spite of some recent improvements [11] in terms of Power Factor (PF) and Total Harmonic Distortion (THD) of the input current, especially in applications operated over a wide input voltage range, as highlighted in [12] voltage-mode control is still more prone to distortion of the input current due to the oscillation of the boost inductor current after the reset of the boost inductor. In fact, depending on the amplitude of this oscillation, the method in [11] may or may not remove this effect. In the end, combining the benefits of voltage-mode control with those of current-mode control appears to be a desirable target.

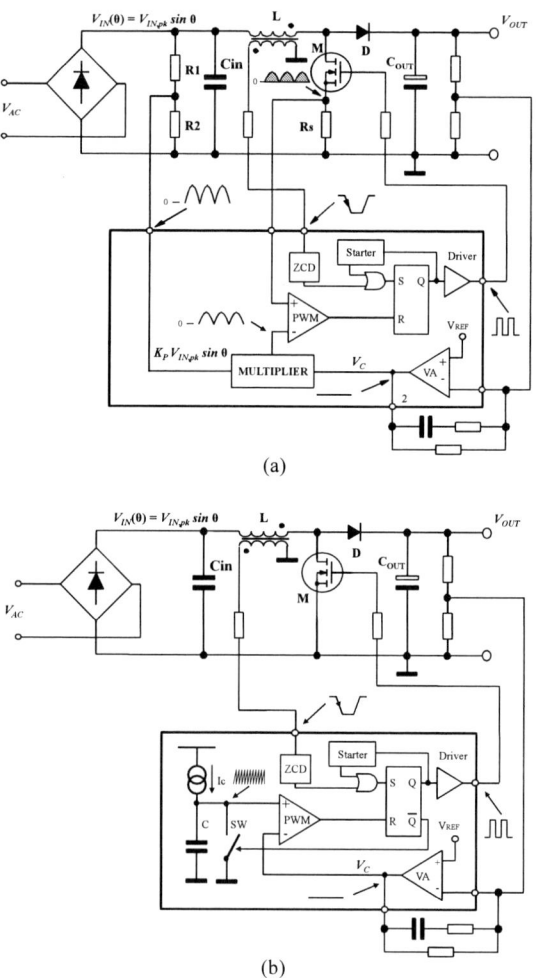

Fig. 1. DCM/CCM Boundary Boost PFC Converter with: (a) current-mode control, (b) voltage-mode control

The novel control method described in this paper enables DCM/CCM Boundary Boost PFC Converters to work with current-mode control without line-sensing circuitry and without the multiplier block. It exploits the volt-second balance across the boost inductor to build the sinusoidal reference of the inner current loop. In this way the resulting current-mode controlled PFC converter features a component count and may target no-load input power consumption equal to those of a voltage-mode controlled converter. Further, the novel method appears to outperform the traditional one also in some respects.

This paper is arranged as follows: in section II the operating principle of the novel method is described along with a circuit implementation; in sections III and IV respectively, the static and the dynamic properties associated to the novel method are derived and compared to those of the traditional multiplier-based one; section V deals with the major nonidealities that affect the operation of the proposed methodology; section VI shows some simulation and experimental results that validate the methodology; section VII provides the conclusions and draws the lines of possible future developments.

II. PRINCIPLE OF OPERATION

Before explaining the novel approach, it is worth reminding that the objective of the current loop in the multiplier-based system shown in fig. 1a can be described mathematically as:

$$V_{CS,REF}(\theta) = K_M V_C \left(K_P V_{IN,pk} \sin\theta \right). \tag{1}$$

$V_{CS,REF}(\theta)$, output of the multiplier, is the reference for the peak primary current, K_M is the multiplier gain, V_C the control voltage (output of the error amplifier that regulates the output voltage), $K_P = R_2/(R_1 + R_2)$ the divider ratio of the line voltage sensing circuitry, $V_{IN,pk}$ the peak value of the line voltage and θ its instantaneous phase ($\theta = 2\pi f_{line} t$). Of course, the control objective of the novel approach will have a similar expression.

Fig. 2 shows a boost PFC converter, along with its control IC embedding the novel control technique and its major functional blocks and most significant signals. It is worth noticing that no divider is sensing the input voltage nor a multiplier is included in the control IC.

The circuit inside the dotted box in fig. 2 is the key element that embodies the novel approach. It generates the reference $V_{CS,REF}(\theta)$ for the peak primary current fed into the inverting input of the PWM comparator, thus it plays the same role as the multiplier in the system in fig. 1a. This functional block will be designated as the Current Reference Generator (CRG) block in the following discussion. Fig. 3 shows its key waveforms. The other functional blocks inside the control IC are the same as in a standard DCM/CCM boundary peak current mode controller.

The CRG includes a voltage-controlled current source delivering a current I_{CH} proportional to the control voltage V_C (output of the error amplifier of the outer voltage loop):

$$I_{CH} = G_M V_C. \tag{2}$$

This generator, which may be realized as exemplarily shown in fig. 4, delivers its current to the circuit made up of the capacitor C_T and the switched resistor R_T during the time interval when the boost inductor current is flowing through the boost diode (FW = high, $\overline{FW}$ = low, see fig. 3). During the rest of the switching cycle (FW = low, $\overline{FW}$ = high) the generator is disconnected from the $R_T C_T$ pair and I_{CH} diverted to ground.

The resistor R_T is connected across the capacitor C_T during the entire switching cycle except for the short time interval T_R from the demagnetization instant of the boost inductor (i.e. when its current touches zero) to the beginning of the next cycle, occurring when the drain voltage has reached the valley of the ringing following the demagnetization. This time interval is identified by the signal ZCD being high ($\overline{ZCD}$ = low).

The ZCD block has therefore a twofold role: in addition to starting a new switching cycle close to the valley of the drain ringing, it has to sense the exact instant of the boost inductor demagnetization. This instant can be identified by detecting the knee in the $V(ZCD)$ voltage (see fig. 3), a technique that is commonly used in commercial products such as [13].

To analyze the operation of the CRG block it is convenient to start from the volt-second balance across the boost inductor. Assuming that the line voltage $V_{AC}(\theta)$ is perfectly sinusoidal and that the rectifier bridge is ideal, it is $V_{IN}(\theta) = V_{IN,pk} \sin\theta$. As a result of the rectification operated by the input bridge, it is possible to assume that $0 \le \theta \le \pi$ (so that $\sin\theta > 0$).

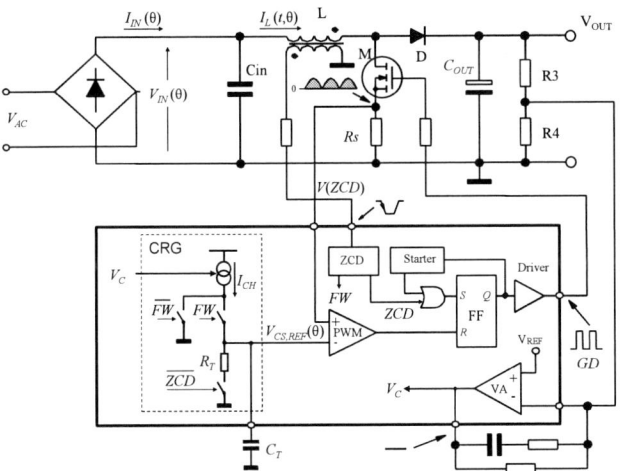

Fig. 2. Principle schematic of a DCM/CCM Boundary Boost PFC Converter using the novel control method

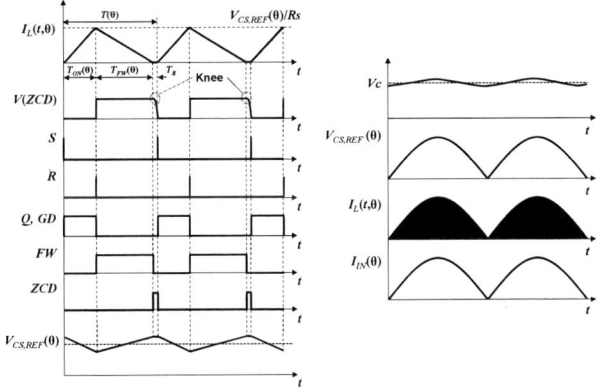

Fig. 3. Key waveforms of the circuit in fig. 2; switching cycle (left) and line cycle time scale (right)

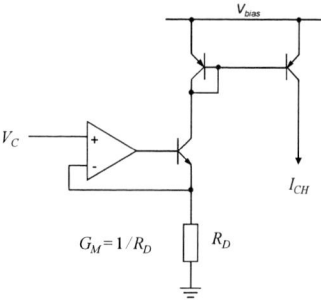

Fig. 4. Exemplary implementation of a voltage-controlled current source I_{CH}.

Referring to figures 2 and 3 for the symbolism:

$$V_{IN}(\theta)\, T_{ON}(\theta) = \left[V_{OUT} - V_{IN}(\theta) \right] T_{FW}(\theta). \qquad (3)$$

Solving (3) for $V_{IN}(\theta)$ and considering that the switching period is $T(\theta) = T_{ON}(\theta) + T_{FW}(\theta) + T_R$, the ratio between input and output voltage is:

$$\frac{V_{IN}(\theta)}{V_{OUT}} = \frac{T_{FW}(\theta)}{T_{ON}(\theta) + T_{FW}(\theta)} = \frac{T_{FW}(\theta)}{T(\theta) - T_R}. \qquad (4)$$

A fundamental assumption for the following analysis is that $T(\theta) \ll R_T C_T \ll 1/2 \cdot f_{line}$, where f_{line} is the line voltage frequency. In this way, on the one hand the switching frequency ripple of the voltage $V(C_T)$ developed across C_T is negligible; on the other hand, $V(C_T)$ may faithfully track a rectified sinusoid at the line frequency. This said, it is possible to find $V(C_T)$, which is used as the current reference $V_{CS,REF}(\theta)$, by charge balance:

$$I_{CH}\, T_{FW}(\theta) = \frac{V(C_T)}{R_T}\left[T(\theta) - T_R \right]. \qquad (5)$$

Solving for $V(C_T)$, considering (2) and that $V(C_T) = V_{CS,REF}(\theta)$:

$$V_{CS,REF}(\theta) = R_T\, G_M\, V_C\, \frac{T_{FW}(\theta)}{T(\theta) - T_R}. \qquad (6)$$

Combining (4) and (6), the result is:

$$V_{CS,REF}(\theta) = V_C\, R_T\, G_M\, \frac{V_{IN,pk}}{V_{OUT}} sin\theta. \qquad (7)$$

Under closed loop steady-state operation V_{OUT} is regulated and, therefore, constant; thus it is possible to state that the expression (7) has the same form as (1), with $R_T\, G_M / V_{OUT}$ in place of $K_M\, K_P$. It is therefore possible to conclude that the circuit in the dotted box in fig. 2 implements a control method enabling DCM/CCM Boundary Boost PFC Converters to draw a sinusoidal current from the power line using current-mode control without line-sensing circuitry and without any multiplier.

III. STATIC CHARACTERISTICS OF THE CURRENT LOOP

To analyze the static characteristics of the current loop a simplifying assumption is helpful: ZCD circuit's delay T_R is negligible, thus the converter works exactly on the CCM/DCM boundary. The objective is to find the operating range for the control variable V_C and its link to the operating conditions of the converter (input voltage $V_{IN}(\theta)$ and output current I_{OUT}).

The boost inductor current $I_L(t, \theta)$ in a switching cycle is triangular-shaped as illustrated in fig. 5. The height of these triangles varies along a line cycle following a sinusoidal law:

$$I_{L,pk}(\theta) = I_L(T_{ON}, \theta) = \frac{1}{L}\left(V_{IN,pk}\, sin\theta \right) T_{ON} = I_{L,PK}\, sin\theta, \quad (8)$$

where L denotes to inductance of the boost inductor. In fact, the peak inductor current is determined by the control loop illustrated in fig. 2. Thus, considering (7) it is possible to write:

$$I_{L,pk}(\theta) = \frac{V_{CS,REF}(\theta)}{Rs} = \frac{1}{Rs} V_C\, R_T\, G_M\, \frac{V_{IN,pk}}{V_{OUT}} sin\theta. \qquad (9)$$

Looking on an "f_{line}" time scale, the input current $I_{IN}(\theta)$ is the average value of each triangle over a switching cycle. Being T_R negligible by hypothesis, the average equals half the peak by geometric considerations:

$$I_{IN}(\theta) = \frac{1}{2} I_{L,pk}(\theta) = \frac{1}{2\,Rs} V_C\, R_T\, G_M\, \frac{V_{IN,pk}}{V_{OUT}} sin\theta, \quad (10)$$

which confirms the sinusoidal shape of the input current. As to the output current $I_{OUT}(\theta)$, on an "f_{line}" time scale it is the average value of the portion of the inductor current that flows for a time $T_{FW}(\theta)$ during the OFF-time of the power switch:

$$I_{OUT}(\theta) = \frac{1}{2} I_{L,pk}(\theta)\, \frac{T_{FW}(\theta)}{T(\theta)}. \qquad (11)$$

Considering (9) and that under the assumption of negligible T_R, it is possible to use (4), it possible to write:

$$I_{OUT}(\theta) = \frac{1}{2\,Rs} V_C\, R_T\, G_M\, \frac{V_{IN,pk}^2}{V_{OUT}^2} sin^2\theta. \qquad (12)$$

Averaging (12) over a line half-cycle yields the dc output current I_{OUT}:

$$I_{OUT} = \frac{1}{4\,Rs} V_C\, R_T\, G_M\, \frac{V_{IN,pk}^2}{V_{OUT}^2}. \qquad (13)$$

By solving this equation for V_C it is possible to find the desired relationships with the operating conditions. Introducing the parameter $Kv = V_{IN,pk} / V_{OUT}$, it is possible to write:

$$V_C = \frac{4}{G_M} \frac{Rs}{R_T\, Kv^2} I_{OUT}. \qquad (14)$$

At this point, it is worth highlighting an interesting property of the novel method that stems from (7) and that can be quantified with (13).

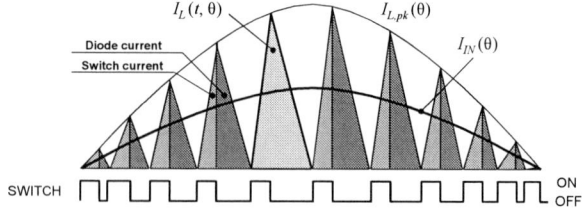

Fig. 5. Inductor current and input current in the boost converter in fig. 2.

From (7), assuming that the outer voltage loop is open (i.e. V_C = constant), it is possible to see that any change in V_{OUT} causes a counteracting change in $V_{CS,REF}(\theta)$. For example, if V_{OUT} drops (e.g. because of a larger load current I_{OUT}), $V_{CS,REF}(\theta)$ will rise and so will do the programmed peak inductor current $V_{CS,REF}(\theta) / Rs$. This tends to compensate the load rise and reduce the V_{OUT} drop. One can get to the same conclusion considering changes in $V_{IN,pk}$. In other words, the CRG features a negative feedback mechanism that acts so as to reduce the variations of the output voltage V_{OUT} in response to changes in the operating conditions (I_{OUT}, V_{IN}).

This mechanism is absent in the multiplier-based approach, as apparent from (1).

Eq. (13) is useful to quantify the effect of this negative feedback. Solving this equation for V_{OUT} provides:

$$V_{OUT} = \frac{1}{2}\sqrt{\frac{R_T\,G_M}{Rs}V_C}\,\frac{V_{IN,pk}}{\sqrt{I_{OUT}}}\,. \tag{15}$$

Starting from (1) and doing the same calculations it is possible to find the relationship equivalent to (15) for the multiplier-based approach:

$$V_{OUT} = \frac{1}{4\,Rs}K_M\,K_p\,V_C\,\frac{V_{IN,pk}^2}{I_{OUT}}\,. \tag{16}$$

From the comparison of (15) and (16), it follows that:

- With the new approach the output voltage changes in inverse proportion to the square root of the output current, whereas with the multiplier-based method it changes in inverse proportion to the output current (see fig. 6a).

- With the new approach the output voltage changes proportionally to the input voltage, whereas with the multiplier-based method it changes with the squared input voltage (see fig. 6b).

The conclusion is that with the new approach the boost stage features lower open-loop output impedance and a higher input voltage rejection ratio. As a consequence, the outer voltage loop is partly relieved of the job needed to keep the output voltage regulated.

Finally, as to the product $R_T\,G_M$ in (14), it is worth noticing that R_T is an integrated resistor while G_M is related to the value of an integrated resistor (precisely, it is the reciprocal of a resistor value, see example in fig. 4).

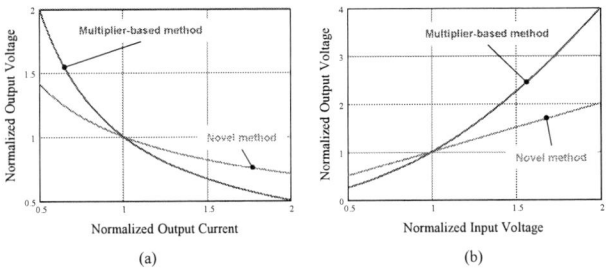

Fig. 6. Comparison of open-loop output voltage variations with the multiplier-based method and the novel method: V_{OUT} vs. I_{OUT} (a) and V_{OUT} vs. V_{IN} (b).

As such, their individual spread due to production tolerance is high (± 25%) but the mismatch in their product can be kept extremely low (< 1%) using resistors of the same type and with some care in laying them out. Therefore, the value of V_C is little affected by the spread of the internal parameters of the control IC. This is another advantage compared to the multiplier-based method, where the value of V_C is directly impacted by the spread of the multiplier gain, which is typically in the range of ±15%.

IV. DYNAMIC CHARACTERISTICS

The block diagram of fig. 7 shows the structure of the overall control loop of the converter in fig. 2, along with the relevant small-signal quantities. It is worth noticing the GRC block, whose transfer function in the Laplace domain is $G_S(s)$, and its feedback loop previously mentioned in section III.

In PFC converters, to achieve low distortion of the input current and nearly unity power factor, the outer feedback loop must have a sufficiently low gain at $2 \cdot f_{line}$ and, then, a narrow bandwidth, typically significantly lower than f_{line}. Therefore, to describe the dynamic properties of the control loop in fig. 7 only frequencies lower than f_{line} are taken into account.

Essentially, there is a double averaging process [14]: a first one is done over the switching period T to remove the switching frequency components and a second one over one half of the line period ($1/2 \cdot f_{line}$). This second averaging removes the components at $2 \cdot f_{line}$ resulting from the power transfer process, which is periodic at that frequency. Only the underlying low-frequency variations are taken into account.

A. Current loop

The small-signal transfer function $G_S(s)$ of the CRG block needs to be determined to characterize its behavior under dynamic conditions. The aim is to ascertain its influence on the overall dynamics of the converter. For simplicity of notation the dependence on θ will not be indicated; the prefix δ will refer to small-signal quantities.

Under dynamic conditions, the net charge change Q_T in the capacitor C_T in a switching cycle, which is zero under steady state conditions as stated by (5), is:

$$Q_T = \left(I_{CH} - \frac{V_{CS,REF}}{R_T}\right)T_{FW} - \frac{V_{CS,REF}}{R_T}T_{ON}\,. \tag{17}$$

Assuming a small change in Q_T due to a small change in I_{CH}, the change $\delta V_{CS,REF}$ in the average value of $V_{CS,REF}$ in a switching cycle can be found by differentiation of (17).

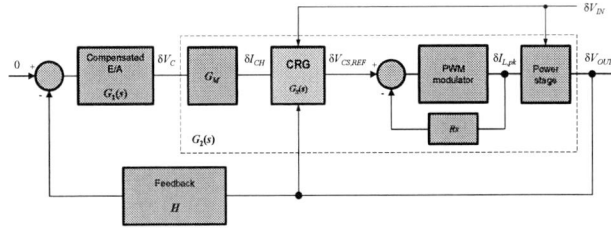

Fig. 7. Block diagram of the overall feedback loop regulating V_{OUT}

Considering that the differential contributions δT_{FW} e δT_{ON} cancel each other:

$$\delta Q_T = \delta I_{ch} T_{FW} - \frac{\delta V_{CS,REF}}{R_T}\left(T_{ON} + T_{FW}\right). \quad (18)$$

It is possible to define an equivalent small-signal charge (discharge) current δI_Q in a switching cycle:

$$\delta I_Q = \frac{\delta Q_T}{T} = \delta I_{CH}\frac{T_{FW}}{T} - \frac{\delta V_{CS,REF}}{R_T}\frac{T_{ON}+T_{FW}}{T}, \quad (19)$$

and determine the rate of change of $\delta V_{CS,REF}$ in a switching cycle as:

$$\frac{d}{dt}\delta V_{CS,REF} = \frac{\delta I_Q}{C_T}. \quad (20)$$

Substituting (19) in (20) and applying Laplace transform:

$$s\,\delta V_{CS,REF} = \frac{1}{C_T}\left(\delta I_{CH}\frac{T_{FW}}{T} - \frac{\delta V_{CS,REF}}{R_T}\frac{T_{ON}+T_{FW}}{T}\right). \quad (21)$$

Finally, after some calculations, considering (4) and introducing the parameter $Kv = V_{IN,pk}\,/\,V_{OUT}$:

$$G_S(s) = \frac{\delta V_{CS,REF}}{\delta I_{CH}} = R_T\,\frac{Kv\sin\theta}{1+s\,R_T C_T\dfrac{T}{T_{ON}+T_{FW}}}. \quad (22)$$

$G_S(s)$ features a variable gain and a moving pole. However, one fundamental assumption was that $R_T C_T << 1\,/\,2\cdot f_{line}$; further, the ratio $T\,/\,(T_{ON} + T_{FW})$ changes little and is always only slightly larger than unity. Consequently, in the $2\cdot f_{line}$ frequency region $G_S(s)$ can be considered independent of frequency:

$$G_S(s) \approx G_S = R_T\,Kv\sin\theta. \quad (23)$$

The component at $2\cdot f_{line}$ in (23) (reminder: θ is defined in the $(0, \pi)$ interval) disappears when averaging over one half of the line period, and just a dc gain remains. The conclusion is that the CRG block adds no significant frequency-dependent contribution to the dynamics of the feedback loop.

B. Control-to-output transfer function $G_2(s)$

With reference to the block diagram in fig. 7, to determine the transfer function $G_1(s)$ of the compensated E/A that achieves the targeted values of phase margin and harmonic distortion, the control-to-output transfer function $G_2(s)$ has to be determined first. As a result of the previous analysis of the CRG block, $G_2(s)$ can be found with the usual quasi-static approach described in [14].

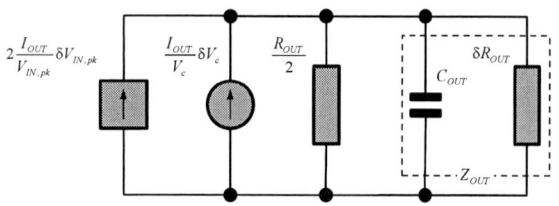

Fig. 8. Equivalent small-signal model used to derive $G_2(s)$

This results in the Norton-type small-signal equivalent schematic depicted in fig. 8. Operatively, this model is derived from (13), which is the result of the aforementioned double averaging process, using the partial derivative method.

The parallel impedance of the Norton-equivalent schematic is equal to $R_{OUT}\,/\,2$, where $R_{OUT} = V_{OUT}\,/\,I_{OUT}$ is the large-signal load resistance. Notice that this value is half that of the small-signal equivalent schematic relevant to the multiplier-based control method, while all the other parameters are the same [14]. This is the small-signal representation of the property of the current loop already highlighted in the previous section.

Z_{OUT} is the impedance of the load seen by the power stage; it includes the parallel of the output capacitor C_{OUT} with the low-frequency incremental load resistance δR_{OUT}. The zero due the ESR of C_{OUT} is normally located at a frequency $>> 2\cdot f_{line}$, thus it is neglected.

The control-to-output transfer function is therefore:

$$G_2(s) = \frac{\delta V_{OUT}}{\delta V_C} = \frac{V_{OUT}}{V_C}\frac{Z_{OUT}}{R_{OUT}+2\,Z_{OUT}}. \quad (24)$$

Considering the case of a boost PFC stage powering a cascaded dc-dc converter, it is worth recalling that a regulated switch-mode converter behaves as a constant power load [14], so that the incremental load resistance δR_{OUT} equals $-R_{OUT}$.

Here one can notice another difference: with the multiplier-based method δR_{OUT} and R_{OUT} cancel out, thus the controlled current generator drives only the capacitor; with the novel method, instead, the parallel combination of δR_{OUT} ($= -R_{OUT}$) and $R_{OUT}\,/\,2$ results in R_{OUT}, so:

$$G_2(s) = \frac{\delta V_{OUT}}{\delta V_C} = \frac{V_{OUT}}{V_C}\frac{1}{1+s\,R_{OUT}\,C_{OUT}}, \quad (25)$$

similarly to the case of the multiplier-based method when the boost converter is loaded with a resistor.

V. Analysis of nonidealities

The theoretical results discussed in sections II and III have been found under some simplifying assumptions. It is expected that nonidealities affecting the real-world operation cause the actual results to deviate from the theoretical ones. Being the nonidealities of the power circuit well-known in the literature and unaffected by the control method, in this context only the impact of those in the control circuit will be analyzed.

A. Switching frequency voltage ripple across C_T

The analysis in section II neglects the voltage ripple across C_T, given that $T(\theta) << R_T C_T$. Actually, because of the ripple – small but existing – the turn-off of the power switch is commanded on the valleys of the voltage ripple across C_T, as illustrated in the key waveforms of fig. 3 on the left-hand side.

Denoting with $\Delta V_{CS,REF}(\theta)$ the peak-to-peak amplitude of the ripple, the turn-off condition (9) becomes:

$$I_{L,pk}(\theta) = \frac{1}{Rs}\left(V_{CS,REF}(\theta) - \frac{1}{2}\Delta V_{CS,REF}(\theta)\right). \quad (26)$$

From the operation of the CRG it is possible to find that:

$$\Delta V_{CS,REF}(\theta) = \frac{V_{CS,REF}(\theta)}{R_T C_T} T_{ON} \; . \tag{27}$$

Combining (8) and (27) into (26) and after a few algebraic manipulations it is possible to show that:

$$I_{L,pk}(\theta) = \frac{1}{Rs} \frac{V_{OUT}}{V_{OUT} + \dfrac{I_{CH}}{2\,Rs\,C_T}} V_{CS,REF}(\theta). \tag{28}$$

Hence, the switching ripple across C_T causes only a reduction of the programmed current and does not introduce significant distortion. Of course, the value of V_C will be affected: it is expected that it will be larger than the value calculated neglecting the ripple. Eq. (14), considering (26) and (7) and after some algebraic manipulations, becomes:

$$V_C = \frac{4}{G_M} \frac{Rs}{R_T\, Kv^2 - 2\dfrac{I_{OUT}}{V_{OUT}}\dfrac{L}{C_T}} I_{OUT} \;. \tag{29}$$

To prevent V_C from exceeding the maximum allowed value V_{Cmax} the capacitor C_T must be selected such that:

$$C_T \geq 2\,\frac{I_{OUT}}{V_{OUT}} \frac{L}{R_T\, Kv^2 - 4\dfrac{I_{OUT}}{G_M}\dfrac{Rs}{V_{Cmax}}} \;. \tag{30}$$

B. Actual shape of the average voltage across C_T

The assumption $R_T C_T \ll 1/2{\cdot}f_{line}$ aims to ensure that $V_{CS,REF}(\theta)$ closely tracks the ideal profile (7). Expressing quantitatively how the time constant $R_T C_T$ impacts on the tracking and, then, on the shape of $I_{IN}(\theta)$ is a complex mathematical task; thus, it has been addressed by simulation. The results are summarized in the graph of fig. 9, showing the THD of $I_{IN}(\theta)$ vs. $2 f_{line} R_T C_T$.

The THD monotonically tends to zero at low values of $R_T C_T$. Although it is possible to minimize the tracking error by choosing a relatively small value of the time constant $R_T C_T$ with no penalty on the THD caused by the switching frequency ripple, the minimum limit for C_T given by (30) must be fulfilled and this sets the minimum useful value for $R_T C_T$.

C. Propagation delay on current sense

Another nonideality neglected in the analysis of section II is the propagation delay T_{PD} of the current sense path (PWM comparator + SR latch + gate driver + turn-off delay). During this time, assumed to be constant for simplicity, the switch is still ON and the input current keeps on ramping up, despite the voltage across the sense resistor has already hit the reference level $V_{CS,REF}(\theta)$. This extra current,

$$\Delta I_{L,pk}(\theta) = \frac{T_{PD}}{L} V_{IN,pk}\, sin\theta \;, \tag{31}$$

affects the peak inductor current $I_{L,pk}(\theta)$ and the input current $I_{IN}(\theta)$. It is possible to find that $I_{IN}(\theta)$ can be expressed as:

$$I_{IN}(\theta) = \frac{1}{2}\left(\frac{G_M\, V_C}{Rs\, V_{OUT}} + \frac{T_{PD}}{L} \right) V_{IN,pk}\, sin\theta \;. \tag{32}$$

Therefore, the propagation delay on current sense causes only the amplitude of the input current to rise and does not introduce significant distortion. The control voltage V_C will be affected in the opposite way the switching ripple across C_T does, so the two effects tend to compensate each other.

D. Low-frequency voltage ripple on V_C and V_{OUT}

The converter output voltage has a substantial ripple ΔV_{OUT} (peak value) at $2{\cdot}f_{line}$ superimposed on the dc value V_{OUT}. This ripple is essentially phase-shifted (lagging) by 180° with respect to the input voltage. The E/A gain $|G_1(s)|$ of the voltage loop at $2{\cdot}f_{line}$ though low is non-zero, so a voltage ripple at $2{\cdot}f_{line}$ with peak amplitude ΔV_C superimposed on the dc value V_C will appear. This ripple, if the E/A is compensated as a type-2 amplifier, will be lagging the output ripple by about 270°, i.e. lagging the input voltage by about 90°.

Based on these considerations, it is possible to assume that the expression of (7) changes into:

$$V_{CS,REF}(\theta) \approx V_C\, R_T\, G_M \frac{V_{IN,pk}}{V_{OUT}} \frac{1 - \dfrac{\Delta V_C}{V_C} cos\,2\theta}{1 - \dfrac{\Delta V_{OUT}}{V_{OUT}} sin\,2\theta} sin\theta . \tag{33}$$

The plot in fig. 10 shows the dependence of the THD of $I_{IN}(\theta)$ on the amplitude of ΔV_{OUT} and ΔV_C calculated with (33). These ripples appear to be the major source of distortion.

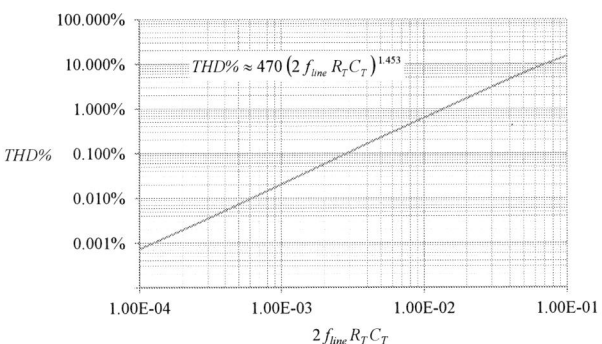

Fig. 9. THD (in %) of the input current caused by the profile tracking error; included is a best-fit approximation of the curve plotted

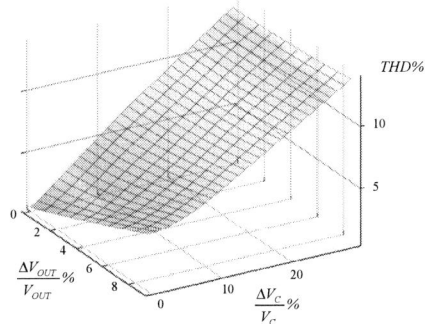

Fig. 10. THD (in %) of the input current caused by the residual ripple ΔV_C in the control voltage V_C and the output voltage ripple (ΔV_{OUT}).

VI. IMPLEMENTATION AND VALIDATION

An 80W DCM/CCM boundary boost PFC converter identical to that described in [15] and embedding the control circuit shown in fig. 2 has been tested with PSIM simulations.

The nonidealities considered in section V are included in the model and so are the components in the power circuit that adversely affect the THD of the input current: the filter capacitor after the bridge rectifier and the overall parasitic capacitance associated to the drain node of the power switch.

The timing diagrams of fig. 11 and the graph of fig. 12 show the results. In fig. 11 the synthesized $V_{CS,REF}(\theta)$ tracks closely the rectified input voltage @ 115 Vac, resulting in a very low THD (2.63%). $V_{CS,REF}(\theta)$ distortion becomes more apparent @ 230 Vac and the THD is higher (7.95%).

Essentially, these results are aligned to those obtained with the traditional multiplier-based method, as visible in fig. 12. This graph compares the THD values resulting from PSIM simulations using both the multiplier-based method (red curve) and the novel method (black curve). To assess the reliability of these results, the experimental data reported in [15] are plotted too (blue curve), showing that the simulation results are in a good agreement with the experiments.

These positive results have encouraged the experimental verification of the novel approach. The CRG has been realized in a simplified way shown in fig. 13 and implanted in the PFC pre-regulator described in [15]. The circuit simplification is based on the same assumption done in section III that considers T_R negligible as compared to $T_{ON}(\theta)$ and $T_{FW}(\theta)$. Therefore, the current I_{CH} feeds the $R_T C_T$ pair during the entire OFF-time of the power switch (and not during $T_{FW}(\theta)$ only) and the resistor R_T is connected in parallel to C_T all the time.

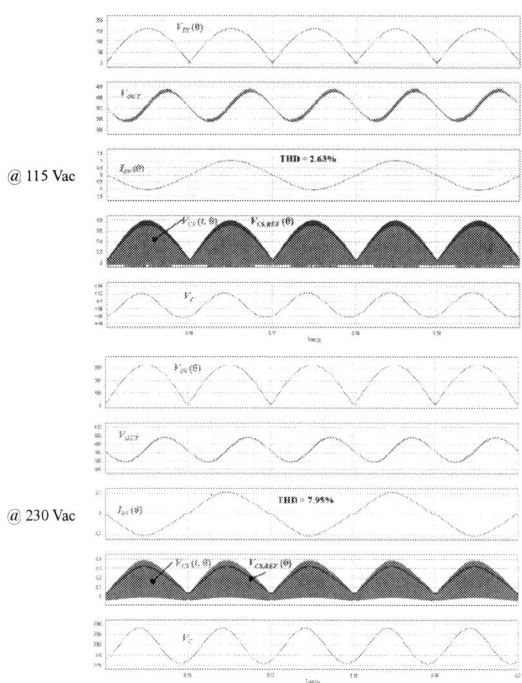

Fig. 11. Simulation results at full load: @115 Vac (upper pic); @230 Vac (lower pic).

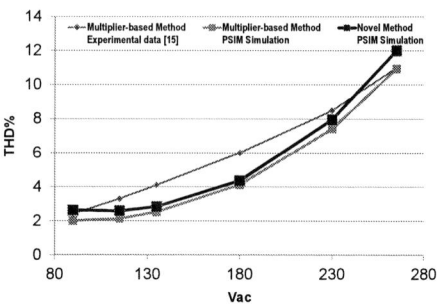

Fig. 12. THD (in %) data comparison at full load on the PFC converter [15].

This is a significant simplification, especially because no "knee detection" circuit – as mentioned in section II - is needed and the gate drive signal of the PFC controller can be used in place of the $\overline{FW}$ signal (see fig. 2 and 3).

In the CRG block it is possible to recognize the structure shown in fig. 4 of the voltage-controlled generator I_{CH}. This is simply switched on and off by means of Q2; Q1 provides the necessary logic inversion. R_D and R_T have been chosen so as to match exactly the gain of the original circuit and have the same range of the control voltage V_C (and the same effects of ΔV_C).

The voltage $V(C_T)$ generated by the CRG block is fed into one input of the multiplier (MULT pin). Notice that the multiplier is used as a simple gain block: in fact, the E/A of the L6562A is configured as a buffer so as to provide a fixed 1V level to the other multiplier input. Its output $V_{CS,REF}(\theta)$, then, equals $K_M V(C_T)$. The op-amp U1A closes the outer loop that regulates the output voltage. Its output is the control voltage V_C.

As a consequence, the operation of the CRG is slightly altered from that described in section II. The reference for the current loop is no longer given by (6) but becomes:

$$V_{CS,REF}(\theta) = R_T\, G_M\, V_C\, \frac{T_{FW}(\theta) + T_R}{T(\theta)}. \tag{34}$$

Essentially, the quantity T_R is added to both the numerator and the denominator of the fraction $T_{FW}(\theta) / [T(\theta) - T_R]$ in (6); since $T_{FW}(\theta) < T(\theta)$, the net result is an increase and a distortion in $V_{CS,REF}(\theta)$. As regards $I_{IN}(\theta)$, however, (10) assumes that T_R is negligible; if T_R is taken into consideration (10) becomes:

$$I_{IN}(\theta) = \frac{1}{2} I_{L,pk}(\theta)\, \frac{T_{ON}(\theta) + T_{FW}(\theta)}{T(\theta)}. \tag{35}$$

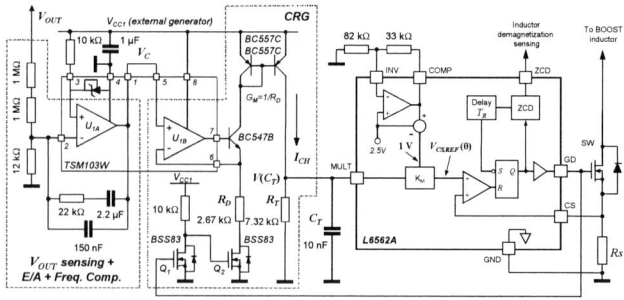

Fig. 13. External circuit added to the PFC pre-regulator described in [15] used to implement and validate the novel control method experimentally.

Thus, irrespective of the control method, the input current $I_{IN}(\theta)$ is not exactly half the peak inductor current programmed by the current loop: this value is further multiplied by a variable term lower than unity that causes a reduction and a distortion in $I_{IN}(\theta)$. Clearly, this reduction in $I_{IN}(\theta)$ tends to compensate the increase in $V_{CS,REF}(\theta)$ stated by (34), though not exactly.

In conclusion, it is expected that the experimental results with this circuit will not be too different from those that would be obtained with a more complex circuit implementing the exact control law (6).

Fig. 14 shows some key waveforms at V_{IN} = 115 Vac (upper pic) and V_{IN} = 230 Vac (lower pic). The voltage $V(C_T)$ generated by the CRG is very close to a rectified sinusoid in both cases and that its amplitude is changed to regulate V_{OUT}. The shape of the input current looks very much like a sinusoid too. The good operation is confirmed by the low THD values shown in fig. 15 at full and half load (black and green curves). On the whole, the performance is comparable to that of the original board reported in [15] (red and blue curves).

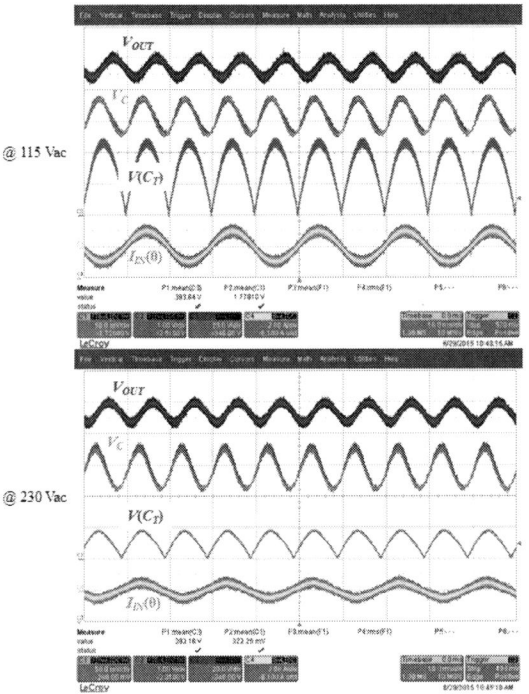

Fig. 14. Experimental waveforms of the prototype at full load; $@V_{IN}$=115 Vac (upper pic), $@V_{IN}$ = 230 Vac (lower pic).

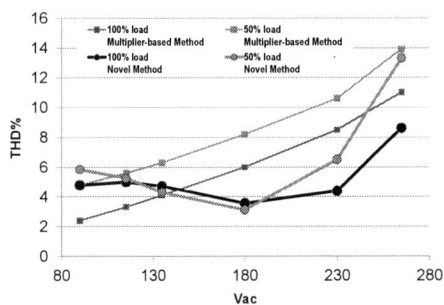

Fig. 15. THD (in %) measurements taken on the prototype at full and half load.

VII. CONCLUSIONS AND FUTURE WORK

A novel control method enabling DCM/CCM Boundary Boost PFC Converters to work with current-mode control without the line-sensing circuitry and without the multiplier has been presented, its operating principle illustrated and then validated by both simulations and experiments.

Both the steady-state and the dynamic characteristics of the novel method have been addressed and the differences with respect to the multiplier-based method have been highlighted. The effect of some significant non-idealities that affect the real-world operation of the novel method and may adversely affect the THD of the input current has been addressed too.

The results of this work demonstrate the viability of the novel control method and pave the way to its successful implementation in a new control IC.

REFERENCES

[1] M. Shen, Z. Qian, M. Chen, "Analysis and average modeling of critical mode boost PFC converter", Power Electronics and Drive Systems (PEDS), 2001, Pages: 138 – 141

[2] L. Rossetto, G. Spiazzi, P. Tenti, "Control techniques for power factor correction converters", Power Electronics, Motion Control (PEMC), 1994, Pages: 1310 – 1318.

[3] European Commission Joint Research Centre, "Code of Conduct on Energy Efficiency of External Power Supplies", version 5, October 2013

[4] D. Maksimovic, Y. Jang, R. Erickson, "Nonlinear-carrier control for high power factor boost rectifiers", Applied Power Electronics Conference and Exposition (APEC), 1995, Pages: 635 – 641

[5] J. Rajagopalan, F. C. Lee, P. Nora, "A general technique for derivation of average current mode control laws for single-phase power-factor-correction circuits without input voltage sensing", IEEE Transactions on Power Electronics, Vol. 14, No. 4, July 1999. Pages: 663 – 672

[6] A. Abramovitz, M. Evzelman, S. Ben-Yaakov, "Investigation of an alternative APFC control with no sensing of line voltage based on a triangular modulation carrier", Applied Power Electronics Conference and Exposition (APEC), 2008, Pages: 709 – 714

[7] R. Brown, M. Soldano "One cycle control IC simplifies PFC designs", Applied Power Electronics Conference and Exposition (APEC), 2005, Volume: 2, Pages: 825 – 829

[8] C. Adragna, "Transition mode operating device for the correction of the power factor in switching power supply units", US Patent # 7,064,527

[9] S.Y.R. Hui, H.S.H. Chung, D.Y. Qiu, "Effective standby power reduction using non-dissipative single-sensor method", Power Electronics Specialists Conference (PESC), 2008, Pages: 678 – 684

[10] A. Stroppa, C. Spini, C. Adragna, "High performance ac-dc notebook PC adapter meets EPA 4 requirements", 13th European Conference on Power Electronics and Applications (EPE), 2009. Pages: 1 – 9

[11] A. Bianco, C. Adragna, G. Scappatura, "Enhanced constant-on-time control for DCM/CCM boundary boost PFC pre-regulators: implementation and performance evaluation", Applied Power Electronics Conference and Exposition (APEC), 2014, Pages: 69 – 75

[12] L. Huber, B. T. Irving, M. Jovanovic, "Effect of valley switching and switching-frequency limitation on line-current distortions of DCM/CCM boundary boost PFC converters", IEEE Transactions on Power Electronics, Vol. 24, No. 2, February 2009. Pages: 339 – 347

[13] "ALTAIR05T-800 Off-line all-primary-sensing switching regulator", STMicroelectronics Datasheet, http://www.st.com

[14] R. W. Erickson, D. Maksimović, "Fundamentals of Power Electronics", 2nd Edition, 2001, ISBN 0-7923-7270-0.

[15] "Solution for designing a transition mode PFC pre-regulator with the L6562A", STMicroelectronics Application Note AN2761, http://www.st.com

Reducing the Switching Frequency Variation range for CRM Buck PFC Converter by Variable On-time Control

Xiaoping WANG, Kai YAO, Junfang Zhang

School of Automation
Nanjing University of Science and Technology
Nanjing, China

Abstract—**Critical conduction mode(CRM) Buck power factor correction(PFC) with peak current-controlled of current-mode control, its traditional control is constant on-time control, the switching frequency varies with the input voltage and load variations, and a relatively large frequency range, lead to a large of switching loss. This paper proposes a variable on-time control strategy for a CRM Buck PFC converter. By injecting a certain amount of third harmonic in the peak current reference signal, obtaining the optimization of the switching frequency range, reduction of the switching loss and the ripple of output voltage. Analyzing the operating principle and performance of CRM Buck PFC converter with constant on-time and variable on-time control strategy respectively, further designing the control circuit. Simulation results show that the variable on-time control strategy can effectively reduce the switching frequency of the switch, improve the performance of CRM Buck PFC converter.**

Keywords—*CRM Buck PFC; third harmonic; variable on-time; optimize the range of switching frequency*

I. INTRODUCTION

Buck PFC converter has high efficiency in the low input, in the case of low voltage input (90V), due to the property of step-down of Buck circuit itself so that the input and output voltages can be close, achieving a higher efficiency. Low output voltage, output voltage of Buck PFC can be set at a low value (typically less than 100V), comparably, output voltage of Boost PFC must be set to be approximately in 400V, lower output voltage can reduce the device stress of stage DC/DC. Common mode Electromagnetic Interference (EMI) noise is small, since the Buck circuit is working, the voltage of V_{ds} change across the dv/dt is small than the Boost PFC, so it has a good property of Common mode EMI. Main inductance is small, due to the Buck PFC main circuit across the inductance volt-second value smaller than the Boost PFC, so the inductance at the same power and the lowest operating frequency it needs to be smaller than the Boost PFC[1-2]. In recent years, more and more researchers began to study the Buck power factor correction [3-13], the efficiency of Buck topology application in AC/DC will not change much with the input voltage, so the

thermal design is relatively simple, it is possible to achieve high power density.

The operation mode of Buck PFC can be divided into continuous conduction mode(CCM), discontinuous conduction mode(DCM) and critical continuous conduction mode(CRM) based on the inductance current is continuous or not. DCM and CRM converters have obtained a wider range of applications because of their control are relatively simple. The advantages of CRM Buck PFC converter are that switch zero current turn on, diode has no reverse recovery, power factor is higher and so on. The switching frequency changes with the input voltage and load changes, inductance and EMI filter design is more complex, suitable for medium power applications.

The literature [14] proposed a method that adjust the on-time of switch to increase the input power factor value, which uses peak current control, when the input voltage is constant, injecting a certain amount of third harmonic to increase the value of PF close to 1.

This paper proposed a varied on-time control method, CRM Buck PFC converter with peak current control, by injecting a certain amount of third harmonic in the peak current reference signal, obtaining the optimization of the switching frequency range. Deriving the relationship between the amount of the third harmonic of the input and output voltage when the switching frequency range is optimization. In meeting the IEC61000-3-2, Class D harmonic standard conditions, it can also reduce the output voltage ripple.

II. OPERATION PRINCIPLE OF CRM BUCK PFC CONVERTER

Fig. 1 shows the main circuit of a buck PFC converter. Fig. 2 shows the inductor current waveform in a switching cycle when the converter operates at CRM.

Supposing that the input voltage is purely sine waveform and it has no distortion, then the input voltage v_{in} and the rectified voltage v_g can be defined as

$$v_{in}(t) = V_m \sin \omega t \qquad (1)$$

$$v_g = V_m |\sin \omega t| \qquad (2)$$

This work was supported by the national natural science foundation of China (51307085).Kai Yao is the corresponding author. Email:13813980876@163.com.

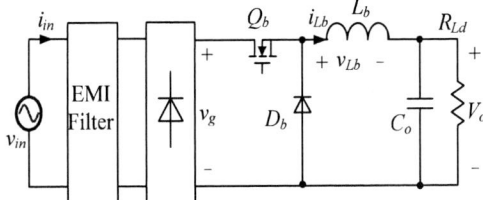

Fig.1. Main circuit of Buck PFC converter

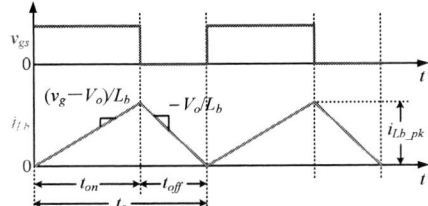

Fig.2. Inductor current waveform in a switching cycle

When Q_b is conducted, D_b is not conducted, the voltage around the boost inductance is $v_g - V_o$, and the current i_{Lb} will rise from zero with the slope of $(v_g - V_o)/L_b$.

When Q_b is turned off, i_{Lb} continuous flow through Db, then the voltage around L_b is V_o, i_{Lb} descending with slope of V_o/L_b. As the Buck converter works on the CRM mode, the switch is conducted again to start a new switching cycle when i_{Lb} was decreased to zero.

In a switching cycle, the inductor peak current i_{Lb_pk} is

$$i_{Lb\_pk} = \frac{v_g - V_o}{L_b} t_{on} = \frac{V_m |\sin \omega t| - V_o}{L_b} t_{on} \quad (3)$$

t_{on} is the conduction time of Q_b.

Within each switching period, the volt-second area of L_b should be balanced, so the turn off time of Q_b can be expressed as

$$t_{off} = \frac{i_{Lb\_pk}}{V_o/L_b} = \frac{V_m |\sin \omega t| - V_o}{V_o} t_{on} \quad (4)$$

Fig. 3 shows the inductance current, peak envelope, average current waveform in half line cycle. Among them $\theta = \arcsin(V_o/V_m)$.

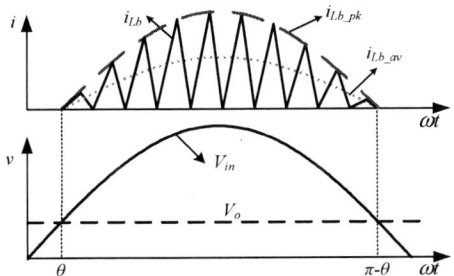

Fig.3. Inductance current waveform in half frequency period

The input current can be defined as

$$i_{in} = \frac{i_{Lb\_pk} \cdot D}{2} = \frac{V_m |\sin \omega t| - V_o}{2L_b} t_{on} \frac{V_o}{V_m |\sin \omega t|} \quad \theta \le \omega t \le \pi - \theta \quad (5)$$

Supposed that the output power is P_o, efficiency is 1, from the balance of input and output power can get that

$$\begin{aligned} P_o = P_{in} &= \frac{1}{\pi} \int_\theta^{\pi-\theta} v_{in} \cdot i_{in} d\omega t \\ &= \frac{1}{\pi} \int_\theta^{\pi-\theta} V_m \sin \omega t \cdot \frac{V_m \sin \omega t - V_o}{2L_b} t_{on} \frac{V_o}{V_m \sin \omega t} d\omega t \\ &= \frac{V_o}{2\pi L_b} t_{on} \int_\theta^{\pi-\theta} (V_m \sin \omega t - V_o) d\omega t \\ &= \frac{2V_o V_m \cos \theta - V_o^2 (\pi - 2\theta)}{2\pi L_b} t_{on} \end{aligned} \quad (6)$$

Combined with (3) can obtain

$$t_{on} = \frac{2\pi P_o L_b}{V_o V_m [2\cos \theta - (\pi - 2\theta)\sin \theta]} \quad \theta \le \omega t \le \pi - \theta \quad (7)$$

Substituting (7) into (3) then the peak current can be expressed as

$$i_{Lb\_pk1} = \frac{2\pi P_o (V_m |\sin \omega t| - V_o)}{2V_m V_o \cos \theta - V_o^2 (\pi - 2\theta)} \quad (8)$$

We can see from (7) and (8) that the traditional control of peak current control, when input and output voltage, output power is constant, the conduction time is constant, combined with the relationship between input and output voltage of Buck, i_{Lb_pk1} can be described as

$$i_{Lb\_pk1} = I_p (\sin \omega t - \sin \theta) \quad (9)$$

Where

$$I_p = \frac{2\pi P_o}{V_o [2\cos \theta - (\pi - 2\theta)\sin \theta]}$$

III. STRATEGY OF VARIABLE ON-TIME CONTROL

A. Traditional Control Method

Combine (4) and (7), the switching frequency of tradition control can be obtained and expressed as

$$f_s = \frac{1}{t_{on} + t_{off}} = \frac{V_o^2 [2\cos \theta - (\pi - 2\theta)\sin \theta]}{2\pi P_o L_b |\sin \omega t|} \quad (10)$$

Within a line cycle, the maximum and minimum switching frequency, the ratio of them respectively is

$$f_{s\_max} = f_{s\_\theta} = \frac{V_o^2 [2\cos \theta - (\pi - 2\theta)\sin \theta]}{2\pi P_o L_b \sin \theta} \quad (11)$$

$$f_{s\_min} = f_{s\_\frac{\pi}{2}} = \frac{V_o^2 [2\cos \theta - (\pi - 2\theta)\sin \theta]}{2\pi P_o L_b} \quad (12)$$

$$\frac{f_{s\_max}}{f_{s\_min}} = \frac{1}{\sin \theta} \quad (13)$$

978-1-4673-9551-9/16 $31.00 © 2016 IEEE

According to (10) and the design index of converter (will be given in Section IV), make the curve of f_s in a half line cycle under various voltages, shown in fig.4 (L_b=428μH, V_o=90V, P_o=100W).

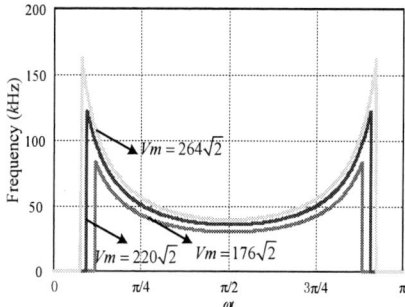

Fig.4. The curve of switching frequency with constant on-time control

B. Variable On-time Control Method

As can be seen from Fig.4, the switching frequency range under this situation is quite large, so envisaged appropriately reduce the switching frequency at $\omega t=\theta$ and increase at $\omega t=\pi/2$, then you can optimize the switching frequency range. So the value of the peak envelope approaching θ and $\pi-\theta$ should be appropriately increased, and reduced close to $\pi/2$. Use graphics to illustrate it as shown in Fig.5, can be achieved by injecting the third harmonic.

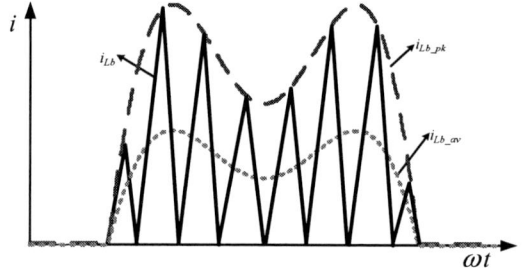

Fig.5. The renderings of inductor current to optimize switching frequency

When inductor current peak i_{Lb_pk} has a certain amount of the third harmonic, express i_{Lb_pk} as

$$i_{Lb\_pk} = I_p(\sin\omega t - \sin\theta + I_3^* \sin3\omega t - I_3^* \sin3\theta) \qquad (14)$$

Supposed that the output power is P_o, efficiency is 1, from the balance of input and output power can get that

$$P_o = \frac{1}{\pi}\int_\theta^{\pi-\theta} V_m \sin\omega t \cdot \frac{I_p\left(\sin\omega t - \sin\theta + I_3^* \sin3\omega t - I_3^* \sin3\theta\right)}{2} \frac{V_o}{V_m \sin\omega t} d\omega t$$

$$= \frac{I_p V_o\left[(6\cos\theta + 2I_3^*\cos3\theta) - 3(\pi - 2\theta)(\sin\theta + I_3^* \sin3\theta)\right]}{6\pi} \qquad (15)$$

$$I_p = \frac{6\pi P_o}{2V_o(3\cos\theta + I_3^*\cos3\theta) - 3V_o(\pi - 2\theta)(\sin\theta + I_3^* \sin3\theta)} \qquad (16)$$

I_p is the fundamental current value of i_{Lb_pk}, I_3^* is standard value that third harmonic of the fundamental current value of i_{Lb_pk}. Combine (3)、(14) and (16) can get

$$t_{on} = \frac{6\pi L_b P_o\left(\sin\omega t - \sin\theta + I_3^* \sin3\omega t - I_3^* \sin3\theta\right)}{2V_o\left(3\cos\theta + I_3^* \cos3\theta\right) - 3V_o(\pi - 2\theta)(\sin\theta + I_3^* \sin3\theta)} \frac{1}{V_m|\sin\omega t| - V_o} \qquad (17)$$

As can be seen from (17), the conduction time of control mode after the third harmonic injection is changed, no longer constant, are variable on-time control.

Combine (17) and (4), the switching frequency f_s can be expressed as

$$f_s = \frac{V_o(V_m|\sin\omega t| - V_o)}{6\pi L_b P_o V_m|\sin\omega t|} \frac{2V_o(3\cos\theta + I_3^* \cos3\theta) - 3V_o(\pi - 2\theta)(\sin\theta + I_3^* \sin3\theta)}{(\sin\omega t - \sin\theta + I_3^* \sin3\omega t - I_3^* \sin3\theta)}$$

$$\theta \le \omega t \le \pi - \theta \qquad (18)$$

From the above equation, the corresponding switching frequency when input voltage is equal to output voltage and the peak ($\omega t = \theta$ and $\omega t = \pi/2$) are

$$f_{s\_\theta} = \frac{2V_o^2(3\cos\theta + I_3^* \cos3\theta) - 3V_o^2(\pi - 2\theta)(\sin\theta + I_3^* \sin3\theta)}{6\pi P_o L_b\left[(1 + 3I_3^*)\sin\theta - 12I_3^* \sin^3\theta\right]} \qquad (19)$$

$$f_{s\_\frac{\pi}{2}} = \frac{2(3\cos\theta + I_3^* \cos3\theta) - 3(\pi - 2\theta)(\sin\theta + I_3^* \sin3\theta)}{1 - \sin\theta - I_3^* - I_3^* \sin3\theta} \frac{V_o^2(V_m - V_o)}{6\pi L_b P_o V_m} \qquad (20)$$

Analyzed (19) and (20), when $f_{s\_\theta} = f_{s\_\pi/2}$, can obtain that

$$I_{3\_\theta=\pi/2}^* = \frac{V_m(1 - \sin\theta) - (V_m - V_o)\sin\theta}{V_m + 3(2V_m - V_o)\sin\theta + 4(3V_o - 4V_m)\sin^3\theta} \qquad (21)$$

When $I_3^* \le I_{3\_\theta=\pi/2}^*$, the switching frequency at $\omega t=\theta$ is higher than at $\omega t=\pi/2$, When $I_3^* \ge I_{3\_\theta=\pi/2}^*$, the switching frequency at $\omega t=\theta$ is lower than at $\omega t=\pi/2$.

According to the design index of converter (Will be given in Section IV), take inductance values under variable on-time control L_b=618μH, output voltage V_o=90V, output power P_o=100W.

In accordance with (18), Made the change curve of f_s in a half frequency cycle under the various voltages with various third harmonic values, as shown in fig.6. Can be seen from the figure, with the gradual increase of I_3^*, the switching frequency at $\omega t=\theta$ is gradually decreasing and the switching frequency at $\omega t=\pi/2$ is gradually increasing, the minimum switching frequency is still at $\omega t=\pi/2$. With the further increases of I_3^*, the lowest switching frequency points appear in the $[\theta,\pi/2]$ and $[\pi/2,\pi-\theta]$ interval, the switching frequency range is gradually reduced. Until $I_3^* = I_{3\_\theta=\pi/2}^*$, the rang is decreased to the minimum. While $I_3^* > I_{3\_\theta=\pi/2}^*$, the switching frequency at $\omega t=\theta$ is lower than at $\omega t=\pi/2$, and the switching frequency range gradually expand.

We obtain the fit function presents the relationship between ωt_0 (the points of minimum frequency) and I_3^* are

$$\omega t_0 = -22.247(I_3^*)^3 + 24.857(I_3^*)^2 - 9.5479 I_3^* + 1.8825 \qquad (22)$$

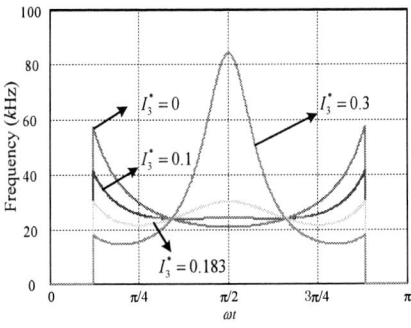

(a) 176 VAC input

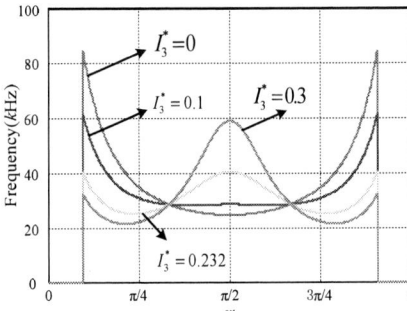

(b) 220VAC input

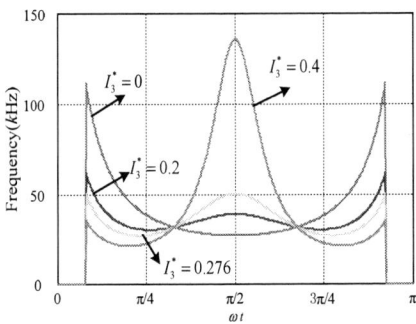

(c) 264 VAC input

Fig.6. The curve of f_s in a half frequency cycle with various third harmonic values

Then get the minimum switching frequency expressed as

$$f_{s\_min} = \frac{2V_o^2(3\cos\theta + I_3^*\cos3\theta) - 3V_o^2(\pi - 2\theta)(\sin\theta + I_3^*\sin3\theta)}{(\sin\omega t_0 - \sin\theta + I_3^*\sin3\omega t_0 - I_3^*\sin3\theta)}$$
$$\cdot \frac{V_m\sin\omega t_0 - V_o}{6\pi P_o L_b V_m \sin\omega t_0} \quad \theta \le \omega t \le \pi - \theta \tag{23}$$

In the analysis, the range of I_3^* is divided into 3 sections, the switching frequency of maximum and minimum value for the ratio of each section is

$$\frac{f_{s\_max}}{f_{s\_min}} = \begin{cases} \dfrac{V_m(1 - I_3^* - \sin\theta - I_3^*\sin3\theta)}{(V_m - V_o)\left[(1 + 3I_3^*)\sin\theta - 12I_3^*\sin^3\theta\right]} \\ \qquad 0 \le I_3^* \le f^{-1}(\omega t_0) \\ \dfrac{2V_o^2(3\cos\theta + I_3^*\cos3\theta) - 3V_o^2(\pi - 2\theta)(\sin\theta + I_3^*\sin3\theta)}{f_s(\omega t_0) \cdot 6\pi \cdot 428 \times 10^{-6} \cdot 100\left[(1 + 3I_3^*)\sin\theta - 12I_3^*\sin^3\theta\right]} \\ \qquad f^{-1}(\omega t_0) \le I_3^* \le I_{3\_\theta = \pi/2}^* \\ \dfrac{V_o^2(V_m - V_o)\left[2(3\cos\theta + I_3^*\cos3\theta) - 3(\pi - 2\theta)(\sin\theta + I_3^*\sin3\theta)\right]}{f_s(\omega t_0) \cdot 6\pi \cdot 428 \times 10^{-6} \cdot 100 V_m\left[1 - \sin\theta - I_3^* - I_3^*\sin3\theta\right]} \\ \qquad I_{3\_\theta = \pi/2}^* \le I_3^* \le 1 \end{cases} \tag{24}$$

According to (24) to make the fig.7, we can see from it, corresponding to any input voltage, you can always find the appropriate I_3^*, achieving the minimum of the ratio of maximum and minimum of the switching frequency, which is the minimum switching frequency range.

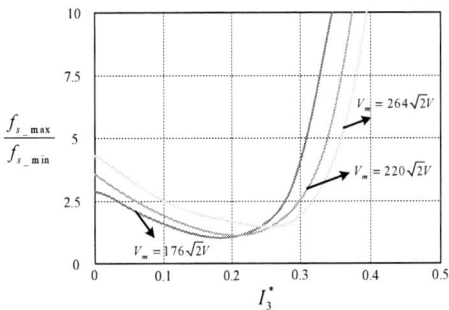

Fig.7. The curve of the relationship of the range of switching frequency and I_3^* under the various input voltage

Derivative I_3^* from (24) can obtain

$$\begin{cases} \dfrac{d\frac{f_{s\_max}}{f_{s\_min}}}{dI_3^*} < 0 & 0 \le I_3^* < I_{3\_\theta = \pi/2}^* \\ \dfrac{d\frac{f_{s\_max}}{f_{s\_min}}}{dI_3^*} = 0 & I_3^* = I_{3\_\theta = \pi/2}^* \\ \dfrac{d\frac{f_{s\_max}}{f_{s\_min}}}{dI_3^*} > 0 & I_{3\_\theta = \pi/2}^* < I_3^* \le 1 \end{cases} \tag{25}$$

Therefore, the optimal third harmonic to attain the minimum of $\frac{f_{s\_max}}{f_{s\_min}}$, also the minimum of switching frequency range within a half of line cycle is

$$I_{3\_optimal}^* = \frac{V_m(1 - \sin\theta) - (V_m - V_o)\sin\theta}{V_m + 3(2V_m - V_o)\sin\theta + 4(3V_o - 4V_m)\sin^3\theta} \tag{26}$$

According to IEC61000-3-2, Class D standards, the input current harmonics and input power ratio should satisfy the (27)

$$\frac{I_3/\sqrt{2}}{P_{in}} = \frac{I_3^* I_1/\sqrt{2}}{\left(V_m/\sqrt{2}\right)\cdot\left(I_1/\sqrt{2}\right)} \le 3.4\cdot10^{-3}$$

$$I_3^* \le I_{3\_limit}^* = 3.4\cdot10^{-3}\cdot\left(V_m/\sqrt{2}\right) \tag{27}$$

where $I_{3\_lim}^*$ is the harmonic limit to meet the standard.

V_{in_rms} varies from 176 V to 264 V, and the output voltage V_o is 90V, in accordance with (26) and (27) make the fig.8. We can see that, under any input voltage situation, the optimal third harmonic are lower than IEC61000-3-2, Class D standard limits.

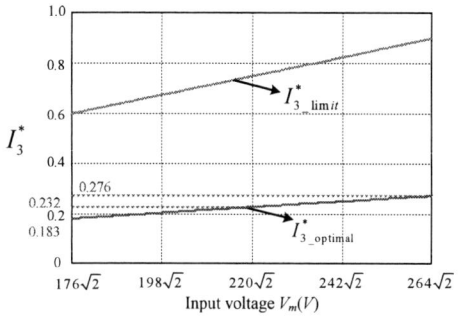

Fig.8.The relationship curves of optimal 3rd harmonic and standard limits with the input voltage

As can be seen from Fig.8, the optimal I_3^* values corresponding to the input voltage 176,220,264VAC were 0.183, 0.232, and 0.276. The optimal values of I_3^* reflected in Fig.7 are the same.

C. Control Circuit

Constant on-time control method and variable on-time control are all peak current control of current-mode control [15-18].

According to (14), (16) and (26) can be obtained the control of inductor current peak envelope expression as (28)

$$i_{Lb\_pk2} = \frac{6\pi P_o\left[\frac{1+3I_{3\_optimal}^*}{V_m}(V_m\sin\omega t - V_o) - \frac{4I_{3\_optimal}^*}{V_m^3}(V_m^3\sin^3\omega t - V_o^3)\right]}{2V_o(3\cos\theta + I_{3\_optimal}^*\cos3\theta) - 3V_o(\pi-2\theta)(\sin\theta + I_{3\_optimal}^*\sin3\theta)} \tag{28}$$

According to (28) can design the control circuit of variable on-time control method, as shown in fig.9. Input voltage v_g get through follower can obtain $v_A = V_m|\sin\omega t|$. Output voltage get through follower can obtain $v_B = V_o$. v_A and v_B get access to a subtraction circuit, then $v_C = k_1(V_m|\sin\omega t|-V_o)$, in which $R_1=R_3$、 $R_2=R_4$、 R_1、 R_3 selected on the basis of proportionality coefficient k_1. v_A and v_B go through multipliers respectively, then get access to a subtraction circuit obtain the potential of F point $v_F = k_2\left(V_m^3\sin^3\omega t - V_o^3\right)$, v_C and v_F get access to another subtraction circuit, and its output is inductor current peak envelope function v_G, in which k_1 and k_2 are the constant of

proportionality that associated with V_m and V_o. Output voltage signal via R_9、 R_{10}、 R_{11} and C_1 form the regulator obtained error signal v_{EA}. v_{EA} get in 2-pin of control chip L6561, v_G get in 3-pin of control chip L6561, the 7-pin output of L6561 drivers switch, make the switch work in accordance with variable on-time variation.

IV. PERFORMANCE COMPARISON

A. Inductance and the reduction of the Switching Frequency

Substituting (26) into (23) obtains the minimum switching frequency when the switching frequency range is optimization as follow

$$f_{s\_min} = \frac{2V_o(3\cos\theta + I_{3\_optimal}^*\cos3\theta) - 3V_o(\pi-2\theta)(\sin\theta + I_{3\_optimal}^*\sin3\theta)}{\left(\sin\omega t_0 - \sin\theta + I_{3\_optimal}^*\sin3\omega t_0 - I_{3\_optimal}^*\sin3\theta\right)}$$
$$\cdot\frac{V_o(V_m\sin\omega t_0 - V_o)}{6\pi L_b P_o V_m\sin\omega t_0} \quad \theta \le \omega t \le \pi-\theta \tag{29}$$

Equation (29) shows that, if define the minimum switching frequency, the maximum inductance value can expressed as

$$L_b = \frac{2V_o(3\cos\theta + I_{3\_optimal}^*\cos3\theta) - 3V_o(\pi-2\theta)(\sin\theta + I_{3\_optimal}^*\sin3\theta)}{\left(\sin\omega t_0 - \sin\theta + I_{3\_optimal}^*\sin3\omega t_0 - I_{3\_optimal}^*\sin3\theta\right)}$$
$$\cdot\frac{V_o(V_m\sin\omega t_0 - V_o)}{6\pi f_{s\_min}P_o V_m\sin\omega t_0} \quad \theta \le \omega t \le \pi-\theta \tag{30}$$

Considering human auditory frequency range, take 20 kHz as the minimum switching frequency, the output voltage V_o=90V, the output power P_o=100W, according to (12) and (30) can make critical inductance values of different input voltages, as shown in fig.10.

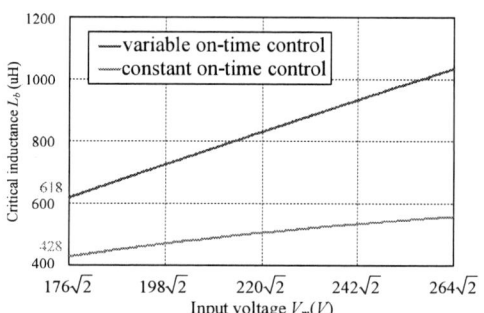

Fig.10. Critical inductance values of different input voltages

Substituting (26) into (18) can obtain

$$f_s = \frac{2V_o(3\cos\theta + I_{3\_optimal}^*\cos3\theta) - 3V_o(\pi-2\theta)(\sin\theta + I_{3\_optimal}^*\sin3\theta)}{\left(\sin\omega t - \sin\theta + I_{3\_optimal}^*\sin3\omega t - I_{3\_optimal}^*\sin3\theta\right)}$$
$$\cdot\frac{V_o(V_m\sin\omega t - V_o)}{6\pi L_b P V_m\sin\omega t} \quad \theta \le \omega t \le \pi-\theta \tag{31}$$

Substituting L_{b1}=428uH into (10), L_{b2}=618uH into (31), can make the change curve of f_s within half line cycle under the two control mode, constant on-time frequency curve was shown in Fig.4, The frequency curve of optimization switching frequency range with variable on-time control is shown in Fig.11. The comparison of frequency range with two control methods is shown in Fig.12.

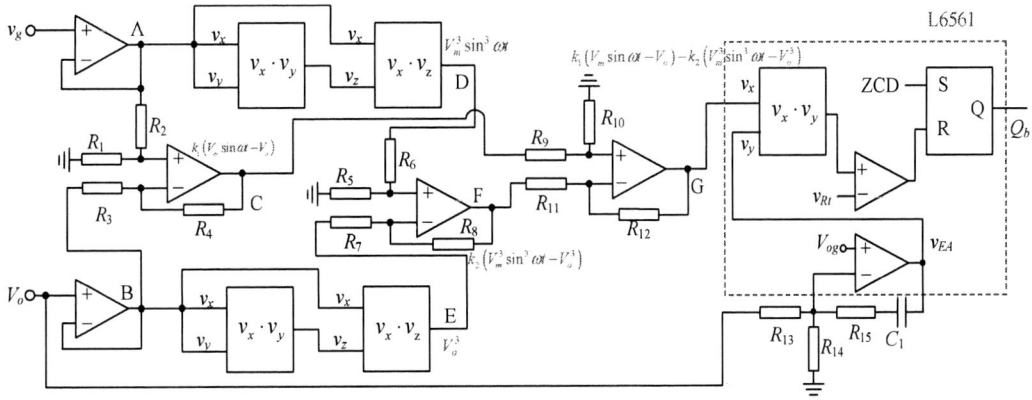

Fig.9. Control circuit of variable on-time control method

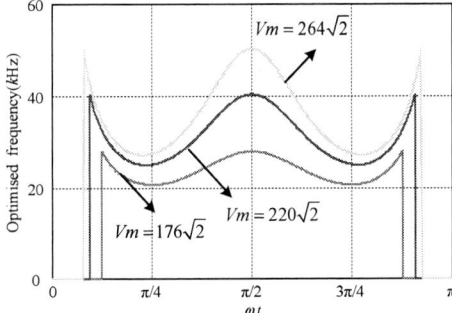

Fig.11. Frequency curve of optimization switching frequency range with variable on-time control

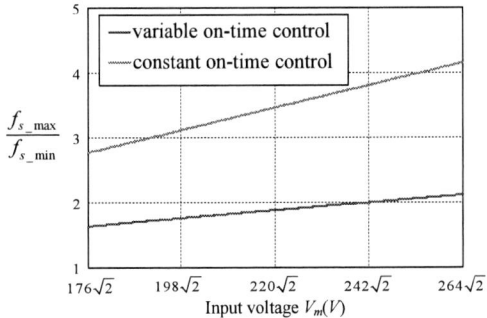

Fig.12.The comparison of frequency range with two control methods

As can be seen from Fig. 11 and 12, with variable on-time control, the switching frequency is significantly reduced under the respective input voltage, and the switching frequency range achieved optimization.

B. The reduction of the Output Voltage Ripple

With constant on-time control, you can get the converter instantaneous input power per unit (reference value of the output power) is

$$p_{in\_1}^*(t) = \frac{v_{in}(t)i_{in}(t)}{P_o} = \frac{\pi(\sin \omega t - \sin \theta)}{2\cos \theta - (\pi - 2\theta)\sin \theta} \quad (32)$$

With variable on-time control, you can get the converter instantaneous input power per unit (reference value of the output power) is

$$p_{in\_2}^*(t) = \frac{3\pi\left[\sin \omega t - \sin \theta + I_{3\_optimal}^*(\sin 3\omega t - \sin 3\theta)\right]}{2(3\cos \theta + I_{3\_optimal}^*\cos 3\theta) - 3(\pi - 2\theta)(\sin \theta + I_{3\_optimal}^*\sin 3\theta)} \quad (33)$$

Through (32) and (33) can make the curve of instantaneous input power per unit in a half frequency cycle with two ways, as shown in Fig.13.

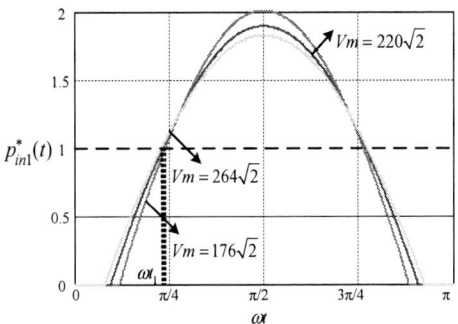

(a) Constant on-time control

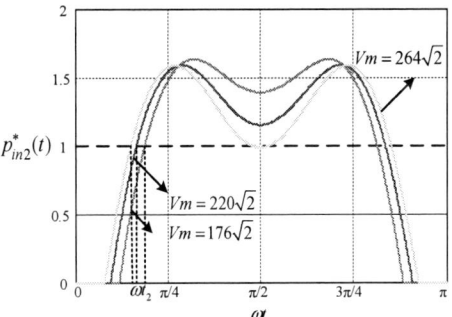

(b) Variable on-time control

Fig.13. Instantaneous input power per unit with two ways

When $p_{in}^*(t) > 1$, the storage capacitor is charging, When $p_{in}^*(t) < 1$, the storage capacitor is discharging. $p_{in}^*(t)$ is assumed from the beginning at $\omega t = 0$, the waveform of $p_{in}^*(t)$ under constant on-time control and variable on-time control intersect with 1, the corresponding time axis coordinates of the first intersection are t_1 and t_2 respectively. The maximum energy per unit (reference value is output power in half frequency cycle) stored of storage capacitor C_o in the half frequency cycle, respectively expressed as

$$\Delta E_1^* = \left\{ 2 \int_0^{t_1} \left[1 - p_{in\_1}^*(t) \right] \cdot dt \right\} \Big/ \left(T_{line}/2 \right) \qquad (34a)$$

$$\Delta E_2^* = \left\{ 2 \int_0^{t_2} \left[1 - p_{in\_2}^*(t) \right] \cdot dt \right\} \Big/ \left(T_{line}/2 \right) \qquad (34b)$$

According to the calculation formula of the capacitance energy storage, ΔE_1^* and ΔE_2^* also can be expressed as

$$\Delta E_1^* \approx \frac{\frac{1}{2} C_o \left(V_o + \frac{\Delta V_{o1}}{2} \right)^2 - \frac{1}{2} C_o \left(V_o - \frac{\Delta V_{o1}}{2} \right)^2}{P_o T_{line}/2} = \frac{2 C_o V_o \cdot \Delta V_{o1}}{P_o T_{line}} \qquad (35a)$$

$$\Delta E_2^* \approx \frac{\frac{1}{2} C_o \left(V_o + \frac{\Delta V_{o2}}{2} \right)^2 - \frac{1}{2} C_o \left(V_o - \frac{\Delta V_{o2}}{2} \right)^2}{P_o T_{line}/2} = \frac{2 C_o V_o \cdot \Delta V_{o2}}{P_o T_{line}}. \qquad (35b)$$

where ΔV_{o1} and ΔV_{o2} are output voltage ripple under constant on-time control and variable on-time control, respectively.

Combine with (34) and (35) can obtain

$$\Delta V_{o1} = 2 P_o \int_0^{t_1} \left[1 - p_{in\_1}^*(t) \right] dt \Big/ C_o V_o \qquad (36a)$$

$$\Delta V_{o2} = 2 P_o \int_0^{t_2} \left[1 - p_{in\_2}^*(t) \right] dt \Big/ C_o V_o \qquad (36b)$$

According to (36), we can compare the output voltage ripple under the two control method as shown in Fig.14. As can be seen, the output voltage ripple with variable on-time control is somewhat reduced when compared to the output voltage ripple with constant on-time control, and the output voltage ripple under the two ways are decreasing with the increasing of the input voltage.

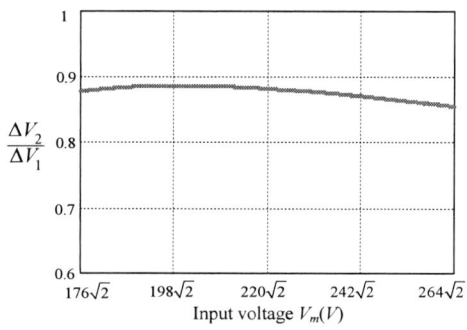

Fig.14.The comparison of output voltage ripple with two ways

V. EXPERIMENTAL VERIFICATION

In order to verify the validity of the proposed variable on-time control scheme, a prototype has been built and tested in the lab. The specifications are given as follows.

- input voltage: v_{in} = 176 ~ 264 VAC / 50 Hz;

- output voltage: V_o = 90 VDC;

- output power: P_o = 100 W;

- Buck inductor: 428μH (constant on-time control), 618μH (variable on-time control);

- output filter capacitor: Co = 470μF;

- control IC: L6561.

The experiment is still going on in the lab，now only simulation result is provided here. Here the simulation results of 176VAC, 220VAC and 264VAC input are presented. Fig.15 shows the waveforms of input voltage, input current (filter) and output voltage in a line cycle with constant on-time control under different input voltage. Fig.16 shows the corresponding waveforms with variable on-time control. The input current contain different number of third harmonic with different input voltage under variable on-time control are consistent with the Fig.5. And the output voltage ripples are decreased. All the waveforms are consistent with the theoretical curves.

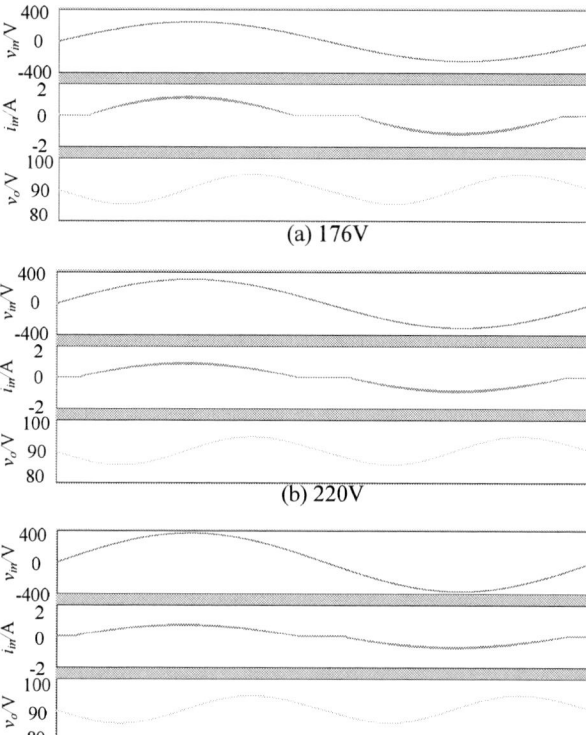

Fig. 15.The waveforms of input voltage, input current, output voltage under constant on-time control

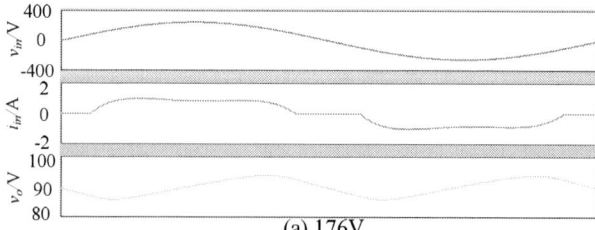

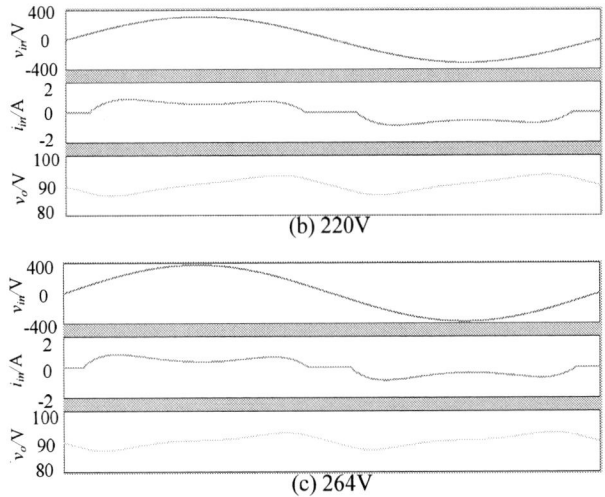

(b) 220V

(c) 264V

Fig.16. The waveforms of input voltage, input current, output voltage under variable on-time control

VI. CONCLUSIONS

This paper proposes a variable on-time control strategy for CRM Buck PFC converter. By injecting a certain amount of third harmonic in the peak current reference signal, can ensure the range of the switching frequency achieves optimization (seen in Fig.12). The control circuit is designed in accordance with the amount of three harmonics. Compared with the constant on-time control, meeting the IEC 61000-3-2, Class D harmonic standards required, can reduce the switching frequency and output voltage ripple (seen in Fig.11 and 14). A 100 W prototype has been built, simulating in Saber based on the control circuits. And the simulation results have verified the correctness of theoretical analysis.

REFERENCES

[1] Jianyou Yang, Xinke Wu, Junming Zhang, Zhaoming Qian, "Constant on-time control of high efficiency buck PFC study." Power Electronics Technology, vol.44, no.41, pp. 33-35, 2010.

[2] Jianyou Yang, "High Efficiency Power Factor Correction Techniques Based on Buck Converter." Hangzhou, Zhejiang University,Jan 2011.

[3] L. Huber, L. Gang, "Design-Oriented Analysis and Performance Evaluation of Buck PFC Front End." IEEE Trans. Power Electron.,vol. 25, no. l, pp.85-94, 2010.

[4] W. Weaver, P. Krein, "Analysis and Applications of a Current-Sourced Buck Converter." in Proc. IEEE APEC, pp.1664-1670, 2007.

[5] A. Mustafa, H. Esam, "Integrated Buck-Boost Quadratic Buck PFC Rectifier for Universal Input Applications."IEEE Trans. Power Electron, vol. 24, no. 12, pp.2886-2896, 2009.

[6] B. Keogh, G Young, H. Wegner, C. Gillmor, "Design Considerations for High Efficiency Buck PFC with Half-Bridge Regulation Stage." in Proc. IEEE APEC, pp.1384-1391, 2010.

[7] J. Yungtaek, M. Jovanovic, "Bridgeless Buck PFC Rectifier." in Proc. IEEE APEC, pp.23-29, 2010.

[8] Bernard Keogh, George Young, Hagen Wegner, Colin Gillmor, " Design considerations for High Efficiency Buck PFC with Half-Bridge Regulation Stage." in Proc IEEE APEC, pp.1384-1391, 2010.

[9] Xinke Wu, Jianyou Yang, Junming Zhang, "Variable On-Time (VOT) Controlled Critical Conduction Mode Buck PFC Converter for High-Input AC/DC HB-LED Lighting Applications." IEEE Transactions on Power factor Electronics, vol.27, no.11, pp.4530-4539, 2012.

[10] Yang Jianyou, Zhang Junming, Wu Xinke, "Performance comparison between buck and boost CRM PFC converter." Control and Modeling for Power Electronics, IEEE, pp.1-5, 2010.

[11] Wu Xinke, Yang Jianyou, Zhang Junming, Xu Ming, "Design Considerations of Soft-Switched Buck PFC Converter with Constant On-Time (COT) Control." IEEE Trans. Power Electron, vol.26, no.11, pp.3144-3152, 2011.

[12] Xiaogao Xie, Chen Zhao, Lingwei Zheng, Shirong Liu, "An Improved Buck PFC Converter With High Power Factor." IEEE Transactions on Power Electronics, vol.28, no.5, May 2013.

[13] Jianyou Yang, Xinke Wu, Junming Zhang, Zhaoming Qian,"Design Considerations of a high efficiency ZVS Buck AC-DC Converter with Constant On-Time Control." In press Telecommunications Energy Conference (INTELEC), 32nd International,pp.1-5,2010.

[14] Schramm D S, Buss M O, "Mathematical analysis of anew harmonic cancellation technique of the input line current in DICM Boost converters." in Proceedings of the IEEE Power Electronics Specialist Conference, pp.1337-1343, 1998.

[15] Yang Fei, "Interleaved Boost PFC converter with coupled inductor." Nanjing, Nanjing University of Aeronautics and Astronautics, 2013.

[16] Hongjia Wu, Yongliang Zhang, Menglian Zhao, Hongfeng Shen, Xiaobo Wu. "A Constant-on-Time Based Buck Controller with Active PFC for Universal Input LED System." In press Power Electronics and Drive Systems (PEDS), pp. 551-556,2015.

[17] Junming Zhang, Siyang Zhao, Hulong Zeng, Xinke Wu. "An Optimal Peak Current Mode Control Scheme for Critical Conduction Mode (CRM) Buck PFC Converter." In press Solid State Lighting (ChinaSSL), pp.182-189,2013.

[18] Hulong Zeng, Junming Zhang. "An Improved Control Scheme for Buck PFC Converter for High Efficiency Adapter Application." In press Energy Conversion Congress and Exposition (ECCE), pp.4569-4576, 2012.

High Efficiency 20-400 MHz PWM Converters using Air-Core Inductors and Monolithic Power Stages in a Normally-Off GaN Process

Alihossein Sepahvand, Yuanzhe Zhang and Dragan Maksimović

Colorado Power Electronics Center
Department of Electrical, Computer and Energy Engineering
University of Colorado, Boulder, CO 80309-425, USA
Email: {ali.sep, yuanzhe.zhang, maksimov}@colorado.edu

Abstract—This paper presents high efficiency dc-dc converters based on monolithic normally-off GaN half-bridge power stages with integrated gate drivers. A new gate driver circuitry is introduced, which enhances both the power stage efficiency and the converter overall efficiency. While using only n-type transistors in the GaN process, the proposed gate driver maintains low quiescent power consumption by emulating the complementary operation commonly employed in CMOS processes. Level shifting is accomplished using a bootstrap technique, with the bootstrap capacitor and the bootstrap diode integrated on the same chip. A family of monolithic GaN chips has been designed, targeting operation from up to 45 V, delivering up to 16 W of output power, and operating at 20-400 MHz switching frequencies. The GaN chips are verified in synchronous buck converters, demonstrating record peak power stage efficiencies of 95.0% at 20 MHz, 94.2% at 50 MHz, 93.2% at 100 MHz, 86.5% at 200 MHz, and 72.5% at 400 MHz.

I. INTRODUCTION

Increasing the switching frequency of power converters to very high frequency (VHF) and ultra high frequency (UHF) levels has the potential to reduce the size of passive components, with a corresponding increase in power density. Furthermore, high frequency operation enables improved transient responses, opening up new application opportunities such as dynamic supply modulation and envelope tracking [1]–[6]. Realizing the aforementioned advantages while maintaining high efficiencies requires advances in circuit topologies and design techniques, high-Q passive components, as well as higher performance semiconductors, such as GaN devices [1]–[19]. In particular, monolithic integration of GaN switching power devices and gate drivers has enabled high efficiency operation of pulse-width modulated (PWM) converters at up to 200 MHz switching frequencies [2]–[4], [7], [8], with 90% power-stage efficiency at 100 MHz switching frequency reported in [2]. These results have been reported for a depletion-mode (normally-on) GaN-on-SiC process, which raises practical concerns related to system start-up issues. Furthermore, pull-up techniques employed in the integrated gate drivers introduce challenging trade-offs between switching speed of

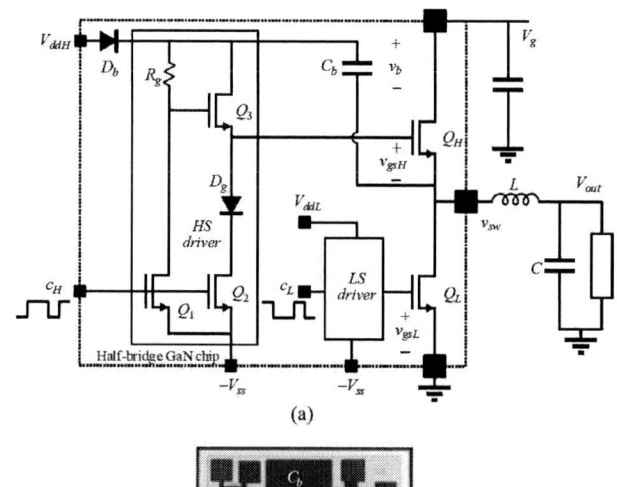

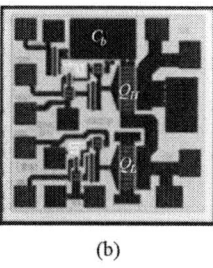

Figure 1: (a) Circuit diagram of the synchronous buck converter using the monolithic GaN half-bridge power stage chip with integrated gate drivers; (b) die photo of the (2mm × 2mm) GaN chip optimized for 100 MHz switching frequency.

power devices, power-stage switching losses, and driver power consumption.

In this paper, the monolithic power stage integration approach is advanced further for an enhancement-mode (normally-off) GaN-on-SiC process using a new gate-driver circuit with improved switching speed versus power consumption trade-off. A synchronous buck converter using the monolithic normally-off GaN half-bridge with the new integrated

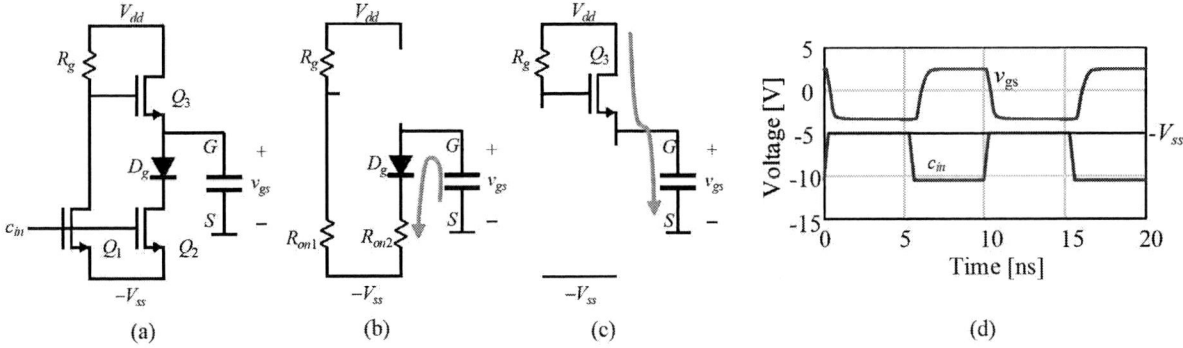

Figure 2: (a) Proposed gate driver circuit with the power-device gate-to-source capacitive load; (b) the driver circuit when the input PWM signal (c_{in}) is high, and (c) when the input PWM signal (c_{in}) is low; (d) simulation waveforms illustrating the gate driver operation at 100 MHz.

gate driver circuitry is shown in Fig. 1(a), and a photograph of the GaN chip die is shown in Fig. 1(b). A family of monolithic GaN chips has been designed, targeting operation from up to 50 V, delivering up to 16 W of output power, and operating at 20-400 MHz switching frequencies.

The paper is organized as follows: the monolithic normally-off GaN power stages including half-bridge power switches and integrated gate drivers are described in Section II. The GaN chip optimization is summarized in Section III. Experimental results, including efficiency measurements and switching waveforms, are presented in Section IV for synchronous buck converter prototypes operating at 20-400 MHz switching frequencies from up to 45 V supply voltage. Finally, Section V concludes the paper.

II. MONOLITHIC NORMALLY-OFF GaN HALF-BRIDGE POWER STAGE WITH INTEGRATED GATE DRIVERS

A circuit diagram of the monolithic GaN power stage chip is shown in Fig. 1(a). This chip includes two power stage transistors in a half-bridge configuration (Q_H and Q_L), with two gate drive circuits, 'HS driver' and 'LS driver', employed for the high-side and the low-side transistors, respectively. The chip is realized in a normally-off GaN process with a threshold voltage $V_{th} \approx +0.1$ V [13]. The positive threshold voltage ensures the leakage current is sufficiently low at zero gate-to-source bias. As a result, the GaN power chip converter can be safely powered up as a normally off power stage. However, since V_{th} is close to zero, the gate driver must be able to supply a positive gate-to-source voltage to turn the power device fully on, and a negative gate-to-source voltage to turn the power device off during normal operation. Using $V_{dd} > 0$ and $V_{ss} < 0$, the gate drivers described in this paper are capable of producing such bipolar gate-to-source drive voltages. Simply by adjusting the gate-driver supply voltages V_{dd} and V_{ss}, the same circuits can also be applied to produce positive or negative unipolar drive voltages to accommodate normally-off (enhancement-mode) or normally-on (depletion mode) processes, respectively.

The logic inputs to the two gate drive circuits are complementary PWM signals with associated dead-times, shown as c_H and c_L in Fig. 1(a). The high-side and the low-side gate driver circuits have identical configurations. Details are shown only for the high-side driver, which further includes a level shifting circuit consisting of the bootstrap capacitor C_b and the bootstrap diode D_b. Note that the bootstrap components are integrated on the same chip. Diodes D_b and D_g are Schottky diodes realized in the same GaN process, hence eliminating losses related to reverse recovery, which would otherwise adversely affect efficiency at very high switching frequencies. The new gate driver circuit is constructed so as to minimize static power consumption while at the same time improving current source and current sink capabilities.

The driver circuit operation is explained with reference to Fig. 2. When the input PWM signal (c_{in}) is logic high, transistors Q_1 and Q_2 are turned on. As Q_1 turns on, the gate of Q_3 is pulled to the negative gate-driver rail ($-V_{ss}$). Furthermore, as Q_2 turns on, a low-impedance path is formed through diode D_g and Q_2 to discharge the power-device gate-to-source capacitance. As a result, v_{gs} is pulled down to

$$v_{gs} = -V_{ss} + V_{Dg} \approx -2.5 \text{ V}, \quad (1)$$

which is sufficient to fully turn off the power device. Due to the voltage drop V_{Dg} across the forward biased diode D_g, the gate-to-source voltage of Q_3 is negative, and Q_3 is off, as shown in Fig. 2(b). In this state, static power consumption in the driver is determined by the power dissipated predominantly in the pull-up resistor R_g taking into account that the on-resistance R_{on1} of the transistor Q_1 is much smaller than R_g. In the half-bridge configuration, neglecting dead times, one of the two gate drivers is in this state. Hence, assuming the same R_g is employed in the high-side and the low-side driver, (3) approximates the total static power consumption in the on-chip integrated gate drivers.

When the input PWM signal (c_{in}) is logic low, the transistors Q_1 and Q_2 are off. The gate of transistor Q_3 is pulled to V_{dd} by R_g. The equivalent circuit of the gate driver in this

978-1-4673-9551-9/16 $31.00 © 2016 IEEE

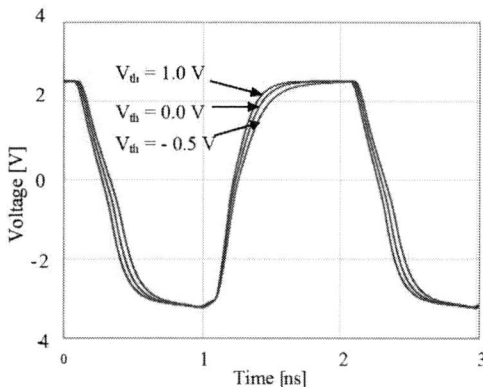

Figure 3: Simulated output voltage of the proposed gate driver at 500 MHz switching frequency, as the threshold voltage of the transistors is changing from $-0.5V$ to $1V$.

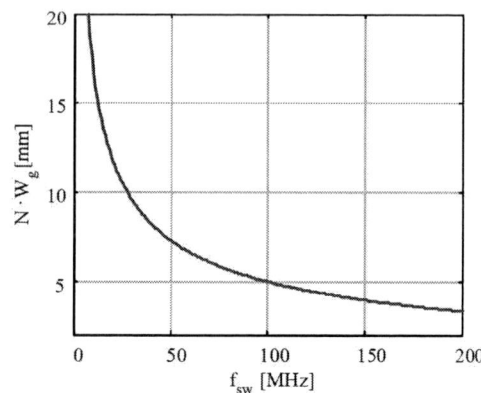

Figure 4: The optimum power-stage device size as a function of switching frequency for a 25 V, 5 W ZVS-QSW synchronous buck converter.

state is shown in Fig. 2(c). The source-follower Q_3 quickly charges the power device gate-to-source capacitance to

$$v_{gs} \approx V_{dd} - V_{th} \approx 2.5 \text{ V}, \tag{2}$$

where $V_{th} \approx +0.1$ V is the device threshold voltage. As a result, the power switch turns on. In this state the static power consumption is zero. It is important to note that the driver current sourcing is greatly enhanced by the source-follower Q_3. As a result, a large R_g can be employed, minimizing the driver static power consumption. Fig. 2(d) shows the waveform of the input PWM signal (c) and the driver output voltage (v_{gs}) for the GaN chip optimized for operation at 100 MHz. One may note that the driver propagation delay, the rise time, and the fall time are all less than 1 ns.

Compared to the previously reported integrated GaN gate drivers [2], [7], the main advantage of the new gate driver is that the pull-up resistor R_g is neither in the source nor in the sink path at the output of the driver. In contrast, Q_3 and Q_2 act as active, low-impedance source and sink paths, respectively, while R_g can be increased in value, thus reducing the quiescent power consumption given by (3 without affecting the driver switching speed. This property resembles the operation of standard complementary gate drivers commonly employed in CMOS processes, even though only n-type devices are available in the considered GaN process.

$$P_{driver} = \frac{(V_{dd} + V_{ss})^2}{R_g}, \tag{3}$$

III. GaN Power Chip and Converter Design Optimization

Another advantage of the proposed gate driver is that the circuit operation is relatively independent of the threshold voltage, which is beneficial as the threshold voltage changes with process variations or with temperature. Simulations have been performed to examine the sensitivity to threshold voltage variations. Fig. 3 shows the output voltage of the proposed

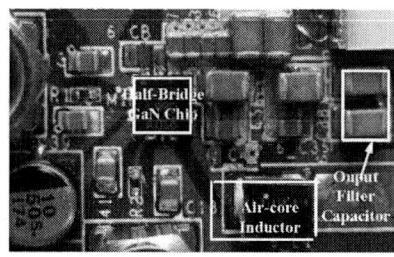

Figure 5: Photograph of the 100 MHz synchronous buck converter prototype using the GaN chip in the 20-pin 4mm × 4mm QFN package and 47 nH air-core inductor.

gate driver as the threshold voltage of the transistors varies from -0.5 V to 1 V. As the threshold voltage varies, only slight changes in the rise and fall times can be observed. It may be noted that even at a switching frequency as high as 500 MHz, the effect of threshold voltage variation is minimal. The monolithic half-bridge GaN power chips presented in the previous section are used to construct synchronous buck converters operated with resonant transitions in the zero-voltage switching quasi-square-wave (ZVS-QSW) mode [20]. The converter circuit utilizing the GaN power chip is shown

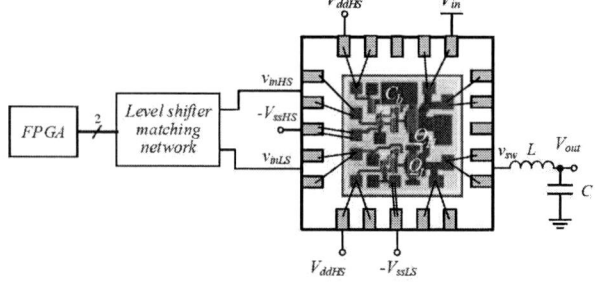

Figure 6: The test setup diagram for a synchronous buck converter using the monolithic GaN power chip.

in Fig. 1(a). Using state-plane analysis, the filter inductance is selected to maintain ZVS operation of the power stage transistors across wide operating ranges in terms of output voltages and power levels. In the GaN power chip, the half-bridge power-stage transistors and the gate-driver circuits are sized to maximize efficiency for a given power level and a target switching frequency, following a loss modeling and optimization approach similar to what has been described in [1]. As an example, Fig. 4 shows the optimum power-stage device size as a function of the switching frequency, for a 25 V, 5 W ZVS-QSW synchronous buck converter operating at 50% duty cycle.

A photograph of one of the GaN power chips, optimized for 100 MHz switching frequency and up to 5 W power, is shown in Fig. 1(b). The size of the chip die is 2mm × 2mm. Depending on intended power level, the fabricated GaN chips have been packaged in 20-pin 4mm × 4mm or 5mm × 5mm QFN packages.

IV. EXPERIMENTAL RESULTS

Based on the operation and design approaches presented in Sections II and III, a family of monolithic GaN half-bridge power stage chips has been designed, fabricated, and tested in ZVS-QSW synchronous buck converters at switching frequencies between 20 MHz and 400 MHz, input voltage levels ranging from 20 V up to 45 V, and output power levels up to 16 W. The power converters use high-Q RF-compatible air-core inductors from Coilcraft (1812SMS and 2222SQ series). Low-ESR ceramic capacitors from American Technical Ceramics are used as input voltage decoupling and output filter capacitors. A photograph of one of the tested buck converter prototypes, using the 100 MHz optimized GaN chip, a 47 nH Coilcraft air-core inductor (1812SMS series), and a 1 µF American Technical Ceramics low-ESR ceramic capacitor is shown in Fig. 5.

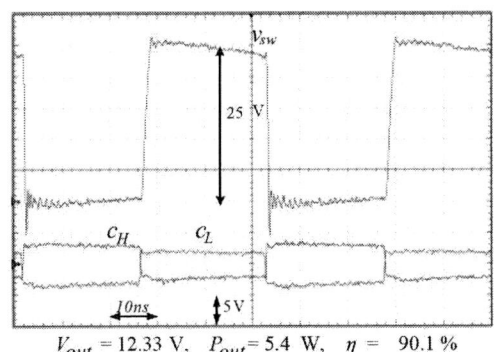

V_{out} = 12.33 V, P_{out} = 5.4 W, η = 90.1 %

Figure 8: Switching node (v_{sw}) and the input control signals (c_H and c_L) at 50% duty cycle for the 20 MHz, 25 V ZVS-QSW synchronous buck converter.

The PWM control signals (c_H and c_L in Fig. 1(a)) for the half-bridge GaN chips are obtained from an Altera Stratix IV FPGA, which provides PWM control signals with 125 ps resolution. A simple level-shifter interfaces the FPGA with the GaN chip gate driver inputs. A diagram of the test setup for the buck converters is shown in Fig. 6. A summary of the power stage and overall efficiency of the converter prototypes is provided in Table I.

Next, measured operation waveforms and efficiency curves are presented for selected converter prototypes.

A. ZVS-QSW buck converter operating at 20 MHz from 25 V input voltage

Power stage efficiency as well as overall efficiency (including the gate drive losses) of the ZVS-QSW synchronous buck converter prototype operating at 20 MHz from 25 V input voltage are shown in Fig. 7(a) and Fig. 7(b) respectively. In this case, the size of the power stage transistors gate periphery (Q_H and Q_L) is 6 mm, and the gate-drive resistor R_g is

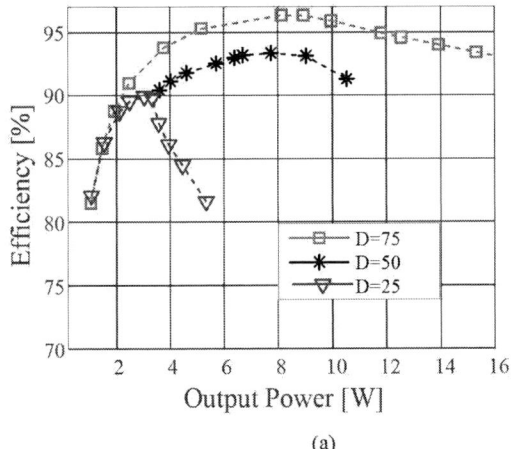

(a)

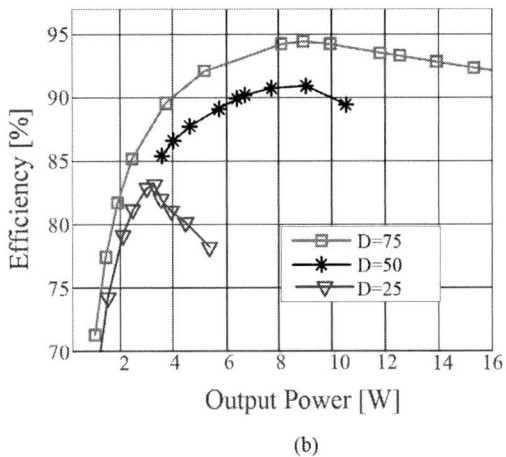

(b)

Figure 7: (a) Power stage efficiency and (b) overall efficiency for different duty cycles as functions of output power for the synchronous buck converter operating at 20 MHz from 25 V supply voltage.

978-1-4673-9551-9/16 $31.00 © 2016 IEEE

Table I: Experimental results for synchronous buck converter prototypes using monolithic GaN power chips with integrated gate drivers

Switching frequency, f_s [MHz]	20	50	100	200	400	100
Input voltage [V]	25	25	25	25	20	45
Maximum output power [W]	16.0	10.1	7.1	3.4	5.0	6.0
Peak power stage efficiency [%]	95.0	94.2	93.2	86.5	72.5	91.7
Peak total efficiency [%]	92.5	91.7	89.2	82.0	67.0	90.2
Inductance (L) [nH]	160	90	47	22	12.5	90
Duty cycle (D) [%]	75	75	75	75	50	50

$950\,\Omega$. A peak power stage efficiency of 95.1% is obtained at 75% duty cycle. The converter maintains greater than 90% power efficiency and greater than 85% overall efficiency over a wide range of output power (4 to 16 W), at 50% and 75% duty cycles. Switching node voltage (v_{sw}) and PWM control signals (c_H and c_L) for 50% duty cycle operation are shown in Fig. 8. Smooth low-to-high and high-to-low transitions typical for ZVS-QSW operation can be observed.

B. ZVS-QSW buck converter operating at 100 MHz from 25 V input voltage

Power stage and overall efficiency for the 100 MHz, 25 V ZVS-QSW synchronous buck converter prototype are shown in Fig. 9. The size of the power stage transistors gate periphery (Q_H and Q_L) is $2.6\,\text{mm}$, and R_g is $950\,\Omega$. This converter achieves a peak efficiency of 93.2% at 75% duty cycle. Switching node voltage and PWM control signals for this prototype, at 50% duty cycle, are shown in Fig. 10, illustrating ZVS operation.

C. ZVS-QSW buck converter operating at 200 MHz from 25 V input voltage

Efficiency plots for the 200 MHz, 25 V ZVS-QSW buck converter prototype are shown in Fig. 11, for operation at 50% duty cycle. The prototype uses the GaN power chip optimized for 100 MHz operation. Peak power stage efficiency greater

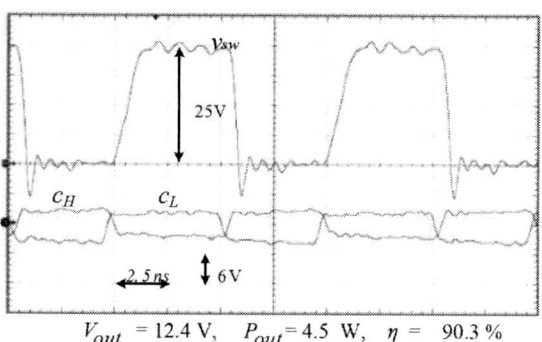

$V_{out} = 12.4$ V, $P_{out} = 4.5$ W, $\eta = 90.3\%$

Figure 10: Switching node voltage (v_{sw}) and the input control signals (c_H and c_L) at 50% duty cycle for the 100 MHz, 25 V ZVS-QSW synchronous buck converter.

than 86% is recorded at high output power levels (> 3 W). The switching node voltage and PWM control signals for this converter at 50% duty cycle are shown in Fig. 12.

D. Buck converter operating at 400 MHz from 20 V input voltage

Power stage efficiency results for the 400 MHz, 20 V ZVS-QSW buck converter prototype are shown in Fig. 13. In this case, the size of the power stage transistors gate periphery

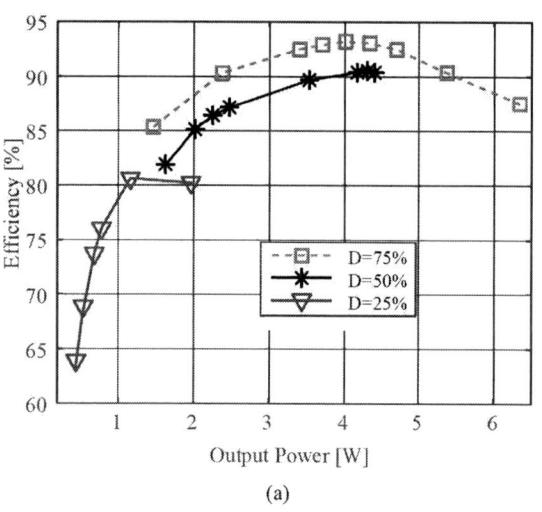

(a)

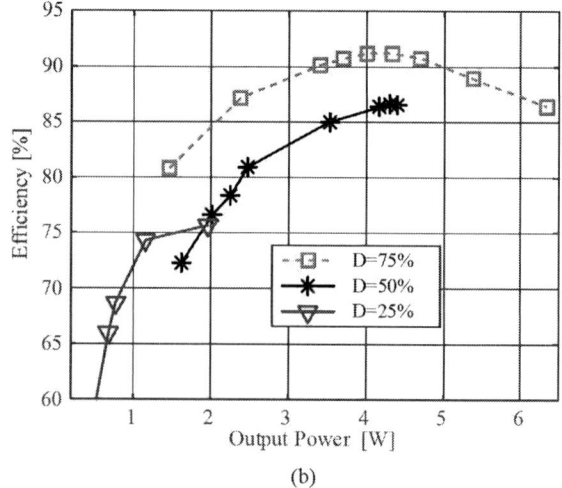

(b)

Figure 9: (a) Power stage efficiency and (b) overall efficiency for different duty cycles as functions of output power for the synchronous buck converter operating at 100 MHz from 25 V supply voltage.

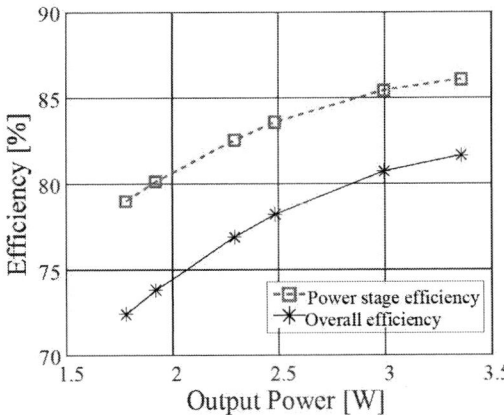

Figure 11: Power stage and overall efficiency (at 50% duty cycle) vs output power for the 200 MHz, 25 V ZVS-QSW synchronous buck converter.

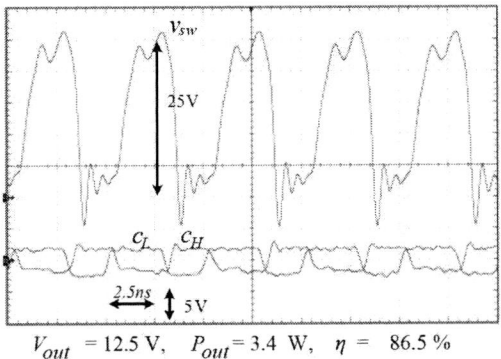

V_{out} = 12.5 V, P_{out} = 3.4 W, η = 86.5 %

Figure 12: Switching node voltage (v_{sw}) and the input control signals (c_H and c_L) at 50% duty cycle for the 200 MHz, 25 V ZVS-QSW synchronous buck converter.

(Q_H and Q_L) is 1.6 mm, and R_g is 150 Ω. Due to practical difficulties of generating complementary input control signals at ultra high switching frequencies, the 400 MHz prototype was operated in the non-synchronous mode. The low side transistor was turned off, and operated as a rectifier in the reverse conduction mode. The reverse conduction behavior of the low-side transistor can be observed in Fig. 14, wherein the switching node (v_{sw}) shows a larger excusrsion to negative voltages. Nevertheless, a greater than 72% peak power stage efficiency has been obtained.

E. ZVS-QSW buck converter operating at 100 MHz *from* 45 V *input voltage*

The power stage and the overall efficiency for the 100 MHz, 45 V ZVS-QSW synchronous buck converter are shown in Fig. 15 as functions of the output power. The size of the power stage transistors gate periphery (Q_H and Q_L) is 1.6 mm, and R_g is 1 kΩ. Note that a larger R_g is employed to support higher-voltage operation. The peak efficiency of 91.7% is

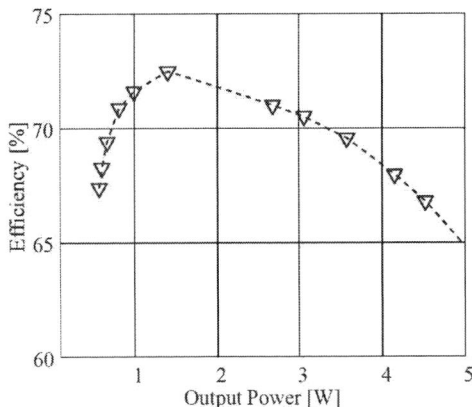

Figure 13: Power stage efficiency (at 50% duty cycle) for the buck converter operating at 400 MHz from 20 V input voltage.

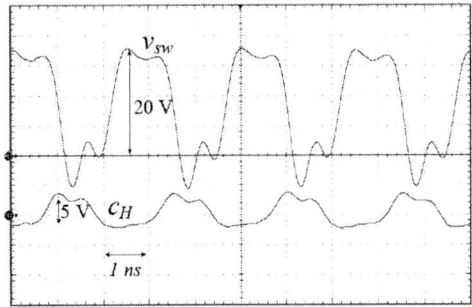

V_{out} = 10.1 V, P_{out} = 1.36 W, η = 72.5 %

Figure 14: Switching node (v_{sw}) and the input control signal (c_H) at 50% duty cycle for the 400 MHz, 20 V buck converter.

obtained at high output power levels ($\approx$ 6 W). Fig. 16 shows the switching node voltage and the control signals for this converter operating at 50% duty ratio.

V. CONCLUSIONS

This paper presents high efficiency dc-dc converters operating at very-to-ultra high frequencies, ranging from the 20 MHz to 400 MHz. The high performance is achieved using monolithic GaN half-bridge power chips with integrated gate drivers in an enhancement-mode (normally-off) GaN-on-SiC process. While using only n-type transistors available in the GaN process, a novel gate driver circuit has been developed to maintain low static power consumption while enabling fast switching by emulating complementary operation similar to approaches commonly employed in processes such as CMOS where complementary devices are available. Level shifting is accomplished using a bootstrap technique, with the bootstrap capacitor and the bootstrap diode integrated on the same GaN power chip. A family of monolithic GaN chips has been designed, targeting operation from up to 45 V, delivering up to 16 W of output power, and operating at 20-400 MHz switching frequencies. The GaN chips are verified in synchronous buck

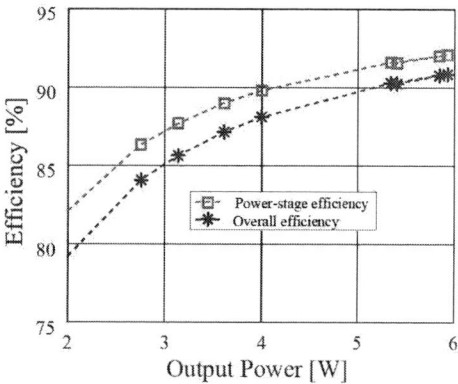

Figure 15: Power stage and overall efficiency vs output power at 50% duty cycle for a 100 MHz and 45 V QSW synchronous buck converter.

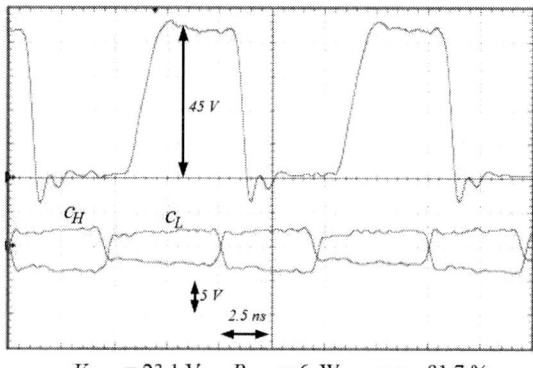

V_{out} = 23.1 V, P_{out} = 6 W, η = 91.7 %

Figure 16: Switching node (v_{sw}) and the input control signals (c_H and c_L) at 50% duty cycle for a 100 MHz and 45 V QSW synchronous buck converter.

converters, demonstrating record peak power stage efficiencies of 95.0% at 20 MHz, 94.2% at 50 MHz, 93.2% at 100 MHz, 86.5% at 200 MHz, and 72.5% at 400 MHz.

REFERENCES

[1] M. Rodriguez, Y. Zhang, and D. Maksimovic, "High-frequency PWM Buck converters using GaN-on-SiC HEMTs," *IEEE Trans. Power Electron.*, vol. 29, no. 5, pp. 2462–2473, May. 2014.

[2] Y. Zhang, M. Rodriguez, and D. Maksimovic, "100 MHz, 20 V, 90% efficient synchronous buck converter with integrated gate driver," in *Proc. IEEE Energy Convers. Congr. Expo.*, 2014, pp. 3664–3671.

[3] D. Maksimovic, Y. Zhang, and M. Rodriguez, "Monolithic very high frequency GaN switched-mode power converters," in *Proc. IEEE Custom Integr. Circuits Conf.*, 2015.

[4] Y.-P. Hong, K. Mukai, H. Gheidi, S. Shinjo, and P. Asbeck, "High efficiency GaN switching converter IC with bootstrap driver for envelope tracking applications," in *Proc. IEEE Radio Freq. Integr. Circuits Symp.*, 2013, pp. 353–356.

[5] J. Garcia, R. Marante, M. Ruiz, and G. Hernandez, "A 1 GHz frequency-controlled class E2 DC/DC converter for efficiently handling wideband signal envelopes," in *Proc. IEEE MTT-S Int. Microw. Symp. Dig.*, 2013, pp. 1–4.

[6] Y. Zhang, J. Strydom, M. de Rooij, and D. Maksimovic, "Envelope tracking GaN power supply for 4G cell phone base stations," in *Proc. IEEE 31th Annu. Appl. Power Electron. Conf. Expo*, 2016.

[7] Y. Zhang, M. Rodriguez, and D. Maksimovic, "High-frequency integrated gate drivers for half-bridge GaN power stage," in *Proc. IEEE 15th Workshop on Control and Modeling for Power Electron.*, 2014, pp. 1–9.

[8] S. Moench, R. Reiner, B. Weiss, P. Waltereit, R. Quay, M. Costa, A. Barner, O. Ambacher, and I. Kallfass, "Monolithic integrated quasi-normally-off gate driver and 600 V GaN-on-Si HEMT," in *IEEE 3rd Workshop on Wide Bandgap Power Devices and Appl.*, 2015.

[9] J. Hu, A. Sagneri, J. Rivas, Y. Han, S. Davis, and D. Perreault, "High-frequency resonant SEPIC converter with wide input and output voltage ranges," *IEEE Trans. Power Electron.*, vol. 27, no. 1, pp. 189–200, Jan. 2012.

[10] J. Garcia, R. Marante, and M. de las Nieves Ruiz Lavin, "GaN HEMT Class E2 resonant topologies for UHF DC/DC power conversion," *IEEE Trans. Microw. Theory Tech.*, vol. 60, no. 12, pp. 4220–4229, Dec. 2012.

[11] M. Roberg, T. Reveyrand, I. Ramos, E. Falkenstein, and Z. Popovic, "High-efficiency harmonically terminated diode and transistor rectifiers," *IEEE Trans. Microw. Theory Tech.*, vol. 60, no. 12, pp. 4043–4052, Dec. 2012.

[12] A. Sagneri, D. Anderson, and D. Perreault, "Optimization of integrated transistors for very high frequency DC-DC converters," *IEEE Trans. Power Electron.*, vol. 28, no. 7, pp. 3614–3626, Jul. 2013.

[13] V. Kumar, D. Fanning, J. Hitt, P. Saunier, M. Rodriguez, F. Zhang, and D. Maksimovic, "GaN power switch development and buck converter demonstrations," in *Proc. GOMAC Tech.*, 2014.

[14] A. Sepahvand, L. Scandola, Y. Zhang, and D. Maksimovic, "Voltage regulation and efficiency optimization in a 100 MHz series resonant DC-DC converter," in *Proc. IEEE 30th Annu. Appl. Power Electron. Conf. Expo*, 2015, pp. 2097–2103.

[15] W. Liang, J. Glaser, and J. Rivas, "13.56 MHz high density DC-DC converter with PCB inductors," *IEEE Trans. Power Electron.*, vol. 30, no. 8, pp. 4291–4301, Aug. 2015.

[16] J. Choi, W. Liang, L. Raymond, and J. Rivas, "A high-frequency resonant converter based on the class phi2 inverter for wireless power transfer," in *Proc. IEEE 79th Veh. Tech. Conf.*, 2014, pp. 1–5.

[17] L. Roslaniec, A. Jurkov, A. Al Bastami, and D. Perreault, "Design of single-switch inverters for variable resistance/load modulation operation," *IEEE Trans. Power Electron.*, vol. 30, no. 6, pp. 3200–3214, Jun. 2015.

[18] J. Santiago-Gonzalez, K. Elbaggari, K. Afridi, and D. Perreault, "Design of class E resonant rectifiers and diode evaluation for VHF power conversion," in *Proc. IEEE Energy Convers. Congr. Expo.*, 2014, pp. 2698–2706.

[19] ——, "Design of class E resonant rectifiers and diode evaluation for VHF power conversion," *Power Electronics, IEEE Transactions on*, vol. 30, no. 9, pp. 4960–4972, Sept 2015.

[20] D. Maksimovic, "Design of the zero-voltage-switching quasi-square-wave resonant switch," in *Proc. IEEE 24th Power Electron. Spec. Conf.*, 1993, pp. 323 –329.

Thermal Evaluation of Chip–Scale Packaged Gallium Nitride Transistors

David Reusch, Johan Strydom, and Alex Lidow

Efficient Power Conversion Corporation
El Segundo, CA USA

Abstract— **With power converters demanding higher power density, transistors must be accommodated in an ever decreasing board space. Beyond gallium nitride based power transistors' ability to improve electrical efficiency, they must also be more thermally efficient. In this paper we will evaluate the thermal performance of chip-scale packaged enhancement-mode GaN field effect transistors (eGaN® FETs) and compare their in-circuit electrical and thermal performance with state-of-art silicon MOSFETs. The paper will conclude with the proposal of a thermal figure of merit for designers to use as a tool to quickly compare the thermal efficiency of device packaging technologies.**

I. INTRODUCTION

Gallium nitride (GaN) transistors have emerged as a replacement for silicon devices in various power conversion applications. GaN transistors are high electron mobility transistors (HEMT) with a higher band gap, critical electric field strength, and electron mobility than silicon devices [1]. These material characteristics allow the GaN transistor to achieve lower on-resistance and lower switching charges in a smaller chip size. This also makes them more suitable for higher frequencies, improving power density in existing applications [2]-[10], and enabling emerging applications not suitable for the silicon (Si) MOSFET [11]-[13].

There are five basic requirements for a better transistor; (1) lower on-resistance, (2) faster switching speeds, (3) better thermal conductivity, (4) smaller size, and (5) lower cost. From time to time different technologies, such as GaAs or SiC, have improved on one or more of these basic requirements. Gallium nitride, grown on a silicon crystal, can improve upon all of the characteristics just listed when compared with the best silicon devices available, and in this paper we will focus on their improved thermal performance in chip-scale packages.

To estimate the thermal performance of a device, the junction temperature (T_J) is modeled by a voltage as shown in figure 1(a) [14]-[16]. The system thermal impedances can be modeled with resistors, with junction-to-board ($R_{\theta JB}$) and junction-to-case ($R_{\theta JC}$) thermal resistances being determined by the device package, and case-to-ambient ($R_{\theta CA}$) and board-to-ambient ($R_{\theta BA}$) thermal impedances determined by

system design. A current source models power loss (P_{LOSS}), and a voltage source models the ambient temperature (T_A). The thermal resistance model can be simplified into a single resistor, shown in figure 1(a), and the effective junction-to-ambient thermal impedance ($R_{\theta JA}$) for the system and the maximum power dissipation (P_{MAX}) can be defined as:

$$P_{MAX} = \frac{T_{JMAX} - T_A}{R_{\theta JA}} \tag{1}$$

Where T_{JMAX} is the maximum junction temperature of the power device.

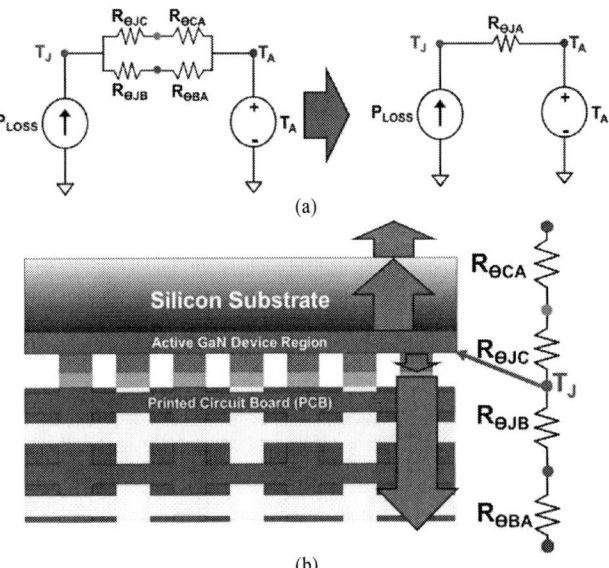

(a)

(b)

Figure 1: (a) Thermal electrical model and (b) side view of chip-scale packaged GaN transistor mounted to a printed circuit board with device and system thermal impedances

The thermal efficiency of a package can be determined by comparing two parameters, $R_{\theta JC}$ and $R_{\theta JB}$, normalized to the package area. $R_{\theta JC}$ is the thermal resistance from junction-to-case; this is the thermal resistance from the active part of the eGaN FET to the top of the silicon substrate, including the sidewalls. $R_{\theta JB}$ is the thermal resistance from junction-to-board; this is the thermal resistance from the active part of the eGaN FET to the printed circuit board (PCB). For this path the heat must transfer through the solder bars/balls to the copper traces on the board. In table I is a compilation of thermally related characteristics from datasheets [17]-[26] of several popular surface mount MOSFET packages as well as two popular eGaN FETs.

TABLE I: COMPARISON OF PACKAGE AREA AND THERMAL RESISTANCE
COMPONENTS $R_{\Theta JC}$ AND $R_{\Theta JB}$

Device Package	$R_{\Theta JC}$ (°C/W)	$R_{\Theta JB}$ (°C/W)	Area (mm²)
CanPAK S [17]	2.9	1	18.2
CanPAK M [18]	1.4	1	30.9
DFN/SON/S3O8 [19]-[21]	-	1-2.7	10.9
DFN/SON/S3O8 Dual Cool [21]	3.5	2.7	10.9
DFN/SON/Super SO8 [22], [23]	20	0.5-0.9	30.0
DFN/SON/Super SO8 Dual Cool [24]	1.2	1.1	30.0
EPC2001C eGaN FET [25]	**1.0**	**2.0**	**6.7**
EPC2021 eGaN FET [26]	**0.5**	**1.4**	**13.9**

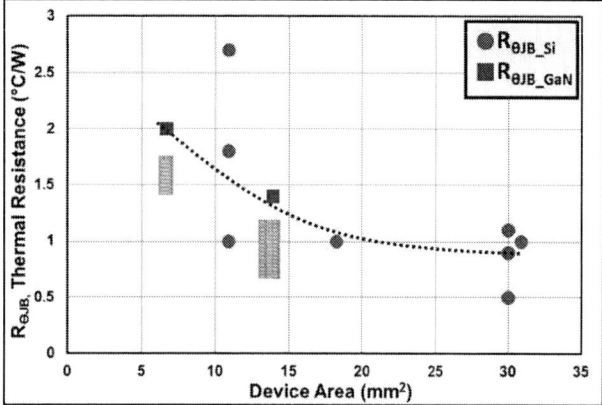

Figure 2: $R_{\Theta JB}$ (Junction-to-Board Thermal Resistance) for several package styles listed in Table I, eGaN FETs represented by blue square dots and Si MOSFETs represented by red circular dots

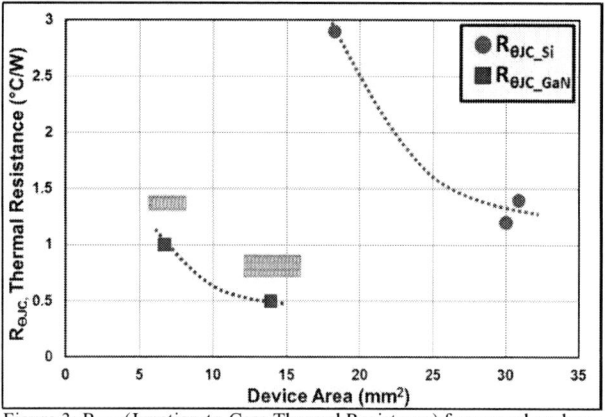

Figure 3: $R_{\Theta JC}$ (Junction-to-Case Thermal Resistance) for several package styles listed in table I, eGaN FETs represented by blue square dots and Si MOSFETs represented by red circular dots

Figure 2 shows a plot of the junction-to-board resistance ($R_{\Theta JB}$) for each of the packages given in Table I. Red circular dots represent the Si MOSFET packages, and blue square dots represent the eGaN FET chip-scale packages. The majority of the sampled packages fall on a single trend line indicating that performance for this element of thermal resistance is determined primarily by package size, and not

technology. It should be noted that the most common Si MOSFET package types, referred to by various manufacturers as Dual Flat No-lead (DFN), Small Outline No-lead (SON), or Small Outline (SO) Super SO8, among other names, while having very similar construction, have a wide range of reported junction-to-board resistances.

In contrast, in figure 3 shows a plot of the thermal resistance from junction to case ($R_{\Theta JC}$). The CanPAK and double-sided cooling SO8 packages are far less efficient at extracting the heat out of the top of the package than the chip-scale packaged eGaN FETs. This makes the chip-scale the most efficient thermal package for double sided cooling and most suitable for high density power designs.

II. TEST SETUP FOR MEASURING IN-CIRCUIT JUNCTION-TO-AMBIENT THERMAL IMPEDANCE

To quantify the in-circuit thermal performance of various packages, a series of thermal evaluation boards were developed. The first, an EPC2021 [26] eGaN FET based buck converter, is shown in figure 4. Matte black thermal measurement points have been added to each board to ensure consistent emissivity for improved measurement accuracy. The thermal evaluation boards have a 1 inch by 1 inch (25.4 mm by 25.4 mm) sized area for copper, outlined in white, and a total size of 2 x 2 inches (50.8 x 50.8 mm) with the copper connections outside of the 1x1 inch (25.4 x 25.4 mm) white border minimized to the input and output power connection points. The circuits are configured as buck converters and the output filter inductor (4.7 µH Coilcraft SER2915L) and output capacitors (5 x TDK C2012X5R1E226M125AC) are located on the backside of the board.

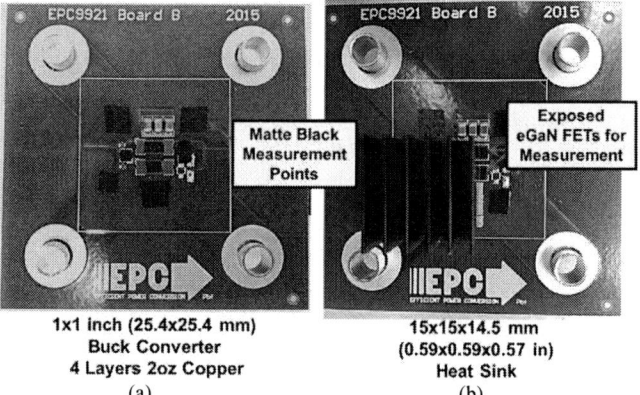

1x1 inch (25.4x25.4 mm)
Buck Converter
4 Layers 2oz Copper
(a)

15x15x14.5 mm
(0.59x0.59x0.57 in)
Heat Sink
(b)

Figure 4: Chip-scale packaged eGaN FET based thermal evaluation boards setup for cases without (a) and with (b) a heat sink

To evaluate the double-sided cooling capability, a heat sink was added to the test circuit, shown in figure 4(b). The matte black measurement points were moved to the edge of the eGaN FET dies, which were left exposed, and outside of the heat sink. While locating the heat sink with a portion of the eGaN FETs exposed was not the best for thermal performance, it allowed for an accurate look at the impact of a heat sink on thermal performance. In an actual design, the heat sink should fully cover all devices to maximize heat transfer. The heat sink used in this evaluation was a 0.59 x 0.59 x 0.57 inches (15 x 15 x 14.5 mm) Advanced Thermal Solutions ATS-54150K-C2-R0, with the thermal interface material removed.

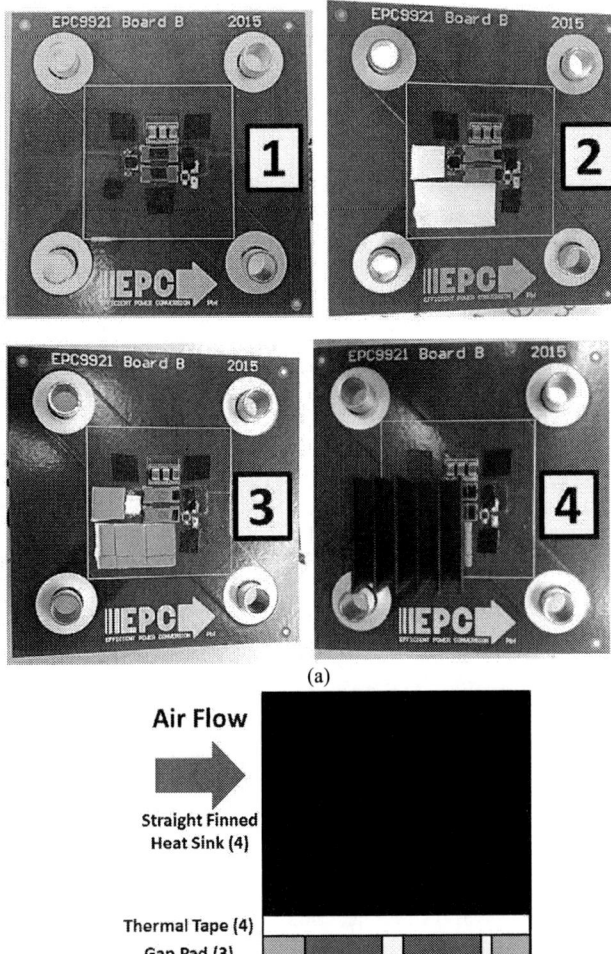

(a)

Air Flow

Straight Finned
Heat Sink (4)

Thermal Tape (4)
Gap Pad (3)
Thermal Tape (2)
PCB (1)

TIM (Optional)

(b)

Figure 5: (a) Heat sink attachment procedure and (b) conceptual cross-section side view with heat sink attachment process

After thermal analysis was completed for a design without heat sinking, following the steps below, the selected heat sink was attached for thermal evaluation:

(1) Starting with the thermal evaluation board without a heat sink
(2) The matte black thermal measurement points from areas where the heat sink will be placed were removed. For the power devices, new thermal measurement points toward the end of the power devices, outside of where the heat sink will be located, were added. A layer of thermal tape (3M 8810) was then added outside of the devices and inside of the 1 x 1 inch (25.4 x 25.4 mm) board area. The main purpose of the thermal tape is to act as a spacer to ensure the heat sink remains level across the power devices.
(3) A layer of gap pad (t-global TG6050) was attached on top of the thermal tape, acting as a spacer. The total height of the thermal tape with gap pad is approximately equal to the height of the power devices. It should be noted that the thermal tape and gap pad are cut to custom fit each design. A layer of thermal tape is also added to the driver to have an approximately uniform height for the heat sink.
(4) A heat sink was added (Advanced Thermal Solutions ATS-54150K-C2-R0 with the thermal interface material removed). The heat sink has a layer of thermal tape (3M 8810) on the bottom to assure good adhesion to the devices and provide electrical isolation. No additional pressure was applied to the heat sink in this evaluation and by using a heat sink

with additional force; a further improvement in double-sided cooling capability could be realized. With additional thermal interface materials (TIM), such as thermal grease, applied between and on the sides of the devices the thermal performance could also be further improved. The use of additional TIMs was not considered in this work.

While the heat sinking in this example is minimal, the objective was to demonstrate the significant thermal performance improvements possible with chip-scale packaged eGaN FETs under a minimal heat sinking setup, and to allow for easily accessible measurement points to accurately characterize the thermal performance of different power packages.

The thermal evaluation boards, meant to accurately represent in-circuit operating conditions, were tested as buck converters with an input voltage of 48 V, an output voltage of 12 V, and a switching frequency of 300 kHz. For the first test case, 80 V EPC2021 eGaN FETs were used [26]. The circuit board image for each test case is shown in the upper left of figures 6 (a) and (b) with corresponding thermal images, taken with a FLIR E40 thermal camera.

With a 200 LFM airflow, the eGaN FET based design achieved an output current of 22 A before reaching a temperature around 100°C at an ambient temperature of approximately 25°C. The selected maximum temperature point for this evaluation was well below the 150°C maximum junction temperature of eGaN FETs. The 100°C point was selected to provide margin for higher ambient temperatures and provide device de-rating, which is a common design practice.

When a heat sink was added to the eGaN FET based design, the output power increased by around 30% while maintaining approximately the same maximum temperature. The heat sink showed a significant temperature rise over ambient, demonstrating that the eGaN FETs are effectively moving heat from the device to the heat sink, allowing for higher output currents and power. This was enabled by the superior backside cooling properties of the eGaN FETs chip-scale packaging.

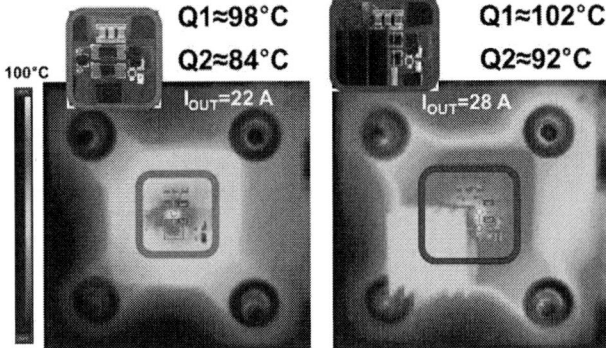

(a)   (b)

Figure 6: Thermal evaluation of eGaN FETs operated as a buck converter with an input voltage of 48 V, an output voltage of 12 V, a switching frequency of 300 kHz, air-flow=200 LFM, ambient temperature=25°C (a) without heat sinking (I_{OUT}=22 A) and (b) with heat sinking (I_{OUT}=28 A)

To create a more complete in-circuit thermal model of the eGaN FET based designs, the thermal evaluation boards were tested at a variety of operating points with various levels of airflow, under two test conditions, with and without a heat sink. A plot of maximum junction temperature versus total system power loss is shown in figure 7. The slope of these lines is equivalent to the junction-to-ambient resistance of the system, as shown in the simplified thermal schematic shown on the right of figure 1.

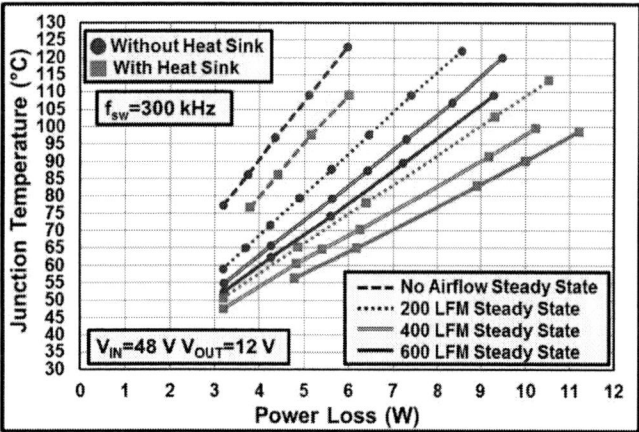

Figure 7: Plot of junction temperature vs. system power loss for a buck converter with an input voltage of 48 V, an output voltage of 12 V, and a switching frequency of 300 kHz

From the data in figure 7, the junction-to-ambient resistances are determined and plotted versus airflow in figure 8. It can be seen that by adding even a small amount of airflow, the thermal performance of eGaN FETs improved significantly. Also, by adding a minimal amount of heat sinking, the thermal impedance of the system decreased by around 30%. With 600 LFM of airflow and a heat sink, the effective system junction-to-ambient resistance is measured to be around 6.5°C/W, which equates to a temperature rise of 65°C for 10 watts of system power loss.

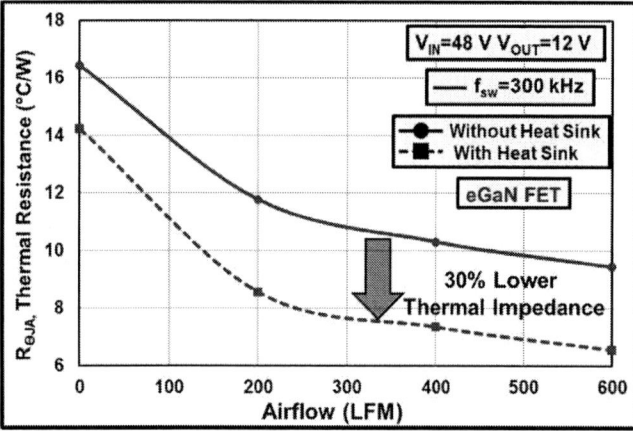

Figure 8: Plot of junction-to-ambient system thermal impedance vs. airflow for a buck converter with an input voltage of 48 V, an output voltage of 12 V, and a switching frequency of 300 kHz

III. COMPARISON OF CHIP-SCALE PACKAGED eGaN FET AND Si MOSFET BASED POWER CONVERTERS

As discussed in [1], [2], and [27], eGaN FETs have universally lower figures of merit and superior electrical packaging. In this section, the electrical and thermal performance of chip-scale packaged eGaN FETs will be compared to Si MOSFETs in conventional and advanced double sided cooling packages in an equivalent test comparison.

To compare the performance of eGaN FETs and state-of-the-art Si MOSFETs, identical buck converter evaluation boards were designed for the three cases, shown in figure 9. All of the boards were two (2) in^2 (2580 mm^2) and had active copper areas of one (1) in^2 (645 mm^2), outlined in white in the figure. The boards, designed on the same PCB panel to be consistent, had four layers of two-ounce (2.8 mils/71 μm) copper and used the optimal layout discussed in [6].

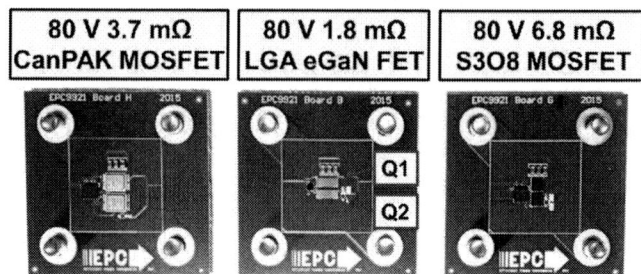

Figure 9: eGaN FET and Si MOSFET thermal evaluation boards

The 80 V Si MOSFETs used for comparison have CanPAK packages (Infineon BSB044N08NN3), shown on the left in figure 9, and the traditional S3O8 packages (Alpha and Omega Semiconductor AON7280), shown on the right in figure 9. For both Si MOSFETs, the lowest commercially available FOM devices for their respective package technologies were selected. A state-of-the-art MOSFET driver (Intersil ISL2111) was used for the MOSFET-based designs and the gate voltage for the Si MOSFETs was 10 V.

For the eGaN FET based design, shown in the center of figure 9, an 80 V Land Grid Array (LGA) chip-scale packaged EPC2021 was used with the Texas Instruments LM5113 driver and the gate voltage for the eGaN FETs was 5 V. When comparing the eGaN FET and Si MOSFET based designs, the eGaN FET based design takes up less than half the active board area of the Si MOSFET based designs. The first reason for the improved power density of the eGaN FET-based design is that the power device is much smaller in size, when compared to the MOSFET in terms of $R_{DS(ON)}$ times the area. The second improvement in density is driven by the gate driver, which is also much smaller than its counterpart, in part from better packaging and in part from the reduced drive requirements of low charge, low drive voltage eGaN FETs.

A. Electrical Performance Comparison

The total system efficiency and power loss comparisons of the eGaN FET and silicon MOSFET based 48 V_{IN} to 12 V_{OUT} buck converters operated at switching frequencies of 300 kHz and 500 kHz are shown in figure 10. This plot takes into account losses of the entire system, including the inductor (4.7 μH Coilcraft SER2915L), capacitors, and PCB losses.

The eGaN FET based design offers similar light load efficiency with the S3O8 packaged Si MOSFET based design, which has an almost 4x higher on-resistance, demonstrating the effect of the low charge of eGaN FETs. At full load, the low eGaN FET on-resistance enables higher efficiency. Combining low conduction losses and low switching losses, the eGaN FET based design has higher efficiency at almost every design point.

978-1-4673-9551-9/16 $31.00 © 2016 IEEE

When comparing the eGaN FET and larger CanPAK MOSFET based design, the performance is improved under all conditions. For the larger, slower switching Si MOSFET the higher switching losses become the major loss mechanism. At a switching frequency of 300 kHz and a load current of 30 A, the eGaN FET based design reduced the total system loss by 35%. At 500 kHz, the eGaN FET based design, with low switching charges, low on-resistance, and improved packaging, had a minimal drop in efficiency when compared to the MOSFET based designs. Furthermore, at 30 A, the total system loss is reduced by almost 40% when compared to the best Si MOSFET based solution.

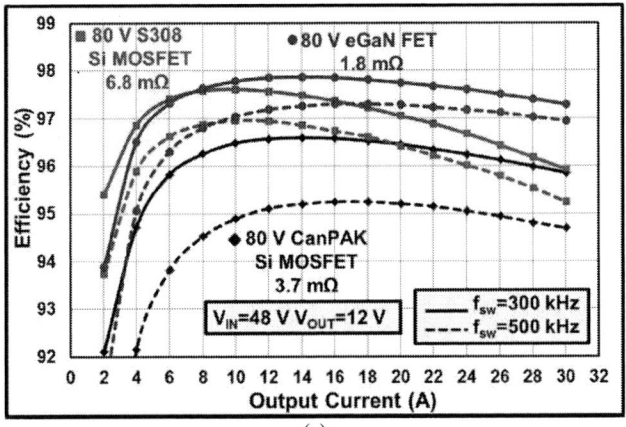

(a)

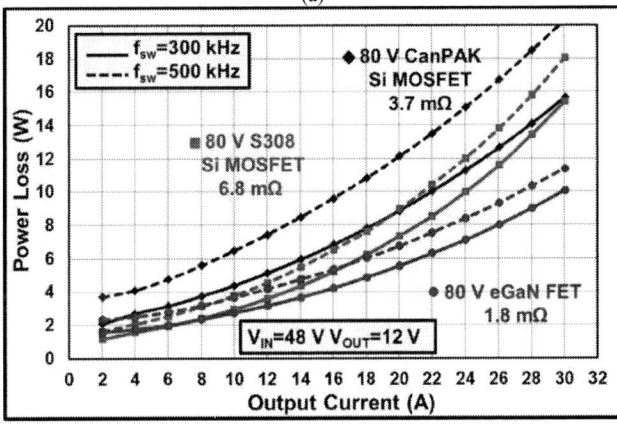

(b)

Figure 10: Experimental electrical comparison between GaN and Si based buck converters (a) efficiency (b) power loss, V_{IN}=48 V to V_{OUT}=12 V, f_{sw}=300 kHz and f_{sw}=500 kHz (L=4.7 µH Coilcraft SER2915L)

B. Thermal Performance Comparison

In figure 11, the thermal performance of chip-scale packaged eGaN FET and S3O8 and CanPAK packaged Si MOSFET based designs are compared. For this paper, the selected maximum temperature point for this evaluation was well below the 150°C maximum junction temperature of power devices. A 100°C maximum temperature was selected to provide margin for higher ambient temperatures and provide device de-rating, which is a common design practice.

At 14 A of output current, the CanPAK based MOSFET design (BSB044N08NN3G), shown on the left of figure 11, reaches 100°C with an airflow of 200 LFM. The Si MOSFET generates higher power loss from lower electrical efficiency

and the package is also less thermally efficient than the chip-scale packaged eGaN FET, shown in the center of figure 11. For the same maximum temperature, the eGaN FET based design is operated to 22 A, where the maximum device temperature matches that of the CanPAK packaged Si MOSFET on the top device (Q1). The eGaN FET based solution can achieve almost 60% more output power while maintaining the same maximum junction temperature and occupying significantly less board space.

At 16 A of output current, the S308 packaged MOSFET design (AON7280), shown on the right of figure 11, approaches 100°C with an airflow of 200 LFM. For the same maximum temperature, the eGaN FET based design is operated to 22 A, where the maximum device temperature matches that of the S308 packaged Si MOSFET on the top device (Q1), while the low side (Q2) eGaN FET runs 13°C cooler. The eGaN FET based solution can push almost 40% more output power while maintaining the same maximum junction temperature.

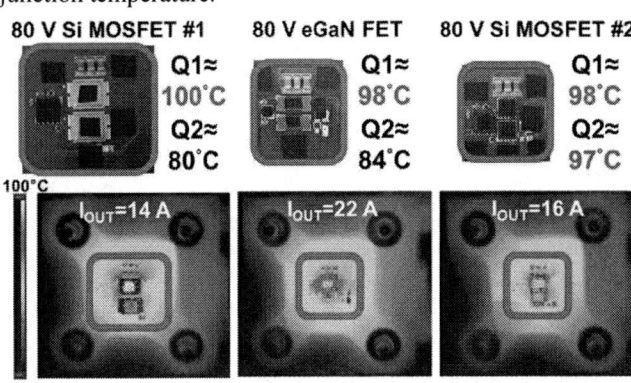

Figure 11: Experimental thermal comparison between chip-scale packaged eGaN FET (center), CanPAK packaged Si MOSFET (left), and S3O8 packaged Si MOSFET (right) based buck converters without heat sinking, V_{IN}=48 V to V_{OUT}=12 V, f_{sw}=300 kHz, air-flow=200 LFM, ambient temperature=25°C with same approximate maximum device temperature

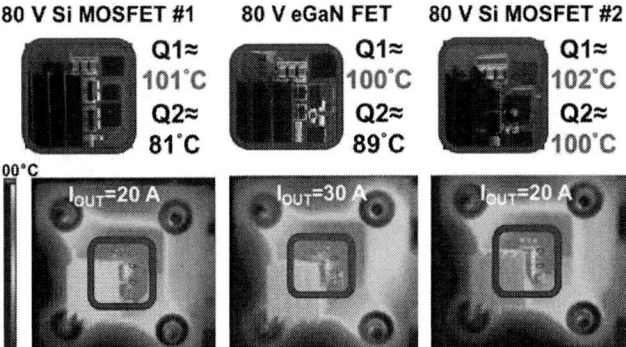

Figure 12: Experimental thermal comparison between chip-scale packaged eGaN FET (center), CanPAK packaged Si MOSFET (left), and S3O8 packaged Si MOSFET (right) based buck converters with heat sinking, V_{IN}=48 V to V_{OUT}=12 V, f_{sw}=300 kHz, air-flow=400 LFM, ambient temperature=25°C with same approximate maximum device temperature

To compare the thermal performance of eGaN FETs and Si MOSFETs with heat sinking, test cases with heat sinks were also evaluated. The heat sink and attachment methods were kept consistent between designs and followed the attachment process outlined earlier in section II.

For the thermal evaluation with heat sinking, the chip-scale packaged eGaN FET based design, shown in the center in figure 12, reaches 100°C at an output current of 30 A with an airflow of 400 LFM. Both Si MOSFET based designs, with the CanPAK and S3O8 packaged Si MOSFET based designs shown on the left and right of figure 12, respectively, reach 100°C at 20 A output current and 400 LFM airflow. The eGaN FET based solution achieves 50% higher overall output power, despite occupying a much smaller board space, while maintaining the same maximum junction temperature. The CanPAK packaged Si MOSFET occupies around three times the board area of the eGaN FET and shows much better junction-to-case cooling than the S3O8 package. The junction-to-case cooling improvement demonstrated by the CanPAK is similar to that of the much smaller eGaN FET, which has larger thermal impedance introduced by the thermal interface used to attach the heat sink. This is due to the much higher transistor density and smaller surface area to connect the heat sink.

Following the same methodology outlined in section II, the equivalent junction-to-ambient system thermal impedances, shown in figure 13, were calculated for the eGaN FET and MOSFET based designs. The eGaN FET based design, occupying the smallest active area, provides the best overall thermal performance. In the case without a heat sink and 400 LFM airflow, the eGaN FET based design has around 9% and 5% lower thermal impedances than the S3O8 and CanPAK packaged Si MOSFET based designs respectively. In the case of a heat sink with 400 LFM airflow, where the junction-to-case thermal path becomes a major heat path, the eGaN FET based design has around 23% and 5% lower thermal impedances than the S3O8 and CanPAK packaged Si MOSFET based designs respectively.

When the superior thermal performance is combined with the improved electrical performance, eGaN FETs can enable improved overall system performance, in a fraction of the occupied board space. This is a combination for game changing performance improvements.

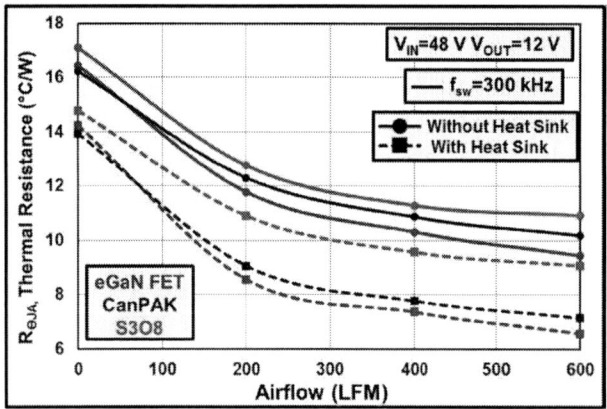

Figure 13: Comparison of junction-to-ambient system thermal impedance vs. airflow for eGaN FET and SI MOSFET based buck converters with an input voltage of 48 V, output voltage of 12 V, and switching frequency of 300 kHz

C. Impact of Thermal Performance on Electrical Performance

The thermal performance of a device will also impact electrical performance. The most notable impact on the device as the temperature rises is an increase in the device on-resistance. The normalized increase in on-resistance for an eGaN FET and Si MOSFETs is very similar [2]. For a device, the larger the initial on-resistance, the greater the additional conduction losses introduced by higher temperatures and increased normalized on-resistance.

Due to the lack of an industry standard on test conditions for thermal steady state, it is difficult to compare such data from different sources. As a result, it is common industry practice for efficiency and power loss measurements to use "instantaneous" efficiency measurements. These are not truly instantaneous since the measurements take several seconds, much longer than time constants for typical semiconductor die. However, they reduce many of the heating effects and form a good baseline measurement.

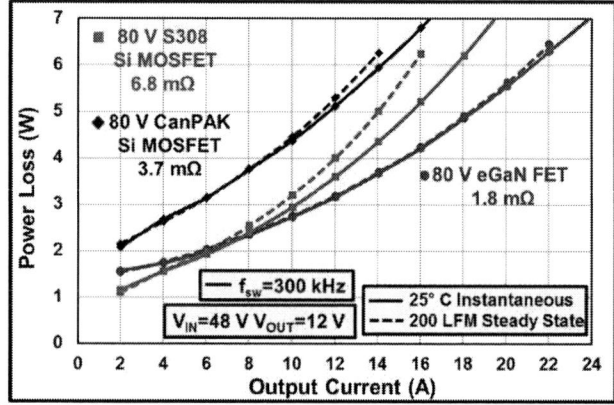

Figure 14: Experimental power loss comparison between eGaN FET and Si based buck converters without heat sinking for instantaneous and thermal steady state operating conditions, V_{IN}=48 V to V_{OUT}=12 V, f_{sw}=300 kHz

Figure 14 shows a comparison of instantaneous and thermal steady-state losses with 200 LFM of airflow and without heat sinking. The eGaN FET based design, which demonstrated lower operating temperatures earlier in the section and uses devices that have the lowest initial on-resistance, shows a very slight increase in power loss in steady state versus instantaneous cases. The S3O8 packaged Si MOSFET based design, which has the highest initial on-resistance, shows a significant degradation in performance in thermal steady state, with power loss increasing almost 20% at 16 A from "instantaneous" to steady state operating conditions. The CanPAK packaged Si MOSFET based design shows less performance degradation in steady state operation than the smaller S3O8 packaged Si MOSFET, but the CanPAK packaged Si MOSFET still displays a larger steady state operation performance penalty than the much smaller chip-scale packaged eGaN FET.

IV. THERMAL PERFORMANCE OF eGaN
MONOLITHIC HALF-BRIDGE IC

As discussed in [28], the monolithic integration of GaN power transistors has the potential to improve electrical performance by allowing for improved die size optimization and providing lower parasitic inductances. In this section, the thermal performance of the eGaN monolithic half-bridge IC will be evaluated. The first eGaN monolithic half-bridge IC is the same size as single discrete eGaN FET and has similar package thermal impedances from junction-to-board and junction-to-case. For half-bridge layouts, thermal evaluation boards were built to the same specifications as discussed in section II and are shown in figure 15. It is anticipated that the eGaN monolithic half-bridge IC, due to its factor of two reduction of chip size, should be at a thermal disadvantage.

Figure 15: Thermal evaluation boards implementation for eGaN FET based half bridges with discrete transistors (left) and an eGaN monolithic half-bridge IC (right)

In figure 16, the thermal performance of the discrete eGaN FET and eGaN monolithic half-bridge IC based designs are compared. At 18 A of output current with an airflow of 200 LFM, the eGaN monolithic half-bridge IC based design approaches 100°C. The discrete eGaN FET based design has 19°C and 28°C lower device temperatures, respectively for the top device (Q1) and low side device (Q2) when compared to the eGaN monolithic half-bridge IC. The thermal performance advantage of the discrete based eGaN FET design is expected here, since the discrete solution has larger chip area to dissipate heat. The eGaN monolithic half-bridge IC, which has the highest power density solution evaluated in this paper, exhibits improved thermal performance when compared to the best Si MOSFET based design, shown on the right of figure 11.

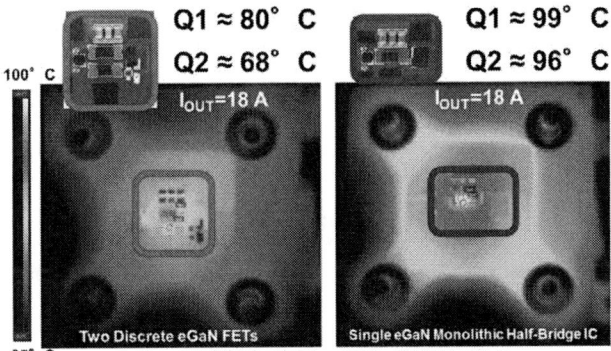

Figure 16: thermal comparison between discrete eGaN FET (left) and eGaN monolithic half-bridge IC (right) based buck converters, V_{IN}=48 V to V_{OUT}=12 V, f_{sw}=300 kHz, I_{OUT}=18 A, air-flow=200 LFM, and ambient temperature=25°C

From the thermal images in figure 16, a major thermal advantage of the monolithic integration can be seen. The advantage is that the monolithic-based solution efficiently distributes heat between the two devices, which are located on a single chip, but with lower thermal impedance between the devices. This will ensure good thermal balancing between the two devices and maximize the overall thermal capability of the system. For the discrete eGaN FET based solution shown on the left in figure 16, the device Q1 is at a much higher temperature than the device Q2 and is the thermally limiting device of the system. For the eGaN monolithic half-bridge IC based design, there is a very small 3°C temperature difference across the eGaN monolithic half-bridge IC. For the monolithic design, the higher loss device, Q1, can use the larger active area (and lower thermal resistance path) of the

device Q2 to spread the heat and achieve much better overall thermal performance.

The equivalent junction-to-ambient thermal impedances, shown in figure 17, were measured for the discrete and eGaN monolithic half-bridge IC based designs, with and without heat sinks. The eGaN monolithic half-bridge IC based design has slightly higher thermal impedance than the discrete eGaN FET based design, despite having only half the chip area.

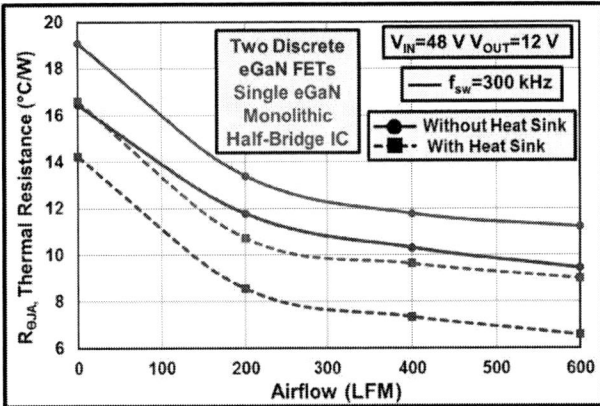

Figure 17: Comparison of junction-to-ambient system thermal impedance vs. airflow for discrete eGaN FET and eGaN monolithic half-bridge IC based buck converters with an input voltage of 48 V, an output voltage of 12 V, and a switching frequency of 300 kHz

V. THERMAL FIGURE OF MERIT (FOM)

Figures of merit have been used for almost half of a century to compare device technologies. FOMs, generally tailored for specific applications, are a simple tool to quickly compare different technologies ability to improve transistor performance [29]-[35]. While the majority of FOMs give valuable insight into the electrical performance achievable with a given device technology, little insight is given into thermal performance at the application level, which is also critical in power converter design. In this section, we will introduce a thermal figure of merit for designers to use as a tool to quickly compare the thermal efficiency of given devices packaging technology. This thermal FOM will be compared for the eGaN FET and Si MOSFET packaging technologies used in this paper.

As shown in figures 2 and 3, the thermal capability of a package can be determined by two parameters, $R_{\theta JB}$ and $R_{\theta JC}$, the junction-to-board and junction-to-case thermal impedances. These thermal impedances scale with the size of a given package technology, with a larger device having lower thermal impedance. The thermal efficiency of a package is related to its ability to dissipate heat based on a given chip area. To compare the thermal efficiency of a given packaging technology a thermal figure of merit is proposed:

$$FOM_{TH} = R_{\theta EQ} \cdot A_{DEVICE} \qquad (2)$$

Where A_{DEVICE} is the size of the packaged device. For design cases where heat can be removed through both the PCB and backside of the device ($R_{\theta BA} \approx R_{\theta CA}$):

$$R_{\theta EQ} = \frac{R_{\theta JB} \cdot R_{\theta JC}}{R_{\theta JB} + R_{\theta JC}} \qquad (3)$$

For cases where the dominant heat removal path is through the printed circuit board (PCB) ($R_{\theta BA} \ll R_{\theta CA}$):

$$R_{\theta EQ} = R_{\theta JB} \qquad (4)$$

And for cases where the dominant heat removal path is the backside of the device ($R_{\theta BA} \gg R_{\theta CA}$):

$$R_{\theta EQ} = R_{\theta JC} \qquad (5)$$

Assuming the device can be effectively cooled through the PCB and the backside of the device, the thermal FOM can be given by:

$$FOM_{TH} = \left(\frac{R_{\theta JB} \cdot R_{\theta JC}}{R_{\theta JB} + R_{\theta JC}} \right) \cdot A_{DEVICE} \qquad (6)$$

Comparing the thermal FOM, where a lower FOM represents improved package thermal efficiency, for eGaN FETs chip-scale packages and the Si MOSFET packages used for comparison in this paper, the thermal efficiency advantages of the chip-scale package can be seen – with the eGaN FET packaging having almost a 2x reduction in FOM_{TH} compared to the best Si MOSFET package. The chip-scale package offers efficient double-sided cooling and the best thermal performance based on the size of the chip.

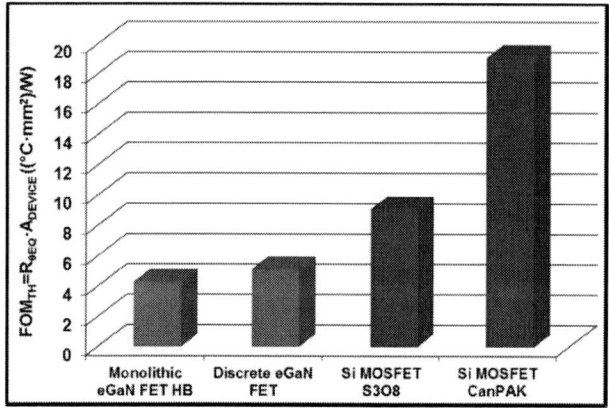

Figure 18: Thermal figure of merit comparison of eGaN FET chip scale packages and state of the art Si MOSFET packages

VI. CONCLUSIONS

In this paper, the thermal performance of chip-scale packaged GaN transistors was evaluated. For high voltage lateral GaN transistors, all of the electrical connections are located on the same side of the die, allowing for the elimination of complex, performance limiting two sided packaging common in vertical Si power MOSFETs. Chip-scale packaging is a more efficient form of packaging that reduces the resistance, inductance, size, thermal impedance, and cost of power transistors, enabling unmatched in-circuit performance.

REFERENCES

[1] A. Lidow, J. Strydom, M. de Rooij, D. Reusch, *GaN Transistors for Efficient Power Conversion*, Second Edition, Wiley, 2014.

[2] D. Reusch and J. Glaser, *DC-DC Converter Handbook, a supplement to GaN Transistors for Efficient Power Conversion*, First Edition, Power Conversion Publications, 2015.

[3] M. Briere, "GaN-based power devices offer game-changing potential in power-conversion electronics," EE Times, 2008.

[4] Y. Wu, M. Jacob-Mitos, M. L. Moore, and S. Heikman, "A 97.8% Efficient GaN HEMT Boost Converter with 300 W Output Power at 1MHz," IEEE Electron Device Letters, Vol. 29, pp. 824-826, 2008.

[5] S. Ji, D. Reusch, and F. C. Lee, "High Frequency High Power Density 3D Integrated Gallium Nitride Based Point of Load Module," Energy Conversion Congress and Exposition (ECCE), pp. 4267-4273, 2012.

[6] D. Reusch and J. Strydom, "Understanding the Effect of PCB Layout on Circuit Performance in a High Frequency Gallium Nitride Based Point of Load Converter," Applied Power Electronics Conference and Exposition (APEC), pp.649-655, 2013.

[7] H. Umeda, Y. Kinoshita, S. Ujita, T. Morita, S. Tamura, M. Ishida and T. Ueda, "Highly Efficient Low-Voltage DC-DC Converter at 2-5 MHz with High Operating Current Using GaN Gate Injection Transistors," International Exhibition and Conference for Power Electronics, Intelligent Motion, Renewable Energy, and Energy Management (PCIM Europe), pp. 1025-1032, 2014.

[8] X. Huang, Z. Liu, Q. Li, and F. C. Lee, "Evaluation and Application of 600V GaN HEMT in Cascode Structure," Applied Power Electronics Conference and Exposition (APEC), pp.1279-1286, 2013.

[9] L. Jenkins, "Optimization of High Power Gallium Nitride Based Point of Load Converters for Data Center Power Supply Chains," PhD Dissertation, Auburn University, 2015.

[10] M. Chen, K. K. Afridi, S. Chakraborty, and D. J. Perreault, "A High-Power-Density Wide-Input-Voltage-Range Isolated Dc-Dc Converter having a MultiTrack Architecture," Energy Conversion Congress and Exposition (ECCE), pp.2017-2026, 2015.

[11] M. A. de Rooij, "Wireless Power Handbook," Second Edition, PCP, El Segundo, October 2015.

[12] D. Cucak, M Vasić, O Garcia, J.A. Oliver, P. Alou, J.A. Cobos, "Application of eGaN FETs for highly efficient Radio Frequency Power Amplifier," Integrated Power Electronics Systems, CIPS 2012, pp.1-6, 2012.

[13] J. Strydom, D. Reusch, "Design and Evaluation of a 10 MHz Gallium Nitride Based 42 V DC-DC Converter," Applied Power Electronics Conference (APEC), pp. 1510-1516. 2014.

[14] Ralph Locher, "Introduction to Power MOSFETs and their Applications," Fairchild Semiconductor Application Note 558, 1998.

[15] "Thermal Design Basics," Analog devices MT-093 Tutorial, 2009.

[16] "DirectFET® Technology Thermal Model and Rating Calculator," International Rectifier Application Note AN-1059, 2010.

[17] Infineon CanPAK S-size BSF134N10NJ3 G datasheet, www.infineon.com

[18] Infineon CanPAK M-size BSB012N03LX3 G datasheet, www.infineon.com

[19] Infineon S3O8 BSZ075N08NS5 datasheet, www.infineon.com

[20] Alpha and Omega AON7280 datasheet, www.aosmd.com.

[21] Texas Instruments S3O8 Dual Cool SON 3.3x3.3mm CSD16323Q3C datasheet, www.TI.com

[22] Super SO8 BSC010N04LS datasheet, www.infineon.com

[23] Super SO8 BSC026N08NS5 datasheet, www.infineon.com

[24] Texas Instruments Super SO8 Dual Cool SON 5x6mm CSD16321Q5C datasheet, www.TI.com

[25] Efficient Power Conversion EPC2001C datasheet, www.epc-co.com

[26] Efficient Power Conversion EPC2021 datasheet, www.epc-co.com

[27] D. Reusch, D. Gilham, Y. Su, and F. C. Lee, "Gallium Nitride Based 3D Integrated Non-Isolated Point of Load Module," Applied Power Electronics Conference and Exposition (APEC), pp. 38 –45, 2012.

[28] D. Reusch, J. Strydom, and J. Glaser "Improving High Frequency DC-DC Converter Performance with Monolithic Half Bridge GaN ICs," Energy Conversion Congress and Exposition (ECCE), 2015.

[29] E. O. Johnson, "Physical Limitations on Frequency and Power Parameters of Transistors," pp. 163-177, 1965.

[30] R. W. Keyes, "Figure of Merit for Semiconductors for High-Speed Switches," Proc. IEEE, p. 225, 1972.

[31] B. J. Baliga, "Semiconductors for High-Voltage, Vertical Channel FET's," J. Appl. Phy. vol. 53, pp. 1759-1764, 1982.

[32] B. J. Baliga, "Power Semiconductor Device Figure-of-Merit for High Frequency Applications," IEEE Electron Device Letters, pp. 455–457, 1989.

[33] I. J. Kim, S. Matsumoto, T. Sakai, and T. Yachi, "New Power Device Figure-of-Merit for High Frequency Applications," in Proc. Int. Symp. Power Semiconductor Devices ICs, Yokohama, Japan, pp. 309–314, 1995.

[34] A. Q. Huang, "New Unipolar Switching Power Device Figures of Merit," IEEE Electron Device Lett., vol. 25, pp. 298-301, 2004.

[35] Y. Ying, "Device Selection Criteria----Based on Loss Modeling and Figure of Merit," Thesis of Master of Science in Electrical Engineering of Virginia Tech, 2008.

Over 300kHz GaN Device Based Resonant Bi-directional DCDC Converter With Integrated Magnetics

Gang Liu[1,2], Dan Li[1], Yungtaek Jang[3], Jianqiu Zhang[1]

[1]Electrical Engineering, Fudan University, Shanghai 200433, People's Republic of China
[2]Delta Power Electronics (Shanghai) Co. Ltd, 201209, People's Republic of China
[3]Power Electronics Laboratory, Delta Products Corporation, 5101 Davis Drive, Research Triangle Park, NC, USA

Abstract—**This paper presents a study of the isolated DC/DC stage of Bi-directional on-board battery charger module (OBCM) for electric vehicle (EV) and plug-in hybrid electric vehicle (PHEV) application. The conventional series resonant converter and secondary-side-switch delay-time control is employed. The delay-time control is used to assist the conventional variable-switching-frequency control of primary switches to reduce the switching-frequency range. By utilizing gallium nitride (GaN) switches, the operating frequency is increased to over 300kHz and the sizes of the passive components are reduced. The new integrated magnetic transformer including two primary windings, one secondary winding, bobbin, and a magnetic core, are designed and implemented. With the magnetic integration, the converter with high power density is achieved while the number of its components and cost are reduced.**

A standard 3.3kW prototype DC/DC converter designed for a high-frequency galvanic isolation of 400Vdc bus is developed with 250V~430V output range. The prototype circuit exhibits the maximum full-load efficiency of over 97% with a switching frequency variation from 300 kHz to 600 kHz over the entire output-voltage range.

I. INTRODUCTION

More and more plug-in hybrid electric vehicles (PHEV) and pure electric vehicles (EV) are being commercialized to reduce fossil-fuel consumption and green house gas emission. Usually a single-phase 3.5kW or 7kW OBCM is installed to charge the EV or PHEV from the power grid. On the other hand, EV and PHEV can be used as energy source, and energy storage in a smart grid is integrated with renewable energy source. Smart grid operation considers smart charging and discharging EV and PHEV [1]. An isolated bidirectional DC/DC converter with high power density and efficiency has attracted increasing attention. Plenty of previous researches focus on the isolated bidirectional DC/DC converters [2]. Generally, resonant converters with variable switching-frequency control are extensively used in state-of-the-art power supplies that offer the highest power densities and efficiencies [3]-[7]. As far as the resonant converters are considered, it has been proved that the LLC resonant converter is good for a unidirectional DC/DC application. Unfortunately, the CLLC resonant structure is requisite for a bidirectional operation converter with big volume, complex

control and high cost [8]-[10]. The conventional series resonant converter (SRC) can only work on a buck mode, which is unsuitable for a wide voltage range OBCM application. A new control technique that significantly improves the performance of a series resonant converter that operates with a wide input-voltage range and/or a wide output-voltage range by substantially reducing their switching-frequency range has been introduced [11]. The boost voltage mode is introduced for conventional SRC by controlling the output voltage with a combination of variable-frequency feedback control and the open-loop delay-time control. Variable-frequency control is used to control the primary switches of the series resonant converter, while delay-time control is used to control secondary-side rectifier switches provided in place of the diode rectifiers. By using the GaN devices as the switches, the proposed SRC can operate at high resonant frequency over 300kHz and keep the good efficiency and thermal performance. As it has high operational frequency, the magnetic components are also optimized to smaller integrated ones. The new integrated magnetic transformer including two primary windings, one secondary winding, bobbin, and a magnetic core, is designed and implemented.

In this paper, the proposed SRC converter is analyzed in section II, the integrated magnetic is designed in section III. Section IV provides complete circuit design, component selections, and extensive experimental results. The performance of the proposed dc-dc converter was verified on a 3.3-kW prototype operating with a 400-V input and an output that varies between 250 V and 430 V.

II. ANALYSES OF PROPOSED CONVERTER

A bidirectional series-resonant converter is proposed in Fig.1. As illustrated in Fig.1 (a), output voltage regulation is achieved using a combination of variable-frequency feedback control and open-loop delay-time control. During charging mode, variable-frequency control is applied to primary switches S_{P1}-S_{P4}, and delay-time control is applied to secondary-side switches S_{S2} and S_{S3}. Figure 1(b) shows ideal gate waveforms of primary switches S_{P1}-S_{P4}, drain and gate waveforms of secondary switches S_{S2} and S_{S3}, and resonant inductor current i_{LR} in the resonant converter of Fig.1(a). As shown in Fig. 1(b) charging mode, switches in the same leg of the primary full-bridge operate in a complementary fashion with a small dead time between their commutations to achieve zero-voltage switching (ZVS).The delay-time control is implemented by delaying the turn-off of switches S_{S2} and S_{S3} with respect to corresponding zero crossings of resonant current i_{LR} so that both switches S_{S2} and S_{S3} conduct during delay-time intervals $[T_0\text{-}T_1]$ and short the secondary of transformer TR. Because of the shorted secondary of the transformer, the voltage across resonant tank $C_R\text{-}L_R$ during delay-time interval $[T_0\text{-}T_1]$ is V_{IN} instead of $V_{IN} - nV_O$ which is the case with no time-delay control. Therefore, with the delay-time control, a higher voltage is applied across the resonant inductor and, consequently, a higher amount of energy is stored in resonant inductor L_R. Therefore, at the same input voltage and switching frequency, secondary-side delay-time control provides a higher output voltage compared to the conventional frequency control. This boost characteristic makes optimizing circuit performance possible by selecting: (i) a higher turns ratio in the transformer to reduce primary conduction losses and (ii) a higher magnetizing inductance to reduce circulating (i.e., magnetizing) current loss. In typical applications, delay-time control is used over the range between the middle and high output voltages so that the switching frequency range is reduced.

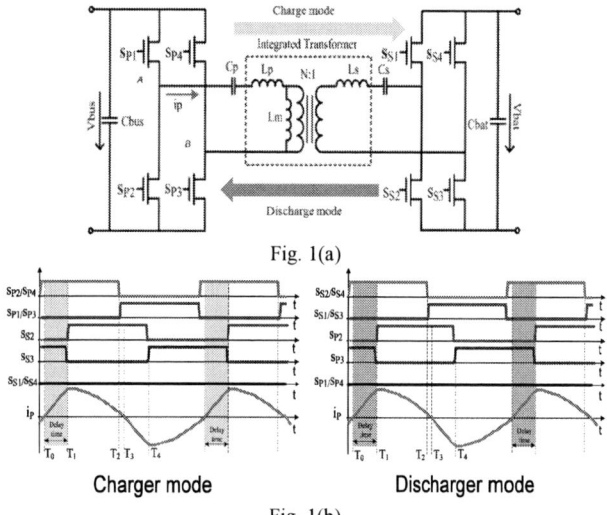

Fig. 1(a)

Fig. 1(b)

Fig. 1 Proposed series resonant converter with additional secondary switch control: (a) circuit diagram, (b) control waveforms.

To derive dc-conversion ratio $M=nV_O/V_{IN}$ of the series-resonant converter with proposed delay-time control, state-plane analysis is utilized and the following relationship is derived [11]:

$$\left(M_{CR_{PK}} + 1\right)^2 + M^2 + M\left(M_{CR_{PK}} + 1\right)(1 - \cos\alpha) +$$
$$\left(M_{CR_{PK}} + 1 + M\right) \times$$
$$\sqrt{\left(M_{CR_{PK}} + 1\right)^2 + M^2 - 2M\left(M_{CR_{PK}} + 1\right)\cos\alpha} \times$$
$$\cos\left(\lambda - \alpha + \beta_P\right) - 2 = 0, \quad (1)$$

where normalized peak resonant capacitor voltage $M_{CR\_PK} = \frac{V_{CR\_PK}}{V_{IN}} = \frac{MQ\lambda + 1 - \cos\alpha}{1 + \cos\alpha}$, $\beta_P = \tan^{-1}\frac{R1\sin\alpha}{R1\cos\alpha - M}$, $\lambda = \alpha + \beta + \gamma = \frac{\omega_0 T_S}{2} = \frac{\pi}{F}$, $R1 = M_{CR\_PK} + 1$, and quality factor $Q = \frac{Z_0}{n^2 R_{LOAD}}$.

For a given Q and duty ratio $D = \alpha/(2\lambda) = T_D/T_S$, dc conversion ratio $M = nV_O/V_{IN}$ can be numerically calculated and plotted as function of normalized frequency $F = f_s/f_O$, as shown in Fig. 2.When duty cycle D is zero ($\alpha = 0°$, $T_D = 0$), the converter characteristic is the same as that of a conventional series resonant converter. As duty cycle D, i.e., delay time T_D increases, dc gain M increases and exhibits a boost characteristic. So during charging mode and discharging mode, different delay time should be selected for different battery voltage, so the frequency change can be reduced and controlled. A microcontroller-based implementation is preferred since delay time T_D that depends on input or output voltage can be easily programmed.

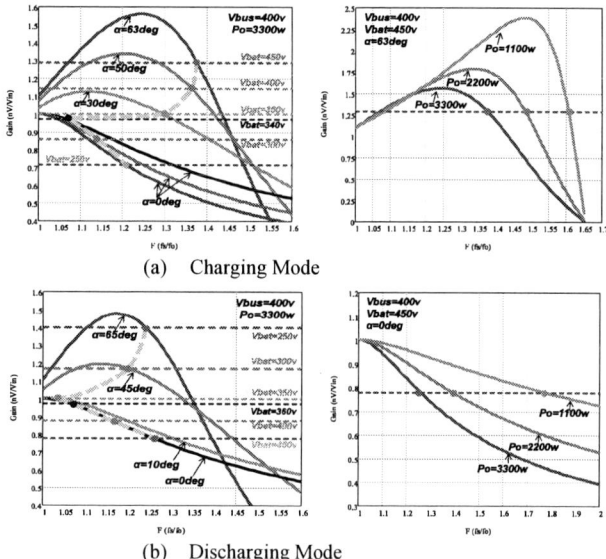

(a) Charging Mode

(b) Discharging Mode

Fig. 2 Input-to-output voltage gain of proposed converter; It should be noted that converter operates as series resonant converter when delay time $T_{D\text{-}N}$ is set to be zero.

Figure 3 shows detailed control waveforms of the proposed series resonant converter with delay-time control shown in Fig. 1(a). During the charging implementation as shown in Fig. 3, primary switches S_{P2} and S_{P4} and secondary switch S_{S3} body diode turn on together at $t=T_0$ and primary switches S_{P1} and S_{P3} and secondary switch S_{S2} body diode turn on together at $t=T_5$ because they use the same internal clock of the microcontroller. To achieve proper delay-time control, the phase shift between primary switch gate transition at $t=T_1$ and zero crossing of resonant inductor current i_{LR} at $t=T_0$ should be detected for gating of switch S_{S3}. The phase shift can be measured by sensing the zero crossing of resonant current i_{LR} by using a current transformer. However, the magnitude of resonant current i_{LR} is too small at light load to detect the zero crossings reliably, the drain voltage waveforms of secondary-side switches S_{S2} and S_{S3} are measured to generate equivalent pulses that represent phase shift. It should be noted that the phase shift duration changes over the input voltage, output voltage, and load ranges. However, it can be assumed that the phase shift over a single switching cycle is near constant since the voltages and load change is much slower than the switching frequency. For gating of switch S_{S2}, the phase shift between primary switch gate transition at $t=T_3$ and zero crossing of resonant inductor current i_{LR} at $t=T_4$ should be detected. The detected duration of phase shift is utilized at the next switching cycle to determine the gate pulse width of switch S_{S2}.

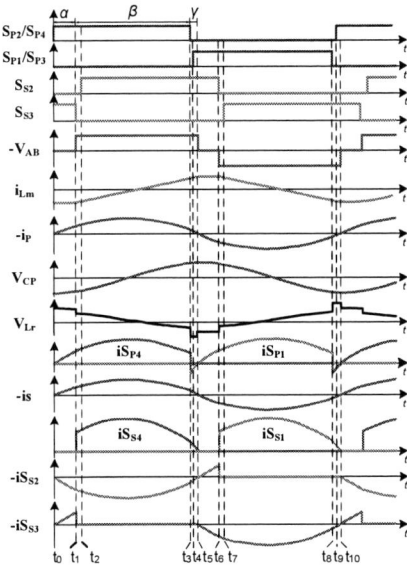

Fig. 3   Detailed control waveforms of proposed series resonant converter shown in Fig. 1.

III. INTEGRTED MAGNETICS DESIGN

The proposed integrated magnetic transformer is illustrated in Fig.4. Its primary and secondary windings have symmetric structure. The core type will be "EE" or "PQ" or others with a symmetric structure. To derive the leakage inductance of the integrated transformer, the size has been represented by letters and marked on them. D and E are related with the core size. X and Z represent the half thickness of the primary and secondary windings, respectively. Y is the space between primary and secondary windings.

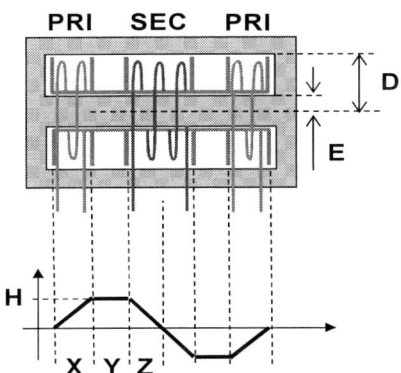

Fig. 4 Structure of integrated magnetic transformer

According to the distribution principle of the electromagnetic field, all electric field intensity of X, Y, and Z regions can be expressed as follows:

$$E_X = \int_0^X \frac{1}{2} u_0 \cdot H_X(x)^2 \cdot A_e dx \qquad (2)$$

$$E_Y = \frac{1}{2} u_0 \cdot H_Y^2 \cdot Y \cdot A_e \qquad (3)$$

$$E_Z = \int_0^z \frac{1}{2} u_0 \cdot H_Z(z)^2 \cdot A_e dz \qquad (4)$$

Where $H_X(x) = H \cdot x/X$, $H_Y = H$, $H_Z(z) = H - H \cdot z/Z$, $H = I \cdot N_p / 2 \cdot G$, $G = (D - E)/2$, $A_e = \pi(D^2 - E^2)/4$, Np is the primary winding turns .

$$E = 2(E_X + E_Y + E_Z) \qquad (5)$$

According to the Faraday's law of induction:

$$E = \frac{1}{2} \cdot L_k \cdot I^2 \qquad (6)$$

Combining (5) and (6), the leak inductance L_k can be expressed as:

$$L_k = \frac{N_p^2}{2G^2} \cdot A_e \cdot u_0 \cdot (\frac{X}{3} + Y + \frac{Z}{3}) . \qquad (7)$$

From (7), it can be found that the value of leakage inductance depends on X, Y, Z, and N_P if the core size is determined. The value of L_k is the leakage inductance which includes both the primary and secondary sides. While the integrated transformer is designed, the winding loss and the core loss are also two important factors. Thus the parameter determination process also needs to be traded off.

Fig. 5 shows that the leakage inductance increases with the increasing of N_P and Y, so the resonant inductance can be determined easily.

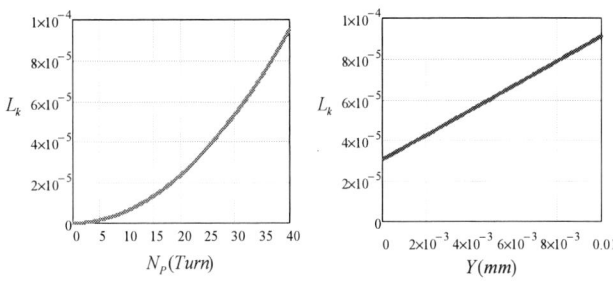

Fig. 5 Integrated transformer leakage inductance (a) L_K vs N_P (b) L_K vs Y

IV. EXPERIMENTAL RESULTS

The performance of the proposed converter shown in Fig. 1 was evaluated on a 3.3-kW prototype circuit that was designed to operate from a 400 V input and deliver power over 250-430V output voltage range. Fig.6 (a) shows a sample of the integrated transformer, and Fig6 (b) shows the side view of the experimental prototype circuit and Fig.7 shows circuit diagram with details of employed power components. All primary and secondary switches are GaN devices TPH3295WS (R_{DS-ON} = 63 mΩ, V_{DS} = 600 V, $I_{D100°C}$ = 24 A) from Transphorm. Table I shows the parameter selection of the resonant tank.

Table I. Resnant Tank Parameters

Parameters	Value
Cp	20nF
Lp	15uH
Lm	1.6mH
Ls	11.5uH
Cs	26nF
N	24:21

During the charging mode, secondary switches S_{S1} and S_{S4} are turned off during the operation so that their body diodes are utilized as output rectifiers, and S_{P1} and S_{P4} do the same way for discharging mode. It should be also noted that the turns ratio (24:21) of transformer TR is chosen to make input-to-output control characteristic M be 1 when the output voltage is approximately 350 V during charging. As a result, the delay time of the secondary switching is set to zero when the converter operates from 250V to approximately 350V battery voltage so that the proposed converter operates as a series resonant converter. However, when the input voltage increases above 350V, the controller starts to increase the delay time monotonically to provide the boost characteristic in order to maintain the output voltage regulation with the selected turns ratio of transformer TR. Table II shows the design parameters of the integrated magnetic

Table II. The Integrated Transformer Parameters

Parameters	Value
D	44mm
E	20mm
X	8mm
Y	5mm
Z	7mm

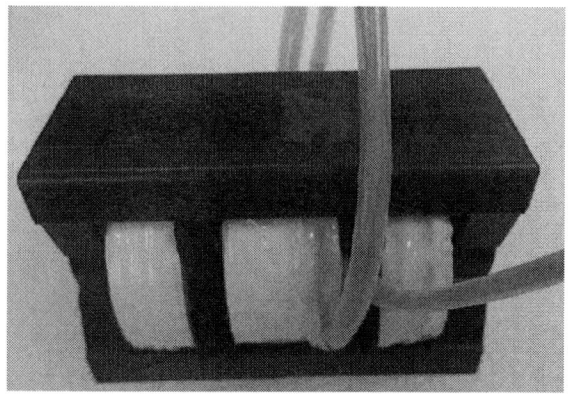

Fig6 (a)

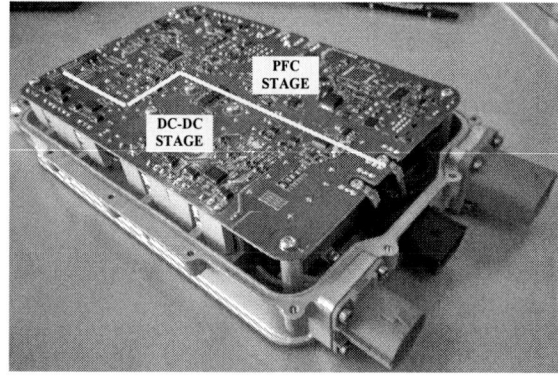

Fig6 (b)

Fig. 6 Experimental prototype circuit (a) integrated transformer (b) side view of the prototype

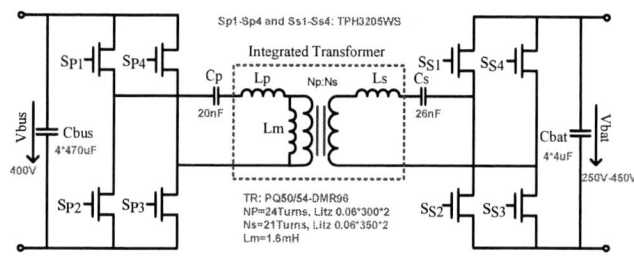

Fig. 7 Prototype circuit diagram with details of employed power components. It should be noted that all primary and secondary switches are GaN devices

Figure 8 shows the measured waveforms of gate and drain voltages of S_{P2} and gate waveforms of S_{S2}, the resonant current i_p, of the experimental circuit when it charges the full power at 430 V, 350 V, and 250 V battery voltages. The waveforms show ZVS of S_{P2}. Although Fig. 8 only shows waveforms of switch S_{P2}, the waveforms of all other primary switches are similar to that of switch S_{P2} and achieve ZVS as well. During the voltage lower than 350V, the converter goes into conventional SRC, it did not have the T_D for the secondary side, so only primary S_{P2} waveforms are shown.

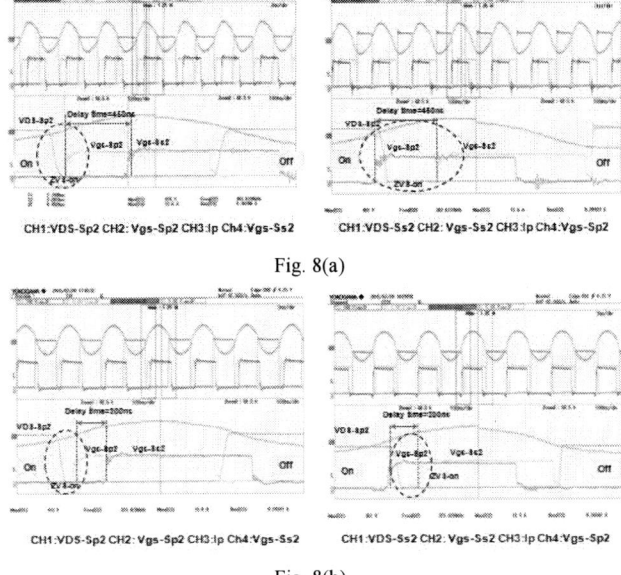

Fig. 8(a)

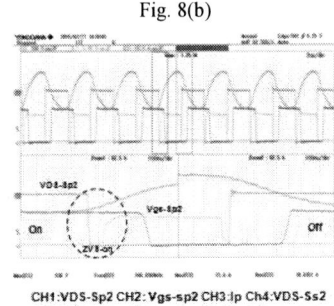

Fig. 8(b)

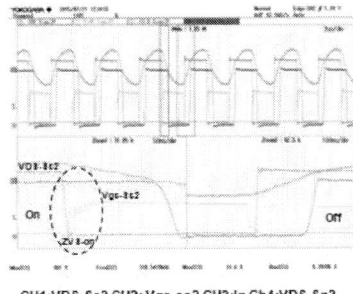

Fig. 8(c)

Fig. 8. Measured waveforms of charging mode for battery voltages :(a) 430 V ;(b) 350 V; and (c) 250 V.

Figure 9 shows the measured waveforms of gate and drain voltages of S_{S2} and gate waveforms of S_{P2} when it discharges full power at 430V, 350V, and 250V battery voltages. As shown in Fig. 9, all the secondary switches operate with ZVS. Moreover, delay time T_D can be observed from the measured drain voltages of switches S_{S2} and S_{P2}. Delay time T_D is the period when both drain voltages of switches S_{S2} and S_{S3} (or S_{P2} and S_{P3}) are zero, i.e., both switches conduct and the winding of TR is short. During the voltage higher than 350V, the converter goes into conventional SRC, it did not have the T_D for the secondary side, so only primary S_{S2} waveforms are shown.

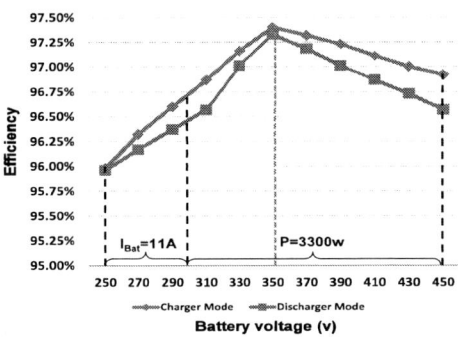

Fig. 9(a)

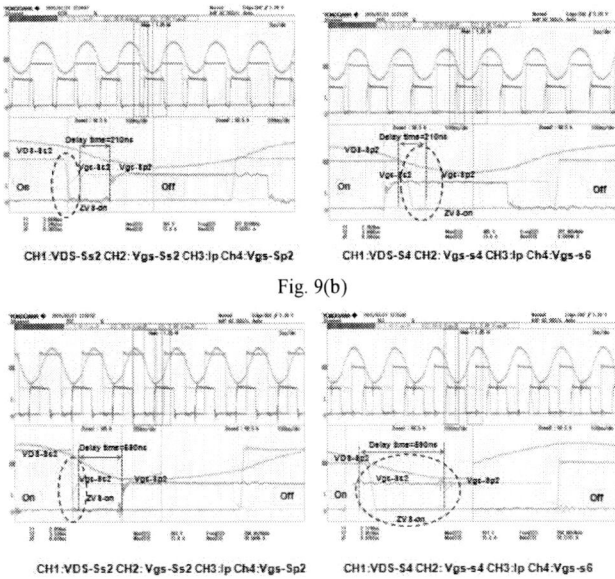

Fig. 9(b)

Fig. 9(c)

Fig. 9 Measured waveforms of discharging mode for battery voltages: (a) 430 V; (b) 350 V; and (c) 250 V.

Figure 10 shows measured efficiency of the prototype converter when it operates over the entire output-voltage range. It should be noted that the converter exhibits the best full-load efficiency when the battery voltage is around 350 V in which the battery for EV or PHEV most frequently operates. Specifically, the converter exhibits the maximum full-load efficiency of 97.2% at 350-V output.

Fig. 10 Measured efficiencies of the experimental prototype as functions of output voltage.

Figure 11 shows the measured full-load switching frequency of the experimental prototype as a function of battery voltage. The measured full-load switching frequencies are in a 300-kHz to 380-kHz range over the entire battery-voltage range, but it will go to higher frequency up to 600-kHz for light load. The converter will goes in to burst mode after the operating frequency is over 600-kHz for better light load efficiency. As seen from Fig.12, the delay-time control is activated around 350V output, which makes converter operates in boost mode and regulates output voltage. The converter has delay-time control from 350V to 430V for charging mode, and the conventional SRC control from 250V to 350V; and it has delay-time control from 250V to 350V for discharging mode, and conventional SRC control from 350V to 430V. With this control strategy, it operates in a narrow frequency range for wide battery voltages, and implements this smaller size integrated magnetic components.

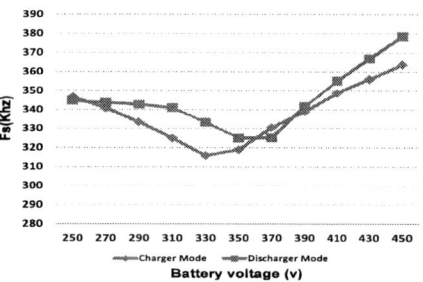

Fig. 11 Measured full load switching frequency of experimental prototype as function of output voltage.

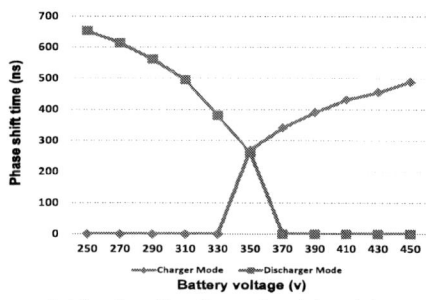

Fig. 12 Measured delay-time T_D of experimental prototype as function of output voltage.

V. SUMMARY

In this paper, an isolated bidirectional series-resonant dc-dc converter that is controlled by a combination of variable-frequency and secondary-side-switch delay-time control has been introduced. The proposed converter is designed as the output stage of an OBCM that operates with a wide battery-voltage range. The delay-time control which is implemented by the modulation of secondary-side switches is used to assist the conventional variable-switching-frequency control of primary switches. The performance evaluation of the proposed series-resonant converter with delay-time control was done on a 3.3-kW prototype delivering energy from 400V bus, which was the output of the PFC/Inverter front end, to a battery operating with voltage range between 250V and 430V. The prototype circuit exhibits the maximum full-load efficiency of 97.2% with a full-load switching-frequency variation from 300-kHz to 380-kHz over the entire output-voltage range.

REFERENCES

[1] Saber, A.Y. and Venayagamoorthy, G.K., "Plug-in Vehicles and Renewable Energy Sources for Cost and Emission Reductions ," IEEE Transactions On Industrial Electronics, vol. 30, no. 6, pp. 1229 – 1238, July 2012

[2] Biao Zhao ; Qiang Song ; Wenhua Liu ; Yandong Sun "Overview of Dual-Active-Bridge Isolated Bidirectional DC–DC Converter for High-Frequency-Link Power-Conversion System" IEEE Trans. Power Electron., vol. 29, no. 8,pp. 4091–4106, Aug. 2014.

[3] B. Yang, R. Chen, and F.C. Lee, "Integrated magnetics for LLC Resonant Converter," in IEEE Applied Power Electronics Conf. Rec. 2002, pp. 346-351.

[4] B. Lu, W. Liu, Y. Liang, F.C. Lee, and J.D. Van Wyk, "Optimal design methodology for LLC resonant converter," in IEEE Applied Power Electronics Conf. Rec. 2006, pp. 533-538.

[5] R. Beiranvand, B. Rashidian, M.R. Zolghadri, S.M.H.Alavi, "A design procedure for optimizing the LLC resonant converter as a wide output range voltage source," IEEE Transactions on Power Electronics, vol. 27, No. 8, pp. 3749-3763, August2012.

[6] F. Musavi, M. Cracium, D.S. Guatam, W. Eberle, and W.G. Dunford, "An LLC resonant DC-DC converter for wide output voltage range battery charging applications," IEEE Transactions on Power Electronics, vol. 28, No. 12, pp. 5437-5445, December2013.

[7] J. Deng, S. Li, S. Hu, C.C. Mi, and R. Ma, "Design Methodology of LLC Resonant Converters for Electric Vehicle Battery Chargers," IEEE Transactions on Vehicular Technology, vol. 63, No. 4, pp. 1581-1592, May2014.

[8] Amit Kumar Jain and Rajapandian Ayyanar, "PWM control of dual active bridge: comprehensive analysis and experimental verification," IEEE Tran. on Power Electronics, vol.26, no.4, pp. 1215-1227, Apr. 2011

[9] Jung, J-.H., Kim, H-.S., Ryu, M-.H. , Baek, J-.W.," Design Methodology of Bidirectional CLLC Resonant Converter for High-Frequency Isolation of DC Distribution Systems", Power Electronics, IEEE Transactions on (Volume:28 , Issue: 4),PP.1741-1755

[10] Wei Chen, Ping Rong , Zhengyu Lu, "Snubberless Bidirectional DC–DC Converter With New CLLC Resonant Tank Featuring Minimized Switching Loss," Industrial Electronics, IEEE Transactions on (Volume:57 , Issue: 9).2010, PP.3075-3086

[11] Y. Jang, M.M. Jovanovic, J.M. Ruiz, and G. Liu, "Series-Resonant Converter with Reduced-Frequency-Range Control," in IEEE Applied Power Electronics Conf. Rec. 2015, pp. 1453-1460.

Effective Control & Software Techniques for High Efficiency GaN FET based flexible Electrical Power System for Cube-Satellites

Ashish Shrivastav
Department of Electrical and Computer Engineering
North Carolina State University
Raleigh, North Carolina, USA
ashriva@ncsu.edu

Shikhar Singh ‡, Anirudha Mahajan*,
Dr. Subhashish Bhattacharya*
*Department of Electrical and Computer Engineering
North Carolina State University
Raleigh, North Carolina, USA.
‡ Systems and Technology Group
IBM, Austin, Texas, USA.
{sbhatta4, aamahaja}@ncsu.edu, shiksing@us.ibm.com

Abstract— This paper investigates an intelligent and configurable electric power system (EPS) for CubeSats and small satellites built using Silicon and GaN FETs at high switching frequency. The EPS is the power source of CubeSat which harnesses power from solar panels and includes battery charging and multi-domain voltage output regulation within the CubeSat. The electrical power system of a Cube-Satellite is developed and is used as a test bed for implementation of the various control algorithms necessary for such photovoltaic – battery based power management systems. The introduction of a digital controller to such power systems provides the system added flexibility and intelligence but also introduces controller design challenges with many control loops running simultaneously and being controlled by a single controller. Efficiency of power converter is increased using pulse frequency modulation (PFM) at light load. For moderate load, higher efficiency can be achieved with comparatively lower Rds-ON based GaN FETs. In this paper we have implemented a control scheme that takes into account various parameters like load transients, control loop update rate and sampling intervals that have an effect on the system performance. Power loss analysis is done for EPS using silicon devices. Simulation and experimental results are presented in this paper using both Silicon and 100V EPC Gallium Nitride Mosfets in PFM and PWM mode for EPS.

Keywords—CubeSat, EPS, GaN, PFM

I. INTRODUCTION

CubeSat is a generic term for a miniature satellite which is designed using commercial off-the-shelf components for development of its electronics framework [1]. A CubeSat EPS should be able to power multiple peripherals with varying power requirements by regulating multiple voltage rails with transients limited to the specifications. This paper discusses the design and development of an effective, scalable electrical power system with efficient control schemes of a CubeSat EPS. This involves addressing issues like solar power harnessing, battery charging mechanism, DC-DC converter design and control, load characterization and code profiling. Growing popularity of wide band gap devices, such as Gallium Nitride (GaN), has led to the development of high

performance power conversion systems. GaN devices can be switched at higher switching frequencies than Silicon (Si) designs and thus size of the filter is reduced [2] and power density of the converter is increased. This also enables higher system efficiency due to reduction in conduction losses with reduced Rds On resistance as compared to silicon devices. These GaN devices with efficient and faster control schemes can be utilized effectively towards building a very powerful power system [3].

Previously, exhaustive work was carried out on EPS prototype using Silicon mosfets at switching frequency of 150 kHz. Figure 1 shows the EPS architecture which has a flexible battery charging module (FBCM) at the input which is fed by solar power [4].

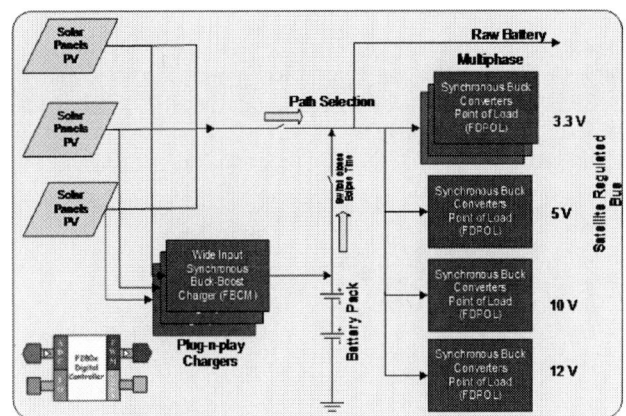

Figure 1: EPS Architecture [5]

FBCM is a buck boost converter which charges lithium ion batteries. The FBCMs perform constant current-constant voltage (CC-CV) charging of the battery and also maximum power point tracking (MPPT) control. Lithium batteries power flexible digital point of load converters (FDPOL) which produce DC voltage rails from 3.3V to 12 V. FDPOLs can be buck or boost DC-DC converters. FDPOLs are operated in automatic pulse frequency/pulse width modulation

978-1-4673-9551-9/16 $31.00 © 2016 IEEE

(PFM/PWM) control for higher efficiency. At lighter loads it is controlled by PFM and PWM control at medium loads. The architecture has a path selection circuitry which connects the output of one of the battery packs to the FDPOL's input. This architecture is controlled effectively by digital controller. Compensator modeling and design are done in software-Code Composer Studio (CCS) for TMS320f28335 DSP [5]. The present implementation for maximum power point tracking of solar power uses perturb and observe (P&O) algorithm because of its simplicity and satisfactory performance for low to medium power systems. In charging mode, the FBCM operates as constant current-constant voltage DC-DC converter with initial trickle charging. To ensure proper functioning of switching circuitry, battery charging, MPPT control at very high switching frequency of GaN Mosfets, it is important to design an optimum state machine model. Different Cubesat EPS were developed using both Silicon and GaN FETs respectively at optimum switching frequency to achieve the maximum efficiency. Figure 2 shows the block diagram of the proposed topology having solar panels to power FBCMs converters. Output of FBCMs are interfaced to path selection circuitry which connects battery pack to FBCMs and FDPOLs.

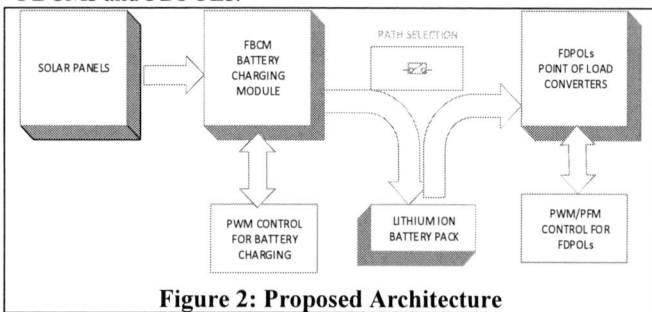

Figure 2: Proposed Architecture

A current sensor card [6] for the EPS is also developed which enables the use of current control techniques. Figure 3 shows the EPS using GaN mosfets. It has one boost FBCM at the input and two buck FDPOLs. The FBCM is programmed to output 7.2V and the FDPOLs are programmed at 3.3V and 5V

Figure 3: EPS using GaN Mosfets

respectively.

Figure 4 shows the Silicon based EPSs which is highly optimized and compact measuring 5 x 4 inches. The silicon

mosfets used are DMN4027S, 40V, 7A, Rds ON 27mΩ, Qg 12.9nC and TPS28226 gate driver is used for gate drive. The FBCM and FDPOL output DC voltages are reconfigurable which adds to the scalability of the system. TI's TMS320f28335 DSP is used for controlling the PWM and closed loop operation of EPS.

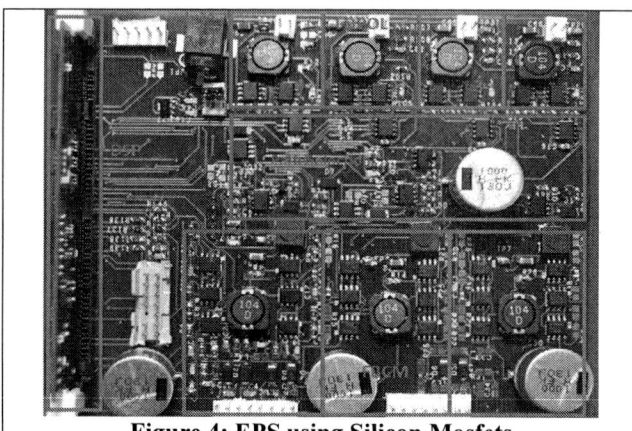

Figure 4: EPS using Silicon Mosfets

II. POWER LOSS ANALYSIS OF EPS -SILICON MOSFETS

The primary focus of this research has been the design of intelligent and efficient EPSs for the CubeSat. To design an efficient converter it is important to analyze the power loss. A single Buck FDPOL of Silicon based EPS is analyzed in detail to benchmark the results with reference to GaN based EPS. Mosfet takes finite time to turn on and off. During the turnon and turnoff transitions, due to the clamping effects of the low side device, the HS device is affected by both high current and high voltage at the same time as they overlap for a very short time, which induces switching losses as seen in Figure 5 [6].

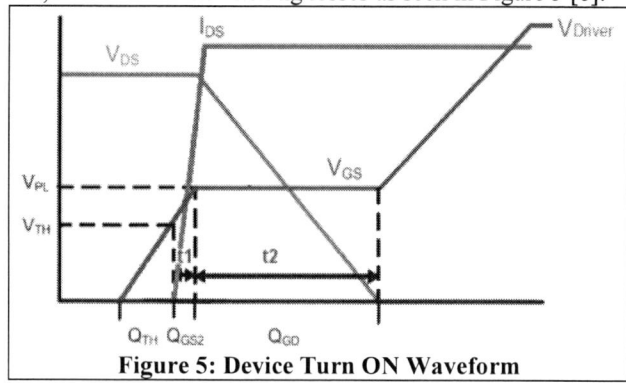

Figure 5: Device Turn ON Waveform

Figure 6 shows gate-source voltage V_{GS}, drain-source voltage V_{DS} and drain current waveforms during turn on for the buck FDPOL on Silicon based EPS. Figure 7 shows V_{GS}, drain current of high side and low side FETs during turn off.

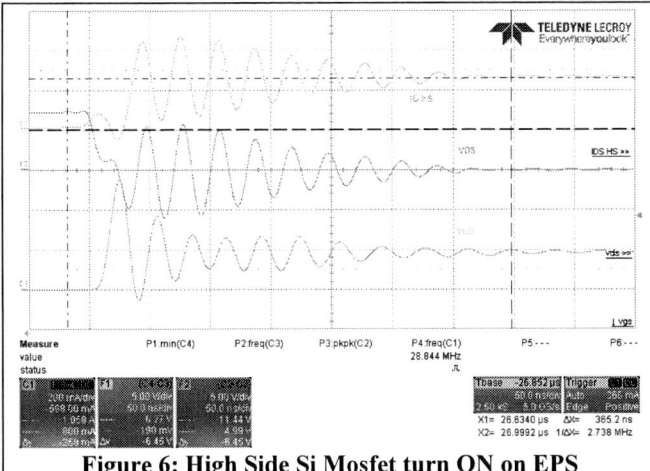

Figure 6: High Side Si Mosfet turn ON on EPS

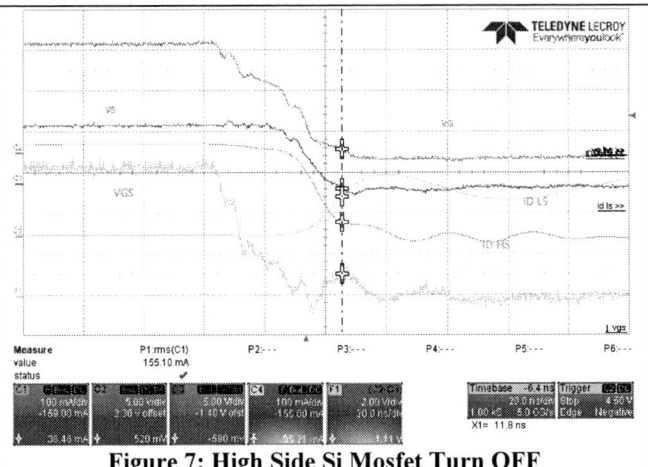

Figure 7: High Side Si Mosfet Turn OFF

In Figure 7 the trace F1(V$_G$-Vs, C2-C3, yellow trace) clearly shows a plateau of about 1.1V after the gate voltage has turned off, as can be seen in C2 (V$_G$). During this plateau, which lasts for about 40 ns, the drain currents of both the high-side (ID HS) and the low-side (LS) mosfets change states. The HS mosfet turns off and the drain current I$_D$ of HS (C1) returns to zero, while the LS mosfet turns on and the drain current I$_D$ of LS (C4) reaches the inductor current. When analyzed it can be seen that as HS gate voltage (V$_G$) goes to ground, drain current will fall to 0 A. If noticed, source voltage C3 (V$_S$) drops to zero in 20 ns but I$_D$ HS goes to 0 A in 40 ns. This rate of change of drain current through PCB tracks will generate a backward voltage as , $L\dfrac{dI_{drain}}{dt}$

This backward voltage pulls the source voltage in a negative direction with respect to the gate hence allowing current to continue to flow though the gate voltage is firmly held to ground [7]. Since the current falls linearly, the backward voltage is given by equation 1 , which creates a voltage plateau of fixed level with time until current completes its transition.

$$L\frac{dI_{drain}}{dt} = \text{constant} \qquad (1)$$

Because of this plateau on V$_{GS}$, the device channel is not completely enhanced and will sustain a definite V$_{DS}$ across it and hence causes losses during switching. Similar argument can be extended during turn ON as well.

Power Loss across HS switching is given by equation 2,

$$P_{HS-ON} = F_{sw} * VDS * I_{out} * \frac{Q_{sw}}{I_g}$$

$$I_g = \frac{V_{driver} - V_{plateau}}{R_g + R_{driver}} \qquad (2)$$

The other source of power loss for switches is conduction loss across HS and LS which is given as equation 3,

$$P_{cond} = I^2_{rms} * R_{ds(on)} \qquad (3)$$

In equation 3, RMS current for HS and LS switches are for their respective duty cycles as shown in Figure 10 and Figure 11.

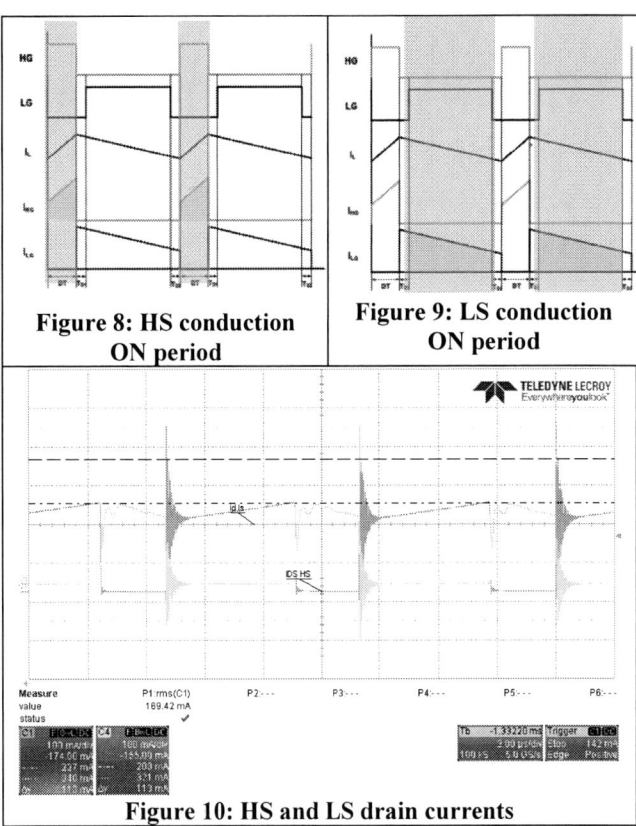

Figure 8: HS conduction ON period

Figure 9: LS conduction ON period

Figure 10: HS and LS drain currents

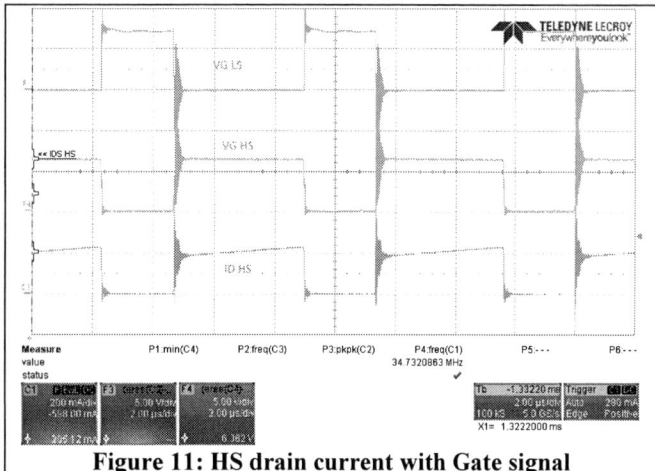

Figure 11: HS drain current with Gate signal

The body diode conduction during dead time also contributes power loss as given in equation 4,

$$P_{deadtime} = VDS * F_{sw} * (I_L * T_{D1} + I_L * T_{D2})$$ (4)

T_{D1}, T_{D2} are deadtime, which is 0.3usec for Fsw=150KHz Figure 12 shows the efficiency vs load profile of Buck FDPOL on Silicon EPS.

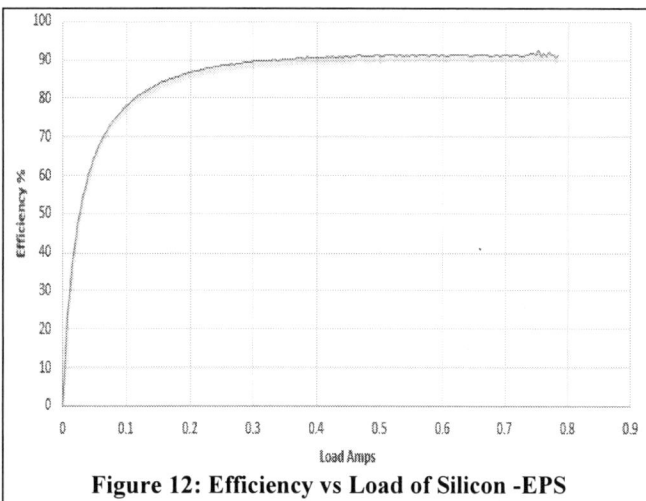

Figure 12: Efficiency vs Load of Silicon -EPS

Hence the total power loss at load 0.2 Amp using DMN4027S Mosfet and TPS28226 gate driver are as follows

Power Loss (W)	Measured (W)	Calculated (W)	Error (W)
	0.142312	0.1229	0.01941

III. GAN FET BASED EPS

As mentioned in the previous sections, the EPS performs power management in the CubeSat. It is responsible for harnessing power, battery management and power rail supply. A good EPS should be scalable, robust, compact and efficient [8]. Figure 13 shows the block diagram of the EPS which is a scaled down version comprising of one FBCM and two buck FDPOLs which produce DC voltages of 3.3V and 5V using

EPC 2016 GaN FET,VDS 100V, ID 11A,Rds ON 16mΩ [9]. Figure 14 shows the efficiency profile of GaN –EPS at different frequencies and based on the optimum efficiency, switching frequencies Fsw 150KHz and 600KHz are selected for further tests. The system uses the EPC2016 FETs. The devices are triggered using TI's LM5113 half bridge gate driver. The design is very compact measuring 4.5 x 2.5 inches. At Fsw 150KHz, efficiency profile is extended till load of 980mA as shown in Figure 15.

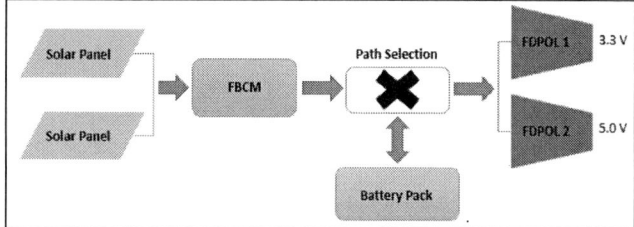

Figure 13 : EPS GaN Architecture

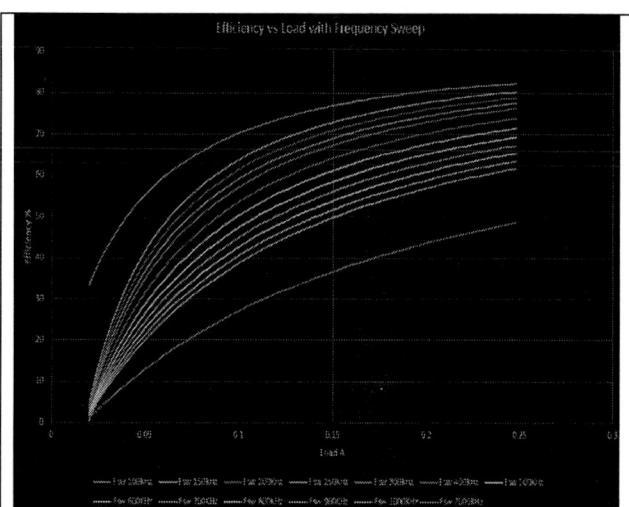

Figure 14 : Efficiency of GaN - EPS with frequency sweep

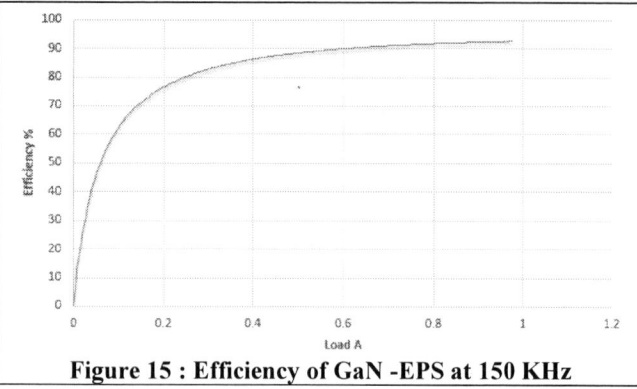

Figure 15 : Efficiency of GaN -EPS at 150 KHz

IV. SOFTWARE AND CONTROL

The software architecture becomes significant for systems like EPS which are scalable, programmable and intelligent. A system that runs simultaneous control loops needs a well-planned software layout so that all the critical functions are completed before the next sampling interval arrives. Issues like control loop update rate, sampling frequency and ADC conversion time need to be considered while laying out the software for the system. The EPS code has a main control loop that runs the compensator, the path switching circuitry and the battery charging code. Figure 16 shows the control flow diagram of the central control loop. The control loop update rate is 100µs which means that the duty cycle is updated every 100µs. A timer triggers the execution of the central control loop. The voltages and current are sensed synchronously every fourth duty period and all the samples before the next update period are averaged to remove sensing aberrations. The central control loop carries out three functions viz. Path Selection, FDPOL Control and FBCM Control. FDPOL Control – The closed loop control of the point of load converters is run as a part of the central control loop. These converters can be enabled or disabled depending on the application. FBCM Control –In this mode the FBCM can run in either battery charging modes or MPPT mode depending on the solar power level at the input.

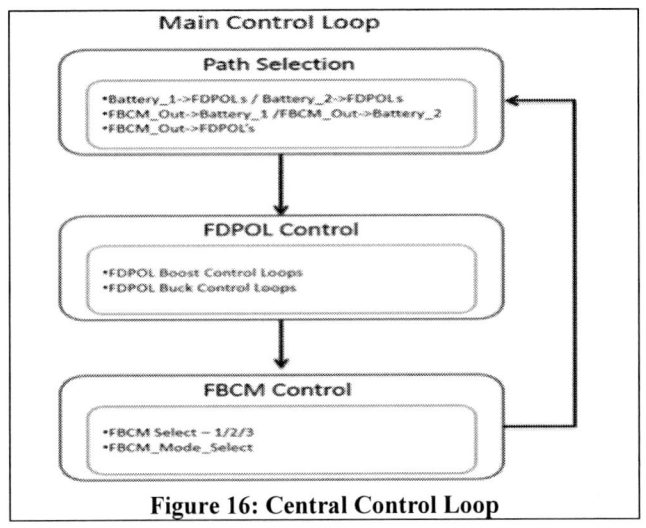

Figure 16: Central Control Loop

In the battery charging mode an FBCM can run in either CC or CV mode depending on the battery charge level pre-charged with trickle mode using PFM control. Figure 17 shows the sequence diagram for the software scheme. The PWM channel triggers ADC start of conversion every 4th positive edge. The voltage and current are recorded and averaged four times every duty cycle. The duty cycle is updated every 100µs by means of timer zero whose ISR triggers the control loop which is responsible for FBCM and FDPOL control. The FBCM can either run in battery mode, track mode or be in idle mode. In battery mode, the FBCM is carrying out CC/CV charging; in track mode, the FBCM runs maximum power point tracking algorithms and the output is unregulated. If the batteries are charged or that particular FBCM is disabled, it is in idle mode. The FBCM module can exist in three states. If the battery voltage is more than the configurable set voltage then it runs in idle mode else it goes

in battery charging mode. Based on the duty ratio and MPPT duty ratio the MPPT mode and battery mode are switched. In our prototype we have used a 7.2V, 2.2 Ah lithium battery. The MPPT algorithm ensures that the solar panels operate at or near the peak efficiency. However, while charging the battery, it might happen that the panels' power decreases and the MPPT algorithm needs to run again. As mentioned before, the control loop runs every 100µs during which the ADC is run four times. Each sweep of the ADC ISR which involves the calculation of current, voltages from ADC values and calculating average and other such functions take around 20µs to execute resulting in a total of around 80µs. The time taken by the control loop is 4.5µs. PI compensators [10] using direct realization techniques are designed for FBCM and FDPOL for optimum phase margin and stability. The analog transfer function is converted into discretized with a sampling period of 100 µs.

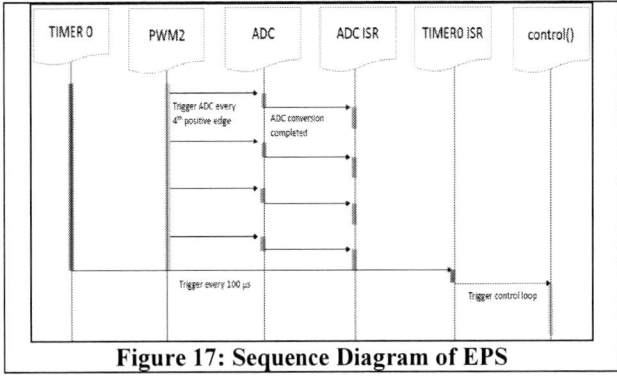

Figure 17: Sequence Diagram of EPS

V. DUAL CONTROL-PWM/PFM

PWM converters suffer from poor conversion efficiency at light-load. Pulse Frequency Modulation (PFM) operating mode used with PWM converters can tackle this problem. A PWM DC/DC uses fixed frequency to transfer energy from the input to the output. The clock frequency is fixed and the pulse width of each clock cycle is adjusted based on compensation loop. Hence, this approach is referred to as "pulse-width modulation," or "PWM." A PFM converter is an alternative DC/DC power-converter architecture that uses a variable frequency clock to drive the power switches and transfer energy from the input to the output. Since frequency is directly controlled to regulate the output voltage, this control scheme is referred to as "pulse-frequency modulation". As the load current decreases, mosfet conduction and passive components' losses decreases while mosfet gate drive and switching losses (dynamic losses) are constant for a PWM converter. Reducing the switching frequency under light load conditions can reduce dynamic losses. Thus, PFM can be used in FDPOLs and at light loads during trickle charging phase.

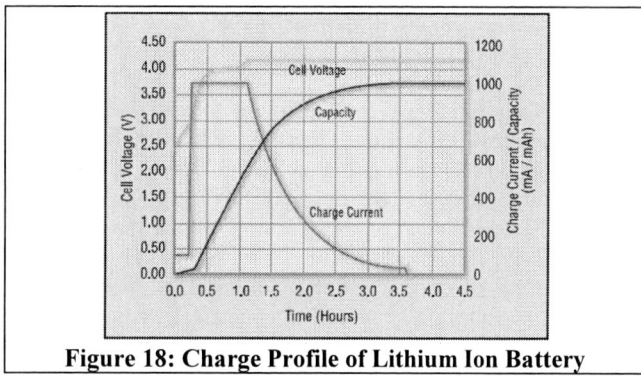

Figure 18: Charge Profile of Lithium Ion Battery

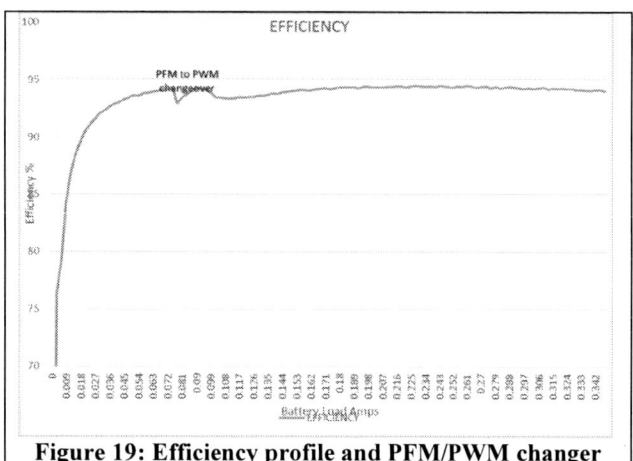

Figure 19: Efficiency profile and PFM/PWM changer

trace shows the current. The current reference is set at 0.45A and current measurement scale is 1Vto 0.34A.

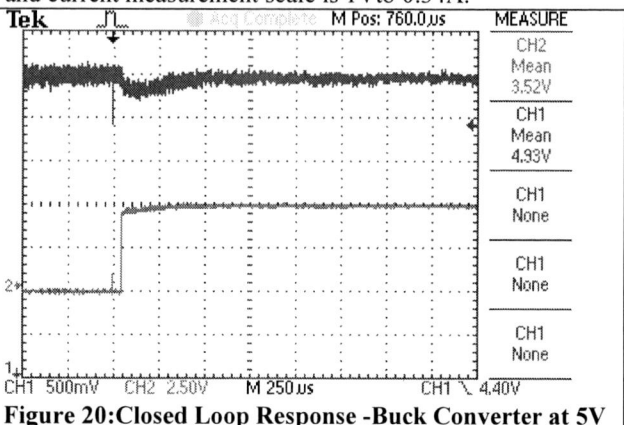

Figure 20:Closed Loop Response -Buck Converter at 5V

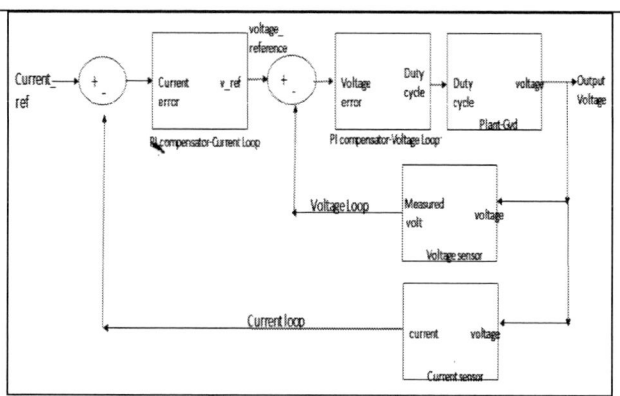

Figure 21: FBCM Constant Current Mode Control Loop Block Diagram

The charge cycle can be broken up into these stages: trickle charge, constant current charge and constant voltage charge as shown in Figure 18[11]. In stage one, trickle charge is employed to restore charge to deeply depleted cells. These are cells in which the cell voltage is below 7.2 V. During this stage, the cell is charged with a constant current rate of 0.1 C to 0.2 C maximum for a 2.2Ah battery used for tests. This is a light load condition for FBCM converter and thus operated in PFM. After the cell voltage has risen above the trickle threshold, the charge current is raised to perform constant current charging. The constant current charge should be in the 0.2 C to 1 C range. In the proposed architecture, FBCM can operate in either PFM or PWM mode depending on constant current amplitude. Simulated efficiency is shown in Figure 19. Closed loop response of Buck FDPOL in PWM mode can be seen in Figure 20. Output voltage is set at 5V, input is 7.2V, 2.2Ah Lithium ion battery. Figure 21 shows the control loop diagram of FBCM running in constant current mode. This scheme employs two control loops –outer current loop and inner voltage loop [12][13]. The voltage loop is similar to the voltage control loop employed in constant voltage control. The outer current loop measures the output current and compares it to the desired current reference. The error term is fed to a PI compensator which generates the voltage reference. The voltage loop runs around 10 times faster than the current loop. Figure 22 shows the FBCM running in constant current mode. The red trace depicts the battery voltage while the blue

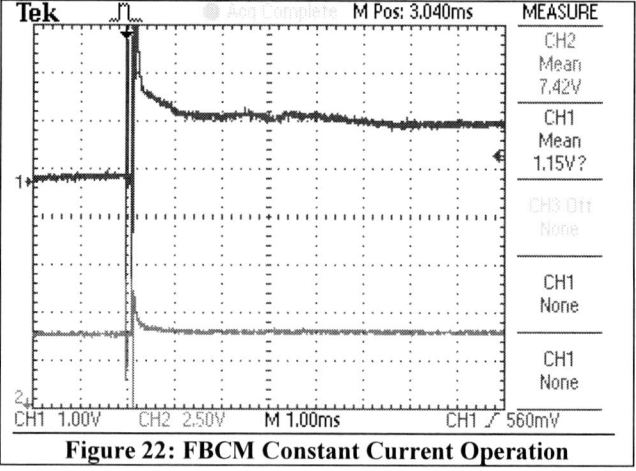

Figure 22: FBCM Constant Current Operation

VI. PFM CONTROLLER DESIGN

In PFM control, HS and LS FETs are controlled by set threshold of output voltage and peak inductor current. Since PFM is useful only for light loads, the efficiency deteriorates at higher loads with PFM control and hence PFM – PWM changer is also required to be a part of control. Much emphasis is laid on the positive inductor current and not allowing it to go below zero. Thus synchronous buck converter is made to operate like an asynchronous buck converter by turning the LS OFF when the inductor current tries to approach less than 0 A as it is seen in Figure 23 [14].

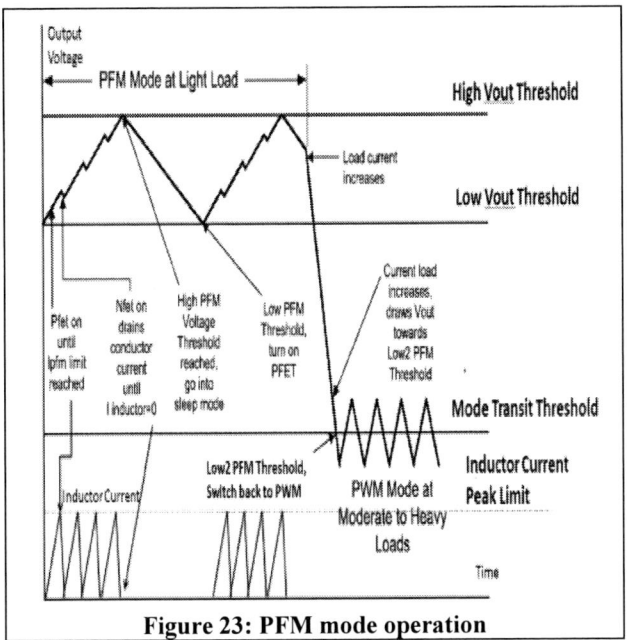

Figure 23: PFM mode operation

The reason for reducing power loss in this control mode is mainly due to the "sleep phase" as the output voltage reaches above high threshold and thus both the switches are turned off till lower threshold is reached and thus capacitor sources power to the load. To maximize this phase, threshold difference must be increased and thus voltage ripple increases. PFM controller involves various blocks for its operation. Hysteritic controller sets the threshold and also sets the limit for PFM to PWM changer as seen in Figure 24. VFB is the feedback voltage and IL is the inductor current. As it can be seen 80mA is the PFM-PWM changer limit set and output voltage threshold is set between 4.98V to 5.02V. These simulations are done in PowerSim. Zero current detection is used to turn off the LS FET as seen in Figure 25. Output voltage and Inductor current can be seen in Figure 26. Since the sampling time for ADC interrupt is 100us, the inductor current goes less than zero only for short time and thus light load efficiency is improved.

The output voltage is regulated between the set thresholds of PFM mode which can be configured depending on the voltage ripple requirements. The inductor current and inductor node voltage during PFM light load condition can be seen in Figure 27.

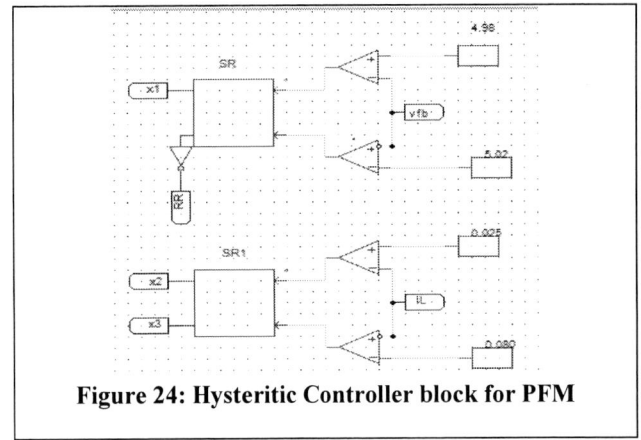

Figure 24: Hysteritic Controller block for PFM

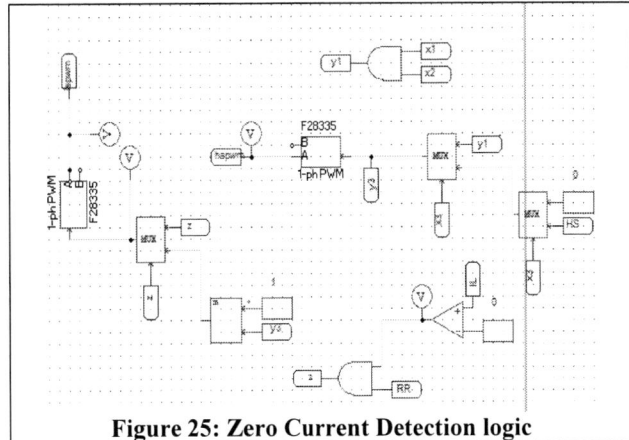

Figure 25: Zero Current Detection logic

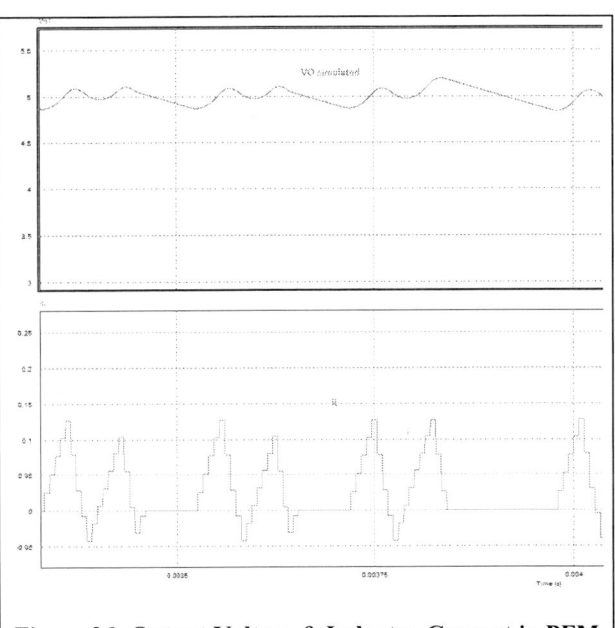

Figure 26: Output Voltage & Inductor Current in PFM Mode

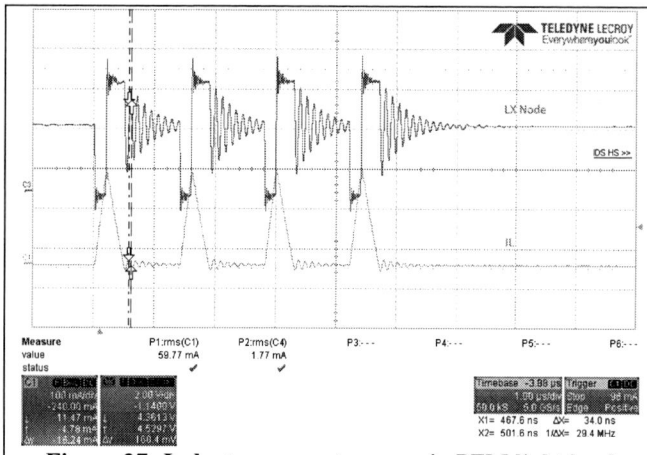

Figure 27: Inductor current as seen in PFM light load

VII. CONCLUSIONS

The development of an intelligent, configurable and efficient power system for CubeSats and small satellites has provided meaningful insights into the feasibility of such power systems which involve digital control and regulation. To achieve higher efficiency it is imperative to understand the power loss at various points of the converter performance like light load and hence efforts should be made in reducing loss of power at these specific and important areas as it was done for the light load management by implementing PFM control. The efficiency can be increased from 10% to 70 % at lighter loads by using dual PFM-PWM control. There is a tradeoff between voltage ripple and efficiency which can be modified depending on the ripple specifications. As the Rds ON resistance for EPC GaN device is less than the Silicon counterpart, its conduction loss is also lower than the Silicon based EPS. Peak efficiency for GaN mosfet is 91.28%while for Silicon mosfet in EPS is 90.7% at 150 KHz. The use of GaN based devices in EPS has led to a reduction in the size and cost of magnetics and has provided an understanding of the performance of these devices when used in systems like the EPS. GaN devices are also radiation hardened which makes them a perfect fit for the Cubesat applications. Design of PCB with WSON packaging of GaN devices is a new challenge at higher switching frequencies. However these devices will play a definite role towards designing of higher power density converters.

VIII. FUTURE WORK

The present implementation of the EPS has one FBCM and two FDPOL modules. These will be increased to provide additional voltage rails and interface to more solar panels. The next step towards improving system performance will be to further increase efficiency of the system by designing even better PCB with less parasitic. The battery charging system will be improved by implementing charge detection algorithms to know the state of charge of batteries.

REFERENCES

[1] M. Wall, "Record-Breaking 33 'Cubesats' to Launch from Space Station," [Online]. Available:
http://www.space.com/24546-cubesat-record-space-station-launch.html.

[2] Shu Ji; Reusch, D.; Lee, F.C., "High-Frequency High Power Density 3-D Integrated Gallium-Nitride-Based Point of Load Module Design," Power Electronics, IEEE Transactions on, vol.28, no.9, pp.4216,4226, Sept. 2013

[3] J. Strydom, "Design and Evaluation of a 10 MHz Gallium Nitride Based 42 V DC-DC Converter," 2014. [Online]. Available: http://epc-co.com/epc/documents/presentations

[4] S. A. Notani, "Develpment of Distributed, Scalable and a Flexible Electrical Power System Module for CubeSat and Small Satellites," North Carolina State University, North Carolina, USA, 2010

[5] M. R. Shah, "Enabling Aggressive Voltage Scaling for Real-Time and Embedded System with Inexpensive and Efficient Power Conversion," North Carolina State University, North Carolina, USA, 2012.

[6] D. Jauregui, B. Wang, and R. Chen (2011, Jun.). Power loss calculation with common source inductance consideration for synchronous buck converters, TI application note [Online]. Available: www.ti.com

[7] "Buck Converter Losses under Microscope" By Alan Elbanhawy, Director, Computing and Telecommunications Segments,Advanced Power Systems Center, Fairchild Semiconductor, SanJose,Ca.

[8] "GaN FET based CubeSat Electrical Power System" Applied Power Electronics Conference and Exposition (APEC), 2015 IEEE, Singh, S. ,Shrivastav,A,Bhattacharya,S.

[9] Efficient Power Conversion, "eGaN® FET DATASHEET EPC2016".

[10] S. Singh, "Development of effective and efficient digital control architectures for a scalable and flexible Electrical Power System of Cube-Satellites and small satellites," North Carolina State University, North Carolina, USA, 2014..

[11] E. Koutroulis and K. Kalaitzakis, "Novel battery charging regulation system for photovoltaic applications," in Electric Power Applications, IEE Proceedings, 2004.

[12] A. Prodic, D. Maksimovic and R. Erickson, "Design and implementation of a digital PWM controller for a high-frequency switching DC-DC power converter," in Industrial Electronics Society, IECON, 2001.

[13] W. Al-Hoor, J. Abu-Qahouq, L. Huang and I. Batarseh, "Design Considerations and Dynamic Technique for Digitally Controlled Variable Frequency DC-DC Converter," in Power Electronics Specialists Conference, 2007.

[14] "An Analysis of Buck Converter Efficiency in PWM/PFM Mode with Simulink" by Cheng Peng, Chia Jiu Wang University of Colorado at Colorado Springs, Department of Electrical and Computer Engineering, Austin Bluffs Parkway, Colorado Springs, USA,Email: cwang@uccs.edu. doi:10.4236/epe.2013.53B013 Published Online May 2013 (http://www.scirp.org/journal/epe)

978-1-4673-9551-9/16 $31.00 © 2016 IEEE

A 98.8% Efficient Bidirectional Full-Bridge Isolated DC-DC GaN Converter

Rakesh Ramachandran
Maersk Mc-Kinney Moller Institute
University of Southern Denmark
Odense, Denmark

Morten Nymand
Maersk Mc-Kinney Moller Institute
University of Southern Denmark
Odense, Denmark

Abstract— **The paper presents the design and development of an ultra-high efficiency bidirectional isolated full bridge dc-dc converter using Gallium Nitride (GaN) devices. To achieve ultra-high efficiency, GaN devices, synchronous rectification and high efficiency magnetics are used. The proposed bidirectional converter allows a power flow in both directions using the same power components; this increases power density and reduce the cost. The performance of a 1.7 kW bidirectional converter is experimentally validated in both forward direction (buck mode) and backward direction (boost mode). The converter operates at a switching frequency of 50 kHz with a voltage of 130V at one side and 52V at the other side of the converter. The fast switching speed of the GaN devices are utilized to achieve extremely high conversion efficiency thus reducing the total volume of the converter. The high power GaN converter has attained an extremely high efficiency of 98.8% in both the directions. This paper demonstrates the highest achievable conversion efficiency with the present technology.**

Keywords— isolated dc-dc converter; high efficiency; magnetics; wide band gap devices; Gallium Nitride; measurement; synchronous rectification

I. INTRODUCTION

Silicon has been used as a power semiconductor material for many decades. The material properties of Silicon have reached its maximum theoretical limit [1, 2]. For power devices, wide band gap materials such as Silicon carbide (SiC) and Gallium Nitride (GaN) are also promising because of their high switching speed and lower switching Figure of Merit (FOM), $Q_{oss}xR_{DS(ON)}$. Compared to conventional silicon devices, for the same breakdown voltage, GaN devices have smaller area for the same on-resistance [3]. This will help in reducing the conduction losses.

The advantages of using GaN devices will include high efficiency, lower cost and high power density [4]. Compared to Si MOSFET, switching losses and conduction losses will be smaller in GaN devices. The lower device loss allows the utilization of smaller and cheaper heat sink components [5]. This reduces operating temperature of the component and hence, increases the reliability of the product.

Bidirectional dc-dc converters are highly desirable in many applications such as battery chargers, uninterruptable power supplies, telecom applications and in smart grid applications. Various bidirectional dc-dc converters are discussed in [6-13]. Some of these bidirectional converters have a dual H-bridge configuration, which is used to provide the galvanic isolation.

In [5], a 10 kW bidirectional converter with dual active bridge is presented. The maximum efficiency achieved is 97.4% excluding drive losses of IGBTs. In [6], a 2 kW, 20 kHz with current doubler topology is presented. The maximum efficiency achieved is 96% at 600W in boost mode and 96% at 1200W in buck mode.

In [7], a 1 kW isolated bidirectional dual bridge with active clamp for electric double layer capacitor (EDLC) is explained. Soft switching is achieved in both buck mode and boost mode of operation. The maximum efficiency achieved during the buck mode is 91% and in boost mode it is 93.3%. A 1 kW resonant bidirectional converter is presented in [8] with an efficiency of 97.5% in forward and 97% in backward direction.

This paper proposes an isolated bidirectional dc-dc converter using GaN devices. High efficiency magnetics design is presented in the paper. Synchronous rectification is also implemented to improve the converter efficiency. The 1.7 kW GaN converter is designed with a switching frequency of 50 kHz. In forward (buck mode) power flow, the input voltage is 130V and output voltage is 52V. The measured maximum efficiency of the bidirectional converter is 98.8% in both forward and backward direction.

II. ISOLATED BIDIRECTIONAL FULL-BRIDGE DC-DC CONVERTER

The circuit diagram for an isolated bidirectional full bridge dc-dc converter is shown in Fig.1.

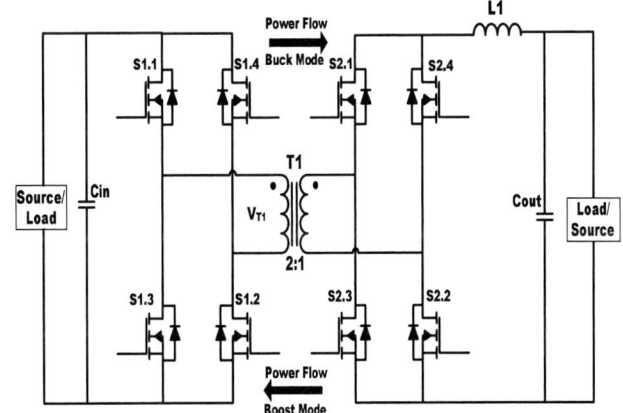

Fig. 1. Isolated full-bridge bidirectional dc-dc converter

978-1-4673-9551-9/16 $31.00 © 2016 IEEE

The converter works as an isolated buck converter in forward direction and isolated boost converter in backward direction.

A. Buck Mode of Operation

The operational waveforms of the converter in buck mode is presented in Fig. 2. The working of the converter is explained in [14].

The output voltage, V_o can be expressed in terms of input voltage V_{in} in a buck mode of operation as,

$$V_0 = 2\, n_c D_c V_{in} \qquad (1)$$

where, n_c is the ratio of number of winding turns in low voltage side to number of winding turns in high voltage side. D_c is the duty cycle of the switches in the high voltage side.

B. Boost Mode of Operation

The operational waveforms of the converter in boost mode is shown in Fig. 3.

The output voltage, V_o can be expressed in terms of input voltage V_{in} in a boost mode of operation can be expressed as,

$$V_0 = \frac{n_d}{1 - D_d} V_{in} \qquad (2)$$

where, $n_c = 1/ n_d$. D_d is the duty cycle of the switches in the low voltage side. $D_d = 1 - D_c$.

The output voltage in boost mode operation can be also written as,

$$V_0 = \frac{1}{n_c D_c} V_{in} \qquad (3)$$

III. CONVERTER DESIGN

The specifications of the converter in forward direction are shown in Table. I.

TABLE I.   CONVERTER SPECIFICATIONS (BUCK MODE)

Parameter	Value
Output Power	1.7 kW
Input Voltage	130 V
Output Voltage	52 V
Switching Frequency	50 kHz
Transformer Core	Ferroxcube EE55/28
Inductance Value	10.2 μH
Inductor Core	Magnetics Kool Mμ E40

A. Transformer

Transformer is designed using Ferroxcube EE55/28 ferrite core. 3C95 material is chosen to reduce the core loss over a wide operating temperature range.

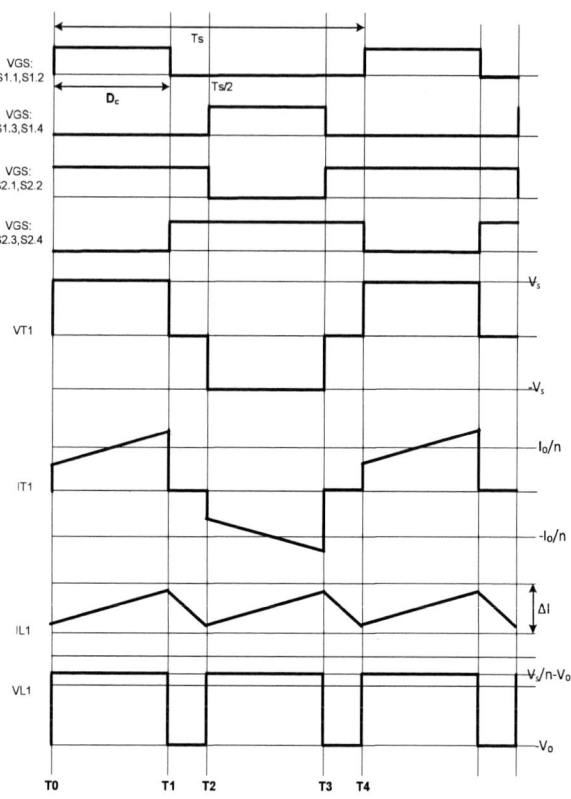

Fig. 2. Operational waveforms of isolated full bridge buck converter

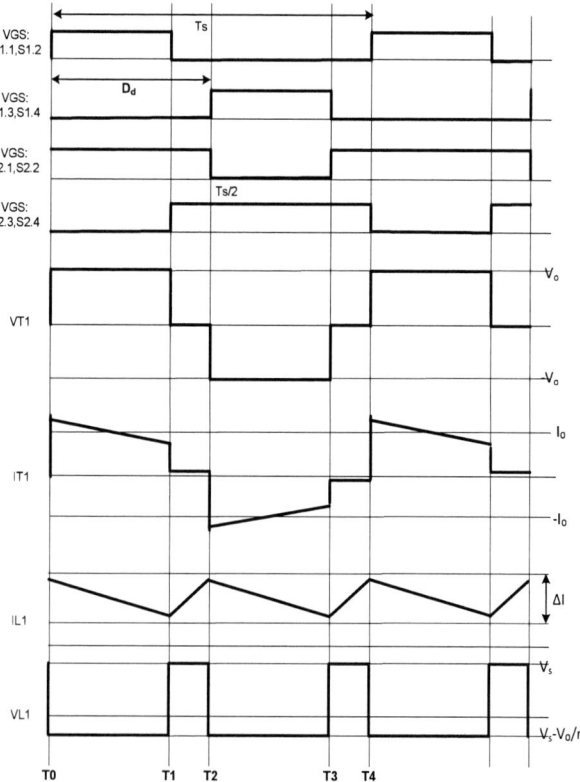

Fig. 3. Operational waveforms of isolated full bridge boost converter

Copper foils are efficiently interleaved to reduce the ac-resistance and thus also reduces the leakage inductance of the transformer. The efficiency of the transformer alone is 99.75%. Eqn. (4) and (5) shows the calculation for the leakage inductance and the ac resistance of the transformer referred to the high voltage side [15].

$$L_{LK} = \mu_0 \frac{N^2 lw}{M^2 bw} \left[\frac{1}{3} \sum hp + \sum h\Delta \right] \qquad (4)$$

where, μ_0 is the permeability of free space, N is the number of turns, M is the number of primary–secondary intersections, lw is the mean turn length, bw is the breadth of winding, hp is the height of pth winding portion and $h\Delta$ is the height of primary-secondary intersection

$$R_{ac} = R_{dc} \left[\varphi \frac{\sinh 2\varphi + \sin 2\varphi}{\cosh 2\varphi - \cos 2\varphi} + \frac{2(m^2-1)}{3} \varphi \frac{\sinh \varphi - \sin \varphi}{\cosh \varphi + \cos \varphi} \right] \quad (5)$$

where, $\varphi = \frac{h}{\delta}$, h is the height of the conductor, δ is the penetration depth, R_{ac} is the ac resistance of the winding, R_{dc} is the dc resistance and m is the number of layers in the winding.

The ac resistance and leakage inductance of the transformer referred to high voltage side is measured using Keysight 4294A precision impedance analyzer. The measurement result is presented in Fig. 4.

B. Inductor

Inductor is wound on Magnetics Kool Mµ E40 core. The inductor is designed using both ac and dc windings separately. A thin ac winding is used to reduce the ac resistance and a thick dc winding to reduce the dc resistance. This helps in improving the efficiency of the inductor and thus increases the overall converter efficiency. The design of high efficiency inductor is explained in [16].

C. GaN Devices

GaN FETs are used at both the high voltage and low voltage sides of the converter. EPC2010C with a breakdown

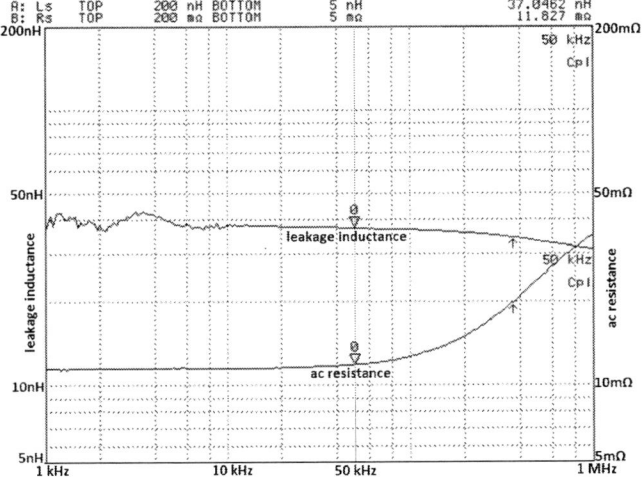

Fig. 4. Measured value of transformer ac-resistance and leakage inductance referred to high voltage side

voltage of 200V is used at the high voltage side. EPC2001C are used at the low voltage sides, they have a breakdown voltage of 100V. Four GaN FETs are paralleled in each switch in the topology. This helps in reducing the conduction losses in the devices.

GaN devices also have a very small gate charge; this reduces the gate drive losses even when four devices are paralleled.

The maximum gate source voltage for the EPC GaN FET is 6V, which is low compared to the conventional Si MOSFET. The gate threshold voltage of these devices is also small in the range of 1 to 1.5V. So driving of these devices requires special consideration in terms of the drivers used and the PCB design layout.

The LM5114, from Texas Instrument's is used as the driver for the GaN FETs. The turn-on and turn-off speed of the devices can be regulated by separate resistors using the LM5114 driver.

Since the threshold voltage of these devices is very small, any small noise can turn on the device. In the PCB design the gate loop inductance is reduced by reducing the distance between the GaN device and the driver IC. Even though the packaging of the GaN FET has negligible impedance, common source inductance has to be considered for PCB design layout. For optimum switching performance, common source inductance should be minimized.

Another major advantage of GaN device, mainly in bidirectional application is the absence of parasitic body diode. Compared to Si MOSFET, GaN devices will be more attractive in synchronous rectification due to zero reverse recovery losses.

D. Filter Capacitors

Ceramic capacitors are used as the filter capacitors at both input and output sides of the converter. Compared to electrolytic capacitors, ceramic capacitors reduce the size of the converter. These capacitors also have a very small ESR value. But the disadvantage is the reduction in capacitance value with the increase in dc bias voltage. To reduce this effect, many ceramic capacitors are paralled. This again reduces the ESR and hence reduces the dielectric loss associated with it.

The diectric loss in the capacitor can be calculated by the equation,

$$P_{di\_cap} = 2\pi f C V^2 \tan \delta \qquad (6)$$

where, f is the frequency, C is the capacitance, V is the voltage across the capacitor and tan δ is the loss tangent or dissipation factor.

IV. LOSS ESTIMATION IN GaN CONVERTER

The proposed GaN converter is hard switched converter. Output charge in the device contributes to a major contribution of switching losses in a hard switched converter. Capacitive switching losses is analysed and presented in [17]. The expression for the capacitive switching losses in the GaN device can be given by the equation,

$$P_{cap(loss)} = \{2Q_{oss,p(V_{in})} - [\frac{5}{8}Q_{oss,p(\frac{3V_{in}}{8})} + \frac{3}{8}Q_{oss,p(\frac{5V_{in}}{8})}] + \frac{2}{n}Q_{oss,s(\frac{V_{in}}{n})}\} V_{in} \ f_{sw} \quad (7)$$

where, $Q_{oss,p(V_{in})}$ is the output charge of four parallel primary switches at V_{in}, $Q_{oss,p(\frac{5V_{in}}{8})}$ is the output charge of four parallel primary switches at $\frac{5V_{in}}{8}$, $Q_{oss,p(\frac{3V_{in}}{8})}$ is the output charge of four parallel primary switches at $\frac{3V_{in}}{8}$, $Q_{oss,s(\frac{V_{in}}{n})}$ is the output charge at $\frac{V_{in}}{n}$ of four parallel secondary switches.

The core loss in the transformer and inductor is calculated using the Steinmetz formula [18].The conduction losses in the magnetics are calculated using the RMS current through the magnetics and the resistance of the copper foil.

Apart from switching loss, GaN devices also have conduction losses and gate drive losses.

The gate drive losses can be calculated as,

$$P_{drive} = V_d \ f_{sw} \ Q_{G(Vd)} \quad (8)$$

where, V_d is the gate drive reference voltage , f_{sw} is the switching frequency and $Q_{G(Vd)}$ is the total gate charge at voltage, V_d.

The conduction losses in GaN devices in buck mode of operation is discussed below,
Losses in the high voltage FET can be calculated by,

$$P_{Hloss} = R_{DS(ON)H}[\frac{P_o\sqrt{D_c}}{n_c V_0}]^2 \quad (9)$$

Similarly, the conduction losses in low voltage FET can be expressed as,

$$P_{Lloss} = R_{DS(ON)L}[\frac{P_o}{V_0}]^2 \quad (10)$$

where, $R_{DS(ON)H}$ is the on-resistance of high voltage FETs, $R_{DS(ON)L}$ is the on-resistance of low voltage FETs, D_c is the duty cycle, n_c is the turns ratio, P_o is the output power, and V_o is the output voltage in buck mode of operation.

The losses in the GaN converter is calculated and presented in Table.II.

The losses in the converter can be both from idle losses and conduction losses. Idle losses in the converter includes the core loss in the magnetics, both transformer and inductor, gate drive losses and capacitive switching losses. Idle losses will be almost constant from no load to full load. Conduction losses in the converter is proportional to the square of output current. They become more dominant at full load condition. The

TABLE II.　ESTIMATED LOSSES AT 1.7 KW (BUCK MODE)

Losses	Idle Loss	Conduction Loss
Transformer	1.56W	2.46W
Inductor	0.55W	1.3W
High Voltage Switches	0.034W	3.4W
Low Voltage Switches	0.055W	3.6W
Dielectric Filter Capacitor Losses	-	0.51W
Capacitive Switching Losses	3.3W	-

estimated losses at 1.7 kW has excluded the PCB losses and the losses from the reverse conduction of GaN FETs.

V.　EXPERIMENTAL RESULTS

The prototype model of the 1.7 kW Isolated bidirectional dc-dc converter is shown in Fig. 5. The 1.7 kW GaN converter is realized with a PCB of size 143mm x 77mm.

When the efficiency becomes higher, for e.g. above 95%, extensive care has to be taken to measure the efficiency accurately and precisely. Compared to precise voltage measurement, measurement of current is always critical. In this efficiency measurement, current at both input and output side of the converter is measured using current sense resistors. They have high temperature stability (< 1 ppm/K) and 0.1% tolerance. Voltage is measured using Keysight 34410, 6½ high performance digital multi-meters.

In order to verify the bidirectional operation of the GaN converter, the converter is operated in both buck (forward) and boost (backward) modes. Experimental results are presented below.

A. Buck Mode

The buck mode operation of the GaN converter is verified by connecting a dc voltage source to the low voltage side of the converter. The high voltage side of the converter is connected to an electronic load. Fig. 6. shows the various waveforms of the converter working in forward direction. The

Fig. 5. 1.7 kW bidirectional Isolated dc-dc GaN converter

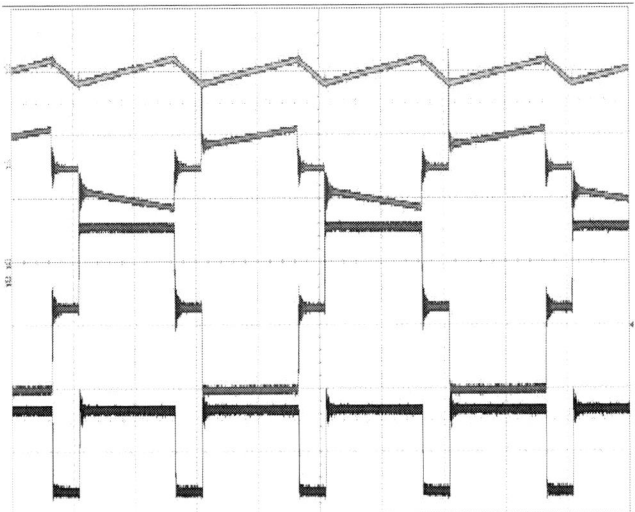

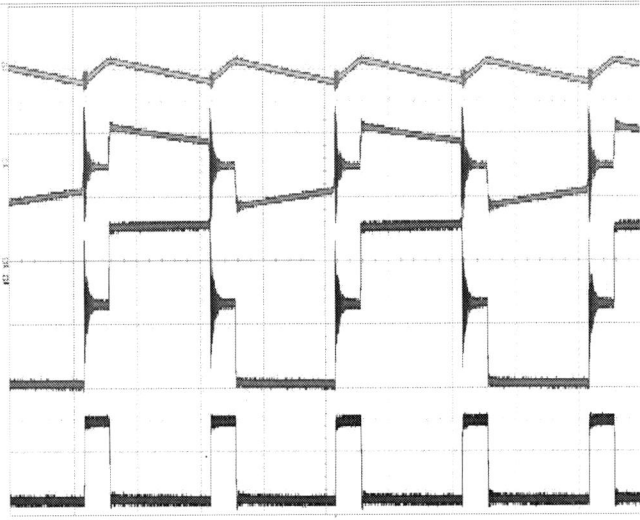

Fig. 6. Measured waveforms of isolated full-bridge dc-dc buck converter (Green: Inductor current 20A/div, Yellow: Transformer primary current 20A/div, Red: Transformer primary voltage 100V/div, Blue: Rectifier voltage 50V/div, Time: 5µS/div)

Fig. 8. Measured waveforms of isolated full-bridge dc-dc boost converter (Green: Inductor current 20A/div, Yellow: Transformer secondary current 20A/div, Red: Transformer secondary voltage 100V/div, Blue: Voltage across Inductor 50V/div, Time: 5µS/div)

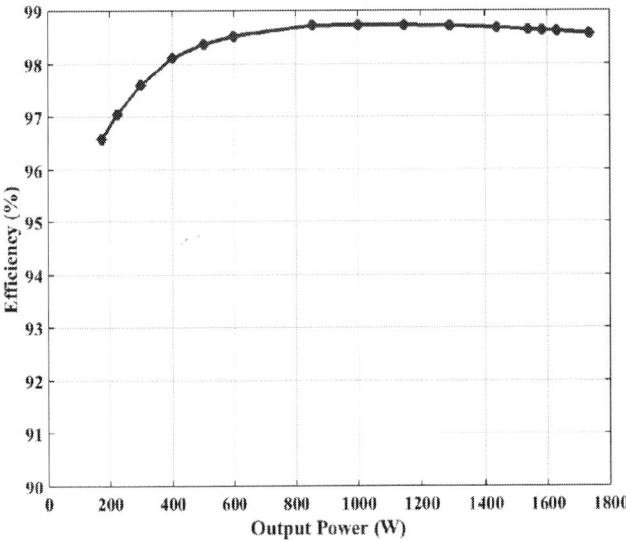

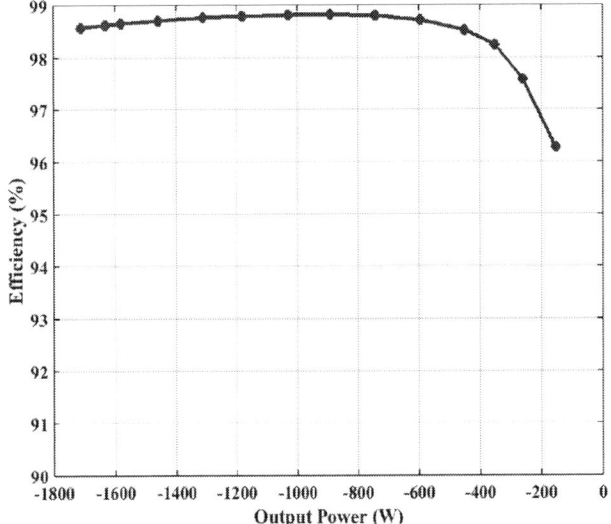

Fig. 7. Measured efficiency curve of the bidirectional dc-dc converter in buck mode

Fig. 9. Measured efficiency curve of the bidirectional dc-dc converter in boost mode

measured efficiency of the converter in buck mode is also presented in Fig. 7. The peak efficiency of the converter in buck mode is measured as 98.8%.

B. Boost Mode

The boost mode operation of the GaN converter is verified by applying an input voltage of 52V. The high voltage side of the converter is connected to an electronic load. Various waveforms of the converter working in backward direction are shown in Fig. 8. The measured efficiency of the converter in boost mode is also shown in Fig. 9. Since the power flow is in reverse direction, the output power is shown as negative in the efficiency curve. The maximum measured efficiency of the converter in backward direction is 98.8%.

VI. CONCLUSION

The paper presents the design and development of an ultra-high efficiency bidirectional isolated dc-dc converter utilizing GaN devices. The prototype model of a 1.7 kW bidirectional GaN converter is presented in the paper. Compared to traditional Silicon bidirectional converter, use of GaN devices reduces the device losses dramatically and improves the converter efficiency. The bidirectional converter maintains an extremely high efficiency of above 98.5% over a wide range of output power in both forward and backward direction. The peak efficiency of the converter measured in both directions is 98.8%. The proposed high power bidirectional GaN converter has achieved a power loss reduction of approximately 40% compared to the recent state of art. This has reduced the heat sink requirement and hence made the converter very compact in size.

ACKNOWLEDGMENT

The project is sponsored by the Danish National Advanced Technology Foundation under Intelligent Efficient Power Electronics (IEPE), strategic research center between the industries and universities in Denmark.

REFERENCES

[1] J. Millan, P. Godignon, X. Perpina, A. Perez-Tomas, J. Rebollo, "A survey of wide bandgap power semiconductor devices," IEEE Trans. on Power Electronics, vol. 29, pp. 2155– 2163, May 2014.

[2] N. Kaminski, "State of the art and the future of wide band-gap devices," in Proc. 13th European Conference on Power Electronics and Applications, pp. 1-9, Sep.2009.

[3] B. J. Baliga, "Semiconductors for high-voltage, vertical channel field-effect transistors," Journal of Applied Physics, vol. 53, pp.1759-1764, Mar 1982

[4] R. Ramachandran, M. Nymand, "Design and Analysis of an Ultra-High Efficiency Phase Shifted Full Bridge GaN Converter," Applied Power Electronics Conference and Exposition, 30th Annual APEC Conference Proceedings 2015, pp.2011-2016, Mar 2015

[5] L. Garcia-Rodriguez, V. Jones, J. Balda, E. Lindstrom, A. Oliva and J. Gonzalez-Llorente, "Design of a GaN-based microinverter for photovoltaic systems,", 2014 IEEE International Symposium in Power Electronics for Distributed Generation Systems (PEDG), June 2014

[6] S. Inoue, H. Akagi, "A bi-directional isolated dc/dc converter as a core circuit of the next-generation medium-voltage power conversion system, " IEEE Trans. Power Electron., vol. 22, no. 2, pp.535 -542, 2007

[7] H. J. Chiu, L. W. Lin, "A bidirectional dc-dc converter for fuel cell electric vehicle driving system", IEEE Trans. Power Electron., vol. 21, no. 4, pp.950 -958, 2006

[8] Y. Miura, M. Kaga, Y. Horita, and T. Ise, "Bidirectional isolated dual full-bridge dc-dc converter with active clamp for EDLC," in Proc. IEEE Energy Convers. Congr. Expo., pp. 1036–1143, Sep. 2010

[9] T.Jiang, J. Zhang, X. Wu, K. Sheng, Y. Wang, "A Bidirectional LLC Resonant Converter With Automatic Forward and Backward Mode Transition, " IEEE Transactions on Power Electronics, vol.30, pp.757-770, 2015

[10] S. Inoue, H. Akagi, "A bidirectional dc–dc converter for an energy storage system with galvanic isolation," IEEE Trans. Power Electron., vol. 22, no. 6, pp.2299 -2306, 2007

[11] X. Yu, P. Yeaman, "A new high efficiency isolated bi-directional DC-DC converter for DC-bus and battery-bank interface," Applied Power Electronics Conference and Exposition (APEC), pp.879-883, 2014

[12] Y. Miura, M. Kaga , Y. Horita and T. Ise "Bidirectional isolated dual full-bridge dc-dc converter with active clamp for EDLC," Proc. IEEE Energy Convers. Congr. Expo, Pp.1036-1143, 2010

[13] L. Zhu, "A novel soft-commutating isolated boost full-bridge ZVS-PWM dc-dc converter for bi-directional high power applications," IEEE Power Electronics Specialists Conf. (PESC'04), pp. 2141-2146, 2004

[14] R. Ramachandran, M. Nymand, N. H. Petersen, "Design of a compact, ultra -high efficient isolated DC-DC converter utilizing GaN devices," Industrial Electronics Society, IECON 2014 - 40th Annual Conference of the IEEE , vol., no., pp.4256-4261, Nov 2014

[15] M. Nymand, M. A. E. Andersen, "High-efficiency isolated boost dc-dc converter for high-power low-voltage fuel cell applications," IEEE Trans. Ind. Electron., vol. 57, no. 2, pp. 505-514, Feb.2010

[16] M. Nymand, U. K. Madawala, M. A. E. Andersen, B. Carsten, O. S. Seiersen, "Reducing ac-winding losses in high-current high-power inductors," IEEE IECON Conf., Portugal, pp. 774-778, Nov. 2009

[17] R. Ramachandran, M. Nymand, "Analysis of Capacitive Losses in GaN Devices for an Isolated Full Bridge DC-DC Converter," International Conf. on Power Electronics and Drive Systems, vol., no., pp.467-472, June 2015

[18] C. P. Steinmetz "On the law of hysteresis", in Proc. IEEE, vol. 72, no. 2, pp.197 -221 1984.

Comparison of Lateral- and Cylindrical-Stator Electrical Machines for High-Speed Direct-Drive Applications in Confined Spaces

Arda Tüysüz and Johann W. Kolar

Power Electronic Systems Laboratory, ETH Zurich

Email: tuysuz@lem.ee.ethz.ch

Abstract—**Lateral-Stator Machine (LSM) topology is presented in earlier works as an unconventional machine that is advantageous for high-speed, direct-drive applications in confined spaces. Owing to its peculiar geometry, LSM makes use of the additional space in a tool head that cannot be utilized by standard Cylindrical-Stator Machines (CSMs). However, a fair and quantitative comparison of LSM and CSM topologies has not been carried out so far. This paper presents a comparative evaluation of the LSM against slotless and slotted permanent-magnet CSMs, not only in terms of torque density but also concerning torque ripple and self-sensing control capability.**

I. INTRODUCTION

The Lateral-Stator Machine (LSM) is introduced in [1], for directly driven high-speed machining applications where the space at the tool head is limited, and where the electrical machine can grow only in one lateral direction. A typical example are dental drills, where ergonomic constraints limit the size of the handpiece.

The LSM topology can be seen in Fig. 1(a). As shown in Fig. 1(b), in a state-of-the-art dental handpiece, a standard, Cylindrical-Stator Machine (CSM) is placed in the handpiece body, where the available space is larger. Several stages of mechanical transmission are used to connect the machine to the drill and to increase the speed from around 40 000 r/min up to around 200 000 r/min. A high-speed CSM placed directly in the head of the handpiece would enable a direct drive and omit the need for mechanical transmissions; however, the space at the tool head is small and a CSM fitting there could potentially not deliver the torque required by the application. In contrast, an LSM can be accommodated in the tool head and drive the drill directly as shown in Fig. 1(c).

Due to the peculiar shape of its stator extending laterally only on one side, the LSM makes use of the space at the tool neck, which would not be used for magnetic parts in case of the conventional CSMs. Therefore, the same tool is able to deliver a higher power output when directly driven by an LSM, compared to a direct drive realized with a CSM. However, so far no quantitative comparison is made between these two types of electrical machines. Thus, this paper deals with the comparative evaluation of the LSM versus two types of permanent-magnet (PM) CSM topologies, namely the slotless and slotted machine types.

Slotless PM machines have a large magnetic air gap and are therefore characterized by weak armature reaction and consequently lower rotor losses caused by the harmonic content

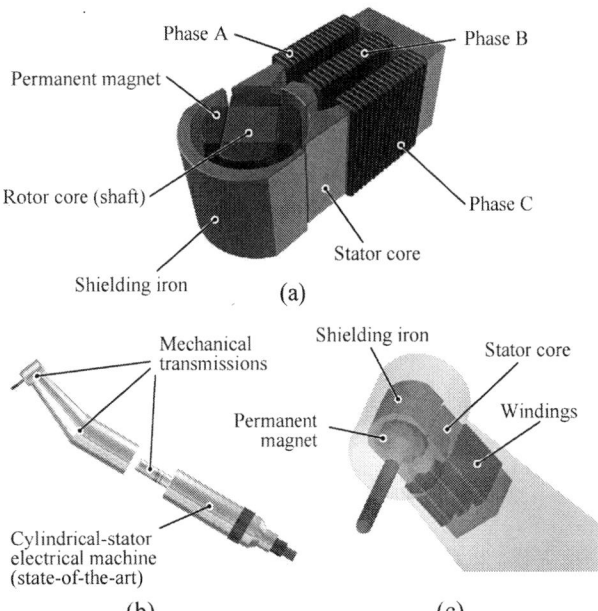

Fig. 1. (a) Conceptual drawing of the Lateral-Stator Machine (LSM). (b) State-of-the-art electric dental handpiece. The rated speed of the electric machine is around 40 000 r/min. Mechanical step-up transmission stages are used to connect the machine to the drill, whose rated speed is 200 000 r/min. (c) An LSM fits in the tool head and can deliver the required torque without using gearboxes. As the stator core is extended on one lateral side, space is gained for the windings outside the tool head [2].

of the armature current [3]. Moreover, this machine type does not suffer from no-load rotor eddy-current losses due to the constant air gap permeance. Thus, it is well suited for high-speed drives. For example, a 500 000 r/min electrical machine is designed in [4], and another one running at 1 000 000 r/min in [5], both with a slotless stator and a one-piece, diametrically magnetized PM rotor.

Even though they are better suited for high-speed applications, the torque density of slotless machines is lower compared to their slotted counterparts [6]. Slotted machines with concentrated (non-overlapping) windings offer higher torque densities due to their shorter end windings compared to machines with overlapping windings [7]. Therefore, they are more commonly used in low-speed, high torque density

978-1-4673-9551-9/16 $31.00 © 2016 IEEE

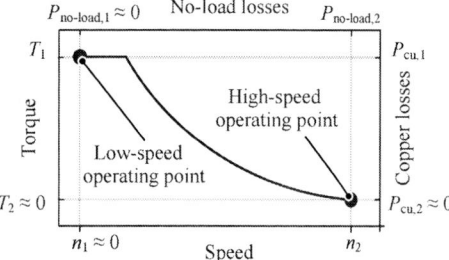

Fig. 2. Typical torque-speed plane of machining applications such as dental drills.

applications such as robotics and power steering [8].

Therefore, this paper compares the LSM topology to the slotless machine type with one pole pair and to the slotted machine type with higher number of poles and concentrated windings. Although the latter is not a very suitable machine topology for high-speed applications due to the higher core losses as a result of the higher electrical frequency and higher no-load rotor losses caused by the non-constant air gap permeability, it is still interesting to compare its torque capability with that of the LSM for applications in confined spaces. Moreover, torque ripple and self-sensing capability of the machines are also discussed.

II. ELECTROMAGNETIC MODELING OF THE MACHINES

A. Target Specifications and Optimization

Various high-speed micro-machining applications ranging from dental drills to high-precision manufacturing tools require electric drives that can deliver high torque at low speeds, while generating low losses at high speeds under low loads. This results in the torque-speed plane illustrated in Fig. 2. Copper (Joule) losses are the only loss component considered in the low-speed operating point whereas they are neglected at the high-speed operating point, where only the no-load losses are considered.

In the LSM optimization presented in [1], the goal is defined as finding the machine geometry that generates the highest torque T_1 for a given $P_{cu,1}$, while generating less no-load losses than a defined $P_{no-load,2}$ at the speed n_2. Based on thermal capacitances and assumptions on the drive cycle, $P_{cu,1}$ is set to 6 W and n_2 is set to 200 000 r/min. In this work, the same specifications are assumed for a direct comparison of the machine types. However, as the main focus is on the torque capabilities of different machine types, the rotor eddy-current losses are neglected.

B. Lateral-Stator Machine

Partial saturation of the stator and the leakage flux between the stator legs play a very important role in determining the performance of the LSM. In order to capture these effects accurately, Two-Dimensional (2-D) Finite-Element Method (FEM) is used for modeling the LSM. The machine geometry is parametrized as shown in Fig. 3(a). Table I shows the discretization of the design space. For each machine in the design space, the winding resistance is calculated assuming

TABLE I. DISCRETIZATION OF THE GEOMETRIC DESIGN SPACE

LSM	w_s	Shaft width	3.5, 4 mm
	r_r	Rotor radius	3.4 mm
	w_n	Tool neck width	8 mm
	w_h	Tool head width	8.8 mm
	l_s	Stator length	10 to 25 mm, 4 steps
	τ	Shoe span	40, 45 deg
	w_l	Leg width	0.8 to 1.4 mm, 4 steps
Slotless	r_s	Stator outer radius	4 mm
	r_r	Rotor radius	0.4 to 2.8 mm, 7 steps[1]
	t_w	Winding thickness	0.4 to 2.8 mm, 7 steps[1]
Slotted	r_s	Stator outer radius	4 mm
	r_b	Stator bore radius	1 to 3.5 mm, 15 steps[2]
	t_m	Magnet thickness	0.5 to 3.5 mm, 15 steps[2]
	$\tau Q/2\pi$	Tooth coverage	0.6 to 0.85, 5 steps
	$\alpha p/\pi$	Magnet coverage	0.6 to 1, 5 steps
	p/Q	Pole-pair/Slot number	2/6, 3/9, 4/12, 5/15, 6/18
All	l_a	Active length[3]	7.4 mm

[1] Excluding designs where the resulting stator core thickness is below 0.2 mm.
[2] Excluding designs where $r_b - t_m < 0.6$ mm.
[3] Axial length of the stator core and the PMs perpendicular to the page plane.

a slot fill factor of 0.3; and a sinusoidal current amplitude is calculated accordingly, such that the total copper losses are $P_{cu,1} = 6$ W. The mean value of the torque is calculated over an electrical period. A second FEM model is run at 200 000 r/min under no load to calculate the stator core losses. Amorphous iron with 23 μm lamination thickness is considered as core material and NdFeB magnets with a remanent flux density of 1.1 T are assumed in the rotor.

A 0.2 mm thick hollow-cylinder-shaped sleeve is assumed on the rotor to hold the permanent magnets in their place under the strong centrifugal stresses occurring at high rotational speeds. The mechanical air gap is constant at 0.2 mm. A 0.2 mm thick plastic hollow cylindrical wall that is coaxial to the rotor separates the mechanical air gap from the lateral stator and the shielding iron. As the sleeve and the plastic wall are made of non-magnetic materials, they are not included in the FEM models. The resulting magnetic air gap is 0.6 mm in the analyzed machines.

Air-friction losses are calculated according to [9]. Further details about the modeling approach can be found in [1].

C. Slotless topology

Fig. 3(b) shows a cross-sectional view of the slotless machine. The stator core is a hollow cylinder with no slots. The air gap windings are usually made of Litz wire (with strand diameters as small as e.g. 50 μm) in order to limit the skin and proximity losses. The rotor consists of a one-piece, diametrically magnetized, cylindrical permanent magnet. As shown in Fig. 4, the torque is transferred to the load using a sleeve that also forms a shaft.

Analytical field models have been presented in literature for analyzing the performance of slotless PM machines [9]. Nevertheless, since only two independent parameters are sufficient to define a unique machine (cf. Table I), 2-D FEM is used in this work for modeling the slotless machine as well. The rotor radius r_r and the winding thickness t_w are both swept from 0.4 mm to 2.8 mm, excluding the designs where the resulting stator core thickness is below 0.2 mm. Core and magnet materials as well as the machine's active length and

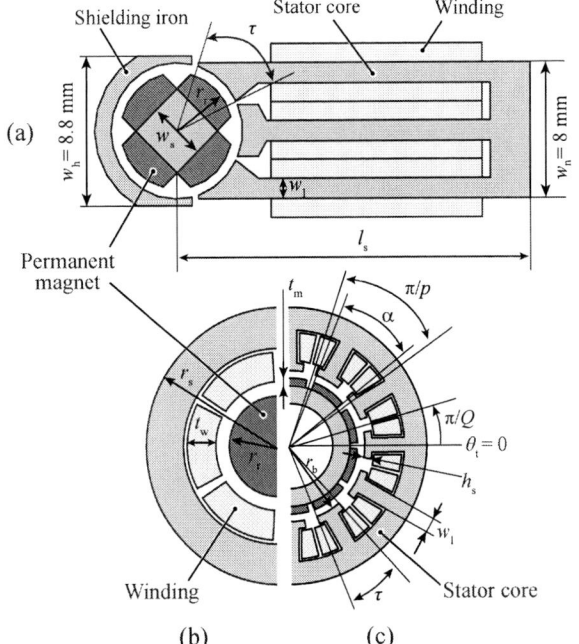

Fig. 3. Cross-sectional view and geometric parameters of (a) LSM, (b) slotless, two-pole CSM, (c) slotted CSM with higher number of poles and concentrated windings. Active (axial, into page plane) length of all machines is set to 7.4 mm. Stator core material is amorphous iron with 23 μm lamination thickness. NdFeB magnets with a remanent flux density of 1.1 T are used in the rotor.

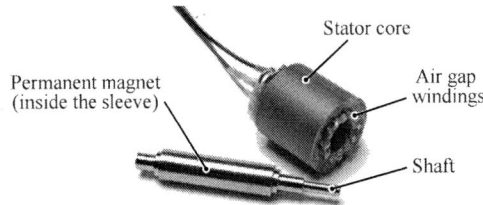

Fig. 4. Photograph of a 150 W, 500 000 r/min slotless PM machine that is described in [10] for a turbocompressor system. Machine's active length is $l_a = 15$ mm and the magnet radius is $r_r = 2.5$ mm [9].

the total copper losses $P_{cu,l}$ are assumed the same as for the LSM.

Straight, overlapping windings are considered with a fill factor of 0.3. The coil length per turn is calculated as

$$l_{coil} = 2(l_a + l_{end}), \qquad (1)$$

where the end winding length l_{end} is approximated as

$$l_{end} = \pi(r_r + 0.6 \text{ mm} + t_w/2). \qquad (2)$$

D. Slotted topology

Fig. 3(c) shows the cross-sectional view of the slotted machine topology. As more independent variables are needed to represent the geometry compared to the slotless machine (cf. Table I), an analytical model is used in order to asses the performance of this machine type in a computationally efficient

way. Firstly, the no-load flux in the air gap is calculated according to [12], by assuming infinite magnetic permeability in the stator core. For this assumption to be realistic, the stator tooth width w_l is dimensioned to avoid saturation as

$$w_l = \frac{r_b}{B_{sat}} \int_{-\pi/Q}^{\pi/Q} B_{ra}(\theta_t) d\theta_t, \qquad (3)$$

where B_{ra} is the radial component of the no-load air gap flux density at the stator bore r_b, B_{sat} is the saturation flux density of the core, Q is the number of slots and θ_t is the integration variable angle in the tangential direction. Evaluating (3) at the rotor position where the considered tooth is aligned with a permanent magnet, the tooth can be dimensioned such that the no-load flux is just enough to saturate it. It has to be noted that the effect of the armature reaction is neglected, resulting in an optimistic machine model in terms of torque capability. The tooth tip height h_s is calculated similarly, such that the tooth tip operates at saturation at no-load

$$h_s = \frac{r_b}{B_{sat}} \int_{w_l/2r_b}^{\pi/Q} B_{ra}(\theta_t) d\theta_t. \qquad (4)$$

Only double-layer windings are considered for minimizing the axial space required by the end windings. The per-turn length of a coil wound around one tooth is calculated as

$$l_{coil} = 2(l_a + l_{end}). \qquad (5)$$

The end winding length is approximated as

$$l_{end} = \frac{1}{2}\left(\frac{2\pi r_{mid}}{Q} - w_l\right) + w_l, \qquad (6)$$

where r_{mid} is the radius of the center of the winding's cross-sectional area.

The flux linkage of one phase is calculated for each rotor position by integrating the radial component of the no-load air gap flux seen by the coils belonging to that phase. The back Electro-Motive Force (EMF) is obtained as the time derivative of the flux linkage. The stator phase currents are calculated such that the resulting copper losses are 6 W, assuming a fill factor of 0.3. Finally, the torque is calculated from the power balance as

$$T = 3\frac{E_{b,rms}I_{rms}}{\omega_m}, \qquad (7)$$

where $E_{b,rms}$ and I_{rms} are the root-mean-square values of the back EMF and the phase current, and ω_m is the the mechanical rotational speed of the rotor.

The stator core losses are calculated based on Steinmetz's equation, assuming as a worst-case scenario that the peak flux is equal to the core material's saturation flux density everywhere in the stator core. Air-friction losses are calculated according to [9], assuming a cylindrical rotor resulting from a 0.2 mm thick sleeve covering the permanent magnets.

(a)

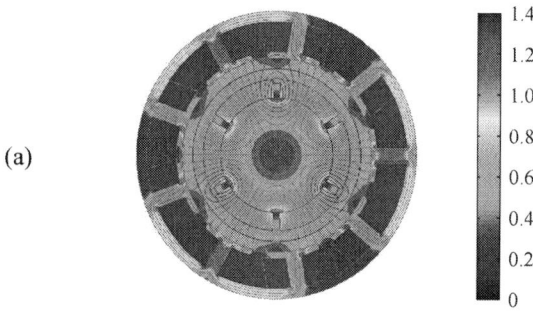

Flux density (T)

(b)

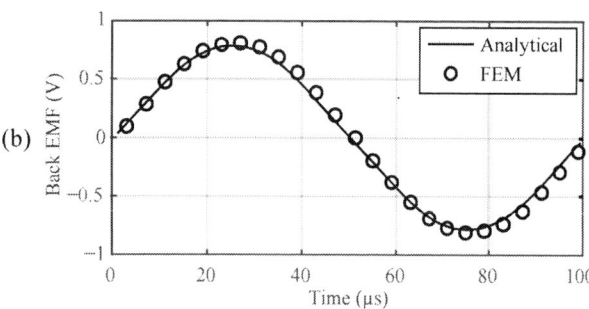

Fig. 5. (a) 2-D FEM simulation results showing the no-load flux distribution of a 3-pole, 9-slot machine with $r_b = 2.7\,\text{mm}$, $t_m = 0.5\,\text{mm}$, $\frac{\alpha p}{\pi} = 0.9$, $\frac{\tau Q}{2\pi} = 0.8$, $w_l = 0.42\,\text{mm}$ and $h_s = 0.17\,\text{mm}$. It can be seen that the maximum flux density in the stator is equal to B_{sat} (saturation flux density of the core material is $1.56\,\text{T}$, and the stacking factor is 84%; hence, the effective saturation flux density is $B_{\text{sat}} = 1.56\,\text{T} \cdot 0.84 = 1.3\,\text{T}$). (b) The back EMF of the same machine at 200 000 r/min. The windings of the machine are double-layer concentrated windings and the number of winding turns is one.

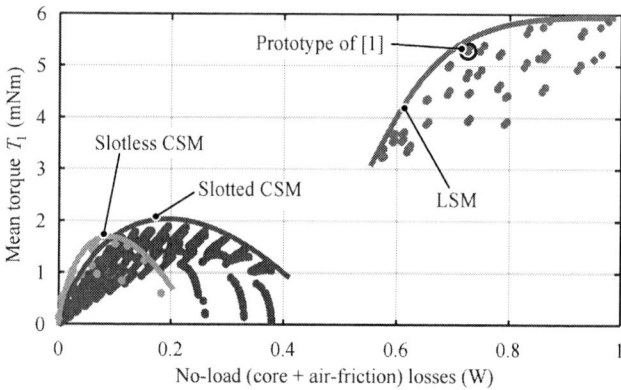

Fig. 6. Comparative evaluation of the performances of two CSMs and the LSM. y-axis shows the mean torque of the machines when producing 6 W of copper losses. x-axis shows the no-load (stator core and air friction) losses at 200 000 r/min.

TABLE II.   SLOTLESS AND SLOTTED CSMs THAT PRODUCE THE HIGHEST TORQUE

Slotless	r_r	Rotor radius	2 mm
	t_w	Winding thickness	0.8 mm
	Ψ	Flux linkage per turn	13.9 μWb
	T_l	Torque	1.55 mNm
	P_c	Stator core losses	54 mW
	P_{air}	Air-friction losses	65 mW
Slotted	r_b	Stator bore radius	2.6 mm
	t_m	Magnet thickness	0.71 mm
	$\tau Q/2\pi$	Tooth coverage	0.6 mm
	$\alpha p/\pi$	Magnet coverage	1 mm
	p/Q	Pole-pair/Slot number	3/9
	w_l	Tooth width	0.42 mm
	h_s	Shoe height	0.16 mm
	Ψ	Flux linkage per turn	12.1 mWb
	T_l	Torque	1.88 mNm
	P_c	Stator core losses	128 mW
	P_{air}	Air-friction losses	65 mW

III.   COMPARSION OF (LOCAL) TORQUE DENSITIES

Fig. 6 shows the mean torque and no-load losses of all the analyzed machines. Each dot in this plot represents a unique machine in the design space. It can be seen clearly that for the same radial space in the tool head, the same active length and the same copper losses, the LSM topology enables significantly higher torque output than both the slotted and the slotless CSMs. Table II describes the slotted and slotless CSMs that produce the highest torque in the analyzed design space.

Certain assumptions made during the modeling process have to be considered for a correct interpretation of Fig. 6. For instance, the same active length is assumed for all the machines. However, due to the straight and overlapping windings with 180° pitch, the slotless machine requires considerably larger space for the end windings compared to the LSM. Consequently, a shorter active length needs to be considered for the slotless machine if it must fit in the same space as the LSM.

As only double-layer concentrated windings are considered in the slotted machine, the space needed for the end windings is less than what is needed for the end windings of the slotless machine. This is illustrated in Fig. 7. However, the end windings still need to be placed in the head of the tool,

Before analyzing all the machines in the parameter range given in Table I, the analytical model described above is verified using 2-D FEM analysis for a single machine geometry. A very high constant relative permeability ($\mu_r = 100\,000$) is assumed in the stator core to approximate the infinite permeability assumed in the analytical approach. Fig. 5(a) shows the no-load flux distribution in this machine. It can be seen that the tooth width and the tooth tip height are calculated properly as the maximum flux density in the stator is equal to the saturation flux density of the core material (saturation flux density of the core material itself is $1.56\,\text{T}$, and the stacking factor is 84%; hence, the effective saturation flux density is $B_{\text{sat}} = 1.56\,\text{T} \cdot 0.84 = 1.3\,\text{T}$). Fig. 5(b) shows the back EMF of this machine for one winding turn and 200 000 r/min, as calculated by analytical and FEM models. These results verify the use of analytical models for the optimization of the slotted machine. Similar to the LSM and the slotless machine types, all slotted machines in the design space are evaluated by first estimating the winding resistance based on the machine geometry, and then calculating the torque for a sinusoidal machine current that generates the copper losses $P_{\text{cu},1} = 6\,\text{W}$. Core and magnet materials as well as the machine's active length are the same as for the LSM and the slotless machine.

978-1-4673-9551-9/16 $31.00 © 2016 IEEE   618

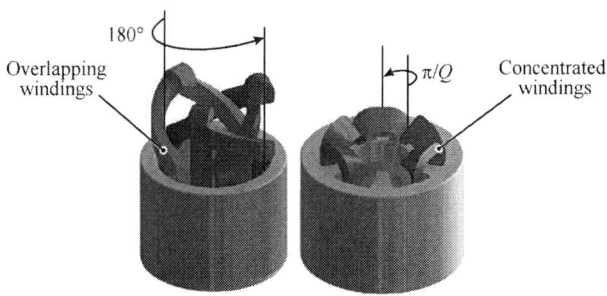

Fig. 7. End windings of the slotless machine with straight and overlapping windings with 180° pitch (left), and slotted machine with double-layer non-overlapping (concentrated) windings. The end winding overhead is smaller for the slotted machine with concentrated windings.

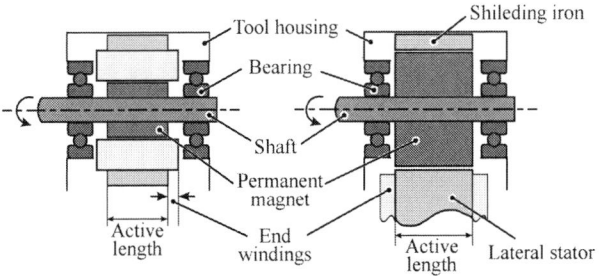

Fig. 8. Positioning of a CSM with end windings (left) and an LSM (right) in the same tool. The end windings of the CSM and the bearings need to fit in the same space, limiting the maximum active (axial) length of the machine. As the windings are taken out of the tool head, the LSM can have not only a radially larger but also an axially longer rotor compared to the CSM.

which is not the case for the LSM, as shown in Fig. 8. Hence, even if the axial length of the end winding is similar for a slotted CSM and an LSM, the LSM can still have a larger active length (axially longer rotor) as its end windings are not in the tool head.

It is clear that incorporating the considerations related to the end windings in the models will change the results shown in Fig. 6 only in favor of the LSM, which can produce significantly higher torque than both types of CSMs even when the same active length is assumed. Therefore, the effect of end windings are not studied any further in this paper.

A direct comparison of the slotted and slotless CSM types, on the other hand, requires a careful consideration of the winding structures. Rotor dynamics may impose a limit on the maximum rotor length in high-speed drives, which makes the end winding overhead a very undesirable feature as it leads to a longer rotor without actually increasing the machine's active length. In practice, ultra-high-speed machines are usually realized using rhombic or skewed air gap windings to shorten or fully avoid the end windings [11]. This, however, reduces the winding factor and the torque-per-current rating of the machine. Nevertheless, this is not studied any further here as the main focus of this paper is not the detailed comparison of the slotted and slotless CSM types to each other.

Moreover, the same copper losses ($P_{\mathrm{cu},1} = 6\,\mathrm{W}$) are as-

sumed in this paper for all the three machine types, for a direct comparison of the machine structures without considerations about cooling. However, a CSM that fits in the tool head has smaller surface for cooling compared to an LSM, hence, its electrical loading may not be as high as that of the LSM. For this reason, the comparison of torque capability is expected to change even more in favor of the LSM following a detailed thermal analysis of both machine types. If Fig. 6 is interpreted considering the remarks about the end windings and the thermal aspects, it can be concluded that the LSM produces *at least* three times the torque that a CSM can offer for a given machining tool.

It has to be noted that so far only applications with a limited space around the rotor have been considered where the stator of the LSM can grow in one lateral direction. Clearly, this gives an advantage to the LSM when comparing the maximum torque that can be generated, as the LSM can utilize more active material (permanent magnets, copper and iron). To be fair, it has to be mentioned that it is the *local* torque density that increases when using an LSM. The *overall* torque density of the LSM prototype of [1] is 3.9 mNm/cm³, whereas the *overall* torque density of the slotless and slotted CSMs of Table II is 4.2 mNm/cm³ and 5.1 mNm/cm³, respectively. Furthermore, it has to be also noted that the volumetric scalability of the LSM is not similar to that of the CSM due to the stray field between the stator legs in an LSM. That is, increasing the stator length l_{s} of an LSM results in higher total slot current for the same copper losses and increases the torque output until an optimum l_{s} value; but beyond that, effect of the stay field becomes visible and the torque output of the machine does not increase further.

IV. LSM versus CSM: Further Aspects

A. Torque ripple

Torque ripple is an undesired effect in applications like machining spindles and dental drills especially at low speeds as it may lead to acoustic noise and vibration. Slotless machines analyzed in this work have a perfectly smooth magnetic circuit and exhibit no cogging torque. The flux linkage is also very sinusoidal, leading to a ripple-free torque when the machine is driven by sinusoidal currents. On the other hand, these machines are also characterized by low phase inductances due to the large magnetic air gap, which means that the drive inverter needs to operate with a high switching frequency and/or a filter needs to be incorporated between the machine and the inverter, in order to drive the machine with sinusoidal currents with low harmonic content. In order to avoid high switching losses and the additional space requirement of filters, Pulse-Amplitude Modulation (PAM) has also been widely applied in inverters driving high-speed slotless machines [13]. However, the block-shaped current waveform resulting from PAM operation leads to torque ripple in machines with sinusoidal back EMF. On the other hand, recent developments in wide-bandgap devices such as Gallium-Nitride (GaN) power switches enable a feasible operation of drive inverters with high switching frequencies, facilitating a clean drive current supply even for very low-inductance machines [14]. For this reason, a purely sinusoidal drive current is assumed in this work for all the analyzed machines.

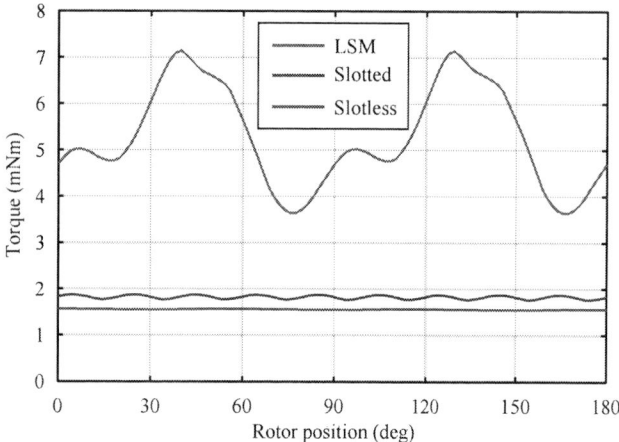

Fig. 9. 2-D FEM simulation results showing the torque of the LSM prototoype [1] and the CSMs designs given in Table II. Copper losses are $P_{cu,1} = 6$ W for all the machines.

Slotted machines suffer from cogging torque due to the attraction between the magnets and the stator tooth tips. The distorted air gap flux distribution also contributes to the torque ripple observed in these machines. On the other hand, the pole-slot combination offers a degree of freedom that can be exploited to minimize the torque ripple [15]. The analytical modeling approach used for the slotted machines in this work only calculates the mean torque. A 2-D FEM model with constant permeability, such as the one used to verify the analytic approach in Section II-D, cannot estimate the torque ripple either, as the influence of the magnetic saturation cannot be captured. Therefore, in this section, a 2-D FEM model that incorporates the nonlinear permeability of the stator core is used –identical to the models used for the lateral-stator and slotless machines– for assessing the torque ripple of the slotted machine.

Fig. 9 shows the torque generated by the LSM described in [1], and the CSMs described in Table II. All three machines are driven by three-phase sinusoidal currents resulting in $P_{cu,1} = 6$ W copper losses. As expected, the slotless machine generates a virtually ripple-free torque, whereas the peak-to-peak ripple is as large as two thirds of the mean torque for the LSM. The slotted machine exhibits a torque ripple whose peak-to-peak value is around 5% of its mean torque. The mean torque of this machine is predicted as 1.88 mNm by the analytic models and as 1.82 mNm by the 2-D FEM simulation incorporating the nonlinear magnetic permeability. Hence, the use of the analytic models is verified once more.

B. Unbalanced magnetic pull

In state-of-the-art permanent-magnet electric machines with cylindrical stators, rotor eccentricity and/or uneven magnetization of permanent magnets may lead to an unbalanced magnetic pull on the rotor [16], [17]. The unbalanced magnetic pull may excite vibrations and shorten the lifetime of the bearings. On the other hand, the asymmetric structure of the LSM results in large magnetic forces acting on the shielding iron, the rotor and the lateral stator, even in the absence of any rotor eccentricity or uneven permanent-magnet magnetization.

Forces up to 7 N pull both the lateral-stator and the shielding iron towards the rotor. As shown in [18], a 3-D printed, 0.2 mm thick plastic wall can withstand these forces. However, long lifetime tests are required to characterize the effect of these forces on the lifetime of the bearings.

C. Position sensing

Rotor position of a PM synchronous machine is needed for a closed-loop operation. A simple way of obtaining the rotor position information is by using dedicated position sensors such as encoders, resolvers or Hall sensors. However, for drive applications in confined spaces as studied in this paper, the space required by an off-the-shelf position sensor may easily be significant compared to the size of the machine, meaning that the machine needs to be built considerably smaller in order to accommodate the position sensor. Consequently, the use of a dedicated position sensor may result in a sizable decrease in the torque that can be generated. For that reason, the possibility of obtaining the rotor position without the use of dedicated position sensors (self-sensing operation) is discussed in the following, for the machines under consideration.

The back EMF of an electrical machine depends on the rotor position, and it can be either measured or estimated using an observer for estimating the rotor position. However, this method is not considered in this paper since it is not applicable for the whole speed range, considering that the back EMF gets more difficult to measure at low speeds and vanishes at standstill. The variation of the machine impedance with the rotor position, on the other hand, can be exploited for position estimation also at lower speeds.

The symmetrical construction of the slotless machine type leads to a virtually non-existent dependency of the machine impedance on the rotor position. Nevertheless, the anisotropic properties of rare-earth magnets introduce a small saliency. Special arrangements of high-frequency signal injection and measurement circuitry have been presented in [19] and [20] for detecting saliencies as small as a few percent. When a metallic sleeve is used on the rotor, alteration of the sleeve geometry may introduce an additional, designer-controlled spatial saliency in the rotor surface resistance [21], which can be detected by a high-frequency signal superimposed on the machine current. However, these methods come with various drawbacks, such as the increased system complexity, the need of using a filter between the machine and the inverter, or the need for having an accessible star-point connection in the machine, or modifying the geometry of the rotor sleeve - a part whose design is subject to very tight tolerances and mechanical constraints. Hence, the reliable and cost-effective self-sensing operation of slotless CSMs in the full speed range remains to be a challenge.

Saliency-tracking-based self-sensing operation of slotted CSMs has been studied extensively in the recent years [22]. Even though surface-mounted PM machines studied in this paper feature a smaller spatial variation of impedance compared to e.g. interior PM machines, observers can be used for increasing the signal-to-noise ratio of the tracked saliency signal, thereby enabling a closed-loop self-sensing operation [23].

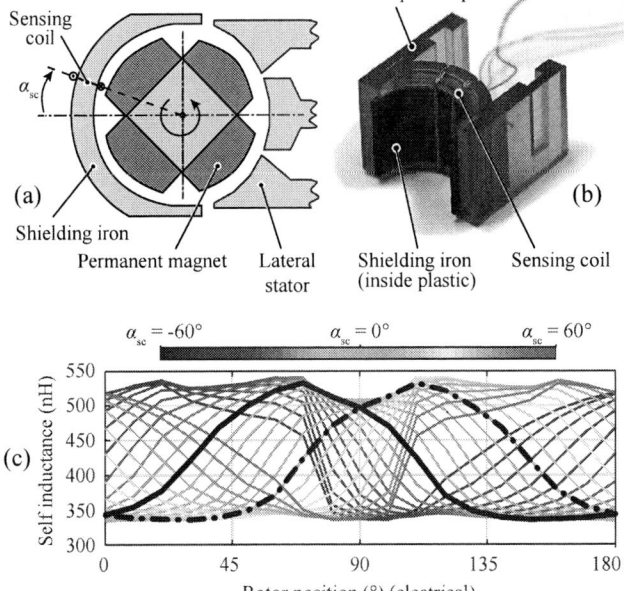

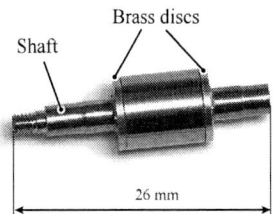

Fig. 11. Rotor of the LSM prototype described in [1]. A long shaft with a thread on one end is used for coupling the LSM to another machine during measurements.

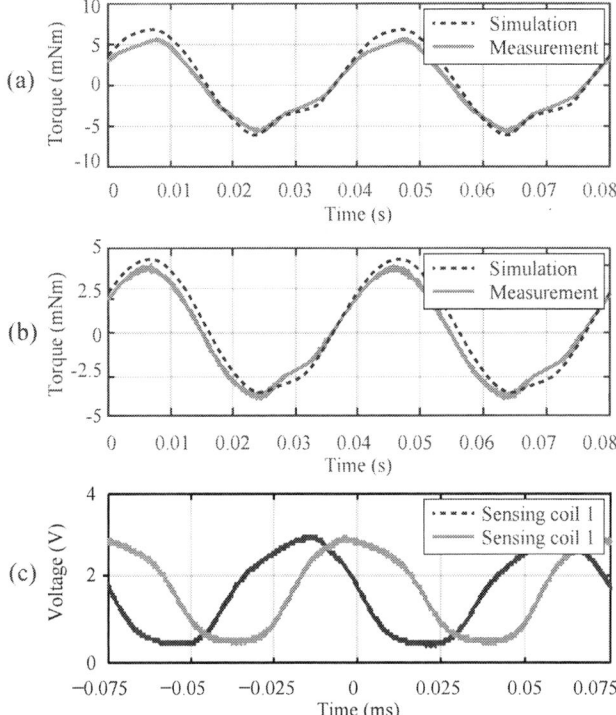

Fig. 10. (a) Illustration of an LSM with a sensing coil wound around the shielding iron. (b) Photograph of a shielding iron with two sensing coils wound on it. The iron is placed inside a 3-D printed plastic case that facilitates the positioning of the shielding iron and the winding of the sensing coil. (c) 2-D FEM analysis results showing the self inductances of the sensing coils placed at positions from $-60°$ to $60°$. Solid and dashed black lines show $\alpha_{sc} = -15°$ and $\alpha_{sc} = 15°$ [2].

In principle, any self-sensing method that can be used for a slotted CSM can also be applied to an LSM, which is by no means a low-saliency machine. On the other hand, an LSM optimized for the highest torque and lowest losses is not necessarily suitable for self-sensing position estimation. For example, all the self and mutual inductances of the LSM optimized in [1], go flat at the same rotor position, making an impedance-tracking-based self-sensing method go blind at these rotor positions [2]. Moreover, the stator core of this machine operates in partial saturation already at no load; hence, the dependency of the impedance on the rotor position is heavily influenced by the load of the machine.

A new position sensing method is presented in [2] for LSMs, where the rotor-position-dependent impedance of the sensing coils wound on the shielding iron is measured by a high-frequency current injection. Even though this is not a self-sensing method, it does not need a large additional space since the sensor is integrated into the machine. Moreover, the the magnetic circuit of the sensing coil is largely decoupled from the stator-field, making the position estimation method much less sensitive to machine's load. Fig. 10(a) and Fig. 10(b) show how the sensing coils are wound on the shielding iron. The rotor-position-dependent variation of the self inductance of a sensing coil wound on the shielding iron at different positions is depicted in Fig. 10(c).

V. HARDWARE AND MEASUREMENTS

Fig. 11 shows a photograph of the rotor of an LSM. Permanent magnets are retained using a titanium sleeve. The

Fig. 12. Simulated and measured torques of an LSM with 33 winding turns when the rotor is fixed for a peak phase current of (a) 5 A, and (b) 3 A. (c) Demodulated and filtered voltage response of the sensing coils measured at 200 000 r/min. The injection frequency is 1 MHz [2].

two brass discs on both axial ends of the sleeve are used for balancing the rotor.

A bearingless standstill torque measurement setup, whose details can be found in [18], is used for verifying the torque capability of the LSM while avoiding bearing friction in the measurements. As seen in Fig. 12(a,b), the FEM models predict the measured torque accurately, which verifies the modeling approach adopted for the optimization of the LSM. Fig. 12(c) shows the filtered and demodulated rotor-position-dependent voltage responses of two sensing coils wound on the shielding iron at $\alpha_{sc} = -15°$ and $\alpha_{sc} = 15°$, for a rotational speed of 200 000 r/min. A digital signal processor sampling these voltages can estimate the rotor position using a simple look-up table [2].

VI. CONCLUSIONS AND OUTLOOK

The Lateral-Stator Machine (LSM) has been proposed in earlier works for high-speed, direct drive applications in confined spaces. However, a quantitative comparison of this machine type to standard, Cylindrical-Stator Machine (CSM) topology has not been carried out so far. This paper compares the LSM to two commonly used CSM types, namely the slotless permanent-magnet machine with one pole pair, and the slotted surface-mount permanent-magnet machine with higher number of pole pairs and concentrated windings. Both analytical and FEM field models are used to compare the torque capabilities of the two CSM types versus the LSM. Due to the peculiar arrangement of its stator, the LSM can make use of the space that cannot be utilized by standard machines, and therefore it can generate at least three times the torque standard machines can offer for the same machining tool. On the other hand, due to the inherent features of the LSM such as the stray field between the legs and the shielding iron that does not contribute to torque generation, the overall torque density of the LSM (3.9 mNm/cm^3) is lower than that of both the slotless (4.2 mNm/cm^3) and the slotted (5.1 mNm/cm^3) CSMs. Moreover, the LSM exhibits a large magnetic pull, whose effect on the bearing lifetime needs to be studied.

By the integration of sensing coils in the machine design, the LSM can be operated closed-loop in the whole speed range without the need for dedicated position sensors.

ACKNOWLEDGMENT

The authors would like to express their sincere appreciation to Celeroton AG for the continuous support they provided during the design, construction and testing of the lateral-stator machine. The authors also acknowledge the support of CADFEM (Suisse) AG concerning the ANSYS software.

REFERENCES

[1] A. Tüysüz, C. Zwyssig, J. W. Kolar, "A novel motor topology for high-speed micro-machining applications," *IEEE Trans. Ind. Electr.*, vol. 61, no. 6, pp. 2960-2968, Jun. 2014.

[2] A. Tüysüz, J. W. Kolar, "New position sensing concept for miniature lateral-stator machines," *IEEE Trans. Ind. Electr.*, accepted for publication in Nov. 2015.

[3] N. Bianchi, S. Bolognani, F. Luise, "Potentials and limits of high-speed PM motors," *IEEE Trans. Ind. Appl.*, vol. 40, no. 6, pp. 1570-1578, Nov./Dec. 2004.

[4] D. Krähenbühl, C. Zwyssig, H. Weser, J. W. Kolar, "A miniature 500 000-r/min electrically driven turbocompressor," *IEEE Trans. Ind. Appl.*, vol. 46, no. 6, pp. 2459-2466, Nov./Dec. 2010.

[5] C. Zwyssig, S. Round, J. W. Kolar, "Megaspeed drive systems: Pushing beyond 1 million r/min," *IEEE/ASME Trans. Mech.*, vol. 14, no. 5, pp. 564-574, Oct. 2009.

[6] J. W. Kolar, T. Friedli, F. Krismer, A. Looser, M. Schweizer, R. Friedemann, P. Steimer, J. Bevirt, "Conceptualization and multi-objective optimization of the electric system of an airborne wind turbine," *IEEE Jour. Emerg. Sel. Top. Pow. Electr.*, vol. 1, no. 2, pp. 72-103, Jun. 2013.

[7] J. Cros, P. Viarouge, "Synthesis of high performance PM motors with concentrated windings," *IEEE Trans. Ene. Conv.*, vol. 17, no. 2, pp. 248-253, Jun. 2002.

[8] K. Wang, Z. Q. Zhu, G. Ombach, M. Koch, S. Zhang, J. Xu, "Electromagnetic performance of an 18-slot/10-pole fractional-slot surface-mounted permanent-magnet machine," *IEEE Trans. Ind. Appl.*, vol. 50, no. 6, pp. 3685-3696, Nov./Dec. 2014.

[9] J. Luomi, C. Zwyssig, A. Looser, J.W. Kolar, "Efficiency optimization of a 100 W 500 000 r/min permanent-magnet machine including air-friction losses," *IEEE Trans. Ind. Appl.*, vol. 45, no. 4, pp. 1368-1377, Jul./Aug. 2009.

[10] C. Zwyssig, D. Krähenbühl, H. Weser, J. W. Kolar, "A miniature turbocompressor system," *Proc. of Smart Ene. Strat.*, (SES), Zurich, Switzerland, Sep. 2008.

[11] A. Looser, T. Baumgartner, J. W. Kolar, C. Zwyssig, "Analysis and measurement of three-dimensional torque and forces for slotless permanent-magnet motors," *IEEE Trans. Ind. Appl.*, vol. 48, no. 4, pp. 1258-1266, Jul. 2012.

[12] Z. Zhu, D. Howe, E. Bolte, and B. Ackermann, "Instantaneous magnetic field distribution in brushless permanent magnet DC motors. I. Open-circuit field," *IEEE Trans. Ind. Appl.*, vol. 29, no. 1, pp. 124-135, Jul. 1993.

[13] L. Schwager, A. Tüysüz, C. Zwyssig, J. W. Kolar, "Modeling and comparison of machine and converter losses for PWM and PAM in high-speed drives," *IEEE Trans. Ind. Appl.*, vol. 50, no. 2, pp. 995-1006, Mar./Apr. 2012.

[14] A. Tüysüz, R. Bosshard, J. W. Kolar, "Performance comparison of a GaN GIT and a Si IGBT for high-speed drive applications," *Proc. of Intl. Pow. Electr. Conf.*, (ECCE Asia, IPEC), Hiroshima, Japan, May 18-21, 2014.

[15] J. A. Güemes, A. M. Iraolagoitia, J. I. Del Hoyo, P. Fernandez, "Torque analysis in permanent-magnet synchronous motors: A comparative study," *IEEE Trans. Ene. Conv.*, vol. 26, no. 1, pp. 55-63, Mar. 2011.

[16] M. Michon, K. Atallah, G. Johnstone, "Effects of unbalanced magnetic pull in large permanent magnet machines," *Proc. of Ene. Conv. Cong. Expo.*, (ECCE), pp. 4815-4820, Pittsburgh, USA, Sep. 2014.

[17] M. Novak, M. Kosek, "Unbalanced magnetic pull induced by the uneven rotor magnetization of permanent magnet synchronous motor," *Proc. of 11th Intl. ELEKTRO Conf.*, pp. 347-351, Rajecke Teplice, Slovakia, May 2014.

[18] A. Tüysüz, D. Koller, A. Looser, J. W. Kolar, C. Zwyssig, "Design of a test bench for a lateral stator electrical machine," *Proc. of 37th IEEE Ann. Conf. Ind. Electr. Soc.*, (IECON), pp. 1801-1806, Melbourne, Australia, Nov. 2014.

[19] A. Tüysüz, M. Schöni, J. W. Kolar, "Novel signal injection methods for high-speed self-sensing electrical drives," *Proc. of Ene. Conv. Cong. Expo.*, (ECCE), pp. 4815-4820, Raleigh, USA, Sep. 2012.

[20] J. Persson, M. Markovic, Y. Perriard, "A new standstill position detection technique for nonsalient permanent-magnet synchronous motors using the magnetic anisotropy method," *IEEE Trans. on Magn.*, vol. 43, no. 2, pp. 554-560, Feb. 2007.

[21] J.-P. Voillat, "Device for controlling an electric motor," *U.S. Patent*, 6 337 554 B1, Jan. 2002.

[22] X. Wang, W. Xie, G. Dajaku, R. Kennel, D. Gerling, R. D. Lorenz, "Position self-sensing evaluation of novel CW-IPMSMs with an HF injection method," *IEEE Trans. Ind. Appl.*, vol. 50, no. 5, pp. 3325-3334, Sep./Oct. 2014.

[23] S.-C. Yang, R. D. Lorenz, "Surface permanent-magnet machine self-sensing at zero and low speeds using improved observer for position, velocity, and disturbance torque estimation," *IEEE Trans. Ind. Appl.*, vol. 48, no. 1, pp. 151-160, Jan./Feb. 2012.

Novel Contactless Axial-Flux Permanent-Magnet Electromechanical Energy Harvester

Michael Flankl[1], Arda Tüysüz[1], Ivan Subotic[2] and Johann W. Kolar[1]

[1]Power Electronic Systems Laboratory
Swiss Federal Institute of Technology (ETH Zurich),
Physikstrasse 3
8092 Zurich, Switzerland

[2]Liverpool John Moores University
School of Engineering,
Byrom Street
Liverpool L3 3AF, UK

Abstract—This paper proposes a novel type of watt-range permanent-magnet energy harvester, which harvests energy from a moving conductive body or surface without mechanical contact, as its operation is purely based on eddy-current coupling. The harvester's main advantage over existing solutions is that it allows energy transfer over atypically large (2 ... 15 mm) air gaps, which are unavoidable in certain industrial applications. The paper provides a detailed description of the system's operating principle, and elaborates its modeling using 3-D Finite-Element Method (FEM) analysis. Two prototypes are built and tested for verifying the models. A power of 2.42 W is harvested from an aluminum surface moving with 10 m/s, over 12 mm air gap using a prototype with $\approx 14\,\mathrm{cm}^3$ magnet volume. Moreover, the effects of the harvester's placement as well as the speed of the moving conductive surface on the maximum harvested power and the system's efficiency are analyzed, both with FEM simulations and measurements.

Index Terms—Kinetic energy harvesting, electromagnetic energy harvesting, watt-range energy harvesting, eddy-current coupling, electromagnetic coupling, air gaps.

NOMENCLATURE

BLDC	Brushless DC
EHS	Energy harvesting system
FEM	Finite-element method
KEH	Kinetic energy harvester
MCS	Moving conductive secondary
MPP	Maximum power point
PM	Permanent magnet
B_r	PM remanent flux density
η_MPP	Efficiency at MPP
φ_1	KEH angle of rotation
$\vec{f}_\mathrm{Lorentz}$	Lorentz force density
g	Air gap
H_cb	PM coercivity
h_m	KEH magnet height
I_1	Generator current
κ	MCS conductivity
k_T1	KEH torque slope
k_P2	MCS power slope
l_ov	Overlap between r_1 and MCS
n_PM	Quantity of PMs
P_1, p_1	Power harvested by KEH (RMS, $p(t)$)
P_2, p_2	Power supplied by MCS (RMS, $p(t)$)
P_ind	Generator input power
P_MPP	Power harvested by KEH at MPP
r_1	KEH PM center radius
r_2	MCS radius
r_m	PM radius
R_gen	Generator phase-to-phase resistance

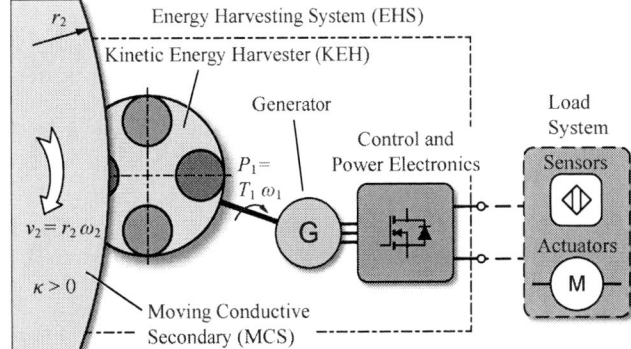

Fig. 1: Overview of the proposed Energy Harvesting System (EHS). Kinetic energy is harvested electromechanically from a moving conductive body (MCS) without mechanical contact and then converted into electrical energy using a generator which supplies a load system through a power electronics interface.

T_1	Electromagnetic torque on KEH
T_2	Torque on MCS
$T_\mathrm{fr+loss}$	Total friction torque and loss torque in generator
T_max	Electromagnetic torque on KEH at $\omega_1 = 0$
v_2	MCS surface speed
ω_1	Angular frequency of KEH
ω_2	Angular frequency of MCS

I. INTRODUCTION

In various fields of industry, remote devices such as wireless sensors [1–3] or actuators require a milliwatt- or watt-range power supply. This is most commonly achieved by employing an energy storage (e.g. batteries), wireless power transfer [4] or a wired connection to a power grid. Nevertheless, occasionally the most practical way of powering such devices is to extract the required power directly from the device's surrounding by using Energy Harvesting Systems (EHS). A diagram of such an EHS with a moving conductive body as energy source is presented in Fig. 1, while Fig. 2 illustrates the realization of the proposed EHS.

Traditionally, the term *energy harvesting* relates to low power levels, ranging from a few microwatts [5–7] to milliwatts [8–12]. However, emerging remote devices such as actuators operating on higher, watt-range power levels cannot be sup-

978-1-4673-9551-9/16 $31.00 © 2016 IEEE

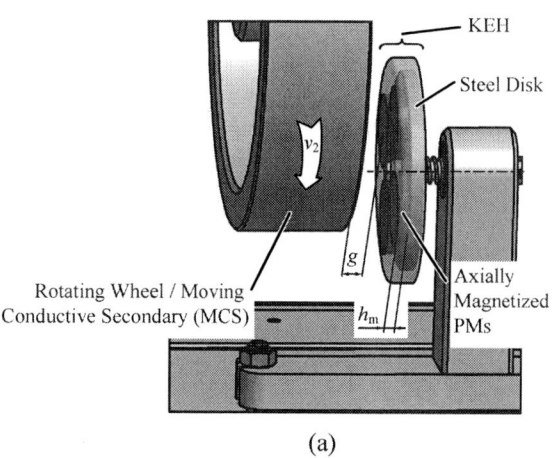
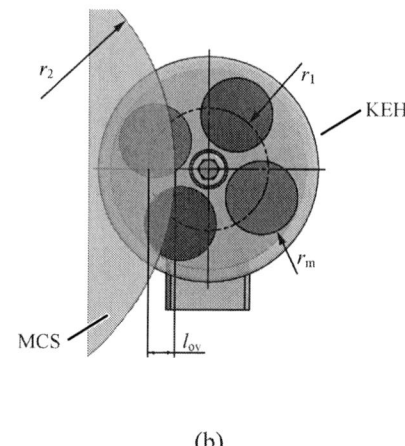

(a)   (b)

Fig. 2: Kinetic Energy Harvester (KEH), comprising axially magnetized PMs and an iron yoke, rotating in close vicinity of a rotating wheel, i.e. a Moving Conductive Secondary (MCS). Kinetic power is transferred electromechanically over an air gap g between MCS and KEH. (a) shows a side view of the system, illustrating air gap (g) and magnet height (h_m). A projection view of the system in given in (b), where radial dimensions r_1, r_2, r_m and the overlap length l_ov are defined.

plied by simple adjustments of existing harvesting topologies. Accordingly, novel energy harvesting methods need to be addressed in research.

A comprehensive overview of well-established low-power kinetic and electromagnetically energy harvesting systems can be found in [13, 14], while a watt-range kinetic EHS for emerging applications has recently been introduced in [15–17] and analyzed in detail in [18]. Power is extracted contactless in an electromagnetic way from a moving conductive body or surface, which will be referred to as Moving Conductive Secondary (MCS) in the following. Fig. 3 shows this EHS, where radially magnetized Permanent Magnets (PMs) are mounted on a magnet wheel (henceforth called the *Kinetic Energy Harvester* (KEH)) that is free to rotate around its axis. The KEH is placed above the MCS, such that its axis of rotation is parallel to the MCS surface. The PM field induces eddy currents in the MCS, and eddy-current coupling takes place between the two mechanical systems, i.e., forces apply both on the KEH and the MCS due to the induced eddy currents. Accordingly, the KEH rotates and drives a generator which converts the kinetic energy of the MCS into electrical energy. The generator may be formed by coils wound directly around the KEH as shown in Fig. 3 [18]; or alternatively, a stand-alone electrical machine can be mechanically coupled to the KEH and be used as a generator. Such an EHS, i.e. the combination of a contactless KEH and a generator could e.g. be used for extracting kinetic energy from a rotating aluminum wheel in order to supply an LED lighting system [15].

In the arrangement of [18] (cf. Fig. 3), the axis of the KEH lies parallel to the MCS surface, which leads to a limited magnetic coupling between PMs and the MCS, as only the PMs that are closest to the MCS interact with it over the narrowest region of the air gap. The rest of the PMs, on the other hand, do not contribute significantly to the magnetic coupling. In order to avoid this drawback, the EHS concept

shown in Fig. 1 is proposed in this paper. The proposed EHS features a new KEH arrangement, which is shown in Fig. 2. This new structure consists of a disk which freely rotates around its axis, and axially magnetized PMs placed on its surface facing the MCS. The disk is made of magnetic steel and provides a low-reluctance return path to the PM flux that induces eddy currents in the MCS across the air gap. The steel disk is dimensioned sufficiently thick, such that the guided flux of PMs does not lead to magnetic saturation in it. As its axis of rotation is perpendicular to the MCS, magnets interact with the MCS over a constant air gap. Therefore, a stronger

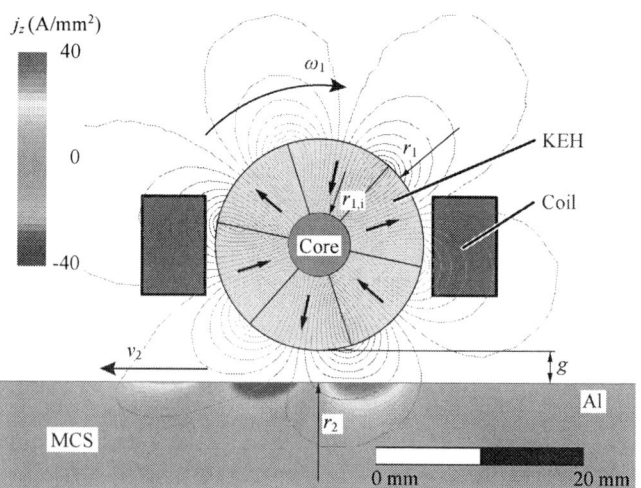

Fig. 3: The EHS analyzed in [18]. Radially magnetized PMs are mounted on a wheel (KEH) that is free to rotate around its axis. The PM field induces eddy currents in the MCS and establishes a contactless eddy-current coupling between the two mechanical systems. The mechanical energy of the KEH wheel may be converted into electrical energy either by coils wound directly around it, or by a mechanical coupling to the shaft of a generator.

978-1-4673-9551-9/16 $31.00 © 2016 IEEE   624

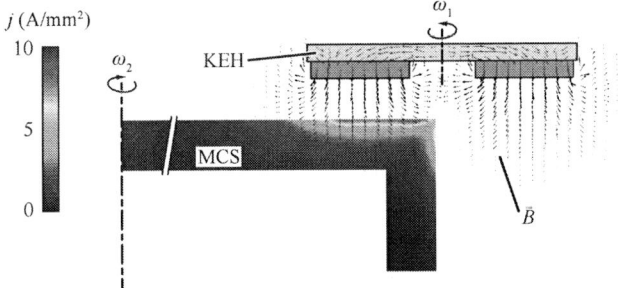

Fig. 4: Cross section of a 3-D FEM simulation. The PM flux, illustrated with an arrow field, induces eddy currents in the moving conductive secondary (MCS). Currents in the MCS are mainly induced in a skin depth in the millimeter-range. Currents and magnetic flux build up Lorentz forces which enable the kinetic energy harvesting.

electromagnetic coupling compared to [18] is achieved.

In order to focus on the new KEH design, in a first step, an off-the-shelf three-phase brushless DC (BLDC) machine is connected mechanically to the disk and employed as the generator.

This paper is about describing the physical operating principle of the presented novel system, its 3-D FEM modeling and the analysis of the influence of the main operation parameters (air gap g, MSC speed v_2 and radial overlap of KEH and MCS l_{ov}) on the system's performance.

The harvester's principle of operation is described and the conducted system modeling is elaborated in Sec. II. Sec. III details the hardware setup with the manufactured harvester prototypes and presents results obtained by measurements and simulations. Sec. III-D focuses on the scaling of harvested power under variation of operation parameters such as air gap (g), MCS speed (v_2) and overlap (l_{ov}), while Sec. IV concludes the paper.

II. Principle of Operation and System Modeling

Eddy-current couplings are a mature technology [19]. They are typically utilized in heavy-duty drivetrains for overload protection and vibration isolation as they are characterized by low maintenance requirements [20]. The functional principle of the KEH shown in Fig. 2 is principally similar to an axial-flux eddy-current coupling. However, in standard eddy-current couplings, the two shafts are usually coaxial in order to maximize the coupling efficiency [21]. On the other hand, in the proposed energy harvesting application, where the energy source is the rotation of a conductive body (e.g. a gear wheel), such coaxial arrangement may not be possible.

Moreover, the MCS movement could be translational instead of rotary, for instance when energy has to be harvested from the translational movement of a conveyor. Therefore, a non-coaxial arrangement with a partial overlap l_{ov} (cf. Fig. 2b) of the KEH and the MCS is considered in this work in order to cover a wider range of energy harvesting applications. For

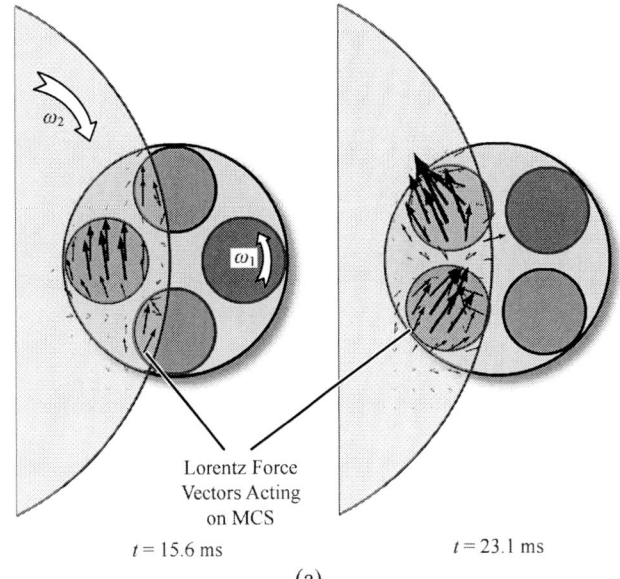

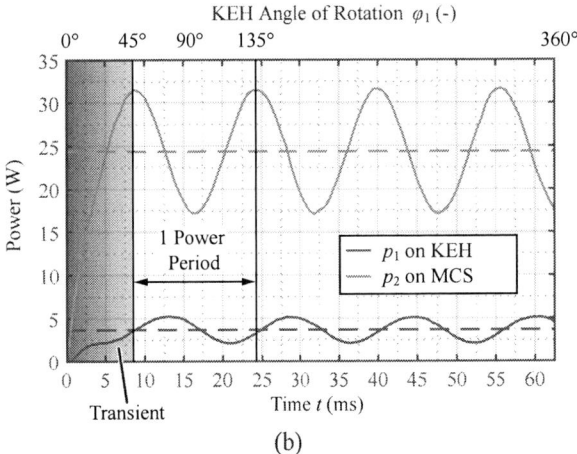

Fig. 5: Build up of torque and power in the energy harvesting system under consideration. The built up Lorentz forces in the MCS are depicted for two different KEH positions in (a) with black arrows as result of a 3-D FEM simulation. Due to the relative speed between KEH magnet field and MCS, eddy currents are induced in the MCS and Lorentz forces are built up. The Lorentz forces describe the power transfer between MCS and KEH. In the figure on the left, a large fraction of the Lorentz force contributes to torque on the KEH. In the right position, the Lorentz force magnitudes are higher, although due to their directions, less effectively contributing to torque on the KEH. The power harvested by the KEH ($p_1(t)$) and the power supplied by the MCS ($p_2(t)$) as outcome of a time-transient 3-D FEM simulation are depicted in (b).

this presented case, computationally efficient analytic models developed for coaxial eddy-current couplings (e.g. [22]) cannot be used, and therefore a FEM-based modeling approach is adopted.

Since the magnetic circuit of Fig. 2 is not continuous in neither a linear nor a rotational axis, a 3-D model is required. Moreover, as the MCS and the KEH are rotating around different axes, the overall system cannot be modeled by only

Fig. 6: Test setup of the proposed EHS. The KEH is positioned in the vicinity of the MCS as illustrated in Fig. 2. A DC machine with speed control is employed as prime mover of the MCS. The KEH extracts kinetic power over an air gap and an off-the-shelf BLDC machine is utilized as generator.

TABLE I: TEST SETUP PARAMETERS.

Parameter	Variable	KEH (1)	KEH (2)
KEH PM center radius	r_1	15 mm	25 mm
PM radius	r_m	7.5 mm	15 mm
PM height	h_m	2 mm	5 mm
Steel disk thickness	h_{Fe}	5 mm	6 mm
Qty. of PMs	n_{PM}	4	
PM grade		N48M	
PM coercivity[a]	H_{cb}	1035 A/mm	
PM remanence[a]	B_r	1.4 T	
PMs volume	V_{mag}	1.4 cm³	14 cm³
MCS radius	r_2	100 mm	
MCS length	l_2	45 mm	
MCS thickness	h_2	15 mm	
MCS material		Al: EN AW-6082 (Ac-112)	
MCS conductivity[a]	κ	24...32 MS/m	
MCS permeability	μ_2	μ_0	
MCS surface speed	v_2	5...15 m/s	
Air gap	g	3...15 mm	
Nominal operation parameters			
Air gap	g	3 mm	12 mm
MCS surface speed	v_2	10 m/s	
Overlap	l_{ov}	10.3 mm	8.7 mm

[a] Datasheet value.

the relative motion of the two mechanical systems, but the true motions of both the MCS and the KEH have to be incorporated in the FEM model by two domains of rotating mesh.

As a first step, only disk-shaped, axially magnetized NdFeB PMs with a remanent flux density of $B_r = 1.4$ T are considered in the KEH. Eddy-currents in the PMs as well as in the steel disk (cf. Fig. 2a) are neglected. An aluminum wheel with 100 mm radius, 45 mm axial length and 15 mm thickness is used as MCS. The conductivity of aluminum is taken as $\kappa_2 = 28$ MS/m for the FEM simulations, which is the mean value of the tolerance band specified by the manufacturer (cf. Table I).

Unlike in the case of the wound or cage rotors of conventional induction machines, separate guides for flux and current (i.e. teeth made of magnetic steel and copper or aluminum conductors in slots therein) do not exist in the analyzed setup. In Fig. 4, the PM flux and the induced current in the MCS are depicted for a cross section of the system as result of a FEM simulation. It can be seen that current is mainly induced in a skin depth in the millimeter-range. Hence, the force acting on an infinitesimal element of the MCS can be calculated based on the volumetric Lorentz force density

$$\vec{f}_{Lorentz} = \vec{j} \times \vec{B}. \qquad (1)$$

Accordingly, the primary (KEH) and secondary (MCS) mechanical powers are obtained by the multiplication of the volume integral of the torque-generating force components

with the specific rotational frequency

$$p_i(t) = \omega_i \, \vec{e}_{ax,i} \cdot \int_{V_{MCS}} (\vec{r} - \vec{r}_{ax,i}) \times \vec{f}_{Lorentz} \, dv, \qquad (2)$$

where $i \in \{1, 2\}$ denotes KEH or MCS, $\vec{e}_{ax,i}$ is the unity coordinate vector of the axis around which the torque is calculated and $(\vec{r} - \vec{r}_{ax,i})$ denotes the vector distance of the integration point to the torque axis.

It has been observed that the volume integration of the Lorentz force density shows better numerical stability compared to a torque computation based on the surface integration of the Maxwell's stress tensor, which is a well-known torque calculation method for FEM simulations. On the other hand, a different torque calculation method may be required in applications where the MCS is made of a magnetic material and the calculation of the cogging torque is of importance.

Fig. 5 shows 3-D FEM simulation results for KEH (2) according to Table I with $n_{PM} = 4$ magnets. Lorentz force vectors acting on the MCS are depicted in Fig. 5a for two different KEH positions. KEH and MCS powers are plotted in Fig. 5b as a function of time. Since a full mechanical rotation covers 4 power periods, it is sufficient to analyze $360°/n_{PM} = 90°$ of KEH rotation only. Moreover, the first $45°$ of KEH rotation after the start of the simulation is disregarded as it is dedicated for allowing the numeric solution to reach its steady state.

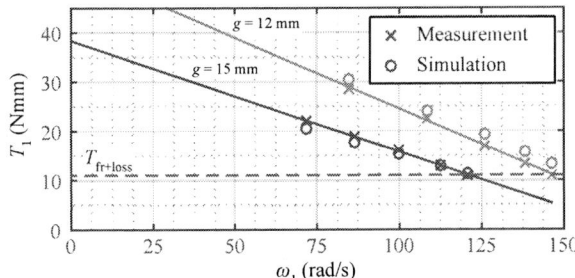

Fig. 7: Torque generation with constant MCS surface speed $v_2 = 10$ m/s. Measurements and simulations show that torque is increasing linearly with decreasing rotational frequency of the KEH ω_1 when the MCS surface speed v_2 is kept constant. An increase in air gap g from 12 mm to 15 mm leads to a more flat torque build up curve and consequently to less harvested power. KEH (2) (cf. Table I) is utilized for the measurements/simulations.

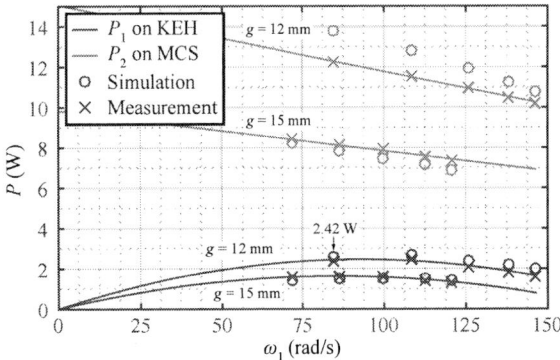

Fig. 8: Measured power extraction with KEH (2) (cf. Table I) for $v_2 = 10$ m/s and $l_{ov} = 8.7$ mm. E.g. a remote sensor could be supplied with the extracted power of ($P > 2.4$ W), which is harvested over an air gap of $g = 12$ mm. 'x' denotes measurement points, solid lines are fitted curves.

III. Test Setup and Measurement Results

Two KEH prototypes are built in order to verify the operating principle qualitatively as well as the 3-D FEM simulations quantitatively. Fig. 6 depicts the hardware test setup with KEH (2) disk (cf. Table I) mounted. Moreover, Table I lists the main parameters of the test setup and the two KEH prototypes. The measurement setup allows to adjust the MCS speed (v_2), the air gap (g) and the overlap (l_{ov}) in wide ranges. An off-the-shelf, three-phase BLDC machine is used as generator. For the measurements conducted in this work, the power electronics interface is omitted and the generator is loaded using a variable resistive load (R_L).

A. Harvested Power and Input Power Measurement

Clearly, the harvested power (P_1) and the input power (P_2) are calculated as

$$P_i = T_i \cdot \omega_i , \qquad (3)$$

with $i \in \{1, 2\}$ denoting KEH and MCS respectively, where rotational frequencies ω_i are measured impulse-based and the torque T_2 on the MCS is measured as reaction torque on the generator with a lever and spring scale system. However, the torque T_1 harvested electromechanically by the KEH cannot be obtained directly in the current setup. Instead, the power and torque harvested by the KEH (P_1 and T_1) are obtained from electrical measurements.

TABLE II: Generator parameters [23].

Parameter	Value
R_{gen}	$4.5\,\Omega$
$U_{1,nom}$	48 V
$I_{1,nom}$	1.4 A
T_{nom}	64.1 Nmm
ω_{nom}	851.4 rad/s

For an accurate estimation, the losses in the generator and in the bearing support of the KEH (cf. Fig. 6) must be taken into account. Given the RMS voltage U_1 and current I_1 at the generator's terminals, the power input to the generator can be expressed as

$$P_{ind} = (U_1 + R_{gen}\, I_1)\, I_1 , \qquad (4)$$

if the generator coil resistance R_{gen} is known (cf. Table II). Moreover, iron (core) and bearing losses of the generator $P_{l,gen1}$ must be accounted for. $P_{l,gen1,nom}$ can be obtained from the generator's datasheet parameters (cf. Table II) for the nominal load point as

$$P_{l,gen1,nom} = P_{ind,nom} - T_{nom}\, \omega_{nom} , \qquad (5)$$

with the nominal torque T_{nom} and nominal rotating frequency ω_{nom}. In a simple, yet accurate approach (as will be confirmed later), $P_{l,gen1}$ is assumed to depend linearly on the rotational frequency. Therefore, an accumulated generator loss torque

$$T_{l,gen} = P_{l,gen1,nom}/\omega_{nom} = 4.5\text{ Nmm} \qquad (6)$$

is estimated.

$T_{fr} = 6.5$ Nmm is assumed as the friction torque for the bearings (two bearings of type *608-2Z*) supporting the KEH, according to the bearing manufacturer's data [24]. Therefore, the electrical power measurements are corrected by a total loss torque of

$$T_{fr+loss} = T_{l,gen} + T_{fr} = 11\text{ Nmm} . \qquad (7)$$

Finally, the harvested torque is calculated as

$$T_1 = \frac{P_{ind}}{\omega_1} + T_{fr+loss} . \qquad (8)$$

B. Torque and Power Build Up

Simulation and measurement results with two different air gaps are presented in Fig. 7 and Fig. 8. Power extraction of $P_1 > 2.42$ W is achieved with KEH (2) (cf. Table I) over an air gap of 12 mm for a MCS speed of $v_2 = 10$ m/s. With the introduced compensation of loss torque on the measurement

978-1-4673-9551-9/16 $31.00 © 2016 IEEE

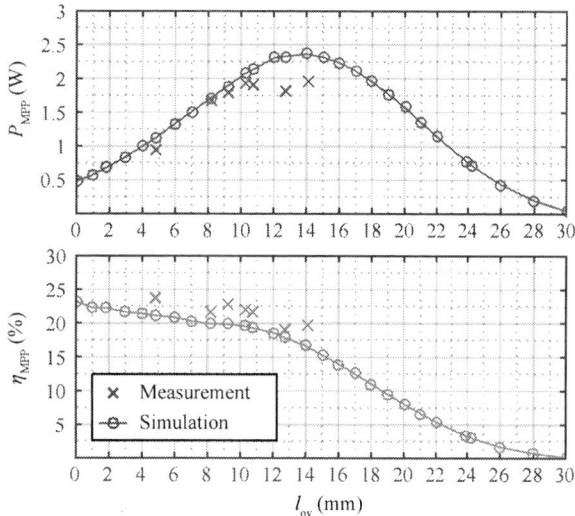

Fig. 9: Influence of overlap (l_{ov}) on the MPP operation. Power and efficiency results for the MPP are extracted according to Sec. III-C from measurements and a total number of 140 3-D FEM simulations. For small overlap values, the coupling between KEH and MCS gradually increases, also P_{MPP} increases, while the efficiency η_{MPP} drops weakly. With the system under consideration, maximum power can be extracted with $l_{ov} \approx 14\,\text{mm}$ and increasing the overlap further is not favorable since both efficiency and power decrease in this region.

data, it can be observed that measurements and results of 3-D FEM simulations are matching for a range of KEH rotational frequencies ω_1 and air gaps g. Therefore, it can be inferred that the time transient 3-D FEM models described above can be used to model the proposed KEH structure accurately.

C. Maximum Power Point (MPP)

For the further analysis of the given energy harvesting system, it is valuable to analyze the Maximum Power Point (MPP) for a given set of operation parameters (g, v_2, l_{ov}), with respect to the KEH rotating frequency ω_1. Based on the results given in Fig. 7, one can infer that the torque T_1 on the KEH can be modeled as linear function

$$T_1 = T_{max} - k_{T1}\,\omega_1\,, \tag{9}$$

where T_{max} and k_{T1} can be identified by at least two measurements and/or simulations with different ω_1. Similar to the maximum power transfer theorem for linear electric networks, the maximum KEH power can be calculated as

$$P_{MPP} = \frac{T_{max}^2}{4\,k_{T1}}\,. \tag{10}$$

Furthermore, the results depicted in Fig. 8 show that the power input at the MCS can be approximated as a linear function in a similar way,

$$P_2 = P_{2,max} - k_{P2}\,\omega_1\,, \tag{11}$$

where $P_{2,max}$ and k_{P2} are again identified by at least two measurement and/or simulation points with varying ω_1. Finally, the

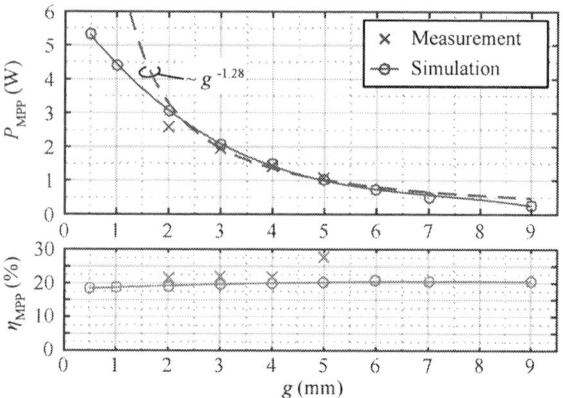

Fig. 10: Influence of air gap on energy harvesting. Clearly, extracted power decreases monotonously with increasing air gap, whereas little influence on efficiency can be observed.

system's efficiency at the MPP can be calculated as

$$\eta_{MPP} = \frac{T_{max}^2}{4\,(P_{2,max}\,k_{T1} - T_{max}\,k_{P2})}\,. \tag{12}$$

D. Impact of Operation Parameters

It is well understood that the width g of the air gap influences the performance of an eddy-current-based electromechanical energy harvester significantly. As shown in [18, 25], the air gap is indeed a limiting factor for electromechanical energy harvesting. As briefly discussed in Sec. II and shown in Fig. 5, different angular rotor positions of the KEH lead to different conditions for torque build up and harvested power. For certain PM positions, it could even occur that insignificant or no power is harvested, while a braking torque acts on the MCS only. Therefore, it is not clear *a priori* how a variation in overlap l_{ov} affects the maximal harvested power P_{MPP} and the efficiency η_{MPP}. In addition, the MCS speed v_2 also influences the harvester operation. In the following, the influence of these operation parameters is analyzed with 3-D FEM simulations and compared to measurements. The smaller prototype KEH (1) with the nominal operation parameters according to Table I is considered in the following, unless otherwise specified.

Both simulations and measurements show that the maximum harvested power P_{MPP} depends on the overlap l_{ov}. The simulation results depicted in Fig. 9 predict a maximum at $l_{ov} \approx 14\,\text{mm}$. Measurement results agree very well with the simulations up to $l_{ov} \approx 11\,\text{mm}$, although a discrepancy can be observed around $l_{ov} \approx 14\,\text{mm}$, which will be investigated in future work.

For $l_{ov} < 14\,\text{mm}$, the coupling between KEH and MCS gradually increases, also increasing P_{MPP}, while the efficiency η_{MPP} reduces weakly. Above the maximum of P_{MPP}, therefore in the region $l_{ov} > 14\,\text{mm}$, P_{MPP} decreases again and efficiency η_{MPP} drops significantly. The drop in the efficiency is caused by increasing Lorentz force components

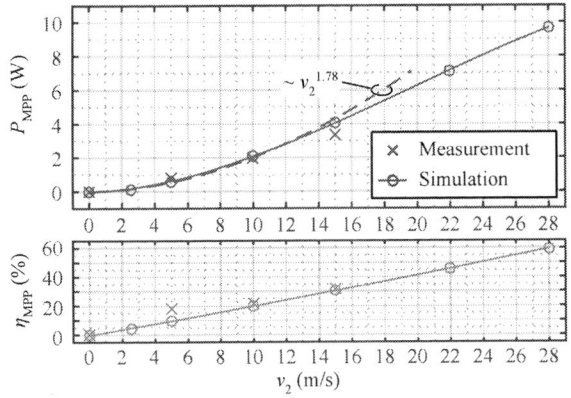

Fig. 11: Influence of MSC surface speed (whose energy sources the harvesting) on the system operation. Extracted power increases stronger than linearly with speed. Both simulations and measurements show that efficiency also increases with speed.

that contribute to the braking torque applying on the MCS, but not to the KEH torque. Consequently, the region slightly below $l_{ov} \leq 14$ mm can be utilized to resolve the power-efficiency trade-off according to the application at hand.

With the provided hardware setup, measurements in the air gap range 2 mm $\leq g \leq 5$ mm, given in Fig. 10, indicate that the proposed electromechanical energy harvesting is capable of operating with a comparably large air gap. A remarkable power of $P_{MPP} \approx 1.5$ W can be harvested electromechanically over an air gap $g = 4$ mm, which is twice the magnet height $h_m = 2$ mm. Simulation results show an estimated power scaling over air gap of $P_{MPP} \propto g^{-1.28}$ in the air gap range 2 mm $\leq g \leq 6$ mm. This power scaling indicates a better performance for large air gaps of the proposed system in comparison to previously discussed electromechanical energy harvesters, where the harvested power scales with approx. $P \propto g^{-2}$ [18, 25]. With decreasing air gap, i.e. values smaller than the magnet height ($g < 2$ mm), P_{MPP} increases less than expected. Air gap reluctance is not dominating the build up of flux for small air gap, but the reluctance path in the MCS limits the generation of Lorentz force and harvested power. For very large air gaps ($g > r_1/2$), harvested power vanishes, since the PM flux closes unfavorably between the magnets in the air gap and not over the MCS. Moreover, it can be observed that air gap hardly affects the efficiency η_{MPP}, as simulations show a variation $< 2\%$ over the investigated air gap range 0.5 mm $\leq g \leq 9$ mm.

As the proposed system harvests the kinetic energy/power of a MCS, its surface speed v_2 clearly affects the level of harvested power. Similar to the previous considerations of overlap l_{ov} and air gap g, simulations and measurements were conducted over a speed range, and results are shown in Fig. 11. For a speed range $v_2 \leq 15$ m/s, a scaling $P_{MPP} \propto v_2^{1.78}$ is estimated based on simulation results. Also the measurement results are in good accordance with the estimated scaling and simulation results. Surprisingly, a major influence of speed v_2 on the efficiency η_{MPP} can be observed. In the

speed range, which is interesting for the proposed harvester, $v_2 \leq 28$ m/s ≈ 100 km/h, a linear scaling of efficiency with speed $\eta_{MPP} \propto v_2$ is suggested by simulation results. Task of future research is to gain further insight how v_2 affects η_{MPP} and to confirm the simulated efficiency over the full speed range with measurements.

IV. CONCLUSION

A novel type of Kinetic Energy Harvester (KEH) of a contact-less axial-flux permanent-magnet electromechanical Energy Harvesting System (EHS) is presented for watt-range power harvesting from a Moving Conductive Secondary (MCS). The proposed new arrangement of the KEH enhances the magnetic coupling between the harvester and the MCS compared to earlier designs [15–18]. 3-D time-transient FEM models are developed for modeling and optimization of the KEH, and two prototypes are manufactured in order to verify the operating principle qualitatively as well as the FEM simulation models quantitatively. Measurements show a power of 2.42 W harvested from an aluminum surface moving with 10 m/s, over 12 mm air gap using prototype KEH (2) with ≈ 14 cm³ magnet volume.

Simulations and measurements demonstrate that the proposed EHS allows to supply watt-range remote sensors and actuators. Especially the ability of the proposed KEH for harvesting over a wide air gap ($g = 12$ mm) allows to encapsulate the system, such that power can be harvested also in harsh industrial environments.

V. ACKNOWLEDGMENT

The authors would like to express their sincere appreciation to Nabtesco Corp., Japan, for the financial and technical support of research on energy harvesting technologies at the Power Electronic Systems Laboratory, ETH Zurich, which provided the basis for achieving the results presented in this paper. In particular, inspiring technical discussions with K. Nakamura and Y. Tsukada are acknowledged. Moreover, the authors acknowledge the support of Comsol Multiphysics GmbH (Switzerland).

REFERENCES

[1] V. Raghunathan, S. Ganeriwal, and M. Srivastava, "Emerging techniques for long lived wireless sensor networks," *IEEE Communications Magazine*, vol. 44, no. 4, pp. 108–114, April 2006.

[2] R. Vullers, R. Schaijk, H. Visser, J. Penders, and C. Hoof, "Energy harvesting for autonomous wireless sensor networks," *IEEE Solid-State Circuits Magazine*, vol. 2, no. 2, pp. 29–38, Spring 2010.

[3] J. Azevedo and F. Santos, "Energy harvesting from wind and water for autonomous wireless sensor nodes," *IET Circuits, Devices & Systems*, vol. 6, no. 6, pp. 413–420, November 2012.

[4] K. O'Brien, G. Scheible, and H. Gueldner, "Analysis of wireless power supplies for industrial automation systems," *Proceedings of the IEEE Industrial Electronics Society Conference (IECON)*, November 2003.

[5] B. Mack, "Elektromagnetische Energiewandler mit dem Potential zur grossflächigen Anwendung (in German)," Ph.D. dissertation, Albert Ludwigs-Univ. Freiburg, 2009.

[6] H. Lhermet, C. Condemine, M. Plissonnier, R. Salot, P. Audebert, and M. Rosset, "Efficient power management circuit: from thermal energy harvesting to above-IC microbattery energy storage," *IEEE Journal of Solid-State Circuits*, vol. 43, no. 1, pp. 246–255, January 2008.

[7] M. Midrio, S. Boscolo, A. Locatelli, D. Modotto, C. De Angelis, and A.-D. Capobianco, "Flared monopole antennas for 10 μm energy harvesting," in *2010 European Microwave Conference (EuMC)*, September 2010.

[8] T. Krupenkin and J. A. Taylor, "Reverse electrowetting as a new approach to high-power energy harvesting," *Nature Communications*, vol. 2, no. 448, August 2011.

[9] D. Brunelli, C. Moser, L. Thiele, and L. Benini, "Design of a solar-harvesting circuit for batteryless embedded systems," *IEEE Transactions on Circuits and Systems I: Regular Papers*, vol. 56, no. 11, pp. 2519–2528, November 2009.

[10] E. Sardini and M. Serpelloni, "Self-powered wireless sensor for air temperature and velocity measurements with energy harvesting capability," *IEEE Transactions on Instrumentation and Measurement*, vol. 60, no. 5, pp. 1838–1844, May 2011.

[11] Q. Sun, S. Patil, N.-X. Sun, and B. Lehman, "Phase/RMS maximum power point tracking for inductive energy harvesting system," *Proceedings of the IEEE Energy Conversion Congress and Exposition (ECCE)*, September 2015.

[12] J. Moon and S. B. Leeb, "Enhancement on energy extraction from magnetic energy harvesters," *Proceedings of the IEEE Energy Conversion Congress and Exposition (ECCE)*, September 2015.

[13] D. Zhu, S. Beeby, "Kinetic energy harvesting," in *Energy harvesting systems : principles, modeling and applications, ch. 1*. T. J. Kazmierski, S. Beeby, editors, Springer, 2011.

[14] S. Beeby, T. O'Donnell, "Electromagnetic energy harvesting," in *Energy Harvesting Technologies, , ch. 5*. S. Priya, D. J. Inman, editors, Springer, 2009.

[15] D. Strothmann, "Device for contactless current generation, in particular bicycle dynamo, vehicle lighting system and bicycle," DE Patent WO 2013/004 320 A1, January 10, 2013.

[16] Magnic Innovations GmbH & Co KG, checked: 19.11.2015. [Online]. Available: http://www.magniclight.com

[17] Shun-Fu Technology Corp., "Induktionsgenerator," DE Gebrauchsmuster DE 202 014 100 380 U1, 04 24, 2014.

[18] M. Flankl, A. Tüysüz, and J. W. Kolar, "Analysis of a watt-range contactless electromechanical energy harvester facing a moving conductive surface," *Proceedings of the IEEE Energy Conversion Congress and Exposition (ECCE)*, September 2015.

[19] E. Davies, "An experimental and theoretical study of eddy-current couplings and brakes," *IEEE Transactions on Power Apparatus and Systems*, vol. 82, no. 67, pp. 401–419, August 1963.

[20] A. S. Erasmus and M. Kamper, "Analysis for design optimisation of double PM-rotor radial flux eddy current couplers," *Proceedings of the IEEE Energy Conversion Congress and Exposition (ECCE)*, September 2015.

[21] T. Lubin and A. Rezzoug, "3-D analytical model for axial-flux eddy-current couplings and brakes under steady-state conditions," *IEEE Transactions on Magnetics*, October 2015.

[22] J. Wang, H. Lin, S. Fang, and Y. Huang, "A general analytical model of permanent magnet eddy current couplings," *IEEE Transactions on Magnetics*, vol. 50, no. 1, pp. 1–9, January 2014.

[23] Maxon Motor AG, "Datasheet Maxon EC-max 30; 272765," checked: 19.11.2015. [Online]. Available: http://www.maxonmotor.com/medias/sys_master/root/8816804986910/15-227-EN.pdf

[24] SKF Group, "SKF bearing calculator," checked: 19.11.2015. [Online]. Available: http://webtools3.skf.com/BearingCalc/

[25] M. Flankl, A. Tüysüz, and J. W. Kolar, "Analysis and power scaling of a single-sided linear induction machine for energy harvesting," *Proceedings of the IEEE Industrial Electronics Society Conference (IECON)*, November 2015.

Design of Rare-Earth Free Five-Phase Outer-Rotor IPM Motor Drive for Electric Bicycle

Md. Zakirul Islam
Student Member, IEEE
Electrical and Computer Engineering
The University of Akron
Ohio, United States 44325
Email: mi15@zips.uakron.edu

Seungdeog Choi
Member, IEEE
Electrical and Computer Engineering
The University of Akron
Ohio, United States 44325
Email: schoi@uakron.edu

Abstract—**This paper presents the design considerations of a five-phase outer rotor interior permanent magnet motor (IPM) for an electric bicycle (e-bicycle). Conventional e-bicycles employ three phase outer rotor surface permanent magnet motors which are less reliable compared to the five-phase IPM motors. Moreover, the traditional electric motors use rare-earth permanent magnet materials such as neodymium which are expensive. Ferrite based magnets can be a better substitute to rare earth materials to make the motor more affordable. In this study, the design of a highly reliable and economical five-phase ferrite based IPM motor to be accommodated in a hub motor drive (HMD) for an e-bicycle is presented. The proposed motor design is optimized using a lumped parameter model (LPM) based optimizer. Detailed analysis on the design parameters and constraints considered for optimization is discussed. Thorough finite element simulations in terms of torque developed, cogging torque, back-EMF, and mechanical stress have been conducted to verify the effectiveness of the proposed motor design.**

Keywords—Permanent magnet machines; interior permanent magnet motor; magnetic circuits; ferrites; design optimization; light electric vehicles; computational efficiency; finite element analysis

I. NOMENCLATURE

L	Length of the permanent magnet
W	Thickness of the permanent magnet
S_h	Slot height
T_w	Tooth width
α_p	Flux barrier span in degree
R_{so}	Outer radius of the stator
g	Air gap length
R_{ro}	Outer radius of the rotor
R_{si}	Inner radius of the stator
A_s	Cross-sectional area of stator tooth pitch
A_m	Cross-sectional area of permanent magnet
λ_{PM}	Flux linkage due to permanent magnet
S_o	Slot opening
T_r	Rated Torque of the motor
R_{ph}	Resistance of phase-A winding
T_r	Rated Torque of the motor
I_m	Equivalent flux of the permanent magnet
I_r	Equivalent leakage flux of the rotor ribs
R_g	Air gap reluctance
R_m	Rotor flux barrier reluctance
I_d	Current of d-axis equivalent circuit
R_t	Rotor tooth reluctance
R_{rb}	Rotor yoke reluctance
R_{sy}	Stator yoke reluctance
I_q	Current in q-axis equivalent circuit
F_q	q-axis MMF
F_d	d-axis MMF
P_{core}	Core loss
P_{out}	Total mechanical output power
P_a	Copper loss

II. INTRODUCTION

The electric bicycle is essentially a mechanical bicycle supplemented with an electric motor. Over the past decade there has been a growing interest in e-bicycles with a number of technologically innovative commercial bicycles available to a growing customer base [1], [2]. With developing technologies, the worldwide sales of e-bicycles are predicted to grow from nearly 32 million units in 2014 to over 40 million units in 2023 [3]. A recent survey conducted indicated that cost and frequent motor fault were among the major obstacles in using e-bicycles while low-price, light-weight, longevity and smart features were among the major incentives in using them [3], [4].

The electric motor is the fundamental and key deciding component in the design of an e-bicycle. The type of electric motor and its performance characteristics would be consequential to the overall performance of the e-bicycle. The higher power density of permanent magnet motors [5] makes them a more promising option as an electric motor to be employed in these e-bicycles. Surface mounted permanent magnet (SPM) motors have been considered in the e-bicycle drives designed in [6]. The drawback of these motors is that the amount of magnet material required is relatively higher and they also requires careful retention of the magnets which tend to pull out. Switched reluctance motors adopted in [7] contribute to a significant amount of mechanical vibration which is not desirable in this application. Additionally, these motors also provide a low torque density. Axial flux motors considered in [8], [9] have higher axial force which could potentially cause an additional stress on the bearing. IPM motors can be potentially applied in e-bicycle which are designed with buried magnets inside the rotor. The use of IPMs with reduced magnet content as an alternative to SPMs has been studied in [10], [11]. These motors provide a more robust and reliable design as they are not prone to magnet pull out. Besides providing a significantly smoother operation, IPM motors generate torque

978-1-4673-9551-9/16 $31.00 © 2016 IEEE

using the hybrid torque generation principle wherein it utilizes both magnetic torque as well as reluctance torque [12] allowing it to have high torque density as well.

The magnet material being utilized in these IPM motors plays a key role in the performance of the machine. Predominantly, there are two types of magnetic materials that are used in electric machines rare earth (Neodymium based) and ceramic (Ferrite based) [13]. The remnant flux density of Neodymium magnets are significantly higher compared to Ceramic magnets. However, these magnet supplies and costs have fluctuated in recent years [14] establishing ferrite based motor designs as a more economical option.

As the motor is directly connected to the wheel of the e-bicycle, it becomes imperative that these motors should have lower torque ripple and mechanical vibrations for smooth operation. Five phase fractional slot concentrated windings provide a lower torque ripple and therefore lower mechanical vibration [15]. In addition to the reduced torque ripple, these windings also provide a higher torque density and better fault tolerant control compared to three phase windings [16], [17].

A major requirement of electric motors being employed in e-bicycles would be the capability of providing higher torque per motor volume [9]. External rotor motors with higher number of pole can provide a higher output torque. These types of motors also allow for relatively easier direct integration of the motor in the wheel without coupling [8].

In this study, the design procedure for a ferrite based five-phase fractional-slot concentrated winding IPM outer rotor motor for an e-bicycle is presented. The multi-objective optimization based on lumped parameter model (LPM) of the design has been implemented with low cost, less weight, and high efficiency as objectives.

III. MOTOR DESIGN

The hub motor system design has been divided into three steps in this study. First step, is the design of the electric motor which incorporates the modeling, optimization and finite element analysis (FEA) of the electric motor. Secondly, design of the control system which includes controlling scheme design and analysis. Finally, the mechanical structure design to ensure sustainable operation for a maximum weight load and for a maximum range of travel. This section discusses only the motor design part.

A. Mathematical Modeling

The IPM motor proposed for this A1-HMD is an outer-rotor motor with permanent magnets inserted in its flux barrier. The motor cross-section of Fig. 1(a) shows the dimension of the motor including the stator coils and rotor permanent magnets. The geometrical parameters to model this motor has been shown in Fig. 1(b). These parameters have been used in the LPM based performance evaluator.

To evaluate the performance of this IPM motor a LPM has been designed which was utilized previously for IPM motor and interior rotor PM assisted Synchronous reluctance motor (PMaSynRM) [19–21]. As IPM motor and PMaSynRM has analogous torque generation principle, similar approach can

TABLE I.   SPECIFICATIONS OF THE PROPOSED MODEL

Parameter	Value
Number stator slot/pole	20/18
Diameter (mm)	190
stack length (mm)	45
Number of phase	5
Rated torque/Speed	11Nm/260rpm
Rated Power	300W
Rated Phase current	4.5A

be followed to for the performance evaluation of the studied IPM motor.

The mentioned LPM uses the d-axis and the q-axis equivalent circuit model of the motor [19–21] to get the equivalent inductances Ld and Lq which are used to measure the torque output as in (5). To get these inductances, the reluctances mentioned in Fig. 2 have to be measured. For measuring these reluctances we need the geometrical parameters of the stator and rotor mentioned in Fig. 1(b).

The reluctance equations for this outer rotor motor can be express as

$$R_m = \frac{2LA_s}{gA_m} \tag{1}$$

$$R_{gq} = \frac{\Delta\alpha_s}{\Delta\alpha_k} \tag{2}$$

$$R_{gd} = \frac{g}{\mu_0 \Delta\alpha_k lr} \tag{3}$$

$$R_{cd} = \frac{l_d}{\mu_0 B_c A_d} \tag{4}$$

Here α is the angular distance between q-axis and rotor ribs. For this single barrier based IPM $\Delta\alpha_k$ refers the angular distance of the ribs to q-axis and ribs to d-axis. The parameter r is the distance between center of the stator and center of the air gap. A_d is the effective cross sectional area in the d-axis flux path and B_c is the relative permeability of the core material. Rest of the parameters are introduced in the nomenclature.

In Fig. 2(a), the reluctance R_g is reluctance in the air gap for d-axis equivalent circuit. The reluctance R_1 and R_{rb} are incorporated in R_{cd} as per (4) . The q-axis equivalent circuit's reluctances are mentioned in the in Fig. 2(b). The reluctance R_{gq} as in (2), is the air gap reluctance for this q-axis circuit.

Equation (5) is used to find I_d and I_q to meet the desired specifications in Table I. Equation (6) and (7) are used to measure the efficiency (η) and cost of the motor ($Cost_M$). These values are used to find the optimized design based on a multi-objective cost function (8). Further discussion about optimization is done in the following subsections.

$$T_r = \frac{m}{2}\frac{p}{2}[\lambda_{PM}I_d + (L_d - L_q)I_dI_q] \tag{5}$$

$$\eta = \frac{P_{out}}{P_{out} + P_{core} + P_a} \tag{6}$$

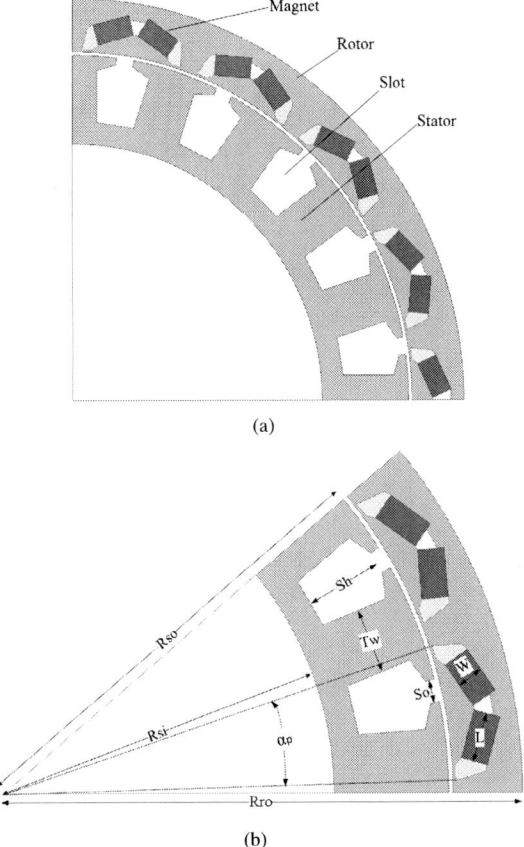

(a)

(b)

Fig. 1. (a) Motor cross-section showing different components of the motor, and (b) geometric parameters for the MEC and optimization.

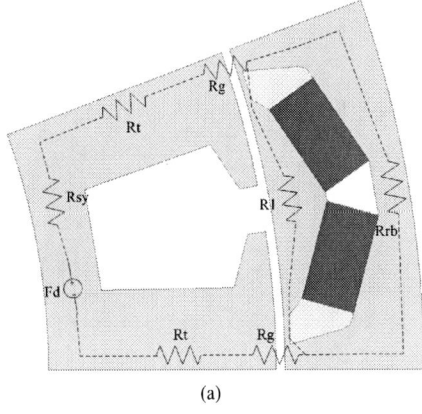

(a)

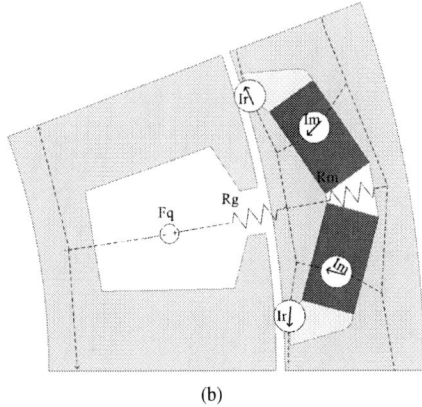

(b)

Fig. 2. Equivalent circuit diagram of (a) d-axis, and (b) q-axis flux path.

$$Cost_M = C_{Cu} + C_{PM} + C_{core} + C_x \qquad (7)$$

Here, P_{core} is total core loss. Production related parameter C_x is dependent on many factors including amount of production, labor cost, and initial inventory cost. Copper cost (C_{Cu}), magnet cost (C_{PM}), steel costs (C_{core}) are mainly dependent on their volume for large scale production.

$$OF = min\left\{ a(1/\eta)^2 + b.MC^2 + c.I_a^2 + d.Weight^2 \right\} \qquad (8)$$

Here, MC which is the machine cost which includes $Cost_M$, and also the cost of other electronic components. I_a is the rms current of phase-A, it represents the per phase current to produce the rated torque of the motor. Weight is the total weight of motor, battery, bearing, housing and all other components in the hub.

The winding configuration for this five-phase motor is shown in Fig. 3. The polarities of different coil sides in this half cross-section describes the relative position of different phase coils.

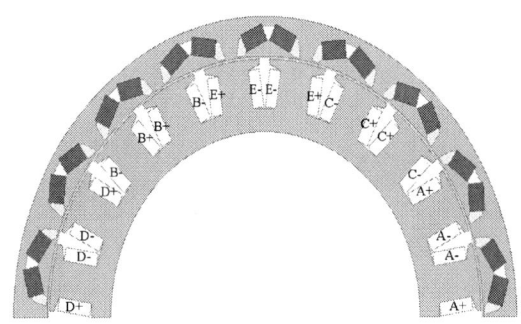

Fig. 3. The polarity of coil sides of the five-phase windings.

B. Design Optimization

This subsection focuses on the optimization of the motor considering multiple objectives including motor cost, weight, efficiency and rated current. Differential evolution optimization procedure has been utilized for this type of multi-objective optimization [19–21].The mathematical modeling defined in the previous subsection has been used for performance evaluation and optimization. There are some constraints and specifications for the designing and optimization of the motor. Table I describes the specifications to be met. Table II shows the list of some constraints which have to be satisfied by the designed motor. These constraints are optional and these are dependent on many conditions such as the features of the geographical terrain on which the user intends to use the bicycle, the

TABLE II.   CONSTRAINTS OF MOTOR DESIGNING

Name of constraint	Level
Terminal voltage	<40V
Outer diameter of the rotor	<200mm
Inner diameter of the stator	>55mm
Maximum current desnity in the stator	<6Amm-2
Cost of Motor	$200
Weight	15lb

TABLE III.   WEIGHT OF HMD COMPONENTS

Component Name	Weight (lb)
Weight of the Battery	4
Weight of Magnet	0.95
Weight of Rotor	3
Weight of Stator	5
Weight of the Motor	9.5
Wight of Whole Hub	15

TABLE IV.   RANGE OF DESIGN VARIABLE

Variable Description	Minimum Value	Maximum Value	Optimum Value
L (in mm)	6.5	10.5	8.8
W (in mm)	3.5	6.5	5.6
Rsi (in mm)	55	85	58
Rro/Rsi	0.95	1.75	1.63
Rso/Rsi	1.2	1.5	1.37
α_p	0.6	0.9	0.85
Sh/(Rso-Rsi)	0.5	0.9	0.54
Tw/(wsa+Tw)	0.45	0.85	0.52

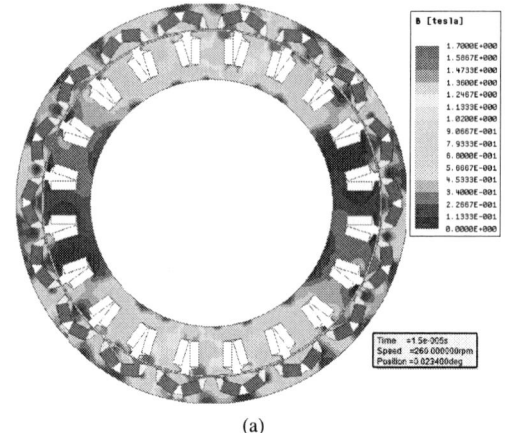

(a)

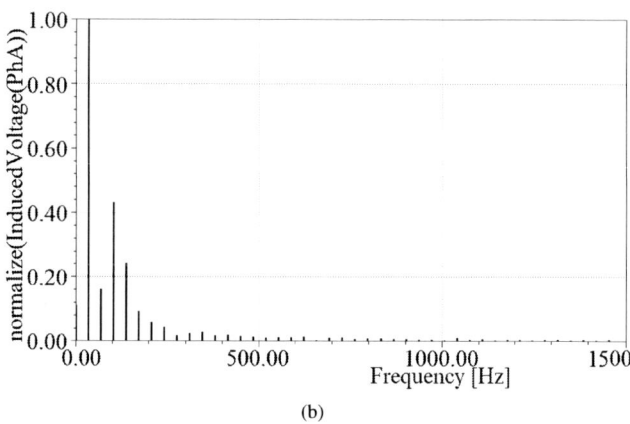

(b)

Fig. 4.   (a) The flux density in the rotor and stator of the motor during rated excitation, and (b) the normalized harmonics components of the back-emf at rated speed.

environmental conditions as well as user specific needs. There can be many other constraints for the design, but these are the most common ones.

The design has to be optimized from the view of cost, efficiency, weight, and torque. The DES optimization procedure has been utilized before for three-phase and five-phase Permanent magnet assisted synchronous reluctance motors [19–21]. The similar strategy will be used for this case. Fig. 5 illustrates the steps for the whole design optimization process.

In the objective function (8), the coefficients a, b, c, and d, are the weight of the corresponding objectives to define the minimizing emphasis on each objective. All the objectives are in per unit value. In this objective function, cost, efficiency, weight, and rated per phase current have been taken to optimize the motor considering the demands and comforts of the customers. The base values for the per unit calculation are determined from an initial optimized design done using an FEA tool.

C. FEA Analysis

The optimized design has been exported from the LPM tool to the FEA tool. The LPM optimized design has been simulated to see the torque, cogging torque, efficiency, inductances and flux linkages. Fig. 4(a) shows the flux density in the rotor and the stator for the rated condition. It shows the motor core is not over saturated during the rated condition of the motor. Fig. 4(b) shows the back-emf harmonics component of the motor. The harmonic components are the result of the fractional slot concentrated winding considered in the design.

Fig.6(a) shows the torque characteristics of the motor with rated current excitation at 260 rpm. This speed corresponds to 20 mile per hour speed of the e-bicycle. The peak to peak torque ripple is 6%. Fig. 6 (b) shows the cogging torque curve of the motor which is the motor torque without any stator excitation while rotating the motor at an angular speed of 1 rad/s. The peak to peak ripple of this torque is as low as 100 mNm, which will ensure very smooth and disturbance free operation during motor turn-off.

The FEA simulations show the efficiency of the motor is 90%. The rated current is 4 A (per phase rms current) to get the rated torque of 11 Nm. The power factor of the motor is 0.6. The eddy current loss was ignored for the magnet as the conductivity of Ceramic magnets are very low. For all the simulation the temperature considered was 24°C.

IV.   INVERTER AND CONTROLLER

The e-bicycles HMD is able to electrically assist the user based on the throttle input or pedal effort generated by user. This A1-HMD has a Bluetooth device embedded inside the hub to communicate between the mobile phone and the motor controller. The Bluetooth device receives commands from the mobile phone and sends reference level of assistance to the controller. Fig. 7(a) shows the phone is communicating with the A1-HMD to command the motor to operate. This

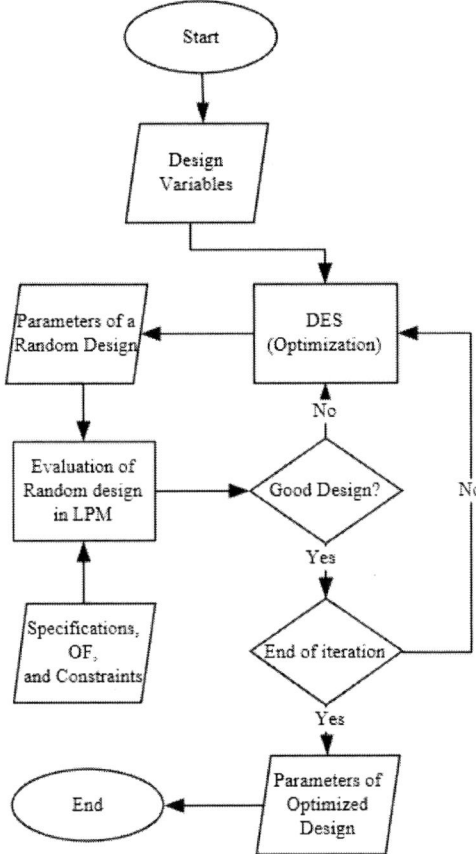

Fig. 5.   Optimization process flow block diagram.

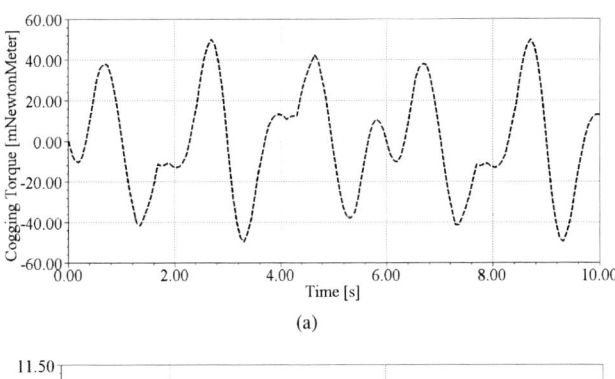

(a)

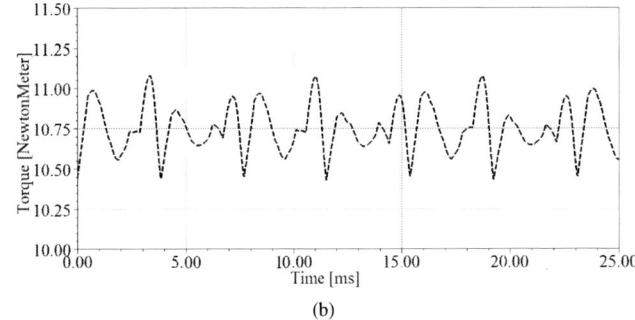

(b)

Fig. 6.   (a) Cogging torque vs time and (b) torque vs time curve for the proposed motor.

communication is bidirectional in type and it provides the mobile with information of the A1-HMD consisting battery state, motor speed, and current.

The assist level reference generated from the mobile phone is based on several factors predefined in the A1-HMD software. These factors will vary based on specific demand of user or battery state or operating condition. If the mobile is not connected to the mobile station because it is out of reach, then the HMD will run with the predefined setup which will also be the default mode of operation of the e-bicycle.

Fig. 7(b) shows the block diagram of the controller. The reference torque is sent to the controller based on required assistance level of the user. It creates the d-axis (I_{dref}) and q-axis (I_{qref}) reference currents. These reference currents are fed to corresponding PI controllers. The PI controller creates it output based on the difference between the reference current and actual motor current. The outputs of the PI controllers are provided as inputs of the inverse park transformation. Five PWM control signals are generated by the inverse park transformation to be used as inverter inputs. Using these control signals, the inverter generates five phase currents to run the IPM motor.

V.   STRUCTURAL AND THERMAL DESIGN

The cross-section of the HMD is given in Fig. 8(a). It has batteries in the inner and the outer part of the motor. FIg. 10(a)

shows the static and rotating part of the motor.

The capability of the wheel hub motor is heavily dependent upon how quickly heat can be dissipated. Heat is produced from several sources for the hub motor including stator, inverter, and batteries. The rotor often is a heat source however with the permanent magnet design chosen, this is not as much of an issue. The proposed design does not seal the stator as shown in Fig. 9 and 10 which helps this model to radiate the generated heat very quickly to the environment. Competitor designs often enclose the entire assembly with minimal vents added to the hub motor which limits their heat dissipation system.

Fig. 9 shows the exploded view of the HMD to illustrate the various components, their arrangement and structure of the motor. A sealed structure is needed to keep particles and moisture out of the rotor. The sealing of the motor is done by two step sealing with hard rubber as shown in Fig. 8(b). Although there will be a certain amount of rolling resistance due to this sealing, it is required for the safety of the electronic components. The motor is supposed to generate a certain amount of torque to prevent this resistance at any moment of operation.

The stress analysis, has been done for the HMD to ensure the rotor operates properly even at twice the rated and allowable maximum speed of the motor. Fig. 11(a) and 11(b) shows the stress generated at the ribs are the maximum stress to be tolerated by the rotor. Fig. 12 shows the total deformation of the rotor structure at that speed. The total deformation is maximum in the middle of the rotor pole piece as expected. The deformation is very low and it predicts the motor will be operating properly without deforming at that speed level.

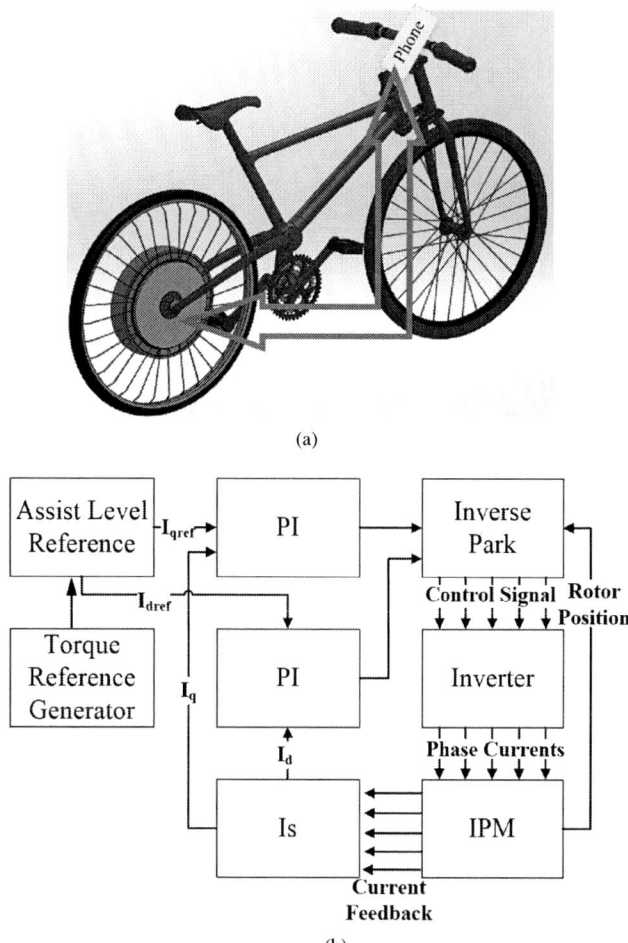

(a)

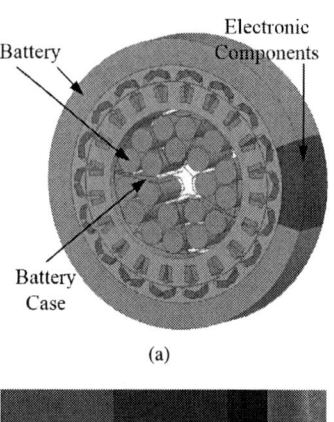

(a)

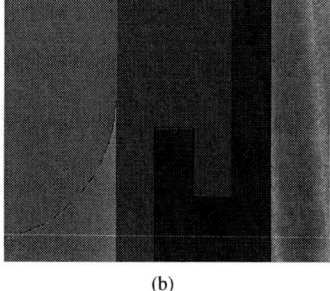

(b)

Fig. 8. (a) Three dimensional cross-section view of hub showing battery, battery casing, and electronics components chamber and (b) seal structure.

(b)

Fig. 7. (a) Controller block diagram, and (b) control implementation in the bicycle.

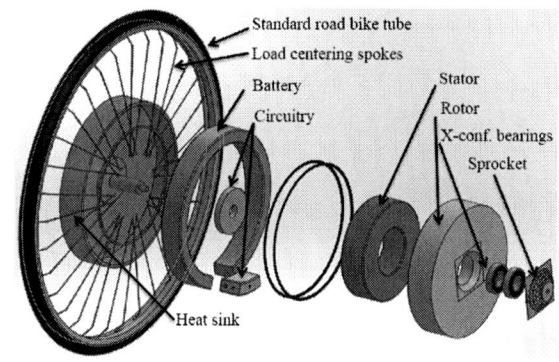

Fig. 9. Exploded view of the HMD.

VI. CONCLUSION AND FUTURE WORKS

In this paper, the design procedure for a ferrite based five-phase fractional-slot concentrated winding IPM outer rotor motor for e-bicycle application is discussed. The designed motor which employs ferrite magnets is found to be economical, reliable and meeting the requirement for e-bicycles. Initially, the mathematical model (LPM) is developed to make the optimized design by LPM based optimizer. Then the adopted process for the optimization procedure using DES has been verified using FEA analysis.

The performance report from the FEA tool shows that the proposed motor design develops 10.8Nm average torque at the rated speed which meets the specification. The peak to peak torque pulsation is around 6%. The cogging torque of the motor at rated speed is 100mNm which is very low ensuring a smooth operation during motor turn-off state. It has been observed that the efficiency of the motor is 90%. The results obtained from these analyses provide a solid framework based on which the production of a prototype A1-HMD can be done.

Apart from the prototype fabrication, additional future work would include thermal analysis to supplement the stress analysis of the proposed design as well as further design

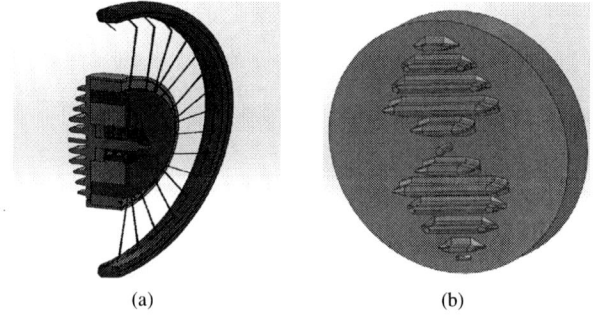

(a)   (b)

Fig. 10. (a) Fixed and rotating components, (b) heat extraction system, and (c) seal structure.

improvement and development of the hub motor and its integration in an e-bicycle.

978-1-4673-9551-9/16 $31.00 © 2016 IEEE

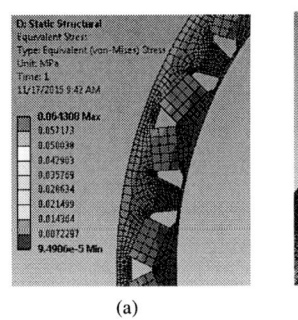
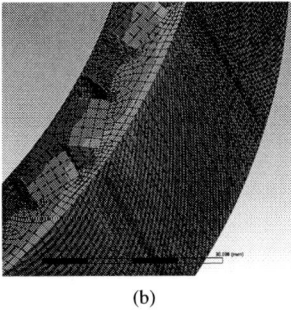

(a)   (b)

Fig. 11.   Equivalent stress (Von-Mises) at the rotor at 600 rpm.

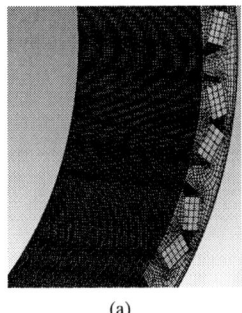
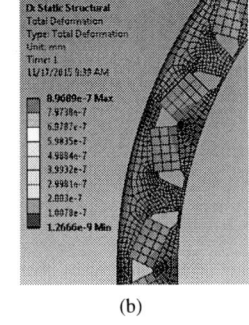

(a)   (b)

Fig. 12.   Total deformation in the rotor at 600 rpm.

ACKNOWLEDGMENT

I-corps site program of the University of Akron Research Foundation has been very helpful to find the actual problems of the customers in order to design this A1-HMD.

REFERENCES

[1] E. Heinen, B. van Wee, and K. Maat, "Commuting by bicycle: an overview of the literature," *Transport reviews*, vol. 30, no. 1, pp. 59–96, 2010.

[2] J. D. Goodman, "An electric boost for bicyclists," *New York Times*, p. B1, 2010.

[3] Throttle-control and pedal-assist e-bicycles, batteries, and motors: Global market opportunities, barriers, technology issues, and demand forecasts. [Online]. Available: http://www.navigantresearch.com/research/electric-bicycles

[4] J. MacArthur, J. Dill, and M. Person, "E-bikes in the north america: Results from an online survey," in *Transportation Research Board 93rd Annual Meeting*, no. 14-4885, 2014.

[5] K. Chau, C. Chan, and C. Liu, "Overview of permanent-magnet brushless drives for electric and hybrid electric vehicles," *Industrial Electronics, IEEE Transactions on*, vol. 55, no. 6, pp. 2246–2257, 2008.

[6] R. K. Dhawan, "Multi-phase multi-pole electric machine," Mar. 29 2012, uS Patent App. 13/433,316.

[7] H. Chen, "The switched reluctance motor drive for application in electric bicycle," in *Industrial Electronics, 2001. Proceedings. ISIE 2001. IEEE International Symposium on*, vol. 2.   IEEE, 2001, pp. 1152–1156.

[8] T. Chan, L.-T. Yan, and S.-Y. Fang, "In-wheel permanent-magnet brushless dc motor drive for an electric bicycle," *Energy Conversion, IEEE Transactions on*, vol. 17, no. 2, pp. 229–233, 2002.

[9] A. Muetze and Y. C. Tan, "Electric bicycles-a performance evaluation," *Industry Applications Magazine, IEEE*, vol. 13, no. 4, pp. 12–21, 2007.

[10] P. B. Reddy, A. M. El-Refaie, K.-K. Huh, J. K. Tangudu, and T. M. Jahns, "Comparison of interior and surface pm machines equipped with fractional-slot concentrated windings for hybrid traction applications,"

Energy Conversion, IEEE Transactions on, vol. 27, no. 3, pp. 593–602, 2012.

[11] M. Barcaro and N. Bianchi, "Interior pm machines using ferrite to substitute rare-earth surface pm machines," in *Electrical Machines (ICEM), 2012 XXth International Conference on*.   IEEE, 2012, pp. 1339–1345.

[12] T. M. Jahns, G. B. Kliman, and T. W. Neumann, "Interior permanent-magnet synchronous motors for adjustable-speed drives," *Industry Applications, IEEE Transactions on*, no. 4, pp. 738–747, 1986.

[13] P. Sekerak, V. Hrabovcova, J. Pyrhonen, S. Kalamen, P. Rafajdus, and M. Onufer, "Comparison of synchronous motors with different permanent magnet and winding types," *Magnetics, IEEE Transactions on*, vol. 49, no. 3, pp. 1256–1263, 2013.

[14] P. C. Dent, "Rare earth elements and permanent magnets," *Journal of applied physics*, vol. 111, no. 7, p. 07A721, 2012.

[15] S. S. R. Bonthu, J. Baek, and S. Choi, "Comparison of optimized permanent magnet assisted synchronous reluctance motors with three-phase and five-phase systems," in *Energy Conversion Congress and Exposition (ECCE), 2014 IEEE*.   IEEE, 2014, pp. 2396–2402.

[16] L. Parsa, "On advantages of multi-phase machines," in *Industrial Electronics Society, 2005. IECON 2005. 31st Annual Conference of IEEE*.   IEEE, 2005, pp. 6–pp.

[17] A. Arafat and S. Choi, "Fault tolerant control of five-phase permanent magnet assisted synchronous reluctance motor based on dynamic current phase advance," in *Energy Conversion Congress and Exposition (ECCE), 2015 IEEE*.   IEEE, 2015, pp. 1208–1214.

[18] The flykly smart wheel for bicycles is now available and first to hit the market in the movement to usher in a new era in urban transportation. [Online]. Available: http://www.businesswire.com/news/home/20150106005432/en/FlyKly-Smart-Wheel-Bicycles-Hit-Market-Movement

[19] E. C. Lovelace, T. M. Jahns, and J. H. Lang, "A saturating lumped-parameter model for an interior pm synchronous machine," *IEEE Transactions on Industry Applications*, vol. 38, no. 3, pp. 645–650, 2002.

[20] J. Baek, M. M. Rahimian, H. Toliyat *et al.*, "Optimal design and comparison of stator winding configurations in permanent magnet assisted synchronous reluctance generator," in *Electric Machines and Drives Conference, 2009. IEMDC'09. IEEE International*.   IEEE, 2009, pp. 732–737.

[21] M. Zakirul, S. S. R. Bonthu, and S. Choi, "Obtaining optimized designs of multi-phase pma-synrm using lumped parameter model based optimizer," in *Electric Machines and Drives Conference, 2015. IEMDC'15. IEEE International*.

Transverse Flux Machines with Rotary Transformer Concept for Wide Speed Operations without using Permanent Magnet Material

Iftekhar Hasan   Md Wasi Uddin   Yilmaz Sozer

Department of Electrical and Computer Engineering,
The University of Akron, Akron, OH, USA

Abstract— **This paper presents a new concept of Transverse Flux Machine (TFM) design that uses rotary transformers to replace the Permanent Magnet (PM) field excitation in the rotor. The rotary transformer has an inductive interface that allows for contactless transfer of energy to the field windings embedded in the rotor core. In order to achieve high magnetic coupling, a highly permeable ferrite core is selected for the transformer which is excited with high electrical frequency of 100 kHz. A field power converter is used to regulate the transformer secondary winding voltage to maintain constant DC field in the rotor. The proposed TFM has a modular structure that is free of PM, which makes it cost-effective, without sacrificing its peak torque and power when compared to a similar sized PM based TFM.**

Keywords— *Transverse flux machine (TFM), rotary transformer, ferrites, finite element analysis (FEA).*

I. INTRODUCTION

Direct-drive applications helps to avoid the drawbacks associated with gearing configurations such as additional space, elasticity and backlash [1]. Direct-drive machines have high torque at low speeds and provide high reliability and low cost due to the elimination of mechanical gearbox [2]. Transverse flux machines (TFM) are renowned for its low speed high power density applications [3-5], thus making it inherently suitable for direct drive applications.

Leakage inductance, high torque ripple [5], low winding utilization [6] and poor power factor especially for TFM in flux concentrating setup [7] are some of the negative attributes that has restricted its industrial applications. However, recent advancements in power electronic converters and 3D finite element software tools has led to the development of better design and analysis procedure to address the complexities and drawbacks associated with the complex 3D geometries of TFM. TFM are compact and complex permanent magnet (PM) machines often designed using rare-earth (RE) materials to achieve high power density. Cost and supply volatility associated with rare earth (RE) magnets has driven the need to develop new machine topologies that either eliminates or reduces the use of RE materials. In [2, 8], a novel ferrite magnet based TFM is presented with flux concentrating setup to achieve air gap flux densities (B > 1T) similar to RE magnets. The use of ferrite magnets addresses the issues related to cost and supply volatility associated with RE magnets. However, demagnetization of ferrite magnets

especially at low temperature is always a concern. In [10-11], the concept of rotary transformer has been used to eliminate brushes and slip rings in doubly fed induction machines. The rotary transformer allowed access to the rotor circuit without any mechanical contact, improving the reliability of the machine. In [12], rotary transformers were used for contactless transfer of energy to establish field current in the rotor, which completely eliminated RE magnets in synchronous motors used for hybrid vehicle applications.

In this paper, a double-sided TFM with embedded rotary transformer is proposed that is modular in structure and free of PM materials. The 3D model of the TFM having quasi-U core stators, rotary transformer and rotor with field coils is shown in Fig. 1. The primary side of the rotary transformer is integrated with the stator and the secondary is embedded on both sides of the rotor. The rotor field windings are excited by using the contactless rotary transformer. The secondary voltage of the rotating transformer is rectified using a bridge rectifier to maintain a constant DC voltage across the field winding. This results in a unidirectional flux being set up in the rotor which replicates the behavior of PM embedded on the rotor.

The Finite Element Analysis (FEA) was carried on the 3D TFM model using Magsoft Flux 3D software. The novel TFM structure and its operating principle is explained in Section II. The rotary transformer principles and its dimensions are discussed in Section III. In Section IV, the FEA results for the rotary transformer based TFM performance are compared against an equivalent PM based TFM.

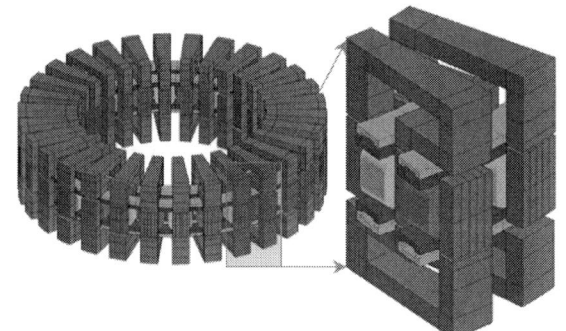

Fig. 1(a). 3D model of the proposed rotary transformer based TFM   Fig. 1(b). One pole pair of the proposed TFM

978-1-4673-9551-9/16 $31.00 © 2016 IEEE

II. TRANSVERSE FLUX MACHINE STRUCTURE

The proposed TFM has a modular structure with ring windings. Modularity is a very important advantage of the proposed TFM topology. Each phase is an independent module which makes the machine very attractive for mass production, since a wide range of power can be covered by placing an adequate number of modules together. Fig. 2 shows a 2D cross-sectional view for a single pole of the proposed TFM where the flux paths are transverse (perpendicular) to the rotating motion of the rotor.

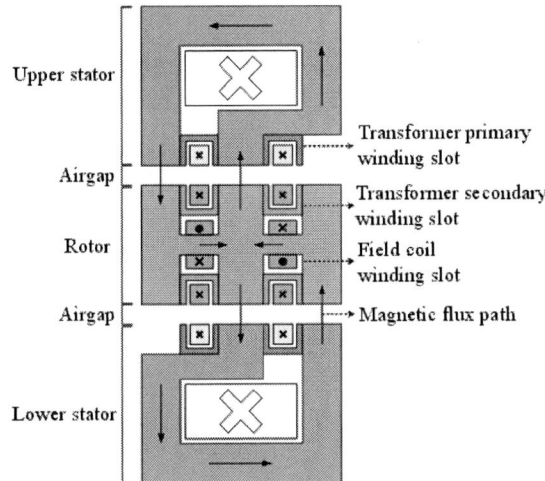

Fig. 2. 2D cross-sectional view of proposed TFM

A. Rotor

Each phase module consists of a rotor in the middle with stator cores located on both sides of the rotor as shown in Fig. 2. The structure of the rotor integrates the transformer secondary winding slots as well as the slots for the field coils as shown in Fig. 3. The rotor windings produce unidirectional field that aids in the buildup of flux in the middle core of the rotor. The flux induced in the field coils are oriented such that they are either pointing to or away from each other. The adjacent rotor cores are arranged to align with the stator cores to the left, middle and the right, respectively, as shown in Fig. 2 to close the magnetic circuit. The rotor pole has the same width to match the stator.

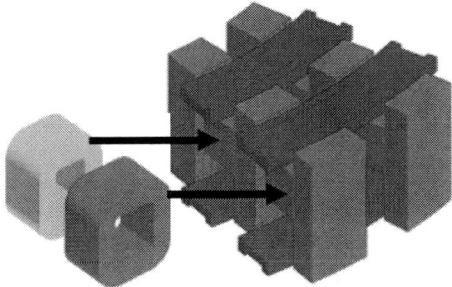

Fig. 3(a). Field coils for rotor   Fig. 3(b). Rotor with embedded transformer secondary slots

B. Stator

The stator consists of two alternate quasi-U cores with double active sides facing the rotor via two axial air gaps and two ring windings in the upper and lower stator slots. The primary winding slots of the rotary transformer are integrated with the stator cores as shown in Fig. 4. The stator orientation minimizes the exposed windings to reduce the leakage fluxes. The double sided windings maximize the stator core utilization and make the machine highly fault tolerant [2]. The 'ring' winding couples each stator core to the entire armature ampere-turns, and thus avoids the limitation of the "BIL" principle of force production [2, 8]. As a result, high torque can be achieved by increasing the pole number without sacrificing the electric loading.

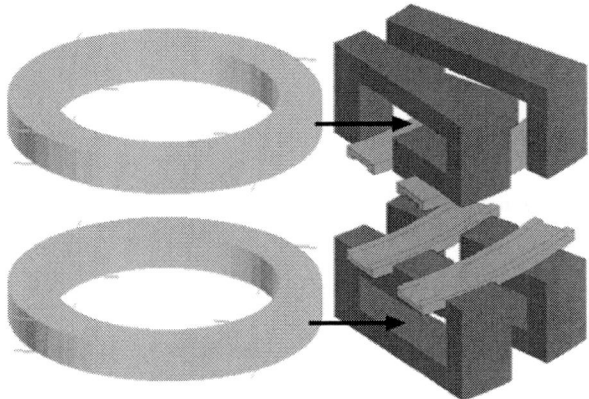

Fig. 4(a). Ring windings for stator upper and lower slots   Fig. 4(b). Stator core with integrated transformer primary slots

In this paper, one module unit is designed and optimized to be used for direct drive wind turbine applications. The motor design parameters along with the design specifications and restrictions are summarized in Table I.

TABLE I. TFM motor parameters

Description	Value	Description	Value
Design and Performance Specifications			
Number of poles	30	Peak Torque	30 Nm
Output power	0.6 kW	DC bus voltage	400 V
Rated speed	400 rpm	Excitation frequency	100 Hz
Current density	5 A/mm²	Peak current	105 A
Air gap length	1 mm	Air gap flux density	1.2 T
Stator outer diameter	255 mm	Axial length	82 mm
Rotor inner diameter	142mm	Rotor outer diameter	225 mm
Stator winding			
Number of phases	1	Winding type	Ring
Number of coils	2	Turns per coil	11
Coil length	585mm	Coil area	270 mm²
Fill factor	0.6	Coil material	Copper
Field winding			
Number of coils	60	Winding type	Ring
Coil length	3 mm	Coil Area	40 mm²
Fill factor	1	Coil material	Copper

III. ROTARY TRANSFORMER AND FIELD POWER CONVERTER DESIGN

Rotary transformers has been used before [10-12] to establish field in the rotor via contactless transfer of energy. However this concept has never been extended to TFM where it could eliminate the use of PM completely from the machine design. In this TFM design, four transformer pairs have been used. The secondary transformer cores are embedded in the rotor, while the primary transformer cores are integrated with the stator cores as shown in Fig. 3 and 4 respectively. The secondary winding of the rotary transformer is physically coupled to the field windings in the rotor through a field power converter.

A. Rotary transformer design

Rotary transformer designs [13-14], are different from conventional transformers due to the presence of an air gap that allows movement between primary (stator) and secondary (rotor) windings as shown in Fig. 5.

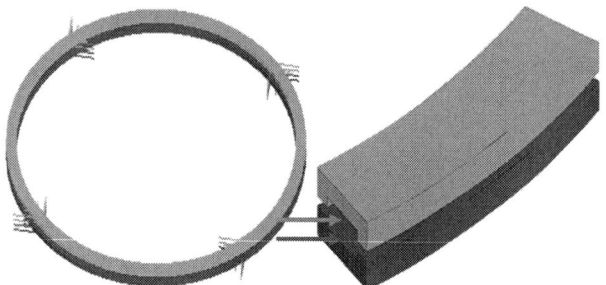

Fig. 5(a). Ring windings for transformer slots   Fig. 5(b). Transformer primary and secondary core

An axial gap rotary transformer with C-shaped ferrite cores are used in this design with a nominal airgap of 1mm. The transformer primary cores are designed in such a way that it fits into the proposed TFM stator core structure used in [8]. The transformer secondary cores are embedded in the rotor of [8] by replacing the magnets sandwiched between the rotor cores. Removing the magnets also provides the slots for the field coils located between the secondary of the transformer cores.

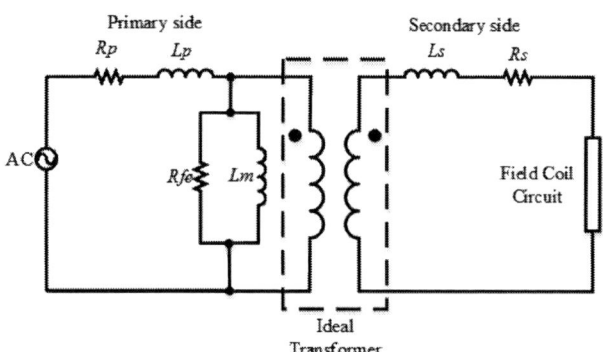

Fig. 6. Equivalent circuit of an ideal transformer

Fig. 6 shows an equivalent circuit of a transformer where Lm is the magnetizing inductance and Rfe is the resistance to represent the core loss. Since the rotary transformer has an airgap between the primary and secondary core, the Lm tends to be high which would require higher current for magnetization. In order to reduce the amount of current going through the magnetizing branch, the magnetizing reactance Xm which depends on the dimensions of the transformer, number of turns in the primary coil and the material properties should be increased.

The expression for Xm can be described as:

$$X_m = \omega L_m = \frac{\omega N^2 \mu_0 A}{g} \qquad (1)$$

where, ω is the applied frequency, N is the number of turns, μ_0 is the air permeability, A is the area of the transformer faces, g is the airgap between the transformer primary and secondary. 'Ring' windings are used in the transformers, where the primary side is excited at $\omega =100$ kHz. Ferrite has been chosen as the material for the transformer cores to be able operate at higher frequencies without having significant core losses.

The primary and secondary winding resistances, magnetizing inductance, leakage inductance of the rotary transformer is calculated through FEA. The parameters of the equivalent circuit shown in Fig 6 are determined and their values summarized in Table II.

TABLE II. Rotary transformer data

Core Material	Ferrite
Primary Number of Turns	4
Primary Resistance Rp	0.014
Primary Leakage inductance Lp	61.1uH
Secondary number of turns	40
Secondary resistance Rs	1.435 Ω
Secondary Leakage inductance Ls	611 mH
Magnetizing inductance	379.74 μH

B. Field power converter circuit

An H bridge inverter is connected to the stationary primary side of the rotary transformer primary with a DC link voltage of 400V as shown in Fig. 7. The rotating secondary side of the transformer is electrically coupled to the field coil through a bridge rectifier. The LC filter ensures constant DC current is maintained at the field coil.

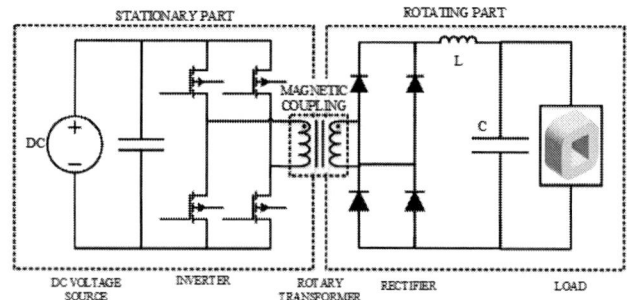

Fig. 7. Field power converter circuit with rotary transformer and field coil

IV. FEA RESULTS

The 3D FEA model was designed for two machines: a PM based TFM and a rotary transformer based TFM as shown in Fig. 8. Both machines have identical outer diameter and axial height. The PM in the rotor of TFM are replaced with field coils (FC) and rotary transformers are integrated to the existing rotor and stator structure of the 3D model without affecting its critical dimensions such as airgap length, pole number, pole length, and pole face area.

The no-load flux linkage and EMF was determined for different rotor positions for both designs. At higher speeds it is important to have the capability to reduce the EMF. For rated operating speed the proposed TFM achieves same EMF with the TFM using PM. This result suggests that proposed machine can achieve wider speed operation as the EMF can be reduced as a function of the field current produced by the rotary transformer.

Torque was also determined for different rotor positions while exciting the coil conductor. The field coil current is controlled such that both of the designs produce similar peak torque and power output at same operating speeds. The no load EMF

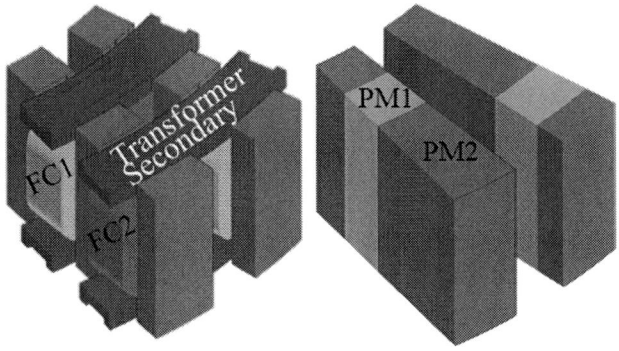

Fig. 8(a). Rotary transformer based TFM rotor

Fig. 8(b). PM based TFM rotor

A. Flux linkage distribution and no-load EMF

Flux linkage under no-load condition at different rotor positions for both of the designs are shown in Fig. 9. Flux linkage in the rotary transformer based TFM is comparable to that of the PM based TFM. No-load EMF for different rotor positions is shown in Fig. 10. The EMF waveform is almost sinusoidal and hence applying sinusoidal current to the main stator winding would ensure a smooth torque. The induced voltage in the transformer secondary winding is rectified to maintain a constant DC field on the rotor as shown in Fig. 11.

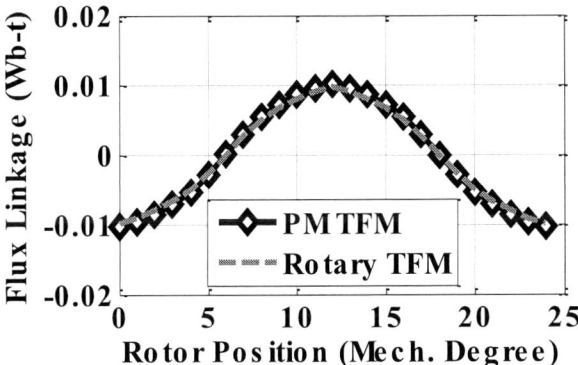

Fig. 9. No-load flux linkage

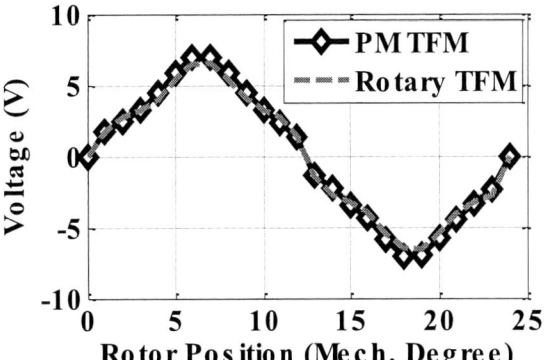

Fig. 10. No-load EMF

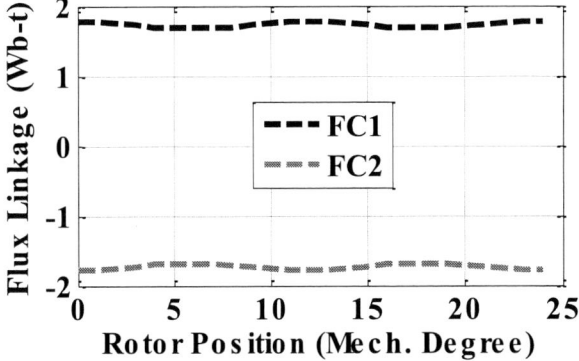

Fig. 11. Flux Linkage in rotor field coils

B. Torque characteristics

The main windings in the stator are driven by single phase 105A peak sinusoidal current. There are two components in the torque generated: a) the electromagnetic torque which comes from the non-sinusoidal no-load EMF, b) cogging torque which comes from the interaction between the rotor field and stator slots. Torque produced at different rotor positions is obtained for both no-load (cogging torque) and loaded conditions as shown in Fig. 12 and 13 respectively.

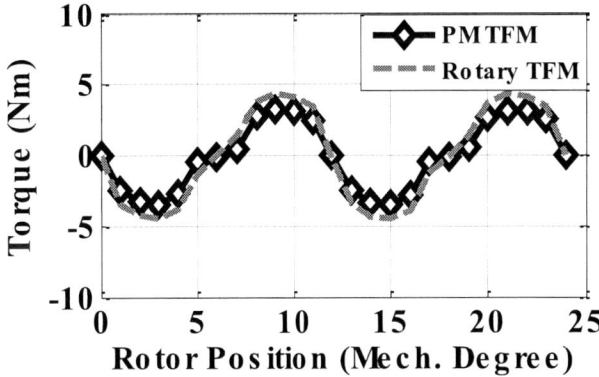

Fig. 12. Cogging Torque (No-load condition)

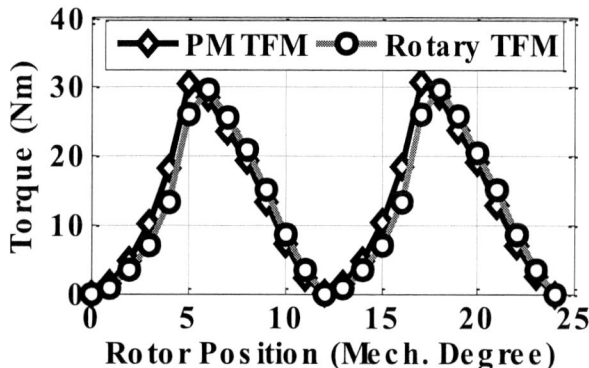

Fig. 13. Electromagnetic Torque (Loaded Condition)

V. CONCLUSION

A single phase, 0.6 kW, 400 rpm TFM for use in wind power applications has been designed using a rotary transformer. The proposed TFM is free of PM materials and does not sacrifice peak power and torque ratings when compared to an equivalent PM based TFM. The designed rotary transformer is retrofitted in an existing TFM geometry without affecting its critical dimensions such as airgap length, pole number, pole length, and pole face area.

REFERENCES

[1] R. Kruse, G. Pfaff, and C. Pfeiffer, "Transverse flux reluctance motor for direct servodrive applications," *IEEE Industry Applications Conference*, vol. 1, pp. 655-662, 1998.

[2] Z. Wan, A. Ahmed, I. Husain, and E. Muljadi, "A novel transverse flux machine for vehicle traction applications," *IEEE Power and Energy Society General Meeting*, Denver, Colorado, July 26-30, 2015.

[3] H. Weh and H. May, "Achievable force densities for permanent magnet excited machines in new configurations," *Proc. Int. Conf. Electrical Machines*, pp. 1107–1111, 1986.

[4] M. R. Harris, G. H. Pajooman, and S. M. A. Sharkh, "Performance and design optimization of electric motors with heteropolar surface magnets and homopolar windings," *Proc. Inst. Elect. Eng., Elect. Power Appl.*, vol. 143, pp. 429–436, 1996.

[5] W. M. Arshad, T. Backstrom, and C. Sadarangani, "Analytical design and analysis procedure for a transverse flux machine," *Proc. IEEE Conf. Elec. Mach. and Drives*, pp. 115-121, 2001.

[6] E. Muljadi, C. P. Butterfield, and Y. H. Wan, "Axial-flux modular permanent-magnet generator with a toroidal winding for wind-turbine applications," *IEEE Trans. Ind. Appl.*, vol. 35, no. 4, pp. 831–836, Jul./Aug. 1999.

[7] L. Strete, L. Tutelea, I. Boldea, C. Martis, and I. Viorel, "Optimal design of a rotating transverse flux motor (TFM) with permanent magnets in rotor," *Proc. Int. Conf. on Electrical Machines (ICEM)*, Sept., 2010

[8] I. Hasan, T. Husain, M. W. Uddin, Y. Sozer, I. Husain, and E. Muljadi, "Analytical modeling of a novel transverse flux machine for direct drive wind turbine applications," *IEEE Energy Conversion Congress and Exposition (ECCE)*, pp. 2161-2168, 2015.

[9] W. M. Arshad, P. Thelin, T. Backstrom, and C. Sadarangani, "Use of Transverse-Flux Machines in a Free-Piston Generator," *IEEE Trans. Ind. Appl.*, vol. 40, no. 4, pp. 1092-1100, July/August 2004.

[10] H. Zhong, Z. Lin, W. Xiuhe, and L. Xiao, "Design and analysis of a three-phase rotary transformer for doubly fed induction generators," *IEEE Industry Applications Society Annual Meeting*, pp. 1-6, 2014.

[11] M. Ruviaro, F. Ruencos, N. Sadowski, and I. M. Borges, "Analysis and test results of a brushless doubly fed induction machine with rotary transformer," *IEEE Trans. Ind. Electron.* vol. 59, no. 6, 2670-2677, 2012.

[12] C. Stancu, T. Ward, K. Rahman, R. Dawsey, and P. Savagian, "Separately excited synchronous motor with rotary transformer for hybrid vehicle application," *IEEE Energy Conversion Congress and Exposition (ECCE)*, pp. 5844-5851, 2014.

[13] K. D. Papastergiou and D. E. Macpherson, "An airborne radar power supply with contactless transfer of energy—Part I: Rotating transformer," *IEEE Trans. Ind. Electron.*, vol. 54, no. 5, pp. 2874–2884, Oct., 2007.

[14] J. P. C. Smeets, D. C. J. Krop, J. W. Jansen, M. A. M. Hedrix, and E. A. Lomonova, "Optimal Design of a Pot Core Rotating Transformer," *Proc. IEEE Energy Conversation Congress and Exposition (ECCE)*, pp. 4390-4397, Sept., 2010.

Field Oriented Modeling and Control of Six Phase, Open-Delta Winding, Interior Permanent Magnet Synchronous Machines considering Current Unbalance and Zero Sequence Currents

Murat Senol, Michael Schubert,
Georges Engelmann and Rik W. De Doncker
Institute for Power Electronics and Electrical Drives (ISEA)
RWTH Aachen University, Germany
e-mail: post@isea.rwth-aachen.de

Thorben Grosse and Kay Hameyer
Institute of Electrical Machines (IEM)
RWTH Aachen University, Aachen, Germany
e-mail: post@iem.rwth-aachen.de

Abstract—Industrial and academic interest on multiphase electric machines have been steadily increasing due to the advantages they provide compared to their three phase counterparts. They offer higher efficiency and better fault tolerance. The power is distributed across a larger number of machine phases and inverter legs, which allows the use of semiconductor devices with lower ratings. These advantages are especially benecifial for electric and hybrid vehicle applications.

Increasing the battery pack voltage of an electric vehicle increases the cost and complexity, while decreasing the energy density of the battery pack. Therefore, electric vehicle system design may benefit on the vehicle level from a more complex drivetrain with higher dc link voltage utilization. H bridge inverters as well as multiphase machines provide increased dc-link voltage utilization compared to three phase inverters and machines.

A low voltage, high power drivetrain is designed for an electric vehicle using an asymmetrical six phase, open-delta interior permanent magnet synchronous machine (IPMSM) and an H Bridge inverter. Field oriented modeling and control of such a system is investigated in this paper. The sources of unbalance and zero sequence current components are explored.

Keywords—*Field oriented control, multiphase machines, asymmetrical six phase, open delta, H Bridge inverter, zero sequence current, resonant controller*

I. INTRODUCTION

Multiphase machine drives provide reduced phase current ratings, increased dc-link utilization, lower dc link current harmonics, higher efficiency, improved fault tolerance, decreased torque ripples, and higher reliability compared to three phase drive systems. Therefore, they are increasingly employed in electric and hybrid electric vehicles, ship propulsion, locomotive traction, aircrafts, and wind power generation.

A large variety of control strategies has been proposed to control six phase machines. Direct torque control (DTC) has low machine parameter dependence and fast dynamic torque response. Predictive control provides similar dynamic torque response as DTC. However, the high computational effort limits its application. Vector control has the advantage of

simplicity, while it is limited in dynamic response by the small bandwidth of the PI controllers. An extensive literature survey about these control methods can be found in [5].

A comprehensive analysis of the machine properties of a six phase IPMSM is presented in this paper to provide insight into the machine-based disturbances which affect the control of six phase open-delta winding machines. This information is used to suggest an accurate and fast machine model which can be used in control simulations of such machines. The control method, which has been presented in [5] for dual three phase, surface-mounted PMSMs, is extended to control six phase, open-delta winding IPMSMs. Simulation results are compared to measurements to verify the validity of the model, and efficiency measurements are provided.

II. FIELD ORIENTED MODELING

In conventional rotor field oriented control, the machine phase variables (current, voltage and flux linkage) are transformed into a two phase orthogonal system which is aligned with the rotor flux linkage. The current component on the direct axis i_d influences the stator flux linkage, while the current component on the quadrature axis i_q produces torque. Therefore, the flux generating and torque generating current components are decoupled and can be controlled with simpler controllers. [1]

Conventionally, two PI controllers are used to control i_d and i_q, as illustrated by the signal flow diagram in Fig. 1. The controller receives information about torque demand T_m^* and maximum allowed stator voltage u_s^{max}. Rotor position and speed are calculated using position sensor data. The data is processed (usually using look-up tables) to obtain reference values $i_{s.d}^*$ and $i_{s.q}^*$. The transformed phase currents are fed back to the PI controllers, which calculate the set voltages u_d^* and u_q^* that should be applied to the machine. These values are then transformed back to the three phase system and applied to the machine through the inverter.

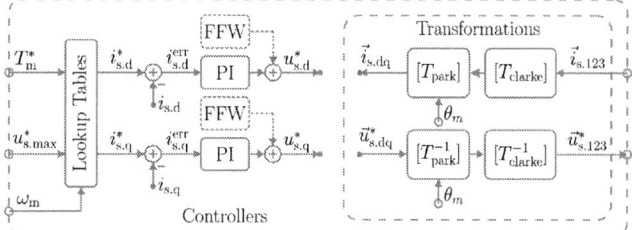

Fig. 1. Control diagram of conventional field oriented control

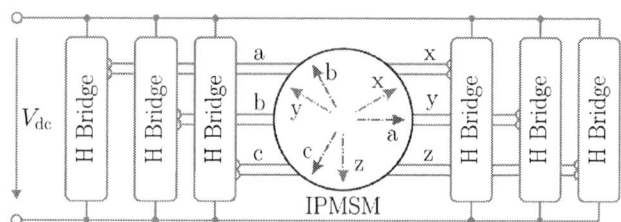

Fig. 2. Six phase open-delta drivetrain topology

A. Field Oriented Transformations

The six phase, open-delta machine can be considered as two three-phase machines with a stator phase shift of $\theta_s = 30°$ in the same stator core, as shown in Fig. 2. To achieve full control of torque and flux linkage, as well as phase currents; a six phase to six orthogonal phase field oriented transformation set is required, as given in (1)–(3). $\vec{F}$ can be current, voltage, or flux linkage vectors in different vector spaces. $\vec{F}_{az}$ represents the vectors in the phase space, $\vec{F}_{\alpha\beta}$ represents the vectors in the stator oriented orthogonal space, and $\vec{F}_{dq}$ represents the vectors in the rotor field oriented orthogonal space. These vectors are given explicitly in (4)–(6).

$$\left[\; \vec{F}_{\alpha\beta} \; \right] = \left[\; T_{\text{clarke}} \; \right] \cdot \left[\; \vec{F}_{az} \; \right] \tag{1}$$

$$\left[\; \vec{F}_{dq} \; \right] = \left[\; T_{\text{park}} \; \right] \cdot \left[\; \vec{F}_{\alpha\beta} \; \right] \tag{2}$$

$$\left[\; \vec{F}_{az} \; \right] = \left[\; T_{\text{clarke}}^{-1} \; \right] \cdot \left[\; T_{\text{park}}^{-1} \; \right] \cdot \left[\; \vec{F}_{dq} \; \right] \tag{3}$$

$$\left[\; \vec{F}_{az} \; \right] = \left[\; F_a \quad F_x \quad F_b \quad F_y \quad F_c \quad F_z \; \right]^T \tag{4}$$

$$\left[\; \vec{F}_{\alpha\beta} \; \right] = \left[\; F_\alpha \quad F_\beta \quad F_{z1} \quad F_{z2} \quad F_{o1} \quad F_{o2} \; \right]^T \tag{5}$$

$$\left[\; \vec{F}_{dq} \; \right] = \left[\; F_d \quad F_q \quad F_{zd} \quad F_{zq} \quad F_{o1} \quad F_{o2} \; \right]^T \tag{6}$$

The power invariant Clarke transformation $[T_{\text{clarke}}]$, given in (11), is constructed considering the harmonic components in the machine variables. This transformation maps the fundamental and $(12k + 1)^{th}$ harmonics to the $\alpha\beta$ plane, $(12k - 6 \pm 1)^{th}$ harmonics to the z_{12} plane, and $(6k - 3)^{rd}$ harmonics to the o_{12} plane, where $k \in \{1, 2, 3, ...\}$. [4]

It should be noted that the z_{12} plane also represents the unbalance between the variables of two machine sets abc and xyz in $\alpha\beta$ plane while the o_{12} plane represents the zero sequence variables of each of these machine sets, as shown in (7)–(10).

$$F_{z1} = F_{\alpha_{abc}} - F_{\alpha_{xyz}} \tag{7}$$

$$F_{z2} = F_{\beta_{xyz}} - F_{\beta_{abc}} \tag{8}$$

$$F_{o1} = \sqrt{2/6} \cdot (F_a + F_b + F_c) \tag{9}$$

$$F_{o2} = \sqrt{2/6} \cdot (F_x + F_y + F_z) \tag{10}$$

The Park transformation $[T_{\text{park}}]$, given in (12), rotates the vectors on $\alpha\beta$ plane to rotor field oriented dq plane. It also rotates the vectors in z_{12} plane to z_{dq} plane such that the z_{dq} plane vectors represents the unbalance between the variables of the two machine sets abc and xyz in dq plane.

It is explained in detail in [5], how the 5th and 7th harmonics in z_{12} plane are converted to 6th harmonics in z_{dq} plane through the Park transformation. Similarly, the 11th and 13th harmonics in $\alpha\beta$ plane are converted to 12th harmonics in dq plane.

The Clarke and inverse Clarke transformations can be calculated and stored numerically instead of variables, for example using matlab, since all elements are constants. For inverse Park transformation, Park transformation can be used with $-\theta_m$ since it is only a rotating transformation.

B. Simplified Machine Model

The stator voltage equation of a six phase, open-delta, interior permanent magnet machine is given in (13). This equation holds for both phase space variables and stator oriented orthogonal space variables, since the Clarke transformation is time and position invariant.

In the following equations, $\vec{u}_s$ is the stator voltage, $\vec{i}_s$ is the stator current, $\vec{\psi}_s$ is the stator flux linkage, $\vec{\psi}_f$ is the magnet flux linkage, $[R_s]$ is the 6×6 diagonal stator resistance matrix, and $[L_s]$ is the stator inductance matrix. The Park transformation is applied to (13) to obtain (14). [1]

$$\vec{u}_{s.\alpha\beta} = [R_s] \vec{i}_{s.\alpha\beta} + \frac{d}{dt} \vec{\psi}_{s.\alpha\beta} \tag{13}$$

$$\begin{bmatrix} u_{s.d} \\ u_{s.q} \\ u_{s.zd} \\ u_{s.zq} \\ u_{s.o1} \\ u_{s.o2} \end{bmatrix} = \begin{bmatrix} R_s i_{s.d} \\ R_s i_{s.q} \\ R_s i_{s.zd} \\ R_s i_{s.zq} \\ R_s i_{s.o1} \\ R_s i_{s.o2} \end{bmatrix} + \frac{d}{dt} \begin{bmatrix} \psi_{s.d} \\ \psi_{s.q} \\ \psi_{s.zd} \\ \psi_{s.zq} \\ \psi_{s.o1} \\ \psi_{s.o2} \end{bmatrix} + \begin{bmatrix} -\omega_m \psi_{s.q} \\ \omega_m \psi_{s.d} \\ -\omega_m \psi_{s.zq} \\ \omega_m \psi_{s.zd} \\ 0 \\ 0 \end{bmatrix} \tag{14}$$

A simplified machine model can be obtained assuming that the machine has sinusoidal back EMF, no iron losses, no mutual leakage inductance, no asymmetries, and no saturation. Considering the assumptions above, it can be further assumed that $\vec{\psi}_{f.dq}$ is a 6 dimensional zero vector with the exception of $\psi_{f.d} = \psi_f$, $[R_s]$ is a 6×6 diagonal matrix with R_s as all non-zero elements, and $[L_{s.dq}]$ is a 6×6 diagonal matrix of constant elements, i.e. independent of rotor position and stator currents.

$$\vec{\psi}_{s.dq} = [L_{s.dq}] \vec{i}_{s.dq} + \vec{\psi}_{f.dq} \tag{15}$$

$$\frac{d}{dt} \vec{\psi}_{s.dq} = [L_{s.dq}] \frac{d}{dt} \vec{i}_{s.dq} + \frac{d}{dt} \vec{\psi}_{f.dq} \tag{16}$$

$$[\,T_{\text{clarke}}\,] = \sqrt{\frac{2}{6}} \cdot \begin{bmatrix} 1 & \cos(\theta_s) & \cos(4\theta_s) & \cos(5\theta_s) & \cos(8\theta_s) & \cos(9\theta_s) \\ 0 & \sin(\theta_s) & \sin(4\theta_s) & \sin(5\theta_s) & \sin(8\theta_s) & \sin(9\theta_s) \\ 1 & \cos(5\theta_s) & \cos(8\theta_s) & \cos(\theta_s) & \cos(4\theta_s) & \cos(9\theta_s) \\ 0 & \sin(5\theta_s) & \sin(8\theta_s) & \sin(\theta_s) & \sin(4\theta_s) & \sin(9\theta_s) \\ 1 & 0 & 1 & 0 & 1 & 0 \\ 0 & 1 & 0 & 1 & 0 & 1 \end{bmatrix} \tag{11}$$

$$[\,T_{\text{park}}\,] = \begin{bmatrix} \cos(\theta_m) & \sin(\theta_m) & 0 & 0 & 0 & 0 \\ -\sin(\theta_m) & \cos(\theta_m) & 0 & 0 & 0 & 0 \\ 0 & 0 & \cos(\theta_m) & \sin(\theta_m) & 0 & 0 \\ 0 & 0 & -\sin(\theta_m) & \cos(\theta_m) & 0 & 0 \\ 0 & 0 & 0 & 0 & 1 & 0 \\ 0 & 0 & 0 & 0 & 0 & 1 \end{bmatrix} \tag{12}$$

The simplified equivalent circuit of the machine, given in Fig. 3, has no disturbances in the z_{dq} and o_{12} planes. It can be used for educational purposes or to observe the isolated effect of the inverter non-idealities on the control. However, it does not sufficiently represent the machine in real life to simulate and investigate a control method. In fact, in this model, the z_{dq} and o_{12} plane currents can be adjusted to zero, simply by setting the voltages to zero.

C. Non-Ideal Properties of the Machine

The effects of current and rotor position dependencies of stator inductances and the magnet flux linkages should be investigated to obtain a more accurate machine model. A set of look-up tables has been created using FEM analysis for this purpose; which provides the stator and magnet flux linkage vectors $\vec{\psi}_{s.az}$ and $\vec{\psi}_{f.az}$, the stator inductance matrix $[L_{s.az}]$, and torque for changing rotor position θ_m, and stator currents $i_{s.d}$ and $i_{s.q}$.

The model can be further improved in precision by adding new dimensions of current dependencies for z_{dq} and o_{12} plane current components. However, this increases the simulation times significantly, both for the FEM analysis and the control simulations where these tables are used. An adequate control that can keep these current components close to zero renders this effort futile in the first place.

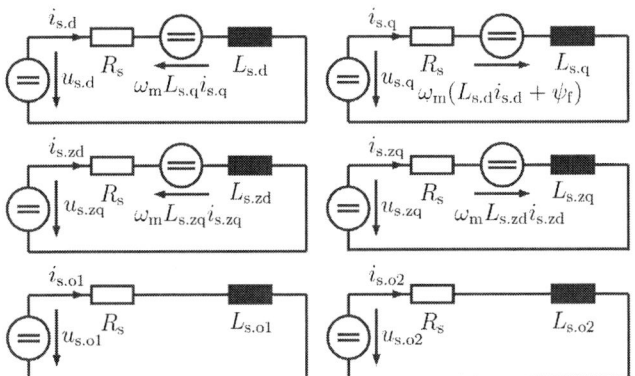

Fig. 3. Simplified equivalent circuit of the machine

1) Stator Current and Rotor Position Dependency of Magnet Flux Linkages:

The magnet flux linkages in each orthogonal dimension as a function of rotor position for different stator currents are given in Fig. 4. The instantaneous values of these flux linkages multiplied by the rotor speed ω_m appear as back EMF within dq and z_{dq} planes, while the derivatives of these components appear as harmonic back EMF in all dimensions of the equivalent circuit.

It can be noted that there are mainly dc components and 12th harmonics in dq plane magnet flux linkages, while there are essentially only 6th harmonics in z_{dq} plane. The o_{12} plane is dominated by 3rd and 9th harmonics.

The transformations in this study have been carried out assuming the magnet flux linkage, i.e. rotor flux linkage, has a constant position on the rotor. Consequently, the rotor position θ_m is used for Park transformations instead of the exact position of rotor flux linkage. The error in this assumption manifests as a dc component in magnet flux linkage $\psi_{f.q}$ when the stator currents increase due to saturation. The magnet flux linkage deviates about $6°$ in the worst case.

However, since the machine is an IPMSM, $i_{s.d} = 0$ control is not employed. Instead, either FEM analysis or measurements are employed to create lookup tables to determine dq current trajectories for torque demand and speed. If rotor position oriented transformations are used throughout this whole process, there will not be any performance degradation.

This issue of changing magnet flux linkage vector position compared to rotor position can also be handled by calculating this deviation as a function of stator currents. It should be noted, however, that this effort will not bring any additional benefits compared to using rotor position and using rotor position oriented lookup tables.

2) Stator Current and Rotor Position Dependency and Non-diagonality of Transformed Inductances:

The transformed stator inductance matrix is easily taken out of the derivative in (16), since it is assumed to be a 6×6 diagonal matrix of constant elements. However, the FEM analysis shows the strong dependence of $[L_{s.dq}]$ on stator currents as well as rotor position, as shown in Fig. 5.

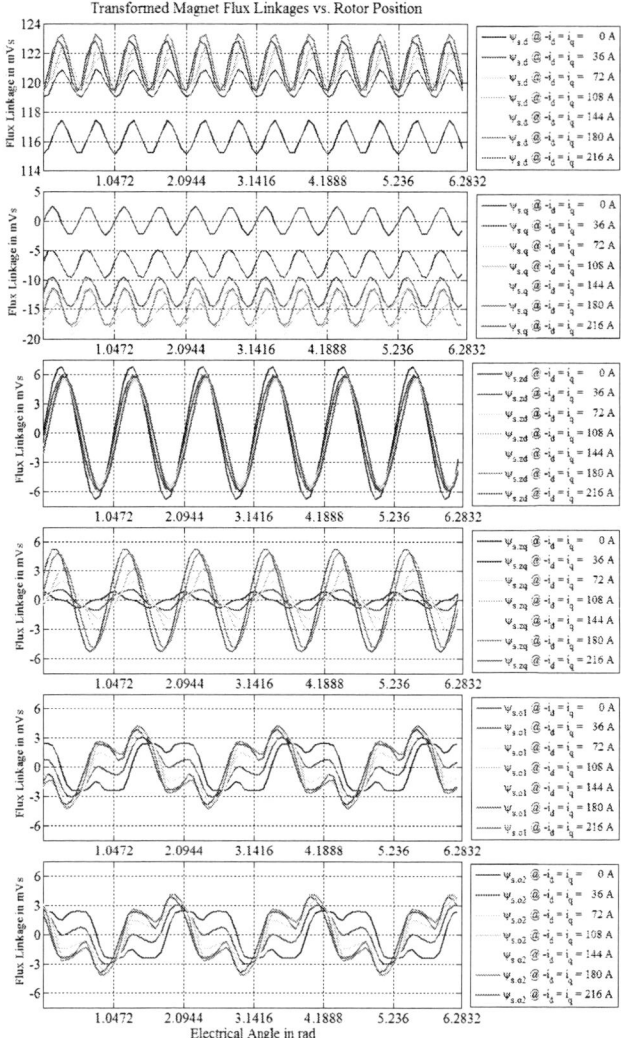

Fig. 4. Magnet flux linkages calculated by FEM analysis

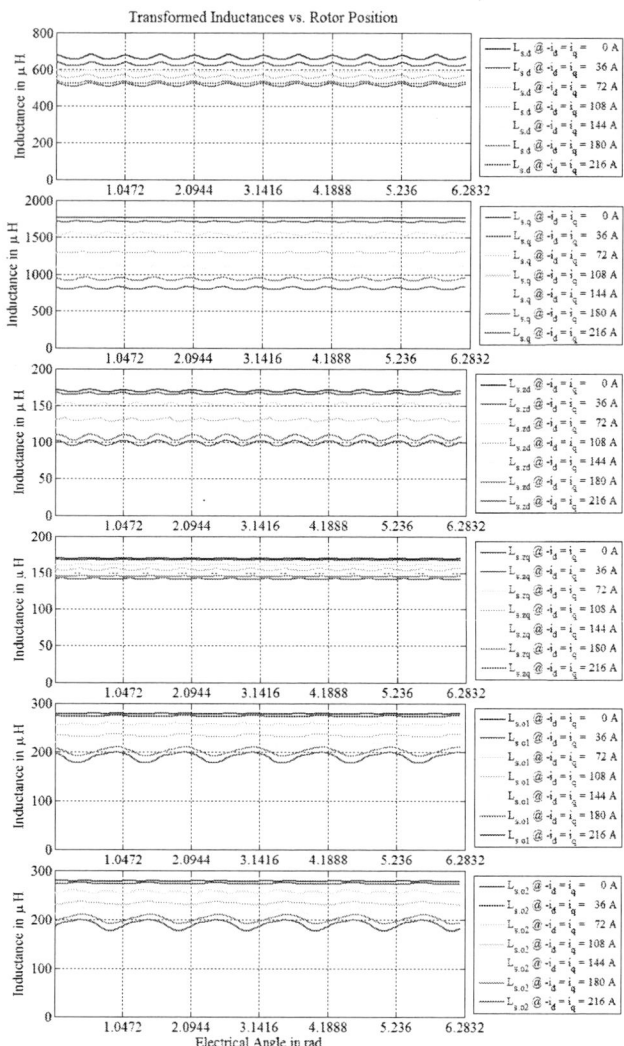

Fig. 5. Transformed self inductances calculated by FEM analysis

Therefore, the stator flux linkage derivative term in (14) takes the more complicated form given in (17) and (18). The first term in (18) is the instantaneous inductance, the second term is saturation, and the last term is the back EMF, induced by the rotor position dependent inductance change.

$$\frac{\mathrm{d}}{\mathrm{d}t}\,\vec{\psi}_{\mathrm{s.dq}} = \frac{\mathrm{d}}{\mathrm{d}t}\,\Big[L_{\mathrm{s.dq}}(\vec{i}_{\mathrm{s.dq}},\theta_{\mathrm{m}})\Big]\,\vec{i}_{\mathrm{s.dq}} + \frac{\mathrm{d}}{\mathrm{d}t}\,\vec{\psi}_{\mathrm{f.dq}} \qquad (17)$$

$$\frac{\mathrm{d}\Big[L_{\mathrm{s.dq}}(\vec{i}_{\mathrm{s.dq}},\theta_{\mathrm{m}})\Big]\,\vec{i}_{\mathrm{s.dq}}(t)}{\mathrm{d}t} =$$

$$\Big[L_{\mathrm{s.dq}}(\vec{i}_{\mathrm{s.dq}},\theta_{\mathrm{m}})\Big]\,\frac{\mathrm{d}\,\vec{i}_{\mathrm{s.dq}}(t)}{\mathrm{d}t} +$$

$$\vec{i}_{\mathrm{s.dq}}(t)\,\frac{\partial\Big[L_{\mathrm{s.dq}}(\vec{i}_{\mathrm{s.dq}},\theta_{\mathrm{m}})\Big]}{\partial\vec{i}_{\mathrm{s.dq}}}\,\frac{\mathrm{d}\,\vec{i}_{\mathrm{s.dq}}(t)}{\mathrm{d}t} + \qquad (18)$$

$$\vec{i}_{\mathrm{s.dq}}(t)\,\frac{\partial\Big[L_{\mathrm{s.dq}}(\vec{i}_{\mathrm{s.dq}},\theta_{\mathrm{m}})\Big]}{\partial\theta_{\mathrm{m}}}\,\frac{\mathrm{d}\,\theta_{\mathrm{m}}(t)}{\mathrm{d}t}$$

The FEM analysis results also indicate that the inductance matrix $\Big[L_{\mathrm{s.dq}}(\vec{i}_{\mathrm{s.dq}},\theta_{\mathrm{m}})\Big]$ is not diagonal, which implies that there is inductive coupling between the orthogonal planes. This coupling also has a significant impact on the machine behaviour and will be shown that it cannot be neglected.

The equivalent circuit model of the machine turns out to be very complex once the non-ideal properties of the machine are also included in the model, especially when the couplings between the supposedly orthogonal dimensions are taken into account. Therefore, the rotor field oriented machine model with an ideal rotating transformer (IRTF) is recommended to model the machine [1].

D. Field Oriented Machine Model

The field oriented machine model with an ideal rotating transformer (IRTF) focuses on the flux linkages in the machine. Since the calculation of the derivative of stator flux linkage is avoided, the model is significantly easier to understand and implement.

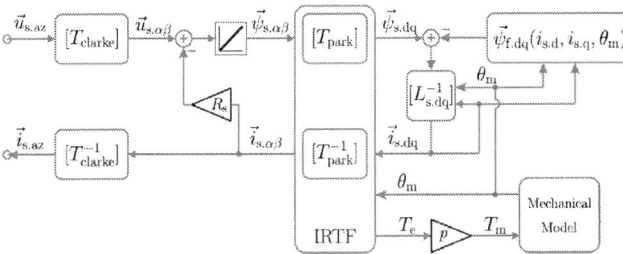

Fig. 6.   IRTF based machine model

The machine model receives the phase voltage information $\vec{u}_{\text{s.az}}$ as input. This vector is transformed using the Clarke transformation to obtain the voltage vector in stator oriented orthogonal space $\vec{u}_{\text{s.}\alpha\beta}$. Once the resistive voltage drop is substracted from $\vec{u}_{\text{s.}\alpha\beta}$, the remaining voltage is integrated to calculate the stator flux linkage vector in stator oriented orthogonal space $\vec{\psi}_{\text{s.}\alpha\beta}$, according to (13).

Rotor field oriented flux linkage vector $\vec{\psi}_{\text{s.dq}}$ can then be calculated by rotating $\vec{\psi}_{\text{s.}\alpha\beta}$ using the Park transformation. The current vector in this vector space $\vec{i}_{\text{s.dq}}$ can be calculated by subtracting the magnet flux linkage $\vec{\psi}_{\text{f.dq}}$ from the stator flux linkage $\vec{\psi}_{\text{s.dq}}$ and multiplying the result with the inverse inductance matrix $\left[L_{\text{s.dq}}^{-1}\right]$, as in (15).

The phase current vector $\vec{i}_{\text{s.az}}$ can finally be calculated by using the inverse Park and inverse Clarke transformations on the rotor field oriented current vector $\vec{i}_{\text{s.dq}}$, according to (3).

The mechanical torque T_{e} produced by the machine can be calculated as the magnitude of the cross product of stator flux linkage vector $\vec{\psi}_{\text{s.}\alpha\beta}$ and stator current vector $\vec{i}_{\text{s.}\alpha\beta}$ multiplied by the pole pair number p. Since the contributions of the other planes are negligibly small, the equation can be simplified to include only the cross product in dq plane, as in (19).

$$T_{\text{m}} = p \cdot T_{\text{e}} = p \cdot (\psi_{\text{s.d}} i_{\text{s.q}} - \psi_{\text{s.q}} i_{\text{s.d}}) \qquad (19)$$

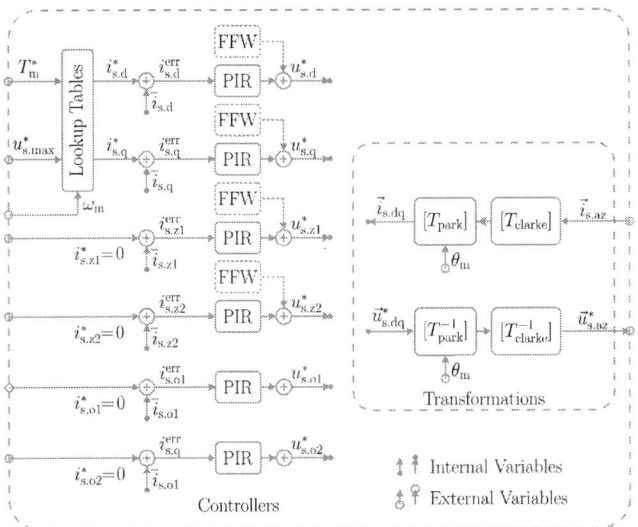

Fig. 7.   Control diagram of a six phase, open-delta IPMSM drivetrain

It is very important to note that $\vec{\psi}_{\text{f.dq}}$ and $\left[L_{\text{s.dq}}^{-1}\right]$ are strongly dependent on the rotor position θ_{m} and the stator currents $i_{\text{s.d}}$ and $i_{\text{s.q}}$. A machine model that neglects these dependencies is likely not to properly represent the machine in real life.

III.   FIELD ORIENTED CONTROL

The rotor field oriented control of six phase, open delta IPMSM is carried out similarly to the conventional field oriented control. Six current controllers are employed for each dimension of the rotor field oriented orthogonal space vector of current $\vec{i}_{\text{s.dq}}$. The set values of the current components in dq plane $i_{\text{s.d}}^*$ and $i_{\text{s.q}}^*$ are calculated using lookup tables, which are created either by using FEM analysis or measurements. Since the current components in z_{dq} plane and o_{12} plane do not contribute to the electromechanical energy conversion, the set values of these current components, $i_{\text{s.zd}}^*$, $i_{\text{s.zq}}^*$, $i_{\text{s.o1}}^*$ and $i_{\text{s.o2}}^*$ are always zero.

The field oriented control diagram is given in Fig. 7, where PIR denotes modified proportional integral resonant controllers, which are explained in detail below,. The optional feed forward components (FFW), which are not explored further in this paper, can be used to improve the dynamic performance of the control.

A.   Proportional Integral Resonant Controllers

PI controllers are not well suited to following alternating references or eliminating alternating errors due to their limited bandwidth. Different kinds of proportional resonant controllers have been proposed to address this control issue. Modified proportional integral resonant controllers (PIR) are used in this paper, since it is known which harmonic disturbance is to be expected in which vector plane.

The transfer function of the resonant component of the PIR controller is given in (20), where n is the order of harmonics that should be suppressed, $X(s)$ is the input and $Y(s)$ is the output of the resonant component of the PIR. Equation (20) can be transformed into (21). The complete PIR controller is shown in Fig. 8.

$$\frac{Y(s)}{X(s)} = \frac{K_{\text{res}} \cdot s}{s^2 + (n\,\omega_{\text{m}})^2} \qquad (20)$$

$$Y(s) = K_{\text{res}} \cdot X(s) \cdot s^{-1} - (n\,\omega_{\text{m}})^2 \cdot Y(s) \cdot s^{-2} \qquad (21)$$

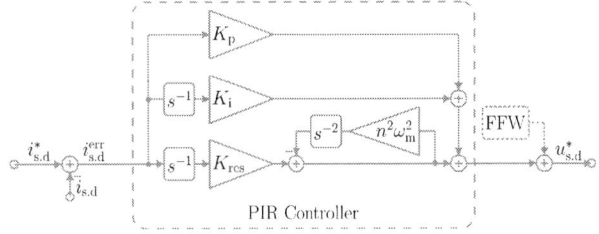

Fig. 8.   PIR controller diagram (exemplarly for $i_{\text{s.d}}$)

978-1-4673-9551-9/16 $31.00 © 2016 IEEE

The resonant frequency $n\,\omega_{\mathrm{m}}$ of the PIR controllers are set to the 12th harmonic frequency in dq plane and the 6th harmonics frequency in z_{dq} plane. A PIR with two resonant components is used in o_{12} plane with resonant frequencies of the 3rd and 9th harmonics.

IV. THE DRIVETRAIN

The drivetrain has been designed as a retrofit replacement for a two-passenger, pure electric vehicle. The specifications of the drivetrain are given in Table I. The project required that the whole drivetrain fits into the space of only the machine of the previous drivetrain. Therefore a modular system, consisting of the machine, the inverter power stage and the control unit, has been designed to form a fully enclosed housing with water cooling to achieve high power density.

TABLE I.   DRIVETRAIN SPECIFICATIONS

Parameter	Variable	Value
Nominal mechanical power	$P_{\mathrm{m.N}}$	31.5 kW
Nominal torque	$T_{\mathrm{m.N}}$	100 Nm
Nominal rotational speed	$n_{\mathrm{m.N}}$	3000 rpm
Maximum rotational speed	$n_{\mathrm{m.max}}$	10000 rpm
Minimum battery voltage	$V_{\mathrm{batt.min}}$	80 V
Nominal battery voltage	$V_{\mathrm{batt.N}}$	108 V
Maximum battery voltage	$V_{\mathrm{batt.max}}$	120 V
Nominal phase voltage	$V_{\mathrm{ph.N}}$	70 V
Nominal phase current	$I_{\mathrm{ph.N}}$	100 A
Power factor	$\cos\varphi$	0.80
Number of pole pairs	p	3
Switching frequency	f_{sw}	50 kHz

The rotor of the machine is designed with a hollow shaft and a rotor mounted fan blade for air cooling. The blade of this cooling system goes through the inverter space, as shown in Fig. 9. This created a significant design challenge for the spacial distribution and cooling of the inverter. Both the rotor cooling system and the inverter spacial distribution led to patent pending ideas, details of which are given in [7] and [8].

A. The Inverter Power Stage and the Control Unit

The inverter consists of a power stage and a control unit, as shown in Fig. 10. The top, left, back isometric view of the inverter is given in Fig. 10-a with the control unit box open. The female plugs to connect the machine to the inverter, which save significant volume compared to a screw solution, can be seen in the top, front, left isometric view in Fig. 10-b.

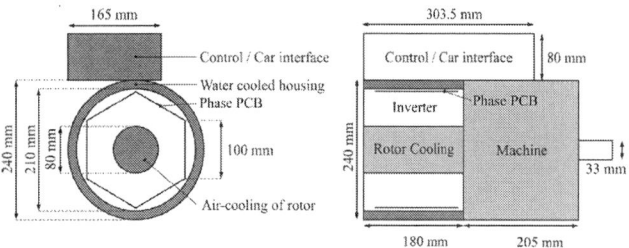

Fig. 9.   The integrated drivetrain design

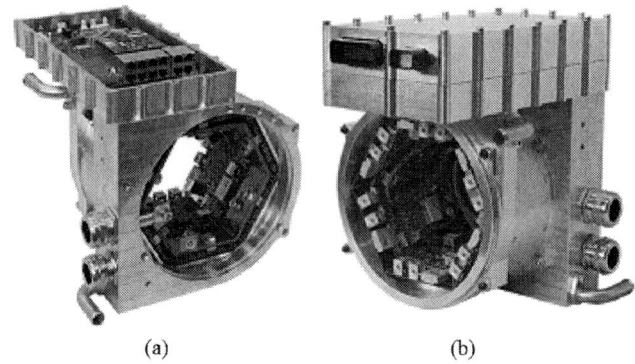

Fig. 10.   The inverter: (a) Back isometric view, (b) Front isometric view

The power unit consists of six individual H bridges, mounted on the inner walls of a hexagonal water cooled housing. Due to the very restricted available mounting space, the H bridges are designed with DirectFET MOSFETs. This allows a very low mounting height as well as low stray inductances in the commutation path, thus allowing operation of the inverter close to the breakdown voltage of the MOSFETs.

Successful operation is verified up to a dc-link voltage of $V_{\mathrm{dc}} = 145\,\mathrm{V}$ using MOSFETs with a rated breakdown voltage of $V_{\mathrm{DS.BD}} = 150\,\mathrm{V}$. To further increase the power density, ceramic SMD capacitors are used in parallel to build up the dc link capacitor. A maximum inverter efficiency of $\eta = 97\,\%$ and power density of 5,4 kW/L is achieved.

The control unit includes the compact control system XCS2100 from the company AixControl. The control algorithm is run on a 150 MHz fixed-point DSP. 12 PWM signals for each half bridge are created by this system and processed by a Xiling CPLD to obtain the 24 gate signals for each MOSFET. Safety functions such as hardware overcurrent and overvoltage protection are also implemented by the CPLD. The control unit handles CAN communication with the vehicle. Further details of the inverter design can be found in [7].

Fig. 11.   The drivetrain installed on the testbench

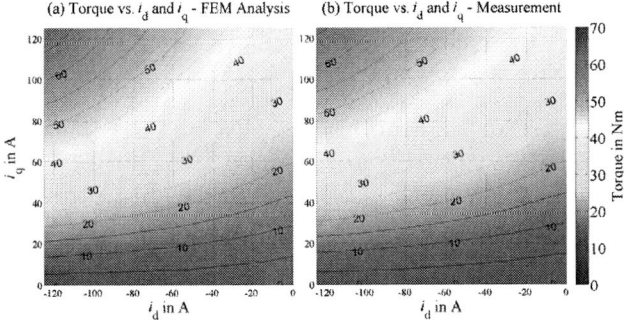

Fig. 12. Torque vs. $i_{\mathrm{s.d}}$ and $i_{\mathrm{s.q}}$: (a) FEM analysis, (b) measurements

B. The Machine

The IPMSM is a 3 pole-pair, distributed winding machine with V-shaped interior magnets. The stator has 36 slots, therefore 1 slot per phase per pole and 8 turns per slot. Reducing the slot pitch of the coil span allows to decrease the harmonic content. However, it is not implemented, since this effect is marginal in a six-phase winding system. In order to reduce the cogging torque, the stator is skewed by one tooth width. The machine has a very compact design, achieving a very high power density of 3 MW/m3. Further details of the machine design can be found in [8].

V. SIMULATIONS AND EXPERIMENTAL RESULTS

A. Machine Model Verification

1) Torque vs. $i_{\mathrm{s.d}}$ and $i_{\mathrm{s.q}}$: As the first step of the experimental model verification, the torque of the machine in FEM analysis and on the test bench have been compared. For this purpose, changing $i_{\mathrm{s.d}}$ and $i_{\mathrm{s.q}}$ values have been applied to the machine and the torque outputs have been measured. The results have shown that the FEM analysis torque calculations successfully represent the actual machine, as illustrated in Fig. 12.

2) Short Circuit Currents: The short circuit currents of machine models with various assumptions are compared to the measurements, in order to observe the impacts of these assumptions on the models. The results are given in stator oriented orthogonal space, since the current measurements do not have corresponding rotor angle data.

The assumption that the rotor field oriented orthogonal space dimensions are inductively decoupled leads to the assumption that the machine can be represented by the machine model, even when only the diagonal inverse inductance matrix is used instead of the complete inverse inductance matrix $\left[L_{\mathrm{s.dq}}^{-1} \right]$. This assumption leads to inaccurate results, as can be seen by comparing (a) and (d) in Fig. 13.

The saturation has a very significant impact on the induced currents in z_{dq} plane as well as o_{12} plane, as the comparison between (b) and (d) in Fig. 13 illustrates. Simulations have also shown that the phase voltage of the machine "triples" at rated speed and torque, if the saturation is not taken into account. Therefore, saturation should not be neglected in order to obtain a representative machine model.

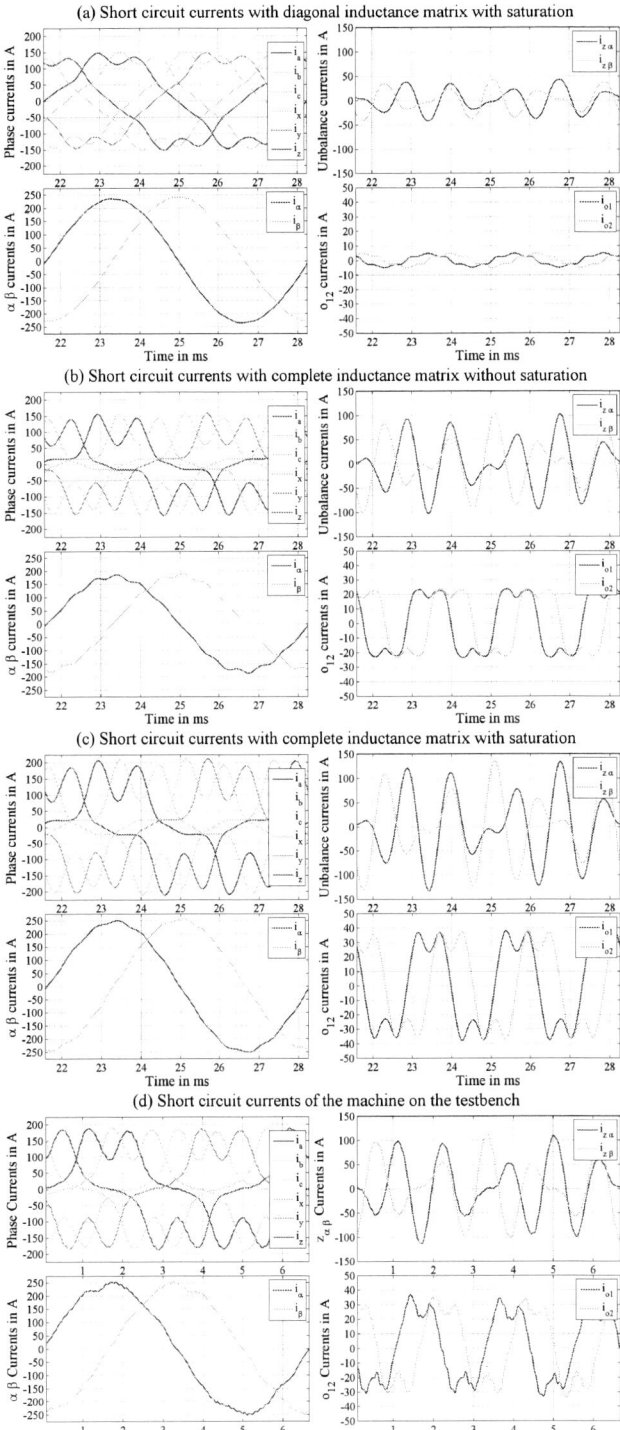

Fig. 13. Short circuit currents of machine models and the real machine

The IRTF based machine model is shown to adequately represent the machine as long as the rotor position and stator current dependencies of inductances and the magnet flux linkages as well as the inductive coupling between the rotor field oriented orthogonal space dimensions are taken into account, as shown in (c) and (d) in Fig. 13.

978-1-4673-9551-9/16 $31.00 © 2016 IEEE

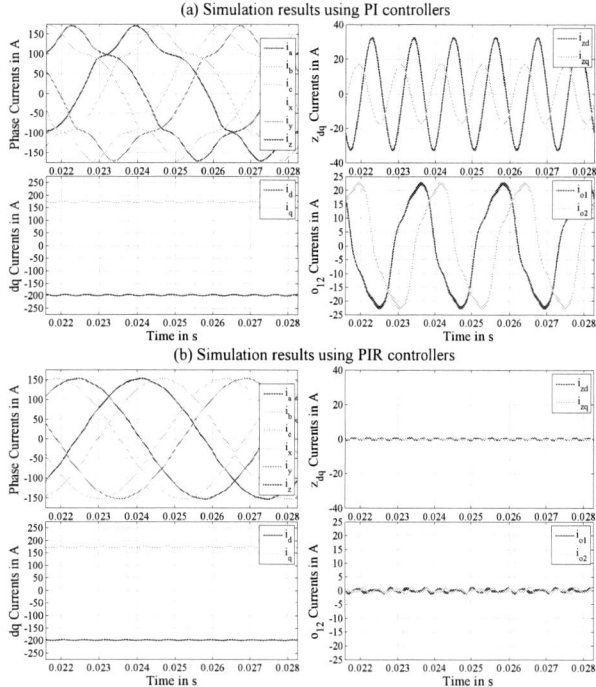

Fig. 14. Control simulations at rated speed and torque

B. Control Simulations

Control simulations have been carried out in MATLAB-Simulink. The controller and the machine models are implemented in Simulink, while the inverter is modeled using PLECS. The current sensor delay and discrete controller operation are implemented, while dead times in the inverter are not implemented.

The current waveforms of the machine at the rated operating point of rotor speed $n = 3000\,\mathrm{rpm}$ and torque $T_m = 100\,\mathrm{Nm}$, using a PI controller and a PIR controller, respectively, are given in Fig. 14. The PIR controllers have successfully damped the harmonics, which could not be damped by the PI controllers.

C. Inverter and Machine Efficiencies

Efficiency maps have been obtained using a power analyzers on the dc link and the six phases of the machine to measure

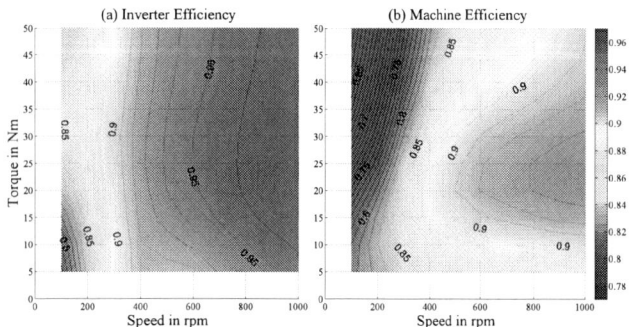

Fig. 15. Efficiency vs. speed and torque of (a) inverter and (b) machine

the electrical power at the input and output of the inverter; and using a torque-speed sensor to measure the mechanical power of the machine. 100 points of measurement have been taken for a varying mechanical torque output T_m between $5\,\mathrm{Nm}$ and $50\,\mathrm{Nm}$, and rotor speed n between $100\,\mathrm{rpm}$ and $1000\,\mathrm{rpm}$ to obtain the efficiency maps given in Fig. 15. An inverter efficiency of $97.1\,\%$ and a machine efficiency of $93.4\,\%$ have been achieved despite partial loading. The full operating range could not be achieved yet as of the date the paper has been written.

VI. Conclusion

A low voltage, high power drivetrain with an asymmetrical six phase, open-delta IPMSM and an H Bridge inverter has been designed for an electric vehicle. Field oriented modeling and control of such a system have been explored. The machine-based sources of unbalance and zero sequence current components have been investigated in further detail. It has been shown how these sources are related to the harmonics induced by the machine. Detailed simulations using machine models with verified FEM analysis are provided, along with efficiency measurements of the drivetrain.

VII. Acknowledgments

The results presented in this paper have been developed during the research project e-MoSys: "Entwicklung und prototypische Umsetzung eines anfordergungsgerechten und modularen Antriebs- und Fahrwerkssystems fuer ein Elektrofahrzeug" (Development and prototype implementation of a specific and modular drive and suspension system for an electric vehicle) granted by the Ministry of Education and Research.

References

[1] R. De Doncker, W. J. D. Pulle, A. Veltman, *Advanced Electrical Drives - Analysis, Modeling, Control*, ISBN 978-94-007-0181-6, Springer, 2011.

[2] E. Levi, *Multiphase Electric Machines for Variable Speed Applications*, IEEE Transactions on Industrial, vol. 55, no. 5, pp.1893-1909, May 2008.

[3] E. Levi, R. Bojoi, F. Profumo, H. A. Toliyat and S. Williamson, *Multiphase Induction Motor Drives A Technology Status Review*, IET Electr. Power Appl., pp. 489-516, 2007.

[4] Y. Zhao, T. A. Lipo, *Space Vector PWM Control of Dual Three Phase Induction Machine Using Vector Space Decomposition*, IEEE Transactions on Industry Applications, vol. 31, no. 5, pp. 1100-1109, September/October 1995.

[5] Y. Hu, Z. Zhu, K. Liu, *Current Control for Dual Three-Phase Permanent Magnet Synchronous Motors Accounting for Current Unbalance and Harmonics*, IEEE Journal Of Emerging And Selected Topics In Power Electronics, Vol. 2, No. 2, June 2014.

[6] M. Neubert, S. Koschik, R.W. De Doncker, *Performance comparison of inverter and drive configurations with open-end and star-connected windings*, Power Electronics Conference (IPEC-Hiroshima 2014 - ECCE-ASIA), 2014.

[7] G. Engelmann, M. Kowal, R.W. De Doncker, *A Highly Integrated Drive Inverter using Directfets and Ceramic Dc-Link Capacitors for Open-End Winding Machines in Electric Vehicles*, Applied Power Electronics Conference and Exposition (APEC), Charlotte, NC, USA, 2015.

[8] T. Grosse, D. Franck, N. Conzelmann, D. Paul, C. Haenelt, A. Stapelmann, K. Hameyer *Compact Machine Design of an Integrated Multiphase VPMSM*, Electric Drives Production Conference (EDPC), Nuremberg, Germany, 2015

Passive Integration Using FMLF Technique for Integrated Boost Resonant Converters

Cheng Deng∗†, *Member, IEEE*, Luciano Andres Garcia Rodriguez∗, *Sudent Member, IEEE*, Juan Zou† and Juan Carlos Balda∗,
Senior Member, IEEE

Email: cdeng@uark.edu, lgarciar@uark.eduVijay, zoujuan@xtu.edu.cn, jbalda@uark.edu
∗Department of Electrical Engineering, University of Arkansas, Fayetteville Arkansas, USA
†College of Information Engineering, Xiangtan University, Xiangtan, Hunan, CHN

Abstract — **In order to further increase the power density of the integrated boost resonant (IBR) converter, a novel passive integrated unit is proposed by combining three inductors, two capacitors and one transformer. For saving the these passive components, the Flexible Multi-Layer Foil (FMLF) integration technique is used.**

To avoid the coupling of two boost inductors and the transformer, the symmetry structure is used in the proposed unit. The electromagnetic analysis of the proposed unit is given, and the calculations of the circuit parameters are provided. Finally, a prototype for a 350-W IBR converter has been built to validate the feasibility of the proposed integrated unit.

Keywords— Integrated Boost Resonant converter; Flexible Multi-Layer Foil integration technique

I. INTRODUCTION

Photovoltaic (PV) systems are a rapidly growing segment in the renewable energy industry [1~3] with the distributed microinverter becoming a popular architecture. In order to improve the efficiency of the microinverter, the IBR converter is presented in [4] and illustrated in Fig. 1. It uses a series-resonant isolation stage to realize the zero-voltage-switching (ZVS). The IBR converter has many passive components, such as two interleaved boost inductors L_{b1} and L_{b2}, one resonant inductor L_r, one transformer T, and two resonant capacitors C_{r1} and C_{r2}. Traditionally, discrete components are used for them. As it is well known, the magnetic energy and the electric energy are stored in their separate spaces respectively, such as the solenoid inductor and the plate capacitor shown in Fig. 2. As a result, these discrete passive components occupy a significant volume in the IBR converter.

FMLF integration technique is an effective approach for passive integration [5~11]. Its essential concept is that a dielectric film is inserted between two flexible conductor films to integrate inductors and capacitors. The basic structure and its lump equivalent model are given in Fig. 3.

In order to reduce the volume of the EMI filter, FMLF integration technique was used to integrate the common-mode (CM) capacitor, CM inductor and differential model (DM) inductor [5]. Compared to the traditional EMI filter, its volume is reduced by about 45%. In [6], an integrated unit is proposed to integrate the two boost inductors and the EMI filter in a PFC converter. The volume of the passive components is reduced by 25% with FMLF integration technique. In [7], another integrated unit is presented to integrate one series resonant

inductor, one parallel resonant inductor, one series resonant capacitor and one transformer for the LLC converter. By using FMLF integration technique, the size of this integrated unit was reduced by 33% compared with the discrete components.

In this paper, a novel integrated unit for realizing three inductors, two capacitors and one transformer with two EE cores by using FMLF integration technique is proposed for the IBR converter. This paper is organized as follows: The design of the symmetry structure used to avoid any coupling of the magnetic structures and a flux analysis of the integrated unit are given in Section II. The calculations of the circuit parameters are provided in Section III. The experimental results on a 350-W IBR converter prototype verifying the effectiveness of the proposed ideas are given in Section IV. Lastly, the conclusions are provided in Section V.

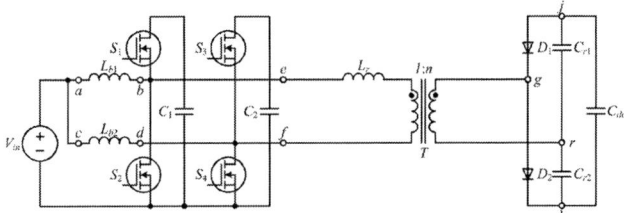

Fig. 1 Integrated boost resonant converter

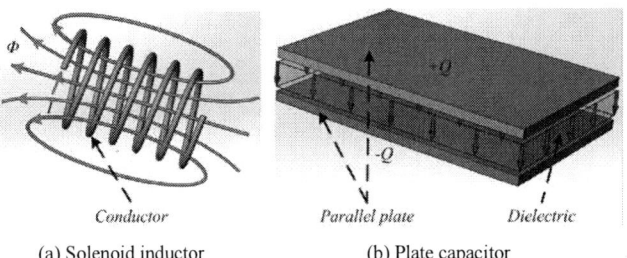

(a) Solenoid inductor (b) Plate capacitor

Fig. 2 Storage modes of magnetic energy and electric energy

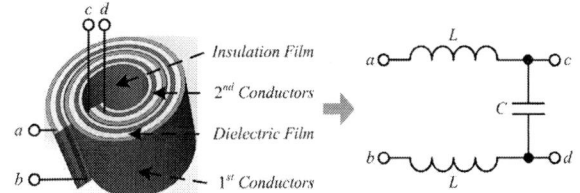

Fig. 3 FMLF and its lumped equivalent model

978-1-4673-9551-9/16 $31.00 © 2016 IEEE

II. PROPOSED INTEGATED UNIT

The symmetry structure is used to realize the six passive components mentioned above into a single integrated unit as described in detail below.

A. Two Types of FMLFs

The flexible single-layer film is presented in Fig. 4 and made up of one copper film and one insulation film can be used as the inductor or the primary winding when its two terminals connect into the circuit.

In this paper, a flexible three-layer film consisting of three conductor films, two dielectric films and one insulation film is proposed as shown in Fig. 5. The 1st dielectric film forms a capacitor since it is inserted between the 1st and the 2nd conductor films. The same result is obtained for the 2nd dielectric as well. Since the two ends q and g of the 2nd conductor film are connected into the circuit, it can operate as an inductor or transformer. But for the 1st and 3rd conductor films, one terminal is connected into the circuit and the other terminal of them is floating. Thus, the 1st and 3rd conductor films can only function as a wire. Fig. 5 presents the distributed electromagnetic component (DEMC) model of the flexible three-layer film.

B. Realization of Two Interleaved Boost Inductors

In Fig. 6, the windings W_1 and W_2 with the flexible single-layer film are rolled on the upper-center and lower-center legs of the cores, respectively. They are used as the interleaved boost inductors when the four terminals a, b, c and d are linked to the IBR converter.

C. Realization of Secondary Winding

With two flexible three-layer films, the 1st and 2nd secondary windings are rolled on the left-side and right-side legs of the cores as shown in Fig. 7(a). The connecting modes between two secondary windings are organized as followed.

- The right terminal of the 1st conductor film of the 1st secondary winding is floating (which is not drawn in Fig. 7(a)). The left terminal is linked to the left terminal of the 1st conductor film of the 2nd secondary winding by the end m. The right terminal of the 1st conductor film of the 2nd secondary winding is connected to the end j in the IBR circuit (see Fig.1).

- The right terminal of the 2nd conductor film of the 1st secondary winding is linked to the end g in IBR circuit. The left terminal connects with the left terminal of the 2nd conductor film of the 2nd secondary winding by the end q. The right terminal of the 2nd conductor film of the 2nd secondary winding is connected to the end r in the IBR circuit.

- The right terminal of the 3rd conductor film of the 1st secondary winding is floating (which is not drawn in Fig. 7(a)). The left terminal connects with the left terminal of the 3rd conductor film of the 2nd secondary winding by the end n. The right terminal of the 3rd conductor film of the 2nd secondary winding is connected to the end k in the IBR circuit.

According to these connecting modes, the DEMC model of two secondary windings is presented in Fig. 7(b). It can be simplified as a lump parameter model as illustrated in Fig. 7(c).

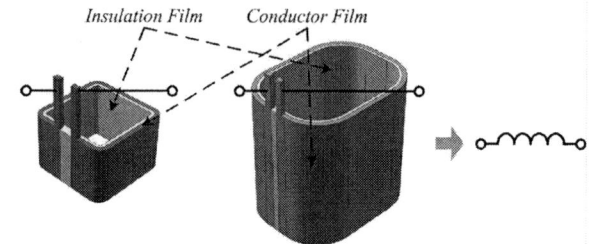

Fig. 4 Flexible single-layer film

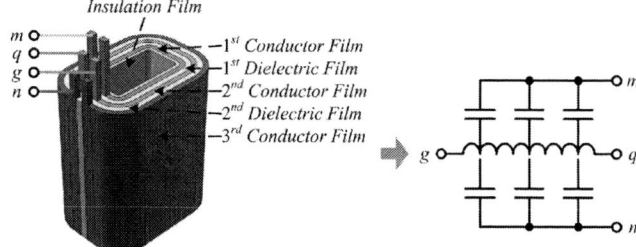

Fig. 5 Flexible three-layer film

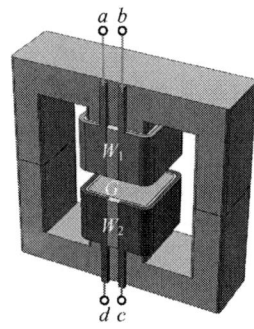

Fig. 6 Realization of two interleaved boost inductors

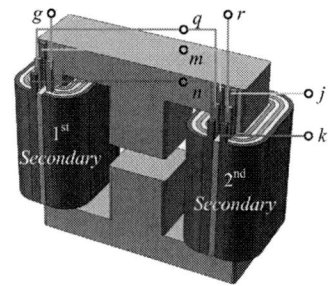

(a) Structure of two secondary windings

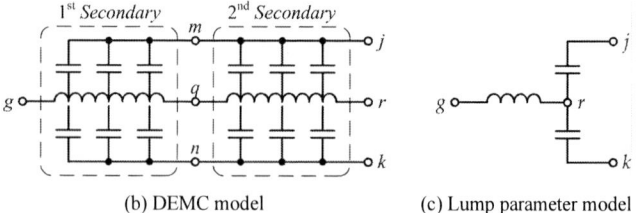

(b) DEMC model (c) Lump parameter model

Fig. 7 Realization of the secondary winding

D. Realization of Transformer

Fig. 8(a) presents an expanded view of the two transformers W_3 and W_4. One leakage material is inserted between one secondary winding and one primary winding. That is the commonly-used method to integrate the series resonant inductor with the transformer [12].

Fig. 8(b) shows the connection modes between W_3 and W_4, where,

- The left terminal of the conductor film of the 1st primary winding is connected to the end e in the IBR circuit. The right terminal connects with the right terminal of the conductor film of the 2nd primary winding at the end p. The left terminal of the conductor film of the 2nd primary winding is connected at the end f in the IBR circuit.

The equivalent circuit model of the two transformers is given in Fig. 8 (c). Since the leakage inductor L_{r_W3} of W_3 is in series with the leakage inductor L_{r_W4} of W_4, the equivalent circuit model can be simplified further.

E. Proposed Integragted Unit

By combined Fig. 6 with Fig. 8 (b), the proposed integrated unit is shown in Fig. 9 which integrates six passive components, namely, L_{b1}, L_{b2}, L_r, T, C_{r1} and C_{r2}.

F. Fluxes Analysis

As show in Fig. 10(a), two fluxes Φ_{W1} and Φ_{W2} are formed when the power current i flows into the terminals a and c respectively. Fluxes Φ_{W1} and Φ_{W2} are superimposed due to having the same magnitude and direction.

In order to avoid flux saturation issues at the maximum power point, the air gap G is inserted into between the upper-center and lower-center legs.

According to Faraday's law, when Φ_{W1} and Φ_{W2} go through the primary windings of W_3 and W_4, two induced electromotive forces V_{M1} and V_{M2} with the same magnitude are generated as shown in Fig. 10(b). Since they are in opposite directions, they counteract each other. That means the fluxes of the two interleaved boost inductors have no impact on the primary windings. The same result occurs on the secondary windings.

As shown in Fig. 11, when the primary current i_{pri} flows into the terminal e and out from the terminal f, the two fluxes

Φ_{W3_pri} and Φ_{W4_pri} are formed. Due to having the same magnitude and direction, Φ_{W3_pri} and Φ_{W4_pri} are superimposed.

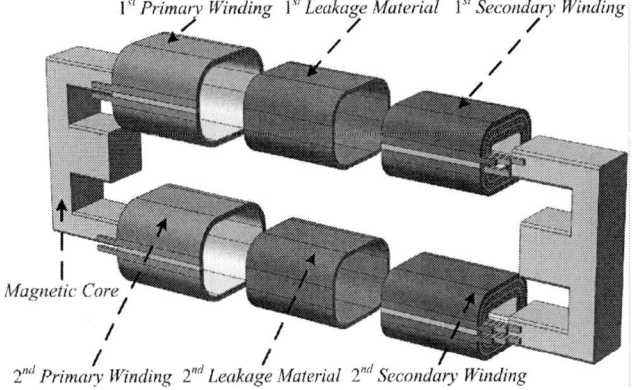

(a) Expanded view of the two transformers

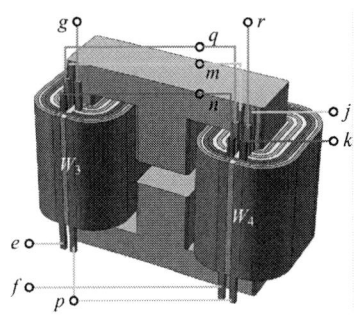

(b) Structure of the two transformers

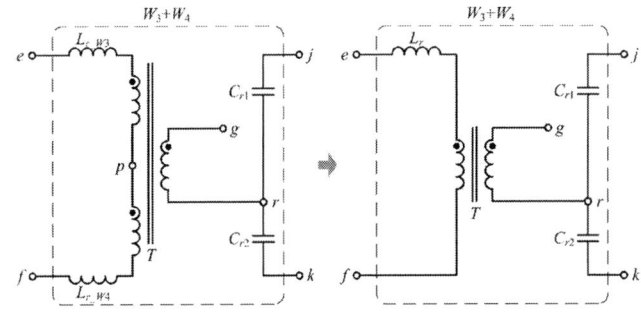

(c) Equivalent circuit model of the two transformers

Fig. 8 Realization of a transformer

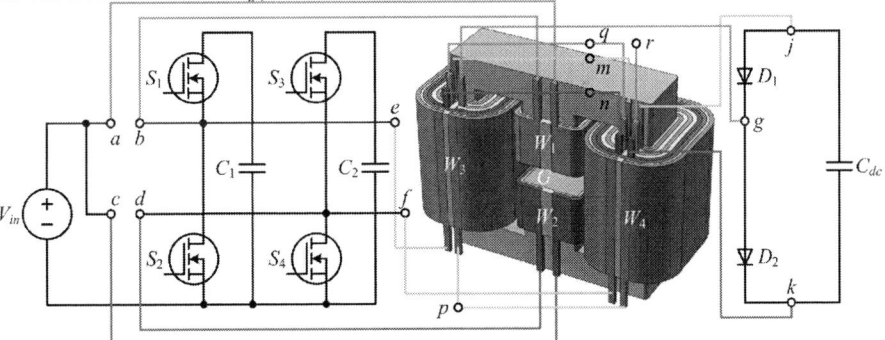

Fig. 9 Proposed integrated unit

978-1-4673-9551-9/16 $31.00 © 2016 IEEE   653

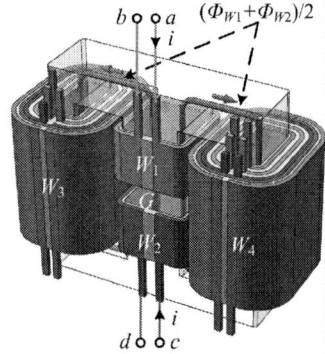

(a) Fluxes of the two interleaved boost inductors

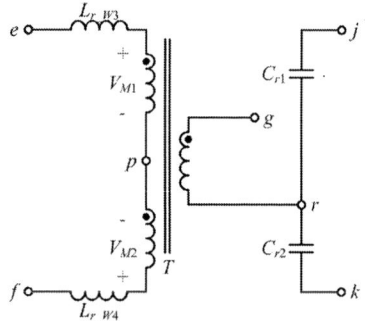

(b) Induced electromotive forces generated by fluxes

Fig. 10 Fluxes of the two interleaved boost inductors and the induced electromotive forces generated by fluxes

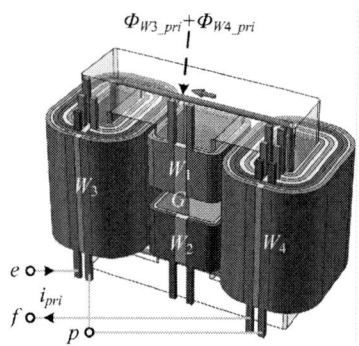

Fig. 11 Fluxes of the two primary windings

Since the reluctance of the air gap G between two center legs is much larger that of the ferrite core, Φ_{W3_pri} and Φ_{W4_pri} flows through the center legs can be neglected.

Thus, the symmetry structure of the integrated unit effectively decouples the fluxes of the two interleaved boost inductors from those of the two transformers.

III. PARAMETER VALUE SELECTION

By specifying L_{B1}/L_{B2}, L_r, C_{r1}/C_{r2}, the output power P_{out}, the primary voltage V_{pir} and the secondary voltage V_{sec}, and the switching frequency f_{sw}, the structure parameters of the integrated unit can be derived as follows.

A. Boost Inductor Value

As shown in Fig. 10(a), the fluxes Φ_{W1} and Φ_{W2} of W_1 and W_2 can be calculated by:

$$\phi_{W1}=\phi_{W2} = \frac{N_b \cdot i_{max}}{R_b} \tag{1}$$

where, N_b is the winding number of turns of W_1 and W_2, R_b is the reluctance of the fluxes of the interleaved boost inductors, and i_{max} is the current at maximum power..

Then, the inductances of W_1 and W_2, L_{b1} and L_{b2} can be deduced as follows:

$$L_{b1} = L_{b2} = \frac{N_b \cdot \left(\phi_{W1} + \phi_{W2}\right)}{i_{max}} = 2 \cdot \frac{N_b^{\,2}}{R_b} \tag{2}$$

Since the air gap G is generated to avoid flux saturation at the maximum power point, the following inequality should be met.

$$B_{sat} \cdot A_{Cen} \ge \phi_{W1} + \phi_{W2} = 2 \cdot \frac{N_b \cdot i_{max}}{R_b} \tag{3}$$

where, B_{sat} is the saturation flux density, and A_{Cen} denotes the effective cross-sectional area of the center leg,

Based on (2) and (3), the winding number of turns N_b of W_1 and W_2 should be chosen by:

$$N_b \ge \frac{L_{b1} \cdot i_{max}}{B_{sat} \cdot A_{e\_Cen}} \tag{4}$$

B. Transformer Number of Turns Values

According to the primary voltage V_{pir} and the secondary voltage V_{sec}, the turn ratio n_T of the transformer can be easily computed as follows:

$$n_T = \frac{V_{pri}}{V_{sec}} \tag{5}$$

Then, the number n_{pri} of the primary winding can be calculated by

$$n_{pri} = \frac{V_{pri}}{k_V \cdot B_{sat} \cdot A_{side} \cdot f_{sw}} \tag{6}$$

where, k_V is the coefficient of the waveform, A_{side} is the effective cross-sectional area of the side leg. Then,

Therefore, the number n_{sec} of the secondary winding can be calculated by:

$$n_{sec} = \frac{n_{pri}}{n_T} \tag{7}$$

C. Series Resonant Capacitor Value

According to Fig. 8, the series resonant capacitors C_{r1} and C_{r2} are the sum of the integrated capacitors C_{W3} and C_{W4} having a single dielectric film in the two secondary windings.

$$C_{r1} = C_{r2} = C_{W3} + C_{W4} = 2 \cdot \frac{\varepsilon_0 \cdot \varepsilon_{die} \cdot l_{die} \cdot h_{die}}{t_{die}} \quad (8)$$

where, ε_0 and ε_{die} are the dielectric constant of the air and the dielectric film, l_{die}, h_{die} and t_{die} denote length, height and thickness of single dielectric film in the 1st/2nd secondary winding, respectively.

D. Series Resonant Inductor Value

In IBR converter [4], the series resonant inductor L_r is selected based on the resonant frequency f_r, the output capacitor C_{dc}, the turn ratio n_T of the transformer, and the resonant capacitors C_{r1} and C_{r2}, namely:

$$f_r = \frac{1}{\pi \sqrt{\dfrac{L_r \cdot C_{dc} \cdot n_T^2 \cdot (C_{r1} + C_{r2})}{C_{dc} + n_T^2 \cdot (C_{r1} + C_{r2})}}} \quad (9)$$

Under the condition without the leakage material, there are three situations for the leakage inductors L_{r_W3} and L_{r_W4} which are used to replace L_r.

Case 1: If,

$$L_{r\_W3} + L_{r\_W4} \gg L_r \quad (10)$$

To meet (9), C_{r1} and C_{r2} should be reduced. This means that the dielectric film has to be a smaller ε_{die} or thicker t_{die}.

Case 2: If, the sum of L_{r_W3} and L_{r_W4} is close to L_r, the leakage inductance of the two transformers can replace L_r. There is no necessary to insert any leakage material between the primary and the secondary windings.

Case 3: Otherwise, a leakage material should be added. Then, the leakage flux distribution is shown in Fig. 12.

By using the method mentioned in [12], L_r can be calculated as follows:

$$L_r = L_{r\_W3} + L_{r\_W4} = 2 \cdot \frac{\mu_0 \cdot \mu_{leak} \cdot n_{pri}^2 \cdot 8 \cdot w_{leak} \cdot t_{leak}}{h_{leak}} \quad (11)$$

where, μ_0 is the permeability of the free space, μ_{leak} is the relative permeability of the leakage material, w_{leak} is the distance between the leakage material and the center of the side leg, t_{leak} and h_{leak} are the thickness and height of the leakage material.

I. EXPERIMENTAL RESULTS

In order to proceed prudently with the passive integration, one prototype for two boost inductors, one resonant inductor and one transformer is built as shown in Fig. 13 according to the working conditions (see Table I). The parameters of the prototype are listed in Table II. In this case, the leakage material can be omitted since the transformer leakage inductance is sufficient to enable the resonant process.

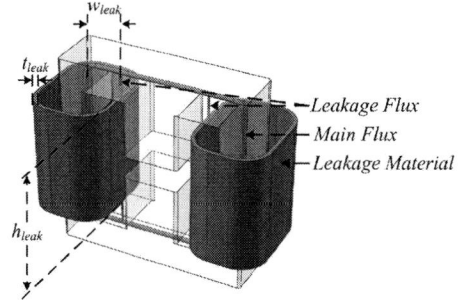

Fig. 12 Leakage fluxes distribution with leakage material in transformer

Fig. 13 Prototype illustrating integration for two boost inductors, one resonant inductor and one transformer

TABLE I.   WORKING CONDITIONS

Item	Value
Maximum Power Pout	350 W
Maximum and Minimum Input Voltage	16 ~ 60V
Rated Output Voltage	400 V
Switching Frequency	100 KHz

TABLE II.   WORKING CONDITIONS

	Item	Value
	Magnetic Core	E35/18/10 (3C95)
Integrated Boost Inductor	Inductance L_{b1} / L_{b2}	22.3 μH
	Turn Number N_b	5 turns
	Thickness of Copper Flim	100 μm
Integrated Transformer	Turn Number of Primary N_{pri}	4 turns
	Turn Number of Secondary N_{sec}	21 turns
	Thickness of Copper Film	50 μm
	Leakage Inductance	10.8 μH
Discrete Resonant Capacitance C_{r1} / C_{r2}		220 nF

II. CONCLUSION

Two interleaved boost inductors, one resonant inductor, one transformer and two resonant capacitors were integrated into an E-E core with two types of FMLF windings. The flux circulations and the equivalent circuits of the integrated structure were analyzed. The procedure for performing a design was presented using the lumped equivalent circuit models. To verify the proposed integrated unit, a prototype for the 350-W IBR converter was built.

ACKNOWLEDGMENT

The authors are grateful for the financial support from the NSF-EPSCoR Vertically Integrated Center for Transformative Energy Research (www.victercenter.com), National Natural Science Foundation of China (51577161, 51277156), Natural Science Foundation of Hunan Province, China (2015JJ2135).

REFERENCES

[1] S. Jiang, D. Cao, Y. Li, and F. Z. Peng, "Grid-Connected Boost-Half-Bridge Photovoltaic Microinverter System Using Repetitive Current Control and Maximum Power Point Tracking," *IEEE Transaction on Power Electronics*, vol. 27, no. 11, pp. 4711–4722, Nov. 2012.

[2] Q. Zhang, C. S. Hu, L. Chen, A. Amirahmadi, N. Kutkut, Z. J. Shen, and I. Batarseh, "A Center Point Iteration MPPT Method With Application on the Frequency-Modulated LLC Microinverter," *IEEE Transaction on Power Electronics*, vol. 29, no. 3, pp. 1262–1274, Mar. 2012.

[3] H. B. Hu, S. Harb, N. Kutkut, I. Batarseh and Z. J. Shen, "A Review of Power Decoupling Techniques for Microinverters With Three Different Decoupling Capacitor Locations in PV Systems," *IEEE Transaction on Power Electronics*, vol. 28, no. 6, pp. 2711–2726, Jun. 2013.

[4] J. B. York, "An Isolated Micro-Converter for Next-Generation Photovoltaic Infrastructure," *PhD Dissertation*, Virginia Polytechnic Institute and State University, 2013.

[5] X. F. Wu, D. H. Xu, Z. W. Wen, Y. Okuma, K. Mino, "Design, modeling and improvement of integrated EMI filter based on FML foils," *IEEE Transaction on Power Electronics*, vol. 26, no. 5, pp. 1344–1354, May 2011.

[6] C. Deng, D. H. Xu, and C. S. Hu, "A PFC Converter with Novel Integration of both EMI Filter and Boost Inductors," *IEEE Transaction on Power Electronics*, volume 29, no 9, pp. 4485-4489, 2014.

[7] Y. J. Zhang, Y. Chen, D. H. Xu, K. Mino, Y. Okuma, "Utilizing flexible printed circuit board (FPCB) to realize passives integration in LLC resonant converter," in *Proc. IEEE Applied Power Electronics Conference* , 2008, pp.1465–1471.

[8] C. Deng, D. H. Xu, P. P. Chen, C. S. Hu, W. P. Zhang, Z. W. Wen, and X. F. Wu, "Integration of Both EMI Filter and Boost Inductor for 1kW PFC Converter," *IEEE Transaction on Power Electronics*, volume 29, no. 11, pp. 5823-5833, 2014.

[9] C. Deng, D. H. Xu, C. S. Hu, Z. W. Wen, "PFC Converter with Novel Integration of both EMI Filter and Boost Inductors," in *Proc. Energy Conversion Congress and Exposition Conference*, Sep. 2013, pp. 3390-3390.

[10] C. Deng, Z. W. Wen, C. S. Hu, and D. H. Xu, "Integration of both EMI filter and Boost inductor for 1 kW PFC converter," in *Proc. Energy Conversion Congress and Exposition Conference*, Sep. 2012, pp. 4600-4607.

[11] C. Deng, D. H. Xu, Y. J. Zhang, Y. Chen, Y. Okuma, K. Mino, "Impact of Dielectric Material on Passive Integration in LLC Resonant Converter," in *Proc. IEEE Power Electronics Specialists Conference*, Apr. 2008, pp. 269-272.

[12] Y. J. Zhang, "Research of High Power Density DC-DC Converters and Passive Components Integration within Them," *PhD Dissertation*, College of Electrical Engineering, Zhengjiang University, China, 2008.

Magnetic Characterization Technique and Materials Comparison for Very High Frequency IVR

Dongbin Hou, Fred C. Lee, Qiang Li
Center for Power Electronics Systems
ECE Department, Virginia Polytechnic and State University
Blacksburg, VA 24061, USA
dongbin@vt.edu

Abstract—To efficiently power multi-core processors in today's computing devices, integrated voltage regulator (IVR) shows significant energy saving ability by dynamic voltage and frequency scaling. One key aspect in developing IVR is to design power inductors with small size and small loss at very high frequency. However, the challenge in very high frequency magnetic characterization is a major obstacle to accurately design and test the IVR inductors. In this work, the magnetic characterization technique at tens of MHz is investigated, and the issue and solution in permeability and loss measurement are demonstrated. The LTCC and NEC flake materials are characterized and compared at very high frequency for IVR inductor design.

I. INTRODUCTION

Multi-core architecture in microprocessors has been widely used in today's computing devices from high-end servers to smartphones. To efficiently power many processor cores, dynamic voltage and frequency scaling (DVFS) was proposed to dramatically reduce power consumption by fast adapting voltage and frequency with respect to the power demand of each core [1]. However, the traditional discrete voltage regulators (VR) are not able to adjust to different voltages at small enough time scales due to their relatively low switching frequency and high parasitic interconnect impedance between VR and processor. Therefore, IVR was proposed to solve this problem with high granularity, small size, near-load integration, and very high switching frequency [2-4]. The primary obstacle facing development of IVR is to design and integrate power inductors with small loss, miniature size, and low stray flux. Despite the recent surge of interest in this area [4-10], the reported measured property of the magnetics is still very limited and indirect. For permeability, although its value at different frequency is often reported [4-6], its saturation property in real DC-biased working condition is still lack of investigation. For loss property, the previous works usually show the equivalent resistance value only [4-8], which is usually measured with small signal excitation from impedance/network analyzer and is not able to represent the real magnetic core loss under large signal excitation in working condition. The lack of magnetic property in real working condition in previous works is caused by the big challenges in the magnetic

characterization technique at very high frequencies, and it is a major obstacle to accurately design and test the IVR inductors.

Recently, 3D integrated non-uniform flux inductors (as shown in Fig. 1) with LTCC and NEC/TOKIN's SENFOLIAGE® flake materials have been reported with superior performance for high frequency, high power density converters[11-14]. To leverage this design methodology for IVR inductors, the properties of LTCC and NEC flake need to be characterized at tens of MHz. Although their permeability and loss have been reported by previous works up to 10MHz [13-17], it is still lack of investigation at higher frequencies for IVR inductor design.

In this work, the magnetic characterization technique at very high frequency is investigated, and the LTCC and NEC flake materials are characterized and compared for IVR application.

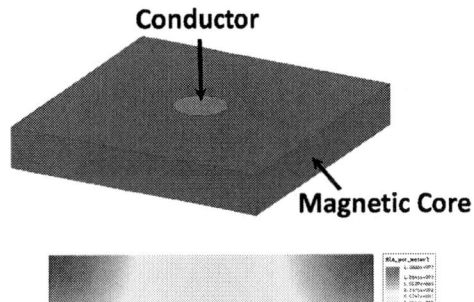

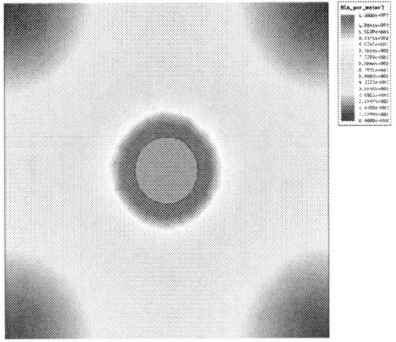

Fig. 1: Unit cell and field distribution (top view) of low profile non-uniform flux inductor.

This work was supported by High Density Integration Consortium of CPES Industry Partnership Program.

978-1-4673-9551-9/16 $31.00 © 2016 IEEE

II. PERMEABILITY CHARACTERIZATION AT VERY HIGH FREQUENCY & DC BIAS

In no DC bias condition, impedance analyzer can be used to test toroid magnetic core sample and provide permeability at very high frequency. Fig. 2 shows the measured relative permeability of LTCC and NEC flake materials up to 80MHz in no DC bias condition with Agilent 4294A impedance analyzer and 16454A magnetic material test fixture. It can be seen that the NEC flake material has higher initial permeability than LTCC materials at tens of MHz for IVR inductor design.

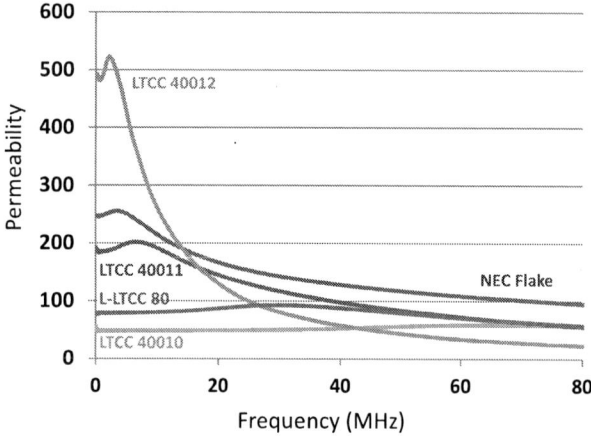

Fig. 2: Permeability in no DC bias condition. LTCC 40010, 40011 & 40012 are from ESL ElectroScience®; L-LTCC 80 are laminated LTCC materials reported in [16].

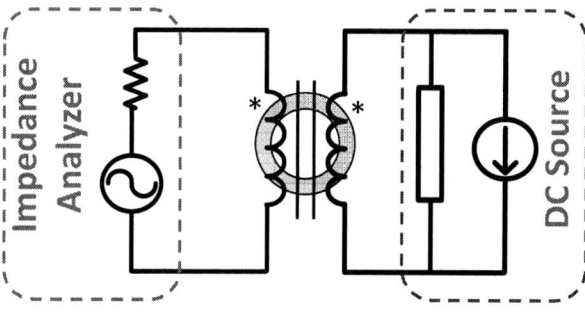

Fig. 3: Permeability measurement setup with impedance analyzer and external DC source.

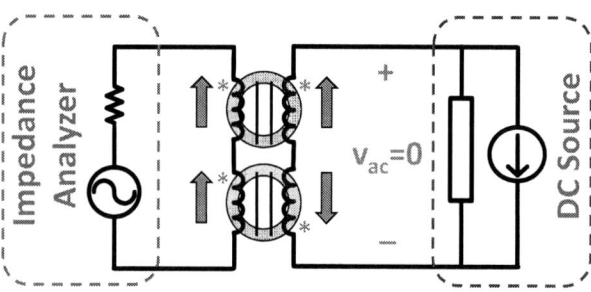

Fig. 4: Improved permeability measurement setup with AC voltage cancellation.

To measure high frequency permeability under DC-biased condition, the internal DC bias of impedance analyzer can be used, but the DC bias ability of impedance analyzer are usually very limited. For example, Agilent 4294A impedance

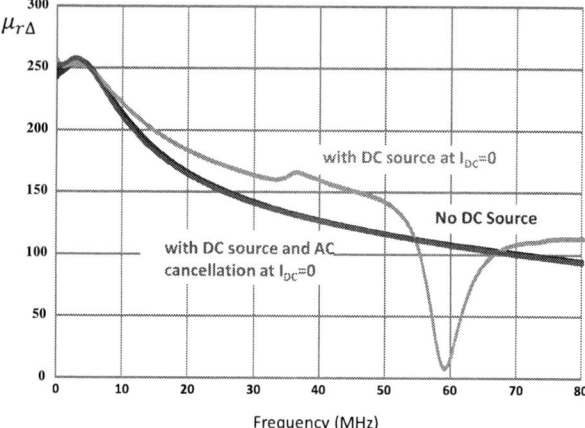

Fig. 5: Measured permeability without DC bias from different setups (Blue: impedance analyzer without external DC source; green: Fig. 3 setup; red: Fig. 4 setup; I_{DC}: output current from DC source).

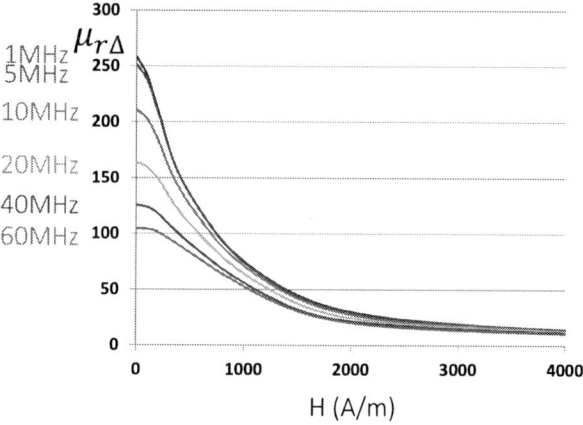

Fig. 6: Permeability under DC bias of NEC flake material. Each curve corresponds to a certain frequency value.

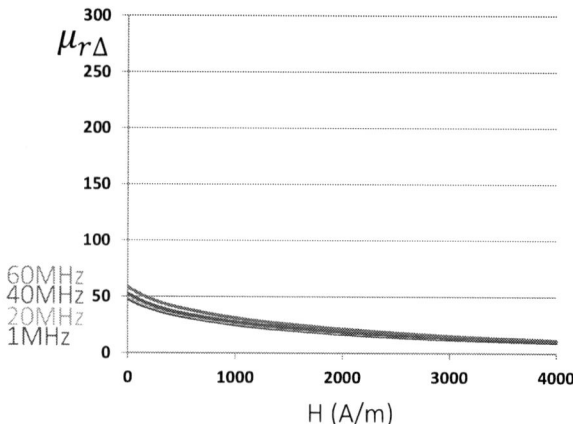

Fig. 7: Permeability under DC bias of LTCC 40010 material at different frequencies.

analyzer can only supply <0.1A DC bias current, which is inadequate to generate the DC bias level in real design. To overcome this problem, an external DC source is adopted to provide required DC bias as shown in Fig. 3. However, the impedance of DC source will introduce big measurement error as shown by the green curve in Fig. 5. To minimize this error, an AC voltage cancellation method applied in [15] is implemented with impedance analyzer as shown in Fig. 4. By using two identical core samples and reversed coupling polarities, the AC voltage generated on the two cores at the DC source side can be cancelled with each other, thus the measurement error caused by the impedance of DC source can be diminished as shown by the red curve in Fig. 5. Using the measurement setup shown in Fig. 4, the DC-biased permeability of NEC flake and LTCC 40010 at different frequency is measured and the results are summarized in Fig. 6 and Fig. 7. Comparing these two materials, NEC flake has higher permeability than LTCC 40010 at very high frequency and DC bias.

III. LOSS CHARACTERIZATION AT VERY HIGH FREQUENCY

A series of high frequency magnetic loss characterization methods has been reported in prior arts [18-20], which have cancellation mechanism to overcome the error caused by phase discrepancy. However, the highest frequency that has been demonstrated in previous works is 10MHz. When pushing to higher frequencies, new problems are observed.

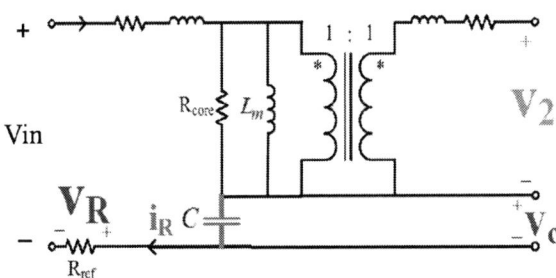

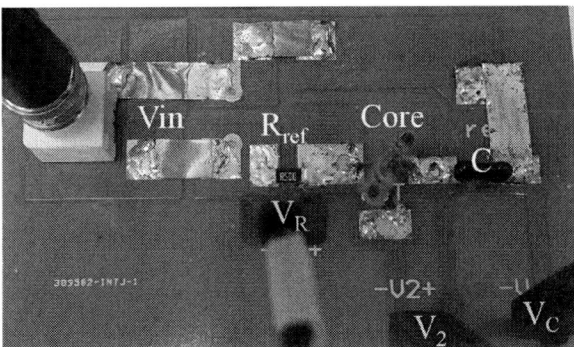

Fig. 8: Core loss measurement schematic and physical setup using partial cancellation method [20]. The magnetic core under test, current sensing resistor (R_{ref}), cancellation capacitor (C), and voltage probes for V_2, V_C and V_R are marked in the figure.

Fig. 8 shows the measurement set up using the partial cancellation method reported recently in [20]. The input port is excited by a power amplifier (Amplifier Research® 25A250A) driven by a function generator (Tektronix AFG3102); the voltage signal V_R, V_2, and V_C are sensed by voltage probes (Tektronix TDP 1000), from which the core loss P_{core} can be extracted by

$$P_{core} = \int v_2 \cdot \frac{v_R}{R_{ref}} + \frac{1}{k} \int v_C \cdot \frac{v_R}{R_{ref}} \qquad (1)$$

where k is cancellation factor [20]. The measurement results of NEC flake material up to 40MHz is shown as the black curve in Fig. 13. It can be seen that the measured core loss density curve is a straight line before 20MHz, but an abnormal variation shows up at higher frequency, which indicates that measurement problem exists at very high frequency. The first step we took to overcome the problem is to minimize the loops in the measurement circuit. As shown in Fig. 9, the area of driving loop, V_2 sensing loop, and V_C sensing loop are reduced by 93%, 95%, and 98%, respectively. The measurement result with the minimized loops is shown as the blue curve in Fig. 13. It can be seen that the abnormal variation is pushed to higher frequency than the original setup's result (black curve). In the next step, three different current sensing methods (current sensing resistor, current shunt, and current probe) are compared at very high frequency. As shown in Fig. 10, the three methods are used to measure the same current simultaneously for comparison.

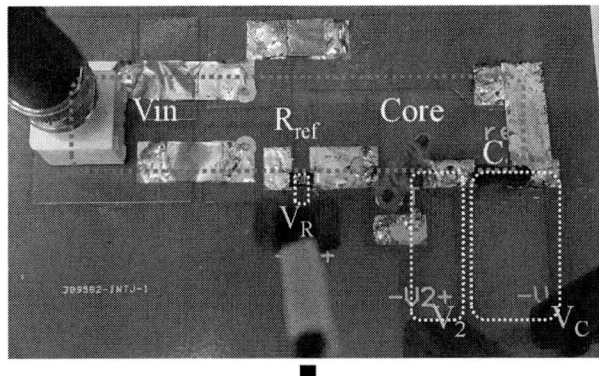

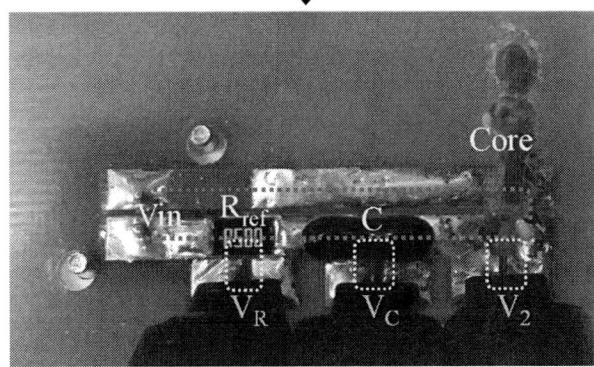

Fig. 9: Minimizing loop area in the measurement setup. The driving loop is marked by red dotted line, and sensing loops are marked by yellow dotted lines.

The magnitude of excitation voltage is kept constant at different frequencies. The current sensing waveforms from the three different methods at different frequencies are summarized in Fig. 11. At 10MHz, the three methods yield almost the same results, but when frequency becomes higher, the results from current sensing resistor and current shunt show abnormal magnitude variation, while the current probe is able to provide the most stable result among the three methods. Therefore, the current probe is adopted in the improved loss measurement setup as shown in Fig. 12, and its measurement result is shown as the red line in Fig. 13, where no abnormal variations is observed.

It is worth mentioning that the loss characterization technique discussed here is applicable not only to the core loss measurement, but also to the total inductor loss measurement. By simply replacing the core under test with an inductor, and using V_2 probe to measure the voltage across the inductor (with everything else kept the same), the measurement result will provide the total inductor loss including core loss and winding loss.

Based on the core loss comparison among LTCC 40010, 40011, 40012, L-LTCC 80 and NEC flake materials in previous works [13-16], the LTCC 40010 and NEC flake have lower core loss density than the other LTCC materials at high frequency. Using the improved measurement setup shown in Fig. 12, the core loss density of LTCC 40010 and NEC flake is measured and compared up to 40MHz as shown in Fig. 14. It shows that the core loss density of LTCC 40010 increases dramatically at higher flux density B_m, especially at very high frequency, while the core loss density change of

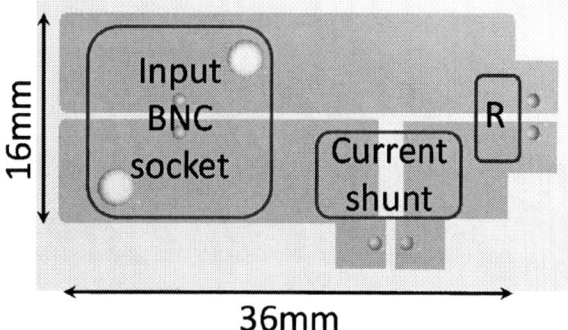

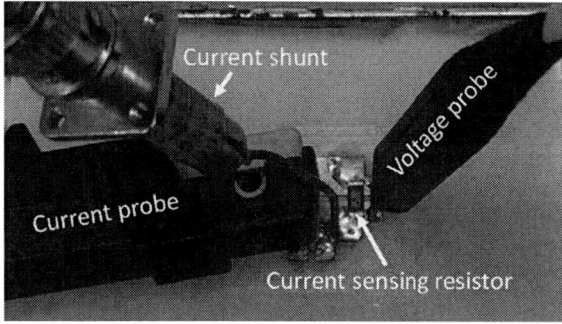

Fig. 10: Board layout (with minimized loops) & physical setup for current sensing methods comparison. Current shunt: T&M Research Products SDN-414-10; current probe: Tektronix TCP0030.

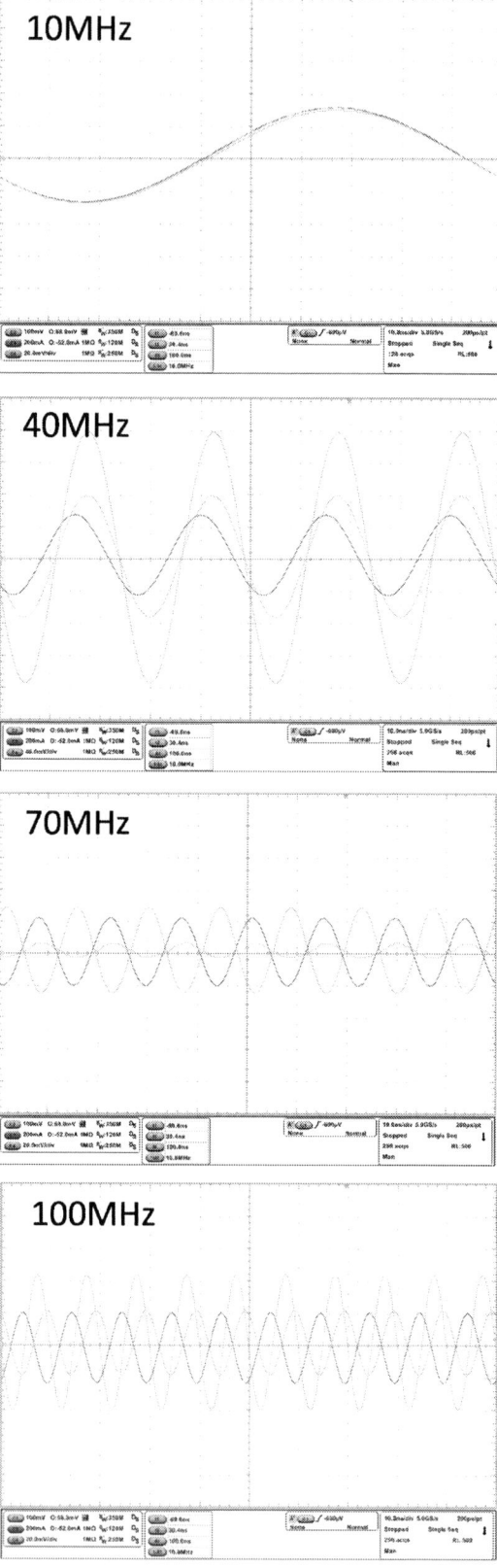

Fig. 11: Current sensing waveforms with different methods. Blue: current sensing resistor; green: current shunt; purple: current probe.

NEC flake keeps the same trend (as indicated by the straight lines in Fig. 14) as the B_m increases. This different behavior can be ascribed to the substantial difference in the magnetic material's composition and structure: ferrite ceramic (LTCC) and metal composite (NEC flake). Comparing NEC flake and LTCC 40010 in tens of MHz and $B_m > 15mT$ range (which is often the IVR inductor design range), NEC flake has lower core loss density than LTCC 40010.

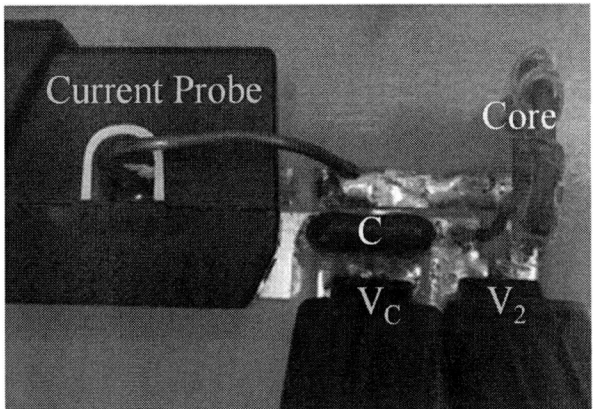

Fig. 12: Improved loss measurement setup with minimized loops and current probe for very high frequency test.

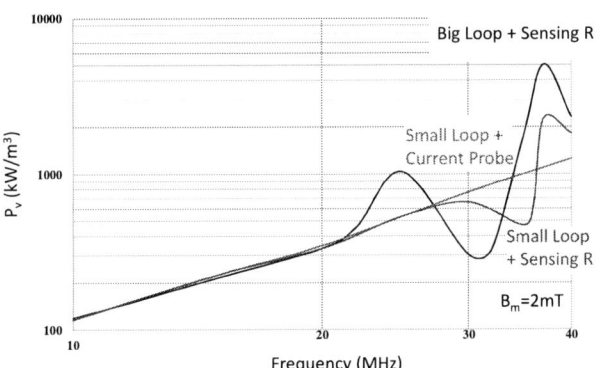

Fig. 13: Measured core loss density of NEC flake material with different measurement setups.

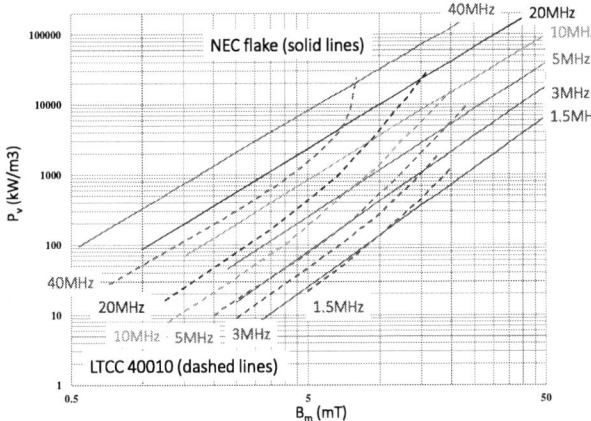

Fig. 14: Measured Core loss density of LTCC 40010 and NEC flake materials at very high frequency.

IV. CONCLUSION

The magnetic characterization technique at very high frequency (tens of MHz) is investigated for IVR inductor design. The issue and solution in the permeability and loss measurement are demonstrated. The LTCC and NEC flake materials are characterized and compared at very high frequency, and the NEC flake shows advantage over LTCC materials from both permeability and core loss density perspective. In future work, non-uniform flux inductors as demonstrated in [11-14] will be designed and tested with NEC flake material for IVR application.

REFERENCES

[1] J. Howard, S. Dighe, S. R. Vangal, G. Ruhl, N. Borkar, S. Jain, *et al.*, "A 48-Core IA-32 Processor in 45 nm CMOS Using On-Die Message-Passing and DVFS for Performance and Power Scaling," *Solid-State Circuits, IEEE Journal of,* vol. 46, pp. 173-183, 2011.

[2] K. Wonyoung, M. S. Gupta, W. Gu-Yeon, and D. Brooks, "System level analysis of fast, per-core DVFS using on-chip switching regulators," in *High Performance Computer Architecture, 2008. HPCA 2008. IEEE 14th International Symposium on,* 2008, pp. 123-134.

[3] E. A. Burton, G. Schrom, F. Paillet, J. Douglas, W. J. Lambert, K. Radhakrishnan, *et al.*, "FIVR — Fully integrated voltage regulators on 4th generation Intel Core SoCs," in *Applied Power Electronics Conference and Exposition (APEC), 2014 Twenty-Ninth Annual IEEE,* 2014, pp. 432-439.

[4] N. Sturcken, E. J. O'Sullivan, N. Wang, P. Herget, B. C. Webb, L. T. Romankiw, *et al.*, "A 2.5D Integrated Voltage Regulator Using Coupled-Magnetic-Core Inductors on Silicon Interposer," *Solid-State Circuits, IEEE Journal of,* vol. 48, pp. 244-254, 2013.

[5] P. R. Morrow, P. Chang-Min, H. W. Koertzen, and J. Dibene, "Design and Fabrication of On-Chip Coupled Inductors Integrated With Magnetic Material for Voltage Regulators," *Magnetics, IEEE Transactions on,* vol. 47, pp. 1678-1686, 2011.

[6] N. Wang, E. J. O'Sullivan, P. Herget, B. Rajendran, L. E. Krupp, L. T. Romankiw, *et al.*, "Integrated on-chip inductors with electroplated magnetic yokes (invited)," *Journal of Applied Physics,* vol. 111, p. 07E732, 2012.

[7] N. Sturcken, R. Davies, C. Cheng, W. E. Bailey, and K. L. Shepard, "Design of coupled power inductors with crossed anisotropy magnetic core for integrated power conversion," in *Applied Power Electronics Conference and Exposition (APEC), 2012 Twenty-Seventh Annual IEEE,* 2012, pp. 417-423.

[8] C. Feeney, W. Ningning, S. C. O Mathuna, and M. Duffy, "A 20-MHz 1.8-W DC-DC Converter With Parallel Microinductors and Improved Light-Load Efficiency," *Power Electronics, IEEE Transactions on,* vol. 30, pp. 771-779, 2015.

[9] W. J. Lambert, M. J. Hill, K. Radhakrishnan, L. Wojewoda, and A. E. Augustine, "Package embedded inductors for integrated voltage regulators," in *Electronic Components and Technology Conference (ECTC), 2014 IEEE 64th,* 2014, pp. 528-534.

[10] C. O. Mathuna, W. Ningning, S. Kulkarni, and S. Roy, "Review of Integrated Magnetics for Power Supply on Chip (PwrSoC)," *Power Electronics, IEEE Transactions on,* vol. 27, pp. 4799-4816, 2012.

[11] D. Hou, Y. Su, Q. Li, and F. C. Lee, "Improving the efficiency and dynamics of 3D integrated POL," in *Applied Power Electronics Conference and Exposition (APEC), 2015 IEEE,* 2015, pp. 140-145.

[12]   Y. Su, Q. Li, F. C. Lee, D. Hou, and S. She, "Planar inductor structure with variable flux distribution - A benefit or impediment?," in *Applied Power Electronics Conference and Exposition (APEC), 2015 IEEE*, 2015, pp. 1169-1176.

[13]   Y. Su, Q. Li, and F. C. Lee, "Design and Evaluation of a High-Frequency LTCC Inductor Substrate for a Three-Dimensional Integrated DC/DC Converter," *Power Electronics, IEEE Transactions on,* vol. 28, pp. 4354-4364, 2013.

[14]   Y. Su, W. Zhang, Q. Li, F. C. Lee, and M. Mu, "High frequency integrated Point of Load (POL) module with PCB embedded inductor substrate," in *Energy Conversion Congress and Exposition (ECCE), 2013 IEEE*, 2013, pp. 1243-1250.

[15]   M. Mu, Y. Su, Q. Li, and F. C. Lee, "Magnetic characterization of low temperature co-fired ceramic (LTCC) ferrite materials for high frequency power converters," in *Energy Conversion Congress and Exposition (ECCE), 2011 IEEE*, 2011, pp. 2133-2138.

[16]   M. Mu, W. Zhang, F. C. Lee, and Y. Su, "Laminated low temperature co-fired ceramic ferrite materials and the applications for high current POL converters," in *Energy*

Conversion Congress and Exposition (ECCE), 2013 IEEE, 2013, pp. 621-627.

[17]   W. Zhang, M. Mu, D. Hou, Y. Su, Q. Li, and F. C. Lee, "Characterization of Low Temperature Sintered Ferrite Laminates for High Frequency Point-of-Load (POL) Converters," *Magnetics, IEEE Transactions on,* vol. 49, pp. 5454-5463, 2013.

[18]   M. Mu, Q. Li, D. J. Gilham, F. C. Lee, and K. D. T. Ngo, "New Core Loss Measurement Method for High-Frequency Magnetic Materials," *Power Electronics, IEEE Transactions on,* vol. 29, pp. 4374-4381, 2014.

[19]   M. Mu, F. C. Lee, Q. Li, D. Gilham, and K. D. T. Ngo, "A high frequency core loss measurement method for arbitrary excitations," in *Applied Power Electronics Conference and Exposition (APEC), 2011 Twenty-Sixth Annual IEEE*, 2011, pp. 157-162.

[20]   D. Hou, M. Mu, F. C. Lee, and Q. Li, "New core loss measurement method with partial cancellation concept," in *Applied Power Electronics Conference and Exposition (APEC), 2014 Twenty-Ninth Annual IEEE*, 2014, pp. 746-751.

Large-Signal Power Circuit Characterization of on-Silicon Coupled Inductors for High Frequency Integrated Voltage Regulation

S. Kulkarni*, Z. Pavlovic*, S. Kubendran*, C. Carretero*†, N. Wang* and C. O'Mathuna*‡

*Microsystems Center, Tyndall National Institute, Cork, Ireland
†Departamento de Fisica Aplicada, Universidad de Zaragoza, Zaragoza, Spain
‡Department of Electrical Engineering, University College Cork, Cork, Ireland

Abstract— This work describes in detail the design, fabrication and characterization of on-silicon coupled inductors. The coupled inductors are fabricated using a double metal layer process for improved performance. Further, the fabricated devices are characterized for small signal, dc bias and large signal testing at high frequencies (up to 15 MHz). The small signal testing includes both impedance analyzer and network analyzer results for inductance, saturation and coupling data on the device. The paper further describes the design and set-up for a large signal testing system allowing for accurate measurement of inductor performance. The parameters are extracted from the ratio between the voltage complex amplitude and the current complex amplitude for sinusoidal excitation. The proposed system is suitable to perform the measurements under different large-signal conditions given by the ac current amplitudes ranging from 0 A to 0.5 A, at frequencies up to 15 MHz. From the impedance analyzer tests, a self-inductance of 50 nH and a dc bias current of 600 mA was measured. From the large signal test, an ac resistance of 1.4 Ω was measured at 15 MHz and ac current amplitude of 0.5 A. The measured data were in line with design parameters.

Keywords—impedance measurement, integrated magnetics, thin-film inductors, large signal testing

I. Introduction

Power Supply on Chip (PwrSoC) constitutes one of the most promising technologies to achieve increased integration, reduced volume and reliability in voltage regulation for low power electronic devices [1]. Silicon integrated magnetic components are a key bottleneck in realization of PwrSoC due to the issues associated with quality and integration of magnetic materials in the device structure. Air-core inductors built over the silicon substrate is the simplest solution, but large surface area is needed to fulfill the inductance requirements for the converter, along with issues of EMI [2, 3]. Thin-film core inductor has been proposed as a promising technology to achieve high efficiency and area effective solutions, [4, 5]. However, the low volume of the magnetic material in these

devices implies a strong dependence on the device behavior with respect to the excitation signal levels. Optimal design of the dc-dc converter therefore requires accurate electrical parameters of the magnetic devices, but the available commercial measurement instruments, at the frequencies involved in the integrated silicon converters, typically provides measurements for current excitation up to several tens of milli-amperes. Presently, small signal impedance analyzers are widely used for characterizing the power devices at high frequencies, which only provide information on certain loss factors which contribute to the device performance i.e, eddy current losses. However, in order to accurately measure the device performance in a 'real circuit' condition would require applying larger voltage signals. With these larger signals, losses, such as hysteresis and excess eddy current losses, can also be measured. With certain magnetic materials, these losses can be more dominant than eddy currents, hence an accurate assessment of the overall device performance is critical before the full circuit implementation. Large-signal measurement system has to be based on the generation of the adequate excitation signal as well as high-precision measurement of the equivalent impedance. The impedance measurements are performed extracting the ratio between the voltage and current waveforms in the device. Voltage measurements can be easily performed by means of a wide-bandwidth oscilloscope. Nevertheless, high-frequency current measurement constitutes a complex task due to the limited bandwidth of the probes.

Several authors have compared the performance of different methods for measuring current [6-8], being the preferred high-frequency current measurement systems: The current monitor Pearson 2877, and the voltage measurement in a shunt resistor [9-12]. Moreover, in [13], the current measurement from the voltage difference between the terminals of a bank resistor is evaluated, but the current redistribution between the resistors implies unacceptable errors.

This work describes the design, fabrication and characterization (small & large signal) of on-silicon coupled inductor devices. The remainder of the paper is organized as follows. In Section II, the on-silicon inductor devices to be

---

The authors would like to acknowledge the support of European Union funded FP7 Project 'Powerswipe' number 318529.

978-1-4673-9551-9/16 $31.00 © 2016 IEEE

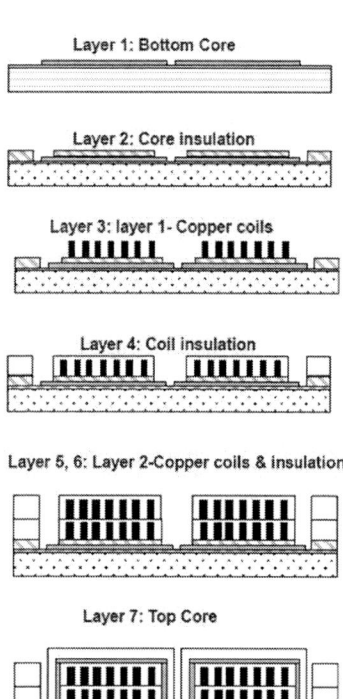

Fig. 1. DLM fabrication process flow.

(a) Top view of low coupling coils.

(b) Cross-section view of low coupling coils.

Fig. 2. Top view and cross-section of a fully fabricated on-silicon coupled inductor

tested in the large signal measurement system are described. In Section IV, details the small signal and coupling measurements of the coupled inductors. Section IV details the design of a large signal testing set-up along with discussion on the measured large signal performance of the on-silicon coupled inductors. Finally, some conclusions are drawn in Section V.

II.   ON-SILICON COUPLED INDUCTOR DEVICES

Several studies proposed using coupled inductors for interleaving buck converter in order to improve the performance of the converter. The device structure considered in this work is a two-phase oppositely coupled inductor with a self-inductance of 47 nH/phase; with a coupling factor of 0.4. The sum of the inductances required by the converter is 90 nH.

The coupled inductor have been built by stacking both coils, therefore, a Double Layer Metal (DLM) is required to fabricate the windings. The inductors have been design to optimize the overall energy efficiency under area limited from 1 to 3 mm². The low coupling 47 nH inductors are two stacked racetrack coils composed of 3 turns with calculated dc resistance of 282 mΩ. Their footprint total area is 1.25 mm². The main characteristics of the coils are detailed in Table I.

TABLE I.   GEOMETRICAL PARAMETERS OF DLM COILS.

Inductor	Core length (mm)	Core thickness (μm)	Cu width (μm)	Cu thickness (μm)	Gap (μm)
Low coupling	0.95	2	40.38	15	15

The 7-mask fabrication procedure for the DLM integrated inductor is shown in Fig. 1, but it is similar to the process described in [14] for integrated transformers. The coupled inductor structure is composed of two sandwiched racetrack coil between two $Ni_{45}Fe_{55}$ magnetic thin-layer cores. The first magnetic layer, whose thickness is 2 μm, is electroplated on native SiO_2 in the wafer. A layer of BCB is deposited for insulating the bottom magnetic layer. The windings are made of two 15 μm tick layers of copper with Su-8 as insulator in-between. Finally, the top magnetic layer (2 μm thick) is electroplated over Su-8. The top views of the coils of a fully fabricated on-silicon coupled inductor is shown in Fig. 2(a). Further, the device cross-section is also shown in Fig. 2(b).

III.   SMALL SIGNAL AND COUPLING MEASUREMENTS FOR ON-SILICON COUPLED INDUCTOR

The small signal characterization was done using HP 4285 LCR meter (75 kHz to 30 MHz) and for bias characterization HP 42841 current source was used. Fig.3 and Fig. 4 show the initial small signal (up to 30 MHz) and low frequency characterization of the two phase on-Silicon Coupled Inductor device, respectively.

From figure the self-inductance measured for each phase is around 50 nH at 1 MHz, this value is within 5% of the design value and this can be explained by the variation in the magnetic core thickness deposited using electroplating. Similarly, the saturation current for coupled inductor is measured at 600 mA.

Further for coupling measurements a two port vector network analyzer was used. Fig. 5 shows measured primary and secondary side inductances of two low coupled sample devices. The two port S-parameters are measured on Agilent E8361A PNA (10 MHz–67 GHz) with GS probes connected to the primary and secondary side terminals of a device. The S-matrix in then converted to Z-matrix and the inductances are extracted according to a standard transformer circuit model.

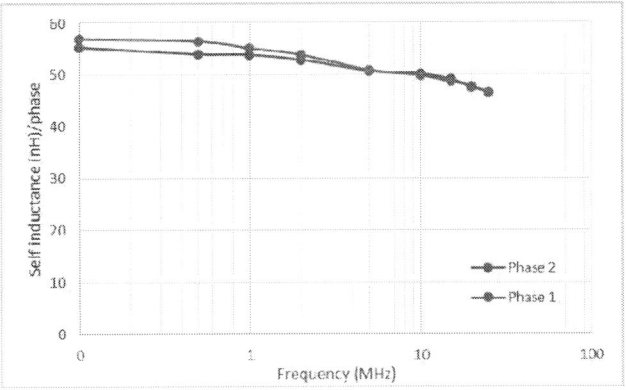

Fig. 3. Self-inductance measurement for two phase coupled inductor using LCR meter.

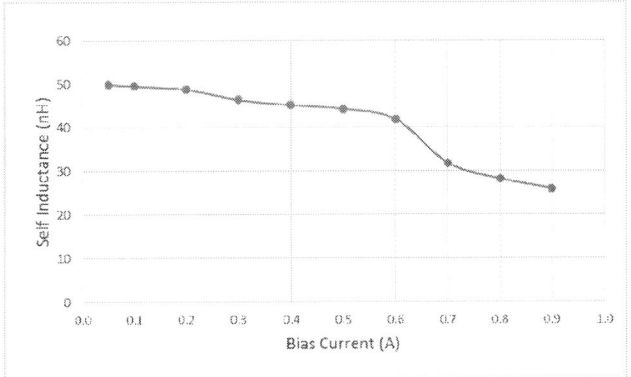

Fig. 4. Self-inductance/phase vs dc bias current for coupled inductor.

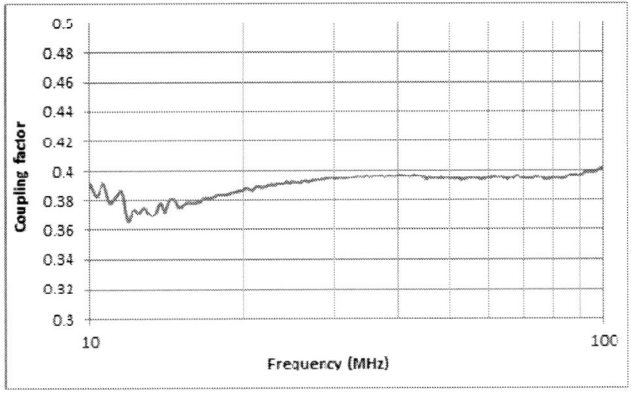

Figure 5. Coupling factor variation with frequency from two-port PNA test of coupled inductor

The extracted coupling factors from the measurements for both devices are $k_1 = 0.37$-0.4. These extracted values are close to the design coupling value of 0.4, the reason for the lower value of measured coupling factor at 10-20 MHz, is due the accuracy of the PNA at low frequencies, as it is designed for high frequency measurements (>1 GHz).

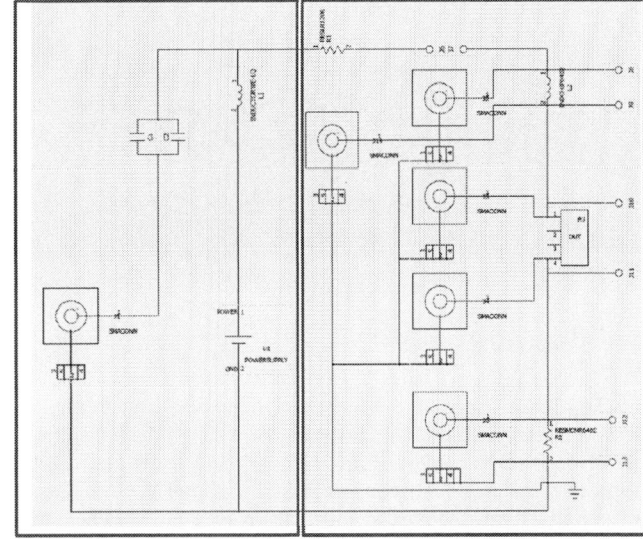

Fig. 6. Schematic of the large signal measurement system for on-silicon inductors.

IV. LARGE SIGNAL MEASUREMENT SYSTEM

The high-frequency measurement system for the characterization of the integrated inductor has been implemented in an optimized Printed Circuit Board (PCB) to reduce the effects of the parasitic elements. The proposed measurement system, shown in Fig. 6, is divided in two blocks: The signal generation system and the measurement system.

The excitation signal is generated by two different branches connected in parallel. A 50 Ω output resistance RF-amplifier model 25A250A from Applied Research, controlled by a signal generator model 8200 from Global Specialties, injects the ac current to the PCB circuit through an SMA connector. The RF Amplifier is suitable to deliver up to 25 W at frequencies fewer than 250 MHz. The RF Amplifier is series-connected to a decoupling bank capacitor composed of 2 SMD 100 nF 25 V devices from Murata, whose purpose is to block the dc current. The dc current is generated by a dc source connected in series with an inductance SMD 1812 3.3 μH from Wurth Elektronics in order to block the ac current. The selected decoupling inductance possesses a high saturation current of 2.7 A, it is suitable to drive up to 1.4 Arms and the self-resonant frequency of 58 MHz is above the maximum frequency to be characterized. Finally, a 10 Ω power resistor is series-connected to both branches with the purpose to limit the total current.

The measurement block consists of several components placed in series, thus, the driven currents are the same in all devices. The main sources of measurements errors are the noise, the error in amplitude measurement and the delay between the signals. PCB has been built with a double layer board, being the bottom layer dedicated to the Signal Ground in order to reduce the noise in the measurements. Trace loops have been minimized in size with the adequate routing in order to reduce the magnetic flux linkage [15].

978-1-4673-9551-9/16 $31.00 © 2016 IEEE

Fig. 7. Large Signal characterization test board with Pearson current probe.

Amplitude errors are minimized by using low tolerance components. However, the delay between the signals becomes the critical parameter, [16], because a small delay implies an important trade-off between the inductance and the resistance measured values, thus, the elements placed in between the measurement points and the input channels of the oscilloscope has to be identical with the purpose to reduce the imbalance of the signals. Voltage measurements in the terminal of the inductor have been taken by using a 4-terminal Kelvin connection, being the driven current traces separated to the voltage measurement traces. Two SMA-connector with the case reference connected to the Signal Ground have been employed to connect the PCB to the oscilloscope with SMA-to-BNC 25 cm RG-68 coaxial cables. The delay of the coaxial cable RG-58 is 5.053 ns per meter, therefore, equal short length measurement cables between probes and oscilloscope input channels are recommended. For calibration and estimating the time lag between current and voltage probes, a reference commercial inductor SMD 0402 with 20 nH inductance and Q-8, 10 at 10 & 15 MHz from CoilCraft is soldered to the circuit. A Pearson Current Monitor model 2877 is selected for the current measurement. This 200 MHz bandwidth probe inserts a low delay to the measured signal compared to the alternative current probes [17].

Fig. 8 and Fig. 9 show the voltage and current waveforms measured for the CoilCraft SMD inductor for calibrations. The first harmonics of the voltage and current waveforms are calculated from the measured data and shown on the plots superimposed on to the real waveforms. From the calculated harmonics the peak voltage and current values and the total phase shift between them are determined. This phase shift is due to the device itself and voltage and current probes. The phase shift inserted by the device is calculated using the SMD inductor's Q value and is equal to $arctgQ$. By subtracting this phase shift from the total one the current probe time lag to the voltage is obtained. At 10 MHz & 15 MHz the time delays of 2.834 ns and 3.714 ns were estimated at the two frequencies.

After calibration, the two-phase coupled inductor was wirebonded on the test board. This test board was then supplied with large voltages at 10 & 15 MHz and the output voltage and current were measured from the device. The voltage and current waveforms for 10 & 15 MHz are shown in Fig. 10 and Fig. 10, respectively.

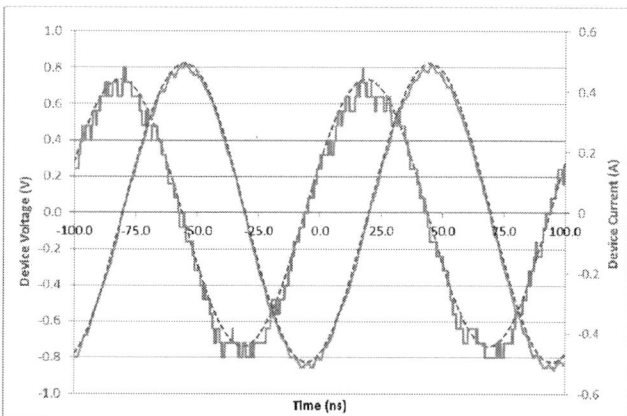

Fig. 8. Device voltage and current waveforms for SMD inductor at 10 MHz.

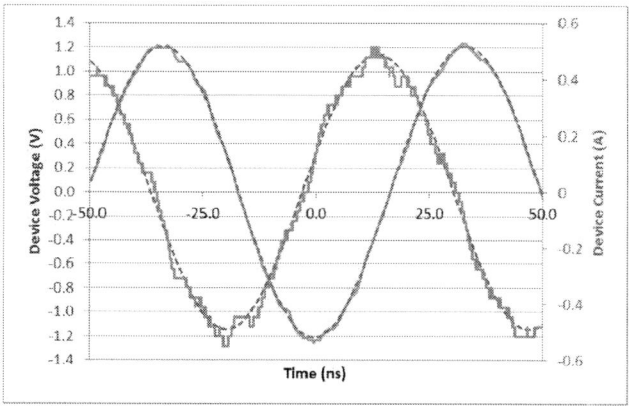

Fig. 9. Device voltage and current waveforms for SMD inductor at 15 MHz.

Again, the peak voltage and current values are calculated from the first harmonics and the magnitude of the device impedance is obtained as a ratio of the two. The phase shift calculated from the first harmonics is corrected for the time delay inserted by the current probe to obtain the device phase angle.

At 10MHz frequency the large signal phase self-inductance and device equivalent series resistance estimations are L=53.41nH and R=672.2mΩ. The ac power loss at approximately 0.5A peak ac current is P_{ac}=86.8mW. The difference between estimated inductance and the measured small signal inductance can be partially attributed to the additional inductance of wirebonds that connect device pads with the PCB. At 15MHz the estimated inductance, resistance ac power loss are L=54.09nH, R=1130.9mΩ and P_{ac}=149.06mW. The slight inductance estimation from large signal testing at 15 MHz can be attributed to the noise in the voltage waveforms measured.

The measured ac resistance which includes the ac winding losses, core losses (eddy current+coercive+anomalous) is almost 70% higher than in case of the 15 MHz measurement compared to 10 MHz measurement.

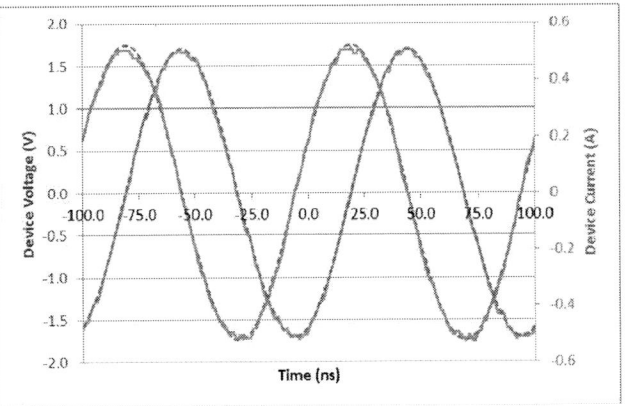

Fig. 10. Device voltage and current waveforms for single phase of two-phase coupled inductor at 10 MHz.

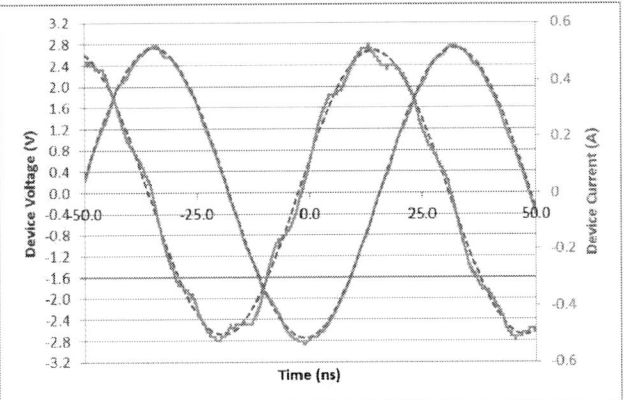

Fig. 11. Device voltage and current waveforms for single phase of two-phase coupled inductor at 15 MHz.

This ac resistance data is not consistent with the design and the exact reason for this variation is still unclear. Additional testing is ongoing to understand this variation and the results will be presented at APEC 2016.

V. CONCLUSION

In this work, we present the design, fabrication and characterization of two phase coupled inductor device on silicon. The coupled inductor uses Tyndall's DLM process for copper deposition. The coupled inductor is designed to have a low coupling factor of 0.4. A detailed characterization of the coupled inductor is presented with both small signal and large signal testing. Data from impedance meter measurement is presented and this matches well with the design parameters. Additional measurements for saturation and coupling is also presented with good agreement with design.

Further, a specific large signal test board was designed and assembled for characterization of the on-silicon coupled inductors. A detailed calibration process for the identifying the time lag between current and voltage waveforms is also explained. Using this calibration technique, large signal testing of one phase of the coupled inductor structure was undertaken. The inductance data for the inductor matches well with the small signal measurements. The ac resistance however, is not consistent with the design data and is currently under investigation.

REFERENCES

[1]   C. O'Mathuna, N. Wang, S. Kulkarni, and S. Roy, "Review of integrated magnetics for power supply on chip (PwrSoC)," *IEEE Transactions on Power Electronics,* vol. 27, pp. 4799-4816, Nov. 2012.

[2]   G. Schrom, P. Hazucha, J. Hahn, D. S. Gardner, B. A. Bloechel, G. Dermer, *et al.,* "A 480-MHz, multi-phase interleaved buck DC-DC converter with hysteretic control," presented at the IEEE Power Electronics Specialists Conference (PESC), 2004.

[3]   C. R. Sullivan, D. V. Harburg, Q. Jizheng, C. G. Levey, and Y. Di, "Integrating magnetics for on-chip power: A perspective," *IEEE Transactions on Power Electronics,* vol. 28, pp. 4342-4353, Sep. 2013.

[4]   N. Wang, R. Miftakhutdinov, S. Kulkarni, and C. O'Mathuna, "High efficiency on Si integrated micro-transformers for isolated power conversion applications," *IEEE Transactions on Power Electronics,* vol. 30, pp. 5746-5754, Oct. 2015.

[5]   N. Wang, T. O'Donnell, S. Roy, S. Kulkarni, P. McCloskey, and C. O'Mathuna, "Thin film microtransformer integrated on silicon for signal isolation," *IEEE Transactions on Magnetics,* vol. 43, pp. 2719-2721, Jun. 2007.

[6]   R. E. Fuja and W. F. Praeg, "Performance of transductors for precision high-current measurement and control," *IEEE Transactions on Nuclear Science,* vol. 20, pp. 411-413, Jun. 1973.

[7]   S. Ziegler, R. C. Woodward, H. H. C. Iu, and L. J. Borle, "Current sensing techniques: A review," *IEEE Sensors Journal,* vol. 9, pp. 354-376, Apr. 2009.

[8]   K. Li, A. Videt, and N. Idir, "Using current surface probe to measure the current of the fast power semiconductors," *IEEE Transactions on Power Electronics,* vol. 30, pp. 2911-2917, Jun. 2015.

[9]   G. Laimer and J. W. Kolar, "Accurate measurement of the switching losses of ultra-high switching speed CoolMOS power transistor/SiC diode combination employed in unity power factor PWM rectifier systems," in *European Power Quality Conference (PCIM),* Nuremberg (Germany), 2002, pp. 14-16.

[10] L. Helong, S. Beczkowski, S. Munk-Nielsen, L. Kaiyuan, and W. Qian, "Current measurement method for characterization of fast switching power semiconductors with Silicon Steel Current Transformer," in *IEEE Applied Power Electronics Conference and Exposition (APEC),* Charlotte (USA), 2015, pp. 2527-2531.

[11] J. Minibock and J. W. Kolar, "Experimental analysis of the appilcation of the latest SiC diode and CoolMOS power transistor technology in a 10 kW three-phase PWM (VIENNA) rectifier," in *European Power Quality Conference (PCIM),* Nuremberg (Germany), 2001, pp. 121-125.

[12] B. Voljc, M. Lindic, and R. Lapuh, "Direct measurement of AC current by measuring the voltage drop on the coaxial current shunt," *IEEE Transactions on Instrumentation and Measurement,* vol. 58, pp. 863-867, Apr. 2009.

[13] M. Danilovic, C. Zheng, W. Ruxi, L. Fang, D. Boroyevich, and P. Mattavelli, "Evaluation of the switching characteristics of a gallium-nitride transistor," in *IEEE Energy Conversion Congress and Exposition (ECCE),* Phoenix (USA), 2011, pp. 2681-2688.

[14] N. Wang, S. Kulkarni, B. Jamieson, J. Rohan, D. Casey, S. Roy, *et al.,* "High efficiency Si integrated micro-transformers using stacked copper windings for power conversion applications," in *IEEE Applied Power Electronics Conference and Exposition (APEC),* 2012, pp. 411-416.

[15] J. A. Ferreira, W. A. Cronje, and W. A. Relihan, "Integration of high frequency current shunts in power electronic circuits," *IEEE Transactions on Power Electronics,* vol. 10, pp. 32-37, Jan. 1995.

[16] J. Lautner and B. Piepenbreier, "Impact of current measurement on switching characterization of GaN transistors," in *IEEE Workshop on Wide Bandgap Power Devices and Applications (WiPDA),* Knoxville (USA), 2014, pp. 98-102.

[17] *Pearson Current Monitor model 2877.* Available: http://www.pearsonelectronics.com/pdf/2877.pdf

[18] *Frequency Response Phase-Shift.* Available: http://www.pearsonelectronics.com/phase-shift

978-1-4673-9551-9/16 $31.00 © 2016 IEEE

Point-of-Load Inductor with High Swinging and Low Loss at Light Load

Ting Ge and Khai Ngo

Bradley Department of Electrical and Computer Engineering
Virginia Polytechnic Institute and State University
Blacksburg, VA 24061, USA
Email: gting@vt.edu

Jim Moss

Texas Instruments, Inc.
Santa Clara, CA 95050, USA

Abstract— Point-of-load converter at light load has low efficiency owing to the "fixed losses" such as core loss and ac winding loss. This paper focuses on two-dimensional (2D) gapping of a ferrite core to shape inductance versus load current to reduce inductor loss at light load. Since the maximum inductance of conventional stepped gap is limited by the cross-sectional area of the thin gap, a 2D gap is formed by joining two orthogonal gaps to gain flexibility. Higher inductance is achieved at light load compared with uniform-gap and stepped-gap geometries having the same volume and dc resistance. Ac resistance is reduced at light load thanks to a magnetic path that steers ac flux away from the winding. Two C-cores with 2D gap were fabricated and tested on a buck converter with 50% reduced total inductor loss at 10% load current.

Keywords—2D gap; Nonlinear inductance; Loss at light load; Swinging inductor

I. INTRODUCTION

The efficiency of a power converter usually drops at light load [1]-[3]. Among the reasons for the degradation are the "fixed losses" in the inductor [4]. Core loss [5] is determined by the inductor's voltage and remains relatively constant as load current varies. Ferrite is employed herein to keep core loss negligibly low. A ferrite core would need a (air) gap to control inductance, but this gap could adversely affect the other fixed loss which is the loss P_{wac} associated with ripple current [6]-[8]. Since P_{wac} is related to ac resistance R_{ac} and ripple current I_{ac} by $P_{wac} = R_{ac}I_{ac}^2$, it can be decreased by reducing R_{ac} or I_{ac}. The paper will address how to shape and design the gaps to control full-load inductance, to swing the inductance up at light load to reduce I_{ac}, and to steer fringing flux from the gaps away from the winding at light load to reduce R_{ac}.

Several methods have been reported to vary inductance with current. A simple way is to use two windings, one sustaining the main current and the other controlling the saturation level [9]-[12]. The extra winding adds loss and volume. Composite material that contains ferrite and powder iron can have current-dependent inductance. Commercial composite inductor in [13] shows high swinging, but the volume almost doubles that of constant inductors with the same current rating and dc resistance. Replacement of the air gap by magnetorheological (MR) fluid can also provide swinging inductance [14]-[15]. Another solution is the use of low-temperature co-fired ceramic (LTCC) with permeability dropping gradually [6]. All these methods have high core loss that originates from powder iron, MR fluid, or LTCC. Several gapping methods, like stepped gap or sloped gap, are available for ferrite core to realize step-shaped inductance with low core loss [16]-[19]. Their maximum inductance is limited by the thin gap.

In this work, a uniform-gap ferrite inductor is retrofitted with a 2D gap to realize high swinging and low loss at light load without modifying the volume or the dc resistance. A thin gap is placed orthogonally to the existent thick gap to take advantage of the significantly larger cross-sectional area in the orthogonal direction. The thick gap is assumed horizontal and the thin gap vertical in the remainder of the paper. Inductance, core loss, and winding loss are optimized simultaneously to minimize the total loss at light load. The trade-off between inductance swinging and saturation is analyzed to optimize the thickness of saturable piece. Finite-element simulation is the primary tool to deal with ac winding loss, bias-dependent and frequency-dependent core loss, and nonlinearity owing to saturation.

In Section II, the material and application circuit are reviewed; the 2D gap inductor is delineated and compared with inductors using 1D gaps (thick gap, thin gap, and stepped gap); simulated inductances and losses versus current are analyzed for all four cases. A design procedure for the inductor with 2D gap is depicted in Section III along with a design example. Experimental results are discussed in Section IV to verify theory and simulation.

II. TWO-DIMENSIONAL GAPPING TO ACHIEVE HIGH SWINGING AND LOW LOSSES

Inductors with 2D gap and 1D gaps (thick gap, thin gap, and stepped gap) are described and compared using the same application circuit, magnetic material, outer core dimensions, dc resistance, rated current, and voltage excitation. These gapping methods are evaluated using the buck converter shown in Fig. 1. The switching frequency is 500 kHz; the input voltage, 10 V; the output voltage, 5 V; and the nominal load current, 5 A. The two MOSFETs are BSC080N03LS with $R_{ds(on)}$ = 8 mΩ and voltage rating of 30 V from Texas Instruments. They are driven synchronously and in continuous-conduction mode. The inductance at rated current is L = 4.7 µH. The capacitors are C_{in} = 10 µF and C_{out} = 10 µF. With these parameters, the current ripple through the inductor is 1 A, and the voltage ripple across the output is 25 mV. Conventional and 2D gap geometries are described next so that their impacts on inductor's loss at light load can be compared in sub-Section II.B.

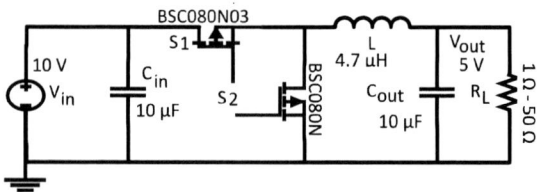

Fig. 1. Buck converter (switched at 500 kHz and outputting 25 W) used to compare gapping methods.

A. Gap Geometries and Inductance Swing

The nonlinear mechanism of a swinging inductor with multiple gaps is explained by the constant reluctance that represents a single thick gap in series with a variable reluctance that represents a single thin gap shown in Fig. 2. A sufficiently thick gap would be able to contain most of the energy and keep the inductance relatively constant over the load range. A thin gap can store only a part of the required energy; the balance of the energy is stored in the ferrite core [20], [21]. The inductance is then controlled by the nonlinear permeability (and the gap length), giving rise to the swinging behavior. To realize high current rating from the thick gap and inductance swinging from the thin gap, the two gaps are employed in one core to behave as a swinging inductor modeled by the series inductances in Fig. 2(a) and the dual parallel reluctances in Fig. 2(b). Light-load inductance L_{max} is determined by the thin gap while the core is not saturated, and inductance at nominal load L_{min} is determined by the thick gap after the thin gap is saturated. Larger dimensions (cross-sectional area and length) of the gap are beneficial for current-handling capability and swinging factor, but the increased volume is undesirable. Alternatives for gap placement to achieve wide inductance range without compromising energy density are studied.

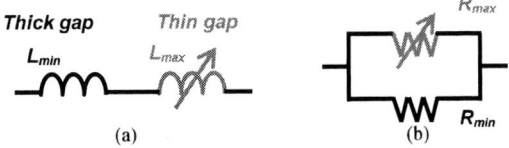

Fig. 2. Swinging inductor modeled by (a) serial inductances and (b) parallel reluctances.

The thick and thin gaps that are in line are termed "1D gap" or "stepped gap" as shown in Fig. 3(b). The gap shown in Fig. 3(a) is a special case of 1D gap where either the thin gap or the thick gap is eliminated. The equivalent circuit for Fig. 3(b) is a constant inductor in series with a swinging inductor as shown in Fig. 2(a). If the width $2A_4$ of the middle leg for the E core in Fig. 3(b) is fixed, an increase in the width A_1 of the thin gap to improve light-load inductance L_{max} causes a corresponding decrease in the width of the thick gap, resulting in loss of nominal-load inductance L_{min}. Another disadvantage of the stepped gap is that the magnetic area around the thin gap would be saturated at heavy load. As the A_1 increases, the whole core would be saturated gradually. Even if the specified nominal-load inductance could be satisfied, e.g., a uniform thin gap with $A_1 = 2A_4$, saturation would bring high bias-dependent core loss, which will be discussed in sub-Section II.B.

In order to overcome the restriction on maximum inductance for stepped gap, and to gain more freedom in controlling the gap dimensions, the inductor with "2D gap" illustrated in Fig. 3(c) places the thin gap perpendicular to, instead of in-line with, the thick gap to take advantage of a larger cross-sectional area ($A_3 > A_1$) to achieve higher inductance at light load with the same thin-gap's length. One I-bar in Fig. 3(c) is inserted between two C-cores to construct the vertical thin gaps, behaving as a swinging reluctance, R_{g3} in Fig. 3(d). The choice of stepped gap or 2D gap depends on the aspect ratio of the core. The 2D gap is advantageous when the core's aspect ratio is close to unity since the volume in two dimensions stores more energy and yields higher swinging than one dimension. As mentioned earlier, the A_1 in the stepped-gap

determines the trade-off between the light-load inductance and saturation level at nominal load. The 2D gap decouples the two effects by using two independent and orthogonal geometrical parameters, the thickness of the I-bar and the cross-sectional area for the thin gap. The light-load inductance then relies only on A_3 and l_{g3}; and the bias level at nominal load can be designed by A_1.

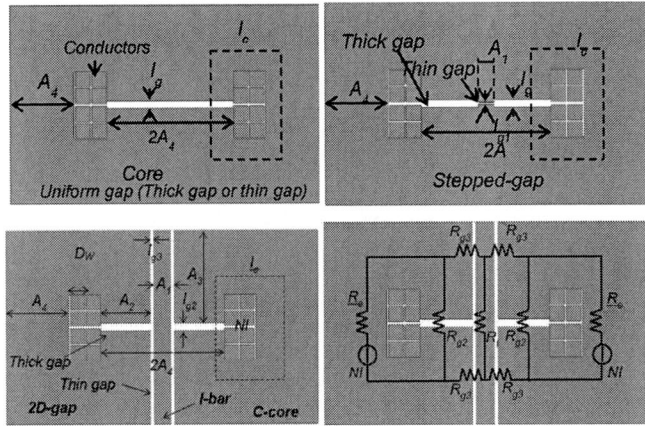

Fig. 3. 2D plot of (a) constant gap design, (b) stepped-gap design, and (c) 2D gap design with (d) equivalent reluctance network. They all have the same outer dimensions (10.8 mm ⊠ 6.5 mm ⊠ 3.5 mm) and dc resistance (11.8 mΩ).

Inductors employing thin gap, stepped gap, and 2D gap exhibit swinging behavior via similar mechanisms. This is explained by the distributions of ac magnetic flux density B_{ac} at light load and nominal load for the 2D gap in Fig. 4(a). The materials used in the simulation are described in sub-Section II.B. At light load where no core saturates, the I-bar is a part of a low-reluctance path (R_{g3}- R_c- R_l in Fig. 3(d)) that not only yields high inductance, but also steers ac flux away from the conductors to lower ac resistance. The higher flux density in the I-bar incurs more core loss, but core loss will be shown negligible to winding-loss reduction in the next section. At nominal current, the I-bar in Fig. 3(c) would be saturated, and has low permeability as well as low ac flux density as shown in Fig. 4(b). The reluctance path then transfers to "R_{g2}- R_c- R_l". The fringing flux in the winding window induces higher ac resistance, but core loss is reduced thanks to lower B_{ac} in the I-bar. Small ac flux caused by the fringing flux remains in the I-bar at the region close to the intersection of the two orthogonal gaps. The equivalent cross-sectional area for the thick gap is increased, and the nominal-load inductance is comparable to that of the inductor with uniform, thick gap. This is the key reason for the inductor with 2D gap achieving swinging inductance without losing heavy-load inductance within the same volume. Since the I-bar absorbs significant fringing flux at nominal load, ac winding resistance is lower than that in 1D-gapped inductor.

The parameters L_{max}, L_{min}, and I_{knee} defined in Fig. 4(c) characterize the inductance swinging of the 2D gap. The light-load inductance L_{max} is defined as the inductance at zero current, derived by R_{g3}, R_l, R_{g2}, and R_c in Fig. 3(d). The reluctances R_l and R_c are negligible at light load since both C-cores and I-bar have high permeability. The dimensions of the gaps are selected such that $lg_3/A_3 << lg_2/A_2$ to make $R_{g3} << R_{g2}$. Fringing effect is neglected since the thin gap's length is much smaller

978-1-4673-9551-9/16 $31.00 © 2016 IEEE

than the height of I-bar in Fig. 3(c). Then L_{max} is approximated by

$$L_{max} = \frac{N^2}{(R_{g3} + R_I)//(\frac{1}{2}R_{g2})//(\frac{1}{2}R_c)} \approx \frac{N^2}{R_{g3}}$$

$$= \frac{N^2 A_3 h \mu_0}{l_{g3}} \quad \left(\frac{l_{g3}}{A_3} \ll \frac{l_{g2}}{A_2}\right) \qquad (1)$$

where N is the number of turns; μ_0 is the air permeability; and h is the core thickness.

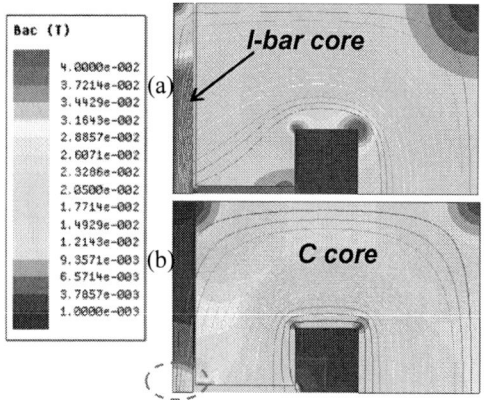

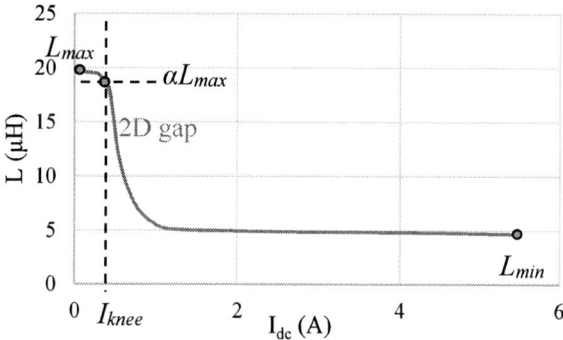

Fig. 4. Simulated ac flux density at (a) I_{dc} = 0.1 A and (b) I_{dc} = 5 A for one-fourth model of 2D gap shown in Fig. 3(c) with l_{g3} = 0.05 mm, l_{g2} = 0.25 mm, A_4 = 2.15 mm and A_1 = 0.7 mm; (c) simulated inductance versus dc current. Core material is P-ferrite described in sub-Section II.B.

The nominal-load inductance L_{min} depends on R_{g2} and R_c. Since $R_c \ll R_{g2}$, the minimum inductance is approximately

$$L_{min} = \frac{N^2}{(\frac{1}{2}R_{g2})//(\frac{1}{2}R_c)} \approx \frac{N^2(2A_2 + l_{g2})(h + l_{g2})\mu_0}{l_{g2}} \qquad (2)$$

where fringing effect is taken into account by extending the cross-sectional dimensions for the thick gap by l_{g2} [22].

Since the inductance at I_{knee} is still high, most of the flux goes through the I-bar and thin gap instead of the thick gap. Let B_I be the average magnetic flux density in the I-bar when the current is I_{knee}. The flux flowing through the cross-sectional areas A_1 and A_3 are $B_I A_1$. The current I_{knee} is given by

$$I_{knee} = \frac{l_{g3} B_I A_1}{\mu_0 A_3 N} \qquad (3)$$

Two candidates to implement the 2D gap are shown in Fig. 5. One is to insert a ferrite plate (I-bar) into the middle region for gapped C cores. Between the I-bar and C-cores are the thin part of the 2D gap. The thick part of the 2D gap is orthogonal to the thin gap and is embedded within the C-cores.

The other structure is a pot core combined with one ferrite cylinder in the center hole. The thick gap is placed between the two core halves, whereas the thin gap surrounds the cylinder core. A drawback of the pot core with 2D gap is the difficulty in controlling the thin gap's length.

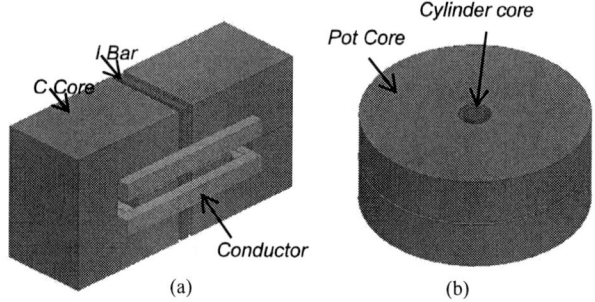

Fig. 5. Two 3D candidates of 2D gap design: (a) two C-cores with saturable bar and (b) pot core with saturable cylinder.

B. Performance Comparison

The material for general gapped inductor is ferrite. Here the material chosen is P-ferrite which has low core loss density at 500 kHz. Measured permeability, flux density, and core loss density are shown in Fig. 6. The saturated magnetic flux density is 0.45 T. Core loss is sensitive to dc bias, especially when the core is saturated [24]-[33]. The core loss versus dc bias in Fig. 6(b) was measured by two U cores (OP41106, P-ferrite) with one winding for the dc excitation (0 – 0.7 A), one winding for ac excitation, and the other winding for flux sensing. Each winding had five turns of 180 strands with a total diameter of 0.72 mm. Measurement procedure followed [34]. A capacitor in series with the tested inductor was used to cancel the passive component of the voltage to improve phase accuracy. This capacitor was adjusted with dc current based on the measured permeability curve in Fig. 6(a).

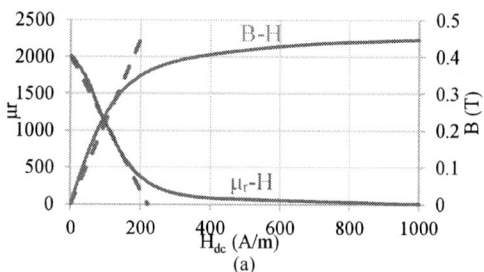

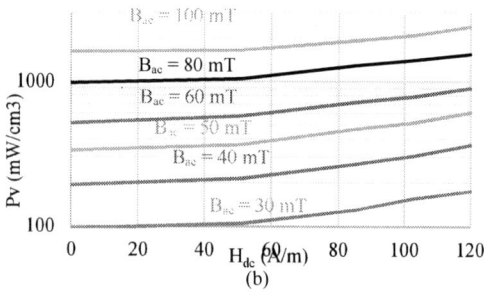

Fig. 6. (a) Relative permeability and (b) measured core loss density of P-ferrite vs. H_{dc} at 500 kHz and 25°C.

978-1-4673-9551-9/16 $31.00 © 2016 IEEE

The dependence of core loss density P_v on dc magnetic field intensity H_{dc} is accounted for by the function $K_{dc}(H_{dc})$ [35]:

$$P_v = P_{v0}K_{dc}(H_{dc}) \tag{4}$$

where P_{v0} is the core loss without dc effect, which can be calculated by such models as MSE [36], GSE [37], iGSE [38], or RESE [39]. Here the model [40] native to the finite-element simulator is used since it deals with the dynamic core loss for 2D or 3D conveniently via the core loss parameters C_m, x, and y extracted from the loss curve associated with sinusoidal excitation. It takes into account minor loops to predict instantaneous hysteresis loss. The model parameters can be extracted from standard loss curves, e.g., $C_m = 6.22 \times 10^{-3}$, $x = 1.93$ and $y = 2.66$ for P-ferrite.

Core loss density was measured as a function of H_{dc} and plotted in Fig. 6(b). The magnitude of the magnetic flux density in the design example is close to 30 mT. Curve-fitting yields

$$K_{dc}(H_{dc}) = 0.000075H_{dc}^2 - 0.00203H_{dc} + 1 \tag{5}$$
$$\text{for } B_{ac} = 30 \text{ mT}$$

where H_{dc} is the dc component of magnetic field intensity in A/m.

The specifications of input voltage equal to 10 V and output voltage equal to 5 V determine the duty cycle D. The case of $D = 0.5$ actually has the highest core loss for a buck converter with a fixed input voltage based on the model in [39]. This paper will confirm that the worst case for a buck converter is at $D = 0.5$ after a 2D gap is designed.

The nonlinear permeability of the core material gives possibility to realize swinging inductance by using only one gap [20]. Two gaps to implement the required inductance at heavy load (e.g., L = 5 μH at 5 A) are shown in Fig. 7(a). The inductance for the thick gap at $l_g = 0.25$ mm is constant, whereas the thin gap at $l_g = 0.114$ mm yields swinging inductance. Simulated inductances for four cases (2D gap, stepped gap, thin gap, and thick gap) are shown in Fig. 7(b). The 2D gap and stepped gap have the same thin gap's length and thick gap's length. The 2D gap has the swinging factor up to four although the knee current is relatively small, which is suitable for point-of-load converter working at two load conditions to improve efficiency as reported in [16]. The inductance for the thin gap is higher than that of stepped gap thanks to its larger effective cross-sectional area at light load. However, the saturation of the thin gap at nominal load would decrease the inductance quickly, and bring higher core loss.

Dc winding loss is dominant at nominal load since ripple current is negligible compared with dc current. The four cases have almost the same winding loss at nominal load as shown in Fig. 8(b) by specifying the same dc resistance. As dc current decreases, the winding loss is dominated by the ac ripple and ac resistance. The 2D gap with the highest inductance at light load shown in Fig. 7(b) has the lowest ac ripple. The ac resistance R_{ac} is calculated from

$$P_{wt} = I_{dc}^2 R_{dc} + (I_{rms}^2 - I_{dc}^2) R_{ac} \tag{6}$$

where P_{wt} is the simulated total winding loss. The ac resistance of 2D gap is small at light load because the flux lines are steered into the I-bar and away from the windings as shown in Fig. 4(a). The ac resistance drops as the current increases as shown in Fig. 8(a) thanks to an increase in equivalent gap length and a decrease in field strength [41]. The ac flux lines transfer to the thick gap after the I-bar saturates as illustrated in

Fig. 4(b), inducing fringing flux in the winding window and high ac resistance. The further increasing current at heavy load would increase bias level of C-cores, and reduce the field intensity in the winding window. The ac resistance drops again with similar mechanism as the first dip.

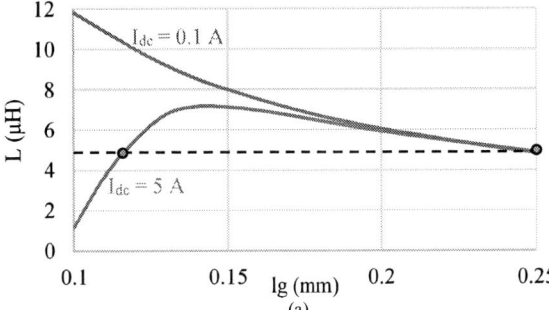

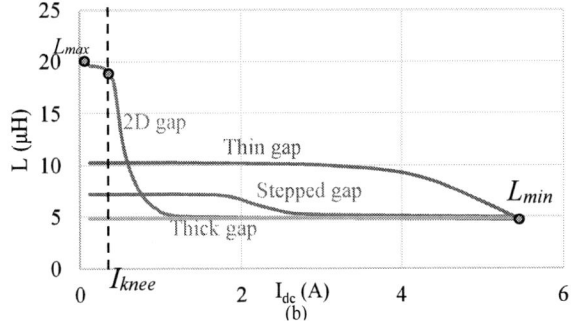

Fig. 7. (a) Simulated L vs. gap length for uniform gap and (b) simulated L vs. I_{dc} for 2D gap with the dimensions in Fig. 4, stepped gap ($l_{g1} = 0.05$ mm and $l_g = 0.25$ mm), thin gap ($l_g = 0.116$ mm), and thick gap ($l_g = 0.25$ mm) with structures in Fig. 3.

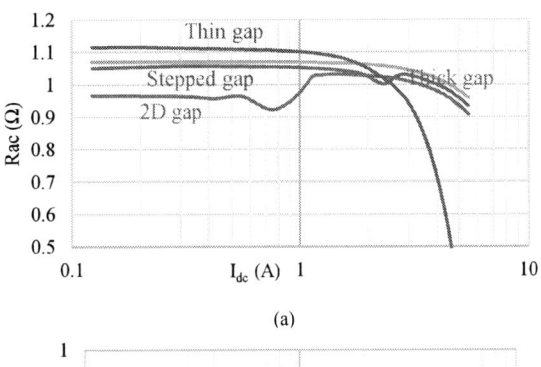

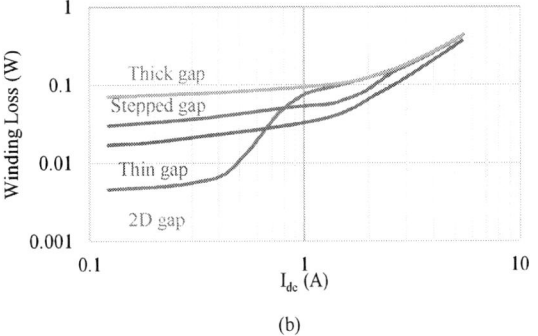

Fig. 8. (a) Simulated ac resistance R_{ac} and (b) total winding loss vs. dc current for 2D gap, stepped gap, thin gap, and thick gap with structures in Fig. 3 and gapping parameters from Fig. 4.

Core loss was simulated for four cases considering the frequency-dependent and bias-dependent nonlinearities. Voltage excitation was an ideal rectangular waveform based on the circuit from Fig. 1. The core loss density P_{v0} and average magnetic field intensity H_{dc} for each element were simulated by FEA. Core loss density with dc effect at each element was then calculated by using the P_{v0} and H_{dc} in the model from (4) and (5). Total core loss was derived by the volume-integration of core loss density. For the stepped gap and thick gap, core loss is almost constant because the distribution of flux density does not change significantly. The high dc bias for thin gap at heavy load leads to saturation shown in Fig. 10(b) and high core loss in Fig. 9(a). The inductor with 2D gap shows high core loss at light load resulted from the I-bar, which is demonstrated by separating the core loss from I-bar and that from the C-cores illustrated in Fig. 9(a). One effective way to reduce the light-load core loss is to utilize the material with low loss density. Another method described in sub-Section II.C is to adjust the thickness of the I-bar to move the peak core loss to higher current. Total loss plotted in Fig. 9(b) is dominated by winding loss at light load for all cases except the 2D gap which has significant reduction of winding loss.

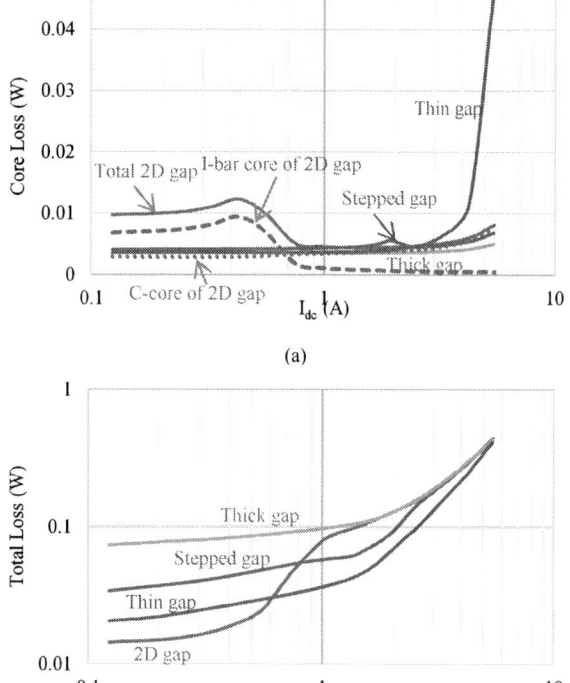

(a)

(b)

Fig. 9. (a) Simulated core loss and (b) total inductor loss vs. dc current for 2D gap, stepped gap, thin gap, and thick gap with structures in Fig. 3 and gapping parameters from Fig. 4.

The dc flux density was also simulated to evaluate the operation at nominal load. Without the thin gap, the whole core has no significant hot points as shown in Fig. 10(a). The magnetic segments adjacent to the thin gaps in Fig. 10(c) and Fig. 10(d) saturate first, and saturation could spread into the magnetic regions in series with the thick gaps unless the core is properly designed (see sub-Section II.C). Once the I-bar in the 2D-gapped inductor saturates, the ac flux prefers to flow through the C-cores and the thick gap. The I-bar then carries negligible ac flux and enjoys low core loss although it is biased by high dc magnetic field intensity. A design priority is to keep

the C-cores away from saturation to avoid high bias-induced core loss. This will be described in sub-Section II.C. The dc flux density for the uniformly thin gap is relatively high as illustrated in Fig. 10(b). If the core were saturated, core loss and temperature would increase, leading to thermal run-away.

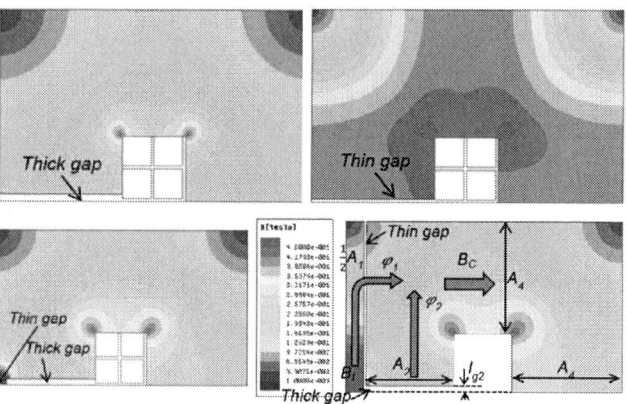

Fig. 10. Simulated dc flux density of (a) thick gap, (b) thin gap, (c) stepped gap, and (d) 2D gap at I_{dc} = 5 A for one-fourth model with structures in Fig. 3 and gapping parameters in Fig. 4.

C. Choice of Thickness for I-bar

The dc flux density B_C in the C-cores under nominal load is derived from the dc flux ϕ_1 through the I-bar, ϕ_2 through the thick gap, and A_4B_c through the C-cores as illustrated in Fig. 10(d). The contribution of the saturated I-bar to nominal-load inductance is negligible, i.e., L_{min} is dominated by the thick gap. The dc flux ϕ_1 is assumed to retain the value $L_{max}I_{knee}/2N$ attained at I_{knee}, whereas ϕ_2 is approximated by $L_{min}I_{load}/2N$. The approximate expression for B_C is then

$$B_C = \frac{\varphi_1 + \varphi_2}{A_4 h} \approx \frac{1}{2NhA_4}(L_{max}I_{knee} + L_{min}I_{load}) \qquad (7)$$

The last equality in (7) constrains the specifications of I_{knee}, L_{max}, and L_{min}, especially since B_C should be lower than saturation flux density to limit core loss. Since I_{knee} is proportional to A_1 via (3), larger A_1 leads to larger B_C and higher bias-dependent core loss.

The B-H curve and μ_r-H curve are approximately linear before saturation as seen in Fig. 6(a), i.e., μ_{rI} is linearly related to B_I:

$$\mu_{rI} = c - k_B B_I \qquad (8)$$

where μ_{Ir} is the relative permeability in the I-bar; c = 2000 and k_B = 3700 [1/T] for the P-ferrite in Fig. 6(a). The current I_{knee} and the corresponding inductance αL_{max}, $0 < \alpha < 1$, characterize the inductance droop caused by saturation of the I-bar. Since the inductance at I_{knee} is still high, most of the flux goes through the I-bar and thin gap instead of the thick gap. Since R_I is no longer negligible compared to R_{g3} at I_{knee},

$$\alpha L_{max} \approx \frac{N^2}{R_{g3} + R_I} \qquad (9)$$

From (1) and (9),

$$R_I = \frac{1 - \alpha}{\alpha} R_{g3} \quad \text{at } I_{knee} \qquad (10)$$

If the flux lines inside the I-bar (see Fig. 4(a)) are approximated to be elliptical [23],

$$R_I = \frac{4A_3 + (\pi - 2)A_1}{\mu_{Ir}\mu_0 h\left(\frac{A_1}{2} + A_3\right) \ln\left(\frac{A_1}{2l_{g3}}\right)} \tag{11}$$

Equations (1), (3), (8), (10), and (11) are combined to yield a relationship from which A_1 is determined:

$$\frac{1-\alpha}{\alpha}$$
$$= \frac{4A_3 + (\pi - 2)A_1}{\left(cl_{g3} - k_B \frac{I_{knee}\mu_0 A_3 N}{A_1}\right)\left(\frac{A_1}{2A_3} + 1\right) \ln\left(\frac{A_1}{2l_{g3}}\right)} \tag{12}$$

III. RETROFIT DESIGN PROCEDURE

An inductor with a uniform thick gap is assumed already available. The retrofit procedure outlined in Fig. 11 adds an I-bar and a 2D gap to swing the inductance up to L_{max} below I_{knee}, keeping inductance L_{min} and dc resistance at nominal load the same. The exterior dimensions, number of turns, material properties, and interior parameters A_3 and A_4 are inherited from the available inductor. Unknown parameters are A_1, l_{g3}, and l_{g2}. The first step is to check the biased condition of B_c. If B_C is lower than maximum flux density B_{satC} of C-cores, the design proceeds. Otherwise, the specifications of L_{max}, L_{min}, I_{knee}, and I_{load} should be revised to meet (7). The gap l_{g3} between the I-bar and the C-core is determined by utilizing the inductance at light load in (1). The thickness A_1 is obtained from specified current I_{knee} in (12). The thick gap's length is calculated using the nominal-load inductance L_{min} and the associated cross-sectional area of the flux path.

To exemplify the design procedure, consider the retrofit of an existing inductor with $A_4 = 2.15$ mm, $A_3 = 3.125$ mm, $N = 8$, $B_{satC} = 0.45$ T, $h = 3.5$ mm, $L_{max} = 18$ µH, $L_{min} = 4.5$ µH, $I_{knee} = 0.27$ A, $\alpha = 0.9$, and $I_{load} = 5$ A. Design results based on Fig. 11 are: $B_C = 0.23$ T $< B_{satC}$, $l_{g3} = 0.049$ mm, $A_1 = 0.71$ mm and $l_{g2} = 0.23$ mm. The results agree with the example in Section II and the simulated results shown in Fig. 7(b).

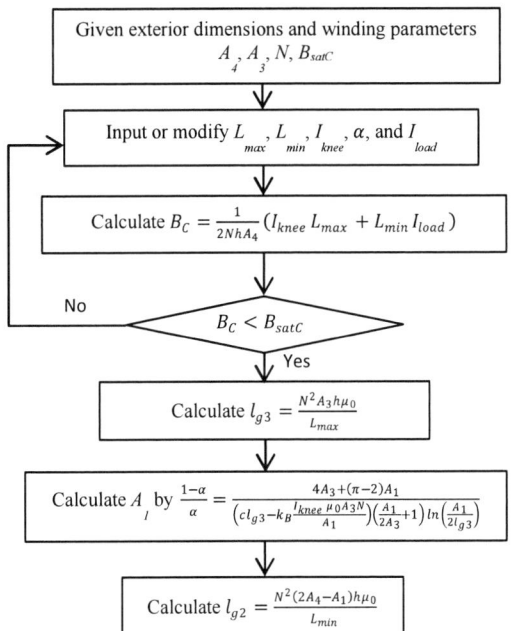

Fig. 11. Flow chart to design 2D gap.

IV. VERIFICATION

The inductor with 2D gap was fabricated by two C-cores and one I-bar with the dimensions shown in Fig. 12(a) and customed hardware in Fig. 12(b). The material for these cores was P-ferrite from Magnetics, Inc. The number of turn was 8; the wire diameter, 0.46 mm; and dc resistance, 18 mΩ. The thin gap was realized by polyester film with thickness of 0.05 mm.

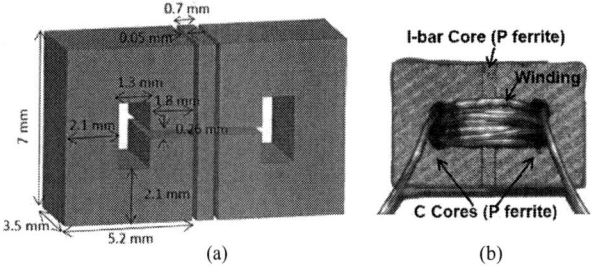

Fig. 12. (a) 3D model and dimensions of (b) 2D-gapped inductor in experiment with N = 8 and winding diameter of 0.46 mm.

The inductor prototype and a reference inductor without the I-bar were tested in a synchronous buck converter with the specifications in Fig. 1. To compare them fairly, the high-side MOSFET was hard-switched within the load range. The current I_L through the inductor and voltage V_L across the inductor were measured to calculate inductor loss. When $I_{dc} \ll I_{knee}$ or $I_{dc} \gg I_{knee}$, I_L is triangular as seen in Fig. 13(a) and Fig. 13(c) since the inductance is constant versus current. When I_{dc} is close to I_{knee}, the current waveform contains harmonics as seen in Fig. 13(b) since inductance varies with instantaneous current in the range of current ripple.

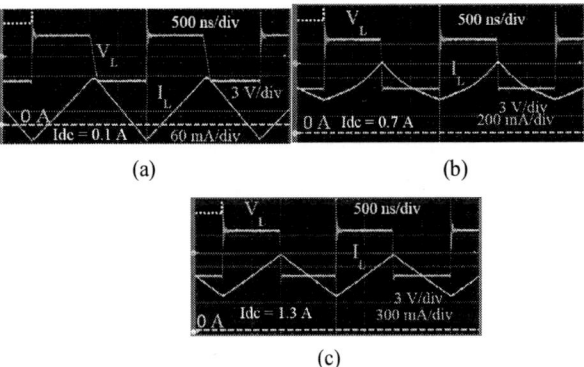

Fig. 13. Measured waveforms of the current through the inductor and the voltage across the inductor for the prototype in Fig. 12 tested in the buck converter in Fig. 1 at (a) $I_{dc} = 0.1$ A, (b) $I_{dc} = 0.7$ A, and (c) $I_{dc} = 1.2$ A.

The inductance is measured by

$$L = \frac{V_{Lp-p}(1-D)D}{I_{Lp-p}f_s} \tag{13}$$

where V_{Lp-p} is the peak-to-peak voltage across the tested inductor; I_{Lp-p} is the peak-to-peak current through the tested inductor; and D is the duty ratio. Error owing to dc resistance of the winding is avoided since L is calculated from the peak-to-peak V_L across the entire switching period.

The inductor power was measured by multiplying the instantaneous V_L by the instantaneous I_L. The waveforms were measured using the mixed-signal oscilloscope MSO5104B set in "HiRes" Mode to increase vertical resolution above 11 bits.

The current bias was calibrated by the Tektronix DMM4020 current meter with resolution of 0.1 mA. The dc voltage was calibrated by multiplying the dc resistance measured by the DMM4020 Ohmmeter by the dc current. The core loss could not be measured separately by the methods described in [42] since the tested inductor had only one winding.

While 2D simulation is expeditious and reliable for nonlinear inductance and core loss, it cannot predict the variation of fringing field and the corresponding winding loss along the length of the conductor. Three-dimensional finite-element simulation was thus employed to check the accuracy of 2D simulation, and to characterize the experiment prototype. Fig. 14 shows a one-eighth model, taking advantage of the symmetrical boundary conditions. The loss density in the conductive regions near the I-bar is lower than that predicted by 2D simulation thanks to reduced fringing field. The maximum mesh length for the core element is 0.5 mm, and the maximum mesh length for the conductor element is 0.1 mm. Each turn is individually closed by a square conductor to save computation time.

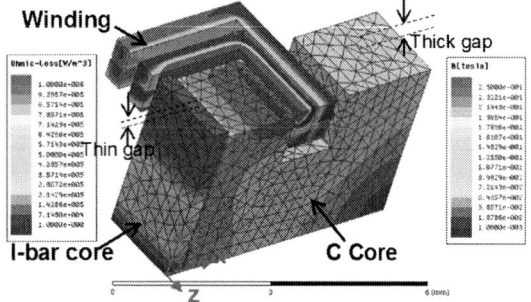

Fig. 14. 3D-simulated ohmic loss and flux density at $I_{dc} = 0.2$ A for one-eighth model of the inductor with 2D gap in Fig. 12.

The measured and simulated inductances are compared in Fig. 15(a), and the inductor losses are compared in Fig. 15(b). The inductor with 2D gap achieves three times the inductance and half the loss of the corresponding conventional inductor employing a 1D thick gap for dc current lower than 0.5 A. Two reasons can explain the loss error in the load range from 0 to 0.5 A for the case with 2D gaps: the voltage ringing caused by the hard switching and phase error between voltage probe and current probe. The total loss at heavy load is dominated by the dc winding loss, which could be measured reliably and simulated with small error.

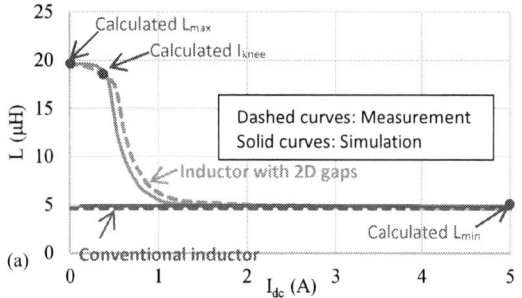

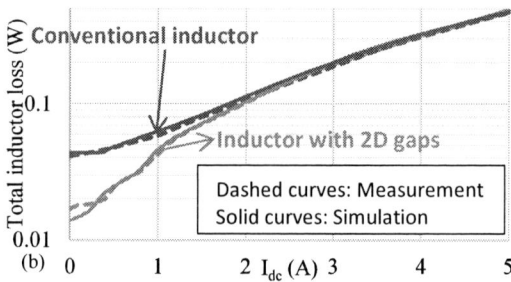

Fig. 15. Measured and simulated (a) inductances and (b) losses for a conventional inductor and an inductor with 2D gaps; calculated L_{max}, L_{min}, and I_{knee} are also shown.

V. CONCLUSION

An inductor employing 2D gap was constructed by two C-cores and one saturable piece, and tested in a point-of-load converter to demonstrate 50% reduction of loss at 10% rated current. In order to achieve a high swinging inductance, a thin gap with small gap length and large cross-sectional area is needed. The gap length is limited by the fabrication tolerance, whereas the area depends on the location of the thin gap. If the thin gap is placed in-line (1D) with the thick gap, as is the case for the "stepped gap", an attempt to increase light-load inductance would cause the core to saturate since an increase of the thin gap's area must be accompanied by a decrease of the thick gap's area. The 2D gap arrangement essentially decouples the two effects by two independent and orthogonal geometrical parameters: the thickness of the I-bar and the cross-sectional area of the thin gap. The thickness can be designed to avoid saturation of the C-cores, whereas the large cross section available in another direction yields a higher light-load inductance.

The sacrifice of the core loss for the inductor with 2D gap at light load is owing to the high ac flux in the I-bar. A peak core loss appears near the knee current where the permeability of the I-bar starts to drop. At nominal load, most of the core loss is incurred in the C-cores because of the redistribution of the ac flux density. The I-bar core would benefit from materials with low core loss density and less bias effect. The C-cores would benefit from materials with high saturation flux density for high power density. The core loss density for the C-cores is not critical since the winding loss and MOSFET loss are dominant in a point-of-load converter at nominal load.

Another approach to diminish the core loss is to shift the peak point to high current by increasing the thickness or flux density of I-bar based on (3). This, however, might induce saturation and bias-dependent core loss and thermal instability at nominal load via (7). The trade-off between the light-load core loss and saturation level at nominal load should be monitored in the design of swinging inductor with 2D gap to avoid additional loss at nominal load.

ACKNOWLEDGMENT

The work is funded by Texas Instruments. The authors would like to thank Dr. Michele Lim for her support.

REFERENCES

[1] Y. Chen, P. Asadi, and P. Parto, "Comparative analysis of power stage losses for synchronous buck converter in diode emulation mode vs. continuous conduction mode at light load condition", in *Proceedings of*

IEEE Applied Power Electronics Conference, Feb. 2010, pp. 1578 - 1583.

[2] A. Ball, M. Lim, D. Gilham, and F. C. Lee, "System design of a 3D integrated non-isolated Point of Load converter," in *Proceedings of IEEE Applied Power Electronics Conference*, Feb. 2008, pp. 181 - 186.

[3] Q. Li and F. C. Lee, "High inductance density low-profile inductor structure for integrated point-of-load converter," in *IEEE Power Electronics Specialists Conference Record*, Feb. 2009, pp. 1011 - 1017.

[4] S. F. Lim and A.M. Khambadkone, "Non-linear inductor design for improving light load efficiency of boost PFC," in *Proceedings of IEEE Energy Conversion Congress and Exposition*, Sept. 2009, pp. 1339 - 1346.

[5] W.G. Hurley and W.H. Wölfle, *Transformers and Inductors for Power Electronics: Theory, Design and Applications*, Chichester: John Wiley & Sons, 1st ed., Feb. 2013, pp. 204 - 212.

[6] M. H. Lim, J. D. van Wyk, F. C. Lee, and Z. Liang, "Internal geometry variation of LTCC inductors to improve light-load efficiency of DC-DC converters," *IEEE Transactions on Components and Packaging Technology*, vol. 32, no. 1, pp. 3 - 11, Mar. 2009.

[7] L. Wang, Y. Pei, X. Yang, and Z. Wang, "Improving Light and Intermediate Load Efficiencies of Buck Converters with Planar Nonlinear Inductors and Variable on Time Control," *IEEE Transactions on Power Electronics*, vol. 27, no. 1, pp. 342 - 353, Jan. 2012.

[8] L. Wang, Z. Hu, Y. Qiu, H. Wang, and Y. Liu, "A new model for designing multi-hole multi-permeability nonlinear LTCC inductors", in *Proceedings of IEEE Applied Power Electronics Conference*, Mar. 2014, pp. 757 - 762.

[9] E. Dallago, M. Passoni, and G. Venchi, "Analysis of High-Frequency IGBT Soft Switching Buck Converter with Saturable Inductors," *IEEE Transactions on Power Electronics*, vol. 22, no. 2, pp. 407 - 416, Mar. 2007.

[10] J. Sun, S. Hamada, J. Yoshitsugu, B. Guo, and M. Nakaoka, "Zero voltage soft-commutation PWM DC-DC converter with saturable reactor switch-cascaded diode rectifier," *IEEE Transactions on Circuits and Systems I*, vol. 45, no. 4, pp. 348 - 354, Apr. 1998.

[11] S. Hamada and M. Nakaoka, "Family of saturable reactor assisted soft-switching PWM DC-DC convertors," in *Proceedings of the Institution of Electrical Engineers*, Jul. 1992, pp. 395 - 401.

[12] M. Stadler and J. Pforr, "Zero-Voltage Switched Multi-Phase Converter utilizing nonlinear and coupled Inductors," in *Proceedings of IEEE Applied Power Electronics Conference*, Feb. 2007, pp. 1038 - 1042.

[13] ST50-267, one of composite cores from Micrometals, Inc., available [Online]:
Http://www.micrometals.com/pcparts/ccore.html?zoom_highlight=ST50-267

[14] H. Ahmed, H. Cha,S. Kim, D. Kim, H. Kim, "Wide Load Range Efficiency Improvement of a High Power-Density Bidirectional DC-DC Converter Using an MR Fluid-Gap Inductor", *IEEE Transactions on Industry Applications*, vol. 51, no. 4, pp. 3216 - 3226, Jul. 2015.

[15] D. W. Kim, H. Cha, S. Lee, and D. H. Kim, "Characteristic of a Variable Inductor Using magnetorheological Fluid for Efficient Power Conversion", *IEEE Transactions on Magnetics*, vol. 49, no. 5, pp. 1901 - 1904, May 2013.

[16] J. Sun, M. Xu, Y. Ren, and F. C. Lee, "Light-load efficiency improvement for buck voltage regulators," *IEEE Transactions on Power Electronics*, vol. 24, no. 3, pp. 742 - 751, Mar. 2009.

[17] W. H. Wölfle, and W. G. Hurley , "Quasi-active power factor correction with a variable inductive filter: theory, design and practice," *IEEE Transactions on Power Electronics*, vol. 18, no. 1, pp. 248 - 255, Jan. 2003.

[18] Magnetics, "Step-gap E core swing chokes", Technical Bulletin, FC-S4, 2001.

[19] ST, "Design tips for L6561 power factor corrector in wide range", application note, AN1214, Dec. 2000.

[20] C.R. Hanna, "Design of reactances and transformers which carry direct current", *Journal of the American Institute of Electrical Engineers*, vol. 46, p. 128, Feb. 1227.

[21] T. Ge, K. Ngo, J. Moss, and M. Lim, "Gap design for nonlinear ferrite cores to maximize inductance", in *Proceedings of IEEE Energy Conversion Congress and Exposition*, Sep. 2014, pp. 5237 - 5242.

[22] N. Mohan, T. M. Undeland, and W. P. Robbins, *Power Electronics-Converters, Applications, and Design*, Chichester: John Wiley & Sons, 2nd ed., 1995.

[23] M. Lu and K. D. T. Ngo, "Model for Electromagnetic Actuator With Significant Fringing Using Minimal Fitting Parameters," *IEEE Transactions on Magnetics*, vol. 51, no. 1, Jul. 2014.

[24] V. C. Valchev, A. P. Van den Bossche, and D. M. Van de Sype, "Ferrite losses of cores with square wave voltage and dc bias", in *Conference of the IEEE Industrial Electronics Society*, Nov. 2005.

[25] C. A. Baguley, B. Carsten, and U. K. Madawala, "An investigation into the impact of DC bias conditions on ferrite core losses," *IEEE Transactions on Magnetics*, vol. 44, no. 2, pp. 246 - 252, Feb. 2008.

[26] C. A. Baguley, U. K. Madawala, and B. Carsten, "The impact of vibration due to agnetostriction on the core losses of ferrite toroidals under DC bias," *IEEE Trans. on Magnetics*, vol. 47, no. 8, pp. 2022 - 2028, Aug. 2011.

[27] J. Mühlethaler, J. Biela, J. W. Kolar, and A. Ecklebe, "Core losses under DC bias condition based on Steinmetz parameters," in *Proc. IEEE/IEEJ International Power Electronics Conference*, Jun. 2010, pp. 2430 - 2437.

[28] K. Venkatachalam, C. R. Sullivan, T. Abdallah, and H. Tacca, "Accurate prediction of ferrite core loss with nonsinusoidal waveforms using only Steinmetz parameters," in *IEEE Workshop on Computers in Power Electronics*, 2002.

[29] J. Mühlethaler, J. Biela, J. W. Kolar, and A. Ecklebe, "Improved core loss calculation for magnetic components employed in power electronic systems," in *Proceedings of IEEE Applied Power Electronnics Conference*, Mar. 2011, pp. 1729 - 1736.

[30] A. Brockmeyer, "Experimental evaluation of the influence of DC premagnetization on the properties of power electronic ferrites", in *Proceedings of IEEE Applied Power Electronics Conference*, Mar. 1996, pp. 454 - 460.

[31] C. A. Baguley, B. Carsten, and U.K. Madawala, "The effect of DC bias conditions on ferrite core losses", *IEEE Transactions on Magnetics*, vol. 44, No. 2, February 2008.

[32] Wai Keung Mo, David K.W. Cheng, and Y. S. Lee, "Simple Approximations of the DC Flux Influence on the Core Loss Power Electronic Ferrites and Their Use in Design of Magnetic Components", *IEEE Transactions on Magnetics*, vol. 44, no. 6, pp. 788 - 799, Dec. 1997.

[33] C. Simão, N. Sadowski, N. J. Batistela, and J. P. A. Bastos, "Evaluation of Hysteresis Losses in Iron Sheets Under DC-biased Inductions," *IEEE Transactions on Magnetics*, vol. 45, no. 3, pp. 1158 - 1161, Mar. 2009.

[34] M. Mu, Q. Li, D. Gilham, F. C. Lee, and K. D.T. Ngo, "New core loss measurement method for high frequency magnetic materials," in *Proceedings of IEEE Energy Conversion Congress and Exposition*, Sep. 2010, pp. 4384 - 4389.

[35] M. Mu, F. Zheng, Q. Li, and F.C. Lee, "Finite Element Analysis of Inductor Core Loss under DC Bias Conditions," *IEEE Transactions on Power Electronics*, vol. 28, no. 9, pp. 4414 - 4421, Sept. 2013.

[36] J. Reinert, A. Brockmeyer, and R. De Doncker, "Calculation of losses in ferro- and ferrimagnetic materials based on the modified Steinmetz equation," *IEEE Transactions on Inst. Appl.*, vol. 37, no. 4, pp. 1055 - 1061, Jul./Aug. 2001.

[37] J. Li, T. Abdallah, and C. R. Sullivan, "Improved calculation of core loss with nonsinusoidal waveforms," in *Proceedings of Industry Applications Conference, Sept./Oct. 2001*, vol. 4, pp. 2203 - 2210.

[38] K. Venkatachalam, C. R. Sullivan, T. Abdallah, and H. Tacca, "Accurate prediction of ferrite core loss with nonsinusoidal waveforms using only Steinmetz parameters," in *Proceedings of IEEE Workshop on Computers in Power Electronics*, Jun. 2002, pp. 36 - 41.

[39] M. Mu and F.C. Lee, "A new core loss model for rectangular AC voltages," in *Proceedings of IEEE Energy Conversion Congress and Exposition, Sept. 2014*, pp. 5214 - 5220.

[40] D. Lin, P. Zhou, W. N. Fu, Z. Badics, and Z. J. Cendes, "A Dynamic-core Loss Model for Soft Ferromagnetic and Power Ferrite Materials in Transient Finite Element Analysis," *IEEE Transactions on Magnetics*, vol. 40, no. 2, pp. 1318 - 1321, Mar. 2004.

[41] J. Hu and C. R. Sullivan, "Ac resistance of planar power inductors and the quasidistributed gap technique", *IEEE Transactions on Power Electronics*, vol. 16, pp. 558 - 567, Jul. 2001.

[42] M. Mu, Q. Li, D. J. Gilham, F. C. Lee, and K. D. T. Ngo, "New core loss measurement method for high-frequency magnetic materials," *IEEE Transactions on Power Electronics*, vol. 29, no. 8, pp. 4374 - 4381, Aug. 2014.

Iron Loss Evaluation of Three-phase Inductor for Three-Phase PWM Inverter

Hiroaki Matsumori and Toshihisa Shimizu
Tokyo Metropolitan University
Tokyo, Japan

Koushi Takano and Ishii Hitoshi
Iwatsu Test Instrument Corporation
Tokyo, Japan

Abstract— The iron losses of three-phase inductor used in a three-phase pulse-width modulation (PWM) inverter are evaluated. First we clarify three-phase inductor design in order to make same inductance value between the each phase. Then based on the design, iron losses for various structure of inductor are simulated. Simulation result shows the iron loss can be reduced about 5% by changing structure from conventional structure. Finally, we create conventional type and newly designed inductor and measure iron loss on a real PWM inverter. In the experiment, iron loss can be reduced about 10% compared with conventional type. Furthermore the calculated agree with measured losses within 10%. The iron loss reduction is verified by simulation and experiment.

Keywords— iron loss, inductor, PWM, three-phase inverter

I. INTRODUCTION

Increasing the power density and conversion efficiency of power converters has been a crucial issue for the growth of future markets. Remarkable advancements in power semiconductor devices, such as those based on SiC and GaN, have enabled the switching frequency of power converters to be increased and the volume of magnetic components, such as AC/DC inductors and transformers, to be reduced. However, the reduced surface area of the magnetic components results in increased thermal resistance and increased temperature. Since the maximum operating temperature of the capacitors is relatively low in general, a tightly packed layout with high-temperature components should be avoided, and as a result, this situation negatively affects the thermal design. Furthermore, since the conduction loss and switching loss of SiC and GaN device are small rather than conventional silicon device, passive components losses is greater than the device losses by the circuit operating conditions. In [1], magnetic loss is dominated half of conversion loss and magnetic material occupy large amount of space. Hence, it is necessary to evaluate the iron loss in order to design converters with high power density.

However, it is difficult to accurately calculate and experimentally verify the core losses in inductive components used in semiconductor power converter because the conventional calculation method is directed to a sine wave and pulse wave excitation [2-10]. In the case of three-phase PWM inverter excitation, the filter inductor excitation voltage is mainly rectangular wave component and the excitation current is contained a bias component. Therefore, the present authors have previously proposed an iron loss calculation method, referred to as the loss map method [10-16]. That method not

only calculates the high-frequency iron loss of an AC filter inductor during one cycle of the low-frequency output current of the PWM inverter, but it also calculates the instantaneous iron loss during each switching period, and its effectiveness can be confirmed. The authors developed an iron loss analyzer (ILA) that measures the instantaneous iron loss during each switching period of the AC filter inductor used in a PWM inverter [11-17]. Though, the authors have yet to evaluate coupled inductor core such as, three-phase inductor [14,15]. Onda et al. have already evaluated and designed symmetrical three-phase inductor [18]. However, above mentioned inductor structure is not commonly used and not suitable for mass production. Therefore, iron loss evaluation of commonly used three-phase inductor having EE core structure with tape winding structure is required.

In this paper, the iron losses of gap-less three-phase inductor on three-phase PWM inverter for various inductor design are evaluated. Firstly, three-phase inductor design method which has same inductance value in each phase is derived. Usually, the three-phase inductors have been designed so that the cross sectional area of center arm is set with the same as the cross sectional area of side arm [19]. But, the design method shows that the core structure can be changed from the conventional structure. Then, iron losses on a various inductor design are calculated. It is clear that the high-frequency iron loss of center arm and that of side arms of three-phase inductor are in the relationship of trade-off. The high-frequency iron loss can be reduced if proper core structure is designed. The calculation result shows that the high-frequency iron loss is reduced about 5% by adjusting the core structure from the conventional structure. Finally, the iron losses of each the conventional and proposed inductors are measured by ILA system. In the experiment, iron loss is reduced about 10% compared with the conventional type. Furthermore, the calculated results coincide well with the measured result within 10%. The iron loss reduction is verified by simulation and experiment.

II. THREE-PHASE INDUCTOR DESIGN

A gap-less three-phase inductor design used for three-phase PWM inverter excitation is introduced. In the general design process of the LC filter, the inductance of %Z (%L) of each phase is set to the same value. And hence, the current ripple flowing through the inductor is also set to the same. In order to satisfy the condition, terminal inductance in each phase should

978-1-4673-9551-9/16 $31.00 © 2016 IEEE

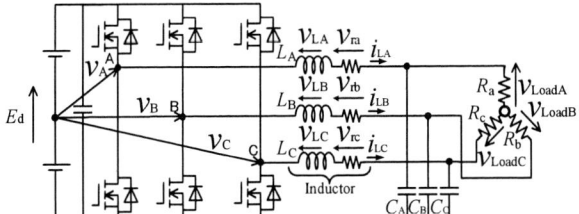

Fig. 1. Three-phase inverter circuit

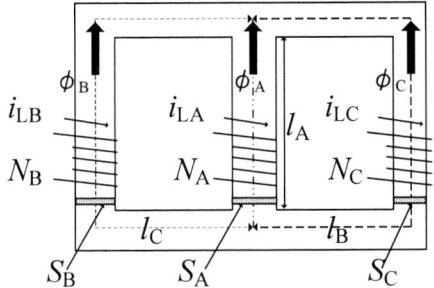

Fig. 2. Three-phase inductor.

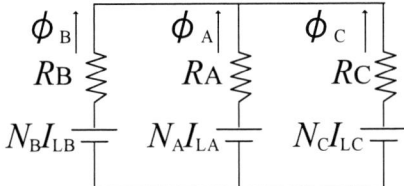

Fig. 3. Equivalent circuit of three-phase inductor.

be the same. However, terminal inductance value is decided not only by the self-inductance, but also by the mutual inductance. By taking account the above mentioned condition, design procedure of the gap-less three-phase inductor is derived.

Firstly, the condition which provides the same ripple current in each phase is derived. Voltage and current of the three-phase inductor on a balanced three-phase load are given by (1) to (3) and (4), respectively.

$$v_{\mathrm{LA}}(t) = \frac{1}{3}\left(2v_{\mathrm{A}}(t) - v_{\mathrm{B}}(t) - v_{\mathrm{C}}(t)\right)$$
$$-v_{\mathrm{ra}}(t) - v_{\mathrm{LoadA}}(t) \quad (1)$$

$$v_{\mathrm{LB}}(t) = \frac{1}{3}\left(2v_{\mathrm{B}}(t) - v_{\mathrm{A}}(t) - v_{\mathrm{C}}(t)\right)$$
$$-v_{\mathrm{rb}}(t) - v_{\mathrm{LoadB}}(t) \quad (2)$$

$$v_{\mathrm{LC}}(t) = \frac{1}{3}\left(2v_{\mathrm{C}}(t) - v_{\mathrm{A}}(t) - v_{\mathrm{B}}(t)\right)$$
$$-v_{\mathrm{rc}}(t) - v_{\mathrm{LoadC}}(t) \quad (3)$$

$$i_{\mathrm{Ln}}(t) = \frac{1}{L}\int_0^T v_{\mathrm{Ln}}(t)dt \ . \ (\mathrm{n = A, B, C}) \quad (4)$$

In order that the ripple current amplitude in each phase is balanced, the terminal inductance, L_{TotalA}, L_{TotalB}, L_{TotalC} should have the same value even though the core structure is changed.

Derivation procedure of the terminal inductance is expressed as follows.

From the equivalent circuit of a three-phase inductor shown in Fig. 2, the fluxes for each magnetic arm are given by (5) to (7) below.

$$\Phi_A(t) = \frac{N_A i_{\mathrm{LA}}(t)}{R_A + \frac{R_B R_C}{R_B + R_C}} - \frac{R_C}{R_A + R_C}\frac{N_B i_{\mathrm{LB}}(t)}{R_B + \frac{R_A R_C}{R_A + R_C}} - \frac{R_B}{R_B + R_A}\frac{N_C i_{\mathrm{LC}}(t)}{R_C + \frac{R_B R_A}{R_B + R_A}} \quad (5)$$

$$\Phi_B(t) = -\frac{R_C}{R_B + R_C}\frac{N_A i_{\mathrm{LA}}(t)}{R_A + \frac{R_B R_C}{R_B + R_C}} + \frac{N_B i_{\mathrm{LB}}(t)}{R_B + \frac{R_A R_C}{R_A + R_C}} - \frac{R_A}{R_B + R_A}\frac{N_C i_{\mathrm{LC}}(t)}{R_C + \frac{R_B R_A}{R_B + R_A}} \quad (6)$$

$$\Phi_C(t) = -\frac{R_B}{R_B + R_C}\frac{N_A i_{\mathrm{LA}}(t)}{R_A + \frac{R_B R_C}{R_B + R_C}} - \frac{R_A}{R_A + R_C}\frac{N_B i_{\mathrm{LB}}(t)}{R_B + \frac{R_A R_C}{R_A + R_C}} + \frac{N_C i_{\mathrm{LC}}(t)}{R_C + \frac{R_B R_A}{R_B + R_A}} \quad (7)$$

Where, $R_A = \frac{l_A}{\mu S_A}$, $R_B = \frac{l_B}{\mu S_B}$, $R_C = \frac{l_C}{\mu S_C}$ are magnetic resistance.

Equation (5) to (7) is expressed by (8) by using the number of flux linkage, $\varphi_A = N_A\Phi_A$, $\varphi_B = N_B\Phi_B$, $\varphi_C = N_C\Phi_C$.

$$\begin{bmatrix} \varphi_A(t) \\ \varphi_B(t) \\ \varphi_C(t) \end{bmatrix} = \begin{bmatrix} L_A & -M_{AB} & -M_{CA} \\ -M_{BA} & L_B & -M_{CB} \\ -M_{AC} & -M_{BC} & L_C \end{bmatrix} \begin{bmatrix} i_{\mathrm{LA}}(t) \\ i_{\mathrm{LB}}(t) \\ i_{\mathrm{LC}}(t) \end{bmatrix} \quad (8)$$

Where, $L_A = \frac{N_A^2}{R_A + \frac{R_B R_C}{R_B + R_C}}$, $L_B = \frac{N_B^2}{R_B + \frac{R_A R_C}{R_A + R_C}}$, $L_C = \frac{N_C^2}{R_C + \frac{R_B R_A}{R_B + R_A}}$ are self-inductance, $M_{AB} = \frac{R_C}{R_A + R_C}\frac{N_B N_A}{R_B + \frac{R_A R_C}{R_A + R_C}}$, $M_{CA} = \frac{R_B}{R_B + R_A}\frac{N_C N_A}{R_C + \frac{R_B R_A}{R_B + R_A}}$, $M_{BA} = \frac{R_C}{R_B + R_C}\frac{N_B N_A}{R_A + \frac{R_B R_C}{R_B + R_C}}$, $M_{CB} = \frac{R_A}{R_B + R_A}\frac{N_C N_B}{R_C + \frac{R_B R_A}{R_B + R_A}}$, $M_{AC} = \frac{R_B}{R_B + R_C}\frac{N_C N_A}{R_A + \frac{R_B R_C}{R_B + R_C}}$, $M_{BC} = \frac{R_A}{R_A + R_C}\frac{N_C N_B}{R_B + \frac{R_A R_C}{R_A + R_C}}$ are mutual inductance.

Furthermore, equation (8) is expressed by (9), since mutual inductances of the diagonal elements have the same value.

$$\begin{bmatrix} \varphi_A(t) \\ \varphi_B(t) \\ \varphi_C(t) \end{bmatrix} = \begin{bmatrix} L_A & -M_{AB} & -M_{CA} \\ -M_{AB} & L_B & -M_{BC} \\ -M_{CA} & -M_{BC} & L_C \end{bmatrix} \begin{bmatrix} i_{\mathrm{LA}}(t) \\ i_{\mathrm{LB}}(t) \\ i_{\mathrm{LC}}(t) \end{bmatrix} \quad (9)$$

Now, the inductor voltage in each phase, v_{LA}, v_{LB} and v_{LC}, are expressed by (10) to (12) by using the self-inductance, L_A, L_B, L_C, the mutual inductance, M_{AB}, M_{BC}, M_{CA}, and inductor current, $i_{\mathrm{LA}}(t)$, $i_{\mathrm{LB}}(t)$, $i_{\mathrm{LC}}(t)$.

$$v_{\mathrm{LA}}(t) = L_A\frac{di_{\mathrm{LA}}(t)}{dt} - M_{AB}\frac{di_{\mathrm{LB}}(t)}{dt} - M_{CA}\frac{di_{\mathrm{LC}}(t)}{dt} \quad (10)$$

$$v_{\mathrm{LB}}(t) = -M_{AB}\frac{di_{\mathrm{LA}}(t)}{dt} + L_B\frac{di_{\mathrm{LB}}(t)}{dt} - M_{BC}\frac{di_{\mathrm{LC}}(t)}{dt} \quad (11)$$

$$v_{\mathrm{LC}}(t) = -M_{CA}\frac{di_{\mathrm{LA}}(t)}{dt} - M_{BC}\frac{di_{\mathrm{LB}}(t)}{dt} + L_C\frac{di_{\mathrm{LC}}(t)}{dt} \quad (12)$$

On the three-phase three-wire system, relationship of each inductor voltage is expressed as, $v_{\mathrm{LA}}(t) + v_{\mathrm{LB}}(t) + v_{\mathrm{LC}}(t) = 0$, and hence the following equation is carried out.

$$v_{\mathrm{LA}}(t) + v_{\mathrm{LB}}(t) + v_{\mathrm{LC}}(t)$$
$$= L_A\frac{di_{\mathrm{LA}}(t)}{dt} - M_{AB}\frac{di_{\mathrm{LB}}(t)}{dt} - M_{CA}\frac{di_{\mathrm{LC}}(t)}{dt}$$
$$+ \left(-M_{AB}\frac{di_{\mathrm{LA}}(t)}{dt} + L_B\frac{di_{\mathrm{LB}}(t)}{dt} - M_{BC}\frac{di_{\mathrm{LC}}(t)}{dt}\right)$$
$$+ \left(-M_{CA}\frac{di_{\mathrm{LA}}(t)}{dt} - M_{BC}\frac{di_{\mathrm{LB}}(t)}{dt} + L_C\frac{di_{\mathrm{LC}}(t)}{dt}\right) = 0. \quad (13)$$

Also, inductor current in each phase is expressed as, $i_{\mathrm{LA}}(t) + i_{\mathrm{LB}}(t) + i_{\mathrm{LC}}(t) = 0$, and hence the following equation is derived:

$$v_{\mathrm{LA}}(t) + v_{\mathrm{LB}}(t) + v_{\mathrm{LC}}(t) = (L_A + M_{BC})\frac{di_{\mathrm{LA}}(t)}{dt}$$

$$+(L_B + M_{CA})\frac{di_{LB}(t)}{dt} + (L_C + M_{AB})\frac{di_{LC}(t)}{dt} = 0. \quad (14)$$

On the balanced three-phase load, each phase of the inductor voltage is expressed by (15) to (17).

$$v_{LA}(t) = (L_A + M_{BC})\frac{di_{LA}(t)}{dt} \quad (15)$$

$$v_{LB}(t) = (L_B + M_{CA})\frac{di_{LB}(t)}{dt} \quad (16)$$

$$v_{LC}(t) = (L_C + M_{AB})\frac{di_{LC}(t)}{dt} \quad (17)$$

In this case, the terminal inductance, $L_{TotalA}, L_{TotalB}, L_{TotalC}$ for A-phase, B-phase, and C-phase are expressed by (18) to (20), respectively.

$$L_{TotalA} = (L_A + M_{BC}) \quad (18)$$
$$L_{TotalB} = (L_B + M_{CA}) \quad (19)$$
$$L_{TotalC} = (L_C + M_{AB}) \quad (20)$$

In order to realize the balanced three-phase current, each phase of the terminal inductance should be the same (e.g. $L_{TotalA} = L_{TotalB} = L_{TotalC}$). In this case, the relationship among magnetic resistance, R, and turn number of winding, N, and terminal inductance, L, are expressed as follows:

$$L = \frac{(R + 2R_A)}{R(R + 2R_A)}N^2. \quad (21)$$

Where, N is turn number of the windings of each phase, L is the terminal inductance of each phase, R_A, $R = R_B = R_C$, are the magnetic resistance of center arm and side arm, respectively.

Usually, the cross sectional area, S_A, have been set with the same as, S_B, and, S_C [19]. However, the design method shows that the core structure can be changed from the conventional structure. Furthermore, the iron loss varies with the inductor core structure because the flux density waveform is a function of the core structure as follows:

$$v_{LA}(t) = L_{TotalA}\frac{di_{LA}(t)}{dt} = N_A\frac{d\Phi_A(t)}{dt} = N_A S_A\frac{dB_A(t)}{dt}. \quad (22)$$

In the case of three-phase PWM inverter, the filter inductor excitation voltage is mainly rectangular wave component and the excitation current is contained a bias component. Therefore, the iron loss calculation is performed by the loss map method.

III. THE LOSS MAP METHOD

In this section, the loss map method is introduced. In order to calculate the high-frequency iron loss, P_{HF}, of the AC filter inductor which is caused by switching action of the PWM inverter, the fundamental data of the iron loss (e.g. the loss map) must be provided beforehand.

A. The loss map

Fig. 4(a) shows the circuit configuration of the iron loss measuring system, and Fig. 4(b) shows the operating waveforms of the buck-chopper circuit shown in Fig. 4(a). The flux density ripple, ΔB, and the premagnetization, H_0, are calculated using the following equations:

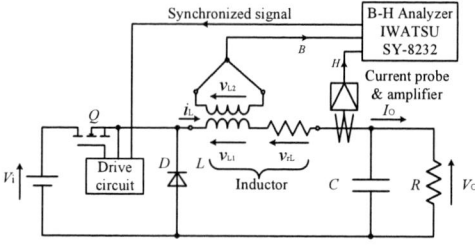

(a) Iron loss measuring circuit

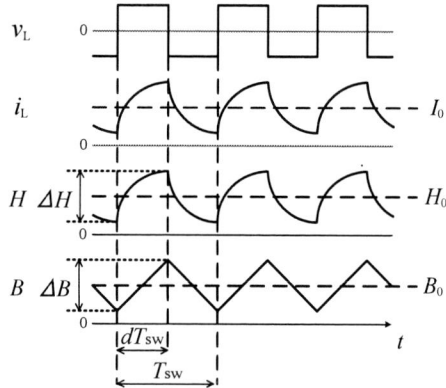

(b) Operation waveforms of the buck chopper
Fig. 4. The iron loss measurement system.

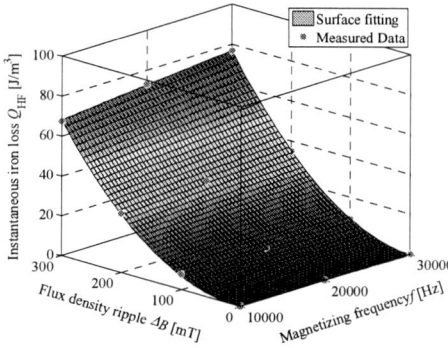

Fig. 5. The loss map (instantaneous iron loss v.s. flux density ripple and magnetizing frequency) at premagnetization $H_0 = 1000$ A/m.

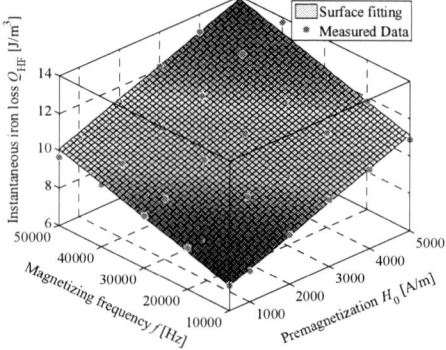

Fig. 6. The loss map (instantaneous iron loss v.s. magnetizing frequency and premagnetization) at flux density ripple $\Delta B = 100$mT.

$$\Delta B = \frac{1}{N_1 S_e} \int_0^{dT_{sw}} v_{L1}\, dt = \frac{1}{N_2 S_e} \int_0^{dT_{sw}} v_{L2}\, dt \, , \quad (23)$$

$$H_0 = \frac{N_1}{l_e\, T_{SW}} \int_0^{T_{sw}} i_L\, dt \, , \quad (24)$$

where N_1 and N_2 are, respectively, the primary and secondly windings of the inductor, l_e is the effective path length of the magnetic core, S_e is the effective cross-sectional area of the magnetic core, and T_{sw} is the switching period.

Fig. 5 and Fig. 6 show the iron loss of the iron powder core (core type SK-14M, Toho Zinc), measured by the circuit shown in Fig. 4(a). The instantaneous iron loss, Q_{HF}, can be expressed as $Q_{HF}(\Delta B, H_0, f)$, and it can be obtained from the loss map tables.

B. Iron loss calculation procedure

Next, the iron loss calculation procedure for the ac filter inductor used in a three-phase PWM inverter is presented. Figure 7 shows the flux density waveform example appeared on the three-phase PWM inverter. The flux density waveform of three-phase PWM inverter is not a simple triangular waveform because of other phase excitation effect [14, 15]. Even in such a case, the iron loss of each part of the flux density waveform can be calculated by referring the loss map, $Q_{HF}(\Delta B, H_0, f)$, when the sign of dB/dt is continuously positive or negative. The instantaneous iron loss, Q_{HF}, which is expressed by Joule, for sections from a to g shown on Fig. 7 is calculated as follows:

$$
\begin{aligned}
Q_{HF(a-g)} &= Q_{(ab)} + Q_{(bc)} + Q_{(cf)} + Q_{(fg)} \\
&= \frac{1}{2} Q_{HF}\big(\Delta B_{(ab)}, H_{0(ab)}, f_{eq(ab)}\big) \\
&\quad + \frac{1}{2} Q_{HF}\big(\Delta B_{(bc)}, H_{0(bc)}, f_{eq(bc)}\big) \\
&\quad + \frac{1}{2} Q_{HF}\big(\Delta B_{(cf)}, H_{0(cf)}, f_{eq(cf)}\big) \\
&\quad + \frac{1}{2} Q_{HF}\big(\Delta B_{(fg)}, H_{0(fg)}, f_{eq(fg)}\big) \quad (25)
\end{aligned}
$$

Where, ΔB is the peak to peak value when dB/dt does not change sign, H_0 is the average value of the magnetic field strength, $H(t)$, in the segment, $f_{eq} = \sum_{k=2}^n \frac{(B_k - B_{k-1})^2}{(B_{max} - B_{min})^2} \cdot \frac{1}{2 \cdot (t_k - t_{k-1})}$ is equivalent frequency in the segment, Q is segmaental iron loss.

The high-frequency iron loss, P_{HF}, which is expressed in Watt, and is the sum of the instantaneous iron losses, Q_{HF}, during one cycle of the low-frequency output current of PWM inverter, is calculated as

$$P_{HF} = \frac{V_e}{T_{LF}} \sum_{n=1}^n Q_{HF(n)} \cdot \quad (26)$$

Where, T_{LF} is one period of the low-frequency output current of the PWM inverter, V_e is volume of inductor.

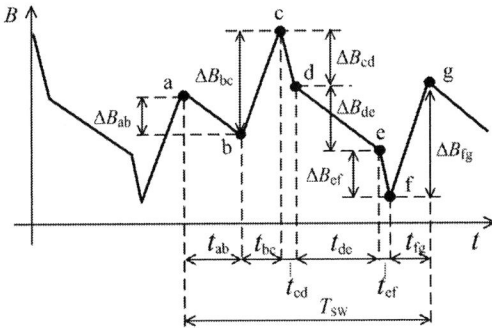

Fig. 7. Examples of flux density waveforms.

C. High-frequency Iron loss estimation

The high-frequency iron loss of gap-less three-phase inductor, shown on Fig. 8, on three-phase PWM inverter is estimated. The three-phase inverter is shown on Fig.1 and this circuit parameters are listed on table 1. An iron powder is used for inductor core materials, and the loss map data of SK-14M core is used for the high-frequency iron loss calculation. Excitation waveforms of inductor are calculated by circuit simulator such as PSIM with JMAG, etc.

In this study, cross sectional area, S_A, and magnetic path length, l_A, of inductor core, shown on Fig. 8, are varied by changing x, and y, respectively, but the inductance, L, cross sectional area, S_B, and, S_C, are kept to 1.65mH, 2cm², 2cm², respectively. And, cross sectional area, S_A, magnetic path length, l_A, magnetic path length, l_B are changed from 1 cm^2 to 2 cm^2, 6.3 cm^2 to 11.3 cm^2, 12.6 cm^2 to 17.6 cm^2, respectively. A turn number, N, for cross sectional area S_A and effective magnetic path length, l_A, is calculated from (11) and the calculated result is shown in Fig. 9. The turn number, N, is increased with the increase of the magnetic path length, l_A, and it is decreased with the increase of the cross sectional area, S_A. Calculated high-frequency iron loss for the one cycle of the output current is shown in Fig. 10. The high-frequency iron loss can be reduced by increase of the magnetic path length, l_A and the adjustment of the cross sectional area, S_A.

The high-frequency iron loss reduction effect due to the cross sectional area, S_A, adjustment is considered. The high-frequency iron loss calculation result for each arm at l_A=6.3cm is shown in Fig. 11. The high-frequency iron loss of arm A and that of arm B plus arm C are in the relationship of trade-off. The iron loss at $S_A = 2\ cm^2$ is 3.30W which is a conventional structure. On the other hand, the iron loss at $S_A = 1 cm^2$ is 3.14W which is the minimum iron loss value. The iron loss can be reduced by about 5% compared with the conventional structure. The reason of high-frequency iron loss changes can be expressed by the flux density waveforms since the iron loss is proportional to the ΔB^2 as shown in Fig. 5. Figs. 12, and 13 show the flux density waveforms when the cross sectional area is $S_A = 2\ cm^2$, $S_A = 1\ cm^2$, respectively. Fig. 12 shows the flux density waveforms in which fundamental and ripple component is contained in the case when cross sectional area in each magnetic arm is $S_A = S_B = S_C = 2\ cm^2$, which is a conventional structure. Fig. 13 shows the flux density waveforms in the case when the cross sectional area of arm A is $S_A = 1\ cm^2$ but that of arm B and C is kept to $S_B = S_C = 2\ cm^2$. It is clear that the reduction of, S_A, increase fundamental and ripple amplitude of the flux density of the arm A and reduce that of the other arm, B and C. As a result, the high-frequency iron loss of arm A is increased and that of arm B and C is reduced. Furthermore, as expressed in (26), the inductor core volume is also one of the iron loss factors. In the cross sectional area S_A adjustment from $2\ cm^2$ to $1\ cm^2$, the high-frequency iron loss decrease ratio of arm B plus C is larger than the high-frequency iron loss increase ratio of arm A because the core volume of the arm, B and C is larger than that of arm A, and the flux density ripple, ΔB, difference between arm A and arm B (or C) is not so large compared with excessively small cross sectional area, S_A, design.

Hence, the total high-frequency iron loss which is the sum of iron loss of arm A, arm B, and arm C can be reduced by optimizing the trade-off relation. In case when magnetic path length, l_A, is l_A =6.3cm, the cross sectional area, S_A, of minimum iron loss design is $S_A = 1\ cm^2$.

In addition, Fig. 13 shows that flux density near the peak amplitude is likely saturated in the case when the cross sectional area of arm A is $S_A = 1\ cm^2$ because peak amplitude of flux density waveform is close to the saturation point of iron powder core about 1.0 to 1.2T [20]. Hence, a design of selecting excessively small cross sectional area should be avoided in order not to cause the magnetic saturation.

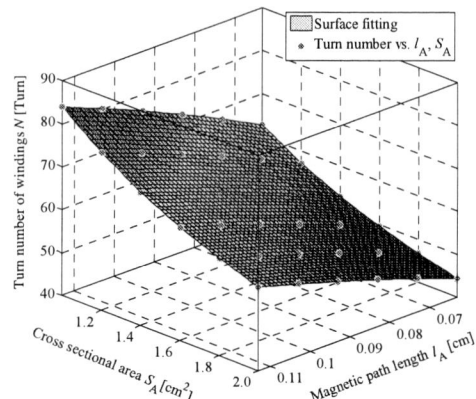

Fig. 9. Estimated turn number of windings

(turn number of windings v.s. cross sectional area and magnetic path length)

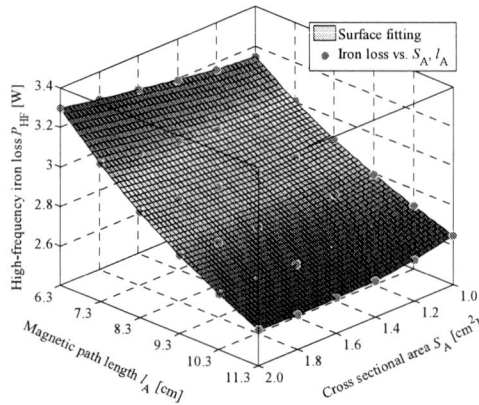

Fig. 10. High-frequency iron loss for various core design

(high-frequency iron loss v.s. cross sectional area and magnetic path length).

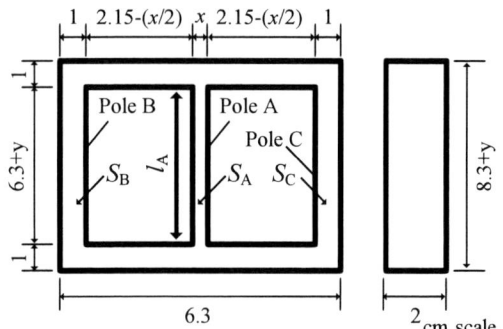

Fig. 8. Three-phase inductor.

Table 1. The circuit parameters of three-phase PWM inverter.

Design parameters	Value
Input voltage: E_d	110 V
Output voltage: V_{Load}	55 V
Output filter inductance: L	1 mH
Output filter capacitance: C	15 μF
Output load resistance: R	25 Ω
Output frequency: f_o	50 Hz
Switching frequency: f_{sw}	10 kHz
Modulation ratio: m_α	0.5

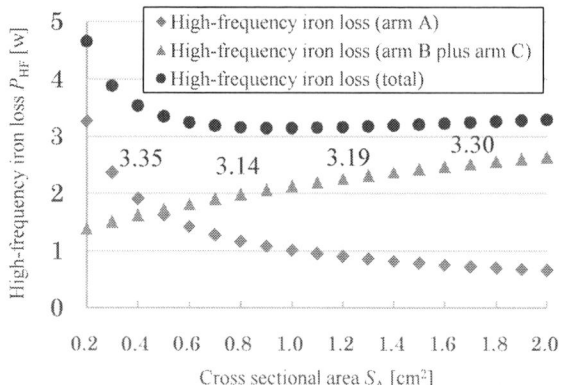

Fig. 11. High-frequency iron loss at l_A=6.3cm.

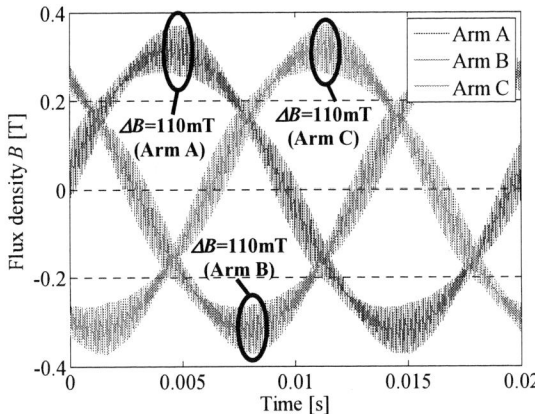

Fig. 12. Flux density waveform at $S_A = 2\ cm^2$

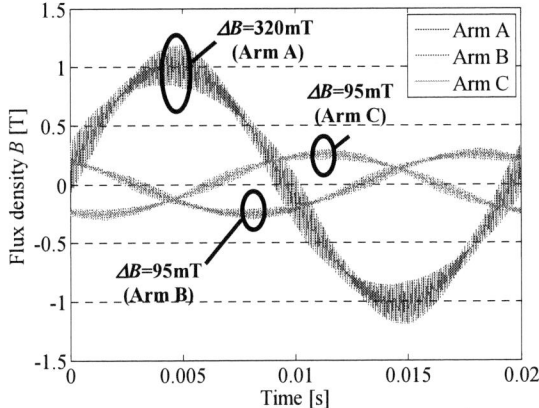

Fig. 13. Flux density waveform at $S_A = 1\ cm^2$

IV. EXPERIMENTAL VERIFICATION

To verify the high-frequency iron loss reduction effect of adjustment method of the cross sectional area, S_A, of arm A, the iron losses of the three-phase inductors made by both the conventional design method and the proposed design method are measured by ILA system [15, 17]. Fig. 14 shows the proposed and conventional inductors. Conventional inductor is designed so that the cross sectional area of arm A is same with that of arm B and C (i.e. $S_A = S_B = S_C = 2\ cm^2$). On the inductor designed by proposed method, the cross sectional area of arm A is $S_A = 1.4 cm^2$, and that of arm B and C is selected to $S_B = S_C = 2\ cm^2$. Magnetic path length of both inductor are set to l_A=6.3cm. In the case of proposed inductor, design of the cross sectional area is selected to $S_A = 1.4 cm^2$. Although the Proposed inditor is not using a minimum iron loss design of the cross sectional area, S_A, obtained by the simulation, the high-frequency iron loss can be reduced without the magnetic saturation risk, expressed in clause III C.

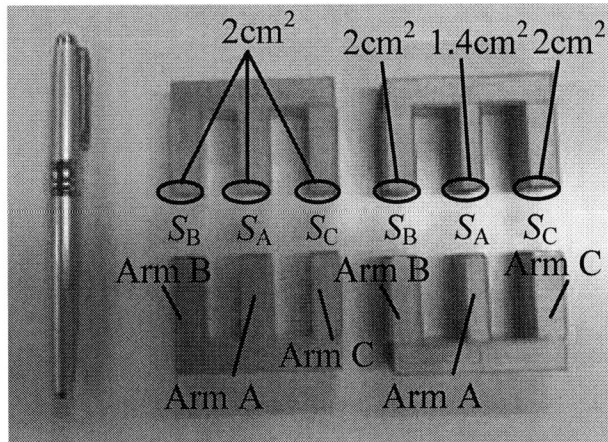

Fig. 14. The conventional and cross sectional area S_A adjustment inductors at $S_A = 2\ cm^2$ (conventional method) and $1.4\ cm^2$ (proposed method) .

A. Iron loss measurment system

Fig. 15 shows the system configurations of the ILA. The ILA system is composed of a PWM signal generator, an inverter with an LC filter, a high-speed sampling digital recorder, and a computer with loss-calculation software. Gate signals that are synchronized with the measurement signal are supplied from the PWM signal generator to the switching devices of the PWM inverter, and the inductor's current and voltage are detected synchronously with the measurement signal. The voltage, v_{L2}, on the secondary winding and the current, i_L, on the primary winding of the AC filter inductors are measured by the voltage and current sensors. In order to obtain an accurate measurement, the phase-shift error between the measured voltage and current should be corrected very precisely.

In the ILA system, the waveforms of both the secondary winding voltage and the primary winding current of the inductor are captured during one cycle of the low-frequency output current in the PWM inverter and are stored into the high-speed sampling digital recorder. The phase-corrected inductor primary current, $i_L^*(t)$, is separated into a low-frequency component, $i_{L(LF)}^*(t)$, and a high-frequency component, $i_{L(HF)}^*(t)$, as shown in the following equations:

$$i_L^*(t) = i_{L(LF)}^*(t) + i_{L(HF)}^*(t)\ , \qquad (27)$$

$$i_{L(LF)}^*(t) = \sum_{n=1}^{k} \{a_n\cos(n\omega t + \Delta\varphi(nf)) \\ + b_n\sin(n\omega t + \Delta\varphi(nf)\}, \qquad (28)$$

$$i_{L(HF)}^*(t) = \sum_{n=k}^{\infty} \{a_n\cos(n\omega t + \Delta\varphi(nf)) \\ + b_n\sin(n\omega t + \Delta\varphi(nf)\}, \qquad (29)$$

978-1-4673-9551-9/16 $31.00 © 2016 IEEE

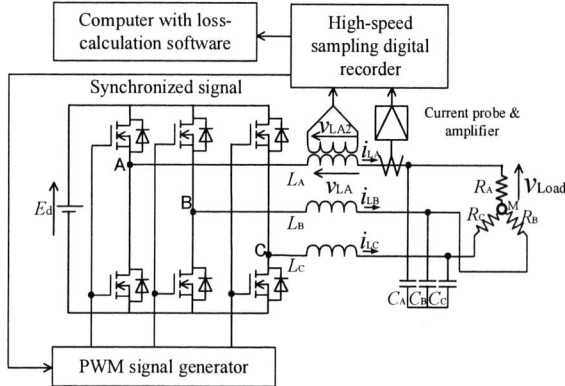

Fig. 15. The ILA system configuration.

Where k (= 3-5, in the present study) is the boundary condition of the low-frequency and high-frequency components.

As in the same manner, the secondary voltage, $v_{L2}(t)$, is separated into a low-frequency component, $v_{L2(LF)}(t)$, and a high-frequency component, $v_{L2(HF)}(t)$, as follows:

$$v_{L2}(t) = v_{L2(LF)}(t) + v_{L2(HF)}(t) \ . \tag{30}$$

During one cycle of the low-frequency output current in the PWM inverter, the high-frequency iron loss, P_{HF}, which is caused by switching action of the PWM inverter, and low-frequency iron loss, P_{LF}, which is caused by fundamental excitation waveform, are calculated as follows:

$$P_{HF} = \frac{N_1/N_2}{T_{LF}} \int_0^{T_{LF}} i^*_{L(HF)}(t) v_{L2(HF)}(t) dt \tag{31}$$

and

$$P_{LF} = \frac{N_1/N_2}{T_{LF}} \int_0^{T_{LF}} i^*_{L(LF)}(t) v_{L2(LF)}(t) dt \ , \tag{32}$$

Where, N_1 and N_2 are turn numbers of the primary and secondary windings of the inductor, respectively. T_{LF} is one period of the low-frequency output current of the PWM inverter, and the unit of P_{HF} and P_{LF} is expressed by Watt.

To the best of our knowledge, there is currently no measurement equipment that can be used to verify the accuracy of the ILA. The overall loss, P_{ALL}, is measured by a high-precision power meter that works over a wide range of frequencies (PPA5530; Newtons 4th), and it is compared with the value measured by the ILA. The maximum deviation is less than 3%. Based on this result, the accuracy of the ILA is expected to be sufficiently high.

B. Iron loss measurement result

The iron loss of the three-phase inductors made by the conventional method and the proposed method are measured. The circuit parameters of three-phase PWM inverter are same as simulation condition which is listed in table 1. Figs. 16 and 17 show the measured results of the magnetic field strength, H, and flux density, B, of two inductors. Amplitude of the magnetic field strength in Arm A is larger than that in arm B, since the magnetic path length, l_A, of arm A is shorter than the

magnetic path length l_B of arm B. In simulation, fundamental and ripple amplitude of flux density of arm A in conventional structure is the same as that of arm B. However, the measured amplitude of that is not the same as that of arm B because the inductor voltages expressed in (7) to (9) are influenced by leakage inductance in each phase. In case of proposed method, even if the flux density waveforms are influenced by leakage inductance in each phase, the fundamental and ripple amplitude of flux density waveforms of arm A is larger than that on the conventional structure and that of arm B is smaller than that on the conventional structure like simulation result. Therefore, the high-frequency iron loss of three-phase inductor can be reduced as in the same manner of simulation.

The measured and calculated high-frequency iron loss values of both inductor are listed in Table 2. In the proposed inductor, the high-frequency iron loss is about 10% smaller than that of the conventional structure. Proposed method improves the trade-off relation of the high-frequency iron loss between arm A and arm B plus C. The high-frequency iron loss reduction effect of the proposed method is confirmed in the experiment. And, the calculated high-frequency iron losses coincide well with measured high-frequency iron loss within 10% deviation.

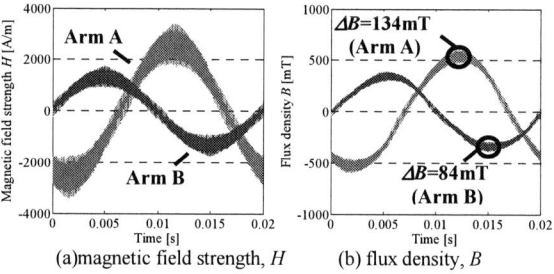

(a) magnetic field strength, H (b) flux density, B

Fig. 16. Magnetizing waveforms at $S_A = 1.4 \ cm^2$ (proposed method).

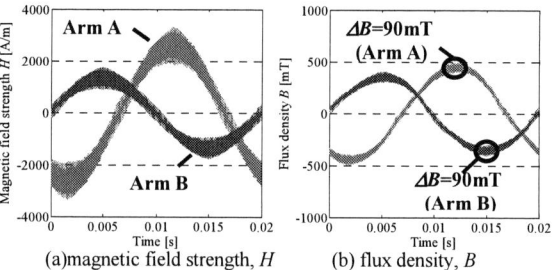

(a) magnetic field strength, H (b) flux density, B

Fig. 17. Magnetizing waveforms at $S_A = 2 \ cm^2$ (conventional method)

Table 2. High-frequency iron loss

Inductor design	Measured iron loss [W]	Calculated iron loss [W]	Deviation [%]
Proposed method ($S_A = 1.4 \ cm^2$)	1.06(arm A) 2.11(arm B+C) 3.17(total)	0.82(arm A) 2.37(arm B+C) 3.19(total)	0.63% (total)
Conventional method ($S_A = 2.0 \ cm^2$)	0.95(arm A) 2.59(arm B+C) 3.54(total)	0.66(arm A) 2.64(arm B+C) 3.30(total)	6.78% (total)

V. CONCLUSION

The iron losses of three-phase inductor used in the three-phase PWM inverter are evaluated. Firstly, we clarify three-phase inductor design method. Usually, the three-phase inductors have been designed so that the cross sectional area of center arm is set with the same as the cross sectional area of side arm. But, the design method shows that the core structure can be changed from the conventional structure. Then based on the design method, iron losses for various inductor structure are calculated by using loss map method. We clarify that the high-frequency iron loss of center arm and that of side arms are in the relationship of trade-off. The high-frequency iron loss can be reduced if proper core structure is designed. The calculation result shows that the high-frequency iron loss can be reduced about 3-5% by optimizing the core structure. Finally, we make proposed and conventional inductor and measure iron losses of both inductor on a real PWM inverter circuit. In the experiment, iron loss can be reduced about 10% compared with conventional type. The calculated high-frequency iron losses coincide well with measured high-frequency iron loss within 10% deviation. The high-frequency iron loss reduction is verified by simulation and experiment.

ACKNOWLEDGEMENT

This work is supported in part by JSPS KAKENHI Grant (Japan) Number 25289077.

REFERENCES

[1] David Reusch, John Glaser, "Improving DC-DC Converter Performance with GaN Transistors", IEEE ECCE 2015.

[2] J. Muhlethaler, J. Biela, J.W. Kolar, and A. Ecklebe, "Core Losses under DC Bias Condition based on Steinmetz Parameters," IEEJ IPEC, pp. 2430-2437, 2010.

[3] J. Muhlethaler, J. W. Kolar and A. Ecklebe, "Loss modeling of inductive components employed in power electronic systems," Proc. 8th Int. Conf. Power Electronics. pp. 945–952, 2011.

[4] J. Muhlethaler, J. Biela, J. W. Kolar and A. Ecklebe, "Improved core loss calculation for magnetic components emplyed in power electronic systems," *Proc, Appl. Power Electronics. Conf. Expo.*, pp. 1729–1736, 2011.

[5] W. A. Roshen, "A practical, accurate and very general core loss model for nonsinusoidal waveforms," IEEE Trans. Power Electronics., vol.22, no.1, pp. 30–40, Jan. 2007.

[6] M. Mu, F. C. Lee, Q. Li, D. Gilham and K. Ngo, "A new high frequency core loss measurement method for arbitrary excitations," Proc, Appl. Power Electronics. Conf. Expo., pp.157-162, 2011.

[7] C. Simao, N. Sadowski, N. J. Batiatela and J. P. A. Bastos, "Analysis of magnetic hysteresis loops under sinusoidal and PWM voltage waveforms," Proc. of Power Electronics Specialists Conference (PESC), pp.1555-1559, 2005.

[8] M. Mu, F. C. Lee, and V. Tech, "A New Core Loss Model for Rectangular AC Voltages," IEEE ECCE, 2014, pp. 5214–5220, 2014.

[9] K. Venkatachalam, C. R. Sullivan, T. Abdallah, and H. Tacca, "Accurate prediction of ferrite core loss with nonsinusoidal waveforms using only Steinmetz parameters," Proc. IEEE Workshop Comput. Power Electronics., pp. 36–41, 2002.

[10] A. Krings and J. Soulard "Overview and Comparison of Iron Loss Models for Electrical Machines," EVER Monaco, March 25-28, 2010

[11] S. Iyasu, T. Shimizu, and K. ishii, "A novel iron loss calculation method on power converters based on dynamic minor loop," *Proc. of European Conference on Power Electronics and Applications*, pp.2016-2022, 2005.

[12] H.Matsumori, T.Shimizu, K.Takano, H.Ishii, "Iron loss calculation of AC filter inductor for three-phase PWM inverters", IEEE ECCE in North Carolina, pp.3271-3279, 2012.

[13] H. Satoh, T. Shimizu, "Iron loss calculation of AC filter inductor for three-phase PWM inverters", ICPE in Korea, 2015.

[14] H. Matsumori, T. Shimizu, K. Takano, H.I shii, "Iron loss calculation of AC filter inductor for three-phase PWM inverters", IEEE ECCE in North Carolina, pp.3271-3279, 2012.

[15] H. Matsumori, T. Shimizu, K. Takano, H.I shii, "Evaluation of Iron Loss of AC Filter Inductor used in Three-Phase PWM inverters Based on an Iron Loss Analyzer (ILA)" IEEE Trans. Power Electronics., Issue:99, DOI: 10.1109/TPEL.2015.2453055, 2015.

[16] K. Kakazu, T. Shimizu, H. Matsumori, T. Takano, and H. Ishii "Iron Loss Evaluation of Filter Inductor used in PWM Inverters," ECCE, pp. 606-613, 2011.

[17] K. Kim, K. Wada, T. Shimizu, T. Takano, and H. Ishii "Dynamic Iron Loss Measurement Method for AC Filter Inductor," EPE, 2007.

[18] N. Kurita, K. Onda, K. Nakanoue, and K. Inagaki, "Loss Estimation Method for Three-Phase AC Reactors of Two Types of Structures Using Amorphous Wound Cores in 400-kVA UPS," vol. 29, no. 7, pp. 3657–3668, 2014.

[19] Colonel Wm. T. McLyman, "Transformer and Inductor Design Handbook Fourth Edition", CRC Press.

[20] T.Maeda, H.Toyoda, N.Igarashi, K.Hirose, K.Mimura, T.Nishioka, "Development of Super Low Iron-loss P/M Soft Magnetic Material" SEI Technical Review, 60, (2005).

CMOS Gate Drive IC with Embedded Cross Talk Suppression Circuitry for SiC Devices

Jeffery Dix, Zheyu Zhang, and Benjamin J. Blalock
Electrical Engineering and Computer Science
University of Tennessee at Knoxville
Knoxville, USA
jdix1@vols.utk.edu

Abstract—**This paper presents a gate driver integrated circuit (IC) for Silicon Carbide (SiC) devices to fully utilize their high switching speed capabilities in a phase-leg configuration. Based upon the intrinsic properties prevalent in SiC devices, gate assist circuitry is integrated into a gate driver IC to control the gate voltages seen by both devices in a phase-leg during different switching transients. Compared to a traditional gate driver IC, the proposed circuit has the potential of suppressing the cross talk seen by both devices thus increasing the overall switching speed of the phase-leg. The replacement of the conventional gate driver with an IC effectively lowers the gate impedance loop by reducing the number of on-board traces and moving essential traces to inside the chip. Therefore, larger transient currents and higher slew rates can be achieved with an IC compared to nominal commercially available gate driver devices. Meanwhile, the added functionality of cross talk suppression, not normally available in other gate drive IC designs, minimizes the spurious gate voltages from cross talk to within the required operating ranges of SiC devices.**

Keywords—Gate Driver, IC, CMOS, SiC, Cross Talk

I. INTRODUCTION

Silicon Carbide (SiC) devices require unique driving characteristics at the gate terminal of the power device. Therefore, SiC devices necessitate different design methodologies to fully utilize their unique characteristics such as faster switching and suppression of the cross talk prevalent in phase-leg devices [1-4]. An intelligent gate driver has been proposed and experimentally tested at the board level [4]. Fig. 1 (a) and (b) show the intelligent gate driver approach and the four voltage levels created by the unique circuitry setup, respectively. The four different voltage levels contribute attributes to the gate driver that will specifically benefit SiC power devices in a phase-leg configuration. The voltage level prevalent between the time periods t_1 and t_2 in Fig. 1b on positive V_{gs_L} allows the lower SiC phase-leg power device to obtain a higher dv/dt allowing for a faster switching speed that SiC devices can achieve. The voltage level seen during time periods t_3 to t_5 in Fig. 1b on negative V_{gs_L} allows the lower device to mitigate spurious voltage induced by cross talk in a phase-leg from the upper device turning on. The intelligent

gate driver design also incorporates different gate resistances for turn-on and turn-off transients. This functionality is achieved by directly connecting the high-side device of the gate driver to the gate resistor optimized for the turn-on transient and likewise connecting the low-side gate driver output device to the turn-off gate resistance [4].

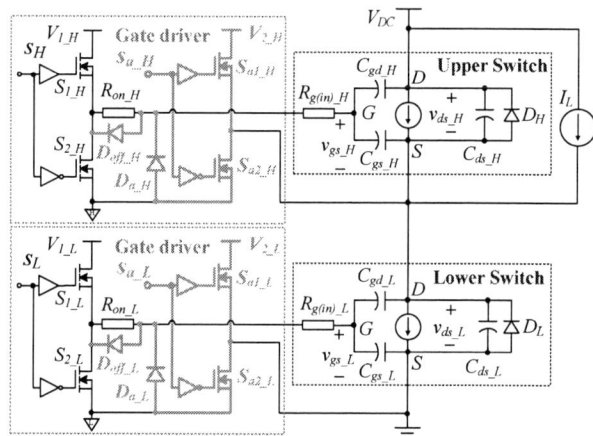

(a) Intelligent gate driver with gate assist circuit.

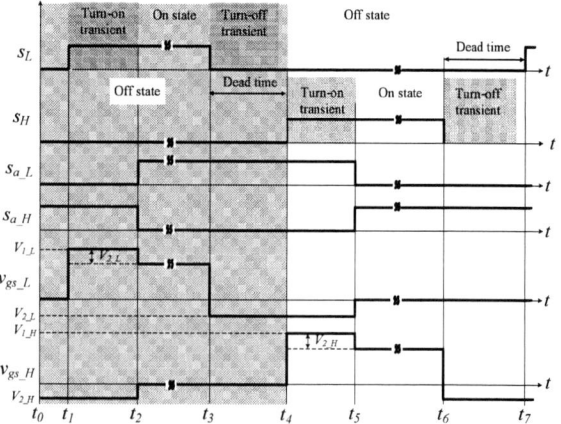

(b) Logic signals of main device and auxiliary transistors.

Fig. 1. Intelligent gate driver for fast switching and cross talk suppression.

978-1-4673-9551-9/16 $31.00 © 2016 IEEE

This intelligent gate driver configuration has been thoroughly tested at the board level against two different conventional gate drivers with the typical two level gate voltage seen in most gate driver applications. The conventional gate drivers have been tested at 20 V / 0 V for the 1st group and 20 V / −5 V for the 2nd group. The results detail significant improvements in total turn-off time reduction (52% for the 1st gate driver and 41% for the 2nd gate driver) and decreased total switching losses (38% for the 1st gate driver and 12% for the 2nd gate driver). The impact of the cross talk on the voltage of the gate node varies greatly across the two conventional gate driver configurations and the intelligent gate driver one. The positive spurious gate voltage of the lower switch induced by cross talk due to the turn-on transient of the upper switch is mitigated down to just 0.6 V, which is better than the 8.2 V of the 1st gate driver and 2.5 V of the 2nd gate driver, but is most importantly lower than the 2.2 V threshold voltage of the SiC power device. During the turn-off transient of the upper switch, the lower switch is subjected to a negative spurious gate voltage from cross talk of only −3.6 V, whereas the voltage is −8.3 V and −10.3 V for the 1st and 2nd gate drivers, respectively [4].

While the previous presented work has significantly benefited the use of SiC power devices, the integration of the intelligent gate driver can improve these numbers even more. The integration of circuitry improves designs by having less parasitics in the important circuit elements and current loops (i.e. gate loop for power devices) while decreasing the complexity in implementation by the end-user. Integration has the opportunity to lower costs as the number of different components is reduced as well as the total area used by phase-leg supporting circuitry. This paper presents the integration of the intelligent gate driver design as well as incorporating more functionality into the gate driver integrated circuit allowing for the driver to accomplish more tasks and increase its overall versatility.

II. GATE VOLTAGE

The Gate Driver integrated circuit (IC) has been developed in a commercially available bulk Silicon CMOS process that allows the utilization of high voltage devices. Figs. 2-4 detail the topology of the Gate Driver IC in three

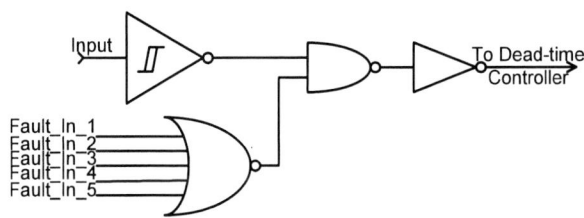

Fig. 2. Gate Driver Input Stage.

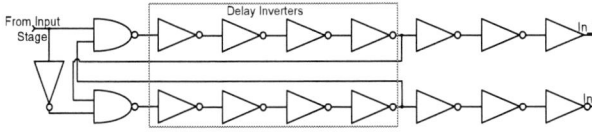

Fig. 3. Gate Driver Dead-time Controller [10].

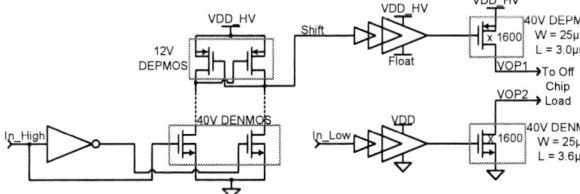

Fig. 4. Gate Driver Output Stage.

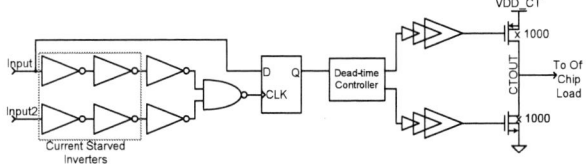

Fig. 5. Cross Talk Mitigation Circuitry Block Diagram.

different stages. The topology is based upon an older generation of gate driver developed previously [9]. Compared with our previous gate driver IC work, the current IC has changed several key components as well as adding in new components to increase the functionality of the chip while opening up a wider market that could potentially use the gate driver. The main addition to the IC is the inclusion of cross talk mitigation circuitry, which is represented in the block diagram in Fig. 5. The Gate Driver's three stages are the input stage, the dead-time controller, and the output stage. The input stage consists of an inverting Schmitt trigger that is combined with five fault inputs through the use of an AND gate. The five faults, when triggered "high", will shut down the main gate driver circuitry thus disabling the output voltage waveform. The dead-time controller utilizes a feedback structure from [10] and a string of delay inverters to create two non-overlapping complementary signals for the output stage. The output stage level shifts one of the complementary signals using high-voltage MOSFETs, diode-connected MOSFETs, and cross-coupled drain extended MOSFETs to drive both the level-shifted and low-side signal through the exponential-horn buffers. After the buffers, the output drivers activate to "push" and "pull" current to and from the off-chip load, respectively.

The same four voltage levels seen in the intelligent gate driver are achieved by the gate driver IC through the use of the incorporated cross talk mitigation circuitry. The turn-on transient voltage level is achieved by the high voltage PMOS output drivers and the NMOS cross talk mitigation drivers being turned on. The turn-off transient voltage level to achieve the −5 V level is fulfilled by the high voltage NMOS output drivers and the PMOS cross talk mitigation drivers being turned on. Fig. 6 shows the four level voltage gate drive signal driving the gate node of a CREE 2nd generation 1200-V/20-A SiC MOSFET while Fig. 7 depicts the conventional gate drive voltage signal using the typical two level method. These figures only utilize a maximum of 15 V due to a design constraint associated with this prototype IC.

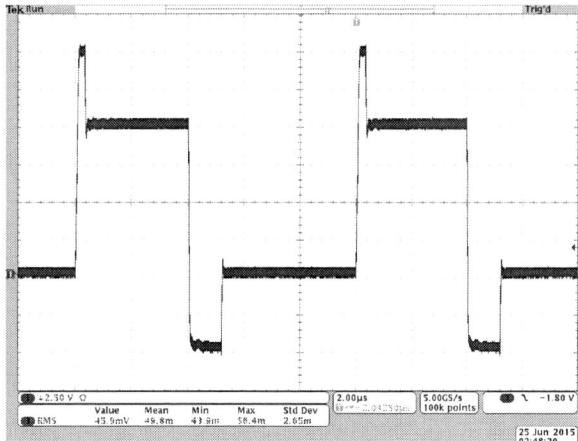

Fig. 6. Intelligent Gate Driver Output.

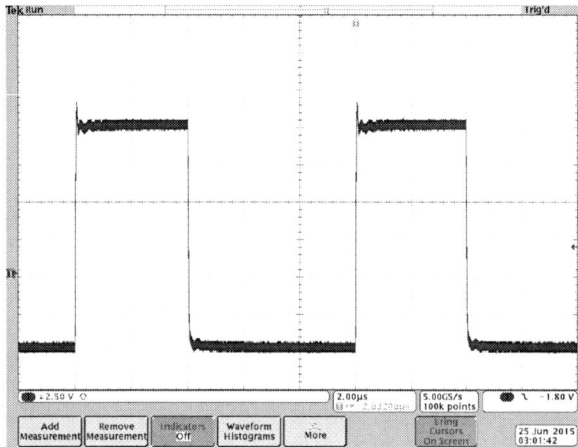

Fig. 7. Conventional Gate Driver Output.

III. TUNABLE DELAY

With the integration of the intelligent gate driver design, additional functionality can be incorporated as well. The first being the ability to dynamically adjust the duration of the maximum gate drive output voltage (15 V level in Fig. 6). Since switching time is highly dependent on the operating conditions and load characteristics [5, 6], tunable delay allows for the maximum gate drive output voltage to follow the change of switching time to minimize the gate stress of SiC devices. This effect is achieved through the incorporation of current starved inverters in the cross talk mitigation circuity in Fig. 5 and can be controlled by the end-user through the current input node. Current starved inverters use current mirrors to power the PMOS and NMOS sections of an inverter instead of directly tying the inverter to VDD and VSS. In this fashion, the current mirrors' bias voltage limits the slew rate of the inverters allowing for a delay to propagate. Fig. 9 shows the results of the tunable delay being adjusted dynamically as the current mirror bias is changed. For this test setup, the delay is managed using a potentiometer but can easily be controlled dynamically in the future through the use of a microcontroller or other circuitry.

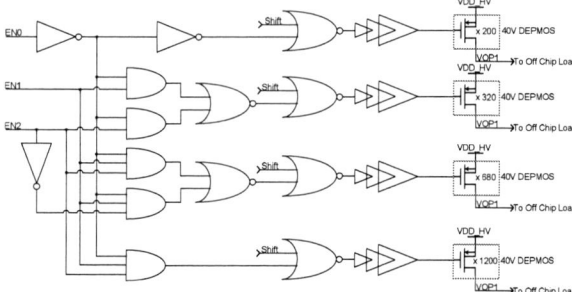

Fig. 8. Decoder to Variable On-Resistance Branches.

TABLE I.   NUMBER OF DEVICES TO ACHIEVE SPECIFIC OHMS.

Ohms	# of NMOS devices	# of PMOS devices
~0	3200	-
2.5	-	3200
4	-	2400
6	-	1200
8	-	520
10	-	200
100	25	-

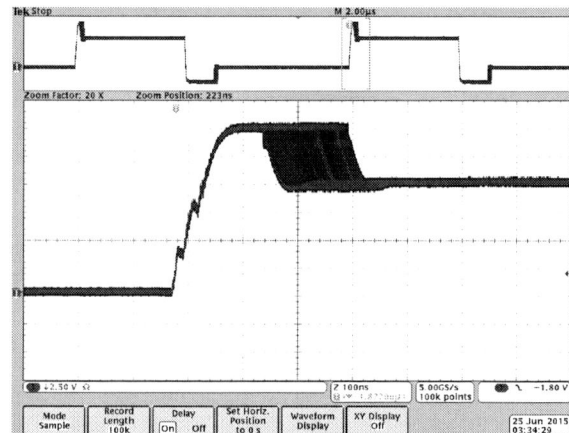

Fig. 9. Gate Driver Output showing Tunable Delay.

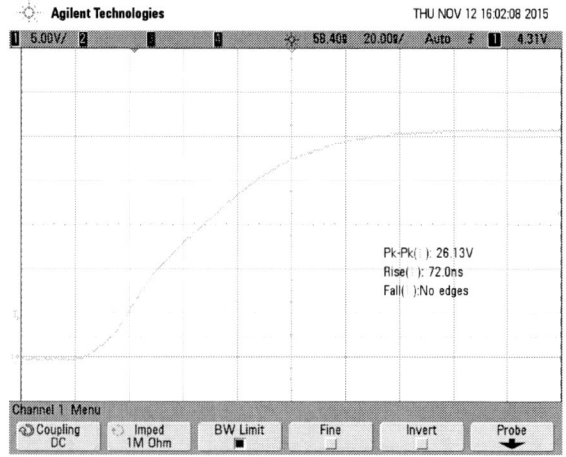

Fig. 10. Variable On-Resistance during Rising Gate Driver IC Edge.

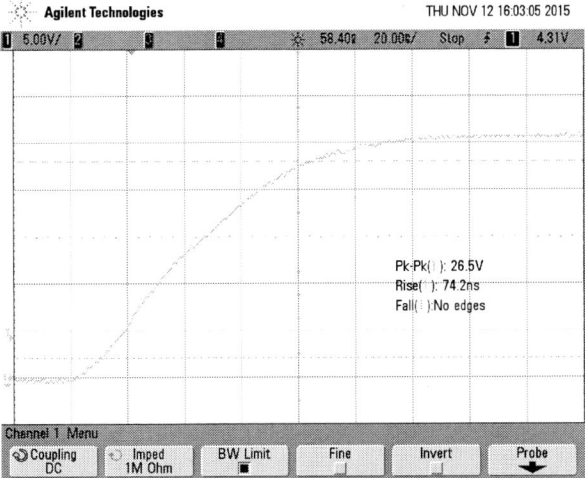

Fig. 11. Conventional Gate Driver Output, 0 V to 10 V.

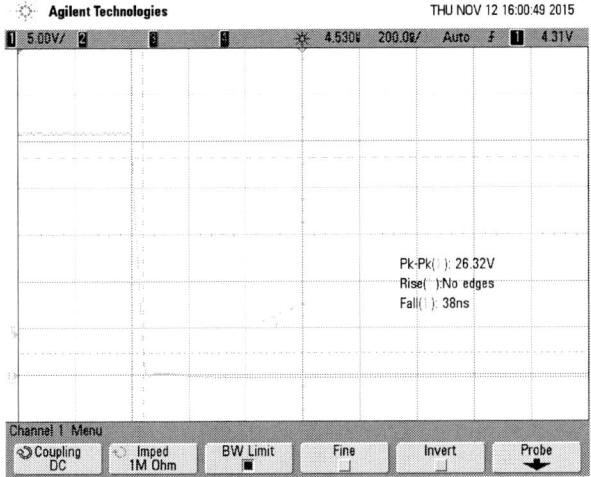

Fig. 12. Conventional Gate Driver Output, −5 V to 10 V.

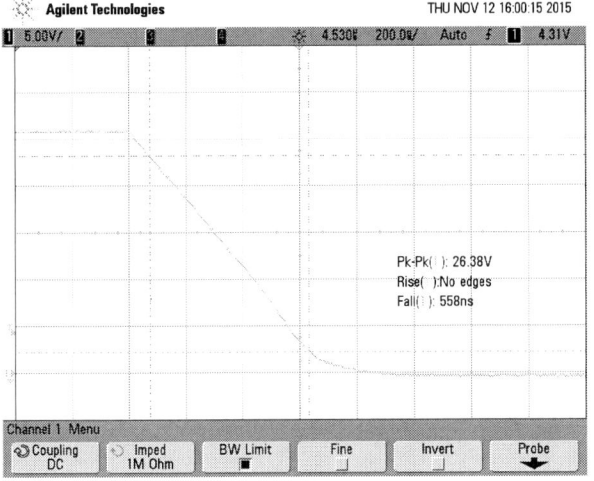

Fig. 13. Gate Driver IC Output with Cross Talk Mitigation.

IV. VARIABLE GATE RESISTANCE

With the goal of maximizing the switching speed of SiC devices without causing a severe over-voltage situation or parasitic ringing [7, 8], the need to design a gate driver with tunable gate resistance is required for a range of different operating conditions. The over-voltage and parasitic ringing are highly dependent upon the operating conditions of the device, which means that a fixed gate resistance cannot achieve the best tradeoff between switching speed and parasitic ringing among a wide operating current range. To incorporate this functionality into the gate driver IC, an on-chip decoder is required to drive "enable" signals to different buffers that pass-through the gate drive signals on to the different sets of output drivers. Each set of enabled output drivers adds devices in parallel to lower the on-resistance of both the high voltage PMOS and NMOS drivers. Fig. 8 details the schematic for the decoder to "enable" pins of the various on-resistance branches. The decoder is only needed for the PMOS branches. Table I details the different resistance values available per number of devices being turned on by the various "enable" signals. In addition, the gate driver IC has independent source and sink outputs that are connected off-chip either through additional gate resistance or jumpers. The separated outputs allow different gate resistances during turn-on and turn-off transients without the need for additional diodes as seen previously in Fig. 1a. Figs. 10 and 11 show the variable on-resistance for the PMOS output drivers for two of the five available states. Fig. 10 is the state representing approximately 4 ohms in Table I whereas Fig. 11 depicts the state for approximately 8 ohms. Figs. 12 and 13 detail the NMOS variable on-resistance as it switches from approximately 0 ohms (Fig. 12) to about 100 ohms (Fig. 13). The testing for Figs. 10-13 was performed on a second prototype chip allowing for a higher gate voltage.

V. DOUBLE PULSE TESTING RESULTS

The double pulse testing was performed by connecting two Gate Driver IC test boards to a power stage board with two CREE 2nd-generation 1200-V/20-A SiC MOSFETs. The setup utilized two isolated power supplies and two signal isolators in order to provide the power rails and signals to the input sections of the Gate Driver IC test boards. The input signals are a double pulse (0-5 V) and a single pulse (0-5 V) for channel 1 and channel 2, respectively. Figs. 14-16 show the V_{GS} drive signals across the gate and source terminals of the SiC devices under test. The three gate drive signal levels tested were 0 V to 10 V (Fig. 14), −5 V to 10 V (Fig. 15), and the four voltage level Gate Driver IC configuration (Fig. 16). These figures detail the double pulse test signals that were used to drive the CREE SiC devices. Each of the three gate drive levels drove the CREE devices that are at a voltage level of 300 V, which is well below the maximum of the devices. Figs. 17-19 depict the interaction of the gate drive signals with the SiC devices as they turn on and off in the same order as previously established. The cross talk seen in phase leg configurations can be seen at the turn-on and turn-off transitions of the lower device in the phase leg.

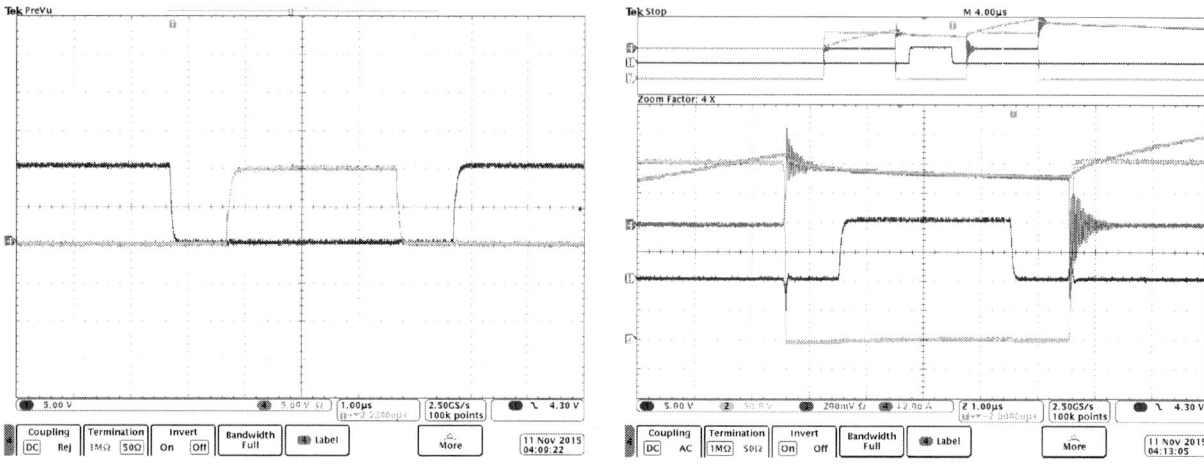

Fig. 14.   Conventional Gate Driver Output, 0 V to 10 V.

Fig. 17.   0 V to 10 V Gate Drive Signal Driving a Phase Leg.

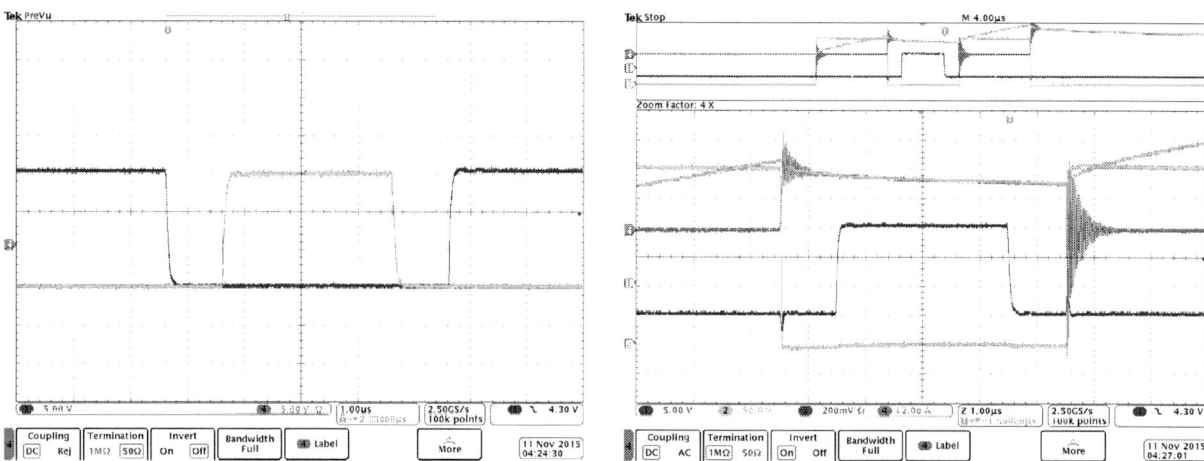

Fig. 15. Conventional Gate Driver Output, −5 V to 10 V.

Fig. 18.   −5 V to 10 V Gate Drive Signal Driving a Phase Leg.

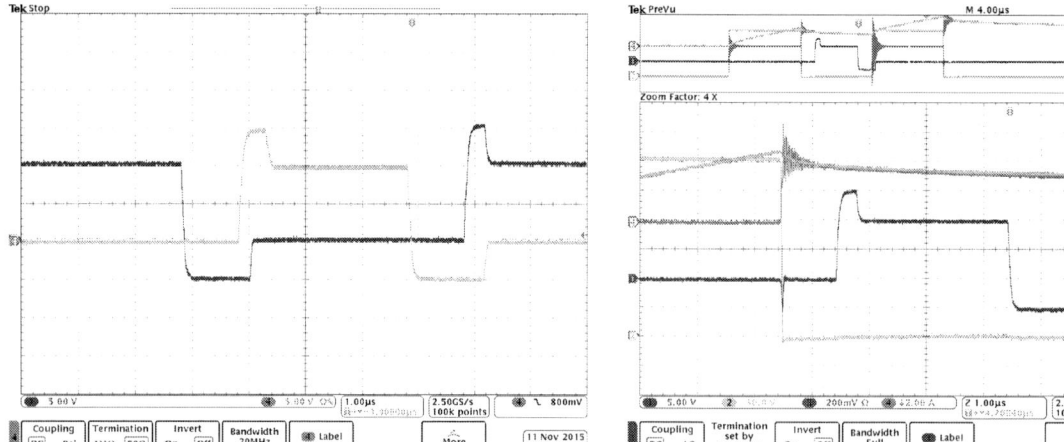

Fig. 16.   Gate Driver IC Output with Cross Talk Mitigation.

Fig. 19.   Gate Driver IC Driving a Phase Leg.

978-1-4673-9551-9/16 $31.00 © 2016 IEEE   688

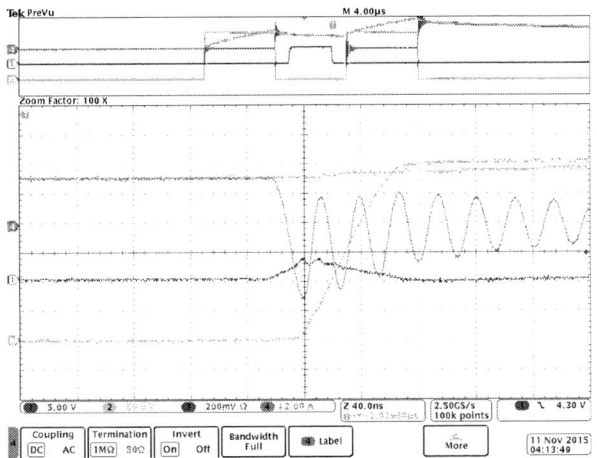

Fig. 20.   0 V to 10 V Gate Drive Signal during Turn-off Transient.

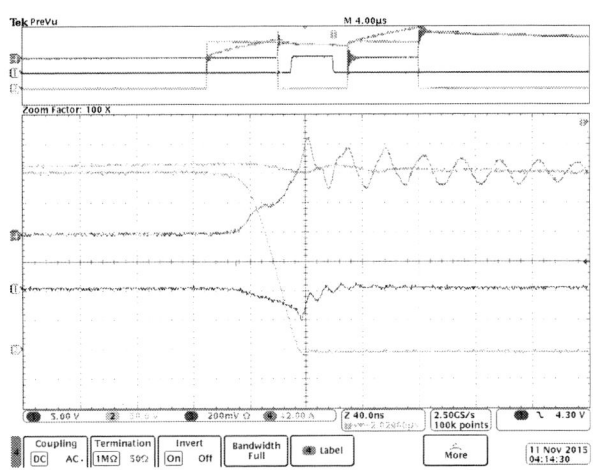

Fig. 23.   0 V to 10 V Gate Drive Signal during Turn-on Transient.

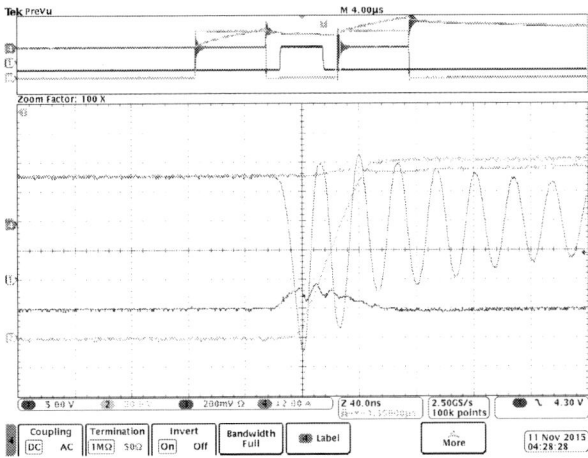

Fig. 21. −5 V to 10 V Gate Drive Signal during Turn-off Transient

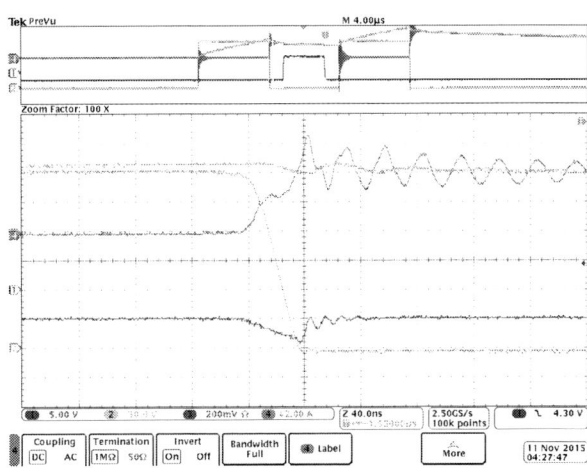

Fig. 24. −5 V to 10 V Gate Drive Signal during Turn-on Transient

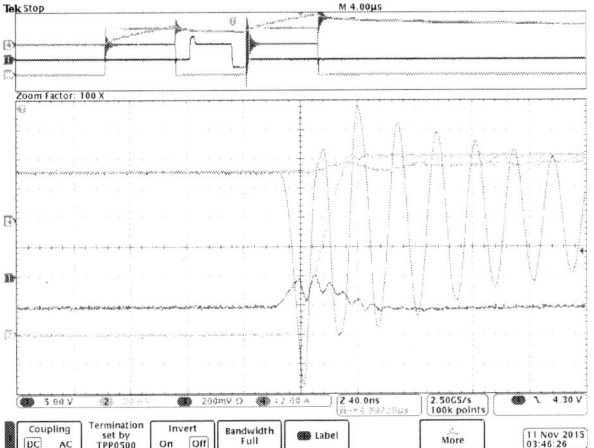

Fig. 22.   Gate Driver IC during Turn-off Transient.

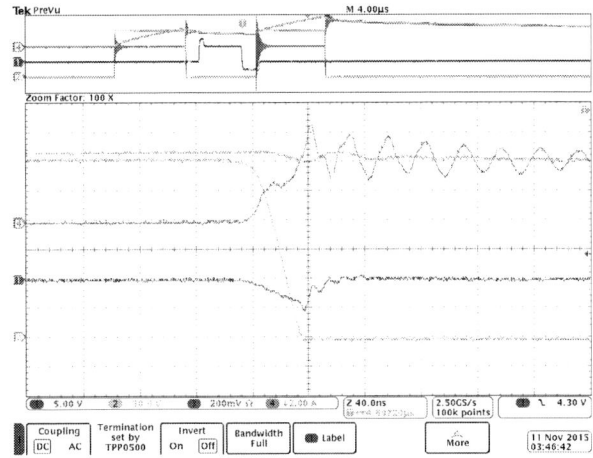

Fig. 25.   Gate Driver IC during Turn-on Transient.

978-1-4673-9551-9/16 $31.00 © 2016 IEEE

Fig. 26. Size Comparison of Gate Driver IC and Circuit Board Gate Driver circuitry.

To explore the effects of the cross talk that occurs in the phase leg configuration, Figs. 20-25 look at the transitioning edges of the signals in order to extract more details from the waveforms. Figs. 20-22 detail the turn-off transition for the lower device. During this period, the cross talk causes a positive voltage spike to occur which can cause a shoot-through current situation to occur if the voltage spike meets or exceeds the threshold voltage of the power device. For the first case with the 0-10 V gate drive signal (Fig. 20), the cross talk in the phase leg causes a voltage spike that is approximately 4 V at its maximum whereas the other two cases mitigate this effect by having the voltage level at −5 V. The −5 V to 10 V case (Fig. 21) appears quite similar in its advantages to the Gate Driver IC case (Fig. 22), but the Gate Driver IC possesses the highest dV/dt seen out of all three cases. This effect is because of the Gate Driver IC's higher voltage level (15 V) during its turn-on transition that allows for a higher dV/dt to be achieved by the power device. The higher dV/dt also can be seen in the larger ringing of the measured current waveform that is caused by the coupling capacitors in conjunction with the faster switching speed.

For the turn-on transition, Figs. 23-25 demonstrate the three gate drive levels in the previously established order. During the turn-on transient, the effect of the cross talk in the phase leg causes a negative voltage spike to occur on the gate drive signal. Starting with Fig. 23, the 0-10 V gate drive level properly mitigates the cross talk in the phase leg. The −5 V to 10 V gate drive level case (Fig. 24) shows that the cross talk causes the gate drive signal to approach the negative V_{GS} maximum for the CREE device (−10 V), which could cause damage to the gate oxide as the device continues to operate at this voltage level. The Gate Driver IC case (Fig. 25) mitigates the cross talk much like the 0-10 V case but has the advantage of a greater dV/dt because of the larger voltage difference seen in the Gate Driver IC during its turn-off transient (10 V to −5 V) as opposed to the 10 V to 0 V, which is seen in the first case. Therefore, the Gate Driver IC performs better during the two transitions of the power devices in the phase leg configuration while mitigating the cross talk that occurs during both transitions, whereas the other two conventional gate driver cases only truly mitigate one cross talk instance and have slower switching speeds. Fig. 26 depicts the volume

Fig. 27. Prototype Gate Driver IC #2 Die Photograph.

advantage that the Gate Driver IC has over the conventional printed circuit board gate driver setup. Fig. 27 shows the prototype die for the second chip run that was used to analyze the variable on-resistance values. The top right corner of the die contains the control circuitry for the entire Gate Driver IC. The large, highly visible structures in the center are the main output drivers with the cross talk output drivers in right center.

ACKNOWLEDGMENT

The authors would like to thank the II-VI Foundation for its support of this research work as well as the support from Texas Instruments. This work made use of the Engineering Research Center Shared Facilities supported by the Engineering Research Center Program of the National Science Foundation and DOE under NSF Award Number EEC-1041877 and the CURENT Industry Partnership Program.

REFERENCES

[1] Z. Chen, D. Boroyevich, and J. Li, "Behavioral comparison of Si and SiC power MOSFETs for high-frequency applications," in *Applied Power Electronics Conference and Exposition (APEC), 2013 Twenty-Eighth Annual IEEE*, 2013, pp. 2453-2460.

[2] Z. Zhang, W. Zhang, F. Wang, L. M. Tolbert, B. J. Blalock, "Analysis of the switching speed limitation of wide band-gap devices in a phase-leg configuration," in *Proc. IEEE Energy Conversion Congress and Exposition*, Sept. 2012, pp. 3950-3955.

[3] Z. Zhang, F. Wang, L. M. Tolbert, B. J. Blalock, "Active gate driver for cross talk suppression of SiC power devices in a phase-leg configuration", in *Proc. IEEE Trans. Power Electronics*, vol. 29, no. 4, April. 2014, pp. 1986-1997.

[4] Z. Zhang, F. Wang, L. M. Tolbert, D. J. Costinett, "Active gate driver for fast switching and cross-talk suppression of SiC devices in a phase-leg configuration", *in Proc. IEEE Applied Power Electronics Conference and Exposition*, March. 2015, pp. 774-781.

[5] Z. Zhang, D. J. Costinett, H. Lu, F. Wang, L. M. Tolbert, B. J. Blalock, "Dead-time optimization of SiC devices for voltage source converter", *in Proc. IEEE Applied Power Electronics Conference and Exposition*, March. 2015, pp. 1145-1152.

[6] Z. Zhang, F. Wang, L. M. Tolbert, B. J. Blalock, D. J. Costinett, "Evaluation of switching performance of SiC devices in PWM inverter fed induction motor drives", *Power Electronics, IEEE Transactions on*, vol.30, no.10, pp.5701,5711, Oct. 2015

[7] Josifovic, J. Popovic-Gerber, and J. A. Ferreira, "Improving SiC JFET Switching Behavior Under Influence of Circuit Parasitics," *Power Electronics, IEEE Transactions on,* vol. 27, pp. 3843-3854, 2012.

[8] Z. Zhang, F. Wang, L. M. Tolbert, B. J. Blalock, D. J. Costinett, "Understanding the limitations and impact factors of wide bandgap devices' high switching-speed capability in a voltage source converter", in *Proc. IEEE Wide Bandgap Power Devices and Applications*, Oct. 2014, pp. 7-12.

[9] Greenwell, Robert Lee, "A Highly Integrated Gate Driver with 100% Duty Cycle Capability and High Output Current Drive for Wide-Bandgap Power Switches in Extreme Environments." PhD diss., University of Tennessee, 2012.

[10] Baker, R. Jacob. *CMOS: circuit design, layout, and simulation.* Vol. 18. John Wiley & Sons, 2011.

Optimal Design of a Voltage Regulator Based Resonant Switched-Capacitor Converter IC

Eli Abramov, *Student Member, IEEE*, Alon Cervera, *Student Member, IEEE*,
and Mor Mordechai Peretz, *Member, IEEE*

The Center for Power Electronics and Mixed-Signal IC, Department of Electrical and Computer Engineering
Ben-Gurion University of the Negev, P.O. Box 653, Beer-Sheva, 8410501 Israel
eliab@post.bgu.ac.il, cervera@bgu.ac.il, and morp@ee.bgu.ac.il
http://www.ee.bgu.ac.il/~pemic

Abstract — This paper details efficiency analysis and characteristics of a gyrator resonant switched-capacitor converter (GRSCC) operating as a voltage regulator. Following the efficiency analysis, this paper introduces an optimal size-efficiency design procedure for IC realization of the converter. In area-sensitive applications, the optimization method combined with the converter's benefits present an attractive approach for better power delivery concepts for point-of-load (PoL) applications. Based on the optimization principles detailed in this study, an on-chip bridge GRSCC topology has been implemented in 0.18μm 5V CMOS process. The analysis has been verified by post-layout analysis and measurements of the fabricated IC. Neglecting the package limitations, the prototype operation is demonstrated with 10 MHz switching frequency, up to 3A, 4.5 W with 3V input voltage, and the efficiency is measured to be 87%. The study has been extended to survey on effects of the package on the performance. The experimental measurements of the manufactured IC have been found to be in very good agreement with the theoretical analysis and optimization process, as well as to accurately estimate the package contribution to the system performance. In addition, a fully monolithic control system to regulate the output voltage is described and implemented on-chip by an automated synthesis process and place-and route tools.

I. INTRODUCTION

Present-day microprocessors and other high-performance ICs require an accurate, dynamically scalable supply voltage in the range of 1V and total current of 10s A/chip. In addition to the tight voltage regulation requirements, the area-efficiency factor of the Point-of-Load (PoL) converter is of key importance to assure the desired performance and to be considered reasonable for commercialization. Improvements of the area-efficiency factor of the PoL converter may enable 3-D power delivery architectures [1] where the converter is integrated with the load, significantly enhancing dynamic power delivery capabilities.

Conventional approaches to reduce the total volume of voltage-regulator modules (VRMs) are carried out by increasing the operating frequency to the 100 MHz range [1]-[2]. By doing so, the integration of magnetics and the decoupling capacitors is more convenient. However, the efficiency and total power that can be processed is limited by the static power consumption at high frequencies. Another area saving concept can be facilitated by resonant-mode converters

[3]-[6]. By employing soft-switching features, the converter's efficiency is not compromised by the high-frequency operation.

Present-day switched-capacitor technology demonstrated superior power density over switched-inductor converters [7]-[8]. However, it lacks the capability of accurate voltage regulation without the penalty of introducing losses, and its transient characteristics are limited [9]-[12].

A solution that overcomes these challenges is presented in [13]. There, an additional switching state has been added to balance the charge difference between the input and output rather than introducing losses for voltage regulation, creating a *gyrator mode resonant switched-capacitor converter (GRSCC) that disengages the efficiency of the system from the voltage gain*. Utilizing such approach allows on-chip integration at operating frequencies in the range of 10MHz without sacrificing the dynamic performance, further improving the power conversion efficiency.

Classical design of the power processing components for VRMs, considers the converter's mode of operation and especially the conversion ratio to optimize the peak efficiency point to the target parameters. E.g., in a 12V to 1.5V buck VRM, the lower transistor of the synchronous rectifier would be much larger in size than the top transistor to assure higher efficiency at the target voltage. In switched-capacitor converter (SCC) technology on the other hand, symmetry of the power transistors is typically assumed since the output voltage is in proximity to the target one. Based on the existing design tools for SCC, the efficiency characteristics of a GRSCC as a voltage regulator will be in-par with other switched-inductor based candidates since it has not been optimized to the target operation [14].

The objective of this study is to present an *optimal size-efficiency design procedure* for the GRSCC when operating as a voltage regulator (VR), and to define the required sizing of the power transistors (and resonant network) based on the target operating point. It is a further objective of this study to present a fully monolithic GRSCC based voltage regulator that is realized by simple constant on-time Pulse Density Modulation (PDM) control (Fig. 1). The new VR scheme with an optimized power converter demonstrate a reasonably sized

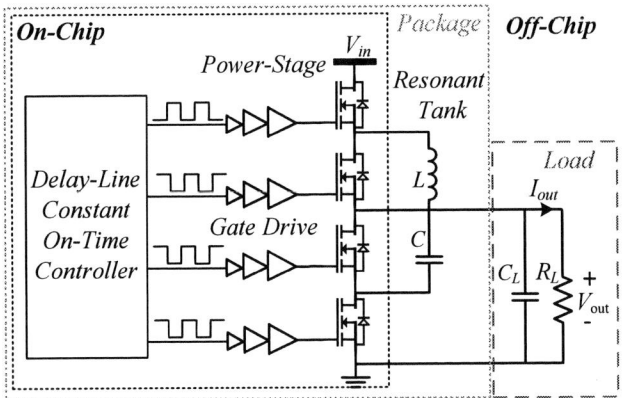

Fig. 1   Circuit diagram of a bridge gyrator mode resonant switched-capacitor converter.

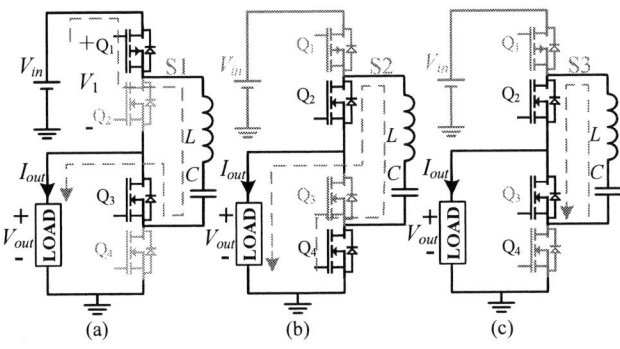

Fig. 2   Bridge GRSCC configuration and operation principle: (a) charge, (b) discharge, and (c) charge balancing.

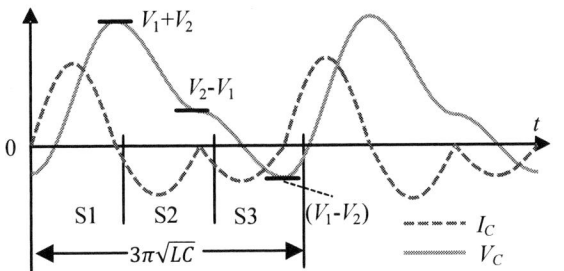

Fig. 3   Typical waveforms of the GRSCC.

solution at much lower operating frequencies and requires virtually no physical inductors, which may be found beneficial for many applications and enable smoother transition towards the 3-D power delivery approach.

The rest of the paper is organized as follows: Section II briefly surveys the GRSCC operation. Section III presents a generalized loss analysis oriented to a desired target conversion ratio. A design procedure and IC implementation are delineated in Section IV. The monolithic delay-line based constant on-time controller is detailed in Section V. Post-layout validation of the converter IC is provided in Section VI. Section VII concludes the paper.

II. BRIDGE GRSCC PRINCIPLE OF OPERATION

The GRSCC-based VR, presented in [14], has evolved from the conventional soft-switched resonant SCC configuration [15]-[17]. As in the common bridge design of a voltage dividing SCC [18], the topology includes four switches and a resonant tank. In addition to the classical complementary switching states, demonstrated by Fig. 2a and 2b, a third state is added which introduces a zero-voltage resonant current path (Fig. 2c). This state is used to balance the residual charge of the flying capacitor, i.e. restore the flying capacitor's voltage to its original state by reversing its polarity. The mechanism of polarity reversal (charge balance) lays the foundations to break the rigid connection of input/output voltage and efficiency dependency. Controlling the sequence of the switches governs power flow direction, hence bidirectional operation.

The operation of the converter shown in Fig. 2 is described for one steady-state charge/discharge/balance cycle and is assisted by Fig. 3 that illustrates the capacitor voltage, V_C, and the resonant tank current, I_C, for an arbitrary case of an uneven voltage ratio. By turning Q_1 and Q_3 on, a charge state (S1) is commenced, in which the resonant tank connects to V_{in} while in series with V_{out}, resonantly charging the flying capacitor from a voltage potential of $V_1 = V_{in} - V_{out}$. After a half-resonant cycle, i.e. at zero current, the switches are turned off and followed by the complementary pair, Q_2 and Q_4 (state S2). At this point, the resonant tank connects in parallel to the output and discharges the flying capacitor onto the potential of $V_2 = V_{out}$. In this example $V_2 > V_1$, so when completing S2 after another half-resonant cycle only a portion of the charge is delivered to the output. This results in V_C that is different from its voltage at the starting point of S1. By turning Q_2, Q_3 on (S3), the resonant tank is short-circuited. This creates the required charge-balance and reverses the flying capacitor voltage polarity such that the voltage at the end of S3 equals the voltage at the beginning of S1.

The relationship between I_{out} and V_{in} follows a gyrator behavior [13] and can be expressed as:

$$I_{out} = 2V_{in}fC, \tag{1}$$

where f is the repetition frequency of S1-S3. Voltage regulation is obtained by introducing time-delay between consecutive sequences, i.e. PDM [19]-[22], meaning that f can be of any value below the resonant limitation, f_{max}, of:

$$f_{max} = \left(3\pi ZC\right)^{-1} \quad , \quad Z = \sqrt{L/C} \ . \tag{2}$$

III. LOSS ANALYSIS

In previous studies, derived from the precursor resonant SCC foundation, symmetrical loop resistances were assumed for all three switching states. While appropriate for most discrete-component realization, in area-sensitive applications and in particular for IC implementation, higher attention should be given to the desired resistance per-loop to obtain the target efficiency. Given a typical operation as described earlier for a case that all three switching phases are dominant, where the inner switches conduct twice per cycle, it may appear that a different organization of the switches' on-resistances results in a higher efficiency from which would be obtained by symmetrical on-resistance organization. An expansion to the loss analysis is detailed in this study.

978-1-4673-9551-9/16 $31.00 © 2016 IEEE   693

Assuming a relatively high ratio between the circuit's characteristic impedance to the loop resistance (i.e., quality factor $Q = R_S/Z \gg 5$) of the resonant network, constant output current I_{out}, and neglecting the output voltage ripple, the relationship between the states' rms currents and the average I_{out} can be expressed as

$$\begin{cases} I_{rms,S1} = \left(\sqrt{A\pi R_L/4Z}\right) I_{out} \\ I_{rms,S2} = \left|\sqrt{A\pi R_L/4Z} - \sqrt{A^{-1}\pi R_L/4Z}\right| I_{out} \\ I_{rms,S3} = \left|2\sqrt{A\pi R_L/4Z} - \sqrt{A^{-1}\pi R_L/4Z}\right| I_{out} \end{cases} , \begin{cases} A = \dfrac{V_{out}}{V_{in}} \end{cases}, \quad (3)$$

where A is the conversion ratio and R_L is the load resistance.

Since the converter operates under ZCS conditions, the dominant contributors to the power dissipation are the conduction losses. To individually identify the per-transistor contribution to these losses, the rms current of the transistors can be written as a function of (3) as:

$$\begin{cases} I_{rms,Q1} = I_{rms,S1} \\ I_{rms,Q2} = \sqrt{I_{rms,S2}^2 + I_{rms,S3}^2} \\ I_{rms,Q3} = \sqrt{I_{rms,S1}^2 + I_{rms,S3}^2} \\ I_{rms,Q4} = I_{rms,S2} \end{cases} \quad (4)$$

As detailed in [23], uniform current distribution results in minimum losses per-area due to even power dissipation. Thus, to minimize losses each transistor should be sized based on the rms current through it. The factor ψ_i defines the required ratio for a transistor's on-resistance (R_{Qi}) with respect to symmetrical operation as:

$$\psi_i = R_{Qi}/R_{sym} = \Sigma I_{rms,Qi}/4I_{rms,Qi} \quad , \quad i = 1, 2, 3, 4 , \quad (5)$$

where R_{sym} is the nominal resistance at symmetrical partition of the switches.

Assuming the transistors' on-resistances, R_{Q1} through R_{Q4}, are the dominant resistances in each loop, the total loss of the converter as a function of the conversion ratio is derived by summation of the losses [24] as presented in (6) below. Substituting (4) and (5) into (6), and after some manipulations, the efficiency of the converter η as a function of A, R_{sym} and ψ can be expressed as (7).

Fig. 4a shows the resulting efficiency versus conversion ratio curves of (7), for several cases of on-resistance selection and compares symmetrical sizing with an optimized one. As can be observed, in the vicinity of the target voltage ($A = 0.5$) where charge-balance of resonant SCC is naturally obtained, the results of both sizing methods coincide. However, as predicted by the initial conjecture, as the conversion ratio deviates from center, a significant efficiency improvement can

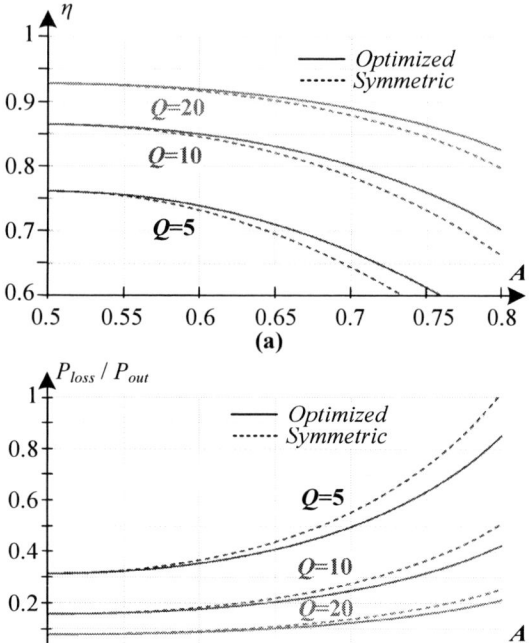

Fig. 4 Theoretical efficiency and losses curves both symmetrical (dotted lines) and optimized (solid lines) partition as a function of A, for various values of Q: (a) efficiency, (b) P_{loss} vs P_{out}.

be observed in favor of the optimized sizing method. An even more interesting is the power losses factor that is depicted in Fig. 4b, showing the possible power saving as a result of better area distribution between the transistors. For example, in conversion ratio of 0.75, $Q = 10$ and output power of 10W, 0.5W of the losses can be trimmed down for the same total area. Alternatively, this implies that if certain power dissipation is allowed, the converter can be further reduced in size.

IV. DESIGN PROCEDURE AND IC IMPLEMENTATION

Using the above analysis and observations, generalized IC design guidelines of a voltage regulator GRSCC are provided and then followed by practical power-stage IC sizing and realization. The procedure is as follows:

a. Given target values for $V_{in,min}$, $I_{out,max}$ and operating frequency, f_{max}.

b. Calculate the resonant network parameters C and L by:

$$C = I_{out,max}/2V_{in,min}f_{max} \quad , \quad L = \left[\left(3\pi f_{max}\right)^2 C\right]^{-1} \quad (8)$$

c. From the desired efficiency η and conversion ratio A,

$$P_{loss} = \frac{\pi R_L}{4Z} I_{out}^2 \left[\left(R_{Q1} + R_{Q2} + 2R_{Q3}\right)\frac{A}{1-A} + \left(R_{Q4} + 2R_{Q2} + R_{Q3}\right)\frac{1-A}{A} - 2\left(R_{Q2} + R_{Q3}\right)\right] \quad (6)$$

$$\eta = \left[1 + \frac{\pi R_{sym}}{4Z}\left(\left(\psi_1 + \psi_2 + 2\psi_3\right)\frac{A}{1-A} + \left(\psi_4 + 2\psi_2 + \psi_3\right)\frac{1-A}{A} - 2\left(\psi_2 + \psi_3\right)\right)\right]^{-1} \quad (7)$$

calculate the symmetrical sized resistance R_{sym}:

$$R_{sym} = \frac{\left(\eta^{-1}-1\right)}{\frac{\pi}{4Z}\left[\left(\psi_1+\psi_2+2\psi_3\right)\frac{A}{1-A}+\left(\psi_4+2\psi_2+\psi_3\right)\frac{1-A}{A}-2\left(\psi_2+\psi_3\right)\right]} \quad (9)$$

d. Calculate ψ_i from (5) and determine the optimized on-resistances by:

$$R_{Qi} = \psi_i R_{sym} \quad (10)$$

e. The individual silicone width per transistor can be directly derived from R_{Qi}:

$$W_i = K'L_g / R_{Qi}, \quad (11)$$

where K' can be found by the values that are given by the vendor's process design kit (PDK) and L_g is the gate length. The total transistors' silicone width is obtained by summing W_i.

A design procedure example is demonstrated by 0.7W GRSCC with target values of: $V_{in,min}$=3V, V_{out}=0.7V, $I_{out,max}$=1A, f_{max}=10MHz, η=80%. The resonant network values are calculated as $C \approx 17$nF and $L \approx 6.7$nH. The required resistances are: $R_{sym}\approx17.5$mΩ; or R_{Q1} through R_{Q4}: 53mΩ, 11mΩ, 18.5mΩ, 14.5mΩ. Given a 0.18μm CMOS process, $K'\cdot L_g\approx3$mΩ, the resulting total width is $W \approx 685,700$μm. Fig. 5 depicts an efficiency prediction for the design example. It can be observed that for the same silicon area efficiency improvement by approximately 3% can be obtained at nominal input voltage of 3.3V. A more noticeable benefit can be viewed in terms of size, where 20% less silicon area is required to obtain the same efficiency.

A. On-Chip Power-Stage Implementation

The selection of the transistor type depends primarily on the driver type. Assuming a conventional ground-referenced driver and that the input voltage is limited to the technology voltage V_{DD}, a pMOS is used for Q_1, while nMOSs are used for Q_2, Q_3, and Q_4, based on the required gate-source threshold voltage to activate the transistor.

To increase current handling capabilities of a 5V CMOS process transistor, paralleled multiple unit cells are realized, as shown in Fig. 6. The size of the device with multiple unit cells is typically defined in terms of the gate boundary $W_i = N_{fi} W_g$, where N_{fi} is the number of fingers for a transistor Q_i, and W_g is the gate width. N_{fi} can be expressed as:

$$N_{fi} = R_{on}A_{eff}/W_g L_g R_{Qi} \quad (12)$$

Given the PDK W_g and L_g constraints, an accurate and efficient quadrilateral layout of the power-stage can be applied. It should be noted that the transistors are connected via a top metal pad, resulting a relatively low parasitic resistance of the conduction and is neglected for the design.

B. Driver Circuitry Implementation

A fully integrated circuit based buffers with the ability to drive transistors with large gate width is implemented. Buffers logic implementation is realized by logical effort technique, including a network of three custom designed buffers for each switch. The ratio between the pull-up network and the pull-

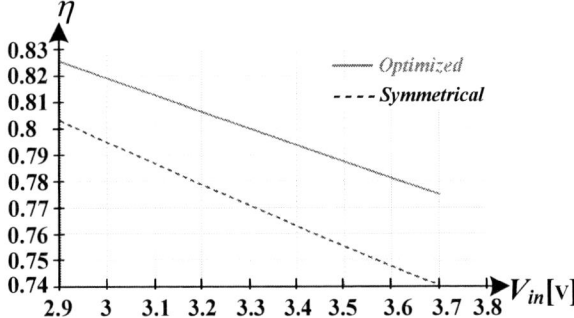

Fig. 5   Efficiency curve of GRSCC as function of V_{in} symmetric and optimized partition, with the target parameters: V_{out}=0.7V, $V_{in,min}$=3V, $I_{out,max}$=1A, f_{max}=10MHz.

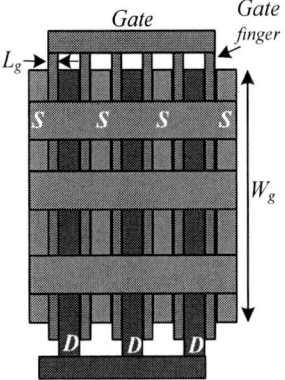

Fig. 6   Simplified structure of a device with paralleled multiple cells.

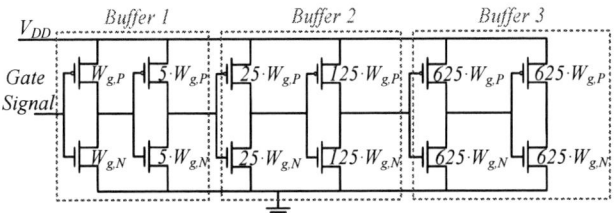

Fig. 7   Schematic of the implemented IC drive circuitry.

down network is $\beta = 1.5$. In the context of driving power transistors, the buffers sizing should also be considered due to the relatively higher input capacitances of the buffers chain. To assure a non-distorted input gate signal, the first inverter of the chain has been matched in size to the analog input unit. Fig. 7 shows in detail a sized example of a driving stage, where $W_{g,P} = 1.5\cdot W_{g,N}$. It should be further emphasized that the sizing characteristics such as gate capacitances are technology-dependent, which should be taken into account to assure proper drive signal.

Explained above, since the driver stage is integrated and the control signals are all ground referenced (with gate voltage swing between 0 to 5V), neither isolation nor level shifters are required. Even for a worst case scenario where $V_{in} = V_{DD}$, the driving circuit is still functional.

C. Package and Bonding Wires Limitations

On-Chip interconnections ultimately connected to the board level via IC packaging. Bond wires technique is extensively used in IC packaging designs, because of its relatively lower cost and simple implementation [25]. Bond wires usually consist of copper, aluminum, gold and silver, while each material has its own electrical and physical characteristics. The bonding is usually farther away from the substrate and therefore has less coupling and loss at high frequencies, still bonding wires can be a major limiting factor in IC design, in particular for high current ICs. The thin and narrow bonding wires are a significant penalty to the conduction losses. In the context of a soft-switched converter, these are the dominant contributors to efficiency reduction. Advances in power semiconductor technology further highlight the limitations of bond wires packaging technology, since the losses contribution by resistances and parasitics of the package and the bond wires exceed those of the silicon die [26]. Performance improvement solutions by means of packaging were detailed in [26]-[29]. However, these technologies require high cost specialized resources and processes. It should be noted that the packaging and bonding wires implementation are beyond the scope of this study.

V. MONOLITHIC DELAY-LINE BASED CONSTANT ON-TIME CONTROLLER

A fully monolithic internal controller has been designed to regulate the output voltage. As described in [14], regulation is facilitated using a single comparator which compares the output voltage to an internal value and triggers the GRSCC when needed. Fig. 8 shows the main components of the control unit. A flip-flop based state-machine (SM) synchronically dictates the active state and out of six options: The three switching states (S1-S3), two intermediate dead-time states (D2, D3) necessary to prevent shoot-through between complementary switches, and an inactivity state (S0). The state-machine is clocked by a configurable delay-line (DL), which enables an individual predefined time length for each state. Once triggered by a comparator (unmasked only at state S0 when inactive for a predefined time value), the DL is activated and triggers the SM to produce the required logic sequence for the GRSCC. Fig. 9 shows a measured experimental gate logic sequence operating at ~3MHz produced from the monolithic DL controller.

VI. POST LAYOUT AND EXPERIMENTAL VERIFICATION

Following the design procedure, an IC prototype that implements the bridge GRSCC voltage regulator has been designed and fabricated. To demonstrate the operation of the IC GRSCC and to verify the theoretical analysis, the Post-Layout design of the power-stage and driver circuitry was verified using Cadence Spectre simulator in 0.18μm 5V CMOS process. R_{on} of the power transistors was designed to value of ~20mΩ (layout shown In Fig. 10). The chip consists of 11-pins and connects to a resonant tank of 2.25nH, 50nF. With the ability to operate at 10MHz, the GRSCC can produce up to 4.5W (1.5V, 3A) from a 3V input. The gate signals are transmitted through an analog input unit and buffers matching network to obtain the desired drive voltage for the converter's power-stage. The timing controller was implemented by an automated synthesis process and place-and route tools, directly from the VHDL representation. Fig. 11 shows the flying capacitor current and voltage waveforms and the gate logic signals that were obtained by cycle-by-cycle post-layout

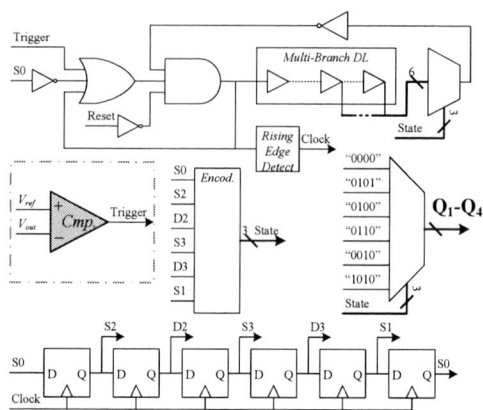

Fig. 8   Schematic of a delay-line based constant on-time controller.

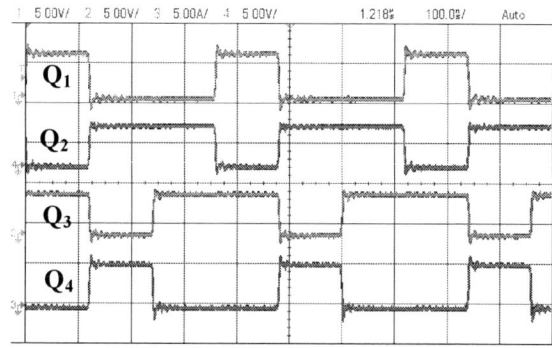

Fig. 9   Experimental gate signals produced from the monolithic controller.

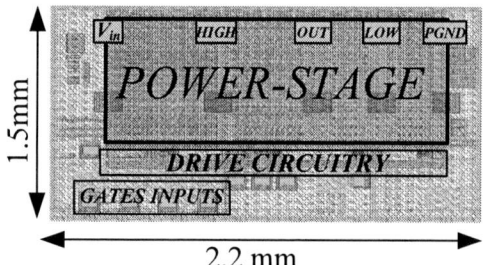

Fig. 10   Chip layout 2 mm x 1.5 mm.

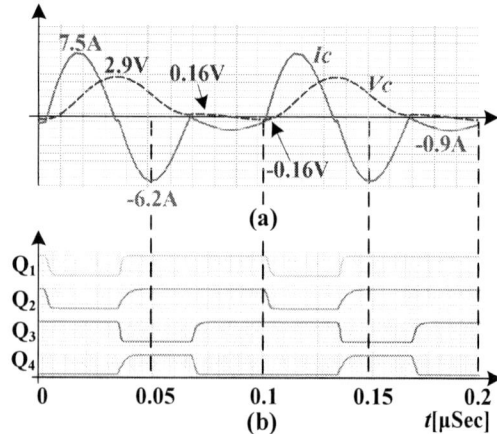

Fig. 11   (a) Flying capacitor voltage and current waveforms. The circuit parameters are V_{in}=3V, V_{out}≈1.5V, R_{loop}≈40mΩ, C=50nF, L=2.25nH; and (b) gate logic signals

simulation. The system has been tested under open-loop conditions and the waveforms verify operability of the design. The efficiency is measured to be 87% (excluding drive losses). Operating under the above specifications, the converter IC is capable to output power up to 4W. These results are in very good agreement with the theoretically predicted efficiency of 85%-90%.

The fabricated IC prototype was packaged using 24-pin 4x4mm QFN package, where 11 pins were in use for the GRSCC. A chip micrograph is depicted in Fig. 12a. The chip connects to a resonant tank of C=100nF and L=6nH. The inductance is realized by stray inductances by the connection and interconnections from the chip to the flying capacitor, Fig. 12b depicts the IC prototype on PCB.

Experimental measurements of the packaged IC converter resistances resulted in approximately 160mΩ per loop, or effective R_{on} of a single transistor is ~80mΩ. Kelvin resistance measurements to cancel out the package effect resulted in resistance of 20mΩ per transistor. These measurements clearly exhibit the bond wires and package limitation, in this case by four times than the targeted R_{on}. Considering these limitations, the best-case experimental results were obtained at maximum switching frequency of 4.5 MHz producing up to 2.25W. The actual quality factor is measured to around 2 and the resulting efficiency is measured to be 67%. Table I summarizes the IC prototype experimental measurements. Fig. 13 shows open loop measured and post-layout simulated results of the capacitor current and voltage.

Efficiency curve of the converter IC as function of V_{in} is depicted in Fig. 14. The experiment was carried out by varying the input voltage and manually compensating with the frequency, such that the output power and the input voltage, 2.25W and 1.5V, respectively, were kept constant. The experimental measurements tightly follow the results obtained by the simulation as well as the theoretical predictions. As can be observed, lower efficiency can be obtained for higher input voltages. This is primarily due to higher conduction losses at higher conversion ratios. Fig. 15 shows the efficiency of the converter IC as a function the output current, I_{out}. As can be seen, in light-load operation there is a discrepancy between the experimental and simulation results from the theoretical analysis, this is due to the resonant characteristics of the three states that are not identical, which was not taken into the theoretical model analysis. The experimental measurements also highlighted the advantage of the GRSCC operating at light loads, at 20mA the efficiency drops only at 4%.

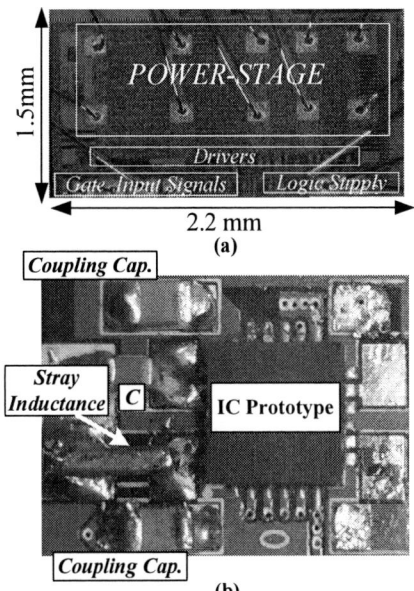

Fig. 12 (a) Chip micrograph of Bridge GRSCC, (b) IC prototype on PCB.

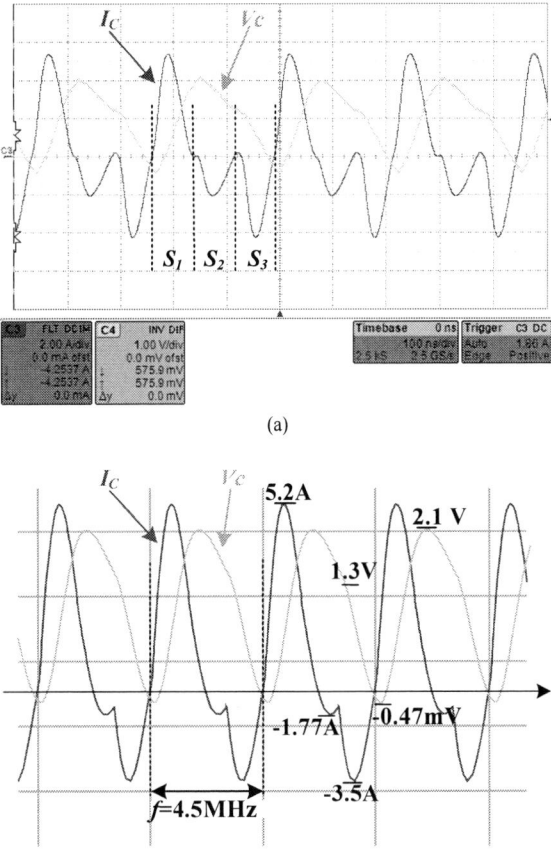

Fig. 13 Flying capacitor voltage and current waveforms including the packaging limitations. (a) Experimental ,(b) Post-Layout.

TABLE I – SUMMARY OF IC PROTOTYPE EXPERIMENTAL MEASUREMENTS

Specifications	Value / Type
Package	4x4 QFN - MLP
V_{in}	3V
R_{on}	~80mΩ
V_{out}	1.5V
I_{out}	1.5A
Off Chip resonant tank	6nH, 100nF
Switching Frequency	4.5MHz
Quality Factor Q	~2
Efficiency	67%

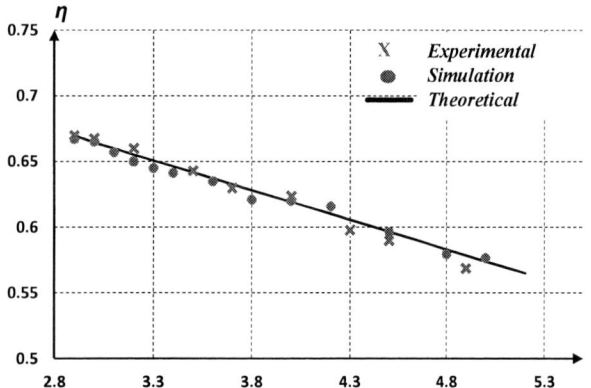

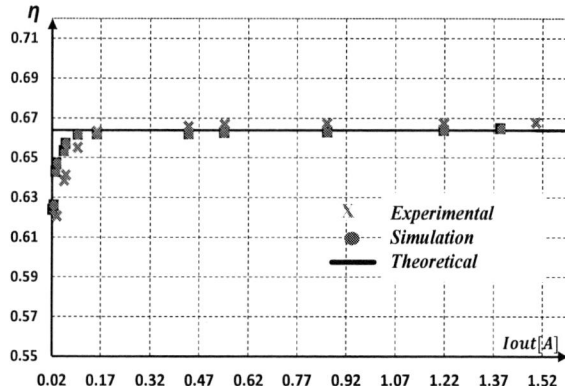

Fig. 14 IC GRSCC efficiency as a function of input voltage, V_{in}, while R_{loop}=160mΩ.

Fig. 15 IC GRSCC efficiency as a function of load current, I_{out}, while R_{loop}=160mΩ.

VII. CONCLUSION

A bridge gyrator resonant-switched capacitor converter IC and optimization methodology have been presented. The detailed efficiency analysis and characteristics of the GRSCC operating as a voltage regulator have been explored. An optimal size-efficiency design procedure and considerations based on the target values of operation are introduced. The effectiveness of the design procedure is verified with a fully monolithic GRSCC fabricated in 0.18µm CMOS process, the analysis was meticulously verified by post-layout simulations and the results are in very good agreement with the theoretical predictions and the efficiency is measured to be 87%. In addition, a fully monolithic control system is described and implemented on-chip to regulate the output voltage of the converter IC.

The experimental results highlighted the significance of packaging and bonding wires limitations. Due to the additional parasitic resistances of the packaging, the experimental tests were slightly different than the targeted. For the given package limitations the obtained measurements from the manufactured converter IC are in very good agreement with the presented optimization procedure. The experimental measurements also further strengthen the advantages of the GRSCC ability to maintain virtually constant efficiency curve for load variations, in particular when operating at light loads. At 20mA loading, the efficiency drops only at 4%. Future work direction would include packaging and wiring optimization to assure absolute certainty of final design.

Significant area saving highlights the benefit of the optimization method, providing a simple and efficient procedure to improve the size-efficiency factor based on the target operating conditions. Combined with the topology benefits, a GRSCC voltage regulation scheme presents an attractive alternative to the switch-inductor converters, in particular in area sensitive application, and establishes the foundations for better power delivery concepts for PoL applications.

VIII. ACKNOWLEDGEMENTS

This research is supported by Vishay Ltd., Siliconix division.

REFERENCES

[1] J. Sun, D. Giuliano, S. Devarajan, J. Lu, T. P. Chow, and R. J. Gutmann, "Fully monolithic cellular Buck converter design for 3-D power delivery," *IEEE Trans. Very Large Scale Integr. Syst.*, vol. 17, no. 3, pp. 447–451, Mar. 2009.

[2] E. A. Burton, G. Schrom, F. Paillet, J. Douglas, W. Lambert, K. Radhakrishnan, and M. Hill, "FIVR-Fully integrated voltage regulators on 4th generation Intel Core SoCs, " in *Proc. IEEE APEC*, pp. 432–439, Mar. 2014.

[3] M. Jabbari, "Unified analysis of switched-resonator converters," *IEEE Tras. Power Electron.*, vol. 26, no. 5, pp. 1364–1376, May 2011.

[4] D. Cao, X. Lu, X. Yu, and F. Peng, "Zero voltage switching double wing multilevel modular switched-capacitor dc-dc converter with voltage regulation," in *Proc. 28th Annu. Appl. Power Electron. Conf. Expo.*, pp. 2029–2036, Mar. 2013.

[5] J. Stauth, M. Seeman, and K. Kesarwani, "Resonant switched-capacitor converters for sub-module distributed photovoltaic power management, " *Power Electronics, IEEE Transactions on*, vol. 28, no. 3, pp. 1189–1198, 2013.

[6] M. Shoyama, T. Naka, and T. Ninomiya, "Resonant switched capacitor converter with high efficiency," in *Proc. IEEE Power Electron. Spec. Conf.*, pp. 3780–3786, Jun. 2004.

[7] T. Santa, M. Auer, C. Sandner, and C. Lindholm, "Switched capacitor DC-DC converter in 65 nm CMOS technology with a peak efficiency of 97%," in *Proc. IEEE Int. Symp. Circuits Syst.*, pp. 1351–1354, 2011.

[8] R. C. N. Pilawa-Podgurski and D. J. Perreault, "Merged two-stage power converter with soft charging switched-capacitor stage in 180 nm CMOS," IEEE J. Solid-State Circuits, vol. 47, no. 7, pp. 1557–1567, Jul. 2012.

[9] Y. P. B.Yeung. Yeung, K. W. E. Cheng, D. Sutanto and S. L. HO, "Zero Current Switching Switched-Capacitor Quasi-resonant Step- Down Converter, " *IEEE Proc. Electr. Power Appl.*, Vol. 149, no.2, pp. 111-121, 2002.

[10] J. M. Henry and J. W. Kimball, "Practical performance analysis of complex switched-capacitor converters," *IEEE Trans. Power Electron.*, vol. 26, no. 1, pp. 127–136, Jun. 2011.

[11] J. W. Kimball and P. T. Krein, "Analysis and design of switched capacitor converters," in *Proc. IEEE Appl. Power Electron. Conf.*, vol. 3, pp. 1473–1477, 2005.

[12] D. Cao, S. Jiang, and F. Z. Peng, "Optimal design of multilevel modular capacitor-clamped DC–DC converter," *IEEE Trans. Power Electron*, vol. 28, no. 8, pp. 3816–3826, Aug. 2013.

[13] A. Cervera, M. Evzelman, M.M. Peretz, and S. Ben-Yaakov, "A High Efficiency Resonant Switched Capacitor Converter with Continuous Conversion Ratio," in *Energy Conversion Congress and Exposition (ECCE), 2013 IEEE*, 2013.

[14] A. Cervera and M. M. Peretz, "Resonant switched-capacitor voltage regulator with ideal transient response," in *Proc. 29th Annu. IEEE Appl. Power Electron. Conf. Exposit. (APEC)*, pp. 867–872, Mar. 2014.

[15] K. Sano and H. Fujita, "A resonant switched-capacitor converter for voltage balancing of series-connected capacitors," in *Proc. Int. Conf. Power Elect. Drive Syst.*, pp. 683–688, 2009.

[16] O. Keiser, P.K. Steimer, and J.W. Kolar, "High power resonant switched-capacitor step-down converter," in *IEEE Power Electronics Specialists Conference, PESC 2008*, pp. 2772-2777, 2008.

[17] Y. P. B.Yeung, K. Cheng, S. Ho,K. Law, and D. Sutanto, "Unified analysis of switched-capacitor resonant converters," *IEEE Trans. Ind. Electron.*, vol. 51, no. 4, pp. 864–873, Aug. 2004.

[18] Jun Chen and A. Ioinovici, "Switching-mode DC-DC converter with switched-capacitor-based resonant circuit," *IEEE Trans. on Circuits and Systems I: Fundamental Theory and Applications*, vol. 43, no. 11, pp. 933-938, 1996.

[19] H. Koizumi, K. Kurokawa, and S. Mori, "Analysis of class D inverter with irregular driving patterns," *IEEE Transactions on Circuits and Systems I: Regular Papers*, vol. 53, no. 3, pp. 677-687 , 2006.

[20] D.J. Tschirhart and P.K. Jain, "Variable frequency pulse density modulation for efficient high frequency operation of series resonant converters operating as voltage regulators," *in IEEE Applied Power Electronics Conference and Exposition, APEC 2010*, pp. 1334-1339, 2010.

[21] Xin Zhang, Yu Pu, K. Ishida, Y. Ryu, Y. Okuma, Po-Hung Chen, K. Watanabe, T. Sakurai, and M. Takamiya, "A 1-V-input switched-capacitor voltage converter with voltage-reference-free pulse-density modulation," *IEEE Trans. on Circuits and Systems II: Express Briefs*, vol. 59, no. 6, pp. 361-365 , 2012.

[22] Y.-H. Liu, S.-C. Wang, and Y.-F. Luo, "Digital dimming control of CCFL drive system using pulse density modulation technique," in *Proc. IEEE Region 10 Conf. TENCON*, pp. 1–4, 2007.

[23] S. Musunuri and P. L. Chapman, "Optimization of CMOS transistors for low power DC-DC converters," *in Proc. IEEE 36th Power Electronics Specialists Conf.*, pp. 2151–2157, 2005.

[24] C.-K. Cheung, S.-C. Tan, C. Tse, and A. Ioinovici, "On energy efficiency of switched-capacitor converters," *IEEE Trans. Power Electron.*, vol. 28, no. 2, pp. 862–876, Feb. 2013.

[25] C.A Plesko, and E.J Vardaman,., "Cost Comparison for Flip Chip, Gold Wire Bond, and Copper Wire Bond Packaging," *Electronic Components and Technology Conference*, Las Vegas, NV, U.S.A., June 1-4, 2010.

[26] L. Xingsheng, and L. Guo-Quan, "Power chip interconnection: from wirebonding to area bonding", *Int Microelectron Packag Soc* 2000; 23:407–13.

[27] S. Haque, K. Xing, R-L. Lin, C. Suchicital, G-Q. Lu, D.J. Nelson, D. Borojevic, and F. C. Lee, "An Innovative Technique for Packaging Power Electronic Building Blocks Using Metal Posts Interconnected Parallel Plate Structures," *IEEE Transactions on Advanced Packaging*, Vol. 22, No. 2, pp. 136 – 144, 1999.

[28] A. Ward and G. Q. Lu, "High Density Power Electronic packaging," in *Virginia Power Electronics Center Seminar proceedings*, pp. 221-224, 1997.

[29] X. Liu, J.N. Calata, J. Wang, and G.Q. Lu, "The Packaging of IPEM Using Flip Chip Technology", *Proceedings 17th Annual VPEC Seminar*, Blacksburg, Virginia, pp. 361-367, September 1999.

Novel Highly Integrated Current Measurement Method for Drive Inverters

N. Langmaack, G. Tareilus, M. Henke
Institute for Electrical Machines, Traction and Drives
Technische Universität Braunschweig
Brunswick, Germany
n.langmaack@tu-braunschweig.de

Abstract— **Measuring the load current of drive inverters is still inevitable for a precise current control. In common industrial and automotive inverters hall effect based current sensors are widely used. [1] The presented novel current measurement method is based on an inductive current sensor like a Rogowski coil. This can be fabricated much smaller and cheaper than hall based current sensors and can be integrated into drive inverter systems easily. Unlike using the standard evaluation electronic for Rogowski coils, AC and DC components of the load current can be measured using a special evaluation scheme. The arrangement of the sensors also allows them to be used for fast short-circuit detection and di/dt measurement, so that some synergies with the gate drive circuitry can be used to create a more compact and reliable system.**

Keywords—current measurement; rogowski coil; drive inverter; integration

I. INTRODUCTION

Rogowski coils are commonly used as current probes for portable measurement equipment as well as permanently installed current sensors in industrial applications. The key disadvantage of Rogowski coils and any other inductive current sensor is that a DC current component cannot be detected but has to be filtered. This usually makes these sensors unsuitable for drive inverter applications.

II. THE PROPOSED CURRENT MEASUREMENT METHOD

With the sensor arrangement and a special evaluation technique proposed in this paper, Rogowski coil sensors can now be used to measure both AC and DC component of the load current of an inverter's half bridge.

A. Basic Principle of Rogowski Coils

The basic principle of the Rogowski coil was published in 1912 [2]. It can be realized as a toroidal air-core coil, which surrounds the wire carrying the current to be measured (see figure 1). The function of this inductive sensor can be understood easily with Ampère's Law in mind:

$$\oint_S \vec{H} \cdot d\vec{s} = I \tag{1}$$

The toroidal coil collects the magnetic field along a closed loop around the current that is to be measured. With the law of induction (2), it can be determined that the induced voltage of the Rogowski coil is proportional to the time derivative of this current.

$$u = \frac{d\Psi}{dt} = N \cdot \frac{d\Psi}{dt} = M \cdot \frac{di}{dt} \tag{2}$$

An integrator is used to obtain the time domain current signal. The DC component of the output signal has to be filtered using a high pass filter, because the integration constant is unknown. [5], [6]

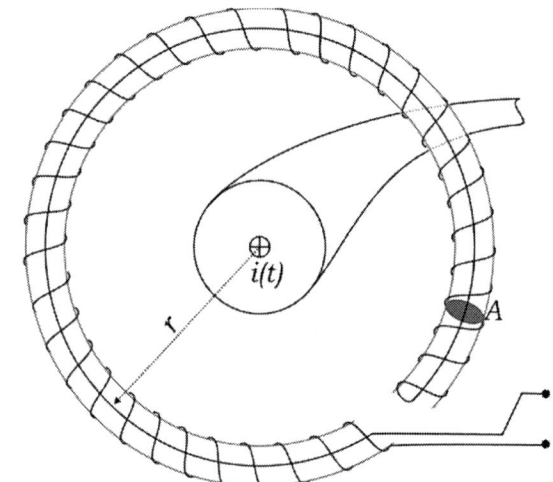

Fig. 1. Classical layout of a Rogowski coil

B. PCB implementation of Rogowski Coils

In the investigations on using Rogowski coils in inverter systems described in the following, the coil itself is realized using PCB technology. This makes it very cheap, robust and highly reproducible. For a sensor which is permanently integrated into an inverter, it is not necessary that it can be opened and closed repeatedly. The coil windings consist of copper traces on the top and bottom layer of the PCB and connecting vias. The air-core in this case consists of the PCB material. Typically this will be FR4, which has a magnetic permeability of one. [3]

978-1-4673-9551-9/16 $31.00 © 2016 IEEE

Figure 2a shows the isolated sensor PCB that was designed for the first tests. Figure 2b shows the sensor of the demonstration inverter including the integrator circuit. Both designs are made to suit the terminals of standard industrial IGBT modules like the EconoDUAL or EconoPACK.

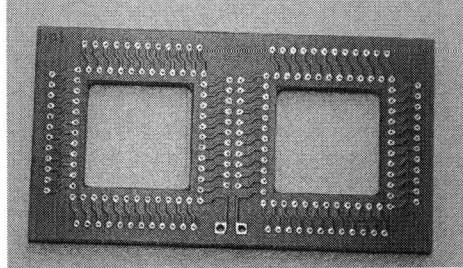

Fig. 2. Dual Rogowski coil on pcb (a), Sensor implementation for demonstrator with active integrator (b)

The implemented Rogowski coils have a number of turns of 48 each. Their cross sectional area is about 4.5 mm², the magnetic path length is 70 mm.

$$M = \frac{\mu_0 \cdot N \cdot A}{l_m} \qquad (3)$$

With the coupling inductance of the Rogowski coil (3), the transfer constant of the sensor can be calculated to be 3.88 nH. With typical current change ratios of 4500 A/µs the induced peak voltage can be expected to be

$$u = M \cdot \frac{di}{dt} = 17.5 \, \text{V} \qquad (4)$$

C. Inverter Leg with Rogowski Coil Current Measurement

To measure the AC and DC component of the load current the sensor is arranged around the DC-link connections of the inverter leg and the measured signal is evaluated according to the switching signals of the half bridge. The information about the switching state of the inverter leg can be used to determine the DC current component from the measured signal by calculating the difference of two consecutive values.

Figure 3 shows a block diagram of the arrangement of the important components.

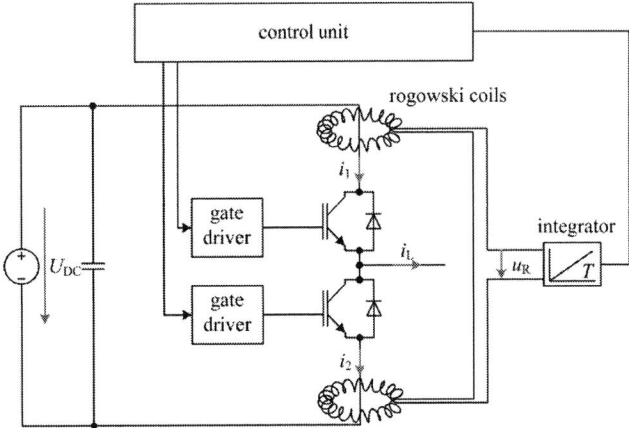

$u_R = M \cdot di_1/dt + M \cdot di_2/dt$

$i_1 = i_L, i_2 = 0$, when high-side switch turned on

$i_1 = 0, i_2 = -i_L$, when low-side switch turned on

Fig. 3. Circuit diagram of one half bridge using the proposed current sensor arrangement

The amplitude of the induced voltage of the sensor can be doubled by using the proposed dual Rogowski coil arrangement with sensors around both DC-link terminals connected in series. In normal operation the di/dt will be the same in both paths (i_1, i_2), so the signal will not change its waveform but only its scaling.

D. Evaluation Method

The output of the Rogowski coil is directly connected to an analog integrator circuit. There are different possible realizations, that are either using only passive or also active components [5]. The best results for the actual application can be achieved using a simple circuit with a single operational amplifier shown in figure 4. The design of the integrator is crucial for the performance of the sensor. Still, determining the component's values and the filter cut-off frequencies is relatively simple, because the maximum di/dt, the typical current range and the sampling rate are all known.

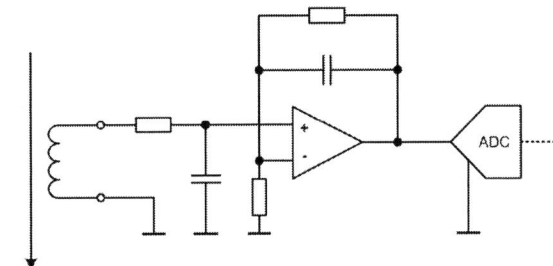

Fig. 4. Active analog integrator

The integrated signal, which is proportional to the DC-link current of the investigated inverter leg, is fed either directly to the analog-digital-converter (ADC) of the inverter's microcontroller or to a high-resolution external ADC.

The load current of one inverter leg can be determined easily from the DC-link current of this inverter leg and the switching state of the half-bridge's power semiconductors [4].

978-1-4673-9551-9/16 $31.00 © 2016 IEEE

One way to do this is sampling the sensor output signal in the middle of the on-state pulse and in the middle of the off-state pulse of the PWM signal. The difference of these two values directly corresponds to the load current. Because it is a differential measurement, the DC offset of the integrator output is negligible. Figure 5 shows examples of the important signals and the described evaluation method.

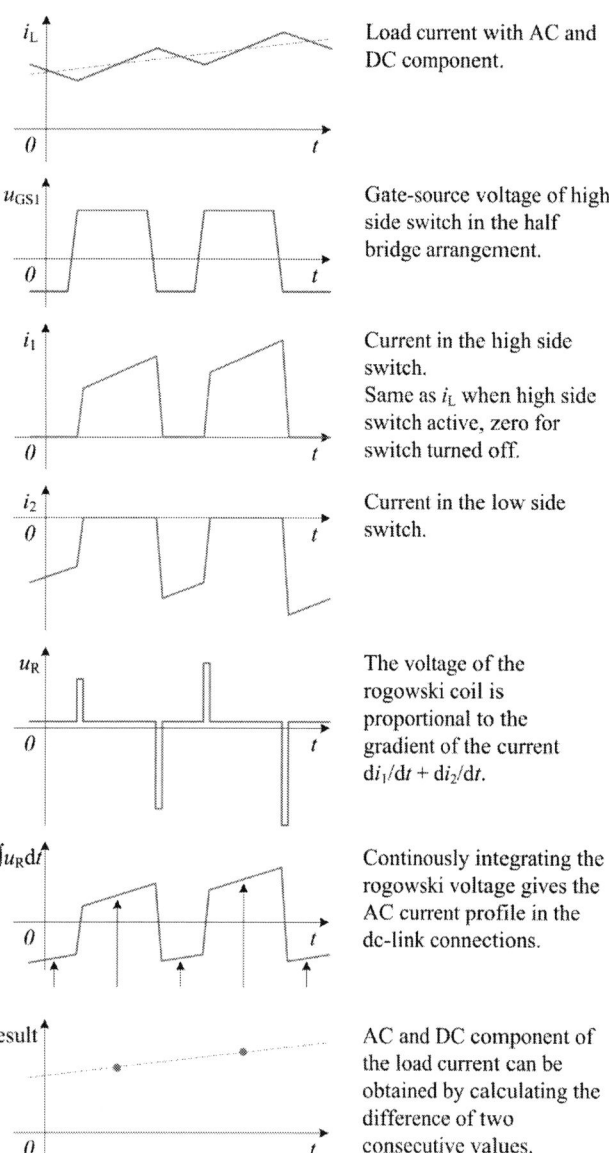

Load current with AC and DC component.

Gate-source voltage of high side switch in the half bridge arrangement.

Current in the high side switch.
Same as i_L when high side switch active, zero for switch turned off.

Current in the low side switch.

The voltage of the rogowski coil is proportional to the gradient of the current $di_1/dt + di_2/dt$.

Continously integrating the rogowski voltage gives the AC current profile in the dc-link connections.

AC and DC component of the load current can be obtained by calculating the difference of two consecutive values.

Fig. 5. Evaluation method step-by-step

III. DEMONSTRATOR

After the first investigations using separate Rogowski coils, a demonstration inverter is built. It uses components which are common for automotive or compact industrial drive inverters. The DC-link voltage is supposed to be 560..700 V, the nominal output power is about 200 kW.

A. Circuit Overview

The demonstration inverter consists of the DC to AC converter only, so it utilizes a single B6-bridge circuit. A standard sixpack IGBT module (EconoPACK, 1200 V / 450 A) is chosen. The DC-link capacitor (1 mF / 900 V, MKP film) is taken from the design of an automotive inverter and fits perfectly on the terminals of the IGBT module resulting in a very small DC-link inductance. Figure 6 shows the circuit diagram of the simple power stage.

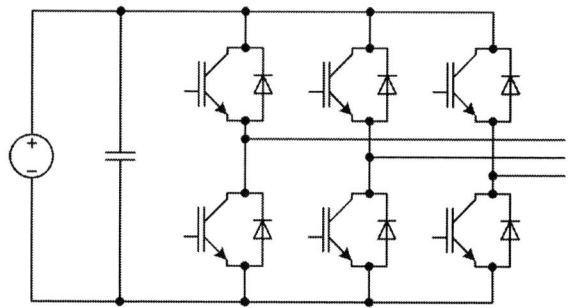

Fig. 6. B6 IGBT inverter

The inverter is completed by the gate driver circuits for the six IGBTs and a controller board with all the necessary interfaces. The implemented PCB Rogowski sensor and its evaluation electronic is designed for the expected peak current of 450 A plus some safety margin.

B. Physical Layout

Figure 7 shows the combined gate driver and current sensor board mounted to the sixpack IGBT module and the connection to the DC-link film capacitor. The analog integrator circuit is placed next to each sensor coil. The connection to the controller board is realized using board-to-board pin headers.

Fig. 7. Gate driver and sensor board on IGBT module and film capacitor

C. Evaluation of the Current Measurement

The functionality of the sensor is investigated by first measurements. The inverter feeds a nearly sinusoidal current to a large inductor. The reference measurement is taken with a high accuracy current probe CP150 by LeCroy. The controller board of the inverter sends the internally measured current value to a digital-analog-converter (DAC) to allow real time analysis.

Figure 8 shows the result in the time domain at a peak current of 150 A. Obviously the current waveform measured with the proposed sensor and evaluation method matches the reference measurement very well.

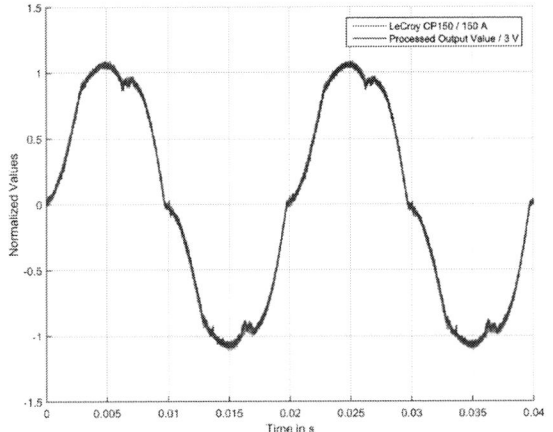

Fig. 8. Comparision in the time domain

Figure 9 shows the x-y-plot of the evaluated sensor signal and the external current probe over 5 periods. The linearity is found to be very good. The total RMS error (linearity and noise) amounts to about 2 % of the nominal current of 450 A. For the first shot, this is already quite satisfactory, since there are still some options for improvements.

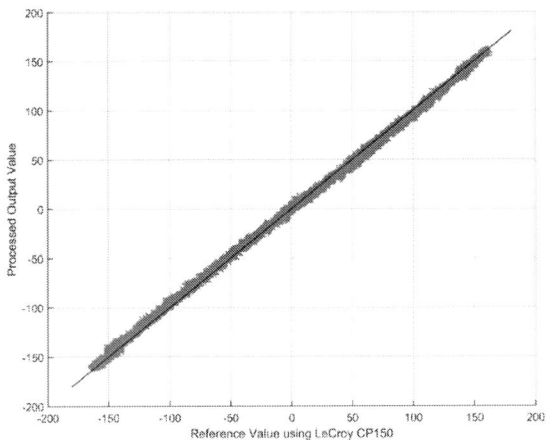

Fig. 9. Measurement of the linearity of the supposed sensor and evaluation principle

IV. CONCLUSIONS

The proposed current measurement method has the potential to replace the commonly used hall effect based current sensors in drive inverter applications, especially when weight and space are important factors. The sensor can be implemented using standard PCB technology which makes it cheap and highly reproducible. The simple construction also makes the sensor interesting for applications with extended thermal requirements [7]. The principle is very scalable. The sensor can be realized for very high current ratings. In the first empirical investigations, the measurement method is found to be highly linear and the variance between the sensors of the three different phases is low. Further investigations on optimizing the Rogowski coil layout, the adaption of the integrator gain to the used ADC and the temperature dependence of the whole measurement are ongoing.

REFERENCES

[1] M. Henke, G. Tareilus, N. Langmaack, "SiC boost converter with high power density for a battery electric sports car", 11th Symposium on Hybrid and Electric Vehicles, pp. 260-273, Braunschweig, February 2014

[2] W. Rogowski, W. Steinhaus, "Die Messung der magnetischen Spannung," Electrical Engineering / Archiv für Elektrotechnik, Volume 1, Issue 4, pp 141-150, 1912

[3] T. Guillod, D. Gerber, J. Biela, A. Müsing, "Design of a PCB Rogowski Coil based on the PEEC Method", CIPS 2012, Paper P14, Nuremberg/Germany, March 2012

[4] S. Rolle, "Measurement of an electrical current in a switching power semiconductor component, using an inductive coil and a sample and hold circuit controlled by a control circuit with sampling triggered by the switch ignition signal", Patent Application DE000010107791A1, 2001

[5] W.F. Ray, "A current measuring device," Patent EP1073908A1, 2001

[6] W.F. Ray et. al., "Developments in Rogowski Current Transducers," EPE '97, 7th European Conference on Power Electronics and Applications, Trondheim, Sept. 8 - 10, 1997, vol. 3, pp. 3.308-3.312

[7] M. Henke, G. Tareilus, N. Langmaack, " High Temperature and High CMR Gate Driver Circuit for Wide-Band-Gap Power Semiconductors ", IEEE PEDS 2015, Sydney/Australia, June 9 - 12, 2015

A Novel DBC Layout for Current Imbalance Mitigation in SiC MOSFET Multichip Power Modules

Helong Li, Stig Munk-Nielsen, Szymon Bęczkowski, Xiongfei Wang

Department of Energy Technology
Aalborg University
Aalborg, Denmark
Email: {hel, smn, sbe, xwa}@et.aau.dk

Abstract— **This paper proposes a novel Direct Bonded Copper (DBC) layout for mitigating the current imbalance among the paralleled SiC MOSFET dies in multichip power modules. Compared to the traditional layout, the proposed DBC layout significantly reduces the circuit mismatch and current coupling effect, which consequently improves the current sharing among the paralleled SiC MOSFET dies in power module. Mathematic analysis and circuit model of the DBC layout are presented to elaborate on the superior features of the proposed DBC layout. Simulation and experimental results further verify the theoretical analysis and current balancing performance of the proposed DBC layout.**

I. Introduction

The limited current capability is challenging the use of single SiC MOSFET die for high current applications. Multichip power modules with the paralleled connection of SiC MOSFET dies are commonly used to increase the current capability [1-3]. However, compared to Si IGBT devices, SiC MOSFET has a faster switching capability, and it is thus more sensitive to the parasitic inductance in the circuit layout [4].

For the parallel connection of the SiC MOSFET dies in multichip power modules, the parasitic circuit mismatch may result in a serious transient current imbalance [5-8]. Also, the current coupling effect in the traditional DBC layout of SiC MOSFET power modules tends to further aggravate this transient current imbalance [9]. Previous research efforts spent on the power modules mainly focus on reducing the total switching parasitic inductance [10, 11], increasing the temperature limitation [12, 13] and improving the reliability of the power module [14]. However, the influence of the DBC layout on mitigating the transient current imbalance in multichip power module is often overlooked.

This paper proposes a split-output DBC layout for multichip SiC MOSFET power modules, in order to mitigate the current imbalance among the paralleled dies. The paper is organized as below. In section II, the mechanism causing the transient current imbalance

between paralleled MOSFETs is first theoretically analyzed. In section III, the influence of the traditional DBC layout on the transient current imbalance in commercial power modules is analyzed. Two effective guidelines are outlined for mitigating the current imbalance. In section IV, a novel DBC layout is proposed following the guidelines in section III. Theoretical modeling and analysis indicate that proposed DBC layout can mitigate the transient current imbalance among the paralleled dies. In section V, simulation and experimental results validate the effectiveness of the proposed guidelines for the mitigation of transient current imbalance and the proposed DBC layout.

II. Mechanism of Transient Current Imbalance

The schematic of paralleling two MOSFETs in a double pulse test circuit are shown in Fig.1. With respect to the common source stray inductance, the gate source voltage (V_{GS}) of the MOSFET can be determined as in (1), during switching transient. L_{s1} and L_{s2} are the common source stray inductance for the paralleled two SiC MOSFETs.

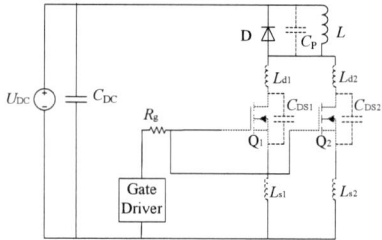

Fig.1. Paralleling two MOSFETs with stray inductance.

$$V_{GS} = V_{driver} - i_G R_G - L_s \frac{di_D}{dt} = V_{driver} - i_G R_G - \Delta V_{LS} \qquad (1)$$

The gate current (i_G) can be neglected compared to the drain-source current (i_D). Therefore, it is reasonable to consider that the paralleled MOSFETs have a simliar gate voltage potential. Consequently, the V_{GS} difference between the paralleled MOSFETs is determined by the MOSFETs source voltage potentials, as shown in (2).

978-1-4673-9551-9/16 $31.00 © 2016 IEEE

$$V_{GS1} - V_{GS2} = (V_{G1} - V_{S1}) - (V_{G2} - V_{S2}) \approx V_{S2S1} \qquad (2)$$

For MOSFETs, the transient drain current (i_D) is controlled by the gate source voltage V_{GS}, as given by (3). V_{th} is the threshold voltage of the MOSFET.

$$i_D = g_{fs}(V_{GS} - V_{th}) \qquad (3)$$

Therefore, the transient drain current imbalance between two paralleled devices can be presented as (4), which indicates that the mechanism of the transient current imbalance between the paralleled devices is the source voltage potential difference.

$$i_{D1} - i_{D2} = g_{fs}(V_{GS1} - V_{th}) - g_{fs}(V_{GS2} - V_{th}) = g_{fs}V_{S2S1} \qquad (4)$$

III. INFLUENCE OF DBC LAYOUT

Two commercial half-bridge multichip power modules with the paralleled connections of SiC MOSFET dies are shown in Fig.2. Power module in Fig.2(a) has the external paralleled Schottky dioes and the auxiliary source connections, while power module in Fig.2(b) does not have. The external Schottky diodes do not affect the transient current distribution among the paralleled MOSFETs. The auxiliary source connection usually has much larger impedance than the power DBC traces. Therefore, most of i_D still goes through the power DBC traces. Consequently, the auxiliary source connection has a limited effect on the transient current distribution and not included in the modeling and analysis in this paper.

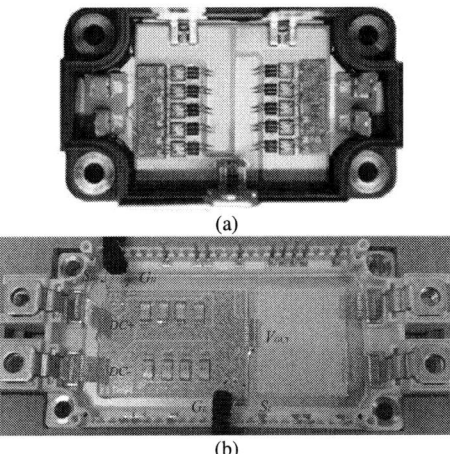

(a)

(b)

Fig.2. Half-bridge SiC MOSFET power modules (a) 1.2kV power module with 5 dies in parallel (b) 1.7kV power module with 4 dies in parallel.

Except for the external Schottky diode and the auxiliary source connections, these two power modules have a very similar DBC layout. This type of DBC layout of multichip power modules is shown in Fig.3.

The common source stray inductance of the DBC traces is depicted with different colors. L_b is the stray inductance of the bond wire for source terminal connection. L_{12}, L_{23} and L_{34} are the stray inductance of the source DBC traces between Q_1 and Q_2, Q_2 and Q_3, Q_3 and Q_4. L_{ss} is the stray inductance of the DBC trace from L_{12} to the DC- terminal.

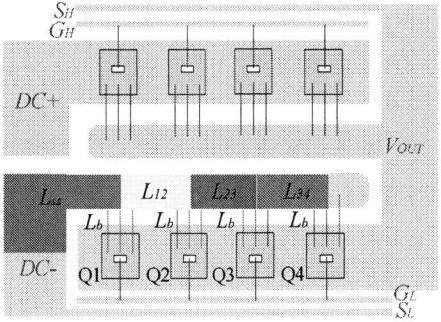

Fig.3. DBC layout of the power module in Fig.2(b).

The modeling for the bottom four paralleled SiC MOSFETs in the DBC layout is shown in Fig.4 with a double pulse test circuit. The common source stray inductance is included in the model in Fig.4.

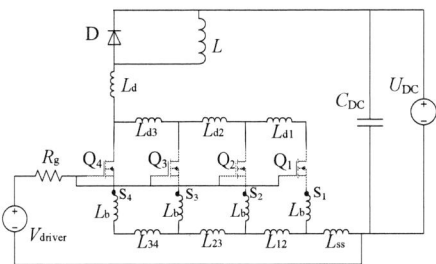

Fig.4. Modeling with traditional DBC layout.

The MOSFETs source voltage potential difference in Fig.4 can be described by (5). The transient current imbalance can be calculated with (4).

$$
\begin{aligned}
V_{S2S1} &= L_b \frac{d(i_{D2} - i_{D1})}{dt} + L_{12}\frac{d(i_{D2} + i_{D3} + i_{D4})}{dt} \\
V_{S3S2} &= L_b \frac{d(i_{D3} - i_{D2})}{dt} + L_{23}\frac{d(i_{D3} + i_{D4})}{dt} \\
V_{S4S3} &= L_b \frac{d(i_{D4} - i_{D3})}{dt} + L_{34}\frac{di_{D4}}{dt} \\
V_{S4S1} &= L_b \frac{d(i_{D4} - i_{D1})}{dt} + L_{12}\frac{d(i_{D2} + i_{D3} + i_{D4})}{dt} + L_{23}\frac{d(i_{D3} + i_{D4})}{dt} L_{34}\frac{di_{D4}}{dt}
\end{aligned}
\qquad (5)
$$

With (4) and (5), it is clear that the transient current imbalance is determined by the common source stray inductance mismatch (L_{12}, L_{23}, L_{34}) and the applied di/dt. Therefore, the design of DBC layout should follow two guidelines to mitigate the current imbalance:

- Reducing the common stray inductance mismatch
- Reducing the di/dt applied to the mismatched stray inductance.

IV. PROPOSED DBC LAYOUT

To mitigate the transient current imbalance in multichip power modules, a novel DBC layout is proposed, as shown in Fig. 5. Vertical connectors are mounted as highlighted in red. In the proposed DBC layout, the MOSFETs and the Schottky diodes are placed based on the concept of switching cell [15], which has been used to reduce the switching loop stray inductance. There are two AC terminals (OUT1 and OUT2) in this DBC layout. If they are directly connected, this is a DBC layout for normal half bridge power module. However, if the two output terminals are separately connected with two leakage inductors, the half bridge power module with the DBC layout will work as a split output half bridge. The operating mechanism of the split-output half bridge has been discussed in [16].

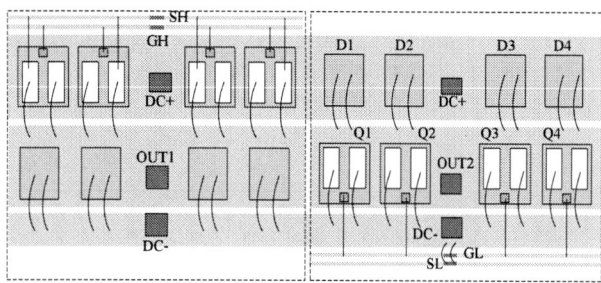

Fig.5. Split-output DBC layout with separate DBC patterns

The model of the proposed DBC layout for Q_1–Q_4 in a double pulse test circuit is shown in Fig.6. The voltage potential difference between the source terminals of the paralleled MOSFETs is given by (6), and transient current imbalance can be calculated with (4).

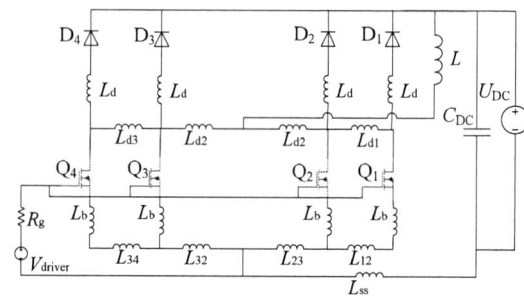

Fig. 6. Modeling with proposed DBC layout for bottom MOSFETs.

$$V_{S2S1}=L_b\frac{d(i_{D2}-i_{D1})}{dt}+L_{12}\frac{di_{D1}}{dt}$$

$$V_{S3S2}=L_b\frac{d(i_{D3}-i_{D2})}{dt}$$

$$V_{S4S3}=L_b\frac{d(i_{D4}-i_{D3})}{dt}+L_{34}\frac{di_{D4}}{dt} \qquad (6)$$

$$V_{S4S1}=L_b\frac{d(i_{D4}-i_{D1})}{dt}$$

Comparing (5) with (6), it is obvious that the mismatch between the common source stray inductances is reduced. The di/dt applied to the mismatched stray inductance is also reduced. Consequently, the transient current imbalance with the proposed DBC layout is mitigated. If there is no device parameters mismatch, there is no transient current imbalance between Q_2 and Q_3, and between Q_1 and Q_4.

V. SIMULATIONS AND EXPERIMENTAL RESULTS

LTspice simulation results are shown in Fig.7.

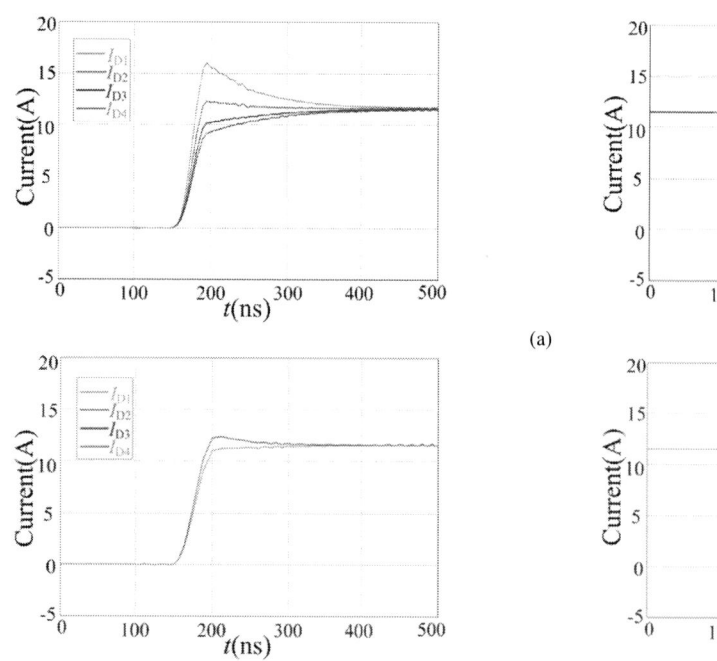

(a)

(b)

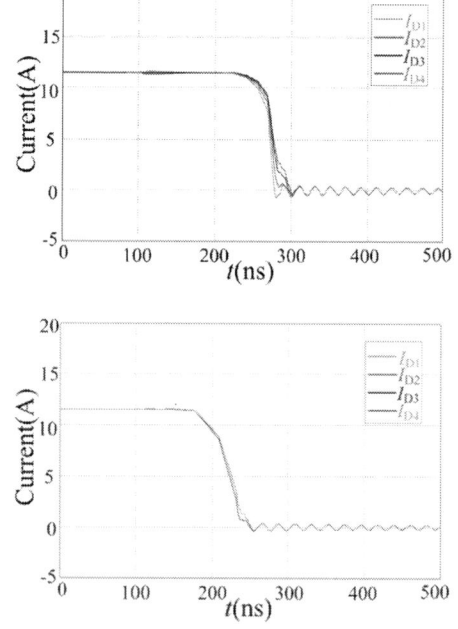

Fig.7. Simulation results of transient current distribution. (a) with traditional DBC layout (b) with proposed DBC layout.

978-1-4673-9551-9/16 $31.00 © 2016 IEEE

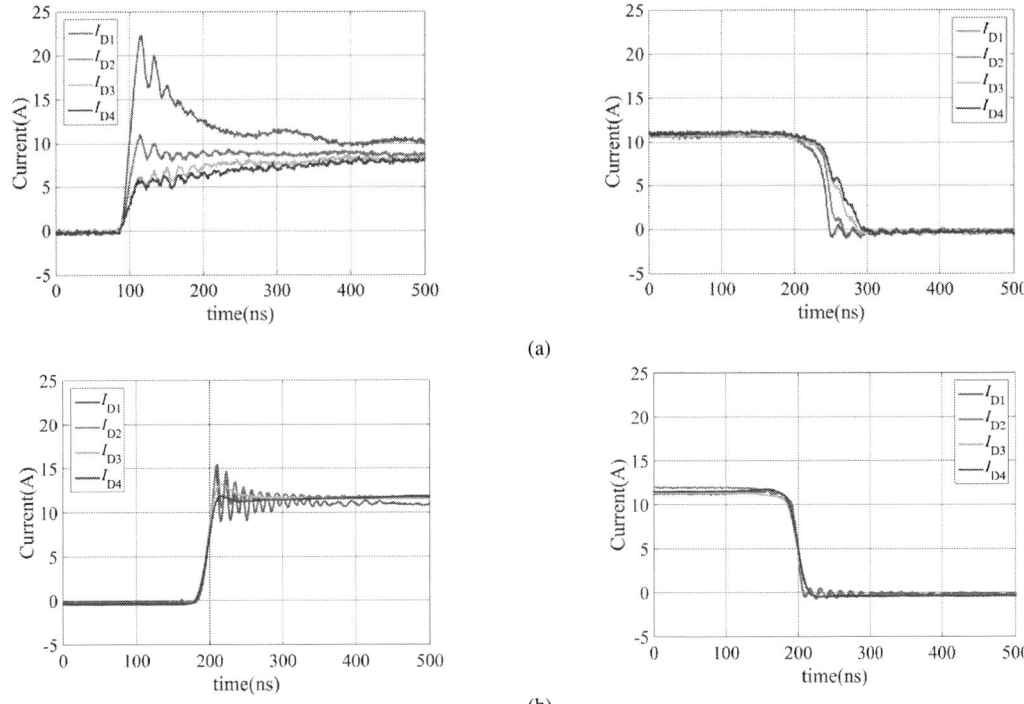

Fig.8. Experimental results of transient current distribution (a) with traditional DBC layout (b) with proposed DBC layout.

In the simulation circuit, $L_{12}=L_{23}=L_{34}=2nH$. It is obvious that the paralleled four SiC MOSFETs currents (I_{D1}, I_{D2}, I_{D3} and I_{D4}) with the proposed DBC layout have a reduced imbalance in Fig.7 (b) compared with that in Fig.7 (a). And with the proposed DBC layout, there is no transient current imbalance between Q_2 and Q_3, and between Q_1 and Q_4, because there is no mismatch of common source stray inductance between them.

Since the die current is hard to be measured accurately, the PCB circuits with similar layouts of the DBC layouts are built in experiments by using discrete SiC MOSFETs. Four paralleled SiC MOSFETs with a close threshold voltage are selected. Fig.8 shows the measured MOSFET drain currents, where Fig.8(a) depicts the current distribution with the traditional layout and Fig.8(b) shows the current distribution with the proposed layout. It is clear that the proposed DBC layout effectively mitigates the transient current imbalance in the SiC MOSFETs compared to the traditional DBC layout. Moreover, comparing Fig.7 and Fig.8, a larger current imbalance than the simulation results is observed in experiments. This is due to the larger stray inductance of the PCB circuits than the DBC in simulations. The simulation results with similar stray inductance as the experiments can be found in [8], which closely match with the experimental results.

VI. CONCLUSION

Traditional DBC layouts of the multichip SiC MOSFET power modules with the paralleled dies have

the large transient current imbalance. The cause of the transient current imbalance has been identified as the MOSFETs source voltage potential difference. Two design guidelines for the transient current imbalance mitigation have been outlined based on mathematic analysis in this paper. A DBC layout which effectively mitigates the transient current imbalance has been developed and validated by simulation and experimental results. The proposed DBC layout is not only for SiC MOSFETs, but also effective for Si IGBTs.

REFERENCES

[1] D. Sadik, J. Colmenares, D. Peftitsis, L. Jang-Kwon, J. Rabkowski and H. Peter-Nee, "Experimental investigations of static and transient current sharing of parallel-connected silicon carbide MOSFETs," in *Proc. IEEE Eur. Conf. Power Electron. Appl.*, 2013, pp. 1-10.

[2] Y. Xue, J. Lu, Z. Wang, L. M. Tolbert, B. J. Blalock and F. Wang, "Active current balancing for parallel-connected silicon carbide MOSFETs," in *Proc. IEEE Energy Convers. Congr. Expo.*, 2013, pp. 1563-1569.

[3] Z. Chen, Y. Yao, D. Boroyevich, K. D. T. Ngo, P. Mattavelli and K. Rajashekara, "A 1200-V, 60-A SiC MOSFET Multichip Phase-Leg Module for High-Temperature, High-Frequency Applications," *IEEE Trans. Power Electron.*, vol. 29, pp. 2307-2320, 2014.

[4] H. Li and S. Munk-Nielsen, "Challenges in switching SiC MOSFET without ringing," in *proc. PCIM Europe*, 2014, pp. 1-6.

[5] H. Li, S. Munk-Nielsen, C. Pham and S. Beczkowski, "Circuit mismatch influence on performance of paralleling silicon carbide MOSFETs," in *Proc. IEEE Eur. Conf. Power Electron. Appl.*, 2014, pp. 1-8.

[6] H. Li, S. Munk-Nielsen, X. Wang, R. Maheshwari, S. Beczkowski, C. Uhrenfeldt and W. Franke, "Influences of Device and Circuit

Mismatches on Paralleling Silicon Carbide MOSFETs," *IEEE Trans. Power Electron.,* pp. 1-1, 2015.

[7] A. Consoli, F. Gennaro, V. John and T. Lipo, "Effects of the internal layout on the performance of IGBT power modules," in *Brazilian Power* Electronics *Conference, Foz do Iguaçu,* 1999, pp.1-9.

[8] J. C. Joyce, "Current sharing and redistribution in high power IGBT modules," Ph.D. dissertation, Dept. Elect. Eng., Cambridge Univ. , MA, 2001.

[9] H. Li, S. Beczkowski, S. Munk-Nielsen, R. Maheshwari and W. Franke, "Circuit mismatch and current coupling effect influence on paralleling SiC MOSFETs in multichip power modules," in *proc. PCIM Europe ,*2015, pp. 1-8.

[10] A. Müsing and J. W. Kolar, "Ultra-Low-Inductance Power Module for Fast Switching Semicon-ductors," in *proc. PCIM Europe ,*2013, pp. 1-8.

[11] S. Li, L. M. Tolbert, F. Wang and F. Z. Peng, "Stray Inductance Reduction of Commutation Loop in the P-cell and N-cell-Based IGBT Phase Leg Module," *IEEE Trans. Power Electron.,* vol. 29, pp. 3616-3624, 2014.

[12] F. Xu, D. Jiang, J. Wang, F. Wang, L. M. Tolbert, T. J. Han, J. Nagashima and S. J. Kim, "High temperature packaging of 50 kW three-phase SiC power module," in *Proc. ECCE Asia,* 2011, pp. 2427-2433.

[13] S. Hazra, S. Madhusoodhanan, S. Bhattacharya, G. K. Moghaddam and K. Hatua, "Design considerations and performance evaluation of 1200 V, 100 a SiC MOSFET based converter for high power density application," in *Proc. IEEE Energy Convers. Congr. Expo.,* 2013, pp. 4278-4285.

[14] P. Ning, R. Lai, D. Huff, F. Wang, K. D. Ngo, V. D. Immanuel and K. J. Karimi, "SiC wirebond multichip phase-leg module packaging design and testing for harsh environment," *IEEE Trans. Power Electron.,* vol. 25, pp. 16-23, 2010.

[15] S. Li, L. M. Tolbert, F. Wang and F. Z. Peng, "Reduction of stray inductance in power electronic modules using basic switching cells," in *Proc. IEEE Energy Convers. Congr. Expo.* 2010, pp. 2686-2691.

[16] H. Li, S. Munk-Nielsen, S. Beczkowski and Xiongfei Wang, "SiC MOSFETs based split output half bridge inverter: Current commutation mechanism and efficiency analysis," in *Proc. IEEE Energy Convers. Congr. Expo.,* 2014, pp. 1581-1588.

A Double-End Sourced Multi-Chip Improved Wire-bonded SiC MOSFET Power Module Design

Miao Wang, Fang Luo, Longya Xu
Department of Electrical and Computer Engineering
The Ohio State University
Columbus, Ohio, United States of America
wang.2965@osu.edu

Abstract— This paper proposes an improved wire-bonded design with a unique double-end sourced (DES) structure for multi-chip paralleled silicon carbide (SiC) power modules. The new structure adopts two pairs of DC bus-bars to source the power module from the two ends, not only shortens the equivalent power loops but also provides a symmetrical structure for the paralleled devices. The proposed design achieved a minimized power-loop inductance of 7.2 nH. In addition, the design improved current sharing among the paralleled devices. A 1200 V, 60 A SiC metal-oxide-semiconductor field-effect-transistor (MOSFET) half-bridge module (3 devices in parallel) is fabricated and tested for verification. Improved performances are observed in both switching and continuous operation. A converter level design is also presented to accommodate this unique module structure.

Keywords— *multi-chip power module; wire-bonded; power-loop inductance; silicon carbide metal-oxide-semiconductor field-effect-transistor; double-end sourced; current-sharing*

I. INTRODUCTION

The silicon carbide (SiC) metal-oxide-semiconductor field-effect-transistor (MOSFET) has attracted a tremendous attention in the past decades due to its superior performances over its silicon (Si) predecessor. The high-speed switching, high-voltage blocking, and high-temperature operating capability of the SiC MOSFET will simplify the requirements for the thermal management and obtain an improved efficiency due to the reduced switching losses. The current ratings of the available commercial power MOSFETs are limited within 100 A. Therefore, in situations where a high-power rating is demanded, power modules are commonly used since the paralleled devices could provide increased current-handling capability. However, the conventional power modules suffer from the large power-loop inductance, which limits the fast switching and threatens the safe-operation of the SiC devices. Therefore, the development of SiC devices and the limitation of the conventional power modules have driven the demand for the next generation of power module packaging.

A great number of studies have been conducted on the power module structures and layouts in order to minimize the module's power-loop inductance, and hence obtain better performances. The concept of the switch cell (SC) was proposed in [1]. It consists of a MOSFET and its freewheeling diode that is placed in close proximity. This arrangement shortens the distance of the current-commuting route between the two devices, and hence the power-loop inductance. Due to the reduced inductances, power modules that adopt SCs have shown superior performances [2-4]. In [5-7], the planar-packaging structure was thoroughly studied. Compared with the conventional power modules, the planar structure gets rid of the wire-bonds and, instead, make the connection on the top pads with another direct bond copper (DBC) board. The absence of the wire-bond can significantly reduce the power-loop inductance and achieve an improved switching performance. In addition, the additional DBC board provides the double-sided cooling capability, which can greatly increase modules' dissipation and enables the devices to operate at higher power ratings under the same junction temperature. However, the planar-packaging structure suffers from complicated fabricating process, and requires special treatment on the bare dice. Therefore, [8, 9] proposed a hybrid-packaging structure, which possesses smaller power-loop inductances compared with the wire-bonded structure and an easier fabrication process compared with the planar packaging. The use of the printed circuit board allows flexible power-loop and gate-loop design, which can shrink the footprint and realized reduced stray inductances. Although the hybrid structure found a compromise between the wire-bonded packaging and planar packaging, the unverified reliability could threatens the safe operation of the power module. Table I summarizes the advantages and disadvantages of the different power module structures presented in [1-9].

The large power-loop inductance of the wire-bonded structure will cause huge overshoot voltage during switching transients and threaten the safe-operation of the devices. However, its high maturity and simplicity make this structure popular in the commercial market. This paper proposes an improved wire-bonded power module design. The new design inherits the advantage of using wire-bond connections, while it significantly reduces power-loop inductances in the module. In addition, this design improves the current sharing between the paralleled devices due to unbalanced power loops [10-13].

II. MODULE LAYOUT DESIGN AND ANALYSIS

SCs have shown improved performances [2-4] and are widely applied in power module packaging. Fig. 1(a) illustrates a baseline design of a parallel-SC module according to [2-4]. As seen, each MOSFET is placed closely to its freewheeling

TABLE I. POWER MODULE STRUCTURE COMPARISON

	Advantages	Disadvantages
Wire-bonded [1-4]	• High maturity, low cost • Easy to fabricate, short fabrication time • High reliability	• Large loop inductance, 12 ~ 25 nH

Planar [5-7]	• Small loop inductance, 4 ~ 12 nH
	• Double-side cooling capability
		• Limited die attachment materials, multi-layer soldering
		• Extremely high fabrication complexity
		  ➤ High coaxial pressure, easy to damage the device
		  ➤ Double-side solder-able device is required
		• Extremely long fabrication time

Hybrid [8, 9]	• Medium fabrication difficulty • Medium loop inductance, 4 ~ 16 nH	• Unverified reliability at raised temperature, the coefficient of thermal expansion (CTE) mismatch might cause interconnection failure • Low maturity

diode to reduce the power-loop inductance, and three MOSFETs are paralleled in order to increase the current-handling capability. The Kelvin-source structure is adopted, whose terminals are arranged at the outer areas of the module. The module is sourced from the DC bus terminals on the right, and a decoupling capacitor C_d is integrated in the module, close to the devices. It closes the power loops for the switching devices so that the stray inductance introduced by the DC-bus terminals would not affect the switching behavior of the module, and as a result it suppresses the voltage overshoot in

the power loop [1].

Fig. 1(a) shows that the distance from the individual SCs to the DC-bus terminals varies. D1-M4, which is placed close to DC^+-DC^-, has the shortest power-loop distance, while M3-D6 at the far end has the longest distance. This varied distances contributes to the unbalanced power-loop inductances seen by the paralleled SCs. In other words, a significantly increasing power-loop inductance will be observed on SCs from D1-M4 to M3-D6. This increased power-loop inductance will generate huge overshoot voltage on the MOSFET, and this unbalanced

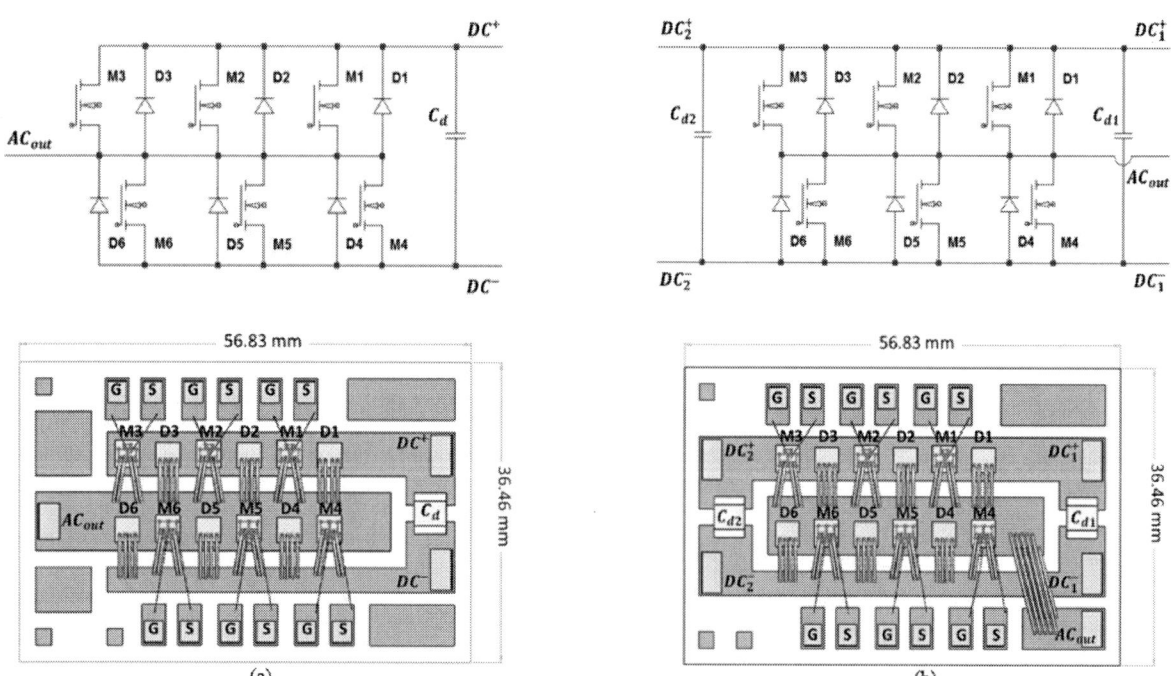

Fig. 1. Power Module Circuit Schematic and Layout. (a) Baseline Structure. (b) DES Structure.

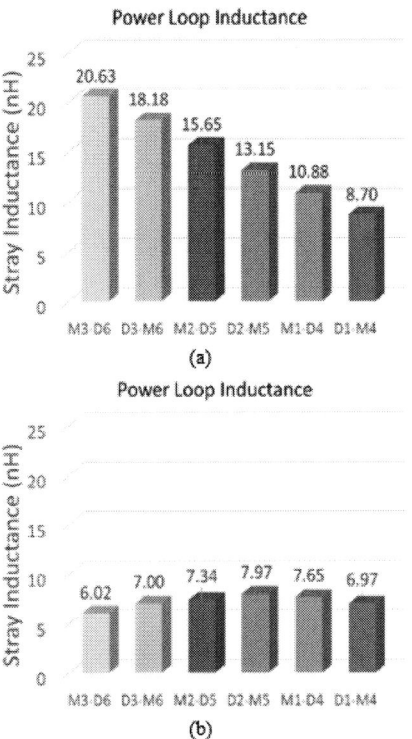

Fig. 2. Power Loop Inductances. (a) Baseline Layout. (b) DES Layout.

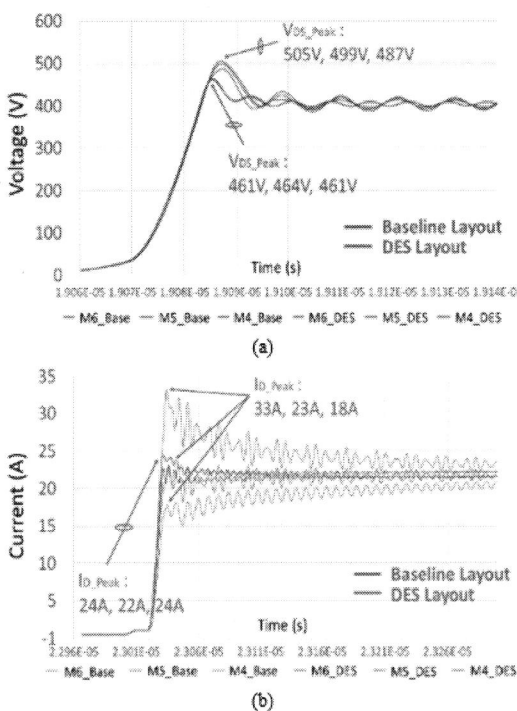

Fig. 3. Switching Transient Performance. (a) Turn-off Overshoot Voltage. (b) Turn-on Current Sharing.

layout will cause unbalanced current distribution within the module.

To suppress the overshoot voltage and improve the current sharing during switching transients, this paper proposes a new double-end sourced power module structure, as shown in Fig. 1(b), where another pair of DC-bus terminals DC_2^+ and DC_2^- was adopted on the left end. This unique structure provides two paralleled power loops for every device, one through C_{d1} and another through C_{d2}, and the equivalent power-loop inductance will be reduced compared with the one with single power loop. For example, in the baseline structure, the SC M3-D6 has a large power-loop inductance due to its long distance to DC^+-DC^-, while, on the other hand, the additional power loop through DC_2^+-DC_2^- in the DES layout shortens its power loop and thus significantly reduces the power-loop inductance. As for D1-M4, the additional power loop can further reduce its power-loop inductance, which would lead to a better switching performance. In addition, this symmetrical structure achieves identical power-loop inductances, from the devices at the two ends to the ones in the middle, and this will distribute currents evenly among the paralleled devices.

The parasitic power-loop inductances of these two designs are illustrated in Fig. 2(a) and 2(b). The baseline layout, as shown in Fig. 2(a), exhibits unbalanced power-loop inductances that vary in a wide range. D1-M4, which is placed close to the DC bus terminals, has the smallest power-loop inductance of 8.7 nH. And as the distance from the SCs to the DC-bus terminals increases, the devices see a rapidly growing

power-loop inductance. As a result, the SC at the opposite end exhibits a large power-loop inductance of 20.63 nH. On the other hand, Fig. 2(b) demonstrates the reduced power-loop inductances obtained with the DES layout, showing that every power-loop inductance is restrained within 8 nH. Furthermore, the inductance difference between devices is minimized to 1.95 nH, compared with 11.93 nH in the Baseline design. In other words, the DES layout effectively reduces the equivalent power loops of the individual MOSFETs, and brings symmetricity and uniformity to the power module.

III. SIMULATION VALIDATION

The switching simulation models for both designs were built in LT-Spice based on the finite element analysis (FEA) results shown in Fig. 2, and the circuit was simulated at 400 V/60 A on the lower-leg devices (M4, M5, and M6). Fig. 3(a) shows the overshoot voltage across each MOSFET during the module's turn-off transient. In the baseline layout, the large power-loop inductances introduce high induced voltages across themselves and are added on the MOSFETs, and large overshoot voltages reaching up to 505 V are generated on the devices. Moreover, due to the unbalanced power-loop inductances, in-consistent turn-off voltages with a difference of 18 V is observed among the paralleled MOSFETs. These performances get improved in the DES layout, where the reduced power-loop inductances lower the peak overshoot voltage from 505 V to 464 V, and the symmetrical structure brings consistency between devices, minimizing the voltage difference to 3 V.

Fig. 3(b) shows the dynamic current sharing comparison of these two designs. It can be seen that M4, M5, and M6 in the baseline layout has a peak dynamic current of 33 A, 23 A, and 18 A, respectively. The MOSFET that is closer to the decoupling capacitor draws more current than other devices because the smaller power-loop inductance imposes less resistance for the rising current. On the other hand, since the symmetricity of the DES layout brings identical power-loop inductances, dynamic currents are shared evenly among the paralleled devices. And the difference is minimized from 15 A to 2 A. This simulation results on the switching behavior of the two layouts reveal that the reduced and balanced power-loop inductances brought by the DES layout contribute to the consistent and improved switching performances among the paralleled MOSFETs.

Fig. 4 illustrates the reduced total switching loss by adopting the DES layout. It shows that the reduced overshoot voltages and balanced sharing currents generate lower switching loss and improve the efficiency of the power module.

IV. EXPERIMENT VALIDATION

Fig. 5(a) shows the internal structure of the fabricated 1200 V, 60 A SiC MOSFET power module with the DES layout, and its assembled product is given in Fig. 5(b). Another power module with the baseline layout was fabricated as well (not shown here). To verify the simulation results shown in Fig. 3(a), the double-pulse test (DPT) was conducted at 300 V/60 A on both designs and the results are given in Fig. 6. In the baseline layout, M4, M5, and M6 obtained a peak overshoot voltage of 352 V, 362 V, and 370 V, respectively. The improved switching performance during the module turn-off is obtained with the DES layout, which reduced the highest overshoot voltage from 370 V to 348 V, and minimized the inconsistency from 18 V to 4 V. The low-frequency component shown in Fig. 6 is caused by the oscillation on the bus-bars connecting the power module.

Measuring the current that goes through the individual device is hard to realize with the proposed structure. Therefore, an alternative approach was adopted to verify the improvement on the current distribution. A continuous power test was conducted on both layouts, and the temperature distribution was recorded as an indicator of the current distribution. Fig. 7 shows the schematic of the continuous power test, where the fabricated power module is configured in a boost converter. In this setup, the lower-leg MOSFETs and the upper-leg Diodes were used to complete the circuit, and the test conditions are given in Fig. 8. Fig. 9 presents the static characterization of the selected MOSFETs, which demonstrates the uniformity of the devices. Fig. 10(a) and 10(b) show the steady status temperature distribution of the baseline layout and DES layout, respectively. The DES layout not only lowered the highest MOSFET temperature from 75.5 °C to 74.2 °C, but also minimized the temperature difference among the MOSFETs from 4.6 °C to 1.7 °C. This more evenly distributed temperature in the DES layout indicates that those devices exhibit consistent performances, and evenly shared current goes through each device. This benefit will become significant when operating the devices at their rated conditions.

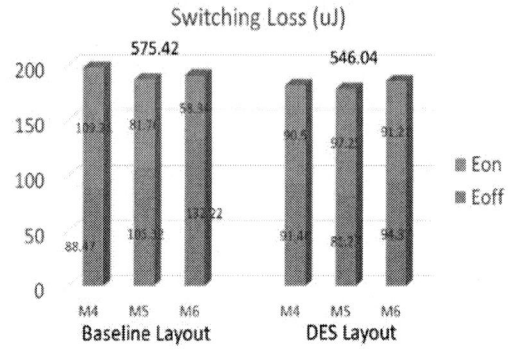

Fig. 4. Switching Loss Comparison.

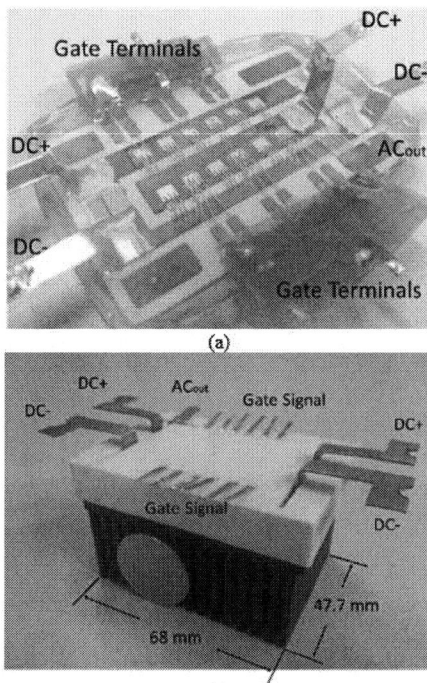

Fig. 5. Fabricated Half-Bridge Module with DES Layout. (a) Internal Structure. (b) Final Product.

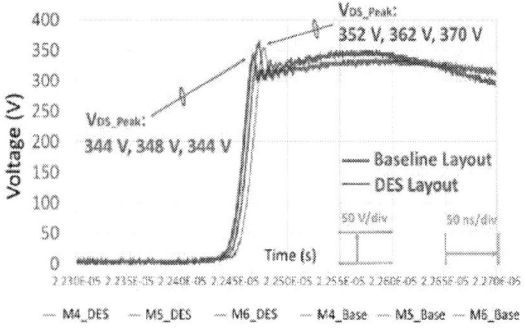

Fig. 6. Experiment DPT Results.

978-1-4673-9551-9/16 $31.00 © 2016 IEEE

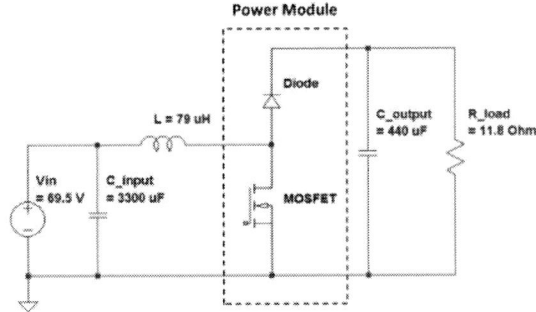

Fig. 7. Continuous Power Test Schematic.

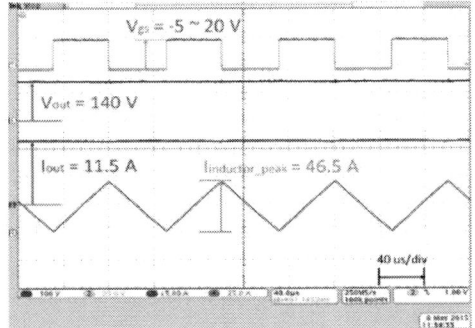

Fig. 8. Continuous Power Test Conditions.

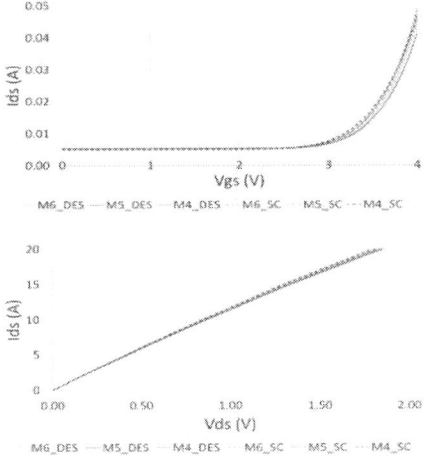

Fig. 9. Vth and I/V Characterization.

Limited by the capability of the facilities in the lab, the module could not be tested at higher ratings under the resistive load condition. Therefore, a full-bridge DC-AC converter was built using the DES modules, and an inductor was selected as the load to absorb the reactive power. Fig. 11 presents the test setup of the full-bridge converter. The DC bus-bars were carefully designed so that the connection did not break the

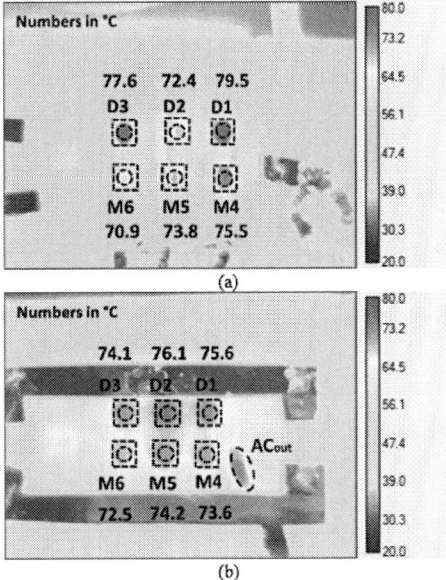

Fig. 10. Steady Status Temperature Distribution. (a) Baseline Layout. (b) DES Layout.

symmetricity of the power modules. Fig. 12(a) shows that the DC voltage was kept at 300 V, and the AC output current was pushed up to 60 A (peak). Fig. 12(b) demonstrates the evenly distributed temperature of the MOSFETs, which indicates a uniform current-sharing within the power module.

V. CONVERTER LEVEL CONSIDERATION

This paper verified the improved performances brought by the unique structure of the DES layout. However, this new double-end bus-bar design does not fit the traditional converter structures. Therefore, this paper proposed a converter level design to accommodate the DES layout power modules. The converter design is presented in Fig. 13. The three half-bridge power modules are placed side-by-side, and a laminated DC bus-bar and the DC capacitors are positioned on top of the modules. The H-shape DC-terminal connectors link the adjacent power modules and connect them to the DC bus-bar. For the individual power module, it is connected to the DC bus from the two ends to ensure a) the symmetricity of the DC source, b) the shortest path from the DC bus to the module. And the gate driver board could be sandwiched between the DC bus-bars and power modules to reduce the gate-loop length and compact the converter system.

This arrangements ensure that each power module is sourced symmetrically from the two ends. And the capacitors that are placed on top of the DC bus-bar could shrink the system size and reduce the stray inductance.

VI. CONCLUSION

This paper proposes a new structure for multi-chip wire-bonded power modules. The featured double-end sourced structure exhibits a reduced power-loop inductance of approximately 7.2 nH, and provides a balanced structure for

978-1-4673-9551-9/16 $31.00 © 2016 IEEE

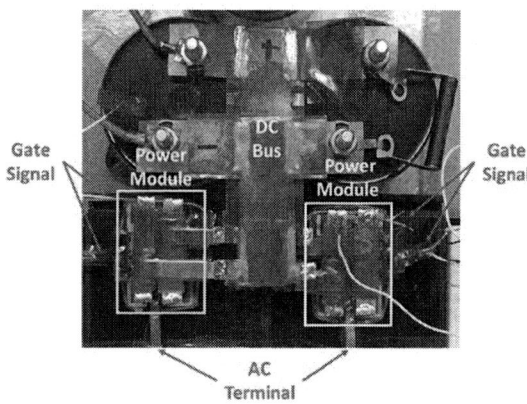

Fig. 11. Full-bridge Test Setup.

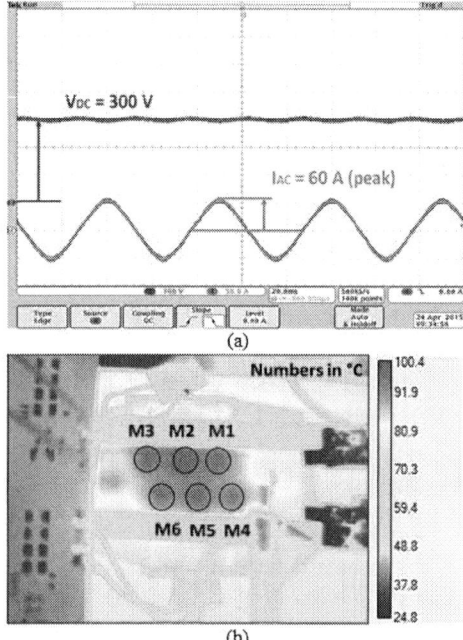

Fig. 12. Full-bridge Test Results. (a) Test Waveforms. (b) Temperature Distribution.

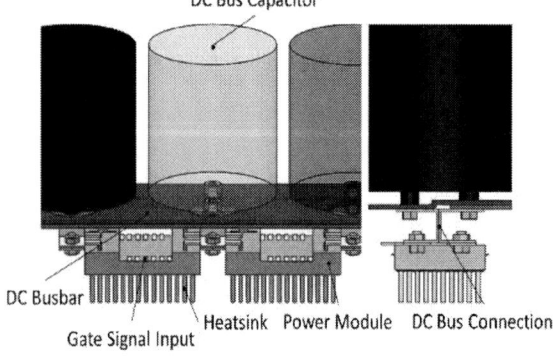

Fig. 13. System Design Diagram.

the paralleled devices. Both the simulation and experiment switching tests were conducted, and the improved transient performances were observed. A continuous power test was conducted, and it verified the improved dynamic current sharing within the proposed power module. In order to accommodate the unique structure, a converter design is provided which ensures a symmetrical structure seen by each power module.

REFERENCES

[1]   S. Li, L. M. Tolbert, F. Wang, and F. Peng, "Reduction of Stray Inductance in Power Electronic Modules Using Basic Switching Cells," IEEE Energy Conversion Congress and Exposition, pp. 2686-2691, 2010

[2]   F. Xu, T. J. Han, D. J, etc., "Development of a SiC JFET-Based Six-Pack Power Module for a Fully Integrated Inverter," IEEE Transactions on Power Electronics, vol. 28, pp. 1464-1478, 2013

[3]   S. Li, L. M. Tolbert, F. Wang, F. Peng, "P-Cell and N-Cell Based IGBT Module: Layout Design, Parasitic Extraction, and Experimental Verification," IEEE Applied Power Electronics Conference and Exposition, pp. 372-378, 2011

[4]   Z. Chen, Y. Yao, D. Boroyevich, etc., "A 1200 V, 60 A SiC MOSFET Multi-Chip Phase-Leg Module for High-Temperature, High-Frequency Applications," IEEE Applied Power Electronics Conference and Exposition, pp 608-615, 2013

[5]   A. A. Bajwa, R. Zeiser, and J. Wilde, "Process Optimization and Characterization of a Novel Micro-Scaled Silver Sintering Paste as a Die-Attach Material for High Temperature High Power Semiconductor Devices," International Spring Seminar on Electronics Technology, pp. 53-58, 2013

[6]   H. Zhang, S. S. Ang, H. A. Mantooth, and S. Krishnamurthy, "A High Temperature, Double-Sided Cooling SiC Power Electronics Module," IEEE Energy Conversion Congress and Exposition, pp. 2877-2883, 2013

[7]   B. J. Grummel, Z. J. Shen, H. A. Mustain, and A. R. Hefner, "Thermo-Mechanical Characterization of Au-In Transient Liquid Phase Bonding Die-Attach," IEEE Transactions on Components, Packaging and Manufacturing Technology, vol. 3, pp. 716-723, 2013

[8]   R. Wang, Z. Chen, D. Boroyevich, etc., "A Novel Hybrid Packaging Structure for High-Temperature SiC Power Modules," IEEE Transactions on Industry Applications, vol. 49, pp. 1609-1618, 2013

[9]   Z. Chen, Y. Yao, D. Boroyevich, etc., "An Ultra-Fast SiC Phase-Leg Module in Modified Hybrid Packaging Structure," IEEE Energy Conversion Congress and Exposition, pp. 2880-2886, 2014

[10] J. Fabre and P. Ladoux, "Parallel Connectioin of SiC MOSFET Modules for Future Use in Traction Converters," International Conference on Electrical Systems for Aircraft, Railway, Ship Propulsion and Road Vehicles, pp.1-6, 2015

[11] J. Rabkowski, D. Peftitsis, and H. P. Nee, "Parallel Operation of Discrete SiC BJTs in a 6-kW/250-kHz DC/DC Boost Converter," IEEE Transactions on Power Electronics, vol. 29, pp. 2482-2491, 2014

[12] J. Colmenares, D. Peftitsis, H. P. Nee, and J. Rabkowski, "Switching Performance of Parallel Connected Power Modules with SiC MOSFETs," International Power Electronics Conference (IPEC Hiroshima), pp. 3712-3717, 2014

[13] H. Li, S. Munk-Nielsen, X. Wang, etc., "Influences of Devices and Circuit Mismatches on Paralleling Silicon Carbide MOSFETs," IEEE Transactions on Power Electronics, vol. PP, 2015

AUTHOR INDEX

A

Abdelmoaty, Ahmed	2437
Abdul Azeez, Najath	3140
Abe, Seiya	1640, 2422
Abedinpour, Siamak	3669
Abramov, Eli	111, 692
Abramson, Rose A.	1138
Abu Qahouq, Jaber A.	1868, 2114, 3611, 3684
Abu-Rub, Haitham	1214, 3663
Acero, J.	3020, 3026, 3566
Achanta, Prasanta K.	3273
Adhikari, Jeevan	9
Adragna, Claudio	564
Afridi, Khurram K.	1138, 1392, 1947, 2395
Afsharian, Jahangir	33, 899, 2312, 2320
Agamy, Mohammed	3403
Agelidis, Vassilios G.	236, 1702
Agostinelli, M.	339, 350
Agostini, Francesco	472
Agrawal, Neeraj	951
Aguilar, A.	2540
Ahmed, Emad M.	1505
Ahmed, Ibrahim	1882
Ahmed, Mahrous	1505
Ahmed, Shamim	1646
Ahmed, Shehab	936
Ahmed-Zaid, Said	2821
Ahn, Jung-Hoon	163, 1273, 2161
Ahsanuzzaman, S.M.	2497
Akagi, Hirofumi	1163
Akamatsu, Keiji	2607
Akin, Bilal	505, 1096, 1176, 2108, 2875
Alatise, Olayiwola	253, 2645
Al-Durra, Ahmed	1941
Alexandrov, Peter	2973
Al-Hallaj, Said	3128
Alharbi, Mahmood	3333
Ali, Kawsar	2491
Allard, Bruno	524
Allen, Scott	979
Allmeling, Jost	1108
Alonso, J. Marcos	1115
Alonso, R.	3020

Alonso, Rafael 3026
Alou, P. 2409
Al-Shyoukh, Mohammad 2437
Am, Sokchea 2401, 2700
Amaro, Mike 66
Ambacher, Oliver 2083
Amirabadi, Mahshid 3704
Amirahmadi, Ahmadreza 3333
Amon, Cristina 1350
Amouzandeh, Maryam S. 329
Andersen, Michael A.E. 1090, 1430, 1541, 1842, 2252, 2473
Andersen, Thomas 1430, 1842
Ando, Masato 2986
Aniruddhan, Sankaran 1878
Anthon, Alexander 1235, 2252
Antunes, Fernando L.M. 2592
Anwar, Saeed 424
Anwar, Usama 1947
Arafat, A.K.M. 1123
Arafat, Akm 2847
Arefifar, Ali 2561
Arias, Andrea 79
Arias, M. 1823
Arias, Manuel 822
Arnold, Cory 1597, 3273
Asa, Erdem 1323, 1756, 1767, 2587
Asensi, R. 1624
Ayers, Curtis 3529
Ayyanar, Raja 432, 3364
Azcondo, F.J. 2389
Azuma, Katsunori 283

B

Badawey, Mohammed 392
Baek, Jeihoon 3004
Bagawade, Snehal 544
Bahman, Amir Sajjad 261, 3012
Bahmani, M.A. 3043
Bai, Hua 529
Bai, Yongjiang 766, 3623
Baier, Thomas 2897
Bak, Claus Leth 3051
Bak, Yeongsu 2764, 3416
Baker, Michael W. 1597
Bakhshai, Alireza 460
Bakker, Cas 2457
Balasubramanian, Bharat 1868
Balda, Juan Carlos 143, 362, 651, 1387, 3712
Ball, Roy 2122, 3038

Bandyopadhyay, Santanu	3286
Banerjee, Arijit	2881
Baranwal, Rohit	2043
Barbosa, A.U.	3231
Bari, Syed	3259
Barlow, Matthew	1646
Barner, Alexander	106
Barrado, A.	2545, 3090
Barth, Christopher	1512
Barthelmebs, Clement	453
Batarseh, Issa	1381, 3333
Bawohl, Melanie	3069
Bayhan, Sertac	3663
Bazzi, Ali M.	2666
Bęczkowski, Szymon	704, 974, 3101
Bede, Lorand	1702, 2264
Beres, Remus	3051
Bergman, Joshua	79
Bermejo, M.	3090
Bermingham, Jack	2794
Berzoy, Alberto	928, 3200
Betz, Vaughn	1882
Bezdenezhnykh, Yevgeny	308
Bhalla, Anup	2973
Bhangu, Bicky	715
Bhardwaj, Manish	505
Bhattachaarjee, Parijat	1344
Bhattacharya, Subhashish	295, 601, 778, 886, 1497, 1632, 2076
Bhattacharya, Tanmoy	199
Bhowmik, Pankaj Kumar	2706
Biglarbegian, Mehrdad	2998
Biswas, Suvankar	1934
Bizjak, Luca	1663
Blaabjerg, Frede	221, 229, 261, 288, 370, 1253, 1872, 1941, 1995, 2011, 2154, 2207, 2215, 2264, 3012, 3051, 3416, 3431, 3500
Blalock, Benjamin J.	684, 893, 1569, 3255
Blasko, Vladimir	2167
Böcker, Joachim	1547
Bodano, Emanuele	1663
Bojarski, Mariusz	1756
Bonthu, Sai Sudheer Reddy	1131
Bonyadi, Roozbeh	253
Bonyadi, Yeganeh	253
Borges, Beatriz	3637
Born, Rachael	1148, 3243
Boroyevich, Dushan	177, 516, 524, 739, 1024, 1315, 1561
Botting, Chris	854
Boynuegri, Ali R.	207
Braga, A.P.S.	3231

Brandt, Tobias 3172
Brar, Berinder 79
Breaz, Elena 3476
Briggs, Roger 2927
Brohlin, Paul 838
Brothers, John A. 990, 1967
Brown, Alan 529
Burdío, J.M. 1040, 1762, 3020, 3026, 3566
Burgos, Rolando 177, 516, 524, 1024, 1561
Buticchi, Giampaolo 2449, 3493, 3629
Buttay, Cyril 524

C

Cai, Wen 1057, 1861, 2599
Campbell, S.L. 1307
Canacsinh, Hiren 3637
Canales, Francisco 472
Cao, Dong 3553
Cao, Jiankun 1371
Cao, Wenchao 2229
Cao, Yuan 1868, 3684
Carlos, Gregory A.A. 3641
Carretero, C. 663, 3020, 3026, 3566
Casady, Jeffrey 979
Castro, Ignacio 25, 822
Castro, Marcus R. 2592
Ceballos, Salvador 236
Cervera, Alon 111, 692, 2298
Cha, Hanju 511, 1708, 3598
Chae, Young-Ho 2801
Chakraborty, Shiladri 1954, 2652, 3389
Challingsworth, Mark 3069
Chang, C.-H. 951
Chatterjee, Urmimala 1183
Chattopadhyay, Ritwik 778
Chattopadhyay, Souvik 1954, 2652, 3389
Chawda, Pradeep 3266
Chee, Seung-Jun 1206, 2370
Chen, Alian 3453, 3465
Chen, Changdong 2981
Chen, Cheng-Po 3255
Chen, Chingchi 1554
Chen, Di 529
Chen, Fang 177
Chen, Guipeng 1450
Chen, Guodong 499
Chen, Guoliang 1227
Chen, Hao 1437
Chen, Hua 1947

Chen, Jie	3453
Chen, Min	138
Chen, Minjie	1138, 1443
Chen, Qianhong	2518
Chen, Runruo	1045
Chen, Weiqiang	2666
Chen, Wenjie	493, 766, 3115, 3623
Chen, Woei-Luen	3471
Chen, Xinwen	1358
Chen, Xuling	1788
Chen, Yang	899, 2304, 2312, 2320
Chen, Yang-Lin	558
Chen, Yaow-Ming	558
Chen, Ying	2071
Chen, Yuxiang	499, 1462
Chen, Zhe	3431
Cheng, Chun Sing	1795
Cheng, Kuang-Yao	118
Cheung, Chun	1616
Chi, Yongning	1462
Chinthavali, M.	1307
Cho, Bo-Hyung	487, 1416
Cho, Shin Young	3690
Choe, Songbaek	2051
Choi, Beomseok	1947
Choi, Byeung G.	1773
Choi, Hee-Su	3153
Choi, Seungdeog	631, 1123, 1131, 1748, 2847, 3004
Choi, Sewan	859
Choi, Sung-Jin	3153
Choi, Wooin	1416
Choi, Wooyoung	2679
Chou, Derek	1512
Chow, Jeff Po Wa	1795
Chowdhury, Md Asif Mahmood	207
Chub, Andrii	2533
Chun, Chang Yoon	3322
Chung, Henry Shu-Hung	1795, 1807, 2154
Chung, Steven	1350
Church, Ron	786
Ci, Song	3189
Ciobotaru, Mihai	1702
Cobos, J.A.	2409
Coelho, Ernane A.A.	3585
Colak, Kerim	1323, 1756, 1767, 2587
Colmenares, Juan	746, 1018
Comanescu, Mihai	2759, 2855
Connaughton, A.M.	355
Conway, Thomas	1670

Cook, M. 3537
Correa, Maurício B.R. 3641, 1032
Corzine, Keith 720, 1191, 1481, 2187, 2840
Cosetin, Marcelo 1115
Costa, Levy F. 2449
Costa, Louelson A. 1032
Costa, Paulo Junior Silva 2376
Costinett, Daniel 424, 872, 893, 1010, 1569, 2441, 3255, 3577
Craciun, Marian 854
Cui, Shenghui 2620
Cui, Yutian 893
Curuvija, Boris 3553
Cuzner, Robert 1577
Czarkowski, Dariusz 1323, 1756, 1767, 2587
Czwickla, Christoph 3069

D

Dahan, Nadav 802
Dai, Ke 1358
Dai, Zhiyong 3134
Dai, Ziwei 1358
Dally, William J. 86
Dang, Zhigang 2114, 3684
Daniel, Michael T. 1695
Dargahi, Vahid 720, 1191, 1481, 2187, 2840
Daryaei, Mohammad 2579
Das, Partha Pratim 2652
Das, Pritam 552, 2491
Dashmiz, Shadi 3297
Davari, Pooya 221, 229
Davletzhanova, Zarina 253
de Almeida Carlos, Gregory A. 2720
de Almeida, Bruno Ricardo 60
De Carne, Giovanni 3493
De Doncker, Rik W. 643
de Oliveira Pacheco, Juliano 3346
de Rooij, Michael 2292
de Souza Oliveira Jr., Demercil 60, 3231, 3346
De, Ankan 295, 1632
Debnath, Suman 1528
Deboy, Gerald 3570
Degner, Michael W. 241
Delhotal, J. 1926
Demerdash, Nabeel A.O. 1065, 2826
Deng, Cheng 143, 362, 651
Deng, Hao 816
Deng, Lu 3521
Deng, Yan 1450
Dias Jr., A.J.S. 3231

Diaz Reigosa, Paula 288
Diaz, Nelson 1227
Dimarino, Christina 516
Ding, Pengling 1371
Ding, Weisheng 440
Dinulovic, Dragan 3097
Ditze, Stefan 864
Divan, Deepak 1437
Dix, Jeffery 684
do Prado, Ricardo N. 1115
Dobmeier, Christian 1741
Domoto, Kazuhide 2422, 2465
Dong, Dong 3403
Dong, Zhou 73, 2518
Doolla, Suryanarayana 2245, 3376
Dorn-Gomba, Lea 453
dos Santos Jr., Euzeli C. 2720
Dos Santos, Gutemberg G. 1032
Dou, Manfeng 3134
Dou, Qingyun 2272
Dousoky, Gamal M. 2735
Driesen, Johan 1183
Driessen, Anton 2457
Drofenik, Uwe 472
Du, Weijing 1002
Du, Weijing 2334
Du, Xiong 2992
Duarte, J.L. 3158
Dujic, Drazen 156, 1108
Dumais, Alex 3219
Dusmez, Serkan 505, 1176, 2108
Dutta, Atanu 3012

E

Eberle, Wilson 1286
Ebrahimi, Mohammad 2579, 3207
Edpuganti, Amarendra 402, 943
Egan, Michael 2794
Ehrlich, Stefan 1741
Einspieler, Sascha 759
Ekhtiari, Marzieh 1430
Elrayyah, Ali 392, 2660, 2806
Elsayed, Ahmed T. 1267
El-Taweel, Nader A. 830
Emadi, Ali 453, 1300
Engelmann, Georges 643
Eni, Emanuel-Petre 974, 3101
Enjeti, Prasad 936, 1695, 2567, 3545
Enshaei, Hossein 2813

Enslin, Johan 2998
Erickson, Robert 1947
Ertl, H. 1
Escobar-Mejía, Andrés 362
Eskandari, Soheila 2127
Essakiappan, Somasundaram 2706
Eum, Hyunchul 2355
Evzelman, Michael 1603
Ezra, Ofer 308

F

Fabricio, Edgard L.L. 3641
Fan, Bo 334
Faraci, Eric 838
Fard, Miad 1403
Farhang, Peyman 733
Farley, Kathleen Blair 1737, 3526
Farnell, Chris 143
Faulkner, Bryan 54
Fayed, Ayman 2437
Fedison, J.B. 247
Fei, Chao 322
Feng, Junjie 1534, 2334
Ferdowsi, Mehdi 1962, 2687
Fernandes, B.G. 2245, 2673, 3376
Fernandes, Darlan A. 1032
Fernandez, A. 1823
Fernandez, C. 2545, 3090
Ferrieux, Jean-Paul 2700
Figge, Heiko 1547
Flankl, Michael 623
Foulkes, Thomas 1512
Francés, A. 1624
Francis, A. Matt 1646
Freitas, Antônio A.A. 2592
Freitas, Luiz C.G. 3585
Frey, David 2401, 2700
Friedrichs, Daniel 3577
Fröhleke, Norbert 1547
Fu, Lixing 1554, 1967
Fu, Shihang 1475
Furukawa, Keita 1336

G

Gaafar, Mahmoud A. 2735
Gafford, James 1577
Gajanayake, C.J. 2942, 3058
Gakhar, Vikram 1878
Galiano Zurbriggen, Ignacio 386

Gan, Yiliang	3560
Gandikota, Srikant	1051
Gao, Fei	3476
Gao, Feng	410, 536, 921, 2259, 2935
Gao, Mingzhi	138
Gao, Rui	3383
Gao, Sugu	2868
Gao, Xieping	3185
Gao, Yabiao	1737, 3526
Gao, Yikai	1861
Garces, Luis	3403
Garcia Rodriguez, Luciano Andres	651
Garcia, Jorge	3508
Garcia, O.	2409, 1624
Garcia, Pablo	3508
Garcia, Virginia	3069
Garcia-Rodriguez, Luciano A.	362, 3712
Gavagsaz-Ghoachani, Roghayeh	446, 3397
Ge, Baoming	1214
Ge, Hongjuan	1424
Ge, Ting	668
Ge, Xiongxuan	3080
Geng, Shengbao	3560
Georgious, Ramy	3508
Gerfer, Alexander	2553, 3097
Gerling, Dieter	215
Ghaffarzadeh, Hooman	3353
Ghandi, Reza	3255
Ghat, Mahendra B.	2342
Ghias, Amer M.Y.M.	236
Giezendanner, Florian	1018
Ginart, Antonio	1737, 3526
Glavanovics, Michael	759
Glover, S.	3537
Goetz, Stefan M.	2349
Gohil, Ghanshyamsinh	1702, 2264
Goktas, Taner	1096, 2875
Gong, Bing	33
Gonnet, Luc	2700
Gonzalez, S.	1926
Gonzalez-Llorente, Jesus	3712
Gorla, Naga Brahmendra Yadav	2491
Gotovac, Ante	1663
Gou, Ruifeng	2071
Gray, C. Thomas	86
Greer III, Thomas H.	86
Gritti, Giovanni	564
Grosse, Thorben	643
Gu, Bin	838

Gu, Dong-Jie	1243
Gu, Lei	2889
Guerrero, Josep M.	398, 1227, 1376, 3459, 3697
Gui, Han-Dong	1243
Guirguis, David	1350
Gulbudak, Ozan	3248
Gunasekaran, Deepak	1045, 2525
Gundel, Paul	3069
Gunter, Samantha J.	1138
Guo, Ben	1010
Guo, Feng	1682
Guo, Suxuan	2063
Guo, Xiaoqiang	398
Gupta, Amit K.	715, 2942, 3058
Gupta, Ankit	1344
Gupta, Mahima	2919
Gupta, Ranjan K.	1520
Gupta, Shalabh	2666
Gurusinghe, Nicoloy	2479

H

Ha, Jung-Ik	193, 487, 1398, 2801, 3717
Hadjidemetriou, Lenos	3500
Hafez, Bahaa	936
Halivni, Bar	111
Hameyer, Kay	643
Han, Di	2861, 2950
Han, Jung Kyu	3690
Han, Xiangyu	2957
Han, Yang	816
Hang, Lijun	3560
Hanrahan, Robert	3266
Hanson, Alex J.	98
Haque, Moinul Shahidul	3004
Hare, James	2666
Harfman-Todorovic, Maja	3403
Hariharan, K.	315
Hariya, Akinori	2430
Harris, Richard Kyle	3255
Harrison, M.J.	247
Hartmann, M.	1
Haryani, Nidhi	1024, 1561
Hasan, Iftekhar	638
Hasegawa, Kazunori	3032
Hata, Katsuhiro	1731
Hata, Yuki	468
Hatae, Shinji	468
Hattori, Yoshiyuki	3146
Haug, Martin	3097

Hayakawa, Seiichi	283
Hayes, John G.	2794
He, Dingyi	2599
He, Haibing	3185
He, Jiangbiao	1065, 1084, 2826
He, Jinwei	1249
He, Ruirui	766
He, Xiangning	499, 1450, 1462, 1475
He, Xiaobin	2182, 2194
Heckel, Thomas	864
Heldwein, Marcelo Lobo	2833
Henke, M.	700
Henkenius, Carsten	1547
Herbert, Joseph	1123, 3004
Hernando, Marta M.	822
Higaki, Yusuke	1713
Hilber, Patrik	746
Hinken, Reiner	303
Hitoshi, Ishii	676
Ho, Carl Ngai-Man	2905
Ho, Kwun Yuan Godwin	2328
Hofmann, Heath	1721, 1726
Hofmann, Wilfried	2175
Holmes, Grahame	2252
Hopkins, Douglas C.	295, 2141
Hori, Yoichi	1731
Hosseini, Rasoul	1577
Hou, Dongbin	657
Hou, Ruoyu	1300
Hsiehu, H.-C.	951
Hu, Haibing	2182, 2194
Hu, Ji	253
Hu, Sheng	3310
Hu, Xiaolei	1071, 3409, 3591
Hu, Zhiyuan	899, 2320
Huang, Alex Q.	132, 269, 983, 2063, 2365, 2727, 2927, 3383, 3648, 3677
Huang, J.-W.	1364
Huang, Kuohsien	2355
Huang, Qingjun	3521
Huang, Qingyun	2727, 3648
Huang, Xiucheng	1002, 1534, 1853, 2334
Huang, Yi	1616
Huang, Ying	1888, 1894
Huang, Yung-Ting	1900
Huang, Zhengrong	1847
Huber, Laszlo	38, 46
Huh, Sungjae	2370
Hui, S.Y. Ron	169, 913, 1169, 1888, 1894, 2328, 3302, 3481
Hui, Zhao	2019

Hull, Brett 979
Husain, Iqbal 2141, 2927, 3383
Hwu, K.I. 2415

I

Iannuzzo, Francesco 288
Ibrahim, Mahmoud 2401, 2700
Iijima, Ryuji 3722
Illa Font, Carlos Henrique 2376
Ilves, Kalle 276
Imura, Takehiro 1731
Inaba, Masamitsu 283
Inokuchi, Seiichiro 468
Inoue, Shuntaro 3146
Ishikuro, Hiroki 1802
Ishizuka, Yoichi 2422, 2430, 2465
Islam, Md. Zakirul 631
Islam, Rakib 3279
Isobe, Takanori 3722
Isurin, Alexander 880
Itagaki, Atsushi 2465
Itakura, Tetsuro 1907
Itoh, Jun-Ichi 1336, 1911
IV, Prasanna 9
Iyer, Vishnu Mahadeva 295
Izuka, Arata 468

J

Jacobina, Cursino B. 2720, 3641
Jahns, Thomas 2167
Jain, Parth 2553
Jain, Praveen 378, 460, 544
Jang, Yujin 3690
Jang, Yungtaek 595, 1292
Jayasinghe, Shantha Gamini 2813
Jedtberg, Holger 3629
Jensen, Scott 1947
Jerinic, Vladan 303
Ji, Junpeng 493, 3115
Ji, Lin 3446
Ji, Shiqi 1456
Jia, Xiaoyu 398
Jiang, Dan 1788
Jiang, Dong 3616
Jiang, Ling 872
Jiang, Qirong 1468
Jiang, W.Z. 2415
Jiang, Xinjian 907
Jiao, Ningfei 2776

Jin, Qian 2409
Jo, Hyunsik 511
Jo, Jongmin 1708
Joffe, Christopher 1741
John, Vinod 951, 2200, 3439
Johnson, J. 1926
Jones, David C. 3273
Jones, Edward A. 1010, 2441
Jones, Vinson 1387
Jourdan, Charlie 3333
Jovanović, Milan M. 38, 46, 1292
Jung, Jae-Jung 2620
Jung, Jee-Hoon 3213
Jung, Kyungsub 17

K

Kakitani, Hisao 2969
Kallfass, Ingmar 2969
Kang, Taeyong 3545
Kang, Yong 3521
Kapat, Santanu 315, 2504, 3224, 3237
Karbalaye Zadeh, Mehdi 446, 3397
Karimi-Ghartemani, Masoud 3165
Karki, Ujjwal 2525
Kashyap, Avinash 3255
Katzir, Liran 3655
Kawai, Yasufumi 2051
Kawajiri, Toru 1802
Kawase, Daisuke 283
Kazama, Taisuke 1585
Ke, Haotao 295
Ke, Xugang 94
Ke, Ziwei 241
Kelly, Anthony 1591, 1670
Kerekes, Tamas 974, 1702, 2264
Khajehoddin, S. Ali 2579, 3207
Khaligh, Alireza 54, 440
Kharezy, M. 3043
Khayat, Joseph 66
Khoshkbar Sadigh, Arash 720, 1191, 1481, 2187, 2840
Khurram, Adil 2782
Kieferndorf, Frederick 472
Kikuchi, Naoto 3146
Kim, Byeong-Heon 1220
Kim, Dong-Hee 1273
Kim, Hyeokjin 1947
Kim, Hyeon-Sik 1206
Kim, Jae-Gu 2161
Kim, Ji-Min 3690

Kim, Jin-Woong 1398
Kim, Jonghoon 1690, 3322
Kim, Minjae 859
Kim, Nari 1273
Kim, Youngjong 2355
Kim, Yun-Sung 163
Kirshenboim, Or 111, 802
Kirtley, James L. 2881
Knott, Arnold 1541, 1842
Ko, Youngjong 3629
Koga, Tomoya 2430
Kolar, Johann W. 615, 623, 1198
Kolluri, Sandeep 552, 2491
Koltsov, H. 339
Kondo, Ryota 1713
Kondo, Takeshi 2788
Konishi, Kyohei 1780
Konrad, Werner 3570
Kostov, Konstantin 1018
Kou, Lei 1489, 2278
Kouchaki, Alireza 2382
Krischan, K. 355, 759
Krishna Moorthy, Radha Sree 794
Krishnamurthy, Mahesh 880, 3128
Kshirsagar, Parag 2027, 3616
Kubendran, S. 663
Kubo, Hajime 2788
Kudva, Sudhir S. 86
Kularatna, Nihal 2479
Kulkarni, Abhijit 2200, 3439
Kulkarni, Onkar Vitthal 2245, 3376
Kulkarni, S. 663
Kulothungan, Gnana Sambandam 402, 943
Kumar, Ashish 1392, 1947, 2395
Kumar, Misha 38, 46, 1292
Kumar, Nikhil 1344
Kumar, V. Inder 3224, 3237
Kumar, V.V.S. Pradeep 2673
Kurokawa, Fujio 754
Kusaka, Keisuke 1336
Kusama, Fumito 2607
Kwak, Sangshin 1748
Kwon, Yong-Cheol 1206
Kyriakides, Elias 3500

L

Lago, Jackson 2833
Lai, Jih-Sheng 1148, 1974
Lai, Jih-Sheng Jason 3243

Lai, Wei-Han	1974
Lam, John	786, 830
Lamar, Diego G.	25, 822, 1823, 2545
Lamo, Paula	2389
Lamoureux, Carl	1882
Langham, Jeff	334
Langmaack, N.	700
Lashway, Christopher R.	1267
Lave, M.	1926
Lazaro, A.	2545, 3090
Lazaro, Orlando	66
Lazzarin, Telles Brunelli	2376
Le, Hoai Nam	1911
Lee, Albert T.L.	169
Lee, Byoung-Kuk	163, 1273, 2161
Lee, C.K.	913
Lee, Eun S.	1773
Lee, Fred C.	322, 343, 657, 1002, 1534, 1608, 1847, 1853, 2334, 3259
Lee, Hyun-jun	1690
Lee, Jaedo	3598
Lee, Jae-Hyun	3086
Lee, Jian-Hsing	1900
Lee, June-Seok	3416
Lee, Junwon	511
Lee, Kevin	2003
Lee, Kun Wang	2875
Lee, Kyo-Beum	2764, 3416
Lee, Kyu-Chan	487
Lee, Moonhyun	1416
Lee, Woongkul	2679, 2861
Lee, Yong-Duk	125
Lee, Yongjae	193
Lee, Younggi	2370
Leeb, Steven B.	2881
Lefranc, Pierre	2401, 2700
Lehman, Brad	417, 2122, 2286, 3038
Lei, Yang	2063
Lei, Yutian	1512
Lemmon, Andrew	1577
Leng, Mingzhi	1554
Leng, Siyu	1941
Lenz, Kevin	303
Leong, K.K.	355
Lequesne, Bruno	2927
Leubner, Martin	2175
Levy, Aron	138
Li, Chendan	3459
Li, Dan	595
Li, David K.W.	1350

Li, Guojie	3560
Li, He	990, 1554, 1967
Li, Helong	704, 3101
Li, Hongxu	1657
Li, Hui	1675, 2237
Li, Jie	2770
Li, Jun	2963
Li, Kai	2613
Li, Kaiyuan	3422
Li, Peide	3697
Li, Qiang	322, 343, 657, 1002, 1534, 1608, 1847, 1853, 2334, 3259
Li, River Tin-Ho	2905
Li, Rui	1675
Li, Sinan	169
Li, Tao	2573
Li, Tengfei	2992
Li, Virginia	343
Li, Wenyu	1450
Li, Wuhua	499, 1462
Li, Xing	728
Li, Xinlei	3623
Li, Xuan	2063
Li, Xueqing	2973
Li, Yalong	2637
Li, Yan	1462
Li, Yan-Cun	1853
Li, Yaohua	3080
Li, Yongdong	3317
Li, Yuan	417, 2525
Li, Yun Wei	1249
Li, Yungui	3560
Li, Yunwei	185
Li, Zhiqing	1329
Li, Zhongxi	2349
Li, Zhongyu	3697
Liang, Beihua	1249
Liang, Lin	2981
Liang, Tsorng-Juu	1900
Liang, Xinyu	2349, 2714
Liao, Yi-Hung	1831
Liao, Zitao	1512
Lidow, Alex	587
Lightbody, Gordon	2794
Liivik, Liisa	2533
Lim, Changjin	2370
Lim, Seungbum	98
Lima, Gustavo B.	3585
Lin, Hua	728, 1078
Lin, L.-C.	951, 1364

Lin, Ni 3189
Lin, P.-H. 1364
Liserre, Marco 2449, 3493, 3629
Lisi, G. 2540
Liu, Baojin 3328, 3370
Liu, Bing 843, 2754
Liu, Bo 966, 2441, 2637
Liu, Fuxin 1788
Liu, Gang 38, 46, 595, 1292
Liu, Haichun 1371
Liu, Haoyan 3180
Liu, Hongpeng 1253
Liu, Jingbo 2147
Liu, Jinjun 739, 2272, 3193, 3328, 3370
Liu, Liming 990
Liu, Pei-Hsin 343
Liu, Pengkun 132, 983
Liu, Sucheng 1489, 2278
Liu, Teng 2272
Liu, Tianshu 899
Liu, Tingting 1410
Liu, Weiguo 1726, 2748, 2776
Liu, Wenbo 899, 2095, 2320
Liu, Wen-Chuen 1512
Liu, Xianzhuo 3317
Liu, Xiaohu 3403
Liu, Xiaokang 2742
Liu, Yan-Fei 899, 1243, 1489, 2087, 2095, 2278, 2304, 2312, 2320
Liu, Yang 959
Liu, Yunting 1045
Liu, Yushan 1214
Liu, Yushi 1392
Liu, Yusi 143
Liu, Zeng 739, 2272, 3328, 3370
Liu, Zhengyang 1847, 1853
Liu, Zhichao 1155
Loh, Poh Chiang 229, 1253, 1872, 1995, 2011, 2207
Lomonova, E.A. 3158
Lope, I. 3020
López del Moral, D. 3090
López, Felipe 2389
Lopez, Ozzie 3065
Lorenz, Robert D. 215, 2055, 2167
Lotfi, Ashraf 1882
Lu, Daorong 2194
Lu, Fei 1721, 1726
Lu, Jie 1392, 1947
Lu, Juncheng 529
Lu, Minghui 1941

Lu, Sizhao 2613
Lu, Ting 1456
Lu, Yong 2215
Lu, Zhengang 2613
Lu, Zhengyu 2003
Lu, Zhigang 398
Lu, Zhouyu 1243
Lucia, Oscar 1040, 1762, 3566
Lukic, Srdjan M. 2349, 2714
Luna, Adriana 1227
Luo, Fang 709, 2981
Luo, Guangzhao 2748
Luo, Haoze 499
Luo, Min 1108
Luo, Tianyi 3065
Lynch, Brian 66

M

Ma, Cong 3279
Ma, Dongsheng 94
Ma, Hongbo 3243
Ma, Jun 499
Ma, Ke 261
Ma, Weizhong 816
Ma, Yingxian 2497
Ma, Yiwei 966, 1261, 2229, 3121
Madhusoodhanan, Sachin 886, 1497, 1632, 2076
Madsen, Mickey P. 1842
Magne, Pierre 453
Mahajan, Anirudh 601
Mahdavikhah, Behzad 329, 3297
Mahmoodzadeh, Zahra 3353
Mainali, Krishna 886, 1497, 1632, 2076
Maitra, Arindam 1974
Makhdoomi Kaviri, Sajjad 378
Makoschitz, M. 1
Maksimović, Dragan 580, 1392, 1947, 2292, 3273
Malcolm, Doug 2087, 2095
Mallik, Ayan 54
Manabe, Shinya 2465
Mandal, Arindam 3237
Mandi, Bipin Chandra 2504
Manjrekar, Madhav 2706
Mansour, Makram 3266
Mantooth, H. Alan 143, 1646, 3012, 3180
Mao, Shuai 2776
Marsili, S. 339
Martín, Kevin 25
Martineau, Donatien 524

Martínez, Gilberto	1115
März, Martin	864, 1741
Mátéfi-Tempfli, Stefan	733
Mathew, Dinto	3286
Matsumori, Hiroaki	676, 3051
Matsuura, Ken	2430
Mattavelli, Paolo	1315
Mauerer, M.	1198
Mawby, Philip	253, 2645
Mazhari, Iman	2998
Mazumder, Paromita	1344
Mazumder, Sudip K.	1344, 1989
Mazzola, Michael	1577
McAmmond, Matt	529
McCann, Roy A.	143
McCue, Benjamin M.	3255
McDonald, Brent	329, 334, 3297
McGrath, Brendan	2252
McHugh, Colin	1947
McIntrye, W.	2540
McKenzie, Craig	3038
McRae, T.	2540
Meder, Dirk	2083
Meere, Ronan	2629
Megyei, George	3255
Mehrizi-Sani, Ali	3353
Mehrotra, Vivek	79
Mekhilef, S.	1163
Méllo, João Paulo R.	2720
Meng, Jinhao	2748
Meng, Peipei	1102
Meng, Tao	2748, 2776
Meola, Marco	1591
Mertens, Axel	3172
Meyer, Jeffrey	1947
Mi, Chris	1721, 1726
Michihira, Masakazu	2607
Mikata, Atsushi	2969
Mikulla, Michael	2083
Miraoui, Abdellatif	3476
Mishima, Tomokazu	1780
Mishra, Richa	2342
Miskovic, Vlatko	2167
Mitra, Rakesh	3279
Miwa, Brett	1597, 3273
Miyazaki, Koutarou	1640
Miyazaki, Takayuki	1907
Modes, Christina	3069
Moeini, Amirhossein	2019

Mohamed, A.A.S. 928
Mohammad, Mostak 1748
Mohammadi, Danyal 2813, 2821
Mohammadi, Mehdi 848
Mohammadpour, Bahador 378
Mohammed, Osama 928, 1267, 3200
Mohan, Ned 1051, 1520, 1934, 1982, 2043
Mojab, Alireza 1989
Molinas, Marta 446, 3397
Mønster, Jakob D. 1842
Moon, Gun-Woo 3690
Moon, Intae 1512
Moon, Seung-Ryul 1974
Morgan, Adam 295, 2141
Moroto, Takahiro 1802
Morris, Casey 2861, 2950
Morsy, Ahmed 2567
Mosesian, Jerry 2122
Moss, Jim 668
Motto, Eric R. 468
Moury, Sanjida 786
Mu, Xianmin 1381
Muetze, A. 355, 759, 3570
Mukherjee, Subhajyoti 2687
Mukhopadhyay, Shayok 2782
Mukhopadhyay, Siddhartha 315
Munk-Nielsen, Stig 288, 704, 974, 1376, 3101
Murmann, Boris 1650
Musavi, Fariborz 772
Musumeci, Salvatore 3669
Muyeen, S.M. 1941

N

Na, Woonki 3322
Nadarajan, Sivakumar 715
Nademi, Hamed 3291
Nagai, Shuichi 2051
Nahid-Mobarakeh, Babak 446, 3397
Nakano, Toshiya 468
Nakao, Hiroshi 754
Nakaoka, Mutsuo 1780
Nakashima, Yoshiyasu 754
Nan, Chenhao 432
Narasimhan, Sneha 2043
Nasr, Miad 1350
Navarro, Angel 3508
Nawaz, Muhammad 276
Nee, Hans-Peter 746, 1018
Neely, J. 1926, 3537

Neft, Charles 79
Negoro, Noboru 2051
Ngo, Khai 668
Nguyen, Duy T. 1773
Ni, Tianheng 2754
Ni, Xijun 983
Niapour, S.A.Kh. Mozaffari 3704
Nikolaidis, Ilias 3069
Ning, Puqi 3080
Ninomiya, Tamotsu 2422, 2430
Nishizawa, Shin-Ichi 3032
Niu, He 2055
Niu, Ying 2770
Noh, Shinyoung 859
Nomura, Katsuya 3146
Nondahl, Thomas A. 2147
Noquil, Jonathan 3065
Norisada, Takaaki 2607
Norum, Lars Einar 3291
Nowak, Torsten 3069
Nymand, Morten 609, 2382

O

O'Donnell, Terence 2629
O'Donovan, Gerard 2794
Ogawa, Taichi 1907
Oh, Chang-Yeol 163
Oh, Jaeyoon 2370
Ojo, Olorunfemi 2035
Okubo, Hiizu 2465
Oliver, J.A. 2409
O'Loughlin, Cathal 2629
O'Mathuna, C. 663
Omura, Ichiro 1640, 3032
Oña, E. 2545
Onal, Yasemin 2693
Onar, O.C. 1307
Orabi, Mohamed 1505
Ordonez, Martin 386, 848, 854, 2561
Orikawa, Koji 1336, 1911
Orr, Ray 1350
Ortiz-Gonzalez, Jose 253
Ortiz-Rivera, Eduardo I. 3712
Otsuka, Masafumi 1350
Ouyang, Ziwei 2473
Ozimek, Patrick E. 1084
Ozpineci, Burak 3529

P

Padhee, Varsha	1982
Pagano, Rosario	3669
Pahlevani, Majid	378
Pala, Vipindas	979
Palaniappan, Vishal	1350
Palmour, John	979
Pam, Srikanth	3266
Pan, Bing	2613
Panda, S.K.	9, 552, 715, 2491
Park, Hwa-Pyeong	3213
Park, Joung-hu	1690
Park, Sung-Yeul	125
Parkhideh, Babak	2998
Parsa, Leila	2573
Parvez, M.	1163
Patel, Ankur	150
Pathan, Abrar Ahmed	2497
Patil, Devendra	2889
Patra, Amit	2504
Pavlick, Stephanie A.	1138
Pavlovic, Z.	663
Paz, Francisco	386
Pedersen, Jeppe A.	1541, 1842
Peixoto, Paulo P.	3346
Peng, Chang	132, 269, 983, 2927
Peng, Fang Z.	959, 1045, 2525
Peng, Hao	1450
Peng, Jichang	2748, 2776
Peng, Kang	2127
Peng, Li	1358
Perales, Mico	1967
Perdigão, Marina	1115
Peretz, Mor Mordechai	111, 308, 692, 802, 2298
Perez, Aday	215
Pérez-Tarragona, Mario	1762
Perreault, David J.	98, 1138
Perrin, Remi	524
Perry, Jeff	3266
Persons, Ryan	3069
Pervaiz, Saad	1947, 2395
Peterchev, Angel V.	2349
Pevere, Alessandro	1183
Phillips, Evan	3684
Piepenbreier, Bernhard	2897
Pierfederici, Serge	446, 3397
Pigazo, Alberto	2389
Pilawa-Podgurski, Robert C.N.	1512
Ping, Dinggang	38, 46

Piya, Prasanna 3165
Pong, M.H. Bryan 2328
Poshtkouhi, Shahab 1350, 1403
Pou, Josep 236
Praça, Paulo P. 60, 3231
Pramod, Prerit 3279
Prasai, Anish 1437
Preciat, Philippe 524
Prieto, R. 1624
Prodić, Aleksandar 329, 1597, 2497, 2540, 2553, 3297
Puukko, Joonas 990

Q

Qi, Feng 2912
Qian, Qiang 1919, 3446
Qiao, Wei 3514
Qin, Jiangchao 1528
Qin, Liang 2525
Qin, Shibin 1512
Qin, Xianhui 843
Qiu, Maohang 138
Qiu, Yajie 2320
Qu, Liyan 3279, 3514
Qu, Xiaohui 2154
Quan, Zhongyi 185
Quay, Rüdiger 2083
Quentin, Nicolas 524

R

Raciti, Angelo 3669
Rahnamaee, Arash 1989
Raizada, Shirish 1344
Rajashekara, Kaushik 2788
Ramachandran, Rakesh 609
Ramadass, Yogesh 838
Ramani, Ramanathan 66
Rambal-Vecino, Andres 3712
Ramezani, Medhi 2035
Ramos, Francisco 215
Ramu, Krishnan 2027
Ran, Li 253, 1443, 2645
Ranstad, Per 1018
Rao, Yuan 1585
Rashkin, L. 3537
Rathore, Akshay K. 402, 794, 943
Ravey, Alexandre 3476
Redondo, Luís M. 3637
Rehman, Habibur 2782
Reiner, Richard 2083

Reitz, Jessica 3069
Remus, Nico 2175
Ren, Hai-Peng 2770
Ren, Ren 2441
Ren, Xiaoyong 73, 2518, 3488
Ren, Xizhou 2912
Ren, Yu 2071, 2102
Rengifo, Johnny 3200
Renjit, Ajit A. 1682
Reusch, David 587
Riazmontazer, Hossein 1989
Rim, Chun T. 1773
Roasto, Indrek 2533
Robbins, William 1934
Roberts II, Charles 3255
Rodrigues, Danillo B. 3585
Rodriguez, A. 1823
Roehrs, Benjamin D. 3255
Rogers, Daniel J. 1650
Romero, David 1350
Rosahl, Thoralf 106
Roßkopf, Andreas 1741
Round, W. Howell 2479
Ruan, Xinbo 1788, 3488
Ruiz, Juan M. 1292

S

Sá Jr., Edilson M. 2592
Saasaa, Raed 1286
Sadik, Diane-Perle 746, 1018
Saeed, Sarah 3508
Saeedifard, Maryam 1528, 2136, 2957
Safaee, Alireza 460
Saha, Aparna 2806
Sahoo, Ashish Kumar 1982
Sahoo, Saroj Kumar 199
Saito, Katsuaki 283
Saito, Shoji 468
Saket, Mohammad Ali 854, 2561
Sakurai, Takayasu 1640
Salameh, Mohamad 3128
Salcines, Cristino 2969
Salem, Ahmed 1505
Sandoval, José Juan 3545
Sangwongwanich, Ariya 370
Sankman, Joseph 94
Santi, Enrico 2127, 3248
Santiago-González, Juan A. 98
Santos de Moura, Diogo Cesar 3553

Santos Guimarães, Jéssica	3346
Sanz, M.	2545, 3090
Sariri, Kouros	3255
Sarlioglu, Bulent	2679, 2861, 2950
Sarnago, Hector	1040, 1762, 3566
Satija, Yudhister	3266
Sato, Masaki	2465
Saublet, Louis-Marie	3397
Saur, Michael	215
Savaghebi, Mehdi	1227, 3697
Scandrett, Brad	417
Schmidt, Peter B.	2147
Schubert, Michael	643
Schweitzer, Ben	3128
Sebastián, Javier	25, 822, 1823
Seeman, Michael	838
Seltzer, Daniel	1947
Sen, Paresh C.	1489, 2278
Senanayake, Thilak	3722
Senol, Murat	643
Seo, Gab-Su	487
Sepahvand, Alihossein	580, 1947
Serrano, J.	3020
Setyawan, Leonardy	3409
Severson, Eric	2043
Shafiei, Navid	848, 854, 2561
Shagar, Viknash	2813
Shah, Neel	2998
Shahbazi, Caitlin	3069
Shamsi, Pourya	1962, 2687
Shang, Fei	880
Shao, Jianwen	3659
Sharkh, Suleiman M.	2223
Sharma, Ratnesh	1682
Shen, Ang	1962
Shen, Guangtong	2003
Shen, Zhiyu	516, 1561
Sheng, Su	417, 2286
Shenoy, Pradeep S.	66
Shi, Baoping	843
Shi, Jianjiang	1475
Shi, Xiaojie	2637
Shi, Yuxiang	1675
Shimizu, Toshihisa	676, 3051
Shiu, T.-H.	1364
Shmilovitz, Doron	3655
Shousha, Mahmoud	3097
Shoyama, Masahito	2735
Shrivastav, Ashish	601

Shu, Zhan 2223
Shukla, Anshuman 2342, 3286
Silva, J. Fernando 3637
Silva, Paulo R. 3585
Silva, Rafael V. 2592
Simanjorang, Rejeki 2942, 3058
Singh, Amandeep 3140
Singh, Shikhar 601
Singh, Surinder P. 1585
Sinha, Sreyam 1947
Siwakoti, Yam P. 1872
Sleik, Roland 759
Sohn, Hoon 3690
Soltani, Hamid 229
Somani, Apurva 1520
Somani, Utsav 3333
Son, Yeongrack 3717
Son, Young-Kwang 2370
Song, Xiaoqing 132, 269, 983
Soni, Harshit 1344
Sozer, Yilmaz 207, 392, 638, 2693, 2806
Srdic, Srdjan 2714
Srinivasan, Dipti 402, 943
Srivastava, Vineet 786
Stack, David 1670
Stamm, Thomas 3659
Steenis, Joel 3219
Steyn-Ross, D. Alistair 2479
Stillwell, Andrew 1512
Strydom, Johan 587, 2292
Stübig, Marc 2175
Styles, Julian 529
Su, Yipeng 118
Subotic, Ivan 623
Suh, Yongsug 17
Sul, Seung-Ki 1206, 1220, 2370, 2620
Sun, Bo 1376
Sun, Kai 1227, 1410, 2182, 2194
Sun, Lei 505
Sun, Lejia 810
Sun, Libing 1227
Sun, Pengju 2992
Sun, Wei 1462
Sung, Won-Yong 163
Sveum, Peter 3128

T

T T, Anandha Ruban 1878
Tabata, Osamu 2051

Tadano, Hiroshi 3722
Tadeparthy, Preetam 1878
Tai, Heng-Ming 2992
Takamiya, Makoto 1640
Takano, Koushi 676
Talebi Khanmiri, Dawood 2122, 3038
Tan, Kai 132, 983
Tan, Linlin 2102
Tan, Nadia M.L. 1163
Tan, Pingan 3185
Tan, Siew-Chong 169, 913, 1169, 1888, 1894, 3302, 3481
Tang, Yi 1071, 3591
Tang, Yichao 440
Tang, Yuan 1443, 2645
Tareilus, G. 700
Tayebi, S. Milad 1381
Teixeira, Carlos 2252
Tekgun, Burak 207
Teodorescu, Remus 974, 1702, 2264
Tewari, Saurabh 1520, 2043
Thiringer, T. 3043
Thone, Jef 3097
Tian, Shuilin 1608
Tian, Ye 499
Ting, Lo Pang-Yen 1900
Tkachov, Sergii 350, 1663
Tolbert, Leon M. 893, 966, 1261, 1307, 1569, 2637, 3121
Tomas-Manez, Kevin 1235
Tomioka, Satoshi 2430
Tong, C.F. 2942, 3058
Trabelsi, Mohamed 3663
Tran, Yan-Kim 156
Trento, Brad 3577
Trescases, Olivier 1350, 1403, 1882
Trintis, Ionut 1376
Tripathi, Awneesh 886, 1497, 1632, 2076
Tse, Zion Tsz Ho 1737, 3526
Tseng, King Jet 1071
Tseng, K.J. 2942, 3058, 3107, 3422, 3591
Tsukuda, Masanori 1640
Tung, Chung-Pui 1807
Tüysüz, Arda 615, 623, 1198

U

Uceda, J. 1624
Uddin, Md Wasi 638
Ueda, Tesuzo 2051
Ueno, Takeshi 1907
Ugur, Enes 1176

Urteaga, Miguel 79

V

Vasić, M. 2409
Vásquez, Juan C. 1227, 3459
Vázquez, A. 25, 2545
Vechalapu, Kasunaidu 295, 886, 1497, 1632, 2076
Vekslender, Timur 308
Venkataramanan, Giri 2919
Venkateswaran, Muthusubramanian 1878
Vermulst, B.J.D. 3158
Vermulst, Bas 2457
Vesti, S. 339
Vilathgamuwa, Mahinda 2813
Villarejo, J.A. 1823
Vinnikov, Dmitri 2533
Vitorino, Montiê A. 1032
Vrankovic, Zoran 1084
Vu, Trong Tue 1835, 2485
Vukadinović, Nenad 1597

W

Wada, Keiji 1640, 2986
Walsh, Ray 2794
Waltereit, Patrick 2083
Wang, Chengshan 1249
Wang, Fan 334
Wang, Feng 810, 2742
Wang, Fred 424, 893, 966, 1010, 1261, 1456, 1569, 2229, 2441, 2637, 3121
Wang, Gangyao 979
Wang, Guo-Xiang 1123
Wang, Haoyu 480, 1280, 1329
Wang, Hongliang 899, 1489, 2278, 2304, 2312, 2320
Wang, Huai 370, 2154
Wang, Jiangfeng 2182
Wang, Jin 990, 1554, 1967
Wang, Jun 516
Wang, Kui 3317
Wang, Kun 1450
Wang, Laili 899, 2087, 2095, 2320
Wang, Li 2365
Wang, Long 2754
Wang, Meilin 1078
Wang, Meng-Jie 3471
Wang, Mengqi 3648
Wang, Miao 709, 2912
Wang, Ming-Hao 3302
Wang, N. 663
Wang, Peng 1071, 3409, 3591

Wang, Qin 1657
Wang, Shike 3328, 3370
Wang, Shiliang 2889
Wang, Shuo 2019, 3603
Wang, Wei 1253
Wang, Xiaoping 572
Wang, Xingwei 1078
Wang, Xiongfei 704, 1253, 1941, 2011, 2207, 2215, 3051, 3101, 3431
Wang, Yanbo 3431
Wang, Yi 1815
Wang, Zhenxiong 3358
Wang, Zhiqiang 2637
Watanabe, Hiroki 1336
Watanabe, Yoshitoshi 3146
Wattes, J.L. 3231
Weber, Bastian 3172
Wei, Chun 3514
Wei, Jiadan 843, 2754
Wei, Lixiang 1065, 2826
Wei, Yingdong 1468
Weise, Nathan 1065
Weiss, Beatrix 2083
Wen, Changyun 3409
Wen, Lucheng 990
Wen, Xuhui 3080
Wens, Mike 3097
Wespel, Matthias 2083
Wicht, Bernhard 106
Wijnands, C.G.E. 3158
Wiktor, Wlodek 66
Williamson, Sheldon S. 3140
Wilson, D. 3537
Winterhalter, Craig 1084
Wittmann, Jürgen 106
Wu, Dalei 3189
Wu, Hongfei 1410, 1424
Wu, John 1967
Wu, Liyao 2136, 2957
Wu, Qunfang 1657
Wu, T.-F. 951, 1364
Wu, Teng 3328, 3370
Wu, Tong 3529

X

Xia, Yinglai 3364
Xiao, Guochun 2215
Xiao, Jianfang 3409
Xiao, Lan 1657
Xiao, Xi 1424, 3338

Xie, Shaojun 1371, 1919, 3446
Xie, Xiaogao 816
Xie, Yicong 2360
Xin, Zhen 1995, 2207, 3697
Xing, Xiangyang 3453, 3465
Xing, Yan 1410, 1424, 2182, 2194
Xiong, Liansong 2742
Xiong, Song 1888, 1894
Xu, Chen 1358
Xu, Dewei David 33
Xu, Dianguo 1253
Xu, Jialin 1657
Xu, Jing 990
Xu, Jinming 1919, 3446
Xu, Longya 709, 2912
Xu, Qianwen 3409
Xu, Tao 921
Xu, Wei 2868
Xu, Yang 2141
Xue, Fei 3677
Xue, Lingxiao 1315

Y

Yadav Gorla, Naga Brahmendra 552
Yadav, Akshat 2076
Yamada, Go 2607
Yamada, Masaki 1713
Yamaguchi, Koji 3075
Yamaguchi, Masahiro 2465
Yamamoto, Keiichi 283
Yamamoto, Yasuhiro 2788
Yan, Shuo 913
Yan, Xingda 2223
Yanagi, Hiroshige 2430
Yang, Ching-Chieh 558
Yang, Enxing 499
Yang, Heya 1462
Yang, Hongbin 1078
Yang, Jianwei 3134
Yang, Liu 1261, 3121
Yang, Pengzhi 1554
Yang, Qichen 2957
Yang, Shuitao 959
Yang, Shunfeng 1071, 3591
Yang, Tao 2629
Yang, Tianbo 913
Yang, Xu 493, 766, 2071, 2102, 3115, 3623
Yang, Yang 1243
Yang, Yong 1788

Yang, Yongheng	221, 370, 1253, 3500
Yang, Yuanyu	843
Yang, Yuchen	1534
Yang, Yun	1169, 3481
Yang, Zhihua	33, 899, 2312, 2320
Yao, Chengcheng	990, 1554
Yao, Jianhui	2182, 2194
Yao, Kai	572, 1815
Yao, Wenxi	2003
Yau, Y.T.	2415
Yazdanian, Mehrdad	3353
Ye, Qing	2237
Ye, Zichao	1512
Yeo, H.L.	3107
Yeong, Lee Meng	3409
Yi, Fan	1057, 1861, 2599
Yi, Hao	3358
Yin, Shan	2942, 3058
Yonezawa, Yu	754
Young, George	1835, 2485
Yu, Hualong	1456
Yu, Ruiyang	3677
Yu, Sheng-Yang	2511
Yu, Wensong	2063, 2141, 2365, 2714, 2727, 3648
Yu, Xinyu	1468
Yu, Yanqi	1286
Yuan, Huawei	907
Yuan, Liqiang	2613

Z

Zafarani, Mohsen	1096, 2875
Zagrodnik, Michael	1071, 3591
Zane, Regan	1603
Zare, Firuz	221, 229
Zefran, Milos	1989
Zeltser, Ilya	802
Zeng, Hulong	1045
Zeng, Xiangjun	2102
Zhak, Serhii M.	3273
Zhan, Xiaohai	1424
Zhan, Xiaoqing	2154
Zhang, Bin	1155
Zhang, Binfeng	1919
Zhang, Canhui	138
Zhang, Chengduo	2349
Zhang, Chenghui	3453, 3465
Zhang, Chi	2714
Zhang, Dan	766, 3623
Zhang, Fan	1947, 2071, 2102

Zhang, Hao 2973

Zhang, Haojiong 2167

Zhang, Hua 1721, 1726

Zhang, Hui 3338

Zhang, Jianqiu 595

Zhang, Julia 241

Zhang, Jun 2992

Zhang, Junfang 572

Zhang, Ke 3476

Zhang, Lanhua 1148, 1974, 3243

Zhang, Li 3488

Zhang, Li-Heng 2770

Zhang, Lin 1868

Zhang, Liqi 269, 2063, 2365

Zhang, Shuoting 966, 3121

Zhang, Weimin 424, 893

Zhang, Wenli 524, 1002

Zhang, Xiangming 1102

Zhang, Xuan 990, 1967, 2229

Zhang, Xuning 177, 1024, 1561

Zhang, Yongchang 2868

Zhang, Yuanzhe 580, 2292

Zhang, Yuzhi 3180

Zhang, Zhe 1090, 1235, 1430, 2252, 3514

Zhang, Zhen 3134

Zhang, Zheyu 684, 1010, 1569, 2441

Zhang, Zhigang 3358

Zhang, Zhiliang 73, 1243, 2518

Zhang, Zhiyu 1475

Zhang, Zicheng 3453, 3465

Zhao, Bo 2912

Zhao, Chongwen 3577

Zhao, Dongdong 3134

Zhao, Hengyang 1475

Zhao, Hui 3603

Zhao, Rende 1995, 2207, 3697

Zhao, Shuze 1882

Zhao, Tao 33

Zhao, Xiaonan 1148, 3243

Zhao, Xin 295

Zhao, Zhengming 1456, 2613

Zheng, Sheng 966

Zheng, Yue 417

Zheng, Zedong 3317

Zhi, Na 3338

Zhou, Bo 843, 2754

Zhou, Daming 3476

Zhou, Jinping 2360

Zhou, Keliang 1155

Zhou, Liwei	410, 2259, 2935
Zhou, Luowei	2992
Zhou, Min	2360
Zhou, Qi	536
Zhou, Sizhan	3193
Zhou, Yong	2567
Zhou, Yuan	73, 2518
Zhou, Zhe	2912
Zhu, Bohang	2788
Zhu, Donghai	3521
Zhu, Guorong	3310
Zhu, Ke	990
Zhu, Qianlai	2365
Zhu, Tianhua	810
Zhuo, Fang	810, 2742, 3358
Zojer, Bernhard	996
Zou, Juan	651
Zou, Ke	1554
Zou, Xudong	3521
Zou, Xuewen	73, 2518
Zou, Zhi-Xiang	3493
Zumel, P.	2545, 3090

IEEE
445 Hoes Lane
Piscataway, NJ 08854-4141

ISBN 978-1-4673-9551-9